Phylonyms

A Companion to the PhyloCode

Phylonyms

A Companion to the PhyloCode

Edited by

Kevin de Queiroz,[1] Philip D. Cantino,[2] and Jacques A. Gauthier[3]

Editorial Board: Sina M. Adl (University of Saskatchewan), Frank E. Anderson (Southern Illinois University), Benoît Dayrat (Pennsylvania State University), Walter G. Joyce (Université de Fribourg), Matjaž Kuntner (National Institute of Biology, Slovenia), Michel Laurin (CNRS, France), Michael S. Y. Lee (Flinders University), Richard G. Olmstead (University of Washington), Greg W. Rouse (University of California, San Diego), Mieczyslaw Wolsan (Polish Academy of Sciences).

[1]Department of Vertebrate Zoology, National Museum of Natural History, Smithsonian Institution, Washington, DC, USA.
[2]Department of Environmental and Plant Biology, Ohio University, Athens, OH, USA.
[3]Department of Geology and Geophysics, Yale University; New Haven, CT, USA.

CRC Press
Taylor & Francis Group
Boca Raton London New York

CRC Press is an imprint of the
Taylor & Francis Group, an **informa** business

CRC Press
Taylor & Francis Group
6000 Broken Sound Parkway NW, Suite 300
Boca Raton, FL 33487-2742

First published in digital version April 30, 2020.

International Standard Book Number-13: 978-1-138-33293-5 (Hardback)

Library of Congress Cataloging-in-Publication Data

Names: De Queiroz, Kevin, editor. | Cantino, Philip D., editor.
Title: Phylonyms / edited by Kevin de Queiroz, Philip Cantino and Jacques
A. Gauthier.
Description: Boca Raton : CRC Press, [2019] | Includes bibliographical
references. | Summary: "Nearly 300 clades - lineages of organisms - will
be defined by reference to hypotheses of phylogenetic history rather
than by taxonomic ranks and types. This volume will document the Real
World uses of PhyloCode and will govern and apply to the names of
clades, while species names will still be governed by traditional
codes"-- Provided by publisher.
Identifiers: LCCN 2019026937 (print) | LCCN 2019026938 (ebook) | ISBN
9781138332935 (hardback) | ISBN 9780429446276 (pdf)
Subjects: LCSH: Cladistic analysis. | Phylogeny--Nomenclature.
Classification: LCC QH83 .P49 2019 (print) | LCC QH83 (ebook) | DDC
578.01/2--dc23
LC record available at https://lccn.loc.gov/2019026937
LC ebook record available at https://lccn.loc.gov/2019026938

Visit the Taylor & Francis Web site at
http://www.taylorandfrancis.com

and the CRC Press Web site at
http://www.crcpress.com

Table of Contents

The table of contents reflects the order of presentation in the volume and is arranged according to the following conventions: The purpose of the arrangement is to provide a linear order for the treatments in the volume, not to represent the phylogenetic relationships of the named clades. Nonetheless, the names are presented as an unranked, indented taxonomy with the names of hypothesized subclades indented beneath those of the clades in which they are included (but see below regarding controversial relationships). With a few exceptions, names that are indented the same amount within a section (see below) designate mutually exclusive clades, but not necessarily sister groups or clades of similar size or age; they are presented in order of the number of subclades whose names are treated in the volume, from smallest to largest. If the numbers of subclades whose names are treated in the volume are equal, the clade names are presented in a manner that is consistent with either phylogenetic relationships or traditional ordering conventions (or both). Names in square brackets are those of clades that are not treated in this volume; they are included because they may be more familiar than the names of their subclades that are included. Entirely extinct taxa are marked with a dagger. The treatments are divided into eight sections, with indentation reset to the left in each. In cases where relationships of a clade are controversial, we have generally followed the primary reference phylogeny cited in the contribution for that clade.

SECTION 4

SECTION 5

SECTION 6

SECTION 7

---

[1] These names are potential synonyms.

SECTION 8

Acknowledgments

The editors would like to thank the following people for reviewing one or more of the contributions to this volume: Sina M. Adl, Frank E. Anderson, Jason S. Anderson, J. David Archibald, Jonathan W. Armbruster, Sandra L. Baldauf, David A. Baum, Bastian Bentlage, Michael J. Benton, Mary L. Berbee, David S. Berman, Cedric Berney, Jonathan I. Bloch, Lynn Bohs, Geoff Boxshall, Heather D. Bracken, Christian Bräuchler, Birgitta Bremer, Daniel Brinkman, Samuel F. Brockington, Harold N. Bryant, Mark A. Buchheim, Frank T. Burbrink, Janet E. Burke, Stephen D. Cairns, Christopher S. Campbell, Nico Cellinese, Ivan Čepička, Stephen G. B. Chester, John C. Clamp, James M. Clark, Julia A. Clarke, Jonathan A. Coddington, Bruce B. Collette, Allen G. Collins, Elena Conti, David Cundall, Marymegan Daly, Charles C. Davis, Benoît Dayrat, Paul De Ley, Frederic Delsuc, Charles F. Delwiche, Thomas A. Deméré, Rob DeSalle, Michael J. Donoghue, Jason P. Downs, James A. Doyle, Gregory D. Edgecombe, Marek Eliáš, Peter K. Endress, Martín D. Ezcurra, Matteo Fabbri, Mark A. Farmer, Kristian Fauchald, Daphne G. Fautin, Daniel J. Fields, William L. Fink, Thomas Friedl, Darrel R. Frost, John Gatesy, Alana Gishlick, Thomas J. Givnish, David J. Gower, Sean W. Graham, Charles E. Griswold, Kenneth M. Halanych, Jason J. Head, Patrick S. Herendeen, Khidir Hilu, John M. Huisman, Pincelli M. Hull, Vladimir D. Ivanov, G. David Johnson, Ulf Jondelius, Walter G. Joyce, Paul Kenrick, Klaus-Dieter Klass, Kathleen A. Kron, Matjaž Kuntner, Conrad C. Labandiera, Gretchen Lambert, Max C. Langer, Allan Larson, Michel Laurin, Michael S. Y. Lee, Frederik Leliaert, Annie R. Lindgren, Aaron Liston, John G. Lundberg, J. Robert Macey, Christopher L. Mah, Jody Martin, Sarah Mathews, Mónica Medina, David P. Mindell, Masaki Miya, Jeffery J. Morawetz, David Moreira, Nancy R. Morin, Cynthia M. Morton, Molly Nepokroeff, Sterling J. Nesbitt, Maureen O'Leary, Richard G. Olmstead, Guillermo Ortí, Kevin Padian, Christopher L. Parkinson, Alan J. Paton, David J. Patterson, David L. Pawson, R. Toby Pennington, Diego Pol, Michael J. Polcyn, Leonid E. Popov, Daniel Potter, Kathleen M. Pryer, Yin-Long Qiu, Rachel A. Racicot, Thomas A. Richards, Stefan Richter, Gar W. Rothwell, Greg W. Rouse, Timothy B. Rowe, Iñaki Ruiz-Trillo, Anthony P. Russell, Leonardo Salgado, George Sangster, Diego San Mauro, Jan Sapp, Eric J. Sargis, Jürg Schönenberger, James A. Schulte II, Jeffrey W. Shultz, Alastair G. B. Simpson, Michael G. Simpson, Jan Šlapeta, Krister T. Smith, Ashleigh B. Smythe, Pamela S. Soltis, Robert J. Soreng, Michelle Spaulding, Erik A. Sperling, Greg S. Spicer, Saša Stefanović, Lena Struwe, Stuart Sumida, Kenneth J. Sytsma, David C. Tank, Ellen Thomas, Susumu Tomiya, Omar Torres-Carvajal, Richard P. Vari, Michael Vecchione, Morgan L. Vis, Sigurd von Boletzkey, Gary D. Wallace, David

W. Weisrock, Lars Werdelin, Wim Willems, Martin F. Wojciechowski, Mieczyslaw Wolsan, Gregory A. Wray, Richard E. Young, Elizabeth A. Zimmer, and Michelle L. Zjhra. We give special thanks to our Editor, Charles R. Crumly, for his patience in seeing this volume through to completion. We also thank Michele Dimont at CRC Press and Rachel Cook at Deanta Global for overseeing copy-editing and production of the book.

The authors of the *Insecta* contribution wish to thank Rolf Beutel, and the author of the *Pan-Nematoda* and *Nematoda* contributions wishes to thank James G. Baldwin, Daniel J. Bumbarger, Paul De Ley, Eric P. Hoberg, W. Duane Hope, and two anonymous reviewers. Author Richard M. McCourt states that his contributions are based upon work done while he was at the National Science Foundation. Any opinion, findings, and conclusions or recommendations expressed in those contributions are those of the author(s) and do not necessarily reflect the views of the National Science Foundation.

Introduction

The publication of this volume marks a turning point in the history of phylogenetic nomenclature. The period prior to this publication can be considered the initial phase in the development of phylogenetic nomenclature during which the general approach was proposed and its basic theory (principles and methods) articulated. The methods of phylogenetic nomenclature were first put into practice during this initial phase, but only unofficially, given that a governing code was still in the process of being developed. This early phase also served as a trial through which the consequences of the methods were examined and evaluated and the methods modified and refined accordingly. Now, with the publication of the *International Code of Phylogenetic Nomenclature (PhyloCode)* and *Phylonyms: A Companion to the PhyloCode*, phylogenetic nomenclature is moving out of this "test" phase to become a formal and fully functional system of nomenclature.

History

The idea behind this book originated early in the development of the *PhyloCode* (*International Code of Phylogenetic Nomenclature*; *ICPN*) and was discussed at the first workshop devoted to the development of such a code (Harvard University Herbaria, August 7–9, 1998). As originally envisioned, this "companion volume" to the *PhyloCode* was to have three functions: (1) provide a starting point for nomenclatural precedence under the *PhyloCode*; (2) establish carefully considered and peer-reviewed phylogenetic definitions for well-known clade names in order to give precedence to definitions that capture the widely understood concepts of important clades; and (3), provide a means for workers who had published definitions of clade names before the implementation of the *PhyloCode* to republish them so that they would have official status under this code.

At the second workshop on phylogenetic nomenclature (Yale Peabody Museum of Natural History, July 28–30, 2002), the scope of the companion volume was reconsidered. As originally conceived, this work would have included phylogenetic definitions of the most widely known names in most or all groups of organisms. It became apparent that accomplishing this goal would require a very lengthy book (or books), that producing this work would be an immense job, and that linking the starting date of the *PhyloCode* to publication of the companion volume would unnecessarily delay implementation of the *PhyloCode*. For this reason, the participants in the second workshop decided to reduce the scope of the companion volume. Instead of attempting a comprehensive treatment of widely known clade names for all major groups of organisms, it was decided that the companion volume would include only treatments of selected taxa. These entries would serve as models for future practitioners of phylogenetic nomenclature, as well as establish precedence for the included names and definitions. Kevin de Queiroz and Jacques A. Gauthier were chosen to edit the companion volume, and Philip Cantino joined as the third editor in 2004.

The inaugural meeting of the International Society for Phylogenetic Nomenclature (ISPN) (July 6–9, 2004, Muséum National d'Histoire Naturelle, Paris) offered an opportunity for the participants to present names and phylogenetic definitions that would form the nucleus of the companion volume. Although this volume would ideally include names for at least some well-known clades in all major groups of organisms, it was inevitable that some groups (e.g., vertebrates and angiosperms) would be more thoroughly covered than others (e.g., insects and archaeans) because these groups have more specialists who are both knowledgeable about their phylogeny and interested in using phylogenetic nomenclature.

In order to expedite completion, a ten-member editorial board was established in 2008 to help review manuscripts. The title *Phylonyms: A Companion to the PhyloCode* was selected by the editors in April 2007. Although the editorial board helped with the review process, completing the companion volume took far longer than anticipated. Because the 285 contributions were accepted over a period of more than ten years, authors were offered an opportunity to revise their contributions in June 2017, but only some did so. Consequently, some entries are more current than are others. The acceptance date (and in some cases, a subsequent revision date) is listed at the end of each contribution, as is the primary editor (or editors) for the contribution.

Objectives of This Book

As noted above, *Phylonyms* serves as the starting point for phylogenetic nomenclature governed by the *PhyloCode*. According to the preamble, "This code will take effect on the publication of *Phylonyms: A Companion to the PhyloCode*, and it

is not retroactive." Thus, names and definitions published here have precedence over any competing names and definitions published either before (or after) the publication of *Phylonyms*.

In addition to serving as the starting point for phylogenetic nomenclature under the *PhyloCode*, this book has two other major objectives. One is to provide well-vetted definitions of many widely known names—definitions that are designed to capture a widely accepted concept of each named clade and avoid unnecessary future changes in the hypothesized composition of those clades.

The other function of *Phylonyms* is to provide examples of well-constructed nomenclatural treatments (protologues), which can serve as models for other users of the *PhyloCode*. All treatments in this book conform to the rules (and in most cases, the recommendations) of the *PhyloCode*, but they vary greatly in length. Some (e.g., *Postciliodesmatophora*, *Convolvulaceae*, *Tunicata*, *Pan-Cryptodira*) are brief but nevertheless exemplify protologues that provide all the information required by the *PhyloCode*. Others (e.g., *Biota*, *Foraminifera*, *Cynodontia*, *Theropoda*) include, for example, lengthy discussions of nomenclatural history, exploration of alternative phylogenetic hypotheses or definitions, or detailed information about composition or diagnostic apomorphies, even though not essential to satisfy the requirements of establishment under the *PhyloCode*.

Every entry includes a brief statement about etymology and a list of phylogenetic synonyms (if any), but neither is required by the *PhyloCode*. The Definition paragraph of every entry includes an abbreviated definition and a sentence naming the kind of definition (e.g., maximum-crown-clade) using the conventions and terminology in the *PhyloCode*. Neither is required by the *PhyloCode*, although use of an abbreviated definition is recommended under certain circumstances (Rec. 9.4A); they are provided in this book to familiarize users with the abbreviations and definitional terminology, some of which differ from previous versions of the *PhyloCode*.

In *Phylonyms*, the emphasis is on well-known names and clades. As a general guideline, a well-known clade is one that might be covered in an introductory course in biology or a major taxon-oriented subdiscipline within biology—e.g., botany, entomology, mammalogy. This volume also emphasizes crown and total clades (crown clades being those that originate in the most recent common ancestor of two or more extant species or organisms; total clades being composed of a crown clade and all organisms that share a more recent common ancestor with that crown clade than with any other non-nested crown clade). However, the book also includes some clades that are neither crown nor total clades.

In addition to being widely known, most of the clades named in this volume are well supported. However, although the authors were expected to address clade support, the editors did not impose a strict criterion for what constitutes strong support. Moreover, useful names need not always be applied to highly corroborated clades. For example, when different data sets yield conflicting results, it may be useful to name the putative clades corresponding to alternative hypotheses.

Organization and Format

Organization of the Volume

Most of this book consists of a set of entries for individual clade names. They are organized into sections based on the more inclusive groups to which the corresponding clades belong in order to facilitate use as a reference. Every entry has the same format, and all discussion relevant to a particular name is included in the entry for that name, as opposed to being hidden in a chapter addressing the names of several internested clades.

Format for Entries

A complete name/clade entry (protologue) is composed of the following parts in this order ("authors" here refers to one or more authors; cited Articles, Recommendations and Notes refer to *PhyloCode* version 6):

1) The defined clade name, authors, dates, and status of the name as new vs. converted (required for establishment; Arts. 9.2, 9.15). In the case of a new clade name, the authors are listed after the name, without enclosing symbols. In the case of a converted clade name, the nominal authors and publication year are listed after the clade name, without enclosing symbols, followed by the definitional authors in square brackets. For more detailed recommendations concerning authorship, see *Authorship of Names* (below).

2) Registration number in RegNum, the ISPN's registration database (required for establishment; Art. 8.1)

3) Phylogenetic definition (required for establishment; Art. 9.3). In most cases, definitions use one of the recommended wordings in the Articles 9.5–9.10. Explicit statements concerning the category of definition (e.g., minimum-clade) and abbreviated definitions are also provided, although neither is required by the *PhyloCode* (abbreviated definitions are suggested when recommended wordings are not used, Rec. 9.4A). Definitions are formulated to capture the spirit of historical (original or current) use of the name, to the degree that this is consistent with monophyly (Rec. 11A). They are also formulated to minimize changes in clade composition

in the context of alternative phylogenetic hypotheses (Recs. 11B-E), although this is not always possible depending on the nature of the differences. For more detailed recommendations concerning definitions, see *Formulating Definitions* (below).

4) Etymology, if known. This is neither a requirement nor a recommendation of the *PhyloCode* but is a convention followed in this volume.

5) Reference phylogeny (required for establishment; Art. 9.13), a published phylogenetic hypothesis that illustrates the context in which the authors of the entry are applying their phylogenetic definition. Additional secondary phylogenies are often included as well. The primary reference phylogeny is expected to contain information sufficient to pinpoint the named clade in the diagram by invoking the definition. If the specifiers (species or organisms) used in the phylogenetic definition are not shown on the primary reference phylogeny, the text of this paragraph or parenthetical information in the definition (which is not considered part of the definition) clarifies which taxa in the reference phylogeny are most closely related to the specifiers.

6) Composition, a list of taxa inferred to be included in the named clade (required for establishment; Art. 9.14). In most cases, this is not a complete list of the known species but, rather, a list of major subclades, sometimes with references to publications discussing the composition of those subclades, or to more comprehensive online databases listing included species.

7) Diagnostic apomorphies for the named clade or a reference to such a list (recommended but not required for establishment; Rec. 9C). If a minimum-clade definition is used, this paragraph lists character states

that are apomorphies for the smallest clade that contains the internal specifiers (but see below for the case of crown clades). If an apomorphy-based definition is used, the character list includes at least the apomorphy upon which the definition is based. If a maximum-clade definition is used, the list includes character states that are apomorphies of the largest clade that contains the internal specifier(s) but does not contain the external specifier(s). For crown clades, there may be two lists of apomorphies: those that distinguish the members of the clade from the members of other crown clades (which may be apomorphies of clades more inclusive than their respective crowns), and those that distinguish the members of the clade from more closely related extinct taxa (members of the stem group of the crown clade in question).

8) Synonyms of the defined name, with references. Synonyms in phylogenetic nomenclature differ from their counterparts in rank-based nomenclature. In both cases, synonyms are names that are spelled differently but refer to the same taxon. However, phylogenetic synonyms are differently spelled names that refer to the same clade, while rank-based synonyms are differently spelled names at the same rank that are based on types that are considered to belong to the same taxon at that rank. Many phylogenetic synonyms are not rank-based synonyms because the names have terminations that indicate different ranks under the latter system (e.g., based on their composition, *Gnetales* and *Gnetopsida* are effectively phylogenetic synonyms of *Gnetophyta*, but they are not rank-based synonyms). Conversely, rank-based synonyms need not be phylogenetic synonyms if the names are given different phylogenetic definitions.

The *PhyloCode* does not require a list of synonyms in order to establish a clade name. However, because a given clade will often have been referred to by several different names, providing a list of synonyms facilitates retrieval of the relevant literature. In this volume, synonyms are classified as unambiguous, approximate, or partial (though some synonyms are both approximate and partial). A synonym is *unambiguous* if it has a published phylogenetic definition or its application to the same clade can be inferred from an explicit published statement about how the taxon was conceptualized. Taxon names labeled on a cladogram or phylogenetic tree can sometimes be considered unambiguous synonyms, but only if the authors clearly distinguished between the relevant clade categories (e.g., crown vs. total vs. apomorphy-based). An *approximate* synonym is one that was apparently applied to the same clade—as inferred from stated composition and/or diagnostic characters and/or labeling of a cladogram or phylogenetic tree—but the authors did not clearly distinguish between the relevant clade categories. A *partial* synonym is a name that has been applied to a group that does not correspond fully to the named clade in terms of composition, including one that was applied to a paraphyletic group originating in the same ancestor as the named clade.

Because a name may be a synonym as applied by one author but not by another, the designation "sensu" is used here to distinguish subsequent uses of the name from the original one.

9) Comments. This section varies in length and content but often includes the following:
a) Historical information concerning recognition of the named group and,

in the case of converted names, use of the defined name, with references. For converted names, the *PhyloCode* (Art. 9.15) requires one or more citations demonstrating prior application of the defined name to the clade approximating the one for which it is being established.

b) Evidence and support for the clade, as well as its existence or lack thereof in the context of alternative phylogenetic hypotheses (i.e., other than the primary reference phylogeny).

c) In the case of converted names, reasons for choosing the selected name over possible alternatives (see Art. 10). If the most widely and consistently used name for the clade of concern is not selected for conversion, a compelling rationale should be provided (Rec. 10.1A).

d) Clarifications concerning the conceptualization of the clade to provide guidance in interpreting the author's intent in formulating the definition. One way to clarify the intended reference of a name is to explain how it is to be applied in the context of alternative phylogenetic hypotheses.

e) Preferred precedence of the defined name relative to other names defined in this volume (or elsewhere), should they become synonymous under an alternative phylogenetic hypothesis.

f) Discussion of alternative phylogenetic hypotheses and how they bear on the application of the name (e.g., if rendered inapplicable by a qualifying clause) or the composition of the clade to which it refers.

g) Information from the stratigraphic archive of body, trace, and/or chemical fossils, and/or the divergence times estimated from them.

General Guidelines for Preparing Phylogenetic Nomenclatural Treatments

Constructing a good phylogenetic definition and associated protologue is not a simple, mechanical procedure. It requires a thorough knowledge of the taxonomic history of the clade, including alternative phylogenetic hypotheses and how they might affect ideas about the composition and diagnostic characters of that clade. Most importantly, it requires a willingness to think carefully about the ramifications of a candidate definition when applied in the context of plausible alternative phylogenies, and how that definition might be modified so that the name will be applied as intended in those contexts.

One reason why this volume took so long to prepare is that the editors were heavily involved in helping authors to formulate robust definitions and carefully consider the ramifications of various possible choices of names and formulations of definitions—a process that entailed more editorial work and more drafts of each contribution than is typical for peer-reviewed publications. As an aid to future authors of phylogenetic nomenclatural treatments, we will summarize here some aspects of the reasoning involved. This is not intended as a literature review, so citations are omitted for the sake of brevity. In omitting them, we do not mean to imply that all of the guidelines presented below are new or uniquely ours, but rather to summarize succinctly what we regard as best practices based on our personal experience. Similarly, we have not cited literature related to the taxa and phylogenies used to exemplify procedures.

Choice of Names

Clade names may be derived from any language but must be Latinized, in the tradition of scientific names. Article 10 of the *PhyloCode* contains a series of rules and recommendations designed to promote nomenclatural continuity. In general, a preexisting name must be used (converted) if possible; the coining of a new name is permitted only in the situations detailed in Article 10.2. Because most preexisting names do not have phylogenetic definitions, inferences about whether a preexisting name has been applied to a particular clade must be based on evidence such as taxon composition, diagnostic characters, statements regarding how the taxon was conceptualized, and the labeling of phylogenetic diagrams. However, the evidence is often ambiguous (or contradictory), and the association of a name with a clade may therefore only be approximate.

If more than one name has been applied to a particular clade, the *PhyloCode* (Rec. 10.1A) recommends selecting the name that is most widely and consistently used in the scientific literature for that clade. As a rule of thumb, if there is less than a twofold difference in the frequency of use of competing names, the choice is left to the discretion of the authors who convert (i.e., establish) the clade's name (Note 10.1A.1). Frequency of use can be assessed using various printed and on-line databases of scientific literature such as *Google Scholar, BIOSIS Previews, Zoological Record, Kew Bibliographic Databases*, and *GeoRef*. If a descriptive name (e.g., *Monocotyledoneae*) and a rank-based name derived from the name of a subordinate taxon (e.g., *Liliopsida*) have roughly similar frequencies of use, we recommend selecting the descriptive name, because its association with the clade may be easier to remember, particularly if it corresponds to a widely used informal name (e.g., monocots). If the only name associated with a particular clade is informal, we recommend selecting a new formal name that corresponds etymologically to the informal name (e.g., *Monilophyta* for "monilophytes"). If no preexisting scientific name has been associated (even loosely) with a particular clade, a preexisting name applied to a paraphyletic group originating from the same ancestor as the clade may be selected, or a new name may be coined. The choice between these two options is left to the discretion of the authors (Note 10.2.1). For crown clades represented in rank-based nomenclature by a single genus (e.g., *Equisetum*) or species (e.g., *Ginkgo biloba*), there is typically a series of roughly equally well-known rank-based names (those associated with the traditional primary ranks) that have been associated, to one degree or another, with the clade of interest— e.g., *Equisetophyta* (Division), *Equisetopsida* (Class), *Equisetales* (Order), *Equisetaceae* (Family), *Equisetum* (Genus). In this situation, it is generally best to select a name associated with one of the lower ranks, which is likely to correspond more closely to the crown clade in traditional use, while the other names are more likely to have been used for larger clades that include extinct taxa outside of the crown.

An Integrated System of Clade Names

In order to advance the development of an integrated system of clade names, the editors of *Phylonyms* have encouraged contributors to adopt some general conventions about how certain categories of clades are named and how certain categories of names are defined. We hope that the advantages of an integrated system (facilitating communication, particularly for those who are not specialists on the group being discussed) will convince future authors of phylogenetic definitions to adopt this approach.

Crown clades. In general, widely known names should be selected for crown clades. Specifically, the name that is most commonly used to refer to a particular crown clade should be defined as referring to that crown, even if it has also been applied (perhaps even more frequently) to one or more clades that include stem taxa (extinct taxa outside of the crown) (Rec. 10.1B). The primary reason for adopting this convention is that the best-known name is used for the clade about which the most can be known (i.e., because we rarely have information about the genetics, development, physiology, behavior, etc. of extinct organisms), and that is how such names are effectively used by people who study aspects of biology that are rarely preserved in fossils. The kinds of definitions that can be used to define crown-clade names are detailed in Art. 9.9.

Total clades. For total clades, we recommend using panclade names (Arts. 10.3– 10.7; names formed by adding the prefix "*Pan-*", including the hyphen, to the name of the corresponding crown clade; e.g., *Pan-Angiospermae*). The primary reason for adopting this convention is that it potentially cuts in half the number of different names that have to be learned, because the names of total clades are immediately recognizable as referring to the total clades of particular crowns. In many cases, no name will have been defined explicitly as the name of a particular total clade (i.e., as opposed to being applied to a slightly less inclusive clade that is neither total nor crown), so that adopting a panclade name will not contradict the recommendation to use the name that has been most widely and consistently applied to that clade. In some cases, however, there may be a preexisting name that has been phylogenetically defined as applying to the total clade. In such cases, the decision whether to convert that name or establish a new name using the prefix "*Pan-*" is left to the author's discretion (Art. 10.6). The *PhyloCode* does not recommend one of these options over the other; however, in the interest of developing an integrated system of clade names, we encourage the use of panclade names. The kinds of definitions that can be used to define total-clade names are detailed in Article 9.10.

Apomorphy-based clades. Clades that are conceptualized in terms of apomorphies should generally be given names that describe those apomorphies. However, some names that describe apomorphies are the most widely known names for crown clades and should therefore be applied to those crowns (see "Crown clades", above). In such cases, the clade originating in the ancestor in which that apomorphy originated may be given a name formed by combining the name of the crown clade with the prefix *Apo-* (e.g., *Apo-Mammalia*), implying that members of the clade possess the apomorphy described by the name (e.g., mammary glands) (see Arts. 10.8 and 10.9). The kinds of definitions that can be used to define the names of clades that are conceptualized in terms of apomorphies are detailed in Art. 9.7.

Authorship of Names

The following conventions follow Recommendation 9.15A, including Notes 9.15A.3 and 9.15A.4. If the clade name is being converted from a preexisting name, the author and date cited are for the spelling that is used in the converted name. If the spelling was "corrected" to a standard ending under one of the rank-based codes, and the corrected spelling is the one being converted, then the citation should be for the first authors to use the corrected spelling (as opposed to the original one) in a publication. However, if the author and date of the corrected spelling are difficult to determine, the author and date of the original spelling may be cited instead, provided that the difference in spelling is noted (e.g., *Cactales* Dumortier 1829, as *Cactarieae*). Similarly, the Principle of Coordination of the *International Code of Zoological Nomenclature* (*ICZN*) is not followed in attributing authorship under the *PhyloCode*. For example, the person who first published a name in the family group is attributed authorship only for the name that he or she actually used, not that of other names in the family group. For those other names, authorship is attributed to the person who first published them. However, if the author and date of the converted spelling are difficult to determine, the author and date under the Principle of Coordination may be cited instead, provided that the difference in spelling is noted (e.g., *Iguaninae* Bell 1825, as *Iguanidae*). We follow the precedent set by the rank-based codes with regard to the starting point for biological nomenclature and thus the authorship of names—that is, there is no need to trace names back farther than the starting point of the appropriate rank-based code (e.g., Linnaeus [1758] for most animal names). In contrast to the *International Code of Nomenclature for Algae, Fungi, and Plants* (*ICNAFP*) requirement for names published between 1935 and 2011, the *PhyloCode* does not require that the description or diagnosis be in Latin (Note 19.1.3). Therefore, the first person who published a name governed by the *ICNAFP* in association with a description or diagnosis and otherwise satisfying the *ICNAFP* rules for a legitimate name is considered the nominal author, regardless of whether the description or diagnosis was in Latin.

Formulating Definitions

When defining converted names, it is particularly important to do so in ways that will minimize disruption of nomenclatural continuity, thus maximizing the utility of names for communication and access to the literature. For this reason, careful consideration should be given to both historical and current use of names, and an attempt should be made to preserve one or, if possible, both. In cases where the two considerations are in conflict, greater weight should be given to current (recent) use.

Consideration should be given not only to prior use but also to the development of a set of names that will be maximally useful in the future. For example, the recommendation to define widely known names as referring to crown clades (Rec. 10.1B; see "Crown clades", above) will often conflict with current use in terms of both taxon composition (i.e., the inclusion of certain fossils) and diagnostic characters. On the other hand, it will often be consistent with the historical composition of taxa (i.e., the groups to which the names were applied before the discovery of fossils outside of the crowns) and at least some diagnostic characters, as well as with some aspects of current use (i.e., the tendency to use these names when discussing characters that apply to all members of particular crowns but not necessarily to various extinct taxa outside of those crowns—e.g., listing the absence of teeth as a diagnostic character of *Aves* or using the name *Aves* when discussing a clade inferred entirely from data on extant species).

One of the most important considerations related to maximizing continuity in nomenclature is that names continue to refer (under a variety of "reasonable" phylogenetic hypotheses) to clades that include certain taxa and exclude others. This is not to say that one should formulate definitions with the goal of maintaining stable taxon composition at all costs. It may be

the case that the inclusion of certain formerly excluded species or clades, or the exclusion of certain formerly included ones, does not violate anything fundamental about the concept of a particular taxon. For example, *Annelida* now includes the formerly excluded *Sipuncula* and *Echiura*, impling that their inclusion does not violate anything fundamental about the concept of *Annelida*. However, in other cases, inclusion of certain traditionally excluded species or clades, or the exclusion of certain traditionally included taxa, may represent violations of the taxon concept. For example, the inference that the traditional taxon *Bryophyta* is paraphyletic relative to vascular plants (*Tracheophyta*) could have led to the application of the name *Bryophyta* to the clade that includes both the traditional "bryophytes" and the previously excluded vascular plants. But that has not been done, suggesting that inclusion of the tracheophytes would have violated the concept of *Bryophyta*. Part of the reason why the name *Bryophyta* was not applied to the clade of "bryophytes" and tracheophytes may have been that this clade already had a widely used name, *Embryophyta*. Another potentially relevant factor is the descriptive meaning of the name. Although the primary purpose of taxon names is simply to refer to clades (*ICPN* Principle 1), so they need not be descriptively accurate, there are advantages to names whose etymologies help users to associate those names with particular clades. For example, some analyses indicate that the clade originating in the immediate common ancestor of all extant gymnosperms also includes angiosperms. Under such a phylogeny, some might consider it confusing to apply the name *Gymnospermae*, which is derived from the Greek *gymnos* (naked) + *sperma* (seed), to a clade that includes plants exhibiting the contrasting character state of seeds enclosed within a fruit (i.e., angiosperms). Nonetheless, these considerations are complicated by unexpected phylogenetic relationships and cases in which the only name closely associated with the ancestor in which the clade to be named originated formerly referred to a paraphyletic taxon (see below). In any case, there are several different mechanisms for restricting the application of phylogenetically defined names so that they do not result in the inclusion or exclusion of taxa that are judged to represent fundamental violations of taxon concepts.

Choice of Specifiers

In general, specifiers should be selected in a way that both (a) captures the spirit of historical use to the degree that it is consistent with monophyly (Rec. 11A), and (b) maintains the association of the name with the intended clade (whether conceptualized in terms of composition, characters, or some other attribute) in the context of a range of plausible phylogenies (Recs. 11B–E).

When selecting internal specifiers, one should avoid taxa whose membership in the clade is questionable. On the other hand, the internal specifiers in a minimum-clade definition should include representatives of all taxa that could plausibly be sister to the rest of the clade (Rec. 11D). Doing so will reduce the chance that, under a new phylogenetic hypothesis, the name will refer to a less inclusive clade than originally intended. If the relationships at the base of the clade are poorly resolved, resulting in there being more than one plausible candidate for sister to the rest of the clade, a minimum-clade definition may require more than two internal specifiers. If there is a conflict between including specifiers representing every plausible sister group candidate and avoiding specifiers whose membership in the clade is questionable (i.e., a taxon that is plausibly one of the two basal subclades of the clade to be named may, alternatively, lie far outside that clade), a qualifying

clause (Art. 11.12; also see below) may be used to specify the inclusion of the problematical subclade under some phylogenies but not others. Alternatively, if the clade to be named is a crown clade, one may be able to deal with poor basal resolution by using a maximum-crown-clade definition or an apomorphy-modified crown-clade definition (Art. 9.9; also see below).

When selecting external specifiers, one should avoid taxa whose position outside the clade is questionable (i.e., some evidence suggests the taxon may instead be part of the clade being named). On the other hand, the external specifiers in a maximum-clade definition should include representatives of all taxa that could plausibly be sister to the clade being named (Rec. 11E). Doing so will reduce the chance that, under a new phylogenetic hypothesis, the name will refer to a more inclusive clade than originally intended. If the close outgroup relationships of the clade being named are poorly resolved, resulting in there being more than one plausible candidate for sister to the clade, a maximum-clade definition may require more than one external specifier.

The above guidelines concern which subclades of the named clade ought to be represented by specifiers but not which species (or specimens) should be selected to represent those subclades. In general, it is advisable to select specifiers from among the species included in the primary reference phylogeny, as that facilitates application of the definition. It is also desirable to select specifiers that are well known biologically, particularly those that are commonly included in phylogenetic studies. In addition, selecting deeply nested rather than early diverging species as internal specifiers has the advantage of reducing the likelihood of compositional changes in the context of alternative phylogenetic hypotheses. Finally, repeated use of the same specifiers for the names of clades in a nested series, or for those of a pair of sister clades, makes it easier to understand the relationships of those clades from the definitions.

In the interest of avoiding confusion, a clade name should not be based on the name of another taxon that is not part of the named clade. Therefore, when a clade name is converted from a preexisting name that is typified under a rank-based code, or is formed from the stem of a typified name, the type (species or specimen) of the preexisting name, or of the genus name from which it is derived, must be used as an internal specifier (Art. 11.10). For example, the definition of *Magnoliophyta* must include the type species or specimen of *Magnolia* as an internal specifier.

Definitions should not necessitate, though they may allow, the inclusion of subtaxa that were historically excluded from the taxon in question (Rec. 11A). This recommendation is particularly relevant in the case of names that formerly referred to paraphyletic taxa but are being converted as clade names. In such cases, species that were traditionally excluded from a paraphyletic taxon should not be chosen as internal specifiers for the definition of the name. For example, no bird species should be used as internal specifiers in the definition of the name *Dinosauria*. This rule does not prohibit the name *Dinosauria* from being defined so that birds are considered a subclade of *Dinosauria* under currently accepted phylogenetic hypotheses; thus, the inclusion or exclusion of birds will be determined by the phylogenetic evidence, as opposed to being necessitated by the definition.

Compositional stability may be jeopardized by an insufficiently precise description of the apomorphy in an apomorphy-based definition. Apomorphy-based definitions should be used only if the apomorphy can be described clearly and precisely, ideally with a clearly labelled image (or reference to one) illustrating the defining apomorphy. For example, if an apomorphy is commonly described by a general term such

as "limbs" or "seeds", the author should provide a detailed description of the properties that a structure must possess to be considered a limb or a seed, preferably by contrasting ancestral and derived states to highlight the distinctions being made. Although similar considerations apply to apomorphy-modified crown-clade definitions, potential ambiguity in the description of the apomorphy is less problematic for reasons discussed below (see *Defining the Names of Crown Clades*).

Qualifying Clauses

Qualifying clauses (Art. 11.12) can be used to restrict the use of the defined name to the context of those phylogenetic hypotheses in which certain specified relationships are present, rendering the name inapplicable in the context of alternative phylogenetic hypotheses. For example, one could define the name *Pinnipedia* as referring to the clade originating in the most recent common ancestor of *Otariidae* (sea lions), *Odobenidae* (walruses), and *Phocidae* (seals), provided that it does not include *Ursidae* (bears), *Procyonidae* (raccoons), or *Mustelidae* (weasels) by inserting specifiers representing these six taxa in their respective positions in this definition. This definition would result in the name *Pinnipedia* not being applicable to any clade under a phylogentic hypothesis in which one or more of the three explicitly excluded taxa (*Ursidae, Procyonidae, Mustelidae*) is considered a descendant of the most recent common ancestor of the three internal specifiers (representatives of *Otariidae, Odobenidae*, and *Phocidae*).

Defining the Names of Crown Clades

A crown clade name may be defined using a minimum-clade definition in which all of the internal specifiers are extant or it is stated explicitly that the named clade is a crown, or using a maximum-crown-clade definition or an apomorphy-modified crown-clade definition (see Art. 9.9 for recommended wordings of all three). Which kind of definition is preferable depends on the degree of well-supported resolution in the reference phylogeny, both the basal relationships within the clade being named and the close outgroup relationships. It is not uncommon for a crown clade to be well supported (at least relative to other crown clades) but to have poorly supported or unresolved basal and/or close-outgroup relationships.

If the basal relationships within a crown clade are well resolved and well supported, but the close outgroup relationships are not, a minimum-clade definition will generally be preferable. It could be as simple as "the smallest crown clade containing A and B", where A and B are specifiers representing the two primary subclades of the crown clade being named. Conversely, if the outgroup relationships are well resolved and well supported but the basal ingroup relationships are not, a maximum-crown-clade definition will generally be preferable. For example, if the monophyly and sister relationship between a crown clade to be named and its sister crown clade are well supported, but the relationships among, say, ten subclades within the former are poorly supported (such that any one of them could be sister to the rest), one would have to include internal specifiers from all ten subclades in a minimum-clade definition of the crown-clade name. By contrast, a maximum-crown-clade definition could be as simple as "the largest crown clade containing A but not Y", where A is a specifier representing the clade being named and Y is a specifier representing its sister crown clade.

If both the basal relationships and the close outgroup relationships of a crown clade are well resolved, these two kinds of definitions work equally well. If both the basal and outgroup relationships are poorly resolved, but

the clade is characterized by a distinctive apomorphy, an apomorphy-modified crown-clade definition may be useful. Such a definition may take the form "the crown clade characterized by apomorphy M (relative to other crown clades) as inherited by A," where A is an internal specifier (see Art. 9.9 for alternative wordings). Because this kind of definition is used only for crown-clade names, potential ambiguities related to the use of apomorphies as specifiers should be less problematic than in the case of standard apomorphy-based definitions. This is true because taxon (specimen) referral is based on whether the taxon lies within the crown (i.e., the clade whose members possess the specified apomorphy), rather than possession of the defining apomorphy itself—which may have arisen much earlier than the ancestor represented by the crown node. For example, if the name *Tetrapoda* were to be defined as the crown clade for which the possession of limbs as inherited by *Homo sapiens* is an apomorphy relative to other crown clades, then even if the precise meaning of "limbs" is left unspecified, fossil stem taxa possessing intermediate morphologies (e.g., *Tiktaalik, Acanthostega*) pose no problem for the definition. Regardless of whether these taxa are considered to possess limbs (because their appendages exhibit some but not all of the derived features shared by extant tetrapods), they clearly lie outside of the tetrapod crown clade.

Alternative Phylogenetic Hypotheses

From the issues discussed in the previous section, it should be clear that considering how a name will be applied (i.e., according to its definition) in the context of alternative phylogenetic hypotheses is *absolutely essential* to formulating robust definitions. We all have our preferred phylogenetic hypotheses, but no matter how strongly one feels that a particular hypothesis is correct, formulating a good definition requires careful consideration of how that definition will cause the name to be applied under a different hypothesis. Conversely, the worst definitions will be those that are strongly dependent on the correctness or acceptance of a particular phylogenetic hypothesis. An exception is when a name is intended to represent a particular phylogenetic hypothesis. In such cases, the definition should be constructed so as to render the name inapplicable under incompatible alternative hypotheses. For example, if the name *Holostei* were to be defined as "the smallest crown clade containing *Amia calva* (representing bowfins) and *Lepisosteus osseus* (representing gars) but not *Perca fluviatilis* (representing teleosts)", then that name would apply to the bowfin-gar crown clade in the context of phylogenies in which bowfins and gars are sisters, but it would not apply to any clade in the context of phylogenies in which bowfins are closer to teleosts than to gars (the alternative *Halecostomi* hypothesis). This is not to say that definitions must result in the name being applied to a clade of identical composition or diagnostic characters no matter which phylogenetic hypothesis is correct (as this will sometimes be impossible). However, knowing the circumstances under which a particular definition will succeed (i.e., the name will be applied to the clade that was intended by the author of the definition), as well as those under which it will fail, will help one to formulate definitions that are robust to changing ideas about phylogeny.

Choice of Primary Reference Phylogeny

A reference phylogeny (Art. 9.13) is a phylogenetic hypothesis that provides a context for applying a clade name by means of its phylogenetic definition. There will often be more than one phylogeny that includes the clade to be named, and it is often worthwhile to cite more than one, but a single diagram should be chosen as the primary reference phylogeny (Rec. 9.13B). The primary reference phylogeny (PRP) may be from the same publication in which the name is defined, or it may be a previously published diagram (as in the case of all names defined in this volume).

The reference phylogeny selected as primary will often be the most recent or the most comprehensive (in terms of taxon and/or character sampling), but this is not necessary (e.g., it could be one that is considered seminal). The PRP should be chosen to show both the clade being named (even if it was not given the same name) and the specifiers used in the definition of its name. If a minimum-clade definition is used, and thus the specifiers are internal, the PRP should show the internal relationships well and should ideally include representatives of all the major subclades of the clade that is being named. It is especially important that the PRP show both branches of the basal split or, if basal relationships are unresolved or are not well supported, all of the relevant subclades.

By contrast, if a maximum-clade definition is used, and thus there are both internal and external specifiers (and often more than one of the latter), it is critical that the PRP show outgroup relationships, even if the clade being named is not as densely sampled as in some other phylogenies. It is especially important that the PRP show the sister of the clade being named or, if outgroup relationships are unresolved or are not well supported, all plausible candidates for sister status.

SECTION 1

Pan-Biota J. Wagner 2004 (as *Panbiota*) [J. Wiemann, K. de Queiroz, T. B. Rowe, N. J. Planavsky, R. P. Anderson, J. P. Gogarten, P. E. Turner, and J. A. Gauthier], converted clade name

Registration Number: 299

Definition: The total clade of the crown clade *Biota*. This a crown-based total-clade definition. Abbreviated definition: total ∇ of *Biota*.

Etymology: Derived from the Greek *pan-* ("all" in reference to a total clade) and *Biota,* the name of a crown clade (see *Biota* entry in this volume for etymology).

Reference Phylogeny: Figure 1 in Hug et al. (2016), in which *Homo sapiens* is part of the clade *Opisthokonta*. *Biota* (this volume) includes all taxa depicted on that tree, and *Pan-Biota* includes not only all members of that crown clade but also all yet-to-be-discovered stem bioentities that share common ancestry with them (see, e.g., Cornish-Bowden and Cardenas, 2017: Fig. 4). There are no phylogenies for early divergences within *Pan-Biota* because no unambiguous stem biotans are known, which is unsurprising given that their potential fossil remains would be very unlikely to preserve substantial morphology or genetic information enabling inferences about their phylogenetic placement. Evolutionary scenarios discussing chemical steps leading to the first ancestral replicator and its organismal assembly giving rise to *Biota* can be found in Comments.

Composition: No unambiguous stem biotans are currently known. Nevertheless, if life on Earth is monophyletic (Theobald, 2010), then *Pan-Biota* necessarily includes all biologically replicating entities (bioentities) that have ever existed on this planet (or spread from it in the past, present, or future), including all extant and extinct bioentities, whether known or not. It thus includes the crown clade *Biota* (this volume) plus all non-biotan cellular and molecular (acellular) entities descended from the first replicator ancestral to that crown clade (= ur-replicator or ur-ancestor), even if that replicator had a deeper extraterrestrial ancestry (viz., Panspermia and derivative hypotheses). Prebiotic molecules do not qualify as pan-biotans unless they are either actively or passively replicated biologically (see Diagnostic Apomorphies and Comments) and that property is homologous with replication in the biotan crown. Well-known replicators considered here as ancestral pan-biotans include DNA single- and double-strands (DNA world), RNA single- and double-strands (RNA world), (hypothetical) nucleic acid-protein complexes (see Comments) and their replicating ancestors (Gilbert, 1986; see critical comments by Orgel, 2003). Additional, protein-only world scenarios predict prion-like protein templating as the ancestral mode of "exchange of biological information" (Pruisner, 1998; Rode et al., 1999; Lupi et al., 2006); however, such protein-protein templating would not fulfill the requirement of biological reproduction. A hypothetical ancestral prion-like protein would be unrelated to extant prions, which are parts of biotans, and emerged multiple times independently in different protein families. Therefore, a hypothetical ancestral prion-like protein would not be part of *Pan-Biota* but would be considered prebiotic chemistry. Other candidate compounds of either hypothetical or in vitro-synthesized nature include (double- and single-stranded) pyranosyl RNA (Eschenmoser, 1999), as well as (single- or double-stranded) peptide nucleic acids (Nielsen and Egholm, 1999). These potential replicators may derive from the biotan stem, or may have formed independently, or may represent only hypothetical replicators. If such artificially synthesized compounds formed naturally, and originated in the biotan stem, they would be considered part of *Pan-Biota*; otherwise, they would be regarded as products of prebiotic chemistry. Ribozymes (a functional term referring to RNA with biocatalytic—including autocatalytic—potential) are bioentities composed of single-stranded RNAs considered central to the currently prevailing RNA-world hypothesis (Gilbert, 1986). Ribozymes are present in all cells (and some viruses), and facilitate, for example, protein biosynthesis. They are therefore parts of *Biota* and *Pan-Biota* as parts of cells, but it is not clear if these extant ribozymes are descendants of an ancestral replicator, and evidence against ribozyme-like pan-biotan replicators is accumulating (Orgel, 2003; see Comments for details). Because most replicator candidates listed here cannot reproduce on their own (except for the aforementioned ribozymes), they require close association with replication catalyzers that can be of inorganic, bioinorganic, or biomolecular origin. While the most popular 'pan-biotan origins' scenarios are compound-exclusive (e.g., RNA world; see Comments), evidence is accumulating for a concurrent emergence of building blocks comprising the ancestral pan-biotan (Patel et al., 2015). The potential coexistence of different molecular building blocks suggests that multiple biological replicators (i.e., RNA and DNA) may have formed simultaneously, while only one of them gave rise to *Biota*. Only this single replicator and all of its descendants would be members of *Pan-Biota*. If other replicators, or their parts, were later to become incorporated into biotans

and thereafter descended along with them, then they would also be parts of *Pan-Biota*.

Diagnostic Apomorphies: The total-clade definition adopted for the name *Pan-Biota* embodies the idea that this name applies to the very largest clade on Earth (containing *Homo sapiens*). To form clades, common ancestry and thus genealogical connectivity via biological replication (an ancestral bioentity giving rise to descendant bioentities) is the only necessary property. That property would then be the fundamental diagnostic apomorphy of *Pan-Biota*. Whether replication occurs through molecular templating, external catalysis, internal catalysis, or autocatalysis does not matter. While we are referring to "biological ancestor-descendant" replication, rather than a chemical educt–product relationship, it should be emphasized that a simpler form of replication likely preceded the complex processes involved in either cellular fission or sexual reproduction. This concept of "biological replication" versus "chemical reactivity" is illustrated below. Given the templating potential of various biomolecules common to *Biota* (this volume), such as DNA, RNA, and even proteins, *Pan-Biota* is expected to contain multiple types of such replicators through sequential integration into an organism derived from the one ancestral replicator (bioentity).

Chemical Reactivity (insufficient for inclusion in *Pan-Biota*)

$$Compound\ A + Compound\ B$$
$$\rightarrow Compound\ C + Compound\ D$$
$$+ Compound\ F$$

$$RNA\ strand + H_2O \xrightarrow{Base} Part\ 1\ of\ RNA\ strand$$
$$+ Part\ 2\ of\ RNA\ strand$$

Biological Replication (creates genealogical relationships required for inclusion in *Pan-Biota*)

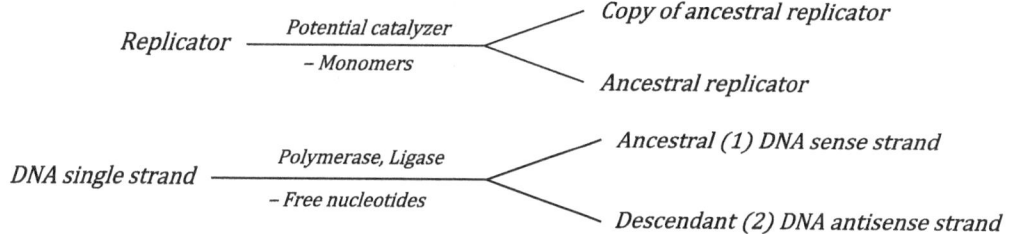

Synonyms: Most of the names listed as synonyms of *Biota* (this volume), as well as *Biota* of Trifonov and Kejnovsky (2015) and others, can also be considered approximate synonyms of *Pan-Biota* in that previous authors seldom distinguished clearly between crown and total clades (some are also partial synonyms). Partial synonyms that refer to paraphyletic taxa that were more clearly considered to have originated in earlier ancestors than the most recent common ancestor of the crown clade include: *Aphanobionta* Novák 1930; *Protobiota* Hu 1965; *Acytota* Jeffrey 1971.

Comments: As in the case of the crown clade (*Biota*), the literature records few instances in which the corresponding total clade has been given a formal taxon name. We take this opportunity to name it using an explicit phylogenetic definition following the general approach of Wagner (2004), which is based on the idea that there is no clade more inclusive than the one being named, although the wording of our definition differs. In the interest of developing an integrated system of clade names (see de Queiroz, 2007), we have chosen *Pan-Biota* as the name for the total clade of *Biota* (this volume). We have attributed the name *Pan-Biota* to Wagner (2004) and treated it as a converted clade name. Although Wagner (2004) used the orthographic variant *Panbiota*, that variant employs the same letters, base name, and prefix as *Pan-Biota*; it differs only in the absence of a hyphen separating the base name and prefix and in the absence of capitalization of the first letter of the base name. Both of those orthographic conventions were adopted in the *PhyloCode* subsequent to Wagner's proposal. Moreover, Article 17.1 implies that the hyphen is not part of the spelling of the name, given that deletion of a hyphen from a preexisting name does not prevent it from being treated as a converted name. Similarly, we do not consider the difference in capitalization of the letter "b" to constitute a difference in spelling (see Art. 17.5).

There are several hypotheses for the nature of the ancestral replicator that gave rise to the clade *Pan-Biota*; we will discuss the more widely accepted alternatives here. Each potential ancestral pan-biotan will be introduced and critically evaluated in light of the defining apomorphy for this clade: biological replication (see Comments in *Biota*, this volume).

The Ancestral Replicator: One of the key problems in diagnosing the clade *Pan-Biota* is that the exact nature of the ancestral replicator remains elusive and the subject of considerable speculation (i.e., Orgel, 2003). In the absence of an informative fossil record, and given the drastic difference between this ur-ancestor and the complex cellular ancestor last shared by all bacteria, archaeans, and eukaryotes (BAEs), theoretical and experimental works have inspired innumerable hypotheses about the features of the ur-ancestral replicator, and how it was initially assembled. The most popular scenarios include the RNA-world (Gilbert, 1986), the DNA-world (Forterre, 2001, 2002), the Protein-only world (Pruisner, 1998; Rode et al., 1999), and further elaborations of these hypotheses that include both terrestrial emergence and extraterrestrial delivery to Earth.

Summarizing the most plausible and established scenarios on the emergence of pan-biotans, it seems reasonable to assume that the ancestral replicator entity was likely composed of (single or double-stranded) RNA and/or DNA (Forterre, 2002; Orgel, 2003). That inference is based on the fact that the "genetic" material found universally among Recent organisms and viruses (or viral derivatives) is composed of either DNA or RNA.

It is often assumed that the ancestral replicator must have been a single compound. This assumption finds expression in the well-established and widely accepted hypothesis that an autocatalytically reproducing ribozyme was the ancestral replicator (RNA-world sensu Gilbert, 1986). Ribozymes are single-stranded RNA with catalytic properties, functionally comparable to enzymes. They are characterized by having both a genotype and phenotype that determines their catalytic activity. Ribozymes perform various functions, including auto-catalytic amplification (Doherty and Doudna, 2001). In vitro, ribozymes have been demonstrated to drastically increase in reaction selectivity and yields when functionally selected (Tsang and Joyce, 1996). Within a few dozens of generations—with generation times that can vary from minutes to decades depending on substrate availability—ribozymes can improve from minimal to maximal reactivity (i.e., Tsang and Joyce, 1996). This renders them suitable candidates for an ancestral replicator, but it also raises a crucial question: if the ancestral replicator was a ribozyme in

an RNA-world, efficiently catalyzing its own amplification, why is the genetic material in all biotans (and candidate bioentities) stored in regular DNA or RNA that is replicated solely by proteins? The enhanced stability of DNA relative to RNA might explain selection for a different information-storage medium (i.e., if RNA was the initial storage medium), but this scenario has little empirical support. We will focus hereafter on the questions that must be answered to better understand the nature and origin of this largest clade on Earth.

The intertwined relationship of DNA, RNA, and proteins in biotans, also known as the central dogma of molecular biology (Crick, 1970), suggests strictly selected molecular coevolution of these compounds. Thus, rather than assuming that the ancestral pan-biotan replicator was formed by just one compound, it seems more plausible to regard it as having originated as a complex composed of a replication substrate and a replication catalyzer. A strand of DNA or RNA would serve as a reasonable substrate for replication that could be catalyzed by RNA (ribozyme) or a protein. In the following we review the key steps culminating in the ur-ancestral pan-biotan replicator and critically assess the requirements necessary and sufficient for the transition from chemistry to biology.

After the initial polymerization of pan-biotan building blocks, two different scenarios can be envisioned that set different stages for the emergence of biological replication: Scenario (1) is compound-exclusive and coincides with the RNA (or DNA)-world hypothesis, according to which only pure nucleic acids (DNA, RNA) formed the ancestral replicator. All other biopolymers (proteins, sugars, lipids) found in biotans are hypothesized to have emerged (spatially and temporally) independently. In this scenario, biological replication could have arisen in the form of a ribozyme catalyzing self-replication, a ribozyme catalyzing the replication of a separate RNA strand, or, a ribozyme catalyzing the replication of a separate DNA strand. Scenario (2) is compound-inclusive; i.e., all single-stranded nucelic acids (DNA, RNA) emerged, as well as peptides with a polymerase function. There is growing evidence for a simultaneous origin of life's building blocks through one-pot chemistry (= all key ingredients react within the same compartment; Patel et al., 2015), which supports the concurrent presence and interaction of different types of short biopolymers during the emergence of biological replication.

All, a few, or only one of these hypothetical ancestral replicators and replicating bioentities may have evolved, but all indications are that only one of them is ancestral to humans (see *Biota*, this volume). Attempts to more accurately identify the ur-ancestral bioentity are plagued by the current lack of an acellular pan-biotan fossil record. Because the ancestral replicator can be thought of as a macromolecule, or macromolecular assembly, in aqueous solution, no morphological residue would be expected (see Briggs and Summons, 2014). DNA and RNA are generally associated with a low fossilization potential, while proteins have been shown to preserve in deep time through chemical transformation into N- and S-heterocyclic polymers (Wiemann et al., 2018). Ancient DNA has been retrieved from samples as old as 1.5 million years but suffers from substantial degradation and alteration (Willerslev et al., 2004). There is thus little reason to expect that there will ever be molecular fossils revealing the nature of the ancestral replicator.

Potential Members of Pan-Biota: The following discussion focuses on the validity of popular candidates as the ur-ancestral replicator.

Ribozymes: The appearance of self-replicating ribonucleic acid (RNA) is widely thought to have been the key innovation in this system

and would fulfil the requirement of biological replication (Gilbert, 1986; Diener, 1989; Maynard Smith and Szathmary, 1995; Neveu et al., 2013; but see Bowman et al., 2015). As long as that RNA replicator directly preceded or contributed to the genetic makeup of biotans, it would be part of *Pan-Biota* (this volume). All extant ribozymes incorporated into biotan cellular systems are parts of *Biota*, even if they evolved independently of the ancestral pan-biotan replicator before biotan cells emerged and subsequently invaded them either as parasites or symbionts.

Viruses: As discussed in *Biota* (this volume), 'viruses' constitute a class of bioentities that parasitize cells (though they can also be viewed as symbionts); there are no data supporting their monophyly but there is considerable evidence indicating that 'viruses' comprise several clades. Two of the three hypotheses for the origins of viruses (Wessner, 2010) propose that they possess genealogical continuity with *Homo sapiens* (Forterre and Prangishvili, 2009). Regardless of whether viruses represent simplified cells, or even less-modified molecular replicators, their survival into the Recent renders them—if not independently evolved—members of crown *Biota* (see extensive discussion in *Biota*). In this case, viruses would also be pan-biotans. Moreover, viral replicators possess the apomorphy required for pan-biotans—the capacity for biological replication (as long as not independently evolved), regardless of the nature of the replication process (i.e., whether autocatalyzed or externally catalyzed).

It remains possible that some viral replicators could have emerged independently on Earth or elsewhere. Hypotheses for independent origins of viral replicators propose that they arose prior to (or in parallel with) the ancestral pan-biotan replicator, in an RNA- or nucleoprotein-world (e.g., Diener, 1989; Forterre, 2006). There, they either coexisted with other nucleic acids (insertions) or used early nucleic acids as hosts, and later infected cellular (pan-)biotans. Under this hypothesis, "viruses" would have been derived from a replicator that evolved reproduction independently of the ancestors of *Biota*, and therefore would not be part of *Pan-Biota*. However, because biological reproduction is a complex process, and is often understood as the key to the success of life on Earth, independent evolution of this trait seems unlikely (see *Biota*, this volume).

DNA, RNA, Their Ancestors and Potential Derivatives: All currently known replicators, or rather replication substrates, present in biotans likely preceded self-sustaining cells or cellular assemblies (= typical organisms). All DNA and RNA bioentities, as well as their ancestors and derivatives, replicate through molecular templating, or under enzyme catalysis, and are therefore—if related to *Homo sapiens*—considered part of *Pan-Biota*.

Dating the Transition from Chemistry to Biology: Accurate dating of the emergence of the clade *Pan-Biota* may prove extremely difficult, given the absence of diagnostic traces, morphological or chemical, of acellular replicators in the fossil record. Recent evidence suggests that by 4.51 billion years ago (from the moon-forming Theia impact onward), the Earth was habitable by pan-biotans, contained liquid water and, contrary to speculation, was unlikely to have been sterilized by heavy bombardments occurring during the Hadean (e.g., Wilde et al., 2001; Abramov and Mojzsis, 2009). There is organic carbon with isotopic signatures consistent with carbon fixation in zircons older than the sedimentary rock record (Bell et al., 2015), and the Earth's oldest sedimentary rocks (ca. 3.8 billion years ago) contain numerous potential if debatable biosignatures (e.g., Schidlowski et al., 1979; Mojzsis et al., 1996; Rosing, 1999; van Zuilen et al., 2002).

However, isotope signatures indicative of carbon fixation have been calibrated based on crown biotan metabolic activity. Simpler, pre-cellular pan-biotans may not have fractionated isotopes in exactly the same way as the compartmentalized anabolic and catabolic systems in cells. Furthermore, pronounced signatures resulting from cellular organisms may not differ among crown and stem members. For this reason, isotopic signatures indicative of carbon fixation cannot reliably distinguish between cellular stem-biotans vs. crown-biotans. There are several generally accepted indications of biological chemistry by 3.5 billion years ago including, most notably, isotopic evidence for carbon and sulfur cycling (e.g., Schopf et al., 2006; Roerdink et al., 2012; Bontognali et al., 2013). These metabolic capacities, however, seem unlikely to have been present in the earliest pre-cellular pan-biotans, as carbon and sulfur cycling requires compartmentalized reaction spaces that can only be generated in a "cellular" environment today.

Nevertheless, minimum age constraints for *Pan-Biota* can be inferred from the fossil record of *Biota*. The body- ("microbial") and trace- (stromatolite) fossil records of this age are difficult to interpret. The "microbial" record certainly reaches back 3.2 billion years with fossils of uncertain phylogenetic affinity (Javaux et al., 2010). However, some degree of cellular organization likely evolved in pan-biotans prior to the origin of *Biota*, and the cellular residues of these putative microbes do not allow an unambiguous assignment to *Biota*. Arguments for a variety of even older body fossils have been made, but their biogenicity is controversial (e.g., Schopf and Packer, 1987; Schopf, 1993; Brasier et al., 2002; Knoll et al., 2016).

In sum, the geologic record indicates the root of *Pan-Biota* may be more than 4 billion years old (but likely younger than 4.51 billion years), is certainly older than 2.3 billion years (conservative estimate for *Biota*), and is probably older than 3.5 billion years ago. The definition of the name *Pan-Biota* may be clear, but as a practical matter, being able to conclusively apply it to potential examples in the rock record billions of years ago may forever lie beyond our grasp.

Literature Cited

Abramov, O., and S. J. Mojzsis. 2009. Microbial habitability of the Hadean Earth during the late heavy bombardment. *Nature* 459:419–422.

Bell, E. A., P. Boehnke, T. M. Harrison, and W. L. Mao. 2015. Potentially biogenic carbon preserved in a 4.1 billion-year-old zircon. *Proc. Natl. Acad. Sci. USA* 112:14518–14521.

Bontognali, T. R. R., A. L. Sessions, A. C. Allwood, W. W. Fischer, J. P. Grotzinger, R. E. Summons, and J. M. Eiler. 2013. Sulfur isotopes or organic matter preserved in 3.45-billion-year-old stromatolites reveal microbial metabolism. *Proc. Natl. Acad. Sci. USA* 109(15):146–151.

Bowman, J. C., N. V. Hud, and L. D. Williams. 2015. The ribosome challenge to the RNA world. *J. Mol. Evol.* 80:143–161.

Brasier, M. D., O. R. Green, A. P. Jephcoat, A. K. Kleppe, M. J. Van Kranendonk, J. F. Lindsey, A. Steele, and N. V. Grassineau. 2002. Questioning the evidence for earth's oldest fossils. *Nature* 416:76–81.

Briggs, D. E., and R. E. Summons. 2014. Ancient biomolecules: their origins, fossilization, and role in revealing the history of life. *BioEssays* 36(5):482–490.

Cornish-Bowden, A., and M. L Cárdenas. 2017. Life before LUCA. *J. Theor. Biol.* 434:68–74.

Crick, F. H. C. 1970. Central dogma of molecular biology. *Nature* 227(5258):561.

de Queiroz, K. 2007. Toward an integrated system of clade names. *Syst. Biol.* 56:956–974.

Diener, T. O. 1989. Circular RNAs: relics of pre-cellular evolution? *Proc. Natl. Acad. Sci. USA* 86(23):9370–9374.

Doherty, E. A., and J. A. Doudna. 2001. Ribozyme structures and mechanisms. *Annu. Rev. Biophys. Biomol. Struct.* 30(1):457–475.

Eschenmoser, A. 1999. Chemical etiology of nucleic acid structure. *Science* 284(5423):2118–2124.

Forterre, P. 2001. Genomics and early cellular evolution. The origin of the DNA world. *C. R. Acad. Sci.* 324(12):1067–1076.

Forterre, P. 2002. The origin of DNA genomes and DNA replication proteins. *Curr. Opin. Microbiol.* 5(5):525–532.

Forterre, P. 2006. The origin of viruses and their possible roles in major evolutionary transitions. *Virus Res.* 117(1):5–16.

Forterre, P., and D. Prangishvili. 2009. The origin of viruses. *Res. Microbiol.* 160(7):466–472.

Gilbert, W. 1986. Origin of life: the RNA world. *Nature* 319:618.

Hu, H. H. 1965. The major groups of living beings: a new classification. *Taxon* 14:254–261.

Hug, L. A., B. B. Baker, K. Anantharaman, C. T. Brown, A. J. Probst, C. J. Castelle, C. N. Butterfield, A. W. Hernsdorf, Y. Y. Amano, K. Ise, Y. Suzuki, N. Dudek, A. A. Relman, K. M. Finstad, R. Amundson, B. C. Thomas, and J. F. Banfield. 2016. A new view of the tree of life. *Nat. Microbiol.* 1:e16048.

Javaux, E. J., C. P. Marshall, and A. Bekker. 2010. Organic-walled microfossils in 3.2 billion-year-old shallow marine siliciclastic deposits. *Nature* 463:934–938.

Jeffrey, C. 1971. Thallophytes and kingdoms: a critique. *Kew Bull.* 25(2):291–299.

Knoll, A. H., K. D. Bergmann, and J. V. Strauss. 2016. Life: the first two billion years. *Philos. Trans. R. Soc. Lond. B Biol. Sci.* 371:e20150493.

Lupi, O., P. Dadalti, L. E. Cruz, P. R. Sanbergd, and Cryopraxis' Task Force for Prion Research. 2006. Are prions related to the emergence of early life? *Med. Hypotheses* 67(5):1027–1033.

Maynard Smith, J., and E. Szathmáry. 1995. *The Major Transitions in Evolution*. W. H. Freeman, New York.

Mojzsis, S. J., G. Arrhenius, K. D. McKeegan, T. M. Harrison, A. P. Nutman, and C. R. L. Friend. 1996. Evidence for life on Earth before 3,800 million years ago. *Nature* 384:55–59.

Neveu, M., H. J. Kim, and S. A. Benner. 2013. The "strong" RNA world hypothesis: fifty years old. *Astrobiology* 13(4):391–403.

Nielsen, P. E., and M. Egholm. 1999. An introduction to peptide nucleic acid. *Curr. Issues Mol. Biol.* 1(1–2):89–104.

Novák, F. A. 1930. *Systematická Botanika.* Aventinum, Praha.

Orgel, L. E. 2003. Some consequences of the RNA world hypothesis. *Origins Life Evol. B.* 33(2):211–218.

Patel, B. H., C. Perciville, D. J. Ritson, C. D. Duffy, and J. D. Sutherland. 2015. Common origins of RNA, protein and lipid precursors in a cyanosulfidic protometabolism. *Nat. Chem.* 7(4):301–307.

Pruisner, S. B. 1998. Prions (Nobel prize lecture). *Proc. Natl. Acad. Sci. USA* 95:13363.

Rode, B. M., W. Flader, C. Sotriffer, and A. Righi. 1999. Are prions a relic of an early stage of peptide evolution? *Peptides* 20(12):1513–1516.

Roerdink, D. L., P. R. D. Mason, J. Farqhuar, and T. Reimer. 2012. Multiple sulfur isotopes in Paleoarchean barites identify an important role for microbial sulfate reduction in the early marine environment. *Earth Planet. Sci. Lett.* 331:177–186.

Rosing, M. T. 1999. ^{13}C-depleted carbon microparticles in >3700-Ma sea-floor sedimentary rocks from west Greenland. *Science* 283:674–676.

Schidlowski, M., P. W. U. Appel, R. Eichmann, and C. E. Junge. 1979. Carbon isotope geochemistry of the 3.7×10^9-yr-old Isua sediments, West Greenland: implications for the Archaean carbon and oxygen cycles. *Geochim. Cosmochim. Acta* 43(2):189–199.

Schopf, J. W. 1993. Microfossils of the early Archean Apex Chert—new evidence for the antiquity of life. *Science* 260:640–646.

Schopf, J. W. 2006. Fossil evidence of Archean life. *Philos. Trans. R. Soc. Lond. B Biol. Sci.* 361:869–885.

Schopf, J. W., and B. M. Packer. 1987. Early Archean (3.3 billion to 3.5 billion-year-old) microfossils from Warrawoona Group, Australia. *Science* 237:70–73.

Theobald, D. L. 2010. A formal test of the theory of universal common ancestry. *Nature* 465:219–222.

Trifonov, E. N., and E. Kejnovsky. 2015. *Acytota* – associated kingdom of neglected life. *J. Biomol. Struct. Dyn.* 34(8):1641–1648.

Tsang, J., and G. F. Joyce. 1996. In vitro evolution of randomized ribozymes. *Methods Enzymol.* 267:410–426.

Van Zuilen, M. A., A. Lepland, and G. Arrhenius. 2002. Reassessing the evidence for the earliest traces of life. *Nature* 418:627–630.

Wagner, J. R. 2004. The general case of phylogenetic definitions, alternate classes of definitions, and the phylogenetic definition of life. P. 8 in *Abstracts of the First International Phylogenetic Nomenclature Meeting,* Paris. Available at https://www.phylocode.org and https://hal.sorbonne-universite.fr/hal-02187647.

Wessner, D. R. 2010. The origins of viruses. *Nat. Educ.* 3(9):37.

Wiemann, J., M. Fabbri, T. R. Yang, K. Stein, P. M. Sander, M. A. Norell, and D. E. Briggs. 2018. Fossilization transforms vertebrate hard tissue proteins into N-heterocyclic polymers. *Nat. Commun.* 9(1):e4741.

Wilde, S. A., J. W. Valley, W. H. Peck, and C. M. Graham. 2001. Evidence from detrital zircons for the existence of continental crust and oceans on the Earth 4.4 Gyr ago. *Nature* 409:175–178.

Willerslev, E., A. J. Hansen, R. Rønn, T. B. Brand, I. Barnes, C. Wiuf, D. A. Gilichinsky, D. Mitchell, and A. Cooper. 2004. Long-term persistence of bacterial DNA. *Curr. Biol.* 14(1):R9–R10.

Authors

Jasmina Wiemann; Department of Geology and Geophysics; Yale University; 210 Whitney Avenue; New Haven, CT 06511, USA. Email: jasmina.wiemann@yale.edu.

Kevin de Queiroz; Department of Vertebrate Zoology; National Museum of Natural History, Smithsonian Institution; Washington, DC 20560, USA. Email: dequeirozk@si.edu.

Timothy B. Rowe; Jackson School of Geosciences, University of Texas at Austin; C1100, Austin, TX 78712, USA. Email: rowe@mail.utexas.edu.

Noah J. Planavsky; Department of Geology and Geophysics; Yale University; 210 Whitney Avenue; New Haven, CT 06511, USA. Email: noah.planavsky@yale.edu.

Ross P. Anderson; Department of Earth Sciences; University of Oxford; Oxford, OX1 3AN, UK. Email: ross.anderson@all-souls.ox.ac.uk.

Johann P. Gogarten; Department of Molecular and Cell Biology; University of Connecticut; Storrs, CT 06269, USA. Email: gogarten@uconn.edu.

Paul E. Turner; Department of Ecology and Evolutionary Biology; Yale University; P. O. Box 208106, New Haven, CT 06520, USA. E-mail: paul.turner@yale.edu.

Jacques A. Gauthier; Department of Geology and Geophysics; Yale University; 210 Whitney Avenue; New Haven, CT 06511, USA. Email: jacques.gauthier@yale.edu.

Date Accepted: 19 July 2019

Primary Editor: Philip Cantino

Biota J. Wagner 2004 [J. Wiemann, K. de Queiroz, T. B. Rowe, N. J. Planavsky, R. P. Anderson, J. P. Gogarten, P. E. Turner, and J. A. Gauthier], converted clade name

Registration Number: 298

Definition: The largest crown clade containing *Homo sapiens* Linnaeus 1758. This is a special case of the maximum-crown-clade definition in that it does not use an external specifier (see Comments); it refers to the crown clade including humans and all other bioentities sharing common ancestry with them (see Comments). Abbreviated definition: max crown ∇ (*Homo sapiens* Linnaeus 1758).

Etymology: Latinized from Ancient Greek βίος (syncope of βίοτος) meaning "life".

Reference Phylogeny: Figure 1 in Hug et al. (2016), in which *Homo sapiens* is part of the clade named *Opisthokonta* (and *Biota* includes all taxa in that tree). See also Lake et al. (1984, "Eocytic Tree" in "Note Added in Proof" on p. 3790), Woese et al. (1990: Fig. 1), Pace (2006), Williams et al. (2012: Fig. 1a), Gouy et al. (2015: Fig. 3), and Hinchliff et al. (2015: Fig. 1).

Composition: Assuming a monophyletic origin of life on Earth (Theobald, 2010), this clade includes the last universal common ancestor (LUCA) of all "living" biological entities (Penny and Poole, 1999) on Earth, and all of its descendants, both extant and extinct, on this planet or anywhere else in the universe. Accumulating evidence for multiple derivations of the building blocks of *Biota* (see *Pan-Biota*, this volume) does not contradict a single origin of *Biota*. Moreover, even should multiple origins of life (on Earth or elsewhere) be demonstrated, if those life-forms do not share a genealogical connection to *Homo sapiens*, then they would not be members of *Biota* (see *Pan-Biota*, this volume).

At our current state of knowledge, *Biota* is generally thought to be composed of three major groups: *Bacteria*, *Archaea*, and *Eukarya* (= BAE), the first two of which might be paraphyla rather than clades (see Comments). In contrast, *Eukarya* (or its synonym *Eukaryota*) has always been associated with a major branch of the tree of life that is now universally regarded as a well-supported clade. Several new taxa of uncertain relations have recently been discovered; they are either deeply imbedded in *Bacteria* (Hug et al., 2016: Suppl. Fig. 2) or join the rest of *Biota* near the putative root of the biotan tree (i.e., some could be sister to all other biotans) (Hug et al., 2016: Suppl. Fig. 1). There are additional sets of (potential) biological entities, or parts of entities, that, while not themselves organisms as traditionally conceived (see Comments), are composed of potentially clade-forming entities that are often discussed when considering the transition from chemistry to biology (i.e., the origin of life), and might thus be parts of *Biota*: viruses, transposons, and nanobacteria (e.g., Trifonov and Kejnovsky, 2015; Kejnovsky and Trifonov, 2016; and see Comments). Extant viruses and transposons represent likely non-monophyletic groups of molecular parasites that seem to be parts of *Pan-Biota*, but not necessarily, or only in part, members of *Biota* (see Comments). Terrestrial and Martian nanobacteria have been suggested as members of *Biota*, posing a significant challenge to understanding the biogeography of early life. Because the phylogeny and composition of most of these candidate biological entities (bioentities) are not well understood, we shall review current insights on each one of them and discuss their significance, composition, and phylogenetic placement in the Comments sections of the contributions on *Biota* and *Pan-Biota*.

If all extant viruses and transposons represent degenerate derivatives of the last common ancestor of bacteria, archaeans, and eukaryotes (BAE), then they are parts of *Biota* as currently understood. If some or all of those bioentities are derivatives of cellular pan-biotans outside of the BAE crown, or even earlier-diverging pre-cellular assemblies that share common ancestry with BAE, they would still be parts of *Biota* as defined here, but that name would apply to a more inclusive clade than the BAE crown. Moreover, if some or all viruses and transposons are found to derive from the other branch stemming from the very earliest divergence of replicators leading to the crown, the name *Biota* would become synonymous with *Pan-Biota*. In each case, the Composition and Diagnosis of *Biota* would change accordingly.

Diagnostic Apomorphies: The definition of the taxon name *Biota* is clear in theory—it refers to the maximum crown clade containing humans, that is, to all descendants of the last common ancestor of all life on Earth today. In practical terms, however, the exact composition of this clade remains elusive, and its diagnosis accordingly unclear. Given that there is still much to be learned about the microorganisms of our planet, there is a possibility that the basal-most crown divergence in the tree of life on Earth has yet to be discovered (see, e.g., "pandoraviruses" sensu Philippe et al., 2013; but see contrary view by Koonin and Yutin, 2018). The diagnosis of *Biota* is further complicated by uncertainty regarding potential phylogenetic relationships of some classes of bioentities, such as viruses and transposons, which may

be secondarily simplified molecular parasites derived from cellular (or even pre-cellular) life-forms (see Comments).

We accordingly focus on the diagnosis of the clade stemming from the last common ancestor that *Homo sapiens* (*Eukarya*) shared with *Halobacterium salinarum* (*Archaea*) and *Escherichia coli* (*Bacteria*), fully cognizant that at least some and probably all of these apomorphies could have arisen well before that ancestor (see Comments below and *Pan-Biota*, this volume). If some or all viruses (or mobile genetic elements) turn out to be members of the BAE clade (see Composition), then they must have lost (most of) these diagnostic features.

The ancestral biotan cell reproduced by binary fission (see *Pan-Biota,* this volume). It was enclosed in an envelope composed of lipids and proteins, containing proteinaceous transmembrane units that facilitate chemosensing of environmental stimuli and endergonic osmoregulation, and chemiosmotic coupling (Oró et al., 1990; Gogarten and Taiz, 1992; Orgel, 1994; Maynard Smith and Szathmáry, 1995; Sára and Sleytr, 2000; Klug et al., 2017; Rodrigues-Oliveira et al., 2017). It stored all vital biomolecules, i.e., amino acid-based proteins (organismic function and structure), fatty acid-based lipids (organismic structure), ribonucleic acids (organismic function), deoxyribonucleic acids (organismic function), and most, if not all, of their monomeric building blocks in a water-based medium (Oró et al., 1990; Orgel, 1994). Unlike some molecular assemblies (e.g., some nanobacteria), the ancestral biotan was capable of growth and reproduction via internal biosynthesis including ribosomes translating genes into proteins, and higher-level compound modification (= metabolism), that responded to chemo-sensed environmental stimuli through intragenerational metabolic adjustments (= physiological adaptation) and intergenerational modifications (= evolution) (Maynard Smith and Szathmáry, 1995, 1997). The ancestral biotan cell stored its DNA wrapped around proteins in chromosomes and contained a set of at least 355 genes (Martin et al., 2016).

Synonyms: Approximate (and partial if descendant viruses, mobile genetic elements, and nanobacteria are excluded) synonyms: *Zotica* Spinola 1850; *Bionta* Walton 1930; *Cellulata* Vorontsov 1965; *Cytobiota* Hu 1965; *Cytota* and *Acytota* Trifonov and Kejnovsky 2015. Numerous partial synonyms refer to a paraphyletic taxon originating in approximately the same ancestor, including *Monera* Haeckel 1866, *Protista* Haeckel 1866 (but not as used by Copeland, 1938, and others), *Bacteria* Haeckel 1894, *Prokaryota* Swain 1969, and *Procytota* Jeffrey 1971 (see https://species.wikimedia.org/wiki/Prokaryota for a more complete list).

Comments: Given the fundamental significance of the most inclusive crown clade of life on Earth, it is surprising that the literature records so few instances in which it has been referred to explicitly by name, rather than being described in terms of the properties thought necessary to its existence—i.e., the often vague concept of "being alive" (see *Pan-Biota*, this volume)—or referred to by implication, for example, through reference to LUCA, the last universal common ancestor (Woese, 1999). By contrast, an inordinate amount of literature has been devoted to how many primary subtaxa should be recognized within this (seldom-named) taxon, and which traditional categorical ranks (e.g., Domain, Kingdom, Phylum) should be assigned to them (e.g., Blackwell, 2004, and references therein). That said, there are a number of taxon names that could arguably be applied to the crown clade that contains all bioentities originating on Earth (related to humans). Most of these candidate names were proposed explicitly for

the paraphylum composed of cells lacking a membrane-bound nucleus (first distinguished by Chatton, 1925; see also Chatton, 1937/1938; though not christened formally with the name "*Prokaryota*" until Swain, 1969). Also, there are a number of candidate names associated with paraphyla that explicitly exclude multicellular organisms (e.g., *Protista* Haeckel 1866; in the twentieth century, this name became associated with unicellular eukaryotic organisms).

There is a much shorter list of taxon names, including *Zotica* Spinola 1850, *Bionta* Walton 1930, *Cytobiota* Hu 1965, *Cellulata* Vorontsov 1965, and *Cytota* Trifonov and Kejnovsky 2015 proposed for a group of bioentities including all cells, whether or not they have a membrane-bound nucleus, as well as both unicellular and multicellular organisms (although still potentially paraphyletic as this group often excluded viruses). Most biologists are unlikely to be familiar with most of these names for they have seldom been used since they were coined.

Biota is an exception, albeit best known as a common noun rather than as a taxon name. Stejneger (1901: 89) first proposed "biota" "as a comprehensive term to include both fauna and flora that will not only designate all animal and plant life of a given region or period, but also any treatise upon the animals and plants of any geographical area or geological period." Following Stejneger, its general use has been as a common noun and "biota" is usually qualified with an adjective (e.g., marine biota, extinct biota), with a prefix (e.g., microbiota, symbiota), or in combination with a possessive (e.g., biota of Brazil). Originally focused on multicellular organisms (i.e., terrestrial flora and fauna), it has since been expanded to include microbes in diverse settings (e.g., marine microbiota or gut biota). Nevertheless, when used as a common noun, "biota" generally refers to some subset of the maximum-crown-clade that is usually not itself a clade.

The term appears only recently to have been employed as a proper noun referring to the taxon composed of all bioentities on Earth. *Biota* was apparently first used in its taxonomic sense in a website (Brands, 1989–2005), then in an abstract volume produced for the Paris meeting of the International Society of Phylogenetic Nomenclature (Wagner, 2004), and later still in a self-published book (Pelletier, 2015). The earliest peer-reviewed journal article using *Biota* as a formal taxon name appears to be that of Trifonov and Kejnovsky (2015; see also Kejnovsky and Trifonov, 2016) (although Hu [1965] also used the term "biota" for "an immense group of living beings" [p. 255; including all viruses regardless of their likely polyphyly]). Given that Wagner (2004) was the first to apply *Biota* as a formal taxon name to this clade in a printed work (which can be downloaded and printed at the phylocode.org website), thereby qualifying it as "in use" and therefore a preexisting name under Article 6.2(b) of the *PhyloCode*, Wagner (2004) is here regarded as the nominal author of *Biota*.

Selecting *Biota* from among the seldom-used names available for this clade takes advantage of the implicit connection between this familiar term for "living beings" and the most inclusive crown clade originating on Earth. Whether used in its ecological or taxonomic sense, expressions such as the "biota" of South America or "biota" of the Burgess Shale would remain perfectly intelligible as referring to all organisms from those times and places. Moreover, as Stejneger noted, "*Biota*" has an obvious (etymological) connection to "biology", the name for the science of life. We accordingly take this opportunity to propose an explicit phylogenetic definition for the taxon name *Biota* using a modification of the form of definition proposed by Wagner (2004), i.e., a maximum-crown-clade definition that does not employ an external specifier. This departure from the standard form of

a maximum-crown-clade definition reflects the exceptional circumstance that the name *Biota* is being applied to *the most* inclusive crown clade on this planet, rendering an external specifier superfluous to limiting the inclusiveness (circumscription) of the clade.

The single internal specifier, *Homo sapiens*, possesses all of the qualities most desirable in (if not required of) a nomenclatural specifier: it is composed of abundant, large organisms that are easily dissected, thoroughly studied and exhaustively illustrated from histological to gross anatomical levels, well-known ontogenetically, physiologically, and behaviorally, and it has a completely sequenced genome and an informative fossil record. It is also deeply imbedded within the named clade, thus buffering the name against unintended taxonomic consequences caused by potential changes in tree topology (e.g., Sereno, 1998). All extant organisms are no less distant in time than *Homo sapiens* is from the most recent common ancestor (= ur-ancestor) of *Biota*, were no less subject to the cumulative contingencies of evolution than we were, and are no less interesting and informative in their own rights (Gee, 2013). Nonetheless, biology is fundamentally and uniquely a human enterprise, as is the practice of coining taxon names to enhance cognitive efficiency (Mervis and Rosch, 1981) as humans communicate about *Biota* and its members. Thus, our specifier choice is not anthropocentric in a prejudicial or pejorative sense; rather, it employs the one and only species in all of *Biota* that is universally known to those creatures who study it, while underscoring that humanity's place in nature derives from a phylogenetic connection with all other life on Earth.

By drawing a clear distinction between the largest crown clade containing *Homo sapiens* (*Biota*) and its corresponding total clade (*Pan-Biota*, this volume), we hope to sharpen the focus of questions regarding the origin of the last common ancestor of all extant replicators related to *H. sapiens*, the relationships of disparate "non-cellular" bioentities (e.g., viruses), and the origin of biological replication itself (sensu *Pan-Biota*).

Tying the taxon name *Biota* to the maximum crown clade containing *Homo sapiens* ensures that all extant organisms and other (less complex) extant bioentities sharing common ancestry with them will be parts of that clade. Nevertheless, ideas about the circumscription of the clade *Biota* may expand stem-ward in the future as new bioentities are discovered, and if some or all debated members of *Biota*, such as viruses and mobile genetic elements, are inferred to have emerged from bioentities that diverged before the last ancestor shared by *E. coli*, *H. salinarum*, and *H. sapiens*, and persisted to the Recent by interacting symbiotically with the descendants of that ancestor (see below). The basalmost possible phylogenetic position of the ancestor in which *Biota* originated would make this clade-name synonymous with *Pan-Biota* (this volume).

In the context of our definition for the name *Biota* and its proposed diagnosis, we now consider more controversial members of *Biota*, such as viruses, transposons, and nanobacteria, which are frequently discussed in modern literature addressing the origin of life. The *PhyloCode* governs clade nomenclature, and clades are typically thought to be composed of integrated bioentities (e.g., organisms) that are self-organizing, self-replicating, and self-sustaining; we accept that none of the controversial candidate members of *Biota* are "organisms" in that sense (Moreira and López-García, 2009). Nevertheless, we see no compelling reason to exclude them *a priori* from *Biota* as a clade (i.e., a common ancestor and *all* of its descendants), especially as they could represent simplified symbiotic offshoots of cellular (or pre-cellular) organisms (e.g., viruses) or their genomes (e.g., transposons and other mobile

genetic elements). The following is a summary of key features of each of these potential members of *Biota*, a critical discussion of their significance to the origin of life, and an overview of current hypotheses concerning their phylogenetic placement, and how their inclusion in *Biota* might affect ideas about the composition of this clade now or in the future.

"Virus" is a functional term for a disparate array of DNA- or RNA-based replicators that are, with few exceptions (see below), enclosed in a proteinaceous sheath called a capsid; the term does not refer to a taxon as such (Van Regenmortel, 2003). These cell-reliant bioentities are the most abundant denizens of Earth, with at least 10 individual virus particles for every cell (Brüssow, 2009) (note also that your body contains about as many bacterial as human cells [Sender et al., 2016]), and have been found everywhere their hosts can survive, even in the most hostile environments (e.g., Sahara Desert; Prestel et al., 2012). Viral genomes typically contain two functional modules: those genes governing genome replication and those regulating capsid self-assembly (Krupovic and Koonin, 2017a,b). Indeed, bacteriophage genomes and capsids are often assembled separately in the cytoplasm of their hosts, where they must encounter one another in order for the former to insert into the latter. In that sense, viruses appear to be composite bioentities, analogous to eukaryotes (or, for that matter, all cells). As Krupovic and Koonin (2017a) argued, the capsid is the key innovation distinguishing viruses from among other "selfish" genetic elements, protecting their replicators in extracellular environments and enabling them to attach to and enter host cells.

Viruses are often explicitly excluded from taxa composed of typical (cellular) organisms bearing the diagnostic features noted above (see Diagnosis), including *Bionta* Walton 1930, *Cellulata* Vorontsov 1965, and *Cytobiota*

Hu 1965 (viruses had yet to be discovered when *Zotica* Spinola 1850 was proposed). In contrast, Trifonov and Kejnovsky (2015; see also Hu, 1965) explicitly included viruses in their *Biota* (as "living beings" if not as "organisms", the latter of which were included in their purportedly less-inclusive *Cytota* Trifonov and Kejnovsky 2015; see also Kejnovsky and Trifonov, 2016). It is not clear whether some or all viruses represent incipient or vestigial organisms, or whether they are rogue bits of genomes that escaped from cellular life-forms (see, e.g., Luria and Darnell, 1967; Podolsky, 1996; Forterre, 2006; Nasir and Caetano-Anollés, 2015), or some combination of these scenarios (e.g., Krupovic and Koonin, 2017a,b).

Viruses are readily distinguished from all other bioentities; for example, their capsids are morphologically and compositionally distinct from cell membranes, and their genome-replicating proteins lack any clear homologs among cellular life-forms (Krupovic and Koonin, 2017b). They are unable to replicate without their hosts, and for that reason it has often been questioned whether viruses are "alive" (see Moreira and López-García, 2009 and López-García and Moreira, 2009, and references therein). Such reasoning seems flawed, however, as many cellular parasites are unable to replicate without their hosts (e.g., *Mycobacterium tuberculosis*), yet they are nonetheless commonly considered to be "alive" (Hegde et al., 2009; Nasir and Caetano-Anollés, 2015). Granted that viruses differ from cells in that they are "unable to transform energy and matter (that is, to actively generate order from disorder)" (López-García and Moreira, 2009:15). But cellular genomes are likewise unable to accomplish those activities on their own and do not differ from viral genomes in that respect; both genomes require access to the elaborate metabolic machinery—including other genes, transcription products, proteins, and metabolites—found only inside membrane-bound cells, in order to reproduce themselves. Although questions about the origin of life remain subject to lively debate (e.g., Root-Bernstein and Root-Bernstein, 2015), our more modest goal is to consider whether one or more viral clades could be descended from uncontroversial "living" bioentities, and thus be part(s) of *Biota*. All viruses sharing ancestry with *Homo sapiens* are by definition part of *Pan-Biota* because they fulfill the requirement of "biological replication" (if that evolved only once), viz., the production of descendants that are themselves able to reproduce (see *Pan-Biota*, this volume).

Inferring phylogenetic relationships among viruses and assessing their place(s) in the tree of life have proven challenging. The viral lifestyle could have originated more than 3.5 billion years ago (see below) and viral "generation times" (from entry to exit from host) can be measured in minutes. As a consequence, viruses can achieve extraordinary rates of evolutionary change that, combined with high mutation rates, can result in remarkably divergent gene sequences, problems for sequence alignment, and a propensity for long-branch attraction, all of which can confound phylogenetic inference. Patterns of viral descent can be obfuscated by additional phenomena, such as horizontal gene transfer from host to virus, and further complicated by horizontal virus transfer among distantly related hosts (e.g., Dolja and Koonin, 2018). Despite these difficulties, several well-supported viral clades have been recognized in recent years, such as bacteriophage DNA viruses (Krupovic and Koonin, 2017a; Yang et al., 2018), eukaryote RNA viruses (Krupovic and Koonin, 2017b; Wolf et al., 2018), and the therian mammal *Lentivirus* clade (Nakano et al., 2017; which includes the subclade SIV and its subclade HIV [Sharp and Hahn, 2011]). No evidence has yet emerged indicating that all viruses comprise a single branch of the tree of life (indeed, no single sequenceable biomolecule is shared by all viruses). Moreover, some viral capsid proteins appear to derive from proteins taken from different major subclades of cellular life-forms. This suggests that viruses—mobile genetic elements that can travel from cell to cell encased in protective sheaths derived from proteins transferred horizontally from preexisting cells—emerged on at least 20 separate occasions in the history of the biotan total clade (Krupovic and Koonin, 2017a; see also Gladyshev and Arkhipova, 2011).

It would be a mistake to regard any of these viruses as not being referable to *Biota* because they are thought, in some sense, not to be "alive". Crown caecilians (vermiform amphibians), for example, are still part of *Tetrapoda* even though they lack any vestiges of this clade's diagnostic four limbs (even as embryos). Similarly, even if viruses lack one, many, or all of the properties described in our Diagnosis, according to our definition, any given Recent virus stemming ultimately from a bioentity that shares ancestry with human cells would still be part of *Biota*, even if it diverged before the origin of the ancestral cell, or before the last cell shared by crown BAE (e.g., Nasir and Caetano-Anollés, 2015). Conversely, if a viral replicator originated independently (e.g., in an "RNA world" sensu Gilbert, 1986; Diener, 1989; Wolf et al., 2018), so that its genome did not share ancestry with *Homo sapiens* (e.g., Koonin and Dolja, 2014), then that virus would not be part of *Biota* (or *Pan-Biota*). That being said, if that independent viral line shared genes acquired from cellular life-forms via horizontal transfer (e.g., Krupovic and Koonin, 2017a), then there are some senses in which it could be considered part of *Biota* (*PhyloCode* Note 2.1.3). If an ancient pan-biotan virus parasitized some non-BAE host, it would only be part of *Biota* if that host was a direct ancestor of BAE, or if the virus later transferred horizontally to a BAE cell (and the virus

survived to the Recent). On that note, some so-called "naked viruses" may well have evolved before the emergence of cellular life, and persisted in solution until they launched an evolutionary arms race with cellular bioentities (that likely predate the complex cell shared by the last common ancestor of BAE). If, as suggested by Koonin and Dolja (2014), viruses emerged from selfish genetic elements (and *vice versa*) on multiple occasions during evolution, then their inclusion in *Biota* would depend on the genealogical connections of those selfish genetic elements: do they ultimately share ancestry with uncontroversial biotan DNA-based genomes or not? Pinpointing the exact phylogenetic placement of the *Biota* node requires a clearer understanding of the origins and evolution of viral bioentities.

Transposons (short for "transposable elements"; also known as "jumping genes") do not comprise a taxon, but instead represent a functional class of DNA- or RNA-based replicators that can change position within the genome (transposition) through excision and insertion into a target DNA sequence (McClintock, 1950). These clade-forming bioentities, like other mobile genetic elements (MGEs, including self-splicing inteins and introns) spread—by both vertical and horizontal transfer—across the tree of life and can be grouped by various criteria, such as the distinctive catalytic mechanisms they employ during transposition (e.g., McDowall, 2006). Their propensity to proliferate within genomes—e.g., they comprise 42% of the human genome (Lander et al., 2001) and 85% of the maize genome (Schnable et al., 2009)—often with deleterious consequences, led to transposons being viewed as "selfish" genome parasites (in contrast to cell-parasitizing viruses). Their abundance within genomes doubtless creates more opportunities for horizontal gene transfer, especially by superabundant retrotransposons in eukaryotes (Kidwell, 1992), that not only jump about within their genomes, but also generate numerous copies of themselves within the genome. Nevertheless, there is growing evidence that transposons may play important roles in the evolution of gene regulation in biotans (e.g., Wagner, 2014; Lanciano and Mirouze, 2018), offering new insights into how the functional morphology of the genome itself evolved (e.g., Britten and Davidson, 1971; Biémont, 2010).

Like viruses, mobile genetic elements are entirely dependent upon their hosts' metabolism and cannot replicate except inside cells (Kidwell, 1992; Villarreal, 2005). Questions about the evolutionary origin(s) of one class of MGEs, transposons, arose when some were found to share aspects of genome form and function with viruses (e.g., Villarreal, 2005). In particular, retroviral gene sequences have been identified in MGEs in eukaryotes that lack capsids and for that reason they have been regarded by some as "retrotransposons" (Koonin and Dolja, 2014). This suggested a possible evolutionary connection between these reverse-transcribing MGEs, with retrotransposons acquiring protein-coding genes from eukaryotic hosts to construct the viral capsids enabling these mobile genetic elements to move not just within cells, but also between them. This transition could have worked both ways, with retrotransposons potentially giving rise to retroviruses and *vice versa* on multiple occasions. However, Krupovic and Koonin (2017b) inferred that the capsidless "retrotransposons" currently known in *Eukarya*, such as non-mobile genetic elements residing exclusively in mitochondria, are actually secondarily simplified retroviruses, as they appear deeply imbedded within a very large eukaryote-wide clade of RNA-based viruses with fully developed viral capsids. An alternative hypothesis by Hickson (1989) suggests that some transposable genetic elements may have emerged from group-II-introns, short genetic elements capable of self-excision from their DNA strands.

Ideas about the evolutionary origins of transposons became intertwined with those of viruses. Nevertheless, whether transposons, or any other mobile genetic element, arose on one or more occasions before, during, or after the ancestral BAE cell first appeared, no one has yet presented any evidence that they are composed of anything but standard genes, such as those encoding transcriptase, common to all biotan cells (however modified these mobile elements might be compared to typical "immobile" genes). Thus, all transposons found in uncontroversial biotan genomes, regardless of their evolutionary origins, must be regarded as parts of *Biota* (and *Pan-Biota*).

Nanobacteria pose another challenge. As with "virus", the term does not apply to a taxon; it refers instead to tiny, spherical assemblies of biomolecules that attract mineral precipitation onto their surfaces. Comparable structures are widely distributed, having been extracted from the lithosphere, human kidney stones and atherosclerotic plaques, as well as from a Martian meteorite (as fossil remains). Whether nanobacteria (originally described as "ultramicrobacteria" by Torrella and Morita, 1981) can be considered "alive", and therefore potentially part of *Biota*, is controversial. Terrestrial and putative Martian nanobacteria (McKay et al., 1996; Çiftçioğlu et al., 2005; Martel and Young, 2008) share a mineral-encapsulated, multilayered organic sphere of very small size (~200 nm circumference), and are allegedly able to replicate this morphology (Sommer et al., 2003; Çiftçioğlu et al., 2005; Martel and Young, 2008) (although this could result from initiation of mineral precipitation, as binary fission has yet to be observed). The idea that nanobacteria are "alive" has been challenged by a fundamental reinterpretation of these structures as organic, nanoparticulate agglomerates that facilitate secondary mineral precipitation via Coulomb effects, or coordination chemistry with surrounding dissolved ions (Cisar et al., 2000; Martel and Young., 2008). Nanobacteria extracted from biotans stain positively for DNA (which might be the product of probe/instrument contamination) (Cisar et al., 2000). However, putative Martian nanobacteria contain only polycyclic aromatic hydrocarbons that could still be fossilization products of original biomolecules (McKay et al., 1996; Wiemann et al., 2018). Therefore, nanobacteria would only be part of *Biota* if they descended from a pre-cellular ancestor of BAE cells, or are derived from secondarily simplified BAE cells. Or they might not be taxa *per se*, but merely represent biotan cellular debris acting as nuclei for mineral deposition, as might be the case for examples extracted from biotan host tissues.

While nanobacteria within human hosts could be part of crown *Biota* (e.g., Çiftçioğlu et al., 1999), putative Martian nanobacteria pose a special problem: their fossil nature suggests that these potentially biological entities existed multiple millions of years before present, well before any possibility of anthropogenic contamination of Earth's adjacent satellite bodies and planets. In that case, Martian nanobacteria might represent either non-anthropogenic contaminants from Earth, transported through whatever means (e.g., by impact debris ejected into space from asteroid collisions). Or they could represent a separate origin of extraterrestrial life, in which case they would not be part of *Biota* as defined here (see *Pan-Biota*, this volume). Alternatively, they could share a common ancestral replicator with *Homo sapiens*—transported either from Earth, from Mars, or to both planets from elsewhere—and thus be members of *Biota* or non-biotan *Pan-Biota*. Or they could just be non-living organic agglomerates arising through abiotic chemistry, and having nothing to do with *Pan-Biota* (although they

could represent non-replicating mineralized biological detritus produced by members of that clade).

One reason for the numerous controversies around these potential candidate members of *Biota* is the fact that the biotan tree is difficult to root (e.g., Hug et al., 2016). It is generally thought that first *Archaea* (whether a clade or paraphylum), and then *Bacteria* (whether a clade or paraphylum), are successive sisters to *Eukarya* (the monophyly of which has never seriously been questioned since Chatton [1925] recognized their foremost shared apomorphies: a membrane-bound nucleus and mitochondria)—in other words, the root of *Biota* is widely held to lie either within (paraphyletic) "*Bacteria*" or between (monophyletic) *Bacteria* on the one hand, and an "*Archaea*" + *Eukarya* clade on the other (e.g., Gogarten et al., 1989; Iwabe et al., 1989; Ciccarelli et al., 2006; Fournier and Gogarten, 2010). As was the case for discovering the clade *Eukarya*, the idea that some "archaeans" are closer to eukaryotes than are others (the "Eocyte Hypothesis") was based initially on (micro)morphology (Lake et al., 1984), and both hypotheses have been corroborated in several recent gene-sequence-based studies (e.g., Williams et al., 2013; Gouy et al., 2015; Hug et al., 2016; Betts et al., 2018). Unfortunately, none of the potential candidate members of *Biota* considered here were included in any of these phylogenetic analyses due to biomolecular incompatibilities, or the current lack of phylogenetically informative molecular residues in their fossil representatives.

Based on our current understanding of *Biota*, it is thought that the ancestral biotan was heterotrophic (Blankenship, 2010; see also Betts et al., 2018) and must have evolved in an aqueous solution containing essential sodium, potassium, magnesium, chloride, phosphate, carbonate, sulfate, nitrate, transition metal ions, and the dissolved organic molecules upon which it fed (e.g., Hunding et al., 2006). This ancestral habitat would not necessarily have to be located in the photic zone, as the primary environmental cue governing metabolic rhythmicity ancestral for *Biota* (e.g., convection—based on temperature change in the upper water layers, and/or UV-light-based chemical reactivity and redox gradients) could be sensed even in the aphotic zone. Biotans can be found today all over Earth's surface and shallow subsurface, and in its hydrosphere and atmosphere. Multiple species can now also be found in some of Earth's adjacent satellite bodies through space-travel-mediated transport of anthropogenic contaminants (Novikova et al., 2006).

Microbial fossils from the 1.9 billion-year-old Belcher Supergroup in Canada (Hofmann, 1976; Golubic and Hofmann, 1976) can unambiguously be identified as (*Pan-*) *Cyanobacteria*, and this clade must be older than the first permanent rise of atmospheric oxygen for which it was responsible—referred to as the Great Oxidation Event—at 2.45–2.32 billion years ago (e.g., Luo et al., 2016). Thus, the basal biotan divergence took place no less than 2.32–2.45 billion years ago based on a literal reading of the body-fossil record. However, the evolution of biotic oxygen production is typically inferred, based on geochemical proxies for divalent oxygen, to have occurred prior to 3.0 billion years ago (e.g., Wang et al., 2018). If isotopic or biomarker signatures correlated with methanogenesis were produced either by methanogenic archaeans, or by the ancestor of archaeans and eukaryotes (e.g., Betts et al., 2018), then *Biota* would be at least 3.5 billion years old. There are multiple claims based on biomarkers and isotope signatures suggesting that *Biota,* or at least *Pan-Biota,* predates 3.5 billion years, but these claims are debated. Divergence-time estimates based on molecular-clock analyses using ancestral biotan genes—those coding for proteins involved in binary fission, ATP metabolism,

protein synthesis (cell division; housekeeping; metabolism), and protein digestion (heterotrophic lifestyle; primitive immune response)—suggest that the basal biotan divergence took place much earlier, more than 3.9 billion years ago (Betts et al., 2018). However, inferred divergence times for organisms passing through mass extinctions—and *Biota* has survived several—can distort such estimates (e.g., Berv and Field, 2018). Moreover, this estimate of divergence time would place the biotan origin before the end of the Hadean Heavy Bombardment, an interval in Earth history unlikely to have been favorable to the proliferation of delicate cellular life-forms. Nevertheless, modeling indicates that this catastrophic episode in Earth history was unlikely to have produced conditions exceeding biotic tolerances planet-wide (Abramov and Mojzsis, 2009). The origin of crown *Biota* is thus likely to have occurred sometime between these two estimates—between 3.9 Ba and 3.5 Ba—and the origin of the biotan total clade is likely to be much older (see *Pan-Biota,* this volume), though doubtless after the moon-forming impact 4.51 billion years ago thought to have sterilized the Earth's surface.

Literature Cited

Abramov, O., and S. J. Mojzsis. 2009. Microbial habitability of the Hadean Earth during the late heavy bombardment. *Nature* 459:419–422.

Berv, J. S., and D. J. Field. 2018. Genomic signature of an avian Lilliput Effect across the K-Pg extinction. *Syst. Biol.* 67(1):1–13.

Betts, H. C., M. N. Puttick, J. W. Clark, T. A. Williams, P. C. Donoghue, and D. Pisani. 2018. Integrated genomic and fossil evidence illuminates life's early evolution and eukaryote origin. *Nat. Ecol. Evol.* 2(10):1556–1562.

Biémont, C. 2010. A brief history of the status of transposable elements: from junk DNA to major players in evolution. *Genetics* 186:1085–1093.

Blackwell, W. H. 2004. Is it kingdoms or domains? Confusion & solutions. *Am. Biol. Teach.* 66(4):268–276.

Blankenship, R. E. 2010. Early evolution of photosynthesis. *Plant Physiol.* 154(2):434–438.

Brands, S. J. 1989–2005. *Systema Naturae 2000.* Amsterdam, the Netherlands. Available at http://sn2000.taxonomy.nl/.

Britten, R. J., and E. H. Davidson. 1971. Repetitive and non-repetitive DNA sequences and a speculation on the origins of evolutionary novelty. *Q. Rev. Biol.* 46(2):111–138.

Brüssow, H. 2009. The not so universal tree of life *or* the place of viruses in the living world. *Philos. Trans. R. Soc. Lond. B Biol. Sci.* 364:2263–2274.

Chatton, E. 1925. *Pansporella perplex*: reflections on the biology and phylogeny of the *Protozoa. Ann. Sci. Nat. Zool.* 8:5–85.

Chatton, E. 1937/1938. *Titres et Travaux Scientifiques (1906–1937).* Sottano Imprimerie, Sète, France.

Ciccarelli, F. D., T. Doerks, C. von Mering, C. J. Creevey, B. Snel, and P. Bork. 2006. Toward automatic reconstruction of a highly resolved tree of life. *Science* 311(5765):1283–1287.

Çiftçioğlu, N., M. Björklund, K. Kuorikoski, K. Bergström, and E. O. Kajander. 1999. Nanobacteria: an infectious cause for kidney stone formation. *Kidney Int.* 56(5):1893–1898.

Çiftçioğlu, N., R. S. Haddad, D. C. Golden, D. R. Morrison, and D. S. McKay. 2005. A potential cause for kidney stone formation during space flights: enhanced growth of nanobacteria in microgravity. *Kidney Int.* 67(2):483–491.

Cisar, J. O., D. Q. Xu, J. Thompson, W. Swaim, L. Hu, and D. J. Kopecko. 2000. An alternative interpretation of nanobacteria-induced biomineralization. *Proc. Natl. Acad. Sci. USA* 97(21):11511–11515.

Copeland, H. F. 1938. The kingdoms of organisms. *Q. Rev. Biol.* 13:383–420.

Diener, T. O. 1989. Circular RNAs: relics of precellular evolution? *Proc. Natl. Acad. Sci. USA* 86(23):9370–9374.

Dolja, V. V., and E. V. Koonin. 2018. Metagenomics reshapes the concepts of RNA virus evolution by extensive horizontal virus transfer. *Virus Res.* 244:36–52.

Forterre, P. 2006. The origin of viruses and their possible roles in major evolutionary transitions. *Virus Res.* 117(1):5–16.

Fournier, G. P., and J. P. Gogarten. 2010. Rooting the ribosomal tree of life. *Mol. Biol. Evol.* 27(8):1792–1801.

Gee, H. 2013. *The Accidental Species: Misunderstandings of Human Evolution.* University of Chicago Press, Chicago, IL.

Gilbert, W. 1986. The RNA world. *Nature* 319:618.

Gladyshev, E. V., and I. R. Arkhipova. 2011. A widespread class of reverse transcriptase-related cellular genes. *Proc. Natl. Acad. Sci. USA* 108(51):20311–20316.

Gogarten, J. P., H. Kibak, P. Dittrich, L. Taiz, E. J. Bowman, B. J. Bowman, M. F. Manolson, R. J. Poole, T. Date, and T. Oshima. 1989. Evolution of the vacuolar H+-ATPase: implications for the origin of eukaryotes. *Proc. Natl. Acad. Sci. USA* 86(17):6661–6665.

Gogarten, J. P., and L. Taiz. 1992. Evolution of proton pumping ATPases: rooting the tree of life. *Photosynth. Res.* 33:137–146.

Golubic, S., and H. J. Hofmann. 1976. Comparison of modern and mid-Precambrian *Entophysalidacae* (*Cyanophyta*) in stromatolitic algal mats: cell division and degradation. *J. Paleontol.* 50:1074–1082.

Gouy, R., D. Baurain, and H. Philippe. 2015. Rooting the tree of life: the phylogenetic jury is still out. *Philos. Trans. R. Soc. Lond. B Biol. Sci.* 370:20140329.

Haeckel, E. 1866. *Generelle Morphologie der Organismen.* Georg Reimer, Berlin.

Haeckel, E. 1894. *Systematische Phylogenie der Protisten und Pflanzen.* Georg Reimer, Berlin.

Hegde, N. R, M. S. Maddur, S. V. Kaveri, and J. Bayry. 2009. Reasons to include viruses in the tree of life. *Nat. Rev. Microbiol.* 7(8):615.

Hickson, R. E. 1989. Self-splicing introns as a source for transposable genetic elements. *J. Theor. Biol.* 141(1):1–10.

Hinchliff, C. E., S. A. Smith, J. F. Allman, J. G. Burleigh, R. Chaudhary, L. M. Coghill, K. A. Crandall, J. Deng, B. T. Drew, R. Gazis, and K. Gude. 2015. Synthesis of phylogeny and taxonomy into a comprehensive tree of life. *Proc. Natl. Acad. Sci. USA* 112(41):12764–12769.

Hofmann, H. J. 1976. Precambrian microflora, Belcher Islands, Canada—significance and systematics. *J. Paleontol.* 50:1040–1073.

Hu, H. H. 1965. The major groups of living beings: a new classification. *Taxon* 14:254–261.

Hug, L. A., B. B. Baker, K. Anantharaman, C. T. Brown, A. J. Probst, C. J. Castelle, C. N. Butterfield, A. W. Hernsdorf, Y. Y. Amano, K. Ise, Y. Susuki, N. Dudek, A. A. Relman, K. M. Finstad, R. Amundson, B. C. Thomas, and J. F. Banfield. 2016. A new view of the tree of life. *Nat. Microbiol.* 1:16048.

Hunding, A., F. Kepes, D. Lancet, A. Minsky, V. Norris, D. Raine, K. Sriram, and R. Root-Bernstein. 2006. Compositional complementarity and prebiotic ecology in the origin of life. *Bioessays* 28(4):399–412.

Iwabe, N., K. Kuma, M. Hasegawa, S. Osawa, and T. Miyata. 1989. Evolutionary relationship of *Archaebacteria*, *Eubacteria*, and eukaryotes inferred from phylogenetic trees of duplicated genes. *Proc. Natl. Acad. Sci. USA* 86(23):9355–9359.

Jeffrey, C. 1971. Thallophytes and kingdoms: a critique. *Kew Bull.* 25(2):291–299.

Kejnovsky, E., and E. N. Trifonov. 2016. Horizontal transfer-imperative mission of acellular life forms, *Acytota. Mob. Genet. Elements* 6(2):1154636.

Kidwell, M. G. 1992. Horizontal transfer of P elements and other short inverted repeat transposons. *Genetica* 86(1–3):275–286.

Klug, Y. A., E. Rotem, R. Schwarzer, and Y. Shai. 2017. Mapping out the intricate relationship of the HIV envelope protein and the membrane environment. *Biochim. Biophys. Acta (BBA)—Biomembr.* 1859(4):550–560.

Koonin, E. V., and V. V. Dolja. 2014. Virus world as an evolutionary network of viruses and capsid-less selfish elements. *Microbiol. Mol. Biol. Rev.* 78(2):278–303.

Koonin, E. V., and N. Yutin. 2018. Multiple evolutionary origins of giant viruses. *F1000Research* 7:1840.

Krupovic, M., and E. V. Kroonin. 2017a. Multiple origins of viral capsid proteins from cellular ancestors. *Proc. Natl. Acad. Sci. USA* 14(12):2401–2410.

Krupovic, M., and E. V. Kroonin. 2017b. Homologous capsid proteins testify to the common ancestry of *Retroviruses*, *Caulimoviruses*, *Pseudoviruses*, and *Metaviruses. J. Virol.* 9(12):e00210–17.

Lake, J. A., E. Henderson, M. Oakes, and M. W. Clark. 1984. Eocytes: a new ribosome structure indicates a kingdom with a close relationship to eukaryotes. *Proc. Natl. Acad. Sci. USA* 81(12):3786–3790.

Lanciano, S., and M. Mirouze. 2018. Transposable elements: all mobile, all different, some stress responsive, some adaptive. *Curr. Opin. Genet. Dev.* 49:106–114.

Lander, E. S., L. M. Linton, B. Birren, C. Nusbaum, M. C. Zody, J. Baldwin, K. Devon, K. Dewar, M. Doyle, W. FitzHugh, R. Funke, D. Gage, K. Harris, A. Heaford, J. Howland, L. Kann, J. Lehoczky, R. LeVine, P. McEwan, K. McKernan, J. Meldrim, J. P. Mesirov, C. Miranda, W. Morris, J. Naylor, C. Raymond, M. Rosetti, R. Santos, A. Sheridan, C. Sougnez, Y. Stange-Thomann, N. Stojanovic, A. Subramanian, D. Wyman, J. Rogers, J. Sulston, R. Ainscough, S. Beck, D. Bentley, J. Burton, C. Clee, N. Carter, A. Coulson, R. Deadman, P. Deloukas, A. Dunham, I. Dunham, R. Durbin, L. French, D. Grafham, S. Gregory, T. Hubbard, S. Humphray, A. Hunt, M. Jones, C. Lloyd, A. McMurray, L. Matthews, S. Mercer, S. Milne, J. C. Mullikin, A. Mungall, R. Plumb, M. Ross, R. Shownkeen, S. Sims, R. H. Waterston, R. K. Wilson, L. W. Hillier, J. D. McPherson, M. A. Marra, E. R. Mardis, L. A. Fulton, A. T. Chinwalla, K. H. Pepin, W. R. Gish, S. L. Chissoe, M. C. Wendl, K. D. Delehaunty, T. L. Miner, A. Delehaunty, J. B. Kramer, L. L. Cook, R. S. Fulton, D. L. Johnson, P. J. Minx, S. W. Clifton, T. Hawkins, E. Branscomb, P. Predki, P. Richardson, S. Wenning, T. Slezak, N. Doggett, J. F. Cheng, A. Olsen, S. Lucas, C. Elkin, E. Uberbacher, M. Frazier, R. A. Gibbs, D. M. Muzny, S. E. Scherer, J. B. Bouck, E. J. Sodergren, K. C. Worley, C. M. Rives, J. H. Gorrell, M. L. Metzker, S. L. Naylor, R. S. Kucherlapati, D. L. Nelson, G. M. Weinstock, Y. Sakaki, A. Fujiyama, M. Hattori, T. Yada, A. Toyoda, T. Itoh, C. Kawagoe, H. Watanabe,

Y. Totoki, T. Taylor, J. Weissenbach, R. Heilig, W. Saurin, F. Artiguenave, P. Brottier, T. Bruls, E. Pelletier, C. Robert, P. Wincker, D. R. Smith, L. Doucette-Stamm, M. Rubenfield, K. Weinstock, H. M. Lee, J. Dubois, A. Rosenthal, M. Platzer, G. Nyakatura, S. Taudien, A. Rump, H. Yang, J. Yu, J. Wang, G. Huang, J. Gu, L. Hood, L. Rowen, A. Madan, S. Qin, R. W. Davis, N. A. Federspiel, A. P. Abola, M. J. Proctor, R. M. Myers, J. Schmutz, M. Dickson, J. Grimwood, D. R. Cox, M. V. Olson, R. Kaul, C. Raymond, N. Shimizu, K. Kawasaki, S. Minoshima, G. A. Evans, M. Athanasiou, R. Schultz, B. A. Roe, F. Chen, H. Pan, J. Ramser, H. Lehrach, R. Reinhardt, W. R. McCombie, M. de la Bastide, N. Dedhia, H. Blöcker, K. Hornischer, G. Nordsiek, R. Agarwala, L. Aravind, J. A. Bailey, A. Bateman, S. Batzoglou, E. Birney, P. Bork, D. G. Brown, C. B. Burge, L. Cerutti, H. C. Chen, D. Church, M. Clamp, R. R. Copley, T. Doerks, S. R. Eddy, E. E. Eichler, T. S. Furey, J. Galagan, J. G. Gilbert, C. Harmon, Y. Hayashizaki, D. Haussler, H. Hermjakob, K. Hokamp, W. Jang, L. S. Johnson, T. A. Jones, S. Kasif, A. Kaspryzk, S. Kennedy, W. J. Kent, P. Kitts, E. V. Koonin, I. Korf, D. Kulp, D. Lancet, T. M. Lowe, A. McLysaght, T. Mikkelsen, J. V. Moran, N. Mulder, V. J. Pollara, C. P. Ponting, G. Schuler, J. Schultz, G. Slater, A. F. Smit, E. Stupka, J. Szustakowki, D. Thierry-Mieg, J. Thierry-Mieg, L. Wagner, J. Wallis, R. Wheeler, A. Williams, Y. I. Wolf, K. H. Wolfe, S. P. Yang, R. F. Yeh, F. Collins, M. S. Guyer, J. Peterson, A. Felsenfeld, K. A. Wetterstrand, A. Patrinos, M. J. Morgan, P. de Jong, J. J. Catanese, and the International Human Genome Sequencing Consortium. 2001. Initial sequencing and analysis of the human genome. *Nature* 409(6822):860–921.

Linnaeus, C. 1758. *Systema Naturae Per Regna Tria Naturae, Secundum Classes, Ordines, Genera, Species, cum Characteribus, Differentiis, Synonymis, Locis*. Tomus I, Editio decima, reformata. Laurentius Salvius, Holmiae (Stockholm).

López-García, P., and D. Moreira. 2009. Yet viruses cannot be included in the tree of life. *Nat. Rev. Microbiol.* 7:615–617.

Luo, G., S. Ono, N. J. Beukes, D. T. Wang, S. Xie, and R. E. Summons. 2016. Rapid oxygenation of Earth's atmosphere 2.33 billion years ago. *Sci. Adv.* 2(5):1600134.

Luria, S. E., and J. Darnell. 1967. *General Virology*. John Wiley, New York.

Martel, J., and J. D. E. Young. 2008. Purported nanobacteria in human blood as calcium carbonate nanoparticles. *Proc. Natl. Acad. Sci. USA* 105(14):5549–5554.

Martin, W. F., M. C. Weiss, S. Neukirchen, S. Nelson-Sathi, and F. L. Sousa. 2016. Physiology, phylogeny and LUCA. *Microbial Cell* 3(12):582–587.

Maynard Smith, J., and E. Szathmáry. 1995. *The Major Transitions in Evolution*. W. H. Freeman, New York.

Maynard Smith, J., and E. Szathmáry. 1997. *The Major Transitions in Evolution*. Oxford University Press, Oxford.

McClintock, B. 1950. The origin and behavior of mutable loci in maize. *Proc. Natl. Acad. Sci. USA* 36(6):344–355.

McDowall, J. 2006. *Protein of the Month: Transposase*. European Bioinformatics Institute. Available at https://www.ebi.ac.uk/interpro/potm/2006_12/Page1.htm.

McKay, D. S., E. K. Gibson, K. L. Thomas-Keprta, H. Vali, C. S. Romanek, S. J. Clemett, X. D. Chillier, C. R. Maechling, and R. N. Zare. 1996. Search for past life on Mars: possible relic biogenic activity in Martian meteorite ALH84001. *Science* 273(5277):924–930.

Mervis, C. B., and E. Rosch. 1981. Categorization of natural objects. *Annu. Rev. Psychol.* 32(1):89–115.

Moreira, D., and P. López-García. 2009. Ten reasons to exclude viruses from the tree of life. *Nat. Rev. Microbiol.* 7(4):306–311.

Nakano, Y., H. Aso, A. Soper, E. Yamada, M. Moriwaki, G. Juarez-Fernandez, Y. Koranagi, and K. Sato. 2017. Conflict of interest: the evolutionary arms race between mammalian APOBEC3 and lentiviral Vif. *Retrovirology* 14(31). doi:10.1186/s12977-017-0355-4.

Nasir, A., and G. Caetano-Anollés. 2015. A phylogenomic data-driven exploration of viral origins and evolution. *Sci. Adv.* 1(8):e1500527.

Novikova, N., P. De Boever, S. Poddubko, E. Deshevaya, N. Polikarpov, N. Rakova, L. Coninx, and M. Mergeay. 2006. Survey of environmental biocontamination on board the International Space Station. *Res. Microbiol.* 157(1):5–12.

Orgel, L. E. 1994. The origin of life on the earth. *Sci. Am.* 271(4):76–83.

Oró, J., S. L. Miller, and A. Lazcano. 1990. The origin and early evolution of life on Earth. *Annu. Rev. Earth Planet. Sci.* 18(1):317–356.

Pace, N. R. 2006. Time for a change. *Nature* 441:289.

Pelletier, B. 2015. *Empire Biota: Taxonomy and Evolution.* 1st edition. Createspace Independent Publishing Platform.

Penny, D., and A. Poole. 1999. The nature of the last universal common ancestor. *Curr. Opin. Genet. Dev.* 9:672–677.

Philippe, N., M. Legendre, G. Doutre, Y. Couté, O. Poirot, M. Lescot, D. Arslan, V. Seltzer, L. Bertaux, C. Bruley, J. Garin, J.-M. Claverie, and C. Aberge. 2013. Pandoraviruses: *Amoeba* viruses with genomes up to 2.5 Mb reaching that of parasitic eukaryotes. *Science* 341(6143):281–286.

Podolsky, S. 1996. The role of the virus in origin-of-life theorizing. *J. Hist. Biol.* 29:79–126.

Prestel, E., C. Regeard, J. Andrews, P. Oger, and M. S. DuBow. 2012. A novel bacteriophage morphotype with a ribbon-like structure at the tail extremity. *Res. J. Microbiol.* 7(1):75–81.

Rodrigues-Oliveira, T., A. Belmok, D. Vasconcellos, B. Schuster, and C. M. Kyaw. 2017. Archaeal s-layers: overview and current state of the art. *Front. Microbiol. Microbial Physiol. Metab.* doi.org/10.3389/fmicb.2017.02597.

Root-Bernstein, M., and R. Root-Bernstein. 2015. The ribosome as a missing link in the evolution of life. *J. Theor. Biol.* 367:130–158.

Sára, M., and U. B. Sleytr. 2000. S-layer proteins. *J. Bacteriol.* 182(4):859–868.

Schnable, P. S., D. Ware, R. S. Fulton, J. C. Stein, F. Wei, S. Pasternak, C. Liang, J. Zhang, L. Fulton, T. A. Graves, P. Minx, A. D. Reily, L. Courtney, S. S. Kruchowski, C. Tomlinson, C. Strong, K. Delehaunty, C. Fronick, B. Courtney, S. M. Rock, E. Belter, F. Du, K. Kim, R. M. Abbott, M. Cotton, A. Levy, P. Marchetto, K. Ochoa, S. M. Jackson, B. Gillam, W. Chen, L. Yan, J. Higginbotham, M. Cardenas, J. Waligorski, E. Applebaum, L. Phelps, J. Falcone, K. Kanchi, T. Thane, A. Scimone, N. Thane, J. Henke, T. Wang, J. Ruppert, N. Shah, K. Rotter, J. Hodges, E. Ingenthron, M. Cordes, S. Kohlberg, J. Sgro, B. Delgado, K. Mead, A. Chinwalla, S. Leonard, K. Crouse, K. Collura, D. Kudrna, J. Currie, R. He, A. Angelova, S. Rajasekar, T. Mueller, R. Lomeli, G. Scara, A. Ko, K. Delaney, M. Wissotski, G. Lopez, D. Campos, M. Braidotti, E. Ashley, W. Golser, H. Kim, S. Lee, J. Lin, Z. Dujmic, W. Kim, J. Talag, A. Zuccolo, C. Fan, A. Sebastian, M. Kramer, L. Spiegel, L. Nascimento, T. Zutavern, B. Miller, C. Ambroise, S. Muller, W. Spooner, A. Narechania, L. Ren, S. Wei, S. Kumari, B. Faga, M. J. Levy, L. McMahan, P. Van Buren, M. W. Vaughn, K. Ying, C. T. Yeh, S. J. Emrich, Y. Jia, A. Kalyanaraman, A. P. Hsia, W. B. Barbazuk, R. S. Baucom, T. P. Brutnell, N. C. Carpita, C. Chaparro, J. M. Chia, J. M. Deragon, J. C. Estill, Y. Fu, J. A. Jeddeloh, Y. Han, H. Lee, P. Li, D. R. Lisch, S. Liu, Z. Liu, D. H. Nagel, M. C. McCann, P. SanMiguel, A. M. Myers, D. Nettleton, J. Nguyen, B. W. Penning, L. Ponnala, K. L. Schneider, D. C. Schwartz, A. Sharma, C. Soderlund, N. M. Springer, Q. Sun, H. Wang, M. Waterman, R. Westerman, T. K. Wolfgruber, L. Yang, Y. Yu, L. Zhang, S. Zhou, Q. Zhu, J. L. Bennetzen, R. K. Dawe, J. Jiang, N. Jiang, G. G. Presting, S. R. Wessler, S. Aluru, R. A. Martienssen, S. W. Clifton, W. R. McCombie, R. A. Wing, and R. K. Wilson 2009. The B73 maize genome: complexity, diversity, and dynamics. *Science* 326(5956):1112–1115.

Sender, R., S. Fuchs, and R. Milo. 2016. Revised estimates for the number of human and bacterial cells in the body. *PLOS Biol.* doi.org/10.1101/036103.

Sereno, P. C. 1998. A rationale for phylogenetic definitions, with application to the higher-level taxonomy of *Dinosauria. Neues Jahrb. Geol. Paläontol.* 210:41–83.

Sharp, P. M., and B. H. Hahn. 2011. Origins of HIV and the AIDS pandemic. *Cold Spring Harb. Perspect. Med.* 1:a006841.

Sommer, A. P., D. S. McKay, N. Çiftçioğlu, U. Oron, A. R. Mester, and E. O. Kajander. 2003. Living nanovesicles chemical and physical survival strategies of primordial biosystems. *J. Proteome Res.* 2(4):441–443.

Spinola, M. M. 1850. Tavola sinottica dei generi spettani all classe degli insetti arthroidignati, *Hemiptera* Linn., Latr.—*Rhyngota* Fab.—*Rhynchota* Burm. *Mem. Soc. Ital. Sci. Nat.* 25(1):1–138.

Stejneger, L. 1901. Scharff's history of the European fauna. *Am. Nat.* 35(410):87–116.

Swain, F. M. 1969. Paleomicrobiology. *Annu. Rev. Microbiol.* 23:455–472.

Theobald, D. L. 2010. A formal test of the theory of universal common ancestry. *Nature* 465:219–222.

Torrella, F., and R. Y. Morita. 1981. Microcultural study of bacterial size changes and microcolony and ultramicrocolony formation by heterotrophic bacteria in seawater. *Appl. Environ. Microbiol.* 41(2):518–527.

Trifonov, E. N., and E. Kejnovsky. 2015. *Acytota*—associated kingdom of neglected life. *J. Biomol. Struct. Dyn.* 34(8):1641–1648.

Van Regenmortel, M. H. 2003. Viruses are real, virus species are man-made, taxonomic constructions. *Arch. Virol.* 148:2481–2488.

Villarreal, L. P. 2005. *Viruses and the Evolution of Life*. ASM Press, Washington, DC.

Vorontsov, N. N. 1965. *Origin of Life and Diversity of Its Forms*. Novosibirsk (in Russian).

Wagner, G. P. 2014. *Homology, Genes, and Evolutionary Innovation*. Princeton University Press, Princeton and Oxford.

Wagner, J. R. 2004. The general case of phylogenetic definitions, alternate classes of definitions, and the phylogenetic definition of Life. P. 18 in *Abstracts of the First International Phylogenetic Nomenclature Meeting*, Paris. Available at https://www.phylocode.org and https://hal.sor bonne-universite.fr/hal-02187647.

Walton, L. B. 1930. Studies concerning organisms occurring in water supplies with particular reference to those founded in Ohio. *Ohio Biol. Surv. Bull.* 24:1–86.

Wang, X., N. J. Planavsky, A. Hofmann, E. E. Saupe, B. P. De Corte, P. Philippot, S. V. LaLonde, N. E. Jemison, H. Zou, F. Ossa, K. Rybacki, N. Alfimova, M. J. Larson, H. Tsikos, P. W. Fralick, T. M. Johnson, A. C. Knudsen, C. T. Reinhard, and K. O. Konhauser. 2018. A Mesoarchean shift in uranium isotope systematics. *Geochim. Cosmochim. Acta* 238:438–452.

Wiemann, J., M. Fabbri, T. R. Yang, K. Stein, P. M. Sander, M. A. Norell, and D. E. Briggs. 2018. Fossilization transforms vertebrate hard tissue proteins into N-heterocyclic polymers. *Nat. Commun.* 9(1):4741.

Williams, T. A., P. G. Foster, C. J. Cox, and T. M. Embley. 2013. An archaeal origin of eukaryotes supports only two primary domains of life. *Nature* 504(7479):231.

Williams, T. A., P. G. Foster, T. M. W. Nye, C. J. Cox, and T. M. Embley. 2012. A congruent phylogenomic signal places eukaryotes with *Archaea*. *Proc. R. Soc. Lond. B Biol. Sci.* 279:4870–4879.

Woese, C. R. 1999. The universal ancestor. *Proc. Natl. Acad. Sci. USA* 95:6854–6859.

Woese, C. R., O. Kandler, and M. L. Wheelis. 1990. Towards a natural system of organisms: proposal for the domains *Archaea*, *Bacteria*, and *Eucarya*. *Proc. Natl. Acad. Sci. USA* 87(12):4576–4579.

Wolf, Y. I., D. Kazlauskas, J. Iranzo, A. Lucia-Sanz, J. H. Kuhn, M. Krupovic, V. V. Dolja, and E. V. Koonin. 2018. Origins and evolution of the global RNA virome. *mBio* 9(6): e02329-18.

Yang, S., G. Xin, J. Meng, A. Zhang, Y. Zhou, M. Long, B. Li, W. Deng, L. Jin, S. Zhao, D. Wu, Y. He, C. Li, S. Liu, Y. Huang, H. Zhang, and L. Zou. 2018. Metagenomic analysis of bacteria, fungi, bacteriophages, and helminths in the gut of Giant Pandas. *Front. Microbiol.* 9:1717.

Authors

Jasmina Wiemann; Department of Geology and Geophysics; Yale University; 210 Whitney Avenue; New Haven, CT 06511, USA. Email: jasmina.wiemann@yale.edu.

Kevin de Queiroz; Department of Vertebrate Zoology; National Museum of Natural History, Smithsonian Institution; Washington, DC 20560, USA. Email: dequeirozk@si.edu.

Timothy B. Rowe; Jackson School of Geosciences,;University of Texas at Austin; C1100, Austin, TX 78712, USA. Email: rowe@mail.utexas.edu.

Noah J. Planavsky; Department of Geology and Geophysics; Yale University; 210 Whitney Avenue; New Haven, CT 06511, USA. Email: noah.planavsky@yale.edu.

Ross P. Anderson; Department of Earth Sciences; University of Oxford; Oxford, OX1 3AN, UK. Email: ross.anderson@all-souls.ox.ac.uk.

Johann P. Gogarten; Department of Molecular and Cell Biology; University of Connecticut; Storrs, CT 06269 USA. Email: gogarten@uconn.edu.

Paul E. Turner; Department of Ecology and Evolutionary Biology; Yale University; P. O. Box 208106, New Haven, CT 06520, USA. Email: paul.turner@yale.edu.

Jacques A. Gauthier; Department of Geology and Geophysics; Yale University; 210 Whitney Avenue; New Haven, CT 06511, USA. Email: jacques.gauthier@yale.edu.

Date Accepted: 11 July 2019

Primary Editor: Philip Cantino

Eukarya R. Creti et al. 1991 [A. G. B. Simpson], converted clade name

Registration Number: 42

Definition: The crown clade for which the possession of nuclei, bounded by nuclear envelopes that include nuclear pores, as inherited by *Homo sapiens* Linnaeus 1758, is an apomorphy relative to other crown clades. This is an apomorphy-modified crown-clade definition. Abbreviated definition: crown ∇ apo nuclei [*Homo sapiens* Linnaeus 1758].

Etymology: Derived from the Greek, *eu-* (true) and *karyon* (kernel, nucleus).

Reference Phylogeny: The reference phylogeny is Ciccarelli et al. (2006: Fig. 2), in which *Eukarya* is called *Eukaryota* (see synonyms) and is colored red (see figure legend). See also Foster et al. (2009: Figs. 2, 4).

Composition: *Archaeplastida*, *Sar* (including *Stramenopila*, *Alveolata*, and *Rhizaria*), *Opisthokonta*, *Discoba*, and *Metamonada* (all as defined in this volume), as well as *Amoebozoa*, *Cryptista*, *Haptophyta*, and several smaller groups (see Adl et al., 2012).

Diagnostic Apomorphies: Lists of features that are universal among eukaryotes, or nearly so, and are generally considered unique to eukaryotes are given by Cavalier-Smith (1987; Table 6) and Roger (1999: Fig. 5). Amongst the most conspicuous or important are:

1) A nucleus, surrounded by a nuclear envelope, with nuclear pores.
2) Genome organized into multiple linear chromosomes.
3) Segregation of the replicated genome through mitosis.
4) Spliceosomal introns, and intron-splicing machinery, including the spliceosome.
5) True sex, involving meiosis (may be secondarily absent in many lineages).
6) A well-developed endomembrane system, including endoplasmic reticulum, and a Golgi apparatus; systems for endocytosis and exocytosis (though see Comments).
7) A complex cytoskeleton including microtubules and actin-based microfilaments as major components (though see Comments).
8) Eukaryotic basal bodies and flagella, and their homologues (e.g., centrioles), constructed partly of microtubules arranged in a 9-fold rotationally symmetrical formation (secondarily absent in several substantial clades).
9) Mitochondria, of endosymbiotic origin (or organelles derived from mitochondria).

Synonyms: *Eucarya* Woese et al. 1990, *Eucaryota* sensu Christensen (1962; also *Eukaryota*), *Eukaryonta* Fott 1959, *Eucaryotae* Traub 1964 (also *Eukaryotae*), *Eucytota* Jeffrey 1971 (all approximate).

Comments: The distinction between extant prokaryotic and eukaryotic cells has often been considered to be the most fundamental discontinuity in the living world (Stanier et al., 1963). There are many profound differences between well-studied eukaryotic and prokaryotic cells (see Diagnostic Apomorphies). Of these, the presence of a membrane-bounded nucleus in eukaryotes is generally seen as the defining feature of these cells. Compartmentalization by membranes of the chromosomal material is extremely rare and phylogenetically restricted in prokaryotes, and is almost certainly analogous, not homologous, to the eukaryote nucleus (Fuerst, 2005). Recent genomic studies of archaea that show evidence of phylogenetic affinity with eukaryotes, especially the 'Asgard archaea', have revealed genes encoding close homologs of some tens of proteins that have previously been treated as eukaryote signatures (albeit in various combinations in different taxa rather than all in one genome; Spang et al., 2015; Eme et al., 2017; Zaremba-Niedzwiedzka et al., 2017). These proteins include homologs of actin and some actin-associated proteins, which appear to be present in all major groups of Asgard archaea, as well as a close homolog of the eukaryotic tubulin families in one case (Zaremba-Niedzwiedzka et al., 2017). All major Asgard groups also encode homologs of several proteins that are involved in membrane trafficking in eukaryotes. One possibility is that the Asgard archaea have cytoskeletons and/or endomembrane systems with some level of special homology to those of extant eukaryotes, though much simpler. As of late 2017, Asgard archaea have not been cultivated, nor examined using high-resolution microscopy, so their fundamental cell structure is unknown. Irrespective, there is no indication from their genome content that Asgard archaea have a nuclear envelope per se, and their genome organization is consistent with that of other prokaryotes (Zaremba-Niedzwiedzka et al. 2017), rather than similar to typical eukaryote genomes.

Acknowledgment of the distinction between what we now refer to as prokaryotic and eukaryotic cells was apparent during the nineteenth century and common by the 1930s (see Copeland, 1938, 1956; Stanier, 1970; Sapp, 2005), and by the 1960s most major structural differences had been documented (see Stanier and van Niel,

1962; Allsopp, 1969; Stanier, 1970). Phylogenies estimated from ribosomal RNA genes and many 'universal' protein-coding genes typically infer an extremely long (and well-supported) branch separating prokaryotic and eukaryotic organisms (e.g., Sogin, 1991; Ciccarelli et al., 2006; Foster et al., 2009). Providing the root of the tree does not lie among eukaryotes, the monophyly of eukaryotes is not seriously doubted. In practice, there has been a widespread, though not universal, belief that eukaryotes are, in fact, a monophyletic group descended from prokaryotic ancestors. The chief reasons, historically, for this view have been forms of 'simple-is-primitive' arguments, but the universal presence of mitochondrial organelles, which are derived from *Proteobacteria*, indicates strongly that the eukaryote crown clade at least is derived (and monophyletic). It is also widely held that strong evidence of prokaryotic life extends earlier into the geological record than reliable evidence of eukaryotic life, consistent with eukaryotes being derived (López-García et al., 2006). The use of universal paralog pairs to root the tree of life usually places the root of the tree among prokaryotes, but is likely unreliable, or at least difficult to interpret (see Philippe and Forterre, 1999, Zhaxybayeva et al., 2005). Fusion hypotheses for the origin of eukaryotic cells (there are several, see Embley and Martin, 2006) generally also imply that eukaryotes are derived with respect to prokaryotic cells and monophyletic.

The term 'eucaryotes' was introduced by Chatton (1925), along with 'procaryotes'. He perhaps intended the term to refer only to unicellular forms, not to 'higher' animals or plants or macroalgae (Chatton, 1925: 76), and the occasional uses of the term for the next two to three decades were apparently mostly in this sense (Katscher, 2004). However, the term 'eucaryotes' then came to be associated unequivocally with all nucleated organisms (e.g., Chadefaud, 1960), and it is in this sense that the term 'eucaryotic' was widely disseminated by Stanier and colleagues (Stanier and van Niel, 1962; Stanier et al., 1963; see Katscher, 2004; Sapp, 2005). Starting around 1960, various taxon names were introduced that were based on the word 'eucaryote' and implicitly or explicitly included multicellular as well as unicellular eukaryotes: *Eukaryonta* was introduced and defined by Fott (1959), while *Eucaryota* was used by Christensen (1962), and Traub (1964) introduced and defined *Eucaryotae*. Stebbins (1964) used *Eukaryota* in passing. A few workers assigned ranks to these taxa, usually considering them 'superkingdoms' or 'kingdoms'. Writing at the end of the 1960s, Allsopp (1969) noted that eukaryotic cells were widely known as '*Eucaryota*'. The alternate transliteration '*Eukaryota*' has now become more popular in works in English. 'Eukaryotes' has also overwhelmingly supplanted 'eucaryotes' in English works (e.g., compare Stanier and van Niel, 1962 and Stanier, 1970). Using a different etymological root, Jeffrey (1971) proposed the name *Eucytota* for eukaryotic organisms, but this has never been widely used.

Much later, Woese, Kandler and Wheelis (Woese et al., 1990) proposed the taxon *Eucarya* for all eukaryotic cells, assigning it the new rank of 'Domain'. It was not made clear why they proposed a new name, rather than using one of the several similar names already in existence. Nonetheless, the name has become rapidly and widely adopted, though with a 'k' instead of 'c' in a large majority of cases, consistent with the now-dominant informal term 'eukaryote'.

Although the name *Eukaryota* is seemingly dominant in general literature, the name *Eukarya* is used much more frequently than *Eukaryota* in scientific literature (with other near-homonyms, e.g.,, *Eukaryotae* and *Eukaryonta*, being little used). It is therefore proposed that *Eukarya* be adopted as the name of this clade, to comply with *PhyloCode* (Cantino and de Queiroz, 2020) Recommendation 10.1A. I do not know who was the first to use *Eukarya* as a taxon name, rather than Woese et al.'s (1990) original *Eucarya*. It was probably introduced by several groups in parallel, in most or all cases unwittingly. To comply with *PhyloCode* Recommendation. 9.15A, the authors of the preexisting name are considered here to be Creti et al. (1991); this is the first published use found that was not an obvious and isolated error. The definition given here follows the recommendation (Rec. 10.1B) that widely used names be attached to crown clades.

Literature Cited

Adl, S. M., A. G. B. Simpson, C. E. Lane, J. Lukeš, D. Bass, S. S. Bowser, M. W. Brown, F. Burki, M. Dunthorn, V. Hampl, A. Heiss, M. Hoppenrath, E. Lara, L. le Gall, D. H. Lynn, H. McManus, E. A. D. Mitchell, S. E. Mozley-Stanridge, L. W. Parfrey, J. Pawlowski, S. Rueckert, L. Shadwick, C. L. Schoch, A. Smirnov, and F. W. Spiegel. 2012. The revised classification of eukaryotes. *J. Eukaryot. Microbiol.* 59:429–493.

Allsopp, A. 1969. Phylogenetic relationships of the *Procaryota* and the origin of the eucaryotic cell. *New Phytol.* 68:591–612.

Cantino, P. D., and K. de Queiroz. 2020. *International Code of Phylogenetic Nomenclature (PhyloCode)*, Version 6. CRC Press, Boca Raton, FL.

Cavalier-Smith, T. 1987. The origin of eukaryote and archaebacterial cells. *Ann. N.Y. Acad. Sci.* 503:17–54.

Chadefaud, M. 1960. *Les Végétaux non Vasculaires (Cryptogamie)*. Masson et Cie, Paris, France.

Chatton, E. 1925. *Pansporella perplexa*. Réflexions sur la biologie et la phylogénie des protozoaires. *Ann. Sci. Nat. Zool.* 10:1–84.

Christensen, T. 1962. *Alger*. Munksgaard, Copenhagen.

Ciccarelli, F. D., T. Doerks, C. von Mering, C. J. Creevey, B. Snel, and P. Bork. 2006. Toward automatic reconstruction of a highly resolved tree of life. *Science* 311:1283–1287.

Copeland, H. F. 1938. The kingdom of organisms. *Q. Rev. Biol.* 13:383–420.

Copeland, H. F. 1956. *The Classification of Lower Organisms.* Pacific Books, Palo Alto, CA.

Creti, R., F. Citarella, O. Tiboni, A. Sanangelantoni, P. Palm, and P. Cammarano. 1991. Nucleotide sequence of a DNA region comprising the gene for elongation factor lα (EF-lα) from the ultrathermophilic archaeote *Pyrococcus woesei*: phylogenetic implications. *J. Mol. Evol.* 33:332–342.

Embley, T. M., and W. Martin. 2006. Eukaryotic evolution, changes and challenges. *Nature* 440:623–630.

Eme, L., A. Spang, J. Lombard, C. W. Stairs, and T. J. Ettema. 2017. *Archaea* and the origin of eukaryotes. *Nat. Rev. Microbiol.* 15:711–723.

Foster, P. G., C. J. Cox, and T. M. Embley. 2009. The primary divisions of life: a phylogenomic approach employing composition-heterogeneous methods. *Philos. Trans. R. Soc. Lond. B Biol. Sci.* 364:2197–2207.

Fott, B. 1959. *Algenkunde.* VEB Gustav Fischer Verlag, Jena, Germany.

Fuerst, J. A. 2005. Intracellular compartmentation in planctomycetes. *Annu. Rev. Microbiol.* 59:299–328.

Jeffrey, C. 1971. Thallophytes and kingdoms: a critique. *Kew Bull.* 25:291–299.

Katscher, F. 2004. The history of the terms prokaryotes and eukaryotes. *Protist* 155:257–263.

López-García, P., D. Moreira, E. Douzery, P. Forterre, M. Van Zuilen, P. Claeys, and D. Prieur. 2006. Ancient fossil record and early evolution. *Earth Moon Planets* 98:247–290.

Philippe, H., and P. Forterre. 1999. The rooting of the universal tree of life is not reliable. *J. Mol. Evol.* 49:509–523.

Roger, A. J. 1999. Reconstructing early events in eukaryotic evolution. *Am. Nat.* 154:S146–S163.

Sapp, J. 2005. The prokaryote-eukaryote dichotomy; meanings and mythology. *Microbiol. Mol. Biol. Rev.* 69:292–305.

Sogin, M. L. 1991. Early evolution and the origin of eukaryotes. *Curr. Opin. Genet. Dev.* 1:457–463.

Spang, A., J. H. Saw, S. L. Jørgensen, K. Zaremba-Niedzwiedzka, J. Martijn, A. E. Lind, R. van Eijk, C. Schleper, L. Guy, and T. J. G. Ettema. 2015. Complex archaea that bridge the gap between prokaryotes and eukaryotes. *Nature* 521:173–179.

Stanier, R. Y. 1970. Some aspects of the biology of cells and their possible evolutionary significance. *Symp. Soc. Gen. Microbiol.* 20:1–38.

Stanier, R. Y., M. Doudoroff, and E. A. Adelberg. 1963. *The Microbial World.* 2nd edition. Prentice Hall Inc., Englewood Cliffs.

Stanier, R. Y., and C. B. van Niel. 1962. The concept of a bacterium. *Arch. Microbiol.* 42:17–35.

Stebbins, G. L. 1964. Four basic questions of plant biology. *Am. J. Bot.* 51:220–230.

Traub, H. P. 1964. *Lineagics.* The American Plant Life Society, La Jolla.

Woese, C. R., O. Kandler, and M. L. Wheelis. 1990. Towards a natural system of organisms: proposal for the domains *Archaea, Bacteria,* and *Eucarya. Proc. Natl. Acad. Sci. USA.* 87:4576–4579.

Zaremba-Niedzwiedzka, K., E. F. Caceres, J. II. Saw, D. Bäckström, L. Juzokaite, E. Vancaester, K. W. Seitz, K. Anantharaman, P. Starnawski, K. U. Kjeldsen, M. B. Stott, T. Nunoura, J. F. Banfield, A. Schramm, B. J. Baker, A. Spang, and T. J. G. Ettema. 2017. Asgard archaea illuminate the origin of eukaryotic cellular complexity. *Nature* 541:353–358.

Zhaxybayeva, O., P. Lapierre, and J. P. Gogarten. 2005. Ancient gene duplications and the root(s) of the tree of life. *Protoplasma* 227:53–64.

Author

Alastair G. B. Simpson; Department of Biology; Dalhousie University; Halifax, NS, B3H 4R2, Canada. Email: alastair.simpson@dal.ca.

Date Accepted: 6 February 2012; updated 20 April 2018

Primary Editor: Philip Cantino

Metamonada T. Cavalier-Smith 1987a [A. G. B. Simpson], converted clade name

Registration Number: 66

Definition: The smallest crown clade containing *Paratrimastix pyriformis* (Klebs 1893) Zhang et al. 2015, *Trichomonas vaginalis* Donné 1836, and *Spironucleus barkhanus* Sterud et al. 1997, provided that the clade does not include any organisms with mitochondria capable of oxidative phosphorylation (e.g., *Malawimonas jakobiformis* O'Kelly and Nerad 1999). This is a minimum-crown-clade definition with a qualifying clause. Abbreviated definition: min crown ∇ (*Paratrimastix pyriformis* (Klebs 1893) Zhang et al. 2015 & *Trichomonas vaginalis* Donné 1836 & *Spironucleus barkhanus* Sterud et al. 1997) | ~ (any organisms with mitochondria capable of oxidative phosphorylation).

Etymology: Derived from the Greek *meta* (next to; in company with) and *monados* (indivisible unit). Historically "monad" was commonly used to refer to various small, simple unicellular microbes, especially small, rounded flagellates with 1–2 flagella. Most members of *Metamonada* are larger and/or have a more complex structure (e.g., more flagella) than typical "monads".

Reference Phylogeny: The primary reference phylogeny is Hampl et al. (2009: Fig. 1, in which *Paratrimastix pyriformis* is referred to as *Trimastix pyriformis*). See also Cavalier-Smith (2003: Fig. 1), Takishita et al. (2012: Fig. S9, in which *Metamonada* is represented by the clade formed by the subclades labeled 'Fornicata' and 'Trimastix'), Leger et al. (2017: Fig. 1, wherein only subtaxa of *Metamonada* are labelled; see Fig. S1 for full taxon sampling amongst outgroups to *Metamonada*), and Yubuki et al.

(2017: Fig. 5, wherein 'caviomonads' refers to *Caviomonadidae*).

Composition: *Preaxostyla* (comprising *Oxymonadida*, e.g., *Monocercomonoides*, *Trimastix* and *Paratrimastix*), *Parabasalia* (e.g., *Trichomonas*), and *Fornicata*. *Fornicata* includes *Diplomonadida* (e.g., *Giardia* and *Spironucleus*), *Retortamonas*, *Chilomastix*, *Caviomonadidae*, *Aduncisulcus*, *Carpediemonas*, *Dysnectes*, *Ergobibamus*, *Hicanonectes*, and *Kipferlia*.

Diagnostic Apomorphies: Mitochondrial organelles lack cristae and lack the ability to perform oxidative phosphorylation.

Synonyms: There are no unambiguous or approximate synonyms. A few names (e.g., *Metamonadina* Grassé 1952, *Metamonadina* sensu Cavalier-Smith [1981, 1983] and *Axostylaria* sensu Grassé [1952]) may be partial synonyms in that they were applied to substantial subsets of *Metamonada* (as that name is defined here; see Comments) that are now known to be non-monophyletic.

Comments: Many eukaryotes, including many 'excavates', lack classical cristae-bearing mitochondria capable of oxidative phosphorylation (i.e., respiration). For example, *Parabasalia* instead have highly modified mitochondrial organelles called hydrogenosomes that perform ATP synthesis, but using an anaerobic, fermentative biochemistry and substrate-level phosphorylation, rather than respiration (Hrdy et al., 2008). *Giardia* (*Diplomonadida*, *Fornicata*), by contrast, has extremely small organelles called mitosomes that appear to have no role in ATP synthesis (Tovar et al., 2003).

Apparently non-respiratory mitochondrion-related organelles have also been reported in *Retortamonadida*, *Carpediemonas*, *Dysnectes*, *Ergobibamus*, *Hicanonectes*, *Kipferlia*, some *Oxymonadida*, *Trimastix*, and *Paratrimastix* (Hampl and Simpson, 2008; Hampl et al., 2008; Park et al., 2009; Yubuki et al., 2013, 2016; Zhang et al., 2015; Leger et al., 2017). Recently, convincing evidence was presented that a strain of *Monocercomonoides* (*Oxymonadida*) has completely lost mitochondrial organelles (Karnkowska et al., 2016).

In the later 1980s and into the 1990s, it was widely thought that many such eukaryotes, including *Parabasalia*, *Diplomonadida*, *Retortamonadida*, and *Oxymonadida*, had diverged prior to the acquisition of mitochondria through the endosymbiotic incorporation of a prokaryotic cell (e.g., Cavalier-Smith, 1987a; Patterson and Sogin, 1992). Molecular phylogenetic studies and the discovery of genes of mitochondrial origin in the nuclear genomes of *Diplomonadida* and *Parabasalia* (amongst others) indicated that all studied organisms without typical mitochondria did in fact once have this organelle (Roger, 1999; Tovar et al, 2003; Embley and Martin, 2006). Some nuclear-encoded proteins inferred to be of mitochondrial origin are actually targeted to hydrogenosomes or mitosomes, providing further evidence that these organelles are homologous to (and likely derived from) respiratory, cristate mitochondria.

Cavalier-Smith (2003) and Simpson and Roger (2004), suggested that the major clades of 'excavate' eukaryotes that lacked respiratory, cristate mitochondria may have descended from a common ancestor, and should then form a clade, since it is unlikely that the complex oxidative phosphorylation machinery of typical mitochondria could re-evolve once lost. Cavalier-Smith (2003) associated the existing taxon name *Metamonada* with this hypothesis,

and this application of the name has become generally adopted (see below).

Molecular phylogenetic studies unite *Diplomonadida*, *Retortamonadida*, *Caviomonadidae*, *Aduncisulcus*, *Carpediemonas*, *Dysnectes*, *Ergobibamus*, *Kipferlia*, and *Hicanonectes* as a clade, named *Fornicata* (Simpson et al., 2002, 2006; Simpson, 2003; Yubuki et al., 2007, 2017; Park et al., 2009; Takishita et al., 2012; Leger et al., 2017), although organisms currently assigned to *Retortamonadida* form two separated subclades within *Fornicata* (Takishita et al., 2012). Multi-protein phylogenies and some shared lateral gene transfers suggest that *Fornicata* is specially related to *Parabasalia* (Henze et al., 2001; Simpson et al., 2006; Hampl et al., 2009), forming a clade *Trichozoa* (Cavalier-Smith, 2003; note that the name *Trichozoa* was earlier used for a different proposed clade of *Parabasalia* plus *Trimastix* and *Paratrimastix* – Cavalier-Smith, 1997). Molecular phylogenies also support a *Trimastix-Paratrimastix-Oxymonadida* clade, called *Preaxostyla* or *Anaeromonada* (Dacks et al., 2001; Cavalier-Smith, 2003; Simpson, 2003; Simpson et al., 2006; Hampl et al., 2009; Zhang et al., 2015; Leger et al., 2017). The relatedness of *Trichozoa* and *Preaxostyla*, and thus the monophyly or otherwise of *Metamonada* sensu Cavalier-Smith (2003), has been debated in the recent past. Phylogenies based on a few genes often place *Paratrimastix* and *Oxymonadida* separately from *Diplomonadida*, *Parabasalia*, etc., and instead (usually) in a clade with *Malawimonadidae*, which have aerobic mitochondria (Simpson et al., 2002, 2006), however this result is likely influenced by long-branch attraction and the possible lateral transfer of tubulin genes (Hampl et al., 2005; Simpson et al., 2006, 2008). When long-branching taxa are excluded, some single gene phylogenies infer a weakly supported *Metamonada* clade (Simpson et al., 2002; Cavalier-Smith, 2003), and phylogenies based on several genes other than alpha tubulin, or on >100 genes, strongly support the monophyly of *Metamonada* to the exclusion of other eukaryotes, including *Malawimonadidae* and *Discoba* (Simpson et al., 2008; Hampl et al., 2009; Takishita et al., 2012; Kamikawa et al., 2014; Leger et al., 2017).

The taxon *Metamonadina* was introduced by Grassé (1952) as a superorder that contained the organisms now assigned to *Retortamonadida*, *Oxymonadida*, *Parabasalia*, *Trimastix*, and *Paratrimastix* but not *Diplomonadida*. Cavalier-Smith (1981, 1983) included within a phylum *Metamonadina* the organisms referred to here as *Retortamonadida*, *Diplomonadida*, and *Oxymonadida* (i.e., it excluded *Parabasalia*, *Trimastix*, and *Paratrimastix*, but included *Diplomonadida*). The same author subsequently referred to the same taxon as phylum *Metamonada* (Cavalier-Smith, 1987a,b). This unexplained change was probably made only to give the name a rank-appropriate suffix (see Cavalier-Smith, 2003: 1745), and Cavalier-Smith continued to consider Grassé as the author of *Metamonada* (Cavalier-Smith, 2003). With this composition, *Metamonada* and 'metamonads' became popular terms when these groups were widely thought to represent very early diverging, primitively amitochondriate eukaryotes (or this idea was being debated), although at that time they were often thought to not form a clade (e.g., Cavalier-Smith, 1987a,b; Brugerolle, 1993; Keeling, 1998). *Metamonada* has since also been used for *Diplomonadida* plus *Retortamonadida* alone (e.g., Cavalier-Smith, 1999). Thus, all of these prior applications of *Metamonada* and its variants excluded at least one of the largest and best-known subclades of *Metamonada* as defined in this entry (i.e., *Diplomonadida*, *Parabasalia*, or *Oxymonadida*) and/or were explicitly intended to apply to a grouping inferred to be paraphyletic.

More recently, however, Cavalier-Smith (2003) proposed an expanded composition

of *Metamonada*, namely *Diplomonadida*, *Retortamonadida*, *Carpediemonas*, *Parabasalia*, *Oxymonadida*, and *Trimastix* (*Aduncisulcus*, *Dysnectes*, *Ergobibamus*, and *Hicanonectes* were unknown at the time, while known members of *Caviomonadidae* were considered to belong to *Diplomonadida*, and *Kipferlia* was not distinguished from *Carpediemonas*), and explicitly argued that they formed a clade that descended from a common ancestor in which mitochondria had evolved into anaerobic organelles (see above). As noted above, recent molecular phylogenies broadly support the monophyly of *Metamonada* sensu Cavalier-Smith (2003) to the exclusion of all studied taxa with mitochondria capable of oxidative phosphorylation. This latest concept of *Metamonada* seems to have superseded the various previous concepts (e.g., Hampl et al., 2005, 2009; Adl et al., 2012; Leger et al., 2017). The supplied definition reflects this most recent concept. The qualifying clause ensures that the name does not apply if the most recent common ancestor of *Metamonada* in fact had mitochondria with a fully functioning oxidative phosphorylation system (i.e., all of respiratory complexes I-V), as would have to be the case if one of its descendants had mitochondria capable of oxidative phosphorylation.

Since the first use of *Metamonada* was seemingly by Cavalier-Smith (1987a), he is considered to have introduced the name (see *PhyloCode* Rec. 9.15A; Cantino and de Queiroz, 2020), even though Cavalier-Smith himself cited Grassé (1952) as the author of *Metamonada* (Cavalier-Smith, 2003), which was apparently derived directly from *Metamonadina* Grassé 1952. The name *Metamonadina* is rarely used today. No other possible partial synonym (i.e., name applied to a non-monophyletic subgroup of *Metamonada*) is in common use today, and there are no unambiguous or approximate synonyms. Grassé (1952) also used the term *Axostylaria*, without assigning a rank, for a grouping that appears to be identical in composition to his *Metamonadina* (i.e., encompassing *Retortamonadida*, *Oxymonadida*, *Parabasalia*, *Trimastix*, and *Paratrimastix* but explicitly excluding *Diplomonadida* – see above), but in recent decades *Axostylaria* has been used predominantly (but still only very occasionally) in a much more restricted sense, as a higher-ranked taxon containing only *Oxymonadida* (Cavalier-Smith, 1993). Chatton and Villeneuve (1937) introduced *Centrosomea*, which was then mentioned by Grassé (1952) in a way that can be interpreted as it represents a group similar in composition to *Metamonada* as defined here (see Cavalier-Smith, 2003). Chatton and Villeneuve's (1937) account of *Centrosomea* itself, however, refers to a substantially more inclusive grouping. Irrespective, this name has not been used for decades.

Literature Cited

Adl, S. M., A. G. B. Simpson, C. E. Lane, J. Lukeš, D. Bass, S. S. Bowser, M. W. Brown, F. Burki, M. Dunthorn, V. Hampl, A. Heiss, M. Hoppenrath, E. Lara, L. le Gall, D. H. Lynn, H. McManus, E. A. D. Mitchell, S. E. Mozley-Stanridge, L. W. Parfrey, J. Pawlowski, S. Rueckert, L. Shadwick, C. L. Schoch, A. Smirnov, and F. W. Spiegel. 2012. The revised classification of eukaryotes. *J. Eukaryot. Microbiol.* 59:429–493.

Brugerolle, G. 1993. Evolution and diversity of amitochondrial zooflagellates. *J. Eukaryot. Microbiol.* 40:616–618.

Cantino, P. D., and K. de Queiroz. 2020. *International Code of Phylogenetic Nomenclature (PhyloCode)*, Version 6. CRC Press, Boca Raton, FL.

Cavalier-Smith, T. 1981. Eukaryote kingdoms: seven or nine? *BioSystems* 14:461–481.

Cavalier-Smith, T. 1983. A 6-kingdom classification and a unified phylogeny. Pp. 265–279 in *Endocytobiology II* (W. Schwemmler and H. E. A. Schenk, eds.). de Gruyter, Berlin.

Cavalier-Smith, T. 1987a. Eukaryotes with no mitochondria. *Nature* 326:332–333.

Cavalier-Smith, T. 1987b. The origin of eukaryote and archaebacterial cells. *Ann. N. Y. Acad. Sci.* 503:17–54.

Cavalier-Smith, T. 1993. Kingdom *Protozoa* and its 18 phyla. *Microbiol. Rev.* 57:953–994.

Cavalier-Smith, T. 1997 ("1996"). Amoeboflagellates and mitochondrial cristae in eukaryote evolution: megasystematics of the new protozoan subkingdoms *Eozoa* and *Neozoa*. *Arch. Protistenkd.* 147:237–258.

Cavalier-Smith, T. 1999. Principles of protein and lipid targeting in secondary symbiogenesis: euglenoid, dinoflagellate, and sporozoan plastid origins and the eukaryote family tree. *J. Eukaryot. Microbiol.* 46:347–366.

Cavalier-Smith, T. 2003. The excavate protozoan phyla *Metamonada* Grasse emend. (*Anaeromonadea, Parabasalia, Carpediemonas, Eopharyngia*) and *Loukozoa* emend. (*Jakobea, Malawimonas*): their evolutionary affinities and new higher taxa. *Int. J. Syst. Evol. Microbiol.* 53:1741–1758.

Chatton, E., and S. Villeneuve. 1937. *Gregarella fabrearum* Chatton et Brachon protiste parasite du cilié *Fabrea salina* Hennéguy. La notion de dépolarisation chez les flagellés et la conception des apomastigines. *Arch. Zool. Exp. Gén.* (Notes et Rev.) 78:216–237.

Dacks, J. B., J. D. Silberman, A. G. Simpson, S. Moriya, T. Kudo, M. Ohkuma, and R. J. Redfield. 2001. Oxymonads are closely related to the excavate taxon *Trimastix*. *Mol. Biol. Evol.* 18:1034–1044.

Embley, T. M., and W. Martin. 2006. Eukaryotic evolution, changes and challenges. *Nature* 440:623–630.

Grassé, P. P. 1952. *Traité de Zoologie. Tome I. Fascicule I. Protozoires (Généralités. Flagellés)*. Masson et Cie, Paris.

Hampl, V., D. S. Horner, P. Dyal, J. Kulda, J. Flegr, P. G. Foster, and T. M. Embley. 2005. Inference of the phylogenetic position of oxymonads based on nine genes: support for *Metamonada* and *Excavata*. *Mol. Biol. Evol.* 22:2508–2518.

Hampl, V., L. Hug, J. W. Leigh, J. B. Dacks, B. F. Lang, A. G. B. Simpson, and A. J. Roger. 2009. Phylogenomic analyses support the monophyly of *Excavata* and resolve relationships among eukaryotic "supergroups". *Proc. Natl. Acad. Sci. USA* 106:3859–3864.

Hampl, V., J. D. Silberman, A. Stechmann, S. Diaz-Trivino, P. J. Johnson, and A. J. Roger. 2008. Genetic evidence for a mitochondriate ancestry in the 'amitochondriate' flagellate *Trimastix pyriformis*. *PLOS ONE* 3:e1383.

Hampl, V., and A. G. B. Simpson. 2008. Possible mitochondria-related organelles in poorly-studied 'amitochondriate' eukaryotes. Pp. 265–282 in *Hydrogenosomes and Mitosomes: Mitochondria of Anaerobic Eukaryotes* (J. Tachezy, ed.). Springer Verlag, Berlin.

Henze, K., D. S. Horner, S. Suguri, D. V. Moore, L. B. Sanchez, M. Müller, and T. M. Embley. 2001. Unique phylogenetic relationships of glucokinase and glucosephosphate isomerase of the amitochondriate eukaryotes *Giardia intestinalis*, *Spironucleus barkhanus* and *Trichomonas vaginalis*. *Gene* 281:123–131.

Hrdy, I., J. Tachezy, and M. Müller. 2008. Metabolism of trichomonad hydrogenosomes. Pp. 113–145 in *Hydrogenosomes and Mitosomes: Mitochondria of Anaerobic Eukaryotes* (J. Tachezy, ed.). Springer Verlag, Berlin.

Kamikawa, R., M. Kolisko, Y. Nishimura, A. Yabuki, M. W. Brown, S. A. Ishikawa, K. Ishida, A. J. Roger, T. Hashimoto, and Y. Inagaki. 2014. Gene content evolution in discobid mitochondria deduced from the phylogenetic position and complete mitochondrial genome of *Tsukubamonas globosa*. *Genome Biol. Evol.* 6:306–315.

Karnkowska, A., V. Vacek, Z. Zubáčová, S. C. Treitli, R. Petrželková, L. Eme, L. Novák, V. Žárský, L. D. Barlow, E. K. Herman, P. Soukal, M. Hroudová, P. Doležal, C. W. Stairs, A. J. Roger, M. Eliáš, J. B. Dacks, Č. Vlček, and V. Hampl. 2016. A eukaryote without a mitochondrial organelle. *Curr. Biol.* 26:1274–1284.

Keeling, P. J. 1998. A kingdom's progress: *Archezoa* and the origin of eukaryotes. *Bioessays* 20:87–95.

Leger, M. M., M. Kolisko, R. Kamikawa, C. W. Stairs, K. Kume, I. Čepička, J. D. Silberman, J. O. Andersson, F. Xu, A. Yabuki, L. Eme, Q. Zhang, K. Takishita, Y. Inagaki, A. G.

B. Simpson, T. Hashimoto, and A. J. Roger. 2017. Organelles that illuminate the origins of *Trichomonas* hydrogenosomes and *Giardia* mitosomes. *Nat. Ecol. Evol.* 1:0092.

Park, J. S., M. Kolisko, A. A. Heiss, and A. G. B. Simpson. 2009. Light microscopic observations, ultrastructure, and molecular phylogeny of *Hicanonectes teleskopos* n. gen., n. sp., a deep-branching relative of diplomonads. *J. Eukaryot. Microbiol.* 56:273–284.

Patterson, D.J., and M. L. Sogin. 1992. Eukaryote origins and protistan diversity. Pp. 14–46 in *The Origin and Evolution of the Cell* (H. Hartmann and K. Matsuno, eds.). World Scientific, Singapore.

Roger, A. J. 1999. Reconstructing early events in eukaryotic evolution. *Am. Nat.* 154:S146–S163.

Simpson, A. G. B. 2003. Cytoskeletal organization, phylogenetic affinities and systematics in the contentious taxon *Excavata (Eukaryota)*. *Int. J. Syst. Evol. Microbiol.* 53:1759–1777.

Simpson, A. G. B., Y. Inagaki, and A. J. Roger. 2006. Comprehensive multigene phylogenies of excavate protists reveal the evolutionary positions of 'primitive' eukaryotes. *Mol. Biol. Evol.* 23:615–625.

Simpson, A. G. B., T. A. Perley, and E. Lara. 2008. Lateral transfer of the gene for a widely used marker, α-tubulin, indicated by a multiprotein study of the phylogenetic position of *Andalucia (Excavata)*. *Mol. Phylogenet. Evol.* 47:366–377.

Simpson A. G. B., and A. J. Roger. 2004. *Excavata* and the origin of amitochondriate eukaryotes. Pp. 27–53 in *Organelles, Genomes, and Eukaryote Phylogeny: An Evolutionary Synthesis in the Age of Genomics* (R. P. Hirt and D. S. Horner, eds.). CRC Press, Boca Raton, FL.

Simpson, A. G. B., A. J. Roger, J. D. Silberman, D. D. Leipe, V. P. Edgcomb, L. S. Jermiin, D. J. Patterson, and M. L. Sogin. 2002. Evolutionary history of 'early-diverging' eukaryotes: the excavate taxon *Carpediemonas* is a close relative of *Giardia*. *Mol. Biol. Evol.* 19:1782–1791.

Takishita, K., M. Kolisko, H. Komatsuzaki, A. Yabuki, Y. Inagaki, I. Čepička, P. Smejkalová, J. D. Silberman, T. Hashimoto, A. J. Roger, and

A. G. B. Simpson. 2012. Multigene phylogenies of diverse *Carpediemonas*-like organisms identify the closest relatives of 'amitochondriate' diplomonads and retortamonads. *Protist* 163:344–355.

Tovar, J., G. León-Avila, L. B. Sánchez, R. Sutak, J. Tachezy, M. van der Giezen, M. Hernández, M. Müller, and J. M. Lucocq. 2003. Mitochondrial remnant organelles of *Giardia* function in iron-sulphur protein maturation. *Nature* 426:172–176.

Yubuki, N., S. S. C. Huang, and B. S. Leander. 2016. Comparative ultrastructure of fornicate excavates, including a novel free-living relative of diplomonads: *Aduncisulcus paluster* gen. et sp. nov. *Protist* 167:584–596.

Yubuki, N., Y. Inagaki, T. Nakayama, and I. Inouye. 2007. Ultrastructure and ribosomal RNA phylogeny of the free-living heterotrophic flagellate *Dysnectes brevis* n. gen., n. sp., a new member of the *Fornicata*. *J. Eukaryot. Microbiol.* 54:191–200.

Yubuki, N., A. G. B. Simpson, and B. S. Leander. 2013. Comprehensive ultrastructure of *Kipferlia bialata* provides evidence for character evolution within the *Fornicata*. *Protist* 164:423–439.

Yubuki, N., E. Zadrobílková, and I. Čepička. 2017. Ultrastructure and molecular phylogeny of *Iotanema spirale* gen. nov. et sp. nov., a new lineage of endobiotic *Fornicata* with strikingly simplified ultrastructure. *J. Eukaryot. Microbiol.* 64:422–433.

Zhang, Q., P. Táborský, J. D. Silberman, T. Pánek, I. Čepička, and A. G. B. Simpson. 2015. Marine isolates of *Trimastix marina* form a plesiomorphic deep-branching lineage within *Preaxostyla*, separate from other known trimastigids (*Paratrimastix* n. gen.). *Protist* 166:468–491.

Author

Alastair G. B. Simpson; Department of Biology; Dalhousie University; Halifax, NS, B3H 4R2, Canada. Email: alastair.simpson@dal.ca.

Date Accepted: 21 January 2014; updated 24 February 2018

Primary Editor: Philip Cantino

Discoba A. G. B. Simpson in Hampl et al. 2009 [A. G. B. Simpson], converted clade name

Registration Number: 37

Definition: The smallest crown clade containing *Jakoba libera* (Ruinen) Patterson 1990 (*Jakobida*), *Andalucia godoyi* Lara et al. 2006 (*Jakobida*), *Euglena gracilis* Klebs 1883 (*Discicristata*), and *Naegleria gruberi* (Schardinger) Alexeieff 1912 (*Discicristata*), but not *Homo sapiens* Linnaeus 1758 (*Opisthokonta*) or *Arabidopsis thaliana* (Linnaeus) Heynhold 1842 (*Archaeplastida*). This is a minimum-crown-clade definition with external specifiers. Abbreviated definition: min crown ∇ (*Jakoba libera* (Ruinen) Patterson 1990 & *Andalucia godoyi*, Lara et al. 2006 & *Euglena gracilis* Klebs 1883 & *Naegleria gruberi* (Schardinger) Alexeieff 1912 ~ *Homo sapiens* Linnaeus 1758 ∇ *Arabidopsis thaliana* (Linnaeus) Heynhold 1842).

Etymology: A blend of the names *Discicristata* and *Jakoba*, representing the two major clades that compose this taxon (*Jakoba* is a member of *Jakobida*).

Reference Phylogeny: The primary reference phylogeny is Hampl et al. (2009: Fig. 1). *Andalucia godoyi* (a specifier in the definition) is closely related to *Stygiella incarcerata*, which is included in the reference phylogeny as *Andalucia incarcerata* (Lara et al., 2006; Simpson et al., 2008). See also Simpson et al. (2008: Figs. 1, 3) and Kamikawa et al. (2014: Fig. 1).

Composition: *Discicristata* (defined in this volume), *Jakobida* Cavalier-Smith 1993, and *Tsukubamonas* Yabuki et al. 2011.

Diagnostic Apomorphies: No non-molecular synapomorphies are known.

Synonyms: Also known informally as 'the JEH group' (e.g., Rodríguez-Ezpeleta et al., 2007), which is an acronym of the three main subclades (*Jakobida*, *Euglenozoa*, and *Heterolobosea*, the latter two composing *Discicristata*). This is an approximate synonym.

Comments: A close relationship between *Jakobida* and *Heterolobosea* and/or *Euglenozoa* (both in *Discicristata*) was proposed tentatively on the basis of (often weakly supported) clades inferred in single gene phylogenies (e.g., Archibald et al., 2002; Cavalier-Smith, 2003). Multigene phylogenies supported this clade more strongly (Simpson et al., 2006, 2008), and the monophyly of the group has now been confirmed solidly by phylogenomic-scale analyses (Rodríguez-Ezpeleta et al., 2007; Burki et al., 2007, 2008; Hampl et al., 2009). Ribosomal RNA and small-scale multigene phylogenies subsequently placed the newly discovered *Tsukubamonas* within the *Jakobida-Heterolobosea-Euglenozoa* clade (Yabuki et al., 2011). This has been more robustly confirmed by phylogenomic-scale analyses, which also generally recover *Tsukubamonas* as the sister to *Discicristata* (Kamikawa et al., 2014; Yang et al., 2017).

Simpson, in Hampl et al. (2009; supporting information), proposed *Discoba* as an unranked taxon, with a phylogenetic definition identical in function to that supplied here (and specifying the same reference phylogeny), but the wording here has been changed, with external specifiers used instead of a qualifying clause.

In isolation, the uniquely ancestral-looking (i.e., bacterial-like) mitochondrial genomes of *Jakobida* (Lang et al., 1997; Gray et al., 2004;

Burger et al., 2013) suggest the possibility that the root of *Eukarya* lies 'within' *Discoba* (see Cavalier-Smith, 2000; Simpson & Roger, 2004). Further, Cavalier-Smith (2010) suggested that the clade *Euglenozoa* is the sister group to all other eukaryotes, which would also place the root of *Eukarya* 'within' *Discoba* (though currently this is not a widely supported hypothesis). The external specifiers included in the definition here (and in the qualifying clause in Hampl et al., 2009) are intended to make the name *Discoba* inapplicable if the root of *Eukarya* lies 'within' *Discoba*.

If *Jakobida* were to be demonstrated to fall within the current composition of *Discicristata* (as defined in this volume) such that the names *Discicristata* and *Discoba* applied to the same clade, the name *Discoba* should have precedence over *Discicristata* (see *Discicristata* for further details).

Literature Cited

Archibald, J. M., C. J. O'Kelly, and W. F. Doolittle. 2002. The chaperonin genes of jakobid and jakobid-like flagellates: implications for eukaryotic evolution. *Mol. Biol. Evol.* 19:422–431.

Burger, G., M. W. Gray, L. Forget, and B. F. Lang. 2013. Strikingly bacteria-like and gene-rich mitochondrial genomes throughout jakobid protists. *Genome Biol. Evol.* 5:418–438.

Burki, F., K. Shalchian-Tabrizi, M. Minge, A. Skjaeveland, S. I. Nikolaev, K. S. Jakobsen, and J. Pawlowski. 2007. Phylogenomics reshuffles the eukaryotic supergroups. *PLOS ONE* 2:e790.

Burki, F., K. Shalchian-Tabrizi, and J. Pawlowski. 2008. Phylogenomics reveals a new 'megagroup' including most photosynthetic eukaryotes. *Biol. Lett.* 4:366–369.

Cavalier-Smith, T. 2000. Flagellate megaevolution: the basis for eukaryote diversification. Pp. 361–390 in *The Flagellates; Unity, Diversity and Evolution* (B. S. C. Leadbeater and J. C. Green, eds.). Taylor & Francis, London.

Cavalier-Smith, T. 2003. The excavate protozoan phyla *Metamonada* Grasse emend. (*Anaeromonadea, Parabasalia, Carpediemonas, Eopharyngia*) and *Loukozoa* emend. (*Jakobea, Malawimonas*): their evolutionary affinities and new higher taxa. *Int. J. Syst. Evol. Microbiol.* 53:1741–1758.

Cavalier-Smith, T. 2010. Kingdoms *Protozoa* and *Chromista* and the eozoan root of the eukaryotic tree. *Biol. Lett.* 6:342–345.

Gray, M. W., B. F. Lang, and G. Burger. 2004. Mitochondria of protists. *Ann. Rev. Genet.* 38:477–524.

Hampl, V., L. Hug, J. W. Leigh, J. B. Dacks, B. F. Lang, A. G. B. Simpson, and A. J. Roger. 2009. Phylogenomic analyses support the monophyly of *Excavata* and resolve relationships among eukaryotic "supergroups". *Proc. Natl. Acad. Sci. USA* 106:3859–3864.

Kamikawa, R., M. Kolisko, Y. Nishimura, A. Yabuki, M. W. Brown, S. A. Ishikawa, K. Ishida, A. J. Roger, T. Hashimoto, and Y. Inagaki. 2014. Gene content evolution in discobid mitochondria deduced from the phylogenetic position and complete mitochondrial genome of *Tsukubamonas globosa*. *Genome Biol. Evol.* 6:306–315.

Lang, B. F., G. Burger, C. J. O'Kelly, R. Cedergren, G. B. Golding, C. Lemieux, D. Sankoff, M. Turmel, and M. W. Gray. 1997. An ancestral mitochondrial DNA resembling a eubacterial genome in miniature. *Nature* 387:493–497.

Lara, E., A. Chatzinotas, and A. G. B. Simpson. 2006. *Andalucia* (gen. nov.): a new taxon for the deepest branch within jakobids (*Jakobida; Excavata*), based on morphological and molecular study of a new flagellate from soil. *J. Eukaryot. Microbiol.* 53:112–120.

Rodríguez-Ezpeleta, N., H. Brinkmann, G. Burger, A. J. Roger, M. W. Gray, H. Philippe, and B. F. Lang. 2007. Toward resolving the eukaryotic tree: the phylogenetic positions of jakobids and cercozoans. *Curr. Biol.* 17:1420–1425.

Simpson, A. G. B., Y. Inagaki, and A. J. Roger. 2006. Comprehensive multigene phylogenies of excavate protists reveal the evolutionary positions of "primitive" eukaryotes. *Mol. Biol. Evol.* 23:615–625.

Simpson, A. G. B., T. A. Perley, and E. Lara. 2008. Lateral transfer of the gene for a widely used marker, α-tubulin, indicated by a multi-protein study of the phylogenetic position of *Andalucia* (*Excavata*). *Mol. Phylogenet. Evol.* 47:366–377.

Simpson A. G. B., and A. J. Roger 2004. *Excavata* and the origin of amitochondriate eukary-otes. Pp. 27–53 in *Organelles, Genomes, and Eukaryote Phylogeny: An Evolutionary Synthesis in the Age of Genomics* (R. P. Hirt and D. S. Horner, eds.). CRC Press, Boca Raton, FL.

Yabuki, A., T. Nakayama, N. Yubuki, T. Hashimoto, K. Ishida, and Y. Inagaki. 2011. *Tsukubamonas globosa* n. gen., n. sp., a novel excavate flagellate possibly holding a key for the early evolution in "Discoba". *J. Eukaryot. Microbiol.* 58:319–331.

Yang, J., T. Harding, R. Kamikawa, A. G. B. Simpson, and A. J. Roger. 2017. Mitochondrial genome evolution and a novel RNA editing system in deep-branching heteroloboseids. *Genome Biol. Evol.* 9:1161–1174.

Author

Alastair G. B. Simpson; Department of Biology; Dalhousie University; Halifax, NS, B3H 4R2, Canada. Email: alastair.simpson@dal.ca.

Date Accepted: 27 July 2013; updated 24 February 2018

Primary Editor: Philip Cantino

Discicristata T. Cavalier-Smith 1998 [A. G. B. Simpson], converted clade name

Registration Number: 36

Definition: The smallest crown clade containing *Euglena gracilis* Klebs 1883 (*Euglenozoa*) and *Naegleria gruberi* (Schardinger) Alexeieff 1912 (*Heterolobosea*). This is a minimum-crown-clade definition. Abbreviated definition: min crown ∇ (*Euglena gracilis* Klebs 1883 & *Naegleria gruberi* (Schardinger) Alexeieff 1912).

Etymology: Refers to the discoidal mitochondrial cristae typical of this group.

Reference Phylogeny: The primary reference phylogeny is Hampl et al. (2009: Fig. 1). See also Yang et al. (2017: Fig. 1).

Composition: *Euglenozoa* (as defined in this volume) and *Heterolobosea* Page and Blanton 1985.

Diagnostic Apomorphies: Discoidal mitochondrial cristae are usually considered an apomorphy for this clade (but see Comments).

Synonyms: *Prodiscea* sensu Patterson (1988) is an approximate synonym (see Comments).

Comments: *Euglenozoa* and *Heterolobosea* have been considered as possible close relatives since the mid-1980s, due mainly to the shared possession of discoidal (disc-shaped) mitochondrial cristae (see Patterson and Brugerolle, 1988). This feature is rare across eukaryotes, though not unique to *Euglenozoa* and *Heterolobosea* (see below).

Some molecular phylogenies, especially those based on inferred amino acid sequences from multiple nuclear genes, united *Euglenozoa* and *Heterolobosea* as a clade, within a moderate sampling of major eukaryote groups (Baldauf et al., 2000). Phylogenomic analyses with a broader sampling of other eukaryotes, including *Jakobida*, *Malawimonadidae* and additional *Metamonada* (i.e., other 'excavates'), demonstrated strongly that *Euglenozoa* and *Heterolobosea* form a clade to the exclusion of other taxa (Rodríguez-Ezpeleta et al., 2007; Hampl et al., 2009). Recently, phylogenomic analyses that also include *Tsukubamonas* have also recovered a strongly supported *Euglenozoa*-*Heterolobosea* clade, with *Tsukubamonas* and then *Jakobida* as its successive sister groups (Kamikawa et al., 2014; Yang et al., 2017). *Stephanopogon*, which has discoidal cristae, and has often been considered as related to *Heterolobosea* and/or *Euglenozoa*, is nested within *Heterolobosea* according to ribosomal RNA gene phylogenies, and is now routinely assigned to the taxon *Heterolobosea* (Cavalier-Smith and Nikolaev, 2008; Yubuki and Leander, 2008; Adl et al., 2012).

The name *Discicristata* was introduced by Cavalier-Smith (1998) as an infrakingdom that contained *Euglenozoa* and *Percolozoa* (an approximate synonym of *Heterolobosea* – see Brugerolle and Simpson, 2004). The name *Discicristata*, and part of Cavalier-Smith's (1998) diagnosis, reflected the shared character of discoidal mitochondrial cristae. The present understanding of eukaryote phylogeny makes it parsimonious that discoidal cristae are indeed an apomorphy for *Discicristata*, since *Tsukubamonas* and *Jakobida* are inferred to be successive sister groups to *Discicristata* in phylogenomic analyses (Kamikawa et al., 2014; Yang et al., 2017) and mostly have tubular cristae (Lara et al., 2006; Yabuki et al., 2011; Strassert et al., 2016). However, all molecular phylogenies so far include only one species of *Tsukubamonas*, making phylogenetic inaccuracy due to poor taxon sampling a possibility, while statistical support for the *Tsukubamonas-Discicristata* clade is modest or weak in some analyses (Yabuki et al., 2011; Kamikawa et al., 2014). Further, the sister group of *Discoba* (i.e., of the *Discicristata-Tsukubamonas-Jakobida* clade*)* is not known at present, and one hypothesis is that *Discoba* is closely related to *Malawimonadidae* and *Metamonada* (i.e., other 'excavates'; Simpson et al., 2006; Hampl et al., 2009). *Metamonada* lack cristae altogether but *Malawimonadidae* have discoidal cristae (O'Kelly and Nerad, 1999). Taken together, these uncertainties make it difficult to rule out the possibility that discoidal cristae are symplesiomorphic for *Discicristata*.

In hierarchical lists of eukaryote groups, Patterson (1988) and Patterson and Sogin (1992) included the name *Prodiscea* for a hypothesized clade composed of *Euglenozoa*, *Heterolobosea* and *Stephanopogon* (which is now considered to be part of *Heterolobosea*; see above) but provided no explicit definition. This name has received little or no subsequent use.

If the root of the eukaryote phylogenetic tree were demonstrated to fall within the current composition of *Discicristata* (see *Discoba*, this volume), such that the names *Discicristata* and *Eukarya* (as defined in this volume) referred to the same clade, *Eukarya* should have precedence over *Discicristata*.

Jakobida falls within the current composition of *Discicristata* in some multigene phylogenetic analyses, especially some with very limited data from *Heterolobosea* (Simpson et al., 2006; Yabuki et al., 2011), such that the names *Discicristata* and *Discoba* (as defined in this volume) apply to identical clades. This is unlikely to represent the true evolutionary relationship; phylogenomic analyses with improved taxon sampling infer a *Euglenozoa-Heterolobosea* clade to the exclusion of *Jakobida* (e.g., Rodríguez-Ezpeleta et al., 2007; Hampl et al., 2009; Kamikawa et al., 2014; Yang et al., 2017), as, in general, do

ribosomal RNA gene phylogenies in which a *Discoba* clade is inferred (e.g., Cavalier-Smith, 2003). Nonetheless, if the names *Discicristata* and *Discoba*, as defined in this volume, end up referring to the same clade, the name *Discoba* should have precedence over *Discicristata*.

Literature Cited

Adl, S. M., A. G. B. Simpson, C. E. Lane, J. Lukeš, D. Bass, S. S. Bowser, M. W. Brown, F. Burki, M. Dunthorn, V. Hampl, A. Heiss, M. Hoppenrath, E. Lara, L. le Gall, D. H. Lynn, H. McManus, E. A. D. Mitchell, S. E. Mozley-Stanridge, L. W. Parfrey, J. Pawlowski, S. Rueckert, L. Shadwick, C. L. Schoch, A. Smirnov, and F. W. Spiegel. 2012. The revised classification of eukaryotes. *J. Eukaryot. Microbiol.* 59:429–493.

Baldauf, S. L., A. J. Roger, I. Wenk-Siefert, and W. F. Doolittle. 2000. A kingdom-level phylogeny of eukaryotes based on combined protein data. *Science* 290:972–977.

Brugerolle, G., and A. G. B. Simpson. 2004. The flagellar apparatus of heteroloboseans. *J. Eukaryot. Microbiol.* 51:96–107.

Cavalier-Smith, T. 1998. A revised six-kingdom system of life. *Biol. Rev.* 73:203–266.

Cavalier-Smith, T. 2003. The excavate protozoan phyla *Metamonada* Grasse emend. (*Anaeromonadea, Parabasalia, Carpediemonas, Eopharyngia*) and *Loukozoa* emend. (*Jakobea, Malawimonas*): their evolutionary affinities and new higher taxa. *Int. J. Syst. Evol. Microbiol.* 53:1741–1758.

Cavalier-Smith, T., and S. Nikolaev. 2008. The zooflagellates *Stephanopogon* and *Percolomonas* are a clade (Class *Percolatea*: Phylum *Percolozoa*). *J. Eukaryot. Microbiol.* 55:501–509.

Hampl, V., L. Hug, J. W. Leigh, J. B. Dacks, B. F. Lang, A. G. B. Simpson, and A. J. Roger. 2009. Phylogenomic analyses support the monophyly of *Excavata* and resolve relationships among eukaryotic "supergroups". *Proc. Natl. Acad. Sci. USA* 106:3859–3864.

Kamikawa, R., M. Kolisko, Y. Nishimura, A. Yabuki, M. W. Brown, S.A. Ishikawa, K. Ishida, A. J. Roger, T. Hashimoto, and Y. Inagaki. 2014. Gene content evolution in discobid mitochondria deduced from the phylogenetic position and complete mitochondrial genome of *Tsukubamonas globosa*. *Genome Biol. Evol.* 6:306–315.

Lara, E., A. Chatzinotas, and A. G. B. Simpson. 2006. *Andalucia* (n. gen.)—the deepest branch within jakobids (*Jakobida*; *Excavata*), based on morphological and molecular study of a new flagellate from soil. *J. Eukaryot. Microbiol.* 53:112–120.

O'Kelly, C. J., and T. A. Nerad. 1999. *Malawimonas jakobiformis* n. gen., n. sp. (*Malawimonadidae* n. fam.): a *Jakoba*-like heterotrophic nanoflagellate with discoidal mitochondrial cristae. *J. Eukaryot. Microbiol.* 46:522–531.

Page, F. C., and R. L. Blanton. 1985. The *Heterolobosea* (*Sarcodina*: *Rhizopoda*), a new class uniting the *Schizopyrenida* and the *Acrasidae* (*Acrasida*). *Protistologica* 12:37–53.

Patterson, D. J. 1988. The evolution of protozoa. *Mem. Inst. Oswaldo Cruz* 83(suppl. 1):580–600.

Patterson, D. J., and G. Brugerolle. 1988. The ultrastructural identity of *Stephanopogon apogon* and the relatedness of the genus to other kinds of protists. *Eur. J. Protistol.* 23:279–290.

Patterson, D.J., and M. L. Sogin 1992. Eukaryote origins and protistan diversity. Pp. 14–46 in *The Origin and Evolution of the Cell* (H. Hartmann and K. Matsuno, eds.). World Scientific, Singapore.

Rodríguez-Ezpeleta, N., H. Brinkmann, G. Burger, A. J. Roger, M. W. Gray, H. Philippe, and B. F. Lang. 2007. Toward resolving the eukaryotic tree: the phylogenetic positions of jakobids and cercozoans. *Curr. Biol.* 17:1420–1425.

Simpson, A. G. B., Y. Inagaki, and A. J. Roger. 2006. Comprehensive multigene phylogenies of excavate protists reveal the evolutionary positions of "primitive" eukaryotes. *Mol. Biol. Evol.* 23:615–625.

Strassert, J. F. H., D. V. Tikhonenkov, J.-F. Pombert, M. Kolisko, V. Tai, A. P. Mylnikov, and P. J. Keeling. 2016. *Moramonas marocensis* gen. nov., sp. nov.: a jakobid flagellate isolated from desert soil with a bacteria-like, but bloated mitochondrial genome. *Open Biol.* 6:150239.

Yabuki, A., T. Nakayama, N. Yubuki, T. Hashimoto, K. Ishida, and Y. Inagaki. 2011. *Tsukubamonas globosa* n. gen., n. sp., a novel excavate flagellate possibly holding a key for the early evolution in "*Discoba*:". *J. Eukaryot. Microbiol.* 58:319–331.

Yang, J., T. Harding, R. Kamikawa, A. G. B. Simpson, and A. J. Roger. 2017. Mitochondrial genome evolution and a novel RNA editing system in deep-branching heteroloboseids. *Genome Biol. Evol.* 9:1161–1174.

Yubuki, N., and B. S. Leander. 2008. Ultrastructure and molecular phylogeny of *Stephanopogon minuta*: an enigmatic microeukaryote from marine interstitial environments. *Eur. J. Protistol.* 44:241–253.

Author

Alastair G.B. Simpson; Department of Biology; Dalhousie University; Halifax, NS, B3H 4R2, Canada. Email: alastair.simpson@dal.ca.

Date Accepted: 24 September 2013; updated 24 February 2018

Primary Editor: Philip Cantino

Euglenozoa T. Cavalier-Smith 1981 [A. G. B. Simpson], converted clade name

Registration Number: 116

Definition: The crown clade for which paraxonemal rods, as inherited by *Entosiphon sulcatum* (Dujardin) Stein 1878, are an apomorphy relative to other crown clades. Heteromorphic in this sense specifically means with a three-dimensional lattice appearance in the posterior/ventral flagellum (flagellum 1) and a tubular/whorled appearance in the anterior/dorsal flagellum (flagellum 2). This is an apomorphy-modified crown-clade definition. Abbreviated definition: crown ∇ apo heteromorphic paraxonemal rods [*Entosiphon sulcatum* (Dujardin) Stein 1878].

Etymology: Derived from *Euglena* (name of an included taxon; from the Greek *eu*, true or original, and *glene*, eyeball) plus the Greek *zoön* (animal).

Reference Phylogeny: The primary reference phylogeny is von der Heyden et al. (2004: Fig. 1). Heteromorphic paraxonemal rods are restricted to the clade labeled *Euglenozoa*. See also Simpson and Roger (2004: Figs. 1–3) and Yubuki et al. (2009: Fig. 10).

Composition: *Euglenida* (referred to as *Euglenoidea* in the primary reference phylogeny), *Kinetoplastea*, *Diplonemea* and *Symbiontida*. *Euglenida* might include *Symbiontida* (see Comments).

Diagnostic Apomorphies: The following are synapomorphies relative to other crown clades:

1) Heteromorphic paraxonemal rods: The paraxonemal rod of the posterior/ventral flagellum (flagellum 1) has a three-dimensional lattice appearance, while the rod of the anterior/dorsal flagellum (flagellum 2) has a tubular/whorled appearance.
2) A distinctive, asymmetric arrangement of three flagellar microtubular roots, with one particular root supporting longitudinally a tubular feeding apparatus.
3) Tubular extrusomes. The extrusomes are tubular, have a thick wall when in the undischarged state, and expand and elongate as a tubular lattice upon discharge. Tubular extrusomes are present in members of all major clades of *Euglenozoa*, but have been secondarily lost in numerous clades.

See Simpson (1997) for further details.

Synonyms: None known

Comments: A close evolutionary relationship between *Euglenida* (e.g., *Euglena*, *Peranema*, *Entosiphon*) and *Kinetoplastea* (e.g., *Trypanosoma*, *Leishmania*, *Bodo*) has been proposed for a long time on the basis of ultrastructural similarities (reviewed in Triemer and Farmer, 1991). This evidence was summarized in the mid-1980s, and at about the same time it became clear that the organisms now referred to as *Diplonemea* were distinct from *Euglenida*, yet related to *Euglenida* and *Kinetoplastea* (Kivic and Walne, 1984; Brugerolle, 1985; Triemer and Farmer, 1991). The members of *Symbiontida* (*Postgaardi*, *Calkinsia*, and *Bihospites*; the clade is also referred to as *Postgaardea*) also share this suite of ultrastructural features (Simpson et al., 1997; Yubuki et al., 2009; Breglia et al., 2010). *Symbiontida* has been considered as an additional major group within *Euglenozoa* (Simpson et al., 1997; Yubuki et al., 2009; Cavalier-Smith, 2016) but could also represent a subclade within *Euglenida*, based on weak molecular phylogenetic results and morphological considerations (Breglia et al., 2010; Yubuki et al., 2013; Lax and Simpson, 2013). Almost all relevant molecular phylogenetic analyses support the monophyly of *Euglenida* plus *Kinetoplastea*, and where included, *Diplonemea* (e.g., Sogin, 1989; Simpson and Roger, 2004; Rodríguez-Ezpeleta et al., 2007; Parfrey et al., 2010; Cavalier-Smith et al., 2016; Yang et al., 2017). Other little-studied organisms that are sometimes proposed to belong within *Euglenozoa* but lack the diagnostic apomorphies were discussed critically by Simpson (1997) and are excluded here.

The name *Euglenozoa* was introduced by Cavalier-Smith (1981) for the groups referred to here as *Euglenida* and *Kinetoplastea* (the distinctiveness of *Diplonemea* from *Euglenida* was not recognized at that time), and in a note added in proof, *Stephanopogon*. *Stephanopogon* has been more-or-less universally excluded from *Euglenozoa* since the mid-1990s (see Simpson, 1997) and is now known to belong to *Heterolobosea* (Cavalier-Smith and Nikolaev, 2008; Yubuki and Leander, 2008). Simpson (1997) explicitly associated the taxon *Euglenozoa* with the diagnostic apomorphies listed above and also defined the name *Euglenozoa* by the apomorphy of heteromorphic paraxonemal rods, in what was essentially an apomorphy-based definition. The definition used here differs in applying to the crown clade associated with the same apomorphy. However, since the defining synapomorphy is not known to fossilize, and fossils for the stem of *Euglenozoa* are unknown, the known compositions of the crown clade and its corresponding apomorphy-based clade are identical. The taxon name *Euglenozoa* is now used very widely, almost always in a manner consistent with this clade composition, and at least implicitly for the crown clade.

For example, molecular phylogenetic studies in which the name *Euglenozoa* was applied to a clade with this composition (e.g., Simpson and Roger, 2004; von der Heyden, 2004; Yubuki et al., 2009; Parfrey et al., 2010; Cavalier-Smith, 2016) implicitly applied the name to the crown clade because the data examined came entirely from extant organisms.

Literature Cited

Breglia, S. A., N. Yubuki, M. Hoppenrath, and B. S. Leander. 2010. Ultrastructure and molecular phylogenetic position of a novel euglenozoan with extrusive episymbiotic bacteria: *Bihospites bacati* n. gen. et sp. (*Symbiontida*). *BMC Microbiol.* 10:e145.

Brugerolle, G. 1985. Des trichocystes chez les bodonides, un caractère phylogénétique supplémentaire entre *Kinetoplastida* et *Euglenida*. *Protistologica* 21:339–348.

Cavalier-Smith, T. 1981. Eukaryote kingdoms: seven or nine? *BioSystems* 14:461–481.

Cavalier-Smith, T. 2016. Higher classification and phylogeny of *Euglenozoa*. *Eur. J. Protistol.* 56:250–276.

Cavalier-Smith, T., E. E. Chao, and K. Vickerman. 2016. New phagotrophic euglenoid species (new genus *Decastava*; *Scytomonas saepesedens*; *Entosiphon oblongum*), Hsp90 introns, and putative euglenoid Hsp90 pre-mRNA insertional editing. *Eur. J. Protistol.* 56:147–170.

Cavalier-Smith, T., and S. Nikolaev. 2008. The zooflagellates *Stephanopogon* and *Percolomonas* are a clade (class *Percolatea*: phylum *Percolozoa*). *J. Eukaryot. Microbiol.* 55:501–509.

Kivic, P. A., and P. L. Walne. 1984. An evaluation of a possible phylogenetic relationship between the *Euglenophyta* and *Kinetoplastida*. *Orig. Life* 13:269–288.

Lax, G., and A. G. B. Simpson. 2013. Combining molecular data with classical morphology for uncultured phagotrophic euglenids (*Excavata*): a single-cell approach. *J. Eukaryot. Microbiol.* 60:615–625.

Parfrey, L. W., J. Grant, Y. I. Tekle, E. Lasek-Nesselquist, H. G. Morrison, M. L. Sogin, D. J. Patterson, and L. A. Katz. 2010. Broadly sampled multigene analyses yield a well-resolved eukaryotic tree of life. *Syst. Biol.* 59:518–533.

Rodríguez-Ezpeleta, N., H. Brinkmann, G. Burger, A. J. Roger, M. W. Gray, H. Philippe, and B. F. Lang. 2007. Toward resolving the eukaryotic tree: the phylogenetic positions of jakobids and cercozoans. *Curr. Biol.* 17:1420–1425.

Simpson, A. G. B. 1997 The identity and composition of the *Euglenozoa*. *Arch. Protistenkd.* 148:318–328.

Simpson, A. G. B., and A. J. Roger. 2004. Protein phylogenies robustly resolve the deep-level relationships within *Euglenozoa*. *Mol. Phylogenet. Evol.* 30:201–212.

Simpson, A. G. B., J. van den Hoff, C. Bernard, H. R. Burton, and D. J. Patterson. 1997. The ultrastructure and systematic position of the euglenozoon *Postgaardi mariagerensis*, Fenchel et al. *Arch. Protistenkd.* 147:213–225.

Sogin, M. L. 1989. Evolution of eukaryotic microorganisms and their small subunit ribosomal RNAs. *Am. Zool.* 29:487–499.

Triemer, R. E., and M. A. Farmer. 1991. An ultrastructural comparison of the mitotic apparatus, feeding apparatus, flagellar apparatus and cytoskeleton in euglenoids and kinetoplastids. *Protoplasma* 164:91–104.

von der Heyden, S., E. E. Chao, K. Vickerman, and T. Cavalier-Smith. 2004. Ribosomal RNA phylogeny of bodonid and diplonemid flagellates and the evolution of *Euglenozoa*. *J. Eukaryot. Microbiol.* 51:402–416.

Yang, J., T. Harding, R. Kamikawa, A. G. B. Simpson, A. J. Roger. 2017. Mitochondrial genome evolution and a novel RNA editing system in deep-branching heteroloboseids. *Genome Biol. Evol.* 9:1161–1174.

Yubuki, N., V. P. Edgcomb, J. M. Bernhard, and B. S. Leander. 2009. Ultrastructure and molecular phylogeny of *Calkinsia aureus*: cellular identity of a novel clade of deep-sea euglenozoans with epibiotic bacteria. *BMC Microbiol.* 9:16–37.

Yubuki, N., and B. S. Leander. 2008. Ultrastructure and molecular phylogeny of *Stephanopogon minuta*: an enigmatic microeukaryote from marine interstitial environments. *Eur. J. Protistol.* 44:241–253.

Yubuki, N., A. G. B. Simpson, and B. S. Leander. 2013. Reconstruction of the feeding apparatus in *Postgaardi mariagerensis* provides evidence for character evolution within the *Symbiontida*. *Eur. J. Protistol.* 49:32–39.

Author

Alastair G. B. Simpson; Department of Biology; Dalhousie University; Halifax, NS, B3H 4R2, Canada. Email: alastair.simpson@dal.ca.

Date Accepted: 24 September 2013; updated 24 February 2018

Primary Editor: Philip Cantino

Sar S. M. Adl et al. 2012 [A. G. B. Simpson and M. Dunthorn], converted clade name

Registration Number: 92

Definition: The smallest crown clade containing *Bigelowiella natans* Moestrup and Sengco 2001 (*Rhizaria*), *Tetrahymena thermophila* Nanney and McCoy 1976 (*Alveolata*), and *Thalassiosira pseudonana* Cleve 1873 (*Stramenopila*) but not *Homo sapiens* Linnaeus 1758 (*Opisthokonta*) or *Dictyostelium discoideum* Raper 1935 (*Amoebozoa*) or *Arabidopsis thaliana* (Linnaeus) Heynhold 1842 (*Archaeplastida*) or *Euglena gracilis* Klebs 1883 (*Discoba*) or *Emiliania huxleyi* (Lohmann) Hay and Mohler in Hay et al. 1967 (*Haptophyta*). This is a minimum-crown-clade definition with external specifiers. Abbreviated definition: min crown ∇ (*Bigelowiella natans* Moestrup and Sengco 2001 & *Tetrahymena thermophila* Nanney and McCoy 1976 & *Thalassiosira pseudonana* Cleve 1873 ~ *Homo sapiens* Linnaeus 1758 ∨ *Dictyostelium discoideum* Raper 1935 ∨ *Arabidopsis thaliana* (Linnaeus) Heynhold 1842 ∨ *Euglena gracilis* Klebs 1883 ∨ *Emiliania huxleyi* (Lohmann) Hay and Mohler in Hay et al. 1967).

Etymology: The name is derived from the acronym of the three groups united in this clade: *Stramenopila*, *Alveolata*, and *Rhizaria* (Burki et al., 2007).

Reference Phylogeny: The primary reference phylogeny is Burki et al. (2008: Fig. 1).

Composition: The clade *Sar* is composed of *Alveolata*, *Rhizaria*, and *Stramenopila* as defined in this volume.

Diagnostic Apomorphies: The clade was identified entirely through molecular phylogenetic analyses. Morphological apomorphies are not currently known. A possible molecular apomorphy (a specific Rab1 paralog) has been proposed (Elias et al., 2009).

Synonyms: *Harosa* Cavalier-Smith 2010 and the informal name 'SAR' (Burki et al., 2007) are approximate synonyms.

Comments: Multi-gene phylogenetic analyses that have included exemplars from the well-known major eukaryotic groups have usually inferred a strongly supported clade consisting of *Stramenopila*, *Alveolata*, and *Rhizaria* (e.g., Burki et al., 2007, 2008, 2012, 2013, 2016; Rodríguez-Ezpeleta et al., 2007; Hackett et al., 2007; Hampl et al., 2009; Brown et al., 2013; Cavalier-Smith et al., 2014; Yabuki et al., 2014; Krabberød et al., 2017). A close relationship between *Stramenopila* and *Alveolata* had been inferred previously in phylogenies of one or a few genes, though usually in the absence of (significant) data from *Rhizaria* (e.g., van de Peer and de Wachter, 1997; Baldauf et al., 2000). More recent phylogenomic analyses often unite *Stramenopila* and *Alveolata* as a monophyletic group within the larger *Stramenopila-Alveolata-Rhizaria* clade (e.g., Burki et al., 2012, 2016; Krabberød et al., 2017), though the interrelationships among *Stramenopila*, *Alveolata*, and *Rhizaria* are not resolved yet (see He et al., 2016).

To date the *Stramenopila-Alveolata-Rhizaria* clade has almost always been referred to by the informal name 'SAR' (or 'SAR clade', or 'sar'), following Burki et al. (2007). More recently Adl et al. (2012) proposed the formal name *Sar*, with a phylogenetic definition identical in function to that supplied here, although the wording here has been changed, with external specifiers used instead of a separate qualifying clause.

Another taxon name, *Harosa*, was also proposed for the *Stramenopila-Alveolata-Rhizaria* clade (Cavalier-Smith, 2010), but this name has been little used, with most researchers instead using 'SAR', 'SAR clade', 'sar' or *Sar*. On this basis, and because of the influential nature of the classification in which it was published (with 25 authors, and under the aegis of the Committee on Systematics and Evolution of the International Society of Protistologists), we believe that the name *Sar* will be more commonly used in the future and selected it for that reason.

The name *Sar* has no relevance outside of the specific phylogenetic hypothesis of a *Stramenopila-Alveolata-Rhizaria* clade that excludes other well-known groups. The use of external specifiers serves to link *Sar* to this phylogenetic hypothesis, making it inapplicable if the minimal *Stramenopila-Alveolata-Rhizaria* clade proves to contain other well-known groups.

Literature Cited

Adl, S. M., A. G. B. Simpson, C. E. Lane, J. Lukeš, D. Bass, S. S. Bowser, M. W. Brown, F. Burki, M. Dunthorn, V. Hampl, A. Heiss, M. Hoppenrath, E. Lara, L. le Gall, D. H. Lynn, H. McManus, E. A. D. Mitchell, S. E. Mozley-Stanridge, L. W. Parfrey, J. Pawlowski, S. Rueckert, L. Shadwick, C. L. Schoch, A. Smirnov, and F. W. Spiegel. 2012. The revised classification of eukaryotes. *J. Eukaryot. Microbiol.* 59:429–493.

Baldauf, S. L., A. J. Roger, I. Wenk-Siefert, and W. F. Doolittle. 2000. A kingdom-level phylogeny of eukaryotes based on combined protein data. *Science* 290:972–977.

Brown, M. W., S. C. Sharpe, J. D. Silberman, A. A. Heiss, B. F. Lang, A. G. B. Simpson, and A. J. Roger. 2013. Phylogenomics demonstrates that breviate flagellates are related to opisthokonts and apusomonads. *Proc. R. Soc. Lond. B Biol. Sci.* 280:20131755.

Burki F., N. Corradi, R. Sierra, J. Pawlowski, G. R. Meyer, C. L. Abbott, and P. J. Keeling. 2013. Phylogenomics of the intracellular parasite *Mikrocytos mackini* reveals evidence for a mitosome in *Rhizaria. Curr. Biol.* 23:1541–1547.

Burki, F., M. Kaplan, D. V. Tikhonenkov, V. Zlatogursky, B. Q. Minh, L. V. Radaykina, A. Smirnov, A. P. Mylnikov, and P. J. Keeling. 2016. Untangling the early diversification of eukaryotes: a phylogenomic study of the evolutionary origins of *Centrohelida*, *Haptophyta* and *Cryptista. Proc. R. Soc. Lond. B Biol. Sci.* 283:20152802.

Burki, F., N. Okamoto, J.-F. Pombert, and P. J. Keeling. 2012. The evolutionary history of haptophytes and cryptophytes: phylogenomic evidence for separate origins. *Proc. R. Soc. Lond. B Biol. Sci.* 279:2246–2254.

Burki, F., K. Shalchian-Tabrizi, M. Minge, A. Skjaeveland, S. I. Nikolaev, K. S. Jakobsen, and J. Pawlowski. 2007. Phylogenomics reshuffles the eukaryotic supergroups. *PLOS ONE* 2:e790.

Burki, F., K. Shalchian-Tabrizi, and J. Pawlowski. 2008. Phylogenomics reveals a new 'megagroup' including most photosynthetic eukaryotes. *Biol. Lett.* 4:366–369.

Cavalier-Smith, T. 2010. Kingdoms *Protozoa* and *Chromista* and the eozoan root of eukaryotes. *Biol. Lett.* 6:342–345.

Cavalier-Smith, T., E. E. Chao, E. A. Snell, C. Berney, A. M. Fiore-Donno, and R. Lewis. 2014. Multigene eukaryote phylogeny reveals the likely protozoan ancestors of opisthokonts (animals, fungi, choanozoans) and *Amoebozoa. Mol. Phylogenet. Evol.* 81:71–85.

Elias, M., N. J. Patron, and P. J. Keeling. 2009. The RAB family GTPase Rab1A from *Plasmodium falciparum* defines a unique paralog shared by chromalveolates and *Rhizaria. J. Eukaryot. Microbiol.* 56:348–356.

Hackett, J. D., H. S. Yoon, S. Li, A. Reyes-Prieto, S. E. Rummele, and D. Bhattacharya. 2007. Phylogenomic analysis supports the monophyly of cryptophytes and haptophytes and the association of *Rhizaria* with chromalveolates. *Mol. Biol. Evol.* 24:1702–1713.

Hampl, V., L. Hug, J. W. Leigh, J. B. Dacks, B. F. Lang, A. G. B. Simpson, and A. J. Roger. 2009. Phylogenomic analyses support the monophyly of *Excavata* and resolve relationships among eukaryotic "supergroups". *Proc. Natl. Acad. Sci. USA* 106:3859–3864.

He, D., R. Sierra, J. Pawlowski, and S. L. Baldauf. 2016. Reducing long-branch effects in multi-protein data uncovers a close relationship between *Alveolata* and *Rhizaria*. *Mol. Phylogenet. Evol.* 101:1–7.

Krabberød, A. K., R. Orr, J. Bråte, T. Kristensen, K. R. Bjørklund, and K. Shalchian-Tabrizi. 2017. Single cell transcriptomics, mega-phylogeny and the genetic basis of morphological innovations in *Rhizaria*. *Mol. Biol. Evol.* 34:1557–1573.

Rodríguez-Ezpeleta, N., H. Brinkmann, G. Burger, A. J. Roger, M. W. Gray, H. Philippe, and B. F. Lang. 2007. Toward resolving the eukaryotic tree: the phylogenetic positions of jakobids and cercozoans. *Curr. Biol.* 17:1420–1425.

van de Peer, Y., and R. de Wachter. 1997. Evolutionary relationships among the eukaryotic crown taxa taking into account site-to-site rate variation in 18S rRNA. *J. Mol. Evol.* 45:619–630.

Yabuki, A., R. Kamikawa, S. A. Ishikawa, M. Kolisko, E. Kim, A. S. Tanabe, K. Kume, K. Ishida, and Y. Inagaki. 2014. *Palpitomonas bilix* represents a basal cryptist lineage: insight into the character evolution in *Cryptista*. *Sci. Rep.* 4:4641.

Authors

Alastair G. B. Simpson; Department of Biology; Dalhousie University; Halifax, NS, B3H 4R2, Canada. Email: alastair.simpson@dal.ca.

Micah Dunthorn; Department of Eukaryotic Microbiology; University of Duisburg-Essen; 45141 Essen, Germany. Email: micah.dunthorn@uni-due.de.

Date Accepted: 21 April 2013; updated 24 February 2018

Primary Editor: Philip Cantino

Stramenopila C. J. Alexopoulos, C. W. Mims, and M. Blackwell 1996 [A. G. B. Simpson and M. Dunthorn], converted clade name

Registration Number: 103

Definition: The crown clade for which flagellar retronemes (tripartite, rigid, hollow flagellar hairs, arrayed in such a way as to cause 'reversed' fluid flow or locomotion), as inherited by *Ochromonas danica* Pringsheim 1955, is an apomorphy relative to other crown clades, provided that it does not include *Tetrahymena thermophila* Nanney and McCoy 1976 (*Alveolata*) and *Bigelowiella natans* Moestrup and Sengco 2001 (*Rhizaria*). This is an apomorphy-modified crown-clade definition with a qualifying clause. Abbreviated definition: crown ∇ apo flagellar retronemes [*Ochromonas danica* Pringsheim 1955] | ~ (*Tetrahymena thermophila* Nanney and McCoy 1976 & *Bigelowiella natans* Moestrup and Sengco 2001).

Etymology: Derived from the Latin *stramen* (straw) and *pilus* (hair) in reference to the structure of the flagellar hairs (the retronemes) in this taxon.

Reference Phylogeny: The primary reference phylogeny is Cavalier-Smith and Chao (2006: Fig. 2), in which the clade is referred to as '*Heterokonta*' (see Comments), and the external specifiers are not included. See Burki et al. (2012: Fig. 1) for the relationships of the external specifiers to stramenopiles. The internal specifier is not included in this latter tree but is most closely related to *Thalassiosira pseudonana*, *Phaeodactylum tricornutum*, *Ectocarpus siliculosus* and *Aureococcus anophagefferens* (as members of the clade labeled *Ochrophyta* in Cavalier-Smith and Chao, 2006: Fig. 2). See also Riisberg et al. (2009: Fig. 2), van de Peer and de Wachter (1997: Figs. 2 and 4), Yubuki et al. (2015: Fig. 5) and Shiratori et al. (2017: Fig. 6).

Composition: Adl et al. (2012) provided a list of most subtaxa known at that time, under the name *Stramenopiles*. See also Andersen (2004) for photosynthetic *Stramenopila* (*Ochrophyta*), under the name 'heterokont algae'. Major groups include: *Bicosoecida, Bacillariophyta (Diatomea), Chrysophyceae* (including *Synurales*), *Developayellaceae, Dictyochophyceae, Eustigmatophyceae* (includes or synonymous with *Eustigmatales*), *Hyphochytidales, Labyrinthulomycetes, Opalinata, Peronosporomycetes* (=*Oomycetes*), *Phaeophyceae, Phaeothamniophyceae, Raphidophyceae* and *Xanthophyceae*. Additional recently described/appended taxa (i.e., not appearing under '*Stramenopiles*' in Adl et al. [2012]) include: *Cantina, Incisomonas, Platysulcus, Pseudophyllomitus* and *Solenicola*.

Diagnostic Apomorphies: The only diagnostic apomorphy is the presence of retronemes (see Definition).

Synonyms: *Stramenopiles, Straminipila, Straminopiles, Stramentopila, Heterokonta* (all approximate) (see Comments).

Comments: In some eukaryotic cells, one flagellum bears stiff, straw-like and generally tripartite hairs, called 'retronemes' by Cavalier-Smith (1989). The retroneme-bearing flagellum is usually directed anteriorly and beats with high-amplitude base-to-tip waves, thereby 'pulling' the cell forwards during swimming. Organisms with retronemes in at least one stage of their lifecycle include most taxa (but often not all) within many major groups of algae with plastids surrounded by four membranes and containing (in most cases) chlorophyll *c*, as well as some entirely heterotrophic groups. These algae include *Chrysophyceae*, *Bacillariophyta* (diatoms), *Phaeophyceae* (brown algae) and several other groups, while the heterotrophs include many small flagellates (e.g., *Bicoseocida*), some fungus-like organisms (e.g., *Peronosporomycetes* [= *Oomycetes*]), and the ectoplasmic net-producing *Labyrinthulomycetes*. Hairs that resemble retronemes in structure, but not function, occur on the cell bodies of some *Opalinata* (*Proteromonas*) and are probably derived from retronemes (Patterson, 1989). As discussed below, scientific opinion, backed by molecular phylogenies, has gradually moved towards a consensus that retronemes are homologous in these diverse eukaryotes and evolved once in a common ancestor of all of them, although they have been secondarily lost or modified multiple times independently, either coincidently with the complete loss of flagella, or leaving the anterior flagellum naked (see, for example, Cavalier-Smith and Chao, 2006). Conversely, it is sometimes proposed that retronemes are ultimately homologous to the non-flow-reversing flagellar hairs with similar substructure in certain other unicellular eukaryotes (e.g., in/within *Cryptophyta* and *Telonemia*; Cavalier-Smith, 1986, 2018; Shalchian-Tabrizi et al., 2006). Nonetheless, recent treatments of this kind still envisage a subsequent, single origin of the flow-reversing property that distinguishes retronemes in *Stramenopila* from other flagellar hairs (Cavalier-Smith, 2018; see below for older ideas).

Close taxonomic or evolutionary relationships between certain of the 'algal' and fungus-like forms mentioned above have been proposed for decades, partly on the basis of having retroneme-bearing flagella (e.g., Bessey, 1950; Manton et al. 1952). Support for a close relationship between photosynthetic and fungus-like stramenopiles was an early result of small subunit ribosomal RNA (SSU rRNA) gene phylogenetic analyses (Gunderson et al., 1987). Subsequent SSU rRNA gene and multi-gene analyses have supported the monophyly of *Stramenopila*, including *Opalinata* (Silberman et al., 1996; Van de Peer and de Wachter, 1997; Kostka et al., 2004; Harper et al., 2005; Burki et al., 2007, 2008; Riisberg et al., 2009).

The name *Stramentopila* (i.e., with a second 't') was introduced in a table by Patterson (1988), without an explicit definition, but with a list of subclades. *Stramentopila* has not been used since; Patterson himself later deliberately used an informal name, 'stramenopiles', for the same taxonomic concept when discussing it in detail (Patterson, 1989, 1999). Patterson (1989) explicitly defined 'stramenopiles' by the apomorphy of hollow, rigid, tripartite hairs that act as retronemes. This essentially apomorphy-based concept has not changed, although some of the organisms suggested tentatively by Patterson (1988, 1989) to fall within this clade do not (*Foraminifera* and some *Alveolata*, neither of which actually have retronemes and whose exclusion is supported by molecular phylogenies). The name *Stramenopila* (without a second 't') seems to have been first introduced by Alexopoulos et al. (1996). While some authors use *Stramenopila*, others use the informal name 'stramenopiles', or formalize it to *Stramenopiles* (see Adl et al., 2005). A few, mainly mycologists, followed an alteration of 'stramenopiles' to 'straminipiles' or 'straminopiles' on the basis that it is more correct Latin, with Dick (2001) formally proposing *Straminipila* as a kingdom, defined in part by the presence of retronemes. However, Dick's citation of the taxon *Chromista* as a subgroup of *Straminipila* implied a much more inclusive composition than that of *Stramenopila*. Irrespective, the name *Straminipila* is rarely used today, other than by some mycologists (e.g., Beakes and Thines, 2017).

Heterokonta is a commonly used alternative to *Stramenopila/Stramenopiles*. *Heterokontae* was first introduced for a small subset of the retroneme-bearing algae (Luther, 1899). It was later expanded in composition. Copeland (1956), for example, included within *Heterokonta* (for which Luther is cited as the authority) a broad collection of algae and heterotrophic organisms, including major groups that are now known to be distantly related to *Stramenopila* (e.g., *Haptophyta* and *Choanoflagellata*), but excluding the best known groups within *Stramenopila*, namely *Bacillariophyta* (diatoms), *Phaeophyceae* (brown algae), and *Peronosporomycetes* [= *Oomycetes*]. With time, however, the term 'heterokont' came to be used to describe organisms with retronemes, especially algae (see Leadbeater, 1989). By the 1970s and 1980s, *Heterokonta* (or 'heterokonts') was being treated as a division/phylum- or kingdom-level grouping that included many (but not all) known retroneme-bearing groups, in various combinations (e.g., Leedale, 1974; Cavalier-Smith, 1981, 1986). Ironically, Cavalier-Smith (1986), who formally proposed *Heterokonta* as a new phylum, argued at the time that flagellar retronemes are a plesiomorphy of *Heterokonta*, not an apomorphy. Nonetheless, since that time, *Heterokonta* has been associated predominantly with the concept of a clade containing all retroneme-bearing taxa (e.g., Cavalier-Smith, 1998). *Heterokontophyta* is sometimes considered an approximate synonym (e.g., van den Hoek et al., 1995), but more usually refers specifically to algal stramenopiles only.

Stramenopila/Stramenopiles and *Heterokonta* are both in wide use to denote approximately the same clade. The frequency of use of *Heterokonta* and *Stramenopila/Stramenopiles* is similar. *Heterokonta* is the older name, but it has changed profoundly in concept and composition and

only recently has become associated predominantly with a clade containing all retroneme-bearing taxa. *Stramenopila* is younger, but from the outset 'stramenopiles' was defined by reference to retronemes as an apomorphy (Patterson, 1989). In this case we prefer to prioritize consistency of concept over the antiquity of the name itself, and we therefore convert *Stramenopila* as the clade name.

In the past it has been proposed that retronemes evolved much earlier than indicated by current phylogenetic trees (e.g., Cavalier-Smith, 1986). While this possibility is unlikely, we recognize that it would be confusing for *Stramenopila* to end up as an extremely inclusive clade. The external specifiers used in the definition in effect prevent *Stramenopila* from being the name of a clade that equals or exceeds the taxon *Sar* (see entry in this volume) in inclusiveness.

Literature Cited

Adl, S. M., A. G. Simpson, M. A. Farmer, R. A. Andersen, O. R. Anderson, J. R. Barta, S. S. Bowser, G. Brugerolle, R. A. Fensome, S. Fredericq, T. Y. James, S. Karpov, P. Kugrens, J. Krug, C. E. Lane, L. A. Lewis, J. Lodge, D. H. Lynn, D. G. Mann, R. M. McCourt, L. Mendoza, O. Moestrup, S. E. Mozley-Standridge, T. A. Nerad, C. A. Shearer, A. V. Smirnov, F. W. Spiegel, and M. F. Taylor. 2005. The new higher level classification of eukaryotes with emphasis on the taxonomy of protists. *J. Eukaryot. Microbiol.* 52:399–451.

Adl, S. M., A. G. B. Simpson, C. E. Lane, J. Lukeš, D. Bass, S. S. Bowser, M. W. Brown, F. Burki, M. Dunthorn, V. Hampl, A. Heiss, M. Hoppenrath, E. Lara, L. le Gall, D. H. Lynn, H. McManus, E. A. D. Mitchell, S. E. Mozley-Stanridge, L. W. Parfrey, J. Pawlowski, S. Rueckert, L. Shadwick, C. L. Schoch, A. Smirnov, and F. W. Spiegel. 2012. The revised classification of eukaryotes. *J. Eukaryot. Microbiol.* 59:429–493.

Alexopoulos, C. J., C. W. Mims, and M. Blackwell. 1996. *Introductory Mycology*. 4th edition. John Wiley & Sons, Inc., New York.

Anderson, R. A. 2004. Biology and systematics of heterokont and haptophyte algae. *Am. J. Bot.* 91:1508–1522.

Beakes G. W., and M. Thines. 2017. *Hyphochytriomycota* and *Oomycota*. Pp. 435–505 in *Handbook of the Protists* (J. Archibald, A. G. B. Simpson and C. H. Slamovits, eds.). Springer, Cham, Switzerland.

Bessey, E. A. 1950. *Morphology and Taxonomy of Fungi*. The Blakiston Company, Philadelphia, PA.

Burki, F., N. Okamoto, J.-F. Pombert, and P. J. Keeling. 2012. The evolutionary history of haptophytes and cryptophytes: phylogenomic evidence for separate origins. *Proc. R. Soc. Lond. B Biol. Sci.* 279:2246–2254.

Burki, F., K. Shalchian-Tabrizi, M. Minge, A. Skjaeveland, S. I. Nikolaev, K. S. Jakobsen, and J. Pawlowski. 2007. Phylogenomics reshuffles the eukaryotic supergroups. *PLOS ONE* 2:e790.

Burki, F., K. Shalchian-Tabrizi, and J. Pawlowski. 2008. Phylogenomics reveals a new 'megagroup' including most photosynthetic eukaryotes. *Biol. Lett.* 4:366–369.

Cavalier-Smith, T. 1981. Eukaryote kingdoms: seven or nine? *BioSystems* 14:461–481.

Cavalier-Smith, T. 1986. The kingdom *Chromista*: origin and systematics. Pp. 309–347 in *Progress in Phycological Research* (F. E. Round and D. J. Chapman, eds.). Biopress, Bristol.

Cavalier-Smith, T. 1989. The kingdom *Chromista*. Pp. 381–407 in *The Chromophyte Algae: Problems and Perspectives* (J. C. Green, B. S. C. Leadbeater, and W. L. Diver, eds.). Oxford University Press, Oxford.

Cavalier-Smith, T. 1998. A revised six-kingdom system of life. *Biol. Rev.* 73:203–266.

Cavalier-Smith, T. 2018. Kingdom *Chromista* and its eight phyla: a new synthesis emphasising periplastid protein targeting, cytoskeletal and periplastid evolution, and ancient divergences. *Protoplasma* 255:297–357.

Cavalier-Smith, T., and E. E. Chao. 2006. Phylogeny and megasystematics of phagotrophic heterokonts (Kingdom *Chromista*). *J. Mol. Evol.* 62:388–420.

Copeland, H. F. 1956. *The Classification of Lower Organisms*. Pacific Books, Palo Alto, CA.

Dick, M. W. 2001. *Straminipilous Fungi*. Kluwer Academic Publishers, Dordrecht.

Gunderson, J. H., H. Elwood, A. Ingold, K. Kindle, and M. L. Sogin. 1987. Phylogenetic relationships between chlorophytes, chrysophytes, and oomycetes. *Proc. Natl. Acad. Sci. USA* 84:5823–5827.

Harper, J. T., E. Waanders, and P. J. Keeling. 2005. On the monophyly of chromalveolates using a six-protein phylogeny of eukaryotes. *Int. J. Syst. Evol. Microbiol.* 55:487–496.

Kostka, M., V. Hampl, I. Cepicka, and J. Flegr. 2004. Phylogenetic position of *Protoopalina intestinalis* based on SSU rRNA gene sequence. *Mol. Phylogenet. Evol.* 33:220–224.

Leadbeater, B. S. C. 1989. The phylogenetic significance of flagellar hairs in the *Chromophyta*. Pp. 145–165 in *The Chromophyte Algae: Problems and Perspectives* (J. C. Green, B. S. C. Leadbeater, and W. L. Diver, eds.). Oxford University Press, Oxford.

Leedale, G. F. 1974. How many are the kingdoms of organisms? *Taxon* 23:261–270.

Luther, A. 1899. Über *Chlorosaccus* eine neue Gattung der Süsswasseralgen. *Bihang til Kongliga Svenska Vetenskaps-Akademiens Handlingar* 24:1–22.

Manton, I., B. Clarke, A. D. Greenwood, and E. A. Flint. 1952. Further observations on the structure of plant cilia, by a combination of visual and electron microscopy. *J. Exp. Bot.* 3:204–215.

Patterson, D. J. 1988. The evolution of protozoa. *Mem. Inst. Oswaldo Cruz* 83(suppl 1):580–600.

Patterson, D. J. 1989. Stramenopiles: chromophytes from a protistan perspective. Pp. 357–379 in *The Chromophyte Algae: Problems and Perspectives* (J. C. Green, B. S. C. Leadbeater, and W. L. Diver, eds.). Oxford University Press, Oxford.

Patterson, D. J. 1999. The diversity of eukaryotes. *Am. Nat.* 154:S96–S124.

Riisberg, I., R. J. Orr, R. Kluge, K. Shalchian-Tabrizi, H. A. Bowers, V. Patil, B. Edvardsen, and K. S. Jakobsen. 2009. Seven gene phylogeny of heterokonts. *Protist* 160:191–204.

Shalchian-Tabrizi, K., W. Eikrem, D. Klaveness, D. Vaulot, M. A. Minge, F. Le Gall, K. Romari, J. Throndsen, A. Botnen, R. Massana, H. A. Thomsen, and K. S. Jakobsen. 2006. *Telonemia*, a new protist phylum with affinity to chromist lineages. *Proc. R. Soc. Lond. B Biol. Sci.* 273:1833–1842.

Shiratori, T., R. Thakur, and K. Ishida. 2017. *Pseudophyllomitus vesiculosus* (Larsen and Patterson 1990) Lee, 2002, a poorly studied phagotrophic biflagellate is the first characterized member of stramenopile environmental clade MAST-6. *Protist* 168:439–451.

Silberman, J. D., M. L. Sogin, D. D. Leipe, and C. G. Clark. 1996. Human parasite finds taxonomic home. *Nature* 380:398.

van den Hoek, C., D. G. Mann, and H. M. Jahns. 1995. *Algae: An Introduction to Phycology.* Cambridge University Press, Cambridge.

van de Peer, Y., and R. De Wachter. 1997. Evolutionary relationships among the eukaryotic crown taxa taking into account site-to-site rate variation in 18S rRNA. *J. Mol. Evol.* 45:619–630.

Yubuki, N., T. Pánek, A. Yabuki, I. Čepička, K. Takishita, Y. Inagaki, and B. S. Leander. 2015. Morphological identities of two different marine stramenopile environmental sequence clades: *Bicosoeca kenaiensis* (Hilliard, 1971) and *Cantina marsupialis* (Larsen and Patterson, 1990) gen. nov., comb. nov. *J. Eukaryot. Microbiol.* 62:532–542.

Authors

Alastair G. B. Simpson; Department of Biology; Dalhousie University; Halifax, NS B3H 4R2, Canada. Email: alastair.simpson@dal.ca.

Micah Dunthorn; Department of Eukaryotic Microbiology; University of Duisburg-Essen; 45141 Essen, Germany. Email: micah.dunthorn@uni-due.de.

Date Accepted: 23 June 2013; updated 9 April 2018

Primary Editor: Philip Cantino

Rhizaria T. Cavalier-Smith 2002 [A. G. B. Simpson], converted clade name

Registration Number: 90

Definition: The smallest crown clade containing *Bigelowiella natans* Moestrup and Sengco 2001 (*Filosa*), *Plasmodiophora brassicae* Woronin 1877 (*Phytomyxea*), *Reticulomyxa filosa* Nauss 1949 (*Retaria*), *Collozoum inerme* (Müller 1855) Haeckel 1887 (*Retaria*), *Gromia sphaerica* Gooday, Bowser, Bett and Smith 2000, and *Bonamia ostreae* Pichot, Comps, Tigé, Grizel and Rabouin 1980 (*Ascetosporea*) but not *Tetrahymena thermophila* Nanney and McCoy 1976 (*Alveolata*) or *Thalassiosira pseudonana* Cleve 1873 (*Stramenopila*). This is a minimum-crown-clade definition with external specifiers. Abbreviated definition: min crown ∇ (*Bigelowiella natans* Moestrup and Sengco 2001 & *Plasmodiophora brassicae* Woronin 1877 & *Reticulomyxa filosa* Nauss 1949 & *Collozoum inerme* (Müller 1855) Haeckel 1887 & *Gromia sphaerica* Gooday, Bowser, Bett and Smith 2000 & *Bonamia ostreae* Pichot, Comps, Tigé, Grizel and Rabouin 1980 ~ *Tetrahymena thermophila* Nanney and McCoy 1976 ∨ *Thalassiosira pseudonana* Cleve 1873).

Etymology: Derived from the Greek, *rhiza* (root), referring to the commonly root-like (thin and branched) nature of the pseudopodia in this group (Cavalier-Smith, 2002).

Reference Phylogeny: The primary reference phylogeny is Sierra et al. (2016: Fig. 1), in which *Collozoum inerme* is represented by '*Collozoum* sp.' See Bass et al. (2009: Fig. 1), Howe et al. (2011: Fig. 1) and Krabberød et al. (2011: Fig. 2) for more complete phylogenetic representations of the diversity of *Rhizaria*.

Composition: The clade *Rhizaria* is composed of *Retaria* (i.e., *Foraminifera*, *Acantharia*, *Sticholonche* and *Polycystinea;* the latter taxon likely not monophyletic), as well as *Filosa*, *Phytomyxea*, *Vampyrellida*, *Gromia*, *Filoreta*, *Tremula* and *Ascetosporea*. *Filosa*, *Phytomyxea* and sometimes others of these taxa have frequently been grouped as a taxon *Cercozoa*, but this is likely to represent a paraphyletic group with respect to *Retaria* (see Comments).

Diagnostic Apomorphies: No unequivocal morphological apomorphies are currently known. See Comments for discussion of pseudopodia. A 1–2 amino acid insertion into polyubiquitin is a likely molecular apomorphy (see Comments).

Synonyms: The informal name 'core *Rhizaria*' sensu Cavalier-Smith (2003a) is an approximate synonym.

Comments: Cavalier-Smith (2002) introduced the taxon *Rhizaria*, including within it *Cercozoa* and *Retaria*, plus *Centrohelida* (a group of 'heliozoan' amoebae) and a heterogeneous collection of heterotrophic flagellates called '*Apusozoa*' (*Apusomonadida*, *Ancyromonadida* and *Hemimastigophora* in that particular publication). *Cercozoa* was a diverse collection of microbial eukaryotes that had been identified shortly before (Cavalier-Smith, 1998) as a probable clade uniting *Filosa* (e.g., *Cercomonadidae* and *Chlorarachniophyceae;* also called 'core *Cercozoa*') and *Phytomyxea* (e.g., *Plasmodiophora*). *Retaria* meanwhile was a proposed clade consisting of two well-known types of amoebae: *Foraminifera* and 'radiolaria'

(Cavalier-Smith, 1999). Cavalier-Smith (2002) did not consider the new taxon *Rhizaria* to be monophyletic, though he did consider *Cercozoa* plus *Retaria* to represent a clade within *Rhizaria*. Cavalier-Smith (2003a) informally denoted *Cercozoa* plus *Retaria* as the 'core *Rhizaria*', and soon thereafter altered *Rhizaria* to be equivalent to this 'core *Rhizaria*' by excluding *Centrohelida* and the remaining apusozoans (Cavalier-Smith, 2003b). In light of molecular evidence supporting its monophyly (see below), this later concept of *Rhizaria* became widely adopted (Nikolaev et al., 2004; Simpson and Roger, 2004; Adl et al., 2005; Keeling et al., 2005; Pawlowski and Burki, 2009). *Cercozoa* has also been widely adopted, although recent phylogenetic analyses indicate that *Cercozoa* sensu lato represents a paraphyletic group within *Rhizaria* (see below).

Phylogenies of some protein-coding genes, some analyses of ribosomal RNA genes, and the possession (usually) of a 1–2 amino acid insertion within the otherwise invariant poly-ubiquitin sequence provided early evidence of a close relationship between *Foraminifera* and taxa assigned to *Cercozoa* (Keeling, 2001; Longet et al., 2003; Berney and Pawlowski, 2003; Archibald et al., 2003). Soon after, phylogenies based largely or entirely on ribosomal RNA (rRNA) genes in turn supported a grouping of *Cercozoa* and *Foraminifera* with various 'radiolarians', although the latter represent a non-monophyletic group within this clade (Nikolaev et al., 2004; Polet et al., 2004; Moreira et al., 2007; see below). These results approximately coincided with the widespread adoption of the name *Rhizaria* (and its application to this clade; see above). Phylogenomic analyses have since strongly supported this *Rhizaria* grouping (Burki et al., 2007, 2008, 2010; Brown et al., 2012; Sierra et al., 2013). Also, multigene molecular phylogenies that include *Centrohelida* and apusozoan groups (i.e., the other original members of *Rhizaria*

– see above), demonstrate that they are not members of *Sar*, a more inclusive clade to which *Rhizaria* belongs (e.g., Parfrey et al., 2010; Burki et al., 2012, 2013, 2016; Brown et al., 2013, 2018; Cavalier-Smith et al. 2015).

The phylogeny within *Rhizaria* is incompletely resolved. *Phaeodarea* were traditionally considered as radiolarians, but instead branch within a robust *Filosa* clade in rRNA gene phylogenies (Nikolaev et al., 2004; Polet et al., 2004). Recent phylogenetic analyses have generally indicated that other radiolarians (traditionally divided into *Acantharia* and *Polycystinea,* but see below) are closely related to *Foraminifera*, together forming the clade *Retaria* (Moreira et al., 2007; Burki et al, 2010; Parfrey et al., 2010; Ishitani et al., 2011; Sierra et al., 2013). Molecular phylogenies, primarily of rRNA genes, indicated that *Rhizaria* also includes taxa such as *Gromia*, *Ascetosporea* (e.g., *Haplosporidia*, *Paramyxida*), *Vampyrellida*, *Filoreta* and *Tremula*, albeit the last of these is often included in *Filosa* (Burki et al., 2002; Cavalier-Smith and Chao, 2003; Tekle et al., 2007; Bass et al., 2009; Howe et al., 2011). These have generally been assigned taxonomically to *Cercozoa*, together with *Filosa* and *Phytomyxea* (e.g., Bass et al., 2009; Adl et al., 2012), but *Cercozoa* in this broad sense is not inferred to be monophyletic in recent phylogenomic analyses (Brown et al., 2012; Burki et al., 2013; Sierra et al., 2013, 2016; Krabberød et al., 2017). In consequence, some authors have used *Cercozoa* as a synonym of the smaller but well supported clade *Filosa* (e.g., Burki et al., 2010, 2013; Sierra et al., 2013; Cavalier-Smith, 2018), while others continue to label this clade as *Filosa* and do not use *Cercozoa* to refer to a clade (e.g., Sierra et al., 2016; Krabberød et al., 2017). The deep-level phylogenetic relationships within *Retaria* are still incompletely resolved as well; for example, most phylogenies with a broad taxon sampling do not infer *Polycystinea* to be a clade (though they all belong within *Retaria*), and there are conflicting inferences as to whether the small taxon *Sticholonche* (*Taxopodida*) is more closely related to a subgroup of *Polycystinea* or to *Acantharia,* or is sister to other *Retaria* (e.g., Kunitomo et al., 2006; Ishitani et al., 2011; Krabberød et al., 2011, 2017).

There are no clear morphological apomorphies for *Rhizaria*. Very fine and often branching pseudopodia are widespread in *Rhizaria* (inspiring the name; Cavalier-Smith, 2002) but differ markedly among different groups. The pseudopodia are microtubule-supported in some taxa, such as *Retaria*; they are actin-microfilament-supported in others, such as *Chlorarachiophyceae*. It is not clear, therefore, whether their fine, branching nature is strictly homologous. The characteristic insertion of 1–2 amino acids into polyubiquitin is likely a molecular apomorphy for *Rhizaria*. This is absent in some *Retaria* (Bass et al., 2005), but currently available phylogenetic evidence strongly suggests that these absences represent secondary loss (Ishitani et al., 2011; Sierra et al., 2016). Krabberød et al. (2017) recently identified a myosin subgroup that may be apomorphic for *Rhizaria*, but data outside *Retaria* are very sparse at present.

The external specifiers used in the definition serve to make the name *Rhizaria* inapplicable if the clade identified by the internal specifiers proves to contain either *Stramenopila* or *Alveolata*.

Literature Cited

Adl, S. M., A. G. Simpson, M. A. Farmer, R. A. Andersen, O. R. Anderson, J. R. Barta, S. S. Bowser, G. Brugerolle, R. A. Fensome, S. Fredericq, T. Y. James, S. Karpov, P. Kugrens, J. Krug, C. E. Lane, L. A. Lewis, J. Lodge, D. H. Lynn, D. G. Mann, R. M. McCourt, L. Mendoza, O. Moestrup, S. E. Mozley-Standridge, T. A. Nerad, C. A. Shearer, A. V. Smirnov, F. W. Spiegel, and M. F. Taylor. 2005. The new higher level classification of eukaryotes with emphasis on the taxonomy of protists. *J. Eukaryot. Microbiol.* 52:399–451.

Adl, S. M., A. G. B. Simpson, C. E. Lane, J. Lukeš, D. Bass, S. S. Bowser, M. W. Brown, F. Burki, M. Dunthorn, V. Hampl, A. Heiss, M. Hoppenrath, E. Lara, L. le Gall, D. H. Lynn, H. McManus, E. A. D. Mitchell, S. E. Mozley-Stanridge, L. W. Parfrey, J. Pawlowski, S. Rueckert, L. Shadwick, C. L. Schoch, A. Smirnov, and F. W. Spiegel. 2012. The revised classification of eukaryotes. *J. Eukaryot. Microbiol.* 59:429–493.

Archibald, J. M., D. Longet, J. Pawlowski, and P. J. Keeling. 2003. A novel polyubiquitin structure in *Cercozoa* and *Foraminifera*: evidence for a new eukaryotic supergroup. *Mol. Biol. Evol.* 20:62–66.

Bass, D., E. E.-Y. Chao, S. Nikolaev, A. Yabuki, K. Ishida, C. Berney, U. Pakzad, C. Wylezich, and T. Cavalier-Smith. 2009. Phylogeny of novel naked filose and reticulose *Cercozoa*: *Granofilosea* cl. n. and *Proteomyxidea* revised. *Protist* 160:75–109.

Bass, D., D. Moreira, P. López-García, S. Polet, E. E. Chao, S. Heyden, J. Pawlowski, and T. Cavalier-Smith. 2005. Polyubiquitin insertions and the phylogeny of *Cercozoa* and *Rhizaria*. *Protist* 156:149–161.

Berney, C., and J. Pawlowski. 2003. Revised small subunit rRNA analysis provides further evidence that *Foraminifera* are related to *Cercozoa*. *J. Mol. Evol.* 57:120–127.

Brown, M. W., A. A. Heiss, R. Kamikawa, Y. Inagaki, A. Yabuki, A. K. Tice, T. Shiratori, K. Ishida, T. Hashimoto, A. G. B. Simpson, and A. J. Roger. 2018. Phylogenomics places orphan protistan lineages in a novel eukaryotic super-group. *Genome Biol. Evol.* 10:427–433.

Brown, M. W., M. Kolisko, J. D. Silberman, and A. J. Roger. 2012. Aggregative multicellularity evolved independently in the eukaryotic super-group *Rhizaria*. *Curr. Biol.* 22:1123–1127.

Brown, M. W., S. C. Sharpe, J. D. Silberman, A. A. Heiss, B. F. Lang, A. G. B. Simpson, and A. J. Roger. 2013. Phylogenomics demonstrates that breviate flagellates are related to opisthokonts and apusomonads. *Proc. R. Soc. Lond. B Biol. Sci.* 280:20131755.

Burki, F., C. Berney, and J. Pawlowski. 2002. Phylogenetic position of *Gromia oviformis* Dujardin inferred from nuclear-encoded small subunit ribosomal DNA. *Protist* 153:251–260.

Burki F., N. Corradi, R. Sierra, J. Pawlowski, G. R. Meyer, C. L. Abbott, and P. J. Keeling. 2013. Phylogenomics of the intracellular parasite *Mikrocytos mackini* reveals evidence for a mitosome in *Rhizaria*. *Curr. Biol.* 23:1541–1547.

Burki, F., M. Kaplan, D. V. Tikhonenkov, V. Zlatogursky, B. Q. Minh, L. V. Radaykina, A. Smirnov, A. P. Mylnikov, and P. J. Keeling. 2016. Untangling the early diversification of eukaryotes: a phylogenomic study of the evolutionary origins of *Centrohelida*, *Haptophyta* and *Cryptista*. *Proc. R. Soc. Lond. B Biol. Sci.* 283:20152802.

Burki, F., A. Kudryavtsev, M. V. Matz, G. V. Aglyamova, S. Bulman, M. Fiers, P. J. Keeling, and J. Pawlowski. 2010. Evolution of *Rhizaria*: new insights from phylogenomic analysis of uncultivated protists. *BMC Evol. Biol.* 10:377.

Burki, F., N. Okamoto, J.-F. Pombert, and P. J. Keeling. 2012. The evolutionary history of haptophytes and cryptophytes: phylogenomic evidence for separate origins. *Proc. R. Soc. Lond. B Biol. Sci.* 279:2246–2254.

Burki, F., K. Shalchian-Tabrizi, M. Minge, A. Skjaeveland, S. I. Nikolaev, K. S. Jakobsen, and J. Pawlowski. 2007. Phylogenomics reshuffles the eukaryotic supergroups. *PLOS ONE* 2:e790.

Burki, F., K. Shalchian-Tabrizi, and J. Pawlowski. 2008. Phylogenomics reveals a new 'megagroup' including most photosynthetic eukaryotes. *Biol. Lett.* 4:366–369.

Cavalier-Smith, T. 1998. A revised six-kingdom system of life. *Biol. Rev.* 73:203–266.

Cavalier-Smith, T. 1999. Principles of protein and lipid targeting in secondary symbiogenesis: euglenoid, dinoflagellate, and sporozoan plastid origins and the eukaryote family tree. *J. Eukaryot. Microbiol.* 46:347–366.

Cavalier-Smith, T. 2002. The phagotrophic origin of eukaryotes and phylogenetic classification of *Protozoa*. *Int. J. Syst. Evol. Microbiol.* 52:297–354.

Cavalier-Smith, T. 2003a. Genomic reduction and evolution of novel genetic membranes and protein-targeting machinery in eukaryote-eukaryote chimaeras (meta-algae). *Philos. Trans. R. Soc. Lond. B Biol. Sci.* 358:109–134.

Cavalier-Smith, T. 2003b. Protist phylogeny and the high-level classification of *Protozoa*. *Eur. J. Protistol.* 39:338–348.

Cavalier-Smith, T. 2018. Kingdom *Chromista* and its eight phyla: a new synthesis emphasising periplastid protein targeting, cytoskeletal and periplastid evolution, and ancient divergences. *Protoplasma* 255:297–357.

Cavalier-Smith, T., and E. E. Y. Chao. 2003. Phylogeny and classification of phylum *Cercozoa* (*Protozoa*). *Protist* 154:341–358.

Cavalier-Smith, T., E. E. Y. Chao, and R. Lewis. 2015. Multiple origins of heliozoa from flagellate ancestors: new cryptist subphylum *Corbihelia*, superclass *Corbistoma*, and monophyly of *Haptista*, *Cryptista*, *Hacrobia* and *Chromista*. *Mol. Phylogenet. Evol.* 93:331–362.

Howe A. T., D. Bass, J. M. Scoble, R. Lewis, K. Vickerman, H. Arndt, and T. Cavalier-Smith. 2011. Novel cultured protists identify deep-branching environmental DNA clades of *Cercozoa*: new genera *Tremula*, *Micrometopion*, *Minimassisteria*, *Nudifila*, *Peregrinia*. *Protist* 162:332–372.

Ishitani, Y., S. A. Ishikawa, Y. Inagaki, M. Tsuchiya, K. Takahashi, and K. Takishita. 2011. Multigene phylogenetic analyses including diverse radiolarian species support the "*Retaria*" hypothesis—the sister relationship of *Radiolaria* and *Foraminifera*. *Mar. Micropaleontol.* 81:32–42.

Keeling, P. J. 2001. *Foraminifera* and *Cercozoa* are related in actin phylogeny: two orphans find a home? *Mol. Biol. Evol.* 18:1551–1557.

Keeling, P. J., G. Burger, D. G. Durnford, B. F. Lang, R. W. Lee, R. E. Pearlman, A. J. Roger, and M. W. Gray. 2005. The tree of eukaryotes. *Trends Ecol. Evol.* 20:670–676.

Krabberød, A. K., J. Bråte, J. K. Dolven, R. F. Ose, D. Klaveness, T. Kristensen, K. R. Bjørklund, and K. Shalchian-Tabrizi. 2011. *Radiolaria* divided into *Polycystina* and *Spasmaria* in combined 18S and 28S rDNA phylogeny. *PLOS ONE* 6:e23526.

Krabberød, A. K., R. Orr, J. Bråte, T. Kristensen, K. R. Bjørklund, and K. Shalchian-Tabrizi. 2017. Single cell transcriptomics, mega-phylogeny and the genetic basis of morphological innovations in *Rhizaria*. *Mol. Biol. Evol.* 34:1557–1573.

Kunitomo, Y., I. Sarashina, M. Iijima, K. Endo, and K. Sashida. 2006. Molecular phylogeny of acantharian and polycystine radiolarians based on ribosomal DNA sequences, and some comparisons with data from the fossil record. *Europ. J. Protistol.* 42:143–153.

Longet, D., J. M. Archibald, P. J. Keeling, and J. Pawlowski. 2003. *Foraminifera* and *Cercozoa* share a common origin according to RNA polymerase II phylogenies. *Int. J. Syst. Evol. Microbiol.* 53:1735–1739.

Moreira, D., S. von der Heyden, D. Bass, P. López-García, E. Chao, and T. Cavalier-Smith. 2007. Global eukaryote phylogeny: combined small-and large-subunit ribosomal DNA trees support monophyly of *Rhizaria*, *Retaria* and *Excavata*. *Mol. Phylogenet. Evol.* 44:255–266.

Nikolaev, S. I., C. Berney, J. F. Fahrni, I. Bolivar, S. Polet, A. P. Mylnikov, V. V. Aleshin, N. B. Petrov, and J. Pawlowski. 2004. The twilight of *Heliozoa* and rise of *Rhizaria*, an emerging supergroup of amoeboid eukaryotes. *Proc. Biol. Sci.* 101:8066–8071.

Parfrey, L. W., J. Grant, Y. I. Tekle, E. Lasek-Nesselquist, H. G. Morrison, M. L. Sogin, D. J. Patterson, and L. A. Katz. 2010. Broadly sampled multigene analyses yield a well-resolved eukaryotic tree of life. *Syst. Biol.* 59:518–533.

Pawlowski, J., and F. Burki. 2009. Untangling the phylogeny of amoeboid protists. *J. Eukaryot. Microbiol.* 56:16–25.

Polet, S., C. Berney, J. Fahrni, and J. Pawlowski. 2004. Small-subunit ribosomal RNA gene sequences of *Phaeodarea* challenge the monophyly of Haeckel's *Radiolaria*. *Protist* 155:53–63.

Sierra, R., S. J. Cañas-Duarte, F. Burki, A. Schwelm, J. Fogelqvist, C. Dixelius, L. N. González-García, G. H. Gile, C. H. Slamovits, C. Klopp, S. Restrepo, I. Arzul, and J. Pawlowski. 2016. Evolutionary origins of rhizarian parasites. *Mol. Biol. Evol.* 33:980–983.

Sierra, R., M. V. Matz, G. Aglyamova, L. Pillet, J. Decelle, F. Not, C. de Vargas, and J. Pawlowski. 2013. Deep relationships of *Rhizaria* revealed by phylogenomics: a farewell to Haeckel's *Radiolaria*. *Mol. Phylogenet. Evol.* 67:53–59.

Simpson, A. G. B., and A. J. Roger. 2004. The real 'kingdoms' of eukaryotes. *Curr. Biol.* 14:693–696.

Tekle, Y. I., J. Grant, J. C. Cole, T. A. Nerad, O. R. Anderson, D. J. Patterson, and L. A. Katz. 2007. A multigene analysis of *Corallomyxa tenera* sp. nov. suggests its membership in a clade that includes *Gromia*, *Haplosporidia* and *Foraminifera*. *Protist* 158:457–472.

Author

Alastair G. B. Simpson; Department of Biology; Dalhousie University; Halifax, NS B3H 4R2, Canada. Email: alastair.simpson@dal.ca.

Date Accepted: 11 November 2014; updated 20 March 2018

Primary Editor: Philip Cantino

Foraminifera C. E. Eichwald 1830 [S. L. Richardson and J. H. Lipps], converted clade name

Registration Number: 297

Definition: The crown clade originating in the most recent common ancestor of *Bulimina marginata* d'Orbigny 1826 (*Globothalamea*) and all extant organisms or species that inherited granuloreticulose pseudopodia synapomorphic with those in *Bulimina marginata* d'Orbigny 1826. This is an apomorphy-modified crown-clade definition. Abbreviated definition: crown ∇ apo granuloreticulose pseudopodia [*Bulimina marginata* d'Orbigny 1826].

Etymology: Derived from the Latin *foramen* (hole) and *-fer* (bearing). The original use of the name refers to the aperture and openings that connect successive chambers (intercameral foramina) in a multi-chambered test, not to the presence of pores in the test wall.

Reference Phylogeny: The primary reference phylogeny is Pawlowski et al. (2013: Fig. 1), where the clade *Foraminifera* is the entire tree (no outgroups are shown). For outgroup relationships, see Sierra et al. (2013: Fig. 2). See also Nikolaev et al. (2004: Fig. 1), Bowser et al. (2006: Fig. 5.3), Pawlowski and Burki (2009: Fig. 2), and Groussin et al. (2011: Fig. 1).

Composition: Recent estimates of the diversity of Recent benthic foraminiferans range from 4,280 species (Murray, 2007) to as many as 15,000 to 100,000 potential species based on the number of unassigned DNA sequences found in environmental samples (Habura et al., 2004a; Pawlowski et al., 2005; Habura et al., 2008), as well as the wide occurrence of cryptic speciation in both planktonic and benthic foraminiferans (Hemleben et al., 1989; Adl et al., 2007; Darling and Wade, 2008; Pawlowski and Holzmann, 2008; Lipps and Finger, 2010). Lipps and Finger (2010) estimated the total number of crown foraminiferan species that have ever lived as less than 5 million. Descriptions of named crown foraminiferan species, both fossil and Recent, are compiled in and accessible through the *Catalogue of* Foraminifera (Ellis and Messina, 1940 et seq.).

Diagnostic Apomorphies: Among crown unicellular eukaryotes, the presence of granuloreticulose pseudopodia in the living cell is a unique and un-reversed synapomorphy regarded as unambiguously diagnostic of crown *Foraminifera* (see Comments). Pseudopodia are fragile and unlikely to be preserved, however, so knowledge of a potentially earlier origin in the extinct foraminiferan stem may remain elusive.

Additional apomorphies that can be used to distinguish *Foraminifera* from other crown clades (such as *Plasmodiophorida*, *Cercozoa*, *Gromida*, *Polycystinea*, and *Acantharea*) within the more inclusive clade *Rhizaria* include the presence of a unique helix in Domain III of the small subunit ribosomal RNA (the novel helix lies between Helices 41 and 39 in Domain III and is up to 350 nucleotides in length; Habura et al. 2004b: Figs. 2, 3; Bowser et al. 2006: Fig. 5.2), and the occurrence of 20 unique spliceosomal introns in the gene sequences that code for the two foraminiferan actin paralogs (Flakowski et al., 2006: Fig. 2).

The presence of a test, or shell, is a traditional foraminiferan apomorphy that can be useful for recognizing foraminiferans in both fossil and Recent sediments (Lipps, 1973; Goldstein, 1999; Hottinger, 2000). Multi-chambered tests, which arose within the crown clade and thus are not an apomorphy of the entire crown, are not constructed by any other single-celled eukaryote, so fossil multi-chambered tests can unequivocally be attributed to *Foraminifera*. Single-chambered tests ancestral for *Foraminifera* are, however, also formed by several other single-celled eukaryotes, including *Gromia* (*Rhizaria*), euglyphids (*Rhizaria: Cercozoa*), testate lobose amoebans (*Amoebozoa*), amphitremids (*Labyrinthulomycetes*), and loricate ciliates (*Alveolata*) (Burki et al., 2002; Wylezich et al., 2002; Longet et al., 2004; Tekle et al., 2007; Nikolaev et al. 2008; Pawlowski and Burki, 2009; Bachy et al., 2012; Gomaa et al., 2013).

Synonyms: The name *Foraminiferida* has been used as an approximate synonym (sensu de Queiroz in Cantino, 2007) of *Foraminifera* as defined here in that it is inferred based on labeling of a phylogenetic tree in which the authors did not clearly indicate whether the name should be applied to a crown clade, a total clade, or to an apomorphy (Pawlowski et al., 1996: Fig. 3; Pawlowski, 2000: Fig. 2).

The following additional names are approximate synonyms (see Cantino and de Queiroz, 2020) of *Foraminifera* as defined here in that their synonymy is inferred based on stated composition and/or diagnostic characters: *Acyttaria* (Haeckel, 1868: 122; Haeckel, 1870: 61); *Asiphonoidea* (de Haan, 1825: 22, in part); *Foraminifères* (d'Orbigny, 1826: 245; De Lanessan, 1882: 58; Granger, 1896: 333); *Foraminiferi* (Schwager, 1877: 18); *Foraminiferiae* (Delage and Hérouard, 1896: 107); *Foraminiferida* (Loeblich and Tappan, 1961: 273; Loeblich and Tappan, 1964: C55; Sleigh, 1973: 283; Loeblich and Tappan, 1984: 2; Loeblich and Tappan, 1988: 7; Sleigh, 1989: 185; Levine et al., 1980: 42; Bock et al., 1985: 252; Culver, 1993: 220; Hart and Williams, 1993: 43); *Granuloreticulosa* (De Saedeleer, 1934: 7; Lee, 1990: 524; Patterson, 1999: S109; Lee et al., 2000: 872);

Granuloreticulosea (Anderson, 1988: 65; Corliss, 1994: 18); *Micropolythalamia* (Zborsewski, 1834: 311); *Polythalamacea* (de Blainville, 1824: 175, in part); *Polythalamia* (Ehrenberg, 1838: 192, 200); *Polythalamiis* (Breyn, 1732, in part: 1); *Reticularia* (Carpenter et al., 1862: 28, in part; Brady, 1881: 43; Brady, 1884: 60); *Reticulariida* (Calkins, 1901: 106, in part); *Reticulosa* (Carpenter et al., 1862: 17, in part); *Rhizopoda* (Owen, 1855: 56, in part; Copeland, 1956: 179); *Rhizopoda imperforata* (Schmarda, 1871: 162, in part); *Rhizopoda reticularia* (Chapman, 1902: 19, in part); *Rhizopoden* (Siebold and Stannius, 1848:11); *Rhizopodes* (Dujardin, 1841: 126, in part); *Testacea* (Schultze, 1854: 52; Pritchard, 1861: 241; Haeckel, 1862: 199; Bütschli in Bronn, 1880: 181); *Thalamophora* (Hertwig, 1876: 53; Haeckel, 1878: 98).

Comments: We choose to convert the name *Foraminifera* instead of other names that could be applied to the clade (see Synonyms) because it has been and still is the most widely used name for the clade in the scientific literature (neontological and palaeontological, ecological and palaeoecological, oceanographic and palaeoceanographic).

The name *Foraminifera* has been applied previously to the same clade to which it is here applied by the following authors as inferred from where they placed that name on their phylogenetic trees, although it was not clearly indicated whether the name applied to a crown- or total-clade or to an apomorphy: Pawlowski et al. (1994: Fig. 5); Pawlowski et al. (1999a: Fig. 1); Pawlowski et al. (1999b: Fig. 2); Lee et al. (2000: Fig. 6b); Keeling (2001: Figs. 1, 2); Pawlowski and Holzmann (2002: Figs. 1, 2); Berney and Pawlowski (2003: Fig. 1); Cavalier-Smith and Chao (2003: Fig. 7); Longet et al. (2003: Fig. 1); Pawlowski and Berney (2003: Fig. 6.1); Pawlowski et al. (2003a: Figs. 1, 2); Archibald and Keeling (2004: Figs. 1, 2); Longet et al. (2004: Figs. 2, 3); Nikolaev et al. (2004: Fig. 1); Flakowski et al. (2005: Fig. 1); Bowser et al. (2006: Fig. 5.3); Moreira et al. (2007: Figs. 1–3); Tekle et al. (2007: Fig. 4); Pawlowski (2008: Fig. 2); Bass et al. (2009: Fig. 1); Pawlowski and Burki (2009: Fig. 2); Groussin et al. (2011: Fig. 1); Pawlowski et al. (2013: Fig. 1); Sierra et al. (2013: Fig. 2).

Similarly, the name *Foraminifera* has been previously applied to the same clade to which it is applied here by the following authors as inferred based on stated composition and/or diagnostic characters: Eichwald (1830: 21); Williamson (1849: 59); Claparède and Lachmann (1858: 434); Williamson (1858: v. 1); Carpenter (1861: 466); Carpenter et al. (1862: 17), in part; Greene (1859: 9); Reuss (1862: 362); Greene (1863: 9); Greene (1868: 9); Greene (1871: 9); Schmarda (1871: 164); Claus (1883: 162); Brady (1884: 60); Doflein (1901: 13); Chapman (1902: 19, in part); Rhumbler (1904: 192), in part; Doflein (1911: 612), in part; Minchin (1912: 231); Calkins (1926: 331); Cushman (1928: 52); Cushman (1933: 62); Galloway (1933: 8); Kudo (1939: 344); Cushman (1948: 65); Moore (1954: 595); Jepps (1956: 20); Banner et al. (1967: 299); Cavalier-Smith (1993: 972); Cavalier-Smith (1998: 232); Sen Gupta (1999: 19); Cavalier-Smith (2002: 326); Cavalier-Smith (2003: 347); Hausmann et al. (2003: 129); Cavalier-Smith (2004: 1252); Adl et al. (2005: 418); Adl et al. (2012: 473); Mikhalevich (2013: 500).

The name *Foraminifera* is a Latinization of the French name *Foraminifères*, first applied to a group of "microscopic cephalopods" by Alcide d'Orbigny in 1826. Earlier researchers also considered foraminiferans to be molluscs and classified them under various names, such as *Polythalamiis* (Breyn, 1732: 1), *Nautilus* (Linnaeus, 1758: 709; Fichtel and Moll, 1798: 12; Fichtel and Moll, 1803: 12), *Polythalamacea* (de Blainville, 1824: 175) and *Asiphonoidea* (de Haan, 1825: 29). Dujardin's (1835abcd) discovery that foraminiferans were not cephalopods,

but relatively unmodified single-celled organisms, led to their reclassification as rhizopods (*Rhizopodes* or *Rhizopoda*) and a reinterpretation of their "tentacles" as pseudopodia, or thread-like extensions of the cell's sarcode or cytoplasm (Dujardin, 1841). Ehrenberg (1838) used the name *Polythalamia* for the group but viewed foraminiferans and other single-celled eukaryotes as having the same organization-systems as multicellular organisms, including miniature circulatory, digestive, excretory, nervous, motor, and reproductive systems (Churchill, 1989).

We use an apomorphy-modified crown-clade definition of the name *Foraminifera* for two reasons: (1) relationships among major subclades within *Foraminifera* are poorly resolved, and (2) estimates of foraminiferan relationships among potentially more inclusive clades remain hampered by poor sampling (e.g., Kosakyan et al., 2016). Following the arguments of Gauthier and de Queiroz (2001), we note that such a definition would fix the name *Foraminifera* with respect to its defining apomorphy—granuloreticulose pseudopodia—that should be robust to changing ideas about ingroup and outgroup tree topologies. In the unlikely event that this 'soft' apomorphy could be observed in stem fossils, our definition stipulates that they would not be part of *Foraminifera*, even though they had the 'defining' synapomorphy, because those fossils are not part of the crown clade. If the 'defining' apomorphy arose convergently among eukaryotes, only those species in which granuloreticulose pseudopodia are homologous with those in the globothalamean foraminiferan *Bulimina marginata* would qualify as *Foraminifera*. Even if some crown foraminiferans were to subsequently lose granuloreticulose pseudopodia, they would still be part of *Foraminifera* because they descended from an ancestor that shared that apomorphy.

Foraminifera is currently thought to be a sub-clade within the more inclusive clade *Rhizaria* (Nikolaev et al., 2004; also see entry in this volume). Microbial diversity in the sea is, however, poorly understood, and the sister clade to *Foraminifera* remains unresolved. Although only a small fraction of the recognized species have been sampled, recent molecular phylogenetic studies indicate that *Foraminifera* is most closely related to one of the following clades of single-celled eukaryotes: *Acantharea* (planktonic marine microbes with a strontium-sulfate skeleton) (Krabberød et al., 2011; Brown et al., 2012; Decelle et al., 2012; Sierra et al., 2013), *Polycystinea* (planktonic marine microbes with siliceous skeletons) (Ishitani et al., 2011; Burki and Keeling, 2014), or *Gromia* (benthic marine microbes with a single-chambered agglutinated test or shell) (Longet et al., 2004; Nikolaev et al, 2004; Flakowski et al., 2005, 2006).

Traditional evolutionary classifications subdivided *Foraminifera* into higher taxa based on wall structure and chamber arrangement (Loeblich and Tappan, 1964, 1988; Culver, 1993; Sen Gupta, 1999). Molecular phylogenetic studies have shown that of the 15 extant foraminiferan taxa traditionally ranked as orders (outlined in Sen Gupta, 1999), three are monophyletic (*Carterinida, Miliolida, Spirillinida*), five are paraphyletic ("*Allogromiida*", "*Astrorhizida*", "*Trochamminida*", "*Textulariida*", "*Rotaliida*"), and three are polyphyletic ("*Globigerinida*", "*Lituolida*", "*Buliminida*")) (Darling et al., 1997; de Vargas et al., 1997; Pawlowski and Holzmann, 2002; Pawlowski et al., 2003a,b; Ertan et al., 2004; Flakowski et al., 2005; Schweizer et al., 2005; Schweizer et al., 2008; Ujiié et al., 2008; Pawlowski et al., 2013; Pawlowski et al., 2014). Information on the phylogenetic relationships based on gene sequences of the remaining four taxa is either equivocal (*Lagenida, Robertinida*) or currently unavailable (*Involutinida, Silicoloculinida*) (Bowser et al., 2006; Pawlowski et al., 2013).

As indicated in the Diagnostic Apomorphies section, only foraminiferans have granuloreticulose pseudopodia, or granulose reticulopodia, characterized by a branching and anastomosing network of filamentous pseudopodia that exhibit rapid bidirectional movement of intracellular granules (Lee, 1990; Lee et al., 2000). The reticulopodial network may spread over a large area relative to the size of the cell body or test; for example, *Reticulomyxa filosa*, a non-testate foraminiferan, can extend a pseudopodial array that covers an area of 1–4 cm^2 (Koonce et al., 1986). Likewise, individual pseudopodia have been observed to extend up to a few centimeters in length (Cushman, 1922; Le Calvez, 1936; Schwab, 1977; Travis et al., 1988).

The granuloreticulose network of pseudopodia extended by foraminiferans is the primary interface through which these single-celled eukaryotes interact with their environment. The network of filose pseudopodia increases the surface area-to-volume ratio of the plasma membrane, allowing more efficient gas exchange, absorption of nutrients, and discharge of wastes (Bowser and Travis, 2002; Travis and Bowser, 1991). Foraminiferans use their pseudopodial network to capture food by actively grazing on unicellular organisms and passively trapping food particles, including small animals and animal larvae (Lipps, 1983). Granuloreticulose pseudopodia are also the primary mechanism used by foraminiferans for locomotion and attachment to biotic and abiotic surfaces (Travis and Bowser, 1991). Pseudopodia play an active role in chamber formation and the biomineralization of the calcite tests (shells) of rotalid foraminiferans by creating a delimited space where the wall is deposited, and by transporting cell components to the site of chamber formation (Hemleben et al., 1986; Erez, 2003). Agglutinated foraminiferans use their pseudopodia to collect sediment particles from their environment for incorporation into their test wall, as well as to create a three-dimensional framework for the formation of new chambers (Angell, 1990; Hemleben and Kaminski, 1990).

The reticulate nature of foraminiferan pseudopodia was first observed by Dujardin, (1835a–d); subsequent researchers described the saltatory bidirectional movement of intracellular granules in the cytoplasm (Carpenter, 1861; Leidy, 1879; Schultze, 1854). The term "granuloreticulose" was first used to describe the pseudopodia of foraminiferans by De Saedeleer (1932), who applied the term to reticulopodia (branched and anastomosing networks of pseudopodia) that could be characterized by the bidirectional motion of intracellular granules (organelles) along pseudopodial filaments. More advanced forms of microscopy have shown that the cytoplasmic granules observed by nineteenth-century and early twentieth-century naturalists were often cell organelles (tubular mitochondria, digestive vacuoles, lysosomes, secretory vacuoles, excretory vacuoles, motility organizing vesicles or elliptical fuzzy-coated vesicles), other unidentified cell cargo, and even membrane-bound endosymbiotic algae (Anderson and Bé, 1978; Anderson and Lee, 1991; Travis and Bowser, 1991; Bowser and Travis, 2002).

Later studies of the cell ultrastructure of foraminiferans using transmission electron microscopy demonstrated that linear arrays of microtubules underlie the granuloreticulose pseudopodia (Hedley et al., 1967; Lengsfeld, 1969a,b; Marszalek, 1969; Schwab, 1969). These parallel bundles of microtubules are not cross-linked into rigid three-dimensional geometric arrays, nor do they nucleate from centrally-located microtubule-organizing centers, as has been observed in the rigid pseudopods (axopods) of polycystines and acantharians (Febvre-Chevalier and Febvre, 1993; Suzuki and Not, 2015). In foraminiferan pseudopodia, the microtubule arrays are dynamic and the microtubules exhibit lateral sliding and bending movements, although adjacent microtubules may also be temporarily connected by dynein-like cross-bridges (Travis and Bowser, 1991; Golz and Hauser, 1994).

Cytoskeletal microtubules form a network of intracellular trackways along which the "granules," or cell organelles, of the interphase cell are transported by molecular motors through the cytoplasm (Lodish et al., 2016). The rate at which organelles are transported in the foraminiferan cell are unusually fast compared to that recorded in the cells of more complex organisms; for example, in *Reticulomyxa filosa*, intracellular organelles have been observed to move from 15 to 25 µm per second (Koonce and Schliwa, 1985), compared to 3 µm per second, the fastest rate of movement of membrane-bound particles in the axon of neurons (Lodish et al., 2016). Bidirectional transport of extracellular particles also takes place along the surface of foraminiferan pseudopodia; material transported includes food particles, sediment particles used in the construction of agglutinated tests and/or feeding cysts, and juvenile foraminiferan tests (dispersed by parent cell after reproduction by multiple fission) (Travis and Bowser, 1991; Bowser and Travis, 2002). Surface transport may be transiently coupled to the transport of intracellular organelles and driven by a common mechanism (Bowser et al., 1984; Bowser and Travis, 2002).

The granuloreticulose pseudopodial network is extremely dynamic and was described by Carpenter (1861: 461) as resembling an "animated spider's web." Pseudopodia are constantly being extended and retracted, and the extended pseudopodia are continually bifurcating and anastomosing. In *Reticulomyxa filosa*, assembly rates of 3.5 µm per sec and disassembly rates of 19.5 µm per sec have been recorded (Chen and Schliwa, 1990). The ability of granuloreticulose pseudopodia to rapidly assemble and disassemble at rates not observed in other organisms or cells is due to the presence of helical filaments, a unique storage form of coiled tubulin, in the cytoplasm of foraminiferan cells (Weinhofer and Travis, 1998). Recent studies have shown that the molecular structure of these unique tubulin helical filaments is due to a non-canonical β-tubulin, referred to as Type 2 β-tubulin or β2-tubulin (Linder et al., 1997; Habura et al., 2005; Bassen et al., 2016). The α-β2-tubulin heterodimers that make up the microtubules, and the helical filaments in the pseudopodia, allow for stronger lateral tubulin heterodimer interactions and weaker longitudinal tubulin heterodimer interactions, a feature which may facilitate the rapid assembly and disassembly of foraminiferan microtubules (Bassen et al., 2016).

Type 2 beta-tubulin, or β2-tubulin, genes have only been found in foraminiferans, polycystines and acantharians, and not in *Gromia* or other rhizarians (Habura et al., 2005; Hou et al., 2013). In phylogenetic analyses, the β2-tubulins of foraminiferans and polycystines-acantharians are genetically distinct from each other and branch in separate clades; that is to say, the foraminiferan β2-tubulins are more closely related to each other than to either the β2-tubulins of polycystines or acantharians (Hou et al., 2013: Fig. 1).

The foraminiferan test is an extracellular covering that encapsulates the cell body and provides protection from the mechanical stress encountered in the interstitial realm of aquatic marine sediments (Giere, 2009). The foraminiferan test has at least one opening, or aperture, through which the granuloreticulose pseudopodia can extrude into the surrounding environment. The aperture lacks an operculum, a feature that can be used to distinguish single-chambered foraminiferan tests from the heavily walled spores of haplosporidians (Perkins, 2000; Tekle et al., 2007).

Tests may also function to protect the cell body from predation and parasitism (Lipps, 1983) and enable epiphytic and other attached foraminiferans to adhere to solid substrates (Langer, 1993). Tubular extensions of the test provide permanent coverings for pseudopodia in *Carterina spiculotesta* (Lipps, 1983); spinose extensions of the test provide support for pseudopodia in planktonic foraminiferans (Hemleben et al., 1989). Root-like structures function to anchor the test of *Notodendrodes antarctikos* in the sediments (Bowser et al., 1995). Specialized brood chambers in soritacean tests protect the developing embryons (products of multiple fission) prior to their release and dispersal (Richardson, 2009). Terminal, gas-filled float chambers in meroplanktonic taxa, such as *Tretomphalus*, allow gametogenic tests to detach from benthic substrates and release their gametes into the water column (Myers, 1943).

The composition of the test wall can be non-mineralized or mineralized. Non-mineralized tests are either organic-walled (secreted by the cell) or agglutinated (adventitious sediment particles bound in a secreted organic cement) (Lipps, 1973; Bender and Hemleben, 1998a). Mineralized tests are composed of calcium carbonate (aragonite, high-magnesium calcite, or low-magnesium calcite) and constructed by two different mechanisms. Mineralized tests in the subclade *Tubothalamea* are built from Golgi-derived vesicles that are formed intracellularly; the membrane-bound crystal packets are transported to, and assembled in, the extracellular environment (Hemleben et al., 1986). Some agglutinated textulariid foraminiferans (early diverging and relatively unmodified *Globothalamea*) use secreted calcareous cement to bind together sediment particles in the test wall (Bender and Hemleben, 1988a,b). Others, including *Carterina* and *Zaninettia*, use an organic cement to bind together secreted calcareous spicules in the test wall (Deutsch and Lipps, 1976; Pawlowski et al., 2014). Biomineralization in rotalid foraminiferans (a later-diverging and more modified subclade of *Globothalamea*) does not involve secreted calcite (or aragonite), but occurs within a protected space formed by the granuloreticulose pseudopodia in the shape of the new chamber (Erez, 2003). Calcite in the test wall is precipitated on both sides of an organic template layer within a microenvironment that is supersaturated in calcium carbonate (Erez, 2003; de Nooijer et al., 2014).

Single-chambered tests with organic or agglutinated walls characterize the earliest-branching clades in *Foraminifera* (Pawlowski et al., 2003a,b; Lecroq et al., 2011). A wide range of test shapes is observed in these single-chambered forms, many of which have spherical, pyriform, ellipsoid, ovoid, hemispherical, tubular, bifurcating tubular, or arborescent test morphologies (Loeblich and Tappan, 1964; 1988). Some xenophyophores have large, complex tests that are composed of an anastomosing network of interconnected tubes (Tendal, 1972; Richardson, 2001). At least two extant species of terrestrial foraminiferans lack a test—*Reticulomyxa filosa* (Nauss, 1949) and *Haplomyxa saranae* (Dellinger et al., 2014). The absence of a test in *R. filosa* has been shown to be an autapomorphy, as this species is most closely related to testate, single-chambered taxa (Pawlowski et al., 1999a,b).

Multi-chambered tests are more variable in shape than single-chambered tests, and chamber arrangement has long been considered an important character in foraminiferal taxonomy (Cifelli, 1990). Some multi-chambered tests in the subclade *Tubothalamea* can be simple two-chambered tests composed of a bulbous proloculus followed by a second, non-septate, tubular chamber that is either coiled or uncoiled. Some multi-chambered tests consist of a linear array of chambers, arranged like a string of interconnected beads. Many multi-chambered tests are arranged in coils. Multi-chambered tests characterize later-branching clades of *Foraminifera*.

In the subclade *Globothalamea*, the test grows by terminal addition of discrete chambers resulting in a multi-chambered test (Pawlowski et al., 2013). In the basal taxa of the subclade *Tubulothalamea*, the test enlarges by accretionary growth whereby new material is added to the test at the apertural margin of a single-chambered tubular test (Pawlowski et al., 2013).

Molecular phylogenies indicate that the earliest-branching clades within crown *Foraminifera* consist of several unnamed taxa of predominantly single-chambered species that construct organic-walled or agglutinated tests (Pawlowski et al., 2003a; Habura et al., 2008); however, some of these taxa, such as the "naked" foraminiferan *Reticulomyxa filosa*, appear to have lost the test secondarily (Pawlowski et al., 1999a,b). In addition, molecular phylogenetic studies of the large agglutinated species *Syringammina corbicula*, *Shinkaiya lindsayi*, and an undescribed xenophyophore, indicate that *Xenophyophora*, formerly classified as a separate phylum (Tendal, 1990), is a clade that is nested within *Foraminifera* and is more closely related to *Globothalamea*, a clade of foraminiferans with multi-chambered calcareous tests, than to other agglutinated foraminiferans with tubular tests (Pawlowski et al., 2003a,b; Lecroq et al., 2009; Hori et al., 2013; Pawlowski et al., 2013).

Biomineralized, multi-chambered tests arose independently in *Tubulothalamea* and *Globothalamea*, two major subclades of *Foraminifera* (Pawlowski et al., 2013). The subclade *Tubulothalamea* is composed of: (1) agglutinated taxa with coiled tubular tests, such as *Ammodiscus*; (2) taxa with coiled, single-crystal calcite tests, such as *Spirillina* and *Patellina*; and (3), miliolid foraminiferans, a diverse clade with multi-chambered tests constructed of haphazardly arranged magnesium calcite crystals (Hansen, 1999; Debenay et al., 2000; Pawlowski et al., 2013). Taxa that unambiguously nest within the subclade *Globothalamea* include: (1) taxa with single- and multi-chambered tests composed of a single-layer of porous calcite, such as *Lenticulina*; (2) taxa with multi-chambered agglutinated tests, such as *Trochammina*; and (3), taxa with multi-chambered tests composed of multiple layers of porous calcite or aragonite, such as *Rosalina* or *Globigerina* (Pawlowski and Holzmann, 2002; Flakowski et al., 2005; Bowser et al., 2006; Schweizer et al., 2008; Pawlowski et al., 2013).

Crown *Foraminifera*, based on taxa sampled so far, is well supported by high non-parametric bootstrap values (maximum likelihood, neighbor joining, and/or maximum parsimony) and/or high Bayesian posterior probabilities in most molecular phylogenetic analyses to date (Pawlowski et al., 1996: Fig. 3; Pawlowski and Holzmann, 2002: Fig. 1; Berney and Pawlowski, 2003: Fig. 1; Cavalier-Smith and Chao, 2003: Fig. 1; Longet et al., 2003: Fig. 1; Pawlowski and Berney, 2003: Fig. 6.1; Longet et al., 2004: Fig. 3; Nikolaev et al., 2004: Fig. 1; Moreira et al., 2007: Fig. 3B; Tekle et al., 2007: Fig. 4; Sierra et al., 2013: Fig. 2).

Molecular estimates of divergence times indicate that the earliest divergence within (crown) *Foraminifera* most likely occurred during the mid-Neoproterozoic between 650–900 Ma (Pawlowski and Berney, 2003; Groussin et al., 2011), a time span that corresponds with palaeontological evidence of a major diversification of marine eukaryotes at ~800 Ma (Porter, 2004; Knoll, 2014). The earliest unambiguous fossil foraminiferans are Neoproterozoic-to-Cambrian-aged species of *Platysolenites* and *Spirosolenites* that have tests characterized by an initial proloculus followed by a second undivided, nonseptate, tubular chamber, much like modern-day species of *Ammodiscus* (Lipps and Rozanov, 1996; McIlroy et al., 2001; Streng et al., 2005; Kontorovich et al., 2008). Ediacaran-aged trace fossils, such as *Palaeopascichnus*, have recently been reinterpreted as foraminiferans (xenophyophores) (Seilacher et al., 2005; Porter,

2006; Dong et al., 2008). Most microfossils with a well-defined proloculus (initial chamber), and test growth by terminal addition, can be unequivocally identified as foraminiferans because those apomorphies appear within the crown (see Diagnostic Apomorphies above). Testate foraminiferans have one of the best fossil records of any organisms, with palaeobiodiversity estimated at 50,000 extinct species (Pawlowski and Holzmann, 2008).

Literature Cited

Adl, S. M., B. S. Leander, A. G. B. Simpson, J. M. Archibald, O. R. Anderson, D. Bass, S. S. Bowser, G. Brugerolle, M. A. Farmer, S. Karpov, M. Kolisko, C. E. Lane, D. J. Lodge, D. G. Mann, R. Meisterfeld, L. Mendoza, Ø. Moestrup, S. E. Mozley-Standridge, A. V. Smirnov, and F. Spiegel. 2007. Diversity, nomenclature, and taxonomy of protists. *Syst. Biol.* 56:684–689.

Adl, S. M., A. G. B. Simpson, M. A. Farmer, R. A. Andersen, O. R. Anderson, J. R. Barta, S. S. Bowser, G. U. Y. Brugerolle, R. A. Fensome, S. Fredericq, T. Y. James, S. Karpov, P. Kugrens, J. Krug, C. E. Lane, L. A. Lewis, J. Lodge, D. H. Lynn, D. G. Mann, R. M. McCourt, L. Mendoza, Ø. Moestrup, S. E. Mozley-Standridge, T. A. Nerad, C. A. Shearer, A. V. Smirnov, F. W. Spiegel, and M. F. J. R. Taylor. 2005. The new higher level classification of eukaryotes with emphasis on the taxonomy of protists. *J. Eukaryot. Microbiol.* 52:399–451.

Adl, S. M., A. G. B. Simpson, C. E. Lane, J. Lukeš, D. Bass, S. S. Bowser, M. W. Brown, F. Burki, M. Dunthorn, V. Hampl, A. Heiss, M. Hoppenrath, E. Lara, L. le Gall, D. H. Lynn, H. McManus, E. A. D. Mitchell, S. E. Mozley-Stanridge, L. W. Parfrey, J. Pawlowski, S. Rueckert, L. Shadwick, C. L. Schoch, A. Smirnov, and F. W. Spiegel. 2012. The revised classification of eukaryotes. *J. Eukaryot. Microbiol.* 59:429–514.

Anderson, O. R. 1988. *Comparative Protozoology: Ecology, Physiology, Life History.* Springer-Verlag, New York.

Anderson, O. R., and A. W. H. Bé. 1978. Recent advances in foraminiferal fine structure research. Pp. 121–202 in Foraminifera, Vol. 3 (R. H. Hedley and C. G. Adams, eds.). Academic Press, London.

Anderson, O. R., and J. J. Lee. 1991. Cytology and fine structure. Pp. 7–40 in *Biology of Foraminifera* (J. J. Lee and O. R. Anderson, eds.). Academic Press, San Diego, CA.

Angell, R. W. 1990. Observations on reproduction and juvenile test building in the foraminifer *Trochammina inflata*. *J. Foramin. Res.* 20:246–247.

Archibald, J. M., and P. J. Keeling. 2004. Actin and ubiquitin protein sequences support a cercozoan/foraminiferan ancestry for the plasmodiophorid plant pathogens. *J. Eukaryot. Microbiol.* 51:113–118.

Bachy, C., F. Gómez, P. López-García, J. R. Dolan, and D. Moreira. 2012. Molecular phylogeny of tintinnid ciliates (*Tintinnida, Ciliophora*). *Protist* 163:873–887.

Banner, F. T., W. J. Clarke, J. L. Cutbill, F. E. Eames, A. J. Lloyd, W. R. Riedel, and A. H. Smout. 1967. *Protozoa*. Pp. 291–332 in *The Fossil Record* (W. B. Harland, C. H. Holland, M. R. House, N. F. Hughes, A. B. Reynolds, M. J. S. Rudwick, G. E. Satterthwaite, L. B. H. Tarlo, and E. C. Willey, eds.). Geological Society of London, London.

Bass, D., E. E. Y. Chao, S. Nikolaev, A. Yabuki, K.-i. Ishida, C. Berney, U. Pakzad, C. Wylezich, and T. Cavalier-Smith. 2009. Phylogeny of novel naked filose and reticulose Cercozoa: *Granofilosea* cl. n. and *Proteomyxidea* revised. *Protist* 160:75–109.

Bassen, D. M., Y. Hou, S. S. Banavali, and N. K. Banavali. 2016. Maintenance of electrostatic stabilization in altered tubulin lateral contacts may facilitate formation of helical filaments in Foraminifera. *Sci. Rep.* 6:31723.

Bender, H., and C. Hemleben. 1988a. Constructional aspects in test formation of some agglutinated Foraminifera. *Abh. Geol. B.-A.* 41:13–21.

Bender, H., and C. Hemleben. 1988b. Calcitic cement secreted by agglutinated foraminifers grown in laboratory culture. *J. Foramin. Res.* 18:42–45.

Berney, C., and J. Pawlowski. 2003. Revised small subunit rRNA analysis provides further evidence that *Foraminifera* are related to *Cercozoa*. *J. Mol. Evol.* 57:S120–S127.

Bock, W., W. Hay, and J. J. Lee. 1985. Order *Foraminiferida* d'Orbigny, 1826. Pp. 253–273 in *An Illustrated Guide to the* Protozoa (J. J. Lee, S. H. Hutner, and E. C. Bovee, eds.). Society of Protozoologists, Lawrence, KS.

Bowser, S. S., A. J. Gooday, S. P. Alexander, and J. M. Bernhard. 1995. Larger agglutinated *Foraminifera* of McMurdo Sound, Antarctica: are *Astrammina rara* and *Notodendrodes antarctikos* allogromiids incognito? *Mar. Micropaleontol.* 26:75–88.

Bowser, S. S., A. Habura, and J. Pawlowski. 2006. Molecular evolution of *Foraminifera*. Pp. 78–93 in *Genomics and Evolution of Microbial Eukaryotes* (L. A. Katz and D. Bhattacharya, eds.). Oxford University Press, Oxford.

Bowser, S. S., H. A. Israel, S. M. McGee-Russell, and C. L. Rieder. 1984. Surface transport properties of reticulopodia: do intracellular and extracellular motility share a common mechanism? *Cell Biol. Int. Rep.* 8:1051–1063.

Bowser, S. S., and J. L. Travis. 2002. Reticulopodia: structural and behavioral basis for the suprageneric placement of granuloreticulosan protists. *J. Foraminif. Res.* 32:440–447.

Brady, H. B. 1881. Notes on some of the reticularian *Rhizopoda* of the "*Challenger*" Expedition. *Q. J. Microsc. Sci.* 21:31–71.

Brady, H. B. 1884. *Report on the* Foraminifera *Collected by H. M. S.* Challenger *during the Years 1873–1876.* Her Majesty's Stationery Office, London.

Breyn, J. P. 1732. *Dissertatio Physica De Polythalamiis, Nova Testaceorum Classe: Cui Quaedam Praemittuntur de Methodo Testacea in Classes Et Genera Distribuendi: Huic Adiicitur Commentatiuncula De Belemnitis Prussicis: Tandemque Schediasma De Echinis Methodice Disponendis.* Cornelium a Beughem, Gedani.

Brown, M. W., M. Kolisko, J. D. Silberman, and A. J. Roger. 2012. Aggregative multicellularity evolved independently in the eukaryotic supergroup *Rhizaria*. *Curr. Biol.* 22:1123–1127.

Burki, F., C. Berney, and J. Pawlowski. 2002. Phylogenetic position of *Gromia oviformis* Dujardin inferred from nuclear-encoded small subunit ribosomal DNA. *Protist* 153:251–260.

Burki, F., and P. J. Keeling. 2014. *Rhizaria. Curr. Biol.* 24:R103–R107.

Bütschli, O. 1880. *Protozoa.* In *Klassen und Ordnungen des Their-Reichs Wissenschaftlich Dargestellt in Wort und Bild* (H. G. Bronn, ed.). C. F. Winter'sche Verlagshandlung, Leipzig.

Calkins, G. N. 1901. *The* Protozoa. Macmillan Company, New York.

Calkins, G. N. 1926. *The Biology of the* Protozoa. Lea and Febiger, New York.

Cantino, P. D., and K. de Queiroz. 2020. *International Code of Phylogenetic Nomenclature (PhyloCode)*, Version 6. CRC Press, Boca Raton, FL.

Cantino, P. D., J. A. Doyle, S. W. Graham, W. S. Judd, R. G. Olmstead, D. E. Soltis, P. S. Soltis, and M. J. Donoghue. 2007. Towards a phylogenetic nomenclature of *Tracheophyta. Taxon* 56:1E–44E.

Carpenter, W. B. 1861. On the systematic arrangement of the *Rhizopoda. Nat. Hist. Rev.* 1:456–472.

Carpenter, W. B., W. K. Parker, and T. R. Jones. 1862. *Introduction to the Study of the* Foraminifera. The Ray Society, London.

Cavalier-Smith, T. 1993. Kingdom *Protozoa* and its 18 phyla. *Microbiol. Rev.* 57:953–994.

Cavalier-Smith, T. 1998. A revised six-kingdom system of life. *Biol. Rev.* 73:203–266.

Cavalier-Smith, T. 2002. The phagotrophic origin of eukaryotes and phylogenetic classification of *Protozoa. Int. J. Syst. Evol. Microbiol.* 52:297–354.

Cavalier-Smith, T. 2003. Protist phylogeny and the high-level classification of *Protozoa. Eur. J. Protistol.* 39:338–348.

Cavalier-Smith, T. 2004. Only six kingdoms of life. *Proc. R. Soc. Lond. B Biol. Sci.* 271:1251–1262.

Cavalier-Smith, T., and E. E. Y. Chao. 2003. Phylogeny of *Choanozoa, Apusozoa*, and other *Protozoa* and early eukaryote megaevolution. *J. Mol. Evol.* 56:540–563.

Chapman, F. 1902. *The* Foraminifera: *An Introduction to the Study of the* Protozoa. Longman, Green and Co., London.

Chen, Y.-T., and M. Schliwa. 1990. Direct observation of microtubule dynamics in *Reticulomyxa*: unusually rapid length changes and microtubule sliding. *Cell Motil. Cytoskel.* 17:214–226.

Churchill, F. B. 1989. The guts of the matter. *Infusoria* from Ehrenberg to Bütschli: 1838–1876. *J. Hist. of Biol.* 22:189–213.

Cifelli, R. 1990. The history of the classification of *Foraminifera* 1826–1933. *Cushman Found. Foramin. Res., Spec. Publ.* 27:1–119.

Claparède, É., and J. Lachmann. 1859. Sur les infusoires et les rhizopodes. *Mém. de l'Inst. Genevois* 6:261–482.

Claus, C. 1872. *Grundzüge der Zoologie.* N. G. Elwert'sche Universitäts-Buchhandlung, Marburg.

Claus, C. 1883. *Lehrbuch der Zoologie.* Elwert'sche Verlagsbuchhandlung, Marburg.

Copeland, H. F. 1956. *The Classification of Lower Organisms.* Pacific Books, Palo Alto, CA.

Corliss, J. O. 1994. An interim utilitarian ("user-friendly") hierarchical classification and characterization of the protists. *Acta Protozool.* 33:1–51.

Culver, S. J. 1993. *Foraminifera.* Pp. 203–247 in *Fossil Prokaryotes and Protists* (J. H. Lipps, ed.). Blackwell Scientific Publications, Boston, MA.

Cushman, J. A. 1922. Shallow-water *Foraminifera* of the Tortugas Region. *Carnegie I. Wash. Publ.* 17:3–85.

Cushman, J. A. 1928. Foraminifera: *Their Classification and Economic Use.* Cushman Laboratory for Foraminiferal Research, Sharon, MA.

Cushman, J. A. 1933. *Foraminifera*; their classification and economic use. *Cushman Lab. Foraminif. Res., Spec. Pub.* 4:1–349 pp.

Cushman, J. A. 1948. Foraminifera: *Their Classification and Economic Use.* Harvard University Press, Cambridge, MA.

Darling, K. F., and C. M. Wade. 2008. The genetic diversity of planktic *Foraminifera* and the global distribution of ribosomal RNA genotypes. *Mar. Micropaleontol.* 67:216–238.

Darling, K. F., C. M. Wade, D. Kroon and A. J. L. Brown. 1997. Planktic foraminiferal molecular evolution and their polyphyletic origins from benthic taxa. *Mar. Micropaleontol.* 30:251–266.

de Blainville, H. M. D. 1824. Mollusques. Pp. 1–414 in *Dictionnaire des Sciences Naturelles* (F. Cuvier, ed.). F. G. Levrault, Strasbourg and Paris.

de Haan, G. 1825. *Monographiae Ammoniteorum et Goniatiteorum*. Lugduni Batavorum, Hazenberg.

De Lanessan, J.-L. 1882. *Traité de Zoologie: Protozoaires*. Octave Doin, Paris.

de Nooijer, L. J., H. J. Spero, J. Erez, J. Bijma, and G. J. Reichart. 2014. Biomineralization in perforate *Foraminifera*. *Earth-Sci. Rev.* 135:48–58.

De Saedeleer, H. 1932. Notes de protistologie. V. Recherches sur les pseudopodes de rhizopodes testacés. Les concepte pseudopodes lobosa, filosa et granulo-reticulosa. *Arch. Zool. Exp. Gén.* 74:597–626.

De Saedeleer, H. 1934. Beitrag zur Kenntnis der Rhizopoden: Morphologische und systematische Untersuchungen und ein Klassifikationsversuch. *Mem. Mus. Hist. Nat. Belgique* 60:1–112.

de Vargas, C., L. Zaninetti, H. Hilbrecht, and J. Pawlowski. 1997. Phylogeny and rates of molecular evolution of planktonic *Foraminifera*: SSU rDNA sequences compared to the fossil record. *J. Mol. Evol.* 45:285–294.

Debenay, J., J.-J. Guillou, E. Geslin, and M. Lesourd. 2000. Crystallization of calcite in foraminiferal tests. Pp. 87–94 in *Advances in the Biology of Foraminifera* (J. J. Lee and P. Hallock, eds.). *Micropaleontology* 46(suppl. 1).

Decelle, J., N. Suzuki, F. Mahé, C. de Vargas, and F. Not. 2012. Molecular phylogeny and morphological evolution of the *Acantharia* (*Radiolaria*). *Protist* 163:435–450.

Delage, Y., and Hérouard, E. 1896. *Traité de Zoologie Concrète. Tome I. La Cellule et Les Protozoaires*. Librairie C. Reinwald, Paris.

Dellinger, M., A. Labat, L. Perrouault, and P. Grellier. 2014. *Haplomyxa saranae* gen. nov. et sp. nov., a new naked freshwater foraminifer. *Protist* 165:317–329.

Deutsch, S., and J. H. Lipps. 1976. Test structure of the foraminifer *Carterina*. *J. Paleontol.* 50:312–317.

Doflein, F. T. 1901. *Die Protozoen als Parasiten und Krankheitserreger Nach Biologischen Gesichtspunkten Dargestellt*. Gustav Fisher, Jena.

Doflein, F. T. 1911. *Lehrbuch der Protozoenkunde*. Gustav Fischer, Jena.

Dong, L., S. Xiao, B. Shen, and C. Zhou. 2008. Silicified *Horodyskia* and *Palaeopascichnus* from upper Ediacaran cherts in South China: tentative phylogenetic interpretation and implications for evolutionary stasis. *J. Geol. Soc. Lond.* 165:367–370.

d'Orbigny, A. D. 1826. Tableau méthodique de la classe des céphalopodes. *Ann. Sci. Nat. (Sér. 1)* 7:96–169, 245–314, pls. 10–15.

Dujardin, F. 1835a. Observations nouvelles sur les céphalopodes microscopiques. *Ann. Sci. Nat. Zool. (Sér. 2)* 3:108–109.

Dujardin, F. 1835b. Observations nouvelles sur les prétendus céphalopodes microscopiques. *Ann. Sci. Nat. Zool. (Sér. 2)* 3:312–316.

Dujardin, F. 1835c. Recherches sur les organisms inférieures. *Ann. Sci. Nat. Zool. (Sér. 2)* 4:343–377.

Dujardin, F. 1835d. Observations sur les rhizopodes et les infusoires. *C. R. Acad. Sci., Paris.* 1835:338–340.

Dujardin, F. 1841. *Histoire Naturelle des Zoophytes. Infusoires, Comprenant la Physiologie et la Classification de ces Animaux et la Maniere de les Étudier à l'Aide du Microscope*. Librairie Encyclopédique de Roret, Paris.

Ehrenberg, C. G. 1838. Dem blossen Auge unsichtbare Kalkthierchen und Kieselthierchen als Hauptbestand theile der Kreidegebirge. Bericht über die zur Bekanntmachung geeigneten. Verhandlungen der König. Preuss. *Akad. Wisse. Berlin, Jahre* 1838:192–200.

Eichwald, C. E. 1830. Zoologia specialis quam expositus animalibus tum vivis, tum fossilibus potissimum Rossiae in universum et Poloniae in species. Josephi Zawadzki, Vilnae.

Ellis, B. F., and A. Messina. *1940 et seq. Catalogue* of *Foraminifera*. Micropaleontology Press, New York.

Erez, J. 2003. The source of ions for biomineralization in *Foraminifera* and their implications for paleoceanographic proxies. *Rev. Mineral. Geochem.* 54:115–149.

Ertan, K. T., V. Hemleben, and C. Hemleben. 2004. Molecular evolution of some selected benthic *Foraminifera* as inferred from sequences of the small subunit ribosomal DNA. *Mar. Micropaleontol.* 53:367–388.

Febvre-Chevalier, C., and J. Febvre. 1993. Structural and physiological basis of axopodial dynamics. *Acta Protozool.* 32:211–228.

Fichtel, L. v., and J. P. C. v. Moll. 1798. *Testacea Microscopica Aliaque Minuta ex Generibus* Argonauta *et* Nautilus *ad Naturam Picta et Descripta.* Anton Pichler, Vienna.

Fichtel, L. v., and J. P. C. v. Moll. 1803. *Testacea Microscopica Aliaque Minuta Ex Generibus* Argonauta *Et* Nautilus *Ad Naturam Picta Et Descripta.* Camesinaischen Buchandlung, Vienna.

Flakowski, J., I. Bolivar, J. Fahrni, and J. Pawlowski. 2005. Actin phylogeny of *Foraminifera. J. Foraminifer. Res.* 35:93–102.

Flakowski, J., I. Bolivar, J. Fahrni, and J. Pawlowski. 2006. Tempo and mode of spliceosomal intron evolution in actin of *Foraminifera. J. Mol. Evol.* 63:30–41.

Galloway, J. J. 1933. *A manual* of Foraminifera. Principia Press, Bloomington, IN.

Gauthier, J. A., and K. de Queiroz. 2001. Feathered dinosaurs, flying dinosaurs, crown dinosaurs, and the name *"Aves"*. Pp. 7–41 in *New Perspectives on the Origin and Evolution of Birds* (J. A. Gauthier and L. F. Gail, eds.). Yale University Press, New Haven, CT.

Giere, O. 2009. *Meiobenthology: The Microscopic Motile Fauna of Aquatic Sediments.* Springer, Hamburg.

Goldstein, S. T. 1999. *Foraminifera*: a biological overview. Pp. 37–55 in *Modern* Foraminifera (B. K. Sen Gupta, ed.). Kluwer Academic Publishers, Dordrecht and Boston.

Golz, R., and M. Hauser. 1994. Spatially separated classes of microtubule bridges in the reticulopodial network of *Allogromia. Eur. J. Protistol.* 30:221–226.

Gomaa, F., E. A. D. Mitchell, and E. Lara. 2013. *Amphitremida* (Poche, 1913) is a new major, ubiquitous labyrinthulomycete clade. *PLOS ONE* 8:e53046.

Granger, A. 1896. *Histoire Naturelle de la France, 17e Partie: Coelentéres, Échinodermes, Protozoaires.* Maisson Émile Deyrolle, Paris.

Greene, J. R. 1859. *A Manual of the Sub-Kingdom* Protozoa. *With a General Introduction on the Principles of Zoology.* Longman, Green, Longman, and Roberts, London.

Greene, J. R. 1863. *A Manual of the Sub-Kingdom* Protozoa. *With a General Introduction on the Principles of Zoology.* Longman, Green, Longman, Roberts & Green, London.

Greene, J. R. 1868. *A Manual of the Sub-Kingdom* Protozoa. *With a General Introduction on the Principles of Zoology.* Longmans, Green, and Co., London.

Greene, J. R. 1871. *A Manual of the Sub-Kingdom* Protozoa. *With a General Introduction on the Principles of Zoology.* Longmans, Green, and Co., London.

Groussin, M., J. Pawlowski, and Z. Yang. 2011. Bayesian relaxed clock estimation of divergence times in *Foraminifera. Mol. Phylogen. Evol.* 61:157–166.

Habura, A., S. T. Goldstein, S, Broderick, and S. S. Bowser. 2008. A bush, not a tree: the extraordinary diversity of cold-water basal foraminiferans extends to warm-water environments. *Limnol. Oceanogr.* 53:1339–1351.

Habura, A., J. A. N. Pawlowski, S. D. Hanes, and S. S. Bowser. 2004a. Unexpected foraminiferal diversity revealed by small-subunit rDNA analysis of Antarctic sediment. *J. Eukarot. Microbiol.* 51:173–179.

Habura, A., D. R. Rosen, and S. S. Bowser. 2004b. Predicted secondary structure of the foraminiferal SSU 3' major domain reveals a molecular synapomorphy for granuloreticulosean protists. *J. Eukaryot. Microbiol.* 51:464–471.

Habura, A., L. Wegener, J. L. Travis, and S. S. Bowser. 2005. Structural and functional implications of an unusual foraminiferal β-tubulin. *Mol. Biol. Evol.* 22:2000–2009.

Haeckel, E. 1862. *Die Radiolarien* (Rhizopoda radiolaria): *Eine Monographie.* Georg Reimer, Berlin.

Haeckel, E. 1868. Monographie der Jena. *Zschr. Med. Naturwiss.* 4:64–137.

Haeckel, E. 1870. *Biologische Studien. I. Studien über Moneren und Andere Protisten, Nebst Einer Ride über Entwickelungsgang und Aufgabe der Zoologie.* Wilhelm Engelmann, Leipzig.

Haeckel, E. 1878. *Das Protistenreich. Eine Populäre Uebersicht über das Formengebiet der Niedersten Lebewesen. Mit Einem Wissenschaftlichen Anhange: System der Protisten.* Wilhelm Engelmann, Leipzig.

Hansen, H. J. 1999. Shell construction in modern calcareous *Foraminifera*. Pp. 57–70 in *Modern* Foraminifera (B. K. Sen Gupta, ed.). Kluwer Academic Publishers, Dordrecht and Boston.

Hart, M. B., and C. L. Williams. 1993. *Protozoa*. Pp. 43–70 in *The Fossil Record 2* (M. J. Benton, ed.). Chapman and Hall, London.

Hausmann, K., N. Hülsmann, and R. Radek. 2003. *Protistology*. E. Schweizerbart'sche Verlagsbuchhandlung, Berlin.

Hedley, R. H., D. M. Parry, and J. S. J. Wakefield. 1967. Fine structure of *Shepheardella taeniformis* (*Foraminifera: Protozoa*). *J. R. Microsc. Soc.* 87:445–456.

Hemleben, C., O. R. Anderson, W.-U. Berthold, and M. Spindler. 1986. Calcification and chamber formation in *Foraminifera*—a brief overview. Pp. 237–249 in *Biomineralization in Lower Plants and Animals* (B. S. C. Leadbeater and R. Riding, eds.). The Clarendon Press, Oxford.

Hemleben, C., and M. A. Kaminski. 1990. Agglutinated *Foraminifera*: an introduction. Pp. 3–11 in *Paleoecology, Biostratigraphy, Paleoceanography and Taxonomy of Agglutinated* Foraminifera (C. Hemleben, ed.). Kluwer Academic Publishers, The Netherlands.

Hemleben, C., M. Spindler, and O. R. Anderson. 1989. *Modern Planktonic* Foraminifera. Springer-Verlag, New York.

Hertwig, R. 1876. Bermerkungen zur Organisation un Systematischen Stellung der Foraminiferen. *Jena. Zeitschr.* 10:41–55.

Hori, S., M. Tsuchiya, S. Nishi, W. Arai, T. Yoshida, and H. Takami. 2013. Active bacterial flora surrounding *Foraminifera* (*Xenophyophorea*) living on the deep-sea floor. *Biosci. Biotech. Biochem.* 77:381–384.

Hottinger, L. C. 2000. Functional morphology of benthic foraminiferal shells, envelopes of cells beyond measure. *Micropaleontology* 46:57–86.

Hou, Y., R. Sierra, D. Bassen, N. K. Banavali, A. Habura, J. Pawlowski, and S. S. Bowser. 2013. Molecular evidence for beta-tubulin neofunctionalization in *Retaria* (*Foraminifera* and radiolarians). *Mol. Biol. Evol.* 30:2487–2493.

Ishitani, Y., S. A. Ishikawa, Y. Inagaki, M. Tsuchiya, K. Takahashi, and K. Takishita. 2011. Multigene phylogenetic analyses including diverse radiolarian species support the "*Retaria*" hypothesis—the sister relationship of *Radiolaria* and *Foraminifera*. *Mar. Micropaleontol.* 81:32–42.

Jepps, M. W. 1956. *The* Protozoa, Sarcodina. Oliver and Boyd, Edinburgh.

Jones, T. R. 1876. Remarks on the *Foraminifera*, with especial reference to their variability of form, illustrated by the cristellarians. *Mon. Microsc. J.* 15:61–92.

Keeling, P. J. 2001. *Foraminifera* and *Cercozoa* are related in actin phylogeny: two orphans find a home? *Mol. Biol. Evol.* 18:1551–1557.

Knoll, A. H. 2014. Paleobiological perspectives on early eukaryotic evolution. *Cold Spring Harb. Perspect. Biol.* 6:a016121.

Kontorovich, A. E., A. I. Varlamov, D. V. Grazhdankin, G. A. Karlova, A. G. Klets, V. A. Kontorovich, S. V. Saraev, A. A. Terleev, S. Y. Belyaev, I. V. Varaksina, A. S. Efimov, B. B. Kochnev, K. E. Nagovitsin, A. A. Postnikov, and Y. F. Filippov. 2008. A section of Vendian in the east of West Siberian Plate (based on data from the Borehole Vostok 3). *Russ. Geol. Geophys.* 49:932–939.

Koonce, M. P., U. Euteneuer, and M. Schliwa. 1986. *Reticulomyxa*: a new model system of intracellular transport. *J. Cell Sci. Suppl.* 5:145–159.

Koonce, M. P., and M. Schliwa. 1985. Bidirectional organelle transport can occur in cell processes that contain single microtubules. *J. Cell Biol.* 100:322–326.

Kosakyan, A., F. Gomaa, E. Lara, and D.J.G. Lahr. 2016. Current and future perspectives on the systematics, taxonomy and nomenclature of testate amoebae. *Eur. J. Protistology* 55:105–117.

Krabberød, A. K., J. Bråte, J. K. Dolven, R. F. Ose, D. Klaveness, T. Kristensen, K. R. Bjørklund, and K. Shalchian-Tabrizi. 2011. *Radiolaria* divided into *Polycystina* and *Spasmaria* in combined 18S and 28S rDNA phylogeny. *PLOS ONE* 6:e23526.

Kudo, R. R. 1939. *Protozoology*. Charles C. Thomas, Springfield, IL.

Langer, M. R. 1993. Epiphytic *Foraminifera*. *Mar. Micropaleontol.* 20:235–265.

Le Calvez, J. 1936. Observations sur le genre *Iridia*. *Arch. Zool. Exp. Gén.* 78:115–131.

Lecroq, B., A. J. Gooday, T. Cedhagen, A. Sabbatini, and J. Pawlowski. 2009. Molecular analyses reveal high levels of eukaryotic richness associated with enigmatic deep-sea protists (*Komokiacea*) *Mar. Biodiv.* 39:45–55.

Lecroq, B., F. Lejzerowicz, D. Bachar, R. Christen, P. Esling, L. Baerlocher, M. Osteras, L. Farinelli, and J. Pawlowski. 2011. Ultra-deep sequencing of foraminiferal microbarcodes unveils hidden richness of early monothalamous lineages. *Proc. Natl. Acad. Sci. USA* 108:13177–13182.

Lee, J. J. 1990. Phylum *Granuloreticulosa* (*Foraminifera*). Pp. 524–548 in *Handbook of Protoctista* (L. Margulis, J. O. Corliss, M. Melkonian, and D. J. Chapman, eds.). Jones and Bartlett Publishers, Boston, MA.

Lee, J. J., J. Pawlowski, J.-P. Debenay, J. Whittaker, F. Banner, A. J. Gooday, O. Tendal, J. Haynes, and W. W. Faber. 2000. Phylum *Granuloreticulosa*. Pp. 872–951 in *The Illustrated Guide to the* Protozoa (J. J. Lee, G. F. Leedale, and P. Bradbury, eds.). Society of Protozoologists, Lawrence, KS.

Leidy, J. 1879. *Fresh-Water Rhizopods of North America*. Government Printing Office, Washington, DC.

Lengsfeld, A. M. 1969a. Zum Feinbau der Foraminifere *Allogromia laticollaris*. *Helgoland Mar. Res.* 19:230–261.

Lengsfeld, A. M. 1969b. Nahrungsaufnahme und Verdauung bei der Foraminifere *Allogromia laticollaris*. *Helgoland Mar. Res.* 19:385–400.

Levine, N. D., J. O. Corliss, F. E. G. Cox, G. Deroux, J. Grain, B. M. Honigberg, G. F. Leedale, A. R. Loeblich, I. J. Lom, D. Lynn, E. G. Merinfeld, F. C. Page, G. Poljansky, V. Sprague, J. Vavra, and F. G. Wallace. 1980. A newly revised classification of the *Protozoa. J. Protozool.* 27:37–58.

Linder, S., M. Schliwa, and E. Kube-Granderath. 1997. Sequence analysis and immunofluorescence study of α- and β-tubulins in *Reticulomyxa filosa*: implications of the high degree of β2-tubulin divergence. *Cell Motil. Cytoskel.* 36:164–178.

Linnaeus, C. 1758. *Systema Naturae Per Regna Tria Naturae: Secundum Classes, Ordines, Genera, Species, cum Characteribus, Differentiis, Synonymis, Locis*. Laurentius Salvius, Holmiae (Stockholm).

Lipps, J. H. 1973. Test structure in *Foraminifera*. *Annu. Rev. Microbiol.* 27:471–486.

Lipps, J. H. 1983. Biotic interactions in benthic *Foraminifera*. Pp. 331–376 in *Biotic Interactions in Recent and Fossil Benthic Communities* (M. J. S. Tevesz and P. L. McCall, eds.). Plenum Press, New York.

Lipps, J. H., and K. L. Finger. 2010. How many *Foraminifera* are there? P. 132 in *FORAMS 2010 International Symposium on* Foraminifera (M. Langer, A. Weinmann, B. Söntgerath, P. Göddertz, and G. Heumann, eds.). Rheinische Freiderich-Wilhelms-Universität Bonn, Germany, September 5–10, 2010.

Lipps, J. H., and A. Y. Rozanov. 1996. The late Precambrian-Cambrian agglutinated fossil *Platysolenites*. *Palaeontol. J.* 30:679–687.

Lodish, H., A. Berk, C. A. Kaiser, M. Krieger, A. Bretscher, H. Ploegh, A. Amon, and K. C. Martin. 2016. *Molecular Cell Biology*. W. H. Freeman and Company, New York.

Loeblich, Jr., A. R., and H. Tappan. 1961. Suprageneric classification of the *Rhizopodea*. *J. Paleontol.* 35:245–330.

Loeblich, Jr., A. R., and H. Tappan. 1964. *Protista 2. Sarcodina* chiefly "thecamoebians" and *Foraminiferida*. Pp. 1–900 in *Treatise on Invertebrate Paleontology* (R. C. Moore, ed.). The Geological Society of America and The University of Kansas Press, Lawrence, KS.

Loeblich, Jr., A. R., and H. Tappan. 1984. Suprageneric classification of the *Foraminiferida* (*Protozoa*). *Micropaleontology* 30:1–70.

Loeblich, Jr., A. R., and H. N. Tappan. 1988. *Foraminiferal Genera and Their Classification*. Van Nostrand Reinhold Co., New York.

Longet, D., J. D. Archibald, P. J. Keeling, and J. Pawlowski. 2003. *Foraminifera* and *Cercozoa* share a common origin according to RNA polymerase II phylogenies. *Int. J. Syst. Evol. Microbiol.* 53:1735–1739.

Longet, D., F. Burki, J. Flakowski, C. Berney, S. Polet, J. F. Fahrni, and J. Pawlowski. 2004. Multigene evidence for close evolutionary relations between *Gromia* and *Foraminifera*. *Acta Protozool.* 43:303–311.

Marszalek, D. S. 1969. Observations on *Iridia diaphana*, a marine foraminifer. *J. Protozool.* 16:599–611.

McIlroy, D., O. R. Green, and M. D. Brasier. 2001. Palaeobiology and evolution of the earliest agglutinated *Foraminifera*: *Platysolenites*, *Spirosolenites* and related forms. *Lethaia* 34:13–29.

Mikhalevich, V. I. 2013. New insight into the systematics and evolution of the *Foraminifera*. *Micropaleontology* 59:493–527.

Minchin, E. A. 1912. *An Introduction to the Study of the* Protozoa, *with Special Reference to the Parasitic Forms*. Edward Arnold, London.

Moore, R. C. 1954. Kingdom of organisms named *Protista*. *J. Paleontol.* 28:588–598.

Moreira, D., S. von der Heyden, D. Bass, P. López-García, E. Chao, and T. Cavalier-Smith. 2007. Global eukaryote phylogeny: combined small- and large-subunit ribosomal DNA trees support monophyly of *Rhizaria*, *Retaria* and *Excavata*. *Mol. Phylogenet. Evol.* 44:255–266.

Murray, J. W. 2007. Biodiversity of living benthic *Foraminifera*: how many species are there? *Mar. Micropaleontol.* 64:163–176.

Myers, E. H. 1943. Biology, ecology and morphogenesis of a pelagic foraminifer. *Stanford Publ. Biol. Sci.* 9:1–30.

Nauss, R. N. 1949. *Reticulomyxa filosa* gen. et sp. nov., a new primitive plasmodium. *B. Torrey Bot. Club* 76:161–173.

Nikolaev, S. I., C. Berney, J. F. Fahrni, I. Bolivar, S. Polet, A. P. Mylnikov, V. V. Aleshin, N. B. Petrov, and J. Pawlowski. 2004. The twilight of *Heliozoa* and rise of *Rhizaria*, an emerging supergroup of amoeboid eukaryotes. *Proc. Natl. Acad. Sci. USA* 101:8066–8071.

Nikolaev, S. I., E. A. D. Mitchell, N. B. Petrov, C. Berney, J. Fahrni, and J. Pawlowski. 2008. The testate lobose *Amoebae* (Order *Arcellinida* Kent, 1880) finally find their home within *Amoebozoa*. *Protist* 156:191–202.

Owen, R. 1855. *Lectures on the Comparative Anatomy and Physiology of the Invertebrate Animals*. Longman, Brown, Green, and Longman, London.

Patterson, D. J. 1999. The diversity of eukaryotes. *Am. Nat.* 154(suppl.):S96–S124.

Pawlowski, J. 2000. Introduction to the molecular systematics of *Foraminifera*. *Micropaleontology* 46(suppl. 1):1–12.

Pawlowski, J. 2008. The twilight of *Sarcodina*: a molecular perspective on the polyphyletic origin of amoeboid protists. *Protistology* 5:281–302.

Pawlowski, J., and C. Berney. 2003. Episodic evolution of nuclear small subunit ribosomal RNA gene in the stem-lineage of *Foraminifera*. Pp. 107–118 in *Telling the Evolutionary Time: Molecular Clocks and the Fossil Record* (P. C. J. Donoghue and M. P. Smith, eds.). CRC Press, Boca Raton, FL.

Pawlowski, J., I. Bolivar, J. F. Fahrni, T. Cavalier-Smith, and M. Gouy. 1996. Early origin of *Foraminifera* suggested by SSU rRNA gene sequences. *Mol. Biol. Evol.* 13:445–450.

Pawlowski, J., I. Bolivar, J. F. Fahrni, C. de Vargas, and S. S. Bowser. 1999a. Naked foraminiferans revealed. *Nature* 399:27.

Pawlowski, J., I. Bolivar, J. F. Fahrni, and C. de Vargas. 1999b. Molecular evidence that *Reticulomyxa filosa* is a freshwater naked foraminifer. *J. Euk. Microbiol.* 46:612–617.

Pawlowski, J., I. Bolivar, J. Guiard-Maffia, and M. Gouy. 1994. Phylogenetic position of *Foraminifera* inferred from LSU rRNA gene sequences. *Mol. Biol. Evol.* 11:929–938.

Pawlowski, J., and F. Burki. 2009. Untangling the phylogeny of amoeboid protists. *J. Euk. Microbiol.* 56:16–25.

Pawlowski, J., J. F. Fahrni, J. Guiard, K. Conlan, J. Hardecker, A. Habura, and S. Bowser. 2005. Allogromiid *Foraminifera* and gromiids from under the Ross Ice Shelf: morphological and molecular diversity. *Polar Biol.* 28:514–522.

Pawlowski, J., and M. Holzmann. 2002. Molecular phylogeny of *Foraminifera* a review. *Eur. J. Protistol.* 38:1–10.

Pawlowski, J., and M. Holzmann. 2008. Diversity and geographic distribution of benthic *Foraminifera*: a molecular perspective. *Biodivers. Conserv.* 17:317–328.

Pawlowski, J., M. Holzmann, C. Berney, J. Fahrni, A. J. Gooday, T. Cedhagen, A. Habura, and S. S. Bowser. 2003a. The evolution of early *Foraminifera*. *Proc. Natl. Acad. Sci. USA* 100:11494–11498.

Pawlowski, J., M. Holzmann, J. Fahrni, and S. L. Richardson. 2003b. Small subunit ribosomal DNA suggests that the xenophyophorean *Syringammina corbicula* is a foraminiferan. *J. Euk. Microbiol.* 50:483–487.

Pawlowski, J., M. Holzmann, and J.-P. Debenay. 2014. Molecular phylogeny of *Carterina spiculotesta* and related species from New Caledonia. *J. Foraminif. Res.* 44:440–450.

Pawlowski, J., M. Holzmann, and J. Tyszka. 2013. New supraordinal classification of *Foraminifera*: molecules meet morphology. *Mar. Micropaleontol.* 100:1–10.

Perkins, F. O. 2000. Phylum *Haplosporidia*. Pp. 1328–1341 in *An Illustrated Guide* to the *Protozoa* (J. J. Lee, S. H. Hutner, and E. C. Bovee, eds.). Society of Protozoologists, Lawrence, KS.

Porter, S. M. 2004. The fossil record of early eukaryotic diversification. *Paleontol. Soc. Paper.* 10:35–50.

Porter, S. M. 2006. The Proterozoic fossil record of heterotrophic eukaryotes. Pp. 1–21 in *Neoproterozoic Geobiology and Paleobiology* (S. Xiao and A. J. Kaufman, eds.). Springer Netherlands.

Pritchard, A. 1861. *A History* of Infusoria, *Including the* Desmidiaceae *and* Diatomaceae, *British and Foreign*. Whittaker and Co., London.

Reuss, A. E. 1862. Entwurf einer systemaischen Zusammenstellung der Foraminiferen. *Akad. Wiss. Wien, Math-Nat. Kl., Sitzb.* 44:355–396.

Rhumbler, L. 1904. Systematische Zusammenstellung der recenten *Reticulosa* (*Nuda* + *Foraminifera.*) I. *Teil. Arch. Protistenkd.* 3:181–294.

Richardson, S. L. 2001. *Syringammina corbicula* sp. nov. (*Xenophyophorea*) from the Cape Verde Plateau, E. Atlantic. *J. Foramin. Res.* 31:201–209.

Richardson, S. L. 2009. An overview of symbiont bleaching in the epiphytic foraminiferan *Sorites dominicensis. Smithson. Contr. Mar. Sci.* 38:429–436.

Schmarda, L. K. 1871. *Zoologie.* Wilhelm Braumüller, Wien.

Schultze, M. S. 1854. *Über den Organismus der Polythalamien (Foraminiferen), Nebst Bermerkungen über die Rhizopoden in Allgemeinen.* Wilhelm Engelmann, Leipzig.

Schwab, D. 1969. Elektronenmikroskopische Untersuchung an der Foraminifere *Myxotheca arenilega* Schaudinn. *Cell Tissue Res.* 96:295–324.

Schwab, D. 1977. Light and electron microscopic investigations on monothalamous foraminifer *Boderia albicollaris* n. sp. *J. Foramin. Res.* 7:188–195.

Schwager, C. 1877. Quadro del proposto sistema di classificazione dei *Foraminiferi* con guscio. *Boll. R. Com. Geol. Ital.* 8:18–24.

Schweizer, M., J. Pawlowski, I. A. P. Duijnstee, T. J. Kouwenhoven, and G. J. van der Zwaan. 2005. Molecular phylogeny of the foraminiferal genus *Uvigerina* based on ribosomal DNA sequences. *Mar. Micropaleontol.* 57:51–67.

Schweizer, M., J. Pawlowski, T. J. Kouwenhoven, J. Guiard, and B. van der Zwaan. 2008. Molecular phylogeny of *Rotaliida (Foraminifera)* based on complete small subunit rDNA sequences. *Mar. Micropaleontol.* 66:233–246.

Seilacher, A., L. A. Buatois, and M. G. Mángano. 2005. Trace fossils in the Ediacaran-Cambrian transition: behavioral diversification, ecological turnover and environmental shift. *Palaeogeogr. Palaeoclimatol. Palaeoecol.* 227:323–356.

Sen Gupta, B. K. 1999. Systematics of modern *Foraminifera*. Pp. 7–36 in *Modern Foraminifera* (B. K. Sen Gupta, ed.). Kluwer Academic Publishers, Boston, MA.

von Siebold, C. T. E., and H. Stannius. 1848. *Lehrbuch der Vergleichenden Anatomie der Wirbellosen Thiere.* Veit and Company, Berlin.

Sierra, R., M. V. Matz, G. V. Aglyamova, L. Pillet, J. Decelle, F. Not, C. de Vargas, and J. Pawlowski. 2013. Deep relationships of *Rhizaria* revealed by phylogenomics: a farewell to Haeckel's *Radiolaria. Mol. Phylogenet. Evol.* 67:53–59.

Sleigh, M. 1973. *The Biology of* Protozoa. Edward Arnold, London.

Sleigh, M. 1989. *Protozoa and Other Protists.* Edward Arnold, London.

Streng, M., L. E. Babcock, and J. S. Hollingsworth. 2005. Agglutinated protists from the Lower Cambrian of Nevada. *J. Paleontol.* 79:1214–1218.

Suzuki, N., and F. Not. 2015. Biology and ecology of *Radiolaria.* Pp. 179–222 in *Marine Protists: Diversity and Dynamics* (S. Ohtsuka, T. Suzuki, T. Horiguchi, N. Suzuki, and F. Not, eds.). Springer, Japan.

Tekle, Y. I., J. Grant, J. C. Cole, T. A. Nerad, O. R. Anderson, D. J. Patterson, and L. A. Katz. 2007. A multigene analysis of *Corallomyxa tenera* sp. nov. suggests its membership in a clade that includes *Gromia, Haplosporidia* and *Foraminifera. Protist* 158:457.

Tendal, Ø. S. 1972. A monograph of the *Xenophyophoria. Galathea Rep.* 12:7–99.

Tendal, Ø. S. 1990. Phylum *Xenophyophora.* Pp. 135–138 in *Handbook* of Protoctista (L. Margulis, J. O. Corliss, M. Melkonian, and D. J. Chapman, eds.). Jones and Bartlett Publishers, Boston, MA.

Travis, J. L., and S. S. Bowser. 1991. The motility of *Foraminifera.* Pp. 91–155 in *Biology of* Foraminifera (J. J. Lee and O. R. Anderson, eds.). Academic Press, San Diego, CA.

Travis, J. L., S. S. Bowser, J. G. Calvin, and J. J. Lee. 1988. Pseudopodial tension in *Amphisorus hemprichii,* a giant Red Sea foraminifer. *Protoplasma Suppl.* 1:64–71.

Ujiié, Y., K. Kimoto, and J. Pawlowski. 2008. Molecular evidence for an independent origin of modern triserial planktonic *Foraminifera* from benthic ancestors. *Mar. Micropaleontol.* 69:334–340.

Weinhofer, E. A., and J. L. Travis. 1998. Evidence for a direct conversion between two tubulin polymers—microtubules and helical filaments—in the foraminiferan, *Allogromia laticollaris. Cell Motil. Cytoskel.* 41:107–116.

Williamson, W. C. 1849. On the structure of the shell and soft animal of *Polystomella crispa*; with some remarks on the zoological position of the *Foraminifera. Trans. Microsc. Soc. London* 2:159–178.

Williamson, W. C. 1858. *On the Recent* Foraminifera *of Great Britain.* Ray Society, London.

Wylezich, C., R. Meisterfeld, S. Meisterfeld, and M. Schlegel. 2002. Phylogenetic analyses of small subunit ribosomal RNA coding regions reveal a monophyletic lineage of euglypid testate amoebae (Order *Euglyphida). J. Eukaryot. Microbiol.* 49:108–118.

Zborsewski, A. 1834. Observations microscopiques sur quelques fossils rares de Podolie et de Volhynie. *Nouv. Mem. Soc. Imp. Mosc.* 3:301–312.

Authors

Susan L. Richardson; Florida Atlantic University; Wilkes Honors College; 5353 Parkside Drive, Jupiter, FL 33458, USA. Email: richards@fau.edu.

Jere H. Lipps; Department of Integrative Biology and Museum of Paleontology; University of California; Berkeley, CA 94720, USA. Email: jlipps@berkeley.edu.

Date Accepted: 20 March 2017

Primary Editor: Jacques A. Gauthier

Alveolata T. Cavalier-Smith 1991 [A. G. B. Simpson,
B. S. Leander, and M. Dunthorn] converted clade name

Registration Number: 3

Definition: The smallest crown clade containing *Alexandrium tamarense* (Lebour 1925) Balech 1992 (*Dinoflagellata*), *Tetrahymena thermophila* Nanney and McCoy 1976 (*Ciliophora*), and *Toxoplasma gondii* (Nicolle and Manceaux 1908) Nicolle and Manceaux 1909 (*Apicomplexa*). This is a minimum-crown-clade definition. Abbreviated definition: min crown ∇ (*Alexandrium tamarense* (Lebour 1925) Balech 1992 & *Tetrahymena thermophila* Nanney and McCoy 1976 & *Toxoplasma gondii* (Nicolle and Manceaux 1908) Nicolle and Manceaux 1909).

Etymology: Refers to the presence of cortical alveoli (see Diagnostic Apomorphies).

Reference Phylogenies: The primary reference phylogeny is van de Peer and de Wachter (1997: Figs 2 and 4, the latter detailing part of the same phylogeny). See also Silberman et al. (2004: Fig. 3), Moore et al. (2008: Fig. 2c), and Park and Simpson (2015: Fig. 5).

Composition: The clade *Alveolata* contains *Ciliophora* as defined in this volume, *Colponema*, *Acavomonas*, *Palustrimonas* and *Myzozoa*. *Myzozoa* in turn contains two major subclades, *Apicomplexa* and *Dinoflagellata*, as well several small groups, a non-exhaustive list of which includes *Perkinsozoa*, *Chromera*, *Vitrella*, *Colpodella*, *Alphamonas*, *Voromonas* and probably *Acrocoelus* (see Comments). *Dinoflagellata* is considered here to include *Oxyrrhis* and 'syndinians', including 'marine alveolate groups 1 and 2', and provisionally, *Ellobiopsidae* and

Psammosa (plus *Colpovora*, if phylogenetically distinguishable from *Psammosa*; see Harada et al., 2007; Okamoto et al., 2012; Cavalier-Smith, 2018).

Diagnostic Apomorphies: The clade *Alveolata* is diagnosed by the presence of cortical alveoli, which are flattened membrane sacs without attached ribosomes that lie immediately underneath the cell membrane (see Patterson, 1999). In *Apicomplexa* these sacs are characteristically pierced in one or many places by small collared invaginations of the cell membrane called micropores, with apparently homologous structures being present in the other major subclades (*Dinoflagellata* and *Ciliophora*; e.g., Appleton and Vickerman, 1996). The homology of alveoli across *Alveolata* is also supported by their association with a distinct family of proteins, alveolins, in representatives of the major subclades (Gould et al., 2008). Superficially similar arrangements of membranes (though without micropores) are found beneath the cell membranes of a few other eukaryotes that are not closely related to *Alveolata*; these are inferred by almost all commentators to have evolved independently (Patterson and Sogin, 1992; Patterson, 1999; Leander and Keeling, 2003; though see Cavalier-Smith, 2002, 2009).

Synonyms: None.

Comments: A close phylogenetic relationship between three major groups of microbial eukaryotes—ciliates (*Ciliophora*), dinoflagellates (*Dinoflagellata*) and apicomplexan parasites (*Apicomplexa*, e.g., *Plasmodium*, *Toxoplasma*; though see below)—was initially identified primarily through analyses of ribosomal RNA (rRNA) gene sequences (Gajadhar et al., 1991; Wolters, 1991; Taylor, 1999). Almost all relevant phylogenetic analyses since have supported this basic relationship (e.g., van de Peer and de Wachter, 1997; Fast et al., 2002; Leander and Keeling, 2004; Burki et al., 2007, 2016; Rodríguez-Ezpeleta et al., 2007; Yoon et al., 2008). Molecular phylogenetic analyses indicate strongly that *Apicomplexa* and *Dinoflagellata* are more closely related to each other than to *Ciliophora* (van de Peer and de Wachter, 1997; Fast et al., 2002; Burki et al., 2007, 2016), and a clade including *Apicomplexa* and *Dinoflagellata* but not *Ciliophora* is now widely known as *Myzozoa*, or at least as 'myzozoans' (Cavalier-Smith and Chao, 2004; Janouškovec et al., 2013; Tikhonenkov et al., 2014; Park and Simpson, 2015; Cavalier-Smith, 2018).

Additional, more obscure, members of *Alveolata* include the predatory or parasitic *Perkinsozoa* (approximate synonym *Perkinsea*) and colpodellids (e.g., *Colpodella*, *Alphamonas*, *Voromonas*), as well as the photosynthetic organisms *Chromera* and *Vitrella* (Brugerolle and Mignot, 1979; Simpson and Patterson, 1996; Siddall et al., 1997; Norén et al., 1999; Moore et al., 2008; Oborník et al., 2012; Reñé et al., 2017; Cavalier-Smith, 2018). There is strong molecular phylogenetic evidence that *Perkinsozoa* is related to *Dinoflagellata*, while colpodellids, *Chromera* and *Vitrella* are more closely related to *Apicomplexa*, likely as a single sister clade wherein colpodellids are not monophyletic with respect to *Chromera* and *Vitrella* (Siddall et al., 1997; van de Peer and de Wachter, 1997; Saldarriaga et al., 2003; Leander et al., 2003; Leander and Keeling, 2004; Silberman et al., 2004; Moore et al., 2008; Janouškovec et al., 2013, 2015; Park and Simpson, 2015; Burki et al., 2016). Some contemporary works include colpodellids, *Chromera* and *Vitrella* in *Apicomplexa* (Votýpka et al., 2017;

Cavalier-Smith, 2018). *Colponema* (a predator of other unicellular eukaryotes; not to be confused with *Colpodella*) and *Acrocoelus* (a parasite) are poorly known organisms with apparent cortical alveoli (Mignot and Brugerolle, 1975, Fernandez et al., 1999) that were often assigned tentatively to *Alveolata*, pending molecular data (Cavalier-Smith, 1998; Patterson, 1999; Cavalier-Smith and Chao, 2003, 2004; Leander and Keeling, 2003). Recent molecular phylogenetic analyses demonstrate that *Colponema* belongs to *Alveolata*, as do the recently characterized predators *Acavomonas* and *Palustrimonas* (Janouškovec et al., 2013; Tikhonenkov et al., 2014; Park and Simpson, 2015); similar data are not available for *Acrocoelus*. In the past *Alveolata* has occasionally been proposed to include some other groups that are now assigned elsewhere; e.g., *Hemimastigophora*, *Haplosporidia* and *Kathablepharidae* (Cavalier-Smith, 1998; Leander and Keeling, 2003).

The name *Alveolata* was first used by Cavalier-Smith (1991), who employed it to refer to what he considered at the time to be a paraphyletic group containing *Apicomplexa*, *Ciliophora* and 'Dinozoa' (i.e., *Dinoflagellata*, and little-studied dinoflagellate-like organisms). Although Cavalier-Smith (1991) cited another manuscript for the introduction of the taxon *Alveolata*, the other manuscript was not actually published. Shortly afterwards Cavalier-Smith (1993) did provide an explicit diagnosis of *Alveolata*, and later explicitly considered the taxon to represent a monophyletic group (Cavalier-Smith, 1998). The name *Alveolata* has been widely adopted and is now consistently used to refer to a clade containing *Apicomplexa*, *Ciliophora*, and *Dinoflagellata* (e.g., Cavalier-Smith, 2002; Leander and Keeling, 2003; Adl et al., 2005, 2012; Cavalier-Smith, 2018). The only names we are aware of that are used for this clade are *Alveolata* and the cognate informal name 'alveolates' (e.g., Patterson and Sogin, 1992;

Patterson, 1999; Baldauf, 2003; Simpson and Roger, 2004; Burki, 2014).

Literature Cited

Adl, S. M., A. G. Simpson, M. A. Farmer, R. A. Andersen, O. R. Anderson, J. R. Barta, S. S. Bowser, G. Brugerolle, R. A. Fensome, S. Fredericq, T. Y. James, S. Karpov, P. Kugrens, J. Krug, C. E. Lane, L. A. Lewis, J. Lodge, D. H. Lynn, D. G. Mann, R. M. McCourt, L. Mendoza, O. Moestrup, S. E. Mozley-Standridge, T. A. Nerad, C. A. Shearer, A. V. Smirnov, F. W. Spiegel, and M. F. Taylor. 2005. The new higher level classification of eukaryotes with emphasis on the taxonomy of protists. *J. Eukaryot. Microbiol.* 52:399–451.

Adl, S. M., A. G. B. Simpson, C. E. Lane, J. Lukeš, D. Bass, S. S. Bowser, M. W. Brown, F. Burki, M. Dunthorn, V. Hampl, A. Heiss, M. Hoppenrath, E. Lara, L. le Gall, D. H. Lynn, H. McManus, E. A. D. Mitchell, S. E. Mozley-Stanridge, L. W. Parfrey, J. Pawlowski, S. Rueckert, L. Shadwick, C. L. Schoch, A. Smirnov, and F. W. Spiegel. 2012. The revised classification of eukaryotes. *J. Eukaryot. Microbiol.* 59:429–493.

Appleton, P. L., and K. Vickerman. 1996. Presence of apicomplexan-type micropores in a parasitic dinoflagellate, *Hematodinium* sp. *Parasitol. Res.* 82:279–282.

Baldauf, S. L. 2003. The deep roots of eukaryotes. *Science* 300:1703–1706.

Brugerolle, G., and J.-P. Mignot. 1979. Observations sur le cycle l'ultrastructure et la position systématique de *Spiromonas perforans* (*Bodo perforans* Hollande 1938), flagellé parasite de *Chilomonas paramecium*: ses relations avec les dinoflagellés et sporozoaires. *Protistologica* 15:183–196.

Burki, F. 2014. The eukaryotic tree of life from a global phylogenomic perspective. *Cold Spring Harb. Perspect. Biol.* 6:a016147.

Burki, F., M. Kaplan, D. V. Tikhonenkov, V. Zlatogursky, B. Q. Minh, L. V. Radaykina, A. Smirnov, A. P. Mylnikov, and P. J. Keeling. 2016. Untangling the early diversification of eukaryotes: a phylogenomic study of the evolutionary origins of *Centrohelida*, *Haptophyta* and *Cryptista*. *Proc. R. Soc. Lond. B Biol. Sci.* 283:20152802.

Burki, F., K. Shalchian-Tabrizi, M. Minge, A. Skjaeveland, S. I. Nikolaev, K. S. Jakobsen, and J. Pawlowski. 2007. Phylogenomics reshuffles the eukaryotic supergroups. *PLOS ONE* 2:e790.

Cavalier-Smith, T. 1991. Cell diversification in heterotrophic flagellates. Pp. 113–131 in *The Biology of Free-Living Heterotrophic Flagellates* (D. J. Patterson and J. Larsen, eds.) The Systematics Association Special Volume No. 45. Clarendon Press, Oxford.

Cavalier-Smith, T. 1993. Kingdom *Protozoa* and its 18 phyla. *Microbiol. Rev.* 57:953–994.

Cavalier-Smith, T. 1998. A revised six-kingdom system of life. *Biol. Rev.* 73:203–256.

Cavalier-Smith, T. 2002. The phagotrophic origin of eukaryotes and phylogenetic classification of *Protozoa*. *Int. J. Syst. Evol. Microbiol.* 52:297–354.

Cavalier-Smith, T. 2009. Megaphylogeny, cell body plans, adaptive zones: causes and timing of eukaryote basal radiations. *J. Eukaryot. Microbiol.* 56:26–33.

Cavalier-Smith, T. 2018. Kingdom *Chromista* and its eight phyla: a new synthesis emphasising periplastid protein targeting, cytoskeletal and periplastid evolution, and ancient divergences. *Protoplasma* 255:297–357.

Cavalier-Smith, T., and E. E. Y. Chao. 2003. Phylogeny and classification of phylum *Cercozoa* (*Protozoa*). *Protist* 154:341–358.

Cavalier-Smith, T., and E. E. Chao. 2004. Protalveolate phylogeny and systematics and the origins of *Sporozoa* and dinoflagellates (phylum *Myzozoa* nom. nov.). *Europ. J. Protistol.* 40:185–212.

Fast, N. M., L. Xue, S. Bingham, and P. J. Keeling. 2002. Re-examining alveolate evolution using multiple protein molecular phylogenies. *J. Eukaryot. Microbiol.* 49:30–37.

Fernandez, I., F. Pardos, J. Benito, and N. L. Arroyo. 1999. *Acrocoelus glossobalani* gen. nov. et sp. nov., a protistan flagellate from the gut of the enteropneust *Glossabalanus minutus*. *Eur. J. Protistol.* 35:55–65.

Gajadhar, A. A., W. C. Marquardt, R. Hall, J. Gunderson, E. V. Ariztia-Carmona, and M. L. Sogin. 1991. Ribosomal RNA sequences of *Sarcocystis muris*, *Theileria annulata* and *Crypthecodinium cohnii* reveal evolutionary relationships among apicomplexans, dinoflagellates, and ciliates. *Mol. Biochem. Parasitol.* 45:147–154.

Gould, S. B., W.-H. Tham, A. F. Cowman, G. I. McFadden, and R. F. Waller. 2008. Alveolins, a new family of cortical proteins that define the protist infrakingdom *Alveolata*. *Mol. Biol. Evol.* 25:1219–1230.

Harada, A., S. Ohtsuka, and T. Horiguchi. 2007. Species of the parasitic genus *Duboscquella* are members of the enigmatic Marine Alveolate Group I. *Protist* 158:337–347.

Janouškovec, J., D. V. Tikhonenkov, F. Burki, A. T. Howe, M. Kolísko, A. P. Mylnikov, and P. J. Keeling. 2015. Factors mediating plastid dependency and the origins of parasitism in apicomplexans and their close relatives. *Proc. Natl. Acad. Sci. USA* 112:10200–10207.

Janouškovec, J., D. V. Tikhonenkov, K. V. Mikhailov, T. G. Simdyanov, V. V. Aleoshin, A. P. Mylnikov, and P. J. Keeling. 2013. Colponemids represent multiple ancient alveolate lineages. *Curr. Biol.* 23:2546–2552.

Leander, B. S., and P. J. Keeling. 2003. Morphostasis in alveolate evolution. *Trends Ecol. Evol.* 18:395–402.

Leander, B. S., and P. J. Keeling. 2004. Early evolutionary history of dinoflagellates and apicomplexans (*Alveolata*) as inferred from hsp90 and actin phylogenies. *J. Phycol.* 40:341–350.

Leander, B. S., O. N. Kuvardina, V. V. Aleshin, A. P. Mylnikov, and P. J. Keeling. 2003. Molecular phylogeny and surface morphology of *Colpodella edax* (*Alveolata*): insights into the phagotrophic ancestry of apicomplexans. *J. Eukaryot. Microbiol.* 50:334–340.

Mignot, J. P., and G. Brugerolle. 1975. Etude ultrastructurale du flagelle phagotrophe *Colponema loxodes* Stein. *Protistologica* 11:429–444.

Moore, R. B., M. Oborník, J. Janouškovec, T. Chrudimský, M. Vancová, D. H. Green, S. W. Wright, N. W. Davies, C. J. S. Bolch, K. Heimann, J. Šlapeta, O. Hoegh-Guldberg, J. M. Logsdon, Jr., and D. A. Carter. 2008. A photosynthetic alveolate closely related to apicomplexan parasites. *Nature* 451:959–963.

Norén, F., O. Moestrup, and A. S. Rehnstam-Holm. 1999. *Parvilucifera infectans* Norén et Moestrup gen. et sp. nov. (*Perkinsozoa* phylum nov.): a parasitic flagellate capable of killing toxic microalgae. *Eur. J. Protistol.* 35:233–254.

Oborník, M., D. Modrý, M. Lukeš, E. Cernotíková-Stříbrná, J. Cihlář, M. Tesařová, E. Kotabová, M. Vancová, O. Prášil, and J. Lukeš. 2012. Morphology, ultrastructure and life cycle of *Vitrella brassicaformis* n. sp., n. gen., a novel chromerid from the Great Barrier Reef. *Protist* 163:306–323.

Okamoto N., A. Horák,, and P. J. Keeling. 2012. Description of two species of early branching dinoflagellates, *Psammosa pacifica* n. g., n. sp. and *P. atlantica* n. sp. *PLOS ONE.* 7:e34900.

Park, J. S., and A. G. B. Simpson. 2015. Diversity of heterotrophic protists from extremely hypersaline habitats. *Protist* 166:422–437.

Patterson, D. J. 1999. The diversity of eukaryotes. *Am. Nat.* 154:S96–S124.

Patterson, D. J., and M. L. Sogin. 1992. Eukaryote origins and protistan diversity. Pp. 13–46 in *The Origin and Evolution of Prokaryotic and Eukaryotic Cells* (H. Hartman and K. Matsuno, eds.). World Scientific, Singapore.

Reñé A., E. Alacid, I. Ferrera, and E. Garcés. 2017. Evolutionary trends of *Perkinsozoa* (*Alveolata*) characters based on observations of two new genera of parasitoids of dinoflagellates, *Dinovorax* gen. nov. and *Snorkelia* gen. nov. *Front. Microbiol.* 8:1594.

Rodríguez-Ezpeleta, N., H. Brinkmann, G. Burger, A. J. Roger, M. W. Gray, H. Philippe, and B. F. Lang. 2007. Toward resolving the eukaryotic tree: the phylogenetic positions of jakobids and cercozoans. *Curr. Biol.* 17:1420–1425.

Saldarriaga, J. F., M. L. McEwan, N. M. Fast, F. J. R. Taylor, and P. J. Keeling. 2003. Multiple protein phylogenies show that *Oxyrrhis marina* and *Perkinsus marinus* are early branches of the dinoflagellate lineage. *Int. J. Syst. Evol. Microbiol.* 53:355–365.

Siddall, M. E., K. S. Reece, J. E. Graves, and E. M. Burreson. 1997. "Total evidence" refutes the inclusion of *Perkinsus* species in the phylum *Apicomplexa*. *Parasitology* 115:165–176.

Silberman J. D., A. G. Collins, L. A. Gershwin, P. J. Johnson, and A. J. Roger. 2004. Ellobiopsids of the genus *Thalassomyces* are alveolates. *J. Eukaryot. Microbiol.* 51:246–252.

Simpson, A. G. B., and D. J. Patterson. 1996. Ultrastructure and identification of the predatory flagellate *Colpodella pugnax* Cienkowski (*Apicomplexa*) with a description of *Colpodella turpis* n. sp. and a review of the genus. *Syst. Parasitol.* 33:187–198.

Simpson, A. G. B., and A. J. Roger. 2004. The real "kingdoms" of eukaryotes. *Curr. Biol.* 14:R693–R696.

Taylor, F. J. R. 1999. Ultrastructure as a control for protistan molecular phylogeny. *Am. Nat.* 154:S125–S136.

Tikhonenkov, D. V., J. Janouškovec, A. P. Mylnikov, K. V. Mikhailov, T. G. Simdyanov, V. V. Aleoshin, and P. J. Keeling. 2014. Description of *Colponema vietnamica* sp. n. and *Acavomonas peruviana* n. gen. n. sp., two new alveolate phyla (*Colponemidia* nom. nov. and *Acavomonidia* nom. nov.) and their contributions to reconstructing the ancestral state of alveolates and eukaryotes. *PLOS ONE* 9:e95467.

van de Peer, Y., and R. de Wachter. 1997. Evolutionary relationships among the eukaryotic crown taxa taking into account site-to-site rate variation in 18S rRNA. *J. Mol. Evol.* 45:619–630.

Votýpka, J., D. Modrý, M. Oborník, J. Šlapeta, and J. Lukeš. 2017. *Apicomplexa*. Pp. 567–624 in *Handbook of the Protists* (J. M. Archibald, A. G. B. Simpson, and C. Slamovits, eds.). Springer International Publishing, Cham, Switzerland.

Wolters, J. 1991. The troublesome parasites: molecular and morphological evidence that *Apicomplexa* belong to the dinoflagellate-ciliate clade. *Biosystems.* 25:75–83.

Yoon, H. S., J. Grant, Y. I. Tekle, M. Wu, B. C. Chaon, J. C. Cole, J. M. Logsdon, Jr., D. J. Patterson, D. Bhattacharya, and L. A. Katz. 2008. Broadly sampled multigene trees of eukaryotes. *BMC Evol. Biol.* 8:14–25.

Author

Alastair G. B. Simpson; Department of Biology; Dalhousie University; Halifax, NS, B3H 4R2, Canada. Email: alastair.simpson@dal.ca.

Brian S. Leander; Departments of Zoology and Botany; The University of British Columbia; Vancouver, BC, V6T 1Z4, Canada. Email: leander@zoology.ubc.ca.

Micah Dunthorn; Department of Eukaryotic Microbiology; University of Duisburg-Essen; 45141 Essen, Germany. Email: micah.dunthorn@uni-due.de.

Date Accepted: 31 September 2012; updated 18 June 2018

Primary Editor: Philip Cantino

Ciliophora F. Doflein 1901 [M. Dunthorn and D. H. Lynn], converted clade name

Registration Number: 29

Definition: The smallest crown clade containing *Tetrahymena thermophila* Nanney and McCoy 1976, *Blepharisma americanum* Suzuki 1954, and *Loxodes striatus* (Engelmann 1862). This is a minimum-crown-clade definition. Abbreviated definition: min crown ∇ (*Tetrahymena thermophila* Nanney and McCoy 1976 & *Blepharisma americanum* Suzuki 1954 & *Loxodes striatus* (Engelmann 1862)).

Etymology: Derived from the Latin *cilium* (eyelash) and Greek *phoreus* (bearer), in reference to the cilia on the cell bodies of all members of this clade (in at least one stage of the life cycle).

Reference Phylogeny: The primary reference phylogeny is Hammerschmidt et al. (1996: Fig. 2). See also Baldauf et al. (2000: Fig. 1) and Yoon et al. (2008: Fig. 2).

Composition: The clade *Ciliophora* contains *Postciliodesmatophora* and *Intramacronucleata* as defined in this volume, which in turn include all of the clades listed by Lynn (2008). Almost all known ciliates are extant; those few fossils found can be placed within previously recognized taxa (Lynn, 2008).

Diagnostic Apomorphies: Ciliates have three apomorphic characters: (1) dimorphic nuclei, with a "germline" micronucleus and a "somatic" macronucleus that are not homologous with those found in *Foraminifera* sensu Lee 1990; (2) cilia, in at least one life-cycle stage, that are derived from a kinetosome (= eukaryotic basal body) that is associated with three fibers (a kinetodesmal fiber, a postciliary microtubular ribbon, and a transverse microtubular ribbon); and (3) sex in the form of conjugation, where there is typically mutual exchange of haploid meiotic products of the micronucleus (Raikov, 1996; Lynn, 2008).

Synonyms: *Ciliata* M. Perty 1852 (approximate), *Infusoria* sensu Bütschli (1887–1889) (approximate); see review by Lynn (2008).

Comments: Ciliates have long been recognized as a group because of their distinctive morphology, and molecular data strongly support this clade (Hammerschmidt et al., 1996; Lynn, 2008). Since the beginning of the twentieth century, *Ciliophora* has been the most widely used name for this clade (e.g., Corliss, 1979; de Puytorac, 1994; Jankowski, 2007; Lynn, 2008).

Literature Cited

Baldauf, S. L., A. J. Roger, I. Wenk-Siefert, and W. F. Doolittle. 2000. A kingdom-level phylogeny of eukaryotes based on combined protein data. *Science* 290:972–977.

Bütschli, O. 1887–1889. *Protozoa*. Abt. III. *Infusoria* und System der *Radiolaria*. Pp. 1098–2035 in *Klassen und Ordnungen des Thier-Reichs* (H. G. Bronn, ed.). C. F. Winter, Leipzig.

Corliss, J. O. 1979. *The Ciliated Protozoa: Characterization, Classification and Guide to the Literature*. 2nd edition. Pergamon Press, Oxford.

de Puytorac, P. 1994. Phylum *Ciliophora* Doflein, 1901. Pp. 1–15 in *Traité de Zoologie, Tome II, Infusoires Ciliés, Fasc. 2, Systématique* (P. de Puytorac, ed.) Masson, Paris.

Doflein, F. 1901. *Die Protozoen als Parasiten und Krankheitserreger nach biologischen Gesichtspunkten dargestellt*. G. Fischer, Jena.

Hammerschmidt, B., M. Schlegel, D. Lynn, D. D. Leipe, M. L. Sogin, and I. B. Raikov. 1996. Insights into the evolution of nuclear dualism in the ciliates revealed by phylogenetic analysis of rRNA sequences. *J. Eukaryot. Microbiol.* 43:225–230.

Jankowski, A. W. 2007. Phylum *Ciliophora* Doflein, 1901. Pp. 415–993 in Protista. *Part 2, Handbook of Zoology* (A. F. Alimov, ed.). Russian Academy of Sciences, Zoological Institure, St. Petersburg.

Lee, J. J. 1990. Class *Foraminifera.* Pp. 877–951 in *An Illustrated Guide to the Protozoa.* 2nd edition (J. J. Lee, G. F. Leedale, and P. Bradbury, eds.). The Society of Protozoologists, Lawrence, KS.

Lynn, D. H. 2008. *The Ciliated Protozoa: Characterization, Classification, and Guide to the Literature.* 3rd edition. Springer, Dordrecht.

Perty, M. 1852. *Zur Kenntniss kleinster Lebensformen Nach Bau, Funktionen, Systematik, Mit Specialverzeichnis der in der Schweiz Beobachteten.* Jent & Reinert, Bern.

Raikov, I. B. 1996. Nuclei of ciliates. Pp. 221–242 in *Ciliates: Cells as Organisms* (K. Hausmann and P. C. Bradbury, eds.). Gustav Fischer, Stuttgart.

Yoon, H. S., J. Grant, Y. I. Tekle, M. Wu, B. C. Chaon, J. C. Cole, J. M. Logsdon, D. J. Patterson, D. Bhattacharya, and L. A. Katz. 2008. Broadly sampled multigene trees of eukaryotes. *BMC Evol. Biol.* 8:14.

Authors

Micah Dunthorn; Department of Eukaryotic Microbiology; University of Duisburg-Essen; 45141 Essen, Germany. Email: micah.dunthorn@uni-due.de.

Denis H. Lynn; Department of Zoology; University of British Columbia; Vancouver, British Columbia V6T 1Z4, Canada. Email: lynn@zoology.ubc.ca.

Date Accepted: 2 September 2011

Primary Editor: Philip Cantino

Postciliodesmatophora Z. P. Gerassimova and L. N. Seravin 1976 [M. Dunthorn and D. H. Lynn], converted clade name

Registration Number: 86

Definition: The smallest crown clade containing *Blepharisma americanum* Suzuki 1954 and *Loxodes striatus* (Engelmann 1862). This is a minimum-crown-clade definition. Abbreviated definition: min crown ∇ (*Blepharisma americanum* Suzuki 1954 & *Loxodes striatus* (Engelmann 1862)).

Etymology: Derived from the Latin *post* (after, behind) and *cilium* (eyelash) and the Greek *desmos* (bond or chain) and *phoreus* (bearer), in reference to the postciliodesmata borne by members of this clade (see Diagnostic Apomorphies).

Reference Phylogeny: The primary reference phylogeny is Hammerschmidt et al. (1996: Fig. 2). See also Hirt et al. (1995: Fig. 2).

Composition: Contains *Karyorelictea* and *Heterotrichea*; see Lynn (2008) for a list of the contents of these clades.

Diagnostic Apomorphies: The members of *Postciliodesmatophora* have postciliodesmata, which are postciliary microtubule ribbons overlapping along the right side of a kinety (integrated somatic file of kinetids). Macronuclei either do not divide (*Karyorelictea*) or divide using extra-macronuclear microtubules (*Heterotrichea*).

Synonyms: None.

Comments: Gerassimova and Seravin (1976) recognized this clade based on the shared postciliodesmata. Subsequent phylogenetic analyses of small subunit ribosomal RNA genes have supported its monophyly (Hammerschmidt et al., 1996; Hirt et al., 1995; Lynn, 2008).

The name is defined here in a manner consistent with Gerassimova and Seravin's (1976) original composition of the group. If a lineage is found that has postciliodesmata but branched off prior to the divergence between *Karyorelictea* and *Heterotrichea*, this larger clade will have to have another name.

Literature Cited

Gerassimova, Z. P., and L. N. Seravin. 1976. Ectoplasmic fibrillar system of *Infusoria* and its role for the understanding of their phylogeny. *Zool. Zh.* 55:645–656.

Hammerschmidt, B., M. Schlegel, D. Lynn, D. D. Leipe, M. L. Sogin, and I. B. Raikov. 1996. Insights into the evolution of nuclear dualism in the ciliates revealed by phylogenetic analysis of rRNA sequences. *J. Eukaryot. Microbiol.* 43:225–230.

Hirt, R. P., P. L. Dyal, M. Wilkinson, B. J. Finlay, D. M. Roberts, and T. M. Embley. 1995. Phylogenetic relationships among karyorelictids and heterotrichs inferred from small subunit rRNA sequences: resolution at the base of the ciliate tree. *Mol. Phylogenet. Evol.* 4:77–87.

Lynn, D. H. 2008. *The Ciliated Protozoa: Characterization, Classification, and Guide to the Literature.* 3rd edition. Springer, Dordrecht.

Authors

Micah Dunthorn; Department of Eukaryotic Microbiology; University of Duisburg-Essen; 45141 Essen, Germany. Email: micah.dunthorn@uni-due.de.

Denis H. Lynn; Department of Zoology; University of British Columbia; Vancouver, BC, V6T 1Z4, Canada. Email: lynn@zoology.ubc.ca.

Date Accepted: 24 August 2011

Primary Editor: Philip Cantino

Intramacronucleata D. H. Lynn 1996 [D. H. Lynn and M. Dunthorn], converted clade name

Registration Number: 53

Definition: The crown clade for which intramacronuclear microtubules (as described under Diagnostic Apomorphies), as inherited by *Tetrahymena thermophila* Nanney and McCoy 1976, is an apomorphy relative to other crown clades. This is an apomorphy-modified crown-clade definition. Abbreviated definition: crown ∇ apo intramacronuclear microtubules [*Tetrahymena thermophila* Nanney and McCoy 1976].

Etymology: Derived from the Latin *intra* (within), Greek *makros* (large), and Latin *nucleus* (kernel), in reference to the presence of intramacronuclear microtubules during cell division (see Diagnostic Apomorphies).

Reference Phylogeny: The primary reference phylogeny is Hammerschmidt et al. (1996: Fig. 2). Intramacronuclear microtubules are inferred to have arisen along the branch leading to the clade that contains (as labeled from top to bottom on the right-hand margin of the figure) *Armophorida* to *Phyllopharyngea*.

Composition: The majority of clades within the *Ciliophora*, as defined in this volume, are within the *Intramacronucleata*. Lynn (2008) lists the following taxa as included in the *Intramacronucleata*: *Armophorea*, *Colpodea*, *Litostomatea*, *Nassophorea*, *Oligohymenophorea*, *Phyllopharyngea*, *Plagiopylea*, *Prostomatea*, and *Spirotrichea*.

Diagnostic Apomorphies: The most distinctive diagnostic apomorphy is the presence of microtubules within the macronuclear envelope during nuclear division. In other *Ciliophora* (i.e. *Postciliodesmotophora*), microtubules are either extra-macronuclear or macronuclei do not divide.

Synonyms: None.

Comments: Lynn (1996) established this taxon, which contains most known ciliates, based on small subunit ribosomal RNA gene phylogenies as well as the presence of intranuclear microtubules in dividing macronuclei. Monophyly of this group is supported (Hirt et al., 1995; Hammerschmidt et al., 1996, Yoon et al., 2008), but internal relationships are unresolved (Lynn, 2008).

Literature Cited

Hammerschmidt, B., M. Schlegel, D. Lynn, D. D. Leipe, M. L. Sogin, and I. B. Raikov. 1996. Insights into the evolution of nuclear dualism in the ciliates revealed by phylogenetic analysis of rRNA sequences. *J. Eukaryot. Microbiol.* 43:225–230.

Hirt, R. P., P. L. Dyal, M. Wilkinson, B. J. Finlay, D. M. Roberts, and T. M. Embley. 1995. Phylogenetic relationships among karyorelictids and heterotrichs inferred from small subunit rRNA sequences: resolution at the base of the ciliate tree. *Mol. Phylogenet. Evol.* 4:77–87.

Lynn, D. H. 1996. My journey in ciliate systematics. *J. Eukaryot. Microbiol.* 43:253–260.

Lynn, D. H. 2008. *The Ciliated Protozoa: Characterization, Classification, and Guide to the Literature.* 3rd edition. Springer, Dordrecht.

Yoon, H. S., J. Grant, Y. I. Tekle, M. Wu, B. C. Chaon, J. C. Cole, J. M. Logsdon, D. J. Patterson, D. Bhattacharya, and L. A. Katz. 2008. Broadly sampled multigene trees of eukaryotes. *BMC Evol. Biol.* 8:14.

Authors

Denis H. Lynn; Department of Zoology; University of British Columbia; Vancouver, BC, V6T 1Z4, Canada. Email: lynn@zoology.ubc.ca.

Micah Dunthorn; Department of Eukaryotic Microbiology; University of Duisburg-Essen; 45141 Essen, Germany. Email: micah. dunthorn@uni-due.de.

Date Accepted: 24 August 2011

Primary Editor: Philip Cantino

Amorphea S. M. Adl et al. 2012. [A. G. B. Simpson, M. Medina, and S. M. Adl], converted clade name

Registration Number: 8

Definition: The smallest crown clade containing *Homo sapiens* Linnaeus 1758 and *Neurospora crassa* Shear and Dodge 1927 and *Dictyostelium discoideum* Raper 1935 but not *Arabidopsis thaliana* (Linnaeus) Heynhold 1842 (*Archaeplastida*) or *Tetrahymena thermophila* Nanney and McCoy 1976 (*Alveolata*) or *Thalassiosira pseudonana* Hasle and Hiemdal 1970 (*Stramenopila*) or *Bigelowiella natans* Moestrup and Sengco 2001 (*Rhizaria*) or *Euglena gracilis* Klebs 1883 (*Discoba*) or *Emiliania huxleyi* (Lohmann) Hay and Mohler 1967 (*Haptophyta*). This is a minimum-crown-clade definition with external specifiers. Abbreviated definition: min crown ∇ (*Homo sapiens* Linnaeus 1758 & *Neurospora crassa* Shear and Dodge 1927 & *Dictyostelium discoideum* Raper 1935 ~ *Arabidopsis thaliana* (Linnaeus) Heynhold 1842 ∨ *Tetrahymena thermophila* Nanney and McCoy 1976 ∨ *Thalassiosira pseudonana* Hasle and Hiemdal 1970 ∨ *Bigelowiella natans* Moestrup and Sengco 2001 ∨ *Euglena gracilis* Klebs 1883 ∨ *Emiliania huxleyi* (Lohmann) Hay and Mohler 1967).

Etymology: From the Greek, *a-* (without) and *morphe* (form, shape), referring to the fact that (the cells of) most members of this taxon do not have fixed form, unless restricted by an external layer, such as a cell wall, lorica, test or extracellular matrix (Adl et al., 2012).

Reference Phylogeny: The primary reference phylogeny is Minge et al. (2009: Fig. 2, in which *Amorphea* is referred to as 'unikonts'). Two external specifiers—*Thalassiosira pseudonana* and *Emiliania huxleyi*—are not present in the primary reference phylogeny but are most closely related to *Phaeodactylum tricornutum* and *Isochrysis galbana*, respectively, in that phylogeny (see Zhao et al., 2012). Note that Figure 2 of Minge et al. (2009) is not intended to communicate that the root of the eukaryote tree falls within *Amorphea*. It is clearly intended that the figure be viewed either as an unrooted tree or as a rooted tree that reflects the hypothesis that the root falls between *Amorphea* and other eukaryotes (Minge et al., 2009; see Richards and Cavalier-Smith, 2005). See also Brown et al. (2018: Fig. 1), where *Amorphea* is labelled, as well as Brown et al. (2009: Fig. 3), Zhao et al. (2012: Fig. 3) and Brown et al. (2013: Fig. 2), in which the clade here named *Amorphea* is inferred but unnamed

Composition: *Opisthokonta* (see entry in this volume), *Amoebozoa*, *Apusomonadida*, and *Breviatea*.

Diagnostic Apomorphies: No non-molecular synapomorphies are known.

Synonyms: *Unikonta*, 'unikonts' (approximate).

Comments: Phylogenies of eukaryotes based on multiple nucleus-encoded proteins usually group *Amoebozoa* and *Opisthokonta* to the exclusion of all other well-studied groups (e.g., Baldauf et al., 2000; Bapteste et al., 2002; Ciccarelli et al, 2006; Steenkamp et al., 2006; Rodríguez-Ezpeleta et al., 2007; Burki et al., 2007, 2008; Hampl et al., 2009; Minge et al., 2009; Zhao et al., 2012; Brown et al., 2013, 2018; Cavalier-Smith et al., 2014). Phylogenies of mitochondrial proteins often show the same relationship (e.g., Lang et al., 2002). These analyses usually do not include prokaryotic outgroups, thereby leaving the tree of eukaryotes unrooted. Even when prokaryotic outgroups are included in the analysis, the inferred position of the root of the eukaryote tree has generally been considered unreliable (Bapteste et al., 2002). These studies therefore leave open the possibility that *Opisthokonta* plus *Amoebozoa* represent a paraphyletic grade relative to the other extant eukaryotes. There have been few explicit proposals to this effect, however (Stechmann and Cavalier-Smith, 2002; Katz et al., 2012), and a *Opisthokonta-Amoebozoa* clade is recovered to the exclusion of the root in recent analyses of selected proteins of (relatively) recent bacterial origin (principally mitochondrial origin; Derelle and Lang, 2012; He et al., 2014; Derelle et al., 2015). The small and poorly studied groups *Apusomonadida* and *Breviatea* usually nest within the *Opisthokonta-Amoebozoa* clade in multigene molecular phylogenies or phylogenomic-scale analyses (Kim et al., 2006, Brown et al., 2009, 2013, 2018; Parfrey et al., 2010; Katz et al., 2011; Derelle and Lang, 2012; Zhao et al., 2012; Cavalier-Smith et al., 2014); very likely they are both specifically related to *Opisthokonta* (Brown et al., 2013). Some phylogenetic analyses infer that this clade also contains other taxa of unicellular eukaryotes, such as *Ancyromonadida* (also known as *Planomonadida*) and *Mantamonas* (Brown et al., 2009; Glücksman et al., 2011; Zhao et al., 2012), but this is contradicted by the most taxon-rich phylogenomic-scale analyses that include these taxa (Cavalier-Smith et al., 2014; Brown et al., 2018).

Stechmann and Cavalier-Smith (2003a, b) used the term the 'unikonts' as an informal name for a putative clade of *Opisthokonta* plus *Amoebozoa*. A few authors subsequently used the name *Unikonta* for this clade, both in specialist literature (e.g., Shatalkin, 2005; Schlegel

and Schmidt, 2007; Tekle et al., 2009) and in biology textbooks and other general works (e.g., Campbell et al., 2008; Freeman, 2008). The name 'unikont' referred to the notion that the common ancestor of *Opisthokonta* and *Amoebozoa* had a single flagellum and only one basal body—the one giving rise to the flagellum (Cavalier-Smith, 2002). This complemented the suggestion by Cavalier-Smith (2002) that the major extant eukaryote groups other than *Opisthokonta* and *Amoebozoa* formed a clade called 'bikonts', identified in part by the apomorphies of having two flagella and undergoing 'flagellar transformation' (see Beech et al., 1991). In fact, the distinction between unikonts and bikonts on the basis of flagellar organization evolution is highly problematic (Kim et al., 2006; Minge et al., 2009; Roger and Simpson, 2009). Flagellated opisthokonts characteristically have two basal bodies, although only one is flagellated. Further, some *Amoebozoa* have a flagellate stage with two flagella, and one species—*Physarum polycephalum*—has been observed to undergo flagellar transformation (Wright et al., 1980; Gely and Wright, 1986). *Apusomonadida* and *Breviatea* also have two basal bodies (both flagellated in *Apusomonadida* and some *Breviatea*), and there is good evidence from *Apusomonadida* in particular that they undergo flagellar transformation (see Walker et al., 2006; Roger and Simpson, 2009; Cavalier-Smith and Chao, 2010; Brown et al., 2013; Heiss et al., 2013a, b). Therefore, it is now almost certain that the most recent common ancestor of *Opisthokonta* and *Amoebozoa* was an organism with two basal bodies, and very likely two flagella (Kim et al., 2006; Roger and Simpson, 2009), and was a bikont-type cell rather than unikont-type cell in the original senses of the terms.

Given this situation, Adl et al. (2012) took the view that using the name 'unikonts' or '*Unikonta*' for the *Opisthokonta-Amoebozoa* clade was unacceptably misleading, and that it would be preferable to introduce a new taxon name. The name *Amorphea* was introduced by Adl et al. (2012) with a phylogenetic definition identical in function to that supplied here, but the wording here has been changed to include external specifiers in the definition itself rather than in a separate qualifying clause. Although the older name *Unikonta* has been more widely used overall than the more recently published name *Amorphea*, the latter is now (year 2017) used as often as the former. We are confident that the trend towards use of *Amorphea* will continue in the future because of the influential nature of the classification in which it was published (with 25 authors, and under the aegis of the Committee on Systematics and Evolution of the International Society of Protistologists), in addition to the misleading nature of the name *Unikonta*. Thus, we feel that the selection of the name *Amorphea* is in the spirit of *PhyloCode* (Cantino and de Queiroz, 2020) Article 10.1 (i.e., the name was chosen to minimize disruption of current usage) even though Recommendation 10.1A was not followed.

The definition of *Amorphea* is intended to reflect the hypothesis that *Opisthokonta* and *Amoebozoa* are related to the exclusion of other well-studied clades of *Eukarya* (emphasis on 'well-studied'). The use of several external specifiers in the definition is intended to make the name *Amorphea* inapplicable should this not be the case, for example, should the root of *Eukarya* prove to fall between *Opisthokonta* and *Amoebozoa*.

Literature Cited

Adl, S. M., A. G. B. Simpson, C. E. Lane, J. Lukeš, D. Bass, S. S. Bowser, M. W. Brown, F. Burki, M. Dunthorn, V. Hampl, A. Heiss, M. Hoppenrath, E. Lara, L. le Gall, D. H. Lynn, H. McManus, E. A. D. Mitchell, S. E. Mozley-Stanridge, L. W. Parfrey, J. Pawlowski, S. Rueckert, L. Shadwick, C. L. Schoch, A. Smirnov, and F. W. Spiegel. 2012. The revised classification of eukaryotes. *J. Eukaryot. Microbiol.* 59:429–493.

Baldauf, S. L., A. J. Roger, I. Wenk-Siefert, and W. F. Doolittle. 2000. A kingdom-level phylogeny of eukaryotes based on combined protein data. *Science* 290:972–977.

Bapteste, E., H. Brinkmann, J. A. Lee, D. V. Moore, C. W. Sensen, P. Gordon, L. Duruflé, T. Gaasterland, P. Lopez, M. Müller, and H. Philippe. 2002. The analysis of 100 genes supports the grouping of three highly divergent amoebae: *Dictyostelium*, *Entamoeba*, and *Mastigamoeba*. *Proc. Natl. Acad. Sci. USA* 99:1414–1419.

Beech, P. L., K. Heimann, and M. Melkonian. 1991. Development of the flagellar apparatus during the cell cycle in unicellular algae. *Protoplasma* 164:23–37.

Brown, M. W., A. A. Heiss, R. Kamikawa, Y. Inagaki, A. Yabuki, A. K. Tice, T. Shiratori, K. Ishida, T. Hashimoto, A. G. B. Simpson, and A. J. Roger. 2018. Phylogenomics places orphan protistan lineages in a novel eukaryotic super-group. *Genome Biol. Evol.* 10:427–433.

Brown, M. W., S. C. Sharpe, J. D. Silberman, A. A. Heiss, B. F. Lang, A. G. B. Simpson, and A. J. Roger. 2013. Phylogenomics demonstrates that breviate flagellates are related to opisthokonts and apusomonads. *Proc. R. Soc. Lond. B Biol. Sci.* 280:20131755.

Brown, M. W., F. W. Spiegel, and J. D. Silberman. 2009. Phylogeny of the "forgotten" cellular slime mold, *Fonticula alba*, reveals a key evolutionary branch within *Opisthokonta*. *Mol. Biol. Evol.* 26:2699–2709.

Burki, F., K. Shalchian-Tabrizi, M. Minge, A. Skjaeveland, S. I. Nikolaev, K. S. Jakobsen, and J. Pawlowski. 2007. Phylogenomics reshuffles the eukaroyotic supergroups. *PLOS ONE* 2:e790.

Burki, F., K. Shalchian-Tabrizi, and J. Pawlowski. 2008. Phylogenomics reveals a new 'megagroup' including most photosynthetic eukaryotes. *Biol. Lett.* 4:366–369.

Campbell, N. A., J. B. Reece, L. A. Urry, M. L. Cain, S. A. Wasserman, P. V. Minorsky, and R. B. Jackson. 2008. *Biology*. 8th edition. Pearson Benjamin Cummings, San Francisco, CA.

Cantino, P. D., and K. de Queiroz. 2020. *International Code of Phylogenetic Nomenclature (PhyloCode)*, Version 6. CRC Press, Boca Raton, FL.

Cavalier-Smith, T. 2002. The phagotrophic origin of eukaryotes and phylogenetic classification of *Protozoa*. *Int. J. Syst. Evol. Microbiol.* 52:297–354.

Cavalier-Smith, T., and E. E. Chao. 2010. Phylogeny and evolution of *Apusomonadida* (*Protozoa: Apusozoa*): new genera and species. *Protist* 161:549–576.

Cavalier-Smith, T., E. E. Chao, E. A. Snell, C. Berney, A. M. Fiore-Donno, and R. Lewis. 2014. Multigene eukaryote phylogeny reveals the likely protozoan ancestors of opisthokonts (animals, fungi, choanozoans) and *Amoebozoa*. *Mol. Phylogenet. Evol.* 81:71–85.

Ciccarelli, F. D., T. Doeks, C. von Mering, C. J. Creevey, B. Snel, and P. Bork. 2006. Toward automatic reconstruction of a highly resolved tree of life. *Science* 311:1283–1287.

Derelle, R., and B. F. Lang. 2012. Rooting the eukaryotic tree with mitochondrial and bacterial proteins. *Mol. Biol. Evol.* 29:1277–1289.

Derelle, R., G. Torruella, V. Klimes, H. Brinkmann, E. Kim, Č. Vlček, B. F. Lang, and M. Eliáš. 2015. Bacterial proteins pinpoint a single eukaryotic root. *Proc. Natl. Acad. Sci. USA* 112:693–699.

Freeman, S. 2008. *Biological Science*. 3rd edition. Pearson Benjamin Cummings, San Francisco, CA.

Gely, C., and M. Wright. 1986. The centriole cycle in the amoebae of the myxomycete *Physarum polycephalum*. *Protoplasma* 132:23–31.

Glücksman, E., E. A. Snell, C. Berney, E. E. Chao, D. Bass, and T. Cavalier-Smith, 2011. The novel marine zooflagellate gliding genus *Mantamonas* (*Mantamonadida* ord. n.: *Apusozoa*). *Protist* 162: 207–221.

Hampl, V., L. Hug, J. W. Leigh, J. B. Dacks, B. F. Lang, A. G. B. Simpson, and A. J. Roger. 2009. Phylogenomic analyses support the monophyly of *Excavata* and resolve relationships among eukaryotic "supergroups". *Proc. Natl. Acad. Sci. USA* 106:3859–3864.

He, D., O. Fiz-Palacios, C. Fu, J. Fehling, C. C. Tsai, and S. L. Baldauf. 2014. An alternative root for the eukaryote tree of life. *Curr. Biol.* 24:465–470.

Heiss, A. A., G. Walker, and A. G. B. Simpson. 2013a. The flagellar apparatus of *Breviata anathema*, a eukaryote without a clear supergroup affinity. *Eur. J. Protistol.* 49:354–372.

Heiss, A. A., G. Walker, and A. G. B. Simpson. 2013b. The microtubular cytoskeleton of the apusomonad *Thecamonas*, a sister lineage to the opisthokonts. *Protist* 164:598–621.

Katz, L. A., J. R. Grant, L. W. Parfrey, A. Grant, C. J. O'Kelly, O. R. Anderson, R. E. Molestina, and T. Nerad. 2011. *Subulatomonas tetraspora* nov. gen. nov. sp. is a member of a previously unrecognized major clade of eukaryotes. *Protist* 162:762–773.

Katz, L. A., J. R. Grant, L. W. Parfrey, and J. G. Burleigh. 2012. Turning the crown upside down: gene tree parsimony roots the eukaryotic tree of life. *Syst. Biol.* 6:653–660.

Kim, E., A. G. B. Simpson, and L. Graham. 2006. Evolutionary relationships of apusomonads inferred from taxon-rich analyses of six nuclear-encoded genes. *Mol. Biol. Evol.* 23:2455–2466.

Lang, B. F., C. J. O'Kelly, T. A. Nerad, M. W. Gray, and G. Burger. 2002. The closest unicellular relatives of animals. *Curr. Biol.* 12:1773–1778.

Minge, M. A., J. D. Silberman, R. J. S. Orr, T. Cavalier-Smith, K. Shalchian-Tabrizi, F. Burki, A. Skjæveland, and K. S. Jakobsen. 2009. Evolutionary position of breviate amoebae and the primary eukaryote divergence. *Proc. R. Soc. Lond. B Biol. Sci.* 276:597–604.

Parfrey, L. W., J. Grant, Y. I. Tekle, E. Lasek-Nesselquist, H. G. Morrison, M. L. Sogin, D. J. Patterson, and L. A. Katz. 2010. Broadly sampled multigene analyses yield a well-resolved eukaryotic tree of life. *Syst. Biol.* 59:518–533.

Richards, T. A., and T. Cavalier-Smith. 2005. Myosin domain evolution and the primary divergence of eukaryotes. *Nature* 436:1113–1118.

Rodríguez-Ezpeleta, N., H. Brinkmann, G. Burger, A. J. Roger, M. W. Gray, H. Philippe, and B. F. Lang. 2007. Toward resolving the eukaryotic tree: the phylogenetic positions of jakobids and cercozoans. *Curr. Biol.* 17:1420–1425.

Roger, A. J., and A. G. B. Simpson. 2009. Revisiting the root of the eukaryotic tree. *Curr. Biol.* 19: 165–167.

Schlegel, M., and S. L. Schmidt. 2007. Evolution und Stammesgeschichte der Eukaryoten. *Denisia* 20:155–164.

Shatalkin, A. I. 2005. Animals (*Animalia*) in system of organisms. 2. Phlyogenetic understanding of animals. *Zh. Obshch. Biol.* 66:389–415 [in Russian].

Stechmann, A., and T. Cavalier-Smith. 2002. Rooting the eukaryote tree by using a derived gene fusion. *Nature* 297:89–91.

Stechmann, A., and T. Cavalier-Smith. 2003a. Phylogenetic analysis of eukaryotes using heat-shock protein hsp90. *J. Mol. Evol.* 57:408–419.

Stechmann, A., and T. Cavalier-Smith. 2003b. The root of the eukaryote tree pinpointed. *Curr. Biol.* 13:R665–R666.

Steenkamp, E. T., J. Wright, and S. L. Baldauf. 2006. The protistan origins of animals and fungi. *Mol. Biol. Evol.* 23:93–106.

Tekle Y. I., L. W. Parfrey, and L. A. Katz. 2009. Molecular data are transforming hypotheses on the origin and diversification of eukaryotes. *Bioscience* 59:471–481.

Walker, G., J. B. Dacks, and T. M. Embley. 2006. Ultrastructural description of *Breviata anathema*, n. gen., n. sp., the organism previously studied as "*Mastigamoeba invertens*". *J. Eukaryot. Microbiol.* 53:65–78.

Wright, M., A. Moisand, and L. Mir. 1980. Centriole maturation in the amoebae of *Physarum polycephalum*. *Protoplasma* 105:149–160.

Zhao, S., F. Burki, J. Bråte, P. J. Keeling, D. Klaveness, and K. Shalchian-Tabrizi. 2012. *Collodictyon*—an ancient lineage in the tree of eukaryotes. *Mol. Biol. Evol.* 29:1557–1568.

Author

Alastair G.B. Simpson; Department of Biology; Dalhousie University; Halifax, NS, B3H 4R2, Canada. Email: alastair.simpson@dal.ca.

Mónica Medina; Department of Biology; Pennsylvania State University; University Park, PA, 16802 USA. Email: mum55@psu.edu.

Sina M. Adl; Department of Soil Science; University of Saskatchewan; Saskatoon, SK S7N 5C5, Canada. Email: sina.adl@usask.ca.

Date Accepted: 12 April 2013; updated 24 February 2018

Primary Editor: Philip Cantino

Opisthokonta T. Cavalier-Smith 1987: Fig. 23.2 [A. G. B. Simpson and M. Medina], converted clade name

Registration Number: 72

Definition: The smallest crown clade containing *Homo sapiens* Linnaeus 1758 (*Metazoa*), *Neurospora crassa* Shear and Dodge 1927 (*Fungi*), and *Monosiga ovata* Saville Kent 1882 (*Choanoflagellata*) but not *Dictyostelium discoideum* Raper 1935 (*Amoebozoa*) or *Arabidopsis thaliana* (Linnaeus) Heynhold 1842 (*Archaeplastida*). This is a minimum-crown-clade definition with external specifiers (see Comments). Abbreviated definition: min crown ∇ (*Homo sapiens* Linnaeus 1758 & *Neurospora crassa* Shear and Dodge 1927 & *Monosiga ovata* Saville Kent 1882 ~ *Dictyostelium discoideum* Raper 1935 ∨. *Arabidopsis thaliana* (Linnaeus) Heynhold 1842).

Etymology: Derived from Greek *opisthen* (posterior) and *kontos* (oar or pole), the latter referring to the flagellum. Flagellated unicellular forms usually have a single emergent flagellum, which is inserted posteriorly (see Diagnostic Apomorphies).

Reference Phylogeny: The primary reference phylogeny is Brown et al. (2009: Figure 3). See also Steenkamp et al. (2006: Fig. 1), Rodríguez-Ezpeleta et al. (2007: Fig. 1), Parfrey et al. (2010: Fig. 2) and Hehenberger et al. (2017: Fig. 2).

Composition: *Metazoa, Fungi, Choanoflagellata, Ichthyosporea* (~*Mesomycetozoea*), *Nucleariidae, Fonticula, Capsaspora, Corallochytrium, Ministeria, Pigoraptor* and *Syssomonas*.

Diagnostic Apomorphies: The identification of apomorphies is complicated by our incomplete

understanding of eukaryote phylogeny. There is good evidence that *Opisthokonta* are closely related to *Apusomonadida* and *Breviatea*, and more distantly to *Amoebozoa* (e.g., Kim et al., 2006; Rodríguez-Ezpeleta et al., 2007; Brown et al., 2009, 2013; Katz et al., 2011; Derelle and Lang, 2012; Derelle et al., 2015), forming the taxon *Amorphea*, however deeper relationships are less well resolved. This makes it difficult to assign many evolutionary transformations to particular branches with certainty. For example, flattened mitochondrial cristae are common to the great majority of opisthokonts, but this feature is present in various other eukaryotic groups as well, including several of the taxa that may be closely related to *Amorphea* (i.e., *Rigifilida* and *Mantamonas* as well as *Ancyromonadida*; see Brown et al., 2018). Thus, while flattened cristae are likely be an apomorphy for *Opisthokonta* (i.e., evolved separately in other groups), they may instead be a plesiomorphy. Here we highlight particularly distinctive features that are unique or nearly unique to *Opisthokonta*.

1. Posterior flagellar insertion. In almost all members of the clade that produce unicellular flagellated stages, there is a single emergent flagellum that is located at the posterior end of the cell, when the anterior-posterior axis is defined by the direction of motion (Cavalier-Smith, 1987). This arrangement is rare among eukaryotes—most flagellated cells, including *Apusomonadida*, *Breviatea* and flagellated *Amoebozoa*, have flagella that insert anteriorly or laterally.

2. Several molecular synapomorphies have been suggested (see Elias, 2008). The

best documented is a large insertion, usually of 12 amino acids, in translation elongation factor 1-α (EF1A). This is present in all studied *Opisthokonta* that encode this protein. The insertion is located in a region that is almost length-invariable across other eukaryotes and *Archaea*, and thus it appears apomorphic (see Baldauf and Palmer, 1993; Steenkamp et al., 2006).

Synonyms: No formal synonyms known. Informally this clade has been referred to as the 'animal-fungal clade'.

Comments: Animals and fungi were treated as separate kingdoms under the pedagogically popular 'five kingdom' systems of life derived from Whittaker (1969). Cavalier-Smith (1987), however, proposed that fungi are closely related to animals and to the unicellular/colonial choanoflagellates. (A relationship between animals, or at least sponges, and choanoflagellates had been proposed earlier, based on the morphological similarities between choanoflagellates and sponge choanocytes – James-Clark, 1868; see King, 2004.) Among the reasons for suspecting a relationship between fungi, animals and choanoflagellates was the distinctive shared combination of flattened but not discoidal mitochondrial cristae, and a single posterior flagellum in motile flagellated cells (Cavalier-Smith, 1987). A few years afterwards, some ribosomal RNA gene phylogenies supported this close relationship (e.g., Wainright et al., 1993), while analyses of several protein-coding genes inferred an animal-fungus clade in the absence of data from choanoflagellates, and also revealed a shared 12-amino-acid insertion in translation elongation factor 1α proteins (Baldauf and Palmer, 1993). Many subsequent molecular phylogenetic analyses employing large samples of genes and/or taxa

confirmed a close relationship between animals, choanoflagellates and fungi (e.g., Lang et al., 2002; Medina et al., 2003; Philippe et al., 2004; Rodríguez-Ezpeleta et al., 2007) but also demonstrated that certain other eukaryotes also fall with or within this clade. These include: the diverse, parasitic *Ichthyosporea* (approximate synonyms *Mesomycetozoea* or 'DRIP's – see Ragan et al., 1996; Mendoza et al., 2002), the filose amoeba group *Nucleariidae* (Amaral Zettler et al., 2001), the unusual and poorly known slime mold *Fonticula* (Brown et al., 2009), the small amoeboid parasites *Corallochytrium* (Cavalier-Smith and Allsopp, 1996) and *Capsaspora* (see Ruiz-Trillo et al., 2004), the small free-living *Ministeria,* which sometimes bears a flagellum (Cavalier-Smith and Chao, 2003), and the recently described free-living flagellates *Pigoraptor* and *Syssomonas* (Hehenberger et al., 2017). Molecular phylogenies of multiple proteins divide the grouping into two well supported subclades: The first subclade is widely referred to as *Holozoa* (Lang et al., 2002; Adl et al., 2012; Hehenberger et al., 2017) and includes *Metazoa*, *Choanoflagellata* and *Ichthyosporea*, plus *Capsaspora*, *Ministeria* and *Pigoraptor* (where studied), as well as *Corallochytrium* and *Syssomonas* (where studied). The second subclade includes *Fungi* (as defined in this volume) and, where studied, *Nucleariidae* and *Fonticula* (Lang et al., 2002; Steenkamp et al., 2006; Ruiz-Trillo et al., 2006, 2008; Shalchian-Tabrizi et al., 2008; Brown et al., 2009; Liu et al., 2009; Parfrey et al., 2010; Hehenberger et al., 2017). This second subclade has been referred to as *Holomycota* (Liu et al., 2009), *Nucletmycea* (Brown et al., 2009), and, possibly in error, '*Holofungi*' (Lara et al., 2010). All of these relatives of *Metazoa* or *Fungi* that include flagellates in their life history produce flagellated cells with a single emergent flagellum (see Cavalier-Smith and Chao, 2003; Mendoza et al., 2002; Hehenberger et al., 2017). Where studied, they also show the characteristic 'animal-fungal' EF1A insertion (Steenkamp et al., 2006). Recent studies based on comparison of whole genomes have uncovered other molecular characters apparently restricted to animals and fungi (and, where studied, choanoflagellates), for example novel genes or shared lateral gene transfers (see Elias, 2008).

Opisthokonta was apparently first proposed by Copeland (1956) as a new high-level taxon for the 'chytrid fungi' (sensu lato), which differ from typical fungi by having a flagellated stage. Copeland included within his phylum *Opisthokonta* one class, the previously created taxon *Archimyces*, which he identified by the same characters as the phylum (i.e., the phylum was a redundant taxon). *Opisthokonta* is not now used to refer to chytrids, and molecular phylogenies indicate that 'chytrids' sensu lato are paraphyletic (e.g., James et al., 2006).

The name '*Opisthokonta*' was subsequently introduced independently by Cavalier-Smith (1987; legend for Figure 23.2) for a very different concept—a hypothesized clade of animals, fungi and choanoflagellates. Because of this independence and the large differences in concept, *Opisthokonta* sensu Copeland (1956) and sensu Cavalier-Smith (1987) are regarded here as homonyms. Despite the capitalization, the name was not proposed as a formal taxon by Cavalier-Smith (1987), as he considered choanoflagellates to be members of a paraphyletic kingdom *Protozoa*, whilst animals and fungi each had their own kingdoms. Nevertheless, the names '*Opisthokonta*' and 'opisthokonts' are now widely accepted as referring to the animal-fungal-choanoflagellate clade (e.g., Cavalier-Smith and Chao, 1995; Baldauf et al., 2000; Baldauf, 2003; Simpson and Roger, 2004; Adl et al., 2005, 2012; Keeling et al., 2005; Parfrey et al., 2006; Steenkamp et al., 2006). *Opisthokonta* was treated as an out-of-code, rankless taxon by Adl et al. (2005), who circumscribed it

with reference to the posterior flagellum and flat cristae (following Cavalier-Smith, 1987). There are no formal synonyms in common use. Notwithstanding Copeland's (1956) earlier homonym (which is dormant, has synonyms, and does not readily refer to a clade – see above), we have selected *Opisthokonta* Cavalier-Smith (1987) for the animal-fungal-choanoflagellate clade, which preserves the contemporary usage of this name.

The stipulation in the definition that *Opisthokonta* not include *Dictyostelium discoideum* (*Amoebozoa*) or *Arabidopsis thaliana* (*Archaeplastida*) is intended to prevent the name from being applied in the (highly unlikely) case that the clade proves to include one or more of the major groups of well-known eukaryotes that have always been excluded from *Opisthokonta*, including the case where the root of the eukaryotic tree falls among the opisthokonts. Non-application of the name in the latter case would be consistent with our view that *Eukarya* should have precedence if *Eukarya* and *Opisthokonta* prove to be synonyms.

Literature Cited

Adl, S. M., A. G. B. Simpson, M. A. Farmer, R. A. Andersen, O. R. Anderson, J. R. Barta, S. S. Bowser, G. Brugerolle, R. A. Fensome, S. Fredericq, T. Y. James, S. Karpov, P. Kugrens, J. Krug, C. E. Lane, L. A. Lewis, J. Lodge, D. H. Lynn, D. G. Mann, R. M. McCourt, L. Mendoza, O. Moestrup, S. E. Mozley-Standridge, T. A. Nerad, C. A. Shearer, A. V. Smirnov, F. W. Spiegel, and M. F. Taylor. 2005. The new higher level classification of eukaryotes with emphasis on the taxonomy of protists. *J. Eukaryot. Microbiol.* 52:399–451.

Adl, S. M., A. G. B. Simpson, C. E. Lane, J. Lukeš, D. Bass, S. S. Bowser, M. W. Brown, F. Burki, M. Dunthorn, V. Hampl, A. Heiss, M. Hoppenrath, E. Lara, L. le Gall, D. H. Lynn, H. McManus, E. A. D. Mitchell, S. E.

Mozley-Stanridge, L. W. Parfrey, J. Pawlowski, S. Rueckert, L. Shadwick, C. L. Schoch, A. Smirnov, and F. W. Spiegel. 2012. The revised classification of eukaryotes. *J. Eukaryot. Microbiol.* 59:429–493.

Amaral Zettler, L. A., T. A. Nerad, C. J. O'Kelly, and M. L. Sogin. 2001. The nucleariid amoebae: more protists at the animal-fungal boundary. *J. Eukaryot. Microbiol.* 48:293–297.

Baldauf, S. L. 2003. The deep roots of eukaryotes. *Science* 300:1703–1706.

Baldauf, S. L., and J. D. Palmer. 1993. Animals and fungi are each other's closest relatives: congruent evidence from multiple proteins. *Proc. Natl. Acad. Sci. USA* 90:11558–11562.

Baldauf, S. L., A. J. Roger, I. Wenk-Siefert, and W. F. Doolittle. 2000. A kingdom-level phylogeny of eukaryotes based on combined protein data. *Science* 290:972–977.

Brown, M. W., A. A. Heiss, R. Kamikawa, Y. Inagaki, A. Yabuki, A. K. Tice, T. Shiratori, K. Ishida, T. Hashimoto, A. G. B. Simpson, and A. J. Roger. 2018. Phylogenomics places orphan protistan lineages in a novel eukaryotic super-group. *Genome Biol. Evol.* 10:427–433.

Brown, M. W., S. C. Sharpe, J. D. Silberman, A. A. Heiss, B. F. Lang, A. G. B. Simpson, and A. J. Roger. 2013. Phylogenomics demonstrates that breviate flagellates are related to opisthokonts and apusomonads. *Proc. R. Soc. Lond. B Biol. Sci.* 280:20131755.

Brown, M. W., F. W. Spiegel, and J. D. Silberman. 2009. Phylogeny of the "forgotten" cellular slime mold, *Fonticula alba*, reveals a key evolutionary branch within *Opisthokonta*. *Mol. Biol. Evol.* 26:2699–2709.

Cavalier-Smith, T. 1987. The origin of fungi and pseudofungi. Pp. 339–353 in *Evolutionary Biology of the Fungi* (A. D. M. Rayner, C. M. Brasier, and D. Moore, eds.). *British Mycological Society Symposium 12*. Cambridge University Press, Cambridge, UK.

Cavalier-Smith, T., and M. T. E. P. Allsopp. 1996. *Corallochytrium*, an enigmatic non-flagellate protozoan related to choanoflagellates. *Eur. J. Protistol.* 32:306–310.

Cavalier-Smith, T., and E. E. Chao. 1995. The opalozoan *Apusomonas* is related to the common ancestor of animals, fungi and choanoflagellates. *Proc. Roy. Soc. Lond. B* 261:1–6.

Cavalier-Smith, T., and E. E. Chao. 2003. Phylogeny of *Choanozoa*, *Apusozoa*, and other protozoa and early eukaryote megaevolution. *J. Mol. Evol.* 56:540–563.

Copeland, H. F. 1956. *The Classification of Lower Organisms*. Pacific Books, Palo Alto, CA.

Derelle, R., and B. F. Lang. 2012. Rooting the eukaryotic tree with mitochondrial and bacterial proteins. *Mol. Biol. Evol.* 29:1277–1289.

Derelle, R., G. Torruella, V. Klimes, H. Brinkmann, E. Kim, Č. Vlček, B. F. Lang, and M. Eliás. 2015. Bacterial proteins pinpoint a single eukaryotic root. *Proc. Natl. Acad. Sci. USA*, 112:693–699.

Elias, M. 2008. The guanine nucleotide exchange factors Sec2 and PRONE: candidate synapomorphies for the *Opisthokonta* and the *Archaeplastida*. *Mol. Biol. Evol.* 25:1526–1529.

Hehenberger, E., D. V. Tikhonenkov, M. Kolisko, J. Del Campo, A. S. Esaulov, A. P. Mylnikov, and P. J. Keeling. 2017. Novel predators reshape holozoan phylogeny and reveal the presence of a two-component signaling system in the ancestor of animals. *Curr. Biol.* 27:2043–2050.

James, T. Y., F. Kauff, C. Schoch, P. B. Matheny, V. Hofstetter, C. Cox, G. Celio, C. Gueidan, E. Fraker, J. Miadlikowska, H. T. Lumbsch, A. Rauhut, V. Reeb, A. E. Arnold, A. Amtoft, J. E. Stajich, K. Hosaka, G.-H. Sung, D. Johnson, B. O'Rourke, M. Crockett, M. Binder, J. M. Curtis, J. C. Slot, Z. Wang, A. W. Wilson, A. Schüßler, J. E. Longcore, K. O'Donnell, S. Mozley-Standridge, D. Porter, P. M. Letcher, M. J. Powell, J. W. Taylor, M. M. White, G. W. Griffith, D. R. Davies, R. A. Humber, J. B. Morton, J. Sugiyama, A. Y. Rossman, J. D. Rogers, D. H. Pfister, D. Hewitt, K. Hansen, S. Hambleton, R. A. Shoemaker, J. Kohlmeyer, B. Volkmann-Kohlmeyer, R. A. Spotts, M. Serdani, P. W. Crous, K. W. Hughes, K. Matsuura, E. Langer, G. Langer, W. A. Untereiner, R. Lücking, B. Büdel, D. M. Geiser, A. Aptroot, P. Diederich, I. Schmitt, M. Schultz, R. Yahr, D. Hibbett, F. Lutzoni,

D. McLaughlin, J. Spatafora, and R. Vilgalys. 2006. Reconstructing the early evolution of *Fungi* using a six-gene phylogeny. *Nature* 443:818–822.

James-Clark, H. 1868. On the spongiae ciliatae as infusoria flagellata; or observations on the structure, animality, and relationship of *Leucosolenia botryroides*. *Ann. Mag. Nat. Hist.* 1:133–142, 188–215, 250–264.

Katz, L. A., J. Grant, L. W. Parfrey, A. Grant, C. J. O'Kelly, O. R. Anderson, R. E. Molestina, and T. Nerad. 2011. *Subulatomonas tetraspora* nov. gen. nov. sp. is a member of a previously unrecognized major clade of eukaryotes. *Protist* 162:762–773.

Keeling, P. J., G. Burger, D. G. Durnford, B. F. Lang, R. W. Lee, R. E. Pearlman, A. J. Roger, and M. W. Gray. 2005. The tree of eukaryotes. *Trends Ecol. Evol.* 20:670–675.

Kim, E., A. G. B. Simpson, and L. E. Graham. 2006. Evolutionary relationships of apusomonads inferred from taxon-rich analyses of six nuclear-encoded genes. *Mol. Biol. Evol.* 23:2455–2466.

King, N., 2004. The unicellular ancestry of animal development. *Dev. Cell* 7:313–325.

Lang, B. F., C. J. O'Kelly, T. A. Nerad, M. W. Gray, and G. Burger. 2002. The closest unicellular relatives of animals. *Curr. Biol.* 12:1773–1778.

Lara, E., D. Moreira, and P. López-García. 2010. The environmental clade LKM11 and *Rozella* form the deepest branching clade of *Fungi*. *Protist* 161:116–121.

Liu, Y., E. T. Steenkamp, H. Brinkmann, L. Forget, H. Philippe, and B. F. Lang. 2009. Phylogenomic analyses predict sistergroup relationship of nucleariids and *Fungi* and paraphyly of zygomycetes with significant support. *BMC Evol. Biol.* 9:e272.

Medina, M., A. Collins, J. Taylor, J. W. Valentine, J. Lipps, L. Amaral-Zettler, and M. L. Sogin. 2003. Phylogeny of *Opisthokonta* and the evolution of multicellularity and complexity in *Fungi* and *Metazoa*. *Int. J. Astrobiot.* 2:203–211.

Mendoza, L., J. W. Taylor, and L. Ajello. 2002. The Class *Mesomycetozoa*: a heterogeneous group of microorganisms at the animal-fungal boundary. *Ann. Rev. Microbiol.* 56:315–344.

Parfrey, L. W., E. Barbero, E. Lasser, M. Dunthorn, D. Bhattacharya, D. J. Patterson, and L. A. Katz. 2006. Evaluating support for the current classification of eukaryotic diversity. *PLOS Genet.* 2:e220.

Parfrey, L. W., J. Grant, Y. I. Tekle, E. Lasek-Nesselquist, H. G. Morrison, M. L. Sogin, D. J. Patterson, and L. A. Katz. 2010. Broadly sampled multigene analyses yield a well-resolved eukaryotic tree of life. *Syst. Biol.* 59:518–533.

Philippe, H., E. A. Snell, E. Bapteste, P. Lopez, P. W. Holland, and D. Casane. 2004. Phylogenomics of eukaryotes: impact of missing data on large alignments. *Mol. Biol. Evol.* 21:1740–1752.

Ragan, M. A., C. L. Goggin, R. J. Cawthorn, L. Cerenius, A. V. Jamieson, S. M. Plourde, T. G. Rand, K. Söderhäll, and R. R. Gutell. 1996. A novel clade of protistan parasites near the animal-fungal divergence. *Proc. Natl. Acad. Sci. USA* 93:11907–11912.

Rodríguez-Ezpeleta, N., H. Brinkmann, G. Burger, A. J. Roger, M. W. Gray, H. Philippe, and B. F. Lang. 2007. Toward resolving the eukaryotic tree: the phylogenetic positions of jakobids and cercozoans. *Curr. Biol.* 17:1420–1425.

Ruiz-Trillo, I., Y. Inagaki, L. A. Davis, S. Sperstad, B. Landfald, and A. J. Roger. 2004. *Capsaspora owczarzaki* is an independent opisthokont lineage. *Curr. Biol.* 14:R946–R947.

Ruiz-Trillo, I., C. E. Lane, J. M. Archibald, and A. J. Roger. 2006. Insights into the evolutionary origin and genome architecture of the unicellular opisthokonts *Capsaspora owczarzaki* and *Sphaeroforma arctica*. *J. Eukaryot. Microbiol.* 53:379–384.

Ruiz-Trillo, I., A. J. Roger, G. Burger, M. W. Gray, and B. F. Lang. 2008. A phylogenomic investigation into the origin of *Metazoa*. *Mol. Biol. Evol.* 25:664–672.

Shalchian-Tabrizi, K., M. A. Minge, M. Espelund, R. Orr, T. Ruden, K. S. Jakobsen, and T. Cavalier-Smith. 2008. Multigene phylogeny of *Choanozoa* and the origin of animals. *PLOS ONE* 3:e2098.

Simpson, A. G. B., and A. J. Roger. 2004. The real 'kingdoms' of eukaryotes. *Curr. Biol.* 14:R693–R696.

Steenkamp, E. T., J. Wright, and S. L. Baldauf. 2006. The protistan origins of animals and fungi. *Mol. Biol. Evol.* 23:93–106.

Wainright, P. O., G. Hinkle, M. L. Sogin, and S. K. Stickel. 1993. Monophyletic orgins of the *Metazoa*: an evolutionary link with *Fungi*. *Science* 260:340–342.

Whittaker, R. H. 1969. New concepts of kingdoms of organisms. *Science* 163:150–160.

Authors

Alastair G. B. Simpson; Department of Biology; Dalhousie University; Halifax, NS, B3H 4R2, Canada. Email: alastair.simpson@dal.ca.

Mónica Medina; Department of Biology; Pennsylvania State University; University Park, PA, 16802, USA. Email: mum55@psu.edu.

Date Accepted: 22 February 2012; updated 8 March 2018

Primary Editor: Philip Cantino

Fungi A. Engler and K. Prantl 1900: iii [D. S. Hibbett, M. Blackwell, T. James, J. W. Spatafora, J. W. Taylor, and R. Vilgalys], converted clade name

Registration Number: 45

Definition: The smallest crown clade containing *Rozella allomycis* F. K. Faust 1937, *Batrachochytrium dendrobatidis* Longcore, Pessier and D. K. Nichols 1999, *Allomyces arbusculus* E. J. Butler 1911, *Entomophthora muscae* (Cohn) Fresen. 1856, *Coemansia reversa* Tiegh. and G. Le Monn. 1873, *Rhizophagus intraradices* (N. C. Schenck and G. S. Sm.) C. Walker and A. Schüßler 2010 [the name *Glomus intraradices* is used on the primary reference phylogeny], *Rhizopus oryzae* Went and Prins. Geerl. 1895, *Saccharomyces cerevisiae* Meyen 1838, and *Coprinopsis cinerea* (Schaeff.) Redhead, Vilgalys and Moncalvo 2001. This is a minimum-crown-clade definition. Abbreviated definition: min crown ∇ (*Rozella allomycis* F. K. Faust 1937 & *Batrachochytrium dendrobatidis* Longcore, Pessier and D. K. Nichols 1999 & *Allomyces arbusculus* E. J. Butler 1911 & *Entomophthora muscae* (Cohn) Fresen. 1856 & *Coemansia reversa* Tiegh. and G. Le Monn. 1873 & *Rhizophagus intraradices* (N. C. Schenck and G. S. Sm.) C. Walker and A. Schüßler 2010 & *Rhizopus oryzae* Went and Prins. Geerl. 1895 & *Saccharomyces cerevisiae* Meyen 1838 & *Coprinopsis cinerea* (Schaeff.) Redhead, Vilgalys and Moncalvo 2001).

Etymology: Derived from the Latin *fungus* (mushroom).

Reference Phylogeny: The primary reference phylogeny is James et al. (2006: Fig. 1). See also James et al. (2013: Fig. 2), Karpov et al. (2013: Fig. 3), Paps et al. (2013: Fig. 1), Chang et al. (2015: Fig. 1), Torruella et al. (2015: Fig. 1), and Spatafora et al. (2016: Fig. 1).

Composition: *Rozella, Microsporidia, Aphelida, Chytridiomycota, Neocallimastigomycota, Blastocladiomycota, Mucoromycota, Zoopagomycota, Ascomycota,* and *Basidiomycota* (Hibbett et al., 2007; Karpov et al., 2014; Spatafora et al., 2016).

Diagnostic Apomorphies: There are no unambiguous morphological, subcellular, or biochemical synapomorphies of *Fungi*. Most *Fungi* are filamentous, have chitinous cell walls, lack flagella, and have intranuclear mitosis with spindle pole bodies (instead of centrioles). However, there are also numerous unicellular forms (yeasts) scattered across the fungal phylogeny, thalli without hyphal growth developing from spores by cell division (*Laboulbeniomycetes*), and forms that develop centrioles and produce flagellated cells that lack cell walls during the motile part of their life cycles (the paraphyletic "chytrids": *Chytridiomycota, Neocallimastigomycota, Blastocladiomycota,* and *Rozella allomycis*). *Rozella, Microsporidia* and *Aphelida* are intracellular parasites of diverse eukaryotes. *Rozella* and *Aphelida* produce both zoosporic stages and endoparasitic amoeboid forms that appear to ingest cytoplasm of their hosts by phagocytosis, whereas *Microsporidia* lack a phagotrophic stage and infect hosts by a polar tube mechanism (James and Berbee, 2012; Corsaro et al., 2014; Karpov et al., 2014; Powell et al., 2017). *Rozella allomycis* may also employ enzymatic degradation to penetrate the host cell wall (Held, 1972). The *R. allomycis* genome encodes four division II chitin synthase genes, which are characteristic of other *Fungi*, including *Microsporidia* (James et al., 2013). However, division II chitin synthase genes have also been found in the holozoan protist *Corallochytrium limacisporum*, suggesting that they may be plesiomorphic in *Opisthokonta* (Torruella et al., 2015).

Synonyms: *Eumycota* sensu Barr (1992) is an approximate synonym.

Comments: Application of the name *Fungi* to this clade, and the choice of this name rather than its approximate synonym *Eumycota*, follows the phylogeny-based classification of Hibbett et al. (2007), which has been adopted in the *Dictionary of the Fungi* (Kirk et al., 2008) and the GenBank taxonomy (http://www.ncbi.nlm.nih.gov/guide/taxonomy). The delimitation of *Fungi* by Hibbett et al. (2007) was based largely on the phylogenetic analysis of James et al. (2006), which used six genes and recovered a clade containing *R. allomycis* and *Microsporidia* as the sister group of all other *Fungi*. Earlier analyses using α-tubulin and β-tubulin genes also placed *Microsporidia* within *Fungi* (Edlind et al., 1996; Keeling and Doolittle, 1996; Keeling, 2003). Recent studies using data derived from whole genomes or transcriptomes have consistently supported monophyly of the clade containing *Rozella* plus *Microsporidia* and have placed it as the sister group to the remaining *Fungi* (James et al., 2013; Torruella et al., 2015; Ren et al., 2016).

Several studies, including combined analyses of genes encoding ribosomal RNA (rRNA) and RNA polymerase II (*rpb*1 and *rpb*2), have suggested that the clade containing *Rozella* and *Microsporidia* also contains the endoparasitic *Aphelida* (Karpov et al., 2013; Corsaro et al., 2014; Letcher et al., 2015), collectively termed the "ARM clade" (Karpov et al., 2014). However, other analyses using rRNA genes only have placed *Aphelida* as the sister group of a clade containing *Rozella*, *Microsporidia*, and all other *Fungi* (Corsaro et al., 2016). The minimum-crown-clade definition of *Fungi*

proposed here employs multiple specifiers, but *R. allomycis* is the only specifier in the ARM clade. *Microsporidia* were not used as specifiers, because they have a dramatically elevated rate of molecular evolution (Corradi, 2015), and *Aphelida* were not used, because they are still represented only by a handful of genes. Nevertheless, current best estimates of the phylogeny suggest that *Microsporidia* and *Aphelida* are members of *Fungi* as defined here.

The sister group of *Fungi* (including *Aphelida*) appears to be a clade containing nucleariids and *Fonticula alba* (Brown et al., 2009; Paps et al., 2013; Torruella et al., 2015). The former are phagotrophic, nonflagellated, amoeboid protists that lack a cell wall, and the latter is a kind of cellular slime mold with aggregative, multicelluar reproductive structures that produces spores with cell walls lacking chitin. Berbee et al. (2017) suggested that the nucleariid-*F. alba* clade should be included in *Fungi*. However, most studies refer to the group containing *Fungi* and the nucleariid-*F. alba* clade as *Holomycota* (Liu et al., 2009; Paps et al., 2013; Corsaro et al., 2014; Karpov et al., 2014; Torruella et al., 2015), or, less often, *Nucletmycea* (Brown et al., 2009; Adl et al., 2012).

Karpov et al. (2014) named the ARM clade *Opisthosporidia* and suggested that it should be excluded from *Fungi*. However, *Rozella* has traditionally been considered a fungus based on morphological and ecological similarities to other chytrids, and *Microsporidia* have been widely regarded as members of *Fungi* ever since the early analyses using tubulin genes (Edlind et al., 1996; Keeling and Doolittle, 1996). Thus, the present definition preserves the composition of *Fungi* as it has come to be understood since the mid-1990s (e.g., James et al., 2006; Hibbett et al., 2007; Kirk et al., 2008; Spatafora et al., 2017), with the likely addition of *Aphelida* and other recently discovered members of the ARM clade (Jones et al., 2011). Moreover, evidence from comparative genomics and ultrastructural studies supports the view that members of the ARM clade are highly reduced and that their common ancestor may have been free-living and possessed many traits typically associated with *Fungi*, including chitinous cell walls and possibly osmoheterotrophy (Held, 1972; Keeling and Corradi, 2011; James et al., 2013; Berbee et al., 2017).

Literature Cited

Adl, S. M., A. G. Simpson, C. E. Lane, Julius Lukeš, D. Bass, S. S. Bowser, M. Brown, F. Burki, M. Dunthorn, V. Hampl, A. Heiss, M. Hoppenrath, E. Lara, L. le Gall, D. H. Lynn, H. McManus, E. A. D. Mitchell, S. E. Mozley-Stanridge, L. Wegener Parfrey, J. Pawlowski, S. Rueckert, L. Shadwick, C. Schoch, A. Smirnov, and F. W. Spiegel. 2012. The revised classification of eukaryotes. *J. Eukaryot. Microbiol.* 59:429–493.

Barr, D. J. S. 1992. Evolution and kingdoms of organisms from the perspective of a mycologist. *Mycologia* 84:1–11.

Berbee, M. L., T. Y. James, and C. Strullu-Derrien. 2017. Early-diverging fungi: diversity and impact at the dawn of terrestrial life. *Annu. Rev. Microbiol.* 71:41–60.

Brown, M. W., F. W. Spiegel, and J. D. Silberman. 2009. Phylogeny of the "forgotten" cellular slime mold, *Fonticula alba*, reveals a key evolutionary branch within *Opisthokonta*. *Mol. Biol. Evol.* 26:2699–2709.

Chang, Y., S. Wang, S. Sekimoto, A. L. Aerts, C. Choi, A. Clum, K. M. LaButti, E. A. Lindquist, C. Y. Ngan, R. A. Ohm, A. A. Salamov, I. V. Grigoriev, J. W. Spatafora, and M. L. Berbee. 2015. Phylogenomic analyses indicate that early fungi evolved digesting cell walls of algal ancestors of land plants. *Genome Biol. Evol.* 7:1590–1601.

Corradi, N. 2015. *Microsporidia*: eukaryotic intracellular parasites shaped by gene loss and horizontal gene transfers. *Annu. Rev. Microbiol.* 69:167–183.

Corsaro, D., R. Michel, J. Walochnik, D. Venditti, K.-D. Müller, B. Hauröder, and C. Wylezich. 2016. Molecular identification of *Nucleophaga terricolae* sp. nov. (*Rozellomycota*), and new insights on the origin of the *Microsporidia*. *Parasitol. Res.* 115:3003–3011.

Corsaro, D., J. Walochnik, D. Venditti, J. Steinmann, K.-D. Müller, and R. Michel. 2014. *Microsporidia*-like parasites of amoebae belong to the early fungal lineage *Rozellomycota*. *Parasitol. Res.* 113:1909–1918.

Edlind, T. D., J. Li, G. S. Visvesvara, M. H. Vodkin, G. L. McLaughlin, and S. K. Katiyar. 1996. Phylogenetic analysis of β-tubulin sequences from amitochondrial protozoa. *Mol. Phylogenet. Evol.* 5:359–367.

Engler, A., and K. Prantl. 1900. *Die natürlichen Pflanzenfamilien.* 1(1). Verlag von Wilhelm Engelmann, Leipzig.

Held, A. A. 1972. Host-parasite relations between *Allomyces* and *Rozella*. *Arch. Mikrobiol.* 82:128–139.

Hibbett, D. S., M. Binder, J. F. Bischoff, M. Blackwell, P. F. Cannon, O. E. Eriksson, S. Huhndorf, T. James, P. M. Kirk, R. Lücking, T. Lumbsch, F. Lutzoni, P. B. Matheny, D. J. Mclaughlin, M. J. Powell, S. Redhead, C. L. Schoch, J. W. Spatafora, J. A. Stalpers, R. Vilgalys, M. C. Aime, A. Aptroot, R. Bauer, D. Begerow, G. L. Benny, L. A. Castlebury, P. W. Crous, Y.-C. Dai, W. Gams, D. M. Geiser, G. W. Griffith, C. Gueidan, D. L. Hawksworth, G. Hestmark, K. Hosaka, R. A. Humber, K. Hyde, J. E. Ironside, U. Kõljalg, C. P. Kurtzman, K.-H. Larsson, R. Lichtwardt, J. Longcore, J. Miądlikowska, A. Miller, J.-M. Moncalvo, S. Mozley-Standridge, F. Oberwinkler, E. Parmasto, V. Reeb, J. D. Rogers, C. Roux, L. Ryvarden, J. P. Sampaio, A. Schüßler, J. Sugiyama, R. G. Thorn, L. Tibell, W. A. Untereiner, C. Walker, Z. Wang, A. Weir, M. Weiß, M. M. White, K. Winka, Y.-J. Yao, and N. Zhang. 2007. A higher-level phylogenetic classification of the *Fungi*. *Mycol. Res.* 111:509–547.

James, T. Y., and M. L. Berbee. 2012. No jacket required—new fungal lineage defies dress code. *Bioessays* 34:94–102.

James, T. Y., F. Kauff, C. Schoch, P. B. Matheny, V. Hofstetter, C. Cox, G. Celio, C. Gueidan, E. Fraker, J. Miadlikowska, H. T. Lumbsch, A. Rauhut, V. Reeb, A. E. Arnold, A. Amtoft, J. E. Stajich, K. Hosaka, G.-H. Sung, D. Johnson, B. O'Rourke, M. Crockett, M. Binder, J. M. Curtis, J. C. Slot, Z. Wang, A. W. Wilson, A. Schüßler, J. E. Longcore, K. O'Donnell, S. Mozley-Standridge, D. Porter, P. M. Letcher, M. J. Powell, J. W. Taylor, M. M. White, G. W. Griffith, D. R. Davies, R. A. Humber, J. B. Morton, J. Sugiyama, A. Y. Rossman, J. D. Rogers, D. H. Pfister, D. Hewitt, K. Hansen, S. Hambleton, R. A. Shoemaker, J. Kohlmeyer, B. Volkmann-Kohlmeyer, R. A. Spotts, M. Serdani, P. W. Crous, K. W. Hughes, K. Matsuura, E. Langer, G. Langer, W. A. Untereiner, R. Lücking, B. Büdel, D. M. Geiser, A. Aptroot, P. Diederich, I. Schmitt, M. Schultz, R. Yahr, D. Hibbett, F. Lutzoni, D. McLaughlin, J. Spatafora, and R. Vilgalys. 2006. Reconstructing the early evolution of *Fungi* using a six-gene phylogeny. *Nature* 443:818–822.

James, T. Y., A. Pelin, L. Bonen, S. Ahrendt, D. Sain, N. Corradi, and J. E. Stajich. 2013. Shared signatures of parasitism and phylogenomics unite *Cryptomycota* and *Microsporidia*. *Curr. Biol.* 23:1548–1553.

Jones, M. D. M., I. Forn, C. Gadelha, M. J. Egan, D. Bass, R. Massana, and T. A. Richards. 2011. Discovery of novel intermediate forms redefines the fungal tree of life. *Nature* 474:200–203.

Karpov, S. A., M. A. Mamkaeva, V. V. Aleoshin, E. Nassonova, O. Lilje, and F. H. Gleason. 2014. Morphology, phylogeny, and ecology of the aphelids (*Aphelidea, Opisthokonta*) and proposal for the new superphylum *Opisthosporidia*. *Front. Microbiol.* 5:1–11.

Karpov, S. A., K. V. Mikhailov, G. S. Mirzaeva, I. M. Mirabdullaev, K. A. Mamkaeva, N. N. Titova, and V. V. Aleoshin. 2013. Obligately phagotrophic aphelids turned out to branch with the earliest-diverging *Fungi*. *Protist* 164:195–205.

Keeling, P. J. 2003. Congruent evidence from alpha-tubulin and beta-tubulin gene phylogenies for a zygomycete origin of microsporidia. *Fungal Genet. Biol.* 38:298–309.

Keeling, P. J., and N. Corradi. 2011. Shrink it or lose it: balancing loss of function with shrinking genomes in the microsporidia. *Virulence* 2:67–70.

Keeling, P. J., and W. F. Doolittle. 1996. Alpha-tubulin from early-diverging eukaryotic lineages and the evolution of the tubulin family. *Mol. Biol. Evol.* 13:1297–1305.

Kirk, P. M., P. F. Cannon, D. W. Minter, and J. A. Stalpers. 2008. *Ainsworth & Bisby's Dictionary of the Fungi*. 10th edition. CAB International, Wallingford, UK.

Letcher, P. M., M. J. Powell, S. Lopez, P. A. Lee, and R. C. McBride. 2015. A new isolate of *Amoeboaphelidium protococcarum*, and *Amoeboaphelidium occidentale*, a new species in phylum *Aphelida* (*Opisthosporidia*). *Mycologia* 107:522–531.

Liu, Y., E. T. Steenkamp, H. Brinkmann, L. Forget, H. Phillipe, and B. F. Lang. 2009. Phylogenomic analyses predict sister group relationship of nucleariids and *Fungi* and paraphyly of zygomycetes with significant support. *BMC Evol. Biol.* 9:272.

Paps, J., L. A. Medina-Chacón, W. Marshall, H. Suga, and I. Ruiz-Trillo. 2013. Molecular phylogeny of unikonts: new insights into the position of apusomonads and ancyromonads and the internal relationships of opisthokonts. *Protist* 164:2–12.

Powell, M. J., P. M. Letcher, and T. Y. James. 2017. Ultrastructural characterization of the host-parasite interface between *Allomyces anomalus* (*Blastocladiomycota*) and *Rozella allomycis* (*Cryptomycota*). *Fungal Biol.* 121:561–572.

Ren, R., Y. Sun, Y. Zhao, D. Geiser, H. Ma, and X. Zhou. 2016. Phylogenetic resolution of deep eukaryotic and fungal relationships using highly conserved low-copy nuclear genes. *Genome Biol. Evol.* 8:2683–2701.

Spatafora, J. W., M. C. Aime, I. V. Grigoriev, F. Martin, J. E. Stajich, and M. Blackwell. 2017. The fungal tree of life: from molecular systematics to genome-scale phylogenies. *Microbiol. Spectr.* 5. doi:10.1128/microbiolspec.FUNK -0053-2016.

Spatafora, J. W., G. L. Benny, K. Lazarus, M. E. Smith, M. L. Berbee, G. Bonito, N. Corradi, I. Grigoriev, A. Gryganskyi, T. Y. James, K. O'Donnell, R. W. Roberson, T. N. Taylor, J.

Uehling, R. Vilgalys, and M. M. White. 2016. A phylum-level phylogenetic classification of zygomycete fungi based on genome-scale data. *Mycologia* 108:1028–1046.

Torruella, G., A. de Mendoza, X. Grau-Bové, M. Antó, M. A. Chaplin, J. del Campo, L. Eme, G. Pérez-Cordón, C. M. Whipps, K. M. Nichols, R. Paley, A. J. Roger, A. Sitjà-Bobadilla, S. Donachie, and I. Ruiz-Trillo. 2015. Convergent evolution of lifestyles in close relatives of animals and fungi. *Curr. Biol.* 25:2404–2410.

Authors

David S. Hibbett; Biology Department; Clark University; Worcester, MA 01610, USA. Email: dhibbett@clarku.edu.

Meredith Blackwell; Department of Biological Sciences; Lousiana State University; Baton Rouge, LA 70803, USA. Email: mblackwell@lsu.edu.

Timothy James; Department of Ecology and Evolutionary Biology; University of Michigan; Ann Arbor, MI 48109, USA. Email: tyjames@umich.edu.

Joseph W. Spatafora; Department of Botany and Plant Pathology; Oregon State University; Corvallis, OR 97331, USA. Email: spatafoj@science.oregonstate.edu.

John W. Taylor; Department of Plant and Microbial Biology; University of California; Berkeley, CA 94720, USA. Email: jtaylor@berkeley.edu.

Rytas Vilgalys; Biology Department; Duke University; Durham NC 27708, USA. Email: fungi@duke.edu.

Date Accepted: 22 July 2011; updated 22 October 2017

Primary Editor: Philip Cantino

Dikarya D. S. Hibbett, T. Y. James, and R. Vilgalys 2007: 518 (in Hibbett et al., 2007) [D. S. Hibbett, M. Blackwell, T. James, J. W. Spatafora, J. W. Taylor, and R. Vilgalys], converted clade name

Registration Number: 35

Definition: The smallest crown clade containing *Coprinopsis cinerea* (Schaeff.) Redhead, Vilgalys and Moncalvo 2001 (*Basidiomycota/Agaricomycotina*), *Saccharomyces cerevisiae* Meyen 1838 (*Ascomycota*), and *Entorrhiza casparyana* (Magnus) Lagerb. 1888 (*Entorrhizomycota*). This is a minimum-crown-clade definition. Abbreviated definition: min crown ∇ (*Coprinopsis cinerea* (Schaeff.) Redhead, Vilgalys and Moncalvo 2001 & *Saccharomyces cerevisiae* Meyen 1838 & *Entorrhiza casparyana* (Magnus) Lagerb. 1888).

Etymology: Derived from the Greek *di-* (two) and *karyon* (nut or kernel, interpreted by biologists to refer to nuclei).

Reference Phylogeny: The primary reference phylogeny is Bauer et al. (2015: Fig. 2). See also James et al. (2006: Fig. 1), Ebersberger et al. (2011: Fig. 3), Chang et al. (2015: Fig. 1), and Ren et al. (2016: Fig. 5).

Composition: *Ascomycota* and *Basidiomycota,* including *Entorrhizomycetes* (Hibbett et al., 2007).

Diagnostic Apomorphies: The dikaryotic condition, which results from cytoplasmic fusion of two haploid, monokaryotic hyphae, is the putative synapomorphy for which the group is named. Clamp connections of *Basidiomycota* and croziers of *Ascomycota*, which are cellular structures that function in the apportioning of nuclei to daughter cells following mitosis in dikaryotic hyphae, may be homologous and could represent an additional synapomorphy. Regularly septate hyphae are also probably a synapomorphy, because members of the candidate sister taxon, *Mucoromycota* (including *Glomeromycotina*; Spatafora et al., 2016, 2017), have predominantly coenocytic hyphae (Benny et al., 2014; Redecker and Schüßler, 2014; Hibbett et al., 2007). If clamps/croziers and septate hyphae of *Basidiomycota* and *Ascomycota* are homologous, then the ancestor of *Dikarya* must have been filamentous, and the unicellular forms (yeasts) that occur in multiple major clades of both *Ascomycota* and *Basidiomycota* were derived by reduction (Nagy et al., 2014).

Synonyms: *Carpomyceteae* Bessey 1907 (approximate), *Dikaryomycota* Kendrick 1985 (approximate), *Neomycota* Cavalier-Smith 1998 (approximate).

Comments: Application of the name *Dikarya* to this clade, and the choice of this name rather than one of the infrequently used synonyms *Dikaryomycota* and *Neomycota*, follow the phylogeny-based classification of Hibbett et al. (2007), which has been adopted in the *Dictionary of the Fungi* (Kirk et al., 2008) and the GenBank taxonomy (http://www.ncbi.nlm.nih.gov/guide/taxonomy). James et al. (2006) used the name *Dikarya* in the same sense as that proposed here, but the name was first validly published (according to the rules of the botanical code [Turland et al., 2018]) by Hibbett et al. (2007). Monophyly of *Dikarya* is strongly supported by independent and combined analyses of nuclear ribosomal RNA genes, RNA polymerase II subunits, and whole genomes

(James et al., 2006; Chang et al., 2015; Ren et al., 2016). The position of *Entorrhizomycetes* within *Dikarya* is not well resolved (see Comments under *Basidiomycota*).

Literature Cited

Bauer, R., S. Garnica, F. Oberwinkler, K. Riess, M. Weiß, and D. Begerow. 2015. *Entorrhizomycota*: a new fungal phylum reveals new perspectives on the evolution of *Fungi. PLOS ONE* 10(7): e0128183.

Benny, G. L., R. A. Humber, and K. Voigt. 2014. Zygomycetous fungi: phylum *Entomophthoromycota* and subphyla *Kickxellomycotina*, *Mortierellomycotina*, *Mucoromycotina*, and *Zoopagomycotina*. Pp. 209–250 in *The* Mycota: *A Comprehensive Treatise on Fungi as Experimental Systems for Basic and Applied Research*, Vol. VII, Part A (D. J. McLaughlin and J. W. Spatafora, eds.). Springer-Verlag, Berlin.

Bessey, C. E. 1907. A synopsis of the plant phyla. *Univ. Stud., University of Nebraska* 7(4):275–373.

Cavalier-Smith, T. 1998. A revised six-kingdom system of life. *Biol. Rev.* 73:203–266.

Chang, Y., S. Wang, S. Sekimoto, A. L. Aerts, C. Choi, A. Clum, K. M. LaButti, E. A. Lindquist, C. Y. Ngan, R. A. Ohm, A. A. Salamov, I. V. Grigoriev, J. W. Spatafora, and M. L. Berbee. 2015. Phylogenomic analyses indicate that early fungi evolved digesting cell walls of algal ancestors of land plants. *Genome Biol. Evol.* 7: 1590–1601.

Ebersberger, I., R. de Matos Simoes, A. Kupczok, M. Gube, E. Kothe, K. Voigt, and A. von Haeseler. 2011. A consistent phylogenetic backbone for the *Fungi. Mol. Biol. Evol.* 29:1319–1334.

Hibbett, D. S., M. Binder, J. F. Bischoff, M. Blackwell, P. F. Cannon, O. E. Eriksson, S. Huhndorf, T. James, P. M. Kirk, R. Lücking, T. Lumbsch, F. Lutzoni, P. B. Matheny, D. J. Mclaughlin, M. J. Powell, S. Redhead, C. L. Schoch, J. W. Spatafora, J. A. Stalpers, R. Vilgalys, M. C. Aime, A. Aptroot, R. Bauer, D. Begerow, G. L. Benny, L. A. Castlebury, P. W. Crous, Y.-C. Dai, W. Gams, D. M. Geiser, G. W. Griffith, C. Gueidan, D. L. Hawksworth, G. Hestmark, K. Hosaka, R. A. Humber, K. Hyde, J. E. Ironside, U. Kõljalg, C. P. Kurtzman, K.-H. Larsson, R. Lichtwardt, J. Longcore, J. Miądlikowska, A. Miller, J.-M. Moncalvo, S. Mozley-Standridge, F. Oberwinkler, E. Parmasto, V. Reeb, J. D. Rogers, C. Roux, L. Ryvarden, J. P. Sampaio, A. Schüßler, J. Sugiyama, R. G. Thorn, L. Tibell, W. A. Untereiner, C. Walker, Z. Wang, A. Weir, M. Weiß, M. M. White, K. Winka, Y.-J. Yao, and N. Zhang. 2007. A higher-level phylogenetic classification of the *Fungi. Mycol. Res.* 111:509–547.

James, T. Y., F. Kauff, C. Schoch, P. B. Matheny, V. Hofstetter, C. Cox, G. Celio, C. Gueidan, E. Fraker, J. Miadlikowska, H. T. Lumbsch, A. Rauhut, V. Reeb, A. E. Arnold, A. Amtoft, J. E. Stajich, K. Hosaka, G.-H. Sung, D. Johnson, B. O'Rourke, M. Crockett, M. Binder, J. M. Curtis, J. C. Slot, Z. Wang, A. W. Wilson, A. Schüßler, J. E. Longcore, K. O'Donnell, S. Mozley-Standridge, D. Porter, P. M. Letcher, M. J. Powell, J. W. Taylor, M. M. White, G. W. Griffith, D. R. Davies, R. A. Humber, J. B. Morton, J. Sugiyama, A. Y. Rossman, J. D. Rogers, D. H. Pfister, D. Hewitt, K. Hansen, S. Hambleton, R. A. Shoemaker, J. Kohlmeyer, B. Volkmann-Kohlmeyer, R. A. Spotts, M. Serdani, P. W. Crous, K. W. Hughes, K. Matsuura, E. Langer, G. Langer, W. A. Untereiner, R. Lücking, B. Büdel, D. M. Geiser, A. Aptroot, P. Diederich, I. Schmitt, M. Schultz, R. Yahr, D. Hibbett, F. Lutzoni, D. McLaughlin, J. Spatafora, and R. Vilgalys. 2006. Reconstructing the early evolution of *Fungi* using a six-gene phylogeny. *Nature* 443:818–822.

Kendrick, B. 1985. *The Fifth Kingdom*. Mycologue Publications, Waterloo.

Kirk, P. M., P. F. Cannon, D. W. Minter, and J. A. Stalpers. 2008. *Ainsworth & Bisby's Dictionary of the* Fungi. 10th edition. CAB International, Wallingford, UK.

Nagy, L. G., R. A. Ohm, G. M. Kovács, D. Floudas, R. Riley, A. Gácser, M. Sipiczki, J. M. Davis, S. L. Doty, G. S. de Hoog, J. W. Spatafora, F. Martin, I. V. Grigoriev, and D. S. Hibbett. 2014. Phylogenomics reveals latent homology behind the convergent evolution of yeast forms. *Nat. Commun.* 5:4471. doi:10.1038/ncomms5471.

Redecker, D., and A. Schüßler. 2014. *Glomeromycota*. Pp. 251–269 in *The Mycota: A Comprehensive Treatise on* Fungi *as Experimental Systems for Basic and Applied Research*, Vol. VII, Part A (D. J. McLaughlin and J. W. Spatafora, eds.). Springer-Verlag, Berlin.

Ren, R., Y. Sun, Y. Zhao, D. Geiser, H. Ma, and X. Zhou. 2016. Phylogenetic resolution of deep eukaryotic and fungal relationships using highly conserved low-copy nuclear genes. *Genome Biol. Evol.* 8:2683–2701.

Spatafora, J. W., M. C. Aime, I. V. Grigoriev, F. Martin, J. E. Stajich, and M. Blackwell. 2017. The fungal tree of life: from molecular systematics to genome-scale phylogenies. *Microbiol. Spectr.* 5. doi:10.1128/microbiolspec.FUNK-0053-2016.

Spatafora, J. W., G. L. Benny, K. Lazarus, M. E. Smith, M. L. Berbee, G. Bonito, N. Corradi, I. Grigoriev, A. Gryganskyi, T. Y. James, K. O'Donnell, R. W. Roberson, T. N. Taylor, J. Uehling, R. Vilgalys, and M. M. White. 2016. A phylum-level phylogenetic classification of zygomycete fungi based on genome-scale data. *Mycologia* 108:1028–1046.

Turland, N. J., J. H. Wiersema, F. R. Barrie, W. Greuter, D. L. Hawksworth, P. S. Herendeen, S. Knapp, W.-H. Kusber, D.-Z. Li, K. Marhold, T. W. May, J. McNeill, A. M. Monro, J. Prado, M. J. Price, and G. F. Smith, eds. 2018. *International Code of Nomenclature for Algae, Fungi, and Plants (Shenzhen Code)*. Adopted by the Nineteenth International Botanical Congress, Shenzhen, China, July 2017. Regnum Vegetabile 159. Koeltz Botanical Books, Glashütten.

Authors

David S. Hibbett; Biology Department; Clark University; Worcester, MA 01610, USA. Email: dhibbett@clarku.edu.

Meredith Blackwell; Department of Biological Sciences; Lousiana State University; Baton Rouge, LA 70803, USA. Email: mblackwell@lsu.edu.

Timothy James; Department of Ecology and Evolutionary Biology; University of Michigan; Ann Arbor, MI 48109, USA. Email: tyjames@umich.edu.

Joseph W. Spatafora; Department of Botany and Plant Pathology; Oregon State University; Corvallis, OR 97331, USA. Email: spatafoj@science.oregonstate.edu.

Rytas Vilgalys; Biology Department; Duke University; Durham, NC 27708, USA. Email: fungi@duke.edu.

Date Accepted: 22 July 2011; updated 22 October 2017

Primary Editor: Philip Cantino

Basidiomycota H. C. Bold 1957: 199 [D. S. Hibbett, T. James, and R. Vilgalys], converted clade name

Registration Number: 19

Definition: The largest crown clade containing *Coprinopsis cinerea* (Schaeff.) Redhead, Vilgalys and Moncalvo 2001, but not *Taphrina wiesneri* (Ráthay) Mix 1954, and *Saccharomyces cerevisiae* Meyen 1838, and *Peziza vesiculosa* Bull 1790. This is a maximum-crown-clade definition. Abbreviated definition: max crown ∇ (*Coprinopsis cinerea* (Schaeff.) Redhead, Vilgalys and Moncalvo 2001 ~ *Taphrina wiesneri* (Ráthay) Mix 1954 & *Saccharomyces cerevisiae* Meyen 1838 & *Peziza vesiculosa* Bull 1790).

Etymology: Derived from the Latin *basis* (base, support) plus diminutive suffix *-idium*, referring to the basidium, a "little pedestal", on which the basidiospores are supported, plus the Greek *mykes* (fungus).

Reference Phylogeny: The primary reference phylogeny is James et al. (2006: Fig. 1). See also Bauer et al. (2015: Fig. 2), Nagy et al. (2016: Fig. 1), and Zhao et al. (2017: Fig. 3).

Composition: *Pucciniomycotina, Ustilaginomycotina, Agaricomycotina* (Hibbett et al., 2007). *Entorrhizomycetes* may also be in *Basidiomycota* (Bauer et al., 2015); see Comments.

Diagnostic Apomorphies: A prolonged, free-living dikaryotic mycelium and the production of meiospores on basidia are putative synapomorphies, although *Basidiomycota* also includes asexual taxa and unicellular forms (yeasts).

Synonyms: *Basidiomycetes* sensu Whittaker (1959) (approximate). *Basidiomycotina* sensu Ainsworth (1973) is a partial synonym because the asexual basidiomycetes were excluded and assigned instead (along with other asexual fungi) to *Deuteromycotina*.

Comments: Application of the name *Basidiomycota* to this clade, and the choice of this name rather than the synonyms *Basidiomycetes* and *Basidiomycotina*, follow the phylogeny-based classification of Hibbett et al. (2007), which has been adopted in the *Dictionary of the Fungi* (Kirk et al., 2008) and the GenBank taxonomy (http://www.ncbi.nlm.nih.gov/guide/taxonomy). Monophyly of *Basidiomycota* has been strongly supported in phylogenetic analyses of multi-locus molecular data (James et al., 2006), including genome-based datasets (Nagy et al., 2016; Zhao et al. 2017), and was also corroborated in an analysis of non-molecular characters (Tehler, 1988). Three major subclades, *Pucciniomycotina* (rusts and relatives), *Ustilaginomycotina* (smuts and relatives), and *Agaricomycotina* (mushrooms, jelly fungi, and relatives), are resolved in most analyses (Aime et al., 2014; Begerow et al., 2014; Hibbett et al., 2014).

The relationship of *Entorrhizomycetes* to *Basidiomycota* is controversial (Matheny et al., 2006; Bauer et al., 2015; Zhao et al., 2017). *Entorrhizomycetes* includes root-gall fungi with similarities to certain *Basidiomycota*, including dolipore septa, dikaryotic vegetative cells, and teliospores with cruciate septation (Bauer et al., 2015). *Entorrhizomycetes* have been classified in *Ustilaginomycotina* (Begerow et al., 2006), but phylogenetic analyses of nuclear ribosomal RNA genes, alone or in combination with RNA polymerase II subunits 1 and 2 (*rpb*1, *rpb*2), suggest that it could be the sister group of all other *Basidiomycota* or of *Dikarya* (Matheny et al., 2006; Bauer et al., 2015; Zhao et al., 2017). Bauer et al. (2015) classified *Entorrhizomycetes* in its own phylum, *Entorrhizomycota*. There are still no whole-genome sequences available for *Entorrhizomycetes*.

Literature Cited

Aime, M. C., M. Toome, and D. J. McLaughlin. 2014. *Pucciniomycotina*. Pp. 271–294 in *The Mycota: A Comprehensive Treatise on* Fungi *as Experimental Systems for Basic and Applied Research*, Vol. VII, Part A (D. J. McLaughlin and J. W. Spatafora, eds.). Springer-Verlag, Berlin.

Ainsworth, G. C. 1973. Introduction and keys to higher taxa. Pp. 1–7 in *The Fungi: An Advanced Treatise*, Vol. 4A (G. C. Ainsworth, F. K. Sparrow, and A. S. Sussman, eds.). Academic Press, New York.

Bauer, R., S. Garnica, F. Oberwinkler, K. Riess, M. Weiß, and D. Begerow. 2015. *Entorrhizomycota*: a new fungal phylum reveals new perspectives on the evolution of *Fungi*. *PLOS ONE* 10(7):e0128183.

Begerow, D., A. M. Schäfer, R. Kellner, A. Yurkov, M. Kemler, F. Oberwinkler, and R. Bauer. 2014. *Ustilaginomycotina*. Pp. 295–329 in *The Mycota: A Comprehensive Treatise on Fungi as Experimental Systems for Basic and Applied Research*, Vol. VII, Part A (D. J. McLaughlin and J. W. Spatafora, eds.). Springer-Verlag, Berlin.

Begerow, D., M. Stoll, and R. Bauer. 2006. A phylogenetic hypothesis of *Ustilaginomycotina* based on multiple gene analyses and morphological data. *Mycologia* 98:906–916.

Bold, H. C. 1957. *Morphology of Plants*. Harper & Row, New York.

Hibbett, D. S., R. Bauer, M. Binder, A. J. Giachini, K. Hosaka, A. Justo, E. Larsson, K. H. Larsson, J. D. Lawrey, O. Miettinen, L. G. Nagy, R. H. Nilsson, M. Weiß, and R. G. Thorn. 2014. *Agaricomycotina*. Pp. 373–429 in *The* Mycota: *A Comprehensive Treatise on Fungi as Experimental Systems for Basic and Applied Research*, Vol. VII. Part A (D. J. McLaughlin and J. W. Spatafora, eds.). Springer-Verlag, Berlin.

Hibbett, D. S., M. Binder, J. F. Bischoff, M. Blackwell, P. F. Cannon, O. E. Eriksson, S. Huhndorf, T. James, P. M. Kirk, R. Lücking, T. Lumbsch, F. Lutzoni, P. B. Matheny, D. J. Mclaughlin, M. J. Powell, S. Redhead, C. L. Schoch, J. W. Spatafora, J. A. Stalpers, R. Vilgalys, M. C. Aime, A. Aptroot, R. Bauer, D. Begerow, G. L. Benny, L. A. Castlebury, P. W. Crous, Y.-C. Dai, W. Gams, D. M. Geiser, G. W. Griffith, C. Gueidan, D. L. Hawksworth, G. Hestmark, K. Hosaka, R. A. Humber, K. Hyde, J. E. Ironside, U. Kõljalg, C. P. Kurtzman, K.-H. Larsson, R. Lichtwardt, J. Longcore, J. Miądlikowska, A. Miller, J.-M. Moncalvo, S. Mozley-Standridge, F. Oberwinkler, E. Parmasto, V. Reeb, J. D. Rogers, C. Roux, L. Ryvarden, J. P. Sampaio, A. Schüßler, J. Sugiyama, R. G. Thorn, L. Tibell, W. A. Untereiner, C. Walker, Z. Wang, A. Weir, M. Weiß, M. M. White, K. Winka, Y.-J. Yao, and N. Zhang. 2007. A higher-level phylogenetic classification of the *Fungi*. *Mycol. Res.* 111:509–547.

James, T. Y., F. Kauff, C. Schoch, P. B. Matheny, V. Hofstetter, C. Cox, G. Celio, C. Gueidan, E. Fraker, J. Miadlikowska, H. T. Lumbsch, A. Rauhut, V. Reeb, A. E. Arnold, A. Amtoft, J. E. Stajich, K. Hosaka, G.-H. Sung, D. Johnson, B. O'Rourke, M. Crockett, M. Binder, J. M. Curtis, J. C. Slot, Z. Wang, A. W. Wilson, A. Schüßler, J. E. Longcore, K. O'Donnell, S. Mozley-Standridge, D. Porter, P. M. Letcher, M. J. Powell, J. W. Taylor, M. M. White, G. W. Griffith, D. R. Davies, R. A. Humber, J. B. Morton, J. Sugiyama, A. Y. Rossman, J. D. Rogers, D. H. Pfister, D. Hewitt, K. Hansen, S. Hambleton, R. A. Shoemaker, J. Kohlmeyer, B. Volkmann-Kohlmeyer, R. A. Spotts, M. Serdani, P. W. Crous, K. W. Hughes, K. Matsuura, E. Langer, G. Langer, W. A. Untereiner, R. Lücking, B. Büdel, D. M. Geiser, A. Aptroot, P. Diederich, I. Schmitt, M. Schultz, R. Yahr, D. Hibbett, F. Lutzoni, D. McLaughlin, J. Spatafora, and R. Vilgalys.

2006. Reconstructing the early evolution of *Fungi* using a six-gene phylogeny. *Nature* 443:818–822.

Kirk, P. M., P. F. Cannon, D. W. Minter, and J. A. Stalpers. 2008. *Ainsworth & Bisby's Dictionary of the* Fungi. 10th edition. CAB International, Wallingford, UK.

Matheny, P. B., J. A. Gossman, P. Zalar, T. K. Arun Kumar, and D. S. Hibbett. 2006. Resolving the phylogenetic position of the *Wallemiomycetes*: an enigmatic major lineage of *Basidiomycota*. *Can. J. Bot.* 84:1794–1805.

Nagy, L. G., R. Riley, A. Tritt, C. Adam, C. Daum, D. Floudas, H. Sun, J. S. Yadav, J. Pangilinan, K.-H. Larsson, K. Matsuura, K. Barry, K. LaButti, R. Kuo, R. Ohm, S. S. Bhattacharya, T. Shirouzu, Y. Yoshinaga, F. M. Martin, I. V. Grigoriev, and D. S. Hibbett. 2016. Comparative genomics of early-diverging mushroom-forming fungi provides insights into the origins of lignocellulose decay capabilities. *Mol. Biol. Evol.* 33:959–970.

Tehler, A. 1988. A cladistic outline of the *Eumycota*. *Cladistics* 4:227.

Whittaker, R. H. 1959. On the broad classifications of organisms. *Q. Rev. Biol.* 34:210–226.

Zhao, R.-L., G.-J. Li, S. Sánchez-Ramírez, M. Stata, Z.-L. Yang, G. Wu, Y.-C. Dai, S.-H. He, B.-K. Cui, J.-L. Zhou, F. Wu, M.-Q. He, J.-M. Moncalvo, and K. D. Hyde. 2017. A six-gene phylogenetic overview of *Basidiomycota* and allied phyla with estimated divergence times of higher taxa and a phyloproteomics perspective. *Fungal Divers.* 84:43–74.

Authors

David S. Hibbett; Biology Department; Clark University; Worcester, MA 01610, USA. Email: dhibbett@clarku.edu.

Timothy James; Department of Ecology and Evolutionary Biology; University of Michigan; Ann Arbor, MI 48109, USA. Email: tyjames@umich.edu.

Rytas Vilgalys; Biology Department; Duke University; Durham NC 27708, USA. Email: fungi@duke.edu.

Date Accepted: 22 July 2011; updated 2 September 2017

Primary Editor: Philip Cantino

Ascomycota H. C. Bold 1957: 180, 616 [J.W. Spatafora, M. Blackwell, and J.W. Taylor], converted clade name

Registration Number: 17

Definition: The largest crown clade containing *Taphrina deformans* (Berk.) Tul. 1866, but not *Puccinia graminis* Pers. 1794, and *Ustilago tritici* (Bjerk.) E. Rostrup 1890, and *Agaricus bisporus* (J.E. Lange) Imbach 1946, and *Entorrhiza casparyana* (Magnus) Lagerb 1888. This is a maximum-crown-clade definition. Abbreviated definition: max crown ∇ (*Taphrina deformans* (Berk.) Tul. 1866 ~ *Puccinia graminis* Pers. 1794 & *Ustilago tritici* (Bjerk.) E. Rostrup 1890 & *Agaricus bisporus* (J.E. Lange) Imbach 1946 & *Entorrhiza casparyana* (Magnus) Lagerb 1888).

Etymology: Derived from the Greek *askos* (sac) + *mykes* (fungus).

Reference Phylogeny: The primary reference phylogeny is Bauer et al. (2015: Fig. 2). See also Lutzoni et al. (2004: Fig. 2), Liu et al. (2008: Fig. 1), James et al. (2006: Fig. 1), Schoch et al. (2009: Fig. S6), Carbone et al. (2017: Fig. 1), and Spatafora et al. (2017: Fig. 1).

Composition: *Taphrinomycotina, Saccharomycotina* and *Pezizomycotina* (Hibbett et al., 2007).

Diagnostic Apomorphies: Morphological synapomorphies of *Ascomycota* include the formation of meiospores (ascospores) within sac-shaped meiosporangia (asci) by the process of free cell formation. Free cell formation involves the production of an enveloping membrane system, which is derived from either the ascus plasmalemma or the nuclear envelope and delimits ascospore initials. Meiotic reproduction is unknown in many species and may have been lost in some. All *Ascomycota* lack flagella and exhibit intranuclear mitosis with spindle pole bodies instead of centrioles (Kumar et al., 2011). Most *Ascomycota* are filamentous with simple septa, but there are numerous yeasts (unicellular forms) especially in the *Taphrinomycotina* (Healy et al., 2013) and *Saccharomycotina* and dimorphic species (capable of both yeast and filamentous growth) in *Pezizomycotina, Taphrinomycotina* and *Saccharomycotina*.

Synonyms: *Ascomycetes* sensu Whittaker (1959) (approximate). *Ascomycotina* sensu Ainsworth (1973) is a partial synonym because the asexual ascomycetes were excluded and assigned instead (along with other asexual fungi) to *Deuteromycotina*.

Comments: Application of the name *Ascomycota* to this clade, and the choice of this name rather than the synonyms *Ascomycetes* and *Ascomycotina*, follow the phylogeny-based classification of Hibbett et al. (2007), which has been adopted in the *Dictionary of the Fungi* (Kirk et al., 2008) and the GenBank taxonomy (http://www.ncbi.nlm.nih.gov/guide/taxonomy). In rank-based classifications (e.g., Kirk et al., 2008; Spatafora et al., 2017), the clade *Ascomycota* is the largest phylum of *Fungi*. It is supported in molecular phylogenetic analyses (Lutzoni et al., 2004; James et al., 2006; Schoch et al., 2009) and comprises three mutually exclusive subclades (Spatafora et al., 2006; Schoch et al., 2009; Carbone et al., 2017). *Taphrinomycotina* is sister group to a well-supported clade comprising *Saccharomycotina* and *Pezizomycotina*. *Pezizomycotina* includes all ascocarp-producing taxa with the exception of *Neolectomycetes* of *Taphrinomycotina*. The monophyly of *Taphrinomycotina* was not supported by early analyses of ribosomal data (reviewed in Sugiyama et al., 2006), but sampling of protein coding loci (RPB1, RPB2, and TEF) and mitochondrial DNA in multi-gene analyses provided support for its monophyly (James et al., 2006; Spatafora et al., 2006; Liu et al., 2008). *Saccharomycotina* (Riley et al., 2016; Shen et al., 2016) and *Pezizomycotina* (Spatafora et al., 2006; Schoch et al., 2009; Kumar et al., 2012; Carbone et al., 2017) are both well-supported clades. The sister group of *Ascomycota* is *Basidiomycota* (James et al., 2006). The fossil record of *Ascomycota* dates to at least the Devonian, with *Paleopyrenomycites* identified as part of the Rhynie Chert fossil fungi (Taylor et al., 2005), but putative ascomycete fossils have been reported from the Silurian (Sherwood-Pike and Gray, 1985). Efforts to fit molecular phylogenies to the fossil record have estimated the origin of *Ascomycota* to be between 0.40 to 1.3 billion years before the present (Heckman et al., 2001; Taylor and Berbee, 2006; Lücking et al., 2009).

Literature Cited

Ainsworth, G. C. 1973. Introduction and keys to higher taxa. Pp. 1–7 in *The* Fungi: *An Advanced Treatise*, Vol. 4A (G. C. Ainsworth, F. K. Sparrow, and A. S. Sussman, eds.). Academic Press, New York.

Bauer, R., S. Garnica, F. Oberwinkler, K. Riess, M. Weiß, and D. Begerow. 2015. *Entorrhizomycota*: a new fungal phylum reveals new perspectives on the evolution of *Fungi*. *PLOS ONE* 10(7):e0128183.

Bold, H. C. 1957. *Morphology of Plants*. Harper & Row, New York.

Carbone, I., J. B. White, J. Miadlikowska, A. E. Arnold, M. A. Miller, F. Kauff, J. M. U'Ren, G. May, and F. Lutzoni. 2017. T-BAS: Tree-Based Alignment Selector toolkit for phylogenetic-based placement, alignment downloads and metadata visualization: an example with the *Pezizomycotina* tree of life. *Bioinformatics* 33:1160–1168.

Healy, R. A., T. K. Kumar, D. A. Hewitt, and D. L. McLaughlin. 2013. Functional and phylogenetic implications of septal pore ultrastructure in the ascoma of *Neolecta vitellina*. *Mycologia* 105:802–813.

Heckman, D. S., D. M. Geiser, B. R. Eidell, R. L. Stauffer, N. L. Kardos, and S. B. Hedges. 2001. Molecular evidence for the early colonization of land by fungi and plants. *Science* 293:1129–1133.

Hibbett, D. S., M. Binder, J. F. Bischoff, M. Blackwell, P. F. Cannon, O. E. Eriksson, S. Huhndorf, T. James, P. M. Kirk, R. Lücking, T. Lumbsch, F. Lutzoni, P. B. Matheny, D. J. Mclaughlin, M. J. Powell, S. Redhead, C. L. Schoch, J. W. Spatafora, J. A. Stalpers, R. Vilgalys, M. C. Aime, A. Aptroot, R. Bauer, D. Begerow, G. L. Benny, L. A. Castlebury, P. W. Crous, Y.-C. Dai, W. Gams, D. M. Geiser, G. W. Griffith, C. Gueidan, D. L. Hawksworth, G. Hestmark, K. Hosaka, R. A. Humber, K. Hyde, J. E. Ironside, U. Kõljalg, C. P. Kurtzman, K.-H. Larsson, R. Lichtwardt, J. Longcore, J. Miądlikowska, A. Miller, J.-M. Moncalvo, S. Mozley-Standridge, F. Oberwinkler, E. Parmasto, V. Reeb, J. D. Rogers, C. Roux, L. Ryvarden, J. P. Sampaio, A. Schüßler, J. Sugiyama, R. G. Thorn, L. Tibell, W. A. Untereiner, C. Walker, Z. Wang, A. Weir, M. Weiß, M. M. White, K. Winka, Y.-J. Yao, and N. Zhang. 2007. A higher-level phylogenetic classification of the *Fungi*. *Mycol. Res.* 111:509–547.

James, T. Y., F. Kauff, C. Schoch, P. B. Matheny, V. Hofstetter, C. Cox, G. Celio, C. Gueidan, E. Fraker, J. Miadlikowska, H. T. Lumbsch, A. Rauhut, V. Reeb, A. E. Arnold, A. Amtoft, J. E. Stajich, K. Hosaka, G.-H. Sung, D. Johnson, B. O'Rourke, M. Crockett, M. Binder, J. M. Curtis, J. C. Slot, Z. Wang, A. W. Wilson, A. Schüßler, J. E. Longcore, K. O'Donnell, S. Mozley-Standridge, D. Porter, P. M. Letcher, M. J. Powell, J. W. Taylor, M. M. White, G. W. Griffith, D. R. Davies, R. A. Humber,

J. B. Morton, J. Sugiyama, A. Y. Rossman, J. D. Rogers, D. H. Pfister, D. Hewitt, K. Hansen, S. Hambleton, R. A. Shoemaker, J. Kohlmeyer, B. Volkmann-Kohlmeyer, R. A. Spotts, M. Serdani, P. W. Crous, K. W. Hughes, K. Matsuura, E. Langer, G. Langer, W. A. Untereiner, R. Lücking, B. Büdel, D. M. Geiser, A. Aptroot, P. Diederich, I. Schmitt, M. Schultz, R. Yahr, D. Hibbett, F. Lutzoni, D. McLaughlin, J. Spatafora, and R. Vilgalys. 2006. Reconstructing the early evolution of *Fungi* using a six-gene phylogeny. *Nature* 443:818–822.

Kirk, P. M., P. F. Cannon, D. W. Minter, and J. A. Stalpers. 2008. *Ainsworth & Bisby's Dictionary of the Fungi*. 10th edition. CAB International, Wallingford, UK.

Kumar, T. K. A., J. A. Crow, T. J. Wennblom, M. Abril, P. M. Letcher, M. Blackwell, R. W. Roberson, and D. J. McLaughlin. 2011. An ontology of fungal subcellular traits. *Am. J. Bot.* 98:1504–1510.

Kumar, T. K. A., R. Healy, J. W. Spatafora, M. Blackwell, and D. J. McLaughlin. 2012. *Orbilia* ultrastructure, character evolution and phylogeny of *Pezizomycotina*. *Mycologia* 104:462–476.

Liu, Y., J. W. Leigh, H. Brinkmann, M. T. Cushion, N. Rodriguez-Ezpeleta, H. Philippe, and B. F. Lang. 2008. Phylogenomic analyses support the monophyly of *Taphrinomycotina*, including *Schizosaccharomyces* fission yeasts. *Mol. Biol. Evol.* 26:27–34.

Lücking, R., S. Huhndorf, D. Pfister, E. R. Plata, and H. Lumbsch. 2009. Fungi evolved right on track. *Mycologia* 101:810–822.

Lutzoni, F., F. Kauff, C. J. Cox, D. McLaughlin, G. Celio, B. Dentinger, M. Padamsee, D. Hibbett, T. Y. James, E. Baloch, M. Grube, V. Reeb, V. Hofstetter, C. Schoch, A. E. Arnold, J. Miadlikowska, J. Spatafora, D. Johnson, S. Hambleton, M. Crockett, R. Shoemaker, G.-H. Sung, R. Lücking, T. Lumbsch, K. O'Donnell, M. Binder, P. Diederich, D. Ertz, C. Gueidan, K. Hansen, R. C. Harris, K. Hosaka, Y.-W. Lim, B. Matheny, H. Nishida, D. Pfister, J. Rogers, A. Rossman, I. Schmitt, H. Sipman, J. Stone, J. Sugiyama, R. Yahr, and R. Vilgalys. 2004. Assembling the fungal tree of life: progress, classification, and evolution of subcellular traits. *Am. J. Bot.* 91:1446–1480.

Riley, R., S. Haridas, K. H. Wolfe, M. R. Lopes, C. T. Hittinger, M. Göker, A. Salamov, J. Wisecaver, T. M. Long, A. L. Aerts, K. Barry, C. Choi, A. Clum, A. Y. Coughlan, S. Deshpande, A. P. Douglass, S. J. Hanson, H.-P. Klenk, K. LaButti, A. Lapidus, E. Lindquist, A. Lipzen, J. P. Meier-Kolthoff, R. A. Ohm, R. P. Otillar, J. P., Y. Peng, A. Rokas, C. A. Rosa, C. Scheuner, A. A. Sibirny, J. C. Slot, J. B. Stielow, H. Sun, C. P. Kurtzman, M. Blackwell, I. V. Grigoriev, and T. W. Jeffries. 2016. Comparative genomics of biotechnologically important yeasts. *Proc. Natl. Acad. Sci. USA* 113:9882–887.

Schoch, C. L., G.-H. Sung, F. L. López-Giráldez, J. P. Townsend, J. Miadlikowska, V. Rie Hofstetter, B. Robbertse, P. B. Matheny, F. Kauff, Z. Wang, C. Gueidan, R. M. Andrie, K. Trippe, L. M. Ciufetti, A. Wynns, E. Fraker, B. P. Hodkinson, G. Bonito, J. Z. Groenewald, M. Arsanlou, G. S. De Hoog, P. W. Crous, D. Hewitt, D. H. Pfister, K. Peterson, M. Grysenhout, M. J. Wingfield, A. Aptroot, S.-O. Suh, M. Blackwell, D. M. Hillis, G. W. Griffith, L. A. Castlebury, A. Y. Rossman, H. T. Lumbsch, R. L. Lücking, B. Büdel, A. Rauhut, P. Diederich, D. Ertz, D. M. Geiser, K. Hosaka, P. Inderbitzin, J. Kohlmeyer, B. Volkmann-Kohlmeyer, L. Mostert, K. O'Donnell, H. Sipman, J. D. Rogers, R. A. Shoemaker, J. Sugiyama, R. C. Summerbell, W. Untereiner, P. R. Johnston, S. Stenroos, A. Zuccaro, P. S. Dyer, P. D. Crittenden, M. S. Cole, K. Hansen, J. M. Trappe, R. Yahr, F. Lutzoni, and J. W. Spatafora. 2009. The *Ascomycota* tree of life: a phylum-wide phylogeny clarifies the origin and evolution of fundamental reproductive and ecological traits. *Syst. Biol.* 58:224–239.

Shen, X.-X., X. Zhou, J. Kominek, C. P. Kurtzman, C. T. Hittinger, and A. Rokas. 2016. Reconstructing the backbone of the *Saccharomycotina* yeast phylogeny using genome-scale data. *G3: Gene. Genom. Genet.* 6:3927–3939.

Sherwood-Pike, M. A., and J. Gray. 1985. Silurian fungal remains: probable records of the Class *Ascomycetes*. *Lethaia* 18:1–20.

Spatafora, J. W., D. Johnson, G.-H. Sung, K. Hosaka, B. O'Rourke, M. Serdani, R. Spotts, F. Lutzoni, V. Hofstette, E. Fraker, C. Gueidan, J. Miadlikowska, V. Reeb, T. Lumbsch, R. Lücking, I. Schmitt, A. Aptroot, C. Roux, A. Miller, D. Geiser, J. Hafellner, G. Hestmark, A. E. Arnold, B. Büdel, A. Rauhut, D. Hewitt, W. Untereiner, M. S. Cole, C. Scheidegger, M. Schultz, H. Sipman, and C. Schoch. 2006. A five-gene phylogenetic analysis of the *Pezizomycotina*. *Mycologia* 98:1018–1028.

Spatafora, J. W., M. C. Aime, I. V. Grigoriev, F. Martin, J. E. Stajich, and M. Blackwell. 2017. The fungal tree of life: from molecular systematics to genome-scale phylogenies. *Microbiol. Spect.* 5(5):FUNK-0053–2016. doi:10.1128/microbiolspec.FUNK-0053-2016.

Sugiyama, J., K. Hosaka, and S.-O. Suh. 2006. Early diverging *Ascomycota*: phylogenetic divergence and related evolutionary enigmas. *Mycologia* 98:998–1007.

Taylor, J. W., and M. L. Berbee. (2006). Dating divergences in the fungal tree of life: review and new analyses. *Mycologia* 98:838–849.

Taylor, T. N., H. Hass, H. Kerp, M. Krings, and R. T. Hanlin. 2005. Perithecial ascomycetes from the 400 million year old Rhynie chert: an example of ancestral polymorphism (Vol. 96, p 1403, 2004). *Mycologia* 97:269–285.

Whittaker, R. H. 1959. On the broad classifications of organisms. *Q. Rev. Biol.* 34:210–226.

Authors

Joseph W. Spatafora; Department of Botany and Plant Pathology; Oregon State University; Corvallis, OR 97331, USA. Email: spatafoj@science.oregonstate.edu.

John W. Taylor; Department of Plant and Microbial Biology; University of California; Berkeley, CA 94720-3102, USA. Email: jtaylor@nature.berkeley.edu.

Meredith Blackwell; Department of Biology; Louisiana State University; Baton Rouge, LA 70803; and Department of Biological Sciences; University of South Carolina; Columbia, SC 29208, USA. Email: mblackwell@lsu.edu.

David S. Hibbett; Biology Department; Clark University; Worcester, MA 01610, USA. Email: dhibbett@clarku.edu.

Date Accepted: 22 July 2011; updated 7 December 2017

Primary Editor: Philip Cantino

SECTION 2

Archaeplastida S. M. Adl et al. 2005 [A. G. B. Simpson], converted clade name

Registration Number: 16

Definition: The crown clade for which possession of primary plastids, as inherited by *Arabidopsis thaliana* (Linnaeus) Heynhold 1842, is an apomorphy relative to other crown clades, provided that it includes *Galdieria sulphuraria* (Galdieri) Merola in Merola et al. 1981 (*Rhodoplantae*) and excludes *Homo sapiens* Linnaeus 1758. In this definition, "plastids" refers to the organelles also known as chloroplasts in the broadest sense, that is, all homologous organelles irrespective of appearance or function; "an apomorphy" means it stems directly from the same event of primary endosymbiosis and does not include later origin through a subsequent event of secondary endosymbiosis (see Comments). This is an apomorphy-modified crown-clade definition with a qualifying clause. Abbreviated definition: crown ∇ apo plastids derived through primary endosymbiosis [*Arabidopsis thaliana* (Linnaeus) Heyhhold 1842] | (*Galdieria sulphuraria* (Galdieri) Merola in Merola et al. 1981) & ~ (*Homo sapiens* Linnaeus 1758).

Etymology: Derived from Greek *arche* (beginning) or *archaios* (old), and *plastida* (from the Greek *plastos*, formed, molded), referring to plastids. The name thus refers to the presence of plastids of primary endosymbiotic origin.

Reference Phylogeny: Burki et al. (2008: Fig. 1), in which the clade is referred to as 'Plants' (see Comments).

Composition: *Viridiplantae* (~*Chloroplastida*) as defined in this volume; *Rhodoplantae* as defined in this volume; *Glaucophyta* (~*Glaucocystophyta*). This list is likely not exhaustive (see Comments).

Diagnostic Apomorphies: Plastids of primary endosymbiotic origin and related genic/genomic features (see Comments).

Synonyms: *Plantae* sensu Cavalier-Smith (1981) and *Primoplantae* Palmer et al. 2004 are approximate synonyms (see Comments).

Comments: The evolutionary history of eukaryotic cells is complicated by the phenomenon of 'endosymbiosis', whereby a cell, hereafter referred to as the nucleocytoplasmic host, acquired as an internal resident a second organism—the endosymbiont, and over evolutionary time this endosymbiont was reduced to the status of a highly integrated organelle, albeit one that still carries a small genome in almost all cases (Archibald, 2015). In a literal sense, all crown eukaryotes are evolutionary chimeras between the nucleocytoplasmic host and the endosymbiont that became the mitochondrion. Photosynthetic eukaryotes, however, are doubly chimeric because their plastids are derived, ultimately, from endosymbiotic cyanobacteria. In the case of plastid-bearing eukaryotes, 'primary endosymbiosis' refers to the situation where the plastid descended directly from an endosymbiosis established between a cyanobacterium and a eukaryotic host cell. The resulting organelle is called a 'primary plastid' and has two bounding membranes. This is distinct from the more complicated phenomena of 'secondary endosymbiosis' and 'tertiary endosymbiosis', in which the plastid descended most immediately from an endosymbiosis involving a eukaryotic endosymbiont that had itself inherited the organelle as a consequence of an earlier primary endosymbiosis. The resulting 'secondary plastids' (or 'tertiary plastids') generally have three or four bounding membranes (Keeling, 2004; Gould et al., 2008; Archibald, 2015).

Three major clades of extant eukaryotes have plastids of primary endosymbiotic origin: *Viridiplantae* (synonyms *Viridaeplantae, Chloroplastida, Chlorobionta*; includes *Embryophyta* and the 'green algae'), *Rhodoplantae* ('red algae') and *Glaucophyta* (also known as *Glaucocystophyta*). Phylogenetic analyses of plastid-encoded genes (e.g., Turner et al., 1999; Rodriguez-Ezpeleta et al., 2005) and the presence of unique gene order organizations on plastid genomes indicate that plastids are a clade, relative to extant cyanobacteria (see McFadden, 2001; Palmer, 2003). These data suggest that there was a single primary endosymbiosis, presumably in a common ancestor of *Viridiplantae, Rhodoplantae* and *Glaucophyta*. An alternative explanation holds that there were multiple primary endosymbioses involving closely related prokaryote endosymbionts but (potentially) distantly related eukaryote hosts. This second model drew support from the fact that *Viridiplantae* and *Rhodoplantae* and/ or *Glaucophyta* did not form a clade in many phylogenies of one or a few nucleus-encoded genes (e.g., Stiller and Hall, 1997). However, phylogenies based on datasets of large numbers of nucleus-encoded proteins have indicated that the nucleocytoplasmic host components of *Viridiplantae, Rhodoplantae* and *Glaucophyta* are closely related to one another, at least to the exclusion of *Sar, Amorphea, Discoba* and *Metamonada* (Rodríguez-Ezpeleta et al., 2005, 2007; Hackett et al., 2007; Burki et al., 2008, 2012, 2016; see Inagaki et al., 2009, and see below). This is broadly consistent with the hypothesis of a single primary endosymbiotic event in their common ancestor. Further, the single primary endosymbiosis hypothesis is strongly supported by the shared presence of apparently derived components of the plastid protein import machinery in *Viridiplantae* and *Rhodoplantae* (see Palmer, 2003; McFadden and van Dooren, 2004) and *Glaucophyta* (Price et al., 2012). This complete system presumably evolved after the establishment of the endosymbiotic association, since its function is to import proteins encoded by the host nucleus. This dates the last common ancestor of the plastids of *Viridiplantae, Rhodoplantae* and *Glaucophyta* to after the establishment of the endosymbiosis, and not before.

The precise phylogenetic positions of *Viridiplantae, Rhodoplantae* and *Glaucophyta* relative to other clades of eukaryotes are not completely resolved, even by recent data-rich phylogenenomic-scale analyses of nucleus-encoded proteins (e.g., Burki et al., 2012, 2016; Yabuki et al., 2014). From these studies, it is unclear whether *Cryptista* (e.g., *Cryptophyta*) is a close sister group to *Archaeplastida* (as defined here) or is actually a member of this clade. A similar uncertainty surrounds *Picozoa*, for which there are few sequence data at present. Some phylogenomic analyses infer that *Haptophyta* branches within *Archaeplastida*, but most recent studies infer that *Haptophyta* is more closely related to *Sar* (e.g., Burki et al., 2016).

One consequence of this uncertainty is that it is difficult to nominate clear apomorphies for *Archaeplastida*, other than the presence of primary plastids, and genic features associated with the plastid. The presence of a photosynthetic organelle derived directly from an endosymbiotic cyanobacterium is all-but-unique to *Archaeplastida*; the only other known case involves a clade within *Paulinella* (*Rhizaria*) as the host, and the symbiosis clearly occurred independently, incorporating a symbiont from a different subclade of *Cyanobacteria* much more recently (Marin et al., 2005; Nowak, 2014). Other distinctive (i.e., rare across other eukaryotes), but not unique, features include the use of starches as primary carbohydrate stores, and flattened mitochondrial cristae (Cavalier-Smith, 1998).

Viridiplantae, Rhodoplantae and *Glaucophyta* have been referred to collectively as the Kingdom *Plantae*. This usage dates back at least as far as one of several alternative schemes proposed by Cavalier-Smith (1981), although he did not at the time consider plastids to be an apomorphy for *Plantae* (see Cavalier-Smith, 1981: 476–477). Subsequently '*Plantae*' or 'plants' has been equated with the primary plastid-bearing eukaryotes by some, especially protistologists (e.g., Cavalier-Smith, 1998; Baldauf et al., 2000; Simpson and Roger, 2004; Keeling et al., 2005; Rodríguez-Ezpeleta et al., 2005; Hackett et al., 2007; Burki et al., 2007, 2008). However, it is much more common, especially among botanists, to associate the name *Plantae* with much less inclusive clades (e.g., Raven et al., 2005). This is due in part to the historical popularity of the Margulis version of the 'five-kingdom' classification scheme for life, in which Kingdom *Plantae* is an approximate synonym of *Embryophyta* (Margulis, 1971). Of course, under earlier 'two-kingdom' systems of life, *Plantae* was an extremely broad and heterogenous grouping that included all photosynthetic organisms (as well as fungi), and it remains very common to use the term 'plants' as a non-taxonomic term for photosynthetic organisms in general (e.g., Judd et al., 2007: chapter 7).

To avoid the confusion associated with *Plantae*, the new name *Archaeplastida* was coined as an outside-of-code taxon name by Adl et al. (2005) to refer to the clade of primary plastid-containing algae. A different name, *Primoplantae*, was introduced independently by Palmer et al. (2004). Of the two, *Archaeplastida* is used far more commonly in the scientific literature, and thus is used here as the name for this clade.

The qualifying clause in the definition addresses two different phylogenetic scenarios: (1) In my view the name *Archaeplastida* should be inapplicable if *Viridiplantae* and *Rhodoplantae* were found to derive from separate events of primary endosymbiosis. The first part of the qualifying clause (necessary inclusion of *Galdieria sulphuraria*) is intended to ensure this. (2) The inference of monophyly of the *Archaeplastida* nucleocytoplasmic component (see above) has been made largely on the basis of unrooted trees, thus strictly speaking, *Archaeplastida* forms a clan (sensu Wilkinson et al., 2007) in these analyses, rather than a clade. The second part of the qualifying clause (necessary exclusion of *Homo sapiens*) is intended make the name *Archaeplastida* inapplicable should the root of a clade consisting of most or all extant eukaryotes fall 'within' *Archaeplastida*. By contrast, in my view it would be appropriate to retain the name *Archaeplastida* should one or more smaller clades without primary plastids prove to lie within the clade identified by the apomorphy of a primary plastid (e.g., *Cryptista* and/or *Picozoa* – see above). Thus, I have not specified a more stringent qualifying clause.

Literature Cited

Adl, S. M., A. G. B. Simpson, M. A. Farmer, R. A. Andersen, O. R. Anderson, J. R. Barta, S. S. Bowser, G. Brugerolle, R. A. Fensome, S. Fredericq, T. Y. James, S. Karpov, P. Kugrens, J. Krug, C. E. Lane, L. A. Lewis, J. Lodge, D. H. Lynn, D. G. Mann, R. M. McCourt, L. Mendoza, O. Moestrup, S. E. Mozley-Standridge, T. A. Nerad, C. A. Shearer, A. V. Smirnov, F. W. Spiegel, and M. F. Taylor. 2005. The new higher level classification of eukaryotes with emphasis on the taxonomy of protists. *J. Eukaryot. Microbiol.* 52:399–451.

Archibald, J. M. 2015. Endosymbiosis and eukaryotic cell evolution. *Curr. Biol.* 25:R911–R921.

Baldauf, S. L., A. J. Roger, I. Wenk-Siefert, and W. F. Doolittle. 2000. A kingdom-level phylogeny of eukaryotes based on combined protein data. *Science* 290:972–977.

Burki, F., M. Kaplan, D. V. Tikhonenkov, V. Zlatogursky, B. Q. Minh, L. V. Radaykina, A. Smirnov, A. P. Mylnikov, and P. J. Keeling. 2016. Untangling the early diversification of eukaryotes: A phylogenomic study of the evolutionary origins of *Centrohelida, Haptophyta* and *Cryptista. Proc. R. Soc. Lond. B Biol. Sci.* 283:20152802.

Burki, F., N. Okamoto, J.-F. Pombert, and P. J. Keeling. 2012. The evolutionary history of haptophytes and cryptophytes: phylogenomic evidence for separate origins. *Proc. R. Soc. Lond. B Biol. Sci.* 279:2246–2254.

Burki, F., K. Shalchian-Tabrizi, M. Minge, A. Skjaeveland, S. I. Nikolaev, K. S. Jakobsen, and J. Pawlowski. 2007. Phylogenomics reshuffles the eukaryotic supergroups. *PLOS ONE* 2:e790.

Burki, F., K. Shalchian-Tabrizi, and J. Pawlowski. 2008. Phylogenomics reveals a new 'megagroup' including most photosynthetic eukaryotes. *Biol. Lett.* 4:366–369.

Cavalier-Smith, T. 1981. Eukaryote kingdoms: seven or nine? *Biosystems* 14:461–481.

Cavalier-Smith, T. 1998. A revised six-kingdom system of life. *Biol. Rev.* 73:203–266.

Gould, S. B., R. F. Waller, and G. I. McFadden. 2008. Plastid evolution. *Annu. Rev. Plant Biol.* 59:491–517.

Hackett, J. D., H. S. Yoon, S. Li, A. Reyes-Prieto, S. E. Rümmele, and D. Bhattacharya. 2007. Phylogenomic analysis supports the monophyly of cryptophytes and haptophytes and the association of *Rhizaria* with chromalveolates. *Mol. Biol. Evol.* 24:1702–1713.

Inagaki, Y., Y. Nakajima, M. Sato, M. Sakaguchi, and T. Hashimoto. 2009. Gene sampling can bias multi-gene phylogenetic inferences: the relationship between red algae and green plants as a case study. *Mol. Biol. Evol.* 26:1171–1178.

Judd, W. S., C. S. Campbell, E. A. Kellogg, P. F. Stevens, and M. J. Donoghue. 2007. *Plant Systematics: A Phylogenetic Approach.* 3rd edition. Sinauer Associates, Sunderland, MA.

Keeling, P. J. 2004. A brief history of plastids and their hosts. *Protist* 155:3–7.

Keeling, P. J., G. Burger, D. G. Durnford, B. F. Lang, R. W. Lee, R. E. Pearlman, A. J. Roger, and M. W. Gray. 2005. The tree of eukaryotes. *Trends Ecol. Evol.* 20:670–675.

Margulis, L. 1971. Whittaker's five kingdoms of organisms: minor revisions suggested by considerations of the origin of mitosis. *Evolution* 25:242–245.

Marin, B., E. C. M. Nowack, and M. Melkonian. 2005. A plastid in the making: primary endosymbiosis. *Protist* 156:425–432.

McFadden, G. I. 2001. Primary and secondary endosymbiosis and the origin of plastids. *J. Phycol.* 37:951–959.

McFadden, G. I., and G. G. van Dooren, 2004. Evolution: red algal genome affirms a common origin of all plastids. *Curr. Biol.* 14:R514–R516.

Nowak, E. C. M. 2014. *Paulinella chromatophora*—rethinking the transition from endosymbiont to organelle. *Acta Soc. Bot. Pol.* 83:387–397.

Palmer, J. D. 2003. The symbiotic birth and spread of plastids: how many times and whodunit? *J. Phycol.* 39:4–11.

Palmer, J. D., D. E. Soltis, and M. W. Chase. 2004. The plant tree of life: an overview and some points of view. *Am. J. Bot.* 91:1437–1445.

Price, D. C., C. X. Chan, H. S. Yoon, E. C. Yang, H. Qiu, A. P. Weber, R. Schwacke, J. Gross, N. A. Blouin, C. Lane, A. Reyes-Prieto, D. G. Durnford, J. A. Neilson, B. F. Lang, G. Burger, J. M. Steiner, W. Löffelhardt, J. E. Meuser, M. C. Posewitz, S. Ball, M. C. Arias, B. Henrissat, P. M. Coutinho, S. A. Rensing, A. Symeonidi, H. Doddapaneni, B. R. Green, V. D. Rajah, J. Boore, and D. Bhattacharya 2012. *Cyanophora paradoxa* genome elucidates origin of photosynthesis in algae and plants. *Science* 335:843–847.

Raven, P. H., R. F. Evert, and S. E. Eichhorn. 2005. *Biology of Plants.* 7th edition. Freeman & Co., New York.

Rodríguez-Ezpeleta, N., H. Brinkmann, S. C. Burey, B. Roure, G. Burger, W. Loffelhardt, H. J. Bohnert, H. Philippe, and B. F. Lang. 2005. Monophyly of primary photosynthetic eukaryotes: green plants, red algae, and glaucophytes. *Curr. Biol.* 15:1325–1330.

Rodríguez-Ezpeleta, N., H. Brinkmann, G. Burger, A. J. Roger, M. W. Gray, H. Philippe, and B. F. Lang. 2007. Toward resolving the eukaryotic tree: the phylogenetic positions of jakobids and cercozoans. *Curr. Biol.* 17:1420–1425.

Simpson, A. G. B., and A. J. Roger. 2004. The real 'kingdoms' of eukaryotes. *Curr. Biol.* 14:R693–696.

Stiller, J. W., and B. D. Hall. 1997. The origins of red algae; implications for plastid evolution. *Proc. Natl. Acad. Sci. USA* 94:4520–4525.

Turner, S., K. M. Pryer, V. P. W. Miao, and J. D. Palmer. 1999. Investigating deep phylogenetic relationships among cyanobacteria and plastids by small submit rRNA sequence analysis. *J. Eukaryot. Microbiol.* 46:327–338.

Wilkinson, M., J. O. McInerney, R. P. Hirt, P. G. Foster, and T. M. Embley. 2007. Of clades and clans: terms for phylogenetic relationships in unrooted trees. *Trends Ecol. Evol.* 22:114–115.

Yabuki, A., R. Kamikawa, S. A. Ishikawa, M. Kolisko, E. Kim, A. S. Tanabe, K. Kume, K. Ishida, and Y. Inagaki. 2014. *Palpitomonas bilix* represents a basal cryptist lineage: insight into the character evolution in *Cryptista*. *Sci. Rep.* 4:4641.

Author

Alastair G. B. Simpson; Department of Biology; Dalhousie University; Halifax, NS, B3H 4R2, Canada. Email: alastair.simpson@dal.ca.

Date Accepted: 25 September 2012; updated 28 February 2018

Primary Editor: Philip Cantino

Rhodoplantae G. W. Saunders and M. H. Hommersand 2004 [G. W. Saunders], converted clade name

Registration Number: 93

Definition: The crown clade originating in the most recent common ancestor of *Cyanidium caldarium* (J. Tilden) Geitler 1933, *Galdieria sulphuraria* (Galdieri) Merola in Merola et al. 1981, and *Compsopogon caeruleus* (C. Agardh) Montagne 1846. This is a minimum-crown-clade definition. Abbreviated definition: min crown ∇ (*Cyanidium caldarium* (J. Tilden) Geitler 1933 & *Galdieria sulphuraria* (Galdieri) Merola 1981 & *Compsopogon caeruleus* (C. Agardh) Montagne 1846).

Etymology: Derived from Greek, *rhodo-* (rosy-red) and Latin, *plantae* (plants).

Reference Phylogeny: Yoon et al. (2006: Fig. 1) is the primary reference phylogeny.

Composition: *Rhodoplantae* includes *Cyanidiales* and *Rhodophyta* as defined in this volume.

Diagnostic Apomorphies: This clade is diagnosed by a combination of putatively apomorphic characters, which nonetheless have analogous states in other clades, rendering diagnosis on the following suite of characters more reliable than diagnosis based on individual features. Attributes include plastids with two bounding membranes lacking interstitial peptidoglycan (likely plesiomorphic), a complete lack of flagellate stages in the life history, and photosynthetic reserve stored as floridean starch (α-1,4-, α-1,6-linked glucan) grains in the cytoplasm (see Gabrielson and Garbary, 1986; Woelkerling, 1990).

Synonyms: *Rhodymeniobiotina* Doweld 2001 and *Rhodophyta* sensu Gabrielson et al. (1985) (as perpetuated in current popular texts; e.g., Raven et al., 2005) are approximate synonyms.

Comments: This clade includes all of the organisms traditionally known as red algae and placed in the subkingdom *Rhodoplantae*, kingdom *Plantae*, under the rank-based system (Saunders and Hommersand, 2004). Among these are all of the macroalgae united by Thuret (1855) under the *Rhodophyceae* and later additions to this taxon. Notably, many unicellular representatives have been subsequently added to this assemblage, including members of the *Cyanidiales* as defined in this volume. Monophyly of *Rhodoplantae* has been established in many studies (Yoon et al., 2006 and references therein).

Saunders and Hommersand (2004) argued that the descriptive (i.e., nontypified) name *Rhodoplantae*, here converted to a clade name, is the best choice for the red algal clade. Saunders and Hommersand based this decision on the fact that *rhodo-* (from Greek for rosy-red) has been associated with a clade including bangiophycean and florideophycean red algae (as *Rhodophyceae*) since Thuret (1855). They thus rejected the typified name *Rhodymeniobiotina* Doweld 2001, which is here considered an approximate synonym of *Rhodoplantae*. The name *Rhodophyta* has a rather convoluted history regarding its constituent taxa, specifically members of the *Cyanidiales*. Although the *Cyanidiales* were considered by a few researchers to be red algae in 1958 (e.g., Hirose, 1958), this was not the consensus view; these algae were variously considered to be cyanobacteria, transitional between cyanobacteria and red algae, green algae, cryptophytes, or a eukaryotic lineage containing an endocyanome (see Chapman, 1974; Kremer, 1982). Indeed, consensus on the red algal affinities of these organisms only solidified in the 1980s (e.g., Gabrielson et al., 1985), which was followed immediately by proposals to recognize cyanidiophytes as distinct at the phylum level (Seckbach, 1987; but see Saunders and Hommersand [2004] for nomenclatural discussion) from the red algae traditionally included in the rank-based taxon *Rhodophyta*. Doweld (2001) subsequently validated this concept, thus restricting the name *Rhodophyta* to all red algae excluding the cyaniodiophytes (see Saunders and Hommersand, 2004). This restricted concept of the name *Rhodophyta* is used here, and thus *Rhodoplantae* has been converted for the larger clade including both *Rhodophyta* and *Cyanidiales*.

Literature Cited

Chapman, D. J. 1974. Taxonomic status of *Cyanidium caldarium*, the *Porphyridiales* and *Goniotrichales*. *Nova Hedwigia* 25:673–682.

Doweld, A. 2001. *Prosyllabus tracheophytorum*. GEOS, Moscow.

Gabrielson, P. W., and D. J. Garbary. 1986. Systematics of red algae (*Rhodophyta*). *Crit. Rev. Plant Sci.* 3:325–366.

Gabrielson, P. W., D. J. Garbary, and R. F. Scagel. 1985. The nature of the ancestral red alga: inferences from a cladistic analysis. *BioSystems* 18:335–346.

Geitler, L. 1933. Diagnose neuer Blaualgen von den Sundalnseln. *Arch. Hydrobiol. Suppl.* 12:622–634.

Hirose, H. 1958. The rearrangement of the systematic position of a thermal alga, *Cyanidium caldarium*. *Bot. Mag. (Tokyo)* 71:347–352.

Kremer, B. P. 1982. *Cyanidium caldarium*: a discussion of biochemical features and taxonomic problems. *Br. Phycol. J.* 17:51–61.

Merola, A., R. Castaldo, P. DeLuca, R. Gambardella, A. Musacchio, and R. Taddei. 1981. Revision of *Cyanidium caldarium*. Three species of acidophilic algae. *G. Bot. Ital.* 115:189–195.

Montagne, C. 1846. Ordo I. *Phyceae* Fries. Pp. 1–197 in *Exploration scientifique de l'Algérie pendant les années 1840, 1841, 1842...Sciences physiques. Botanique. Cryptogamie* (M. C. Durieu De Maisonneuve, ed.). Imprimerie Royale, publiée par ordre du Gouvernement et avec le concours d'une Commission Académique, Paris.

Raven, P. H., R. F. Evert, and S. E. Eichhorn. 2005. *Biology of Plants*. 7th edition. W.H. Freeman & Co., New York.

Saunders, G. W., and M. H. Hommersand. 2004. Assessing red algal supraordinal diversity and taxonomy in the context of contemporary systematic data. *Am. J. Bot.* 91:1494–1507.

Seckbach, J. 1987. Evolution of eukaryotic cells via bridge algae, the cyanidia connection. *Ann. N.Y. Acad. Sci.* 503:424–437.

Thuret, G. 1855. Note sur un nouveau genre d'algues, de la famille des Floridées. Feuardent, Cherbourg. Reprinted as *Mem. Soc. Sci. Nat., Cherbourg* 3:155–160, 2 plates.

Woelkerling, W. J. 1990. An introduction. Pp. 1–6 in *Biology of the Red Algae* (K. M. Cole and R. G. Sheath, eds.). Cambridge University Press, Cambridge, UK.

Yoon, H. S., K. M. Müller, R. G. Sheath, F. D. Ott, and D. Bhattacharya. 2006. Defining the major lineages of red algae (*Rhodophyta*). *J. Phycol.* 42:482–492.

Author

Gary W. Saunders; Centre for Environmental & Molecular Algal Research; Dept. of Biology; University of New Brunswick; Fredericton, NB, E3B 5A3, Canada. Email: gws@unb.ca

Date Accepted: 19 July 2010; updated 24 February 2018

Primary Editor: Philip Cantino

Cyanidiales T. Christensen 1962 [G. W. Saunders], converted clade name

Registration Number: 119

Definition: The largest crown clade containing *Cyanidium caldarium* (J. Tilden) Geitler 1933 and *Galdieria sulphuraria* (Galdieri) Merola in Merola et al. 1981 but not *Compsopogon caeruleus* (C. Agardh) Montagne 1846 (*Rhodophyta*). This is a maximum-crown-clade definition. Abbreviated definition: max crown ∇ (*Cyanidium caldarium* (J. Tilden) Geitler 1933 & *Galdieria sulphuraria* (Galdieri) Merola in Merola et al. 1981 ~ *Compsopogon caeruleus* (C. Agardh) Montagne 1846).

Etymology: Based on the genus name *Cyanidium*, which is derived from Greek, *cyan-* (blue) and *-idium* (small) in reference to the small size and bluish-green color of these species.

Reference Phylogeny: Yoon et al. (2006: Fig. 1) is the primary reference phylogeny.

Composition: Includes *Cyanidium* Geitler 1933, *Cyanidioschyzon* P. DeLuca, R. Taddei and L. Varano 1978 and *Galdieria* Merola in Merola et al. 1981 (see Yoon et al., 2006, and Schneider and Wynne, 2007).

Diagnostic Apomorphies: This clade is diagnosed by a combination of putatively apomorphic characters, which nonetheless have analogous states in other clades of red algae, rendering diagnosis on the following suite of characters more reliable than considering individual attributes. Features include Golgi associated with endoplasmic reticulum, cell walls thick and proteinaceous, endospores produced, and heterotrophic capacity (see Saunders and Hommersand (2004), Yoon et al. (2006), and references therein).

Synonyms: *Cyanidiophyceae* Merola et al. 1981, *Cyanidiophytina* Yoon et al. 2006, and *Cyanidiophyta* Doweld 2001 are approximate synonyms.

Comments: The *Cyanidiales* are unicellular red algae, frequently inhabiting extreme acidophilic/thermophilic environments. Yoon et al. (2006) provided the most comprehensive molecular phylogenetic analyses of this clade, which they also named *Cyanidiales*, with regards to its sister relationship to the remaining red algae (*Rhodophyta* as defined in this volume). Although support for this clade is weak in the reference phylogeny, there are no published phylogenies that question strongly its monophyly (see Yoon et al., 2006). However, the inclusion of a second internal specifier ensures that the name will not apply to any clade if *Cyanidium caldarium* (J. Tilden) Geitler 1933 and *Galdieria sulphuraria* (Galdieri) Merola in Merola et al. 1981 are not part of a clade that excludes *Rhodophyta* as defined in this volume.

The taxon has had a highly volatile taxonomic history with regard to the rank at which it should be recognized in traditional nomenclature (Saunders and Hommersand, 2004). It had been designated by that time as a family, order, subclass, class and phylum, and subsequently (Yoon et al., 2006) as a subphylum. The *PhyloCode* (Cantino and de Queiroz, 2020) has the advantage of terminating this seemingly never-ending taxonomic shuffle in the literature and of assigning this clade a single name. The name traditionally associated with the rank of order (*Cyanidiales*), which has been widely applied in contemporary research, is selected here for consistency with published literature on the group.

Literature Cited

Cantino, P. D., and K. de Queiroz. 2020. *International Code of Phylogenetic Nomenclature (PhyloCode)*, Version 6. CRC Press, Boca Raton, FL.

Christensen, T. 1962. Alger. Pp. 1–178 in *Botanik. Bind II. Systematisk Botanik Nr. 2. I Kommission hos Munksgaard* (T. W. Böcher, M. Lange, and T. Sørensen (Redig.), eds.). København, Denmark.

DeLuca, P., R. Taddei, and L. Varano. 1978. '*Cyanidioschyzon merolae*': a new alga of thermal acidic environments. *Weebia* 33:37–44.

Doweld, A. 2001. *Prosyllabus tracheophytorum*. GEOS, Moscow.

Geitler, L. 1933. Diagnose neuer Blaualgen von den Sundalnseln. *Arch. Hydrobiol. Suppl.* 12:622–634.

Merola, A., R. Castaldo, P. DeLuca, R. Gambardella, A. Musacchio, and R. Taddei. 1981. Revision of *Cyanidium caldarium*. Three species of acidophilic algae. *Giornale Botanico Italiano* 115:189–195.

Montagne, C. 1846. Ordo I. *Phyceae* Fries. Pp. 1–197 in *Exploration scientifique de l'Algérie pendant les années 1840, 1841, 1842 . . . Sciences physiques. Botanique. Cryptogamie* (M.C. Durieu De Maisonneuve, ed.). Imprimerie Royale, publiée par ordre du Gouvernement et avec le concours d'une Commission Académique, Paris.

Saunders, G. W., and M. H. Hommersand. 2004. Assessing red algal supraordinal diversity and taxonomy in the context of contemporary systematic data. *Am. J. Bot.* 91:1494–1507.

Schneider, C. W., and M. J. Wynne. 2007. A synoptic review of the classification of red algal genera a half century after Kylin's "Die Gattungen der Rhodophyceen". *Bot. Mar.* 50:197–249.

Yoon, H. S., K. M. Müller, R. G. Sheath, F. D. Ott, and D. Bhattacharya. 2006. Defining the major lineages of red algae (*Rhodophyta*). *J. Phycol.* 42:482–492.

Author

Gary W. Saunders; Centre for Environmental & Molecular Algal Research; Dept. of Biology; University of New Brunswick; Fredericton, NB, E3B 5A3, Canada. Email: gws@unb.ca.

Date Accepted: 19 July 2010; updated 24 February 2018

Primary Editor: Philip Cantino

Rhodophyta A. Wettstein 1901 [G. W. Saunders], converted clade name

Registration Number: 123

Definition: The largest crown clade containing *Compsopogon caeruleus* (C. Agardh) Montagne 1846 but not *Cyanidium caldarium* (J. Tilden) Geitler 1933 and *Galdieria sulphuraria* (Galdieri) Merola in Merola et al. 1981. This is a maximum-crown-clade definition. Abbreviated definition: max crown ∇ (*Compsopogon caeruleus* (C. Agardh) Montagne 1846 ~ *Cyanidium caldarium* (J. Tilden) Geitler 1933 & *Galdieria sulphuraria* (Galdieri) Merola 1981).

Etymology: Derived from Greek, *rhodo-* (rosy-red) and *phyton* (plant).

Reference Phylogeny: Yoon et al. (2006: Fig. 1).

Composition: *Rhodophyta* includes *Proteorhodophytina* and *Eurhodophytina* as defined in this volume.

Diagnostic Apomorphies: This clade is diagnosed by a combination of putatively apomorphic characters, which nonetheless have analogous states in other clades of red algae, rendering diagnosis on the following suite of characters more reliable than considering individual attributes. Features include Golgi association usually with the ER plus the mitochondrion, or rarely just the nucleus or ER alone; cell walls composed of various polysaccharides; production of spores by conversion of complete cell contents to sporangia or by the formation of packets through successive divisions (see Gabrielson and Garbary, 1986; Saunders and Hommersand, 2004).

Synonyms: *Rhodophyceae* sensu Thuret (1855), *Rhodymeniophyta* Doweld 2001 and *Rhodophytina* Yoon et al. 2006 are approximate synonyms.

Comments: The clade *Rhodophyta* consists of unicellular and multicellular red algae in freshwater and marine habitats. Monophyly of *Rhodophyta* has been established in many studies (Yoon et al. (2006) and references cited therein). This clade contains the bulk of the red algal species traditionally assigned to the *Bangiophyceae* and *Florideophyceae* under the rank-based system (for a review, see Saunders and Hommersand, 2004). The name *Rhodophyta* is applied here in the restricted sense excluding the highly distinctive *Cyanidiales* (see Comments under *Rhodoplantae*, this volume).

The pre-existing name *Rhodophyta*, here converted as a clade name, has long been associated with the phylum level in the rank-based system. The application of a typified name, *Rhodymeniophyta*, by Doweld (2001) was considered a superfluous act that was rejected by Saunders and Hommersand (2004). *Rhodymeniophyta* has rarely been used and is here considered an approximate synonym of *Rhodophyta*. Thuret (1855) was the first to unite the bangiophycean and florideophycean algae into a taxon, for which he used the name *Rhodophyceae* Ruprecht (Ragan and Gutell, 1995). As the name *Rhodophyceae* has a rather convoluted taxonomic history (Dixon, 1973), and is less commonly used than *Rhodophyta* in the literature, the latter name is converted here for this clade, and *Rhodophyceae* is an approximate synonym.

Literature Cited

Dixon, P. S. 1973. *Biology of the* Rhodophyta. Oliver and Boyd, Edinburgh.

Doweld, A. 2001. *Prosyllabus tracheophytorum*. GEOS, Moscow.

Gabrielson, P. W., and D. J. Garbary. 1986. Systematics of red algae (*Rhodophyta*). *Crit. Rev. Plant Sci.* 3:325–366.

Geitler, L. 1933. Diagnose neuer Blaualgen von den Sundalnseln. *Arch. Hydrobiol. Suppl.* 12:622–634.

Merola, A., R. Castaldo, P. DeLuca, R. Gambardella, A. Musacchio, and R. Taddei. 1981. Revision of *Cyanidium caldarium*. Three species of acidophilic algae. *G. Bot. Ital.* 115:189–195.

Montagne, C. 1846. Ordo I. *Phyceae* Fries. Pp. 1–197 in *Exploration scientifique de l'Algérie pendant les années 1840, 1841, 1842...Sciences physiques. Botanique. Cryptogamie* (M. C. Durieu De Maisonneuve, ed.). Imprimerie Royale, publiée par ordre du Gouvernement et avec le concours d'une Commission Académique, Paris.

Ragan, M. A., and R. R. Gutell. 1995. Are red algae plants? *Bot. J. Linn. Soc.* 118:81–105.

Saunders, G. W., and M. H. Hommersand. 2004. Assessing red algal supraordinal diversity and taxonomy in the context of contemporary systematic data. *Am. J. Bot.* 91:1494–1507.

Thuret, G. 1855. Note sur un nouveau genre d'algues, de la famille des Floridées. Feuardent, Cherbourg. Reprinted as *Mem. Soc. Sci. Nat., Cherbourg* 3:155–160, 2 plates.

Wettstein, A. 1901. *Handbuch der Systematischen Botanik.* Leipzig, Vienna.

Yoon, H. S., K. M. Müller, R. G. Sheath, F. D. Ott, and D. Bhattacharya. 2006. Defining the major lineages of red algae (*Rhodophyta*). *J. Phycol.* 42:482–492.

Author

Gary W. Saunders; Centre for Environmental & Molecular Algal Research; Dept. of Biology; University of New Brunswick; Fredericton, NB, E3B 5A3, Canada. Email: gws@unb.ca.

Date Accepted: 19 July 2010; updated 24 February 2018

Primary Editor: Philip Cantino

Proteorhodophytina S. A. Muñoz-Gómez , F. G. Mejía-Franco, K. Durnin, M. Colp, C. J. Grisdale, J. M. Archibald and C. H. Slamovits 2017 [G. W. Saunders], converted clade name

Registration Number: 124

Definition: The crown clade originating in the most recent common ancestor of *Bangiopsis subsimplex* (Montagne) F. Schmitz 1896, *Corynoplastis japonica* Yokoyama, J. L. Scott, G. C. Zuccarello, M. Kajikawa, Y. Hara and J. A. West 2009, *Porphyridium cruentum* (S.F. Gray) Nägeli 1849 (= *Porphyridium purpureum* (Bory de Saint-Vincent) K. M. Drew and R. Ross 1965), and *Rhodochaete pulchella* Thuret ex Bornet 1892. This is a minimum-crown-clade definition. Abbreviated definition: min crown ∇ (*Bangiopsis subsimplex* (Montagne) F. Schmitz 1896 & *Corynoplastis japonica* Yokoyama, J. L. Scott, G. C. Zuccarello, M. Kajikawa, Y. Hara and J. A. West 2009 & *Porphyridium purpureum* (Bory de Saint-Vincent) *Porphyridium cruentum* (S. F. Gray) Nägeli 1849 and R. Ross 1965 & *Rhodochaete pulchella* Thuret ex Bornet 1892).

Etymology: Derived from Greek god of the sea, *Proteus*, and from the Greek *rhodo-* (rosy-red), and *phyton* (plant). The reference to Proteus, who was "able to display many different forms, … alludes to the vast ('protean') phenotypic and genotypic diversity exhibited by the members" of this clade (Muñoz-Gómez et al., 2017).

Reference Phylogeny: Muñoz-Gómez et al. (2017: Fig. 3; in which *Rhodochaete pulchella* is referred to by the incorrect name *Rhodochaete parvula* [Guiry and Guiry, 2017]).

Composition: *Proteorhodophytina* includes *Compsopogonophyceae*, *Porphyridiophyceae*, *Rhodellophyceae*, and *Stylonematales* as defined in this volume.

Diagnostic Apomorphies: Plastid genomes with their large numbers of introns and quadripartite organization around the rRNA operon, which have relatively small single copy regions, may best define this lineage, but more study is needed (Muñoz-Gómez et al., 2017).

Synonyms: None.

Comments: Qiu et al. (2016) and Muñoz-Gómez et al. (2017) provided the first solid evidence for a clade comprising *Compsopogonophyceae*, *Porphyridiophyceae*, *Rhodellophyceae*, and *Stylonematales* as defined in this volume, relationships among which had remained elusive (see Saunders and Hommersand (2004), Yoon et al. (2006), and references therein). This clade contains mesophylic red algae displaying a wide diversity of vegetative types—unicellular, pseudofilamentous, filamentous, and pseudoparenchymatous—as well as a wide variety of organellar arrangements and plastid morphologies.

This clade name is converted from the subphylum name *Proteorhodophytina* Muñoz-Gómez, Mejía-Franco, Durnin, Colp, Grisdale, J.M. Archibald and Slamovits 2017 in the rank-based system.

Literature Cited

Guiry, M. D., and G. M. Guiry. 2017. *AlgaeBase.* World-wide electronic publication, National University of Ireland, Galway. Available at http://www.algaebase.org, accessed on 2 October 2017.

Muñoz-Gómez, S. A., F. G. Mejía-Franco, K. Durnin, M. Colp, C. J. Grisdale, J. M. Archibald, and C. H. Slamovits. 2017. The new red algal subpylum *Proteorhodophytina* comprises the largest and most divergent plastid genomes known. *Curr. Biol.* 27:1–8.

Qiu, H., H. S. Yoon, and D. Bhattacharya. 2016. Red algal phylogenomics provides a robust framework for inferring evolution of key metabolic pathways. *PLOS Curr. Tree Life.* 2 December 2016. 1st edition. doi:10.1371/currents.tol.7b03 7376e6d84a1be34af756a4d90846.

Saunders, G. W., and M. H. Hommersand. 2004. Assessing red algal supraordinal diversity and taxonomy in the context of contemporary systematic data. *Am. J. Bot.* 91:1494–1507.

Yoon, H. S., K. M. Müller, R. G. Sheath, F. D. Ott, and D. Bhattacharya. 2006. Defining the major lineages of red algae (*Rhodophyta*). *J. Phycol.* 42:482–492.

Author

Gary W. Saunders; Centre for Environmental & Molecular Algal Research; Dept. of Biology; University of New Brunswick; Fredericton, NB, E3B 5A3, Canada. Email: gws@unb.ca.

Date Accepted: 24 February 2018

Primary Editor: Philip Cantino

Compsopogonophyceae G. W. Saunders and M. H. Hommersand 2004 [G. W. Saunders], converted clade name

Registration Number: 125

Definition: The largest crown clade containing *Compsopogon caeruleus* (C. Agardh) Montagne 1846 and *Rhodochaete pulchella* Thuret ex Bornet 1892 but not *Bangia fuscopurpurea* (Dillwyn) Lyngbye 1819 and *Porphyridium aerugineum* Geitler 1923 and *Rhodella maculata* L. Evans 1970 and *Stylonema cornu-cervi* Reinsch 1875. This is a maximum-crown-clade definition. Abbreviated definition: max crown ∇ (*Compsopogon caeruleus* (C. Agardh) Montagne 1846 & *Rhodochaete pulchella* Thuret ex Bornet 1892 ~ *Bangia fuscopurpurea* (Dillwyn) Lyngbye 1819 & *Porphyridium aerugineum* Geitler 1923 & *Rhodella maculata* L. Evans 1970 & *Stylonema cornu-cervi* Reinsch 1875).

Etymology: Based on the genus name *Compsopogon*, which is derived from Greek, *kompsos* (elegant) and *pogon* (beard) in reference to the gross morphology of the type species. The ending *phyceae* is the standard termination under the botanical code (Turland et al., 2018) for algal taxa at the class rank and is derived from the Greek *phykos*, meaning seaweed or alga.

Reference Phylogeny: Zuccarello et al. (2000) provided the first molecular evidence in support of this clade, but the strongest support was realized in subsequent phylogenetic analyses of Müller et al. (2001: Fig. 1), which serve as the primary reference phylogeny. See also Yoon et al. (2006: Figs. 1–3). *Rhodochaete pulchella* is incorrectly labeled *R. parvula* in all of these publications (Guiry and Guiry, 2017).

Composition: *Compsopogonophyceae* includes *Compsopogonales*, *Erythropeltidales* and *Rhodochaetales* (see Silva, 1996: 915 for nomenclatural discussion) (see Saunders and Hommersand, 2004, and references therein). See Schneider and Wynne (2007) and Guiry and Guiry (2017) for comprehensive listings of taxa traditionally ranked as genera in these rank-based taxa.

Diagnostic Apomorphies: This clade is diagnosed by a combination of putatively apomorphic characters, which nonetheless have analogous states in other clades of red algae, rendering diagnosis on the following suite of characters more reliable than considering individual attributes: where known (secondary loss in some taxa?), pit plugs naked (lacking caps and membranes) (possibly plesiomorphic); monosporangia and spermatangia usually cut out by curved walls from ordinary vegetative cells; Golgi-ER association; encircling thylakoids in the plastid; sexual life history, where known, is biphasic (likely plesiomorphic) (see Garbary et al., 1980; Zuccarello et al., 2000).

Synonyms: *Metarhodophytina* Saunders and Hommersand 2004 is an approximate synonym, as is the name *Metarhodophycidae* Magne 1989.

Comments: Although support for the monophyly of *Compsopogonophyceae* is variable and weak in molecular phylogenies to date (e.g., Yoon et al., 2006), it is one of the few groups of red algae diagnosed by shared anatomical features, including the type of pit plug (which may or may not represent a derived state; Scott and Broadwater, 1989) and the division pattern in the production of reproductive structures

(Garbary et al., 1980). A study by Nelson et al. (2003) indicates that at least some of the taxa now included in this clade have lost the diagnostic pattern of sporangial and spermatangial development and have converged on the mode of division more typical of the *Bangiales*. This is an interesting discovery that requires more investigation, but not one that would alter the composition of the clade *Compsopogonophyceae*. The inclusion of a second internal specifier in the definition ensures that the name will not apply to any clade if *Compsopogonales*, *Rhodochaetales*, and *Erythropeltidales* do not form a clade.

The clade name *Compsopogonophyceae* is converted from *Compsopogonophyceae* Saunders and Hommersand 2004, a class name under the rank-based system, which they applied to the same clade to which it is applied here. It was the only class in the subphylum *Metarhodophytina* Saunders and Hommersand 2004 (an elevation in rank of Magne's (1989) *Metarhodophycidae*). The name *Compsopogonophyceae* is the more widely used of the three names and is therefore selected here for conversion to a clade name.

Literature Cited

Bornet, E. 1892. Les algues de P.-K.-A. Schousboe. *Mém. Soc. Sci. Nat., Cherbourg* 28: 165–376, 3 plates.

Evans, L. V. 1970. Electron microscopical observations on a new red algal unicell, *Rhodella maculata* gen. nov., sp. nov. *Br. Phycol. J.* 5:1–13.

Garbary, D. J., G. K. Hansen, and R. F. Scagel. 1980. A revised classification of the *Bangiophyceae* (*Rhodophyta*). *Nova Hedwigia* 33:145–166.

Geitler, L. 1923. *Porphyridium aerugineum* nov. sp. *Osterreichische Botanische Zeitschrift* 72:4.

Guiry, M. D., and G. M. Guiry. 2017. *AlgaeBase*. World-wide Electronic Publication, National University of Ireland, Galway. Available at http://www.algaebase.org, searched on 2 October 2017.

Lyngbye, H. C. 1819. *Tentamen Hydrophytologiae Danicae*. Schultz, Copenhagen.

Magne, F. 1989. Classification et phylogénie des Rhodophycées. *Cryptgam. Algol.* 10:101–115.

Montagne, C. 1846. Ordo I. *Phyceae* Fries. Pp. 1–197 in *Exploration Scientifique de l'Algérie Pendant les Années 1840, 1841, 1842...Sciences Physiques. Botanique. Cryptogamie.* (M.C. Durieu De Maisonneuve, ed.). Imprimerie Royale, publiée par ordre du Gouvernement et avec le concours d'une Commission Académique, Paris.

Müller, K. M., M. C. Oliveira, R. G. Sheath, and D. Bhattacharya. 2001. Ribosomal DNA phylogeny of the *Bangiophycidae* (*Rhodophyta*) and the origin of secondary plastids. *Am. J. Bot.* 88:1390–1400.

Nelson, W. A., J. E. Broom, and T. J. Farr. 2003. *Pyrophyllon* and *Chlidophyllon* (*Erythropeltidales, Rhodophyta*): two new genera for obligate epiphytic species previously placed in *Porphyra*, and a discussion of the orders *Erythropeltidales* and *Bangiales*. *Phycologia* 42:308–315.

Reinsch, P. F. 1875. *Contributiones Ad Algologiam et Fungologiam*, Vol. 1, pp. XII + 103, 131 plates [I–III, IIIa, IV–VI, VIa, VII–XII, XIIa, XIII–XX, XXa, XXI–XXXV, XXXVa, XXXVI (*Melanophyceae*); I–XLII, XLIIa, XLIII–XLVII, XLVIIa, XLVIII–LXI (*Rhodophyceae*); I–XVIII (*Chlorophyllophyceae*); I–IX (*Fungi*)]. Nürnberg.

Saunders, G. W., and M. H. Hommersand. 2004. Assessing red algal supraordinal diversity and taxonomy in the context of contemporary systematic data. *Am. J. Bot.* 91: 1494–1507.

Schneider, C. W., and M. J. Wynne. 2007. A synoptic review of the classification of red algal genera a half century after Kylin's "Die Gattungen der Rhodophyceen". *Bot. Mar.* 50: 197–249.

Scott, J. L., and S. Broadwater. 1989. Ultrastructure of vegetative organization and cell division in the freshwater red alga *Compsopogon*. *Protoplasma* 152:112–122.

Silva, P. C. 1996. Taxonomic and nomenclatural notes. Pp. 910–937 in *Catalogue of the Benthic Marine Algae of the Indian Ocean* (P. C. Silva,

P. W. Basson, and R. L. Moe, eds.). University of California Press, Berkeley, CA.

Turland, N. J., J. H. Wiersema, F. R. Barrie, W. Greuter, D. L. Hawksworth, P. S. Herendeen, S. Knapp, W.-H. Kusber, D.-Z. Li, K. Marhold, T. W. May, J. McNeill, A. M. Monro, J. Prado, M. J. Price, and G. F. Smith, eds. 2018. *International Code of Nomenclature for Algae, Fungi, and Plants (Shenzhen Code)*. Adopted by the Nineteenth International Botanical Congress, Shenzhen, China, July 2017. Regnum Vegetabile 159. Koeltz Botanical Books, Glashütten.

Yoon, H. S., K. M. Müller, R. G. Sheath, F. D. Ott, and D. Bhattacharya. 2006. Defining the major lineages of red algae (*Rhodophyta*). *J. Phycol.* 42:482–492.

Zuccarello, G., J. West, A. Bitans, and G. Kraft. 2000. Molecular phylogeny of *Rhodochaete parvula* (*Bangiophycidae*, *Rhodophyta*). *Phycologia* 39:75–81.

Author

Gary W. Saunders; Centre for Environmental & Molecular Algal Research; Dept. of Biology; University of New Brunswick; Fredericton, NB, E3B 5A3, Canada. Email: gws@unb.ca.

Date Accepted: 19 July 2010; updated 24 February 2018

Primary Editor: Philip Cantino

Porphyridiophyceae H. S. Yoon, K. M. Müller, R. G. Sheath, F. D. Ott and D. Bhattacharya 2006 [G.W. Saunders], converted clade name

Registration Number: 155

Definition: The largest crown clade containing *Porphyridium cruentum* (S. F. Gray) Nägeli 1849 (= *Porphyridium purpureum* (Bory de Saint-Vincent) K. M. Drew and R. Ross 1965) but not *Bangia fuscopurpurea* (Dillwyn) Lyngbye 1819 and *Compsopogon caeruleus* (C. Agardh) Montagne 1846 and *Rhodella maculata* L. Evans 1970 and *Rhodochaete pulchella* Thuret ex Bornet 1892 and *Stylonema alsidii* (Zanardini) K. M. Drew 1956. This is a maximum-crown-clade definition. Abbreviated definition: max crown ∇ (*Porphyridium cruentum* (S. F. Gray) Nägeli 1849 ~ *Bangia fuscopurpurea* (Dillwyn) Lyngbye 1819 & *Compsopogon caeruleus* (C. Agardh) Montagne 1846 & *Rhodella maculata* L. Evans 1970 & *Rhodochaete pulchella* Thuret ex Bornet 1892 & *Stylonema alsidii* (Zanardini) K. M. Drew 1956).

Etymology: Based on the genus name *Porphyridium*, which is derived from Greek, *porphyry-* (purple) and *-idium* (small) in reference to the small purplish cells. The ending *phyceae* is the standard termination under the botanical code (Turland et al., 2018) for algal taxa at the class rank and is derived from the Greek *phykos*, meaning seaweed or alga.

Reference Phylogeny: Yoon et al. (2006: Fig. 3, where this clade is labelled "*Porphyridiales* (3)") is the primary reference phylogeny. See also Müller et al. (2001: Fig. 1). *Rhodochaete pulchella* is incorrectly labeled *R. parvula* in both of these phylogenies (Guiry and Guiry, 2017).

Composition: Includes *Porphyridium* Nägeli 1849, *Erythrolobus* Baca, Wolf and Cox nom.

illeg. and *Flintiella* F.D. Ott in Bourrelly 1970 according to the reference phylogeny (Yoon et al. (2006; "*Porphyridiales* (3)"). There is some confusion surrounding the other included taxa, and the reader is directed to Schneider and Wynne (2007) and Guiry and Guiry (2017) for comprehensive listings of genera and a discussion of the controversies.

Diagnostic Apomorphies: This clade is diagnosed by a combination of putatively apomorphic characters, which nonetheless have analogous states in other clades of red algae, rendering diagnosis on the following suite of characters more reliable than considering individual attributes. Features include unicellular body (possibly a plesiomorphic state), a single branched or stellate plastid, Golgi associated with mitochondrion and ER, and with floridoside as a low molecular weight carbohydrate (see Gabrielson and Garbary, 1986; Yoon et al., 2006).

Synonyms: *Porphyridiales* H. Skuja 1939 and *Porphyridiaceae* H. Skuja 1939 (as applied by some authors) are approximate synonyms.

Comments: This clade was included along with the clades *Rhodellophyceae* and *Stylonematales* (as defined in this volume) in the taxon *Rhodellophyceae* (circumscribed more broadly than in this volume) by Saunders and Hommersand (2004). Their classification was recognized as an interim step until new observations were generated to resolve relationships among these three clades and the other major clades of *Rhodophyta* (*Compsopogonophyceae* and *Eurhodophytina* as designated here). The recent studies by Qiu et al. (2016) and Muñoz-Gómez et al. (2017) have resolved many of these uncertainties.

This clade name is converted from the class name *Porphyridiophyceae* H.S. Yoon, K.M. Müller, R.G. Sheath, F.D. Ott and D. Bhattacharya in the rank-based system; these authors applied the name to the same clade to which it is applied here. The names *Porphyridiales* and *Porphyridiaceae* (see Synonyms) are much more widely used in the literature but have not been applied consistently to this clade, having wide and variable applications in various treatments over the decades. In contrast, the name *Porphyridiophyceae* applies unambiguously to this clade. Silva (1996: 913) provides critical insights into the nomenclature of this clade under the rank-based system.

Literature Cited

Bornet, E. 1892. Les algues de P.-K.-A. Schousboe. *Mém. Soc. Sci. Nat., Cherbourg* 28: 165–376, 3 plates.

Bourrelly, P. 1970. *Les Algues d'Eau Douce, Initiation á la Systématique. Tome III. Les Algues Blues et Rouges, Les Eugléniens, Peridiniens et Cryptomonadines.* Boubée and Cie, Paris.

Drew, K. M. 1956. *Conferva ceramicola* Lyngbye. *Bot. Tidssk.* 53:67–74.

Drew, K. M., and R. Ross. 1965. Some generic names in the *Bangiophycidae. Taxon* 14:93–99.

Evans, L. V. 1970. Electron microscopical observations on a new red algal unicell, *Rhodella maculata* gen. nov., sp. nov. *Br. Phycol. J.* 5:1–13.

Gabrielson, P. W., and D. J. Garbary. 1986. Systematics of red algae (*Rhodophyta*). *Crit. Rev. Plant Sci.* 3:325–366.

Guiry, M. D., and G. M. Guiry. 2017. *AlgaeBase.* World-wide Electronic Publication, National University of Ireland, Galway. Available at http://www.algaebase.org, searched on 2 October 2017.

Lyngbye, H. C. 1819. *Tentamen Hydrophytologiae Danicae.* Schultz, Copenhagen.

Montagne, C. 1846. Ordo I. *Phyceae* Fries. Pp. 1–197 in *Exploration Scientifique de l'Algérie Pendant Les Années 1840, 1841, 1842...Sciences Physiques.* Botanique. *Cryptogamie* (M.C. Durieu De Maisonneuve, ed.). Imprimerie Royale, Publiée Par Ordre Du Gouvernement et Avec le Concours d'une Commission Académique, Paris.

Müller K. M., M. C. Oliveira, R. G. Sheath, and D. Bhattacharya. 2001. Ribosomal DNA phylogeny of the *Bangiophycidae* (*Rhodophyta*) and the origin of secondary plastids. *Am. J. Bot.* 88:1390–1400.

Muñoz-Gómez, S. A., F. G. Mejía-Franco, K. Durnin, M. Colp, C. J. Grisdale, J. M. Archibald, and C. H. Slamovits. 2017. The new red algal subpylum *Proteorhodophytina* comprises the largest and most divergent plastid genomes known. *Curr. Biol.* 27:1–8.

Nägeli, C. 1849. Gattungen einzelliger Algen, physiologisch und systematisch bearbeitet. *Neue Denkschriften der Allg. Schweizerischen Gesellschaft für die Gesammten Naturwissenschaften* 10(7):VIII+139, VIII plates.

Qiu, H., H. S. Yoon, and D. Bhattacharya. 2016. Red algal phylogenomics provides a robust framework for inferring evolution of key metabolic pathways. *PLOS Curr. Tree Life.* 2 December 2016. 1st edition. doi:10.1371/currents.tol. 7b037376e6d84a1be34af756a4d90846.

Saunders, G. W., and M. H. Hommersand. 2004. Assessing red algal supraordinal diversity and taxonomy in the context of contemporary systematic data. *Am. J. Bot.* 91:1494–1507.

Schneider, C. W., and M. J. Wynne. 2007. A synoptic review of the classification of red algal genera a half century after Kylin's "Die Gattungen der Rhodophyceen". *Bot. Mar.* 50:197–249.

Silva, P. C. 1996. Taxonomic and nomenclatural notes. Pp. 910–937 in *Catalogue of the Benthic Marine Algae of the Indian Ocean* (P. C. Silva, P. W. Basson, and R. L. Moe, eds.). University of California Press, Berkeley, CA.

Skuja, H. 1939. Versuch einer systematischen Einteilung der Bangioideen oder Protoflorideen. *Acta Hort. Bot. Univ. Latv.* 11/12:23–38.

Turland, N. J., J. H. Wiersema, F. R. Barrie, W. Greuter, D. L. Hawksworth, P. S. Herendeen, S. Knapp, W.-H. Kusber, D.-Z. Li, K. Marhold, T. W. May, J. McNeill, A. M. Monro, J. Prado, M. J. Price, and G. F. Smith, eds. 2018. *International Code of Nomenclature for Algae, Fungi, and Plants (Shenzhen Code)*. Adopted by the Nineteenth *International Botanical Congress*, Shenzhen, China, July 2017. Regnum Vegetabile 159. Koeltz Botanical Books, Glashütten.

Yoon, H. S., K. M. Müller, R. G. Sheath, F. D. Ott, and D. Bhattacharya. 2006. Defining the major lineages of red algae (*Rhodophyta*). *J. Phycol.* 42:482–492.

Author

Gary W. Saunders; Centre for Environmental & Molecular Algal Research; Dept. of Biology; University of New Brunswick; Fredericton, NB, E3B 5A3, Canada. Email: gws@unb.ca.

Date Accepted: 19 July 2010; updated 24 February 2018

Primary Editor: Philip Cantino

Rhodellophyceae Cavalier-Smith 1998 [G. W. Saunders], converted clade name

Registration Number: 126

Definition: The largest crown clade containing *Rhodella maculata* L. Evans 1970 but not *Bangia fuscopurpurea* (Dillwyn) Lyngbye 1819 and *Compsopogon caeruleus* (C. Agardh) Montagne 1846 and *Porphyridium aerugineum* Geitler 1923 and *Rhodochaete pulchella* Thuret ex Bornet 1892 and *Stylonema cornu-cervi* Reinsch 1875. This is a maximum-crown-clade definition. Abbreviated definition: max crown ∇ (*Rhodella maculata* L. Evans 1970 ~ *Bangia fuscopurpurea* (Dillwyn) Lyngbye 1819 & *Compsopogon caeruleus* (C. Agardh) Montagne 1846 & *Porphyridium aerugineum* Geitler 1923 & *Rhodochaete pulchella* Thuret ex Bornet 1892 & *Stylonema cornu-cervi* Reinsch 1875).

Etymology: Based on the genus name *Rhodella*, which is derived from Greek, *rhodo-* (rosy-red) and Latin, *-ellus* (small). The ending *phyceae* is the standard termination under the botanical code (Turland et al., 2018) for algal taxa at the class rank and is derived from the Greek *phykos*, meaning seaweed or alga.

Reference Phylogeny: The primary reference phylogeny is Müller et al. (2001: Fig. 1), where this clade is labelled "*Porphyridiales* (1)". See also Yoon et al. (2006: Fig. 3). *Rhodochaete pulchella* is incorrectly labeled *R. parvula* in both of these phylogenies (Guiry and Guiry, 2017).

Composition: Includes *Rhodella* L. Evans 1970, *Dixoniella* J. L. Scott, S. T. Broadwater, B. D. Saunders, J. P. Thomas and P. W. Gabrielson 1992, and *Glaucosphaera* Korshikov 1930 (see Saunders and Hommersand (2004); Yoon et al. (2006; "*Porphyridiales* (1)")). See Schneider and Wynne (2007) and Guiry and Guiry (2017) for comprehensive listings of taxa traditionally ranked as genera in this clade (under the name *Rhodellophyceae*).

Diagnostic Apomorphies: This clade is diagnosed by a combination of putatively apomorphic characters, which nonetheless have analogous states in other clades of red algae, rendering diagnosis on the following suite of characters more reliable than considering individual attributes. Features include unicellular body (possibly a plesiomorphic state), a single highly lobed plastid with a pyrenoid, Golgi associated with the nucleus and ER, and mannitol as a low molecular weight carbohydrate (see Gabrielson and Garbary, 1986; Yoon et al., 2006).

Synonyms: The phylogenetically equivalent *Rhodellales* Yoon et al. 2006 and *Rhodellaceae* Yoon et al. 2006 are both approximate synonyms.

Comments: This clade was included along with *Porphyridiophyceae* and *Stylonematales* (as defined in this volume) in a single taxon (called *Rhodellophyceae* but circumscribed more broadly than here) by Saunders and Hommersand (2004). This taxon was recognized as an interim step until new observations were generated to try and resolve relationships among these three clades and the other major clades of *Rhodophyta* (*Compsopogonophyceae* and *Eurhodophytina* as defined in this volume). The recent studies by Qiu et al. (2016) and Muñoz-Gómez et al. (2017) have resolved many of these uncertainties. The clade name *Rhodellophyceae* is converted from the class name *Rhodellophyceae*

Cavalier-Smith 1998 in the rank-based system and is applied in the restricted sense of Yoon et al. (2006: 490). It is preferred over the later names *Rhodellales* and *Rhodellaceae* in being more widely used in the literature.

Literature Cited

Bornet, E. 1892. Les algues de P.-K.-A. Schousboe. *Mém. Soc. Sci. Nat., Cherbourg* 28:165–376, 3 plates.

Cavalier-Smith, T. 1998. A revised six-kingdom system of life. *Biol. Rev.* 73:203–266.

Evans, L. V. 1970. Electron microscopical observations on a new red algal unicell, *Rhodella maculata* gen. nov., sp. nov. *Br. Phycol. J.* 5–13.

Gabrielson, P. W., and D. J. Garbary. 1986. Systematics of red algae (*Rhodophyta*). *Crit. Rev. Plant Sci.* 3:325–366.

Geitler, L. 1923. *Porphyridium aerugineum* nov. sp. *Österr. Bot. Z.* 72:4.

Guiry, M. D., and G. M. Guiry. 2017. *AlgaeBase*. World-wide Electronic Publication, National University of Ireland, Galway. Available at http://www.algaebase.org, searched on 2 October 2017.

Korshikov, A. A. 1930. *Glaucosphaera vacuolata*, a new member of the *Glaucophyceae*. *Arch. Protistenk*. 70: 217–222.

Lyngbye, H. C. 1819. *Tentamen Hydrophytologiae Danicae*. Schultz, Copenhagen.

Montagne, C. 1846. Ordo I. *Phyceae* Fries. Pp. 1–197 in *Exploration Scientifique de l'Algérie Pendant Les Années 1840, 1841, 1842... Sciences Physiques. Botanique. Cryptogamie* (M. C. Durieu De Maisonneuve, ed.). Imprimerie Royale, publiée par ordre du Gouvernement et avec le concours d'une Commission Académique, Paris.

Müller K. M., M. C. Oliveira, R. G. Sheath, and D. Bhattacharya. 2001. Ribosomal DNA phylogeny of the *Bangiophycidae* (*Rhodophyta*) and the origin of secondary plastids. *Am. J. Bot.* 88:1390–1400.

Muñoz-Gómez, S. A., F. G. Mejía-Franco, K. Durnin, M. Colp, C. J. Grisdale, J. M. Archibald, and C. H. Slamovits. 2017. The new red algal subpylum *Proteorhodophytina* comprises the largest and most divergent plastid genomes known. *Curr. Biol.* 27:1–8.

Qiu, H., H. S. Yoon, and D. Bhattacharya. 2016. Red algal phylogenomics provides a robust framework for inferring evolution of key metabolic pathways. *PLOS Curr. Tree Life*. 2 December 2016. doi:10.1371/currents.tol.7b037376e6d84a1be34af756a4d90846.

Reinsch, P. F. 1875. *Contributiones Ad Algologiam et Fungologiam*, Vol. 1, pp. XII + 103, 131 plates [I–III, IIIa, IV–VI, VIa, VII–XII, XIIa, XIII–XX, XXa, XXI–XXXV, XXXVa, XXXVI (Melanophyceae); I–XLII, XLIIa, XLIII–XLVII, XLVIIa, XLVIII–LXI (*Rhodophyceae*); I–XVIII (*Chlorophyllophyceae*); I–IX (Fungi)]. Nürnberg.

Saunders, G. W., and M. H. Hommersand. 2004. Assessing red algal supraordinal diversity and taxonomy in the context of contemporary systematic data. *Am. J. Bot.* 91:1494–1507.

Schneider, C. W., and M. J. Wynne. 2007. A synoptic review of the classification of red algal genera a half century after Kylin's "Die Gattungen der Rhodophyceen". *Bot. Mar.* 50:197–249.

Scott, J. L., S. T. Broadwater, B. D. Saunders, J. P. Thomas, and P. W. Gabrielson. 1992. Ultrastructure of vegetative organization and cell division in the unicellular red alga *Dixoniella griesa* gen. nov. (*Rhodophyta*) and a consideration of the genus *Rhodella*. *J. Phycol.* 28:649–660.

Turland, N. J., J. H. Wiersema, F. R. Barrie, W. Greuter, D. L. Hawksworth, P. S. Herendeen, S. Knapp, W.-H. Kusber, D.-Z. Li, K. Marhold, T. W. May, J. McNeill, A. M. Monro, J. Prado, M. J. Price, and G. F. Smith, eds. 2018. *International Code of Nomenclature for Algae, Fungi, and Plants (Shenzhen Code)*. Adopted by the Nineteenth International Botanical Congress, Shenzhen, China, July 2017. Regnum Vegetabile 159. Koeltz Botanical Books, Glashütten.

Yoon, H. S., K. M. Müller, R. G. Sheath, F. D. Ott, and D. Bhattacharya. 2006. Defining the major lineages of red algae (*Rhodophyta*). *J. Phycol.* 42:482–492.

Author

Gary W. Saunders; Centre for Environmental & Molecular Algal Research; Dept. of Biology; University of New Brunswick; Fredericton, NB, E3B 5A3, Canada. Email: gws@unb.ca.

Date Accepted: 19 July 2010; updated 24 February 2018

Primary Editor: Philip Cantino

Stylonematales K. M. Drew 1956 [G. W. Saunders], converted clade name

Registration Number: 127

Definition: The largest crown clade containing *Stylonema cornu-cervi* Reinsch 1875 but not *Bangia fuscopurpurea* (Dillwyn) Lyngbye 1819 and *Compsopogon caeruleus* (C. Agardh) Montagne 1846 and *Porphyridium aerugineum* Geitler 1923 and *Rhodella maculata* L. Evans 1970 and *Rhodochaete pulchella* Thuret ex Bornet 1892. This is a maximum-crown-clade definition. Abbreviated definition: max crown ∇ (*Stylonema cornu-cervi* Reinsch 1875 ~ *Bangia fuscopurpurea* (Dillwyn) Lyngbye 1819 & *Compsopogon caeruleus* (C. Agardh) Montagne 1846 & *Porphyridium aerugineum* Geitler 1923 & *Rhodella maculata* L. Evans 1970 & *Rhodochaete pulchella* Thuret ex Bornet 1892).

Etymology: Derived from Greek, *stylus* (pillar or column) and *nema* (thread) in reference to the fine, thread-like thalli of some members.

Reference Phylogeny: The primary reference phylogeny is Müller et al. (2001: Fig. 1), where this clade is labelled "*Porphyridiales* (2)". See also Yoon et al. (2006: Fig. 3). *Rhodochaete pulchella* is incorrectly labeled *R. parvula* in both of these phylogenies (Guiry and Guiry, 2017).

Composition: Includes a number of unicellular and pseudofilamentous red algal taxa as outlined in Yoon et al. (2006; "*Porphyridiales* (2)"). Owing to some uncertainty as to the composition of this taxon, only those genera included in Yoon et al. (2006) are listed here: *Bangiopsis* F. Schmitz 1896, *Chroodactylon* Hansgirg 1885, *Chroothece* Hansgirg in Wittrock and Nordstedt 1884, *Kyliniella* H. Skuja 1926, *Purpureofilum* J. A. West, G. C. Zuccarello and J. L. Scott in West et al. 2005, *Rhodosorus* Geitler 1930, *Rhodospora* Geitler 1927, *Rufusia* D. E. Wujek and P. Timpano 1986 and *Stylonema* Reinsch 1875. See Schneider and Wynne (2007) and Guiry and Guiry (2017) for comprehensive listings of taxa traditionally ranked as genera in this clade (under the name *Stylonematophyceae*), as well as discussions regarding taxonomic controversies in this clade.

Diagnostic Apomorphies: This clade is diagnosed by the presence of putative group II introns in the *psa*A gene at conserved locations (see Yoon et al., 2006). In addition, the taxa in this clade are generally characterized by sorbitol/digeneaside or only sorbitol as a low molecular weight carbohydrate, unicellular, pseudofilamentous or filamentous construction, and the Golgi associated with the mitochondrion and ER (see Gabrielson and Garbary, 1986; Yoon et al., 2006).

Synonyms: *Stylonematophyceae* Yoon et al. 2006 and the phylogenetically equivalent name *Stylonemataceae* Drew 1956 are both approximate synonyms.

Comments: This clade was included along with the clades *Porphyridiophyceae* and *Rhodellophyceae* (as defined in this volume) in the taxon *Rhodellophyceae* (circumscribed more broadly than in this volume) by Saunders and Hommersand (2004). This action was recognized as an interim step until new observations were generated to resolve relationships among these clades and the other major lines of *Rhodophyta* (*Compsopogonophyceae* and *Eurhodophytina* as defined in this volume). The recent studies by Qiu et al. (2016) and

Muñoz-Gómez et al. (2017) have resolved many of these uncertainties.

The clade name *Stylonematales* is based on the name *Stylonematales* K. M. Drew in the rank-based system and is applied here in the sense of Yoon et al. (2006: 490). It is preferred over the names *Stylonematophyceae* and *Stylonemataceae* in being more widely and consistently used in the literature.

Literature Cited

Bornet, E. 1892. Les algues de P.-K.-A. Schousboe. *Mém. Soc. Sci. Nat., Cherbourg* 28:165–376, 3 plates.

Drew, K. M. 1956. *Conferva ceramicola* Lyngbye. *Bot. Tidsskr.* 53:67–74.

Evans, L. V. 1970. Electron microscopical observations on a new red algal unicell, *Rhodella maculata* gen. nov., sp. nov. *Br. Phycol. J.* 5:1–13.

Gabrielson, P. W., and D. J. Garbary. 1986. Systematics of red algae (*Rhodophyta*). *Crit. Rev. Plant Sci.* 3:325–366.

Geitler, L. 1923. *Porphyridium aerugineum* nov. sp. *Österr. Bot. Z.* 72:4.

Geitler, L. 1927. *Rhodospora sordida*, nov. gen. et n. sp., eine neue "Bangiacee" des Süsswassers. *Österr. Bot. Z.* 76:25–28.

Geitler, L. 1930. Ein grünes Filarplasmodium und andere neue Protisten. *Arch. Protistenk.* Bd. 69, Jena.

Guiry, M. D., and G. M. Guiry. 2017. *AlgaeBase.* World-wide Electronic Publication, National University of Ireland, Galway. Available at http://www.algaebase.org, searched on 2 October 2017.

Hansgirg, A. 1885. Ein Beitrag zur Kenntniss von der Verbreitung der Chromatophoren und Zellkernen bei den Schizophyceen (Phycochromaceen). *Ber. Deutsch. Bot. Ges.* 3:14–22.

Lyngbye, H. C. 1819. *Tentamen Hydrophytologiae Danicae.* Schultz, Copenhagen.

Montagne, C. 1846. Ordo I. *Phyceae* Fries. Pp. 1–197 in *Exploration Scientifique de l'Algérie Pendant Les Années 1840, 1841, 1842...Sciences Physiques. Botanique. Cryptogamie* (M. C. Durieu De Maisonneuve, ed.). Imprimerie Royale, Publiée Par Ordre du Gouvernement et Avec le Concours d'une Commission Académique, Paris.

Müller K. M., M. C. Oliveira, R. G. Sheath, and D. Bhattacharya. 2001. Ribosomal DNA phylogeny of the *Bangiophycidae* (*Rhodophyta*) and the origin of secondary plastids. *Am. J. Bot.* 88:1390–1400.

Muñoz-Gómez, S. A., F. G. Mejía-Franco, K. Durnin, M. Colp, C. J. Grisdale, J. M. Archibald, and C. H. Slamovits. 2017. The new red algal subpylum *Proteorhodophytina* comprises the largest and most divergent plastid genomes known. *Curr. Biol.* 27:1–8.

Qiu, H., H. S. Yoon, and D. Bhattacharya. 2016. Red algal phylogenomics provides a robust framework for inferring evolution of key metabolic pathways. *PLOS Curr. Tree Life.* 2 December 2016. doi:10.1371/currents.tol.7b03 7376e6d84a1be34af756a4d90846.

Reinsch, P. F. 1875. *Contributiones Ad Algologiam et Fungologiam*, Vol. 1, pp. XII + 103, 131 plates [I–III, IIIa, IV–VI, VIa, VII–XII, XIIa, XIII–XX, XXa, XXI–XXXV, XXXVa, XXXVI (*Melanophyceae*); I–XLII, XLIIa, XLIII-XLVII, XLVIIa, XLVIII–LXI (*Rhodophyceae*); I–XVIII (*Chlorophyllophyceae*); I–IX (*Fungi*)]. Nürnberg.

Saunders, G. W., and M. H. Hommersand. 2004. Assessing red algal supraordinal diversity and taxonomy in the context of contemporary systematic data. *Am. J. Bot.* 91:1494–1507.

Schmitz, F. 1896. *Bangiaceae*. Pp. 307–316 in *Die Natürlichen Pflanzenfamilien* I. Teil, Abt. 2. (A. Engler and K. Prantl eds.). Leipzig.

Schneider, C. W., and M. J. Wynne. 2007. A synoptic review of the classification of red algal genera a half century after Kylin's "Die Gattungen der Rhodophyceen". *Bot. Mar.* 50:197–249.

Skuja, H. 1926. Eine neue Süsswasserbangiacee *Kyliniella* n. g., n. sp. *Acta Hort. Bot. Univ. Latv.* Bd. 1, Riga.

West, J. A., G. C. Zuccarello, J. Scott, J. Pickett-Heaps, and G. H. Kim. 2005. Observations on *Purpureofilum apyrenoidigerum* gen. et sp. nov. from Australia and *Bangiopsis subsimplex* from India (*Stylonematales, Bangiophyceae, Rhodophyta*). *Phycol. Res.* 53:49–66.

Wittrock, V. B., and O. Nordstedt. 1884. Algae aquae dulcis exsiccatae. Fasc. 14: 651–700. Stockholm. (Exsiccata with printed labels. Index and descriptions of new taxa reprinted in *Botaniska Notiser* 1884:121–128.)

Wujek, D. E., and P. Timpano. 1986. *Rufusia* (*Porphyridiales, Phragmonemataceae*), a new red alga from sloth hair. *Brenesia* 25/26:163–168.

Yoon, H. S., K. M. Müller, R. G. Sheath, F. D. Ott, and D. Bhattacharya. 2006. Defining the major lineages of red algae (*Rhodophyta*). *J. Phycol.* 42:482–492.

Author

Gary W. Saunders; Centre for Environmental & Molecular Algal Research; Dept. of Biology; University of New Brunswick; Fredericton, NB, E3B 5A3, Canada. Email: gws@unb.ca.

Date Accepted: 19 July 2010; updated 24 February 2018

Primary Editor: Philip Cantino

Eurhodophytina G. W. Saunders and M. H. Hommersand 2004 [G. W. Saunders], converted clade name

Registration Number: 128

Definition: The crown clade originating in the most recent common ancestor of *Bangia fuscopurpurea* (Dillwyn) Lyngbye 1819 and *Palmaria palmata* (Linnaeus) Kuntze 1891. This is a minimum-crown-clade definition. Abbreviated definition: min crown ∇ (*Bangia fuscopurpurea* (Dillwyn) Lyngbye 1819 & *Palmaria palmata* (Linnaeus) Kuntze 1891).

Etymology: Derived from Greek, *eu-* (true), *rhodo-* (rosy-red), and *phyton* (plant).

Reference Phylogeny: The primary reference phylogeny is Yoon et al. (2006: Fig. 3). See also Müller et al. (2001: Fig. 1).

Composition: Contains *Bangiales* and *Florideophyceae* as defined in this volume.

Diagnostic Apomorphies: This clade is diagnosed by a combination of putatively apomorphic characters, which nonetheless have analogous states in other clades of red algae, rendering diagnosis on the following suite of characters more reliable than considering individual attributes: Golgi in an endoplasmic reticulum/mitochondrial association; sexual life histories, where known, biphasic or triphasic (see Gabrielson and Garbary, 1986); pit plugs present in at least one stage of sexual life histories and displaying caps and or membranes at their cytoplasmic faces (Pueschel and Cole, 1982) (secondarily lost in *Ahnfeltia*; Le Gall and Saunders, 2007).

Synonyms: *Eurhodophycidae* Magne (1989) is an approximate synonym.

Comments: This clade is best characterized by the complexity of representative pit-plug ultrastructure (Pueschel and Cole, 1982) and is resolved in nearly every publication to date that addresses red algal phylogeny (e.g., Yoon et al., 2006; Müller et al., 2001). The majority of the red macroalgae are included in *Eurhodophytina* as defined here and in the rank-based construct of Saunders and Hommersand (2004). This includes all taxa of the clades *Bangiales* and *Florideophyceae* as defined in this volume. The reader is directed to Saunders and Hommersand (2004) and references therein for an in-depth review of this clade.

Eurhodophycidae Magne (1989) was the base name for the subphylum name *Eurhodophytina* Saunders and Hommersand (2004). The latter is nonetheless better known in the literature and is selected here as the source for the converted clade name. *Macrorhodophytina* Cavalier-Smith (1998) was a subphylum erected to contain all of *Eurhodophytina* as defined here, but also species of *Rhodochaetales*, thus the name referred to a group that we now consider to be polyphyletic. Saunders and Hommersand (2004) rejected Cavalier-Smith's construct as "too broad to be applied" to any of the groups they recognized and instead applied the name *Eurhodophytina* to the same clade to which it is applied here.

Literature Cited

Cavalier-Smith, T. 1998. A revised six-kingdom system of life. *Biol. Rev.* 73:203–266.

Gabrielson, P. W., and D. J. Garbary. 1986. Systematics of red algae (*Rhodophyta*). *Crit. Rev. Plant Sci.* 3:325–366.

Kuntze, O. 1891. *Revisio Generum Plantarum*, Pars 2, pp. 375–1011. Arthur Felix, Leipzig.

Le Gall, L., and G. W. Saunders. 2007. A nuclear phylogeny of the *Florideophyceae* (*Rhodophyta*) inferred from combined EF2, small subunit and large subunit ribosomal DNA: establishing the new red algal subclass *Corallinophycidae*. *Mol. Phylogenet. Evol.* 43: 1118–1130.

Lyngbye, H. C. 1819. *Tentamen Hydrophytologiae Danicae*. Schultz, Copenhagen.

Magne, F. 1989. Classification et phylogénie des Rhodophycées. *Cryptgam. Algol.* 10: 101–115.

Müller, K. M., M. C. Oliveira, R. G. Sheath, and D. Bhattacharya. 2001. Ribosomal DNA phylogeny of the *Bangiophycidae* (*Rhodophyta*) and the origin of secondary plastids. *Am. J. Bot.* 88:1390–1400.

Pueschel, C. M., and K. M. Cole. 1982. Rhodophycean pit plugs: an ultrastructural survey with taxonomic implications. *Am. J. Bot.* 69:703–720.

Saunders, G. W., and M. H. Hommersand. 2004. Assessing red algal supraordinal diversity and taxonomy in the context of contemporary systematic data. *Am. J. Bot.* 91:1494–1507.

Yoon, H. S., K. M. Müller, R. G. Sheath, F. D. Ott, and D. Bhattacharya. 2006. Defining the major lineages of red algae (*Rhodophyta*). *J. Phycol.* 42:482–492.

Author

Gary W. Saunders; Centre for Environmental and Molecular Algal Research; Dept. of Biology; University of New Brunswick; Fredericton, NB, E3B 5A3, Canada. Email: gws@unb.ca.

Date Accepted: 19 July 2010; updated 24 February 2018

Primary Editor: Philip Cantino

Bangiales Nägeli 1847 [G. W. Saunders], converted clade name

Registration Number: 129

Definition: The largest crown clade containing *Bangia fuscopurpurea* (Dillwyn) Lyngbye 1819 but not *Palmaria palmata* (Linnaeus) Kuntze 1891. This is a maximum-crown-clade definition. Abbreviated definition: max crown ∇ (*Bangia fuscopurpurea* (Dillwyn) Lyngbye 1819 ~ *Palmaria palmata* (Linnaeus) Kuntze 1891).

Etymology: Based on the genus name *Bangia*, which was named in honour of Niels Hofmann-Bang (1776–1855).

Reference Phylogeny: The primary reference phylogeny is Müller et al. (2001: Fig. 1). See also Yoon et al. (2006: Fig. 3).

Composition: This clade contains the following red algal taxa traditionally ranked as genera: *Bangia* Lyngbye 1819 and *Porphyra* C. Agardh 1824 (Müller et al., 2003; Saunders and Hommersand, 2004), *Bangiadulcis* W. A. Nelson 2007, *Dione* W. A. Nelson 2005 in Nelson et al. 2005, *Minerva* W. A. Nelson 2005 in Nelson et al. 2005, and *Pseudobangia* K. M. Müller and R. G. Sheath 2005 in Müller et al. 2005. Species of *Porphyra* sensu lato have recently been redistributed into a number of genera (see Sutherland et al., 2011). Consult Guiry and Guiry (2017) for a comprehensive list.

Diagnostic Apomorphies: Species in this group are diagnosed by pit plugs with a single cap layer only at the cytoplasmic faces and a characteristic biphasic life history in which parenchymatous gametophytes lacking pit plugs form reproductive structures in packets by successive divisions alternating with a filamentous sporophyte with pit plugs and conchospore production (Pueschel and Cole, 1982; Saunders and Hommersand, 2004).

Synonyms: As used by some recent authors (e.g., Saunders and Hommersand, 2004; Yoon et al., 2006), the names *Bangiophyceae* Wettstein 1901 and *Bangiophycidae* Wettstein 1901 are approximate synonyms of *Bangiales*, as is the phylogenetically equivalent *Bangiaceae* Engler 1892.

Comments: *Bangiales* was selected as the source of the converted name because it has generally been applied unambiguously to this clade since 1980 (Garbary et al., 1980), in contrast to the names *Bangiophyceae* and *Bangiophycidae*, which have been more broadly applied (e.g., Dixon, 1973; Gabrielson and Garbary, 1986). *Bangiales* is more widely used in the literature than *Bangiaceae*. Although monophyly of the *Bangiales* is well supported (cf. Müller et al., 2001), some of the component taxa traditionally ranked as genera are not monophyletic (cf. Müller et al., 2003; and references therein) and are characterized by genetic divergence levels among species that dwarf some ordinal level distances within the *Florideophyceae* (Oliveira et al., 1995). Considerable taxonomic work is required in this clade, which will undoubtedly lead to the recognition of many more clades among these morphologically conservative species. For examples of recent improvements, see Müller et al. (2005), Nelson et al. (2005), Nelson (2007) and significantly Sutherland et al. (2011).

Literature Cited

Agardh, C. A. 1824. *Systema Algarum*, pp. [i]–xxxviii, [1]–312. Berling, Lund.

Dixon, P. S. 1973. *Biology of the* Rhodophyta. Oliver and Boyd, Edinburgh.

Engler, A. 1892. *Syllabus der Vorlesungen Über Specielle und Medicinisch–Pharmaceutische Botanik... Grosse Ausgabe. Gebr.* Borntraeger, Berlin.

Gabrielson, P. W., and D. J. Garbary. 1986. Systematics of red algae (*Rhodophyta*). *Crit. Rev. Plant Sci.* 3:325–366.

Garbary, D. J., G. K. Hansen, and R. F. Scagel. 1980. A revised classification of the *Bangiophyceae* (*Rhodophyta*). *Nova Hedwigia* 33:145–166.

Guiry, M. D., and G. M. Guiry. 2017. *AlgaeBase.* World-wide Electronic Publication, National University of Ireland, Galway. Available at http://www.algaebase.org, searched on 2 October 2017.

Kuntze, O. 1891. *Revisio Generum Plantarum*, Pars 2. Leipzig. 375–1011.

Lyngbye, H. C. 1819. *Tentamen Hydrophytologiae Danicae.* Schultz, Copenhagen.

Müller, K. M., J. J. Cannone, and R. G. Sheath. 2005. A molecular phylogenetic analysis of the *Bangiales* (*Rhodophyta*) and description of a new genus and species, *Pseudobangia kaycoleia.* *Phycologia* 44:146–155.

Müller, K. M., K. M. Cole, and R. G. Sheath. 2003. Systematics of *Bangia* (*Bangiales, Rhodophyta*) in North America. II. Biogeographical trends in karyology: chromosome numbers and linkage with gene sequence phylogenetic trees. *Phycologia* 42:209–219.

Müller, K. M., M. C. Oliveira, R. G. Sheath, and D. Bhattacharya. 2001. Ribosomal DNA phylogeny of the *Bangiophycidae* (*Rhodophyta*) and the origin of secondary plastids. *Am. J. Bot.* 88:1390–1400.

Nägeli, C. 1847. *Die Neuern Algensysteme und Versuch zur Begründung Eines Eigenen Systems der Algae und Florideen.* Schulthess, Zurich.

Nelson, W. A. 2007. *Bangiadulcis* gen. nov.: a new genus for freshwater filamentous *Bangiales* (*Rhodophyta*). *Taxon* 56:883–886.

Nelson, W. A., T. J. Farr, and J. E. S. Broom. 2005. *Dione* and *Minerva*, two new genera from New Zealand circumscribed for basal taxa in the *Bangiales* (*Rhodophyta*). *Phycologia* 44:139–145.

Oliveira, M. C., J. Kurniawan, C. J. Bird, E. L. Rice, C. A. Murphy, R. K. Singh, R. R. Gutell, and M. A. Ragan. 1995. A preliminary investigation of the order *Bangiales* (*Bangiophycidae, Rhodophyta*) based on sequences of nuclear small-subunit ribosomal RNA genes. *Phycol. Res.* 43:71–79.

Pueschel, C. M., and K. M. Cole. 1982. Rhodophycean pit plugs: an ultrastructural survey with taxonomic implications. *Am. J. Bot.* 69:703–720.

Saunders, G. W., and M. H. Hommersand. 2004. Assessing red algal supraordinal diversity and taxonomy in the context of contemporary systematic data. *Am. J. Bot.* 91:1494–1507.

Sutherland, J. E., S. C. Lindstrom, W. A. Nelson, J. Brodie, M. D. Lynch, M. S. Hwang, H.-G. Choi, M. Miyata, N. Kikuchi, M. C. Oliveira, T. Farr, C. Neefus, A. Mols-Mortensen, D. Milstein, and K. M. Müller. 2011. A new look at an ancient order: generic revision of the *Bangiales* (*Rhodophyta*). *J. Phycol.* 47(5):1131–1151.

Wettstein, A. 1901. *Handbuch der Systematischen Botanik.* Leipzig, Vienna.

Yoon, H. S., K. M. Müller, R. G., Sheath, F. D. Ott, and D. Bhattacharya. 2006. Defining the major lineages of red algae (*Rhodophyta*). *J. Phycol.* 42:482–492.

Author

Gary W. Saunders; Centre for Environmental & Molecular Algal Research; Dept. of Biology; University of New Brunswick; Fredericton, NB, E3B 5A3, Canada. Email: gws@unb.ca.

Date Accepted: 19 July 2010; updated 24 February 2018

Primary Editor: Philip Cantino

Florideophyceae A. Cronquist 1960 [G. W. Saunders], converted clade name

Registration Number: 130

Definition: The crown clade originating in the most recent common ancestor of *Ahnfeltia plicata* (Hudson) Fries 1836 (*Ahnfeltiophycidae*), *Chondrus crispus* Stackhouse 1797 (*Rhodymeniophycidae*), *Corallina officinalis* Linnaeus 1758 (*Corallinophycidae*), *Hildenbrandia rubra* (Sommerfelt) Meneghini 1841 (*Hildenbrandiales*) and *Palmaria palmata* (Linnaeus) Kuntze 1891(*Nemaliophycidae*). This is a minimum-crown-clade definition. Abbreviated definition: min crown ∇ (*Ahnfeltia plicata* (Hudson) Fries 1836 & *Chondrus crispus* Stackhouse 1797 & *Corallina officinalis* Linnaeus 1758 & *Hildenbrandia rubra* (Sommerfelt) Meneghini 1841 & *Palmaria palmata* (Linnaeus) Kuntze 1891).

Etymology: Derived from Latin, *floridus* (profusely flowering) and Greek, *phykos* (seaweed or alga). The ending *phyceae* is the standard termination under the botanical code (Turland et al., 2018) for algal taxa at the class rank.

Reference Phylogeny: The primary reference phylogeny is Saunders and Bailey (1997: Fig. 1). See also Müller et al. (2001: Fig. 2), Saunders and Bailey (1999: Fig. 1) and Yoon et al. (2006: Fig. 3).

Composition: Contains *Ahnfeltiophycidae*, *Corallinophycidae*, *Hildenbrandiales*, *Nemaliophycidae*, and *Rhodymeniophycidae* as defined in this volume.

Diagnostic Apomorphies: Florideophyte red algae are diagnosed by the presence of tetrasporangia, although this feature has been lost secondarily in some clades (e.g., in *Batrachospermales*).

In addition, these algae are characterized generally by apical vegetative growth and apical production of reproductive structures, life histories that are triphasic (including a carposporophyte), the presence of pit plugs in both gametophytic and sporophytic generations, and trichogynes (hair-like extensions) on the carpogonia (female gametangia) (see Gabrielson and Garbary, 1986; Saunders and Hommersand, 2004).

Synonyms: *Rhodymeniophyceae* Doweld 2001 is an approximate synonym.

Comments: This clade has long been recognized as monophyletic owing to a shared suite of unusual and unique reproductive structures and patterns (see Saunders and Hommersand, 2004). Its monophyly is supported by the initial SSU phylogenies of Ragan et al. (1994) and all subsequent phylogenies to consider red algal-wide relationships (see Saunders and Hommersand 2004).

The clade name *Florideophyceae* is converted from *Florideophyceae* A. Cronquist 1960, a class name under the rank-based system, and is applied to the same clade as by Saunders and Hommersand (2004). Doweld (2001) introduced the typified name *Rhodymeniophyceae* for this taxon (assigned to the rank of class under the rank-based system), but it was rejected by Saunders and Hommersand (2004) because the name *Florideophyceae* traces back to the vernacular name Floridées Lamouroux (1813)—the first time red algae were grouped together as a distinct taxon—and also because this name has been applied for over a century to an unambiguous taxonomic construct. *Florideophyceae* is much more frequently used in the scientific literature than its approximate synonym

(*Rhodymeniophyceae*) and thus is selected for the converted clade name.

Literature Cited

Cronquist, A. 1960. The divisions and classes of plants. *Bot. Rev.* 26:425–482.

Doweld, A. 2001. *Prosyllabus tracheophytorum.* GEOS, Moscow.

Fries, E. M. 1836. *Corpus Floranum Provinciallum Sueciae. I. Floram Scanicam Scripsit Elias Fries.* Uppsala.

Gabrielson, P. W., and D. J. Garbary. 1986. Systematics of red algae (*Rhodophyta*). *Crit. Rev. Plant Sci.* 3:325–366.

Kuntze, O. 1891. *Revisio generum plantarum*, Pars 2. Arthur Felix, Leipzig, pp. 375–1011.

Lamouroux, J. V. F. 1813. Essai sur les genres de la famille de Thalassiophytes non articulées. *Ann. Mus. Hist. Nat. Paris* 20:21–47, 115–139, 267–293, seven plates.

Linnaeus, C. 1758. *Systema Naturae Per Regna Tria Naturae.* Laurentii Salvii, Holmiae (Stockholm).

Meneghini, G. 1841. Algologia dalmatica. *Atti del Terza Riunione Degli Scienziati Italiani Tenuta in Firenze* 3:424–431.

Müller K. M., M. C. Oliveira, R. G. Sheath, and D. Bhattacharya. 2001. Ribosomal DNA phylogeny of the *Bangiophycidae* (*Rhodophyta*) and the origin of secondary plastids. *Am. J. Bot.* 88:1390–1400.

Ragan, M. A., C. J. Bird, E. L. Rice, R. R. Gutell, C. A. Murphy, and R. K. Singh. 1994. A molecular phylogeny of the marine red algae (*Rhodophyta*) based on the nuclear small-subunit rRNA gene. *Proc. Natl. Acad. Sci. USA* 91:727–7280.

Saunders, G. W., and J. C. Bailey. 1997. Phylogenesis of pit-plug-associated features in the *Rhodophyta*: inferences from molecular systematic data. *Can. J. Bot.* 75:1436–1447.

Saunders, G. W., and J. C. Bailey. 1999. Molecular systematic analyses indicate that the enigmatic *Apophlaea* is a member of the *Hildenbrandiales* (*Rhodophyta*, *Florideophycidae*). *J. Phycol.* 35:171–175.

Saunders, G. W., and M. H. Hommersand. 2004. Assessing red algal supraordinal diversity and taxonomy in the context of contemporary systematic data. *Am. J. Bot.* 91:1494–1507.

Stackhouse, J. 1797. *Nereis Britannica; Continens Species Omnes Fucorum in Insulis Britannicis Crescentium: Descriptione Latine et Anglico, Necnon Iconibus Ad Vivum Depictis...* Fasc. 2, pp. ix–xxiv, 31–70, pls IX–XIII. J. White, London.

Turland, N. J., J. H. Wiersema, F. R. Barrie, W. Greuter, D. L. Hawksworth, P. S. Herendeen, S. Knapp, W.-H. Kusber, D.-Z. Li, K. Marhold, T. W. May, J. McNeill, A. M. Monro, J. Prado, M. J. Price, and G. F. Smith, eds. 2018. *International Code of Nomenclature for Algae, Fungi, and Plants (Shenzhen Code).* Adopted by the Nineteenth International Botanical Congress, Shenzhen, China, July 2017. Regnum Vegetabile 159. Koeltz Botanical Books, Glashütten.

Yoon, H. S., K. M. Müller, R. G. Sheath, F. D. Ott, and D. Bhattacharya. 2006. Defining the major lineages of red algae (*Rhodophyta*). *J. Phycol.* 42:482–492.

Author

Gary W. Saunders; Centre for Environmental and Molecular Algal Research; Dept. of Biology; University of New Brunswick; Fredericton, NB, E3B 5A3, Canada. Email: gws@unb.ca.

Date Accepted: 19 July 2010; updated 24 February 2018

Primary Editor: Philip Cantino

Hildenbrandiales C. M. Pueschel and K. M. Cole 1982
[G. W. Saunders], converted clade name

Registration Number: 133

Definition: The largest crown clade containing *Hildenbrandia prototypus* Nardo 1834 (= *Hildenbrandia rubra* (Sommerfelt) Meneghini 1841) but not *Ahnfeltia plicata* (Hudson) Fries 1836 and *Chondrus crispus* Stackhouse 1797 and *Corallina officinalis* Linnaeus 1758, and *Palmaria palmata* (Linnaeus) Kuntze 1891. This is a maximum-crown-clade definition. Abbreviated definition: max crown ∇ (*Hildenbrandia prototypus* Nardo 1834 ~ *Ahnfeltia plicata* (Hudson) Fries 1836 & *Chondrus crispus* Stackhouse 1797 & *Corallina officinalis* Linnaeus 1758 & *Palmaria palmata* (Linnaeus) Kuntze 1891).

Etymology: Derived from the genus name *Hildenbrandia*, which was named in honour of Professor F. E. Hildenbrandt.

Reference Phylogeny: The primary reference phylogeny is Saunders and Bailey (1999: Fig. 1). See also Le Gall and Saunders (2007: Fig. 4).

Composition: The clade *Hildenbrandiales* is composed of those species traditionally included in the order *Hildenbrandiales* (confined to the two genera *Apophlaea* J.D. Hooker and Harvey 1845 and *Hildenbrandia* Nardo 1834) under the rank-based system (see Saunders and Bailey (1999) and Saunders and Hommersand (2004) for references and details).

Diagnostic Apomorphies: This clade is diagnosed by a series of putatively apomorphic characters, which nonetheless have analogous states in other clades of red algae, rendering diagnosis on the following suite of characters more reliable than considering individual attributes. Diagnostic features include a thallus that is crustose and smooth to tubercular or with erect branches; anatomically consisting of a basal layer of laterally adhering branched filaments and laterally adhering simple or branched erect filaments. Reproduction is by direct cycling of apomeiotic tetrasporangia; sexual life histories are unknown. Tetrasporangia are zonately or irregularly divided, borne in ostiolate conceptacles, and division is initially successive but completed simultaneously. Secondary pit connections of the direct type are common between vegetative filaments. Pit plugs have a membrane and single cap layer at cytoplasmic faces (Pueschel and Cole, 1982; Pueschel 1982, 1988; Hawkes, 1983; Saunders and Hommersand, 2004).

Synonyms: *Hildenbrandiophycidae* G. W. Saunders and M. H. Hommersand 2004 and *Hildenbrandiaceae* Rabenhorst 1868 are approximate synonyms of *Hildenbrandiales*.

Comments: Saunders and Bailey (1999) provided the first molecular support for a close phylogenetic relationship between *Hildenbrandia* and *Apophlaea*. Hawkes (1983) had earlier suggested this as a likely association based on a series of morphological and anatomical similarities between species of these two genera. *Hildenbrandiales* is a poorly understood clade of red algae considered to have very low species diversity (possibly an artifact of the simple crustose morphology common to this group) and a complete absence of sexual life histories. As traditional red algal taxonomy was based almost exclusively on the latter, the taxonomic placement of this group has been ambiguous, with the current perspectives based almost exclusively on molecular phylogenies (see Saunders and Bailey, 1999). In a rank-based system, the use of *Hildenbrandiophycidae* G. W. Saunders and M. H. Hommersand would be preferred to signify that this clade is equivalent in rank (subclass) to other major clades included in the clade *Florideophyceae*. However, the older name *Hildenbrandiales* (traditionally associated with the rank of order) is more extensively used in contemporary systematic literature than either of its phylogenetic synonyms (*Hildenbrandiophycidae* and *Hildenbrandiaceae*) and is therefore the preferred source for a converted clade name to establish consistency between the phylogenetic nomenclature and previous literature on this group of algae.

Literature Cited

Fries, E. M. 1836. *Corpus Floranum Provinciallum Sueciae. I. Floram Scanicam Scripsit Elias Fries.* Uppsala.

Hawkes, M. W. 1983. Anatomy of *Apophlaea sinclairii* – an enigmatic red alga endemic to New Zealand. *Jpn. J. Phycol.* 31:51–64.

Hooker, J. D., and W. H. Harvey. 1845. Algae Novae Zelandiae; being a catalogue of all of the species of algae yet recorded as inhabiting the shores of New Zealand, with characters and brief descriptions of the new species discovered during the voyage of H.M. discovery ships "Erebus" and "Terror" and of others communicated to Sir W. Hooker by D. Sinclair, the Rev. Colenso, and M. Raoul. *London J. Bot.* 4:521–551.

Kuntze, O. 1891. *Revisio Generum Plantarum,* Pars 2, pp. 375–1011. Leipzig.

Le Gall, L., and G. W. Saunders. 2007. A nuclear phylogeny of the *Florideophyceae* (*Rhodophyta*) inferred from combined EF2, small subunit and large subunit ribosomal DNA: establishing the new red algal subclass *Corallinophycidae. Mol. Phylogenet. Evol.* 43:1118–1130.

Linnaeus, C. 1758. *Systema Naturae Per Regna Tria Naturae.* Laurentii Salvii, Holmiae (Stockholm).

Meneghini, G. 1841. Algologia dalmatica. *Atti Del Terza Riunione Degli Scienziati Italiani Tenuta in Firenze* 3:4–431.

Nardo, J. D. 1834. *De Novo Genere Algarum Cul Nomen Est* Hildenbrandia prototypus. Isis, Oken.

Pueschel, C. M. 1982. Ultrastructural observations of tetrasporangia and conceptacles in *Hildenbrandia* (*Rhodophyta, Hildenbrandiales*). *Br. Phycol. J.* 17:333–341.

Pueschel, C. M. 1988. Secondary pit connections in *Hildenbrandia* (*Rhodophyta, Hildenbrandiales*). *Br. Phycol. J.* 23:25–32.

Pueschel, C. M., and K. M. Cole. 1982. Rhodophycean pit plugs: an ultrastructural survey with taxonomic implications. *Am. J. Bot.* 69:703–720.

Rabenhorst, L. 1868. *Flora Europaea Algarum Aquae Dulcis et Submarinae.* Kummer, Leipzig.

Saunders, G. W., and J. C. Bailey. 1999. Molecular systematic analyses indicate that the enigmatic *Apophlaea* is a member of the *Hildenbrandiales* (*Rhodophyta, Florideophycidae*). *J. Phycol.* 35:171–175.

Saunders, G. W., and M. H. Hommersand. 2004. Assessing red algal supraordinal diversity and taxonomy in the context of contemporary systematic data. *Am. J. Bot.* 91:1494–1507.

Stackhouse, J. 1797. *Nereis Britannica; Continens Species Omnes Fucorum in Insulis Britannicis Crescentium: Descriptione Latine et Anglico, Necnon Iconibus ad Vivum Depictis...* Fasc. 2, pp. ix–xxiv, 31–70, pls IX–XIII. J. White, London.

Author

Gary W. Saunders; Centre for Environmental & Molecular Algal Research; Dept. of Biology; University of New Brunswick; Fredericton, NB, E3B 5A3, Canada. Email: gws@unb.ca.

Date Accepted: 19 July 2010; updated 24 February 2018

Primary Editor: Philip Cantino

Nemaliophycidae T. Christensen 1978 [G. W. Saunders], converted clade name

Registration Number: 134

Definition: the largest crown clade containing *Nemalion lubricum* Duby 1830 (=*Nemalion helminthoides* (Velley) Batters 1902) but not *Ahnfeltia plicata* (Hudson) Fries 1836 and *Chondrus crispus* Stackhouse 1797 and *Corallina officinalis* Linnaeus 1758 and *Hildenbrandia rubra* (Sommerfelt) Meneghini 1841. This is a maximum-crown-clade definition. Abbreviated definition: max crown ∇ (*Nemalion lubricum* Duby 1830 ~ *Ahnfeltia plicata* (Hudson) Fries 1836 & *Chondrus crispus* Stackhouse 1797 & *Corallina officinalis* Linnaeus 1758 & *Hildenbrandia rubra* (Sommerfelt) Meneghini 1841).

Etymology: Based on the genus name *Nemalion*, which is derived from Greek, *nema* (thread) referring to the threadlike thalli. The ending *phycidae* is the standard termination under the botanical code (Turland et al., 2018) for algal taxa at the subclass rank.

Reference Phylogeny: The primary reference phylogeny is Le Gall and Saunders (2007: Fig. 4, where this clade is labelled as the "APB complex"). See also Harper and Saunders (2001: Fig. 1, the clade labeled L2B).

Composition: *Nemaliophycidae* includes species assigned to *Nemaliales* and related taxa of the ANP complex (*Acrochaetiales*, *Colaconematales* and *Palmariales*; Harper and Saunders, 2002), *Balbianiales*, *Balliales*, *Batrachospermales*, *Rhodachlyales* and *Thoreales* (see Harper and Saunders (2002), Saunders and Hommersand (2004) and West et al. (2008) for additional references). Most recently the orders *Corynodactylales* (Saunders et al. 2017) and *Entwisleiales* (Scott et al. 2013) have been added. See Guiry and Guiry (2017) for a comprehensive list.

Diagnostic Apomorphies: *Nemaliophycidae* is diagnosed by an insert of two amino acid residues (a serine and an asparagine) at position 98 in the EF2 gene (Le Gall and Saunders, 2007). Additionally, this clade is characterized by pit plugs with two cap layers at cytoplasmic faces, an absence of auxiliary cells in postfertilization development and, where present, cruciate tetrasporangia (all three possibly plesiomorphic) (see Pueschel and Cole, 1982; Gabrielson and Garbary, 1986; Saunders and Hommersand, 2004).

Synonyms: There are no known synonyms.

Comments: This clade includes a wide assortment of species with highly variable vegetative and reproductive attributes and life history strategies (Saunders and Kraft, 1997). Additionally, most freshwater florideophyte red algae are included in this clade. The monophyly of *Nemaliophycidae* was first demonstrated in a molecular study by Saunders et al. (1995) and has subsequently been corroborated in several other studies, most recently in the comprehensive molecular investigations of Harper and Saunders (2001, 2002) and Le Gall and Saunders (2007).

The converted clade name *Nemaliophycidae* is here defined so that it corresponds in composition to *Nemaliophycidae* sensu Le Gall and Saunders (2007). In contrast, *Nemaliophycidae* sensu Saunders and Hommersand (2004) applied to a larger group, which also contained species that are included here within *Corallinophycidae*.

Literature Cited

Batters, E. A. L. 1902. A catalogue of the British marine algae. *J. Bot., Lond.* 40(suppl.):1–107.

Christensen, T. 1978. Annotations to a textbook of phycology. *Bot. Tidsskr.* 73:65–70.

Duby, J. É. 1830. *Aug. Pyrami de Candolle Botanicon Gallicum Sen Synopsis Plantarum in Flora Gallica Descriptarum. Editio Secunda. Ex Herbariis et Schedis Candollianis Propriisque Digestum a J. É. Duby. Pars Secunda Plantas Cellulares Continens*, pp. [i–vi], [545]–1068, [i]–lviii. Mme Ve Bouchard-Huzard, Paris.

Fries, E. M. 1836. *Corpus Floranum Provinciallum Sueciae. I. Floram Scanicam Scripsit Elias Fries.* Uppsala.

Gabrielson, P. W., and D. J. Garbary. 1986. Systematics of red algae (*Rhodophyta*). *CRC Crit. Rev. Plant Sci.* 3:325–366.

Guiry, M. D., and G. M. Guiry. 2017. *AlgaeBase*. World-wide Electronic Publication, National University of Ireland, Galway. Available at http://www.algaebase.org, searched on 2 October 2017.

Harper, J. T., and G. W. Saunders. 2001. Molecular systematics of the *Florideophyceae* (*Rhodophyta*) using nuclear large and small subunit rDNA sequence data. *J. Phycol.* 37:1073–1082.

Harper, J. T., and G. W. Saunders. 2002. A re-classification of the *Acrochaetiales* based on molecular and morphological data, and establishment of the *Colaconematales* ord. nov. (*Florideophyceae, Rhodophyta*). *Eur. J. Phycol.* 37:463–476.

Le Gall, L., and G. W. Saunders. 2007. A nuclear phylogeny of the *Florideophyceae* (*Rhodophyta*) inferred from combined EF2, small subunit and large subunit ribosomal DNA: establishing the new red algal subclass *Corallinophycidae*. *Mol. Phylogenet. Evol.* 43:1118–1130.

Linnaeus, C. 1758. *Systema Naturae Per Regna Tria Naturae.* Laurentii Salvii, Holmiae (Stockholm).

Meneghini, G. 1841. Algologia dalmatica. *Atti Del Terza Riunione degli Scienziati Italiani Tenuta in Firenze* 3:424–431.

Pueschel, C. M., and K. M. Cole. 1982. Rhodophycean pit plugs: an ultrastructural survey with taxonomic implications. *Am. J. Bot.* 69:703–720.

Saunders, G. W., C. J. Bird, M. A. Ragan, and E. L. Rice. 1995. Phylogenetic relationships of species of uncertain taxonomic position within the *Acrochaetiales/Palmariales* complex (*Rhodophyta*): inferences from phenotypic and 18S rDNA sequence data. *J. Phycol.* 31:601–611.

Saunders, G. W., and G. T. Kraft. 1997. A molecular perspective on red algal evolution: focus on the *Florideophycidae. Plant Syst. Evol.* 11(suppl.):115–138.

Saunders, G. W., K. L. Wadland, E. D. Salomaki, and C. E. Lane. 2017. A contaminant DNA barcode sequence reveals a new red algal order, *Corynodactylales* (*Nemaliophycidae, Florideophyceae*). *Botany* 95:561–566.

Saunders, G. W., and M. H. Hommersand. 2004. Assessing red algal supraordinal diversity and taxonomy in the context of contemporary systematic data. *Am. J. Bot.* 91:1494–1507.

Scott, F. J., G. W. Saunders, and G. T. Kraft. 2013. *Entwisleia bella*, gen. et sp. nov., a novel marine "batrachospermaceous" red alga from southeastern Tasmania representing a new family and order in the *Nemaliophycidae. Eur. J. Phycol.* 48:398–410.

Stackhouse, J. 1797. *Nereis Britannica; Continens Species Omnes Fucorum in Insulis Britannicis Crescentium: Descriptione Latine et Anglico, Necnon Iconibus Ad Vivum Depictis...* Fasc. 2, pp. ix–xxiv, 31–70, pls IX–XIII. J. White, London.

Turland, N. J., J. H. Wiersema, F. R. Barrie, W. Greuter, D. L. Hawksworth, P. S. Herendeen, S. Knapp, W.-H. Kusber, D.-Z. Li, K. Marhold, T. W. May, J. McNeill, A. M. Monro, J. Prado, M. J. Price, and G. F. Smith, eds. 2018. *International Code of Nomenclature for Algae,*

Fungi, and Plants (Shenzhen Code). Adopted by the Nineteenth International Botanical Congress, Shenzhen, China, July 2017. Regnum Vegetabile 159. Koeltz Botanical Books, Glashütten.

West, J. A., J. L. Scott, K. A. West, U. Karsten, S. L. Clayden, and G. W. Saunders. 2008. *Rhodachlya madagascarensis* gen et sp. nov.: a distinct acrochaetioid represents a new order and family (*Rhodachlyales* ord. nov., *Rhodachlyaceae* fam. nov.) of the *Florideophyceae* (*Rhodophyta*). *Phycologia* 47:203–212.

Author

Gary W. Saunders; Centre for Environmental & Molecular Algal Research; Dept. of Biology; University of New Brunswick; Fredericton, NB, E3B 5A3, Canada. Email: gws@unb.ca.

Date Accepted: 19 July 2010; updated 24 February 2018

Primary Editor: Philip Cantino

Corallinophycidae L. Le Gall and G. W. Saunders 2007
[G. W. Saunders], converted clade name

Registration Number: 132

Definition: The crown clade originating in the most recent common ancestor of *Corallina officinalis* Linnaeus 1758, *Rhodogorgon ramosissima* J. N. Norris and Bucher 1989, and *Sporolithon ptychoides* Heydrich 1897. This is a minimum-crown-clade definition. Abbreviated definition: min crown ∇ (*Corallina officinalis* Linnaeus 1758 & *Rhodogorgon ramosissima* J. N. Norris and Bucher 1989 & *Sporolithon ptychoides* Heydrich 1897).

Etymology: Based on the genus name *Corallina*, which was derived from Greek, *korallion* (coral), referring to the coral-like habit of these algae. The ending *phycidae* is the standard termination under the botanical code (Turland et al., 2018) for algal taxa at the subclass rank.

Reference Phylogeny: Le Gall et al. (2010: Fig. 1).

Composition: The clade *Corallinophycidae* is composed of those species traditionally included in the taxa *Corallinales*, *Hapalidiales* (Nelson et al., 2015), *Sporolithales* (Le Gall et al., 2010), and *Rhodogorgonales* (Fredericq and Norris, 1995) under the rank-based system (see Saunders and Hommersand (2004) for references and details). The reader is directed to Schneider and Wynne (2007) and Guiry and Guiry (2017) for comprehensive lists of taxa traditionally ranked as genera in this clade.

Diagnostic Apomorphies: This clade is characterized by calcite deposition in an organic matrix in the cell walls (Pueschel et al., 1992). Additional possible apomorphies are the presence of two-celled carpogonial branches and of two cap layers, the outer distinctly inflated, on the cytoplasmic faces of the pit plugs (see Le Gall and Saunders, 2007).

Synonyms: There are no known formal synonyms, but the clade has been referred to informally as the "CR complex" (Le Gall and Saunders, 2007).

Comments: Saunders and Bailey (1997) used nuclear small-subunit ribosomal DNA to establish a relationship between the taxa *Corallinales* and *Rhodogorgonales*, which together were subsequently recognized as a major clade within *Nemaliophycidae* of the *Florideophyceae* (Saunders and Hommersand, 2004). More recently, detailed analyses of molecular data (Le Gall and Saunders, 2007) have resulted in this clade being granted subclass status under the rank-based system as *Corallinophycidae*, which is converted here as the clade name. In a study by Harvey et al. (2002), a relationship between *Corallinales* and *Rhodogorgonales* was similarly inferred, but the authors questioned the molecular results because they considered these taxa too divergent in aspects of their vegetative and reproductive anatomy to be closely related. I am swayed more heavily by the molecular results in this regard and contend that these two clades are not as divergent as perceived by some taxonomists (i.e., the changes in vegetative and reproductive anatomy are not necessarily as significant as currently argued in the literature). Subsequently two segregate orders were recognized for coralline algae, *Hapalidiales* (Nelson et al. 2015) and *Sporolithales* (Le Gall et al. 2010), in the rank-based system.

Literature Cited

Fredericq, S., and J. N. Norris. 1995. A new order (*Rhodogorgonales*) and family (*Rhodogorgonaceae*) of red algae composed of two tropical calciferous genera, *Renouxia* gen. nov. and *Rhodogorgon. Crypt. Bot.* 5:316–331.

Guiry, M. D., and G. M. Guiry. 2017. *AlgaeBase.* World-wide Electronic Publication, National University of Ireland, Galway. Available at http://www.algaebase.org, searched on 2 October 2017.

Harvey, A. S., W. J. Woelkerling, and A. J. K. Millar. 2002. The *Sporolithaceae* (*Corallinales, Rhodophyta*) in south-eastern Australia: taxonomy and 18S rRNA phylogeny. *Phycologia* 41:207–227.

Heydrich, F. 1897. *Corallinaceae,* insbesondere *Melobesieae. Ber. Deutsch. Bot. Ges.* 15:34–70, 3 figs., pl. III.

Le Gall, L., and G. W. Saunders. 2007. A nuclear phylogeny of the *Florideophyceae* (*Rhodophyta*) inferred from combined EF2, small subunit and large subunit ribosomal DNA: establishing the new red algal subclass *Corallinophycidae. Mol. Phylogenet. Evol.* 43:1118–1130.

Le Gall, L., C. Payri, L. Bittner, and G. W. Saunders. 2010. Multigene phylogenetic analyses support recognition of the *Sporolithales* ord. nov. *Mol. Phylogenet. Evol.* 54:302–305.

Linnaeus, C. 1758. *Systema Naturae Per Regna Tria Naturae.* Laurentii Salvii, Holmiae (Stockholm).

Nelson, W. A., J. E. Sutherland, T. J. Farr, D. R. Hart, K. F. Neill, H. J. Kim, and H. S. Yoon. 2015. Multi-gene phylogenetic analyses of New Zealand coralline algae: *Corallinapetra novaezealndiae* gen. et sp. nov. and recognition of the *Hapalidiales* ord. nov. *J. Phycol.* 51:454–468.

Norris, J. N., and K. E. Bucher. 1989. *Rhodogorgon,* an anomalous new red algal genus from the Caribbean Sea. *Proc. Biol. Soc. Wash.* 102:1050–1066.

Pueschel, C. M., H. H. Eichelberger, and H. N. Trick. 1992. Specialized calcareous cells in the marine alga *Rhodogorgon carriebowensis* and their implications for models of red algal calcification. *Protoplasma* 166:89–98.

Saunders, G. W., and J. C. Bailey. 1997. Phylogenesis of pit-plug-associated features in the *Rhodophyta*: inferences from molecular systematic data. *Can. J. Bot.* 75:1436–1447.

Saunders, G. W., and M. H. Hommersand. 2004. Assessing red algal supraordinal diversity and taxonomy in the context of contemporary systematic data. *Am. J. Bot.* 91:1494–1507.

Schneider, C. W., and M. J. Wynne. 2007. A synoptic review of the classification of red algal genera a half century after Kylin's "Die Gattungen der Rhodophyceen". *Bot. Mar.* 50:197–249.

Turland, N. J., J. H. Wiersema, F. R. Barrie, W. Greuter, D. L. Hawksworth, P. S. Herendeen, S. Knapp, W.-H. Kusber, D.-Z. Li, K. Marhold, T. W. May, J. McNeill, A. M. Monro, J. Prado, M. J. Price, and G. F. Smith, eds. 2018. *International Code of Nomenclature for Algae, Fungi, and Plants (Shenzhen Code).* Adopted by the Nineteenth International Botanical Congress, Shenzhen, China, July 2017. Regnum Vegetabile 159. Koeltz Botanical Books, Glashütten.

Author

Gary W. Saunders; Centre for Environmental & Molecular Algal Research; Dept. of Biology; University of New Brunswick; Fredericton, NB, E3B 5A3, Canada. E-mail: gws@unb.ca.

Date Accepted: 19 July 2010; updated 24 February 2018

Primary Editor: Philip Cantino

Ahnfeltiophycidae G. W. Saunders and M. H. Hommersand 2004 [G. W. Saunders], converted clade name

Registration Number: 131

Definition: The largest crown clade containing *Ahnfeltia plicata* (Hudson) Fries 1836 but not *Chondrus crispus* Stackhouse 1797 and *Corallina officinalis* Linnaeus 1758 and *Hildenbrandia rubra* (Sommerfelt) Meneghini 1841 and *Palmaria palmata* (Linnaeus) Kuntze 1891. This is a maximum-crown-clade definition. Abbreviated definition: max crown ∇ (*Ahnfeltia plicata* (Hudson) Fries 1836 ~ *Chondrus crispus* Stackhouse 1797 & *Corallina officinalis* Linnaeus 1758 & *Hildenbrandia rubra* (Sommerfelt) Meneghini 1841 & *Palmaria palmata* (Linnaeus) Kuntze 1891).

Etymology: Based on the genus name *Ahnfeltia*, which was named in honour of Nicolaus Otto Ahnfelt (1801–1837). The ending *phycidae* is the standard termination under the botanical code (Turland et al., 2018) for algal taxa at the subclass rank.

Reference Phylogeny: The primary reference phylogeny is Huisman et al. (2003: Fig. 4). See also Saunders and Bailey (1997), Saunders and Kraft (1997), Harper and Saunders (2001), and Le Gall and Saunders (2007).

Composition: The clade *Ahnfeltiophycidae* comprises *Ahnfeltia* Fries 1836 and *Pihiella* Huisman, A. R. Sherwood and I. A. Abbott 2003 (in the monogeneric orders *Ahnfeltiales* and *Pihiellales* in the rank-based system); see Maggs and Pueschel (1989) and Huisman et al. (2003).

Diagnostic Apomorphies: This clade is diagnosed by a series of putatively apomorphic characters, which nonetheless have analogous states in other clades of red algae, rendering diagnosis on the following suite of characters more reliable than considering individual attributes: carpogonia sessile on vegetative filaments with postfertilization development directly from the fertilized carpogonium (both possibly plesiomorphic states), carposporophyte (where present) external, pit plugs naked at cytoplasmic faces (likely through the loss of membranes and/or caps) (Maggs and Pueschel, 1989; Huisman et al., 2003; Le Gall and Saunders, 2007).

Synonyms: None.

Comments: The monophyly of this group has been demonstrated in a series of molecular studies (see Reference Phylogeny). Maggs and Pueschel (1989) presented a series of meticulous life history and ultrastructural observations, which resulted in recognition of the genus *Ahnfeltia* as distinct at the ordinal level in the rank-based system. The molecular investigation of Saunders and Bailey (1997) supported the previous taxonomic conclusion and further indicated that this taxon was quite distinct from others in the *Florideophyceae*. More recently, Huisman et al. (2003) added a novel taxon, *Pihiellales*, as sister to the *Ahnfeltiales*, and these two clades formed Saunders and Hommersand's (2004) *Ahnfeltiophycidae*, which is here converted as a clade name.

Literature Cited

Fries, E. M. 1836. *Corpus Floranum Provinciallum Sueciae. I. Floram Scanicam Scripsit Elias Fries.* Uppsala.

Harper, J. T., and G. W. Saunders. 2001. Molecular systematics of the *Florideophyceae* (*Rhodophyta*) using nuclear large and small subunit rDNA sequence data. *J. Phycol.* 37:1073–1082.

Huisman, J. M., A. R. Sherwood, and I. A. Abbott. 2003. Morphology, reproduction, and the 18S rRNA gene sequence of *Pihiella liagoraciphila* gen. et sp. nov. (*Rhodophyta*), the so-called 'monosporangial discs' associated with members of the *Liagoraceae* (*Rhodophyta*) and proposal of the *Pihiellales* ord. nov. *J. Phycol.* 39:978–987.

Kuntze, O. 1891. *Revisio Generum Plantarum.* Pars 2, pp. 375–1011. Arthur Felix, Leipzig.

Le Gall, L., and G. W. Saunders. 2007. A nuclear phylogeny of the *Florideophyceae* (*Rhodophyta*) inferred from combined EF2, small subunit and large subunit ribosomal DNA: establishing the new red algal subclass *Corallinophycidae*. *Mol. Phylogenet. Evol.* 43:11–1130.

Linnaeus, C. 1758. *Systema Naturae Per Regna Tria Naturae.* Laurentii Salvii, Holmiae (Stockholm).

Maggs, C. A., and C. M. Pueschel. 1989. Morphology and development of *Ahnfeltia plicata* (*Rhodophyta*): proposal of *Ahnfeltiales* ord. nov. *J. Phycol.* 25:333–351.

Meneghini, G. 1841. Algologia dalmatica. *Atti Del Terza Riunione Degli Scienziati Italiani Tenuta in Firenze* 3:424–431.

Saunders, G. W., and J. C. Bailey. 1997. Phylogenesis of pit-plug-associated features in the *Rhodophyta*: inferences from molecular systematic data. *Can. J. Bot.* 75:1436–1447.

Saunders, G. W., and M. H. Hommersand. 2004. Assessing red algal supraordinal diversity and taxonomy in the context of contemporary systematic data. *Am. J. Bot.* 91:1494–1507.

Saunders, G. W., and G. T. Kraft. 1997. A molecular perspective on red algal evolution: focus on the *Florideophycidae*. *Plant Syst. Evol.* 11(suppl.):115–138.

Stackhouse, J. 1797. *Nereis Britannica; Continens Species Omnes Fucorum in Insulis Britannicis Crescentium: Descriptione Latine et Anglico, Necnon Iconibus Ad Vivum Depictis...* Fasc. 2, pp. ix–xxiv, 31–70, pls IX–XIII. J. White, London.

Turland, N. J., J. H. Wiersema, F. R. Barrie, W. Greuter, D. L. Hawksworth, P. S. Herendeen, S. Knapp, W.-H. Kusber, D.-Z. Li, K. Marhold, T. W. May, J. McNeill, A. M. Monro, J. Prado, M. J. Price, and G. F. Smith, eds. 2018. *International Code of Nomenclature for Algae, Fungi, and Plants (Shenzhen Code).* Adopted by the Nineteenth International Botanical Congress, Shenzhen, China, July 2017. Regnum Vegetabile 159. Koeltz Botanical Books, Glashütten.

Author

Gary W. Saunders; Centre for Environmental & Molecular Algal Research; Dept. of Biology; University of New Brunswick; Fredericton, NB, E3B 5A3, Canada. Email: gws@unb.ca.

Date Accepted: 19 July 2010; updated 24 February 2018

Primary Editor: Philip Cantino

Rhodymeniophycidae G. W. Saunders and M. H. Hommersand 2004 [G. W. Saunders], converted clade name

Registration Number: 135

Definition: The largest crown clade containing *Rhodymenia palmetta* (Stackhouse) Greville 1830 (= *Rhodymenia pseudopalmata* (J.V. Lamouroux) P. C. Silva 1952) but not *Ahnfeltia plicata* (Hudson) Fries 1836 and *Corallina officinalis* Linnaeus 1758 and *Hildenbrandia rubra* (Sommerfelt) Meneghini 1841 and *Palmaria palmata* (Linnaeus) Kuntze 1891. This is a maximum-crown-clade definition. Abbreviated definition: max crown ∇ (*Rhodymenia palmetta* (Stackhouse) Greville 1830 ~ *Ahnfeltia plicata* (Hudson) Fries 1836 & *Corallina officinalis* Linnaeus 1758 & *Hildenbrandia rubra* (Sommerfelt) Meneghini 1841 & *Palmaria palmata* (Linnaeus) Kuntze 1891).

Etymology: Based on the genus name *Rhodymenia*, which was derived from Greek, *rhodo-* (rosy-red) and *hymen-* (membranous), referring to the blade-like habit typical of species in this genus. The ending *phycidae* is the standard termination under the botanical code (Turland et al., 2018) for algal taxa at the subclass rank.

Reference Phylogeny: The primary reference phylogeny is Saunders and Bailey (1997: Fig. 1, where the internal specifier and generic type, *Rhodymenia pseudopalmata*, is represented by its presumed close relative *R. linearis* J. Agardh). See also Le Gall and Saunders (2007: Fig. 4, where *Rhodymenia pseudopalmata* is represented by five related taxa of the *Rhodymeniales* (Saunders et al., 1999)).

Composition: The clade *Rhodymeniophycidae* includes *Rhodymeniales* and allied taxa (*Halymeniales* and *Sebdeniales*), as well as the florideophyte taxa *Acrosymphytales*, *Bonnemaisoniales*, *Ceramiales*, *Gelidiales*, *Gigartinales*, *Gracilariales*, *Nemastomatales*, and *Plocamiales* (see Saunders and Hommersand (2004) and Withall and Saunders (2006)). More recently the orders *Atractophorales*, *Catenellopsidales* (Saunders et al., 2016), and *Peyssonneliales* (Krayesky et al., 2009), have been added. See Schneider and Wynne (2007) and Guiry and Guiry (2017) for comprehensive listings of taxa traditionally ranked as genera in this clade.

Diagnostic Apomorphies: This clade is diagnosed by the presence of only membranes at the cytoplasmic faces of pit plugs (with the exception of the *Gelidiales*, which has independently evolved a single cap layer adjacent to the membrane; see Pueschel and Cole, 1982). Additionally, sexual life histories are generally triphasic with the carposporophytes developing directly from the carpogonium or carpogonial fusion cell, or indirectly from an auxiliary cell that has received the postfertilization diploid nucleus (Saunders and Hommersand, 2004).

Synonyms: There are no known synonyms.

Comments: This clade was first resolved in a molecular study by Saunders and Bailey (1997), and its monophyly has subsequently been supported in several studies, most recently in the comprehensive molecular investigations of Le Gall and Saunders (2007). Saunders et al.

(2004), Withall and Saunders (2006), and, most recently, Saunders et al. (2016) have completed comprehensive molecular systematic surveys of *Rhodymeniophycidae*, including a listing of orders and accepted families under the rank-based system (also see Saunders and Hommersand, 2004). The clade name *Rhodymeniophycidae* is converted from *Rhodymeniophycidae* Saunders and Hommersand 2004, a subclass name under the rank-based system, which they applied to the same clade to which it is applied here.

Literature Cited

Fries, E. M. 1836. *Corpus Floranum Provinciallum Sueciae. I. Floram Scanicam Scripsit Elias Fries.* Uppsala.

Greville, R. K. 1830. Algae *Britannicae, or Descriptions of the Marine and Other Inarticulated Plants of the British Islands, Belonging to the Order Algae; With Plates Illustrative of the Genera*, pp. [i*–iii*], [i]–lxxxviii, [1]–218, pl. 1–19. McLachlan & Stewart; Baldwin & Cradock, Edinburgh and London.

Guiry, M. D., and G. M. Guiry. 2017. *AlgaeBase.* World-wide Electronic Publication, National University of Ireland, Galway. Available at http://www.algaebase.org, searched on 2 October 2017.

Krayesky, D. M., J. N. Norris, P. W. Gabrielson, D. Gabriel, and S. Fredericq. 2009. A new order of red algae based on the *Peyssonneliaceae*, with an evaluation of the ordinal classification of the *Florideophyceae* (*Rhodophyta*). *Proc. Biol. Soc. Wash.* 122:364–391.

Kuntze, O. 1891. *Revisio Generum Plantarum*, Pars 2, pp. 375–1011. Arthur Felix, Leipzig.

Le Gall, L., and G. W. Saunders. 2007. A nuclear phylogeny of the *Florideophyceae* (*Rhodophyta*) inferred from combined EF2, small subunit and large subunit ribosomal DNA: establishing the new red algal subclass *Corallinophycidae*. *Mol. Phylogenet. Evol.* 43:1118–1130.

Linnaeus, C. 1758. *Systema Naturae Per Regna Tria Naturae.* Laurentii Salvii, Holmiae (Stockholm).

Meneghini, G. 1841. Algologia dalmatica. *Atti Del Terza Riunione Degli Scienziati Italiani Tenuta in Firenze* 3:424–431.

Pueschel, C. M., and K. M. Cole. 1982. Rhodophycean pit plugs: an ultrastructural survey with taxonomic implications. *Am. J. Bot.* 69:703–720.

Saunders, G. W., and J. C. Bailey. 1997. Phylogenesis of pit-plug-associated features in the *Rhodophyta*: inferences from molecular systematic data. *Can. J. Bot.* 75:1436–1447.

Saunders, G. W., A. Chiovitti, and G. T. Kraft. 2004. Small-subunit rDNA sequences from representatives of selected families of the *Gigartinales* and *Rhodymeniales* (*Rhodophyta*). 3. recognizing the *Gigartinales* sensu stricto. *Can. J. Bot.* 82:43–74.

Saunders, G. W., G. Filloramo, K. Dixon, L. Le Gall, C. A. Maggs, and G. T. Kraft. 2016. Multigene analyses resolve early diverging lineages in the *Rhodymeniophycidae* (*Florideophyceae, Rhodophyta*). *J. Phycol.* 52:505–522.

Saunders, G. W., and M. H. Hommersand. 2004. Assessing red algal supraordinal diversity and taxonomy in the context of contemporary systematic data. *Am. J. Bot.* 91:1494–1507.

Saunders, G. W., I. Strachan, and G. T. Kraft. 1999. The families of the order *Rhodymeniales* (*Rhodophyta*): a molecular-systematic investigation with a description of *Faucheaceae* fam. nov. *Phycologia* 38:23–40.

Schneider, C. W., and M. J. Wynne. 2007. A synoptic review of the classification of red algal genera a half century after Kylin's "Die Gattungen der Rhodophyceen". *Bot. Mar.* 50:197–249.

Silva, P. C. 1952. A review of nomenclatural conservation in the algae from the point of view of the type method. *Univ. CA Pub. Bot.* 25:241–323.

Turland, N. J., J. H. Wiersema, F. R. Barrie, W. Greuter, D. L. Hawksworth, P. S. Herendeen, S. Knapp, W.-H. Kusber, D.-Z. Li, K. Marhold, T. W. May, J. McNeill, A. M. Monro, J. Prado, M. J. Price, and G. F. Smith, eds. 2018. *International Code of Nomenclature for Algae, Fungi, and Plants (Shenzhen Code).* Adopted by the Nineteenth International Botanical Congress, Shenzhen, China, July 2017. Regnum Vegetabile 159. Koeltz Botanical Books, Glashütten.

Withall, R. D., and G. W. Saunders. 2006. Combining small and large subunit ribosomal DNA genes to resolve relationships among orders of *Rhodymeniophycidae* (*Rhodophyta*): recognition of the *Acrosymphytales* ord. nov. and *Sebdeniales* ord. nov. *Eur. J. Phycol.* 41:379–394.

Author

Gary W. Saunders; Centre for Environmental & Molecular Algal Research; Dept. of Biology; University of New Brunswick; Fredericton, NB, E3B 5A3, Canada. Email: gws@unb.ca.

Date Accepted: 19 July 2010; updated 24 February 2018

Primary Editor: Philip Cantino

Viridiplantae T. Cavalier-Smith 1981 [B. D. Mishler, J. D. Hall, R. M. McCourt, K. G. Karol, C. F. Delwiche, and L. A. Lewis], converted clade name

Registration Number: 110

Definition: The smallest crown clade containing *Arabidopsis thaliana* (Linnaeus) Heynh. 1842, *Chlamydomonas reinhardtii* P. A. Dangeard 1888, *Ulva intestinalis* Linnaeus 1753, *Palmophyllum umbracola* W. A. Nelson and K. G. Ryan 1986, *Nephroselmis olivacea* F. Stein 1878, and *Mesostigma viride* Lauterborn 1894. This is a minimum-crown-clade definition. Abbreviated definition: min crown ∇ (*Arabidopsis thaliana* (Linnaeus) Heynh 1842 & *Chlamydomonas reinhardtii* P. A. Dangeard 1888 & *Ulva intestinalis* Linnaeus 1753 & *Palmophyllum umbracola* W. A. Nelson and K. G. Ryan 1986 & *Nephroselmis olivacea* F. Stein 1878 & *Mesostigma viride* Lauterborn 1894).

Etymology: Derived from Latin, *viridis* (green) and *planta* (plant).

Reference Phylogeny: The primary reference phylogeny is Figure 2A in Zechman et al. (2010), a phylogeny of two plastid genes (*rbc*L and *atp*B). See also Zechman et al. (2010: Fig. 2B) and Leliaert et al. (2016: Figs. 4, 5).

Composition: This is one of the largest and most structurally disparate eukaryotic clades, containing perhaps 500,000 named species ranging from unicells such as *Chlamydomonas* and *Chlorella*, to coenocytes such as *Codium*, colonial forms such as *Volvox*, and multicellular organisms such as *Ulva*, *Spirogyra*, mosses, ferns, conifers, and flowering plants. It includes two major subclades: the total clades of *Charophyta* (called *Streptophyta* in the reference phylogeny) and *Chlorophyta*. In addition, the small clade *Palmophyllales*, which was classified with *Prasinococcales* into *Palmophyllophyceae* (Leliaert et al., 2016), may either be part of *Chlorophyta* or sister to *Chlorophyta* plus *Charophyta* (Zechman et al., 2010; Leliaert et al., 2016).

Diagnostic Apomorphies: Chlorophyll b, thylakoids stacked into grana, "stellate" structure of the flagellae (as seen under TEM in cross-section), and true starch stored in plastids (Pickett-Heaps, 1982; Mishler and Churchill, 1985).

Synonyms: *Chlorobionta* sensu Simpson (2006), *Chlorobiota* sensu Kenrick and Crane (1997), *Chloroplastida* Adl et al. 2005, *Viridaeplantae* sensu Cavalier-Smith (2004) (a linguistically incorrect alternative spelling), and *Plantae* (in some uses of that name; see Comments).

Comments: We use *Viridiplantae* instead of *Chlorobionta*, *Chlorobiota*, *Chloroplastida*, or *Viridaeplantae* because it has much wider use. *Plantae* has often been used for this clade (e.g., Copeland, 1956), but is also used widely in several other senses: e.g., for just the embryophytes (e.g., Margulis, 1974; Raven et al., 1999), or for green plants (i.e., *Viridiplantae* as defined here) plus red plants (*Rhodoplantae*), or both green and red plants plus glaucophytes (e.g., Rodríguez-Ezpeleta et al., 2005; Weber et al., 2006; Moustafa et al., 2008), or for a highly polyphyletic group consisting of all photosynthetic, multicellular organisms minus their unicellular relatives (e.g., Whittaker and Margulis, 1978; and many popular biology textbooks). Because of this ambiguity in application, we choose not to use *Plantae* for this clade so as to avoid confusion.

It was difficult to decide whether to use a minimum- or maximum-clade definition because of two empirical problems. One, which cautions against a minimum-clade definition, is that there remain a number of understudied unicellular organisms that have been called micromonadophytes (Mattox and Stewart, 1984; Mishler et al., 1994) or prasinophytes (Lewis and McCourt, 2004), some of which branch near the base of *Viridiplantae* (see Guillou et al., 2004; Turmel et al., 2009; Zechman et al., 2010). Some of these may later be found to be sister to the clade comprising *Charophyta* and *Chlorophyta* (or these two subclades plus *Palmophyllales*, if the latter is not part of *Chlorophyta*), and thus not be members of *Viridiplantae* under the minimum-crown-clade definition proposed here. The other empirical problem, which cautions against a maximum-clade definition, is that there is too little certainty about the sister group of *Viridiplantae*. It has been argued that the sister group is either the glaucophytes or the *Rhodoplantae* (Rodríguez-Ezpeleta et al., 2005, 2007; Hackett et al., 2007), but this may be an artifact of gene phylogenies derived from the plastid (which certainly is closely related in *Viridiplantae*, glaucophytes, and *Rhodoplantae*). The host organisms that gave rise to these three clades are possibly only distantly related once the chloroplast signature is removed completely—it may be a case of one clade of endosymbionts having become established in a polyphyletic set of hosts, as seen for example in modern dinoflagellates, which occur as endosymbionts in such disparate hosts as sea anemones, sponges, corals, jellyfish, clams, radiolarians, and foraminiferans (Taylor et al., 2008). We are uncertain enough about this issue to choose not to use a member of *Rhodoplantae* as an external specifier for a maximum-crown-clade definition, because the name might end up applying to a more inclusive clade than intended. We conclude that the future risks of the minimum-clade definition are less than those of a maximum-clade definition. We include multiple internal specifiers representing all the major known extant subclades, to help ensure that the name applies to the crown clade originating in their most recent common ancestor regardless of future changes in relationships among them.

Literature Cited

Adl, S. M., A. G. Simpson, M. A. Farmer, R. A. Andersen, O. R. Anderson, J. R. Barta, S. S. Bowser, G. Brugerolle, R. A. Fensome, S. Fredericq, T. Y. James, S. Karpov, P. Kugrens, J. Krug, C. E. Lane, L. A. Lewis, J. Lodge, D. H. Lynn, D. G. Mann, R. M. McCourt, L. Mendoza, O. Moestrup, S. E. Mozley-Standridge, T. A. Nerad, C. A. Shearer, A. V. Smirnov, F. W. Spiegel, and M. F. Taylor. 2005. The new higher level classification of eukaryotes with emphasis on the taxonomy of protists. *J. Eukaryot. Microbiol.* 52:399–451.

Cavalier-Smith, T. 1981. Eukaryote kingdoms: seven or nine? *Biosystems* 14:461–481.

Cavalier-Smith, T. 2004. Only six kingdoms of life. *Proc. R. Soc. Lond. B Biol. Sci.* 271:1251–1262.

Copeland, H. F. 1956. *The Classification of the Lower Organisms.* Pacific Books, Palo Alto, CA.

Guillou, L., W. Eikrem, M.-J. Chrétiennot-Dinet, F. Le Gall, R. Massana, K. Romari, C. Pedrós-Alió, and D. Vaulot. 2004. Diversity of picoplanktonic prasinophytes assessed by direct nuclear SSU rDNA sequencing of environmental samples and novel isolates retrieved from oceanic and coastal marine ecosystems *Protist* 155:193–214.

Hackett, J. D., H. S. Yoon, S. Li, A. Reyes-Prieto, S. E. Rümmele, and D. Bhattacharya. 2007. Phylogenomic analysis supports the monophyly of cryptophytes and haptophytes and the association of *Rhizaria* with chromalveolates. *Mol. Biol. Evol.* 24:1702–1713.

Kenrick, P., and P. R. Crane. 1997. *The Origin and Early Diversification of Land Plants. A Cladistic Study.* Smithsonian Institution Press, Washington, DC.

Leliaert, F., A. Tronholm, C. Lemieux, M. Turmel, M. S. DePriest, D. Bhattacharya, K. G. Karol, S. Fredericq, F. W. Zechman, and J. M. Lopez-Bautista. 2016. Chloroplast phylogenomic analyses reveal the deepest-branching lineage of the *Chlorophyta, Palmophyllophyceae* class. nov. *Sci. Rep.* 6:25367.

Lewis, L. A., and R. M. McCourt. 2004. Green algae and the origin of land plants. *Am. J. Bot.* 91:1535–1556.

Margulis L. 1974. Five-kingdom classification and the origin and evolution of cells. *Evol. Biol.* 7:45–78.

Mattox, K. R., and K. D. Stewart. 1984. Classification of the green algae: a concept based on comparative cytology. Pp. 29–72 in *Systematics of the Green Algae* (D. Irvine and D. John, eds.). Systematics Association Special Volume No. 27. Academic Press, London and Orlando.

Mishler, B. D., and S. P. Churchill. 1985. Transition to a land flora: phylogenetic relationships of the green algae and bryophytes. *Cladistics* 1:305–328.

Mishler, B. D., L. A. Lewis, M. A. Buchheim, K. S. Renzaglia, D. J. Garbary, C. F. Delwiche, F. W. Zechman, T. S. Kantz, and R. L. Chapman. 1994. Phylogenetic relationships of the "green algae" and "bryophytes". *Ann. Mo. Bot. Gard.* 81:451–483.

Moustafa A., A. Reyes-Prieto, and D. Bhattacharya. 2008. *Chlamydiae* has contributed at least 55 genes to *Plantae* with predominantly plastid functions. *PLOS ONE* 3(5):e2205. doi:10.1371/journal.pone.0002205.

Pickett-Heaps, J. D. 1982. *New Light on the Green Algae.* Carolina Biology Reader 115, Carolina Biological Supply Company, Burlington, NC.

Raven, P. H., R. F. Evert, and S. E. Eichhorn, 1999. *Biology of Plants.* 6th edition. W. H. Freeman and Company, New York.

Rodríguez-Ezpeleta N., H. Brinkmann, S. C. Burey, B. Roure, G. Burger, W. Loffelhardt, H. J. Bohnert, H. Philippe, and B. F. Lang. 2005. Monophyly of primary photosynthetic eukaryotes: green plants, red algae, and glaucophytes. *Curr. Biol.* 15:1325–1330.

Rodríguez-Ezpeleta, N., H. Brinkmann, G. Burger, A. J. Roger, M. W. Gray, H. Philippe, and B. F. Lang. 2007. Toward resolving the eukaryotic tree: the phylogenetic positions of jakobids and cercozoans. *Curr. Biol.* 17:1420–1425.

Simpson, M. G. 2006. *Plant Systematics.* Elsevier Press. Burlington, MA.

Taylor, F. J. R., M. Hoppenrath, and J. F. Saldarriaga. 2008. Dinoflagellate diversity and distribution. *Biodivers. Conserv.* 17:407–418.

Turmel, M., M.-C. Gagnon, C. J. O'Kelly, C. Otis, and C. Lemieux. 2009. The chloroplast genomes of the green algae *Pyramimonas, Monomastix,* and *Pycnococcus* shed new light on the evolutionary history of prasinophytes and the origin of the secondary chloroplasts of euglenids. *Mol. Biol. Evol.* 26:631–648.

Weber, A. P. M., M. Linka, and D. Bhattacharya. 2006. Single, ancient origin of a plastid metabolite translocator family in *Plantae* from an endomembrane-derived ancestor. *Eukaryot. Cell* 5:609–612.

Whittaker, R. H., and L. Margulis. 1978. Protist classification and kingdoms of organisms. *BioSystems* 10:3–18.

Zechman, F. W., H. Verbruggen, F. Leliaert, M. Ashworth, M. A. Buchheim, M. W. Fawley, H. Spalding, C. M. Pueschel, J. A. Buchheim, B. Verghese, and M. D. Hanisak. 2010. An unrecognized ancient lineage of green plants persists in deep marine waters. *J. Phycol.* 46:1288–1295.

Authors

Brent D. Mishler; University Herbarium, Jepson Herbarium, and Department of Integrative Biology; University of California; Berkeley, CA 94720-2465, USA. Email: bmishler@berkeley.edu.

John D. Hall; Department of Plant Science and Landscape Architecture, University of Maryland, College Park, MD 20742, USA. Email: jdhall@umd.edu.

Richard M. McCourt; Department of Botany; Academy of Natural Sciences of Drexel University; Philadelphia, PA 19103, USA. Email: rmm45@drexel.edu

Kenneth G. Karol; Cullman Program for Molecular Systematics; The New York Botanical Garden; Bronx, NY 10458, USA. Email: kkarol@nybg.org.

Charles F. Delwiche; Department of Cell Biology and Molecular Genetics; University of Maryland; College Park, MD 20740, USA. Email: delwiche@umd.edu.

Louise A. Lewis; Department of Ecology and Evolutionary Biology; University of Connecticut; Storrs, CT 06269-3043, USA. Email: louise.lewis@uconn.edu.

Date Accepted: 28 July 2016

Primary Editor: Philip Cantino

Chlorophyta Reichenbach 1834: 5 [J. D. Hall, L. A. Lewis, R. M. McCourt, C. F. Delwiche, B. Mishler and K. G. Karol], converted clade name

Registration Number: 27

Definition: The largest crown clade containing *Chlamydomonas reinhardti* P. C. A. Dangeard 1891 but not *Chara vulgaris* Linnaeus 1753. This is a maximum-crown-clade definition. Abbreviated definition: max crown ∇ (*Chlamydomonas reinhardti* P. C. A. Dangeard 1891 ~ *Chara vulgaris* Linnaeus 1753).

Etymology: Derived from Greek *chloros* (green) and *phyton* (plant).

Reference Phylogeny: Cocquyt et al. (2009: Fig. 1). The green algae have been the subject of numerous phylogenetic studies, but very few studies have included a broad sampling of *Chlorophyta* and outgroup taxa. Cocquyt et al. (2009) presented a phylogeny with broad sampling of *Chlorophyta* and other green algae with strong support for many branches. *Chara vulgaris* is used as a specifier because it is the type species of *Chara*, upon which the name *Charophyta* (this volume) is based (*ICPN* Art. 11.10; Cantino and de Queiroz, 2020). Because we intend the definitions of *Chlorophyta* and *Charophyta* to be reciprocal, *Chara vulgaris* must be the external specifier for *Chlorophyta*. *Chara vulgaris* is not included in the primary reference phylogeny; however, for the purpose of applying the definition, *Chara vulgaris* should be considered most closely related to *Chara connivens* in the reference phylogeny.

Composition: *Chlorophyta* is a species-rich clade that includes many green algae but not land plants or their close algal relatives (*Charophyta*). Most classifications divide the *Chlorophyta* into several major groups: *Chlorophyceae* (including *Chlamydomonas reinhardti*), *Trebouxiophyceae*, *Ulvophyceae*, *Chlorodendrophyceae* and *Pedinophyceae*, together comprising the core *Chlorophyta*, plus several clades of prasinophytes (small planktonic unicells), and *Palmophyllophyceae*, a lineage of unicellular and gelatinous thalloid forms (e.g., Lewis and McCourt, 2004; Becker and Marin, 2009; Leliaert et al., 2012; Fučíková et al., 2014; Leliaert et al., 2016). The prasinophytes and *Palmophyllophyceae* form a paraphyletic grade with respect to the core *Chlorophyta* (e.g., Guillou et al., 2004; Turmel et al., 2009; Leliaert et al., 2016) (see Comments).

Diagnostic Apomorphies: *Chlorophyta* is an ancient and structurally heterogeneous clade. There are no universally present structural synapomorphies for the clade. However, most *Chlorophyta* with a swimming stage have an even number of apically inserted flagella with flagellar roots arranged in a cruciate pattern and no multilayered structure (Pickett-Heaps and Marchant, 1972; Mattox and Stewart, 1984; van den Hoek et al., 1995). This contrasts to the *Charophyta*, which typically have paired, laterally inserted flagella in a unilateral arrangement and an associated multilayer structure (see *Charophyta*, this volume).

Synonyms: *Chlorophycophyta* Papenfuss 1946 (approximate); *Isophyta* Jeffrey 1967 (approximate); *Chlamydomonadophyta* (as a synonym of *Isophyta*) Jeffrey 1967 (approximate).

Comments: *Chlorophyta* contains diverse macroscopic and microscopic green algae including flagellate and coccoid unicells, motile and non-motile colonies, as well as filamentous, siphonous and thalloid multicellular species. This clade includes many microscopic freshwater and marine species, all marine green seaweeds, as well as some terrestrial species. We limit the name *Chlorophyta* to the most inclusive crown clade of green algae that excludes land plants (and their closest algal relatives), which is consistent with its modern use in the taxonomic literature (e.g., Bremer et al., 1987; Lewis and McCourt, 2004; Becker and Marin, 2009).

According to the *Index Nominum Algarum* (viewed in 2012) the name *Chlorophyta* was originally used to classify algae, mosses and ferns, presumably excluding seed plants (Reichenbach, 1834) (original not seen). The name was later limited to green algae (those containing chlorophylls *a* and *b*) and explicitly or implicitly excluded land plants (e.g., Pascher, 1914; Smith, 1933). To be consistent with the modern literature, we have applied the name *Chlorophyta* to the largest clade of green algae that excludes land plants and their closest algal relatives (Bremer et al., 1987; Guillou et al., 2004; Lewis and McCourt, 2004; Turmel et al., 2009; Cocquyt et al., 2009). The name *Chlorophyta* was chosen for this clade because it is the only name that has been widely used.

It is now well established that there was substantial diversification within the green plant clade (*Viridiplantae*) long before the origin of land plants (*Embryophyta*) (reviewed in Lewis and McCourt, 2004; Becker and Marin, 2009; Leliaert et al., 2012). Molecular and morphological data both support an ancient division in green algae distinguishing *Chlorophyta* from *Charophyta* (Pickett-Heaps and Marchant, 1972; Mattox and Stewart, 1984; Mishler and Churchill, 1985; Bremer et al., 1987; Bhattacharya et al., 1998; Karol et al., 2001; Turmel et al., 2002; Cocquyt et al., 2009) (see *Charophyta*, this volume). *Chlorophyta* was traditionally divided into several major groups: *Chlorophyceae*, *Trebouxiophyceae*, *Ulvophyceae* and several groups of prasinophytes

(planktonic unicells). Two prasinophyte groups, *Pedinophyceae* and *Chlorodendrophyceae* were recently allied into the core *Chlorophyta* with *Chlorophyceae*, *Trebouxiophyceae*, and *Ulvophyceae* (Fučíková et al., 2014). Many names have been proposed for the remaining prasinophytes, including *Mamiellophyceae* and *Nephroselmidophyceae*. Almost all prasinophytes belong to the *Chlorophyta* (Guillou et al., 2004; Turmel et al., 2009). Only *Mesostigma* is thought to belong to another clade (see *Charophyta*, this volume). However, the exact branching relationship remains uncertain and new lineages are still being discovered. For example, an early-branching clade of *chlorophytes*, *Palmophyllophyceae*, was recently described that contains gelatinous thalloid forms (Zechman et al., 2010; Leliaert et al., 2016).

Literature Cited

Becker, B., and B. Marin. 2009. Streptophyte algae and the origin of embryophytes. *Ann. Bot.* 103:999–1004.

Bhattacharya, D., K. Weber, S. S. An, and W. Berning-Koch. 1998. Actin phylogeny identifies *Mesostigma viride* as a flagellate ancestor of the land plants. *J. Mol. Evol.* 47:544–550.

Bremer, K., C. J. Humphries, B. D. Mishler, and S. P. Churchhill. 1987. On cladistic relationships in green plants. *Taxon* 36:339–349.

Cantino, P. D., and K. de Queiroz. 2020. *International Code of Phylogenetic Nomenclature (PhyloCode)*, Version 6. CRC Press, Boca Raton, FL.

Cocquyt, E., H. Verbruggen, F. Leliaert, F. W. Zechman, K. Sabbe, and O. De Clerck. 2009. Gain and loss of elongation factor genes in green algae. *BMC Evol. Biol.* 9:39.

Fučíková, K., F. Leliaert, E. D. Cooper, P. Škaloud, S. D'Hondt, O. De Clerck, C. F. D. Gergel, L. A. Lewis, P. O. Lewis, J. M. Lopez-Bautista, C. F. Delwiche, and H. Verbruggen. 2014. New phylogenetic hypotheses for the core *Chlorophyta* based on chloroplast sequence data. *Front. Ecol. Evol.* 2: 63.

Guillou, L., W. Eikrem, M.-J. Chrétiennot-Dinet, F. Le Gall, R. Massana, K. Romari, C. Pedrós-Alió, and D. Vaulot. 2004. Diversity of pico-planktonic prasinophytes assessed by direct nuclear SSU rDNA sequencing of environmental samples and novel isolates retrieved from oceanic and coastal marine ecosystems. *Protist* 155:193–214.

Index Nominum Algarum. University Herbarium, University of California, Berkeley, CA. Compiled by Paul Silva. Available at http://ucjeps.berkeley.edu/INA.html.

Jeffrey, C. 1967. The origin and differentiation of the archegoniate land plants: a second contribution. *Kew Bull.* 21:335–349.

Karol, K. G., R. M. McCourt, M. T. Cimino, and C. F. Delwiche. 2001. The closest living relatives of land plants. *Science* 294:2351–2353.

Leliaert, F., D. R. Smith, H. Moreau, M. D. Herron, H. Verbruggen, C. F. Delwiche, and O. De Clerck. 2012. Phylogeny and molecular evolution of the green algae. *Crit. Rev. Plant Sci.* 31:1–46.

Lewis, L. A., and R. M. McCourt. 2004. Green algae and the origin of land plants. *Am. J. Bot.* 91:1535–1556.

Leliaert, F., A. Tronholm, C. Lemieux, M. Turmel, M. S. DePriest, D. Bhattacharya, K. G. Karol, S. Fredericq, F. W. Zechman, and J. M. Lopez-Bautista. 2016. Chloroplast phylogenomic analysis reveal the deepest-branching lineage of the *Chlorophyta*, *Palmophyllophyceae* class. nov. *Sci. Rep.* 6:25367.

Mattox, K. R., and K. D. Stewart. 1984. Classification of the green algae: a concept based on comparative cytology. Pp. 29–72 in *Systematics of the Green Algae* (D. Irvine and D. John, eds.). Systematics Association Special Volume No. 27. Academic Press, London and Orlando.

Mishler, B. D., and S. P. Churchill. 1985. Transition to a land flora: phylogenetic relationships of the green algae and bryophytes. *Cladistics* 1:305–328.

Papenfuss, G. F. 1946. Proposed names for the phyla of algae. *Bull. Torrey Bot. Club.* 73:217–218.

Pascher, A. 1914. Über Flagellaten und Algen. *Ber. Deutsch. Bot. Ges.* 32:136–160.

Pickett-Heaps, J. D., and H. J. Marchant. 1972. The phylogeny of the green algae: a new proposal. *Cytobios* 6:255–264.

Reichenbach, H. G. L. 1834. *Das Pflanzenreich in Seinen Natürlichen Classen und Familien Entwickelt und Durch Mehr Als Tausend in Kupfer Gestochene Übersichtlich-Bildliche Darstellungen fur Anfänger und Freunde der Botanik*. Verlag der Expedition des Naturfreundes, Leipzig.

Smith, G. M. 1933. *The Freshwater Algae of the United States*. McGraw-Hill, New York.

Turmel, M., M. Ehara, C. Otis, and C. Lemieux. 2002. Phylogenetic relationships among streptophytes as inferred from chloroplast small and large subunit rRNA gene sequences. *J. Phycol.* 38:364–375.

Turmel, M., M.-C. Gagnon, C. J. O'Kelly, C. Otis, and C. Lemieux. 2009. Chloroplast genomes of the green algae *Pyramimonas*, *Monomastix* and *Pycnococcus* shed new light on the evolutionary history of prasinophytes and origin of the secondary chloroplasts of euglenids. *Mol. Biol. Evol.* 26:631–648.

van den Hoek, C., D. G. Mann, and H. M. Jahns. 1995. *Algae. An Introduction to Phycology*. Cambridge University Press, Cambridge, New York , and Melbourne.

Zechman, F. W., H. Verbruggen, F. Leliaert, M. Ashworth, M. A. Bucheim, M. W. Fawley, H. Spalding, C. M. Pueschel, J. A. Bucheim, B. Verghese, and M. D. Hanisak. 2010. An unrecognized ancient lineage of green plants persists in deep marine waters. *J. Phycol.* 46:1288–1295.

Authors

John D. Hall; Botany; Academy of Natural Sciences of Drexel University; Philadelphia, PA 19103, USA. Email: jdhall2@gmail.com.

Louise A. Lewis; Department of Ecology and Evolutionary Biology; University of Connecticut; Storrs, CT 06269, USA. Email:louise.lewis@uconn.edu.

Richard M. McCourt; Botany; Academy of Natural Sciences of Drexel University; Philadelphia, PA 19103, USA. Email: rmm45@drexel.edu.

Charles F. Delwiche; Department of Cell Biology and Molecular Genetics; University of Maryland; College Park, MD 20740, USA. Email: delwiche@umd.edu.

Brent Mishler; University and Jepson Herbaria; 1001 Valley Life Sciences Building, # 2465; Berkeley, CA 94720-2465, USA. Email: bmishler@berkeley.edu.

Kenneth G. Karol; Cullman Program for Molecular Systematics; The New York Botanical Garden; Bronx, NY 10458, USA. Email: kkarol@nybg.org.

Date Accepted: 7 January 2013; updated 15 January 2020

Primary Editor: Philip Cantino

Charophyta E. F. A. W. Migula 1897: 94 [K. G. Karol, R. M. McCourt, B. D. Mishler, C. F. Delwiche and J. D. Hall], converted clade name

Registration Number: 26

Definition: The largest crown clade containing *Chara vulgaris* Linnaeus 1753 but not *Chlamydomonas reinhardti* P. C. A. Dangeard 1891 (*Chlorophyta*). This is a maximum-crown-clade definition. Abbreviated definition: max crown ∇ (*Chara vulgaris* Linnaeus 1753 ~ *Chlamydomonas reinhardti* P. C. A. Dangeard 1891).

Etymology: Derived from *Chara* Linnaeus 1753, the name of an included taxon, plus Greek *phyton* (plant). See *Charophyceae* (this volume) regarding the etymology of *Chara*.

Reference Phylogeny: Cocquyt et al. (2009: Fig. 1), a molecular phylogenetic study including representatives of most of the major subclades of *Charophyta* (under the name *Streptophyta* in that paper) with broad outgroup sampling. However, the reference phylogeny lacked a representative of one of the major subclades of *Charophyta*, *Coleochaetophyceae* (for a phylogeny that includes *Coleochaetophyceae*, see Karol et al., 2001: Fig. 1).

Composition: The clade *Charophyta* contains many thousands of species. The forms range in size and structural complexity from unicellular flagellates to multicellular (and massive) sequoia trees. This clade contains diverse organisms belonging to the subclades here named *Charophyceae*, *Coleochaetophyceae*, *Embryophyta*, *Klebsormidiophyceae* and *Zygnematophyceae* as well as *Chlorokybus atmophyticus* Geitler and probably *Mesostigma viride* Lauterborn (see Comments).

Diagnostic Apomorphies: Motile cells of *Charophyta* have subapically inserted flagella and an associated multilayer structure (MLS) (Stewart and Mattox, 1975; Mattox and Stewart, 1984). Vegetative cells have a microbody associated with the chloroplast, nucleus and a branched mitochondrion (Mikhailyuk et al., 2008), as well as a persistent inter-zonal spindle during cell division (Stewart and Mattox, 1975).

Synonyms: *Charophyceae* sensu Mattox and Stewart (1984) (approximate); *Streptophyta* Jeffrey 1971 (approximate); *Streptobionta* Kenrick and Crane 1997 (approximate).

Comments: The names *Charophyceae*, *Streptophyta* and *Streptobionta* have also been used to refer to this clade of green algae and land plants. We consider it more appropriate to apply *Charophyceae* to a smaller clade (that of muskgrasses). The names *Charophyta* and *Streptophyta* are both widely used in recent works. We prefer the former because the latter was originally used for a clade including just *Embryophyta* and *Charophyceae* (Jeffrey, 1967, 1971) – the organisms that have twisted sperm (*strepto* means twisted). Kenrick and Crane (1997) proposed the name *Streptobionta* but it has not been widely used.

The name *Charophyta* has sometimes been used to refer to a paraphyletic assemblage of algal lineages related to but excluding land plants. Our definition describes a crown clade that includes land plants. The inclusion of land plants in the *Charophyta* is consistent with the use of that name in the recent literature (reviewed in Lewis and McCourt, 2004).

The core group of *Charophyta* (*Charophyceae, Coleochaetophyceae, Chlorokybus atmophyticus, Embryophyta, Klebsormidiophyceae* and *Zygnematophyceae*) is not controversial. Their close relationship was determined by shared structural characteristics (Pickett-Heaps and Marchant, 1972; Stewart and Mattox, 1975; Mattox and Stewart, 1984; Mishler and Churchill, 1985) and later supported by molecular phylogenetic data (e.g., Kranz et al., 1995; Karol et al., 2001; Cocquyt et al., 2009; Wickett et al., 2014). The inclusion of *Mesostigma viride* in *Charophyta* is less certain. *Mesostigma* was considered a close relative of charophytic algae because it possesses a multilayer structure in the flagellar apparatus (Rogers et al., 1981; Melkonian, 1989). The flagella in *Mesostigma* are laterally or subapically inserted. However, mitochondrial sequence data and chloroplast ribosomal operon data suggest that *Mesostigma viride* may be sister to all green plants (Turmel et al., 2002a,b). If this is correct, *Mesostigma* is not part of *Charophyta*, but rather is a remnant of a lineage that originated before the divergence of *Chlorophyta* and *Charophyta*. However, most phylogenetic studies suggest that *Mesostigma* belongs within *Charophyta* (Bhattacharya et al., 1998; Marin and Melkonian, 1999; Karol et al., 2001; Delwiche et al., 2002; Kim et al., 2006). Inclusion of *Mesostigma* is supported by the fact that it shares a unique gene family and gene duplication with land plants and other charophytes (Nedelcu et al., 2006; Peterson et al., 2006).

Literature Cited

Bhattacharya, D., K. Weber, S. S. An, and W. Berning-Koch. 1998. Actin phylogeny identifies *Mesostigma viride* as a flagellate ancestor of the land plants. *J. Mol. Evol.* 47:544–550.

Cocquyt, E., H. Verbruggen, F. Leliaert, F. W. Zechman, K. Sabbe, and O. De Clerck. 2009. Gain and loss of elongation factor genes in green algae. *BMC Evol. Biol.* 9:1–39.

Delwiche, C. F., K. G. Karol, and M. T. Cimino. 2002. Phylogeny of the genus *Coleochaete* (*Coleochaetales, Charophyta*) and related taxa inferred by analysis of the chloroplast gene *rbcL*. *J. Phycol.* 38:394–403.

Jeffrey, C. 1967. The origin and differentiation of the archegoniate land plants: a second contribution. *Kew Bull.* 21(2):335–349.

Jeffrey, C. 1971. Thallophytes and kingdoms—a critique. *Kew Bull.* 25:291–299.

Karol, K. G., R. M. McCourt, M. T. Cimino, and C. F. Delwiche. 2001. The closest living relatives of land plants. *Science* 294:2351–2353.

Kenrick, P., and P. R. Crane. 1997. *The Origin and Early Diversification of Land Plants. A Cladistic Study.* Smithsonian Institution Press, Washington, DC and London.

Kim, E., L. W. Wilcox, M. W. Fawley, and L. E. Graham. 2006. Phylogenetic position of the green flagellate *Mesostigma viride* based on α-tubulin and β-tubulin gene sequences. *Int. J. Plant Sci.* 167:873–883.

Kranz, H. D., D. Miks, M.-L. Siegler, I. Capesius, C. W. Sensen, and V. A. R. Huss. 1995. The origin of land plants: phylogenetic relationships among charophytes, bryophytes, and vascular plants inferred from complete small-subunit ribosomal RNA gene sequences. *J. Mol. Evol.* 41:74–84.

Lewis, L. A., and R. M. McCourt. 2004. Green algae and the origin of land plants. *Am. J. Bot.* 91:1535–1556.

Marin, B., and M. Melkonian. 1999. *Mesostigmatophyceae*, a new class of streptophyte green algae revealed by SSU rRNA sequence comparisons. *Protist* 150:399–417.

Mattox, K. R., and K. D. Stewart. 1984. Classification of the green algae: a concept based on comparative cytology. Pp. 29–72 in *Systematics of the Green Algae* (D. Irvine and D. John, eds.). Systematics Association Special Volume No. 27. Academic Press, London and Orlando.

Melkonian, M. 1989. Flagellar apparatus ultrastructure in *Mesostigma viride* (*Prasinophyceae*). *Plant Syst. Evol.* 163:93–122.

Migula, E. F. A. W. 1897. Die Characeen. Pp. 1–765 in *Kryptogamen-Flora von Deutschland, Oesterreich und der Schweiz,* Bd. 5 (Rabenhorst, ed.). Edward Kummer. Leipzig.

Mikhailyuk, T. I., H. J. Sluiman, A. Massalski, O. Mudimu, E. M. Demchenko, S. Y. Kondratyuk, and T. Friedl. 2008. New streptophyte green algae from terrestrial habitats and an assessment of the genus *Interfilum* (*Klebsormidiophyceae, Streptophyta*). *J. Phycol.* 44:1586–1603.

Mishler, B. D., and S. P. Churchill. 1985. Transition to a land flora: phylogenetic relationships of the green algae and bryophytes. *Cladistics* 1:305–328.

Nedelcu, A. M., T. Borza, and R. W. Lee. 2006. A land-plant specific multigene family in the unicellular *Mesostigma* argues for its close relationship to *Streptophyta. Mol. Biol. Evol.* 23:1011–1015.

Peterson, J., R. Teich, B. Becker, R. Cerff, and H. Brinkmann. 2006. The *GapA/B* gene duplication marks the origin of *Streptophyta* (charophytes and land plants). *Mol. Biol. Evol.* 23:1109–1118.

Pickett-Heaps, J. D., and H. J. Marchant. 1972. The phylogeny of the green algae: a new proposal. *Cytobios* 6:255–264.

Rogers, C. E., D. S. Domozych, K. D. Stewart, and K. R. Mattox. 1981. The flagellar apparatus of *Mesostigma viride* (*Prasinophyceae*): multilayered structures in a scaly green flagellate. *Plant Syst. Evol.* 138:247–258.

Stewart, K. G., and K. R. Mattox. 1975. Comparative cytology, evolution and classification of the green algae with some consideration of the origins of other organisms with chlorophylls A and B. *Bot. Rev.* 41:104–135.

Turmel, M., M. Ehara, C. Otis, and C. Lemieux. 2002a. Phylogenetic relationships among streptophytes as inferred from chloroplast small and large subunit rRNA gene sequences. *J. Phycol.* 38:364–375.

Turmel, M., C. Otis, and C. Lemieux. 2002b. The complete mitochondrial DNA sequence of *Mesostigma viride* indentifies this green alga as the earliest green plant divergence and predicts a highly compact mitochondrial genome in the ancestor of all green plants. *Mol. Biol. Evol.* 19:24–38.

Wickett, N. J., S. Mirarab, N. Nguyen, T. Warnow, E. Carpenter, N. Matasci, S. Ayyampalayam, M. S. Barker, J. G. Burleigh, M. A. Gitzendanner, B. R. Ruhfel, E. Wafula, J. P. Der, S. W. Graham, S. Mathews, M. Melkonian, D. E. Soltis, P. S. Soltis, N. W. Miles, C. J. Rothfels, L. Pokorny, A. J. Shaw, L. DeGironimo, D. W. Stevenson, B. Surek, J. C. Villarreal, B. Roure, H. Philippe, C. W. dePamphilis, T. Chen, M. K. Deyholos, R. S. Baucom, T. M. Kutchan, M. M. Augustin, J. Wang, Y. Zhang, Z. Tian, Z. Yan, X. Wu, X. Sun, G. K.-S. Wong, and J. Leebens-Mack. 2014. Transcriptomic analysis of the original and early diversification of land plants. *PNAS* 111(45):E4859–E4868.

Authors

Kenneth G. Karol; Cullman Program for Molecular Systematics; The New York Botanical Garden; Bronx, NY 10458, USA. Email: kkarol@nybg.org.

Richard M. McCourt; Department of Botany; Academy of Natural Sciences of Drexel University; Philadelphia, PA 19103, USA. Email: rmm45@drexel.edu.

Brent D. Mishler; University and Jepson Herbaria; 1001 Valley Life Sciences Building, # 2465; Berkeley, CA 94720-2465, USA. Email: bmishler@berkeley.edu.

Charles F. Delwiche; Department of Cell Biology and Molecular Genetics; University of Maryland; College Park, MD 20740, USA. Email: delwiche@umd.edu.

John D. Hall; Department of Botany; Academy of Natural Sciences of Drexel University; Philadelphia, PA 19103, USA. Email: jdhall2@gmail.com.

Date Accepted: 21 October 2013; updated 6 March 2018

Primary Editor: Philip Cantino

Klebsormidiophyceae C. Jeffrey 1982: 411 [J. D. Hall, C. F. Delwiche and K. G. Karol], converted clade name

Registration Number: 153

Definition: The largest crown clade containing *Klebsormidium flaccidum* (Kützing) P. C. Silva, K. R. Mattox and W. H. Blackwell 1972 but not *Coleochaete scutata* Brébisson 1844 (*Coleochaetophyceae*) and *Spirogyra porticalis* (O. F. Müller) Dumort. 1822 and *Zygnema cruciatum* (Vaucher) C. Agardh 1817 (the latter two specifiers representing *Zygnematophyceae*). This is a maximum-crown-clade definition. Abbreviated definition: max crown ∇ (*Klebsormidium flaccidum* (Kützing) P. C. Silva, K. R. Mattox and W. H. Blackwell 1972 ~ *Coleochaete scutata* Brébisson 1844 & *Spirogyra porticalis* (O. F. Müller) Dumort. 1822 & *Zygnema cruciatum* (Vaucher) C. Agardh 1817).

Etymology: Name derived from *Klebsormidium*, the name of one of the included taxa. *Klebsormidium* is so named because it approximately agrees with Klebs' (1896) concept of *Hormidium* (Silva et al., 1972).

Reference Phylogeny: Mikhailyuk et al. (2008: Fig. 6). This phylogeny included most of the known members of *Klebsormidiophyceae*. *Coleochaete scutata* is presumed to be closest to *Coleochaete orbicularis* (Delwiche et al., 2002) in the reference phylogeny. *Spirogyra porticalis* and *Zygnema cruciatum* are thought to belong to the smallest clade including *Cosmarium botrytis* and *Cylindrocystis brebissonii* in the reference phylogeny (Gontcharov et al., 2003; Hall et al., 2008). These external specifiers were chosen because they are internal specifiers of *Coleochaetophyceae* and *Zygnematophyceae* (this volume).

Composition: *Klebsormidiophyceae* contains species of *Interfilum* and *Klebsormidium* as well as *Hormidiella attenuata* and probably *Entransia fimbriata* (Mikhailyuk et al., 2008; see Comments regarding *Entransia*). Two strains identified as *Geminella* and *Microspora* are also included in this clade according to the primary reference phylogeny, but we suspect these to be misidentifications. *Geminella* and *Microspora* are members of *Chlorophyta* (Mikhailyuk et al., 2008) and thus distantly related to *Klebsormidiophyceae*.

Diagnostic Apomorphies: Members of this clade can be diagnosed by a combination of character states, none of which is a clear synapomorphy of the clade as a whole. They are coccoid or uniseriate in the vegetative stage, reproduce by zoospores with two subapically inserted flagella, and lack a phragmoplast (Mattox and Stewart, 1984; Cook, 2004; Mikhailyuk et al., 2008). Most members of this clade have a saddle-shaped (selliform) parietal chloroplast.

Synonyms: *Klebsormidiales* Stewart and Mattox 1975 (approximate); *Klebsormidiophyta* Kenrick and Crane 1997 (approximate).

Comments: The names *Klebsormidiophyceae* and *Klebsormidiales* have both been commonly used to refer to a group approximating the named clade; the name *Klebsormidiophyta* has not. The first two names are usually used for taxa of the same composition that are assigned to different ranks. We selected the name *Klebsormidiophyceae* because the names selected for closely related mutually exclusive clades have the same ending (e.g., *Charophyceae*, *Coleochaetophyceae*,

Zygnematophyceae). It is also the name used in the reference phylogeny (Mikhailyuk et al., 2008).

The clade *Klebsormidiophyceae* was only recently recognized. The taxon *Klebsormidium* was not named until 1972 (Silva et al., 1972), and the relationships among *Klebsormidium*, *Interfilum* and *Entransia* were not known until molecular phylogenetic studies suggested that they form a clade (Karol et al., 2001; Turmel et al., 2002; Mikhailyuk et al., 2008; Sluiman et al., 2008). Only a few dozen species are currently known. Most species are filamentous, but the clade also includes some coccoid forms. *Klebsormidiophyceae* almost certainly contains unrecognized and undiscovered species. Support for this clade is generally high, but the relationships among the subclades remain uncertain (Karol et al., 2001; Mikhailyuk et al., 2008; Sluiman et al., 2008). Because of the poor support for its basal relationships, a maximum-crown-clade definition is preferable to one with a minimum-clade structure.

Entransia fimbriata seems to be distantly related to *Klebsormidium* and in some phylogenies appears as sister to *Coleochaete* (McCourt et al., 2000; Rindi et al., 2008). It should be noted that phylogenies suggesting a relationship between *Entransia* and *Klebsormidium* include nuclear ribosomal data (Karol et al., 2001; Mikhailyuk et al., 2008; Sluiman et al., 2008), while the relationship between *Entransia* and *Coleochaete* is suggested by the chloroplast gene *rbcL* (McCourt et al., 2000; Rindi et al., 2008).

Literature Cited

Cook, M. E. 2004. Structure and asexual reproduction of the enigmatic charophycean green alga *Entransia fimbriata* (*Klebsormidiales, Charophyceae*). *J. Phycol.* 40:424–431.

Delwiche, C. F., K. G. Karol, and M. T. Cimino. 2002. Phylogeny of the genus *Coleochaete* (*Coleochaetales, Charophyta*) and related taxa inferred by analysis of the chloroplast gene *rbcL*. *J. Phycol.* 38:394–403.

Gontcharov, A. A., B. Marin, and M. Melkonian. 2003. Molecular phylogeny of conjugating green algae (*Zygnemophyceae, Streptophyta*) inferred from SSU rDNA sequence comparisons. *J. Mol. Evol.* 56:89–104.

Hall, J. D., K. G. Karol, R. M. McCourt, and C. F. Delwiche. 2008. Phylogeny of conjugating green algae based on chloroplast and mitochondrial nucleotide sequence data. *J. Phycol.* 44:467–477.

Jeffrey, C. 1982. Kingdoms, codes and classification. *Kew Bull.* 37:403–416.

Karol, K. G., R. M. McCourt, M. T. Cimino, and C. F. Delwiche. 2001. The closest living relatives of land plants. *Science* 294:2351–2353.

Klebs, G. 1896. *Die Bedingung der Fortpflanzung Bei Einigen Algen und Pilzen*. Jena.

Mattox, K. R., and K. D. Stewart. 1984. Classification of the green algae: a concept based on comparative cytology. Pp. 29–72 in *Systematics of the Green Algae* (D. Irvine and D. John, eds.). Systematics Association Special Volume No. 27. Academic Press, London and Orlando.

McCourt, R. M, K. G. Karol, J. Bell, M. Helm-Bychowski, A. Grajewska, M. F. Wojciechowski, and R. W. Hoshaw. 2000. Phylogeny of the conjugating green algae (*Zygnematophyceae*) based on *rbcL* sequences. *J. Phycol.* 36:747–758.

Mikhailyuk, T. I., H. J. Sluiman, A. Massalski, O. Mudimu, E. M. Demchenko, S. Y. Kondratyuk, and T. Friedl. 2008. New streptophyte green algae from terrestrial habitats and an assessment of the genus *Interfilum* (*Klebsormidiophyceae, Streptophyta*). *J. Phycol.* 44:1586–1603.

Rindi, F., M. D. Guiry, and J. M. López-Bautista. 2008. Distribution, morphology and phylogeny of *Klebsormidium* (*Klebsormidiales, Charophyceae*) in urban environments in Europe. *J. Phycol.* 44:1529–1540.

Silva, P. C., K. R. Mattox, and W. H. Blackwell, Jr. 1972. The generic name *Hormidium* as applied to green algae. *Taxon* 21:639–645.

Sluiman, H. J., C. Guihal, and O. Mudimu. 2008. Assessing phylogenetic affinities and species delimitations in *Klebsormidiales* (*Streptophyta*): nuclear-encoded rDNA phylogenies and its secondary structure models in *Klebsormidium, Hormidiella,* and *Entransia. J. Phycol.* 44:183–195.

Turmel, M., M. Ehara, C. Otis, and C. Lemieux. 2002. Phylogenetic relationships among streptophytes as inferred from chloroplast small and large subunit rRNA gene sequences. *J. Phycol.* 38:364–375.

Author

John D. Hall; Department of Botany; Academy of Natural Sciences of Drexel University; Philadelphia, PA 19103, USA. Email: jdhall2@gmail.com.

Charles F. Delwiche; Department of Cell Biology and Molecular Genetics; University of Maryland; College Park, MD 20740, USA. Email: delwiche@umd.edu.

Kenneth G. Karol; Cullman Program for Molecular Systematics; The New York Botanical Garden; Bronx, NY 10458, USA. Email: kkarol@nybg.org.

Date Accepted: 20 December 2013; updated 16 July 2018

Primary Editor: Philip Cantino

Phragmoplastophyta G. Lecointre and H. Le Guyader 2006: 158 [J. D. Hall, R. M. McCourt, C. F. Delwiche, B. D. Mishler and K. G. Karol], converted clade name

Registration Number: 150

Definition: The crown clade originating in the most recent common ancestor of *Arabidopsis thaliana* (Linnaeus) Heynh. 1842 (*Embryophyta*), *Chara vulgaris* Linnaeus 1753 (*Charophyceae*), *Coleochaete scutata* Brébisson 1844 (*Coleochaetophyceae*), and *Zygnema cruciatum* (Vaucher) C. Agardh 1817 (*Zygnematophyceae*). This is a minimum-crown-clade definition. Abbreviated definition: min crown ∇ (*Arabidopsis thaliana* (Linnaeus) Heynh. 1842 & *Chara vulgaris* Linnaeus 1753 & *Coleochaete scutata* Brébisson 1844 & *Zygnema cruciatum* (Vaucher) C. Agardh 1817).

Etymology: From the Greek *phragmos* (fence) and *plastos* (formed), referring to the characteristic array of microtubules (phragmoplast) formed during cell division in many members of this clade, and *phyton* (plant).

Reference Phylogeny: Karol et al. (2001: Fig. 1). Although phylogenetic studies present several different hypotheses with regard to which algal clade is most closely related to land plants, they collectively indicate that either *Charophyceae, Coleochaetophyceae, Zygnematophyceae* or some combination of those taxa is sister to land plants (Huss and Kranz, 1995; Kranz et al., 1995; McCourt et al., 2000; Karol et al., 2001; Turmel et al., 2002b, 2003). Because of these competing hypotheses, a phylogeny with broad taxon and molecular sampling (Karol et al., 2001: Fig. 1) was chosen as the reference phylogeny. *Chara vulgaris, Coleochaete scutata*, and *Zygnema cruciatum* are represented by or assumed to be closely related to *Chara, Coleochaete orbicularis*, and *Zygnema*, respectively, in the reference phylogeny. The former three taxa were chosen as internal specifiers for consistency with the definitions of the names of three of the four major subclades of *Phragmoplastophyta*. Because these species are the type species of their respective genera under rank-based nomenclature, they must be used as internal specifiers for names based on those genera (*ICPN* Art. 11.10; Cantino and de Queiroz, 2020): *Charophyceae, Coleochaetophyceae*, and *Zygnematophyceae* as defined in this volume.

Composition: *Phragmoplastophyta* contains those organisms belonging to the clades *Embryophyta, Charophyceae, Coleochaetophyceae*, and *Zygnematophyceae* (all covered in this volume)—i.e., land plants and their three closest algal relatives (according to the reference phylogeny). See Comments regarding *Klebsormidiophyceae*.

Diagnostic Apomorphies: Although this group includes structurally diverse species, its members share some fundamental characteristics that distinguish them from other *Charophyta*. Most organisms in this clade display some degree of phragmoplastic cell division (Mattox and Stewart, 1984). The phragmoplasts of *Coleochaetophyceae, Charophyceae*, and *Embryophyta* are very similar (Graham et al., 1991). The phragmoplast in *Zygnematophyceae* (known only from some species of *Spirogyra* and *Mougeotia*) is similar in location, but different in composition and function (Fowke and Pickett-Heaps, 1969; Sawitzky and Grolig, 1995).

Additionally, some distantly related species in *Chlorophyta* contain a phragmoplast (Chapman et al., 2001). Members of *Phragmoplastophyta* are also unique within *Charophyta* in that many are oogamous. Some land plants (including all relatively basal embryophytes) and all extant *Charophyceae* and *Coleochaetophyceae* are undisputedly oogamous. Some members of *Zygnematophyceae* are anisogamous with a non-motile female gamete and a motile, non-flagellate male gamete, although the female gamete is not always larger than the male. Sexual reproduction is not known in the related taxa *Klebsormidiophyceae*, *Chlorokybus atmophyticus* or *Mesostigma viride*. In view of the non-universal representation of both apomorphies (i.e., phragmoplasts and oogamy) in *Zygnematophyceae*, and their questionable homology with the corresponding features in the other three major clades of *Phragmoplastophyta*, phragmoplasts and oogamy may be synapomorphies of a smaller clade rather than *Phragmoplastophyta* as a whole. The same concern applies if *Klebsormidiophyceae* is part of *Phragmoplastophyta* (see Comments), since the members of *Klebsormidiophyceae* lack phragmoplasts and are not known to reproduce sexually.

Synonyms: None.

Comments: Nearly every phylogenetic study that included land plants and algal members of *Charophyta* has inferred a clade that corresponds closely in composition to the clade to which the name *Phragmoplastophyta* is applied here (Mishler and Churchill, 1985; Graham et al., 1991; Mishler et al., 1994; Huss and Kranz, 1995; Kranz et al., 1995; Bhattacharya and Medlin, 1998; Bhattacharya et al., 1998; McCourt et al., 2000; Karol et al., 2001; Turmel et al., 2002a,b, 2003; Wickett et al., 2014). Surprisingly, this clade was left unnamed in all of these studies.

However, Lecointre and Le Guyader (2006) applied the name *Phragmoplastophyta* to the same clade to which it is applied here, although without a phylogenetic definition.

Based on some published phylogenies (e.g., McCourt et al., 2000), this clade may also include *Klebsormidiophyceae*, which our definition allows to be either included or excluded.

Literature Cited

Bhattacharya, D., and L. Medlin. 1998. Algal phylogeny and the origin of land plants. *Plant Physiol.* 116:9–15.

Bhattacharya, D., K. Weber, S. S. An, and W. Berning-Koch. 1998. Actin phylogeny identifies *Mesostigma viride* as a flagellate ancestor of the land plants. *J. Mol. Evol.* 47:544–550.

Cantino, P. D., and K. de Queiroz. 2020. *International Code of Phylogenetic Nomenclature (PhyloCode)*, Version 6. CRC Press, Boca Raton, FL.

Chapman, R., O. Borkhsenious, R. C. Brown, M. C. Henk, and D. C. Waters. 2001. Phragmoplast-mediated cytokinesis in *Trentepohlia*: results of TEM and immunofluorescense cytochemistry. *Int. J. Syst. Evol. Microbiol.* 51:759–765.

Fowke, L. C., and J. D. Pickett-Heaps. 1969. Cell division in *Spirogyra*. II. Cytokinesis. *J. Phycol.* 5:273–281.

Graham, L. E., Delwiche, C. F., and B. D. Mishler. 1991. Phylogenetic connections between the green algae and the bryophytes. *Adv. Bryol.* 4:213–244.

Huss, V. A. R., and H. D. Kranz. 1995. Charophyte evolution and the origin of land plants. Pp. 103–114 in *Origins of Algae and Their Plastids* (D. Bhattacharya, ed.). Springer, Wien and New York.

Karol, K. G., R. M. McCourt, M. T. Cimino, and C. F. Delwiche. 2001. The closest living relatives of land plants. *Science* 294:2351–2353.

Kranz, H. D., D. Miks, M.-L. Siegler, I. Capesius, C. W. Sensen, and V. A. R. Huss. 1995. The origin of land plants: phylogenetic relationships among charophytes, bryophytes, and vascular plants inferred from complete small-subunit ribosomal RNA gene sequences. *J. Mol. Evol.* 41:74–84.

Lecointre G., and H. Le Guyader. 2006. *The Tree of Life.* Harvard University Press, Cambridge, MA.

Mattox, K. R., and K. D. Stewart. 1984. Classification of the green algae: a concept based on comparative cytology. Pp. 29–72 in *Systematics of the Green Algae* (D. Irvine and D. John, eds.). Systematics Association Special Volume No. 27. Academic Press, London and Orlando.

McCourt, R. M, K. G. Karol, J. Bell, M. Helm-Bychowski, A. Grajewska, M. F. Wojciechowski, and R. W. Hoshaw. 2000. Phylogeny of the conjugating green algae (*Zygnematophyceae*) based on *rbcL* sequences. *J. Phycol.* 36:747–758.

Mishler, B. D., and S. P. Churchill. 1985. Transition to a land flora: phylogenetic relationships of the green algae and bryophytes. *Cladistics* 1:305–328.

Mishler, B. D., L. A. Lewis, M. A. Buchheim, K. S. Renzaglia, D. J. Garbary, C. F. Delwiche, F. W. Zechman, T. S. Kantz, and R. L. Chapman. 1994. Phylogenetic-relationships of the green algae and bryophytes. *Ann. Mo. Bot. Gard.* 81:451–483.

Sawitzky, H., and F. Grolig. 1995. Phragmoplast of the green alga *Spirogyra* is functionally distinct from the higher plant phragmoplast. *J. Cell Biol.* 130:1359–1371.

Turmel, M., M. Ehara, C. Otis, and C. Lemieux. 2002a. Phylogenetic relationships among streptophytes as inferred from chloroplast small and large subunit rRNA gene sequences. *J. Phycol.* 38:364–375.

Turmel, M., C. Otis, and C. Lemieux. 2002b. The complete mitochondrial DNA sequence of *Mesostigma viride* identifies this green alga as the earliest green plant divergence and predicts a highly compact mitochondrial genome in the ancestor of all green plants. *Mol. Biol. Evol.* 19:24–38.

Turmel, M., C. Otis, and C. Lemieux. 2003. The mitochondrial genome of *Chara vulgaris*: insights into the mitochondrial DNA architecture of the last common ancestor of green algae and land plants. *Plant Cell* 15:1888–1903.

Wickett, N. J., S. Mirarab, N. Nguyen, T. Warnow, E. Carpenter, N. Matasci, S. Ayyampalayam, M. S. Barker, J. G. Burleigh, M. A. Gitzendanner, B. R. Ruhfel, E. Wafula, J. P. Der, S. W. Graham, S. Mathews, M. Melkonian, D. E. Soltis, P. S. Soltis, N. W. Miles, C. J. Rothfels, L. Pokorny, A. J. Shaw, L. DeGironimo, D. W. Stevenson, B. Surek, J. C. Villarreal, B. Roure, H. Philippe, C. W. dePamphilis, T. Chen, M. K. Deyholos, R. S. Baucom, T. M. Kutchan, M. M. Augustin, J. Wang, Y. Zhang, Z. Tian, Z. Yan, X. Wu, X. Sun, G. K.-S. Wong, and J. Leebens-Mack. 2014. Transcriptomic analysis of the original and early diversification of land plants. *PNAS* 111(45):E4859–E4868.

Authors

John D. Hall; Department of Botany; Academy of Natural Sciences of Drexel University; Philadelphia, PA 19103, USA. Email: jdhall2@gmail.com.

Richard M. McCourt; Department of Botany; Academy of Natural Sciences of Drexel University; Philadelphia, PA 19103, USA. Email: rmm45@drexel.edu.

Charles F. Delwiche; Department of Cell Biology and Molecular Genetics; University of Maryland; College Park, MD 20740, USA. Email: delwiche@umd.edu.

Brent D. Mishler; University and Jepson Herbaria; 1001 Valley Life Sciences Building, # 2465; Berkeley, CA 94720-2465, USA. Email: bmishler@berkeley.edu.

Kenneth G. Karol; Cullman Program for Molecular Systematics; The New York Botanical Garden; Bronx, NY 10458, USA. Email: kkarol@nybg.org.

Date Accepted: 21 October 2013; updated 6 March 2018

Primary Editor: Philip Cantino

Zygnematophyceae C. van den Hoek, D. G. Mann, and H. M. Jahns 1995: 461 [J. D. Hall, R. M. McCourt and K. G. Karol], converted clade name

Registration Number: 154

Definition: The largest crown clade containing *Zygnema cruciatum* (Vaucher) C. Agardh 1817 but not *Marchantia polymorpha* Linnaeus 1753 (*Embryophyta*) and *Chara vulgaris* Linnaeus 1753 (*Charophyceae*) and *Coleochaete scutata* Brébisson 1844 (*Coleochaetophyceae*) and *Entransia fimbriata* E. O. Hughes 1943 and *Klebsormidium flaccidum* (Kützing) Silva, Mattox and Blackwell 1972 (*Klebsormidiophyceae*). This is a maximum-crown-clade definition. Abbreviated definition: max crown ∇ (*Zygnema cruciatum* (Vaucher) C. Agardh 1817 ~ *Marchantia polymorpha* Linnaeus 1753 & *Chara vulgaris* Linnaeus 1753 & *Coleochaete scutata* Brébisson 1844 & *Entransia fimbriata* E. O. Hughes 1943 & *Klebsormidium flaccidum* (Kützing) Silva, Mattox and Blackwell 1972).

Etymology: The name is derived from *Zygnema*, the name of an included taxon, and *phykos*, Greek for seaweed. *Zygnema* is derived from the Greek *zygos* (yoke or pair) and *nema* (thread).

Reference Phylogeny: McCourt et al. (2000: Fig. 1) was selected as the reference phylogeny for the *Zygnematophyceae*, but see also Gontcharov et al. (2003: Fig. 1, 2004: Fig. 3) and Hall et al. (2008: Fig. 1), which included more zygnematophycean taxa. The specifiers *Chara vulgaris*, *Coleochaete scutata* and *Zygnema cruciatum* are presumed to belong to the subclades containing species of *Chara*, *Coleochaete* and *Zygnema*, respectively, in the reference phylogeny. *Klebsormidium flaccidum* is presumed to be most closely related to *Entransia fimbriata*

in the reference phylogeny (Karol et al., 2001). Most of the external specifiers used in the definition have been selected because they are internal specifiers in the definitions of *Charophyceae*, *Coleochaetophyceae* and *Klebsormidiophyceae* (this volume).

Composition: The clade *Zygnematophyceae* (conjugating green algae) includes about 4,000 species of mostly freshwater microalgae. The clade includes two well-known groups—the species-rich desmids (*Desmidiales* in modern classifications) and the paraphyletic '*Zygnematales*.'

Diagnostic Apomorphies: Members of *Zygnematophyceae* are unique among the green algae in their mode of sexual reproduction, conjugation. This process involves the fusion of non-flagellate gametes. There are no flagella in their life history and they have no centrioles (Hoshaw et al., 1990).

Synonyms: *Akontae* Blackman and Tansley 1902 (approximate); *Conjugaphyta* sensu Hoshaw et al. (1990) (approximate); *Conjugatophyceae* Engler 1892 (approximate); *Zygnematales* sensu Mattox and Stewart (1984) (approximate); *Zygnemophyceae* Round 1971 (approximate); *Zygnemophyta* Kenrick and Crane 1997; *Zygnematophyta* Lecointre and Le Guyader 2006 (approximate); *Zygophyceae* sensu Widder (1960) (approximate).

Comments: Phylogenetically, the core group of conjugating green algae is easily identified (Bhattacharya et al., 1994; McCourt et al., 2000; Gontcharov et al., 2004; Hall et al., 2008).

However, in molecular phylogenetic analyses, some taxa appear on long branches, including some forms currently assigned to *Mesotaenium* (Gontcharov et al., 2004; Hall et al., 2008) and all species of *Spirotaenia* (Gontcharov and Melkonian, 2004). Although a strain identified as *Spirotaenia* was included in the reference phylogeny (McCourt et al., 2000), subsequent study revealed that this was probably a contaminant (Gontcharov and Melkonian, 2004) and that *Spirotaenia* species are only distantly related to *Zygnematophyceae*. If future molecular phylogenetic studies find *Spirotaenia* to be sister to *Zygnematophyceae* then *Spirotaenia* should be included.

Zygnematophyceae in its modern conception—one largely consistent with molecular phylogenetic studies as regards inclusiveness—has been known for a very long time. Ralfs (1848) acknowledged the relationship, and the affinity of the filamentous and unicellular forms of the conjugating green algae was acknowledged by nearly every subsequent publication on these organisms. That the smooth-walled unicellular and filamentous forms do not form natural groups—i.e., that the *Zygnemataceae* (as traditionally circumscribed) and *Mesotaeniaceae* (as traditionally circumscribed) are each polyphyletic—was not known until molecular phylogenetic analyses including many unicellular taxa were performed (McCourt et al., 1995, 2000).

Zygnematophyceae is the name most commonly used for this clade in modern classifications (Gontcharov et al., 2003, 2004; Gontcharov and Melkonian, 2004; Hall et al., 2008). Many other names proposed for this clade were based on antiquated nomenclature or were previously applied to polyphyletic groups. Although sometimes applied to the clade *Zygnematophyceae*, the form names *Akontae* and *Zygophyceae* were both used in classifications that considered diatoms and desmids as close relatives (e.g., Oltmanns, 1904;

Bessey, 1914), whereas they are actually very distantly related clades. The names *Conjugatophyceae* and *Conjugaphyta* are not descriptive names but rather based on the now rejected genus name *Conjugata* (a synonym of *Spirogyra*) (Silva, 1980).

Literature Cited

Bessey, C. E. 1914. Synopsis of the conjugate algae: *Zygophyceae*. *Trans. Am. Microsc. Soc.* 33:11–49.

Bhattacharya, D., B. Surek, M. Rusing, S. Damberger, and M. Melkonian. 1994. Group I introns are inherited through common ancestry in the nuclear-encoded rRNA of *Zygnematales* (*Charophyceae*). *Proc. Nat. Acad. Sci. USA* 91:9916–9920.

Gontcharov, A. A., B. Marin, and M. Melkonian. 2003. Molecular phylogeny of conjugating green algae (*Zygnemophyceae, Streptophyta*) inferred from SSU rDNA sequence comparisons. *J. Mol. Evol.* 56:89–104.

Gontcharov, A. A., B. Marin, and M. Melkonian. 2004. Are combined analyses better than single gene phylogenies? A case study using SSU rDNA and *rbcL* sequence comparisons in the *Zygnematophyceae* (*Streptophyta*). *Mol. Biol. Evol.* 21:612–624.

Gontcharov, A. A., and M. Melkonian. 2004. Unusual position of the genus *Spirotaenia* (*Zygnematophyceae*) among streptophytes revealed by SSU rDNA and *rbcL* sequence comparisons. *Phycologia* 43:105–113.

Hall, J. D., K. G. Karol, R. M. McCourt, and C. F. Delwiche. 2008. Phylogeny of conjugating green algae based on chloroplast and mitochondrial nucleotide sequence data. *J. Phycol.* 44:467–477.

Hoek, C. van den, D. G. Mann, and H. M. Jahns. 1995. *Algae. An Introduction to Phycology*. Cambridge University Press, Cambridge, UK.

Hoshaw, R. W., R. M. McCourt, and J.-C. Wang. 1990. Phylum *Conjugaphyta*. Pp. 119–131 in *Handbook of* Protoctista (L. Margulis, J. Corliss, M. Melkonian and D. Chapman, eds.). Jones and Bartlett, Boston, MA.

Karol, K. G., R. M. McCourt, M. T. Cimino, and C. F. Delwiche. 2001. The closest living relatives of land plants. *Science* 294:2351–2353.

Mattox, K. R., and K. D. Stewart. 1984. Classification of the green algae: a concept based on comparative cytology. Pp. 29–72 in *Systematics of the Green Algae* (D. Irvine and D. John, eds.). Systematics Association Special Volume No. 27. Academic Press, London and Orlando.

McCourt, R. M., K. G. Karol, J. Bell, M. Helm-Bychowski, A. Grajewska, M. F. Wojciechowski, and R. W. Hoshaw. 2000. Phylogeny of the conjugating green algae (*Zygnematophyceae*) based on *rbcL* sequences. *J. Phycol.* 36:747–758.

McCourt, R. M., K. G. Karol, S. Kaplan, and R. W. Hoshaw. 1995. Using *rbcL* sequences to test hypotheses of chloroplast and thallus evolution in the conjugating green algae (*Zygnematales, Charophyceae*). *J. Phycol.* 31:989–995.

Oltmanns, F. 1904. *Morphologie und Biologie der Algen.* Verlag von Gustav Fischer, Jena.

Ralfs, J. 1848. *British* Desmidieae. Richard and John E. Taylor, London.

Silva, P. C. 1980. *Names of Classes and Families of Living Algae.* W. Junk, Hague, Netherlands.

Widder, F. 1960. Review: Fott, B. 1959. Algenkunde. *Phyton (Horn)* 9:167–168.

Authors

John D. Hall; Department of Botany; Academy of Natural Sciences of Drexel University; Philadelphia, PA 19103, USA. Email: jdhall2@gmail.com.

Richard M. McCourt; Department of Botany; Academy of Natural Sciences of Drexel University; Philadelphia, PA 19103, USA. Email: rmm45@drexel.edu.

Kenneth G. Karol; Cullman Program for Molecular Systematics; The New York Botanical Garden; Bronx, NY 10458, USA. Email: kkarol@nybg.org.

Date Accepted: 11 December 2013; updated 16 July 2018

Primary Editor: Philip Cantino

Coleochaetophyceae C. Jeffrey 1982: 411
[C. F. Delwiche, K. G. Karol and J. D. Hall], converted clade name

Registration Number: 152

Definition: The largest crown clade containing *Coleochaete scutata* Brébisson 1844 but not *Marchantia polymorpha* Linnaeus 1753 (*Embryophyta*) and *Chara vulgaris* Linnaeus 1753 (*Charophyceae*) and *Zygnema cruciatum* (Vaucher) C. Agardh 1817 (*Zygnematophyceae*). This is a maximum-crown-clade definition. Abbreviated definition: max crown ∇ (*Coleochaete scutata* Brébisson 1844 ~ *Marchantia polymorpha* Linnaeus 1753 & *Chara vulgaris* Linnaeus 1753 & *Zygnema cruciatum* (Vaucher) C. Agardh 1817.)

Etymology: Derived from the Greek *coleos* (sheath) and *chaete* (long hair) and *phykos* (seaweed), referring to the sheathed hairs characteristic of members of the clade.

Reference Phylogeny: Delwiche et al. (2002: Fig. 3), which was based on a broad sampling within *Coleochaetophyceae*. The name *Coleochaetophyceae* refers to the clade in the reference phylogeny that is labeled *Coleochaetales*. *Zygnema cruciatum* and *Chara vulgaris* are not included in the reference phylogeny but are used here as specifiers because they are internal specifiers of *Zygnematophyceae* and *Charophyceae*, respectively (this volume). They are thought to be most closely related to *Zygnema peliosporum* and *Chara connivens*, respectively, in the reference phylogeny.

Composition: *Coleochaetophyceae* contains about 25 currently recognized species belonging to *Coleochaete* and *Chaetosphaeridium* (Delwiche et al., 2002). See Comments for other possible members of this clade.

Diagnostic Apomorphies: There are no clear non-molecular synapomorphies of this clade. Members of this clade can be diagnosed by a combination of character states, none of which can be used independently to separate *Coleochaetophyceae* from all other green algae. Specifically, coleochaetophytes form branched filaments, have phragmoplastic cell division and possess specialized cells with long protruding sheathed hairs (Delwiche et al., 2002).

Synonyms: *Coleochaetales* Stewart and Mattox 1975 (approximate). See Comments regarding the names *Coleochaetophyta*, *Coleochaetophytina* and *Coleochaetobiotes* of Kenrick and Crane (1997).

Comments: The members of *Coleochaetophyceae* were recognized as a group based on their sheathed hairs. However, it was not until ultrastructural methods were employed that *Coleochaete* and *Chaetosphaeridium* were removed from the distantly related *Chaetophorales* (*Chlorophyta*) (Pickett-Heaps and Marchant, 1972). The name *Coleochaetophyceae* is equivalent to *Coleochaetales* in some phylogenetic studies (e.g., Cimino and Delwiche, 2002; Delwiche et al., 2002). The names *Coleochaetophyta*, *Coleochaetophytina* and *Coleochaetobiotes* were published by Kenrick and Crane (1997); however, their concept of this taxon, which they named at three different ranks, explicitly excluded *Chaetosphaeridium*; thus these names are synonyms of *Coleochaete* rather than *Coleochaetophyceae*.

One molecular study suggested that *Chaetosphaeridium* is more closely related to *Mesostigma* than to *Coleochaete* (Marin and Melkonian, 1999). However, subsequent molecular phylogenetic studies have corroborated the relationship between *Chaetosphaeridium* and *Coleochaete* (Karol et al., 2001; Cimino and Delwiche, 2002; Delwiche et al., 2002; Turmel et al., 2002; Wickett et al., 2014). It is not known if species currently assigned to other taxa with sheathed hairs (e.g., *Awahdiella* and *Chaetotheke*) belong to this clade (Hall and Delwiche, 2007).

Literature Cited

Cimino, M. T., and C. F. Delwiche. 2002. Molecular and morphological data identify a cryptic species complex in endophytic members of the genus *Coleochaete* Brébisson (*Charophyta: Coleochaetaceae*). *J. Phycol.* 38:1213–1221.

Delwiche, C. F., K. G. Karol, and M. T. Cimino. 2002. Phylogeny of the genus *Coleochaete* (*Coleochaetales, Charophyta*) and related taxa inferred by analysis of the chloroplast gene *rbcL*. *J. Phycol.* 38:394–403.

Hall, J. D., and C. F. Delwiche. 2007. In the shadow of giants; systematics of the charophyte green algae. Pp. 155–169 in *Unraveling the Algae: The Past, Present, and Future of Algal Systematics* (J. Brodie and J. Lewis, eds.). Systematics Association.

Jeffrey, C. 1982. Kingdoms, codes and classification. *Kew Bull.* 37:403–416.

Karol, K. G., R. M. McCourt, M. T. Cimino, and C. F. Delwiche. 2001. The closest living relatives of land plants. *Science* 294:2351–2353.

Kenrick, P., and P. R. Crane. 1997. *The Origin and Early Diversification of Land Plants. A Cladistic Study.* Smithsonian Institution Press, Washington, DC and London.

Marin, B., and M. Melkonian. 1999. *Mesostigmatophyceae*, a new class of streptophyte green algae revealed by SSU rRNA sequence comparisons. *Protist* 150:399–417.

Pickett-Heaps, J. D., and H. J. Marchant. 1972. The phylogeny of the green algae: a new proposal. *Cytobios* 6:255–264.

Turmel, M., M. Ehara, C. Otis, and C. Lemieux. 2002. Phylogenetic relationships among streptophytes as inferred from chloroplast small and large subunit rRNA gene sequences. *J. Phycol.* 38:364–375.

Wickett, N. J., S. Mirarab, N. Nguyen, T. Warnow, E. Carpenter, N. Matasci, S. Ayyampalayam, M. S. Barker, J. G. Burleigh, M. A. Gitzendanner, B. R. Ruhfel, E. Wafula, J. P. Der, S. W. Graham, S. Mathews, M. Melkonian, D. E. Soltis, P. S. Soltis, N. W. Miles, C. J. Rothfels, L. Pokorny, A. J. Shaw, L. DeGironimo, D. W. Stevenson, B. Surek, J. C. Villarreal, B. Roure, H. Philippe, C. W. dePamphilis, T. Chen, M. K. Deyholos, R. S. Baucom, T. M. Kutchan, M. M. Augustin, J. Wang, Y. Zhang, Z. Tian, Z. Yan, X. Wu, X. Sun, G. K.-S. Wong and J. Leebens-Mack. 2014. Transcriptomic analysis of the original and early diversification of land plants. *PNAS* 111(45):E4859–E4868.

Authors

Charles F. Delwiche; Department of Cell Biology and Molecular Genetics; University of Maryland; College Park, MD 20740, USA. Email: delwiche@umd.edu.

Kenneth G. Karol; Cullman Program for Molecular Systematics; The New York Botanical Garden; Bronx, NY 10458, USA. Email: kkarol@nybg.org.

John D. Hall; Department of Botany; Academy of Natural Sciences of Drexel University; Philadelphia, PA 19103, USA. Email: jdhall2@gmail.com.

Date Accepted: 21 October 2013; updated 16 July 2018

Primary Editor: Philip Cantino

Charophyceae G. M. Smith 1938: 127 [K. G. Karol, R. M. McCourt, and J. D. Hall], converted clade name

Registration Number: 151

Definition: The largest crown clade containing *Chara vulgaris* Linnaeus 1753 but not *Arabidopsis thaliana* (Linnaeus) Heynh. 1842 (*Embryophyta*) and *Coleochaete scutata* Brébisson 1844 (*Coleochaetophyceae*) and *Entransia fimbriata* E. O. Hughes 1943 and *Klebsormidium flaccidum* (Kützing) Silva, Mattox and Blackwell 1972 (*Klebsormidiophyceae*) and *Zygnema cruciatum* (Vaucher) C. Agardh 1817 (*Zygnematophyceae*). This is a maximum-crown-clade definition. Abbreviated definition: max crown ∇ (*Chara vulgaris* Linnaeus 1753 ~ *Arabidopsis thaliana* (Linnaeus) Heynh. 1842 & *Coleochaete scutata* Brébisson 1844 & *Entransia fimbriata* E. O. Hughes 1943 & *Klebsormidium flaccidum* (Kützing) Silva, Mattox and Blackwell 1972 & *Zygnema cruciatum* (Vaucher) C. Agardh 1817).

Etymology: Derived from *Chara* Linnaeus 1753, the name of an included taxon, and *phykos*, Greek for seaweed. The etymology of *Chara* is uncertain (see Robinson, 1906 for discussion).

Reference Phylogeny: Karol et al. (2001: Fig 1). *Chara vulgaris*, *Coleochaete scutata* and *Zygnema cruciatum* are represented by or assumed to be closely related to *Chara*, *Coleochaete orbicularis* and *Zygnema*, respectively, in the reference phylogeny. The former three taxa were chosen as internal specifiers for consistency with the definitions of three of the four major subclades of *Phragmoplastophyta*. Because these species are the type species of their respective genera under rank-based nomenclature, they must be used as internal specifiers for names based on those genera (*ICPN* Art. 11.10; Cantino and de Queiroz,

2020): *Charophyceae*, *Coleochaetophyceae*, and *Zygnematophyceae* as defined in this volume.

Composition: This clade includes those species currently assigned to the following living taxa traditionally ranked as genera: *Chara* Linnaeus 1753, *Lamprothamnium* J. Groves 1916, *Lychnothamnus* (Ruprecht) A. Braun 1856, *Nitella* C. Agardh 1824, *Nitellopsis* Hy 1889, and *Tolypella* (A. Braun) A. Braun 1857 as well as those fossils that originated after the most recent common ancestor of the extant species. A nearly comprehensive list of extant species was provided by Wood and Imahori (1965).

Diagnostic Apomorphies: Members of *Charophyceae* are unique among the green algae in having a thallus composed of whorls of branchlets at nodes connected by a single internodal cell. They also have structurally complex reproductive organs: the oogonium comprising an egg cell and jacket cells, and the antheridium composed of several shield cells, each with a manubrium that gives rise to spermatogenous filaments (summarized in Wood and Imahori, 1965). The presence of a thick-walled spore with various surface ornamentations is also unique to this group and allows for the referral of many species known only from the fossil record to the total clade of *Charophyceae* (Feist et al., 2005a).

Synonyms: *Charophyta* Migula 1897 (approximate); *Charophycophyta* Papenfuss 1946 (approximate); *Charales* sensu Mattox and Stewart (1984) (approximate); *Characeae* S. F. Gray 1821 (approximate); *Charophytina* Bremer and Wanntorp 1981 (approximate).

Comments: The name *Charophyceae* has sometimes been used to refer to a broader group of algae than those commonly called stoneworts or muskgrasses. It is applied here in its strict sense—to the crown clade that includes *Chara vulgaris* and all extant species that share a more recent common ancestor with *Chara* than with *Embryophyta*, *Coleochaetophyceae*, *Zygnematophyceae*, *Klebsormidiophyceae*, and *Entransia*.

The inferred composition of the crown clade of *Charophyceae* (see Composition) has been strongly supported by molecular phylogenetic and structural evidence (McCourt et al., 1996, 1999; Meiers et al., 1999; Karol et al., 2001; Feist et al., 2005a). The relationships among some of the subclades of *Charophyceae* remains moderately or weakly supported in molecular phylogenetic studies, but the inclusion of all living species of *Charophyceae* is unambiguous. The inclusion of several fossil taxa is less certain. Feist et al. (2005a) proposed that more than 40 fossil taxa traditionally ranked as genera (i.e., those included in her family *Characeae*) belong to *Charophyceae* as defined here, thus greatly expanding the diversity thought to belong to this crown clade.

The names *Charophyceae* and *Charophyta* have been used in the past in both a narrow sense, restricted to the stoneworts, and a broad sense, referring to a larger assemblage of green algae that are closely related to land plants (with or without inclusion of the land plants, depending on the author, thus the name variably applied to a clade or to a paraphyletic group). In this volume, we follow Lewis and McCourt (2004) in applying the name *Charophyta* to the larger crown clade containing not only stoneworts and land plants but also several other algal relatives, and restricting *Charophyceae* to the smaller crown clade originating in the most recent common ancestor of the living stoneworts (Martín-Closas et al., 1999; Feist et al., 2005a,b).

Literature Cited

Cantino, P. D., and K. de Queiroz. 2020. *International Code of Phylogenetic Nomenclature* (*PhyloCode*), Version 6. CRC Press, Boca Raton, FL.

Feist, M., N. Grambast-Fessard, M. Guerlesquin, K. Karol, R. M. McCourt, H. Lu, S. Zhang, and Q. Wang. 2005a. Pp. 1–170 in *Treatise on Invertebrate Paleontology*, Part C, Protista 1, Vol. I (R. L. Kaesler, ed.). Charophyta. Geological Society of America and the University of Kansas Press, Lawrence, Kansas.

Feist, M., J. Liu, and P. Tafforeau. 2005b. New insights into Paleozoic charophyte morphology and phylogeny. *Am. J. Bot.* 92:1152–1160.

Karol, K. G., R. M. McCourt, M. T. Cimino, and C. F. Delwiche. 2001. The closest living relatives of land plants. *Science* 294:2351–2353.

Lewis, L. A., and R. M. McCourt. 2004. Green algae and the origin of land plants. *Am. J. Bot.* 91:1535–1556.

Martín-Closas, C., R. Bosch and J. Serra-Kiel. 1999. Biomechanics and evolution of spiralization in charophyte fructifications. Pp. 399–421 in *The Evolution of Plant Architecture* (M. H. Kurmann and A. Hemsley, eds.). Royal Botanic Gardens Kew, London.

Mattox, K. R., and K. D. Stewart. 1984. Classification of the green algae: a concept based on comparative cytology. Pp. 29–72 in *Systematics of the Green Algae* (D. Irvine and D. John, eds.). Systematics Association Special Volume No. 27. Academic Press, London and Orlando.

McCourt, R. M., K. G. Karol, M. T. Casanova and M. Feist. 1999. Monophyly of genera and species of *Characeae* based on *rbcL* sequences with special reference to Australian and European *Lychnothamnus barbatus* (*Characeae*; *Charophyceae*). *Aust. J. Bot.* 47: 361–369.

McCourt, R. M., K. G. Karol, M. Guerlesquin, and M. Feist. 1996. Phylogeny of extant genera in the family *Characeae* (*Charales*, *Charophyceae*) based on *rbcL* sequences and morphology. *Am. J. Bot.* 83:125–131.

Meiers, S. T., V. W. Proctor, and R. L. Chapman. 1999. Phylogeny and biogeography of *Chara* (*Charophyta*) inferred from 18S rDNA sequences. *Aust. J. Bot.* 47:347–360.

Robinson, C. B. 1906. The *Characeae* of North America. *Bull. N. Y. Bot. Gard.* 4:244–308.

Smith, G. M. 1938. *Cryptogamic Botany, Vol. 1, Algae and Fungi.* McGraw-Hill, New York.

Wood, R. D., and K. Imahori. 1965. *A Revision of the* Characeae. Verlag von J. Cramer, Weinheim, Germany.

Authors

Kenneth G. Karol; Cullman Program for Molecular Systematics; The New York Botanical Garden; Bronx, NY 10458, USA. Email: kkarol@nybg.org.

Richard M. McCourt; Department of Botany; Academy of Natural Sciences of Drexel University; Philadelphia, PA 19103, USA. Email: rmm45@drexel.edu.

John D. Hall; Department of Botany; Academy of Natural Sciences of Drexel University; Philadelphia, PA 19103, USA. Email: jdhall2@gmail.com.

Date Accepted: 27 December 2013; updated 6 March 2018

Primary Editor: Philip Cantino

SECTION 3

Embryophyta A. Engler and K. Prantl 1889 [B. D. Mishler and Y.-L. Qiu], converted clade name

Registration Number: 136

Definition: The smallest crown clade containing *Magnolia tripetala* (L.) L. 1759 (*Tracheophyta*), *Anthoceros punctatus* L. 1753 (*Anthocerotae*), *Marchantia polymorpha* L. 1753 (*Hepaticae*), and *Sphagnum palustre* L. 1753 (*Musci*). This is a minimum-crown-clade definition. Abbreviated definition: min crown ∇ (*Magnolia tripetala* (L.) L. 1759 & *Anthoceros punctatus* L. 1753 & *Marchantia polymorpha* L. 1753 & *Sphagnum palustre* L. 1753).

Etymology: From Greek *embryon* (young one) and *phyta* (plants). This refers to one of the most important synapomorphies of the group, the retention of the fertilized egg inside the female gametangium and its development into a multicellular diploid life phase attached to the mother haploid plant.

Reference Phylogeny: Figure 1 in Qiu et al. (2007), a multigene, 3-genomic compartment (chloroplast, mitochondrion, nucleus) analysis of DNA sequence data. *Anthoceros punctatus*, the type of *Anthocerotae*, was not included in the reference phylogeny but is closely related to *A. agrestis* (Duff et al., 2007), one of the species used in the analysis through which that phylogeny was inferred.

Composition: The clade *Embryophyta* consists of four major subclades: vascular plants, liverworts, mosses, and hornworts (*Pan-Tracheophyta* and total clades of *Hepaticae*, *Musci*, and *Anthocerotae*, respectively).

Diagnostic Apomorphies: Cuticle; embryo; archegonia, antheridia, and sporangia with an outer jacket of sterile cells (see discussion in Mishler and Churchill, 1984, 1985; Kenrick and Crane, 1997b; Qiu et al. 2012); a number of aspects of sperm architecture (Duckett et al., 1982); and probably gravitropism (see discussion by Qiu [2008] and Doyle [2013]).

Synonyms: The following are approximate synonyms: land plants, *Cormophyta* Endlicher 1836–1840, *Cormobionta* Rothmaler 1948, *Euplanta* Barkley 1949, *Embryobionta* Cronquist, Takhtajan and Zimmerman 1966, *Metaphyta* sensu Whittaker (1969), *Plantae* sensu Margulis (1974), *Embryophytina* Bremer and Wanntorp 1981, *Embryobiota* Kenrick and Crane 1997b, *Embryobiotes* Kenrick and Crane 1997b, *Embryophyceae* sensu Lewis and McCourt (2004), *Equisetopsida* sensu Chase and Reveal (2009), and *Embryopsida* Pirani and Prado 2012. *Acrobrya* Endlicher 1836–1840 and *Bryophyta* Engler and Prantl 1909 are partial synonyms, but *Bryophyta* sensu Crum (2001) is a synonym of *Musci*.

Comments: Of the many names that have been applied to this clade (see Synonyms), *Embryophyta* and *Plantae* are much more frequently used than the rest. We have chosen the former because it refers unambiguously to this clade. In contrast, although *Plantae* has often been used for this clade (e.g., Margulis, 1974; Raven et al., 1999), it is also used widely in several other senses: e.g., for all the green plants (i.e., *Viridiplantae*; e.g., Copeland, 1956), or for greens plus red plants (*Rhodophyta*) plus glaucophytes (e.g., Rodríguez-Ezpeleta, 2005; Weber et al., 2006; Moustafa et al., 2008), or for a highly polyphyletic group consisting of all photosynthetic, multicellular organisms minus their unicellular relatives (e.g., Whittaker and Margulis, 1978 and many popular biology textbooks).

The name *Embryophyta* has been applied to this clade for more than 100 years. The first use of this name we can find (see also Gundersen, 1918) for the modern clade is in the monumental work *Die natürlichen Pflanzenfamilien*, founded by Engler and Prantl with many contributors. We attribute it to Engler and Prantl (1889: Teil 2, page 1) themselves rather than any of the contributing authors because it is part of the structural classification of the overall work. Engler and Prantl used both *Embryophyta zoidiogama* (for *Bryophyta* and *Pteridophyta*), and *Embryophyta siphonogama* (for *Gymnospermae* and *Angiospermae*) but clearly intended "*Embryophyta*" *per se* to refer to the taxon that includes all of these plants.

The crown here named *Embryophyta* is an extremely well-supported clade, recognized in all published analyses, including early cladistic studies using morphological data of extant taxa alone (Mishler and Churchill, 1984, 1985) and with fossils added (Kenrick and Crane, 1997a,b), as well as studies using DNA sequence data (e.g., Nickrent et al., 2000; Qiu et al., 2006, 2007; Chang and Graham, 2011; Ruhfel et al., 2014; Liu et al., 2014) and genome structural data such as gene rearrangements and intron gain/loss (e.g., Qiu et al., 1998, 2006; Kelch et al, 2004; Mishler and Kelch, 2009). Because the monophyly of each of the four principal subclades (*Tracheophyta, Hepaticae, Musci, Anthocerotae*) is well supported (Mishler et al., 1994; Kenrick and Crane, 1997b; Qiu et al., 2006, 2007; Chang and Graham, 2011; Liu et al., 2014), but disagreements remain about the relationships among them (Nickrent et al., 2000; Wickett et al., 2014; Cox et al., 2014), a single internal specifier was used to represent each.

Literature Cited

Barkley, F. A. 1949. Un esbozo de clasificación de los organismos. *Rev. Fac. Nac. Agron.* 10:83–103.

Bremer, K., and H.-E. Wanntorp. 1981. A cladistic classification of green plants. *Nord. J. Bot.* 1:1–3.

Chang, Y., and S. W. Graham. 2011. Inferring the higher-order phylogeny of mosses (*Bryophyta*) and relatives using a large, multigene plastid data set. *Am. J. Bot.* 98:839–849.

Chase, M. W., and J. L. Reveal. 2009. A phylogenetic classification of the land plants to accompany APG III. *Bot. J. Linn. Soc.* 161:122–127.

Copeland, H. F. 1956. *The Classification of the Lower Organisms.* Pacific Books, Palo Alto, CA.

Cox, C. J., B. Li, P. G. Foster, T. M. Embley, and P. Civáň. 2014. Conflicting phylogenies for early land plants are caused by composition biases among synonymous substitutions. *Syst. Biol.* 63:272–279.

Cronquist, A., A. Takhtajan, and W. Zimmerman. 1966. On the higher taxa of *Embryobionta*. *Taxon* 15:129–134.

Crum, H. 2001. *Structural Diversity of Bryophytes.* The University of Michigan Herbarium, Ann Arbor, MI.

Doyle, J. A. 2013. Phylogenetic analyses and morphological innovations in land plants. Pp. 1–50 in *The Evolution of Plant Form* (B.A. Ambrose and M. Purugganan, eds.). Wiley-Blackwell, Oxford.

Duckett, J. G., Z. B. Carothers, and C. C. J. Miller. 1982. Comparative spermatology and bryophyte phylogeny. *J. Hattori Bot. Lab.* 53:107–125.

Duff, R. J., J. C. Villareal, D. C Cargill, and K. S. Renzaglia. 2007. Progress and challenges toward developing a phylogeny and classification of the hornworts. *Bryologist* 110:214–243.

Endlicher, S. 1836–1840. *Genera Plantarum Secundum Ordines Naturales Disposita.* Vindobonae, Apud Fr. Beck Universitatis Bibliopolam.

Engler, A., and K. Prantl. 1889. *Die natürlichen Pflanzenfamilien,* Teil 2. Leipzig. Available at http://www.biodiversitylibrary.org/item/56456-page/5/mode/1up.

Engler, A., and K. Prantl. 1909. *Die Natürlichen Pflanzenfamilien, I. Teil, Abteilung 3, I. Hälfte.* Verlag von Wilhelm Engelmann, Leipzig.

Gundersen, A. 1918. A sketch of plant classification from Theophrastus to the present (continued). *Torreya* 18:231–239.

Kelch, D. G., A. Driskell, and B. D. Mishler. 2004. Inferring phylogeny using genomic characters: a case study using land plant plastomes. Pp. 3–12 in *Molecular Systematics of Bryophytes* [Monographs in Systematic Botany 98] (B. Goffinet, V. Hollowell, and R. Magill, eds.). Missouri Botanical Garden Press, St. Louis.

Kenrick, P., and P. R. Crane. 1997a. The origin and early evolution of plants on land. *Nature* 389:33–39.

Kenrick, P., and P. R. Crane. 1997b. *The Origin and Early Diversification of Land Plants. A Cladistic Study.* Smithsonian Institution Press, Washington, DC.

Lewis, L. A., and R. M. McCourt. 2004. Green algae and the origin of land plants. *Am. J. Bot.* 91:1535–1556.

Liu, Y., C. J. Cox, W. Wang, and B. Goffinet. 2014. Mitochondrial phylogenomics of early land plants: mitigating the effects of saturation, compositional heterogeneity, and codon-usage bias. *Syst. Biol.* 63:862–878.

Margulis, L. 1974. Five-kingdom classification and the origin and evolution of cells. *Evol. Biol.* 7:45–78.

Mishler, B. D., and S. P. Churchill. 1984. A cladistic approach to the phylogeny of the "bryophytes". *Brittonia* 36:406–424.

Mishler, B. D., and S. P. Churchill. 1985. Transition to a land flora: phylogenetic relationships of the green algae and bryophytes. *Cladistics* 1:305–328.

Mishler, B. D., and D. G. Kelch. 2009. Phylogenomics and early land plant evolution. Pp. 173–197 in *Bryophyte Biology* (B. Goffinet and A. J. Shaw, eds.). Cambridge University Press, Cambridge, UK.

Mishler, B. D., L. A. Lewis, M. A. Buchheim, K. S. Renzaglia, D. J. Garbary, C. F. Delwiche, F. W. Zechman, T. S. Kantz, and R. L. Chapman. 1994. Phylogenetic relationships of the "green algae" and "bryophytes". *Ann. Mo. Bot. Gard.* 81:451–483.

Moustafa A., A. Reyes-Prieto, and D. Bhattacharya 2008. *Chlamydiae* has contributed at least 55 genes to *Plantae* with predominantly plastid functions. *PLOS ONE* 3(5):e2205. doi:10.1371/journal.pone.0002205.

Nickrent, D. L., C. L. Parkinson, J. D. Palmer, and R. J. Duff. 2000. Multigene phylogeny of land plants with special reference to bryophytes and the earliest land plants. *Mol. Biol. Evol.* 17:1885–1895.

Pirani, J. R., and J. Prado. 2012. *Embryopsida*, a new name for the class of land plants. *Taxon* 61:1096–1098.

Qiu, Y.-L. 2008. Phylogeny and evolution of charophytic algae and land plants. *J. Syst. Evol.* 46:287–306.

Qiu, Y.-L., Y. Cho, J. C. Cox and J. D. Palmer. 1998. The gain of three mitochondrial introns identifies liverworts as the earliest land plants. *Nature* 394:671–674.

Qiu, Y.-L., L. Li, B. Wang, Z. Chen, O. Dombrovska, J. Lee, L. Kent, R. Li, R. W. Jobson, T. A. Hendry, D. W. Taylor, C. M. Testa, and M. Ambros. 2007. A nonflowering land plant phylogeny inferred from nucleotide sequences of seven chloroplast, mitochondrial, and nuclear genes. *Int. J. Plant Sci.* 168:691–708.

Qiu, Y.-L., L. Li, B. Wang, Z. Chen, V. Knoop, M. Groth-Malonek, O. Dombrovska, J. Lee, L. Kent, J. Rest, G. F. Estabrook, T. A. Hendry, D. W. Taylor, C. M. Testa, M. Ambros, B. Crandall-Stotler, R. J. Duff, M. Stech, W. Frey, D. Quandt, and C. C. Davis. 2006. The deepest divergences in land plants inferred from phylogenomic evidence. *Proc. Natl. Acad. Sci. USA* 103:15511–15516.

Qiu, Y.-L., A. B. Taylor, and H. A. McManus. 2012. Evolution of the life cycle in land plants. *J. Syst. Evol.* 50:171–194.

Raven, P. H., R. F. Evert, and S. E. Eichhorn. 1999. *Biology of Plants.* 6th edition. W. H. Freeman and Company, New York.

Rodríguez-Ezpeleta N., H. Brinkmann, S. C. Burey, B. Roure, G. Burger, W. Loffelhardt, H. J. Bohnert, H. Philippe, and B. F. Lang. 2005. Monophyly of primary photosynthetic eukaryotes: green plants, red algae, and glaucophytes. *Curr. Biol.* 15:1325–1330.

Rothmaler, W. 1948. Über das natürliche System der Organismen. *Biolog. Zentralbl.* 67:242–250.

Ruhfel, B. D., M. A. Gitzendanner, P. S. Soltis, D. E. Soltis, and J. G. Burleigh. 2014. From algae to angiosperms – inferring the phylogeny of green plants (*Viridiplantae*) from 360 plastid genomes. *BMC Evol. Biol.* 14:23.

Weber, A. P. M., M. Linka, and D. Bhattacharya. 2006. Single, ancient origin of a plastid metabolite translocator family in *Plantae* from an endomembrane-derived ancestor. *Eukaryot. Cell* 5:609–612.

Whittaker, R. H. 1969. New concepts of kingdoms of organisms. *Science* 163:150–160.

Whittaker, R. H., and L. Margulis. 1978. Protist classification and kingdoms of organisms. *BioSystems* 10:3–18.

Wickett, N. J., S. Mirarab, N. Nguyen, T. Warnow, E. Carpenter, N. Matasci, S. Ayyampalayam, M. S. Barker, J. G. Burleigh, M. A. Gitzendanner, B. R. Ruhfel, E. Wafula, J. P. Der, S. W. Graham, S. Mathews, M. Melkonian, D. E. Soltis, P. S. Soltis, N. W. Miles, C. J. Rothfels, L. Pokorny, A. J. Shaw, L. DeGironimo, D. W. Stevenson, B. Surek, J. C. Villarreal, B. Roure, H. Philippe, C. W. dePamphilis, T. Chen, M. K. Deyholos, R. S. Baucom, T. M. Kutchan, M. M. Augustin, J. Wang, Y. Zhang, Z. Tian, Z. Yan, X. Wu, X. Sun, G. K. Wong, and J. Leebens-Mack. 2014. Phylotranscriptomic analysis of the origin and early diversification of land plants. *Proc. Natl. Acad. Sci. USA* 111:E4859–E4868.

Authors

Brent D. Misher; University & Jepson Herbaria, and Department of Integrative Biology; University of California; Berkeley, CA, 94720-2465, USA. Email: bmishler@berkeley.edu.

Yin-Long Qiu; Department of Ecology & Evolutionary Biology, and The University Herbarium; University of Michigan; Ann Arbor, MI 48109-1048, USA. Email: ylqiu@umich.edu.

Date Accepted: 29 September 2014; updated 1 July 2017

Primary Editor: Philip Cantino

Hepaticae A. L. Jussieu 1789 [B. D. Mishler, L. Forrest, B. Crandall-Stotler, and R. Stotler], converted clade name

Registration Number: 50

Definition: The largest crown clade containing *Marchantia polymorpha* L. 1753 but not *Anthoceros punctatus* L. 1753 (*Anthocerotae*), and *Sphagnum palustre* L. 1753 (*Musci*), and *Equisetum telmateia* Ehrh. 1783 (*Tracheophyta*). This is a maximum-crown-clade definition. Abbreviated definition: max crown ∇ (*Marchantia polymorpha* L. 1753 ~ *Anthoceros punctatus* L. 1753 & *Sphagnum palustre* L. 1753 & *Equisetum telmateia* Ehrh. 1783).

Etymology: From the Greek *hepar* (liver), referring to the perceived resemblance of some liverworts such as *Concephalum* and *Marchantia* to the lobes that are common in vertebrate livers, which according to the ancient doctrine of signatures meant that these plants were supposedly useful for treating liver ailments in humans (Gibson and Gibson, 2007).

Reference Phylogeny: Figure 1 (plus more detailed Figs. 3–9) in Forrest et al. (2006). In the reference phylogeny, *Anthocerotae*, *Musci*, and *Tracheophyta* are labeled hornworts, mosses, and tracheophytes, respectively. Also see Qiu et al. (2006: Fig. 1).

Composition: This is a morphologically disparate clade consisting of about 5,000 recognized species (Crandall-Stotler et al., 2009), with a range of thalloid to leafy growth forms. The three deepest-branching subclades (Forrest et al., 2006) are *Haplomitriaceae* + *Treubiaceae*, *Blasiales*+complex thalloids, and a large clade containing the simple thalloids and the true leafy liverworts.

Diagnostic Apomorphies: Elaters (unicellular sterile cells among the spores, which are morphologically and developmentally distinct from the pseudoelaters of the hornworts), along with oil bodies (membrane-bound organelles), are unique synapomorphies of the liverworts (Mishler and Churchill, 1985).

Synonyms: *Marchantiophyta* sensu Kenrick and Crane (1997), *Hepaticophytina* sensu Frey and Stech (2005), *Hepaticopsida* sensu Crum (1991), and *Hepatophyta* sensu Crum (2001) are approximate synonyms.

Comments: The name *Hepaticae* was first used by Jussieu (1789: 7) for the liverworts, including what is now known to be a separate clade, the hornworts (*Anthocerotae*). We apply the name here to the liverworts alone, in accordance with the universal modern usage (Schuster, 1966; Lewis et al., 1997). We use *Hepaticae* instead of *Marchantiophyta*, *Hepaticophytina*, *Hepaticopsida*, or *Hepatophyta* because *Hepaticae* has much wider usage.

The name *Hepaticae* is applied here to the crown clade, in accordance with *PhyloCode* (Cantino and de Queiroz, 2020) Recommendation 10.1B. Palaeobotanists and bryologists have often viewed all fossil liverworts, some of which may lie outside the crown clade, as members of *Hepaticae* (e.g., Boureau et al., 1967; Taylor et al., 2009). Following the definition of *Hepaticae* here, liverwort fossils shown to be outside the crown would instead be considered part of the informally named total clade pan-Hepaticae in accordance with Recommendation 10.3A.

This is an extremely well-supported clade that has been inferred from very different sources of data (Mishler and Churchill, 1985; Lewis et al., 1997; Qiu et al., 2006; Forrest et al., 2006). There is a consensus across many studies (e.g., Mishler and Churchill, 1984; Mishler et al., 1994; Kenrick and Crane, 1997; Qiu et al. 1998, 2006; Forrest et al. 2006) that the liverworts are sister to the rest of *Embryophyta* (i.e., *Musci + Anthocerotae + Tracheophyta*). However, some studies have found *Hepaticae* to be sister to *Musci* (Nickrent et al., 2000; Wickett et al., 2014). Given that the hypothesized relationships among the four principal extant embryophyte clades (i.e., *Hepaticae, Musci, Anthocerotae,* and *Tracheophyta*) may be subject to change in the future (e.g., Cox et al., 2014), we have selected external specifiers representing each of the latter three.

Literature Cited

Boureau, É., S. Jovet-Ast, O. A. Höeg, and W. G. Chaloner. 1967. *Traité de Paléobotanique: Tome II: Bryophyta, Psilophyta, Lycophyta*. Masson et Cie Editeurs, Paris.

Cantino, P. D., and K. de Queiroz. 2020. *International Code of Phylogenetic Nomenclature (PhyloCode)*, Version 6. CRC Press, Boca Raton, FL.

Cox, C. J., B. Li, P. G. Foster, T. Embley, and P. Civan. 2014. Conflicting phylogenies for early land plants are caused by composition biases among synonymous substitutions. *Syst. Biol.* 63:272–279.

Crandall-Stotler, B., R. E. Stotler, and D. G. Long. 2009. Morphology and classification of the *Marchantiophyta*. Pp. 1–54 in *Bryophyte Biology*. 2nd edition (A. J. Shaw and B. Goffinet. eds.). Cambridge University Press, Cambridge, UK.

Crum, H. A. 1991. *Liverworts and Hornworts of Southern Michigan*. University of Michigan Herbarium, Ann Arbor, MI.

Crum, H. A. 2001. *Structural Diversity of Bryophytes*. University of Michigan Herbarium, Ann Arbor, MI.

Forrest, L. L., E. C. Davis, D. G. Long, B. J. Crandall-Stotler, A. Clarke, and M. L. Hollingsworth. 2006. Unraveling the evolutionary history of the liverworts (*Marchantiophyta*): multiple taxa, genomes and analyses. *Bryologist* 109:303–334.

Frey, W., and M. Stech. 2005. A morpho-molecular classification of the liverworts (*Hepaticophytina, Bryophyta*). *Nova Hedwigia* 81:55–78.

Gibson, J. P., and T. R. Gibson. 2007. *Plant Diversity*. Infobase Publishing, New York.

Jussieu, A. L. 1789. *Genera Plantarum*. Herissant & Barrois, Paris.

Kenrick, P., and P. R. Crane. 1997. *The Origin and Early Diversification of Land Plants: A Cladistic Study*. Smithsonian Institution Press, Washington, DC.

Lewis, L. A., B. D. Mishler, and R. Vilgalys. 1997. Phylogenetic relationships of the liverworts (*Hepaticae*), a basal embryophyte lineage, inferred from nucleotide sequence data of the chloroplast gene *rbcL*. *Mol. Phylog. Evol.* 7:377–393.

Mishler, B. D., and S. P. Churchill. 1984. A cladistic approach to the phylogeny of the "bryophytes". *Brittonia* 36:406–424.

Mishler, B. D., and S. P. Churchill. 1985. Transition to a land flora: phylogenetic relationships of the green algae and bryophytes. *Cladistics* 1:305–328.

Mishler, B. D., L. A. Lewis, M. A. Buchheim, K. S. Renzaglia, D. J. Garbary, C. F. Delwiche, F. W. Zechman, T. S. Kantz, and R. L. Chapman. 1994. Phylogenetic relationships of the "green algae" and "bryophytes". *Ann. Mo. Bot. Gard.* 81:451–483.

Nickrent, D. L., C. L. Parkinson, J. D. Palmer, and R. J. Duff. 2000. Multigene phylogeny of land plants with special reference to bryophytes and the earliest land plants. *Mol. Biol. Evol.* 17:1885–1895.

Qiu, Y.-L., Y. Cho, J. C. Cox, and J. D. Palmer, 1998. The gain of three mitochondrial introns identifies liverworts as the earliest land plants. *Nature* 394:671–674.

Qiu, Y.-L., L. Li, B. Wang, Z. Chen, V. Knoop, M. Groth-Malonek, O. Dombrovska, J. Lee, L. Kent, J. Rest, G. F. Estabrook, T. A. Hendry, D. W. Taylor, C. M. Testa, M. Ambros, B. Crandall-Stotler, R. J. Duff, M. Stech, W. Frey, D. Quandt, and C. C. Davis. 2006. The deepest divergences in land plants inferred from phylogenomic evidence. *Proc. Natl. Acad. Sci. USA* 103:15511–15516.

Schuster, R. M. 1966. *The* Hepaticae *and* Anthocerotae *of North America East of the Hundredth Meridian*. Columbia University Press, New York.

Taylor, T. N., E. L. Taylor, and M. Krings. 2009. *Paleobotany: The Biology and Evolution of Fossil Plants*. 2nd edition. Elsevier, Oxford.

Wickett, N. J., S. Mirarab, N. Nguyen, T. Warnow, E. Carpenter, N. Matasci, S. Ayyampalayam, M. S. Barker, J. G. Burleigh, M. A. Gitzendanner, B. R. Ruhfel, E. Wafula, J. P. Der, S. W. Graham, S. Mathews, M. Melkonian, D. E. Soltis, P. S. Soltis, N. W. Miles, C. J. Rothfels, L. Pokorny, A. J. Shaw, L. DeGironimo, D. W. Stevenson, B. Surek, J. C. Villarreal, B. Roure, H. Philippe, C. W. dePamphilis, T. Chen, M. K. Deyholos, R. S. Baucom, T. M. Kutchan, M. M. Augustin, J. Wang, Y. Zhang, Z. Tian, Z. Yan, X. Wu, X. Sun, G. K. Wong, and J. Leebens-Mack. 2014. Phylotranscriptomic analysis of the origin and early diversification of land plants. *Proc. Natl. Acad. Sci. USA* 111:E4859–E4868.

Authors

Brent D. Mishler; University & Jepson Herbaria, and Department of Integrative Biology; University of California; Berkeley, CA 94720-2465, USA. Email: bmishler@berkeley.edu.

Laura L. Forrest; Royal Botanic Garden, 20A Inverleith Row, Edinburgh EH3 5LR, Scotland. Email: l.forrest@rbge.ac.uk.

Barbara J. Crandall-Stotler; Department of Plant Biology; University of Southern Illinois; Carbondale, IL 62901, USA. Email: crandall@.siu.edu.

Raymond Stotler[†]; Department of Plant Biology; University of Southern Illinois; Carbondale, IL 62901, USA.

Date Accepted: 30 August 2016

Primary Editor: Philip Cantino

[†] Our esteemed colleague, Ray Stotler, an expert on liverwort and hornwort classification, contributed to early versions of this manuscript but sadly passed away December 4, 2013.

Musci A. L. Jussieu 1789. [B. D. Mishler, A. E. Newton, N. Bell, B. Crandall-Stotler, and R. Stotler], converted clade name

Registration Number: 69

Definition: The largest crown clade containing *Sphagnum palustre* L. 1753 but not *Marchantia polymorpha* L. 1753 (*Hepaticae*), and *Anthoceros punctatus* L. 1753 (*Anthocerotae*), and *Zea mays* L. 1753 (*Tracheophyta*). This is a maximum-crown-clade definition. Abbreviated definition: max crown ∇ (*Sphagnum palustre* L. 1753 ~ *Marchantia polymorpha* L. 1753 & *Anthoceros punctatus* L. 1753 & *Zea mays* L. 1753).

Etymology: From the Latin *muscus* (moss).

Reference Phylogeny: Figure 3 in Newton et al. (2000), a combined analysis of molecular and morphological data. Also see Qiu et al. (2006: Fig. 1).

Composition: This is a diverse clade consisting of nearly 13,000 recognized species. The main subclades are *Takakia* and *Sphagnum* (these two possibly sister within one clade or two separate clades), *Andreales*, *Oedipodium*, and the peristomate mosses (see Comments).

Diagnostic Apomorphies: Leaves on the gametophyte (not homologous with leaves in leafy liverworts or *Tracheophyta*); multicellular rhizoids (Mishler and Churchill, 1984).

Synonyms: The following are approximate synonyms: *Bryopsida* sensu Vitt (1984) and *Bryophyta* (in some applications of this name— e.g., Kenrick and Crane, 1997; Goffinet and Buck, 2004; Chang and Graham, 2011). *Bryophyta* has been used by other authors (e.g., Cavers, 1910; Haskell, 1949; Hiesel et al., 1994) to include the liverworts and hornworts as well as the mosses; see Comments).

Comments: We use *Musci* instead of *Bryopsida* or *Bryophyta* because it has much wider usage as a name for mosses alone. *Bryophyta* has a common and long-standing use in the literature for a most-likely paraphyletic group comprising the liverworts, mosses, and hornworts (e.g., Cavers, 1910; Haskell, 1949; Hiesel et al., 1994).

The name *Musci* was first used by Jussieu (1789: 10) for the mosses alone (as applied here and by most modern authors); he recognized liverworts as *Hepaticae*. Hedwig (1801: 341), in the starting point publication for mosses under the botanical code (Turland et al., 2018), applied *Musci* more broadly. He had two subcategories in his classification: *Musci frondosi* (mosses) and *Musci hepatici* (liverworts).

The name *Musci* is applied here to the crown clade, in accordance with *PhyloCode* (Cantino and de Queiroz, 2020) Recommendation 10.1B. Palaeobotanists and bryologists have often viewed all fossil mosses, some of which may lie outside the crown clade, as *Musci* (e.g., Boureau et al., 1967; Taylor et al., 2009). Following the definition of the name *Musci* proposed here, however, fossils shown to be outside the crown would instead be considered part of the informally named total clade pan-Musci in accordance with Recommendation 10.3A.

The clade *Musci* is extremely well supported, and our understanding of its internal cladistic relationships has remained remarkably consistent throughout the history of study (e.g., Mishler and Churchill, 1984; Newton et al, 2000; Cox et al., 2004; Bell et al., 2007; Wahrmund et al., 2010; Cox et al., 2010; Chang and Graham, 2011). On the other hand, the relationship of mosses to other land plants has been subject to considerable revision. They have been inferred in some studies to be the sister of *Tracheophyta* (Mishler and Churchill, 1984, 1985; Kenrick and Crane, 1997), and in others to be sister to *Hepaticae* (Nickrent et al., 2000; Wickett et al., 2014). The dominant hypothesis in the last decade has been that mosses are sister to hornworts plus tracheophytes (Kelch et al., 2004; Qiu et al., 2006). Given that the hypothesized relationships among the four principal extant embryophyte clades (i.e., *Musci, Anthocerotae, Hepaticae*, and *Tracheophyta*) may be subject to change in the future (e.g., Cox et al., 2014), we have selected external specifiers representing each of the latter three.

The diverse clade of mosses that is sister to *Oedipodium* in the reference phylogeny is characterized by the presence of a peristome (teeth surrounding the mouth of the capsule). There are two types of peristome: nematodontous (i.e., composed of remnants of whole cells), characteristic of *Polytrichales* and *Tetraphis*, and arthrodontous (i.e., composed of remnants of adjoining cell walls only), characteristic of the remainder of this clade. There is some uncertainty as to whether these two peristome types are homologous (or indeed whether the nematodontous peristomes of *Polytrichales* and *Tetraphis* are homologous to each other), but as all these peristomes develop from the same fundamental tissue layers in the capsule, it is likely that they are homologous at this level in the phylogeny, with modifications occurring subsequently during evolution of the descendant lineages (see Discussion and Fig. 29 in Goffinet et al., 1999).

Literature Cited

Bell, N. E., D. Quandt, T. J. O'Brien, and A. E. Newton. 2007. Taxonomy and phylogeny in the earliest diverging pleurocarps: square holes and bifurcating pegs. *Bryologist* 110:533–560.

Boureau, É., S. Jovet-Ast, O. A. Höeg, and W. G. Chaloner. 1967. *Traité de Paléobotanique: Tome II: Bryophyta, Psilophyta, Lycophyta*. Masson et Cie Editeurs, Paris.

Cantino, P. D., and K. de Queiroz. 2020. *International Code of Phylogenetic Nomenclature (PhyloCode)*, Version 6. CRC Press, Boca Raton, FL.

Cavers, F. 1910. The inter-relationships of the *Bryophyta* IV. Acrogynous *Jungermanniales. New Phytol.* 9:269–304.

Chang, Y., and S. W. Graham. 2011. Inferring the higher-order phylogeny of mosses (*Bryophyta*) and relatives using a large, multigene plastid data set. *Am. J. Bot.* 98:839–849.

Cox, C. J., B. Goffinet, A. J. Shaw, and S. B. Boles. 2004. Phylogenetic relationships among the mosses based on heterogeneous Bayesian analysis of multiple genes from multiple genomic compartments. *Syst. Bot.* 29:234–250.

Cox, C. J., B. Goffinet, N. J. Wickett, S. B. Boles, and A. J. Shaw. 2010. Moss diversity: a molecular phylogenetic analysis of genera. *Phytotaxa* 9:175–195.

Cox, C. J., B. Li, P. G. Foster, T. Embley, and P. Civan. 2014. Conflicting phylogenies for early land plants are caused by composition biases among synonymous substitutions. *Syst. Biol.* 63:272–279.

Goffinet, B., and W. R. Buck. 2004. Systematics of the *Bryophyta* (mosses): from molecules to a revised classification. Pp. 205–239 in *Molecular Systematics of Bryophytes* (B. Goffinet, V. Hollowell, and R. Magill, eds.). Missouri Botanical Garden Press, St. Louis.

Goffinet, B., J. Shaw, L. E. Anderson, and B. D. Mishler. 1999. Peristome development in mosses in relation to systematics and evolution. V. *Diplolepideae: Orthotrichaceae. Bryologist* 102:581–594.

Haskell, G. 1949. Some evolutionary problems concerning the *Bryophyta. Bryologist* 52:50–57.

Hedwig, J. 1801. *Species Muscorum Frondosorum*. J. A. Barthii, Leipzig.

Hiesel, R., B. Combettes, and A. Brennicke. 1994. Evidence for RNA editing in mitochondria of all major groups of land plants except the *Bryophyta*. *Proc. Natl. Acad. Sci., USA* 91:629–633.

Jussieu, A. L. de. 1789. *Genera Plantarum*. Herissant & Barrois, Paris.

Kelch, D. G., A. Driskell, and B. D. Mishler. 2004. Inferring phylogeny using genomic characters: a case study using land plant plastomes. Pp. 3–12 in *Molecular Systematics of Bryophytes* (B. Goffinet, V. Hollowell, and R. Magill, eds.). Missouri Botanical Garden Press, St. Louis, MO.

Kenrick, P., and P. R. Crane. 1997. *The Origin and Early Diversification of Land Plants. A Cladistic Study*. Smithsonian Institution Press, Washington, DC.

Mishler, B. D., and S. P. Churchill. 1984. A cladistic approach to the phylogeny of the "bryophytes". *Brittonia* 36:406–424.

Mishler, B. D., and S. P. Churchill. 1985. Transition to a land flora: phylogenetic relationships of the green algae and bryophytes. *Cladistics* 1:305–328.

Newton, A. E., C. J. Cox, J. G. Duckett, J. Wheeler, B. Goffinet, T. A. J. Hedderson, and B. D. Mishler. 2000. Evolution of the major moss lineages: phylogenetic analyses based on multiple gene sequences and morphology. *Bryologist* 103:187–211.

Nickrent, D. L., C. L. Parkinson, J. D. Palmer, and R. J. Duff. 2000. Multigene phylogeny of land plants with special reference to bryophytes and the earliest land plants. *Mol. Biol. Evol.* 17:1885–1895.

Qiu, Y.-L., L. Li, B. Wang, Z. Chen, V. Knoop, M. Groth-Malonek, O. Dombrovska, J. Lee, L. Kent, J. Rest, G. F. Estabrook, T. A. Hendry, D. W. Taylor, C. M. Testa, M. Ambros, B. Crandall-Stotler, R. J. Duff, M. Stech, W. Frey, D. Quandt, and C. C. Davis. 2006. The deepest divergences in land plants inferred from phylogenomic evidence. *Proc. Natl. Acad. Sci., USA* 103:15511–15516.

Taylor, T. N., E. L. Taylor, and M. Krings. 2009. *Paleobotany: The Biology and Evolution of Fossil Plants*. 2nd edition. Elsevier, Oxford.

Turland, N. J., J. H. Wiersema, F. R. Barrie, W. Greuter, D. L. Hawksworth, P. S. Herendeen, S. Knapp, W.-H. Kusber, D.-Z. Li, K. Marhold, T. W. May, J. McNeill, A. M. Monro, J. Prado, M. J. Price, and G. F. Smith, eds. 2018. *International Code of Nomenclature for Algae, Fungi, And Plants (Shenzhen Code)*. Adopted by the Nineteenth International Botanical Congress, Shenzhen, China, July 2017. Regnum Vegetabile 159. Koeltz Botanical Books, Glashütten.

Vitt, D. H. 1984. Classification of the *Bryopsida*. Pp. 696–759 in *New Manual of Bryology*, Vol. 2. (R. M. Schuster, ed.). Hattori Botanical Laboratory, Nichinan, Japan.

Wahrmund, U., D. Quandt, and V. Knoop. 2010. The phylogeny of mosses—addressing open issues with a new mitochondrial locus: group I intron cobi420. *Mol. Phy. Evol.* 54:417–442.

Wickett, N. J., S. Mirarab, N. Nguyen, T. Warnow, E. Carpenter, N. Matasci, S. Ayyampalayam, M. S. Barker, J. G. Burleigh, M. A. Gitzendanner, B. R. Ruhfel, E. Wafula, J. P. Der, S. W. Graham, S. Mathews, M. Melkonian, D. E. Soltis, P. S. Soltis, N. W. Miles, C. J. Rothfels, L. Pokorny, A. J. Shaw, L. DeGironimo, D. W. Stevenson, B. Surek, J. C. Villarreal, B. Roure, H. Philippe, C. W. dePamphilis, T. Chen, M. K. Deyholos, R. S. Baucom, T. M. Kutchan, M. M. Augustin, J. Wang, Y. Zhang, Z. Tian, Z. Yan, X. Wu, X. Sun, G. K. Wong, and J. Leebens-Mack. 2014. Phylotranscriptomic analysis of the origin and early diversification of land plants. *Proc. Natl. Acad. Sci. USA* 111:E4859–E4868.

Authors

Brent D. Mishler; University & Jepson Herbaria, and Department of Integrative Biology; University of California; Berkeley, CA 94720-2465, USA. Email: bmishler@berkeley.edu.

Angela E. Newton; 1 Cayzer Court, Gartmore, Stirling FK8 3RE, UK. Email: anangienewton38@googlemail.com.

Neil E. Bell; Royal Botanic Garden Edinburgh; 20A Inverleith Row, Edinburgh EH3 5LR, UK. Email: n.bell@rbge.ac.uk.

Barbara J. Crandall-Stotler; Department of Plant Biology; University of Southern Illinois; Carbondale, IL 62901, USA. Email: crandall@siu.edu.

Raymond Stotler[†]; Department of Plant Biology; University of Southern Illinois; Carbondale, IL 62901, USA.

Date Accepted: 7 September 2016

Primary Editor: Philip Cantino

[†] Our esteemed colleague, Ray Stotler, an expert on liverwort and hornwort classification, contributed to early versions of this manuscript but sadly passed away December 4, 2013.

Anthocerotae W. Mitten 1855 [B. D. Mishler, D. C. Cargill, J. C. Villarreal, K. Renzaglia, B. Crandall-Stotler, and R. Stotler], converted clade name

Registration Number: 12

Definition: The largest crown clade containing *Anthoceros punctatus* L. 1753 but not *Plagiochila adianthoides* (Sw.) Lindenb. 1840 (*Hepaticae*), and *Sphagnum palustre* L. 1753 (*Musci*), and *Osmunda cinnamomea* L. 1753 (*Tracheophyta*). This is a maximum-crown-clade definition. Abbreviated definition: max crown ∇ (*Anthoceros punctatus* L. 1753 ~ *Plagiochila adianthoides* (Sw.) Lindenb. 1840 & *Sphagnum palustre* L. 1753 & *Osmunda cinnamomea* L. 1753).

Etymology: From the Greek *anthos* (flower) and *keras* (horn), in reference to the most distinctive synapomorphy of the clade, the elongated and progressively maturing sporangium.

Reference Phylogeny: Figure 2 in Duff et al. (2007).

Composition: A relatively small clade with only about 200–250 named species, it nonetheless contains much structural disparity (Renzaglia, 1978; Renzaglia et al., 2007; Villarreal et al., 2010). There appear to be four main subclades: *Leiosporoceros*, *Anthocerotidae*, *Notothylatidae*, and *Dendrocerotidae* (Duff et al., 2007).

Diagnostic Apomorphies: Pseudoelaters in the sporangium, non-synchronous production of spores in an elongated and progressively maturing sporangium, pyrenoid in the chloroplast, antheridia endogenous, bilaterally symmetric spermatozoids with basal bodies similar and inserted side-by-side, and vertical division of the zygote (see discussion in Mishler and Churchill, 1984, 1985).

Synonyms: *Anthoceroteae* (alternative spelling of Nees (1833); see Comments), *Anthocerotales* sensu Proskauer (1951), *Anthocerotopsida* sensu Bell and Hemsley (2000), *Anthocerotophyta* sensu Frey and Stech (2005), and *Anthocerophyta* sensu Crum (2001) are approximate synonyms.

Comments: We attribute the name *Anthocerotae* to Mitten (1855) because he was the first to name the hornworts (albeit as a subgroup of liverworts) using this spelling. There are some orthographic complications, however, because Nees (1833) used the slightly different spelling *Anthoceroteae* for the hornworts as a subgrouping of liverworts. Janczewski (1872) also used *Anthoceroteae* but was the first to recognize the hornworts as a taxon distinct from the liverworts. *Anthocerotae* is the spelling that has been used in all twentieth and twenty-first century literature. We use *Anthocerotae* instead of one of its synonyms because it has much wider usage.

The name *Anthocerotae* is applied here to the crown clade, in accordance with *PhyloCode* (Cantino and de Queiroz, 2020) Recommendation 10.1B. Palaeobotanists and bryologists have often viewed all fossil hornworts, some of which might lie outside the crown clade, as *Anthocerotae* (e.g., Boureau et al., 1967; Taylor et al., 2009). Following the definition of *Anthocerotae* proposed here, hornwort fossils shown to be outside the crown would instead be considered part of the informally named total clade pan-Anthocerotae in accordance with Recommendation 10.3A.

The clade *Anthocerotae* is extremely well supported and is recognized in all published phylogenetic analyses (e.g., Mishler and Churchill, 1984; Kenrick and Crane, 1997; Nickrent et al., 2000; Kelch et al., 2004; Qiu et al., 2006; Duff et al., 2007; Villarreal et al., 2010). The relationship of *Anthocerotae* to other land plants remains uncertain. In earlier cladistic analyses, the clade was either inferred to be sister to mosses plus tracheophytes (Mishler and Churchill, 1984; Kenrick and Crane, 1997) or sister to all other embryophytes (Nickrent et al., 2000; Duff et al., 2007), but other recent studies have inferred it to be sister to tracheophytes (Kelch et al., 2004; Qiu et al., 2006). Given that the hypothesized relationships among the four principal embryophyte crown clades (i.e., *Anthocerotae*, *Hepaticae*, *Musci*, and *Tracheophyta*) may be subject to change in the future (e.g., Cox et al., 2014; Wickett et al., 2014), we have selected external specifiers representing each of the latter three.

Literature Cited

Bell, P. R., and A. R. Hemsley. 2000. *Green Plants: Their Origin and Diversity*. Cambridge University Press, Cambridge, UK.

Boureau, É., S. Jovet-Ast, O. A. Höeg, and W. G. Chaloner. 1967. *Traité de Paléobotanique: Tome II: Bryophyta, Psilophyta, Lycophyta*. Masson et Cie Editeurs, Paris.

Cantino, P. D., and K. de Queiroz. 2020. *International Code of Phylogenetic Nomenclature (PhyloCode)*, Version 6. CRC Press, Boca Raton, FL.

Cox, C. J., B. Li, P. G. Foster, T. Embley, and P. Civan. 2014. Conflicting phylogenies for early land plants are caused by composition biases among synonymous substitutions. *Syst. Biol.* 63:272–279.

Crum, H. A. 2001. *Structural Diversity of Bryophytes*. University of Michigan Herbarium, Ann Arbor, MI.

Duff, R. J., J. C. Villareal, D. C. Cargill, and K. S. Renzaglia. 2007. Progress and challenges toward developing a phylogeny and classification of the hornworts. *Bryologist* 110:214–243.

Frey, W., and M. Stech, 2005. A morpho-molecular classification of the *Anthocerotophyta* (hornworts). *Nova Hedwigia* 80:541–546.

Janczewski, E. 1872. Vergleichende Untersuchungen über die Entwickelungsgeschichte des Archegoniums. *B. Z.* (Berlin) 30:377–394, 401–420, 440–443.

Kelch, D. G., A. Driskell, and B. D. Mishler. 2004. Inferring phylogeny using genomic characters: a case study using land plant plastomes. Pp. 3–12 in *Molecular Systematics of Bryophytes* (B. Goffinet, V. Hollowell, and R. Magill, eds.). Missouri Botanical Garden Press, St. Louis.

Kenrick, P., and P. R. Crane. 1997. *The Origin and Early Diversification of Land Plants: A Cladistic Study*. Smithsonian Institution Press, Washington, DC.

Mishler, B. D., and S. P. Churchill. 1984. A cladistic approach to the phylogeny of the "bryophytes". *Brittonia* 36:406–424.

Mishler, B. D., and S. P. Churchill. 1985. Transition to a land flora: phylogenetic relationships of the green algae and bryophytes. *Cladistics* 1:305–328.

Mitten, W. 1855. *Hepaticae*. Pp. 125–172 in *Flora Novae-Zelandiae. Part II* (J. D. Hooker, ed.). Reeve Brothers, London.

Nees von Esenbeck, C. G. D. 1833. *Naturgeschichte der Europäischen Lebermoose* 1:86. Rücker, Berlin.

Nickrent, D. L., C. L. Parkinson, J. D. Palmer, and R. J. Duff. 2000. Multigene phylogeny of land plants with special reference to bryophytes and the earliest land plants. *Mol. Biol. Evol.* 17:1885–1895.

Proskauer, J. 1951. Studies on *Anthocerotales*. III. *Bull. Torrey Bot. Club* 78:331–349.

Qiu, Y.-L., L. Li, B. Wang, Z. Chen, V. Knoop, M. Groth-Malonek, O. Dombrovska, J. Lee, L. Kent, J. Rest, G. F. Estabrook, T. A. Hendry, D. W. Taylor, C. M. Testa, M. Ambros, B. Crandall-Stotler, R. J. Duff, M. Stech, W. Frey, D. Quandt, and C. C. Davis. 2006. The deepest divergences in land plants inferred from phylogenomic evidence. *Proc. Natl. Acad. Sci. USA* 103:15511–15516.

Renzaglia, K. S. 1978. A comparative morphology and developmental anatomy of the *Anthocerophyta*. *J. Hattori Bot. Lab.* 44:31–90.

Renzaglia, K. S., S. Schuette, R. J. Duff, R. Ligrone, A. J. Shaw, B. D. Mishler, and J. G. Duckett. 2007. Bryophyte phylogeny: advancing the molecular and morphological frontiers. *Bryologist* 110:179–213.

Taylor, T. N., E. L. Taylor, and M. Krings. 2009. *Paleobotany: The Biology and Evolution of Fossil Plants*. 2nd edition. Elsevier, Oxford.

Villarreal, J. C., D. C. Cargill, A. Hagborg, L. Söderström, and K. S. Renzaglia. 2010. A synthesis of hornwort diversity: patterns, causes and future work. *Phytotaxa* 9:150–166.

Wickett, N. J., S. Mirarab, N. Nguyen, T. Warnow, E. Carpenter, N. Matasci, S. Ayyampalayam, M. S. Barker, J. G. Burleigh, M. A. Gitzendanner, B. R. Ruhfel, E. Wafula, J. P. Der, S. W. Graham, S. Mathews, M. Melkonian, D. E. Soltis, P. S. Soltis, N. W. Miles, C. J. Rothfels, L. Pokorny, A. J. Shaw, L. DeGironimo, D. W. Stevenson, B. Surek, J. C. Villarreal, B. Roure, H. Philippe, C. W. dePamphilis, T. Chen, M. K. Deyholos, R. S. Baucom, T. M. Kutchan, M. M. Augustin, J. Wang, Y. Zhang, Z. Tian, Z. Yan, X. Wu, X. Sun, G. K. Wong, and J. Leebens-Mack. 2014. Phylotranscriptomic analysis of the origin and early diversification of land plants. *Proc. Natl. Acad. Sci. USA* 111:E4859–E4868.

Authors

Brent D. Mishler; University & Jepson Herbaria, and Department of Integrative Biology; University of California; Berkeley, CA 94720-2465, USA. Email: bmishler@berkeley.edu.

D. Christine Cargill; Centre for Australian National Biodiversity Research, GPO Box 1600, Canberra. ACT 2601, Australia. Email: Chris.Cargill@environment.gov.au.

Juan Carlos Villarreal; Département de Biologie; Université Laval; Québec (Québec), G1V 0A6, Canada. Email: juan-carlos.villarreal-aguilar@bio.ulaval.ca.

Karen Renzaglia; Department of Plant Biology; University of Southern Illinois; Carbondale, IL 62901, USA. Email: renzaglia@cos.siu.edu.

Barbara J. Crandall-Stotler; Department of Plant Biology; University of Southern Illinois; Carbondale, IL 62901, USA. Email: crandall@siu.edu.

Raymond Stotler[†]; Department of Plant Biology; University of Southern Illinois; Carbondale, IL 62901, USA.

Date Accepted: 28 July 2016

Primary Editor: Philip Cantino

[†] Our esteemed colleague, Ray Stotler, an expert on liverwort and hornwort classification, contributed to early versions of this manuscript but sadly passed away December 4, 2013.

Pan-Tracheophyta P. D. Cantino and M. J. Donoghue in P. D. Cantino et al. (2007): 830 [P. D. Cantino and M. J. Donoghue], converted clade name

Registration Number: 83

Definition: The total clade of the crown clade *Tracheophyta*. This is a crown-based total-clade definition. Abbreviated definition: total ∇ of *Tracheophyta*.

Etymology: From the Greek *pan-* or *pantos* (all, the whole), indicating that the name refers to a total clade, and *Tracheophyta* (see entry in this volume for etymology), the name of the corresponding crown clade.

Reference Phylogeny: Crane et al. (2004: Fig. 1; the clade stemming from the base of the branch labeled "Polysporangiophytes"). See also Kenrick and Crane (1997: Fig. 4.31).

Composition: *Tracheophyta* (this volume) and all extinct plants (e.g., *Aglaophyton, Horneophytopsida*, and *Rhyniopsida* sensu Kenrick and Crane 1997) that share more recent ancestry with *Tracheophyta* than with extant *Musci* (mosses), *Hepaticae* (liverworts), and *Anthocerotae* (hornworts).

Diagnostic Apomorphies: An independent sporophyte, multiple sporangia, and sunken archegonia are possible synapomorphies of *Pan-Tracheophyta*. All three were listed by Kenrick and Crane (1997: Table 7.2) as synapomorphies of *Polysporangiomorpha* ("polysporangiophytes" of Crane et al., 2004; a slightly less inclusive clade than *Pan-Tracheophyta*; see Comments). The order in which the three features evolved is not known. Moreover, sunken archegonia also occur in *Anthocerotae* (Kenrick and Crane, 1997: Fig. 3.33, pp. 63–64) and thus may be

a synapomorphy of a more inclusive clade if *Anthocerotae* is the extant sister group of tracheophytes, a hypothesis that has received some molecular support (Kelch et al., 2004; Wolf et al., 2005; Qiu et al., 2006, 2007; however, see Wickett et al., 2014). Another possible synapomorphy of *Pan-Tracheophyta* is the production of a sporophyte stem, if Kato and Akiyama (2005) are correct that the stem is not homologous with the bryophyte seta.

Synonyms: The phylogenetically defined names *Polysporangiomorpha* (Kenrick and Crane, 1997) and *Polysporangiophyta* (Crane and Kenrick, 1997) have the same known composition as *Pan-Tracheophyta* but potentially refer to different clades (see Comments).

Comments: The definition of *Pan-Tracheophyta* is identical to the one we used when we first published the name (Cantino et al., 2007). The name *Polysporangiomorpha* (polysporangiophytes) sensu Kenrick and Crane (1997: Table 7.2, Fig. 4.31) has an apomorphy-based definition and thus is unlikely to be fully synonymous with *Pan-Tracheophyta*. Its currently known composition is the same as that of *Pan-Tracheophyta*, but there may have been pan-tracheophytes that preceded the evolution of multiple sporangia. Similarly, *Polysporangiophyta* sensu Crane and Kenrick (1997) has the same known composition as both *Pan-Tracheophyta* and *Polysporangiomorpha*, but its name has a node-based definition and thus the clade *Polysporangiophyta* may be less inclusive than either of the other two. The three names would no longer be synonyms if a fossil were to be found in the future that possesses an intermediate combination of ancestral and derived states of the three characters mentioned above as possible synapomorphies (for example, a sporophyte that is dependent on the gametophyte throughout its life, as is the case in bryophytes, but has more than one sporangium).

Literature Cited

Cantino, P. D., J. A. Doyle, S. W. Graham, W. S. Judd, R. G. Olmstead, D. E. Soltis, P. S. Soltis, and M. J. Donoghue. 2007. Towards a phylogenetic nomenclature of *Tracheophyta*. *Taxon* 56:822–846 and E1–E44.

Crane, P. R., P. Herendeen, and E. M. Friis. 2004. Fossils and plant phylogeny. *Am. J. Bot.* 91:1683–1699.

Crane, P. R., and P. Kenrick. 1997. Problems in cladistic classification: higher-level relationships in land plants. *Aliso* 15:87–104.

Kato, M., and H. Akiyama. 2005. Interpolation hypothesis for origin of the vegetative sporophyte of land plants. *Taxon* 54:443–450.

Kelch, D. G., A. Driskell, and B. D. Mishler. 2004. Inferring phylogeny using genomic characters: a case study using land plant plastomes. Pp. 3–11 in *Molecular Systematics of Bryophytes* (B. Goffinet, V. Hollowell, and R. Magill, eds.). Missouri Botanical Garden Press, St. Louis, MO.

Kenrick, P., and P. R. Crane. 1997. *The Origin and Early Diversification of Land Plants—A Cladistic Study*. Smithsonian Institution Press, Washington, DC.

Qiu, Y.-L., L. Li, B. Wang, Z. Chen, O. Dombrovska, J. Lee, L. Kent, R. Li, R. W. Jobson, T. A. Hendry, D. W. Taylor, C. M. Testa, and M. Ambros. 2007. A nonflowering land plant phylogeny inferred from nucleotide sequences of seven chloroplast, mitochondrial, and nuclear genes. *Int. J. Plant Sci.* 168:691–708.

Qiu, Y.-L., L. Li, B. Wang, Z. Chen, V. Knoop, M. Groth-Malonek, O. Dombrovska, J. Lee, L. Kent, J. Rest, G. F. Estabrook, T. A. Hendry, D. W. Taylor, C. M. Testa, M. Ambros, B. Crandall-Stotler, R. J. Duff, M. Stech, W. Frey, D. Quandt, and C. C. Davis. 2006. The deepest divergences in land plants inferred from phylogenomic evidence. *Proc. Natl. Acad. Sci. USA* 103:15511–15516.

Wickett, N. J., S. Mirarab, N. Nguyen, T. Warnow, E. Carpenter, N. Matasci, S. Ayyampalayam, M. S. Barker, J. G. Burleigh, M. A. Gitzendanner, B. R. Ruhfel, E. Wafula, J. P. Der, S. W. Graham, S. Mathews, M. Melkonian, D. E. Soltis, P. S. Soltis, N. W. Miles, C. J. Rothfels, L. Pokorny, A. J. Shaw, L. DeGironimo, D. W. Stevenson, B. Surek, J. C. Villarreal, B. Roure, H. Philippe, C. W. dePamphilis, T. Chen, M. K. Deyholos, R. S. Baucom, T. M. Kutchan, M. M. Augustin, J. Wang, Y. Zhang, Z. Tian, Z. Yan, X. Wu, X. Sun, G. K. Wong, and J. Leebens-Mack. 2014. Phylotranscriptomic analysis of the origin and early diversification of land plants. *Proc. Natl. Acad. Sci. USA* 111:E4859–E4868.

Wolf, P. G., K. G. Karol, D. F. Mandoli, J. Kuehl, K. Arumuganathan, M. W. Ellis, B. D. Mishler, D. G. Kelch, R. G. Olmstead, and J. L. Boore. 2005. The first complete chloroplast genome sequence of a lycophyte, *Huperzia lucidula* (*Lycopodiaceae*). *Gene* (Amst.) 350:117–128.

Authors

Philip D. Cantino; Department of Environmental and Plant Biology; Ohio University; Athens, OH 45701, USA. Email: cantino@ohio.edu.

Michael J. Donoghue; Department of Ecology and Evolutionary Biology; Yale University; P.O. Box 208106; New Haven, CT 06520, USA. Email: michael.donoghue@yale.edu.

Date Accepted: 1 April 2011; updated 9 February 2018

Primary Editor: Kevin de Queiroz

Apo-Tracheophyta P. D. Cantino and M. J. Donoghue in P. D. Cantino et al. (2007): E10 [P. D. Cantino and M. J. Donoghue], converted clade name

Registration Number: 15

Definition: The clade characterized by the apomorphy tracheids (i.e., differentially thickened water-conducting cells) as inherited by *Pinus sylvestris* Linnaeus 1753. This is an apomorphy-based definition. Abbreviated definition: ∇ apo tracheids [*Pinus sylvestris* Linnaeus 1753].

Etymology: From the Greek *apo-*, indicating that the name refers to an apomorphy-based clade, and *Tracheophyta* (see entry in this volume for etymology).

Reference Phylogeny: The primary reference phylogeny is Crane et al. (2004: Fig. 1; the clade labeled "Tracheophytes"). See also Kenrick and Crane (1997: Fig. 4.31; the clade labeled *Tracheophyta*).

Composition: Assuming that tracheids with S-type and G-type cell walls (see Kenrick and Crane, 1997: Fig. 4.26) are homologous, *Apo-Tracheophyta* includes *Tracheophyta* sensu Cantino and Donoghue (this volume) and *Rhyniopsida* sensu Kenrick and Crane (1997) (*Rhyniaceae* in the primary reference phylogeny). Under the alternative hypothesis that these tracheid types evolved independently, *Rhyniopsida* would not be part of *Apo-Tracheophyta*, and the currently known membership of *Apo-Tracheophyta* and *Tracheophyta* would be the same.

Diagnostic Apomorphies: Tracheids (see Definition), and possibly lignin deposition on the inner surface of the tracheid cell wall (Kenrick and Crane, 1997: Table 7.2, under *Tracheophyta*).

Synonyms: Based on composition, *Tracheophyta* sensu Kenrick and Crane (1997: Tables 7.1, 7.2, p. 236) is an approximate synonym. Although Kenrick and Crane (1997: 236) listed *Tracheidatae* Bremer (1985) as a synonym of their "Eutracheophytes," implying that that *Tracheidatae* referred to the crown group, it is clear from Bremer's comments (p. 382) that he considered rhyniopsids to be part of *Tracheidatae*; thus, based on composition, *Tracheidatae* is an approximate synonym of *Apo-Tracheophyta*. *Pteridophyta* of some earlier authors (e.g., Haupt, 1953) is a partial synonym; the pteridophytes originated from the same ancestor as *Apo-Tracheophyta* but are paraphyletic because they exclude *Apo-Spermatophyta* sensu Cantino and Donoghue (this volume).

Comments: Although the name *Tracheophyta* is often applied to this apomorphy-based clade, we opted here and previously (Cantino et al., 2007) to apply that name instead to the crown clade, following *PhyloCode* (version 6) Recommendation 10.1B, and a new name to the apomorphy-based clade.

Literature Cited

Bremer, K. 1985. Summary of green plant phylogeny and classification. *Cladistics* 1:369–385.

Cantino, P. D., J. A. Doyle, S. W. Graham, W. S. Judd, R. G. Olmstead, D. E. Soltis, P. S. Soltis, and M. J. Donoghue. 2007. Towards a phylogenetic nomenclature of *Tracheophyta*. *Taxon* 56:822–846 and E1–E44.

Crane, P. R., P. Herendeen, and E. M. Friis. 2004. Fossils and plant phylogeny. *Am. J. Bot.* 91:1683–1699.

Haupt, A. W. 1953. *Plant Morphology*. McGraw-Hill, New York.

Kenrick, P., and P. R. Crane. 1997. *The Origin and Early Diversification of Land Plants—A Cladistic Study*. Smithsonian Institution Press, Washington, DC.

Authors

Philip D. Cantino; Department of Environmental and Plant Biology; Ohio University; Athens, OH 45701, USA. Email: cantino@ohio.edu.

Michael J. Donoghue; Department of Ecology and Evolutionary Biology; Yale University; P.O. Box 208106; New Haven, CT 06520, USA. Email: michael.donoghue@yale.edu.

Date Accepted: 1 April 2011; updated 9 February 2018

Primary Editor: Kevin de Queiroz

Tracheophyta E. W. Sinnott 1935: 441
[P. D. Cantino and M. J. Donoghue], converted clade name

Registration Number: 107

Definition: The smallest crown clade containing *Magnolia tripetala* (Linnaeus) Linnaeus 1759 (*Euphyllophyta*) and *Lycopodium clavatum* Linnaeus 1753 (*Lycopodiophyta*). This is a minimum-crown-clade definition. Abbreviated definition: min crown ∇ (*Magnolia tripetala* (Linnaeus) Linnaeus 1759 & *Lycopodium clavatum* Linnaeus 1753).

Etymology: From the Latin *tracheida*, refering to the presence of tracheids (differentially thickened water conducting cells in the xylem), and the Greek *phyton* (plant).

Reference Phylogeny: The primary reference phylogeny is Qiu et al. (2007: Fig. 1). See also Kenrick and Crane (1997: Fig. 4.31; as "Eutracheophytes"), Pryer et al. (2001: Fig. 1), Wolf et al. (2005: Fig. 3), Qiu et al. (2006: Fig. 1), and Ruhfel et al. (2014: Figs. 5–7).

Composition: All extant vascular plants and their extinct relatives that fall within the crown clade. The crown clade is composed of two primary subclades, *Pan-Lycopodiophyta* and *Pan-Euphyllophyta* (see entries in this volume).

Diagnostic Apomorphies: The walls of the water-conducting cells in the xylem have a thick, lignified, decay-resistant layer. A free-living sporophyte and multiple sporangia per sporophyte are synapomorphies relative to other crown clades; however, when fossils are considered, these traits are synapomorphies at a more inclusive level (see *Pan-Tracheophyta*). Sterome (a well-developed peripheral zone of the stem consisting of thick-walled, decay-resistant cells) and pitlets in the tracheid wall are listed by Kenrick and Crane (1997: Table 7.2, pp. 114, 120) as synapomorphies of "eutracheophytes" (= *Tracheophyta* as defined here), but the extent of missing data for fossils combined with the apparent loss of these traits in all extant tracheophytes reduces confidence in their inferred originations.

Synonyms: "Eutracheophytes" sensu Kenrick and Crane (1997: 236) was described as "the tracheophyte crown group" and is thus an unambiguous synonym. Crane and Kenrick (1997) formalized this name as *Eutracheophyta* and defined it (node-based) using species of *Huperzia* and *Nymphaea* as specifiers; this is an unambiguous synonym. *Cormatae* Jeffrey (1982) is an approximate synonym; all listed subordinate taxa are extant.

Comments: The vascular plants, or tracheophytes, have long been recognized as a clade based on their possession of tracheids as a synapomorphy, and the crown clade is strongly supported by DNA analyses (Nickrent et al., 2000; Pryer et al., 2001; Qiu et al., 2006, 2007; Ruhfel et al., 2014; Wickett et al., 2014). Here and previously (Cantino et al., 2007: 829) we have defined the name *Tracheophyta* to refer to the crown clade, following *PhyloCode* Recommendation 10.1B. This application is somewhat unconventional in that this name has more often been applied to the slightly more inclusive clade originating with the evolution of tracheids (i.e., *Apo-Tracheophyta* in this volume). Sinnott (1935) introduced the name *Tracheophyta* for the vascular plants but the Latin diagnosis required by the botanical code (Turland et al., 2018) was first provided by Cavalier-Smith (1998: 251).

We earlier (Cantino et al., 2007) used a maximum-crown-clade definition with external specifiers representing liverworts, mosses, and hornworts, because we thought this definition would provide greater compositional stability than a simple node-based definition. However, in view of the very strong support for the basal dichotomy within *Tracheophyta* in recent molecular analyses (Qiu et al., 2006, 2007; Wickett et al., 2014), we are opting here for a simpler minimum-crown-clade definition with the two internal specifiers representing the two clades originating from the basal split: *Pan-Lycopodiophyta* and *Pan-Euphyllophyta*.

Literature Cited

Cantino, P. D., J. A. Doyle, S. W. Graham, W. S. Judd, R. G. Olmstead, D. E. Soltis, P. S. Soltis, and M. J. Donoghue. 2007. Towards a phylogenetic nomenclature of *Tracheophyta*. *Taxon* 56:822–846 and E1–E44.

Cavalier-Smith, T. 1998. A revised six-kingdom system of life. *Biol. Rev. Camb. Philos. Soc.* 73:203–266.

Crane, P. R., and P. Kenrick. 1997. Problems in cladistic classification: higher-level relationships in land plants. *Aliso* 15:87–104.

Jeffrey, C. 1982. Kingdoms, codes and classification. *Kew Bull.* 37:403–416.

Kenrick, P., and P. R. Crane. 1997. *The Origin and Early Diversification of Land Plants—A Cladistic Study*. Smithsonian Institution Press, Washington, DC.

Nickrent, D. L., C. L. Parkinson, J. D. Palmer, and R. J. Duff. 2000. Multigene phylogeny of land plants with special reference to bryophytes and the earliest land plants. *Mol. Phylogenet. Evol.* 17:1885–1895.

Pryer, K. M., H. Schneider, A. R. Smith, R. Cranfill, P. G. Wolf, J. S. Hunt, and S. D. Sipes. 2001. Horsetails and ferns are a monophyletic group and the closest living relatives to seed plants. *Nature* 409:618–622.

Qiu, Y.-L., L. Li, B. Wang, Z. Chen, O. Dombrovska, J. Lee, L. Kent, R. Li, R. W. Jobson, T. A. Hendry, D. W. Taylor, C. M. Testa, and M. Ambros. 2007. A nonflowering land plant phylogeny inferred from nucleotide sequences of seven chloroplast, mitochondrial, and nuclear genes. *Int. J. Plant Sci.* 168:691–708.

Qiu, Y.-L., L. Li, B. Wang, Z. Chen, V. Knoop, M. Groth-Malonek, O. Dombrovska, J. Lee, L. Kent, J. Rest, G. F. Estabrook, T. A. Hendry, D. W. Taylor, C. M. Testa, M. Ambros, B. Crandall-Stotler, R. J. Duff, M. Stech, W. Frey, D. Quandt, and C. C. Davis. 2006. The deepest divergences in land plants inferred from phylogenomic evidence. *Proc. Natl. Acad. Sci. USA* 103:15511–15516.

Ruhfel, B. R., M. A. Gitzendanner, P. S. Soltis, D. E. Soltis, and J. G. Burleigh. 2014. From algae to angiosperms—inferring the phylogeny of green plants (*Viridiplantae*) from 360 plastid genomes. *BMC Evol. Biol.* 14:23.

Sinnott, E. W. 1935. *Botany: Principles and Problems*. 3rd edition. McGraw-Hill, New York.

Turland, N. J., J. H. Wiersema, F. R. Barrie, W. Greuter, D. L. Hawksworth, P. S. Herendeen, S. Knapp, W.-H. Kusber, D.-Z. Li, K. Marhold, T. W. May, J. McNeill, A. M. Monro, J. Prado, M. J. Price, and G. F. Smith, eds. 2018. *International Code of Nomenclature for Algae, Fungi and Plants (Shenzhen Code)*. Adopted by the Nineteenth International Botanical Congress, Shenzhen, China, July 2017. Regnum Vegetabile 159. Koeltz Botanical Books, Glashütten.

Wickett, N. J., S. Mirarab, N. Nguyen, T. Warnow, E. Carpenter, N. Matasci, S. Ayyampalayam, M. S. Barker, J. G. Burleigh, M. A. Gitzendanner, B. R. Ruhfel, E. Wafula, J. P. Der, S. W. Graham, S. Mathews, M. Melkonian, D. E. Soltis, P. S. Soltis, N. W. Miles, C. J. Rothfels, L. Pokorny, A. J. Shaw, L. DeGironimo, D. W. Stevenson, B. Surek, J. C. Villarreal, B. Roure, H. Philippe, C. W. dePamphilis, T. Chen, M. K. Deyholos, R. S. Baucom, T. M. Kutchan, M. M. Augustin, J. Wang, Y. Zhang, Z. Tian, Z. Yan, X. Wu, X. Sun, G. K. Wong, and J. Leebens-Mack. 2014. Phylotranscriptomic analysis of the origin and early diversification of land plants. *Proc. Natl. Acad. Sci. USA* 111:E4859–E4868.

Wolf, P. G., K. G. Karol, D. F. Mandoli, J. Kuehl, K. Arumuganathan, M. W. Ellis, B. D. Mishler, D. G. Kelch, R. G. Olmstead, and J. L. Boore. 2005. The first complete chloroplast genome sequence of a lycophyte, *Huperzia lucidula* (*Lycopodiaceae*). *Gene* (Amst.) 350:117–128.

Authors

Philip D. Cantino; Department of Environmental and Plant Biology; Ohio University; Athens, OH 45701, USA. Email: cantino@ohio.edu.

Michael J. Donoghue; Department of Ecology and Evolutionary Biology; Yale University; P.O. Box 208106; New Haven, CT 06520, USA. Email: michael.donoghue@yale.edu.

Date Accepted: 23 March 2011; updated 9 February 2018, 11 July 2018

Primary Editor: Kevin de Queiroz

Pan-Lycopodiophyta P. D. Cantino and M. J. Donoghue in P. D. Cantino et al. (2007): E11 [P. D. Cantino and M. J. Donoghue], converted clade name

Registration Number: 81

Definition: The total clade of the crown clade *Lycopodiophyta*. This is a crown-based total-clade definition. Abbreviated definition: total ∇ of *Lycopodiophyta*.

Etymology: From the crown clade name, *Lycopodiophyta* (see entry in this volume for its etymology), and the Greek *pan-* or *pantos* (all, the whole), indicating reference to a total clade.

Reference Phylogeny: Doyle (2013: Fig. 1.1). See also Crane et al. (2004: Fig. 1). In both of these phylogenies, the clade here named *Pan-Lycopodiophyta* is not labeled as such but is the most inclusive clade containing the lycophytes but not the euphyllophytes.

Composition: *Lycopodiophyta* (this volume) and all extinct plants that share more recent ancestry with *Lycopodiophyta* than with *Euphyllophyta* (this volume). According to the phylogenies of Kenrick and Crane (1997), Crane et al. (2004), and Doyle (2013), extinct taxa outside of the crown include (not an exhaustive list): the relatively apical stem groups *Asteroxylon*, *Drepanophycus*, *Baragwanathia*; the zosterophylls; and the relatively basal stem groups *Renalia*, *Yunia*, *Uskiella*, *Sartilmania*, and *Cooksonia cambrensis*. The inclusion of some other species of *Cooksonia* is uncertain because of their unresolved position on these trees.

Diagnostic Apomorphies: A possible synapomorphy is sporangium dehiscence by a transverse, apical slit. Doyle (1998, 2013) showed this character as arising at or near the base of the (unnamed) lycopodiophyte total clade. Kenrick and Crane (1997: Table 4.6) cited it as a possible synapomorphy of their node 52, which is near the base of the total clade. Crane and Kenrick (1997) listed the following additional potential synapomorphies for their *Lycophytina* (which is somewhat less inclusive than the total clade but much more inclusive than the crown; see Comments): reniform sporangia; marked sporangial dorsiventrality; inconspicuous cellular thickening of the dehiscence line; sporangia on short, laterally inserted stalks; and exarch xylem differentiation. However, the more taxonomically comprehensive analysis of Kenrick and Crane (1997: Fig. 4.32 and Table 4.6) suggests that some of these characters may be synapomorphic for more inclusive clades than *Lycophytina* within the total clade (also see Doyle, 2013).

Synonyms: Based on its composition, the name *Lycopodiobiotina* Doweld (2001) is an approximate synonym. *Lycophytina* (Kenrick and Crane, 1997; Crane and Kenrick, 1997; Bateman et al., 1998) applies to a clade that is less inclusive than *Pan-Lycopodiophyta* (see Comments).

Comments: Our definition of *Pan-Lycopodiophyta* here is identical to that used when we published the name (Cantino et al. 2007). We chose the panclade name rather than its approximate synonym *Lycopodiobiotina* (Doweld, 2001), as permitted by *PhyloCode* Article 10.6, so that its application to a total clade would be evident. The name *Lycophytina* sensu Kenrick and Crane (1997: Fig. 4.31 and Table 7.2) has a "synapomorphy-based definition" (not a formal definition; perhaps better described as an apomorphy-based conceptualization of the taxon, with six synapomorphies cited) and is somewhat less inclusive than *Pan-Lycopodiophyta*. The name *Lycophytina* sensu DiMichele and Bateman (1996) and Bateman et al. (1998) appears to be applied to a clade that, based on its synapomorphies and composition, is circumscribed similarly to *Lycophytina* sensu Kenrick and Crane (1997). *Lycophytina* sensu Crane and Kenrick (1997) has a minimum-clade definition and approximates *Pan-Lycopodiophyta* in composition if only the taxa in their cladogram are considered, but it is less inclusive than the total clade in the context of trees that include more stem fossils (e.g., Crane et al., 2004). For example, *Yunia* and *Renalia*, which were not included in Crane and Kenrick's (1997) tree, are part of *Pan-Lycopodiophyta* but outside of *Lycophytina* when Crane and Kenrick's (1997) definition of *Lycophytina* is applied in the context of the tree of Crane et al. (2004).

Literature Cited

Bateman, R. M., P. R. Crane, W. A. DiMichele, P. R. Kenrick, N. P. Rowe, T. Speck, and W. E. Stein. 1998. Early evolution of land plants: phylogeny, physiology, and ecology of the primary terrestrial radiation. *Annu. Rev. Ecol. Syst.* 29:263–292.

Cantino, P. D., J. A. Doyle, S. W. Graham, W. S. Judd, R. G. Olmstead, D. E. Soltis, P. S. Soltis, and M. J. Donoghue. 2007. Towards a phylogenetic nomenclature of *Tracheophyta*. *Taxon* 56:822–846 and E1–E44.

Crane, P. R., P. Herendeen, and E. M. Friis. 2004. Fossils and plant phylogeny. *Am. J. Bot.* 91:1683–1699.

Crane, P. R., and P. Kenrick. 1997. Problems in cladistic classification: higher-level relationships in land plants. *Aliso* 15:87–104.

DiMichele, W. A., and R. M. Bateman. 1996. The rhizomorphic lycopsids: a case-study in paleobotanical classification. *Syst. Bot.* 21:535–552.

Doweld, A. 2001. *Prosyllabus Tracheophytorum— Tentamen Systematis Plantarum Vascularum* (Tracheophyta). Institutum Nationale Carpologiae, Moscow.

Doyle, J. A. 1998. Phylogeny of vascular plants. *Annu. Rev. Ecol. Syst.* 29:567–599.

Doyle, J. A. 2013. Phylogenetic analyses and morphological innovations in land plants. *Ann. Plant Rev.* 45:1–50.

Kenrick, P., and P. R. Crane. 1997. *The Origin and Early Diversification of Land Plants—A Cladistic Study*. Smithsonian Institution Press, Washington, DC.

Authors

Philip D. Cantino; Department of Environmental and Plant Biology; Ohio University; Athens, OH 45701, USA. Email: cantino@ohio.edu.

Michael J. Donoghue; Department of Ecology and Evolutionary Biology; Yale University; P.O. Box 208106; New Haven, CT 06520, USA. Email: michael.donoghue@yale.edu.

Date Accepted: 1 April 2011; updated 9 February 2018

Primary Editor: Kevin de Queiroz

Lycopodiophyta A. Cronquist, A. Takhtajan, and W. Zimmermann 1966: 133 [P. D. Cantino and M. J. Donoghue], converted clade name

Registration Number: 160

Definition: The smallest crown clade containing *Lycopodium clavatum* Linnaeus 1753, *Huperzia selago* (Linnaeus) Schrank & Martius 1829 (originally described as *Lycopodium selago* Linnaeus 1753), *Isoetes lacustris* Linnaeus 1753, and *Selaginella apoda* (Linnaeus) Spring 1840 (originally described as *Lycopodium apodum* Linnaeus 1753). This is a minimum-crown-clade definition. Abbreviated definition: min crown ∇ (*Lycopodium clavatum* Linnaeus 1753 & *Huperzia selago* (Linnaeus) Schrank & Martius 1829 & *Isoetes lacustris* Linnaeus 1753 & *Selaginella apoda* (Linnaeus) Spring 1840).

Etymology: Based on the name *Lycopodium*, which is derived from Greek *lycos* and *podus*, meaning "wolf's foot", from a fancied resemblance (Fernald, 1970), and Greek *phyton* (plant) (Stearn, 1973).

Reference Phylogeny: The primary reference phylogeny is Korall et al. (1999: Fig. 2). See also Rydin and Wikström (2002: Fig. 2), Pryer et al. (2004b: Fig. 3; labeled "lycophytes"), Qiu et al. (2007: Fig. 1), Ruhfel et al. (2014: Fig. 6), and Wickett et al. (2014: Fig. 2).

Composition: The total clades of *Lycopodiaceae* (including *Huperziaceae*) and *Isoëtopsida* (*Isoëtes* + *Selaginella*; Cantino et al., 2007). According to current understanding of phylogeny (Doyle, 1998; Pryer et al., 2004a; Judd et al., 2016), the total clade of *Isoëtes* includes Palaeozoic arborescent forms such as *Lepidodendron*.

Diagnostic Apomorphies: Kenrick and Crane (1997: Table 6.3 and Fig. 6.19; node 35), Crane and Kenrick (1997: Table 9; "*Lycopsida*"), Doyle (1998), and Gensel (1992) listed the following synapomorphies for the crown clade: close developmental association of a single axillary or adaxial sporangium with a sporophyll; absence of vasculature in the sporangium; metaxylem tracheids pitted; root stele bilaterally symmetrical, with phloem located on only one side of the stele (but there are a lot of missing data for fossils outside the crown, so this trait may be synapomorphic for a more inclusive clade); crescent-shaped root xylem (but there are a lot of missing data for fossils outside the crown). The following are synapomorphies of this crown clade relative to other crowns but are known to be apomorphic at a more inclusive level when stem fossils are considered (Kenrick and Crane, 1997: Fig. 6.18 and Table 7.2; Doyle, 2013: Fig. 1.1): microphylls ("lycophylls"; Schneider et al., 2002; Pryer et al., 2004a); exarch xylem differentiation in stem (Kenrick and Crane, 1997; Doyle, 1998; Schneider et al., 2002); stellate xylem strand in stem; reniform sporangia with transverse dehiscence (Doyle, 1998). This list is not exhaustive; see Kenrick and Crane (1997: Table 7.2) and DiMichele and Bateman (1996) for other synapomorphies listed under *Lycophytina* and *Lycopsida*.

Synonyms: *Lycophyta*, *Lycopsida*, and *Lycopodiopsida* are approximate synonyms (see Comments below). *Lycopsida* sensu Crane and Kenrick (1997) is an unambiguous synonym in that it was phylogenetically defined as applying to the same crown clade, but the definition conflicts with the labeling of the reference phylogeny (see Comments).

Comments: The monophyly of this clade is strongly supported by molecular data (Korall et al., 1999; Pryer et al., 2001, 2004b; Qiu et al., 2006, 2007; Ruhfel et al., 2014; Wickett et al., 2014) as well as by numerous morphological synapomorphies (detailed above). One study (Garbary et al., 1993) based solely on male gametogenesis characters found this group to be polyphyletic, with *Selaginella* sharing closer ancestry with bryophytes than with *Lycopodium*, but this hypothesis has not been supported by any other analysis.

While the monophyly of the morphologically distinct taxa *Selaginella* and *Isoëtes* is not in question, the monophyly of *Lycopodiaceae* (including *Huperziaceae*) is supported by only one possible morphological synapomorphy (foveolate-fossulate microspore wall; Kenrick and Crane, 1997). Although there is molecular support for its monophyly, those studies either were based solely on *rbc*L (Korall et al., 1999) or included too small a sample of *Lycopodiaceae* to be convincing (Qiu et al., 2006, 2007; Wickett et al. 2014). We remain concerned that *Huperzia* could end up being sister to the rest of *Lycopodiophyta*. Consequently, the clade *Lycopodiaceae* is represented by two specifiers (*Lycopodium clavatum* and *Huperzia selago*) in the definition of *Lycopodiophyta*.

The names *Lycophyta*, *Lycopodiophyta*, *Lycopsida*, and *Lycopodiopsida* have historically applied to the same set of clades (from crown to total), but most phylogenetic studies have instead used the informal equivalents, "lycophytes" and "lycopsids". Since *Lycophyta* and *Lycopsida* are apparently based on the name *Lycopodium*, they should be corrected to *Lycopodiophyta* and *Lycopodiopsida* under the botanical code (Turland et al., 2018; Art.

16.1). We are narrowing our choice to the latter pair of names in the interest of promoting consistency between rank-based and phylogenetic nomenclature. We prefer the *-phyta* ending because it means "plant" while *-opsida* is solely an indicator of (class) rank, and we have therefore chosen *Lycopodiophyta*, as in our previous paper (Cantino et al., 2007). We also follow that paper (p. E11) in applying the name explicitly to the lycopodiophyte crown clade. The clade *Lycopsida* sensu Kenrick and Crane (1997) is somewhat larger in that it includes fossils such as *Asteroxylon* and *Baragwanathia* that are shown (p. 239; see also Pryer et al., 2004a: Fig. 10.3) as being outside the crown clade. The same is true of *Microphyllophyta* sensu Bold (1957) and Bold et al. (1980), *Lepidophyta* sensu Smith (1955), and *Lycopodiopsida* sensu Bierhorst (1971). Crane and Kenrick (1997) gave *Lycopsida* a minimum-clade definition with extant species of *Huperzia* and *Isoetes* as the specifiers, thereby effectively (but not explicitly) applying the name to the crown clade. However, in the accompanying cladogram and classification, they applied this same name to a larger clade by including an extinct stem group (*Drepanophycales*); thus, their conceptualization of the name was ambiguous.

Following the *PhyloCode* (Version 6, Rec. 9.15A), we attribute the preexisting name *Lycopodiophyta* to Cronquist et al. (1966), but under the botanical code (Turland et al., 2018; Art. 16.3), the name is attributed to Scott (1909), who spelled it *Lycopsida* (Hoogland and Reveal, 2005).

Literature Cited

Bierhorst, D. W. 1971. *Morphology of Vascular Plants*. Macmillan Co., New York.

Bold, H. C. 1957. *Morphology of Plants*. Harper & Row, New York.

Bold, H. C., C. J. Alexopoulos, and T. Delevoryas. 1980. *Morphology of Plants and Fungi*. 4th edition. Harper & Row, New York.

Cantino, P. D., J. A. Doyle, S. W. Graham, W. S. Judd, R. G. Olmstead, D. E. Soltis, P. S. Soltis, and M. J. Donoghue. 2007. Towards a phylogenetic nomenclature of *Tracheophyta*. *Taxon* 56:822–846 and E1–E44.

Crane, P. R., and P. Kenrick. 1997. Problems in cladistic classification: higher-level relationships in land plants. *Aliso* 15:87–104.

Cronquist, A., A. Takhtajan, and W. Zimmermann. 1966. On the higher taxa of *Embryobionta*. *Taxon* 15:129–168.

DiMichele, W. A., and R. M. Bateman. 1996. The rhizomorphic lycopsids: a case-study in paleobotanical classification. *Syst. Bot.* 21: 535–552.

Doyle, J. A. 1998. Phylogeny of vascular plants. *Annu. Rev. Ecol. Syst.* 29:567–599.

Doyle, J. A. 2013. Phylogenetic analyses and morphological innovations in land plants. Pp. 1–50 in *The Evolution of Plant Form* (B. A. Ambrose and M. Purugganan, eds.). Annual Plant Reviews 45. Blackwell, Oxford.

Fernald, M. L. 1970. *Gray's Manual of Botany*. 8th edition, corrected printing. Van Nostrand, New York.

Garbary, D. J., K. S. Renzaglia, and J. G. Duckett. 1993. The phylogeny of land plants: a cladistic analysis based on male gametogenesis. *Plant Syst. Evol.* 188:237–269.

Gensel, P. G. 1992. Phylogenetic relationships of the zosterophylls and lycopsids: evidence from morphology, paleoecology, and cladistic methods of inference. *Ann. Mo. Bot. Gard.* 79:450–473.

Hoogland, R. D., and J. L. Reveal. 2005. Index nominum familiarum plantarum vascularium. *Bot. Rev.* 71:1–291.

Judd, W. S., C. S. Campbell, E. A. Kellogg, P. F. Stevens, and M. J. Donoghue. 2016. *Plant Systematics: A Phylogenetic Approach*. 4th edition. Sinauer Associates, Sunderland, MA.

Kenrick, P., and P. R. Crane. 1997. *The Origin and Early Diversification of Land Plants—A Cladistic Study*. Smithsonian Institution Press, Washington, DC.

Korall, P., P. Kenrick, and J. P. Therrien. 1999. Phylogeny of *Selaginellaceae*: evaluation of generic/subgeneric relationships based on *rbcL* gene sequences. *Int. J. Plant Sci.* 160:585–594.

Pryer, K. M., H. Schneider, and S. Magallón. 2004a. The radiation of vascular plants. Pp. 138–153 in *Assembling the Tree of Life* (J. Cracraft and M. J. Donoghue, eds.). Oxford University Press, Oxford.

Pryer, K. M., H. Schneider, A. R. Smith, R. Cranfill, P. G. Wolf, J. S. Hunt, and S. D. Sipes. 2001. Horsetails and ferns are a monophyletic group and the closest living relatives to seed plants. *Nature* 409:618–622.

Pryer, K. M., E. Schuettpelz, P. G. Wolf, H. Schneider, A. R. Smith, and R. Cranfill. 2004b. Phylogeny and evolution of ferns (monilophytes) with a focus on the early leptosporangiate divergences. *Am. J. Bot.* 91: 1582–1598.

Qiu, Y.-L., L. Li, B. Wang, Z. Chen, O. Dombrovska, J. Lee, L. Kent, R. Li, R. W. Jobson, T. A. Hendry, D. W. Taylor, C. M. Testa, and M. Ambros. 2007. A nonflowering land plant phylogeny inferred from nucleotide sequences of seven chloroplast, mitochondrial, and nuclear genes. *Int. J. Plant Sci.* 168:691–708.

Qiu, Y.-L., L. Li, B. Wang, Z. Chen, V. Knoop, M. Groth-Malonek, O. Dombrovska, J. Lee, L. Kent, J. Rest, G. F. Estabrook, T. A. Hendry, D. W. Taylor, C. M. Testa, M. Ambros, B. Crandall-Stotler, R. J. Duff, M. Stech, W. Frey, D. Quandt, and C. C. Davis. 2006. The deepest divergences in land plants inferred from phylogenomic evidence. *Proc. Natl. Acad. Sci. USA* 103:15511–15516.

Ruhfel, B. R., M. A. Gitzendanner, P. S. Soltis, D. E. Soltis, and J. G. Burleigh. 2014. From algae to angiosperms—inferring the phylogeny of green plants (*Viridiplantae*) from 360 plastid genomes. *BMC Evol. Biol.* 14:23.

Rydin, C., and N. Wikström. 2002. Phylogeny of *Isoëtes* (*Lycopsida*): resolving basal relationships using *rbcL* sequences. *Taxon* 51:83–89.

Schneider, H., K. M. Pryer, R. Cranfill, A. R. Smith, and P. G. Wolf. 2002. Evolution of vascular plant body plans: a phylogenetic perspective. Pp. 330–364 in *Developmental Genetics and Plant Evolution* (Q. C. B. Cronk, R. M. Bateman, and J. A. Hawkins, eds.). Taylor & Francis, London.

Scott, D. H. 1909. *Studies in Fossil Botany*. 2nd edition. Adam & Charles Black, London.

Smith, G. M. 1955. *Cryptogamic Botany*. 2nd edition, Vol. 2. McGraw-Hill, New York.

Stearn, W. T. 1973. *Botanical Latin*. David & Charles, Newton Abbot, Devon.

Turland, N. J., J. H. Wiersema, F. R. Barrie, W. Greuter, D. L. Hawksworth, P. S. Herendeen, S. Knapp, W.-H. Kusber, D.-Z. Li, K. Marhold, T. W. May, J. McNeill, A. M. Monro, J. Prado, M. J. Price, and G. F. Smith, eds. 2018. *International Code of Nomenclature for Algae, Fungi and Plants (Shenzhen Code)*. Adopted by the Nineteenth International Botanical Congress, Shenzhen, China, July 2017. Regnum Vegetabile 159. Koeltz Botanical Books, Glashütten.

Wickett, N. J., S. Mirarab, N. Nguyen, T. Warnow, E. Carpenter, N. Matasci, S. Ayyampalayam, M. S. Barker, J. G. Burleigh, M. A. Gitzendanner, B. R. Ruhfel, E. Wafula, J. P. Der, S. W. Graham, S. Mathews, M. Melkonian, D. E. Soltis, P. S. Soltis, N. W. Miles, C. J. Rothfels, L. Pokorny, A. J. Shaw, L. DeGironimo, D. W. Stevenson, B. Surek, J. C. Villarreal, B. Roure, H. Philippe, C. W. dePamphilis, T. Chen, M. K. Deyholos, R. S. Baucom, T. M. Kutchan, M. M. Augustin, J. Wang, Y. Zhang, Z. Tian, Z. Yan, X. Wu, X. Sun, G. K. Wong, and J. Leebens-Mack. 2014. Phylotranscriptomic analysis of the origin and early diversification of land plants. *Proc. Natl. Acad. Sci., USA*. 111:E4859–E4868.

Authors

Philip D. Cantino; Department of Environmental and Plant Biology; Ohio University; Athens, OH 45701, USA. Email: cantino@ohio.edu.

Michael J. Donoghue; Department of Ecology and Evolutionary Biology; Yale University; P.O. Box 208106; New Haven, CT 06520, USA. Email: michael.donoghue@yale.edu.

Date Accepted: 1 April 2011; updated 21 May 2018, 11 July 2018

Primary Editor: Kevin de Queiroz

Pan-Euphyllophyta P. D. Cantino and M. J. Donoghue in P. D. Cantino et al. (2007): E12 [P. D. Cantino and M. J. Donoghue], converted clade name

Registration Number: 78

Definition: The total clade of the crown clade *Euphyllophyta*. This is a crown-based total-clade definition. Abbreviated definition: total ∇ of *Euphyllophyta*.

Etymology: From the Greek *eu-* (true, original, primitive), *phyllon* (leaf), and *phyton* (plant), plus *pan-* or *pantos* (all, the whole), indicating that this is a total clade. Euphylls (also known as megaphylls; i.e., leaves that usually have more than one vein and are thought to have evolved from branch systems) are characteristic of the crown clade *Euphyllophyta*.

Reference Phylogeny: Crane and Kenrick (1997: Fig. 1, the clade labeled *Euphyllophytina*) (see Synonyms).

Composition: *Euphyllophyta* (this volume) and all extinct plants that share more recent ancestry with *Euphyllophyta* than with *Lycopodiophyta*. The stem euphyllophytes are often classified as trimerophytes, a paraphyletic group that has been named at various ranks (see Synonyms). The trimerophytes include species of *Psilophyton* and *Eophyllophyton*, which lie outside the euphyllophyte crown (Crane and Kenrick, 1997; Kenrick and Crane, 1997; Doyle, 2013), but *Pertica* (which is also classified as a trimerophyte; Stewart and Rothwell, 1993; Taylor, 1981) is thought to lie within the crown (Kenrick and Crane, 1997: Figs. 4.31, 7.10).

Diagnostic Apomorphies: Several synapomorphies were listed by Crane and Kenrick (1997: Table 9) and Kenrick and Crane (1997: 240,

Table 7.2, and pages listed below), most of which have been lost or modified in some or all extant members of the clade: pseudomonopodial or monopodial branching (Kenrick and Crane, 1997: 109, 359; Doyle, 2013) (although if the fernlike leaves of early seed plants were derived from pseudomonopodial branch systems of more basal lignophytes (Doyle, 1998, 2013), the axillary monopodial branching of seed plants and the pseudomonopodial branching of more basal lignophytes may not be homologous; Doyle, 2013); helical arrangement of branches (Kenrick and Crane, 1997: 110, 360); dichotomous appendages (Kenrick and Crane, 1997: 113, 361); recurved branch apexes (Kenrick and Crane, 1997: 112–113, 360); paired sporangia grouped into terminal trusses (Kenrick and Crane, 1997: 121–122, 364); sporangial dehiscence along one side through a single longitudinal slit (Kenrick and Crane, 1997: 125, 366); Doyle, 2013; and radially aligned xylem in larger axes (Crane and Kenrick, 1997). Kenrick and Crane also cited scalariform bordered pitting of metaxylem cells as a synapomorphy, but it does not occur in *Eophyllophyton* and therefore is synapomorphic for a less inclusive group than the total clade (Kenrick and Crane, 1997: 120, 363, Fig. 7.10).

Synonyms: *Euphyllophytina* Kenrick and Crane (1997: Table 7.1) and Crane and Kenrick (1997) may be an approximate synonym (see Comments). *Trimerophytophyta* sensu Bold et al. (1980), *Trimerophytina* Banks 1968, *Trimerophytopsida* Foster and Gifford 1974 have been applied to a paraphyletic group originating in approximately the same ancestor as *Pan-Euphyllophyta* according to the phylogenies of

Kenrick and Crane (1997: Figs. 4.31, 4.32) and Crane and Kenrick (1997: Fig. 1) and therefore are partial (and approximate) synonyms.

Comments: *Euphyllophytina* (Crane and Kenrick, 1997; Kenrick and Crane, 1997) referred to a clade that is similar in composition to *Pan-Euphyllophyta*. However, there is conflict within and between these two papers regarding whether the name *Euphyllophytina* applies to a node-based, apomorphy-based, or branch-based clade. Kenrick and Crane (1997) gave the name a "synapomorphy-based definition" in Table 7.2, but they described the clade (p. 240) as the sister group of *Lycophytina*, suggesting that both of these clades were conceptualized as originating from their point of divergence rather than from the origin of a particular apomorphy. In contrast, the same authors (Crane and Kenrick, 1997) used a "node-based" (minimum-clade) definition for *Euphyllophytina* with *Eophyllophyton bellum* and *Nymphaea odorata* as the specifiers. In any case, the composition of *Euphyllophytina*, as shown on the reference trees in these two papers by Crane and Kenrick, seems to approximate that of *Pan-Euphyllophyta* as defined here. However, there could be undiscovered fossil pan-euphyllophytes that lie outside the node-based *Euphyllophytina* of Crane and Kenrick.

Literature Cited

Banks, H. P. 1968. The early history of land plants. Pp. 73–107 in *Evolution and Environment* (E. T. Drake, ed.). Yale University Press, New Haven.

Bold, H. C., C. J. Alexopoulos, and T. Delevoryas. 1980. *Morphology of Plants and Fungi*. 4th edition. Harper & Row, New York.

Cantino, P. D., J. A. Doyle, S. W. Graham, W. S. Judd, R. G. Olmstead, D. E. Soltis, P. S. Soltis, and M. J. Donoghue. 2007. Towards a phylogenetic nomenclature of *Tracheophyta*. *Taxon* 56:822–846 and E1–E44.

Crane, P. R., and P. Kenrick. 1997. Problems in cladistic classification: higher-level relationships in land plants. *Aliso* 15:87–104.

Doyle, J. A. 1998. Phylogeny of vascular plants. *Annu. Rev. Ecol. Syst.* 29:567–599.

Doyle, J. A. 2013. Phylogenetic analyses and morphological innovations in land plants. Pp. 1–50 in *The Evolution of Plant Form* (B. A. Ambrose and M. Purugganan, eds.). Annual Plant Reviews 45. Blackwell, Oxford.

Foster, A. S., and E. M. Gifford. 1974. *Comparative Morphology of Vascular Plants*. 2nd edition. W. H. Freeman, San Francisco, CA.

Kenrick, P., and P. R. Crane. 1997. *The Origin and Early Diversification of Land Plants—A Cladistic Study*. Smithsonian Institution Press, Washington, DC.

Stewart, W. N., and G. W. Rothwell. 1993. *Paleobotany and the Evolution of Plants*. 2nd edition. Cambridge University Press, Cambridge, UK.

Taylor, T. N. 1981. *Paleobotany*. McGraw-Hill, New York.

Authors

Philip D. Cantino; Department of Environmental and Plant Biology; Ohio University; Athens, OH 45701, USA. Email: cantino@ohio.edu.

Michael J. Donoghue; Department of Ecology and Evolutionary Biology; Yale University; P.O. Box 208106; New Haven, CT 06520, USA. Email: michael.donoghue@yale.edu.

Date Accepted: 31 January 2013; updated 9 February 2018

Primary Editor: Kevin de Queiroz

Euphyllophyta G. Lecointre and H. Le Guyader 2006: 181
[P. D. Cantino and M. J. Donoghue], converted clade name

Registration Number: 43

Definition: The largest crown clade containing *Ginkgo biloba* Linnaeus 1771 (*Spermatophyta*) and *Pteridium aquilinum* (Linnaeus) Kuhn 1879 (originally described as *Pteris aquilina* Linnaeus 1753) (*Leptosporangiatae*), but not *Selaginella apoda* (Linnaeus) Spring 1840 (originally described as *Lycopodium apodum* Linnaeus 1753) (*Lycopodiophyta*). This is a maximum-crown-clade definition. Abbreviated definition: max crown ∇ (*Ginkgo biloba* Linnaeus 1771 & *Pteridium aquilinum* (Linnaeus) Kuhn 1879 ~ *Selaginella apoda* (Linnaeus) Spring 1840).

Etymology: From the Greek *eu* (true), *phyllon* (leaf), and *phyton* (plant). Megaphylls (sometimes referred to as euphylls; e.g., Simpson, 2006; Schneider et al., 2009), which are leaves that usually have more than one vein and are thought to have evolved either from dichotomously forking lateral branches or from branch systems bearing such units, are characteristic of this clade though probably not a synapomorphy (see Diagnostic Apomorphies).

Reference Phylogeny: The primary reference phylogeny is Pryer et al. (2001: Fig. 1, the clade labeled *Euphyllophytina*). See also Kenrick and Crane (1997: Fig. 7.10), Pryer et al. (2004: Fig. 3), Qiu et al. (2007: Fig. 1), Schneider et al. (2009: Figs. 1 and 2), Ruhfel et al. (2014: Figs. 6 and 7), and Wickett et al. (2014: Figs. 2 and 3).

Composition: All extant seed plants (*Spermatophyta*), ferns (*Leptosporangiatae, Ophioglossales,* and *Marattiales*), horsetails (*Equisetum*), and whisk ferns (*Psilotaceae*), as well as their extinct relatives that fall within the crown clade. The crown clade is composed of two primary subclades, *Pan-Spermatophyta* (see entry in this volume) and the unnamed total clade of *Monilophyta*.

Diagnostic Apomorphies: Apomorphies relative to other crown clades include roots with monopodial branching and endogenous lateral roots (Schneider et al., 2002); sporangia terminal on lateral branches (Pryer et al., 2004a) and dehiscing longitudinally (Doyle, 1998, 2013) (these features characterize the earliest members of *Pan-Euphyllophyta* and were modified in most extant representatives); lobed, mesarch primary xylem strand (Stein, 1993; Kenrick and Crane, 1997: Fig. 7.10 and p. 241; Doyle, 1998, 2013), which has been modified in the stems of most extant members; multiflagellate spermatozoids (apparently convergent in *Isoëtes*) (Garbary et al., 1993; Kenrick and Crane, 1997: 240, 275); a 30-kb inversion in the chloroplast genome (Raubeson and Jansen, 1992). Megaphylls are sometimes cited as a synapomorphy of this clade (Schneider et al., 2002), but analyses that include fossils suggest that megaphylls of monilophytes and seed plants evolved independently (Stewart and Rothwell, 1993; Kenrick and Crane, 1997; Doyle, 1998, 2013; Boyce and Knoll, 2002; Friedman et al., 2004; Galtier, 2010). Even within *Lignophyta* (i.e., *Spermatophyta* and their stem relatives that have a bifacial vascular cambium; phylogenetically defined by Donoghue and Doyle in Cantino et al., 2007), the small, wedge-shaped leaves of *Archaeopteris* may not be homologous with the whole fernlike fronds of "seed ferns" (a non-monophyletic assortment of extinct seed plants with more or less fernlike leaves), but rather with individual leaflets of such fronds (Doyle and Donoghue, 1986;

Doyle, 1998, 2013). The single-veined leaves of *Equisetum* may be characterized as microphylls or megaphylls, depending on whether one's definition of these terms is based on structure (specifically, presence of only one vein in the case of microphylls) or evolutionary origin (specifically, from dichotomously forking branches or branch systems in the case of megaphylls versus from unvascularized enations [or possibly from sterilized sporangia; Kenrick and Crane, 1997] in the case of microphylls). Some fossil relatives of *Equisetum* such as *Sphenophyllum* had multiple-veined leaves that apparently evolved from dichotomous branches (reviewed by Doyle, 2013), suggesting that the one-veined leaves of *Equisetum* would qualify as megaphylls if the definition is based on origin.

Synonyms: There are no synonyms; *Euphyllophytina* (Kenrick and Crane, 1997; Crane and Kenrick, 1997) refers to a more inclusive clade (see Comments).

Comments: A group composed of 'ferns' (excluding *Equisetum* and *Psilotophyta*) and seed plants has been recognized on morphological grounds for many years (Banks, 1968), and the monophyly of this group, when *Equisetum* and *Psilotophyta* are included, is strongly supported by molecular data (Pryer et al., 2001, 2004; Qiu et al., 2007; Wickett et al., 2014). However, there was no scientific name for the crown clade until we explicitly applied the name *Euphyllophyta* to it (Cantino et al., 2007). We incorrectly referred to it as a new name, rather than converted, because we were unaware of its use by Lecointre and Le Guyader (2006). The latter authors were not explicit about the clade to which the name *Euphyllophyta* was intended to apply, but their inclusion of *Psilophyton* indicates that it was more inclusive than the crown (see Kenrick and Crane, 1997: Fig. 4.32).

Euphyllophytina Kenrick and Crane (1997: Table 7.1 and Fig. 7.10) referred to a more inclusive clade than the crown, though it is unclear whether the name was applied to an apomorphy-based or a total clade. The name was given a "synapomorphy-based definition" in Table 7.2, but the clade was described (p. 240) as the sister group of *Lycophytina*, suggesting that both of these clades were conceptualized as originating from their point of divergence rather than from the origin of a particular apomorphy. In contrast, the same authors (Crane and Kenrick, 1997) used a "node-based" (minimum-clade) definition for *Euphyllophytina*. Although there is conflict within and between these two papers as to whether the name *Euphyllophytina* applies to a minimum, apomorphy-based, or total clade, all three are more inclusive than the crown clade to which we apply the name *Euphyllophyta* here.

A maximum-crown-clade definition requires only one internal specifier, but two are used here in order to disqualify the name under certain conditions. In the context of a phylogeny in which either 'ferns' or seed plants share more recent ancestry with lycophytes than they do with each other (e.g., Rothwell and Nixon [2006: Fig. 6]; Ruhfel et al. [2014: Fig. 5]), the name *Euphyllophyta* would not apply to any clade. In our earlier definition (Cantino et al., 2007), we included a species of *Equisetum* as a third internal specifier, but we have omitted it here because it is unnecessary. To the best of our knowledge, *Euphyllophyta* has the same composition with either definition in the context of all recently published phylogenies.

Literature Cited

Banks, H. P. 1968. The early history of land plants. Pp. 73–107 in *Evolution and Environment* (E. T. Drake, ed.). Yale University Press, New Haven, CT.

Boyce, C. K., and A. H. Knoll. 2002. Evolution of developmental potential and the multiple independent origins of leaves in Paleozoic vascular plants. *Paleobiology* 28:70–100.

Cantino, P. D., J. A. Doyle, S. W. Graham, W. S. Judd, R. G. Olmstead, D. E. Soltis, P. S. Soltis, and M. J. Donoghue. 2007. Towards a phylogenetic nomenclature of *Tracheophyta*. *Taxon* 56:822–846 and E1–E44.

Crane, P. R., and P. Kenrick. 1997. Problems in cladistic classification: higher-level relationships in land plants. *Aliso* 15:87–104.

Doyle, J. A. 1998. Phylogeny of vascular plants. *Annu. Rev. Ecol. Syst.* 29:567–599.

Doyle, J. A. 2013. Phylogenetic analyses and morphological innovations in land plants. Pp. 1–50 in *The Evolution of Plant Form* (B. A. Ambrose, and M. Purugganan, eds.). Annual Plant Reviews 45. Blackwell, Oxford.

Doyle, J. A., and M. J. Donoghue. 1986. Seed plant phylogeny and the origin of angiosperms: an experimental cladistic approach. *Bot. Rev.* 52:321–431.

Friedman, W. E., R. C. Moore, and M. D. Purugganan. 2004. The evolution of plant development. *Am. J. Bot.* 91:1726–1741.

Galtier, J. 2010. The origins and early evolution of the megaphyllous leaf. *Int. J. Plant Sci.* 171:641–661.

Garbary, D. J., K. S. Renzaglia, and J. G. Duckett. 1993. The phylogeny of land plants: a cladistic analysis based on male gametogenesis. *Plant Syst. Evol.* 188:237–269.

Kenrick, P., and P. R. Crane. 1997. *The Origin and Early Diversification of Land Plants: A Cladistic Study*. Smithsonian Institution Press, Washington, DC.

Lecointre, G., and H. Le Guyader. 2006. *The Tree of Life—A Phylogenetic Classification*. Belknap Press, Cambridge, UK.

Pryer, K. M., H. Schneider, A. R. Smith, R. Cranfill, P. G. Wolf, J. S. Hunt, and S. D. Sipes. 2001. Horsetails and ferns are a monophyletic group and the closest living relatives to seed plants. *Nature* 409:618–622.

Pryer, K. M., E. Schuettpelz, P. G. Wolf, H. Schneider, A. R. Smith, and R. Cranfill. 2004. Phylogeny and evolution of ferns (monilophytes) with a focus on the early leptosporangiate divergences. *Am. J. Bot.* 91:1582–1598.

Qiu, Y.-L., L. Li, B. Wang, Z. Chen, O. Dombrovska, J. Lee, L. Kent, R. Li, R. W. Jobson, T. A. Hendry, D. W. Taylor, C. M. Testa, and M. Ambros. 2007. A nonflowering land plant phylogeny inferred from nucleotide sequences of seven chloroplast, mitochondrial, and nuclear genes. *Int. J. Plant Sci.* 168:691–708.

Raubeson, L. A., and R. K. Jansen. 1992. Chloroplast DNA evidence on the ancient evolutionary split in vascular land plants. *Science* 255:1697–1699.

Rothwell, G. W., and K. C. Nixon. 2006. How does the inclusion of fossil data change our conclusions about the phylogenetic history of euphyllophytes? *Int. J. Plant Sci.* 167:737–749.

Ruhfel, B. R., M. A. Gitzendanner, P. S. Soltis, D. E. Soltis, and J. G. Burleigh. 2014. From algae to angiosperms—inferring the phylogeny of green plants (*Viridiplantae*) from 360 plastid genomes. *BMC Evol. Biol.* 14:23.

Schneider, H., K. M. Pryer, R. Cranfill, A. R. Smith, and P. G. Wolf. 2002. Evolution of vascular plant body plans: a phylogenetic perspective. Pp. 330–364 in *Developmental Genetics and Plant Evolution* (Q. C. B. Cronk, R. M. Bateman, and J. A. Hawkins, eds.). Taylor & Francis, London.

Schneider, H., A. R. Smith, and K. M. Pryer. 2009. Is morphology really at odds with molecules in estimating fern phylogeny? *Syst. Bot.* 34:455–475.

Simpson, M. G. 2006. *Plant Systematics*. Elsevier Academic Press, Burlington, MA.

Stein, W. 1993. Modeling the evolution of stelar architecture in vascular plants. *Int. J. Plant Sci.* 154:229–263.

Stewart, W. N., and G. W. Rothwell. 1993. *Paleobotany and the Evolution of Plants*. 2nd edition. Cambridge University Press, Cambridge, UK.

Wickett, N. J., S. Mirarab, N. Nguyen, T. Warnow, E. Carpenter, N. Matasci, S. Ayyampalayam, M. S. Barker, J. G. Burleigh, M. A. Gitzendanner, B. R. Ruhfel, E. Wafula, J. P. Der, S. W. Graham, S. Mathews, M. Melkonian, D. E. Soltis, P. S. Soltis, N. W. Miles, C. J. Rothfels, L. Pokorny, A. J. Shaw, L. DeGironimo, D. W. Stevenson, B. Surek, J. C. Villarreal, B. Roure, H. Philippe, C. W.

dePamphilis, T. Chen, M. K. Deyholos, R. S. Baucom, T. M. Kutchan, M. M. Augustin, J. Wang, Y. Zhang, Z. Tian, Z. Yan, X. Wu, X. Sun, G. K. Wong, and J. Leebens-Mack. 2014. Phylotranscriptomic analysis of the origin and early diversification of land plants. *Proc. Natl. Acad. Sci. USA.* 111:E4859–E4868.

Authors

Philip D. Cantino; Department of Environmental and Plant Biology; Ohio University; Athens, OH 45701, USA. Email: cantino@ohio.edu.

Michael J. Donoghue; Department of Ecology and Evolutionary Biology; Yale University; P.O. Box 208106; New Haven, CT 06520, USA. Email: michael.donoghue@yale.edu.

Date Accepted: 21 January 2013; updated 9 February 2018

Primary Editor: Kevin de Queiroz

Monilophyta P. D. Cantino and M. J. Donoghue in P. D. Cantino et al. (2007): E13 [J. A. Doyle, P. D. Cantino, and M. J. Donoghue], converted clade name

Registration Number: 67

Definition: The largest crown clade containing *Pteridium aquilinum* (Linnaeus) Kuhn 1879 (originally *Pteris aquilina*) (*Leptosporangiatae*) and *Equisetum hyemale* Linnaeus 1753 but not *Oryza sativa* Linnaeus 1753 (*Spermatophyta*) or *Huperzia lucidula* (Michaux) Trevisan de Saint-Léon 1875 (originally *Lycopodium lucidulum*) (*Lycopodiophyta*). This is a maximum-crown-clade definition. Abbreviated definition: max crown ∇ (*Pteridium aquilinum* (Linnaeus) Kuhn 1879 & *Equisetum hyemale* Linnaeus 1753 ~ *Oryza sativa* Linnaeus 1753 ∨ *Huperzia lucidula* (Michaux) Trevisan de Saint-Léon 1875).

Etymology: From the Latin *monile*, meaning necklace, in reference to the "position and ontogeny of protoxylem in the lobed primary xylem of early fossil groups" (Kenrick and Crane, 1997: 248), and the Greek *phyton* (plant).

Reference Phylogeny: The primary reference phylogeny is Knie et al. (2015: Fig. 4). See also Rothfels et al. (2015: Fig. 1, the clade labeled "*Polypodiopsida*"), Pryer et al. (2001: Fig. 1, as "*Moniliformopses*"), Pryer et al. (2004: Fig. 3, as "ferns (monilophytes)"), Qiu et al. (2007: Fig. 1), and Wickett et al. (2014: Figs. 2, 3).

Composition: *Equisetum, Psilotophyta, Ophioglossales, Marattiales,* and *Leptosporangiatae,* their most recent common ancestor, and all of its other descendants. In addition to fossil representatives of the five listed crown subgroups, fossil members presumably include various taxa considered to be their stem relatives, such as *Sphenophyllales, Archaeocalamites,* and *Calamites* in the case of *Equisetum; Psaronius* in the case of *Marattiales;* and *Ankyropteris* in the case of *Leptosporangiatae* (see Doyle, 2013). Less can be said about other fossil members because most of the analyses that inferred the existence of this clade included only extant plants, and some analyses that included both fossils and extant plants did not infer the existence of this clade. Kenrick and Crane (1997: Table 7.1) included the fossil groups *Cladoxyliidae, Zygopteridae* (which may include additional stem relatives of *Leptosporangiatae*: see Galtier, 2010), and *Stauropteridae* within *Moniliformopses* (a clade that may be either equivalent to or slightly more inclusive than *Monilophyta;* see Comments). In contrast, Rothwell (1999) inferred that *Cladoxyliidae* and *Zygopteridae,* along with *Equisetum,* are more closely related to seed plants than to extant ferns (thus the clade *Monilophyta* as defined here does not exist on that phylogeny) and stauropterids are even more distant from extant ferns. Some of the analyses of Rothwell and Nixon (2006) inferred the existence of *Monilophyta* (as defined here) and others did not, but in those trees in which there is a clade that fits our definition of *Monilophyta,* all three of the above-mentioned extinct groups included in *Moniliformopses* by Kenrick and Crane lie outside *Monilophyta.*

Diagnostic Apomorphies: In the analysis of Kenrick and Crane (1997), the main synapomorphy is mesarch protoxylem confined to the outer lobed ends of the xylem strand (Crane and Kenrick, 1997); typically the protoxylem is parenchymatous, so that the metaxylem forms a conspicuous "peripheral loop" around a spongy protoxylem area. This feature occurs in fossil cladoxylopsids, zygopterids, and *Ankyropteris*, and it has been assumed that the spongy protoxylem areas were modified into the protoxylem canals of *Equisetum* and related fossil calamites; however, homologous structures have not been recognized in living ferns. Schneider et al. (2009), considering extant groups only, reported four unambiguous apomorphies: sporangia arranged in a sorus, presence of a pseudoendospore, centrifugal spore wall formation (also reported by Schneider et al., 2002), and plasmodial tapetum. However, it is not clear that the groups of sporangia in *Equisetum*, *Psilotophyta*, *Ophioglossales*, and many presumed fossil monilophytes are more comparable to the typical sori of leptosporangiate ferns than are the groups of sporangia and microsynangia of seed plant stem relatives ("progymnosperms," "seed ferns"). It is also not certain that the spore characters of monilophytes are derived relative to the pollen of living seed plants, which have more complex exine development, or to the more fern-like spores and "pre-pollen" of seed plant stem relatives, in which exine development is largely unknown.

Synonyms: *Moniliformophyta* Crane and Kenrick 1997 is an unambiguous heterodefinitional synonym in the context of some phylogenetic hypotheses but not others (see Comments).

Moniliformopses sensu Lecointre and Guyader (2006) is a possible synonym (see Comments). The names *Filicophyta*, *Filicopsida*, *Polypodiophyta*, *Pterophyta*, and *Pteropsida* are partial synonyms of *Monilophyta* in the context of some phylogenies (e.g., Pryer et al., 2001) in that they have been commonly applied to the paraphyletic group ("ferns") originating from the same ancestor as the clade *Monilophyta* but excluding *Equisetum* and usually (but not always;

e.g., Bierhorst, 1971) *Psilotophyta*. However, in the context of other phylogenies (e.g., Wickett et al., 2014; Knie et al., 2015; Rothfels et al., 2015), where *Equisetum* is sister to all other monilophytes, names such as *Filicophyta* are not synonyms of *Monilophyta* but instead apply to a major subclade comprising the "ferns" and *Psilotophyta*.

Comments: Kenrick and Crane (1997) first proposed the existence of a clade that includes ferns and *Equisetum* (represented in the tree shown in their Figure 4.32 by the presumed fossil ferns *Pseudosporochnus*, a "cladoxylopsid," and *Rhacophyton*, a "zygopterid," and the presumed fossil sphenopsid *Ibyka*) but excludes *Lycopodiophyta* and seed plants. Ferns and *Equisetum* do not form a monophyletic group in several other morphological studies (Bremer et al., 1987; Rothwell, 1999; Rothwell and Nixon, 2006), but the two groups (also including *Psilotophyta* or "whisk ferns") are strongly supported as a clade by molecular analyses (Nickrent et al., 2000; Pryer et al., 2001, 2004; Wikström and Pryer, 2005; Schuettpelz et al., 2006; Qiu et al., 2007; Ruhfel et al., 2014; Wickett et al., 2014; Knie et al., 2015; Rothfels et al., 2015) and weakly supported by a morphological analysis of extant taxa by Schneider et al. (2009).

A maximum-crown-clade definition usually has only one internal specifier, but a second internal specifier is included here in order to disqualify the name under certain conditions. In the context of a phylogenetic hypothesis in which extant ferns share more recent ancestry with seed plants than with *Equisetum* (Bremer et al., 1987: Fig. 1), or one in which *Equisetum* shares more recent ancestry with seed plants than with extant ferns (Rothwell, 1999: Fig. 2; Rothwell and Nixon, 2006: Fig. 3A), the name *Monilophyta* would not apply to any clade. Abandonment of the name would be appropriate in such cases because it is universally associated with the hypothesis that ferns (including *Psilotophyta*) and horsetails form a clade exclusive of seed plants and lycopodiophytes. The maximum-crown-clade definition that we have adopted here and the slightly different definition of Cantino et al. (2007) both capture the essence of this hypothesis. The two definitions differ in the specifiers chosen to represent *Equisetum*, *Spermatophyta*, and *Lycopodiophyta*, with each definition using species included in the respective primary reference phylogeny.

When we phylogenetically defined the name *Monilophyta* as referring to this crown clade (Cantino et al., 2007), we selected this name because it closely approximates the informal name "monilophytes," which is often used for this clade (e.g., Judd et al., 2002, 2008; Simpson, 2006). The name *Monilophyta* was applied to this clade in a field guide (Cobb et al., 2005) but without a description or diagnosis, so it did not qualify as a preexisting name. Another candidate name, *Moniliformopses* Kenrick and Crane (1997: Table 7.1), was apparently apomorphy-based (1997: Table 7.2), and it is unclear whether subsequent uses of this name (e.g., Lecointre and Guyader, 2006) refer to the crown or an apomorphy-based clade. A third candidate name, *Moniliformophyta* (Crane and Kenrick, 1997), was given a "node-based" (minimum-clade) definition using a species of *Equisetum* and a leptosporangiate fern as the specifiers. In the context of those authors' reference phylogeny, our primary reference phylogeny (Knie et al., 2015), and others (e.g., Wickett et al., 2014; Rothfels et al., 2015) in which *Equisetum* is sister to the rest of the clade, *Moniliformophyta* refers to the same crown clade as does *Monilophyta*, but in the context of other molecular phylogenies (e.g., Pryer et al., 2001; Wikström and Pryer, 2005; Schuettpelz et al., 2006; Qiu et al., 2007), in which *Equisetum* is nested deeper within the clade, *Moniliformophyta* applies to a less inclusive crown clade than *Monilophyta*;

i.e., the ophioglossoid ferns and whisk ferns (*Psilotophyta*) are part of *Monilophyta* but lie outside of *Moniliformophyta*.

One might argue that a name based on *Filico-*, *Ptero-*, or *Polypodio-* (see Synonyms) should have been chosen for this clade given that these partial synonyms are widely applied to the plants that are commonly called ferns. However, the name *Monilophyta* has already been phylogenetically defined for the clade of concern here (i.e., including *Equisetum*), unlike any of the alternatives, and it avoids the suggestion that the common ancestor of *Equisetum* and ferns was fernlike in having compound leaves, when it more likely had branch systems with dichotomous ultimate appendages, as in "cladoxylopsids" (which may include stem relatives of the whole clade and some of its subgroups). Names such as *Filicophyta* and *Polypodiophyta* are better reserved for a clade that excludes *Equisetum*, such as "fern clade 3" of Rothwell (1999: Fig. 2), or the clade including all monilophytes except *Equisetum* in Knie et al. (2015) and Rothfels et al. (2015). Indeed, the name *Polypodiophyta* has already been phylogenetically defined to apply in precisely that way (Cantino et al., 2007: E14); the three specifiers were species of *Ophioglossales*, *Marattiales*, and *Leptosporangiatae*, and if the sister-group relationship of *Psilotophyta* and *Ophioglossales* found in most molecular analyses is correct, this clade also includes *Psilotophyta*.

Literature Cited

Bierhorst, D. W. 1971. *Morphology of Vascular Plants*. MacMillan, New York.

Bremer, K., C. J. Humphries, B. D. Mishler, and S. P. Churchill. 1987. On cladistic relationships in green plants. *Taxon* 36:339–349.

Cantino, P. D., J. A. Doyle, S. W. Graham, W. S. Judd, R. G. Olmstead, D. E. Soltis, P. S. Soltis, and M. J. Donoghue. 2007. Towards a phylogenetic nomenclature of *Tracheophyta*. *Taxon* 56:822–846 and E1–E44.

Cobb, B., E. Farnsworth, and C. Lowe. 2005. *A Field Guide to Ferns and Their Related Families.* 2nd edition. Houghton-Mifflin, Boston, MA.

Crane, P. R., and P. Kenrick. 1997. Problems in cladistic classification: higher-level relationships in land plants. *Aliso* 15:87–104.

Doyle, J. A. 2013. Phylogenetic analyses and morphological innovations in land plants. Pp. 1–50 in *The Evolution of Plant Form* (B. A. Ambrose, and M. Purugganan, eds.). Annual Plant Reviews 45. Blackwell, Oxford.

Galtier, J. 2010. The origins and early evolution of the megaphyllous leaf. *Int. J. Plant Sci.* 171:641–661.

Judd, W. S., C. S. Campbell, E. A. Kellogg, P. F. Stevens, and M. J. Donoghue. 2002. *Plant Systematics—A Phylogenetic Approach.* 2nd edition. Sinauer Associates, Sunderland, MA.

Judd, W. S., C. S. Campbell, E. A. Kellogg, P. F. Stevens, and M. J. Donoghue. 2008. *Plant Systematics—A Phylogenetic Approach.* 3rd edition. Sinauer Associates, Sunderland, MA.

Kenrick, P., and P. R. Crane. 1997. *The Origin and Early Diversification of Land Plants—A Cladistic Study.* Smithsonian Institution Press, Washington, DC.

Knie, N., S. Fischer, F. Grewe, M. Polsakiewicz, and V. Knoop. 2015. Horsetails are the sister group to all other monilophytes and *Marattiales* are sister to leptosporangiate ferns. *Mol. Phylogenet. Evol.* 90:140–149.

Lecointre, G., and H. Le Guyader. 2006. *The Tree of Life—A Phylogenetic Classification.* Belknap Press, Cambridge, MA.

Nickrent, D. L., C. L. Parkinson, J. D. Palmer, and R. J. Duff. 2000. Multigene phylogeny of land plants with special reference to bryophytes and the earliest land plants. *Mol. Biol. Evol.* 17:1885–1895.

Pryer, K. M., H. Schneider, A. R. Smith, R. Cranfill, P. G. Wolf, J. S. Hunt, and S. D. Sipes. 2001. Horsetails and ferns are a monophyletic group and the closest living relatives to seed plants. *Nature* 409:618–622.

Pryer, K. M., E. Schuettpelz, P. G. Wolf, H. Schneider, A. R. Smith, and R. Cranfill. 2004. Phylogeny and evolution of ferns (monilophytes) with a focus on the early leptosporangiate divergences. *Am. J. Bot.* 91:1582–1598.

Qiu, Y.-L., L. Li, B. Wang, Z. Chen, O. Dombrovska, J. Lee, L. Kent, R. Li, R. W. Jobson, T. A. Hendry, D. W. Taylor, C. M. Testa, and M. Ambros. 2007. A nonflowering land plant phylogeny inferred from nucleotide sequences of seven chloroplast, mitochondrial, and nuclear genes. *Int. J. Plant Sci.* 168:691–708.

Rothfels, C. J., F.-W. Li, E. M. Sigel, L. Huiet, A. Larsson, D. O. Burge, M. Ruhsam, M. Deyholos, D. E. Soltis, C. N. Stewart, Jr., S. W. Shaw, L. Pokorny, T. Chen, C. dePamphilis, L. DeGironimo, L. Chen, X. Wei, X. Sun, P. Korall, D. W. Stevenson, S. W. Graham, G. K.-S. Wong, and K. M. Pryer. 2015. The evolutionary history of ferns inferred from 25 low-copy nuclear genes. *Am. J. Bot.* 102:1089–1107.

Rothwell, G. W. 1999. Fossils and ferns in the resolution of land plant phylogeny. *Bot. Rev.* 65:188–218.

Rothwell, G. W., and K. C. Nixon. 2006. How does the inclusion of fossil data change our conclusions about the phylogenetic history of euphyllophytes? *Int. J. Plant Sci.* 167:737–749.

Ruhfel, B. R., M. A. Gitzendanner, P. S. Soltis, D. E. Soltis, and J. G. Burleigh. 2014. From algae to angiosperms—inferring the phylogeny of green plants (*Viridiplantae*) from 360 plastid genomes. *BMC Evol. Biol.* 14:23.

Schneider, H., K. M. Pryer, R. Cranfill, A. R. Smith, and P. G. Wolf. 2002. Evolution of vascular plant body plans: a phylogenetic perspective. Pp. 330–364 in *Developmental Genetics and Plant Evolution* (Q. C. B. Cronk, R. M. Bateman, and J. A. Hawkins, eds.). Taylor & Francis, London.

Schneider, H., A. R. Smith, and K. M. Pryer. 2009. Is morphology really at odds with molecules in estimating fern phylogeny? *Syst. Bot.* 34:455–475.

Schuettpelz, E., P. Korall, and K. M. Pryer. 2006. Plastid *atpA* data provide improved support for deep relationships among ferns. *Taxon* 55:897–906.

Simpson, M. G. 2006. *Plant Systematics.* Elsevier, Amsterdam.

Wickett, N. J., S. Mirarab, N. Nguyen, T. Warnow, E. Carpenter, N. Matasci, S. Ayyampalayam, M. S. Barker, J. G. Burleigh, M. A. Gitzendanner, B. R. Ruhfel, E. Wafula, J. P. Der, S. W. Graham, S. Mathews, M. Melkonian, D. E. Soltis, P. S. Soltis, N. W. Miles, C. J. Rothfels, L. Pokorny, A. J. Shaw, L. DeGironimo, D. W. Stevenson, B. Surek, J. C. Villarreal, B. Roure, H. Philippe, C. W. dePamphilis, T. Chen, M. K. Deyholos, R. S. Baucom, T. M. Kutchan, M. M. Augustin, J. Wang, Y. Zhang, Z. Tian, Z. Yan, X. Wu, X. Sun, G. K. Wong, and J. Leebens-Mack. 2014. Phylotranscriptomic analysis of the origin and early diversification of land plants. *Proc. Natl. Acad. Sci. USA* 111:E4859–E4868.

Wikström, N., and K. M. Pryer. 2005. Incongruence between primary sequence data and the distribution of a mitochondrial *atp1* group II intron among ferns and horsetails. *Mol. Phylogenet. Evol.* 36:484–493.

Authors

James A. Doyle; Department of Evolution and Ecology; University of California; Davis, CA 95616, USA. Email: jadoyle@ucdavis.edu.

Philip D. Cantino; Department of Environmental and Plant Biology; Ohio University; Athens, OH 45701, USA. Email: cantino@ohio.edu.

Michael J. Donoghue; Department of Ecology and Evolutionary Biology; Yale University; P.O. Box 208106; New Haven, CT 06520, USA. Email: michael.donoghue@yale.edu.

Date Accepted: 29 September 2011; updated 17 April 2018

Primary Editor: Kevin de Queiroz

Pan-Spermatophyta P. D. Cantino and M. J. Donoghue in P. D. Cantino et al. (2007): 831 [P. D. Cantino, J. A. Doyle, and M. J. Donoghue], converted clade name

Registration Number: 82

Definition: The total clade of the crown clade *Spermatophyta*. This is a crown-based total-clade definition. Abbreviated definition: total ∇ of *Spermatophyta*.

Etymology: From the Greek *pan-* or *pantos* (all, the whole), indicating that the name refers to a total clade, and *Spermatophyta* (see entry in this volume for etymology), the name of the corresponding crown clade.

Reference Phylogeny: The primary reference phylogeny is Kenrick and Crane (1997: Fig. 7.10), where the clade *Pan-Spermatophyta* originates at the base of the branch leading to *Pertica*, *Tetraxylopteris*, and seed plants. For more detailed representation of the composition of *Pan-Spermatophyta* (but no outgroups), see Hilton and Bateman (2006: Fig. 10). For a broader view of outgroup relationships, see Kenrick and Crane (1997: Fig. 4.31), where *Pan-Spermatophyta* is the clade originating at the base of the branch that leads to *Pertica* and *Tetraxylopteris* (no extant pan-spermatophytes are shown).

Composition: *Spermatophyta* (this volume) and all extinct plants that share more recent ancestry with *Spermatophyta* than with any extant plants that do not bear seeds (e.g., *Monilophyta*, *Lycopodiophyta*). According to the reference phylogenies, stem spermatophytes that lacked seeds include *Tetraxylopteris*, *Archaeopteris*, *Cecropsis*, and possibly *Pertica*; for stem spermatophytes that possessed seeds, see *Apo-Spermatophyta* (this volume).

Diagnostic Apomorphies: Synapomorphies of most *Pan-Spermatophyta*, including

"progymnosperms" (a paraphyletic group) such as *Aneurophyton*, *Tetraxylopteris*, *Archaeopteris*, and *Cecropis* (Beck, 1960), include a bifacial vascular cambium that produces both secondary xylem and secondary phloem and probably a cork cambium that produces periderm (Scheckler and Banks, 1971). Kenrick and Crane (1997) listed three synapomorphies supporting inclusion of *Pertica* in this clade, but these are problematical. Synapomorphies proposed in their Table 4.6 were tetrastichous branching and presence of xylem rays (indicative of secondary xylem). However, tetrastichous branching is questionable because it occurs in *Tetraxylopteris* but not in other "progymnosperms," and branching in *Pertica* is highly variable (Hotton et al., 2001). Xylem rays and other anatomical characters are actually unknown in *Pertica*, because its stems are not preserved anatomically, and Kenrick and Crane (1997) correctly scored these characters as unknown; their identification of rays as a synapomorphy appears to be a result of using acctran character optimization. However, secondary xylem with apparent rays has been described in a possibly related fossil, *Armoricaphyton* (Gerrienne and Gensel, 2016). In Table 7.2, Kenrick and Crane (1997) indicated that the clade including *Pertica*, *Tetraxylopteris*, and seed plants, designated *Radiatopses*, is united by tetrastichous branching and "a distinctive form of protoxylem ontogeny with multiple strands occurring along the midplanes of the primary xylem ribs" ("radiate protoxylem" of Stein, 1993), but again anatomical characters are unknown in *Pertica*, and *Armoricaphyton* has a presumably more plesiomorphic *Psilophyton*-like centrarch stele with unlobed xylem (Gerrienne and Gensel, 2016). In any case, it is equivocal whether radiate protoxylem is ancestral or derived within the monilophyte-spermatophyte clade relative to the "permanent protoxylem" of early monilophytes.

Synonyms: The name *Radiatopses* (Kenrick and Crane, 1997: Tables 7.1, 7.2) is an approximate synonym. It has a "synapomorphy-based definition," but its presumed composition may be identical to that of *Pan-Spermatophyta* if the latter includes *Pertica*. The name *Progymnospermopsida*, proposed for relatives of seed plants that have secondary xylem and phloem but lack seeds (Beck, 1960), may be a partial and approximate synonym in that it refers to a paraphyletic group that originated in approximately the same ancestor as *Pan-Spermatophyta* (see Comments).

Comments: Uncertainty about the status of *Progymnospermopsida* and the corresponding informal name "progymnosperms" as (partial) synonyms of *Pan-Spermatophyta* is related to the distinction between *Pan-Spermatophyta* and *Lignophyta*, an apomorphy-based clade (Donoghue and Doyle in Cantino et al., 2007) characterized by a bifacial vascular cambium that produces secondary xylem (wood) and phloem. If a close relationship to seed plants is more critical to the concept of "progymnosperms" than is the possession of secondary xylem and phloem, then the name "progymnosperms" is appropriately interpreted as a partial and approximate synonym of *Pan-Spermatophyta*. However, if the possession of secondary xylem and phloem is more critical to the concept, then the name is more appropriately interpreted as a partial and approximate synonym of *Lignophyta*.

Literature Cited

Beck, C. B. 1960. The identity of *Archaeopteris* and *Callixylon*. *Brittonia* 12:351–368.

Cantino, P. D., J. A. Doyle, S. W. Graham, W. S. Judd, R. G. Olmstead, D. E. Soltis, P. S. Soltis, and M. J. Donoghue. 2007. Towards a phylogenetic nomenclature of *Tracheophyta*. *Taxon* 56:822–846 and E1–E44.

Gerrienne, P., and P. G. Gensel. 2016. New data about anatomy, branching, and inferred growth patterns in the Early Devonian plant *Armoricaphyton chateaupannense*, Montjean-sur-Loire, France. *Rev. Palaeobot. Palynol.* 224:38–53.

Hilton, J., and R. M. Bateman. 2006. Pteridosperms are the backbone of seed plant evolution. *J. Torrey Bot. Soc.* 133:119–168.

Hotton, C. L., F. M. Hueber, D. H. Griffing, and J. S. Bridge. 2001. Early terrestrial plant environments: An example from the Emsian of Gaspé, Canada. Pp. 179–212 in *Plants Invade the Land: Evolutionary and Environmental Perspectives* (P. G. Gensel and D. Edwards, eds.). Columbia University Press, New York.

Kenrick, P., and P. R. Crane. 1997. *The Origin and Early Diversification of Land Plants—A Cladistic Study*. Smithsonian Institution Press, Washington, DC.

Scheckler, S. E., and H. P. Banks. 1971. Anatomy and relationships of some Devonian progymnosperms from New York. *Am. J. Bot.* 58:737–751.

Stein, W. 1993. Modeling the evolution of stelar architecture in vascular plants. *Int. J. Plant Sci.* 154:229–263.

Authors

Philip D. Cantino; Department of Environmental and Plant Biology; Ohio University; Athens, OH 45701, USA. Email: cantino@ohio.edu.

James A. Doyle; Department of Evolution and Ecology; University of California; Davis, CA 95616, USA. Email: jadoyle@ucdavis.edu.

Michael J. Donoghue; Department of Ecology and Evolutionary Biology; Yale University; P.O. Box 208106; New Haven, CT 06520, USA. Email: michael.donoghue@yale.edu.

Date Accepted: 31 January 2013; updated 09 February 2018

Primary Editor: Kevin de Queiroz

Apo-Spermatophyta P. D. Cantino and M. J. Donoghue in P. D. Cantino et al. (2007): 831 [P. D. Cantino, J. A. Doyle, and M. J. Donoghue], converted clade name

Registration Number: 14

Definition: The clade characterized by seeds as inherited by *Magnolia tripetala* (Linnaeus) Linnaeus 1759 (*Angiospermae*), *Podocarpus macrophyllus* (Thunberg) Sweet 1818 (*Coniferae*), *Ginkgo biloba* Linnaeus 1771, *Cycas revoluta* Thunberg 1782 (*Cycadophyta*), and *Gnetum gnemon* Linnaeus 1767 (*Gnetophyta*). This is an apomorphy-based definition. A seed is a fertilized ovule, the ovule being an indehiscent megasporangium surrounded by an integument (represented by unfused or partially fused integumentary lobes in the earliest members). Presence of integument(s) (fused or unfused) and megasporangium indehiscence are fully correlated in all known seed plants, with the exception of some parasitic angiosperms (e.g., derived *Santalales*; Brown et al., 2010) in which the integuments have been lost. If only one of the two features is present, the presence of an integument rather than indehiscence will determine whether a structure is an ovule according to the definition used here. Abbreviated definition: ∇ apo seeds [*Magnolia tripetala* (Linnaeus) Linnaeus 1759 & *Podocarpus macrophyllus* (Thunberg) Sweet 1818 & *Ginkgo biloba* Linnaeus 1771 & *Cycas revoluta* Thunberg 1782 & *Gnetum gnemon* Linnaeus 1767].

Etymology: From the Greek *apo-*, indicating that the name refers to an apomorphy-based clade, and *Spermatophyta* (see entry in this volume for etymology).

Reference Phylogeny: Hilton and Bateman (2006: Fig. 10), where *Apo-Spermatophyta* includes the entire illustrated tree except the "progymnosperms." See also Rothwell and Serbet (1994: Fig. 3).

Composition: The crown clade *Spermatophyta* (this volume) and all extinct seed-bearing plants that lie outside the crown (e.g., Palaeozoic seed ferns such as *Elkinsia*, *Lyginopteris*, and *Medullosa*; for additional members, see Rothwell and Serbet, 1994; Hilton and Bateman, 2006).

Diagnostic Apomorphies: Ovules and seeds (see Definition and Comments for more detailed descriptions). Some associated apomorphies are listed under Comments.

Synonyms: The name *Spermatophytata* Kenrick and Crane (1997: Table 7.2) is an unambiguous synonym. It is defined based on two apomorphies—presence of an integument and only one functional megaspore—both of which characterize the ovule and the former of which we have used as the defining feature of an ovule for the purpose of our definition (see Comments). The name *Spermatophyta* as used by many earlier authors is also implicitly apomorphy-based, but we prefer to use it for the crown clade (see *Spermatophyta*, this volume). The name *Gymnospermae*, as used for living and fossil seed plants other than angiosperms, is a partial synonym; gymnosperms originated from the same immediate ancestor as *Apo-Spermatophyta* but are paraphyletic with respect to angiosperms. The informal names "pteridosperms" and "seed ferns" and corresponding formal names (e.g., *Cycadofilicales*, *Pteridospermales*, *Pteridospermopsida*) refer to a multiply paraphyletic group of extinct seed plants that originated in the same ancestor as *Apo-Spermatophyta* but includes only a small fraction of its total diversity. Seed ferns, so named because they retained the fernlike foliage that appears to have been ancestral for *Apo-Spermatophyta*, include both stem relatives and plesiomorphic members of the crown clade *Spermatophyta*.

Comments: Some earlier workers (e.g., Arnold, 1948; Beck, 1966) suggested that seeds evolved more than once (ignoring the seedlike structures of some fossil lycophytes), specifically in "coniferophytes" (including cordaites, ginkgophytes, and conifers) and "cycadophytes" (including cycads, "seed ferns," and bennettites). This hypothesis has not been supported by modern phylogenetic analyses, which nest "coniferophytes" among "seed ferns" (Crane, 1985; Doyle and Donoghue, 1986; Rothwell and Serbet, 1994; Doyle, 2006; Hilton and Bateman, 2006). However, our definition of *Apo-Spermatophyta* is constructed such that the name would not apply to any clade in the context of a phylogeny in which the seeds of the five extant subgroups are not homologous.

The definition of "ovule" adopted here includes what some authors (e.g., Stewart and Rothwell, 1993) have referred to as preovules, with a ring of more or less free integumentary lobes surrounding the megasporangium, so that a micropyle is lacking, as in *Elkinsia* and *Genomosperma*. Such fossils have been described as seed plants. Various features are closely associated in the reproductive biology of seed plants (Stewart and Rothwell, 1993): e.g., an indehiscent megasporangium, an integument, pollination, and one functional megaspore (with derived exceptions in *Angiospermae* and *Gnetophyta*; Gifford and Foster, 1989). However, in order to apply the name *Apo-Spermatophyta* unambiguously, it is best for the definition to focus on one feature to determine whether a particular structure is an ovule (and thus whether the plant that bears it is a member of *Apo-Spermatophyta*). Cantino et al. (2007) chose indehiscence of the megasporangium, on the grounds that it is "fundamental to the reproductive biology of seed plants," but we prefer to use the presence of an integument because it is more consistent with the common definition of an ovule as an integumented megasporangium (e.g., Gifford and Foster, 1989; Stewart and Rothwell, 1993) and more readily applied in practice. Presence of an integument has been widely used to classify fossilized structures as ovules when the megasporangium character is unknown (Stewart and Rothwell, 1993). The two features are associated in all known fossils in which the characters are preserved, but if a fossil were to be found with an indehiscent megasporangium and no integument it would not be a member of *Apo-Spermatophyta*.

Literature Cited

Arnold, C. A. 1948. Classification of gymnosperms from the viewpoint of paleobotany. *Bot. Gaz.* 110:2–12.

Beck, C. B. 1966. On the origin of gymnosperms. *Taxon* 15:337–339.

Brown, R. H., D. L. Nickrent, and C. S. Gasser. 2010. Expression of ovule and integument-associated genes in reduced ovules of *Santalales*. *Evol. Dev.* 12:231–240.

Cantino, P. D., J. A. Doyle, S. W. Graham, W. S. Judd, R. G. Olmstead, D. E. Soltis, P. S. Soltis, and M. J. Donoghue. 2007. Towards a phylogenetic nomenclature of *Tracheophyta*. *Taxon* 56:822–846 and E1–E44.

Crane, P. R. 1985. Phylogenetic relationships in seed plants. *Cladistics* 1:329–348.

Doyle, J. A. 2006. Seed ferns and the origin of angiosperms. *J. Torrey Bot. Soc.* 133:169–209.

Doyle, J. A., and M. J. Donoghue. 1986. Seed plant phylogeny and the origin of angiosperms: an experimental cladistic approach. *Bot. Rev.* 52:321–431.

Gifford, E. M., and A. S. Foster. 1989. *Morphology and Evolution of Vascular Plants*. 3rd edition. W. H. Freeman, New York.

Hilton, J., and R. M. Bateman. 2006. Pteridosperms are the backbone of seed plant evolution. *J. Torrey Bot. Soc.* 133:119–168.

Kenrick, P., and P. R. Crane. 1997. *The Origin and Early Diversification of Land Plants—A Cladistic Study Smithsonian* . Institution Press, Washington, DC.

Rothwell, G. W., and R. Serbet. 1994. Lignophyte phylogeny and the evolution of spermatophytes: A numerical cladistic analysis. *Syst. Bot.* 19:443–482.

Stewart, W. N., and G. W. Rothwell. 1993. *Paleobotany and the Evolution of Plants*. 2nd edition. Cambridge University Press, Cambridge, UK.

Authors

Philip D. Cantino; Department of Environmental and Plant Biology; Ohio University; Athens, OH 45701, USA. Email: cantino@ohio.edu.

James A. Doyle; Department of Evolution and Ecology; University of California; Davis, CA 95616, USA. Email: jadoyle@ucdavis.edu.

Michael J. Donoghue; Department of Ecology and Evolutionary Biology; Yale University; P.O. Box 208106; New Haven, CT 06520, USA. Email: michael.donoghue@yale.edu.

Date Accepted: 31 January 2013; updated 09 February 2018

Primary Editor: Kevin de Queiroz

Spermatophyta N. L. Britton and A. Brown 1896: 49
[P. D. Cantino, J. A. Doyle, and M. J. Donoghue],
converted clade name

Registration Number: 100

Definition: The smallest crown clade containing *Magnolia tripetala* (Linnaeus) Linnaeus 1759 (*Angiospermae*), *Podocarpus macrophyllus* (Thunberg) Sweet 1818 (*Coniferae/Cupressophyta*), *Ginkgo biloba* Linnaeus 1771, *Cycas revoluta* Thunberg 1782 (*Cycadophyta*), and *Gnetum gnemon* Linnaeus 1767 (*Gnetophyta*). This is a minimum-crown-clade definition. Abbreviated definition: min crown ∇ (*Magnolia tripetala* (Linnaeus) Linnaeus 1759 & *Podocarpus macrophyllus* (Thunberg) Sweet 1818 & *Ginkgo biloba* Linnaeus 1771 & *Cycas revoluta* Thunberg 1782 & *Gnetum gnemon* Linnaeus 1767).

Etymology: Derived from Greek, *sperma* (seed) and *phyton* (plant) (Stearn, 1973).

Reference Phylogeny: The primary reference phylogeny is Qiu et al. (2007: Fig. 1). See also Burleigh and Matthews (2004: Fig. 5), Xi et al. (2013: Fig. 2), and Ruhfel et al. (2014: Fig. 5).

Composition: All extant seed plants and any fossils that are part of the crown, including *Pan-Angiospermae*, *Pan-Coniferae*, and the total clades (not named in this volume) of *Ginkgo*, *Cycadophyta*, and *Gnetophyta*. In the reference phylogeny and some other published trees, the total clade of *Gnetophyta* is contained within *Coniferae*, but there are other phylogenies in which these two clades are non-overlapping (see *Coniferae*, this volume).

Diagnostic Apomorphies: Possible apomorphies of this crown clade are endarch primary xylem in the stem (assuming that the mesarch primary xylem in *Callistophyton* is a reversal, if this taxon is nested in the crown clade; Doyle, 2006), distal aperture position in the pollen (assuming that absence of a distal aperture in *Cordaitales* and *Emporia* represents a reversal to the state in early fossil seed plants (*Apo-Spermatophyta*), as implied by trees of Rothwell and Serbet, 1994 and Doyle, 2006), a linear tetrad of megaspores (Doyle and Donoghue, 1986; Doyle, 2006), honeycomb-alveolar infratectal structure of the pollen exine, and platyspermic ovules (Doyle, 2006). Because ovules of *Cycas* are platyspermic (biradial) but ovules of other cycads are radiospermic, the level at which platyspermy is synapomorphic is uncertain if the clade *Cycadophyta* is the extant sister group of the rest of *Spermatophyta* (e.g., Doyle, 2006: Fig. 6). Furthermore, Rothwell and Serbet (1994) questioned the distinction between radiospermic and platyspermic and divided ovule symmetry into four states.

The following are apomorphies of *Spermatophyta*, as defined here, relative to other crown clades (lycophytes and monilophytes) but are apomorphic at a more inclusive level when fossils are considered (not an exhaustive list): heterospory (which evolved independently in some monilophytes and lycophytes), ovule (i.e., an integumented, indehiscent megasporangium that develops after fertilization into a seed; Stewart, 1983), axillary branching (reversed in living cycads), eustele, cataphylls, and a free-nuclear stage followed by alveolar cellularization in embryogeny (Doyle and Donoghue, 1986; Rothwell and Serbet, 1994; Doyle, 1998, 2006).

Synonyms: The name *Spermatophytatinae* sensu Jeffrey (1982) is an approximate synonym in that all listed subordinate taxa are extant. The "platyspermic clade" of Doyle and Donoghue (1986: 354) is an approximate synonym based on composition (independent of whether platyspermic ovules are a synapomorphy), but the name "platysperms" of Crane (1985) is not (because that taxon excludes *Cycadophyta*).

Comments: Although the clade *Spermatophyta* has very strong support from both morphological and molecular analyses, relationships among its principal extant subgroups (*Angiospermae*, *Ginkgo*, *Cycadophyta*, *Coniferae*, and *Gnetophyta*) remain unresolved in spite of intensive study (Doyle and Donoghue, 1986; Rothwell and Serbet, 1994; Bowe et al., 2000; Chaw et al., 2000; Magallón and Sanderson, 2002; Rydin et al., 2002; Soltis et al., 2002; Burleigh and Mathews, 2004; Doyle, 2006, 2008; Hilton and Bateman, 2006; Qiu et al., 2007; Mathews, 2009; Mathews et al., 2010; Lee et al., 2011; Xi et al, 2013; Ruhfel et al., 2014). Consequently, all five of these subgroups are represented by internal specifiers in our definition. Although there is disagreement as to whether the clade *Pinaceae* is more closely related to *Gnetophyta* or to *Cupressophyta*, both are represented in the reference phylogeny, so there is no need to include a specifier to represent *Pinaceae*.

Here and previously (Cantino et al., 2007) we have defined the name *Spermatophyta* to refer to the seed plant crown clade, following *PhyloCode* Recommendation 10.1B. This application is somewhat unconventional, since this name has more often been applied to the apomorphy-based clade originating with the origin of the seed. However, it has been implicitly applied to the crown in molecular studies. For example, Jager et al. (2003: 843) discussed the need for data from cycads and *Ginkgo* "to infer the MADS-box gene content of the last common ancestor of *Spermatophyta*." Since there is currently no way to study the MADS-box genes of extinct plants, their statement implies application of the name *Spermatophyta* to the crown (see *PhyloCode* Note 10.1B.1). We are aware of only one other scientific name having been applied to the crown (see Synonymy), and only one use of it.

Literature Cited

Bowe, L. M., G. Coat, and C. W. dePamphilis. 2000. Phylogeny of seed plants based on all three genomic compartments: Extant gymnosperms are monophyletic and *Gnetales'* closest relatives are conifers. *Proc. Natl. Acad. Sci. USA* 97:4092–4097.

Britton, N. L., and A. Brown. 1896. *An Illustrated Flora of the Northern United States, Canada and the British Possessions*. Scribner's Sons, New York.

Burleigh, J. G., and S. Mathews. 2004. Phylogenetic signal in nucleotide data from seed plants: Implications for resolving the seed plant tree of life. *Am. J. Bot.* 91:1599–1613.

Cantino, P. D., J. A. Doyle, S. W. Graham, W. S. Judd, R. G. Olmstead, D. E. Soltis, P. S. Soltis, and M. J. Donoghue. 2007. Towards a phylogenetic nomenclature of *Tracheophyta*. *Taxon* 56:822–846 and E1–E44.

Chaw, S.-M., C. L. Parkinson, Y. Cheng, T. M. Vincent, and J. D. Palmer. 2000. Seed plant phylogeny inferred from all three plant genomes: Monophyly of extant gymnosperms and origin of *Gnetales* from conifers. *Proc. Natl. Acad. Sci. USA* 97:4086–4091.

Crane, P. R. 1985. Phylogenetic relationships in seed plants. *Cladistics* 1:329–348.

Doyle, J. A. 1998. Phylogeny of vascular plants. *Annu. Rev. Ecol. Syst.* 29:567–599.

Doyle, J. A. 2006. Seed ferns and the origin of angiosperms. *J. Torrey Bot. Soc.* 133:169–209.

Doyle, J. A. 2008. Integrating molecular phylogenetic and paleobotanical evidence on origin of the flower. *Int. J. Plant Sci.* 169:816–843.

Doyle, J. A., and M. J. Donoghue. 1986. Seed plant phylogeny and the origin of angiosperms: an experimental cladistic approach. *Bot. Rev.* 52:321–431.

Hilton, J., and R. M. Bateman. 2006. Pteridosperms are the backbone of seed plant evolution. *J. Torrey Bot. Soc.* 133:119–168.

Jager, M., A. Hassanin, M. Manuel, H. Le Guyader, and J. Deutsch. 2003. MADS-box genes in *Ginkgo biloba* and the evolution of the AGAMOUS family. *Mol. Biol. Evol.* 20:842–854.

Jeffrey, C. 1982. Kingdoms, codes and classification. *Kew Bull.* 37:403–416.

Lee, E. K., A. Cibrian-Jaramillo, S. Kolokotronis, M. S. Katari, A. Stamatakis, M. Ott, J. C. Chiu, D. P. Little, D. W. Stevenson, W. R. McCombie, R. A. Martienssen, G. Coruzzi, and R. DeSalle. 2011. A functional phylogenomic view of the seed plants. *PLOS Genet.* 7:e1002411.

Magallón, S., and M. J. Sanderson. 2002. Relationships among seed plants inferred from highly conserved genes: Sorting conflicting phylogenetic signals among ancient lineages. *Am. J. Bot.* 89:1991–2006.

Mathews, S. 2009. Phylogenetic relationships among seed plants: Persistent questions and the limits of molecular data. *Am. J. Bot.* 96:228–236.

Mathews, S., M. D. Clements, and M. A. Beilstein. 2010. A duplicate gene rooting of seed plants and the phylogenetic position of flowering plants. *Philos. Trans. R. Soc. Lond. B Biol. Sci.* 365:383–395.

Qiu, Y.-L., L. Li, B. Wang, Z. Chen, O. Dombrovska, J. Lee, L. Kent, R. Li, R. W. Jobson, T. A. Hendry, D. W. Taylor, C. M. Testa, and M. Ambros. 2007. A nonflowering land plant phylogeny inferred from nucleotide sequences of seven chloroplast, mitochondrial, and nuclear genes. *Int. J. Plant Sci.* 168:691–708.

Rothwell, G. W., and R. Serbet. 1994. Lignophyte phylogeny and the evolution of spermatophytes: A numerical cladistic analysis. *Syst. Bot.* 19:443–482.

Ruhfel, B. R., M. A. Gitzendanner, P. S. Soltis, D. E. Soltis, and J. G. Burleigh. 2014. From algae to angiosperms—inferring the phylogeny of green plants (*Viridiplantae*) from 360 plastid genomes. *BMC Evol. Biol.* 14:23.

Rydin, C., M. Källersjö, and E. M. Friis. 2002. Seed plant relationships and the systematic position of *Gnetales* based on nuclear and chloroplast DNA: conflicting data, rooting problems, and the monophyly of conifers. *Int. J. Plant Sci.* 163:197–214.

Soltis, D. E., P. S. Soltis, and M. J. Zanis. 2002. Phylogeny of seed plants based on evidence from eight genes. *Am. J. Bot.* 89:1670–1681.

Stearn, W. T. 1973. *Botanical Latin.* David & Charles, Newton Abbot, Devon.

Stewart, W. N. 1983. *Paleobotany and the Evolution of Plants.* Cambridge University Press, Cambridge, UK.

Xi, Z, J. S. Rest, and C. C. Davis. 2013. Phylogenomics and coalescent analyses resolve extant seed plant relationships. *PLOS ONE* 8:e80870.

Authors

Philip D. Cantino; Department of Environmental and Plant Biology; Ohio University; Athens, OH 45701, USA. Email: cantino@ohio.edu.

James A. Doyle; Department of Evolution and Ecology; University of California; Davis, CA 95616, USA. Email: jadoyle@ucdavis.edu.

Michael J. Donoghue; Department of Ecology and Evolutionary Biology; Yale University; P.O. Box 208106; New Haven, CT 06520, USA. Email: michael.donoghue@yale.edu.

Date Accepted: 31 January 2013; updated 09 February 2018

Primary Editor: Kevin de Queiroz

Pan-Gnetophyta J. A. Doyle, M. J. Donoghue, and P. D. Cantino in Cantino et al. (2007): E22 [J. A. Doyle, M. J. Donoghue, and P. D. Cantino], converted clade name

Registration Number: 80

Definition: The total clade of the crown clade *Gnetophyta*. This is a crown-based total-clade definition. Abbreviated definition: total ∇ of *Gnetophyta*.

Etymology: Derived from the Greek *pan*, meaning "all," in reference to the total clade, and *Gnetophyta*, the name of the corresponding crown clade (see *Gnetophyta* in this volume for the etymology of that name).

Reference Phylogeny: The primary reference phylogeny is Doyle (1996: Fig. 5). See also Rothwell and Serbet (1994: Fig. 8) and Doyle (2008: Fig. 3).

Composition: The crown clade *Gnetophyta* (this volume) and all extinct plants that share more recent ancestry with *Gnetophyta* than with any other extant seed plants. In Doyle's (1996) analysis, this total clade includes *Piroconites* and, on some (but not other) trees, *Bennettitales*, *Pentoxylon*, and *Glossopteridales*. However, the striate pollen character that suggested *Piroconites* was related to *Gnetophyta* was based on a misinterpretation (Osborn, 2000). *Gnetophyta* are nested within *Coniferae* in some trees in the morphological analysis of Doyle (2008), as found in many molecular analyses (e.g., Bowe et al., 2000; Chaw et al., 2000); in these trees, *Bennettitales*, *Pentoxylon*, and *Glossopteridales* are linked more closely with *Angiospermae* than with *Gnetophyta*, and no other fossil taxa are related to *Gnetophyta* (*Piroconites* was not included). A better candidate for a stem taxon is *Dechellyia* (Ash, 1972; Crane, 1996), represented by shoots bearing opposite, linear leaves and strobili containing striate pollen. An analysis by Friis et al. (2007) identified *Bennettitales*, *Erdtmanithecales*, and Cretaceous charcoalified seeds (subsequently described by Friis et al., 2009) as stem relatives of *Gnetophyta*; the charcoalified seeds were also associated with *Gnetophyta* by Rothwell et al. (2009). Rothwell and Stockey (2013) interpreted the female strobilus *Protoephedrites*, which differs from extant *Gnetophyta* in having two ovules rather than one per fertile short shoot, as a probable stem gnetophyte.

Diagnostic Apomorphies: The treatment of *Gnetophyta* in this volume lists many synapomorphies relative to other crown clades. It is not known where on the gnetophyte stem these synapomorphies evolved. Striate (polyplicate) pollen similar to that of *Ephedra* and *Welwitschia* occurs in the earliest fossils that have been interpreted as possible stem gnetophytes (Crane, 1988, 1996), so this character may have arisen near the base of the total clade. *Dechellyia* (Late Triassic), one of the earliest macrofossils that is associated with striate pollen, has opposite phyllotaxy and possibly terminal ovules (Ash, 1972; Crane, 1996), suggesting that these apomorphies may also have arisen near the base of *Pan-Gnetophyta*, but because the ovules are usually borne in pairs it is also possible that the structure bearing them is homologous with a single short shoot with two lateral ovules in *Protoephedrites* (Rothwell and Stockey, 2013). The ovule of *Protoephedrites* has only a very short micropylar tube, which could mean that the long micropylar tube of extant *Gnetophyta* originated late on the stem lineage.

Synonyms: None that unambiguously apply to the total clade.

Literature Cited

Ash, S. R. 1972. Late Triassic plants from the Chinle Formation in north-eastern Arizona. *Palaeontology* 15:598–618.

Bowe, L. M., G. Coat, and C. W. dePamphilis. 2000. Phylogeny of seed plants based on all three genomic compartments: extant gymnosperms are monophyletic and *Gnetales'* closest relatives are conifers. *Proc. Natl. Acad. Sci. USA* 97:4092–4097.

Cantino, P. D., J. A. Doyle, S. W. Graham, W. S. Judd, R. G. Olmstead, D. E. Soltis, P. S. Soltis, and M. J. Donoghue. 2007. Towards a phylogenetic nomenclature of *Tracheophyta*. *Taxon* 56:822–846 and E1–E44.

Chaw, S. -M., C. L. Parkinson, Y. Cheng, T. M. Vincent, and J. D. Palmer. 2000. Seed plant phylogeny inferred from all three plant genomes: monophyly of extant gymnosperms and origin of *Gnetales* from conifers. *Proc. Natl. Acad. Sci. USA* 97:4086–4091.

Crane, P. R. 1988. Major clades and relationships in the "higher" gymnosperms. Pp. 218–272 in *Origin and Evolution of Gymnosperms* (C. B. Beck, ed.). Columbia University Press, New York.

Crane, P. R. 1996. The fossil history of the *Gnetales*. *Int. J. Plant Sci.* 157:S50–S57.

Doyle, J. A. 1996. Seed plant phylogeny and the relationships of *Gnetales*. *Int. J. Plant Sci.* 157 (suppl.):S3–S39.

Doyle, J. A. 2008. Integrating molecular phylogenetic and paleobotanical evidence on origin of the flower. *Int. J. Plant Sci.* 169:816–843.

Friis, E. M., P. R. Crane, K. R. Pedersen, S. Bengtson, P. C. J. Donoghue, G. W. Grimm, and M. Stampanoni. 2007. Phase-contrast X-ray microtomography links Cretaceous seeds with *Gnetales* and *Bennettitales*. *Nature* 450:549–552.

Friis, E. M., K. R. Pedersen, and P. R. Crane. 2009. Early Cretaceous mesofossils from Portugal and eastern North America related to the *Bennettitales-Erdtmanithecales-Gnetales* group. *Am. J. Bot.* 96:252–283.

Osborn, J. M. 2000. Pollen morphology and ultrastructure of gymnospermous anthophytes. Pp. 163–185 in *Pollen and Spores: Morphology and Biology* (M. M. Harley, C. M. Morton and S. Blackmore, eds.). Royal Botanic Gardens, Kew.

Rothwell, G. W., W. L. Crepet, and R. A. Stockey. 2009. Is the anthophyte hypothesis alive and well? New evidence from the reproductive structures of *Bennettitales*. *Am. J. Bot.* 96:296–322.

Rothwell, G. W., and R. Serbet. 1994. Lignophyte phylogeny and the evolution of spermatophytes: a numerical cladistic analysis. *Syst. Bot.* 19:443–482.

Rothwell, G. W., and R. A. Stockey. 2013. Evolution and phylogeny of gnetophytes: evidence from the anatomically preserved seed cone *Protoephedrites eamesii* gen. et sp. nov. and the seeds of several bennettitalean species. *Int. J. Plant Sci.* 174:511–529.

Authors

James A. Doyle; Department of Evolution and Ecology; University of California; Davis, CA 95616, USA. Email: jadoyle@ucdavis.edu.

Michael J. Donoghue; Department of Ecology and Evolutionary Biology; Yale University; P.O. Box 208106; New Haven, CT 06520, USA. Email: michael.donoghue@yale.edu.

Philip D. Cantino; Department of Environmental and Plant Biology; Ohio University; Athens, OH 45701, USA. Email: cantino@ohio.edu.

Date Accepted: 29 April 2011; updated 09 February 2018

Primary Editor: Kevin de Queiroz

Gnetophyta C. E. Bessey 1907: 323 (as "*Gnetales*"; Hoogland and Reveal, 2005) [M. J. Donoghue, J. A. Doyle, and P. D. Cantino], converted clade name

Registration Number: 47

Definition: The smallest crown clade containing *Gnetum gnemon* Linnaeus 1767, *Ephedra distachya* Linnaeus 1753, and *Welwitschia mirabilis* J. D. Hooker 1862. This is a minimum-crown-clade definition. Abbreviated definition: min crown ∇ (*Gnetum gnemon* Linnaeus 1767 & *Ephedra distachya* Linnaeus 1753 & *Welwitschia mirabilis* J. D. Hooker 1862).

Etymology: Derived from *Gnetum* (the name of an included genus) and the Greek *phyton*, meaning "plant." The name *Gnetum* is apparently derived from the same word as the specific epithet *gnemon* (i.e., *genemo*, the vernacular name for *Gnetum gnemon* in the Molucca Islands; Markgraf, 1951).

Reference Phylogeny: The primary reference phylogeny is Rydin et al. (2002: Fig. 1). See also Hou et al. (2015: Fig. 2).

Composition: *Ephedra* (40 species), *Gnetum* (30 species), and *Welwitschia* (1 species) (species numbers from Mabberley, 2008). For descriptions of these three taxa, see Kubitzki (1990).

Diagnostic Apomorphies: Synapomorphies relative to other crown clades include multiple axillary buds, opposite phyllotaxy, terminal ovules with one or two outer envelopes derived from opposite primordia, basally fused microsporophylls with terminal microsporangia, vessels in xylem (assuming these are not homologous with angiosperm vessels, as inferred even in analyses in which angiosperms and gnetophytes are extant sister groups: Doyle and Donoghue, 1986; Doyle, 1996), compound microsporangiate strobili, striate pollen (modified to echinate in *Gnetum*; Yao et al., 2004), micropylar tube, and apical meristem with one tunica layer (Doyle and Donoghue, 1986, 1992; Crane, 1988; Rothwell and Serbet, 1994; Doyle, 2006).

Synonyms: *Gnetopsida* and *Gnetales* (see Comments); also *Gnetidae* (Chase and Reveal, 2009) and *Gnetatae* (Kubitzki, 1990). All four are approximate synonyms with the same composition.

Comments: The monophyly of *Gnetophyta* is very strongly supported by both molecular and morphological analyses (Crane, 1985, 1988; Doyle and Donoghue, 1986, 1992; Doyle et al., 1994; Rothwell and Serbet, 1994; Doyle, 1996, 2006; Rydin et al., 2002; Soltis et al., 2002; Burleigh and Mathews, 2004; Hilton and Bateman, 2006; Zhong, 2010, 2011; Lu et al., 2014; Wickett et al., 2014). The names *Gnetophyta*, *Gnetopsida*, and *Gnetales* are widely applied to this clade (*Gnetidae* and *Gnetatae* much less frequently). Here and previously (Cantino et al., 2007), we have opted for the *-phyta* ending even though the name *Gnetales* is more frequently used, because, although *-phyta* is used to designate the rank of phylum (division) in botanical rank-based nomenclature, it also means "plants," whereas *-opsida* and *-ales* have no meaning other than indicating rank.

Literature Cited

Bessey, C. E. 1907. A synopsis of plant phyla. *Univ. Stud.,* University of Nebraska, 7:275–373.

Burleigh, J. G., and S. Mathews. 2004. Phylogenetic signal in nucleotide data from seed plants: implications for resolving the seed plant tree of life. *Am. J. Bot.* 91:1599–1613.

Cantino, P. D., J. A. Doyle, S. W. Graham, W. S. Judd, R. G. Olmstead, D. E. Soltis, P. S. Soltis, and M. J. Donoghue. 2007. Towards a phylogenetic nomenclature of *Tracheophyta*. *Taxon* 56:822–846 and E1–E44.

Chase, M. W., and J. L. Reveal. 2009. A phylogenetic classification of the land plants to accompany APG III. *Bot. J. Linn. Soc.* 161:122–127.

Crane, P. R. 1985. Phylogenetic analysis of seed plants and the origin of angiosperms. *Ann. Mo. Bot. Gard.* 72:716–793.

Crane, P. R. 1988. Major clades and relationships in the "higher" gymnosperms. Pp. 218–272 in *Origin and Evolution of Gymnosperms* (C. B. Beck, ed.). Columbia University Press, New York.

Doyle, J. A. 1996. Seed plant phylogeny and the relationships of *Gnetales*. *Int. J. Plant Sci.* 157 (suppl.):S3–S39.

Doyle, J. A. 2006. Seed ferns and the origin of angiosperms. *J. Torrey Bot. Soc.* 133:169–209.

Doyle, J. A., and M. J. Donoghue. 1986. Seed plant phylogeny and the origin of angiosperms: an experimental cladistic approach. *Bot. Rev.* 52:321–431.

Doyle, J. A., and M. J. Donoghue, 1992. Fossils and seed plant phylogeny reanalyzed. *Brittonia* 44:89–106.

Doyle, J. A., M. J. Donoghue, and E. A. Zimmer. 1994. Integration of morphological and ribosomal RNA data on the origin of angiosperms. *Ann. Mo. Bot. Gard.* 81:419–450.

Hilton, J., and R. M. Bateman. 2006. Pteridosperms are the backbone of seed-plant phylogeny. *J. Torrey Bot. Soc.* 133:119–168.

Hou, C., A. M. Humphreys, O. Thureborn, and C. Rydin. 2015. New insights into the evolutionary history of *Gnetum* (*Gnetales*). *Taxon* 64:239–252.

Kubitzki, K. 1990. Gnetatae. Pp. 378–391 in *Families and Genera of Vascular Plants*, Vol. 1 (K. U. Kramer and P. S. Green, vol. eds.). Springer-Verlag, Berlin.

Lu, Y., J.-H. Ran, D.-M. Guo, Z.-Y. Yang, and X.-Q. Wang. 2014. Phylogeny and divergence times of gymnosperms inferred from single-copy nuclear genes. *PLOS ONE* 9(9):e107679.

Mabberley, D. J. 2008. *Mabberley's Plant-Book*. 3rd edition. Cambridge University Press, Cambridge, UK.

Markgraf, F. 1951. *Gnetaceae*. Pp. 336–347 in *Flora Malesiana*, Series 1, Vol. 4(3). (C. G. G. J. van Steenis, ed.). Noordhoff-Kolff, Djakarta.

Rothwell, G. W., and R. Serbet. 1994. Lignophyte phylogeny and the evolution of spermatophytes: a numerical cladistic analysis. *Syst. Bot.* 19:443–482.

Rydin, C., M. Källersjö, and E. Friis. 2002. Seed plant relationships and the systematic position of *Gnetales* based on nuclear and chloroplast DNA: conflicting data, rooting problems, and the monophyly of conifers. *Int. J. Plant Sci.* 163:197–214.

Soltis, D. E., P. S. Soltis, and M. J. Zanis. 2002. Phylogeny of seed plants based on evidence from eight genes. *Am. J. Bot.* 89:1670–1681.

Wickett, N. J., S. Mirarab, N. Nguyen, T. Warnow, E. Carpenter, N. Matasci, S. Ayyampalayam, M. S. Barker, J. G. Burleigh, M. A. Gitzendanner, B. R. Ruhfel, E. Wafula, J. P. Der, S. W. Graham, S. Mathews, M. Melkonian, D. E. Soltis, P. S. Soltis, N. W. Miles, C. J. Rothfels, L. Pokorny, A. J. Shaw, L. DeGironimo, D. W. Stevenson, B. Surek, J. C. Villarreal, B. Roure, H. Philippe, C. W. dePamphilis, T. Chen, M. K. Deyholos, R. S. Baucom, T. M. Kutchan, M. M. Augustin, J. Wang, Y. Zhang, Z. Tian, Z. Yan, X. Wu, X. Sun, G. K. Wong, and J. Leebens-Mack. 2014. Phylotranscriptomic analysis of the origin and early diversification of land plants. *Proc. Natl. Acad. Sci. USA* 111:E4859–E4868.

Yao, Y., Y. Xi, B. Geng, and C. Li. 2004. The exine ultrastructure of pollen grains in *Gnetum* (*Gnetaceae*) from China and its bearing on the relationship with the ANITA group. *Bot. J. Linn. Soc.* 146:415–425.

Zhong, B., O. Deusch, V. V. Goremykin, D. Penny, P. J. Biggs, R. A. Atherton, S. V. Nikiforova, and P. J. Lockhart. 2011. Systematic error in seed plant phylogenomics. *Genome Biol. Evol.* 3:1340–1348.

Zhong, B., T. Yonezawa, Y. Zhong, and M. Hasegawa. 2010. The position of *Gnetales* among seed plants: overcoming pitfalls of chloroplast phylogenomics. *Mol. Biol. Evol.* 27:2855–2863.

Authors

Michael J. Donoghue; Department of Ecology and Evolutionary Biology; Yale University; P.O. Box 208106; New Haven, CT 06520, USA. Email: michael.donoghue@yale.edu.

James A. Doyle; Department of Evolution and Ecology; University of California; Davis, CA 95616, USA. Email: jadoyle@ucdavis.edu.

Philip D. Cantino; Department of Environmental and Plant Biology; Ohio University; Athens, OH 45701, USA. Email: cantino@ohio.edu.

Date Accepted: 29 August 2011; updated 09 February 2018

Primary Editor: Kevin de Queiroz

Pan-Coniferae P. D. Cantino, M. J. Donoghue, and J. A. Doyle in P. D. Cantino et al. (2007): E20 [J. A. Doyle, P. D. Cantino, and M. J. Donoghue], converted clade name

Registration Number: 77

Definition: The total clade of the crown clade *Coniferae*. This is a crown-based total-clade definition. Abbreviated definition: total ∇ of *Coniferae*.

Etymology: From the Greek *pan-* or *pantos* (all, the whole), indicating that the name refers to a total clade, and *Coniferae* (see entry in this volume for etymology), the name of the corresponding crown clade.

Reference Phylogeny: The reference phylogeny is Doyle (2006: Fig. 8).

Composition: *Coniferae* (this volume) and all extinct plants that share more recent ancestry with *Coniferae* than with any other extant seed plants. Morphological homologies between extant and Palaeozoic conifers and *Cordaitales* proposed by Florin (1951) could be evidence that the cordaites belong in *Pan-Coniferae*, but this is uncertain, because Florin (1949) adduced similar evidence that *Ginkgo* and related fossils are also related to *Cordaitales*. In phylogenetic analyses that include fossils, depending in part on the position of *Ginkgo*, *Pan-Coniferae* may include *Cordaitales*, Palaeozoic conifers such as *Emporia* and *Lebachia*, both, or neither (Crane, 1985; Doyle and Donoghue, 1986; Rothwell and Serbet, 1994; Doyle, 1996, 2006, 2008; Hilton and Bateman, 2006). Because of uncertainty whether the clade *Gnetophyta* is part of *Coniferae* (see *Coniferae* in this volume), *Pan-Coniferae* may, similarly, include or exclude *Gnetophyta*.

Diagnostic Apomorphies: Based on a phylogenetic tree that included *Gnetophyta* within *Coniferae*, Doyle (2006: Fig. 11) listed the following synapomorphies for the clade labeled "Conifers" (including *Emporia* but not *Cordaitales* and corresponding roughly to *Pan-Coniferae* as defined here): leaves with a single vein (modified to several-veined within some groups such as *Agathis* and *Nageia*), compound female strobili but simple male strobili, bilateral (dorsiventral) symmetry in ovuliferous short shoots (modified into cone scales in extant taxa), inverted ovules, and sarcotesta absent or uniseriate. Of these, the sarcotesta character is questionable because it is unclear whether a sarcotesta is present or absent in the immature ovules of *Emporia* (Mapes and Rothwell, 1984).

Synonyms: The names *Pinopsida*, *Coniferophyta*, *Coniferopsida*, and *Coniferales*, which are listed as approximate synonyms of *Coniferae* (this volume), as well as *Coniferae* itself, could equally well be considered to be approximate synonyms of *Pan-Coniferae*, since authors have rarely made it clear whether these names apply to the crown, the total clade, or an intermediate clade.

Comments: We prefer the name *Pan-Coniferae* rather than its approximate synonyms because the name *Coniferae* (as defined in this volume) is unambiguous and neither *Coniferae* nor *Pan-Coniferae* has any implication of rank. As discussed in the entry on *Coniferae* (this volume), the names *Pinopsida*, *Coniferophyta*, and *Coniferopsida* have been used in broader senses. Moreover, the endings *-opsida* and *-ales* imply class and ordinal rank, respectively.

Literature Cited

Cantino, P. D., J. A. Doyle, S. W. Graham, W. S. Judd, R. G. Olmstead, D. E. Soltis, P. S. Soltis, and M. J. Donoghue. 2007. Towards a phylogenetic nomenclature of *Tracheophyta*. *Taxon* 56:822–846 and E1–E44.

Crane, P. R. 1985. Phylogenetic analysis of seed plants and the origin of angiosperms. *Ann. Mo. Bot. Gard.* 72:716–793.

Doyle, J. A. 1996. Seed plant phylogeny and the relationships of *Gnetales*. *Int. J. Plant Sci.* 157 (suppl.):S3–S39.

Doyle, J. A. 2006. Seed ferns and the origin of angiosperms. *J. Torrey Bot. Soc.* 133: 169–209.

Doyle, J. A. 2008. Integrating molecular phylogenetic and paleobotanical evidence on origin of the flower. *Int. J. Plant Sci.* 169:816–843.

Doyle, J. A., and M. J. Donoghue. 1986. Seed plant phylogeny and the origin of angiosperms: an experimental cladistic approach. *Bot. Rev.* 52:321–431.

Florin, R. 1949. The morphology of *Trichopitys heteromorpha* Saporta, a seed-plant of Palaeozoic age, and the evolution of the female flowers in the *Ginkgoinae*. *Acta Horti Bergiani* 15:79–109.

Florin, R. 1951. Evolution in cordaites and conifers. *Acta Horti Bergiani* 15:285–388.

Hilton, J., and R. M. Bateman. 2006. Pteridosperms are the backbone of seed-plant phylogeny. *J. Torrey Bot. Soc.* 133:119–168.

Mapes, G., and G. W. Rothwell. 1984. Permineralized ovulate cones of *Lebachia* from late Palaeozoic limestones of Kansas. *Palaeontology* 27:69–94.

Rothwell, G. W., and R. Serbet. 1994. Lignophyte phylogeny and the evolution of spermatophytes: a numerical cladistic analysis. *Syst. Bot.* 19:443–482.

Authors

James A. Doyle; Department of Evolution and Ecology; University of California; Davis, CA 95616, USA. Email: jadoyle@ucdavis.edu.

Philip D. Cantino; Department of Environmental and Plant Biology; Ohio University; Athens, OH 45701, USA. Email: cantino@ohio.edu.

Michael J. Donoghue; Department of Ecology and Evolutionary Biology; Yale University; P.O. Box 208106; New Haven, CT 06520, USA. Email: michael.donoghue@yale.edu.

Date Accepted: 29 August 2011; updated 09 February 2018

Primary Editor: Kevin de Queiroz

Coniferae A. L. Jussieu 1789: 411 [J. A. Doyle, P. D. Cantino, and M. J. Donoghue], converted clade name

Registration Number: 31

Definition: The smallest crown clade containing *Pinus strobus* Linnaeus 1753, *Cupressus sempervirens* Linnaeus 1753, *Podocarpus macrophyllus* (Thunberg) Sweet 1818 (originally described as *Taxus macrophylla* Thunberg 1783), and *Taxus baccata* Linnaeus 1753. This is a minimum-crown-clade definition. Abbreviated definition: min crown ∇ (*Pinus strobus* Linnaeus 1753 & *Cupressus sempervirens* Linnaeus 1753 & *Podocarpus macrophyllus* (Thunberg) Sweet 1818 & *Taxus baccata* Linnaeus 1753).

Etymology: Derived from Latin, *conus* (cone) and *ferre* (to carry or bear).

Reference Phylogeny: The reference phylogeny is Rydin et al. (2002: Fig. 1). See also Rai et al. (2008: Figs. 1, 2), Leslie et al. (2012: Fig. S4), Lu et al. (2014: Fig. 1), and Ruhfel et al. (2014: Fig. 5).

Composition: *Araucariaceae*, *Cephalotaxaceae*, *Cupressaceae* (including "*Taxodiaceae*"), *Pinaceae*, *Podocarpaceae*, *Taxaceae*, and *Sciadopitys*. The clade *Gnetophyta* (this volume) is included within *Coniferae* in some trees based on molecular data (Bowe et al., 2000; Chaw et al., 2000; Sanderson et al., 2000; Gugerli et al., 2001; Magallón and Sanderson, 2002; Burleigh and Mathews, 2004; Zhong et al., 2010, 2011; Xi et al., 2013; Lu et al., 2014; Ruhfel et al., 2014), but not in others (Sanderson et al., 2000; Magallón and Sanderson, 2002; Rydin et al., 2002; Burleigh and Mathews, 2004; Rai et al., 2008). The chloroplast-DNA structural mutation that characterizes conifers (see Diagnostic Apomorphies) is absent in *Gnetophyta* (Raubeson and Jansen, 1992). Most analyses based on morphology (see Comments) have not supported inclusion of *Gnetophyta*, but the clade *Gnetophyta* is nested in *Coniferae* in some most-parsimonious trees in the morphological analysis of Doyle (2008). Although *Emporia* (a Palaeozoic conifer) or *Emporia* and *Cordaitales* are part of *Coniferae*, as defined here, in some of the trees of Rothwell and Serbet (1994: Figs. 2a,b), this is not the case in their other equally parsimonious trees (Figs. 1, 2c,d), or in those of other authors (Miller, 1999; Doyle, 1996, 2006, 2008; Hilton and Bateman, 2006).

Diagnostic Apomorphies: Synapomorphies relative to other crown clades include one-veined needlelike leaves (modified to several-veined in some groups), resin canals, compound female strobili but simple male strobili, tiered proembryos, siphonogamy (Doyle and Donoghue, 1986; Rothwell and Serbet, 1994; Doyle, 2006, 2008), and loss or extreme reduction of one copy of the inverted repeat in the chloroplast genome (Raubeson and Jansen, 1992; Wakasugi et al., 1994). Siphonogamy in *Coniferae* is not homologous with that in angiosperms and gnetophytes if phylogenies such as those of Crane (1985), Doyle and Donoghue (1986), and Doyle (1996) are correct, but it may be homologous with siphonogamy in the latter groups in phylogenies that link *Coniferae* with angiosperms and gnetophytes (Nixon et al., 1994; Rothwell and Serbet, 1994; Doyle, 2006: Fig. 6), and it is presumably homologous with siphonogamy in *Gnetophyta* if the latter are nested in *Coniferae*.

Synonyms: *Pinopsida, Coniferophyta, Coniferopsida,* and *Coniferales* are approximate synonyms (see Comments).

Comments: Many molecular analyses (e.g., Stefanovic et al., 1998; Bowe et al., 2000; Chaw et al., 2000; Gugerli et al., 2001; Magallón and Sanderson, 2002; Rydin et al., 2002; Soltis et al., 2002; Rai et al., 2008; Zhong et al., 2010, 2011; Leslie et al., 2012; Lu et al., 2014; Ruhfel et al., 2014) and a morphological analysis (Hart, 1987) of extant conifers agreed that the clade *Pinaceae* or a clade comprising *Pinaceae* and *Gnetophyta* (see below) is sister to the rest of the conifers. However, phylogenetic analyses based on morphology have variously suggested that the clade *Taxaceae* is the extant sister to the rest (Miller, 1988, 1999; Rothwell and Serbet, 1994), that a clade comprising *Podocarpaceae* and *Pinaceae* occupies this position (Doyle, 1996; Hilton and Bateman, 2006), or that the position of *Podocarpaceae* is unresolved relative to *Pinaceae* and the rest of the conifers (Doyle, 2006: Fig. 6, 2008)—hence our inclusion of species of *Taxus* and *Podocarpus* as internal specifiers. Because no member of *Gnetophyta* is an internal or external specifier, our definition permits application of the name *Coniferae* in the context of the "gnepine hypothesis" (Bowe et al., 2000; Zhong et al., 2011), or of the "gnecup hypothesis" (Ruhfel et al., 2014), in which case the clade *Gnetophyta* is nested within (and therefore part of) *Coniferae*, but it does not require inclusion of *Gnetophyta*.

The names *Coniferae, Pinopsida, Coniferophyta, Coniferopsida,* and *Coniferales* are all widely applied to this clade. The name *Pinophyta* is ambiguous because it is often applied to the paraphyletic group that includes all gymnosperms (Cronquist et al., 1972; Jones and Luchsinger, 1986; Meyen, 1987; Fedorov, 1999; Woodland, 2000). In accordance with our preference for names that are descriptive or end in *-phyta* (meaning plants) rather than having an ending that is meaningless except in indicating rank, *Coniferae* and *Coniferophyta* are the best candidate names for this clade. Here and previously (Cantino et al., 2007), we have chosen *Coniferae* over *Coniferophyta* because the informal names "coniferophytes" and "coniferopsids" (generally used synonymously) traditionally referred to a hypothesized larger group that includes *Ginkgo* and *Cordaitales* as well as conifers (e.g., Coulter and Chamberlain, 1910; Chamberlain, 1935; Foster and Gifford, 1974). The name *Coniferophyta*, which is not defined in this volume, is best reserved for this larger group in the context of phylogenies in which it is a clade (e.g., Crane, 1985; Doyle and Donoghue, 1986; Doyle, 1996, 2008).

Literature Cited

Bowe, L. M., G. Coat, and C. W. dePamphilis. 2000. Phylogeny of seed plants based on all three genomic compartments: extant gymnosperms are monophyletic and *Gnetales'* closest relatives are conifers. *Proc. Natl. Acad. Sci. USA* 97:4092–4097.

Burleigh, J. G., and S. Mathews. 2004. Phylogenetic signal in nucleotide data from seed plants: implications for resolving the seed plant tree of life. *Am. J. Bot.* 91:1599–1613.

Cantino, P. D., J. A. Doyle, S. W. Graham, W. S. Judd, R. G. Olmstead, D. E. Soltis, P. S. Soltis, and M. J. Donoghue. 2007. Towards a phylogenetic nomenclature of *Tracheophyta*. *Taxon* 56:822–846 and E1–E44.

Chamberlain, C. J. 1935. *Gymnosperms. Structure and Evolution*. University of Chicago Press, Chicago, IL.

Chaw, S.-M., C. L. Parkinson, Y. Cheng, T. M. Vincent, and J. D. Palmer. 2000. Seed plant phylogeny inferred from all three plant genomes: monophyly of extant gymnosperms and origin of *Gnetales* from conifers. *Proc. Natl. Acad. Sci. USA* 97:4086–4091.

Coulter, J. M., and C. J. Chamberlain. 1910. *Morphology of Gymnosperms.* University of Chicago Press, Chicago, IL.

Crane, P. R. 1985. Phylogenetic analysis of seed plants and the origin of angiosperms. *Ann. Mo. Bot. Gard.* 72:716–793.

Cronquist, A., A. H. Holmgren, N. H. Holmgren, and J. L. Reveal. 1972. *Intermountain Flora,* Vol. 1. Hafner, New York.

Doyle, J. A. 1996. Seed plant phylogeny and the relationships of *Gnetales. Int. J. Plant Sci.* 157 (suppl.):S3–S39.

Doyle, J. A. 2006. Seed ferns and the origin of angiosperms. *J. Torrey Bot. Soc.* 133:169–209.

Doyle, J. A. 2008. Integrating molecular phylogenetic and paleobotanical evidence on origin of the flower. *Int. J. Plant Sci.* 169:816–843.

Doyle, J. A., and M. J. Donoghue. 1986. Seed plant phylogeny and the origin of angiosperms: an experimental cladistic approach. *Bot. Rev.* 52:321–431.

Fedorov, A. 1999. *Flora of Russia,* Vol. 1. A. A. Balkema, Rotterdam.

Foster, A. S., and E. M. Gifford. 1974. *Comparative Morphology of Vascular Plants.* 2nd edition. W. H. Freeman, San Francisco, CA.

Gugerli, F., C. Sperisen, U. Büchler, I. Brunner, S. Brodbeck, J. D. Palmer, and Y. Qiu. 2001. The evolutionary split of *Pinaceae* from other conifers: evidence from an intron loss and a multigene phylogeny. *Mol. Phylogenet. Evol.* 21:167–175.

Hart, J. A. 1987. A cladistic analysis of conifers: preliminary results. *J. Arnold Arboretum* 68:269–307.

Hilton, J., and R. M. Bateman. 2006. Pteridosperms are the backbone of seed-plant phylogeny. *J. Torrey Bot. Soc.* 133:119–168.

Jones, Jr., S. B., and A. E. Luchsinger. 1986. *Plant Systematics.* 2nd edition. McGraw-Hill, New York.

Jussieu, A. L. de. 1789. *Genera Plantarum.* Herissant et Barrois, Paris.

Leslie, A. B., J. M. Beaulieu, H. S. Rai, P. R. Crane, M. J. Donoghue, and S. Mathews. 2012. Hemisphere-scale differences in conifer evolutionary dynamics. *Proc. Natl. Acad. Sci. USA* 109:16217–16221.

Lu, Y., J.-H. Ran, D.-M. Guo, Z.-Y. Yang, and X.-Q. Wang. 2014. Phylogeny and divergence times of gymnosperms inferred from single-copy nuclear genes. *PLOS ONE* 9(9):e107679.

Magallón, S., and M. J. Sanderson. 2002. Relationships among seed plants inferred from highly conserved genes: sorting conflicting phylogenetic signals among ancient lineages. *Am. J. Bot.* 89:1991–2006.

Meyen, S. V. 1987. *Fundamentals of Palaeobotany.* Chapman & Hall, London.

Miller, Jr., C. N. 1988. The origin of modern conifer families. Pp. 448–486 in *Origin and Evolution of Gymnosperms* (C. B. Beck, ed.). Columbia University Press, New York.

Miller, Jr., C. N. 1999. Implications of fossil conifers for the phylogenetic relationships of living families. *Bot. Rev.* 65:239–277.

Nixon, K. C., W. L. Crepet, D. Stevenson, and E. M. Friis. 1994. A reevaluation of seed plant phylogeny. *Ann. Mo. Bot. Gard.* 81:484–533.

Rai, H. S., P. A. Reeves, R. Peakall, R. G. Olmstead, and S. W. Graham. 2008. Inference of higher-order conifer relationships from a multi-locus plastid data set. *Botany* 86:658–669.

Raubeson, L. A., and R. K. Jansen. 1992. A rare chloroplast-DNA structural mutation is shared by all conifers. *Biochem. Syst. Ecol.* 20:17–24.

Rothwell, G. W., and R. Serbet. 1994. Lignophyte phylogeny and the evolution of spermatophytes: a numerical cladistic analysis. *Syst. Bot.* 19:443–482.

Ruhfel, B. R., M. A. Gitzendanner, P. S. Soltis, D. E. Soltis, and J. G. Burleigh. 2014. From algae to angiosperms—inferring the phylogeny of green plants (*Viridiplantae*) from 360 plastid genomes. *BMC Evol. Biol.* 14:23.

Rydin, C., M. Källersjö, and E. M. Friis. 2002. Seed plant relationships and the systematic position of *Gnetales* based on nuclear and chloroplast DNA: conflicting data, rooting problems, and the monophyly of conifers. *Int. J. Plant Sci.* 163:197–214.

Sanderson, M. J., M. F. Wojciechowski, J.-M. Hu, T. Sher Khan, and S. G. Brady. 2000. Error, bias, and long-branch attraction in data for two chloroplast photosystem genes in seed plants. *Mol. Biol. Evol.* 17:782–797.

Soltis, D. E., P. S. Soltis, and M. J. Zanis. 2002. Phylogeny of seed plants based on evidence from eight genes. *Am. J. Bot.* 89:1670–1681.

Stefanovic, S., M. Jager, J. Deutsch, J. Broutin, and M. Masselot. 1998. Phylogenetic relationships of conifers inferred from partial 28S rRNA gene sequences. *Am. J. Bot.* 85:688–697.

Wakasugi, T., J. Tsudzuki, S. Ito, K. Nakashima, T. Tsudzuki, and M. Sugiura. 1994. Loss of all *ndh* genes as determined by sequencing the entire chloroplast genome of the black pine *Pinus thunbergii. Proc. Natl. Acad. Sci. USA* 91:9794–9798.

Woodland, D. W. 2000. *Contemporary Plant Systematics.* 3rd edition. Andrews University Press, Berrien Springs, MI.

Xi, Z., J. S. Rest, and C. C. Davis. 2013. Phylogenomics and coalescent analyses resolve extant seed plant relationships. *PLOS ONE* 8 (11):e80870.

Zhong, B., O. Deusch, V. V. Goremykin, D. Penny, P. J. Biggs, R. A. Atherton, S. V. Nikiforova, and P. J. Lockhart. 2011. Systematic error in seed plant phylogenomics. *Genome Biol. Evol.* 3:1340–1348.

Zhong, B., T. Yonezawa, Y. Zhong, and M. Hasegawa. 2010. The position of *Gnetales* among seed plants: overcoming pitfalls of chloroplast phylogenomics. *Mol. Biol. Evol.* 27:2855–2863.

Authors

James A. Doyle; Department of Evolution and Ecology; University of California; Davis, CA 95616, USA. Email: jadoyle@ucdavis.edu.

Philip D. Cantino; Department of Environmental and Plant Biology; Ohio University; Athens, OH 45701, USA. Email: cantino@ohio.edu.

Michael J. Donoghue; Department of Ecology and Evolutionary Biology; Yale University; P.O. Box 208106; New Haven, CT 06520, USA. Email: michael.donoghue@yale.edu.

Date Accepted: 29 August 2011; updated 09 February 2018

Primary Editor: Kevin de Queiroz

Cupressophyta P. D. Cantino and M. J. Donoghue, in P. D. Cantino et al. 2007:832 [S. W. Graham, P. D. Cantino and M. J. Donoghue], converted clade name

Registration Number: 249

Definition: The smallest crown clade containing *Cupressus sempervirens* L. 1753 and *Podocarpus macrophyllus* (Thunb.) Sweet 1818, but not *Gnetum gnemon* L. 1767 or *Pinus strobus* L. 1753. This is a minimum-crown-clade definition with external specifiers. Abbreviated definition: min crown ∇ (*Cupressus sempervirens* L. 1753 & *Podocarpus macrophyllus* (Thunb.) Sweet 1818 ~ *Gnetum gnemon* L. 1767 ∨ *Pinus strobus* L. 1753).

Etymology: Derived from *Cupressus* (name of an included taxon), which is the Latin name of the Mediterranean cypress (Eckenwalder, 2009).

Reference Phylogeny: The primary reference phylogeny is Rai et al. (2008: Fig. 2); see also Quinn et al. (2002: Fig. 3), Rydin et al. (2002: Fig. 1), and Leslie et al. (2012: Fig. 1). The species of *Podocarpus* listed as having been used in the primary reference phylogeny, *Podocarpus chinensis* Wall. ex J. Forbes, is a synonym of *P. macrophyllus* (Fu et al., 1999). *Cupressus sempervirens*, the type species of *Cupressus* and used here as an internal specifier, is not present in the primary reference phylogeny but is most closely related to *Juniperus* in that phylogeny (e.g., Mao et al., 2012: Fig. S1; Yang et al., 2012).

Composition: *Araucariaceae, Cupressaceae* (including "*Taxodiaceae*"), *Podocarpaceae, Sciadopitys verticillata,* and *Taxaceae* (including *Cephalotaxus*) and extinct descendents of their most recent common ancestor.

Diagnostic Apomorphies: A likely synapomorphy is the existence of secondary phloem fibers in the form of regular, uniseriate, tangential bands, which are modified to irregular masses or bands in *Araucariaceae* (Doyle, 2006). Hart (1987: Fig. 2) inferred three additional apomorphies for this clade, but Cantino et al. (2007: E21) questioned their validity in the context of current molecular phylogenetic trees.

Synonyms: No scientific names. The informal name "conifer II" has been used as an approximate synonym for this clade in some publications (e.g., Bowe et al., 2000; Gugerli et al., 2001; Rydin et al., 2002).

Comments: The name *Cupressophyta* was formally applied to this clade, with an explicit phylogenetic definition, by Cantino and Donoghue in Cantino et al. (2007: 832 and E21). Most molecular analyses infer this clade with strong support (e.g., Chaw et al., 1997; Stefanovic et al., 1998; Bowe et al., 2000; Gugerli et al., 2001; Quinn et al., 2002; Rydin et al., 2002; Burleigh and Mathews, 2004; Qiu et al., 2006, 2007; Rai et al., 2008; Graham and Iles, 2009; Mao et al., 2012: Fig. S1; Ruhfel et al., 2014; Wickett et al., 2014). Contrary to these results, *Pinaceae* or *Gnetophyta* are nested inside *Cupressophyta* in some optimal or nearly optimal trees in some morphology-based analyses (e.g., Rothwell and Serbet, 1994; Doyle, 1996, 2006, 2008; and Hilton and Bateman, 2006; but not Hart, 1987, who inferred a clade with the composition of *Cupressophyta* as inferred here). We use *Pinus strobus* (*Pinaceae*) and *Gnetum gnemon* (*Gnetaceae*) as external specifiers, equivalent to a qualifying clause, to render the name *Cupressophyta* inapplicable in the context of any phylogeny in which *Pinaceae* or *Gnetophyta* is more closely related to any of the internal specifiers than the latter are to each other. The definition used here differs from that of Cantino and Donoghue (in Cantino et al., 2007: 832 and E21) by the addition of *Gnetum gnemon* as an external specifier and the removal of *Araucaria araucana* as an internal specifier. Because recent molecular phylogenies (Quinn et al., 2002; Rai et al., 2008; Mao et al., 2012: Fig. S1; Leslie et al., 2012: Fig. 1) have provided very strong support for a sister-group relationship between *Araucariaceae* and *Podocarpaceae*, the inclusion of specifiers representing both taxa is unnecessary.

Literature Cited

Bowe, L. M., G. Coat, and C. W. dePamphilis. 2000. Phylogeny of seed plants based on all three genomic compartments: extant gymnosperms are monophyletic and *Gnetales'* closest relatives are conifers. *Proc. Natl. Acad. Sci. USA* 97:4092–4097.

Burleigh, J. G., and S. Mathews. 2004. Phylogenetic signal in nucleotide data from seed plants: implications for resolving the seed plant tree of life. *Am. J. Bot.* 91:1599–1613.

Cantino, P. D., J. A. Doyle, S. W. Graham, W. S. Judd, R. G. Olmstead, D. E. Soltis, P. S. Soltis, and M. J. Donoghue. 2007. Towards a phylogenetic nomenclature of *Tracheophyta*. *Taxon* 56:822–846 and E1–E44.

Chaw, S.-M., A. Zharkikh, H.-M. Sung, T.-C. Lau, and W.-H. Li. 1997. Molecular phylogeny of extant gymnosperms and seed plant evolution: analysis of nuclear 18S rRNA sequences. *Mol. Biol. Evol.* 14:56–68.

Doyle, J. A. 1996. Seed plant phylogeny and the relationships of *Gnetales*. *Int. J. Plant Sci.* 157:S3–S39.

Doyle, J. A. 2006. Seed ferns and the origin of angiosperms. *J. Torrey Bot. Soc.* 133:169–209.

Doyle, J. A. 2008. Integrating molecular phylogenetic and paleobotanical evidence on origin of the flower. *Int. J. Plant Sci.* 169:816–843.

Eckenwalder, J. E. 2009. *Conifers of the World.* Timber Press, Inc., Portland, OR.

Fu, L., Y. Li, and R. R. Mill. 1999. *Podocarpaceae. Flora of China* 4:78–84. Available at http://www.efloras.org/florataxon.aspx?flora_id=2&taxon_id=200005472.

Graham, S. W., and W. J. D. Iles. 2009. Different gymnosperm outgroups have (mostly) congruent signal regarding the root of flowering plant phylogeny. *Am. J. Bot.* 96:216–227.

Gugerli, F., C. Sperisen, U. Büchler, I. Brunner, S. Brodbeck, J. D. Palmer, and Y.-L. Qiu. 2001. The evolutionary split of *Pinaceae* from other conifers: evidence from an intron loss and a multigene phylogeny. *Mol. Phylogenet. Evol.* 21:167–175.

Hart, J. A. 1987. A cladistic analysis of conifers: preliminary results. *J. Arnold Arboretum* 68:269–307.

Hilton, J., and R. M. Bateman. 2006. Pteridosperms are the backbone of seed-plant phylogeny. *J. Torrey Bot. Soc.* 133:119–168.

Leslie, A. B., J. M. Beaulieu, H. Rai, P. R. Crane, M. J. Donoghue, and S. Mathews. 2012. Hemisphere-scale differences in conifer evolutionary dynamics. *Proc. Nat. Acad. Sci. USA* 109:16217–16221.

Mao, K., R. I. Milne, L. Zhang, Y. Peng, J. Liu, P. Thomas, R. R. Mill, and S. S. Renner. 2012. Distribution of living *Cupressaceae* reflects the breakup of Pangea. *Proc. Natl. Acad. Sci. USA* 109:7793–7798.

Qiu, Y.-L., L. Li, B. Wang, Z. Chen, O. Dombrovska, J. Lee, L. Kent, R. Li, R. W. Jobson, T. A. Hendry, D. W. Taylor, C. M. Testa, and M. Ambros. 2007. A nonflowering land plant phylogeny inferred from nucleotide sequences of seven chloroplast, mitochondrial, and nuclear genes. *Int. J. Plant Sci.* 168:691–708.

Qiu, Y.-L., L. Li, B. Wang, Z. Chen, V. Knoop, M. Groth-Malonek, O. Dombrovska, J. Lee, L. Kent, J. Rest, G. F. Estabrook, T. A. Hendry, D. W. Taylor, C. M. Testa, M. Ambros, B. Crandall-Stotler, R. J. Duff, M. Stech, W. Frey, D. Quandt, and C. C. Davis. 2006. The deepest divergences in land plants inferred from phylogenomic evidence. *Proc. Natl. Acad. Sci. USA* 103:15511–15516.

Quinn, C. J., R. A. Price, and P. A. Gadek. 2002. Familial concepts and relationships in the conifers based on *rbcL* and *matK* sequence comparisons. *Kew Bull.* 57:513–531.

Rai, H. S., P. A. Reeves, R. Peakall, R. G. Olmstead, and S. W. Graham. 2008. Inference of higher-order conifer relationships from a multi-locus plastid data set. *Botany* 86:658–669.

Rothwell, G. W., and R. Serbet. 1994. Lignophyte phylogeny and the evolution of spermatophytes: a numerical cladistic analysis. *Syst. Bot.* 19:443–482.

Ruhfel, B. R., M. A. Gitzendanner, P. S. Soltis, D. E. Soltis, and J. G. Burleigh. 2014. From algae to angiosperms—inferring the phylogeny of green plants (*Viridiplantae*) from 360 plastid genomes. *BMC Evol. Biol.* 14:23.

Rydin, C., M. Källersjö, and E. M. Friis. 2002. Seed plant relationships and the systematic position of *Gnetales* based on nuclear and chloroplast DNA: conflicting data, rooting problems, and the monophyly of conifers. *Int. J. Plant Sci.* 163:197–214.

Stefanovic, S., M. Jager, J. Deutsch, J. Broutin, and M. Masselot. 1998. Phylogenetic relationships of conifers inferred from partial 28S rDNA gene sequences. *Am. J. Bot.* 85:688–697.

Wickett, N. J., S. Mirarab, N. Nguyen, T. Warnow, E. Carpenter, N. Matasci, S. Ayyampalayam, M. S. Barker, J. G. Burleigh, M. A. Gitzendanner, B. R. Ruhfel, E. Wafula, J. P. Der, S. W. Graham, S. Mathews, M. Melkonian, D. E. Soltis, P. S. Soltis, N. W. Miles, C. J. Rothfels, L. Pokorny, A. J. Shaw, L. DeGironimo, D. W. Stevenson, B. Surek, J. C. Villarreal, B. Roure, H. Philippe, C. W. dePamphilis, T. Chen, M. K. Deyholos, R. S. Baucom, T. M. Kutchan, M. M. Augustin, J. Wang, Y. Zhang, Z. Tian, Z. Yan, X. Wu, X. Sun, G. K. Wong, and J. Leebens-Mack. 2014. Phylotranscriptomic analysis of the origin and early diversification of land plants. *Proc. Natl. Acad. Sci. USA* 111:E4859–E4868.

Yang, Z. -Y., J.- H. Ran, and X. Q. Wang. 2012. Three genome phylogeny of *Cupressaceae* s.l.: further evidence for the evolution of gymnosperms and Southern Hemisphere biogeography. *Mol. Phylogenet. Evol.* 64:452–470.

Authors

Sean W. Graham; Department of Botany, and UBC Botanical Garden & Centre for Plant Research; University of British Columbia; Vancouver, BC, V6T 1Z4, Canada. Email: swgraham@interchange.ubc.ca.

Philip D. Cantino; Department of Environmental and Plant Biology; Ohio University; Athens, OH 45701, USA. Email: cantino@ohio.edu.

Michael J. Donoghue; Department of Ecology and Evolutionary Biology; Yale University; P.O. Box 208106; New Haven, CT 06520, USA. Email: michael.donoghue@yale.edu.

Date Accepted: 6 September 2013; updated 14 July 2018

Primary Editor: Kevin de Queiroz

Pan-Angiospermae P. D. Cantino and M. J. Donoghue in Cantino et al. (2007): 833 [J. A. Doyle, P. D. Cantino, and M. J. Donoghue], converted clade name

Registration Number: 75

Definition: The total clade of the crown clade *Angiospermae*. This is a crown-based total-clade definition. Abbreviated definition: total ∇ of *Angiospermae*.

Etymology: Derived from the Greek *pan-* or *pantos* (all, the whole), indicating that this is a total clade, and *Angiospermae*, the name of the crown (see *Angiospermae*, this volume, for etymology of that name).

Reference Phylogeny: The primary reference phylogeny is Doyle (2008: Fig. 3C). See also Doyle (2006: Figs. 6, 8), Doyle (2008: Figs. 3A, 4), Rothwell et al. (2009: Fig. 30), and Rothwell and Stockey (2016: Fig. 28).

Composition: *Angiospermae* (this volume) and all extinct plants that share more recent ancestry with *Angiospermae* than with *Gnetophyta*, *Coniferae*, *Cycadales*, or *Ginkgo*. In the primary reference phylogeny and in the analysis of Hilton and Bateman (2006), *Caytonia*, *Bennettitales*, *Pentoxylon*, and *Glossopteridales* are pan-angiosperms. However, in some other phylogenies, in which *Gnetophyta* are more closely related to *Angiospermae*, only *Bennettitales* and *Pentoxylon* are stem relatives of angiosperms (Rothwell et al., 2009: Fig. 30), only *Bennettitales* (Rothwell and Serbet, 1994: Fig. 2a; Doyle, 2006: Fig. 6; Doyle, 2008: Fig. 3A), only *Caytonia* (Doyle, 1996), or only *Petriellales* (Rothwell and Stockey, 2016: Fig. 28), or there are no known stem relatives of angiosperms (Crane, 1985; Doyle and Donoghue, 1986, 1992; Rothwell and Serbet, 1994: Fig. 1).

Diagnostic Apomorphies: The crown clade has many apomorphies relative to other crown clades (see *Angiospermae*, this volume), some of which also occur in fossil plants (listed under Composition) that may be part of *Pan-Angiospermae*. Because of uncertainty about which fossils lie within *Pan-Angiospermae* and incomplete information on characters in fossils, it is difficult to identify apomorphies that arose near the base of this total clade. With the reference phylogeny and dataset of Doyle (2008), two unequivocally optimized apomorphies uniting putative stem relatives of angiosperms with members of the crown are absence of a lagenostome at the apex of the nucellus (a condition that arose independently in *Coniferae*) and a thick nucellar cuticle. Other apomorphies that are largely or entirely restricted to *Pan-Angiospermae* (not including those known only in extant angiosperms) but are equivocally optimized are ovule(s) borne on the adaxial surface of a foliar structure (potentially ancestral in crown *Spermatophyta*) and absence of an exinous megaspore membrane (a megaspore membrane is absent in *Pentoxylon*, *Bennettitales*, *Caytonia*, and *Angiospermae*, but present in *Glossopteridales*).

Synonyms: *Magnoliophyta* sensu Doweld (2001) is an approximate synonym; its inclusion of extinct, non-carpel-bearing seed plants such as *Caytonia* and *Leptostrobus* suggests that it was conceptualized as a total clade. Although not a scientific name, "angiophytes" (Doyle and Donoghue, 1993: 146) is an unambiguous synonym.

Literature Cited

Cantino, P. D., J. A. Doyle, S. W. Graham, W. S. Judd, R. G. Olmstead, D. E. Soltis, P. S. Soltis, and M. J. Donoghue. 2007. Towards a phylogenetic nomenclature of *Tracheophyta*. *Taxon* 56:822–846 and E1–E44.

Crane, P. R. 1985. Phylogenetic analysis of seed plants and the origin of angiosperms. *Ann. Mo. Bot. Gard.* 72:716–793.

Doweld, A. 2001. *Prosyllabus Tracheophytorum. Tentamen Systematis Plantarum Vascularum* (Tracheophyta). Institutum Nationale Carpologiae, Moscow.

Doyle, J. A. 1996. Seed plant phylogeny and the relationships of *Gnetales*. *Int. J. Plant Sci.* 157 (suppl.):S3–S39.

Doyle, J. A. 2006. Seed ferns and the origin of angiosperms. *J. Torrey Bot. Soc.* 133:169–209.

Doyle, J. A. 2008. Integrating molecular phylogenetic and paleobotanical evidence on origin of the flower. *Int. J. Plant Sci.* 169:816–843.

Doyle, J. A., and M. J. Donoghue. 1986. Seed plant phylogeny and the origin of angiosperms: an experimental cladistic approach. *Bot. Rev.* 52:321–431.

Doyle, J. A., and M. J. Donoghue. 1992. Fossils and seed plant phylogeny reanalyzed. *Brittonia* 44:89–106.

Doyle, J. A., and M. J. Donoghue. 1993. Phylogenies and angiosperm diversification. *Paleobiology* 19:141–167.

Hilton, J., and R. M. Bateman. 2006. Pteridosperms are the backbone of seed-plant phylogeny. *J. Torrey Bot. Soc.* 133:119–168.

Rothwell, G. W., W. L. Crepet, and R. A. Stockey. 2009. Is the anthophyte hypothesis alive and well? New evidence from the reproductive structures of *Bennettitales*. *Am. J. Bot.* 96:296–322.

Rothwell, G. W., and R. Serbet. 1994. Lignophyte phylogeny and the evolution of spermatophytes: a numerical cladistic analysis. *Syst. Bot.* 19:443–482.

Rothwell, G. W., and R. A. Stockey. 2016. Phylogenetic diversification of Early Cretaceous seed plants: the compound seed cone of *Doylea tetrahedrasperma*. *Am. J. Bot.* 103:923–937.

Authors

James A. Doyle; Department of Evolution and Ecology; University of California; Davis, CA 95616, USA. Email: jadoyle@ucdavis.edu.

Philip D. Cantino; Department of Environmental and Plant Biology; Ohio University; Athens, OH 45701, USA. Email: cantino@ohio.edu.

Michael J. Donoghue; Department of Ecology and Evolutionary Biology; Yale University; P.O. Box 208106; New Haven, CT 06520, USA. E-mail: michael.donoghue@yale.edu.

Date Accepted: 1 April 2011; updated 9 February 2018

Primary Editor: Kevin de Queiroz

SECTION 4

Angiospermae J. Lindley 1830: xxxvi [P. D. Cantino, J. A. Doyle, and M. J. Donoghue], converted clade name

Registration Number: 11

Definition: The largest crown clade containing *Liquidambar styraciflua* Linnaeus 1753 but not *Cycas revoluta* Thunberg 1782 (*Cycadophyta*) and *Ginkgo biloba* Linnaeus 1771 and *Gnetum gnemon* Linnaeus 1767 (*Gnetophyta*) and *Pinus strobus* Linnaeus 1753 (*Coniferae*). This is a maximum-crown-clade definition. Abbreviated definition: max crown ∇ (*Liquidambar styraciflua* Linnaeus 1753 ~ *Cycas revoluta* Thunberg 1782 & *Ginkgo biloba* Linnaeus 1771 & *Gnetum gnemon* Linnaeus 1767 & *Pinus strobus* Linnaeus 1753).

Etymology: Derived from Greek, *angeion* (vessel or case) and *sperma* (seed), referring to the enclosure of the ovules inside a carpel and, consequently, the seeds inside a fruit.

Reference Phylogeny: The primary reference phylogeny is Doyle (2008: Fig. 3C). See also Doyle (2008: Figs. 3A, 4), Soltis et al. (2011: Fig. 2), Ruhfel et al. (2014: Fig. 5), and Rothwell and Stockey (2016: Fig. 28).

Composition: The total clades corresponding to *Amborella*, *Nymphaeales* (including *Hydatellaceae*), *Austrobaileyales*, *Ceratophyllum*, *Chloranthaceae*, *Magnoliidae* (as defined in this volume), *Monocotyledoneae* (this volume), and *Eudicotyledoneae* (this volume); for summary of internal relationships, see Soltis et al. (2011). *Angiospermae* is a huge clade, with an estimated 271,500 extant species (Mabberley, 2008), a figure that is certain to increase as many new species continue to be described, particularly from the tropics. For information about the composition of subgroups whose names are not defined in this volume, see Stevens (2001 onwards) and Mabberley (2008).

Diagnostic Apomorphies: The following are apomorphies relative to other crown clades, some of which also occur in fossil plants that may be stem relatives of *Angiospermae* (these are noted parenthetically): closed carpel, which develops into a fruit; ovule with two integuments; lack of an exinous megaspore membrane (also in *Caytonia*, *Bennettitales*, and *Pentoxylon*); highly reduced female gametophyte, most commonly with eight nuclei (potentially based on a module consisting of an egg, two synergids, and a polar nucleus, as seen in the four-celled, four-nucleate female gametophyte of *Nymphaeales* and *Austrobaileyales*; Friedman and Williams, 2003; Friedman and Ryerson, 2009); endosperm (either diploid or triploid) resulting from double fertilization; microgametophyte with three nuclei; scalariform pitting or perforations in secondary xylem (also in *Bennettitales*); apical meristem with tunica of two cell layers; more than two orders of leaf venation; poles of stomatal guard cells level with aperture (also in *Caytonia*); axially aligned companion cells derived from the same mother cells as the sieve elements; anther wall with endothecium; pollen with unlaminated endexine; stamen with two pairs of pollen sacs (Crane, 1985; Doyle and Donoghue, 1986a, 1992; Rothwell and Serbet, 1994; Doyle, 1996, 2006, 2008; P. Soltis et al., 2004).

Synonyms: Approximate synonyms include *Magnoliophyta* sensu Cronquist (1981) and many other authors (see Comments), *Anthophyta* sensu Bold (1957) and *Magnoliopsida* sensu Jeffrey (1982), Scagel et al. (1984), and Thorne and Reveal (2007). However, the name *Magnoliopsida* is more widely applied to the paraphyletic group "dicots" (e.g., Takhtajan, 1987, 1997; Cronquist, 1981; and many texts that adopted Cronquist's system).

Comments: The monophyly of *Angiospermae* is very strongly supported by both molecular and morphological analyses (Magallón and Sanderson, 2002; Rydin et al., 2002; Soltis et al., 2002; Doyle, 2006, 2008; Xi et al., 2014), and the clade has many morphological apomorphies (see Diagnostic Apomorphies). In various phylogenetic analyses, the extant sister group of the angiosperm crown clade has been inferred to be either *Gnetophyta* (Crane, 1985; Doyle and Donoghue, 1986a,b, 1992; Loconte and Stevenson, 1990; Doyle et al., 1994; Rothwell and Serbet, 1994; Doyle, 1996, 2006, 2008: Fig. 3A; Stefanovic et al., 1998; Rydin et al., 2002: Fig. 3; Hilton and Bateman, 2006), a clade comprising *Gnetum* and *Welwitschia* (Nixon et al., 1994), all extant gymnosperms (Bowe et al., 2000; Chaw et al., 2000; Gugerli et al., 2001; Magallón and Sanderson, 2002; Soltis et al., 2002: Figs. 2, 4–6; Qiu et al., 2006, 2007; Xi et al., 2013; Ruhfel et al., 2014; Wickett et al., 2014), a clade comprising conifers, cycads, and *Ginkgo* (Hamby and Zimmer, 1992; Magallón and Sanderson, 2002; Rydin et al., 2002: Figs. 1, 2; Soltis et al., 2002: Fig. 3; Rai et al., 2003), a clade comprising conifers and *Gnetophyta* (Hill and Crane, 1982; Soltis et al., 2002: Fig. 1), or *Cycadophyta* (Doyle, 2006: Fig. 7; Doyle, 2008: Fig. 3C; Mathews et al., 2010). Because of this disagreement about outgroup relationships, four external specifiers are used here. A minimum-clade definition with three specifiers, two of which would be *Amborella trichopoda* and any species of *Nymphaeales* or *Hydatellaceae*, would be simpler and is consistent with the results of recent molecular analyses (e.g., Soltis et al., 2011; Wickett et al., 2014; Xi et al., 2014; Zeng et al., 2014). However, the immensity of *Angiospermae* and the recency of the discovery that *Amborella* or a clade comprising *Amborella* and *Nymphaeales/Hydatellaceae* is (apparently) sister to the rest of the angiosperms argue against this kind of definition. Regardless of how confident one may currently feel about the position of *Amborella*, one must consider the possibility that some other species that has to date not been included in a molecular analysis may turn out to be sister to the rest of *Angiospermae*. The discovery (Saarela et al., 2007) that the taxon *Hydatellaceae*, formerly thought to be part of the monocot clade, is related to *Nymphaeales* near the base of the angiosperm tree illustrates this point. Furthermore, some genome-based phylogenies (Goremykin et al., 2003, 2004; but see D. Soltis et al., 2004) suggest that monocots (represented only by grasses in Goremykin's papers), rather than *Amborella*, are sister to the rest of the angiosperms (i.e., a basal split between monocots and dicots). Compositional stability is better served by a maximum-crown-clade definition with the relatively few candidates for the extant sister group represented by external specifiers.

Angiospermae and *Magnoliophyta* are the principal names for this clade. Here and previously (Cantino et al., 2007), we have adopted the name *Angiospermae* because we prefer to avoid names with rank-based endings if there is a reasonable alternative, and it appears to be the more widely used of the two names. We attribute the name *Angiospermae* to Lindley (1830). Although Lindley published this name as a tribe that contains orders, and thus it was not validly published by Lindley according to the botanical code (Art. 37.6; Turland et al., 2018), this does not disqualify Lindley as the earliest author of the preexisting name *Angiospermae* under the *PhyloCode* (Rec. 9.15A).

Literature Cited

Bold, H. C. 1957. *Morphology of Plants.* Harper & Row, New York.

Bowe, L. M., G. Coat, and C. W. dePamphilis. 2000. Phylogeny of seed plants based on all three genomic compartments: extant gymnosperms are monophyletic and *Gnetales'* closest relatives are conifers. *Proc. Natl. Acad. Sci. USA* 97:4092–4097.

Cantino, P. D., J. A. Doyle, S. W. Graham, W. S. Judd, R. G. Olmstead, D. E. Soltis, P. S. Soltis, and M. J. Donoghue. 2007. Towards a phylogenetic nomenclature of *Tracheophyta. Taxon* 56:822–846 and E1–E44.

Chaw, S.-M., C. L. Parkinson, Y. Cheng, T. M. Vincent, and J. D. Palmer. 2000. Seed plant phylogeny inferred from all three plant genomes: monophyly of extant gymnosperms and origin of *Gnetales* from conifers. *Proc. Natl. Acad. Sci. USA* 97:4086–4091.

Crane, P. R. 1985. Phylogenetic analysis of seed plants and the origin of angiosperms. *Ann. Missouri Bot. Gard.* 72:716–793.

Cronquist, A. 1981. *An Integrated System of Classification of Flowering Plants.* Columbia University Press, New York.

Doyle, J. A. 1996. Seed plant phylogeny and the relationships of *Gnetales. Int. J. Plant Sci.* 157 (suppl.):S3–S39.

Doyle, J. A. 2006. Seed ferns and the origin of angiosperms. *J. Torrey Bot. Soc.* 133:169–209.

Doyle, J. A. 2008. Integrating molecular phylogenetic and paleobotanical evidence on origin of the flower. *Int. J. Plant Sci.* 169:816–843.

Doyle, J. A., and M. J. Donoghue. 1986a. Seed plant phylogeny and the origin of angiosperms: an experimental cladistic approach. *Bot. Rev.* 52:321–431.

Doyle, J. A., and M. J. Donoghue. 1986b. Relationships of angiosperms and *Gnetales*: a numerical cladistic analysis. Pp. 177–198 in *Systematic and Taxonomic Approaches to Paleobotany* (B. A. Thomas and R. A. Spicer, eds.). Oxford University Press, Oxford.

Doyle, J. A., and M. J. Donoghue. 1992. Fossils and seed plant phylogeny reanalyzed. *Brittonia* 44:89–106.

Doyle, J. A., M. J. Donoghue, and E. A. Zimmer. 1994. Integration of morphological and ribosomal RNA data on the origin of angiosperms. *Ann. Mo. Bot. Gard.* 81:419–450.

Friedman, W. E., and K. C. Ryerson. 2009. Reconstructing the ancestral female gametophyte of angiosperms: insights from *Amborella* and other ancient lineages of flowering plants. *Am. J. Bot.* 96:129–143.

Friedman, W. E., and J. H. Williams. 2003. Modularity in the angiosperm female gametophyte and its bearing on the early evolution of endosperm in flowering plants. *Evolution* 57:216–230.

Goremykin, V. V., K. I. Hirsch-Ernst, S. Wölfl, and F. H. Hellwig. 2003. Analysis of the *Amborella trichopoda* chloroplast genome sequence suggests that *Amborella* is not a basal angiosperm. *Mol. Biol. Evol.* 20:1499–1505.

Goremykin, V. V., K. I. Hirsch-Ernst, S. Wölfl, and F. H. Hellwig. 2004. The chloroplast genome of *Nymphaea alba*: Whole-genome analyses and the problem of identifying the most basal angiosperm. *Mol. Biol. Evol.* 21:1445–1454.

Gugerli, F., C. Sperisen, U. Büchler, I. Brunner, S. Brodbeck, J. D. Palmer, and Y. Qiu. 2001. The evolutionary split of *Pinaceae* from other conifers: evidence from an intron loss and a multigene phylogeny. *Mol. Phylogenet. Evol.* 21:167–175.

Hamby, R. K., and E. A. Zimmer. 1992. Ribosomal RNA as a phylogenetic tool in plant systematics. Pp. 50–91 in *Molecular Systematics of Plants* (P. S. Soltis, D. E. Soltis and J. J. Doyle, eds.). Chapman and Hall, New York.

Hill, C. R., and P. R. Crane. 1982. Evolutionary cladistics and the origin of angiosperms. Pp. 269–361 in *Problems of Phylogenetic Reconstruction* (K. A. Joysey and A. E. Friday, eds.). Academic Press, London.

Hilton, J., and R. M. Bateman. 2006. Pteridosperms are the backbone of seed-plant phylogeny. *J. Torrey Bot. Soc.* 133:119–168.

Jeffrey, C. 1982. Kingdoms, codes and classification. *Kew Bull.* 37:403–416.

Lindley, J. 1830. *Introduction to the Natural System of Botany.* Longman, Rees, Orme, Brown, and Green, London.

Loconte, H., and D. W. Stevenson. 1990. Cladistics of *Spermatophyta. Brittonia* 42:197–211.

Mabberley, D. J. 2008. *Mabberley's Plant-book.* 3rd edition. Cambridge University Press, Cambridge, UK.

Magallón, S., and M. J. Sanderson. 2002. Relationships among seed plants inferred from highly conserved genes: sorting conflicting phylogenetic signals among ancient lineages. *Am. J. Bot.* 89:1991–2006.

Mathews, S., M. D. Clements, and M. A. Beilstein. 2010. A duplicate gene rooting of seed plants and the phylogenetic position of flowering plants. *Philos. Trans. R. Soc. Lond. B Biol. Sci.* 365:383–395.

Nixon, K. C., W. L. Crepet, D. Stevenson, and E. M. Friis. 1994. A reevaluation of seed plant phylogeny. *Ann. Mo. Bot. Gard.* 81:484–533.

Qiu, Y.-L., L. Li, B. Wang, Z. Chen, O. Dombrovska, J. Lee, L. Kent, R. Li, R. W. Jobson, T. A. Hendry, D. W. Taylor, C. M. Testa, and M. Ambros. 2007. A nonflowering land plant phylogeny inferred from nucleotide sequences of seven chloroplast, mitochondrial, and nuclear genes. *Int. J. Plant Sci.* 168:691–708.

Qiu, Y.-L., L. Li, B. Wang, Z. Chen, V. Knoop, M. Groth-Malonek, O. Dombrovska, J. Lee, L. Kent, J. Rest, G. F. Estabrook, T. A. Hendry, D. W. Taylor, C. M. Testa, M. Ambros, B. Crandall-Stotler, R. J. Duff, M. Stech, W. Frey, D. Quandt, and C. C. Davis. 2006. The deepest divergences in land plants inferred from phylogenomic evidence. *Proc. Natl. Acad. Sci. USA* 103:15511–15516.

Rai, H. S., H. E. O'Brien, P. A. Reeves, R. G. Olmstead, and S. W. Graham. 2003. Inference of higher-order relationships in the cycads from a large chloroplast data set. *Mol. Phylogenet. Evol.* 29:350–359.

Rothwell, G. W., and R. Serbet. 1994. Lignophyte phylogeny and the evolution of spermatophytes: a numerical cladistic analysis. *Syst. Bot.* 19:443–482.

Rothwell, G. W., and R. A. Stockey. 2016. Phylogenetic diversification of Early Cretaceous seed plants: The compound seed cone of *Doylea tetrahedrasperma. Am. J. Bot.* 103:923–937.

Ruhfel, B. R., M. A. Gitzendanner, P. S. Soltis, D. E. Soltis, and J. G. Burleigh. 2014. From algae to angiosperms—inferring the phylogeny of green plants (*Viridiplantae*) from 360 plastid genomes. *BMC Evol. Biol.* 14:23.

Rydin, C., M. Källersjö, and E. Friis. 2002. Seed plant relationships and the systematic position of *Gnetales* based on nuclear and chloroplast DNA: conflicting data, rooting problems, and the monophyly of conifers. *Int. J. Plant Sci.* 163:197–214.

Saarela, J. M., H. S. Rai, J. A. Doyle, P. K, Endress, S. Mathews, A. D. Marchant, B. G. Briggs, and S. W. Graham. 2007. *Hydatellaceae* identified as a new branch near the base of the angiosperm phylogenetic tree. *Nature* 446: 312–315.

Scagel, R. F., R. J. Bandoni, J. R. Maze, G. E. Rouse, W. B. Schofield, and J. R. Stein. 1984. *Plants: An Evolutionary Survey.* Wadsworth, Belmont, CA.

Soltis, D. E., V. A. Albert, V. Savolainen, K. Hilu, Y-L. Qiu, M. W. Chase, J. S. Farris, S. Stefanovic, D. W. Rice, J. D. Palmer, and P. S. Soltis. 2004. Genome-scale data, angiosperm relationships, and 'ending incongruence': a cautionary tale in phylogenetics. *Trends Plant Sci.* 9:477–483.

Soltis, D. E., S. A. Smith, N. Cellinese, K. J. Wurdack, D. C. Tank, S. F. Brockington, N. F. Refulio-Rodriguez, J. B. Walker, M. J. Moore, B. S. Carlsward, C. D. Bell, M. Latvis, S. Crawley, C. Black, D. Diouf, Z. Xi, C. A. Rushworth, M. A. Gitzendanner, K. J. Sytsma, Y-L. Qiu, K. H. Hilu, C. C. Davis, M. J. Sanderson, R. S. Beaman, R. G. Olmstead, W. S. Judd, M. J. Donoghue, and P. S. Soltis. 2011. Angiosperm phylogeny: 17 genes, 640 taxa. *Am. J. Bot.* 98:704–730.

Soltis, D. E., P. S. Soltis, and M. J. Zanis. 2002. Phylogeny of seed plants based on evidence from eight genes. *Am. J. Bot.* 89:1670–1681.

Soltis, P. S., D. E. Soltis, M. W. Chase, P. K. Endress, and P. R. Crane. 2004. The diversification of flowering plants. Pp. 154–167 in *Assembling the Tree of Life* (J. Cracraft and M. J. Donoghue, eds.). Oxford University Press, Oxford.

Stefanovic, S., M. Jager, J. Deutsch, J. Broutin, and M. Masselot. 1998. Phylogenetic relationships of conifers inferred from partial 28S rRNA gene sequences. *Am. J. Bot.* 85:688–697.

Stevens, P. F. 2001 onwards. Angiosperm Phylogeny Website, Version 14, July 2017 [and more or less continuously updated since]. Available at http://www.mobot.org/MOBOT/research/APweb/.

Takhtajan, A. 1987. *Systema Magnoliophytorum.* Nauka, Leningrad.

Takhtajan, A. 1997. *Diversity and Classification of Flowering Plants.* Columbia University Press, New York.

Thorne, R. F. 2007, and J. L. Reveal. 2007. An updated classification of the Class *Magnoliopsida* ("*Angiospermae*"). *Bot. Rev.* 73:67–182.

Turland, N. J., J. H. Wiersema, F. R. Barrie, W. Greuter, D. L. Hawksworth, P. S. Herendeen, S. Knapp, W.-H. Kusber, D.-Z. Li, K. Marhold, T. W. May, J. McNeill, A. M. Monro, J. Prado, M. J. Price, and G. F. Smith, eds. 2018. *International Code of Nomenclature for Algae, Fungi, and Plants (Shenzhen Code).* Adopted by the Nineteenth International Botanical Congress, Shenzhen, China, July 2017. Regnum Vegetabile 159. Koeltz Botanical Books, Glashütten.

Wickett, N. J., S. Mirarab, N. Nguyen, T. Warnow, E. Carpenter, N. Matasci, S. Ayyampalayam, M. S. Barker, J. G. Burleigh, M. A. Gitzendanner, B. R. Ruhfel, E. Wafula, J. P. Der, S. W. Graham, S. Mathews, M. Melkonian, D. E. Soltis, P. S. Soltis, N. W. Miles, C. J. Rothfels, L. Pokorny, A. J. Shaw, L. DeGironimo, D. W. Stevenson, B. Surek, J. C. Villarreal, B. Roure, H. Philippe, C. W. dePamphilis, T. Chen, M. K. Deyholos, R. S. Baucom, T. M. Kutchan, M. M. Augustin, J. Wang, Y. Zhang, Z. Tian, Z. Yan, X. Wu, X. Sun, G. K. Wong, and J. Leebens-Mack. 2014. Phylotranscriptomic analysis of the origin and early diversification of land plants. *Proc. Natl. Acad. Sci. USA* 111:E4859–E4868.

Xi, Z., L. Liu, J. S. Rest and C. C. Davis. 2014. Coalescent versus concatenation methods and the placement of *Amborella* as sister to water lilies. *Syst. Biol.* 63:919–932.

Xi, Z., J. S. Rest, and C. C. Davis. 2013. Phylogenomics and coalescent analyses resolve extant seed plant relationships. *PLOS ONE* 8 (11):e80870.

Zeng, L., Q. Zhang, R. Sun, H. Kong, N. Zhang, and H. Ma. 2014. Resolution of deep angiosperm phylogeny using conserved nuclear genes and estimates of early divergence times. *Nat. Commun.* 5:4956.

Authors

Philip D. Cantino; Department of Environmental and Plant Biology; Ohio University; Athens, OH 45701, USA. Email: cantino@ohio.edu.

James A. Doyle; Department of Evolution and Ecology; University of California; Davis, CA 95616, USA. Email: jadoyle@ucdavis.edu.

Michael J. Donoghue; Department of Ecology and Evolutionary Biology; Yale University; P.O. Box 208106; New Haven, CT 06520, USA. Email: michael.donoghue@yale.edu.

Date Accepted: 01 April 2011; updated 4 April 2018, 11 July 2018

Primary Editor: Kevin de Queiroz

Mesangiospermae M. J. Donoghue, J. A. Doyle, and P. D. Cantino in
P. D. Cantino et al. (2007): 834 [M. J. Donoghue, J. A.
Doyle, and P. D. Cantino], converted clade name

Registration Number: 65

Definition: The largest crown clade containing *Platanus occidentalis* Linnaeus 1753 but not *Amborella trichopoda* Baillon 1869 and *Nymphaea odorata* Aiton 1789 (*Nymphaeales*) and *Austrobaileya scandens* C. T. White 1933 (*Austrobaileyales*). This is a maximum-crown-clade definition. Abbreviated definition: max crown ∇ (*Platanus occidentalis* Linnaeus 1753 ~ *Amborella trichopoda* Baillon 1869 & *Nymphaea odorata* Aiton 1789 & *Austrobaileya scandens* C. T. White 1933).

Etymology: *Mesangiospermae* is a rough translation of "core angiosperms," the informal name applied to this clade by Judd et al. (2002). The prefix *mes-* (Greek) means "middle" (Stearn, 1973), and *Angiospermae* (this volume) is the name of a larger clade.

Reference Phylogeny: The primary reference phylogeny is Qiu et al. (2005: Fig. 2). See also Soltis et al. (2011: Figs. 1, 2) and Wickett et al. (2014: Figs. 2, 3).

Composition: *Chloranthaceae, Ceratophyllum, Magnoliidae, Monocotyledoneae,* and *Eudicotyledoneae* (see entries in this volume for the last three clades).

Diagnostic Apomorphies: Unambiguous morphological apomorphies for *Mesangiospermae* are not yet known. One possibility is plicate carpels sealed by postgenital fusion of the margins (see Endress and Igersheim, 2000), but this interpretation depends on the placement of two of the groups of *Mesangiospermae, Chloranthaceae,* and *Ceratophyllum,* which have ascidiate carpels sealed by secretion, comparable to those of *Amborella trichopoda* and other members of what Qiu et al. (1999) called the "ANITA" grade (i.e., *Amborella, Nymphaeales,* and *Austrobaileyales,* later extended to include *Hydatellaceae*: Saarela et al., 2007). Most molecular analyses have indicated that *Chloranthaceae* and *Ceratophyllum* are nested within *Mesangiospermae,* and with some topologies (Endress and Doyle, 2009: Fig. 8B) it is equally parsimonious to assume either that plicate carpels and sealing by postgenital fusion are apomorphies at the level of *Mesangiospermae,* with reversals in some lines, or that these features originated more than once within the clade. However, with other topologies (Soltis et al., 2005: Fig. 3.17) the latter scenario is more parsimonious. Combined molecular and morphological analyses (Doyle and Endress, 2000; Endress and Doyle, 2009) and some molecular analyses (Qiu et al., 2005, 2010) supported the placement of *Chloranthaceae* or a clade consisting of *Chloranthaceae* and *Ceratophyllum* as sister to all remaining *Mesangiospermae,* in which case it could be inferred that the ascidiate carpels of *Chloranthaceae* (with or without *Ceratophyllum*) are plesiomorphic and plicate carpels sealed by postgenital fusion evolved once within *Mesangiospermae.* In all these cases, some homoplasy would remain (e.g., reversals to ascidiate carpels in *Nelumbo* and *Berberidaceae*; convergent origins of partially plicate carpels in *Illicium*; Doyle and Endress, 2000: Fig. 7; Soltis et al., 2005: Fig. 3.17). Finally, recent embryological studies (Williams and Friedman, 2002; Friedman, 2006) raise the possibility that the typical 7-celled, 8-nucleate *Polygonum*-type embryo sac is an apomorphy of *Mesangiospermae,* assuming that the 9-nucleate embryo sac of *Amborella* was independently derived from the 4-nucleate type found in *Nymphaeales* and *Austrobaileyales*.

Synonyms: The informal names "euangiosperms" (Qiu et al., 2000) and "core angiosperms" (Judd et al., 2002) are approximate synonyms.

Comments: Until we published the name *Mesangiospermae* (Cantino et al., 2007), there was no preexisting scientific name for this large and well-supported clade, which includes the vast majority of living angiosperms. In most recent analyses of the basal angiosperm problem (e.g., Mathews and Donoghue, 1999; Doyle and Endress, 2000; Qiu et al., 2000; Zanis et al., 2003; Xi et al., 2014), which have focused on resolving relationships among *Amborella trichopoda*, *Nymphaeales*, and *Austrobaileyales*, the clade comprising the remaining angiosperms has not been named, with the exception that Qiu et al. (2000: S7) referred to it as "euangiosperms." It did not receive even an informal name in several phylogenetic studies of the angiosperms as a whole (e.g., Soltis et al., 2000; Hilu et al., 2003) or in summary treatments (e.g., APG II, 2003; Soltis et al., 2005) despite rather high levels of support, though one text (Judd et al., 2002: 178) referred to the clade as the "core angiosperms." More recent editions of this text (e.g., Judd et al., 2008, 2016) use the name *Mesangiospermae* (citing Cantino et al., 2007) as well as "core angiosperms." The Angiosperm Phylogeny Group (APG IV, 2016) used the informal name "Mesangiosperms" in its appendix.

Because outgroup relationships are better resolved than basal relationships within *Mesangiospermae*, compositional stability can be achieved more simply with a maximum-crown-clade definition than a minimum-crown-clade definition. Relationships among five clades at the base of *Mesangiospermae* (*Chloranthaceae*, *Ceratophyllum*, *Magnoliidae*, *Monocotyledoneae*,

and *Eudicotyledoneae*) remain unclear. Some analyses have suggested that *Chloranthaceae* (e.g., Doyle and Endress, 2000; Qiu et al., 2005: Fig. 1) or a clade comprising *Chloranthaceae* and *Magnoliidae* (Saarela et al., 2007: Fig. 2) is the sister group of the rest of *Mesangiospermae*. Others have supported *Ceratophyllum* alone (e.g., Zanis et al., 2003: Fig. 4), *Monocotyledoneae* alone (Qiu et al., 2005: Fig. 2), or a clade consisting of *Ceratophyllum* and monocots (Qiu et al., 2005: Fig. 3C; Zanis et al., 2003: Fig. 3) as sister to the rest (see Soltis et al., 2005, for discussion). In still other analyses *Ceratophyllum* has been linked instead with eudicots (Hilu et al., 2003; Qiu et al., 2005: Fig. 2; Graham et al., 2006; Jansen et al., 2007; Moore et al., 2007; Saarela et al., 2007) or with *Chloranthaceae* (Qiu et al., 2005: Fig. 3A,B; Qiu et al., 2006: Fig. 3; Qiu et al., 2010; Endress and Doyle, 2009; Moore et al., 2011; Zhang et al., 2012; Zeng et al., 2014). By using a maximum-crown-clade definition, and citing all plausible candidates for the extant sister group among the external specifiers, we ensure that all of the major clades of *Mesangiospermae* will be included regardless of their interrelationships. This definition also ensures that the name *Mesangiospermae* will still apply to a clade that includes the three major subclades *Magnoliidae*, *Monocotyledoneae*, and *Eudicotyledoneae* in the unlikely event that *Chloranthaceae*, *Ceratophyllum*, or both are shown to be linked with one of the more basal angiosperm clades.

Literature Cited

APG II (Angiosperm Phylogeny Group II). 2003. An update of the Angiosperm Phylogeny Group classification for the orders and families of flowering plants: APG II. *Bot. J. Linn. Soc.* 141:399–436.

APG IV (Angiosperm Phylogeny Group IV). 2016. An update of the Angiosperm Phylogeny Group classification for the orders and families of flowering plants: APG IV. *Bot. J. Linn. Soc.* 181:1–20.

Cantino, P. D., J. A. Doyle, S. W. Graham, W. S. Judd, R. G. Olmstead, D. E. Soltis, P. S. Soltis, and M. J. Donoghue. 2007. Towards a phylogenetic nomenclature of *Tracheophyta*. *Taxon* 56:822–846 and E1–E44.

Doyle, J. A., and P. K. Endress. 2000. Morphological phylogenetic analysis of basal angiosperms: comparison and combination with molecular data. *Int. J. Plant Sci.* 161:S121–S153.

Endress, P. K., and J. A. Doyle. 2009. Reconstructing the ancestral angiosperm flower and its initial specializations. *Am. J. Bot.* 96:22–66.

Endress, P. K., and A. Igersheim. 2000. Gynoecium structure and evolution in basal angiosperms. *Int. J. Plant Sci.* 161:S211–S223.

Friedman, W. E. 2006. Embryological evidence for developmental lability during early angiosperm evolution. *Nature* 441:337–340.

Graham, S. W., J. M. Zgurski, M. A. McPherson, D. M. Cherniawsky, J. M. Saarela, E. S. C. Horne, S. Y. Smith, W. A. Wong, H. E. O'Brien, V. L. Biron, J. C. Pires, R. G. Olmstead, M. W. Chase, and H. S. Rai. 2006. Robust inference of monocot deep phylogeny using an expanded multigene plastid data set. Pp. 3–20 in *Monocots: Comparative Biology and Evolution (Excluding* Poales*)* (J. T. Columbus, E. A. Friar, J. M. Porter, L. M. Prince, and M. G. Simpson, eds.). Rancho Santa Ana Botanic Garden, Claremont, CA.

Hilu, K. W., T. Borsch, K. Müller, D. E. Soltis, P. S. Soltis, V. Savolainen, M. W. Chase, M. P. Powell, L. A. Alice, R. Evans, H. Sauquet, C. Neinhuis, T. A. B. Slotta, J. G. Rohwer, C. S. Campbell, and L. W. Chatrou. 2003. Angiosperm phylogeny based on *matK* sequence information. *Am. J. Bot.* 90:1758–1776.

Jansen, R. K., Z. Cai, L. A. Raubeson, H. Daniell, C. W. dePamphilis, J. Leebens-Mack, K. F. Müller, M. Guisinger-Bellian, R. C. Haberle, A. K. Hansen, T. W. Chumley, S.-B. Lee, R. Peery, J. R. McNeal, J. V. Kuehl, and J. L. Boore. 2007. Analysis of 81 genes from 64 plastid genomes resolves relationships in angiosperms and identifies genome-scale evolutionary patterns. *Proc. Natl. Acad. Sci. USA* 104:19369–19374.

Judd, W. S., C. S. Campbell, E. A. Kellogg, P. F. Stevens, and M. J. Donoghue. 2002. *Plant Systematics—A Phylogenetic Approach.* 2nd edition. Sinauer Associates, Sunderland, MA.

Judd, W. S., C. S. Campbell, E. A. Kellogg, P. F. Stevens, and M. J. Donoghue. 2008. *Plant Systematics—A Phylogenetic Approach.* 3rd edition. Sinauer Associates, Sunderland, MA.

Judd, W. S., C. S. Campbell, E. A. Kellogg, P. F. Stevens, and M. J. Donoghue. 2016. *Plant Systematics—A Phylogenetic Approach.* 4th edition. Sinauer Associates, Sunderland, MA.

Mathews, S., and M. J. Donoghue. 1999. The root of angiosperm phylogeny inferred from duplicate phytochrome genes. *Science* 286:947–949.

Moore, M. J., C. D. Bell, P. S. Soltis, and D. E. Soltis. 2007. Using plastid genome-scale data to resolve enigmatic relationships among basal angiosperms. *Proc. Natl. Acad. Sci. USA* 104:19363–19368.

Moore, M. J., N. Hassan, M. A. Gitzendanner, R. A. Bruenn, M. Croley, A. Vandeventer, J. W. Horn, A. Dhingra, S. F. Brockington, M. Latvis, J. Ramdial, R. Alexandre, A. Piedrahita, Z. Xi, C. C. Davis, P. S. Soltis, and D. E. Soltis. 2011. Phylogenetic analysis of the plastid inverted repeat for 244 species: insights into deeper-level angiosperm relationships from a long, slowly evolving sequence region. *Int. J. Plant Sci.* 172:541–558.

Qiu, Y.-L., O. Dombrovska, J. Lee, L. Li, B. A. Whitlock, F. Bernasconi-Quadroni, J. S. Rest, C. C. Davis, T. Borsch, K. W. Hilu, S. S. Renner, D. E. Soltis, P. S. Soltis, M. J. Zanis, J. J. Cannone, R. R. Gutell, M. Powell, V. Savolainen, L. W. Chatrou, and M. W. Chase. 2005. Phylogenetic analyses of basal angiosperms based on nine plastid, mitochondrial, and nuclear genes. *Int. J. Plant Sci.* 166:815–842.

Qiu, Y.-L., J.-Y. Lee, F. Bernasconi-Quadroni, D. E. Soltis, P. S. Soltis, M. Zanis, Z. Chen, V. Savolainen, and M. W. Chase. 1999. The earliest angiosperms: evidence from mitochondrial, plastid and nuclear genomes. *Nature* 402:404–407.

Qiu, Y.-L., J.-Y. Lee, F. Bernasconi-Quadroni, D. E. Soltis, P. S. Soltis, M. Zanis, E. Zimmer, Z. Chen, V. Savolainen, and M. W. Chase. 2000. Phylogeny of basal angiosperms: analyses of five genes from three genomes. *Int. J. Plant Sci.* 161:S3–S27.

Qiu, Y.-L., L. Li, T. A. Hendry, R. Li, D. W. Taylor, M. J. Issa, A. J. Ronen, M. L. Vekaria, and A. M. White. 2006. Reconstructing the basal angiosperm phylogeny: evaluating information content of mitochondrial genes. *Taxon* 55:837–856.

Qiu, Y.-L., L. Li, B. Wang, J.-Y. Xue, T. A. Hendry, R.-Q. Li, J. W. Brown, Y. Liu, G. T. Hudson, and Z.-D. Chen. 2010. Angiosperm phylogeny inferred from sequences of four mitochondrial genes. *J. Syst. Evol.* 48:391–425.

Saarela, J. M., H. S. Rai, J. A. Doyle, P. K. Endress, S. Mathews, A. D. Marchant, B. G. Briggs, and S. W. Graham. 2007. *Hydatellaceae* identified as a new branch near the base of the angiosperm phylogenetic tree. *Nature* 446:312–315.

Soltis, D. E., S. A. Smith, N. Cellinese, K. J. Wurdack, D. C. Tank, S. F. Brockington, N. F. Refulio-Rodriguez, J. B. Walker, M. J. Moore, B. S. Carlsward, C. D. Bell, M. Latvis, S. Crawley, C. Black, D. Diouf, Z. Xi, C. A. Rushworth, M. A. Gitzendanner, K. J. Sytsma, Y.-L. Qiu, K. H. Hilu, C. C. Davis, M. J. Sanderson, R. S. Beaman, R. G. Olmstead, W. S. Judd, M. J. Donoghue, and P. S. Soltis. 2011. Angiosperm phylogeny: 17 genes, 640 taxa. *Am. J. Bot.* 98:704–730.

Soltis, D. E., P. S. Soltis, M. W. Chase, M. E. Mort, D. C. Albach, M. Zanis, V. Savolainen, W. H. Hahn, S. B. Hoot, M. F. Fay, M. Axtell, S. M. Swensen, L. M. Prince, W. J. Kress, K. C. Nixon, and J. S. Farris. 2000. Angiosperm phylogeny inferred from 18S rDNA, *rbcL*, and *atpB* sequences. *Bot. J. Linn. Soc.* 133:381–461.

Soltis, D. E., P. S. Soltis, P. K. Endress, and M. W. Chase. 2005. *Phylogeny and Evolution of Angiosperms*. Sinauer Associates, Sunderland, MA.

Stearn, W. T. 1973. *Botanical Latin*. David & Charles, Newton Abbot, Devon.

Wickett, N. J., S. Mirarab, N. Nguyen, T. Warnow, E. Carpenter, N. Matasci, S. Ayyampalayam, M. S. Barker, J. G. Burleigh, M. A. Gitzendanner, B. R. Ruhfel, E. Wafula, J. P. Der, S. W. Graham, S. Mathews, M. Melkonian, D. E. Soltis, P. S. Soltis, N. W. Miles, C. J. Rothfels, L. Pokorny, A. J. Shaw, L. DeGironimo, D. W. Stevenson, B. Surek, J. C. Villarreal, B. Roure, H. Philippe, C. W. dePamphilis, T. Chen, M. K. Deyholos, R. S. Baucom, T. M. Kutchan, M. M. Augustin, J. Wang, Y. Zhang, Z. Tian, Z. Yan, X. Wu, X. Sun, G. K. Wong, and J. Leebens-Mack. 2014. Phylotranscriptomic analysis of the origin and early diversification of land plants. *Proc. Natl. Acad. Sci. USA* 111:E4859–E4868.

Williams, J. H., and W. E. Friedman. 2002. Identification of diploid endosperm in an early angiosperm lineage. *Nature* 415:522–526.

Xi, Z., L. Liu, J. S. Rest, and C. C. Davis. 2014. Coalescent versus concatenation methods and the placement of *Amborella* as sister to water lilies. *Syst. Biol.* 63:919–932.

Zanis, M., P. S. Soltis, Y.-L. Qiu, E. Zimmer, and D. E. Soltis. 2003. Phylogenetic analyses and perianth evolution in basal angiosperms. *Ann. Mo. Bot. Gard.* 90:129–150.

Zeng, L., Q. Zhang, R. Sun, H. Kong, N. Zhang, and H. Ma. 2014. Resolution of deep angiosperm phylogeny using conserved nuclear genes and estimates of early divergence times. *Nat. Commun.* 5:4956.

Zhang, N., L. Zeng, H. Shan, and H. Ma. 2012. Highly conserved low-copy nuclear genes as effective markers for phylogenetic analyses in angiosperms. *New Phytol.* 195:923–937.

Authors

Michael J. Donoghue; Department of Ecology and Evolutionary Biology; Yale University; P.O. Box 208106; New Haven, CT 06520, USA. Email: michael.donoghue@yale.edu.

James A. Doyle; Department of Evolution and Ecology; University of California; Davis, CA 95616, USA. Email: jadoyle@ucdavis.edu.

Philip D. Cantino; Department of Environmental and Plant Biology; Ohio University; Athens, OH 45701, USA. Email: cantino@ohio.edu.

Date Accepted: 29 April 2011; updated 09 February 2018

Primary Editor: Kevin de Queiroz

Magnoliidae Novák ex Takhtajan 1967: 51
[W. S. Judd, P. S. Soltis, and D. E. Soltis],
converted clade name

Registration Number: 64

Definition: The smallest crown clade containing *Canella winterana* (L.) Gaertn. 1788 (*Canellales*), *Magnolia virginiana* L. 1753 (*Magnoliales*), *Cinnamomum camphora* (L.) T. Nees and C. H. Eberm. 1831 (*Laurales*), and *Piper betle* L. 1753 (*Piperales*). This is a minimum-crown-clade definition. Abbreviated definition: min crown ∇ (*Canella winterana* (L.) Gaertn. 1788 & *Magnolia virginiana* L. 1753 & *Cinnamomum camphora* (L.) T. Nees and C. H. Eberm. 1831 & *Piper betle* L. 1753).

Etymology: From the Latinized name of Pierre Magnol (1638–1715), French physician and botanist, and professor of botany and director of the Royal Botanic Garden at Montpellier.

Reference Phylogeny: Qiu et al. (2006: Fig. 1). *Magnolia virginiana* is used as a specifier because it is the type species of *Magnoliales*; although it is not included in the primary reference phylogeny, its close relationship to *Magnolia tripetala*, which was employed in the analysis of Qiu et al. (2006), is supported by Kim et al. (2001). See also Mathews and Donoghue (1999), Qiu et al. (1999, 2000, 2005), Soltis et al. (1999, 2000), Graham and Olmstead (2000), Nickrent et al. (2002), Zanis et al. (2002, 2003), and Hilu et al. (2003).

Composition: *Canellales, Laurales, Magnoliales,* and *Piperales.*

Diagnostic Apomorphies: Non-DNA synapomorphies are problematic; possible synapomorphies may include the phenylpropane compound asarone, the lignans galbacin and veraguensin, and the neolignan licarin (Hegnauer, 1962–1994; Soltis et al., 2005). Stevens (2001) suggested that ovules with a raphal bundle branching at the chalaza and a nucellar cap may also be synapomorphic.

Synonyms: Thorne and Reveal (2007) recognized a subclass (*Magnoliidae*) and superorder (*Magnolianae*) of identical circumscription, which correspond to the clade here named *Magnoliidae*. The name *Magnolianae*, which was also used by Chase and Reveal (2009) for the same clade, is therefore an approximate synonym of *Magnoliidae* as here defined (see also Comments).

Comments: *Magnoliidae*, as circumscribed by Takhtajan (1997) and Cronquist (1988), is significantly different in circumscription from the clade given the informal name "magnoliids" or "eumagnoliids" in many recent publications (Soltis et al., 2000, 2005; Judd et al., 2002, 2008; APG II, 2003; Hilu et al., 2003; Soltis and Soltis, 2004; Simpson, 2006, APG III, 2009). However, the name *Magnoliidae* was formally linked with the latter clade by Thorne and Reveal (2007), who provided a comprehensive list of its subgroups. We phylogenetically defined the name *Magnoliidae* (in Cantino et al. 2007), using the same definition that is used here.

Literature Cited

APG II (Angiosperm Phylogeny Group II). 2003. An update of the Angiosperm Phylogeny Group classification for the orders and families of flowering plants: APG II. *Bot. J. Linn. Soc.* 141:399–436.

APG III (Angiosperm Phylogeny Group III). 2009. An update of the Angiosperm Phylogeny Group classification for the orders and families of flowering plants: APG III. *Bot. J. Linn. Soc.* 161:105–121.

Cantino, P. D., J. A. Doyle, S. W. Graham, W. S. Judd, R. G. Olmstead, D. E. Soltis, P. S. Soltis, and M. J. Donoghue. 2007. Towards a phylogenetic nomenclature of *Tracheophyta*. *Taxon* 56:822–846 and E1–E44.

Chase, M. W., and J. L. Reveal. 2009. A phylogenetic classification of the land plants to accompany APG III. *Bot. J. Linn. Soc.* 161:122–127.

Cronquist, A. 1988. *The Evolution and Classification of Flowering Plants.* 2nd edition. New York Botanical Garden, Bronx, New York.

Graham, S. W., and R. G. Olmstead. 2000. Utility of 17 chloroplast genes for inferring the phylogeny of the basal angiosperms. *Am. J. Bot.* 87:1712–1730.

Hegnauer, R. 1962–1994. *Chemotaxonomie der Pflanzen*. Birkhäusen, Basel.

Hilu, K. W., T. Borsch, K. Müller, D. E. Soltis, P. S. Soltis, V. Savolainen, M. W. Chase, M. P. Powell, L. A. Alice, R. Evans, H. Sauquet, C. Neinhuis, T. A. B. Slotta, J. G. Rohwer, C. S. Campbell, and L. W. Chatrou. 2003. Angiosperm phylogeny based on *matK* sequence information. *Am. J. Bot.* 90:1758–1776.

Judd, W. S., C. S. Campbell, E. A. Kellogg, P. F. Stevens, and M. J. Donoghue. 2002. *Plant Systematics—A Phylogenetic Approach.* 2nd edition. Sinauer Associates, Sunderland, MA.

Judd, W. S., C. S. Campbell, E. A. Kellogg, P. F. Stevens, and M. J. Donoghue. 2008. *Plant Systematics—A Phylogenetic Approach.* 3rd edition. Sinauer Associates, Sunderland, MA.

Kim, S., C.-W. Park, Y.-D. Kim, and Y. Suh. 2001. Phylogenetic relationships in family *Magnoliaceae* inferred from *ndhF* sequences. *Am. J. Bot.* 88:717–728.

Mathews, S., and M. J. Donoghue. 1999. The root of angiosperm phylogeny inferred from duplicate phytochrome genes. *Science* 286:947–949.

Nickrent, D. L., A. Blarer, Y.-L. Qiu, D. E. Soltis, P. S. Soltis, and M. Zanis. 2002. Molecular data place *Hydnoraceae* with *Aristolochiaceae*. *Am. J. Bot.* 89:1809–1817.

Qiu, Y.-L., O. Dombrovska, J. Lee, L. Li, B. A. Whitlock, F. Bernasconi-Quadroni, J. S. Rest, C. C. Davis, T. Borsch, K. W. Hilu, S. S. Renner, D. E. Soltis, P. S. Soltis, M. J. Zanis, J. J. Cannone, R. R. Gutell, M. Powell, V. Savolainen, L. W. Chatrou, and M. W. Chase. 2005. Phylogenetic analyses of basal angiosperms based on nine plastid, mitochondrial, and nuclear genes. *Int. J. Plant Sci.* 166:815–842.

Qiu, Y.-L., J.-Y. Lee, F. Bernasconi-Quadroni, D. E. Soltis, P. S. Soltis, M. Zanis, Z. Chen, V. Savolainen, and M. W. Chase. 1999. The earliest angiosperms: evidence from mitochondrial, plastid and nuclear genomes. *Nature* 402:404–407.

Qiu, Y.-L., J.-Y. Lee, F. Bernasconi-Quadroni, D. E. Soltis, P. S. Soltis, M. Zanis, E. Zimmer, Z. Chen, V. Savolainen, and M. W. Chase. 2000. Phylogeny of basal angiosperms: analyses of five genes from three genomes. *Int. J. Plant Sci.* 161:S3–S27.

Qiu, Y.-L., L. Li, T. A. Hendry, R. Li, D. W. Taylor, M. J. Issa, A. J. Ronen, M. L. Vekaria, and A. M. White. 2006. Reconstructing the basal angiosperm phylogeny: evaluating information content of the mitochondrial genes. *Taxon* 55:837–856.

Simpson, M. G. 2006. *Plant Systematics.* Elsevier, Amsterdam.

Soltis, D. E., P. S. Soltis, M. W. Chase, M. E. Mort, D. C. Albach, M. Zanis, V. Savolainen, W. H. Hahn, S. B. Hoot, M. F. Fay, M. Axtell, S. M. Swensen, L. M. Prince, W. J. Kress, K. C. Nixon, and J. S. Farris. 2000. Angiosperm phylogeny inferred from 18S rDNA, *rbcL*, and *atpB* sequences. *Bot. J. Linn. Soc.* 133:381–461.

Soltis, D. E., P. S. Soltis, P. K. Endress, and M. W. Chase. 2005. *Phylogeny and Evolution of Angiosperms*. Sinauer Associates, Sunderland, MA.

Soltis, P. S., and D. E. Soltis. 2004. The origin and diversification of angiosperms. *Am. J. Bot.* 91:1614–1626.

Soltis, P. S., D. E. Soltis, and M. W. Chase. 1999. Angiosperm phylogeny inferred from multiple genes: a research tool for comparative biology. *Nature* 402:402–404.

Takhtajan, A. 1967. *Sistema I Filogeniia Tesvetkovykh Rastenii (Systema et Phylogenia Magnoliophytorum)*. Bauka, Moscow. [Dated 1966, but published 4 Feb. 1967; J. Reveal, pers. comm.]

Takhtajan, A. 1997. *Diversity and Classification of Flowering Plants*. Columbia University Press, New York.

Thorne, R. F., and J. L. Reveal. 2007. An updated classification of the class *Magnoliopsida* ("*Angiospermae*"). *Bot. Rev.* 73:67–182.

Stevens, P. F. 2001 onwards. Angiosperm Phylogeny Website, Version 9, June 2008 [and more or less continuously updated since]. Available at http://www.mobot.org/MOBOT/research/APweb/.

Zanis, M., D. E. Soltis, P. S. Soltis, S. Mathews, and M. J. Donoghue. 2002. The root of the angiosperms revisited. *Proc. Natl. Acad. Sci. USA* 99:6848–6853.

Zanis, M., P. S. Soltis, Y.-L. Qiu, E. Zimmer, and D. E. Soltis. 2003. Phylogenetic analyses and perianth evolution in basal angiosperms. *Ann. Mo. Bot. Gard.* 90:129–150.

Authors

Walter S. Judd; Department of Biology; University of Florida; Gainesville, FL 32611-8526, USA. Email: lyonia@ufl.edu.

Douglas E. Soltis; Department of Biology; University of Florida; Gainesville, Florida 32611-8526, USA. Email: dsoltis@ufl.edu.

Pamela S. Soltis; Florida Museum of Natural History; University of Florida, Gainesville, Florida 32611-7800, USA. Email: psoltis@flmnh.ufl.edu.

Date Accepted: 22 August 2010

Primary Editor: Philip Cantino

Monocotyledoneae de Candolle 1817: 122
[W. S. Judd, P. S. Soltis, D. E. Soltis and S. W. Graham],
converted clade name

Registration Number: 68

Definition: The smallest crown clade containing *Acorus calamus* Linnaeus 1753, *Gymnostachys anceps* Robert Brown 1810 (*Alismatales*), and *Lilium superbum* Linnaeus 1762 (*Liliales*). This is a minimum-crown-clade definition. Abbreviated definition: min crown ∇ (*Acorus calamus* Linnaeus 1753 & *Gymnostachys anceps* Robert Brown 1810 & *Lilium superbum* Linnaeus 1762).

Etymology: From Greek *monos* (one) and *kotyledon* (cup-shaped cavity), referring to the presence of only one cotyledon in the embryo.

Reference Phylogeny: The primary reference phylogeny is Chase et al. (2006: Fig. 2). See also Chase et al. (1995a,b, 2000), Soltis et al. (2000), Stevenson et al. (2000), Hilu et al. (2003), Davis et al. (2004, 2006), Graham et al. (2006), and Givnish et al. (2006).

Composition: *Acorus, Alismatales, Asparagales, Commelinidae, Dioscoreales, Liliales, Pandanales,* and *Petrosaviales.*

Diagnostic Apomorphies: Clear synapomorphies, at least relative to other crown clades, include an embryo with a single cotyledon, parallel-veined leaves (see Givnish et al., 2005 for secondary evolution of net venation, and Doyle et al., 2008, for additional discussion of this condition), stem with scattered vascular bundles, and sieve tube plastids with cuneate proteinaceous crystalloids (Dahlgren et al., 1985). Monocot-like sieve tube plastids also occur in some *Piperales*, where they apparently evolved

independently. Other possible synapomorphies (Stevens, 2001; Judd et al., 2008) include sheathing leaf base, mature plant with adventitious root system, the loss of a vascular cambium, and sympodial growth. All of these character states occur in other angiosperms, and some of them do not occur in all monocots, but they may still be apomorphies of *Monocotyledoneae*, depending on outgroup and ingroup tree topology.

Synonyms: *Monocotyledones* Jussieu 1789, *Liliopsida* (e.g., Cronquist, 1981), and *Lilianae* sensu Chase and Reveal (2009) are approximate synonyms. The name *Monocotyledonae* appears in the titles of a few papers and floras (22 found in *Kew Bibliographic Databases*), but we are unaware of any major taxonomic work that has used this name, which might best be viewed as a misspelling of *Monocotyledoneae*.

Comments: The monocotyledons have been recognized as a taxon for over 300 years. John Ray, in his *Methodus Plantarum* (1703), divided *Florifera* (which included gymnosperms as well as angiosperms) into *Dicotyledones* and *Monocotyledones*, and he thus was the first botanist to make a major distinction between monocots and other seed plants. They constitute one of the best supported major clades of angiosperms. In addition to several morphological and ultrastructural apomorphies (listed above), the clade is very strongly supported by molecular analyses (see Reference Phylogeny).

There are five names that are commonly applied to this clade: *Monocotyledoneae, Monocotyledonae, Monocotyledones, Lilianae,* and *Liliopsida.* The *Kew Bibliographic Databases* (www.kew.org/kbd/searchpage.do) yielded far more links to

Monocotyledoneae and *Monocotyledones* than the other three names. We also prefer descriptive names based on distinctive apomorphies over nondescriptive, rank-based names unless a name of the latter sort is much more widely used. The corresponding informal names "monocots" and "monocotyledons" have been applied to this clade in nearly all recent phylogenetic treatments of angiosperms (e.g., Chase, 2004; Soltis and Soltis, 2004; Soltis et al., 2005; Simpson, 2006; Judd et al., 2008; APG III, 2009). Our choice of *Monocotyledoneae* over *Monocotyledones* is somewhat arbitrary, but the former appears to have been used in more post-1900 classifications, floras and textbooks. The name *Monocotyledoneae* was phylogenetically defined by us (in Cantino et al., 2007), using nearly the same definition as used here. The only difference is that there was one additional specifier in the 2007 definition. We no longer think it necessary to include *Tofieldia glutinosa* as a specifier because the clade *Alismatales*, which is already represented by another specifier, *Gymnostachys anceps*, is very strongly supported (Hilu et al., 2003; Chase et al., 2006; Givnish et al, 2006; Graham et al., 2006).

Literature Cited

APG III (Angiosperm Phylogeny Group II). 2009. An update of the Angiosperm Phylogeny Group classification for the orders and families of flowering plants: APG III. *Bot. J. Linn. Soc.* 161:105–121.

de Candolle, A. P.. 1817. *Regni Vegetabilis Systema Naturale*. Treuttel & Würtz, Paris.

Cantino, P. D., J. A. Doyle, S. W. Graham, W. S. Judd, R. G. Olmstead, D. E. Soltis, P. S. Soltis, and M. J. Donoghue. 2007. Towards a phylogenetic nomenclature of *Tracheophyta*. *Taxon* 56:822–846 and E1–E44.

Chase, M. W. 2004. Monocot relationships: an overview. *Am. J. Bot.* 91:1645–1655.

Chase, M. W., M. R. Duvall, H. G. Hills, J. G. Conran, A. V. Cox, L. E. Eguiarte, J. Hartwell, M. F. Fay, L. R. Caddick, K. M. Cameron, and S. Hoot. 1995a. Molecular phylogenetics of *Lilianae*. Pp. 109–137 in *Monocotyledons: Systematics and Evolution* (P. J. Rudall, P. J. Cribb, D. F. Cutler, and C. J. Humphries, eds.). Royal Botanic Gardens, Kew.

Chase, M. W., M. F. Fay, D. S. Devey, O. Maurin, N. Rønsted, J. Davies, Y. Pillon, G. Petersen, O. Seberg, M. N. Tamura, C. B. Asmussen, K. Hilu, T. Borsch, J. I. Davis, D. W. Stevenson, J. C. Pires, T. J. Givnish, K. J. Sytsma, M. M. McPherson, S. W. Graham, and H. S. Rai. 2006. Multigene analyses of monocot relationships: a summary. Pp. 62–74 in *Monocots: Comparative Biology and Evolution (Excluding Poales)* (J. T. Columbus, E. A. Friar, J. M. Porter, L. M. Prince, and M. G. Simpson, eds.). Rancho Santa Ana Botanic Garden, Claremont, CA.

Chase, M. W., and J. L. Reveal. 2009. A phylogenetic classification of the land plants to accompany APG III. *Bot. J. Linn. Soc.* 161:122–127.

Chase, M. W., D. W. Stevenson, P. Wilkin, and P. J. Rudall. 1995b. Monocot systematics: a combined analysis. Pp. 685–730 in *Monocotyledons: Systematics and Evolution* (P. J. Rudall, P. J. Cribb, D. F. Cutler, and C. J. Humphries, eds.). Royal Botanic Gardens, Kew.

Chase, M. W., D. E. Soltis, P. S. Soltis, P. J. Rudall, M. F. Fay, W. H. Hahn, S. Sullivan, J. Joseph, M. Molvray, P. J. Kores, T. J. Givnish, K. J. Sytsma, and J. C. Pires. 2000. Higher-level systematics of the monocotyledons: an assessment of current knowledge and a new classification. Pp. 3–16 in *Monocots: Systematics and Evolution* (K. L. Wilson and D. A. Morrison, eds.). CSIRO, Melbourne.

Cronquist, A. 1981. *An Integrated System of Classification of Flowering plants*. Columbia University Press, New York.

Dahlgren, R. M. T., H. T. Clifford, and P. F. Yeo. 1985. *The Families of Monocotyledons: Structure, Evolution, and Taxonomy*. Springer-Verlag, Berlin.

Davis, J. I., G. Petersen, O. Seberg, D. W. Stevenson, C. R. Hardy, M. P. Simmons, F. A. Michelangeli, D. H. Goldman, L. M. Campbell, C. D. Specht, and J. I. Cohen.

2006. Are mitochondrial genes useful for the analysis of monocot relationships? *Taxon* 55:857–870.

Davis, J. I, D. W. Stevenson, G. Petersen, O. Seberg, L. M. Campbell, J. V. Freudenstein, D. H. Goldman, C. R. Hardy, F. A. Michelangeli, M. P. Simmons, C. D. Specht, F. Vergara-Silva, and M. Gandolfo. 2004. A phylogeny of the monocots, as inferred from *rbcL* and *atpA* sequence variation, and a comparison of methods for calculating jackknife and bootstrap values. *Syst. Bot.* 29:467–510.

Doyle, J. A., P. K. Endress, and G. R. Upchurch, Jr. 2008. Early Cretaceous monocots: a phylogenetic evaluation. *Acta Mus. Nat. Pragae, Ser. B Hist. Nat.* 64(2–4):59–87.

Givnish, T. J., J. C. Pires, S. W. Graham, M. A. McPherson, L. M. Prince, T. B. Patterson, H. S. Rai, E. H. Roalson, T. M. Evans, W. J. Hahn, K. C. Millam, A. W. Meerow, M. Molvray, P. J. Kores, H. E. O'Brien, J. C. Hall, W. J. Kress, and K. J. Sytsma. 2005. Repeated evolution of net venation and fleshy fruits among monocots in shaded habitats confirms *a priori* predictions: evidence from an *ndhF* phylogeny. *Proc. Roy. Soc. Lond. B, Biol. Sci.* 272:1481–1490.

Givnish, T. J., J. C. Pires, S. W. Graham, M. A. McPherson, L. M. Prince, T. B. Patterson, H. S. Rai, E. H. Roalson, T. M. Evans, W. J. Hahn, K. C. Millam, A. W. Meerow, M. Molvray, P. J. Kores, H. E. O'Brien, J. C. Hall, W. J. Kress, and K. J. Sytsma. 2006. Phylogenetic relationships of monocots based on the highly informative plastid gene *ndhF*: evidence for widespread concerted convergence. Pp. 27–50 in *Monocots: Comparative Biology and Evolution (Excluding Poales)* (J. T. Columbus, E. A. Friar, J. M. Porter, L. M. Prince, and M. G. Simpson, eds.). Rancho Santa Ana Botanic Garden, Claremont, CA.

Graham, S. W., J. M. Zgurski, M. A. McPherson, D. M. Cherniawsky, J. M. Saarela, E. S. C. Horne, S. Y. Smith, W. A. Wong, H. E. O'Brien, V. L. Biron, J. C. Pires, R. G. Olmstead, M. W. Chase, and H. S. Rai. 2006. Robust inference of monocot deep phylogeny using an expanded multigene plastid data set.

Pp. 3–20 in *Monocots: Comparative Biology and Evolution (Excluding Poales)* (J. T. Columbus, E. A. Friar, J. M. Porter, L. M. Prince, and M. G. Simpson, eds.). Rancho Santa Ana Botanic Garden, Claremont, CA.

Hilu, K. W., T. Borsch, K. Müller, D. E. Soltis, P. S. Soltis, V. Savolainen, M. W. Chase, M. P. Powell, L. A. Alice, R. Evans, H. Sauquet, C. Neinhuis, T. A. B. Slotta, J. G. Rohwer, C. S. Campbell, and L. W. Chatrou. 2003. Angiosperm phylogeny based on *matK* sequence information. *Am. J. Bot.* 90:1758–1776.

Judd, W. S., C. S. Campbell, E. A. Kellogg, P. F. Stevens, and M. J. Donoghue. 2008. *Plant Systematics—A Phylogenetic Approach*. 3rd edition. Sinauer Associates, Sunderland, MA.

de Jussieu, A. L.. 1789. *Genera Plantarum*. Herissant and Barrois, Paris.

Ray, J. 1703. *Methodus Plantarum Emendata*. Samuelis Smith and Benjamini Walford, London.

Simpson, M. G. 2006. *Plant Systematics*. Elsevier, Amsterdam.

Soltis, D. E., P. S. Soltis, M. W. Chase, M. E. Mort, D. C. Albach, M. Zanis, V. Savolainen, W. H. Hahn, S. B. Hoot, M. F. Fay, M. Axtell, S. M. Swensen, L. M. Prince, W. J. Kress, K. C. Nixon, and J. S. Farris. 2000. Angiosperm phylogeny inferred from 18S rDNA, *rbcL*, and *atpB* sequences. *Bot. J. Linn. Soc.* 133:381–461.

Soltis, D. E., P. S. Soltis, P. K. Endress, and M. W. Chase. 2005. *Phylogeny and Evolution of Angiosperms*. Sinauer Associates, Sunderland, MA.

Soltis, P. S., and D. E. Soltis. 2004. The origin and diversification of angiosperms. *Am. J. Bot.* 91:1614–1626.

Stevens, P. F. 2001 and onwards. Angiosperm Phylogeny Website, Version 9, June 2008 [and more or less continuously updated since]. Available at http://www.mobot.org/MOBOT/research/APweb/.

Stevenson, D. W., J. I. Davis, J. V. Freudenstein, C. R. Hardy, M. P. Simmons, and C. D. Specht. 2000. A phylogenetic analysis of the monocotyledons based on morphological and molecular character sets, with comments on the placement of *Acorus* and *Hydatellaceae*.

Pp. 17–24 in *Monocots: Systematics and Evolution* (K. L. Wilson and D. A. Morrison, eds.). CSIRO, Melbourne.

Authors

Walter S. Judd; Department of Biology; University of Florida; Gainesville, FL 32611, USA. E-mail: lyonia@ufl.edu.

Pamela S. Soltis; Florida Museum of Natural History; University of Florida; Gainesville, FL 32611-7800, USA. E-mail: psoltis@flmnh.ufl.edu.

Douglas E. Soltis; Department of Biology; University of Florida; Gainesville, FL 32611, USA. Email: dsoltis@ufl.edu.

Sean W. Graham; UBC Botanical Garden & Centre for Plant Research, and Dept. of Botany; University of British Columbia; Vancouver, BC, V6T 1Z4, Canada. Email: swgraham@ interchange.ubc.ca.

Date Accepted: 13 September 2010

Primary Editor: Philip Cantino

Petrosaviidae S. W. Graham and W. S. Judd in Cantino et al., 2007:834 [S. W. Graham, W. S. Judd and W. J. D. Iles], converted clade name

Registration Number: 84

Definition: The smallest crown clade containing *Lilium candidum* L. 1753 and *Petrosavia stellaris* Becc. 1871 but not *Acorus calamus* L. 1753 or *Alisma plantago-aquatica* L. 1753. This is a minimum-crown-clade definition with external specifiers. Abbreviated definition: min crown ∇ (*Lilium candidum* L. 1753 & *Petrosavia stellaris* Becc. 1871 ~ *Acorus calamus* L. 1753 ∨ *Alisma plantago-aquatica* L. 1753).

Etymology: Derived from *Petrosavia* Becc. 1871, the name of an included taxon, named in honor of the Italian botanist Pietro Savi, 1811–1871 (Cameron et al., 2003).

Reference Phylogeny: The primary reference phylogeny is Tamura et al. (2004: Fig 1). The specifiers *Lilium candidum* and *Petrosavia stellaris* are, respectively, most closely related to *Lilium regale* and *Petrosavia sakuraii* in the reference phylogeny (see Comments). The genus *Alisma* in this reference phylogeny is a composite taxon comprising two species, one of which is the relevant specifier species. See also Chase et al. (2006: Fig. 2) and Graham et al. (2006: Fig. 1B).

Composition: The crown clades *Asparagales*, *Commelinidae* (this volume), *Dioscoreales*, *Liliales*, *Pandanales*, and *Petrosaviales*.

Diagnostic Apomorphies: Stevens (2001+) suggested several possible synapomorphies for the clade, including simple, amylophobic starch grains, and a lack of colleters and cyanogenic glycosides, but data on the distributions of these characters are incomplete, and these hypotheses have not been formally examined in a phylogenetic analysis.

Synonyms: None.

Comments: The name *Petrosaviidae* was formally applied to this clade, with an explicit phylogenetic definition, by Graham and Judd in Cantino et al. (2007: 834 and E25). The definition used here is a modification of our earlier one. Specifically, one internal specifier, *Typha latifolia*, is deleted here because it is unnecessary; that is, both it and *Lilium candidum* represent the same well supported clade, the sister group of *Petrosaviales*. Secondly, we now use type species to represent *Liliales* (internal specifier) and *Alismatales* (external specifier). (The other specifiers were used in our 2007 definition and are also type species.) The clade *Petrosaviidae* is very well supported in all recent molecular phylogenetic analyses (Tamura et al., 2004; Davis et al., 2004, 2006; Chase et al., 2006; Graham et al., 2006; see also Cameron et al., 2003). To be cautious, however, we have included two external specifiers in case extreme rate elevation found in some genes in *Acorus* and some members of *Alismatales* (see examples in Davis et al, 2004; Petersen et al., 2006) has contributed to inaccurate phylogenetic inference near the base of *Monocotyledoneae* (this volume). The name *Petrosaviidae* would not be applicable if *Acorus* or *Alismatales* were found to be part of the least inclusive clade containing *Petrosavia* and *Lilium*.

Literature Cited

Cameron, K. M., M. W. Chase, and P. J. Rudall. 2003. Recircumscription of the monocotyledonous family *Petrosaviaceae* to include *Japonolirion*. *Brittonia* 55:214–225.

Cantino, P. D., J. A. Doyle, S. W. Graham, W. S. Judd, R. G. Olmstead, D. E. Soltis, P. S. Soltis, and M. J. Donoghue. 2007. Towards a phylogenetic nomenclature of *Tracheophyta*. *Taxon* 56:822–846 and E1–E44.

Chase, M. W., M. F. Fay, D. S. Devey, O. Maurin, N. Rønsted, J. Davies, Y. Pillon, G. Petersen, O. Seberg, M. N. Tamura, C. B. Asmussen, K. Hilu, T. Borsch, J. I Davis, D. W. Stevenson, J. C. Pires, T. J. Givnish, K. J. Sytsma, M. A. McPherson, S. W. Graham, and H. S. Rai. 2006. Multigene analyses of monocot relationships: a summary. Pp. 62–74 in *Monocots: Comparative Biology and Evolution (Excluding* Poales) (J. T. Columbus, E. A. Friar, J. M. Porter, L. M. Prince, and M. G. Simpson, eds.). Rancho Santa Ana Botanic Garden, Claremont, CA.

Davis, J. I., G. Petersen, O. Seberg, D. W. Stevenson, C. R. Hardy, M. P. Simmons, F. A. Michelangeli, D. H. Goldman, L. M. Campbell, C. D. Specht, and J. I. Cohen. 2006. Are mitochondrial genes useful for the analysis of monocot relationships? *Taxon* 55:857–870.

Davis, J. I., D. W. Stevenson, G. Petersen, O. Seberg, L. M. Campbell, J. V. Freudenstein, D. H. Goldman, C. R. Hardy, F. A. Michelangeli, M. P. Simmons, C. D. Specht, F. Vergara-Silva, and M. Gandolfo. 2004. A phylogeny of the monocots, as inferred from *rbcL* and atpA sequence variation, and a comparison of methods for calculating jackknife and bootstrap values. *Syst. Bot.* 29:467–510.

Graham, S. W., J. M. Zgurski, M. A. McPherson, D. M. Cherniawsky, J. M. Saarela, E. S. C. Horne, S. Y. Smith, W. A. Wong, H. E. O'Brien, V. L. Biron, J. C. Pires, R. G. Olmstead, M. W. Chase, and H. S. Rai. 2006. Robust inference of monocot deep phylogeny using an expanded multigene plastid data set. Pp. 3–20 in *Monocots: Comparative Biology and Evolution (Excluding* Poales) (J. T. Columbus, E. A. Friar, J. M. Porter, L. M. Prince, and M. G. Simpson, eds.). Rancho Santa Ana Botanic Garden, Claremont, CA.

Petersen, G., O. Seberg, J. I. Davis, D. H. Goldman, D. W. Stevenson, L. M. Campbell, F. A. Michelangeli, C. D. Specht, M. W. Chase, M. F. Fay, J. C. Pires, J. V. Freudenstein, C. R. Hardy, and M. P. Simmons. 2006. Mitochondrial data in monocot phylogenetics. Pp. 52–64 in *Monocots: Comparative Biology and Evolution (Excluding* Poales) (J. T. Columbus, E. A. Friar, J. M. Porter, L. M. Prince, and M. G. Simpson, eds.). Rancho Santa Ana Botanic Garden, Claremont, CA.

Stevens, P. F. 2001 onwards. Angiosperm Phylogeny Website. Available at http://www.mobot.org/mobot/research/apweb/, accessed on 25 September 2010.

Tamura, M. N., J. Yamashita, S. Fuse, and M. Haraguchi. 2004. Molecular phylogeny of monocotyledons inferred from combined analysis of plastid *matK* and *rbcL* gene sequences. *J. Plant Res.* 117:109–5120.

Authors

Sean W. Graham; UBC Botanical Garden & Centre for Plant Research, and Dept. of Botany; University of British Columbia; Vancouver, BC, V6T 1Z4, Canada. Email: swgraham@interchange.ubc.ca.

Walter S. Judd; Department of Biology; University of Florida; Gainesville, FL 32611, USA. Email: yonia@ufl.edu.

William J. D. Iles; UBC Botanical Garden & Centre for Plant Research, and Dept. of Botany; University of British Columbia; Vancouver, BC, V6T 1Z4, Canada. Email: will.iles@botany.ubc.ca.

Date Accepted: 28 September 2011

Primary Editor: Philip Cantino

Commelinidae A. Takhtajan 1967: 171
[S. W. Graham, J. M. Saarela and W. S. Judd],
converted clade name

Registration Number: 30

Definition: The smallest crown clade containing *Commelina communis* L. 1753 (*Commelinales*), *Dasypogon hookeri* J. Drumm. 1843 (*Dasypogonaceae*), *Oryza sativa* L. 1753 (*Poales*) and *Roystonea princeps* (Becc.) Burret 1929 (*Arecales*). This is a minimum-crown-clade definition. Abbreviated definition: min crown ∇ (*Commelina communis* L. 1753 & *Dasypogon hookeri* J. Drumm. 1843 & *Oryza sativa* L. 1753 & *Roystonea princeps* (Becc.) Burret 1929).

Etymology: Derived from *Commelina* L. 1753, an included genus named in honor of the Dutch botanists Jan Commelijn (1629–1692) and Caspar(us) Commelijn (1667/1668–1731) (Stearn, 2002).

Reference Phylogeny: The primary reference phylogeny is Saarela et al. (2008: Fig. 2); see also Chase et al. (2006: Fig. 3), Givnish et al. (2006: Fig. 1B, 2010: Fig. 3) and Graham et al. (2006: Fig. 1B). We include the type species of *Commelina* according to rank-based systems of classification as an internal specifier in our definition, as *Commelina* provides the stem of the clade name. *Commelina communis* is most closely related to *Tradescantia ohiensis* or *Palisota bogneri* in the reference phylogeny (accounting for uncertainty in the phylogeny of *Commelinaceae*; Evans et al., 2003).

Composition: *Arecales*, *Commelinales*, *Dasypogonaceae*, *Poales*, and *Zingiberales*.

Diagnostic Apomorphies: UV-fluorescent ferulic acid in cell walls is an unreversed synapomorphy (Dahlgren and Rasmussen, 1983; Clark et al., 1993; Givnish et al., 1999; Harris and Trethewey, 2010; the latter authors also note that ferulic acid is ester-linked to a non-cellulosic polysaccharide, glucurono-arabinoxylan, in the primary cell wall in this clade). "*Strelitzia*-type" epicuticular wax sculpturing, with wax crystalloids aggregated as rod-like projections (Dahlgren et al., 1985: 65) is a probable synapomorphy, but there were multiple losses within the clade and presumed convergences outside it (Dahlgren and Rasmussen, 1983; Clark et al., 1993; Givnish et al., 1999). Other possible synapomorphies include starchy pollen and endosperm, silica bodies and bracteate inflorescences (but see the caveats noted by Graham and Judd in Cantino et al., 2007: E25).

Synonyms: There are no synonymous scientific names, but the informal names "commelinoids" and "commelinids" (or "commelinoid monocots," etc.) have been applied to this clade in many publications (e.g., APG, 1998; APG II, 2003; APG III, 2009; Chase, 2004; Chase et al., 1995, 2000, 2006; Graham et al., 2006; Zona, 2001).

Comments: The circumscription of *Commelinidae* in the classification by Takhtajan (2009) is very similar to the composition noted here, although he excluded *Arecales*, in addition to more minor differences. The name *Commelinidae* was applied formally to the clade of concern here by Givnish et al. (1999), and was given an explicit phylogenetic definition by Graham and Judd in Cantino et al. (2007: E25).

The same definition is adopted here. The clade is well supported in Givnish et al. (1999) and large-scale phylogenetic studies (e.g., Chase et al., 2006; Davis et al., 2006; Graham et al., 2006; Saarela et al., 2008; Givnish et al., 2010). The reference phylogeny cited here includes closer relatives of *Commelina* than those in the reference phylogeny for the earlier phylogenetic definition. We include four specifiers because of persistent uncertainty about relationships among most of the taxa in the composition list. For example, while a whole-plastid genome study by Givnish et al. (2010) yielded a moderately to strongly supported set of relationships for the major clades represented by the specifiers, the authors noted that these relationships should be considered to be tentative, because conflicting relationships were recovered using different phylogenetic methods. One major component clade, *Zingiberales*, is not represented by a specifier here because there is consistent and substantial support in multigene phylogenetic analyses (Chase et al., 2006; Davis et al., 2006; Graham et al., 2006; Saarela et al., 2008; Givnish et al., 2010) for a sister-group relationship between *Zingiberales* and *Commelinales*, and the latter is represented by a specifier.

Literature Cited

APG (Angiosperm Phylogeny Group). 1998. An ordinal classification for the families of flowering plants. *Ann. Mo. Bot. Gard.* 85:531–553.

APG II (Angiosperm Phylogeny Group II). 2003. An update of the Angiosperm Phylogeny Group classification for the orders and families of flowering plants: APG II. *Bot. J. Linn. Soc.* 141:399–436.

APG III (Angiosperm Phylogeny Group III). 2009. An update of the Angiosperm Phylogeny Group classification for the orders and families of flowering plants: APG III. *Bot. J. Linn. Soc.* 161:105–121.

Cantino, P. D., J. A. Doyle, S. W. Graham, W. S. Judd, R. G. Olmstead, D. E. Soltis, P. S. Soltis, and M. J. Donoghue. 2007. Towards a phylogenetic nomenclature of *Tracheophyta*. *Taxon* 56:E1–E44.

Chase, M. W. 2004. Monocot relationships: an overview. *Am. J. Bot.* 91:1645–1655.

Chase, M. W., M. R. Duvall, H. G. Hills, J. G. Conran, A. V. Cox, L. E. Eguiarte, J. Hartwell, M. F. Fay, L. R. Caddick, K. M. Cameron, and S. Hoot. 1995. Molecular phylogenetics of *Lilianae*. Pp. 109–137 in *Monocotyledons: Systematics and Evolution* (P. J. Rudall, P. J. Cribb, D. F. Cutler and C. J. Humphries, eds.). Royal Botanic Gardens, Kew.

Chase, M. W., M. F. Fay, D. Devey, N. Rønsted, J. Davies, Y. Pillon, G. Petersen, O. Seberg, M. N. Tamura, C. B. Asmussen, K. Hilu, T. Borsch, J. I. Davis, D. W. Stevenson, J. C. Pires, T. J. Givnish, K. J. Sytsma, M. A. McPherson, S. W. Graham, and H. S. Rai. 2006. Multigene analyses of monocot relationships: a summary. *Aliso* 22:63–75.

Chase, M. W., D. E. Soltis, P. S. Soltis, P. J. Rudall, M. F. Fay, W. H. Hahn, S. Sullivan, J. Joseph, M. Molvray, P. J. Kores, T. J. Givnish, K. J. Sytsma, and J. C. Pires. 2000. Higher-level systematics of the monocotyledons: an assessment of current knowledge and a new classification. Pp. 3–16 in *Monocots: Systematics and Evolution* (K. L. Wilson and D. A. Morrison, eds.). CSIRO, Melbourne.

Clark, W. D., B. S. Gaut, M. R. Duvall, and M. T. Clegg. 1993. Phylogenetic relationships of the *Bromeliiflorae-Commeliniflorae-Zingiberiflorae* complex of monocots based on *rbc*L sequence comparisons. *Ann. Mo. Bot. Gard.* 80:987–998.

Dahlgren, R. M. T., and F. N. Rasmussen. 1983. Monocotyledon evolution: characters and phylogenetic estimation. Pp. 255–395 in *Evolutionary Biology*, Vol. 16 (M. K. Hecht, B. Wallace, and G. T. Prance, eds.). Plenum Publ. Corp., New York, U.S.A.

Dahlgren, R. M. T., H. T. Clifford, and P. F. Yeo. 1985. *The Families of Monocotyledons: Structure, Evolution and Taxonomy*. Springer-Verlag, Berlin.

Davis, J. I., G. Petersen, O. Seberg, D. W. Stevenson, C. R. Hardy, M. P. Simmons, F. A. Michelangeli, D. H. Goldman, L. M. Campbell, C. D. Specht, and J. I. Cohen. 2006. Are mitochondrial genes useful for the analysis of monocot relationships? *Taxon* 55:857–870.

Evans, T. M., K. J. Sytsma, R. B. Faden, and T. J. Givnish. 2003. Phylogenetic relationships in the *Commelinaceae*: II. A cladistic analysis of *rbcL* sequences and morphology. *Syst. Bot.* 28:270–212.

Givnish, T. J., T. M. Evans, J. C. Pires, and K. J. Sytsma. 1999. Polyphyly and convergent morphological evolution in *Commelinales* and *Commelinidae*: evidence from *rbcL* sequence data. *Mol. Phylogenet. Evol.* 12:360–385.

Givnish, T. J., J. C. Pires, S. W. Graham, M. A. McPherson, L. M. Prince, T. B. Patterson, H. S. Rai, E. H. Roalson, T. M. Evans, W. J. Hahn, K. C. Millam, A. W. Meerow, M. Molvray, P. J. Kores, H. E. O'Brien, J. C. Hall, W. J. Kress, and K. J. Sytsma. 2006. Phylogenetic relationships of monocots based on the highly informative plastid gene *ndhF*: evidence for widespread concerted convergence. *Aliso* 22:28–51.

Givnish, T. J., M. A. Sevillano, J. R. McNeal, M. R. McKain, P. R. Steele, C. W. dePamphilis, S. W. Graham, J. C. Pires, D. W. Stevenson, W. B. Zomlefer, B. G. Briggs, M. L. Duvall, M. J. Moore, D. E. Soltis, P. S. Soltis, K. Thiele, and J. H. Leebens-Mack. 2010. Assembling the tree of the monocotyledons: plastome sequence phylogeny and evolution of Poales. *Ann. Mo. Bot.* 97:584–616.

Graham, S. W., J. M. Zgurski, M. A. McPherson, D. M. Cherniawsky, J. M. Saarela, E. F. C. Horne, S. Y. Smith, W. A. Wong, H. E. O'Brien, V. L. Biron, J. C. Pires, R. G. Olmstead, M. W. Chase, and H. S. Rai. 2006. Robust inference of monocot deep phylogeny using an expanded multigene plastid data set. *Aliso* 22:3–21.

Harris, P. J., and J. A. K. Trethewey. 2010. The distribution of ester-linked ferulic acid in the cell walls of angiosperms. *Phytochem. Rev.* 9:19–33.

Saarela, J. M., P. J. Prentis, H. S. Rai, and S. W. Graham. 2008. Phylogenetic relationships in the monocots order *Commelinales*, with a focus on *Philydraceae*. *Botany* 86:719–731.

Stearn, W. T. 2002. *Dictionary of Plant Names for Gardeners*. Timber Press, Inc., Portland, OR.

Takhtajan, A. 1967. *Sistema I Filogeniia Tsvetkovykh Rastenii (Systema et Phylogenia Magnoliophytorum)*. Nauka, Moscow. [Dated 1966, but published 4 Feb. 1967.]

Takhtajan, A. 2009. *Flowering Plants*. Springer, New York.

Zona, S. 2001. Starchy pollen in commelinoid monocots. *Ann. Bot.* 87:109–116.

Authors

Sean W. Graham; UBC Botanical Garden & Centre for Plant Research, and Dept. of Botany; University of British Columbia; Vancouver, BC, V6T 1Z4, Canada. Email: swgraham@interchange.ubc.ca.

Jeffery M. Saarela; Canadian Museum of Nature; P.O. Box 3443, Station D; Ottawa, ON, K1P 6P4, Canada. Email: jsaarela@mus-nature.ca.

Walter S. Judd; Department of Biology; University of Florida; Gainesville, FL 32611, USA. Email: lyonia@ufl.edu.

Date Accepted: 30 September 2011

Primary Editor: Philip Cantino

Poineae J. M. Saarela and S. W. Graham, new clade name

Registration Number: 85

Definition: The smallest crown clade containing *Poa pratensis* Linnaeus 1753 (*Poaceae*), *Flagellaria indica* Linnaeus 1753 (*Flagellariaceae*), and *Restio paludosus* Pillans 1922 (*Restionaceae*). This is a minimum-crown-clade definition. Abbreviated definition: min crown ∇ (*Poa pratensis* L. 1753 & *Flagellaria indica* Linnaeus 1753 & *Restio paludosus* Pillans 1922).

Etymology: Derived from *Poa*, the name of the largest included genus, which is an ancient Greek word for grass or fodder (Fernald, 1970).

Reference Phylogeny: The reference phylogeny is Bremer (2002: Fig. 1, where *Poineae* is labeled "graminoid clade"). *Poa pratensis* is used as a specifier because in rank-based nomenclature it is the type of the genus *Poa*, from which the name *Poineae* is derived. Of the taxa of *Poaceae* included in the reference phylogeny, *Poa* is most closely related to *Bambusa* (e.g., see Bouchenak-Khelladi et al., 2008).

Composition: *Anarthriaceae, Centrolepidaceae, Ecdeiocoleaceae, Flagellariaceae, Joinvilleaceae, Poaceae,* and *Restionaceae.*

Diagnostic Apomorphies: Possible apomorphies include presence of girdle-like endothecial wall thickenings, presence of scrobiculi, pollen aperture margins annulate, branched stigmas, orthotropous ovules, and shoot apices not sunken, although all of these characters may be homoplasious in the broader context of *Commelinidae* phylogeny (see Table 3 in Michelangeli et al., 2003).

Synonyms: The formal name *Poales* and informal names "graminoid clade", "graminids", and "core *Poales*" are approximate synonyms (see Comments).

Comments: The clade here named *Poineae* was first inferred in a combined analysis of morphological and *rbc*L data by Linder and Kellogg (1995), and later by Michelangeli et al. (2003) in an analysis of monocot phylogeny based on morphology and two genes. The clade has generally been inferred with strong support in large-scale molecular studies that have sampled different combinations of genes and taxa in *Poales* and relatives (i.e., Bremer 2002; Chase et al., 2006; Givnish et al, 2010; Graham et al., 2006; Saarela and Graham, 2010), although a weakly conflicting relationship was recovered in some analyses in Davis et al. (2004). While the composition of the clade seems clear, inferences about relationships among its components vary, including the identity of the sister group of the grasses, and the affinities of *Flagellariaceae* and *Centrolepidaceae* (e.g., above references; Doyle et al., 1992; Katayama and Ogihara, 1996; Briggs et al., 2000; Linder et al., 2000; Davis et al., 2004, 2006; Linder and Rudall, 2005; Marchant and Briggs, 2007; Briggs et al., 2010).

Multiple names, both formal and informal, have been applied to the clade here named *Poineae*. *Poales* was used widely for this clade in morphological and early molecular-based classifications (e.g., Dahlgren, 1980; Dahlgren and Clifford, 1982; Dahlgren et al. 1985; Campbell and Kellogg, 1987; Linder, 1987; Kellogg and Linder, 1995; Kubitzki, 1998), whereas the Angiosperm Phylogeny Group circumscribed *Poales* more broadly (APG, 1998; APG II, 2003; APG III, 2009). The clade *Poineae* has since been referred to as "core *Poales*" or "*Poales* s.s.*" (e.g., Michelangeli et al., 2003; Ramos et al., 2005; Marchant and Briggs, 2007), though some authors have construed "core *Poales*" more broadly (e.g., Linder and Rudall, 2005) or narrowly (e.g., Leitch et al., 2010). The informal names "graminids" and "graminoids" are widely used, particularly in the non-scientific literature, to refer to grasses and grass-like plants, including sedges and rushes, which are not included in *Poineae*. However, "graminoid clade" (e.g., Bremer, 2002; Trethewey et al., 2005), "graminoid Poales" (Marchant and Briggs, 2007) and "graminids" (e.g., Chase, 2004) have been used for *Poineae*. "Graminids" has also been used for a less inclusive clade comprising *Ecdeiocoleaceae*, *Flagellariaceae*, *Joinvilleaceae*, and *Poaceae*, with the clade comprising *Anarthriaceae*, *Centrolepidaceae*, and *Restionaceae* referred to as "restiids", and *Poineae* referred to as the "graminid-restiid clade" (Linder and Rudall, 2005; Rudall et al., 2005; Givnish et al. 2010).

The name *Poineae* was used by Thorne (1968, 1992) in his earlier classifications for a suborder that included only *Poaceae*. Shipunov (2003) used *Poineae* for a group with nearly the same composition as *Poineae* here but including the distantly related *Dasypogonaceae*; he later removed *Dasypogonaceae* from *Poineae* (Shipunov, 2007, 2010). *Poineae* is designated here as a new clade name because it has not been validated (i.e., with a description or diagnosis, or with reference to one that is validly published) under the botanical code (Turland et al., 2018; see Reveal, 2010), and thus does not qualify as a preexisting name under the *PhyloCode* (Art. 6.2; Cantino and de Queiroz, 2020).

Literature Cited

APG (Angiosperm Phylogeny Group). 1998. An ordinal classification for the families of flowering plants. *Ann. Mo. Bot. Gard.* 85:531–553.

APG II (Angiosperm Phylogeny Group II). 2003. An update of the Angiosperm Phylogeny Group classification for the orders and families of flowering plants: APG II. *Bot. J. Linn. Soc.* 141:399–436.

APG III (Angiosperm Phylogeny Group III). 2009. An update of the Angiosperm Phylogeny Group classification for the orders and families of flowering plants: APG III. *Bot. J. Linn. Soc.* 161:105–121.

Bouchenak-Khelladi, Y., N. Salamin, V. Savolainen, F. Forest, M. van der Bank, M. W. Chase, and T. R. Hodkinson. 2008. Large multi-gene phylogenetic trees of the grasses (*Poaceae*): progress towards complete tribal and generic level sampling. *Mol. Phylogenet. Evol.* 47:488–505.

Bremer, K. 2002. Gondwanan evolution of the grass alliance of families (*Poales*). *Evolution* 56:1374–1387.

Briggs, B. G., A. D. Marchant, S. Gilmore, and C. L. Porter. 2000. A molecular phylogeny of *Restionaceae* and allies. Pp. 661–671 in *Monocots: Systematics and Evolution* (K. L. Wilson and D. A. Morrisson, eds.). CSIRO, Collingwood, Victoria.

Briggs, B.G., A. D. Marchant, and A. J. Perkins 2010. Phylogeny and features in *Restionaceae*, *Centrolepidaceae* and *Anarthriaceae* (restiid clade of *Poales*). Pp. 357–388 in *Diversity, Phylogeny, and Evolution in the Monocotyledons* (O. Seberg, G. Petersen, A. Barfod and J. I Davis, eds.). Aarhus University Press, Aarhus.

Campbell, C. S., and E. A. Kellogg. 1987. Sister group relationships of the *Poaceae*. Pp. 217–224 in *Grass Systematics and Evolution* (T. R. Soderstrom, K. W. Hilu, C. S. Campbell, and M.E. Barkworth, eds.). Smithsonian Institution Press, Washington, DC.

Cantino, P. D., and K. de Queiroz. 2020. *International Code of Phylogenetic Nomenclature (PhyloCode)*, Version 6. CRC Press, Boca Raton, FL.

Chase, M. W. 2004. Monocot relationships: an overview. *Am. J. Bot.* 91:1645–1655.

Chase, M. W., M. F. Fay, D. Devey, N. Rønsted, J. Davies, Y. Pillon, G. Petersen, O. Seberg, M. N. Tamura, C. B. Asmussen, K. Hilu, T. Borsch, J. I. Davis, D. W. Stevenson, J. C. Pires, T. J.

Givnish, K. J. Sytsma, M. A. McPherson, S. W. Graham, and H. S. Rai. 2006. Multigene analyses of monocot relationships: a summary. *Aliso* 22:63–75.

Dahlgren, R. M. T. 1980. A revised system of classification of the angiosperms. *Bot. J. Linn. Soc.* 80:91–124.

Dahlgren, R. M. T., and H. T. Clifford. 1982. *The Monocotyledons: A Comparative Study.* Academic Press, London.

Dahlgren, R. M. T., H. T. Clifford, and P. F. Yeo. 1985. *The Families of Monocotyledons: Structure, Evolution and Taxonomy.* Springer-Verlag, Berlin.

Davis, J. I, G. Petersen, O. Seberg, D. W. Stevenson, C. R. Hardy, M. P. Simmons, F. A. Michelangeli, D. H. Goldman, L. M. Campbell, C. D. Specht, and J. I. Cohen 2006. Are mitochondrial genes useful for the analysis of monocot relationships? *Taxon* 55:857–870.

Davis J. I., D. W. Stevenson, G. Peterson, O. Seberg, L. M. Campbell, J. V. Freudenstein, D. H. Goldman, C. R. Hardy, F. A. Michelangeli, M. P. Simmons, C. D. Specht, F. Vergara-Silva, and M. A. Gandolfo. 2004. A phylogeny of the monocots, as inferred from *rbcL* and *atpA* sequence variation. *Syst. Bot.* 29: 467–510.

Doyle J. J., J. I Davis, R. J. Soreng, D. Garvin, and M. J. Anderson. 1992. Chloroplast DNA inversions and the origin of the grass family (*Poaceae*). *Proc. Natl. Acad. Sci. USA* 89:7722–7726.

Fernald, M. L. 1970. *Gray's Manual of Botany.* 8th edition, corrected printing. Van Nostrand Co., New York.

Givnish, T. J., M. A. Sevillano, J. R. McNeal, M. R. McKain, P. R. Steele, C. W. dePamphilis, S. W. Graham, J. C. Pires, D. W. Stevenson, W. B. Zomlefer, B. G. Briggs, M. L. Duvall, M. J. Moore, D. E. Soltis, P. S. Soltis, K. Thiele, and J. H. Leebens-Mack. 2010. Assembling the tree of the monocotyledons: plastome sequence phylogeny and evolution of *Poales*. *Ann. Mo. Bot. Gard.* 97:584–616.

Graham, S. W., J. M. Zgurski, M. A. McPherson, D. M. Cherniawsky, J. M. Saarela, E. F. C. Horne, S. Y. Smith, W. A. Wong, H. E. O'Brien, V. L. Biron, J. C. Pires, R. G. Olmstead, M. W. Chase, and H. S. Rai. 2006. Robust inference of monocot deep phylogeny using an expanded multigene plastid data set. *Aliso* 22:3–21.

Katayama, H., and Y. Ogihara. 1996. Phylogenetic affinities of the grasses to other monocots as revealed by molecular analysis of chloroplast DNA. *Curr. Genet.* 29:572–581.

Kellogg, E. A., and H. P. Linder. 1995. Phylogeny of *Poales*. Pp. 511–542 in *Monocotyledons: Systematics and Evolution* (P. J. C. Rudall, D. F. Cutler, and C. J. Humphries, eds.). Royal Botanic Gardens, Kew.

Kubitzki, K., ed. 1998. *The Families and Genera of Vascular Plants, Vol. 3. Flowering Plants, Monocotyledons*: Alismatanae *and* Commelinanae *(Except* Gramineae*).* Springer-Verlag, Berlin.

Leitch, I. J., J. M. Beaulieu, M. W. Chase, A. R. Leitch, and M. F. Fay. 2010. Genome size dynamics and evolution in monocots. *J. Bot.*, 2010:862516 (18 pp.).

Linder, H. P. 1987. The evolutionary history of the *Poales/Restionales* – a hypothesis. *Kew Bull.* 42:297–318.

Linder, H. P., B. G. Briggs, and L. A. S. Johnson. 2000. *Restionaceae*—a morphological phylogeny. Pp. 653–660 in *Monocots: Systematics and Evolution* (K. L. Wilson and D. A. Morrison, eds.). CSIRO, Melbourne.

Linder, H. P., and E. A. Kellogg. 1995. Phylogenetic patterns in the commelinid clade. Pp. 473–497 in *Monocotyledons: Systematics and Evolution* (P. J. C. Rudall, D. F. Cutler, and C. J. Humphries, eds.). Royal Botanic Gardens Kew.

Linder, H. P., and P. J. Rudall. 2005. Evolutionary history of *Poales*. *Ann. Rev. Ecol. Evol. Syst.* 36:107–124.

Marchant, A. D., and B. G. Briggs. 2007. *Ecdeiocoleaceae* and *Joinvilleaceae*, sisters of *Poaceae* (*Poales*): evidence from *rbcL* and *matK* data. *Telopea* 11:437–450.

Michelangeli, F. A., J. I. Davis, and D. W. Stevenson. 2003. Phylogenetic relationships among *Poaceae* and related families as inferred from morphology, inversions in the plastid genome, and sequence data from the mitochondrial and plastid genomes. *Am. J. Bot.* 90:93–106.

Ramos, C., E. Borba, and L. Funch. 2005. Pollination in Brazilian *Syngonanthus* (*Eriocaulaceae*) species: evidence for entomophily instead of anemophily. *Ann. Bot.* 96:387–397.

Reveal, J. L. 2010. A checklist of familial and suprafamilial names for extant vascular plants. *Phytotaxa* 6:1–402.

Rudall, P. J., W. Stuppy, J. Cunniff, E. A. Kellogg, and B. G. Briggs. 2005. Evolution of reproductive structures in grasses (*Poaceae*) inferred by sister-group comparison with their putative closest living relatives, *Ecdeiocoleaceae*. *Am. J. Bot.* 92:1432–1443.

Saarela, J. M., and S. W. Graham. 2010. Inference of phylogenetic relationships among the subfamilies of grasses (*Poaceae*: *Poales*) using meso-scale exemplar-based sampling of the plastid genome. *Botany* 88:65–84.

Shipunov, A. B. 2003. The system of flowering plants: synthesis of classical and molecular approaches. *Zhurnal Obshchei Biologii* 64:501–509 [in Russian].

Shipunov, A. B. 2007. *Systema Angiospermarum*, v. 4.804 (January 28, 2007). Available at http://herba.msu.ru/shipunov/ang/v4.8/syang.pdf, accessed on 10 November 2010 (archived by WebCite® at http://www.webcitation.org/5u8lBBkCu).

Shipunov, A. B. 2010. *Systema Angiospermarum*, v. 5.0 (11 August 2010). Available at http://herba.msu.ru/shipunov/ang/current/syang.pdf, accessed on 10 November 2010 (archived by WebCite® at http://www.webcitation.org/5u8kHpnOA).

Thorne, R. F. 1968. Synopsis of a putatively phylogenetic classification of the flowering plants. *Aliso* 6:57–66.

Thorne, R. F. 1992. An updated phylogenetic classification of the flowering plants. *Aliso* 13:365–389.

Trethewey, J. A. K., L. M. Campbell, and P. J. Harris. 2005. (1->3),(1->4)-{beta}-d-Glucans in the cell walls of the *Poales* (sensu lato): an immunogold labeling study using a monoclonal antibody. *Am. J. Bot.* 92:1660–1674.

Turland, N. J., J. H. Wiersema, F. R. Barrie, W. Greuter, D. L. Hawksworth, P. S. Herendeen, S. Knapp, W.-H. Kusber, D.-Z. Li, K. Marhold, T. W. May, J. McNeill, A. M. Monro, J. Prado, M. J. Price, and G. F. Smith, eds. 2018. *International Code of Nomenclature for Algae, Fungi, and Plants (Shenzhen Code)*. Adopted by the Nineteenth International Botanical Congress, Shenzhen, China, July 2017. Regnum Vegetabile 159. Koeltz Botanical Books, Glashütten.

Authors

Jeffery M. Saarela; Canadian Museum of Nature; P.O. Box 3443, Station D; Ottawa, ON, K1P 6P4, Canada. Email: jsaarela@mus-nature.ca.

Sean W. Graham; UBC Botanical Garden & Centre for Plant Research, and Dept. of Botany; University of British Columbia; Vancouver, BC, V6T 1Z4, Canada. Email: swgraham@interchange.ubc.ca.

Date Accepted: 30 November 2011

Primary Editor: Philip Cantino

Tricolpatae M. J. Donoghue, J. A. Doyle, and P. D. Cantino in P. D. Cantino et al. (2007): E26 [M. J. Donoghue, J. A. Doyle, and P. D. Cantino], converted clade name

Registration Number: 108

Definition: The most inclusive clade exhibiting tricolpate (or derivative) pollen grains synapomorphic with those found in *Platanus occidentalis* Linnaeus 1753 (*Eudicotyledoneae*). A tricolpate pollen grain is one having three elongate, furrow-like apertures (colpi) located at and oriented perpendicular to the equator. This is an apomorphy-based definition. Abbreviated definition: ∇ apo tricolpate pollen [*Platanus occidentalis* Linnaeus 1753].

Etymology: From Greek, *tri-* (three) and *kolpos* (a fold, lap, hollow, or bay, here an elongate longitudinal aperture in a pollen grain).

Reference Phylogeny: The primary reference phylogeny is Doyle (2005: Fig. 4). See also Doyle and Endress (2000: Fig. 4).

Composition: *Eudicotyledoneae* (this volume) and stem taxa with tricolpate pollen. So far, all well-reconstructed Early Cretaceous (Albian) fossil taxa with tricolpate pollen (e.g., Friis et al., 1988; Drinnan et al., 1991; Crane et al., 1993; Mendes et al., 2014; Friis et al., 2017) appear to be part of the crown group, *Eudicotyledoneae* (see Doyle and Endress, 2010), but some dispersed tricolpate pollen types may represent stem taxa that are part of *Tricolpatae* but not of *Eudicotyledoneae*.

Diagnostic Apomorphies: Tricolpate pollen (see Definition).

Synonyms: None (but see Comments).

Comments: Until we published the name *Tricolpatae* (Cantino et al., 2007), there was no scientific name for this clade. Published uses of the terms "eudicots" and "tricolpates" have not clearly distinguished whether they refer to the crown clade or to a more inclusive clade originating with the evolution of tricolpate pollen. Here, as in our 2007 paper, we separate the meanings associated with these names by explicitly applying *Eudicotyledoneae* to the crown clade and *Tricolpatae* to the apomorphy-based clade. We think that this distinction will be helpful in view of the substantial fossil record of pollen and the possibility of discovering plants within the tricolpate clade that fall outside of the eudicot crown. However, if only extant plants are considered, *Eudicotyledoneae* and *Tricolpatae* have the same composition.

The reference phylogenies and definition require comment. The pollen of *Illicium* and *Schisandraceae* was scored as tricolpate by Donoghue and Doyle (1989) and then inferred to have evolved separately from the grains of the tricolpate clade. However, these grains differ from standard tricolpate grains in that the colpi are located 60 degrees from those of the latter grains (Huynh, 1976; Doyle et al., 1990) and usually fused at the distal pole (syntricolpate). Accordingly, they were scored as representing a separate state by Doyle and Endress (2000) and again (defined somewhat differently) by Doyle (2005). In any case, all relevant phylogenetic analyses clearly indicate that the three apertures of *Illicium* and *Schisandraceae* are not homologous with those of *Tricolpatae*. Many different forms of pollen grains have evolved (in most cases multiple times) from the first tricolpate grains of the clade *Tricolpatae*. These modifications include increases and decreases in the number of colpi (di-, tetra-, penta-, hexa-, and polycolpate forms). Compound-aperturate and porate forms, especially tricolporate and triporate grains, appear to have originated frequently, and in some cases the position and/or orientation of the colpi or pores has shifted away from the equator of the grain (e.g., polyrugate and polyforate grains). In other cases, apertures have been lost completely (inaperturate pollen). The resulting multitude of pollen forms all appear to be modifications of the original grains of *Tricolpatae*.

Literature Cited

Cantino, P. D., J. A. Doyle, S. W. Graham, W. S. Judd, R. G. Olmstead, D. E. Soltis, P. S. Soltis, and M. J. Donoghue. 2007. Towards a phylogenetic nomenclature of *Tracheophyta*. *Taxon* 56:822–846 and E1–E44.

Crane, P. R., K. R. Pedersen, E. M. Friis, and A. N. Drinnan. 1993. Early Cretaceous (early to middle Albian) platanoid inflorescences associated with *Sapindopsis* leaves from the Potomac Group of eastern North America. *Syst. Bot.* 18:328–344.

Donoghue, M. J., and J. A. Doyle. 1989. Phylogenetic analysis of angiosperms and the relationship of *Hamamelidae*. Pp. 17–45 in *Evolution, Systematics and Fossil History of the* Hamamelidae, Vol. 1 (P. Crane and S. Blackmore, eds.). Clarendon Press, Oxford.

Doyle, J. A. 2005. Early evolution of angiosperm pollen as inferred from molecular and morphological phylogenetic analyses. *Grana* 44:227–251.

Doyle, J. A., and P. K. Endress. 2000. Morphological phylogenetic analysis of basal angiosperms: comparison and combination with molecular data. *Int. J. Plant Sci.* 161:S121–S153.

Doyle, J. A., and P. K. Endress. 2010. Integrating Early Cretaceous fossils into the phylogeny of living angiosperms: *Magnoliidae* and eudicots. *J. Syst. Evol.* 48:1–35.

Doyle, J. A., C. L. Hotton, and J. V. Ward. 1990. Early Cretaceous tetrads, zonasulculate pollen, and *Winteraceae*. II. Cladistic analysis and implications. *Am. J. Bot.* 77:1558–1568.

Drinnan, A. N., P. R. Crane, E. M. Friis, and K. R. Pedersen. 1991. Angiosperm flowers and tricolpate pollen of buxaceous affinity from the Potomac Group (mid-Cretaceous) of eastern North America. *Am. J. Bot.* 78:153–176.

Friis, E. M., P. R. Crane, and K. R. Pedersen. 1988. Reproductive structures of Cretaceous *Platanaceae*. *Biol. Skr. Dan. Vid. Selsk.* 31:1–55.

Friis, E. M., K. R. Pedersen, and P. R. Crane. 2017. *Kenilanthus*, a new eudicot flower with tricolpate pollen from the Early Cretaceous (early-middle Albian) of eastern North America. *Grana* 56:161–173.

Huynh, K.-L. 1976. L'arrangement du pollen du genre *Schisandra* (*Schisandraceae*) et sa signification phylogénique chez les Angiospermes. *Beitr. Biol. Pflanz.* 52:227–253.

Mendes, M. M., G. W. Grimm, J. Pais, and E. M. Friis. 2014. Fossil *Kajanthus lusitanicus* gen. et sp. nov. from Portugal: floral evidence for Early Cretaceous *Lardizabalaceae* (*Ranunculales*, basal eudicot). *Grana* 53:283–301.

Authors

Michael J. Donoghue; Department of Ecology and Evolutionary Biology; Yale University; P.O. Box 208106; New Haven, CT 06520, USA. Email: michael.donoghue@yale.edu.

James A. Doyle; Department of Evolution and Ecology; University of California; Davis, CA 95616, USA. Email: jadoyle@ucdavis.edu.

Philip D. Cantino; Department of Environmental and Plant Biology; Ohio University; Athens, OH 45701, USA. Email: cantino@ohio.edu.

Date Accepted: 27 January 2013; updated 09 February 2018

Primary Editor: Kevin de Queiroz

Eudicotyledoneae M. J. Donoghue, J. A. Doyle, and P. D. Cantino in
P. D. Cantino et al. (2007): 835 [M. J. Donoghue,
J. A. Doyle, and P. D. Cantino], converted clade name

Registration Number: 250

Definition: The smallest crown clade containing *Ranunculus acris* Linnaeus 1753 (*Ranunculales*) and *Helianthus annuus* Linnaeus 1753 (*Asterales*). This is a minimum-crown-clade definition. Abbreviated definition: min crown ∇ (*Ranunculus acris* Linnaeus 1753 & *Helianthus annuus* Linnaeus).

Etymology: From the Greek *eu-* (true) and the name *Dicotyledoneae* (the name of a more inclusive taxon that is now understood to be paraphyletic). *Dicotyledoneae* is derived from the Greek *di-* (two) and *kotyledon* (seed leaf).

Reference Phylogeny: The primary reference phylogeny is Soltis et al. (2011: Figs. 1, 2). See also Doyle and Endress (2000: Fig. 4), Soltis et al. (2000: Figs. 1, 5), Hilu et al. (2003: Fig. 2), Soltis et al. (2003: Fig. 2), Kim et al. (2004: Fig. 4), Zeng et al., (2014: Fig. 3), and Sun et al. (2016: Fig. 2).

Composition: *Ranunculales* (sensu APG II, 2003) and its presumed sister clade, the latter including *Proteales* (*Proteaceae, Platanus*, and *Nelumbo*), *Sabiaceae, Trochodendraceae* (including *Tetracentron*), *Buxaceae* (including *Didymeles*), and *Gunneridae* (as defined in this volume).

Diagnostic Apomorphies: Members of *Eudicotyledoneae* are distinguished from other extant angiosperms by having pollen grains that are tricolpate or have aperture conditions that were derived from the tricolpate condition. Tricolpate pollen appears to have originated on the line leading to crown eudicots from the monosulcate (and globose, columellar) grains that appear to be ancestral in angiosperms (Doyle, 2005). Loss of oil cells in the mesophyll and dry fruit wall have also been identified as apomorphies of *Eudicotyledoneae* (Doyle and Endress, 2000: Fig. 4), but this inference is sensitive to outgroup relationships.

Synonyms: The informal names "eudicots" and "tricolpates" are approximate synonyms (see Comments).

Comments: Until we phylogenetically defined the name *Eudicotyledoneae* (Cantino et al., 2007), there was no scientific name for this crown clade, which has been referred to informally as either "eudicots" or "tricolpates." It was originally recognized based on morphology by Dahlgren and Bremer (1985) and Donoghue and Doyle (1989), although only equivocally supported in the latter study, and subsequently strongly supported by molecular data (cited under Reference Phylogeny). This clade was originally referred to as the "tricolpates" (Donoghue and Doyle, 1989). Doyle and Hotton (1991) later coined the name "eudicots" ("true dicots") to signify that this very large subset of "dicots" (dicotyledonous angiosperms—a paraphyletic group) forms a clade. Since that time, the name "eudicots" has been used most frequently, and it has been adopted in widely cited phylogenetic studies and classification schemes (e.g., APG, 1998; Doyle and Endress, 2000; Hilu et al., 2003; APG II, 2003; Soltis et al., 2003, 2005; Soltis and Soltis, 2004; APG III, 2009; APG IV, 2016), as well as in textbooks (e.g., Judd et al., 2002, 2008, 2016; Soltis et al., 2005, 2018; Simpson, 2006).

Although cogent arguments have been made in favor of reverting to use of the name tricolpates (Judd and Olmstead, 2004), we chose *Eudicotyledoneae* for the crown clade owing to the widespread use of the name eudicots, which now extends well beyond the plant systematics literature. We have defined the name *Tricolpatae* (this volume and Cantino et al., 2007) for the more inclusive clade based on the apomorphy of tricolpate pollen. Pollen grains are well represented in the fossil record, and the appearance of tricolpate grains has taken on great importance in assessing the timing of angiosperm evolution (see Soltis et al., 2005). It is not clear that all early fossil tricolpate pollen grains represent members of *Eudicotyledoneae*, but they can be confidently assigned to *Tricolpatae*.

Cantino et al. (2007) defined the name *Eudicotyledoneae* using six internal specifiers (*Ranunculus trichophyllus* Villars 1786, *Platanus occidentalis* Linnaeus 1753, *Sabia swinhoei* W. B. Hemsley 1886, *Trochodendron aralioides* P. F. Siebold and Zuccarini 1838, *Buxus sempervirens* Linnaeus 1753, and *Helianthus annuus* Linnaeus 1753). Six specifiers were used because at the time measures of support for basal and near-basal eudicot relationships were variable, and in some cases low (e.g., see Soltis et al., 2000; Soltis et al., 2003; Hilu et al., 2003; and Kim et al., 2004). However, more recent analyses have provided strong support for the hypothesis that the members of *Ranunculales* form a clade that is sister to a clade containing the remaining eudicots (Soltis et al., 2011; Zeng et al., 2014; Sun et al., 2016). Consequently, a simpler minimum-crown-clade definition with only two specifiers can be used.

Literature Cited

APG (Angiosperm Phylogeny Group). 1998. An ordinal classification for the families of flowering plants. *Ann. Mo. Bot. Gard.* 85:531–553.

APG II (Angiosperm Phylogeny Group). 2003. An update of the Angiosperm Phylogeny Group classification for the orders and families of flowering plants: APG II. *Bot. J. Linn. Soc.* 141:399–436.

APG III (Angiosperm Phylogeny Group). 2009. An update of the Angiosperm Phylogeny Group classification for the orders and families of flowering plants: APG III. *Bot. J. Linn. Soc.* 161:105–121.

APG IV (Angiosperm Phylogeny Group). 2016. An update of the Angiosperm Phylogeny Group classification for the orders and families of flowering plants: APG IV. *Bot. J. Linn. Soc.* 181:1–20.

Cantino, P. D., J. A. Doyle, S. W. Graham, W. S. Judd, R. G. Olmstead, D. E. Soltis, P. S. Soltis, and M. J. Donoghue. 2007. Towards a phylogenetic nomenclature of *Tracheophyta*. *Taxon* 56:822–846 and E1–E44.

Dahlgren, R., and K. Bremer. 1985. Major clades of angiosperms. *Cladistics* 1:349–368.

Donoghue, M. J., and J. A. Doyle. 1989. Phylogenetic analysis of angiosperms and the relationships of *Hamamelidae*. Pp. 17–45 in *Evolution, Systematics, and Fossil History of the* Hamamelidae, Vol. 1 (P. Crane and S. Blackmore, eds.). Clarendon Press, Oxford.

Doyle, J. A. 2005. Early evolution of angiosperm pollen as inferred from molecular and morphological phylogenetic analyses. *Grana* 44:227–251.

Doyle, J. A., and P. K. Endress. 2000. Morphological phylogenetic analysis of basal angiosperms: comparison and combination with molecular data. *Int. J. Plant Sci.* 161:S121–S153.

Doyle, J. A., and C. L. Hotton. 1991. Diversification of early angiosperm pollen in a cladistic context. Pp. 165–195 in *Pollen and Spores: Patterns of Diversification* (S. Blackmore and S. H. Barnes, eds.). Clarendon Press, Oxford.

Hilu, K. W., T. Borsch, K. Müller, D. E. Soltis, P. S. Soltis, V. Savolainen, M. W. Chase, M. P. Powell, L. A. Alice, R. Evans, H. Sauquet, C. Neinhuis, T. A. B. Slotta, J. G. Rohwer, C. S. Campbell, and L. W. Chatrou. 2003. Angiosperm phylogeny based on *matK* sequence information. *Am. J. Bot.* 90:1758–1776.

Judd, W. S., C. S. Campbell, E. A. Kellogg, P. F. Stevens, and M. J. Donoghue. 2002. *Plant Systematics—A Phylogenetic Approach.* 2nd edition. Sinauer Associates, Sunderland, MA.

Judd, W. S., C. S. Campbell, E. A. Kellogg, P. F. Stevens, and M. J. Donoghue. 2008. *Plant Systematics—A Phylogenetic Approach.* 3rd edition. Sinauer Associates, Sunderland, MA.

Judd, W. S., C. S. Campbell, E. A. Kellogg, P. F. Stevens, and M. J. Donoghue. 2016. *Plant Systematics—A Phylogenetic Approach.* 4th edition. Sinauer Associates, Sunderland, MA.

Judd, W. S., and R. G. Olmstead. 2004. A survey of tricolpate (eudicot) phylogenetic relationships. *Am. J. Bot.* 91:1627–1644.

Kim, S., D. E. Soltis, P. S. Soltis, M. J. Zanis, and Y. Suh. 2004. Phylogenetic relationships among early-diverging eudicots based on four genes: were the eudicots ancestrally woody? *Mol. Phylogenet. Evol.* 31:16–30.

Simpson, M. G. 2006. *Plant Systematics.* Elsevier, Amsterdam.

Soltis, D. E., A. E. Senters, M. Zanis, S. Kim, J. D. Thompson, P. S. Soltis, L. P. Ronse De Craene, P. K. Endress, and J. S. Farris. 2003. *Gunnerales* are sister to other core eudicots: implications for the evolution of pentamery. *Am. J. Bot.* 90:461–470.

Soltis, D. E., S. A. Smith, N. Cellinese, K. J. Wurdack, D. C. Tank, S. F. Brockington, N. F. Refulio-Rodriguez, J. B. Walker, M. J. Moore, B. S. Carlsward, C. D. Bell, M. Latvis, S. Crawley, C. Black, D. Diouf, Z. Xi, C. A. Rushworth, M. A. Gitzendanner, K. J. Sytsma, Y.-L. Qiu, K. H. Hilu, C. C. Davis, M. J. Sanderson, R. S. Beaman, R. G. Olmstead, W. S. Judd, M. J. Donoghue, and P. S. Soltis. 2011. Angiosperm phylogeny: 17 genes, 640 taxa. *Am. J. Bot.* 98:704–730.

Soltis, D. E., P. S. Soltis, M. W. Chase, M. E. Mort, D. C. Albach, M. Zanis, V. Savolainen, W. H. Hahn, S. B. Hoot, M. F. Fay, M. Axtell, S. M. Swensen, L. M. Prince, W. J. Kress, K. C. Nixon, and J. S. Farris. 2000. Angiosperm phylogeny inferred from 18S rDNA, *rbcL*, and *atpB* sequences. *Bot. J. Linn. Soc.* 133:381–461.

Soltis, D. E., P. S. Soltis, P. K. Endress, and M. W. Chase. 2005. *Phylogeny and Evolution of Angiosperms.* Sinauer Associates, Sunderland, MA.

Soltis, D. E., P. S. Soltis, P. Endress, M. Chase, S. Manchester, W. Judd, L. Majure, and E. Mavrodiev. 2018. *Phylogeny and Evolution of the Angiosperms: Revised and Updated Edition.* University of Chicago Press, Chicago, IL.

Soltis, P. S., and D. E. Soltis. 2004. The origin and diversification of angiosperms. *Am. J. Bot.* 91:1614–1626.

Sun, Y., M. J. Moore, S. Zhang, P. S. Soltis, D. E. Soltis, T. Zhao, A. Meng, X. Li, J. Li, and H. Wang. 2016. Phylogenomic and structural analyses of 18 complete plastomes across nearly all families of early-diverging eudicots, including an angiosperm-wide analysis of IR gene content evolution. *Mol. Phylogenet. Evol.* 96:93–101.

Zeng, L., Q. Zhang, R. Sun, H. Kong, N. Zhang, and H. Ma. 2014. Resolution of deep angiosperm phylogeny using conserved nuclear genes and estimates of early divergence times. *Nat. Commun.* 5:4956.

Authors

Michael J. Donoghue; Department of Ecology and Evolutionary Biology; Yale University; P.O. Box 208106; New Haven, CT 06520, USA. Email: michael.donoghue@yale.edu.

James A. Doyle; Department of Evolution and Ecology; University of California; Davis, CA 95616, USA. Email: jadoyle@ucdavis.edu.

Philip D. Cantino; Department of Environmental and Plant Biology; Ohio University; Athens, OH 45701, USA. Email: cantino@ohio.edu.

Date Accepted: 31 January 2013; updated 16 May 2018

Primary Editor: Kevin de Queiroz

Gunneridae D. E. Soltis, P. S. Soltis, and W. S. Judd (in Cantino et al., 2007: 835) [D. E. Soltis, P. S. Soltis, and W. S. Judd], converted clade name

Registration Number: 48

Definition: The smallest crown clade containing _Gunnera perpensa_ L. 1767 (_Gunnerales_), _Viscum album_ L. 1753 (_Santalales_), _Berberidopsis corallina_ Hook. f. 1862 (_Berberidopsidales_), _Stellaria media_ (L.) Cirillo 1784 (_Caryophyllales_), _Dillenia indica_ L. 1753 (_Dilleniaceae_), _Saxifraga mertensiana_ Bong. 1835 (_Saxifragales_), _Vitis aestivalis_ Michx. 1803 (_Vitaceae_), _Photinia x fraseri_ Dress 1961 (_Rosidae_), and _Helianthus annuus_ L. 1753 (_Asteridae_). This is a minimum-crown-clade definition. Abbreviated definition: min crown ∇ (_Gunnera perpensa_ L. 1767 & _Viscum album_ L. 1753 & _Berberidopsis corallina_ Hook. f. 1862 & _Stellaria media_ (L.) Cirillo 1784 & _Dillenia indica_ L. 1753 & _Saxifraga mertensiana_ Bong. 1835 & _Vitis aestivalis_ Michx. 1803 & _Photinia x fraseri_ Dress 1961 & _Helianthus annuus_ L. 1753).

Etymology: Based on the genus name _Gunnera_, which was named after Johan E. Gunnerus (1718–1773), a Norwegian bishop and botanist.

Reference Phylogeny: Soltis et al. (2003: Fig. 2). _Gunnera perpensa_ is used as a specifier because it is the type species of _Gunnerales_; its close relationship to _G. hamiltonii_ Kirk. ex W. S. Ham. (included in the reference phylogeny) is supported by the monophyly of _Gunnera_ (see Wanntorp et al., 2001). See also Burleigh et al. (2009), Hoot et al. (1999), Savolainen et al. (2000a,b), Soltis et al. (2000, 2007), Hilu et al. (2003), and Zhu et al. (2007).

Composition: _Gunnerales_ and _Pentapetalae_ (defined in this volume). The latter includes _Asteridae_, _Berberidopsidales_, _Caryophyllales_, _Dilleniaceae_, _Rosidae_, _Santalales_, _Saxifragales_, and _Vitaceae_.

Diagnostic Apomorphies: Possible gene duplications in a number of gene families that underlie flower development, including MADS-box and TCP genes (Lamb and Irish, 2003; Litt and Irish, 2003; Howarth and Donoghue, 2006), could be synapomorphic for this clade. The presence of ellagic acid is also synapomorphic (Soltis et al., 2005).

Synonyms: There are no synonymous scientific names, but the informal names "core eudicots" and "core tricolpates" have been applied to this clade (APG, 1998; APG II, 2003; APG III 2009; Hilu et al., 2003; Soltis et al., 2003, 2005; Judd and Olmstead, 2004; Soltis and Soltis, 2004). If _Gunneridae_ and _Pentapetalae_ become synonymous in the context of a future phylogeny, _Pentapetalae_ should have precedence (see Comments).

Comments: This clade is strongly supported by several molecular analyses (Savolainen et al., 2000; Hilu et al., 2003; Soltis et al., 2000, 2003, 2007; Burleigh et al., 2009), though it was not recognized previous to the advent of DNA sequence data. There was no scientific name for the clade until the name _Gunneridae_ was introduced by Soltis et al. (in Cantino et al., 2007), using a phylogenetic definition similar to the one employed here, except that only _Gunnera perpensa_ and _Helianthus annuus_ were included as specifiers. The inclusion of _Photinia_ × _fraseri_ as a specifier (a member of the _Rosidae_) ensures compositional stability even if _Gunnera_ is part of a clade that includes *Asteridae* but excludes *Rosidae* (as in Zhu et al., 2007: Fig. 4), which we consider unlikely. The inclusion of the additional specifiers *Viscum album, Berberidopsis corallina, Stellaria media, Dillenia indica, Saxifraga mertensiana,* and *Vitis aestivalis* (representing *Santalales, Berberidopsidales, Caryophyllales, Dilleniaceae, Saxifragales,* and *Vitaceae,* respectively) ensures compositional stability even if *Gunnera* turned out to fall among these subclades of *Gunneridae,* instead of in its currently best supported position as sister to the remaining members of *Gunneridae.* In such cases, *Gunneridae* and *Pentapetalae* would become synonymous (see definition of *Pentapetalae*), and we intend that *Pentapetalae* would then have precedence. We prefer *Pentapetalae* in such situations because it is a descriptive name for this major clade in which most species have highly synorganized, pentamerous flowers, and thus the name will be easily remembered. We note that the relationships among the above-listed subclades of *Gunneridae* are only moderately supported. The addition of specifiers representing each of them thus ensures that *Gunneridae* will not inadvertently refer to a clade within *Pentapetalae.*

Literature Cited

APG (Angiosperm Phylogeny Group). 1998. An ordinal classification for the families of flowering plants. *Ann. Mo. Bot. Gard.* 85:531–553.

APG II (Angiosperm Phylogeny Group II). 2003. An update of the Angiosperm Phylogeny Group classification for the orders and families of flowering plants: APG II. *Bot. J. Linn. Soc.* 141:399–436.

APG III (Angiosperm Phylogeny Group III). 2009. An update of the Angiosperm Phylogeny Group classification for the orders and families of flowering plants: APG III. *Bot. J. Linn. Soc.* 161:105–121.

Burleigh, J. G., K. W. Hilu, and D. E. Soltis. 2009. Inferring phylogenies with incomplete data sets: a 5-gene, 567-taxon analysis of angiosperms. *BMC Evol. Biol.* 9:61 (11 pp.). Available at http://www.biomedcentral.com/1 471-2148/9/61.

Cantino, P. D., J. A. Doyle, S. W. Graham, W. S. Judd, R. G. Olmstead, D. E. Soltis, P. S. Soltis, and M. J. Donoghue. 2007. Towards a phylogenetic nomenclature of *Tracheophyta. Taxon* 56:822–846 and E1–E44.

Hilu, K. W., T. Borsch, K. Müller, D. E. Soltis, P. S. Soltis, V. Savolainen, M. W. Chase, M. P. Powell, L. A. Alice, R. Evans, H. Sauquet, C. Neinhuis, T. A. B. Slotta, J. G. Rohwer, C. S. Campbell, and L. W. Chatrou. 2003. Angiosperm phylogeny based on *matK* sequence information. *Am. J. Bot.* 90:1758–1776.

Hoot, S. B., S. Magallón, and P. R. Crane. 1999. Phylogeny of basal eudicots based on three molecular data sets: *atpB, rbcL,* and 18S nuclear ribosomal DNA sequences. *Ann. Mo. Bot. Gard.* 86:1–32.

Howarth, D. G., and M. J. Donoghue. 2006. Phylogenetic analyses of the "ECE" (CYC/TB1) clade reveal duplications that predate the core eudicots. *Proc. Natl. Acad. Sci. USA* 103:9101–9106.

Judd, W. S., and R. G. Olmstead. 2004. A survey of tricolpate (eudicot) phylogenetic relationships. *Am. J. Bot.* 91:1627–1644.

Lamb, R. S., and V. F. Irish. 2003. Functional divergence within the APETALA3/PISTILLATA floral homeotic gene lineages. *Proc. Natl. Acad. Sci. USA* 100:6558–6563.

Litt, A., and V. F. Irish. 2003. Duplication and diversification in the APETALA/FRUITFULL floral homeotic gene lineage: implications for the evolution of floral development. *Genetics* 165:821–833.

Savolainen, V., M. W. Chase, S. B. Hoot, C. M. Morton, D. E. Soltis, C. Bayer, M. F. Fay, A. Y. de Bruijn, S. Sullivan, and Y.-L. Qiu. 2000a. Phylogenetics of flowering plants based on combined analysis of plastid *atpB* and *rbcL* gene sequences. *Syst. Biol.* 49:306–362.

Savolainen, V., M. F. Fay, D. C. Albach, A. Bachlund, M. van der Bank, K. M. Cameron, S. A. Johnson, M. D. Lledó, J.-C. Pintaud, M. Powell, M. C. Sheahan, D. E. Soltis, P. S. Soltis, P. Weston, M. W. Whitten,

K. J. Wurdack, and M. W. Chase. 2000b. Phylogeny of the eudicots: a nearly complete familial analysis based on *rbcL* gene sequences. *Kew Bull.* 55:257–309.

Soltis, D. E., M. A. Gitzendanner, and P. S. Soltis. 2007. A 567-taxon data set for angiosperms: The challenges posed by Bayesian analyses of large data sets. *Int. J. Plant Sci.* 168:137–157.

Soltis, D. E., A. E. Senters, M. Zanis, S. Kim, J. D. Thompson, P. S. Soltis, L. P. Ronse De Craene, P. K. Endress, and J. S. Farris. 2003. *Gunnerales* are sister to other core eudicots: implications for the evolution of pentamery. *Am. J. Bot.* 90:461–470.

Soltis, D. E., P. S. Soltis, M. W. Chase, M. E. Mort, D. C. Albach, M. Zanis, V. Savolainen, W. H. Hahn, S. B. Hoot, M. F. Fay, M. Axtell, S. M. Swensen, L. M. Prince, W. J. Kress, K. C. Nixon, and J. S. Farris. 2000. Angiosperm phylogeny inferred from 18S rDNA, *rbcL*, and *atpB* sequences. *Bot. J. Linn. Soc.* 133:381–461.

Soltis, D. E., P. S. Soltis, P. K. Endress, and M. W. Chase. 2005. *Phylogeny and Evolution of Angiosperms*. Sinauer Associates, Sunderland, MA.

Soltis, P. S., and D. E. Soltis. 2004. The origin and diversification of angiosperms. *Am. J. Bot.* 91:1614–1626.

Wanntorp, L., H.-E. Wanntorp, B. Oxelman, and M. Källersjö. 2001. Phylogeny of *Gunnera*. *Plant Syst. Evol.* 226:85–107.

Zhu, X.-Y., M. W. Chase, Y.-L. Qiu, H. Z. Kong, D. L. Dilcher, J.-H. Li, and Z. D. Chen. 2007. Mitochondrial *matR* sequences help to resolved deep phylogenetic relationships in rosids. *BMC Evol. Biol.* 7:217. Available at http://www.biomedcentral.com/1471-2148/7/217.

Authors

Douglas E. Soltis; Florida Museum of Natural History and Department of Biology; University of Florida; Gainesville, FL 32611, USA. Email: dsoltis@botany.ufl.edu.

Pamela S. Soltis; Florida Museum of Natural History; University of Florida; Gainesville, FL 32611-7800, USA. Email: psoltis@flmnh.ufl.edu.

Walter S. Judd; Department of Biology; University of Florida; Gainesville, FL 32611, USA. Email: wjudd@botany.ufl.edu.

Date Accepted: 21 June 2011

Primary Editor: Philip Cantino

Pentapetalae D. E. Soltis, P. S. Soltis, and W. S. Judd
(in Cantino et al., 2007: 835) [D. E. Soltis, P. S. Soltis, and W. S. Judd],
converted clade name

Registration Number: 117

Definition: The smallest crown clade containing *Viscum album* L. 1753 (*Santalales*), *Berberidopsis corallina* Hook. f. 1862 (*Berberidopsidales*), *Stellaria media* (L.) Cirillo 1784 (*Caryophyllales*), *Dillenia indica* L. 1753 (*Dilleniaceae*), *Saxifraga mertensiana* Bong. 1835 (*Saxifragales*), *Vitis aestivalis* Michx. 1803 (*Vitaceae*), *Photinia* × *fraseri* Dress 1961 (*Rosidae*), and *Helianthus annuus* L. 1753 (*Asteridae*). This is a minimum-crown-clade definition. Abbreviated definition: min crown ∇ (*Viscum album* L. 1753 & *Berberidopsis corallina* Hook. f. 1862 & *Stellaria media* (L.) Cirillo 1784 & *Dillenia indica* L. 1753 & *Saxifraga mertensiana* Bong. 1835 & *Vitis aestivalis* Michx. 1803 & *Photinia* × *fraseri* Dress 1961 & *Helianthus annuus* L. 1753).

Etymology: From the Greek *pente* (five) and *petalon* (petal), in reference to the typically pentamerous flowers of this clade.

Reference Phylogeny: Soltis et al. (2003: Fig. 2). See also Burleigh et al. (2009), Hilu et al. (2003), Hoot et al. (1999), Savolainen et al. (2000), and Soltis et al. (2007).

Composition: *Asteridae, Berberidopsidales, Caryophyllales, Dilleniaceae, Rosidae, Santalales, Saxifragales*, and *Vitaceae*.

Diagnostic Apomorphies: A possible synapomorphy is the duplication of the AP1/FUL pair of MADS-box floral regulatory genes (Litt and Irish, 2003). There may have been related changes in developmental mechanisms that are correlated with the fixation of floral structures characteristic of this clade, i.e., the evolution of a pentamerous, highly synorganized flower with a differentiated perianth composed of distinct calyx and corolla (Soltis et al., 2003).

Synonyms: None. If *Gunneridae* and *Pentapetalae* become synonymous in the context of a revised estimate of phylogeny, *Pentapetalae* should be given precedence (see Comments under *Gunneridae* in this volume).

Comments: This is a recently discovered clade that has been strongly supported only in recent molecular analyses (Soltis et al., 2003, 2007; Wang et al., 2009), although its existence was supported (without strong statistical support) in some other analyses (e.g., Savolainen et al., 2000; Hilu et al., 2003). Its circumscription and characterization have been discussed by Judd and Olmstead (2004), Soltis and Soltis (2004), and Soltis et al. (2005). There was no scientific name for the clade until the name *Pentapetalae* was introduced by Soltis et al. (in Cantino et al., 2007), using the same phylogenetic definition as used here.

Literature Cited

Burleigh, J. G., K. W. Hilu, and D. E. Soltis. 2009. Inferring phylogenies with incomplete data sets: a 5-gene, 567-taxon analysis of angiosperms. *BMC Evol. Biol.* 9:61 (11 pp.). Available at http://www.biomedcentral.com/1471-2148/9/61.

Cantino, P. D., J. A. Doyle, S. W. Graham, W. S. Judd, R. G. Olmstead, D. E. Soltis, P. S. Soltis, and M. J. Donoghue. 2007. Towards a phylogenetic nomenclature of *Tracheophyta*. *Taxon* 56:822–846 and E1–E44.

Hilu, K. W., T. Borsch, K. Müller, D. E. Soltis, P. S. Soltis, V. Savolainen, M. W. Chase, M. P. Powell, L. A. Alice, R. Evans, H. Sauquet, C. Neinhuis, T. A. B. Slotta, J. G. Rohwer, C. S. Campbell, and L. W. Chatrou. 2003. Angiosperm phylogeny based on *matK* sequence information. *Am. J. Bot.* 90:1758–1776.

Hoot, S. B., S. Magallón, and P. R. Crane. 1999. Phylogeny of basal eudicots based on three molecular data sets: *atpB*, *rbcL*, and 18S nuclear ribosomal DNA sequences. *Ann. Mo. Bot. Gard.* 86:1–32.

Judd, W. S., and R. G. Olmstead. 2004. A survey of tricolpate (eudicot) phylogenetic relationships. *Am. J. Bot.* 91:1627–1644.

Litt, A., and V. F. Irish. 2003. Duplication and diversification in the APETALA/FRUITFULL floral homeotic gene lineage: implications for the evolution of floral development. *Genetics* 165:821–833.

Savolainen, V., M. W. Chase, S. B. Hoot, C. M. Morton, D. E. Soltis, C. Bayer, M. F. Fay, A. Y. de Bruijn, S. Sullivan, and Y.-L. Qiu. 2000. Phylogenetics of flowering plants based on combined analysis of plastid *atpB* and *rbcL* gene sequences. *Syst. Biol.* 49:306–362.

Soltis, D. E., M. A. Gitzendanner, and P. S. Soltis. 2007. A 567-taxon data set for angiosperms: The challenges posed by Bayesian analyses of large data sets. *Int. J. Plant Sci.* 168:137–157.

Soltis, D. E., A. E. Senters, M. Zanis, S. Kim, J. D. Thompson, P. S. Soltis, L. P. Ronse De Craene, P. K. Endress, and J. S. Farris. 2003. *Gunnerales* are sister to other core eudicots: implications for the evolution of pentamery. *Am. J. Bot.* 90:461–470.

Soltis, D. E., P. S. Soltis, P. K. Endress, and M. W. Chase. 2005. *Phylogeny and Evolution of Angiosperms.* Sinauer Associates, Sunderland, MA.

Soltis, P. S., and D. E. Soltis. 2004. The origin and diversification of angiosperms. *Am. J. Bot.* 91:1614–1626.

Wang, H.-C., M. J. Moore, P. S. Soltis, C. D. Bell, S. F. Brockington, R. Alexandre, C. C. Davis, M. Latvis, S. R. Manchester, and D. E. Soltis. 2009. Rosid radiation and the rapid rise of angiosperm-dominated forests. *Proc. Nat. Acad. Sci. USA.* 106:3853–3858.

Authors

Douglas E. Soltis; Florida Museum of Natural History and Department of Biology; University of Florida; Gainesville, FL 32611, USA. Email: dsoltis@botany.ufl.edu.

Pamela S. Soltis; Florida Museum of Natural History; University of Florida; Gainesville, FL 32611-7800, USA. Email: psoltis@flmnh.ufl.edu.

Walter S. Judd; Department of Biology; University of Florida; Gainesville, FL 32611, USA. Email: wjudd@botany.ufl.edu.

Date Accepted: 21 June 2011

Primary Editor: Philip Cantino

Superrosidae W. S. Judd, D. E. Soltis, and P. S. Soltis in Soltis et al., 2011 [W. S. Judd, D. E. Soltis, and P. S. Soltis], converted clade name

Registration Number: 105

Definition: The largest crown clade containing *Rosa cinnamomea* L. 1753 (*Rosidae/Rosales*) but not *Aster amellus* L. 1753 (*Asteridae/Asterales*). This is a maximum-crown-clade definition. Abbreviated definition: max crown ∇ (*Rosa cinnamomea* L. 1753 ~ *Aster amellus* L. 1753).

Etymology: From the Latin *super* (above, over, or on top) and *Rosidae,* a converted clade name based on the included taxon *Rosa*, the Latin name for rose (and probably originally from the Greek, *rhodon*), in reference to the fact that the *Superrosidae* is intended to apply to a crown clade more inclusive than *Rosidae*.

Reference Phylogeny: The primary reference phylogeny is Soltis et al. (2011: Figs. 1–2). See also Wang et al. (2009: Fig. 1) and Moore et al. (2010: Fig. 1). *Rosa cinnamomea* is used as a specifier because it is the type species of *Rosidae*, which forms part of the defined clade name (see *PhyloCode* (Cantino and de Queiroz, 2020), Art. 11.10). The close relationship of *Rosa cinnamomea* to the species of *Rosaceae* included in the primary reference phylogeny (i.e., *Spiraea betulifolia*, *S. vanhouttei* and *Prunus persica*) is supported by the analyses of Evans et al. (2000) and Potter et al. (2002, 2007), which found *Rosaceae* to be monophyletic. Similarly, *Aster amellus* is used as a specifier because it is the type species of *Asteridae*, which forms part of the name *Superasteridae* (see entry in this volume). Its close relationships to the species of *Asteraceae* included in the primary reference phylogeny (i.e., *Barnadesia arborea*, *Gerbera jamesonii*, *Echinops*

bannaticus, *Tragopogon dubius*, *T. porrifolius*, *Cichorium intybus*, *Lactuca sativa*, *Tagetes erecta*, *Guizotia abyssinica* and *Helianthus annuus*) is supported by the series of phylogenetic studies of *Asteraceae* presented in Funk et al. (2009).

Composition: *Rosidae* (incl. *Vitaceae*) *Saxifragales*, and possibly *Dilleniaceae* (see Comments).

Diagnostic Apomorphies: No non-DNA synapomorphies are known.

Synonyms: There are no synonymous scientific names, but the informal name "super-rosids" (Moore et al., 2010; Stevens, 2001 and onwards, version 9) has been used for this clade.

Comments: This is a recently discovered clade that is strongly supported by the molecular analyses of Soltis et al. (2011). A clade with similar composition was also well supported in both the 12-gene ML analysis of Wang et al. (2009: Fig. 1), a study that focused only on relationships within the *Rosidae* and thus included many fewer sampled taxa, especially among non-rosid taxa, and the 83-gene ML and MP analyses of Moore et al. (2010: Fig. 1), which also were based on a much smaller array of sampled taxa. An unnamed clade with the same composition was inferred (but without strong support) in all most-parsimonious trees in the 3-gene analyses of Soltis et al. (2000: see Figs. 1(B), 5 and 6) and in some of the most parsimonious trees resulting from the analysis of *atpB* and *rbcL* sequences in Savolainen et al. (2000). The clade received weak support in the ML analyses of angiosperms based on 5 genes (Burleigh et al., 2009:

see Fig. 3 and the "full tree" included with the on-line supplemental information). The clade was informally named "the superrosids" by Moore et al. (2010) and formally named *Superrosidae* by Soltis et al. (2011) with the same phylogenetic definition that we use here.

There is disagreement among recent phylogenetic analyses regarding the position of *Dilleniaceae*, which has been variably inferred to be part of *Superasteridae*, part of *Superrosidae*, or lie outside both (see Comments on *Superasteridae* in this volume). For both *Superrosidae* and *Superasteridae*, the use of maximum-crown-clade definitions that do not include any members of *Dilleniaceae* as specifiers accommodates placement of *Dilleniaceae* in either *Superrosidae* or *Superasteridae*, or positioned as the sister taxon to a *Superrosidae* + *Superasteridae* clade within the *Pentapetalae*.

The essential feature of our concept of *Superrosidae* that we have tried to capture in our definition is its inclusion of every extant species that is closer to *Rosidae* than to *Asteridae*. Furthermore, when this definition is used in conjunction with our reciprocal definition of *Superasteridae* (this volume), it ensures that *Superrosidae* and *Superasteridae* are always sister crown clades, regardless of the placement of *Dilleniaceae*.

Literature Cited

Burleigh, J. G., K. W. Hilu, and D. E. Soltis. 2009. Inferring phylogenies with incomplete data sets: a 5-gene, 567-taxon analysis of angiosperms. *BMC Evol. Biol.* 9:61 (11 pp.). Available at http://www.biomedcentral.com/1471-2148/9/61.

Cantino, P. D., and K. de Queiroz. 2020. *International Code of Phylogenetic Nomenclature* (*PhyloCode*), Version 6. CRC Press, Boca Raton, FL.

Evans, R. C., L. A. Alice, C. S. Campbell, T. A. Dickinson, and E. A. Kellogg. 2000. The granule-bound starch synthase (GBSSI) gene in the *Rosaceae*: multiple loci and phylogenetic utility. *Mol. Phylog. Evol.* 17:388–400.

Funk, V. A., A. Susanne, T. F. Stuessy, and R. J. Bayer, eds. 2009. *Systematics, Evolution, and Biogeography of* Compositae. International Association for Plant Taxonomy, Institute of Botany, University of Vienna, Vienna.

Moore, M. J., P. S. Soltis, C. D. Bell, J. G. Burleigh, and D. E. Soltis. 2010. Phylogenetic analysis of 83 plastid genes further resolves the early diversification of eudicots. *Proc. Nat. Acad. Sci. USA* 107:4623–4628.

Potter, D., F. Gao, P. E. Bortiri, S.-H. Oh, and S. Baggett. 2002. Phylogenetic relationships in *Rosaceae* inferred from chloroplast *matK* and *trnL-trnF* nucleotide sequence data. *Plant Syst. Evol.* 231:77–89.

Potter, D., T. Eriksson, R. C. Evans, S.-H. Oh, J. Smedmark, D. Morgan, M. Kerr, K. R. Robertson, M. Arsenault, T. A. Dickinson, and C. S. Campbell. 2007. Phylogeny and classification of *Rosaceae*. *Plant Syst. Evol.* 266:5–43.

Savolainen, V., M. W. Chase, S. B. Hoot, C. M. Morton, D. E. Soltis, C. Bayer, M. F. Fay, A. Y. de Bruijn, S. Sullivan, and Y.-L. Qiu. 2000. Phylogenetics of flowering plants based on combined analysis of plastid *atpB* and *rbcL* gene sequences. *Syst. Biol.* 49:306–382.

Soltis, D. E., S. Smith, N. Cellinese, K. J. Wurdack, D. C. Tank, S. F. Brockington, N. F. Refulio-Rodriguez, J. B. Walker, M. J. Moore, B. S. Carlsward, C. D. Bell, M. Latvis, S. Crawley, C. Black, D. Diouf, Z.-X. Xi, C. A. Rushworth, M. A. Gitzendanner, K. J. Sytsma, Y.-L. Qiu, K. W. Hilu, C. C. Davis, M. J. Sanderson, R. S. Beaman, R. G. Olmstead, W. S. Judd, M. J. Donoghue, and P. S. Soltis. 2011. Angiosperm phylogeny: 17-genes, 640 taxa. *Am. J. Bot.* 98:704–730.

Soltis, D. E., P. S. Soltis, M. W. Chase, M. E. Mort, D. C. Albach, M. Zanis, V. Savolainen, W. H. Hahn, S. B. Hoot, M. F. Fay, M. Axtell, S. M. Swensen, L. M. Prince, W. J. Kress, K. C. Nixon, and J. S. Farris. 2000. Angiosperm phylogeny inferred from 18S rDNA, *rbcL*, and *atpB* sequences. *Bot. J. Linn. Soc.* 133:381–461.

Stevens, P. F. 2001 onwards. Angiosperm Phylogeny Website, Version 9, June 2008 [and more or less continuously updated since]. Available at http://www.mobot.org/MOBOT/research/APweb/.

Wang, H.-C., M. J. Moore, P. S. Soltis, C. D. Bell, S. F. Brockington, R. Alexandre, C. C. Davis, M. Latvis, S. R. Manchester, and D. E. Soltis. 2009. Rosid radiation and the rapid rise of angiosperm-dominated forests. *Proc. Nat. Acad. Sci. USA* 106:3853–3858.

Authors

Walter S. Judd; Department of Biology; University of Florida; Gainesville, FL 32611-8526, USA. Email: lyonia@ufl.edu.

Douglas E. Soltis; Department of Biology; University of Florida; Gainesville, FL 32611-8526, USA. Email: dsoltis@ufl.edu.

Pamela S. Soltis; Florida Museum of Natural History; University of Florida, Gainesville, FL 32611-7800, USA. Email: psoltis@flmnh.ufl.edu.

Date Accepted: 24 December 2011

Primary Editor: Philip Cantino

Rosidae A. L. Takhtajan 1967: 264
[W. S. Judd, P. D. Cantino, D. E. Soltis, and P. S. Soltis],
converted clade name

Registration Number: 251

Definition: The largest crown clade containing *Rosa cinnamomea* L. 1753 (*Fabidae*/*Rosaceae*) but not *Saxifraga mertensiana* Bong. 1835 (*Saxifragales*) and *Helianthus annuus* L. 1753 (*Asteridae*) and *Berberidopsis corallina* Hook. f. 1862 (*Berberidopsidales*) and *Stellaria media* (L.) Cirillo 1784 (*Caryophyllales*) and *Viscum album* L. 1753 (*Santalales*) and *Dillenia indica* L. 1753 (*Dilleniaceae*) and *Gunnera perpensa* L. 1753 (*Gunneraceae*). This is a maximum-crown-clade definition. Abbreviated definition: max crown ∇ (*Rosa cinnamomea* L. 1753 ~ *Saxifraga mertensiana* Bong. 1835 & *Helianthus annuus* L. 1753 & *Berberidopsis corallina* Hook. f. 1862 & *Stellaria media* (L.) Cirillo 1784 & *Viscum album* L. 1753 & *Dillenia indica* L. 1753 & *Gunnera perpensa* L. 1753).

Etymology: Based on *Rosa*, the Latin name for rose (and probably originally from the Greek, *rhodon*).

Reference Phylogeny: Soltis et al. (2011: Figs. 1, overview, and 2, detailed). See also Wang et al. (2009: Fig. 1). *Rosa cinnamomea* is used as a specifier because it is the type species of *Rosidae* (under the Botanical Code; Turland et al., 2018); its close relationship to the species of *Spiraea* and *Prunus* used in the reference phylogeny is supported by the analyses of Evans et al. (2000) and Potter et al. (2002, 2007).

Composition: *Fabidae* (this volume), *Malvidae* (this volume), and probably *Vitaceae* (see Comments).

Diagnostic Apomorphies: No non-DNA synapomorphies are known.

Synonyms: There are no synonymous scientific names, but the informal names "rosids" and "eurosids," which have been used in many molecular phylogenetic studies and classifications, are approximate synonyms except that some of the clades to which they are applied exclude *Vitaceae* (see Comments).

Comments: The taxon *Rosidae*, as circumscribed by Takhtajan (1997) and Cronquist (1988), differed significantly in composition from the clade given the informal name "rosids" or "eurosids" in many recent phylogenetic studies (Chase et al., 1993; Savolainen et al., 2000a,b; Soltis et al., 2000, 2003, 2005, 2007; Judd et al., 2002, 2008; Hilu et al., 2003; Judd and Olmstead, 2004; Soltis and Soltis, 2004) and classifications (APG, 1998; APG II, 2003; APG III, 2009; APG IV, 2016). The name *Rosidae* was linked with this clade by Fukuda et al. (2003: 589) and Soltis and Soltis (2003: 1793), and *Rosidae* was subsequently defined by us (in Cantino et al., 2007), with the same definition used here. It is the only scientific name for this clade.

The concept of "rosids" has changed gradually with regard to inclusion of *Vitaceae*. In earlier phylogenetic studies (e.g., Savolainen et al., 2000a,b; Soltis et al., 2000: Fig. 6; Hilu et al., 2003) and a few more recent ones (e.g., Worberg et al., 2009), "eurosids" and "rosids" tended to exclude *Vitaceae*, whether or not *Vitaceae* was inferred to be sister to the clade labeled as "rosids" or "eurosids". The inclusion of *Vitaceae* within the "rosids" in APG II (2003) was an exception. However, there has been recent movement towards inclusion of *Vitaceae* (e.g., APG III, 2009; Wang et al., 2009, in the text but not labeled on the figure; Soltis et al., 2011; APG IV, 2016), as the evidence for its sister group relationship to the rest of the rosids has become progressively stronger. In a few older phylogenies (e.g., Savolainen et al., 2000b; Hilu et al., 2003), *Vitaceae* was inferred to be somewhat distant from the rosids, and other studies (e.g., Savolainen et al., 2000a: Fig. 6; Zhu et al., 2007) found a poorly supported sister group relationship between *Vitaceae* and the other rosids. In contrast, three recent studies (Soltis et al., 2007, 2011; Wang et al., 2009) and one older one (Soltis et al., 2000: Fig. 6) provided moderate to strong support for this sister-group relationship. Our definition of *Rosidae* does not explicitly include or exclude *Vitaceae*, but *Vitaceae* is part of *Rosidae* when our definition is applied in the context of the most comprehensive phylogenies (Soltis et al., 2007, 2011; Wang et al., 2009).

The definition of *Superrosidae* (Soltis et al., 2011 and this volume), when applied to recent phylogenies, includes *Rosidae* (as defined here), *Saxifragales*, and possibly *Dilleniaceae* (Moore et al., 2010: Fig. 1). In the context of a phylogeny in which *Rosidae* and *Superrosidae* are synonyms (e.g., Hilu et al., 2003: Fig. 3), we intend that *Rosidae* have precedence.

Literature Cited

APG (Angiosperm Phylogeny Group). 1998. An ordinal classification for the families of flowering plants. *Ann. Missouri Bot. Gard.* 85:531–553.

APG II (Angiosperm Phylogeny Group II). 2003. An update of the Angiosperm Phylogeny Group classification for the orders and families of flowering plants: APG II. *Bot. J. Linn. Soc.* 141:399–436.

APG III (Angiosperm Phylogeny Group III). 2009. An update of the Angiosperm Phylogeny Group classification for the orders and families of flowering plants: APG III. *Bot. J. Linn. Soc.* 161:105–121.

APG IV (Angiosperm Phylogeny Group IV). 2016. An update of the Angiosperm Phylogeny Group classification for the orders and families of flowering plants: APG IV. *Bot. J. Linn. Soc.* 181:1–20.

Cantino, P. D., J. A. Doyle, S. W. Graham, W. S. Judd, R. G. Olmstead, D. E. Soltis, P. S. Soltis, and M. J. Donoghue. 2007. Towards a phylogenetic nomenclature of *Tracheophyta*. *Taxon* 56:822–846 and E1–E44.

Chase, M. W., D. E. Soltis, R. G. Olmstead, D. Morgan, D. H. Les, B. D. Mishler, M. R. Duvall, R. A. Price, H. G. Hills, Y.-L. Qiu, K. A. Kron, J. H. Rettig, E. Conti, J. D. Palmer, J. R. Manhart, K. J. Sytsma, H. J. Michaels, W. John Kress, K. G. Karol, W. Dennis Clark, M. Hedren, B. S. Gaut, R. K. Jansen, K.-J. Kim, C. F. Wimpee, J. F. Smith, G. R. Furnier, S. H. Strauss, Q.-Y. Xiang, G. M. Plunkett, P. S. Soltis, S. M. Swensen, S. E. Williams, P. A. Gadek, C. J. Quinn, L. E. Eguiarte, E. Golenberg, G. H. Learn, Jr., S. W. Graham, S. C. H. Barrett, S. Dayanandan, and V. A. Albert. 1993. Phylogenetics of seed plants: an analysis of nucleotide sequences from the plastid gene *rbcL*. *Ann. Mo. Bot. Gard.* 80:528–580.

Cronquist, A. 1988. *The Evolution and Classification of Flowering Plants*. 2nd edition. New York Botanical Garden, Bronx, NY.

Evans, R. C., L. A. Alice, C. S. Campbell, T. A. Dickinson, and E. A. Kellogg. 2000. The granule-bound starch synthase (GBSSI) gene in the *Rosaceae*: multiple loci and phylogenetic utility. *Mol. Phylogenet. Evol.* 17:388–400.

Fukuda, T., J. Tokoyama, and M. Maki. 2003. Molecular evolution of cycloidea-like genes in *Fabaceae*. *J. Mol. Evol.* 57:588–597.

Hilu, K. W., T. Borsch, K. Müller, D. E. Soltis, P. S. Soltis, V. Savolainen, M. W. Chase, M. P. Powell, L. A. Alice, R. Evans, H. Sauquet, C. Neinhuis, T. A. B. Slotta, J. G. Rohwer,

C. S. Campbell, and L. W. Chatrou. 2003. Angiosperm phylogeny based on *matK* sequence information. *Am. J. Bot.* 90:1758–1776.

Judd, W. S., C. S. Campbell, E. A. Kellogg, P. F. Stevens, and M. J. Donoghue. 2002. *Plant Systematics—A Phylogenetic Approach.* 2nd edition. Sinauer Associates, Sunderland, MA.

Judd, W. S., C. S. Campbell, E. A. Kellogg, P. F. Stevens, and M. J. Donoghue. 2008. *Plant Systematics—A Phylogenetic Approach.* 3rd edition. Sinauer Associates, Sunderland, MA.

Judd, W. S., and R. G. Olmstead. 2004. A survey of tricolpate (eudicot) phylogenetic relationships. *Am. J. Bot.* 91:1627–1644.

Moore, M. J., P. S. Soltis, C. D. Bell, J. G. Burleigh, and D. E. Soltis. 2010. Phylogenetic analysis of 83 plastid genes further resolves the early diversification of eudicots. *Proc. Natl. Acad. Sci. USA* 107:4623–4628.

Potter, D., T. Eriksson, R. C. Evans, S.-H. Oh, J. Smedmark, D. Morgan, M. Kerr, K. R. Robertson, M. Arsenault, T. A. Dickinson, and C. S. Campbell. 2007. Phylogeny and classification of *Rosaceae. Plant Syst. Evol.* 266: 5–43.

Potter, D., F. Gao, P. E. Bortiri, S.-H. Oh, and S. Baggett. 2002. Phylogenetic relationships in *Rosaceae* inferred from chloroplast *matK* and *trnL-trnF* nucleotide sequence data. *Plant Syst. Evol.* 231:77–89.

Savolainen, V., M. W. Chase, S. B. Hoot, C. M. Morton, D. E. Soltis, C. Bayer, M. F. Fay, A. Y. de Bruijn, S. Sullivan, and Y.-L. Qiu. 2000a. Phylogenetics of flowering plants based on combined analysis of plastid *atpB* and *rbcL* gene sequences. *Syst. Biol.* 49:306–362.

Savolainen, V., M. F. Fay, D. C. Albach, A. Bachlund, M. van der Bank, K. M. Cameron, S. A. Johnson, M. D. Lledó, J.-C. Pintaud, M. Powell, M. C. Sheahan, D. E. Soltis, P. S. Soltis, P. Weston, M. W. Whitten, K. J. Wurdack, and M. W. Chase. 2000b. Phylogeny of the eudicots: a nearly complete familial analysis based on *rbcL* gene sequences. *Kew Bull.* 55:257–309.

Soltis, D. E., M. A. Gitzendanner, and P. S. Soltis. 2007. A 567-taxon data set for angiosperms: The challenges posed by Bayesian analyses of large data sets. *Int. J. Plant Sci.* 168: 137–157.

Soltis, D. E., A. E. Senters, M. Zanis, S. Kim, J. D. Thompson, P. S. Soltis, L. P. Ronse De Craene, P. K. Endress, and J. S. Farris. 2003. *Gunnerales* are sister to other core eudicots: implications for the evolution of pentamery. *Am. J. Bot.* 90:461–470.

Soltis, D. E., S. A. Smith, N. Cellinese, K. J. Wurdack, D. C. Tank, S. F. Brockington, N. F. Refulio-Rodriguez, J. B. Walker, M. J. Moore, B. S. Carlsward, C. D. Bell, M. Latvis, S. Crawley, C. Black, D. Diouf, Z. Xi, C. A. Rushworth, M. A. Gitzendanner, K. J. Sytsma, Y-L. Qiu, K. H. Hilu, C. C. Davis, M. J. Sanderson, R. S. Beaman, R. G. Olmstead, W. S. Judd, M. J. Donoghue, and P. S. Soltis. 2011. Angiosperm phylogeny: 17 genes, 640 taxa. *Am. J. Bot.* 98: 704–730.

Soltis, D. E., and P. S. Soltis. 2003. The role of phylogenetics in comparative genetics. *Plant Physiol.* 132:1790–1800.

Soltis, D. E., P. S. Soltis, M. W. Chase, M. E. Mort, D. C. Albach, M. Zanis, V. Savolainen, W. H. Hahn, S. B. Hoot, M. F. Fay, M. Axtell, S. M. Swensen, L. M. Prince, W. J. Kress, K. C. Nixon, and J. S. Farris. 2000. Angiosperm phylogeny inferred from 18S rDNA, *rbcL*, and *atpB* sequences. *Bot. J. Linn. Soc.* 133: 381–461.

Soltis, D. E., P. S. Soltis, P. K. Endress, and M. W. Chase. 2005. *Phylogeny and Evolution of Angiosperms.* Sinauer Associates, Sunderland, MA.

Soltis, P. S., and D. E. Soltis. 2004. The origin and diversification of angiosperms. *Am. J. Bot.* 91:1614–1626.

Takhtajan, A. 1967. *Sistema i Filogeniia Tesvetkovyhk Rastenii (Systema et Phylogenia Magnoliophytorum).* Bauka, Moscow. [Dated 1966, but published 4 Feb. 1967; J. Reveal, pers. comm.]

Takhtajan, A. 1997. *Diversity and Classification of Flowering Plants.* Columbia University Press, New York.

Turland, N. J., J. H. Wiersema, F. R. Barrie, W. Greuter, D. L. Hawksworth, P. S. Herendeen, S. Knapp, W.-H. Kusber, D.-Z. Li, K. Marhold, T. W. May, J. McNeill, A. M. Monro, J. Prado, M. J. Price, and G. F. Smith, eds. 2018. *International Code of Nomenclature for Algae, Fungi, and Plants (Shenzhen Code)*. Adopted by the Nineteenth International Botanical Congress, Shenzhen, China, July 2017. Regnum Vegetabile 159. Koeltz Botanical Books, Glashütten.

Wang, H.-C., M. J. Moore, P. S. Soltis, C. D. Bell, S. F. Brockington, R. Alexandre, C. C. Davis, M. Latvis, S. R. Manchester, and D. E. Soltis. 2009. Rosid radiation and the rapid rise of angiosperm-dominated forests. *Proc. Natl. Acad. Sci. USA* 106:3853–3858.

Worberg, A., M. H. Alford, D. Quandt, and T. Borsch. 2009. *Huerteales* sister to *Brassicales* plus *Malvales*, and newly circumscribed to include *Dipentodon*, *Gerrardina*, *Huertea*, *Perrottetia*, and *Tapiscia*. *Taxon* 58:468–478.

Zhu, X.-Y., M. W. Chase, Y.-L. Qiu, H.-Z. Kong, D. L. Dilcher, J.-H. Li, and Z.-D. Chen. 2007. Mitochondrial *matR* sequences help to resolve deep phylogenetic relationships in rosids. *BMC Evol. Biol.* 7:217 (15 pp.). Available at http://www.biomed-central.com/1471-2148/7/217.

Authors

Walter S. Judd; Department of Biology; University of Florida; Gainesville, FL 32611-8525, USA. Email: lyonia@ufl.edu.

Philip Cantino; Department of Environmental and Plant Biology; Ohio University; Athens, OH 45701, USA. Email: cantino@ohio.edu.

Douglas E. Soltis; Department of Biology; University of Florida, Gainesville, FL 32611-8525, USA. Email: dsoltis@ufl.edu.

Pamela S. Soltis; Florida Museum of Natural History; University of Florida; Gainesville, FL 32611-7800, USA. Email: psoltis@flmnh.ufl.edu.

Date Accepted: 27 January 2013; updated 13 July 2018

Primary Editor: Kevin de Queiroz

Malvidae W. S. Judd, D. E. Soltis, and P. S. Soltis in Cantino et al. 2007: 836 [W. S. Judd, D. E. Soltis, P. S. Soltis, and P. D. Cantino], converted clade name

Registration Number: 161

Definition: The largest crown clade containing *Malva sylvestris* L. 1753 (*Malvales*) but not *Vicia faba* L. 1753 (*Fabidae*). This is a maximum-crown-clade definition. Abbreviated definition: max crown ∇ (*Malva sylvestris* L. 1753 ~ *Vicia faba* L. 1753).

Etymology: Based on the Latin *malva* and the Greek *malache* (mallow).

Reference Phylogeny: The primary reference phylogeny is Wang et al. (2009: Fig. 1). See also Burleigh et al. (2009: Figs. 2, 3), Soltis et al. (2011: Figs. 1, 2), Ruhfel et al., (2014: Fig. 7), Sun et al. (2015: Figs. 1–3), and Sun et al. (2016: Fig. 2). The specifiers used in our definition, *Malva sylvestris* and *Vicia faba,* are not included in the primary reference phylogeny. *Malva sylvestris* is used as a specifier because it is the type species of *Malva*, upon which the name *Malvidae* is based (see *ICPN*, Art. 11.10). For the purpose of applying the definition, *Malva sylvestris* should be considered most closely related to *Gossypium* on the reference phylogeny, based on the combined results of Bayer et al., (1999) and Escobar García et al. (2009). Similarly, *Vicia faba* is used as a specifier because it is the type species of *Faba*, upon which the name *Fabidae* is based, and it is the internal specifier in the definition of *Fabidae* (this volume); *Malvidae* and *Fabidae* have reciprocal definitions using the same two specifiers. For the purpose of applying the definition, *Vicia faba* should be considered most closely related to *Medicago* on the reference phylogeny, based on the results of Wojciechowski et al. (2004).

Composition: *Brassicales, Malvales, Sapindales, Huerteales*, probably also *Crossosomatales* and *Picramniaceae*, and possibly *Geraniales* and *Myrtales* (see Comments). The COM clade (*Celastrales-Oxalidales-Malpighiales*) may also be part of this clade, based on analysis by Sun et al. (2015) (see Comments). The circumscription for these clades follows APG IV (2016).

Diagnostic Apomorphies: No non-DNA apomorphies are known.

Synonyms: There are no synonymous scientific names, but the informal names "eurosids II" and "malvids" have been used for clades with more or less similar compositions (see Comments).

Comments: *Malvidae*, as the name is defined here, corresponds roughly to the clade that has been informally named "eurosids II" (e.g., Soltis et al., 2000, 2005; Judd et al., 2002; APG II, 2003; Hilu et al., 2003; Soltis and Soltis, 2004; Burleigh et al., 2009) or "malvids" (e.g., Judd and Olmstead, 2004; Soltis et al., 2007; Zhu et al., 2007; Judd et al., 2008; APG III, 2009; Worberg et al., 2009; APG IV, 2016). The core groups—*Brassicales, Malvales*, and *Sapindales*—are part of eurosids II (or malvids) in the context of all well-resolved phylogenies, and *Huerteales* are part of this clade in all studies in which representatives were included. The monophyly of the core *Malvidae* has been strongly supported in several phylogenetic analyses (Soltis et al., 2007, 2011; Burleigh et al., 2009; Wang et al., 2009; Worberg et al., 2009). However, these studies do not fully agree on the composition of this clade beyond this core. *Crossosomatales, Geraniales, Myrtales*, and *Picramniaceae* were included in eurosids II (or malvids) by some authors but not others. In some phylogenies, the exclusion of these taxa from *Malvidae* was due simply to lack of resolution, absence from the study, or application of the name to a smaller clade than we do here (e.g., *Myrtales*, *Geraniales*, and *Crossosomatales* were excluded from "eurosid II" by Burleigh et al. [2009: Fig. 3] but are part of the clade in that figure that corresponds to *Malvidae* as defined here). However, there are also some true disagreements in the positions of these taxa relative to *Malvidae* and *Fabidae*, as defined in this work. In contrast to the reference phylogeny (Wang et al., 2009), Ruhfel et al. (2014), and Gitzendanner et al. (2018), which found *Geraniales* and *Myrtales* to be part of *Malvidae*, other molecular analyses have inferred *Geraniales* and *Myrtales* to lie outside of both *Fabidae* and *Malvidae* (Zhu et al., 2007; Sun et al., 2016), and Worberg et al. (2009) inferred *Geraniales* to be part of *Fabidae* (as defined in this work). Early molecular analyses inferred *Picramniaceae* and *Crossosomatales* to be part of *Fabidae* (Savolainen et al., 2000; Hilu et al., 2003), but both were found in more recent studies to be part of *Malvidae* (Wang et al., 2009; Ruhfel et al., 2014; Sun et al., 2016). The position of the COM clade (*Celastrales-Oxalidales-Malpighiales*) is an especially interesting situation. Sun et al. (2015) reported a putative ancient hybridization event, resulting in the COM clade being part of *Fabidae* based on plastid genes and part of *Malvidae* based on nuclear and mitochondrial genes (see *Fabidae* entry for further discussion).

Our preliminary node-based (minimum-clade) definition for *Malvidae* (in Cantino et al., 2007) applied that name to a particular composition by including specifiers representing the four major subclades of eurosids II in Soltis et al. (2000: Fig. 9). However, in view of the varying compositions of eurosids II (and "malvids" in more recent literature), the maximum-crown-clade definition adopted here seems preferable. This definition delimits a composition identical or close to that of the clade labeled eurosids II, malvids or *Malvidae* when applied to most of the relevant phylogenies. For example, when applied to the phylogeny of Wang et al. (2009: Fig. 1; see also Gitzendanner et al., 2018: Appendix S2), *Malvidae* as defined here includes all of the taxa within the bracket labeled *Malvidae* in those figures (namely, *Brassicales*, *Malvales*, *Huerteales*, *Sapindales*, *Geraniales*, *Myrtales*, *Crossosomatales*, and *Picramniaceae*). However, when applied to the topology of Soltis et al. (2000: Fig. 9A), the definite membership of *Malvidae* would be restricted to *Brassicales*, *Malvales*, *Huerteales*, and *Sapindales* (the groups included within "eurosid II" on that tree); the other taxa might or might not be part of *Malvidae* because the phylogeny (in Soltis et al., 2000) is equivocal on their placement relative to both *Malvidae* and *Fabidae*.

The concept of this clade has been changing as the understanding of phylogeny improves, but the feature that we are trying to capture in the definition is its inclusion of everything that is closer to the original core groups of eurosids II (viz., *Brassicales*, *Malvales*, and *Sapindales*) than to the core groups of eurosids I (*Fabidae*) (viz., *Fabales*, *Rosales*, *Fagales*, and *Cucurbitales*). When this definition is used in conjunction with our reciprocal definition of *Fabidae* (this volume), it ensures that *Fabidae* and *Malvidae* are always sister groups regardless of their exact composition in the context of a particular phylogeny. This sister-group relationship seems to us to be a more critical part of the current concept of these two clades, as represented in the highly influential APG classification (APG III, 2009; APG IV, 2016), than the inclusion or exclusion of particular subgroups (e.g., *Geraniales*, *Myrtales*, or *Malpighiales*), hence our preference for the definition proposed here rather than one that uses multiple internal specifiers to guarantee the inclusion of particular subgroups (other than *Malvales*). However, the primary reference phylogeny (Wang et al., 2009: Fig. 1), which is based on two nuclear genes and ten plastid genes, provides very strong support for the compositions of *Fabidae* and *Malvidae* as hypothesized here (see Composition sections for both entries). These compositions are also strongly supported by the 17-gene phylogenetic analysis of angiosperms (Soltis et al., 2011) as well as other recent large-scale analyses of plastid genomes (Ruhfel et al., 2014; Gitzendanner et al., 2018).

Literature Cited

APG II (Angiosperm Phylogeny Group II). 2003. An update of the Angiosperm Phylogeny Group classification for the orders and families of flowering plants: APG II. *Bot. J. Linn. Soc.* 141:399–436.

APG III (Angiosperm Phylogeny Group III). 2009. An update of the Angiosperm Phylogeny Group classification for the orders and families of flowering plants: APG III. *Bot. J. Linn. Soc.* 161:105–121.

APG IV (Angiosperm Phylogeny Group IV). 2016. An update of the Angiosperm Phylogeny Group classification for the orders and families of flowering plants: APG IV. *Bot. J. Linn. Soc.* 181:1–20.

Bayer, C., M. F. Fay, A. Y. de Bruijn, V. Savolainen, C. M. Morton, K. Kubitzki, W. S. Alverson, and M. W. Chase. 1999. Support for an expanded family concept of *Malvaceae* within a re-circumscribed order *Malvales*: combined analysis of plastid *atpB* and *rbcL* DNA sequences. *Bot. J. Linn. Soc.* 129:267–303.

Burleigh, J. G., K. W. Hilu, and D. E. Soltis. 2009. Inferring phylogenies with incomplete data sets: A 5-gene, 567-taxa analysis of angiosperms. *BMC Evol. Biol.* 9:61 (11 pp.). Available at http://www/biomedcentral.com/1471-2148/9/61.

Cantino, P. D., J. A. Doyle, S. W. Graham, W. S. Judd, R. G. Olmstead, D. E. Soltis, P. S. Soltis, and M. J. Donoghue. 2007. Towards a phylogenetic nomenclature of *Tracheophyta*. *Taxon* 56:822–846 and E1–E44.

Escobar García, P., P. Schönswetter, J. Fuertes Aguilar, G. Nieto Feliner, and G. M. Schneeweiss. 2009. Five molecular markers reveal extensive morphological homoplasy and reticulate evolution in the *Malva* alliance (*Malvaceae*). *Mol. Phylogenet. Evol.* 50:226–239.

Gitzendanner, M. A., P. S. Soltis, G. K.S. Wong, B. R. Ruhfel, and D. E. Soltis. 2018. Plastid phylogenomic analysis of green plants: A billion years of evolutionary history. *Am. J. Bot.* 105:291–301.

Hilu, K. W., T. Borsch, K. Müller, D. E. Soltis, P. S. Soltis, V. Savolainen, M. W. Chase, M. P. Powell, L. A. Alice, R. Evans, H. Sauquet, C. Neinhuis, T. A. B. Slotta, J. G. Rohwer, C. S. Campbell, and L. W. Chatrou. 2003. Angiosperm phylogeny based on *matK* sequence information. *Am. J. Bot.* 90:1758–1776.

Judd, W. S., C. S. Campbell, E. A. Kellogg, P. F. Stevens, and M. J. Donoghue. 2002. *Plant Systematics—A Phylogenetic Approach*. 2nd edition. Sinauer Associates, Sunderland, MA.

Judd, W. S., C. S. Campbell, E. A. Kellogg, P. F. Stevens, and M. J. Donoghue. 2008. *Plant Systematics—A Phylogenetic Approach*. 3rd edition. Sinauer Associates, Sunderland, MA.

Judd, W. S., and R. G. Olmstead. 2004. A survey of tricolpate (eudicot) phylogenetic relationships. *Am. J. Bot.* 91:1627–1644.

Ruhfel, B. R., M. A. Gitzendanner, P. S. Soltis, D. E. Soltis, and J. G. Burleigh. 2014. From algae to angiosperms—inferring the phylogeny of green plants (*Viridiplantae*) from 360 plastid genomes. *BMC Evol. Biol.* 14:23.

Savolainen, V., M. W. Chase, S. B. Hoot, C. M. Morton, D. E. Soltis, C. Bayer, M. F. Fay, A. Y. de Bruijn, S. Sullivan, and Y.-L. Qiu. 2000. Phylogenetics of flowering plants based on combined analysis of plastid *atpB* and *rbcL* gene sequences. *Syst. Biol.* 49:306–362.

Soltis, D. E., M. A. Gitzendanner, and P. S. Soltis. 2007. A 567-taxon data set for angiosperms: the challenges posed by Bayesian analyses of large data sets. *Int. J. Plant Sci.* 168:137–157.

Soltis, D. E., S. A. Smith, N. Cellinese, K. J. Wurdack, D. C. Tank, S. F. Brockington, N. F. Refulio-Rodriguez, J. B. Walker, M. J. Moore, B. S. Carlsward, C. D. Bell, M. Latvis, S. Crawley, C. Black, D. Diouf, Z. Xi, C. A. Rushworth, M. A. Gitzendanner, K. J. Sytsma, Y.-L. Qiu, K. H. Hilu, C. C. Davis, M. J. Sanderson, R. S. Beaman, R. G. Olmstead, W. S. Judd, M. J. Donoghue, and P. S. Soltis. 2011. Angiosperm phylogeny: 17 genes, 640 taxa. *Am. J. Bot.* 98:704–730.

Soltis, D. E., P. S. Soltis, M. W. Chase, M. E. Mort, D. C. Albach, M. Zanis, V. Savolainen, W. H. Hahn, S. B. Hoot, M. F. Fay, M. Axtell, S. M. Swensen, L. M. Prince, W. J. Kress, K. C. Nixon, and J. S. Farris. 2000. Angiosperm phylogeny inferred from 18S rDNA, *rbcL*, and *atpB* sequences. *Bot. J. Linn. Soc.* 133:381–461.

Soltis, D. E., P. S. Soltis, P. K. Endress, and M. W. Chase. 2005. *Phylogeny and Evolution of Angiosperms*. Sinauer Associates, Sunderland, MA.

Soltis, P. S., and D. E. Soltis. 2004. The origin and diversification of angiosperms. *Am. J. Bot.* 91:1614–1626.

Sun, M., R. Naeem, J.-X. Su, Z.-Y. Cao, J. G. Burleigh, P. S. Soltis, D. E. Soltis, and Z.-D. Chen. 2016. Phylogeny of the *Rosidae*: a dense taxon sampling analysis. *J. Syst. Evol.* 54:363–391.

Sun, M., D. E. Soltis, P. S. Soltis, X. Zhu, J. G. Burleigh, and Z. Chen. 2015. Deep phylogenetic incongruence in the angiosperm clade *Rosidae*. *Mol. Phylogenet. Evol.* 83:156–166.

Wang, H.-C., M. J. Moore, P. S. Soltis, C. D. Bell, S. F. Brockington, R. Alexandre, C. C. Davis, M. Latvis, S. R. Manchester, and D. E. Soltis. 2009. Rosid radiation and the rapid rise of angiosperm-dominated forests. *Proc. Natl. Acad. Sci. USA* 106:3853–3858.

Wojciechowski, M. F., M. Lavin, and M. J. Sanderson. 2004. A phylogeny of legumes (*Leguminosae*) based on analysis of the plastid *matK* gene resolves many well-supported subclades within the family. *Am. J. Bot.* 91:1846–1862.

Worberg, A., M. H. Alford, D. Quandt, and T. Borsch. 2009. *Huerteales* sister to *Brassicales* plus *Malvales*, and newly circumscribed to include *Dipentodon*, *Gerrardina*, *Huertea*, *Perrottetia*, and *Tapiscia*. *Taxon* 58:468–478.

Zhu, X.-Y., M. W. Chase, Y.-L. Qiu, H.-Z. Kong, D. L. Dilcher, J.-H. Li, and Z.-D. Chen. 2007. Mitochondrial *matR* sequences help to resolve deep phylogenetic relationships in rosids. *BMC Evol. Biol.* 7:217 (15 pp.). Available at http://www.biomedcentral.com/1471-2148/7/217.

Authors

Walter S. Judd; Department of Biology; University of Florida; Gainesville, FL 32611, USA. Email: lyonia@ufl.edu.

Douglas E. Soltis; Department of Biology; University of Florida; Gainesville, FL 32611, USA. Email: dsoltis@ufl.edu.

Pamela S. Soltis; Florida Museum of Natural History; University of Florida; Gainesville, FL 32611-7800, USA. Email: psoltis@flmnh.ufl.edu.

Philip D. Cantino; Department of Environmental and Plant Biology; Ohio University; Athens, OH 45701, USA. Email: cantino@ohio.edu.

Date Accepted: 18 January 2012; updated 30 September 2018

Primary Editor: Kevin de Queiroz

Myrtales Jussieu ex Berchtold and J. Presl. 1820: 233 [W. S. Judd], converted clade name

Registration Number: 70

Definition: The smallest crown clade containing *Terminalia catappa* Linnaeus 1771 (*Combretaceae*), *Lythrum salicaria* Linnaeus 1753 (*Lythraceae*), *Clidemia rubra* (Aublet) Martius 1832 (*Melastomataceae*), and *Myrtus communis* Linnaeus 1753 (*Myrtaceae*). This is a minimum-crown-clade definition. Abbreviated definition: min crown ∇ (*Terminalia catappa* Linnaeus 1771 & *Lythrum salicaria* Linnaeus 1753 & *Clidemia rubra* (Aublet) Martius 1832 & *Myrtus communis* Linnaeus 1753).

Etymology: Based on *Myrtus*, an ancient Latin name for myrtle.

Reference Phylogeny: The primary reference phylogeny is Sytsma et al. (2004: Fig. 2). See also Conti et al. (1996: Fig. 1, 1997: Fig. 2), Savolainen et al. (2000a: Fig. 6), Soltis et al. (2000: Fig. 9), and Wang et al. (2009: Fig. 1).

Composition: Major clades within *Myrtales* include *Combretaceae*, *Onagraceae*, *Lythraceae* (including *Punicaceae*, *Sonneratiaceae*, *Trapaceae*), *Vochysiaceae*, *Myrtaceae* (including *Heteropyxidaceae*, *Psiloxylaceae*), *Melastomataceae*, *Memecylaceae*, and the "CAROP clade" (*Crypteroniaceae*, *Alzateaceae*, *Rhynchocalycaceae*, *Oliniaceae*, and *Penaeaceae*).

Diagnostic Apomorphies: *Myrtales* are easily diagnosed; morphological synapomorphies include: vessel elements with vestured pits; stems with internal phloem; flowers with a short to elongate hypanthium, the inner surface of which is nectariferous; and stamens incurved in bud (Dahlgren and Thorne, 1984; Johnson and Briggs, 1984; van Vliet and Baas, 1984; Jansen et al., 2001; Stevens, 2001; Judd and Olmstead, 2004; Judd et al., 2008). Other possible synapomorphies include: bark that tends to shed in plates or strips (Dahlgren and Thorne, 1984; Johnson and Briggs, 1984); the presence of libriform septate fibers in the wood; unilacunar nodes, the twigs with stratified secondary phloem (van Vliet and Baas, 1984); stipules that are characteristically minute, and frequently are dissected into several small, finger-like projections (or entirely absent; Dahlgren and Thorne, 1984); and pollen with pseudocolpi (Patel et al., 1984).

Synonyms: None.

Comments: This broadly distributed clade includes over 11,000 species. *Combretaceae*, *Lythraceae*, *Melastomataceae*, *Myrtaceae*, and *Onagraceae* are large, ecologically important groups, which often dominate plant communities. The clade, as a result of its morphological distinctiveness, has long been recognized in approximately its current circumscription (see discussion in Dahlgren and Thorne, 1984). However, it was not until the advent of phylogenetic methodology that our knowledge of the limits of the group began to achieve its current form, both through the removal of clades that are now thought to be only distantly related (e.g., *Lecythidaceae* (*Ericales*), *Rhizophoraceae* (*Malpighiales*), and *Thymelaeaceae* (*Malvales*)), and through inclusion of clades that were formerly excluded, most notably *Vochysiaceae*. The results of the morphology-based phylogenetic analysis of Johnson and Briggs (1984) are remarkably similar to those of more recent molecular analyses (e.g., compare with Sytsma et al., 2004), differing mainly in the absence of *Vochysiaceae*, which Johnson and Briggs did not include in their analyses. Phylogenetic analyses that support the monophyly of *Myrtales* include Johnson and Briggs (1984), Conti et al. (1996, 1997), Savolainen et al. (2000a,b), Soltis et al. (2000), Sytsma et al. (2004), and Wang et al. (2009). All these authors except Johnson and Briggs (1984), who did not treat *Vochysiaceae*, adopted a circumscription of *Myrtales* identical to the one proposed here. The circumscription of *Myrtales* in the classification of the Angiosperm Phylogeny Group (1998, 2003, 2009) is also identical. Summaries of phylogenetic relationships within *Myrtales* are provided by Johnson and Briggs (1984), Stevens (2001), Judd and Olmstead (2004), Soltis et al. (2005), and Judd et al. (2008).

Relationships within *Myrtales* are quite well understood. An entire issue of the *Annals of the Missouri Botanical Garden* (1984: vol. 71, no. 3) was dedicated to the clade, summarizing aspects of morphology, chemistry, anatomy, and palynology (e.g., Behnke, 1984; Dahlgren and Thorne, 1984; Johnson and Briggs, 1984; Patel et al., 1984; van Vliet and Baas, 1984); embryology has also been investigated (Tobe and Raven, 1983). *Vochysiaceae* and *Myrtaceae* sensu *lato* are sister clades (Wilson et al., 2001; Sytsma et al., 2004). *Lythraceae* and *Onagraceae* are sister clades, sharing a valvate calyx and grouped vessels in the wood (Johnson and Briggs, 1984; Conti et al., 1996, 1997; Soltis et al., 2000). *Melastomataceae* and *Memecylaceae* (Conti et al., 1996, 1997; Clausing and Renner, 2001; Renner, 2004; Sytsma et al., 2004) are sister clades, and both lack nectaries. The CAROP clade (i.e., *Crypteroniaceae, Alzateaceae, Rhynchocalycaceae, Oliniaceae, Penaeaceae*; Conti et al., 1996, 1999; Clausing and Renner, 2001; Schönenberger and Conti, 2003; Renner, 2004) is characterized by square stems with more or less swollen nodes and stamens equal in number to and opposite the petals; it is sister to the *Melastomataceae + Memecylaceae* clade (Sytsma et al., 2004). The position of *Combretaceae* has varied in these analyses, although this clade is often inferred to be sister to *Lythraceae + Onagraceae* (Conti et al., 1996, 1997; Savolainen et al., 2000a; B. Berger & K. J. Sytsma, unpublished data).

The minimum-crown-clade definition used here is based on the clear division of *Myrtales* into (1) the *Lythraceae + Onagraceae* clade, (2) the *Melastomataceae + Memecylaceae + CAROP* clade, (3) the *Myrtaceae* sensu *lato + Vochysiaceae* clade, and (4) *Combretaceae*. Thus, a specifier has been selected from each of these clades. Although the support for clades 1 and 3 is only moderate in the primary reference phylogeny, these clades have 100% bootstrap support (though far fewer representatives) in the 12-gene analysis of Wang et al. (2009).

The position of the clade *Myrtales* in the *Rosidae* (see Cantino et al., 2007: Fig. 1) is still problematic. It was placed in *Malvidae* (eurosids II) in Angiosperm Phylogeny Group (1998, 2009) but was left unplaced within *Rosidae* in Angiosperm Phylogeny Group (2003). It has been associated with *Malvidae* in several studies (Savolainen et al., 2000a; Soltis et al., 2003; Hilu et al., 2003; Wang et al., 2009), but it grouped with *Fabidae* in Soltis et al. (2000) and was positioned as an early diverging lineage within *Rosidae* in Soltis et al. (2007).

Literature Cited

Angiosperm Phylogeny Group. 1998. An ordinal classification for the families of flowering plants. *Ann. Mo. Bot. Gard.* 85:531–553.

Angiosperm Phylogeny Group. 2003. An update of the Angiosperm Phylogeny Group classification for the orders and families of flowering plants: APG II. *Bot. J. Linn. Soc.* 141:399–436.

Angiosperm Phylogeny Group. 2009. An update of the Angiosperm Phylogeny Group classification for the orders and families of flowering plants: APG III. *Bot. J. Linn. Soc.* 161:105–121.

Behnke, H.-D. 1984. Ultrastructure of sieve-element plastids of *Myrtales* and allied groups. *Ann. Mo. Bot. Gard.* 71:824–831.

Berchtold, F., and J. S. Presl. 1820. *Přirozenosti Rostlin.* Krala Wiljma Endevsa, Praha.

Cantino, P. D., J. A. Doyle, S. W. Graham, W. S. Judd, R. G. Olmstead, D. E. Soltis, P. S. Soltis, and M. J. Donoghue. 2007. Towards a phylogenetic nomenclature of *Tracheophyta. Taxon* 56:822–846 and E1–E44.

Clausing, G., and S. S. Renner. 2001. Molecular phylogenetics of *Melastomataceae* and *Memecylaceae*: implications for character evolution. *Am. J. Bot.* 88:486–498.

Conti, E., D. Baum, and K. J. Sytsma. 1999. Phylogeny of *Crypteroniaceae* and related families: implications for morphology and biogeography. In *XVI International Botanical Congress*, abstracts, 250. Missouri Botanical Garden, St. Louis, MO.

Conti, E., A. Litt, and J. K. Sytsma. 1996. Circumscription of *Myrtales* and their relationship to other rosids: evidence from *rbcL* sequence data. *Am. J. Bot.* 83:221–233.

Conti, E., A. Litt, P. G. Wilson, S. A. Graham, B. G. Briggs, L. A. S. Johnson, and K. J. Sytsma. 1997. Interfamilial relationships in *Myrtales*: molecular phylogeny and patterns of morphological evolution. *Syst. Bot.* 22:629–647.

Dahlgren, R., and R. F. Thorne. 1984. The order *Myrtales*: circumscription, variation, and relationships. *Ann. Mo. Bot. Gard.* 71:633–699.

Hilu, K. W., T. Borsch, K. Müller, D. E. Soltis, P. S. Soltis, V. Savolainen, M. W. Chase, M. P. Powell, L. A. Alice, R. Evans, H. Sauquet, C. Neinhuis, T. A. B. Slotta, J. G. Rohwer, C. S. Campbell, and L. W. Chatrou. 2003. Angiosperm phylogeny based on *matK* sequence information. *Am. J. Bot.* 90:1758–1776.

Jansen, S., P. Baas, and E. Smets. 2001. Vestured pits: their occurrence and systematic importance in eudicots. *Taxon* 50:135–167.

Johnson, L. A. S., and B. G. Briggs. 1984. *Myrtales* and *Myrtaceae*: a phylogenetic analysis. *Ann Mo. Bot. Gard.* 71:700–756.

Judd, W. S., C. S. Campbell, E. A. Kellogg, P. F. Stevens, and M. J. Donoghue. 2008. *Plant Systematics—A Phylogenetic Approach.* 3rd edition. Sinauer Association, Sunderland, MA.

Judd, W. S., and R. G. Olmstead. 2004. A survey of tricolpate (eudicot) phylogenetic relationships. *Am. J. Bot.* 91:1627–1644.

Patel, V. C., J. J. Skvarla, and P. H. Raven. 1984. Pollen characters in relation to the delimitation of *Myrtales. Ann. Mo. Bot. Gard.* 71:858–969.

Renner, S. S. 2004. Bayesian analysis of combined chloroplast loci, using multiple calibrations, supports the recent arrival of *Melastomataceae* in Africa and Madagascar. *Am. J. Bot.* 91:1427–1435.

Savolainen, V., M. W. Chase, S. B. Hoot, C. M. Morton, D. E. Soltis, C. Bayer, M. F. Fay, A. Y. de Bruijn, S. Sullivan, and Y.-L. Qiu. 2000a. Phylogenetics of flowering plants based on combined analysis of plastid *atpB* and *rbcL* gene sequences. *Syst. Biol.* 49:306–362.

Savolainen, V. M., M. F. Fay, D. C. Albach, A. Backlund, M. van der Bank, K. M. Cameron, S. A. Johnson, M. D. Lledo, J.-C. Pintaud, M. Powell, M. C. Sheahan, D. E. Soltis, P. S. Soltis, P. Weston, M. W. Whitten, K. J. Wurdack, and M. W. Chase. 2000a. Phylogeny of eudicots: a nearly complete familial analysis based on *rbcL* gene sequences. *Kew Bull.* 55:257–309.

Schönenberger, J., and E. Conti. 2003. Molecular phylogeny and floral evolution of *Pennaeaceae, Oliniaceae, Rhynchocalycaceae,* and *Alzateaceae* (*Myrtales*). *Am. J. Bot.* 90:293–309.

Soltis, D. E., M. A. Gitzendanner, and P. S. Soltis. 2007. A 567-taxon data set for angiosperms: the challenges posed by Bayesian analyses of large data sets. *Int. J. Plant Sci.* 168:137–157.

Soltis, D. E., A. E. Senters, M. J. Zanis, S. Kim, J. D. Thompson, P. S. Soltis, L. P. Rosnse de Craene, P. K. Endress, and J. S. Farris. 2003. *Gunnerales* are sister to other core eudicots: implications for the evolution of pentamery. *Am. J. Bot.* 90:416–470.

Soltis, D. E., P. S. Soltis, M. W. Chase, M. E. Mort, D. C. Albach, M. Zanis, V. Savolainen, W. H. Hahn, S. B. Hoot, M. F. Fay, M. Axtell, S. M. Swensen, L. M. Prince, W. J. Kress, K. C. Nixon, and J. S. Farris. 2000. Angiosperm phylogeny inferred from 18S rDNA, *rbcL*, and *atpB* sequences. *Bot. J. Linn. Soc.* 133:381–461.

Soltis, D. E., P. S. Soltis, P. K. Endress, and M. W. Chase. 2005. *Phylogeny and Evolution of Angiosperms*. Sinauer Association, Sunderland, MA.

Stevens, P. F. 2001 onwards. Angiosperm Phylogeny Website, Version 9, June 2008 [and more or less continuously updated since]. Available at http://www.mobot.org/MOBOT/research/APweb/.

Sytsma, K. J., A. Litt., M. L. Zjhra, C. Pires, M. Nepokroeff, E. Conti, J. Walker, and P. G. Wilson. 2004. Clades, clocks, and continents: historical and biogeographical analysis of *Myrtaceae*, *Vochysiaceae*, and relatives in the Southern Hemisphere. *Int. J. Plant Sci.* 165(4, suppl.): S85–S105.

Tobe, H., and P. H. Raven. 1983. An embryological analysis of *Myrtales*: its definition and characteristics. *Ann. Mo. Bot. Gard.* 70:71–94.

van Vliet, G. J. C. M., and P. Baas. 1984. Wood anatomy and classification of the *Myrtales*. *Ann. Mo. Bot. Gard.* 71:783–800.

Wang, H.-C., M. J. Moore, P. S. Soltis, C. D. Bell, S. F. Brockington, R. Alexandre, C. C. Davis, M. Latvis, S. R. Manchester, and D. E. Soltis. 2009. Rosid radiation and the rapid rise of angiosperm-dominated forests. *Proc. Nat. Acad. Sci. USA.* 106:3853–3858.

Wilson, P. G., M. M. O'Brien, P. A. Gadek, and C. J. Quinn. 2001. *Myrtaceae* revisited: a reassessment of infrafamilial groups. *Am. J. Bot.* 88:2013–2025.

Author

Walter S. Judd; Department of Biology; University of Florida; Gainesville, FL 32611, USA. Email: lyonia@ufl.edu.

Date Accepted: 20 August 2010

Primary Editor: Philip Cantino

Fabidae W. S. Judd, D. E. Soltis, and P. S. Soltis in Cantino et al., 2007: 836 [W. S. Judd, D. E. Soltis, P. S. Soltis, and P. D. Cantino], converted clade name

Registration Number: 159

Definition: The largest crown clade containing *Vicia faba* L. 1753 (*Fabales*) but not *Malva sylvestris* L. 1753 (*Malvidae*). This is a maximum-crown-clade definition. Abbreviated definition: max crown ∇ (*Vicia faba* L. 1753 ~ *Malva sylvestris* L. 1753).

Etymology: Based on the Latin *faba* (bean).

Reference Phylogeny: The primary reference phylogeny is Wang et al. (2009: Fig. 1). See also Burleigh et al. (2009: Figs. 2, 3), Soltis et al. (2011: Figs. 1, 2), Sun et al. (2015: Figs. 1–3), and Sun et al. (2016: Fig. 2). The specifiers used in our definition, *Vicia faba* and *Malva sylvestris,* are not included in the primary reference phylogeny. *Vicia faba* is used as a specifier because it is the type species of *Faba,* upon which the name *Fabidae* is based (see *PhyloCode*, Art. 11.10). For the purpose of applying the definition, *Vicia faba* should be considered most closely related to *Medicago* on the reference phylogeny, based on the results of Wojciechowski et al. (2004). Similarly, *Malva sylvestris* is used as a specifier because it is the type species of *Malva,* upon which the name *Malvidae* is based, and it is the internal specifier in the definition of *Malvidae* (this volume); *Fabidae* and *Malvidae* have reciprocal definitions using the same two specifiers. For the purpose of applying the definition, *Malva sylvestris* should be considered most closely related to *Gossypium* on the reference phylogeny based on the combined results of Bayer et al. (1999) and Escobar García et al. (2009).

Composition: *Cucurbitales, Fabales, Fagales, Huaceae, Rosales,* and probably *Zygophyllales* (see Comments). The COM clade (*Celastrales-Oxalidales-Malpighiales*) is probably part of *Fabidae,* but see Comments. The circumscription for these clades follows APG IV (2016).

Diagnostic Apomorphies: No non-DNA apomorphies are known.

Synonyms: There are no synonymous scientific names, but the informal names "eurosids I" and "fabids" have been used for clades with more or less similar compositions (see Comments).

Comments: The clade *Fabidae* is large and economically important, including over 51,000 species and a diversity of well-known plants such as oaks, birches, roses, figs, elms, legumes, willows, spurges, violets, and squashes. The monophyly of *Fabidae* has been strongly supported in several phylogenetic analyses (Soltis et al., 2007, 2011; Burleigh et al., 2009; Wang et al., 2009; Ruhfel et al., 2014) and moderately supported by Soltis et al. (2000) and Sun et al. (2016). Taxa that are part of *Fabidae* (as defined here) in nearly all analyses in which they were represented (Savolainen et al., 2000a,b; Soltis et al., 2000, 2003, 2007, 2011; Hilu et al., 2003; Zhu et al., 2007: Fig. 4; Burleigh et al., 2009; Wang et al., 2009: Fig. 1; Worberg et al., 2009; Ruhfel et al., 2014; Sun et al., 2016; Gitzendanner et al., 2018) are *Celastrales, Cucurbitales, Fabales, Fagales, Huaceae, Oxalidales, Malpighiales, Rosales,* and *Zygophyllales* (but see discussion below regarding *Zygophyllales* and the *Celastrales-Oxalidales-Malpighiales* (COM) clade).

Zygophyllales groups with *Malvidae* in the combined MP analysis of Wang et al. (2009: Suppl. tree 1), but is strongly supported in *Fabidae* in the combined ML analysis of Wang et al. (2009: Fig. 1) and in MP analyses of other data partitions in Wang et al. (2009), in agreement with several other phylogenetic analyses (e.g., Savolainen et al., 2000a,b; Soltis et al., 2000, 2003, 2007, 2011; Hilu et al., 2003; Burleigh et al., 2009; Ruhfel et al., 2014; Sun et al., 2016). We consider it unlikely that *Zygophyllales* are part of *Malvidae*. The combined MP topology (in Wang et al., 2009) may be the result of long-branch attraction, because when sequences from just the slowly evolving Inverted Repeat region of the chloroplast are used, both MP and ML analyses place *Zygophyllales* within *Fabidae* (Wang et al., 2009). Finally, we note that an analysis of 83 plastid genes (Moore et al., 2010) places *Bulnesia* (*Zygophyllaceae*) in *Fabidae* with 100% bootstrap support.

All plastid-based trees to date (e.g., Ruhfel et al., 2014; Gitzendanner et al., 2018) support the inclusion of the COM clade in *Fabidae* (as in Soltis et al., 2011). However, trees reconstructed from nuclear and mitochondrial genes place the COM clade in *Malvidae* (see analyses and references in Sun et al., 2015; nuclear data from Lee et al., 2011). Morphological evidence also supports the latter placement (Endress and Matthews, 2006; Endress et al., 2013). Sun et al. (2015) attribute these results to a possible ancient hybridization event that resulted in the plastid genome of a fabid ancestor in the nuclear background of malvids. If this hypothesis is correct, the COM clade is correctly placed in both *Fabidae*, based on its plastid genome, and *Malvidae*, based on its nuclear genome. Phylogenetic nomenclature can accommodate the inclusion of a subclade in two named clades that are otherwise mutually exclusive.

The positions of *Geraniales*, *Myrtales*, *Crossosomatales*, and *Picramniaceae* are currently unresolved relative to *Fabidae* and *Malvidae*. In contrast to the reference phylogeny (Wang et al., 2009), Ruhfel et al. (2014), and Gitzendanner et al. (2018), which found *Geraniales* and *Myrtales* to be part of *Malvidae*, other molecular analyses inferred *Geraniales* and *Myrtales* to lie outside of both *Fabidae* and *Malvidae* (Zhu et al., 2007; Sun et al., 2016); moreover, Worberg et al. (2009) inferred *Geraniales* to be part of *Fabidae* (as defined here). Early molecular analyses inferred *Fabidae* (as defined here) to include *Picramniaceae* and *Crossosomatales*, respectively (Savolainen et al., 2000a; Hilu et al., 2003), but both were inferred in more recent studies to be part of *Malvidae* (Wang et al., 2009; Sun et al., 2016). The bulk of the evidence, however, suggests that none of these taxa are part of *Fabidae*.

This clade was informally named "eurosids I" in several phylogenetic treatments of angiosperms (e.g., Savolainen et al., 2000b; Soltis et al., 2000, 2005; Judd et al., 2002; APG II, 2003; Soltis and Soltis, 2004) and "fabids" in others (e.g., Judd and Olmstead, 2004; Judd et al., 2008; APG III, 2009; APG IV, 2016). There being no previous scientific name for this clade, we (in Cantino et al. 2007) proposed and phylogenetically defined the name *Fabidae*, using a node-based (minimum-clade) definition with internal specifiers representing *Celastrales*, *Cucurbitales*, *Fabales*, *Fagales*, *Huaceae*, *Oxalidales*, *Malpighiales*, *Rosales*, and *Zygophyllales*. However, we here propose a maximum-crown-clade definition as a way to accommodate the varying positions of the COM clade and *Zygophyllales*. The essential feature of the concept of *Fabidae* that we are trying to capture in the definition is its inclusion of everything that is closer to the core groups of eurosids I (viz., *Fabales*, *Rosales*, *Fagales*, and *Cucurbitales*, which constitute the nitrogen-fixing clade; see Soltis et al., 2000, 2005; Zhu et al., 2007) than to the core groups of eurosids II (*Malvidae*) (viz., *Brassicales*, *Malvales*, and *Sapindales*). When this definition is used in conjunction with our reciprocal definition of *Malvidae* (this volume), it ensures that *Fabidae* and *Malvidae* are always sister groups regardless of their exact composition in the context of a particular phylogeny. This sister-group relationship seems to us to be a more critical part of the accepted concept of these two clades, as represented in the highly influential APG classification (APG III, 2009; APG IV, 2016), than the inclusion or exclusion of particular subgroups.

Literature Cited

APG II (Angiosperm Phylogeny Group II). 2003. An update of the Angiosperm Phylogeny Group classification for the orders and families of flowering plants: APG II. *Bot. J. Linn. Soc.* 141:399–436.

APG III (Angiosperm Phylogeny Group III). 2009. An update of the Angiosperm Phylogeny Group classification for the orders and families of flowering plants: APG III. *Bot. J. Linn. Soc.* 161:105–121.

APG IV (Angiosperm Phylogeny Group IV). 2016. An update of the Angiosperm Phylogeny Group classification for the orders and families of flowering plants: APG IV. *Bot. J. Linn. Soc.* 181:1–20.

Bayer, C., M. F. Fay, A. Y. de Bruijn, V. Savolainen, C. M. Morton, K. Kubitzki, W. S. Alverson, and M. W. Chase. 1999. Support for an expanded family concept of *Malvaceae* within a re-circumscribed order *Malvales*: combined analysis of plastid *atpB* and *rbcL* DNA sequences. *Bot. J. Linn. Soc.* 129:267–303.

Burleigh, J. G., K. W. Hilu, and D. E. Soltis. 2009. Inferring phylogenies with incomplete data sets: a 5-gene, 567-taxon analysis of angiosperms. *BMC Evol. Biol.* 9:61 (11 pp.). Available at http://www.biomedcentral.com/1471-2148/9/61.

Cantino, P. D., J. A. Doyle, S. W. Graham, W. S. Judd, R. G. Olmstead, D. E. Soltis, P. S. Soltis, and M. J. Donoghue. 2007. Towards a phylogenetic nomenclature of *Tracheophyta*. *Taxon* 56:822–846 and E1–E44.

Endress, P. K., C. C. Davis, and M. L. Matthews. 2013. Advances in the floral structural characterization of the major subclades of *Malpighiales*, one of the largest orders of flowering plants. *Ann. Bot.* 111:969–985.

Endress, P. K., and M. L. Matthews. 2006. Floral structure and systematics in four orders of rosids, including a broad survey of floral mucilage cells. *Plant Syst. Evol.* 260:223–251.

Escobar García, P., P. Schönswetter, J. Fuertes Aguilar, G. Nieto Feliner, and G. M. Schneeweiss. 2009. Five molecular markers reveal extensive morphological homoplasy and reticulate evolution in the *Malva* alliance (*Malvaceae*). *Mol. Phylogenet. Evol.* 50:226–239.

Gitzendanner, M. A., P. S. Soltis, G. K.S. Wong, B. R. Ruhfel, and D. E. Soltis. 2018. Plastid phylogenomic analysis of green plants: a billion years of evolutionary history. *Am. J. Bot.* 105:291–301.

Hilu, K. W., T. Borsch, K. Müller, D. E. Soltis, P. S. Soltis, V. Savolainen, M. W. Chase, M. P. Powell, L. A. Alice, R. Evans, H. Sauquet, C. Neinhuis, T. A. B. Slotta, J. G. Rohwer, C. S. Campbell, and L. W. Chatrou. 2003. Angiosperm phylogeny based on *matK* sequence information. *Am. J. Bot.* 90:1758–1776.

Judd, W. S., C. S. Campbell, E. A. Kellogg, P. F. Stevens, and M. J. Donoghue. 2002. *Plant Systematics—A Phylogenetic Approach.* 2nd edition. Sinauer Associates, Sunderland, MA.

Judd, W. S., C. S. Campbell, E. A. Kellogg, P. F. Stevens, and M. J. Donoghue. 2008. *Plant Systematics—A Phylogenetic Approach.* 3rd edition. Sinauer Associates, Sunderland, MA.

Judd, W. S., and R. G. Olmstead. 2004. A survey of tricolpate (eudicot) phylogenetic relationships. *Am. J. Bot.* 91:1627–1644.

Lee, E. K., A. Cibrian-Jaramillo, S.-K. Kolokotronis, M. S. Katari, A. Stamatakis, M. Ott, J. C. Chiu, D. P. Little, D. W. Stevenson, W. R. McCombie, R. A. Martienssen, G. Coruzzi, and R. DeSalle. 2011. A functional phylogenomic view of the seed plants. *PLOS Genet.* 7 (12):e1002411.

Moore, M. J., P. S. Soltis, C. D. Bell, J. G. Burleigh, and D. E. Soltis. 2010. Phylogenetic analysis of 83 plastid genes resolves the rapid origin of eudicot diversity. *Proc. Natl. Acad. Sci. USA* 107:4623–4628.

Ruhfel, B. R., M. A. Gitzendanner, P. S. Soltis, D. E. Soltis, and J. G. Burleigh. 2014. From algae to angiosperms—inferring the phylogeny of green plants (*Viridiplantae*) from 360 plastid genomes. *BMC Evol. Biol.* 14:23.

Savolainen, V., M. W. Chase, S. B. Hoot, C. M. Morton, D. E. Soltis, C. Bayer, M. F. Fay, A. Y. de Bruijn, S. Sullivan, and Y.-L. Qiu. 2000a. Phylogenetics of flowering plants based on combined analysis of plastid *atpB* and *rbcL* gene sequences. *Syst. Biol.* 49:306–362.

Savolainen, V., M. F. Fay, D. C. Albach, A. Bachlund, M. van der Bank, K. M. Cameron, S. A. Johnson, M. D. Lledó, J.-C. Pintaud, M. Powell, M. C. Sheahan, D. E. Soltis, P. S. Soltis, P. Weston, M. W. Whitten, K. J. Wurdack, and M. W. Chase. 2000b. Phylogeny of the eudicots: a nearly complete familial analysis based on *rbcL* gene sequences. *Kew Bull.* 55:257–309.

Soltis, D. E., M. A. Gitzendanner, and P. S. Soltis. 2007. A 567-taxon data set for angiosperms: the challenges posed by Bayesian analyses of large data sets. *Int. J. Plant Sci.* 168:137–157.

Soltis, D. E., A. E. Senters, M. Zanis, S. Kim, J. D. Thompson, P. S. Soltis, L. P. Ronse De Craene, P. K. Endress, and J. S. Farris. 2003. *Gunnerales* are sister to other core eudicots: implications for the evolution of pentamery. *Am. J. Bot.* 90:461–470.

Soltis, D. E., S. A. Smith, N. Cellinese, K. J. Wurdack, D. C. Tank, S. F. Brockington, N. F. Refulio-Rodriguez, J. B. Walker, M. J. Moore, B. S. Carlsward, C. D. Bell, M. Latvis, S. Crawley, C. Black, D. Diouf, Z. Xi, C. A. Rushworth, M. A. Gitzendanner, K. J. Sytsma, Y.-L. Qiu, K. H. Hilu, C. C. Davis, M. J. Sanderson, R. S. Beaman, R. G. Olmstead, W. S. Judd, M. J. Donoghue, and P. S. Soltis. 2011. Angiosperm phylogeny: 17 genes, 640 taxa. *Am. J. Bot.* 98:704–730.

Soltis, D. E., P. S. Soltis, M. W. Chase, M. E. Mort, D. C. Albach, M. Zanis, V. Savolainen, W. H. Hahn, S. B. Hoot, M. F. Fay, M. Axtell, S. M. Swensen, L. M. Prince, W. J. Kress, K. C. Nixon, and J. S. Farris. 2000. Angiosperm phylogeny inferred from 18S rDNA, *rbcL*, and *atpB* sequences. *Bot. J. Linn. Soc.* 133:381–461.

Soltis, D. E., P. S. Soltis, P. K. Endress, and M. W. Chase. 2005. *Phylogeny and Evolution of Angiosperms*. Sinauer Associates, Sunderland, MA.

Soltis, P. S., and D. E. Soltis. 2004. The origin and diversification of angiosperms. *Am. J. Bot.* 91:1614–1626.

Sun, M., R. Naeem, J.-X. Su, Z.-Y. Cao, J. G. Burleigh, P. S. Soltis, D. E. Soltis, and Z.-D. Chen. 2016. Phylogeny of the *Rosidae*: a dense taxon sampling analysis. *J. Syst. Evol.* 54:363–391.

Sun, M., D. E. Soltis, P. S. Soltis, X. Zhu, J. G. Burleigh, and Z. Chen. 2015. Deep phylogenetic incongruence in the angiosperm clade *Rosidae*. *Mol. Phylogenet. Evol.* 83:156–166.

Wang, H.-C., M. J. Moore, P. S. Soltis, C. D. Bell, S. F. Brockington, R. Alexandre, C. C. Davis, M. Latvis, S. R. Manchester, and D. E. Soltis. 2009. Rosid radiation and the rapid rise of angiosperm-dominated forests. *Proc. Natl. Acad. Sci. USA* 106:3853–3858.

Wojciechowski, M. F., M. Lavin, and M. J. Sanderson. 2004. A phylogeny of legumes (*Leguminosae*) based on analysis of the plastid *matK* gene resolves many well-supported subclades within the family. *Am. J. Bot.* 91:1846–1862.

Worberg, A., M. H. Alford, D. Quandt, and T. Borsch. 2009. *Huerteales* sister to *Brassicales* plus *Malvales*, and newly circumscribed to include *Dipentodon*, *Gerrardina*, *Huertea*, *Perrottetia*, and *Tapiscia*. *Taxon* 58:468–478.

Zhu, X.-Y., M. W. Chase, Y.-L. Qiu, H.-Z. Kong, D. L. Dilcher, J.-H. Li, and Z.-D. Chen. 2007. Mitochondrial *matR* sequences help to resolve deep phylogenetic relationships in rosids. *BMC Evol. Biol.* 7:217 (15 pp.). Available at http://www.biomedcentral.com/1471-2148/7/217.

Authors

Walter S. Judd; Department of Biology; University of Florida; Gainesville, FL 32611, USA. Email: lyonia@ufl.edu.

Douglas E. Soltis; Department of Biology; University of Florida; Gainesville, FL 32611, USA. Email: dsoltis@ufl.edi.

Pamela S. Soltis; Florida Museum of Natural History; University of Florida; Gainesville, FL 32611-7800, USA. Email: psoltis@flmnh.ufl.edu.

Philip D. Cantino; Department of Environmental and Plant Biology; Ohio University; Athens, OH 45701, USA. Email: cantino@ohio.edu.

Date Accepted: 21 May 2012; updated 30 September 2018

Primary Editor: Kevin de Queiroz

Leguminosae Jussieu 1789: 345 [M. J. Sanderson, M. F. Wojciechowski, M. M. McMahon and M. Lavin], converted clade name

Registration Number: 60

Definition: The crown clade originating in the most recent common ancestor of *Vicia faba* Linnaeus 1753 and all extant organisms or species that share a more recent common ancestor with *Vicia faba* than with *Polygala californica* Nuttall 1840, *Quillaja saponaria* Molina 1782, and *Suriana maritima* Linnaeus 1753. This is a maximum-crown-clade definition. Abbreviated definition: max crown ∇ (*Vicia faba* Linnaeus 1753 ~ *Polygala californica* Nuttall 1840 & *Quillaja saponaria* Molina 1782 & *Suriana maritima* Linnaeus 1753).

Etymology: Derived from French *légume* (legume), and ultimately from the Latin *legumen*, referring to clade members' characteristic fruit.

Reference Phylogeny: Wojciechowski et al. (2004: Figs. 1–5) is the primary reference phylogeny. Important others include Kajita et al. (2001: Figs. 1–3), Hilu et al. (2003: Fig. 9), Forest et al. (2007: Fig. 1), Bruneau et al. (2008: Figs. 1–4), Soltis et al. (2011: Fig. 2), and Legume Phylogeny Working Group (2017: Figs. 1, S1; tree file at https:// doi.org/10.5061/ dryad.61pd6).

Composition: Lewis et al. (2005) comprehensively reviewed this diverse clade of ~19,500 species, which historically have been classified among three subfamilies: *Caesalpinioideae* (paraphyletic), *Mimosoideae*, and *Papilionoideae* (all three established at various ranks), whereas the most recent classification proposes six monophyletic subfamilies (LPWG, 2017).

Diagnostic Apomorphies: The legume, which is a fruit arising from a superior ovary consisting of a single carpel, with seeds disposed along marginal placentae, in two adjacent rows, and dorsiventrally dehiscent along two sutures, is typical of most *Leguminosae*, and has usually been regarded as a diagnostic apomorphy of the clade (Judd et al., 2018). Innumerable modifications of the legume have evolved, including fleshy or nutlike indehiscent fruits, winged fruits, and fruits breaking into parts transversely or rarely, longitudinally. Some question about whether the legume is diagnostic has arisen with the finding that *Quillaja* might be the sister group of *Leguminosae*, because its fruit, although composed of five nearly free carpels, also dehisces dorsiventrally, suggesting that a single carpel of *Quillaja* could be homologous with the legume fruit.

Pinnately (or bi-pinnately) compound leaves with stipules and pulvini are present in the vast majority of *Leguminosae* and, in combination, represent a diagnostic syndrome (Judd et al., 2018), but variation among outgroups and rarely within *Leguminosae* makes assessment of it as a synapomorphy complicated. Among the likely outgroups, all have simple leaves except a few species of *Surianaceae*. The presence of simple leaves in some *Cercideae* (*Cercis* and relatives), which may be the sister group to the rest of the *Leguminosae*, raises the possibility that compound leaves are derived *within* the group and are thus not diagnostic. However, even if these taxa are correctly positioned in the tree, there is morphological (i.e., position of pulvini) and molecular genetic evidence (Champagne et al., 2007) supporting the hypothesis that these "simple leaves" are reductions of compound leaves to a single leaflet, a trend seen elsewhere in *Leguminosae*.

The seeds of *Leguminosae* have a unique exotestal structure consisting of a usually hard outer layer of elongate epidermal cells with spiral thickenings inside that render the testa watertight upon drying. In most legumes, excepting *Cercis* and close relatives, the hypodermal cells are characteristically hour-glass shaped (Gunn, 1981).

Synonyms: *Fabaceae* Lindley 1836 (e.g., Hickman, 1993), *Fabales* Bromhead 1838, and *Leguminales* Jones 1955 are approximate synonyms. See Comments.

Comments: *Leguminosae* is a large clade of great economic and ecological importance. Members of all six subfamilies are dominant or otherwise ecologically important throughout many biomes of the world. Perhaps this is due in part to a high nitrogen metabolism in legumes, which is promoted by various mechanisms including their ability to fix atmospheric nitrogen via symbiotic interactions with rhizobial bacteria in their specialized root nodules. Papilionoids comprise most of the economically important legume crops on the planet, used for food (peas, beans, soybeans, lentils, peanuts, etc.), forage (alfalfa, clovers, vetches, etc.) and many other purposes (Lewis et al., 2005).

Despite the highly stable content of *Leguminosae*, arguments about rank have dogged its taxonomic history and led to acceptance of the names *Leguminosae*, *Fabaceae*, *Leguminales*, and *Fabales* in various works. Especially problematic is the ambiguous use of *Fabaceae* for both the clade here named *Leguminosae* and the subclade of *Leguminosae* comprising the papilionoid legumes only (e.g., *Fabaceae* sensu Cronquist (1981)). Recent treatments (e.g., Lewis et al., 2005; LPWG, 2017) tend to recognize a single family *Leguminosae*, equivalent to our usage. Lewis and Schrire (2003) discussed the arguments over the choice of *Fabaceae* or *Leguminosae*, both of which are acceptable under the botanical code (Turland et al., 2018). In addition, we note that the suffix "*-osae*" has no rank-dependent meaning in classical botanical nomenclature, which seems to tip the balance in favor of *Leguminosae* for phylogenetic nomenclature.

The monophyly of *Leguminosae* is unambiguously supported by all molecular phylogenetic analyses (see those cited above under Reference Phylogeny). Evidence from several molecular markers now points to *Polygalaceae* (950 spp.), *Surianaceae* (8 spp.), and *Quillaja* (4 spp.) as the closest relatives within *Fabales*, clearly nested in the *Rosidae* clade (LPWG, 2017). Molecular phylogenies sampling multiple representatives of *Polygalaceae* and *Surianaceae* (Forest et al., 2007) support each of these taxa as monophyletic, thus allowing us to use a single exemplar of each as an external specifier to exclude them from the crown clade designated by our definition.

Most recent phylogenetic analyses resolve representatives of the former, but paraphyletic subfamily *Caesalpinioideae* as spanning the earliest nodes in the *Leguminosae* and including the sister group of all other legumes (Wojciechowski et al., 2004; Forest et al., 2007; LPWG, 2017). An alternative to the maximum-crown-clade definition adopted here is a minimum-crown-clade definition based on the six strongly supported subfamilies now recognized (LPWG, 2017). However, ongoing phylogenomic research has not yet resolved these relationships, and consequently such a definition would entail using specifiers from all six clades.

Although we do not define it formally here, the name *Apo-Leguminosae* may ultimately prove useful for the clade stemming from the ancestor in which the legume fruit originated. The earliest unambiguous fossil *Leguminosae* are from the Palaeocene, but typically these have been assigned to subclades of the crown group *Leguminosae* as defined herein (Herendeen and Dilcher, 1992). Only fossils possessing legumes but indisputably residing outside the crown group would require reference to *Apo-Leguminosae*.

Literature Cited

Bromhead, E. F. 1838. An attempt to ascertain characters of the botanical alliances. *Edinburgh New Philos. J.* 25:123–134.

Bruneau, A., M. Mercure, G. P. Lewis, and P. S. Herendeen. 2008. Phylogenetic patterns and diversification in the caesalpinioid legumes. *Botany* 86:697–718.

Champagne, C. E. M., T. E. Goliber, M. F. Wojciechowski, R. W.-B. Mei, B. T. Townsley, K. Wang, M. M. Paz, R. Geeta, and N. R. Sinha. 2007. Compound leaf development and evolution in the legumes. *Plant Cell* 19:3369–3378.

Cronquist, A. 1981. *An Integrated System of Classification of Flowering Plants*. Columbia University Press, New York.

Forest, F., M. W. Chase, C. Persson, P. R. Crane, and J. A. Hawkins. 2007. The role of biotic and abiotic factors in evolution of ant dispersal in the milkwort family (*Polygalaceae*). *Evolution* 61:1675–1694.

Gunn, C. R. 1981. Seeds of *Leguminosae*. Pp. 913–925 in *Advances in Legume Systematics: Part 2* (R. Polhill and P. Raven, eds.). Royal Botanic Gardens, Kew.

Herendeen, P. S., and D. L. Dilcher, eds. 1992. *Advances in Legume Systematics, Part 4: The Fossil Record*. Royal Botanic Gardens, Kew.

Hickman, J. C., ed. 1993. *The Jepson Manual: Higher Plants of California*. University of California Press, Berkeley, CA.

Hilu, K. W., T. Borsch, K. Muller, D. E. Soltis, P. S. Soltis, V. Savolainen, M. W. Chase, M. P. Powell, L. A. Alice, R. Evans, H. Sauquet, C. Neinhuis, T. A. B. Slotta, J. G. Rohwer, C. S. Campbell, and L. W. Chatrou. 2003. Angiosperm phylogeny based on *matK* sequence information. *Am. J. Bot.* 90:1758–1776.

Jones, G. N. 1955. *Leguminales*: a new ordinal name. *Taxon* 4:188–189.

Judd, W. S., C. S. Campbell, E. A. Kellogg, P. F. Stevens, and M. J. Donoghue. 2018. *Plant Systematics*. 4th edition. Sinauer Associates, Sunderland, MA.

Jussieu, A. L. 1789. *Genera Plantarum*. Apud Viduam Herissant et Theophilum Barrois, Parisii.

Kajita, T., H. Ohashi, Y. Tateishi, C. Bailey, and J. Doyle. 2001. *RbcL* and legume phylogeny, with particular reference to *Phaseoleae, Millettieae*, and allies. *Syst. Bot.* 26:515–536.

Legume Phylogeny Working Group (LPWG; N. Azani, M. Babineau, C. D. Bailey, et al. [92 authors]). 2017. A new subfamily classification of the *Leguminosae* based on a taxonomically comprehensive phylogeny. *Taxon* 66:44–77.

Lewis, G., and B. D. Schrire. 2003. *Leguminosae* or *Fabaceae*? Pp. 1–3 in *Advances in Legume Systematics: Part 10* (B. Klitgaard and A. Bruneau, eds.). Royal Botanic Gardens, Kew.

Lewis, G., B. Schrire, B. Mackinder, and M. Lock. 2005. *Legumes of the World*. Royal Botanic Gardens, Kew.

Lindley, J. 1836. *A Natural System of Botany*. Longman, Rees, Orme, Brown, Green, and Longman, London.

Soltis, D. E., S. A. Smith, N. Cellinese, K. J. Wurdack, D. C. Tank, S. F. Brockington, N. F. Refulio-Rodriquez, J. B. Walker, M. J. Moore, B. S. Carlsward, C. D. Bell, M. Latvis, S. Crawley, C. Black, D. Diouf, Z. Xi, C. A. Rushworth, M. A. Gitzendanner, K. J. Sytsma, Y-L. Qiu, K. W. Hilu, C. C. Davis, M. J. Sanderson, R. S. Beeman, R. G. Olmstead, W. S. Judd, M. J. Donoghue, and P. S. Soltis. 2011. Angiosperm phylogeny: 17 genes, 640 taxa. *Am. J. Bot.* 98:704–730.

Turland, N. J., J. H. Wiersema, F. R. Barrie, W. Greuter, D. L. Hawksworth, P. S. Herendeen, S. Knapp, W.-H. Kusber, D.-Z. Li, K. Marhold, T. W. May, J. McNeill, A. M. Monro, J. Prado, M. J. Price, and G. F. Smith, eds. 2018. *International Code of Nomenclature for Algae, Fungi, and Plants (Shenzhen Code)*. Adopted by the Nineteenth International Botanical Congress, Shenzhen, China, July 2017. Regnum Vegetabile 159. Koeltz Botanical Books, Glashütten.

Wojciechowski, M., M. Lavin, and M. Sanderson. 2004. A phylogeny of legumes (*Leguminosae*) based on analyses of the plastid *matK* gene resolves many well-supported subclades within the family. *Am. J. Bot.* 91:1846–1862.

Authors

Michael J. Sanderson; Department of Ecology and Evolutionary Biology; University of Arizona; Tucson, AZ 85721, USA. Email: sanderm@email.arizona.edu.

Martin F. Wojciechowski; School of Life Sciences; Arizona State University; Tempe, AZ 85287, USA. Email: mfwojciechowski@asu.edu.

Michelle M. McMahon; Department of Plant Sciences; University of Arizona; Tucson, AZ 85721, USA. Email: mcmahonm@email.arizona.edu.

Matt Lavin; Department of Plant Sciences and Plant Pathology; Montana State University; Bozeman, MT 59717, USA. Email: mlavin@montana.edu.

Date Accepted: 13 February 2009; updated 3 October 2019

Primary Editor: Philip Cantino

Superasteridae W. S. Judd, D. E. Soltis, and P. S. Soltis in Soltis et al., 2011 [W. S. Judd, D. E. Soltis, and P. S. Soltis], converted clade name

Registration Number: 104

Definition: The largest crown clade containing *Aster amellus* L. 1753 (*Asteridae/Asterales*) but not *Rosa cinnamomea* L. 1753 (*Rosidae/Rosales*). This is a maximum-crown-clade definition. Abbreviated definition: max crown ∇ (*Aster amellus* L. 1753 ~ *Rosa cinnamomea* L. 1753).

Etymology: From the Latin *super* (above, over, or on top) and *Asteridae*, a converted clade name based on the name of the included taxon *Aster* (derived from the Latin, *aster*, meaning star, so called because of the form of the radiate floral heads of these plants), in reference to the fact that *Superasteridae* is intended to refer to a crown clade more inclusive than *Asteridae*.

Reference Phylogeny: The primary reference phylogeny is Soltis et al. (2011: Figs. 1–2). See also Burleigh et al. (2009: Fig. 2, and online supplemental file 7) and Moore et al. (2010: Fig. 1). *Aster amellus* is used as a specifier because it is the type species of *Asteridae*, which forms part of the defined clade name (see *PhyloCode* (Cantino and de Queiroz, 2020), Art. 11.10). The close relationship of *Aster amellus* to the species of *Asteraceae* included in the primary reference phylogeny (i.e., *Barnadesia arborea, Gerbera jamesonii, Echinops bannaticus, Tragopogon dubius, T. porrifolius, Cichorium intybus, Lactuca sativa, Tagetes erecta, Guizotia abyssinica,* and *Helianthus annuus*) is supported by the series of phylogenetic studies of *Asteraceae* presented in Funk et al. (2009). Similarly, *Rosa cinnamomea* is used as a specifier because it is the type species of *Rosidae*, which forms part of the name *Superrosidae* (this volume). The close relationship of *Rosa cinnamomea* to the species of *Rosaceae* included in the primary reference phylogeny (i.e., *Spiraea betulifolia, S. vanhouttei,* and *Prunus persica*) is supported by the analyses of Evans et al. (2000) and Potter et al. (2002, 2007), which found *Rosaceae* to be monophyletic.

Composition: *Santalales, Berberidopsidales, Caryophyllales, Asteridae,* and possibly *Dilleniaceae* (see Comments).

Diagnostic Apomorphies: No non-DNA synapomorphies are known.

Synonyms: There are no synonymous scientific names, but the informal name "superasterids" (Moore et al., 2010; Stevens, 2001 and onwards, version 9) has been used for this clade.

Comments: This is a recently discovered clade that is strongly supported by the molecular analyses of Soltis et al. (2011). It was also strongly supported in the 83-gene ML and MP analyses of Moore et al. (2010: Fig. 1), which were based on a much smaller array of sampled taxa (and did not include any representative of *Berberidopsidales*). An unnamed clade with the same composition received weak support in the ML analyses of angiosperms based on 5 genes (Burleigh et al., 2009: see Fig. 3 and the "full tree" included with the on-line supplemental information in supplemental file 7). The clade was informally named "the superasterids" by Moore et al. (2010) and formally named *Superasteridae* by Soltis et al. (2011) with the same phylogenetic definition that we use here.

There is some disagreement among recent phylogenetic analyses regarding the position of *Dilleniaceae*. In the 17-gene analysis of Soltis et al. (2011), the representatives of *Dilleniaceae* (*Tetracera*, *Hibbertia*, and *Dillenia*) were found to form a clade within *Superasteridae*, where they were strongly supported as sister to a clade comprising *Santalales*, *Caryophyllales*, *Berberidopsidales*, and *Asteridae*. The analyses of Burleigh (2009, including online supplemental files 6 and 7) also found *Superasteridae* to include *Dilleniaceae*; i.e., *Dillenia* and *Tetracera* were sister to *Caryophyllales* using the 3-gene data (supplemental file 6), while these two taxa were sister to the rest of the *Superasteridae* using the 5-gene data (supplemental file 7). In contrast, in an analysis of complete plastid genome sequence data (Moore et al., 2010), the sole representative of *Dilleniaceae* (*Dillenia*) was found to lie outside *Superasteridae* as sister to a clade comprising *Saxifragales* and *Rosidae*. In an analysis of IR (inverted repeat) sequences (Moore et al., unpublished), the representative of *Dilleniaceae* (*Dillenia*) was also found to lie outside *Superasteridae* as the sister group to a large clade comprising *Rosidae*, *Saxifragales*, *Asteridae*, *Berberidopsidales*, *Caryophyllales*, and *Santalales*. For both *Superasteridae* and *Superrosidae*, the use of maximum-crown-clade definitions that do not include any members of *Dilleniaceae* as specifiers accommodates placement of *Dilleniaceae* in either *Superasteridae* or *Superrosidae*, or positioned as the sister taxon to a *Superrosidae* + *Superasteridae* clade, within the *Pentapetalae*.

The essential feature of our concept of *Superasteridae* that we have tried to capture in our definition is its inclusion of every extant species that is closer to *Asteridae* than to *Rosidae*. Furthermore, when this definition is used in conjunction with our reciprocal definition of *Superrosidae* (this volume), it ensures that *Superasteridae* and *Superrosidae* are always sister crown clades, regardless of the placement of *Dilleniaceae*.

Literature Cited

Burleigh, J. G., K. W. Hilu, and D. E. Soltis. 2009. Inferring phylogenies with incomplete data sets: a 5-gene, 567-taxon analysis of angiosperms. *BMC Evol. Biol.* 9:61 (11 pp.). Available at http://www.biomedcentral.com/1471-2148/9/61.

Cantino, P. D., and K. de Queiroz. 2020. *International Code of Phylogenetic Nomenclature* (*PhyloCode*), Version 6. CRC Press, Boca Raton, FL.

Evans, R. C., L. A. Alice, C. S. Campbell, T. A. Dickinson, and E. A. Kellogg. 2000. The granule-bound starch synthase (GBSSI) gene in the *Rosaceae*: multiple loci and phylogenetic utility. *Mol. Phylogenet Evol.* 17:388–400.

Funk, V. A., A. Susanne, T. F. Stuessy, and R. J. Bayer, eds. 2009. *Systematics, Evolution, and Biogeography of* Compositae. International Association for Plant Taxonomy, Institute of Botany, University of Vienna, Vienna.

Moore, M. J., P. S. Soltis, C. D. Bell, J. G. Burleigh, and D. E. Soltis. 2010. Phylogenetic analysis of 83 plastid genes further resolves the early diversification of eudicots. *Proc. Nat. Acad. Sci. USA.* 107:4623–4628.

Potter, D., T. Eriksson, R. C. Evans, S.-H. Oh, J. Smedmark, D. Morgan, M. Kerr, K. R. Robertson, M. Arsenault, T. A. Dickinson, and C. S. Campbell. 2007. Phylogeny and classification of *Rosaceae*. *Plant Syst. Evol.* 266:5–43.

Potter, D., F. Gao, P. E. Bortiri, S.-H. Oh, and S. Baggett. 2002. Phylogenetic relationships in *Rosaceae* inferred from chloroplast *matK* and *trnL-trnF* nucleotide sequence data. *Plant Syst. Evol.* 231:77–89.

Soltis, D. E., S. A. Smith, N. Cellinese, K. J. Wurdack, D. C. Tank, S. F. Brockington, N. F. Refulio-Rodriguez, J. B. Walker, M. J. Moore, B. S. Carlsward, C. D. Bell, M. Latvis, S. Crawley, C. Black, D. Diouf, Z.-X. Xi, C. A. Rushworth, M. A. Gitzendanner, K. J. Sytsma, Y.-L Qiu, K. W. Hilu, C. C. Davis, M. J. Sanderson, R. S. Beaman, R. G. Olmstead, W. S. Judd, M. J. Donoghue, and P. S. Soltis. 2011. Angiosperm phylogeny: 17 genes, 640 taxa. *Am. J. Bot.* 98:704–730.

Stevens, P. F. 2001 onwards. Angiosperm Phylogeny Website. Version 9, June 2008 [and more or less continuously updated since]. Available at http://www.mobot.org/MOBOT/research/APweb/.

Authors

Walter S. Judd; Department of Biology; University of Florida; Gainesville, FL 32611-8526, USA. Email: lyonia@ufl.edu.

Douglas E. Soltis; Department of Biology; University of Florida; Gainesville, FL 32611-8526, USA. Email: dsoltis@ufl.edu.

Pamela S. Soltis; Florida Museum of Natural History; University of Florida, Gainesville, FL 32611-7800, USA. Email: psoltis@flmnh.ufl.edu.

Date Accepted: 3 December 2011

Primary Editor: Philip Cantino

Caryophyllales K. J. Perleb 1826: 312 (as "Caryophylleae") [P. S. Soltis, W. S. Judd, and D. E. Soltis, 2009], converted clade name

Registration Number: 24

Definition: the smallest crown clade containing *Dianthus caryophyllus* L. 1753 (*Caryophyllaceae*), *Polygonum sachalinense* F. Schmidt ex Maxim. 1859 (*Polygonaceae*), *Simmondsia chinensis* (Link) C. K. Schneid. 1907 (*Simmondsiaceae*), and *Rhabdodendron amazonicum* (Spruce ex Benth.) Huber 1909 (*Rhabdodendraceae*). This is a minimum-crown-clade definition. Abbreviated definition: min crown ∇ (*Dianthus caryophyllus* L. 1753 & *Polygonum sachalinense* F. Schmidt ex Maxim. 1859 & *Simmondsia chinensis* (Link) C. K. Schneid. 1907 & *Rhabdodendron amazonicum* (Spruce ex Benth.) Huber 1909).

Etymology: Based on *Caryophyllus* P. Miller (which included the carnation, now *Dianthus caryophyllus* L.), not *Caryophyllus* L. (a rejected name for the clove, now *Syzygium aromaticum* (L.) Merr. & L. M. Perry), and derived from the Greek *karyon* (nut, or kernel) and *phyllon* (leaf), referring to the smell of walnut leaves, which led to the use of the name *karyophyllon* for the clove (Brown, 1956), and thence the clove pink (or carnation) (Stearn, 1992) due to its distinctive odor. The name *Caryophyllus* P. Miller is not in current use because it is a later homonym of *Caryophyllus* L.

Reference Phylogeny: Soltis et al. (2000: Fig. 5). *Dianthus caryophyllus*, which is used as a specifier because it is the type species of *Caryophyllales* under the botanical code (Turland et al., 2018), is not included in the primary reference phylogeny, but its closest relative there is *Stellaria*. See also Cuénoud et al.

(2002: Fig. 2), Hilu et al. (2003: Fig. 8), and Brockington et al. (2009: Fig. 1).

Composition: See APG III (2009) for taxa included in *Caryophyllales*. Some included clades that are particularly large or economically important are *Aizoaceae*, *Amaranthaceae*, *Cactaceae*, *Caryophyllaceae*, and *Polygonaceae*.

Diagnostic Apomorphies: More studies are needed, but possible synapomorphies include anthers with outer parietal cells developing directly into the endothecium and gynoecia with elongate style branches (Judd et al., 2008; Stevens, 2001 onwards).

Synonyms: *Caryophyllanae* sensu Thorne and Reveal (2007) and sensu Chase and Reveal (2009) are approximate synonyms, based in the former case on composition and in the latter on labeling of a cladogram in which it is not specified whether the name applies to a node or a branch. *Caryophyllidae* sensu Takhtajan (1967, 1987, 1997) and Cronquist (1981, 1988) is, in each case, a partial synonym; some taxa placed by those authors in *Dillenidae*, *Rosidae*, and *Hamamelidae* are part of *Caryophyllales*. However, *Caryophyllidae* sensu Judd et al. (2002), Soltis and Soltis (2003: 1793), and Judd and Olmstead (2004) is an approximate synonym (see Comments).

Comments: This clade is strongly supported in all multi-gene molecular phylogenetic analyses of the angiosperms as a whole (e.g., Savolainen et al., 2000; Soltis et al., 1999, 2000, 2007) and approximates the group that has often been called *Caryophyllidae* (or informally, the caryophyllids; see Chase et al., 1993). The name *Caryophyllales* was first applied to this clade by APG (1998), a phylogenetic system in which many broad ordinal circumscriptions were proposed, and this name was retained in APG II (2003) and APG III (2009) and thus adopted in many recent phylogenetic papers (e.g., Savolainen et al., 2000; Soltis et al., 2000; Cuénoud et al., 2002; Hilu et al., 2003; Brockington et al., 2009) and textbooks (Judd et al., 2008; Simpson, 2006). It was phylogenetically defined by Soltis et al. (in Cantino et al., 2007), using the same definition as is used here. However, this leaves the less inclusive clade that was long associated with the name *Caryophyllales* (e.g., Takhtajan, 1967, 1987, 1997; Cronquist, 1981, 1988)—a group characterized by P-type plastids, seeds with perisperm and a curved embryo, and betalain pigments—in need of a new name. This less inclusive clade was also commonly referred to in mid-twentieth century literature (e.g., Lawrence, 1951; Engler and Harms, 1960; Melchior, 1964; Rendle, 1967) as *Centrospermae*, a name that could be phylogenetically defined in the future to apply to this smaller clade. There is yet a third clade that has also been called *Caryophyllales* in relatively recent papers. The clear support (e.g., Soltis et al., 2000; Cuénoud et al., 2002; APG II, 2003; Brockington et al., 2009) for two large clades within *Caryophyllales* sensu APG has led some (Judd et al., 2002; Judd & Olmstead, 2004) to propose that the name *Caryophyllidae* be used for the more inclusive clade (i.e., *Caryophyllales* sensu APG), and that its two subclades be named *Caryophyllales* (composed of the P-type plastid clade that has traditionally been called *Caryophyllales* plus *Rhabdodendraceae*, *Simmondsiaceae*, *Asteropeiaceae*, and *Physenaceae*) and *Polygonales*. After careful consideration, we have provided a phylogenetic definition linking the name *Caryophyllales* to the most inclusive of the three clades to which it has historically been applied, in agreement with APG III (2009; see also Stevens, 2001). Because the APG system has proven to be very influential, we feel that defining *Caryophyllales* in this manner will best facilitate communication within the systematic community.

It is noteworthy that *Caryophyllales* includes a number of highly unusual subgroups, some of which exhibit morphological, anatomical, and physiological specializations to arid environments (e.g., *Aizoaceae, Cactaceae, Didiereaceae,* and *Portulacaceae*), while others are carnivorous (e.g., *Droseraceae, Drosophyllaceae,* and *Nepenthaceae*).

Literature Cited

APG (Angiosperm Phylogeny Group). 1998. An ordinal classification for the families of flowering plants. *Ann. Mo. Bot. Gard.* 85:531–553.

APG II (Angiosperm Phylogeny Group II). 2003. An update of the Angiosperm Phylogeny Group classification for the orders and families of flowering plants: APG II. *Bot. J. Linn. Soc.* 141:399–436.

APG III (Angiosperm Phylogeny Group III). 2009. An update of the Angiosperm Phylogeny Group classification for the orders and families of flowering plants: APG III. *Bot. J. Linn. Soc.* 161:105–121.

Brockington, S. F., R. Alexandre, J. Ramdial, M. J. Moore, S. Crawley, A. Dhingra, K. Hilu, D. E. Soltis, and P. S. Soltis. 2009. Phylogeny of the *Caryophyllales* sensu lato: revisiting hypotheses on pollination biology and perianth differentiation in the core *Caryophyllales*. *Int. J. Plant Sci.* 170:627–643.

Brown, R. W. 1956. *Composition of Scientific Words.* Smithsonian Books, Washington, D.C.

Cantino, P. D., J. A. Doyle, S. W. Graham, W. S. Judd, R. G. Olmstead, D. E. Soltis, P. S. Soltis, and M. J. Donoghue. 2007. Towards a phylogenetic nomenclature of *Tracheophyta*. *Taxon* 56:822–846 and E1–E44.

Chase, M. W., and J. L. Reveal. 2009. A phylogenetic classification of the land plants to accompany APG III. *Bot. J. Linn. Soc.* 161:122–127.

Chase, M. W., D. E. Soltis, R. G. Olmstead, D. Morgan, D. H. Les, B. D. Mishler, M. R. Duvall, R. A. Price, H. G. Hills, Y.-L. Qiu, K. A. Kron, J. H. Rettig, E. Conti, J. D. Palmer, J. R. Manhart, K. J. Sytsma, H. J. Michaels, W. J. Kress, K. G. Karol, W. D. Clark, M. Hedrén, B. S. Gaut, R. K. Jansen, K.-J. Kim, C. F. Wimpee, J. F. Smith, G. R. Furnier, S. H. Strauss, Q.-Y. Xiang, G. M. Plunkett, P. S. Soltis, S. M. Swensen, S. E. Williams, P. A. Gadek, C. J. Quinn, L. E. Equiarte, E. Golenberg, G. H. Learn, Jr., S. W. Graham, S. C. H. Barrett, S. Dayanandan, and V. A. Albert. 1993. Phylogeny of seed plants: an analysis of nucleotide sequences from the plastid gene rbcL. *Ann. Mo. Bot. Gard.* 80:528–580.

Cronquist, A. 1981. *An Integrated System of Classification of Flowering Plants.* Columbia University Press, New York.

Cronquist, A. 1988. *The Evolution and Classification of Flowering Plants.* 2nd edition. New York Botanical Garden, Bronx, New York.

Cuénoud, P., V. Savolainen, L. W. Chatrou, M. Powell, R. J. Grayer, and M. W. Chase. 2002. Molecular phylogenetics of *Caryophyllales* based on nuclear 18S rDNA and plastid *rbcL*, *atpB*, and *matK* DNA sequences. *Am. J. Bot.* 89:132–144.

Engler, A., and H. Harms. 1960. *Die Natürlichen Pflanzenfamilien.* 2nd edition, Vol. 16c. Duncker & Humblot, Berlin.

Hilu, K. W., T. Borsch, K. Müller, D. E. Soltis, P. S. Soltis, V. Savolainen, M. W. Chase, M. P. Powell, L. A. Alice, R. Evans, H. Sauquet, C. Neinhuis, T. A. B. Slotta, J. G. Rohwer, C. S. Campbell, and L. W. Chatrou. 2003. Angiosperm phylogeny based on *matK* sequence information. *Am. J. Bot.* 90:1758–1776.

Judd, W. S., C. S. Campbell, E. A. Kellogg, P. F. Stevens, and M. J. Donoghue. 2002. *Plant Systematics—A Phylogenetic Approach.* 2nd edition. Sinauer Associates, Sunderland, MA.

Judd, W. S., C. S. Campbell, E. A. Kellogg, P. F. Stevens, and M. J. Donoghue. 2008. *Plant Systematics—A Phylogenetic Approach.* 3rd edition. Sinauer Associates, Sunderland, MA.

Judd, W. S., and R. G. Olmstead. 2004. A survey of tricolpate (eudicot) phylogenetic relationships. *Am. J. Bot.* 91:1627–1644.

Lawrence, G. H. M. 1951. *Taxonomy of Vascular Plants.* MacMillan, New York.

Melchior, H. 1964. A. Engler's *Syllabus der Pflanzenfamilien*, Vol. 2. Gebruder Borntraeger, Berlin.

Perleb, K. J. 1826. *Lehrbuch der Naturgeschicte des Pflanzenreichs.* Freiburg im Breisgau, Verlag von Friedrich Wagner.

Rendle, A. B. 1967. *The Classification of Flowering Plants*, Vol. 2. Cambridge University Press, Cambridge, UK.

Savolainen, V., M. W. Chase, S. B. Hoot, C. M. Morton, D. E. Soltis, C. Bayer, M. F. Fay, A. Y. de Bruijn, S. Sullivan, and Y.-L. Qiu. 2000. Phylogenetics of flowering plants based on combined analysis of plastid *atpB* and *rbcL* gene sequences. *Syst. Biol.* 49:306–362.

Simpson, M. G. 2006. *Plant Systematics.* Elsevier Academic Press, Amsterdam.

Soltis, D. E., M. A. Gitzendanner, and P. S. Soltis. 2007. A 567-taxon data set for angiosperms: the challenges posed by Bayesian analyses of large data sets. *Int. J. Plant Sci.* 168:137–157.

Soltis, D. E., and P. S. Soltis. 2003. The role of phylogenetics in comparative genetics. *Plant Phys.* 132:1790–1800.

Soltis, D. E., P. S. Soltis, M. W. Chase, M. E. Mort, D. C. Albach, M. Zanis, V. Savolainen, W. H. Hahn, S. B. Hoot, M. F. Fay, M. Axtell, S. M. Swensen, L. M. Prince, W. J. Kress, K. C. Nixon, and J. S. Farris. 2000. Angiosperm phylogeny inferred from 18S rDNA, *rbcL*, and *atpB* sequences. *Bot. J. Linn. Soc.* 133:381–461.

Soltis, P. S., D. E. Soltis, and M. W. Chase. 1999. Angiosperm phylogeny inferred from multiple genes: a research tool for comparative biology. *Nature* 402:402–404.

Stearn, W. T. 1992. *Stearn's Dictionary of Plant Names for Gardeners: A Handbook on the Origin and Meaning of the Botanical Names of Some Cultivated Plants.* Timber Press, Portland, OR.

Stevens, P. F. 2001 and onwards. Angiosperm Phylogeny Website, Version 9, June 2008 [and more or less continuously updated since]. Available at http://www.mobot.org/MOBOT/research/APweb/.

Takhtajan, A. 1967. *Sistema i Filogeniia Tsvetkovykh Rastenii (Systema et Phylogenia Magnoliophytorum)*. Nauka, Moscow. [Dated 1966, but published 4 Feb. 1967; J. Reveal, pers. comm.]

Takhtajan, A. 1987. *Systema Magnoliophytorum*. Nauka, Leningrad.

Takhtajan, A. 1997. *Diversity and Classification of Flowering Plants*. Columbia University Press, New York.

Thorne, R. F., and J. L. Reveal. 2007. An updated classification of the class *Magnoliopsida* ("*Angiospermae*"). *Bot. Rev.* 73:67–182.

Turland, N. J., J. H. Wiersema, F. R. Barrie, W. Greuter, D. L. Hawksworth, P. S. Herendeen, S. Knapp, W.-H. Kusber, D.-Z. Li, K. Marhold, T. W. May, J. McNeill, A. M. Monro, J. Prado, M. J. Price, and G. F. Smith, eds. 2018. *International Code of Nomenclature for Algae, Fungi, and Plants (Shenzhen Code)*. Adopted by the Nineteenth International Botanical Congress, Shenzhen, China, July 2017. Regnum Vegetabile 159. Koeltz Botanical Books, Glashütten.

Authors

Pamela S. Soltis; Florida Museum of Natural History; University of Florida; Gainesville, FL 32611-7800, USA. Email: psoltis@flmnh.ufl.edu.

Walter S. Judd; Department of Biology; University of Florida; Gainesville, FL 32611, USA. Email: lyonia@ufl.edu.

Douglas E. Soltis; Florida Museum of Natural History and Department of Biology, University of Florida; Gainesville, FL 32611, USA. Email: dsoltis@ufl.edu.

Date Accepted: 21 August 2011

Primary Editor: Philip Cantino

Asteridae Takhtajan 1967: 405 [R. G. Olmstead and W. S. Judd], converted clade name

Registration Number: 18

Definition: The smallest crown clade containing *Lamium purpureum* Linnaeus 1753 (*Lamiidae/Lamiales*), *Cornus mas* Linnaeus 1753 (*Cornales*), *Aster amellus* Linnaeus 1753 (*Campanulidae/Asterales*), and *Arbutus unedo* Linnaeus 1753 (*Ericales*). This is a minimum-crown-clade definition. Abbreviated definition: min crown ∇ (*Lamium purpureum* Linnaeus 1753 & *Cornus mas* Linnaeus 1753 & *Aster amellus* Linnaeus 1753 & *Arbutus unedo* Linnaeus 1753).

Etymology: Derived from *Aster* (name of a genus), which is Greek and Latin for "star" (Brown, 1954), referring to the shape of the flowering head.

Reference Phylogeny: Soltis et al. (2011: Figs. 1–2). *Aster amellus*, which is used as a specifier because it is the type species of *Asteridae* in rank-based nomenclature, is most closely related to the clade comprising *Helianthus*, *Guizotia*, and *Tagetes* in this reference phylogeny (Funk et al., 2009: Figs. 44.3–44.8). See also Olmstead et al. (2000: Fig. 1) and Bremer et al. (2002: Figs. 1–2).

Composition: The crown clades *Cornales*, *Ericales* (this volume), and *Gentianidae* (this volume), and extinct descendants of their most recent common ancestor.

Diagnostic Apomorphies: Possible synapomorphies include tenuinucellate and unitegmic ovules, sympetaly, and iridoid compounds, but all of these traits may be synapomorphic at a less inclusive level (Olmstead et al., 1993;

Albach et al., 2001a; Judd and Olmstead, 2004; Judd et al., 2007).

Synonyms: *Asteranae* sensu Chase and Reveal (2009) has the same composition and is therefore an approximate synonym. The informal name "asterids" has been used for this clade in many publications (e.g., Savolainen et al., 2000; Soltis et al., 2000; Albach et al., 2001b; Bremer et al., 2002; Judd and Olmstead, 2004).

Comments: *Asteridae* was first used by Takhtajan (1967) for a group of plants that mostly shared a suite of floral characters including sympetalous corollas, stamens adnate to the corolla and arranged alternately with the corolla lobes, and two fused carpels. Circumscription of that group included *Asterales* and *Dipsacales* sensu APG II (2003) and *Lamianae* (as defined in this volume) with some minor differences. Takhtajan (1987, 1997) later recognized a much-reduced *Asteridae*, which approximated *Asterales* sensu APG II (2003). The name *Asteridae* was applied formally to the clade of concern here by Olmstead et al. (1992: Fig. 2) and, using an explicit phylogenetic definition, by Olmstead and Judd in Cantino et al. (2007: E29). This expanded concept of *Asteridae* has been accepted (as *Asteridae* or "asterids") in all recent phylogenetic analyses and classifications (e.g., Olmstead et al., 2000; Soltis et al., 2000; Albach et al. 2001b; Bremer et al., 2002; Hilu et al. 2003; APG II 2003; Judd and Olmstead, 2004; Soltis et al., 2005, 2011; APG IV, 2016).

Asteridae, as defined here, were first recognized as a clade by Olmstead et al. (1992) based on *rbcL* sequences, although with weak bootstrap support. Subsequent studies using other genes (e.g., *ndhF* – Olmstead et al., 2000; *matK* – Hilu et al., 2003) and multiple gene datasets (e.g., Savolainen et al., 2000; Soltis et al., 2000; Albach et al., 2001b; Bremer et al., 2002) infer this clade with very high support (>90%).

All of the possible synapomorphies occur elsewhere in angiosperms, although rarely outside of *Asteridae*, and are variable within *Asteridae*. Unitegmic ovules probably are the most consistent synapomorphy. Both tenuinucellate and crassinucellate ovules are found in *Cornales*, which are sister to the remaining *Asteridae*, but the distribution and frequency with which tenuinucellate ovules occur in *Cornales* and their rarity outside of *Asteridae* suggests they arose in the common ancestor of *Asteridae*. Sympetaly is likewise variable in *Cornales* and *Ericales* and may represent a case of multiple origins within *Asteridae*, although developmental evidence of early corolla initiation in polypetalous clades within *Asteridae* often shows a fused primordial ring before development of unfused petals at maturity (Erbar, 1991; Erbar and Leins, 1996).

Literature Cited

Albach, D. C., P. S. Soltis, and D. E. Soltis. 2001a. Patterns of embryological and biochemical evolution in the asterids. *Syst. Bot.* 2:242–262.

Albach, D. C., P. S. Soltis, D. E. Soltis, and R. G. Olmstead. 2001b. Phylogenetic analysis of asterids based on sequences of four genes. *Ann. Mo. Bot. Gard.* 88:163–212.

APG II (Angiosperm Phylogeny Group II). 2003. An update of the Angiosperm Phylogeny Group classification for the orders and families of flowering plants: APG II. *Bot. J. Linn. Soc.* 141:399–436.

APG IV (Angiosperm Phylogeny Group IV). 2016. An update of the Angiosperm Phylogeny Group classification for the orders and families of flowering plants: APG IV. *Bot. J. Linn. Soc.* 181:1–20.

Bremer, B., K. Bremer, N. Heidari, P. Erixon, R. G. Olmstead, A. A. Anderberg, M. Källersjö, and E. Barkhordarian. 2002. Phylogenetics of asterids based on 3 coding and 3 non-coding chloroplast DNA markers and the utility of non-coding DNA at higher taxonomic levels. *Mol. Phylogenet. Evol.* 24:274–301.

Brown, R. W. 1954. *Composition of Scientific Words*. Smithsonian Books, Washington, D.C.

Cantino, P. D., J. A. Doyle, S. W. Graham, W. S. Judd, R. G. Olmstead, D. E. Soltis, P. S. Soltis, and M. J. Donoghue. 2007. Towards a phylogenetic nomenclature of *Tracheophyta*. *Taxon* 56:822–846 and E1–E44.

Chase, M. W., and J. L. Reveal. 2009. A phylogenetic classification of the land plants to accompany APG III. *Bot. J. Linn. Soc.* 161:122–127.

Erbar, C. 1991. Sympetaly – a systematic character? *Bot. Jahrb. Syst.* 112:417–451.

Erbar, C., and P. Leins. 1996. Distribution of the character states "early" and "late sympetaly" within the "Sympetalae tetracyclicae" and presumably related groups. *Bot. Acta* 109:429–440.

Funk, V. A., and 52 coauthors. 2009. *Compositae metatrees: the next generation. Pp. 747–777 in Systematics, Evolution, and Biogeography of* Compositae (V. A. Funk, A. Susanna, T. F. Stuessy, and R. J. Bayer, eds.). International Association for Plant Taxonomy, Vienna, Austria.

Hilu, K. W., T. Borsch, K. Müller, D. E. Soltis, P. S. Soltis, V. Savolainen, M. W. Chase, M. P. Powell, L. A. Alice, R. Evans, H. Sauquet, C. Neinhuis, T. A. B. Slotta, J. G. Rohwer, C. S. Campbell, and L. W. Chatrou. 2003. Angiosperm phylogeny based on *matK* sequence information. *Am. J. Bot.* 90:1758–1776.

Judd, W. S., C. S. Campbell, E. A. Kellogg, P. F. Stevens, and M. J. Donoghue. 2007. *Plant Systematics—A Phylogenetic Approach*. 3rd edition. Sinauer Associates, Sunderland, MA.

Judd, W. S., and R. G. Olmstead. 2004. A survey of tricolpate (eudicot) phylogenetic relationships. *Am. J. Bot.* 91:1627–1644.

Olmstead, R. G., B. Bremer, K. M. Scott, and J. D. Palmer. 1993. A parsimony analysis of the *Asteridae* sensu lato based on *rbcL* sequences. *Ann. Mo. Bot. Gard.* 80:700–722.

Olmstead, R. G., K. Kim, R. K. Jansen, and S. J. Wagstaff. 2000. The phylogeny of the *Asteridae* sensu lato based on chloroplast *ndhF* gene sequences. *Mol. Phylogenet. Evol.* 16:96–112.

Olmstead, R. G., H. J. Michaels, K. M. Scott, and J. D. Palmer. 1992. Monophyly of the *Asteridae* and identification of their major lineages inferred from DNA sequences of *rbcL*. *Ann. Mo. Bot. Gard.* 79:249–265.

Savolainen, V., M. W. Chase, S. B. Hoot, C. M. Morton, D. E. Soltis, C. Bayer, M. F. Fay, A. Y. de Bruijn, S. Sullivan, and Y.-L. Qiu. 2000. Phylogenetics of flowering plants based on combined analysis of plastid *atpB* and *rbcL*. *Syst. Biol.* 49:306–362.

Soltis, D. E., S. A. Smith, N. Cellinese, K. J. Wurdack, D. C. Tank, S. F. Brockington, N. F. Refulio-Rodriguez, J. B. Walker, M. J. Moore, B. S. Carlsward, C. D. Bell, M. Latvis, S. Crawley, C. Black, D. Diouf, Z. Xi., M. A. Gitzendanner, K. J. Sytsma, Y.-L. Qiu, K. W. Hilu, S. R. Manchester, C. C. Davis, M. J. Sanderson, R. G. Olmstead, W. S. Judd, M. J. Donoghue, and P. S. Soltis. 2011. Angiosperm phylogeny: 17 genes, 640 taxa. *Am. J. Bot.* 98:704–730.

Soltis, D. E., P. S. Soltis, M. W. Chase, M. E. Mort, D. C. Albach, M. Zanis, V. Savolainen, W. H. Hahn, S. B. Hoot, M. F. Fay, M. Axtell, S. M. Swensen, L. M. Prince, W. J. Kress, K. C. Nixon, and J. S. Farris. 2000. Angiosperm phylogeny inferred from 18S rDNA, *rbcL*, and *atpB* sequences. *Bot. J. Linn. Soc.* 133:381–461.

Soltis, D. E., P. S. Soltis, P. K. Endress, and M. W. Chase. 2005. *Phylogeny and Evolution of Angiosperms.* Sinauer Associates, Sunderland, MA.

Takhtajan, A. 1967. *Sistema i Filogeniia Tsvetkovykh Rastenii (Systema et Phylogenia Magnoliophytorum).* Nauka, Moscow. [Dated 1966, but published 4 Feb. 1967; J. Reveal, pers. comm.]

Takhtajan, A. 1987. *Systema Magnoliophytorum.* Nauka, Leningrad.

Takhtajan, A. 1997. *Diversity and Classification of Flowering Plants.* Columbia University Press, New York.

Authors

Richard G. Olmstead; Department of Biology and Burke Museum; University of Washington; Seattle, WA 98195, USA. Email: olmstead@u.washington.edu.

Walter S. Judd; Department of Biology; University of Florida; Gainesville, FL 32611, USA. Email: lyonia@ufl.edu.

Date Accepted: 24 August 2010; updated 18 July 2017

Primary Editor: Philip Cantino

Ericales Berchtold and Presl 1820: 251 [W. S. Judd and K. A. Kron], converted clade name

Registration Number: 41

Definition: The smallest crown clade containing *Impatiens capensis* Meerb. 1794 and *Erica cinerea* L. 1753. This is a minimum-crown-clade definition. Abbreviated definition: min crown ∇ (*Impatiens capensis* Meerb. 1794 & *Erica cinerea* L. 1753).

Etymology: Based on *erice*, an ancient Latin name for heather.

Reference Phylogeny: The primary reference phylogeny is Schönenberger et al. (2005: Fig. 1). *Erica cinerea* is used as a specifier because it is the type of *Ericales*; although not included in the primary reference phylogeny, its close relationship to those species of *Ericaceae* that are included is supported by Kron et al. (2002) and McGuire and Kron (2005). See also Soltis et al. (2000: Fig. 10), Savolainen et al. (2000b: Fig. 8), Albach et al. (2001: Fig. 2), Anderberg et al. (2002: Fig.1), and Bremer et al. (2002: Fig. 1).

Composition: Major clades within *Ericales* include *Balsaminaceae*, *Marcgraviaceae*, *Tetrameristaceae* (incl. *Pellicieraceae*), *Polemoniaceae*, *Fouquieriaceae*, *Lecythidaceae*, *Sladeniaceae*, *Pentaphylacaceae* (incl. *Ternstroemiaceae*), *Sapotaceae*, *Ebenaceae*, *Primulaceae* sensu lato (incl. *Maesaceae*, *Theophrastaceae*, and *Myrsinaceae*, although many workers maintain these at the familial level, e.g., Anderberg and Ståhl, 1995; Anderberg et al., 2000), *Theaceae*, *Symplocaceae*, *Styracaceae*, *Diapensiaceae*, *Actinidiaceae*, *Roridulaceae*, *Sarraceniaceae*, *Clethraceae*, *Cyrillaceae*, *Ericaceae* (defined in this volume), and *Mitrastemonaceae* (see Barkman et al., 2004, and Nickrent et al.,

2004, for discussion of phylogenetic placement of this enigmatic parasitic taxon).

Diagnostic Apomorphies: Morphological support for *Ericales* is weak, but a possible synapomorphy is the presence of theoid leaf teeth, i.e., a condition in which a single vein enters the tooth and ends in an opaque deciduous cap or gland; such teeth are found in at least some members of most of the included families (Hickey and Wolfe, 1974; Stevens, 2001; Judd et al., 2008), but entire margins have frequently evolved. In *Ericaceae*, the tooth condition is somewhat modified, each tooth being associated with a multicellular, often gland-headed hair. Placentas that protrude into the ovary locule/locules are a second possible synapomorphy (Nandi et al., 1998).

Synonyms: None.

Comments: This nearly cosmopolitan clade includes over 11,000 species, comprising 5.9% of eudicot diversity (Magallón et al., 1999), and three of its component clades (*Ebenaceae*, *Ericaceae*, and *Sapotaceae*) are major components of many ecosystems. In the latter part of the twentieth century, *Ericales* were considered merely as *Ericaceae* and a few close relatives. For example, Cronquist (1981) included only *Cyrillaceae*, *Clethraceae*, and *Ericaceae* (with *Empetraceae*, *Epacridaceae*, *Pyrolaceae*, and *Monotropaceae* segregated) along with the misplaced *Grubbiaceae*; Takhtajan (1980) included *Actinidiaceae*, *Clethraceae*, *Ericaceae* (with *Empetraceae* and *Epacridaceae* segregated) and also *Diapensiaceae*, *Cyrillaceae*, and *Grubbiaceae*; and Thorne (1992) included only

Ericaceae, although he segregated *Empetraceae* and *Epacridaceae*. Morphology-based phylogenetic analyses by Anderberg (1992, 1993) and Judd and Kron (1993) supported the close phylogenetic relationship of *Ericaceae* sensu *lato*, *Cyrillaceae*, *Clethraceae*, *Actinidiaceae*, and *Diapensiaceae*, and suggested that clades such as *Theaceae*, *Sarraceniaceae*, and *Rorulidaceae* also were related to *Ericaceae*, recognizing the distinctiveness of the core *Ericales*. Hufford (1992), in a morphology-based cladistic analysis focusing on "rosid" angiosperms, also found some support for a clade containing *Ericaceae*, *Actinidiaceae*, *Clethraceae*, *Fouquieriaceae*, and possibly also *Sarraceniaceae*. Soon, numerous molecular analyses (Chase et al., 1993; Kron and Chase, 1993; Olmstead et al., 1993; Morton et al., 1997; Soltis et al., 2000; Savolainen et al., 2000a,b; Albach et al., 2001; Bremer et al., 2002; Hilu et al., 2003) supported the monophyly of an expanded *Ericales*, including not only the familial clades mentioned above, but also many others, resulting in the *Ericales*, as here circumscribed. Statistical support for the group in multigenic analyses is generally strong; e.g., *Ericales* are supported by jackknife values ranging from 98–100% in Soltis et al. (2000: Fig. 10), Bremer et al. (2002: Fig. 1), and Hilu et al. (2003: Fig. 11). Summaries of phylogenetic relationships within *Ericales* are provided by APG (1998), Stevens (2001), APG II (2003), Judd and Olmstead (2004), Soltis et al. (2005), and Judd et al. (2008), all of whom applied the name *Ericales* in the same way as it is applied here.

Relationships within this large clade are still rather poorly understood, although the position of the *Balsaminaceae* + *Marcgraviaceae* + *Tetrameristaceae* clade as sister to the remaining taxa is clear (Anderberg et al., 2002; Schönenberger et al., 2005). Other species-rich, strongly supported subclades within *Ericales* (see Schönenberger et al., 2005, Fig. 1A) include (1)

the *Polemoniaceae* + *Fouquieriaceae* clade (Clade II in the primary reference phylogeny), (2) the *Primulaceae* sensu *lato* + *Ebenaceae* + *Sapotaceae* clade (Clade III in the primary reference phylogeny, but *Sapotaceae* is only placed with *Primulaceae s.l.* and *Ebenaceae* in the Bayesian analysis; see also the phylogeny presented by Källersjö et al., 2000), (3) the *Symplocaceae* + *Styracaceae* + *Diapensiaceae* clade (Clade IV in the reference phylogeny), and (4) Clade V in the reference phylogeny (sometimes called the core *Ericales*), which includes *Roridulaceae*, *Actinidiaceae*, *Sarraceniaceae*, *Clethraceae*, *Cyrillaceae*, and *Ericaceae*. The core *Ericales* are morphologically distinctive because their inflorescences are racemose, their anthers usually invert in development and open by pores or short slits, and the syncarpous gynoecia have an impressed style (or styles), which is usually hollow (Anderberg, 1992, 1993; Judd and Kron, 1993).

Members of *Ericales* can usually be distinguished from *Gentianidae* (defined in this volume) by having at least twice as many stamens as petals (vs. stamens as many or fewer than petals). However, staminal reductions have occurred in a few clades, e.g., in *Primulaceae* sensu lato (but in some members of this clade, the outer staminal whorl is represented by staminodia), *Balsaminaceae*, and *Polemoniaceae*. Patterns of character variation within Ericales are discussed by Schönenberger et al. (2005).

A minimum-clade definition is preferred based on the clear position of *Balsaminaceae* + *Marcgraviaceae* + *Tetrameristaceae* as sister to the remaining members of *Ericales*. The relationship between *Ericales* and its two closest relatives (*Cornales* and *Gentianidae*) remains uncertain. The topology (*Cornales* (*Ericales*, *Gentianidae*)) is supported by most analyses (e.g., Chase et al., 1993; Olmstead et al., 2000; Savolainen et al., 2000a; Bremer et al., 2002), but the topology

(*Ericales* (*Cornales, Gentianidae*)) was found by Hilu et al. (2003), and the topology ((*Ericales, Cornales*) *Gentianidae*) was found by Savolainen et al. (2000a).

Literature Cited

Albach, D. C., P. S. Soltis, D. E. Soltis, and R. G. Olmstead. 2001. Phylogenetic analysis of asterids based on sequences of four genes. *Ann. Mo. Bot. Gard.* 88:162–212.

Anderberg, A. A. 1992. The circumscription of the *Ericales*, and their cladistic relationships to other families of "higher" dicotyledons. *Syst. Bot.* 17:660–675.

Anderberg, A. A. 1993. Cladistic relationships and major clades of the *Ericales*. *Plant Syst. Evol.* 184:207–231.

Anderberg, A. A., C. Rydin, and M. Källersjö. 2002. Phylogenetic relationships in the order *Ericales* s.l.: analyses of molecular data from five genes from the plastid and mitochondrial genomes. *Am. J. Bot.* 89:677–689.

Anderberg, A. A., and B. Ståhl. 1995. Phylogenetic interrelationships in the order *Primulales*, with special emphasis on the family circumscriptions. *Can. J. Bot.* 73:1699–1730.

Anderberg, A. A., B. Ståhl, and M. Källersjö. 2000. *Maesaceae*, a new primuloid family in the order *Ericales* s.l. *Taxon* 49:183–187.

APG (Angiosperm Phylogeny Group). 1998. An ordinal classification for the families of flowering plants. *Ann. Mo. Bot. Gard.* 85:531–553.

APG II (Angiosperm Phylogeny Group II). 2003. An update of the Angiosperm Phylogeny Group classification for the orders and families of flowering plants: APG II. *Bot. J. Linn. Soc.* 141:399–436.

Barkman, T. J., S.-H. Lim, K. M. Salleh, and K. Nais. 2004. Mitochondrial DNA sequences reveal the photosynthetic relatives of *Rafflesia*, the world's largest flower. *Proc. Natl. Acad. Sci. USA* 101:787–792.

Berchtold, F., and J. S. Presl. 1820. *Přirozenosti Rostlin*. Krala Wiljma Endevsa, Praha.

Bremer, B., K. Bremer, N. Heidari, P. Erixon, R. G. Olmstead, A. A. Anderberg, M. Källersjö,

and E. Barkhordarian. 2002. Phylogenetics of asterids based on 3 coding and 3 non-coding chloroplast DNA markers and utility of non-coding DNA at higher taxonomic levels. *Mol. Phylogenet. Evol.* 24:274–301.

Chase, M. W., S. Mirarab, N. Nguyen, T. Warnow, E. Carpenter, N. Matasci, S. Ayyampalayam, M. S. Barker, J. G. Burleigh, M. A. Gitzendanner, B. R. Ruhfel, E. Wafula, J. P. Der, S. W. Graham, S. Mathews, M. Melkonian, D. E. Soltis, P. S. Soltis, N. W. Miles, C. J. Rothfels, L. Pokorny, A. J. Shaw, L. DeGironimo, D. W. Stevenson, B. Surek, J. C. Villarreal, B. Roure, H. Philippe, C. W. dePamphilis, T. Chen, M. K. Deyholos, R. S. Baucom, T. M. Kutchan, M. M. Augustin, J. Wang, Y. Zhang, Z. Tian, Z. Yan, X. Wu, X. Sun, G. K. Wong, and J. Leebens-Mack. 1993. Phylogenetics of seed plants: an analysis of nucleotide sequences from the plastid gene *rbcL*. *Ann. Mo. Bot. Gard.* 80:528–580.

Cronquist, A. 1981. *An Integrated System of Classification of Flowering Plants*. Columbia University Press, New York.

Hickey, L. J., and J. A. Wolfe. 1974. The bases of angiosperm phylogeny: vegetative morphology. *Ann. Mo. Bot. Gard.* 62:538–589.

Hilu, K. W., T. Borsch, K. Müller, D. E. Soltis, P. S. Soltis, V. Savolainen, M. W. Chase, M. P. Powell, L. A. Alice, R. Evans, H. Sauquet, C. Neinhuis, T. A. B. Slotta, J. G. Rohwer, C. S. Campbell, and L. W. Chatrou. 2003. Angiosperm phylogeny based on *matK* sequence information. *Am. J. Bot.* 90:1758–1776.

Hufford, L. 1992. *Rosidae* and their relationships to other nonmagnoliid dicotyledons: a phylogenetic analysis using morphological and chemical data. *Ann. Mo. Bot. Gard.* 79:218–248.

Judd, W. S., C. S. Campbell, E. A. Kellogg, P. F. Stevens, and M. J. Donoghue. 2008. *Plant Systematics: A Phylogenetic Approach*. Sinauer Association, Sunderland, MA.

Judd, W. S., and K. A. Kron. 1993. Circumscription of *Ericaceae* (*Ericales*) as determined by preliminary cladistic analyses based on morphological, anatomical, and embryological features. *Brittonia* 45:99–114.

Judd, W. S., and R. G. Olmstead. 2004. A survey of tricolpate (eudicot) phylogenetic relationships. *Am. J. Bot.* 91:1627–1644.

Källersjö, M., G. Bergqvist, and A. A. Anderberg. 2000. Generic realignment in primuloid families of the *Ericales* s.l.: a phylogenetic analysis based on DNA sequences from three chloroplast genes and morphology. *Am. J. Bot.* 87:1325–1341.

Kron, K. A., and M. W. Chase. 1993. Systematics of the *Ericaceae, Empetraceae, Epacridaceae* and related taxa based on *rbcL* sequence data. *Ann. Mo. Bot. Gard.* 80:735–741.

Kron, K. A., W. S. Judd, P. F. Stevens, D. M. Crayn, A. A. Anderberg, P. A. Gadek, C. J. Quinn, and J. L. Luteyn. 2002. Phylogenetic classification of *Ericaceae*: molecular and morphological evidence. *Bot. Rev.* 68:1–426.

Magallón, S. A., P. R. Crane, and P. S. Herendeen. 1999. Phylogenetic pattern, diversity, and diversification of eudicots. *Ann. Mo. Bot. Gard.* 86:297–372.

McGuire, A. F., and K. A.Kron. 2005. Phylogenetic relationship of European and African ericas. *Int. J. Plant Sci.* 166:311–318.

Morton, C. M., K. A. Kron, and M. W. Chase. 1997. A molecular evaluation of the monophyly of the order *Ebenales* based upon *rbcL* sequence data. *Syst. Bot.* 21:567–586.

Nandi, O. I., M. W. Chase, and P. K. Endress. 1998. A combined cladistic analysis of angiosperms using *rbcL* and non-molecular data. *Ann. Mo. Bot. Gard.* 85:137–212.

Nickrent, D. L., A. Blarer, Y.-L. Qiu, R. Vidal-Russell, and F. E. Anderson. 2004. Phylogenetic inference in *Rafflesiales*: the influence of rate heterogeneity and horizontal gene transfer. *BMC Evol. Biol.* 4:40. Available at http://www.biomedcentral.com/1471-2148/4/40.

Olmstead, R. G., B. Bremer, K. M. Scott, and J. D. Palmer. 1993. A parsimony analysis of the *Asteridae* sensu lato based on *rbcL* sequences. *Ann. Mo. Bot. Gard.* 80:700–722.

Olmstead, R. G., K.-J. Kim, R. K. Jansen, and S. J. Wagstaff. 2000. The phylogeny of the *Asteridae* sensu lato based on chloroplast *ndhF* gene sequences. *Mol. Phylogenet. Evol.* 16:96–112.

Savolainen, V., M. W. Chase, S. B. Hoot, C. N. Morton, D. E. Soltis, C. Bayer, M. F. Fay, A. Y. de Bruijn, S. Sullivan, and Y.-L. Qiu. 2000a. Phylogenetics of flowering plants based on combined analysis of plastid *atpB* and *rbcL* gene sequences. *Syst. Bot.* 49:306–362.

Savolainen, V., M. F. Fay, D. C. Albach, A. Backlund, M. van der Bank, K. M. Cameron, S. A. Johnson, M. D. Lledo, J.-C. Pintaud, M. Powell, M. C. Sheahan, D. E. Soltis, P. S. Soltis, P. Weston, W. M. Whitten, K. J. Wurdack, and M. W. Chase. 2000b. Phylogeny of the eudicots: a nearly complete familial analysis based on *rbcL* gene sequences. *Kew Bull.* 55:257–309.

Schönenberger, J., A. A. Anderberg, and K. J. Sytsma. 2005. Molecular phylogenetics and patterns of floral evolution in the *Ericales*. *Int. J. Plant Sci.* 166:265–288.

Soltis, D. E., P. S. Soltis, M. W. Chase, M. E. Mort, D. C. Albach, M. Zanis, V. Savolainen, W. H. Hahn, S. B. Hoot, M. F. Fay, M. Axtell, S. M. Swensen, L. M. Prince, W. J. Kress, K. C. Nixon, and J. S. Farris. 2000. Angiosperm phylogeny inferred from 18S rDNA, *rbcL*, and *atpB* sequences. *Bot. J. Linn. Soc.* 133:381–461.

Soltis, D. E., P. S. Soltis, P. K. Endress, and M. W. Chase. 2005. *Phylogeny and Evolution of Angiosperms*. Sinauer Associates, Sunderland, MA.

Stevens, P. F. 2001, and onwards. Angiosperm Phylogeny Website, Version 9, June 2008 [and more or less continuously updated since]. Available at http://www.mobot.org/MOBOT/research/APweb/.

Takhtajan, A. L. 1980. Outline of the classification of flowering plants (*Magnoliophyta*). *Bot. Rev.* 46:225–359.

Thorne, R. F. 1992. Classification and geography of the flowering plants. *Bot. Rev.* 58:225–348.

Authors

Walter S. Judd; Department of Biology; University of Florida; Gainesville, FL 32611, USA. Email: lyonia@ufl.edu.

Kathleen A. Kron; Department of Biology; Wake Forest University; Winston-Salem, FL 27109, USA. Email: kronka@wfu.edu.

Date Accepted: 14 July 2010

Primary Editor: Philip Cantino

Ericaceae Jussieu 1789: 159 [W. S. Judd and K. A. Kron], converted clade name

Registration Number: 40

Definition: The smallest crown clade containing *Enkianthus campanulatus* G. Nicholson 1884, *Pyrola rotundifolia* L. 1753, *Hypopitys monotropa* Crantz 1766, *Arctostaphylos uva-ursi* (L.) Spreng. 1825, and *Erica cinerea* L. 1753. This is a minimum-crown-clade definition. Abbreviated definition: min crown ∇ (*Enkianthus campanulatus* G. Nicholson 1884 & *Pyrola rotundifolia* L. 1753 & *Hypopitys monotropa* Crantz 1766 & *Arctostaphylos uva-ursi* (L.) Spreng. 1825 & *Erica cinerea* L. 1753).

Etymology: Based on *Erica*, an ancient Latin name for heather.

Reference Phylogeny: The primary reference phylogeny is Kron et al. (2002a: Fig. 7, but also see Figs. 1–6). *Erica cinerea* is used as a specifier because it is the type of *Ericaceae*; although it is not included in the primary reference phylogeny, its close relationship to the three species of *Erica* that are included is supported by a recent phylogenetic analysis of *Erica* (McGuire and Kron, 2005). Another specifier, *Hypopitys monotropa*, is also not present on the primary reference phylogeny, but two possible positions for this species (labeled as *Monotropa hypopithys*; see Comments) are shown in Figures 3 and 4 of the same paper (Kron et al., 2002a). See also Schönenberger et al. (2005: Fig. 1), Anderberg et al. (2002: Fig. 1), and Kron and Chase (1993: Fig. 2), and for the mycoparasitic taxa, Bidartondo and Bruns (2001: Fig. 3) and Tsukaya et al. (2008: Fig. 3).

Composition: Major clades within *Ericaceae* include *Enkianthoideae*, *Monotropoideae* (incl. *Pyroloideae*), *Arbutoideae*, *Ericoideae* (incl. *Rhododendroideae*), *Cassiopoideae*, *Harrimanelloideae*, *Epacridoideae* (incl. *Styphelioideae*), and *Vaccinioideae* (including both inferior-ovaried species, i.e., *Vaccinieae*, and superior-ovaried species, i.e., *Oxydendreae*, *Lyonieae*, *Gaultherieae*, and *Andromedeae*).

Diagnostic Apomorphies: Probable synapomorphies of *Ericaceae* are the pendulous flowers and an urceolate to more or less cylindrical corolla, although reversals to erect or horizontal flowers occur (e.g., in *Bejaria*, *Elliottia*, *Phyllodoceae*, *Empetreae*, *Rhodoreae*, *Cosmelieae*, and many *Monotropeae*), and reversals to distinct petals (or tepals) also occur, e.g., in *Empetreae*, *Bejaria*, *Elliottia*, the *Bryanthus* + *Ledothamnus* clade, *Pyroleae*, and many *Monotropeae* (Kron et al. 2002a). Another possible synapomorphy is the distinctive form of the marginal leaf teeth, i.e., each tooth is associated with a multicellular hair, but entire leaves have evolved repeatedly within the clade.

Synonyms: None.

Comments: This nearly cosmopolitan clade includes about 4,200 species and was recently reviewed by Kron et al. (2002a) and Stevens et al. (2004); the former provides detailed phylogenetic analyses with extensive intra-clade sampling, employing 18S, ITS, *matK*, *rbcL* sequences, and morphology. Although recent authors (Anderberg et al., 2002; Kron et al., 2002a; APG II, 2003; Stevens et al., 2004; Schönenberger et al., 2005; Heywood et al., 2007; Judd et al., 2008) apply the name *Ericaceae* in the same way it is applied here, the name has historically been applied to a non-monophyletic group originating with the same ancestor (Hooker, 1876; Drude, 1897; Stevens, 1971; Takhtajan, 1980; Cronquist, 1981), i.e., *Ericaceae* was formerly divided into *Ericaceae* sensu *stricto*, *Epacridaceae*, and *Empetraceae*. In addition, the highly mycorrhizal, ± herbaceous (in contrast to most *Ericaceae*, which are woody), photosynthetic taxa frequently have been segregated as *Pyrolaceae*, and the non-chlorophyllous, mycoparasitic taxa have been segregated as *Monotropaceae*. Sometimes the inferior-ovaried taxa have been segregated as *Vacciniaceae*.

Enkianthus is sister to the rest of *Ericaceae* (Kron et al., 2002a) and differs from them in having anthers with a fibrous endothecium and opening by two longitudinal slits, monadinous pollen, and seeds with vascular bundles in the raphe (all of which are putative plesiomorphies). Anthers with an exothecium, but more or less lacking a fibrous endothecium, and seeds lacking a vascular bundle in the raphe probably are synapomorphic for the clade comprising all *Ericaceae* except for *Enkianthus*. Pollen released in tetrahedral tetrads may also be a synapomorphy at this level, although the condition is variable within *Monotropoideae*, with pollen monads occurring in *Orthilia*, *Chimaphila* (some), and the mycoparasitic genera, so it may be a synapomorphy for the *Arbutoideae* + "early anther inversion clade." "Pseudomonads," i.e., tetrads in which three of the four grains abort, occur in some *Epacridoideae* (especially *Styphelieae*) (Furness, 2009). The clade comprising *Vaccinioideae*, *Epacridoideae*, *Harrimanelloideae*, *Cassiopoideae*, and *Ericoideae*, i.e., referred to as the "early anther inversion clade" by Kron et al. (2002a) and constituting the vast majority of the species of *Ericaceae*, is supported by the synapomorphy of anthers inverting early in the development of the flower. *Arbutoideae*, *Monotropoideae*, and *Enkianthus* retain the plesiomorphic condition, i.e., anthers inverting just before anthesis, but the anthers of

Hemitomes, *Pleuricospora*, and *Cheilotheca* (in the *Monotropeae*) do not invert.

The monophyly of *Ericaceae* is strongly supported. See especially Schönenberger et al. (2005), in which the clade was supported in a parsimony analysis by a bootstrap value of 98% and by a Bayesian posterior probability of 1.0, in analyses utilizing 18S and 26S nrDNA sequences, the mitochondrial genes *atp1* and *matR*, and the chloroplast regions *rbcL*, *atpB*, *ndhF*, *rps16*, *matK*, *trnT-F* spacer, and *trnV-atpE* spacer. Its monophyly has also been supported by several other phylogenetic analyses based on morphological and/or molecular characters (Anderberg, 1993; Judd and Kron, 1993; Kron and Chase, 1993; Kron, 1996; Anderberg et al., 2002; Kron et al., 2002a). These analyses, along with those focused on relationships within the clade (Anderberg, 1994a,b; Crayn et al., 1996, 1998; Kron and King, 1996; Powell et al. 1996, 1997; Kron and Judd, 1997; Kron, 1997; Freudenstein, 1999; Kron et al., 1999a,b, 2002b; Crayn and Quinn, 2000; Oliver, 2000; Bidartondo and Bruns, 2001; Hileman et al., 2001; Powell and Kron, 2001, 2002; Taaffe et al., 2001; Quinn et al., 2003; Goetsch et al., 2005; McGuire and Kron, 2005; Bush and Kron, 2008; Tsukaya et al., 2008; Waselkov and Judd, 2008; Bush et al., 2009; Pedraza-Peñalosa, 2009) have resulted in a fairly well-resolved phylogeny for the group. Remaining problems mainly concern relationships within the *Vaccinieae* and among the mycoparasitic species.

A minimum-clade definition is preferred based on the well resolved phylogenetic structure within *Ericaceae*, especially the position of *Enkianthus* as sister to the remaining taxa and the monophyly of the "early anther inversion clade," although relationships are still somewhat unclear among the mycoparasitic species.

The specifier *Hypopitys monotropa* Crantz has long been recognized as *Monotropa hypopithys* L., but phylogenetic analyses (Bidartondo and

Bruns, 2001; Tsukaya et al., 2008) indicate that the genus *Monotropa*, as traditionally circumscribed, is non-monophyletic, and that genus is now restricted to *Monotropa uniflora*.

Literature Cited

Anderberg, A. A. 1993. Cladistic interrelationships and major clades of the *Ericales*. *Plant Syst. Evol.* 184:207–231.

Anderberg, A. A. 1994a. Cladistic analysis of *Enkianthus* with notes on the early diversification of the *Ericaceae*. *Nord. J. Bot.* 14:385–401.

Anderberg, A. A. 1994b. Phylogeny of the *Empetraceae*, with special emphasis on character evolution in the genus *Empetrum*. *Syst. Bot.* 19:35–46.

Anderberg, A. A., C. Rydin, and M. Källersjö. 2002. Phylogenetic relationships in the order *Ericales* s.l.: analyses of molecular data from five genes from the plastid and mitochondrial genomes. *Am. J. Bot.* 89:677–687.

APG II (Angiosperm Phylogeny Group II). 2003. An update of the Angiosperm Phylogeny Group classification for the orders and families of flowering plants: APG II. *Bot. J. Linn. Soc.* 141:399–436.

Bidartondo, M. I., and T. D. Bruns. 2001. Extreme specificity in epiparasitic *Monotropoideae* (*Ericaceae*): widespread phylogenetic and geographical structure. *Mol. Ecol.* 10:2285–2295.

Bush, C. M., and K. A. Kron. 2008. A phylogeny of *Bejaria* (*Ericaceae*: *Ericoideae*) based on molecular data. *J. Bot. Res. Inst. Texas* 2:1193–1205.

Bush, C. M., L. Lu, P. W. Fritsch, D.-Z. Li, and K. A. Kron. 2009. Phylogeny of *Gaultherieae* (*Ericaceae*: *Vaccinioideae*) based on DNA sequence data from *matK*, *ndhF*, and nrITS. *Int. J. Plant Sci.* 170:355–364.

Crayn, D. M., K. A. Kron, P. A. Gadek, C. J. Quinn. 1996. Delimitation of *Epacridaceae*: preliminary molecular evidence. *Ann. Bot. (Lond.)* 77:317–321.

Crayn, D. M., K. A. Kron, P. A. Gadek, and C. J. Quinn. 1998. Phylogenetics and evolution of epacrids: a molecular analysis using the plastid gene *rbcL*, with a reappraisal of the position of *Lebetanthus*. *Aust. J. Bot.* 46:187–200.

Crayn, D. M., and C. J. Quinn. 2000. The evolution of the *atpB-rbcL* intergenic spacer in the epacrids (*Ericales*) and its systematic and evolutionary implications. *Mol. Phyogenet. Evol.* 16:238–252.

Cronquist, A. 1981. *An Integrated System of Classification of Flowering Plants*. Columbia University Press, New York.

Drude, O. 1897. *Ericaceae*. Pp. 15–65 in *Die natürlichen Pflanzenfamilien*, Vol. 4, Part 1 (A. Engler and K. Prantl, eds.). Engelmann, Leipzig.

Freudenstein, J. 1999. Relationships and character transformation in *Pyroloideae* (*Ericaceae*) based on ITS sequences, morphology, and development. *Syst. Bot.* 24:398–408.

Furness, C. A. 2009. Pollen evolution and development in *Ericaceae*, with particular reference to pseudomonads and variable pollen sterility in *Styphelioideae*. *Int. J. Plant Sci.* 170:476–490.

Goetsch, L., A. J. Eckert, and B. D. Hall. 2005. The molecular systematics of *Rhododendron* (*Ericaceae*): a phylogeny based upon *RPB2* gene sequences. *Syst. Bot.* 30:616–626.

Heywood, V. H., R. K. Brummitt, A. Culman, and O. Seberg. 2007. *Flowering Plant Families of the World*. Firefly Books, Ontario.

Hileman, L. C., M. C. Vasey, and V. T. Parker. 2001. Phylogeny and biogeography of the *Arbutoideae* (*Ericaceae*): implications for the Madrean-Tethyan hypothesis. *Syst. Bot.* 26:131–143.

Hooker, J. D. 1876. *Vacciniaceae*, Pp. 564–577, and *Ericaceae*, Pp. 577–604 in *Genera Plantarum*, Vol. 2 (G. Bentham and J. D. Hooker, eds.). London.

Judd, W. S., C. S. Campbell, E. A. Kellogg, P. F. Stevens, and M. J. Donoghue. 2008. *Plant Systematics: A Phylogenetic Approach*. Sinauer Association, Sunderland, MA.

Judd, W. S., and K. A. Kron. 1993. Circumscription of *Ericaceae* (*Ericales*) as determined by preliminary cladistic analyses based on morphological, anatomical, and embryological features. *Brittonia* 45:99–114.

de Jussieu, A. L. 1789. *Genera Plantarum*. Herissant and Barrois, Paris.

Kron, K. A. 1996. Phylogenetic relationships of *Empetraceae, Epacridaceae, Ericaceae, Monotropaceae, Pyrolaceae*: evidence from nuclear ribosomal 18s sequence data. *Ann. Bot. (Lond.)* 77:293–303.

Kron, K. A. 1997. Phylogenetic relationships of *Rhododendroideae* (*Ericaceae*). *Am. J. Bot.* 84:973–980.

Kron, K. A., and M. W. Chase. 1993. Systematics of the *Ericaceae, Empetraceae, Epacridaceae*, and related taxa based upon *rbcL* sequence data. *Ann. Mo. Bot. Gard.* 80:735–741.

Kron, K. A., R. Fuller, D. M. Crayn, P. A. Gadek, and C. J. Quinn. 1999a. Phylogenetic relationships of epacrids and vaccinioids (*Ericaceae* s.l.) based on *matK* sequence data. *Plant Syst. Evol.* 218:55–65.

Kron, K. A., and W. S. Judd. 1997. Systematics of the *Lyonia* group (*Andromedeae, Ericaceae*) and the use of species as terminals in higher-level cladistic analyses. *Syst. Bot.* 22:479–492.

Kron, K. A., W. S. Judd, and D. M. Crayn. 1999b. Phylogenetic analyses of *Andromedeae* (*Ericaceae* subfam. *Vaccinioideae*). *Am. J. Bot.* 86:1290–1300.

Kron, K. A., W. S. Judd, P. F. Stevens, D. M. Crayn, A. A. Anderberg, P. A. Gadek, C. J. Quinn, and J. L. Luteyn. 2002a. Phylogenetic classification of *Ericaceae*: molecular and morphological evidence. *Bot. Rev.* 68:335–423.

Kron, K. A., and J. M. King. 1996. Cladistic relationships of *Kalmia, Leiophyllum*, and *Loiseleuria* (*Phyllodoceae, Ericaceae*) based on *rbcL* and nrITS data. *Syst. Bot.* 21:17–29.

Kron, K. A., E. A. Powell, and J. L. Luteyn. 2002b. Phylogenetic relationships within the blueberry tribe (*Vaccinieae, Ericaceae*) based on sequence data from *matK* and nuclear ribosomal ITS regions, with comments on the placement of *Satyria. Am. J. Bot.* 89:327–336.

McGuire, A. F., and K. A. Kron. 2005. Phylogenetic relationship of European and African ericas. *Int. J. Plant Sci.* 166:311–318.

Oliver, E. G. H. 2000. *Systematics of Ericeae* (*Ericaceae: Ericoideae*): *Species with Indehiscent and Partially Dehiscent Fruits. Contributions from the Bolus Herbarium*, No. 19. Bolus Herbarium, University of Cape Town, Cape Town..

Pedraza-Peñalosa, P. 2009. Systematics of the Neotropical blueberry genus *Disterigma* (*Ericaceae*). *Syst. Bot.* 34:406–413.

Powell, E. A., and K. A. Kron. 2001. An analysis of the phylogenetic relationships in the wintergreen group (*Diplycosia, Gaultheria, Pernettya, Tepuia; Ericaceae*). *Syst. Bot.* 26: 808–817.

Powell, E. A., and K. A. Kron. 2002. Hawaiian blueberries and their relatives—a phylogenetic analysis of *Vaccinium* sections *Macropelma, Myrtillus*, and *Hemimyrtillus* (*Ericaceae*). *Syst. Bot.* 27:768–779.

Powell, J. M., D. M. Crayn, P. A. Gadek, C. J. Quinn, D. A. Morrison, and A. R. Chapman. 1996. A reassessment of relationships within *Epacridaceae. Ann. Bot. (Lond.)* 77: 305–315.

Powell, J. M., D. A. Morrison, P. A. Gadek, D. M. Crayn, and C. J. Quinn. 1997. Relationships and generic concepts within *Styphelieae* (*Epacridaceae*). *Aust. Syst. Bot.* 10:15–29.

Quinn, C. J., D. M. Crayn, M. M. Heslewood, E. A. Brown, and P. A. Gadek. 2003. A molecular estimate of the phylogeny of *Styphelieae* (*Ericaceae*). *Aust. Syst. Bot.* 16:581–594.

Schönenberger, J., A. A. Anderberg, and K. J. Sytsma. 2005. Molecular phylogenetics and patterns of floral evolution in the *Ericales. Int. J. Plant Sci.* 166:265–288.

Stevens, P. F. 1971. A classification of the *Ericaceae*: subfamilies and tribes. *Bot. J. Linn. Soc.* 64:1–53.

Stevens, P. F., J. L. Luteyn, E. G. H. Oliver, T. L. Bell, E. A. Brown, R. K. Crowden, A. S. George, G. J. Jordan, P. Ladd, K. Lemson, C. B. McLean, Y. Menadue, J. S. Pate, H. M. Stace, and C. M. Weiller. 2004. *Ericaceae*. Pp. 145–194 in *The Families and Genera of Vascular Plants: VI. Flowering Plants: Dicotyledons*: Celastrales, Oxalidales, Rosales, Cornales, Ericales (K. Kubitzki, ed.). Springer-Verlag, Berlin.

Taaffe, G., E. A. Brown, D. M. Crayn, P. A. Gadek, and C. J. Quinn. 2001. Generic concepts in *Styphelieae*: resolving the limits of *Leucopogon*. *Aust. J. Bot.* 49:107–120.

Takhtajan, A. L. 1980. Outline of the classification of flowering plants (*Magnoliophyta*). *Bot. Rev.* 46:225–359.

Tsukaya, H., J. Yokoyama, R. Imaichi, and H. Ohba. 2008. Taxonomic status of *Monotropastrum humile*, with special reference to *M. humile* var. *glaberrimum* (*Ericaceae, Monotropoideae*). *J. Plant Res.* 121:271–278.

Waselkov, K., and W. S. Judd. 2008. A phylogenetic analysis of *Leucothoe* s.l. (*Ericaceae*; tribe *Gaultherieae*) based on phenotypic characters. *Brittonia* 60:382–397.

Authors

Walter S. Judd; Department of Biology; University of Florida; Gainesville, FL 32611, USA. E-mail: lyonia@ufl.edu

Kathleen A. Kron; Department of Biology; Wake Forest University; Winston-Salem, FL 27109, USA. Email: kronka@wfu.edu.

Date Accepted: 18 July 2010

Primary Editor: Philip Cantino

Gentianidae R. G. Olmstead, W. S. Judd, and P. D. Cantino in P. D. Cantino et al. 2007: E29-30 [R. G. Olmstead, W. S. Judd and P. D. Cantino], converted clade name

Registration Number: 217

Definition: The smallest crown clade containing *Gentiana lutea* Linnaeus 1753 (*Lamiidae*) and *Campanula elatines* Linnaeus 1759 (*Campanulidae*). This is a minimum-crown-clade definition. Abbreviated definition: min crown ∇ (*Gentiana lutea* Linnaeus 1753 & *Campanula elatines* Linnaeus 1759).

Etymology: Based on *Gentiana* (name of an included taxon), which is derived from the Greek *Gentius*, the name of the last king of Illyria (second century BC), who used the roots to treat his troops for malaria (Charters, 2008; Gledhill, 1989).

Reference Phylogeny: Soltis et al. (2011: Figs. 1–2). See also Olmstead et al. (2000: Fig. 1), Savolainen et al. (2000: Fig. 5), Soltis et al. (2000: Fig. 11), Albach et al. (2001b: Fig. 2), Bremer et al. (2002: Fig. 1), and Hilu et al. (2003: Fig.12).

Composition: *Lamiidae* and *Campanulidae* (see entries in this volume).

Diagnostic Apomorphies: Possible synapomorphies include epipetalous stamens arising from a sympetalous corolla and equal in number to (or fewer than) the number of corolla lobes, unitegmic ovules, and two fused carpels (Albach et al., 2001a; Judd and Olmstead, 2004; Soltis et al., 2005; Stevens, 2006; Judd et al., 2008).

Synonyms: No formal synonyms; however, this clade has been referred to informally as "euasterids" (Olmstead et al., 2000; Savolainen et al., 2000; Bremer et al., 2002; APG II, 2003; Stevens, 2006) and "core asterids" (Judd et al., 2008; Hilu et al., 2003; Judd and Olmstead, 2004).

Comments: Composition of *Gentianidae* is similar to that of *Asteridae* sensu Takhtajan (1980) and Cronquist (1981) but also includes *Apiales, Aquifoliales, Garryales,* and *Icacinaceae.* These groups were excluded from traditional circumscriptions of *Asteridae* (sensu Cronquist or Takhtajan), because they lack one or more of the morphological traits cited above as possible synapomorphies. Because *Asteridae* has come to be used for a more inclusive clade in recent classifications (APG III, 2009) and texts (Soltis et al., 2005; Judd et al., 2008), *Gentianidae* was proposed for this clade by Olmstead et al. (in Cantino et al., 2007: E29–30). Good support for this clade comes from molecular phylogenetic studies (e.g., Olmstead et al., 2000; Soltis et al., 2000; Bremer et al., 2002; Hilu et al., 2003).

Literature Cited

Albach, D. C., P. S. Soltis, and D. E. Soltis. 2001a. Patterns of embryological and biochemical evolution in the asterids. *Syst. Bot.* 26:242–262.

Albach, D. C., P. S. Soltis, D. E. Soltis, and R. G. Olmstead. 2001b. Phylogenetic analysis of asterids based on sequences of four genes. *Ann. Mo. Bot. Gard.* 88:163–212.

APG II (Angiosperm Phylogeny Group II). 2003. An update of the Angiosperm Phylogeny Group classification for the orders and families of flowering plants: APG II. *Bot. J. Linn. Soc.* 141:399–436.

APG III (Angiosperm Phylogeny Group III). 2009. An update of the Angiosperm Phylogeny Group classification for the orders and families of flowering plants: APG III. *Bot. J. Linn. Soc.* 161:105–121.

Bremer, B., K. Bremer, N. Heidari, P. Erixon, R. G. Olmstead, A. A. Anderberg, M. Källersjö, and E. Barkhordarian. 2002. Phylogenetics of asterids based on 3 coding and 3 non-coding chloroplast DNA markers and the utility of non-coding DNA at higher taxonomic levels. *Mol. Phylogenet. Evol.* 24:274–301.

Cantino, P. D., J. A. Doyle, S. W. Graham, W. S. Judd, R. G. Olmstead, D. E. Soltis, P. S. Soltis, and M. J. Donoghue. 2007. Towards a phylogenetic nomenclature of *Tracheophyta*. *Taxon* 56:822–846 and E1–E44.

Charters, M. L. 2008. *California Plant Names: Latin and Greek Meanings and Derivations.* Available at www.calflora.net/botanicalnames, accessed on 2 November 2011.

Cronquist, A. 1981. *An Integrated System of Classification of Flowering Plants.* Columbia University Press, New York.

Gledhill, D. 1989. *The Names of Plants.* 2nd edition. Cambridge University Press, Cambridge, UK.

Hilu, K. W., T. Borsch, K. Müller, D. E. Soltis, P. S. Soltis, V. Savolainen, M. W. Chase, M. P. Powell, L. A. Alice, R. Evans, H. Sauquet, C. Neinhuis, T. A. B. Slotta, J. G. Rohwer, C. S. Campbell, and L. W. Chatrou. 2003. Angiosperm phylogeny based on *matK* sequence information. *Am. J. Bot.* 90:1758–1776.

Judd, W. S., C. S. Campbell, E. A. Kellogg, P. F. Stevens, and M. J. Donoghue. 2008. *Plant Systematics—A Phylogenetic Approach.* 3rd edition. Sinauer Associates, Sunderland, MA.

Judd, W. S., and R. G. Olmstead. 2004. A survey of tricolpate (eudicot) phylogenetic relationships. *Am. J. Bot.* 91:1627–1644.

Olmstead, R. G., K. Kim, R. K. Jansen, and S. J Wagstaff. 2000. The phylogeny of the *Asteridae* sensu lato based on chloroplast *ndhF* gene sequences. *Mol. Phylogenet. Evol.* 16:96–112.

Savolainen, V., M. W. Chase, S. B. Hoot, C. M. Morton, D. E. Soltis, D. Bayer, M. F. Fay, A. Y. de Bruijn, S. Sullivan, and Y.-L. Qiu. 2000. Phylogenetics of flowering plants based on combined analysis of plastid *atpB* and *rbcL* gene sequences. *Syst. Biol.* 49:306–362.

Soltis, D. E., S. A. Smith, N. Cellinese, K. J. Wurdack, D. C. Tank, S. F. Brockington, N. F. Refulio-Rodriguez, J. B. Walker, M. J. Moore, B. S. Carlsward, C. D. Bell, M. Latvis, S. Crawley, C. Black, D. Diouf, Z. Xi. M. A. Gitzendanner, K. J. Systma, Y.-L. Qiu, K. W. Hilu, S. R. Manchester, C. C. Davis, M. J. Sanderson, R. G. Olmstead, W. S. Judd, M. J. Donoghue, and P. S. Soltis. 2011. Angiosperm phylogeny: 17-genes, 640 taxa. *Am. J. Bot.* 98:704–730.

Soltis, D. E., P. S. Soltis, M. W. Chase, M. E. Mort, D. C. Albach, M. Zanis, V. Savolainen, W. H. Hahn, S. B. Hoot, M. F. Fay, M. Axtell, S. M. Swensen, L. M. Prince, W. J. Kress, K. C. Nixon, and J. S. Farris. 2000. Angiosperm phylogeny inferred from 18S rDNA, *rbcL*, and *atpB* sequences. *Bot. J. Linn. Soc.* 133:381–461.

Soltis, D. E., P. S. Soltis, P. K. Endress, and M. W. Chase. 2005. *Phylogeny and Evolution of Angiosperms.* Sinauer Associates, Sunderland, MA.

Stevens, P. F. 2006. Angiosperm Phylogeny Website, Version 7. Available at http://www.mobot.org/mobot/research/apweb/.

Takhtajan, A. 1980. Outline of the classification of flowering plants (*Magnoliophyta*). *Bot. Rev.* 46:225–359.

Authors

Richard G. Olmstead; Department of Biology and Burke Museum; University of Washington; Seattle, WA 98195, USA. Email: olmstead@u.washington.edu.

Walter S. Judd; Department of Botany; University of Florida; Gainesville, FL 32611, USA. Email: wjudd@botany.ufl.edu.

Philip D. Cantino; Department of Environmental and Plant Biology; Ohio University; Athens, OH 45701, USA. Email: cantino@ohio.edu.

Date Accepted: 27 January 2013

Primary Editor: Kevin de Queiroz

Campanulidae M. J. Donoghue and P. D. Cantino in P. D. Cantino et al. (2007): 837 [M. J. Donoghue and P. D. Cantino], converted clade name

Registration Number: 248

Definition: The largest crown clade containing *Campanula latifolia* Linnaeus 1753 (*Asterales*) but not *Lamium purpureum* Linnaeus 1753 (*Lamiidae/Lamiales*) and *Cornus mas* Linnaeus 1753 (*Cornales*) and *Erica carnea* Linnaeus 1753 (*Ericales/Ericaceae*). This is a maximum-crown-clade definition. Abbreviated definition: max crown ∇ (*Campanula latifolia* Linnaeus 1753 ~ *Lamium purpureum* Linnaeus 1753 & *Cornus mas* Linnaeus 1753 & *Erica carnea* Linnaeus 1753).

Etymology: Derived from *Campanula* (name of an included taxon), which is Latin for "little bell" (Gledhill, 1989).

Reference Phylogeny: The primary reference phylogeny is Soltis et al. (2011: Figs. 1, 2e–g). See also Soltis et al. (2000: Figs. 1, 12), Kårehed (2001: Figs. 1, 2), Bremer et al. (2002: Fig. 1), Winkworth et al. (2008: Fig. 1), and Tank and Donoghue (2010: Figs. 1, 3).

Composition: *Apiidae* (this volume) and probably *Aquifoliales* sensu APG II (2003). There is a slight possibility that some or all of the taxa that are currently included in *Aquifoliales* are not part of *Campanulidae* (see Comments).

Diagnostic Apomorphies: We know of no unambiguous non-molecular synapomorphies. Stevens (2011) cited several characters for this clade, including vessel elements with scalariform perforations, small flowers, short styles, copious endosperm, and short embryos. Several of these characters are poorly sampled; others are ill-defined or highly variable both within and outside of this clade (e.g., flower size, style length). Erbar and Leins (1996) showed that "early sympetaly" is largely restricted to this clade, but its correlation with inferior ovary and reduced calyx should be explored further (Endress, 2001), and its placement on the tree remains uncertain. For example, it may be an apomorphy of the less inclusive clade *Apiidae* (defined in this volume), as suggested by Stevens (2011).

Synonyms: The informal names "asterid II", "euasterid(s) II", and "campanulids" are approximate synonyms (see Comments).

Comments: Until we published the name *Campanulidae* (Cantino et al., 2007), there was no preexisting scientific name for this clade, which is strongly supported in molecular analyses (Soltis et al., 2000; Bremer et al., 2002; Tank and Donoghue, 2010; Soltis et al., 2011) and in an analysis that combined molecular and morphological data (Kårehed, 2001). It has been referred to informally as "asterid II" (Chase et al., 1993), "euasterid(s) II" (APG, 1998; Olmstead et al., 2000; Savolainen et al., 2000; Soltis et al., 2000; Albach et al., 2001a,b; Lundberg, 2001; Judd et al., 2002; APG II, 2003), and "campanulids" (Bremer et al., 2002; Judd and Olmstead, 2004; APG III, 2009). The definition used here differs slightly from our earlier one (Cantino et al., 2007) in that we no longer use *Garrya elliptica* as an external specifier. With 100% bootstrap support for the grouping of *Garryales* with the rest of *Lamiidae* (Soltis et al., 2011), there is no longer any need to include two external specifiers representing *Lamiidae*.

There is a slight possibility that *Ilex* (*Aquifoliaceae*) is a member of *Lamiidae* (as defined in this volume), rather than being closely related to *Apiidae* (this volume) as in the reference phylogeny. *Ilex* was linked with *Lamiidae* in an analysis of *RPB2* duplications (Oxelman et al., 2004) and in an analysis of *matK* sequences (Hilu et al., 2003). Because these studies did not include any members of *Helwingia*, *Phyllonoma*, *Cardiopteridaceae* or *Stemonuraceae*, which have been linked strongly with *Ilex* (Kårehed, 2001; Bremer et al., 2002; Tank and Donoghue, 2010; Soltis et al., 2011) in *Aquifoliales*, these taxa presumably could also be related to *Lamiidae*. Our definition of *Campanulidae* is designed to include *Ilex* and these relatives (*Aquifoliales*) if they are more closely related to *Apiidae* than to *Lamiidae* and to exclude them if that is not the case. If *Ilex* and its relatives were to be found to be more closely related to *Lamiidae* than to *Apiidae*, then *Campanulidae* and *Apiidae* would become synonyms. As we stated previously (Cantino et al., 2007), it is our intent that *Campanulidae* have precedence over *Apiidae* in the unlikely event that both names refer to the same clade.

Literature Cited

Albach, D. C., P. S. Soltis, and D. E. Soltis. 2001a. Patterns of embryological and biochemical evolution in the asterids. *Syst. Bot.* 26:242–262.

Albach, D. C., P. S. Soltis, D. E. Soltis, and R. G. Olmstead. 2001b. Phylogenetic analysis of asterids based on sequences of four genes. *Ann. Mo. Bot. Gard.* 88:163–212.

APG (Angiosperm Phylogeny Group). 1998. An ordinal classification for the families of flowering plants. *Ann. Mo. Bot. Gard.* 85:531–553.

APG II (Angiosperm Phylogeny Group II). 2003. An update of the Angiosperm Phylogeny Group classification for the orders and families of flowering plants: APG II. *Bot. J. Linn. Soc.* 141:399–436.

APG III (Angiosperm Phylogeny Group III). 2009. An update of the Angiosperm Phylogeny Group classification for the orders and families of flowering plants: APG III. *Bot. J. Linn. Soc.* 161:105–121.

Bremer, B., K. Bremer, N. Heidari, P. Erixon, R. G. Olmstead, A. A. Anderberg, M. Källersjö, and E. Barkhordarian. 2002. Phylogenetics of asterids based on 3 coding and 3 non-coding chloroplast DNA markers and the utility of non-coding DNA at higher taxonomic levels. *Mol. Phylogenet. Evol.* 24:274–301.

Cantino, P. D., J. A. Doyle, S. W. Graham, W. S. Judd, R. G. Olmstead, D. E. Soltis, P. S. Soltis, and M. J. Donoghue. 2007. Towards a phylogenetic nomenclature of *Tracheophyta*. *Taxon* 56:822–846 and E1–E44.

Chase, M. W., M. W. Chase, D. E. Soltis, R. G. Olmstead, D. Morgan, D. H. Les, B. D. Mishler, M. R. Duvall, R. A. Price, H. G. Hills, Y.-L. Qiu, K. A. Kron, J. H. Rettig, E. Conti, J. D. Palmer, J. R. Manhart, K. J. Sytsma, H. J. Michaels, W. J. Kress, K. G. Karol, W. D. Clark, M. Hedren, B. S. Gaut, R. K. Jansen, K.-J. Kim, C. F. Wimpee, J. F. Smith, G. R. Furnier, S. H. Strauss, Q.-Y. Xiang, G. M. Plunkett, P. S. Soltis, S. M. Swensen, S. E. Williams, P. A. Gadek, C. J. Quinn, L. E. Eguiarte, E. Golenberg, G. H. Learn, Jr., S. W. Graham, S. C. H. Barrett, S. Dayanandan, and V. A. Albert. 1993. Phylogenetics of seed plants: an analysis of nucleotide sequences from the plastid gene *rbcL*. *Ann. Mo. Bot. Gard.* 80:528–580.

Endress, P. K. 2001. Origins of flower morphology. *J. Exp. Zool.* 291:105–115.

Erbar, C., and P. Leins. 1996. Distribution of the character states "early sympetaly" and "late sympetaly" within the "Sympetalae Tetracyclicae" and presumably allied groups. *Bot. Acta* 109:427–440.

Gledhill, D. 1989. *The Names of Plants.* 2nd edition. Cambridge University Press, Cambridge, UK.

Hilu, K. W., T. Borsch, K. Müller, D. E. Soltis, P. S. Soltis, V. Savolainen, M. W. Chase, M. P. Powell, L. A. Alice, R. Evans, H. Sauquet, C. Neinhuis, T. A. B. Slotta, J. G. Rohwer,

C. S. Campbell, and L. W. Chatrou. 2003. Angiosperm phylogeny based on *matK* sequence information. *Am. J. Bot.* 90:1758–1776.

Judd, W. S., C. S. Campbell, E. A. Kellogg, P. F. Stevens, and M. J. Donoghue. 2002. *Plant Systematics—A Phylogenetic Approach.* 2nd edition. Sinauer Associates, Sunderland, MA.

Judd, W. S., and R. G. Olmstead. 2004. A survey of tricolpate (eudicot) phylogenetic relationships. *Am. J. Bot.* 91:1627–1644.

Kårehed, J. 2001. Multiple origin of the tropical forest tree family *Icacinaceae. Am. J. Bot.* 88:2259–2274.

Lundberg, J. 2001. Phylogenetic studies in the Euasterids II, with particular reference to *Asterales* and *Escalloniaceae.* Ph.D. dissertation, Uppsala University. Available at http://publications.uu.se/theses/abstract.xsql?dbid=1597.

Olmstead, R. G., K. Kim, R. K. Jansen, and S. J. Wagstaff. 2000. The phylogeny of the *Asteridae* sensu lato based on chloroplast *ndhF* gene sequences. *Mol. Phylogenet. Evol.* 16: 96–112.

Oxelman, B., N. Yoshikawa, B. L. McConaughy, J. Luo, A. L. Denton, and B. D. Hall. 2004. *RPB2* gene phylogeny in flowering plants, with particular emphasis on asterids. *Mol. Phylogenet. Evol.* 32:462–479.

Savolainen, V., M. W. Chase, S. B. Hoot, C. M. Morton, D. E. Soltis, D. Bayer, M. F. Fay, A. Y. de Bruijn, S. Sullivan, and Y.-L. Qiu. 2000. Phylogenetics of flowering plants based on combined analysis of plastid *atpB* and *rbcL* gene sequences. *Syst. Biol.* 49: 306–362.

Soltis, D. E., S. A. Smith, N. Cellinese, K. J. Wurdack, D. C. Tank, S. F. Brockington, N. F. Refulio-Rodriguez, J. B. Walker, M. J. Moore, B. S. Carlsward, C. D. Bell, M. Latvis, S. Crawley, C. Black, D. Diouf, Z. Xi, C. A. Rushworth, M. A. Gitzendanner, K. J. Sytsma, Y.-L. Qiu, K. H. Hilu, C. C. Davis, M. J. Sanderson, R. S. Beaman, R. G. Olmstead, W. S. Judd, M. J. Donoghue, and P. S. Soltis. 2011. Angiosperm phylogeny: 17 genes, 640 taxa. *Am. J. Bot.* 98: 704–730.

Soltis, D. E., P. S. Soltis, M. W. Chase, M. E. Mort, D. C. Albach, M. Zanis, V. Savolainen, W. H. Hahn, S. B. Hoot, M. F. Fay, M. Axtell, S. M. Swensen, L. M. Prince, W. J. Kress, K. C. Nixon, and J. S. Farris. 2000. Angiosperm phylogeny inferred from 18S rDNA, *rbcL*, and *atpB* sequences. *Bot. J. Linn. Soc.* 133:381–461.

Stevens, P. F. 2011. Angiosperm Phylogeny Website, Version 9. Available at http://www.mobot.org/mobot/research/apweb/, last updated on 25 June 2011.

Tank, D. C., and M. J. Donoghue. 2010. Phylogeny and phylogenetic nomenclature of the *Campanulidae* based on an expanded sample of genes and taxa. *Syst. Bot.* 35:425–441.

Winkworth, R. C., J. Lundberg, and M. J. Donoghue. 2008. Toward a resolution of campanulid phylogeny, with special reference to the placement of *Dipsacales. Taxon* 57:1–13.

Authors

Michael J. Donoghue; Department of Ecology and Evolutionary Biology; Yale University; P.O. Box 208106; New Haven, CT 06520, USA. Email: michael.donoghue@yale.edu.

Philip D. Cantino; Department of Environmental and Plant Biology; Ohio University; Athens, OH 45701, USA. Email: cantino@ohio.edu.

Date Accepted: 31 January 2013

Primary Editor: Kevin de Queiroz

Apiidae M. J. Donoghue and P. D. Cantino in P. D. Cantino et al. (2007): E31 [M. J. Donoghue and P. D. Cantino], converted clade name

Registration Number: 247

Definition: The largest crown clade containing *Apium graveolens* Linnaeus 1753 (*Apiales*), *Helianthus annuus* Linnaeus 1753 (*Asterales*), and *Dipsacus sativus* (Linnaeus) Honckeny 1782 (originally described as *Dipsacus fullonum* var. *sativus* Linnaeus 1763) (*Dipsacales*), but not *Ilex crenata* Thunberg 1784 (*Aquifoliales*) or *Lamium purpureum* Linnaeus 1753 (*Lamiidae*). This is a maximum-crown-clade definition with multiple internal and external specifiers. Abbreviated definition: max crown ∇ (*Apium graveolens* Linnaeus 1753 & *Helianthus annuus* Linnaeus 1753 & *Dipsacus sativus* (Linnaeus) Honckeny 1782 ~ *Ilex crenata* Thunberg 1784 v *Lamium purpureum* Linnaeus 1753).

Etymology: Derived from *Apium*, the name of a subclade to which celery belongs and a name used by Pliny "for a celery-like plant" (Gledhill, 1989).

Reference Phylogeny: The primary reference phylogeny is Soltis et al. (2011: Figs. 1, 2e–g). See also Soltis et al. (2000: Figs. 1, 12), Kårehed (2001: Figs. 1, 2), Bremer et al. (2002: Fig. 1), Winkworth et al. (2008: Fig. 1), and Tank and Donoghue (2010: Figs. 1, 3).

Composition: The clade *Apiidae* includes three major subclades (*Apiales*, *Asterales* and *Dispsacales*) and three smaller ones (*Escalloniaceae*, *Paracryphiaceae* and *Bruniales*) (all taxa sensu Tank and Donoghue, 2010).

Diagnostic Apomorphies: There are no clear non-molecular synapomorphies. Possible synapomorphies cited by Stevens (2011) include early sympetaly (see Erbar and Liens, 1996; Leins and Erbar, 2003), a gynoecium of two or three carpels, and an inferior ovary. In addition, polyacetylenes are mentioned by Judd and Olmstead (2004). However, corolla tube development and polyacetylenes are still poorly sampled, and the gynoecial characters are widespread in *Asteridae* and may thus be plesiomorphic. A noteworthy tendency within *Apiidae* is the aggregation of small flowers into more conspicuous, head-like or umbellate inflorescences.

Synonyms: None.

Comments: Traditionally, the taxon *Apiales* was considered distantly related to *Asterales* and *Dipsacales*, and even to *Asteridae* (e.g., Cronquist, 1981, placed *Apiales* in *Rosidae*). However, recent analyses based on molecular data have indicated that *Apiales* (and several smaller taxa; see Composition) are closely related to *Asterales* and *Dipsacales*. To highlight the relationship of *Apiales* to *Asterales* and *Dipsacales*, Cantino et al. (2007) proposed the name *Apiidae* for a clade composed primarily of these three taxa.

Although the monophyly of *Apiidae* has very strong molecular support (Olmstead et al., 2000; Soltis et al., 2000; Albach et al., 2001; Lundberg, 2001; Bremer et al., 2002; Hilu et al., 2003; Tank and Donoghue, 2010; Soltis et al., 2011), basal relationships remain somewhat uncertain, and a minimum-clade definition would consequently require a long list of internal specifiers. We therefore prefer a maximum-crown-clade definition with two external specifiers. Two additional external specifiers that we used in our earlier, otherwise-similar definition (Cantino et al., 2007)—*Cardiopteris quinqueloba* and

Garrya elliptica—are excluded here because recent analyses (Tank and Donoghue, 2010; Soltis et al., 2011) have clarified their positions. Specifically, there is now strong support for the monophyly of *Aquifoliales*, eliminating the need to represent this clade by two external specifiers (*Ilex* and *Cardiopteris*). Similarly, the 100% boostrap support for the inclusion of *Garryales* within *Lamiidae* (Soltis et al., 2011) has eliminated the need to include species of both *Garrya* and *Lamium* as external specifiers.

A maximum-crown-clade definition normally has only one internal specifier, but three are used here to render the name inapplicable to any clade in the context of certain phylogenies. In the unlikely event that *Apiales*, *Asterales* and *Dipsacales* turn out not to be closely related, the name *Apiidae* may not apply to any clade.

Under any phylogenetic hypothesis in which *Campanulidae* and *Apiidae* are synonyms, we intend *Campanulidae* to have precedence; see *Campanulidae* (this volume).

Literature Cited

Albach, D. C., P. S. Soltis, D. E. Soltis, and R. G. Olmstead, 2001. Phylogenetic analysis of asterids based on sequences of four genes. *Ann. Mo. Bot. Gard.* 88:163–212.

Bremer, B., K. Bremer, N. Heidari, P. Erixon, R. G. Olmstead, A. A. Anderberg, M. Källersjö, and E. Barkhordarian. 2002. Phylogenetics of asterids based on 3 coding and 3 non-coding chloroplast DNA markers and the utility of non-coding DNA at higher taxonomic levels. *Mol. Phylogenet. Evol.* 24:274–301.

Cantino, P. D., J. A. Doyle, S. W. Graham, W. S. Judd, R. G. Olmstead, D. E. Soltis, P. S. Soltis, and M. J. Donoghue. 2007. Towards a phylogenetic nomenclature of *Tracheophyta*. *Taxon* 56:822–846 and E1–E44.

Cronquist, A. 1981. *An Integrated System of Classification of Flowering Plants*. Columbia University Press, New York.

Erbar, C., and P. Leins. 1996. Distribution of the character states "early sympetaly" and "late sympetaly" within the "Sympetalae Tetracyclicae" and presumably allied groups. *Bot. Acta* 109:427–440.

Gledhill, D. 1989. *The Names of Plants*. 2nd edition. Cambridge University Press, Cambridge, UK.

Hilu, K. W., T. Borsch, K. Müller, D. E. Soltis, P. S. Soltis, V. Savolainen, M. W. Chase, M. P. Powell, L. A. Alice, R. Evans, H. Sauquet, C. Neinhuis, T. A. B. Slotta, J. G. Rohwer, C. S. Campbell, and L. W. Chatrou. 2003. Angiosperm phylogeny based on *matK* sequence information. *Am. J. Bot.* 90:1758–1776.

Judd, W. S., and R. G. Olmstead. 2004. A survey of tricolpate (eudicot) phylogenetic relationships. *Am. J. Bot.* 91:1627–1644.

Kårehed, J. 2001. Multiple origin of the tropical forest tree family *Icacinaceae*. *Am. J. Bot.* 88:2259–2274.

Leins, P., and C. Erbar. 2003. Floral developmental features and molecular data in plant systematics. Pp. 81–105 in *Deep Morphology: Toward a Renaissance of Morphology in Plant Systematics* (T. F. Stuessy, V. Mayer, and E. Hörandl, eds.). Regnum Vegetabile, Vol. 141. A. R. G. Gantner Verlag, Liechtenstein.

Lundberg, J. 2001. Phylogenetic studies in the Euasterids II, with particular reference to *Asterales* and *Escalloniaceae*. Ph.D. dissertation, Uppsala University. Available at http://publications.uu.se/theses/abstract.xsql?dbid=1597.

Olmstead, R. G., K. Kim, R. K. Jansen, and S. J. Wagstaff. 2000. The phylogeny of the *Asteridae* sensu lato based on chloroplast *ndhF* gene sequences. *Mol. Phylogenet. Evol.* 16:96–112.

Soltis, D. E., S. A. Smith, N. Cellinese, K. J. Wurdack, D. C. Tank, S. F. Brockington, N. F. Refulio-Rodriguez, J. B. Walker, M. J. Moore, B. S. Carlsward, C. D. Bell, M. Latvis, S. Crawley, C. Black, D. Diouf, Z. Xi, C. A. Rushworth, M. A. Gitzendanner, K. J. Sytsma, Y-L. Qiu, K. H. Hilu, C. C. Davis, M. J. Sanderson, R. S. Beaman, R. G. Olmstead, W. S. Judd, M. J. Donoghue, and P. S. Soltis. 2011. Angiosperm phylogeny: 17 genes, 640 taxa. *Am. J. Bot.* 98:704–730.

Soltis, D. E., P. S. Soltis, M. W. Chase, M. E. Mort, D. C. Albach, M. Zanis, V. Savolainen, W. H. Hahn, S. B. Hoot, M. F. Fay, M. Axtell, S. M. Swensen, L. M. Prince, W. J. Kress, K. C. Nixon, and J. S. Farris. 2000. Angiosperm phylogeny inferred from 18S rDNA, *rbcL*, and *atpB* sequences. *Bot. J. Linn. Soc.* 133:381–461.

Stevens, P. F. 2011. Angiosperm Phylogeny Website, Version 9. Available at http://www.mobot.org/mobot/research/apweb/, last updated on 25 June 2011.

Tank, D. C., and M. J. Donoghue. 2010. Phylogeny and phylogenetic nomenclature of the *Campanulidae* based on an expanded sample of genes and taxa. *Syst. Bot.* 35:425–441.

Winkworth, R. C., J. Lundberg, and M. J. Donoghue. 2008. Toward a resolution of campanulid phylogeny, with special reference to the placement of *Dipsacales*. *Taxon* 57:1–13.

Authors

Michael J. Donoghue; Department of Ecology and Evolutionary Biology; Yale University; P.O. Box 208106; New Haven, CT 06520, USA. Email: michael.donoghue@yale.edu.

Philip D. Cantino; Department of Environmental and Plant Biology; Ohio University; Athens, OH 45701, USA. Email: cantino@ohio.edu.

Date Accepted: 31 January 2013

Primary Editor: Kevin de Queiroz

Campanulaceae Jussieu 1789: 163 [N. Cellinese], converted clade name

Registration Number: 20

Definition: The crown clade originating in the most recent common ancestor of *Campanula latifolia* Linnaeus 1753 and all extant organisms or species that share a more recent common ancestor with *Campanula latifolia* than with *Roussea simplex* J. E. Smith 1789, *Pentaphragma ellipticum* Poulsen 1903, and *Stylidium graminifolium* Swartz ex Willdenow 1805. This is a maximum-crown-clade definition. Abbreviated definition: max crown ∇ (*Campanula latifolia* Linnaeus 1753 ~ *Roussea simplex* J. E. Smith 1789 & *Pentaphragma ellipticum* Poulsen 1903 & *Stylidium graminifolium* Swartz ex Willdenow 1805).

Etymology: Derived from Latin *campana* (bell), referring to clade members' characteristic bell-shaped flowers.

Reference Phylogeny: Tank and Donoghue (2010: Fig. 3). *Campanula latifolia*, the internal specifier and the type of *Campanula*, is not in the primary reference phylogeny but is closely related to *Campanula elatines*, which was included in that analysis (Cellinese et al., 2009: Figs. 2–3). See also Cosner et al. (1994: Fig. 1), Gustafsson et al. (1996: Figs. 2–3), Gustafsson and Bremer (1995: Figs. 1–3), Bremer and Gustafsson, (1997: Fig. 1), Bremer et al. (2002: Figs. 2–C), Lundberg and Bremer (2003: Figs. 3–4), Winkworth et al. (2008: Fig. 1).

Composition: This cosmopolitan clade includes approximately 2,400 species and was recently reviewed by Lammers (2007). Major groups include *Campanuloideae* and *Lobelioideae*, as defined in this volume, in addition to *Nemacladoideae, Cyphioideae* and *Cyphocarpoideae* as delimited by Lammers (2007).

Diagnostic Apomorphies: Non-DNA synapomorphies of *Campanulaceae* include laticifers with milky sap, stamens attached to disk at the apex of the ovary (Judd et al., 2008), and secondary pollen presentation mechanisms with a late style elongation phase (with the exception of *Cyphia*) (Erbar and Leins, 1995).

Secondary pollen presentation mechanisms occur in scattered clades of flowering plants. Within *Asterales*, they occur only in *Campanulaceae* and the clade that includes *Asteraceae, Calyceraceae,* and *Goodeniaceae* (including *Brunoniaceae*). Based on the phylogeny generated by Winkworth et al. (2008), this set of traits has evolved independently at least twice. This finding contradicts previous understandings that secondary pollen presentation mechanisms evolved only once in the *Asterales* and were subsequently lost in several lineages (Leins and Erbar, 1997, 2006). In *Campanulaceae,* a few different types of secondary pollen presentation mechanisms can be observed and include a deposition mechanism, a brushing mechanism, and a pump mechanism (Erbar and Leins, 1989, 1995; Leins and Erbar, 1990). These mechanisms differ from similar syndromes present in other clades.

Synonyms: None

Comments: *Campanulaceae* are mainly terrestrial, but occasionally aquatic or epiphytic. Species are primarily herbaceous perennials, rarely annuals or biennials, but also include woody and herbaceous lianas, shrubs, and a few trees.

The monophyly of *Campanulaceae*, as the name is applied here, is well supported by morphological and molecular phylogenetic analyses (Cosner et al., 1994; Gustafsson and Bremer, 1995; Gustafsson et al., 1996; Bremer and Gustafsson, 1997; Kårehed et al., 1999; Bremer et al., 2002; Lundberg and Bremer, 2003; Winkworth et al., 2008; Tank and Donoghue, 2010). All taxa traditionally included in rank-based *Lobeliaceae* (= *Lobelioideae*), *Nemacladaceae* (= *Nemacladoideae*), *Cyphocarpaceae* (= *Cyphocarpoideae*), and *Cyphiaceae* (= *Cyphioideae*) are consistently members of *Campanulaceae* (as delimited here), whereas other taxa such as *Pentaphragma* and *Sphenoclea*, which have been included in *Campanulaceae* by some authors (see below), consistently fall outside this clade. The studies with the most extensive molecular dataset to date (Tank and Donoghue, 2010; Winkworth et al., 2008) show that *Campanulaceae* are sister to *Rousseaceae*, and the clade comprising these two taxa is sister to a larger clade that comprises all other *Asterales*, with *Pentaphragma* sister to the rest of the latter clade.

The name *Campanulaceae* has been used very broadly to include *Campanulaceae* sensu *stricto*, *Lobeliaceae*, *Nemacladaceae*, *Cyphocarpaceae*, *Cyphiaceae*, *Pentaphragmaceae*, and *Sphenocleaceae* (Schönland, 1889; Takhtajan, 1980), and very narrowly to include only the immediate relatives of *Campanula* (*Campanulaceae* sensu *stricto*; Shetler and Morin, 1986; Kolakovsky, 1994; Takhtajan, 1997). Wagenitz (1964), Cronquist (1981), and Lammers (2007) proposed an intermediate circumscription of *Campanulaceae* that corresponds to the clade named here—i.e., including *Campanulaceae* sensu *stricto*, *Lobeliaceae* (as *Lobelioideae*), *Nemacladaceae* (as *Nemacladoideae*), *Cyphocarpaceae* (as *Cyphocarpoideae*), and *Cyphiaceae* (as *Cyphioideae*). No other names have been applied to this clade.

Because outgroup relationships of *Campanulaceae* are well supported, while its basal internal relationships are unclear, a maximum-crown-clade definition is used here.

Literature Cited

Bremer, B., K. Bremer, N. Heidari, P. Erixon, R. G. Olmstead, A. A. Anderberg, M. Kallersjo, and E. Barkhordarian. 2002. Phylogenetics of asterids based on 3 coding and 3 non-coding chloroplast DNA markers and the utility of non-coding DNA at higher taxonomic levels. *Mol. Phylogenet. Evol.* 24:274–301.

Bremer, K., and M. H. G. Gustafsson. 1997. East Gondwana ancestry of the sunflower alliance of families. *Proc. Natl. Acad. Sci. USA.* 94:9188–9190.

Cellinese, N., S. A. Smith, E. J. Edwards, S.-T. Kim, R. C. Haberle, M. Avramakis, and M. J. Donoghue. 2009. Historical biogeography of the endemic *Campanulaceae* of Crete. *J. Biogeogr.* 53:125–1269.

Cosner, M. E., R. K. Jansen, and T. G. Lammers. 1994. Phylogenetic relationships in the *Campanulales* based on *rbcL* sequences. *Plant Syst. Evol.* 190:79–95.

Cronquist, A. 1981. *An Integrated System of Classification of Flowering Plants.* Columbia University Press, New York.

Erbar, C., and P. Leins. 1989. On the early floral development and the mechanisms of secondary pollen presentation in *Campanula*, *Jasione*, and *Lobelia*. *Bot. Jahrb. Syst. Pflanzengesch. Pflanzengeogr.* 111:29–55.

Erbar, C., and P. Leins. 1995. Portioned pollen release and the syndromes of secondary pollen presentation in the *Campanulales-Asterales* complex. *Flora* 190:323–338.

Gustafsson, M. H. G., A. Backlund, and B. Bremer. 1996. Phylogeny of the *Asterales* sensu lato based on *rbcL* sequences with particular reference to the *Goodeniaceae*. *Plant Syst. Evol.* 199:217–242.

Gustafsson, M. H. G., and K. Bremer. 1995. Morphology and phylogenetic interrelationships of the *Asteraceae*, *Calyceraceae*, *Campanulaceae*, *Goodeniaceae*, and related families (*Asterales*). *Am. J. Bot.* 82:250–265.

Judd, W. S., C. S. Campbell, E. A. Kellogg, P. F. Stevens, and M. J. Donoghue. 2008. *Plant Systematics: A Phylogenetic Approach.* 3rd edition. Sinauer, Sunderland, MA.

Jussieu, A. L. 1789. *Genera Plantarum.* Apud Viduam Herissant et Theophilum Barrois, Parisii.

Kårehed, J., J. Lundberg, B. Bremer, and K. Bremer. 1999. Evolution of the Australasian families *Alseuosmiaceae, Argophyllaceae,* and *Phellinaceae. Syst. Bot.* 24:660–682.

Kolakovsky, A. A. 1994. The conspectus of the system of the Old World *Campanulaceae. Bot. Zh. (St. Petersbg.)* 79:109–124.

Lammers, T. G. 2007. *Campanulaceae.* Pp. 26–56 in *The Families and Genera of Vascular Plants* (K. Kubitzki, ed.). Vol. 8 (J. W. Kadereit, and C. Jeffrey, eds.). Springer, Berlin.

Leins, P., and C. Erbar. 1990. On the mechanisms of secondary pollen presentation in the *Campanulales-Asterales*-complex. *Botanica Acta* 103:87–92.

Leins, P., and C. Erbar. 1997. Floral developmental studies: some old and new questions. *Int. J. Plant Sci.* 158:S3–S12.

Leins, P., and C. Erbar. 2006. Secondary pollen presentation syndromes of the *Asterales*—a phylogenetic perspective. *Bot. Jahrb. Syst. Pflanzengesch. Pflanzengeogr.* 127:83–103.

Lundberg, J., and K. Bremer. 2003. Phylogenetic study of the order *Asterales* using one morphological and three molecular data dets. *Int. J. Plant Sci.* 164:553–578.

Schönland, S. 1889. *Campanulaceae.* Pp. 40–70 in *Die Natürlichen Planzenfamilien* (A. Engler, and K. Prantl, eds.). Engelmann, Leipzig.

Shetler, S. G., and N. R. Morin. 1986. Seed morphology in North American *Campanulaceae. Ann. Mo. Bot. Gard.* 73:653–688.

Takhtajan, A. 1980. Outline of the classification of flowering plants (*Magnoliophyta*). *Bot. Rev.* 46:225–359.

Takhtajan, A. 1997. *Diversity and Classification of Flowering Plants.* Columbia University Press, New York.

Tank, D. C., and M. J. Donoghue. 2010. Phylogeny and phylogenetic nomenclature of *Campanulideae* based on an expanded sample of genes and taxa. *Syst. Bot.* 35:425–441.

Wagenitz, G. 1964. Reihe *Campanulales.* Pp. 478–497 in A. Engler's *Syllabus der Planzenfamilien* (H. Melchior, ed.). Gebrüder-Bornträger, Berlin.

Winkworth, R. C., J. Lundberg, and M. J. Donoghue. 2008. Toward a resolution of campanulid phylogeny, with special reference to the placement of *Dipsacales. Taxon* 57:53–65.

Author

Nico Cellinese; University of Florida; Florida Museum of Natural History; 354 Dickinson Hall, Museum Rd, Gainesville, FL 32611, USA. Email: ncellinese@flmnh.ufl.edu.

Date Accepted: 5 December 2011

Primary Editor: Philip Cantino

Campanuloideae Burnett 1835: 942, 1094, 1110 [N. Cellinese], converted clade name

Registration Number: 21

Definition: The crown clade originating in the most recent common ancestor of *Campanula latifolia* Linnaeus 1753, *Wahlenbergia linifolia* A. de Candolle 1830, and *Platycodon grandiflorus* (Jacquin) A. de Candolle 1830. This is a minimum-crown-clade definition. Abbreviated definition: min crown ∇ (*Campanula latifolia* Linnaeus 1753 & *Wahlenbergia linifolia* A. de Candolle 1830 & *Platycodon grandiflorus* (Jacquin) A. de Candolle 1830).

Etymology: Derived from Latin *campana* (bell), referring to clade members' characteristic bell-shaped flowers.

Reference Phylogeny: Cellinese et al. (2009: Figs. 2–3). See also Eddie et al. (2003: Fig. 1), Haberle et al. (2009: Fig. 3).

Composition: The clade *Campanuloideae* comprises mainly temperate taxa, with approximately 1,050 species, mostly occurring in the Old World. Two large taxa within the *Campanuloideae* are *Campanula* (approximately 420 species) and *Wahlenbergia* (approximately 260 species).

Diagnostic Apomorphies: Possible non-DNA synapomorphies for *Campanuloideae* include scalariform vessel perforations and verrucose pollen surface (Gustafsson and Bremer, 1995), and stylar hair invagination (Erbar and Leins, 1989, 1995; Leins and Erbar, 1990; Judd et al., 2008), although basal taxa such as *Platycodon* have not been thoroughly investigated.

The invagination of stylar hairs is a process occurring during a late stage of the secondary pollen presentation mechanism. In *Campanuloideae*, this syndrome involves pollen being deposited onto stylar hairs, which subsequently invaginate, leaving a glabrous, pitted style (Erbar and Leins, 1989, 1995). Rarely (e.g., *Phyteuma*), a brushing mechanism is present, but it is different from the typical brushing mechanism present in *Asteraceae* and *Lobelioideae* in that the tube of fused anthers is lacking (Erbar and Leins, 1989, 1995).

Synonyms: *Campanulaceae* sensu Shetler and Morin (1986), Kolakovsky (1994), and Takhtajan (1997) is an approximate synonym. In the present work, the name *Campanulaceae* is applied to a larger clade, as by Wagenitz (1964), Cronquist (1981), and Lammers (2007).

Comments: The monophyly of *Campanuloideae* is strongly supported by molecular phylogenies (Eddie et al., 2003; Cellinese et al., 2009; Haberle et al., 2009). These studies also revealed that the species of *Campanuloideae* fall into three clades: one including the campanuloids, another the wahlenbergioids, and a third including the platycodonoids (Eddie et al., 2003; Cellinese et al., 2009; Haberle et al., 2009). Because these three clades are well supported and include all of the *Campanuloideae*, a minimum-crown-clade definition with three specifiers works well.

The name *Campanulaceae* was applied to this clade by Shetler and Morin (1986), Kolakovsky (1994), and Takhtajan (1997). However, following Wagenitz (1964) and Cronquist (1981), most authors have used the name *Campanuloideae* for this clade, applying the name *Campanulaceae* to a more inclusive clade. The latter application of these two names is adopted here.

Literature Cited

Burnett, G. T. 1835. *Outlines of Botany*. Renshaw, London.

Cellinese, N., S. A. Smith, E. J. Edwards, S.-T. Kim, R. C. Haberle, M. Avramakis, and M. J. Donoghue. 2009. Historical biogeography of the endemic *Campanulaceae* of Crete. *J. Biogeogr.* 53:1254–1269.

Cronquist, A. 1981. *An Integrated System of Classification of Flowering Plants*. Columbia University Press, New York.

Eddie, W. M. M., T. Shulkina, J. Gaskin, R. C. Haberle, and R. K. Jansen. 2003. Phylogeny of *Campanulaceae* s. str. inferred from ITS Sequences of Nuclear Ribosomal DNA. *Ann. Mo. Bot. Gard.* 90:554–575.

Erbar, C., and P. Leins. 1989. On the early floral development and the mechanisms of secondary pollen presentation in *Campanula, Jasione,* and *Lobelia. Bot. Jahrb. Syst. Pflanzengesch. Pflanzengeogr.* 111:29–55.

Erbar, C., and P. Leins. 1995. Portioned pollen release and the syndromes of secondary pollen presentation in the *Campanulales-Asterales* complex. *Flora* 190:323–338.

Gustafsson, M. H. G., and K. Bremer. 1995. Morphology and phylogenetic interrelationships of the *Asteraceae, Calyceraceae, Campanulaceae, Goodeniaceae,* and related families (*Asterales*). *Am. J. Bot.* 82:250–265.

Haberle, R. C., A. Dang, T. Lee, C. Penaflor, H. Cortes-Burn, A. Oestreich, L. Raubenson, N. Cellinese, E. J. Edwards, S.-T. Kim, W. M. M. Eddie, and R. K. Jansen. 2009. Taxonomic and biogeographic implications of a phylogenetic analysis of the *Campanulaceae* based on three chloroplast genes. *Taxon* 58:715–734.

Judd, W. S., C. S. Campbell, E. A. Kellogg, P. F. Stevens, and M. J. Donoghue. 2008. *Plant Systematics—A Phylogenetic Approach*. 3rd edition. Sinauer, Sunderland, MA.

Kolakovsky, A. A. 1994. The conspectus of the system of the Old World *Campanulaceae. Bot. Zh.* (*St. Petersbg.*) 79:109–124.

Lammers, T. G. 2007. *Campanulaceae*. Pp. 26–56 in *The Families and Genera of Vascular Plants* (K. Kubitzki, ed.). Vol. 8 (J. W. Kadereit, and C. Jeffrey, eds.). Springer, Berlin.

Leins, P., and C. Erbar. 1990. On the mechanisms of secondary pollen presentation in the *Campanulales-Asterales*-complex. *Botanica Acta* 103:87–92.

Shetler, S. G., and N. R. Morin. 1986. Seed morphology in North American *Campanulaceae. Ann. Mo. Bot. Gard.* 73:653–688.

Takhtajan, A. 1997. *Diversity and Classification of Flowering Plants*. Columbia University Press, New York.

Wagenitz, G. 1964. Reihe *Campanulales*. Pp. 478–497 in *A. Engler's Syllabus der Planzenfamilien* (H. Melchior, ed.). Gebrüder-Bornträger, Berlin.

Author

Nico Cellinese; University of Florida; Florida Museum of Natural History; 354 Dickinson Hall, Museum Rd, Gainesville, FL 32611, USA. Email: ncellinese@flmnh.ufl.edu.

Date Accepted: 5 December 2011

Primary Editor: Philip Cantino

Lobelioideae Burnett 1835: 942, 1094, 1110 [N. Cellinese], converted clade name

Registration Number: 2

Definition: The crown clade originating in the most recent common ancestor of *Lobelia cardinalis* Linnaeus 1753 and *Lobelia coronopifolia* Linnaeus 1753. This is a minimum-crown-clade definition. Abbreviated definition: min crown ∇ (*Lobelia cardinalis* Linnaeus 1753 & *Lobelia coronopifolia* Linnaeus 1753).

Etymology: Derived from the genus name *Lobelia*, which honors the Flemish botanist Matthias de L'Obel (1538–1616).

Reference Phylogeny: Antonelli (2008: Fig. 2).

Composition: The clade *Lobelioideae* includes approximately 1,200 species, mostly occurring in the New World. The clade comprises *Lobelia* with about 400 species, and its closest relatives (see Comments).

Diagnostic Apomorphies: Non-DNA synapomorphies for *Lobelioideae* include resupinate flowers, connate anthers forming a tube, and a pump mechanism of secondary pollen presentation. Pump mechanisms are also present in *Asteraceae*. However, in *Lobelioideae*, where it evolved independently, the anther tube is markedly oblique, and the style is much shorter in late developmental stages (Erbar and Leins, 1989).

Synonyms: *Lobeliaceae* Jussieu 1813 (sensu Presl 1836) is an approximate synonym.

Comments: The monophyly of *Lobelioideae* is strongly supported by molecular data (Antonelli, 2008). Lammers (2007) provided the most recent comprehensive synopsis, but the only formal taxonomic treatment dates back to Wimmer (1943, 1953). The names *Lobeliaceae* and *Lobelioideae* have been applied to the clade that includes *Lobelia* (which is paraphyletic) and its offshoots. However, following Wimmer (1943, 1953), most authors have treated *Lobelioideae* as a subfamily of *Campanulaceae*. Therefore, based on frequency and stability of usage, the name *Lobelioideae* is adopted here. Given the well-supported phylogeny, a minimum-crown-clade definition is appropriate.

According to Antonelli (2008), *Lobelia dortmanna* is the type species of *Lobelia*. However, the correct type species is *Lobelia cardinalis* L., which is chosen here as an internal specifier.

Literature Cited

Antonelli, A. 2008. Higher level phylogeny and evolutionary trends in *Campanulaceae* subfam. *Lobelioideae*: molecular signal overshadows morphology. *Mol. Phylogenet. Evol.* 46:1–18.

Burnett, G.T. 1835. *Outlines of Botany*. Renshaw, London.

Erbar, C., and P. Leins. 1989. On the early floral development and the mechanisms of secondary pollen presentation in *Campanula*, *Jasione*, and *Lobelia*. *Bot. Jahrb. Syst. Pflanzengesch. Pflanzengeogr.* 111:29–55.

Jussieu, A. L. 1813. *Description des Plantes Rares Cultivées à Malmaison et à Navarre*, by Bonpland. P. Didot l'aîné, Paris.

Lammers, T. G. 2007. *Campanulaceae*. Pp. 26–56 in *The Families and Genera of Vascular Plants* (K. Kubitzki, ed.). Vol. 8 (J. W. Kadereit, and C. Jeffrey, eds.). Springer, Berlin.

Presl, C. B. 1836. *Prodromusmonographiae Lobeliacearum*. Theophilus Haase, Prague.

Wimmer, F. E. 1943. *Campanulaceae-Lobelioideae*, I Teil. Pp. 1–260 in *Das Pflanzenreich, IV.276b* (R. Mansfeld, ed.). W. Engelmann, Leipzig.

Wimmer, F. E. 1953. *Campanulaceae-Lobelioideae*, II Teil. Pp. 261–813 in *Das Pflanzenreich, IV.276b* (H. Stubbe and K. Noack, eds.). Akademie Verlag, Berlin.

Author

Nico Cellinese; University of Florida; Florida Museum of Natural History; 354 Dickinson Hall, Museum Rd, Gainesville, FL 32611, USA. Email: ncellinese@flmnh.ufl.edu.

Date Accepted: 5 December 2011

Primary Editor: Philip Cantino

Lamiidae A. Takhtajan 1987:228 [R. G. Olmstead, W. S. Judd, and P. D. Cantino], converted clade name

Registration Number: 59

Definition: The largest crown clade containing *Lamium album* Linnaeus 1753 (*Lamiales*) but not *Campanula latifolia* Linnaeus 1753 (*Campanulidae*) and *Cornus mas* Linnaeus 1753 (*Cornales*) and *Erica cinerea* Linnaeus 1753 (*Ericales*). This is a maximum-crown-clade definition. Abbreviated definition: max crown ∇ (*Lamium album* Linnaeus 1753 ~ *Campanula latifolia* Linnaeus 1753 & *Cornus mas* Linnaeus 1753 & *Erica cinerea* Linnaeus 1753).

Etymology: Derived from *Lamium* (name of an included taxon), referring to the throat-like corolla tube (DeFelice, 2005). The genus name was presumably derived from the Greek *laimos* (throat or gullet) (Brown, 1954).

Reference Phylogeny: Soltis et al. (2011: Figs. 1–2), where the closest relative of *Lamium album* is *Lamium*, of *Campanula latifolia* is *Campanula*, of *Cornus mas* is *Cornus*, and of *Erica cinerea* is *Arbutus* (*Ericaceae*) (terminal taxa in this paper were only identified to genus). See also Olmstead et al. (2000: Fig. 1), Soltis et al. (2000: Fig. 11), Kårehed (2001: Figs. 1–2), Bremer et al. (2002: Fig. 1), and Refulio and Olmstead (2014: Fig 1).

Composition: *Garryales*, *Icacinaceae*, *Oncothecaceae*, *Metteniusiaceae* (González et al., 2007), and *Lamianae* (this volume).

Diagnostic Apomorphies: No non-DNA synapomorphies have yet been identified.

Synonyms: *Garryidae* (Cantino et al., 2007) is an unambiguous synonym. This clade has been referred to informally as "asterid I" (Chase et al., 1993), "euasterids I" (Olmstead et al., 2000; Soltis et al., 2000; Savolainen et al., 2000a; Albach et al., 2001; Hilu et al., 2003; APG II, 2003) and "lamiids" (Bremer et al., 2002; Judd and Olmstead, 2004; Judd et al., 2008; APG III, 2009; APG IV, 2016).

Comments: The name *Lamiidae* was first applied to this clade by Olmstead et al. (1992), although they did not sample *Garryales*. In that study, the four principal clades of *Asteridae* recognized today were identified (*Cornales*, *Ericales*, *Campanulidae*, *Lamiidae*). In a more extensively sampled study based on *rbcL*, Olmstead et al. (1993) included *Garryales* but, due to lack of resolution, used the name *Lamiidae* for the smaller clade (named *Lamianae* in this volume) that excludes *Garryales*. Subsequent studies (e.g., those cited under Reference Phylogeny) have confirmed the monophyly of the larger clade (i.e., the one treated in this entry). Olmstead et al. proposed *Garryidae* for this clade in Cantino et al. (2007: 836), but that name is inconsistent with the frequent use of "lamiids" (as an informal version of "*Lamiidae*") for the same clade (e.g., Judd et al., 2008; APG IV, 2016). Therefore, we herein adopt *Lamiidae* rather than *Garryidae* for this clade.

The name *Lamiidae* was first used by Takhtajan (1987) for a group that differs in circumscription in some details. Takhtajan included a description of *Lamiidae* in Russian, but the name was not validly published according to the botanical code (Turland et al., 2018) until Reveal (1993) provided a Latin description. Takhtajan is considered the nominal author of *Lamiidae* according to the *PhyloCode* (see Arts. 6.2 and 19.1).

Most molecular phylogenetic studies, beginning with Chase et al. (1993), have inferred this clade, although mostly with weak support (e.g., Olmstead et al., 2000; Savolainen et al., 2000a; Soltis et al., 2000; Albach et al., 2001; Hilu et al., 2003). However, several multi-gene analyses with more extensive sampling (Kårehed, 2001; Bremer et al., 2002; Soltis et al., 2011) provided strong support for this clade. In some early trees based on the single gene *rbcL*, *Garryales* either were not associated with the rest of this clade (Savolainen et al., 2000b), or were unresolved with respect to the rest of *Asteridae* (Olmstead et al., 1993). In the unlikely event that *Garryales* turn out in the future to have quite a different position than is currently believed (for example, if *Garryales* were found to be related to *Cornaceae*, as proposed by Cronquist, 1981), the name *Lamiidae* could (depending on the positions of *Icacinaceae*, *Oncothecaceae*, *Metteniusiaceae*) end up applying to the same clade that is herein named *Lamianae*. If this were to occur, *Lamiidae* should have precedence over *Lamianae*.

Literature Cited

Albach, D. C., P. S. Soltis, D. E. Soltis, and R. G. Olmstead. 2001. Phylogenetic analysis of asterids based on sequences of four genes. *Ann. Mo. Bot. Gard.* 88:163–212.

APG II (Angiosperm Phylogeny Group II). 2003. An update of the Angiosperm Phylogeny Group classification for the orders and families of flowering plants: APG II. *Bot. J. Linn. Soc.* 141:399–436.

APG III (Angiosperm Phylogeny Group III). 2009. An update of the Angiosperm Phylogeny Group classification for the orders and families of flowering plants: APG III. *Bot. J. Linn. Soc.* 161:105–121.

APG IV (Angiosperm Phylogeny Group IV). 2016. An update of the Angiosperm Phylogeny Group classification for the orders and families of flowering plants: APG IV. *Bot. J. Linn. Soc.* 181:1–20.

Bremer, B., K. Bremer, N. Heidari, P. Erixon, R. G. Olmstead, A. A. Anderberg, M. Källersjö, and E. Barkhordarian. 2002. Phylogenetics of asterids based on 3 coding and 3 non-coding chloroplast DNA markers and the utility of non-coding DNA at higher taxonomic levels. *Mol. Phylogenet. Evol.* 24:274–301.

Brown, R. W. 1954. *Composition of Scientific Words*. Smithsonian Books, Washington, DC.

Cantino, P. D., J. A. Doyle, S. W. Graham, W. S. Judd, R. G. Olmstead, D. E. Soltis, P. S. Soltis, and M. J. Donoghue. 2007. Towards a phylogenetic nomenclature of *Tracheophyta*. *Taxon* 56:822–846 and E1–E44.

Chase, M. W., M. W. Chase, D. E. Soltis, R. G. Olmstead, D. Morgan, D. H. Les, B. D. Mishler, M. R. Duvall, R. A. Price, H. G. Hills, Y.-L. Qiu, K. A. Kron, J. H. Rettig, E. Conti, J. D. Palmer, J. R. Manhart, K. J. Sytsma, H. J. Michaels, W. J. Kress, K. G. Karol, W. D. Clark, M. Hedren, B. S. Gaut, R. K. Jansen, K.-J. Kim, C. F. Wimpee, J. F. Smith, G. R. Furnier, S. H. Strauss, Q.-Y. Xiang, G. M. Plunkett, P. S. Soltis, S. M. Swensen, S. E. Williams, P. A. Gadek, C. J. Quinn, L. E. Eguiarte, E. Golenberg, G. H. Learn, Jr., S. W. Graham, S. C. H. Barrett, S. Dayanandan, and V. A. Albert. 1993. Phylogenetics of seed plants: an analysis of nucleotide sequences from the plastid gene *rbcL*. *Ann. Mo. Bot. Gard.* 80:528–580.

Cronquist, A. 1981. *An Integrated System of Classification of Flowering Plants*. Columbia University Press, New York.

DeFelice, M. S. 2005. Henbit and the deadnettles, *Lamium* spp.—archangels or demons? *Weed Tech.* 19:768–774.

González, F., J. Betancur, O. Maurin, J. V. Freudenstein, and M. W. Chase. 2007. *Metteniusaceae*, an early-diverging family in the lamiid clade. *Taxon* 56:795–800.

Hilu, K. W., T. Borsch, K. Müller, D. E. Soltis, P. S. Soltis, V. Savolainen, M. W. Chase, M. P. Powell, L. A. Alice, R. Evans, H. Sauquet, C. Neinhuis, T. A. B. Slotta, J. G. Rohwer, C. S. Campbell, and L. W. Chatrou. 2003. Angiosperm phylogeny based on *matK* sequence information. *Am. J. Bot.* 90:1758–1776.

Judd, W. S. C. S. Campbell, E. A. Kellogg, P. F. Stevens, and M. J. Donoghue. 2008. *Plant Systematics—A Phylogenetic Approach*. 3rd edition. Sinauer Associates, Sunderland, MA.

Judd, W. S., and R. G. Olmstead. 2004. A survey of tricolpate (eudicot) phylogenetic relationships. *Am. J. Bot.* 91:1627–1644.

Kårehed, J. 2001. Multiple origin of the tropical forest tree family *Icacinaceae. Am. J. Bot.* 88:2259–2274.

Olmstead, R. G., B. Bremer, K. M. Scott, and J. D. Palmer. 1993. A parsimony analysis of the *Asteridae* sensu lato based on *rbcL* sequences. *Ann. Missouri Bot. Gard.* 80:700–722.

Olmstead, R. G., K. Kim, R. K. Jansen, and S. J Wagstaff. 2000. The phylogeny of the *Asteridae* sensu lato based on chloroplast *ndhF* gene sequences. *Mol. Phylogenet. Evol.* 16:96–112.

Olmstead, R. G., H. J. Michaels, K. Scott, and J. D. Palmer. 1992. Monophyly of the *Asteridae* and identification of their major lineages inferred from DNA sequences of *rbcL. Ann. Mo. Bot. Gard.* 79:249–265.

Refulio, N. F., and R. G. Olmstead. 2014. Phylogeny of *Lamiidae. Amer. J. Bot.* 101 287–299.

Reveal, J. L. 1993. New subclass and superordinal names for extant vascular plants. *Phytologia* 74:178–179.

Savolainen, V., M. W. Chase, S. B. Hoot, C. M. Morton, D. E. Soltis, D. Bayer, M. F. Fay, A. Y. de Bruijn, S. Sullivan, and Y.-L. Qiu. 2000a. Phylogenetics of flowering plants based on combined analysis of plastid *atpB* and *rbcL* gene sequences. *Syst. Biol.* 49:306–362.

Savolainen, V., M. F. Fay, D. C. Albach, A. Bachlund, M. van der Bank, K. M. Cameron, S. A. Johnson, M. D. Lledó, J.-C. Pintaud, M. Powell, M. C. Sheahan, D. E. Soltis, P. S. Soltis, P. Weston, W. M. Whitten, K. J. Wurdack, and M. W. Chase. 2000b. Phylogeny of the eudicots: a nearly complete familial analysis based on *rbcL* gene sequences. *Kew Bull.* 55:257–309.

Soltis, D. E., S. A. Smith, N. Cellinese, K. J. Wurdack, D. C. Tank, S. F. Brockington, N. F. Refulio-Rodriguez, J. B. Walker, M. J. Moore, B. S. Carlsward, C. D. Bell, M. Latvis, S. Crawley, C. Black, D. Diouf, Z. Xi. M. A. Gitzendanner, K. J. Systma, Y.-L. Qiu, K. W. Hilu, S. R. Manchester, C. C. Davis, M. J. Sanderson, R. G. Olmstead, W. S. Judd, M. J. Donoghue, and P. S. Soltis. 2011. Angiosperm phylogeny: 17-genes, 640 taxa. *Am. J. Bot.* 98:704–730.

Soltis, D. E., P. S. Soltis, M. W. Chase, M. E. Mort, D. C. Albach, M. Zanis, V. Savolainen, W. H. Hahn, S. B. Hoot, M. F. Fay, M. Axtell, S. M. Swensen, L. M. Prince, W. J. Kress, K. C. Nixon, and J. S. Farris. 2000. Angiosperm phylogeny inferred from 18S rDNA, *rbcL*, and *atpB* sequences. *Bot. J. Linn. Soc.* 133:381–461.

Takhtajan, A. 1987. *Systema Magnoliophytorum*. Nauka, Leningrad.

Turland, N. J., J. H. Wiersema, F. R. Barrie, W. Greuter, D. L. Hawksworth, P. S. Herendeen, S. Knapp, W.-H. Kusber, D.-Z. Li, K. Marhold, T. W. May, J. McNeill, A. M. Monro, J. Prado, M. J. Price, and G. F. Smith, eds. 2018. *International Code of Nomenclature for Algae, Fungi, and Plants (Shenzhen Code)*. Adopted by the Nineteenth International Botanical Congress, Shenzhen, China, July 2017. Regnum Vegetabile 159. Koeltz Botanical Books, Glashütten.

Authors

Richard G. Olmstead; Department of Biology and Burke Museum; University of Washington; Seattle, WA 98195, USA. Email: olmstead@u.washington.edu.

Walter S. Judd; Department of Botany; University of Florida; Gainesville, FL 32611, USA. Email: wjudd@botany.ufl.edu.

Philip D. Cantino; Department of Environmental and Plant Biology; Ohio University; Athens, OH 45701, USA. Email: cantino@ohio.edu.

Date Accepted: 27 January 2013; updated 21 July 2017, 11 July 2018

Primary Editor: Kevin de Queiroz

Lamianae A. L. Takhtajan 1967: 230 [R. G. Olmstead and W. S. Judd], converted clade name

Registration Number: 58

Definition: The smallest crown clade containing *Lamium album* Linnaeus 1753 (*Lamiales*), *Solanum lycopersicum* Linnaeus 1753 (*Solanales*), *Gentiana procera* T. Holm 1901 (*Gentianales*), *Borago officinalis* Linnaeus 1753 (*Boraginales*), and *Vahlia capensis* (Linnaeus f.) Thunberg 1782 (*Vahliaceae*). This is a minimum-crown-clade definition. Abbreviated definition: min crown ∇ (*Lamium album* Linnaeus 1753 & *Solanum lycopersicum* Linnaeus 1753 & *Gentiana procera* T. Holm 1901 & *Borago officinalis* Linnaeus 1753 & *Vahlia capensis* (Linnaeus f.) Thunberg 1782).

Etymology: Derived from *Lamium* (name of an included genus), referring to the throat-like corolla tube (DeFelice, 2005). The genus name was presumably derived from the Greek *laimos* (throat or gullet) (Brown, 1954).

Reference Phylogeny: Refulio and Olmstead (2014: Fig 1), where the closest relative of the specifier *Lamium album* is *Lamium purpureum*. See also Olmstead et al. (2000: Fig. 1), Bremer et al. (2002: Fig. 1), and Soltis et al. (2011: Fig. 2).

Composition: *Gentianales* (this volume), *Solanales* (this volume), *Lamiales* (this volume), *Boraginales*, and *Vahliaceae*.

Diagnostic Apomorphies: vessels with simple perforations (Baas et al., 2003; Stevens, 2001 onwards); perhaps "corolla tube initiation late" (i.e., in floral development the corolla lobes are initiated first, followed by development of a ring primordium that forms the tube), but sampling very limited (Leins and Erbar, 2003; Stevens, 2001 onwards).

Synonyms: *Lamiidae* sensu Olmstead et al. (1993) is an approximate synonym, and *Lamiidae* sensu Olmstead and Judd (in Cantino et al., 2007) is an unambiguous synonym; see Comments.

Comments: The name *Lamianae* was first used by Takhtajan (1967) for a group that differs in circumscription in relatively minor details from the clade named here. In subsequent revisions to his classification, Takhtajan (1987, 1997) restricted *Lamianae* to a group corresponding more closely to what is now called *Lamiales* (this volume). The clade that is herein named *Lamianae* was referred to as *Lamiidae* by Olmstead et al. (1993: Fig. 3) in a study based on *rbcL* sequences. That study failed to resolve the larger clade that includes *Garryales* and which in almost all subsequent studies has been recognized as "lamiids" or "*Lamiidae*." Following the lead of the 1993 paper, Olmstead and Judd (in Cantino et al., 2007: E30) phylogenetically defined the name *Lamiidae* to apply to the smaller clade (the one named *Lamianae* here). However, in recognition of the fact that the informal clade name "lamiids" was clearly based on *Lamiidae* and has been applied to the larger clade in more studies (e.g., Bremer et al., 2002; Judd and Olmstead, 2004; Judd et al., 2007; APG III, 2009) than to the smaller one, we prefer to apply *Lamiidae* to the larger clade (this volume), thus necessitating a new name for the smaller one.

This clade was first recognized by Olmstead et al. (1992, 1993), based on *rbcL* sequences with strong support, and has been inferred in analyses of 18S rDNA (Albach et al., 2001) and many subsequent molecular studies (e.g., Olmstead et al., 2000; Savolainen et al., 2000; Soltis et al, 2000; Albach et al., 2001; Bremer et al., 2002; Hilu et al., 2003).

Literature Cited

Albach, D. C., P. S. Soltis, D. E. Soltis, and R. G. Olmstead. 2001. Phylogenetic analysis of asterids based on sequences of four genes. *Ann. Mo. Bot. Gard.* 88:163–212.

APG III (Angiosperm Phylogeny Group II). 2009. An update of the Angiosperm Phylogeny Group classification for the orders and families of flowering plants: APG III. *Bot. J. Linn. Soc.* 161:105–121.

Baas, P., S. Jansen, and E. A. Wheeler. 2003. Ecological adaptations and deep phylogenetic splits—evidence and questions from the secondary xylem. Pp. 221–239 in *Deep Morphology: Toward a Renaissance of Morphology in Plant Systematics* (T. F. Stuessy, V. Mayer, and E. Hörandl, eds.). Regnum Vegetabile, Vol. 141. A.R.G. Gantner Verlag, Liechtenstein.

Bremer, B., K. Bremer, N. Heidari, P. Erixon, R. G. Olmstead, A. A. Anderberg, M. Källersjö, and E. Barkhordarian. 2002. Phylogenetics of asterids based on 3 coding and 3 non-coding chloroplast DNA markers and the utility of non-coding DNA at higher taxonomic levels. *Mol. Phylogenet. Evol.* 24:274–301.

Brown, R. W. 1954. *Composition of Scientific Words.* Smithsonian Books, Washington, DC.

Cantino, P. D., J. A. Doyle, S. W. Graham, W. S. Judd, R. G. Olmstead, D. E. Soltis, P. S. Soltis, and M. J. Donoghue. 2007. Towards a phylogenetic nomenclature of *Tracheophyta.* *Taxon* 56:822–846 and E1–E44.

DeFelice, M. S. 2005. Henbit and the deadnettles, *Lamium* spp.—archangels or demons? *Weed Tech.* 19:768–774.

Hilu, K. W., T. Borsch, K. Müller, D. E. Soltis, P. S. Soltis, V. Savolainen, M. W. Chase, M. P. Powell, L. A. Alice, R. Evans, H. Sauquet, C. Neinhuis, T. A. B. Slotta, J. G. Rohwer, C. S. Campbell, and L. W. Chatrou. 2003. Angiosperm phylogeny based on *matK* sequence information. *Am. J. Bot.* 90:1758–1776.

Judd, W. S., C. S. Campbell, E. A. Kellogg, P. F. Stevens, and M. J. Donoghue. 2007. *Plant Systematics—A Phylogenetic Approach.* 3rd edition. Sinauer Associates, Sunderland, MA.

Judd, W. S., and R. G. Olmstead. 2004. A survey of tricolpate (eudicot) phylogenetic relationships. *Am. J. Bot.* 91:1627–1644.

Leins, P., and C. Erbar. 2003. Floral developmental features and molecular data in plant systematics. Pp. 81–105 in *Deep Morphology: Toward a Renaissance of Morphology in Plant Systematics* (T. F. Stuessy, V. Mayer, and E. Hörandl, eds.). Regnum Vegetabile, Vol. 141. A.R.G. Gantner Verlag, Liechtenstein.

Olmstead, R. G., B. Bremer, K. M. Scott, and J. D. Palmer. 1993. A parsimony analysis of the *Asteridae* sensu lato based on *rbcL* sequences. *Ann. Mo. Bot. Gard.* 80:700–722.

Olmstead, R. G., K. Kim, R. K. Jansen, and S. J. Wagstaff. 2000. The phylogeny of the *Asteridae* sensu lato based on chloroplast *ndhF* gene sequences. *Mol. Phylogenet. Evol.* 16:96–112.

Olmstead, R. G., H. J. Michaels, K. M. Scott, and J. D. Palmer. 1992. Monophyly of the *Asteridae* and identification of their major lineages inferred from DNA sequences of *rbcL.* *Ann. Mo. Bot. Gard.* 79:249–265.

Refulio, N. F., and R. G. Olmstead. 2014. Phylogeny of *Lamiidae. Am. J. Bot.* 101: 287–299.

Savolainen, V., M. W. Chase, S. B. Hoot, C. M. Morton, D. E. Soltis, C. Bayer, M. F. Fay, A. Y. de Bruijn, S. Sullivan, and Y.-L. Qiu. 2000. Phylogenetics of flowering plants based on combined analysis of plastid *atpB* and *rbcL.* *Syst. Biol.* 49:306–362.

Soltis, D. E., S. A. Smith, N. Cellinese, K. J. Wurdack, D. C. Tank, S. F. Brockington, N. F. Refulio-Rodriguez, J. B. Walker, M. J. Moore, B. S. Carlsward, C. D. Bell, M. Latvis, S. Crawley, C. Black, D. Diouf, Z. Xi, M. A. Gitzendanner, K. J. Sytsma, Y.-L. Qiu, K. W.

Hilu, S. R. Manchester, C. C. Davis, M. J. Sanderson, R. G. Olmstead, W. S. Judd, M. J. Donoghue, and P. S. Soltis. 2011. Angiosperm phylogeny: 17 genes, 640 taxa. *Am. J. Bot.* 98:704–730.

Soltis, D. E., P. S. Soltis, M. W. Chase, M. E. Mort, D. C. Albach, M. Zanis, V. Savolainen, W. H. Hahn, S. B. Hoot, M. F. Fay, M. Axtell, S. M. Swensen, L. M. Prince, W. J. Kress, K. C. Nixon, and J. S. Farris. 2000. Angiosperm phylogeny inferred from 18S rDNA, *rbcL*, and *atpB* sequences. *Bot. J. Linn. Soc.* 133:381–461.

Stevens, P. F. 2001 onwards. Angiosperm Phylogeny Website, Version 7. Available at http://www.mobot.org/mobot/research/apweb/.

Takhtajan, A. 1967. *A System and Phylogeny of the Flowering Plants.* Moscow and Leningrad. (In Russian.)

Takhtajan, A. 1987. *Systema Magnoliophytorum.* Nauka, Leningrad.

Takhtajan, A. 1997. *Diversity and Classification of Flowering Plants.* Columbia University Press, New York.

Authors

Richard G. Olmstead; Department of Biology and Burke Museum; University of Washington; Seattle, WA 98195, USA. Email: olmstead@u.washington.edu.

Walter S. Judd; Department of Botany; University of Florida; Gainesville, FL 32611, USA. Email: lyonia@ufl.edu.

Date Accepted: 21 August 2011; updated 18 July 2017

Primary Editor: Philip Cantino

Gentianales J. Lindley 1833: 248 [L. Struwe, C. Frasier, and R. G. Olmstead], converted clade name

Registration Number: 46

Definition: The smallest crown clade containing *Gentiana lutea* Linnaeus 1753 (*Gentianaceae*) and *Cinchona pubescens* Vahl 1790 (*Rubiaceae*). This is a minimum-crown-clade definition. Abbreviated definition: min crown ∇ (*Gentiana lutea* Linnaeus 1753 & *Cinchona pubescens* Vahl 1790).

Etymology: Based on *Gentiana* (name of a genus), which is derived from the Greek "Gentius," the last king of Illyria (ruled 180–168 BC), who was credited by Pliny the Elder with discovering the pharmacological uses of the plant (Fernald, 1970).

Reference Phylogeny: The primary reference phylogeny is Backlund et al. (2000: Fig. 1; see Fig. 2 and Table 4 for support values). See also Olmstead et al. (1993, 2000), Savolainen et al. (2000), Soltis et al. (2000, 2007), Albach et al. (2001), Bremer et al. (2002), Struwe et al. (2002), and Frasier (2008).

Composition: *Apocynaceae* (including *Asclepiadaceae* and *Periplocaceae*; Endress and Stevens, 2001), *Gelsemiaceae* (including *Pteleocarpaceae*; Jiao and Li, 2007; Struwe et al., 2014; Struwe, 2019), *Gentianaceae* (including *Saccifoliaceae* and *Potaliaceae*; Struwe et al., 2002; Struwe and Pringle, 2019), *Loganiaceae* (incl. *Antoniaceae, Spigeliaceae, Strychnaceae*; Frasier, 2008; Struwe et al., 2019), *Rubiaceae* (incl. *Dialypetalanthaceae*; Robbrecht and Manen, 2006; Bremer and Eriksson, 2009).

Diagnostic Apomorphies: Possible synapomorphies include vestured pits; opposite leaves with interpetiolar stipules or lines; and the presence of colleters (short, multicellular glands) on the inner, adaxial surfaces of interpetiolar stipules, leaf axils, and/or calyces (Albert and Struwe, 2002).

Synonyms: *Rubiales* sensu Thorne and Reveal (2007) is an approximate synonym based on a list of included taxa.

Comments: The name *Gentianales* has been associated with this clade in traditional classifications (Stebbins, 1974; Cronquist, 1981; Takhtajan, 1987, 1997; Thorne, 1992, but not Thorne and Reveal, 2007) and molecular phylogenetic studies, where it has received strong support (Olmstead et al., 1993, 2000; Bremer et al., 1994, 2002; Backlund et al., 2000; Soltis et al., 2000, 2007; Struwe et al., 2002; Hilu et al., 2003). Several taxa included in *Gentianales* in one or more of the traditional classifications cited above have been found not to belong to the clade, including *Buddleja, Carlemanniaceae, Desfontainia, Menyanthaceae, Oleaceae, Plocosperma,* and *Retzia.* Molecular evidence supports the inclusion of *Pteleocarpa* in *Gelsemiaceae* (Struwe et al., 2014). *Dialypetalanthus* occasionally was placed outside this group in traditional classifications (e.g., Cronquist, 1981), but belongs in *Gentianales* (in *Rubiaceae*; Fay et al., 2000). *Rubiaceae* are supported as the sister group to the rest of *Gentianales* (Frasier, 2008; Soltis et al., 2007; Yang et al., 2016). We have selected the name *Gentianales* over the synonym *Rubiales* (sensu Thorne and Reveal, 2007) because the former is much more frequently used.

Literature Cited

Albach, D. C., P. S. Soltis, D. E. Soltis, and R. G. Olmstead. 2001. Phylogenetic analysis of asterids based on sequences of four genes. *Ann. Mo. Bot. Gard.* 88:163–212.

Albert, V. A., and L. Struwe. 2002. *Gentianaceae* in context. Pp. 1–20 in *Gentianaceae – Systematics and Natural History* (L. Struwe and V. A. Albert, eds.). Cambridge University Press, Cambridge, UK.

Backlund, M., B. Oxelman, and B. Bremer. 2000. Phylogenetic relationships within the *Gentianales* based on *ndhF* and *rbcL* sequences, with particular reference to the *Loganiaceae*. *Am. J. Bot.* 87:1029–1043.

Bremer, B., K. Bremer, N. Heidari, P. Erixon, R. G. Olmstead, A. A. Anderberg, M. Källersjö, and E. Barkhordarian. 2002. Phylogenetics of asterids based on 3 coding and 3 non-coding chloroplast DNA markers and the utility of non-coding DNA at higher taxonomic levels. *Mol. Phylogenet. Evol.* 24:274–301.

Bremer, B., and T. Eriksson. 2009. Time tree of *Rubiaceae*: phylogeny and dating of the family, subfamilies, and tribes. *Int. J. Plant Sci.* 170:766–793.

Bremer, B., R. G. Olmstead, L. Struwe, and J. A. Sweere. 1994. *rbcL* sequences support exclusion of *Retzia*, *Desfontainia*, and *Nicodemia* from the *Gentianales*. *Plant Syst. Evol.* 190:213–230.

Cronquist, A. 1981. *An Integrated System of Classification of Flowering Plants*. Columbia University Press, New York.

Endress, M. E., and W. D. Stevens. 2001. The renaissance of the *Apocynaceae* s.l.: recent advances in systematics, phylogeny, and evolution. *Ann. Mo. Bot. Gard.* 88:517–522.

Fay, M. F., B. Bremer, G. T. Prance, M. van der Bank, D. Bridson, and M. W. Chase. 2000. Plastid *rbcL* sequence data show *Dialypetalanthus* to be a member of *Rubiaceae*. *Kew Bull.* 55:853–864.

Fernald, M. L. 1970. *Gray's Manual of Botany*. 8th edition, corrected printing. Van Nostrand, New York.

Frasier, C. L. 2008. Evolution and systematics of the angiosperm order *Gentianales* with an in-depth focus on *Loganiaceae* and its species-rich and toxic genus *Strychnos*. Ph.D. dissertation, Rutgers University, New Brunswick, NJ.

Hilu, K. W., T. Borsch, K. Müller, D. E. Soltis, P. S. Soltis, V. Savolainen, M. W. Chase, M. P. Powell, L. A. Alice, R. Evans, H. Sauquet, C. Neinhuis, T. A. B. Slotta, J. G. Rohwer, C. S. Campbell, and L. W. Chatrou. 2003. Angiosperm phylogeny based on *matK* sequence information. *Am. J. Bot.* 90:1758–1776.

Jiao, Z., and J. Li. 2007. Phylogeny of intercontinental disjunct *Gelsemiaceae* inferred from chloroplast and nuclear DNA sequences. *Syst. Bot.* 32:617–627

Lindley, J. 1833. *Nixus Plantarum*. Ridgway and Sons, London.

Olmstead, R. G., B. Bremer, K. M. Scott, and J. D. Palmer. 1993. A parsimony analysis of the *Asteridae* sensu lato based on *rbcL* sequences. *Ann. Mo. Bot. Gard.* 80:700–722.

Olmstead, R. G., K. Kim, R. K. Jansen, and S. J. Wagstaff. 2000. The phylogeny of the *Asteridae* sensu lato based on chloroplast *ndhF* gene sequences. *Mol. Phylogenet. Evol.* 16:96–112.

Robbrecht, E., and J.-F. Manen. 2006. The major evolutionary lineages of the coffee family (*Rubiaceae*, angiosperms). Combined analysis (nDNA and cpDNA) to infer the position of *Coptosapelta* and *Luculia*, and supertree construction based on *rbcL*, *rps16*, *trnL-trnF* and *atpB-rbcL* data. A new classification in two subfamilies, *Cinchonoideae* and *Rubioideae*. *Syst. Geogr. Plants* 76:85–146.

Savolainen, V., M. F. Fay, D. C. Albach, A. Bachlund, M. van der Bank, K. M. Cameron, S. A. Johnson, M. D. Lledó, J.-C. Pintaud, M. Powell, M. C. Sheahan, D. E. Soltis, P. S. Soltis, P. Weston, W. M. Whitten, K. J. Wurdack, and M. W. Chase. 2000. Phylogeny of the eudicots: a nearly complete familial analysis based on *rbcL* gene sequences. *Kew Bull.* 55:257–309.

Soltis, D. E., M. A. Gitzendanner, and P. S. Soltis. 2007. A 567-taxon data set for angiosperms: the challenges posed by Bayesian analyses of large data sets. *Int. J. Plant Sci.* 168:137–157.

Soltis, D. E., P. S. Soltis, M. W. Chase, M. E. Mort, D. C. Albach, M. Zanis, V. Savolainen, W. H. Hahn, S. B. Hoot, M. F. Fay, M. Axtell, S. M. Swensen, L. M. Prince, W. J. Kress, K. C. Nixon, and J. S. Farris. 2000. Angiosperm phylogeny inferred from 18S rDNA, *rbcL*, and *atpB* sequences. *Bot. J. Linn. Soc.* 133:381–461.

Stebbins, G. L. 1974. *Flowering Plants: Evolution Above the Species Level.* Harvard University Press, Cambridge, MA.

Struwe, L. 2019. *Gelsemiaceae.* In *The Families and Genera of Flowering Plants* (K. Kubitzki, ed.). Vol. 15: *Apiales, Gentianales* (*Except* Rubiaceae) (J. W. Kadereit and V. Bittrich, eds.). Springer Verlag, Berlin.

Struwe, L., K. L. Gibbons, B. J. Conn, and T. Motley. 2019. *Loganiaceae.* In *The Families and Genera of Flowering Plants* (K. Kubitzki, ed.). Vol. 15: *Apiales, Gentianales* (*Except* Rubiaceae) (J. W. Kadereit and V. Bittrich, eds.). Springer Verlag, Berlin.

Struwe, L., J. Kadereit, J. Klackenberg, S. Nilsson, M. Thiv, K. B. von Hagen, and V. A. Albert. 2002. Systematics, character evolution, and biogeography of *Gentianaceae*, including a new tribal and subtribal classification. Pp. 2–309 in Gentianaceae – *Systematics and Natural History* (L. Struwe and V. A. Albert, eds.). Cambridge University Press, Cambridge, UK.

Struwe, L., and J. Pringle. 2019. *Gentianaceae.* In *The Families and Genera of Flowering Plants* (K. Kubitzki, ed.). Vol. 15: *Apiales, Gentianales* (*except* Rubiaceae) (J. W. Kadereit and V. Bittrich, eds.). Springer Verlag, Berlin.

Struwe, L., V. L. Soza, S. Manickam, and R. G. Olmstead. 2014. *Gelsemiaceae* (*Gentianales*) expanded to include the enigmatic Asian genus *Pteleocarpa. Bot. J. Linn. Soc.* 175:482–496.

Takhtajan, A. 1987. *Systema Magnoliophytorum.* Nauka, Leningrad.

Takhtajan, A. 1997. *Diversity and Classification of Flowering Plants.* Columbia University Press, New York.

Thorne, R. F. 1992. Classification and geography of the flowering plants. *Bot. Rev.* 58:225–348.

Thorne, R. F., and J. L. Reveal. 2007. An updated classification of the class *Magnoliopsida* ("*Angiospermae*"). *Bot. Rev.* 73:67–182.

Yang, L.-L., H.-L. Li, L. Wei, T. Yang, D.-Y. Kuang, M.-H. Li, Y.-Y. Liao, Z.-D. Chen, H. Wu, and S.-Z. Zhang. 2016. A supermatrix approach provides a comprehensive genus-level phylogeny for *Gentianales. J. Syst. Evol.* 54:400–415.

Authors

Lena Struwe; Dept. of Ecology, Evolution, and Natural Resources/Dept. of Plant Biology; Rutgers University; New Brunswick, NJ 08901-8551, USA. Email: lena.struwe@rutgers.edu.

Cynthia Frasier; Department of Conservation Genetics; Omaha's Henry Doorly Zoo and Aquarium; Omaha, NE 68107, USA. Email: cfrasier77@gmail.com.

Richard G. Olmstead; Department of Biology and Burke Museum; University of Washington; Seattle, WA 98195, USA. Email: olmstead@u.washington.edu.

Date Accepted: 26 September 2011; updated 24 July 2017

Primary Editor: Philip Cantino

Rubiaceae A. L. de Jussieu 1789: 196 [B. Bremer and T. Eriksson], converted clade name

Registration Number: 91

Definition: The largest crown clade containing *Rubia tinctorum* Linnaeus 1753 but not *Alstonia scholaris* (Linnaeus) Robert Brown 1810 and *Gentiana procera* Holm 1901 and *Gelsemium sempervirens* (Linnaeus) Saint-Hilaire 1805 and *Logania vaginalis* Ferdinand Mueller 1868. This is a maximum-crown-clade definition. Abbreviated definition: max crown ∇ (*Rubia tinctorum* Linnaeus 1753 ~ *Alstonia scholaris* (Linnaeus) Robert Brown 1810 & *Gentiana procera* Holm 1901 & *Gelsemium sempervirens* (Linnaeus) Saint-Hilaire 1805 & *Logania vaginalis* Ferdinand Mueller 1868.)

Etymology: Derived from *Rubia* (the name of an included taxon), which in turn is derived from the Latin *ruber* (red), referring to the red dye obtained from the roots of *Rubia tinctorum*. According to Linnaeus (1737), the name *Rubia* had previously been used by Plinius and others.

Reference Phylogeny: The primary reference phylogeny is Bremer et al. (1999: Fig. 5). *Rubia tinctorum*, which is the type of *Rubiaceae* under the botanical code (Turland et al., 2018) and the sole internal specifier in our definition, is not present in the primary reference phylogeny, but its position within a monophyletic *Rubioideae* (most closely related to *Anthospermum* among the sampled taxa within the "RUBI" clade in the reference phylogeny) is well supported (Bremer and Eriksson, 2009: Figs. 5, 6). See also Rydin et al. (2017: Fig. 1) for an alternative hypothesis of relationships within *Rubiaceae*.

Composition: *Rubiaceae* has been considered (Bremer et al., 1999; Wikström et al., 2015) to contain three major clades (*Rubioideae*, *Cinchonoideae*, and *Ixoroideae*) and two other small taxa ranked as genera (*Luculia* and *Coptosapelta*), mainly based on chloroplast DNA, but a recent analysis of mitochondrial DNA (Rydin et al., 2017) has called into question the monophyly of *Cinchonoideae* and *Ixoroideae* as traditionally delimited. *Rubiaceae* includes many well-known ornamentals such as gardenias (*Gardenia*) and West Indian jasmine (*Ixora*), as well as providing medicines—most notably quinine (*Cinchona*), but the most economically important members are the two coffee species, *Coffea arabica* and *C. canephora* (syn. *C. robusta*).

Diagnostic Apomorphies: Synapomorphies relative to the closely related clades include leaves with distinct interpetiolar stipules, inferior ovary (few exceptions), absence of internal phloem, and alkaloids (when present) never with C-16 skeleton as reaction center (Andersson and Persson, 1991; Bremer and Struwe, 1992). Distinct interpetiolar stipules are in some cases found in *Spigelia* (*Loganiaceae*), but as a result of convergent evolution (Bremer and Struwe, 1992).

Synonyms: None.

Comments: The monophyly of *Rubiaceae* is strongly supported by molecular data (Bremer and Jansen, 1991; Bremer et al. 1995, 1999, 2002; Rova et al., 2002; Wikström et al., 2015; Rydin et al., 2017) and morphology (see Diagnostic Apomorphies). The delimitation of this clade has been stable for many years, and the name *Rubiaceae* is consistently applied to it (Bremer et al., 1995 and many previous works).

Literature Cited

Andersson, L., and C. Persson. 1991. Circumscription of the tribe *Cinchoneae* (*Rubiaceae*) – a cladistic approach. *Plant Syst. Evol.* 178:65–94.

Bremer, B., K. Andreasen, and D. Olsson. 1995. Subfamilial and tribal relationships in the *Rubiaceae* based on rbcL sequence data. *Ann. Mo. Bot. Gard.* 82:383–397.

Bremer, B., K. Bremer, N. Heidari, P. Erixon, R. G. Olmstead, A. A. Anderberg, M. Källersjö, and E. Barkhordarian. 2002. Phylogenetics of asterids based on 3 coding and 3 non-coding chloroplast DNA markers and the utility of non-coding DNA at higher taxonomic levels. *Mol. Phylogenet. Evol.* 24:274–301.

Bremer, B., and T. Eriksson. 2009. Time tree of *Rubiaceae*: phylogeny and dating the family, subfamilies, and tribes. *Int. J. Plant. Sci.* 170:766–793.

Bremer, B., and R. K. Jansen. 1991. Comparative restriction site mapping of chloroplast DNA implies new phylogenetic relationships within *Rubiaceae*. *Am. J. Bot.* 78:198–213.

Bremer, B., R. K. Jansen, B. Oxelman, M. Backlund, H. Lantz, and K. J. Kim. 1999. More characters or more taxa for a robust phylogeny – case study from the coffee family (*Rubiaceae*). *Syst. Biol.* 48:413–435.

Bremer, B., and L. Struwe. 1992. Phylogeny of the *Rubiaceae* and the *Loganiaceae* – congruence or conflict between morphological and molecular data. *Am. J. Bot.* 79:1171–1184.

de Jussieu, A. L. 1789. *Genera plantarum secundum Ordines Naturales Disposita*. Herissant & Barrois, Paris.

Linnaeus, C. 1737. *Critica Botanica*. Wishoff, Leiden, (English translation: Hort, A. 1938. Ray Society, London.)

Rova, J. H. E., P. G. Delprete, L. Andersson, and V. A. Albert. 2002. A trnL-F cpDNA sequence study of the *Condamineeae-Rondeletieae-Sipaneeae* complex with implications on the phylogeny of the *Rubiaceae*. *Am. J. Bot.* 89:145–159.

Rydin, C., N. Wikström, and B. Bremer. 2017. Conflicting results from mitochondrial genomic data challenge current views of *Rubiaceae* phylogeny. *Am. J. Bot.* 104:1522–1532.

Turland, N. J., J. H. Wiersema, F. R. Barrie, W. Greuter, D. L. Hawksworth, P. S. Herendeen, S. Knapp, W.-H. Kusber, D.-Z. Li, K. Marhold, T. W. May, J. McNeill, A. M. Monro, J. Prado, M. J. Price, and G. F. Smith, eds. 2018. *International Code of Nomenclature for Algae, Fungi, and Plants (Shenzhen Code)*. Adopted by the Nineteenth International Botanical Congress, Shenzhen, China, July 2017. Regnum Vegetabile 159. Koeltz Botanical Books, Glashütten.

Wikström, N., K. Kainulainen, S. G. Razafimandimbison, J. E. E. Smedmark, and B. Bremer. 2015. A revised time tree of the asterids: establishing a temporal framework for evolutionary studies of the coffee family (*Rubiaceae*). *PLOS ONE* 10:e0126690.

Authors

Birgitta Bremer; The Bergius Foundation at the Royal Swedish Academy of Sciences and Department of Ecology, Environment and Plant Sciences; Stockholm University; SE-106 91 Stockholm, Sweden. Email: Bremer.Birgitta@gmail.com.

Torsten Eriksson; IT Department; University of Bergen; Postboks 7800, NO-5020 Bergen, Norway. Email: Torsten.Eriksson@adm.uib.no.

Date Accepted: 21 December 2012; updated 7 November 2017

Primary Editor: Philip Cantino

Solanales Dumortier 1829:180 [R. G. Olmstead and S. Stefanović], converted clade name

Registration Number: 99

Definition: The smallest crown clade containing *Solanum nigrum* L. 1753 (*Solanaceae*), *Convolvulus arvensis* L. 1753 (*Convolvulaceae*), and *Hydrolea ovata* Nutt. ex Choisy 1833 (*Hydroleaceae*). This is a minimum-crown-clade definition. Abbreviated definition: min crown ∇ (*Solanum nigrum* L. 1753 & *Convolvulus arvensis* L. 1753 & *Hydrolea ovata* Nutt. ex Choisy 1833).

Etymology: Derived from *Solanum* (name of a genus), which in turn comes from the Latin for "quieting" due to the narcotic effect of alkaloids in some species (Charters, 2008). The Latin word in question is probably *solor* (comfort, soothe) (Brown, 1954).

Reference Phylogeny: Refulio and Olmstead (2014: Fig. 1). See also Olmstead et al. (2000: Fig. 1), Bremer et al. (2002: Fig. 1), and Soltis et al. (2011: Fig. 2). The specifiers *Solanum nigrum* and *Convolvulus arvensis* are represented by *Solanum lycopersicum* and *Convolvulus cneorum*, respectively, in the primary reference phylogeny (see Comments).

Composition: *Solanaceae* (this volume), *Convolvulaceae* (this volume), *Montiniaceae*, *Hydroleaceae*, and *Sphenocleaceae* (APG IV, 2016).

Diagnostic Apomorphies: No non-molecular synapomorphies are known.

Synonyms: *Solanineae* sensu Thorne and Reveal (2007) is an approximate synonym based on composition.

Comments: *Solanales* in most traditional classifications comprised a non-monophyletic group of asterid families with radiate corolla symmetry and superior ovaries, including, for example, *Polemoniaceae*, *Boraginaceae*, and *Hydrophyllaceae* (Cronquist, 1981; Thorne, 1992), and excluded *Montiniaceae* and *Sphenocleaceae*. The restriction of *Solanales* to a clade centered on *Solanaceae* and *Convolvulaceae* was suggested by Olmstead et al. (1992) based on *rbcL* sequences. Subsequently, the small families *Montiniaceae*, *Hydroleaceae*, and *Sphenocleaceae* were added as evidence became available (Olmstead et al., 1993, 2000; Cosner et al., 1994; Fay et al., 1998; Bremer et al., 2002).

Some of the early molecular studies based on single genes found weak support for this clade (e.g., Olmstead et al., 1993, 2000; Albach et al., 2001) or did not infer a clade with this composition (e.g., Savolainen et al., 2000b excluded *Sphenocleaceae* and *Hydroleaceae*). Other studies inferred a clade comprising *Solanales* as defined here and *Boraginaceae* (Olmstead et al., 1992, 2000; Cosner et al., 1994; Savolainen et al., 2000a,b); the name *Solanales* has been used for this larger clade (Savolainen et al., 2000a,b). However, studies with multiple genes in the data set (Soltis et al., 2000; Bremer et al., 2002; Refulio and Olmstead, 2014) do not infer a clade comprising *Solanales* and *Boraginaceae* and instead provide strong support for *Solanales* as defined here.

Thorne and Reveal (2007) designated a group with identical composition at the rank of suborder, *Solanineae*. The name *Solanales* is adopted here following the Angiosperm Phylogeny Group (APG IV, 2016) and in keeping with the preponderance of recent use (Olmstead et al., 1993, 2000; Cosner et al., 1994; Soltis et al., 2000; Albach et al., 2001; Bremer et al., 2002).

Two species that are not included in the primary reference phylogeny were used as specifiers: *Solanum nigrum* because it is the type species of *Solanales* under rank-based nomenclature, and *Convolvulus arvensis* because it is used as specifier in the definition of *Convolvulaceae* (this volume). *Solanum nigrum* is closely related to *S. lycopersicum* in the primary reference phylogeny (Olmstead et al., 2008), and *Convolvulus arvensis* is closely related to *C. cneorum* in the reference phylogeny (Stefanović et al., 2002; Carine et al., 2004).

Literature Cited

Albach, D. C., P. S. Soltis, D. E. Soltis, and R. G. Olmstead. 2001. Phylogenetic analysis of asterids based on sequences of four genes. *Ann. Mo. Bot. Gard.* 88:163–212.

APG IV (Angiosperm Phylogeny Group IV). 2016. An update of the Angiosperm Phylogeny Group classification for the orders and families of flowering plants: APG IV. *Bot. J. Linn. Soc.* 181:1–20.

Bremer, B., K. Bremer, N. Heidari, P. Erixon, R. G. Olmstead, A. A. Anderberg, M. Källersjö, and E. Barkhordarian. 2002. Phylogenetics of asterids based on 3 coding and 3 non-coding chloroplast DNA markers and the utility of non-coding DNA at higher taxonomic levels. *Mol. Phylogenet. Evol.* 24:274–301.

Brown, R. W. 1954. *Composition of Scientific Words*. Smithsonian Books, Washington, DC.

Carine, M. A., S. J. Russell, A. Santos-Guerra, and J. Francisco-Ortega. 2004. Relationships of the Macaronesian and Mediterranean floras: molecular evidence for multiple colonizations into Macarenesia and back-colonization of the continent in *Convolvulus* (*Convolvulaceae*). *Am. J. Bot.* 91:1070–1085.

Charters, M. L. (compiler). 2008. California Plant Names: Latin and Greek Meanings and Derivations. Available at http://www.calflora.net/botanicalnames.

Cosner, M. E., R. K. Jansen, and T. G. Lammers. 1994. Phylogenetic relationships in the *Campanulales* based on rbcL sequences. *Plant Syst. Evol.* 190:79–95.

Cronquist, A. 1981. *An Integrated System of Classification of Flowering Plants*. Columbia University Press, New York.

Dumortier, B.-C. 1829. *Analyse des Familles des Plantes*. Tournay, Paris.

Fay, M. F., R. G. Olmstead, J. E. Richardson, E. Santiago, G. T. Prance, and M. W. Chase. 1998. Molecular data support the inclusion of *Duckeodendron cestroides* in *Solanaceae*. *Kew Bull.* 53:203–212.

Olmstead, R. G., L. Bohs, H. Abdel Migid, E. Santiago-Valentin, V. F. Garcia, and S. M. Collier. 2008. A molecular phylogeny and classification of the *Solanaceae*. *Taxon* 57:1159–1181.

Olmstead, R. G., B. Bremer, K. M. Scott, and J. D. Palmer. 1993. A parsimony analysis of the *Asteridae* sensu lato based on *rbcL* sequences. *Ann. Mo. Bot. Gard.* 80:700–722.

Olmstead, R. G., K. Kim, R. K. Jansen, and S. J. Wagstaff. 2000. The phylogeny of the *Asteridae* sensu lato based on chloroplast *ndhF* gene sequences. *Mol. Phylogenet. Evol.* 16:96–112.

Olmstead, R. G., H. J. Michaels, K. M. Scott, and J. D. Palmer. 1992. Monophyly of the *Asteridae* and identification of their major lineages inferred from DNA sequences of *rbcL*. *Ann. Mo. Bot. Gard.* 79:249–265.

Refulio, N. F., and R. G. Olmstead. 2014. Phylogeny of *Lamiidae*. *Am. J. Bot.* 101:287–299.

Savolainen, V., M. W. Chase, S. B. Hoot, C. M. Morton, D. E. Soltis, D. Bayer, M. F. Fay, A. Y. de Bruijn, S. Sullivan, and Y.-L. Qiu. 2000a. Phylogenetics of flowering plants based on combined analysis of plastid *atpB* and *rbcL* gene sequences. *Syst. Biol.* 49:306–362.

Savolainen, V., M. F. Fay, D. C. Albach, A. Bachlund, M. van der Bank, K. M. Cameron, S. A. Johnson, M. D. Lledó, J.-C. Pintaud, M. Powell, M. C. Sheahan, D. E. Soltis, P. S. Soltis, P. Weston, W. M. Whitten, K. J. Wurdack, and M. W. Chase. 2000b. Phylogeny of the eudicots: a nearly complete familial analysis based on *rbcL* gene sequences. *Kew Bull.* 55:257–309.

Soltis, D. E., S. A. Smith, N. Cellinese, K. J. Wurdack, D. C. Tank, S. F. Brockington, N. F. Refulio-Rodriguez, J. B. Walker, M. J. Moore, B. S. Carlsward, C. D. Bell, M. Latvis, S. Crawley, C. Black, D. Diouf, Z. Xi., M. A. Gitzendanner, K. J. Sytsma, Y.-L. Qiu, K. W. Hilu, S. R. Manchester, C. C. Davis, M. J. Sanderson, R. G. Olmstead, W. S. Judd, M. J. Donoghue, and P. S. Soltis. 2011. Angiosperm phylogeny: 17 genes, 640 taxa. *Am. J. Bot.* 98:704–730.

Soltis, D. E., P. S. Soltis, M. W. Chase, M. E. Mort, D. C. Albach, M. Zanis, V. Savolainen, W. H. Hahn, S. B. Hoot, M. F. Fay, M. Axtell, S. M. Swensen, L. M. Prince, W. J. Kress, K. C. Nixon, and J. S. Farris. 2000. Angiosperm phylogeny inferred from 18S rDNA, *rbcL*, and *atpB* sequences. *Bot. J. Linn. Soc.* 133:381–461.

Stefanović, S., L. E. Krueger, and R. G. Olmstead. 2002. Monophyly of the *Convolvulaceae* and circumscription of their major lineages based on DNA sequences of multiple chloroplast loci. *Am. J. Bot.* 89:1510–1522.

Thorne, R. F. 1992. Classification and geography of the flowering plants. *Bot. Rev.* 58:225–348.

Thorne, R. F., and J. L. Reveal. 2007. An updated classification of the class *Magnoliopsida* ("*Angiospermae*"). *Bot. Rev.* 73:67–182.

Authors

Richard G. Olmstead; Department of Biology; University of Washington; Seattle, WA 98195, USA. Email: olmstead@u.washington.edu.

Saša Stefanović; Department of Biology; University of Toronto Mississauga; 3359 Mississauga Rd N; Mississauga, ON, Canada, L5L 1C6. Email: sasa.stefanovic@utoronto.ca.

Date Accepted: 30 September 2010; updated 18 July 2017

Primary Editor: Philip Cantino

Solanaceae Jussieu 1789: 124 [R. G. Olmstead], converted clade name

Registration Number: 98

Definition: The smallest crown clade containing *Solanum nigrum* Linnaeus 1753, *Schizanthus pinnatus* Ruiz et Pavon 1794, *Goetzea elegans* Wydler 1830, *Duckeodendron cestroides* Kuhlmann 1925, *Schwenckia lateriflora* (Vahl) Carvalho 1969, *Salpiglossis sinuata* Ruiz et Pavon 1794, *Benthamiella skottsbergii* Soriano, and *Petunia axillaris* (Lam.) Britton, Stearns, et Poggenb. 1888. This is a minimum-crown-clade definition. Abbreviated definition: min crown ∇ (*Solanum nigrum* Linnaeus 1753 & *Schizanthus pinnatus* Ruiz et Pavon 1794 & *Goetzea elegans* Wydler 1830 & *Duckeodendron cestroides* Kuhlmann 1925 & *Schwenckia lateriflora* (Vahl) Carvalho 1969 & *Salpiglossis sinuata* Ruiz et Pavon 1794 & *Benthamiella skottsbergii* Soriano & *Petunia axillaris* (Lam.) Britton, Stearns, et Poggenb. 1888).

Etymology: Derived from *Solanum* (name of a genus), which in turn comes from the Latin for "quieting" due to the narcotic effect of alkaloids in some species (Charters, 2008). The Latin word in question is probably *solor* (comfort, soothe) (Brown, 1954).

Reference Phylogeny: Olmstead et al. (2008: Fig. 1). See also Olmstead and Palmer (1992: Fig. 2), Olmstead et al. (1999: Figs. 2, 4), and Särkinen et al. (2013: Fig. 1).

Composition: *Solanoideae, Nicotianoideae, Schwenckieae, Petunieae, Benthamielleae, Cestroideae, Goetzeoideae, Duckeodendron,* and *Schizanthus* (Olmstead et al., 2008).

Diagnostic Apomorphies: A bicarpellate ovary oriented obliquely to the median plane of the flower is often cited as a diagnostic trait for *Solanaceae* (Cronquist, 1981; Hunziker, 2001; Judd et al., 2007). Careful ontogenetic study has shown that this is a long-standing error in interpretation of the *Solanaceae* flower (Ampornpan and Armstrong, 2002). However, Ampornpan and Armstrong (2002) suggested that the orientation of the flower with a calyx lobe (and staminode in zygomorphic flowers) in the abaxial position (inverted relative to other asterid flowers) may be a synapomorphy of *Solanaceae*. A three amino acid deletion in the coding sequence for the chloroplast gene *ndhF* is a molecular synapomorphy for the clade (Olmstead et al., 2000).

Synonyms: None.

Comments: The name *Solanaceae* has been applied consistently to a paraphyletic group originating with the same ancestor as this clade (ranked as a family) in traditional angiosperm classifications (e. g., Cronquist, 1981; Takhtajan, 1997). However, a series of small groups included within this clade (*Duckeodendron, Nolana, Sclerophylax, Goetzeaceae*), but having atypical fruit morphology, have been excluded from *Solanaceae* in those classifications.

Molecular phylogenetic studies provide strong support for the monophyly of *Solanaceae* (Olmstead et al., 2000; Soltis et al., 2000; Albach et al., 2001, Refulio and Olmstead, 2014). Most studies identify *Schizanthus* as sister to the remaining taxa (Olmstead and Palmer, 1992; Olmstead and Sweere, 1994; Martins and Barkman, 2005; Hu and Saedler, 2007; Olmstead et al., 2008), with a deletion in the chloroplast genome providing a molecular synapomorphy for *Solanaceae* exclusive of *Schizanthus* (Olmstead and Palmer, 1992). However, a few studies have identified *Schwenckia* as sister to all others (Olmstead et al., 1999; Santiago-Valentin and Olmstead, 2003), or a clade consisting of *Goetzea* and *Metternichia* (Fay et al., 1998). The most complete study to date (Särkinen et al., 2013) was unable to provide clear resolution at the base of the tree. In consideration of the inconsistency of signal in some studies at the base of the tree and the relatively low support for other deep nodes, I have selected representatives of several early-diverging lineages as specifiers.

Literature Cited

Albach, D. C., P. S. Soltis, D. E. Soltis, and R. G. Olmstead, 2001. Phylogenetic analysis of asterids based on sequences of four genes. *Ann. Mo. Bot. Gard.* 88:163–212.

Ampornpan, L., and J. E. Armstrong. 2002. Floral ontogeny of *Salpiglossis* (*Solanaceae*) and the oblique gynoecium. *Bull. Torr. Bot. Soc.* 129:85–95.

Brown, R. W. 1954. *Composition of Scientific Words*. Smithsonian Books, Washington, DC.

Charters, M. L. 2008. California plant names: Latin and Greek meanings and derivations, 9 July, 2010. Available at www.calflora.net/botanicalnames.

Cronquist, A. 1981. *An Integrated System of Classification of Flowering Plants*. Columbia University Press, New York.

Fay, M. F., R. G. Olmstead, J. E. Richardson, E. Santiago, G. T. Prance, and M. W. Chase. 1998. Molecular data support the inclusion of *Duckeodendron cestroides* in *Solanaceae*. *Kew Bull.* 53:203–212.

Hu, J.-Y., and H. Saedler. 2007. Evolution of the inflated calyx syndrome in *Solanaceae*. *Mol. Biol. Evol.* 24:2443–2453.

Hunziker, A. T. 2001. *Genera Solanacearum: The Genera of* Solanaceae *Illustrated, Arranged According to a New System*. A. R. G. Gantner Verlag, Ruggell, Liechtenstein.

Judd, W. S., C. S. Campbell, E. A. Kellogg, P. F. Stevens, and M. J. Donoghue. 2007. *Plant Systematics—A Phylogenetic Approach*. 3rd edition. Sinauer Associates, Sunderland, MA.

Jussieu, A. L. 1789. *Genera Plantarum*. Paris.

Martins, T. R., and T. J. Barkman. 2005. Reconstruction of *Solanaceae* phylogeny using the nuclear gene SAMT. *Syst. Bot.* 30:435–447.

Olmstead, R. G., L. Bohs, H. Abdel Migid, E. Santiago-Valentin, V. F. Garcia, and S. M. Collier. 2008. A molecular phylogeny and classification of the *Solanaceae*. *Taxon* 57:1159–1181.

Olmstead, R. G., K. Kim, R. K. Jansen, and S. J. Wagstaff. 2000. The phylogeny of the *Asteridae* sensu lato based on chloroplast *ndhF* gene sequences. *Mol. Phylogenet. Evol.* 16:96–112.

Olmstead, R. G., and J. D. Palmer. 1992. A chloroplast DNA phylogeny of the *Solanaceae*: subfamilial relationships and character evolution. *Ann. Mo. Bot. Gard.* 79:346–360.

Olmstead, R. G., and J. A. Sweere. 1994. Combining data in phylogenetic systematics: an empirical approach using three molecular data sets in the Solanaceae. *Syst. Biol.* 43:467–481.

Olmstead, R. G., J. A. Sweere, R. E. Spangler, L. Bohs, and J. D. Palmer. 1999. Phylogeny and provisional classification of the *Solanaceae* based on chloroplast DNA. Pp. 111–137 in Solanaceae *IV: Advances in Biology and Utilization* (M. Nee, D. Symon, R. N. Lester, and J. Jessop, eds.). Royal Botanic Gardens, Kew.

Refulio, N. F., and R. G. Olmstead. 2014. Phylogeny of *Lamiidae*. *Am. J. Bot.* 101:287–299.

Santiago-Valentin, E., and R. G. Olmstead. 2003. Phylogenetics of the Antillean *Goetzeoideae* (*Solanaceae*) and their relationships within the *Solanaceae* based on chloroplast and ITS DNA sequence data. *Syst. Bot.* 28:452–460.

Särkinen, T., L. Bohs, R. G. Olmstead, and S. D. Knapp. 2013. A phylogenetic framework for evolutionary study of the nightshades (*Solanaceae*): a dated 1000-tip tree. *BMC Evol. Biol.* 13:214

Soltis, D. E., P. S. Soltis, M. W. Chase, M. E. Mort, D. C. Albach, M. Zanis, V. Savolainen, W. H. Hahn, S. B. Hoot, M. F. Fay, M. Axtell, S. M. Swensen, L. M. Prince, W. J. Kress, K. C. Nixon, and J. S. Farris. 2000. Angiosperm phylogeny inferred from 18S rDNA, *rbcL*, and *atpB* sequences. *Bot. J. Linn. Soc.* 133:381–461.

Takhtajan, A. 1997. *Diversity and Classification of Flowering Plants.* Columbia University Press, New York.

Author

Richard G. Olmstead; Department of Biology; University of Washington; Seattle, WA 98195, USA. Email: olmstead@u.washington.edu.

Date Accepted: 21 May 2011; updated 18 July 2017

Primary Editor: Philip Cantino

Convolvulaceae Jussieu 1789: 132 [S. Stefanović and R. G. Olmstead], converted clade name

Registration Number: 32

Definition: The smallest crown clade containing *Convolvulus arvensis* L. 1753 and *Humbertia madagascariensis* Lam. 1786. This is a minimum-crown-clade definition. Abbreviated definition: min crown ∇ (*Convolvulus arvensis* L. 1753 & *Humbertia madagascariensis* Lam. 1786).

Etymology: Derived from *Convolvulus* (name of an included genus), which is from the Latin *convolvere*, to twine (Fernald, 1970).

Reference Phylogeny: Stefanović et al. (2002: Fig. 1). See also Stefanović and Olmstead (2004: Figs. 1, 3).

Composition: *Humbertieae, Cardiochlamyeae, Erycibeae, Dichondreae, Cresseae, Maripeae, Jacquemontieae, Cuscuteae, Aniseieae, Convolvuleae,* "*Merremieae,*" and *Ipomoeeae* (Stefanović et al., 2003).

Diagnostic Apomorphies: The deletion of an intron in the *rpl2* gene of the chloroplast genome is an unambiguous molecular synapomorphy for the clade (Stefanović et al., 2002), although this deletion also is found in other distantly related angiosperms. The presence of articulated latex canals and associated latex cells and milky sap is a probable anatomical synapomorphy (Stefanović et al., 2003).

Synonyms: None.

Comments: In traditional classifications (e.g., Cronquist, 1981; Takhtajan, 1997), the name *Convolvulaceae* has been applied to this group with the exclusion of *Cuscuta*, which is commonly placed in its own family *Cuscutaceae*. The monophyly of *Convolvulaceae* including *Cuscuta* is strongly supported by molecular studies based on chloroplast and mitochondrial DNA sequences (Stefanović et al., 2002; Stefanović and Olmstead, 2004). The clade includes three groups, sometimes segregated as families in traditional classifications (*Humbertiaceae, Dichondraceae,* and *Cuscutaceae*). A phylogenetic definition identical to the one adopted here was proposed by Stefanović et al. (2003). The position of *Cuscuta* is still not clearly resolved within *Convolvulaceae,* although tests reject its placement outside *Convolvulaceae* (Stefanović and Olmstead, 2004).

Literature Cited

Cronquist, A. 1981. *An Integrated System of Classification of Flowering Plants.* Columbia University Press, New York.

Fernald, M. L. 1970. *Gray's Manual of Botany.* 8th edition, corrected printing. Van Nostrand, New York.

Jussieu, A. L. 1789. *Genera Plantarum.* Apud Viduam Herissant et Theophilum Barrois, Paris.

Stefanović, S., D. F. Austin, and R. G. Olmstead. 2003. Classification of *Convolvulaceae*: a phylogenetic approach. *Syst. Bot.* 28:791–806.

Stefanović, S., L. E. Krueger, and R. G. Olmstead. 2002. Monophyly of the *Convolvulaceae* and circumscription of their major lineages based on DNA sequences of multiple chloroplast loci. *Am. J. Bot.* 89:1510–1522.

Stefanović, S., and R. G. Olmstead. 2004. Testing the phylogenetic position of a parasitic plant (*Cuscuta, Convolvulaceae, Asteridae*): Bayesian inference and the parametric bootstrap on data drawn from three genomes. *Syst. Biol.* 53:384–399.

Takhtajan, A. 1997. *Diversity and Classification of Flowering Plants*. Columbia University Press, New York.

Authors

Saša Stefanović; Department of Biology; University of Toronto Mississauga; 3359 Mississauga Rd N; Mississauga, ON, Canada, L5L 1C6. Email: sasa.stefanović@utoronto.ca.

Richard G. Olmstead; Department of Biology; University of Washington; Seattle, WA 98195; USA. Email: olmstead@u.washington.edu.

Date Accepted: 15 October 2010; updated 17 July 2017

Primary Editor: Philip Cantino

Lamiales E. F. Bromhead 1838: 210 [R. G. Olmstead and D. C. Tank], converted clade name.

Registration Number: 57

Definition: The smallest crown clade containing *Lamium album* L. 1753, *Olea europaea* L. 1753, and *Plocosperma buxifolium* Benth. 1876. This is a minimum-crown-clade definition. Abbreviated definition: min crown ∇ (*Lamium album* L. 1753 & *Olea europaea* L. 1753 & *Plocosperma buxifolium* Benth. 1876).

Etymology: Derived from *Lamium* (name of an included genus), referring to the throat-like corolla tube (DeFelice, 2005). The genus name was presumably derived from the Greek *laimos* (throat or gullet) (Brown, 1954).

Reference Phylogeny: Refulio and Olmstead (2014: Fig. 1), where the closest relative of the specifier *Lamium album* is *Lamium purpureum* See also Schäferhoff et al. (2010: Fig. 2) and Soltis et al. (2011: Figs. 1–2).

Composition: *Acanthaceae* (Scotland and Vollesen, 2000; Schwarzbach and McDade, 2002), *Bignoniaceae* (Fischer et al., 2004b), *Byblidaceae* (Conran and Carolin, 2004), *Calceolariaceae* (Tank et al., 2006), *Carlemanniaceae* (Thiv, 2004), *Gesneriaceae* (Weber, 2004), *Labiatae* (Harley et al., 2004), *Lentibulariaceae* (Fischer et al., 2004a, *Linderniaceae* (Tank et al., 2006), *Martyniaceae* (Ihlenfeldt, 2004a), *Oleaceae* (Green, 2004), *Orobanchaceae* (Tank et al., 2006), *Paulowniaceae* (Nakai, 1949), *Pedaliaceae* (Ihlenfeldt, 2004b), *Phrymaceae* (Tank et al., 2006), *Plantaginaceae* (synonym: *Veronicaceae*) (Tank et al., 2006), *Plocospermataceae* (Struwe and Jensen, 2004), *Schlegeliaceae* (as tribe *Schlegelieae*, Gentry, 1980), *Scrophulariaceae* (Tank et al., 2006), *Stilbaceae* (Tank et al., 2006), *Tetrachondraceae* (Wagstaff, 2004), *Thomandersiaceae* (Wortley et al., 2007), and *Verbenaceae* (Atkins, 2004).

Diagnostic Apomorphies: No non-DNA synapomorphies discovered.

Synonyms: *Scrophulariales* sensu Thorne (1992), but not Cronquist (1981) or others, is an approximate synonym based on a list of included families, which is very similar to the composition described here but excludes *Plocospermataceae* and *Tetrachondraceae*.

Comments: The name *Lamiales* often has been narrowly applied to a smaller group of plants characterized by a reduced number of ovules (1–2) per carpel (e.g., Cronquist 1981; Takhtajan 1997), which has been shown through molecular phylogenetic study to be a polyphyletic assemblage (e.g., Wagstaff and Olmstead, 1997). *Scrophulariales* also has been applied to a non-monophyletic assemblage of varying circumscription, but only once (Thorne, 1992) to a group circumscribed similarly to *Lamiales* as defined here. Use of *Scrophulariales* in this context never came into common use, and Thorne has used *Lamiales* in subsequent classifications (Thorne, 2000; Thorne and Reveal, 2007).

This clade was recognized by Olmstead et al. (1992, 1993) based on plastid *rbcL* sequences and given the name *Lamiales* (Olmstead et al., 1993). Numerous subsequent molecular phylogenetic studies have corroborated this clade with strong support (Olmstead et al., 2000, 2001; Savolainen et al., 2000a,b; Soltis et al., 2000; Albach et al., 2001; Kårehed, 2001; Bremer et al., 2002; Hilu et al., 2003; Lee et al., 2007; Schäferhoff et al., 2010).

Literature Cited

Albach, D. C., P. S. Soltis, D. E. Soltis, and R. G. Olmstead. 2001. Phylogenetic analysis of asterids based on sequences of four genes. *Ann. Mol. Bot. Gard.* 88:163–212.

Atkins, S. 2004. *Verbenaceae*. Pp. 449–468 in *The Families and Genera of Vascular Plants* (K. Kubitzki, ed.). Vol. 7 (J. W. Kadereit, ed.). Springer, Berlin.

Bremer, B., K. Bremer, N. Heidari, P. Erixon, R. G. Olmstead, A. A. Anderberg, M. Källersjö, and E. Barkhordarian. 2002. Phylogenetics of asterids based on 3 coding and 3 non-coding chloroplast DNA markers and the utility of non-coding DNA at higher taxonomic levels. *Mol. Phylogenet. Evol.* 24:274–301.

Bromhead, E. F. 1838. Remarks on the affinities of *Lythraceae* and *Vochyaceae. Mag. Nat. Hist., N.S.* 2:210–214.

Brown, R. W. 1954. *Composition of Scientific Words.* Smithsonian Books, Washington, DC.

Conran, J. G., and R. Carolin. 2004. *Byblidaceae.* Pp. 45–49 in *The Families and Genera of Vascular Plants* (K. Kubitzki, ed.). Vol. 7 (J. W. Kadereit, ed.). Springer, Berlin.

Cronquist, A. 1981. *An Integrated System of Classification of Flowering Plants.* Columbia University Press, New York.

DeFelice, M. S. 2005. Henbit and the Deadnettles, *Lamium* spp.—archangels or demons? *Weed Technol.* 19:768–774.

Fischer, E., W. Barthlott, R. Seine, and I. Theisen. 2004a. *Lentibulariaceae.* Pp. 276–282 in *The Families and Genera of Cascular Plants* (K. Kubitzki, ed.). Vol. 7 (J. W. Kadereit, ed.). Springer, Berlin.

Fischer, E., I. Theisen, and L. G. Lohmann. 2004b. *Bignoniaceae.* Pp. 9–38 in *The Families and Genera of Vascular Plants* (K. Kubitzki, ed.). Vol. 7 (J. W. Kadereit, ed.). Springer, Berlin.

Gentry, A. H. 1980. *Bignoniaceae* Part I—Tribes *Crescentieae* and *Tourretieae. Flora Neotrop. Monogr.* 25:1–131.

Green, P. S. 2004. *Oleaceae.* Pp. 296–306 in *The Families and Genera of Vascular Plants* (K. Kubitzki, ed.). Vol. 7 (J. W. Kadereit, ed.). Springer, Berlin.

Harley, R. M., S. Atkins, A. L. Budantsev, P. D. Cantino, B. J. Conn, R. Grayer, M. M. Harley, R. de Kok, T. Krestovskaja, R. Morales, A. J. Paton, O. Ryding, and T. Upson. 2004. *Labiatae.* Pp. 167–275 in *The Families and Genera of Vascular Plants* (K. Kubitzki, ed.). Vol. 7 (J. W. Kadereit, ed.). Springer, Berlin.

Hilu, K. W., T. Borsch, K. Müller, D. E. Soltis, P. S. Soltis, V. Savolainen, M. W. Chase, M. P. Powell, L. A. Alice, R. Evans, H. Sauquet, C. Neinhuis, T. A. B. Slotta, J. G. Rohwer, C. S. Campbell, and L. W. Chatrou. 2003. Angiosperm phylogeny based on *matK* sequence information. *Am. J. Bot.* 90:1758–1776.

Ihlenfeldt, H.-D. 2004a. *Martyniaceae.* Pp. 283–288 in *The Families and Genera of Vascular Plants* (K. Kubitzki, ed.). Vol. 7 (J. W. Kadereit, ed.). Springer, Berlin.

Ihlenfeldt, H.-D. 2004b. *Pedaliaceae.* Pp. 307–322 in *The Families and Genera of Vascular Plants* (K. Kubitzki, ed.). Vol. 7 (J. W. Kadereit, ed.). Springer, Berlin.

Kårehed, J. 2001. Multiple origin of the tropical forest tree family *Icacinaceae. Am. J. Bot.* 88:2259–2274.

Lee, H.-L., R. K. Jansen, T. W. Chumley, and K.-J. Kim. 2007. Gene relocations within chloroplast genomes of *Jasminum* and *Menodora (Oleaceae)* are due to multiple, overlapping inversions. *Mol. Biol. Evol.* 24: 1161–1180.

Nakai, T. 1949. Classes, ordinae, familiae, subfamileae, tribus, genera nova quae attinent ad plantas Koreanas. *J. Jpn. Bot.* 24: 8–14.

Olmstead, R. G., B. Bremer, K. M. Scott, and J. D. Palmer. 1993. A parsimony analysis of the *Asteridae* sensu lato based on *rbcL* sequences. *Ann. Mo. Bot. Gard.* 80:700–722.

Olmstead, R. G., C. W. dePamphilis, A. D. Wolfe, N. D. Young, W. J. Elisons, and P. A. Reeves. 2001. Disintegration of the *Scrophulariaceae. Am. J. Bot.* 88:348–361.

Olmstead, R. G., K. Kim, R. K. Jansen, and S. J Wagstaff. 2000. The phylogeny of the *Asteridae* sensu lato based on chloroplast *ndhF* gene sequences. *Mol. Phylogenet. Evol.* 16: 96–112.

Olmstead, R. G., H. J. Michaels, K. M. Scott, and J. D. Palmer. 1992. Monophyly of the *Asteridae* and identification of their major lineages inferred from DNA sequences of *rbcL*. *Ann. Mo. Bot. Gard.* 79:249–265.

Refulio, N. F., and R. G. Olmstead. 2014. Phylogeny of *Lamiidae*. *Am. J. Bot.* 101:287–299.

Savolainen, V., M. W. Chase, S. B. Hoot, C. M. Morton, D. E. Soltis, D. Bayer, M. F. Fay, A. Y. de Bruijn, S. Sullivan, and Y.-L. Qiu. 2000a. Phylogenetics of flowering plants based on combined analysis of plastid *atpB* and *rbcL* gene sequences. *Syst. Biol.* 49:306–362.

Savolainen, V., M. F. Fay, D. C. Albach, A. Bachlund, M. van der Bank, K. M. Cameron, S. A. Johnson, M. D. Lledó, J.-C. Pintaud, M. Powell, M. C. Sheahan, D. E. Soltis, P. S. Soltis, P. Weston, W. M. Whitten, K. J. Wurdack, and M. W. Chase. 2000b. Phylogeny of the eudicots: a nearly complete familial analysis based on *rbcL* gene sequences. *Kew Bull.* 55:257–309.

Schäferhoff, B., A. Fleischmann, E. Fischer, D. C. Albach, T. Borsch, G. Heubl, and K. Müller. 2010. Towards resolving *Lamiales* relationships: insights from rapidly evolving chloroplast sequences. *BMC Evol. Biol.* 10:352.

Schwarzbach, A. E., and L. A. McDade. 2002. Phylogenetic relationships of the mangrove family *Avicenniaceae* based on chloroplast and nuclear ribosomal DNA sequences. *Syst. Bot.* 27:84–98.

Scotland, R. W., and K. Vollesen. 2000. Classification of *Acanthaceae*. *Kew Bull.* 55:513–589.

Soltis, D. E., S. A. Smith, N. Cellinese, K. J. Wurdack, D. C. Tank, S. F. Brockington, N. F. Refulio-Rodriguez, J. B. Walker, M. J. Moore, B. S. Carlsward, C. D. Bell, M. Latvis, S. Crawley, C. Black, D. Diouf, Z. Xi., M. A. Gitzendanner, K. J. Sytsma, Y.-L. Qiu, K. W. Hilu, S. R. Manchester, C. C. Davis, M. J. Sanderson, R. G. Olmstead, W. S. Judd, M. J. Donoghue, and P. S. Soltis. 2011. Angiosperm phylogeny: 17 genes, 640 taxa. *Am. J. Bot.* 98:704–730.

Soltis, D. E., P. S. Soltis, M. W. Chase, M. E. Mort, D. C. Albach, M. Zanis, V. Savolainen, W. H. Hahn, S. B. Hoot, M. F. Fay, M. Axtell, S. M. Swensen, L. M. Prince, W. J. Kress, K. C. Nixon, and J. S. Farris. 2000. Angiosperm phylogeny inferred from 18S rDNA, *rbcL*, and *atpB* sequences. *Bot. J. Linn. Soc.* 133:381–461.

Struwe, L., and S. R. Jensen. 2004. *Plocospermataceae*. Pp. 330–333 in *The Families and Genera of Vascular Plants* (K. Kubitzki, ed.). Vol. 7 (J. W. Kadereit, ed.). Springer, Berlin.

Takhtajan, A. 1997. *Diversity and Classification of Flowering Plants*. Columbia University Press, New York.

Tank, D. C., P. Beardsley, S. Kelchner, and R. G. Olmstead. 2006. L.A.S. Johnson Review No, 7. Review of the systematics of *Scrophulariaceae* s.l. and their current disposition. *Aust. Syst. Bot.* 19:289–307.

Thiv, M. 2004. *Carlemanniaceae*. Pp. 57–59 in *The Families and Genera of Vascular Plants* (K. Kubitzki, ed.). Vol. 7 (J. W. Kadereit, ed.). Springer, Berlin.

Thorne, R. F. 1992. Classification and geography of the flowering plants. *Bot. Rev.* 58:225–348.

Thorne, R. F. 2000. The classification and geography of the flowering plants: dicotyledons of the class *Angiospermae* (subclasses *Magnoliidae*, *Ranunculidae*, *Caryophyllidae*, *Dilleniidae*, *Rosidae*, *Asteridae*, and *Lamiidae*). *Bot. Rev.* 66:441–647.

Thorne, R. F., and J. L. Reveal. 2007. An updated classification of the class *Magnoliopsida* ("*Angiospermae*"). *Bot. Rev.* 73:67–182.

Wagstaff, S. J. 2004. *Tetrachondraceae*. Pp. 441–444 in *The Families and Genera of Vascular Plants* (K. Kubitzki, ed.). Vol. 7 (J. W. Kadereit, ed.). Springer, Berlin.

Wagstaff, S. J., and R. G. Olmstead. 1997. Phylogeny of *Labiatae* and *Verbenaceae* inferred from *rbcL* sequences. *Syst. Bot.* 22:165–179.

Weber, A. 2004. *Gesneriaceae*. Pp. 63–158 in *The Families and Genera of Vascular Plants* (K. Kubitzki, ed.). Vol. 7 (J. W. Kadereit, ed.). Springer, Berlin.

Wortley, A. H., D. J. Harris, and R. W. Scotland. 2007. On the taxonomy and phylogenetic position of *Thomandersia. Syst. Bot.* 32:415–444.

Authors

Richard G. Olmstead; Department of Biology; University of Washington; Seattle, WA 98195; USA. Email: olmstead@u.washington.edu.

David C. Tank; Department of Forest Resources; College of Natural Resources; University of Idaho; PO Box 441133; Moscow, ID 83844-1133; USA. Email: dtank@uidaho.edu.

Date Accepted: 27 December 2011; updated 18 July 2017

Primary Editor: Philip Cantino

Bignoniaceae Jussieu 1789: 137 [R. G. Olmstead and L. G. Lohmann], converted clade name

Registration Number: 22

Definition: The smallest crown clade containing *Bignonia capreolata* Linnaeus 1753, *Jacaranda mimosifolia* D. Don 1822, and *Eccremocarpus scaber* Ruiz and Pavon 1794. This is a minimum-crown-clade definition. Abbreviated definition: min crown ∇ (*Bignonia capreolata* Linnaeus 1753 & *Jacaranda mimosifolia* D. Don 1822 & *Eccremocarpus scaber* Ruiz and Pavon 1794).

Etymology: Derived from *Bignonia* (name of an included genus), named in honor of Jean-Paul Bignon (1662–1743), the French royal librarian (Fernald, 1970).

Reference Phylogeny: Olmstead et al. (2009: Fig. 1). See also Spangler and Olmstead (1999) and Grose and Olmstead (2007).

Composition: *Bignonieae, Catalpeae, Coleeae, Crescentieae, Jacarandeae, Oroxyleae, Tecomeae,* and *Tourretieae* as circumscribed in Zjhra et al. (2004), Lohmann (2006), Grose and Olmstead (2007), and Olmstead et al. (2009), plus several taxa assigned to the non-monophyletic *Tecomeae* sensu Fischer et al. (2004).

Diagnostic Apomorphies: A suite of vegetative and reproductive traits help characterize *Bignoniaceae*. Of these, the existence of four didynamous stamens and a staminode, winged seeds, seeds with reduced endosperm, two placentae per carpel, and bilamellate and sensitive stigmas are probable synapomorphies (Gentry, 1980; Armstrong, 1985; Manning, 2000; Fischer et al., 2004; Judd et al., 2007). A six base pair repeat in the non-coding spacer between the *trnL* and *trnF* genes in the plastid genome provides a molecular synapomorphy for *Bignoniaceae* (Olmstead et al., 2009).

Synonyms: None.

Comments: The name *Bignoniaceae* long has been associated with this group in traditional classifications (e.g., Cronquist, 1981; Takhtajan, 1997). The four small genera comprising *Schlegeliaceae* (*Exarata, Gibsoniothamnus, Schlegelia, Synapsis*) often have been included in *Bignoniaceae* (e.g., Cronquist, 1981; Takhtajan, 1987, but not 1997). However, molecular evidence for a relationship between these clades is equivocal. One study found weak support for a sister group relationship (Olmstead et al., 2001), but most studies (e.g., Spangler and Olmstead, 1999; Oxelman et al., 1999, 2005; Bremer et al., 2002; Olmstead et al., 2009) do not place *Schlegelia* with *Bignoniaceae*. Similarly, *Paulownia* has been assigned to *Bignoniaceae* in some classifications (e.g., Cronquist, 1981), but phylogenetic studies reject that relationship (e.g., Spangler and Olmstead, 1999; Olmstead et al., 2001). Parsimony bootstrap support for monophyly of *Bignoniaceae* based on cpDNA (Spangler and Olmstead, 1999; Olmstead et al., 2009) is modest, but Bayesian analysis of the same data, along with the suite of morphological synapomorphies and the repeat in the *trn*LF spacer, provide additional confidence in their monophyly.

Literature Cited

Armstrong, J. E. 1985. The delimitation of *Bignoniaceae* and *Scrophulariaceae* based on floral anatomy, and the placement of problem genera. *Am. J. Bot.* 72:755–766.

Bremer, B., K. Bremer, N. Heidari, P. Erixon, R. G. Olmstead, A. A. Anderberg, M. Källersjö, and E. Barkhordarian. 2002. Phylogenetics of asterids based on 3 coding and 3 non-coding chloroplast DNA markers and the utility of non-coding DNA at higher taxonomic levels. *Mol. Phylogenet. Evol.* 24:274–301.

Cronquist, A. 1981. *An Integrated System of Classification of Flowering Plants.* Columbia University Press, New York.

Fernald, M. L. 1970. *Gray's Manual of Botany.* 8th edition, corrected printing. Van Nostrand, New York.

Fischer, E., I. Theisen, and L. G. Lohmann. 2004. *Bignoniaceae.* Pp. 9–38 in *The Families and Genera of Vascular Plants* (K. Kubitzki, ed.). Vol. 7 (J. W. Kadereit, ed.). Springer, Berlin.

Gentry, A. H. 1980. *Bignoniaceae* part I—tribes *Crescentieae* and *Tourretieae. Flora Neotrop. Monogr.* 25:1–131.

Grose, S. O., and R. G. Olmstead. 2007. Evolution of a charismatic neotropical clade: molecular phylogeny of *Tabebuia* s.l., *Crescentieae*, and allied genera (*Bignoniaceae*). *Syst. Bot.* 32:650–659.

Judd, W. S., C. S. Campbell, E. A. Kellogg, P. F. Stevens, and M. J. Donoghue. 2007. *Plant Systematics—A Phylogenetic Approach.* 3rd edition. Sinauer Associates, Sunderland, MA.

Jussieu, A. L. 1789. *Genera Plantarum.* Paris.

Lohmann, L. G. 2006. Untangling the phylogeny of neotrpropical lianas (*Bignonieae*, *Bignoniaceae*). *Am. J. Bot.* 93:304–318.

Manning, S. D. 2000. The genera of *Bignoniaceae* in the southeastern United States. *Harv. Papers Bot.* 5:1–77.

Olmstead, R. G., C. W. de Pamphilis, A. D. Wolfe, N. D. Young, W. J. Elisens, and P. A. Reeves. 2001. Disintegration of the *Scrophulariaceae. Am. J. Bot.* 88:348–361.

Olmstead, R. G., M. L. Zjhra, L. G. Lohmann, S. O. Grose, and A. J. Eckert. 2009. A molecular phylogeny and classification of *Bignoniaceae. Am. J. Bot.* 96:1731–1743.

Oxelman, B., M. Backlund, and B. Bremer. 1999. Relationships of the *Buddlejaceae* s.l. investigated using parsimony jackknife and branch support analysis of chloroplast *ndhF* and *rbcL* sequence data. *Syst. Bot.* 24:164–182.

Oxelman, B., P. Kornhall, R. G. Olmstead, and B. Bremer. 2005. Further disintegration of *Scrophulariaceae. Taxon* 54:411–425.

Spangler, R. E., and R. G. Olmstead. 1999. Phylogenetic analysis of *Bignoniaceae* based on the cpDNA gene sequences of *rbcL* and *ndhF. Ann. Mo. Bot. Gard.* 86:33–46.

Takhtajan, A. 1987. *Systema Magnoliophytorum.* Nauka, Leningrad.

Takhtajan, A. 1997. *Diversity and Classification of Flowering Plants.* Columbia University Press, New York.

Zjhra, M. L., K. J. Sytsma, and R. G. Olmstead. 2004. Delimitation of tribe *Coleeae* and implications for fruit evolution in *Bignoniaceae* inferred from a chloroplast DNA phylogeny. *Plant Syst. Evol.* 245:55–67.

Authors

Richard G. Olmstead; Department of Biology and Burke Museum; University of Washington; Seattle, WA 98195; USA. Email: olmstead@u.washington.edu.

Lúcia G. Lohmann; Universidade de São Paulo; Instituto de Biociências; Departmamento de Botânica; Rua do Matão, 277, 05508-090; São Paulo, SP; Brazil. Email: llohmann@usp.br.

Date Accepted: 23 August 2010; updated 17 July 2017

Primary Editor: Philip Cantino

Orobanchaceae E. P. Ventenant 1799: 292 [D. C. Tank, A. D. Wolfe, S. Mathews, and R. G. Olmstead], converted clade name

Registration Number: 73

Definition: The largest crown clade containing *Orobanche major* L. 1753, but not *Phryma leptostachya* L. 1753 and *Paulownia tomentosa* (Thunb.) Steud. 1841 and *Mazus japonicus* (Thunb.) Kuntze 1891. This is a maximum-crown-clade definition. Abbreviated definition: max crown ∇ (*Orobanche major* L. 1753 ~ *Phryma leptostachya* L. 1753 & *Paulownia tomentosa* (Thunb.) Steud. 1841 & *Mazus japonicus* (Thunb.) Kuntze 1891).

Etymology: From the Greek *orobos*, a name for a kind of vetch, and *anchein*, "to strangle," referring to its parasitic nature (Quattrocchi, 2000).

Reference Phylogeny: Xia et al. (2009: Fig. 4), where *Orobanchaceae* is labeled "Clade G". See also McNeal et al. (2013: Fig. 3) for additional detail of relationships within *Orobanchaceae*, Albach et al. (2009: Fig. 3) and Refulio-Rodriguez and Olmstead (2014: Fig. 1) for the inclusion of *Rehmanniaceae* (*Rehmannia* and *Triaenophora*), and Morawetz et al. (2010: Fig. 2) for the inclusion of *Cyclocheilaceae* (*Asepalum* and *Cyclocheilon*) as the sister group of the remainder of the *Buchnereae* clade of McNeal et al. (2013). The specifier *Orobanche major* is not included in the primary reference phylogeny but is related to other species of *Orobanche* that are included (see Comments).

Composition: This clade includes the following taxa: *Orobanchaceae* (sensu Table 1 in Wolfe et al., 2005, which is slightly less inclusive than the clade defined here), *Rehmanniaceae* (Albach et al., 2009; Xia et al., 2009; Reveal, 2011; Refulio-Rodriguez and Olmstead, 2014; see Li et al. [2011] for taxonomic treatment of *Rehmannia*; not considered part of *Orobanchaceae* sensu Wolfe et al., [2005]), *Cyclocheilaceae* (Bremer et al., 2002; Oxelman et al., 2005; Morawetz et al., 2010; see Demissew [2004] for taxonomic treatment; not considered part of *Orobanchaceae* sensu Wolfe et al., [2005]), *Nesogenes* (Bennett and Mathews, 2006; Morawetz et al., 2010; see Harley [2004] for taxonomic treatment; not considered part of *Orobanchaceae* sensu Wolfe et al., [2005]), and *Brandisia* (Oxelman et al., 2005; Bennett and Mathews, 2006; not considered part of *Orobanchaceae* sensu Wolfe et al., [2005]).

Diagnostic Apomorphies: The rhinanthoid corolla organization, in which the abaxial corolla lobes are outside the adaxial lobes, capsules that are half or partly exserted from the persistent calyx tubes, iridoid glycosides lacking harpagide and 6-rhamnopyranosyl-catalpol and their esters, and the presence of two lateral bracteoles borne at the flower pedicel just above the leaf-like subtending bract may be synapomorphies for the entire clade (Olmstead et al., 2001; Wolfe et al., 2005; Xia et al., 2009) but require more thorough investigation throughout *Orobanchaceae* to confirm.

Synonyms: None (but see Comments regarding *Rhinanthoideae* sensu Bellini).

Comments: In most traditional classifications, the holoparasitic, nonphotosynthetic species that comprise a subset of the clade named here have been recognized as the family *Orobanchaceae* (Stebbins, 1974; Cronquist,

1981) or as a subfamily (*Orobanchoideae*) within *Scrophulariaceae* (Thorne, 1992; Takhtajan, 1997, 2009). Molecular phylogenetic studies have shown that the clade whose name is defined here is separate from *Scrophulariaceae*, although it includes elements assigned to *Scrophulariaceae* in traditional classifications. Bellini (1907) recognized the relationship among the known hemiparasitic members of *Scrophulariaceae* and the holoparasitic species in his *Rhinanthoideae* of *Scrophulariaceae*. However, *Rhinanthoideae* sensu Bellini differed from *Orobanchaceae* (as defined here) by excluding the nonparasitic *Lindenbergia, Rehmannia,* and *Triaenophora,* as well as several hemiparasitic taxa that are now known to belong to this clade (e.g., *Brandisia, Cyclocheilaceae, Nesogenes*; Bremer et al., 2002; Oxelman et al., 2005; Bennett and Mathews, 2006; Morawetz et al., 2010). In contrast, molecular studies have used the name *Orobanchaceae* consistently for the clade containing all hemi- and holoparasitic lineages and the nonparasitic *Lindenbergia* (e.g., Olmstead et al., 2001; Oxelman et al., 2005; Refulio-Rodriguez and Olmstead, 2014), and the clade *Rehmanniaceae* (*Rehmannia* and *Triaenophora*) is consistently sister to this clade (Albach et al., 2009; Xia et al., 2009; Schäferhoff et al., 2010; Reveal, 2011; Refulio-Rodriguez and Olmstead, 2014). Young et al. (1999) provided a minimum-crown-clade definition for *Orobanchaceae* that used species of *Orobanche, Lindenbergia,* and *Schwalbea* as specifiers. However, we prefer to apply a maximum-crown-clade definition in light of the strength of support for close outgroup relationships (Xia et al., 2009; Refulio-Rodriguez and Olmstead, 2014) and to accommodate phylogenetic uncertainty within *Orobanchaceae* among molecular systematic studies (e.g., dePamphilis et al., 1997; Wolfe and dePamphilis, 1998; Young et al., 1999; Olmstead et al., 2001; Wolfe et al., 2005; Bennett and Mathews, 2006; Park et al., 2008; Morawetz et al., 2010; McNeal et al., 2013).

Phylogenetic studies have shown that the loss of photosynthesis (i.e., holoparasitism) in this clade followed the origin of parasitism and has occurred at least five times independently, thus the group circumscribed in traditional classifications under the name *Orobanchaceae* or *Orobanchoideae* is polyphyletic (dePamphilis et al., 1997; Wolfe and dePamphilis, 1998; Young et al., 1999; McNeal et al., 2013). Molecular studies using plastid DNA regions (*matK* and *rps2,* Young et al., 1999; *rbcL, rps2, ndhF,* Olmstead et al., 2001; *rps2,* Park et al., 2008) support the monophyly of both a parasitic clade and a more inclusive clade within which *Lindenbergia* is sister to the parasitic clade. However, support for this more inclusive clade including *Lindenbergia* was stronger in all studies (97% bootstrap support, Young et al., 1999; 99% bootstrap support, Olmstead et al., 2001; 86% bootstrap support, Park et al., 2008) than the support for the parasitic clade alone (92% bootstrap support, Young et al., 1999; 75% bootstrap support, Olmstead et al., 2001; 65% bootstrap support, Park et al., 2008). In a study of the nuclear Phytochrome A gene (Bennett and Mathews, 2006), a group of hemiparasitic taxa, including *Schwalbea,* forms a clade with *Lindenbergia* instead of with the other parasites, albeit with low bootstrap support (67%). Likewise, relationships among the major subclades of *Orobanchaceae* as defined here (including both *Schwalbea* and *Lindenbergia*) received little to no bootstrap support (< 50%) using nuclear rDNA ITS sequences (Wolfe et al., 2005). McNeal et al. (2013) expanded the sampling of Bennett and Mathews (2006) and Wolfe et al. (2005) with a five-gene study using the plastid DNA regions *matK* and *rps2,* nuclear rDNA ITS sequences, and the nuclear Phytochrome A and Phytochrome B genes. Phylogenetic analysis of this five-gene dataset recovered the same major subclades of *Orobanchaceae* as defined here, but with increased nodal support throughout, including resolving the non-parasitic *Lindenbergia* as the sister group of the parasitic members of the clade with strong bootstrap support. Unfortunately, none of these studies included the non-parasitic *Rehmannia* or *Triaenophora* in their phylogenetic analyses, despite the strongly supported sister-group relationship of this lineage to the remainder of *Orobanchaceae* as defined here (Albach et al., 2009; Xia et al., 2009). Furthermore, several putative synapomorphies including corolla organization, secondary chemistry, and inflorescence morphology (Olmstead et al., 2001; Wolfe et al., 2005; Xia et al., 2009) support the inclusion of *Rehmanniaceae* in *Orobanchaceae* as defined here.

Orobanche major was not included in the sampling of Xia et al. (2009) or McNeal et al. (2013), but it is used as a specifier in accordance with the *ICPN* (Cantino and de Queiroz, 2020; Art. 11.10), because it is the type of *Orobanchaceae* in rank-based nomenclature. It has been shown to be closely related to the species of *Orobanche* included in the primary reference phylogeny (Schneeweiss et al., 2004; Park et al., 2008).

Literature Cited

Albach, D. C., K. Yan, S. R. Jensen, and H.-Q. Li. 2009. Phylogenetic placement of *Triaenophora* (formerly *Scrophulariaceae*) with some implications for the phylogeny of *Lamiales*. *Taxon* 58:749–756.

Bellini, R. 1907. Criteri per una nuova classificazione delle *Personatae* (*Scrophulariaceae* et *Rhinanthaceae*). *Ann. Bot. (Rome)* 6:131–145.

Bennett, J. R., and S. Mathews. 2006. Phylogeny of the parasitic plant family *Orobanchaceae* inferred from phytochrome A. *Am. J. Bot.* 93:1039–1051.

Bremer, B., K. Bremer, N. Heidari, P. Erixon, R. G. Olmstead, A. A. Anderberg, M. Källersjö, and E. Barkhordarian. 2002. Phylogenetics of asterids based on 3 coding and 3 non-coding chloroplast DNA markers and the utility of non-coding DNA at higher taxonomic levels. *Mol. Phylogenet. Evol.* 24:274–301.

Cantino, P. D., and K. de Queiroz. 2020. *International Code of Phylogenetic Nomenclature (PhyloCode)*, Version 6. CRC Press, Boca Raton, FL.

Cronquist, A. 1981. *An Integrated System of Classification of Flowering Plants*. Columbia University Press, New York.

Demissew, S. 2004. *Cyclocheilaceae*. Pp. 60–62 in *The Families and Genera of Vascular Plants* (K. Kubitzki, ed.). Vol. 7 (J. W. Kadereit, ed.). Springer, Berlin.

dePamphilis, C. W., N. D. Young, and A. D. Wolfe. 1997. Evolution of plastid gene *rps2* in a lineage of hemiparasitic and holoparasitic plants: many losses of photosynthesis and complex patterns of rate variation. *Proc. Nat. Acad. Sci. USA* 94:7362–7372.

Harley, R. M. 2004. *Nesogenaceae*. Pp. 293–295 in *The Families and Genera of Vascular Plants* (K. Kubitzki, ed.). Vol. 7 (J. W. Kadereit, ed.). Springer, Berlin.

Li, X.-D., Y.-Y. Zan, M.-M. Luo, H.-T. Liu, and J.-Q. Li. 2011. Taxonomic revision of the genus *Rehmannia*. *Plant Sci. J.* 29:423–431.

McNeal, J. R., J. R. Bennett, A. D. Wolfe, and S. Mathews. 2013. Phylogeny and origins of holoparasitism in *Orobanchaceae*. *Am. J. Bot.* 100:971–983.

Morawetz, J. J., C. P. Randle, and A. D. Wolfe. 2010. Phylogenetic relationships within the tropical clade of *Orobanchaceae*. *Taxon* 59:416–426.

Olmstead, R. G., C. W. dePamphilis, A. D. Wolfe, N. D. Young, W. J. Elisons, and P. A. Reeves. 2001. Disintegration of the *Scrophulariaceae*. *Am. J. Bot.* 88:348–361.

Oxelman, B., P. Kornhall, R. G. Olmstead, and B. Bremer. 2005. Further disintegration of *Scrophulariaceae*. *Taxon* 54:411–425.

Park, J.-M., J.-F. Manen, A. E. Colwell, and G. M. Schneeweiss. 2008. A plastid gene phylogeny of the non-photosynthetic parasitic *Orobanche* (*Orobanchaceae*) and related genera. *J. Plant Res.* 121:365–376.

Quattrocchi, U. 2000. *CRC World Dictionary of Plant Names: Common Names, Scientific Names, Eponyms, Synonyms, and Etymology*. CRC Press, Boca Raton, FL.

Refulio-Rodriguez, N. F., and R. G. Olmstead. 2014. Phylogeny of *Lamiidae. Am. J. Bot.* 101:287–299.

Reveal, J. L. 2011. Summary of recent systems of angiosperm classification. *Kew Bull.* 66:5–48.

Schäferhoff, B., A. Fleischmann, E. Fischer, D. C. Albach, T. Borsch, G. Heubl, and K. F. Müller. 2010. Towards resolving *Lamiales* relationships: insights from rapidly evolving chloroplast sequences. *BMC Evol. Biol.* 10:352.

Schneeweiss, G. M., A. E. Colwell, J.-M. Park, C.-G. Jang, and T. Stuessy. 2004. Phylogeny of holoparasitic *Orobanche (Orobanchaceae)* inferred from nuclear ITS sequences. *Mol. Phylogenet. Evol.* 30:465–478.

Stebbins, G. L. 1974. *Flowering Plants: Evolution above the Species Level*. Belknap Press of Harvard University Press, Cambridge, MA.

Takhtajan, A. 1997. *Diversity and Classification of Flowering Plants*. Columbia University Press, New York.

Takhtajan, A. 2009. *Flowering Plants*. 2nd edition. Springer.

Thorne, R. F. 1992. Classification and geography of the flowering plants. *Bot. Rev.* 58:225–348.

Ventenant, E. P. 1799. *Orobanchaceae*. Pp. 292 in *Tableau du Règne Végétal Selon la Méthode de Jussieu*, Vol. 2. J. Drisonnier, Paris.

Wolfe, A. D., and C. dePamphilis. 1998. The effect of relaxed functional constraints on the photosynthetic gene *rbcL* n photosynthetic and nonphotosynthetic parasitic plants. *Mol. Biol. Evol.* 15:1243–1258.

Wolfe, A. D., C. P. Randle, L. Liu, and K. E. Steiner. 2005. Phylogeny and biogeography of *Orobanchaceae. Fol. Geobot.* 40:115–134.

Xia, Z., Y.-Z. Wang, and J. F. Smith. 2009. Familial placement and relations of *Rehmannia* and *Triaenophora* (*Scrophulariaceae* s.l.) inferred from five gene regions. *Am. J. Bot.* 96:519–530.

Young, N. D., K. E. Steiner, and C. W. dePamphilis. 1999. The evolution of parasitism in *Scrophulariaceae/Orobanchaceae*: plastid gene sequences refute an evolutionary transition series. *Ann. Mo. Bot. Gard.* 86:876–893.

Authors

David C. Tank; Department of Biological Sciences and Stillinger Herbarium; University of Idaho; Moscow, ID 83844; USA. Email: dtank@uidaho.edu.

Andrea Wolfe; Department of Evolution, Ecology, and Organismal Biology; Ohio State University; Columbus, OH 43210; USA. Email: wolfe.205@osu.edu.

Sarah Mathews; Department of Biological Sciences; Louisiana State University, Baton Rouge, LA 70803, USA. Email: sarahmathews@lsu.edu.

Richard G. Olmstead; Department of Biology and Burke Museum; University of Washington; Seattle, WA 98195; USA. E-mail: olmstead@uw.edu.

Date Accepted: 25 July 2012; updated 13 September 2018

Primary Editor: Philip Cantino

Labiatae M. Adanson 1763:180 [P. D. Cantino and R. G. Olmstead], converted clade name

Registration Number: 55

Definition: The smallest crown clade containing *Lamium album* L. 1753 (*Perolamiina/Lamioideae/ Lamieae*), *Ajuga reptans* L. 1753 (*Ajugoideae*), *Premna microphylla* Turcz. 1863 (*Premnoideae*), *Tectona grandis* L.f. 1782, *Glechoma hederacea* L. 1753 (*Nepetoideae*), *Vitex agnus-castus* L. 1753 (*Viticoideae*), *Congea tomentosa* Roxb. 1820 (*Symphorematoideae*), *Callicarpa dichotoma* (Lour.) K. Koch 1872, and *Prostanthera rotundifolia* R. Br. 1810 (*Prostantheroideae*). This is a minimum-crown-clade definition. Abbreviated definition: min crown ∇ (*Lamium album* L. 1753 & *Ajuga reptans* L. 1753 & *Premna microphylla* Turcz. 1863 & *Tectona grandis* L.f. 1782 & *Glechoma hederacea* L. 1753 & *Vitex agnuscastus* L. 1753 & *Congea tomentosa* Roxb. 1820 & *Callicarpa dichotoma* (Lour.) K. Koch 1872 & *Prostanthera rotundifolia* R. Br. 1810).

Etymology: From Latin *labiatus* (lipped), referring to the organization of the corolla lobes into two lips (or less frequently one)—a characteristic of many members of this clade, but quite possibly a plesiomorphy. When Adanson (1763) coined the name *Labiatae*, he Latinized the French, les Labiées, which he attributed to Tournefort, noting that "les découpures inégales & irrégulières de leur corolle imitent comunément les deux levres de la bouche d'un animal."

Reference Phylogeny: Li et al. (2016: Figs. 1–3). For outgroup relationships, see Schäferhoff et al. (2010: Fig. 2) and Refulio-Rodriguez and Olmstead (2014: Fig. 1). *Lamium album* is used as a specifier because it is the type species of *Labiatae* (see *PhyloCode* Art. 11.10); it is part of the clade labeled *Lamieae* on Figure 2 of the reference phylogeny.

Composition: The following clades (Li et al., 2016: Fig. 1): *Perolamiina* (contains *Lamioideae*, *Cymarioideae*, *Scutellarioideae*, and *Peronematoideae*), *Ajugoideae*, *Premnoideae*, *Tectona*, *Nepetoideae* (see entry in this volume), *Viticisymphorina* (contains *Viticoideae* and *Symphorematoideae*), and *Calliprostantherina* (includes *Callicarpa* and *Prostantheroideae*). For descriptions of all the genera of *Labiatae*, see Harley et al. (2004).

Diagnostic Apomorphies: Careful studies of ovary structure (Junell, 1934) revealed a probable synapomorphy (Cantino, 1992): the carpel walls recurve into the interior of the carpel, with the ovules borne short of the carpel margins. This is the only known morphological synapomorphy, but the clade is very well supported by molecular data (see Comments).

Synonyms: *Lamiaceae* sensu Cantino and Olmstead (2004) is an unambiguous synonym and is an equally correct alternative name for the same taxon under the botanical code (Turland et al., 2018) (see Comments).

Comments: The names *Labiatae* and *Lamiaceae* are widely applied to the same taxon (originally to a non-monophyletic group but more recently to a clade; see below) and are accepted in the botanical code as alternative family names with the same type (Arts. 18.5 and 18.6; Turland et al., 2018). The *PhyloCode* does not permit alternative names for the same clade, so a choice must be made.

We have previously defined (phylogenetically) both names as applying to this clade (Cantino et al., 1997; Cantino and Olmstead, 2004) in the course of experimenting with phylogenetic nomenclature, but precedence does not apply to these names and definitions because the *PhyloCode* was not yet in effect. We prefer the name *Labiatae* because it is descriptive of the plants (and thus easier to remember) and does not have a rank-based ending. Furthermore, the informal equivalent "labiates" is widely used for this group. However, although most species of *Labiatae* have bilabiate (or at least bilaterally symmetrical) corollas, there are a few subgroups such as *Callicarpa* and *Tectona* in which the corolla is radially symmetrical and thus not "lipped".

The taxon *Labiatae*, as traditionally delimited (except by Junell [1934]), is now known to have been non-monophyletic. This discovery was based initially on a phylogenetic analysis of morphological data (Cantino, 1992) and was later strongly corroborated by a series of molecular phylogenetic analyses in which genera such as *Callicarpa* and *Clerodendrum* that were traditionally included in *Verbenaceae* were instead nested within *Labiatae* (Wagstaff and Olmstead, 1997; Wagstaff et al., 1998; Lindqvist and Albert, 2002; Steane et al., 2004; Scheen et al., 2010; Schäferhoff et al., 2010; Bendiksby et al., 2011; Refulio-Rodriguez and Olmstead, 2014; Li et al., 2016). Consequently, the taxon *Labiatae* was expanded so that it would be monophyletic (Cantino et al., 1992; Harley et al., 2004), and it is that application of the name *Labiatae* that is used here. It is unclear whether that expanded circumscription of *Labiatae* applies the name to all of the descendants of the same ancestor as the paraphyletic group to which the name had traditionally been applied, or whether the name has been moved down the tree to an older ancestor. The latter hypothesis is supported by the

cpDNA tree of Bendiksby et al (2011: Fig. 1) in which the traditionally verbenaceous *Callicarpa* is sister to the rest of *Labiatae*, but the former hypothesis is supported by more recent molecular analyses (Refulio-Rodriguez and Olmstead, 2014; Li et al., 2016) in which a *Callicarpa-Prostantheroideae* clade (which includes some taxa that have always been considered *Labiatae*) is sister to the rest of *Labiatae*. Although the monophyly of this expanded *Labiatae* and that of its major subgroups have strong molecular support, the relationships among many of those subgroups—the "backbone" of the clade *Labiatae*—are still relatively poorly supported (Li et al., 2016). The same is true of the closest outgroup relationships (Refulio-Rodriguez and Olmstead, 2014). Consequently, achieving compositional stability requires a definition with many specifiers, regardless whether a minimum-crown-clade or maximum-crown-clade definition is used. An apomorphy-modified crown-clade definition would be much simpler but is not advisable in this case because the single putative morphological synapomorphy is a microscopic character that is difficult to observe and therefore has been examined in relatively few species.

The definition used here is very similar to our 2004 definition of *Lamiaceae* (see above), but that definition included *Petraeovitex multiflora* as an additional internal specifier. There is now strong molecular support for the position of *Petraeovitex* within *Perolamiina* (Li et al., 2016), which is represented by the specifier *Lamium album*, so there is no longer a need to include *Petraeovitex multiflora* as a specifier. Similar reasoning led us to exclude here several other internal specifiers that were used in our earlier definition of *Labiatae* (Cantino et al., 1997). For example, *Physopsis spicata* and *Chloanthes parviflora*, which were used as internal specifiers in our 1997 definition, are not used here because more recent analyses (Li et al., 2016) have demonstrated that they are part of a clade (*Prosantheroideae*) that is represented by *Prostanthera rotundifolia* in the present definition.

Literature Cited

Adanson, M. 1763. *Familles des Plantes*, Part 2. Vincent, Paris.

Bendiksby, M., L. Thorbek, A. Scheen, C. Lindqvist, and O. Ryding. 2011. An updated phylogeny and classification of *Lamiaceae* subfamily *Lamioideae*. *Taxon* 60:471–484.

Cantino, P. D. 1992. Evidence for a polyphyletic origin of the *Labiatae. Ann. Mo. Bot. Gard.* 79:361–379.

Cantino, P. D., R. M. Harley, and S. J. Wagstaff. 1992. Genera of *Labiatae*: status and classification. Pp. 511–522 in *Advances in Labiate Science* (R. M. Harley and T. Reynolds, eds.). Royal Botanic Gardens, Kew.

Cantino, P. D., and R. G. Olmstead. 2004. Phylogenetic nomenclature of *Lamiaceae*. P. 13 in *Abstracts of the First International Phylogenetic Nomenclature Meeting* (M. Laurin, ed.). Paris. Available at https://hal.sorbonne-universite.fr/hal-02187647.

Cantino, P. D., R. G. Olmstead, and S. J. Wagstaff. 1997. A comparison of phylogenetic nomenclature with the current system: a botanical case study. *Syst. Bot.* 46:313–331.

Harley, R. M., S. Atkins, A. L. Budantsev, P. D. Cantino, B. J. Conn, R. Grayer, M. M. Harley, R. de Kok, T. Krestovskaja, R. Morales, A. J. Paton, O. Ryding, and T. Upson. 2004. *Labiatae*. Pp. 167–275 in *The Families and Genera of Vascular Plants* (K. Kubitzki, ed.). Vol. 7 (J. W. Kadereit, ed.). Springer, Berlin.

Junell, S. 1934. Zur Gynäceummorphologie und Systematik der Verbenaceen und Labiaten. *Symb. Bot. Upsal.* 4:1–219.

Li, B., P. D. Cantino, R. G. Olmstead, G. L. C. Bramley, C.-L. Xiang, Z.-H. Ma, Y.-H. Tan, and D.-X. Zhang. 2016. A large-scale chloroplast phylogeny of the *Lamiaceae* sheds new light on its subfamilial classification. *Sci. Rep.* 6:34343.

Lindqvist, C., and V. A. Albert. 2002. Origin of the Hawaiian endemic mints within North American *Stachys* (*Lamiaceae*). *Am. J. Bot.* 89:1709–1724.

Refulio-Rodriguez, N. F., and R. G. Olmstead. 2014. Phylogeny of *Lamiidae. Am. J. Bot.* 101:287–299.

Schäferhoff, B., A. Fleischmann, E. Fischer, D. C. Albach, T. Borsch, G. Heubl, and K. Müller. 2010. Towards resolving *Lamiales* relationships: insights from rapidly evolving chloroplast sequences. *BMC Evol. Biol.* 10, 352.

Scheen, A., M. Bendiksby, O. Ryding, C. Mathiesen, V. A. Albert, and C. Lindqvist. 2010. Molecular phylogenetics, character evolution, and suprageneric classification of *Lamioideae* (*Lamiaceae*). *Ann. Mo. Bot. Gard.* 97:191–217.

Steane, D. A., R. P. J. de Kok, and R. G. Olmstead. 2004. Phylogenetic relationships between *Clerodendrum* (*Lamiaceae*) and other ajugoid genera inferred from nuclear and chloroplast DNA sequence data. *Mol. Phylogenet. Evol.* 32:39–45.

Turland, N. J., J. H. Wiersema, F. R. Barrie, W. Greuter, D. L. Hawksworth, P. S. Herendeen, S. Knapp, W.-H. Kusber, D.-Z. Li, K. Marhold, T. W. May, J. McNeill, A. M. Monro, J. Prado, M. J. Price, and G. F. Smith, eds. 2018. *International Code of Nomenclature for Algae, Fungi, and Plants (Shenzhen Code).* Adopted by the Nineteenth International Botanical Congress, Shenzhen, China, July 2017. Regnum Vegetabile 159. Koeltz Botanical Books, Glashütten.

Wagstaff, S. J., L. Hickerson, R. Spangler, P. A. Reeves, and R. G. Olmstead. 1998. Phylogeny in *Labiatae* s.l., inferred from cpDNA sequences. *Plant Syst. Evol.* 209:265–274.

Wagstaff, S. J., and R. G. Olmstead. 1997. Phylogeny of *Labiatae* and *Verbenaceae* inferred from *rbcL* sequences. *Syst. Bot.* 22:165–179.

Authors

Philip D. Cantino; Department of Environmental and Plant Biology; Ohio University; Athens, OH 45701, USA. Email: cantino@ohio.edu.

Richard G. Olmstead; Department of Biology and Burke Museum; University of Washington; Seattle, WA 98195, USA. Email: olmstead@u.washington.edu.

Date Accepted: 29 January 2013; updated 23 October 2017, 11 July 2018.

Primary Editor: Kevin de Queiroz.

Nepetoideae G. Erdtman 1945: 280-281 [P. D. Cantino and R. G. Olmstead], converted clade name

Registration Number: 71

Definition: The smallest crown clade containing *Nepeta cataria* Linnaeus 1753 (*Mentheae*), *Ocimum basilicum* Linnaeus 1753 (*Ocimeae*), and *Elsholtzia fruticosa* Rehder 1916 (*Elsholtzieae*). This is a minimum-crown-clade definition. Abbreviated definition: min crown ∇ (*Nepeta cataria* Linnaeus 1753 & *Ocimum basilicum* Linnaeus 1753 & *Elsholtzia fruticosa* Rehder 1916).

Etymology: Derived from *Nepeta*, the name of an included taxon, plus –*oideae*, an ending associated with the rank of subfamily under the botanical code (Turland et al., 2018). *Nepeta* is thought to be named after the Italian city of Nepi, which was called Nepete in Roman times (Fernald, 1970; Hyam and Pankhurst, 1995).

Reference Phylogeny: Y.-P. Chen et al. (2016: Fig. 4). See Comments for a list of papers focusing on relationships within the three principal subclades of *Nepetoideae*.

Composition: Approximately 3400 species in 117 genera, which are divided among three major clades: *Elsholtzieae*, *Mentheae*, and *Ocimeae*. Harley et al. (2004) provided descriptions for 105 genera. Based on a review of subsequent literature, Li et al. (2016: 10) recognized 118 genera but overlooked the synonymization of *Leocus* with *Plectranthus* by Pollard and Paton (2009), reducing the number to 117.

Diagnostic Apomorphies: The only clear synapomorphy is hexazonocolpate pollen, usually shed at a three-celled stage. This character is associated to varying degrees with nine other morphological and phytochemical characters (Cantino and Sanders, 1986), but spotty information about the distribution of these features, particularly in the outgroups, precludes stating with confidence that any of them are apomorphies of *Nepetoideae*. The best candidates are production of rosmarinic acid, an "investing" embryo (one in which the large cotyledons encase the rest of the embryo; Martin, 1946), and myxocarpy.

The investing embryo was inferred to be an apomorphy for *Nepetoideae* in Cantino's (1992) morphological phylogenetic analysis of *Labiatae* (which did not use rosmarinic acid or myxocarpy as characters), but only a few representatives of *Nepetoideae* were included. Exalbuminous seeds also were apomorphic for *Nepetoideae* in that analysis, but this character is widespread in *Labiatae* and thus may be apomorphic for a more inclusive clade in the context of other inferred phylogenies. Ryding (1992, 2001) documented that myxocarpy (mucilage production by the pericarp) occurs in about 60% of the species of *Nepetoideae* and has not been found in any other *Labiatae*. He suggested that it is another apomorphy for *Nepetoideae* (the absences in some species of *Nepetoideae* being due to secondary loss).

In molecular phylogenies (e.g., Wagstaff et al., 1998; Li et al., 2016), *Nepetoideae* and *Lamioideae* are widely separated; in this context, the distinctive gynobasic style that they share is most parsimoniously hypothesized to have arisen independently in these two clades and thus is apomorphic for each of them (Cantino et al., 2005). However, in the context of the topology yielded by a phylogenetic analysis of morphological data (Cantino, 1992), the gynobasic style is apomorphic for a clade comprising *Nepetoideae* and *Lamioideae*. Similarly, a fruit consisting of four one-seeded nutlets is probably apomorphic for *Nepetoideae* in the context of the molecular phylogenies of Wagstaff et al. (1998) and Li et al. (2016), but it evolved at a more inclusive level on the morphological tree.

There are also many molecular apomorphies for *Nepetoideae*, including an *ndhF* deletion (Wagstaff et al., 1998) and two unspecified *trnL-F* indel events (Walker et al., 2004).

Synonyms: None.

Comments: The preexisting name *Nepetoideae* is credited here to Erdtman (1945), who applied it to the same clade that we have. This is the earliest use of this orthography that we are aware of, but there may well be earlier uses. The name *Nepetoideae* is attributed under the botanical code (Art. 19.7 of the Shenzhen Code; Turland et al., 2018) to Burnett (1835), who spelled it "*Nepetidae*" but ranked it as a subfamily (Reveal, 2012). In contrast, under the *PhyloCode* (Rec. 9.15A), a preexisting name should be attributed to the first person who published that orthography. Erdtman did not publish a formal nomenclatural protologue for *Nepetoideae*, but he used this name on p. 281 to refer to a group that he tentatively circumscribed based on diagnostic pollen features that he described on p. 280. We consider this to constitute a diagnosis (which, under the *PhyloCode*, is not required to be in Latin). We have twice phylogenetically defined *Nepetoideae* as applying to this clade in the course of experimenting with phylogenetic nomenclature (Cantino et al., 1997; Cantino and Olmstead, 2004), but precedence does not apply to these names and definitions because the *PhyloCode* was not yet in effect. The differences between those previous definitions and the one provided here are explained below, but all three result in the name being applied to the same clade.

Nepetoideae is the largest well-supported clade within *Labiatae* (= *Lamiaceae* in the publication containing the reference phylogeny), including nearly half of its species (approximately 3,400 of 7,200 species; estimates from Harley et al, 2004). It was first recognized by Erdtman (1945) based on a tight correlation between the number of pollen colpi and cells. Wunderlich (1967) provided additional support through a more extensive pollen survey and a co-occurrence of Erdtman's pollen characters with several embryological features. Cantino and Sanders (1986) further expanded the pollen survey and tabulated some additional associated characters. Based on outgroup comparison, Cantino and Sanders (1986) hypothesized that *Nepetoideae* sensu Erdtman is a clade, based on the synapomorphy of hexazonocolpate pollen. The monophyly of *Nepetoideae* was strongly supported by subsequent molecular studies (Wagstaff et al., 1995, 1998; Wagstaff and Olmstead, 1997; Paton et al., 2004; Walker et al., 2004; Bräuchler et al., 2010; Drew and Sytsma, 2012; Y.-P. Chen et al., 2016; Li et al., 2016), with 100% bootstrap support or a Bayesian posterior probability of 1.0 in every case.

Internal relationships within *Nepetoideae* are now well known due to a series of molecular phylogenetic analyses (Wagstaff et al., 1995; Paton et al., 2004; Walker et al., 2004; Bräuchler et al., 2005, 2010; Edwards et al., 2006; Moon et al., 2010; Zhong et al., 2010; Drew and Sytsma, 2011, 2012; Pastore et al., 2011; Y.-P. Chen et al., 2016; Drew et al., 2017; Li et al., 2017; Will and Classen-Bockhoff, 2017). Only three of the 117 genera have never been included in a molecular analysis: *Eriothymus* (1 species, possibly extinct, Brazil, assigned by Harley et al. [2004] to *Mentheae*), *Benguellia* (1 species, Angola, *Ocimeae*), and *Madlabium* (1 species, Madagascar, *Ocimeae*). All other species of *Nepetoideae* have been demonstrated through molecular analyses to fall into three strongly supported clades: *Elsholtzieae, Mentheae,* and *Ocimeae*. Since the phylogenetic position of the three genera that have never been included in a molecular analysis can be inferred with confidence from their morphology, it is highly unlikely that any of them will turn out to lie outside these three principal clades. Therefore, a minimum-crown-clade definition with just three specifiers can be used with very little risk that it could end up excluding from *Nepetoideae* a taxon that has not yet been included in a molecular phylogenetic analysis. Three specifiers are needed because it remains unclear which one is sister to the other two, with some analyses inferring *Elsholtzieae* to be sister to an *Ocimeae-Mentheae* clade (Bräuchler et al., 2005, 2010; Zhong et al., 2010; Y.-P. Chen et al., 2016; Li et al., 2016) but others finding *Mentheae* to be sister to an *Elsholtzieae-Ocimeae* clade (Wagstaff et al., 1998; Paton et al., 2004; Zhong et al., 2010; Drew and Sytsma, 2011, 2012) or *Ocimeae* to be sister to an *Elsholtzieae-Mentheae* clade (Wagstaff et al., 1995; Z.-D. Chen et al., 2016).

Our previous definitions of *Nepetoideae* (Cantino et al., 1997; Cantino and Olmstead, 2004) are more complex than the one provided here because knowledge of phylogeny was far less complete than it is now. Our 1997 definition has nine internal specifiers to accommodate uncertainty about the basal topology. In 2004, we used an apomorphy-modified crown-clade definition with hexacolpate pollen and *Nepeta cataria* as the specifiers because we felt that there was still too little known about the topology within *Nepetoideae* to choose internal specifiers with confidence. Fortunately, the explosion of molecular phylogenetic studies since 2004

has made it possible to use a relatively simple minimum-crown-clade definition.

Phylogenetic relationships within *Mentheae* have been documented by Walker et al. (2004), Bräuchler et al. (2005, 2010), Edwards et al. (2006), Moon et al. (2010), Drew and Sytsma (2011, 2012), Drew et al. (2014, 2017), Li et al. (2017), and Will and Classen-Bockhoff (2017). For *Ocimeae*, see Paton et al. (2004), Zhong et al. (2010), and Pastore et al. (2011). For *Elsholtzieae*, see Y.-P. Chen et al. (2016).

Literature Cited

Bräuchler, C., H. Meimberg, T. Abele, and G. Heubl. 2005. Polyphyly of the genus *Micromeria* (*Lamiaceae*)—evidence from cpDNA sequence data. *Taxon* 54:639–650.

Bräuchler, C., H. Meimberg, and G. Heubl. 2010. Molecular phylogeny of *Menthinae* (*Lamiaceae, Nepetoideae, Mentheae*)—taxonomy, biogeography and conflicts. *Mol. Phylogenet. Evol.* 55:501–523.

Burnett, G. T. 1835. *Outlines of Botany.* John Churchill, London.

Cantino, P. D. 1992. Evidence for a polyphyletic origin of the *Labiatae. Ann. Mo. Bot. Gard.* 79:361–379.

Cantino, P. D., and R. G. Olmstead. 2004. Phylogenetic nomenclature of *Lamiaceae.* P. 13 in *Abstracts of the First International Phylogenetic Nomenclature Meeting* (M. Laurin, ed.). Paris. Available at https://hal.sorbonne-universite.fr/hal-02187647.

Cantino, P. D., R. G. Olmstead, and S. J. Wagstaff. 1997. A comparison of phylogenetic nomenclature with the current system: a botanical case study. *Syst. Biol.* 46:313–331.

Cantino, P. D., and R. W. Sanders. 1986. Subfamilial classification of *Labiatae. Syst. Bot.* 11:163–185.

Cantino, P. D., Y. Yuan, and R. G. Olmstead. 2005. Phylogeny and floral evolution in *Lamiaceae.* P. 316 in Abstract P0485. XVII International Botanical Congress.

Chen, Y.-P., B. T. Drew, B. Li, D. E. Soltis, P. S. Soltis, and C.-L. Xiang. 2016. Resolving the phylogenetic position of *Ombrocharis* (*Lamiaceae*), with reference to the molecular phylogeny of tribe *Elsholtzieae*. *Taxon* 65:123–136.

Chen, Z.-D., T. Yang, L. Lin, L.-M. Lu, H.-L. Li, M. Sun, B. Liu, M. Chen, Y.-T. Niu, J.-F. Ye, Z.-Y. Cao, H.-M. Liu, X.-M. Wang, W. Wang, J.-B. Zhang, Z. Meng, W. Cao, J.-H. Li, S.-D. Wu, H.-L. Zhao, Z.-J. Liu, Z.-Y. Du, Q.-F. Wang, J. Guo, X.-X. Tan, J.-X. Su, L.-J. Zhang, L.-L. Yang, Y.-Y. Liao, M.-H. Li, G.-Q. Zhang, S.-W. Chung, J. Zhang, K.-L. Xiang, R.-Q. Li, D. E. Soltis, P. S. Soltis, S.-L. Zhou, J.-H. Ran, X.-Q. Wang, X.-H. Jin, Y.-S. Chen, T.-G. Gao, J.-H. Li, S.-Z. Zhang, A.-M. Lu, and China Phylogeny Consortium. 2016. Tree of life for the genera of Chinese vascular plants. *J. Syst. Evol.* 54:277–306.

Drew, B. T., J. G. González-Gallegos, C.-L. Xiang, R. Kriebel, C. P. Drummond, J. B. Walker, and K. J. Sytsma. 2017. *Salvia* united: the greatest good for the greatest number. *Taxon* 66:133–145.

Drew, B. T., N. Ivalú Cacho, and K. J. Sytsma. 2014. The transfer of two rare monotypic genera, *Neoeplingia* and *Chaunostoma*, to *Lepechinia* (*Lamiaceae*), and notes on their conservation. *Taxon* 63:831–842.

Drew, B. T., and K. J. Sytsma. 2011. Testing the monophyly and placement of *Lepechinia* in the tribe *Mentheae* (*Lamiaceae*). *Syst. Bot.* 36:1038–1049.

Drew, B. T., and K. J. Sytsma. 2012. Phylogenetics, biogeography, and staminal evolution in the tribe *Mentheae* (*Lamiaceae*). *Am. J. Bot.* 99:933–953.

Edwards, C. E., D. E. Soltis, and P. S. Soltis. 2006. Molecular phylogeny of *Conradina* and other scrub mints (*Lamiaceae*) from the southeastern USA: evidence for hybridization in Pleisotcene refugia? *Syst. Bot.* 31:193–207.

Erdtman, G. 1945. Pollen morphology and plant taxonomy. IV. *Labiatae, Verbenaceae,* and *Avicenniaceae. Sven. Bot. Tidskr.* 39:279–285.

Fernald, M. L. 1970. *Gray's Manual of Botany.* 8th edition, corrected printing. Van Nostrand, New York.

Harley, R. M., S. Atkins, A. L. Budantsev, P. D. Cantino, B. J. Conn, R. Grayer, M. M. Harley, R. de Kok, T. Krestovskaja, R. Morales, A. J. Paton, O. Ryding, and T. Upson. 2004. *Labiatae.* Pp. 167–275 in *The Families and Genera of Vascular Plants* (K. Kubitzki, ed.). Vol. 7 (J. W. Kadereit, ed.). Springer, Berlin.

Hyam, R., and R. Pankhurst. 1995. *Plants and Their Names: A Concise Dictionary.* Oxford University Press, Oxford.

Li, B., P. D. Cantino, R. G. Olmstead, G. L. C. Bramley, C.-L. Xiang, Z.-H. Ma, Y.-H. Tan, and D.-X. Zhang. 2016. A large-scale chloroplast phylogeny of the *Lamiaceae* sheds new light on its subfamilial classification. *Sci. Rep.* 6:34343.

Li, J.-C., J.-W. Zhang, D.-G. Zhang, T. Deng, S. Volis, H. Sun, and Z.-M. Li. 2017. Phylogenetic position of the Chinese endemic genus *Heterolamium*: a close relative of subtribe *Nepetinae* (*Lamiaceae*). *J. Jpn. Bot.* 92:12–19.

Martin, A. C. 1946. The comparative internal morphology of seeds. *Am. Midl. Nat.* 36:513–660.

Moon, H.-K., E. Smets, and S. Huysmans. 2010. Phylogeny of tribe *Mentheae* (*Lamiaceae*): the story of molecules and micromorphological characters. *Taxon* 59:1065–1076.

Pastore, J. F. B., R. M. Harley, F. Forest, A. Paton, and C. van den Berg. 2011. Phylogeny of the subtribe *Hyptidinae* (*Lamiaceae* tribe *Ocimeae*) as inferred from nuclear and plastid DNA. *Taxon* 60:1317–1329.

Paton, A. J., D. Springate, S. Duddee, D. Otieno, R. J. Grayer, M. M. Harley, F. Willis, M. S. J. Simmonds, M. P. Powell, and V. Savolainen. 2004. Phylogeny and evolution of basils and allies (*Ocimeae, Labiatae*) based on three plastid DNA regions. *Mol. Phylogenet. Evol.* 31:277–299.

Pollard, B. J., and A. Paton. 2009. The African *Plectranthus* (*Lamiaceae*) expansion continues. Vale *Leocus*! *Kew Bull.* 64:259–261.

Reveal, J. L. 2012. *Index Nominum Suprageneri-corum Plantarum Vascularium.* Available at www.plantsystematics.org/reveal/pbio/fam/allspgnames.html, version of 9 April 2012.

Ryding, O. 1992. The distribution and evolution of myxocarpy in *Lamiaceae.* Pp. 85–96 in *Advances in Labiate Science* (R. M. Harley and T. Reynolds, eds.). Royal Botanic Gardens, Kew.

Ryding, O. 2001. Myxocarpy in the *Nepetoideae* (*Lamiaceae*) with notes on myxodiaspory in general. *Syst. Geogr. Pl.* 71:503–514.

Turland, N. J., J. H. Wiersema, F. R. Barrie, W. Greuter, D. L. Hawksworth, P. S. Herendeen, S. Knapp, W.-H. Kusber, D.-Z. Li, K. Marhold, T. W. May, J. McNeill, A. M. Monro, J. Prado, M. J. Price, and G. F. Smith, eds. 2018. *International Code of Nomenclature for Algae, Fungi, and Plants (Shenzhen Code).* Adopted by the Nineteenth International Botanical Congress, Shenzhen, China, July 2017. Regnum Vegetabile 159. Koeltz Botanical Books, Glashütten.

Wagstaff, S. J., L. Hickerson, R. Spangler, P. A. Reeves, and R. G. Olmstead. 1998. Phylogeny in *Labiatae* s.l., inferred from cpDNA sequences. *Plant Syst. Evol.* 209:265–274.

Wagstaff, S. J., and R. G. Olmstead. 1997. Phylogeny of *Labiatae* and *Verbenaceae* inferred from *rbcL* sequences. *Syst. Bot.* 22:165–179.

Wagstaff, S. J., R. G. Olmstead, and P. D. Cantino. 1995. Parsimony analysis of cpDNA restriction site variation in subfamily *Nepetoideae* (*Labiatae*). *Am. J. Bot.* 82:886–892.

Walker, J. B., K. J. Sytsma, J. Treutlein, and M. Wink. 2004. *Salvia* (*Lamiaceae*) is not mono-phyletic: implications for the systematics, radiation, and ecological specializations of *Salvia* and tribe *Mentheae. Am. J. Bot.* 91:1115–1125.

Will, M., and R. Classen-Bockhoff. 2017. Time to split *Salvia* s.l. (*Lamiaceae*)—new insights from Old World *Salvia* phylogeny. *Mol. Phyloget. Evol.* 109:33–58.

Wunderlich, R. 1967. Ein Vorschlag zu einer natürlichen Gliederung der Labiaten auf Grund der Pollen-körner, der Samenentwicklung und des reifen Samens. *Oesterr. Bot. Z.* 114:383–483.

Zhong, J.-S., J. Li, L. Li, J. G. Conran, and H.-W. Li. 2010. Phylogeny of *Isodon* (Schrad. Ex Benth.) Spach (*Lamiaceae*) and related genera inferred from nuclear ribosomal ITS, *trnL-trnF* region, and *rps16* intron sequences and morphology. *Syst. Bot.* 35:207–219.

Authors

Philip D. Cantino; Department of Environmental and Plant Biology; Ohio University; Athens, OH 45701, USA. Email: cantino@ohio.edu.

Richard G. Olmstead; Department of Biology; University of Washington; Seattle, WA 98195, USA. Email: olmstead@u.washington.edu.

Date Accepted: 31 January 2013; updated 23 October 2017, 11 July 2018

Primary Editor: Kevin de Queiroz

SECTION 5

Metazoa E. Haeckel 1874 [J. R. Garey and K. M. Halanych], converted clade name

Registration Number: 192

Definition: The smallest crown clade containing *Strongylocentrotus purpuratus* (Stimpson 1857) (*Deuterostomia/Echinodermata*), *Drosophila melanogaster* Meigen 1830 (*Ecdysozoa/Arthropoda*), *Capitella teleta* Blake, Grassle and Eckelbarger 2009 (*Lophotrochozoa/Annelida*), *Convolutriloba longifissura* Bartolomaeus and Balzer 1997 (*Acoelomorpha/Acoela*), *Hydra vulgaris* Pallas 1766 (*Cnidaria*), *Trichoplax adhaerens* Schulze 1883 (*Placozoa*), *Oscarella carmela* Muricy and Pearse 2004 (*Porifera*), *Amphimedon queenslandica* Hooper and van Soest 2006 (*Porifera*), and *Mnemiopsis leidyi* Agassiz 1865 (*Ctenophora*), but not *Monosiga ovata* Kent 1881 (*Choanoflagellata*). This is a minimum-crown-clade definition with an external qualifier. Abbreviated definition: min crown ∇ (*Strongylocentrotus purpuratus* (Stimpson 1857) & *Drosophila melanogaster* Meigen 1830 & *Capitella teleta* Blake, Grassle and Eckelbarger 2009 & *Convolutriloba longifissura* Bartolomaeus and Balzer 1997 & *Hydra vulgaris* Pallas 1766 & *Trichoplax adhaerens* Schulze 1883 & *Oscarella carmela* Muricy and Pearse 2004 & *Amphimedon queenslandica* Hooper and van Soest 2006 & *Mnemiopsis leidyi* Agassiz 1865 ~ *Monosiga ovata* Kent 1881).

Etymology: From the Greek *meta* (after) and *zoa* (animal), referring to the fact that *Metazoa* came after *Protozoa* (*proto*—first and *zoa*—animal). Metazoans were recognized as having evolved later as they had tissue layers and went through the developmental process of gastrulation.

Reference Phylogeny: Hejnol et al. (2009) Figure 2 is the primary reference phylogeny.

Note, *Hydra magnipapillata* Itô 1947 in the reference topology is a junior synonym of *Hydra vulgaris* Pallas 1766, as explained by Martínez et al. (2010). Secondary reference phylogenies include Philippe et al. (2009: Fig. 1), Pick et al. (2010: Fig. 1), Chang et al. (2015: Fig. 2), Whelan et al. (2015: Fig. 3), and Cannon et al. (2016: Fig. 2).

Composition: All species within the total clades of *Protostomia* and *Deuterostomia* (see entries in this volume), the total clade of *Xenacoelomorpha* (sensu Cannon et al. 2016), the total clades of *Ctenophora*, *Porifera*, *Placozoa* and *Cnidaria*, and all descendants of the most recent common ancestor of *Protostomia*, *Deuterostomia*, *Ctenophora* and *Porifera*.

Diagnostic Apomorphies: Multicellularity and motility are the most general apomorphies for *Metazoa*. The presence of septate junctions and an extracellular matrix (related to multicellularity), and details of gametogenesis and gamete structure are additional apomorphies (reviewed in Ax, 1996).

Synonyms: *Animalia* Linnaeus 1758, *Blastozoa* Haeckel 1874, *Gastraeae* Huxley 1875, *Enterozoa* Lankester 1877, *Polyzoa* Scimkewitsch 1890, and *Polyplastida* Goette 1902 are approximate synonyms.

Comments: The composition of *Metazoa* has not changed substantially since the taxon was named by Haeckel (1874). Historically, "protozoans" and metazoans were all considered to be animals (*Animalia*), and *Metazoa* referred specifically to the multicellular animals. Occasionally some single-celled or colonial eukaryote lineages were included in *Metazoa*, although in the modern sense, the single-celled organisms archaically known as "protozoans" by Haeckel (1874) and Hyman (1940) are no longer considered animals. Some organisms, for example, dicyemids, orthonectids (known as *Mesozoa*) and *Trichoplax* (known as *Placozoa*) now known to be within *Metazoa*, were considered intermediate between "*Protozoa*" and *Metazoa* (see Hyman, 1940, chapter IV for discussion). Although *Porifera* was traditionally considered as the earliest diverging metazoan clade, several molecular phylogenetic studies based on independent sources of data place *Ctenophora* as the earliest diverging metazoan clade (Dunn et al., 2008; Hejnol et al., 2009; Whelan et al., 2015; Borowiec et al., 2015; Cannon et al., 2016; Shen et al., 2017; but see Philippe et al., 2009; Pick et al., 2010).

All of the synonyms of *Metazoa* (see Synonyms), with the exception *Animalia*, are rarely used and therefore are not appropriate names for this clade. Use of *Animalia* for the named clade differs from its use by Linnaeus (1758–1759), who excluded sponges, and later authors (e.g., Haeckel 1874), who included "protozoans". In contrast, *Metazoa* has been used in a more consistent manner since the name was first proposed and is therefore selected as the name for this clade.

Literature Cited

Ax, P. 1996. *Multicellular Animals. A New Approach to the Phylogenetic Order in Nature*, Vol. I. Springer, Berlin.

Borowiec M. L., E. K. Lee, J. C. Chiu, and D. C. Plachetzki. 2015. Extracting phylogenetic signal and accounting for bias in whole-genome datasets supports the *Ctenophora* as sister to remaining *Metazoa*. *BMC Genomics* 16:987.

Cannon, J. T., B. C. Vellutini, J. Smith, III, F. Ronquist, U. Jondelius, and A. Hejnol. 2016. *Xenacoelomorpha* is sister to *Nephrozoa*. *Nature* 530:89–93.

Chang, E. S., M. Neuhof, N. D. Rubinstein, A. Diamant, H. Philippe, D. Huchon, and P. Cartwright. 2015. Genomic insights into the evolutionary origin of *Myxozoa* within *Cnidaria*. *Proc. Natl. Acad. Sci. USA* 112:14912–14917.

Dunn, C. W., A. Hejnol, D. Q. Matus, D. Pang, W. E. Browne, S. A. Smith, E. Seaver, G. W. Rouse, M. Obst, G. D. Edgecombe, M. V. Sørensen, S. H. D. Haddock, A. Schmidt-Rhaesa, A. Okusu, R. M. Kristensen, W. C. Wheeler, M. Q. Martindale, and G. Giribet. 2008. Broad phylogenomic sampling improves resolution of the animal tree of life. *Nature* 452:745–749.

Haeckel E., 1874. Die Gastrea-Theorie, die phylogenetische Classification des Tierreichs und die Homologie der Keimblatter. *Jenaische Zeitschrift für Naturwissenschaft* 8:1–55. Note: an English translation exists: Haeckel, E. transl. by E. P. Wright. 1874. The gastraea-theory, the phylogenetic classification of the animal kingdom and the homology of the germ-lamellae. *Q. J. Microsc. Sci.* 14:142–165 & 223–247.

Hejnol, A., M. Obst, A. Stamatakis, M. Ott, G. W. Rouse, G. D. Edgecombe, P. Martinez, J. Baguñà, X. Bailly, U. Jondelius, M. Wiens, W. E. G. Müller, E. Seaver, W. C. Wheeler, M. Q. Martindale, G. Giribet, and C. W. Dunn. 2009. Assessing the root of bilaterian animals with scalable phylogenomic methods. *Proc. R. Soc. Lond. B Biol. Sci.* 276:4261–4270.

Hyman, L. H. 1940. *The Invertebrates*: Protozoa *through* Ctenophora. 1st edition, Vol. I. McGraw-Hill, New York and London.

Linnaeus, C. 1758–1759. *Systema Naturæ Per Regna Tria Naturæ, Secundum Classes, Ordines, Genera, Species, cum Characteribus, Differentiis, Synonymis, Locis*. Tomi I–II, Editio decima, reformata. Laurentii Salvii, Holmiæ (Stockholm).

Martínez D. E., A. R. Iñiguez, K. M. Percell, J. B. Willner, J. Signorovitch, and R. D. Campbell. 2010. Phylogeny and biogeography of *Hydra* (*Cnidaria*: *Hydridae*) using mitochondrial and nuclear DNA sequences. *Mol. Phylogenet. Evol.* 57:403–410.

Philippe, H., R. Derelle, P. Lopez, K. Pick, C. Borchiellini, N. Boury-Esnault, J. Vacelet, E. Renard, E. Houliston, E. Quéinnec, and C. Da Silva. 2009. Phylogenomics revives traditional views on deep animal relationships. *Curr. Biol.* 19:706–712.

Pick, K. S., H. Philippe, F. Schreiber, D. Erpenbeck, K. J. Jackson, P. Wrede, M. Wiens, A. Alié, B. Morgenstern, M. Manuel, and G. Wörheide. 2010. Improved phylogenomic taxon sampling noticeably affects nonbilaterian relationships. *Mol. Biol. Evol.* 27:1983–1987.

Shen, X.-X., C. T. Hittinger, and A. Rokas. 2017. Contentious relationships in phylogenomic studies can be driven by a handful of genes. *Nat. Ecol. Evol.* 1:0126.

Whelan, N. V., K. M. Kocot, L. L. Moroz, and K. M. Halanych. 2015. Error, signal, and the placement of *Ctenophora* sister to all other animals. *Proc. Natl. Acad. Sci. USA* 112:5773–5778.

Authors

James R. Garey; Department of Cell Biology, Microbiology and Molecular Biology; University of South Florida; 4202 East Fowler Avenue ISA2015; Tampa, FL 33620, USA. Email: garey@usf.edu.

Kenneth M. Halanych; Department of Biological Sciences; 101 Rouse Building; Auburn University; Auburn, AL 36849, USA. Email: ken@auburn.edu.

Date Accepted: 6 November 2017

Primary Editors: Jacques A. Gauthier, Kevin de Queiroz

Porifera R. E. Grant 1836 [M. Manuel and N. Boury-Esnault], converted clade name

Registration Number: 201

Definition: The least inclusive crown clade containing *Carteriospongia foliacens* (Pallas 1766) (*Demospongiae*), *Oopsacas minuta* Topsent 1927 (*Hexactinellida*), *Oscarella lobularis* (Schmidt 1862) (*Homoscleromorpha*), and *Sycon raphanus* Schmidt 1862 (*Calcispongia*). This is a minimum-crown-clade definition. Abbreviated definition: min crown ∇ (*Carteriospongia foliacens* (Pallas 1766) & *Oopsacas minuta* Topsent 1927 & *Oscarella lobularis* (Schmidt 1862) & *Sycon raphanus* Schmidt 1862).

Etymology: The name *Porifera* is derived from latin *porus*, which means "small hole," or "pore" and the verb *ferre*, which means "to bear," thus "pore-bearing".

Reference Phylogeny: The primary reference phylogeny is Pick et al. (2010: Fig. 1). See also Philippe et al. (2009: Fig. 1) and Nosenko et al. (2013: Fig. 1).

Composition: *Porifera* is currently believed to contain about 9200 recognized extant species but the true number may be much higher (van Soest, 2007). All (currently known) extant species of *Porifera* belong to one of the following four clades: *Demospongiae, Hexactinellida, Calcispongia* and *Homoscleromorpha* (see entries in this volume). For an updated list of extant sponge species see van Soest et al. (2020).

Diagnostic Apomorphies: An aquiferous system of internal canals and choanocyte chambers through which water flows, and a thin epithelium covering the surface and the canals called the pinacoderm. Water enters the system through small pores (ostia) and leaves through one or several large hole(s) (osculum, oscula) (Hooper et al., 2002; Philippe et al., 2009: Fig. 2). Members of the demosponge subclade *Cladorhizidae* have evolved carnivory, and this adaptation has resulted in the loss of the aquiferous system and choanocytes (Vacelet and Boury-Esnault, 1995).

Synonyms: *Spongiae* Linnaeus 1759; *Poriphora* Sommerville 1859; *Spongida* Huxley 1864 (all approximate).

Comments: In the beginning of the history of their classification, sponges were designated under the names *Spongiae, Spongida,* and *Poriphora* by different authors, both before and after the name *Porifera* was proposed by Grant (1836). However, since the end of nineteenth century *Porifera* has been most widely and consistently used for that clade (e.g., Zittel, 1878; Ridley and Dendy, 1887; Laubenfels, 1936; Hyman, 1940; Brusca and Brusca, 2003) and was selected for that reason.

The monophyly of *Porifera* has been the subject of debate from the nineteenth century until the present. There are three main hypotheses:

1. All sponges share an exclusive common ancestor and thus the group is monophyletic (Ax, 1995, 1996; Dohrmann et al., 2008; Hooper et al., 2002; Philippe et al., 2009; Philippe et al., 2011; Pick et al., 2010; Wörheide et al., 2012; Simion et al., 2017). Philippe et al. (2009) found a sister group relationship between a clade *Demospongiae/Hexactinellida* and a clade *Calcispongia/Homoscleromorpha*. The monophyly of sponges is supported by

the largest amount of phylogenomic data, in terms of amino acid positions and ingroup taxon sampling, and by morphological characters such as an aquiferous system of internal canals and choanocyte chambers through which water flows, and an epithelium covering the surface and the canals (Leys and Riesgo, 2012; Philippe et al., 2009; Philippe et al., 2011; Wörheide et al., 2012).

2. *Hexactinellida* constitutes a separate clade from other sponges (Bidder 1929; Bergquist 1985) based on their partly syncytial organization and is the sister group of the clade (*Demospongiae + Calcispongiae + Homoscleromorpha + Eumetazoa*). According to this hypothesis the syncytial organisation of *Hexactinellida* is considered a plesiomorphic character, and the cellular organisation is regarded as a synapomorphy of a clade composed of *Demospongiae*, *Calcispongia* and *Eumetazoa*. However, if the choanoflagellates are the sister group of *Metazoa* and if metazoans are derived from colonial unicellular ancestors (Alié and Manuel, 2010; Carr et al., 2008; King et al., 2008; Nielsen, 2008), then a (multi)cellular organization is a synapomorphy of all *Metazoa* and the syncytial organization becomes an autapomorphy of *Hexactinellida*.

3. *Hexactinellida* and *Demospongiae* are sister-groups, and *Calcispongia* and/or *Homoscleromorpha* is/are the sister group of *Eumetazoa* (e.g., Zrzavý et al., 1998; Borchiellini et al., 2001; Medina et al., 2001; Peterson and Eernisse, 2001; Sperling et al., 2007). The morphological synapomorphies for the *Hexactinellida/Demospongiae* clade are the siliceous nature of the skeleton and the intracellular secretion of siliceous spicules around an axial filament. The sister-group relationship between *Calcispongia* plus *Homoscleromorpha* and *Eumetazoa* was inferred by Erwin et al. (2011), but without firm morphological synapomorphies (Nielsen, 2008; Wörheide et al., 2012). Other results place *Calcispongia* and *Homoscleromorpha* successive sister groups of *Eumetazoa* (Sperling et al., 2007, 2010).

These two last hypotheses imply the paraphyly of *Porifera*. Sponge paraphyly has been rejected by Ax (1995), Hooper et al. (2002) and many other sponge taxonomists, although the debate has been flourishing (e.g., Nielsen, 2008; Sperling et al., 2009; Erwin et al., 2011). In hypothesis 3, inferred paraphyly is mainly based on nuclear housekeeping genes. These results are hampered by insufficient taxon sampling and provide weak support for the hypothesized relationships. With molecular phylogenies based on multiple genes and an increasing number of taxa, the monophyly of *Porifera* is the more widely accepted hypothesis (Philippe et al., 2009; Pick et al., 2010; Philippe et al., 2011; Wörheide et al., 2012; Simion et al., 2017). Nonetheless, in the context of either hypothesis of sponge paraphyly, the name *Porifera* would become a synonym of *Metazoa* according to our definition. If so, *Metazoa* should have precedence.

Literature Cited

Alié, A., and Manuel, M. 2010. The backbone of the post-synaptic density originated in a unicellular ancestor of choanoflagellates and metazoans. *BMC Evol. Biol.* 10:34.

Ax, P. 1995. *Das System der Metazoa. ein Lehrbuch der Phylogenetischen Systematik.* Gustav Fischer Verlag, Stuttgart.

Ax, P. 1996. *Multicellular Animals: A New Approach to the Phylogenetic Order in Nature*, Vol. 1. Springer Verlag, Berlin.

Bergquist, P. 1985. Poriferan relationships. Pp. 14–27 in *The Origin and Relationships of Lower Invertebrates* (S. C. Morris, J. D. George, R. Gibson, and H. M. Platt, eds.). The Systematics Association, 28. Clarendon Press, London.

Bidder, G. P. 1929. Sponges. Pp. 254–261 in *Encyclopedia Britannica*. London.

Borchiellini, C., M. Manuel, E. Alivon, N. Boury-Esnault, J. Vacelet, and Y. Le Parco. 2001. Sponge paraphyly and the origin of *Metazoa*. *J. Evol. Biol.* 14:171–179.

Brusca, R. C., and G. J. Brusca. 2003. *Invertebrates*. Sinauer Associates Inc., Sunderland, MA.

Carr, M., B. S. C. Leadbeater, R. Hassan, M. Nelson, and S. L. Baldauf. 2008. Molecular phylogeny of choanoflagellates, the sister group to *Metazoa. Proc. Natl. Acad. Sci. USA* 105:16641–16646.

Dohrmann, M., D. Janussen, J. Reitner, A. G. Collins, and G. Wörheide. 2008. Phylogeny and evolution of glass sponges (*Porifera, Hexactinellida*). *Syst. Biol.* 57:388–405.

Erwin, D. H., M. Laflamme, S. M. Tweedt, E. A. Sperling, D. Pisani, and K. J. Peterson. 2011. The Cambrian Conundrum: early divergence and later ecological success in the early history of animals. *Science* 334:1091–1095.

Grant, R. E. 1836. Animal kingdom. Pp. 107–118 in *The Cyclopaedia of Anatomy and Physiology*, Vol. 1 (R. B. Todd, ed.). Sherwood, Gilbert, and Piper, London.

Hooper, J. N. A., R. W. M. van Soest, and F. Debrenne. 2002. Phylum *Porifera* Grant, 1836. Pp. 9–14 in *Systema* Porifera: *A Guide to the Classification of Sponges* (J. N. A. Hooper and R. W. M van Soest, eds.). Kluwer Academic and Plenum Publishers, Dordrecht.

Huxley, T. H. 1864. Lectures on the elements of comparative anatomy. On the classification of animals: the *Gregarinida, Rhizopoda, Spongida* and *Infusoria. Q. J. Microsc. Soc. Lond.* 4:63–81.

Hyman, L. H. 1940. *The Invertebrates*, Vol. 1. McGraw-Hill, New York.

King, N., M. Westbrook, S. Young, A. Kuo, M. Abedin, J. Chapman, S. Fairclough, U. Hellsten, Y. Isogai, I. Letunic, M. Marr, D. Pincus, N. Putnam, A. Rokas, K. J. Wright, R. Zuzow, W. Dirks, M. Good, D. Goodstein, D. Lemons, W. Li, J. B. Lyons, A. Morris, S. Nichols, D. J. Richter, A. Salamov, J. G. Sequencing, P. Bork, W. A. Lim, G. Manning, W. T. Miller, W. McGinnis, H. Shapiro, R. Tjian, I. V. Grigoriev, and D. Rokhsar. 2008. The genome of the choanoflagellate *Monosiga brevicollis* and the origin of metazoans. *Nature* 451:783–788.

de Laubenfels, M. W. 1936. *A Discussion of the Sponge Fauna of the Dry Tortugas in particular and the West Indies in general, with Material for a Revision of the Families and Orders of the* Porifera. Papers from the Tortugas Laboratory, Vol. 30. Carnegie Institution of Washington, Washington, DC.

Leys, S. P., and A. Riesgo. 2012. Epithelia, an evolutionary novelty of metazoans. *J. Exp. Zool. Part B Mol. Dev. Evol.* doi:10.1002/jez.b.21442.

Linnaeus, C. 1759. *Species Plantarum*. Laurentii Salvii, Stockholm.

Medina, M., A. G. Collins, J. D. Silberman, and M. L. Sogin. 2001. Evaluating hypotheses of basal animal phylogeny using complete sequences of large and small subunit rRNA. *Proc. Natl. Acad. Sci. USA* 98:9707–9712.

Nielsen, C. 2008. Six major steps in animal evolution: are we derived sponge larvae? *Evol. Dev.* 10:241–257.

Nosenko, T., F. Schreiber, M. Adamska, M. Adamski, M. Eitel, J. Hammel, M. Maldonado, W. E. G. Müller, M. Nickel, B. Schierwater, J. Vacelet, M. Wiens, and G. Wörheide. 2013. Deep metazona phylogeny: when different genes tell different stories. *Mol. Phylogenet. Evol.* 67:223–233.

Pallas, P. S. 1766. *Elenchus Zoophytorum Sistens Generum Adumbrations Generaliores et Specierum Cognitarum Succinctas Descriptiones cum Selectis Auctorum Synonymis*. P. van Cleef, The Hague.

Peterson, K. J., and D. J. Eernisse. 2001. Animal phylogeny and the ancestry of bilaterians: inferences from morphology and 18S rDNA sequences. *Evol. Dev.* 3:170–205.

Philippe, H., H. Brinkmann, D. V. Lavrov, D. T. J. Littlewood, M. Manuel, G. Wörheide, and D. Baurain. 2011. Resolving difficult phylogenetic questions: why more sequences are not enough. *PLOS Biol.* 9:e1000602.

Philippe, H., R. Derelle, P. Lopez, K. Pick, C. Borchiellini, N. Boury-Esnault, J. Vacelet, E. Renard, E. Houliston, E. Quéinnec, C. Da Silva, P. Wincker, H. Le Guyader, S. Leys, D.J. Jackson, F. Schreiber, D. Erpenbeck, B. Morgenstern, W. Wörheide, and M. Manuel. 2009. Phylogenomics revives traditional views on deep animal relationships. *Curr. Biol.* 19:1–7.

Pick, K. S., H. Philippe, F. Schreiber, D. Erpenbeck, D. J. Jackson, P. Wrede, M. Wiens, A. Alié, B. Morgenstern, M. Manuel, and G. Wörheide. 2010. Improved phylogenomic taxon sampling noticeably affects nonbilaterian relationships. *Mol. Biol. Evol.* 27:1983–1987.

Ridley, S. O., and A. Dendy. 1887. Report on the *Monaxonida* collected by H.M.S. 'Challenger' during the years 1873–1876. Report on the Scientific Results of the Voyage of H.M.S. 'Challenger', 1873–1876. *Zoology* 20(59):1–275

Schmidt, O. 1862. *Die Spongien des adriatischen Meeres*. Wilhelm Engelmann, Leipzig.

Simion, P., H. Philippe, D. Baurain, M. Jager, D.J. Richter, A. Di Franco, B. Roure, N. Satoh, E. Quéinnec, A. Ereskovsky, P. Lapébie, E. Corre, F. Delsuc, N. King, G. Wörheide, and M. Manuel. 2017. A large and consistent phylogenomic dataset supports sponges as the sister group to all other animals. *Curr. Biol.* 27:958–967.

Sommerville, J. M. 1859. *Ocean Life*. Barnard and Jones, Philadelphia, PA.

Sperling, E. A., K. J. Peterson, and D. Pisani. 2009. Phylogenetic-signal dissection of nuclear housekeeping genes supports the paraphyly of sponges and the monophyly of *Eumetazoa*. *Mol. Biol. Evol.* 26(10):2261–2274.

Sperling, E. A, D. Pisani, and K. J. Peterson. 2007. Poriferan paraphyly and its implications for Precambrian palaeobiology. *Geol. Soc., Lond. Spec. Publ.* 286:355–368.

van Soest, R. W. M. 2007. Sponge biodiversity. *J. Mar. Biol. Assoc. UK* 87:1345–1348.

van Soest, R. W. M. N. Boury-Esnault J. N. A. Hooper, K. Rützler, N. J. de Voogd, B. Alvarez, E. Hajdu, A. B. Pisera, R. Manconi, C. Schœnberg, M. Klautau, M. Kelly, J. Vacelet, M. Dohrmann, M.-C. Díaz, P. Cárdenas, J. L. Carballo, P. Ríos, R. Downey, and C. C Morrow. 2020. World *Porifera* Database. Available at http://www.marinespecies.org/porifera. doi:10.14284/359.

Topsent, E. 1927. Diagnoses d'Éponges nouvelles recueillies par le Prince Albert ler de Monaco. *Bull. Inst. Océanogr. Monaco* 502:1–19.

Vacelet, J., and N. Boury-Esnault. 1995. Carnivorous sponges. *Nature* 373:333–335.

Wörheide, G.., M. Dohrmann, D. Erpenbeck, C. Larroux, M. Maldonado, O. Voigt, C. Borchiellini, and D. Lavrov. 2012. Deep phylogeny and evolution of sponges (Phylum *Porifera*). *Adv. Mar. Biol.* 61:3–78.

Zittel, K. A., 1878. *Zur Stammes-Geschichte der Spongien*. Ludwig-Maximilian's Universität, München.

Zrzavý, Y., S. Mihulka, P. Kepka, A. Bezděk, and D. Tietz. 1998. Phylogeny of the *Metazoa* Based on Morphological and 18S Ribosomal DNA Evidence. *Cladistics* 14:249–285.

Authors

Michaël Manuel; Sorbonne Université; Institut de Systématique, Evolution, Biodiversité (ISYEB, UMR 7205 MNHN SU CNRS EPHE); bâtiment A, 4o étage; 7, quai St Bernard, 75005; Paris, France. Email: michael.manuel@upmc.fr.

Nicole Boury-Esnault; IMBE CNRS; Station marine d'Endoume; Université de la Méditerranée; rue de la Batterie des Lions 13007-Marseille, France. Email: nicole.boury-esnault@orange.fr.

Date Accepted: 18 March 2013

Primary Editor: Kevin de Queiroz.

Demospongiae W. J. Sollas 1885 [N. Boury-Esnault and M. Manuel], converted clade name

Registration Number: 202

Definition: The least inclusive crown clade containing *Spongia officinalis* Linnaeus 1759 (*Keratosa*), *Chondrosia reniformis* Nardo 1847 (*Myxospongiae*), *Haliclona mediterranea* Griessinger 1971 (*Haploscleromorpha)* and *Axinella polypoides* Schmidt 1862 (*Democlavia*). This is a minimum-crown-clade definition. Abbreviated definition: min crown ∇ (*Spongia officinalis* Linnaeus 1759 & *Chondrosia reniformis* Nardo 1847 & *Haliclona mediterranea* Griessinger 1971 & *Axinella polypoides* Schmidt 1862).

Etymology: The name *Demospongiae* is derived from the Greek *dēmos* "people" and *spongiá* "sponge" and means "common sponges".

Reference Phylogeny: The primary reference phylogeny is Borchiellini et al. (2004: Fig. 2). See also Erwin et al. (2011: Fig. 1).

Composition: *Demospongiae* contains about 85% of all extant sponge species, with about 7600 living species described in the literature. This estimate is probably conservative, and new species are continually being described thanks to the investigation of new areas and to new techniques that allow the discrimination of sibling species. Four subclades have been recently recognized among *Demospongiae*: *Keratosa*, *Myxospongiae*, *Haploscleromorpha* and *Democlavia* or *Heteroscleromorpha* (Borchiellini et al., 2004; Sperling et al., 2009; Cárdenas et al., 2012). For an updated classification and species list see van Soest et al. (2020).

Diagnostic Apomorphies: A skeleton composed of monaxonic and/or tetraxonic siliceous spicules and/or spongin fiber. The axial filament of the spicules is triangular or hexagonal in section (Hooper and van Soest, 2002). Siliceous spicules with a triangular or hexagonal axial filament and spongin fibers are apomorphies of *Demospongiae*. However, one or both characters, and sometimes even the skeleton itself, have been lost in some members of the clade.

Synonyms: *Demospongiaria* Topsent 1928; *Demospongida* Dendy 1905 (both approximate).

Comments: All the names used for this clade in the literature (see Synonyms) have the same etymological basis. *Demospongiae* is both the oldest name and by far the most widely used (Hooper and van Soest, 2002) and was selected for those reasons. In addition, the name *Demospongiae* has previously been defined phylogenetically as applying to the same crown clade (Borchiellini et al., 2004), although the previous definition used *Geodia cydonium* rather than *Axinella polypoides* to represent *Democlavia*.

The composition of this clade has been subject to debate (by contrast, *Hexactinellida* and *Calcispongia* have been clearly delimited since the end of the nineteenth century). Traditionally, all taxa that were not hexactinellid or calcareous sponges were considered demosponges (Lévi, 1999; Boury-Esnault, 2006). *Demospongiae*, defined by either a combination of non-unique characters or negative characters, had over the years a fluctuating content (Erpenbeck and Wörheide, 2007; Cárdenas et al., 2012; Wörheide et al., 2012). Sponges without a skeleton or those with a skeleton composed only of spongin fibers have been often excluded from this group (see Lévi, 1956 for a discussion). In the 1970s the discovery of species with hypercalcified skeletons led Hartman and Goreau (1970) to exclude these species from *Demospongiae* and to erect a new class *Sclerospongiae*. This position was rejected by Vacelet (1964, 1979, 1985), who contended that species with hypercalcified skeletons share characters with different subgroups of *Calcispongia* or *Demospongiae*. He was followed by many other authors (e.g., van Soest, 1984; Engeser et al., 1986; Wood et al., 1989; Wood, 1990). Molecular data have confirmed that *Sclerospongiae* is polyphyletic (Chombard et al., 1997). Recently, Borchiellini et al. (2004) have shown that *Homoscleromorpha* does not belong to *Demospongiae*, and this inference has been confirmed by several subsequent studies (Dohrmann et al., 2008; Philippe et al., 2009; Sperling et al., 2009; Pick et al., 2010; Sperling et al., 2010; Erwin et al., 2011; Philippe et al., 2011). Although some studies have found evidence for the placement of hexactinellids within *Desmospongiae* (e.g., Cavalier-Smith and Chao, 2003; Nichols, 2005; Haen et al., 2007), most recent studies have supported the monophyly of *Desmospongiae* excluding hexactinellids (Dohrmann et al., 2008; Philippe et al., 2009; Philippe et al., 2011).

Literature Cited

Borchiellini, C., C. Chombard, M. Manuel, E. Alivon, J. Vacelet, and N. Boury-Esnault. 2004. Molecular phylogeny of *Demospongiae*: implications for classification and scenarios of character evolution. *Mol. Phylogenet. Evol.* 32:823–837.

Boury-Esnault, N. 2006. Systematics and evolution of *Demospongiae*. *Can. J. Zool.* 84:205–224.

Cárdenas, P., T. Pérez, and N. Boury-Esnault, 2012. Sponge systematics facing new challenges. *Adv. Mar. Biol.* 61:79–209.

Cavalier-Smith, T., and E. E. Y. Chao. 2003. Phylogeny of *Choanozoa*, *Apusozoa*, and other *Protozoa* and early eukaryote megaevolution. *J. Mol. Evol.* 56:540–563.

Chombard, C., A. Tillier, N. Boury-Esnault, and J. Vacelet. 1997. Polyphyly of "sclerosponges" (*Porifera*, *Demospongiae*) supported by 28S ribosomal sequences. *Biol. Bull.* (*Woods Hole*), 193:359–367.

Dendy, A. 1905. Report on the sponges collected by Professor Herdman, at Ceylon, in 1902. Pp. 57–246 in *Report to the Government of Ceylon on the Pearl Oyster Fisheries of the Gulf of Manaar*, Vol. 3(suppl. 18) (W. A. Herdman, ed.). Royal Society London.

Dohrmann, M., D. Janussen, J. Reitner, A. Collins, and G. Wörheide. 2008. Phylogeny and evolution of glass sponges (*Porifera*, *Hexactinellida*). *Syst. Biol.* 57:388–405.

Engeser, T., M. Floquet, and J. Reitner, 1986. *Acanthochaetetidae* (*Hadromerida*, *Demospongiae*) from the Coniacian of Vera de Bidasoa (Basque Pyrénées, Northern Spain). *Geobios (Lyon)* 19:849–854.

Erpenbeck, D., and G. Wörheide. 2007. On the molecular phylogeny of sponges (*Porifera*). Pp. 107–126 in *Linnaeus Tercentenary: Progress in Invertebrate Taxonomy* (Z.-Q. Zhang and W. A. Shear, eds.). *Zootaxa* 1668.

Erwin, D. H., M. Laflamme, S. M. Tweedt, E. A. Sperling, D. Pisani, and K. J. Peterson. 2011. The Cambrian Conundrum: early divergence and later ecological success in the early history of animals. *Science* 334:1091–1095.

Griessinger, M. 1971. Etude des Réniérides de Méditerranée (Démosponges, Haplosclérides). *Bull. Mus. Natl. Hist. Nat. Paris* 3:97–182.

Haen, K. M., B. F. Lang, S. A. Pomponi, and D. V. Lavrov. 2007. Glass sponges and bilaterian animals share derived mitochondrial genomic features: a common ancestry or parallel evolution? *Mol. Biol. Evol.* 24:1518–1527.

Hartman, W. D., and T. F. Goreau. 1970. Jamaican coralline sponges: their morphology, ecology and fossil relatives. Pp. 205–243 in *The Biology of the* Porifera (W. G. Fry, ed.). *Symp. Zool. Soc. Lond.*, 25. Academic Press, London.

Hooper, J. N. A., and R. W. M. van Soest. 2002. Class *Demospongiae* Sollas, 1885. Pp. 15–18 in *Systema* Porifera: *A Guide to the Classification of Sponges* (J. N. A. Hooper and R. W. M. van Soest, eds.). Kluwer Academic and Plenum Publishers, Dordrecht.

Lévi, C. 1956. Etude des *Halisarca* de Roscoff. Embryologie et systématique des démosponges. *Arch. Zool. Exp. Gén.* 93:1–184.

Lévi, C. 1999. Sponge science from origin to outlook. *Mem. Queensl. Mus.* 44:1–7.

Linnaeus, C. 1759. *Species Plantarum*. Laurentii Salvii, Stockholm.

Nardo, G. D. 1847. Prospetto della fauna marina volgare del Veneto Estuario con cenni sulle principali specie commestibili dell'Adriatico, sulle venete pesche, sulle valli, ecc. Pp. 113–156 (1–45 in reprint) in *Venezia e le sue Lagune*. Giovanni Correr. Antonelli, Venezia.

Nichols, S. A. 2005. An evaluation of support for order-level monophyly and interrelationships within the class *Demospongiae* using partial data from the large subunit rDNA and cytochrome oxidase subunit I. *Mol. Phylogenet Evol.* 34:81–96.

Philippe, H., H. Brinkmann, D. V. Lavrov, D. T. J. Littlewood, M. Manuel, G. Wörheide, and D. Baurain. 2011. Resolving difficult phylogenetic questions: why more sequences are not enough. *PLOS Biol.* 9:e1000602.

Philippe, H., R. Derelle, P. Lopez, K. Pick, C. Borchiellini, N. Boury-Esnault, J. Vacelet, E. Renard, E. Houliston, E. Quéinnec, C. Da Silva, P. Wincker, H. Le Guyader, S. Leys, D.J. Jackson, F. Schreiber, D. Erpenbeck, B. Morgenstern, W. Wörheide, and M. Manuel. 2009. Phylogenomics revives traditional views on deep animal relationships. *Curr. Biol.* 19:1–7.

Pick, K. S., H. Philippe, F. Schreiber, D. Erpenbeck, D. J. Jackson, P. Wrede, M. Wiens, A. Alié, B. Morgenstern, M. Manuel, and G. Wörheide. 2010. Improved phylogenomic taxon sampling noticeably affects nonbilaterian relationships. *Mol. Biol. Evol.* 27:1983–1987.

Schmidt, O. 1862. Die Spongien des adriatischen Meeres. Wilhelm Engelmann, Leipzig.

Sperling, E. A., K. V. Peterson, and D. Pisani. 2009. Phylogenetic-signal dissection of nuclear housekeeping genes supports the paraphyly of sponges and the monophyly of *Eumetazoa*. *Mol. Biol. Evol.* 26(10):2261–2274.

Sperling, E. A., J. M. Robinson, D. Pisani, and K. J. Peterson. 2010. Where's the glass? Biomarkers, molecular clocks, and microRNAs suggest a 200-Myr missing Precambrian fossil record of siliceous sponge spicules. *Geobiology* 8:24–36.

van Soest, R. W. M. 1984. Deficient *Merlia normani* Kirkpatrick, 1908, from the Curaçao Reefs with a discussion on the phylogenetic interpretation of sclerosponges. *Bijdr. Dierkd.* 54:211–219.

van Soest, R. W. M., N. Boury-Esnault, J. N. A. Hooper, K. Rützler, N. J. de Voogd, B. Alvarez, E. Hajdu, A. B. Pisera, R. Manconi, C. Schoenberg, M. Klautau, M. Kelly, J. Vacelet, M. Dohrmann, M.-C. Díaz, P. Cárdenas, J. L. Carballo, P. Ríos, R. Downey, and C. C. Morrow. 2020. World *Porifera* Database. Available at http://www.marinespecies.org/porifera. doi:10.14284/359.

Sollas, W. J. 1885. A Classification of the Sponges. *Ann. Mag. Nat. Hist.* 16(95):395.

Topsent, E. 1928. Spongiaires de l'Atlantique et de la Méditerranée provenant des croisières du Prince Albert ler de Monaco. *Résultats des Campagnes Scientifiques Accomplies par le Prince Albert I, Prince Souverain de Monaco* 74:1–376.

Vacelet, J. 1964. Etude monographique de l'éponge calcaire pharétronide de Méditerranée, *Petrobiona massiliana* Vacelet et Lévi. Les pharétronides actuelles et fossiles. *Recl. Trav. Stn. Mar. Endoume* 34:1–125.

Vacelet, J. 1979. Description et affinités d'une éponge Sphinctozoaire actuelle. Pp. 483–493 in *Biologie des Spongiaires* (C. Lévi and N. Boury-Esnault, eds.). Editions du C.N.R.S., Paris.

Vacelet, J. 1985. Coralline sponges and the evolution of the *Porifera*. Pp. 1–13 in *The Origins and Relationships of Lower Invertebrates* (S. Conway Morris, J. D. George, R. Gibson and H. M. Platt, eds.). Clarendon Press, Oxford.

Wood, R. 1990. Reef-building sponges. *Am. Sci.* 78:224–235.

Wood, R., J. Reitner, and R. R. West. 1989. Systematics and phylogenetic implications of the haplosclerid stromatoporoid *Newellia mira* nov. gen. *Lethaia* 22:85–93.

Wörheide, G., M. Dohrmann, D. Erpenbeck, C. Larroux, M. Maldonado, O. Voigt, C. Borchiellini, and D. Lavrov. 2012. Deep phylogeny and evolution of sponges (Phylum *Porifera*). *Adv. Mar. Biol.* 61:3–78.

Authors

Nicole Boury-Esnault; IMBE-CNRS; Station marine d'Endoume; Université de la Méditerranée; rue de la Batterie des Lions 13007-Marseille, France. Email: nicole.boury-esnault@orange.fr.

Michaël Manuel; Université Paris 06, UMR 7138 CNRS UPMC MNHN IRD, bâtiment A, 4° étage; 7, quai St Bernard, 75005; Paris, France. E-mail: michael.manuel@snv.jussieu.fr.

Date Accepted: 18 March 2013

Primary Editor: Kevin de Queiroz

Hexactinellida O. Schmidt 1870 [M. Manuel and N. Boury-Esnault], converted clade name

Registration Number: 205

Definition: The most inclusive crown clade containing *Oopsacas minuta* Topsent 1927 but not *Oscarella lobularis* (Schmidt 1862) (*Homoscleromorpha*) and *Spongia officinalis* Linnaeus 1759 (*Demospongiae*) and *Grantia compressa* (Fabricius 1780) (*Calcispongia* ≈ *Calcarea* in the reference phylogeny). This is a maximum-crown-clade definition. Abbreviated definition: max crown ∇ (*Oopsacas minuta* Topsent 1927 ~ *Oscarella lobularis* (Schmidt 1862) & *Spongia officinalis* Linnaeus 1759 & *Grantia compressa* (Fabricius 1780)).

Etymology: The name *Hexactinellida* is derived from the Greek *hexas* "the number six" and *aktinos* "rays", referring to the six-rayed symmetry of the skeletal elements (spicules).

Reference Phylogeny: The primary reference phylogeny is Dohrmann et al. (2008: Fig. 2). See also Dohrmann et al. (2012: Fig. 3).

Composition: *Hexactinellida* presently contains about 600–700 recognized extant species (ca. 7% of all described species of Recent *Porifera*). Two subclades are widely accepted, *Amphidiscophora* Schulze 1886 and *Hexasterophora* Schulze 1886, one composed of species with amphidiscs and the other composed of species with hexasters (Dohrmann et al., 2008: 2012; Reiswig, 2002). For an updated species list see van Soest et al. (2020).

Diagnostic Apomorphies: Spicules with three-axes (six-rays), or modifications thereof, with an axial filament that is square in section (Reiswig 2002, 2006), and with a partially syncytial organization of the soft parts (Leys et al., 2007).

Synonyms: *Hexactinellidae* Schmidt 1870; *Coralliospongia* Gray 1867 (in part); *Vitrea* Thomson 1868; *Hyalospongiae* Claus 1872 (in part); *Triaxonia* Schulze 1886 (all approximate).

Comments: Some of the names given to this group of sponges refer to the vitreous aspect of its members (*Vitrea* and *Hyalospongiae*), while others refer to their spicules, which have three axes (*Triaxonia*) and 6 rays (*Hexactinellida*). This last name is the most frequently used (e.g., Schulze, 1887; Ijima, 1901; Lévi and Lévi, 1982; Reiswig, 2002; 2006) and was selected for that reason.

The molecular phylogenetic analysis of *Hexactinellida* supports its monophyly and largely corroborates previous morphology-based hypotheses (Dohrmann et al., 2008, 2012). Hexactinellids are characterized by a rich suite of morphological characters, especially an "amazing array of spicules of various shapes and sizes" (Leys et al., 2007: 59), providing taxonomists and phylogeneticists with sufficient information to delineate natural groups (Dohrmann et al., 2008). The molecular phylogeny is also largely consistent with results of a phylogenetic analysis of morphological data (Mehl, 1992). Hexactinellids have been collected from depths of 5–6,770 m and are unknown in freshwater habitats. They constitute important members of deep-water marine communities, which are under-sampled.

Literature Cited

Claus, C. F. W. 1872. *Grundzuge der Zoologie.* 2nd edition. N. G. Elwert, Marburg.

Dohrmann, M, D. Janussen, J. Reitner, A. G. Collins, and G. Wörheide. 2008. Phylogeny and evolution of glass sponges (*Porifera, Hexactinellida*). *Syst. Biol.* 57:388–405.

Dohrmann, M., K. M. Haen, D. Lavrov and G. Wörheide. 2012. Molecular phylogeny of glass sponges (*Porifera, Hexactinellida*): increased taxon sampling and inclusion of the mitochondrial protein-coding gene, cytochrome oxidase subunit I. *Hydrobiologia* 687:11–20.

Fabricius, O. 1780. *Fauna Groenlandica: Systematice Sistens Animalia Groenlandiae Occidentalis Hactenus Indagata, Quod Nomen Specificium.* Hafniae et Lipsiae, Copenhagen, Denmark.

Gray, J. E. 1867. Notes on the arrangement of sponges with the descriptions of some new genera. *Proc. Zool. Soc. Lond.* 2:492–558.

Ijima, I. 1901. Studies on the *Hexactinellida.* Contribution. I. *Euplectellidae. J. Coll. Sci. Imp. Univ. Tokyo* 15:1–299.

Lévi, C., and P. Lévi. 1982. Spongiaires hexactinellides du Pacifique sud-ouest (Nouvelle-Calédonie). *Bull. Mus. Natl. Hist. Nat. Paris* 4:288–317.

Leys, S. P., G. O. Mackie, and H. M. Reiswig. 2007. The biology of glass sponges. *Adv. Mar. Biol.* 52:1–145.

Linnaeus, C. 1759. *Species Plantarum.* Laurentii Salvii, Stockholm.

Mehl, D. 1992. Die Entwicklung der *Hexactinellida* seit dem Mesozoikum Paläobiologie, Phylogenie und Evolutionsökologie. *Berl. Geowiss. Abh. Reihe E* 2:1–164.

Reiswig, H. M. 2002. Class *Hexactinellida* Schmidt, 1870. Pp. 1201–1202 in *Systema* Porifera: *A Guide to the Classification of Sponges* (J. N. A. Hooper and R. W. M. van Soest, eds.). Kluwer Academic and Plenum, Dordrecht.

Reiswig, H. M. 2006. Classification and phylogeny of *Hexactinellida* (*Porifera*). *Can. J. Zool.* 84:195–204.

Schmidt, O. 1862. *Die Spongien des adriatischen Meeres.* Wilhelm Engelmann, Leipzig.

Schmidt, O. 1870. *Grundzüge einer Spongien-fauna des Atlantischen Gebietes.* Engelmann, Leipzig.

Schulze, F. E. 1886. Über den Bau und das System der Hexactinelliden. *Abh. K. Akad. wiss. Berlin, Phys.-Math. Kl.* 1886:1–97.

Schulze, F. E. 1887. Report on the *Hexactinellida* collected by H. M. S. Challenger during the years 1873–1876. *Zoology* 21:1–514.

van Soest, R. W. M., N. Boury-Esnault, J. N. A. Hooper, K. Rützler, N. J. de Voogd, B. Alvarez, E. Hajdu, A. B. Pisera, R. Manconi, C. Schoenberg, M. Klautau, and M. Kelly, J. Vacelet, M. Dohrmann, M.-C. Díaz, P. Cárdenas, J. L. Carballo, P. Ríos, R. Downey, and C. C Morrow. 2020. World *Porifera* Database. Available at http://www.marinespecies.org/porifera.

Thomson, C. W. 1868. On the "vitreous" sponges. *Ann. Mag. Nat. Hist.* 1(4):114–132.

Topsent, E. 1927. Diagnoses d'Éponges nouvelles recueillies par le Prince Albert ler de Monaco. *Bull. Inst. Océanogr. Monaco* 502:1–19.

Authors

Michaël Manuel; Sorbonne Université; Institut de Systématique, Evolution, Biodiversité (ISYEB, UMR 7205 MNHN SU CNRS EPHE); bâtiment A, 4o étage; 7, quai St Bernard; Paris 75005, France. Email: michael.manuel@upmc.fr.

Nicole Boury-Esnault; IMEB-CNRS Station marine d'Endoume; Université de la Méditerranée; rue de la Batterie des Lions; Marseille 13007, France. Email: nicole.boury-esnault@orange.fr.

Date Accepted: 13 April 2013

Primary Editor: Kevin de Queiroz

Homoscleromorpha P. Bergquist 1978 [N. Boury-Esnault and M. Manuel], converted clade name

Registration number: 203

Definition: The most inclusive crown clade containing *Oscarella lobularis* (Schmidt 1862) but not *Spongia officinalis* Linnaeus 1759 (*Demospongiae*) and *Oopsacas minuta* Topsent 1927 (*Hexactinellida*) and *Grantia compressa* (Fabricius 1780) (*Calcispongia*). This is a maximum-crown-clade definition. Abbreviated definition: max crown ∇ (*Oscarella lobularis* (Schmidt 1862) ~ *Spongia officinalis* Linnaeus 1759 & *Oopsacas minuta* Topsent 1927 & *Grantia compressa* (Fabricius 1780)).

Etymology: The name *Homoscleromorpha* is derived from the Greek *homos* "similar" plus *skleros* "hard" plus *morphe* "form", referring to the existence of a single size class of spicules (no distinction between micro- and macroscleres).

Reference Phylogeny: The primary reference phylogeny is Borchiellini et al. (2004: Fig. 2). See also Dorhmann et al. (2008: Fig. 2) and, for more extensive ingroup sampling, Gazavé et al. (2010: Fig. 2). Although *Oscarella lobularis* (the type of *Oscarella*) is not included in the reference phylogeny, it is considered to be more closely related to *Oscarella tuberculata* than to *Plakortis simplex* (Gazave et al., 2010). Similarly, although *Grantia comprsessa* is not included in the reference phylogeny, it is considered to be more closely related to *Leucosolenia sp.* and *Petrobiona massiliana* than to *Leucaltis clathria* and *Soleniscus radovani* (Dohrmann et al., 2006). In order to ensure the exclusion of hexactinellids from *Homoscleromorpha* but because no hexactinellid sponges are included in the reference phylogeny, a representative

of *Hexactinellida* (*Oopsacas minuta*) is also included as an external specifier (see Dorhmann et al., 2008, for the relationships between *Hexactinellida* and *Homoscleromorpha*).

Composition: *Homoscleromorpha* contains about 123 described extant species. Of the subclades of *Porifera* named in this volume, it is the one with the smallest number of species. It is also the clade for which new species are being described at the highest rate, about half of the species having been described in the last 20 years. Two subclades of *Homoscleromorpha* are recognized, one composed of species with spicules (*Plakinidae*) and the other composed of species without spicules (*Oscarellidae*) (Schulze, 1880; Muricy, 1999; Gazave et al., 2010; Ivanišević et al., 2011). For an updated species list, see van Soest et al. (2020).

Diagnostic Apomorphies: Flagellated exo- and endopinacocytes, a basement membrane lining choanoderm and pinacoderm, and a cinctoblastula larva (Muricy and Diaz, 2002).

Synonyms: *Microsclerophora* Sollas 1887; *Megasclerophora* Lendenfeld 1903; *Homosclerophora*, *Homosclerophorida* Dendy 1905 (all approximate).

Comments: In sponges, siliceous spicules are generally classified into two size classes: 1) megascleres, which are relatively simple and large (>~100 µm); and 2) microscleres, which are considerably smaller (<~100 µm) and more complex. Sollas (1887) considered the spicules of the group here named *Homoscleromorpha* similar to the microscleres of other *Demospongiae* and named the group *Microsclerophora*, whereas Lendenfeld (1903) considered these spicules similar to megascleres and named the group *Megasclerophora*. Dendy (1905) decided that it is impossible to allocate the spicules of plakinid sponges to one of these categories and named the group *Homosclerophorida*. Lévi (1973: 591) upgraded the group to the rank of subclass of *Demospongia* with the name "Homosclérophorides". Later Bergquist (1978), on the same basis, named the group *Homoscleromorpha*. The names *Homoscleromorpha* and *Homosclerophorida* are currently used with roughly equal frequency, and we have selected the former because it was previously defined phylogenetically (Borchiellini et al., 2004). However, in contrast to Borchiellini et al. (2004), who defined the name as applying to a total clade, we have applied it to the corresponding crown clade.

The monophyly of *Homoscleromorpha* has been accepted for many years, and the group has been considered a subgroup of *Demospongiae* in previous classifications (Lévi, 1957, 1973; Bergquist, 1978; Muricy and Diaz, 2002). However, recent molecular phylogenies have shown that *Homoscleromorpha* is not part of *Demospongiae* but has its closest relationships with *Calcispongia* (Borchiellini et al., 2004; Dohrmann et al., 2008; Sperling et al., 2009, 2010; Philippe et al., 2009, 2011; Pick et al., 2010; Gazave et al., 2010, 2012).

Literature Cited

Bergquist, P. R. 1978. *Sponges.* Hutchinson, London; University of California Press, Berkeley, CA and Los Angeles, CA.

Borchiellini, C., C. Chombard, M. Manuel, E. Alivon, J. Vacelet, and N. Boury-Esnault. 2004. Molecular phylogeny of *Demospongiae*: implications for classification and scenarios of character evolution. *Mol. Phylogenet. Evol.* 32:823–837.

Dendy, A. 1905. Report on the sponges collected by Prof. Herdman at Ceylon, in 1902. Pp. 57–246 in *Report to the Government of Ceylon on the Pearl Oyster Fisheries of the Gulf of Manaar* (W. A. Herdman, ed.). Royal Society, London.

Dohrmann, M., D. Janussen, J. Reitner, A. G. Collins, and G. Wörheide. 2008. Phylogeny and evolution of glass sponges (*Porifera, Hexactinellida*). *Syst. Biol.* 57:388–405.

Dohrmann, M., O. Voigt, D. Erpenbeck, and G. Wörheide. 2006. Non-monophyly of most supraspecfic taxa of calcareous sponges (*Porifera, Calcarea*) revealed by increased taxon sampling and partitioned Bayesian analysis of ribosomal DNA. *Mol. Phylogenet. Evol.* 40:830–843.

Fabricius, O. 1780. *Fauna Groenlandica, Systematice Sistens, Animalia Groenlandiae Occidentalis Hactenus Indagata, Quoad Nomen Specificum, Triviale, Vernaculumque; Synonym Auctorum Plurium, Descriptionem, Locum, Victum, Generationem. Mores, Usum, Capturamque Singuli; Prout Detegendi Occasio Fuit, Maximaque Parti Secundum Proprias Observationes.* Rothe, Hafniae & Lipsiae.

Gazave, E., P. Lapébie, A. V. Ereskovsky, J. Vacelet, E. Renard, P. Cárdenas, and C. Borchiellini. 2012. No longer *Demospongiae*: *Homoscleromorpha* formal nomination as a fourth class of *Porifera*. *Hydrobiologia* 687:3–10.

Gazave, E., P. Lapébie, E. Renard, J. Vacelet, C. Rocher, A. V. Ereskovsky, D. V. Lavrov, and C. Borchiellini. 2010. Molecular phylogeny restores the supra-generic subdivision of homoscleromorph sponges (*Porifera, Homoscleromorpha*). *PLOS ONE* 5:1–15.

Ivanišević, J., O. P. Thomas, C. Lejeusne, P. Chevaldonné, and T. Pérez. 2011. Metabolic fingerprinting as an indicator of biodiversity: towards understanding inter-specific relationships among *Homoscleromorpha* sponges. *Metabolomics* 7:289–304.

Von Lendenfeld, R. 1903. *Porifera. Tetraxonia.* Pp. 1–168 in *Das Tierreich. 19* (F. E. Schulze, ed.). Friedländer, Berlin.

Lévi, C. 1957. Ontogeny and systematics in sponges. *Syst. Zool.* 6(4):174–183.

Lévi, C. 1973. Systématique de la classe des *Demospongiaria* (Démosponges). Pp. 577–631 in *Traité de Zoologie. Anatomie, Systématique, Biologie. Spongiaires*, Vol. 3(1) (P.-P. Grassé, ed.). Masson et Cie, Paris.

Linnaeus, C. 1759. *Species Plantarum*. Laurentii Salvii, Stockholm.

Muricy, G. 1999. An evaluation of morphological and cytological data sets for the phylogeny of *Homosclerophorida* (*Porifera*: *Demospongiae*). *Mem. Queensl. Mus.* 44:399–409.

Muricy, G., and C. Diaz. 2002. Order *Homosclerophorida* Dendy, 1905, Family *Plakinidae* Schluze, 1880. Pp. 71–82 in *Systema Porifera: A Guide to the Classification of Sponges* (J. N. A. Hooper and R. W. M. van Soest, eds.). Kluwer Academic and Plenum Publishers, Dordrecht.

Philippe, H., R. Derelle, P. Lopez, K. Pick, C. Borchiellini, N. Boury-Esnault, J. Vacelet, E. Renard, E. Houliston, E. Quéinnec, C. Da Silva, P. Wincker, H. Le Guyader, S. Leys, D. J. Jackson, F. Schreiber, D. Erpenbeck, B. Morgenstern, W. Wörheide, and M. Manuel. 2009. Phylogenomics revives traditional views on deep animal relationships. *Curr. Biol.* 19:1–7.

Philippe, H., H. Brinkmann, D. V. Lavrov, D. T. J. Littlewood, M. Manuel, G. Wörheide, and D. Baurain. 2011. Resolving difficult phylogenetic questions: why more sequences are not enough. *PLOS Biol.* 9(3):e1000602.

Pick, K. S., H. Philippe, F. Schreiber, D. Erpenbeck, D. J. Jackson, P. Wrede, M. Wiens, A. Alié, B. Morgenstern, M. Manuel, and G. Wörheide. 2010. Improved phylogenomic taxon sampling noticeably affects nonbilaterian relationships. *Mol. Biol. Evol.* 27:1983–1987.

Schmidt, O. 1862. *Die Spongien des Adriatischen Meeres*. Wilhelm Engelmann, Leipzig.

Schulze, F. E. 1880. Untersuchungen über den Bau und die Entwicklung der Spongien. Neunte Mittheilung. Die Plakiniden. *Zeits. wiss. Zool.* 34(2):407–451.

Sollas, W. J. 1887. Sponges. Pp. 412–429 in *Encyclopaedia Britannica*. 9th edition, Vol. 22 (W. R. Smith, ed.). A. Black and C. Black, Edinburgh.

Sperling, E. A., K. Peterson, and D. Pisani. 2009. Phylogenetic-signal dissection of nuclear housekeeping genes supports the paraphyly of sponges and the monophyly of *Eumetazoa*. *Mol. Biol. Evol.* 26:2261–2274.

Sperling, E. A., J. M. Robinson, D. Pisani, and K. J. Peterson. 2010. Where's the glass? Biomarkers, molecular clocks, and microRNAs suggest a 200-Myr missing Precambrian fossil record of siliceous sponge spicules. *Geobiology* 8:24–36.

van Soest, R. W. M., N. Boury-Esnault, J. N. A. Hooper, K. Rützler, N. J. de Voogd, B. Alvarez, E. Hajdu, A. B. Pisera, R. Manconi, C. Schœnberg, M. Klautau, M. Kelly, J. Vacelet, M. Dohrmann, M.-C. Díaz, P. Cárdenas, J. L. Carballo, P. Ríos, R. Downey, and C. C Morrow. 2020. World *Porifera* Database. Available at http://www.marinespecies.org/porifera. doi:10.14284/359.

Topsent, E. 1927. Diagnoses d'Éponges nouvelles recueillies par le Prince Albert ler de Monaco. *Bull. Inst. Océanogr. Monaco* 502:1–19.

Authors

Nicole Boury-Esnault; IMEB-CNRS Station marine d'Endoume; Université de la Méditerranée; rue de la Batterie des Lions 13007 Marseille, France. Email: nicole.boury-esnault@orange.fr.

Michaël Manuel; Université Paris 06; UMR 7138 CNRS UPMC MNHN IRD, bâtiment A, 4° étage; 7, quai St Bernard, 75005 Paris, France. Email: michael.manuel@snv.jussieu.fr.

Date Accepted: 15 May 2013

Primary Editor: Kevin de Queiroz

Calcispongia M. H. D. de Blainville 1830 [M. Manuel and N. Boury-Esnault], converted clade name

Registration Number: 204

Definition: The least inclusive crown clade containing *Clathrina clathrus* (Schmidt 1864) (*Calcinea*) and *Grantia compressa* (Fabricius 1780) (*Calcaronea*). This is a minimum-crown-clade definition. Abbreviated definition: min crown ∇ (*Clathrina clathrus* (Schmidt 1864) & *Grantia compressa* (Fabricius 1780)).

Etymology: The name *Calcispongia* is derived from the Latin *calcis* "lime, chalk" referring to the calcium carbonate composition of the spicules in members of this taxon, and *spongia* "sponge".

Reference Phylogeny: The primary reference phylogeny is Manuel et al. (2003: Fig. 4). See also Manuel et al. (2004: Fig. 1); Dohrmann et al. (2006: Fig. 1); Voigt et al. (2012: Fig. 3).

Composition: *Calcispongia* is composed of approximately 785 currently recognized species, which represent about 9 % of known sponge species. Two sister clades *Calcinea* and *Calcaronea* proposed by Bidder (1898) have been shown to be monophyletic and are now widely accepted (Manuel et al., 2003, 2004; Manuel, 2006; Dohrmann et al., 2006). For an extant species list see van Soest et al. (2020).

Diagnostic Apomorphies: The skeleton is composed of monocrystalline calcareous spicules (Haeckel, 1872; Poléjaeff, 1883; Borojević et al., 1990, 2000; Manuel et al., 2002; Manuel, 2006).

Synonyms: *Calcarea* Bowerbank 1864 (approximate).

Comments: Two names have been widely used for calcareous sponges: *Calcarea* Bowerbank 1864 and *Calcispongia* Blainville 1830 (including the variant *Calcispongiae* Schmidt 1862). The latter name was originally coined as a genus name but was subsequently elevated to the class level by Johnston (1842) and used in this sense by, for example, Haeckel (1872). Manuel (2006) proposed *Calcispongia* for the crown clade (the smallest clade comprising all extant calcareous sponge species) and the name *Calcarea* for the corresponding total clade (*Calcispongia* + all fossil species closer to *Calcispongia* than to any other crown clade). The main fossil calcareous sponge taxon, *Heteractinida*, belongs to *Calcarea* but not to *Calcispongia*, according to Pickett (2002).

The possession of a skeleton made of calcium carbonate spicules makes the *Calcispongia* unique relative to all other sponges. Over the past two centuries, the monophyletic origin of calcareous sponges has never been seriously doubted. Recently, molecular phylogenetic studies using small- and large-subunit rDNA sequences (Manuel et al., 2003, 2004; Manuel, 2006; Dohrmann et al., 2006) have confirmed, with strong support, their monophyly as well as that of *Calcinea* and *Calcaronea*. Members of *Calcinea* have equiangular triactine spicules, a basal nucleus in the choanocytes, a flagellum arising independently from the nucleus, a coeloblastula larva, and triactines as the first spicules to appear during ontogenesis, whereas those of *Calcaronea* possess inequiangular triactines, an apical nucleus in the choanocytes, a flagellum arising from the nucleus, a stomoblastula larva which after eversion becomes an amphiblastula, and diactines as the first spicules to appear during ontogenesis. However, the knowledge of the world fauna of *Calcispongia* is very fragmentary. The total number of described species represents less than 9% of all described extant sponges, partially due to a bias in taxonomic effort, although recently there has been a revival of taxonomic interest in *Calcispongia* due to the interest in this clade as a model for evolutionary developmental studies.

Literature Cited

de Blainville, M. H. D. 1830. Zoologie. Vers et zoophytes. Pp. 1–546 in *Dictionnaire des Sciences Naturelles* (F. G. Levrault, ed.). Le Normant, Paris.

Bidder, G. P. 1898. The skeleton and classification of calcareous sponges. *Proc. R. Soc. Lond. B Biol. Sci.* 64:61–76.

Borojević, R., N. Boury-Esnault, and J. Vacelet. 1990. A revision of the supraspecific classification of the subclass *Calcinea* (*Porifera*, class *Calcarea*). *Bull. Mus. Natl. Hist. Nat.* (4, A) 12(2):243–276.

Borojević, R., N. Boury-Esnault, and J. Vacelet. 2000. A revision of the supraspecific classification of the subclass *Calcaronea* (*Porifera*, class *Calcarea*). *Zoosystema* 22(2):203–263.

Bowerbank, J. S. 1864. *A Monograph of the British Spongiadae*, Vol. 1, Pp. i–xx, 1–290. Ray Society, London.

Dohrmann, M., O. Voigt, D. Erpenbeck, and G. Wörheide. 2006. Non-monophyly of most supraspecific taxa of calcareous sponges (*Porifera*, *Calcarea*) revealed by increased taxon sampling and partitioned Bayesian analysis of ribosomal DNA. *Mol. Phylogenet. Evol.* 40:830–843.

Fabricius, O. 1780. *Fauna Groenlandica: Systematice Sistens Animalia Groenlandiae Occidentalis Hactenus Indagata, quod Nomen Specificium.* Hafniae et Lipsiae, Copenhagen, Denmark.

Haeckel, E. 1872. *Die Kalkschwämme. Eine Monographie in zwei Bänden Text und einem Atlas mit 60 Tafeln.* G. Reimer, Berlin.

Johnston, G. 1842. *A History of British Sponges and Lithophytes*, Pp. i–xii, 1–264 W. H. Lizars, Edinburgh.

Manuel, M. 2006. Phylogeny and evolution of calcareous sponges. *Can. J. Zool.* 84:225–241.

Manuel, M., C. Borchiellini, E. Alivon, and N. Boury-Esnault. 2004. Molecular phylogeny of calcareous sponges using 18S rRNA and 28S rRNA sequences. *Boll. Mus. Ist. Biol. Genova* 68:449–461.

Manuel, M., C. Borchiellini, E. Alivon, Y. Le Parco, J. Vacelet, and N. Boury-Esnault. 2003. Phylogeny and evolution of calcareous sponges: monophyly of *Calcinea* and *Calcaronea*, high level of morphological homoplasy, and the primitive nature of axial symmetry. *Syst. Biol.* 52:311–333.

Manuel, M., R. Borojević, N. Boury-Esnault, and J. Vacelet. 2002. Class *Calcarea* Bowerbank, 1864. Pp. 1103–1110 in *Systema* Porifera: *A Guide to the Classification of Sponges* (J. N. A. Hooper and R. W. M. van Soest, eds.). Kluwer Academic and Plenum Publishers, Dordrecht.

Pickett, J. 2002. Fossil *Calcarea*. An overview. Pp. 1117–1119 in *Systema* Porifera: *A Guide to the Classification of Sponges.* (J. N. A. Hooper and R. W. M.van Soest, eds.). Kluwer Academic and Plenum Publishers, Dordrecht.

Poléjaeff, N. 1883. Report on the *Calcarea* dredged by H.M.S. 'Challenger', during the years 1873–1876. Report on the Scientific Results of the Voyage of H.M.S. 'Challenger', 1873–1876. *Zoology* 8(2):1–76.

Schmidt, O. 1864. *Supplement der Spongien des Adriatischen Meeres. Enthaltend die Histologie und Systematische Ergänzungen.* Wilhelm Engelmann, Leipzig.

Soest, R. W. M. van, N. Boury-Esnault, J. N. A. Hooper, K. Rützler, N. J. de Voogd, B. Alvarez, E. Hajdu, A. B. Pisera, R. Manconi, C. Schoenberg, M. Klautau, M. Kelly, J. Vacelet, M. Dohrmann, M.-C. Díaz, P. Cárdenas, J. L. Carballo, P. Ríos, R. Downey, and C. C Morrow. 2020. World *Porifera* Database. Available at http://www.marinespecies.org/porifera. doi:10.14284/359.

Voigt, O., E. Wülfing, and G. Wörheide. 2012. Molecular phylogenetic evaluation of classification and scenarios of character evolution in calcareous sponges (*Porifera*, Class *Calcarea*). *PLOS ONE* 7(3):e33417.

Authors

Michaël Manuel; Sorbonne Université, Institut de Systématique, Evolution, Biodiversité (ISYEB, UMR 7205 MNHN SU CNRS EPHE), bâtiment A, 4o étage; 7, quai St Bernard, 75005 Paris, France. Email: michael.manuel@upmc.fr.

Nicole Boury-Esnault; IMBE-CNRS; Station marine d'Endoume; Université de la Méditerranée; rue de la Batterie des Lions 13007 Marseille, France. Email: nicole.boury-esnault@orange.fr.

Date Accepted: 18 March 2013

Primary Editor: Kevin de Queiroz

Cnidaria A. E. Verrill 1865 [A. G. Collins, M. Daly, and C. W. Dunn], converted clade name

Registration Number: 165

Definition: The crown clade originating in the most recent common ancestor of *Hydra vulgaris* (Pallas 1766) and *Nematostella vectensis* Stephenson 1935. This is a minimum-crown-clade definition. Abbreviated definition: min crown ∇ (*Hydra vulgaris* (Pallas 1766) & *Nematostella vectensis* Stephenson 1935).

Etymology: Derived from the Greek word *knide* (χνιδη) for nettle, in reference to the stinging sensation caused on contact with some species. Its use in reference to cnidarian taxa can be traced back 2,350 years to Aristotle's *Historia Animalium* (Balme, 2002).

Reference Phylogeny: Collins (2002: Fig. 1), showing *Cnidaria* as a clade consisting of *Anthozoa* and *Medusozoa*, is chosen as the reference phylogeny because of its relatively strong sampling of both outgroup and ingroup taxa. For the purpose of applying our definition, *Hydra vulgaris* (representing *Medusozoa*) is regarded as closely related to *Hydra littoralis* and *H. circumcincta*, and *Nematostella vectensis* (representing *Anthozoa*) is regarded as closely related to *Anthopleura midori* and *A. kuogane*.

Composition: *Anthozoa* and *Medusozoa* as defined in this volume; i.e., all sea anemones (*Actiniaria*, *Zoanthidea*, *Corallimorpharia*), jellyfishes (some *Cubozoa*, *Hydrozoa*, and *Scyphozoa*), soft corals (some *Octocorallia*), stony corals (*Scleractinia*), hydroids (some *Hydrozoa*), sea fans (some *Octocorallia*), sea pens (*Pennatulacea*), black corals (*Antipatharia*), and their relatives, and probably *Myxozoa* (see Comments). *Cnidaria* presently contains more than 13,000 extant species, including approximately 11,000 anthozoans and medusozoans (Daly et al., 2007) and roughly 2,200 myxozoans (Lom and Dyková, 2006). Given the relative lack of taxonomic effort on the group, the true species richness of *Cnidaria* is likely to be significantly higher.

Diagnostic Apomorphies: The production of cnidae, organelle-like capsules with eversible tubules (Watson, 1988), is the diagnostic apomorphy of *Cnidaria*. Of the three types of cnidae, nematocysts – ptychocysts – and spirocysts, only nematocysts are found in all anthozoan and medusozoan species surveyed. The latter two types are restricted to ceriantharian tube anemones and *Anthozoa*, respectively, so they are likely to be derived. If *Myxozoa* is part of *Cnidaria* (see below), then myxozoan polar capsules would likely be derived cnidae as well, as suggested previously (Weill, 1938; Siddall et al., 1995). Thus, cnidae of the nematocyst form, though this varies greatly, might be considered a shared derived character of the group. There are no known reversals in this character across the clade; that is, no cnidarians are known to lack cnidae. Three other features that are sometimes considered to be diagnostic of *Cnidaria* are radial symmetry and planula and polyp stages in development, but all are problematic. Although many cnidarians exhibit radial symmetry, some are directionally asymmetric (Dunn and Wagner, 2006), and many have a bilateral organization, leading some to conclude that bilateral symmetry is the ancestral condition for the clade (Salvini-Plawen, 1978; Matus et al., 2006). The other two features are difficult to define. For example, the motile stage between embryo and settled juvenile in any given cnidarian's life cycle is typically termed a planula, and, although this stage is usually ciliated, sausage-shaped, and non-feeding, deviations from this pattern—e.g., *Haliclystus* (*Medusozoa*/*Staurozoa*), *Hydra* (*Hydrozoa*/*Hydroidolina*), *Zoanthidea* (*Anthozoa*/*Hexacorallia*)—have been well documented (Marques and Collins, 2004). Polyp forms are even more variable than planulae. They may be solitary or colonial, with or without polymorphic zooids (if colonial), with or without mineralized skeletons, benthic or pelagic, with or without tentacles, and even absent (Hyman, 1940). Thus, cnidae are the one clear derived feature of the group.

Synonyms: *Coelenterata* Frey and Leuckart 1847 (approximate).

Comments: Milne-Edwards (1857) introduced the name *Cnidaires* to describe a group that more or less corresponds to our present concept of *Anthozoa*. Nonetheless, his *Cnidaires* included members of both *Anthozoa* and *Medusozoa*. This usage was adopted by Verrill (1865), who was the first to use the name *Cnidaria*. Hatschek (1888) is often cited as the first to use the name in its modern sense because he separated Frey and Leuckart's (1847) *Coelenterata* into *Cnidaria*, *Ctenophora*, and sponges (as *Spongiaria*). However, Hatschek's *Cnidaria* contained mesozoans (e.g., *Dyciemidae* and *Orthonectidae*). The earliest use of *Cnidaria* referring exclusively to anthozoans and medusozoans in a text covering other metazoan taxa (including sponges, ctenophores, and mesozoans) appears to have been that of McMurrich (1894). The earliest tree-like phylogenetic hypotheses including the name *Cnidaria* are presented by Haeckel (e.g., 1896). Although Haeckel's *Cnidaria* included *Ctenophora*, he made it clear that the origin of *Ctenophora* was problematic and difficult to determine with any certainty (Haeckel, 1896).

The name *Coelenterata* (Frey and Leuckart, 1847) is occasionally used synonymously with *Cnidaria*. *Coelenterata* originally encompassed cnidarians and ctenophores, but later included sponges as well. After Hatschek (1888) broke up *Coelenterata* into the three distinct groups, *Coelenterata* has been inconsistently used to refer either to *Cnidaria* plus *Ctenophora* or to just *Cnidaria*.

The reference phylogeny (Collins, 2002), based on 18S rDNA data, follows a series of papers (Cavalier-Smith et al., 1996; Collins, 1998; Berntson et al., 1999; Kim et al., 1999; Won et al., 2001) in which sampling of 18S from cnidarian species increased, starting from the seminal work of Wainright et al. (1993: Fig. 1), which showed a well-supported monophyletic *Cnidaria* with *Anthozoa* and *Medusozoa* each represented by a single species. Phylogenetic topologies including a monophyletic *Cnidaria* have also been derived on morphological grounds (Salvini-Plawen, 1978; Bridge et al., 1995) and 28S rDNA sequences (Odorico and Miller, 1997; Medina et al., 2001).

Myxozoa is a group of enigmatic parasites that have sometimes been placed within *Cnidaria*. Though various analyses of its phylogenetic position have come to conflicting conclusions (Siddall et al., 1995; Zrzavý and Hypša, 2003; Jiménez-Guri et al., 2007), most available evidence suggests that *Myxozoa* is part of *Cnidaria*. If it were found instead that *Cnidaria* and *Myxozoa* are sister groups, an additional name would be required to designate the clade including both groups. No fossils appear to have been hypothesized to be stem members outside the crown clade *Cnidaria* as defined here, although it is difficult to imagine what evidence could be used to make or test such a conjecture given that the relevant features (e.g., presence/absence of life cycle stages, nematocyst types, etc.) have little fossilization potential.

Literature Cited

Balme, D. M. 2002. *Aristotle, Historia Animalium. Cambridge Classical Texts and Commentaries* (No. 38). Cambridge University Press, Cambridge, UK.

Berntson, E. A., S. C. France, and L. S. Mullineaux. 1999. Phylogenetic relationships within the Class *Anthozoa* (Phylum *Cnidaria*) based on nuclear 18S rDNA sequences. *Mol. Phylogenet. Evol.* 13:417–433.

Bridge, D., C. W. Cunningham, R. Desalle, and L. W. Buss. 1995. Class-level relationships in the phylum *Cnidaria*: molecular and morphological evidence. *Mol. Biol. Evol.* 12:679–689.

Cavalier-Smith, T., M. T. E. P. Allsopp, E. E. Chao, N. Boury-Esnault, and J. Vacelet. 1996. Sponge phylogeny, animal monophyly, and the origin of the nervous system: 18S rRNA evidence. *Can. J. Zool.* 74:2031–2045.

Collins, A. G. 1998. Evaluating multiple alternative hypotheses for the origin of *Bilateria*: an analysis of 18S rRNA molecular evidence. *Proc. Natl. Acad. Sci. USA* 95:15458–15463.

Collins, A. G. 2002. Phylogeny of *Medusozoa* and the evolution of cnidarian life cycles. *J. Evol. Biol.* 15:418–432.

Daly, M., M. R. Brugler, P. Cartwright, A. G. Collins, M. N. Dawson, D. G. Fautin, S. C. France, C. S. McFadden, D. M. Opresko, E. Rodriguez, S. L. Romano, and J. L. Stake. 2007. The phylum *Cnidaria*: a review of phylogenetic patterns and diversity 300 years after Linnaeus. *Zootaxa* 1668:127–182.

Dunn, C. W., and G. P. Wagner. 2006. The evolution of colony-level development in the *Siphonophora* (*Cnidaria*: *Hydrozoa*). *Dev. Genes Evol.* doi:10.1007/s00427-006-0101-8.

Frey, H., and R. Leuckart. 1847. *Beiträgezur Kenntniss wirbelloser Thiere mit besonderer Berücksichtigung der Fauna des Norddeutschen Meeres.* Friedrich Vieweg und sohn, Braunschweig.

Haeckel E., 1896. *Systematische Phylogenie. Zweiter Theil: Systematische Phylogenie der Wirbellosen Thiere* (Invertebrata). Georg Reimer, Berlin.

Hatschek 1888. *Lehrbuch der Zoologie*, Vol. 1. G. Fischer, Jena.

Hyman, L. H. 1940. *The Invertebrates. Vol. 1:* Protozoa *through* Ctenophora. McGraw-Hill, New York.

Jiménez-Guri, E., H. Philippe, B. Okamura, and P. W. H. Holland. 2007. *Buddenbrockia* is a cnidarian worm. *Science* 317:116–118.

Kim, J. H., W. Kim, and C. W. Cunningham. 1999. A new perspective on lower metazoan relationships from 18S rDNA sequences. *Mol. Biol. Evol.* 16:423–427.

Lom, J., and I. Dyková. 2006. Myxozoan genera: definition and notes on taxonomy, life-cycle terminology and pathogenic species. *Folia Parasitol.* 53:1–36.

Marques, A. C., and A. G. Collins. 2004. Cladistic analysis of *Medusozoa* and cnidarian evolution. *Invertebr. Biol.* 123:23–42.

Matus. D. Q., K. Pang, H. Marlow, C. W. Dunn, G. H. Thomsen, and M. Q. Martindale. 2006. Molecular evidence for deep evolutionary roots of bilaterality in animal development. *Proc. Natl. Acad. Sci. USA* 103:11195–11200.

McMurrich, J. P. 1894. *A Textbook of Invertebrate Morphology.* Henry Holt and Company, New York.

Medina, M., A. G. Collins, J. D. Silberman, and M. L. Sogin. 2001. Evaluating hypotheses of basal animal phylogeny using complete sequences of large and small subunit rRNA. *Proc. Natl. Acad. Sci. USA* 98:9707–9712.

Milne-Edwards, H. 1857. *Histoire Naturelle des Coralliares ou Polypes Proporement Dits,* Vol. 1. Libraire Encyclopedique de Roret, Paris.

Odorico, D. M., and D. J. Miller. 1997. Internal and external relationships of the *Cnidaria*: implications of primary and predicted secondary structure of the 5′-end of the 23S-like rDNA. *Proc. R. Soc. Lond. B Biol. Sci.* 264:77–82.

Salvini-Plawen. 1978. On the origin and evolution of the lower *Metazoa. Z. Zool. Syst. Evol.-Forsch.* 16:40–88.

Siddall, M. E., D. S. Martin, D. Bridge, D. M. Cone, and S. S. Desser. 1995. The demise of a phylum of protists: *Myxozoa* and other parasitic *Cnidaria. J. Parasitol.* 81:961–967.

Verrill, A. E. 1865. Classification of polyps. *Proc. Essex Inst.* 4:145–152.

Wainright, P. O., G. Hinkle, M. L. Sogin, and S. K. Stickel. 1993. Monophyletic origins of the *Metazoa*: an evolutionary link with fungi. *Science* 260:340–342.

Watson, G. M. 1988. Ultrastructure and cytochemistry of developing nematocysts. Pp. 143–164 in *The Biology of Nematocysts* (D. A. Hessinger and H. M. Lenhoff, eds.). Academic Press, San Diego and other cities.

Weill, R. 1938. L'interpretation des Cnidosporidies et la valeur taxonomique de leur cnidome. Leur cycle comparé à la phase larvaire des Narcomeduses Cuninides. *Trav. Stat. Zool. Wimer.* 13:727–744.

Won, J., B. Rho, and J. Song. 2001. A phylogenetic study of the *Anthozoa* (phylum *Cnidaria*) based on morphological and molecular characters. *Coral Reefs* 20:39–50.

Zrzavý, J., and V. Hypša. 2003. *Myxozoa, Polypodium*, and the origin of the *Bilateria*: the phylogenetic position of "*Endocnidozoa*" in light of the rediscovery of *Buddenbrockia*. *Cladistics* 19:91–180.

Authors

Allen G. Collins; National Systematics Laboratory of NOAA's Fisheries Service & National Museum of Natural History, MRC-153, Smithsonian Institution, P.O. Box 37012, Washington, DC 20013-7012, USA. Email: collinsa@si.edu.

Marymegan Daly; Department of Evolution; Ecology and Organismal Biology; The Ohio State University; 1315 Kinnear Road; Columbus, OH 43212, USA. Email: daly.66@osu.edu.

Casey W. Dunn; Department of Ecology and Evolutionary Biology; Yale University; 165 Prospect St.; New Haven, CT 06511, USA. E-mail: casey.dunn@yale.edu.

Date Accepted: 31 March 2011

Primary Editor: Kevin de Queiroz

Anthozoa C. G. Ehrenberg 1834 [M. Daly], converted clade name

Registration Number: 164

Definition: The most inclusive crown clade containing both *Nematostella vectensis* Stephenson 1935 (*Hexacorallia*) and *Alcyonium digitatum* Linnaeus 1758 (*Octocorallia*) but not *Hydra vulgaris* (Pallas 1766) (*Hydrozoa*). This is a maximum-crown-clade definition with two internal specifiers. Abbreviated definition: max crown ∇ (*Nematostella vectensis* Stephenson 1935 & *Alcyonium digitatum* Linnaeus 1758 ~ *Hydra vulgaris* (Pallas 1766)).

Etymology: Derived from *anthos* (Gk., flower) + *zoön* (Gk., animal) referring to the flower-like external appearance of polyps.

Reference Phylogeny: The primary reference phylogeny is France et al. (1996: Fig. 3), showing *Anthozoa* as a clade consisting of *Octocorallia* and *Hexacorallia* (the latter represented in the reference phylogeny by *Ceriantharia*, *Corallimorpharia*, *Scleractinia*, *Antipatharia*, and *Actiniaria*). This analysis is based on mitochondrial 16S rDNA data and samples the ingroup more densely than previous studies (e.g., Song et al., 1994; Bridge et al., 1995; Cavalier-Smith et al., 1996). For the purpose of applying our definition in the context of the reference phylogeny, *Nematostella vectensis* is to be considered part of *Actiniaria* (see Daly et al., 2003, 2008), *Alcyonium digitatum* is to be considered closely related to *Alcyonium* sp. (see McFadden et al., 2006), and *Hydra vulgaris* is to be considered part of *Hydrozoa* (see Bridge et al., 1995).

Composition: *Hexacorallia* and *Octocorallia* as defined in this volume; i.e., all sea anemones (*Actiniaria*), stony corals (*Scleractinia*), soft corals and sea fans (*Alcyonacea*), sea pens (*Pennatulacea*), black corals (*Antipatharia*), and their relatives. All members of *Anthozoa* are exclusively polypoid; they may be colonial, clonal, or solitary, and either naked or with a mineralic or proteinaceous skeleton. *Anthozoa* presently contains about 6,000 extant species (Dunn, 1982; Daly et al., 2007).

Fossils attributed to *Anthozoa* generally have been associated with particular subclades, rather than inferred to be stem anthozoans. For example, members of *Tetracorallia* (rugose corals) are inferred to belong to *Hexacorallia*. Neoproterozoic fossils that had been interpreted as belonging to *Anthozoa*, such as the Ediacaran *Charnia*, have had this placement seriously questioned (Antcliffe and Brasier, 2007). Fossil embryos from the Neoproterozoic of China have been interpreted as belonging to *Anthozoa* (see Chen et al., 2002), but the particular subclade or relationship has not been specified.

Diagnostic Apomorphies: *Anthozoa* has at least 2 diagnostic apomorphies: the actinopharynx, and mesenteries. The actinopharynx (= stomadeum, gullet) is an ectodermally lined tube that projects into the gastrovascular cavity; this structure is found in all *Anthozoa*, with no known reversals. A densely ciliated, often more highly glandular region of the actinopharynx, called siphonoglyph (= sulcus) is characteristic of many *Anthozoa*, though it is absent in *Antipatharia* and some *Actiniaria*. The siphonoglyph helps to determine the bilateral symmetry for the polyp (Finnerty et al., 2004). Bilateral symmetry is further defined by the mesenteries (= septa, gastric septa), sheets of tissue that extend all or part of the way from the body wall to the actinopharynx. The longitudinal epidermal musculature of mesenteries allows a polyp to shorten. The homology of mesenteries and the gastric septa of some medusozoans, e.g., *Stauromedusae*, is not clear; the two differ morphologically in that at least some of the mesenteries span the gastrovascular cavity, attaching on one side to the body wall and on the other to the actinopharynx. Unlike gastric septa, mesenteries bear the gametogenic tissue. Furthermore, the free edge of a mesentery is typically elaborated into a mesenterial filament that may contain numerous gland cells, nematocysts, and cilia. The exclusively polypoid nature of the anthozoan life cycle is sometimes considered a synapomorphy for the group (e.g., Hyman, 1940; Brusca and Brusca, 1990), but this attribute is shared with at least some medusozoans (e.g., *Hydra vulgaris*), and may be a plesiomorphy (Collins et al., 2006).

Synonyms: *Actinoidea* Dana 1846 (approximate); *Cnidaria* Milne-Edwards 1857 (approximate).

Comments: *Anthozoa* is preferred over alternative names because it is the most widely used of all previously proposed names (see Synonyms) for this clade. Older names fail to include all relevant lineages, or erroneously include members of *Medusozoa* or other taxa. For example, Rapp (1829) subdivided polyps based on the ways in which reproductive cells were formed, and thus considered members of *Anthozoa* part of the group he called *Endoairer*; Hertwig and Hertwig (1879) followed this nomenclature and taxonomic composition but called the group *Endocarpe* rather than *Endoairer*. De Blainville (1834) grouped members of *Anthozoa* (and several *Medusozoa*) into his "Type" *Actinozoaires*; most *Anthozoa* placed in *Actinozoaires* were grouped in the "Class" *Zoanthaires*. Because the names *Zoanthaires* and *Actinozoaires* are associated with *Hexacorallia* or clades within it, it is inappropriate to adopt either of these names for the clade here named. The later name *Actinoidea* Dana 1846 closely parallels the modern concept of *Anthozoa*, including the distinction between *Octocorallia* and *Hexacorallia*

(as *Alcyonaria* and *Actiniaria*, respectively); in addition to being slightly younger than Ehrenberg's (1834) name, *Actinoidea* has the disadvantage of having been used to refer to *Actiniaria*, a clade within *Hexacorallia*. Similarly, Milne-Edwards (1857) used the name *Cnidaria* (as *Cnidaries*) for a group that closely corresponds to the current circumscription of *Anthozoa*. The name *Cnidaria* has been used most commonly as the name for the clade encompassing *Anthozoa* and its sister clade, *Medusozoa*; thus, restricting it to one of these two clades would generate unnecessary confusion.

An early phylogeny based on "alignment-stable" regions of nuclear 18S rDNA (=stems) of a relatively small subset of *Cnidaria* found *Anthozoa* to be paraphyletic with respect to a monophyletic *Medusozoa* (Bridge et al., 1995: Fig. 2). However, this result was inferred by the authors to be an artifact of the analytical methods and the taxon sample because *Anthozoa* was recovered as a clade under some analytical parameters and when additional data and taxa were considered (Bridge et al., 1995). Indeed, the monophyly of *Anthozoa* has been corroborated in subsequent studies of nuclear 18S rDNA (Song and Won, 1997; Berntson et al., 1999; Collins, 2002) and 28S rDNA (Odorico and Miller, 1997). Morphological data have long been offered in support of anthozoan monophyly (e.g., Hyman, 1940; Wells and Hill, 1956; Brusca and Brusca, 1990; Schuchert, 1993), and phylogenetic analyses of morphological data (e.g., Won et al., 2001: Fig. 2) have corroborated this hypothesis. The only exceptions to this pattern are phylogenetic analyses based on mitochondrial genomes (Kayal and Lavrov, 2008; Kayal et al., 2013), which studied a taxon sample similar to that of Bridge et al. (1995) and came to similar conclusions. Genome-scale analyses (Zapata et al., 2015; Kayal et al., 2018) consistently find a monophyletic *Anthozoa*. A maximum-crown-clade definition with both hexacoral and octocoral internal specifiers was chosen so that the name *Anthozoa* will not apply to any clade in the context of phylogenies in which a group composed of hexacorals and octocorals is paraphyletic relative to hydrozoans (and thus also medusozoans).

The scenario depicted in Bridge et al. (1995) and Kayal and Lavrov (2008) corresponds to a hypothesis in the historical literature (e.g., Hadzi, 1963; Werner, 1973), namely, that *Anthozoa* gave rise to *Medusozoa*. Although most recent phylogenetic analyses have supported reciprocal monophyly of *Medusozoa* and *Anthozoa*, because anthozoans lack the specialized life-history of medusozoans, and because most larger-scale phylogenetic analyses sample *Medusozoa* more densely, *Anthozoa* is often interpreted as the "basal lineage" of *Cnidaria* (e.g., Wikramanayake et al., 2003; Extavour et al., 2005; Kamm et al., 2006). However, because *Anthozoa* and *Medusozoa* are reciprocally monophyletic sister clades, characterization of *Anthozoa* as the basal lineage of *Cnidaria* cannot be justified phylogenetically.

Literature Cited

Antcliffe, J. B., and M.D. Brasier. 2007. *Charnia* and sea pens are poles apart. *J. Geol. Soc. Lond.* 164:49–51.

Berntson, E. A., S. C. France, and L. S. Mullineaux. 1999. Phylogenetic relationships within the Class *Anthozoa* (Phylum *Cnidaria*) based on nuclear 18S rDNA sequences. *Mol. Phylogenet. Evol.* 13:417–433.

Bridge, D., C. W. Cunningham, R. Desalle, and L. W. Buss. 1995. Class-level relationships in the phylum *Cnidaria*: molecular and morphological evidence. *Mol. Biol. Evol.* 12:679–689.

de Blainville, H. M. D. 1834. *Manuel de Actinologie ou de Zoophytologie.* F.G. Levrault, Paris.

Brusca, R. C., and G. J. Brusca. 1990. *The Invertebrates.* Sinauer Associates, Sunderland, MA.

Cavalier-Smith, T., M. T. E. P. Allsopp, E. E. Chao, N. Boury-Esnault, and J. Vacelet. 1996. Sponge phylogeny, animal monophyly, and the origin of the nervous system: 18S rRNA evidence. *Can. J. Zool.* 74:2031–2045.

Chen, J.-Y., P. Oliveri, F. Gao, S. Q. Dornbos, C.-W. Li, D. J. Bottjer, and E. H. Davidson. 2002. Precambrian animal life: probable developmental and adult cnidarian forms from southwest China. *Dev. Biol.* 248:182–196.

Collins, A. G. 2002. Phylogeny of *Medusozoa* and the evolution of cnidarian life cycles. *J. Evol. Biol.* 15:418–432.

Collins, A. G., P. Schuchert, A. C. Marques, T. Jankowski, M. Medina, and B. Schierwater. 2006. Medusozoan phylogeny and character evolution clarified by new large and small subunit rDNA data and an assessment of the utility of phylogenetic mixture models. *Syst. Biol.* 55:97–115.

Daly, M., M. Brugler, P. Cartwright, A. G. Collins, M. N. Dawson, S. C. France, D. G. Fautin, C. S. McFadden, D. M. Opresko, E. Rodríguez, S. L. Romano, and J. L. Stake. 2007. The phylum *Cnidaria*: a review of phylogenetic patterns and diversity three hundred years after Linnaeus. *Zootaxa* 1668:127–186.

Daly, M., A. Chaudhuri, L. Gusmão, and E. Rodriguez. 2008. Phylogenetic relationships among sea anemones (*Cnidaria*: *Anthozoa*: *Actiniaria*). *Mol. Phylogenet. Evol.* 48:292–301.

Daly, M., D. G. Fautin, and V. A. Cappola. 2003. Systematics of the *Hexacorallia* (*Cnidaria*: *Anthozoa*). *Zool. J. Linn. Soc.* 139:419–437.

Dana, J. D. 1846. *Structure and Classification of Zoophytes.* Lea and Blanchard, Philadelphia, PA.

Dunn, D. F. 1982. *Cnidaria.* Pp. 669–706 in *Synopsis and Classification of Living Orgasnisms* (S. P. Parker, ed.). McGraw-Hill, New York.

Ehrenberg, C. G. 1834. *Die Corallienthiere des Rothen Meeres.* Königlichen Akademie der Wissenschaften, Berlin.

Extavour, C. G., K. Pang, D. Q. Matus, and M. Q. Martindale. 2005. *vasa* and *nanos* expression patterns in a sea anemone and the evolution of bilaterian germ cell specification mechanisms. *Evol. Dev.* 7:201–215.

Finnerty, J. R., K. Pang, P. Burton, D. Paulson, and M. Q. Martindale. 2004. Origins of bilateral symmetry: hox and dpp expression in a sea anemone. *Science* 304:1335–1337.

France, S. C., P. E. Rosel, J. E. Agenbroad, L. S. Mullineaux, and T. D. Kocher. 1996. DNA sequence variation of mitochondrial large-subunit rRNA provides support for a two-sublass organization of the *Anthozoa* (*Cnidaria*). *Mol. Mar. Biol. Biotech.* 5:15–28.

Hadzi, J. 1963. *The Evolution of the* Metazoa. Macmillan, New York.

Hertwig, O., and R. Hertwig. 1879. *Die Actinien Anatomisch und Histologisch mit Besonderer Berücksichtigung des Nervenmuskelsystems Untersucht*. Gustav Fischer, Jena.

Hyman, L. 1940. *The Invertebrates*, Vol. 1. McGraw-Hill, New York.

Kamm, K., B. Schierwater, W. Jakob, S. L. Dellaporta, and D. J. Miller. 2006. Axial patterning and diversification in the *Cnidaria* predate the Hox system. *Curr. Biol.* 16:920–926.

Kayal, E., and D. V. Lavrov. 2008. The mitochondrial genome of *Hydra oligactis* (*Cnidaria, Hydrozoa*) sheds new light on animal mtDNA evolution and cnidarian phylogeny. *Gene* 410:177–180.

Kayal, E., B. Roure, H. Philippe, A. G. Collins, and D. V. Lavrov. 2013. Cnidarian phylogenetic relationships as revealed by mitogenomics. *BMC Evol. Biol.* 13:5.

Kayal, E., B. Bentlage, M. Sabrina Pankey, A. H. Ohdera, M. Medina, D. C. Plachetzki, A. G. Collins, and J. F. Ryan. 2018. Phylogenomics provides a robust topology of the major cnidarian lineages and insights on the origins of key organismal traits. *BMC Evol. Biol.* 18:68.

McFadden, C. S., S. C. France, J. A. Sánchez, and P. Alderslade. 2006. A molecular phylogenetic analysis of the *Octocorallia* (*Cnidaria*: *Anthozoa*) based on mitochondrial protein-coding sequences. *Mol. Phylogenet. Evol.* 41:513–527.

Milne-Edwards, H. 1857. *Histoire Naturelle des Coralliares ou Polypes Proprement Dits*, Vol. 1. Libraire Encyclopedique de Roret, Paris.

Odorico, D. M., and D. J. Miller. 1997. Internal and external relationships of the *Cnidaria*: implications of primary and predicted secondary structure of the 5′-end of the 23S-like rDNA. *Proc. R. Soc. Lond. B Biol. Sci.* 264:77–82.

Rapp, W. 1829. *Uber die Polypen im Allgemeinen und die Actinien Insbesondere*. Grossherzogl. Sächs., Weimar.

Schuchert, P. 1993. Phylogenetic analysis of the *Cnidaria*. *Z. Zool. Syst. Evolutionsforsch.* 31:161–173.

Song, J.-I., W. Kim, E. K. Kim, and J. Kim. 1994. Molecular phylogeny of anthozoans (phylum *Cnidaria*) based on the nucleotide sequences of 18S rRNA gene. *Korean J. Zool.* 37:343–351.

Song, J.-I., and J. H. Won. 1997. Systematic relationships of the anthozoan orders based on the partial nuclear 18S rDNA sequences. *Korean J. Biol. Sci.* 1:43–52.

Stephenson, T. A. 1935. *The British Sea Anemones*. The Ray Society, London.

Wells, J. W., and D. Hill. 1956. *Anthozoa*. Pp. F161–F166 in *Treatise on Invertebrate Paleontology: Part F*. Coelenterata (R. C. Moore, ed.). University Kansas Press, Lawrence, KS.

Werner, B. 1973. New investigations on systematics and evolution of the class *Scyphozoa* and the phylum *Cnidaria*. *Publ. Seto Mar. Biol. Lab.* 20:35–61.

Wikramanayake, A. H., M. Hong, P. N. Lee, K. Pang, C. A. Byrum, J. M. Bince, R. Xu and Mark Q. Martindale. 2003. An ancient role for nuclear b-catenin in the evolution of axial polarity and germ layer segregation. *Nature* 426:446–450.

Won, J., B. Rho, and J. Song. 2001. A phylogenetic study of the *Anthozoa* (phylum *Cnidaria*) based on morphological and molecular characters. *Coral Reefs* 20:39–50.

Zapata, F., F. E. Goetz, S. A. Smith, M. Howison, S. Siebert, S. H. Church, S. M. Sanders, C. L. Ames, C. S. McFadden, S. C. France, and M. Daly. 2015. Phylogenomic analyses support traditional relationships within *Cnidaria*. *PLOS ONE* 10:10.

Author

Marymegan Daly; Department of Evolution; Ecology and Organismal Biology; The Ohio State University; 1315 Kinnear Road, Columbus, OH 43212, USA. Email: daly.66@osu.edu.

Date Accepted: 18 March 2011

Primary Editor: Kevin de Queiroz

Hexacorallia E. Haeckel 1866 [M. Daly], converted clade name

Registration Number: 167

Definition: The crown clade originating in the most recent common ancestor of *Nematostella vectensis* Stephenson 1935 and all extant species that share a more recent common ancestor with *Nematostella vectensis* Stephenson 1935 than with *Alcyonium digitatum* Linnaeus 1758 (*Octocorallia*). This is a maximum-crown-clade definition. Abbreviated definition: max crown ∇ (*Nematostella vectensis* Stephenson 1935 ~ *Alcyonium digitatum* Linnaeus 1758).

Etymology: Derived from *hex* (Gk., six) + *korallion* (Gk., coralline) referring to the hexamerous arrangement of mesenteries and tentacles in a typical hexacorallian polyp.

Reference Phylogeny: The reference phylogeny is France et al. (1996: Fig. 3), whose analysis of 16S mitochondrial rDNA found that *Anthozoa* comprises two mutually exclusive subclades, *Octocorallia* and *Hexacorallia*. For the purpose of applying our definition in the context of the reference phylogeny, *Nematostella vectensis* is to be considered part of *Actiniaria* (see Daly et al., 2003, 2008), and *Alcyonium digitatum* is to be considered closely related to *Alcyonium* sp. (see McFadden et al., 2006).

Composition: *Hexacorallia* includes all black corals (*Antipatharia*), sea anemones in the broadest sense, (i.e., members of *Actiniaria*, *Ceriantharia*, *Corallimorpharia*, and *Zoanthidea*), and stony corals (*Scleractinia*). *Hexacorallia* presently contains about 4,300 extant species (Doumenc, 1987; Daly et al., 2007). Fossil "corals" belonging to groups such as *Cothoniida*, *Heliolitida*, *Heterocorallia*, *Kilbuchophyllida*, *Numidiaphyllida*, *Rugosa*, *Tabulaconida*, and *Tabulata* are typically included within *Hexacorallia*, based on the symmetry and morphology inferred from the fossilized skeletons (Oliver, 1996). The arrangement of the calcareous septa demonstrates that these groups have attributes of *Hexacorallia*: for example, *Tetracorallia* and *Heterocorallia* have serially-inserted, coupled mesenteries (Oliver, 1980, 1996; Semenoff-Tian-Chansky, 1987), and *Tabulata* has hexamerously-arranged mesenteries (Scrutton, 1979), as do the few fossils attributed to *Kilbuchophyllia* (Scrutton and Clarkson, 1991). However, the arrangement of septa in these fossil groups is unlike that of any extant group, and so association of the fossil lineages with modern lineages is not possible. The relationship of the fossil groups to extant forms is not clear (see Fautin et al., 2000, and below).

Diagnostic Apomorphies: All hexacorallians have spirocysts, a type of cnida with a single-walled capsule into which a tubule bearing tiny, hollow threads is coiled (reviewed in Daly et al., 2003). Spirocysts are primarily adhesive; the threads solublize in seawater and form a meshwork that enwraps prey and aids in attachment. The other two diagnostic apomorphies of *Hexacorallia* pertain to the arrangement and morphology of the mesenteries. In hexacorallians, some of the mesenteries are incomplete (= imperfect), spanning only part of the distance between the body wall and the actinopharynx (Hyman, 1940; see Daly et al., 2003). Additionally, in all hexacorallians, mesenteries develop as couples, a set of two mesenteries symmetrical with respect to the sulcal axis (see Daly et al., 2003). The six-part symmetry that gives the clade its name is not characteristic of all *Hexacorallia*, being notably absent in *Ceriantharia* and several lineages of *Actiniaria* (Hyman, 1940).

Synonyms: *Zoantharia* de Blainville 1830 (approximate); *Zoocorallia Polyactinia* Ehrenberg 1834 (partial).

Comments: de Blainville (1830) grouped soft-bodied actiniarians with skeletonized scleractinians into *Zoantharia* (in French, as *Zoanthaires*). This is the most commonly encountered synonym of *Hexacorallia*, being used by Hyman (1940) and in synoptic classifications (e.g., Dunn, 1982). This name has been used to refer to a group consisting of *Actiniaria*, *Corallimorpharia*, *Scleractinia*, and *Zoanthidea*, excluding *Ceriantharia* and *Antipatharia*, which together have been called *Ceriantipatharia* Van Beneden 1889 (see, e.g., Hyman, 1940). Furthermore, *Zoantharia* is a homonym, having been used by Milne-Edwards (1857) for the group now called *Zoanthidea*, one of the constituent subclades of *Hexacorallia*. Other authors have named groups that contain some members of *Hexacorallia*, but most of these, like Ehrenberg's (1834) *Zoocorallia Polyactinia*, exclude certain lineages of *Scleractinia*. The name *Hexacorallia* avoids nomenclatural confusion, is descriptive in reference to its sister group *Octocorallia*, and is widely used, including in the reference phylogeny of France et al. (1996).

Relationships within *Hexacorallia* remain unclear. In particular, the relationship between *Actiniaria* and *Zoanthidea*, and between *Scleractinia* and *Corallimorpharia* are controversial, with nuclear (Won et al., 2001; Daly et al., 2002, 2003) and mitochondrial (Medina et al., 2006) DNA suggesting different patterns of relationship. Nonetheless, most analyses seem to agree that these four taxa plus *Antipatharia* constitute a clade, with *Ceriantharia* as its sister group.

Ceriantharia has been postulated to represent a separate clade from *Hexacorallia*, based on its unique developmental pathway and arrangement of mesenteries (McMurrich, 1891; Hyman, 1940). This scenario bolstered the creation of the subclass *Ceriantipatharia*, which included ceriantharians and antipatharians (Van Beneden, 1889). Schmidt's (1974) study of the distribution and morphology of cnidae provided the first character-based assessment of phylogenetic relationships in *Anthozoa* and concluded that *Ceriantharia* and *Antipatharia* were not sister taxa, although both were part of *Hexacorallia*: *Ceriantharia* is its basal-most extant member, and *Antipatharia* is a more deeply nested clade. The topology proposed by Schmidt (1974) has not been replicated in subsequent analyses, although most molecular and morphological phylogenies support his more general conclusions that *Ceriantharia* and *Antipatharia* are not sister taxa. France et al. (1996) addressed the placement of *Ceriantharia* and *Antipatharia* explicitly and concluded that *Ceriantharia* and *Antipatharia* are not sister taxa; *Ceriantharia* is sister to the rest of the group, and *Antipatharia* is nested within the other hexacorallian taxa. Subsequent studies using other data have confirmed the conclusions of France et al. (18S rDNA: Berntson et al., 1999, 2001; Daly et al., 2002; Won et al., 2001; 28S rDNA: Chen et al., 1995; morphology: Won et al., 2001; 16S, 18S, 28S, morphology: Daly et al., 2003). Some phylogenies have found that *Ceriantharia* lies outside of *Hexacorallia* (Song et al., 1994; Song and Won, 1997; Stampar et al., 2014), but this seems to reflect some differences in rate of evolution and sensitivity to taxon sampling (reviewed in Zapata et al., 2015; Stampar et al., 2019). Because *Ceriantharia* is an unstable lineage in terms of its relationship to other *Anthozoa*, the definition of the name *Hexacorallia* proposed here allows *Ceriantharia* to be part of *Hexacorallia* or not depending on its phylogenetic relationships.

This treatment of *Hexacorallia* does not fully address the placement of several enigmatic fossil groups, such as *Tetracorallia* (= *Rugosa*) and *Tabulata*, which are typically placed within *Hexacorallia*. These groups are known only from fossilized calcium carbonate skeletons, and so are difficult to associate with modern hexacorallian clades, only one of which has a carbonate skeleton. Tabulate and rugose corals have been hypothesized to be ancestors to modern scleractinians, but the differences between them in terms of skeletal organization and mineralization argue against close ancestry (Oliver, 1980; Stanley and Fautin, 2002).

Literature Cited

Berntson, E. A., S. C. France, and L. S. Mullineaux. 1999. Phylogenetic relationships within the Class *Anthozoa* (Phylum *Cnidaria*) based on nuclear 18S rDNA sequences. *Mol. Phylogenet Evol.* 13:417–433.

Chen, C. A., D. M. Odorico, M. ten Lohuis, J. E. N. Veron, and D. J. Miller. 1995. Systematic relationships within the *Anthozoa* using the 5′-end of the 28S rDNA. *Mol. Phylogenet Evol.* 4:175–183.

Daly, M., M. Brugler, P. Cartwright, A. G. Collins, M. N. Dawson, S. C. France, D. G. Fautin, C. S. McFadden, D. M. Opresko, E. Rodríguez, S. L. Romano, and J. L. Stake. 2007. The phylum *Cnidaria*: a review of phylogenetic patterns and diversity three hundred years after Linnaeus. *Zootaxa* 1668:127–186.

Daly, M., A. Chaudhuri, L. Gusmão, and E. Rodriguez. 2008. Phylogenetic relationships among sea anemones (*Cnidaria*: *Anthozoa*: *Actiniaria*). *Mol. Phylogenet Evol.* 48, 292.

Daly, M., D. G. Fautin, and V. A. Cappola. 2003. Systematics of the *Hexacorallia* (*Cnidaria*: *Anthozoa*). *Zool. J. Linn. Soc.* 139:419–437.

Daly, M., D. L. Lipsomb, and M. W. Allard. 2002. A simple test: evaluating explanations for the relative simplicity of the *Edwardsiidae*. *Evolution* 56:502–510.

Doumenc, D. 1987. Sous-classe des hexacoralliares. Pp. 187–188 in *Traite de Zoologie, Tome III, Fasicule 3. Cnidaries Anthozoaires* (P.-P. Grassé, ed.). Masson, Paris.

Dunn, D. F. 1982. *Cnidaria*. Pp. 669–706 in *Synopsis and Classification of Living Things* (S. P. Parker, ed.). McGraw-Hill, New York.

Ehrenberg, C. G. 1834. *Die Corallenthiere des rothen Meeres*. Königlichen Akademie der Wissenschaften, Berlin.

Fautin, D. G., S. L. Romano, and W. A. Oliver. 2000. *Zoantharia*. Sea Anemones and Corals, Version 4 October 2000. Available at http://tolweb.org/Zoantharia/17643/2000. 10.04 in The Tree of Life Web Project, http://tolweb.org/.

France, S. C., P. E. Rosel, J. E. Agenbroad, L. S. Mullineaux, and T. D. Kocher. 1996. DNA sequence variation of mitochondrial large-subunit rRNA provides support for a two-sublass organization of the *Anthozoa* (*Cnidaria*). *Mol. Mar. Biol. Biotech.* 5:15–28.

Haeckel, E. 1866. *Generale Morphologie der Organismen*. Verlag von Georg Reimer, Berlin.

Hyman, L. 1940. *The Invertebrates*, Vol. 1. McGraw-Hill, New York.

McFadden, C. S., S. C. France, J. A. Sánchez, and P. Alderslade. 2006. A molecular phylogenetic analysis of the *Octocorallia* (*Cnidaria*: *Anthozoa*) based on mitochondrial protein-coding sequences. *Mol. Phylogenet. Evol.* 41:513–527.

McMurrich, J. P. 1891. Contributions to the morphology of the *Actinozoa* III. On the phylogeny of the *Actinozoa. J. Morphol.* 5:125–164.

Medina, M., A. G. Collins, T. L. Takaoka, J. V. Kuehl, and J. L. Boore. 2006. Naked corals: skeleton loss in *Scleractinia*. *Proc. Natl. Acad. Sci. USA* 103:9096–9100.

Milne-Edwards, H. 1857. *Histoire Naturelle des Coralliares ou Polypes Proprement Dits*, Vol. 1. Librairie Encyclopedique de Roret, Paris.

Oliver, W. A. 1980. The relationship of the scleractinian corals to the rugose corals. *Paleobiology* 6:146–160.

Oliver, W. A. 1996. Origins and relationships of Paleozoic coral groups and the origin of *Scleractinia. Paleontol. Soc. Pap.* 1:107–134.

Schmidt, H. 1974. On the evolution in the *Anthozoa*. *Proc. 2nd Intl. Coral Reef Symp.* 1:533–560.

Scrutton, C. T. 1979. Early fossil cnidarians. Pp. 161–207 in *The Origin of Major Invertebrate Groups* (M. R. House, ed.). Academic Press, London.

Scrutton, C. T., and E. N. K. Clarkson. 1991. A new scleractinian-like coral from the Ordovician of the southern uplands Scotland. *Paleontology* 34:179–194.

Semenoff-Tian-Chansky, P. 1987. Sous-classe des Tetracoralliares. Pp. 765–781 in *Traite de Zoologie, Tome III, Fasicule 3. Cnidaries Anthozoaires* (P.-P. Grassé, ed.). Masson, Paris.

Song, J.-I., W. Kim, E. K. Kim, and J. Kim. 1994. Molecular phylogeny of anthozoans (Phylum *Cnidaria*) based on the nucleotide sequences of 18S rRNA gene. *Korean J. Zool.* 37:343–351.

Song, J.-I., and J. H. Won. 1997. Systematic relationships of the anthozoan orders based on the partial nuclear 18S rDNA sequences. *Korean J. Biol. Sci.* 1:43–52.

Stampar, S. N., M. B. Broe, J. Macrander, A. M. Reitzel, M. R. Brugler, and M. Daly. 2019. Linear mitochondrial genome in *Anthozoa* (*Cnidaria*): a case study in *Ceriantharia*. *Sci. Rep.* 9:1–12.

Stanley, Jr., G. D., and D. G. Fautin. 2002. The origins of modern corals. *Science* 291:1913–1914.

Van Bendeden, E. 1889. Recherches sur le développement des *Arachnactis*. *Arch. Biol.* 11:115–146.

Won, J., B. Rho, and J. Song. 2001. A phylogenetic study of the *Anthozoa* (phylum *Cnidaria*) based on morphological and molecular characters. *Coral Reefs* 20:39–50.

Zapata, F., F. E. Goetz, S. A. Smith, M. Howison, S. Siebert, S. H. Church, S. M. Sanders, C. L. Ames, C. S. McFadden, S. C. France, M. Daly, A. G. Collins, S. H. D. Haddock, C. W. Dunn, and P. Cartwright. 2015. Phylogenomic analyses support traditional relationships within *Cnidaria*. *PLOS ONE* 10(10).

Author

Marymegan Daly; Department of Evolution, Ecology and Organismal Biology, The Ohio State University, 1315 Kinnear Road, Columbus, OH 43212, USA. Email: daly.66@osu.edu.

Date Accepted: 18 March 2011

Primary Editor: Kevin de Queiroz

Octocorallia E. Haeckel 1866 [M. Daly and C. S. McFadden], converted clade name

Registration Number: 172

Definition: The crown clade originating in the most recent common ancestor of *Alcyonium digitatum* Linnaeus 1758 and all extant species that share a more recent common ancestor with *Alcyonium digitatum* Linnaeus 1758 than with *Nematostella vectensis* Stephenson 1935 (*Hexacorallia*). This is a maximum-crown-clade definition. Abbreviated definition: max crown ∇ (*Alcyonium digitatum* Linnaeus 1758 ~ *Nematostella vectensis* Stephenson 1935).

Etymology: Derived from *okto* (Gk., eight) + *korallion* (Gk., coralline) referring to the octamerous arrangement of tentacles and mesenteries in an octocoral polyp.

Reference Phylogeny: France et al. (1996: Fig. 3) is the primary reference phylogeny. See also Berntson et al. (1999: Fig. 4), Berntson et al. (2001: Fig. 1) and Won et al. (2001: Fig. 6, as *Alcyonaria*). For the purpose of applying our definition in the context of the reference phylogeny, *Nematostella vectensis* is to be considered part of *Actiniaria* (see Daly et al., 2003, 2008), and *Alcyonium digitatum* is to be considered closely related to *Alcyonium sp.* (see McFadden et al., 2006).

Composition: *Octocorallia* includes all extant species of soft corals and sea fans (*Alcyonacea*), sea pens (*Pennatulacea*), and blue corals (*Helioporacea*). *Octocorallia* is currently estimated to contain approximately 3,000 extant species (Daly et al., 2007).

Diagnostic Apomorphies: The presence of polyps with eight tentacles and eight unpaired, complete mesenteries is the most obvious apomorphy of *Octocorallia* and distinguishes it unequivocally from *Hexacorallia*. Pinnate tentacles, in which hollow projections or pinnules are typically arrayed in a single plane along the margin of the tentacle, and coloniality have also long been considered to be apomorphies of *Octocorallia*. Although a few species that lack these characters are known (e.g., the pinnuleless *Acrossota amboinensis* (Burchard 1902) and the solitary taxon *Taiaroa tauhou* Bayer and Muzik 1976), the current phylogenetic understanding of *Octocorallia* suggests that loss of pinnules and loss of coloniality are apomorphies for those small subclades (McFadden and Ofwegen, 2012). Other attributes common to many *Octocorallia* but of unknown historical origin include characters associated with coloniality, such as coenenchyme (colonial tissue surrounding the polyps) and solenia (networks of gastrodermal canals within the coenenchyme), as well as sclerites (= spicules), free calcareous skeletal elements found within the coenenchyme and polyps (Bayer, 1956). Whether these attributes represent symplesiomorphies, synapomorphies, or convergences can only be assessed in the context of a stable, robust phylogeny of *Octocorallia*.

Synonyms: *Alcyonaria* Dana 1846 (approximate sensu Dana, 1846; partial sensu Kükenthal, 1925); *Octactinia* Ehrenberg 1828 (approximate); *Zoophytaria* de Blainville 1834 (approximate).

Comments: *Alcyonaria* is the most commonly used synonym of *Octocorallia*, perpetuated in part because of its use in Hyman's (1940) influential treatise on invertebrates. Kükenthal (1925) restricted this name to a subset of

Octocorallia. Octocorallia is in wider use and is preferred over *Alcyonaria*, as the latter is easily confused with *Alcyonacea*, the name of one of its subgroups, and is less descriptive than *Octocorallia*, which brings to mind the eight-part organization of the polyp. Furthermore, this is the name used in the reference phylogeny of France et al. (1996).

Because of the uniformity of polyp morphology across the group, *Octocorallia* has long been considered to be monophyletic (e.g., Dana, 1846; Haeckel, 1866; Kükenthal, 1925), and all molecular phylogenetic analyses of *Anthozoa* conducted to date also provide robust support for octocorallian monophyly (16S rDNA: France et al., 1996; 18S rDNA: Song and Won, 1997; Berntson et al., 1999, 2001; Won et al., 2001; 28S rDNA: Chen et al., 1995; mitochondrial protein-coding sequences: Park et al., 2012; Kayal et al. 2013; transcriptomes: Zapata et al., 2015; Pratlong et al., 2016). Interpretations of relationships within *Octocorallia* vary widely, however (e.g., Kükenthal, 1925; Hickson, 1930; Bayer, 1981; McFadden et al., 2006, 2010), and at present there is no consensus among octocoral taxonomists about higher order relationships within the clade. Phylogenetic relationships among the six to eight mutually exclusive groups (traditionally ranked as orders or suborders) into which the clade has historically been subdivided (e.g., Hickson, 1930; Bayer, 1981) remain unclear, and few, if any, of those groups appear to be monophyletic (Berntson et al., 2001; Sánchez et al., 2003; McFadden et al., 2006, 2010; McFadden and Ofwegen, 2012, 2013; Cairns and Wirshing, 2015).

Literature Cited

Bayer, F. M. 1956. *Octocorallia*. Pp. F166–F230 in *Treatise on Invertebrate Paleontology: Part F. Coelenterata* (R. C. Moore, ed.). University of Kansas Press, Lawrence, KS.

Bayer, F. M. 1981. Key to the genera of *Octocorallia* exclusive of *Pennatulacea* (*Coelenterata: Anthozoa*), with diagnoses of new taxa. *Proc. Biol. Soc. Wash.* 94:902–947.

Berntson, E. A., F. M. Bayer, A. G. MacArthur, and S. C. France. 2001. Phylogenetic relationships within the *Octocorallia* (*Cnidaria: Anthozoa*) based on nuclear 18S rRNA sequences. *Mar. Biol.* 138:235–246.

Berntson, E. A., S. C. France, and L. S. Mullineaux. 1999. Phylogenetic relationships within the Class *Anthozoa* (Phylum *Cnidaria*) based on nuclear 18S rDNA sequences. *Mol. Phylogenet. Evol.* 13:417–433.

Cairns, S. D., and H. A. Wirshing. 2015. Phylogenetic reconstruction of scleraxonian octocorals supports the resurrection of the family *Spongiodermidae* (*Cnidaria, Alcyonacea*). *Invertebr. Syst.* 29: 345–368.

Chen, C. A., D. M. Odorico, M. ten Lohuis, J. E. N. Veron, and D. J. Miller. 1995. Systematic relationships within the *Anthozoa* using the 5′-end of the 28S rDNA. *Mol. Phylogenet. Evol.* 4:175–183.

Daly, M., M. Brugler, P. Cartwright, A. G. Collins, M. N. Dawson, S. C. France, D. G. Fautin, C. S. McFadden, D. M. Opresko, E. Rodriguez, S. L. Romano, and J. L. Stake. 2007. The phylum *Cnidaria*: a review of phylogenetic patterns and diversity three hundred years after Linnaeus. *Zootaxa* 1668:127–186.

Daly, M., A. Chaudhuri, L. Gusmão, and E. Rodriguez. 2008. Phylogenetic relationships among sea anemones (*Cnidaria: Anthozoa: Actiniaria*). *Mol. Phylogenet. Evol.* 48:292–301.

Daly, M., D. G. Fautin, and V. A. Cappola. 2003. Systematics of the *Hexacorallia* (*Cnidaria: Anthozoa*). *Zool. J. Linn. Soc.* 139:419–437.

Dana, J. D. 1846. *Structure and Classification of Zoophytes*. Lea and Blanchard, Philadelphia, PA.

France, S. C., P. E. Rosel, J. E. Agenbroad, L. S. Mullineaux, and T. D. Kocher. 1996. DNA sequence variation of mitochondrial large-subunit rRNA provides support for a two-sublass organization of the *Anthozoa* (*Cnidaria*). *Mol. Mar. Biol. Biotech.* 5:15–28.

Haeckel, E. 1866. *Generale Morphologie der Organismen*. Georg Reimer, Berlin.

Hickson, S. J. 1930. On the classification of the *Alcyonaria. Proc. Zool. Soc. Lond.* 1:229–252.

Hyman, L. H. 1940. *The Invertebrates*: Protozoa *through* Ctenophora. McGraw-Hill, New York.

Kayal, E., B. Roure, H. Philippe, A. G. Collins, and D. V. Lavrov. 2013. Cnidarian phylogenetic relationships as revealed by mitogenomics. *BMC Evol. Biol.* 13:5.

Kükenthal, W. 1925. *Handbuch der Zoologie*, Vol. 1. Walter de Gruyter, Berlin.

McFadden, C. S., S. C. France, J. A. Sánchez, and P. Alderslade. 2006. A molecular phylogenetic analysis of the *Octocorallia* (*Cnidaria*: *Anthozoa*) based on mitochondrial protein-coding sequences. *Mol. Phylogenet. Evol.* 41:513–527.

McFadden, C. S., J. A. Sánchez, and S. C. France. 2010. Molecular phylogenetic insights into the evolution of *Octocorallia*: a review. *Integr. Comp. Biol.* 50:389–410.

McFadden, C. S., and L. P. van Ofwegen. 2012. Stoloniferous octocorals (*Anthozoa, Octocorallia*) from South Africa, with descriptions of a new family of *Alcyonacea*, a new genus of *Clavulariidae*, and a new species of *Cornularia* (*Cornulariidae*). *Invertebr. Syst.* 26:331–356.

McFadden, C. S., and L. P. van Ofwegen. 2013. Molecular phylogenetic evidence supports a new family of octocorals and a new genus of *Alcyoniidae* (*Octocorallia, Alcyonacea*). *ZooKeys* 346:59–83.

Park, E., D.-S. Hwang, J.-S. Lee, J.-I. Song, T.-K. Seo, and Y.-J. Won. 2012. Estimation of divergence times in cnidarian evolution based on mitochondrial protein-coding genes and the fossil record. *Mol. Phylogenet. Evol.* 62:329–345.

Pratlong, M., C. Rancurel, P. Pontarotti, and D. Aurelle. 2016. Monophyly of *Anthozoa* (*Cnidaria*): why do nuclear and mitochondrial phylogenies disagree? *Zool. Scr.* 46:363–371.

Sánchez, J. A., H. R. Lasker, and D. J. Taylor. 2003. Phylogenetic analyses among octocorals (*Cnidaria*): mitochondrial and nuclear DNA sequences (lsu-rRNA, 16S and ssu-rRNA, 18S) support two convergent clades of branching gorgonians. *Mol. Phylogenet. Evol.* 29:31–42.

Song, J.-I., and J. H. Won. 1997. Systematic relationships of the anthozoan orders based on the partial nuclear 18S rDNA sequences. *Korean J. Biol. Sci.* 1:43–52.

Won, J., B. Rho, and J. Song. 2001. A phylogenetic study of the *Anthozoa* (phylum *Cnidaria*) based on morphological and molecular characters. *Coral Reefs* 20:39–50.

Zapata, F. et al. 2015. Phylogenomic analyses support traditional relationships within *Cnidaria*. *PLOS ONE*: e0139068.

Authors

Marymegan Daly; Department of Evolution, Ecology and Organismal Biology; The Ohio State University; 1315 Kinnear Road; Columbus, OH 43212, USA. Email: daly.66@osu.edu.

Catherine S. McFadden; Department of Biology; Harvey Mudd College; 1250 N. Dartmouth Ave.; Claremont, CA 91711, USA. Email: mcfadden@hmc.edu.

Date Accepted: 18 March 2011

Primary Editor: Kevin de Queiroz

Medusozoa K. Peterson 1979 [A. G. Collins and C. W. Dunn], converted clade name

Registration Number: 171

Definition: The least inclusive crown clade containing *Chironex fleckeri* Southcott 1956, *Hydra vulgaris* (Pallas 1766), *Atolla vanhoeffeni* Russell 1957, and *Craterolophus convolvulus* (Johnston 1835). This is a minimum-crown-clade definition. Abbreviated definition: min crown ∇ (*Chironex fleckeri* Southcott 1956 & *Hydra vulgaris* (Pallas 1766) & *Atolla vanhoeffeni* Russell 1957 & *Craterolophus convolvulus* (Johnston 1835)).

Etymology: Derived from the Greek *Medousa* (Μέδουσα), the name of a mythological character who had snakes for hair, + *zoön*, animal, in reference to the free-swimming, tentacled medusa or jellyfish phase in the life cycle of many of its species.

Reference Phylogeny: Collins (2002: Fig. 1)— showing *Medusozoa* as a clade within *Cnidaria* consisting of *Cubozoa*, *Hydrozoa*, *Scyphozoa*, and *Staurozoa*, as represented respectively by *Chironex fleckeri*, *Hydra littoralis*, *Atolla vanhoeffeni*, and *Craterolophus convolvulus*, but distinct from *Anthozoa* (as defined in this volume)—is chosen as the reference phylogeny because of its relatively extensive sampling of both cnidarian and non-cnidarian taxa. For the purpose of applying our definition, *Hydra vulgaris* (which is not shown on the reference phylogeny) is regarded as closely related to *Hydra littoralis* (Martinez et al., 2010).

Composition: *Cubozoa* (cubopolyps, cubomedusae, box jellyfishes, sea wasps), *Hydrozoa* (hydropolyps including hydroids and hydro-corals, hydromedusae), *Scyphozoa* (scyphopolyps, scyphomedusae, true jellyfishes), and *Staurozoa* (stauromedusae, stalked jellyfishes). *Myxozoa*, a group of enigmatic parasites, may be derived from within *Cnidaria* (see entry for *Cnidaria*, this volume; Siddall et al., 1995; Zrzavý and Hypša, 2003; Jiménez-Guri et al., 2007) and could be either part of or the sister group to *Medusozoa*. Roughly 3,800 extant species of *Medusozoa*, exclusive of *Myxozoa*, are known (Daly et al., 2007). See Comments for a discussion of possible fossil members of this clade.

Diagnostic Apomorphies: Identifying diagnostic apomorphies for *Medusozoa* is hampered by lack of study in diverse species and character reversals. Lists of putative apomorphies for *Medusozoa* can be found in the analyses of Schuchert (1993), Bridge et al. (1995), Marques and Collins (2004), and Van Iten et al. (2006). All analyses agree in suggesting that linear mitochondrial genomes, cnidocils (cilia of cnidocytes lacking basal rootlets), and microbasic eurytele nematocysts are shared derived features of *Medusozoa*. Other potential synapomorphies include podocyst resting and pelagic adult (i.e., medusae) stages in the life cycle. Of the characters mentioned here, reversals are not known for cnidocils and linear mitochondrial genomes, but these characters have not been documented in representatives of all major medusozoan lineages (Marques and Collins, 2004).

Synonyms: *Tesserazoa* Salvini-Plawen 1978 (approximate).

Comments: Salvini-Plawen (1978) was the first to provide a name covering *Medusozoa* in the sense that it is defined here. He used the name *Tesserazoa*, in reference to the tetraradial symmetry that many medusozoans exhibit in cross section. Petersen (1979) is the originator of the name *Medusozoa*, which refers to the medusa stage constituting the sexual adult of many species within this clade. Despite the priority of *Tesserazoa* over *Medusozoa*, the latter name is by far the more widely used (e.g., Bridge et al., 1995; Odorico and Miller, 1997; Collins, 2002) and was selected as the name of the clade for this reason.

The reference phylogeny (Collins, 2002: Fig. 1) is based on 18S rDNA data and follows earlier studies (Collins, 1998; Kim et al., 1999) with weaker taxon sampling that also showed *Medusozoa* as a clade. Phylogenetic topologies showing a monophyletic *Medusozoa* within *Cnidaria* have also been derived on morphological grounds (Werner, 1973; Salvini-Plawen, 1978; Schuchert, 1993; Bridge et al., 1995) and 28S rDNA sequences (Odorico and Miller, 1997; Medina et al., 2001; Collins et al., 2006). To date, analyses based on the most complete taxon sampling for *Cnidaria*, including multiple exemplars from *Cubozoa*, *Hydrozoa*, *Scyphozoa*, and *Staurozoa*, are presented by Collins et al. (2006: Figs. 3, 6 and Fig. 2, for 18S and 28S respectively), but these analyses do not contain any non-cnidarian outgroup taxa and are rooted on a sampling of anthozoans. Conflicting phylogenetic hypotheses with *Anthozoa* nested within medusozoan taxa (Haeckel, 1896; Hyman, 1940; Brusca and Brusca, 1990) appear to be incorrect. Should one of these hypotheses unexpectedly turn out to be correct, according to our definitions of *Medusozoa* and *Cnidaria*, these names would become synonyms, in which case *Cnidaria* should be granted precedence because it would not be associated with a change in composition.

Conulariids, an extinct group of animals with long-debated phylogenetic origins, is most likely to have originated within *Medusozoa* (Van Iten et al., 2006). Conulariids are interpreted to have lived attached to the substrate by a flexible stalk and possessed a steeply pyramidal apatitic exoskeleton that is square in cross section. This four-parted organization appears to be plesiomorphic for *Medusozoa* (Collins et al., 2006) and is manifest in the benthic polypoid stages of members of *Scyphozoa* and *Staurozoa*. The cladistic analysis of Van Iten et al. (2006) suggests that conulariids are part of *Scyphozoa*, with an especially close relationship to *Coronatae*.

Literature Cited

Bridge, D., C. W. Cunningham, R. Desalle, and L. W. Buss. 1995. Class-level relationships in the phylum *Cnidaria*: molecular and morphological evidence. *Mol. Biol. Evol.* 12:679–689.

Brusca, R. C., and G. J. Brusca. 1990. *Invertebrates*. Sinauer Associates, Sunderland, MA.

Collins, A. G. 1998. Evaluating multiple alternative hypotheses for the origin of *Bilateria*: an analysis of 18S rRNA molecular evidence. *Proc. Natl. Acad. Sci. USA* 95:15458–15463.

Collins, A. G. 2002. Phylogeny of *Medusozoa* and the evolution of cnidarian life cycles. *J. Evol. Biol.* 15:418–432.

Collins, A. G., P. Schuchert, A. C. Marques, T. Jankowski, M. Medina, and B. Schierwater. 2006. Medusozoan phylogeny and character evolution clarified by new large and small subunit rDNA data and an assessment of the utility of phylogenetic mixture models. *Syst. Biol.* 55:97–115.

Daly, M., M. R. Brugler, P. Cartwright, A. G. Collins, M. N. Dawson, D. G. Fautin, S. C. France, C. S. McFadden, D. M. Opresko, E. Rodriguez, S. L. Romano, and J. L. Stake. 2007. The phylum *Cnidaria*: a review of phylogenetic patterns and diversity 300 years after Linnaeus. *Zootaxa* 1668:127–182.

Haeckel, E., 1896. *Systematische Phylogenie. Zweiter Theil: Systematische Phylogenie der wirbellosen Thiere* (Invertebrata). Georg Reimer, Berlin.

Hyman, L. H. 1940. *The Invertebrates. Vol. 1:* Protozoa *through* Ctenophora. McGraw-Hill, New York.

Jiménez-Guri, E., H. Philippe, B. Okamura, and P. W. H. Holland. 2007. *Buddenbrockia* is a cnidarian worm. *Science* 317:116–118.

Kim, J. H., W. Kim, and C. W. Cunningham. 1999. A new perspective on lower metazoan relationships from 18S rDNA sequences. *Mol. Biol. Evol.* 16:423–427.

Marques, A. C., and A. G. Collins. 2004. Cladistic analysis of *Medusozoa* and cnidarian evolution. *Invertebr. Biol.* 123:23–42.

Martinez, D. E., A. R. Iniguez, K. M. Percell, J. B. Willner, J. Signorovitch, and R. D. Campbell. 2010. Phylogeny and biogeography of *Hydra* (*Cnidaria: Hydridae*) using mitochondrial and nuclear DNA sequences. *Mol. Phylogenet. Evol.* 57:403–410.

Medina, M., A. G. Collins, J. D. Silberman, and M. L. Sogin. 2001. Evaluating hypotheses of basal animal phylogeny using complete sequences of large and small subunit rRNA. *Proc. Natl. Acad. Sci. USA* 98:9707–9712.

Odorico, D. M., and D. J. Miller. 1997. Internal and external relationships of the *Cnidaria*: implications of primary and predicted secondary structure of the 5′-end of the 23S-like rDNA. *Proc. R. Soc. Lond. B Biol. Sci.* 264:77–82.

Petersen, K. W. 1979. Development of coloniality in *Hydrozoa*. Pp. 105–139 in *Biology and Systematics of Colonial Organisms* (G. Larwood and B. R. Rosen, eds.). Academic Press, New York.

Salvini-Plawen. 1978. On the origin and evolution of the lower *Metazoa*. *Z. Zool. Syst. Evolut.-forsch.* 16:40–88.

Schuchert, P. 1993. Phylogenetic analysis of the *Cnidaria*. *Z. Zool. Syst. Evolut.-forsch.* 31:161–173.

Siddall, M. E., D. S. Martin, D. Bridge, D. M. Cone, and S. S. Desser. 1995. The demise of a phylum of protists: *Myxozoa* and other parasitic *Cnidaria*. *J. Parasitol.* 81:961–967.

Van Iten, H., J. De Moraes Leme, M. G. Simões, A. C. Marques, and A. G. Collins. 2006. Reassessment of the phylogenetic position of conulariids (?Ediacaran-Triassic) within the subphylum *Medusozoa* (phylum *Cnidaria*). *J. Syst. Palaeontol.* 4:109–118.

Werner, B. 1973. New investigations on systematics and evolution of the class *Scyphozoa* and the phylum *Cnidaria*. *Publ. Seto Mar. Biol. Lab.* 20:35–61.

Zrzavý, J., and V. Hypša. 2003. *Myxozoa, Polypodium,* and the origin of the *Bilateria*: the phylogenetic position of *"Endocnidozoa"* in light of the rediscovery of *Buddenbrockia*. *Cladistics* 19:91–180.

Authors

Allen G. Collins; National Systematics Laboratory of NOAA's Fisheries Service and National Museum of Natural History; MRC-153, Smithsonian Institution; P.O. Box 37012, Washington, DC 20013-7012, USA. Email: collinsa@si.edu.

Casey W. Dunn; Department of Ecology and Evolutionary Biology; Yale University; 165 Prospect St.; New Haven, CT 06511, USA. E-mail: casey.dunn@yale.edu.

Date Accepted: 31 March 2011

Primary Editor: Kevin de Queiroz

Cubozoa B. Werner 1973 [A. G. Collins], converted clade name

Registration Number: 166

Definition: The crown clade originating in the most recent common ancestor of *Carybdea rastonii* Haacke 1886 and *Chironex fleckeri* Southcott 1956. This is a minimum-crown-clade definition. Abbreviated definition: min crown ∇ (*Carybdea rastonii* Haacke 1886 & *Chironex fleckeri* Southcott 1956).

Etymology: Derived from the Greek *kubos* (κύβος), meaning cube, + *zoön* (animal), presumably in reference to the cuboidal or box-like shape of the medusa phase of the life cycle of its members.

Reference Phylogeny: Collins et al. (2006: Fig. 3) serves as the reference phylogenetic hypothesis.

Composition: All cubopolyps and cubomedusae (box jellyfishes, sea wasps, irukandjis, fire jellies). Some 40 species are currently recognized (Daly et al., 2007), including those listed by Kramp (1961) and recent additions (e.g., Gershwin, 2005; Gershwin and Alderslade, 2005; Lewis and Bentlage, 2009), classified in two reciprocally monophyletic groups, *Carybdeida* and *Chirodropida* (Bentlage et al., 2010).

Diagnostic Apomorphies: Lists of putative synapomorphies for *Cubozoa* can be found in the cladistic analyses of Schuchert (1993), Bridge et al. (1995), Marques and Collins (2004), and Van Iten et al. (2006). In the medusa stage, which all species are thought to possess, cubozoans exhibit several characters that readily distinguish them from other cnidarians. Most distinctive are the four perradial sensory rhopalia, which contain strikingly complex eyes involving ocelli, vitreous bodies, lenses, and retinas (Pearse and Pearse, 1978), in addition to statocysts. The velarium, a piece of tissue entirely of subumbrellar origin that projects into and serves to narrow the subumbrellar opening, is also unique to *Cubozoa*. The velarium is infused by canals and supported by perpendicularly arranged structures known as frenulae. The tentacles of cubomedusae are concentrated at the four interradial corners and have thickened muscular bases termed pedalia. Less readily observed but diagnostic features of *Cubozoa* include the organization of the nervous system of the polyps and the process by which the solitary polyp entirely metamorphoses into a single juvenile medusa (Werner et al., 1971; Werner, 1976).

Synonyms: Marsupialées Lesson 1837 (partial), *Marsupialae* Lesson 1843 (partial), *Charybdeidae* Gegenbaur 1856, *Marsupialidae* L. Aggasiz 1862, *Conomedusae* Haeckel 1879, *Cubomedusae* Haeckel 1880, *Marsupialida* Claus 1886, *Carybdeidae* (of Mayer, 1910), and *Carybdeida* Cockerell 1911 (all approximate).

Comments: *Carybdea marsupialis* (Linnaeus, 1758) was the first described cubozoan species. As additional species were described, they began to be classified together. *Marsupialées* of Lesson (1837) and *Marsupialae* of Lesson (1843) are partial synonyms of *Cubozoa* as they united all cubozoan species known at the time, but also included disparate hydrozoans. In contrast, Lesson's taxa *Carybdées* (1837) and *Carybdeae* (1843) are not synonyms (approximate or partial) because they did not contain any cubozoan species. Gegenbaur's *Charybdeidae* (1856) was the earliest higher taxon that united all known and only cubozoan species. Agassiz's *Marsupialidae* (1862) also exclusively contained cubozoans and is notable because it was the first cubozoan taxon to contain representatives of its two primary clades (see Composition). Haeckel (1880) was the first to recognize them as fundamentally distinct from other major groups of medusozoans by recognizing *Cubomedusae* as one of four mutually exclusive medusozoan taxa (ranked as orders).

Even though Gegenbaur (1856) based his name *Charybdeidae* on Péron and Lesueur's (1810) genus name *Carybdea*, he spelled both names with a "Ch", probably stemming from a difference in opinion about how to correctly transliterate the Greek root (χάρυβδις). Subsequently, Gegenbaur's name has been variously spelled beginning with a "C" or "Ch." Mayer (1910) appears to be the first to have used the spelling *Carybdeidae* in reference to the clade here named *Cubozoa*, but the same spelling had been applied earlier to a subset of cubozoan species (Leuckart, 1879). Cockerell (1911) seems to have been the first to use the spelling *Carybdeida* for a taxon referring to all known cubozoans. Of all the names used to refer to the group herein named *Cubozoa*, Haeckel's *Cubomedusae* has been most commonly used in recent decades (e.g., Kramp, 1961). Werner (1973) reformulated the name as *Cubozoa* and removed the taxon (which he ranked as a class) from *Scyphozoa* (also ranked as a class), when he and others observed that the life cycle and polyps of cubmomedusae were rather distinct from those of scyphomedusae. *Cubozoa* is chosen over *Cubomedusae* because it is in more common current use.

Although there have been a number of phylogenetic analyses that have addressed the position of *Cubozoa* within *Cnidaria* (Werner, 1973; Salvini-Plawen, 1978; Schuchert 1993; Bridge et al., 1995), few analyses have contained sufficient taxon sampling to provide stringent tests of the monophyly of *Cubozoa*. The group is so distinctive that such tests were probably not conceived as necessary. Collins (2002) used 18S rDNA data from nine different cubozoan species and found robust support for cubozoan monophyly. Collins et al. (2006) and Bentlage et al. (2010), who increased taxon sampling and collected complete 28S data from disparate cubozoans, found further support for cubozoan monophyly while refining understanding of evolution within this fascinating group of animals. No study has suggested non-monophyly of *Cubozoa*.

Literature Cited

Agassiz, L. 1862. *Contributions to the Natural History of the United States*, Vol. IV. Little, Brown and Company, Boston, MA.

Bentlage, B., P. Cartwright, A. A. Yanagihara, C. Lewis, G. S. Richards, and A. G. Collins. 2010. Evolution of box jellyfish (*Cnidaria*: *Cubozoa*), a group of highly toxic invertebrates. *Proc. R. Soc. B Biol. Sci.* 277:493–501.

Bridge, D., C. W. Cunningham, R. Desalle, and L. W. Buss. 1995. Class-level relationships in the phylum *Cnidaria*: molecular and morphological evidence. *Mol. Biol. Evol.* 12:679–689.

Cockerell, T. D. A. 1911. The nomenclature of the *Scyphomedusae*. *Proc. Biol. Soc. Wash.* 24:7–12.

Collins, A. G. 2002. Phylogeny of *Medusozoa* and the evolution of cnidarian life cycles. *J. Evol. Biol.* 15:418–432.

Collins, A. G., P. Schuchert, A. C. Marques, T. Jankowski, M. Medina, and B. Schierwater. 2006. Medusozoan phylogeny and character evolution clarified by new large and small subunit rDNA data and an assessment of the utility of phylogenetic mixture models. *Syst. Biol.* 55:97–115.

Daly, M., M. R. Brugler, P. Cartwright, A. G. Collins, M. N. Dawson, D. G. Fautin, S. C. France, C. S. McFadden, D. M. Opresko, E. Rodriguez, S. L. Romano, and J. L. Stake. 2007.

The phylum *Cnidaria*: a review of phylogenetic patterns and diversity 300 years after Linnaeus. *Zootaxa* 1668:127–182.

Gegenbaur, C. 1856. Versuch eines Systemes der Medusen, mit Beschreibung neuer oder wenig gekannter Formen; zugleich ein Beitrag zur Kenntnis der Fauna des Mittelmeeres. *Z. wiss. Zool. Leipzig* 8:202–273.

Gershwin, L. 2005. *Carybdea alata* auct. and *Manokia stiasnyi*, reclassification to a new family with description of a new genus and two new species. *Mem. Queensl. Mus.* 51:501–523.

Gershwin, L., and P. Alderslade 2005. A new genus and species of box jellyfish (*Cubozoa*: *Carybdeida*) from tropical Australian waters. *Beagle* 21:27–36.

Haeckel, E. 1879. Uber das System der Medusen. *Jena. Zeit. Naturwiss.* 14:68–90.

Haeckel, E. 1880. *System der Acraspeden. Zweite Halfte des Systems der Medusen.* G. Fischer, Jena.

Kramp, P. L. 1961. Synopsis of the medusae of the world. *J. Mar. Biol. Assoc. UK* 40:1–469.

Lesson, R. P. 1837. *Prodrome d'une Monographie des Méduses.* Rochefort, Paris.

Lesson, R. P. 1843. *Histoire Naturelle des Zoophytes. Acalèphes.* Librairie Encyclopédique de Roret, Paris.

Lewis, C., and B. Bentlage. 2009. Clarifying the identity of the Japanese Habu-kurage, *Chironex yamaguchii*, sp. nov. (*Cnidaria*: *Cubozoa*: *Chirodropida*). *Zootaxa* 2030:59–65.

Marques, A. C., and A. G. Collins. 2004. Cladistic analysis of *Medusozoa* and cnidarian evolution. *Invertebr. Biol.* 123:23–42.

Mayer, A. G. 1910. *Medusae of the World.* Carnegie Institution, Washington, DC.

Pearse, J. S., and V. B. Pearse. 1978. Vision in cubomedusan jellyfishes. *Science* 199:458.

Péron, F., and C. A. Lesueu. 1810. Tableau des caractères génériques et spécifiques de toutes les espèces de méduses connues jusqu'à ce jour. *Ann. Mus. Hist. Nat. Paris* 14:325–366.

Salvini-Plawen, L. V. 1978. On the origin and evolution of the lower *Metazoa. Z. Zool. Syst. Evolut.-forsch.* 16:40–88.

Schuchert, P. 1993. Phylogenetic analysis of the *Cnidaria. Z. Zool. Syst. Evolut.-forsch.* 31:161–173.

Van Iten, H., J. De Moraes Leme, M. G. Simões, A. C. Marques, and A. G. Collins. 2006. Reassessment of the phylogenetic position of conulariids (?Ediacaran-Triassic) within the subphylum *Medusozoa* (phylum *Cnidaria*). *J. Syst. Palaeontol.* 4:109–118.

Werner, B. 1973. New investigations on systematics and evolution of the class *Scyphozoa* and the phylum *Cnidaria. Pub. Seto Mar. Biol. Lab.* 20:35–61.

Werner, B. 1976. Muscular and nervous systems of the cubopolyp (*Cnidaria*). *Experientia* 32:1047–1049.

Werner, B., D. M. Chapman, and C. E. Cutress. 1971. Life cycle of *Tripedalia cystophora* Conant (*Cubomedusae*). *Nature* 232:582–583.

Authors

Allen G. Collins; National Systematics Laboratory of NOAA's Fisheries Service & National Museum of Natural History; MRC-153, Smithsonian Institution; P.O. Box 37012, Washington, DC 20013-7012, USA. Email: collinsa@si.edu.

Date Accepted: 20 March 2011

Primary Editor: Kevin de Queiroz

Hydrozoa R. Owen 1843 [C. W. Dunn and A. G. Collins], converted clade name

Registration Number: 168

Definition: The crown clade originating in the most recent common ancestor of *Hydra vulgaris* (Pallas 1766) (*Hydroidolina/Aplanulata*), *Solmissus marshalli* A. Agassiz and Mayer 1902 (*Trachylina/Narcomedusae*), and *Craspedacusta sowerbii* Lankester 1908 (*Limnomedusae*). This is a minimum-crown-clade definition. Abbreviated definition: min crown ∇ (*Hydra vulgaris* (Pallas 1766) & *Solmissus marshalli* A. Agassiz and Mayer 1902 & *Craspedacusta sowerbii* Lankester 1908).

Etymology: The name *Hydrozoa* is presumably derived from the Greek *hydra* (υδρα), a many-headed mythical creature, + *zoön* (animal), though Owen (1843) gave no explicit etymology.

Reference Phylogeny: Collins et al. (2006a: Fig. 6). For the purpose of applying our definition, *Hydra vulgaris* is to be regarded as a close relative of *Hydra littoralis*.

Composition: The most recent common ancestor of *Hydroidolina* and *Trachylina*, as defined in this volume, and all of its descendants. Schuchert (1997) listed 3,260 extant species belonging to *Hydrozoa*. *Polypodiozoa* has been hypothesized to be within *Hydrozoa* (Bouillon and Boero, 2000), but its phylogenetic position within *Cnidaria* remains unknown (Evans et al., 2008).

Diagnostic Apomorphies: Despite strong support from molecular data (including that underlying the reference phylogeny) and wide consensus among authors regarding the monophyly and composition of the group (Schuchert, 1993), there are few unambiguous morphological apomorphies that are diagnostic of *Hydrozoa*. Lists of putative apomorphies for *Hydrozoa* can be found in the cladistic analyses of Schuchert (1993), Bridge et al. (1995), Marques and Collins (2004), and Van Iten et al. (2006). The medusae, when present, possess a velum (with the exception of *Obelia*) and two nerve rings. When the lifecycle includes both polyps and medusae, the medusae are usually budded laterally from the polyps (rather than arising by strobilation, as in *Scyphozoa*, or by complete transformation of the polyp, as in *Cubozoa*). Within *Cnidaria*, intercellular gap junctions have only been documented in hydrozoans. Polyps, when present, lack septa or mesenteries. In contrast with non-hydrozoan cnidarians, which possess gonads of endodermal origin and location, most hydrozoans have gonads that are located in the ectoderm. Gonad origins in different hydrozoan species can be either endodermal or ectodermal (Bouillon et al., 2004), and it is unclear how variable this character is.

Synonyms: Bouillon et al. (1992) provide a partial list of many other names that closely "define the group of hydropolyps and hydromedusae" (including *Hydrophorae* Huxley 1856, *Hydromedusae* Claus 1880 (= *Polypomedusae*), and *Aphacellae* Haeckel 1879), most of which can be interpreted as approximate synonyms of *Hydrozoa*. *Hydromedusae* McMurrich (1894) is also an approximate synonym of the group including hydropolyps and hydromedusae. *Hydroidomedusae* was proposed by Bouillon et al. (1992) as a name for all *Hydrozoa* exclusive of *Siphonophora*. This is a partial synonym of

Hydrozoa in that it refers to a paraphyletic group originating in the same ancestor. *Hydroidomedusae* has also been used to designate a group comprising *Capitata*, *Filifera*, *Laingiomedusae*, *Leptomedusae*, *Limnomedusae*, and *Siphonophora* (Bouillon and Boero, 2000), another partial synonym of *Hydrozoa* that refers to a paraphyletic group originating in the same ancestor, but this time lacking *Narcomedusae*, *Trachymedusae*, and other taxa (Collins, 2002). Owen (1843) listed *Dimorphoea* Ehernberg, *Sertulariens* Milne Edwards (perhaps a French vernacular name), and *Nudibranchiata* Farre as synonyms of *Hydrozoa*.

Comments: The taxonomy of *Hydrozoa* has been greatly complicated by parallel sets of names employed for the polyps and medusae of the same life cycle. This has resulted in many synonyms both for subclades within *Hydrozoa* and for the group as a whole. For instance, *Hydromedusae* is sometimes used as a synonym of *Hydrozoa* (e.g., Bigelow, 1919) but other times it is used only to designate medusae of *Hydrozoa* (Bouillon et al., 1992). In this latter sense "hydromedusae" is not a taxon name at all, but rather a reference to particular stages in the lifecycle of some members of *Hydrozoa*. Indeed, even in its original usage by Owen (1843), *Hydrozoa* referred only to the benthic polyp stages of *Hydrozoa*. By the end of the 1800s, it was understood that many hydrozoan species possessed both polyps and medusae and a single taxon name—*Hydrozoa* (Haeckel, 1896) or *Hydromedusae* (McMurrich, 1894)—encompassing all life stages came into common use. For unknown reasons, *Hydrozoa* became the dominant name and we use it here for that reason. There has been just a single explicit phylogenetic analysis of *Hydrozoa* using morphology and life history data (Marques and Collins, 2004) and this supported hydrozoan monophyly. Molecular studies have also supported hydrozoan monophyly (Collins, 2002;

Collins et al., 2006a). Although descent of *Limnomedusae* from the most recent common ancestor of *Hydra vulgaris* and *Solmissus marshalli* is strongly supported by molecular data (Collins et al., 2006a, 2008), *Craspedacusta sowerbii* is used as an internal specifier in the definition of the name *Hydrozoa* to ensure inclusion of *Limnomedusae* in the event that it branched off earlier, a possibility raised by cladistic analysis of morphological and life history data (Marques and Collins, 2004).

Literature Cited

Bigelow, H. B. 1919. *Hydromedusae*, siphonophores, and ctenophores of the "Albatross" Philippine Expedition. *Bull U.S. Natl. Mus.* 100:279–362.

Bouillon, J., and F. Boero. 2000. The *Hydrozoa*: a new classification in the light of old knowledge. *Thalass. Salentina* 24:3–45.

Bouillon, J., F. Boero, F. Cicogna, J. M. Gili, and R. G. Hughes. 1992. Non-siphonophoran *Hydrozoa*: what are we talking about? *Sci. Mar.* 56:279–284.

Bouillon, J., M. D. Medel, F. Pagès, J. M. Gili, F. Boero, and C. Gravili. 2004. Fauna of the Mediterranean *Hydrozoa*. *Sci. Mar.* 68(suppl. 2):5–438.

Bridge, D., C. W. Cunningham, R. Desalle, and L. W. Buss. 1995. Class-level relationships in the phylum *Cnidaria*—molecular and morphological evidence. *Mol. Biol. Evol.* 12:679–689.

Claus, C. 1880. *Grundzüge der Zoologie. Zum wissenschaftlichen Gebrauche.* 4 Auflage, N. G. Elwert, Marburg.

Collins, A. G. 2002. Phylogeny of *Medusozoa* and the evolution of cnidarian life cycles. *J. Evol. Biol.* 15:418–432.

Collins, A. G., B. Bentlage, A. Lindner, D. Lindsay, S. H. D. Haddock, G. Jarms, J. L. Norenburg, T. Jankowski, and P. Cartwright. 2008. Phylogenetics of *Trachylina* (*Cnidaria*: *Hydrozoa*) with new insights on the evolution of some problematical taxa. *J. Mar. Biol. Assoc. UK* 88:1673–1685.

Collins, A. G., P. Schuchert, A. C. Marques, T. Jankowski, M. Medina, and B. Schierwater. 2006. Medusozoan phylogeny and character evolution clarified by new large and small subunit rDNA data and an assessment of the utility of phylogenetic mixture models. *Syst. Biol.* 55:97–115.

Evans, N. M., A. Lindner, E. V. Raikova, A. G. Collins, and P. Cartwright. 2008. Phylogenetic placement of the enigmatic parasite, *Polypodium hydriforme*, within the Phylum *Cnidaria*. *BMC Evol. Biol.* 8:139.

Haeckel, E. 1879. *Das System der Medusen: Erster Theil einer Monographie der Medusen*. G. Fischer, Jena.

Haeckel, E. 1896. *Systematische Phylogenie. Zweiter Theil: Systematische Phylogenie der wirbellosen Thiere (Invertebrata)*. Georg Reimer, Berlin.

Huxley, T. H. 1856. Lectures on general natural history. *Med. Times Gazette* 12–13.

Marques, A. C., and A. G. Collins. 2004. Cladistic analysis of *Medusozoa* and cnidarian evolution. *Invertebr. Biol.* 123:23–42.

McMurrich, J. P. 1894. *A Textbook of Invertebrate Morphology*. Henry Holt and Company, New York.

Owen, R. 1843. *Lectures in Comparative Anatomy and Physiology of the Invertebrate Animals, Delivered at the Royal College of Surgeons, in 1843*. Longman, Brown, Green, and Longmans, London.

Schuchert, P. 1993. Phylogenetic analysis of the *Cnidaria*. *Z. Zool. Syst. Evolut.-forsch.* 31:161–173.

Schuchert, P. 1997. How many hydrozoan species are there? *Zool. Verh. Leiden* 323:209–219.

Van Iten, H., J. De Moraes Leme, M. G. Simões, A. C. Marques, and A. G. Collins. 2006. Reassessment of the phylogenetic position of conulariids (?Ediacaran-Triassic) within the subphylum *Medusozoa* (phylum *Cnidaria*). *J. Syst. Palaeontol.* 4:109–118.

Authors

Casey W. Dunn; Department of Ecology and Evolutionary Biology; Yale University; 165 Prospect St., New Haven, CT 06511, USA. Email: casey.dunn@yale.edu.

Allen G. Collins; National Systematics Laboratory of NOAA's Fisheries Service and National Museum of Natural History; MRC-153. Smithsonian Institution; P.O. Box 37012, Washington, DC 20013-7012, USA. Email: collinsa@si.edu.

Date Accepted: 27 July 2011

Primary Editor: Kevin de Queiroz

Trachylina E. Haeckel 1879 as *Trachylinae* [C. W. Dunn and A. G. Collins], converted clade name

Registration Number: 244

Definition: The most inclusive crown clade containing *Solmissus marshalli* A. Agassiz and Mayer 1902 but not *Hydra vulgaris* (Pallas 1766) (*Hydroidolina*). This is a maximum-crown-clade definition. Abbreviated definition: max crown ∇ (*Solmissus marshalli* A. Agassiz and Mayer 1902 ~ *Hydra vulgaris* (Pallas 1766)).

Etymology: Derived from *trachys* (Greek for rough) and λίνον (*linon*) which means "thread" or "anything made of flax." Haeckel (1879) stated that *linon* could mean "thread" or "sail" and could therefore refer to tentacles or the velum, features of both of which he felt distinguished the medusae of *Trachylina* from those of his *Leptolinae* (approximates *Hydroidolina*; see entry in this volume).

Reference Phylogeny: Collins et al. (2008: Fig. 4). For the purpose of applying our definition, *Hydra vulgaris* is to be regarded as a representative of *Hydroidolina*.

Composition: *Limnomedusae, Trachymedusae, Narcomedusae, Actinulida* (Collins et al., 2008: Fig. 8), and potentially some members of *Laingiomedusae* (Collins et al., 2006a: Fig. 6). *Actinulida* contains two relatively disparate groups traditionally ranked as genera (*Otohydra* and *Halammohydra*) that live between sand grains and only one of these (*Halammohydra*) was included in the reference phylogeny (Collins et al., 2008). The monophyly of *Actinulida* has not specifically been tested. *Tetraplatia*, a medusozoan whose phylogenetic position has been quite controversial, has been shown to belong to *Trachylina* within *Narcomedusae* (Collins et al., 2006b).

Diagnostic Apomorphies: While molecular data provide strong support for the monophyly of *Trachylina* (Collins et al., 2006a, 2008), there appear to be no diagnostic apomorphies of the clade. However, trachyline medusae are unique within *Hydrozoa* in possessing statocysts of ecto-endodermal origin (Haeckel, 1879). This character appears to be plesiomorphic for the group (Collins et al., 2006a).

Synonyms: *Trachylinae*, Haeckel's (1879) original spelling of the name, is an approximate synonym, given that he did not explicitly deal with species of *Actinulida*, which were unknown at the time. Harrington and Moore's (1956) *Trachylinida* is another approximate synonym, often used in the palaeontological community, that also did not explicitly address *Actinulida*. *Automedusa* Bouillon and Boero 2000 is a partial synonym for *Trachylina*, differing in its exclusion of *Limnomedusae*. *Limnomedusae* (Kramp, 1938) appears to be a partial synonym in that it refers to a paraphyletic group originating in the same ancestor (see Comments).

Comments: Haeckel's (1879) *Trachylinae* included two main groups, *Narcomedusae* and *Trachymedusae*, whose members so far as was known at the time developed directly from planula to medusa. Indeed, most trachylines are open-ocean, direct-developing species that lack polyp stages (Hyman, 1940). While this general description fits the species of *Narcomedusae* and *Trachymedusae*, members of *Limnomedusae* (which has a complicated taxonomic history; see Collins et al., 2008) differ in possessing a polyp phase in the lifecycle, and members of *Actinulida* differ in inhabiting the interstices of sand grains. Remane (1927) was the first to recognize that the directly developing interstitial *Halammohydra* (representative of *Actinulida*) was probably derived from trachyline ancestors. Indeed, analyses based on morphological characters have consistently concluded that *Actinulida*, *Narcomedusae* and *Trachymedusae* form a clade (Bouillon and Boero, 2000; Marques and Collins, 2004; Van Iten et al., 2006). However, these same studies were either inconclusive about the relationship of the polyp-possessing *Limnomedusae* to the other trachyline taxa (Marques and Collins, 2004; Van Iten et al., 2006) or placed *Limnomedusae* with the taxa of *Hydroidolina* (Bouillon and Boero, 2000). When molecular data first suggested that *Limnomedusae* is part of *Trachylina* (Collins, 2002), the result was somewhat surprising. On the other hand, the medusa stages of *Limnomedusae* strongly resemble those of *Trachymedusae*, and prior to the discovery of the polyp stages of *Limnomedusae*, they were classified within *Trachymedusae* (e.g., Hyman, 1940). Increased taxon sampling within *Trachylina* suggests that *Limnomedusae* is a paraphyletic grade that gave rise to other trachyline lineages (Collins et al., 2006a, 2008). The name *Trachylina* was selected over other possible names for this clade (see Synonyms) because it is more common in current use.

Literature Cited

Bouillon, J., and F. Boero. 2000. The *Hydrozoa*: a new classification in the light of old knowledge. *Thalass. Salentina* 24:3–45.

Collins, A. G. 2002. Phylogeny of *Medusozoa* and the evolution of cnidarian life cycles. *J. Evol. Biol.* 15:418–432.

Collins, A. G., B. Bentlage, A. Lindner, D. Lindsay, S. H. D. Haddock, G. Jarms, J. L. Norenburg, T. Jankowski, and P. Cartwright. 2008. Phylogenetics of *Trachylina* (*Cnidaria*: *Hydrozoa*) with new insights on the evolution of some problematical taxa. *J. Mar. Biol. Assoc. UK* 88:1673–1685.

Collins, A. G., B. Bentlage, G. I. Matsumoto, S. H. D. Haddock, K. J. Osborn, and B. Schierwater. 2006b. Solution to the phylogenetic enigma of *Tetraplatia*, a worm-shaped cnidarian. *Biol. Lett.* 2:120–124.

Collins, A. G., P. Schuchert, A. C. Marques, T. Jankowski, M. Medina, and B. Schierwater. 2006a. Medusozoan phylogeny and character evolution clarified by new large and small subunit rDNA data and an assessment of the utility of phylogenetic mixture models. *Syst. Biol.* 55:97–115.

Haeckel, E. 1879. *Das System der Medusen: Erster Theil einer Monographie der Medusen*. G. Fischer, Jena.

Harrington, H. J., and R. C. Mooore. 1956. *Trachylinida*. Pp. F68–F76 in *Treatise on Invertebrate Palaeontology, Part F*. Coelenterata (R. C. Moore, ed.). Geological Society of America, Boulder, and University of Kansas Press, Lawrence, KS.

Hyman, L. H. 1940. *The Invertebrates: Volume 1*. Protozoa *through* Ctenophora. McGraw-Hill, New York.

Kramp, P. L. 1938. Die meduse von *Ostroumovia inkermanica* (Pal.-Ostr.) und die systematische Stellung der Olindiiden. *Zool. Anz.* 122:103–108.

Marques, A. C., and A. G. Collins. 2004. Cladistic analysis of *Medusozoa* and cnidarian evolution. *Invertebr. Biol.* 123:23–42.

Remane, A. 1927. *Halammohydra*, ein eigenartiges Hydrozoon der Nordund Ostsee. *Z. Morphol. Ökol. Tiere* 7:643–677.

Van Iten, H., J. de Moraes Leme, M. G. Simões, A. C. Marques, and A. G. Collins. 2006. Reassessment of the phylogenetic position of conulariids (?Ediacaran-Triassic) within the subphylum *Medusozoa* (phylum *Cnidaria*). *J. Syst. Palaeontol.* 4:109–118.

Authors

Casey W. Dunn; Department of Ecology and Evolutionary Biology; Yale University; 165 Prospect St., New Haven, CT 06511, USA. Email: casey.dunn@yale.edu.

Allen G. Collins; National Systematics Laboratory of NOAA's Fisheries Service and National Museum of Natural History; MRC-153, Smithsonian Institution; P.O. Box 37012; Washington, DC 20013-7012, USA. Email: collinsa@si.edu.

Date Accepted: 18 March 2011

Primary Editor: Kevin de Queiroz

Hydroidolina A. G. Collins 2000 [A. G. Collins and C. W. Dunn], converted clade name

Registration Number: 169

Definition: The most inclusive crown clade containing *Hydra vulgaris* (Pallas 1766) but not *Solmissus marshalli* A. Agassiz and Mayer 1902 (*Trachylina*). This is a maximum-crown-clade definition. Abbreviated definition: max crown ∇ (*Hydra vulgaris* (Pallas 1766) ~ *Solmissus marshalli* A. Agassiz and Mayer 1902).

Etymology: Derived from "hydroid" (a name used for the polyp stages of hydrozoans) + λίνον (*linon*) which means "thread" or "anything made of flax." Haeckel (1879) stated that *linon* could mean "thread" or "sail" and could therefore refer to tentacles or the velum, features of both of which he felt distinguished the medusae of his *Leptolinae* (approximates *Hydroidolina*) from those of *Trachylina* (see Comments).

Reference Phylogeny: Collins et al. (2006: Fig. 6). For the purpose of applying our definition, *Hydra vulgaris* is to be regarded as a close relative of *Hydra littoralis*.

Composition: *Leptothecata*, *Siphonophora* (this volume), *Aplanulata*, at least one member of *Laingiomedusae*, and the various other taxa assigned to the paraphyletic group *Anthoathecata* (*Capitata* and *Filifera*). *Hydroidolina* includes almost all hydrozoans whose lifecycle includes a polyp stage (the exception being *Limnomedusae*, which is part of *Trachylina*).

Diagnostic Apomorphies: Five apomorphies for *Hydroidolina* were implied by the cladistic analysis of Marques and Collins (2004), but all of these were homoplastic. An apomorphic state of medusae lacking statocysts of ecto-endodermal origin was identified by Collins et al. (2006), but the majority of hydroidolinan species do not possess a medusa stage. Statocysts, when present, are exclusively ectodermal in origin. Colonial hydroid stages within *Hydroidolina*, especially siphonophores, tend to have greater functional specialization between zooids than do other colonial members of *Cnidaria* (Hyman, 1940).

Synonyms: *Leptolinae* Haeckel 1879 is a partial synonym that differs from *Hydroidolina* in excluding *Siphonophora*. An alternative spelling, *Leptolina*, has sometimes been used (e.g., Mayer, 1900). *Hydroidomedusae* Bouillon and Boero 2000 is also a partial synonym that differs from *Hydroidolina* in including *Limnomedusae*. This name also has a variant spelling, *Hydroidomedusa* (e.g., within Bouillion and Boero, 2000).

Comments: Collins (2000) was the first to publish the name *Hydroidolina*, but its origin stems from the dissertation work of Marques (partly published as Marques and Collins, 2004). Collins (2000) described the clade *Hydroidolina* as comprising "Capitata, Filifera, Hydridae, Leptomedusae, and Siphonophora", which approximates the composition of the clade to which the name is applied here. *Hydroidolina* is chosen over *Leptolinae* as the name of this clade because the former included *Siphonophora* in its original conception. Despite the lack of clear morphological synapomorphies, the clade was supported by the morphology-based phylogenetic analyses of Bouillon and Boero (2000) and Marques and Collins (2004). Molecular corroboration of *Hydroidolina* is strong in multigene analyses with broad taxon sampling (Collins et al., 2006a; Cartwright et al., 2008).

Literature Cited

Bouillon, J., and F. Boero. 2000. The *Hydrozoa*: a new classification in the light of old knowledge. *Thalass. Salentina* 24:3–45.

Cartwright, P., N. M. Evans, C. W. Dunn, A. C. Marques, M. P. Miglietta, P. Schuchert, and A. G. Collins. 2008. Phylogenetics of *Hydroidolina* (*Hydrozoa*: *Cnidaria*). *J. Mar. Biol. Assoc. UK* 88(8):1663–1672.

Collins, A. G. 2000. Towards understanding the phylogenetic history of *Hydrozoa*: hypothesis testing with 18S gene sequence data. *Sci. Mar.* 64:5–22.

Collins, A. G., P. Schuchert, A. C. Marques, T. Jankowski, M. Medina, and B. Schierwater. 2006. Medusozoan phylogeny and character evolution clarified by new large and small subunit rDNA data and an assessment of the utility of phylogenetic mixture models. *Syst. Biol.* 55:97–115.

Haeckel, E. 1879. *Das System der Medusen: Erster Theil einer Monographie der Medusen.* G. Fischer, vormals F. Mauke, Jena.

Hyman, L. H. 1940. *The Invertebrates. Vol. 1: Protozoa* through *Ctenophora.* McGraw-Hill, New York.

Marques, A. C., and A. G. Collins. 2004. Cladistic analysis of *Medusozoa* and cnidarian evolution. *Invertebr. Biol.* 123:23–42.

Mayer, A. G. 1900. Some medusae from Tortugas, Florida. *Bull. Mus. Comp. Zool.* 37:13–82.

Authors

Allen G. Collins; National Systematics Laboratory of NOAA's Fisheries Service and National Museum of Natural History; MRC-153, Smithsonian Institution; P.O. Box 37012, Washington, DC 20013-7012, USA. Email: collinsa@si.edu.

Casey W. Dunn; Department of Ecology and Evolutionary Biology; Yale University, 165 Prospect St.; New Haven CT 06511, USA. Email: casey.dunn@yale.edu.

Date Accepted: 18 March 2011

Primary Editor: Kevin de Queiroz

Siphonophora F. Eschscholtz 1829, as *Siphonophorae*
[C. W. Dunn], converted clade name

Registration Number: 170

Definition: The crown clade originating in the most recent common ancestor of *Physalia physalis* Linnaeus 1758, *Physophora hydrostatica* Forskål 1775, *Apolemia uvaria* (Lesueur ?1811), and *Praya dubia* Quoy and Gaimard 1833. This is a minimum-crown-clade definition. Abbreviated definition: min crown ∇ (*Physalia physalis* Linnaeus 1758 & *Physophora hydrostatica* Forskål 1775 & *Apolemia uvaria* (Lesueur ?1811) & *Praya dubia* Quoy and Gaimard 1833).

Etymology: Derived from *siphon* (Gr., tube) + *-phore* (Gr., bearing); these organisms were known as the röhrenquallen ("tube jelly-fish") in the early German literature, in reference to the tube-like stem and polyps.

Reference Phylogeny: Dunn et al. (2005: Fig. 6) is designated as the primary reference phylogeny. Collins et al. (2006a: Figs. 3, 6) includes more outgroups but has fewer siphonophores. For the purpose of applying our definition, *Apolemia uvaria* is to be regarded as a close relative of *Apolemia* species 1–4, as sampled in the reference phylogeny.

Composition: *Siphonophora* contains about 185 currently recognized species. These include all representatives of *Cystonectae* and *Codonophora*, the latter including all members of the *Calycophorae* and the paraphyletic grade *Physonectae* (Dunn et al., 2005).

Diagnostic Apomorphies: Siphonophores are polymorphic colonies, bearing highly specialized polyps and medusae (Totton, 1965). Siphonophores differ from other hydrozoans in the following derived characters. Feeding polyps bear a single basal tentacle. The colony is organized along a single linear stem (which is reduced to a bulbous corm in several taxa and is entirely absent in *Physalia physalis* Linnaeus, 1758) that develops from the elongated body column of the protozooid. Siphonophore colonies bear asexual propulsive medusae (nectophores), which can be attached directly to the stem or attached to clusters of reproductive medusae (gonodendra).

Synonyms: *Syphonophorae* Eschscholtz 1829 and *Siphonophorae* Eschscholtz 1829 (see Comments).

Comments: Eschscholtz (1829) spelled the name as both *Syphonophorae* and *Siphonophorae*. The former seems to have been a misspelling. The spelling *Siphonophorae* was changed to *Siphonophora* by different authors at different times in the first half of the twentieth century (e.g., Garstang, 1946) and is maintained here. Eschscholtz (1829) considered *Siphonophora* to include *Chondrophora* (represented by *Porpita* and *Vellela* in the reference phylogeny). Totton (1954) abandoned this scheme, and since then *Chondrophora* has not been placed within *Siphonophora* (Mackie et al., 1987; Pugh, 1999). Consequently, the definition provided here for *Siphonophora* does not include any taxa belonging to *Chondrophora* as internal specifiers.

Literature Cited

Collins, A. G., P. Schuchert, A. C. Marques, T. Jankowski, M. Medina, and B. Schierwater. 2006. Medusozoan phylogeny and character evolution clarified by new large and small subunit rDNA data and an assessment of the utility of phylogenetic mixture models. *Syst. Biol.* 55:97–115.

Dunn, C. W., P. R. Pugh, and S. H. D. Haddock. 2005. Molecular phylogenetics of the *Siphonophora* (*Cnidaria*), with implications for the evolution of functional specialization. *Syst. Biol.* 54:916–935.

Eschscholtz, F. 1829. *System der Acalephen*. F. Dümmier, Berlin.

Garstang, W. 1946 The morphology and relations of the *Siphonophora*. *Q. J. Microsc. Sci.* 87:103–193.

Mackie, G. O., P. R. Pugh, and J. E. Purcell. 1987. Siphonophore biology. *Adv. Mar. Biol.* 24:97–262.

Pugh, P. R. 1999. *Siphonophorae*. Pp. 467–511 in *South Atlantic Zooplankton* (D. Boltovskoy, ed.). Backhuys Publishers, Leiden.

Totton, A. K. 1954. *Siphonophora* of the Indian Ocean together with systematic and biological notes on related specimens from other oceans. *Discovery Rep.* 27:1–162.

Totton, A. K. 1965. *A Synopsis of the* Siphonophora. British Museum of Natural History, London.

Authors

Allen G. Collins; National Systematics Laboratory of NOAA's Fisheries Service and National Museum of Natural History; MRC-153, Smithsonian Institution; P.O. Box 37012, Washington, DC 20013-7012, USA. Email: collinsa@si.edu.

Casey W. Dunn; Department of Ecology and Evolutionary Biology; Yale University; 165 Prospect St., New Haven, CT 06511, USA. Email: casey.dunn@yale.edu.

Date Accepted: 27 July 2011

Primary Editor: Kevin de Queiroz

Bilateria B. Hatschek 1888 [J. A. Lake, J. R. Garey, J. A. Servin, and K. M. Halanych], converted clade name

Registration Number: 188

Definition: The largest crown clade containing *Strongylocentrotus purpuratus* (Stimpson 1857) (*Deuterostomia/Echinodermata*), *Drosophila melanogaster* Meigen 1830 (*Ecdysozoa/Arthropoda*), *Capitella teleta* Blake, Grassle and Eckelbarger 2009 (*Lophotrochozoa/Annelida*), and *Paratomella rubra* Rieger and Ott 1971 (*Acoelomorpha/Acoela*), but not *Hydra vulgaris* Pallas 1766 (*Cnidaria*) and *Pleurobrachia bachei* A. Agassiz 1860 (*Ctenophora*). This is a maximum-crown-clade definition. Abbreviated definition: max crown ∇ (*Strongylocentrotus purpuratus* (Stimpson 1857) & *Drosophila melanogaster* Meigen 1830 & *Capitella teleta* Blake, Grassle and Eckelbarger 2009 & *Paratomella rubra* Rieger and Ott 1971 ~ *Hydra vulgaris* Pallas 1766 & *Pleurobrachia bachei* A. Agassiz 1860).

Etymology: From the Latin *bi* (two) and *lateralis* (side), referring to the apomorphy of having left/right symmetry at some point in the life cycles of most animals in, and presumably the last common ancestor of, this clade.

Reference Phylogeny: Paps et al. (2009) Figure 1 is the primary reference phylogeny. Secondary reference phylogenies include Peterson and Eernisse (2001: Figs. 2 and 3), Halanych (2004: Fig. 2), Hejnol et al. (2009: Fig. 2), Ryan et al. (2013: Fig. 3), and Whelan et al. (2015: Fig. 3).

Composition: All species within the total clades of *Protostomia* + *Deuterostomia* (see entries in this volume), and *Xenacoelomorpha* (sensu Cannon et al., 2016). See Comments concerning the inclusion of *Orthonectida* and *Dicyemida*.

Diagnostic Apomorphies: Bilateral symmetry and the presence of three tissue layers in adult animals have long been considered apomorphies for *Bilateria* (Hatschek, 1888; Hyman, 1940). The distribution of symmetry across animals is more complex than previously thought. For example, cnidarians and ctenophores often exhibit bilateral features (see Hyman, 1940, Nielsen, 2012). An additional apomorphy diagnosing *Bilateria* is the addition of central-class genes in the Hox-cluster (de Rosa et al., 1999).

Synonyms: There are no unambiguous synonyms. The name *Triploblastica* Lankester 1873 is an approximate synonym, which is often used informally as "triploblast". *Triploblastica* sensu Nielsen (2012) has a different meaning as it includes *Ctenophora* and *Bilateria*.

Comments: The composition of *Bilateria* has, for the most part, not been controversial, but placement of *Myxozoa* and *Ctenophora* were subject to debate. *Myxozoa* had been considered to be within "protozoans" (see Siddall et al., 1995) and among bilateral animals (Anderson et al., 1998), but current evidence suggests they are highly reduced cnidarians (Chang et al., 2015). Nielsen (1995) suggested that ctenophores are allied to deuterostomes, but current analyses support their divergence prior to the origin of *Bilateria* (Nielsen, 2012; Ryan et al., 2013; Whelan et al., 2015). Cannon et al. (2016) placed *Xenacoelomorpha* (i.e., *Xenoturbella*, *Acoela*, and *Nemertodermatida*) as sister to the rest of *Bilateria* (i.e., *Protostomia* +

Deuterostomia). *Orthonectida* and *Dicyemida* are likely within *Protostomia* but placement is not certain (reviewed by Kocot, 2016).

The taxon *Bilateria* was first recognized by Hatschek (1888). There is strong support for the clade based on both morphological (Peterson and Eernisse, 2001) and molecular data (Aguinaldo et al., 1997; Dunn et al., 2008; Paps et al., 2009; Whelan et al., 2015). Although the primary synapomorphy for the clade is presence of a symmetrical body with right and left halves, these features are also found in some ctenophores (e.g., with respect to their lobes and tentacles) and cnidarians (e.g., the siphonogylph of anemones) (see Diagnostic Apomorphies), which diverged prior to the origin of *Bilateria*. Although presence of three tissue layers has been associated and discussed with the origins of *Bilateria* (e.g., Hyman, 1940), mesoderm and nervous systems clearly evolved near the base of the animal tree (Moroz et al., 2014). Thus, while the character of bilateral symmetry may not be restricted to members of the clade here named *Bilateria*, the name is adopted here because of precedence and utility due to common use.

Literature Cited

Aguinaldo, A. M. A., J. M. Turbeville, L. S. Linford, M. C. Rivera, J. R. Garey, R. A. Raff, and J. A. Lake. 1997. Evidence for a clade of nematodes, arthropods and other moulting animals. *Nature* 387:489–493.

Anderson, C. L., E. U. Canning, and B. Okamura. 1998. A triploblast origin of *Myxozoa*? *Nature* 392:346–347.

Cannon, J. T., B. C. Vellutini, J. Smith III, F. Ronquist, U. Jondelius, and A. Hejnol. 2016. *Xenacoelomorpha* is sister to *Nephrozoa*. *Nature* 530:89–93.

Chang, E. S., M. Neuhof, N. D. Rubinstein, A. Diamant, H. Philippe, D. Huchon, and P. Cartwright. 2015. Genomic insights into the evolutionary origin of *Myxozoa* within *Cnidaria*. *Proc. Natl. Acad. Sci. USA.* 112:14912–14917.

Dunn, C. W., A. Hejnol, D. Q. Matus, D. Pang, W. E. Browne, S. A. Smith, E. Seaver, G. W. Rouse, M. Obst, G. D. Edgecombe, M. V. Sørensen, S. H. D. Haddock, A. Schmidt-Rhaesa, A. Okusu, R. M. Kristensen, W. C. Wheeler, M. Q. Martindale, and G. Giribet. 2008. Broad phylogenomic sampling improves resolution of the animal tree of life. *Nature* 452:745–749.

Halanych, K. M. 2004. The new view of animal phylogeny. *Annu. Rev. Ecol. Evol. Syst.* 35:229–256.

Hatschek, B. 1888. *Lehrbuch der Zoologie 1. Liefrung*. Gustav Fischer, Jena.

Hejnol, A., M. Obst, A. Stamatakis, M. Ott, G. W. Rouse, G. D. Edgecombe, P. Martinez, J. Baguñà, X. Bailly, U. Jondelius, M. Wiens, W. E. G. Müller, E. Seaver, W. C. Wheeler, M. Q. Martindale, G. Giribet, and C. W. Dunn. 2009. Assessing the root of bilaterian animals with scalable phylogenomic methods. *Proc. R. Soc. Lond. B Biol. Sci.* 276:4261–4270.

Hyman, L. H. 1940. *The Invertebrates*: Protozoa *through* Ctenophora. 1st edition, Vol. I. McGraw-Hill, New York and London.

Kocot, K. M. 2016. On 20 years of *Lophotrochozoa*. *Org. Divers. Evol.* 16:329–343.

Lankester, E. R. 1873. On the primitive cell-layers of the embryo as the basis of genealogical classification of animals, and on the origin of vascular and lymph systems. *Ann. Mag. Nat. Hist.*, 4th series 11:321–338.

Moroz, L. L., K. M. Kocot, M. R. Citarella, S. Dosung, T. P. Norekian, I. S. Povolotskaya, A. P. Grigorenko, C. Dailey, E. Berezikov, K. Buckley, D. Reshetov, A. Ptytsyn, K. Mukherjee, T. P. Moroz, E. Dabe, F. Yu, V. V. Kapitonov, J. Jurka, Y. Bobkov, J. J. Swore, D. O. Girardo, R. Bruders, C. E. Mills, V. V. Solovyev, J. Rast, B. J. Swalla, F. A. Kondrashov, J. V. Sweedler, E. I. Rogaev, K. M. Halanych, and A. B. Kohn. 2014. The ctenophore genome and the emergence of neural systems. *Nature* 510:109–114.

Nielsen, C. 1995. *Animal Evolution: Interrelationships of the Living Phyla.* 1st edition. Oxford University Press, Oxford.

Nielsen, C. 2012. *Animal Evolution: Interrelationships of the Living Phyla.* 3rd edition. Oxford University Press, Oxford.

Paps, J., J. Baguñà, and M. Riutort. 2009. *Lophotrochozoa* internal phylogeny: new insights from an up-to-date analysis of nuclear ribosomal genes. *Proc. R. Soc. Lond. B Biol. Sci.* 276:1245–1254.

Peterson, K. J., and D. J. Eernisse. 2001. Animal phylogeny and the ancestry of bilaterians: inferences from morphology and 18S rDNA gene sequences. *Evol. Dev.* 3:170–205.

de Rosa, R., J. K. Grenier, T. Andreeva, C. E. Cook, A. Adoutte, M. Akam, S. B. Carroll, and G. Balavoine. 1999. Hox genes in brachiopods and priapulids and protostome evolution. *Nature* 399:772–776.

Ryan, J. F., K. Pang, C. E. Schnitzler, A. D. Nguyen, R. T. Moreland, D. K. Simmons, B. J. Koch, W. R. Francis, P. Havlak, NISC Comparative Sequencing Program, S. A. Smith, N. H. Putnam, S. H. D. Haddock, C. W. Dunn, T. G. Wolfsberg, J. C. Mullikin, M. Q. Martindale, and A. D. Baxevanis. 2013. The genome of the ctenophore *Mnemiopsis leidyi* and its implications for cell type evolution. *Science* 342:1242592.

Siddall, M. E., D. S. Martin, D. Bridge, S. S. Desser, and D. K. Cone. 1995. The demise of a phylum of protists: phylogeny of *Myxozoa* and other parasitic *Cnidaria. J. Parasitol.* 81:961–967.

Whelan, N. V., K. M. Kocot, L. L. Moroz, and K. M. Halanych. 2015. Error, signal, and the placement of *Ctenophora* sister to all other animals. *Proc. Natl. Acad. Sci. USA* 112:5773–5778.

Authors

James Lake; Molecular Biology Institute; 232 Boyer Hall; University of California Los Angeles; Los Angeles, CA 90095, USA. Email: Lake@mbi.ucla.edu.

James R. Garey; Department of Cell Biology, Microbiology and Molecular Biology; University of South Florida; 4202 East Fowler Avenue ISA2015; Tampa, FL 33620, USA. Email: garey@usf.edu.

Jacqueline Servin; Molecular Biology Institute; 232 Boyer Hall; University of California Los Angeles; Los Angeles, CA 90095, USA. Email: jacquelineservin@gmail.com.

Kenneth M. Halanych; Department of Biological Sciences; 101 Rouse Building; Auburn University; Auburn, AL 36849, USA. Email: ken@auburn.edu.

Date Accepted: 24 July 2017

Primary Editors: Jacques A. Gauthier, Kevin de Queiroz

Protostomia K. Grobben 1908 [J. R. Garey, K. M. Halanych, J. A. Servin, and J. A. Lake], converted clade name

Registration number: 193

Definition: The largest crown clade containing *Drosophila melanogaster* Meigen 1830 (*Arthropoda/Hexapoda*) and *Capitella teleta* Blake, Grassle and Eckelbarger 2009 (*Annelida*) but not *Strongylocentrotus purpuratus* (Stimpson 1857) (*Echinodermata/Echinoidea*) and *Homo sapiens* Linnaeus 1758 (*Chordata/Vertebrata*). This is a maximum-crown-clade definition. Abbreviated definition: max crown ∇ (*Drosophila melanogaster* Meigen 1830 & *Capitella telata* Blake, Grassle and Eckelbarger 2009 ~ *Strongylocentrotus purpuratus* (Stimpson 1857) & *Homo sapiens* Linnaeus 1758).

Etymology: From the Greek *proto-* (first) and *stoma* (mouth), referring to a mouth derived from a primary embryonic opening (see Comments).

Reference Phylogeny: Dunn et al. (2008: Fig. 1) is the primary reference phylogeny. Secondary reference phylogenies include Aguinaldo et al. (1997: Fig. 2), Halanych (2004: Fig. 2), Mallatt et al. (2004: Fig. 2), Philippe et al. (2005: Fig. 4), and Pick et al. (2010: Fig. 1). Also see Peterson and Eernisse (2001: Table 1).

Composition: *Protostomia* includes all taxa within *Lophotrochozoa* (including *Mollusca*, *Annelida* [contains *Echiura Siboglinidae*, *Myzostomida*, and *Sipuncula*], *Phoronida*, *Brachiopoda*, *Nemertea*, *Gastrotricha*, *Cycliophora*, *Entoprocta*, *Gnathostomulida*, *Micrognathozoa*, *Rotifera*, *Acanthocephala*, *Bryozoa*, *Platyhelminthes*) and within *Ecdysozoa* (including *Arthropoda*, *Tardigrada*, *Onychophora*, *Nematoda*, *Nematomorpha*, *Kinorhyncha*, *Priapulida*, *Loricifera*) and all extant organisms that share a more recent common ancestor with these organisms than with *Deuterostomia*. See Comments regarding the potential inclusion of *Acoelomorpha* and *Chaetognatha*.

Diagnostic Apomorphies: The monophyly of *Protostomia* is based primarily on ribosomal RNA sequences and large nuclear protein coding gene datasets (see references cited in Reference Phylogeny). The Ubd-A peptide within the Hox gene cluster also appears to support the monophyly of protostomes (Balavoine et al., 2002). Spiral cleavage has historically been considered an apomorphy for *Protostomia* (Hyman, 1940; Ruppert and Barnes, 1994), but ecdysozoans and many lophotrochozoans do not exhibit spiral cleavage (Schmidt-Rhaesa et al., 1998).

Synonyms: There are several approximate synonyms: *Zygoneura* Hatschek 1888, *Hypogastrica* Goette 1902, *Ecterocoelia* Hatschek 1911, *Spiralia* as used by Siewing (1976, 1980) and Willmer (1990), although the original use by Schleip (1929) was more restrictive, and *Gastroneuralia* Ulrich 1951.

Comments: The name *Protostomia* has been used to represent various collections of taxa and refers to the fate of the blastopore as being retained to form the mouth (as opposed to deuterostomy, in which the blastopore becomes the anus). However, this fate of the blastopore is not an apomorphy for the group, as it is the plesiomorphic condition (Ax, 1996), and variation in this character (including deuterostomy or loss) has been reported in lophotrochozoans (Nielsen, 2012: 89; Halanych, 2004) and ecydsozoans (Wennberg et al., 2008; Martín-Durán et al., 2012). Halanych et al. (1995) and Aguinaldo et al. (1997) were two of the primary studies to reshape understanding of protostome phylogeny. Halanych et al. (1995) showed that lophophorates (*Brachiopoda*, *Bryozoa*, *Phoronida*) with some developmental features similar to deuterostomes were closely related to the protostomous annelids and mollusks. Aguinaldo et al. (1997) debunked the idea that the segmented annelids and arthropods where closely related, and instead showed that molting, which placed nematodes next to arthropods, was a phylogenetically informative character.

Four approximate synonyms of *Protostomia* (*Zygoneura*, *Ecterocoelia*, *Hypogastrica*, and *Gastroneuralia*) have fallen out of use or were never widely accepted. The name *Spiralia* was coined by Schleip (1929) to include all taxa with spiral embryonic cleavage patterns (although not precisely defined), and the taxon has a long history of including *Arthropoda* (Anderson, 1973; Costello and Henley, 1976; Siewing, 1976, 1980; Willmer, 1990). *Spiralia* is still occasionally used, though less frequently than *Protostomia*. *Spiralia* is not adopted for the clade here named *Protostomia* because its meaning is interpreted differently by various authors (see Halanych, 2016). For example, Ax (1996) included *Arthropoda* within *Spiralia*, but excluded lophophorates; Nielsen (1995) used *Spiralia* for a subset of *Protostomia*, excluding lophophorates and most ecdysozoans but including *Arthropoda*, while later using *Spiralia* as a synonym of *Lophotrochozoa* (Nielsen, 2012).

The monophyly of *Protostomia* (Grobben, 1908) is supported by both molecular (Dunn et al., 2008) and morphological analyses (Nielsen, 2012). The phylogenetic positions of two taxa are problematic relative to the composition of *Protostomia*. Molecular studies suggest uncertainty in the phylogenetic position of *Acoelomorpha* (*Acoela* + *Nemertodermatida*), which has been placed in *Lophotrochozoa* (Dunn et al., 2008), as early diverging bilaterians (Ruiz-Trillo et al., 1999, Hejnol et al., 2009), or in *Deuterostomia* (Philippe et al., 2011). However subsequent genomic analyses support their placement as early branching bilaterians (Simakov et al., 2015; Cannon et al., 2016; Rouse et al., 2016). Another problematic group is *Chaetognatha*, which historically was included in *Deuterostomia* (e.g., Hyman, 1940; Ruppert, 1997), but molecular analyses indicate that chaetognaths are early diverging protostomes (Halanych, 2004; Helfenbein et al., 2004; Marlétaz et al., 2006; Dunn et al., 2008), lophotrochozoans (Papillon et al., 2004; Matus et al., 2006) or ecdysozoans (Peterson and Eernisse, 2001; Paps et al., 2009). If either *Acoelomorpha* or *Chaetognatha* is conclusively identified as diverging prior to the bifurcation that led to *Lophotrochozoa* and *Ecdysozoa* but still more closely related to those clades than to *Deuterostomia*, it will be included within *Protostomia* under our definition. If either of our internal specifiers (*Drosophila melanogaster* or *Capitella telata*) were found to be more closely related to deuterostomes than they are to each other, the name *Protostomia* would not apply to any clade.

Literature Cited

Aguinaldo, A. M. A., J. M. Turbeville, L. S. Linford, M. C. Rivera, J. R. Garey, R. A. Raff, and J. A. Lake. 1997. Evidence for a clade of nematodes, arthropods and other moulting animals. *Nature* 387:489–493.

Anderson, D. T. 1973. *Embryology and Phylogeny of Annelids and Arthropods*. Pergamon Press, Oxford.

Ax, P. 1996. *Multicellular Animals: A New Approach to the Phylogenetic Order in Nature*, Vol. I. Springer, Berlin.

Balavoine, G., R. de Rosa, and A. Adoutte. 2002. Hox clusters and bilaterian phylogeny. *Mol. Phylogenet. Evol.* 24:366–373.

Cannon, J. T., B. C. Velutini, J. Smith, III, F. Ronquist, U. Jondelius, and A. Henjol. 2016. *Xenacoelomorpha* is the sister group to *Nephrozoa. Nature* 530:89–93.

Costello, D. P., and C. Henley. 1976. Spiralian development: a perspective. *Am. Zool.* 16:277–291.

Dunn, C. W., A. Hejnol, D. Q. Matus, D. Pang, W. E. Browne, S. A. Smith, E. Seaver, G. W. Rouse, M. Obst, G. S. Edgecombe, M. V. Sørensen, S. H. D. Haddock, A. Schmidt-Rhaesa, A. Okusu, R. M. Kristensen, W. C. Wheeler, M. Q. Martindale, and G. Giribet. 2008. Broad phylogenomic sampling improves resolution of the animal tree of life. *Nature* 452:745–749.

Goette, A. 1902. *Lehrbuch der Zoologie*. Verlag won Wilhelm Engelmann, Leipzig.

Grobben, K. 1908. Die systematische Einteilung des Tierreichs. *Verh. Zool. Bot. Ges. Wien* 58:491–511.

Halanych, K. M. 2004. The new view of animal phylogeny. *Annu. Rev. Ecol. Evol. Syst.* 35:229–56.

Halanych, K. M. 2016. How our view of animal phylogeny was reshaped by molecular approaches: lessons learned. *Org. Divers. Evol.* 16:319–328.

Halanych, K. M., J. D. Bacheller, A. M. A. Aguinaldo, S. M. Liva, D. M. Hillis, and J. A. Lake. 1995. Evidence from 18S ribosomal DNA that the lophophorates are protostome animals. *Science* 267:1641–1643.

Hatschek, B. 1888–1891. *Lehrbuch der Zoologie: eine morphologische Übersicht des Thierreiches zur Einführung in das Studium dieser Wissenschaft*. Gustav Fischer, Jena.

Hatschek, B. 1911. *Das neue Zoologisches System*. W. Engelmann, Leipzig.

Hejnol, A., M. Obst, A. Stamatakis, M. Ott, G. W. Rouse, G. D. Edgecombe, P. Martinez, J. Baguña, X. Bailly, U. Joneelius, M. Wiens, W. E. G. Müller, E. Seaver, W. C. Wheeler, M. Q. Martindale, G. Giribet and C. Dunn. 2009.

Assessing the root of bilaterian animals with scalable phylogenomic methods. *Proc. R. Soc. Lond. B Biol. Sci.* 276:4261–4270.

Helfenbein, K. G., H. M. Fourcade, R. G. Vanjani and J. L. Boore. 2004. The mitochondrial genome of *Paraspadella gotoi* is highly reduced and reveals that chaetognaths are a sister group to protostomes. *Proc. Natl. Acad. Sci. USA* 101:10639–10643.

Hyman, L. 1940. *The Invertebrates*: Protozoa *through* Ctenophora. McGraw-Hill, New York and London.

Mallatt, J. M., J. R. Garey, and J. W. Shultz. 2004. Ecdysozoan phylogeny and Bayesian inference: first use of nearly complete 28S and 18S rRNA gene sequences to classify the arthropods and their kin. *Mol. Phylogenet. Evol.* 31:178–191.

Marlétaz, F., E. Martin, Y. Perez, D. Papillon, X. Caubit, C. J. Lowe, B. Freeman, L. Fasano, C. Dossat, P. Wincker, J. Weissenbach, and Y. Le Parco. 2006. Chaetognath phylogenomics: a protostome with deuterostome-like development. *Curr. Biol.* 16:R577–R578.

Martín-Durán, J. M., R. Janssen, S. Wennberg, G. E. Budd, and A. Hejnol. 2012. Deuterostomic development in the protostome *Priapulus caudatus. Curr. Biol.* 22:2161–2166.

Matus, D. Q., R. R. Copley, C. W. Dunn, A. Hejnol, H. Eccleston, K. M. Halanych, M. Q. Martindale, and M. J. Telford. 2006. Broad taxon and gene sampling indicate that chaetognaths are protostomes. *Curr. Biol.* 16:R575–R576.

Nielsen, C. 1995. *Animal Evolution: Interrelationships of the Living Phyla*. 1st edition. Oxford University Press, Oxford.

Nielsen, C. 2012. *Animal Evolution: Interrelationships of the Living Phyla*. 3rd edition. Oxford University Press, Oxford.

Papillon, D., Y. Perez, X. Caubit and Y. Le Parco. 2004. Identification of chaetognaths as protostomes is supported by the analysis of their mitochondrial genome. *Mol. Biol. Evol.* 21: 2122–2129.

Paps, J., J. Baguñà and M. Riutort. 2009. Bilaterian phylogeny: a broad sampling of 13 nuclear genes provides a new *Lophotrochozoa* phylogeny and supports a paraphyletic basal *Acoelomorpha. Mol. Biol. Evol.* 26:2397–2406.

Peterson, K. J., and D. J. Eernisse. 2001 Animal phylogeny and the ancestry of bilaterians: inferences from morphology and 18S rDNA gene sequences. *Evol. Dev.* 3:170–205.

Pick, K. S., H. Philippe, F. Schreiber, D. Erpenbeck, K. J. Jackson, P. Wrede, M. Wiens, A. Alié, B. Morgenstern, M. Manuel, and G. Wörheide. 2010. Improved phylogenomic taxon sampling noticeably affects nonbilaterian relationships. *Mol. Biol. Evol.* 27:1983–1987.

Philippe, H., H. Brinkmann, R. R. Copley, L. L. Moroz, H. Nakano, A. J. Poustka, A. Wallberg, K. J. Peterson, and M. J. Telford. 2011. Acoelomorph flatworms are deuterostomes related to *Xenoturbella*. *Nature* 470:255–258.

Philippe, H., N. Lartillot, and H. Brinkmann. 2005. Multigene analyses of bilaterian animals corroborate the monophyly of *Ecdysozoa*, *Lophotrochozoa*, and *Protostomia*. *Mol. Biol. Evol.* 22:1246–1253.

Rouse, G. W., N. G. Wilson, J. I. Carvajal, and R. C. Vrijenhoek. 2016. New deep-sea species of *Xenoturbella* and the position of *Xenocoelomorpha*. *Nature* 530:94–97.

Ruiz-Trillo, I., M. Riutort, D. T. J. Littlewood, E. A. Herniou and J. Baguña. 1999. Acoel flatworms: earliest extant bilaterian metazoans, not members of *Platyhelminthes*. *Science* 283:1919–1923.

Ruppert, E. E. 1997. Introduction: microscopic anatomy of the notochord, heterochrony, and chordate evolution. Pp. 1–13 in *Microscopic Anatomy of Invertebrates* (F. W. Harrison, ed.). Vol. 15: *Hemichordata, Chaetognatha, and the Invertebrate Chordates*. Wiley-Liss, New York.

Ruppert, E. E., and R. D. Barnes. 1994. *Invertebrate Zoology*. 6th edition. Saunders Publishing, New York.

Schleip, W. 1929. *Die Determination der Primitiventwicklung: Eine Zusammenfassende Darstellung der Ergebnisse über das Determinationsgeschehen in den ersten Entwicklungsstadien der Tiere.* Akademische Verlagsgesellschaft, Leipzig.

Schmidt-Rhaesa, A., U. Ehlers, T. Bartolomaeus, C. Lemburg, and J. R. Garey. 1998. The phylogenetic position of the *Arthropoda*. *J. Morphol.* 238:263–285.

Siewing, R. 1976. Probleme und neuere Erkenntnisse in der Grosssystematik der Wirbellosen. *Verhandl. Dtsch. Zool. Ges. Stuttgart* 1976:59–83.

Siewing, R. 1980. Das Archicoelomatenkonzept. *Zool. Jb. Syst.* 103:439–482.

Simakov, O., T. Kawaashima, F. Marlétaz, J. Jenkins, R. Koyanagi, T. Mitros, K. Hisata, J. Bredeson, E. Shoguchi, F. Gyoja, J. X. Yue, Y. C. Chen, R. M. Freeman, Jr., A. Sasaki, T. Hikosaka-Katayama, A. Sato, M. Fujie, K. W. Baughman, J. Levine, P. Gonzalez, C. Cameron, J. H. Fritzenwanker, A. M. Pani, H. Goto, M. Kanda, N. Arakaki, S. Yamasaki, J. Qu, A. Cree, Y. Ding, H. H. Dinh, S. Dugan, M. Holder, S. N. Jhangiani, C. L. Kovar, S. L. Lee, L. R. Lewis, D. Morton, L. V. Nazareth, G. Okwuonu, J. Santibanez, R. Chen, S. Richards, D. M. Muzny, A. Gillis, L. Peshkin, M. Wu, T. Humphreys, Y. H. Su, N. H. Putnam, J. Schmutz, A. Fujiyama, J. K. Yu, K. Tagawa, K. C. Worley, R. A. Gibbs, M. W. Kirschner, C. J. Lowe, N. Satoh, D. S. Rokhsar, and J. Gerhart. 2015. Hemichordate genomes and deuterostome origins. *Nature* 527:459–465.

Ulrich, W. 1951. Vorschläge zu einer Revision der Grosseinteilung des Tierreichs. *Zool. Anz. Suppl.* 15:244–271.

Wennberg, S. A., R. Janssen, and G. E. Budd. 2008. Early development of the priapulid worm *Priapulus caudatus*. *Evol. Dev.* 10:326–338.

Willmer, P. 1990. *Invertebrate Relationships, Patterns in Animal Evolution*. Cambridge University Press, New York.

Authors

James R. Garey; Department of Cell Biology, Microbiology and Molecular Biology; University of South Florida; 4202 East Fowler Avenue ISA2015; Tampa, FL 33620; USA. Email: garey@usf.edu.

Kenneth M. Halanych; Department of Biological Sciences; 101 Rouse Building; Auburn University; Auburn, AL 36849, USA. Email: ken@auburn.edu

Jacqueline Servin; Molecular Biology Institute; 232 Boyer Hall; University of California Los Angeles; Los Angeles, CA 90095, USA. Email: jacquelineservin@gmail.com.

James Lake; Molecular Biology Institute; 232 Boyer Hall; University of California Los Angeles; Los Angeles, CA 90095, USA. Email: Lake@mbi.ucla.edu.

Date Accepted: 16 September 2016

Primary Editor: Kevin de Queiroz

Lophotrochozoa K. M. Halanych, J. D. Bacheller, A. M. A. Aguinaldo, S. M. Liva, D. M. Hillis, and J. A. Lake 1995 [K. M. Halanych, J. R. Garey, J. A. Servin, and J. A. Lake], converted clade name

Registration Number: 191

Definition: The largest crown clade containing *Glycera americana* Leidy 1855 (*Annelida*), *Placopecten magellanicus* (Gmelin 1791) (*Mollusca*), *Phoronis vancouverensis* Pixell 1912 (*Phoronida*), *Terebratalia transversa* (Sowerby 1846) (*Brachiopoda*), and *Plumatella repens* (Linnaeus 1758) (*Bryozoa*), but not *Drosophila melanogaster* Meigen 1830 (*Arthropoda/ Insecta*), *Caenorhabditis elegans* (Maupas 1900) (*Nematoda*) and *Strongylocentrotus purpuratus* (Stimpson 1857) (*Deuterostomia/Echinodermata*). This is a maximum-crown-clade definition. Abbreviated definition: max crown ∇ (*Glycera americana* Leidy 1855 & *Placopecten magellanicus* (Gmelin 1791) & *Phoronis vancouverensis* Pixell 1912 & *Terebratalia transversa* (Sowerby 1846) & *Plumatella repens* (Linnaeus 1758) ~ *Drosophila melanogaster* Meigen 1830 & *Caenorhabditis elegans* (Maupas 1900) & *Strongylocentrotus purpuratus* (Stimpson 1857)).

Etymology: Derived from the Greek *lophos* (crest), *trochos* (wheel), and *zōia* (animals), referring to the observation that this clade includes animals that possess either a lophophore feeding structure or a trochozoan larva (Halanych et al., 1995).

Reference Phylogeny: Halanych et al. (1995) Figure 1 is the primary reference phylogeny (the external specifiers are not included in this phylogeny, although arthropods and deuterostomes are represented by other species). In this phylogeny, articulate and inarticulate are brachiopods, chiton and bivalve are mollusks, and polychaete is an annelid. Secondary reference phylogenies include Mackey et al. (1996) Figure 1, Aguinaldo et al. (1997) Figures 2 and 3, Peterson and Eernisse (2001) Figures 3 and 6, Ruiz-Trillo et al. (2002) Figures 2 and 3, Anderson et al. (2004) Figure 1, Halanych (2004) Figure 2, Philippe et al. (2005) Figure 4, Dunn et al. (2008) Figures 1 and 2, Pick et al. (2010) Figure 1, and Kocot et al. (2017) Figures 1 and 2.

Composition: All species within *Mollusca*, *Annelida* (including *Echiura*, *Siboglinidae*, *Myzostomida*, and *Sipuncula*), *Phoronida*, *Brachiopoda*, *Nemertea*, *Gastrotricha*, *Cycliophora*, *Entoprocta*, *Gnathostomulida*, *Micrognathozoa*, *Rotifera*, *Acanthocephala*, *Platyhelminthes*, and *Bryozoa* (see articles cited under Reference Phylogeny). See Comments regarding the relationships of *Acoelomorpha* and *Chaetognatha*.

Diagnostic Apomorphies: The Hox genes Lox 2, Lox 5, Lox 4, Post-1, and Post-2 have Lophotrochozoan-specific peptides as described in Balavoine et al. (2002).

Synonyms: *Spiralia* sensu Giribet (2009), Hejnol et al. (2009), and Nielsen (2012) is an approximate synonym; however, *Spiralia* sensu Schleip (1929) and others (e.g., Nielsen, 1995; Ax, 1996) is not (see Comments).

Comments: Halanych et al. (1995) first recognized that lophophorates (*Brachiopoda*, *Phoronida*, *Bryozoa*) formed a subclade of protostome taxa along with animals possessing trochophore larvae (e.g., *Annelida*, *Mollusca*, *Nemertea*, *Platyhelminthes*) and named the clade *Lophotrochozoa*. Subsequent studies have shown other taxa (e.g., *Platyhelminthes*) to be within *Lophotrochozoa* (Aguinaldo et al., 1997; de Rosa et al., 1999; Kocot et al., 2017) and support the *Lophotrochozoa* hypothesis (Erber et al., 1998; Valentine and Collins, 2000; Kocot et al., 2017). Large nuclear protein coding gene datasets also support the clade (Dunn et al., 2008, Pick et al., 2010).

Historically, spiral cleavage has been considered an apomorphy for *Protostomia* (Hyman, 1940; Ruppert and Barnes, 1994), but ecdysozoans and many other protostomes, including some lophotrochozoans, do not exhibit spiral cleavage (Ax, 1996; Schmidt-Rhaesa et al., 1998). The name *Spiralia* has sometimes been used for the clade here named *Lophotrochozoa*. The name *Spiralia* was coined by Schleip (1929) to include only taxa with spiral embryonic cleavage patterns. He included five groups: *Nemertina*, *Annelida*, *Scaphopoda*, *Gastropoda* and *Lamellibranchia* but excluded *Cephalopoda*. *Spiralia*'s meaning has been interpreted differently by various authors. For example, Ax (1996) included *Arthropoda*, but excluded lophophorates. Nielsen (1995) used *Spiralia* for a clade approximating *Protostomia* (see entry in this volume), except that he excluded lophophorates, which he considered more closely related to deuterostomes. Later, *Spiralia* was applied to the clade here named *Lophotrochozoa* (Giribet, 2008; Nielsen, 2012). Given the varied definitions, imprecise meaning, and less common use of the name *Spiralia*, *Lophotrochozoa* is therefore selected as the name for this clade.

The name *Lophotrochozoa* was originally defined by Halanych et al. (1995) for the clade originating in the most recent common ancestor of lophophorates, mollusks, and annelids—a node-based (minimum-clade) definition. However, it was later adopted by some of the same authors (Aguinaldo et al., 1997), who implicitly applied it as either a maximum-clade (branch-based) or a maximum-crown-clade (branch-modified node-based) definition by including rotifers and platyhelminths, which diverged prior to the most recent common ancestor of lophophorates, mollusks, and annelids in their tree. Other authors have adopted the original definition, thus applying the name *Lophotrochozoa* to a subclade of the one to which that name is applied here and using *Spiralia* for the larger clade (e.g., Hausdorf et al., 2007; Giribet et al., 2009). We take this opportunity to clear up this ambiguity by establishing the name *Lophotrochozoa* using a maximum-crown-clade definition that employs as internal specifiers only members of clades originally included in *Lophotrochozoa* by Halanych et al. (1995) but results in the inclusion of other extant taxa that are more closely related to the originally included taxa than to ecdysozoans, as the name was applied by some of the original authors in subsequent publications (e.g., Aguinaldo et al., 1997).

Acoelomorpha has been placed in a variety of locations in the metazoan tree but is likely not a member of *Lophotrochozoa* (Ruiz-Trillo et al., 1999; Paps et al., 2009; Philippe et al., 2011; Cannon et al., 2016). *Chaetognatha* may be a member of *Lophotrochozoa*, but its position is uncertain (e.g., Matus et al., 2006; Marlétaz et al., 2006).

Literature Cited

Aguinaldo, A. M. A., J. M. Turbeville, L. S. Linford, M. C. Rivera, J. R. Garey, R. A. Raff, and J. A. Lake. 1997. Evidence for a clade of nematodes, arthropods and other moulting animals. *Nature* 387:489–493.

Anderson, F. E., A. J. Cordoba, and M. Thollesson. 2004. Bilaterian phylogeny based on analyses of a region of the sodium-potassium ATPase α-subunit gene. *J. Mol. Evol.* 58:252–268.

Ax, P. 1996. *Multicellular Animals I. A New Approach to the Phylogenetic Order in Nature.* Springer, Berlin.

Balavoine, G., R. de Rosa, and A. Adoutte. 2002. Hox clusters and bilaterian phylogeny. *Mol. Phylogenet. Evol.* 24:366–373.

Cannon, J. T., B. C. Vellutini, J. Smith, F. Ronquist, U. Jondelius, and A. Henjol. 2016. *Xenacoelomorpha* is the sister group to *Nephrozoa*. *Nature* 530:89–93.

Dunn, C. W., A. Hejnol, D. Q. Matus, D. Pang, W. E. Browne, S. A. Smith, E. Seaver, G. W. Rouse, M. Obst, G. D. Edgecombe, M. V. Sørensen, S. H. D. Haddock, A. Schmidt-Rhaesa, A. Okusu, R. M. Kristensen, W. C. Wheeler, M. Q. Martindale, and G. Giribet. 2008. Broad phylogenomic sampling improves resolution of the animal tree of life. *Nature* 452:745–749.

Erber, A., D. Riemer, M. Bovenschulte, and K. Weber. 1998. Molecular phylogeny of metazoan intermediate filament proteins. *J. Mol. Evol.* 47:751–762.

Giribet, G. 2008. Assembling the lophotrochozoan (=spiralian) tree of life. *Philos. Trans. R. Soc. Lond. B Biol. Sci.* 363:1513–1522.

Giribet, G., C. W. Dunn, G. D. Edgecombe, A. Hejnol, M. Q. Martindale, and G. W. Rouse. 2009. Assembling the spiralian tree of life. Pp. 52–64 in *Animal Evolution: Genomes, Fossils, and Trees* (M. J. Telford and D. T. J. Littlewood, eds.). Oxford University Press, Oxford.

Halanych, K. M. 2004. The new view of animal phylogeny. *Annu. Rev. Ecol. Evol. Syst.* 35:229–56.

Halanych, K. M., J. D. Bacheller, A. M. A. Aguinaldo, S. M. Liva, D. M. Hillis, and J. A. Lake. 1995. Evidence from 18S ribosomal DNA that the lophophorates are protostome animals. *Science* 267:1641–1643.

Hausdorf, B., M. Helmkampf, A. Meyer, A. Witek, H. Herlyn, I. Bruchhaus, T. Hankeln, T. H. Struck, and B. Lieb. 2007. Spiralian phylogenomics supports the resurrection of *Bryozoa* comprising *Ectoprocta* and *Entoprocta*. *Mol. Biol. Evol.* 24:2723–2729.

Hejnol, A., M. Obst., A. Stamatakis, M. Ott, G. W. Rouse, G. D. Edgecombe, P. Martinez, J. Baguña, X. Bailly, U. Jondelius, M. Wiens, W. E. G. Müller, E. Seaver, W. C. Wheeler, M. Q. Martindale, G. Giribet, and C. W. Dunn. 2009. Assessing the root of bilaterian animals with scalable phylogenomic methods. *Proc. R. Soc. Lond. B Biol. Sci.* 276:4261–4270.

Hyman, L. H. 1940. *The Invertebrates*: Protozoa *through* Ctenophora. McGraw-Hill, New York.

Kocot, K. M., T. H. Struck, J. Merkel, D. S. Waits, C. Todt, P. M. Brannock, D. A. Weese, J. T. Cannon, L. L. Moroz, B. Leib, and K. M. Halanych. 2017. Phylogenomics of *Lophotrochozoa* with consideration of systematic error. *Syst. Biol.* 66:256–282.

Mackey, L. Y., B. Winnepennickx, R. De Wachter, T. Backeljau, P. Emschermann, and J. R. Garey. 1996. 18S rRNA suggests that *Entoprocta* are protostomes, unrelated to *Ectoprocta*. *J. Mol. Evol.* 42:552–559.

Marlétaz, F., E. Martin, Y. Perez, D. Papillion, X. Caubit, C. J. Lowe, B. Freeman, L. Fasano, C. Dossat, P. Wincker, J. Weissenbach, and Y. Le Parco. 2006. Chaetognath phylogenomics: a protostome with deuterostome-like development. *Curr. Biol.* 16:R577–R578.

Matus, D. Q., R. R. Copley, C. W. Dunn, A. Hejnol, H. Eccleston, K. M. Halanych, M. Q. Martindale, and M. J. Telford. 2006. Broad taxon and gene sampling indicate that chaetognaths are protostomes. *Curr. Biol.* 16:R575–R576.

Nielsen, C. 1995. *Animal Evolution. Interrelationships of the Living Phyla.* Oxford University Press, Oxford.

Nielsen, C. 2012. *Animal Evolution. Interrelationships of the Living Phyla.* 3rd edition. Oxford University Press, Oxford.

Paps, J., J. Baguña, and M. Riutort. 2009. Bilaterian phylogeny: a broad sampling of 13 nuclear genes provides a new *Lophotrochozoa* phylogeny and supports a paraphyletic basal *Acoelomorpha*. *Mol. Biol. Evol.* 26:2397–2406.

Peterson, K. J., and D. J. Eernisse. 2001. Animal phylogeny and the ancestry of bilaterians: inferences from morphology and 18S rDNA gene sequences. *Evol. Dev.* 3:170–205.

Philippe, H., H. Brinkmann, R. R. Copley, L. L. Moroz, H. Nakano, A. J. Poustka, A. Wallberg, K. J. Peterson, and M. J. Telford. 2011. Acoelomorph flatworms are deuterostomes related to *Xenoturbella. Nature* 470: 255–258.

Philippe, H., N. Lartillot, and H. Brinkmann. 2005. Multigene analyses of bilaterian animals corroborate the monophyly of *Ecdysozoa, Lophotrochozoa,* and *Protostomia. Mol. Biol. Evol.* 22:1246–1253.

Pick, K. S., H. Philippe, F. Schreiber, D. Erpenbeck, K. J. Jackson, P. Wrede, M. Wiens, A. Alié, B. Morgenstern, M. Manuel, and G. Wörheide. 2010. Improved phylogenomic taxon sampling noticeably affects nonbilaterian relationships. *Mol. Biol. Evol.* 27:1983–1987.

de Rosa, R., J. K. Grenier, T. Andreeva, C. E. Cook, A. Adoutte, M. Akam, S. B. Carroll, and G. Balavoine. 1999. HOX genes in brachiopods and priapulids and protostome evolution. *Nature* 399:772–776.

Ruiz-Trillo, I., J. Paps, M. Loukota, C. Ribera, U. Jondelius, J. Baguña, and M. Riutort. 2002. A phylogenetic analysis of myosin heavy chain type II sequences corroborates that *Acoela* and *Nemertodermatida* are basal bilaterians. *Proc. Natl. Acad. Sci. USA* 99:11246–11251.

Ruiz-Trillo, I., M. Riutort, T. J. Littlewood, E. A. Herniou, and J. Baguña. 1999. Acoel flatworms: earliest extant bilaterian metazoans, not members of *Platyhelminthes. Science* 283:1919–23.

Ruppert, E. E., and R. D. Barnes. 1994. *Invertebrate Zoology.* 6th edition. Saunders College Publishing, Fort Worth, TX and Harcourt Brace and Company, Orlando, FL.

Schleip, W. 1929. *Die Determination der Primitiventwicklung.* Akademische Verlagsgesellschaft, Leipzig.

Schmidt-Rhaesa, A., T. Bartolomaeus, C. Lemburg, U. Ehlers, and J. R. Garey. 1998. The position of the *Arthropoda* in the phylogenetic system. *J. Morphol.* 238:263–285

Valentine, J. W., and A. G. Collins. 2000. The significance of moulting in ecdysozoan evolution. *Evol. Dev.* 2:152–156.

Authors

Kenneth M. Halanych; Department of Biological Sciences; 101 Rouse Building; Auburn University; Auburn, AL 36849, USA. Email: ken@auburn.edu

James R. Garey; Department of Cell Biology, Microbiology and Molecular Biology; University of South Florida; 4202 East Fowler Avenue ISA2015; Tampa, FL 33620, USA. Email: garey@usf.edu.

Jacqueline Servin; Molecular Biology Institute; 232 Boyer Hall; University of California Los Angeles; Los Angeles, CA 90095, USA. Email: jacquelineservin@gmail.com.

James Lake; Molecular Biology Institute; 232 Boyer Hall; University of California Los Angeles; Los Angeles, CA 90095, USA. Email: Lake@mbi.ucla.edu.

Date Accepted: 6 November 2017

Primary Editors: Jacques A. Gauthier, Kevin de Queiroz

Annelida Lamarck 1802 [F. Pleijel and G. W. Rouse], converted clade name

Registration Number: 218

Definition: The least inclusive crown clade containing (in order of appearance on the reference phylogeny) *Eulalia viridis* (Linnaeus 1767) (*Phyllodocidae*), *Riftia pachyptila* Jones 1981 (*Siboglinidae)*, *Aeolosoma hemprichi* Ehrenberg 1828, *Chaetopterus variopedatus* (Renier 1804), *Eurythoe complanata* (Pallas 1766) (*Amphinomidae*), *Notomastus latericeus* M. Sars 1851 (*Capitellidae*), *Ophelina acuminata* Ørsted 1843, (*Opheliidae*), *Cirratulus cirratus* (O. F. Müller 1776), *Pectinaria regalis* Verrill 1901, *Arenicola marina* (Linnaeus 1758), *Terebellides stroemi* M. Sars 1835 (*Terebelliformia*), *Flabelligera affinis* (M. Sars 1829), *Scoloplos armiger* (O. F. Müller 1776) (*Orbiniidae*), *Sabellaria alveolata* (Linnaeus 1767), *Harmothoe imbricata* (Linnaeus 1767) (*Aphroditiformia*), *Scalibregma inflatum* Rathke 1843, *Nereis pelagica* Linnaeus 1758, *Eunice pennata* (O. F. Müller 1776), *Sabella pavonina* Savigny 1822, *Polydora ciliata* (Johnston 1838) (*Spionidae*), *Capilloventer australis* Erséus 1993 (incertae sedis), *Lumbricus terrestris* Linnaeus 1758 (*Clitellata*), and *Hirudo medicinalis* Linnaeus 1758. This is a minimum-crown-clade definition. Abbreviated definition: min crown ∇ (*Eulalia viridis* (Linnaeus 1767) & *Riftia pachyptila* Jones 1981 & *Aeolosoma hemprichi* Ehrenberg 1828 & *Chaetopterus variopedatus* (Renier 1804) & *Eurythoe complanata* (Pallas 1766) & *Notomastus latericeus* M. Sars 1851 & *Ophelina acuminata* Ørsted 1843 & *Cirratulus cirratus* (O. F. Müller 1776) & *Pectinaria regalis* Verrill 1901 & *Arenicola marina* (Linnaeus 1758) & *Terebellides stroemi* M. Sars 1835 & *Flabelligera affinis* (M. Sars 1829) & *Scoloplos armiger* (O. F. Müller 1776) & *Sabellaria alveolata* (Linnaeus 1767) & *Harmothoe imbricata* (Linnaeus 1767) & *Scalibregma inflatum* Rathke 1843 & *Nereis pelagica* Linnaeus 1758 & *Eunice pennata* (O. F. Müller 1776) & *Sabella pavonina* Savigny 1822 & *Polydora ciliata* (Johnston 1838) & *Capilloventer australis* Erséus 1993 & *Hirudo medicinalis* Linnaeus 1758 & *Lumbricus terrestris* Linnaeus 1758).

Etymology: Derived from Latin, *anellus*, a little ring, diminutive of *annulus*, a ring.

Reference Phylogeny: Figure 2 in Rousset et al. (2007).

Composition: *Annelida* currently includes ca. 14,000 described species, with 9,000 polychaetes, 650 leeches, 150 branchiobdellids and 4,000 oligochaetes (e.g., Rouse and Pleijel, 2005). Note, however, that leeches and branchiobdellids are nested among oligochaetes (Siddall et al., 2001), together referred to as *Clitellata*, and that this latter group in turn may be nested among the polychaetes (e.g., McHugh, 1997; Rouse and Pleijel, 2001, 2003; Struck et al., 2002; Bleidorn et al., 2003a,b; Jördens et al., 2004). Today there is also strong support that *Vestimentifera* and *Pogonophora* (now referred to as *Siboglinidae*) are annelids (Bartolomaeus, 1995, 1998; McHugh, 1997; Rouse and Fauchald, 1997; Kojima, 1998; Rousset et al., 2004), and also that *Echiura* and possibly *Sipuncula* are members of *Annelida* (McHugh, 1997; Hessling, 2002; Rousset et al., 2007; Struck et al., 2007; Dunn et al., 2008). Currently there is debate on the overall position of *Myzostomida*, with authors variously favouring a position among annelids (Rouse and Pleijel, 2001; Bleidorn et al., 2007; Struck et al., 2011) or outside annelids and closer to platyhelminths (Eeckhaut et al., 2000) or rotifers (Zrzavy et al., 2001).

Diagnostic Apomorphies: Sequentially formed segments established during development from growth zones located at the posterior end of the body (Seaver, 2003). The apparent absence of segmentation in *Echiura* and *Sipuncula* has been argued to be a loss in each case (Hessling, 2002; Kristof et al., 2008). Resolution of the phylogenetic placement of these groups is required to fully determine whether "segmentation" is an apomorphy for *Annelida*, but currently this would appear to be the case. Chaetae formed by microvillar borders of invaginated epidermal cells (O'Clair and Cloney, 1974) may represent an annelid apomorphy, with a loss in some taxa (e.g., *Sipuncula*, *Polygordius*, and *Hirudinea*). Very similar chaetae do appear in non-annelids such as *Brachiopoda* (see Lüter, 2000) and further study on this is required. Nuchal organs may constitute another apomorphy for *Annelida* (Rouse and Fauchald, 1995); however, they appear to be absent for *Clitellata* and some other annelids, and thus the level at which this feature constitutes an apomorphy depends on the relationships between clitellates and polychaetes.

Synonyms: *Polychaeta* (partial and ambiguous); see Comments.

Comments: Lamarck (1802:56, 65) originally introduced the name *"Annelides"* for a group identified by a series of characters (e.g., presence of branchiae and segmentation) but made no reference to any included taxa. However, he later (1818) specified the content in including a series of polychaetes, some (but not all) oligochaetes, and echiurans and leeches. Echiurans were removed from annelids and referred to a separate phylum by Newby (1940), but more recent studies indicate that Lamarck's original inclusion was correct (e.g., McHugh, 1997).

Annelida has often been applied for a group that includes *Polychaeta*, *Oligochaeta* and *Hirudinea* (≈ *Hirudinida*). If *Clitellata* (*Oligochaeta* and *Hirudinea*) is nested within *Polychaeta* (as in the reference tree) then *Annelida* and *Polychaeta* become synonymous. We have chosen to apply the name *"Annelida"* rather than *"Polychaeta"* because this is in close agreement with current and previous usage regarding composition. We suggest that the vernacular name "polychaete" can be applied for non-clitellate annelids, and "oligochaete" for non-hirudinean clitellates.

The phylogenetic analysis by Rousset et al. (2007) is not the most recent one on annelids (see Zrzavy et al., 2009 and Struck et al., 2011), but was chosen because it has the most complete taxon sampling until now. We have opted for a minimum-clade definition with many specifiers since both the relationships between major clades within annelids, and the sister group relationships of annelids, are poorly understood. We have selected specifiers that are present in the reference phylogeny, and, as far as possible, also from taxa that are types for genera and families under the *ICZN*, and which are easily re-collected.

In the reference phylogeny of Rousset et al. (2007), *Apistobranchidae* and *Oweniidae* are no more closely related to the remaining annelids than are molluscs and some nemerteans. Following this phylogeny the inclusion of *Apistobranchidae* and *Oweniidae* in the definition would mean that both molluscs and some nemerteans also should be included in *Annelida*. Since this application of the name goes against both traditional and current usage of the name we have chosen not to include any apistobranchids or oweniids as specifiers. It should be noted, however, that the positions of apistobranchids and oweniids have very low support and we believe them to be spurious. Nevertheless, should future analyses support this topology then apistobranchids and oweniids will indeed remain outside the *Annelida*.

Literature Cited

Bartolomaeus, T. 1995. Structure and formation of the uncini in *Pectinaria koreni, Pectinaria auricoma* (*Terebellida*) and *Spirorbis spirorbis* (*Sabellida*): implications for annelid phylogeny and the position of the *Pogonophora. Zoomorphology* (*Berl.*) 115:161–177.

Bartolomaeus, T. 1998. Chaetogenesis in polychaetous *Annelida*—significance for annelid systematics and the position of the *Pogonophora. Zoology* (*Jena*) 100:348–364.

Bleidorn, C., I. Eeckhaut, L. Podsiadlowski, N. Schult, D. McHugh, K. Halanych, C. Milinkovitsch, and R. Tiedemann. 2007. Mitochondrial genome and nuclear sequence data support *Myzostomida* as part of the annelid radiation. *Mol. Biol. Evol.* 24:1690–1701.

Bleidorn, C., L. Vogt, and T. Bartolomaeus. 2003a. New insights into polychaete phylogeny (*Annelida*) inferred from 18S rDNA sequences. *Mol. Phylogenet. Evol.* 29:279–288.

Bleidorn, C., L. Vogt, and T. Bartolomaeus. 2003b. A contribution to sedentary polychaete phylogeny using 18S rRNA sequence data. *J. Zool. Syst. Evol. Res.* 41:186–195.

Dunn, C. W., A. Hejnol, D. Q. Matus, K. Pang, W. E. Browne, S. A. Smith, E. Seaver, G. W. Rouse, M. Obst, G. D. Edgecombe, M. V. Sørensen, S. H. D. Haddock, A. Schmidt-Rhaesa, A. Okusu, R. M. Kristensen, W. C. Wheeler, M. Q. Martindale, and G. Giribet. 2008. Broad phylogenomic sampling improves resolution of the animal tree of life. *Nature* 452:745–749.

Eeckhaut, I., D. McHugh, P. Mardulyn, R. Tiedemann, D. Monteyne, M. Jangoux, and C. Milinkovitch. 2000. *Myzostomida*: a link between trochozoans and flatworms? *Proc. R. Soc. Lond. B Biol. Sci.* 267:1383–1392.

Hessling, R. 2002. Metameric organisation of the nervous system in developmental stages of *Urechis caupo* (*Echiura*) and its phylogenetic implications. *Zoomorphology* (*Berl.*) 121:221–234.

Jördens, J., T. Struck, and G. Purschke. 2004. Phylogenetic inference regarding *Parergodrilidae* and *Hrabeiella periglandulata* ("*Polychaeta*", *Annelida*) based on 18S rDNA, 28S rDNA and COI sequences. *J. Zool. Syst. Evol. Res.* 42:270–280.

Kojima, S. 1998. Paraphyletic status of *Polychaeta* suggested by phylogenetic analysis based on the amino acid sequences of elongation factor 1-alpha. *Mol. Phylogenet. Evol.* 9:255–261.

Kristof, A., T. Wollesen, and A. Wanninger. 2008. Segmental mode of neural patterning in *Spinucula. Curr. Biol.* 18:1129–1132.

Lamarck, J.-B. 1802. *Discours d'Ouverture, Prononcé le 27 Floréal an 10, au Muséum d'Histoire Naturelle.* Maillard, Paris. (Reprinted in 1907 in *Bull. Scient. Fr. Belg.* 40:45–105.)

Lamarck, J.-B. 1818. *Histoire Naturelle des Animaux sans Vertébres.* Tome cinquième. Deterville, Paris.

Lüter, C. 2000. Ultrastructure of larval and adult setae of *Brachiopoda. Zool. Anz.* 239:75–90.

McHugh, D. 1997. Molecular evidence that echiurans and pogonophorans are derived annelids. *Proc. Natl. Acad. Sci. USA* 94:8006–8009.

Newby, W. W. 1940. The embryology of the echiuroid worm *Urechis caupo. Mem. Am. Philos. Soc.* 16:1–213.

O'Clair, R. M., and R. A. Cloney. 1974. Patterns of morphogenesis mediated by dynamic microvilli: chaetogenesis in *Nereis vexillosa. Cell Tissue Res.* 151:141–157.

Rouse, G. W., and K. Fauchald. 1995. The articulation of annelids. *Zool. Scr.* 24:269–301.

Rouse, G. W., and K. Fauchald. 1997. Cladistics and polychaetes. *Zool. Scr.* 26:139–204.

Rouse, G. W., and F. Pleijel. 2001. *Polychaetes.* Oxford University Press, Oxford.

Rouse, G. W., and F. Pleijel. 2003. Problems in polychaete systematics. *Hydrobiologia* 496:175–189.

Rouse, G. W., and F. Pleijel. 2005. Annelid phylogeny and systematics. Pp. 3–21 in *Reproductive Biology and Phylogeny of Annelida* (G. W. Rouse, and F. Pleijel, eds.). Science Publishers Inc., Enfield, NH.

Rousset, V., F. Pleijel, G. W. Rouse, C. Erséus, and M. Siddall. 2007. A molecular phylogeny of annelids. *Cladistics* 23:41–63.

Rousset, V., G. W. Rouse, M. E. Siddall, A. Tillier, and F. Pleijel. 2004. The phylogenetic position of *Siboglinidae* (*Annelida*), inferred from 18S rRNA, 28S rRNA, and morphological data. *Cladistics* 20:518–533.

Seaver, E. C. 2003. Segmentation: mono- or polyphyletic? *Int. J. Dev. Biol.* 47:583–595.

Siddall, M. E., K. Apakupakul, E. M. Burreson, K. A. Coates, C. Erséus, S. R. Gelder, M. Källersjö, and H. Trapido-Rosenthal. 2001. Validating Livanow: molecular data agree that leeches, branchiobdellidans, and *Acanthobdella peledina* form a monophyletic group of oligochaetes. *Mol. Phylogenet. Evol.* 21:346–351.

Struck, T., R. Hessling, and G. Purschke. 2002. The phylogenetic position of the *Aeolosomatidae* and *Parergodrilidae*, two enigmatic oligochaete-like taxa of the "*Polychaeta*", based on molecular data from 18S rDNA sequences. *J. Zool. Syst. Evol. Res.* 40:155–163.

Struck T., C. Paul, N. Hill, S. Hartmann, C. Hösel, M. Kube, B. Lieb, A. Meyer, R. Tiedemann, G. Purschke, and C. Bleidorn. 2011. Phylogenomic analyses unravel annelid evolution. *Nature* 471:95–98.

Struck, T., N. Schult, T. Kusen, E. Hickman, C. Bleidorn, D. McHugh, and K. Halanych. 2007. Annelid phylogeny and the status of *Sipuncula* and *Echiura*. *BMC Evol. Biol.* 7:57.

Zrzavy, J., V. Hypsa, and D. F. Tietz. 2001. *Myzostomida* are not annelids: molecular and morphological support for a clade of animals with anterior sperm flagella. *Cladistics* 17:170–198.

Zrzavy, J., P. Riha, L. Pialek, and J. Janouskovec. 2009. Phylogeny of *Annelida* (*Lophotrochozoa*): total-evidence analysis of morphology and six genes. *BMC Evol. Biol.* 9:189.

Authors

Fredrik Pleijel; University of Gothenburg; Department of Biological and Environmental Sciences – Tjärnö; SE-452 96 Strömstad, Sweden. Email: fredrik.pleijel@gu.se.

Greg W. Rouse; Scripps Institution of Oceanography; University of California, San Diego; 9500 Gilman Drive; La Jolla, CA 92093-0202, USA. Email: grouse@ucsd.edu.

Date Accepted: 07 May 2012

Primary Editor: Kevin de Queiroz

Rhabdocoela C. G. Ehrenberg 1831 [T. Artois], converted clade name

Registration Number: 89

Definition: the smallest crown clade containing *Polycystis naegelii* Kölliker 1845 and *Provortex balticus* (Schultze 1851) Graff 1882 and *Mariplanella frisia* Ax and Heller 1970. This is a minimum-crown-clade definition. Abbreviated definition: min crown ∇ (*Polycystis naegelii* Kölliker 1845 & *Provortex balticus* (Schultze 1851) Graff 1882 & *Mariplanella frisia* Ax and Heller 1970).

Etymology: From Greek, *rhabdo-* (rod, stick, staff) and *coel* or *koilos* (hollow). The name refers to the bar-shaped gut, as opposed to the branched gut of some other turbellarians.

Reference Phylogeny: The primary reference phylogeny is Willems et al. (2006: Fig. 1). See also Littlewood et al. (1999b), Norén and Jondelius (2002) and Joffe and Kornakova (2001).

Composition: *Rhabdocoela* contains *Kalyptorhynchia* and *Dalytyphloplanida* as defined in this volume. Approximately 1,500 species of rhabdocoels have been described (estimate based on Tyler et al., 2006–2011).

Diagnostic Apomorphies: A number of possible synapomorphies were cited by Willems et al. (2006), although they stated that "the *Rhabdocoela* is still not supported by a clear morphological apomorphy". However, a re-evaluation of all of the data in the literature indicates that two of the possible synapomorphies discussed by Willems et al. (2006) are indeed unique features of the *Rhabdocoela*, and therefore can be considered synapomorphies. One is the presence of a pharynx of the bulbosus-type, either directed forwards (pharynx doliiformis) or directed ventrally (pharynx rosulatus). Such a pharynx can be recognised by the presence of a septum that completely separates the pharynx from the parenchyma, with the pharyngeal glands completely situated within the pharynx proper. Moreover, the ciliation of the pharynx lumen is reduced (if cilia are present they are only found on the proximal part of the prepharyngeal cavity). The pharyngeal circular muscles are situated externally from the longitudinal muscles near the pharynx lumen, underneath the longitudinal muscles at the side of the septum. Bulbosus-like pharynges are also found in some lecithoepitheliates and prolecithophorans, but these never combine all the characteristics mentioned above. The different nature of the pharynx in lecithoepitheliates and prolecitophorans was already acknowledged by Meixner (1938), who coined the name pharynx variabilis for the type of pharynx present in these taxa. The same goes for the bulbosus-like pharynx found in the otoplanid *Bulbotoplana acephala* Ax 1956 and in *Ciliopharyngiella intermedia* Ax 1956. A detailed discussion is provided by Rieger et al. (1991). Ehlers (1985) considered the pharynx of some species of *Neodermata* homologous with the pharynx doliiformis (and thus included the *Neodermata* within the *Rhabdocoela*), but this was criticised by Joffe (1987). A second synapomorphy is the fact that the terminal cell of the protonephridia forms a weir consisting of only a single row of ribs containing microtubules, as was recognised by Littlewood et al. (1999b) (also see Rohde, 2001). A terminal cell similar to that found in rhabdocoels is found in the protonephridial system of *Lecithoepitheliata*, but here the ribs have a different form in transverse section and do not contain microtubules. Apart from these two apomorphies, Culioli et al. (2004) proposed the disto-proximal rotation of the centrioles during spermiogenesis as a synapomorphy of the rhabdocoels. However, this character was studied only in a very limited number of species, and therefore it is still not entirely clear whether it is indeed a synapomorphy for all of the rhabdocoels.

Synonyms: *Neorhabdocoela* Meixner 1938 is an approximate synonym of *Rhabdocoela* as defined above (see Comments).

Comments: The name *Rhabdocoela* was first introduced by Ehrenberg (1831) to denote all turbellarians with an unbranched intestinal tract. As such, *Rhabdocoela* encompassed all turbellarians known at that time except for the triclads and polyclads. In 1882, Graff removed *Acoela* and *Alloiocoela* from *Rhabdocoela* and assigned the three resulting "tribi" to the ordo *Rhabdocoelida*. In 1905, Graff split *Rhabdocoela* further into the taxa *Rhabdocoela Hysterophora* and *Rhabdocoela Lecithophora*. The latter corresponds approximately to *Rhabdocoela* as we define the name here. Meixner (1938) coined the name *Neorhabdocoela* for a group with the same composition as the *Rhabdocoela Lecithophora*, and indicated the pharynx bulbosus to be a diagnostic feature for this taxon. Subsequently, the terms *Rhabdocoela* and *Neorhabdocoela* were used interchangeably, often by the same authors (e.g., Luther, 1950, 1955; Karling, 1963, 1974). Ehlers (1985) used the name *Rhabdocoela* for a group that included *Neodermata* (parasitic flatworms) as well as Meixner's *Neorhabdocoela*. In later molecular phylogenies, it became clear that the *Neodermata* are not part of a monophyletic *Rhabdocoela*, and that *Rhabdocoela* is a very well supported clade (Littlewood et al., 1999a,b; Baguña et al., 2001; Littlewood and Olson, 2001; Joffe and Kornakova, 2001; Norén and

Jondelius, 2002; Lockyer et al., 2003; Willems et al., 2006). In all these papers (and the majority of the other recent papers on this taxon), the name *Rhabdocoela* is used to denote the clade for which the name is defined here. That name was previously defined phylogenetically by Willems et al. (2006), who used a maximum-clade definition. However, the name is being applied here to a crown clade using a minimum-crown-clade definition.

Of the approximately 1500 species of rhabdocoels that have been described, only a very limited number have been included in phylogenetic analyses. The most comprehensive study (Willems et al., 2006) included 56 species.

Literature Cited

Baguña, J., S. Carranza, J. Paps, I. Ruiz-Trillo, and M. Riutort. 2001. Molecular taxonomy and phylogeny of the *Tricladida*. Pp. 49–73 in *Interrelationships of the* Platyhelminthes (D. T. J. Littlewood and R. A. Bray, eds.). Taylor & Francis, London.

Culioli, J.-L., J. Foata, C. Mori, A. Orsini, and B. Marchand. 2004. Ultrastructure of spermiogenesis and the spermatozoon in *Castrada cristatispina* Papi, 1951 (*Platyhelminthes, Rhabdocoela, Typhloplanida*). *Acta Zool.* 85:245–256.

Ehlers, U. 1985. *Das Phylogenetische System der Plathelminthes.* Gustav Fisher Verlag, Stuttgart.

Ehrenberg, C. G. 1831. Animalia evertebrata exclusis insectis recensuit Dr. C. G. Ehrenberg. Series prima cum tabularum decade prima. Pp. 1–15 in *Symbolae Physicae II.* Phytozoa Turbellaria (W. F. Hemprich and C. G. Ehrenberg, eds.). Berolini.

von Graff, L. 1882. *Monographie der Turbellarien. I.* Rhabdocoelida. Verlag von Wilhelm Engelman, Leipzig.

von Graff, L. 1905. Marine Turbellarien Orotavas und der Küsten Europas. Ergebnisse einiger, mit Unterstützung der kaiserlichen Akademie der Wissenschaften in Wien (aus dem Legate

Wedl) in den Jahren 1902 und 1903 unternommenStudienreise. II. *Rhabdocoela. Z. wiss. Zool.* 83:68–148.

Joffe, B. I. 1987. On the evolution of the pharynx in *Plathelminthes. Proc. Zool. Inst., Leningrad* 167:34–71.

Joffe, B. I., and E. E. Kornakova. 2001. Flatworm phylogeneticist: between molecular hammer and morphological anvil. Pp. 279–291 in *Interrelationships of the* Platyhelminthes (D. T. J. Littlewood and R. A. Bray, eds.). Taylor & Francis, London.

Karling, T. G. 1963. Die Turbellarien ostfennoscandiens. V. *Neorhabdocoela. 3. Kalyptorhynchia. Soc. Flora Fauna Fenn.* 17:6–59.

Karling, T. G. 1974. Turbellarian fauna of the Baltic proper. Identification, ecology and biogeography. *Fauna Fenn.* 27:4–101.

Littlewood, D. T. J., and P. D. Olson. 2001. Small subunit rDNA and the *Platyhelminthes*: signal, noise, conflict and compromise. Pp. 262–278 in *Interrelationships of the* Platyhelminthes (D. T. J. Littlewood and R. A. Bray, eds.). Taylor & Francis, London.

Littlewood, D. T. J., K. Rohde, R. A. Bray, and E. A. Herniou. 1999a. Phylogeny of the *Platyhelminthes* and the evolution of parasitism. *Biol. J. Linn. Soc.* 68:257–287.

Littlewood, D. T. J., K. Rohde, and K. A. Clough. 1999b. The interrelationships of all major groups of *Platyhelminthes*: phylogenetic evidence from morphology and molecules. *Biol. J. Linn. Soc.* 66:75–114.

Lockyer, A. E., P. D. Olson, and D. T. J. Littlewood. 2003. Utility of complete large and small subunit rRNA genes in resolving the phylogeny of the *Neodermata (Platyhelminthes)*: implications and a review of the cercomer theory. *Biol. J. Linn. Soc.* 78:155–171.

Luther, A. 1950. Untersuchungen an rhabdocoelen Turbellarien. IX. Zur Kenntnis einiger *Typhoplanida. Acta Zool. Fenn.* 60:3–40.

Luther, A., 1955. Die Dalyelliiden (*Turbellaria Neorhabdocoela*), eine Monographie. *Acta Zool. Fenn.* 87:1–337.

Meixner, J. 1938. *Turbellaria* (Strudelwürmer) I. (Allgemeiner Teil). *Die Tierwelt der Nord- und Ostsee* 33:1–146.

Norén, M., and U. Jondelius. 2002. The phylogenetic position of the *Prolecithophora (Rhabditophora, 'Platyhelminthes'). Zool. Scr.* 31:403–414.

Rieger, R. M., S. Tyler, J. P. S. Smith III, and G. Rieger. 1991. *Platyhelminthes*: *Turbellaria*. Pp. 7–140 in *Microscopic Anatomy of Invertebrates* (F. W. Harrison and B. J. Bogitsch, eds.). Wiley-Liss, New York.

Rohde, K. 2001. Protonephridia as phylogenetic characters. Pp. 203–216 in *Interrelationships of the* Platyhelminthes (D. T. J. Littlewood and R. A. Bray, eds.). Taylor & Francis, London.

Tyler, S., S. Schilling, M. Hooge, and L. F. Bush. 2006–2011. Turbellarian Taxonomic Database, Version 1.7. Available at http://turbellaria.umaine.edu.

Willems, W. R., A. Wallberg, U. Jondelius, D. T. J. Littlewood, T. Backeljau, E. R. Schockaert, and T. J. Artois. 2006. Filling a gap in the phylogeny of flatworms: relationships within the *Rhabdocoela (Platyhelminthes)*, inferred from 18S ribosomal DNA sequences. *Zool. Scr.* 35:1–17.

Author

Tom J. Artois; Research Group Zoology; Biodiversity and Toxicology; Centre for Environmental Sciences; Hasselt University; Universitaire campus Gebouw D; 3590 Diepenbeek, Belgium. Email: tom.artois@uhasselt.be.

Date Accepted: 16 January 2012

Primary Editor: Philip Cantino

Dalytyphloplanida W. R. Willems et al. 2006 [T. Artois], converted clade name

Registration Number: 33

Definition: The largest crown clade containing *Provortex balticus* (Schultze 1851) Graff 1882 but not *Polycystis naegelii* Kölliker 1845. This is a maximum-crown-clade definition. Abbreviated definition: max crown ∇ (*Provortex balticus* (Schultze 1851) Graff 1882 ~ *Polycystis naegelii* Kölliker 1845).

Etymology: The name is a contraction of "*Dalyellioida*" and "*Typhloplanoida*", as *Dalytyphloplanida* contains all of the species of these two non-monophyletic groups.

Reference Phylogeny: Willems et al. (2006: Fig. 1); see also Figure 4, where *Dalytyphloplanida* is labeled.

Composition: This clade contains all species formerly placed within the taxa "*Dalyellioida*", *Temnocephalida*, and "*Typhloplanoida*" (except *Ciliopharyngiella constricta* Martens and Schockaert 1981, as discussed by Willems et al. [2006]). For an overview of the sub-taxa included in these three taxa, see Cannon (1986). This work does not list the following taxa: *Achenella* Cannon 1993, *Aegira* Willems et al. 2005, *Archopistomum* An der Lan 1939, *Austradenopharynx* Willems et al. 2005, *Austrodalyellia* Hochberg and Cannon 2002, *Caridinicola* Annandale 1912, *Cephalopharynx* Hochberg 2004, *Decadidymus* Cannon 1991, *Eldenia* Ax 2008, *Feanora* De Clerck and Schockaert 1995, *Gandalfia* Willems et al. 2005, *Gaziella* De Clerck and Schockaert 1995, *Gelasinella* Sewell and Cannon 1998, *Haplodidymos* Hochberg and Cannon 2002, *Hartogia* Mack-Fira 1968, *Heptacraspedella* Cannon and Sewell 1995, *Kaitalugia* Willems et al. 2005, *Luriculus* Faubel et al. 1994, *Magnetia* Hochberg and Cannon 2003, *Mahurubia* Willems et al. 2005, *Marcomesostoma* Janssen and Faubel 1992, *Microcalyptorhynchus* Kepner and Ruebush 1935, *Moevenbergia* Armonies and Hellwig 1987, *Monticellina* Westblad 1953, *Papiella* Mack-Fira 1970, *Parafallacohospes* Shinn 1987, *Parapharyngiella* Willems et al. 2005, *Paraproboscifer* De Clerck 1994, *Pilamonila* Willems et al. 2004, *Polliculus* Van Steenkiste et al. 2008, *Poseidoplanella* Willems et al. 2005, *Protandrella* Ehlers et al. 1994, *Protoascus* Hayes 1941, *Protopharyngiella* Schwank 1980, *Pseudophaenocora* Gilbert 1938, *Rhomboplanilla* Schwank 1980, *Scoliopharyngia* Ehlers et al. 1994, *Sphagnella* Sekera 1912, *Syringoplana* Artois et al. 2005, *Taborella* Sekera 1912, *Temnohaswellia* Pereira and Cuoccolo 1941, *Temnomonticellia* Pereira and Cuoccolo 1941, *Temnosewellia* Damborenea and Cannon 2001, *Vauclusia* Willems et al. 2004, *Westbladiella* Luther 1943, and *Zygopella* Cannon and Sewell 1995. On the other hand, Cannon (1986) lists *Fecampiidae* Graff 1905 as *Rhabdocoela*, which they are not (see Norén and Jondelius, 2002). An overview of all taxa can be found in the Turbellarian taxonomic database (Tyler et al., 2006). "*Dalyellioida*" and "*Typhloplanoida*" are both polyphyletic, but all of their species can be found in the clades called *Neodalyeliida* and *Neotyphloplanida* by Willems et al. (2006). *Mariplanella frisia* may also be part of *Dalytyphloplanida* (see Comments).

Diagnostic Apomorphies: According to Littlewood et al. (1999) and Willems et al. (2006), possible synapomorphies are found in the ultrastructure of the sperm: the presence of longitudinal rows of small dense granules; a spur-shaped structure at the base of the sperm axoneme; longitudinal microtubules lying between the embedded ends of the sperm axonemes; and a dense heel on the basal bodies during spermiogenesis. However, these characters have a mosaic-like distribution within the taxon, with many species lacking some or all of them. It is presently impossible to infer whether some (or all) of these characters are apomorphies for *Dalytyphloplanida*, with multiple losses within the clade, or whether these characters arose in parallel in different subclades within the *Dalytyphloplanida*.

Synonyms: No known synonyms.

Comments: The clade name *Dalytyphloplanida* was first defined by Willems et al. (2006) with a maximum-clade definition, using *Mariplanella frisia* Ax and Heller 1970 as one of two external specifiers. This species was included as an external specifier because the sister-group relationship between *M. frisia* and the rest of the non-kalyptorhynch rhabdocoels was supported by a jackknife value of only 81%. In the phylogeny depicted by Willems et al. (2006: Fig. 4), in which all nodes with less than 90% jackknife value are collapsed, *M. frisia* is part of a trichotomy with two clades: *Kalyptorhynchia* and a clade consisting of the rest of the non-kalyptorhynch rhabdocoels (i.e., *Dalytyphloplanida*). However, I think it is best not to use *M. frisia* as an external specifier, so that the content of the name *Dalytyphloplanida* is not dependent on the position of this species. If the sister-group relationship between *M. frisia* and the other non-kalyptorhynch rhabdocoels is corroborated, *M. frisia* will be part of *Dalytyphloplanida*. In spite of this difference in the two definitions, the clade to which Willems et al. (2006: Fig. 4) applied the name *Dalytyphloplanida* at least approximates the one specified by the definition used here, the only likely difference being the inclusion or exclusion of *Mariplanella frisia*.

Although there is no clear synapomorphy for all *Dalytyphloplanida*, members of this clade can very easily be recognised. They all have a pharynx bulbosus, an apomorphy of *Rhabdocoela*, either of the rosulatus-type or of the doliiformis-type, but they lack the proboscis, the presence of which is a synapomorphy of *Kalyptorhynchia* (this volume).

Literature Cited

Cannon, L. 1986. Turbellaria *of the World: A Guide to Families and Genera*. Queensland Museum, Brisbane.

Littlewood, D. T. J., K. Rohde, and K. A. Clough. 1999. The interrelationships of all major groups of *Platyhelminthes*: phylogenetic evidence from morphology and molecules. *Biol. J. Linn. Soc.* 66:75–114.

Norén, M., and U. Jondelius. 2002. The phylogenetic position of the *Prolecithophora* (*Rhabditophora*, 'Platyhelminthes'). *Zool. Scr.* 31:403–414.

Tyler, S., S. Schilling, M. Hooge, and L. F. Bush. 2006. Turbellarian Taxonomic Database, Version 1.5. Available at http://turbellaria. umaine.edu.

Willems, W. R., A. Wallberg, U. Jondelius, D. T. J. Littlewood, T. Backeljau, E. R. Schockaert, and T. J. Artois. 2006. Filling a gap in the phylogeny of flatworms: relationships within the *Rhabdocoela* (*Platyhelminthes*), inferred from 18S ribosomal DNA sequences. *Zool. Scr.* 35:1–17.

Author

Tom J. Artois; Research Group Zoology; Biodiversity and Toxicology; Centre for Environmental Sciences; Hasselt University; Universitaire campus Gebouw D; 3590 Diepenbeek, Belgium. Email: tom.artois@uhasselt.be.

Date Accepted: 23 June 2011

Primary Editor: Philip Cantino

Kalyptorhynchia L. Graff 1905 [T. Artois], converted clade name

Registration Number: 54

Definition: the largest crown clade containing *Polycystis naegelii* Kölliker 1845 but not *Provortex balticus* (Schultze 1851) Graff 1882 and *Mariplanella frisia* Ax and Heller 1970. This is a maximum-crown-clade definition. Abbreviated definition: max crown ∇ (*Polycystis naegelii* Kölliker 1845 ~ *Provortex balticus* (Schultze 1851) Graff 1882 & *Mariplanella frisia* Ax and Heller 1970).

Etymology: From Greek, *kalyptos* (covered) and *rhynchos* (snout). The name refers to the presence of an anterior proboscis that can only be seen when viewed under a microscope and therefore is considered hidden.

Reference Phylogeny: Willems et al. (2006: Fig. 1).

Composition: Contains the clade *Schizorhynchia* (this volume) and "*Eukalyptorhynchia*", a group that is probably not monophyletic (my unpublished data). For an overview of the genera within "*Eukalyptorhynchia*", see Cannon (1986). This work does not list following taxa: *Acirrostylus* Van Steenkiste et al. 2008, *Alchoides* Willems et al. 2006, *Ametochus* Willems et al. 2006, *Arrawaria* Willems et al. 2006, *Cohenella* Timoshkin 2004, *Coulterella* Timoshkin 2004, *Cystirete* Brunet 1965, *Duplexostylus* Willems et al. 2006, *Elvertia* Noldt 1989, *Galapagorhynchus* Artois and Schockaert 1998, *Gnorimorhynchus* Brunet 1972, *Kawanabella* Timoshkin 2004, *Lagenopolycystis* Artois and Schockaert 2000, *Marcusia* Artois and Schockaert 1998, *Mariareuterella* Timoshkin and Grygier 2004, *Marirhynchus* Schilke 1970, *Mityusha* Timoshkin 2004, *Myobulla* Artois and Schockaert 2000, *Obolkinaella* Timoshkin 2004, *Paragnatorhynchus* Meixner 1938, *Pygmorhynchus* Artois and Schockaert 1998, *Rhynchokarlingia* Timoshkin and Mamkaev 2004, *Riedelella* Timoshkin 2004, *Sabulirhynchus* Artois and Schockaert 2000, *Sitnikovaella* Timoshkin 2004, *Smithsoniarhynches* Hochberg 2004, *Stradorhynchus* Willems et al. 2006, *Syatkinella* Timoshkin 2004, *Syltorhynchus* Noldt 1989, *Triaustrorhynchus* Willems et al. 2006, and *Wadaella* Timoshkin 2004. An overview of all taxa can be found in the Turbellarian taxonomic database (Tyler et al., 2006).

Diagnostic Apomorphies: The clade *Kalyptorhynchia* is characterised by two morphological synapomorphies. All kalyptorhynchids have an anterior proboscis, as was described by Rieger et al. (1991). The proboscis is a muscular organ that is completely separated from the rest of the parenchyma by a septum consisting of extracellular matrix and muscles. It is pierced by gland necks. In the other rhabdocoels with proboscides, the proboscis is actually a permanent invaginated pit, not separated from the rest of the parenchyma by a septum. The kalyptorhynch proboscis can either be of one piece (the so-called conorhynch), as in the "*Eukalyptorhynchia*", or it consists of a dorsal and ventral half (the so-called schizorhynch of the *Schizorhynchia*). A second synapomorphy of *Kalyptorhynchia* is the complete incorporation of the axonemes in the body of the mature sperm, where they are located within the ring of cortical microtubules (Ehlers, 1985; Watson, 2001).

Synonyms: *Acrorhynchina* is an approximate synonym. This name was proposed by Graff (1882) for all species with a true proboscis (i.e., as in kalyptorhynchs) that were known at that time.

Comments: When Graff first introduced the name *Kalyptorhynchia* in 1905, he used it to denote a much larger group than the one to which the name now refers. He also included all species with a permanently invaginated anterior pit (species within *Trigonostomidae*). Meixner (1924) removed *Trigonostomidae* from *Kalyptorhynchia*, and since that time, the name *Kalyptorhynchia* always has been used for a group with the composition covered by our definition (e.g., Schilke, 1970; Evdonin, 1977; Karling, 1992). This group was shown to be monophyletic in the first cladistic analyses of turbellarians (Ehlers, 1985; Littlewood 1999a,b). The name was phylogenetically defined by Willems et al. (2006) using a maximum-clade definition with the same internal and external specifiers as used here.

Literature Cited

Cannon, L. 1986. Turbellaria *of the World: A Guide to Families and Genera*. Queensland Museum, Brisbane.

Ehlers, U. 1985. *Das Phylogenetische System der* Plathelminthes. Gustav Fisher Verlag, Stuttgart.

Evdonin, L. A. 1977. *Turbellaria Kalyptorhynchia* in the fauna of the USSR and adjacent areas. *Fauna USSR* 115:1–400.

Graff, L. von. 1882. *Monographie der Turbellarien. I.* Rhabdocoelida. Verlag von Wilhelm Engelman, Leipzig.

von Graff, L. 1905. Marine Turbellarien Orotavas und der Küsten Europas. Ergebnisse einiger, mit Unterstützung der kaiserlichen Akademie der Wissenschaften in Wien (aus dem Legate Wedl) in den Jahren 1902 und 1903 unternommenStudienreise. II. *Rhabdocoela. Z. wiss. Zool.* 83:68–148.

Karling, T. G. 1992. Identification of the *Kalyptorhynchia* (*Plathelminthes*) in Meixner's 'Turbellaria' 1938 with remarks on the morphology and distribution of the species in the North Sea and the Baltic Sea. *Zool. Scr.* 21:103–118.

Littlewood, D. T. J., K. Rohde, R. A. Bray, and E. A. Herniou. 1999a. Phylogeny of the *Platyhelminthes* and the evolution of parasitism. *Biol. J. Linn. Soc.* 68:257–287.

Littlewood, D. T. J., K. Rohde, and K. A. Clough. 1999b. The interrelationships of all major groups of *Platyhelminthes*: phylogenetic evidence from morphology and molecules. *Biol. J. Linn. Soc.* 66:75–114.

Meixner, J. 1924. Studien zu einer Monographie der *Kalyptorhynchia* und zum system der *Turbellaria Rhabdocoela. Zool. Anz.* 60:113–125.

Rieger, R. M., S. Tyler, J. P. S. Smith III, and G. Rieger. 1991. *Platyhelminthes: Turbellaria*. Pp. 7–140 in *Microscopic Anatomy of Invertebrates* (F. W. Harrison and B. J. Bogitsch, eds.). Wiley-Liss, New York.

Schilke, K. 1970. *Kalyptorhynchia* (*Turbellaria*) aus dem Eulitoral der deutschen Nordseeküste. *Helgoländer wiss. Meeresunters.* 21:143–265.

Tyler, S., S. Schilling, M. Hooge, and L. F. Bush. 2006. Turbellarian Taxonomic Database, Version 1.5. Available at http://turbellaria.umaine.edu.

Watson, N. 2001. Insights from comparative spermatology in the 'turbellarian' *Rhabdocoela*. Pp. 217–230 in *Interrelationships of the* Platyhelminthes (D. T. J. Littlewood and R. A. Bray, eds.). Taylor & Francis, London.

Willems, W. R., A. Wallberg, U. Jondelius, D. T. J. Littlewood, T. Backeljau, E. R. Schockaert, and T. J. Artois. 2006. Filling a gap in the phylogeny of flatworms: relationships within the *Rhabdocoela* (*Platyhelminthes*), inferred from 18S ribosomal DNA sequences. *Zool. Scr.* 35:1–17.

Author

Tom J. Artois; Research Group Zoology; Biodiversity and Toxicology; Centre for Environmental Sciences; Hasselt University; Universitaire campus Gebouw D; 3590 Diepenbeek, Belgium. Email: tom.artois@uhasselt.be.

Date Accepted: 3 August 2011

Primary Editor: Philip Cantino

Schizorhynchia J. Meixner 1928 [T. Artois], converted clade name

Registration Number: 95

Definition: The crown clade for which a proboscis split into two halves (dorsal and ventral), as inherited by *Schizorhynchus coecus* Hallez 1894, is an apomorphy relative to other crown clades. This is an apomorphy-modified crown-clade definition. Abbreviated definition: crown ∇ apo split proboscis [*Schizorhynchus coecus* Hallez 1894].

Etymology: From Greek, *schizo-* (to split) and *rhynchos* (snout), referring to the fact that in all species the proboscis is split into a ventral and a dorsal half.

Reference Phylogeny: Willems et al. (2006: Fig. 1). Because the name *Schizorhynchia* is based on the genus name *Schizorhynchus* Hallez 1894, the type of *Schizorhynchus* must be used as an internal specifier (*PhyloCode* Art. 11.10; Cantino and de Queiroz, 2020). However, both the type species (*S. coecus*) and the only other species of *Schizorhynchus* (*S. tataricus* Graff 1905) are known only from their original collections and therefore have not been included in any molecular analysis. Based on their morphology, they appear to be related to the exemplars of the clade that is labeled as *Schizorhynchia* in the reference phylogeny.

Composition: *Schizorhynchia* currently consists of about 140 species. For an overview of the genera see Cannon (1986). This work does not list *Proschizorhynchella* Schilke 1970 and *Serratorhynchus* Noldt 1988, which also are part of *Schizorhynchia*. An overview of all taxa can be found in the Turbellarian taxonomic database (Tyler et al., 2006), where *Schizorhynchia* can be found within *Kalyptorhynchia*.

Diagnostic Apomorphies: The only clear synapomorphy is the one used in the definition: the proboscis consists of a dorsal and a ventral half, which gives it a pincer-like appearance. Willems et al. (2006) hypothesised that the presence of only one axoneme in the mature sperm also characterizes *Schizorhynchia*. However, this feature also occurs in the eukalyptorhynch *Nannorhynchides herdlaensis* Karling 1956 (see Watson, 2001), so the level at which it is synapomorphic is unclear.

Synonyms: No known synonyms.

Comments: Meixner (1928) introduced the name *Schizorhynchia* for a taxon consisting of all kalyptorhynch flatworms with a split proboscis, and since then it has always been applied in this way (e.g., Schilke, 1970; Karling, 1983; Noldt, 1989). The monophyly of *Schizorhynchia* was weakly supported in earlier publications (e.g., Norén and Jondelius, 2002), but the latest published analysis on rhabdocoel flatworms, including a larger sample of species within and close to *Schizorhynchia*, showed very high support for its monophyly (Willems et al., 2006). The name was phylogenetically defined by Willems et al. (2006) using a maximum-clade definition, but to ensure that the name will always refer to the clade intended by Meixner (1928), an apomorphy-modified crown-clade definition is used here.

Whereas the proboscis is split in all schizorhynch kalyptorhynchs that have a proboscis, two species within the taxon *Typhlorhynchus* Laidlaw 1902 do not have a proboscis at all. According to Karling (1981) and Noldt (1985), *Typhlorhynchus* is the sister clade to *Proschizorhynchus* Meixner 1928 and has secondarily lost the proboscis. A thorough phylogenetic analysis based also on molecular data is needed to assess whether *Typhlorhynchus* is indeed part of a monophyletic *Schizorhynchia*.

Literature Cited

Cannon, L. 1986. Turbellaria *of the World: A Guide to Families and Genera*. Queensland Museum, Brisbane.

Cantino, P. D., and K. de Queiroz. 2020. *International Code of Phylogenetic Nomenclature (PhyloCode)*, Version 6. CRC Press, Boca Raton, FL.

Karling, T. 1981. *Typhlorhynchus nanus* Laidlaw, a kalyptorhynch turbellarian without a proboscis (*Platyhelminthes*). *Ann. Zool. Fenn.* 18:169–177.

Karling, T. 1983. Structural and systematic studies on *Turbellaria Schizorhynchia* (*Platyhelminthes*). *Zool. Scr.* 12:77–89.

Meixner, J. 1928. Aberrante *Kalyptorhynchia* (*Turbellaria Rhabdocoela*) aus dem Sande der Kieler Bucht. *Zool. Anz.* 60:113–125.

Norén, M., and U. Jondelius. 2002. The phylogenetic position of the *Prolecithophora* (*Rhabditophora*, '*Platyhelminthes*'). *Zool. Scr.* 31:403–414.

Noldt, U. 1985. *Typhlorhynchus syltenis* n. sp. (*Schizorhynchia*, *Plathelminthes*) and the adelphotaxa-relationship of *Typhlorhynchus* and *Prochizorhynchus*. *Microfauna Mar.* 2:347–370.

Noldt, U. 1989. *Kalyptorhynchia* (*Plathelminthes*) from sublittoral coastal areas near the Island of Sylt (North Sea) 1. *Schizorhynchia*. *Microfauna Mar.* 5:7–85.

Schilke, K. 1970. Zur morphologie und phylogenie der *Schizorhynchia* (*Turbellaria, Kalyptorhynchia*). *Z. Morph. Tiere* 65:118–171.

Tyler, S., S. Schilling, M. Hooge, and L. F. Bush. 2006. Turbellarian Taxonomic Database, Version 1.5. Available at http://turbellaria.umaine.edu.

Watson, N. 2001. Insights from comparative spermatology in the 'turbellarian' *Rhabdocoela*. Pp. 217–230 in *Interrelationships of the* Platyhelminthes (D. T. J. Littlewood and R. A. Bray, eds.). Taylor & Francis, London.

Willems, W. R., A. Wallberg, U. Jondelius, D. T. J. Littlewood, T. Backeljau, E. R. Schockaert, and T. J. Artois. 2006. Filling a gap in the phylogeny of flatworms: relationships within the *Rhabdocoela* (*Platyhelminthes*), inferred from 18S ribosomal DNA sequences. *Zool. Scr.* 35:1–17.

Author

Tom J. Artois; Research Group Zoology; Biodiversity and Toxicology; Centre for Environmental Sciences; Hasselt University; Universitaire campus Gebouw D; 3590 Diepenbeek, Belgium. Email: tom.artois@uhasselt.be.

Date Accepted: 23 June 2011

Primary Editor: Philip Cantino

Pan-Brachiopoda S. J. Carlson and B. L. Cohen, new clade name

Registration Number: 146

Definition: The total clade of the crown clade *Brachiopoda*. This is a crown-based total-clade definition. Abbreviated definition: total ∇ of *Brachiopoda*.

Etymology: *Pan-* is derived from the Greek Παντος (*pantos*) meaning "all" (a standard prefix for total clades; see *PhyloCode* Art. 10.3). See *Brachiopoda* (this volume) for the etymology of that name.

Reference Phylogeny: Dunn et al. (2008: Fig. 2).

Composition: The crown clade *Brachiopoda* (this volume) and all extinct organisms or species that share more recent common ancestry with *Brachiopoda* than with any other mutually exclusive (non-nested) crown clade in *Lophotrochozoa*; according to the reference phylogeny, those mutually exclusive crown clades are those of *Platyhelminthes, Mollusca, Annelida* (including *Echiura* and *Sipuncula*), *Nemertea*, and possibly *Phoronida*. In the context of the reference phylogeny, the inclusion of *Phoronida* in *Brachiopoda* (and *Pan-Brachiopoda*) is unclear because only a single brachiopod and a single phoronid were included in the analysis; it is therefore possible that some brachiopods are more closely related to phoronids than are others (see entry for *Brachiopoda* in this volume, as well as Comments, below).

Several Lower Cambrian fossils have been considered possible "stem-group" brachiopods including: *Mickwitzia* (Skovsted and Holmer, 2003, 2005; Skovsted et al., 2008); *Micrina* (Holmer et al., 2002, 2008; Williams and Holmer, 2002; Li and Xiao, 2004; Conway Morris and Caron, 2007; Skovsted et al., 2008; Vendrasco et al., 2009); *Tannuolina* (Skovsted et al., 2008); *Paterimitra* (Skovsted et al., 2009; Vendrasco et al., 2009); *Heliomedusa* (Zhang et al., 2009); and *Sunnaginia* (Murdock et al., 2012). *Acanthotretella* (Holmer and Caron, 2006), a Middle Cambrian fossil, and *Drakozoon* (Sutton et al., 2010), a Silurian fossil, have also been suggested as belonging to the brachiopod stem-group.

Halkieria and related halkieriids were thought to be possible stem-group brachiopods (Conway Morris and Peel, 1995; Ushatinskaya, 2002; Williams and Holmer, 2002; Cohen et al., 2003; Li and Xiao, 2004), but are more likely to be crown molluscs (Bengtson, 1992; Runnegar, 2000; Vendrasco et al., 2004; Vinther and Nielsen, 2005; Caron et al., 2007; Conway Morris and Caron, 2007; Vinther et al., 2008; Vinther, 2009; Paterson et al., 2009; see also Porter, 2008), or possibly stem-group polychaetes (Butterfield, 2006).

The phylogenetic position of each of the above-mentioned fossil taxa, whether within or outside the crown clade *Brachiopoda* (or related to a different crown clade), is strongly contentious, due largely to uncertainty about the homology and polarity of features preserved in these early fossils, and the lack of a rigorous phylogenetic analysis of those features (see Carlson and Cohen, 2009).

Diagnostic Apomorphies: Uncertainties about both the relationship of *Brachiopoda* to other *Lophotrochozoa* and the relationship of numerous fossil taxa (listed above) to *Brachiopoda* make it impossible at this time to identify apomorphies of *Pan-Brachiopoda*. The presence or absence of a mineralized skeleton, number of mineralized elements, number of types of mineralized elements, symmetry of mineralized elements, chemical composition of mineralized elements, identification of a lophophore, "attachment organ," and other potentially diagnostic features of "soft" anatomy in fossils, have all been discussed as possible apomorphies of a group that might approximate the *Pan-Brachiopoda* clade (e.g., Skovsted et al., 2009, 2011; Murdock et al., 2012), but no analyses of phylogenetic relationships have been published to test these assertions (see Carlson, 1995, 2007; Williams and Carlson, 2007 and papers cited in the Comments section for further discussion).

Synonyms: There are no unambiguous synonyms of *Pan-Brachiopoda* as applied to the total clade of *Brachiopoda*. *Brachiopoda* in the sense of Williams et al. (1996, 2000) and many earlier authors is considered to be an approximate synonym of *Pan-Brachiopoda* because it is unclear if the name refers to the total clade, the crown clade, or something in between. Note that the informal term "pan-Brachiopoda", introduced by Santagata and Cohen (2009) for the crown clade composed of phoronids and a paraphyletic *Brachiopoda*, refers to a different clade than *Pan-Brachiopoda* as defined here (see *Brachiopoda* in this volume).

Comments: Many recent studies have attempted to determine phylogenetic relationships among metazoans; lophotrochozoan relationships have been particularly difficult to establish with confidence. Each of the following studies included at least one representative of *Brachiopoda* in the analyses: Giribet et al. (2000); Mallatt and Winchell (2002); Glenner et al. (2004); Telford et al. (2005); Baguna et al. (2008); Bourlat et al. (2008); Colgan et al. (2008); Giribet (2008); Helmkampf et al. (2008); Yokobori et al. (2008); Jang and Hwang (2009); Paps et al. (2009); Podsiadlowski et al. (2009); Dordel et al. (2010); Hausdorf et al. (2010); Mallatt et al.

(2010); Nesnidal et al. (2010); Edgecombe et al. (2011); and Sperling et al. (2011). These studies produced a minimum of six different hypotheses concerning the sister group of *Brachiopoda*, which therefore remains uncertain, as does the sister group of *Pan-Brachiopoda*.

Recognition of the brachiopod stem group requires the recognition of the corresponding crown clade (*Brachiopoda* sensu Carlson and Cohen in this volume). Prior to explicit nomenclatural distinction between the crown and total clades, numerous studies labeled a range of early Palaeozoic fossil organisms as "stem-group" brachiopods (Conway Morris and Peel, 1995; Holmer et al., 2002, 2008, 2009; Williams and Holmer, 2002; Cohen et al., 2003; Li and Xiao, 2004; Skovsted and Holmer, 2005; Holmer and Caron, 2006; Conway Morris and Caron, 2007; Caron et al., 2007; Skovsted et al., 2008, 2009, 2011; Zhang et al., 2008; Paterson et al., 2009; Vendrasco et al., 2009; Sutton et al., 2010; Murdock et al., 2012). Several of these "small shelly fossils" (Matthews and Missarzhevsky, 1975) had been classified previously in *Tommotiida* Missarzhevsky 1970. Some of these fossils appear to possess some morphological characters in common with brachiopods that may (Balthasar et al., 2009), or may not (Vinther and Nielsen, 2005; Carlson, 2007; Williams and Carlson, 2007), be homologous with those in brachiopods. If homologous, the polarity of the evolutionary transformation of these characters can be difficult to determine because: the brachiopod sister group within *Lophotrochozoa* has not been identified (see also Jenner and Littlewood, 2008); too little is known of the ontogeny of these extinct organisms to be able to rely confidently on ontogenetic polarity criteria; and there are definite crown group *Brachiopoda* that are almost as old, thus calling stratigraphic polarity criteria into question. Diagenetic and other preservational artifacts (e.g., Balthasar, 2007; see also Porter,

2004; Donoghue and Purnell, 2009) can also hinder the accurate characterization of these features as they occurred in the once-living animals, as can the dearth of morphological characters for which homology can be hypothesized.

Of the many named extinct species (more than 10,000) that are considered to be brachiopods on the basis of morphology (see Williams et al., 2000 and subsequent volumes of the *Treatise on Invertebrate Paleontology*, Part H), all appear to nest within the crown clade, because of the broad phylogenetic distribution that is considered to reflect ancient basal divergences of the extant species (Carlson, 2007). Some early Palaeozoic species that look quite different from typical brachiopods may be stem-group brachiopods; they have been discussed in the preceding paragraph. Our use of the names *Brachiopoda* and *Pan-Brachiopoda* highlights the distinction between crown and total clades, thus more clearly delineating the paraphyletic "stem-group" between them (even though the membership of that stem group is currently unclear).

Literature Cited

Baguna, J., P. Martinez, J. Paps, and M. Riutort. 2008. Back in time: a new systematic proposal for the *Bilateria*. *Philos. Trans. R. Soc. Lond. B Biol. Sci.* 363:1481–1491.

Balthasar, U. 2007. An Early Cambrian organophosphatic brachiopod with calcitic granules. *Palaeontology* 50:1319–1325.

Balthasar, U., C. B. Skovsted, L. E. Holmer, and G. E. Brock. 2009. Homologous skeletal secretion in tommotiids and brachiopods. *Geology* 37:1143–1146.

Bengtson, S. 1992. The cap-shaped Cambrian fossil *Maikhanella* and the relationship between coeloscleritophorans and molluscs. *Lethaia* 25:401–420.

Bourlat, S. J., C. Nielsen, A. D. Economou, and M. J. Telford. 2008. Testing the new animal phylogeny: a phylum level molecular analysis of the animal kingdom. *Mol. Phylogenet. Evol.* 49:23–31.

Butterfield, N. J. 2006. Hooking some stem-group "worms": fossil lophotrochozoans in the Burgess Shale. *BioEssays* 28:1161–1166.

Carlson, S. J. 1995. Phylogenetic relationships among extant brachiopods. *Cladistics* 11:131–197.

Carlson, S. J. 2007. Recent research on brachiopod evolution. Pp. 2878–2900 in *Treatise on Invertebrate Paleontology, Part H*, Brachiopoda, *Revised*, Vol. 6 (P. A. Selden, ed.). Geological Society of America, Boulder, CO; University of Kansas Press, Lawrence, KS.

Carlson, S. J., and B. L. Cohen. 2009. Separating the crown from the stem: defining *Brachiopoda* and *Pan-Brachiopoda* delineates stem-brachiopods. *Geol. Soc. Am., Abstracts with Programs* 41(7):562.

Caron, J.-B., A. Scheltema, C. Schander, and D. Rudkin. 2007. Reply to Butterfield on stem-group "worms": fossil lophotrochozoans in the Burgess Shale. *BioEssays* 29:200–202.

Cohen, B. L., L. E. Holmer, and C. Luter. 2003. The brachiopod fold: a neglected body plan hypothesis. *Palaeontology* 46:59–65.

Colgan, D. J., P. A. Hutchings, and E. Beacham. 2008. Multi-gene analyses of the phylogenetic relationships among the *Mollusca, Annelida*, and *Arthropoda. Zool. Stud.* 47:338–351.

Conway Morris, S., and J.-B. Caron. 2007. Halwaxiids and the early evolution of the lophotrochozoans. *Science* 315:1255–1258.

Conway Morris, S., and J. S. Peel. 1995. Articulated halkieriids from the Lower Cambrian of North Greenland and their role in early protostome evolution. *Philos. Trans. R. Soc. Lond. B Biol. Sci.* 347:305–358.

Donoghue, P. C. J., and M. A. Purnell. 2009. Distinguishing heat from light in debate over controversial fossils. *BioEssays* 31:178–189.

Dordel, J., F. Fisse, G. Purschke, and T. H. Struck. 2010. Phylogenetic position of *Sipuncula* derived from multi-gene and phylogenomic data and its implications for the evolution of segmentation. *J. Zool. Syst. Evol. Res.* 48:197–207.

Dunn, C. W., A. Hejnol, D. Q. Matus, K. Pang, W. E. Browne, S. A. Smith, E. Seaver, G. W. Rouse, M. Obst, G. D. Edgecombe, M. V. Sørensen, S. H. D. Haddock, A. Schmidt-Rhaesa, A. Okusu, R. Møbjerg Kristensen, W. C. Wheeler, M. Q. Martindale, and G. Giribet. 2008. Broad phylogenomic sampling improves resolution of the animal tree of life. *Nature* 452:745–749.

Edgecombe, G. D., G. Giribet, C. W. Dunn, A. Hejnol, R. M. Kristensen, R. C. Neves, G. W. Rouse, K. Worsaae, and M. V. Sorensen. 2011. Higher-level metazoan relationships: recent progress and remaining questions. *Org. Divers. Evol.* 11:151–172.

Giribet, G. 2008. Assembling the lophotrochozoan (=spiralian) tree of life. *Philos. Trans. R. Soc. Lond. B Biol. Sci.* 363:1513–1522.

Giribet, G., D. L. Distel, M. Polz, W. Sterrer, and W. C. Wheeler. 2000. Triploblastic relationships with emphasis on the acoelomates and the position of *Gnathostomulida, Cycliophora, Plathelminthes,* and *Chaetognatha*: a combined approach of 18S rDNA sequences and morphology. *Syst. Biol.* 49:539–562.

Glenner, H., A. J. Hansen, M. V. Sørensen, F. Ronquist, J. P. Huelsenbeck, and E. Willerslev. 2004. Bayesian inference of the metazoan phylogeny: a combined molecular and morphological approach. *Curr. Biol.* 14:1644–1649.

Hausdorf, B., M. Helmkampf, M. P. Nesnidal, and I. Bruchhaus. 2010. Phylogenetic relationships within the lophophorate lineages (*Ectoprocta, Brachiopoda* and *Phoronida*). *Mol. Phylogenet. Evol.* 55:1121–1127.

Helmkampf, M., I. Bruchhaus, and B. Hausdorf. 2008. Phylogenomic analyses of lophophorates (brachiopods, phoronids and bryozoans) confirm the *Lophotrochozoa* concept. *Proc. R. Soc. Lond. B Biol. Sci.* 275:1927–1933.

Holmer, L. E., and J.-B. Caron. 2006. A spinose stem group brachiopod with pedicle from the Middle Cambrian Burgess Shale. *Acta Zool. (Stockh.)* 87:273–290.

Holmer, L. E., C. B. Skovsted, G. A. Brock, J. L. Valentine, and J. R. Paterson. 2008. The Early Cambrian tommotiid *Micrina*, a sessile bivalved stem group brachiopod. *Biol. Lett.* 4:724–728.

Holmer, L. E., C. B. Skovsted, and A. Williams. 2002. A stem-group brachiopod from the Lower Cambrian—support for *Micrina* (halkieriid) ancestry. *Palaeontology* 45:875–882.

Holmer, L. E., S. P. Stolk, C. B. Skovsted, U. Balthasar, and L. Popov. 2009. The enigmatic Early Cambrian *Salanygolina*—a stem group of rhynchonelliform chileate brachiopods? *Palaeontology* 52:1–10.

Jang, K. H., and U. W. Hwang. 2009. Complete mitochondrial genome of *Bugula neritina* (*Bryozoa, Gymnolaemata, Cheilostomata*): phylogenetic position of *Bryozoa* and phylogeny of lophophorates within the *Lophotrochozoa. BMC Genomics* 10:167 (18 pp.).

Jenner, R. A., and D. T. J. Littlewood. 2008. Problematica old and new. *Philos. Trans. R. Soc. Lond. B Biol. Sci.* 363:1503–1512.

Li, G., and S. Xiao. 2004. *Tannuolina* and *Micrina* (*Tannuolinidae*) from the Lower Cambrian of Eastern Yunnan, South China, and their scleritome reconstruction. *J. Paleontol.* 78:900–913.

Mallatt, J., C. W. Craig, and M. Y. Yoder. 2010. Nearly complete rRNA genes assembled from across the metazoan animals: effects of more taxa, a structure-based alignment, and paired-sites evolutionary models on phylogeny reconstruction. *Mol. Phylogenet. Evol.* 55:1–17.

Mallatt, J., and C. J. Winchell. 2002. Testing the new animal phylogeny: first use of combined large-subunit and small-subunit rRNA gene sequences to classify the protostomes. *Mol. Bio. Evol.* 19(30):289–301.

Matthews, S. C., and V. V. Missarzhevsky. 1975. Small shelly fossils of Late Precambrian and Early Cambrian age: a review of recent work. *J. Geol. Soc.* 131:289–304.

Missarzhevsky, V. V. 1970 (1969). Description of hyolithids, gastropods, hyolithelminths, camenides and forms of an obscure taxonomic position. Pp. 103–175 in *The Tommotian Stage and the Cambrian Lower Boundary Problem* (M. R. Raben, ed.). Akad. Nauk. SSSR, Moscow.

Murdock, D. J. E., P. C. J. Donoghue, S. Bengtson, and F. Marone. 2012. Ontogeny and microstructure of the enigmatic Cambrian tommotiid *Sunnaginia* Missarzhevsky, 1969. *Palaeontology* 55:661–676.

Nesnidal, M. P., M. Helmkampf, I. Bruchhaus, and B. Hausdorf. 2010. Compositional heterogeneity and phylogenomic inference of metazoan relationships. *Mol. Biol. Evol.* 27:2095–2104.

Paps, J., J. Baguna, and M. Riutort. 2009. *Lophotrochozoa* internal phylogeny: new insights from an up-to-date analysis of nuclear ribosomal genes. *Proc. R. Soc. Lond. B Biol. Sci.* 276:1245–1254.

Paterson, J. R., G. A. Brock, and C. B. Skovsted. 2009. *Oikozetetes* from the early Cambrian of South Australia: implications for halkieriid affinities and functional morphology. *Lethaia* 42:199–203.

Podsiadlowski, L., A. Braband, T. H. Struck, J. von Dohren, and T. Bartolomaeus. 2009. Phylogeny and mitochondrial gene order variation in *Lophotrochozoa* in the light of new mitogenomic data from *Nemertea*. *BMC Genomics* 10:364.

Porter, S. M. 2004. Halkieriids in Middle Cambrian phosphatic limestones from Australia. *J. Paleontol.* 78:574–590.

Porter, S. M. 2008. Skeletal microstructure indicates chancelloriids and halkieriids are closely related. *Palaeontology* 51:865–879.

Runnegar, B. N. 2000. Body building in *Halkieria* and comparisons with chitons and other molluscs. *Geo. Soc. Am., Abstracts with Programs* 32:A72.

Santagata, S., and B. L. Cohen. 2009. Phoronid phylogenetics (*Brachiopoda*; *Phoronata*): evidence from morphological cladistics, small and large subunit rDNA sequences, and mitochondrial *cox*1. *Zool. J. Linn. Soc.* 157:34–50.

Skovsted, C. B., G. A. Brock, J. R. Paterson, L. E. Holmer, and G. E. Budd. 2008. The scleritome of *Eccentrotheca* from the Lower Cambrian of South Australia: lophophorate affinities and implications for tommotiid phylogeny. *Geology* 36:171–174.

Skovsted, C. B., G. A. Brock, T. P. Topper, J. R. Paterson, and L. E. Holmer. 2011. Scleritome construction, biofacies, biostratigraphy and systematics of the tommotiid *Eccentrotheca helenia* sp. nov. from the Early Cambrian of South Australia. *Palaeontology* 54:253–286.

Skovsted, C. B., and L. E. Holmer. 2003. The Early Cambrian (Botomian) stem group brachiopod *Mickwitzia* from Northeast Greenland. *Acta Palaeontol. Pol.* 48:1–20.

Skovsted, C. B., and L. E. Holmer. 2005. Early Cambrian brachiopods from North-east Greenland. *Palaeontology* 48:325–345.

Skovsted, C. B., L. E. Holmer, C. M. Larsson, A. E. S. Hogstrom, G. A. Brock, T. P. Popper, U. Balthasar, S. P. Stolk, and J. R. Paterson. 2009. The scleritome of *Paterimitra*: an Early Cambrian stem group brachiopod from South Australia. *Proc. R. Soc. Lond. B Biol. Sci.* 276:1651–1656.

Sperling, E. A., D. Pisani, and K. J. Peterson. 2011. Molecular paleobiological insights into the origin of the *Brachiopoda*. *Evol. Dev.* 13:290–303.

Sutton, M. D., D. E. G. Briggs, D. J. Siveter, and D. J. Siveter. 2010. A soft-bodied lophophorate from the Silurian of England. *Biol. Lett.* 7:146–149.

Telford, M. J., M. J. Wise, and V. Gowri-Shankar. 2005. Consideration of RNA secondary structure significantly improves likelihood-based estimates of phylogeny: examples from the *Bilateria*. *Mol. Biol. Evol.* 22:1129–1136.

Ushatinskaya, G. T. 2002. Genus *Micrina* (small shelly problematics) from the Lower Cambrian of South Australia: morphology, microstructures, and possible relation to halkieriids. *Paleontol. Zh.* 2002:11–21 [in Russian].

Vendrasco, M. J., G. Li, S. M. Porter, and C. Z. Fernandez. 2009. New data on the enigmatic *Ocruranus-Eohalobia* group of Early Cambrian small skeletal fossils. *Palaeontology* 52:1373–1396.

Vendrasco, M. J., T. E. Wood, and B. N. Runnegar. 2004. Articulated Paleozoic fossil with 17 plates greatly expands disparity of early chitons. *Nature* 429:288–291.

Vinther, J. 2009. The canal system in sclerites of Lower Cambrian *Sinosachites* (*Halkieriidae*: *Sachitida*): significance for the molluscan affinities of the sachitids. *Palaeontology* 52:689–712.

Vinther, J., and C. Nielsen. 2005. The Early Cambrian *Halkieria* is a mollusc. *Zool. Scr.* 34:81–89.

Vinther, J., P. Van Roy, and D. E. G. Briggs. 2008. Machaeridians are Palaeozoic armoured annelids. *Nature* 451:185–188.

Williams, A., and S. J. Carlson. 2007. Affinities of brachiopods and trends in their evolution. Pp. 2822–2877 in *Treatise on Invertebrate Paleontology, Part H*, Brachiopoda, *Revised*, Vol. 6. (R. L. Kaesler, ed.). Geological Society of America, Boulder, CO; University of Kansas Press, Lawrence, KS.

Williams, A., S. J. Carlson, and C. H. C. Brunton. 2000. *Brachiopoda*. Pp. 28–29 in *Treatise on Invertebrate Paleontology, Part H*, Brachiopoda, *Revised*, Vol. 2. (R. L. Kaesler, ed.). Geological Society of America, Boulder, CO; University of Kansas Press, Lawrence, KS.

Williams, A., S. J. Carlson, C. H. C. Brunton, L. E. Holmer, and L. Popov. 1996. A supra-ordinal classification of the *Brachiopoda. Philos. Trans. R. Soc. Lond. B Biol. Sci.* 351:1171–1193.

Williams, A., and L. E. Holmer. 2002. Shell structure and inferred growth, function and affinities of the sclerites of the problematic *Micrina. Palaeontology* 45:845–873.

Yokobori, S., T. Iseto, S. Asakawa, T. Sasaki, N. Shimizu, A. Yamagishi, T. Oshima, and E. Hirose. 2008. Complete nucleotide sequences of mitochondrial genomes of two solitary entoprocts, *Loxocorone allax* and *Loxosomella aloxiata*: implications for lophotrochozoan phylogeny. *Mol. Phylogenet. Evol.* 47: 612–628.

Zhang, Z., G. Li, C. C. Emig, J. Han, L. E. Holmer, and D. Shu. 2009. Architecture and function of the lophophore in the problematic brachiopod *Heliomedusa orienta* (Early Cambrian, South China). *Geobios* 42:649–661.

Zhang, Z., S. P. Robson, C. Emig, and D. Shu. 2008. Early Cambrian radiation of brachiopods: a perspective from South China. *Gondwana Res.* 14:241–254.

Authors

Sandra J. Carlson; Department of Geology; University of California; One Shields Avenue; Davis, CA 95616-8605, USA. Email: sjcarlson@ucdavis.edu.

Bernard L. Cohen; Faculty of Biomedical and Life Sciences; University of Glasgow; G12 8QQ, Scotland, UK. Email: b.l.cohen@bio.gla.ac.uk.

Date Accepted: 12 October 2012

Primary Editor: Kevin de Queiroz

Brachiopoda A. Duméril 1805 (1806): 154 [S. J. Carlson and B. L. Cohen], converted clade name

Registration number: 144

Definition: The least inclusive crown clade containing *Lingula anatina* Lamarck 1801 (*Linguloidea*), *Discinisca* (originally *Orbicula*) *tenuis* (Sowerby 1847) (*Discinoidea*), *Novocrania* (originally *Patella*) *anomala* (Müller 1776) (*Craniida*), *Notosaria* (originally *Terebratula*) *nigricans* (Sowerby 1846) (*Rhynchonellida*), *Terebratalia* (originally *Terebratula*) *transversa* (Sowerby 1846) (*Terebratulida*), and *Thecidellina blochmanni* Dall 1920 (*Thecideida*). This is a minimum-crown-clade definition. Abbreviated definition: min crown ∇ (*Lingula anatina* Lamarck 1801 & *Discinisca tenuis* (Sowerby 1847) & *Novocrania anomala* (Müller 1776) & *Notosaria nigricans* (Sowerby 1846) & *Terebratalia transversa* (Sowerby 1846) & *Thecidellina blochmanni* Dall 1920).

Etymology: Derived from the Greek βραχ|ων (*brachion*) meaning "arm" and πΟυσ (*pous*) meaning "foot," referring to the lophophore, which resembles two arms with tentacles, but was thought, in error, to function like the foot of bivalve mollusks.

Reference Phylogeny: Cohen and Weydmann (2005: Fig. 2). Although not included in the reference phylogeny, *Thecidellina blochmanni* Dall 1920 is included as an internal specifier in the definition of the name *Brachiopoda* because it represents a clade of micromorphic brachiopods whose relationship to other extant brachiopods is unclear and controversial.

Composition: The total clade of *Linguliformea* Williams et al. 1996, the total clade of

Craniiformea Popov et al. 1993, and *Pan-Neoarticulata* (this volume). If one accepts the phylogeny hypothesized by Cohen and Weydmann (2005: Fig. 2), *Phoronida* Hatschek 1888 is part of *Brachiopoda* (see Comments). Because of the broad phylogenetic distribution of extant species within *Brachiopoda* (Carlson, 2007: Fig. 1908), most named extinct species that have been referred to *Brachiopoda* (see Williams et al., 2000 and subsequent volumes of the *Treatise on Invertebrate Paleontology*, Part H), appear to nest within the crown clade.

Diagnostic Apomorphies: *Brachiopoda* has traditionally been defined (e.g., Williams et al., 2000) as those organisms possessing both a lophophore and a bivalved shell that encloses the lophophore, with the plane of bilateral symmetry normal to the surface of separation between the valves. Apposition of the valve margins (Holmer et al., 2002) and the ability to seal the mantle cavity along the commissure by the closure of the two valves (Rudwick, 1970), which often, but not exclusively (see Nielsen, 1991), grow in coordination with one another during ontogeny (Carlson and Cohen, herein), are also considered to be diagnostic features, if not previously identified as apomorphies. See Comments for a discussion of possible complications concerning the interpretation of the bivalved shell as a diagnostic apomorphy.

Synonyms: The informal term "pan-Brachiopoda" was introduced by Santagata and Cohen (2009) for the crown clade of brachiopods including phoronids and is thus an unambiguous synonym (under the reference phylogeny). The name *Brachiozoa* (Cavalier-Smith, 1998) includes

Brachiopoda and *Phoronida* and is considered to be an approximate synonym of *Brachiopoda*, as defined here. It was diagnosed "with lophophore and vascular system" but was not explicitly defined as referring to the crown clade. *Brachiopia* Rafinesque 1831, *Brachiopodidae* Broderip 1839, *Brachionopoda* Agassiz 1848, *Polymaria* King 1850, *Brachionacephala* Bronn 1862, and *Branchionoconchae* Bronn 1862 are also approximate synonyms of *Brachiopoda*. *Spirobrachiophora* Gray 1821 and *Palliobranchiata* de Blainville 1824 are partial (and approximate) synonyms of *Brachiopoda* in that they include some taxa now considered to be molluscan.

Comments: We include six internal specifiers in our minimum-crown-clade definition of the name *Brachiopoda* to acknowledge uncertainty expressed by some about the relationships among subclades within *Brachiopoda* (Cohen and Weydmann, 2005; Cohen, 2007, 2013) that have been classified into three taxa of uncertain monophyly (Williams et al., 1996; Carlson, 2007): *Linguliformea*, *Craniiformea*, and *Rhynchonelliformea*. The names *Brachiopoda* Duméril 1805 (1806) and "brachiopod" have been used more commonly in the literature for a much longer period of time than *Brachiozoa* and brachiozoan. For this reason, the clade name *Brachiopoda* was selected over *Brachiozoa*.

The composition of the crown clade *Brachiopoda* as conceptualized here corresponds, for the most part, to the traditional composition of *Brachiopoda* (Williams and Rowell, 1965; Williams et al., 2000). Despite the fact that approximately 95% of brachiopod species are extinct, virtually all extinct species nest within the crown clade (See Composition). The main difference from the traditional interpretation of *Brachiopoda* is the possible inclusion of *Phoronida*, suggesting that bivalved shells (and other features) were lost secondarily in *Phoronida*. Because of the high species diversity in *Brachiopoda* (more than 10,000 species named), the majority of which are extinct, and its long evolutionary history (more than 530 million years since the basal internal divergence), morphological evidence necessarily plays an important role in the diagnosis of this clade and the referral of fossil species to it.

The inclusion of *Phoronida* within *Brachiopoda*, as the sister group to *Craniiformea* + *Linguliformea*, is supported by some analyses of molecular evidence (Cohen and Gawthrop, 1996; Cohen et al., 1998a,b; Cohen, 2000; Cohen and Weydmann, 2005; Cohen, 2007, 2013) but not supported by others (Giribet et al., 2000; Helmkampf et al., 2008; Sperling et al., 2011). Following Recommendation 11A of the *PhyloCode* (Cantino and de Queiroz, 2020), no species of *Phoronida* is used as a specifier in our definition of *Brachiopoda*; consequently, that definition permits but does not require the inclusion of *Phoronida* within *Brachiopoda*. Some analyses with much more limited species sampling within *Brachiopoda* have suggested that *Phoronida* may be the sister group to *Brachiopoda* (Giribet et al., 2000; Mallatt and Winchell, 2002; Glenner et al., 2004; Dunn et al., 2008; Paps et al., 2009; Helmkampf et al., 2008; Mallatt et al., 2010; Nesnidal et al., 2010; Dordel et al., 2010; Sperling et al., 2011; see also Zrzavy et al., 1998) or more distantly related to *Brachiopoda* within *Lophotrochozoa* (Bourlat et al., 2008; Dunn et al., 2008; Yokobori et al., 2008; Podsiadlowski et al., 2009; Jang and Hwang, 2009; Edgecombe et al., 2011). Greater species sampling of both brachiopods and phoronids is necessary to test the hypotheses of relationship referred to above. Brachiopod polyphyly (Passamaneck and Halanych, 2006; Colgan et al., 2008), using molecular evidence now thought to be less robust than when originally published, is not consistently supported by any subsequent studies of metazoan phylogeny.

If future data provide additional support for the nesting of *Phoronida* within *Brachiopoda*, the bivalved condition of *Brachiopoda* appears to have been lost, as a secondarily derived condition, in phoronids. It is also possible that the absence of two valves in *Phoronida* represents the retention of the ancestral condition for *Brachiopoda*, requiring the independent origin of the bivalved condition at least twice (possibly more than twice, depending on the genetic regulation of mineralization in the Early Cambrian) within *Brachiopoda* (see Valentine, 1975; Williams and Hurst, 1977; Wright, 1979; Gorjansky and Popov, 1985, 1986; Williams and Carlson, 2007; Zhang et al., 2008). No unambiguous evidence exists currently to allow the rejection of this possibility (Carlson, 1995, 2007; see also Santagata and Cohen, 2009).

Although Cuvier used the name "Brachiopodes" in 1805 for the (extant) acephalous "mollusks" *Lingula*, *Orbicula* (now *Discinisca*), and *Terebratula* (now *Terebratalia* and numerous other genera), the name was not formalized until Duméril proposed *Brachiopoda* as the fifth order of *Mollusca* in 1805 (1806). *Brachiopoda* was then characterized (in French) as "skeletonized mollusks without a head, with ciliated tentacles that retract to the interior of the sessile shell." Brachiopods were thought to be closely related to mollusks, most commonly lamellibranchs (bivalves), as well as to ascidians (tunicates: Lamarck, 1809), for much of the rest of the nineteenth century. Morse (1902) suggested that brachiopods enjoy a closer phylogenetic affinity to annelids than to molluscs, based on the possession of setae, considered to be inherited from a close common ancestor of the two groups (see also Glenner et al., 2004; Helmkampf et al., 2008; Podsiadlowski et al., 2009).

Tentaculata (*Lophophorata* sensu Hyman, 1959; see also Emig, 1977, 1984) was proposed by Hatschek (1888) for a group that included brachiopods, bryozoans, and phoronids based on the possession of a lophophore, with or without a bivalved shell. Hyman (1959) further considered lophophorates to be "intermediate" between protostomes and deuterostomes, as had been suggested indirectly by Lamarck (1809), because of the mosaic of mostly developmental features of protostomes and deuterostomes exhibited by brachiopods in particular (see Carlson, 1995, for further discussion). Despite the common possession of a lophophore, Nielsen (1985) proposed that bryozoans shared close common ancestry with protostomes, while brachiopods and phoronids were considered more closely related to each other, and together, to the deuterostomes. Molecular evidence, however, (e.g., Field et al., 1988; Halanych et al., 1995; Cohen et al., 1998b; Giribet, 2008; Sperling et al., 2011) has consistently supported the protostome affinities of brachiopods, as well as generally supporting the closer relationship between brachiopods and phoronids than either one to bryozoans (Helmkampf et al., 2008; Paps et al., 2009; Mallatt et al., 2010; Nesnidal et al., 2010; Sperling et al., 2011; but see Bourlat et al., 2008; Yokobori et al., 2008; and Podsiadlowski et al., 2009, for alternative hypotheses of relationships among brachiopods, phoronids, and bryozoans).

Brachiopod monophyly (excluding phoronids) was tested and supported with morphological data in studies by Rowell (1981, 1982), Carlson (1991, 1995) and Williams et al. (1996, 2000). The first serious challenge to a traditional morphological interpretation of *Brachiopoda*, diagnosed as bivalved lophophorates, was Cohen and Gawthrop (1996), in which molecular evidence was presented that supported the nesting of phoronids, non-mineralized tube-dwelling lophophorates, within *Brachiopoda*, as the sister-group to the inarticulated brachiopods (the clade of *Linguliformea* and *Craniiformea*). Cohen and Gawthrop (1996) sampled many more species within *Brachiopoda* than previous molecular phylogenetic analyses involving brachiopods. As more species are sampled in the future, it is possible that the inclusion of *Phoronida* within *Brachiopoda* could become widely accepted, as *Dinosauria* is now widely considered to include *Aves*.

Literature Cited

Agassiz, L. 1848. *Nomenclatoris Zoologici Index Universalis: Continens Nomina Systematica Classium, Ordinum, Familiarum et Generum Animalium Omnium ... Homonymiis Plantarum.* Sumptibus Jent et Gassman, Soloduri.

Blainville, H. de. 1824. Mollusques. *Dict. Sci. Nat.* 32:1–567.

Bourlat, S. J., C. Nielsen, A. D. Economou, and M. J. Telford. 2008. Testing the new animal phylogeny: a phylum level molecular analysis of the animal kingdom. *Mol. Phylogenet. Evol.* 49:23–31.

Broderip, W. J. 1839. Article on "Malacology." *Penny Cyclopaeia, London* 14:314–325.

Bronn, H. G. 1862. *Die Klassen und Ordnungen der Weichthiere (Malacozoa),* Vol. 3, Part I. Kopflose Weichthiere.

Cantino, P. D., and K. de Queiroz. 2020. *International Code of Phylogenetic Nomenclature (PhyloCode),* Version 6. CRC Press, Boca Raton, FL.

Carlson, S. J. 1991. Phylogenetic relationships among brachiopod higher taxa. Pp. 3–10 in *Brachiopods through Time* (D. I. MacKinnon, D. E. Lee, and J. D. Campbell, eds.). A. A. Balkema, Rotterdam.

Carlson, S. J. 1995. Phylogenetic relationships among extant brachiopods. *Cladistics* 11:131–197.

Carlson, S. J. 2007. Recent research on brachiopod evolution. Pp. 2878–2900 in *Treatise on Invertebrate Paleontology, Part H,* Brachiopoda, *Revised,* Vol. 6 (P. A. Selden, ed.). Geological Society of America, Boulder, CO; University of Kansas Press, Lawrence, KS.

Cavalier-Smith, T. 1998. A revised six-kingdom system of life. *Biol. Rev.* 73:203–266.

Cohen, B. L. 2000. Monophyly of brachiopods and phoronids: reconciliation of molecular evidence with Linnaean classification (the subphylum *Phoroniformea* nov.). *Proc. R. Soc. Lond. B Biol. Sci.* 267:225–231.

Cohen, B. L. 2007. The brachiopod genome. Pp. 2356–2372 in *Treatise on Invertebrate Paleontology, Part H,* Brachiopoda, *Revised,* Vol. 6 (P. A. Selden, ed.). Geological Society of America, Boulder, CO; University of Kansas Press, Lawrence, KS.

Cohen, B. L. 2013. Rerooting the rDNA gene tree reveals phoronids to be 'brachiopods without shells'; dangers of wide taxon samples in metazoan phylogenetics (*Phoronida*; *Brachiopoda*). *Zool. J. Linn. Soc.* 167:82–92.

Cohen, B. L., and A. Weydmann. 2005. Molecular evidence that phoronids are a subtaxon of brachiopods (*Brachiopoda: Phoronata*) and that genetic divergence of metazoan phyla began long before the early Cambrian. *Org. Divers. Evol.* 5:253–273.

Cohen, B. L., and A. B. Gawthrop. 1996. Brachiopod molecular phylogeny. Pp. 73–80 in *Brachiopods: Proceedings of the Third International Brachiopod Congress, Sudbury, Ontario,* 1995 (P. Copper and J. Jin, eds.). A. A. Balkema, Rotterdam.

Cohen, B. L., A. B. Gawthrop, and T. Cavalier-Smith. 1998a. Molecular phylogeny of brachiopods and phoronids based on nuclear-encoded small subunit ribosomal RNA sequences. *Philos. Trans. R. Soc. Lond. B Biol. Sci.* 1378:2039–2061.

Cohen, B. L., S. Stark, A. B. Gawthrop, M. E. Burke, and C. W. Thayer. 1998b. Comparison of articulate brachiopod nuclear and mitochondrial gene trees leads to a clade-based redefinition of protostomes (*Protostomozoa*) and deuterostomes (*Deuterostomozoa*). *Proc. R. Soc. Lond. B Biol. Sci.* 265:475–482.

Colgan, D. J., P. A. Hutchings, and E. Beacham. 2008. Multi-gene analyses of the phylogenetic relationships among the *Mollusca, Annelida,* and *Arthropoda. Zool. Stud.* 47(3):338–351.

Cuvier, G. L. C. F. D. 1805. *Leçons d'Anatomie Comparée de G. Cuvier Recueillies et Publiées sous ses yeux par G. L. Duvernoy*, Vols. 3–5. Baudouin, Paris.

Dall, W. H. 1920. Annotated list of the recent *Brachiopoda* in the collection of the United States National Museum, with descriptions of thirty-three new forms. *Proc. U.S. Natl. Mus.* 57:261–377.

Dordel, J., F. Fisse, G. Purschke, and T. H. Struck. 2010. Phylogenetic position of *Sipuncula* derived from multi-gene and phylogenomic data and its implications for the evolution of segmentation. *J. Zool. Syst. Evol. Res.* 48:197–207.

Duméril, A. M. C. 1805 (1806). *Zoologie Analytique, ou Méthode Naturelle de Classification des Animaux*, P. 171 in particular. Allais, Paris.

Dunn, C. W., A. Hejnol, D. Q. Matus, K. Pang, W. E. Browne, S. A. Smith, E. Seaver, G. W. Rouse, M. Obst, G. D. Edgecombe, M. V. Sørensen, S. H. D. Haddock, A. Schmidt-Rhaesa, A. Okusu, R. Møbjerg Kristensen, W. C. Wheeler, M. Q. Martindale, and G. Giribet. 2008. Broad phylogenomic sampling improves resolution of the animal tree of life. *Nature* 452:745–749.

Edgecombe, G. D., G. Giribet, C. W. Dunn, A. Hejnol, R. M. Kristensen, R. C. Neves, G. W. Rouse, K. Worsaae, and M. V. Sorensen. 2011. Higher-level metazoan relationships: recent progress and remaining questions. *Org. Divers. Evol.* 11:151–172.

Emig, C. C. 1977. Un nouvel embranchement: les lophophorates. *Bull. Soc. Zool. Fr.* 102:341–344.

Emig, C. C. 1984. On the origin of the *Lophophorata*. *Z. Zool. Syst. Evolutionforsch.* 22:91–94.

Field, K. G., G. J. Olsen, D. J. Lane, S. J. Giovannoni, M. T. Ghiselin, E. C. Raff, N. R. Pace, and R. A. Raff. 1988. Molecular phylogeny of the animal kingdom. *Science* 239:748–753.

Giribet, G. 2008. Assembling the lophotrochozoan (=spiralian) tree of life. *Philos. Trans. R. Soc. Lond. B Biol. Sci.* 363:1513–1522.

Giribet, G., D. L. Distel, M. Polz, W. Sterrer, and W. C. Wheeler. 2000. Triploblastic relationships with emphasis on the acoelomates and the position of *Gnathostomulida, Cycliophora,* *Plathelminthes*, and *Chaetognatha*: a combined approach of 18S rDNA sequences and morphology. *Syst. Biol.* 49:539–562.

Glenner, H., A. J. Hansen, M. V. Sørensen, F. Ronquist, J. P. Huelsenbeck, and E. Willerslev. 2004. Bayesian inference of the metazoan phylogeny: a combined molecular and morphological approach. *Curr. Biol.* 14:1644–1649.

Gorjansky, V. Y., and L. E. Popov. 1985. The morphology, systematic position, and origin of inarticulate brachiopods with carbonate shells. *Paleontol. Zh.* 3:3–13.

Gorjansky, V. Y., and L. E. Popov. 1986. On the origin and systematic position of the calcareous-shelled inarticulate brachiopods. *Lethaia* 19:233–240.

Gray, J. E. 1821. A natural arrangement of *Mollusca* according to their internal structure. *London Med. Repos.* 15:229–239.

Halanych, K. M., J. D. Bacheller, A. M. A. Aguinaldo, S. M. Liva, D. M. Hillis, and J. A. Lake. 1995. Evidence from 18S ribosomal DNA that the lophophorates are protostome animals. *Science* 267:1641–1643.

Hatschek, B. 1888–1891. *Lehrbuch der Zoologie.* Fischer, Jena.

Helmkampf, M., I. Bruchhaus, and B. Hausdorf. 2008. Phylogenomic analyses of lophophorates (brachiopods, phoronids and bryozoans) confirm the *Lophotrochozoa* concept. *Proc. R. Lond. Soc. B* 275:1927–1933.

Holmer, L. E., C. B. Skovsted, and A. Williams. 2002. A stem-group brachiopod from the Lower Cambrian—support for *Micrina* (halkieriid) ancestry. *Palaeontology* 45:875–882.

Hyman, L. H. 1959. *The Invertebrates: Smaller Coelomate Groups*, Vol. 5. McGraw-Hill, New York.

Jang, K. H., and U. W. Hwang. 2009. Complete mitochondrial genome of *Bugula neritina* (*Bryozoa, Gymnolaemata, Cheilostomata*): phylogenetic position of *Bryozoa* and phylogeny of lophophorates within the *Lophotrochozoa*. *BMC Genomics* 10:167.

King, W. 1850. *A Monograph of the Permian Fossils of England.* Palaeontological Society of London.

Lamarck, J. B. P. 1801. *Systeme des Animaux sans Vertebres.* Deterville, Paris.

Lamarck, J. B. P. 1809. *Philosophie Zoologique, ou Eexposition des Considérations Relatives à l'Histoire Naturelle des Animaux.* Macmillan, London.

Mallatt, J., and C. J. Winchell. 2002. Testing the new animal phylogeny: first use of combined large-subunit and small-subunit rRNA gene sequences to classify the protostomes. *Mol. Bio. Evol.* 19:289–301.

Mallatt, J., C. W. Craig, and M. Y. Yoder. 2010. Nearly complete rRNA genes assembled from across the metazoan animals: effects of more taxa, a structure-based alignment, and paired-sites evolutionary models on phylogeny reconstruction. *Mol. Phylogenet. Evol.* 55:1–17.

Morse, E. S. 1902. Observations on living *Brachiopoda. Mem. Boston Soc. Nat. Hist.* 5(8):313–386.

Müller, O. F. 1776. *Zoologiae Danicae Prodromus, seu Animalium Daniae et Norvegiae Indigenarum Characteres, Nomina, et Synonyma Imprimis Popularium.* Typis Hallageriis, Havniae (Copenhagen).

Nesnidal, M. P., M. Helmkampf, I. Bruchhaus, and B. Hausdorf. 2010. Compositional heterogeneity and phylogenomic inference of metazoan relationships. *Mol. Biol. Evol.* 27(9):2095–2104.

Nielsen, C. 1985. Animal phylogeny in the light of the trochaea theory. *Biol. J. Linn. Soc.* 25:243–299.

Nielsen, C. 1991. The development of the brachiopod *Crania* (*Neocrania*) *anomala* (O. F. Müller) and its phylogenetic significance. *Acta Zool.* 72(1):7–28.

Paps, J., J. Baguna, and M. Riutort. 2009. *Lophotrochozoa* internal phylogeny: new insights from an up-to-date analysis of nuclear ribosomal genes. *Proc. R. Lond. Soc. B* 276:1245–1254.

Passamaneck, Y., and K. M. Halanych. 2006. Lophotrochozoan phylogeny assessed with LSU and SSU data: evidence of lophophorate polyphyly. *Mol. Phylogenet. Evol.* 40(1):20–28.

Podsiadlowski, L., A. Braband, T. H. Struck, J. von Dohren, and T. Bartolomaeus. 2009. Phylogeny and mitochondrial gene order variation in *Lophotrochozoa* in the light of new mitogenomic data from *Nemertea. BMC Genomics* 10:364.

Popov, L., M. G. Bassett, L. E. Holmer, and J. R. Laurie. 1993. Phylogenetic analysis of higher taxa of *Brachiopoda. Lethaia* 26:1–5.

Rafinesque, C. S. 1831. On the fossil bivalve shells of the western region, supplement to A monograph of the bivalve shells of the River Ohio. Philadelphia, PA.

Rowell, A. J. 1981. The Cambrian radiation: monophyletic or polyphyletic origins? In *Short Papers for the Second International Symposium on the Cambrian System. U.S. Geol. Surv. Open File Rep.* 81–743: 184–187.

Rowell, A. J. 1982. The monophyletic origin of the *Brachiopoda. Lethaia* 15:299–307.

Rudwick, M. J. S. 1970. *Living and Fossil Brachiopods.* Hutchinson University Library, London.

Santagata, S., and B. L. Cohen. 2009. Phoronid phylogenetics (*Brachiopoda*; *Phoronata*): evidence from morphological cladistics, small and large subunit rDNA sequences, and mitochondrial *cox*1. *Zool. J. Linn. Soc.* 157:34–50.

Sowerby, G. B. 1846. Description of thirteen new species of brachiopods. *Proc. Zool. Soc. Lond.* 14:91–95.

Sowerby, G. B. 1847. The recent *Brachiopoda. Thesaurus Conchyliorum* 1:362.

Sperling, E. A., D. Pisani, and K. J. Peterson. 2011. Molecular paleobiological insights into the origin of the *Brachiopoda. Evol. Dev.* 13:290–303.

Valentine, J. W. 1975. Adaptive strategies and the origin of grades and ground-plans. *Am. Zool.* 15:391–404.

Williams, A., and A. J. Rowell. 1965. Classification. Pp. H214–H237 in *Treatise on Invertebrate Paleontology, Part H, Brachiopoda*, Vol. 1 (R. C. Moore, ed.). Geological Society of America, Boulder, CO; University of Kansas Press, Lawrence, KS.

Williams, A., and S. J. Carlson. 2007. Affinities of brachiopods and trends in their evolution. Pp. 2822–2877 in *Treatise on Invertebrate Paleontology, Part H*, Brachiopoda, *Revised*, Vol. 6 (R. L. Kaesler, ed.). Geological Society of America, Boulder, CO; University of Kansas Press, Lawrence, KS.

Williams, A., S. J. Carlson, C. H. C. Brunton, L. E. Holmer, and L. Popov. 1996. A supra-ordinal classification of the *Brachiopoda. Philos. Trans. R. Soc. Lond. B Biol. Sci.* 351:1171–1193.

Williams, A., S. J. Carlson, and C. H. C. Brunton. 2000. *Brachiopoda.* Pp. 28–29 in *Treatise on Invertebrate Paleontology, Part H,* Brachiopoda, *Revised,* Vol. 2 (R. L. Kaesler, ed.). Geological Society of America, Boulder, CO; University of Kansas Press, Lawrence, KS.

Williams, A., and J. M. Hurst. 1977. Brachiopod evolution. Pp. 75–121 in *Patterns of Evolution* (A. Hallam, ed.). Elsevier, Amsterdam and New York.

Wright, A. 1979. Brachiopod radiation. Pp. 235–252 in *The Origin of Major Invertebrate Groups* (M. R. House, ed.). Academic Press, New York and London.

Yokobori, S., T. Iseto, S. Asakawa, T. Sasaki, N. Shimizu, A. Yamagishi, T. Oshima, and E. Hirose. 2008. Complete nucleotide sequences of mitochondrial genomes of two solitary entoprocts, *Loxocorone allax* and *Loxosomella aloxiata*: implications for lophotrochozoan phylogeny. *Mol. Phylogenet. Evol.* 47:612–628.

Zhang, Z., S. P. Robson, C. Emig, and D. Shu. 2008. Early Cambrian radiation of brachiopods: a perspective from South China. *Gondwana Res.* 14:241–254.

Zrzavy, J., S. Mihulka, P. Kepka, A. Bezdek, and D. Tietz. 1998. Phylogeny of the *Metazoa* based on morphological and 18S ribosomal DNA evidence. *Cladistics* 14:249–285.

Authors

Sandra J. Carlson; Department of Geology; University of California; One Shields Avenue; Davis, CA 95616-8605, USA. Email: sjcarlson@ucdavis.edu.

Bernard L. Cohen; Faculty of Biomedical and Life Sciences; University of Glasgow; G12 8QQ, Scotland, UK. Email: b.l.cohen@bio.gla.ac.uk.

Date Accepted: 12 October 2012; subsequent revisions accepted 20 May 2013

Primary Editor: Kevin de Queiroz

Pan-Neoarticulata S. J. Carlson and B. L. Cohen,
new clade name

Registration Number: 147

Definition: The total clade of the crown clade *Neoarticulata*. This is a crown-based total-clade definition. Abbreviated definition: total ∇ of *Neoarticulata*.

Etymology: *Pan-* is derived from the Greek Παντος meaning "all" (a standard prefix for total clades; see *PhyloCode* Art. 10.3). See *Neoarticulata* (this volume) for the etymology of that name.

Reference Phylogeny: Carlson (2007: Fig. 1908). See Comments for further discussion on the reference phylogeny. See also Carlson (1991: Fig. 4); Carlson (1995: Fig. 3); Williams et al. (1996: Fig. 6); Holmer and Popov (2000: Fig. 50); Popov et al. (2000: Fig. 88); Williams et al. (2000a: Fig. 5); Carlson and Leighton (2001a: Fig. 1); Carlson and Leighton (2001b: Fig. 26.1); Cohen and Weydmann (2005: Fig. 2); Cohen (2007: Fig. 1530); Sperling et al. (2011: Fig. 2C); Cohen (2013: Fig. 3).

Composition: *Pan-Neoarticulata* is composed of the crown clade *Neoarticulata* (this volume) and all extinct organisms or species that share more recent common ancestry with *Neoarticulata* than with any other mutually exclusive (non-nested) crown clade in *Brachiopoda;* according to the reference phylogeny, those mutually exclusive crown clades are those of *Linguliformea* and *Craniiformea* (and possibly *Phoronida;* see entry for *Brachiopoda* in this volume, and Cohen, 2013). *Pan-Neoarticulata* is currently thought to include the following five non-overlapping groups whose monophyletic status

and relationships to one another within *Pan-Neoarticulata* are uncertain (Williams et al., 1996): *Rhynchonellata* Williams et al. 1996 (which includes the crown clade *Neoarticulata*), and four extinct groups, namely *Strophomenata* Williams et al. 1996, *Kutorginata* Williams et al. 1996, *Obolellata* Williams et al. 1996, *Chileata* Williams et al. 1996 (see Comments).

Several Lower Cambrian fossils have been considered possible "stem" rhynchonelliforms (and thus implicitly stem neoarticulates), including *Longtancunella* (Zhang et al., 2011) and *Paterimitra* (Holmer et al., 2011). Their status as stem neoarticulates is tentative, though, because none of the putative phylogenetic affinities hypothesized in these studies has been corroborated by an explicit phylogenetic analysis of the characters present in these taxa.

Diagnostic Apomorphies: Apomorphies of *Pan-Neoarticulata* are thought to be the calcitic (organocarbonate), not organophosphate, bivalved and articulated shell with a fibrous, not tabular laminated, secondary layer (Williams et al., 1996, 2000a). These characters distinguish living *Neoarticulata* from *Linguliformea* and *Craniiformea* (and *Phoronida*) and are thought to have arisen earliest in the history of the total clade because of their combined occurrence in several groups of Cambrian fossils (see Comments).

Features related to articulation between the two valves have been considered to be apomorphies by some (Williams et al., 1996, 2000a), but the homology of teeth and sockets and their manner of articulation among crown group *Neoarticulata* and all basal members of *Pan-Neoarticulata* has not yet been clearly demonstrated (Williams and Rowell, 1965; Rudwick, 1970; Carlson, 1991; Popov, 1992; Popov et al., 1996, 2000; Carlson and Leighton, 2001a,b; Holmer, 2001; Bassett et al., 2001; Carlson, 2007; Williams and Carlson, 2007). The putative apomorphy is valve articulation consisting of a pair of teeth or tooth-like structures (typically ventral) and sockets or socket-like structures (typically dorsal) that lie on either side of a postero-median indentation or notch on each valve, with a hinge axis formed by the posterior margins of valve interareas, or by a hinge axis that is less clearly identifiable on the valves themselves (see discussion in Williams and Carlson, 2007). Movement about the hinge axis is effected by the contraction of adductor (opening) and diductor (closing) muscles (Eshleman et al., 1982), as demonstrated from the location of muscle scars indicating origin and insertion of these muscles on the valves.

Synonyms: *Rhynchonelliformea* Williams et al. 1996; *Articulata* Huxley 1869 (both approximate).

Comments: *Pan-Neoarticulata* as conceptualized here corresponds in known composition with *Rhynchonelliformea* Williams et al. 1996, as traditionally conceptualized. However, *Pan-Neoarticulata*, referring to *Neoarticulata* and its entire stem lineage, is more inclusive than *Rhynchonelliformea*, which was diagnosed by mineralized features that could be preserved in fossils (calcitic shell with fibrous microstructure), at least some of which likely originated after the initial divergence of the lineage.

The reference phylogeny chosen represents the most recent hypothesis of phylogenetic relationships among all the major named groups in *Brachiopoda*, extant and extinct. It is a composite hypothesis of topological elements extracted from seven different, more restricted, analyses, and it was constructed "by hand" following a consensus method. Nodes present in the reference phylogeny are those in agreement with the majority of analyses evaluated; conflicts are collapsed to polytomies and uncertainties are represented by question marks, both of which require more comprehensive testing than has been attempted thus far. A large component of the uncertainty surrounding these groups and their phylogenetic identities and affinities is the fact that taxonomic names were often applied to groups before their monophyly was even investigated, much less determined.

Three brachiopod taxa established prior to the present definition of *Neoarticulata* as the name of a crown clade correspond to an internested series of clades including various extinct brachiopods that are successively more distantly related to the crown clade *Neoarticulata*. None of these taxa (*Rhynchonelliformea*, *Articulata*, and *Rhynchonellata*) corresponds exactly to *Pan-Neoarticulata* as conceptualized here. *Rhynchonellata* Williams et al. 1996 (see Williams and Carlson, 2000) includes the crown clade *Neoarticulata* as well as *Pentamerida* Schuchert and Cooper 1931 (see Carlson et al., 2002). *Orthida* Schuchert and Cooper 1932 (Williams and Harper, 2000b) and *Protorthida* Schuchert and Cooper 1931 (Williams and Harper, 2000a) are also included in *Rhynchonellata* Williams et al. 1996, outside of *Neoarticulata*, but relationships among extinct taxa outside of the crown clade are unclear (see Carlson and Leighton, 2001a,b; Carlson, 2007). *Rhynchonellata* is diagnosed by the presence of a pedicle developed from a larval rudiment rather than as an outgrowth of the posterior body wall, as evidenced by the presence and location of pedicle adjustor muscle scars visible on valves of extinct forms. Most rhynchonellates also possess brachiophores, blades of secondary shell that project from the inner socket ridges and may have functioned to support the mouth region of the lophophore (Williams et al., 1997);

their homology with crura, loops, and spiralia is not entirely clear, however.

Articulata is a well-supported and commonly recognized clade that has always included extinct taxa now though to be outside of the crown clade *Neoarticulata* (Huxley, 1869; Ager et al., 1965; Williams and Rowell, 1965; Rudwick, 1970). *Articulata* has been diagnosed by many apomorphies (Carlson, 1995), including most notably the presence of non-interlocking ventral valve teeth and dorsal valve sockets that articulate along a hinge axis located between the valves (referred to as deltidiodont articulation; Jaanusson, 1971). Several soft anatomical or embryological characters are sometimes listed as apomorphies of *Articulata*, but given that their presence cannot be confirmed in extinct members of the clade, this practice seems to represent an imprecise application of the name to both the crown and a more inclusive clade. Several hard-part features that have traditionally been considered diagnostic for *Articulata* (Ager et al., 1965) have since been interpreted to be plesiomorphies rather than apomorphies of the clade (e.g., valves invariably calcareous with well-developed primary and secondary shell layers; see Carlson, 1995). *Articulata* includes *Rhynchonellata* and the extinct *Strophomenata* Williams et al. 1996, which includes *Strophomenida* Opik 1934 (see Cocks and Rong, 2000), *Productida* Sarytcheva and Sokolskaya 1959 (see Brunton et al., 2000), *Billingsellida* Schuchert 1893 (see Williams and Harper, 2000c), and *Orthotetida* Waagen 1884 (see Williams et al., 2000c). The evidence supporting the sister-group relationship of *Kutorginata* Kuhn 1949 and *Strophomenata* depicted in the reference phylogeny is weak; the relationships of *Kutorginata* Kuhn 1949 to *Rhynchonellata* and *Strophomenata* are controversial (see Carlson, 2007).

Rhynchonelliformea Williams et al. 1996 (see also 2000a) is a clade diagnosed by a fibrous secondary layer in a calcitic shell; the name is an approximate synonym of *Pan-Neoarticulata* in that it includes all known putative members of the stem-group of *Neoarticulata*. *Rhynchonelliformea* includes *Articulata* as well as *Kutorginata* Kuhn 1949 (see Williams et al., 1996; Popov and Williams, 2000), *Chileida* Popov and Tikhonov 1990 (see Popov and Holmer, 2000a), *Dictyonellida* Cooper 1956 (see Holmer, 2000), *Obolellida* Rowell 1965a (see Popov and Holmer, 2000b), and *Naukatida* Popov and Tikhonov 1990 (see Popov and Holmer, 2000b). All except for crown clade *Neoarticulata* are extinct (see Carlson, 2007). These five mostly short-lived, Cambrian calcitic taxa are thought to share more recent common ancestry with *Neoarticulata* than with any other crown clade in *Brachiopoda*. The evidence to support this conclusion is weak, however, because it is not clear if the shell structure and primitive form of valve articulation (see Williams and Carlson, 2007) present in these five extinct groups are homologous with those present in the crown clade *Neoarticulata*; if they are homologous, it is not clear that they are apomorphic and not plesiomorphic. Lingering doubts about the homology of articulation in *Articulata* and these five extinct groups prompted the naming of a new taxon (*Rhynchonelliformea* Williams et al. 1996) that does not focus on the character of valve articulation, but rather on the articulated group containing extant representatives with the earliest first appearance in the stratigraphic record, *Rhynchonellida*.

The traditional concept of *Articulata* (e.g., Ager et al., 1965) was revised and expanded under the name *Rhynchonelliformea* (Williams et al., 1996, 2000a,b) to include five extinct, poorly known, primarily early Cambrian groups exhibiting a variety of diverse "articulatory" types (as outlined in Williams and Carlson, 2007); they are considered to be more closely related to *Articulata* (sensu Ager et al., 1965) than to *Linguliformea* or *Craniiformea* (see Williams et al., 1996, 2000a). In the definitive reference on brachiopod classification prior to the revision (Williams et al., 1965), *Kutorginida* was in "Class Uncertain" (Rowell, 1965b), *Dictyonellidina* was in "Order Uncertain" (Rowell, 1965a) in Class *Articulata*, and *Obolellida* was named as a new order (Rowell, 1965c) in Class *Inarticulata*. Both *Naukatida* Popov and Tikhonov 1990 and *Chileida* Popov and Tikhonov 1990 were named subsequently. These five groups, while all possessing calcitic bivalved shells, each exhibit a mosaic of characters of shell structure and valve articulation; for example, chileates and some obolellates lack articulatory structures, while obolellates (and strophomenates) possess laminar not fibrous shell structure. In comparison, early members of *Articulata* (sensu Ager et al., 1965) are characterized by deltidiodont (Jaanusson, 1971) hinge structures, while members of the crown clade are characterized by cyrtomatodont hinge structures (see *Neoarticulata* in this volume). Monophyly of each of these five groups has not been established; relationships of the five groups to one another and to *Articulata* (sensu Ager et al., 1965) are uncertain; homology and polarity of often poorly preserved structures (particularly articulatory structures; see discussion in Williams et al., 2000a and Bassett et al., 2001) are unclear. These groups are, however, thought to be more closely related to *Neoarticulata* than to than any other crown clade in *Brachiopoda* (Williams et al., 1996) because of at least some features of shell structure or valve articulation, and, as such, are tentatively considered to be included in the total clade *Pan-Neoarticulata*. See Comments under *Neoarticulata* (this volume) and Bassett et al. (2001) for further discussion of some issues surrounding the homology of articulatory structures.

In addition to valve articulation, the homology and polarity of shell mineralogy has long been a contentious issue in discussions of brachiopod evolution (see Carlson, 1995; Williams

et al., 1996, 2000a). Mineralization among lophotrochozoans is predominantly calcareous, suggesting that a calcitic mineralogy may be a retained primitive feature within *Brachiopoda*, and thus also within *Pan-Neoarticulata* (Carlson, 1995; although see discussion in Williams et al., 2000a). If homologous among lophotrochozoans, a calcitic shell would be plesiomorphic for *Pan-Neoarticulata* regardless of whether craniiforms are more closely related to linguliforms (Rowell, 1981a,b, 1982; Carlson, 1995; Cohen and Weydmann, 2005; Cohen, 2007; Sperling et al., 2011; Cohen, 2013) or rhynchonelliforms (Gorjansky and Popov, 1985, 1986; Popov et al., 1993; Holmer et al., 1995). However, if mineralization evolved independently in multiple clades from non-mineralized metazoans (see Wright, 1979), then calcareous mineralization is not necessarily homologous in all metazoans in which it occurs (see discussion in Williams and Carlson, 2007; Carlson, 2007). Among lophotrochozoans, therefore, polarity assessment of biomineral type would be more difficult to establish. However, if calcitic mineralization can be demonstrated to be a plesiomorphy of *Brachiopoda*, then the organophosphatic shell mineralogy would be a derived condition. Conversely, if organophosphatic mineralization can be demonstrated to be a plesiomorphy of *Brachiopoda* (consistent with the hypothesis that organophosphatic tommotiids are in the brachiopod stem-group, and not merely the linguliform stem-group; see discussion in *Pan-Brachiopoda*, this volume), then the calcitic shell mineralogy of *Pan-Neoarticulata* would be derived within *Brachiopoda*.

The shell structure of *Pan-Neoarticulata* consists of an organic periostracum underlain by a thin outer granular or acicular "primary" shell layer and an inner fibrous secondary layer (Williams, 1997). The presence of a fibrous secondary layer distinguishes the total clade *Pan-Neoarticulata* from all other brachiopods. The cross-bladed laminar shell fabric characteristic of many strophomenates can be distinguished from the tabular lamination of craniiforms by examination of the microstructure with scanning electron microscopy. It appears that cross-bladed lamination has arisen more than once from the fibrous condition within strophomenates (Brunton, 1972; Williams, 1997; Williams et al., 2000a; Carlson and Leighton, 2001a) and independently in obolellates (Popov and Holmer, 2000b).

Literature Cited

Ager, D. V., T. W. Amsden, G. Biernat, A. J. Boucot, G. F. Elliott, R. E. Grant, K. Hatai, J. G. Johnson, D. J. McLaren, H. M. Muir-Wood, C. W. Pitrat, A. J. Rowell, H. Schmidt, R. D. Staton, F. G. Stehli, A. Williams, and A. D. Wright. 1965. *Articulata*. Pp. H297–H299 in *Treatise on Invertebrate Paleontology, Part H, Brachiopoda*, Vol. 1 (R. L. Kaesler, ed.). Geological Society of America, Boulder, CO; University of Kansas Press, Lawrence, KS.

Bassett, M. G., L. E. Popov, and L. E. Holmer. 2001. Functional morphology of articulatory structures and implications for patterns of musculature in Cambrian rhynchonelliform brachiopods. Pp. 163–17 in *Brachiopods Past and Present* (C. H. C. Brunton, L. R. M. Cocks and S. L. Long, eds.). Taylor & Francis, London and New York.

Brunton, C. H. C. 1972. The shell structure of chonetacean brachiopods and their ancestors. *Bull. Br. Mus. Nat. Hist.* 21:1–26.

Brunton, C. H. C., S. S Lazarev, R. E. Grant, and J. Yu-gan. 2000. *Productida*. Pp. 424–609 in *Treatise on Invertebrate Paleontology, Part H, Brachiopoda, Revised*, Vol. 3 (R. L. Kaesler, ed.). Geological Society of America, Boulder, CO; University of Kansas Press, Lawrence, KS.

Carlson, S. J. 1991. Phylogenetic relationships among brachiopod higher taxa. Pp. 3–10 in *Brachiopods through Time* (D. I. MacKinnon, D. E. Lee and J. D. Campbell, eds.). A. A. Balkema, Rotterdam.

Carlson, S. J. 1995. Phylogenetic relationships among extant brachiopods. *Cladistics* 11:131–197.

Carlson, S. J. 2007. Recent research on brachiopod evolution. Pp. 2878–2900 in *Treatise on Invertebrate Paleontology, Part H,* Brachiopoda, *Revised*, Vol. 6 (P. A. Selden, ed.). Geological Society of America, Boulder, CO; University of Kansas Press, Lawrence, KS.

Carlson, S. J., A. J. Boucot, Rong Jia-yu, and R. B. Blodgett. 2002. *Pentamerida*. Pp. 921–928 in *Treatise on Invertebrate Paleontology, Part H,*, Brachiopoda, *revised*, Vol. 4 (R. L. Kaesler, ed.). Geological Society of America, Boulder, CO; University of Kansas Press, Lawrence, KS.

Carlson, S. J., and L. R. Leighton. 2001a. The phylogeny and classification of *Rhynchonelliformea*. Pp. 27–51 in *Brachiopods Ancient and Modern* (S. J. Carlson and M. R. Sandy, eds.). The Paleontological Society, New Haven, CT.

Carlson, S. J., and L. R. Leighton. 2001b. Incorporating stratigraphic data in the phylogenetic analysis of the *Rhynchonelliformea*. Pp. 248–258 in *Brachiopods Past and Present* (C. H. C. Brunton, L. R. M. Cocks and S. L. Long, eds.). Taylor & Francis, London and New York.

Cocks, L. R. M., and Rong Jia-yu. 2000. *Strophomenida*. Pp. 216–348 in *Treatise on Invertebrate Paleontology, Part H,* Brachiopoda, *Revised*, Vol. 2 (R. L. Kaesler, ed.). Geological Society of America, Boulder, CO; University of Kansas Press, Lawrence, KS.

Cohen, B. L. 2007. The brachiopod genome. Pp. 2356–2372 in *Treatise on Invertebrate Paleontology, Part H,* Brachiopoda, *Revised*, Vol. 6 (P. A. Selden, ed.). Geological Society of America, Boulder, CO; University of Kansas Press, Lawrence, KS.

Cohen, B. L. 2013. Rerooting the rDNA gene tree reveals phoronids to be 'brachiopods without shells'; dangers of wide taxon samples in metazoan phylogenetics (*Phoronida*; *Brachiopoda*). *Zool. J. Linn. Soc.* 167:82–92.

Cohen, B. L., and A. Weydmann. 2005. Molecular evidence that phoronids are a subtaxon of brachiopods (*Brachiopoda: Phoronata*) and that genetic divergence of metazoan phyla began long before the early Cambrian. *Org. Divers. Evol.* 5(4):253–273.

Cooper, G. A. 1956. Chazyan and related brachiopods. *Smithson. Misc. Collect.* 127(I–II):1–1245.

Eshleman, W. P., J. L. Wilkens, and M. J. Cavey. 1982. Electrophoretic and electron microscopic examination of the adductor and diductor muscles of an articulate brachiopod, *Terebratalia transversa. Can. J. Zool.* 60(4):550–559.

Gorjansky, V. Y., and L. E. Popov. 1985. The morphology, systematic position, and origin of inarticulate brachiopods with carbonate shells. *Paleontol. Zh.* 3:3–13.

Gorjansky, V. Y., and L. E. Popov. 1986. On the origin and systematic position of the calcareous-shelled inarticulate brachiopods. *Lethaia* 19:233–240.

Holmer, L. E. 2000. *Dictyonellida.* Pp. 196–200 in *Treatise on Invertebrate Paleontology, Part H,* Brachiopoda, *Revised,* Vol. 2 (R. L. Kaesler, ed.). Geological Society of America and University of Kansas Press, Boulder and Lawrence.

Holmer, L. E. 2001. Phylogeny and classification: *Linguliformea* and *Craniiformea.* Pp. 11–26 in *Brachiopods Ancient and Modern* (S. J. Carlson and M. R. Sandy, eds.). The Paleontological Society, New Haven, CT.

Holmer, L. E., and L. E. Popov. 2000. *Lingulata.* Pp. 30–146 in *Treatise on Invertebrate Paleontology, Part H,* Brachiopoda, *Revised,* Vol. 2 (R. L. Kaesler, ed.). Geological Society of America, Boulder, CO; University of Kansas Press, Lawrence, KS.

Holmer, L. E., L. E. Popov, M. G. Bassett, and J. Laurie. 1995. Phylogenetic analysis and classification of the *Brachiopoda. Palaeontology* 38:713–741.

Holmer, L. E., C. B. Skovsted, C. Larsson, G. A. Brock and Z. Zhang. 2011. First record of a bivalved larval shell in Early Cambrian tommotiids and its phylogenetic significance. *Palaeontology* 54:235–239.

Huxley, T. H. 1869. *An Introduction to the Classification of Animals.* John Churchill and Sons, London.

Jaanusson, V. 1971. Evolution of the brachiopod hinge. *Smithson. Contrib. Paleobiol.* 3:33–46.

Kuhn, O. 1949. *Lehrbuch der Palaozoologie.* E. Schweizerbart, Stuttgart.

Opik, A. A. 1934. Uber Klitamboniten. Universitatis Tartuensis (Dorpatensis). *Acta Comment. (Ser. A)* 26:1–239.

Popov, L. E. 1992. The Cambrian radiation of brachiopods. Pp. 399–423 in *Origin and Early Evolution of the* Metazoa (J. H. Lipps and P. W. Signor, eds.). Plenum Press, New York.

Popov, L. E., M. G. Bassett, and L. E. Holmer. 2000. *Craniata.* Pp. 158–192 in *Treatise on Invertebrate Paleontology, Part H,* Brachiopoda, *Revised,* Vol. 2 (R. L. Kaesler, ed.). Geological Society of America, Boulder, CO; University of Kansas Press, Lawrence, KS.

Popov, L. E., M. G. Bassett, L. E. Holmer, and J. R. Laurie. 1993. Phylogenetic analysis of higher taxa of *Brachiopoda. Lethaia* 26:1–5.

Popov, L. E., and L. E. Holmer. 2000a. *Chileata.* Pp. 193–196 in *Treatise on Invertebrate Paleontology, Part H,* Brachiopoda, *Revised,* Vol. 2 (R. L. Kaesler, ed.). Geological Society of America, Boulder, CO; University of Kansas Press, Lawrence, KS.

Popov, L. E., and L. E. Holmer. 2000b. *Obolellata.* Pp. 200–207 in *Treatise on Invertebrate Paleontology, Part H,* Brachiopoda, *Revised,* Vol. 2 (R. L. Kaesler, ed.). Geological Society of America, Boulder, CO; University of Kansas Press, Lawrence, KS.

Popov, L. E., L. E. Holmer, and M. G. Bassett. 1996. Radiation of the earliest calcareous brachiopods. Pp. 209–214 in *Brachiopods: Proceedings of the Third International Brachiopod Congress, Sudbury, Ontario,* 1995 (P. Copper and J. Jin, eds.). A. A. Balkema, Rotterdam.

Popov, L. E., and Y. A. Tikhonov. 1990. Rannekembriiskie brakhiopody iz iuzhnoi Kirgizii (Early Cambrian brachiopods from south Kirgizii). *Paleontol. Zh.* 1990(3):33–44. [In Russian].

Popov, L. E., and A. Williams. 2000. *Kutorginata.* Pp. 208–215 in *Treatise on Invertebrate Paleontology, Part H,* Brachiopoda, *Revised,* Vol. 2 (R. L. Kaesler, ed.). Geological Society of America, Boulder, CO; University of Kansas Press, Lawrence, KS.

Rowell, A. J. 1965a. Order *Obolellida*. Pp. H291–293 in *Treatise on Invertebrate Paleontology, Part H,* Brachiopoda, Vol. 1 (R. C. Moore, ed.). Geological Society of America, Boulder, CO; University of Kansas Press, Lawrence, KS.

Rowell, A. J. 1965b. Order *Kutorginida*. Pp. H296–297 in *Treatise on Invertebrate Paleontology, Part H,* Brachiopoda, Vol. 1 (R. C. Moore, ed.). Geological Society of America, Boulder, CO; University of Kansas Press, Lawrence, KS.

Rowell, A. J. 1965c. Order Uncertain—*Dictyonellidina*. Pp. H359–H361 in *Treatise on Invertebrate Paleontology, Part H,* Brachiopoda, Vol. 1 (R. C. Moore, ed.). Geological Society of America, Boulder, CO; University of Kansas Press, Lawrence, KS.

Rowell, A. J. 1981a. The Cambrian radiation: monophyletic or polyphyletic origins? In: M. E. Taylor (ed.). Short papers for the Second International Symposium on the Cambrian System. *U.S. Geol. Surv. Open File Rep.* 81–743:184–187.

Rowell, A. J. 1981b. The origin of the brachiopods. Pp. 97–109 in *Lophophorates, Notes for a Short Course* (T. W. Broadhead, ed.). University of Tennessee, Knoxville, TN.

Rowell, A. J. 1982. The monophyletic origin of the Brachiopoda. *Lethaia* 15:299–307.

Rudwick, M. J. S. 1970. *Living and Fossil Brachiopods.* Hutchinson University Library, London.

Sarytcheva, T. G., and A. N. Sokolskaya. 1959. O klassifikatsin lozhnoporistykh brakhiopod (On the classification of pseudopunctate brachiopods). *Dokl. Akad. Nauk SSSR (Moscow)* 125:181–184.

Schuchert, C. 1893. Classification of the Brachiopoda. *Am. Geol.* 11:141–167.

Schuchert, C., and G. A. Cooper. 1931. Synopsis of the brachiopod genera of the suborders *Orthoidea* and *Pentameroidea,* with notes on the *Telotremata. Am. J. Sci.* (Ser. 5) 22:241–255.

Schuchert, C., and G. A. Cooper. 1932. Brachiopod genera of the suborders *Orthoidea* and *Pentameroidea. Mem. Peabody Mus. Nat. Hist.* 4:1–270.

Sperling, E. A., D. Pisani, and K. J. Peterson. 2011. Molecular paleobiological insights into the origin of the Brachiopoda. *Evol. Dev.* 13:290–303.

Waagen, W. 1884. Salt Range fossils, vol. 1, part 4. *Productus* Limestone fossils, *Brachiopoda. Mem. Geol. Surv. India, Palaeontol. Indica* (Ser. 13) 3–4:547–728.

Williams, A. 1997. Shell structure. Pp. 267–320 in *Treatise on Invertebrate Paleontology, Part H,* Brachiopoda, *Revised,* Vol. 1 (R. L. Kaesler, ed.). Geological Society of America, Boulder, CO; University of Kansas Press, Lawrence, KS.

Williams, A., C. H. C. Brunton, and D. I. MacKinnon. 1997. Morphology. Pp. 321–422 in *Treatise on Invertebrate Paleontology, Part H,* Brachiopoda, *Revised,* Vol. 1 (R. L. Kaesler, ed.). Geological Society of America, Boulder, CO; University of Kansas Press, Lawrence, KS.

Williams, A., C. H. C. Brunton and A. D. Wright. 2000c. *Orthotetida.* Pp. 644 in *Treatise on Invertebrate Paleontology, Part H,* Brachiopoda, *Revised,* Vol. 3. (R. L. Kaesler, ed.). Geological Society of America, Boulder, CO; University of Kansas Press, Lawrence, KS.

Williams, A., and S. J. Carlson. 2000. *Rhynchonellata.* Pp. 708–709 in *Treatise on Invertebrate Paleontology, Part H,* Brachiopoda, *Revised,* Vol. 3 (R. L. Kaesler, ed.). Geological Society of America, Boulder, CO; University of Kansas Press, Lawrence, KS.

Williams, A., and S. J. Carlson. 2007. Affinities of brachiopods and trends in their evolution. Pp. 2822–2877 in *Treatise on Invertebrate Paleontology, Part H,* Brachiopoda, *Revised,* Vol. 6 (P. A. Selden, ed.). Geological Society of America, Boulder, CO; University of Kansas Press, Lawrence, KS.

Williams, A., S. J. Carlson, and C. H. C. Brunton. 2000a. Brachiopod classification. Pp. 1–29 in *Treatise on Invertebrate Paleontology, Part H,* Brachiopoda, *Revised,* Vol. 2 (R. L. Kaesler, ed.). Geological Society of America, Boulder, CO; University of Kansas Press, Lawrence, KS.

Williams, A., S. J. Carlson, and C. H. C. Brunton. 2000b. *Rhynchonelliformea*. P. 193 in *Treatise on Invertebrate Paleontology, Part H,* Brachiopoda, *Revised,* Vol. 2 (R. L. Kaesler, ed.). Geological Society of America, Boulder, CO; University of Kansas Press, Lawrence, KS.

Williams, A., S. J. Carlson, C. H. C. Brunton, L. E. Holmer, and L. Popov. 1996. A supra-ordinal classification of the *Brachiopoda. Philos. Trans. R. Soc. Lond. B Biol. Sci.* 351:1171–1193.

Williams, A., and D. A. T. Harper. 2000a. *Protorthida*. Pp. 709–714 in *Treatise on Invertebrate Paleontology, Part H,* Brachiopoda, *Revised,* Vol. 3 (R. L. Kaesler, ed.). Geological Society of America, Boulder, CO; University of Kansas Press, Lawrence, KS.

Williams, A., and D. A. T. Harper. 2000b. *Orthida*. Pp. 714–846 in *Treatise on Invertebrate Paleontology, Part H,* Brachiopoda, *Revised,* Vol. 3 (R. L. Kaesler, ed.). Geological Society of America, Boulder, CO; University of Kansas Press, Lawrence, KS.

Williams, A., and D. A. T. Harper. 2000c. *Billingsellida*. Pp. 689–708 in *Treatise on Invertebrate Paleontology, Part H,* Brachiopoda, *Revised,* Vol. 3 (R. L. Kaesler, ed.). Geological Society of America, Boulder, CO; University of Kansas Press, Lawrence, KS.

Williams, A., A. J. Rowell and 17 others. 1965. Pp. H1–H927 in *Treatise on Invertebrate Paleontology, Part H,* Brachiopoda, Vol. 1 and 2 (R. C. Moore, ed.). Geological Society of America, Boulder, CO; University of Kansas Press, Lawrence, KS.

Williams, A., and A. J. Rowell. 1965. Evolution and phylogeny. Pp. H164–H199 in *Treatise on Invertebrate Paleontology, Part H,* Brachiopoda, Vol. 1 (R. C. Moore, ed.). Geological Society of America, Boulder, CO; University of Kansas Press, Lawrence, KS.

Wright, A. 1979. Brachiopod radiation. Pp. 235–252 in *The Origin of Major Invertebrate Groups* (M. R. House, ed.). Academic Press, London and New York.

Zhang, Z., L. E. Holmer, Q. Ou, J. Han, and D. Shu. 2011. The exceptionally preserved early Cambrian stem rhynchonelliform brachiopod *Longtancunella* and its implications. *Lethaia* 44:490–495.

Authors

Sandra J. Carlson; Department of Geology; University of California; One Shields Avenue; Davis, CA, 95616-8605, USA. Email: sjcarlson@ucdavis.edu.

Bernard L. Cohen; Faculty of Biomedical and Life Sciences; University of Glasgow; G12 8QQ, UK. Email: b.l.cohen@bio.gla.ac.uk.

Date Accepted: 7 May 2013

Primary Editor: Kevin de Queiroz

Neoarticulata S. J. Carlson and B. L. Cohen, new clade name

Registration Number: 145

Definition: The least inclusive crown clade containing *Notosaria* (originally *Terebratula*) *nigricans* (Sowerby 1846) (*Rhynchonellida*), *Terebratalia* (originally *Terebratula*) *transversa* (Sowerby 1846) (*Terebratellidina*), *Terebratulina* (originally *Anomia*) *retusa* (Linnaeus 1758) (*Terebratulidina*) and *Thecidellina blochmanni* Dall 1920 (*Thecideida*). This is a minimum-crown-clade definition. Abbreviated definition: min crown ∇ (*Notosaria nigricans* (Sowerby 1846) & *Terebratalia transversa* (Sowerby 1846) & *Terebratulina retusa* (Linnaeus 1758) & *Thecidellina blochmanni* Dall 1920).

Etymology: *Neoarticulata* is named in reference to *Articulata* Huxley, 1869, traditionally ranked as a class (Ager et al., 1965) and referring to brachiopods with "valves connected along a hinge-line, often provided with teeth and sockets." The name *Articulata* is derived from the Latin *articulatus* meaning "to divide into distinct parts" and "furnished with joints" in reference to the two valves that are connected by teeth and sockets and can rotate relative to one another about a hinge. The prefix *neo-*, from the Greek *néos* meaning "new," is meant to indicate that the name applies to the crown clade of *Articulata* Huxley 1869.

Reference Phylogeny: Cohen (2007: Fig. 1530). See also Carlson (1995: Fig. 3); Williams et al. (1996: Fig. 6); Carlson and Leighton (2001a: Fig. 1); Carlson and Leighton (2001b: Fig. 26.1); Cohen and Weydmann (2005: Fig. 2); Carlson (2007: Fig. 1908); Sperling et al. (2011: 2Fig. 2C).

Composition: *Terebratulida* Waagen 1883, *Thecideida* Elliott 1958, and the extant (and possibly all extinct) *Rhynchonellida* Kuhn 1949. The extinct groups *Atrypida* Rzhonsnitskaia 1960 and *Athyridida* Boucot et al. 1964 are also likely to be within *Neoarticulata*, and possibly also the extinct groups *Spiriferida* Waagen 1883 and *Spiriferinida* Ivanova 1972. See Comments for further discussion. For a list of extant species, see Emig (2013) under *Rhynchonellata*.

Diagnostic Apomorphies: Anatomical and embryological apomorphies of *Articulata* Huxley 1869 (Carlson, 1995: Table 5) are diagnostic apomorphies of *Neoarticulata*, relative to the other crown clades, because both taxa include the same groups of extant brachiopods. Apomorphies that distinguish *Neoarticulata* from other crown clades within *Brachiopoda* include: presence of mantle lobes that are fused posteriorly and mineralize two valves (of fibrous calcite) possessing interareas, the posterior cardinal margins of which form a hinge between the valves that articulates in a pair of interlocking (cyrtomatodont; see Jaanusson, 1971; Carlson, 1989) ventral teeth and dorsal sockets; an outer mantle lobe indented by a periostracal slot located between vesicular and lobate cells; a larval mantle rudiment that reverses at settlement on the substrate; a tissue-filled epistome lacking a protocoel-like cavity; a curved C-shaped alimentary canal that ends blindly, without an anus; a lophophore supported by calcitic extensions from the dorsal hingeline in the form of crura, spiralia, or loops; relatively simple muscle system composed of adductor, diductor, and adjustor muscles; an anterior shift of the attachment area of the ventral muscle field (indicative of larval mantle reversal), creating a well-defined chamber for the pedicle base, which fills the posterior portion of the ventral valve; the presence of a non-muscular pedicle that develops from a larval rudiment, rather than from the ventral mantle, that resides within the delthyrial (ventral posterior) area, is without a coelomic core (filled with connective tissue), and is controlled by adjustor muscles (see Williams and Rowell, 1965; Carlson, 1995; Williams et al., 1996, 2000e; Williams and Carlson, 2007; Carlson, 2007).

Synonyms: There are no unambiguous synonyms of *Neoarticulata*. *Rhynchonelliformea* Williams et al. 1996, *Rhynchonellata* Williams et al. 1996, and *Articulata* Huxley 1869 include *Neoarticulata* as conceptualized herein, but have always included taxa that are now considered members of its stem group (see Comments under *Pan-Neoarticulata*). Because there has been no name applied specifically to the crown, some of these names have sometimes been used as if they applied to the crown (e.g., Cohen and Gawthrop, 1996). *Telotremata* Beecher 1891 and *Kampylopegmata* Waagen 1882–5 are partial and approximate synonyms of *Neoarticulata*, as each refers to a paraphyletic group that originated in approximately the same ancestor (see Comments).

Comments: There is no pre-existing scientific name that has been applied unambiguously to this crown clade. *Telotremata* Beecher 1891 was named as one of four brachiopod orders characterized by the nature of the pedicle opening and its relationship to the valves; it included taxa here considered members of *Neoarticulata* that were later named *Rhynchonellida* Kuhn 1949 and *Terebratulida* Waagen 1883 (both with extant members), and *Atrypida* Rzhonsnitskaia 1960, *Athyridida* Boucot et al. 1964, *Spiriferida* Waagen 1883, and *Spiriferinida* Ivanova 1972 (all extinct), all of which have a pedicle opening that is shared by the two valves early in ontogeny, but is later restricted to the ventral valve only in adults. *Telotremata* excluded *Thecideida* Elliott 1958 (with extant members) lacking a pedicle as adults, but with a pedicle lobe thought to exist in the larvae (Lacaze-Duthiers, 1861), and now thought to lie within the crown clade (Cohen, 2007). *Telotremata* thus refers to a paraphyletic group originating in (approximately) the same ancestor as *Neoarticulata* (according to the phylogenies of, e.g., Williams et al., 1996 and Carlson, 2007). *Kampylopegmata* Waagen 1882-5 was named as one of three brachiopod suborders characterized by the presence and form of the calcareous brachidia; it included taxa here considered members of *Neoarticulata* that were later named *Rhynchonellida* Kuhn 1949 (with crura), *Terebratulida* Waagen 1883 (with loops), and *Thecideida* Elliott 1958 (with brachial ridges), all of which are represented by extant species, but excluded the extinct taxa *Atrypida* Rzhonsnitskaia 1960, *Athyridida* Boucot et al. 1964, *Spiriferida* Waagen 1883, and *Spiriferinida* Ivanova 1972 (with spiralia), which are now thought likely to lie within the crown clade (Williams et al., 1996; Carlson, 2007). If mineralized lophophore supports arose once (see Fig. 617 in Carlson et al., 2002, and discussion therein) and if curved supports (crura, in *Rhynchonellida*, *Camerelloidea*, and *Pentameridina*) are ancestral relative to spiral supports (spiralia, in *Atrypida*, *Athyridida*, *Spiriferida*, and *Spiriferinida*), then *Kampylopegmata* appears to be a partial synonym in the sense that it refers to a paraphyletic taxon originating in approximately the same ancestor, though it included some taxa now thought to lie outside of the crown clade. Neither *Telotremata* nor *Kampylopegmata* are deemed appropriate as names for the crown clade, as neither name has been used for most of the preceding century, and because each refers to only a portion of the crown clade (see Muir-Wood, 1955, for further discussion).

We chose the name *Neoarticulata* because it makes direct reference to *Articulata* Huxley 1869 and applies to its crown clade. This seems preferable to selecting *Articulata*, which has always included extinct taxa now thought to be outside of the crown. Because a large proportion (approximately 95%) of known brachiopod diversity is extinct, the majority of brachiopod workers are palaeontologists. In this context, and in contrast to the situation with many other widely known names, restricting use of the name *Articulata* to the crown clade only would be more disruptive than would retaining the more inclusive use that is traditional among palaeontologists. Another long-standing difficulty with the name *Articulata* for a clade within *Brachiopoda* is that it is also the name of a taxon within *Crinoidea* (*Echinodermata*): *Articulata* Zittel 1879. When Huxley (1869) named *Articulata*, he included the extant *Terebratulidae* and *Rhynchonellidae*, and also the extinct *Spiriferidae* and *Orthidae*. *Orthidae* is now widely considered to lie outside the crown clade (Williams and Rowell, 1965; Williams et al., 1996, 2000; Carlson, 2007), thus the crown clade of *Articulata* Huxley 1869 has been left unnamed. *Rhynchonellata* Williams et al. 1996 is more restricted in its composition than *Articulata*, but it also includes taxa (*Orthida*, *Protorthida*, and *Pentamerida*) now thought to lie outside the crown clade *Neoarticulata* (see further comments below). *Rhynchonelliformea* Williams et al. 1996 is even more inclusive in its composition than *Articulata* and includes taxa (*Kutorginata*, *Chileida*, *Dictyonellida*, *Obolellida*, and *Naukatida*) thought to lie well outside the crown clade *Neoarticulata*. *Rhynchonelliformea* was named for the articulated group containing extant representatives with the earliest first appearance in the stratigraphic record, *Rhynchonellida*. *Rhynchonellata*, *Articulata*, and *Rhynchonelliformea* were each proposed, and are best retained, for more inclusive clades that contain several extinct taxa whose valve articulation

is of questionable homology with crown clade articulated brachiopods and that almost certainly lie outside the crown clade *Neoarticulata* (see Carlson, 2007: Fig. 1908; Afanasjeva, 2008).

We include four internal specifiers in our minimum-crown-clade definition representing *Rhynchonellida*, *Terebratulida* (*Terebratulidina and Terebratellidina*), and *Thecideida*. Those four taxa are more closely related to each other than any one is to any other crown clade in *Brachiopoda*, although the relationship of *Thecideida* to *Rhynchonellida* (see Carlson, 1995) and *Terebratulida* (see Cohen, 2007) is somewhat uncertain. *Thecideida* is monophyletic (Baker, 1990, 2006; Baker and Carlson, 2010). *Thecideida* is composed of miniaturized brachiopods that have long been considered to be paedomorphic (Elliott, 1948; Pajaud, 1970; Williams, 1973; Baker, 1984, 1990, 1991, 2006; Carlson, 1995; Carlson and Jaecks, 2001; Carlson and Leighton, 2001a, b; Cohen, 2007); possibly for this reason, their sister group relationships are controversial. *Terebratulida* is monophyletic (Carlson, 1995; Cohen, 2007). *Rhynchonellida* appears to be monophyletic when considering only extant taxa (Cohen, 2007), but some extinct groups (*Atrypida* and possibly *Athyridida*) are likely to share common ancestry with different taxa in *Rhynchonellida* thus rendering the group paraphyletic (Carlson et al., 2002). Also, uncertainty regarding phylogenetic relationships among all extinct and extant *Rhynchonellida* render the composition of the crown clade of *Rhynchonellida* uncertain.

The crown clade here named *Neoarticulata* was previously an unnamed clade nested within *Rhynchonelliformea* as proposed by Williams et al. (1996; see also Carlson and Leighton, 2001a,b; Carlson, 2007). *Rhynchonelliformea* was diagnosed by mineralized features that could be preserved in fossils–brachiopods having a calcitic shell with fibrous microstructure. Phylogenetic analyses involving both molecular data (Cohen and Gawthrop, 1996; Cohen and Weydmann, 2005; Cohen, 2007; Sperling et al., 2011) and morphological data (Rowell, 1982; Carlson, 1995; Williams and Carlson, 2007) have begun to clarify relationships (Carlson, 2007) within *Neoarticulata*, although numerous uncertainties remain, as discussed below.

Uncertainty about the homology and polarity of a spiral brachidium and shell endopunctation in extinct and extant forms complicates interpretation of relationships within (and thus the composition of) *Neoarticulata* (Williams and Carlson, 2007; Carlson, 2007). The possession of a calcified lophophore support in the shape of two spiral coils characterizes *Atrypida*, *Athyridida*, *Spiriferida*, and *Spiriferinida*; all are extinct, and none is definitively monophyletic (Carlson and Leighton, 2001a). It is not clear if a spiralium is a primitive feature that evolved once from crura (Copper, in Copper and Gourvennec, 1996), and from which loops (*Terebratulida*) and possibly brachial ridges (*Thecideida*) later evolved paedomorphically, or if spiralia evolved two or more times independently (Gourvennec, in Copper and Gourvennec, 1996). *Atrypida* and *Athyridida* almost certainly cluster within the crown clade *Neoarticulata* (see Carlson and Leighton, 2001a; Carlson, 2007). Gourvennec has suggested that *Atrypida* and *Athyridida* evolved from *Rhynchonellida*, and *Spiriferida* and *Spiriferinida* from *Orthida*, which is not in crown clade *Neoarticulata* (see Carlson, 2007: Fig. 1908). If spiralia evolved once, then *Spiriferida* and *Spiriferinida* may also cluster within *Neoarticulata*; if spiralia evolved more than once, it is not clear that either group would cluster within *Neoarticulata*. *Spiriferida* and *Spiriferinida* may be sister groups; it is also possible that *Spiriferinida* nests within *Spiriferida*.

It is not clear whether the presence or absence of endopunctation is primitive within *Neoarticulata* and if the derived state evolved once (Cooper, 1944) or more than

once (see Muir-Wood, 1955, and discussion in Williams and Carlson, 2007). Given the distribution of an impunctate shell structure among outgroups to *Neoarticulata* within *Rhynchonelliformea* (see Carlson, 2007: Fig. 1908), and in some extant *Rhynchonellida* (see Cohen, 2007: Fig. 1530), the lack of any form of endopunctation is likely to be the primitive condition, with endopunctae evolving independently more than once (up to four or five times; Williams and Carlson, 2007; Carlson, 2007). Because the relationship of *Thecideida* to other neoarticulates is controversial, uncertainty surrounds the identity of extinct groups that might nest within *Neoarticulata* (Carlson, 2007: Fig. 1908). *Thecideida* may be the sister group to extinct *Spiriferida* (with impunctate shells; Baker, 1990), extinct *Spiriferinida* (with endopunctate shells; Williams, 1973), extant *Terebratulida* (Copper in Copper and Gourvennec, 1996; Cohen, 2007), or the clade of *Terebratulida* and *Rhynchonellida* together (Carlson, 1995).

Literature Cited

Afanasjeva, G. A. 2008. Supraordinal brachiopod classification. *Paleontol. J.* 42:792–802.

Ager, D. V., T. W. Amsden, G. Biernat, A. J. Boucot, G. F. Elliott, R. E. Grant, K. Hatai, J. G. Johnson, D. J. McLaren, H. M. Muir-Wood, C. W. Pitrat, A. J. Rowell, H. Schmidt, R. D. Staton, F. G. Stehli, A. Williams, and A. D. Wright. 1965. *Articulata*. Pp. H297–H299 in *Treatise on Invertebrate Paleontology, Part H*, Brachiopoda, Vol. 1 (R. L. Kaesler, ed.). Geological Society of America, Boulder, CO; University of Kansas Press, Lawrence, KS.

Baker, P. G. 1984. New evidence of a spiriferide ancestor for the *Thecideidina* (*Brachiopoda*). *Palaeontology* 27:857–866.

Baker, P. G. 1990. The classification, origin and phylogeny of thecideidine brachiopods. *Palaeontology* 33:175–191.

Baker, P. G. 1991. Morphology and shell microstructure of Cretaceous thecideidine brachiopods and their bearing on thecideidine phylogeny. *Palaeontology* 34:815–836.

Baker, P. G. 2006. *Thecideida*. Pp. 1938–1964 in *Treatise on Invertebrate Paleontology, Part H*, Brachiopoda, *Revised*, Vol. 5 (R. L. Kaesler, ed.). Geological Society of America, Boulder, CO; University of Kansas Press, Lawrence, KS.

Baker, P. G., and S. J. Carlson. 2010. The early ontogeny of Jurassic thecideoid brachiopods and its contribution to the understanding of thecideoid ancestry. *Palaeontology* 53:645–667.

Beecher, C. E. 1891. Development of the *Brachiopoda*. Part I. Introduction. *Am. J Sci.* 41:343–357.

Boucot, A. J., J. G. Johnson, and R. D. Staton. 1964. On some atrypoid, retzioid, and athyridoid *Brachiopoda*. *J. Paleontol.* 38:805–822.

Carlson, S. J. 1989. The articulate brachiopod hinge mechanism: morphological and functional variation. *Paleobiology* 15(4):364–386.

Carlson, S. J. 1995. Phylogenetic relationships among extant brachiopods. *Cladistics* 11:131–197.

Carlson, S. J. 2007. Recent research on brachiopod evolution. Pp. 2878–2900 in *Treatise on Invertebrate Paleontology, Part H*, Brachiopoda, *Revised*, Vol. 6 (P. A. Selden, ed.). Geological Society of America, Boulder, CO; University of Kansas Press, Lawrence, KS.

Carlson, S. J., A. J. Boucot, R. Jia-Yu, and R. B. Blodgett. 2002. Introduction: order *Pentamerida*. Pp. 921–928 in *Treatise on Invertebrate Paleontology, Part H*, Brachiopoda, *Revised*, Vol. 4 (R. L. Kaesler, ed.). Geological Society of America, Boulder, CO; University of Kansas Press, Lawrence, KS.

Carlson, S. J., and G. S. Jaecks. 2001. How phylogenetic inference can shape our view of heterochrony: examples from thecideide brachopods. *Paleobiology* 27:205–225.

Carlson, S. J., and L. R. Leighton. 2001a. The phylogeny and classification of *Rhynchonelliformea*. Pp. 27–51 in *Brachiopods Ancient and Modern* (S. J. Carlson and M. R. Sandy, eds.). The Paleontological Society, New Haven, CT.

Carlson, S. J., and L. R. Leighton. 2001b. Incorporating stratigraphic data in the phylogenetic analysis of the *Rhynchonelliformea*. Pp. 248–258 in *Brachiopods Past and Present* (C. H. C. Brunton, L.R. M. Cocks, and S. L. Long, eds.). Taylor & Francis, London and New York.

Cohen, B. L. 2007. The brachiopod genome. Pp. 2356–2372 in *Treatise on Invertebrate Paleontology, Part H*, Brachiopoda, *Revised*, Vol. 6 (P. A. Selden, ed.). Geological Society of America, Boulder, CO; University of Kansas Press, Lawrence, KS..

Cohen, B. L., and A. B. Gawthrop. 1996. Brachiopod molecular phylogeny. Pp. 73–80 in *Brachiopods: Proceedings of the Third International Brachiopod Congress, Sudbury, Ontario, 1995* (P. Copper and J. Jin, eds.). A. A. Balkema, Rotterdam.

Cohen, B. L., and A. Weydmann. 2005. Molecular evidence that phoronids are a subtaxon of brachiopods (*Brachiopoda: Phoronata*) and that genetic divergence of metazoan phyla began long before the early Cambrian. *Org. Divers. Evol.* 5: 253–273.

Cooper, G. A. 1944. Phylum *Brachiopoda*. Pp. 277–365 in *Index Fossils of North America*. (H. W. Shimer and R. R. Shrock, eds.). Wiley & Sons, New York.

Copper, P., and R. Gourvennec. 1996. Evolution of the spire-bearing brachiopods (Ordovician – Jurassic). Pp. 81–88 in *Brachiopods: Proceedings of the Third International Brachiopod Congress, Sudbury, Ontario, 1995* (P. Copper and J. Jin, eds.). A. A. Balkema, Rotterdam.

Dall, W. H. 1920. Annotated list of the recent *Brachiopoda* in the collection of the United States National Museum, with descriptions of thirty-three new forms. *Proc. U.S. Natl. Mus.* 57:261–377.

Elliott, G. F. 1948. Palingenesis in *Thecidea* (*Brachiopoda*). *Ann. Mag. Natl. Hist. (Ser. 12)* 1:1–30.

Elliott, G. F. 1958. Classification of thecidean brachiopods. *J. Paleontol.* 32:373.

Emig, C. C. 2013. *Rhynchonellata*. In World *Brachiopoda* Database (C.C. Emig, F. Alvarez, and M.A. Bitner, eds.). Accessed through:

World Register of Marine Species at http://www.marinespecies.org/aphia.php?p=taxdetails&id=104020, accessed on 18 May 2013.

Huxley, T. H. 1869. *An Introduction to the Classification of Animals.* John Churchill and Sons, London.

Ivanova, E. A. 1972. Osnovnyye zakonomernosti evolyutsii spirifer (*Brachiopoda*). [Main features of spiriferid evolution (*Brachiopoda*)]. *Paleontol. Zh.* 1972:28–42 [in Russian].

Jaanusson, V. 1971. Evolution of the brachiopod hinge. *Smithson. Contrib. Paleobiol.* 3:33–46.

Kuhn, O. 1949. *Lehrbuch der Palaozoologie.* E. Schweizerbart'sche Verlagsbuchhandlung, Stuttgart.

de Lacaze-Duthiers, F. J. H. 1861. Histoire naturelle des brachiopods vivants de la Mediterranee. I. Histoire naturelle de la Thecidie (*Thecidium mediterraneum*). *Ann. Sci. Nat. Zool. (Ser. 4)* 15:259–330.

Linnaeus [Linne], C. 1758. *Systema Naturae, sive Regna tria Naturae Systematicae Proposita per Classes, Ordines, Genera et Species.* 10th edition, Vol. 1. Laurentius Salvius, Holmiae (Stockholm).

Muir-Wood, H. M. 1955. *A History of the Classification of the Phylum* Brachiopoda. British Museum (Natural History), London.

Pajaud, D. 1970. Monographie des Thecidees (*Brachiopodes*). *Mem. Soc. Geol. France, Nouvelle Serie* 49:1–349.

Rowell, A. J. 1982. The monophyletic origin of the *Brachiopoda. Lethaia* 15:299–307.

Rzhonsnitskaia, M. A. 1960. Order *Atrypida.* Pp. 257–264 in *Osnovy Paleontologii, Mshanki, Brakhiopody (Bryozoa, Brachiopoda)*, Vol. 7 (Y. A. Orlov, ed.). Akademiia Nauk SSSR, Moscow.

Sowerby, G. B. 1846. Descriptions of thirteen new species of brachiopods. *Proc. Zool. Soc. Lond.* 14:91–95.

Sperling, E. A., D. Pisani, and K. J. Peterson. 2011. Molecular paleobiological insights into the origin of the *Brachiopoda. Evol. Dev.* 13:290–303.

Waagen, W. H. 1882–1885. Salt Range fossils. *Productus* limestone fossils, *Brachiopoda. Mem. Geol. Surv. India, Palaeontol. Indica, Ser. 13* 1:329–770.

Williams, A. 1973. The secretion and structural evolution of the shell of thecideidine brachiopods. *Philos. Trans. R. Soc. Lond. B Biol. Sci.* 264:439–478.

Williams, A., and S. J. Carlson. 2007. Affinities of brachiopods and trends in their evolution. Pp. 2822–2877 in *Treatise on Invertebrate Paleontology, Part H,* Brachiopoda, *Revised,* Vol. 6 (P. A. Selden, ed.). Geological Society of America, Boulder, CO; University of Kansas Press, Lawrence, KS.

Williams, A., S. J. Carlson, and C. H. C. Brunton. 2000. Brachiopod classification. Pp. 1–27 in *Treatise on Invertebrate Paleontology, Part H,* Brachiopoda, *Revised,* Vol. 2 (R. L. Kaesler, ed.). Geological Society of America, Boulder, CO; University of Kansas Press, Lawrence, KS.

Williams, A., S. J. Carlson, C. H. C. Brunton, L. E. Holmer, and L. Popov. 1996. A supra-ordinal classification of the *Brachiopoda. Philos. Trans. R. Soc. Lond. B Biol. Sci.* 351:1171–1193.

Williams, A., and A. J. Rowell. 1965. Evolution and phylogeny. Pages H164–H199 in *Treatise on Invertebrate Paleontology, Part H,* Brachiopoda, Vol. 1. (R. C. Moore, ed.). Geological Society of America, Boulder, CO; University of Kansas Press, Lawrence, KS.

von Zittel, K. A. 1879. *Echinodermata.* Pp. 308–560 in *Handbuch der Palaontologie: Palaozoologie,* Vol. 1, Part 1. R. Oldenbourg, Munchen & Leipzig.

Authors

Sandra J. Carlson; Department of Geology; University of California; One Shields Avenue; Davis, CA 95616-8605, USA. Email: sjcarlson@ucdavis.edu.

Bernard L. Cohen; Faculty of Biomedical and Life Sciences; University of Glasgow; G12 8QQ, UK. Email: b.l.cohen@bio.gla.ac.uk.

Date Accepted: 21 May 2013

Primary Editor: Kevin de Queiroz

Cephalopoda J. Macartney 1802 [F. Anderson], converted clade name

Registration Number: 138

Definition: The crown clade originating in the most recent common ancestor of *Nautilus pompilius* Linnaeus 1758 and *Sepia officinalis* Linnaeus 1758 (*Neocoleoidea*). This is a minimum-crown-clade definition. Abbreviated definition: min crown ∇ (*Nautilus pompilius* Linnaeus 1758 & *Sepia officinalis* Linnaeus 1758).

Etymology: Derived from the Greek *kephalos* (head) plus *podos* (foot).

Reference Phylogeny: Figure 6 in Lindgren et al. (2004).

Composition: *Cephalopoda* is composed of two primary crown clades—*Nautilus* and *Neocoleoidea*, as defined in this volume—and comprises about 800 described extant species (Sweeney, 2001; Sweeney and Roper, 1998). In addition, thousands of extinct cephalopod forms, most notably ammonites and belemnites, have been described (Teichert, 1988; Nishiguchi and Mapes, 2008).

Diagnostic Apomorphies: Several apomorphies that distinguish cephalopods from all other mollusks were listed by Berthold and Engeser (1987) and Engeser and Bandel (1988). These include:

1. A calcified shell comprising a body chamber and a septate, gas-filled region (phragmocone) with a single siphuncle (a strand of tissue) passing through each septum. This structure is present in many extinct fossil cephalopods, but among extant cephalopods it is seen only in nautiloids, cuttles and *Spirula*. Septation of the shell is not unique to cephalopods; some extant gastropods and numerous fossil forms (believed to be monoplacophorans) have septate regions of the shell (Clarkson, 1986). Penetration of the septa by a siphuncle, however, is only known in cephalopods.
2. Muscular arms around the oral opening.
3. A funnel (or siphon). The funnel is a muscular tube on the ventral side of the body used for locomotion; this structure is derived from the foot (Bandel and Boletzky, 1988; Boletzky, 2003; Shigeno et al., 2008) and thus is not homologous with the siphons of bivalve mollusks and many gastropods, which are derived from the mantle (Yonge, 1982; Geiger, 2006).
4. A pair of beak-like jaws composed of chitin-protein complexes.
5. Nidamental glands (in females; used to produce a protective egg sheath).
6. Pericardial glands (branchial heart appendages that play a role in ultrafiltration of the blood and secretion).
7. A unique spermatophore storage organ (Needham's sac) in the distal region of the gonoduct in males.

These features are known from extant representatives of *Nautilus* and *Neocoleoidea*, and a few (including circumoral arms and a beak) are known from some extinct groups (e.g., *Ammonoidea* and *Belemnoidea*). These features therefore constitute diagnostic apomorphies of *Cephalopoda* as defined here relative to other extant mollusks. Some (if not all) of these features are likely to have occurred in stem cephalopods, but the soft-part anatomy of many early cephalopod fossils is unknown (Teichert, 1988).

Synonyms: *Siphonopoda* E. R. Lankester 1877 not G. O. Sars 1878 (approximate).

Comments: The term *Céphalopodes*—and the modern concept of the group—first appeared in Cuvier (1795). As noted by Donovan (1996) and Nixon and Young (2003), the formal name *Cephalopoda* is a Latinized translation of *Céphalopodes* that appeared in an English translation (Macartney, 1802) of Cuvier's *Leçons d'Anatomie Comparée* (Cuvier, 1800). Strong support for *Cephalopoda* monophyly has been seen in several phylogenetic analyses, including those of Salvini-Plawen and Steiner (1996; morphological data, with *Archaeogastropoda* as an outgroup), Carlini and Graves (1999: Fig. 2; partial mitochondrial cytochrome oxidase I data, one member of *Polyplacophora* as an outgroup), Passamaneck et al. (2004: Fig. 2; combined small- and large-subunit ribosomal RNA sequences, several molluscan outgroups), Giribet et al. (2006) and Lindgren et al. (2004; which includes several figures depicting analyses of morphological as well as separate and combined molecular data sets, with eighteen outgroup taxa representing most other major molluscan groups). Relatively few phylogenies include both extant and extinct taxa, and most of these use the name *Cephalopoda* for a more inclusive clade that includes at least some members of the stem group (e.g., Engeser and Bandel, 1988: Fig. 1).

There are some reasonable alternative phylogenetic definitions for *Cephalopoda*. A definition based on presence of a siphunculate phragmocone would include several early Palaeozoic fossils that may have arisen prior to the divergence between *Nautilus* and *Neocoleoidea* (Berthold and Engeser, 1987; Engeser, 1996; Flower, 1988). *Cephalopoda* has been defined by the presence of a siphunculate phragmocone by some previous authors (e.g., Clarkson, 1986), and the use of an apomorphy-based definition would minimize nomenclatural disruption for palaeontologists. However, there are obvious disadvantages of a definition for *Cephalopoda* based on the siphunculate phragmocone. First, the definitive apomorphy is not present in most living cephalopods, because the shell has been reduced or lost. Second, very little is known about the soft-part anatomy and general biology of stem-group taxa, minimizing the biological inferences that could be made for many extinct members of a clade that is defined by possession of a siphunculate phragmocone.

By contrast, a crown-clade definition means that *Cephalopoda* will be applied to a group about which the most biological information can be known. For instance, all but one of the diagnostic apomorphies listed by Berthold and Engeser (1987) and Engeser and Bandel (1988)—the features that most biologists and laypeople associate with cephalopods, such as muscular appendages around the mouth—are known only from members of the crown group. Furthermore, many recent papers on cephalopod phylogeny include only crown-group cephalopods. Finally, a minimum-crown-clade definition for *Cephalopoda* using *Nautilus pompilius* and *Sepia officinalis* as internal specifiers designates a clade that includes the two main extinct cephalopod subclades—*Ammonoidea* and *Belemnoidea*—under recent phylogenetic hypotheses (Berthold and Engeser, 1987; Engeser, 1990; Engeser, 1996; Flower, 1988; Haas, 2002). Stem-group taxa could be accommodated with a new name, *Pan-Cephalopoda*.

The name *Siphonopoda* has also been used in reference to this clade (Salvini-Plawen, 1980) but is rejected here because (1) it is a senior homonym of a taxon name within another molluscan clade (*Scaphopoda*) (Steiner and Kabat, 2001), which could lead to confusion, (2) *Cephalopoda* is by far the most widely used name for this group, and (3) Boletzky (2006) has argued that the name *Siphonopoda* is based on an incorrect interpretation of the development of the brachial arm crown and siphon.

Literature Cited

Bandel, K., and S. v. Boletzky. 1988. Features of development and functional morphology required in the reconstruction of early coleoid cephalopods. Pp. 229–246 in *Cephalopods Past and Present* (J. Wiedmann and J. Kullmann, eds.). E. Schweizerbart'sche Verlagsbuchhandlung, Stuttgart.

Berthold, T., and T. Engeser. 1987. Phylogenetic analysis and systematization of the *Cephalopoda* (*Mollusca*). *Verh. Naturwiss. Ver. Hamburg* 29:187–220.

Boletzky, S.v. 2003. Biology of early life stages in cephalopod molluscs. *Adv. Mar. Biol.* 44:143–203.

Boletzky, S. v. 2006. From head to foot—and back again: brachial crown development in the *Coleoidea* (*Mollusca, Cephalopoda*). Pp. 33–42 in *Proceedings of the Second International Symposium "Coleoid Cephalopods Through Time"* (M. Kostak, and J. Marek, eds.). Acta Universitatis Carolinae, Prague.

Carlini, D. B., and J. E. Graves. 1999. Phylogenetic analysis of cytochrome *c* oxidase I sequences to determine higher-level relationships within the coleoid cephalopods. *Bull. Mar. Sci.* 64:57–76.

Clarkson, E. N. K. 1986. *Invertebrate Paleontology and Evolution*. Allen & Unwin, Boston, MA.

Cuvier, G. L. C. F. D. 1795. Second mémoire sur l'organisation et les rapports des animaux á sang blanc, dans lequel on traite de la structure des Mollusques et de leur division en ordre. *Magazin Encycl. (Anné 1)* 2:433–449.

Cuvier, G. L. C. F. D. 1800. *Leçons d'Anatomie Comparée de G. Cuvier. Recueillies et Publiées sous ses yeux, par C. Duméril. 1. Contenant les Organes du Mouvement*. Baudouin, Paris.

Donovan, D. T. 1996. Origins of the terms cephalopod, *Cephalopoda* and *Gastropoda*, and early subdivisions of the *Mollusca Bull. Zool. Nomencl.* 53:247–252.

Engeser, T. 1990. Phylogeny of the fossil coleoid *Cephalopoda* (*Mollusca*). *Berliner Geowiss. Abh. A* 124:123–191.

Engeser, T. 1996. The position of the *Ammonoidea* within the *Cephalopoda*. Pp. 3–19 in *Ammonoid Paleobiology* (N. H. Landman, K. Tanabe, and R. A. Davis, eds.). Plenum Press, New York.

Engeser, T., and K. Bandel. 1988. Phylogenetic classification of coleoid cephalopods. Pp. 105–116 in *Cephalopods Past and Present* (J. Wiedmann, and J. Kullmann, eds.). E. Schweizerbart'sche Verlagsbuchhandlung, Stuttgart.

Flower, R. H. 1988. Progress and changing concepts in cephalopod and particularly nautiloid phylogeny and distribution. Pp. 17–24 in *Cephalopods Past and Present* (J. Wiedmann, and J. Kullmann, eds.). E. Schweizerbart'sche Verlagsbuchhandlung, Stuttgart.

Geiger, D. L. 2006. Marine *Gastropoda*. Pp. 295–312 in *The Mollusks: A Guide to their Study, Collection and Preservation* (C. F. Sturm, T. A. Pearce and A. Valdés, eds.). Universal Publishers, Inc., Boca Raton, FL.

Giribet, G., A. Okusu, A. R. Lindgren, S. W. Huff, M. Schrodl, and M. K. Nishiguchi. 2006. Evidence for a clade composed of molluscs with serially repeated structures: monoplacophorans are related to chitons. *Proc. Natl. Acad. Sci. USA* 103:7723–8.

Haas, W. 2002. The evolutionary history of the eight-armed *Coleoidea*. *Abh. Geol. Bundesanstalt Wien* 57:341–351.

Lankester, E. R. 1877. Notes on the embryology and classification of the animal kingdom: comprising a revision of speculations relative to the origin and significance of the germ layers. *Q. J. Microsc. Sci.* (new series) 17:399–454.

Lindgren, A. R., G. Giribet, and M. K. Nishiguchi. 2004. A combined approach to the phylogeny of *Cephalopoda* (*Mollusca*). *Cladistics.* 20:454–486.

Linnaeus, C. v. 1758. *Systema Naturae Per Regna Tria Naturae, Secundum Classes, Ordines, Genera, Species, cum Characteribus, Differentiis, Synonymis, Locis*. 10th edition. Laurentii Salvii, Holmiae (Stockholm).

Macartney, J. 1802. *Lectures on Comparative Anatomy, Translated from The French of G. Cuvier. 1. On the Organs of Motion*. Oriental Press, London.

Nishiguchi, M. K., and Mapes, R. H. 2008. *Cephalopoda*. Pp. 163–199 in *Phylogeny and Evolution of the* Mollusca (W. F. Ponder and D. R. Lindberg, eds.). University of California Press, Berkeley, CA.

Nixon, M., and J. Z. Young. 2003. *The Brains and Lives of Cephalopods*. Oxford University Press, New York.

Passamaneck, Y. J., C. Schander, and K. M. Halanych. 2004. Investigation of molluscan phylogeny using large-subunit and small-subunit nuclear rRNA sequences. *Mol. Phylogenet. Evol.* 32:25–38.

Salvini-Plawen, L. v. 1980. A reconsideration of systematics in *Mollusca* (phylogeny and higher classification). *Malacologia* 19:247–278.

Salvini-Plawen, L. v., and G. Steiner. 1996. Synapomorphies and plesiomorphies in higher classification of *Mollusca*. Pp. 29–52 in *Origin and Evolutionary Radiation of the* Mollusca (J. Taylor, ed.). Oxford University Press, Oxford.

Sars, G. O. 1878. *Bidrag til Kundskaben om Norges Arktiske Fauna. I:* Mollusca *regionis Arcticae Norvegiae. Oversigt over de i Norges arktiske region forkommende Bløddyr*. A. W. Brøgger, Christiania.

Shigeno, S., T. Sasaki, T. Moritaki, T. Kasugai, M. Vecchione, and K. Agata. 2008. Evolution of the cephalopod head complex by assembly of multiple molluscan body parts: evidence from *Nautilus* embryonic development. *J. Morphol.* 269:1–17.

Steiner, G., and A. R. Kabat. 2001. Catalogue of supraspecific taxa of *Scaphopoda* (*Mollusca*). *Zoosystema* 23:433–460.

Sweeney, M. J. 2001. Current Classification of Recent *Cephalopoda*. Available at http://www .mnh.si.edu/cephs/newclass.pdf.

Sweeney, M. J., and C. F. E. Roper. 1998. Classification, type localities and type repositories of recent *Cephalopoda*. Pp. 561–599 in *Systematics and Biogeography of Cephalopods*, Vol. II (N. A. Voss, M. Vecchione, R. B. Toll, and M. J. Sweeney, eds.). Smithsonian Institution Press, Washington, DC.

Teichert, C. 1988. Main features of cephalopod evolution. Pp. 11–79 in *The* Mollusca*: Paleontology and Neontology of Cephalopods* (M. R. Clarke, and E. R. Trueman, eds.). Academic Press, New York.

Yonge, C. M. 1982. Mantle margins with a revision of siphonal types in the *Bivalvia*. *J. Mollus. Stud.* 48:102–103.

Author

Frank E. Anderson; Department of Zoology; Southern Illinois University; Carbondale, IL 62901, USA. Email: feander@siu.edu.

Date Accepted: 05 October 2011; updated 29 May 2012

Primary Editor: Kevin de Queiroz

Nautilus C. Linnaeus 1758 [F. Anderson], converted clade name

Registration Number: 140

Definition: The crown clade originating in the most recent common ancestor of *Nautilus pompilius* Linnaeus 1758, *Nautilus perforatus* Conrad 1847, and *Nautilus scrobiculatus* Lightfoot 1786. This is a minimum-crown-clade definition. Abbreviated definition: min crown ∇ (*Nautilus pompilius* Linnaeus 1758 & *Nautilus perforatus* Conrad 1847 & *Nautilus scrobiculatus* Lightfoot 1786).

Etymology: From the Greek *nautes* (sailor).

Reference Phylogeny: Figure 2 in Ward and Saunders (1997) and Figures 1–3 in Harvey et al. (1999) are consensus trees depicting relationships within *Nautilus* based on parsimony analyses of morphological data, and all include representatives of extinct taxa. Three molecular data sets have also been used to address relationships within extant members of *Nautilus* (Bonacum et al., 2011; Woodruff et al., 1987; Wray et al., 1995), but these studies did not include extinct taxa. For the purposes of this definition, Figure 2 in Ward and Saunders (1997) can be considered the primary reference phylogeny.

Composition: The most recent common ancestor of the extant nautiluses and all of its descendants. According to the primary reference phylogeny, this clade includes *Allonautilus* (two putative species traditionally referred to *Nautilus*, see Comments) and the extinct taxa *Aturia* and *Hercoglossa*. However, according to the reanalysis of Harvey et al. (1999: Fig. 3), *Aturia* and *Hercoglossa* are excluded. There is considerable disagreement concerning the number of extant species of *Nautilus* (see Comments).

Diagnostic Apomorphies: Ward and Saunders (1997) did not list any apomorphies for the clade originating in the most recent common ancestor of *Nautilus scrobiculatus*, *Nautilus perforatus* and *Nautilus pompilius*. Examination of their data matrix shows that two shell characters—absence of a simple external lobe and absence of an umbilical lobe—may be apomorphies of *Nautilus* as defined here, as these characters have one state in *Cenoceras* (the outgroup) and another state in *Nautilus* (including *Allonautilus*, *Hercoglossa*, and *Aturia*). Harvey et al. (1999) listed a different apomorphy for a clade originating in the most recent common ancestor of *Nautilus scrobiculatus*, *Nautilus perforatus* and *Nautilus pompilius*—seven septa at hatching—when four characters were recoded to treat similarities between *Nautilus scrobiculatus* and *Cenoceras* as convergences, as suggested by Ward and Saunders. However, their version of this clade (Harvey et al., 1999: Fig. 3) has a different composition than that seen in the reference phylogeny (*Hercoglossa* and *Aturia* are excluded). This conflict highlights the uncertainty regarding the composition of *Nautilus* as defined here with respect to certain fossil taxa. Several major features distinguish *Nautilus* from all other extant cephalopods (Neocoleoidea as defined in this volume). For example, members of *Nautilus* possess (1) an external shell (neocoleoids have an internal calcified shell, a shell remnant, or lack a shell entirely), (2) several dozen arms (neocoleoids have eight arms, eight arms and two dorsolateral filaments or eight arms and two ventrolateral tentacles), (3) pinhole-type eyes that lack lenses (neocoleoids have camera-type eyes with lenses) and (4) two pairs of gills (neocoleoids have a single pair of gills). Retention of an external shell in *Nautilus* is certainly a plesiomorphy (Teichert, 1988), and many of the remaining features may also be plesiomorphies whose alternative states are coleoid or neocoleoid apomorphies.

Synonyms: *Nautarius* Duméril 1806 (approximate).

Comments: Teichert and Matsumoto (1987) provide an overview of the possible ancestry of extant nautiluses. Though most nautiloid fossils were initially placed in *Nautilus*, over time the use of the name was restricted to Mesozoic and Tertiary forms (Kummel, 1956). Spath (1927) stated that *Nautilus* includes some Eocene forms, but Miller (1951) ultimately restricted *Nautilus* to include only extant species, stating that he knew of no Tertiary forms that belong to the genus. This view was supported by Kummel (1956), but Figures 2 and 3 in Harvey et al. (1999) support the inclusion of two Tertiary fossil taxa (*Kummelonautilus cookanus* and *N. praepompilius*) in *Nautilus*.

Both molecular (Wray et al., 1995) and morphological (Ward and Saunders, 1997; Harvey et al., 1999) phylogenies suggest that a clade comprising the most recent common ancestor of *N. scrobiculatus*, *N. perforatus* and *N. pompilius* and all of its descendants includes all known extant nautiluses (plus, perhaps, several extinct species). Ward and Saunders (1997) argued that two extant species of *Nautilus*—*N. scrobiculatus* Lightfoot 1786 and *N. perforatus* Conrad 1847—should be placed in a separate genus, *Allonautilus* Ward and Saunders 1997, based on differences in shell, gill and male reproductive system anatomy as well as the results of a phylogenetic analysis of morphological data. However, Harvey et al. (1999) argued against this interpretation, noting that neither the original analysis of Ward and Saunders (1997), nor any of the reanalyses performed by

Harvey et al., support reciprocal monophyly of *Allonautilus* and *Nautilus*.

All of the phylogenies in Ward and Saunders (1997) and Harvey et al. (1999) are poorly resolved consensus trees that leave many questions—including the position of the root—unanswered. Furthermore, the number of extant *Nautilus* species remains unclear. A total of eleven species and seven variants have been described (Saunders, 1987), but there is disagreement regarding how many species are valid. Different authors have recognized six (Teichert and Matsumoto, 1987), four (Saunders, 1987) or even fewer (Miller, 1951; Kummel, 1956) valid *Nautilus* species. Levels of mitochondrial 16S sequence divergence among representatives of several extant populations representing five conchological forms are quite low (less than 5%) (Wray et al., 1995). However, *N. scrobiculatus* appears to be highly divergent from other sampled nautiluses based on mitochondrial 16S (Wray et al., 1995), nuclear 18S (Bonnaud et al., 2004) and allozyme (Woodruff et al., 1987) data (though the significance of this magnitude of divergence is difficult to interpret, given that neither study included outgroups). Wray et al. (1995) suggested that there may be only two valid species: *Nautilus pompilius* and *Nautilus scrobiculatus*. Harvey et al. (1999) also noted this possibility, stating that "... the pertinent question is not whether there is more than one genus represented by these species, but rather whether there are more than two species represented by these specimens" (p. 1216).

In light of this uncertainty, a phylogenetic definition of *Nautilus* that closely matches traditional usage (i.e., prior to the proposal of *Allonautilus*) has been adopted here. However, it must be noted that this use of the name *Nautilus* does not preclude the use of the name *Allonautilus* for the *Nautilus* subclade containing *N. scrobiculatus* and *N. perforatus,* if further research supports its monophyly.

Literature Cited

Bonacum, J., N. H. Landman, R. H. Mapes, M. M. White, A. J. White, and J. Irlam. 2011. Evolutionary radiation of present-day *Nautilus* and *Allonautilus*. *Am. Malacol. Bull.* 29:77–93.

Bonnaud, L., C. Ozouf-Costaz, and R. Boucher-Rodoni. 2004. A molecular and karyological approach to the taxonomy of *Nautilus*. *C. R. Biol.* 327:133–138.

Duméril, A. M. C. 1806. *Zoologie Analytique, ou Methode Naturelle de Classification des Animaux, Rendue plus Facile a l'aide de Tableaux Synoptiques*. Allais, Paris.

Harvey, A. W., R. Mooi, and T. M. Gosliner. 1999. Phylogenetic taxonomy and the status of *Allonautilus* Ward and Saunders, 1997. *J. Paleontol.* 73:1214–1217.

Kummel, B. 1956. Post–Triassic nautiloid genera. *Bull. Mus. Comp. Zool.* 114:319–494.

Linnaeus, C. 1758. *Systema Naturae Per Regna Tria Naturae, Secundum Classes, Ordines, Genera, Species, cum Characteribus, Differentiis, Synonymis, Locis*. 10th edition. Laurentii Salvii, Holmiae (Stockholm).

Miller, A. K. 1951. Tertiary nautiloids of west-coastal Africa. *Ann. Mus. Roy. Congo Belge, Sér. 8°, Sci. Géol.* 8:1–88.

Saunders, W. B. 1987. The species of *Nautilus*. Pp. 35–52 in Nautilus: *The Biology and Paleobiology of a Living Fossil* (W. B. Saunders, and N. H. Landman, eds.). Plenum Press, New York.

Spath, L. F. 1927. On the classification of the Tertiary nautili. *Ann. Mag. Nat. Hist.* 20:424–428.

Teichert, C. 1988. Main features of cephalopod evolution. Pp. 11–79 in *The* Mollusca: *Paleontology and Neontology of Cephalopods* (M. R. Clarke, and E. R. Trueman, eds.). Academic Press, New York.

Teichert, C., and T. Matsumoto. 1987. The ancestry of the genus *Nautilus*. Pp. 25–32 in Nautilus: *The Biology and Paleobiology of a Living Fossil* (W. B. Saunders, and N. H. Landman, eds.). Plenum Press, New York.

Ward, P. D., and W. B. Saunders. 1997. *Allonautilus*: A new genus of living nautiloid cephalopod and its bearing on phylogeny of the *Nautilida*. *J. Paleontol.* 71:1054–1064.

Woodruff, D. S., M. P. Carpenter, W. B. Saunders, and P. D. Ward. 1987. Genetic variation and phylogeny in *Nautilus*. Pp. 65–83 in Nautilus: *The Biology and Paleobiology of a Living Fossil* (W. B. Saunders, and N. H. Landman, eds.). Plenum Press, New York.

Wray, C. G., N. H. Landman, and J. Bonacum. 1995. Genetic divergence and geographic diversification in *Nautilus*. *Paleobiology* 21:220–228.

Author

Frank E. Anderson; Department of Zoology; Southern Illinois University; Carbondale, IL 62901, USA. Email: feander@siu.edu.

Date Accepted: 05 October 2011; updated 29 May 2012

Primary Editor: Kevin de Queiroz

Neocoleoidea W. Haas 1997 [F. Anderson], converted clade name

Registration Number: 141

Definition: The crown clade originating in the most recent common ancestor of *Argonauta argo* Linnaeus 1758 (*Octopoda/Incirrata*), *Stauroteuthis syrtensis* Verrill 1879 (*Octopoda/Cirrata*), *Sepia officinalis* Linnaeus 1758 (*Decapodiformes*), and *Vampyroteuthis infernalis* Chun 1903, provided that this clade does not include *Passaloteuthis paxillosa* Schlotheim 1820 (*Belemnoidea*). This is a minimum-crown-clade definition with a qualifying clause. Abbreviated definition: min crown ∇ (*Argonauta argo* Linnaeus 1758 & *Stauroteuthis syrtensis* Verrill 1879 & *Sepia officinalis* Linnaeus 1758 & *Vampyroteuthis infernalis* Chun 1903) | ~ (*Passaloteuthis paxillosa* Schlotheim 1820).

Etymology: From the Greek *neo* (new) + *koleos* (sheath) + *eidos* (form). The name refers to a sheath protecting the protoconch seen in the development of the internal shells (or shell remnants) of *Belemnoidea* (an extinct cephalopod group) and all extant cephalopods except *Nautilus*; *neo* refers to the crown clade (all extant cephalopods except *Nautilus*, which does not include belemnoids).

Reference Phylogeny: Figure 6 in Lindgren et al. (2004). For the purposes of applying the definition, *Argonauta argo* is inferred to be more closely related to *A. nodosa* than to any other species in the reference phylogeny (Strugnell and Allcock, 2010).

Composition: The most recent common ancestor of *Decapodiformes*, *Octopoda* and *Vampyroteuthis*, as defined in this volume, and all of its descendants. A list of valid extant species within *Neocoleoidea* can be found in Sweeney (2001).

Diagnostic Apomorphies: Berthold and Engeser (1987) listed thirteen diagnostic apomorphies for *Neocoleoidea* (referred to therein as *Dibranchiata*). These include a ventrally closed funnel, chromatophores, suckers, a buccal membrane, and gill (branchial) hearts. Unfortunately, the states of most of the proposed apomorphies listed by Berthold and Engeser are unknown in *Ammonoidea*, *Belemnoidea*, or both. The ink sac is an apomorphy of *Neocoleoidea* relative to *Nautilus*, but it is known to have been present in belemnoids (i.e., stem neocoleoids) (Owen, 1845). Of all of these characters, only suckers are known to be present in *Vampyroteuthis*, *Octopoda* and *Decapodiformes* but absent in *Nautilus* and the relatively few representatives of *Belemnoidea* and *Ammonoidea* for which presence or absence of this character could be determined (Haas, 1997).

Synonyms: *Dibranchiata* Owen 1832 (approximate).

Comments: The internal specifiers represent the three main subclades of *Neocoleoidea* among which relationships are currently unresolved—*Decapodiformes* (*Sepia officinalis*), *Octopoda* (*Argonauta argo*) and *Vampyroteuthis* (*Vampyroteuthis infernalis*). Two of these taxa were chosen because they were described by Linnaeus (1758), and the third is the only currently recognized extant species of *Vampyroteuthis*. The external specifier *Passaloteuthis paxillosa* Schlotheim 1820 was chosen as a representative belemnoid because it is one of the few belemnoids for which information about the soft-part anatomy (including arm, hook, jaw, mantle muscle and ink sac morphology) is available (Reitner and Ulrichs, 1983; Riegraf and Hauff, 1983). Furthermore, it is a member of *Belemnitida*, which Engeser (1990) argues is deeply nested within *Belemnoidea*. Berthold and Engeser (1987) and Hass (1997) noted that only a single character supports monophyly of *Neocoleoidea*—the presence of suckers. However, Fuchs et al. (2010) described an exceptionally well-preserved belemnoid that appears to possess functional suckers, and argued that this calls *Neocoleoidea* into question. Inclusion of a belemnoid as an external specifier ensures that the name *Neocoleoidea* cannot be applied to any clade that includes *Belemnoidea*, in which case it would have the same composition as *Coleoidea* (see following discussion).

Berthold and Engeser (1987) clarified the distinction between *Coleoidea* Bather 1888 (which explicitly included *Belemnoidea*, an extinct group now thought to lie outside of the crown clade) and *Dibranchiata* Owen 1832 (which was intended for the clade originating with the most recent common ancestor of the Recent endocochleate cephalopods, i.e., the crown clade). Engeser (1990) and Berthold and Engeser (1987) supported the use of the name *Dibranchiata* for the crown clade, following Owen's original intent. However, Engeser (1990) acknowledged that this contradicted the usage of *Dibranchiata* in much of the palaeontological literature, where *Dibranchiata* has been used for a clade comprising *Belemnoidea* and the crown clade. Haas (1997) proposed an alternative name—*Neocoleoidea*—for the crown clade.

Thus, two names for the crown clade comprising *Vampyroteuthis*, *Octopoda* and *Decapodiformes* exist in the literature: *Dibranchiata* and *Neocoleoidea*. However, many recent studies of cephalopod phylogeny ignore both of these names, and instead refer to members of the clade comprising *Decapodiformes*, *Vampyroteuthis* and *Octopoda* as coleoids (Akasaki et al., 2006; Boletzky, 1999; Bonnaud et al., 1997; Carlini and Graves, 1999; Lindgren et al., 2004; Strugnell et al., 2005). This is unfortunate, given that (as noted above) *Coleoidea* was originally proposed as the name of a group that includes *Belemnoidea*, and it refers to a character that is shared among belemnoids, *Vampyroteuthis*, octopods and decapodiforms. The name *Neocoleoidea* will help with some of this confusion, as it is clearly not just a synonym of *Coleoidea* (as *Dibranchiata* has often been used), but has always referred explicitly to the crown clade of *Coleoidea*. Furthermore, *Neocoleoidea* has been used by some neontologists (Young et al., 1998; Vecchione et al., 2000) and molecular phylogeneticists (Strugnell and Nishiguchi, 2007) for the crown clade.

Literature Cited

Akasaki, T., M. Nikaido, K. Tsuchiya, S. Segawa, M. Hasegawa, and N. Okada. 2006. Extensive mitochondrial gene arrangements in coleoid *Cephalopoda* and their phylogenetic implications. *Mol. Phylogenet. Evol.* 38:648–658.

Bather, F. A. 1888. Shell-growth in *Cephalopoda* (*Siphonopoda*). *Ann. Mag. Nat. Hist.* 6:298–310.

Berthold, T., and T. Engeser. 1987. Phylogenetic analysis and systematization of the *Cephalopoda* (*Mollusca*). *Verh. Naturwiss. Ver. Hamburg* 29:187–220.

Boletzky, S. v. 1999. Brève mise au point sur la classification des cephalopodes actuels. *Bull. Soc. Zool. Fr.* 124:271–278.

Bonnaud, L., R. Boucher-Rodoni, and M. Monnerot. 1997. Phylogeny of cephalopods inferred from mitochondrial DNA sequences. *Mol. Phylogenet. Evol.* 7:44–54.

Carlini, D. B., and J. E. Graves. 1999. Phylogenetic analysis of cytochrome *c* oxidase I sequences to determine higher-level relationships within the coleoid cephalopods. *Bull. Mar. Sci.* 64:57–76.

Chun, C. 1903. *Aus den Tiefen des Weltmeeres*. 2nd edition. Verlag Gustav Fischer, Jena.

Engeser, T. 1990. Phylogeny of the fossil coleoid *Cephalopoda* (*Mollusca*). *Berliner Geowiss. Abh. A* 124:123–191.

Fuchs, D., S. von Boletzky, and H. Tischlinger. 2010. New evidence of functional suckers in belemnoid coleoids (*Cephalopoda*) weakens support for the 'Neocoleoidea' concept. *J. Mollus. Stud.* 76:404–406.

Haas, W. 1997. The evolutionary history of the *Decabrachia* (*Cephalopoda, Coleoidea*). *Palaeontogr. Abt. A* 245:63–81.

Lindgren, A. R., G. Giribet, and M. K. Nishiguchi. 2004. A combined approach to the phylogeny of *Cephalopoda* (*Mollusca*). *Cladistics* 20:454–486.

Linnaeus, C. v. 1758. *Systema Naturae Per Regna Tria Naturae, Secundum Classes, Ordines, Genera, Species, cum Characteribus, Differentiis, Synonymis, Locis.* 10th edition. Laurentii Salvii, Holmiae (Stockholm).

Owen, R. 1832. *Memoir on the Pearly Nautilus (Nautilus pompilius, Linn.) with Illustrations of its External Form and Internal Structure.* Council of the Royal College of Surgeons, London.

Owen, R. 1845. A description of certain belemnites, preserved, with a great proportion of their soft parts, in the Oxford Clay at Christian Malford, Wilts. *Q. J. Geol. Soc.* 1:119–125.

Reitner, J., and M. Ulrichs. 1983. Echte Weichteilbelemniten aus dem Untertoarcium (Posidonnienschiefer) Südwestdeutschlands. *Neues Jahrb. Geol. Paläontol. Abh.* 165:450–465.

Riegraf, W., and R. Hauff. 1983. Belemnitenfund mit Weichkörper, Fangarmen und Gladius aus dem Untertoarcium (Posidenienschiefer) und Unteraalenium (Opalinuston) Südwestdeutschlands. *Neues Jahrb. Geol. Paläontol Abh.* 165:466–483.

von Schlotheim, E. F. 1820. *Die Petrefactenkunde auf Ihrem Jetzigen Standpunkte Durch Die Beschreibung Seiner Sammlung Versteinerter und Fossilier Uberrerste des Thier- und Pflanzenreichs der Vorwelt Erläutert.* Beckersche Buchhandlung, Gotha.

Strugnell, J., and A. L. Allcock. 2010. Co-estimation of phylogeny and divergence times of *Argonautoidea* using relaxed phylogenetics. *Mol. Phylogenet. Evol.* 54:701–708.

Strugnell, J., and M. Nishiguchi. 2007. Molecular phylogeny of coleoid cephalopods (*Mollusca: Cephalopoda*) inferred from three mitochondrial and six nuclear loci: a comparison of alignment, implied alignment and analysis methods. *J. Mollus. Stud.* 73:399–410.

Strugnell, J., M. Norman, J. Jackson, A. J. Drummond, and A. Cooper. 2005. Molecular phylogeny of coleoid cephalopods (*Mollusca: Cephalopoda*) using a multigene approach; the effect of data partitioning on resolving phylogenies in a Bayesian framework. *Mol. Phylogenet. Evol.* 37:426–441.

Sweeney, M. J. 2001. Current Classification of Recent *Cephalopoda*. Available at http://www.mnh.si.edu/cephs/newclass.pdf.

Vecchione, M., R. E. Young, and D. B. Carlini. 2000. Reconstruction of ancestral character states in neocoleoid cephalopods based on parsimony. *Am. Malacol. Bull.* 15:179–193.

Young, R. E., M. Vecchione, and D. T. Donovan. 1998. The evolution of coleoid cephalopods and their present biodiversity and ecology. *S. Afr. J. Mar. Sci.* (*Suid-Afrikaanse Tydskrif Vir Seewetenskap*) 20:393–420.

Author

Frank E. Anderson; Department of Zoology; Southern Illinois University; Carbondale, IL 62901, USA. Email: feander@siu.edu.

Date Accepted: 05 October 2011; updated 29 May 2012

Primary Editor: Kevin de Queiroz

Decapodiformes R. E. Young, M. Vecchione and D. T. Donovan 1998 [F. Anderson], converted clade name

Registration Number: 139

Definition: The crown clade for which a pair of tentacles, as inherited by *Sepia officinalis* Linnaeus 1758 (*Sepiidae*), is an apomorphy relative to other crown clades. Tentacles are modified fourth arms that consist of an elongate elastic stalk (circular or tetrahedral in cross-section) with a distal oval, usually widened, club (Nesis, 1987; Roper et al., 1984). The tentacle stalk is either suckerless or bears only minute suckers; the club bears suckers and occasionally hooks. This is an apomorphy-modified-crown-clade definition that is intended to apply to the most inclusive crown clade whose members exhibit all components just described of the complex apomorphy "tentacles". Abbreviated definition: crown ∇ apo tentacles [*Sepia officinalis* Linnaeus 1758].

Etymology: From the Greek *deca* (ten) plus *podos* (foot) plus *formes*, reflecting the presence of ten circumoral appendages (eight arms and two tentacles) in members of this clade.

Reference Phylogeny: Figure 6 in Lindgren et al. (2004); tentacles are thought to have arisen on the branch subtending the least inclusive clade that contains both *Enoploteuthis leptura* and *Histioteuthis corona* in this figure.

Composition: The common ancestor of all extant cephalopods that possess eight arms and two tentacles at some point in their life cycle, and all of its descendants. This includes *Spirulida* (ram's horn squids), *Sepiidae* (cuttles), *Idiosepiidae* (pygmy squids), *Sepiolida* (bobtail squids), *Loliginidae*, and *Oegopsida* (squids lacking corneal membranes); note that *Oegopsida* may be paraphyletic.

Diagnostic Apomorphies: Haas (1997, 2003) listed three apomorphies for *Decapodiformes* (though he referred to this group as *Decabrachia*): (1) fourth arm pair modified as tentacles, (2) possession of pedunculate suckers bearing chitinous rings and (3) fused renal sacs (kidneys). Berthold and Engeser (1987) cited these three characters and two others—(4) a hectocotylized ventral arm pair (though it must be noted that the location of the hectocotylized arms varies widely among decapodiforms, and some lack hectocotyli altogether) and (5) giant nerve fibers in the mantle. Young and Vecchione (1996) found nine morphological character states that are only found in decapodiforms (tentacles, horny sucker rings, sucker stalks with a base and neck, fin cartilages at proximal end of fins, bilaterally symmetrical oviducal glands, photosensitive vesicles within the cephalic cartilage only, subfrontal brain lobes, and digestive gland duct apparatuses within the nephridial coelom). However, Young and Vecchione argued that only one of these characters could be polarized—Arms IV ("unmodified" or "tentacles")—and the presence of tentacles was the only character state that they considered to be an unambiguous apomorphy for *Decapodiformes*. All of these apomorphies distinguish *Decapodiformes* from other crown clades within *Neocoleoidea*.

Synonyms: *Decapoda* W. Leach 1817 not *Decapoda* Latreille 1802–1803, *Decabrachia* E. Haeckel 1866, *Decembrachiata* R. Winckworth 1932, *Decabrachiomorpha* Haas 2003 (all approximate).

Comments: An apomorphy-modified-crown-clade definition was chosen because the crown clade to be named is characterized by a distinct apomorphy, but basal relationships within the clade and its outgroup relationships are poorly resolved. Monophyly of *Decapodiformes* has been inferred in many phylogenetic analyses of both morphological and molecular data. Morphology-based phylogenies that support decapodiform monophyly can be found in Donovan (1977), Berthold and Engeser (1987), Engeser and Bandel (1988), Engeser (1990), Haas (1997) and Haas (2003) (these include representatives of both extant and extinct subgroups, but they were not generated using explicit, reproducible analytical methods). Other morphology-based phylogenies that support decapodiform monophyly have been published by Young and Vecchione (1996), Young et al. (1998) and Salvini-Plawen and Steiner (1996) (these include only subclades with extant representatives). Molecular phylogenies that support decapodiform monophyly include those published by Bonnaud et al. (1997), Carlini and Graves (1999), Carlini et al. (2000), Anderson (2000a,b), Lindgren et al. (2004), Strugnell et al. (2005), and Giribet et al. (2006). Relationships among the decapodiform subclades, however, are poorly understood, and the sister taxon of *Decapodiformes* is also unclear. Most researchers have favored a sister group relationship between a *Vampyroteuthis* + *Octopoda* clade and *Decapodiformes* (Young and Vecchione, 1996; Bonnaud et al., 1997; Fig. 2b in Carlini and Graves, 1999; Fig. 5b and 6b in Carlini et al., 2000), but there is some support for a *Vampyroteuthis* + *Decapodiformes* sister-group arrangement, along with possible octopod paraphyly (Figs. 4–6 in Lindgren et al., 2004).

There is disagreement in the recent systematic literature regarding the appropriate name for the crown clade that includes cuttles and all squids. Several names have been used for this group: *Decapoda, Decabrachia, Decabrachiomorpha, Decembrachiata* and *Decapodiformes*. Leach (1817) erected both *Octopoda* and *Decapoda* as names of cephalopod groups, but *Decapoda* is a junior homonym of a crustacean taxon name (Boettger, 1952; Engeser, 1990; Young et al., 1998; Boletzky, 1999) and is rejected herein for that reason. Haeckel (1866) proposed the name *Decabrachia* and Boettger (1952) also used this name (without reference to Haeckel). Both Haeckel and Boettger included *Belemnoidea*—an extinct group now generally believed to belong to the neocoleoid stem group (Berthold and Engeser, 1987; Doyle et al., 1994; Engeser, 1990; Engeser and Bandel, 1988; Fuchs et al., 2003)—in *Decabrachia*. Fioroni (1981) also used the name *Decabrachia* (without reference to either Haeckel or Boettger) but included *Vampyroteuthis* within *Decabrachia*. Finally, Young et al. (1998) proposed the name *Decapodiformes* for this clade, arguing that this name acknowledges the long history of referring to these animals as "decapods", but is clearly distinct from that of the crustacean taxon. In short, *Decapodiformes, Decabrachia, Decembrachiata* and *Decabrachiomorpha* and *Decabrachia* would all be suitable names for this clade. Of these names, *Decabrachia* has priority, but the confusion regarding its composition is problematic. *Decapodiformes* has been the most widely used and was selected for that reason.

Literature Cited

Anderson, F. E. 2000a. Phylogenetic relationships among loliginid squids (*Cephalopoda: Myopsida*) based on analyses of multiple data sets. *Zool. J. Linn. Soc.* 130:603–633.

Anderson, F. E. 2000b. Phylogeny and historical biogeography of the loliginid squids (*Mollusca: Cephalopoda*) based on mitochondrial DNA sequence data. *Mol. Phylogenet. Evol.* 15:191–214.

Berthold, T., and T. Engeser. 1987. Phylogenetic analysis and systematization of the *Cephalopoda* (*Mollusca*). *Verh. Naturwiss. Ver. Hamburg* 29:187–220.

Boettger, C. R. 1952. Die Stämme des Tierreichs in ihrer systematischen Gliederung. *Abh. Braunschw. Wiss. Ges.* 4:238–300.

Boletzky, S. v. 1999. Brève mise au point sur la classification des cephalopodes actuels. *Bull. Soc. Zool. France* 124:271–278.

Bonnaud, L., R. Boucher-Rodoni, and M. Monnerot. 1997. Phylogeny of cephalopods inferred from mitochondrial DNA sequences. *Mol. Phylogenet. Evol.* 7:44–54.

Carlini, D. B., and J. E. Graves. 1999. Phylogenetic analysis of cytochrome *c* oxidase I sequences to determine higher-level relationships within the coleoid cephalopods. *Bull. Mar. Sci.* 64:57–76.

Carlini, D. B., K. S. Reece, and J. E. Graves. 2000. Actin gene family evolution and the phylogeny of coleoid cephalopods (*Mollusca*: *Cephalopoda*). *Mol. Biol. Evol.* 17:1353–1370.

Donovan, D. T. 1977. Evolution of the dibranchiate *Cephalopoda*. Pp. 15–48 in *The Biology of Cephalopods* (M. Nixon, and J. B. Messenger, eds.). Academic Press, London.

Doyle, P., D. T. Donovan, and M. Nixon. 1994. Phylogeny and systematics of the *Coleoidea*. *Univ. Kansas Paleontol. Contr. New Ser.* 5:1–15.

Engeser, T. 1990. Phylogeny of the fossil coleoid *Cephalopoda* (*Mollusca*). *Berliner Geowiss. Abh., A* 124:123–191.

Engeser, T., and K. Bandel. 1988. Phylogenetic classification of coleoid cephalopods. Pp. 105–116 in *Cephalopods Past and Present* (J. Wiedmann, and J. Kullmann, eds.). E. Schweizerbart'sche Verlagsbuchhandlung, Stuttgart.

Fioroni, P. 1981. Die Sonderstellung der Sepioliden, ein Vergleich der Ordnungen der rezenten Cephalopoden. *Zool. Jahrb.* 108:178–228.

Fuchs, D., H. Keupp, and T. Engeser. 2003. New records of soft parts of *Muensterella scutellaris* Muenster, 1842 (*Coleoidea*) from the late Jurassic plattenkalks of Eichstätt and their significance for octobrachian relationships. *Berliner Paläobiol. Abh.* 3:101–111.

Giribet, G., A. Okusu, A. R. Lindgren, S. W. Huff, M. Schrodl, and M. K. Nishiguchi. 2006. Evidence for a clade composed of molluscs with serially repeated structures: monoplacophorans are related to chitons. *Proc. Natl. Acad. Sci. USA* 103:7723–7728.

Haas, W. 1997. The evolutionary history of the *Decabrachia* (*Cephalopoda*, *Coleoidea*). *Palaeontogr. Abt. A* 245:63–81.

Haas, W. 2003. Trends in the evolution of the *Decabrachia*. *Berliner Paläobiol. Abh.* 3:113–129.

Haeckel, E. H. P. A. 1866. *Generelle Morphologie der Organismen: Allgemeine Grundzüge der Organischen Formen-Wissenschaft, Mechanisch Begründet Durch Die von Charles Darwin Reformirte Descendenz-Theorie*. G. Reimer, Berlin.

Latreille, P. A. 1802–1803. *Histoire Naturelle, Général et Particulière, des Crustacés et des Insects*. F. Dufart, Paris.

Leach, W. E. 1817. Synopsis of the orders, families and genera of the class *Cephalopoda*. *Zool. Miscell. Being Descript. New Interest. Anim.* 3:137–141.

Lindgren, A. R., G. Giribet, and M. K. Nishiguchi. 2004. A combined approach to the phylogeny of *Cephalopoda* (*Mollusca*). *Cladistics* 20:454–486.

Linnaeus, C. v. 1758. *Systema Naturae Per Regna Tria Naturae, Secundum Classes, Ordines, Genera, Species, cum Characteribus, Differentiis, Synonymis, Locis*. 10th edition. Laurentii Salvii, Holmiae (Stockholm).

Nesis, K. N. 1987. *Cephalopods of the World*. T. F. H. Publications, Neptune City, NJ.

Roper, C. F. E., M. J. Sweeney, and C. E. Nauen. 1984. *Cephalopods of the World: An Annotated and Illustrated Catalogue of Species of Interest to Fisheries*. United Nations Development Programme; Food and Agricultural Organization of the United Nations, New York.

Salvini-Plawen, L. v., and G. Steiner. 1996. Synapomorphies and plesiomorphies in higher classification of *Mollusca*. Pp. 29–52 in *Origin and Evolutionary Radiation of the* Mollusca (J. Taylor, ed.). Oxford University Press, Oxford.

Strugnell, J., M. Norman, J. Jackson, A. J. Drummond, and A. Cooper. 2005. Molecular phylogeny of coleoid cephalopods (*Mollusca*: *Cephalopoda*) using a multigene approach; the effect of data partitioning on resolving phylogenies in a Bayesian framework. *Mol. Phylogenet. Evol.* 37:426–441.

Winckworth, R. 1932. The British *Mollusca*. *J. Conchol.* 19:211–252.

Young, R. E., and M. Vecchione. 1996. Analysis of morphology to determine primary sister-taxon relationships within coleoid cephalopods. *Am. Malacol. Bull.* 12:91–112.

Young, R. E., M. Vecchione, and D. T. Donovan. 1998. The evolution of coleoid cephalopods and their present biodiversity and ecology. *S. Afr. J. Mar. Sci.* (*Suid-Afrikaanse Tydskrif Vir Seewetenskap*) 20:393–420.

Author

Frank E. Anderson; Department of Zoology; Southern Illinois University; Carbondale, IL 62901, USA. Email: feander@siu.edu.

Date Accepted: 05 October 2011; updated 29 May 2012

Primary Editor: Kevin de Queiroz

Vampyroteuthis C. Chun 1903 [F. Anderson], converted clade name

Registration Number: 143

Definition: The most inclusive crown clade containing *Vampyroteuthis infernalis* Chun 1903 but not *Argonauta argo* Linnaeus 1758 (*Octopoda/Incirrata*) and *Stauroteuthis syrtensis* Verrill 1879 (*Octopoda/Cirrata*) and *Sepia officinalis* Linnaeus 1758 (*Decapodiformes*). This is a maximum-crown-clade definition. Abbreviated definition: max crown ∇ (*Vampyroteuthis infernalis* Chun 1903 ~ *Argonauta argo* Linnaeus 1758 & *Stauroteuthis syrtensis* Verrill 1879 & *Sepia officinalis* Linnaeus 1758).

Etymology: Derived from the Slavic *vampir* (vampire)—referring to the dark color and cloak-like appearance of these animals—plus the Greek *teuthis* (squid).

Reference Phylogeny: Figure 1 in Lindgren et al. (2004). For the purposes of applying the definition, *Argonauta argo* is inferred to be more closely related to *A. nodosa* than to any other species in the reference phylogeny (Strugnell and Allcock, 2010).

Composition: *Vampyroteuthis infernalis* Chun 1903 (see Comments).

Diagnostic Apomorphies: The most notable apomorphy for *Vampyroteuthis* is the transformation of the second arm pair into retractile filaments. Additional apomorphies (Young, 2008) include:

1. Spermatangia receptacles of females form deep pockets, one anterior to each eye.
2. Photosensitive vesicles located immediately dorsal to the funnel.

3. Two pairs of fins in the young.
4. Presence of a posterior duct of the visceropericardial coelom.
5. Suckers present only on distal half of arms, where they alternate with cirri.

Synonyms: *Vampyroteuthidae* J. Thiele 1915, *Vampyroteuthinae* G. Grimpe 1917, *Eurytreta* G. Grimpe 1917, *Vampyromorphae* G. Grimpe 1917, *Watasellidae* M. Sasaki 1920, *Odontoglossa* A. Naef 1922, *Protopinnata* M. Sasaki 1929, *Dicerata* L. Joubin 1929, *Vampyromorpha* G. C. Robson 1929, *Pseudoctobrachia* A. Guerra 1992, *Vampyromorphida* M. Sweeney and C. Roper 1998, along with several genera listed in Robson (1932) (all approximate; Haas [2002] argued that *Vampyromorpha* includes only *Vampyroteuthis infernalis*—rendering *Vampyromorpha* an approximate synonym of *Vampyroteuthis*—but there is substantial disagreement regarding the placement of several fossils [see Comments]).

Comments: Several vampyromorph species and genera were described in the early twentieth century (Robson, 1932), typically on the basis of damaged and fragmentary specimens. Pickford (1939) referred all of these animals (except *Laetmoteuthis lugubris*, which she argued was not a vampyromorph) to a single species—*Vampyroteuthis infernalis*—and this view prevails today. However, there are some morphological data (e.g., beak size differences between Atlantic and Pacific specimens; Young, 1972) that suggest there may be more than one extant species of *Vampyroteuthis*.

The phylogenetic positions of several fossil coleoids relative to *Vampyroteuthis infernalis* are unclear. There has been substantial debate regarding the phylogenetic positions of several so-called "fossil teuthids" (summarized in Klug et al., 2005), which have been placed in *Vampyromorpha* (Engeser, 1990), either *Vampyromorpha* or *Decapodiformes* (Young et al., 1998), as a stem group of either *Decapodiformes* or *Vampyropoda* (a proposed clade comprising *Vampyroteuthis* + *Octopoda*; Boletzky, 1999; Doyle et al., 1994; Young et al., 1998), or in a paraphyletic taxon dubbed "*Trachyteuthimorpha*" along the stem of *Vampyropoda* (Haas, 2002).

The phylogenetic position of *Vampyroteuthis* among extant cephalopods is also uncertain. Most phylogenetic analyses have supported *Octopoda* as the sister taxon of *Vampyroteuthis* (Carlini and Graves, 1999; Carlini et al., 2000; Young and Vecchione, 1996; Fig. 1 of Lindgren et al., 2004), but some phylogenetic hypotheses presented by Lindgren et al. (2004) suggest a sister-group relationship between *Decapodiformes* and *Vampyromorpha*. A maximum-crown-clade definition is adopted here because it will allow referral of other species to *Vampyroteuthis* if they are more closely related to *V. infernalis* than to other extant cephalopods (e.g., if specimens currently referred to *V. infernalis* turn out to represent more than one species).

Approximate synonyms for *Vampyroteuthis* include *Pseudoctobrachia* and several permutations of *Vampyroteuthis* or *Vampyromorpha*, reflecting attempts to accommodate this single species in a system with mandatory ranks. *Vampyroteuthis* is preferred as a name for the crown group over *Vampyromorpha*, *Vampyroteuthidae* and *Vampyroteuthinae* as well as *Pseudoctobrachia* and several older names (*Dicerata*, *Eurytreta*, *Protopinnata* and *Odontoglossa*) simply because it has been used more frequently. *Vampyromorpha* would, however, be a suitable name for a more inclusive clade containing *Vampyroteuthis* and some or all of its stem forms.

Literature Cited

Boletzky, S. v. 1999. Brève mise au point sur la classification des cephalopodes actuels. *Bull. Soc. Zool. France* 124:271–278.

Carlini, D. B., and J. E. Graves. 1999. Phylogenetic analysis of cytochrome *c* oxidase I sequences to determine higher-level relationships within the coleoid cephalopods. *Bull. Mar. Sci.* 64:57–76.

Carlini, D. B., K. S. Reece, and J. E. Graves. 2000. Actin gene family evolution and the phylogeny of coleoid cephalopods (*Mollusca*: *Cephalopoda*). *Mol. Biol. Evol.* 17:1353–1370.

Chun, C. 1903. *Aus den Tiefen des Weltmeeres*. 2nd edition. Verlag Gustav Fischer, Jena.

Doyle, P., D. T. Donovan, and M. Nixon. 1994. Phylogeny and systematics of the *Coleoidea*. *Univ. Kansas Paleontol. Contr., New Ser.* 5:1–15.

Engeser, T. 1990. Phylogeny of the fossil coleoid *Cephalopoda* (*Mollusca*). *Berliner Geowiss. Abh. A* 124:123–191.

Guerra, A. 1992. *Mollusca: Cephalopoda*. Pp. 1–327 in *Fauna Ibérica*, Vol. 1 (M. A. Ramos, J. Alba, X. Bellés, J. Gosálbez, A. Guerra, E. Macpherson, F. Martin, J. Serrano, and J. Templado, eds.). Museo Nacional de Ciencias Naturales (CSIC), Madrid.

Haas, W. 2002. The evolutionary history of the eight-armed *Coleoidea*. *Abh. Geol. Bundesanst. Wien* 57:341–351.

Joubin, L. 1929. Notes preliminaires sur les Céphalopodes des croisières du DANA (1921–1922). *Ann. Inst. Océanograph.* 6:363–394.

Klug, C., G. Schweigert, G. Dietl, and D. Fuchs. 2005. Coleoid beaks from the Nusplingen Lithographic Limestone (Upper Kimmeridgian, SW Germany). *Lethaia* 38:173–191.

Lindgren, A. R., G. Giribet, and M. K. Nishiguchi. 2004. A combined approach to the phylogeny of *Cephalopoda* (*Mollusca*). *Cladistics* 20:454–486.

Linnaeus, C. 1758. *Systema Naturae Per Regna Tria Naturae, Secundum Classes, Ordines, Genera, Species, cum Characteribus, Differentiis, Synonymis, Locis*. 10th edition. Laurentii Salvii, Holmiae (Stockholm).

Pickford, G. E. 1939. A re-examination of the types of *Melanoteuthis lucens* Joubin. *Bull. Inst. Oceanograph.* Monaco 777:1–12.

Robson, G. C. 1929. On the rare abyssal octopod *Melanoteuthis beebei* (sp. n.): a contribution to the phylogeny of the *Octopoda. Proc. Zool. Soc. Lond.* 1929:469–486.

Robson, G. C. 1932. *A Monograph of the Recent* Cephalopoda. *Part 2. The* Octopoda (*Excluding the* Octopodinae). The Trustees of the British Museum, London.

Sasaki, M. 1920. Report of cephalopods collected during 1906 by the United States Bureau of Fisheries Steamer "Albatross" in the Northwestern Pacific. *Proc. U.S. Natl. Mus.* 52:163–203.

Sasaki, M. 1929. A monograph of the dibranchiate cephalopods of the Japanese and adjacent waters. *J. Coll. Agric. Hokkaido Imp. Univ.* 20 (suppl.):1–357.

Strugnell, J., and A. L. Allcock. 2010. Co-estimation of phylogeny and divergence times of *Argonautoidea* using relaxed phylogenetics. *Mol. Phylogenet. Evol.* 54:701–708.

Sweeney, M. J., and C. F. E. Roper. 1998. Classification, type localities and type repositories of recent *Cephalopoda*. Pp. 561–599 in *Systematics and Biogeography of Cephalopods*, Vol. II (N. A. Voss, M. Vecchione, R. B. Toll, and M. J. Sweeney, eds.). Smithsonian Institution Press, Washington, DC.

Verrill, A. E. 1879. Notice of recent additions to the marine fauna of the eastern coast of North America, No. 7. Brief Contributions to Zoology from the Museum of Yale College No. XLIV. *Am. J. Sci. Arts* 18:468–470.

Young, R. E. 1972. The systematics and areal distribution of pelagic cephalopods from the seas off Southern California. *Smithsonian Contrib. Zool.* 97:1–159.

Young, R. E. 2008. *Vampyromorpha* Robson, 1929. *Vampyroteuthis infernalis* Chun, 1903. The Vampire Squid, Version 30 May 2008. Available at http://tolweb.org/Vampyroteuthis_infernalis/20084/2008.05.30 in The Tree of Life Web Project, http://tolweb.org/.

Young, R. E., and M. Vecchione. 1996. Analysis of morphology to determine primary sister-taxon relationships within coleoid cephalopods. *Am. Malacol. Bull.* 12:91–112.

Young, R. E., M. Vecchione, and D. T. Donovan. 1998. The evolution of coleoid cephalopods and their present biodiversity and ecology. *S. Afr. J. Mar. Sci. (Suid-Afrikaanse Tydskrif Vir Seewetenskap)* 20:393–420.

Author

Frank E. Anderson; Department of Zoology; Southern Illinois University; Carbondale, IL 62901, USA. Email: feander@siu.edu.

Date Accepted: 05 October 2011; updated 29 May 2012

Primary Editor: Kevin de Queiroz

Octopoda W. Leach 1817 [F. Anderson], converted clade name

Registration Number: 142

Definition: The crown clade originating in the most recent common ancestor of *Octopus vulgaris* Cuvier 1797 (*Incirrata*) and *Stauroteuthis syrtensis* Verrill 1879 (*Cirrata*), provided that this clade does not include *Sepia officinalis* Linnaeus 1758. This is a minimum-crown-clade definition with a qualifying clause. Abbreviated definition: min crown ∇ (*Octopus vulgaris* Cuvier 1797 & *Stauroteuthis syrtensis* Verrill 1879) | ~ (*Sepia officinalis* Linnaeus 1758).

Etymology: From the Greek *octo* (eight) plus *podos* (foot), reflecting the presence of eight circumoral arms in all known members of the clade.

Reference Phylogeny: Figure 1 in Lindgren et al. (2004), where the clade is named *Octobrachia*. *Octopus vulgaris* is not included in the reference phylogeny but is considered more closely related to *Japetella diaphana*, *Thaumeledone gunteri*, and *Argonauta nodosa* than to *Opisthoteuthis sp.* (Uribe and Zardoya, 2017).

Composition: *Incirrata* Grimpe 1916 (the finless octopuses) and *Cirrata* Grimpe 1916 (the finned octopuses), along with their extinct relatives that fall within the crown clade. A list of currently recognized extant octopod species can be found in Sweeney (2001).

Diagnostic Apomorphies: Berthold and Engeser (1987) listed the following apomorphic characters for this clade:

1. Ink sac embedded in the digestive gland.
2. Coelomic cavity reduced.

3. Unilobed digestive gland (however, this is also seen in some members of *Decapodiformes*).
4. Reduction of the magnocellular lobe of the brain (Young, 1977).
5. Presence of a suprabrachial commissure.
6. Possession of a sac-like body.

Young and Vecchione (1996) list several other apomorphies for this clade, including:

1. Head/mantle fusion without nuchal cartilage.
2. Dorsal mantle cavity.
3. Photosensitive vesicles located on stellate ganglia.
4. Central nervous system with inferior frontal lobe system present; superior buccal and posterior buccal lobes fused; suprabrachial commisure separate from brain.
5. Suckers with cuticular lining.

Haas (2002) proposed two additional apomorphies for *Octopoda* as defined here: reduction of the middle field of the gladius and loss of the second arm pair.

Synonyms: *Octobrachia* E. Haeckel 1866, *Octobrachiata* R. Winckworth 1932, *Octopodida* M. Sweeney and C. Roper 1998 (all approximate).

Comments: The monophyly of *Octopoda* is supported in most published phylogenies, including Figure 6 in Berthold and Engeser (1987), Figure 5 of Engeser (1990), Figure 1 of Doyle et al. (1994), Figure 2 of Kluessendorf and Doyle (2000), and Figures 1 and 11 of Haas (2002) (all derived from morphological data, including fossils, but not based on reproducible analyses of data matrices); Figure 15 of Young and Vecchione (1996), Figure 3 of Young et al. (1998), and Figure 1 of Lindgren et al. (2004) (based on morphological data for extant taxa only); and Figure 2b of Carlini and Graves (1999), Figures 2, 5 and 6b of Carlini et al. (2000), Figure 5 of Lindgren et al. (2004), and Figures 3 and 4 of Strugnell et al. (2005) (molecular data from several mitochondrial and nuclear gene regions).

A number of taxa representing *Incirrata* and *Cirrata* would be suitable internal specifiers. *Octopus vulgaris* Cuvier 1797 (*Incirrata*) is the type species of *Octopus*. Representatives of *Stauroteuthis syrtensis* Verrill 1879 (*Cirrata*) are often included in cephalopod phylogenetic studies (Carlini and Graves, 1999; Carlini et al., 2000; Giribet et al., 2006; Lindgren et al., 2004); furthermore, this species appears to be nested deeply within *Cirrata* (Piertney et al., 2003).

A few molecular phylogenetic studies have cast some doubt on the monophyly of *Octopoda*. Parsimony analysis of an equally weighted data set of partial mitochondrial cytochrome oxidase I sequences supports octopod paraphyly (Fig. 2a in Carlini and Graves, 1999), although an alternative analysis of the same data set supported octopod monophyly (Fig. 2b). Furthermore, Figures 2–4, 6 and 8 in Lindgren et al. (2004) depict octopods as paraphyletic with respect to *Vampyroteuthis* and/or *Decapodiformes*, though their analysis of a morphological data set supported both octopod monophyly and a sister-group relationship between *Octopoda* and *Vampyroteuthis* (Fig. 1), and their analysis of a combined molecular data set supported octopod monophyly (Fig. 5). Because of this lingering uncertainty regarding monophyly of *Octopoda*, a qualifying clause has been added to the phylogenetic definition, insuring that the name *Octopoda* will not be applied to any clade that includes *Decapodiformes*. It must be noted, however, that *Vampyroteuthis* is not included as an external specifier in this definition. Thus, if future analyses find that *Vampyroteuthis* is a member of *Octopoda* as defined here, the name will not be nullified. If *Vampyroteuthis* is instead found to be the extant sister taxon to *Octopoda*, either *Octopodiformes* Young et al., 1998 or *Vampyropoda* Boletzky, 1992 could be used for the clade comprising *Vampyroteuthis* plus *Octopoda*.

There are three approximate synonyms of *Octopoda* as defined here—*Octobrachia*, *Octobrachiata* and *Octopodida*. Although *Octobrachiata* and *Octopodida* have been rarely used, *Octobrachia* has been used by several authors (e.g., Boettger, 1952; Fioroni, 1981; Boletzky, 1999; Lindgren et al., 2004) to denote a group including *Incirrata* Grimpe 1916 and *Cirrata* Grimpe 1916. Haeckel (1866) originally proposed the name *Octobrachia* to refer to a group comprising *Cirroteuthida* (i.e., *Cirrata*) as well as *Eledonelida* and *Philonexida* (= *Tremoctopodidae*) (members of *Incirrata*), but the name has been used differently by subsequent authors. Boettger (1952) and two recent cephalopod classifications (Nixon and Young, 2003; Sweeney and Roper, 1998) included *Vampyroteuthis* within *Octobrachia*. However, *Vampyroteuthis* was unknown to Haeckel, and Fioroni (1981) (the source cited by Sweeney and Roper, 1998, for the name *Octobrachia*) included vampyromorphs in *Decabrachia*, not *Octobrachia* (Boletzky, 1999).

Octopoda is the oldest name for this group, and it has been used to refer to the crown clade comprising incirrate and cirrate octopuses in the recent scientific literature far more frequently than have any of the alternatives (*Octobrachia*, *Octobrachiata* or *Octopodida*). For these reasons, *Octopoda* is adopted herein.

As might be expected due to their dearth of hard parts, octopuses do not fossilize well. However, a few fossils that may represent octopods have been found, including *Palaeoctopus newboldi* (Woodward, 1896), *Proteroctopus ribeti* (Fischer and Riou, 1982) and several fossil brood chambers of argonautids (Engeser, 1990). *Palaeoctopus newboldi* is regarded by some researchers as a member of *Octopoda* as defined here. Despite possessing fins, it may represent a stem lineage of the incirrate octopuses (Berthold and Engeser, 1987; Engeser, 1988; Engeser, 1990; Haas, 2002), but others believe it represents the stem group of a larger clade comprising *Vampyroteuthis* and *Octopoda* (Doyle et al., 1994). Engeser (1988) argued that *Proteroctopus ribeti* is either a stem octopod or a "fossil teuthid" (i.e., not an octopod as defined here), and Young et al. (1998) suggested that *Proteroctopus ribeti* may not even be a cephalopod.

Literature Cited

Berthold, T., and T. Engeser. 1987. Phylogenetic analysis and systematization of the *Cephalopoda* (*Mollusca*). *Verh. Naturwiss. Ver. Hamburg* 29:187–220.

Boettger, C. R., 1952. *Die Stämme des Tierreichs in Ihrer Systematischen Gliederung.* Vieweg+ Teubner Verlag, Wiesbaden.

Boletzky, S. v. 1992. Evolutionary aspects of development, life style and reproductive mode in incirrate octopods (*Mollusca, Cephalopoda*). *Rev. Suisse Zool.* 99:755–770.

Boletzky, S. v. 1999. Brève mise au point sur la classification des cephalopodes actuels. *Bull. Soc. Zool. Fr.* 124:271–278.

Carlini, D. B., and J. E. Graves. 1999. Phylogenetic analysis of cytochrome *c* oxidase I sequences to determine higher-level relationships within the coleoid cephalopods. *Bull. Mar. Sci.* 64:57–76.

Carlini, D. B., K. S. Reece, and J. E. Graves. 2000. Actin gene family evolution and the phylogeny of coleoid cephalopods (*Mollusca: Cephalopoda*). *Mol. Biol. Evol.* 17:1353–1370.

Doyle, P., D. T. Donovan, and M. Nixon. 1994. Phylogeny and systematics of the *Coleoidea*. *Univ. Kansas Paleontol. Contr. New Ser.* 5:1–15.

Engeser, T. 1988. Fossil "octopods"—a critical review. Pp. 81–88 in *The* Mollusca: *Paleontology and Neontology of Cephalopods* (M. R. Clarke, and E. R. Trueman, eds.). Academic Press, New York.

Engeser, T. 1990. Phylogeny of the fossil coleoid *Cephalopoda* (*Mollusca*). *Berliner Geowiss. Abh., A* 124:123–191.

Fioroni, P. v. 1981. Die Sonderstellung der Sepioliden, ein Vergleich der Ordnungen der rezenten Cephalopoden. *Zool. Jahrb. Abt. Syst. Okol. Geogr. Tiere* 108:178–228.

Fischer, J.-C., and B. Riou. 1982. Le plus ancien Octopode connu (*Cephalopoda, Dibranchiata*): *Proteroctopus ribeti* nov. ben., nov. sp., du Callovien de l'Ardèche (France). *C. R. Seances Acad. Sci.* 295:277–280.

Giribet, G., A. Okusu, A. R. Lindgren, S. W. Huff, M. Schrodl, and M. K. Nishiguchi. 2006. Evidence for a clade composed of molluscs with serially repeated structures: monoplacophorans are related to chitons. *Proc. Natl. Acad. Sci. USA* 103:7723–7728.

Grimpe, G. 1916. *Chunioteuthis*: Eine neue Cephalopoden-gattung. *Zool. Anz.* 52:297–305.

Haas, W. 2002. The evolutionary history of the eight-armed *Coleoidea*. *Abh. Geol. Bundesanst. Wien* 57:341–351.

Haeckel, E. H. P. A. 1866. *Generelle Morphologie der Organismen: Allgemeine Grundzüge der Organischen Formen-Wissenschaft, Mechanisch Begründet Durch Die von Charles Darwin Reformirte Descendenz-Theorie.* G. Reimer, Berlin.

Kluessendorf, J., and P. Doyle. 2000. *Pohlsepia mazonensis*, an early 'octopus' from the Carboniferous of Illinois, USA. *Paleontology* 43:919–926.

Leach, W. E. 1817. Synopsis of the orders, families and genera of the class *Cephalopoda*. *The Zoological Miscellany: Being Descriptions of New, or Interesting Animals* 3:137–141.

Lindgren, A. R., G. Giribet, and M. K. Nishiguchi. 2004. A combined approach to the phylogeny of *Cephalopoda* (*Mollusca*). *Cladistics* 20:454–486.

Linnaeus, C. v. 1758. *Systema Naturae Per Regna Tria Naturae, Secundum Classes, Ordines, Genera, Species, cum Characteribus, Differentiis, Synonymis, Locis.* 10th edition. Laurentii Salvii, Holmiae (Stockholm).

Nixon, M., and J. Z. Young. 2003. *The Brains and Lives of Cephalopods*. Oxford University Press, New York.

Piertney, S. B., C. Hudelot, F. G. Hochberg, and M. A. Collins. 2003. Phylogenetic relationships among cirrate octopods (*Mollusca*: *Cephalopoda*) resolved using mitochondrial 16S ribosomal sequences. *Mol. Phylogenet. Evol.* 27:348–353.

Strugnell, J., M. Norman, J. Jackson, A. J. Drummond, and A. Cooper. 2005. Molecular phylogeny of coleoid cephalopods (*Mollusca*: *Cephalopoda*) using a multigene approach; the effect of data partitioning on resolving phylogenies in a Bayesian framework. *Mol. Phylogenet. Evol.* 37:426–441.

Sweeney, M. J. 2001. Current Classification of Recent *Cephalopoda*. Available at http://www.mnh.si.edu/cephs/newclass.pdf.

Sweeney, M. J., and C. F. E. Roper. 1998. Classification, type localities and type repositories of recent *Cephalopoda*. Pp. 561–599 in *Systematics and Biogeography of Cephalopods*, Vol. II (N. A. Voss, M. Vecchione, R. B. Toll, and M. J. Sweeney, eds.). Smithsonian Institution Press, Washington, DC.

Uribe, J. E., and R. Zarodya. 2017. Revisiting the phylogeny of *Cephalopoda* using complete mitochondrial genomes. *J. Mollus. Stud.* 83:133–144.

Verrill, A. E. 1879. Notice of recent additions to the marine fauna of the eastern coast of North America, No. 7. Brief Contributions to Zoology from the Museum of Yale College No. XLIV. *Am. J. Sci. Arts* 18:468–470.

Winckworth, R. 1932. The British *Mollusca*. *J. Conchol.* 19:211–252.

Woodward, H. 1896. On a fossil octopus (*Calais newboldi*, J. de C. Sby. MS) from the Cretaceous of the Lebanon. *Q. J. Geol. Soc. Lond.* 52:229–234.

Young, J. Z., 1977. Brain, behaviour and evolution of cephalopods. In *Symp. Zool. Soc. Lond.* 38:377–434.

Young, R. E., and M. Vecchione. 1996. Analysis of morphology to determine primary sister-taxon relationships within coleoid cephalopods. *Am. Malacol. Bull.* 12:91–112.

Young, R. E., M. Vecchione, and D. T. Donovan. 1998. The evolution of coleoid cephalopods and their present biodiversity and ecology. *S. Afr. J. Mar. Sci.* (*Suid-Afrikaanse Tydskrif Vir Seewetenskap*) 20:393–420.

Author

Frank E. Anderson; Department of Zoology; Southern Illinois University; Carbondale, IL 62901, USA. Email: feander@siu.edu.

Date Accepted: 05 October 2011; updated 29 May 2012

Primary Editor: Kevin de Queiroz

Ecdysozoa A. M. A. Aguinaldo, J. M. Turbeville, L. S. Linford, M. C. Rivera, J. A. Garey, R. A. Raff, and J. A. Lake 1997 [J. A. Lake, K. M. Halanych, J. A. Servin, and J. R. Garey], converted clade name

Registration Number: 190

Definition: The largest crown clade containing *Drosophila melanogaster* Meigen 1830 (*Arthropoda/Insecta*), *Priapulus caudatus* Lamarck 1816 (*Priapula*) and *Caenorhabditis elegans* (Maupas 1900) (*Nematoda*), but not *Capitella teleta* Blake, Grassle and Eckelbarger 2009 (*Annelida*) and *Girardia* (formerly *Dugesia*) *tigrina* (Girard 1850) (*Platyhelminthes*). This is a maximum-crown-clade definition. Abbreviated definition: max crown ∇ (*Drosophila melanogaster* Meigen 1830 & *Priapulus caudatus* Lamarck 1816 & *Caenorhabditis elegans* (Maupas 1900) ~ *Capitella teleta* Blake, Grassle and Eckelbarger 2009 & *Girardia tigrina* (Girard 1850)).

Etymology: From the Greek *ekdysis* (act of getting out) and *zōia* (animals), referring to the apomorphy of ecdysis (i.e., molting) exhibited by the animals in this clade.

Reference Phylogeny: Aguinaldo et al. (1997: Fig. 2) is the primary reference phylogeny; see Halanych (2004: Fig. 2) for the relationships of the external specifiers. Secondary reference phylogenies include Garey (2001: Fig. 2), Peterson and Eernisse (2001: Fig. 6), Anderson et al. (2004: Fig. 1), Halanych (2004: Fig. 2), Mallatt et al. (2004: Fig. 2), Philippe et al. (2005: Fig. 4), Dunn et al. (2008: Fig. 2), and Yamasaki et al. (2015: Fig. 2).

Composition: *Ecdysozoa* consists of all species within *Arthropoda*, *Tardigrada*, *Onychophora*, *Nematoda*, *Nematomorpha*, *Kinorhyncha*, *Loricifera*, and *Priapulida*.

Diagnostic Apomorphies: Apomorphies of *Ecdysozoa* are molting of the exoskeleton cuticle and loss of locomotory cilia (Aguinaldo et al., 1997). For a detailed discussion on the significance of molting in ecdysozoan evolution, see Valentine and Collins (2000). The Hox genes Ultrabithorax, Abdominal-A and Abdominal-B have ecdysozoan-specific amino acid residues as described by Balavoine et al. (2002). A tissue-specific marker of *Ecdysozoa* (the presence of anti-HRP-reactive glycoproteins in neural tissue) was described by Haase et al. (2001).

Synonyms: There are no synonyms.

Comments: Aguinaldo et al. (1997) first recognized a clade of molting animals, based on phylogenetic analysis of 18S ribosomal DNA sequences, and named it *Ecdysozoa*. Large nuclear-protein-coding gene datasets also support the clade (e.g., Dunn et al., 2008; Pick et al., 2010). In addition to phylogenetic analyses, other evidence supports *Ecdysozoa* (Valentine and Collins, 2000; Budd, 2002; Roy and Gilbert, 2005), including developmental Hox genes when optimized on a tree (de Rosa et al., 1999).

Literature Cited

Aguinaldo, A. M. A., J. M. Turbeville, L. S. Linford, M. C. Rivera, J. R. Garey, R. A. Raff, and J. A. Lake. 1997. Evidence for a clade of nematodes, arthropods and other moulting animals. *Nature* 387:489–493.

Anderson, F. E., A. J. Cordoba, and M. Thollesson. 2004. Bilaterian phylogeny based on analyses of a region of the sodium-potassium ATPase α-subunit gene. *J. Mol. Evol.* 58:252–268.

Balavoine, G., R. de Rosa, and A. Adoutte. 2002. Hox clusters and bilaterian phylogeny. *Mol. Phylogenet. Evol.* 24:366–73.

Budd, G. E. 2002. A palaeontological solution to the arthropod head problem. *Nature* 417:271–275.

Dunn, C. W., A. Hejnol, D. Q. Matus, D. Pang, W. E. Browne, S. A. Smith, E. Seaver, G. W. Rouse, M. Obst, G. D. Edgecombe, M. V. Sørensen, S. H. D. Haddock, A. Schmidt-Rhaesa, A. Okusu, R. M. Kristensen, W. C. Wheeler, M. Q. Martindale, and G. Giribet. 2008. Broad phylogenomic sampling improves resolution of the animal tree of life. *Nature* 452:745–749.

Garey, J. R. 2001. *Ecdysozoa*: The relationship between *Cycloneuralia* and *Panarthropoda*. *Zool. Anz.* 240:321–330.

Haase, A., M. Stern, K. Wachtler, and G. Bicker. 2001. A tissue-specific marker of *Ecdysozoa*. *Dev. Genes Evol.* 211:428–433.

Halanych, K. M. 2004. The new view of animal phylogeny. *Annu. Rev. Ecol. Evol. Syst.* 35:229–56.

Mallatt, J. M., J. R. Garey, and J. W. Shultz. 2004. Ecdysozoan phylogeny and Bayesian inference: first use of nearly complete 28S and 18S rRNA gene sequences to classify the arthropods and their kin. *Mol. Phylogenet. Evol.* 31:178–191.

Peterson, K. J., and D. J. Eernisse. 2001. Animal phylogeny and the ancestry of bilaterians: inferences from morphology and 18S rDNA gene sequences. *Evol. Dev.* 3:170–205.

Philippe, H., N. Lartillot, and H. Brinkmann. 2005. Multigene analyses of bilaterian animals corroborate the monophyly of *Ecdysozoa*, *Lophotrochozoa*, and *Protostomia*. *Mol. Biol. Evol.* 22:1246–1253.

Pick, K. S., H. Philippe, F. Schreiber, D. Erpenbeck, K. J. Jackson, P. Wrede, M. Wiens, A. Alié, B. Morgenstern, M. Manuel, and G. Wörheide. 2010. Improved phylogenomic taxon sampling noticeably affects nonbilaterian relationships. *Mol. Biol. Evol.* 27:1983–1987.

de Rosa, R., J. K. Grenier, T. Andreeva, C. E. Cook, A. Adoutte, M. Akam, S. B. Carroll, and G. Balavoine. 1999. HOX genes in brachiopods and priapulids and protostome evolution. *Nature* 399:772–776.

Roy, S. W., and W. Gilbert. 2005. Resolution of a deep animal divergence by the pattern of intron conservation. *Proc. Natl. Acad. Sci. USA* 102:4403–4408.

Valentine, J. W., and A. G. Collins. 2000. The significance of moulting in ecdysozoan evolution. *Evol. Dev.* 2:152–156.

Yamasaki, H., S. Fujimoto, and K. Miyazaki. 2015. Phylogenetic position of *Loricifera* inferred from nearly complete 18S and 28S rRNA gene sequences. *Zool. Lett.* 1:18.

Authors

James Lake; Molecular Biology Institute; 232 Boyer Hall; University of California Los Angeles; Los Angeles, CA 90095, USA. Email: Lake@mbi.ucla.edu.

Kenneth M. Halanych; Department of Biological Sciences; 101 Rouse Building; Auburn University; Auburn, AL 36849, USA. Email: ken@auburn.edu.

Jacqueline Servin; Molecular Biology Institute; 232 Boyer Hall; University of California Los Angeles; Los Angeles, CA 90095, USA. Email: jacquelineservin@gmail.com.

James R. Garey; Department of Cell Biology, Microbiology and Molecular Biology; University of South Florida; 4202 East Fowler Avenue ISA2015; Tampa, FL 33620, USA. Email: garey@usf.edu.

Date Accepted: 22 June 2017

Primary Editors: Jacques A. Gauthier and Kevin de Queiroz

Nematomorpha F. Vejdovsky 1886 [A. Schmidt-Rhaesa], converted clade name

Registration Number: 227

Definition: The crown clade originating in the most recent common ancestor of *Nectonema agile* Verrill 1879, *Gordius aquaticus* Linnaeus 1758 (*Gordiida*), and *Paragordius varius* Leidy 1851 (*Gordiida*). This is a minimum-crown-clade definition. Abbreviated definition: min crown ∇ (*Nectonema agile* Verrill 1879 & *Gordius aquaticus* Linnaeus 1758 & *Paragordius varius* Leidy 1851).

Etymology: Derived from the Greek "*nematos*", thread, plus "*morphe*" form, shape, meaning "shaped as a thread".

Reference Phylogeny: Figure 3 of Bleidorn et al. (2002).

Composition: *Nematomorpha* includes about 360 currently recognized species. European species are listed in Schmidt-Rhaesa (1997), Nearctic species in Schmidt-Rhaesa et al. (2003), Central American species in Schmidt-Rhaesa and Menzel (2005) and Neotropical species in de Miralles & de Villalobos (1993a,b). Five species of the genus *Nectonema* are marine and parasitize crustaceans; all others live in freshwater and parasitize insects, millipeds or chelicerates.

Diagnostic Apomorphies: Autapomorphies of *Nematomorpha* were listed in Schmidt-Rhaesa (1998):

1. Circumpharyngeal brain with subpharyngeal main portion and fine suprapharyngeal commissure.
2. Larva with introvert-like preseptum, cuticular hooks and stylets.

Other characters such as a life cycle including a parasitic and a free-living phase, adult body size in the cm-range, adult cuticle with crossed fiber layers and musculature polymyar and coelomyar have been considered to be apomorphies of *Nematomorpha* (Schmidt-Rhaesa, 1998) but also occur in some *Nematoda*.

Synonyms: *Gordiacea* of some authors (e.g., Carvalho, 1942) (approximate).

Comments: Members of *Nematomorpha* were first known from the free-living phase of freshwater species and were included, e.g., by Linnaeus (1758), in "*Vermes intestina*". Later, they were included in *Nematoda* (this volume) and sometimes united with mermithid nematodes into a group "*Gordiacea*" (see Schmidt-Rhaesa, 1997). The name *Nematomorpha* was introduced by Vejdovsky (1886), shortly after the discovery of marine species by Verrill (1879), to include marine and the longer-known freshwater species and has since then been applied to a group corresponding to the clade treated here (e.g., Hyman, 1951; Bresciani, 1991). Although the name *Gordiacea* has sometimes been applied to a group corresponding to the clade here named *Nematomorpha* (e.g., Carvalho, 1942), it has more commonly been applied (after removal of the mermithid nematodes) to a subclade of nematomorphs composed of the freshwater species (other names for this subclade are *Gordiida*, *Gordioida* and *Gordioidea*). Monophyly of *Nematomorpha* is generally not doubted, but there are alternative phylogenetic hypotheses concerning its sister group (see Schmidt-Rhaesa, 1998; Sørensen et al., 2008). The selection of the name *Nematomorpha* is straightforward, because the only alternative name has been applied more commonly to a less inclusive clade.

Literature Cited

Bleidorn, C., A. Schmidt-Rhaesa, and J. R. Garey. 2002. Systematic relationships of *Nematomorpha* based on molecular and morphological data. *Invertebr. Biol.* 121:357–364.

Bresciani, J. 1991. *Nematomorpha.* Pp. 197–218 in *Microscopic Anatomy of Invertebrates*, Vol. 4 (F. W. Harrisson, and E. E. Ruppert, eds.). Wiley-Liss, Inc., New York.

Carvalho, J. C. M. 1942. Studies on some *Gordiacea* of North and South America. *J. Parasitol.* 28:213–222.

De Miralles, D. A. B., and L. C. de Villalobos. 1993a. Distribucion geografica de los gordiaceos en la Republica Argentina. Pp. 1–16 in *Fauna de Agua Dulce de la Republica Argentina*, Vol. 13 (Z. A. de Castellanos, ed.). PROFADU, CONICET, La Plata, Argentina.

De Miralles, D. A. B., and L. C. de Villalobos. 1993b. Distribucion geografica de los gordiaceos en la region Neotropical. Pp. 17–32 in *Fauna de Agua Dulce de la Republica Argentina*, Vol. 13 (Z. A. de Castellanos, ed.). PROFADU, CONICET, La Plata, Argentina.

Hyman, L. H. 1951. *The Invertebrates:* Acanthocephala, Aschelminthes, and Entoprocta. *The Pseudocoelomate* Bilateria, Vol. 3. McGraw-Hill, New York.

von Linnaeus, C. 1758. *Systema Naturae Per Regna Tria Naturae, Secundum Classes, Ordines, Genera, Species, cum Characteribus, Differentiis, Synonymis, Locis.* 10th edition, Tome 1, Part 6. Laurentii Salvii, Holmiae (Stockholm).

Schmidt-Rhaesa, A. 1997. *Nematomorpha.* Pp. 1–124 in *Süßwasserfauna Mitteleuropas*, Vol. 4/4 (J. Schwoerbel, and P. Zwick, eds.). Gustav Fischer Verlag, Stuttgart.

Schmidt-Rhaesa, A. 1998. Phylogenetic relationships of the *Nematomorpha*—a discussion of current hypotheses. *Zool. Anz.* 236:203–216.

Schmidt-Rhaesa, A., B. Hanelt, and W. Reeves. 2003. Redescription and compilation of Nearctic freshwater *Nematomorpha* (*Gordiida*), with the description of two new species. *Proc. Acad. Nat. Sci. Phila.* 153:77–117.

Schmidt-Rhaesa, A., and L. Menzel. 2005. Central American and Caribbean species of horsehair worms (*Nematomorpha*), with the description of three new species. *J. Nat. Hist.* 39:515–529.

Sørensen, M. V., M. B. Hebsgaard, I. Heiner, H. Glenner, E. Willerslev, and R. M. Kristensen 2008. New data from an enigmatic phylum: evidence from molecular sequence data supports a sister-group relationship between *Loricifera* and *Nematomorpha. J. Zool. Syst. Evol. Res.* 46:231–239.

Vejdovsky, F. 1886. Zur Morphologie der Gordiiden. *Z. wiss. Zool.* 43:369–433.

Verrill, A. E. 1879. Notice of recent additions to the marine invertebrates of the Northern coast of America. *Proc. U.S. Natl. Mus.* 2:165–205.

Author

Andreas Schmidt-Rhaesa; Zoological Museum of the University Hamburg; Martin-Luther-King-Platz 3, 20146 Hamburg, Germany. Email: andreas.schmidt-rhaesa@uni-hamburg.de.

Date Accepted: 21 March 2011

Primary Editor: Kevin de Queiroz

Pan-Nematoda A.B. Smythe, new clade name

Registration Number: 242

Definition: The total clade of the crown clade *Nematoda*. This is a crown-based total-clade definition. Abbreviated definition: total ∇ of *Nematoda*.

Etymology: Derived from the Greek *pan*, meaning "all", in reference to the total clade, and *Nematoda*, the name of the corresponding crown clade (see *Nematoda* in this volume for the etymology of that name).

Reference Phylogeny: The primary reference phylogeny is Smythe et al. (2006: Fig. 7).

Composition: *Pan-Nematoda* includes the crown clade *Nematoda* (this volume) and all extinct nematodes that share more recent common ancestry with *Nematoda* than with any other mutually exclusive crown clade. No currently known fossil nematodes retain ancestral features suggesting that they diverged prior to the last common ancestor of all extant nematodes; therefore, there are no currently known taxa that are thought to be unambiguous stem nematodes. Several fossils unassignable to extant families (e.g., *Vetus* Taylor 1935 and *Captivonema cretacea* Manum, Bose, Sayer and Boström 1994) may represent stem nematodes but are missing informative structures needed to make this determination (Taylor, 1935; Manum et al., 1994).

Diagnostic Apomorphies: Referral of an extinct organism to the stem of *Nematoda* presumably would be based on its possession of some, but not all, of the diagnostic apomorphies of *Nematoda*; however, as unambiguous stem nematodes are currently unknown (see Composition), the order of acquisition of those apomorphies is also unknown.

Synonyms: *Nematoda* Diesing 1861 (approximate); *Nemates* Cobb 1919 (approximate); *Nemata* Chitwood 1958 (approximate); *Adenophorea* von Linstow 1905 (partial). As no name has previously been applied explicitly to the total clade, there are no unambiguous synonyms.

Comments: Due to the relatively recent age of most known nematode fossils and their lack of ancestral features, the fossil record is uninformative regarding nematode origins (Andrássy, 1976; Conway Morris, 1981). The majority of nematode fossils are known from amber (Poinar, 1992) and few are older than 40 million years. None of the amber fossils, nor *Palaeonema phyticum*, recently described from Devonian chert (Poinar et al., 2008), are thought to be stem nematodes. Recent molecular clock estimates of divergence dates for members of *Bilateria* range from 600–1300 million years ago (Hedges et al., 2004; Peterson et al., 2004), suggesting that the total clade of nematodes originated at least 600 million years ago. See *Nematoda* (this volume) for a brief discussion of the closest living relatives of nematodes.

Literature Cited

Andrássy, I. 1976. *Evolution as a Basis for the Systematization of Nematodes*. Pitman Publishers, London.

Conway Morris, S. 1981. Parasites and the fossil record. *Parasitology* 82:489–509.

Hedges, S., J. Blair, M. Venturi, and J. Shoe. 2004. A molecular timescale of eukaryote evolution and the rise of complex multicellular life. *BMC Evol. Biol.* 4:2.

Manum, S., M. Bose, R. Sayer, and S. Boström. 1994. A nematode (*Captivonema cretacea* gen. et sp. n.) preserved in a clitellate cocoon wall from the Early Cretaceous. *Zool. Scr.* 23: 27–31.

Peterson, K., J. Lyons, K. Nowak, C. Takacs, M. Wargo, and M. McPeek. 2004. Estimating metazoan divergence times with a molecular clock. *Proc. Natl. Acad. Sci. USA* 101:6536–6541.

Poinar, G. O. 1992. *Life in Amber*. Stanford University Press, Stanford, CA.

Poinar, J. O., George, H. Kerp, and H. Hass. 2008. *Palaeonema phyticum* gen. n., sp. n. (*Nematoda: Palaeonematidae* fam. n.), a Devonian nematode associated with early land plants. *Nematology* 10:9–14.

Smythe, A. B., M. J. Sanderson, and S. A. Nadler. 2006. Nematode small subunit phylogeny correlates with alignment parameters. *Syst. Biol.* 55:972–992.

Taylor, A. 1935. A review of the fossil nematodes. *Proc. Helminthol. Soc. Wash.* 2:47–49.

Author

Ashleigh B. Smythe; Department of Invertebrate Zoology; National Museum of Natural History; Smithsonian Institution; Washington, DC. Current address: Department of Biology; Virginia Military Institute; 301B Maury-Brooke Hall; Lexington, VA 24450, USA. Email: smytheab@vmi.edu.

Date Accepted: 9 April 2011

Primary Editor: Kevin de Queiroz

Nematoda C. Diesing 1861 [A.B. Smythe], converted clade name

Registration Number: 241

Definition: The most inclusive crown clade containing *Enoplus brevis* Bastian 1865 but not *Gordius aquaticus* Linnaeus 1758 (*Nematomorpha*), *Priapulus caudatus* Lamarck 1816 (*Priapulida*), *Echinoderes horni* Higgins 1983 (*Kinorhyncha*), *Nanaloricus mysticus* Kristensen 1983 (*Loricifera*), *Brachionus plicatilis* Mueller 1786 (*Rotifera*), *Lepidodermella squamata* (Dujardin 1841) (*Gastrotricha*), *Echiniscus viridissimus* Péterfi 1959 (*Tardigrada*), *Peripatopsis capensis* (Grube 1866) (*Onychophora*), and *Drosophila melanogaster* Meigen 1830 (*Arthropoda*). This is a maximum-crown-clade definition. Abbreviated definition: max crown ∇ (*Enoplus brevis* Bastian 1865 ~ *Gordius aquaticus* Linnaeus 1758 & *Priapulus caudatus* Lamarck 1816 & *Echinoderes horni* Higgins 1983 & *Nanaloricus mysticus* Kristensen 1983 & *Brachionus plicatilis* Mueller 1786 & *Lepidodermella squamata* (Dujardin 1841) & *Echiniscus viridissimus* Péterfi 1959 & *Peripatopsis capensis* (Grube 1866) & *Drosophila melanogaster* Meigen 1830).

Etymology: Derived from the Greek, *nema*, meaning "thread" and *odes*, meaning "like" or "having the form of".

Reference Phylogeny: The primary reference phylogeny is Smythe et al. (2006: Fig. 7). *Echinoderes horni*, *Nanaloricus mysticus*, *Lepidodermella squamata*, *Echiniscus viridissimus*, *Peripatopsis capensis*, and *Drosophila melanogaster* are not included in the reference phylogeny but are considered to lie outside of the clade originating in the last common ancestor of *Enoplus brevis* and *Trichinella spiralis* (Mallatt and Giribet, 2006; Meldal et al., 2007; Sørensen et al., 2008).

Composition: *Nematoda* currently consists of approximately 25,000 extant described species (Baldwin et al., 2000), but estimates of the true number of extant species vary radically, from 500,000 to 100,000,000 species (Platt and Warwick, 1980; Lambshead, 1993, 2004). Extinct members found as fossils include *Oligaphelenchoides atrebora* (fungal-feeding nematodes in *Aphelenchoididae*), *Oligoplectus* (microbivores in *Plectidae*), *Heydenius* and *Heleidomermis libani* (both insect parasites in *Mermithidae*), and *Palaeonema phyticum* (plant-associated nematodes in the newly erected *Palaeonematidae*) (Taylor, 1935; Poinar, 1977; Poinar et al., 1994, 2008).

Diagnostic Apomorphies:

1. Three circlets of sensilla surrounding the stoma (with six, six, and four sensilla per circlet respectively) (Coomans, 1979; Maggenti, 1991).
2. Amphids (paired anteriolateral organs, each with an external opening leading to a modified cilium and combined sensory-secretory functions) (Nielsen, 2001; Baldwin and Perry, 2004).
3. Cuticle molted four times during post-embryonic development (Lorenzen, 1985; Hodda, 2007).
4. Lateral epidermal cords containing the nuclei of epidermal cells (Ax, 2003; Hodda, 2007).
5. Locomotion through dorsoventral undulation (Ax, 2003).

6. Somatic musculature having separate contractile and non-contractile regions (Baldwin and Perry, 2004).

7. Ovaries (if didelphic) opposed to each other and open through a vulva placed at or near mid-body (Lorenzen, 1985).

8. Sperm non-flagellate, with cytoskeleton composed of major sperm protein and lacking an acrosome (Nielsen, 2001; Justine, 2002).

9. Male with paired copulatory spicules and cloaca (cloaca absent in females) (Andrássy, 1976; Hodda, 2007).

See Comments for exceptions to numbers 1, 7, and 9.

Synonyms: *Nemates* Cobb 1919 (approximate); *Nemata* Chitwood 1958 (approximate); *Adenophorea* von Linstow 1905 (partial); see Comments.

Comments: *Nematoidea* Rudolphi 1808 was first used for "rundwurms"—nematodes and nematomorphs (Maggenti et al., 1987). Diesing (1861) first proposed *Nematoda*, but still included the nematomorphs. Cobb (1919) removed the nematomorphs and left only nematodes when he proposed the name *Nemates*. Potts (1932) and Chitwood and Chitwood (1950) agreed with the removal of *Nematomorpha* but did not accept the name *Nemates* and synonymized it with *Nematoda*, a modification of *Nematodes* Burmeister 1837 (which was applied to a group that included nematomorphs). Chitwood (1958) proposed renaming the taxon *Nemata*, but that name failed to gain widespread acceptance. *Nematoda* was chosen as the preferred name due to its older and more common present use for a group corresponding to the named clade (De Ley and Blaxter, 2002).

Lorenzen (1985) listed three apomorphies ("holapomorphies") for *Nematoda*: three circlets of sensilla anteriorly; if two ovaries, they

are opposed and open through a vulva situated at mid-body; cuticle molted four times. The basic arrangement of sensilla surrounding the stoma (considered primitive by many authors, e.g., Chitwood and Chitwood, 1974; Maggenti, 1981) is a circlet of six inner labial sensilla, a circlet of six outer labial sensilla, and a more posterior circlet of four cephalic (or paralabial) sensilla (De Coninck, 1965). Sensilla can protrude externally as setae or papillae, or occur hidden below the cuticular surface. Sometimes they are repositioned or absent relative to the basic formula (see Coomans, 1979, and references therein). Chitwood and Chitwood (1974) considered two ovaries to be the primitive condition, but many nematodes have just one (e.g., *Cephalobidae*). Lorenzen's (1985) ovarian character best applies to free-living nematodes with two ovaries, as several plant-parasites (e.g., *Meloidogyne*) and animal-parasites (e.g., *Ascaris*) have parallel (overlapping and extending in the same direction) ovaries (Maggenti, 1981).

Nielsen (2001) included the circlets of sensilla and four molts as apomorphies, but disregarded Lorenzen's (1985) ovarian character, including instead the "highly specialized sperm": non-flagellate, usually amoeboid (but elongate to rod-like in some taxa, e.g., *Nippostrongylus* [Jamuar, 1966], *Deontostoma* [Wright et al., 1973]) and characterized by a cytoskeleton protein found only in nematodes, the major sperm protein, which likely takes the place of actin in amoeboid movement (King et al., 1992; Justine, 2002).

Nielsen (2001) also suggested that the amphids may be a nematode apomorphy, with one located on either side of the head or cervical region of all nematodes (Filipjev and Schuurmans Stekhoven Jr., 1941). Steiner (1919, 1920) suggested that the amphids and accompanying organs are homologous with the retrocerebral organs and subcerebral glands of rotifers, particularly *Callidina*. Ultrastructural examination indicates that these organs in rotifers consist entirely of glands (Clément and Wurdak, 1991) and lack the modified cilia of amphids. It has also been suggested that amphids are homologous with the cephalic sense organs of gastrotrichs (Remane, 1936; Teuchert, 1977). The primary distinguishing feature between nematode amphids and gastrotrich cephalic sense organs is that the former are open to the environment while the latter are closed by a layer of cuticle (Lorenzen, 1994). An exception to this are the blister-shaped amphids of the *Desmoscolecida*, which are covered by a thin membrane. Ultrastructural studies are needed to determine if this membrane is composed of cuticle or the corpus gelatum (Riemann, 1972; Lorenzen, 1994). Lee (1965) pointed out that many animals, including insects, have sensory structures with similar features (e.g., modified ciliated neurons surrounded by specialized accessory cells), and Ward et al. (1975) and Wright (1983) suggested that the similarities among different taxa reflect convergence.

Several authors have considered the lack of circular somatic muscles to be diagnostic for nematodes (e.g., Andrássy, 1976; Hodda, 2007) but nematomorphs also lack circular muscles (Ax, 2003). Somatic musculature consisting of only longitudinal muscles results in a characteristically sinusoidal, undulatory motion, which occurs in the lateral dimension in nematomorphs and dorsoventrally in nematodes (Ax, 2003). While a number of invertebrates swim with a dorsoventral undulatory motion (e.g., leeches [Kristan Jr. et al., 1974]), only nematodes also travel in this way while lying on their sides on a hard substrate. Another widely known feature of nematodes is the separate contractile and noncontractile regions of the muscle cells (Baldwin and Perry, 2004), but this has never previously been proposed as an apomorphy. This type of muscle cell structure has not been demonstrated in other animals and is therefore proposed as an apomorphy for *Nematoda*.

Andrássy (1976) suggested that the copulatory spicules of the male nematode are an apomorphy. The majority of nematodes have 2 spicules, but a single spicule does occur in many groups (e.g., *Trichuroidea, Oxyuridae, Desmodoridae* [Chitwood and Chitwood, 1974]). Several taxa, usually in groups whose other members have single or fused spicules, are devoid of spicules entirely (e.g., *Trichinella, Hystrignathus* [Chitwood and Chitwood, 1974]; *Myolaimus* [De Ley and Blaxter, 2002]). Ax (2003) considered the presence of a cloaca only in males to be an apomorphy for nematodes, and contrasted this with the presence of a cloaca in both male and female nematomorphs.

While the monophyly of nematodes is widely accepted by contemporary taxonomists (e.g., Hodda, 2007; Meldal et al., 2007; but see Inglis, 1983; Lorenzen, 1996), the relationship of nematodes to other animals remains controversial. The primary hypotheses of the relationships of nematodes to other metazoans are termed the "*Coelomata*" and "*Ecdysozoa*" hypotheses. In the "*Coelomata*" hypothesis, nematodes and other "pseudocoelomates" (animals lacking a true coelom, e.g., *Rotifera, Gastrotricha*) are early-branching metazoans while arthropods and chordates (coelomates) form a clade (e.g., Blair et al., 2002; Wolf et al., 2004; Philip et al., 2005). In the "*Ecdysozoa*" hypothesis, nematodes are grouped with arthropods in a clade of molting animals (e.g., Aguinaldo et al., 1997; Peterson and Eernisse, 2001; Philippe et al., 2005; Sørensen et al., 2008). Most morphological analyses and many of the early molecular phylogenies suggest that *Nematomorpha* is the sister taxon to nematodes (e.g., Aleshin et al., 1998; Adoutte et al., 2000; Nielsen, 2001; Ax, 2003), and *Nematoidea* Rudolfi 1808 would seem an appropriate name for this clade, although a recent molecular phylogeny found *Nematoda* to be sister to a clade formed by *Loricifera* and *Nematomorpha* (Sørensen et al., 2008).

A maximum-crown-clade definition with external specifiers from all likely sister taxa is provided here to promote compositional stability regardless of the eventual determination of the sister taxon to nematodes. External specifiers not shown in the reference phylogeny were included in the definition because the reference phylogeny did not include all of the potential sister taxa of nematodes. Within nematodes, relationships among the earliest-branching lineages are also unclear, with major clades at the base of the tree either unresolved or poorly supported (e.g., Blaxter et al., 1998; De Ley and Blaxter, 2002; Smythe et al., 2006; Meldal et al., 2007; Bik et al., 2010). Blaxter and Koutsovoulos (2015) constructed a phylogeny using 181 protein coding genes that placed *Enoplus brevis* as the earliest branching lineage, but suggested this may be an artifact of that taxon's relatively short branch length. The definition used here will apply to a clade composed of the same set of species in the context of any potential resolution of the relationships within *Nematoda*.

The potential discovery of a currently unknown taxon that is more closely related to all of the taxa that have traditionally been called nematodes than to any listed external specifier would pose a problem for this definition. If this occurs, the morphology of the new taxon and common sense should determine its placement. If the new taxon is sufficiently worm-like to be considered a nematode, it should be placed in *Nematoda*. If the animal is not worm-like, the definition should be emended (*PhyloCode* Art. 15) to include a species of the new taxon as an external specifier.

Literature Cited

Adoutte, A., G. Balavoine, N. Lartillot, O. Lespinet, B. Prud'homme, and R. de Rosa. 2000. The new animal phylogeny: reliability and implications. *Proc. Natl. Acad. Sci. USA* 97:4453–4456.

Aguinaldo, A. M. A., J. M. Turbeville, L. S. Linford, M. C. Rivera, J. R. Garey, R. A. Raff, and J. A. Lake. 1997. Evidence for a clade of nematodes, arthropods and other moulting animals. *Nature* 387:489–493.

Aleshin, V., I. Milyutina, O. Kedrova, N. Vladychenskaya, and N. Petrov. 1998. Phylogeny of *Nematoda* and *Cephalorhyncha* derived from 18S rDNA. *J. Mol. Evol.* 47:597–605.

Andrássy, I. 1976. *Evolution as a Basis for the Systematization of Nematodes*. Pitman Publishers, London.

Ax, P. 2003. *Multicellular Animals: Order in Nature—System Made by Man*. Springer, Berlin.

Baldwin, J. G., S. A. Nadler, and D. H. Wall. 2000. Nematodes: pervading the earth and linking all life. Pp. 176–191 in *Nature and Human Society: The Quest for a Sustainable World: Proceedings of the 2000 Forum on Biodiversity* (P. Raven and T. Williams, eds.). National Academy Press, Washington, D.C.

Baldwin, J. G., and R. N. Perry. 2004. Nematode morphology, sensory structure and function. Pp. 175–257 in *Nematology: Advances and Perspectives*, Vol. 1 (Z. X. Chen, S. Y. Chen, and D. W. Dickson, eds.). CABI Publishing, Beijing.

Bik, H. M., P. J. D. Lambshead, W. K. Thomas, and D. H. Lunt. 2010. Moving towards a complete molecular framework of the *Nematoda*: a focus on the *Enoplida* and early-branching clades. *BMC Evol. Biol.* 10:353.

Blair, J., K. Ikeo, T. Gojobori, and S. B. Hedges. 2002. The evolutionary position of nematodes. *BMC Evol. Biol.* 2:7.

Blaxter, M. L., P. De Ley, J. R. Garey, L. X. Liu, P. Scheldeman, A. Vierstraete, J. R. Vanfleteren, L. Y. Mackey, M. Dorris, L. M. Frisse, J. T. Vida, and W. K. Thomas. 1998. A molecular evolutionary framework for the phylum *Nematoda*. *Nature* 392:71–75.

Blaxter, M., and G. Koutsovoulos, 2015. The evolution of parasitism in *Nematoda*. *Parasitology* 142(suppl. 1):S26–S39.

Burmeister, H. 1837. *Handbuch der Naturgeschichte. Zum Gebrauch bei Vorlesungen. Zweite Abtheilung, Zoologie*. T. C. F. Enslin, Berlin.

Chitwood, B. G. 1958. The designation of official names for higher taxa of invertebrates. *Bull. Zool. Nomen.* 15:860–895.

Chitwood, B. G., and M. B. Chitwood. 1950. *Introduction to Nematology.* University Park Press, Baltimore, MD.

Chitwood, B. G., and M. B. Chitwood. 1974. *Introduction to Nematology.* Revised edition. University Park Press, Baltimore, MD.

Clément, P., and E. Wurdak. 1991. *Rotifera.* Pp. 219–297 in *Microscopic Anatomy of Invertebrates. Vol. 4:* Aschelminthes (F. W. Harrison and E. E. Ruppert, eds.). Wiley-Liss, Inc., New York.

Cobb, N. A. 1919. The orders and classes of nemas. Pp. 213–216 in *Contributions to a Science of Nematology,* Vol. 8. Waverly Press, Baltimore, MD.

Coomans, A. 1979. The anterior sensilla of nematodes. *Revue Nématologie* 2:259–283.

De Coninck, L. A. 1965. Classe des nématodes—systématique des nématodes et sous-classe des *Adenophorea. Traité Zool.* (Grassé, ed.). 4:586–681.

De Ley, P., and M. L. Blaxter. 2002. Systematic position and phylogeny. Pp. 1–30 in *The Biology of Nematodes* (D. Lee, ed.). Taylor & Francis, London.

Diesing, K. M. 1861. Revision der Nematoden. *Sitzungber. Kaiserl. Akad. Wissensch. Wien* 42:595–736.

Filipjev, I. N., and J. H. Schuurmans Stekhoven, Jr. 1941. *A Manual of Agricultural Helminthology.* E.J. Brill, Leiden.

Hodda, M. 2007. Phylum *Nematoda. Zootaxa* 1668:265–293.

Inglis, G. W. 1983. An outline classification of the phylum *Nematoda. Aust. J. Zool.* 31:243–255.

Jamuar, M. P. 1966. Studies of spermiogenesis in a nematode, *Nippostrongylus brasiliensis. J. Cell Biol.* 31:381–396.

Justine, J. 2002. Male and female gametes and fertilisation. Pp. 73–119 in *The Biology of Nematodes* (D. Lee, ed.). Taylor & Francis, London.

King, K. L., M. Stewart, T. M. Roberts, and M. Seavy. 1992. Structure and macromolecular assembly of two isoforms of the major sperm protein (MSP) from the amoeboid sperm of the nematode, *Ascaris suum. J. Cell Sci.* 101:847–857.

Kristan, Jr., W., G. Stent, and C. Ort. 1974. Neuronal control of swimming in the medicinal leech. I. Dynamics of the swimming rhythm. *J. Comp. Physiol.* 94:97–119.

Lambshead, P. 1993. Recent developments in marine benthic biodiversity research. *Oceanis* 19:5–24.

Lambshead, P. 2004. Marine nematode biodiversity. Pp. 438–468 in *Nematology: Advances and Perspectives,* Vol. 1 (Z. Chen, S. Chen, and D. Dickson, eds.). CABI Publishing, Beijing.

Lee, D. L. 1965. *The Physiology of Nematodes.* Oliver and Boyd, London.

Lorenzen, S. 1985. Phylogenetic aspects of pseudocoelomate evolution. Pp. 211–223 in *The Origins and Relationships of Lower Invertebrates* (S. C. Morris, J. George, R. Gibson, and H. Platt, eds.). The Systematics Association Special Volume No. 28. Clarendon Press, Oxford.

Lorenzen, S. 1994. *The Phylogenetic Systematics of Freeliving Nematodes.* The Ray Society, London.

Lorenzen, S. 1996. *Nemathelminthes (Aschelminthes).* Pp. 682–732 in *Spezielle Zoologie—Erster Teil* (W. Westheide and R. Rieger, eds.). G. Fischer, Jena.

Maggenti, A. 1981. *General Nematology.* Springer-Verlag, New York.

Maggenti, A. 1991. *Nemata:* higher classification. Pp. 147–187 in *Manual of Agricultural Nematology* (W. Nickle, ed.). Marcel Dekker, New York.

Maggenti, A. R., M. Luc, D. J. Raski, R. Fortuner, and E. Geraert. 1987. A reappraisal of *Tylenchina (Nemata).* 2. Classification of the suborder *Tylenchina (Nemata: Diplogasteria). Revue Nématol.* 10:135–142.

Mallatt, J., and G. Giribet. 2006. Further use of nearly complete 28S and 18S rRNA genes to classify *Ecdysozoa:* 37 more arthropods and a kinorynch. *Mol. Phylogenet. Evol.* 40:772–794.

Meldal, B. H. M., N. J. Debenham, P. De Ley, I. T. De Ley, J. R. Vanfleteren, A. R. Vierstraete, W. Bert, G. Borgonie, T. Moens, P. A. Tyler, M. C. Austen, M. L. Blaxter, A. D. Rogers, and P. J. D. Lambshead. 2007. An improved molecular phylogeny of the *Nematoda* with special emphasis on marine taxa. *Mol. Phylogenet. Evol.* 42:622–636.

Nielsen, C. 2001. *Animal Evolution: The Interrelationships of the Living Phyla.* 2nd edition. Oxford University Press, Oxford.

Peterson, K. J., and D. J. Eernisse. 2001. Animal phylogeny and the ancestry of bilaterians: Inferences from morphology and 18S rDNA gene sequences. *Evol. Dev.* 3:170–205.

Philip, G. K., C. J. Creevey, and J. O. McInerney. 2005. The *Opisthokonta* and the *Ecdysozoa* may not be clades: stronger support for the grouping of plant and animal than for animal and fungi and stronger support for the *Coelomata* than *Ecdysozoa. Mol. Biol. Evol.* 22:1175–1184.

Philippe, H., N. Lartillot, and H. Brinkmann. 2005. Multigene analyses of bilatarian animals corroborate the monophyly of *Ecdysozoa*, *Lophotrochozoa*, and *Prostomia. Mol. Biol. Evol.* 22:1246–1253.

Platt, H. M., and R. M. Warwick. 1980. The significance of free-living nematodes to the littoral ecosystem. Pp. 729–759 in *The Shore Environment. Vol. 2: Ecosystems* (H. Price, D. Irvine, and W. Farnham, eds.). Academic Press, London.

Poinar, Jr., G. O. 1977. Fossil nematodes from Mexican amber. *Nematologica* 23:232–238.

Poinar, Jr., G. O., A. Acra, and F. Acra. 1994. Earliest fossil nematode (*Mermithidae*) in Cretaceous Lebanese amber. *Fund. Appl. Nematol.* 17:475–477.

Poinar, Jr., G. O. H. Kerp, and H. Hass. 2008. *Palaeonema phyticum* gen. n., sp. n. (*Nematoda*: *Palaeonematidae* fam. n.), a Devonian nematode associated with early land plants. *Nematology* 10:9–14.

Potts, F. A. 1932. The phylum *Nematoda*. Pp. 214–227 in *The Invertebrata: A Manual for the Use of Students* (L. A. Borradaile, F. A. Potts, L. E. S. Eastham, and J. T. Saunders, eds.). Cambridge University Press, Cambridge, UK.

Remane, A. 1936. *Gastrotricha* und *Kinorhyncha*. Pp. 1–242 in *Klassen und Ordnungen des Tierreichs* (Vermes), Vol. 4 (H. Bronn, ed.). Akademische Verlagsgesellschaft, Leipzig.

Riemann, F. 1972. Corpus gelatum und ciliäre Strukturen als lichtmikroskopisch sichtbare Bauelemente des Seitenorgans freilebender Nematoden. *Z. Morphol. Tiere* 72:46–76.

Smythe, A. B., M. J. Sanderson, and S. A. Nadler. 2006. Nematode small subunit phylogeny correlates with alignment parameters. *Syst. Biol.* 55:972–992.

Sørensen, M. V., M. B. Hebsgaard, I. Heiner, H. Glenner, E. Willerslev, and R. M. Kristensen. 2008. New data from an enigmatic phylum: evidence from molecular sequence data supports a sister-group relationship between *Loricifera* and *Nematomorpha. J. Zool. Syst. Evol. Res.* 46:231–239.

Steiner, G. 1919. Zur Kenntnis der *Kinorhyncha*, nebst Bemerkungen über ihr Verwandtschaftsverhältnis zu den Nematoden. *Zool. Anz.* 50:177–187.

Steiner, G. 1920. Betrachtungen zur Frage des Verwandtschafts-verhältnisses der Rotatorien und Nematoden. *Festschrift Zschokke, Basel* 31:1–15.

Taylor, A. 1935. A review of the fossil nematodes. *Proc. Helminthol. Soc. Wash.* 2:47–49.

Teuchert, G. 1977. The ultrastructure of the marine gastrotrich *Turbanella cornuta* Remane (*Macrodasyoidea*) and its functional and phylogenetical importance. *Zoomorphologie* 88:189–246.

Ward, S., N. Thomson, J. G. White, and S. Brenner. 1975. Electron microscopical reconstruction of the anterior sensory anatomy of the nematode *Caenorhabditis elegans. J. Comp. Neurol.* 160:313–338.

Wolf, Y. I., I. B. Rogozin, and E. V. Koonin. 2004. *Coelomata* and not *Ecdysozoa*: evidence from genomewide phylogenetic analysis. *Genome Res.* 14:29–36.

Wright, K., W. Hope, and N. Jones. 1973. The ultrastructure of the sperm of *Deontostoma californicum*, a free-living marine nematode. *Proc. Helminthol. Soc. Wash.* 40:30–36.

Wright, K. A. 1983. Nematode chemosensilla: form and function. *J. Nematol.* 15:151–158.

Author

Ashleigh B. Smythe; Department of Invertebrate Zoology; National Museum of Natural History; Smithsonian Institution; Washington, DC. Current address: Department of Biology; Virginia Military Institute; 301B Maury-Brooke Hall; Lexington, VA 24450, USA. Email: smytheab@vmi.edu.

Date Accepted: 18 March 2012; updated 17 July 2017

Primary Editor: Kevin de Queiroz

Branchiopoda J.-B. Lamarck 1801 [T. A. Hegna and J. Olesen], converted clade name

Registration Number: 289

Definition: The smallest crown clade containing *Artemia salina* (originally *Cancer salinus*) (Linnaeus 1758) and *Daphnia magna* Straus 1820. This is a minimum-crown-clade definition. Abbreviated definition: min crown ∇ (*Artemia salina* (Linnaeus 1758) & *Daphnia magna* Straus 1820).

Etymology: Derived from the Greek βραγχΕα (*branchia*; gill) and πΟυς (*pous*; foot) (see Baird 1850: p. 17).

Reference Phylogeny: The primary reference phylogeny is Figure 1 in Regier et al. (2010). See also Richter et al. (2007: Fig. 8).

Composition: All extant anostracans, notostracans, 'conchostracans' (including spinicaudatans, laevicaudatans, and *Cyclestheria*), and cladocerans, plus all extinct taxa that are descended from their most recent common ancestor. Estimates of number of described living branchiopod species vary from 839 (Martin and Davis, 2006) to 1,180 (Adamowicz and Purvis, 2005). Adamowicz and Purvis (2005) estimated that the total number of living branchiopods (described and undescribed) is around 2,500 species.

Diagnostic Apomorphies: Apomorphies for the group were reviewed by Olesen (2004, 2007, 2009). The morphological heterogeneity of the clade makes it difficult to specify apomorphies that are not secondarily lost in one or more derived subclades. Among these are the reduction of the first and second maxillae in adults to coxae only (Olesen, 2004, 2007, 2009). However, males of the Devonian putative stem anostracan *Lepidocaris rhyniensis* Scourfield 1926 seem to have a first maxilla modified for clasping. Adult branchiopod trunk limbs have 5–6 endites and an unsegmented endopod (modified in cladocerans) (Olesen, 2004, 2007, 2009). Mandibles in adults consist only of a large *pars molaris* gnathal edge of the rolling-grinding type (Richter, 2004; Olesen, 2007), except in *Notostraca*, *Laevicaudata*, and various predatory/raptorial branchiopods such as onychopods and haplopods.

There are also a significant number of larval characters that unite branchiopods. The larval first antenna is unsegmented with setation restricted to the distal end (Olesen, 2007, 2009). The larval second antenna possesses a long protopod and a coxal masticatory spine that becomes branched after one of the first molts (Olesen, 2004, 2007, 2009). The larval mandible is a four-segmented, uniramous limb lacking an endopod (Olesen, 2009; Fritsch et al., 2013).

Synonyms: *Phyllopoda* Latreille 1802b sensu Sars (1867), *Euphyllopoda* Grobben 1892, and *Eubranchiopoda* Pennak 1953 are partial and approximate synonyms (see Comments for further discussion of these and other names associated with branchiopods).

Comments: The concept of *Branchiopoda* was refined during the first one hundred years of its use and has been essentially stable since Calman (1909; see Dumont and Negrea, 2002, for a review). The name '*Branchiopoda* = Branchiopode' was originally coined by Lamarck

(1801) as a genus name that was applied to then-known anostracans. Its use was continued by Latreille (1802a), whose more precise diagnosis would extend to all anostracans (p. 19, Tome Troisième). The taxon was elevated to the rank of order by Latreille (1817), and all subsequent authors have credited *Branchiopoda* to Latreille (1817), not to Lamarck (1801). Despite this confusion, there is no doubt that Lamarck (1801) was the original author of the name—he was credited as such by Latreille (1802b: p. 300, Tome Quatrième,) and we accordingly follow Latreille in assigning nominal authorship to Lamarck.

Branchiopoda, as applied by Latreille (1817), was polyphyletic and consequently differed markedly from current concepts and usage. Latreille's *Branchiopoda* included all branchiopods known at the time plus several maxillopod crustaceans (branchiurans, copepods, and ostracods) as well as non-crustacean xiphosurans (see Olesen, 2007). Both Gerstaeker's (1866) and Sars's (1867) formulations of *Branchiopoda* were closer to the modern concept. Both encompassed all known branchiopods plus branchiurans and either ostracods (Gerstaeker) or leptostracans. Both formulations were still polyphyletic. Calman (1909) is credited for formulating the modern concept of *Branchiopoda* (Dumont and Negrea, 2002). Calman's (1909) character support, however, has proven insufficient, and his characters cannot be translated directly into synapomorphies (see Olesen, 2007). But, as Olesen (2007) showed, this shortcoming of Calman's diagnosis does not invalidate the taxon, as numerous apomorphies provide robust support for its monophyly. As the name is defined here, *Branchiopoda* does not have any full synonyms, but it has three partial synonyms: *Phyllopoda* Latreille 1802b sensu Sars (1867), *Euphyllopoda* Grobben 1892, and *Eubranchiopoda* Pennak 1953.

Phyllopoda Latreille 1802b has been used in a couple of senses (taxonomic and descriptive), as discussed by Fryer (1987) and Martin and Davis (2006). It was first used as a taxon name by Latreille (1802b), but it only contained one species: *Limulus apus* (presumably the notostracan species currently known as *Lepidurus* (originally *Monoculus*) *apus* Linnaeus 1758). Sars (1867) used *Phyllopoda* for a taxon within *Branchiopoda* that encompassed the non-cladoceran branchiopods; in Sars's concept, *Phyllopoda* could be considered a partial synonym of *Branchiopoda* as it seems to refer to a paraphyletic group originating in approximately the same ancestor. *Phyllopoda* is often credited to Preuss (1951), who did not coin the name but instead substantially reinterpreted the taxon to contain all non-anostracan branchiopods; because *Anostraca* is sister to other branchiopods (Regier et al., 2010), *Phyllopoda* sensu Preuss would not be synonymous with *Branchiopoda*, as the former would apply to a subtaxon of the latter. This has been the most common taxonomic use of *Phyllopoda* over the last twenty years and is generally accepted currently. Schram (1986), following a phylogenetic analysis, recast *Phyllopoda* to include leptostracans and cephalocarids. His analysis was criticized for this suggestion (e.g., Dahl, 1987; Fryer, 1987), and these criticisms inspired successively more detailed analyses with better-supported results indicating the polyphyly of this grouping (Wilson, 1992; Spears and Abele, 1997, 1999; Giribet et al., 2005; Richter et al., 2007; and see below). In a paper critical of Schram's proposal, Martin and Christiansen (1995) advocated using the name *Phyllopoda* strictly as a descriptive term—not as a phylogenetic one—citing the tortured history associated with that taxon name.

Euphyllopoda Grobben 1892 has been used in field guides and faunal lists, but not in any explicit taxonomic studies. In modern usage, it encompasses the large branchiopods (*Anostraca*, *Notostraca*, *Laevicaudata*, *Spinicaudata*, and *Cyclestheria*)—i.e., all branchipods except the cladocerans. This association with a paraphyletic assemblage is consistent with Grobben's (1892) original concept as interpreted from his schematic tree (Fig. 1); as a group originating in approximately the same ancestor, it thus qualifies as a partial synonym of *Branchiopoda*.

Eubranchiopoda (neither an author nor a date have conventionally been assigned to this name—Pennak's 1953 text book contains the earliest reference to the name that the authors have been able to find) includes the 'large branchiopods' (all branchiopods excluding cladocerans). It has been used in several faunal guides (Pennak, 1953; Saunders, 1980, 1981; Dong et al., 1982; Saunders and Wu, 1984). However, it has never been used in any explicit taxonomic studies, and a recent textbook (Saxena, 2005) urged its abandonment because of non-monophyly.

The following names associated with branchiopod taxonomy are worth discussing, although they are not synonyms of the taxon name *Branchiopoda* as defined here: *Entomostraca* Müller 1785, *Branchipodiodes* Starobogatov 1984, *Gnathostraca* Dahl 1956, and *Pan-Branchiopoda* Olesen 2004. *Entomostraca* Müller 1785 was an early name used for a group that included many freshwater crustaceans; Baird's (1850) usage in his monograph on British entomostracans included branchiopods as we recognize them today plus branchiurans, copepods, leptostracans, and ostracods. The name was essentially abandoned during the twentieth century (see Boxshall, 2007). However, Walossek (1999) has attempted to resurrect the taxon—emending it to include *Branchiopoda*, *Cephalocarida*, and *Maxillopoda* based on evidence derived from exceptionally preserved fossil crustaceans from the Upper Cambrian of Sweden (the 'Orsten' deposits). Walossek's (1999) revised version of *Entomostraca* may, however, rest on crustacean plesiomorphies, thereby casting its monophyly into doubt (Boxshall, 2007; Wolfe and Hegna, 2014). That name has in any case been applied uniformly to a more inclusive group than

Branchiopoda and thus cannot be considered a synonym.

Branchipodiodes Starobogatov 1986 (English translation 1988) was proposed as part of his somewhat idiosyncratic system of classification for crustaceans and has not been widely used. Besides a rigorous application of his own system of taxonomic suffixes reflecting hierarchical rank (Starobogatov, 1984), the composition of his Class *Branchipodiodes* contains all members of *Branchiopoda* as recognized herein, as well as *Remipedia*, *Cephalocarida*, and a range of fossil taxa. Now regarded as referring to a polyphylum (e.g., Regier et al., 2010), *Branchipodiodes* could be used for a more inclusive clade, but it would not be a synonym of *Branchiopoda*.

Gnathostraca Dahl 1956 was originally proposed to include the *Anostraca*, *Cephalocarida*, and *Phyllopoda* (sensu Preuss, 1951) with the anostracans hypothesized to be more closely related to cephalocarids than to phyllopods. This proposal followed closely on the discovery of cephalocarids (Sanders, 1955), and it was one of the first attempts to infer their phylogenetic position. *Gnathostraca* has not seen wide usage, with Dahl (1963) himself retreating from its use as more became known about cephalocarids. Because cephalocarids (xenocaridan mircocrustaceans) are only distantly related to vericrustacean branchiopods (Regier et al., 2010), *Gnathostraca* was applied to a polyphyletic group and is not a synonym of *Branchiopoda*.

Pan-Branchiopoda Olesen 2004 was proposed to encompass stem-group branchiopods as well as crown-group branchiopods.

The tree topology of *Branchiopoda* has been fairly stable in terms of the monophyly of its major subclades—whether analyzed using morphology (Olesen, 2009; Richter et al., 2007) or molecules (Brabrand et al., 2002; deWaard et al., 2006; Richter et al., 2007; Stenderup et al., 2006; Sun et al., 2006, 2016). Nearly all analyses infer *Anostraca* as sister to the rest, with *Cyclestherida* and *Cladocera* as sisters. *Rehbachiella kinnekullensis* is the earliest putative branchiopod fossil (Walossek, 1993)—living nearly 500 million years ago in the late Cambrian. Its phylogenetic position as either a stem anostracan (i.e., thus a crown branchiopod; Walossek, 1993) or as a stem branchiopod (Olesen, 2007) is a matter of debate. This conflict is complicated by the fact that *Rehbachiella kinnekullensis* is only represented by larval stages—its adult morphology is unknown. This complicates interpreting relevant anatomical features and discerning their phylogenetic significance (Boxshall, 2007; Wolfe and Hegna, 2014). Additional putative branchiopod fossils are known from the Ordovician (Caster and Brooks, 1956; Young et al., 2007) and the Silurian (Mikulic et al., 1985a,b; Schram, 1986; von Bitter et al., 2007), but all of these fossils are in need of closer study. The oldest unambiguous branchiopod remains are anostracan-like mandibles from Middle/Late Cambrian (Harvey et al., 2012). These mandibles are consistent in age with the divergence-time estimates of Oakley et al. (2013). Other unambiguous branchiopod fossils are known from the Devonian Rhynie Chert in Scotland, i.e., the stem anostracan *Lepidocaris rhyniensis* Scourfield 1926 and the stem diplostracan *Castracollis wilsonae* Fayers and Trewin 2002. These fossils are almost universally thought to represent crown branchiopods (Olesen, 2007, 2009; Wolfe et al., 2016; but see Schram, 1986, for an alternate interpretation of *Lepidocaris rhyniensis's*). In addition, fossil 'conchostracan' (i.e., ?spinicaudatan) carapaces are known from the Devonian (e.g., Novozhilov, 1961; Brummer, 1980; Shen, 1983).

Literature Cited

Adamowicz, S. J., and A. Purvis. 2005. How many branchiopod crustacean species are there? Quantifying the components of underestimation. *Glob. Ecol. Biogeogr.* 14:455–468.

Baird, W. 1850. *The Natural History of the British Entomostraca.* Ray Society, London.

von Bitter, P. H., M. A. Purnell, D. K. Tetreault, and C. A. Stott. 2007. Eramosa Lagerstätte—exceptionally preserved soft-bodied biotas with shallow-marine shelly and bioturbating organisms (Silurian, Ontario, Canada). *Geology* 35:879–882.

Boxshall, G. A. 2007. Crustacean classification: ongoing controversies and unresolved problems. *Zootaxa* 1668:313–325.

Braband, A., S. Richter, R. Hiesel, and G. Scholtz. 2002. Phylogenetic relationships within the *Phyllopoda* (*Crustacea, Branchiopoda*) based on mitochondrial and nuclear markers. *Mol. Phylo. Evol.* 25:229–244.

Brummer, G. J. 1980. Midden devonische conchostracenmergel, een Zuidoost-Baltisch gesteente uit de rode "schollen"-keileem van Groningen (Middle Devonian conchostracan marl, southeastern Baltic rock from the red shell marls of Groningen). *Grondboor Hamer* 6:186–190.

Calman, W. T. 1909. *Crustacea.* Pp. 1–136 in *Treatise on Zoology Part VII Appendiculata*, (R. Lancaster, ed.). Adam and Charles Black, London.

Caster, K. E., and H. K. Brooks. 1956. New fossils from the Canadian-Chazyan (Ordovician) hiatus in Tennessee. *Bul. Am. Paleo.* 36:157–199.

Dahl, E. 1956. On the differentiation of the topography of the crustacean head. *Acta Zool.* 37:123–192.

Dahl, E. 1963. Main evolutionary lines among recent crustaceans. Pp. 1–26 in *Phylogeny and Evolution of Crustacea* (H. B. Whittington and W. D. I. Rolfe, eds.). Museum of Comparative Zoology Special Publication, Cambridge, MA.

Dahl, E. 1987. *Malacostraca* maltreated—the case of the *Phyllocarida. J. Crust. Biol.* 7:721–726.

deWaard, J. R., V. Sacherova, M. E. A. Cristescu, E. A. Remigio, T. J. Crease, and P. D. N. Hebert. 2006. Probing the relationships of the branchiopod crustaceans. *Mol. Phylo. Evol.* 39:491–502.

Dong, J.-M., A.-Y. Dai, Y.-Z. Jiang, S.-Z. Chen, Y.-S. Chen, and R.-X. Cai 1982 (in Chinese). Class *Crustacea.* Pp. 1–114 in *Illustrated Animals of China.* 2nd edition, Vol. 1. Science and Technology Press, Beijing.

Dumont, H. J., and S. V. Negrea. 2002. *Branchiopoda*. Pp. 1–398 in *Guides to the Identification of the Microinvertebrates of the Continental Waters of the World* (H. J. Dumont, ed.). Backhuys Publishers, Leiden.

Fayers, S. R., and N. H. Trewin. 2002. A new crustacean from the Early Devonian Rhynie chert, Aberdeenshire, Scotland. *Trans. R. Soc. Edinb. Earth Sci.* 93:355–382.

Fritsch, M., O. R. Bininda-Emonds, and S. Richter. 2013. Unraveling the origin of *Cladocera* by identifying heterochrony in the developmental sequences of *Branchiopoda*. *Front. Zool.* 10:35.

Fryer, G. 1987. A new classification of the branchiopod *Crustacea*. *Zool. J. Linn. Soc.* 91:357–383.

Gerstaecker, A. 1866. *Crustacea* (Erste Hälfte). Pp. 1–1320 in *Die Klassen und Ordnungen der Thier-Reichs, Bd. 5* (Arthropoda). Abt. 1. Akademie Verlag, Leipzig.

Giribet, G., S. Richter, G. D. Edgecombe, and W. C. Wheeler. 2005. The position of crustaceans within *Arthropoda*—evidence from nine molecular loci and morphology. Pp. 307–353 in *Crustacea and Arthropod Relationships* (S. Koenenmann and R. A. Jenner, eds.). CRC Press, Boca Raton, FL.

Grobben, K. 1892. Zur kenntniss des stammbaumes und des systems der Crustaceen. Sitzungsberichte der Kaiserlichen Akademie der Wissenschaften. *Math. Natur. Classe* 101:237–274.

Harvey, T. H., M. I. Velez and N. J. Butterfield. 2012. Exceptionally preserved crustaceans from western Canada reveal a cryptic Cambrian radiation. *Proc. Natl. Acad. Sci. USA* 109:1589–1594.

Lamarck, J. B. 1801. *Système des animaux sans vertèbres*. Déterville, Paris.

Latreille, P. A. 1802a. *Familles Naturelles des Genres, Tome Troisème. Histoire Naturelle, Générale et Particulière des Crustacés et Des Insectes*. F. Dufart, Paris.

Latreille, P.A. 1802b. *Familles Naturelles des Genres, Tome Quatrième. Histoire Naturelle, Générale et Particulière des Crustacés et Des Insectes*. F. Dufart, Paris.

Latreille, P. A. 1817. Les crustacés, les arachnides, les insectes. Pp. 1–653 in *Le Regne Animal Distribue d'Apres Son Organisation, Pour Servir de Base a l'Histoire Naturelle des Animaux et d'Introduction a l'Anatomie*, Vol. 3. Deterville, Paris.

Linnaeus, C. 1758. *Systema Naturae Per Regna Tria Naturae, Secundum Classes, Ordines, Genera, Species, cum Characteribus, Differentiis, Synonymis, Locis*. Tomus I, Editio decima, reformata. Laurentius Salvius, Holmiae (Stockholm).

Martin, J. W., and J. C. Christiansen. 1995. A morphological comparison of the phyllopodous thoracic limbs of a leptostracan (*Nebalia* sp.) and a spinicaudate conchostracan (*Leptestheria* sp.), with comments on the use of *Phyllopoda* as a taxonomic category. *Can. J. Zool.* 73:2283–2291.

Martin, J. W., and G. E. Davis. 2006. Historical trends in crustacean systematics. *Crustaceana* 79:1347–1368.

Mikulic, D. G., D. E. G. Briggs, and J. Kluessendorf. 1985a. A new exceptionally preserved biota from the Lower Silurian of Wisconsin, U.S.A. *Philos. Trans. R. Soc. Lond. B Biol. Sci.* 311:75–85.

Mikulic, D. G., D. E. G. Briggs, and J. Kluessendorf. 1985b. A Silurian soft-bodied biota. *Science* 228:715–717.

Müller, O. F. 1785. *Entomostraca seu Insecta Testacea, Quae in Aquis Daniae et Norvegiae Reperit, Descripsit et Iconobus Illustravit*. J. G. Mülleriani, Lipsiae et Havniae.

Novozhilov, N. I. 1961. Dvustvorcatye Listonogie Devona (Devonian Conchostracans). *Akad. Nauk SSSR Trudy Paleontogiceskogo Inst.* 81:1–132.

Oakley, T. H., J. M. Wolfe, A. R. Lindgren and A. K. Zaharoff. 2013. Phylotranscriptomics to bring the understudied into the fold: monophyletic *Ostracoda*, fossil placement, and pancrustacean phylogeny. *Mol. Biol. Evo.* 30:215–233.

Olesen, J. 2004. On the ontogeny of the *Branchiopoda* (*Crustacea*): contribution of development to the phylogeny and classification. Pp. 217–269 in *Evolutionary Developmental Biology of Crustacea* (G. Scholtz, ed.). A.A. Balkema Publishers, Lisse.

Olesen, J. 2007. Monophyly and phylogeny of the *Branchiopoda*, with focus on morphology and homologies of the branchiopod phyllopodous limbs. *J. Crust. Biol.* 27:165–183.

Olesen, J. 2009. Phylogeny of *Branchiopoda* (*Crustacea*)—character evolution and contribution of uniquely preserved fossils. *Arthro. Syst. Phylo.* 67:3–39.

Pennak, R. W. 1953. *Fresh-Water Invertebrates of the United States.* The Ronald Press Company, New York.

Preuss, G. 1951. Die Verwandtschaft der *Anostràca* und *Phyllopoda. Zool. Anz.*147:49–64.

Regier, J. C., J. W. Shultz, A. Zwick, A. Hussey, B. Ball, R. Wetzer, J. W. Martin, and C. W. Cunningham. 2010. Arthropod relationships revealed by phylogenomic analysis of nuclear protein-coding sequences. *Nature* 463:1079–1083.

Richter, S. 2004. A comparison of the mandibular gnathal edges in branchiopod crustaceans: implications for the phylogenetic position of the *Laevicaudata. Zoomorphology.* 123:31–44.

Richter, S., J. Olesen, and W. C. Wheeler. 2007. Phylogeny of *Branchiopoda* (*Crustacea*) based on a combined analysis of morphological data and six molecular loci. *Cladistics* 23:301–336.

Sanders, H. L. 1955. The *Cephalocarida*, a new subclass of *Crustacea* from Long Island Sound. *Proc. Nat. Acad. Sci. USA* 41:61–66.

Saunders, III, J. F. 1980. *Eubranchiopoda* of Colorado, part 1, introduction and *Notostraca. Nat. Hist. Invent. Colorado* 5:11–24.

Saunders, III, J. F. 1981. *Eubranchiopoda* of Colorado, part 2, *Anostraca. Nat. Hist. Invent. Colorado* 6:1–23.

Saunders, III, J. F., and S.-K. Wu. 1984. *Eubranchiopoda* of Colorado, part 3, *Conchostraca. Natural History Inventory of Colorado* 8:1–23.

Sars, G. O. 1867. *Histoire Naturelle des Crustacés d'Eau Douce de Norvège.* Les Malacostracés. Chr. Johnsen, Christiania.

Saxena, A. 2005. *Text Book of Crustacea.* Discovery Publishing House, New Delhi.

Schram, F. R. 1986. *Crustacea.* Oxford University Press, New York.

Scourfield, D. J. 1926. On a new type of crustacean from the Old Red Sandstone (Rhynie Chert Bed, Aberdeenshire)—*Lepidocaris rhyniensis,* gen et sp. nov. *Philos. Trans. R. Soc. Lond. B Biol. Sci.* 214:153–187.

Shen, Y.-B. 1983. Restudy of Devonian leaiid conchostracans from Hunan and Guangdong Provinces. *Bull. Nanjing Inst. Geol. Palaeontol.* 6:185–207.

Spears, T., and L. G. Abele. 1997. Crustacean phylogeny inferred from 18S rDNA. Pp. 169–187 in *Arthropod Relationships* (R. A. Fortey and R. H. Thomas, eds.). Chapman & Hall, London.

Spears, T., and L. G. Abele. 1999. Phylogenetic relationships of crustaceans with foliaceous limbs: an 18S rDNA study of *Branchiopoda, Cephalocarida,* and *Phyllocarida. J. Crust. Biol.* 19:825–843.

Starobogatov, Ya. I. 1984. O problemakh nomenklatury vysshikh taksonomischeskikh kategoriy (On the problem of the nomenclature of higher taxonomic categories). Pp. 174–187 in *Spravochnik po Sistematike Iskopayemykh Organizmov* (Handbook on the systematics of fossil organisms) (L. P. Tatarinov and V. N. Shimanskiy, eds.). Nauka, Moscow.

Starobogatov, Ya. I. 1986 (Systematics of *Crustacea*). *Zool. Zhurn.* 65:1769–1781.

Starobogatov, Ya. I. 1988. Systematics of *Crustacea. J. Crust. Biol.* 8:300–311.

Straus, E. H. 1820. Mémoire sur les *Daphnia* de la classe des Crustacés. *Mém. Mus. d'Hist. Nat., Paris* 5:380–425.

Stenderup, J. T., J. Olesen, and H. Glenner. 2006. Molecular phylogeny of the *Branchiopoda* (*Crustacea*)—multiple approaches suggest a "diplostracan" ancestry of the *Notostraca. Mol. Phylo. Evol.* 41:182–194.

Sun, X., Q. Yang, and Y. Shen. 2006. Jurassic radiation of large *Branchiopoda* (*Arthropoda: Crustacea*) using secondary structure-based phylogeny and relaxed molecular clocks. *Prog. Nat. Sci.* 16:292–302.

Sun, X.-Y., X. Xia, and Q. Yang. 2016. Dating the origin of the major lineages of *Branchiopoda. Palaeoworld* 25:303–317.

Walossek, D. 1993. The Upper Cambrian *Rehbachiella* and the phylogeny of *Branchiopoda* and *Crustacea. Fossils Strata* 32:1–202.

Walossek, D. 1999. On the Cambrian diversity of *Crustacea.* Pp. 3–27 in *Crustaceans and the Biodiversity Crisis. Proceedings of the Fourth*

International Crustacean Congress (F. R. Schram and J. C. von Vaupel Klein, eds.). Brill Academic Publishers, Leiden.

Wilson, G. D. F. 1992. Computerized analysis of crustacean relationships. *Acta Zool.* 73:383–389.

Wolfe, J. M., A. C. Daley, D. A. Legg, and G. D. Edgecombe. 2016. Fossil calibrations for the arthropod tree of life. *Earth-Sci. Rev.* 160:43–110.

Wolfe, J. M., and T. A. Hegna. 2014. Testing the phylogenetic position of Cambrian pancrustacean larval fossils by coding ontogenetic stages. *Cladistics* 30:366–390.

Young, G. A., D. M. Rudkin, E. P. Dobrzanski, S. P. Robson, and G. S. Nowlan. 2007. Exceptionally preserved Late Ordovician biotas from Manitoba, Canada. *Geology* 35:883–886.

Authors

Thomas A. Hegna; Department of Geology and Environmental Sciences; SUNY Fredonia; 203 Jewett Hall; 280 Central Ave.; Fredonia, NY. 14063, USA; hegna@fredonia.edu.

Jørgen Olesen; Natural History Museum of Denmark; University of Copenhagen; Universitetsparken 15; DK-2100 Copenhagen Ø, Denmark. Email: jolesen@snm.ku.dk

Date Accepted: 02 May 2017

Primary Editor: Jacques A. Gauthier

Insecta C. Linnaeus 1758 [K. M. Kjer and R. W. Holzenthal], converted clade name

Registration Number: 177

Definition: The crown clade originating in the most recent common ancestor of *Petrobius brevistylis* Carpenter 1913 (*Archaeognatha*), *Tricholepidion gertschi* Wygodzinsky 1961 (*Dicondylia/Zygentoma*), and *Anax junius* (Drury 1773) (*Dicondylia/Pterygota/Odonata*). This is a minimum-crown-clade definition. Abbreviated definition: min crown ∇ (*Petrobius brevistylis* Carpenter 1913 & *Tricholepidion gertschi* Wygodzinsky 1961 & *Anax junius* (Drury 1773)).

Etymology: From the Latin *insectum*, translated from the Greek *entomon*, meaning segmented, or cut into.

Reference Phylogeny: The primary reference phylogeny is Beutel and Gorb (2001: Fig. 12).

Composition: *Insecta* includes the ancestrally wingless *Archaeognatha* and *Zygentoma*, as well as the winged *Pterygota*. The first major dichotomy within *Insecta* is between *Archaeognatha* and *Dicondylia* (possessing dicondylic mandibles). *Dicondylia* includes *Zygentoma* and *Pterygota*. Most insects belong to *Pterygota*, which is a name that refers to the presence of a pair of wings on both the second and third thoracic segments (the mesothorax and metathorax). Within *Pterygota*, there are two major subdivisions, *Palaeoptera*, and *Neoptera*. The monophyly of *Palaeoptera*, which includes *Odonata* (dragonflies and damselflies) and *Ephemeroptera* (mayflies), is still debatable. Members of *Neoptera* have a series of apomorphies that allow the wings to be folded over the dorsal thorax and abdomen. Within *Neoptera*, there are three major subgroups, *Polyneoptera*, *Acercaria*, and *Holometabola*. *Polyneoptera* is of equivocal monophyly, and includes *Orthoptera* sensu *stricto* (grasshoppers and crickets), *Dermaptera* (earwigs), *Plecoptera* (stoneflies), *Phasmatodea* (stick insects), *Dictyoptera* (mantids and roaches [including termites]), and other taxa traditionally ranked as orders (e.g., *Zoraptera*) whose members have neopterous wings, and chewing mouthparts, but do not have a pupal stage. *Acercaria* includes *Hemiptera* (aphids, hoppers, cicadas, true bugs), along with *Psocodea* (barklice, booklice, parasitic lice), and *Thysanoptera* (thrips). *Holometabola* (or *Endopterygota* [lacking external wing buds]) includes the insects that possess a quiescent pupal stage, during which a major transition from a distinct larva to the winged reproductive adult takes place. *Holometabola* contains *Neuropterida* (raphidiopterans, megalopterans, lacewings), *Strepsiptera*, *Trichoptera* (caddisflies), *Mecoptera* and *Siphonaptera* (fleas), and the extremely species-rich clades (traditionally ranked as orders) *Hymenoptera* (wasps, bees, ants), *Coleoptera* (beetles), *Diptera* (flies), and *Lepidoptera* (butterflies, moths). These last 4 clades comprise over 80% of insect species.

Diagnostic Apomorphies: The synapomorphies recognized by Hennig (1981), Kristensen (1975), and Beutel and Gorb (2006) include:

1. Antennae with muscles confined to the most basal antennomere (scape, which is followed by a pedicel and an annulated flagellum).
2. "Johnston's organ" (Schmidt, 1975) in pedicel, a group of chordotonal organs that sense motion of the flagellum.
3. Posterior tentorial arms fused medially to form a transverse bar, the metatentorium or tentorial bridge.

4. Coxae of thoracic legs primarily without sternal articulations.

5. Tarsi subsegmented (although probably not in the †*Monura*, see Comments).

6. Pretarsus not developed as a distinct distal leg segment, modified as claw-bearing unguitractor plate.

7. Ovipositor formed by gonapophyses (lengthened and subsegmented) of abdominal segments 8 and 9, which are bound to each other by a sliding interlock (olistheter) (Klass, 2008).

8. Long, annulated "terminal filament" on dorsum of segment 11 (secondarily reduced or absent in winged insects, except for *Ephemeroptera*).

9. Antennal vessels separated from the aorta (Gereben-Krenn and Pass, 1999).

10. Embryonic membrane fold (serosal or amnioserosal fold to secrete cuticle ventral to the embryo; Machida, 2006).

Synonyms: Hennig (1969, 1981) and others used the named *Ectognatha* (approximate synonym) and Boudreaux (1979) and others used the name *Ectognathata* (approximate synonym) for the clade here named *Insecta*; these authors used the name *Insecta* for a broader clade (now commonly called *Hexapoda*) that also contains collembolans, proturans, and diplurans. Kluge (2010) used the name *Amyocerata* Remington 1955 (approximate synonym) for *Insecta* as it is used here. *Thysanura s.l.* (e.g., Snodgrass, 1938, 1951) included the current taxa *Archaeognatha* and *Zygentoma*. Their most recent common ancestor is the same as for *Insecta*, as that name is defined here, so *Thysanura s. l.* is an approximate and partial synonym of *Insecta* (*Thysanura s.l.* was a precladistic term for a paraphyletic group, which is why many authors prefer the name *Zygentoma* over *Thysanura* for the monophyletic group that includes *Zygentoma* but not *Archaeognatha*).

Comments: Most animals are insects. There are over a million described species (Zhang, 2011), but that figure represents a great underestimate of insect diversity (Erwin, 1983; Gaston, 1991; Basset, 2001). Over 99% of insects are pterygotes (members of the clade of winged insects) and over 80% of pterygotes are holometabolous (having complete metamorphosis), but these features are not present in the earliest insects. Insects exist on every continent, and in almost every terrestrial and freshwater habitat.

The name *Insecta* was used by Linnaeus (1758), but he applied the name to a group that included taxa that are no longer considered insects, such as mites, spiders, scorpions, crabs, and even a polychaete. Linnaeus' *Insecta* most closely corresponds to what is now commonly known as *Arthropoda*. Leach (1815) applied the name *Insecta* to what we now call *Hexapoda*, which was named by Blainville (1816). For over 150 years, there was no general consensus on which of the wingless hexapods were included in *Insecta*. *Insecta*, as defined here, gained wide support with the work of Kristensen (1975, 1981, 1991, 1995, 1997) who restricted the name to a clade composed of *Archaeognatha*, *Zygentoma*, and *Pterygota*. "Insect" is the English common name for the members of this clade, but in common usage, little attention is given to whether other small and seldom-noticed hexapods (*Collembola*, *Protura*, and *Diplura*) are to be considered insects.

Although resolution of the phylogenetic relationships among the traditionally recognized orders of insects remains a difficult problem (Whitfield and Kjer, 2007), there is remarkable consensus regarding the monophyly of *Insecta*. Hennig (1969) provided a well-resolved phylogeny of higher-level insect phylogeny (published in English in 1981). Kristensen (1975, 1981, 1991, 1995, 1997) provided authoritative reviews and commentary on insect phylogeny, reconciling differences among workers, and in general providing less-resolved phylogenies than earlier workers (which appropriately reflect our uncertainty to this day, even with the advent of molecular data). However, all credible morphological analyses have yielded a monophyletic *Insecta*, including more recent analyses (Beutel and Gorb, 2001, 2006; Bitsch and Bitsch, 2004; Bitsch et al., 2004).

The choice of which clade should be called *Insecta*, as well as why the clade to which the name is applied in this contribution should be called *Insecta* (as opposed to *Ectognatha*), is based on current trends. Most recent works (e.g., Kristensen, 1975; Grimaldi and Engel, 2005; Zhang, 2011; Trautwein et al., 2012) use the name *Insecta* as it is defined here. Kristensen (1975) preferred the names *Hexapoda* and *Insecta* rather than *Insecta* and *Ectognatha* because *Ectognatha* was not widely used, and *Hexapoda* alluded to a characteristic apomorphy.

The inclusion or exclusion of the extinct *Monura* (*Dasyleptidae*) is problematic, but under the definition adopted here, it is dependent on the phylogenetic placement of *Monura* relative to the specified common ancestor. Monurans retain the apparently plesiomorphic condition of non-subsegmented tarsi (see Diagnostic Apomorphies). In addition, the possibility that *Protura* and *Diplura* form a monophyletic group (*Nonoculata*), found in many recent molecular studies (Kjer, 2004; Kjer et al., 2006; Giribet et al., 2004; Luan et al., 2005; von Reumont et al., 2009; Meusemann et al., 2010) would indicate that paired claws could be another apomorphy for *Insecta*, independently derived in *Diplura* and lacking in *Monura*. Both characters suggest that *Monura* may be part of the stem group of the crown clade to which the name *Insecta* is applied here.

Molecular analyses have also supported the monophylum *Insecta*. Before transcriptomes, the dominant marker used in higher-level insect phylogenetics has been the nuclear rRNA, and the analyses of nuclear rRNA by Giribet and Ribera (2000), Kjer (2004), Luan et al. (2005), Kjer et al. (2006), Mallatt and Giribet (2006), and von Reumont et al. (2009) all support a monophyletic *Insecta*. Many other studies, with other sources of data, also support insect monophyly, including combined morphological and molecular data (Edgecombe et al., 2000; Wheeler et al., 2001; Giribet et al., 2004; Kjer et al., 2006), mitochondrial data (Kjer et al., 2006; Carapelli et al., 2007), the Elongation Factor 1-alpha gene (EF-1a) (Kjer et al., 2006), and combined nuclear protein-coding genes (Regier et al., 2004, 2005). Carapelli et al. (2006) reviewed the phylogenetic relationships among *Protura*, *Collembola*, *Diplura*, and *Insecta*. Transcriptomic work confirms the monophyly of *Insecta* (Misof et al., 2014). Grimaldi and Engel (2005) provided the most comprehensive recent review of insect diversity and evolution, including extensive discussions of fossil species.

Literature Cited

Basset, Y. 2001. Invertebrates in the canopy of tropical rain forests. How much do we really know? *Plant Ecol.* 153:87–107.

Beutel, R. G., and S. Gorb. 2001. Ultrastructure of attachment specializations of hexapods (*Arthropoda*): evolutionary patterns inferred from a revised ordinal phylogeny. *J. Zool. Syst. Evol. Res.* 39:177–207.

Beutel, R. G., and Gorb, S. 2006. A revised interpretation of the evolution of attachment structures in *Hexapoda* (*Arthropoda*), with special emphasis on *Mantophasmatodea*. *Arthropod Syst. Phylo.* 64:3–25.

Bitsch, C., and J. Bitsch. 2004. Phylogenetic relationships of basal hexapods among the mandibulate arthropods: a cladistic analysis based on comparative morphological characters. *Zool. Scr.* 33:511–550.

Bitsch, J., C. Bitsch, T. Bourgoin, and C. D'Haese. 2004. The phylogenetic position of early hexapod lineages: morphological data contradict molecular data. *Syst. Entomol.* 29:433–440.

de Blainville, H. 1816. Prodrome d'une nouvelle distribution systématique du règne animal. *Bull. Sci. Soc. Philom. Paris* 1816:105–112.

Boudreaux, H. B. 1979. *Arthropod Phylogeny, with Special Reference to Insects.* John Wiley & Sons, New York, Chichester, Brisbane, Toronto.

Carapelli, A., F. Nardi, R. Dallai, and F. Frati. 2006. A review of molecular data for the phylogeny of basal hexapods. *Pedobiologia* 50:191–204.

Carapelli, A., P. Liòl, F. Nardi, E. van der Wath, and F. Frati. 2007. Phylogenetic analysis of mitochondrial protein coding genes confirms the reciprocal paraphyly of *Hexapoda* and *Crustacea. BMC Evol. Biol.* 7(suppl. 2):S8.

Edgecombe, G. D., G. D. F. Wilson, D. J. Colgan, M. R. Gray, and G. Cassis. 2000. Arthropod cladistics: combined analysis of Histone H3 and U2 snRNA sequences and morphology. *Cladistics* 16:155–203.

Erwin, T. L. 1983. Tropical forest canopies: the last biotic frontier. *Bull. Entomol. Soc. Am.* 29:14–19.

Gaston, K. J. 1991. The magnitude of global insect species richness. *Conserv. Biol.* 5:283–296.

Gereben-Krenn, B.-A., and G. Pass. 1999. Circulatory organs of *Diplura* (*Hexapoda*): the basic design in *Hexapoda? Int. J. Insect Morphol. Embryol.* 28:71–79.

Giribet, G., and C. Ribera. 2000. A review of arthropod phylogeny: new data based on ribosomal DNA sequences and direct character optimization. *Cladistics* 16:204–231.

Giribet, G., G. D. Edgecombe, J. M. Carpenter, C. A. D'Haese, and W. C. Wheeler. 2004. Is *Ellipura* monophyletic? A combined analysis of basal hexapod relationships with emphasis on the origin of insects. *Org. Divers. Evol.* 4:319–340.

Grimaldi, D., and M. S. Engel. 2005. *Evolution of the Insects.* Cambridge University Press, Cambridge, New York, etc.

Hennig, W. 1969. *Die Stammesgeschichte der Insekten.* Waldemar Kramer, Frankfurt.

Hennig, W. 1981. *Insect Phylogeny.* John Wiley & Sons, New York.

Kjer, K. M. 2004. Aligned 18S and insect phylogeny. *Syst. Biol.* 53:506–514.

Kjer, K. M., F. L. Carle, J. Litman, and J. Ware. 2006. A molecular phylogeny of *Hexapoda. Arthropod Syst. Phylo.* 64:35–44.

Klass, K. 2008. The female abdomen of ovipositor-bearing *Odonata* (*Insecta: Pterygota*) *Arthropod Syst. Phylo.* 66:45–142.

Kluge, N. J. 2010. Circumscriptional names of higher taxa in *Hexapoda. Bionomina* 1:15–55.

Kristensen, N. P. 1975. The phylogeny of hexapod "orders": A critical review of recent accounts. *Z. Zool. Syst. Evolutionsforsch.* 13:1–44.

Kristensen, N. P. 1981. Phylogeny of insect orders. *Annu. Rev. Entomol.* 26:135–157.

Kristensen, N. P. 1991. Phylogeny of extant hexapods. Pp. 125–140 in *The Insects of Australia* (C.S.I.R.O., ed.). Cornell University Press, Ithaca, NY.

Kristensen, N. P. 1995. Forty years' insect phylogenetic systematics. *Zool. Beitr. NF* 36:83–124.

Kristensen, N. P. 1997. The groundplan and basal diversification of the hexapods. Pp. 282–293 in *Arthropod Relationships* (R. A. Fortey and R. H. Thomas, eds.). Systematics Association Special, Volume No. 55. Springer, Dordrecht.

Leach, W. E. 1815. Classification (Entomology). Pp. 76–162 in *Edinburgh Encyclopedia,* Vol. 9 (D. Brewster, ed.). Blackwood, Edinburgh.

Linnaeus, C. 1758. *Systema Naturae Per Regna Tria Naturae, Secundum Classes, Ordines, Genera, Species, cum Characteribus, Differentiis, Synonymis, Locis.* 10th edition. Vol. 1: Regnum *Animalia.* Laurentii Salvii, Holmiae (Stockholm).

Luan, Y.-X., J. M. Mallat, R.-D. Xie, Y.-M. Yang, and W.-Y. Yin. 2005. The phylogenetic positions of the three basal-hexapod groups (*Protura, Diplura,* and *Collembola*) based on ribosomal RNA gene sequences. *Mol. Biol. Evol.* 22:1579–1592.

Machida, R. 2006. Evidence from embryology for reconstructing the relationships of hexapod basal clades. *Arthropod Syst. Phylo.* 64: 95–104.

Mallatt, J., and Giribet, G. 2006. Further use of nearly complete 28S and 18S rRNA genes to classify *Ecdysozoa*: 37 more arthropods and a kinorhynch. *Mol. Phylogenet. Evol.* 40:772–794.

Meusemann, K., B. M. von Reumont, S. Simon, F. Roeding, S. Strauss, P. Patrick Kück, I. Ebersberger, M. Walzl, G. Pass, S. Breuers, V. Achter, A. von Haeseler, T. Burmester, H. Hadrys, J. W. Wägele, and B. Misof. 2010. A phylogenomic approach to resolve the arthropod tree of life. *Mol. Biol. Evol.* 27:2451–2464.

Misof, B., S. Liu, K. Meusemann, R. S. Peters, A. Donath, C. Mayer, P. B. Frandsen, J. Ware, T. Flouri, R. G. Beutel, O. Niehuis, M. Petersen, F. Izquierdo-Carrasco, T. Wappler, J. Rust, A. J. Aberer, U. Aspöck, H. Aspöck, D. Bartel, A. Blanke, S. Berger, A. Böhm, T. R. Buckley, B. Calcott, J. Chen, F. Friedrich, M. Fukui, M. Fujita, C. Greve, P. Grobe, S. Gu, Y. Huang, L. S. Jermiin, A. Y. Kawahara, L. Krogmann, M. Kubiak, R. Lanfear, H. Letsch, Y. Li, Z. Li, J. Li, H. Lu, R. Machida, Y. Mashimo, P. Kapli, D. D. McKenna, G. Meng, Y. Nakagaki, J. L. Navarrete-Heredia, M. Ott, Y. Ou, G. Pass, L. Podsiadlowski, H. Pohl, B. M. von Reumont, K. Schütte, K. Sekiya, S. Shimizu, A. Slipinski, A. Stamatakis, W. Song, X. Su, N. U. Szucsich, M. Tan, X. Tan, M. Tang, J. Tang, G. Timelthaler, S. Tomizuka, M. Trautwein, X. Tong, T. Uchifune, M. G. Walzl, B. M. Wiegmann, J. Wilbrandt, B. Wipfler, T. K. Wong, Q. Wu, G. Wu, Y. Xie, S. Yang, Q. Yang, D. K. Yeates, K. Yoshizawa, Q. Zhang, R. Zhang, W. Zhang, Y. Zhang, J. Zhao, C. Zhou, L. Zhou, T. Ziesmann, S. Zou, Y. Li, X. Xu, Y. Zhang, H. Yang, J. Wang, J. Wang, K. M. Kjer, and X. Zhou. 2014. Phylogenomics resolves the timing and pattern of insect evolution. *Science* 346:763–767.

Regier, J. C., J. W. Shultz, and R. E. Kambic. 2004. Phylogeny of basal hexapod lineages and estimates of divergence times. *Ann. Entomol. Soc. Am.* 97:411–419.

Regier, J. C., J. W. Shultz, and R. E. Kambic. 2005. Pancrustacean phylogeny: hexapods are terrestrial crustaceans and maxillopods are not monophyletic. *Proc. R. Soc. Lond. B Biol. Sci.* 272:395–401.

Remington, C. L. 1955. The *"Apterygota"*. Pp. 495–505 in *A Century of Progress in the Natural Sciences* (E. L. Kessel, ed.). California Academy of Sciences, San Francisco, CA.

von Reumont, B. M., K. Meusemann, N. U. Szucsich, E. Dell'Ampio, V. Gowri-Shankar, D. Bartel, S. Simon, H. O. Letsch, R. R. Stocsits, Y.-X. Luan, J. W. Wägele, G. Pass, H. Hadrys, and B. Misof. 2009. Can comprehensive background knowledge be incorporated into substitution models to improve phylogenetic analyses? A case study on major arthropod relationships. *BMC Evol. Biol.* 9:119.

Schmidt, K. 1975. Das Johnstonsche Organ der primär flügellosen *Ectognatha* (*Lepisma*, *Zygentoma*; *Machilis*, *Archaeognatha*). *Cytobiologie* 11:153–171.

Snodgrass, R. E. 1938. Evolution of the *Annelida*, *Onychophora*, and *Arthropoda*. *Smithson. Misc. Collect.* 97:1–159.

Snodgrass, R. E. 1951. *Comparative Studies of the Head of Mandibulate Arthropods*. Comstock Publishing Co., Ithaca, NY.

Trautwein, M. D., B. M. Wiegmann, R. Beutel, K. M. Kjer, and D. K. Yeates. 2012. Advances in insect phylogeny at the dawn of the postgenomic era. *Annu. Rev. Entomol.* 57:449–468.

Wheeler, W. C., M. Whiting, Q. D. Wheeler and J. M. Carpenter. 2001. The phylogeny of the extant hexapod orders. *Cladistics* 17:113–169.

Whitfield, J. B., and K. M. Kjer. 2007. Ancient rapid radiations of insects: challenges for phylogenetic analysis. *Annu. Rev. Entomol.* 53:1–23.

Zhang, Z.-Q. 2011. Phylum *Arthropoda* von Siebold, 1848. In *Animal Biodiversity: An Outline of Higher-Level Classification and Survey of Taxonomic Richness*. (Z.-Q. Zhang, ed.). *Zootaxa* 3148:99–103.

Authors

Karl M. Kjer; 3 New Brunswick Ave, Lavallette, NJ 08735, USA. Email: karl.kjer@gmail.com.

Ralph W. Holzenthal; Department of Entomology; 1980 Folwell Ave.; 219 Hodson Hall; University of Minnesota; St. Paul, MN 55108, USA. Email: holze001@umn.edu.

Date Accepted: 10 July 2013

Primary editor: Kevin de Queiroz

Trichoptera W. Kirby 1813 [R. W. Holzenthal and K. M. Kjer], converted clade name

Registration Number: 178

Definition: The crown clade originating in the most recent common ancestor of *Phyrganea grandis* Linnaeus 1758 (*Integripalpia, Plenitentoria* [*=Limnephiloidea*], *Phryganeidae*) and all extant species that inherited both apneustic aquatic larval stages and the prelabio-hypopharyngeal lobe forming a "haustellum" in adults synapomorphic with those in *P. grandis* Linnaeus. This is an apomorphy-modified crown-clade definition. Abbreviated definition: crown ∇ apo both apneustic aquatic larval stages & haustellate adult mouthparts [*Phryganea grandis* Linnaeus 1758].

Etymology: Derived from the Greek *trichos* for "hair" and *pteron* for "wing," referring to the dense clothing of hairs on the wings of adult *Trichoptera*.

Reference Phylogeny: The primary reference phylogeny is that of Holzenthal et al. (2007a: Figs. 1–10), where the name applies to the clade at the right end of the branch labeled *Trichoptera* (the branch on which apneutic aquatic larvae and haustellate adult mouthparts are inferred to have arisen). See also Kjer et al. (2001: Fig. 11) and Kjer et al. (2002: Fig. 4). The specifier *Phryganea grandis* is more closely related to *Phyrganea cinerea* Walker 1852 than to any other species included in the reference phylogeny (Holzenthal et al., 2007a: Fig. 6); although not included in the reference phylogeny, it is the type species of *Phryganea* (Wiggins, 1998).

Composition: *Trichoptera*, or caddisflies, the English common name for members of the clade, contains approximately 14,000 extant species found in freshwater habitats in all faunal regions except the Antarctic (de Moor and Ivanov, 2008; Holzenthal et al., 2011); larvae of a few species are marine, while a handful of others are semi-terrestrial. Holzenthal et al. (2007b) reviewed the diversity of included taxa and species. Morse (2009) provided an online checklist of world species and Holzenthal et al. (2009) provided an online database of world literature. The clade is especially species-diverse in the tropics (notably the Neotropics and Southeast Asia), where many new species are being discovered and described. Ivanov and Sukatcheva (2002) reviewed both total and crown clade fossil species. There is a rich Baltic (Ulmer, 1912) and Dominican (Wichard, 2007) amber fauna. *Trichoptera* includes two primary subclades, *Annulipalpia*, the retreat-making caddisflies, and *Integripalpia*, the cocoon-making and case-making caddisflies (Holzenthal et al., 2011; Wiggins, 2004).

Diagnostic Apomorphies: Most of the synapomorphies of *Trichoptera* recognized by Hennig (1981) still stand and have been variously listed, discussed, or expanded by other authors, including Weaver (1984), Kristensen (1991, 1997), and Ivanov and Sukatcheva (2002).

1. Larvae aquatic, apneustic (no open spiracles), respiration epidermal, often by filamentous abdominal gills.
2. "Larvae hiding in silken retreats in stream bottom (in aerobic conditions)" (Ivanov and Sukatcheva, 2002: 209) (not present in all *Trichoptera*, although all extant species construct a pupal shelter, larval retreat, or larval case; production of silken larval and pupal structures also occurs in *Lepidoptera*).

3. Larval tentorium reduced, delicate.

4. Larval antennae greatly reduced (at most 2-segmented) and without extrinsic muscles.

5. Larval abdominal segments I–IX without prolegs (possibly apomorphic for a more inclusive clade, such as *Amphiesmenoptera* or *Endopterygota*, N. P. Kristensen, pers. comm.).

6. Larval anal prolegs well developed.

7. Larval abdominal segment IX with dorsal tergite.

8. "Pupation underwater in special dome-shaped pupal retreat (made of sand and attached to the solid ground), in osmotically active semipermeable cocoon filled with special liquid ensuring ion-friendly environment and protecting from mechanical stress" (Ivanov and Sukatcheva, 2002: 209). (Not all *Trichoptera* have semipermeable fluid-fill cocoons, although almost all extant species pupate underwater in a pupal structure, except for a few semiterrestial species.)

9. Adult mandibles reduced, with loss of mandibular articulation.

10. Adult prelabium joined with hypopharynx to form a unique protrusible/eversible "haustellum" which serves as a lapping/sponging organ (Crichton, 1957; Kristensen, 1997; Holzenthal et al., 2007a,b: Fig. 12).

11. "Maxillary and labial palps well developed, with delimited subterminal sensory complex" (Ivanov and Sukatcheva, 2002: 209).

12. "Mesonotum with distinct warts along lateral sutures" (Ivanov and Sukatcheva, 2002: 209).

13. Forewing vein CuP with strong apical bend (listed as a potential autapomorphy by Kristensen, 1997, but several extant species have straight CuP apex).

Synonyms: (All approximate): *Phryganeida* Rohdendorf and Rasnitsyn 1980, derived from *Phryganites* Latreille 1810. The name was used in some Russian literature related to fossils and was listed as a synonym of *Trichoptera* by Ivanov and Sukatcheva (2002), but it has not been widely adopted. Fischer (1960: 1) and Morse (2009) listed other names used for the clade in the nineteenth century, including *Phryganides* Latreille 1805, and *Phryganites* Latreille 1810, which predate Kirby's (1813) first use of the name *Trichoptera* (in a footnote in his paper on *Strepsiptera*).

Comments: The monophyly of *Trichoptera* is strongly supported by morphology (Kristensen, 1991) and by molecular and combined data (Kjer et al., 2001; Holzenthal et al., 2007a). There are no credible published studies that do not infer a monophyletic *Trichoptera*. *Trichoptera* has long been recognized as a distinct and well-supported clade within the winged insects and is inferred to be the sister group to *Lepidoptera* (among extant species) in phylogenetic assessments based on morphological (Kristensen, 1984, 1991, 1997), molecular (Kjer, 2004), and combined data (Kjer et al., 2006). Ulmer (1912) presented the first phylogenetic tree for the clade, and Ross (1956, 1964, 1967: Fig. 1) presented the first modern evolutionary assessment of caddisfly phylogeny. More recent treatments include those of Weaver (1984), Weaver and Morse (1986), Wiggins and Wichard (1989), Frania and Wiggins (1997), Ivanov (2002), and Ivanov and Sukatcheva (2002). Morse (1997) reviewed phylogenetic studies within the clade.

Trichopterans appear in Linnaeus (1758, 1761) under his genus *Phryganea* in his order *Neuroptera*, but *Phryganea* also included other, non-trichopteran species, now referred to *Megaloptera* and *Plecoptera*. The name *Trichoptera* was first applied to a taxon corresponding to the then-known members of the clade to which the name is applied here by Kirby (1813) and has been in continuous use ever since (Fischer, 1960). Given this continuous use and that none of the synonyms have been commonly used since the early nineteenth century (except in some Russian literature on fossil insects; see Synonyms), the name *Trichoptera* is clearly the most appropriate name for the clade in question. The name *Trichoptera* was used implicitly for the crown clade, as defined here, by Kristensen (1991) and Holzenthal et al. (2007a).

According to molecular phylogenies (Kjer et al., 2001, 2002; Holzenthal et al., 2007a), the branch leading to extant *Trichoptera* is unusually long and distinct from *Lepidoptera*. It is not impossible to imagine an extant insect that possesses some but not all of the apomorphies listed here (see Diagnostic Apomorphies), subdividing the branch leading to extant, described *Trichoptera*, but such an insect is yet to be discovered. Some Mesozoic fossil impressions have been referred to the total clade of *Trichoptera*. These fossils are distinguished from *Lepidoptera* by the presence of macrotrichia, a long stalk of the forewing anal loop, a pattern of setose warts on the adult body, long adult legs, and/or larval cases (Ivanov and Sukatcheva, 2002).

Literature Cited

Crichton, M. I. 1957. The structure and function of the mouth parts of adult caddis flies (*Trichoptera*). *Trans. R. Entomol. Soc. Lond., Ser. B Biol. Sci.* 241:45–91.

Fischer, F. C. J. 1960. Necrotaulidae, Prosepididontidae, Rhyacophilidae. *Trichopterorum Catalogus 1.* Nederlandsche Entomologische Vereeniging (Netherlands Entomological Society), Amsterdam.

Frania, H. E., and G. B. Wiggins. 1997. Analysis of morphological and behavioural evidence for the phylogeny and higher classification of *Trichoptera* (*Insecta*). *R. Ont. Mus. Life Sci. Contrib.* 160:1–67.

Hennig, W. 1981. *Insect Phylogeny.* John Wiley & Sons, New York.

Holzenthal, R. W., R. J. Blahnik, K. M. Kjer, and A. P. Prather. 2007a. An update on the phylogeny of caddisflies (*Trichoptera*). Pp. 143–153 in *Proceedings of the 12th International Symposium on* Trichoptera (J. Bueno-Soria, R. Barba-Alvarez, and B. Armitage, eds.). The Caddis Press, Columbus, OH.

Holzenthal, R. W., R. J. Blahnik, A. L. Prather, and K. M. Kjer. 2007b. Order *Trichoptera* Kirby, 1813 (*Insecta*), caddisflies. *Zootaxa* 1668:639–698.

Holzenthal, R. W., P. K. Mendez, and J. W. H. Steiner. 2009. *Trichoptera* literature database: a collaborative bibliographic resource for world caddisfly research. Available at http://www.trichopteralit.umn.edu/.

Holzenthal, R. W., J. C. Morse, and K. M. Kjer. 2011. Order *Trichoptera* Kirby, 1813. Pp. 209–211 in *Animal Biodiversity: An Outline of Higher-Level Classification and Survey of Taxonomic Richness* (Z.-Q. Zhang, ed.). *Zootaxa* 3148:209–211.

Ivanov, V. D. 2002. Contribution to the *Trichoptera* phylogeny: new family tree with considerations of *Trichoptera-Lepidoptera* relations. *Nova Suppl. Entomol.* (*Proc. 10th Int. Symp.* Trichoptera) 15:277–292.

Ivanov, V. D., and I. D. Sukatcheva. 2002. Order *Trichoptera* Kirby, 1813. The caddisflies (=*Phryganeida* Latreille, 1810). Pp. 199–222 in *History of Insects* (A. P. Rasnitsyn and D. L. J. Quicke, eds.). Kluwer Academic Publishers, Dordrecht.

Kirby, W. 1813. *Strepsiptera*, a new order of insects proposed; and the characters of the order, with those of its genera, laid down. *Trans. Linn. Soc. Lond.* 11:86–122.

Kjer, K. M. 2004. Aligned 18S and insect phylogeny. *Syst. Biol.* 53:506–514.

Kjer, K. M., R. J. Blahnik, and R. W. Holzenthal. 2001. Phylogeny of *Trichoptera* (caddisflies): Characterization of signal and noise within multiple datasets. *Syst. Biol.* 50:781–816.

Kjer, K. M., R. J. Blahnik, and R. W. Holzenthal. 2002. Phylogeny of caddisflies (*Insecta, Trichoptera*). *Zool. Scr.* 31:83–91.

Kjer, K. M., F. L. Carle, J. Litman, and J. Ware. 2006. A molecular phylogeny of *Hexapoda*. *Arthropod Syst. Phylo.* 64:35–44.

Kristensen, N. P. 1984. Studies on the morphology and systematics of primitive *Lepidoptera*. *Steenstrupia* 10:141–191

Kristensen, N. P. 1991. Phylogeny of extant hexapods. Pp. 125–140 in *The Insects of Australia* (C.S.I.R.O., ed.). Cornell University Press, Ithaca, NY.

Kristensen, N. P. 1997. Early evolution of the *Lepidoptera* + *Trichoptera* lineage: phylogeny and the ecological scenario. Pp. 253–271 in *The Origin of Biodiversity in Insects: Phylogenetic Tests of Evolutionary Scenarios* (P. Grandcolas, ed.). Mem. Mus. Natl. Hist. Nat., Éditions du Muséum, Paris.

Latrielle, P. A. 1805. *Histoire Naturelle, Générale et Particulière des Crustacés et des Insects, XIII*. F. Dufart, Paris.

Latreille, P. A. 1810. *Considérations Générales sur l'ordre Naturel des Animaux Composant les Classes des Crustacés, des Arachnides, et des Insectes; avec un Tableau Méthodique de Leurs Genres, Disposés en Familles*. F. Schoell, Paris.

Linnaeus, C. 1758. *Systema Naturae Per Regna Tria Naturae, Secundum Classes, Ordines, Genera, Species, cum Characteribus, Differentiis, Synonymis, Locis*. 10th edition. Vol. 1: Regnum *Animalia*. Laurentii Salvii, Holmiae (Stockholm).

Linnaeus, C. 1761. *Fauna Svecica, Sistens Animalia Sveciae Regni:* Mammalia, Aves, Amphibia, Pisces, Insecta, Vermes, *Distributa per Classes & Ordines, Genera & Species. Cum Differentiis Specierum, Synonymis Auctorum, Nominibus Incolarum, Locis Natalium, Descriptionibus Insectorum. Editio altera auctior*. Laurentii Salvii, Holmiae (Stockholm).

de Moor, F. C., and V. D. Ivanov. 2008. Global diversity of caddisflies (*Trichoptera: Insecta*) in freshwater. *Hydrobiologia* 595:393–407.

Morse, J. C. 1997. Phylogeny of *Trichoptera*. *Annu. Rev. Entomol.* 42:427–450.

Morse, J. C. 2009. *Trichoptera* World Checklist. Available at http://entweb.clemson.edu/data base/trichopt/.

Rohdendorf, B. B., and A. P. Rasnitsyn, eds. 1980. Historical Development of the Class *Insecta*. *Trans. Paleontol. Inst. Acad. Sci. USSR* 175. Nauka Press, Moscow (in Russian).

Ross, H. H. 1956. *Evolution and Classification of the Mountain Caddisflies*. University of Illinois Press, Urbana, IL.

Ross, H. H. 1964. Evolution of caddisworm cases and nets. *Am. Zool.* 4:209–220.

Ross, H. H. 1967. The evolution and past dispersal of the *Trichoptera*. *Annu. Rev. Entomol.* 12:169–206.

Ulmer, G. 1912. Die Trichopteren des Baltischen Bernsteins. *Beitr. Naturkd. Preussens* 10:1–380.

Weaver, J. S., III. 1984. The evolution and classification of *Trichoptera*. Part I: the groundplan of *Trichoptera*. Pp. 413–419 in *Proceedings of the 4th International Symposium on* Trichoptera (J. C. Morse, ed.). Dr. W. Junk, The Hague.

Weaver, J. S., III, and J. C. Morse. 1986. Evolution of feeding and case-making behavior in *Trichoptera*. *J. North. Am. Benthol. Soc.* 5:150–158.

Wichard, W. 2007. Overview and descriptions of caddisfiles (*Insecta, Trichoptera*) in Dominican amber (Miocene). *Stuttg. Beitr. Natkd. Ser. B* (*Geol. Palaeontol.*) 366:1–51.

Wiggins, G. B. 1998. *The Caddisfly Family* Phryganeidae (Trichoptera). University of Toronto Press, Toronto.

Wiggins, G. B. 2004. *Caddisflies: The Underwater Architects*. University of Toronto Press, Toronto.

Wiggins, G. B., and W. Wichard. 1989. Phylogeny of pupation in *Trichoptera*, with proposals on the origin and higher classification of the order. *J. N. Am. Benthol. Soc.* 8:260–276.

Authors

Ralph W. Holzenthal; Department of Entomology; 1980 Folwell Ave.; 219 Hodson Hall; University of Minnesota; St. Paul, MN 55108, USA. Email: holze001@umn.edu.

Karl M. Kjer; 3 New Brunswick Ave, Lavallette, NJ 08735, USA. Email: karl.kjer@gmail.com

Date Accepted: 16 June 2013

Primary Editor: Kevin de Queiroz

Polycarpidea C. S. Bate 1888 [M. L. Christoffersen], converted clade name

Registration Number: 162

Definition: The crown clade originating in the most recent common ancestor of *Plesionika martia* (A. Milne Edwards 1883) (*Pandaloidea*), *Philocheras gorei* (Dardeau 1980) (*Crangonoidea*), and *Merhippolyte agulhasensis* Bate 1888 (*Alpheoidea*), provided that the multiarticulate second pereiopod shared by those taxa represents a single evolutionary origin. This is a minimum-crown-clade definition with a qualifying clause that allows the name to be applied only in the context of phylogenies in which there was a single origin of the multiarticulate second pereiopods (but it does not preclude secondary losses). Abbreviated definition: min crown ∇ (*Plesionika martia* (A. Milne Edwards 1883) & *Philocheras gorei* (Dardeau 1980) & *Merhippolyte agulhasensis* Bate 1888) | the multiarticulate second pereiopod shared by those taxa had a single evolutionary origin.

Etymology: Derived from the Greek *polys* (many) and *karpos* (wrist), referring to the morphological condition of the carpus of the second pereiopod, which is multiarticulate in most representatives of this clade (being secondarily undivided only in *Thalassocaris* and *Crangonidae*).

Reference Phylogeny: Unnumbered figure on p. 102 of Christoffersen (1990); see also Fig. 1 of Christoffersen (1987), and further comment in Christoffersen (2001: 110), although the clade is not named on those phylogenies.

Composition: Contains 1,504 extant and 4 fossil species, which represent 45% of the species of *Caridea* Dana 1852 according to the latest estimate of 3,268 Recent and 57 fossil species of *Caridea* by De Grave et al. (2009). In their classification, these species belong to the superfamilies *Pandaloidea* Haworth 1825, *Physetocaridoidea* Chace 1940, *Crangonoidea* Haworth 1825, and *Alpheoidea* Rafinesque 1915. This is the same clade identified by Christoffersen (1987, 1988, 1989a, 1990) as *Pandaloidea* + *Crangonoidea* + *Alpheoidea* (where *Physetocarididae* is part of *Pandaloidea*), although there are several other differences in the attribution of the genera and families among the respective superfamilies when the traditional classification of De Grave et al. (2009) is compared to the phylogenetic classification of Christoffersen (1987, 1990).

Diagnostic Apomorphy: Carpus of second pereiopod multiarticulated. In the stem-lineage of this clade, the second pereiopods are hypothesized to have become very elongate and slightly unequal, the carpus, merus and ischium having become uniquely multiarticulated from post-larval stages to adults. Bauer (2004) suggested that this innovation enhanced the exploration of the surroundings by the shrimp. Conceivably the multiarticulated carpi have then been used for more efficient cleaning of their own bodies. A subsequent evolutionary trend within the clade is for the second pereiopods to become reduced in length, to become symmetrical, to lose the articles of the ischium and merus, and to gradually reduce the number of articles of the carpus, from about 75 in *Pandalus* (Christoffersen, 1989a) to a single article in *Thalassocaris* (Christoffersen, 1989a) and *Crangonidae* (Christoffersen, 1988).

Synonym: *Polycarpinea* Stebbing 1893 contains the very same taxa included by Bate (1888) in his *Polycarpidea*: *Nikidae* Bate 1888

(= *Processidae* Ortmann 1896), *Alpheidae* Rafinesque 1915, *Hippolytidae* Bate 1888, and *Pandalidae* Haworth 1825. The name is an approximate synonym.

Comments: Pandaloids, crangonoids, and alpheoids share a remarkable synapomorphy, the multiarticulation of the second pereiopod, which conceivably represents a unique adaptation within the caridean shrimps for body autocleaning. Although the diagnostic value of this single synapomorphy becomes compromised in the only two polycarpideans that have a secondarily undivided carpus, *Thalassocaris* and *Crangonidae*, morphological and molecular evidence discussed below support a single origin for this apomorphy with reversals in *Thalassocaris* and *Crangonidae*. In the nematocarcinoid sistergroup of the *Polycarpidea*, the carpus of the second pereiopod is relatively elongate, that is, its length is equal to the merus + ischium, while in the plesiomorphic condition of more distantly related carideans the carpus is shorter than the merus + ischium (Christoffersen, 1990: character 9). Furthermore, *Thalassocaris* is well nested within pandaloid clades that have the carpus progressively reduced to 3, 2, and finally to an undivided carpus (Christoffersen, 1989a: Fig. 1).

Regarding *Crangonidae*, the carpus is multiarticulate in its sister-clade *Glyphocrangon*, and the second pereiopods are characteristically elongate, asymmetrical and with ischium, merus and carpus multiarticulate in the next most closely related taxon, *Processidae* (Christoffersen, 1987: Fig. 2). Furthermore, the general trend for reducing the second pereiopod continues further within some lineages of the *Crangonidae*. The second pereiopods become reduced in *Sabinea* and the *Pontophilus - Parapontophilus* clade, and totally lost in *Paracrangon* (Christoffersen, 1988: 45, Fig. 1).

Bate's (1888) tribe *Polycarpidea* was originally established to include *Nikidae* Yokoya

1933 (= *Processidae* Ortmann 1896), *Alpheidae* Rafinesque 1815, *Hippolytidae* Bate 1888 and *Pandalidae* Haworth 1825, and was subsequently referred to as (Legion) *Polycarpinea* by Stebbing (1893). In both classifications, *Thalassocaris* and *Crangonidae* were excluded because of their undivided carpi. Neither *Polycarpidea* nor *Polycarpinea* has been widely used, so Bate's original name *Polycarpidea* was selected for conversion.

Ortmann (1896) was the first to suggest that the "*Crangonidae* are connected with the *Nikidae* (= *Processidae*) by the genus *Glyphocrangon*". Coutière (1899: 250) also clearly perceived the evolutionary trend affecting the second pereiopods: "on peut suivre, dans la tribu *Polycarpidea*, la diminution progressive des segments du carpe". Finally, Kemp (1925: 272) was not deceived about the phylogenetic relationships of *Thalassocaris* Stimpson 1860 with undivided carpi on the second pereiopods: "the genus *Chlorotocoides* is in some respects intermediate between *Chlorotocus* and *Thalassocaris* and it indicates, in my opinion, that the subfamily Thalassocarinae (sometimes even exalted to the rank of a family) should be abandoned and the genus placed without distinction in the Pandalidae".

Subsequent outlooks on caridean classification (Holthuis, 1955, 1993; Thompson, 1967; Bowman and Abele, 1982; Chace, 1992; Martin and Davis, 2001; De Grave et al., 2009), based mostly on convenience, authority or overall resemblance of taxa, have prevented, in my opinion, the evolutionary insights obtained in the nineteenth century from becoming accepted by the modern carcinological community. If such evolutionary insights had been logically extended by recent authors to a modern phylogenetic system, the phylogenetic importance of segmented second pereiopods would not have been discarded so lightly without detailed consideration (e.g., Thompson, 1967), and the absence of articulations in the carpi of *Thalassocaris* and *Crangonidae* would not have clouded for so long the recognition of a close relationship between *Pandaloidea*, *Crangonoidea*, and *Alpheoidea*.

The phylogenetic classifications of Christoffersen (1987, 1988, 1989a, 1989b, 1990) were based on a limited number of morphological characters, but these were selected at the time as being the best transformation series available for producing phylogenetic inferences. Although these works were not accepted at first, they are slowly gaining some recognition. Some suggestions were incorporated by Martin and Davies (2001), and recently some elements have been corroborated by Li et al. (2011).

After the first phylogenetic reconstruction of the *Caridea* based on morphological characters (Christoffersen, 1990), three molecular phylogenies have focused recently on the *Caridea* (Xu et al., 2005; Bracken et al., 2009; Li et al., 2011). Unfortunately, the pioneer molecular analysis of Xu et al. (2005), based on 16S data, lacked sufficient taxon sampling (20 species) and showed little support for the resulting phylogeny.

Bracken et al. (2009) inferred the most extensive caridean phylogeny to date, using 104 species and a genetic coverage of 1835 base pairs of mitochondrial and nuclear genes. Bracken et al. (2009) did not corroborate the monophyly of the polycarpideans, but then their results provided only weak support for deep relationships within the *Caridea*, due to a shortage of phylogenetically informative markers at deeper nodes. Yet even their phylogeny does not imply that polycarpidean lineages arose multiple times throughout caridean history, as suggested initially by Thompson (1967). Multiarticulated carpi may have arisen only once, and then been lost several times. However, if that is the case, then the phylogeny of Bracken et al. (2009) implies that *Palaemonoidea* is part of *Polycarpidea* and represents an additional loss of the multiarticulate carpus.

Li et al. (2011) used five nuclear protein-coding genes to produce a dataset of 3,819 base pairs for 35 carideans species that was capable of resolving phylogenetic relationships among taxa at higher levels within the *Caridea*. A clade composed of *Pandaloidea*, *Crangonoidea* and *Alpheoidea* was supported in this analysis (with a 0.97 Bayesian posterior probability, though parsimony and likelihood bootstrap values were less than 50%). The results confirm that *Crangonidae* belongs to the clade despite the secondary loss of multiple articulations on the second pereiopods. This is the first time that morphology and molecules have shown congruence at a deep level of caridean phylogeny.

Literature Cited

Bate, C. S. 1888. Report on the *Crustacea Macrura* collected by H. M. S. Challenger during the years 1873–1876. *Rep. Sci. Res. Explor. Voy. Challenger, Zool.* 24:1–942.

Bauer, R. T. 2004. *Remarkable Shrimp: Adaptations and Natural History of the Caridens*. University of Oklahoma Press, Norman, OK.

Bowman, T. E., and L. G. Abele. 1982. Classification of Recent *Crustacea*. Pp. 1–27 in *The Biology of* Crustacea, *Systematics, the Fossil Record, and Biogeography* (L. G. Abele, ed.). Academic Press, New York.

Bracken, H. D., S. De Grave, and D. L. Felder. 2009. Phylogeny of the infraorder *Caridea* based on nuclear and mitochondrial genes (*Crustacea*: *Decapoda*). Pp. 274–300 in *Decapod Crustacean Phylogenetics* (Crustacean Issues 18) (J. W. Martin, K. A. Crandall, and D. L. Felder, eds.). CRC Press, Boca Raton, FL.

Chace, Jr., F. A. 1992. On the classifiction of the *Caridea* (Decapoda). *Crustaceana* 63:70–80.

Christoffersen, M. L. 1987. Phylogenetic relationships of hippolytid genera, with an assignment of new families for the *Crangonoidea* and *Alpheoidea* (Crustacea, Decapoda, Caridea). *Cladistics* 3:348–362.

Christoffersen, M. L. 1988. Genealogy and phylogenetic classification of the world *Crangonidae* (*Crustacea, Caridea*), with a new species and new records for the south western Atlantic. *Rev. Nordest. Biol.* 6:43–59.

Christoffersen, M. L. 1989a. Phylogeny and classification of the *Pandaloidea* (*Crustacea, Caridea*). *Cladistics* 5:259–274.

Christoffersen, M. L. 1989b. Phylogenetic relationships between *Oplophoridae, Atyidae, Pasiphaeidae, Alvinocarididae* fam. n., *Bresiliidae, Psalidopodidae* and *Disciadidae* (*Crustacea Caridea Atyoidea*). *Bol. Zool.* 10:273–281.

Christoffersen, M. L. 1990. A new superfamily classification of the *Caridea* (*Crustacea*: *Pleocyemata*) based on phylogenetic pattern. *Z. Zool. Syst. Evolutions-forsch.* 28:94–106.

Christoffersen, M. L. 2001. *Decapoda: Caridea*. Pp. 110 in *An Updated Classification of the Recent* Crustacea (J. W. Martin, and G. E. Davis, eds.). Science Series, 39. Natural History Museum of Los Angeles County, Los Angeles, CA.

Coutière, H. 1899. Les *"Alpheidae"*, morphologie externe et interne, formes larvaires, bionomie. *Ann. Sci. Nat. Zool., Ser. 8* 9:1–559, pls. 1–6.

De Grave, S., N. D. Pentcheff, S. T. Ahyong, T.-Y. Chan, K. A. Crandall, P. C. Dworschak, D. L. Felder, R. M. Feldmann, Fransen, C. H. J. M.; L. Y. D. Goulding, R. Lemaitre, M. E. Y. Low, J. W. Martin, P. K. L. Ng, C. E. Schweitzer, S. H. Tan, D. Tshudy, and R. Wetzer. 2009. A classification of living and fossil genera of decapod crustaceans. *Raffles Bull. Zool. Suppl.* 21:1–109.

Holthuis, L. B. 1955. The Recent genera of the caridean and stenopodidean shrimps (Class *Crustacea*, Order *Decapoda*, Supersection *Natantia*) with keys for their determination. *Zool. Verh.* 26:1–157.

Holthuis, L. B. 1993. *The Recent Genera of the Caridean and Stenopodidean Shrimps* (Crustacea, Decapoda) *with an Appendix on the Order* Amphionidacea. Nationaal Natuurhistorisch Museum, Leiden.

Kemp, S. 1925. On various *Caridea*. Notes on *Crustacea Decapoda* in the Indian Museum. XVII. *Rec. Ind. Mus.* 27:249–343.

Li, C. P., S. De Grave, T. Y. Chan, H. C. Lei, and K. H . Chu. 2011. Molecular systematics of caridean shrimps based on five nuclear genes: implications for superfamily classification. *Zool. Anz.* 4(Spec. Iss.):270–279.

Martin, J. W., and G. E. Davis, eds. 2001. *An Updated Classification of the Recent* Crustacea. Science Series, 39. Natural History Museum of Los Angeles County, Los Angeles, CA.

Ortmann, A. 1896. A study of the systematics and geographical distribution of the decapod family *Crangonidae* Bate. *Proc. Acad. Nat. Sci. Philad.* 47:173–197.

Stebbing, T. R. R. 1893. *A History of* Crustacea. *Recent* Malacostraca. The International Scientific Series, 71. Kegan Paul, Trench, Trübner, and Co., London.

Thompson, J. R. 1967. Comments on phylogeny of section *Caridea* (*Decapoda Natantia*) and the phylogenetic importance of the *Oplophoroidea*. Pp. 314–326 in *Proceedings of the Symposium on Crustacea held at Ernakulam from January 12 to 15, 1965. Part 1.* Symposium Series 2. Marine Biological Association of India, Mandapam Camp, India.

Xu, Y., L.-S. Song, and X.-Z. Li. 2005. The molecular phylogeny of infraoder *Caridea* based on 16S rDNA sequences. *Mar. Sci.* (Beijing) 29:36–41.

Author

Martin Lindsey Christoffersen; Departamento de Sistemática e Ecologia; Universidade Federal da Paraíba; 58.059-900, João Pessoa, Paraíba, Brazil. Email: mlchrist@dse.ufpb.br.

Date Accepted: 15 May 2012

Primary Editor: Kevin de Queiroz

Prochelata M. L. Christoffersen, new clade name

Registration Number: 163

Definition: The most inclusive crown clade containing *Crangon crangon* (Linnaeus 1758) (*Crangonoidea* Haworth 1825 sensu Christoffersen, 1987, 1988, 1990) and *Alpheus armillatus* H. Milne Edwards 1837 (*Alpheoidea* Rafinesque 1815 sensu Christoffersen, 1987, 1990), but not *Pandalus paucidens* Miers 1881 (*Pandaloidea* Haworth 1825). This is a maximum-crown-clade definition with two internal specifiers. Abbreviated definition: max crown ∇ (*Crangon crangon* (Linnaeus 1758) & *Alpheus armillatus* H. Milne Edwards 1837 ~ *Pandalus paucidens* Miers 1881).

Etymology: Derived from the Latin *pro-* (before) + *chela* (claw) + *-ata* (provided with), referring to the fact that members of the named clade have the chelae of the first chelipeds more developed than those of the second cheliped, a unique condition within those carideans with an elongate carpus on the second pereiopod (i.e., *Nematocarcinoidea* + *Pandaloidea* + *Crangonoidea* + *Alpheoidea*) (Christoffersen, 1990).

Reference Phylogeny: Unnumbered figure on p. 102 of Christoffersen (1990); see also Fig. 1 of Christoffersen (1987), although the clade of interest is not named on those phylogenies.

Composition: Contains 1,302 extant and 4 fossil species, which represent 87% of the species of the polycarpidean carideans (see entry for *Polycarpidea*, this volume), according to the latest estimate of 3,268 Recent and 57 fossil species of *Caridea* by De Grave et al. (2009). In their classification, these species belong to superfamilies *Crangonoidea* Haworth 1825 and *Alpheoidea* Rafinesque 1915. These correspond to the subclades identified by Christoffersen (1987, 1988, 1990) as *Crangonoidea* and *Alpheoidea*, but there are several differences in the assignments of the genera and families to the respective superfamilies when the traditional classification of De Grave et al. (2009) is compared to the phylogenetic classification of Christoffersen (1990).

Diagnostic Apomorphy: First pereiopod more robust than second pereiopod in post-larval stages (Christoffersen, 1987: character 3; 1990: character 14).

Synonyms: None.

Comments: Relatively robust first chelipeds apparently evolved convergently in several carideans belonging to more basal lineages (e.g., *Caridina*, *Bresilia*, *Discias*, *Eugonatonotus*, *Rhynchocinetes*; Holthuis, 1955, 1993). As a consequence, this character is generally perceived as unreliable within the *Caridea*. On the other hand, it should not be assumed that this character is convergent within the large clade composed at least of *Palaemonoidea* + *Pandaloidea* + *Physetocaridoidea* + *Processoidea* + *Crangonoidea* + *Alpheoidea*, where it appears to be a synapomophy of *Crangonoidea* plus *Alpheoidea* within *Polycarpidea* Bate 1888 (see entry in this volume). The *Crangonoidea* plus *Alpheoidea* represented an unrecognized clade until Christoffersen (1987) postulated this relationship based the robust first chelipeds. Although the clade has thus been recognized previously, I am naming it for the first time herein.

In the very first system proposed for the *Caridea*, Dana (1852) included both *Nika*

Risso 1816 (= *Processa* Leach 1815) and *Lysmata* Risso 1816 in the *Crangonidae*. Although later recognized as a separate family, the *Processidae* continued to be classified together with *Glyphocrangon* and the *Crangonidae* by other early authors (Borradaile, 1907; Balss, 1927, 1957). Later Holthuis (1955, 1993) transferred *Processa* and *Lysmata* to *Alpheoidea*. Martin and Davis (2001) and De Grave and Fransen (2011) established a monotypic superfamily *Processoidea*, but retained *Lysmata* within the *Alpheoidea*.

In my cladistic analyses of the *Crangonoidea* and *Alpheoidea* (Christoffersen, 1987, 1988, 1990), taxa included in the *Alpheoidea* by Holthuis (1955, 1993) (*Ligur, Barbouria, Janicea, Parhippolyte,* and *Somersiella* [these 5 taxa = *Barbouriidae* Christoffersen 1987], *Calliasmata, Lysmata, Exhippolysmata,* and *Mimocaris* [these 4 taxa = *Lysmatidae* of Christoffersen, 1987], *Merguia* [= *Merguiidae* Christoffersen 1990] and *Processidae* Ortmann 1896), were transferred back from the *Alpheoidea* to the *Crangonoidea*, thus eliminating two notably paraphyletic taxa, *Hippolytidae* and *Alpheoidea* (sensu Holthuis, 1955, 1993; and Chace, 1992).

Two recent molecular phylogenies bear on the proposed name and definition. According to the results of Bracken et al. (2009) it would seem that *Prochelata* would include palaemonoids and *Thalassocarididae* (formerly referred to *Pandaloidea*), in addition to crangonoids and alpheoids, and would possibly exclude some alpheoids (*Hippolytidae*). On the other hand, according to the phylogeny of Li et al. (2011), it would seem that the name *Prochelata* does not apply to any clade, because there is no clade in that phylogeny that includes both alpheoids and crangonoids but not pandaloids (which therein are closer to crangonoids). Yet the critical relationships are poorly supported in both cases (<50% bootstrap values in parsimony and likelihood analyses). In addition, the placement of *Pandalidae* rather than *Glyphocrangonidae* as the sister of *Crangonidae* + *Barbouridae* + *Lysmatidae* in Li et al. (2011) and the placement of *Thalassocarididae* within the clade originating in the most recent common ancestor of *Glyphocrangonidae, Processidae* and *Crangonidae* in Bracken et al. (2009), do not agree with the concept of *Crangonoidea* of Christoffersen (1987, 1990).

On the other hand, both recent molecular analyses provide support for some relationships within *Prochelata* hypothesized by Christoffersen (1987). Bracken et al. (2009) placed processids (yet together with thalassocaridids) as the sister group of crangonoids, and *Ogyrididae* as the sister group of *Alpheidae*, in concordance with Christoffersen (1987). Interestingly, Li et al. (2011) also confirmed that portions of the former hippolytids containing *Lysmata* cluster with *Barbouridae* and *Crangonidae*, while other portions of the former *Hippolytidae* containing *Saron* group with the *Alpheidae*, thus at least partially corroborating the concepts of *Crangonoidea* and *Alpheoidea* sensu Christoffersen (1987, 1990). Another partial corroboration of the concept of *Crangonoidea* sensu Christoffersen (1987, 1990) is that *Crangonidae* and *Glyphocrangonidae* form a clade with *Processidae* in both molecular analyses.

Literature Cited

Balss, H. 1927. *Decapoda*. Pp. 840–1038 in *Handbuch der Zoologie*, Vol. 3, Part 1 (W. Kükenthal, and T. Krumbach, eds.). De Gruyter, Berlin.

Balss, H., ed. 1957. *Decapoda*, VIII, Systematik. Pp. 1505–1672 in *Dr. H. G. Bronns Klassen und Ordnungen des Tierreichs*, Vol. 5, Section 1, Book 7, Issue 12. Akademische Verlagsgesellschaft Geest und Portig, Leipzig.

Borradaile, L. A. 1907. On the classification of the decapod Crustaceans. *Ann. Mag. Nat. Hist., Ser. 7* 19:457–486.

Bracken, H. D., S. De Grave, and D. L. Felder. 2009. Phylogeny of the infraorder *Caridea* based on nuclear and mitochondrial genes (*Crustacea: Decapoda*). Pp. 274–300 in *Decapod Crustacean Phylogenetics* (Crustacean Issues 18) (J. W. Martin, K. A. Crandall, and D. L. Felder, eds.). CRC Press, Bocan Raton, FL.

Chace, Jr., F. A., 1992. On the classification of the *Caridea* (*Decapoda*). *Crustaceana* 63:70–80.

Christoffersen, M. L. 1987. Phylogenetic relationships of hippolytid genera, with an assignment of new families for the *Crangonoidea* and *Alpheoidea* (*Crustacea, Decapoda, Caridea*). *Cladistics* 3:348–362.

Christoffersen, M. L. 1988. Genealogy and phylogenetic classification of the world *Crangonidae* (*Crustacea, Caridea*), with a new species and new records for the south western Atlantic. *Rev. Nordest. Biol.* 6:43–59.

Christoffersen, M. L. 1990. A new superfamily classification of the *Caridea* (*Crustacea: Pleocyemata*) based on phylogenetic pattern. *Z. zool. Syst. Evolutions-forsch.* 28:94–106.

Dana, J. D. 1852. *Crustacea*. Pp. 1–1620 in *United States Exploring Expedition during the Years 1838, 1839, 1840, 1841, 1842 Under the Command of Charles Wilkes, U. S. N.*, Vol. 13. C. Sherman, Philadelphia, PA.

De Grave, S., and C. H. J. M. Fransen. 2011. Carideorum catalogus: the Recent species of the dendrobranchiate, stenopodidean, procarididean and caridean shrimps (*Crustacea: Decapoda*). *Zool. Med.* 85:195–589.

De Grave, S., N. D. Pentcheff, and S. T. Ahyong, et al. 2009. A classification of living and fossil genera of decapod crustaceans. *Raffles Bull. Zool. Suppl.* 21:1–109.

Holthuis, L. B. 1955. The Recent genera of the caridean and stenopodidean shrimps (Class *Crustacea*, Order *Decapoda*, Supersection *Natantia*) with keys for their determination. *Zool. Verh.* 26:1–157.

Holthuis, L. B. 1993. *The Recent Genera of the Caridean and Stenopodidean Shrimps* (Crustacea, Decapoda) *with an Appendix on the Order* Amphionidacea. Nationaal Natuurhistorisch Museum, Leiden.

Leach, W. E. 1815. Malacostraca podophthalmata Brittanniae; *or, Description of Such British Species of the Linnean Genus* Cancer *as have their Eyes Elevated on Footstalks.* James Sowerby, London.

Li, C. P., S. De Grave, T.-Y. Chan, H. C. Lei, and K. H. Chu. 2011. Molecular systematics of caridean shrimps based on five nuclear genes: Implications for superfamily classification. *Zool. Anz.* 4(Spec. Iss.):270–279.

Martin, J. W., and G. E. Davis, eds. 2001. *An Updated Classification of the Recent* Crustacea. Science Series, 39. Natural History Museum of Los Angeles County, Los Angeles, CA.

Ortmann, A. 1896. Das System der Decapoden-Krebse. *Zool. Jahrb. Abt. Syst. Geog. Biol. Tiere* 9:409–453.

Author

Martin Lindsey Christoffersen; Departamento de Sistemática e Ecologia; Universidade Federal da Paraíba; 58.059-900, João Pessoa, Paraíba, Brazil. Email: mlchrist@dse.ufpb.br.

Date Accepted: 15 May 2012

Primary Editor: Kevin de Queiroz

Araneae C. Linnaeus 1758 [I. Agnarsson and M. Kuntner], converted clade name

Registration Number: 182

Definition: The crown clade for which male pedipalps modified as secondary sperm transfer organs (e.g., Foelix 2011: Fig. 7.10), as inherited by *Araneus angulatus* Clerck 1757 (*Araneidae*), is an apomorphy relative to other crown clades. This is an apomorphy-modified crown-clade definition. Abbreviated definition: crown ∇ apo male pedipalps modified as secondary sperm transfer organs [*Araneus angulatus* Clerck 1757].

Etymology: *Araneae* is the plural of the Latin feminine noun *aranea* (spider) (Cameron, 2005).

Reference Phylogeny: Coddington (2005: Fig. 2.2), a manually built super tree placing nearly all taxa traditionally ranked as families, is the primary reference phylogeny. On that phylogeny, pedipalps modified as secondary sperm transfer organs are inferred to have originated along the branch at the base of the tree (i.e., the one subtending the node that unites *Mesothelae* and *Opistothelae*). Recent analyses have suggested alternative placements of some taxa such as palpimanoids (Schutt, 2003; Blackledge et al., 2009), nicodamids (Blackledge et al., 2009), nephilids (Kuntner et al., 2008), and others (Agnarsson et al., 2013). Selden and Penney (2010) reviewed fossil spiders and proposed an alternative summary phylogeny including extant and extinct taxa. Earlier phylogenies were also summary trees that are near-identical in structure, at least concerning the deep relationships within the clade (Coddington et al., 2004: Fig. 18.3; Coddington and Colwell, 2001: Fig. 2). The first comprehensive summary phylogeny of spiders was a standard reference for more than a decade (Coddington and Levi, 1991: Fig. 1), but included fewer taxa than does the primary reference phylogeny.

Composition: To date 43,678 extant (Platnick, 2013) and 1,106 extinct crown species (Dunlop et al., 2011) of spiders are currently recognized. Spiders are distributed in three well-diagnosed major clades (see entries in this volume): *Araneomorphae* (the "modern" or "true" spiders), *Mygalomorphae* (trapdoor and tarantula spiders), and the *Mesothelae* (burrowing spiders from Asia that retain abdominal segmentation).

Diagnostic Apomorphies: Apomorphies relative to other crown clades include abdominal spinnerets, male pedipalpi modified as secondary sperm transfer organs, cheliceral venom glands and ducts, trochanter-femur depressor muscles absent, sternum unsegmented, pygidium absent (for reviews and references see e.g., Shultz, 1990, 2007; Coddington and Levi, 1991; Foelix and Erb, 2010), and possibly, abdominal pedicillate setae (Agnarsson et al., 2007). All these features are putative apomorphies of this clade (Shultz, 1990, 2007; Coddington and Levi, 1991; Coddington, 2005), and at least the modified pedipalpi, cheliceral venom glands, and the abdominal spinnerets and associated spigots and glands are unique (not present in any other animals), universal, and unreversed across the entire clade (Coddington and Levi, 1991).

Synonyms: *Aranei* Clerck 1757 (approximate) (see Comments).
Araneides Latreille 1806 (approximate).
Araneidea Leach 1817 (approximate).

Araneadae Leach 1819 in Samouelle (1819) (approximate).

Diechisoma Billberg 1820 (approximate).

Tetrapneumones Latreille 1825 (partial and approximate)

Araneida (e.g., Blackwall, 1844, 1853) (approximate).

Maripalpi Cook 1899 (approximate).

Orthognatha Berland 1932 (partial and approximate).

Comments: There are two words for "spider" in Latin, the masculine "araneus" (plural "aranei") and the feminine "aranea" (plural "araneae"). Clerck (1757) used the masculine word, while Linnaeus (1758) used the feminine. Subsequently, the use of the feminine form has been overwhelmingly favored, hence *Araneae* for spiders, rather than *Aranei*. The name *Araneae* (plural of *Aranea*) has always been attributed to Clerck (1757), although he used *Aranei*. However, Linnaeus (1758: 619, footnote) in his *Systema Naturae* was the first to use *"Araneae"*, so spelled, and according to the *PhyloCode* (see Rec. 9.15A), he is the author of this clade name.

The names *Tetrapneumones* Latreille 1825 and *Orthognatha* Berland 1932 are partial synonyms in the sense that they refer to paraphyletic groups including *Mesothelae* and some *Opisthothelae* spiders (*Mygalomorphae*), thus originating in approximately the same ancestor as *Araneae*.

Of the remaining synonyms, only *Araneida* is widely used. That name was created in an effort to unify the endings of arachnid order names with "-ida", in accordance with those of some other animal orders. Although forms ending with "-ida" are grammatically incorrect (see below), they were intended to be neuter plural nouns, companions to forms ending in "-idae." Petrunkevitch (1955) formally used the name *Araneida* in the *Treatise on Invertebrate Paleontology*, but first use of

that name goes back much further, at least to Blackwall (1844, 1853). Although *Araneida* is fairly widely used, the authors of most recent systematic studies on spiders prefer *Araneae* (see Coddington and Levi, 1991; Coddington, 2005; Agnarsson et al., 2013 for reviews), as we do here. Savory (1972: 125) offered a justification for disfavoring *Araneida* "First, it is a hybrid of Greek and Latin; and secondly it is foolish, because it can only mean "like a spider". To describe the order of spiders as one that contains animals that are "like spiders" is as intelligent and as helpful as saying that the shape of an egg is ovoid."

Phylogenetic analyses of mitochondrial genomes (Masta et al., 2009) provide limited support for the monophyly of spiders.

Literature Cited

Agnarsson, I., J. A. Coddington, and M. Kuntner. 2013. Systematics: progress in the study of spider diversity and evolution. Pp. 58–111 in *Spider Research in the 21st Century* (D. Penney, ed.). Siri Scientific Press, Manchester.

Agnarsson, I., J. A. Coddington, and L. J. May-Collado. 2007. Elongated pedicillate setae—a putative sensory system and synapomorphy of spiders. *J. Arachnol.* 35:411–426.

Billberg, G. J. 1820. *Enumeratio Insectorum in Museo Gust. Joh. Billberg.* Typus Gadelianus, Stockholm.

Blackledge, T. A., N. Scharff, J. A. Coddington, T. Szuts, J. W. Wenzel, C. Y. Hayashi, and I. Agnarsson. 2009. Reconstructing web evolution and spider diversification in the molecular era. *Proc. Natl. Acad. Sci. USA* 106:5229–5234.

Blackwall, J. 1844. Descriptions of some newly discovered species of *Araneida. Ann. Mag. Nat. Hist.* 13:179–188.

Blackwall, J. 1853. Descriptions of some newly discovered species of *Araneida. Ann. Mag. Nat. Hist.* 11:14–25.

Cameron, H. D. 2005. An etymological dictionary of North American genus names. Pp. 274–330 in *Spiders of North America: An*

Identification Manual (D. Ubick, P. Paquin, P. E. Cushing, and V. Roth, eds.). The American Arachnological Society.

Clerck, C. 1757. *Svenska Spindlar uti sina Hufvud-slägter Indelte Samt under Några och Sextio Särskildte arter Beskrefne och med Illuminerade Figurer Uplyste. Aranei Svecici, Descriptionibus et Figuris æneis Illustrati, ad Genera Subalterna Redacti, Speciebus Ultra LX Determinati.* L. Salvii, Holmiae (Stockholm).

Coddington, J. A. 2005. Phylogeny and classification of spiders. Pp. 18–24 in *Spiders of North America: An Identification Manual* (D. Ubick, P. Paquin, P. E. Cushing, and V. Roth, eds.). The American Arachnological Society.

Coddington, J. A., and R. K. Colwell. 2001. *Arachnida.* Pp. 199–218 in *Encyclopedia of Biodiversity,* Vol. 1 (S. C. Levin, ed.). Academic Press, New York.

Coddington, J. A., G. Giribet, M. S. Harvey, L. Prendini, and D. E. Walter. 2004. *Arachnida.* Pp. 296–318 in *Assembling the Tree of Life* (J. Cracraft and M. J. Donoghue, eds.). Oxford University Press, New York.

Coddington, J. A., and H. W. Levi. 1991. Systematics and evolution of spiders (*Araneae*). *Annu. Rev. Ecol. Syst.* 22:565–592.

Cook, O. F. 1899. *Hubbardia,* a new genus of *Pedipalpi. Proc. Entomol. Soc. Wash.* 4:249–261.

Dunlop, J. A., D. Penney, and D. Jekel. 2011. A summary list of fossil spiders and their relatives. Pp. 1–253 in *The World Spider Catalog,* Version 11.5 (N. I. Platnick, ed.). American Museum of Natural History. Available at http://research.amnh.org/iz/spiders/catalog. doi:10.5531/db.iz.0001.

Foelix, R. F. 2011. *Biology of Spiders.* Oxford University Press, New York.

Foelix, R. F., and B. Erb. 2010. *Mesothelae* have venom glands. *J. Arachnol.* 38:596–598.

Kuntner, M., J. A. Coddington, and G. Hormiga. 2008. Phylogeny of extant nephilid orb-weaving spiders (*Araneae, Nephilidae*): testing morphological and ethological homologies. *Cladistics* 24:147–217.

Latreille, P. A. 1806. *Genera Crustaceorum et Insectorum,* Tome 1, 302 pp. (*Araneae,* pp. 82–127). Amand Koenig, Parisiis et Argentorati.

Leach, W. E. 1817. *The Zoological Miscellany: Being Descriptions of New, or Interesting Animals.* McMillan, London.

Linnaeus, C. 1758. *Systema Naturae Per Regna Tria Naturae, Secundum Classes, Ordines, Genera, Species, cum Characteribus, Differentiis, Synonymis, Locis.* Tomus I, Editio decima, reformata. Laurentii Salvii, Holmiae (Stockholm).

Masta, S. E., S. J. Longhorn, and J. L. Boore. 2009. Arachnid relationships based on mitochondrial genomes: asymmetric nucleotide and amino acid bias affects phylogenetic analyses. *Mol. Phylogenet. Evol.* 50:117–128.

Petrunkevitch, A. 1955. *Arachnida.* Pp. 42–162 in *Treatise on Invertebrate Paleontology, Pt. P,* Arthropoda 2 (R. C. Moore, ed.). Geological Society of America, Boulder, CO; University of Kansas Press, Lawrence, KS.

Platnick, N. I. 2013. *The World Spider Catalog,* Version 13.5. American Museum of Natural History. Available at http://research.amnh. org/iz/spiders/catalog/ARANEIDAE.html.

Samouelle, G. 1819. *The Entomologist's Useful Compendium; Or, An Introduction to the Knowledge of British Insects.* T. Boys, London.

Savory, T. 1972. On the names of the orders of *Arachnida. Syst. Zool.* 21:122–125.

Schuett, K. 2003. Phylogeny of *Symphytognathidae* s.l. (*Araneae, Araneoidea*). *Zool. Scr.* 32:129–151.

Selden, P. A., and D. Penney. 2010. Fossil spiders. *Biol. Rev.* 85:171–206.

Shultz, J. W. 1990. Evolutionary morphology and phylogeny of *Arachnida. Cladistics* 6:1–38.

Shultz, J. W. 2007. A phylogenetic analysis of the arachnid orders based on morphological characters. *Zool. J. Linn. Soc.* 150:221–265.

Authors

Ingi Agnarsson; Department of Biology; University of Vermont; Burlington, VT 05405, USA. Email: iagnarsson@gmail.com.

Matjaž Kuntner; Department of Organisms and Ecosystems Research; National Institute of Biology; Večna pot 111, 1000 Ljubljana, Slovenia, and Institute of Biology;

Scientific Research Centre of the Slovenian Academy of Sciences and Arts; Novi trg 2, P. O. Box 306, SI-1001 Ljubljana, Slovenia. Email: kuntner@gmail.com.

Date Accepted: 2 July 2013

Primary Editor: Kevin de Queiroz

Mesothelae R. I. Pocock 1892 [I. Agnarsson and M. Kuntner], converted clade name

Registration Number: 184

Definition: The most inclusive crown clade containing *Liphistius desultor* Schiödte 1849 but not *Araneus angulatus* (Clerck 1757) (*Opisthothelae/Araneomorphae/Araneidae*) and *Theraphosa blondi* (Latreille 1804) (*Opisthothelae/Mygalomorphae/Theraphosidae*). This is a maximum-crown-clade definition. Abbreviated definition: max crown ∇ (*Liphistius desultor* Schiödte 1849 ~ *Araneus angulatus* (Clerck 1757) & *Theraphosa blondi* (Latreille 1804)).

Etymology: *Meso* is Greek for "middle" and *thele* is Greek for "nipple", which has become a conventional substitute for "spinneret." This refers to the more anterior position of the spinnerets in the members of this clade.

Reference Phylogeny: Coddington (2005: Fig. 2.2) is the primary reference phylogeny.

Composition: *Mesothelae* contains the extant *Liphistiidae* (89 species) (Platnick, 2013), and putatively the extinct *Arthrolycosidae* and *Arthromygalidae* (Dunlop et al., 2011).

Diagnostic Apomorphies: Apomorphies relative to other crown clades include a narrow ventral plate at the prosoma, the "sternite" (Raven, 1985), and tibial spurs specialized as sense organs (Platnick and Goloboff, 1985; Haupt, 2003).

Synonyms: *Liphistiomorphae* Petrunkevitch 1923 (approximate).

Comments: Pocock (1892) named two major groups (suborders) containing all spiders: "For these I propose the names Mesothelae and Opisthothelae, the terms being derived from the position of the spinning-organs". Petrunkevitch (1923: 150–152) disliked Pocock's name, which he considered "misleading and therefore objectionable", and proposed *Liphistiomorphae* instead. Nonetheless, *Mesothelae* is preferred on the basis of both priority and usage (Coddington, 2005).

The members of *Mesothelae* are also readily recognized based on plesiomorphic characters: they are the only spiders with external marks of opisthosoma segmentation, bearing tergite plates on the dorsal side, and possess eight segmented spinnerets, which are positioned on the mid venter (Coddington and Colwell, 2001) rather than terminally (see *Opisthothelae*); however, contrary to previous reports (Haupt, 2003), members of this clade do have venom glands and ducts synapomorphic for spiders (Foelix and Erb, 2010). Discussion of the evidence for the monophyly of *Mesothelae* is presented by Platnick and Goloboff (1985) and Platnick and Gertsch (1976: Fig. 6); however, no modern phylogenetic analysis has been performed to date.

Literature Cited

Coddington, J. A. 2005. Phylogeny and classification of spiders. Pp. 18–24 in *Spiders of North America: An Identification Manual* (D. Ubick, P. Paquin, P. E. Cushing, and V. Roth, eds.). The American Arachnological Society.

Coddington, J. A., and R. K. Colwell. 2001. *Arachnida*. Pp. 199–218 in *Encyclopedia of Biodiversity*, Vol. 1 (S. C. Levin, ed.). Academic Press, New York.

Dunlop, J. A., D. Penney, and D. Jekel. 2011. A summary list of fossil spiders and their relatives. Pp. 1–253 in *The World Spider Catalog*, Version 11.5 (N. I. Platnick, ed.). American Museum of Natural History. Available at http://research.amnh.org/iz/spiders/catalog. doi:10.5531/db.iz.0001.

Foelix, R., and B. Erb. 2010. *Mesothelae* have venom glands. *J. Arachnol.* 38:596–598.

Haupt, J. 2003. The *Mesothelae* – a monograph of an exceptional group of spiders (*Araneae*: *Mesothelae*). *Zoologica* 154:1–102.

Petrunkevitch, A. 1923. On families of spiders. *Ann. NY. Acad. Sci.* 29:145–180.

Platnick, N. I. 2013. *The World Spider Catalog*, Version 13.5. American Museum of Natural History. Available at http://research.amnh.org/iz/spiders/catalog/ARANEIDAE.html.

Platnick, N. I., and W. J. Gertsch. 1976. The suborders of spiders: a cladistic analysis (*Arachnida*, *Araneae*). *Am. Mus. Novit.* 2607:1–15.

Platnick, N. I., and P. A. Goloboff. 1985. On the monophyly of the spider suborder *Mesothelae* (*Arachnida*, *Araneae*). *NY. Entomol. Soc.* 93:1265–1270.

Pocock, R. I. 1892. *Liphistius* and its bearing upon the classification of spiders. *Ann. Mag. Nat. Hist.* 6:306–314.

Raven, R. J. 1985. The spider infraorder *Mygalomorphae* (*Araneae*): cladistics and systematics. *Bull. Am. Mus. Nat. Hist.* 182:1–180.

Authors

Ingi Agnarsson; Department of Biology; University of Vermont; Burlington, VT 5405, USA. Email: iagnarsson@gmail.com.

Matjaž Kuntner; Department of Organisms and Ecosystems Research; National Institute of Biology; Večna pot 111, 1000 Ljubljana, Slovenia, and Institute of Biology; Scientific Research Centre of the Slovenian Academy of Sciences and Arts; Novi trg 2, P. O. Box 306, SI-1001 Ljubljana, Slovenia. Email: kuntner@gmail.com.

Date Accepted: 20 August 2013

Primary Editor: Kevin de Queiroz

Opisthothelae R. I. Pocock 1892 [M. Kuntner and I. Agnarsson], converted clade name

Registration Number: 186

Definition: The crown clade for which a posteriorly positioned set of six or fewer spinnerets, as inherited by *Araneus angulatus* Clerck 1757 (*Araneidae*), is an apomorphy relative to other crown clades. This is an apomorphy-modified crown-clade definition designating a crown clade. Abbreviated definition: crown ∇ apo posteriorly positioned spinnerets [*Araneus angulatus* Clerck 1757].

Etymology: A Latinized feminine plural compound adjective formed by combining *Ophisho-* (Greek), "rear," with *thele* (Greek), "nipple." The name refers to the more posterior position of the spinnerets in these spiders relative to that in their sister group.

Reference Phylogeny: Coddington (2005: Fig. 2.2) is the primary reference phylogeny.

Composition: *Opisthothelae* contains all extant spiders to the exclusion of *Liphistiidae* (see *Mesothelae* in this volume), in total over 43,000 currently recognized species (Platnick, 2013). Within *Opisthothelae* the main clades are *Mygalomorphae* and *Araneomorphae* (see entries in this volume). *Opisthothelae* also contains over 1,000 known extinct species (Dunlop et al., 2011).

Diagnostic Apomorphies: Apomorphies relative to other crown clades (see Platnick and Gertsch, 1976; Raven, 1985; Coddington and Colwell, 2001; Coddington, 2005) include posteriorly positioned spinnerets on the opisthosoma, the anterior median spinnerets modified into cribellum or colulus (ancestrally in *Araneomorphae*) or lost (see *Mygalomorphae*), and the loss of external segmentation on the opisthosoma (Platnick and Gertsch, 1976; Raven, 1985).

Synonyms: To our knowledge there are no true synonyms, but *Araneae* and many older clade names that are considered synonyms of *Araneae* (see entry in this volume) were applied to *Opisthothelae* (in terms of composition) prior to the discovery of *Mesothelae*.

Comments: Pocock (1892) divided spiders into two major groups, *Opisthothelae* and *Mesothelae*. This division was accepted by many subsequent authors and first supported with an explicit phylogenetic analysis by Platnick and Gertsch (1976). *Mesothelae* (see entry in this volume) contains a small number of Southeast Asian spiders retaining external abdominal segmentation; *Opisthothelae* contains the majority of spiders, which lack external segmentation (Coddington, 2005).

Literature Cited

Coddington, J. A. 2005. Phylogeny and classification of spiders. Pp. 18–24 in *Spiders of North America: An Identification Manual* (D. Ubick, P. Paquin, P. E. Cushing, and V. Roth, eds.). The American Arachnological Society.

Coddington, J. A., and R. K. Colwell. 2001. *Arachnida*. Pp. 199–218 in *Encyclopedia of Biodiversity*, Vol. 1 (S. C. Levin, ed.). Academic Press, New York.

Dunlop, J. A., D. Penney, and D. Jekel. 2011. A summary list of fossil spiders and their relatives. Pp. 1–253 in *The World Spider Catalog*, Version 11.5 (N. I. Platnick, ed.). American

Museum of Natural History. Available at http://research.amnh.org/iz/spiders/catalog. doi:10.5531/db.iz.0001.

Platnick, N. I. 2013. *The World Spider Catalog,* Version 13.5. American Museum of Natural History. Available at http://research.amnh.org/iz/spiders/catalog/ARANEIDAE.html.

Platnick, N. I., and W. J. Gertsch. 1976. The suborders of spiders: a cladistic analysis (*Arachnida, Araneae*). *Am. Mus. Novit.* 2607:1–15.

Pocock, R. I. 1892. *Liphistius* and its bearing upon the classification of spiders. *Ann. Mag. Nat. Hist., Ser.* 6 10:306–314.

Raven, R. J. 1985. The spider infraorder *Mygalomorphae* (*Araneae*): cladistics and systematics. *Bull. Am. Mus. Nat. Hist.* 182:1–180.

Authors

Matjaž Kuntner; Department of Organisms and Ecosystems Research; National Institute of Biology; Večna pot 111, 1000 Ljubljana, Slovenia, and Institute of Biology; Scientific Research Centre of the Slovenian Academy of Sciences and Arts; Novi trg 2, P. O. Box 306, SI-1001 Ljubljana, Slovenia. Email: kuntner@gmail.com.

Ingi Agnarsson; Department of Biology; University of Vermont; Burlington, VT 5405, USA. Email: iagnarsson@gmail.com.

Date Accepted: 2 July 2013

Primary Editor: Kevin de Queiroz

Mygalomorphae R. I. Pocock 1892 [M. Kuntner and I. Agnarsson], converted clade name

Registration Number: 185

Definition: The most inclusive crown clade containing *Theraphosa blondi* (Latreille 1804) (*Theraphosidae*) but not *Liphistius desultor* Schiödte 1849 (*Mesothelae*) and *Araneus angulatus* (Clerck 1757) (*Araneomorphae*). This is a maximum-crown-clade definition. Abbreviated definition: max crown ∇ (*Theraphosa blondi* (Latreille 1804) ~ *Liphistius desultor* Schiödte 1849 & *Araneus angulatus* (Clerck 1757)).

Etymology: *Mygalomorphae* is a feminine plural compound adjective composed of the Greek word *mygale* (shrew, mouse) and the Greek *morphe* (form, shape). The combination thus literally means "mouse-shaped", although Pocock apparently intended it to mean "shaped like a tarantula" as it is based on the genus name *Mygale* Latreille 1802 (a junior homonym of *Mygale* Cuvier 1800).

Reference Phylogeny: Bond et al. (2012: Fig. 2) is the primary reference phylogeny; see also Coddington (2005: Fig. 2.2). *Araneus angulatus*, an external specifier, was not included in the primary reference phylogeny; however, that species is considered to be most closely related to *Hypochilus* of the species included in the reference phylogeny (see Coddington, 2005). Earlier primary data analyses with discussions of the evidence for monophyly have been published by Raven (1985) and Hedin and Bond (2006).

Composition: *Mygalomorphae* is a clade of mostly ground-dwelling spiders. It contains the tarantulas (*Theraphosidae*) with the largest spider species, *Theraphosa blondi* (Latreille 1804), trap-door spiders (*Ctenizidae*, *Migidae*, *Idiopidae*, *Actinopodidae*, *Nemesiidae*), purse web spiders (*Atypidae*) and others. Most construct only ground or arboreal retreats and do not rely on webs to catch and entangle prey. Exceptions are the web-building funnel-web spiders (*Dipluridae*, *Hexathelidae*), the latter containing several Australian species that are dangerously venomous to humans (*Atrax*, *Hadronyche*). *Mygalomorphae* contains a little over 2,600 extant species (Platnick, 2013) and about 20 known extinct species (Dunlop et al. 2011).

Diagnostic Apomorphies: Apomorphies relative to proximal clades include the absence of the anterior median spinnerets and their homologs: cribellum or colulus (Platnick and Gertsch, 1976; Raven 1985).

Synonyms: *Avicularidae* Simon 1874 (approximate).
Avicularoidea sensu Comstock 1913 (approximate).
Mygaloidea Berland 1932 (approximate).
Theraphosomorphae Caporiacco 1938 (approximate).

Comments: The names *Tetrapneumones* Latreille 1825 and *Orthognatha* Berland 1932 are not synonyms as they both have been used for taxa that include *Mesothelae*. Rather, these two names are partial synonyms of *Araneae*. *Avicularidae*, *Avicularoidea* and *Mygaloidea* are rarely used names derived from the genus names *Avicularia* and *Mygale*, respectively. *Theraphosomorphae* derives from the tarantula genus name *Theraphosa* and "morpha" or form and means the same as the older name *Mygalomorphae*. The name *Mygalomorphae* has been predominantly used in systematic literature (e.g., Coddington and Levi, 1991; Coddington, 2005; Bond et al., 2012; Agnarsson et al., 2013) and was selected for that reason.

Relative to the *Araneomorphae* (the so-called "true spiders"), members of *Mygalomorphae* share plesiomorphies with those of *Mesothelae*, including four booklungs and paraxial (orthognathous) cheliceral fangs (Raven, 1985).

Literature Cited

Agnarsson, I., J. A. Coddington, and M. Kuntner. 2013. Systematics: progress in the study of spider diversity and evolution. Pp. 58–111 in *Spider Research in the 21st Century* (D. Penney, ed.). Siri Scientific Press, Manchester.

Berland, L. 1932. Les arachnides (scorpions, araignées, etc.). *Encyclopédie Entomol.* 16:1–485.

Bond, J. E., B. E. Hendrixson, C. A. Hamilton, and M. Hedin. 2012. A reconsideration of the classification of the spider infraorder *Mygalomorphae* (*Arachnida: Araneae*) based on three nuclear genes and morphology. *PLOS ONE* 7:e38753–e38753.

di Caporiacco, L. 1938. Il sistema degli Araneidi. *Arch. Zool. Ital.* 25:35–155.

Coddington, J. A. 2005. Phylogeny and classification of spiders. Pp. 18–24 in *Spiders of North America: An Identification Manual* (D. Ubick, P. Paquin, P. E. Cushing, and V. Roth, eds.). The American Arachnological Society.

Coddington, J. A., and H. W. Levi. 1991. Systematics and evolution of spiders (*Araneae*). *Ann. Rev. Ecol. Syst.* 22:565–592.

Comstock, J. H. 1913. *The Spider Book*. Doubleday, Page, and Company, Garden City, NY.

Dunlop, J. A., D. Penney, and D. Jekel. 2011. A summary list of fossil spiders and their relatives. Pp. 1–253 in *The World Spider Catalog*, Version 11.5. (N. I. Platnick, ed.). American Museum of Natural History. Available at http://research.amnh.org/iz/spiders/catalog. doi:10.5531/db.iz.0001.

Hedin, M., and J. E. Bond. 2006. Molecular phylogenetics of the spider infraorder *Mygalomorphae* using nuclear rRNA genes (18S and 28S): Conflict and agreement with the current system of classification. *Mol. Phylogenet. Evol.* 41:454–471.

Platnick, N. I. 2013. *The World Spider Catalog*, Version 13.5. American Museum of Natural History. Available at http://research.amnh.org/iz/spiders/catalog/ARANEIDAE.html.

Platnick N. I., and W. J. Gertsch. 1976. The suborders of spiders: a cladistic analysis (*Arachnida, Araneae*). *Am. Mus. Novit.* 2607:1–15.

Pocock, R. I. 1892. *Liphistius* and its bearing upon the classification of spiders. *Ann. Mag. Nat. Hist., Ser. 6* 10:306–314.

Raven, R. J. 1985. The spider infraorder *Mygalomorphae* (*Araneae*): cladistics and systematics. *Bull. Am. Mus. Nat. Hist.* 182:1–180.

Simon, E. 1874. *Les Arachnides de France*. Tome Premier. Roret, Paris.

Authors

Matjaž Kuntner; Department of Organisms and Ecosystems Research; National Institute of Biology; Večna pot 111, 1000 Ljubljana, Slovenia, and Institute of Biology; Scientific Research Centre of the Slovenian Academy of Sciences and Arts; Novi trg 2, P. O. Box 306, SI-1001 Ljubljana, Slovenia. Email: kuntner@gmail.com.

Ingi Agnarsson; Department of Biology; University of Vermont; Burlington, VT 05405, USA. Email: iagnarsson@gmail.com.

Date Accepted: 20 August 2013

Primary Editor: Kevin de Queiroz

Araneomorphae F. P. Smith 1902 [I. Agnarsson and M. Kuntner], converted clade name

Registration Number: 183

Definition: The crown clade originating in the most recent common ancestor of *Hypochilus thorelli* Marx 1888 (*Paleocribellatae/Hypochilidae*), *Gradungula sorenseni* Forster 1955, *Austrochilus manni* Gertsch and Zapfe 1955 (*Austrochiloidea/Austrochilidae*), *Filistata insidiatrix* (Forskål 1775) (*Araneoclada/Filistatidae*) and *Araneus angulatus* Clerck 1757 (*Araneoclada/Araneidae*). This is a minimum-crown-clade definition. Abbreviated definition: min crown ∇ (*Hypochilus thorelli* Marx 1888 & *Gradungula sorenseni* Forster 1955 & *Austrochilus manni* Gertsch and Zapfe 1955 & *Filistata insidiatrix* (Forskål 1775) & *Araneus angulatus* Clerck 1757).

Etymology: *Araneomorphae* is a feminine plural compound Latin adjective combining the Latin word *araneus* (spider) and the Latinized Greek word *morphe* (form, shape). Thus, the name means "spider-shaped."

Reference Phylogeny: Coddington (2005: Fig. 2.2), a manually built super tree including nearly all taxa traditionally ranked as families, is the primary reference phylogeny.

Composition: *Araneomorphae*, sometimes called the "true" spiders, contains the majority of spiders—all except *Mesothelae* and *Mygalomorphae*—including over 32,000 extant species (Platnick, 2013) and over 1,000 known fossil species (Dunlop et al., 2011).

Diagnostic Apomorphies: Apomorphies relative to other clades of *Araneae* include labidognath chelicerae (Platnick and Gertsch, 1976), elongate 3rd abdominal entapophyses (Ramirez, 2000), anterior median spinnerets modified into a cribellum with associated calamistrum on leg 4 (Platnick and Gertsch, 1976), posterior lateral spinnerets consisting of only one or two segments (Platnick and Gertsch, 1976), and a single pair of coxal glands (Millot, 1949).

Synonyms: *Arachnomorphae* Pocock 1892 (approximate).
> *Araneae Verae* Simon 1892 (approximate).
> *Antiodontes* Thorell 1895 (approximate).
> *Labidognatha* auct. (approximate).
> *Dipneumones* Latreille 1825 (approximate).
> *Dipneumomorphae* Petrunkevitch 1946 (approximate).

Comments: *Araneomorphae* is a substitute name for *Arachnomorphae* Pocock 1892 (see Platnick and Gertsch, 1976). Although Simon's *Araneae Verae* was rarely used subsequently, the directly translated term "true spiders" has continued to be associated with this group. It is not clear who first coined the term *Labidognatha*; the first use of it, so spelled, we have encountered is by Gerhardt and Kästner (1938). Berland (1932) introduced its counterpart—*Orthognatha* Berland 1932, and he referred to the other half of the dichotomy in French as "araignees labidognathes". Platnick and Gertsch (1976) advocated abandoning *Labidognatha* and *Orthognatha* as the latter is paraphyletic and the former poorly diagnosed.

Although the name *Labidognatha* still commonly appears in literature, the authors of most recent systematic analyses use the name

Araneomorphae for this clade (e.g., Coddington and Levi, 1991; Coddington, 2005; Agnarsson et al., 2013), and we prefer it here for that reason (note that *Labidognatha* Ehlers, is the name of a group of *Polychaeta*). *Dipneumones* (also used for a group of fishes; the author of the spider taxon name is uncertain, with the first use known to us being that of Latreille, 1825) and *Dipneumomorphae* Petrunkevitch 1946, are approximate synonyms referring to spiders with one pair of booklungs.

A recent molecular analysis of spider phylogenetic relationships (Agnarsson et al., 2013) failed to find support for *Araneomorphae*.

Literature Cited

Agnarsson, I., J. A. Coddington, and M. Kuntner. 2013. Systematics: progress in the study of spider diversity and evolution. Pp. 58–111 in *Spider Research in the 21st Century* (D. Penney, ed.). Siri Scientific Press, Manchester.

Berland, L. 1932. Les Arachnides (Scorpions, Araignées, etc.). *Encyclopédie Entomologique*, Tome 16. Lechevalier, Paris.

Coddington, J. A. 2005. Phylogeny and classification of spiders. Pp. 18–24 in *Spiders of North America: An Identification Manual* (D. Ubick, P. Paquin, P. E. Cushing, and V. Roth, eds.). The American Arachnological Society.

Coddington, J. A., and H. W. Levi. 1991. Systematics and evolution of spiders (*Araneae*). *Annu. Rev. Ecol. Syst.* 22:565–592.

Dunlop, J. A., D. Penney, and D. Jekel. 2011. A summary list of fossil spiders and their relatives. Pp. 1–253 in *The World Spider Catalog, Version 11.5* (N. I. Platnick, ed.). American Museum of Natural History. Available at http://research.amnh.org/iz/spiders/catalog. doi:10.5531/db.iz.0001.

Gerhardt, U., and A. Kästner. 1938. *Araneae*— Echte Spinnen—Webspinnen. Pp. 497–656 in *Handbuch der Zoologie* (W. Kükenthal and T. Krumbach, eds.). DeGruyter, Berlin.

Latreille, M. 1825. *Familles Naturelles du Règne Animal*. J.-B. Baillière, Paris.

Millot, J. 1949. Classe des arachnides: morphologie generale et anatomie inteme; ordre des *Araneides*. Pp. 263–320 and 589–743 in *Traite de Zoologie*, Vol. 6. P. P. Grasse, Paris.

Petrunkevitch, A. 1946. Palaeozoic *Arachnida* of Illinois: An inquiry into their evolutionary trends. *Sci. Pap. Illinois State Mus.* 3 (2):1–72.

Platnick, N. I. 2013. *The World Spider Catalog*, Version 13.5. American Museum of Natural History. Available at http://research.amnh.org/iz/spiders/catalog/ARANEIDAE.html.

Platnick, N. I., and W. J. Gertsch. 1976. The suborders of spiders: a cladistic analysis (*Arachnida, Araneae*). *Am. Mus. Novit.* 2607:1–15.

Pocock, R. I. 1892. *Liphistius* and its bearing upon the classification of spiders. *Ann. Mag. Nat. Hist.* 6:306–314.

Ramírez, M. 2000. Respiratory system morphology and the phylogeny of haplogyne spiders (*Araneae, Araneomorphae*). *J. Arachnol.* 28:149–157.

Simon, E. 1874. *Les Arachnides de France*. Tome Premier. Roret, Paris.

Smith, F. P. 1902. The spiders of Epping Forest. *Essex Nat.* 12:181–201.

Thorell, T. 1895. *Descriptive Catalogue of the Spiders of Burma*. Trustees of the British Museum (Natural History), London.

Authors

Ingi Agnarsson; Department of Biology; University of Vermont; Burlington VT, 5405, USA. Email: iagnarsson@gmail.com.

Matjaž Kuntner; Department of Organisms and Ecosystems Research; National Institute of Biology; Večna pot 111, 1000 Ljubljana, Slovenia, and Institute of Biology; Scientific Research Centre of the Slovenian Academy of Sciences and Arts; Novi trg 2, P. O. Box 306, SI-1001 Ljubljana, Slovenia. Email: kuntner@gmail.com.

Date Accepted: 2 July 2013

Primary Editor: Kevin de Queiroz

Deuterostomia K. Grobben 1908 [J. R. Garey, C. B. Cameron, K. M. Halanych, J. A. Servin and J. A. Lake], converted clade name

Registration Number: 34

Definition: The largest crown clade containing *Strongylocentrotus purpuratus* Stimpson 1857 (*Echinodermata/Echinoidea*) and *Homo sapiens* Linnaeus 1758 (*Chordata/Vertebrata*) but not *Drosophila melanogaster* Meigen 1830 (*Arthropoda/Insecta*) and *Capitella teleta* Blake, Grassle and Eckelbarger 2009 (*Annelida/Polychaeta*). This is a maximum-crown-clade definition. Abbreviated definition: max crown ∇ (*Strongylocentrotus purpuratus* Stimpson 1857 & *Homo sapiens* Linnaeus 1758 ~ *Drosophila melanogaster* Meigen 1830 & *Capitella teleta* Blake, Grassle and Eckelbarger 2009).

Etymology: From the Greek *deutero-* (second) and *stoma* (mouth), referring to a mouth derived from a secondary opening (see Diagnostic Apomorphies).

Reference Phylogeny: Dunn et al. (2008: Fig. 1) is the primary reference phylogeny. Secondary reference phylogenies include Cameron et al. (2000: Fig. 2), Turbeville et al. (1994: Fig. 2a), Wada and Satoh (1994: Fig. 1), and Halanych (1995: Fig. 2).

Composition: *Deuterostomia* is currently thought to include *Hemichordata*, *Echinodermata*, *Chordata*, and *Xenoturbellida* (Bourlat et al., 2006; Dunn et al., 2008).

Diagnostic Apomorphies: Ribosomal RNA sequences are the primary apomorphies of this clade, as documented in the references above (see Reference Phylogeny). *Deuterostomia*-specific peptides in Hox genes (Balavoine et al., 2002) and other nuclear protein-coding gene datasets also support the clade (e.g., Pick et al., 2010). Other apomorphies may include metanephridial system and monociliated cells, the use of creatine phosphate as the phosphate store, and the endo-mesoderm is generally derived from the gut (Cameron, 2005). Traditionally, deuterostomes were diagnosed by embryological features such as radial cleavage, indeterminate cell fates, a mouth derived from a secondary opening rather than the blastopore, and a coelom that develops by enterocoely (Hyman, 1940). There are numerous exceptions to these developmental characteristics (Halanych, 2004; Nielsen, 2011), and they may be more correlated with the volume of the egg than with shared ancestry (Galis and Sinervo, 2002).

Synonyms: There are several approximate synonyms: *Enterocölia* Schneider 1902, *Pleurogastrica* Goette 1902, *Enterocoelia* Hatschek 1911, and *Neorenalia* Nielsen 1995.

Comments: The monophyly of *Deuterostomia* (Grobben, 1908) is supported by both molecular and morphological analyses (Peterson and Eernisse, 2001; Winchell et al., 2002; Halanych, 2004; Webster et al., 2006; Nielsen, 2011). The name *Deuterostomia* has priority over *Enterocoelia*, which has fallen from common usage. The earlier names *Enterocölia* and *Pleurogastrica* have also fallen from common usage. The name *Neorenalia*, proposed by Nielsen (1995) for the traditional deuterostomes (*Ambulacraria* + *Chordata*) when he expanded *Deuterostomia* to include two protostome taxa, *Phoronida* and *Brachiopoda*, was never widely used and was later abandoned by Nielsen (2011).

The name *Deuterostomia* is defined here in a manner that is congruent with its historical use (Horst, 1939; Hyman, 1959), that is, for a clade composed of *Ambulacraria* + *Chordata* with the more recent addition of *Xenoturbellida* (Bourlat et al., 2006) and the absence of lophophorates (Halanych et al., 1995).

Several taxa once thought to be members of *Deuterostomia* have been shown to be members of *Protostomia* instead. Halanych et al. (1995) conclusively demonstrated that the lophophorate taxa (*Brachiopoda*, *Phoronida*, and *Bryozoa*) are nested within a subset of protostome taxa that includes animals with trochophore larvae, despite having strong similarities to deuterostomes in their patterns of early embryological development. *Chaetognatha*, also traditionally considered deuterostomes, has also been removed (Telford and Holland, 1993; Wada and Satoh, 1994; Marlétaz et al., 2006; Matus et al., 2006).

Several taxa not historically viewed as members of *Deuterostomia* have been inferred to belong to this clade, primarily due to molecular sequence analyses. For example, *Xenoturbella* has been included in *Deuterostomia* based on large concatenated alignments of multiple protein coding genes (Bourlat et al., 2003; Bourlat et al, 2006; Dunn et al., 2008; Pick et al., 2010). Phillipe et al. (2011) suggested that *Acoelomorpha* are deuterostomes closely allied with *Xenoturbella*, although organisms with rapidly evolving or highly divergent genomes such as *Acoelomorpha* are difficult to place in molecular trees.

Although a maximum-crown-clade definition only requires one internal specifier, two internal specifiers are used in our definition to represent chordates and echinoderms. In the unlikely event that either of these clades turned out to be more closely related to protostomes than to one another, the name *Deuterostomia* would not apply to any clade.

Literature Cited

Balavoine, G., R. de Rosa, and A. Adoutte. 2002. Hox clusters and bilaterian phylogeny. *Mol. Phylogenet. Evol.* 24:366–373.

Bourlat, S. J., T. Juliusdottir, C. J. Lowe, R. Freeman, J. Aronowicz, M. Kirschner, E. S. Lander, M. Thorndyke, H. Nakano, A. B. Kohn, A. Heyland, L. L. Moroz, R. R. Copley, and M. J. Telford. 2006. Deuterostome phylogeny reveals monophyletic chordates and the new phylum *Xenoturbellida*. *Nature* 444:85–88.

Bourlat, S. J., C. Nielsen, A. E. Lockyer, D. T. Littlewood, and M. J. Telford. 2003. *Xenoturbella* is a deuterostome that eats molluscs. *Nature* 424:885–886.

Cameron, C. B. 2005. A phylogeny of the hemichordates based on morphological characters. *Can. J. Zool.* 83:196–215.

Cameron, C. B., J. R. Garey, and B. J. Swalla. 2000. Evolution of the chordate body plan: new insights from phylogenetic analyses of deuterostome phyla. *Proc. Natl. Acad. Sci. USA* 97:4469–4474.

Dunn, C. W., A. Hejnol, D. Q. Matus, D. Pang, W. E. Browne, S. A. Smith, E. Seaver, G. W. Rouse, M. Obst, G. D. Edgecombe, M. V. Sørensen, S. H. D. Haddock, A. Schmidt-Rhaesa, A. Okusu, R. M. Kristensen, W. C. Wheeler, M. Q. Martindale, and G. Giribet. 2008. Broad phylogenomic sampling improves resolution of the animal tree of life. *Nature* 452:745–749.

Galis, F., and B. Sinervo. 2002. Divergence and convergence in early embryonic stages of metazoans. *Contrib. Zool.* 71(1/3):101–113.

Goette, A. 1902. *Lehrbuch der Zoologie*. Wilhelm Engelmann, Leipzig.

Grobben, K. 1908. Die systematische Einteilung des Tierreichs. *Verh. Zool. Bot. Ges. Wien* 58:491–511.

Halanych, K. M. 1995. The phylogenetic position of the pterobranch hemichordates based on 18S rDNA sequence data. *Mol. Phylogenet. Evol.* 4:72–76.

Halanych, K. M. 2004. The new view of animal phylogeny. *Annu. Rev. Ecol. Evol. Syst.* 35:229–56.

Halanych, K. M., J. D. Bacheller, A. M. A. Aguinaldo, S. M. Liva, D. M. Hillis, and J. A. Lake. 1995. Evidence from 18S ribosomal DNA that the lophophorates are protostome animals. *Science* 267:1641–1643.

Hatschek, B. 1911. *Das neue zoologisches System* Leipzig (reference cited in Hyman, 1940).

van der Horst, C. J. 1939. *Hemichordata*. P. 737 in *Klassen und Ordnungen des Tier-Reichs wissenschaftlich dargestellt in Wort und Bild*, Band 4, Abt. 4, Buch 2, Tiel 2 (H. G. Bronns, ed.). Akademische Verlagsgesellschaft, Leipzig.

Hyman, L. 1940. *The Invertebrates*: Protozoa *through* Ctenophora. McGraw-Hill, New York and London.

Hyman, L. 1959. *The Invertebrates: Smaller Coelomate Groups*. 5th edition. McGraw-Hill.

Marlétaz, F., E. Martin, Y. Perez, D. Papillon, X. Caubit, C. J. Lowe, B. Freeman, L. Fasano, C. Dossat, P. Wincker, J. Weissenback, and Y. L. Parco. 2006. Chaetognath phylogenomics: a protostome with deuterostome-like development. *Curr. Biol.* 16:R577–R578.

Matus, D. Q., R. R. Copley, C. W. Dunn, A. Hejnol, H. Eccleston, K. M. Halanych, M. Q. Martindale, and M. J. Telford. 2006. Broad taxon and gene sampling indicate that chaetognaths are protostomes. *Curr. Biol.* 16:R575–R576.

Nielsen, C. 1995. *Animal Evolution: Interrelationships of the Living Phyla*. Oxford University Press, Oxford.

Nielsen, C. 2011. *Animal Evolution: Interrelationships of the Living Phyla*. 3rd edition. Oxford University Press, Oxford.

Peterson, K. J., and D. J. Eernisse. 2001. Animal phylogeny and the ancestry of bilaterians: inferences from morphology and 18S rDNA gene sequences. *Evol. Dev.* 3:170–205.

Phillipe, H., H. Brinkmann, R. R. Copley, L. L. Moroz, H. Nakano, A. J. Poustka, A. Wallberg, K. J. Peterson, and M. J. Telford. 2011. Acoelomorph flatworms are deuterostomes related to *Xenoturbella*. *Nature* 470:255–260.

Pick, K. S., H. Philippe, F. Schreiber, D. Erpenbeck, K. J. Jackson, P. Wrede, M. Wiens, A. Alié, B. Morgenstern, M. Manuel, and G. Wörheide. 2010. Improved phylogenomic taxon sampling noticeably affects nonbilaterian relationships. *Mol. Biol. Evol.* 27:1983–1987.

Schneider, K. C. 1902. *Lehrbuch der Vergleichenden Histologie der Tiere*. Gustav Fischer, Jena.

Telford, M. J., and P. W. Holland. 1993. The phylogenetic affinities of the chaetognaths: a molecular analysis. *Mol. Biol. Evol.* 3:660–676.

Turbeville, J. M., J. R. Schulz, and R. A. Raff. 1994. Deuterostome phylogeny and the sister group of the chordates: evidence from molecules and morphology. *Mol. Biol. Evol.* 4:648–655.

Wada, H., and N. Satoh. 1994. Details of the evolutionary history from invertebrates to vertebrates, as deduced from the sequences of 18S rRNA. *Proc. Nat. Acad. Sci. USA* 5:1801–1804.

Webster, B. L., R. R. Copley, R. A. Jenner, J. A. Mackenzie-Dodds, S. J. Bourlat, O. Rota-Stabelli, D. T. Littlewood, and M. J. Telford. 2006. Mitogenomics and phylogenomics reveal priapulid worms as extant models of the ancestral ecdysozoan. *Evol. Dev.* 8:502–510.

Winchell, C. J., J. Sullivan, C. B. Cameron, B. J. Swalla, J. Mallat. 2002. Evaluating hypotheses of deuterostome phylogeny and chordate evolution with new LSU and SSU ribosomal DNA data. *Mol. Biol. Evol.* 19:762–776.

Authors

James R. Garey; Department of Cell Biology, Microbiology and Molecular Biology; University of South Florida; 4202 East Fowler Avenue ISA2015; Tampa, FL 33620, USA. Email: garey@usf.edu.

Christopher B. Cameron; Département de Sciences Biologiques, F-208-8, Pavillon Marie-Victorin, Montréal, Québec, H2V 2S9, Canada. Email: c.cameron@umontreal.ca.

Kenneth M. Halanych; Department of Biological Sciences; 101 Rouse Building; Auburn University; Auburn, AL 36849, USA. Email: ken@auburn.edu.

Jacqueline Servin; Molecular Biology Institute; 232 Boyer Hall; University of California Los Angeles; Los Angeles, CA 90095, USA. Email: jacquelineservin@gmail.com.

James Lake; Molecular Biology Institute; 232 Boyer Hall; University of California Los Angeles; Los Angeles, CA 90095, USA. Email: Lake@mbi.ucla.edu.

Date Accepted: 8 April 2014

Primary Editor: Philip Cantino

Ambulacraria V. E. Metschnikoff 1881 [C. B. Cameron and K. Nanglu], converted clade name

Registration Number: 6

Definition: The crown clade originating in the most recent common ancestor of the hemichordates *Cephalodiscus gracilis* Harmer 1905, *Saccoglossus pusillus* Ritter 1902, and *Ptychodera flava* Eschscholtz 1825, and the echinoderms *Florometra serratissima* A. H. Clark 1907, *Asterias amurensis* Lutken 1871, *Arbacia punctulata* Lamarck 1816, and *Cucumeria salva* Lampert 1885. This is a minimum-crown-clade definition. Abbreviated definition: min crown ∇ (*Cephalodiscus gracilis* Harmer 1905 & *Saccoglossus pusillus* Ritter 1902 & *Ptychodera flava* Eschscholtz 1825 & *Florometra serratissima* A. H. Clark 1907 & *Asterias amurensis* Lutken 1871 & *Arbacia punctulata* Lamarck 1816 & *Cucumeria salva* Lampert 1885).

Etymology: Derived from the Latin *ambulare* (to walk), referring to the podia, or tube feet, with which echinoderms grip the substrate and move.

Reference Phylogeny: Mallatt and Winchell (2007: Fig. 1) is the primary reference phylogeny. See also Winchell et al. (2002), Cameron (2005), and Telford et al. (2014).

Composition: *Ambulacraria* includes *Hemichordata* and *Echinodermata* and their stem groups.

Diagnostic Apomorphies: Apomorphies of *Ambulacraria* include a larval preoral, monociliated feeding band (Morgan, 1891; Strathmann and Bonar, 1976); a perioral, ciliated band (Lacalli and Gilmour, 2001); unique larval multipolar neurons with apical processes (Lacalli, 1993); and coelomic sacs that are organized anterior to posterior as paired protocoels (echinoderm axocoels), mesocoels (echinoderm hydrocoels) and metacoels (echinoderm somatocoels) (Gemmill, 1914; Gislén, 1930). The anterior left coelom has a ciliated duct that is lined with nephridia (mesothelial monociliated podocytes and myocytes) and opens to the exterior via a left dorsal lateral pore (Balser and Ruppert, 1990). The larval pore canal – hydropore complex is retained in adult hemichordates as a heart – kidney (echinoderms axial) complex (Goodrich, 1917; Dilly et al., 1986; Balser and Ruppert, 1990).

Synonyms: *Ambulacralia* Hatschek 1888 (approximate); *Coelomopora* Marcus 1958 (approximate).

Comments: The monophyly of *Ambulacraria* is strongly supported by both molecular and morphological analyses (Cameron, 2005; Mallatt and Winchell, 2007). The clade name *Ambulacraria* was selected over its synonyms both because it has priority and because it is the most commonly used. Should *Tunicata*, *Cephalochordata* or *Vertebrata* be found to be part of *Ambulacraria*, in which case *Ambulacraria* would become synonymous with *Deuterostomia* (as defined in this volume), *Deuterostomia* should have precedence over *Ambulacraria*.

Literature Cited

Balser, E. J., and E. E. Ruppert. 1990. Structure, ultrastructure, and function of the preoral heart-kidney in *Saccoglossus kowalevskii* (*Hemichordata, Enteropneusta*) including new data on the sto-mochord. *Acta Zool.* (Stockh.) 71:23–249.

Cameron, C. B. 2005. A phylogeny of the hemi-chordates based on morphological characters. *Can. J. Zool.* 83:196–215.

Dilly, P. N., U. Welsch, and G. Rehkamper. 1986. Fine structure of heart, pericardium and glomerular vessel in *Cephalodiscus gracilis* M'Intosh, 1882 (*Pterobranchia, Hemichordata*). *Acta Zool.* (Stockh.) 67(3):173–179.

Gemmill, J. F. 1914. The development and certain points in the adult structure of the starfish *Asterias rubens*. *Philos. Trans. R. Soc. Lond. B Biol. Sci.* 205:213–294.

Gislén, T. 1930. Affinities between the *Echinodermata*, *Enteropneusta* and *Chordonia*. *Zoologiska Bidrag fran Uppsala* 12:199–304.

Goodrich, E. S. 1917. Proboscis pores in craniate vertebrates, a suggestion concerning the pre-mandibular somites and hypophysis. *Q. J. Microsc. Sci.* 62:539–553.

Hatschek, B. 1888. Lehrbuch der Zoologie. *Lieferung* 1:1–144.

Lacalli, T. C. 1993. Ciliary bands in echinoderm lar-vae: Evidence for structural homologies and a common plan. *Acta Zool.* (Stockh.) 74:127–133.

Lacalli, T. C., and T. H. J. Gilmour. 2001. Locomotory and feeding effectors of the tor-naria larva of *Balanoglossus biminiensis*. *Acta Zool.* (Stockh.) 82:117–126.

Mallatt, J., and C. J. Winchell. 2007. Ribosomal RNA genes and deuterostome phylogeny revis-ited: More cyclostomes, elasmobranchs, rep-tiles, and a brittle star. *Mol. Phylogenet. Evol.* 43:1005–1022.

Marcus, E. 1958. On the evolution of the animal phyla. *Q. Rev. Biol.* 33:24–58.

Metschnikoff, V. E. 1881. Uber die system-atische Stellung von *Balanoglossus*. *Zool. Anz.* 4:153–157.

Morgan, T. H. 1891. The growth and metamorpho-sis of tornaria. *J. Morphol.* 5:407–458.

Strathmann, R. R., and D. Bonar. 1976. Ciliary feeding of tornaria larvae of *Ptychodera flava* (*Hemichordata: Enteropneusta*). *Mar. Biol.* (Berl.) 34:317–324.

Telford, M.J., C. J. Lowe, C. B. Cameron, O. Ortega-Martinez, J. Aronowicz, P. Oliveri, and R. R. Copley. 2014. Phylogenomic analy-sis of echinoderm class relationships supports Asterozoa. *Proc. R. Soc. Lond. B Biol. Sci.* 281(1786). doi:10.1098/rspb.2014.0479.

Winchell, C. J., J. Sullivan, C. B. Cameron, B. J. Swalla, and J. Mallatt. 2002. Evaluating hypothesis of deuterostome phylogeny and chordate evolution with new LSU and SSU ribosomal DNA data. *Mol. Biol. Evol.* 19:762–776.

Author

Chris B. Cameron; Sciences biologiques; Universite de Montreal; C.P. 6128, Succ. Centre-ville; Montreal, QC, Canada H3C 3J7. Email: c.cameron@umontreal.ca.

Karma Nanglu; Department of Paleobiology; Smithsonian Institution; Washington, DC 20560, USA. Email: nangluk@si.edu.

Date Accepted: 22 August 2010; updated 27 November 2019

Primary Editor: Philip Cantino

Hemichordata W. Bateson 1885 [C. B. Cameron and E. Bell], converted clade name

Registration Number: 49

Definition: The smallest crown clade containing *Cephalodiscus gracilis* Harmer 1905 and *Ptychodera flava* Eschscholtz 1825. This is a minimum-crown-clade definition. Abbreviated definition: min crown ∇ (*Cephalodiscus gracilis* Harmer 1905 & *Ptychodera flava* Eschscholtz 1825).

Etymology: Derived from the Greek *hemi* (half) and *chorde* (string), referring to the stomochord. Like the chordate notochord, the stomochord is derived from the endoderm and is composed of turgid cells, but unlike the notochord, it is restricted to the anterior part of the body and thus is a 'half-chord.

Reference Phylogeny: The primary reference phylogeny is Cameron (2005: Fig. 5). See also Cameron et al. (2000), Winchell et al. (2002), and Cannon et al. (2009).

Composition: The clade *Hemichordata* is currently thought to contain some 106 extant species of *Enteropneusta*, *Pterobranchia*, and *Planctosphaeridae*, and the extinct *Graptolithina*. An up-to-date comprehensive list of living species may be found in Cameron (2009).

Diagnostic Apomorphies: *Hemichordata* apomorphies include the prosoma: a muscular–secretory–locomotory preoral organ (enteropneust proboscis and pterobranch cephalic shield) that encloses a heart–kidney coelomic complex, including a stomochord. Further apomorphies include the paired valved mesocoel ducts and pores; and a ventral postanal extension of the metacoels (Schepotieff, 1909; Horst, 1939; Hyman, 1959).

Synonyms: *Stomochordata* Dawydoff 1948 (approximate).

Comments: The monophyly of *Hemichordata* is supported by both molecular and morphological analyses (Cameron et al. 2000; Winchell et al. 2002; Cameron, 2005). The hypothesis that the pterobranchs are the sister group of the echinoderms and the enteropneusts are the sister group of the chordates (Bateson, 1885; Nielsen, 2001) is no longer widely accepted. Instead, hemichordates are widely regarded as monophyletic and the sister group of *Echinodermata* (Horst, 1939; Dawydof, 1948; Hyman, 1959; Halanych, 1996; Cameron et al., 2000). Dawydoff (1948) created a taxon named *Stomochordata* to unambiguously include *Enteropneusta* and *Pterobranchia*; however, the name *Hemichordata*, originally applied to enteropneusts alone (Bateson, 1885), had already been applied to this group in two major works (Harmer, 1904; Horst, 1939). *Hemichordata* is the name most commonly applied to the clade composed of pterobranchs and enteropneusts (e.g., Cameron et al., 2000; Winchell et al., 2002; Cannon et al., 2009) and was selected here for that reason. *Deuterostomia* should have precedence over *Hemichordata* in the context of phylogenies in which the two names are synonyms. Members of the monotypic taxon *Planctosphaeridae* may be long-lived enteropneust larvae (Hadfield and Young, 1983).

Literature Cited

Bateson, W. 1885. The later stages in the development of *Balanoglossus kowalevskii*, with a suggestion as to the affinities of the *Enteropneusta*. *Q. J. Microsc. Sci.* 25(suppl.):81–122.

Cameron, C. B. 2005. A phylogeny of the hemichordates based on morphological characters. *Can. J. Zool.* 83:196–215.

Cameron, C. B. 2009. A Comprehensive List of Extant Hemichordate Species with Links to Images. Available at https://www.webdepot.umontreal.ca/Usagers/cameroc/MonDepotPublic/Cameron/Species.html.

Cameron, C. B., B. J. Swalla, and J. R. Garey. 2000. Evolution of the chordate body plan: new insights from phylogenetic analysis of deuterostome phyla. *Proc. Natl. Acad. Sci. USA* 97:4469–4474.

Cannon, J. T., A. L. Rychel, H. Eccleston, K. M. Halanych, and B. J. Swalla. 2009. Molecular phylogeny of *Hemichordata*, with updated status of deep-sea enteropneusts. *Mol. Phylogenet. Evol.* 52:17–24.

Dawydoff, C. 1948. Classe des entéropneustes. Pp. 369–453 in *Traité de Zoologie* (P. P. Grassé, ed.). Masson, Paris.

Hadfield, M. G., and R. E. Young. 1983. *Planktospaera* (Hemichordata: Enteropneusta) in the Pacific Ocean. *Mar. Biol.* 73:151–153.

Halanych, K. M. 1996. Convergence in the feeding apparatuses of lophophorates and pterobranch hemichordates revealed by 18S rDNA: an interpretation. *Biol. Bull. (Woods Hole)* 190:1–5.

Harmer, S. F. 1904. *Hemichordata*. Pp. 3–111 in *The Cambridge Natural History*, Vol. 7 (S. F. Harmer and A. E. Shipley, eds.). Macmillan & Co., London.

van der Horst, C. J. 1939. *Hemichordata*. In *Klassen und Ordnungen des Tier-Reichs wissenschaftlich dargestellt in Wort und Bild*, Band 4, Abt. 4, Buch 2, Tiel 2 (H. G. Bronns, ed.). Akademische Verlagsgesellschaft, Leipzig.

Hyman, L. H. 1959. *The Invertebrates: Smaller Coelomate Groups*, Vol. 5. McGraw-Hill, New York.

Nielsen, C. 2001. *Animal Evolution: Interrelationships of the Living Phyla.* 2nd edition. Oxford University Press, Oxford.

Schepotieff, A. 1909. Die Pterobranchier des Indischen Ozeans. *Zool. Jahrb. Abt. Syst. Oekol. Geogr. Tiere* 28:429–448, plates 7–8.

Winchell, C. J., J. Sullivan, C. B. Cameron, B. J. Swalla, and J. Mallatt. 2002. Evaluating hypothesis of deuterostome phylogeny and chordate evolution with new LSU and SSU ribosomal DNA data. *Mol. Biol. Evol.* 19:762–776.

Author

Chris B. Cameron; Sciences biologiques; Universite de Montreal; C.P. 6128, Succ. Centre-ville; Montreal, Québec, H3C 3J7, Canada. Email: c.cameron@umontreal.ca.

Elena Bell; Sciences biologiques; Université de Montréal; C.P. 6128, Succ. Centre-ville; Montreal, Québec, H3C 3J7, Canada. Email: elena.bell@umontreal.ca.

Date Accepted: 8 March 2012

Primary Editor: Philip Cantino

Enteropneusta C. Gegenbaur 1870 [C. B. Cameron and N. Jabr], converted clade name

Registration Number: 39

Definition: The crown clade originating in the most recent common ancestor of *Ptychodera flava* Eschscholtz 1825, *Harrimania kupfferi* von Willemoes-Suhm 1871, *Saccoglossus kowalevskii* Agassiz 1873, *Glandiceps hacksi* Marion 1885, *Stereobalanus canadensis* Spengel 1893, *Protoglossus koehleri* Caullery and Mesnil 1900, *Xenopleura vivipara* Gilchrist 1925, *Saxipendium coronatum* Woodwick and Sensenbaugh 1985, and *Tergivelum baldwinae* Holland et al. 2009. This is a minimum-crown-clade definition. Abbreviated definition: min crown ∇ (*Ptychodera flava* Eschscholtz 1825 & *Harrimania kupfferi* von Willemoes-Suhm 1871 & *Saccoglossus kowalevskii* Agassiz 1873 & *Glandiceps hacksi* Marion 1885 & *Stereobalanus canadensis* Spengel 1893 & *Protoglossus koehleri* Caullery and Mesnil 1900 & *Xenopleura vivipara* Gilchrist 1925 & *Saxipendium coronatum* Woodwick and Sensenbaugh 1985 & *Tergivelum baldwinae* Holland et al. 2009).

Etymology: Derived from the Greek *enteron* (intestine) and *pneustikos* (breathing), in reference to the gut that is perforated by gill slits, forming a pharynx that is used in filter-feeding food particles from seawater.

Reference Phylogeny: The primary reference phylogeny is Cameron (2005: Fig. 5). See also Winchell et al. (2002: Fig. 1) and Osborn et al. (2011: Fig. 4). *Tergivelum baldwinae* is used as an internal specifier despite being absent from the primary reference phylogeny. However, there is evidence (Osborn et al., 2011) suggesting a close relationship between *Tergivelum* and *Ptychoderidae* (the latter is shown on the primary reference phylogeny).

Composition: The clade *Enteropneusta* currently contains some 99 extant species in four taxa ranked as families: *Harrimaniidae*, *Spengelidae*, *Ptychoderidae*, and *Torquaratoridae*. An up-to-date comprehensive list of living species may be found in Cameron (2012).

Diagnostic Apomorphies: Enteropneust autapomorphies include the preoral ciliary organ (Brambell and Cole, 1939); the hepatic/branchial pharynx (Cameron, 2002); a Y-shaped nuchal skeleton; perihaemal coeloms associated with a dorsal blood vessel (Pardos and Benito, 1990); mesocoel ducts that open into the first pair of gill pores; a larval locomotory ciliated band (telotroch); and a larval apical plate retractor muscle (Morgan, 1891).

Synonyms: *Hemichordata* Bateson 1885 (approximate); see Comments.

Comments: The monophyly of *Enteropneusta* is supported by molecular and morphological analyses (Winchell et al., 2002; Cameron, 2005; Osborn et al., 2011). An alternative phylogeny inferred from the 18S rDNA gene suggests that *Pterobranchia* is sister to the enteropneusts *Saccoglossus* and *Harrimania*, and the grouping of *Saccoglossus*, *Harrimania*, and pterobranchs is sister to the enteropneust taxon *Ptychoderidae* (Cameron et al., 2000; Mallatt and Winchell, 2007; Cannon et al., 2009). However the 18S rDNA sequence does not code for protein and so it can tolerate extensive insertion, deletion and point mutations (Hassouna et al., 1984). Ribosomal DNA genes are also multicopy loci that can undergo unusual patterns of sequence evolution, and they can undergo radically different rates of evolution in different lineages (Abouheif et al., 1998). The pterobranch 18S rDNA displays rapid rates of evolution resulting in long branches, and their grouping with the enteropneusts *Saccoglossus* and *Harrimania* may be a consequence of long-branch attraction (Cameron, 2005). In the event that *Enteropneusta* is found to be paraphyletic, and thus *Enteropneusta* would become a phylogenetic synonym of *Hemichordata*, *Hemichordata* should have precedence over *Enteropneusta* to maintain compositional stability, because *Pterobranchia* is traditionally included in *Hemichordata* but not in *Enteropneusta*.

Bateson (1885) proposed the name *Hemichordata* as a synonym of *Enteropneusta* and suggested it be included in *Chordata*. However, subsequent authors also included *Pterobranchia* in *Hemichordata*, so that what is now the traditional concept of *Hemichordata* includes both *Enteropneusta* and *Pterobranchia*, an arrangement used in three major treatises on zoology (Horst, 1939; Dawydoff, 1948; Hyman, 1959). Consequently, although Bateson (1885) applied the name *Hemichordata* to *Enteropneusta* alone, it is currently applied to a more inclusive clade (Winchell et al., 2002; Cameron, 2005). Thus, the name *Enteropneusta* was selected because it is the name most commonly applied to the clade in question.

Literature Cited

Abouheif, E., R. Zardoya, and A. Meyer. 1998. Limitations of metazoan 18S rRNA sequence data: implications for reconstructing a phylogeny of the animal kingdom and inferring the reality of the Cambrian explosion. *J. Mol. Evol.* 47:394–405.

Bateson, W. 1885. The later stages in the development of *Balanoglossus kowalevskii*, with a suggestion as to the affinities of the *Enteropneusta*. *Q. J. Microsc. Sci.* 25(suppl.):81–122.

Brambell, F. W. R., and H. A. Cole. 1939. The preoral ciliary organ of the *Enteropneusta*: its occurrence, structure, and possible phylogenetic significance. *Proc. Zool. Soc. Lond.* 109:181–193.

Cameron, C. B. 2002. Particle retention and flow in the pharynx of the enteropneust worm *Harrimania planktophilus*: the filter feeding pharynx may have evolved prior to the chordates. *Biol. Bull. (Woods Hole)* 202:192–200.

Cameron, C. B. 2005. A phylogeny of the hemichordates based on morphological characters. *Can. J. Zool.* 83:196–215.

Cameron, C. B. 2012. A Comprehensive List of Extant Hemichordate Species with Links to Images. Available at https://www.webdepot.umontreal.ca/Usagers/cameroc/MonDepotPublic/Cameron/Species.html.

Cameron, C. B., B. J. Swalla, and J. R. Garey. 2000. Evolution of the chordate body plan: new insights from phylogenetic analysis of deuterostome phyla. *Proc. Nat. Acad. Sci. USA* 97:4469–4474.

Cannon, J. T., A. L. Rychel, H. Eccleston, K. M. Halanych, and B. J. Swalla. 2009. Molecular phylogeny of *Hemichordata*, with updated status of deep-sea enteropneusts. *Mol. Phylogenet. Evol.* 52:17–24.

Dawydoff, C. 1948. Classe des entéropneustes. Pp. 369–453 in *Traité de Zoologie* (P. P. Grassé, ed.). Masson, Paris.

Gegenbaur, C. 1870. *Grundzüge der Vergleichenden Anatomie. 2. Umgearb. Auflage. Mit 319 Holzschnitten.* Wilhelm Engelmann, Leipzig.

Hassouna, N., B. Michot, and J.-P. Bachellerie. 1984. The complete nucleotide sequence of mouse 28S rRNA gene. Implications for the process of size increase of the large subunit rRNA in higher eukaryotes. *Nucleic Acids Res.* 12:3563–3583.

van der Horst, C. J. 1939. *Hemichordata.* Pp. 1–737 in *Klassen und Ordnungen des Tier-Reichs Wissenschaftlich Dargestellt in Wort und Bild*, Band 4, Abt. 4, Buch 2, Teil 2. (H. G. Bronns, ed.). Akademische Verlagsgesellschaft, Leipzig.

Hyman, L. H. 1959. *The Invertebrates: Smaller Coelomate Groups*, Vol. 5. McGraw-Hill, New York.

Mallatt, J., and C. J. Winchell. 2007. Ribosomal RNA genes and deuterostome phylogeny revisited: more cyclostomes, elasmobranchs, reptiles, and a brittle star. *Mol. Phylogenet. Evol.* 43:1005–1022.

Morgan, T. H. 1891. The growth and metamorphosis of *Tornaria. J. Morphol.* 5:407–458.

Osborn, K. J., L. A. Kuhnz, I. G. Priede, M. Urata, A. V. Gebruk, and N. D. Holland. 2011. Diversification of acorn worms (*Hemichordata, Enteropneusta*) revealed in the deep sea. *Proc. R. Soc. Lond. B Biol. Sci.* 279:1646–1654.

Pardos, F., and J. Benito. 1990. The main trunk vessels and blood components of *Glossobalanus minutus* (Enteropneusta). *Eur. Arch. Biol.* (Bruxelles) 101:455–468.

Winchell, C. J., J. Sullivan, C. B. Cameron, B. J. Swalla, and J. Mallatt. 2002. Evaluating hypothesis of deuterostome phylogeny and chordate evolution with new LSU and SSU ribosomal DNA data. *Mol. Biol. Evol.* 19:762–776.

Author

Chris B. Cameron; Sciences biologiques; Université de Montréal; C.P. 6128, Succ. Centre-ville; Montreal, Québec, H3C 3J7, Canada. Email: ccameron@umontreal.ca.

Noura Jabr; Sciences biologiques; Université de Montréal; C.P. 6128, Succ. Centre-ville; Montreal, Québec, H3C 3J7, Canada. Email: noura.jabr@umontreal.ca.

Date Accepted: 27 November 2012

Primary Editor: Philip Cantino

Pterobranchia E. R. Lankester 1877 [M. J. Melchin and C. B. Cameron], converted clade name

Registration Number: 87

Definition: The smallest crown clade containing *Rhabdopleura normani* Allman 1869, *Cephalodiscus dodecalophus* M'Intosh 1882, and *Atubaria heterolopha* Sato 1935. This is a minimum-crown-clade definition. Abbreviated definition: min crown ∇ (*Rhabdopleura normani* Allman 1869 & *Cephalodiscus dodecalophus* M'Intosh 1882 & *Atubaria heterolopha* Sato 1935).

Etymology: Derived from the Greek *pteron* (feather) and *brachion* (arm), referring to the ciliated arms and tentacles used to capture particles of food from the plankton.

Reference Phylogeny: The primary reference phylogeny is Cameron (2005: Fig. 5). See also Winchell et al. (2002). *Atubaria heterolopha* (Sato, 1935) is used as an internal specifier despite being absent from the reference phylogeny. Its distinctness from *Cephalodiscus* is questionable (see Comments), but there is a chance that *Atubaria* is indeed a distinct taxon and that it represents the first branch within *Pterobranchia*. To accommodate this possibility, it is included here as an internal specifier to ensure that it remains in *Pterobranchia*.

Composition: *Pterobranchia* currently contains 25 extant species, belonging to 3 genera: *Rhabdopleura* (Allman, 1869), *Cephalodiscus* (M'Intosh, 1882), and *Atubaria* (Sato, 1935). An up-to-date, comprehensive list of living species may be found in Cameron (2009).

Diagnostic Apomorphies: Pterobranch apomorphies include the collagenous/proteinaceous tubes that they secrete using special cells of the locomotory cephalic shield (protosome); a pigmented band of ciliated cells on the ventral cephalic shield (Horst, 1939); a nonmigratory mesenchymal, pulsatile pericardium (Lester, 1988); a U-shaped gut; and two or more mesocoelic arms and tentacles.

Synonyms: *Axobranches* Delage and Hérouard 1897, *Graptolithoidea* Lapworth 1875 (sensu Beklemishev, 1951), and *Pterobranchea* Kinman 1994; all of these are approximate synonyms.

Comments: The monophyly of *Pterobranchia* is supported by both molecular (Winchell et al., 2002; Cannon et al., 2009) and morphological (Cameron, 2005) analyses. The tentaculated arms, which are used in suspension feeding, are sometimes referred to collectively as a lophophore, but the feeding structure is convergent with the feeding structure of the protostome "lophophorates": bryozoans, phoronids and brachiopods (Horst, 1939; Hyman, 1959; Halanych, 1996). Pterobranchs are unique in producing asexual zooids that are linked by a common stolon (*Rhabdopleura*) or germinal disk (*Cephalodiscus*) (Lester, 1985), and any one colony may be dioecious or monoecious (Hyman, 1959). *Atubaria*, which was brought up by a deep-sea dredge off the coast of Japan, has no secreted tubes and may be a *Cephalodiscus* that had abandoned its tubes on collection (Dilly, 1986). The collagenous tubes of the living *Rhabdopleura* have very strong similarities to those of members of the extinct *Graptolithina* (Rigby, 1994), a group for which details of soft part anatomy are not known. A phylogenetic analysis based on tubarium morphology (Mitchell et al., 2013) suggested that the *Graptolithina* form a clade within the *Pterobranchia* and that the living genus *Rhabdopleura* is an extant member of the *Graptolithina*. The name *Pterobranchia* was selected over other names that have been applied to a group approximating this clade (see Synonyms) because it is by far the most commonly used in the classic and recent literature (e.g., Delage and Hérouard, 1897; Horst, 1939; Hyman, 1959; Cameron, 2005).

Literature Cited

Allman, G. 1869. On *Rhabdopleura. Q. J. Microsc. Sci.* 9:57.

Beklemishev, V. N. 1951. To the building of animals. *Deuterostomia*, their origin and composition. *Uspehi Sovremennoy Biologii* 32:256–270.

Cameron, C. B. 2005. A phylogeny of the hemichordates based on morphological characters. *Can. J. Zool.* 83:196–215.

Cameron, C. B. 2009. A Comprehensive List of Extant Hemichordate Species with Links to Images. Available at www.webdepot.umontreal.ca/Usagers/cameroc/MonDepotPublic/Cameron/Species.html.

Cannon, J. T., A. L. Rychel, H. Eccleston, K. M. Halanych, and B. J. Swalla. 2009. Molecular phylogeny of *Hemichordata*, with updated status of deep-sea enteropneusts. *Mol. Phylogenet. Evol.* 52:17–24.

Delage, Y., and E. Hérouard. 1897. *Traité de Zoologie Concrète. Vol. 5: Les Vermidiens*. Paris.

Dilly, P. N. 1986. Modern pterobranchs: observations on their behaviour and tube building. Pp. 261–269 in *Palaeoecology and Biostratigraphy of Graptolites* (C.P. Hughes and R.B. Rickards, eds.). Geological Society Special Publication No. 20. Blackwell, Oxford.

Halanych, K. M. 1996. Convergence in the feeding apparatuses of lophophorates and pterobranch hemichordates revealed by 18S rDNA: an interpretation. *Biol. Bull. (Woods Hole)* 190:1–5.

van der Horst, C. J. 1939. *Hemichordata*. In *Klassen und Ordnungen des Tier-Reichs wissenschaftlich dargestellt in Wort und Bild*, Band 4, Abt. 4, Buch 2, Tiel 2 (H.G. Bronns, ed.). Akademische Verlagsgesellschaft, Leipzig.

Hyman, L. H. 1959. *The Invertebrates: Smaller Coelomate Groups*, Vol. 5. McGraw-Hill, New York.

Kinman, K. E. 1994. *The Kinman System: Toward a Stable Cladisto-Eclectic Classification of Organisms*. K. E. Kinman, Hays, KS

Lankester, E. R. 1877. Notes on the embryology and classification of the animal kingdom; comprising a revision of speculations relative to the origin and significance of the germlayers. *Q. Jour. Microsc. Sci.* 17:339–454.

Lester, S. M. 1985. *Cephalodiscus* sp. (Hemichordata: Pterobranchia): observations of functional morphology, behavior and occurrence in shallow water around Bermuda. *Mar. Biol.* (Berl.) 85:263–268.

Lester, S. M. 1988. Fine structure of the heart vesicle and pericardium in the pterobranch *Rhabdopleura* (Hemichordata). *Am. Zool.* 22:938A.

M'Intosh, W. C. 1882. Preliminary notice of *Cephalodiscus*, a new type allied to Prof. Allman's *Rhabdopleura*, dredged from the H.M.S. Challenger. *Ann. Mag. Nat. Hist.* 5(10):337–348.

Mitchell, C. E., M. J. Melchin, C. B. Cameron, and J. Maletz. 2013. Phylogenetic analysis reveals that *Rhabdopleura* is an extant graptolite. *Lethaia* 46:34–56.

Rigby, S. 1994. Hemichordate skeletal growth: shared patterns in *Rhabdopleura* and graptoloids. *Lethaia* 27:317–324.

Sato, T. 1935. Über *Atubaria heterolopha. Zool. Anz.* 115:105.

Winchell, C. J., J. Sullivan, C. B. Cameron, B. J. Swalla, and J. Mallatt. 2002. Evaluating hypothesis of deuterostome phylogeny and chordate evolution with new LSU and SSU ribosomal DNA data. *Mol. Biol. Evol.* 19:762–776.

Authors

Michael Melchin, Department of Earth Sciences, St. Francis Xavier University, P.O. Box 5000, Antigonish, NS, B2G 2W5, Canada. Email: mmelchin@stfx.ca.

Chris B. Cameron; Sciences biologiques; Université de Montréal; C.P. 6128, Succ. Centre-ville; Montreal, Québec, H3C 3J7, Canada. Email: c.cameron@umontreal.ca.

Date Accepted: 7 March 2012

Primary Editor: Philip Cantino

Pan-Echinodermata C. D. Sumrall, new clade name

Registration Number: 228

Definition: The total clade of the crown clade *Echinodermata*. This is a crown-based total-clade definition. Abbreviated definition: total ∇ of *Echinodermata*.

Etymology: Derived from the Greek *pan-* (all), indicating that the name refers to a total clade or pan-monophyllum, plus *Echinodermata*, the name of the corresponding crown clade (see the entry for *Echinodermata* in this volume for the etymology of that name).

Reference Phylogeny: Janies et al. (2011: Fig. 3). Although many extinct clades of pan-echinoderms were not included in this phylogeny, it shows the relationships of *Echinodermata* (*Echinoidea*, *Ophiuroidea*, *Holothuroidea*, *Asteroidea*, *Crinoidea*) with its putative most closely related extant outgroups (*Enteropneusta*, *Pterobranchia*, *Urochordata*) and is, therefore, appropriate as the reference phylogeny for *Pan-Echinodermata*.

Composition: *Echinodermata* (this volume) and all extinct animals that share more recent ancestry with *Echinodermata* than with any other extant animals (*Hemichordata* according to most current phylogenies; Janies et al., 2011). Extinct animals sometimes considered to be stem echinoderms include *Helicoplacoidea* Durham and Caster 1963, *Edrioasteroidea* Jaekel 1918, *Eocrinoidea* Jaekel 1918, *Cyclocystoidea* Miller and Gurley 1895, *Soluta* Jaekel 1901 (= *Homoiostelea* Gill and Caster 1960), *Cincta* Jaekel 1918 (= *Homostelea* Gill and Caster 1960), *Ctenocystoidea* Robison and Sprinkle 1969, and *Stylophora* Gill and Caster 1960. Several alternative phylogenies

for *Pan-Echinodermata* have been presented over the last several years (Sumrall, 1997; David and Mooi, 1998, 1999; David et al., 2000; Smith, 2005). Sumrall (1997) only inferred three Early Cambrian taxa—*Helicoplacoidea*, *Imbricata*, and *Camptostroma*—as stem echinoderms. Mooi (2001), based on earlier work by himself and collaborators (David and Mooi 1998, 1999; David et al., 2000), presented a phylogeny in which nearly all extinct clades of pan-echinoderms listed above except for *Stylophora* were outside the crown clade. Smith (2005) presented a phylogeny in which "homalozoan" clades (*Cincta*, *Ctenocystoidea*, *Soluta* and *Stylophora*) were shown as stem echinoderms, but no explicit statement was made concerning other pentaradial extinct pan-echinoderm clades.

Diagnostic Apomorphies: Deuterostomes with a skeleton composed of unicrystalline ossicles of Mg-calcite that bear a stereom microstructure. Although these apomorphies probably did not arise simultaneously with the origin of the echinoderm stem lineage, they appear to have arisen early in the history of that lineage. Pentaradiate symmetry appears to be derived within *Pan-Echinodermata*, but its placement is unclear because of instability in the tree topology (Sumrall, 1997; Mooi, 2001; Smith, 2005).

Synonyms: The name *Echinodermata* as generally used in the palaeontological literature is based on the presence of diagnostic traits (typically, a skeleton composed of unicrystalline ossicles of Mg-calcite that bear a stereom microstructure). This usage is effectively apomorphy-based but has the same known composition as *Pan-Echinodermata* (Sprinkle, 1973, 1976; Paul and Smith, 1984; Smith, 1984, 1988a,b, 1990,

2005; David et al., 2000) and is in that sense an approximate synonym. Sumrall (1997) used the name *Echinodermoformes* for the total clade that is here named *Pan-Echinodermata* (i.e., the name is an unambiguous synonym).

Comments: The internal relationships among members of *Pan-Echinodermata* are not generally agreed upon. Three main scenarios are described largely centering on the position of an assemblage of unusual pan-echinoderms named "*Homalozoa*" including *Cincta*, *Ctenocystoidea*, *Soluta* and *Stylophora*. Sumrall (1997) viewed *Homalozoa* as monophyletic (though with reservations) and within crown *Echinodermata*. According to this view, only a few groups of early pan-echinoderms lie outside of crown *Echinodermata*, and these were placed there *a priori* as outgroups. David et al. (2000) viewed *Homalozoa* as polyphyletic with *Stylophora* as sister taxon to *Crinoidea* and this clade as sister taxon to *Eleutherozoa* (i.e., *Asteroidea*, *Ophiuroidea*, *Echinoidea*, and *Holothuroidea*). The remaining *Homalozoa* were variously placed in *Blastozoa* (i.e., as stem echinoderms). According to this hypothesis, (crown) *Echinodermata* is a relatively small clade, as many fossil taxa traditionally considered echinoderms are placed in the echinoderm stem group. Smith (2005) viewed *Homalozoa* as paraphyletic with respect to crown *Echinodermata*, though no specific indication was made concerning the position of most clades of extinct echinoderms within this scenario. The calcichordate theory (see Jefferies, 1986), though largely discredited (Peterson, 1995; David et al., 2000; Lefebvre, 2000; Smith, 2005), places different members of *Homalozoa* in the stem groups of crown *Chordata*, crown *Cephalochordata*, crown *Urochordata*, and crown *Echinodermata*.

The clade name *Pan-Echinodermata* is used here over *Echinodermoformes* in the interest of developing an integrated system of crown and total clade names (de Queiroz, 2007). *Pan-Echinodermata* is roughly equivalent to *Echinodermata* as the latter name has been used in the palaeontological literature (Sprinkle, 1973, 1976; Smith 1984, 1988b, Sumrall, 1997; Mooi, 2001).

Literature Cited

David, B., B. Lefebvre, R. Mooi, and R. L. Parsley. 2000. Are homalozoans echinoderms? An answer from the extraxial-axial theory. *Paleobiology* 26:529–555.

David, B., and R. Mooi. 1998. Major events in the evolution of echinoderms viewed by the light of embryology. Pp. 21–28 in *Echinoderms San Francisco* (R. Mooi and M. Telford, eds.). A. A. Balkema, Rotterdam.

David, B., and R. Mooi. 1999. Comprendre les échinodermes: la contribution du modèle extraxial-axial. *Bull. Soc. Geol. Fr.* 170:91–101.

Durham, J. W., and K. E. Caster. 1963. *Helicoplacoidea*: a new class of echinoderms. *Science* 140:97–102.

Gill, E. D., and K. E. Caster. 1960. Carpoid echinoderms from the Silurian and Devonian of Australia. *Bull. Am. Paleontol.* 41:7–43.

Janies, D. A., J. R. Voight, and M. Daly. 2011. Echinoderm phylogeny including *Xyloplax*, a progenetic asteroid. *Syst. Biol.* 60:420–438.

Jaekel, O. 1901. Ueber *Carpoideen*; eine neue Classe von Pelmatozoen. *Zeit. Deuts. Geo. Gesell.* 52:661–677.

Jaekel, O. 1918. Phylogenie und system der Pelmatozoen. *Paläont. Zeit.* 3:1–128.

Jefferies, R. P. S. 1986. *The Ancestry of the Vertebrates*. British Museum (Natural History) and Cambridge University Press, London.

Lefebvre, B. 2000. Homologies in *Stylophora*: a test of the "Calcichordate Theory". *GEOBIOS* 33:359–364.

Miller, S. A., and W. F. E. Gurley. 1895. Description of new species of Palaeozoic *Echinodermata*. *Bull. Ill. St. Nat. Hist. Mus.* 6:1–62.

Mooi, R. 2001. Not all written in stone: interdisciplinary syntheses in echinoderm paleontology. *Can. J. Zool.* 79:1209–1231.

Paul, C. R. C., and A. B. Smith. 1984. The early radiation and phylogeny of echinoderms. *Biol. Rev.* 59:443–481.

Peterson, K. J. 1995. A phylogenetic test of the calcichordate scenario. *Lethaia* 28:25–37.

de Queiroz, K. 2007. Toward an integrated system of clade names. *Syst. Biol.* 56:956–974.

Robison, R. A., and J. Sprinkle. 1969. *Ctenocystoidea*: new class of primitive echinoderms. *Science* 166:1512–1514.

Smith, A. B. 1984. Classification of the *Echinodermata. Palaeontology* 27:431–459.

Smith, A. B. 1988a. Fossil evidence for the relationships of extant echinoderm classes and their times of divergence. Pp. 85–101 in *Echinoderm Phylogeny and Evolution* (C. R. C. Paul and A. B. Smith, eds.). Clarendon Press, Oxford.

Smith, A. B. 1988b. Patterns of diversification and extinction in Early Palaeozoic echinoderms. *Palaeontology* 31:799–828.

Smith, A. B. 1990. Evolutionary diversification of echinoderms during the Early Palaeozoic. Pp. 256–286 in *Major Evolutionary Radiations* (P. D. Taylor and G. P. Larwood, eds.). Clarendon Press, Oxford.

Smith, A. B. 2005. The pre-radial history of echinoderms. *Geol. J.* 40:255–280.

Sprinkle, J. 1973. *Morphology and Evolution of Blastozoan Echinoderms*. Harvard University Museum of Comparative Zoology, Special Publication.

Sprinkle, J. 1976. Classification and phylogeny of pelmatozoan echinoderms. *Syst. Zool.* 25:83–91.

Sumrall, C. D. 1997. The role of fossils in the phylogenetic reconstruction of *Echinodermata*. Pp. 267–288 in *Geobiology of Echinoderms* (J. A. Waters and C. G. Maples eds.). *Paleontol. Soc. Pap.* 3.

Author

Colin D. Sumrall; Department of Earth and Planetary Sciences; The University of Tennessee; Knoxville, TN 37996-1526, USA. Email: csumrall@utk.edu.

Date Accepted: 16 September 2013

Primary Editor: Kevin de Queiroz

Echinodermata T. Klein 1754 [C. D. Sumrall], converted clade name

Registration Number: 243

Definition: The crown clade originating in the most recent common ancestor of *Asterias forbesi* (Desor 1848) (*Asteroidea*), *Ophioderma brevispinum* (Say 1825) (*Ophiuroidea*), *Strongylocentrotus purpuratus* (Stimpson 1857) (*Echinoidea*), *Cucumaria miniata* (Brandt 1835) (*Holothuroidea*), *Metacrinus rotundus* Carpenter 1885 (*Crinoidea*) and *Xyloplax janetae* Mah 2006 (*Concentricycloidea*). This is a minimum-crown-clade definition. Abbreviated definition: min crown ∇ (*Asterias forbesi* (Desor 1848) & *Ophioderma brevispinum* (Say 1825) & *Strongylocentrotus purpuratus* (Stimpson 1857) & *Cucumaria miniata* (Brandt 1835) & *Metacrinus rotundus* Carpenter 1885 & *Xyloplax janetae* Mah 2006).

Etymology: Derived from the Greek *echinos* (spiny) and *derma* (skin).

Reference Phylogeny: Janies et al. (2011: Fig. 3). Janies et al. (2011) provided a thorough analysis of all major extant clades of echinoderms. This analysis also included *Xyloplax* (*Concentriclycloidea*) nested within *Asteroidea*, and addressed monophyly and relationships of the five main clades of *Echinodermata*.

Composition: *Echinodermata* is currently believed to contain some 7,000 extant species, placed in five mutually exclusive groups: the total clades of *Echinoidea* (sea urchins), *Holothuroidea* (sea cucumbers), *Asteroidea* (sea stars), *Ophiuroidea* (brittle stars and sea baskets), and *Crinoidea* (sea lilies and feather stars). A sixth group known from three species, once considered a separate class *Concentriclycloidea* (Baker et al., 1986; Rowe, et al, 1988), is nested within *Asteroidea* (Smith 1988b; Belyaev, 1988; Janies and Mooi, 1999; Janies, 2001; Mah, 2006; Janies et al., 2011). Furthermore, two extinct clades are nested within *Echinodermata*: *Ophiocistioidea* Sollas 1899, which forms the sister taxon to either echinoids, holothuroids or both (Sumrall, 1997; Reich and Haude, 2004); and *Somasteroidea* Spencer 1951, which forms the sister taxon to either asteroids, ophiuroids or both (Blake, 1982, Smith and Jell, 1990).

Diagnostic Apomorphies: Deuterostomes with water vascular system originating from the left mesocoel, and mutable collagenous tissue. These two synapomorphies are known in all living *Echinodermata*, but because they relate to soft tissue, they cannot clearly be established outside of (crown) *Echinodermata*. Based on conflicting phylogenies for *Pan-Echinodermata*, it is unclear whether pentaradiate symmetry is apomorphic or plesiomorphic for *Echinodermata*.

Synonyms: *Echinodermaria* Blainville 1816 and *Echinoderma* (e.g., of Bather, 1900) are approximate synonyms of *Echinodermata*.

Comments: The monophyly of *Echinodermata* as a group including asteroids, ophiuroids, echinoids, holothuroids and crinoids has been generally accepted. The definition given here is general by design because the relationships within *Echinodermata* and *Pan-Echinodermata* are not universally agreed upon. *Eleutherozoa*, which contains *Asteroidea*, *Ophiuroidea*, *Holothuroidea*, and *Echinoidea*, is nearly universally inferred as sister taxon to *Crinoidea* (but see Smiley, 1986, 1988). However, relationships among the members of *Eleutherozoa* vary among analyses. Many analyses have inferred *Asterozoa* (*Asteroidea* + *Ophiuroidea*) (Wada and Satoh, 1994; Sumrall,

1997; Sumrall and Sprinkle, 1998; Pearse and Pearse, 1994; Janies, 2001). Others have inferred *Ophiuroidea* closer to *Echinozoa* (*Echinoidea* + *Holothuroidea*) (Paul and Smith, 1984; Smith, 1988a; Cameron, 2005), *Asteroidea* closer to *Echinozoa* (Raff et al., 1988; Field et al., 1988), or *Asteroidea* (*Echinoidea* (*Ophiuroidea* + *Holothuroidea*)) (Janies et al., 2011). Smiley (1986, 1988) placed *Holothuroidea* as sister taxon to (*Crinoidea* (*Asteroidea* (*Ophiuroidea*, *Echinoidea*))). This uncertainty precludes the formal naming of the largest subclades within *Eleutherozoa* at this time. Regardless, the extinct clades *Ophiocistioidea* Sollas 1899 and *Somasteroidea* Spencer 1951 are considered to be *Echinodermata*—that is, members of the crown clade (Blake, 1982; Smith and Jell, 1990; Sumrall, 1997; Reich and Haude, 2004).

The clade name *Echinodermata* is used here for the crown clade in the interest of developing an integrated system of crown and total clade names (de Queiroz, 2007). This is the implicit usage of *Echinodermata* by the zoological community (Janies et al., 2011). *Echinodermata* as has been typically used in the palaeontological literature (e.g., Sprinkle, 1973, 1976; Smith 1984, 1988b; Sumrall, 1997; Mooi, 2001) is roughly equivalent to the total clade referred to in this volume as *Pan-Echinodermata*.

Literature Cited

Baker, A. N., F. W. E. Rowe, and H. E. S. Clark. 1986. A new class of *Echinodermata* from New Zealand. *Nature* 321:862–864.

Bather, F. A. 1900. *Echinoderma. A Treatise on Zoology, Part III* (E. R. Lankester, ed.). Adam and Charles Black, London.

Belyaev, G. M. 1988. Is it valid to isolate the genus *Xyloplax* as an independent class of echinoderms? *Zool. Z.* 69:83–96.

de Blainville, H. M. D. 1816. Prodrome d'une nouvelle distribution systématique du règne animal. *Bull. Soc. Philom. Paris* 8:105–112.

Blake, D. B. 1982. *Somasteroidea, Asteroidea*, and the Affinities of *Luidia* (*Plasterias*) *latiradiata*. *Paleontology* 25:167–191.

Brandt, J. F. 1835. Prodromus descriptionis animalium ab H. Mertensio in orbis terrarum circumnavigatione observatorum. Fascic. I. Polypos, Acalephas Discophoras et Siphonophoras, nec non *Echinodermata* continens. *Recl. Act. Acad. Imp. Sci. St. Petersb.* 1834:201–275.

Cameron, C. B. 2005. A phylogeny of the hemichordates based on morphological characters. *Can. J. Zool.* 83:196–215.

Carpenter, P. H. 1885. On three new species of *Metacrinus. Trans. Linn. Soc. Ser. 2 (Zool.)* 2:435–446.

Desor, E. 1848. Zoölogical investigation among the shoals of Nantucket. *Proc. Bost. Soc. Nat. Hist.* 3:65–68.

Field, K. J., M. T. Ghislen, D. J. Lane, G. J. Olson, N. R. Pace, E. C. Raff, and R. A. Raff. 1988. Molecular phylogeny of the animal kingdom. *Science* 239:748–753.

Janies, D. 2001. Phylogenetic relationships of extant echinoderm classes. *Can. J. Zool.* 79:1232–1250.

Janies, D. A., J. R. Voight, and M. Daly. 2011. Echinoderm phylogeny including *Xyloplax*, a progenetic asteroid. *Syst. Biol.* 60:420–438.

Janies, D., and R. Mooi. 1999. *Xyloplax* is an asteroid. Pp. 311–316 in *Echinoderm Research 1998* (C. Carnevali and F. Bonsoro, eds.). Balkema, Rotterdam.

Klein, T. 1754. *Ordre Naturel des Oursins de Mer et Fossiles, avec des Observations sur les Piquants des Oursins de Mer et Quelques Remarques sur les Bélemnites.* C. J. B. Bauche, Paris.

Mah, C. L. 2006. A new species of *Xyloplax* (*Echinodermata: Asteroidea: Concentricycloidea*) from the northeast Pacific: comparative morphology and a reassessment of phylogeny. *Invertebr. Biol.* 125:136–153.

Mooi, R. 2001. Not all written in stone: interdisciplinary syntheses in echinoderm paleontology. *Can. J. Zool.* 79:1209–1231.

Paul, C. R. C., and A. B. Smith. 1984. The early radiation and phylogeny of echinoderms. *Biol. Rev.* 59:443–481.

Pearse, V. B., and J. S. Pearse. 1994. Echinoderm phylogeny and the place of concentricycloids.

Pp. 121–126 in *Echinoderms through Time* (D. Bruno, A. Guille, J. Féral, and M. Roux, eds.). Balkema, Rotterdam.

de Queiroz, K. 2007. Toward an integrated system of clade names. *Syst. Biol.* 56:956–974.

Raff, R. A., K. G. Field, M. T. Ghislen, D. J. Lane, G. J. Olson, N. R. Pace, A. L. Parks, B. A. Parr, and E. C. Raff. 1988. Molecular analysis of distant phylogenetic relationships in echinoderms. Pp. 29–41 in *In Echinoderm Phylogeny and Evolution* (C. R. C. Paul and A. B. Smith, eds.). Clarendon Press, Oxford.

Reich, M., and R. Haude. 2004. *Ophiocistioidea* (fossil *Echinodermata*): an overview. Pp. 489–494 in *Echinoderms München* (T. Heinzeller and J. H. Nebelsick, eds.). Balkema, Rotterdam.

Rowe, F. W. E., A. N. Baker, and H. E. S. Clark. 1988. The morphology, development and taxonomic status of *Xyloplax* Baker, Rowe and Clark, 1986 (*Echinodermata*: *Concentricycloidea*) with the description of a new species. *Proc. R. Soc. Lond. B Biol. Sci.* 233:431–459.

Say, T. 1825. On the species of the Linnaean genus *Asterias* inhabiting the coast of the U.S. *J. Acad. Nat. Sci. Phila.* 5:141–154.

Smiley, S. 1986. Metamorphosis of *Stichopus californicus* (*Echinodermata*: *Holothuroidea*) and its phylogenetic implications. *Biol. Bull. Mar. Bio. Lab., Woods Hole* 171:671–691.

Smiley, S. 1988. The phylogenetic relationships of holothurians: a cladistic analysis of the extant echinoderm classes. Pp. 69–84 in *Echinoderm Phylogeny and Evolution* (C. R. C. Paul and A. B. Smith, eds.). Clarendon Press, Oxford.

Smith, A. B. 1984. Classification of the *Echinodermata*. *Palaeontology* 27:431–459.

Smith, A. B. 1988a. Patterns of diversification and extinction in Early Palaeozoic echinoderms. *Palaeontology* 31:799–828.

Smith, A. B. 1988b. To group or not to group: The taxonomic position of *Xyloplax*. Pp. 17–23 in *Echinoderm Biology* (R. D. Burke, P. V. Mlandenov, P. Lambert, and R. L. Parsley, eds.). A. A. Balkema, Rotterdam.

Smith, A. B., and P. A. Jell. 1990. Cambrian edrioasteroids from Australia and the origin of starfishes. *Mem. Queens. Mus.* 28:715–778.

Sollas, W. J. 1899. Fossils in the university museum, Oxford. I. On Silurian *Echinoidea* and *Ophiuroidea*. *Q. J. Geol. Soc. Lond.* 55:692–715.

Spencer, W. K. 1951. Early Paleozoic starfish. *Philos. Trans. R. Soc. Lond. B Biol. Sci.* 235:87–129.

Sprinkle, J. 1973. *Morphology and Evolution of Blastozoan Echinoderms*. Harvard University Museum of Comparative Zoology, Special Publication, Cambridge, MA.

Sprinkle, J. 1976. Classification and phylogeny of pelmatozoan echinoderms. *Syst. Zool.* 25:83–91.

Stimpson, W. 1857. Prodromus descriptionis animalium evertebratorum in expeditione ad Oceanum Pacificum Septentrionalem missa C. Ringgold et Johanne Rodgers, ducibus, observatorum et descriptorum. *Proc. Acad. Nat. Sci. Phila.* 9:31–40, 93–110, 159–163, 225–252.

Sumrall, C. D. 1997. The role of fossils in the phylogenetic reconstruction of *Echinodermata*. Pp. 267–288 in *Geobiology of Echinoderms* (J. A. Waters and C. G. Maples, eds.). Paleontol. Soc. Pap. 3.

Sumrall, C. D., and J. Sprinkle. 1998. Phylogenetic analysis of living *Echinodermata* based on primitive fossil taxa. Pp. 81–87 in *Echinoderm Biology* (R. D. Burke, P. V. Mlandenov, P. Lambert, and R. L. Parsley, eds.). A. A. Balkema, Rotterdam.

Wada, H., and N. Satoh. 1994. Phylogenetic relationships among extant classes of echinoderms, as inferred from sequences of 18S rDNA, coincide with relationships deduced from the fossil record. *J. Mol. Evol.* 38:41–49.

Author

Colin D. Sumrall; Department of Earth and Planetary Sciences; The University of Tennessee; Knoxville, TN 37996-1526, USA. Email: csumrall@utk.edu.

Date Accepted: 25 August 2013

Primary Editor: Kevin de Queiroz

Edrioasterida B. M. Bell 1976 [C. D. Sumrall], converted clade name

Registration Number: 229

Definition: The clade originating in the most recent common ancestor of *Edrioaster bigsbyi* (Billings 1857), *Cyathotheca suecica* Jaekel 1927, *Lampteroblastus hintzei* Guensburg and Sprinkle 1994 and *Rhenopyrgus piojoensis* Sumrall et al. 2013. This is a minimum-clade definition. Abbreviated definition: min ∇ (*Edrioaster bigsbyi* (Billings 1857) & *Cyathotheca suecica* Jaekel 1927 & *Lampteroblastus hintzei* Guensburg and Sprinkle 1994 & *Rhenopyrgus piojoensis* Sumrall et al. 2013).

Etymology: Derived from the Greek *hedra* (diminutive *hedrion*) (seated) and *aster* (star).

Reference Phylogeny: Sumrall et al. (2013: Fig. 2A). Although not included in the reference phylogeny, *Edrioaster bigsbyi* is thought to be closely related to *Paredriophus elongatus* Guensburg and Sprinkle 1994 (Guensburg and Sprinkle, 1994).

Composition: *Edrioasterida* is extinct and ranges minimally from the Early Ordovician through Late Devonian (Bell, 1976; Guensburg and Sprinkle, 1994). *Edrioasterida* includes four clades: *Edrioasteridae* Bell 1976, including *Edrioaster*, *Edriophus* and *Paredriophus*; *Cyathocystidae* Bather 1899, including *Cyathocystis* and *Cyathotheca*; *Astrocystitidae* Bassler 1935 (= *Edrioblastoidea* Fay 1962), including *Astrocystites*, *Cambroblastus* and *Lampteroblastus*; and *Rhenopyrgidae* Holloway and Jell 1983, including *Rhenopyrgus* (Smith and Jell, 1990; Guensburg and Sprinkle, 1994; Sumrall et al., 2013).

Diagnostic Apomorphies: Floor plates biserial with sutural podial pores; cover plates biserial and in line with floor plates that lack intrathecal and intrambulacral extensions; theca offset from pedunculate zone; aboral ring bordering interambulacral plating (derived mainly from Bell, 1976; Guensburg and Sprinkle, 1994; Sumrall et al., 2013).

Synonyms: There are no synonyms for *Edrioasterida*.

Comments: The name *Edrioasterida* as defined here follows the use of several authors whose phylogenetic analyses inferred a clade including *Edrioaster*, *Astrocystitidae* (edrioblastoids), *Cyathocystidae* and *Rhenopyrgidae* (Smith and Jell, 1990; Guensburg and Sprinkle, 1994; Sumrall et al., 2013). This is the common use in the literature. Bell (1976) erected *Edrioasterida* to include only *Edrioaster* and *Edriophus*. This taxon was expanded by Bell and Sprinkle (1978) to include the Cambrian taxon *Totiglobus* because of similarities in the shape of the theca. In the context of later phylogenetic analyses, the reference of this name was expanded to include *Astrocystitidae* (edrioblastoids), *Cyathocystidae* and *Rhenopyrgidae* (Smith and Jell, 1990; Guensburg and Sprinkle, 1994; Sumrall et al., 2013). As defined here, *Edrioasterida* excludes *Totiglobus* though that taxon may be the closest presently known outgroup. Guensburg and Sprinkle (2007) suggested a close relationship between *Astrocystitidae* and *Crinoidea* though this remains highly controversial. *Dinocystis barroisi* Bather 1898, thought to be an edrioasterid by several authors (Bather, 1914; Regnéll, 1966; Bell, 1976), though poorly preserved, shows the small peripheral rim and organized pedunculate zone characteristic of *Discocystinae* (see entry in this volume) in *Isorophida*.

Much confusion exists in the literature about the name *Edrioasteroidea* (not to be confused with *Edrioasterida*). *Edrioasteroidea* is applied to an assemblage of discoidal to globular Palaeozoic taxa that share recumbent ambulacra that lack accessory feeding appendages (Bell, 1976). It is unclear whether edrioasteroids represent a clade or are a plesiomorphic grade on the stem of *Asteroidea* (see Smith and Jell, 1990) or of *Crinoidea* (see Guensburg and Sprinkle, 2007). Although *Edrioasterida* and *Isorophida* (see entry this volume) are clades, how they relate within the broader context of *Edrioasteroidea* has not been determined because of limited taxon sampling. For this reason, the name *Edrioasteroidea* has not been defined here.

Literature Cited

Bassler, R. S. 1935. The classification of the *Edrioasteroidea. Smithson. Misc. Coll.* 93(8):1–11.

Bather, F. A. 1898. *Dinocystis barroisi*, n. g. et sp., Psammites du Condroz. *Geol. Mag.* 5:543–548.

Bather, F. A. 1899. A phylogenetic classification of the *Pelmatozoa. Rep. Br. Ass. Adv. Sci.* 68:916–923.

Bather, F. A. 1914. Edrioasters in the Trenton Limestone (parts 1 and 2) *Geol. Mag.* 1:115–125, 162–171.

Bell, B. M. 1976. A Study of North American *Edrioasteroidea. New York St. Mus. Mem.* 21: 476 pp.

Bell, B. M., and J. Sprinkle. 1978. *Totiglobus*, an unusual new edrioasteroid from the Middle Cambrian of Nevada. *J. Paleontol.* 52:243–266.

Billings, E. 1857. Report for the year 1856, Fossils from Anicosti and new species of fossils from the Lower Silurian rocks of Canada. *Geol. Surv. Canada, Rept. Prog.* 1853–1856:247–345.

Fay, R. O. 1962. *Edrioblastoidea*, a new class of Echinodermata. *J. Paleontol.* 36:201–205.

Guensburg, T. E., and J. Sprinkle. 1994. Revised phylogeny and functional interpretation of the *Edrioasteroidea* based on new taxa from the Early and Middle Ordovician of western Utah. *Fieldiana (Geology)* 29:1–43.

Guensburg, T. E., and Sprinkle, J. 2007. Phylogenetic implications of the *Protocrinoida*: blastozoans are not ancestral to crinoids. *Ann. Paléontol.* 93:277–290.

Holloway, D. J., and P. A. Jell. 1983. Silurian and Devonian edrioasteroids from Australia. *J. Paleontol.* 57:1001–1016.

Jaekel, O. 1927. *Cyathotheca suecica* n.g. n.sp., eine Thecoidee des schwedischen Ordoviciums. *Ark. Zool.* 19A, 4:1–15.

Regnéll, G. 1966. Edrioasteroids. Pp. U135–U173 in *Treatise on Invertebrate Paleontology, Part U* (Echinodermata 3) (R. C. Moore, ed.). Geological Society of America, New York; University of Kansas, Lawrence, KS.

Smith, A. B., and P. A. Jell. 1990. Cambrian edrioasteroids from Australia and the origin of starfishes. *Mem. Queensl. Mus.* 28:715–778.

Sumrall, C. D., S. Heredia, C. M. Rodríguez, and A. I. Mestre. 2013. The first report of South American edrioasteroids and the paleoecology and ontogeny of rhenopyrgid echinoderms. *Acta Palaeontol. Pol.* 58:763–776.

Author

Colin D. Sumrall; Department of Earth and Planetary Sciences; The University of Tennessee; Knoxville, TN 37996-1526, USA. Email: csumrall@utk.edu.

Date Accepted: 25 August 2013

Primary Editor: Kevin de Queiroz

Isorophida B. M. Bell 1976 [C. D. Sumrall], converted clade name

Registration Number: 230

Definition: The clade originating in the most recent common ancestor of *Streptaster vorticellatus* (Hall 1866), *Argodiscus espilezorum* Sumrall and Zamora 2011, *Carneyella pilea* (Hall 1866), *Isorophus cincinnatiensis* (Roemer 1851), and *Agelacrinites hamiltonensis* Vanuxem 1842. This is a minimum-clade definition. Abbreviated definition: min ∇ (*Streptaster vorticellatus* (Hall 1866) & *Argodiscus espilezorum* Sumrall and Zamora 2011 & *Carneyella pilea* (Hall 1866) & *Isorophus cincinnatiensis* (Roemer 1851) & *Agelacrinites hamiltonensis* Vanuxem 1842).

Etymology: Derived from the Greek *isos* (equal), and *orophe* (roof).

Reference Phylogeny: Figure 6 of Sumrall and Zamora (2011). Analyses of Guensburg and Sprinkle (1994) and Sumrall and Gahn (2006) inferred largely congruent phylogenetic hypotheses.

Composition: *Isorophida* includes approximately 80 named species (Bell, 1976). This clade is extinct and ranges minimally from the Early Ordovician through the Late Permian and includes nearly all of the taxa regularly thought of as edrioasteroids (see comments on *Edrioasterida* in this volume). Many animals, especially from the Cambrian, included in *Edrioasteroidea* in the older literature lie outside the clade here named *Isorophida* (this volume). Some of these taxa have been argued to be within *Isorophida* (Zamora and Smith, 2010), whereas other isorophid-like animals have been argued to be within the stem groups of specific subclades of crown echinoderms,

such as *Asteroidea* (Smith, 1985; Smith and Jell, 1990). *Isorophida* includes *Lebetodiscina* Bell 1976, *Pyrgocystidae* Kesling 1967, *Isorophina* (this volume), *Agelacrinitidae* (this volume), *Lepidodiscina* (this volume) and *Discocystinae* (this volume).

Diagnostic Apomorphies: Floor plates hidden, uniserial, lacking podial pores; hydropore and gonopore united into shared structure along the sutures between modified cover plates and hydropore oral plate, cover plates with intrathecal extensions (reversed in some members of the clade) (derived from Bell, 1976; Sumrall, 1993; Guensburg and Sprinkle, 1994; Sumrall and Gahn, 2006). Other characters cited by Bell (1976) as isorophid characters are plesiomorphic, including: discoidal theca bearing recumbent ambulacra structurally embedded into the thecal wall; theca bordered by a peripheral rim (Guensburg and Sprinkle, 1994).

Synonyms: There are no synonyms for *Isorophida*.

Comments: The definition of the name *Isorophida* adopted here closely approximates the traditional usage in the literature (e.g., Bell, 1976; Sumrall, 1993; Guensburg and Sprinkle, 1994). The presence of a domal theca and peripheral rim were traditionally considered diagnostic of *Isorophida* but here are regarded as plesiomorphic. Therefore, such taxa as *Walcottidiscus* and *Cambraster,* while similar to *Isorophida* based on shared plesiomorphic characters, are excluded from *Isorophida* based on their phylogenetic position (Smith and Jell, 1990; Guensburg and Sprinkle, 1994; Zhao et al., 2010; Zamora et al., 2012).

The name *Isorophida* was chosen for this clade because it closely approximates the original use by Bell (1976) and follows recent usage (Sumrall, 1993; Guensburg and Sprinkle, 1994; Sumrall and Zamora, 2011).

Literature Cited

Bell, B. M. 1976. A study of North American *Edrioasteroidea*. *New York St. Mus. Mem.* 21: 476 pp.

Guensburg, T. E., and J. Sprinkle. 1994. Revised phylogeny and functional interpretation of the *Edrioasteroidea* based on new taxa from the Early and Middle Ordovician of western Utah. *Fieldiana (Geology)* 29:1–43.

Hall, J. 1866. Descriptions of some new species of *Crinoidea* and other fossils. *New York St. Mus. 20th Ann. Rep.* (Adv. Pub.):1–17.

Kesling, R. V. 1967. Edrioasteroid with unique shape from Mississippian strata of Alberta. *J. Paleontol.* 41:197–202.

Roemer, F. (C. F. von). 1851. Beiträge zur Kenntniss der fossilen Fauna des devonischen Gebirges am Rhein. *Decheniana (Verh. Naturhist. Ver. Rheinlands Westfalens)* 8:357–376.

Smith, A. B. 1985. Cambrian eleutherozoan echinoderms and the early diversification of edrioasteroids. *Palaeontology* 28:715–756.

Smith, A. B., and P. A. Jell. 1990. Cambrian edrioasteroids from Australia and the origin of starfishes. *Mem. Queens. Mus.* 28: 715–778.

Sumrall, C. D. 1993. Thecal designs in isorophinid edrioasteroids. *Lethaia* 26:289–302.

Sumrall, C. D., and F. J. Gahn. 2006. Morphological and systematic reinterpretation of two enigmatic edrioasteroids (*Echinodermata*) from Canada. *Can. J. Earth Sci.* 43:497–507.

Sumrall, C. D., and S. Zamora. 2011. Ordovician edrioasteroids from Morocco: faunal exchanges across the Rheic Ocean. *J. Syst. Palaeontol.* 9:425–454.

Vanuxem, L. 1842. Survey of the third geological district. *Nat. Hist. NY. Geo.* 3:1–307.

Zamora, S., and A. B. Smith. 2010. The oldest isorophid edrioasteroid (*Echinodermata*) and the evolution of attachment strategies in Cambrian edrioasteroids. *Acta Palaeontol. Pol.* 55:487–494.

Zamora S., C. D. Sumrall, and D. Vizcaïno. 2012. Morphology and ontogeny of the Cambrian edrioasteroid (*Echinodermata*) *Cambraster cannati* Miquel from western Gondwana. *Acta Palaeontol. Pol.* doi:10.4202/app.2011.0152.

Zhao Y., C. D. Sumrall, R. L. Parsley, J. Peng. 2010. *Kailidiscus*, a new plesiomorphic edrioasteroid from the basal middle Cambrian Kaili Biota of Guizhou Province, China. *J. Paleontol.* 84:668–680.

Author

Colin D. Sumrall; Department of Earth and Planetary Sciences; The University of Tennessee; Knoxville, TN 37996-1526, USA. Email: csumrall@utk.edu.

Date Accepted: 25 August 2013

Primary Editor: Kevin de Queiroz

Isorophina B. M. Bell 1976 [C. D. Sumrall], converted clade name

Registration Number: 245

Definition: The clade originating in the most recent common ancestor of *Isorophus cincinnatiensis* (Roemer 1851), *Agelacrinites hamiltonensis* Vanuxem 1842, and *Hypsiclavus huntsvillensis* Sumrall 1996. This is a minimum-clade definition. Abbreviated definition: min ∇ (*Isorophus cincinnatiensis* (Roemer 1851) & *Agelacrinites hamiltonensis* Vanuxem 1842 & *Hypsiclavus huntsvillensis* Sumrall 1996).

Etymology: Derived from the Greek *isos* (equal) and *orophe* (roof).

Reference Phylogeny: Sumrall and Gahn (2006: Fig. 3.1).

Composition: *Isorophina* is extinct and includes approximately 50 named species that range minimally from the Early Ordovician through Late Permian (Bell, 1976; Guensburg and Sprinkle, 1994; Sumrall, 1996; Sumrall and Zamora, 2011). *Isorophina* includes *Agelacrinitidae* (this volume) and *Isorophidae*, which is here applied to a less inclusive clade than by Sumrall and Zamora (2011).

Diagnostic Apomorphies: Secondary ambulacral cover plates present; four primary oral cover plates (reduced to two in *Agelacrinitidae*) cover plates with intrambulacral extensions; valvular anal structure; floor plates with vertical sutures (becoming imbricate in a few taxa) (derived from Bell, 1976; Smith and Arbizu, 1987; Sumrall, 1993; Sumrall and Gahn, 2006). The presence of intrathecal cover plate extensions mentioned by Bell (1976) for some *Isorophina* is plesiomorphic as it is apomorphic for the more inclusive clade *Isorophida* (see entry in this volume).

Synonyms: *Agelacrinitinae* as used by Smith and Arbizu (1987) and *Agelacrinitidae* as used by Basler (1936), Regnéll (1966), Kesling (1967) and Guensburg and Sprinkle (1994) are approximate synonyms of *Isorophina*.

Comments: The name *Isorophina* as defined here closely approximates the original use of the name by Bell (1976). This clade was first inferred by Smith and Arbizu (1987, therein referred to as *Agelacrinitinae*). Later analyses by Sumrall (1993) and Sumrall and Gahn (2006) inferred a largely congruent monophyletic group, which they referred to as *Isorophina* following Bell (1976). *Isorophina* as conceptualized here includes most of the well-known and all of the post-Ordovician edrioasteroids. The characteristic discoidal theca with well-defined peripheral rim that is often considered diagnostic is plesiomorphic for the clade (Smith and Arbizu, 1987; Sumrall, 1993; Sumrall and Gahn, 2006).

Isorophina was selected as the name for this clade because it conforms in composition and diagnostic characters to the original use of Bell (1976) that was followed in most of the subsequent literature (Sumrall 1993, 1996, 2001; Sumrall and Gahn, 2006; Sumrall and Zamora, 2011; Sumrall et al., 2000, 2006).

Literature Cited

Bassler, R. S. 1936. New species of American *Edrioasteroidea*. *Smithson. Misc. Coll.* 95(6):1–33.

Bell, B. M. 1976. A study of North American *Edrioasteroidea*. *New York St. Mus. Mem.* 21:476 pp.

Guensburg, T. E., and J. Sprinkle. 1994. Revised phylogeny and functional interpretation of the *Edrioasteroidea* based on new taxa from the

Early and Middle Ordovician of western Utah. *Fieldiana (Geology)* 29:1–43.

Kesling, R. V. 1967. Edrioasteroid with unique shape from Mississippian strata of Alberta. *J. Paleontol.* 41:197–202.

Regnéll, G. 1966. Edrioasteroids. Pp. U135–U173 in *Treatise on Invertebrate Paleontology, Part U* (Echinodermata 3) (R. C. Moore, ed.). Geological Society of America, New York; University of Kansas, Lawrence, KS.

Roemer, F. (C. F. von). 1851. Beiträge zur Kenntniss der fossilen Fauna des devonischen Gebirges am Rhein. *Decheniana (Verh. Naturhist. Ver. Rheinlands und Westfalens)* 8:357–376.

Smith, A. B., and M. A. Arbizu. 1987. Inverse development in a Devonian edrioasteroid from Spain and the phylogeny of *Agelacrinitinae*. *Lethaia*. 20:49–62.

Sumrall, C. D. 1993. Thecal designs in isorophinid edrioasteroids. *Lethaia* 26:289–302.

Sumrall, C. D. 1996. Late Paleozoic edrioasteroids from the North American mid-continent. *J. Paleontol.* 70:969–985.

Sumrall, C. D. 2001. Paleoecology of two new edrioasteroids from a Mississippian hardground in Kentucky. *J. Paleontol.* 75:136–146.

Sumrall, C. D., and F. J. Gahn. 2006. Morphological and systematic reinterpretation of two enigmatic edrioasteroids (*Echinodermata*) from Canada. *Can. J. Earth Sci.* 43:497–507.

Sumrall, C. D., J. Garbisch, and J. P. Pope. 2000. The systematics of postbullinid edrioasteroids. *J. Paleontol.* 74:72–83.

Sumrall, C. D., J. Sprinkle, and R. M. Bonem. 2006. An edrioasteroid-dominated echinoderm assemblage from a Lower Pennsylvanian marine conglomerate in Oklahoma. *J. Paleontol.* 80:229–244.

Sumrall, C. D., and S. Zamora. 2011. Ordovician edrioasteroids from Morocco: Faunal exchanges across the Rheic Ocean. *J. Syst. Palaeontol.* 9:425–454.

Vanuxem, L. 1842. Survey of the Third Geological District. *Nat. Hist. NY. Geol.* 3:1–307.

Author

Colin D. Sumrall; Department of Earth and Planetary Sciences; The University of Tennessee; Knoxville, TN 37996-1526, USA. Email: csumrall@utk.edu.

Date Accepted: 16 September 2013

Primary Editor: Kevin de Queiroz

Agelacrinitidae E. J. Chapman 1860 [C. D. Sumrall, this volume], converted clade name

Registration Number: 231

Definition: The clade originating in the most recent common ancestor of *Agelacrinites hamiltonensis* Vanuxem 1842, *Lepidodiscus squamosus* (Meek and Worthen 1868), and *Hypsiclavus huntsvillensis* Sumrall 1996 (referred to as "Chester *L. laudoni*" in the reference phylogeny). This is a minimum-clade definition. Abbreviated definition: min ∇ (*Agelacrinites hamiltonensis* Vanuxem 1842 & *Lepidodiscus squamosus* (Meek and Worthen 1868) & *Hypsiclavus huntsvillensis* Sumrall 1996).

Etymology: Derived from the Greek *agelaios* (gregarious) and *krinon* (lily).

Reference Phylogeny: Sumrall (1993: Fig. 10A).

Composition: *Agelacrinitidae* is extinct and includes approximately 35 named species that range minimally from the Late Silurian through Late Pennsylvanian (Bell, 1976; Sumrall, 1996; Sumrall et al., 2006a). Traditional genera include: *Agelacrinites* Chapman 1860, *Bostryclavus* Sumrall 1996, *Clavidiscus* Sumrall 1996, *Cooperidiscus* Bassler 1935, *Dinocystis* Bather 1898, *Discocystis* Gregory 1897, *Eopostibulla* Müller and Hahn 2010, *Giganticlavus* Sumrall and Bowsher 1996, *Hypsiclavus* Sumrall 1996, *Krama* Bell 1976, *Lepidodiscus* Meek and Worthen 1868, *Lispidecodus* Kesling 1967, *Parakrama* Müller and Hahn 2010, *Parapostibulla* Sumrall et al. 2000, *Postibulla* Bell 1976, *Pyrgopostibulla* Sumrall et al. 2006b, *Spiraclavus* Sumrall 1992, *Stalticodiscus* Smith 1983, *Timeischytes* Ehlers and Kesling 1958, *Torquerisediscus* Sumrall 2001, and *Ulrichidiscus* Bassler 1935.

Diagnostic Apomorphies: Peristome covered by numerous oral cover plates; primary oral plates poorly differentiated; hydro-gonopore offset from oral rise; cover plates with an externally-exposed cyclic pattern; intrathecal cover plate extensions lacking (derived from Bell, 1976; Smith and Arbizu, 1987; Sumrall, 1993; Sumrall and Gahn, 2006).

Synonyms: None.

Comments: Although *Agelacrinitidae* was first recognized by Chapman (1860), its modern conceptualization did not begin to take shape until Bassler (1935). Bassler (1936) provided further clarification of *Agelacrinitidae* by excluding the Cambrian taxa *Walcottidiscus* and *Stromatocystites* and including only those taxa previously referred to *Agelacrinitidae* (Bassler, 1935) that bear small oral cover plates, as opposed to *Hemicystitidae*, whose members bear large oral cover plates. This use was subsequently followed by other authors (Regnéll, 1966; Kesling, 1967), until Bell (1976) further refined the concept of *Agelacrinitidae* by excluding taxa that he assigned to *Isorophidae*.

Agelacrinitidae was chosen as the name for this clade because it follows the dominant usage in the recent literature (Sumrall, 1992, 1993, 1996, 2001; Sumrall and Bowsher, 1996; Sumrall et al., 2000, 2006a). Guensburg and Sprinkle (1994) conceptualized *Agelacrinitidae* more broadly, approximating the use of Bassler (1936) by applying the name *Agelacrinitidae* to the clade here named *Isorophina* (this volume).

Literature Cited

Bassler, R. S. 1935. The classification of the *Edrioasteroidea. Smithson. Misc. Coll.* 93(8):1–11.

Bassler, R. S. 1936. New species of American *Edrioasteroidea. Smithson. Misc. Coll.* 95(6):1–33.

Bather, F. A. 1898. *Dinocystis barroisi,* n. g. et sp., Psammites du Condroz. *Geol. Mag.* 5:543–548.

Bell, B. M. 1976. A study of North American *Edrioasteroidea. New York St. Mus. Mem.* 21:476 pp.

Chapman, E. J. 1860. On a new species of *Agelacrinites,* and on the structural relations of that genus. *Can. J. Ind. Sci. Art* 5:358–365.

Ehlers, G. M., and R. V. Kesling. 1958. *Timeischityes,* a new genus of hemicystitid edrioasteroid from the Middle Devonian Four Mile Dam Limestone of Michigan. *J. Paleontol.* 32:933–936.

Gregory, J. W. 1897. On *Echinocystis* and *Palaeodiscus*—two genera of *Echinoidea. Q. J. Geol. Soc. Lond.* 53:123–136.

Guensburg, T. E., and J. Sprinkle. 1994. Revised phylogeny and functional interpretation of the *Edrioasteroidea* based on new taxa from the Early and Middle Ordovician of western Utah. *Fieldiana (Geology)* 29:1–43.

Kesling, R. V. 1967. Edrioasteroid with unique shape from Mississippian strata of Alberta. *J. Paleontol.* 41:197–202.

Meek, F. B., and A. H. Worthen. 1868. Remarks on some types of Carboniferous *Crinoidea* with descriptions of new genera and species of the same, and one echinoid. *Proc. Acad. Nat. Sci. Phila.* 5:335–359.

Müller, P., and G. Hahn. 2010. *Edrioasteroidea* aus den Seifen-Schichten im Westerwald, Rheinisches Schiefer-Gebirge (Unter-Devon, Deutschland) Part 1. *Geol. Palaeont.* 43:9–47.

Regnéll, G. 1966. Edrioasteroids. Pp. U135–U173 in *Treatise on Invertebrate Paleontology, Part U* (Echinodermata 3) (R. C. Moore, ed.). Geological Society of America, New York; University of Kansas, Lawrence, KS.

Smith, A. B. 1983. British Carboniferous edrioasteroids (*Echinodermata*). *Bull. Brit. Mus. Nat. Hist. (Geol.)* 37:113–138.

Smith, A. B., and M. A. Arbizu. 1987. Inverse development in a Devonian edrioasteroid from Spain and the phylogeny of *Agelacrinitinae. Lethaia* 20:49–62.

Sumrall, C. D. 1992. *Spiraclavus nacoensis,* a new clavate agelacrinitid edrioasteroid from central Arizona. *J. Paleontol.* 66:90–98.

Sumrall, C. D. 1993. Thecal designs in isorophinid edrioasteroids. *Lethaia* 26:289–302.

Sumrall, C. D. 1996. Late Paleozoic edrioasteroids from the North American mid-continent. *J. Paleontol.* 70:969–985.

Sumrall, C. D. 2001. Paleoecology of two new edrioasteroids from a Mississippian hardground in Kentucky. *J. Paleontol.* 75:136–146.

Sumrall, C. D., and A. L. Bowsher. 1996. *Giganticlavus,* a new genus of Pennsylvanian edrioasteroid from North America. *J. Paleontol.* 70:986–993.

Sumrall, C. D., C. E. Brett, and S. Cornell. 2006b. The systematics and ontogeny of *Pyrgopostibulla belli,* a new edrioasteroid (*Echinodermata*) from the Lower Devonian of New York. *J. Paleontol.* 80:187–192.

Sumrall, C. D., and F. J. Gahn. 2006. Morphological and systematic reinterpretation of two enigmatic edrioasteroids (*Echinodermata*) from Canada. *Can. J. Earth Sci.* 43:497–507.

Sumrall, C. D., J. Garbisch, and J. P. Pope. 2000. The systematics of postibullinid edrioasteroids. *J. Paleontol.* 74:72–83.

Sumrall, C. D., J. Sprinkle, and R. M. Bonem. 2006a. An edrioasteroid-dominated echinoderm assemblage from a Lower Pennsylvanian marine conglomerate in Oklahoma. *J. Paleontol.* 80:229–244.

Vanuxem, L. 1842. Survey of the Third Geological District. *Nat. Hist. New York Geol.* 3:1–307.

Author

Colin D. Sumrall; Department of Earth and Planetary Sciences; The University of Tennessee; Knoxville, TN 37996-1526, USA. Email: csumrall@utk.edu.

Date Accepted: 25 August 2013

Primary Editor: Kevin de Queiroz

Lepidodiscina C. D. Sumrall, new clade name

Registration Number: 232

Definition: The clade originating in the most recent common ancestor of *Lepidodiscus squamosus* (Meek and Worthen 1868), *Ulrichidiscus pulaskiensis*, (Miller and Gurley 1894) and *Hypsiclavus huntsvillensis* Sumrall 1996. This is a minimum-clade definition. Abbreviated definition: min ∇ (*Lepidodiscus squamosus* (Meek and Worthen 1868) & *Ulrichidiscus pulaskiensis* (Miller and Gurley 1894) & *Hypsiclavus huntsvillensis* Sumrall 1996).

Etymology: Derived from the Greek *lepidos* (scale) and *diskos* (circular plate).

Reference Phylogeny: Sumrall and Gahn (2006: Fig. 3.1).

Composition: *Lepidodiscina* is extinct and includes approximately 20 named species that range minimally from the Late Devonian through Late Pennsylvanian (Bell, 1976; Smith and Arbizu, 1987; Sumrall, 1993). Traditional genera include: *Bostryclavus* Sumrall 1996, *Clavidiscus* Sumrall 1996, *Cooperidiscus* Bassler 1935, *Dinocystis* Bather 1898, *Discocystis* Gregory 1897, *Giganticlavus* Sumrall and Bowsher 1996, *Hypsiclavus* Sumrall 1996, *Lepidodiscus* Meek and Worthen 1868, *Lispidecodus* Kesling 1967, *Spiraclavus* Sumrall 1992, *Stalticodiscus* Smith 1983, and *Ulrichidiscus* Bassler 1935.

Diagnostic Apomorphies: Three or more sets of shared cover plates; floor plates with lateral extensions; complex hydro-gonopore with numerous hydropore oral plates; great expansion of the pedunculate zone between oral surface and peripheral rim (derived from Smith and Arbizu, 1987; Sumrall, 1993; Sumrall and Gahn, 2006).

Synonyms: None.

Comments: *Lepidodiscina* has been inferred by every analysis that included diverse members of *Agelacrinitidae* (Smith and Arbizu, 1987; Sumrall, 1993; Sumrall and Gahn, 2006). This clade includes a number of taxa that share generally large thecae with an extendable pedunculate zone. All of these taxa bear complex cyclic patterns of ambulacral cover plates, differences in which form the basis of species delimitation (Bell, 1976; Sumrall, 1993, 1996). *Agelacrinites hamiltonensis* (an outgroup to *Lepidociscina*) and *Discocystis kaskaskiensis* have these cycles expressed on the interior of the cover plates, yet they are not exposed on the external surface (Sumrall, 1993).

Literature Cited

Bassler, R. S. 1935. The classification of the *Edrioasteroidea. Smithson. Misc. Coll.* 93(8):1–11.

Bather, F. A. 1898. *Dinocystis barroisi*, n. g. et sp., Psammites du Condroz. *Geol. Mag.* 5:543–548.

Bell, B. M. 1976. A study of North American *Edrioasteroidea. New York State Mus. Mem.* 21:476 pp.

Gregory, J. W. 1897. On *Echinocystis* and *Palaeodiscus* – two genera of *Echinoidea. Q. J. Geol. Soc. Lond.* 53:123–136.

Kesling, R. V. 1967. Edrioasteroid with unique shape from Mississippian strata of Alberta. *J. Paleontol.* 41:197–202.

Meek, F. B., and A. H. Worthen. 1868. Remarks on some types of Carboniferous *Crinoidea* with descriptions of new genera and species of the same, and one echinoid. *Proc. Acad. Nat. Sci. Phila.* 5:335–359.

Miller, S. A., and F. E. Gurley. 1894. New genera and species of *Echinodermata. Ill. State Mus. Bull.* 5:5–53.

Smith, A. B. 1983. British Carboniferous edrioasteroids (*Echinodermata*). *Bull. Brit. Mus. Nat. Hist. (Geol.)* 37:113–138.

Smith, A. B., and M. A. Arbizu. 1987. Inverse development in a Devonian edrioasteroid from Spain and the phylogeny of *Agelacrinitinae. Lethaia* 20:49–62.

Sumrall, C. D. 1992. *Spiraclavus nacoensis*, a new clavate agelacrinitid edrioasteroid from central Arizona. *J. Paleontol.* 66:90–98.

Sumrall, C. D. 1993. Thecal designs in isorophinid edrioasteroids. *Lethaia* 26:289–302.

Sumrall, C. D. 1996. Late Paleozoic edrioasteroids from the North American mid-continent. *J. Paleontol.* 70:969–985.

Sumrall, C. D., and A. L. Bowsher. 1996. *Giganticlavus*, a new genus of Pennsylvanian edrioasteroid from North America. *J. Paleontol.* 70:986–993.

Sumrall, C. D., and F. J. Gahn. 2006. Morphological and systematic reinterpretation of two enigmatic edrioasteroids (*Echinodermata*) from Canada. *Can. J. Earth Sci.* 43:497–507.

Author

Colin D. Sumrall; Department of Earth and Planetary Sciences; The University of Tennessee; Knoxville, TN 37996-1526, USA. Email: csumrall@utk.edu.

Date Accepted: 25 August 2013

Primary Editor: Kevin de Queiroz

Discocystinae C. D. Sumrall 1996 [C. D. Sumrall], converted clade name

Registration Number: 233

Definition: The clade originating in the most recent common ancestor of *Discocystis kaskaskiensis* (Hall 1858), *Clavidiscus laudoni* (Bassler 1936), *Hypsiclavus huntsvillensis* Sumrall 1996, and *Giganticlavus bennisoni* Sumrall and Bowsher 1996. This is a minimum-clade definition. Abbreviated definition: min ∇ (*Discocystis kaskaskiensis* (Hall 1858) & *Clavidiscus laudoni* (Bassler 1936) & *Hypsiclavus huntsvillensis* Sumrall 1996 & *Giganticlavus bennisoni* Sumrall and Bowsher 1996).

Etymology: Derived from the Greek *diskos* (circular plate) and *kystis* (sac).

Reference Phylogeny: Sumrall and Gahn (2006: Fig. 3.1) should be considered the reference phylogeny. Although not included in the reference phylogeny, *Discocystis kaskaskiensis* (Hall, 1858) is thought to be more closely related to *Hypsiclavus huntsvillensis* than to other taxa in the reference phylogeny (Sumrall, 1996).

Composition: *Discocystinae* is extinct and includes approximately 12 named species that range minimally from the Late Devonian through Late Pennsylvanian including the largest of the isorophids (Sumrall, 1996; Sumrall and Bowsher, 1996). Traditional genera include: *Bostryclavus* Sumrall 1996, *Clavidiscus* Sumrall 1996, *Dinocystis* Bather 1898, *Discocystis* Gregory 1897, *Giganticlavus* Sumrall and Bowsher 1996, *Hypsiclavus* Sumrall 1996, *Lispidecodus* Kesling 1967, and *Spiraclavus* Sumrall 1992.

Diagnostic Apomorphies: Adjacent interambulacral plates, peduncular plates aligned into columns, floor plates thick, peripheral rim much smaller than thecal diameter (derived from Sumrall, 1993, 1996; Sumrall and Gahn, 2006).

Synonyms: None.

Comments: *Discocystinae* as used here follows the original use by Sumrall (1996) and subsequently followed by other authors (Sumrall and Bowsher, 1996; Sumrall and Parsley, 2003; Sumrall and Gahn, 2006). It is a distinctive clade of isorophidans bearing clavate thecae. *Dinocystis barroisi* Bather 1898, thought to be an edrioasterid by several authors (Bather, 1914; Regnéll, 1966; Bell, 1976), though poorly preserved, shows the small peripheral rim and organized pedunculate zone characteristic of discocystines. Though highly morphologically unusual and incomplete, *Lispidecodus plinthotus* Kesling 1967 has been inferred to be a member of *Discocytinae* in three analyses (Smith and Arbizu, 1987; Sumrall, 1993; Sumrall and Gahn, 2006).

Literature Cited

Bassler, R. S. 1936. New species of American *Edrioasteroidea. Smithson. Misc. Coll.* 95(6):1–33.

Bather, F. A. 1898. *Dinocystis barroisi*, n. g. et sp., Psammites du Condroz. *Geol. Mag.* 5:543–548.

Bather, F. A. 1914. Edrioasters in the Trenton Limestone (parts 1 and 2). *Geol. Mag.* 1:115–125, 162–171.

Bell, B. M. 1976. A study of North American *Edrioasteroidea. New York State Mus. Mem.* 21:476 pp.

Gregory, J. W. 1897. On *Echinocystis* and *Palaeodiscus* – two genera of *Echinoidea. Q. J. Geol. Soc. Lond.* 53:123–136.

Hall, J. 1858. Palaeontology of Iowa. Pp. 473–724 in *Report on the Geological Survey of the State of Iowa: Embracing the Results of Investigations Made During Portions of the Years 1855, 56 & 57, Vol. 1, Pt. II. Palaeonotology.* Legislature of Iowa, Des Moines.

Kesling, R. V. 1967. Edrioasteroid with unique shape from Mississippian strata of Alberta. *J. Paleontol.* 41:197–202.

Regnéll, G. 1966. Edrioasteroids. Pp. U135–U173 in *Treatise on Invertebrate Paleontology, Part U* (Echinodermata 3) (R. C. Moore, ed.). Geological Society of America, New York; University of Kansas, Lawrence, KS.

Smith, A. B., and M. A. Arbizu. 1987. Inverse development in a Devonian edrioasteroid from Spain and the phylogeny of *Agelacrinitinae.* *Lethaia* 20:49–62.

Sumrall, C. D. 1992. *Spiraclavus nacoensis*, a new clavate agelacrinitid edrioasteroid from central Arizona. *J. Paleontol.* 66:90–98.

Sumrall, C. D. 1993. Thecal designs in isorophinid edrioasteroids. *Lethaia* 26:289–302.

Sumrall, C. D. 1996. Late Paleozoic edrioasteroids from the North American mid-continent. *J. Paleontol.* 70:969–985.

Sumrall, C. D., and A. L. Bowsher. 1996. *Giganticlavus*, a new genus of Pennsylvanian edrioasteroid from North America. *J. Paleontol.* 70:986–993.

Sumrall, C. D., and F. J. Gahn. 2006. Morphological and systematic reinterpretation of two enigmatic edrioasteroids (*Echinodermata*) from Canada. *Can. J. Earth Sci.* 43:497–507.

Sumrall, C. D., and R. L. Parsley. 2003. Morphology and biomechanical implications of isolated discocystinid plates (*Edrioasteroidea, Echinodermata*) from the Carboniferous of North America. *Palaeontology* 46:113–138.

Author

Colin D. Sumrall; Department of Earth and Planetary Sciences; The University of Tennessee; Knoxville, TN 37996-1526, USA. Email: csumrall@utk.edu.

Date Accepted: 16 September 2013

Primary Editor: Kevin de Queiroz

Chordata E. Haeckel 1874 [C. B. Cameron], converted clade name

Registration Number: 28

Definition: The largest crown clade containing *Petromyzon marinus* Linnaeus 1758 but not *Strongylocentrotus purpuratus* Stimpson 1857. This is a maximum-crown-clade definition. Abbreviated definition: max crown ∇ (*Petromyzon marinus* Linnaeus 1758 ~ *Strongylocentrotus purpuratus* Stimpson 1857).

Etymology: Derived from the Greek *chorde* (string), a reference to the notochord (see Diagnostic Apomorphies).

Reference Phylogeny: The primary reference phylogeny is Delsuc et al. (2008: Fig. 1). See also Winchell et al. (2002), Zeng and Swalla (2005), Bourlat et al. (2006), Delsuc et al. (2006), and Putnam et al. (2008).

Composition: *Chordata* is currently hypothesized to include *Cephalochordata*, *Tunicata*, *Cyclostomata* (often included in *Vertebrata*), and *Vertebrata* (Delsuc et al., 2008).

Diagnostic Apomorphies: *Chordata* apomorphies (Cameron, 2005) include a notochord that extends to the end of the body using serially arranged, discoid shaped cordal cells; a dorsal hollow nerve cord (with an anterior to posterior flow in the neurocoel); a postanal tail that is dorsal to the anus; and an endostyle (thought to be homologous to the thyroid gland of vertebrates) that binds iodine.

Synonyms: *Protochordata* (for *Cephalochordata* plus *Tunicata*) is a partial synonym in that it refers to a paraphyletic group associated with the same ancestor but does not include *Vertebrata*.

Comments: The monophyly of *Chordata*—that is, of a group composed of cephalochordates, tunicates, cyclostomes, and vertebrates—is supported by molecular and morphological analyses (Winchell et al., 2002; Zeng and Swalla, 2005; Bourlat et al., 2006; Delsuc et al., 2006; Delsuc et al., 2008; and Putnam et al., 2008). The name *Chordata* derives from the notochord, a structure composed of serially arranged discoidal cordal cells that extends to the end of the body. Bourlat et al. (2006) and Delsuc et al. (2006) positioned the cephalochordates closer to echinoderms than to vertebrates or tunicates. Bourlat et al. (2006) tested this unusual hypothesis for systematic error by (1) adding more taxa, thereby enabling better detection of misleading multiple substitutions at the same site, and (2) removing problematic taxa with high rate of substitutions (and therefore susceptible to long branch attraction); in both tests, a monophyletic *Chordata* was inferred. In a later paper, Delsuc et al. (2008) reanalyzed their data using an improved model of sequence evolution and using more genes and more taxa, and they inferred a monophyletic *Chordata* with good statistical support.

Literature Cited

Bourlat, S. J., T. Juliusdotti, C.J. Lowe, R. Freeman, J. Aronowicz, M. Kirschner, E.S. Lander, M. Thorndyke, H. Nakano, A.B. Kohn, A. Heyland, L.L. Moroz, R.R. Copley, and M.J. Telford. 2006. Deuterostome phylogeny reveals monophyletic chordates and the new phylum *Xenoturbellida*. *Nature* 444: 85–88.

Cameron, C. B. 2005. A phylogeny of the hemichordates based on morphological characters. *Can. J. Zool.* 83:196–215.

Delsuc F., G. Tsagkogeorga, N. Lartillot, and H. Philippe. 2008. Additional molecular support for the new chordate phylogeny. *Genesis* 46:592–604.

Delsuc, F., H. Brinkmann, D. Chourrout, and H. Philippe. 2006. Tunicates and not cephalochordates are the closest living relatives of vertebrates. *Nature* 439:965–968.

Haeckel, E. 1874. *Anthropogenie oder Entwickelungsgeschichte des Menschen.* Engelmann, Leipzig.

Putnam N. H., T. Butts, D. E. Ferrier, R. F. Furlong, U. Hellsten, T. Kawashima, M. Robinson-Rechavi, E. Shoguchi, A. Terry, J. K. Yu, E. L. Benito-Gutierrez, I. Dubchak, J. Garcia-Fernandez, J. J. Gibson-Brown, I. V. Grigoriev, A. C. Horton, P. J. de Jong, J. Jurka, V. V. Kapitonov, Y. Kohara, Y. Kuroki, E. Lindquist, S. Lucas, K. Osoegawa, L. A. Pennacchio, A. A. Salamov, Y. Satou, T. Sauka-Spengler, J. Schmutz, I. T. Shin, A. Toyoda, M. Bronner- Fraser, A. Fujiyama, L. Z. Holland, P. W. Holland, N. Satoh, and D. S. Rokhsar. 2008. The amphioxus genome and the evolution of the chordate karyotype. *Nature* 453:1064–1071.

Winchell, C. J., J. Sullivan, C. B. Cameron, B. J. Swalla, and J. Mallatt. 2002. Evaluating hypothesis of deuterostome phylogeny and chordate evolution with new LSU and SSU ribosomal DNA data. *Mol. Biol. Evol.* 19:762–776.

Zeng, L., and B. J. Swalla. 2005. Molecular phylogeny of the protochordates: chordate evolution. *Can. J. Zool.* 83:24–33.

Author

Chris B. Cameron; Sciences biologiques; Université de Montréal; C.P. 6128, Succ. Centre-ville; Montreal, Québec, H3C 3J7, Canada. Email: c.cameron@umontreal.ca.

Date Accepted: 28 February 2012

Primary Editor: Philip Cantino

Cephalochordata R. Owen 1846 [C. B. Cameron], converted clade name

Registration Number: 25

Definition: The largest crown clade containing *Branchiostoma lanceolatum* Pallas 1744 but not *Petromyzon marinus* Linnaeus 1758 (*Vertebrata*) and *Ciona intestinalis* Linnaeus 1767 (*Tunicata*) and *Strongylocentrotus purpuratus* Stimpson 1857 (*Echinodermata*) and *Ptychodera flava* Eschscholtz 1825 (*Hemichordata*). This is a maximum-crown-clade definition. Abbreviated definition: max crown ∇ (*Branchiostoma lanceolatum* Pallas 1744 ~ *Petromyzon marinus* Linnaeus 1758 & *Ciona intestinalis* Linnaeus 1767 & *Strongylocentrotus purpuratus* Stimpson 1857 & *Ptychodera flava* Eschscholtz 1825).

Etymology: Derived from the Greek *kephale* (head) and *chorde* (string), referring to the extension of the notochord to the front of the body.

Reference Phylogeny: Kon et al. (2007: Fig. 3). Although not included in the reference phylogeny, *Ciona intestinalis* is considered more closely related to *Petromyzon marinus* than to *Branchiostoma lanceolatum* (Delsuc et al., 2008); *Strongylocentrotus purpuratus* is considered more closely related to *Balanoglussus carnosus* than to *Branchiostoma lanceolatum* (Cameron, 2005; Mallatt and Winchell, 2007); and *Ptychodera flava* is considered most closely related to *Balanoglussus carnosus* (Cameron, 2005).

Composition: The clade *Cephalochordata* currently contains some 29 extant species assignable to the genera *Branchiostoma*, *Epigonichthys* (Poss and Boschung, 1996) and *Asymmetron* (Nishikawa, 2004).

Diagnostic Apomorphies: Characters uniting the cephalochordates include a laterally compressed body; a ciliated wheel organ before the opening to the mouth (Ruppert, 1997a); a notochord that extends the entire length of the body (Ruppert, 1997b); bilateral myomeres that are out-of-register (rather than in-register, as in vertebrates) mononucleate cells with slender tails that arise from each muscle cell and extend to the nerve cord (Holland, 1996); a Hatschek's nephridium (the anterior-most nephridia, which open into the pharynx inside the dorsal lip); a ventral atrium enclosing the pharynx (Northcutt, 1996; Ruppert, 1997a); and coelomic cavities that form fin rays in series along the dorsal and ventral midline (Goodsear, 1844).

Synonyms: Gill (1883) listed the following approximate synonyms: *Amphioxidae* Gray 1842, *Amphioxi* Bonaparte 1846, *Amphioxini* Müller 1846, *Branchiostomidae* Bonaparte 1846, *Cirrostomi* Owen 1846, *Leptocardii* Müller 1846, *Myelozoa* Bonaparte 1856, *Amphioxoidei* Bleeker 1859, *Branchiostomoidae* Gill 1860, *Acrania* Haeckel 1866, *Leptocardia* Haeckel 1866, *Cirrhostomi* Günther 1870, and *Entomocrania* Huxley 1875 (date and publication incorrectly attributed by Gill). Other approximate synonyms are *Branchiostomiformes* and *Amphioxiformes* (authors not determined) and the vernacular name "lancelets".

Comments: Traditional classification schemes of cephalochordates separate *Branchiostoma*, with a bilateral series of gonads, from *Epigonichthys* and *Asymmetron*, with a single asymmetric series of gonads. In contrast, the mitochondrial tree in Kon et al. (2007) suggests that *Asymmetron* is sister to a clade comprising

Epigonichthys and *Branchiostoma*. Nohara et al. (2005) provided further evidence for the monophyly of the cephalochordates using complete mitochondrial DNA gene sequences. The clade name *Cephalochordata* is preferred to its various synonyms because it is most commonly used.

Literature Cited

Cameron, C. B. 2005. A phylogeny of the hemichordates based on morphological characters. *Can. J. Zool.* 83:196–215.

Delsuc F., G. Tsagkogeorga, N. Lartillot, and H. Philippe. 2008. Additional molecular support for the new chordate phylogeny. *Genesis* 46:592–604.

Gill, T. 1883. Note on the leptocardians. *Proc. U. S. Natl. Mus.* 5:515–516.

Goodsear, J. 1844. On the anatomy of *Amphioxus lanceolatus*; Lancelet, Yarrell. *Trans. Edinb. R. Soc.* 15:247–264.

Holland, L. Z. 1996. Muscle development in amphioxus: morphology, biochemistry, and molecular biology. *Isr. J. Zool.* 42(suppl.):235–246.

Huxley, T. H. 1875. On the classification of the animal kingdom. *J. Linn. Soc. Lond.* 12:199–226.

Kon, T., M. Nohara, Y. Yamanoue, Y. Fujiwara, M. Nishida, and T. Nishikawa. 2007. Phylogenetic position of a whale-fall lancelet (*Cephalochordata*) inferred from whole mitochondrial genome sequences. *BMC Evol. Biol.* 7:127.

Mallatt, J., and C. J. Winchell. 2007. Ribosomal RNA genes and deuterostome phylogeny revisited: more cyclostomes, elasmobranchs, reptiles, and a brittle star. *Mol. Phylogenet. Evol.* 43:1005–1022.

Nishikawa, T. 2004. A new deep-water lancelet (*Cephalochordata*) from off Cape Nomamisaki, SW Japan, with a proposal of the revised system recovering the genus *Asymmetron*. *Zool. Sci.* 21:1131–1136.

Nohara, M., M. Nishida, M. Miya, and T. Nishikawa. 2005. Evolution of the mitochondrial genome in *Cephalochordata* as inferred from complete nucleotide sequences from two *Epigonichthys* species. *J. Mol. Evol.* 60: 526–537.

Northcutt, R. G. 1996. The origin of craniates: neural crest, neurogenic placodes, and homeobox genes. *Isr. J. Zool.* 42(suppl.):273–313

Owen, R. 1846. *On the Archetype and Homologies of the Vertebrate Skeleton*. Report of the British Association for the Advancement of Science, Southampton, 169–340.

Poss, S. G., and H. T. Boschung. 1996. Lancelets (*Cephalochordata*: *Branchiostomatidae*): how many species are valid? *Isr. J. Zool.* 42(suppl.):S13–S66.

Ruppert, E. E. 1997a. *Cephalochordata*. Pp. 349–504 in *Microscopic Anatomy of Invertebrates*, Vol. 15 (F. W. Harrison and E. E. Ruppert, eds.). Wiley-Liss, Inc., New York.

Ruppert, E. E. 1997b. Introduction: microscopic anatomy of the notochord, heterochrony, and chordate evolution. Pp. 1–13 in *Microscopic Anatomy of Invertebrates*, Vol. 15 (F. W. Harrison and E. E. Ruppert, eds.). Wiley-Liss, Inc., New York.

Author

Chris B. Cameron; Sciences biologiques; Université de Montréal; C.P. 6128, Succ. Centre-ville; Montreal, Québec, H3C 3J7, Canada. Email: c.cameron@umontreal.ca.

Date Accepted: 28 February 2012

Primary Editor: Philip Cantino

Tunicata J. B. Lamarck 1816 [C. B. Cameron and B. J. Swalla], converted clade name

Registration Number: 109

Definition: The crown clade for which a tunic, as inherited by *Ciona intestinalis* Linnaeus 1767, is an apomorphy relative to other crown clades. The tunic is a covering that surrounds the mantle and organs of the animal and is composed of cellulose and protein. This is an apomorphy-modified crown-clade definition. Abbreviated definition: crown ∇ apo tunic [*Ciona intestinalis* Linnaeus 1767].

Etymology: Derived from latin *tunica* (garment, skin or husk).

Reference Phylogeny: The primary reference phylogeny is Swalla et al. (2000: Fig. 1), in which the tunic is inferred to have arisen along the branch subtending the node representing the last common ancestor of *Oikopleura dioica* and *Thalia democratica*. See also Stach and Turbeville (2002), Turon and López-Legentil (2004), Zeng and Swalla (2005) Yokobori et al. (2006), and Zeng et al. (2006).

Composition: *Tunicata* includes the sessile "*Ascidiacea*" and the planktonic *Thaliacea* and *Appendicularia* (*Larvacea*).

Diagnostic Apomorphies: The defining apomorphy of *Tunicata* relative to other crown clades is a tunic that is composed of cellulose and protein. The following additional characters are synapomorphic relative to other crown clades: During early development, tunicates develop a single pair of coelomic cavities that secondarily fuse on the mid-ventral line to form the pericardium; otherwise they lack coelomic cavities (Berrill, 1950). They are unlike any other invertebrate deuterostome in having a closed nephridial system, and except for larvaceans, they have periodic heartbeat reversals that result in a change of the direction of blood flow (Ruppert et al., 2004). The larval tadpole is lecithotrophic with a trunk and a locomotory tail (except salps, pyrosomids and some molgulid ascidians, which have lost the larval stage) (Hadfield et al., 1995; Jeffery and Swalla, 1997; Huber et al., 2000). Tunicates have a perforate pharynx that usually takes up most of the body volume.

Synonyms: *Urochordata* Balfour 1881 is an approximate synonym.

Comments: The monophyly of *Tunicata* is strongly supported by morphological (Cameron, 2005) and molecular analyses (Swalla et al., 2000; Stach and Turbeville, 2002; Turon and López-Legentil, 2004; Zeng and Swalla, 2005; Yokobori et al., 2006; Zeng et al., 2006). The name *Urochordata* was not used until Balfour (1881) created it as a replacement name for *Tunicata*, presumably to emphasize the chordate affinity. A name already existed for this taxon, so the name *Urochordata* was not needed. The name *Tunicata* is almost universally used to refer to this group of organisms in the major monographic works, including Alder (1863), Van Name (1945), Berrill (1950), and Bone (1998), and it is more frequently used. It was selected for the clade named here for those reasons.

Literature Cited

Alder, J. 1863. Observations on the British *Tunicata* with descriptions of several new species *Ann. Mag. Nat. Hist.* 3(11):153–173.

Balfour, F. M. 1881. *A Treatise of Comparative Embryology*, Vol. 2. Macmillan, London.

Berrill, N. J. 1950. *The* Tunicata, *with an Account of the British Species*. Ray Society, London.

Bone, Q. 1998. *The Biology of Pelagic Tunicates*. Oxford University Press, Oxford.

Cameron, C. B. 2005. A phylogeny of the hemichordates based on morphological characters. *Can. J. Zool.* 83:196–215

Hadfield, K. A., B. J. Swalla, and W. R. Jeffery. 1995. Multiple origins of anural development in ascidians inferred from rDNA sequences. *J. Mol. Evol.* 40:413–427.

Huber, J., K. Burke da Silva, W. Bates, and B. J. Swalla. 2000. The evolution of anural larvae in molgulid ascidians. *Semin. Cell Dev. Biol.* 11:419–426.

Jeffery, W. R., and B. J. Swalla. 1997. Tunicates. Pp. 331–364 in *Embryology: Constructing the Organism* (S. F. Gilbert and A. M. Raunio, eds.). Sinauer Associates, Sunderland, MA.

Lamarck, J. B. 1816. *Histoire Naturelle des Animaux sans Vertebrates. Tome III. Tuniciers*, pp. 80–130. Déterville, Paris

Ruppert, E. E., R. S. Fox, and R. D. Barnes. 2004. *Invertebrate Zoology: A Functional Evolutionary Approach*. 7th edition. Thomas Brooks/Cole.

Stach, T., and J. M. Turbeville. 2002. Phylogeny of *Tunicata* inferred from molecular and morphological characters. *Mol. Phyl. Evol.* 25:408–428.

Swalla, B. J., C. B. Cameron, L. S. Corley, and J. R. Garey. 2000. Urochordates are monophyletic within the deuterostomes. *Syst. Biol.* 49:52–64.

Turon, X., and S. López-Legentil. 2004. Ascidian molecular phylogeny inferred from mtDNA data with emphasis on the *Aplousobranchiata*. *Mol. Phylogenet. Evol.* 33:309–320.

Van Name, W. G. 1945. The North and South American Ascidians. *Bull. Am. Mus. Nat. Hist.* 84:1–476.

Yokobori, S., A. Kurabayashi, B. A. Neilan, T. Maruyama, and E. Hirose. 2006. Multiple origins of the ascidian-*Prochloron* symbiosis: molecular phylogeny of photosymbiotic and non-symbiotic colonial ascidians inferred from 18S rDNA sequences. *Mol. Phylogenet. Evol.* 40:8–19.

Zeng, L., M. J. Jacobs, and B. J. Swalla. 2006. Coloniality and sociality has evolved only once in stolidobranch ascidians. *Integr. Comp. Biol.* 46:255–268.

Zeng, L., and B. J. Swalla. 2005. Molecular phylogeny of the protochordates: chordate evolution. *Can. J. Zool.* 83:24–33.

Authors

Chris B. Cameron; Sciences biologiques; Université de Montréal; C.P. 6128, Succ. Centre-ville; Montreal, Québec, H3C 3J7, Canada. Email: ccameron@umontreal.ca.

Billie J. Swalla; Department of Biology; University of Washington; Seattle, WA 98195-1800, USA. Email: bjswalla@u.washington.edu.

Date Accepted: 28 February 2012

Primary Editor: Philip Cantino

SECTION 6

Pan-Gnathostomata J. P. Downs, new clade name

Registration Number: 253

Definition: The total clade of the crown clade *Gnathostomata*. This is a crown-based total-clade definition. Abbreviated definition: total ∇ of *Gnathostomata*.

Etymology: Derived from the Greek *pan-* (all, the whole), here referring to the fact that the name designates a total clade or pan-monophyllum, and *Gnathostomata*, the name of the corresponding crown clade. See *Gnathostomata* (this volume) for the etymology of that name.

Reference Phylogeny: Donoghue and Smith (2001: Fig. 8B), in which *Pan-Gnathostomata* includes jawed vertebrates and all taxa that are more closely related to them than to *Petromyzontida* (lampreys).

Composition: *Gnathostomata* (this volume) and all extinct vertebrates that share a more recent common ancestor with *Gnathostomata* than with any other mutually exclusive crown clade. In addition to the crown clade, it includes many extinct stem taxa, including the paraphyletic "ostracoderms" *Anaspida*, *Astraspida*, *Pteraspidomorphi*, *Galeaspida*, and *Osteostraci*— all of which lack jaws—and the paraphyletic "placoderms"—all of which have jaws. It is currently thought to include *Conodonta* as well (Donoghue and Smith, 2001).

Diagnostic Apomorphies: Total clades need not have any apomorphies, but those currently diagnosing *Pan-Gnathostomata* could include pharyngeal denticles (presence of the odontode sensu Donoghue and Sansom, 2002), the tissue dentine (sensu Donoghue and Sansom, 2002),

and the loss of keratinized oral "horny teeth" (Donoghue et al., 2000).

Synonyms: *Gnathostomata* sensu Donoghue et al. (2000) is an unambiguous synonym.

Comments: Morphological and combined morphological-molecular phylogenetic studies suggest that *Petromyzontida* (lampreys) is the extant sister clade of *Gnathostomata* (e.g., Løvtrup, 1977; Janvier and Blieck, 1979; Janvier, 1981; Hardisty, 1982; Forey, 1984; Maisey, 1986; Forey and Janvier, 1993; Forey, 1995; Janvier, 1996; Donoghue et al., 2000; Donoghue and Smith, 2001; Donoghue and Sansom, 2002; Near, 2009). Analyses of molecular data yield a sister relationship between *Petromyzontida* and *Myxini* (hagfishes) (e.g., Stock and Whitt, 1992; Mallatt and Sullivan, 1998; Kuraku et al., 1999; Furlong and Holland, 2002; Yu et al., 2008). Regardless of which hypothesis is accepted, the phylogenetic position of *Petromyzontida* will determine the composition of *Pan-Gnathostomata*, as stated above. The composition of *Pan-Gnathostomata* inferred above follows from the phylogenetic hypotheses of Donoghue et al. (2000), Donoghue and Smith (2001), Donoghue and Sansom (2002), and Sansom et al. (2005). Other phylogenetic analyses exclude various "ostracoderms" from *Pan-Gnathostomata* by inferring a closer relationship between *Petromyzontida* and *Gnathostomata*. Taxa that have at one time or another been proposed for exclusion from *Pan-Gnathostomata* (as that name is defined here) include *Heterostraci* (e.g., Janvier and Blieck, 1979; Janvier, 1981; Forey, 1984; Maisey, 1986; Forey and Janvier, 1993), *Astraspida* (e.g., Forey and Janvier, 1993), *Arandaspida* (e.g., Forey and Janvier, 1993),

Anaspida (e.g., Janvier and Blieck, 1979; Janvier, 1981; Forey and Janvier, 1993; Forey, 1995), the "thelodonts" (e.g., Forey, 1984), *Osteostraci* (e.g., Janvier and Blieck, 1979), and *Galeaspida* (e.g., Janvier, 1981).

Literature Cited

Donoghue, P. C. J., P. L. Forey, and R. J. Aldridge. 2000. Conodont affinity and chordate phylogeny. *Biol. Rev.* 75:191–251.

Donoghue, P. C. J., and I. J. Sansom. 2002. Origin and early evolution of vertebrate skeletonization. *Microsc. Res. Techniq.* 59:352–372.

Donoghue, P. C. J., and M. P. Smith. 2001. The anatomy of *Turinia pagei* (Powrie), and the phylogenetic status of the *Thelodonti*. *Trans. R. Soc. Edinb. Earth Sci.* 92:15–37.

Forey, P. L. 1984. Yet more reflections on agnathan-gnathostome relationships. *J. Vertebr. Paleontol.* 4(3):330–343.

Forey, P. L. 1995. Agnathans recent and fossil, and the origin of jawed vertebrates. *Re. Fish Biol. Fisher.* 5:267–303.

Forey, P. L., and P. Janvier. 1993. Agnathans and the origin of jawed vertebrates. *Nature* 361:129–134.

Furlong, R. F., and P. W. H. Holland. 2002. Bayesian phylogenetic analysis supports monophyly of *Ambulacraria* and of cyclostomes. *Zool. Sci.* 19:593–599.

Hardisty, M. W. 1982. Lampreys and hagfishes: analysis of cyclostome relationships. Pp. 165–259 in *The Biology of Lampreys*, Vol. 4B (M. W. Hardisty and I. C. Potter, eds.). Academic Press, London.

Janvier, P. 1981. The phylogeny of the *Craniata*, with particular reference to the significance of fossil "agnathans." *J. Vertebr. Paleontol.* 1(2):121–159.

Janvier, P. 1996. *Early Vertebrates*. Clarendon Press, Oxford.

Janvier, P., and A. Blieck. 1979. New data on the internal anatomy of the *Heterostraci* (*Agnatha*), with general remarks on the phylogeny of the *Craniata*. *Zool. Scr.* 8:287–296.

Kuraku, S., D. Hoshiyama, K. Katoh, H. Sug, and T. Miyata. 1999. Monophyly of lampreys and hagfishes supported by nuclear DNA-coded genes. *J. Mol. Evol.* 49:729–735.

Løvtrup, S. 1977. *The Phylogeny of Vertebrata*. John Wiley & Sons, New York.

Maisey, J. G. 1986. Heads and tails: a chordate phylogeny. *Cladistics* 2:201–256.

Mallatt, J., and J. Sullivan. 1998. 28S and 18S rDNA sequences support the monophyly of lampreys and hagfishes. *Mol. Biol. Evol.* 15:1706–1718.

Near, T. J. 2009. Conflict and resolution between phylogenies inferred from molecular and phenotypic data sets for hagfish, lampreys, and gnathostomes. *J. Exp. Zool. (Mol. Dev. Evol.)* 312B:749–761.

Sansom, I. J., P. C. J. Donoghue, and G. Albanesi. 2005. Histology and affinity of the earliest armoured vertebrate. *Biol. Lett.* 1:446–449.

Stock, D. W., and G. S. Whitt. 1992. Evidence from 18S ribosomal RNA sequence that lampreys and hagfishes form a natural group. *Science* 257:787–789.

Yu, S. Y., W. W. Zhang, L. Li, H. F. Huang, F. Ma, and Q. W. Li. 2008. Phylogenetic analysis of 48 gene families revealing relationships between hagfishes, lampreys, and *Gnathostomata*. *J. Genet. Genom.* 35:285–290.

Author

Jason P. Downs; Delaware Valley University; 700 East Butler Avenue, Doylestown, PA 18901, USA. Email: jason.downs@delval.edu.

Date Accepted: 29 April 2015

Primary Editor: Jacques A. Gauthier

Apo-Gnathostomata J. P. Downs, new clade name

Registration Number: 254

Definition: The clade originating in the first species of *Pan-Gnathostomata* to exhibit jaws synapomorphic with those in *Scyliorhinus canicula* Linnaeus 1758 (*Chondrichthyes*). 'Jaws' are here defined as opposing upper (palatoquadrate) and lower (Meckel's cartilage) endoskeletal mouth parts (homologous with components of the first pharyngeal arch in jawless vertebrates), and any associated dermoskeletal elements (e.g., gnathals), that are capable of producing an oral bite. This is an apomorphy-based definition. Abbreviated definition: ∇ apo jaws [*Scyliorhinus canicula* Linnaeus 1758].

Etymology: Derived from the Greek *apo-* (away, after), in reference to the apomorphy-based definition, and *Gnathostomata*. See *Gnathostomata* (this volume) for the etymology of that name.

Reference Phylogeny: Brazeau (2009: Fig. 3A), in which jaws are hypothesized to have originated along the branch subtending the node marked with a black dot.

Composition: In addition to *Gnathostomata* (a crown clade; see entry in this volume), *Apo-Gnathostomata* includes the extinct paraphylum "placoderms".

Diagnostic Apomorphies: Diagnostic apomorphies of *Apo-Gnathostomata* include jaws as defined above (see Definition), segmented branchial arches lying internally to the blood and nerve supply and the gill lamellae (Donoghue et al., 2000), paired nasal sacs (Young, 2008), a third (horizontal) semicircular canal (Janvier and Blieck, 1979), a mineralized axial endoskeleton (Donoghue et al., 2000), paired pelvic appendages (Young, 2008), and separate endoskeletal pectoral and pelvic girdles and fin skeletons (Donoghue et al., 2000).

Synonyms: *Gnathostomata* and the informal name 'gnathostomes,' though not explicitly defined phylogenetically, have been used as synonyms to the extent that the character 'jaws' was conceptualized similarly (e.g., Janvier and Blieck, 1979; Forey, 1980, 1984, 1995; Schaeffer and Thomson, 1980; Janvier 1981, 1996a,b, 1998; Gardiner, 1984; Maisey, 1986; Young, 1986; Forey and Janvier, 1993; Goujet, 2001; Goujet and Young, 2004).

Comments: In the interests of promoting an integrated system of taxonomy (de Queiroz, 2007), *Apo-Gnathostomata* replaces the original (Gegenbaur, 1874) and enduring (e.g., Janvier and Blieck, 1979; Forey, 1980, 1984, 1995; Schaeffer and Thomson, 1980; Janvier, 1981, 1996a, 1996b, 1998; Gardiner, 1984; Maisey, 1986; Young, 1986; Forey and Janvier, 1993; Goujet, 2001; Goujet and Young, 2004) use of *Gnathostomata* to refer to all vertebrates with jaws. The taxon name *Gnathostomata*, as defined in this volume, refers to the crown clade only. According to both contemporary (Brazeau, 2009) and traditional (Schaeffer, 1975, 1981; Miles and Young, 1977; Schaeffer and Williams, 1977; Denison, 1978; Rosen et al., 1981; Zangerl, 1981; Reif, 1982; Young, 1986; Goujet and Young, 1995, 2004) views of vertebrate phylogeny, *Apo-Gnathostomata* refers to a more inclusive clade.

Literature Cited

Brazeau, M. D. 2009. The braincase and jaws of a Devonian 'acanthodian' and modern gnathostome origins. *Nature* 457:305–308.

Denison, R. H. 1978. *Placodermi. Handbook of Paleoichthyology*, Vol. 2 (H.-P. Schultze, ed.). Gustav Fischer Verlag, Stuttgart.

Donoghue, P. C. J., P. L. Forey, and R. J. Aldridge. 2000. Conodont affinity and chordate phylogeny. *Biol. Rev.* 75:191–251.

Forey, P. L. 1980. *Latimeria*: a paradoxical fish. *Proc. R. Soc. Lond. B Biol. Sci.* 208:369–384.

Forey, P. L. 1984. Yet more reflections on agnathan-gnathostome relationships. *J. Vertebr. Paleontol.* 4(3):330–343.

Forey, P. L. 1995. Agnathans recent and fossil, and the origin of jawed vertebrates. *Rev. Fish Biol. Fisher.* 5:267–303.

Forey, P. L., and P. Janvier. 1993. Agnathans and the origin of jawed vertebrates. *Nature* 361:129–134.

Gardiner, B. G. 1984. The relationships of placoderms. *J. Vertebr. Paleontol.* 4(3):379–195.

Gegenbaur, C. 1874. *Grundriss der Vergleichenden Anatomie*. Wilhelm Engelmann, Leipzig.

Goujet, D. 2001. Placoderms and basal gnathostome apomorphies. Pp. 209–222 in *Major Events in Early Vertebrate Evolution: Palaeontology, Phylogeny, Genetics, and Development. Systematics Association Series*, Vol. 61 (P. E. Ahlberg, ed.). Taylor & Francis, London.

Goujet, D., and G. C. Young. 1995. Interrelationships of placoderms revisited. *Geobios Mém. Spéc.* 19:89–95.

Goujet, D., and G. C. Young. 2004. Placoderm anatomy and phylogeny: new insights. Pp. 109–126 in *Recent Advances in the Origin and Early Radiation of Vertebrates* (G. Arratia, M. V. H. Wilson, and R. Clouthier, eds.). Verlag Dr. Friedrich Pfeil, München.

Janvier, P. 1981. The phylogeny of the *Craniata*, with particular reference to the significance of fossil "agnathans." *J. Vertebr. Paleontol.* 1(2):121–159.

Janvier, P. 1996a. The dawn of the vertebrates: characters versus common ascent in the rise of current vertebrate phylogenies. *Palaeontology* 39(2):259–287.

Janvier, P. 1996b. *Early Vertebrates*. Clarendon Press, Oxford.

Janvier, P. 1998. Les Vertébrés avant le Silurien. *Geobios* 30(7):931–950.

Janvier, P., and A. Blieck. 1979. New data on the internal anatomy of the *Heterostraci* (*Agnatha*), with general remarks on the phylogeny of the *Craniata*. *Zool. Scr.* 8:287–296.

Linnaeus, C. 1758. *Systema Naturae Per Regna Tria Naturae, Secundum Classes, Ordines, Genera, Species, cum Characteribus, Differentiis, Synonymis, Locis*. 10th edition, Tome 1–2. Laurentii Salvii, Holmiae (Stockholm).

Maisey, J. G. 1986. Heads and tails: a chordate phylogeny. *Cladistics* 2:201–256.

Miles, R. S., and G. C. Young. 1977. Placoderm interrelationships reconsidered in the light of new ptyctodontids from Gogo, Western Australia. Pp. 123–198 in *Problems in Vertebrate Evolution* (S. M. Andrews, R. S. Miles, and A. D. Walker, eds.). Academic Press, London.

de Queiroz, K. 2007. Toward an integrated system of clade names. *Syst. Biol.* 56:956–974.

Reif, W. E. 1982. Evolution of dermal skeleton and dentition in vertebrates. The odontode regulation theory. *Evol. Biol.* 15:287–368.

Rosen, D. E., P. L. Forey, B. G. Gardiner, and C. Patterson. 1981. Lungfishes, tetrapods, paleontology, and plesiomorphy. *Bull. Am. Mus. Nat. Hist.* 167:159–276.

Schaeffer, B. 1975. Comments on the origin and basic radiation of the gnathostome fishes with particular reference to the feeding mechanism. *Colloq. Int. CNRS* 218:101–109.

Schaeffer, B. 1981. The xenacanth shark neurocranium, with comment on elasmobranch monophyly. *Bull. Am. Mus. Nat. Hist.* 169:1–66.

Schaeffer, B., and K. S. Thomson. 1980. Reflections on agnathan-gnathostome relationships. Pp. 19–33 in *Aspects of Vertebrate History* (L. L. Jacobs, ed.). Museum of Northern Arizona Press, Flagstaff.

Schaeffer, B., and M. Williams. 1977. Relationships of fossil and living elasmobranchs. *Am. Zool.* 17:293–302.

Young, G. C. 1986. The relationships of placoderm fishes. *Zool. J. Linn. Soc.* 88:1–57.

Young, G. C. 2008. Number and arrangement of extraocular muscles in primitive gnathostomes: evidence from extinct placoderm fishes. *Biol. Lett.* 4:110–114.

Zangerl, R. 1981. *Chondrichthyes* I. Paleozoic *Elasmobranchii. Handbook of Paleoichthyology*, Vol. 3A (H.-P. Schultze, ed.). Gustav Fischer Verlag, Stuttgart.

Date Accepted: 29 April 2015

Primary Editor: Jacques A. Gauthier

Author

Jason P. Downs; Delaware Valley University; 700 East Butler Avenue; Doylestown, PA 18901, USA. Email: jason.downs@delval.edu.

Gnathostomata C. Gegenbaur 1874 [J. P. Downs], converted clade name

Registration Number: 255

Definition: The most inclusive crown clade whose members exhibit jaws synapomorphic with those in *Scyliorhinus canicula* Linnaeus 1758 (*Chondrichthyes*). This is an apomorphy-modified crown-clade definition. Abbreviated definition: crown ∇ apo jaws [*Scyliorhinus canicula* Linnaeus 1758].

Etymology: Derived from the Greek *gnathos* (jaw), *stoma* (mouth), and -*ata* (having, provided with).

Reference Phylogeny: Brazeau (2009: Fig. 3A). *Scyliorhinus canicula* is not shown on the reference phylogeny, which focused on extinct species, but is part of the clade labeled *Chondrichthyes*. *Gnathostomata* applies to the node labeled 5–92 and includes *Osteichthyes* and *Chondrichthyes*. *Culmacanthus, Gladiobranchus, Diplacanthus*, and *Tetanopsyrus* belong to *Gnathostomata* only if they fall within the total groups of either *Osteichthyes* or *Chondrichthyes*.

Composition: All members of the crown clade *Chondrichthyes* and its stem group and *Osteichthyes* and its stem group (see *Osteichthyes* and *Pan-Osteichthyes*, this volume). With nearly 60,000 species, *Gnathostomata* contains 99% of all Recent vertebrates.

Diagnostic Apomorphies: Because of the long stem, an enormous number of apomorphies diagnose the crown clade of *Gnathostomata* relative to other vertebrate crown clades. It would be difficult to present all of them here, but the most comprehensive current list can be compiled from the dataset of Maisey (1986).

Even an abbreviated list of skeletal apomorphies that distinguish the members of *Gnathostomata* from those of other crown-clade vertebrates (the presence of a dermal skeleton; a mineralized endoskeleton; paired pectoral and pelvic appendages; the tissues bone, dentine, and enameloid) suggests the enormity of an exhaustive list (Donoghue and Sansom, 2002).

Jaws are traditionally associated with *Gnathostomata*. However, as discussed in *Apo-Gnathostomata* below, jaws in the gnathostome crown constitute a complex apomorphy whose components were acquired along the gnathostome stem. Even jaws bearing teeth with dentine and a centripetally-filled pulp cavity, once considered a synapomorphy of the gnathostome crown, appear prior to the origin of the crown clade (Rücklin et al., 2012; Rücklin and Donoghue, 2015). Indeed, it remains possible that, at some level, homology may be drawn between the teeth of crown gnathostomes and the oral denticles of conodonts (Donoghue et al., 2006) or even the keratinized horny pharyngeal teeth that are apparently ancestral for *Craniata*.

Unambiguous synapomorphies of crown clade *Gnathostomata* relative to known members of the stem group include: the presence of multiple, bony hyoidean gill covers (branchiostegals of Brazeau, 2009); the presence of a large otic process on the palatoquadrate (a "cleaver-shaped" palatoquadrate; Schaeffer, 1975; Schaeffer, 1981; Brazeau, 2009); the presence of a suprapterygoid articulation of the palatoquadrate (Brazeau, 2009); insertion of the rectus eye muscle in a caudal position in the orbit (Young, 1986); insertion of the superior oblique eye muscle in a rostral part of the orbit (Young, 1986); and fusion of the nasal capsule to the rest of the braincase (Young, 1986).

Synonyms: The name *Eugnathostomata* Turner 1970 has been used as an unambiguous synonym of *Gnathostomata* as defined here (Wagner and Larsson, 2007; Turner et al., 2010; Blais et al., 2011). It has also been used as an apomorphy-based name. Robertson (1970) used *Eugnathostomata* to refer to vertebrates with a Meckelian lower jaw, which, according to the author, excluded "placoderms". Blieck et al. (2010) and Blieck (2011) used the name to refer to the clade of vertebrates with both jaws and teeth. As an apomorphy-based name, *Eugnathostomata* of those authors would not explicitly exclude members of the stem group and therefore is not a synonym of the taxon name *Gnathostomata* as defined here.

Comments: By several orders of magnitude, *Gnathostomata* is a far more commonly used name than *Eugnathostomata* and so it is selected for the crown clade over its approximate synonym (3,090 vs. 19 in Google Scholar as of 29 July 2015). In its traditional use, the name *Gnathostomata* has referred to all vertebrates with jaws, a usage that includes "placoderms", currently regarded as a paraphylum of jawed forms on the gnathostome stem (Brazeau, 2009). Examples of the use of *Gnathostomata* to refer to all jawed vertebrates abound in phylogenetic treatments of vertebrates (e.g., Janvier and Blieck, 1979; Forey, 1980, 1984, 1995; Schaeffer and Thomson, 1980; Janvier 1981, 1996a, 1996b, 1998; Gardiner, 1984; Maisey, 1986; Young, 1986; Forey and Janvier, 1993; Goujet, 2001; Goujet and Young, 2004). *Gnathostomata* has also been used to refer to the total clade (Donoghue et al., 2000). The crown clade definition of *Gnathostomata* presented above follows the convention of applying widely known, traditional names to crown clades (de Queiroz and Gauthier, 1992; de Queiroz, 2007). Its use in this context is not novel: Donoghue and

Sansom (2002) explicitly used *Gnathostomata* to refer to the crown clade, noting that the name does not equate to "jawed vertebrates" due to the presence of jaws in near-crown stem gnathostomes.

A contemporary view of vertebrate phylogeny regards "placoderms" as a paraphylum composed of several clades on the gnathostome stem (Johanson, 2002; Brazeau, 2009), and regards "acanthodians" as a polyphylum spread across the bases of the total groups of *Osteichthyes* and *Chondrichthyes*, with others having unresolved relations at the base of *Gnathostomata* (Brazeau, 2009). This differs from the traditional view which regards "placoderms" as a clade on the *Gnathostomata* stem (Schaeffer, 1975, 1981; Miles and Young, 1977; Schaeffer and Williams, 1977; Denison, 1978; Rosen et al., 1981; Zangerl, 1981; Reif, 1982; Young, 1986; Goujet and Young, 1995, 2004), and regards "acanthodians" as a clade on the *Osteichthyes* stem (Miles, 1973; Schaeffer and Williams, 1977; Reif, 1982). Historically, numerous other configurations have been proposed, many of which placed all jawed vertebrates within the crown. Early phylogenetic reconstructions placed "placoderms" within the total group of *Chondrichthyes* (Quenstedt, 1838; Holmgren, 1942; Jarvik, 1955, 1980; Ørvig, 1957, 1962, 1967; Stensiö, 1959, 1963; Moy-Thomas and Miles, 1971; Miles and Young, 1977; Goujet, 1982; Janvier, 1996b) or within the total group of *Osteichthyes* (McCoy, 1848; Huxley, 1861; Newberry, 1875; Woodward, 1891; Jaeckel, 1911, 1919; Forey, 1980; Gardiner, 1984; Maisey, 1986). "Acanthodians" have been placed on the *Gnathostomata* stem (Forey, 1980; Rosen et al., 1981) or within the total clade of *Chondrichthyes* (Holmgren, 1942; Quenstedt, 1852; Woodward, 1891; Jarvik, 1955, 1977; Ørvig, 1957; Nelson, 1969). In some of these alternative configurations, the name *Gnathostomata* referred exclusively to members of the crown clade simply because all vertebrates that possessed jaws were hypothesized to belong to the crown.

Competing theories regarding the phylogenetic status and position of the extinct groups "acanthodians" and "placoderms" have no effect on the above-stated theoretical composition of crown clade *Gnathostomata* (although they do affect whether fossils referred to those taxa are inferred to be included in the crown, and alter its diagnosis accordingly). Recognition of a basal split between *Chondrichthyes*-line (panchondrichthyan) and *Osteichthyes*-line (panosteichthyan) gnathostomes has a long history (Goodrich, 1909). By using the above crownclade definition of *Gnathostomata*, the composition of the clade under all phylogenetic schemes includes only these two total crown clades.

Literature Cited

Blais, S. A., L. A. MacKenzie, and M. V. Wilson. 2011. Tooth-like scales in Early Devonian eugnathostomes and the 'outside-in' hypothesis for the origins of teeth in vertebrates. *J. Vertebr. Paleontol.* 31(6):1189–1199.

Blieck, A. 2011. The André Dumont medallist lecture: from adaptive radiations to biotic crises in Palaeozoic vertebrates: a geobiological approach. *Geol. Belg.* 14(3–4):203–227.

Blieck, A., S. Turner, C. J. Burrow, H.-P. Schultze, C. B. Rexroad, P. Bultynck, and G. S. Nowlan. 2010. Fossils, histology, and phylogeny: why conodonts are not vertebrates. *Episodes* 33(4):234–241.

Brazeau, M. D. 2009. The braincase and jaws of a Devonian 'acanthodian' and modern gnathostome origins. *Nature* 457:305–308.

Denison, R. H. 1978. *Placodermi*. In *Handbook of Paleoichthyology*, Vol. 2 (H.-P. Schultze, ed.). Gustav Fischer Verlag, Stuttgart.

Donoghue, P. C. J., P. L. Forey, and R. J. Aldridge. 2000. Conodont affinity and chordate phylogeny. *Biol. Rev.* 75:191–251.

Donoghue, P. C. J., and I. J. Sansom. 2002. Origin and early evolution of vertebrate skeletonization. *Microsc. Res. Techn.* 59:352–372.

Donoghue, P. C., I. J. Sansom, and J. P. Downs. 2006. Early evolution of vertebrate skeletal tissues and cellular interactions, and the canalization of skeletal development. *J. Exp. Zool. (Mol. Dev. Evol.)* 306:278–94.

Forey, P. L. 1980. *Latimeria*: a paradoxical fish. *Proc. R. Soc. Lond. B Biol. Sci.* 208:369–384.

Forey, P. L. 1984. Yet more reflections on agnathan-gnathostome relationships. *J. Vertebr. Paleontol.* 4(3):330–343.

Forey, P. L. 1995. Agnathans recent and fossil, and the origin of jawed vertebrates. *Rev. Fish Biol. Fisher.* 5:267–303.

Forey, P. L., and P. Janvier. 1993. Agnathans and the origin of jawed vertebrates. *Nature* 361:129–134.

Gardiner, B. G. 1984. The relationships of placoderms. *J. Vertebr. Paleontol.* 4(3):379–195.

Gegenbaur, C. 1874. *Grundriss der Vergleichenden Anatomie*. Wilhelm Engelmann, Leipzig.

Goodrich, E. S. 1909. Part IX. *Vertebrata Craniata* (First Fascicle: Cyclostomes and Fishes). P. 518 in *A Treatise on Zoology* (R. Lankester, ed.). Adam and Charles Black, London.

Goujet, D. 1982. Les affinités des placodermes, une revue des hypothèses actuelles. *Geobios Mém. Spéc.* 6:27–38.

Goujet, D. 2001. Placoderms and basal gnathostome apomorphies. Pp. 209–222 in *Major Events in Early Vertebrate Evolution: Palaeontology, Phylogeny, Genetics, and Development. Systematics Association Series*, Vol. 61 (P. E. Ahlberg, ed.). Taylor & Francis, London.

Goujet, D., and G. C. Young. 1995. Interrelationships of placoderms revisited. *Geobios, Mémoire Spécial* 19:89–95.

Goujet, D., and G. C. Young. 2004. Placoderm anatomy and phylogeny: new insights. Pp. 109–126 in *Recent Advances in the Origin and Early Radiation of Vertebrates* (G. Arratia, M. V. H. Wilson, and R. Clouthier, eds.). Verlag Dr. Friedrich Pfeil, Munich.

Holmgren, N. 1942. Studies on the head of fishes: an embryological, morphological, and phylogenetic study. Part III. The phylogeny of elasmobranch fishes. *Acta Zool.* 23:129–261.

Huxley, T. H. 1861. Preliminary essay upon the systematic arrangement of the fishes of the

Devonian epoch. *Mem. Geol. Surv. UK* 10:1–40.

Jaeckel, O. 1911. *Die Wirbeltiere, Eine Übersicht Über die Fossilen und Lebenden Formen.* Borntraeger, Berlin.

Jaeckel, O. 1919. Die Mundbildung der Placodermen. *Sitzungsber. d. Ges. Naturforsch. Fr. Berlin* 1919:73–110.

Janvier, P. 1981. The phylogeny of the *Craniata*, with particular reference to the significance of fossil "agnathans." *J. Vertebr. Paleontol.* 1(2):121–159.

Janvier, P. 1996a. The dawn of the vertebrates: characters versus common ascent in the rise of current vertebrate phylogenies. *Palaeontology* 39(2):259–287.

Janvier, P. 1996b. *Early Vertebrates.* Clarendon Press, Oxford.

Janvier, P. 1998. Les Vertébrés avant le Silurien. *Geobios* 30(7):931–950.

Janvier, P., and A. Blieck. 1979. New data on the internal anatomy of the *Heterostraci* (*Agnatha*), with general remarks on the phylogeny of the *Craniata. Zool. Scr.* 8:287–296.

Jarvik, E. 1955. The oldest tetrapods and their forerunners. *Sci. Mon.* 80:141–154.

Jarvik, E. 1977. The systematic position of acanthodian fishes. Pp. 199–225 in *Problems in Vertebrate Evolution* (S. M. Andrews, R. S. Miles, and A. D. Walker, eds.). Academic Press, London.

Jarvik, E. 1980. *Basic Structure and Evolution of the Vertebrates*, Vol. 1. Academic Press, London.

Johanson, Z. 2002. Vascularization of the osteostracan and antiarch (*Placodermi*) pectoral fin: similarities and implications for placoderm relationships. *Lethaia* 35:169–186.

Linnaeus, C. 1758. *Systema Naturae Per Regna Tria Naturae, Secundum Classes, Ordines, Genera, Species, Cum Characteribus, Differentiis, Synonymis, Locis.* 10th edition, Tomes 1–2. Laurentii Salvii, Holmiae (Stockholm).

Maisey, J. G. 1986. Heads and tails: a chordate phylogeny. *Cladistics* 2:201–256.

McCoy, F. 1848. On some new fossil fish of the Carboniferous Period. *Ann. Mag. Nat. Hist., Second Series* 2:1–10.

Miles, R. S. 1973. Relationships of acanthodians. Pp. 63–104 in *Interrelationships of Fishes* (P. H. Greenwood, R. S. Miles, and C. Patterson, eds.). Academic Press, London.

Miles, R. S., and G. C. Young. 1977. Placoderm interrelationships reconsidered in the light of new ptyctodontids from Gogo, Western Australia. Pp. 123–198 in *Problems in Vertebrate Evolution* (S. M. Andrews, R. S. Miles, and A. D. Walker, eds.). Academic Press, London.

Moy-Thomas, J. A., and R. S. Miles. 1971. *Paleozoic Fishes.* 2nd edition. Chapman and Hall, London.

Nelson, G. J. 1969. Gill arches and the phylogeny of fishes, with notes on the classification of vertebrates. *Bull. Am. Mus. Nat. Hist.* 141:475–552.

Newberry, J. S. 1875. Descriptions of fossil fishes. *Rep. Geol. Surv. Ohio* 2(2):1–64.

Ørvig, T. 1957. Notes on some Palaeozoic lower vertebrates from Spitzbergen and North America. *Nor. Geol. Tidsskr.* 37(3–4):285–353.

Ørvig, T. 1962. Y a-t-il une relation directs entre les arthrodires ptyctodontides et les holocéphales? *Colloq. Int. CNRS* 104:49–61.

Ørvig, T. 1967. Phylogeny of tooth tissues: evolution of some calcified tissues in early vertebrates. Pp. 45–110 in *Structural and Chemical Organization of Teeth*, Vol. 1. (A. E. W. Miles, ed.). Academic Press, New York.

de Queiroz, K. 2007. Toward an integrated system of clade names. *Syst. Biol.* 56:956–974.

de Queiroz, K., and J. Gauthier. 1992. Phylogenetic taxonomy. *Annu. Rev. Ecol. Evol. Syst.* 23:449–480.

Quenstedt, F. A. 1838. Über die fossilen knochen im rothen Sandsteine Livland's und Esthland's. *Neues. Jahrb. Mineral. Geog. Geol. Petref.* 1838:12–16.

Quenstedt, F. A. 1852. *Handbuch der Petrefaktenkunde.* Laupp & Siebeck, Tübingen.

Reif, W. E. 1982. Evolution of dermal skeleton and dentition in vertebrates. The odontode regulation theory. *Evol. Biol.* 15:287–368.

Robertson, G. M. 1970. The oral region of ostracoderms and placoderms: possible phylogenetic significance. *Am. J. Sci.* 269:39–64.

Rosen, D. E., P. L. Forey, B. G. Gardiner, and C. Patterson. 1981. Lungfishes, tetrapods, paleontology, and plesiomorphy. *Bull. Am. Mus. Nat. Hist.* 167:159–276.

Rücklin, M., P. C. J. Donoghue, Z. Johanson, K. Trinajstic, F. Marone, and M. Stampanoni. 2012. Development of teeth and jaws in the earliest jawed vertebrates. *Nature* 491:748–754.

Rücklin, M., and P. C. J. Donoghue. 2015. *Romundina* and the evolutionary origin of teeth. *Biol. Lett.* 11:20150326.

Schaeffer, B. 1975. Comments on the origin and basic radiation of the gnathostome fishes with particular reference to the feeding mechanism. *Colloq. Int. CNRS* 218:101–109.

Schaeffer, B. 1981. The xenacanth shark neurocranium, with comment on elasmobranch monophyly. *Bull. Am. Mus. Nat. Hist.* 169:1–66.

Schaeffer, B., and K. S. Thomson. 1980. Reflections on agnathan-gnathostome relationships. Pp. 19–33 in *Aspects of Vertebrate History* (L. L. Jacobs, ed.). Museum of Northern Arizona Press, Flagstaff.

Schaeffer, B., and M. Williams. 1977. Relationships of fossil and living elasmobranchs. *Am. Zool.* 17:293–302.

Stensiö, E. A. 1959. On the pectoral fin and shoulder girdle of the arthrodires. *K. Sven. Vetenskakad. Handl., Serie 4* 8:1–229.

Stensiö, E. A. 1963. Anatomical studies on the arthrodiran head. Part 1. Preface, geological and geographical distribution, the organisation of the head in the *Dolichothoraci, Coccosteomorphi* and *Pachyosteomorphi.* Taxonomic appendix. *K. Sven. Vetensk.akad. Handl., Serie 4* 9:1–419.

Turner, S., C. J. Burrow, H.-P. Schultze, A. Blieck, W.-E. Reif, C. B. Rexroad, P. Bultynck, and G. S. Nowlan. 2010. False teeth: conodont vertebrate phylogenetic relationships revisited. *Geodiversitas* 32(4):545–594.

Wagner, G. P., and H. C. Larsson. 2007. Fins and limbs in the study of evolutionary novelties. Pp. 49–61 in *Fins into Limbs: Evolution, Development, and Transformation* (B. K. Hall, ed.). The University of Chicago Press, Chicago, IL.

Woodward, A. S. 1891. *Catalogue of the Fossil Fishes in the British Museum (Natural History), Part II.* Taylor & Francis, London.

Young, G. C. 1986. The relationships of placoderm fishes. *Zool. J. Linn. Soc.* 88:1–57.

Zangerl, R. 1981. *Chondrichthyes* I. Paleozoic *Elasmobranchii.* In *Handbook of Paleoichthyology*, Vol. 3A (H.-P. Schultze, ed.). Gustav Fischer Verlag, Stuttgart.

Author

Jason P. Downs; Delaware Valley University; 700 East Butler Avenue, Doylestown, PA 18901, USA. Email: jason.downs@delval.edu.

Date Accepted: 29 April 2015

Primary Editor: Jacques A. Gauthier

Pan-Osteichthyes T. Rowe 2004 [J. A. Moore and T. J. Near], converted clade name

Registration Number: 215

Definition: The total clade of the crown clade *Osteichthyes*. This is a crown-based total-clade definition. Abbreviated definition: total ∇ of *Osteichthyes*.

Etymology: Derived from the Greek *pantos* (all), indicating that the name refers to a total clade or pan-monophylum, plus *Osteichthyes*, the name of the corresponding crown clade (see the entry for *Osteichthyes* in this volume for the etymology of that name).

Reference Phylogeny: Brazeau (2009: Fig. 3, where *Pan-Osteichthyes* is designated as *Osteichthyes* Stem + Crown) is the primary reference phylogeny, see also Zhu and Schultze (2001: Fig. 17.4c, where *Pan-Osteichthyes* corresponds roughly with the clade designated as *Osteichthyes*).

Composition: *Osteichthyes* (this volume) and all extinct fishes that share a more recent ancestry with *Osteichthyes* than with *Chondrichthyes*. According to the results of Brazeau (2009), many of the acanthodians (including *Acanthodes*) are stem osteichthyans. The anomalous bony fish *Psarolepis romeri* may represent a member of the osteichthyan stem group as shown in some phylogenies (Ahlberg, 1999; Zhu et al., 1999; Zhu and Schultze, 2001). *Dialipina, Ligulalepis, Andreosteus,* and *Lophosteus* have also been proposed as stem *Osteichthyes* (Friedman, 2007; Brazeau, 2009; Zhu et al. 2009; Johanson et al. 2010). See Comments regarding the possibility that placoderms are stem osteichthyans.

Diagnostic Apomorphies: The characters listed here did not necessarily arise immediately at the origin of the *Pan-Osteichthyes*, but instead arose early in the history of the stem lineage leading to the crown *Osteichthyes*. These features are useful in recognizing fossils as members of the total clade. Janvier (1996) proposed the presence of a narrow-based braincase and 3 pairs of otoliths as uniting acanthodians and *Osteichthyes*. Miles (1973) proposed the presence of dermal branchiostegal rays, pattern of perichondral ossification of the neurocranium, and several other features in common between the acanthodian *Acanthodes* and osteichthyans. Maisey (1986, 1994) pointed out the structure of the hemibranchs and interbranchial septum as additional evidence allying acanthodians with osteichthyans. Friedman and Brazeau (2010) recently listed 7 characters that united *Acanthodes* with crown *Osteichthyes*.

Synonyms: The more inclusive version of *Teleostomi* (Gardiner, 1973; Nelson, 2006; Helfman et al., 2009; Long, 2011; see Comments on *Osteichthyes* in this volume), which includes acanthodians, is an approximate synonym, as is *Osteichthyes* sensu Zhu and Schultze (2001: Fig 17.4c), Brazeau (2009: Fig. 3), and Zhu et al. (2009: Fig. 5). *Ganoidei* of Agassiz (1833–1844) is a partial and approximate synonym of *Pan-Osteichthyes* (as it included all of the early osteichthyans and their presumed relatives, but excluded tetrapods and most teleosts). Similarly, *Ganoidei* of Huxley (1861) is partial and approximate synonyms given the inclusion of acanthodians. *Osteichthyes* of Friedman and Brazeau (2010), who explicitly adopted a total-clade definition, is an unambiguous synonym.

Comments: We have chosen to follow Rowe (2004) in using the name *Pan-Osteichthyes* for total clade composed of the crown clade *Osteichthyes* plus its stem group. Rowe (2004) included the placoderms within *Pan-Osteichthyes* following Gardiner (1984). Most other workers (e.g., Burrow and Turner, 1999; Goujet, 2001; Brazeau, 2009; Friedman and Brazeau, 2010) have followed Young (1986, 2010) in placing placoderms as a monophyletic group of stem gnathostomes, although Young (2010) also considered the possibilities that placoderms are a monophyletic group of stem chondrichthyans or a monophyletic group of stem osteichthyans. As another alternative, Brazeau (2009) showed placoderms as a paraphyletic assemblage making up much of the stem group of *Gnathostomata*.

The name *Osteichthyes* has previously been applied to the total clade whose name is defined in this entry (see Synonyms), but we have chosen to apply that name to the corresponding crown clade (to which it has also been previously applied). Another approximate synonym, *Teleostomi*, has been used to denote the clade *Osteichthyes* plus *Acanthodii* (Gardiner, 1973; Nelson, 2006; Helfman et al., 2009; Long, 2011) and could be defined phylogenetically as applying to that clade (a subclade of *Pan-Osteichthyes*). In any case, it seems useful to have a name for the (total) clade that includes all extinct forms that are more closely related to (crown) osteichthyans than to (crown) chondrichthyans, especially given that acanthodians may not be a single monophyletic group (Brazeau, 2009).

Literature Cited

Agassiz, J. L. R. 1833–1844. *Recherches sur les Poissons Fossiles*. Neuchatel, Switzerland.

Ahlberg, P. E. 1999. Something fishy in the family tree. *Nature* 397:564–565.

Brazeau, M. D. 2009. The braincase and jaws of a Devonian 'acanthodian' and modern gnathostome origins. *Nature* 457:305–308.

Burrow, C. J., and S. Turner. 1999. A review of placoderm scales, and their significance in placoderm phylogeny. *J. Vertebr. Paleontol.* 19(2):204–219.

Friedman, M. 2007. *Styloichthys* as the oldest coelacanth: implications for early osteichthyan interrelationships. *J. Syst. Palaeontol.* 5, 289–343.

Friedman, M., and M. D. Brazeau. 2010. A reappraisal of the origin and basal radiation of the *Osteichthyes*. *J. Vertebr. Paleontol.* 30:36–56.

Gardiner, B. G. 1973. Interrelationships of teleostomes. Pp. 105–135 in *Interrelationships of Fishes* (P. H. Greenwood, R. S. Miles, and C. Patterson, eds.). Academic Press, London.

Gardiner, B. G. 1984. The relationship of placoderms. *J. Vertebr. Paleontol.* 4(3):379–395.

Goujet, D. 2001. Placoderms and basal gnathostome apomorphies. Pp. 209–222 in *Major Events in Early Vertebrate Evolution: Paleontology, Phylogeny, Genetics, and Development* (P. E. Ahlberg, ed.). Taylor & Francis, London.

Helfman, G. S., B. B. Collette, D. E. Facey, and B. W. Bowen. 2009. *The Diversity of Fishes. Biology, Evolution, and Ecology.* 2nd edition. Wiley-Blackwell, Hoboken, NJ.

Huxley, T. H. 1861. Preliminary essay upon the systematic arrangement of the fishes of the Devonian Epoch. Pp. 1–40 in *Memoirs of the Geological Survey of the United Kingdom*, Decade X. Longman, Green, Longman, and Roberts, London.

Janvier, P. 1996. *Early Vertebrates.* Oxford University Press, Oxford.

Johanson, Z., A. Kearsley, J. den Blaauwen, M. Newman, and M. M. Smith. 2010. No bones about it: an enigmatic Devonian fossil reveals a new skeletal framework—a potential role of loss of gene regulation. *Semin. Cell Dev. Biol.* 21:414–423.

Long, J. A. 2011. *The Rise of Fishes: 500 Million Years of Evolution.* Johns Hopkins Press, Baltimore, MD.

Maisey, J. G. 1986. Heads and tails: a chordate phylogeny. *Cladistics* 2:201–256.

Maisey, J. G. 1994. Gnathostomes (jawed fishes). Pp. 38–56 in *Major Features of Vertebrate Evolution* (D. R. Prothero and R. M. Schoch, eds.). Short Courses in Paleontology, No. 7. Paleontological Society, University of Tennessee, Knoxville, TN.

Miles, R. S. 1973. Relationships of acanthodians. Pp. 63–103 in *Interrelationships of Fishes* (P. H. Greenwood, R. S. Miles, and C. Patterson, eds.). Academic Press, London.

Nelson, J. 2006. *Fishes of the World*. 4th edition. John Wiley & Sons, New York.

Rowe, T. 2004. Chordate phylogeny and development. Pp. 384–409 in *Assembling the Tree of Life* (J. Cracraft and M. J. Donoghue, eds.). Oxford University Press, Oxford.

Young, G. C. 1986. The relationships of placoderm fishes. *Zool. J. Linn. Soc.* 88:1–56.

Young, G. C. 2010. Placoderms (armored fishes): dominant vertebrates of the Devonian period. *Annu. Rev. Earth. Planet. Sci.* 38:523–550.

Zhu, M., and H. P. Schultze. 2001. Interrelationships of basal osteichthyans. Pp. 289–314 in *Major Events in* Early *Vertebrate Evolution. Paleontology, phylogeny, genetics, and development* (P. E. Ahlberg, ed.). Taylor & Francis, London.

Zhu, M., X. Yu, and P. Janvier. 1999. A primitive fossil fish sheds light on the origin of bony fishes. *Nature* 397:607–610.

Zhu, M., W. Zhao, L.Jia, J. Lu, T. Qiao, and Q. Qu. 2009. The oldest articulated osteichthyan reveals mosaic gnathostome characters. *Nature* 458:469–474.

Authors

Jon A. Moore; Wilkes Honors College; Florida Atlantic University; Jupiter, FL 33458, USA. Email: jmoore@fau.edu.

Thomas J. Near; Department of Ecology and Evolutionary Biology; Yale University; New Haven, CT 06520, USA. Email: Thomas. Near@yale.edu.

Date Accepted: 12 August 2013

Primary Editor: Kevin de Queiroz

Osteichthyes T. H. Huxley 1880 [J. A. Moore and T. J. Near], converted clade name

Registration Number: 214

Definition: The least inclusive crown clade that contains *Latimeria chalumnae* Smith 1939, *Neoceratodus forsteri* (Krefft 1870), *Polypterus bichir* Lacepede 1803, *Acipenser sturio* Linnaeus 1758, *Lepisosteus osseus* (Linnaeus 1758), *Amia calva* Linnaeus 1766, and *Perca fluviatilis* Linnaeus 1758. This is a minimum-crown-clade definition. Abbreviated definition: min crown ∇ (*Latimeria chalumnae* Smith 1939 & *Neoceratodus forsteri* (Krefft 1870) & *Polypterus bichir* Lacepede 1803 & *Acipenser sturio* Linnaeus 1758 & *Lepisosteus osseus* (Linnaeus 1758) & *Amia calva* Linnaeus 1766 & *Perca fluviatilis* Linnaeus 1758).

Etymology: Derived from the Greek *osteon* (bone) and *ichthys* (fishes).

Reference Phylogeny: Diogo (2007: Figs. 3 and 4) is the primary reference phylogeny, but see also Brazeau (2009: Fig. 3), Stiassny et al. (2004: Fig. 24.1), Rowe (2004: Fig. 23.1) Rosen et al. (1981: Fig. 62), Maisey (1986: Fig. 5), Patterson (1994: Fig. 1), and Zhu and Schultze (2001: Fig. 17.4). Although *Perca fluviatilis* is not included in the primary reference phylogeny, available evidence indicates that it is deeply nested within the clade originating in branch 6 on that phylogeny (Dettai and Lecointre, 2008; Li et al., 2009; Wiley and Johnson, 2010; Near et al., 2012).

Composition: *Osteichthyes* is currently thought to include about 53,000 living species, roughly equally divided between its two primary sub-clades persisting today as the crown clades *Actinopterygii* (this volume) and *Sarcopterygii* (including *Tetrapoda*) (Gardiner, 1984; Stiassny et al., 2004). *Osteichthyes* thus includes all extant reptiles (including birds) listed in Cogger and Zweifel (1998) and Clements (2007), mammals listed in Macdonald (2009), amphibians listed by Frost (2009), and all living actinopterygian and sarcopterygian fishes listed by Eschmeyer et al. (1998) and Eschmeyer (2011). Partial lists of extinct *Osteichthyes* can be found in Janvier (1996) and in Benton (2005).

Diagnostic Apomorphies: Maisey (1986) listed and discussed 18 characters for *Osteichthyes*. Some of the more notable apomorphies include: presence of a swim bladder or lung, endochondral bone (however Maisey considered the possibility that this feature might instead be apomorphic for *Gnathostomata*), lepidotrichia in fins (lost in tetrapods), and the oral margin made of dentary, premaxillary and maxillary bones. Springer and Johnson (2004) stated that the presence of gill arch levator muscles is an apomorphy for *Osteichthyes* (their *Teleostomi*). Rowe (2004) highlighted several characters, particularly of the lungs, as features of *Osteichthyes*. Friedman and Brazeau (2010) listed and discussed 17 characters that appear to be apomorphies of the crown *Osteichthyes*.

Synonyms: *Teleostomi* of Gill (1892), Gregory (1906), and Berg (1947), *Teleostomes* of Gill (1893), and *Pisces* of Jordan (1923) and Regan (1929) are partial (and approximate) synonyms of *Osteichthyes* in that they were used for a paraphyletic group excluding tetrapods that originated in approximately the same ancestor. *Teleostomi* of Owen (1866) and Woodward (1891) and *Hyopomata* of Cope (1877) are also partial synonyms that excluded both dipnoans and tetrapods. Inclusion of tetrapods and therefore a monophyletic *Teleostomi* is found in

Nelson (1969: Fig. 25), Wiley (1979), and Springer and Johnson (2004) so that the name as used by those authors is a full (if still approximate) synonym. *Crossopterygidae* of Huxley (1861) is a partial and approximate synonym. Nelson (1994, 2006) proposed using *Euteleostomi* instead of *Osteichthyes*, because the historical, pre-cladistic use of the name *Osteichthyes* excluded the tetrapods; *Euteleostomi* is thus another approximate synonym of *Osteichthyes*.

Comments: Many early authors used the name *Osteichthyes* in a paraphyletic sense that excluded tetrapods (Garstang, 1931; Schaeffer, 1968). Exclusion of tetrapods was justified in the context of an older paradigm that emphasized degree of difference. We have chosen to use the name *Osteichthyes* for a monophyletic group originating in the same ancestor, because of its historical and widespread use among many scientists and the public. Granted, some of those individuals still conceive of *Osteichthyes* as the paraphyletic group including only the "bony fishes," but as Hennig's principle of monophyly has become more pervasive, more individuals are becoming familiar with the concept of *Osteichthyes* that includes tetrapods (Shubin, 2008). Forey (1980) and Rosen et al. (1981) first used *Osteichthyes* in the monophyletic sense, explicitly including the tetrapods within the sarcopterygians.

Several authors have used the name *Teleostomi* as a synonym of *Osteichthyes* (see Synonyms); however, other authors have used *Teleostomi* for a more inclusive group of acanthodians + osteichthyans (Moy-Thomas and Miles, 1971; Gardiner, 1973; Nelson, 2006; Helfman et al., 2009; Long, 2011), and the name could be defined as applying to that clade (but see Brazeau, 2009, and Davis et al., 2012, concerning the non-monophyly of "acanthodians"). Other groups have sometimes been included within *Osteichthyes*. For example, Regan (1904) and Goodrich (1909) included not just the

non-tetrapod bony fishes, but also the placoderms (including ostracoderms, in the case of Regan 1904). The name *Euteleostomi* has seen more limited use and remains an unfamiliar term. The more inclusive application of the name *Teleostomi* has continued into the modern literature, whereas the more inclusive application of *Osteichthyes* faded from use over 100 years ago and the name *Osteichthyes* has come to mean the combined group of actinopterygians and sarcopterygians (including tetrapods) in present day literature. For these reasons, we selected *Osteichthyes* as the name of the clade composed of actinopterygians and sarcopterygians.

The name *Osteichthyes* historically referred to those "fishes" with the presence of endochondral ossification within the braincase and axial skeleton, i.e., "bony fishes". An apomorphy-based definition might therefore seem appropriate for the name of such a clade. But if Maisey's (1986) suggestion that endochondral bone is found in chondrichthyans holds true, then an apomorphy-based definition would make *Osteichthyes* a more inclusive clade that would include chondrichthyans. We have elected to restrict *Osteichthyes* to a crown clade (using a minimum-crown-clade definition) because such clades have ancestors that we can know the most about (de Queiroz and Gauthier, 1992; Gauthier and de Queiroz, 2001).

Noack et al. (1996), Zardoya and Meyer (2001), Meyer and Zardoya (2003), Kikugawa et al. (2004), and Alfaro et al. (2009) have provided molecular evidence supporting the monophyly of the crown clade *Osteichthyes*.

Literature Cited

Alfaro, M. E., F. Santini, C. Brock, H. Alamillo, A. Dornburg, D. L. Rabosky, G. Carnevale, and L. J. Harmon. 2009. Nine exceptional radiations plus high turnover explain species diversity in jawed vertebrates. *Proc. Natl. Acad. Sci. USA* 106:13410–13414 (and supplemental info pp. 1–20).

Benton, M. J. 2005. *Vertebrate Paleontology.* 3rd edition. Blackwell Publishing, Malden, MA.

Berg, L. S. 1947. *Classification of Fishes Both Recent and Fossil.* J. W. Edwards, Ann Arbor, MI.

Brazeau, M. D. 2009. The braincase and jaws of a Devonian 'acanthodian' and modern gnathostome origins. *Nature* 457:305–308.

Clements, J. F. 2007. *Birds of the World: A Checklist.* 6th edition. Cornell University Press, Ithaca, NY.

Cogger, H. G., and R. G. Zweifel. 1998. *Encyclopedia of Reptiles and Amphibians.* Academic Press, San Diego, CA.

Cope, E. D. 1877. Synopsis of the cold blooded *Vertebrata*, procured by Prof. James Orton during his exploration of Peru in 1876–77. *Proc. Am. Philos. Soc.* 17:33–48.

Davis, S. P., J. A. Finarelli, and M. I. Coates. 2012. *Acanthodes* and shark-like conditions in the last common ancestor of modern gnathostomes. *Nature* 486:247–250.

Dettai, A., and G. Lecointre. 2008. New insights into the organization and evolution of vertebrate IRBP genes and utility of IRBP gene sequences for the phylogenetic study of the *Acanthomorpha* (*Actinopterygii: Teleostei*). *Mol. Phylogenet. Evol.* 48:258–269.

Diogo, R. 2007. *The Origin of Higher Clades.* Science Publishers, Enfield, NH.

Eschmeyer, W. N., C. J. Ferraris, Jr., M. Hoang, and D. J. Long. 1998. *Catalog of Fishes.* Special Publication. California Academy of Sciences, San Francisco, CA.

Eschmeyer, W. N., ed. 2011. *Catalog of Fishes*, electronic version 14 July 2011. Available at http://research.calacademy.org/ichthyology/catalog/fishcatmain.asp.

Forey, P. L. 1980. *Latimeria*: a paradoxical fish. *Proc. R. Soc. Lond. B Biol. Sci.* 208:369–384.

Friedman, M., and M. D. Brazeau. 2010. A reappraisal of the origin and basal radiation of the *Osteichthyes. J. Vertebr. Paleontol.* 30:36–56.

Frost, Darrel R. 2009. Amphibian species of the world: an online reference, Version 5.3 (12 February 2009). Available at http://research.amnh.org/herpetology/amphibia/.

Gardiner, B. G. 1973. Interrelationships of teleostomes. Pp. 105–135 in *Interrelationships of Fishes* (P. H. Greenwood, R. S. Miles, and C. Patterson, eds.). Academic Press, London.

Gardiner, B. G. 1984. The relationships of the paleoniscid fishes, a review based on new specimens of *Mimia* and *Moythomasia* from the Upper Devonian of Western Australia. *Bull. Br. Mus. (Nat. Hist.) Geol.* 37:173–428.

Garstang, W. 1931. The phyletic classification of *Teleostei. Proc. Leeds Philos. Lit. Soc. Sci. Sect.* 2:240–261.

Gauthier, J. A., and K. de Queiroz, 2001. Feathered dinosaurs, flying dinosaurs, crown dinosaurs, and the name "*Aves.*" Pp. 7–41 in *New Perspectives on the Origin and Early Evolution of Birds: Proceeding of the International Symposium in Honor of John H. Ostrom* (J. Gauthier and L. F. Gall, eds.). Special Publication of the Peabody Museum of Natural History, New Haven, CT.

Gill, T. N. 1892. Arrangement of the families of fishes. *Smithson. Misc. Coll.* 247:1–95.

Gill, T. N. 1893. Families and subfamilies of fishes. *Mem. Natl. Acad. Sci.* 6:127–138.

Goodrich, E. S. 1909. *Vertebrata Craniata* (first fascicle: cyclostomes and fishes). Pp. 1–518 in *A Treatise on Zoology, Part IX* (E. R. Lankester, ed.). A and C Black, London.

Gregory, W. K. 1906. The orders of teleostomous fishes. *Ann. N. Y. Acad. Sci.* 17:437–508.

Helfman, G. S., B. B. Collette, D. E. Facey, and B. W. Bowen. 2009. *The Diversity of Fishes. Biology, Evolution, and Ecology.* 2nd edition. Wiley-Blackwell, Hoboken, NJ.

Huxley, T. H. 1861. Preliminary essay upon the systematic arrangement of the fishes of the Devonian epoch. Pp. 1–40 in *Memoirs of the Geological Survey of the United Kingdom,* Decade X. Longman, Green, Longman, and Roberts, London.

Huxley, T. H. 1880. On the application of the laws of evolution to the arrangement of the *Vertebrata* and more particularly of the *Mammalia. Proc. Zool. Soc. Lond.* 43:649–662.

Janvier, P. 1996. *Early Vertebrates.* Oxford University Press, Oxford.

Jordan, D. S. 1923. *A Classification of Fishes: Including Families and Genera as Far as Known.* Stanford University Press, Palo Alto, CA.

Kikugawa, K., K. Katoh, S. Kuraku, H. Sakurai, O. Ishida, N. Iwabe, and T. Miyata. 2004. Basal jawed vertebrate phylogeny inferred from multiple nuclear DNA-coded genes. *BMC Biol.* 2:3.

Li, B., A. Dettai, C. Cruaud, A. Coiloix, M. Desoutter-Meniger, and G. Lecointre. 2009. RNF213, a new nuclear marker for acanthomorph phylogeny. *Mol. Phylogenet. Evol.* 50:345–363.

Long, J. A. 2011. *The Rise of Fishes: 500 Million Years of Evolution.* Johns Hopkins Press, Baltimore, MD.

MacDonald, D. W. 2009. *The Princeton Encyclopedia of Mammals.* Princeton University Press, Princeton, NJ.

Maisey, J. G. 1986. Heads and tails: a chordate phylogeny. *Cladistics* 2:201–256.

Meyer, A., and R. Zardoya. 2003. Recent advances in the (molecular) phylogeny of vertebrates. *Annu. Rev. Ecol. Syst.* 34:311–338.

Moy-Thomas, J. A., and R. S. Miles. 1971. *Paleozoic Fishes.* Saunders, Philadelphia, PA.

Near, T. J., R. I. Eytan, A. Dornburg, K. L. Kuhn, J. A. Moore, M. P. Davis, P. C. Wainwright, M. Friedman, and W. L. Smith. 2012. Resolution of ray-finned fish phylogeny and timing of diversification. *Proc. Natl. Acad. Sci. USA* 109:13698–13703.

Nelson, G. 1969. Gill arches and the phylogeny of fishes, with notes on the classification of vertebrates. *Bull. Am. Mus. Nat. Hist.* 141:475–568.

Nelson, J. 1994. *Fishes of the World.* 3rd edition. John Wiley & Sons, New York.

Nelson, J. 2006. *Fishes of the World.* 4th edition. John Wiley & Sons, New York.

Noack, K., R. Zardoya, and A. Meyer. 1996. The complete mitochondrial DNA sequence of the bichir (*Polypterus ornatipinnis*), a basal ray-finned fish: ancient establishment of the consensus vertebrate gene order. *Genetics* 144:1165–1180.

Owen, R. 1866. *The Anatomy of Vertebrates. Vol. 1: Fishes and Reptiles.* Longmans, Green and Co., London.

Patterson, C. 1994. Bony fishes. Pp. 57–84 in *Major Features of Vertebrate Evolution* (D. R. Prothero and R. M. Schoch, eds.). Short Courses in Paleontology, No. 7. Paleontological Society, University of Tennessee, Knoxville, TN.

de Queiroz, K., and J. A. Gauthier. 1992. Phylogenetic taxonomy. *Annu. Rev. Ecol. Syst.* 23:449–480.

Regan, C. T. 1904. The phylogeny of the *Teleostomi. Ann. Mag. Nat. Hist.* (*Ser. 7*) 13:329–349, pl. 7.

Regan, C. T. 1929. Fishes. *Encyclopedia Brittanica.* 14th edition. London

Rosen, D. E., P. L. Forey, B. G. Gardiner, and C. Patterson. 1981. Lungfish, tetrapods, paleontology and plesiomorphy. *Bull. Am. Mus. Nat. Hist.* 167:159–276.

Rowe, T. 2004. Chordate phylogeny and development. Pp. 384–409 in *Assembling the Tree of Life* (J. Cracraft and M. J. Donoghue, eds.). Oxford University Press, Oxford.

Schaeffer, B. 1968. The origin and basic radiation of the *Osteichthyes.* Pp. 207–222 in *Current Problems of Lower Vertebrate Phylogeny* (T. Orvig, ed.). Proceedings of the Fourth Nobel Symposium. Almquist and Wiksell, Stockholm.

Shubin, N. 2008. *Your Inner Fish: A Journey into the 3.5-Billion-Year History of the Human Body.* Pantheon, New York.

Springer, V. G., and G. D. Johnson. 2004. Study of the dorsal gill-arch musculature of teleostome fishes, with special reference to the *Actinopterygii. Bull. Biol. Soc. Wash.* 11:1–236.

Stiassny, M. L. J., E. O. Wiley, G. D. Johnson, and M. R. de Carvalho. 2004. Gnathostome fishes. Pp. 410–429 in *Assembling the Tree of Life* (M. J. Donoghue and J. Cracraft, eds.). Oxford University Press, Oxford.

Wiley, E. O. 1979. Ventral gill arch muscles and the interrelationships of gnathostomes, with a new classification of the *Vertebrata. Zool. J. Linn. Soc.* 67:149–179.

Wiley, E. O., and G. D. Johnson. 2010. A teleost classification based on monophyletic groups. Pp. 123–182 in *Origin and Phylogenetic Interrelationships of Teleosts* (J. S. Nelson, H.-P. Schultze, and M. V. H. Wilson, eds.). Verlag Dr. Friedrich Pfeil, Munich.

Woodward, A. S. 1891. *Catalogue of Fossil Fishes in the British Museum (Natural History)*, Vol. 2. British Museum (Natural History), London.

Zardoya, R., and A. Meyer. 2001. Vertebrate phylogeny: limits of inference of mitochondrial genome and nuclear rDNA sequence data due to an adverse phylogenetic signal/noise ratio. Pp. 106–118 in *Major Events in Early Vertebrate Evolution: Paleontology, Phylogeny, Genetics, and Development* (P. E. Ahlberg, ed.). Taylor & Francis, London.

Zhu, M., and H. P. Schultze. 2001. Interrelationships of basal osteichthyans. Pp. 289–314 in *Major Events in Early Vertebrate Evolution: Paleontology, Phylogeny, Genetics, and Development* (P. E. Ahlberg, ed.). Taylor & Francis, London.

Authors

Jon A. Moore; Wilkes Honors College; Florida Atlantic University; Jupiter, FL 33458, USA. Email: jmoore@fau.edu.

Thomas J. Near; Department of Ecology and Evolutionary Biology; Yale University; New Haven, CT 06520, USA. E-mail: Thomas.Near@yale.edu.

Date Accepted: 22 August 2013

Primary Editor: Kevin de Queiroz

Pan-Actinopterygii T. Rowe 2004 [J. A. Moore and T. J. Near], converted clade name

Registration Number: 207

Definition: The total clade of the crown clade *Actinopterygii*. This is a crown-based total-clade definition. Abbreviated definition: total ∇ of *Actinopterygii*.

Etymology: Derived from the Greek *pantos* (all), indicating that the name refers to a total clade or pan-monophyllum, plus *Actinopterygii*, the name of the corresponding crown clade (see the entry for *Actinopterygii* in this volume for the etymology of that name).

Reference Phylogeny: Gardiner (1984: Fig. 146), where *Pan-Actinopterygii* includes all of the taxa in the tree, is the primary reference phylogeny; see also Janvier (1996: Fig. 4.70), Gardiner et al. (2005: Fig. 1), Schultze and Cumbaa (2001: Fig. 18.5), and Zhu et al. (2009: Fig. 5).

Composition: *Actinopterygii* (this volume) and all extinct fishes that share a more recent ancestry with *Actinopterygii* than with *Sarcopterygii* (lobe-finned fishes). *Cheirolepis, Andreolepis, Lophosteus, Ligulalepis*, and *Dialipina* are often cited as stem actinopterygians (Patterson, 1982; Lauder and Liem, 1983: Fig 6; Gardiner and Schaeffer, 1989; Schultze, 1992; Patterson, 1994; Janvier, 1996; Schultze and Cumbaa, 2001; Cloutier and Arratia, 2004; Rowe, 2004; Benton, 2005: Fig. 7.7; Gardiner et al., 2005; Zhu et al., 2006, 2009). *Cheirolepis* is most consistently inferred to be a stem actinopterygian, while *Dialipina, Ligulalepis, Andreolepis, and Lophosteus* have sometimes been proposed as stem osteichthyans (Basden et al., 2000; Friedman, 2007; Brazeau, 2009; Friedman and Brazeau, 2010).

Diagnostic Apomorphies: The characters listed here did not necessarily arise immediately at the origin of the *Pan-Actinopterygii* but instead arose early in the history of the stem taxa leading to the crown *Actinopterygii*. These features are useful in recognizing fossils as members of the total clade. One feature often cited for taxa within the *Pan-Actinopterygii* is ganoine on the scales and dermal bones (Lauder and Liem, 1983; Janvier, 1996; Schultze and Cumbaa, 2001; Cloutier and Arratia, 2004). Janvier (1996) also cited the presence of fringing fulcral scales on the anterior edge of the caudal fin and also the presence of notches in the nasal bone for the nares. The loss of separate splenials, postsplenials, and splenialpostsplenials, was inferred as a feature arising very early within the *Pan-Actinopterygii* (see Cloutier and Arratia, 2004: character 26). Friedman and Brazeau (2010) listed and discussed 9 characters that appear to place fossil taxa within the *Pan-Actinopterygii*.

Synonyms: Use of the name *Actinopterygii* by various palaeontologists (Gardiner, 1984; Gardiner and Schaeffer, 1989; Schultze and Cumbaa, 2001; Cloutier and Arratia, 2004; Gardiner et al., 2005; Zhu et al., 2006) to include increasingly more stem taxa makes the name effectively synonymous with *Pan-Actinopterygii* as defined here. Though not a formal taxon name, the term "total-group *Actinopterygii*" (e.g., Friedman and Blom, 2006) is an unambiguous synonym. *Chondrostei* of Woodward (1891, 1895) and Goodrich (1909, 1930) is a partial (and approximate) synonym: it refers to a paraphyletic assemblage of crown chondrosteans (Goodrich's *Acipenseroidei*) and fossil "paleoniscoids" originating in approximately the same ancestor as our *Pan-Actinopterygii*. Similarly, Goodrich's (1909, 1930) *Palaeoniscoidei* is also a partial (and approximate) synonym, given that it includes fossil stem taxa to *Actinopterygii* (e.g., *Cheirolepis*), *Actinopteri* (e.g., *Boreosomus* and *Cheirodus*), and *Neopterygii* (e.g., *Palaeoniscum* and *Bergeria*). *Palaeonisciformes* (Hay, 1929; Gardiner 1966, 1967) is also a partial (and approximate) synonym.

Comments: We selected the name *Pan-Actinopterygii* for this total clade in keeping with our selection of the name *Actinopterygii* for the corresponding crown clade and our desire to contribute to the development of an integrated approach to naming crown and total clades (e.g., de Queiroz, 2007). As a total clade, *Pan-Actinopterygii* will accommodate all presently known and yet to be discovered fossils that diverged from (or were part of) the ancestral lineage of *Actinopterygii* after its most recent common ancestor with *Sarcopterygii*.

Literature Cited

Basden, A. M., G. C. Young, M. I. Coates, and A. Ritchie. 2000. The most primitive osteichthyan braincase? *Nature* 403:185–188.

Benton, M. J. 2005. *Vertebrate Paleontology*. 3rd edition. Blackwell Publishing, Malden, MA.

Brazeau, M. D. 2009. The braincase and jaws of a Devonian 'acanthodian' and modern gnathostome origins. *Nature* 457:305–308.

Cloutier, R., and G. Arratia. 2004. Early diversification of actinopterygians. Pp. 217–270 in *Recent Advances in the Origin and Early Radiation of Vertebrates* (G. Arratia, M. V. H. Wilson, R. Cloutier, eds.). Verlag Dr. Friedrich Pfeil, Munich.

Friedman, M. 2007. *Styloichthys* as the oldest coelacanth: implications for early osteichthyan interrelationships. *J. Syst. Palaeontol.* 5:289–343

Friedman, M., and H. Blom. 2006. A new actinopterygian from the Famennian of East Greenland and the interrelationships of Devonian ray-finned fishes. *J. Paleontol.* 80:1186–1204.

Friedman, M., and M. D. Brazeau. 2010. A reappraisal of the origin and basal radiation of the Osteichthyes. *J. Vertebr. Paleontol.* 30:36–56.

Gardiner, B. G. 1966. A catalogue of Canadian fossil fishes. *Roy. Ont. Mus. Univ. Toronto Life Sci. Contrib.* 68:1–154.

Gardiner, B. G. 1967. Further notes on palaeoniscoid fishes with a classification of the Chondrostei. *Bull. Brit. Mus. (Nat. Hist.) Geol.* 14:143–206.

Gardiner, B. G. 1984. The relationships of the paleoniscid fishes, a review based on new specimens of *Mimia* and *Moythomasia* from the Upper Devonian of Western Australia. *Bull. Brit. Mus. (Nat. Hist.), Geol.* 37:173–428.

Gardiner, B. G., and B. Schaffer. 1989. Interrelationships of the lower actinopterygian fishes. *Zool. J. Linn. Soc.* 97:135–187.

Gardiner, B. G., B. Schaeffer, and J. A. Masserie. 2005. A review of the lower actinopterygian phylogeny. *Zool. J. Linn. Soc.* 144:511–525.

Goodrich, E. S. 1909. *Vertebrata Craniata* (first fascicle: cyclostomes and fishes). Pp. 1–518 in *A Treatise on Zoology*, Part IX (E. R. Lankester, ed.). A. and C. Black, London.

Goodrich, E. S. 1930. *Studies on the Structure and Development of Vertebrates*. Macmillan and Co., London.

Hay, O. P. 1929. *Second Bibliography and Catalogue of the Fossil* Vertebrata *of North America*, Vol. I. Publ. no. 390, Carnegie Institute, Washington, DC.

Janvier, P. 1996. *Early Vertebrates*. Oxford University Press, Oxford.

Lauder, G. V., and K. F. Liem. 1983. The evolution and interrelationships of the actinopterygian fishes. *Bull. Mus. Comp. Zool.* 150(3):95–197.

Patterson, C. 1982. Morphology and interrelationships of primitive actinopterygian fishes. *Am. Zool.* 22:241–259.

Patterson, C. 1994. Bony fishes. Pp. 57–84 in *Major Features of Vertebrate Evolution* (D. R. Prothero and R. M. Schoch, eds.). Short Courses in Paleontology, No. 7. Paleontological Society, University of Tennessee, Knoxville, TN.

de Queiroz, K. 2007. Toward an integrated system of clade names. *Syst. Biol.* 56:956–974.

Rowe, T. 2004. Chordate phylogeny and development. Pp. 384–409 in *Assembling the Tree of Life* (J. Cracraft and M. J. Donoghue, eds.). Oxford University Press, Oxford.

Schultze, H-. P. 1992. Early Devonian actinopterygians (*Osteichthyes*, *Pisces*) from Siberia. Pp. 233–242 in *Fossil Fishes as Living Animals* (E. Mark-Kurik, ed.). Academy of Sciences of Estonia, Tallinn.

Schultze, H.-P., and S. L. Cumbaa. 2001. *Dialipina* and the characters of basal actinopterygians. Pp. 315–332 in *Major Events in Early Vertebrate Evolution: Paleontology, Phylogeny, Genetics and Development* (P. E. Ahlberg, ed.). Taylor & Francis, London.

Woodward, A. S. 1891. *Catalogue of Fossil Fishes in the British Museum (Natural History)*, Vol. 2. British Museum (Natural History), London.

Woodward, A. S. 1895. *Catalogue of Fossil Fishes in the British Museum (Natural History)*, Vol. 3. British Museum (Natural History), London.

Zhu, M., X. Yu, W. Wang, W. Zhao, and L. Jia. 2006. A primitive fish provides key characters bearing on deep osteichthyan phylogeny. *Nature* 441:77–80.

Zhu, M., W. Zhao, L. Jia, T. Quio, and Q. Qu. 2009. The oldest articulated osteichthyan reveals mosaic gnathostome characters. *Nature* 458:469–474.

Authors

Jon A. Moore; Wilkes Honors College; Florida Atlantic University; Jupiter, FL 33458, USA. Email: jmoore@fau.edu.

Thomas J. Near; Department of Ecology and Evolutionary Biology; Yale University; New Haven, CT 06520, USA. Email: Thomas.Near@yale.edu.

Date Accepted: 13 August 2013

Primary Editor: Kevin de Queiroz

Actinopterygii A. S. Woodward 1891 [J. A. Moore and T. J. Near], converted clade name

Registration Number: 206

Definition: The least inclusive crown clade that contains *Polypterus bichir* Lacepede 1803, *Acipenser sturio* Linnaeus 1758, *Psephurus gladius* (Martens 1862), *Lepisosteus osseus* (Linnaeus 1758), *Amia calva* Linnaeus 1766, and *Perca fluviatilis* Linnaeus 1758. This is a minimum-crown-clade definition. Abbreviated definition: min crown ∇ (*Polypterus bichir* Lacepede 1803 & *Acipenser sturio* Linnaeus 1758 & *Psephurus gladius* (Martens 1862) & *Lepisosteus osseus* (Linnaeus 1758) & *Amia calva* Linnaeus 1766 & *Perca fluviatilis* Linnaeus 1758).

Etymology: Derived from the Greek *aktinos* (rays) and *pterygion* (little fins).

Reference Phylogeny: Diogo (2007: Figs. 3 and 4) is the primary reference phylogeny; see also Patterson (1994: Fig. 1), Maisey (1986: Fig. 10), Nelson (2006: 87), Stiassny et al. (2004: Fig. 24.1) and Near et al. (2012: Figs. 1 and s1). Although *Perca fluviatilis* is not included in the primary reference phylogeny, available evidence indicates that it is deeply nested within the clade originating in branch 22 (*Neoteleostei*) on that phylogeny (Dettai and Lecointre, 2008; Li et al., 2009; Wiley and Johnson, 2010; Near et al., 2012).

Composition: *Actinopterygii* is currently thought to include 31,012 living species (Froese and Pauly, 2013), divided between its two primary subclades persisting today as the crown clades *Cladistia* (polypterids) and *Actinopteri* (*Acipenseroidei* sensu Grande and Bemis [1996] and *Neopterygii*). *Actinopterygii* thus includes all living polypterids, sturgeons and paddlefishes, gars, amiids and teleosts listed in Nelson (2006) and Eschmeyer et al. (1998, 2011). Partial lists of extinct *Actinopterygii* can be found in Janvier (1996), Benton (2005), and Nelson (2006).

Lund et al. (1995) and Lund and Poplin (2002) inferred a phylogeny that placed *Polypterus* as the sister group of all known fossil and living ray-finned fishes, which resulted in the placement of all known fossil actinopterygians within the crown clade. Recent phylogenies by most other palaeontologists (Gardiner, 1984; Gardiner and Schaeffer, 1989; Schultze and Cumbaa, 2001; Cloutier and Arratia, 2004; Gardiner et al., 2005; Zhu et al., 2006) place most of those fossil taxa outside the crown clade, so that they are here considered members of *Pan-Actinopterygii* (this volume) but not of *Actinopterygii*.

Diagnostic Apomorphies: Løvtrup (1977) discussed a number of soft anatomical features of actinopterygians (including polypterids) including particular lens proteins, isotocin as a neurohypophyseal hormone, and large lagenar and saccular otoliths made of vaterite. Patterson (1982) listed several additional apomorphies, such as a valvula in the cerebellum, pectoral proterygium, and an ascending process of the parasphenoid. Maisey (1986) listed 12 apomorphies, but some (e.g., fulcral scales on leading edge of upper caudal fin) are actually features of larger clades within *Pan-Actinopterygii* as defined in this volume. Diogo (2007) listed two apomorphies for the crown clade *Actinopterygii*: presence of a single intermandibularis section and presence of a peculiar branchiomandibularis muscle (see Wiley, 1979). The shortened midpiece of the sperm was given as yet another putative apomorphy (Jamieson, 1991: Fig. 9.5). Another apomorphy proposed for this crown clade is an acrodin cap on the teeth (Schultze and Cumbaa, 2001). Gardiner et al. (2005: Fig.1 Node B) confirmed the acrodin cap as an apomorphy and added a particular maxilla-premaxilla relationship for crown *Actinopterygii*. Janvier (1996) listed the loss of the pineal plate as an apomorphy of crown *Actinopterygii*.

Synonyms: Though not a formal taxon name, the term "crown-group *Actinopterygii*" (e.g., Friedman and Blom, 2006) is an unambiguous synonym.

Comments: Klein (1885) is sometimes given as the original author of the name *Actinopterygii* (see Gardiner, 1993), but a reading of Klein's paper finds several references to *Acanthopterygii* with no reference at all to *Actinopterygii*. All the genera and species mentioned in Klein (1885) are teleostean fishes; no chondrostean or holostean fishes are mentioned in either the text or plates.

Cope (1871a,b,c) originally proposed the taxon *Actinopteri* for paddlefishes, sturgeons, gars, bowfins, and teleosts. When he later (Cope, 1877) redefined *Actinopteri* to exclude *Chondrostei* (paddlefishes and sturgeons), he created a group that contained just the gars, bowfins and teleosts, which later became the widely accepted composition of *Neopterygii* (Regan 1923; see *Neopterygii* in this volume). Cope (1887) still later replaced his modified version of *Actinopteri* with the name *Actinopterygia*, without changing the composition of the group. Cope's (1887, 1891) *Actinopterygia* included neither the *Cladistia* (polypterids), which he placed within the *Crossopterygia*, nor *Chondrostei*, which he placed in the *Podopterygia*. Woodward (1891) modified the spelling of the name to *Actinopterygii* and added the *Chondrostei* back to the taxon. Subsequent authors added the polypterids to *Actinopterygii* (Goodrich, 1928, 1930; Save-Soderbergh, 1934; White, 1939; Gardiner, 1967, 1973, 1984; Wiley, 1979; Rosen et al., 1981; Patterson, 1982). This composition of polypterids, chondrosteans, gars, bowfin, and teleosts has since become entrenched in the literature and for that reason we choose the name *Actinopterygii*, as spelled by Woodward (1891), for this clade. Rosen et al. (1981: Fig 62) showed an early cladogram with synapomorphies for this group and Patterson (1982: Fig. 3A) added a significant number of synapomorphies.

The name *Actinopterygii* has previously been applied to a group characterized by the presence of ganoine on the scales and dermal bones (Cloutier and Arratia, 2004), which has resulted in the inclusion of many fossil lineages outside of the crown clade (see Gardiner and Schaeffer, 1989; Schultze and Cumbaa, 2001). We have nevertheless elected to restrict *Actinopterygii* to a crown clade because such clades have ancestors that we can know the most about, therefore maximizing justified inferences regarding the ancestor of *Actinopterygii* (sensu de Queiroz and Gauthier, 1992; Gauthier and de Queiroz, 2001). We use the name *Pan-Actinopterygii* (see entry in the volume) for the total clade including fossils outside of the crown.

Lê et al. (1993), Noack et al. (1996), Venkatesh et al. (2001), Zardoya and Meyer (2001), Meyer and Zardoya (2003), Inoue et al. (2003), Kikugawa et al. (2004), Santini et al. (2009), Alfaro et al. (2009) and Near et al. (2012) have provided molecular evidence supporting the monophyly of the crown clade *Actinopterygii* and Meyer and Zardoya (2003), Inoue et al. (2003), Kikugawa et al. (2004), Santini et al. (2009), Alfaro et al. (2009) and Near et al. (2012) all implicitly applied the name *Actinopterygii* to this crown clade.

Literature Cited

Alfaro, M. E., F. Santini, C. Brock, H. Alamillo, A. Dornburg, D. L. Rabosky, G. Carnevale, and L. J. Harmon. 2009. Nine exceptional radiations plus high turnover explain species diversity in jawed vertebrates. *Proc. Natl. Acad. Sci. USA* 106(32):13410–13414 (and supplemental info pp. 1–20).

Benton, M. J. 2005. *Vertebrate Paleontology.* 3rd edition. Blackwell Publishing, Malden, MA.

Cloutier, R., and G. Arratia. 2004. Early diversification of actinopterygians. Pp. 217–270 in *Recent Advances in the Origin and Early Radiation of Vertebrates* (G. Arratia, M. V. H. Wilson, R. Cloutier, eds.). Verlag Dr. Friedrich Pfeil, Munich, Germany.

Cope, E. D. 1871a. Observations on the systematic relations of the fishes. *Am. Nat.* 5:579–593.

Cope, E. D. 1871b. Contribution to the ichthyology of the Lesser Antilles. *Trans. Am. Philos. Soc.* 14:445–460.

Cope, E. D. 1871c. Observations on the systematic relations of the fishes. *Proc. Am. Assoc. Adv. Sci.* 20:317–343.

Cope, E. D. 1877. On the classification of the extinct fishes of the lower types. *Proc. Am. Assoc. Adv. Sci.* 26:292–300.

Cope, E. D. 1887. Geology and paleontology. Zittel's Manual of Palaeontology. *Am. Nat.* 21:1014–1020. .

Cope, E. D. 1891. On the characters of some Paleozoic fishes. *Proc. U. S. Natl. Mus.* 14:447–463

de Queiroz, K., and J. Gauthier. 1992. Phylogenetic taxonomy. *Annu. Rev. Ecol. Syst.* 23:449–480.

Dettai, A., and G. Lecointre. 2008. New insights into the organization and evolution of vertebrate IRBP genes and utility of IRBP gene sequences for the phylogenetic study of the Acanthomorpha (*Actinopterygii: Teleostei*). *Mol. Phylogenet. Evol.* 48:258–269.

Diogo, R. 2007. *The Origin of Higher Clades.* Science Publishers, Enfield, NH.

Eschmeyer, W. N., ed. 2011. Catalog of Fishes, electronic version 14 July 2011. Available at http://research.calacademy.org/ichthyology/catalog/fishcatmain.asp.

Eschmeyer, W. N., C. J. Ferraris, Jr., M. Hoang, and D. J. Long. 1998. *Catalog of Fishes.* Special Publication. California Academy of Sciences, San Francisco, CA.

Friedman, M., and H. Blom. 2006. A new actinopterygian from the Famennian of East Greenland and the interrelationships of Devonian ray-finned fishes. *J. Paleontol.* 80:1186–1204.

Froese, R., and D. Pauly, eds. 2013. *FishBase: Fishes Used by Humans* (based on FishBase 2/2013) [under Tools/Fish statistics]. Available at http://fishbase.org/Report/FishesUsedByHumans.php, accessed on 17 March 2013.

Gardiner, B. G. 1967. Further notes on paleoniscoid fishes with a classification of the Chondrostei. *Bull. Brit. Mus. (Nat. Hist.) Geol.* 14:146–206.

Gardiner, B. G. 1973. Interrelationships of teleostomes. Pp. 105–135 in *Interrelationships of Fishes* (P. H. Greenwood, R. S. Miles, and C. Patterson, eds.). Academic Press, London.

Gardiner, B. G. 1984. The relationships of the paleoniscid fishes, a review based on new specimens of *Mimia* and *Moythomasia* from the Upper Devonian of Western Australia. *Bull. Brit. Mus. (Nat. Hist.) Geol.* 37:173–428.

Gardiner, B. G. 1993. *Osteichthyes*: Basal *Actinopterygii.* Pp. 611–619 in *Fossil Record 2* (M. J. Benton, ed.). Chapman and Hall, London.

Gardiner, B. G., and B. Schaffer. 1989. Interrelationships of the lower actinopterygian fishes. *Zool. J. Linn. Soc.* 97:135–187.

Gardiner, B. G., B. Schaeffer, and J. A. Masserie. 2005. A review of the lower actinopterygian phylogeny. *Zool. J. Linn. Soc.* 144:511–525.

Gauthier, J. A., and K. de Queiroz, 2001. Feathered dinosaurs, flying dinosaurs, crown dinosaurs, and the name *"Aves."* Pp. 7–41 in *New Perspectives on the Origin and Early Evolution of Birds: Proceeding of the International Symposium in Honor of John H. Ostrom* (J. Gauthier and L.F. Gall, eds.). Special Publication of the Peabody Museum of Natural History, New Haven, CT.

Goodrich, E.S. 1928. *Polypterus* a palaeoniscid? *Palaeobiologica* 7:87–92.

Goodrich, E. S. 1930. *Studies on the Structure and Development of Vertebrates*. Macmillan and Co., London.

Grande, L., and W. E. Bemis, 1996. Interrelationships of *Acipenseriformes*, with comments on "*Chondrostei*". Pp. 85–115 in *Interrelationships of Fishes* (M. L. J. Stiassny, L. R. Parenti, and G. D. Johnson, eds.). Academic Press, New York.

Inoue, J. G., M. Miya, K. Tsukamoto, and M. Nishida. 2003. Basal actinopterygian relationships: a mitogenomic perspective on the phylogeny of the "ancient fish". *Mol. Phylogenet. Evol.* 26:110–120.

Jamieson, B. G. M. 1991. *Fish Evolution and Systematics: Evidence from Spermatozoa*. Cambridge University Press, Cambridge, UK.

Janvier, P. 1996. *Early Vertebrates*. Oxford University Press, Oxford.

Kikugawa, K., K. Katoh, S. Kuraku, H. Sakurai, O. Ishida, N. Iwabe, and T. Miyata. 2004. Basal jawed vertebrate phylogeny inferred from multiple nuclear DNA-coded genes. *BMC Biol.* 2:3.

Klein, E. F. 1885. Beitrage zur Bildung des Schadels der Knochenfische II. *Jahresh. Ver. Vaterl. Naturkd. Wurttemb.* 41:107–261.

Lê, H. L., G. Lecointre, and R. Perasso. 1993. A 28S rRNA-based phylogeny of the gnathostomes: first steps in the analysis of conflict and congruence with morphologically based cladograms. *Mol. Phylogenet. Evol.* 2:31–51.

Li, B., A. Dettai, C. Cruaud, A. Coiloix, M. Desoutter-Meniger, and G. Lecointre. 2009. RNF213, a new nuclear marker for acanthomorph phylogeny. *Mol. Phylogenet. Evol.* 50:345–363.

Løvtrup, S. 1977. *The Phylogeny of Vertebrata*. Wiley Interscience, New York.

Lund, R., and C. Poplin. 2002. Cladistic analysis of the relationships of the tarrasiids (Lower Carboniferous actinopterygians). *J. Vertebr. Paleontol.* 22:480–486.

Lund, R., C. Poplin, and K. McCarthy. 1995. Preliminary analysis of the interrelationships of some Paleozoic *Actinopterygii*. *Geobios* 28(suppl. 2):215–220.

Maisey, J. G. 1986. Heads and tails: a chordate phylogeny. *Cladistics* 2:201–256.

Meyer, A., and R. Zardoya. 2003. Recent advances in the (molecular) phylogeny of vertebrates. *Annu. Rev. Ecol. Syst.* 34:311–338

Near, T. J., R. I. Eytan, A. Dornburg, K. L. Kuhn, J. A. Moore, M. P. Davis, P. C. Wainwright, M. Friedman, and W. L. Smith. 2012. Resolution of ray-finned fish phylogeny and timing of diversification. *Proc. Natl. Acad. Sci. USA* 109(34):13698–13703.

Nelson, J. 2006. *Fishes of the World*. 4th edition. John Wiley & Sons, New York.

Noack, K., R. Zardoya, and A. Meyer. 1996. The complete mitochondrial DNA sequence of the bichir (*Polypterus ornatipinnis*), a basal ray-finned fish: ancient establishment of the consensus vertebrate gene order. *Genetics* 144:1165–1180

Patterson, C. 1982. Morphology and interrelationships of primitive actinopterygian fishes. *Am. Zool.* 22:241–259.

Patterson, C. 1994. Bony fishes. Pp. 57–84 in *Major Features of Vertebrate Evolution* (D. R. Prothero and R. M. Schoch, eds.). Short Courses in Paleontology, No. 7. Paleontological Society, University of Tennessee, Knoxville, TN.

Regan, C. T. 1923. The skeleton of *Lepidosteus*, with remarks on the origin and evolution of the lower neopterygian fishes. *Proc. Zool. Soc.* 93:445–461.

Rosen, D. E., P. L. Forey, B. G. Gardiner, and C. Patterson. 1981. Lungfish, tetrapods, paleontology and plesiomorphy. *Bull. Am. Mus. Nat. Hist.* 167:159–276.

Santini, F., L. J. Harmon, G. Carnevale, and M. E. Alfaro, 2009. Did genome duplication drive the origin of teleosts? A comparative study of diversification in ray-finned fishes. *BMC Evol. Biol.* 9:194–209.

Säve-Söderbergh, G. 1934. Some points of view concerning the evolution of the vertebrates and the classification of this group. *Ark. Zool.* 26:1–20.

Schultze, H.-P., and S. L. Cumbaa. 2001. *Dialipina* and the characters of basal actinopterygians. Pp. 315–332 in *Major Events in Early*

Vertebrate Evolution. Paleontology, Phylogeny, Genetics, and Development (P. E. Ahlberg, ed.). Taylor & Francis, London.

Stiassny, M. L. J., E. O. Wiley, G. D. Johnson, and M. R. de Carvalho. 2004. Gnathostome fishes. Pp. 410–429 in *Assembling the Tree of Life* (M. J. Donoghue and J. Cracraft, eds.). Oxford University Press, Oxford.

Venkatesh, B., M. V. Erdmann, and S. Brenner. 2001. Molecular synapomorphies resolve evolutionary relationships of extant jawed vertebrates. *Proc. Natl. Acad. Sci. USA* 98:11382–11387.

White, E. I. 1939. A new type of palaeoniscoid fish, with remarks on the evolution of the actinopterygian pectoral fins. *Proc. Zool. Soc. Lond.* 109:41–61.

Wiley, E. O. 1979. Ventral gill arch muscles and the interrelationships of gnathostomes, with a new classification of the *Vertebrata. Zool. J. Linn. Soc.* 67:149–179.

Wiley, E. O., and G. D. Johnson. 2010. A teleost classification based on monophyletic groups. Pp. 123–182 in *Origin and Phylogenetic Interrelationships of Teleosts* (J. S. Nelson, H.-P. Schultze, and M. V. H. Wilson, eds.). Verlag Dr. Friedrich Pfeil, Munich.

Woodward, A. S. 1891. *Catalogue of Fossil Fishes in the British Museum (Natural History)*, Vol. 2. British Museum (Natural History), London.

Zardoya, R., and A. Meyer. 2001. Vertebrate phylogeny: limits of inference of mitochondrial genome and nuclear rDNA sequence data due to an adverse phylogenetic signal/noise ratio. Pp. 106–118 in *Major Events in Early Vertebrate Evolution. Paleontology, Phylogeny, Genetics, and Development* (P. E. Ahlberg, ed.). Taylor & Francis, London.

Zhu, M., X. Yu, W. Wang, W. Zhao, and L. Jia. 2006. A primitive fish provides key characters bearing on deep osteichthyan phylogeny. *Nature* 441:77–80.

Authors

Jon A. Moore; Wilkes Honors College; Florida Atlantic University; Jupiter, FL 33458, USA. Email: jmoore@fau.edu.

Thomas J. Near; Department of Ecology and Evolutionary Biology; Yale University; New Haven, CT 06520, USA. Email: Thomas. Near@yale.edu.

Date Accepted: 8 September 2013

Primary Editor: Kevin de Queiroz

Pan-Actinopteri J. A. Moore and T. J. Near, new clade name

Registration Number: 209

Definition: The total clade of the crown clade *Actinopteri*. This is a crown-based total-clade definition. Abbreviated definition: total ∇ of *Actinopteri*.

Etymology: Derived from the Greek *pantos* (all), indicating that the name refers to a total clade or pan-monophyllum, plus *Actinopteri*, the name of the corresponding crown clade (see the entry for *Actinopteri* in this volume for the etymology of that name).

Reference Phylogeny: Coates (1999: Fig 9a), where *Pan-Actinopteri* applies to branch C and all of its descendants, is the primary reference phylogeny; see also Janvier (1996: Fig 4.70, nodes 3–9), Taverne (1997: Fig. 14) and Gardiner et al. (2005: Fig. 1).

Composition: *Actinopteri* (this volume) and all extinct fishes that share more recent ancestry with *Actinopteri* than with polypterids (*Cladistia*). Gardiner (1984, 1993), Janvier (1996: Fig. 4.70), Coates (1999: Fig. 9a) list or illustrate several fossil taxa in the stem group of *Actinopteri* (including *Mimia*, *Moythomasia*, *Kentuckia*, *Redfieldiidae*, and others). The *Tarrasiidae* (Lund and Poplin, 2002) and the *Aeduelliformes* (Poplin and Dutheil, 2005) have also been included in the stem group of *Actinopteri*. Lund and Poplin (2002) and Poplin and Dutheil (2005) also included *Cheirolepis* in the stem group of *Actinopteri*, contrary to its placement by most authors in the stem group of *Actinopterygii*. Taverne (1997) and Swartz (2009) inferred several additional fossil forms as stem actinopterans, including well-known

species such as *Howqualepis rostridens*, *Tegeolepis clarki*, and *Stegotrachelus finlayi*, as well as newly described species such as *Donnrosenia schaefferi* (Long et al., 2008) and *Cuneognathus gardineri* (Friedman and Blom, 2006). Stem actinopterans have sometimes been placed in a paraphyletic group *Palaeoniscimorpha* (Lund et al., 1995; Poplin and Dutheil, 2005), although that group also contains taxa that are not considered stem actinopterans by other authors (e.g., Gardiner et al., 2005).

Diagnostic Apomorphies: The characters listed here did not necessarily arise immediately at the origin of the *Pan-Actinopteri*, but instead arose early in the history of the stem lineage leading to the crown *Actinopteri*. These features are useful in recognizing fossils as members of the total clade. One feature cited for taxa within the *Pan-Actinopteri* is a perforated proterygium with bases of rays embracing the proterygium (Gardiner, 1984). Fringing fulcral scales on the leading edges of all the fins (not just the caudal fin) was also proposed as a apomorphy for *Pan-Actinopteri*, although this feature is lost in *Amia* and in most teleosts (although *Megalops* and some fossil elopiforms have fringing fulcral scales in the caudal fin) (Janvier, 1996; Arratia, 2009). Janvier (1996) also listed the lateral cranial canal in the braincase as a feature that arose early in the stem lineage of *Actinopteri*.

Synonyms: *Actinopteri* of Gregory (1906) included palaeoniscids and is therefore an approximate synonym of our *Pan-Actinopteri*. *Actinopterygii* as used by earlier authors (Woodward, 1891) prior to the addition of polypterids to that group (Goodrich, 1909) encompassed fossil and living *Actinopteri*, including several stem taxa, and is thus an approximate synonym. *Actinopteri* of some more recent vertebrate palaeontologists (e.g., Gardiner, 1993; Janvier, 1996; Benton, 2005: Fig. 7.7) also constitutes an approximate synonym of *Pan-Actinopteri*. *Palaeoniscimorpha* of Lund et al. (1995) and Lund and Poplin (2002) is a partial and approximate synonym of *Pan-Actinopteri* that refers to a paraphyletic assemblage of stem and early crown actinopterans. If Lund et al. (1995) and Lund and Poplin (2002) are correct in including *Cheirolepis* in the stem group of *Actinopteri*, then *Palaeonisciformes* (e.g., Hay, 1929; Gardiner, 1966, 1967) and *Palaeoniscoidei* (e.g., Goodrich, 1909, 1930) should be considered partial (and approximate) synonyms of *Pan-Actinopteri*.

Comments: We selected the name *Pan-Actinopteri* for this total clade because the names *Actinopterygii* and *Actinopteri* were selected for crown clades. As a total clade, *Pan-Actinopteri* will accommodate all presently known and yet to be discovered fossils that diverged from (or were part of) the ancestral lineage of *Actinopteri* after its most recent common ancestor with *Cladistia*.

Literature Cited

Arratia, G. 2009. Identifying patterns of diversity of the actinopterygian fulcra. *Acta Zool.* 90(suppl. 1):220–235.

Benton, M. J. 2005. *Vertebrate Paleontology.* 3rd edition. Blackwell Publishing, Malden, MA.

Coates, M. I. 1999. Endocranial preservation of a Carboniferous actinopterygian from Lancashire, U.K., and the interrelationships of primitive actinopterygians. *Philos. Trans. R. Soc. Lond. B Biol. Sci.* 354:435–462.

Friedman, M., and H. Blom. 2006. A new actinopterygian from the Famennian of East Greenland and the interrelationships of Devonian ray-finned fishes. *J. Paleontol.* 80:1186–1204.

Gardiner, B. G. 1966. A catalogue of Canadian fossil fishes. *Roy. Ont. Mus. Univ. Toronto Life Sci. Contrib.* 68:1–154.

Gardiner, B. G. 1967. Further notes on palaeoniscoid fishes with a classification of the *Chondrostei. Bull. Brit. Mus. (Nat. Hist.) Geol.* 14:143–206.

Gardiner, B. G. 1984. The relationships of the paleoniscid fishes, a review based on new specimens of *Mimia* and *Moythomasia* from the Upper Devonian of Western Australia. *Bull. Brit. Mus. (Nat. Hist.) Geol.* 37:173–428.

Gardiner, B. G. 1993. *Osteichthyes:* basal *Actinopterygii.* Pp. 611–619 in *Fossil Record 2.* (M. J. Benton, ed.). Chapman and Hall, London.

Gardiner, B. G., B. Schaeffer, and J. A. Masserie. 2005. A review of the lower actinopterygian phylogeny. *Zool. J. Linn. Soc.* 144:511–525.

Goodrich, E. S. 1909. *Vertebrata Craniata* (first fascicle: cyclostomes and fishes). Pp. 1–518 in *A Treatise on Zoology* (E. R. Lankester, ed.). A. and C. Black, London.

Goodrich, E. S. 1930. *Studies on the Structure and Development of Vertebrates.* Macmillan and Co., London.

Gregory, W. K. 1906. The orders of teleostomous fishes. *Ann. N. Y. Acad. Sci.* 17:437–508.

Hay, O. P. 1929. *Second Bibliography and Catalogue of the Fossil* Vertebrata *of North America,* Vol. I. Publ. no. 390. Carnegie Institute, Washington, DC.

Janvier, P. 1996. *Early Vertebrates.* Oxford University Press, Oxford.

Long, J. A., B. Choo, and G. C. Young. 2008. A new basal actinopterygian fish from the Middle Devonian Aztec Siltstone of Antarctica. *Antarct. Sci.* 20:393–412.

Lund, R., and C. Poplin. 2002. Cladistic analysis of the relationships of the tarrasiids (Lower Carboniferous actinopterygians). *J. Vertebr. Paleontol.* 22:480–486.

Lund, R., C. Poplin, and K. McCarthy. 1995. Preliminary analysis of the interrelationships of some Paleozoic *Actinopterygii. Geobios* 19:215–220.

Poplin, C., and D. B. Dutheil. 2005. Les *Aeduellidae* (*Pisces, Actinopterygii*) carbonifères et permiens: systématique et étude phylogénétique préliminaire. *Geodiversitas* 27(1):17–33.

Swartz, B. A. 2009. Devonian actinopterygian phylogeny and evolution based on a redescription of *Stegotrachelus finlayi*. *Zool. J. Linn. Soc.* 156:750–784.

Taverne, L. 1997. *Osorioichthys marginis* "Paléonisciforme" du Framennian de Belgique, et la phylogénie des actinoptérygiens dévoniens (*Pisces*). *Bull. Inst. Roy. Soc. Nat. Belgique, Sci. Terre* 67:57–78.

Authors

Jon A. Moore; Wilkes Honors College; Florida Atlantic University; Jupiter, FL 33458, USA. Email: jmoore@fau.edu.

Thomas J. Near; Department of Ecology and Evolutionary Biology; Yale University; New Haven, CT 06520, USA. Email: Thomas. Near@yale.edu.

Date Accepted: 13 August 2013

Primary Editor: Kevin de Queiroz

Actinopteri E. D. Cope 1871a [J. A. Moore and T. J. Near], converted clade name

Registration Number: 208

Definition: The least inclusive crown clade that contains *Acipenser sturio* Linnaeus 1758, *Psephurus gladius* (Martens 1862), *Lepisosteus osseus* (Linnaeus 1758), *Amia calva* Linnaeus 1766, and *Perca fluviatilis* Linnaeus 1758. This is a minimum-crown-clade definition. Abbreviated definition: min crown ∇ (*Acipenser sturio* Linnaeus 1758 & *Psephurus gladius* (Martens 1862) & *Lepisosteus osseus* (Linnaeus 1758) & *Amia calva* Linnaeus 1766 & *Perca fluviatilis* Linnaeus 1758).

Etymology: Derived from the Greek *aktinos* (rays) and *pteron* (fin).

Reference Phylogeny: Diogo (2007: Figs. 3 and 4) is the primary reference phylogeny; see also Patterson (1982: Fig. 3A and 1994: Fig.1), Maisey (1986: Fig. 10), Stiassny et al. (2004: Fig. 24.1) and Near et al. (2012: Figs. 1 and s1). Although *Perca fluviatilis* is not included in the primary reference phylogeny, available evidence indicates that it is deeply nested within the clade originating in branch 22 (*Neoteleostei*) on that phylogeny (Dettai and Lecointre, 2008; Li et al., 2009; Wiley and Johnson, 2010; Near et al., 2012).

Composition: *Actinopteri* is currently thought to include 30,998 living species (Froese and Pauly, 2013), divided between its two primary subclades persisting today as the crown clades *Acipenseroidei* (sensu Grande and Bemis, 1996) and *Neopterygii* (this volume). *Actinopteri* thus includes all sturgeons and paddlefishes, gars, bowfins, and teleosts listed in Nelson (2006) and Eschmeyer et al. (1998, 2011). Partial lists of extinct *Actinopteri* can be found in Gardiner (1984), Janvier (1996), Benton (2005), and Nelson (2006).

Diagnostic Apomorphies: Patterson (1982: Fig. 3A) listed a number of apomorphies, including a spiracular canal that opens into the fossa bridgei and three ossifications or cartilages in the hyoid bar beneath the interhyal. Maisey (1986) provided 7 apomorphies and an additional 5 that have limited distribution among basal actinopterans, although some of these were later found to have a more inclusive distribution within *Pan-Actinopteri* (Gardiner and Schaeffer, 1989; Gardiner et al., 2005). Maisey (1986) listed one soft character—the swimbladder connects dorsally to the foregut. Gardiner et al. (2005: Fig.1, Node G) cited three unambiguous apomorphies for *Actinopteri* as defined here: bone-enclosed spiracular canal; lateral cranial canal and fossa bridgei; and an unpaired myodome. Janvier (1996) gave a keystone-shaped dermosphenotic and rudimentary post-temporal fossa as two apomorphies. The presence of a micropyle in the egg is yet another apomorphy (Jamieson 1991: Fig. 9.5).

Synonyms: *Actinopterygii* of Woodward (1891), Boulenger (1892), Dean (1895), McAllister (1968), G. Nelson (1969), J. Nelson (1976, 1984), Løvtrup (1977), and Forey (1980) in each case excluded *Cladistia* (= *Brachiopterygii*) and is thus an approximate synonym of *Actinopteri*. *Epipneusta* Garstang 1931, apparently coined as a substitute for *Actinopterygii* in the sense excluding polypterids (see Comments on *Actinopterygii* in this volume), is also an approximate synonym.

Comments: Cope (1871a,b,c) originally included chondrosteans, holosteans, and teleosteans within *Actinopteri*, but excluded cladistians. When Cope (1877, 1887) revised *Actinopteri* to exclude the chondrosteans, he made *Actinopteri* essentially equivalent in content to what was later called *Neopterygii* (Regan, 1923). The name *Actinopteri* fell out of use for a long period after Jordan (1905, 1923) and Gregory (1906), until Patterson (1982) resurrected *Actinopteri* to encompass the original composition as Cope (1871a,b,c) first proposed (all actinopterygians except cladistians).

We have elected to restrict *Actinopteri* to a crown clade because such clades have ancestors that we can know the most about, therefore maximizing justified inferences regarding the ancestor of *Actinopteri* (sensu de Queiroz and Gauthier, 1992; Gauthier and de Queiroz, 2001). When Cope (1871a,b,c) originally proposed the *Actinopteri*, he based that concept on extant taxa.

We chose the name *Actinopteri* for this particular crown clade because of its prior historic use for this group (Cope, 1871a,b,c; Jordan, 1905; Gregory, 1906; Patterson, 1982; Maisey, 1986; Janvier, 1996). A broader discussion of the history of the taxon *Actinopteri* and its composition can be found in Patterson (1982).

Lê et al. (1993), Venkatesh et al. (2001), Inoue et al. (2003), Kikugawa et al. (2004), Hurley et al. (2007), Santini et al. (2009), Alfaro et al. (2009), and Near et al. (2012) provided molecular support for the monophyly of the crown clade *Actinopteri*.

Literature Cited

Alfaro, M. E., F. Santini, C. Brock, H. Alamillo, A. Dornburg, D. L. Rabosky, G. Carnevale, and L. J. Harmon. 2009. Nine exceptional radiations plus high turnover explain species diversity in jawed vertebrates. *Proc. Natl. Acad. Sci. USA* 106:13410–13414 (and supplemental info pp. 1–20).

Benton, M. J. 2005. *Vertebrate Paleontology*. 3rd edition. Blackwell Publishing, Malden, MA.

Boulenger, G. A. 1892. *Pisces. Zool. Rec.* 28:1–41.

Cope, E. D. 1871a. Observations on the systematic relations of the fishes. *Am. Nat.* 5:579–593.

Cope, E. D. 1871b. Contribution to the ichthyology of the Lesser Antilles. *Trans. Am. Philos. Soc.* 14:445–460.

Cope, E. D. 1871c. Observations on the systematic relations of the fishes. *Proc. Am. Assoc. Adv. Sci.* 20:317–343.

Cope, E. D. 1877. On the classification of the extinct fishes of the lower types. *Proc. Am. Assoc. Adv. Sci.* 26:292–300.

Cope, E. D. 1887. Geology and paleontology. Zittel's Manual of Paleontology. *Am. Nat.* 21:1014–1020.

Dean, B. 1895. *Fishes, Living and Fossil*. Columbia University Press, New York.

de Queiroz, K., and J. Gauthier. 1992. Phylogenetic taxonomy. *Annu. Rev. Ecol. Syst.* 23:449–480.

Dettai, A., and G. Lecointre. 2008. New insights into the organization and evolution of vertebrate IRBP genes and utility of IRBP gene sequences for the phylogenetic study of the *Acanthomorpha* (*Actinopterygii: Teleostei*). *Mol. Phylogenet. Evol.* 48:258–269.

Diogo, R. 2007. *The Origin of Higher Clades*. Science Publishers, Enfield, NH.

Eschmeyer, W. N., ed. 2011. *Catalog of Fishes*, electronic version 14 July 2011. Available at http://research.calacademy.org/ichthyology/catalog/fishcatmain.asp.

Eschmeyer, W. N., C. J. Ferraris, Jr., M. Hoang, and D. J. Long. 1998. *Catalog of Fishes*. Special Publication. California Academy of Sciences, San Francisco, CA.

Forey, P. L. 1980. *Latimeria*: a paradoxical fish. *Proc. R. Soc. Lond. B Biol. Sci.* 208:369–384.

Froese, R., and D. Pauly, eds. 2013. *FishBase: Fishes Used by Humans* (based on FishBase 2/2013) [under Tools/Fish statistics]. Available at http://fishbase.org/Report/FishesUsedByHumans.php, accessed on 17 March 2013.

Gardiner, B. G. 1984. The relationships of the pale-oniscid fishes, a review based on new specimens of *Mimia* and *Moythomasia* from the Upper Devonian of Western Australia. *Bull. Brit. Mus. (Nat. Hist.) Geol.* 37:173–428.

Gardiner, B. G., and B. Schaffer. 1989. Interrelationships of the lower actinopterygian fishes. *Zool. J. Linn. Soc.* 97:135–187.

Gardiner, B. G., B. Schaeffer, and J. A. Masserie. 2005. A review of the lower actinopterygian phylogeny. *Zool. J. Linn. Soc.* 144:511–525.

Garstang, W. 1931. The phyletic classification of *Teleostei. Proc. Leeds Philos. Lit. Soc. Sci. Sect.* 2:240–261.

Gauthier, J. A., and K. de Queiroz, 2001. Feathered dinosaurs, flying dinosaurs, crown dinosaurs, and the name "*Aves.*" Pp. 7–41 in *New Perspectives on the Origin and Early Evolution of Birds: Proceeding of the International Symposium in Honor of John H. Ostrom* (J. Gauthier and L.F. Gall, eds.). Special Publication of the Peabody Museum of Natural History New Haven, CT.

Grande, L., and W. E. Bemis. 1996. Interrelationships of *Acipenseriformes*, with comments on "*Chondrostei*". Pp. 85–115 in *Interrelationships of Fishes* (M. L. J. Stiassny, L. R. Parenti and G. D. Johnson, eds.). Academic Press, New York.

Gregory, W. K. 1906. The orders of teleostomous fishes. *Ann. N. Y. Acad. Sci.* 17:437–508.

Hurley, I. A., R. L. Mueller, K. A. Dunn, E. J. Schmidt, M. Friedman, R. K. Ho, V. E. Prince, Z. Yang, M. G. Thomas, and M. I. Coates. 2007. A new time scale for ray-finned fish evolution. *Proc. R. Soc. Lond. B Biol. Sci.* 274:489–498.

Inoue, J. G., M. Miya, K. Tsukamoto, and M. Nishida. 2003. Basal actinopterygian relationships: a mitogenomic perspective on the phylogeny of the "ancient fish". *Mol. Phylogenet. Evol.* 26:110–120.

Jamieson, B. G. M. 1991. *Fish Evolution and Systematics: Evidence From Spermatozoa.* Cambridge University Press, Cambridge, UK.

Janvier, P. 1996. *Early Vertebrates.* Oxford University Press, Oxford.

Jordan, D. S. 1905. *A Guide to the Study of Fishes.* H. Holt, New York.

Jordan, D. S. 1923. *A Classification of Fishes, Including Families and Genera as Far as Known.* Stanford University, Palo Alto, CA.

Kikugawa, K., K. Katoh, S. Kuraku, H. Sakurai, O. Ishida, N. Iwabe, and T. Miyata. 2004. Basal jawed vertebrate phylogeny inferred from multiple nuclear DNA-coded genes. *BMC Biol.* 2:3.

Lê, H. L., G. Lecointre, and R. Perasso. 1993. A 28S rRNA-based phylogeny of the gnathostomes: first steps in the analysis of conflict and congruence with morphologically based cladograms. *Mol. Phylogenet. Evol.* 2:31–51.

Li, B., A. Dettai, C. Cruaud, A. Coiloix, M. Desoutter-Meniger, and G. Lecointre. 2009. RNF213, a new nuclear marker for acanthomorph phylogeny. *Mol. Phylogenet. Evol.* 50:345–363.

Løvtrup, S. 1977. *The Phylogeny of* Vertebrata. Wiley Interscience, New York

Maisey, J. G. 1986. Heads and tails: a chordate phylogeny. *Cladistics* 2:201–256.

McAllister, D. E. 1968. Evolution of branchiostegals and classification of teleostome fishes. *Bull. Natl. Mus. Can.* 221:1–239.

Near, T. J., R. I. Eytan, A. Dornburg, K. L. Kuhn, J. A. Moore, M. P. Davis, P. C. Wainwright, M. Friedman, and W. L. Smith. 2012. Resolution of ray-finned fish phylogeny and timing of diversification. *Proc. Natl. Acad. Sci. USA* 109:13698–13703.

Nelson, G. 1969. Gill arches and the phylogeny of fishes, with notes on the classification of vertebrates. *Bull. Am. Mus. Nat. Hist.* 141:475–568.

Nelson, J. 1976. *Fishes of the World.* 1st edition. John Wiley & Sons, New York

Nelson, J. 1984. *Fishes of the World.* 2nd edition. John Wiley & Sons, New York

Nelson, J. 2006. *Fishes of the World.* 4th edition. John Wiley & Sons, New York

Patterson, C. 1982. Morphology and interrelationships of primitive actinopterygian fishes. *Am. Zool.* 22:241–259.

Regan, C. T. 1923. The skeleton of *Lepidosteus*, with remarks on the origin and evolution of the lower neopterygian fishes. *Proc. Zool. Soc.* 93:445–461.

Santini, F., L. J. Harmon, G. Carnevale and M. E. Alfaro. 2009. Did genome duplication drive the origin of teleosts? A comparative study of diversification in ray-finned fishes. *BMC Evol. Biol.* 9:194–209.

Stiassny, M. L. J., E. O. Wiley, G. D. Johnson, and M. R. de Carvalho. 2004. Gnathostome fishes. Pp. 410–429 in *Assembling the Tree of Life* (M. J. Donoghue and J. Cracraft, eds.). Oxford University Press, Oxford.

Venkatesh, B., M. V. Erdmann, and S. Brenner. 2001. Molecular synapomorphies resolve evolutionary relationships of extant jawed vertebrates. *Proc. Natl. Acad. Sci. USA* 98:11382–11387.

Wiley, E. O., and G. D. Johnson. 2010. A teleost classification based on monophyletic groups. Pp. 123–182 in *Origin and Phylogenetic Interrelationships of Teleosts* (J. S. Nelson, H.-P. Schultze, and M. V. H. Wilson, eds.). Verlag Dr. Friedrich Pfeil, Munich.

Woodward, A. S. 1891. *Catalogue of Fossil Fishes in the British Museum (Natural History)*, Vol. 2. British Museum (Natural History), London.

Authors

Jon A. Moore; Wilkes Honors College; Florida Atlantic University; Jupiter, FL 33458, USA. Email: jmoore@fau.edu.

Thomas J. Near; Department of Ecology and Evolutionary Biology; Yale University; New Haven, CT 06520, USA. Email: Thomas. Near@yale.edu.

Date Accepted: 13 August 2013

Primary Editor: Kevin de Queiroz

Pan-Neopterygii J. A. Moore and T. J. Near, new clade name

Registration Number: 211

Definition: The total clade of the crown clade *Neopterygii*. This is a crown-based total-clade definition. Abbreviated definition: total ∇ of *Neopterygii*.

Etymology: Derived from the Greek *pantos* (all), indicating that the name refers to a total clade or pan-monophyllum, plus *Neopterygii*, the name of the corresponding crown clade (see the entry for *Neopterygii* in this volume for the etymology of that name).

Reference Phylogeny: Gardiner and Schaeffer (1989: Fig. 12, in which the name *Pan-Neopterygii* applies to the clade composed of all branches to the right of node O) is the primary reference phylogeny; see also Janvier (1996: Fig. 4.71, in which the name applies to the clade composed of all branches below node 1).

Composition: *Neopterygii* (this volume) and all extinct fishes that share a more recent ancestry with *Neopterygii* than with living paddle-fishes and sturgeons (*Chondrostei*). Gardiner et al. (2005: Fig. 12) and Janvier (1996: Fig. 4.71 and Table 4.9) inferred several fossils to be stem neopterygians, including *Palaeoniscum*, *Birgeria*, *Australosomus*, *Cleithrolepis*, *Perleidus*, *Peltopleurus*, and *Luganoia*. Hurley et al. (2007: Fig. 2) presented a phylogeny that places a number of fossils outside of the crown *Neopterygii*, including *Discoserra*, *Peltopleurus*, *Luganoia*, and *Ebenaqua*.

Diagnostic Apomorphies: The characters listed here did not necessarily arise immediately at the origin of the *Pan-Neopterygii*, but instead arose early in the history of the stem lineage leading to the crown *Neopterygii*. These features are useful in recognizing fossils as members of the total clade. Patterson (1973) listed at least one "neopterygian" feature that was actually present in several of the *Pan-Neopterygii* stem lineages: the upper caudal fin rays elongate. Janvier (1996) proposed numerous anamestic supraorbitals as a feature that arose early in the stem lineage of *Neopterygii*. Gardiner (1984) gave the presence of a myodome as the first character to arise in the stem lineage of *Neopterygii*, but Gardiner et al. (2005) stated that the unpaired myodome is actually a feature of the more inclusive clade composed of *Actinopteri* and the stem actinopteran *Kentuckia*.

Synonyms: Gardiner's (1984, 1993) *Neopterygii* included several fossil groups (e.g., *Palaeonisciformes*, *Pholidopleuriformes*, *Perleidiiformes*, *Peltopleuriformes*) outside of the crown, which implies that his *Neopterygii* is an approximate synonym of *Pan-Neopterygii*. Patterson (1994) implicitly used the name *Neopterygii* for a total clade, although he also associated it with the corresponding crown node (pp. 71–72). Ignoring that inconsistency, Patterson's *Neopterygii* could be considered an unambiguous synonym of *Pan-Neopterygii*. The synonyms listed for *Neopterygii* (this volume) can also be considered approximate synonyms of *Pan-Neopterygii*.

Comments: We selected the name *Pan-Neopterygii* for this total clade, because the name *Neopterygii* was selected for the crown clade. As a total clade, *Pan-Neopterygii* will accommodate all presently known and yet to be discovered fossils that diverged from (or were part of) the ancestral lineage of *Neopterygii* after its most recent common ancestor with *Chondrostei*.

Literature Cited

Gardiner, B. G. 1984. The relationships of the paleoniscid fishes, a review based on new specimens of *Mimia* and *Moythomasia* from the Upper Devonian of Western Australia. *Bull. Brit. Mus. (Nat. Hist.) Geol.* 37:173–428.

Gardiner, B. G. 1993. *Osteichthyes:* basal *Actinopterygii.* Pp. 611–619 in *Fossil Record 2.* (M. J. Benton, ed.). Chapman and Hall, London.

Gardiner, B. G., and B. Schaffer. 1989. Interrelationships of the lower actinopterygian fishes. *Zool. J. Linn. Soc.* 97:135–187.

Gardiner, B. G., B. Schaeffer, and J. A. Masserie. 2005. A review of the lower actinopterygian phylogeny. *Zool. J. Linn. Soc.* 144:511–525.

Hurley, I. A., R. L. Mueller, K. A. Dunn, E. J. Schmidt, M. Friedman, R. K. Ho, V. E. Prince, Z. Yang, M. G. Thomas, and M. I. Coates. 2007. A new time-scale for ray-finned fish evolution. *Proc. R. Soc. Lond. B Biol. Sci.* 274:489–498.

Janvier, P. 1996. *Early Vertebrates.* Oxford University Press, Oxford.

Patterson, C. 1973. Interrelationships of holosteans. Pp. 233–305 in *Interrelationships of Fishes* (P. H. Greenwood, R. S. Miles, and C. Patterson, eds.). Academic Press, London.

Patterson, C. 1994. Bony fishes. Pp. 57–84 in *Major Features of Vertebrate Evolution* (D. R. Prothero and R. M. Schoch, eds.). Short Courses in Paleontology, No. 7. Paleontological Society, University of Tennessee, Knoxville, TN.

Authors

Jon A. Moore; Wilkes Honors College; Florida Atlantic University; Jupiter, FL 33458, USA. Email: jmoore@fau.edu.

Thomas J. Near; Department of Ecology and Evolutionary Biology; Yale University; New Haven, CT 06520, USA. Email: Thomas. Near@yale.edu.

Date Accepted: 13 August 2013

Primary Editor: Kevin de Queiroz

Neopterygii C. T. Regan 1923 [J. A. Moore and T. J. Near], converted clade name

Registration Number: 210

Definition: The least inclusive crown clade containing *Lepisosteus osseus* (Linnaeus 1758), *Amia calva* Linnaeus 1766, and *Perca fluviatilis* Linnaeus 1758. This is a minimum-crown-clade definition. Abbreviated definition: min crown ∇ (*Lepisosteus osseus* (Linnaeus 1758) & *Amia calva* Linnaeus 1766 & *Perca fluviatilis* Linnaeus 1758).

Etymology: Derived from the Greek *neo* (new) and *pterygion* (little fin).

Reference Phylogeny: Diogo (2007: Figs. 3 and 4) is the primary reference phylogeny; see also Rosen et al. (1981: Fig 62), Maisey (1986: Fig. 10), Patterson (1994: Fig. 1), Stiassny et al. (2004: Fig. 24.1), Nelson (2006: 87), and Near et al. (2012: Figs. 1 and s1). Although *Perca fluviatilis* is not included in the primary reference phylogeny, available evidence indicates that it is deeply nested within the clade *Neoteleostei* originating in branch 22 on that phylogeny (Dettai and Lecointre, 2008; Li et al., 2009; Wiley and Johnson, 2010; Near et al., 2012).

Composition: *Neopterygii* is currently thought to include 30,971 living species (Froese and Pauly, 2013), persisting today as the crown clades *Ginglymodi* (gars) and *Halecostomi* (= *Halecomorphi* [bowfins] + *Teleostei*). *Neopterygii* thus includes the gars, bowfins, and teleosts listed in Nelson (2006), Eschmeyer et al. (1998), and Eschmeyer (2011). Partial lists of extinct *Neopterygii* can be found in Gardiner et al. (1996), Janvier (1996), Benton (2005), and Nelson (2006).

Diagnostic Apomorphies: Patterson (1973) provided the first analysis of apomorphies for *Neopterygii*, with a list including the reduction or loss of clavicles, presence of a symplectic, and a coronoid process on the articular bone. Gardiner (1984) reiterated several of Patterson's characters and added more apomorphies for the *Neopterygii*, including a hyomandibula with opercular process, palatoquadrate disconnected from cheek bones posteriorly and dorsally, and quadratojugal that braces the quadrate. Janvier (1996) also cited the reduction or loss of the clavicle as a neopterygian apomorphy. Springer and Johnson (2004) stated that the presence of obliquus dorsalis and transversus dorsalis muscles represent two apomorphies supporting the monophyly of the *Neopterygii*. Jamieson (1991: Fig 10.3) proposed that the ectaquasperm lacking an acrosome characterizes *Neopterygii*. Diogo (2007) listed five apomorphies.

Synonyms: *Pisces* of Linnaeus (1758) is a partial (and approximate) synonym of *Neopterygii* in that it included most of the then-known teleosts (except *Lophius* and *Hypostomus plecostomus*, the latter of which was included in *Acipenser*) and gars (as *Esox osseus*) but excluded sturgeons (*Acipenser*); *Amia* had not yet been described. *Actinopteri* sensu Cope (1877, 1887), *Teleostei* sensu Regan (1904, 1909), and *Holostei* sensu Goodrich (1930) all included what are now considered amioids, gars, and teleosts, and therefore are approximate synonyms of *Neopterygii* as defined here. *Holostei* sensu J. Nelson (1976) is a partial (and approximate) synonym in that it applied to a paraphyletic group of amioids and gars that excluded teleosts.

Comments: The name *Neopterygii* has consistently been used for a taxon including the holostean and teleostean fishes. We chose this name for the crown clade because of this prior history of consistent use and because all of its synonyms are, given their histories, more appropriately applied to different clades. *Actinopteri* and *Teleostei* are both used elsewhere in this volume for other crown clades, and *Holostei* has been used to denote a group including the extant gars and bowfins, which is paraphyletic according to some inferences (Patterson, 1973) or monophyletic according to others (G. Nelson, 1969; Santini et al., 2009; Wiley and Johnson, 2010; Grande, 2010). The name *Neopterygii* has been implicitly applied to the crown clade by Diogo (2007: Fig. 4), Santini et al. (2009), and Near et al. (2012). We have elected to restrict *Neopterygii* to a crown clade because such clades have ancestors that we can know the most about (de Queiroz and Gauthier, 1992; Gauthier and de Queiroz, 2001).

If amiids and gars do not form a monophyletic group, there is also considerable debate over whether amiids or gars are the sister-group to the *Teleostei* (Arratia, 2000, 2001). Patterson (1973, 1994) and Gardiner et al. (1996) discussed the debate on morphological grounds, and concluded that amiids (*Halecomorphi*) are the closest clade. However, Olsen (1984) and Olsen and McCune (1991) argued that lepisosteids are closer. The molecular data are also equivocal, with Inoue et al. (2003) and Venkatesh et al. (2001) supporting an "ancient fish clade" composed of chondrostean and holosteans together. Lê et al. (1993), Kikugawa et al. (2004), Hurley et al. (2007), Li et al. (2008), Santini et al. (2009), Broughton (2010: Fig. 2), and Near et al. (2012) found strong molecular support for *Neopterygii*, but often with the amiids and lepisosteids together in a monophyletic *Holostei* (Normark et al., 1991; Lê et al., 1993; Kikugawa et al., 2004; Li et al., 2008; Santini

et al., 2009; Near et al., 2012). Molecular support for the *Halecostomi* (amiids + teleosts) is weaker (Hurley et al., 2007; Alfaro et al., 2009).

Literature Cited

Alfaro, M. E., F. Santini, C. Brock, H. Alamillo, A. Dornburg, D. L. Rabosky, G. Carnevale, and L. J. Harmon. 2009. Nine exceptional radiations plus high turnover explain species diversity in jawed vertebrates. *Proc. Natl. Acad. Sci. USA* 106:13410–13414 (and supplemental info pp. 1–20).

Arratia, G. 2000. Phylogenetic relationships of *Teleostei*: past and present. *Estud. Oceanol.* 19:19–51.

Arratia, G. 2001. The sister-group of *Teleostei*: consensus and disagreements. *J. Vertebr. Paleontol.* 21:767–773.

Benton, M. J. 2005. *Vertebrate Paleontology.* 3rd edition. Blackwell Publishing, Malden, MA.

Broughton, R. E. 2010. Phylogeny of teleosts based on mitochondrial genome sequences. Pp. 61–76 in *Origin and Phylogenetic Interrelationships of Teleosts* (J. S. Nelson, H.-P. Schultze, and M. V. H. Wilson, eds.). Verlag Dr. Friedrich Pfeil, Munich.

Cope, E. D. 1877. On the classification of the extinct fishes of the lower types. *Proc. Am. Assoc. Adv. Sci.* 26:292–300.

Cope, E. D. 1887. Geology and paleontology. Zittel's Manual of Paleontology. *Am. Nat.* 21:1014–1020.

Dettai, A., and G. Lecointre. 2008. New insights into the organization and evolution of vertebrate IRBP genes and utility of IRBP gene sequences for the phylogenetic study of the *Acanthomorpha* (*Actinopterygii*: *Teleostei*). *Mol. Phylogenet. Evol.* 48:258–269.

Diogo, R. 2007. *The Origin of Higher Clades.* Science Publishers, Enfield, NH.

Eschmeyer, W. N., ed. 2011. *Catalog of Fishes*, electronic version 14 July 2011. Available at http://research.calacademy.org/ichthyology/catalog/fishcatmain.asp.

Eschmeyer, W. N., C. J. Ferraris, Jr., M. Hoang, and D. J. Long. 1998. *Catalog of Fishes*. Special Publication. California Academy of Sciences, San Francisco, CA.

Froese, R., and D. Pauly, eds. 2013. *FishBase: Fishes Used by Humans* (based on FishBase 2/2013) [under Tools/Fish statistics]. Available at http://fishbase.org/Report/FishesUsedByHumans.php, accessed on 17 March 2013.

Gardiner, B. G. 1984. The relationships of the paleoniscid fishes, a review based on new specimens of *Mimia* and *Moythomasia* from the Upper Devonian of Western Australia. *Bull. Brit. Mus. (Nat. Hist.) Geol.* 37:173–428.

Gardiner, B. G., J. G. Maisey, and D. T. J. Littlewood. 1996. Interrelationships of basal neopterygians. Pp. 117–146 in *Interrelationships of Fishes* (M. L. J. Stiassny, L. R. Parenti, and G. D. Johnson, eds.). Academic Press, New York.

Gauthier, J. A., and K. de Queiroz. 2001. Feathered dinosaurs, flying dinosaurs, crown dinosaurs, and the name *"Aves."* Pp. 7–41 in *New Perspectives on the Origin and Early Evolution of Birds: Proceeding of the International Symposium in Honor of John H. Ostrom* (J. Gauthier and L.F. Gall, eds.). Special Publication of the Peabody Museum of Natural History., New Haven, CT.

Goodrich, E. S. 1930. *Studies on the Structure and Development of Vertebrates*. Macmillan and Co., London.

Grande, L. 2010. Empirical synthetic pattern study of gars (*Lepisosteidae*) and closely related species, based mostly on skeletal anatomy: the resurrection of *Holostei*. *Am. Soc. Ichthyol. Herpetol. Spec. Publ.* 6:1–871.

Hurley, I. A., R. L. Mueller, K. A. Dunn, E. J. Schmidt, M. Friedman, R. K. Ho, V. E. Prince, Z. Yang, M. G. Thomas, and M. I. Coates. 2007. A new time scale for ray-finned fish evolution. *Proc. R. Soc. Lond. B Biol. Sci.* 274:489–498.

Inoue, J. G., M. Miya, K. Tsukamoto, and M. Nishida. 2003. Basal actinopterygian relationships: a mitogenomic perspective on the phylogeny of the "ancient fish". *Mol. Phylogenet. Evol.* 26:110–120.

Jamieson, B. G. M. 1991. *Fish Evolution and Systematics: Evidence from Spermatozoa*. Cambridge University Press, Cambridge, UK.

Janvier, P. 1996. *Early Vertebrates*. Oxford University Press, Oxford.

Kikugawa, K., K. Katoh, S. Kuraku, H. Sakurai, O. Ishida, N. Iwabe, and T. Miyata. 2004. Basal jawed vertebrate phylogeny inferred from multiple nuclear DNA-coded genes. *BMC Biol.* 2:3.

Lê, H. L., G. Lecointre, and R. Perasso. 1993. A 28S rRNA-based phylogeny of the gnathostomes: first steps in the analysis of conflict and congruence with morphologically based cladograms. *Mol. Phylogenet. Evol.* 2(1):31–51.

Li, B., A. Dettai, C. Cruaud, A. Coiloix, M. Desoutter-Meniger, and G. Lecointre. 2009. RNF213, a new nuclear marker for acanthomorph phylogeny. *Mol. Phylogenet. Evol.* 50:345–363.

Li, C., G. Lu, and G. Orti. 2008. Optimal data partitioning and a test case for ray-finned fishes (*Actinopterygii*) based on ten nuclear loci. *Syst. Biol.* 57:519–539.

Linnaeus, C. 1758. *Systema Naturae*. 10th edition. Laurentii Salvii, Holmiae (Stockholm).

Maisey, J. G. 1986. Heads and tails: a chordate phylogeny. *Cladistics* 2:201–256.

Near, T. J., R. I. Eytan, A. Dornburg, K. L. Kuhn, J. A. Moore, M. P. Davis, P. C. Wainwright, M. Friedman, and W. L. Smith. 2012. Resolution of ray-finned fish phylogeny and timing of diversification. *Proc. Natl. Acad. Sci. USA* 109:13698–13703.

Nelson, G. 1969. Gill arches and the phylogeny of fishes, with notes on the classification of vertebrates. *Bull. Am. Mus. Nat. Hist.* 141:475–568.

Nelson, J. 1976. *Fishes of the World*. 1st edition. John Wiley & Sons, New York.

Nelson, J. 2006. *Fishes of the World*. 4th edition. John Wiley & Sons, New York.

Normark, B. B., A. R. McCune, and R. G. Harrison. 1991. Phylogenetic relationships of neopterygian fishes inferred from mitochondrial DNA sequences. *Mol. Biol. Evol.* 8:819–834.

Olsen, P. E. 1984. The skull and pectoral girdle of the parasemionotid fish *Watsonulus eugnathoides* from the Early Triassic Sakamena Group of Madagascar, with comments on the relationships of the holostean fishes. *J. Vertebr. Paleontol.* 4:481–499.

Olsen, P. E., and A. M. McCune. 1991. Morphology of the *Semionotus elegans* species group from the Early Jurassic part of the Newark supergroup of eastern North America with comments on the family *Semionotidae* (*Neopterygii*). *J. Vertebr. Paleontol.* 11:269–292.

Patterson, C. 1973. Interrelationships of holosteans. Pp. 233–305 in *Interrelationships of Fishes* (P. H. Greenwood, R. S. Miles, and C. Patterson, eds.). Academic Press, London.

Patterson, C. 1994. Bony fishes. Pp. 57–84 in *Major Features of Vertebrate Evolution* (D. R. Prothero and R. M. Schoch, eds.). Short Courses in Paleontology, No. 7. Paleontological Society, University of Tennessee, Knoxville, TN.

de Queiroz, K., and J. A. Gauthier. 1992. Phylogenetic taxonomy. *Annu. Rev. Ecol. Syst.* 23:449–480.

Regan, C. T. 1904. The phylogeny of the *Teleostomi*. *Ann. Mag. Nat. Hist.* (*Ser. 7*) 13:329–349, pl. 7.

Regan, C. T. 1909. The classification of teleostean fishes. *Ann. Mag. Nat. Hist.* (*Ser. 8*) 3:75–86.

Regan, C. T. 1923. The skeleton of *Lepidosteus*, with remarks on the origin and evolution of the lower neopterygian fishes. *Proc. Zool. Soc.* 93:445–461.

Rosen, D. E., P. L. Forey, B. G. Gardiner, and C. Patterson. 1981. Lungfish, tetrapods, paleontology, and plesiomorphy. *Bull. Am. Mus. Nat. Hist.* 167:159–276.

Santini, F., L. J. Harmon, G. Carnevale, and M. E. Alfaro. 2009. Did genome duplication drive the origin of teleosts? A comparative study of diversification in ray-finned fishes. *BMC Evol. Biol.* 9:194–209.

Springer, V. G., and G. D. Johnson. 2004. Study of the dorsal gill-arch musculature of teleostome fishes, with special reference to the *Actinopterygii*. *Bull. Biol. Soc. Wash.* 11:1–236

Stiassny, M. L. J., E. O. Wiley, G. D. Johnson, and M. R. de Carvalho. 2004. Gnathostome fishes. Pp. 410–429 in *Assembling the Tree of Life* (M. J. Donoghue and J. Cracraft, eds.). Oxford University Press, Oxford.

Venkatesh, B., M. V. Erdmann, and S. Brenner. 2001. Molecular synapomorphies resolve evolutionary relationships of extant jawed vertebrates. *Proc. Natl. Acad. Sci. USA* 98:11382–11387.

Wiley, E. O., and G. D. Johnson. 2010. A teleost classification based on monophyletic groups. Pp. 123–182 in *Origin and Phylogenetic Interrelationships of Teleosts* (J. S. Nelson, H.-P. Schultze, and M. V. H. Wilson, eds.). Verlag Dr. Friedrich Pfeil, Munich, Germany.

Authors

Jon A. Moore; Wilkes Honors College; Florida Atlantic University; Jupiter, FL 33458, USA. Email: jmoore@fau.edu.

Thomas J. Near; Department of Ecology and Evolutionary Biology; Yale University; New Haven, CT 06520, USA. Email: Thomas. Near@yale.edu.

Date Accepted: 8 September 2013

Primary Editor: Kevin de Queiroz

Pan-Teleostei J. A. Moore and T. J. Near, new clade name

Registration Number: 213

Definition: The total clade of the crown clade *Teleostei*. This is a crown-based total-clade definition. Abbreviated definition: total ∇ of *Teleostei*.

Etymology: Derived from the Greek *pantos* (all), indicating that the name refers to a total clade or pan-monophyllum, plus *Teleostei*, the name of the corresponding crown clade (see the entry for *Teleostei* in this volume for the etymology of that name).

Reference Phylogeny: De Pinna (1996: Fig. 1, where the clade here named *Pan-Teleostei* is called *Teleostei*) is the primary reference phylogeny, but see also Janvier (1996: Fig. 4.73) and Arratia (2000a: Fig. 20, 2000b: Fig. 21) who also use the name *Teleostei* for a clade approximating the one here named *Pan-Teleostei*.

Composition: *Teleostei* (this volume) and all extinct fishes that share a more recent ancestry with *Teleostei* than with any holostean fishes (amiids and lepisosteids). Patterson (1973, 1993) inferred several fossils as members of the teleost stem group, and Patterson and Rosen (1977: Fig. 54), Lauder and Liem (1983: Fig 14), Arratia (1996: Fig. 5; 1997: Fig. 97; 2000a: Fig. 20, 2000b: Fig. 21, 2008a: Fig. 7), and Janvier (1996: Fig 4.73) presented cladograms identifying pholidophorid, ichthyodectiform, crossognathiform, and leptolepid fishes as stem teleosts. Patterson (1977), de Pinna (1996), and Arratia (2000c) included *Pachycormiformes*, *Aspidorhynchiformes*, and *Pleuropholidae* as stem teleosts as well. Gardiner et al. (1996) and Arratia (2000c) indicated that

both *Dapedium* and pycnodonts may also be stem teleosts. However, Arratia (2001) pointed out that *Aspidorhynchiformes*, *Pachycormiformes*, *Dapedium*, and pycnodonts frequently change their positions on phylogenetic trees and sometimes end up associated with holostean, rather than teleostean, taxa.

Diagnostic Apomorphies: The characters listed here did not necessarily arise immediately at the origin of *Pan-Teleostei* but instead arose early in the history of the stem lineage leading to crown *Teleostei*. These features are useful in recognizing fossils as members of the total clade. We used the distributions of characters and character states given in De Pinna (1996) to determine that the following six features originated near the base of the teleost stem lineage (i.e., early in the history of *Pan-Teleostei*): the presence of uroneural bones in the caudal fin, hypurals divided into two groups, seven or fewer epurals, urohyal formed as an unpaired tendon bone, foramen in the hypohyals for the hyoidean artery, and lower jaw without coronoid bones. DePinna (1996) listed another 17 features that diagnose various nodes along the teleost stem lineage prior to the crown node. In addition, presence of an elongate posteroventral process on the quadrate, the single feature defining Arratia's (1999, 2000b, 2001) apomorphy-based *Teleostei*, occurs about midway up the teleost stem lineage.

Synonyms: *Teleostei* as used by several authors (Jordan, 1905; Patterson, 1977, 1994; Patterson and Rosen 1977, Arratia, 1991, 1997; Janvier 1996; Wiley and Johnson, 2010) encompasses both living teleosts and fossils representing the teleost stem group and is therefore an approximate synonym for *Pan-Teleostei*. However, de Pinna (1996) provided a maximum (and implicitly a total clade) definition for *Teleostei* that makes the name in his use an unambiguous synonym of *Pan-Teleostei*. Arratia (2001, 2008b) used the name *Teleosteomorpha* for the clade composed of an apomorphy-based *Teleostei* plus its stem group, which makes that name another unambiguous synonym.

Comments: *Teleostei* is unavailable as a name for this total clade, because we have used it for the crown clade, which, as we argued (see entry for *Teleostei* in this volume), is consistent with the original definition of Müller (1845). We choose to use the name *Pan-Teleostei* over the existing name *Teleosteomorpha*, because *Teleosteomorpha* is not a widely used name and the prefix "*Pan-*" will readily identify our name as designating the total clade of the crown clade *Teleostei*.

Arratia (1999, 2000b,c, 2001, 2008b, 2010) used *Teleostei* as an apomorphy-based name for the clade composed of pholidophorids + *Leptolepis* + varasichthyids + ichthyodectiforms + crown teleosts. Although Wiley and Johnson (2010) followed Arratia's use of *Teleostei*, they indicated (p.128) that Johnson preferred to use the name *Teleostei* for the crown clade.

Patterson (1973) recognized the inherent problem with applying the name *Teleostei* to a more inclusive clade containing a series of fossil stem taxa. He stated, "I realize that by expanding the content of the *Teleostei* in this way I have drastically altered the meaning of the word 'teleost'" (Patterson 1973: 300). This has precisely been the problem as new research by various authors has moved the node at which *Teleostei* is applied to include more fossil stem taxa. The composition of *Teleostei* has been ever expanding as the name *Teleostei* has crept from one node to another down the tree. By restoring the name *Teleostei* to the crown clade, as we have done above, we are avoiding the ever-expanding

composition and altered meaning that has characterized the history of this name. The utility of *Pan-Teleostei* is that it will accommodate all presently known and yet to be discovered fossils that diverged from (or were part of) the ancestral lineage of teleosts after its most recent common ancestor with holosteans (amiids and/or gars) while clearly indicating their relationship to the teleost crown clade.

Literature Cited

Arratia, G. 1991. The caudal skeleton of Jurassic teleosts; a phylogenetic analysis. Pp. 249–340 in *Early Vertebrates and Related Problems in Evolutionary Biology* (Chang, M.-M., Y.-H. Liu, and G.-R. Zhang, eds.). Science Press, Beijing.

Arratia, G. 1996. Reassessment of the phylogenetic relationships of certain Jurassic teleosts and their implications on teleostean phylogeny. Pp. 219–242 in *Mesozoic Fishes – Systematics and Paleoecology* (G. Arratia and G. Viohl, eds.). Verlag Dr. Friedrich Pfeil, Munich, Germany.

Arratia, G. 1997. Basal teleosts and teleostean phylogeny. *Palaeoichthyologica* 7:5–168.

Arratia, G. 1999. The monophyly of *Teleostei* and stem-group teleosts. Consensus and disagreements. Pp. 265–334 in *Mesozoic Fishes 2: Systematics and Fossil Record* (G. Arratia and H.-P. Schultze, eds.). Dr. Friedrich Pfeil, Munich, Germany.

Arratia, G. 2000a. New teleostean fishes from the Jurassic of southern Germany and the systematic problems concerning the 'pholidophoriforms'. *Palaeontol. Zeit.* 74:113–143.

Arratia, G. 2000b. Remarkable teleostean fishes from the Late Jurassic of southern Germany and their phylogenetic relationships. *Mitt. Mus. Nat.kd. Berlin, Geowiss. Reihe* 3:137–179.

Arratia, G. 2000c. Phylogenetic relationships of *Teleostei*: past and present. *Estud. Oceanol.* 19:19–51.

Arratia, G. 2001. The sister-group of *Teleostei*: consensus and disagreements. *J. Vertebr. Paleontol.* 21:767–773.

Arratia, G. 2008a. The varasichthyid and other crossognathiform fishes, and the break-up of Pangea. Pp.71–92 in *Fishes and the Break-up of Pangea* (L. Cavin, A. Longbottom, and M. Richter, eds.). Geological Society, London.

Arratia, G. 2008b. Actinopterygian postcranial skeleton with special reference to the diversity of fin ray elements, and the problem of identifying homologies. Pp. 49–191 in *Mesozoic Fishes 4 – Homology and Phylogeny* (G. Arratia, H.- P. Schultze, and M. V. H. Wilson, eds.). Verlag Dr. Friedrich Pfeil, Munich.

Arratia, G. 2010. Critical analysis of the impact of fossils on teleostean phylogenies, especially that of basal teleosts. Pp. 247–274 in *Morphology, Phylogeny and Paleobiogeography of Fossil Fishes* (D. K. Elliot, J. G. Maisey, X. Yu, and D. Miao, eds.). Verlag Dr. Friedrich Pfeil, Munich.

de Pinna, M. C. C. 1996. Teleostean monophyly. Pp. 147–162 in *Interrelationships of Fishes* (M. L. J. Stiassny, L. R. Parenti, and G. D. Johnson, eds.). Academic Press, New York.

Gardiner, B. G., J. G. Maisey, and D. T. J. Littlewood. 1996. Interrelationships of basal neopterygians. Pp. 117–146 in *Interrelationships of Fishes* (M. L. J. Stiassny, L. R. Parenti and G. D. Johnson, eds.). Academic Press, New York.

Janvier, P. 1996. *Early Vertebrates*. Oxford University Press, Oxford.

Jordan, D. S. 1905. *A Guide to the Study of Fishes*. H. Holt, New York.

Lauder, G. V., and K. F. Liem. 1983. The evolution and interrelationships of the actinopterygian fishes. *Bull. Mus. Comp. Zool.* 150:95–197.

Müller, J. 1845. Über den Bau und die Grenzen der Ganoiden, und über das natürliche System der Fische. *Arch. Naturgesch.* 11:91–141.

Patterson, C. 1973. Interrelationships of holosteans. Pp. 207–226 in *Interrelationships of Fishes* (P. H. Greenwood, R. S. Miles, and C. Patterson, eds.). Academic Press, London

Patterson, C. 1977. The contribution of paleontology to teleostean phylogeny. Pp. 579–643 in *Major Patterns in Vertebrate Evolution* (P. C. Hecht, P. C. Goody, and B. M. Hecht, eds.). Plenum Press, New York.

Patterson, C. 1993. *Osteichthyes: Teleostei*. Pp. 622–656 in *Fossil Record* 2. (M. J. Benton, ed.). Chapman and Hall, London.

Patterson, C. 1994. Bony fishes. Pp. 57–84 in *Major Features of Vertebrate Evolution* (D. R. Prothero and R. M. Schoch, eds.). Short Courses in Paleontology, No. 7. Paleontological Society, University of Tennessee, Knoxville, TN.

Patterson, C., and D. E. Rosen. 1977. Review of ichthyodectiform and other Mesozoic teleost fishes and the theory and practice of classifying fossils. *Bull. Am. Mus. Nat. Hist.* 158:81–172.

Wiley, E. O., and G. D. Johnson. 2010. A teleost classification based on monophyletic groups. Pp. 123–182 in *Origin and Phylogenetic Interrelationships of Teleosts* (J. S. Nelson, H.-P. Schultze, and M. V. H. Wilson, eds.). Verlag Dr. Friedrich Pfeil, Munich.

Authors

Jon A. Moore; Wilkes Honors College; Florida Atlantic University; Jupiter, FL 33458, USA. Email: jmoore@fau.edu.

Thomas J. Near; Department of Ecology and Evolutionary Biology; Yale University; New Haven, CT 06520, USA. Email: Thomas.Near@yale.edu.

Date Accepted: 13 August 2013

Primary Editor: Kevin de Queiroz

Teleostei J. Müller 1845 [J. A. Moore and T. J. Near], converted clade name

Registration Number: 212

Definition: The least inclusive crown clade that contains *Hiodon tergisus* Lesueur 1818 (*Osteoglossomorpha*), *Elops saurus* Linnaeus 1766 (*Elopomorpha*), *Engraulis encrasicolus* (Linnaeus 1758) (*Otocephala/Clupeomorpha*), and *Perca fluviatilis* Linnaeus 1758 (*Euteleostei*). This is a minimum-crown-clade definition. Abbreviated definition: min crown ∇ (*Hiodon tergisus* Lesueur 1818 & *Elops saurus* Linnaeus 1766 & *Engraulis encrasicolus* (Linnaeus 1758) & *Perca fluviatilis* Linnaeus 1758).

Etymology: Derived from the Greek *teleos* (perfect, complete) and *osteon* (bone) in reference to the more complete ossification of the skeleton in fishes of this clade relative to chondrosteans and holosteans.

Reference Phylogeny: Diogo (2007: Figs. 3 and 4) is the primary reference phylogeny; see also Patterson (1994: Fig. 1), de Pinna (1996: Fig. 1), Nelson (2006: 87), Stiassny et al. (2004: Fig. 24.1), and Near et al. (2012: Figs. 1 and s1). Although *Perca fluviatilis* is not included in the primary reference phylogeny, available evidence indicates that it is deeply nested within the clade *Neoteleostei* originating in branch 22 on that phylogeny (Dettai and Lecointre, 2008; Li et al., 2009; Wiley and Johnson, 2010; Near et al., 2012).

Composition: *Teleostei* is currently thought to include 30,963 living species (Froese and Pauly, 2013), divided between several subclades persisting today as *Osteoglossomorpha*, *Elopomorpha*, *Otocephala* (*Clupeomorpha* + *Ostariophysi*), and *Euteleostei*. *Teleostei* thus includes all bony tongues, elopiforms, eels, herrings, ostariophysans, protacanthopterygians, esocoids, and neoteleosts listed in Nelson (2006), Eschmeyer et al. (1998), and Eschmeyer (2011). Partial lists of extinct *Teleostei* can be found in Patterson (1993), Janvier (1996), Benton (2005), and Nelson (2006).

Diagnostic Apomorphies: De Pinna (1996) provided seven features that are synapomorphies for the crown group *Teleostei* as defined here (= his *Teleocephala*): seven or fewer hypurals, presence of a basihyal, four pharyngobranchials, presence of a craniotemporal muscle, myelinated fibers in the lateral forebrain bundle, eminentiae granulares of the cerebellum, and presence of accessory nasal sacs. Janvier (1996: Fig. 4.73) cited the sharply upturned caudal axis at the level of the first preural centrum as an apomorphy of crown-group teleosts. Arratia (1999) also listed five or fewer uroneurals and bases of the dorsal fin rays aligned with the hypurals. Springer and Johnson (2004) stated that the insertion of obliquus dorsalis muscle on epibranchial 3 represents a synapomorphy supporting the monophyly of the *Teleostei* relative to other extant actinopterygians. Wiley and Johnson (2010) reiterated several of the same characters listed above, but added uroneural 1 reaching anterior to preural centrum 2. Diogo (2007) gave 19 unambiguous characters, most of which are muscular features, diagnostic for the living members of the crown group *Teleostei* relative to other extant actinopterygians. Hurley et al. (2007) demonstrated that *Teleostei* is characterized by a whole-genome duplication.

Synonyms: De Pinna (1996) named the group of Recent teleosts *Teleocephala*, which is an approximate synonym of *Teleostei* as defined here, and he used the name *Teleostei* for the total clade (see also Arratia, 2001).

Comments: Müller's (1845) original description of *Teleostei* was based on two soft body features: the fibrous nature of the ventral aorta and presence of only two valves in the conus arteriosus, so that the name was implicitly applied to a crown clade. Heckel (1850a,b, 1851) assigned the fossils *Leptolepis*, *Thrissops*, and *Tharsis*, now generally considered members of the teleost stem group, to the teleosts, and Woodward (1895, 1901) and Boulenger (1904) added pholidophorids. While those earlier authors did not explicitly indicate the phylogenetic relationships of those fossils, Nelson (1969), Patterson and Rosen (1977), and Arratia (1991) unambiguously placed pholidophorids and several other fossil taxa in the stem group of *Teleostei*. Patterson (1977) put *Pachycormiformes*, *Aspidorhynchiformes*, *Pleuropholidae*, and *Ichthyokentema* outside of the paraphyletic "pholidophorids" as early branches within his *Teleostei*. Gardiner et al. (1996) added the clade of *Dapedium* + pycnodonts as a still earlier branch. This series of additions continually changed the node at which the name *Teleostei* was placed, as well as the composition of the taxon, to the point where a new name (*Teleocephala*) had to be created for the crown group, because the name *Teleostei* had crept so far down the tree. Up to the introduction of cladistics, these expansions were effectively apomorphy-based changes. Patterson's (1973, 1994) advocacy of a total clade definition for *Teleostei* influenced other ichthyologists to accept the ever-expanding composition as more stem taxa were recognized. Different uses of the name *Teleostei* (i.e., total clade versus crown clade) by palaeontologists and neontologists can have a huge impact on molecular divergence time analyses if those different uses are confused. Biologists performing such analyses could potentially make the mistake of using the fossil age estimates for the total clade to calibrate the crown node. Alfaro et al. (2009) and Near et al. (2012) are among the few studies using molecular data to estimate divergence times in fishes to make this distinction between ages derived from stem vs. crown fossils.

Müller (1845) originally used soft features that can be assessed only in living fishes to diagnose *Teleostei*. Some palaeontologists (e.g., Arratia, 1997, 2000, 2001) have been critical of this decision, because such features are not applicable to fossils. However, the fact that soft features can only be assessed in living fishes also means that a much larger body of information is available for inferences about the crown node relative to those along the stem (de Queiroz and Gauthier, 1992). We choose *Teleostei* for the crown clade name, rather than *Teleocephala*, for this reason and because of the pervasive use and understanding of *Teleostei* as the more ossified, ray-finned fishes as evidenced by its widespread use for the crown clade among neontologists (e.g., Inoue et al., 2003, 2004; Freeman et al., 2013; Betancur-R. et al., 2013). It is appropriate to return the name to the node to which it originally applied—that representing the last common ancestor of the living taxa. If the scientific community would agree to adopt this use of the name, communication among neontologists and palaeontologists would be facilitated, which would help to avoid inappropriate fossil calibrations for divergence time estimates and other errors related to the distinction between crown and total clades.

There is still some debate as to which subclade of *Teleostei* is the sister group of the remaining crown teleosts. Several studies proposed *Osteoglossomorpha* as the sister group to the rest (Patterson and Rosen, 1977; Forey et al., 1996; Inoue et al., 2001, 2004) whereas others advocated *Elopomorpha* (Arratia, 1991, 1997, 2000, 2010; Diogo, 2007; Arratia and Tischlinger, 2010; Alfaro et al., 2009; Santini et al., 2009).

Venkatesh et al. (2001), Inoue et al. (2001, 2003, 2004), Kikugawa et al. (2004), Hurley et al. (2007), Li et al. (2008), Alfaro et al. (2009) Broughton (2010), and Near et al. (2012) provided molecular support for the monophyly of the crown clade *Teleostei*.

Literature Cited

Alfaro, M. E., F. Santini, C. Brock, H. Alamillo, A. Dornburg, D. L. Rabosky, G. Carnevale, and L. J. Harmon. 2009. Nine exceptional radiations plus high turnover explain species diversity in jawed vertebrates. *Proc. Natl. Acad. Sci. USA* 106:13410–13414 (and supplemental info pp. 1–20).

Arratia, G. 1991. The caudal skeleton of Jurassic teleosts; a phylogenetic analysis. Pp. 249–340 in *Early Vertebrates and Related Problems in Evolutionary Biology* (Chang, M.-M., Y.-H. Liu, and G.-R. Zhang, eds.). Science Press, Beijing.

Arratia, G. 1997. Basal teleosts and teleostean phylogeny. *Palaeoichthyologica* 7:5–168.

Arratia, G. 1999. The monophyly of *Teleostei* and stem-group teleosts. Consensus and disagreements. Pp. 265–334 in *Mesozoic Fishes 2: Systematics and Fossil Record* (G. Arratia and H.-P. Schultze, eds.). Dr. Friedrich Pfeil, Munich.

Arratia, G. 2000. Phylogenetic relationships of *Teleostei*: past and present. *Estud. Oceanol.* 19:19–51.

Arratia, G. 2001. The sister-group of *Teleostei*: consensus and disagreements. *J. Vertebr. Paleontol.* 21:767–773.

Arratia, G. 2010. Critical analysis of the impact of fossils on teleosten phylogenies, especially that of basal teleosts. Pp. 247–274 *in Morphology, Phylogeny and Paleobiogeography of Fossil Fishes* (D. K. Elliott, J. G. Maisey, X. Yu, and D. Miao, eds.). Verlag Dr. Friedrich Pfeil, Munich.

Arratia, G., and H. Tischlinger. 2010. The first record of Late Jurassic crossognathiform fishes from Europe and their phylogenetic importance for teleostean phylogeny. *Foss. Rec.* 13:317–341.

Benton, M. J. 2005. *Vertebrate Paleontology.* 3rd edition. Blackwell Publishing, Malden, MA.

Betancur-R., R., R. E. Broughton, E. O. Wiley, K. Carpenter, J. A. López, C. Li, N. I. Holcroft, D. Arcila, M. Sanciangco, J. C. Cureton II, F. Zhang, T. Buser, M. A. Campbell, J. A. Ballesteros, A. Roa-Varon, S. Willis, W. C. Borden, T. Rowley, P. C. Reneau, D. J. Hough, G. Lu, T. Grande, G. Arratia, and G. Ortí. 2013. The tree of life and a new classification of bony fishes. *PLOS Curr. Tree Life* 2013 April 18:5.

Boulenger, G. A. 1904. A synopsis of the suborders and families of teleostean fishes. *Ann. Mag. Nat. Hist. (Ser. 7)* 13:161–190.

Broughton, R. E. 2010. Phylogeny of teleosts based on mitochondrial genome sequences. Pp. 61–76 in *Origin and Phylogenetic Interrelationships of Teleosts* (J. S. Nelson, H.-P. Schultze, and M. V. H. Wilson, eds.). Verlag Dr. Friedrich Pfeil, Munich.

Diogo, R. 2007. *The Origin of Higher Clades.* Science Publishers, Enfield, NH.

Dettai, A., and G. Lecointre. 2008. New insights into the organization and evolution of vertebrate IRBP genes and utility of IRBP gene sequences for the phylogenetic study of the *Acanthomorpha (Actinopterygii: Teleostei). Mol. Phylogenet. Evol.* 48:258–269.

Eschmeyer, W. N., ed. 2011. *Catalog of Fishes*, electronic version 14 July 2011. Available at http://research.calacademy.org/ichthyology/catalog/fishcatmain.asp.

Eschmeyer, W. N., C. J. Ferraris, Jr., M. Hoang, and D. J. Long. 1998. *Catalog of Fishes.* Special Publication. California Academy of Sciences, San Francisco, CA.

Forey, P., D. T. J. Littlewood, P. Ritchie, and A. Meyer. 1996. Interrelationships of elopomorph fishes. Pp. 175–191 in *Interrelationships of Fishes* (M. L. J. Stiassny, L. R. Parenti, and G. D. Johnson, eds.). Academic Press, New York.

Freeman, S., K. Quillin, and L. Allison. 2013. *Biological Science.* 5th edition. Benjamin Cummings, Upper Saddle River, NJ.

Froese, R., and D. Pauly, eds. 2013. *FishBase: Fishes Used by Humans* (based on FishBase 2/2013) [under Tools/Fish statistics]. Available at http://fishbase.org/Report/FishesUsedByHumans.php, accessed on 17 March 2013.

Gardiner, B. G., J. G. Maisey, and D. T. J. Littlewood. 1996. Interrelationships of basal neopterygians. Pp. 117–146 in *Interrelationships of Fishes* (M. L. J. Stiassny, L. R. Parenti, and G. D. Johnson, eds.). Academic Press, New York.

Heckel, J. J. 1850a. Ueber das Wirbelsäulen-Ende bei Ganoiden und Teleostiern. *Sitzungsber. Kaiserl. Akad. Wiss. Math.-Naturwiss. Cl.* 5:143–148.

Heckel, J. J. 1850b. Ueber die Wirbelsäule fossiler Ganoiden. *Sitzungsber. Kaiserl. Akad. Wiss. Math.-Naturwiss. Cl.* 5:358–368.

Heckel, J. J. 1851. Über die Ordnung der *Chondrostei* und die Gattungen *Amia, Cyclurus, Notaeus. Sitzungsber. Kaiserl. Akad. Wiss. Math.-Naturwiss. Cl.* 6:219–224.

Hurley, I. A., R. L. Mueller, K. A. Dunn, E. J. Schmidt, M. Friedman, R. K. Ho, V. E. Prince, Z. Yang, M. G. Thomas, and M. I. Coates. 2007. A new time-scale for ray-finned fish evolution. *Proc. R. Soc. Lond. B Biol. Sci.* 274:489–498.

Inoue, J. G., M. Miya, K. Tsukamoto, and M. Nishida. 2001. A mitogenomic perspective on the basal teleostean phylogeny: resolving higher-level relationships with longer DNA sequences. *Mol. Phylogenet. Evol.* 20:275–285.

Inoue, J. G., M. Miya, K. Tsukamoto, and M. Nishida. 2003. Basal actinopterygian relationships: a mitogenomic perspective on the phylogeny of the "ancient fish". *Mol. Phylogenet. Evol.* 26:110–120.

Inoue, J. G., M. Miya, K. Tsukamoto, and M. Nishida. 2004. Mitogenomic evidence for the monophyly of elopomorph fishes (*Teleostei*) and the evolutionary origin of the leptocephalus larva. *Mol. Phylogenet. Evol.* 32:274–286.

Janvier, P. 1996. *Early Vertebrates.* Oxford University Press, Oxford.

Kikugawa, K., K. Katoh, S. Kuraku, H. Sakurai, O. Ishida, N. Iwabe, and T. Miyata. 2004. Basal jawed vertebrate phylogeny inferred from multiple nuclear DNA-coded genes. *BMC Biol.* 2:3.

Li, B., A. Dettai, C. Cruaud, A. Coiloix, M. Desoutter-Meniger, and G. Lecointre. 2009. RNF213, a new nuclear marker for acanthomorph phylogeny. *Mol. Phylogenet. Evol.* 50:345–363.

Li, C., G. Lu, and G. Orti. 2008. Optimal data partitioning and a test case for ray-finned fishes (*Actinopterygii*) based on ten nuclear loci. *Syst. Biol.* 57:519–539.

Müller, J. 1845. Über den Bau und die Grenzen der Ganoiden, und über das natürliche System der Fische. *Arch. Naturgesch.* 11:91–141.

Near, T. J., R. I. Eytan, A. Dornburg, K. L. Kuhn, J. A. Moore, M. P. Davis, P. C. Wainwright, M. Friedman, and W. L. Smith. 2012. Resolution of ray-finned fish phylogeny and timing of diversification. *Proc. Natl. Acad. Sci. USA* 109:13698–13703.

Nelson, G. 1969. Gill arches and the phylogeny of fishes, with notes on the classification of vertebrates. *Bull. Am. Mus. Nat. Hist.* 141:475–568.

Nelson, J. 2006. *Fishes of the World.* 4th edition. John Wiley & Sons, New York.

Patterson, C. 1973. Interrelationships of holosteans. Pp. 233–305 in *Interrelationships of Fishes* (P. H. Greenwood, R. S. Miles, and C. Patterson, eds.). Academic Press, London.

Patterson, C. 1977. The contribution of paleontology to teleostean phylogeny. Pp. 579–643 in *Major Patterns in Vertebrate Evolution* (P. C. Hecht, P. C. Goody, and B. M. Hecht, eds.). Plenum Press, New York.

Patterson, C. 1993. *Osteichthyes: Teleostei.* Pp. 622–656 in *Fossil Record 2* (M. J. Benton, ed.). Chapman and Hall, London.

Patterson, C. 1994. Bony fishes. Pp. 57–84 in *Major Features of Vertebrate Evolution* (D. R. Prothero and R. M. Schoch, eds.). Short Courses in Paleontology, No. 7. Paleontological Society, University of Tennessee, Knoxville, TN.

Patterson, C., and D. E. Rosen. 1977. Review of ichthyodectiform and other Mesozoic teleost fishes and the theory and practice of classifying fossils. *Bull. Am. Mus. Nat. Hist.* 158:81–172.

de Pinna, M. C. C. 1996. Teleostean monophyly. Pp. 147–162 in *Interrelationships of Fishes* (M. L. J. Stiassny, L. R. Parenti, and G. D. Johnson, eds.). Academic Press, New York.

de Queiroz, K., and J. Gauthier. 1992. Phylogenetic taxonomy. *Annu. Rev. Ecol. Syst.* 23:449–480.

Santini, F., L. J. Harmon, G. Carnevale, and M. E. Alfaro. 2009. Did genome duplication drive the origin of teleosts? A comparative study of diversification in ray-finned fishes. *BMC Evol. Biol.* 9:194–209.

Springer, V. G., and G. D. Johnson. 2004. Study of the dorsal gill-arch musculature of teleostome fishes, with special reference to the *Actinopterygii. Bull. Biol. Soc. Wash.* 11:1–236.

Stiassny, M. L. J., E. O. Wiley, G. D. Johnson, and M. R. de Carvalho. 2004. Gnathostome fishes. Pp. 410–429 in *Assembling the Tree of Life* (M. J. Donoghue and J. Cracraft, eds.). Oxford University Press, Oxford.

Venkatesh, B., M. V. Erdmann, and S. Brenner. 2001. Molecular synapomorphies resolve evolutionary relationships of extant jawed vertebrates. *Proc. Natl. Acad. Sci. USA* 98:11382–11387.

Wiley, E. O., and G. D. Johnson. 2010. A teleost classification based on monophyletic groups. Pp. 123–182 in *Origin and Phylogenetic Interrelationships of Teleosts* (J. S. Nelson, H.-P. Schultze, and M. V. H. Wilson, eds.). Verlag Dr. Friedrich Pfeil, Munich.

Woodward, A. S. 1895. *Catalogue of Fossil Fishes in the British Museum (Natural History)*, Vol. 3. British Museum (Natural History), London.

Woodward, A. S. 1901. *Catalogue of Fossil Fishes in the British Museum (Natural History)*, Vol. 4. British Museum (Natural History), London.

Authors

Jon A. Moore; Wilkes Honors College; Florida Atlantic University; Jupiter, FL 33458, USA. Email: jmoore@fau.edu.

Thomas J. Near; Department of Ecology and Evolutionary Biology; Yale University; New Haven, CT 06520, USA. Email: Thomas.Near@yale.edu.

Date Accepted: 22 September 2013

Primary Editor: Kevin de Queiroz

Ostariophysi M. Sagemehl 1885 [J. G. Lundberg], converted clade name

Registration Number: 196

Definition: The crown clade originating in the most recent common ancestor of *Gonorynchus* (originally *Cyprinus*) *gonorynchus* (Linnaeus 1766) (*Gonorynchiformes*), *Cyprinus carpio* Linnaeus 1758 (*Cypriniformes*), *Charax* (originally *Salmo*) *gibbosus* (Linnaeus 1758) (*Characiformes*), *Gymnotus carapo* Linnaeus 1758 (*Gymnotiformes*; *Gymnotoidei* on the reference phylogeny) and *Silurus glanis* Linnaeus 1758 (*Siluriformes*; *Siluroidei* on the reference phylogeny). This is a minimum-crown-clade definition. Abbreviated definition: min crown ∇ (*Gonorynchus gonorynchus* (Linnaeus 1766) & *Cyprinus carpio* Linnaeus 1758 & *Charax gibbosus* (Linnaeus 1758) & *Gymnotus carapo* Linnaeus 1758 & *Silurus glanis* Linnaeus 1758).

Etymology: Derived from Greek *ostarion* meaning "little bone" and *physinx* meaning "bladder" in reference to bones linking the swim bladder to the inner ear.

Reference Phylogeny: For the purposes of applying the definition above, Figure 1 of Fink and Fink (1981: 302) is the primary reference phylogeny. Although the terminal taxa in the reference phylogeny are large clades rather than species, the five specifier species used in the definition are the type species associated with the names *Gonorynchus* (and therefore also *Gonorynchiformes*), *Cyprinus* (and therefore also *Cypriniformes*), *Charax* (and therefore also *Characiformes*), *Gymnotus* (and therefore also *Gymnotiformes*) and *Silurus* (and therefore also *Siluriformes*) under the rank-based *ICZN*.

Composition: *Ostariophysi* is a crown clade that contains the five mutually exclusive crown subclades *Gonorynchiformes* (milkfishes), *Cypriniformes* (carps), *Characiformes* (characins), *Gymnotiformes* (American knifefishes) and *Siluriformes* (catfishes), with about 11,144 currently recognized extant species (Fricke, Eschmeyer and Fong, 2020). Several extinct species have been referred to *Ostariophysi* (Fink and Fink, 1996; Grande and Poyato-Ariza, 1999; Chang and Chen, 2008; Malabarba and Malabarba, 2010; Ferraris, 2007; Gayet and Meunier, 2003; Albert and Fink, 2007; Azpelicueta and Cione, 2011). Filleul and Maisey (2004) suggested that the Early Cretaceous *Santanichthys diasii* is a stem characiform, but Malabarba and Malabarba (2010) argued that this taxon is more likely a stem member of *Ostariophysi* or *Otophysi* (this volume).

Diagnostic Apomorphies: Ostariophysan characters and discussions of homologies are given by Rosen and Greenwood (1970), Roberts (1973, 1982), Fink and Fink (1981, 1996), Hoffmann and Britz (2006) and Wiley and Johnson (2010). Salient apomorphies that exhibit no or relatively little variation within the clade are emphasized here.

1. Basisphenoid bone absent.
2. Dermopalatine bone absent.
3. Saccular and lagenar otoliths in a relatively posteromedial position in the floor of the otic capsule.
4. Gasbladder anterior chamber enveloped by a silvery sheet of the peritoneum that also attaches to first and second pleural ribs.
5. Gasbladder dorsally suspended by a thickened sheet of mesentery.
6. No first supraneural or accessory neural arch present anterior to first vertebra.
7. Neural arches of anterior vertebrae form roof over neural canal.

8. Possession of a pheromone, the alarm substance (Schreckstoff), produced by epidermal cells that stimulates a fright reaction behavior in conspecifics (Pfeiffer, 1977).

Synonyms: None.

Comments: Greenwood et al. (1966) first presented phylogenetic evidence relating *Gonorynchiformes* to the clade comprising *Cypriniformes*, *Characiformes*, *Gymnotiformes* and *Siluriformes*, which was then named *Ostariophysi*. Following the lead of Rosen and Greenwood (1970) and subsequent authors (e.g., Fink and Fink, 1981, 1996; Nelson, 2006), *Ostariophysi* was expanded to include *Gonorynchiformes*, and the name *Otophysi* was resurrected for the clade to which the name *Ostariophysi* had previously been applied (i.e., the clade comprising *Cypriniformes*, *Characiformes*, *Gymnotiformes* and *Siluriformes*). The relationship between *Gonorynchiformes* and *Otophysi* (this volume) has been supported by subsequent phylogenetic analyses using morphological and molecular data (Patterson, 1984; Grande et al., 2010; Wiley and Johnson, 2010; Saitoh et al., 2003; Lavoué et al., 2005; Li et al., 2008; Poulsen et al., 2009; Chen et al., 2013; Betancur-R. et al., 2013). In all of these works, authors applied the name *Ostariophysi* to the clade including just *Gonorynchiformes* and *Otophysi*.

Literature Cited

Albert, J. S., and W. L. Fink. 2007. Phylogenetic relationships of fossil Neotropical electric fishes (*Osteichthyes*: *Gymnotiformes*) from the Upper Miocene of Bolivia. *J. Vertebr. Paleontol.* 27:17–25.

Azpelicueta, M. M., and A. L. Cione. 2011. Redescription of the Eocene catfish *Bachmannia chubutensis* (*Teleostei*: *Bachmanniidae*) of southern South America. *J. Vertebr. Paleontol.* 31:258–269.

Betancur-R. R., R. E. Broughton, E. O. Wiley, K. Carpenter, J. A. López, C. Li, N. I. Holcroft, D. Arcila, M. Sanciangco, J. C. Cureton II, F. Zhang, T. Buser, M. A. Campbell, J. A. Ballesteros, A. Roa-Varon, S. Willis, W. C. Borden, T. Rowley, P. C. Reneau, D. J. Hough, G. Lu, T. Grande, G. Arratia and G. Ortí. 2013. The tree of life and a new classification of bony fishes. *PLOS Curr. Tree Life* 2013 April 18:1–41.

Chang, M., and G. Chen. 2008. Fossil *Cypriniformes* from China and its adjacent areas and their palaeobiogeographical implications. Pp. 337–350 in *Fishes and the Break-up of Pangaea* (L. Cavin, A. Longbottom, M. Richter, eds.). Geological Society, Special Publications, London.

Chen, W. J., S. Lavoué and R. L. Mayden. 2013. Evolutionary origin and early biogeography of otophysan fishes (*Ostariophysi*: *Teleostei*). *Evolution* 67:1–22.

Ferraris, Jr., C. J. 2007. Checklist of catfishes, recent and fossil (*Osteichthys*: *Siluriformes*), and catalogue of siluriform primary types. *Zootaxa* 1418:1–628.

Filleul, A., and J. Maisey. 2004. Redescription of *Santanichthys diasii* (*Otophysi*, *Characiformes*) from the Albian of the Santana Formation and comments on its implications for otophysan relationships. *Am. Mus. Novit.* 3455:1–21.

Fink, S. V., and W. L. Fink. 1981. Interrelationships of the ostariophysan fishes (*Teleostei*). *Zool. J. Linn. Soc.* 72:297–353.

Fink, S. V., and W. L. Fink. 1996. Interrelationships of the *Ostariophysi*. Pp. 491–522 in *Interrelationships of Fishes* (M. Stiassney, L. Parenti and D. Johnson, eds.). Academic Press, San Diego, CA.

Fricke, R., W. N. Eschmeyer, and J. D. Fong. 2020. Species by Family/Subfamily in the Catalog of Fishes. Electronic version. Available at http://research.calacademy.org/research/ichthyology/catalog/SpeciesByFamily.asp, accessed on 14 January 2020.

Gayet, M., and F. J. Meunier. 2003. Paleontology and palaeobiogeography of catfishes. Pp. 491–522 in *Catfishes* (G. Arratia, B. G. Kapoor, M. Chardon and R. Diogo, eds.). Science Publishers, Enfield, NH.

Grande, T., and F. Poyato-Ariza. 1999. Phylogenetic relationships of fossil and Recent gonorynchiform fishes (*Teleostei: Ostariophysi*). *Zool. J. Linn. Soc.* 125:197–238.

Grande, T., F. Poyato-Ariza and R. Diogo, eds. 2010. Gonorynchiformes *and Ostariophysan Relationships: A Comprehensive Review*. Science Publishers, Enfield, NH.

Greenwood, P. H., D. E. Rosen, S. H. Weitzman and G. S. Myers. 1966. Phyletic studies of teleostean fishes, with a provisional classification of living forms. *Bull. Am. Mus. Nat. Hist.* 131:341–455.

Hoffmann, M., and R. Britz. 2006. Ontogeny and homology of the neural complex of otophysan *Ostariophysi. Zool. J. Linn. Soc.* 147:301–330.

Lavoué, S., M. Miya, J. G. Inoue, K. Saitoh, N. B. Ishiguro, and M. Nishida. 2005. Molecular systematics of the gonorynchiform fishes (*Teleostei*) based on whole mitogenome sequences: implications for higher-level relationships within the *Otocephala. Mol. Phylogenet. Evol.* 37:165–177.

Li, C., G. Lu, and G. Ortí. 2008. Optimal data partitioning and a test case for ray-finned fishes (*Actinopterygii*) based on ten nuclear loci. *Syst. Biol.* 57:519–539.

Linnaeus, C. 1758. *Systema Naturae Per Regna Tria Naturae, Secundum Classes, Ordines, Genera, Species, cum Characteribus, Differentiis, Synonymis, Locis.* 10th edition, Tomes 1–2. Laurentii Salvii, Holmiae (Stockholm).

Linnaeus, C. 1766. *Systema Naturae Per Regna Tria Naturae, Secundum Classes, Ordines, Genera, Species, cum Characteribus, Differentiis, Synonymis, Locis.* 12th edition. Laurentii Salvii, Holmiae (Stockholm).

Malabarba, M. C., and L. R. Malabarba. 2010. Pp. 317–336 in *Origin and Phylogenetic Interrelationships of Teleosts* (J. S. Nelson, H.-P. Schultze, and M. V. H. Wilson, eds.). Verlag Dr. Friedrich Pfeil, Munich.

Nelson, J. S. 2006. *Fishes of the World*. 4th edition. John Wiley & Sons, New York.

Patterson, C. 1984. *Chanoides*, a marine Eocene otophysan fish (*Teleostei: Ostariophysi*). *J. Vertebr. Paleontol.* 4:430–456.

Pfeiffer, W. 1977. The distribution of fright reaction and alarm substance cells in fishes. *Copeia* 1977:653–665.

Poulsen, J. Y., P. R. Møller, S. Lavoué, S. W. Knudsen, M. Nishida, and M. Miya. 2009. Higher and lower-level relationships of the deep-sea fish order *Alepocephaliformes* (*Teleostei: Otocephala*) inferred from whole mitogenome sequences. *Biol. J. Linn. Soc.* 98:923–936.

Roberts, T. R. 1973. Interrelationships of ostariophysans. Pp. 373–395 in *Interrelationships of Fishes* (P. H. Greenwood, R. S. Miles, and C. Patterson, eds.). Academic Press, London.

Roberts, T. R. 1982. Unculi (horny projections arising from single cells), an adaptive feature of the epidermis of ostariophysan fishes. *Zool. Scr.* 11:55–76.

Rosen, D., and P. H. Greenwood. 1970. Origin of the Weberian apparatus and relationships of the ostariophysan and gonorynchiform fishes. *Am. Mus. Novit.* 2468:1–49.

Sagemehl, M. 1885. Beitrage zur vergleichenden Anatomie der Fische: III. Das Cranium der Characiniden nebst allgemeinen Bemerkungen über die mit einen Weber'schen Apparat versehenen Physostomenfamilien. *Gegenbauers Morphol. Jahrb.* 10:1–119.

Saitoh, K., M. Miya, J. G. Inoue, N. B. Ishiguro, and M. Nishida. 2003. Mitochondrial genomics of ostariophysan fishes: perspectives on phylogeny and biogeography. *J. Mol. Evol.* 56:464–472.

Wiley, E. O., and G. D. Johnson. 2010. A teleost classification based on monophyletic groups. Pp. 123–182 in *Origin and Phylogenetic Interrelationships of Teleosts* (J. S. Nelson, H. P. Schultze, M. V. H. Wilson, eds.). Verlag, Dr. Friedrich Pfeil, Munich.

Author

John G. Lundberg, Department of Ichthyology, Academy of Natural Sciences, 1900 Benjamin Franklin Parkway, Philadelphia, PA 19103, USA. Email: lundberg@ansp.org.

Date Accepted: 9 September 2013

Primary Editor: Kevin de Queiroz

Otophysi W. Garstang 1931 [J. G. Lundberg], converted clade name

Registration Number: 197

Definition: The crown clade originating in the most recent common ancestor of *Cyprinus carpio* Linnaeus 1758 (*Cypriniformes*), *Charax* (originally *Salmo*) *gibbosus* (Linnaeus 1758) (*Characiformes*), *Gymnotus carapo* Linnaeus 1758 (*Gymnotiformes*; *Gymnotoidei* on the reference phylogeny), *Silurus glanis* Linnaeus 1758 (*Siluriformes*; *Siluroidei* on the reference phylogeny). This is a minimum-crown-clade definition. Abbreviated definition: min crown ∇ (*Cyprinus carpio* Linnaeus 1758 & *Charax gibbosus* (Linnaeus 1758) & *Gymnotus carapo* Linnaeus 1758 & *Silurus glanis* Linnaeus 1758).

Etymology: Derived from the Greek *oto* meaning "ear" and *physinx* meaning "bladder," in reference to a link between the swim bladder and inner ear.

Reference Phylogeny: For the purposes of applying the definition above, Figure 1 of Fink and Fink (1981: 302) is the primary reference phylogeny. Although the terminal taxa in the reference phylogeny are large clades rather than species, the four specifier species used in the definition are the type species associated with the names *Cyprinus* (and therefore also *Cypriniformes*), *Charax* (and therefore also *Characiformes*), *Gymnotus* (and therefore also *Gymnotiformes*) and *Silurus* (and therefore also *Siluriformes*) under the rank-based *ICZN*.

Composition: *Otophysi* is a crown clade that contains the four mutually exclusive crown subclades *Cypriniformes* (carps), *Characiformes* (characins), *Gymnotiformes* (American knife-fishes) and *Siluriformes* (catfishes; this volume), with over 11,107 currently recognized extant species (Fricke, Eschmeyer and Fong, 2020). Several extinct species of *Otophysi* are known that represent *Cypriniformes* (Chang and Chen, 2008), *Characiformes* (Malabarba and Malabarba, 2010) and *Siluriformes* (Ferraris, 2007; Gayet and Meunier, 2003; Azpelicueta and Cione, 2011) and there is one named fossil *gymnotiform* (Albert and Fink, 2007). The Eocene *Chanoides macropoma* is a stem group otophysan (Patterson, 1984). Filleul and Maisey (2004) suggested that the Early Cretaceous *Santanichthys diasii* is a stem characiform but (Malabarba and Malabarba, 2010) argued that this taxon is more likely a stem member of *Ostariophysi* (this volume) or *Otophysi*.

Diagnostic Apomorphies: Otophysan characters and discussions of homologies are given by Rosen and Greenwood (1970), Roberts (1973, 1982), Fink and Fink (1981, 1996), Hoffmann and Britz (2006) and Wiley and Johnson (2010). Salient apomorphies inferred to have originated in the most recent common ancestor of *Otophysi* are listed here, although some are further modified in subgroups.

1. Development of the Weberian apparatus including parts of the anterior four vertebrae and adjacent bones that link the swim bladder to inner ear (Chranilov, 1927; Chardon et al., 2003). The Weberian apparatus functions in sound reception, transmission and perception and includes the following components:
   a. centra 1–4 are relatively foreshortened,
   b. anteriormost parapophyses 1–2 are fused to their centra,
   c. neural arch of vertebra 1 is modified to form the first (claustrum) and

second (scaphium) paired Weberian ossicles,

  d. neural arch of vertebra 2 is modified to form the third paired Weberian ossicle or intercalarium,

  e. rib and parapophysis of vertebra 3 are modified to form the fourth paired Weberian ossicles or tripus that contacts the gas bladder wall,

  f. rib and parapophysis of vertebra 4 form a process, the os suspensorium, to which the gas bladder is attached.

  g. a posteromedial extension (sinus impar) of the perilymph system of the ear passes out of the skull through a median canal ventral to the foramen magnum to contact the anterior Weberian ossicles.

2. No supraneural is present anterior to neural arch of second vertebra.

3. Supraneurals 2 and/or 3 are fused with supradorsals 3 and 4 and expanded to articulate with neural arches of vertebra 3 and/or 4.

4. The caudal-fin skeleton has a compound terminal centrum formed by fusion of preural centrum 1, ural centra 1 and 2, and paired uroneural bones.

5. The caudal-fin skeleton has hypural 2 fused to the compound centrum.

6. The pelvic girdle is bifurcated anteriorly.

7. Unicellular unculi are developed on lips and fins (Roberts, 1982).

Synonyms: *Otophysa* Wiley and Johnson 2010 is an approximate synonym, as are *Ostariophysi* Sagemehl 1885 and *Cypriniformes* sensu Goodrich (1909).

Comments: The selection here of the name *Otophysi* for the clade including *Cypriniformes*, *Characiformes*, *Gymnotiformes*, and *Siluriformes*

follows Rosen and Greenwood (1970) and subsequent authors (e.g., Fink and Fink, 1981, 1996; Nelson, 2006). Prior to the inclusion of *Gonorynchiformes* in *Ostariophysi* by Rosen and Greenwood (1970), the name *Ostariophysi* was applied to the clade that was renamed *Otophysi*. Garstang (1931), who coined the name *Otophysi*, also included hiodontoids and clupeoids in the group. Goodrich's (1909) *Cypriniformes* was applied to the clade now named *Otophysi*, and *Cypriniformes* is now the name applied to a subgroup of *Otophysi*.

In addition to the morphological synapomorphies noted above, monophyly of *Otophysi* is well supported by molecular evidence (Dimmick and Larson, 1996; Saitoh et al., 2003, Lavoué et al., 2005; Li et al., 2008; Poulsen et al., 2009; Natakani et al., 2011; Betancur-R. et al., 2013).

Literature Cited

Albert, J. S., and W. L. Fink. 2007. Phylogenetic relationships of fossil Neotropical electric fishes (*Osteichthyes*: *Gymnotiformes*) from the Upper Miocene of Bolivia. *J. Vertebr. Paleontol.* 27:17–25.

Azpelicueta, M. M., and A. L. Cione. 2011. Redescription of the Eocene catfish *Bachmannia chubutensis* (*Teleostei*: *Bachmanniidae*) of southern South America. *J. Vertebr. Paleontol.* 31:258–269.

Betancur-R. R., R. E. Broughton, E. O. Wiley, K. Carpenter, J. A. López, C. Li, N. I. Holcroft, D. Arcila, M. Sanciangco, J. C. Cureton II, F. Zhang, T. Buser, M. A. Campbell, J. A. Ballesteros, A. Roa-Varon, S. Willis, W. C. Borden, T. Rowley, P. C. Reneau, D. J. Hough, G. Lu, T. Grande, G. Arratia, and G. Ortí. 2013. The tree of life and a new classification of bony fishes. *PLOS Curr. Tree Life* 2013 April 18:1–41.

Chang, M., and G. Chen. 2008. Fossil *Cypriniformes* from China and its adjacent areas and their palaeobiogeographical implications. Pp. 337–350 in *Fishes and the Break-up of Pangaea* (L. Cavin, A. Longbottom, and M. Richter, eds.). Geological Society, Special Publications, London.

Chardon, M., E. Parmentier, and P. Vandewalle. 2003. Morphology, development and evolution of the Weberian apparatus in catfishes. Pp. 71–120 in *Catfishes* (G. Arratia, B. G. Kapoor, M. Chardon, and R. Diogo, eds.). Science Publishers, Enfield, NH.

Chranilov, N. S. 1927. Beitrage zur kenntnis desWeber'schen apparatus der *Ostariophysi*. I. Vergleichend-anatomishe ubersicht der knochenelemente der Weber'schen apparates bei *Cypriniformes. Zool. Jahrb.* 49:501–597.

Dimmick, W. W., and A. Larson. 1996. A molecular and morphological perspective on the phylogenetic relationships of the otophysan fishes. *Mol. Phylogenet. Evol.* 6:120–133.

Ferraris, Jr., C. J. 2007. Checklist of catfishes, recent and fossil (*Osteichthys: Siluriformes*), and catalogue of siluriform primary types. *Zootaxa* 1418:1–628.

Filleul, A., and J. Maisey. 2004. Redescription of *Santanichthys diasii* (*Otophysi, Characiformes*) from the Albian of the Santana Formation and comments on its implications for otophysan relationships. *Am. Mus. Novit.* 3455:1–21.

Fink, S. V., and W. L. Fink. 1981. Interrelationships of the ostariophysan fishes (*Teleostei*). *Zool. J. Linn. Soc.* 72:297–353.

Fink, S. V., and W. L. Fink. 1996. Interrelationships of the *Ostariophysi*. Pp. 209–249 in *Interrelationships of Fishes* (M. Stiassney, L. Parenti, and D. Johnson, eds.). Academic Press, San Diego, CA.

Fricke, R., W. N. Eschmeyer, and J. D. Fong. 2020. Species by Family/Subfamily in the Catalog of Fishes. Electronic version. Available at http://research.calacademy.org/research/ichthyology/catalog/SpeciesByFamily.asp, accessed on 14 January 2020.

Garstang, W. 1931. The phyletic classification of *Teleostei. Proc. Leeds Philos. Lit. Soc., Sci. Sect.,* 2:240–260.

Gayet, M., and F. J. Meunier. 2003. Paleontology and palaeobiogeography of catfishes. Pp. 491–522 in *Catfishes* (G. Arratia, B. G. Kapoor, M. Chardon, and R. Diogo, eds.). Science Publishers, Enfield, NH.

Goodrich, E. S. 1909. *Vertebrata Craniata* (first fascicle: cyclostomes and fishes). Part IX in *A Treatise on Zoology* (E. R. Lankester, ed.). Adam and Charles Black, London.

Hoffmann, M., and R. Britz. 2006. Ontogeny and homology of the neural complex of otophysan *Ostariophysi. Zool. J. Linn. Soc.* 147:301–330.

Lavoué, S., M. Miya, J. G. Inoue, K. Saitoh, N. B. Ishiguro, and M. Nishida. 2005. Molecular systematics of the gonorynchiform fishes (*Teleostei*) based on whole mitogenome sequences: implications for higher-level relationships within the *Otocephala. Mol. Phylogenet. Evol.* 37:165–177.

Li, C., G. Lu, and G. Ortí. 2008 Optimal data partitioning and a test case for ray-finned fishes (*Actinopterygii*) based on ten nuclear loci. *Syst. Biol.* 57:519–539.

Linnaeus, C. 1758. *Systema Naturae Per Regna Tria Naturae, Secundum Classes, Ordines, Genera, Species, cum Characteribus, Differentiis, Synonymis, Locis.* 10th edition, Tome 1. Laurentii Salvii, Holmiae (Stockholm).

Malabarba, M. C., and L. R. Malabarba. 2010. Biogeography of *Characiformes*: an evaluation of the available information of fossil and extant taxa. Pp. 317–336 in *Origin and Phylogenetic Interrelationships of Teleosts.* (J. S. Nelson, H.-P. Schultze, and M. V. H. Wilson, eds.). Dr. Friedrich Pfeil, Munich.

Nakatani, M., M. Miya, K. Mabuchi, K. Saitoh, and M. Nishida. 2011. Evolutionary history of *Otophysi* (*Teleostei*), a major clade of the modern freshwater fishes: Pangaean origin and Mesozoic radiation. *BMC Evol. Biol.* 11:177.

Nelson, J. S. 2006. *Fishes of the World.* 4th edition. John Wiley & Sons.

Patterson, C. 1984. *Chanoides*, a marine Eocene otophysan fish (*Teleostei: Ostariophysi*). *J. Vertebr. Paleontol.* 4:430–456.

Poulsen, J. Y., P. R. Møller, S. Lavoué, S. W. Knudsen, M. Nishida, and M. Miya. 2009. Higher and lower-level relationships of the deep-sea fish order *Alepocephaliformes* (*Teleostei: Otocephala*) inferred from whole mitogenome sequences. *Biol. J. Linn. Soc.* 98: 923–936.

Roberts, T. R. 1973. Interrelationships of ostariophysans. Pp. 373–396 in *Interrelationships of fishes* (P. H. Greenwood, R. S. Miles, and C. Patterson, eds.). Academic Press, London.

Roberts, T. R. 1982. Unculi (horny projections arising from single cells), an adaptive feature of the epidermis of ostariophysan fishes. *Zool. Scr.* 11:55–76.

Rosen, D., and P. H. Greenwood. 1970. Origin of the Weberian apparatus and relationships of the ostariophysan and gonorynchiform fishes. *Am. Mus. Novit.* 2468:1–49.

Sagemehl, M. 1885. Beitrage zur vergleichenden Anatomie der Fische: III. Das Cranium der Characiniden nebst allgemeinen Bemerkungen uber die mit einen Weber'schen Apparat versehenen Physostomenfamilien. *Morphol. Jahrb.* 10:1–119.

Saitoh, K., M. Miya, J. G. Inoue, N. B. Ishiguro, and M. Nishida. 2003. Mitochondrial genomics of ostariophysan fishes: perspectives on phylogeny and biogeography. *J. Mol. Evol.* 56:464–472.

Wiley, E. O., and G. D. Johnson. 2010. A teleost classification based on monophyletic groups. Pp. 123–182 in *Origin and Phylogenetic Interrelationships of Teleosts* (J. S. Nelson, H. P. Schultze, M. V. H. Wilson, eds.). Verlag Dr. Friedrich Pfeil, Munich.

Author

John G. Lundberg; Department of Ichthyology; Academy of Natural Sciences; 1900 Benjamin Franklin Parkway, Philadelphia, PA 19103, USA. Email: lundberg@ansp.org.

Date Accepted: 9 September 2013

Primary Editor: Kevin de Queiroz

Pan-Siluriformes J. G. Lundberg, new clade name

Registration Number: 198

Definition: The total clade of the crown clade *Siluriformes*. This is a crown-based total-clade definition. Abbreviated definition: total ∇ of *Siluriformes*.

Etymology: Derived from the Greek *pantos* (all), indicating that the name refers to a total clade or pan-monophyllum, plus *Siluriformes*, the name of the corresponding crown clade (see the entry for *Siluriformes* in this volume for the etymology of that name).

Reference Phylogeny: The primary reference phylogeny is Figure 2 in Lundberg et al. (2007), where *Pan-Siluriformes* is the clade originating in the branch above and to the right of Node 8.

Composition: *Pan-Siluriformes* contains *Siluriformes* (this volume) and all extinct species that share a more recent common ancestor with *Siluriformes* than with any other mutually exclusive (non-nested) crown clade. The fragmentary and thus poorly known members of *Andinichthyidae* Gayet (1988, 1991) are the only known fossil catfishes that might be part of the stem group of the crown group *Siluriformes*. However, the incomplete fossil crania on which this taxon is based preserve the apomorphies of the skull roof and posttemporo-supracleithrum that are diagnostic for crown group catfishes (Arratia and Gayet, 1995). Whether they exhibit plesiomorphic states for other diagnostic characters of crown group catfishes is currently unknown.

Diagnostic Apomorphies: See *Siluriformes* (this volume) for characters that arose along its stem lineage and may thus permit reference of yet-to-be discovered stem fossils to *Pan-Siluriformes*. Because of the lack of fossil taxa that can be confidently referred to the catfish stem group (see Composition), it is not currently possible to determine the order in which those characters arose in the stem lineage.

Synonyms: Because most previous authors did not distinguish between crown and total clades, most or all of the names listed as synonyms of *Siluriformes* Hay 1929 are likely also approximate synonyms of *Siluriformes*: *Oplophoria* Rafinesque 1815, *Siluroideae* Richardson 1836, *Siluridae* of Swainson (1838), *Siluri* of Bleeker (1858) (*Siluri* of Bleeker [1858] was applied to catfishes only, whereas Bonaparte [1841] used the name *Siluri* earlier but applied it to a phylogenetically mixed assemblage of fishes including catfishes, gars, grenadiers and squaretails), *Nematognathi* Gill 1861, *Siluroidei* of Goodrich (1909) (*Siluroidei* of Bleeker [1847] was applied to a subgroup of catfishes), *Siluroidea* of Regan (1911). Greenwood et al. (1966) also listed *Siluroidiformes* as a synonym but did not give an author.

Comments: The panclade name was selected over various alternatives (see Synonyms) in the interest of developing an integrated approach for the names of crown and total clades.

Literature Cited

Arratia, G., and M. Gayet. 1995. Sensory canals and related bones of tertiary siluriform crania from Bolivia and North America and comparison with recent forms. *J. Verebrt. Palaeontol.* 15:482–505.

Bleeker, P. 1847. Pharyngognathorum Siluroideorumque species novae Javanenses. *Nat. Geneesk. Arch. N. I.* 4:155–169.

Bleeker, P. 1858. Ichthyologiae archipelagi Indici prodromus, Vol 1. *Siluri. Act. Soc. Sci. Indo-Neerl.* 4:1–370, i–xii.

Bonaparte, C. L. 1841. A new systematic arrangement of vertebrated animals. *Trans. Linn. Soc. London* 18:247–304.

Gayet, M. 1988. Le plus ancien crane de siluriforme: *Andinichthys bolivianensis* nov. gen., nov. sp. (*Andinichthyidae* nov. fam.) du Maastrichtien de Tiupampa (Bolivie). *C. R. Acad. Sci. Paris Ser. II* 307:833–836.

Gayet, M. 1991. Holostean and teleostean fossils from Bolivia. Pp. 453–494 in *Fósiles y Facies de Bolivia* (R. Suárez-Sourco, ed.). Revista Tech. YPFB Cochabamba, Bolivia.

Gill, T. N. 1861. Catalogue of the fishes of the eastern coast of North America, from Greenland to Georgia. *Proc. Acad. Nat. Sci., Phila.* 13 (suppl.):1–63.

Goodrich, E. S. 1909. Part IX. *Vertebrata Craniata* (first fascicle: cyclostomes and fishes). In *A Treatise on Zoology* (E. R. Lankester, ed.). Adam and Charles Black, London.

Greenwood, P. H., D. E. Rosen, S. H. Weitzman and G. S. Myers. 1966. Phyletic studies of teleostean fishes, with a provisional classification of living forms. *Bull. Am. Mus. Nat. Hist.* 131:341–455.

Lundberg, J. G., J. P. Sullivan, R. Rodiles-Hernández and D. A. Hendrickson. 2007. Discovery of African roots for the Mesoamerican Chiapas catfish, *Lacantunia enigmatica*, requires an ancient intercontinental passage. *Proc. Acad. Nat. Sci. Phila.* 156:39–53.

Rafinesque, C. S. 1815. *Analyse de la Nature, ou Tableau de l'Univers et des Corps Organisés.* Palerme, privately published.

Regan, C. T. 1911. The classification of the teleostean fishes of the order *Ostariophysi.* 2. *Siluroidea. Ann. Mag. Nat. Hist.* 8:553–577.

Richardson, J. 1836. *Fauna Boreali-Americana; or the Zoology of the Northern Parts of British America: Containing Descriptions of the Objects of Natural History Collected on the Late Northern Land Expeditions under Command of Captain Sir John Franklin, R.N. Part Third. The Fish.* Richard Bentley, London.

Swainson, W. 1838. *The Natural History and Classification of Fishes, Amphibians, and Reptiles, or Monocardian Animals.* Longman, Orme, Brown, Green, and Longmans, London.

Author

John G. Lundberg; Department of Ichthyology; Academy of Natural Sciences; 1900 Benjamin Franklin Parkway, Philadelphia, PA 19103, USA. Email: lundberg@ansp.org.

Date Accepted: 8 August 2013

Primary Editor: Kevin de Queiroz

Siluriformes O. P. Hay 1929 [J. G. Lundberg], converted clade name

Registration Number: 199

Definition: The crown clade originating in the most recent common ancestor of *Loricaria cataphracta* Linnaeus 1758 (*Loricarioidei*), *Diplomystes* (originally *Silurus*) *chilensis* (Molina 1782) (*Diplomystidae*), and *Silurus glanis* Linnaeus 1758 (*Siluroidei*). This is a minimum-crown-clade definition. Abbreviated definition: min crown ∇ (*Loricaria cataphracta* Linnaeus 1758 & *Diplomystes chilensis* (Molina 1782) & *Silurus glanis* Linnaeus 1758).

Etymology: Derived from the Greek *silouros* or Latin *silurus*, the vernacular name for the members of a subgroup of catfishes that includes *Silurus glanis*, the type species of *Silurus* under the rank-based *International Code of Zoological Nomenclature* (*ICZN*).

Reference Phylogeny: For the purposes of applying our definition, Sullivan et al. (2006: Figs. 1 and 2) is the primary reference phylogeny. Although not included in the reference phylogeny, the three specifiers used in the definition are the type species associated with the names *Loricaria* (and therefore also *Loricarioidei*), *Diplomystes* (and *Diplomystidae*), and *Silurus* (and *Siluroidei*) under the rank-based *ICZN*.

Composition: *Siluriformes*, the crown clade of catfishes, currently contains the three mutually exclusive subclades *Loricarioidei*, *Diplomystidae*, and *Siluroidei*, comprising more than 3,972 currently recognized extant species (Fricke, Eschmeyer and Fong, 2020). The most up-to-date lists of extinct species of *Siluriformes* are by Ferraris (2007) and Gayet and Meunier (2003).

Diagnostic Apomorphies: Much of Regan's (1911) diagnosis of *Siluriformes* (as *Siluroidea*) still serves, and these features are apomorphies inferred to have originated in the most recent common ancestor of *Siluriformes*. Some of these characters are further changed in subclades. Additional characters and discussions of homologies are described by Fink and Fink (1981, 1996), Arratia (2003a,b), Wiley and Johnson (2010), and by authors cited in the rest of this section and in Comments.

1. Skin naked (scale-less). A few catfish families have bony plates in the skin but these are neomorphic and not homologous to the thin, bony-ridge scales present in most teleosts.
2. Parietal bones absent. Based on evidence from the position of sensory canals (Arratia and Gayet, 1995) and unconfirmed development of ossification centers (Bamford, 1948), the paired parietal bones are hypothesized to fuse with the supraoccipital.
3. Autopalatine separate from remaining suspensorium skeleton and moveably articulating with lateral ethmoid and maxilla.
4. Ectopterygoid and endopterygoid reduced and free from metapterygoid, quadrate and hyomandibular. Some catfishes have lost one or both of the anterior pterygoid bones.
5. Metapterygoid located anterodorsal to quadrate.
6. Symplectic bone and posterior projection of quadrate absent.
7. Preopercle and interopercle bones relatively small and anteroposteriorly shortened.

8. Subopercle bone absent.

9. Maxillary bone bears a fleshy barbel with an elastic cartilage core that extends from its posterior end.

10. Basihyal absent.

11. Second, third, and fourth vertebrae fused to form complex centrum of the Weberian apparatus.

12. Third and fourth neural arches fused to each other and to complex centrum.

13. Parapophysis of second vertebral centrum absent.

14. Parapophysis of fourth vertebral centrum broadly expanded and articulating with expanded facet of posttemporo-supracleithrum. Some catfish subgroups have lost the articulation between the fourth parapophysis and shoulder girdle.

15. Intermuscular bones (epipleurals and epineurals) absent.

16. Upper element of pectoral girdle (post-temporo-supracleithrum) triradiate, its dorsal limb articulating with skull, ventral limb articulating with cleithrum, and medial limb (ossified Baudelot's or transcapular ligament) articulating medially with basioccipital and exoccipital, and posteriorly with expanded parapophysis of fourth vertebra. Some catfish subgroups have the posttemporo-supracleithrum fused with the pterotic bone.

17. Dorsal fin with two anterior spines (anteriormost two lepidotrichia), first spine short and tightly bound to longer second spine. First dorsal-fin spine articulating with 4th supraneural; second dorsal-fin spine articulating with first dorsal-fin pterygiophore, together this anatomical cluster forms a spine-locking joint. Some catfishes have reduced or lost the spine-locking joint and some have lost the entire dorsal fin.

18. Principal caudal-fin rays plesiomorphically 9 in both upper and lower lobes, but most commonly 8 in upper lobe and 9 in lower lobe, and often fewer.

19. Pectoral fin with single spine (anteriormost lepidotrichium) articulating through a rotating and locking joint with cleithrum. Some catfishes have reduced or lost the spinous structure of the anterior lepidotrichium and locking joint.

20. Postcleithra absent.

Synonyms: *Oplophoria* Rafinesque 1815, *Siluroideae* Richardson 1836, *Siluridae* of Swainson (1838), *Siluri* of Bleeker (1858) (*Siluri* of Bleeker [1858] was applied to catfishes only whereas Bonaparte [1841] used the name *Siluri* earlier but applied it to a phylogenetically mixed assemblage of fishes including catfishes, gars, grenadiers and squaretails), *Nematognathi* Gill 1861, *Siluroidei* of Goodrich (1909) (*Siluroidei* of Bleeker [1847] was applied to a subgroup of catfishes), *Siluroidea* of Regan (1911); all of the foregoing are approximate synonyms. Greenwood et al. (1966) also listed *Siluroidiformes* as a synonym, but did not give an author.

Comments: The name *Siluriformes* has been attributed to Cuvier (1816) based on his Siluroïdes, which included all of catfishes in his work (Ferraris and de Pinna, 1999). Although this is the oldest published name for catfishes as a group, it is a French vernacular name.

Our choice of the name *Siluriformes*, attributed to Hay (1929), for the crown clade of catfishes is based on its essentially universal usage since 1966 following the classification of teleosts by Greenwood et al. *Siluriformes* is widely used in textbooks (e.g., Nelson, 2006), checklists (e.g., Ferraris, 2007) and comprehensive phylogenetic studies (see below). The older names applied to catfishes collectively that are not based on the name *Silurus* were never (*Oplophoria*) or rarely (*Nematognathi*) used subsequent to their original proposals.

Members of *Siluriformes* are highly distinctive morphologically and are globally distributed in tropical and temperate fresh and coastal marine waters. Many species are important food, ornamental or sport fish. It is thus not surprising that catfishes are among the most widely known groups of fishes, and they were named early in both folk and biological taxonomies. Under rank-based nomenclature, the taxonomic rank bestowed on the catfish clade has steadily increased from family to order. Since the detailed and broadly comparative work of Fink and Fink on major ostariophysan clades (1981, 1996), catfishes are generally considered to be the sister clade of electric knife-fishes (*Gymnotiformes*). There are no known living or fossil fishes that are morphologically intermediate between the members of this pair of clades, or between *Siluriformes* and any other group of *Ostariophysi*. Two of the earliest arrangements of *Siluriformes* based on phylogenetic concepts are those of Eigenmann and Eigenmann (1890, South American taxa) and Regan (1911, all taxa). Other systematic studies of catfishes of broad taxonomic scope have provided more explicit and comprehensive phylogenetic classifications or trees: Chardon (1968), Mo (1991), de Pinna (1998), Diogo (2003), Hardman (2005), Sullivan et al. (2006), and Lundberg et al. (2007).

The earliest descriptions of fossil catfishes were published in the mid-nineteenth century; for examples see Marck (1868), Cope (1872), and Leidy (1873). Since then several unusual and some relatively plesiomorphic fossil catfishes have come to light, but as yet there are no known extinct catfishes that are unequivocally excluded from the catfish crown group.

Hypsidoris Lundberg and Case 1970 from the Eocene of North America plesiomorphically retains teeth on the maxilla, as does the extant taxon *Diplomystidae*, but otherwise *Hypsidoris* shares derived characters with the members of *Siluroidei* (Grande, 1987). Among the oldest known catfishes, *Andinichthys* Gayet (1988) from the Late Cretaceous of South America is known only from fragmentary skulls showing the apparently fused parieto-supraoccipital region characteristic of all members of *Siluriformes*. Thus, all that can be concluded of *Andinichthys* is that it belongs to *Pan-Siluriformes* (this volume).

Literature Cited

Arratia, G. 2003a. Catfish head skeleton: an overview. Pp. 3–46 in *Catfishes* (G. Arratia, A. S. Kapoor, M. Chardon, R. Diogo, eds.). Science Publishers, Enfield, NH.

Arratia G. 2003b. The siluriform postcranial skeleton. Pp. 121–158 in *Catfishes* (G. Arratia, A. S. Kapoor, M. Chardon, R. Diogo, eds.). Science Publishers, Enfield, NH.

Arratia, G., and M. Gayet. 1995. Sensory canals and related bones of tertiary siluriform crania from Bolivia and North America and comparison with recent forms. *J. Vertebr. Palaeontol.* 15:482–505.

Bamford, T. W. 1948. The cranial development of *Galeichthys felis*. *Proc. Zool. Soc. Lond.* 118:364–391.

Bleeker, P. 1847. Pharyngognathorum Siluroideorumque species novae Javanenses. *Nat. Geneesk. Arch. N. I.* 4:155–169.

Bleeker, P. 1858. Ichthyologiae archipelagi Indici prodromus. Vol. 1. *Siluri. Act. Soc. Sci. Indo-Neerl.* 4:1–370, i–xii.

Bonaparte, C. L. 1841. A new systematic arrangement of vertebrated animals. *Trans. Linn. Soc. Lond.* 18:247–304.

Chardon, M. 1968. Anatomie comparée de l'appareil de Weber et structures connexes chez les *Siluriformes. Ann. Mus. R. Afr. Cent. Ser. 8 Sci. Zool.* 169:1–277.

Cope, E. D. 1872. Notices of new *Vertebrata* from the upper waters of Bitter Creek, Wyoming Territory. *Proc. Am. Philos. Soc.* 7:483–486.

Cuvier, G. 1816. *Le Règne Animal Distribué d'Après son Organisation, pour Servir de Base à l'Histoire Naturelle des Animaux et d'Introduction à l'Anatomie Comparée. Tome 2. Les Reptiles, les Poissons, les Mollusques et les Annélides.* Deterville, Paris.

Diogo, R. 2003. Higher-level phylogeny of *Siluriformes*—an overview. Pp. 353–384 in *Catfishes* (G. Arratia, A. S. Kapoor, M. Chardon, R. Diogo, eds.). Science Publishers, Enfield, NH.

Eigenmann, C. H., and Eigenmann, R. S. 1890. A revision of the South American *Nematognathi* or cat-fishes. *Occ. Pap. Calif. Acad. Sci.* 1:1–508.

Ferraris, Jr., C. J. 2007. Checklist of catfishes, recent and fossil (*Osteichthys:Siluriformes*), and catalogue of siluriform primary types. *Zootaxa* 1418:1–628.

Ferraris, Jr., C. J., and M. C. C. de Pinna. 1999. Higher-level names for catfishes (*Actinopterygii*: *Ostariophysi*: *Siluriformes*). *Proc. Calif. Acad. Sci.* 15:1–17.

Fink, S. V., and W. L. Fink. 1981. Interrelationships of the ostariophysan fishes (*Teleostei*). *Zool. J. Linn. Soc.* 72:297–353.

Fink, S. V., and W. L. Fink. 1996. Interrelationships of the *Ostariophysi*. Pp. 209–249 in *Interrelationships of Fishes* (M. Stiassny, L. Parenti and D. Johnson, eds.). Academic Press, San Diego and London.

Fricke, R., W. N. Eschmeyer, and J. D. Fong. 2020. Species by Family/Subfamily in the Catalog of Fishes. Electronic version. Available at http://research.calacademy.org/research/ichthyology/catalog/SpeciesByFamily.asp, accessed on 14 January 2020.

Gayet, M. 1988. Le plus ancien crane de siluriforme: *Andinichthys bolivianensis* nov. gen., nov. sp. (*Andinichthyidae* nov. fam.) du Maastrichtien de Tiupampa (Bolivie). *C. R. Acad. Sci. Paris Ser. II* 307:833–836.

Gayet, M., and F. J. Meunier. 2003. Paleontology and palaeobiogeography of catfishes. Pp. 491–522 in *Catfishes* (G. Arratia, A. S. Kapoor, M. Chardon, R. Diogo, eds.). Science Publishers, Enfield, NH.

Gill, T. N. 1861. Catalogue of the fishes of the eastern coast of North America, from Greenland to Georgia. *Proc. Acad. Nat. Sci. Phila.* 13 (suppl.):1–63.

Goodrich, E. S. 1909. Part IX. *Vertebrata Craniata* (first fascicle: cyclostomes and fishes). In *A Treatise on Zoology* (E. R. Lankester, ed.). London, Adam and Charles Black.

Grande, L. 1987. Redescription of *Hypsidoris farsonensis* (*Teleostei*: *Siluriformes*), with a reassessment of its phylogenetic relationships. *J. Vertebr. Palaeontol.* 7:24–54.

Greenwood, P. H., D. E. Rosen, S. H. Weitzman, and G. S. Myers. 1966. Phyletic studies of teleostean fishes, with a provisional classification of living forms. *Bull. Am. Mus. Nat. Hist.* 131:341–455.

Hardman, M. 2005. The phylogenetic relationships among non-diplomystid catfishes as inferred from mitochondrial cytochrome b sequences; the search for the ictalurid sister taxon (*Otophysi*: *Siluriformes*). *Mol. Phylogenet. Evol.* 37:700–720.

Hay, O. P. 1929. *Second Bibliography and Catalogue of the Fossil Vertebrata of North America.* Carnegie Institution of Washington, Washington, DC.

Leidy, J. 1873. Notice of remains of fishes in the Bridger Tertiary Formation of Wyoming. *Proc. Acad. Nat. Sci. Phila.* 1873:97–99.

Linnaeus, C. 1758. *Systema Naturae Per Regna Tria Naturae, Secundum Classes, Ordines, Genera, Species, cum Characteribus, Differentiis, Synonymis, Locis.* 10th edition, Tomes 1–2. Laurentii Salvii, Holmiae (Stockholm).

Lundberg, J. G., and G. R. Case. 1970. A new catfish from the Eocene Green River Formation, Wyoming. *J. Palaeontol.* 44:451–457.

Lundberg, J. G., J. P. Sullivan, R. Rodiles-Hernández, and D. A. Hendrickson. 2007. Discovery of African roots for the Mesoamerican Chiapas catfish, *Lacantunia enigmatica*, requires an ancient intercontinental passage. *Proc. Acad. Nat. Sci. Phila.* 156:39–53.

von der Marck, W., and C. Schlüter. 1868. Neue Fische und Krebse aus der Kreide von Westphalen. *Palaeontographica* 15:269–305.

Mo, T., 1991. Anatomy, relationships and systematics of the Bagridae (*Teleostei, Siluroidei*) with a hypothesis of siluroid phylogeny. *Theses Zool.* 17:1–216.

Molina, G. I. 1782. *Saggio sulla Storia Naturale del Chili.* S. Tommaso d'Aquino, Bologna.

de Pinna, M. C. C. 1998. Phylogenetic relationships of Neotropical *Siluriformes* (*Teleostei: Ostariophysi*): historical overview and synthesis of hypotheses. Pp. 279–330 in *Phylogeny and Classification of Neotropical Fishes* (L. R. Malabarba, R. E. Reis, R. P. Vari, C. A. S. Lucena, Z. M. S. Lucena, eds.). Museu de Ciencias e Tecnologia, Edipucrs, Porto Alegre.

Nelson, J. S. 2006. *Fishes of the World.* 4th edition. John Wiley & Sons, Hoboken, NJ.

Rafinesque, C. S. 1815. *Analyse de la Nature, ou Tableau de l'Univers et des Corps Organisés.* Palerme, Privately Published.

Regan, C. T. 1911. The classification of the teleostean fishes of the order *Ostariophysi.* 2. *Siluroidea. Ann. Mag. Nat. Hist.* 8:553–577.

Richardson, J. 1836. *Fauna Boreali-Americana; Or The Zoology of the Northern Parts of British America: Containing Descriptions of the Objects of Natural History Collected on the Late Northern Land Expeditions under Command of Captain Sir John Franklin, R.N. Part Third. The Fish.* Richard Bentley, London.

Sullivan, J. P., J. G. Lundberg, and M. Hardman. 2006. A phylogenetic analysis of the major groups of catfishes (*Teleostei: Siluriformes*) using nuclear *rag1* and *rag2* gene sequences. *Mol. Phylogenet. Evol.* 41:636–662.

Swainson, W. 1838. *The Natural History and Classification of Fishes, Amphibians, and Reptiles, or Monocardian Animals.* Longman, Orme, Brown, Green, and Longmans, London.

Wiley, E. O., and G. D. Johnson. 2010. A teleost classification based on monophyletic groups. Pp. 123–182 in *Origin and Phylogenetic Interrelationships of Teleosts* (J. S. Nelson, H. P. Schultze, M. V. H. Wilson, eds.). Verlag Dr. Friedrich Pfeil, Munich.

Author

John G. Lundberg; Department of Ichthyology; Academy of Natural Sciences; 1900 Benjamin Franklin Parkway, Philadelphia, PA 19103, USA. Email: lundberg@ansp.org.

Date Accepted: 9 July 2013

Primary Editor: Kevin de Queiroz

Stegocephali E. D. Cope 1868 [M. Laurin], converted clade name

Registration Number: 102

Definition: The largest clade that includes *Eryops megacephalus* Cope 1877 (*Temnospondyli*) but not *Tiktaalik roseae* Daeschler et al. 2006, *Panderichthys rhombolepis* Gross 1930 (*Panderichthyidae*), and *Eusthenopteron foordi* Whiteaves 1881 (*Osteolepiformes*). This is a maximum-clade definition. Abbreviated definition: max ∇ (*Eryops megacephalus* Cope 1877 ~ *Tiktaalik roseae* Daeschler et al. 2006 & *Panderichthys rhombolepis* Gross 1930 & *Eusthenopteron foordi* Whiteaves 1881).

Etymology: Presumably (Cope did not specify) from the Greek *stege* (roof) and *kephale* (head), which presumably refers to the solid skull roof, devoid of fenestration.

Reference Phylogeny: The primary reference phylogeny is Vallin and Laurin (2004: Fig. 6). In that tree, *Tiktaalik roseae* would be located immediately crownward of *Panderichthys rhombolepis* (Daeschler et al., 2006). Alternative phylogenies that differ substantially from the reference phylogeny with regard to the relationships of temnospondyls, "lepospondyls", embolomeres, and seymoriamorphs to amniotes and lissamphibians (e.g., Ruta et al., 2003: Fig. 4) result in application of the name *Stegocephali* to a clade of similar composition. For the earliest (and earliest-branching) members of the taxon, see the phylogenies of Coates (1996: Figs. 27–29), Ahlberg and Clack (1998: Figs. 20–22) and Marjanovic and Laurin (2019: Figs. 10, 11, 14, 20, 21). All these studies focused on stegocephalians and included few other sarcopterygians. Phylogenies that show the position of stegocephalians within sarcopterygians include Ahlberg (1991a: Fig. 14), Schultze (1994: Fig. 1), Zhu and Schultze (1997: Fig. 8), and Zhu et al. (2001: Fig. 3a).

Composition: *Stegocephali* includes all known limbed vertebrates and perhaps some that had paired fins, because the paired appendages of its most basal and earliest (Devonian) members, such as *Obruchevichthys gracilis*, *Elginerpeton pancheni*, *Ventastega curonica*, and *Metaxygnathus denticulatus* (Clack et al., 2012), are poorly known (Campbell and Bell, 1977; Ahlberg, 1991b, 1995; Ahlberg et al., 1994; Laurin, 1998a; Laurin et al., 2000). In the Palaeozoic, this taxon included the better known Devonian forms *Acanthostega gunnari*, *Ichthyostega stensioei*, and *Tulerpeton curtum*, along with a much greater diversity of Permo-Carboniferous taxa, such as *Crassigyrinus scoticus*, baphetids, temnospondyls, embolomeres, and seymouriamorphs (stem-tetrapods according to the primary reference phylogeny), as well as stem-lissamphibians (represented by the paraphyletic group known as "lepospondyls"), stem-amniotes (diadectomorphs) and (crown) amniotes. Extant stegocephalians include about 5,300 species of lissamphibians and about 22,000 species of amniotes (Pough et al., 2004).

Diagnostic Apomorphies: Several apomorphies of this taxon may reflect the acquisition of a slightly more terrestrial lifestyle, even though the first stegocephalians seem to have been primarily and primitively aquatic (Coates and Clack, 1991; Clack and Coates, 1995; Clack, 2002; Laurin et al., 2004). Characters that reflect the appearance of limbs with digits may be diagnostic of a slightly less inclusive clade, such as *Labyrinthodontia* (this volume), and are not listed here; only characters that can be inferred to have been present in the most recent common ancestor of *Eryops megacephalus* and *Acanthostega gunnari* are listed here. The list of apomorphies is relatively short because *Stegocephali* is a maximum-clade name and its earliest diverging members are known from fragmentary, disarticulated material. The apomorphies of its basalmost node include (Ahlbergh et al., 1994; Ahlberg and Clack, 1998; Laurin, 1998a,b): a contact between the lacrimal and nasal bone that reflects the smaller size of the prefrontal (Ahlberg et al., 1994); a contact between left and right pterygoid that hides the contact (which may nevertheless be present) between vomer and parasphenoid in palatal view; fangs on middle coronoid more or less in line with marginal tooth row and not much larger than marginal teeth; posterior coronoid fangs not much larger than marginal teeth; meckelian bone does not floor precoronoid fossa; mesial parasymphysial foramen; a tooth row rather than a tooth field on the parasymphysial plate; a longitudinal ridge on the dorsal surface of the prearticular; a sacrum composed of the ilium that articulates with at least one pair of ribs; a femoral adductor blade (Ahlberg, 1998). Several other characters that were until recently considered diagnostic of stegocephalians are now thought to diagnose a more inclusive clade because *Tiktaalik roseae* possesses the derived condition (Daeschler et al., 2006). This includes the loss of the opercular bones that cover the gill chamber and the loss of the extrascapular bones that link the dermal shoulder girdle with the skull.

Synonyms: *Amphibia* Linnaeus 1758 (partial and approximate); *Tetrapoda* Fischer 1808 (approximate); *Ganocephala* Owen 1859 (partial and approximate); *Labyrinthodontia* Owen 1859 (approximate); and *Stegocephalia* Woodward 1898 (approximate).

Comments: *Stegocephali* was erected by Cope (1868) for a taxon that included all then-known early limbed vertebrates, the oldest of which dated from the Carboniferous, such as *Amphibamus grandiceps* (a Carboniferous temnospondyl), stereospondyls (from the Mesozoic), *Baphetes spp.*, and *Microsauria*. The latter included at that time various "lepospondyls" that are no longer considered "microsaurs" (e.g., the nectridean *Sauropleura longidentata*, the adelogyrinid *Molgophis macrurus*, and even *Hylonomus lyelli*, which has since been shown to be an amniote). Cope (1881) retained a similar concept, although he added embolomeres. Cope's (1868, 1881) *Stegocephali* excluded *Reptilia*. Later, Cope (1888) removed from *Stegocephali* most of the taxa that it had previously included, but this was ignored by most subsequent authors (e.g., Broili, 1904: 48–49; Säve-Soderbergh, 1932; Case, 1946). When Devonian limbed vertebrates (*Ichthyostega stensioei* and *Acanthostega gunnari*) were discovered, they were immediately included in *Stegocephalia* (Säve-Soderbergh, 1932). *Seymouriamorpha* was first included in *Stegocephali* by Case (1946), who considered that it retained primitive characters from embolomeres, and by Romer (1947), who recognized larval seymouriamorphs. These conclusions were accepted in most subsequent studies. Thus, since its origin, *Stegocephali* encompassed all known limbed vertebrates except for amniotes and lissamphibians (Moodie, 1916: 44). This meaning was retained by Goodrich (1930: xxv) and Case (1946), who included all known Palaeozoic limbed vertebrates except amniotes in *Stegocephalia* (sic). Thus, the name *Stegocephali* is here applied to a clade that includes all limbed vertebrates and any currently unknown forms that are more closely related to them than to forms that are currently known to have retained paired fins. It is preferred over the more frequently used variant *Stegocephalia*, because the latter represents a subsequent emendation of *Stegocephali*.

The origins of stegocephalians have been sought in the finned sarcopterygians since the nineteenth century. Most early palaeontologists suggested that among the finned sarcopterygians, the osteolepiforms, now known to be a paraphyletic group, were the closest known relatives of stegocephalians (Gregory, 1915: 326; Westoll 1943), although rhizodontids were also considered potential close relatives in early studies (Gregory, 1915), as were dipnoans (lungfishes) in some later ones (Rosen et al. 1981). Most early authors considered stegocephalians monophyletic in the pre-Hennigian sense (Gregory, 1915: 326; Westoll, 1943; Szarski, 1962: 238), with the exception of Holmgren (1933), Säve-Soderbergh (1934), and Jarvik (1955, 1962), who suggested that stegocephalians were polyphyletic. This latter view was rejected by most contemporaneous and more recent studies (e.g., Thomson, 1964: 353; Schultze, 1994; Zhu and Schutlze, 1997; Ahlberg and Johanson, 1998; Zhu et al., 2001; Laurin, 1998a). *Panderichthys rhombolepis* and *Tiktaalik roseae* are now viewed by nearly all systematists as the nearest known relatives of limbed vertebrates that retained paired fins (Shultze and Arsenault, 1985; Vorobyeva and Schultze, 1991; Laurin, 1998a; Ahlberg et al., 2000; Ruta et al., 2003; Daeschler et al., 2006), with the possible exception of some poorly known Devonian forms such as *Elpistostege watsoni* and *Livoniana multidentata*, whose paired appendages are either unknown or are represented only by fragmentary remains of the proximal segments (Laurin et al., 2000).

Several names could be proposed for the clade that includes *Acanthostega gunnari*, *Ichthyostega stensioei*, and other limbed vertebrates but excludes *Tiktaalik roseae*, *Panderichthys rhombolepis*, and taxa that diverged earlier from the tetrapod stem lineage. The best known are *Tetrapoda*, *Stegocephali*, and *Labyrinthodontia*. However, the only names that have been unambiguously and consistently associated with approximately the same ancestor are *Labyrinthodontia* and *Stegocephali* because *Tetrapoda* and *Amphibia* have sometimes been associated with the crown clade (Laurin and Anderson, 2004). *Labyrinthodontia* has usually been considered to be part of *Stegocephali,* so *Labyrinthodontia* is defined (this volume) as the name of a less inclusive clade.

The definition of the name *Stegocephali* uses extinct species as internal and external specifiers because this taxon traditionally included only fossils (e.g., *Eusthenopteron foordi, Panderichthys rhombolepis, Elpistostege watsoni, Livoniana multidentata,* and *Tiktaalik roseae* are also extinct).

The conversion of names such as *Stegocephali* has raised objections because it has been claimed that such names have been "largely or completely out of use among cladists" (Marjanović, 2004: 5). Whether the systematists that have used these names are cladists is of little relevance; what matters is frequency of use in the literature dealing with the relevant taxa. Internet searches show that these terms have not fallen into disuse; a Google Scholar search for the word *Stegocephalia* during the period 1980–2014 found 93 records for *Stegocephali*, 171 for *Stegocephalia*, and 292 for stegocephalian.

Literature Cited

Ahlberg, P. E. 1991a. A re-examination of sarcopterygian interrelationships, with special reference to the *Porolepiformes. Zool. J. Linn. Soc.* 103:241–287.

Ahlberg, P. E. 1991b. Tetrapod or near-tetrapod fossils from the Upper Devonian of Scotland. *Nature* 354:298–301.

Ahlberg, P. E. 1995. *Elginerpeton pancheni* and the earliest tetrapod clade. *Nature* 373:420–425.

Ahlberg, P. E. 1998. Postcranial stem tetrapod remains from the Devonian of Scat Craig, Morayshire, Scotland. *Zool. J. Linn. Soc.* 122:99–141.

Ahlberg, P. E., and J. A. Clack. 1998. Lower jaws, lower tetrapods – a review based on the Devonian genus *Acanthostega*. *Trans. R. Soc. Edinb.* 89:11–46.

Ahlberg, P. E., and Z. Johanson. 1998. Osteolepiforms and the ancestry of tetrapods. *Nature* 395:792–794.

Ahlberg, P. E., E. Luksevics, and O. Lebedev. 1994. The first tetrapod finds from the Devonian (Upper Famennian) of Latvia. *Philos. Trans. R. Soc. Lond. B Biol. Sci.* 343:303–328.

Ahlberg, P. E., E. Luksevics, and E. Mark-Kurik. 2000. A near-tetrapod from the Baltic Middle Devonian. *Palaeontology* 43:533–548.

Broili, F. 1904. Permische Stegocephalen und Reptilien aus Texas. *Palaeontogr. Abt. A* 51:1–120.

Campbell, K. S. W., and M. W. Bell. 1977. A primitive amphibian from the Late Devonian of New South Wales. *Alcheringa* 1:369–381.

Case, E. C. 1946. A census of the determinable genera of the *Stegocephalia*. *Trans. Am. Philos. Soc.* 35:325–420.

Clack, J. A. 2002. *Gaining Ground: The Origin and Evolution of Tetrapods*. Indiana University Press, Bloomington, IN.

Clack J. A., P. E. Ahlberg, H. Blom, and S. M. Finney. 2012. A new genus of Devonian tetrapod from North-East Greenland, with new information on the lower jaw of *Ichthyostega*. *Palaeontology* 55:73–86.

Clack, J. A., and M. I. Coates. 1995. *Acanthostega gunnari*, a primitive, aquatic tetrapod? *Bull. Mus. Natl. Hist. Nat., Paris, 4ème Sér.* 17:359–372.

Coates, M. I. 1996. The Devonian tetrapod *Acanthostega gunnari* Jarvik: postcranial anatomy, basal tetrapod interrelationships and patterns of skeletal evolution. *Trans. R. Soc. Edinb.* 87:363–421.

Coates, M. I., and J. A. Clack. 1991. Fish-like gills and breathing in the earliest known tetrapod. *Nature* 352:234–236.

Cope, E. D. 1868. Synopsis of the extinct *Batrachia* of North America. *Proc. Acad. Nat. Sci. Phila.* 20:208–221.

Cope, E. D. 1877. Descriptions of extinct *Vertebrata* from the Permian and Triassic formations of the United States. *Proc. Am. Philos. Soc.* 17:182–193.

Cope, E. D. 1881. Catalogue of *Vertebrata* of the Permian formation of the United States. *Am. Nat.* 15:162–164.

Cope, E. D. 1888. Systematic catalogue of the species of *Vertebrata* in the beds of the Permian epoch in North America with notes and descriptions. *Trans. Am. Philos. Soc.* 16:285–297.

Daeschler, E. B., N. H. Shubin, and F. A. Jenkins, Jr. 2006. A Devonian tetrapod-like fish and the evolution of the tetrapod body plan. *Nature* 440:757–763.

Fischer, G. 1808. *Tableaux Synoptiques de Zoognosie*. Imprimerie de l'Université impériale, Moscou.

Goodrich, E. S. 1930. *Studies on the Structure and Development of Vertebrates*. Macmillan, London.

Gregory, W. K. 1915. Present status of the problem of the origin of the *Tetrapoda*, with special reference to the skull and paired limbs. *Ann. N. Y. Acad. Sci.* 26:317–383.

Gross, W. 1930. Die Fische des mittleren Old Red Süd-Livlands. *G. Pal. Abh.* 18:123–156.

Holmgren, N. 1933. On the origin of the tetrapod limb. *Acta Zool. Stockh.* 14:185–295.

Jarvik, E. 1955. The oldest tetrapods and their forerunners. *Sci. Monthly* 80:141–154.

Jarvik, E. 1962. Les porolépiformes et l'origine des urodèles. Problèmes Actuels de Paléontologie—Evolution des Vertébrés. *Coll. Internat. Cent. Nat. Rech. Sci., Paris* 104:87–101.

Laurin, M. 1998a. The importance of global parsimony and historical bias in understanding tetrapod evolution. Part I. Systematics, middle ear evolution, and jaw suspension. *Ann. Sci. Nat., Zool., 13 Ser.* 19:1–42.

Laurin, M. 1998b. The importance of global parsimony and historical bias in understanding tetrapod evolution. Part II. Vertebral centrum, costal ventilation, and paedomorphosis. *Ann. Sci. Nat., Zool., 13 Ser.* 19:99–114.

Laurin, M., and J. S. Anderson. 2004. Meaning of the name *Tetrapoda* in the scientific literature: an exchange. *Syst. Biol.* 53:68–80.

Laurin, M., M. Girondot, and M.-M. Loth. 2004. The evolution of long bone microanatomy and lifestyle in lissamphibians. *Paleobiology* 30:589–613.

Laurin, M., M. Girondot, and A. de Ricqlès. 2000. Early tetrapod evolution. *Trends Ecol. Evol.* 15:118–123.

Linnaeus, C. 1758. *Systema Naturae.* 10th edition. Laurentii Salvii, Holmiae (Stockholm).

Marjanović, D. 2004. How to preserve historical usage in phylogenetic definitions? Selfdestructive definitions for names of grades. Pp. 1–5 in *Abstracts of the First International Phylogenetic Nomenclature Meeting* (M. Laurin, ed.). Paris. Available at https://hal.sorbonne-universite.fr/hal-02187647.

Marjanović, D., and M. Laurin. 2019. Phylogeny of Paleozoic limbed vertebrates reassessed through revision and expansion of the largest published relevant data matrix. *PeerJ* 6:e5565.

Moodie, R. L. 1916. The coal measures *Amphibia* of North America. *Publ. Carnegie Inst. Wash.* 238:1–222.

Owen, R. 1859. On the orders of fossil and recent *Reptilia* and their distribution in time. *Rept. Brit. Assn. Adv. Sci.* 153–166.

Pough, F. H., R. M. Andrews, J. E. Cadle, M. L. Crump, A. H. Savitzky, and K. Wells. 2004. *Herpetology.* 3rd edition. Prentice Hall, Upper Saddle River, NJ.

Romer, A. S. 1947. Review of the *Labyrinthodontia*. *Bull. Mus. Comp. Zool. Harv.* 99:1–368.

Rosen, D. E., P. L. Forey, B. G. Gardiner, and C. Patterson 1981. Lungfishes, tetrapods, paleontology, and plesiomorphy. *Bull. Am. Mus. Nat. Hist.* 167:163–275.

Ruta, M., M. I. Coates, and D. D. L. Quicke. 2003. Early tetrapod relationships revisited. *Biol. Rev.* 78:251–345.

Säve-Söderbergh, G. 1932. Preliminary note on Devonian stegocephalians from East Greenland. *Medd. Grønl.* 94:1–105.

Säve-Söderbergh, G. 1934. Some points of view concerning the evolution of the vertebrates and the classification of this group. *Ark. Zool.* 26A:1–20.

Schultze, H.-P. 1994. Comparison of hypotheses on the relationships of sarcopterygians. *Syst. Biol.* 43:155–173.

Schultze, H.-P., and M. Arsenault. 1985. The panderichthyid fish *Elpistostege*: a close relative of tetrapods? *Palaeontology* 28:293–309.

Szarski, H. 1962. The origin of the *Amphibia*. *Q. Rev. Biol.* 37:189–241.

Thomson, K. S. 1964. The comparative anatomy of the snout in rhipidistian fishes. *Bull. Mus. Comp. Zool. Harv.* 131:315–357.

Vallin, G., and M. Laurin. 2004. Cranial morphology and affinities of *Microbrachis*, and a reappraisal of the phylogeny and lifestyle of the first amphibians. *J. Vertebr. Paleontol.* 24:56–72.

Vorobyeva, E., and H.-P. Schultze. 1991. Description and systematics of panderichthyid fishes with comments on their relationship to tetrapods. Pp. 68–109 in *Origins of the Higher Groups of Tetrapods—Controversy and Consensus* (H.-P. Schultze and L. Trueb, eds.). Cornell University Press, Ithaca, NY.

Westoll, T. S. 1943. The origin of the tetrapods. *Biol. Rev.* 18:78–98.

Whiteaves, J. F. 1881. On some remarkable fossil fishes from the Devonian rocks of Scaumenac Bay, in the province of Quebec. *Ann. Mag. Nat. Hist.*, 5th Ser., 8:159–162.

Woodward, A. S. 1898. *Outlines of Vertebrate Palaeontology for Students of Zoology.* Cambridge University Press, Cambridge, UK.

Zhu, M., and H.-P. Schultze. 1997. The oldest sarcopterygian fish. *Lethaia* 30:293–304.

Zhu, M., X. Yu, and E. P. Ahlberg. 2001. A primitive sarcopterygian fish with an eyestalk. *Nature* 410:81–84.

Author

Michel Laurin; CR2P, CNRS/MNHN/SU; Muséum National d'Histoire Naturelle; CP 48; 43 rue Buffon, 75005 Paris, France. Email: michel.laurin@mnhn.fr.

Date Accepted: 1 May 2014

Editors: Philip Cantino, Jacques A. Gauthier, Kevin de Queiroz

Labyrinthodontia R. Owen 1859 [M. Laurin], converted clade name

Registration Number: 56

Definition: The smallest clade containing *Acanthostega gunnari* Jarvik 1952, *Ichthyostega stensiöi* Säve-Söderbergh 1932, *Crassigyrinus scoticus* (Lydekker 1890), *Mastodonsaurus giganteus* (Jaeger 1828), *Baphetes kirkbyi* Owen 1854 (*Baphetidae*), and *Anthracosaurus russelli* Huxley 1863. This is a minimum-clade definition. Abbreviated definition: min ∇ (*Acanthostega gunnari* Jarvik 1952 & *Ichthyostega stensiöi* Säve-Söderbergh 1932 & *Crassigyrinus scoticus* (Lydekker 1890) & *Mastodonsaurus giganteus* (Jaeger 1828) & *Baphetes kirkbyi* Owen 1854 & *Anthracosaurus russelli* Huxley 1863).

Etymology: Presumably derived (Owen did not specify) from the Greek *labyrinthos* (maze) and *odontos* (tooth), in reference to the deep infolding of the dentine of the teeth, which is best seen in histological cross-sections but also leaves grooves on the surface of the enamel.

Reference Phylogeny: The primary reference phylogeny is Vallin and Laurin (2004: Fig. 6). *Anthracosaurus russelli* and *Mastodonsaurus giganteus* are not included in the reference phylogeny but are thought to be part of *Embolomeri* and *Temnospondyli*, respectively (Yates and Warren, 2000; Ruta and Coates, 2007).

Composition: *Labyrinthodontia* includes all vertebrates that are currently known to have possessed limbs, that is, (crown) *Tetrapoda* as well as part of its stem group, including seymouriamorphs, gephyrostegids, embolomeres, temnospondyls, colosteids, baphetids, *Crassigyrinus*, *Tulerpeton*, *Ichthyostega*, and *Acanthostega* (see *Tetrapoda* in this volume regarding alternative views about which of these taxa are inside and outside of the crown group). It differs from *Stegocephali* by excluding, under the reference phylogeny, some poorly known Devonian taxa that may have had paired fins, such as *Obruchevichthys gracilis*, *Elginerpeton pancheni*, and possibly other poorly known Devonian species such as *Ventastega curonica* and *Metaxygnathus denticulatus* (Campbell and Bell, 1977; Ahlberg, 1991, 1995; Ahlberg et al., 1994; Laurin, 1998c; Laurin et al., 2000; Marjanović and Laurin, 2019).

Diagnostic Apomorphies: Several apomorphies of this taxon reflect the appearance of the limb with digits, which may be diagnostic of a slightly more inclusive clade. However, given the fragmentary nature of the remains of several potential sister-taxa of *Labyrinthodontia*, an objective method for dealing with missing data had to be found. Thus, for the purpose of the diagnostic list of apomorphies, characters have been treated as if optimized under a DELTRAN optimization (Swofford, 2003), which assigns to apomorphies the most recent possible date of appearance. These apomorphies include (Laurin, 1998a,b): contact between skull table and cheek at the level of the tabular; a paroccipital process; basioccipital and exoccipital remain suturally distinct throughout ontogeny (reversed in most extant taxa); fangs on anterior coronoid more or less in line with vertical lamina and marginal tooth row (Ahlberg and Clack, 1998); neural arch halves firmly fused to each other; transverse processes for articulation with the ribs; ilium expanded dorsally to form a distinct iliac blade; digits (defined here as unbranched rows of endoskeletal elements articulating proximally to metapodial elements); loss of the lepidotrichia in the paired appendages.

Synonyms: *Stegocephali* Cope 1868 and the variant *Stegocephalia* (partial, approximate); *Stegocephali* sensu Laurin (1998a) (approximate); *Tetrapoda* Hatschek and Cori 1896 (approximate); *Tetrapoda* sensu Goodrich (1930) (approximate); *Amphibia* Linnaeus 1758 (partial, approximate); *Amphibia* sensu Goodrich (1930) (partial, approximate).

Comments: The proposed definition uses as internal specifiers only taxa that were members of *Labyrinthodontia* in numerous older studies, in which this name designated a paraphyletic group. The taxon *Labyrinthodon* is a junior synonym of *Mastodonsaurus* (Romer, 1947: 227), so the type species of *Mastodonsaurus* (Damiani, 2001) is included as a specifier. All major clades that were part of the old, paraphyletic concept of *Labyrinthodontia* in most studies are represented. However, a few stegocephalians that may have paired fins (*Elginerpeton*, *Obruchevichthys*, *Ventastega*, and *Metaxygnathus*) are deliberately excluded from the specifiers. Thus, *Labyrinthodontia* is the smallest clade that includes all currently known limbed, or more precisely digitate, vertebrates.

The choice of this name must be put into a historical perspective to be understood. The name *Labyrinthodontia* was erected by Owen (1859). However, the very similar name *Labyrinthodontes* was proposed by von Meyer in 1842 (Moodie, 1916: 40). Labyrinthodonts (as well as amphibians, which Owen called "batrachians") were initially classified among the reptiles, and they continued to be so classified, by at least some authors, for nearly a quarter of a century after the discovery of *Mastodonsaurus*, the first taxon traditionally attributed to *Labyrinthodontia* to be described (Moodie, 1916). The first suggestions that labyrinthodonts were amphibians were made in the 1850s, but they were largely ignored (Moodie, 1916). Initially, this taxon included

only forms that are now considered stereospondyls, a clade of Triassic temnospondyls (stem-tetrapods, according to the reference phylogeny). This concept of *Labyrinthodontia* was retained by Owen (1860). The Permo-Carboniferous temnospondyls, which formed the taxon *Ganocephala*, remained outside *Labyrinthodontia* until Huxley (1863) included them. *Labyrinthodontia* was further expanded by Miall (1874), who also included embolomeres and all known "lepospondyls" (nectrideans, aistopods, "microsaurs"). This taxonomy, and the very similar scheme proposed by Lydekker (1890), probably assigned more taxa to *Labyrinthodontia* than most previous or subsequent taxonomies in that among limbed vertebrates, only lissamphibians and (recognized) amniotes were excluded. Most authors from the late nineteenth century onwards included in *Labyrinthodontia* some or all early limbed vertebrates of medium to large body size that possessed dentine infolding. Cope (1875) included in this taxon baphetids and embolomeres but neither "lepospondyls" nor colosteids (it is not clear whether stereospondyls were included). Watson (1917, 1919: 64–68) included *Embolomeri*, *Rachitomi*, and *Stereospondyli* in *Labyrinthodontia*. Romer (1947: 310–319) similarly included *Temnospondyli* (which included baphetids, rhachitomes, trematosaursn and *Ichthyostega*) and *Anthracosauria* (for embolomeres and seymouriamorphs). Lehman (1955) adopted a similar composition for *Labyrinthodontia*, which included limbed vertebrates except amniotes, lissamphibians, and "phyllospondyls" (branchiosaurs; now usually considered larval and immature temnospondyls). Romer (1966) and Carroll (1988) similarly included Devonian stem-tetrapods such as *Ichthyostega*, *Acanthostega*, and temnospondyls, and stem-amniotes such as limnoscelids, in *Labyrinthodontia*. Only "lepospondyls", lissamphibians and amniotes were excluded. The composition of *Labyrinthodontia* differed from *Stegocephali* mostly in the exclusion of "lepospondyls" in most recent taxonomies. In the context of phylogenetic nomenclature, which would include "lepospondyls", lissamphibians, and amniotes in any clade that contains the taxa traditionally included in *Labyrinthodontia*, the name *Labyrinthodontia* could be considered redundant with *Stegocephali*, as it was by some earlier authors (Fritsch, 1883: 68; Goodrich, 1930: XXV). Another possibility is to use the name *Labyrinthodontia* for a slightly more or less inclusive clade, and the second option is adopted here (see *Stegocephali* in this volume). Should both names apply to the same clade due to future changes in our understanding of the affinities of the relevant taxa, the name *Stegocephali*, which historically referred to a more inclusive set of taxa than *Labyrinthodontia*, should have precedence. The name *Labyrinthodontia* has been widely used in palaeontology as well as reasonably often in the scientific literature since 1980. It is here applied to a clade originating in approximately the same ancestor as the paraphyletic group with which it was formerly associated.

Literature Cited

Ahlberg, P. E. 1991. Tetrapod or near-tetrapod fossils from the Upper Devonian of Scotland. *Nature* 354:298–301.

Ahlberg, P. E. 1995. *Elginerpeton pancheni* and the earliest tetrapod clade. *Nature* 373:420–425.

Ahlberg, P. E., and J. A. Clack. 1998. Lower jaws, lower tetrapods—a review based on the Devonian genus *Acanthostega*. *Trans. R. Soc. Edinb.* 89:11–46.

Ahlberg, P. E., E. Luksevics, and O. Lebedev. 1994. The first tetrapod finds from the Devonian (Upper Famennian) of Latvia. *Philos. Trans. R. Soc. Lond. B Biol. Sci.* 343:303–328.

Campbell, K. S. W., and M. W. Bell. 1977. A primitive amphibian from the Late Devonian of New South Wales. *Alcheringa* 1:369–381.

Carroll, R. L. 1988. *Vertebrate Paleontology and Evolution*. W. H. Freeman, New York.

Cope, E. D. 1868. Synopsis of the extinct *Batrachia* of North America. *Proc. Acad. Nat. Sci. Phila.*:208–221.

Cope, E. D. 1875. Check-list of North American *Batrachia* and *Reptilia*, with a systematic list of the higher groups and an essay on geographical distribution based on the specimens contained in the U. S. National Museum. *Bull. U. S. Nat. Mus.* 1:7–12.

Damiani, R. J. 2001. A systematic revision and phylogenetic analysis of Triassic mastodonsauroids (*Temnospondyli: Stereospondyli*). *Zool. J. Linn. Soc.* 133:379–482.

Fritsch, A. 1883. *Fauna der Gaskohle und der Kalksteine der Permformation* Böhmens, Prague.

Goodrich, E. S. 1930. *Studies on the Structure and Development of Vertebrates*. Macmillan, London.

Hatschek, B., and C. J. Cori. 1896. *Elementarcus der Zootomie in fünfzen Vorlesungen*. Gustav Fischer, Jena.

Huxley, T. H. 1863. Description of *Anthracosaurus russelli*, a new labyrinthodont from the Lanarkshire coal field. *Q. J. Geol. Soc. Lond.* 19:56–68.

Jaeger, G. 1828. *Über die fossile Reptilien, welche in Württemberg aufgefunden worden sind*. J. B. Metzler, Stuttgart.

Jarvik, E. 1952. On the fish-like tail in the ichthyostegid stegocephalians with descriptions of a new stegocephalian and a new crossopterygian from the Upper Devonian of East Greenland. *Meddr Grønland* 114:1–90.

Laurin, M. 1998a. The importance of global parsimony and historical bias in understanding tetrapod evolution. Part I. Systematics, middle ear evolution, and jaw suspension. *Ann. Sci. Nat., Zool., 13 Ser.* 19:1–42.

Laurin, M. 1998b. The importance of global parsimony and historical bias in understanding tetrapod evolution. Part II. Vertebral centrum, costal ventilation, and paedomorphosis. *Ann. Sci. Nat., Zool., 13 Ser.* 19:99–114.

Laurin, M. 1998c. A reevaluation of the origin of pentadactyly. *Evolution* 52:1476–1482.

Laurin, M., M. Girondot, and A. de Ricqlès. 2000. Early tetrapod evolution. *Trends Ecol. Evol.* 15:118–123.

Lehman, J.-P. 1955. Les amphibiens: généralités. Pp. 2–52 in *Traité de Paléontologie* (J. Piveteau, ed.) Masson & Cie, Paris.

Linnaeus, C. 1758. *Systema Naturae*, 10th edition. Laurentii Salvii, Holmiae (Stockholm).

Lydekker, R. 1890. On two new species of labyrinthodonts. *Q. J. Geol. Soc. Lond.* 46:289–294.

Marjanović, D., and M. Laurin. 2019. Phylogeny of Paleozoic limbed vertebrates reassessed through revision and expansion of the largest published relevant data matrix. *PeerJ* 6:e5565.

Miall, L. C. 1874. On the classification of the *Labyrinthodontia. Rept. Brit. Assn. Adv. Sci.*:149–192.

Moodie, R. L. 1916. The coal measures *Amphibia* of North America. *Publ. Carnegie Inst. Wash.* 238:1–222.

Owen, R. 1854. On a fossil reptilian skull embedded in a mass of Pictou coal from Nova Scotia. *Q. J. Geol. Soc. Lond.* 10:207–208.

Owen, R. 1859. On the orders of fossil and recent *Reptilia* and their distribution in time. *Rept. Brit. Assn. Adv. Sci. 29th meeting*:153–166.

Owen, R. 1860. On the orders of fossil and recent *Reptilia*, and their distribution in time. *Can. J. Ind. Sci. Art* 25:73–86.

Romer, A. S. 1947. Review of the *Labyrinthodontia. Bull. Mus. Comp. Zool. Harv.* 99:1–368.

Romer, A. S. 1966. *Vertebrate Paleontology.* 3rd edition. University of Chicago Press, Chicago, IL.

Ruta, M., and M. I. Coates. 2007. Dates, nodes and character conflict: addressing the lissamphibian origin problem. *J. Syst. Palaeontol.* 5:69–122.

Säve-Söderbergh, G. 1932. Preliminary note on Devonian stegocephalians from East Greenland. *Medd. Grønl.* 94:1–105.

Swofford, D. L. 2003. *PAUP* Phylogenetic Analysis Using Parsimony (*and Other Methods)*, Version 4.0b10. Sinauer Associates.

Vallin, G., and M. Laurin. 2004. Cranial morphology and affinities of *Microbrachis*, and a reappraisal of the phylogeny and lifestyle of the first amphibians. *J. Vertebr. Paleontol.* 24:56–72.

Watson, D. M. S. 1917. A sketch classification of the pre-Jurassic tetrapod vertebrates. *Proc. Zool. Soc. Lond.* 1917:167–186.

Watson, D. M. S. 1919. The structure, evolution and origin of the *Amphibia.* The "Orders' *Rachitomi* and *Stereospondyli. Philos. Trans. R. Soc. Lond. B Biol. Sci.* 209:1–73.

Yates, A.M., and A. Warren. 2000. The phylogeny of the 'higher' temnospondyls (*Vertebrata: Choanata*) and its implications for the monophyly and origins of the *Stereospondyli. Zool. J. Linn. Soc.* 128:77–121.

Author

Michel Laurin; CR2P, CNRS/MNHN/SU; Muséum National d'Histoire Naturelle; CP 48; 43 rue Buffon, 75005 Paris, France. Email: michel.laurin@mnhn.fr.

Date Accepted: 5 April 2014

Editors: Philip Cantino, Jacques A. Gauthier, Kevin de Queiroz

Anthracosauria G. Säve-Söderbergh 1934 [M. Laurin and T. R. Smithson], converted clade name

Registration Number: 13

Definition: The largest clade containing *Anthracosaurus russelli* Huxley 1863 (*Embolomeri*) but not *Ascaphus truei* Stejneger 1899 (closest to *Discoglossidae* and *Pipidae* in the reference phylogeny) and *Eryops megacephalus* Cope 1877 (*Temnospondyli*). This is a maximum-clade definition. Abbreviated definition: max ∇ (*Anthracosaurus russelli* Huxley 1863 ~ *Ascaphus truei* Stejneger 1899 & *Eryops megacephalus* Cope 1877).

Etymology: Säve-Söderbergh (1934) did not specify it, but the derivation is apparently from the name of the included taxon *Anthracosaurus* Huxley 1863, which is itself apparently derived from the Greek *anthrakos*, coal, and *sauros*, lizard.

Reference Phylogeny: The primary reference phylogeny is Vallin and Laurin (2004: Fig. 6), which shows the relationships between embolomeres (the core group of anthracosaurs) and other stegocephalians.

Composition: *Anthracosauria* includes *Embolomeri* Cope 1884 and, depending on the phylogeny, other taxa. Under the reference phylogeny and some other phylogenies (e.g., Laurin, 1998a: Fig. 3; Laurin and Reisz, 1999: Fig. 5; Marjanović and Laurin, 2019: Figs. 10, 11, 14, 19–21), *Anthracosauria* includes only *Embolomeri*, but under other traditional phylogenies (e.g., Panchen and Smithson, 1988: Fig. 1.1; Gauthier et al., 1988: Fig. 4.3; Lombard and Sumida, 1992: Fig. 6), *Anthracosauria* may also include *Seymouriamorpha* Watson 1917 and/or *Crassigyrinus scoticus* Lydekker 1890,

diadectomorphs, and amniotes. When it includes the latter, it refers to the amniote total group and becomes synonymous with *Pan-Amniota* Laurin and Smithson (this volume). Under the phylogeny of Ruta et al. (2003), *Anthracosauria* includes all of those taxa except *Crassigyrinus scoticus*, but it also includes lepospondyls (as stem amniotes). Many older works included gephyrostegids (=*Gephyrostegoidea* of Carroll 1969) in *Anthracosauria* (e.g., Carroll, 1970; Panchen, 1980; Godfrey and Reisz, 1991), and this is congruent with the phylogeny of Laurin and Reisz (1999). This possibility is not ruled out by the primary reference phylogeny (Vallin and Laurin, 2004: Fig. 6), which is based on a more recent version of the same data set, in which the position of gephyrostegids is unresolved relative to embolomeres and a large clade that includes the tetrapod crown. Detailed phylogenies of embolomeres have been published by Holmes (1984: Fig. 39), Smithson (1985: Fig. 34), and Clack (1987: Fig. 36),

Diagnostic Apomorphies: The list of apomorphies is taken from Laurin (1998a,b), Laurin and Reisz (1999), Ruta et al. (2003), and Vallin and Laurin (2004), assuming the phylogeny of Vallin and Laurin (2004), and not considering *Gephyrostegidae* part of *Anthracosauria* (see Composition). Diagnostic characters include: dermal sculpturing consisting of shallow pits; septomaxilla absent; lacrimal excluded from orbit by a contact between prefrontal and nasal; jugal extending anterior to anterior orbital rim; subdermal tabular horn; presence of an incomplete lateral palatal tooth row; absence of denticles on palatine; paroccipital process absent; posttemporal fossae absent; exoccipitals separated from dermatocranium by opisthotic; diameter of pleurocentrum greater than its length; internal trochanter confluent with proximal articular surface of femur. Additionally, the following characters suggest that *Gephyrostegoidea* is part of *Anthracosauria*: posterior part of joint between skull table and cheek smooth (not serrated); presence of a ridge near posterior edge of flexor surface of fibula; presence of five phalanges in fifth toe.

Synonyms: In the context of the primary reference phylogeny (Vallin and Laurin, 2004: Fig. 6), *Embolomeri* Cope 1884 and *Anthracosauroideae* Watson 1929 are approximate synonyms. However, under traditional phylogenies (Gauthier et al., 1988: Fig. 4.3; Lombard and Sumida, 1992: Fig. 6) and that of Ruta et al. (2003), *Pan-Amniota* Rowe 2004 (see Laurin and Smithson, this volume) is an unambiguous synonym, and *Reptiliomorpha* Säve-Söderbergh 1934, although not given an explicit phylogenetic definition, is also considered an unambiguous synonym.

Comments: The taxon *Anthracosauria* was erected by Säve-Soderbergh (1934) for a taxon then thought to be on the amniote stem. It included embolomeres (and possibly seymouriamorphs, but they are not explicitly mentioned). Subsequently, the taxon *Anthracosauria* came to encompass an increasingly large proportion of the presumed extinct relatives of amniotes, until it included all of them (Carroll, 1988). Thus, in the taxonomy of Panchen (1970: 2), *Anthracosauria* included only embolomeres, but Panchen (1975: 609) included embolomeres, proterogyrinids, and gephyrostegids, and Panchen (1980: 343) included seymouriamorphs in addition to all these taxa. Finally, Carroll (1988: 613) expanded *Anthracosauria* to include all the taxa already included by Panchen (1980), as well as *Solenodonsaurus janenschi*,

diadectomorphs, and two taxa that are now considered amniotes, *Nycteroleter ineptus* and *Macroleter poezicus* (Reisz and Laurin, 2001). When these taxa were included in an explicit phylogenetic analysis along with amniotes (Gauthier et al., 1988), they were found to correspond to the amniote stem group according to the inferred phylogeny. Consequently, when Gauthier et al. (1988, 1989) proposed a taxonomy adopting Hennig's principle of monophyly, they further expanded the formerly paraphyletic *Anthracosauria* to include *Amniota*, defining the name *Anthracosauria* phylogenetically as the sister group of *Amphibia*. Similarly, the taxonomy of de Queiroz and Gauthier (1992: 474) used a maximum-clade definition of *Anthracosauria* that applied the name to the total clade comprising amniotes and their extinct relatives.

The definition proposed by Gauthier et al. (1988, 1989) became problematic when subsequent studies (Laurin 1998a,b; Laurin and Reisz, 1999; Vallin and Laurin, 2004) suggested that all embolomeres (including *Anthracosaurus*), seymouriamorphs and several other taxa traditionally included in *Anthracosauria* were stem-tetrapods rather than stem amniotes. When the definition proposed by Gauthier et al. (1988, 1989) is applied in the context of these phylogenies, *Anthracosauria* no longer includes *Anthracosaurus russelli* or any other taxon that was part of the original composition of *Anthracosauria*, which is contrary to Article 11.10 of the *PhyloCode* (ver. 6; Cantino and de Queiroz, 2020). Therefore, Laurin (2001: 208) proposed an alternative phylogenetic definition based on *Anthracosaurus russelli*, in conformity with the *PhyloCode*: "The largest clade that includes *Anthracosaurus russelli* but not *Ascaphus truei*". The definition published here adds a temnospondyl as an external specifier because temnospondyls have always been excluded from *Anthracosauria,* and the relationships among embolomeres, lissamphibians, and temnospondyls are debated (so that using only a lissamphibian as an external specifier could result in the inclusion of temnospondyls in *Anthracosauria*). This definition ensures that the taxon *Anthracosauria* will always include *Anthracosaurus russelli*, on whose name it is based, and will always exclude temnospondyls (and lissamphibians), whether lissamphibians or temnospondyls are most closely related to embolomeres. It ensures that the maximal number of taxa traditionally considered anthracosaurs are part of this taxon, to the extent that this is compatible with the criterion of monophyly.

As mentioned above, under some phylogenies (e.g., Panchen and Smithson, 1988: Fig. 1.1, both trees; Gauthier et al., 1988: Fig. 4.3; Lombard and Sumida, 1992: Fig. 6), *Anthracosauria* as defined here is synonymous with *Pan-Amniota* (see Laurin and Smithson, this volume). In this case, we wish for *Anthracosauria* to have precedence because it has been used more frequently, as suggested by a Google Scholar search (April 23, 2014), which yielded only one result for *Pan-Amniota*, versus 319 for *Anthracosauria*. Although some of the uses of *Anthracosauria*, particularly those prior to 1975, are neither for the amniote total clade nor for a paraphyletic group composed of part or all of the amniote stem, occurrences of *Anthracosauria* in papers published in 1975 or later account for 261 of the results for that name.

Literature Cited

Cantino, P. D., and K. de Queiroz. 2020. *International Code of Phylogenetic Nomenclature (PhyloCode)*, Version 6. CRC Press, Boca Raton, FL.

Carroll, R. L. 1969. Problems of the origin of reptiles. *Biol. Rev.* 44:393–432.

Carroll, R. L. 1970. The ancestry of reptiles. *Philos. Trans. R. Soc. Lond. B Biol. Sci.* 257:267–308.

Carroll, R. L. 1988. *Vertebrate Paleontology and Evolution*. W. H. Freeman, New York.

Clack, J. A. 1987. *Pholiderpeton scutigerum* Huxley, an amphibian from the Yorkshire coal measures. *Philos. Trans. R. Soc. Lond. B Biol. Sci.* 318:1–107.

Cope, E. D. 1877. Descriptions of extinct *Vertebrata* from the Permian and Triassic formations of the United States. *Proc. Am. Philos. Soc.* 17:182–193.

Cope, E. D. 1884. The *Batrachia* of the Permian period of North America. *Am. Nat.* 18:26–39.

Gauthier, J. A., D. C. Cannatella, K. De Queiroz, A. G. Kluge, and T. Rowe. 1989. Tetrapod phylogeny. Pp. 337–353 in *The Hierarchy of Life* (B. Fernholm, K. Bremer, and H. Jornvall, eds.). Elsevier Science Publishers B. V. (Biomedical Division), New York.

Gauthier, J. A., A. G. Kluge, and T. Rowe. 1988. The early evolution of the *Amniota*. Pages 103–155 in *The Phylogeny and Classification of the Tetrapods, Vol. 1: Amphibians, Reptiles, Birds* (M. J. Benton, ed.). Clarendon Press, Oxford.

Godfrey, S. J., and R. R. Reisz. 1991. The vertebral morphology of *Gephyrostegus bohemicus* Jaekel 1902, with comments on the atlas-axis complex in primitive tetrapods. *Hist. Biol.* 5:27–36.

Holmes, R. 1984. The Carboniferous amphibian *Proterogyrinus scheelei* Romer, and the early evolution of tetrapods. *Philos. Trans. R. Soc. Lond. B Biol. Sci.* 306:431–527.

Huxley, T. H. 1863. Description of *Anthracosaurus russelli*, a new labyrinthodont from the Lanarkshire coal field. *Q. J. Geol. Soc. Lond.* 19:56–68.

Laurin, M. 1998a. The importance of global parsimony and historical bias in understanding tetrapod evolution. Part I. Systematics, middle ear evolution, and jaw suspension. *Ann. Sci. Nat., Zool., 13 Ser.* 19:1–42.

Laurin, M. 1998b. The importance of global parsimony and historical bias in understanding tetrapod evolution. Part II. Vertebral centrum, costal ventilation, and paedomorphosis. *Ann. Sci. Nat., Zool., 13 Ser.* 19:99–114.

Laurin, M. 2001. L'utilisation de la taxonomie phylogénétique en paléontologie: avantages et inconvénients. *Biosystema* 19:197–211.

Laurin, M., and R. R. Reisz. 1999. A new study of *Solenodonsaurus janenschi*, and a reconsideration of amniote origins and stegocephalian evolution. *Can. J. Earth Sci.* 36:1239–1255.

Lombard, R. E., and S. S. Sumida. 1992. Recent progress in understanding early tetrapods. *Amer. Zool.* 32:609–622.

Lydekker, R. 1890. On two new species of labyrinthodonts. *Q. J. Geol. Soc. Lond.* 46:289–294.

Marjanović, D., and M. Laurin. 2019. Phylogeny of Paleozoic Limbed Vertebrates Reassessed through Revision and Expansion of the Largest Published Relevant Data Matrix. *PeerJ* 6:e5565.

Panchen, A. L. 1970. *Anthracosauria* Pp. 1–83 in *Encyclopedia of Paleoherpetology* (O. Kuhn, ed.). Gustav Fischer, Stuttgart.

Panchen, A. L. 1975. A new genus and species of anthracosaur amphibian from the Lower Carboniferous of Scotland and the status of *Pholidogaster pisciformis* Huxley. *Philos. Trans. R. Soc. Lond. B Biol. Sci.* 269:581–637.

Panchen, A. L. 1980. The origin and relationships of the anthracosaur *Amphibia* from the late Paleozoic. Pp. 319–350 in *The Terrestrial Environment and the Origin of Land Vertebrates* (A. L. Panchen, ed.). Academic Press, London.

Panchen, A. L., and T. R. Smithson. 1988. The relationships of the earliest tetrapods. Pp. 1–32 in *The Phylogeny and Classification of the Tetrapods, Vol. 1: Amphibians, Reptiles, Birds* (M. J. Benton, ed.). Clarendon Press, Oxford.

de Queiroz, K., and J. A. Gauthier. 1992. Phylogenetic taxonomy. *Annu. Rev. Ecol. Syst.* 23:449–480.

Reisz, R. R., and M. Laurin. 2001. The reptile *Macroleter*: first vertebrate evidence for correlation of Upper Permian continental strata of North America and Russia. *GSA Bulletin* 113:1229–1233.

Rowe, T. 2004. Chordate phylogeny and development. Pp. 384–409 in *Assembling the Tree of Life* (J. Cracraft and M. J. Donoghue, eds.). Oxford University Press, Oxford.

Ruta, M., M. I. Coates, and D. D. L. Quicke. 2003. Early tetrapod relationships revisited. *Biol. Rev.* 78:251–345.

Säve-Söderbergh, G. 1934. Some points of view concerning the evolution of the vertebrates and the classification of this group. *Ark. Zool.* 26A:1–20.

Smithson, T. R. 1985. The morphology and relationships of the Carboniferous amphibian *Eoherpeton watsoni* Panchen. *Zool. J. Linn. Soc.* 85:317–410.

Stejneger, L. 1899. Description of a new genus and species of discoglossoid toad from North America. *Proc. U. S. Natl. Mus.* 21:899–901.

Vallin, G., and M. Laurin. 2004. Cranial morphology and affinities of *Microbrachis*, and a reappraisal of the phylogeny and lifestyle of the first amphibians. *J. Vertebr. Paleontol.* 24:56–72.

Watson, D. M. S. 1917. A sketch classification of the pre-Jurassic tetrapod vertebrates. *Proc. Zool. Soc. Lond.* 1917:167–186.

Watson, D. M. S. 1929. The Carboniferous *Amphibia* of Scotland. *Palaeont. Hung.* 1:219–252.

Authors

Michel Laurin; CR2P, CNRS/MNHN/SU; Muséum National d'Histoire Naturelle; CP 48; 43 rue Buffon, 75005 Paris, France. Email: michel.laurin@mnhn.fr.

Tim Smithson; University Museum of Zoology; Downing Street; Cambridge; CB2 3EJ, UK. Email: ts556@cam.ac.uk.

Date Accepted: 1 May 2014

Editors: Philip Cantino, Jacques A. Gauthier, Kevin de Queiroz

Seymouriamorpha D. M. S. Watson 1917 [M. Laurin and J. Klembara], converted clade name

Registration Number: 97

Definition: The largest clade containing *Seymouria baylorensis* Broili 1904 but not *Homo sapiens* Linnaeus 1758 (*Synapsida*) and *Procolophon trigoniceps* Owen 1876 and *Anthracosaurus russelli* Huxley 1863 (*Embolomeri*) and *Eryops megacephalus* Cope 1877 (*Temnospondyli*) and *Gephyrostegus bohemicus* Jaekel 1902 (*Gephryostegidae*) and *Solenodonsaurus janenschi* Broili 1924 and *Diadectes sideropelicus* Cope 1878 (*Diadectomorpha*). This is a maximum-clade definition. Abbreviated definition: max ∇ (*Seymouria baylorensis* Broili 1904 ~ *Homo sapiens* Linnaeus 1758 & *Procolophon trigoniceps* Owen 1876 & *Anthracosaurus russelli* Huxley 1863 & *Eryops megacephalus* Cope 1877 & *Gephyrostegus bohemicus* Jaekel 1902 & *Solenodonsaurus janenschi* Broili 1924 & *Diadectes sideropelicus* Cope 1878).

Etymology: Watson 1917 did not specify it, but the derivation is apparently from *Seymouria*, the name of an included taxon, and from the Greek *morphe* (external appearance).

Reference Phylogeny: Vallin and Laurin (2004: Fig. 6) is the primary reference phylogeny, but the position of *Seymouriamorpha* in this phylogeny is controversial (see Comments). The phylogeny within *Seymouriamorpha* has been inferred by Klembara and Ruta (2005b: Fig. 10C), and is fairly similar to that of Laurin (1996a: Fig. 6). Both include only a fraction of known seymouriamorphs (the Permo-Carboniferous forms) and are partly incompatible with each other.

Composition: *Seymouriamorpha* includes extinct taxa from the Lower Permian of: North America, such as *Seymouria* (White, 1939; Berman et al., 1987; Laurin, 1995,b; Klembara et al. 2005, 2007); central Europe, such as *Discosauriscus* (Klembara and Meszároš, 1992; Klembara, 1997), *Makowskia* (Klembara, 2005), *Spinarerpeton* (Klembara, 2009), and *Seymouria* (Berman and Martens, 1993; Berman et al., 2000; Klembara et al. 2005); and central Asia, on the former Kazakhstan microcontinent, such as *Ariekanerpeton* (Ivakhnenko, 1981; Laurin, 1996c; Klembara and Ruta, 2005a,b), *Utegenia* (Kuznetsov and Ivakhnenko, 1981; Laurin, 1996a; Klembara and Ruta, 2004a,b), and *Urumqia* (Zhang et al., 1984; synonymized with *Utegenia* by Ivakhnenko, 1987). It also includes taxa from the Middle or Upper Permian of European Russia, such as *Karpinskiosaurus* (Bulanov, 2002, Klembara, 2011), *Kotlassia* (Amalitzky, 1921), and *Microphon* (Bulanov, 2003). A useful review of the seymouriamorphs found in Russia and central Asia can be found in Bulanov (2003), although the phylogeny presented in that paper (Fig. 55) was not based on an explicit phylogenetic analysis of a data matrix and proposes that nycteroleteromorphs are derived from a paraphyletic *Seymouriamorpha*. Other recent phylogenetic analyses have concluded that nycteroleteromorphs are parareptiles and, thus, amniotes (Lee, 1995: Fig. 22; deBraga and Reisz, 1996: Fig. 3; deBraga and Rieppel, 1997: Fig. 1; Lee, 1997: Fig. 1; Rieppel and Reisz, 1999: Fig. 1; Modesto, 2000: Fig. 3; Laurin and Piñeiro, 2018: Fig. 2b, c). As such, they are not derived from seymouriamorphs.

Diagnostic Apomorphies: Seymouriamorph apomorphies include: posterior edge of parasphenoid forms a posteromedian process, in some species more or less split in the median plane; otic tube formed by prootic and opisthotic; stapes has no ossified contact with quadrate and appears to have lost its role in jaw suspension (perhaps transformed into an auditory ossicle, as has occurred independently in several tetrapod taxa); stapedial footplate represented by a gentle proximal swelling; anterior dorsal neural arches swollen (Laurin, 1998a,b).

Synonyms: none.

Comments: *Seymouriamorpha* was erected by Watson (1917: 171), who included only *Seymouria* and considered it part of *Reptilia*. Other taxa from Russia were subsequently included in *Seymouriamorpha*, such as *Kotlassia* (Case, 1946: 415) and *Rhinosaurus* (Romer, 1947: 286). Case (1946) even included *Solenodonsaurus* in this taxon, but this was not accepted in most subsequent studies (e.g., Carroll, 1969). He also recognized a "dominantly reptilian nature of the Seymouriamorpha", but included this taxon in *Stegocephalia*, probably because "The group retains many characters stemming directly from the early Embolomeri" (Case, 1946: 418). When larval seymouriamorphs with external gills and grooves for a lateral-line organ were recognized (Romer, 1947: 281), *Seymouriamorpha* was removed from *Reptilia* and further expanded (i.e., by the addition of *Discosauriscus* and later *Ariekanerpeton* and *Utegenia*). The taxon has generally been restricted to *Seymouria* and closely related forms, such as *Kotlassia, Discosauriscus, Utegenia*, and *Ariekanerpeton* (Laurin, 2000; Klembara, 2005b). The proposed definition respects this usage and is similar to the definition proposed by Laurin (1998a). However, in contrast to Laurin (1998a), multiple external specifiers are used here because the affinities of

seymouriamorphs are debated (Ruta et al., 2003; Marjanović and Laurin, 2019).

There is no general consensus as to the phylogenetic position of *Seymouriamorpha*. Under the phylogeny of Vallin and Laurin (2004: Fig. 6) and Marjanovic and Laurin (2009, 2019: Figs. 10, 11, 14, 19–21), seymouriamorphs are stem tetrapods. According to Klembara and Ruta (2005b), Klembara (2011) and Klembara et al. (2020) seymouriamorphs are stem amniotes. However, these uncertainties have no impact on the composition of *Seymouriamorpha* as its name is defined herein.

Literature Cited

Amalitzky, A. 1921. *Seymouriidae*, North Dvina excavations of Professor V. P. Amalitzky, vol. II. *Petrograd: Ros. Gos. Akad.*:1–14.

Berman, D. S., A. C. Henrici, S. S. Sumida, and T. Martens. 2000. Redescription of *Seymouria sanjuanensis* (*Seymouriamorpha*) from the Lower Permian of Germany based on complete, mature specimens with a discussion of paleoecology of the Bromacker locality assemblage. *J. Vertebr. Paleontol.* 20:253–268.

Berman, D. S., and T. Martens. 1993. First occurrence of *Seymouria* (*Amphibia: Batrachosauria*) in the Lower Permian Rotliegend of central Germany. *Carnegie Mus. Nat. Hist.* 62:63–79.

Berman, D. S., R. R. Reisz, and D. A. Eberth. 1987. *Seymouria sanjuanensis* (*Amphibia, Batrachosauria*) from the Lower Permian Cutler Formation of north-central New Mexico and the occurrence of sexual dimorphism in that genus questioned. *Can. J. Earth Sci.* 24:1769–1784.

deBraga, M., and R. R. Reisz. 1996. The Early Permian reptile *Acleistorhinus pteroticus* and its phylogenetic position. *J. Vertebr. Paleontol.* 16:384–395.

deBraga, M., and O. Rieppel. 1997. Reptile phylogeny and the interrelationships of turtles. *Zool. J. Linn. Soc.* 120:281–354.

Broili, F. 1904. Permische Stegocephalen und Reptilien aus Texas. *Palaeontogr. Abt. A* 51:1–120.

Broili, F. 1924. Ein Cotylosaurier aus der oberkarbonischen Gaskohle von Nürschan in Böhmen. *Sber. Bayer. Akad. Wiss.* 1924:3–11.

Bulanov, V. V. 2002. *Karpinskiosaurus ultimus* (*Seymouriamorpha, Parareptilia*) from the Upper Permian of European Russia. *Paleont. Jour.* 36:72–79.

Bulanov, V. V. 2003. Evolution and systematics of seymouriamorph parareptiles. *Paleont. Jour.* 37:1–105.

Carroll, R. L. 1969. Problems of the origin of reptiles. *Biol. Rev.* 44:393–432.

Case, E. C. 1946. A census of the determinable genera of the *Stegocephalia*. *Trans. Am. Philos. Soc.* 35:325–420.

Cope, E. D. 1877. Descriptions of extinct *Vertebrata* from the Permian and Triassic formations of the United States. *Proc. Am. Philos. Soc.* 17:182–193.

Cope, E. D. 1878. Descriptions of extinct *Batrachia* and *Reptilia* from the Permian formation of Texas. *Proc. Am. Philos. Soc.* 17:505–530.

Huxley, T. H. 1863. Description of *Anthracosaurus russelli*, a new labyrinthodont from the Lanarkshire coal field. *Q. J. Geol. Soc. Lond.* 19:56–68.

Ivakhnenko, M. F. 1981. *Discosauriscidae* from the Permian of Tadzhikistan. *Paleont. Jour.* 1981:90–102.

Ivakhnenko, M. F. 1987. *Permskie Parareptilii SSSR* Nauka, Moskva (in Russian).

Jaekel, O. 1902. Ueber *Gephyrostegus bohemicus* n.g. n.sp. *Z. Dtsch. Geol. Ges.* 54:127–132.

Klembara, J. 1997. The cranial anatomy of *Discosauriscus* Kuhn, a seymouriamorph tetrapod from the Lower Permian of the Boskovice Furrow (Czech Republic). *Philos. Trans. R. Soc. Lond. B Biol. Sci.* 352: 257–302.

Klembara, J. 2005. A new discosauriscid seymouriamorph tetrapod from the Lower Permian of Moravia, Czech Republic. *Acta Palaeont. Polonica* 50:25–48.

Klembara, J. 2009 (for 2008). The skeletal anatomy and relationships of a new discosauriscid seymouriamorph from the Lower Permian of Moravia (Czech Republic). *Ann. Carnegie Mus.* 77:451–484.

Klembara, J. 2011. The cranial anatomy, ontogeny and relationships of *Karpinskiosaurus secundus* (Amalitzky) (*Seymouriamorpha, Karpinskiosauridae*) from the Upper Permian of the European Russia. *Zool. J. Linn. Soc. (London).* 161:184–212.

Klembara, J., D. S. Berman, A. Henrici, and A. Čerňanský. 2005. New structures and reconstructions of the skull of the seymouriamorph *Seymouria sanjuanensis*, Vaughn. *Ann. Carnegie Mus.* 74(4):217–224.

Klembara, J., M. Hain, M. Ruta, D. S. Berman, S. E. Pierce, and A. C. Henrici. 2020. Inner ear morphology of diadectomorphs and seymouriamorphs (*Tetrapoda*) uncovered by high-resolution x-ray microcomputed tomography, and the origin of the amniote crown group. *Palaeontology* 63: 131–154.

Klembara, J., D. S. Berman, A. Henrici, A. Čerňanský, R. Werneburg, and T. Martens. 2007. First description of skull of Lower Permian *Seymouria sanjuanensis* (*Seymouriamorpha: Seymouriidae*) at an early juvenile stage. *Ann. Carnegie Mus.* 76(1):53–72.

Klembara, J., and S. Meszáros. 1992. New finds of *Discosauriscus austriacus* (Makowsky 1876) from the Lower Permian of Boskovice furrow (Czecho-Slovakia). *Geol. Carpathica (Bratislava)* 43:305–312.

Klembara, J., and M. Ruta. 2004a. The seymouriamorph tetrapod *Utegenia shpinari* from the ?Upper Carboniferous – Lower Permian of Kazakhstan. Part I: cranial anatomy and ontogeny. *Trans. R. Soc. Edinb.* 94:45–74.

Klembara, J., and M. Ruta. 2004b. The seymouriamorph tetrapod *Utegenia shpinari* from the ?Upper Carboniferous – Lower Permian of Kazakhstan. Part II: postcranial anatomy and relationships. *Trans. R. Soc. Edinb.* 94:75–93.

Klembara, J., and M. Ruta. 2005a. The seymouriamorph tetrapod *Ariekanerpeton sigalovi* from the Lower Permian of Tadzhikistan. Part I: cranial anatomy and ontogeny. *Trans. R. Soc. Edinb.* 96:43–70.

Klembara, J., and M. Ruta. 2005b. The seymouriamorph tetrapod *Ariekanerpeton sigalovi* from the Lower Permian of Tadzhikistan. Part II: postcranial anatomy and relationships. *Trans. R. Soc. Edinb.* 96:71–93.

Kuznetsov, V. V., and M. F. Ivakhnenko. 1981. Discosauriscids from the Upper Paleozoic in Southern Kazakhstan. *Paleont. Jour.* 1981:101–108.

Laurin, M. 1995. Comparative cranial anatomy of *Seymouria sanjuanensis* (*Tetrapoda: Batrachosauria*) from the Lower Permian of Utah and New Mexico. *PaleoBios* 16:1–8.

Laurin, M. 1996a. A reappraisal of *Utegenia*, a Permo-Carboniferous seymouriamorph (*Tetrapoda: Batrachosauria*) from Kazakhstan. *J. Vertebr. Paleontol.* 16:374–383.

Laurin, M. 1996b. A redescription of the cranial anatomy of *Seymouria baylorensis*, the best known seymouriamorph (*Vertebrata: Seymouriamorpha*). *PaleoBios* 17:1–16.

Laurin, M. 1996c. A reevaluation of *Ariekanerpeton*, a Lower Permian seymouriamorph (*Tetrapoda: Seymouriamorpha*) from Tadzhikistan. *J. Vertebr. Paleontol.* 16:653–665.

Laurin, M. 1998a. The importance of global parsimony and historical bias in understanding tetrapod evolution. Part I. Systematics, middle ear evolution, and jaw suspension. *Ann. Sci. Nat., Zool., 13 Ser.* 19:1–42.

Laurin, M. 1998b. The importance of global parsimony and historical bias in understanding tetrapod evolution. Part II. Vertebral centrum, costal ventilation, and paedomorphosis. *Ann. Sci. Nat., Zool., 13 Ser.* 19:99–114.

Laurin, M. 2000. Seymouriamorphs. Pp. 1064–1080 in *Amphibian Biology* (H. Heatwole and R. L. Carroll, eds.). Surrey Beatty & Sons, Chipping Norton.

Laurin, M., and G. Piñeiro. 2018. Response: commentary: a reassessment of the taxonomic position of mesosaurs, and a surprising phylogeny of early amniotes. *Front. Earth Sci.* 6:220.

Lee, M. S. Y. 1995. Historical burden in systematics and the interrelationships of 'Parareptiles'. *Biol. Rev.* 70:459–547.

Lee, M. S. Y. 1997. Reptile relationships turn turtles. *Nature* 389:245–246.

Linnaeus, C. 1758. *Systema Naturae*. 10th edition. Laurentii Salvii, Holmiae (Stockholm).

Marjanović, D., and M. Laurin. 2019. Phylogeny of Paleozoic limbed vertebrates reassessed through revision and expansion of the largest published relevant data matrix. *PeerJ* 6:e5565.

Marjanović, D., and M. Laurin. 2009. The origin(s) of modern amphibians: a commentary. *Evol. Biol.* 36:336–338.

Modesto, S. P. 2000. *Eunotosaurus africanus* and the Gondwanan ancestry of anapsid reptiles. *Palaeont. Afr.* 36:15–20.

Owen, R. 1876. *Descriptive and Illustrated Catalogue of the Fossil* Reptilia *of South Africa in the Collections of the British Museum*. London.

Rieppel, O., and R. R. Reisz. 1999. The origin and early evolution of turtles. *Annu. Rev. Ecol. Syst.* 30:1–22.

Romer, A. S. 1947. Review of the *Labyrinthodontia*. *Bull. Mus. Comp. Zool. Harv.* 99:1–368.

Ruta, M., M. I. Coates, and D. D. L. Quicke. 2003. Early tetrapod relationships revisited. *Biol. Rev.* 78:251–345.

Vallin, G., and M. Laurin. 2004. Cranial morphology and affinities of *Microbrachis*, and a reappraisal of the phylogeny and lifestyle of the first amphibians. *J. Vertebr. Paleontol.* 24:56–72.

Watson, D. M. S. 1917. A sketch classification of the pre-Jurassic tetrapod vertebrates. *Proc. Zool. Soc. Lond.* 1917:167–186.

White, T. E. 1939. Osteology of *Seymouria baylorensis* Broili. *Bull. Mus. Comp. Zool. Harv.* 85:325–409.

Zhang, F., Y. Li, and X. Wan. 1984. A new occurrence of Permian seymouriamorphs in Xinjiang, China. *Vertebr. PalAsiatica* 22:294–304.

Authors

Michel Laurin; CR2P, CNRS/MNHN/SU; Muséum National d'Histoire Naturelle; CP 48; 43 rue Buffon, 75005 Paris, France. Email: michel.laurin@mnhn.fr.

Jozef Klembara; Comenius University in Bratislava, Faculty of Natural Sciences, Ilkovičova 6, Mlynská dolina, Department of Ecology, 842 15 Bratislava, Slovakia. Email: jozef.klembara@uniba.sk

Date Accepted: 21 March 2014

Editors: Philip Cantino, Jacques A. Gauthier, Kevin de Queiroz

Tetrapoda B. Hatschek and C. J. Cori 1896 [M. Laurin], converted clade name

Registration Number: 106

Definition: *Tetrapoda* is the smallest crown clade containing *Homo sapiens* Linnaeus 1758 (*Synapsida*), *Caecilia tentaculata* Linnaeus 1758 (*Ichthyophiidae* is the closest relative in the reference phylogeny), *Siren lacertina* Österdam 1766 (*Sirenidae*), and *Pipa pipa* Linnaeus 1758 (*Pipidae*). This is a minimum-crown-clade definition. Abbreviated definition: min crown ∇ (*Homo sapiens* Linnaeus 1758 & *Caecilia tentaculata* Linnaeus 1758 & *Siren lacertina* Österdam 1766 & *Pipa pipa* Linnaeus 1758).

Etymology: Presumably from Greek *tetra* (four) and *podos* (foot). Hatschek and Cori (1896) did not specify, although they provided the German translation "Vierfûßer", which is compatible with these roots.

Reference Phylogeny: The primary reference phylogeny is Vallin and Laurin (2004: Fig. 6). However, early tetrapod phylogeny is currently controversial (see Comments).

Composition: The extant composition of *Tetrapoda* is highly stable: lissamphibians and amniotes. The taxon also includes all extinct taxa that constitute the stems of these two crown clades, but the composition of those stems is much more unstable because the relationships of early extinct limbed vertebrates are currently debated (see Comments). Most phylogenies imply that diadectomorphs and *Solenodonsaurus* are stem amniotes, and that "lepospondyls" are either stem lissamphibians (Vallin and Laurin, 2004; Marjanović and Laurin, 2019) or stem amniotes (Ruta et al., 2003) and therefore part of *Tetrapoda*. Other extinct taxa may or may not be part of *Tetrapoda*, depending on which reference phylogeny is adopted. Under the primary reference phylogeny (Vallin and Laurin, 2004), only taxa listed above are part of *Tetrapoda*. However, under the phylogeny of Ruta et al. (2003), seymouriamorphs and embolomeres are stem amniotes, and temnospondyls are stem lissamphibians, and therefore parts of *Tetrapoda*. Baphetids may also be tetrapods under some phylogenies (Clack, 1998), but not under most recently published trees (Marjanović and Laurin, 2019).

Diagnostic Apomorphies: As a crown clade, several soft anatomical, physiological, behavioral, and molecular apomorphies can be attributed to this taxon relative to other extant vertebrates (Laurin and Anderson, 2004). These and several osteological characters distinguish extant tetrapods from the nearest crown clade, dipnoans (lungfishes). These characters include (among others): an autopod including five or fewer digits and lacking lepidotrichia (or any other dermal fin rays, such as actinotrichia); an articulation between the pelvis and at least one pair of ribs; loss of the internal gills and of the opercular bones; a mobile neck resulting from the loss of bony contact between skull and shoulder girdle and including at least one specialized anterior cervical vertebra (atlas) articulating with the occipital condyle; a more complex lung with many alveoli; a fenestra ovalis in the braincase; and a better ossified axial endoskeleton, with a cylindrical pleurocentrum that fuses to the neural arch.

However, only osteological characters distinguish *Tetrapoda* from stem-tetrapods. These characters include, according to the reference phylogeny (Vallin and Laurin, 2004; see also Laurin, 1998a,b): absence of the intertemporal; loss of the temporal emargination; a convex occipital flange of the squamosal; braincase endochondral roof ossified; reduction or loss of labyrinthine infolding of the teeth; loss of vomerine and palatine fangs; epipterygoid suturally distinct from pterygoid throughout ontogeny; and loss of anterior coronoid (no more than two coronoids are normally present).

Synonyms: *Stegocephali* Cope 1868 (approximate, partial); *Labyrinthodontia* Owen 1859 (approximate, partial); *Reptilia* of Owen (1859) (approximate, partial); *Amphibia* Linnaeus 1758 (approximate, partial); *Neotetrapoda* Gaffney 1979 (approximate).

Comments: Several authors have credited Goodrich (1930) with erecting the taxon *Tetrapoda* (e.g., Ahlberg, 1995; Laurin 1998a: 10; Laurin and Anderson, 2004: 74). However, this is clearly not the case. Several earlier works used *Tetrapoda* in the same sense as Goodrich (1930), for the smallest clade that includes all limbed vertebrates. These include von Huene (1913), but the earliest occurrence I found is Hatschek and Cori (1896), a book designed for teaching that unfortunately does not cite its sources. I was not able to trace this usage of the name *Tetrapoda* (in the sense of all limbed vertebrates) further back into the past, despite having searched in several papers in the period 1860–1920. Several older uses of the name *Tetrapoda* exist, but these refer to other groups. For instance, the name *Tetrapoda* was used much earlier by Fischer (1808) for a group of mammals that possess four limbs (whales were excluded), by Haworth (1825) for a group of squamates that possess four limbs, and by Grant (1861) for a group of perennibranchiate urodeles. These uses are considered homonyms.

The first phylogenetic definition of *Tetrapoda* (Gauthier et al., 1988: 106) was formulated as follows: "Tetrapoda is restricted to the most recent common ancestor of extant Lissamphibia and Amniota, and all of its descendants." This definition was then already known to exclude at least some limbed vertebrates, such as *Ichthyostega* (Gauthier et al., 1988) and *Crassigyrinus* (Gauthier et al., 1989), even though those taxa had been included in *Tetrapoda* by previous authors. This departure from tradition was done to "ensure maximum informativeness" (i.e., that most statements about tetrapods, which are based on observations in extant organisms, can be inferred to apply to their most recent common ancestor). The choice to define the name *Tetrapoda* and other popular names (*Aves, Mammalia, Reptilia, Diapsida*, etc.) as crown clades has led to a controversy because some authors prefer to tie such names to apomorphies and thus to retain in them various extinct species that are outside the crown. For instance, Anderson (in Laurin and Anderson, 2004) preferred an apomorphy-based definition of the name *Tetrapoda* that delimits this taxon using the appearance of limbs with digits. Other authors have agreed that restricting widely used taxon names to crown clades is a good idea, and several, including some palaeontologists, have adopted a crown-clade definition of the name *Tetrapoda* (Gauthier et al., 1989; de Queiroz and Gauthier, 1992; Laurin, 1998a; Laurin and Reisz, 1999; Laurin et al., 2000; Vallin and Laurin, 2004).

Several arguments favor a crown-clade definition of the name *Tetrapoda*. First, the most common use of the name *Tetrapoda* in the literature overall (which is dominated by neontological works) is not "limbed vertebrates" but, rather, "the crown clade of limbed vertebrates". An earlier survey (Laurin and Anderson, 2004: Fig. 2) reviewed the various uses of the name *Tetrapoda*, which has often been explicitly or implicitly restricted to the crown clade of limbed vertebrates. Most often in the literature, the name *Tetrapoda* is implicitly used to refer to the crown clade in that many characters that cannot fossilize are attributed to it (Laurin and Anderson, 2004). Such characters that are present in all extant tetrapods can reasonably be inferred to have been present in their most recent common ancestor (i.e., the ancestor in which the crown originated) but not in more remote ancestors (e.g., that of all vertebrates that possess limbs).

Second, other preexisting names (see Synonyms) have been more consistently associated with the most recent common ancestor of all known limbed vertebrates, if not with the entire clade, as some descendant clades were often explicitly excluded. In fact, the only names that have been consistently associated with that ancestor are *Labyrinthodontia* and *Stegocephali*, both of which were traditionally used for paraphyletic groups containing the earliest known limbed vertebrates and which are defined elsewhere in this volume for clades originating close to the origin of limbs. The same is not true for *Amphibia*, which has been used for at least two different clades as well as a paraphyletic group associated with the origin of limbs and is applied in this volume to the total clade of lissamphibians. Thus, the names *Labyrinthodontia*, *Stegocephali*, and *Amphibia* are all arguably more appropriately applied to clades other than the crown, leaving the name *Tetrapoda* available for the crown clade. Another possible name for this clade, *Neotetrapoda*, has infrequently been used in the literature, as shown by the fact that a Google Scholar (performed on March 26, 2014) found only 21 pages (versus 4070 for *Tetrapoda*). Furthermore, it is not clear that Gaffney (1979), who first published the name *Neotetrapoda*, meant it to designate a crown group (Laurin, 2002: 367).

Finally, it should be noted that all known limbed vertebrates (except for the Devonian taxa, under some hypotheses) were thought to be part of the crown *Tetrapoda* until the late 1980s because according to the prevailing (almost universally accepted) phylogenies (Romer, 1947; Carroll, 1970; Gauthier et al.,

1988; Panchen and Smithson, 1988; Milner, 1993), seymouriamorphs and embolomeres were stem amniotes (reptiliomorphs), whereas "lepospondyls" and temnospondyls were considered stem lissamphibians (amphibians). Even Devonian limbed vertebrates such as *Tulerpeton* and *Ichthyostega* were usually considered stem amniotes (Lebedev and Coates, 1995) or stem lissamphibians (Panchen and Smithson, 1988), although they were sometimes excluded (Gaffney, 1979). Thus, there is no major contradiction of palaeontological tradition in using a crown-clade definition of the name *Tetrapoda*; however, under all recent phylogenies (e.g., Anderson, 2001; Ruta et al., 2003; Ruta and Coates, 2007), Devonian and, in some cases, Early Carboniferous, limbed vertebrates are excluded from *Tetrapoda* as conceptualized here, and in a few others (e.g., Laurin, 1998a; Marjanović and Laurin, 2019; Laurin et al., 2019), several Late Carboniferous and Permian limbed vertebrates also fall outside the crown.

As noted above (see Composition), the relationships of early fossil limbed vertebrates are controversial. Traditionally, lissamphibians (or at least anurans) have been thought to be derived from temnospondyls since the works of Cope (1888). This hypothesis has been supported by numerous subsequent studies (Watson, 1940; Bolt, 1969, 1977; Bolt and Lombard, 1985; Trueb and Cloutier, 1991; Milner, 1993; Schoch, 1995; Ruta et al., 2003). Similarly, most authors have followed Cope's (1880, 1884) suggestion that embolomeres were stem amniotes (e.g., Carroll, 1970; Gauthier et al., 1988; Panchen and Smithson, 1988; Lombard and Sumida, 1992; Lebedev and Coates, 1995). Although Broili's (1904) suggestion that *Seymouria* was an amniote (in an apomorphy-based sense) was contradicted by the discovery of larvae in it and related forms (Watson, 1942; Spinar, 1952), until recently, most authors have considered seymouriamorphs stem amniotes (Gauthier et al., 1988;

Panchen and Smithson, 1988; Lombard and Sumida, 1992; Ruta et al., 2003). Other taxa usually considered to be stem amniotes include *Gephyrostegus*, *Solenodonsaurus* (Broili, 1924; Carroll, 1970), and diadectomorphs (Case, 1911; Watson, 1917; Laurin and Reisz, 1995), although diadectomorphs have been placed within *Amniota* by a few authors (Berman et al., 1992). More recently, *Westlothiana* has also been argued to be a stem amniote (Smithson et al., 1994), although more recent studies raise doubts about the affinities of this taxon and suggest that it is either a stem tetrapod or a stem lissamphibian (Laurin and Reisz, 1999; Ruta et al., 2003; Vallin and Laurin, 2004; Marjanović and Laurin, 2019: Figs. 10, 11, 14, 18–21).

The position of "lepospondyls" is more controversial and less often discussed. Most "lepospondyls" have been considered tetrapods by most authors (Dawson, 1863; Romer, 1950; Gregory, 1965; Vaughn, 1962; Carroll and Currie, 1975; Carroll and Holmes, 1980). However, some early studies have suggested that "lepospondyls" were so different from other early stegocephalians (labyrinthodonts or "apsidospondyls" in older literature) that they might have arisen separately from the same stock of sarcopterygians (Thomson, 1967). More recently, adelogyrinids have been argued to be stem-tetrapods (Ruta and Coates, 2007).

The advent of computer-assisted phylogenetics did not immediately call into question these well-established ideas (Gauthier et al., 1988; Trueb and Cloutier, 1991; Milner, 1993); however, those analyses were initially restricted to presumed reptiliomorphs (stem and crown amniotes) or to presumed amphibians (temnospondyls and lissamphibians). When both putative groups were analyzed together, substantially different relationships were inferred (Laurin and Reisz, 1997, 1999; Laurin, 1998a,b; Anderson, 2001; Vallin and Laurin, 2004; Pawley, 2006; Germain, 2008; Marjanović and Laurin, 2009, 2019),

and this had important implications about the composition of *Tetrapoda*. According to these analyses, temnospondyls, embolomeres, and seymouriamorphs are excluded from *Tetrapoda*. However, analyses by Ruta et al. (2003) and Ruta and Coates (2007) supported the traditional ideas that temnospondyls are stem lissamphibians, and that "lepospondyls", along with diadectomorphs, seymouriamorphs and embolomeres, are stem amniotes. This situation leaves considerable uncertainty about the composition of *Tetrapoda*.

Literature Cited

Ahlberg, P. E. 1995. *Elginerpeton pancheni* and the earliest tetrapod clade. *Nature* 373:420–425.

Anderson, J. S. 2001. The phylogenetic trunk: maximal inclusion of taxa with missing data in an analysis of the *Lepospondyli* (*Vertebrata*, *Tetrapoda*). *Syst. Biol.* 50:170–193.

Berman, D. S., S. S. Sumida, and R. E. Lombard. 1992. Reinterpretation of the temporal and occipital regions in *Diadectes* and the relationships of diadectomorphs. *J. Paleont.* 66:481–499.

Bolt, J. R. 1969. Lissamphibian origins: possible protolissamphibian from the Lower Permian of Oklahoma. *Science, N. Y.* 166:888–891.

Bolt, J. R. 1977. Dissorophoid relationships and ontogeny, and the origin of the *Lissamphibia*. *J. Paleont.* 51:235–249.

Bolt, J. R., and R. E. Lombard. 1985. Evolution of the amphibian tympanic ear and the origin of frogs. *Zool. J. Linn. Soc.* 24:83–99.

Broili, F. 1904. Permische Stegocephalen und Reptilien aus Texas. *Palaeontogr. Abt. A* 51:1–120.

Broili, F. 1924. Ein Cotylosaurier aus der oberkarbonischen Gaskohle von Nürschan in Böhmen. *Sber. Bayer. Akad. Wiss.* 1924:3–11.

Carroll, R. L. 1970. The ancestry of reptiles. *Philos. Trans. R. Soc. Lond. B Biol. Sci.* 257:267–308.

Carroll, R. L., and P. J. Currie. 1975. Microsaurs as possible apodan ancestors. *Zool. J. Linn. Soc.* 57:229–247.

Carroll, R. L., and R. Holmes. 1980. The skull and jaw musculature as guides to the ancestry of salamanders. *Zool. J. Linn. Soc.* 68:1–40.

Case, E. C. 1911. A revision of the *Cotylosauria* of North America. *Carnegie Inst. Washington* 145:1–122.

Clack, J. A. 1998. A new Early Carboniferous tetrapod with a melange of crown-group characters. *Nature* 394:66–69.

Cope, E. D. 1880. Extinct *Batrachia*. *Am. Nat.* 14:609–610.

Cope, E. D. 1884. The *Batrachia* of the Permian period of North America. *Am. Nat.* 18:26–39.

Cope, E. D. 1888. On the intercentrum of the terrestrial *Vertebrata*. *Trans. Am. Philos. Soc.* 16:243–253.

Dawson, J. W. 1863. Air-breathers of the coal period. *Am. J. Sci.* 36:430–432.

Fischer, G. 1808. *Tableaux Synoptiques de Zoognosie.* Imprimerie de l"Universite Imperiale, Moscou, Moscow.

Gaffney, E. S. 1979. Tetrapod monophyly: a phylogenetic analysis. *Bull. Carnegie Mus. Nat. Hist.* 13:92–105.

Gauthier, J. A., D. C. Cannatella, K. De Queiroz, A. G. Kluge, and T. Rowe. 1989. Tetrapod phylogeny. Pp. 337–353 in *The Hierarchy of Life* (B. Fernholm, K. Bremer, and H. Jornvall, eds.). Elsevier Science Publishers B. V. (Biomedical Division), New York.

Gauthier, J. A., A. G. Kluge, and T. Rowe. 1988. The early evolution of the *Amniota*. Pp. 103–155 in *The Phylogeny and Classification of the Tetrapods, Vol. 1: Amphibians, Reptiles, Birds* (M. J. Benton, ed.). Clarendon Press, Oxford.

Germain, D. 2008. *Anatomie des Lépospondyles et Origine des Lissamphibiens.* Doctoral thesis, Muséum National d'Histoire Naturelle, Paris.

Goodrich, E. S. 1930. *Studies on the Structure and Development of Vertebrates.* Macmillan, London.

Grant, R. E. 1861. *Tabular View of the Primary Divisions of the Animal Kingdom, Intended to Serve as an Outline of an Elementary Course of Recent Zoology (Cainozoology), or the Natural History of Existing Animals.* Walton and Maberly, London.

Gregory, J. T. 1965. Microsaurs and the origin of captorhinomorph reptiles. *Am. Zool.* 5:277–286.

Hatschek, B., and C. J. Cori. 1896. *Elementarcus der Zootomie in fünfzen Vorlesungen.* Gustav Fischer, Jena.

Haworth, A. H. 1825. A binary arrangement of the class *Amphibia*. *Philos. Mag.* 65:372–373.

von Huene, F. 1913. The skull elements of the Permian *Tetrapoda* in the American Museum of Natural History, New York. *Bull. Am. Mus. Nat. Hist.* 32:315–386.

Laurin, M. 1998a. The importance of global parsimony and historical bias in understanding tetrapod evolution. Part I. Systematics, middle ear evolution, and jaw suspension. *Ann. Sci. Nat., Zool., 13 Ser.* 19:1–42.

Laurin, M. 1998b. The importance of global parsimony and historical bias in understanding tetrapod evolution. Part II. Vertebral centrum, costal ventilation, and paedomorphosis. *Ann. Sci. Nat., Zool., 13 Ser.* 19:99–114.

Laurin, M. 2002. Tetrapod phylogeny, amphibian origins, and the definition of the name *Tetrapoda*. *Syst. Biol.* 51:364–369.

Laurin, M., and J. S. Anderson. 2004. Meaning of the name *Tetrapoda* in the scientific literature: an exchange. *Syst. Biol.* 53:68–80.

Laurin, M., M. Girondot, and A. de Ricqlès. 2000. Early tetrapod evolution. *Trends Ecol. Evol.* 15:118–123.

Laurin, M., O. Lapauze, and D. Marjanović. 2019. What do ossification sequences tell us about the origin of extant amphibians? *PCI Paleontol.* doi:10.1101/352609.

Laurin, M., and R. R. Reisz. 1995. A reevaluation of early amniote phylogeny. *Zool. J. Linn. Soc.* 113:165–223.

Laurin, M., and R. R. Reisz. 1997. A new perspective on tetrapod phylogeny. Pp. 9–59 in *Amniote Origins—Completing the Transition to Land* (S. Sumida and K. Martin, eds.). Academic Press, San Diego.

Laurin, M., and R. R. Reisz. 1999. A new study of *Solenodonsaurus janenschi*, and a reconsideration of amniote origins and stegocephalian evolution. *Can. J. Earth Sci.* 36:1239–1255.

Lebedev, O. A., and M. I. Coates. 1995. The post-cranial skeleton of the Devonian tetrapod *Tulerpeton curtum* Lebedev. *Zool. J. Linn. Soc.* 114:307–348.

Linnaeus, C. 1758. *Systema Naturae*. 10th edition. Laurentii Salvii, Holmiae (Stockholm).

Lombard, R. E., and S. S. Sumida. 1992. Recent progress in understanding early tetrapods. *Amer. Zool.* 32:609–622.

Marjanović, D., and M. Laurin. 2009. The origin(s) of modern amphibians: a commentary. *Evol. Biol.* 36:336–338.

Marjanović, D., and M. Laurin. 2019. Phylogeny of Paleozoic limbed vertebrates reassessed through revision and expansion of the largest published relevant data matrix. *PeerJ* 6:e5565.

Milner, A. R. 1993. The Paleozoic relatives of lissamphibians. *Herpetol. Monogr.* 7:8–27.

Österdam, A. 1766. *Siren lacertina*. PhD thesis, Uppsala University University, Uppsala.

Owen, R. 1859. On the orders of fossil and recent *Reptilia* and their distribution in time. *Rept. Brit. Assn. Adv. Sci.*:153–166.

Panchen, A. L., and T. R. Smithson. 1988. The relationships of the earliest tetrapods. Pp. 1–32 in *The Phylogeny and Classification of the Tetrapods, Vol. 1: Amphibians, Reptiles, Birds* (M. J. Benton, ed.). Clarendon Press, Oxford.

Pawley, K. 2006. *The Postcranial Skeleton of Temnospondyls* (Tetrapoda: Temnospondyli). La Trobe University, Melbourne.

de Queiroz, K., and J. A. Gauthier. 1992. Phylogenetic taxonomy. *Annu. Rev. Ecol. Syst.* 23:449–480.

Romer, A. S. 1947. Review of the *Labyrinthodontia*. *Bull. Mus. Comp. Zool. Harv.* 99:1–368.

Romer, A. S. 1950. The nature and relationships of the Paleozoic microsaurs. *Am. J. Sci.* 248:628–654.

Ruta, M., and M. I. Coates. 2007. Dates, nodes and character conflict: addressing the lissamphibian origin problem. *J. Syst. Paleontol.* 5:69–122.

Ruta, M., M. I. Coates, and D. D. L. Quicke. 2003. Early tetrapod relationships revisited. *Biol. Rev.* 78:251–345.

Schoch, R. 1995. Heterochrony in the development of the amphibian head. Pp. 107–124 in *Evolutionary Change and Heterochrony* (K. J. McNamara, ed.). John Wiley & Sons, New York.

Smithson, T. R., R. L. Carroll, A. L. Panchen, and S. M. Andrews. 1994. *Westlothiana lizziae* from the Viséan of East Kirkton, West Lothian, Scotland, and the amniote stem. *Trans. R. Soc. Edinb.* 84:383–412.

Spinar, Z. 1952. Revision of some Moravian *Discosauriscidae*. *Sborník Ustr. Ust. Geol.* 15:1–159.

Thomson, K. S. 1967. Notes on the relationships of the rhipidistian fishes and the ancestry of the tetrapods. *J. Paleont.* 41:660–674.

Trueb, L., and R. Cloutier. 1991. A phylogenetic investigation of the inter- and intrarelationships of the *Lissamphibia* (Amphibia: Temnospondyli). Pp. 223–313 in *Origins of the Higher Groups of Tetrapods—Controversy and Consensus* (H.-P. Schultze and L. Trueb, eds.). Cornell University Press, Ithaca, NY.

Vallin, G., and M. Laurin. 2004. Cranial morphology and affinities of *Microbrachis*, and a reappraisal of the phylogeny and lifestyle of the first amphibians. *J. Vertebr. Paleontol.* 24:56–72.

Vaughn, P. P. 1962. The Paleozoic microsaurs as close relatives of reptiles, again. *Am. Midl. Nat.* 67:79–84.

Watson, D. M. S. 1917. A sketch classification of the pre-Jurassic tetrapod vertebrates. *Proc. Zool. Soc. Lond.* 1917:167–186.

Watson, D. M. S. 1940. The origin of frogs. *Trans. R. Soc. Edinb.* 60:195–231.

Watson, D. M. S. 1942. On Permian and Triassic tetrapods. *Geol. Mag.* 79:81–116.

Author

Michel Laurin; CR2P, CNRS/MNHN/SU; Muséum National d'Histoire Naturelle; CP 48; 43 rue Buffon, 75005 Paris, France. Email: michel.laurin@mnhn.fr.

Date Accepted: 5 April 2014

Editors: Philip Cantino, Jacques A. Gauthier, Kevin de Queiroz

Amphibia C. Linnaeus 1758 [M. Laurin, J. W. Arntzen, A. M. Báez, A. M. Bauer, R. Damiani, S. E. Evans, A. Kupfer, A. Larson, D. Marjanović, H. Müller, L. Olsson, J.-C. Rage and D. Walsh], converted clade name

Registration Number: 9

Definition: The largest total clade containing *Caecilia tentaculata* Linnaeus 1758 (*Gymnophiona*), *Siren lacertina* Österdam 1766 (*Sirenidae*), *Andrias japonicus* (Temminck 1836), *Proteus anguinus* Laurenti 1768 (*Proteidae*), and *Rana temporaria* Linnaeus 1758 (*Anura*) but not *Homo sapiens* Linnaeus 1758 (*Synapsida*). This is a maximum-total-clade definition with multiple internal specifiers (see Comments). Abbreviated definition: max total ∇ (*Caecilia tentaculata* Linnaeus 1758 & *Siren lacertina* Österdam 1766 & *Andrias japonicus* (Temminck 1836) & *Proteus anguinus* Laurenti 1768 & *Rana temporaria* Linnaeus 1758 ~ *Homo sapiens* Linnaeus 1758).

Etymology: Derived from the ancient Greek *amphi-* (both) and *bios* (life). This refers to the biphasic life history that characterizes most extant species of this taxon, with aquatic larvae and more terrestrial (amphibious to truly terrestrial) adults.

Reference Phylogeny: The reference phylogeny is Vallin and Laurin (2004: Fig. 6). However, amphibian phylogeny is currently extremely controversial (Laurin et al., 2019; Marjanović and Laurin, 2019), and alternative phylogenies are associated with major differences in the hypothesized composition of this taxon (see Composition). The internal specifiers *Caecilia tentaculata*, *Andrias japonicus*, and *Rana temporaria* are not shown in the reference phylogeny but are most closely related to *Ichthyophiidae*, *Hynobiidae*, and *Pipidae*, respectively, in that phylogeny (Pyron and Wiens, 2011).

Composition: *Amphibia* includes lissamphibians (members of the crown clade) as well as all extinct taxa that are more closely related to lissamphibians than to amniotes. Under the primary reference phylogeny, such taxa (stem lissamphibians) include only the "lepospondyls" (adelogyrinids, aïstopods, nectrideans, lysorophians, and "microsaurs"). Under the phylogeny of Ruta et al. (2003: Fig. 4) and Ruta and Coates (2007: Figs. 5, 6), amphibians include only lissamphibians and "temnospondyls". Under traditional phylogenies (e.g., Panchen and Smithson, 1988: Fig. 1.1; Trueb and Cloutier, 1991: Fig. 4; Lombard and Sumida, 1992: Fig. 6; Milner, 1993: Fig. 5A), *Amphibia* includes "lepospondyls", "temnospondyls", colosteids, and sometimes ichthyostegids.

Diagnostic Apomorphies: Because *Amphibia* has a maximum-clade definition, only a provisional list of apomorphies that arose along that branch, and thus diagnose the basalmost node within *Amphibia*, can be given. The characters depend strongly on the reference phylogeny adopted; the list below is based on the phylogeny of Vallin and Laurin (2004: Fig. 6) and on the apomorphies given by Laurin (1988a,b). These include: transverse flange of pterygoid absent, vertebral centrum composed of a large pleurocentrum, and in some cases also a small intercentrum; neurocentral sutures fused early in ontogeny; atlantal intercentrum absent; four digits (or fewer) present in manus (derived independently in all "temnospondyls" and in some amniote clades).

Synonyms: *Pan-Amphibia* Rowe (2004) (unambiguous); *Batrachomorpha* Säve-Söderbergh

1934 (partial and approximate); *Batrachomorpha* sensu Benton (1997) (approximate).

Comments: *Amphibia* of Linnaeus (1758) originally included all extant taxa now considered lissamphibians, as well as reptiles (birds excluded), some actinopterygians (*Acipenser*), chondrichthyans (*Squalus, Raja, Chimaera*), and the jawless vertebrate *Petromyzon*. Subsequently, all these extant taxa except lissamphibians were removed from *Amphibia*, but many extinct, limbed tetrapods were added to this taxon. This reorganization was done in a stepwise manner and its history is complicated. A few key steps may be mentioned here. Linnaeus and Gmelin (1788) removed from *Amphibia* the taxa that are not tetrapods, giving it a composition that remained stable for several decades, as in Haworth (1825), in which *Amphibia* included only extant amphibians and "reptiles" (although anurans, urodeles, and gymnophionans were not considered each other's closest relatives). Anurans and urodeles were first considered close relatives by Brongniart (1800). This was an important step in the evolution of the concept of *Amphibia*, even though this name does not appear in Brongniart (1800), who used the name *Batrachia* instead. Subsequently, other authors used the name *Amphibia*, or variants, such as the French "Amphibies" for the same group (e.g., Latreille, 1825; Duméril and Bibron, 1834). Thus, *Amphibia* was reduced further, to include only anurans and urodeles. Gymnophionans were first argued to be closely related by Oppel (1811), who placed them (as *Apoda*) in the taxa *Nuda* and *Batracii*, along with anurans (*Ecaudata*) and urodeles (*Caudata*).

Extinct taxa have long been included in *Amphibia*, although their affinities were not always clear (and controversy persists to this day). The first long-extinct amphibian form was described by Jaeger (1828) under the name *Salamandroides giganteus*; these remains in fact pertain to the Triassic temnospondyl *Mastodonsaurus giganteus*. Such early stegocephalians were long considered "reptiles", but Quenstedt (1850) argued that they were batrachians, a suggestion that was quickly accepted by Vogt (1854), who proposed that batrachians (the extant forms) and stegocephalians together formed *Amphibia*. From then on, *Amphibia* has included long-extinct taxa from the Palaeozoic and Triassic, at least in the palaeontological literature. In Haeckel (1866: fig. 7), *Amphibia* included *Lissamphibia* (the anurans, caudatans and closely related extinct forms) and *Phractamphibia* (extinct, Palaeozoic and Mesozoic limbed vertebrates, and gymnophionans because, among other reasons, of the presence of dermal bony scales). Thus, *Amphibia* has long included all limbed vertebrates except amniotes. A few studies that used such a paraphyletic concept of *Amphibia* include Lydekker (1890), Case (1911: 14), Moodie (1916), Goodrich (1930), Romer (1947, 1966), Lehman (1955: 30), Duellman and Trueb (1986), Carroll (1988), and Benton (1997).

Among the early authors who advocated a monophyletic use of this name are Hennig (1983), who adopted both a total and a crown-clade concept of *Amphibia*; Gardiner (1983), who included in *Amphibia* only lissamphibians and the "lepospondyls" that figure in his taxonomy, namely adelogyrinids, aïstopods and nectrideans; and Ax (1987: 91), who included in this taxon at least extant amphibians (extinct groups often attributed to *Amphibia* are not mentioned). Gauthier et al. (1989) proposed the first phylogenetic definition of the name *Amphibia*, applying that name to a total clade, and Laurin (1998a, 2001), Laurin and Reisz (1999), de Ricqlès and Laurin (1999), and Vallin and Laurin (2004), adopted the same definition. By contrast, de Queiroz and Gauthier (1992), Cannatella and Hillis (1993), Rowe (2004), and Frost et al. (2006) explicitly advocated defining *Amphibia* as the name of a crown clade. In this contribution, we propose a definition consistent with the use of the name *Amphibia* in most of the recent literature that adopts phylogenetic nomenclature; this definition is also the most consistent with palaeontological literature because it maximizes the number of early, limbed vertebrates (traditionally called amphibians) that are included in a monophyletic *Amphibia* that does not include *Amniota*.

Unlike the case of *Tetrapoda*, it is probably best not to restrict the definition of *Amphibia* to a crown-group because extinct taxa that were generally thought to fit well outside the crown ("lepospondyls" under our reference phylogeny; these and temnospondyls under other phylogenies) form an important component of *Amphibia*. This is obvious from a survey of the main herpetology textbooks that discuss this taxon; in nearly every case, *Amphibia* includes Palaeozoic taxa that are thought to be excluded from the crown, even though the authors are not palaeontologists (Noble, 1931; Goin et al., 1978; Duellman and Trueb, 1986; Zug et al., 2001; Pough et al., 2004). Furthermore, they use the name *Lissamphibia* for crown amphibians. General textbooks on vertebrates also indicate that *Amphibia* includes Palaeozoic members that are not part of the crown (Goodrich, 1930; Pirlot, 1969; Romer and Parsons, 1977; Liem et al., 2001). Similarly, several recent papers written by herpetologists and phylogeneticists working only on extant amphibians make a distinction between *Lissamphibia* (for the crown) and *Amphibia,* which is a more inclusive taxon (either a grade or a total group). A few examples include Zhang et al. (2003a,b) and San Mauro et al. (2005). Even neontologists who do not explicitly discuss Palaeozoic amphibians use the name *Lissamphibia* for the crown-clade (e.g., Feller and Hedges, 1998). Thus, palaeontologists are not the only systematists who use a

concept of *Amphibia* that extends well beyond the crown delimited by anurans, gymnophionans, and urodeles.

We have selected the name *Amphibia* for the clade of interest because it is much more widely used than its synonyms. Based on a Google Scholar search (April 25, 2014), the results are as follows: 131,000 for *Amphibia*, 66 for *Batrachomorpha*, and 1 for *Pan-Amphibia*.

Our definition differs from standard maximum-clade definitions by having multiple internal specifiers rather than only one. This is done deliberately to prevent use of the name *Amphibia* under phylogenies (e.g., Carroll, 2007: Fig. 78; Anderson, 2007: Fig. 5.5; Anderson et al., 2008: Fig. 4) in which the least inclusive clade containing extant amphibians also includes amniotes, because the name *Tetrapoda* is more appropriate for that clade. However, most of these hypotheses (Carroll and Currie, 1975; Carroll and Holmes, 1980; Anderson, 2001; Schoch and Carroll, 2003) were not accompanied by a taxon/character data matrix, are incompatible with recent palaeontological and molecular phylogenies, and were not based on phylogenetic analyses (Laurin, 2002). Notable exceptions are the analyses of Anderson (2007: Fig. 5.5) and Anderson et al. (2008), which included large datasets, although rescoring of the matrix supports a phylogeny in which *Amphibia* as defined above can be recognized (Marjanović and Laurin, 2009).

Literature Cited

Anderson, J. S. 2001. The phylogenetic trunk: maximal inclusion of taxa with missing data in an analysis of the *Lepospondyli* (*Vertebrata, Tetrapoda*). *Syst. Biol.* 50:170–193.

Anderson, J. S. 2007. Incorporating ontogeny into the matrix: a phylogenetic evaluation of developmental evidence for the origin of modern amphibians. Pp. 182–227 in *Major Transition in Vertebrate Evolution* (J. S. Anderson, and H.-D. Sues, eds.). Indiana University Press, Bloomington, IN.

Anderson J. S., R. R. Reisz, D. Scott, N. B. Fröbisch, and S. S. Sumida. 2008. A stem batrachian from the Early Permian of Texas and the origin of frogs and salamanders. *Nature* 453:515–518.

Ax, P. 1987. *The Phylogenetic System: The Systematization of Organisms on the Basis of their Phylogenesis.* 1st edition. John Wiley & Sons, Toronto.

Benton, M. J. 1997. *Vertebrate Palaeontology.* 2nd edition. Blackwell Science Ltd., London.

Brongniart, A. 1800. Essai d'une classification naturelle des Reptiles. 1ère partie. Établissement des ordres. *Bull. Sci. Soc. Philom.* 2:81–82.

Cannatella, D. C., and D. M. Hillis. 1993. Amphibian relationships: phylogenetic analysis of morphology and molecules. *Herpetol. Monogr.* 7:1–7.

Carroll, R. L. 1988. *Vertebrate Paleontology and Evolution.* W. H. Freeman, New York.

Carroll, R.L. 2007. The Palaeozoic ancestry of salamanders, frogs and caecilians. *Zool. J. Linn. Soc.* 150(suppl.):1–140.

Carroll, R. L., and P. J. Currie. 1975. Microsaurs as possible apodan ancestors. *Zool. J. Linn. Soc.* 57:229–247.

Carroll, R. L., and R. Holmes. 1980. The skull and jaw musculature as guides to the ancestry of salamanders. *Zool. J. Linn. Soc.* 68:1–40.

Case, E. C. 1911. Revision of the *Amphibia* and *Pisces* of the Permian of North America. *Publ. Carnegie Inst. Wash.* 146:1–179.

Duellman, W. E., and L. Trueb. 1986. *Biology of Amphibians.* McGrawHill, New York.

Duméril, A. M. C., and G. Bibron. 1834. *Erpétologie Générale ou Histoire Naturelle des Reptiles.* Librairie Encyclopédique de Roret, Paris.

Feller, A. E., and S. B. Hedges. 1998. Molecular evidence for the early history of living amphibians. *Mol. Phyl. Evol.* 9:509–516.

Frost, D. R., T. Grant, J. Faivovich, R. H. Bain, A. Haas, C. F. B. Haddad, R. O. de Sá, A. Channing, M. Wilkinson, S. C. Donnellan, C. J. Raxworthy, J. A. Campbell, B. Blotto, P. Moler, R. C. Drewes, R. A. Nussbaum, J.

D. Lynch, D. M. Green, and W. C. Wheeler. 2006. The amphibian tree of life. *Bull. Am. Mus. Nat. Hist.* 297:1–370.

Gardiner, B. G. 1983. Gnathostome vertebrae and the classification of the *Amphibia*. *Zool. J. Linn. Soc.* 79:1–59.

Gauthier, J., D. C. Cannatella, K. de Queiroz, A. G. Kluge, and T. Rowe. 1989. Tetrapod phylogeny. Pp. 337–353 in *The Hierarchy of Life* (B. Fernholm, K. Bremer, and H. Jørnvall, eds.). Elsevier Science Publishers B. V. (Biomedical Division), New York.

Goin, C. J., O. B. Goin, and G. R. Zug. 1978. *Introduction to Herpetology.* 3rd edition. W. H. Freeman and Company, New York.

Goodrich, E. S. 1930. *Studies on the Structure and Development of Vertebrates.* Macmillan, London.

Haeckel, E. 1866. *Generelle Morphologie der Organismen. Allgemeine Grundzüge der organischen Formen-Wissenschaft, Mechanisch Begründet Durch die von Charles Darwin Reformirte Descendenz-Theorie. Zweiter Band: Allgemeine Entwicklungsgeschichte der Organismen.* Georg Reimer, Berlin.

Haworth, A. H. 1825. A binary arrangement of the class *Amphibia*. *Philos. Mag.* 65:372–373.

Hennig, W. 1983. *Stammesgeschichte der Chordaten.* P. Parey, Hamburg.

Jaeger, G. 1828. *Über die Fossile Reptilien, Welche in Württemberg Aufgefunden Worden Sind.* J. B. Metzler, Stuttgart.

Latreille, P.-A. 1825. *Familles Naturelles du Règne Animal, Exposées Succintement et Dans un Ordre Analytique, Avec L'indication de Leurs Genres.* J. B. Baillière, Paris.

Laurenti, J. N. 1768. *Specimen Medicum, Exhibens Synopsin Reptilium Emendatam cum Experimentis Circa Venena et Antidota Reptilium Austriacorum.* PhD thesis, University of Vienna, Vienna.

Laurin, M. 1998a. The importance of global parsimony and historical bias in understanding tetrapod evolution. Part I. Systematics, middle ear evolution, and jaw suspension. *Ann. Sci. Nat., Zool., 13 Ser.* 19:1–42.

Laurin, M. 1998b. The importance of global parsimony and historical bias in understanding tetrapod evolution. Part II. Vertebral centrum, costal ventilation, and paedomorphosis. *Ann. Sci. Nat., Zool., 13 Ser.* 19:99–114.

Laurin, M. 2001. L'utilisation de la taxonomie phylogénétique en paléontologie: avantages et inconvénients. *Biosystema* 19:197–211.

Laurin, M. 2002. Tetrapod phylogeny, amphibian origins, and the definition of the name *Tetrapoda*. *Syst. Biol.* 51:364–369.

Laurin, M., Lapauze O., and D. Marjanović. 2019. What do ossification sequences tell us about the origin of extant amphibians? *PCI Paleontol.* doi:10.1101/352609.

Laurin, M., and R. R. Reisz. 1999. A new study of *Solenodonsaurus janenschi*, and a reconsideration of amniote origins and stegocephalian evolution. *Can. J. Earth Sci.* 36:1239–1255.

Lehman, J.-P. 1955. Les amphibiens: généralités. Pp. 2–52 in *Traité de Paléontologie* (J. Piveteau, ed.). Masson & Cie, Paris.

Liem, K. F., W. E. Bemis, W. F. Walker, Jr., and L. Grande. 2001. *Functional Anatomy of the Vertebrates, an Evolutionary Perspective.* 3rd edition. Harcourt College Publishers, Orlando, FL.

Linnaeus, C. 1758. *Systema Naturae Per Regna Tria Naturae, Secundum Classes, Ordines, Genera, Species, cum Characteribus, Differentiis, Synonymis, Locis.* Tomus I, Editio decima, reformata. Laurentius Salvius, Holmiae (Stockholm).

Linnaeus, C., and J. F. Gmelin. 1788. *Systema Naturae.* 13th edition. Georg. Emanuel. Beer, Lipsiae.

Lombard, R. E., and S. S. Sumida. 1992. Recent progress in understanding early tetrapods. *Amer. Zool.* 32:609–622.

Lydekker, R. 1890. *Catalogue of the Fossil* Reptilia *and* Amphibia *in the British Museum (Natural History), Part IV. Containing the Orders* Anomodontia, Ecaudata, Caudata, and Labyrinthodontia; *and Supplement.* Taylor & Francis, London.

Marjanović, D., and M. Laurin. 2009. The origin(s) of modern amphibians: a commentary. *Evol. Biol.* 36:336–338.

Marjanović, D., and M. Laurin. 2019. Phylogeny of Paleozoic limbed vertebrates reassessed through revision and expansion of the largest published relevant data matrix. *PeerJ* 6:e5565.

Milner, A. R. 1993. The Paleozoic relatives of lissamphibians. *Herpetol. Monogr.* 7:8–27.

Moodie, R. L. 1916. The coal measures *Amphibia* of North America. *Carnegie Instit. Wash., Publ.* 238:1–222.

Noble, G. K. 1931. *The Biology of the* Amphibia. 1st edition. McGrawHill, New York.

Oppel, M. 1811. *Die Ordnungen, Familien und Gattungen der Reptilien als Prodrom einer Naturgeschichte derselben.* Königlich Baierische Akademie der Wissenschaften, Munich.

Österdam, A. 1766. *Siren lacertina.* PhD thesis, Uppsala University, Uppsala.

Panchen, A. L., and T. R. Smithson. 1988. The relationships of the earliest tetrapods. Pp. 1–32 in *The Phylogeny and Classification of the Tetrapods. Vol. 1: Amphibians, Reptiles, Birds* (M. J. Benton, ed.). Clarendon Press, Oxford.

Pirlot, P. 1969. *Morphologie Évolutive des Chordés.* Les Presses de l'Université de Montréal, Montréal.

Pough, F. H., R. M. Andrews, J. E. Cadle, M. L. Crump, A. H. Savitzky, and K. Wells. 2004. *Herpetology.* 3rd edition. Prentice Hall, Upper Saddle River, NJ.

Pyron, R. A., and J. J. Wiens. 2011. A large-scale phylogeny of *Amphibia* including over 2800 species, and a revised classification of extant frogs, salamanders, and caecilians. *Mol. Phyl. Evol.* 61:543–583.

de Queiroz, K., and J. A. Gauthier. 1992. Phylogenetic taxonomy. *Annu. Rev. Ecol. Syst.* 23:449–480.

Quenstedt, F. A. 1850. *Die Mastodonsaurier im grünen Keupersandsteine Württemberg's sind Batrachie*r. Verlag der H. Laupp'schen Buchhandlung, Tübingen.

de Ricqlès, A., and M. Laurin. 1999. The origin of tetrapods. Pp. 23–33 in *9th Ordinary General Meeting, Societas Europaea Herpetologica* (C. Miaud and R. Guyétant, eds.). Société Herpétologique de France, Le Bourget du Lac, France.

Romer, A. S. 1947. Review of the *Labyrinthodontia*. *Bull. Mus. Comp. Zool.* Harv. 99:1–368.

Romer, A. S. 1966. *Vertebrate Paleontology.* 3rd edition. University of Chicago Press, Chicago, IL.

Romer, A. S., and T. S. Parsons. 1977. *The Vertebrate Body.* 5th edition. W. B. Saunders, Philadelphia, PA.

Rowe, T. 2004. Chordate phylogeny and development. Pp. 384–409 in *Assembling the Tree of Life* (J. Cracraft and M. J. Donoghue, eds.). Oxford University Press, Oxford.

Ruta, M., and M. I. Coates. 2007. Dates, nodes and character conflict: addressing the lissamphibian origin problem. *J. Syst. Paleontol.* 5:69–122.

Ruta, M., M. I. Coates, and D. D. L. Quicke. 2003. Early tetrapod relationships revisited. *Biol. Rev.* 78:251–345.

San Mauro, D., M. Vences, M. Alcobendas, R. Zardoya, and A. Meyer. 2005. Initial diversification of living amphibians predated the breakup of Pangaea. *Am. Nat.* 165:590–599.

Säve-Söderbergh, G. 1934. Some points of view concerning the evolution of the vertebrates and the classification of this group. *Ark. Zool.* 26A:1–20.

Schoch, R. R., and R. L. Carroll. 2003. Ontogenetic evidence for the Paleozoic ancestry of salamanders. *Evol. Dev.* 5:314–324.

Temminck, C.J. 1836. *Coup d'oeil sur la Faune des îles de la Sonde et de l'Empire du Japon*, A. Arnz, Leiden.

Trueb, L., and R. Cloutier. 1991. A phylogenetic investigation of the inter- and intrarelationships of the *Lissamphibia* (*Amphibia: Temnospondyli*). Pp. 223–313 in *Origins of the Higher Groups of Tetrapods—Controversy and Consensus* (H.P. Schultze and L. Trueb, eds.). Cornell University Press, Ithaca, CA.

Vallin, G., and M. Laurin. 2004. Cranial morphology and affinities of *Microbrachis*, and a reappraisal of the phylogeny and lifestyle of the first amphibians. *J. Vertebr. Paleontol.* 24:56–72.

Vogt, K. C. 1854. *Archegosaurus* is kein Batrachier, doch ein Amphibium. *Neues Jahrb. Miner., Geogno., Geol. Petref.-Kunde* 676–677.

Zhang, P., Y. Q. Chen, Y. F. Liu, H. Zhou, and L. H. Qu. 2003a. The complete mitochondrial genome of the Chinese giant salamander, *Andrias davidianus* (*Amphibia: Caudata*). *Gene* 311:93–98.

Zhang, P., Y. Q. Chen, H. Zhou, X. L. Wang, and L. H. Qu. 2003b. The complete mitochondrial genome of a relic [sic] salamander, *Ranodon sibiricus* (*Amphibia: Caudata*) and implications for amphibian phylogeny. *Mol. Phyl. Evol.* 28:620–626.

Zug, G. R., L. J. Vitt, and J. P. Caldwell. 2001. *Herpetology: An Introductory Biology of Amphibians and Reptiles*. 2nd edition. Academic Press, San Diego, CA.

Authors

Michel Laurin; CR2P, CNRS/MNHN/SU; Muséum National d'Histoire Naturelle; CP 48; 43 rue Buffon, 75005 Paris, France. Email: michel.laurin@mnhn.fr.

Jan W. Arntzen; Naturalis Biodiversity Center; P. O. Box 9517; NL-2300 RA Leiden, the Netherlands. Email: pim.arntzen@naturalis.nl.

Ana María Báez; Department of Geological Sciences, Universidad de Buenos Aires Pabellón II, Ciudad Universitaria 1428 Ciudad de Buenos Aires, Argentina. Email: baez@gl.fcen.uba.ar.

Aaron M. Bauer; Department of Biology, Villanova University; 800 Lancaster Avenue, Villanova, PA 19085, USA. Email: aaron.bauer@villanova.edu.

Ross Damiani; Staatliches Museum für Naturkunde Stuttgart; Rosenstein 1; D-70191 Stuttgart, Germany. Email: rossano1973@gmail.com.

Susan E. Evans; Department of Cell and Developmental Biology; UCL – University College London; Gower Street; London WC1E 6BT, UK. Email: ucgasue@ucl.ac.uk.

Alexander Kupfer; Department of Zoology; Herpetology; State Museum of Natural History Stuttgart; Rosenstein 1; 70191 Stuttgart, Germany. Email: alexander.kupfer@smns-bw.de.

Allan Larson; Department of Biology; Washington University; St. Louis, MO 63130, USA. Email: larson@wustl.edu.

David Marjanović; Museum für Naturkunde; 10115 Berlin, Germany. Email: david.marjanovic@gmx.at.

Hendrik Müller; Institut für Spezielle Zoologie und Evolutionsbiologie mit Phyletischem Museum; Friedrich-Schiller-Universität Jena;

Erbertstraße 1; D-07743 Jena, Germany. Email: hendrik.mueller@uni-jena.de.

Lennart Olsson; Institut für Spezielle Zoologie und Evolutionsbiologie mit Phyletischem Museum Friedrich-Schiller-Universität Jena; Erbertstraße 1; D-07743 Jena, Germany. Email: Lennart.Olsson@uni-jena.de.

Jean-Claude Rage[†]; Département Histoire de la Terre, Muséum National d'Histoire Naturelle, CP 38, 8 rue Buffon; F-75231 Paris Cedex 05, France.

Denis Walsh; Department of Philosophy and Institute for the History and Philosophy of Science and Technology; University of Toronto; Victoria College; 91 Charles Street; Toronto, ON, M5S 1K7, Canada. Email: denis.walsh@utoronto.ca.

Date Accepted: 30 March 2014

Editors: Philip Cantino, Jacques A. Gauthier, Kevin de Queiroz

---

[†] Deceased.

Lissamphibia E. Haeckel 1866 [M. Laurin, J. W. Arntzen, A. M. Báez, A. M. Bauer, R. Damiani, S. E. Evans, A. Kupfer, A. Larson, D. Marjanović, H. Müller, L. Olsson, J.-C. Rage, and D. Walsh], converted clade name

Registration Number: 62

Definition: The smallest crown clade containing *Caecilia tentaculata* Linnaeus 1758, *Andrias japonicus* (Temminck 1836), *Siren lacertina* Österdam 1766, and *Rana temporaria* Linnaeus 1758 but not *Homo sapiens* Linnaeus 1758 or *Eryops megacephalus* Cope 1877 or *Diplocaulus salamandroides* Cope 1877. This is a minimum-crown-clade definition with external specifiers. Abbreviated definition: min crown ∇ (*Caecilia tentaculata* Linnaeus 1758 & *Andrias japonicus* (Temminck 1836) & *Siren lacertina* Österdam 1766 & *Rana temporaria* Linnaeus 1758 ~ *Homo sapiens* Linnaeus 1758 ∨ *Eryops megacephalus* Cope 1877 ∨ *Diplocaulus salamandroides* Cope 1877).

Etymology: Presumably derived (Haeckel did not specify except for mentioning the absence of dermal scales) from the Ancient Greek *lissos* (smooth) and *Amphibia*, which is probably derived from the Ancient Greek *amphi-* (both) and *bios* (life). These last two roots refer to the fact that many lissamphibians have a biphasic life cycle, with aquatic larvae and more terrestrial adults.

Reference Phylogeny: The primary reference phylogeny is Frost et al. (2006: Fig. 50) for the phylogeny within *Lissamphibia*, and Vallin and Laurin (2004: Fig. 6) for the relationships between extant amphibians and other limbed vertebrates. In Vallin and Laurin (2004), *Diplocaulus* is not represented directly, but it is deeply nested within *Nectridea* (Milner and Ruta, 2009). Other recent, generally compatible phylogenies of major clades of lissamphibians include San Mauro et al. (2005: Fig. 1) and Roelants et al. (2007: Fig. 1).

Composition: In addition to anurans, caudatans, gymnophionans, and their stem groups, this clade possibly includes the extinct albanerpetontids, which have been hypothesized to be stem-caudatans (Trueb and Cloutier, 1991), stem-batrachians (McGowan and Evans, 1995; Gardner, 2001: Fig. 8), stem-gymnophionans (Ruta and Coates, 2007), and the sister-group of *Lissamphibia* (mentioned by McGowan and Evans (1995) as being only slightly less parsimonious than a position as the sister-group of *Batrachia* Latreille 1800; found in at least one of 64 most-parsimonious trees by Ruta et al., 2003). There is a controversy about the relationships among anurans, caudatans and gymnophionans, but many recent analyses suggest that these three groups, along with the enigmatic *Albanerpetontidae*, form a clade that excludes all known Palaeozoic taxa (Marjanović and Laurin, 2013, 2019; Laurin et al., 2019). A few studies have suggested that at least some dissorophoids (e.g., Rage, 1985: Fig. 3; Schoch and Carroll, 2003) and some lepospondyls (McGowan, 2002; Carroll, 2007) might be part of *Lissamphibia*, but only a few of these were supported by data matrices (e.g., McGowan, 2002; Carroll, 2007; Anderson et al., 2008).

Diagnostic Apomorphies: Skeletal apomorphies (seen in adults) include (Parsons and Williams, 1963; Laurin, 1998a,b; Marjanović and Laurin, 2019): bicuspid, pedicellate teeth; an operculum-plectrum complex (Jenkins et al., 2007); loss of tabular, postorbital, postparietal, one of the two splenials, and the surangular; absence of occipital flange of squamosal; a broad interpterygoid vacuity, which is also present in most temnospondyls as well as the "lepospondyls" *Diplocaulus salamandroides* (Williston, 1909), *Diploceraspis burkei* (Beerbower, 1963), *Ptyonius marshii* (Bossy and Milner, 1998), and probably *Carrolla craddocki* (Maddin et al., 2011); palatine participates in edge of interpterygoid vacuities that represent at least half the width of the palate (also present in some temnospondyls, such as *Doleserpeton annectens*); basioccipital never present as a discrete bone in adults; interclavicle absent; intertrochanteric fossa and femoral adductor blade absent.

Synonyms: *Amphibia* of de Queiroz and Gauthier (1992), Cannatella and Hillis (1993), and Frost et al. (2006) is an unambiguous synonym. The following are approximate synonyms: Batraciens Brongniart 1800, *Nuda* and *Batrachii* Oppel 1811, *Batrachia* (Merrem 1820), *Amphibia nuda* Müller 1831, *Dipnoa* Bonaparte 1838, *Diplopnoa* Bonaparte 1841, *Amphibia Dipnoa* Stannius 1856.

Comments: Four lissamphibians are used as internal specifiers. A single anuran and a single gymnophionan are sufficient because numerous characters indicate the monophyly of *Anura* and *Gymnophiona*, but the monophyly of *Caudata* is more difficult to demonstrate; molecular studies generally corroborate monophyly of caudatans (e.g., Hay et al., 1995; Feller and Hedges, 1998; Zhang et al., 2003a,b, 2005; San Mauro et al., 2005; Wiens et al., 2005; Frost et al., 2006; Roelants et al., 2007), but morphological studies do not always support this (Laurin and Reisz, 1997, 1999; Laurin, 1998a; Vallin and Laurin, 2004; Marjanović and Laurin, 2019). Three of the seven thorough analyses of molecular and morphological data of Wiens et al. (2005: Figs. 2, 5–6) failed to infer caudatan monophyly. *Siren* and other (extant and extinct) sirenids have

often been considered to be distantly related to other caudatans, and the taxa *Meantes* Palisot de Beauvois 1799 and *Trachystomata* Cope 1866 (not nested in *Caudata*) had been erected just for sirenids (Goin et al., 1978: 216). Some molecular studies placed *Siren* at the base of caudatans (Hay et al., 1995; Wiens et al., 2005: Figs. 6–7), but they have always been considered lissamphibians; thus, the type species, *Siren lacertina*, is included among the specifiers. Cryptobranchoids are usually considered the sister-group of all other extant caudatans (e.g., Wiens et al., 2005: Fig. 8); thus, a cryptobranchoid (*Andrias japonicus*) is also used as an internal specifier. This choice ensures that *Lissamphibia* always includes all extant amphibians. Under phylogenies in which extant amphibians are paraphyletic (or polyphyletic) relative to *Amniota*, *Temnospondyli* or *Nectridea* (e.g., Carroll, 2007: Fig. 78; Anderson, 2007: Fig. 5.5; Anderson et al., 2008: Fig. 4), the definition proposed above does not apply to any clade.

Haeckel (1866) coined *Lissamphibia* for caudatans and anurans because they typically lack dermal scales (although some anurans are now known to have osteoderms). Gadow (1898) was the first to include gymnophionans in *Lissamphibia*. After Gadow, the name *Lissamphibia* fell into disuse until Parsons and Williams (1963) brought it back. Starting with Gadow (1898), the taxon *Lissamphibia* has always included the three major clades of extant amphibians (anurans, gymnophionans and caudatans) and has always excluded all known Palaeozoic amphibians. Use of this name is pervasive in the literature; palaeontologists use it to refer specifically to crown-amphibians, and every major herpetology textbook published after 1950 that we have examined (Goin et al., 1978: 66; Duellman and Trueb, 1986: 494; Milner, 1993; Zug et al., 2001: 12; Pough et al., 2004: 28) has made the distinction between *Lissamphibia* (clearly the crown) and *Amphibia* (a more inclusive clade or a grade). This is also true of most textbooks on vertebrates (Romer and Parsons, 1977: 63; Hildebrand, 1982: 58; Liem et al., 2001: 80). Older textbooks (Noble, 1931: 463) did not always name the clade of extant amphibians, although this clade was usually discussed and distinguished from Palaeozoic amphibians (in the paraphyletic sense). The recent primary literature on extant amphibians also includes numerous uses of the name *Lissamphibia* (see Comments on *Amphibia*, this volume). *Amphibia* is much more widely used (140,000 pages in Google Scholar for *Amphibia*, but only 1,510 for *Lissamphibia*, according to a search carried out on May 19, 2012), but *Amphibia* is often used for a more inclusive clade than the crown lissamphibians, even though several authors have used *Amphibia* for the crown (e.g., de Queiroz and Gauthier, 1992; Cannatella and Hillis, 1993; Frost et al., 2006). Thus, the most appropriate name for the crown of amphibians is *Lissamphibia*.

Literature Cited

Anderson, J. S. 2007. Incorporating ontogeny into the matrix: a phylogenetic evaluation of developmental evidence for the origin of modern amphibians. Pp. 182–227 in *Major Transition in Vertebrate Evolution* (J. S. Anderson, and H.-D. Sues, eds.). Indiana University Press, Bloomington, IN.

Anderson, J. S., R. R. Reisz, D. Scott, N. B. Fröbisch, and S. S. Sumida. 2008. A stem batrachian from the Early Permian of Texas and the origin of frogs and salamanders. *Nature* 453:515–518.

Beerbower J. R. 1963. Morphology, paleoecology, and phylogeny of the Permo-Pennsylvanian amphibian *Diploceraspis*. *Bull. Mus. Comp. Zool. Harv.* 130:33–108.

Bonaparte, C. L. 1838. *Iconographia Della Fauna Italica per le Quattro Classi Degli Animali Vertebrati*. Salviucci, Rome.

Bonaparte, C. L. 1841. A new systematic arrangement of vertebrated animals. *Proc. Linn. Soc. London* 18:247–304.

Bossy K. A., and A. C. Milner. 1998. Order *Nectridea*. Pp. 73–131 in *Encyclopedia of Paleoherpetology*—Lepospondyli (R. L. Carroll, K. A. Bossy, A. C. Milner, S. M. Andrews, and C. F. Wellstead, eds.). Gustav Fischer Verlag, Stuttgart.

Brongniart, A. 1800. Essai d'une classification naturelle des reptiles. Tome Ier: Des règles à suivre dans cette classification. Formation des ordres. *Lect. l'Inst. Natl.*:1–53.

Cannatella, D. C., and D. M. Hillis. 1993. Amphibian relationships: phylogenetic analysis of morphology and molecules. *Herpetol. Monogr.* 7:1–7.

Carroll, R. L. 2007. The Palaeozoic ancestry of salamanders, frogs and caecilians. *Zool. J. Linn. Soc.* 150:1–140.

Cope, E. D. 1866. On the structure and distribution of the genera of the arciferous *Anura*. *J. Acad. Nat. Sci. Phila.* 6:67–112.

Cope, E. D. 1877. Descriptions of extinct *Vertebrata* from the Permian and Triassic formations of the United States. *Proc. Am. Philos. Soc.* 17:182–193.

Duellman, W. E., and L. Trueb. 1986. *Biology of Amphibians*. McGraw-Hill, New York.

Feller, A. E., and S. B. Hedges. 1998. Molecular evidence for the early history of living amphibians. *Mol. Phyl. Evol.* 9:509–516.

Frost, D. R., T. Grant, J. Faivovich, R. H. Bain, A. Haas, C. F. B. Haddad, R. O. de Sá, A. Channing, M. Wilkinson, S. C. Donnellan, C. J. Raxworthy, J. A. Campbell, B. Blotto, P. Moler, R. C. Drewes, R. A. Nussbaum, J. D. Lynch, D. M. Green, and W. C. Wheeler. 2006. The amphibian tree of life. *Bull. Am. Mus. Nat. Hist.* 297:1–370.

Gadow, H. 1898. A *Classification of* Vertebrata, *Recent and Extinct*. Adam and Charles Black, London.

Gardner, J. D. 2001. Monophyly and affinities of albanerpetontid amphibians (*Temnospondyli*; *Lissamphibia*). *Zool. J. Linn. Soc.* 131:309–352.

Goin, C. J., O. B. Goin, and G. R. Zug. 1978. *Introduction to Herpetology*. 3rd edition. W. H. Freeman and Company, New York.

Haeckel, E. 1866. *Generelle Morphologie der Organismen. Allgemeine Grundzüge der Organischen Formen-Wissenschaft, Mechanisch*

Begründet Durch die von Charles Darwin reformirte Descendenz-Theorie. Zweiter Band, *Allgemeine Entwicklungsgeschichte der Organismen.* Georg Reimer, Berlin.

Hay, J. M., I. Ruvinsky, S. B. Hedges, and L. R. Maxson. 1995. Phylogenetic relationships of amphibian families inferred from DNA sequences of mitochondrial 12S and 16S ribosomal RNA genes. *Mol. Biol. Evol.* 12:928–937.

Hildebrand, M. 1982. *Analysis of Vertebrate Structure.* 2nd edition. John Wiley & Sons, New York.

Jenkins, Jr., F. A., D. M. Walsh, and R. L. Carroll. 2007. Anatomy of *Eocaecilia micropodia*, a limbed caecilian of the Early Jurassic. *Bull. Mus. Comp. Zool. Harv.* 158:285–365.

Latreille, P. A. 1800. *Histoire Naturelle des Salamandres de France, Précédée d'un Tableau Méthodique des Autres Reptiles Indigènes.* Imprimerie de Crapelet, Paris.

Laurin, M. 1998a. The importance of global parsimony and historical bias in understanding tetrapod evolution. Part I. Systematics, middle ear evolution, and jaw suspension. *Ann. Sci. Nat., Zool., 13 Ser.* 19:1–42.

Laurin, M. 1998b. The importance of global parsimony and historical bias in understanding tetrapod evolution. Part II. Vertebral centrum, costal ventilation, and paedomorphosis. *Ann. Sci. Nat., Zool., 13 Ser.* 19:99–114.

Laurin, M., O. Lapauze, and D. Marjanović. 2019. What do ossification sequences tell us about the origin of extant amphibians? *PCI Paleontol.* doi:10.1101/352609.

Laurin, M., and R. R. Reisz. 1997. A new perspective on tetrapod phylogeny. Pp. 9–59 in *Amniote Origins—Completing the Transition to Land* (S. Sumida and K. Martin, eds.). Academic Press, San Diego, CA.

Laurin, M., and R. R. Reisz. 1999. A new study of *Solenodonsaurus janenschi*, and a reconsideration of amniote origins and stegocephalian evolution. *Can. J. Earth Sci.* 36:1239–1255.

Liem, K. F., W. E. Bemis, W. F. Walker, Jr., and L. Grande. 2001. *Functional Anatomy of the Vertebrates: An Evolutionary Perspective.* 3rd edition. Harcourt College Publishers, Orlando, FL.

Linnaeus, C. 1758. *Systema Naturae Per Regna Tria Naturae, Secundum Classes, Ordines, Genera, Species, cum Characteribus, Differentiis, Synonymis, Locis.* Tomus I, Editio decima, reformata. Laurentius Salvius, Holmiae (Stockholm).

Maddin, H. C., J. C. Olori, and J. S. Anderson. 2011. A redescription of *Carrolla craddocki* (*Lepospondyli: Brachystelechidae*) based on high-resolution CT, and the impacts of miniaturization and fossoriality on morphology. *J. Morph.* 272:722–743.

Marjanović, D., and M. Laurin. 2013. The origin(s) of extant amphibians: a review with emphasis on the "lepospondyl hypothesis". *Geodiversitas* 35:207–272.

Marjanović, D., and M. Laurin. 2019. Phylogeny of Paleozoic limbed vertebrates reassessed through revision and expansion of the largest published relevant data matrix. *PeerJ* 6:e5565.

McGowan, G. J. 2002. Albanerpetontid amphibians from the Lower Cretaceous of Spain and Italy: a description and reconsideration of their systematics. *Zool. J. Linn. Soc.* 135:1–32.

McGowan, G., and S. E. Evans. 1995. Albanerpetontid amphibians from the Cretaceous of Spain. *Nature* 373:143–145.

Merrem, B. 1820. *Versuch eines Systems der Amphibien/Tentamen Systematis Amphibiorum.* J. C. Krieger, Marburg.

Milner, A. R. 1993. Amphibian-grade *Tetrapoda.* Pp. 665–679 in *The Fossil Record* (M. J. Benton, ed.). Chapman & Hall, London.

Milner, A. C., and M. Ruta. 2009. A revision of *Scincosaurus* (*Tetrapoda, Nectridea*) from the Moscovian of Nyrany, Czech Republic, and the phylogeny and interrelationships of nectrideans. *Spec. Pap. Palaeontol.* 81:71–89.

Müller, J. 1831. Beiträge zur Anatomie und Naturgeschichte der Amphibien. Pp. 190–275 in *Zeitschrift für Physiologie* (F. Tiedemann, G. R. Treviranus, and L. C. Treviranus, eds.). Karl Groos, Heidelberg.

Noble, G. K. 1931. *The Biology of the Amphibia.* 1st edition. McGraw-Hill, New York.

Oppel, M. 1811. Die *Ordnungen, Familien und Gattungen der Reptilien als Prodrom einer Naturgeschichte derselben.* Königlich Baierische Akademie der Wissenschaften, Munich.

Österdam, A. 1766. *Siren lacertina*. PhD thesis, Uppsala University, Uppsala.

Palisot de Beauvois, A. M. F. J. 1799. Translation of a memoir on a new species of *Siren*. *Trans. Am. Philos. Soc.* 4:277–281.

Parsons, T. S., and E. E. Williams. 1963. The relationships of the modern *Amphibia*: a re-examination. *Q. Rev. Biol.* 38:26–53.

Pough, F. H., R. M. Andrews, J. E. Cadle, M. L. Crump, A. H. Savitzky, and K. Wells. 2004. *Herpetology*. 3rd edition. Prentice Hall, Upper Saddle River, NJ.

de Queiroz, K., and J. A. Gauthier. 1992. Phylogenetic taxonomy. *Annu. Rev. Ecol. Syst.* 23:449–480.

Rage, J. -C. 1985. Origine et phylogénie des amphibiens. *Bull. Soc. Herp. Fr.* 34:1–19.

Roelants, K., D. J. Gower, M. Wilkinson, S. P. Loader, S. D. Biju, K. Guillaume, L. Moriau, and F. Bossuyt. 2007. Global patterns of diversification in the history of modern amphibians. *Proc. Natl. Acad. Sci. USA* 104:887–892.

Romer, A. S., and T. S. Parsons. 1977. *The Vertebrate Body*. 5th edition. W. B. Saunders, Philadelphia, PA.

Ruta, M., and M. I. Coates. 2007. Dates, nodes and character conflict: addressing the lissamphibian origin problem. *J. Syst. Paleontol.* 5:69–122.

Ruta, M., M. I. Coates, and D. D. L. Quicke. 2003. Early tetrapod relationships revisited. *Biol. Rev.* 78:251–345.

San Mauro, D., M. Vences, M. Alcobendas, R. Zardoya, and A. Meyer. 2005. Initial diversification of living amphibians predated the breakup of Pangaea. *Am. Nat.* 165:590–599.

Schoch, R. R., and R. L. Carroll. 2003. Ontogenetic evidence for the Paleozoic ancestry of salamanders. *Evol. Dev.* 5:314–324.

Stannius, H. 1856. *Handbuch der Zootomie*. 2nd edition. Veit & Comp., Berlin.

Temminck, C. J. 1836. *Coup d'œil sur la Faune des îles de la Sonde et de l'Empire du Japon*. Leiden.

Trueb, L., and R. Cloutier 1991. A phylogenetic investigation of the inter- and intrarelationships of the *Lissamphibia* (*Amphibia*: *Temnospondyli*). Pp. 223–313 in *Origins of the Higher Groups of Tetrapods—Controversy and Consensus* (H.-P. Schultze and L. Trueb, eds.). Cornell University Press, Ithaca, NY.

Vallin, G., and M. Laurin. 2004. Cranial morphology and affinities of *Microbrachis*, and a reappraisal of the phylogeny and lifestyle of the first amphibians. *J. Vertebr. Paleontol.* 24:56–72.

Wiens, J. J., R. M. Bonett, and P. T. Chippindale. 2005. Ontogeny discombobulates phylogeny: paedomorphosis and higher-level salamander relationships. *Syst. Biol.* 54:91–110.

Williston, S. W. 1909. The skull and extremities of *Diplocaulus*. *Trans. Kansas Acad. Sci.* 22:122–131.

Zhang, P., Y.-Q. Chen, H. Zhou, X.-L. Wang, and L.-H. Qu. 2003a. The complete mitochondrial genome of a relic [sic] salamander, *Ranodon sibiricus* (*Amphibia: Caudata*) and implications for amphibian phylogeny. *Mol. Phyl. Evol.* 28:620–626.

Zhang, P., Y.-Q. Chen, Y.-F. Liu, H. Zhou, and L.-H. Qu. 2003b. The complete mitochondrial genome of the Chinese giant salamander, *Andrias davidianus* (*Amphibia: Caudata*). *Gene* 311:93–98.

Zhang, P., H. Zhou, Y.-Q. Chen, Y.-F. Liu, and L.-H. Qu. 2005. Mitogenomic perspectives on the origin and phylogeny of living amphibians. *Syst. Biol.* 54:391–400.

Zug, G. R., L. J. Vitt, and J. P. Caldwell. 2001. *Herpetology: An Introductory Biology of Amphibians and Reptiles*. 2nd edition. Academic Press, San Diego, CA.

Authors

Michel Laurin; CR2P, CNRS/MNHN/SU; Muséum National d'Histoire Naturelle; CP 48; 43 rue Buffon, 75005 Paris, France. Email: michel.laurin@mnhn.fr.

Jan W. Arntzen; Naturalis Biodiversity Center; P. O. Box 9517; 2300 RA Leiden, the Netherlands. Email: pim.arntzen@naturalis.nl.

Ana María Báez; Department of Geological Sciences; Universidad de Buenos Aires Pabellón II; Ciudad Universitaria 1428 Ciudad de Buenos Aires, Argentina. Email: baez@gl.fcen.uba.ar.

Aaron M. Bauer; Department of Biology, Villanova University; 800 Lancaster Avenue, Villanova, PA 19085, USA. Email: aaron.bauer@villanova.edu.

Ross Damiani; Staatliches Museum für Naturkunde Stuttgart; Rosenstein 1; D-70191 Stuttgart, Germany. Email: rossano1973@gmail.com.

Susan E. Evans; Department of Anatomy and Developmental Biology; UCL – University College London; Gower Street; London WC1E 6BT, UK. Email: ucgasue@ucl.ac.uk.

Alexander Kupfer; Department of Zoology; Herpetology; State Museum of Natural History Stuttgart; Rosenstein 1; 70191 Stuttgart, Germany. Email: alexander.kupfer@smns-bw.de.

Allan Larson; Department of Biology; Washington University; St. Louis, MO 63130, USA. Email: larson@wustl.edu.

David Marjanović; Museum für Naturkunde; D-10115 Berlin, Germany. Email: david.marjanovic@gmx.at.

Hendrik Müller; Institut für Spezielle Zoologie und Evolutionsbiologie mit Phyletischem Museum; Friedrich-Schiller-Universität Jena; Erbertstraße 1; D-07743 Jena, Germany. Email: hendrik.mueller@uni-jena.de.

Lennart Olsson; Institut für Spezielle Zoologie und Evolutionsbiologie mit Phyletischem Museum Friedrich-Schiller-Universität Jena; Erbertstraße 1; D-07743 Jena, Germany. Email: Lennart.Olsson@uni-jena.de.

Jean-Claude Rage[†]; Département Histoire de la Terre, Muséum National d'Histoire Naturelle, CP 38, 8 rue Buffon; F-75231 Paris Cedex 05, France.

Denis Walsh; Department of Philosophy and Institute for the History and Philosophy of Science and Technology; University of Toronto; Victoria College; 91 Charles Street; Toronto, ON, M5S 1K7, Canada. Email: denis.walsh@utoronto.ca.

Date Accepted: 2 May 2014

Editors: Philip Cantino, Jacques A. Gauthier, Kevin de Queiroz

---

[†] Deceased.

Gymnophiona J. Müller 1832 [M. H. Wake], converted clade name

Registration Number: 235

Definition: The crown clade originating in the most recent common ancestor of *Caecilia tentaculata* Linnaeus 1758 *(Caeciliidae), Ichthyophis glutinosus* (Linnaeus 1758) *(Ichthyophiidae),* and *Rhinatrema bivittatum* (Cuvier *in* Guérin-Méneville 1829) *(Rhinatrematidae).* This is a minimum-crown-clade definition. Abbreviated definition: min crown ∇ (*Caecilia tentaculata* Linnaeus 1758 & *Ichthyophis glutinosus* (Linnaeus 1758) & *Rhinatrema bivittatum* (Cuvier *in* Guérin-Méneville 1829)).

Etymology: Derived from the Greek with reference to the snake-like, elongate and limbless body form, and the lack of external scales, characteristic of all modern members (*gymno-* = naked, *ophios* = snake).

Reference Phylogeny: Wilkinson and Nussbaum (2006: Fig. 2.2) (see also Wilkinson et al., 2011: Fig. 1).

Composition: *Gymnophiona* currently contains 205 extant species, (AmphibiaWeb, 2017; accessed 9 July 2017). In addition, the extinct taxon *Apodops pricei* was referred to the *Caeciliidae* (Estes and Wake, 1972), and is therefore inferred to be within the crown. The fossil taxa *Eocaecilia* (Jenkins and Walsh, 1993) and *Rubricacaecilia* (Evans and Sigogneau-Russell, 2001) have been assigned to *Gymnophiona*; however, because the name *Gymnophiona* is here applied to the caecilian crown clade, those fossils are not included assuming they have been correctly interpreted as members of its stem group (e.g., Trueb and Cloutier, 1991; Evans and Sigogneau-Russell, 2001; Schoch and Milner, 2004). Similarly, *Chinlestegophis jenkinsi*, regarded as the third stem caecilian by Pardo et al. (2017), is not included in the crown clade. See Comments for a discussion of some of the named clades within *Gymnophiona*.

Diagnostic Apomorphies: Caecilians (as the members of the named clade are commonly known) are the only crown group amphibians that lack limbs and girdles, have extreme body elongation (typically 90–280 body vertebrae, depending on the species), and have short (6–~20 postcloacal vertebrae) tails or lack them entirely (though some have 'pseudotails') (see Taylor, 1968). Additional diagnostic characters include eyes covered by skin (also seen in some salamanders) or skin and bone (Wake, 1985), and a pair of chemosensory tentacles that passes odorants from the environment to the vomeronasal chamber (Billo and Wake, 1987; Schmidt and Wake, 1990) and may be tactile (Fox, 1985), a dual jaw-closing mechanism (Bemis et al., 1983; Nussbaum, 1983) and an eversible cloacal phallodeum in males. For detailed information, see Nussbaum and Wilkinson (1989), Trueb and Cloutier (1991), Wake (2003), and Wilkinson and Nussbaum (2006). Based on fossils referred to the caecilian stem group (see Composition), Pardo et al. (2017) presented a proposed sequence of evolution of caecilian apomorphies.

Synonyms: A plethora of approximate synonyms exists, largely in the literature of the nineteenth century, when very few taxa of caecilians were recognized, and they were often included with snakes. These include, among many, *Gymnophia* (Rafinesque-Schmaltz, 1814) serpens nus (Cuvier, 1817), pseudophydiens, coeciles (de Blainville, 1816, 1822), *Batrachophidies* (Latreille, 1825), *Nuda, Coecilioidea, Ceciloides* (Fitzinger (1826), *Caeciliae* (Wagler, 1830), *Batrachophides* (Bonaparte, 1831), *Gymnophidia* (Müller, 1831), *Abranchia* (Hogg, 1839), *Celatibranchia* (Hogg, 1841), *Pseudophidia* (Gray, 1850), *Batrachia repentia* or *Peromeles* (Duméril, 1863). Several reflect the view that caecilians are 'smooth' or 'nude' ophidians. In addition, the name *Apoda* (Oppel, 1811a,b) has been used in the literature for more than 150 years as a synonym for *Gymnophiona*, rarely defined but referring to a clade that includes the caecilian crown clade and possibly some or all members of its stem group (see Comments).

Comments: The clade here named *Gymnophiona* is strongly supported by both morphological (Nussbaum and Wilkinson, 1989; Wilkinson and Nussbaum, 2006) and molecular data (Cannatella and Hillis, 1993; Frost, 2006; Wilkinson et al., 2011). The history of the name of the caecilian clade is complex, so only the most widely used names will be mentioned in the following discussion. *Apoda* was substituted in 1879 for *Pseudophidia* (*Reptilia*) in the Zoological Record (O'Shaughnessy, 1880). Boulenger (1882) retained the name *Apoda* when he transferred caecilians to *Batrachia*, maintaining that usage for many years. *Gymnophiona* replaced *Apoda* in the Systematic Index for *Amphibia* of the Zoological Record in 1954 (pers. obs.); the basis for the change was not stated. Dubois (2004) pointed out reasons that *Apoda* Oppel (1811a,b) is not a valid name, such as homonymy and priority of the name for eels; Wilkinson and Nussbaum (2006) also argued against the use of *Apoda*. A few authors continue to use *Apoda*, but most recent formal taxonomies and phylogenies do not. Trueb and Cloutier (1991) used the name *Apoda* for a clade approximating the caecilian crown and *Gymnophiona* for a larger clade including the crown and at least *Eocaecilia*.

However, de Quieroz and Gauthier (1992) and Cannatella and Hillis (1993) explicitly defined *Gymnophiona* as the name of the crown clade of caecilians (see also Frost et al., 2006). This author concurs with Wilkinson's and Nussbaum's (2006) suggestion that *Eocaecilia* not be included in *Gymnophiona* because the literature is filled with comments about *Gymnophiona* and changing the name to include putative fossil forms would invalidate many of the current generalizations based exclusively on extant taxa. Given this history, *Gymnophiona* is chosen here as the name of the crown clade. De Quieroz and Gauthier (1992) and Cannatella and Hillis (1993) used the name *Apoda* for the caecilian total clade, although the former authors now prefer the name *Pan-Gymnophiona* (Gauthier and de Quieroz, 2001). Carroll (2007) used the name *Caecilia* for a clade that includes part but not all of the caecilian stem, but that name is most commonly used for a much smaller taxon (traditionally ranked as a genus).

Cannatella and Hillis (1993) used the name *Stegokrotaphia* for the sister taxon of the *Rhinatrematidae* (see also Frost et al., 2006), representing the basal divergence within *Gymnophiona*. Wilkinson and Nussbaum (2006) named the same clade *Neocaecilia*; within it they named the clade originating in the last common ancestor of *Ichthyophis, Caudacaecilia*, and *Uraeotyphlus* (*Diatriata*), and the clade originating in the last common ancestor of the rest of the extant taxa (*Teresomata*). Recognizing that some recent molecular phylogenies have presented evidence that the previously recognized taxon *Caeciliidae* is paraphyletic, Wilkinson et al. (2011) have proposed restricting that name to the clade of all caecilians more closely related to *Caecilia* than to *Typhlonectes* and resurrecting five names (*Dermophiidae, Herpelidae, Indotyphlidae, Siphonopidae*, and *Typhlonectidae*, the last of which had been treated as a synonym of *Caeciliidae* by Hedges et al. [1993] and Frost et al. [2006]) for clades formerly included in *Caeciliidae* based largely on molecular phylogenetic studies and better knowledge of diversity. This author considers these taxonomic proposals (as well as that of Kamel et al. [2012]) useful but provisional, still awaiting more conclusive evidence for relationships among extant as well as extinct taxa.

Literature Cited

AmphibiaWeb: Information on Amphibian Biology and Conservation [web application]. 2017. University of California, Berkeley, CA. Available at http://amphibiaweb.org, accessed on 9 July 2017.

Bemis, W. E., K. Schwenk, and M. H. Wake. 1983. Morphology and function of the feeding apparatus in *Dermophis mexicanus* (*Amphibia*: Gymnophiona). *Zool. J. Linn. Soc.* 77:75–96.

Billo, R., and M. H. Wake. 1987. Tentacle development in *Dermophis mexicanus* (*Amphibia*: Gymnophiona: Caeciliidae). *J. Morphol.* 192:101–111.

de Blainville, H. M. D.. 1816. Prodrome d'une nouvelle distribution systématique du règne animal. *Bull. Sci. Soc. Philom. Paris*: 111.

de Blainville, H. M. D. 1822. *De l'Organisation des Animaux: ou Principes d'Anatomie Comparée*, Vol. 1. F. G. Levrault, Paris.

Bonaparte, C. L. 1831. *Saggio di una Distribuzione Methodica Degli Animali Vertebrati a Sanuge Freddo*. A. Boulzaler, Rome.

Boulenger, G. A. 1882. *Catalogue of the* Batrachia Gradientia S. Caudata *and* Batrachia Apoda *in the Collection of the British Museum*. Taylor & Francis, London.

Cannatella, D. C., and D. M. Hillis. 1993. Amphibian relationships: phylogenetic analysis of morphology and molecules. *Herpetol. Monogr.* 7:1–7.

Carroll, R. L. 2007. The Paleozoic ancestry of salamanders, frogs and caecilians. *Zool. J. Linn. Soc.* 150(suppl.):1–140.

Cuvier, G. 1817. *Le Règne Animal Distribué après son Organization pour Servir de Base a l'Histoire Naturelle des Animaux et d'Introduction à l'Anatomie Comparée*. Deterville, Paris.

Dubois, A. 2004. The higher nomenclature of recent amphibians. *Alytes* 22:1–14.

Duméril, A. 1863. Catalogue méthodique de la collection des batraciens du Muséum d'Histoire naturelle de Paris. *Mém. Soc. Imp. Sci. Nat. Cherbourg* 9:293–321.

Estes, R., and M. H. Wake. 1972. Caecilian amphibians: their first fossil record. *Nature* 239:228–231.

Evans, S. E., and D. Sigogneau-Russell. 2001. A stem-group caecilian (*Lissamphibia*: Gymnophiona) from the Lower Cretaceous of North Africa. *Paleontology* 44:259–273.

Fitzinger, L. J. 1826. *Neue Classification der Reptilien Nach Ihren Natürlichen Verwandtschaften*. Heubner, Vienna.

Fox, H. 1985. The tentacles of *Ichthyophis* (*Amphibia: Caecilia*) with special reference to the skin. *J. Zool. (London)* 205:223–234.

Frost, D. R. 2006. *Amphibian Species of the World*, an online reference, Version 5.5. American Museum of Natural History, New York. Available at http://research.amnh.org/herpetology'amphibia/index.php.

Frost, D. R., T. Grant, J. Faivovich, R. H. Bain, A. Haas, C. F. B. Haddad, R. O. de Sa, A. Channing, M. Wilkinson, S. C. Donnellan, C. J. Raxworthy, J. A. Campbell, B. Blotto, P. Moler, R. C. Drewes, R. A. Nussbaum, J., D. Lynch, D. M. Green, and W. C. Wheeler. 2006. The amphibian tree of life. *Bull. Am. Mus. Nat. Hist.* 297:1–370.

Gauthier, J. A., and K. de Queiroz. 2001. Feathered dinosaurs, flying dinosaurs, crown dinosaurs, and the name *"Aves."* Pp. 7–41 in *New Perspectives on the Origin and Early Evolution of Birds*: *Proceedings of the International Symposium in Honor of John H. Ostrom* (J. A. Gauthier and L. F. Gall, eds.). Peabody Museum of Natural History, Yale University, New Haven, CT.

Gray, J. E. 1850. *Catalogue of the Specimens of* Amphibia *in the Collection of the British Museum: Part II*. Batrachia Gradientia. Trustees of the British Museum, London.

Guérin-Méneville, F. E. 1829. *Icongraphie du Règne Animal de G. Cuvier.* 3 Vols. J. B. Bailliere, Paris.

Hedges, S. B., R. A. Nussbaum, and L. R. Maxson 1993. Caecilian phylogeny and biogeography inferred from mitochondrial DNA sequences of the 12S rRNA and 16S rRNA genes (*Amphibia*: *Gymnophiona*). *Herpetol. Monogr.* 7:64–76.

Hogg, J. 1839. On the classification of the *Amphibia*. *Mag. Nat. Hist.* (n. s.) 3:265–274 and 367–378.

Hogg, J. 1841. On the existence of branchiae in the young *Caeciliae*, and on a modification and extension of the branchial classification of the *Amphibia*. *Ann. Mag. Nat. Hist.* 45:353–363.

Jenkins, Jr., F. A., and D. M. Walsh. 1993. An Early Jurassic caecilian with limbs. *Nature* 365:246–250.

Kamel, R. G., D. San Mauro, D. J. Gower, I. Van Bocxlaer, E. Sherratt, A. Thomas, S. Babu, F. Bossuyt, M. Wilkinson, and S. D. Biju. 2012. Discovery of a new family of amphibians from northeast India with ancient links to Africa. *Proc. R. Lond. Soc. B Biol. Sci.* 279:2396–2401.

Latreille, P.A. 1825. *Families Naturelles du Règne Animal, Exposées Succinctement et dans un Ordre Analytique, Avec L'indication de Leurs Genres.* J. B. Baillere, Paris.

Linnaeus, C. 1758. *Systema Naturae Per Regna Tria Naturae, Secundum Classes, Ordines, Genera, Species, cum Characteribus, Differentiis, Synonymis, Locis.* 10th edition, Vol. I. Laurentii Salvii, Holmiae (Stockholm).

Müller, J. 1831. Kiemenlöcher an einer jungen *Caecilia hypocyanae*, in Museum der Naturgeschichte zu Leiden beobachtet. *Isis* 24:709–711.

Müller, J. 1832. Beiträge zur Anatomie und Naturgeschichte der Amphibien. I. Ueber die natürliche Eintheilung der Amphibien. *Z. Physiol.* 4:190–275.

Nussbaum, R. A. 1983. The evolution of a unique dual jaw-closing mechanism in caecilians (*Amphibia*: *Gymnophiona*) and its bearing on caecilian ancestry. *J. Zool.* (*London*) 199:545–554.

Nussbaum, R. A., and M. Wilkinson. 1989. On the classification and phylogeny of caecilians (*Amphibia*: *Gymnophiona*), a critical review. *Herpetol. Monogr.* 3:1–42.

Oppel, M. 1811a. Mémoire sur la classification des reptiles. *Ann. Mus. Hist. Nat.* 16:254–295, 376–393, 394–418.

Oppel, M. 1811b. *Die Ordnungen, Familien und Gattungen der Reptilien als Prodrom einer Naturgeschichte derselben.* Lindauer, Munich.

O'Shaughnessy, A. W. E. 1880. *Reptilia. Zool. Record* 1879(1880):1–19.

Pardo, J. D., B. J. Small, and A. K. Huttenlocker. 2017. Stem caecilian from the Triassic of Colorado sheds light on the origins of *Lissamphibia*. *Proc. Natl. Acad. Sci. USA.* doi:10.1073/pnas.1706752114.

de Queiroz, K., and J. A. Gauthier. 1992. Phylogenetic taxonomy. *Annu. Rev. Ecol. Syst.* 23:449–480.

Rafinesque-Schmaltz, C. S. 1814. Fine del prodromo d'erpetologia Siciliana. *Speccio Sci.* 2:102–104.

Schmidt, A., and M. H. Wake. 1990. The olfactory and vomeronasal system of caecilians (*Amphibia*: *Gymnophiona*). *J. Morphol.* 205:255–268.

Schoch, R. R., and R. A. Milner. 2004. Structure and implications of theories on the origin of lissamphibians. Pp. 345–377 in *Recent Advances in the Origin and Early Radiation of Vertebrates* (G. Arratia, M. V. H. Wilson, and R. Cloutier eds.). Verlag Dr. Friedrich Pfeil, Munchen, Germany.

Taylor, E. H. 1968. *Caecilians of the World: A Taxonomic Review.* University of Kansas, Lawrence, KA.

Trueb, L., and R. Cloutier. 1991. A phylogenetic investigation of the inter- and intrarelationships of the *Lissamphibia* (*Amphibia*: *Temnospondyli*). Pp. 223–313 in *Origins of the Major Groups of Tetrapods: Controversies and Consensus* (H. P. Schultze and L. Trueb eds.). Cornell University Press, Ithaca, NY.

Wagler, J. 1830. *Natürliches System der Amphibien mit voranhender Classification der Saügethiere und Vogel.* Cotta, Munich.

Wake, M. H. 1985. The comparative morphology and evolution of the eyes of caecilians (*Amphibia, Gymnophiona*). *Zoomorphology* 105:277–295.

Wake, M. H. 2003. Osteology of caecilians. Pp. 1809–1876 in *Biology of Amphibians. Vol. 5: Osteology* (H. Heatwole and M. Davies, eds.). Surrey Beatty and Sons, Chipping Norton, Australia.

Wilkinson, M., and R. A. Nussbaum, 2006. Caecilian phylogeny and classification. Pp. 39–78 in *Reproductive Biology and Phylogeny of* Gymnophiona (*Caecilians*) (J.-M. Exbrayat and B. G. M. Jamieson eds.). Science Publishers, Enfield, NH.

Wilkinson, M., D. San Mauro, E. Sherratt, and D. Gower. 2011. A nine-family classification of caecilians (*Amphibia: Gymnophiona*). *Zootaxa* 2874:41–64.

Author

Marvalee H. Wake; Department of Integrative Biology and Museum of Vertebrate Zoology; University of California; Berkeley, CA 94720-3140, USA. Email: mhwake@berkeley.edu.

Date Accepted: 09 November 2012; updated 13 July 2017

Primary Editor: Kevin de Queiroz

Caudata J. A. Scopoli 1777 [D. Wake], converted clade name

Registration Number: 234

Definition: The crown clade originating in the most recent common ancestor of *Salamandra* (originally *Lacerta*) *salamandra* (Linnaeus 1758) (*Salamandridae*), *Siren lacertina* Österdam 1766 (*Sirenidae*) and *Cryptobranchus* (originally *Salamandra*) *alleganiensis* (Daudin 1803) (*Cryptobranchidae*). This is a minimum-crown-clade definition. Abbreviated definition: min crown ∇ (*Salamandra salamandra* (Linneaus 1758) & *Siren lacertina* Österdam 1766 & *Cryptobranchus alleganiensis* (Daudin 1803)).

Etymology: Derived from the Latin *cauda* with reference to the presence of a tail, characteristic of all contained taxa.

Reference Phylogeny: Figure 1 of Roelants et al. (2007). This tree includes representatives of all taxa traditionally ranked as families and is based on combined nuclear and mitochondrial gene sequences.

Composition: *Caudata* currently contains 739 extant species, based on lists maintained on AmphibiaWeb (2020) and Amphibian Species of the World (Frost, 2020). Dubois and Raffaëlli (2012) listed all known fossil taxa.

Diagnostic Apomorphies: Salamanders are more generalized in morphology than other living tetrapods and most of their obvious traits are plesiomorphic. They are unique in limb structure and development; the manus (maximum of four digits) and pes show preaxial dominance during development, the metapterygial axis extends between digits one and two, and in both manus and pes the first and second distal mesopodials are represented by a single basale commune (Shubin and Wake, 2003). For a detailed list of morphological characters see Wiens et al. (2005).

Synonyms: *Gradientia* Laurenti 1768 included salamanders and diverse tetrapods but was restricted to salamanders by Merrem (1820). The name subsequently was used in catalogues of the British Museum (Natural History) by Gray (1850) and Boulenger (1882); Gray did not include all then-known taxa currently recognized as taxa of *Caudata*, but Boulenger did. *Urodela* was recommended as the name of the salamander taxon by Dubois (2004; see also Dubois and Raffaëlli, 2012), who attributed the name to Duméril (1806), who, however, used the French spelling Urodèles. *Urodela* first was used with the current spelling and content by either Lichtenstein (1856) or possibly Stannius (1856). Both *Gradientia* sensu Merrem (1820) and *Urodela* sensu Lichtenstein (1856) and Stannius (1856) are thus approximate synonyms of *Caudata* as defined here. Many other approximate synonyms exist but none included all then-known taxa currently referred to *Caudata* and none were used after the mid-late nineteenth century (Frost, 2020).

Comments: Wiens et al. (2005) performed a Bayesian analysis of combined morphological and molecular data that inferred *Cryptobranchidae* + *Hynobiidae* to be sister to the remaining salamander taxa. In turn, *Sirenidae* is sister to the remaining taxa. *Plethodontidae* and *Amphiumidae* are sister taxa, and the combined clade is sister to *Rhyacotritonidae*. *Dicamptodontidae* and *Ambystomatidae* are sister taxa, and the combined clade is sister to

Salamandridae. Roelants et al. (2007) used only molecular data but more than used by Wiens et al., (2005). They inferred similar relationships except that *Proteidae* is the sister group of *Rhyachotritonidae* + *Amphiumidae* + *Plethodontidae* rather than of *Ambystomatidae* + *Dicamptodontidae* + *Salamandridae.* Some nodes were weakly or not supported. Zhang and Wake (2009), using complete mitochondrial genomes, obtained a tree identical regarding the taxa mentioned above to that of Roelants et al. (2007), except that *Sirenidae* is sister to all remaining salamanders, and a *Cryptobranchidae* + *Hynobiidae* clade is sister to the crown clade of salamanders with internal fertilization (all remaining salamanders). All major nodes are well supported. Pyron and Wiens (2011) combined all available molecular data for all available taxa and obtained a tree with a topology identical to that of Roelants et al. (2007) with regard to the taxa mentioned above (i.e., the traditional families). Because of uncertainty concerning whether *Sirenidae* or *Cryptobranchidae* + *Hynobiidae* is the sister group of the remaining salamanders, representatives of both taxa are used as internal specifiers (see Definition).

Frost et al. (2006) attributed the names *Caudata* and *Urodela* to Fischer von Waldheim (or, Fischer) 1813 (as *Urodeli*, a synonym of his *Caudati*, both unranked names), and selected *Caudata.* Dubois (2004; see also Dubois and Raffaëlli, 2012) rejected *Caudata* and argued in favor of *Urodela*, in part because the original use of *Caudata* by Scopoli (1777) was for a heterogeneous group including not only salamanders, and instead accepted the Latinized version of Duméril's *Urodéles. Urodela* was first used to include all current taxa and nothing else by Stannius (1856), but later authors often failed to include all taxa (e.g., Cope, 1889, who excluded *Proteidae* and *Sirenidae*). *Urodela* continues to be widely used, but most recent authors use *Caudata* for the clade in question (e.g., Roelants

et al., 2007; Zhang and Wake, 2009; Pyron and Wiens, 2011), which was selected for that reason. Authorship is attributed to Scopoli (1977) according to Article 19.1 of the *PhyloCode.* One suggestion to eliminate conflict between the names *Caudata* and *Urodela* (de Queiroz and Gauthier, 1992; Cannatella and Hillis, 1993; Trueb and Cloutier, 1991) refers living and crown-group fossil taxa of salamanders to *Caudata* and extinct salamander taxa that fall outside the crown clade to a more inclusive total clade *Urodela,* which includes *Caudata* as a subordinate taxon. The convention of forming the name of the total clade by combining the prefix *Pan-* with the name of the crown clade (e.g., de Queiroz, 2007; *PhyloCode* Art. 10.3), in this case *Pan-Caudata,* would also allow the name *Urodela* to be used for a larger clade that includes *Caudata,* though not for the total clade.

Literature Cited

AmphibiaWeb: Information on Amphibian Biology and Conservation. [web application]. 2020. AmphibiaWeb, Berkeley, CA. Available at http://amphibiaweb.org/, accessed on 14 January, 2020.

Boulenger, G. A. 1882. *Catalogue of the* Batrachia Gradientias. Caudata *of the British Museum.* 2nd edition. Taylor & Francis, London.

Cannatella, D. C., and D. M. Hillis. 1993. Amphibian relationships: phylogenetic analysis of morphology and molecules. *Herpetol. Monogr.* 7:1–7.

Cope, E. D. 1889. The *Batrachia* of North America. *Bull. U.S. Natl. Mus.* 34:1–525.

Dubois, A. 2004. The higher nomenclature of recent amphibians. *Alytes* 22:1–14.

Dubois, A., and J. Raffaëlli. 2012. A new ergotaxonomy of the order *Urodela* Duméril, 1805 (*Amphibia, Batrachia*). *Alytes* 28:77–161.

Duméril, A. M. C., 1806 (1805). *Zoologie Analytique, ou Méthode Naturelle de Classification des Animaux, Rendue plus Facile à l'Aide de Tableaux Synoptiques.* Allais, Paris.

Fischer, G. 1813. *Zoognosia Tabulis Synopticis Illustrate. Editio Tertia, Classium, Ordinum, Generum Illustration Perpetua Aucta.* Typis Nicolai Sergeidis Vsevolozsky, Moscow.

Frost, D. R. 2013. *Amphibian Species of the World: An Online Reference,* Version 6.0. Electronic database 14 January 2013. American Museum of Natural History, New York. Available at http://research.amnh.org/herpetology/amphibia/index.php.

Frost, D. R., T. Grant, J. Faivovich, R. H. Bain, A. Haas, C. F. B. Haddad, R. O. de Sá. A. Channing, M. Wilkinson, S. C. Donnellan, C. J. Raxworthy, J. A. Campbell, B. L. Blotto, P. Moler, R. C. Drewes, R. A. Nussbaum, J. D. Lynch, D. M. Green, and W. C. Wheeler. 2006. The amphibian tree of life. *Bull. Am. Mus. Nat. Hist.* 297:1–370.

Gray, J. E. 1850. *Catalogue of the Specimens of* Amphibia *in the Collection of the British Museum. Part II.* Batrachia Gradientia, *etc.* Spottiswoodes & Shaw, London.

Laurenti, J. N. 1768. *Specimen Medicum, Exhibens Synopsin* Reptilium *Emendatam com Experimentis circa Venena et Antidota* Reptilium *Austriacorum.* Joan. Thom. Nob. De Trattnern, Vienna.

Lichtenstein, H. 1856. *Nomenclator Reptilium et Amphibiorum Musei Zoologici Berolinensis.* Königlichen Akademie der Wissenschaften, Berlin.

Merrem, B. 1820. *Versuch eines Systems der Amphibien. Tentamen Systematis Amphibiorum.* Johann Christian Krieger, Marburg.

de Queiroz, K. 2007. Toward and integrated system of clade names. *Syst. Biol.* 56:956–974.

de Queiroz, K., and J. Gauthier. 1992. Phylogenetic taxonomy. *Annu. Rev. Ecol. Syst.* 23:449–480.

Pyron, R. A., and J. J. Wiens. 2011. A large-scale phylogeny of *Amphibia* including over 2,800 species, and a revised classification of extant frogs, salamanders, and caecilians. *Mol. Phylogenet. Evol.* 61:543–583.

Roelants, K., D. J. Gower, M. Wilkinson, S. P. Loader, S. D. Biju, K. Guillaume, L. Moriau, and F. Bossuyt. 2007. Global patterns of diversification in the history of modern amphibians. *Proc. Natl. Acad. Sci. USA* 104:887–892.

Scopoli, J. A. 1777. *Introductio ad Historiam Naturelam, Sistens Genera Lapidum, Planarum, et Animalium Hactenus Detecta, Caracteribus Essentialibus Donate, in Tribus Divisa, Subinde ad Leges Naturae.* Gerle, Prague.

Shubin, N. H., and D. B. Wake. 2003. Morphological variation, development, and evolution of the limb skeleton of salamanders. Pp.1782–1808 in *Amphibian Biology,* Vol. 5 (H. Heatwole and M. Davies, eds.). Surrey Beatty and Sons, Chipping Norton, Australia.

Stannius, H. 1856. *Zootomie der Amphibien, 2nd Buch, Handbuch der Zootomie,* 2nd Theil, Wirbelthiere (Von Siebold and Stannius, eds.). Verlag Von Veit & Comp., Berlin.

Trueb, L., and R. Cloutier. 1991. A phylogenetic investigation of the inter- and intrarelationships of the *Lissamphibia* (*Amphibia:* *Temnospondyli*). Pp. 223–313 in *Origins of the Higher Groups of Tetrapods: Controversy and Consensus* (H.-P. Schultze and L. Trueb, eds.). Cornell University Press, Ithaca, NY.

Wiens, J. J., R. M. Bonett, and P. T. Chippindale. 2005. Ontogeny discombobulates phylogeny: paedomorphosis and higher-level salamander relationships. *Syst. Biol.* 54:91–110.

Zhang, P., and D. B. Wake. 2009. Higher-level salamander relationships and divergence dates inferred from complete mitochondrial genomes. *Mol. Phylogenet. Evol.* 53:492–508.

Author

David B. Wake; Museum of Vertebrate Zoology; University of California; Berkeley, CA 94720-3160, USA. Email: wakelab@berkeley.edu.

Date Accepted: 08 May 2013

Primary Editor: Kevin de Queiroz

Pan-Amniota T. Rowe 2004 [M. Laurin and T. R. Smithson], converted clade name

Registration Number: 74

Definition: The largest total clade containing *Homo sapiens* Linnaeus 1758 (*Amniota*) but not *Pipa pipa* Linnaeus 1758 (*Pipidae*) and *Caecilia tentaculata* Linnaeus 1758 and *Siren lacertina* Österdam 1766 (*Sirenidae*). This is a maximum-total-clade definition. Abbreviated definition: max total ∇ (*Homo sapiens* Linnaeus 1758 ~ *Pipa pipa* Linnaeus 1758 & *Caecilia tentaculata* Linnaeus 1758 & *Siren lacertina* Österdam 1766).

Etymology: *Pan* (prefix indicating that the name refers to a total clade; derived from the Greek *Pantos*, all, the whole) + *Amniota* (See etymology for *Amniota* in this volume).

Reference Phylogeny: The reference phylogeny is Vallin and Laurin (2004: Fig. 6). The external specifier *Caecilia tentaculata* is not included in the reference phylogeny but is most closely related to *Ichthyophiidae* (Pyron and Wiens, 2011: Fig. 2).

Composition: *Pan-Amniota* includes amniotes (the crown-group) as well as all extinct taxa that are more closely related to amniotes than to lissamphibians. Under the main reference phylogeny, pan-amniotes include *Diadectomorpha* Watson 1917 and *Solenodonsaurus* Broili 1924, which are stem-amniotes, and possibly *Westlothiana* (the position of which is unresolved on the reference phylogeny), as well as the amniote crown. Under the phylogeny of Ruta et al. (2003: Fig. 4), pan-amniotes also include "lepospondyls", *Seymouriamorpha* Watson 1917, *Gephyrostegus bohemicus* Jaekel 1902, *Brukterpeton fiebigi* Boy and Bandel 1973, and *Embolomeri* Cope 1884. Under traditional phylogenies (e.g., Panchen and Smithson, 1988: Fig. 1.1; Lombard and Sumida, 1992: Fig. 6), *Pan-Amniota* includes the same taxa as under the phylogeny of Ruta et al. (2003) except that "lepospondyls" are excluded. Thus, the composition and diagnosis of *Pan-Amniota* is highly dependent on the choice of the reference phylogeny, which is controversial (Laurin et al., 2019; Marjanović and Laurin, 2019).

Diagnostic Apomorphies: The list of apomorphies is taken from Laurin (1998a,b), with modifications from Laurin and Reisz (1999). The basal position of *Solenodonsaurus* and the fragmentary nature of the holotype and only known specimen limit the list of apomorphies. Most of the characters listed below may diagnose a slightly less inclusive clade that excludes *Soleondonsaurus*, and all are formulated in the context of the reference phylogeny. Diagnostic pan-amniote characters include: lateral-line sulci (and the organ itself) absent throughout ontogeny; tabular located on occiput rather than on skull table; exoccipital does not contact dermal skull roof; convex occipital condyle; basioccipital excluded from edge of foramen magnum; stapes with a dorsal process separated from the footplate by a deep notch; atlantal pleurocentrum fused to axial intercentrum; and three scapulocoracoid ossifications.

Synonyms: *Anthracosauria* Säve-Söderbergh 1934 sensu Panchen (1980) (partial and approximate) and sensu Gauthier et al. (1988) and Laurin (1998a) (unambiguous); *Reptiliomorpha* Säve-Söderbergh 1934, although not given an explicit phylogenetic definition, is considered an unambiguous synonym; *Reptiliomorpha* sensu Vallin and Laurin (2004) is also an unambiguous synonym.

Comments: The taxon *Reptiliomorpha* was erected by Säve-Soderbergh (1934) for amniotes and embolomeres, which he believed to be closely related. Säve-Soderbergh believed that reptiliomorphs were the sister-group of *Batrachomorpha* (which he also erected), which included *Ichthyostegalia, Labyrinthodontia, Phyllospondyli* (the last two taxa included forms now considered temnospondyls), and *Anura.* Urodeles were thought to have evolved limbs independently and to be most closely related to lungfishes. Thus, both *Reptiliomorpha* and *Batrachomorpha* approximated total clades and may have been conceptualized as such by Säve-Soderbergh (1934). *Reptiliomorpha* has been used for a clade approximating the total clade of *Amniota* by several subsequent authors (e.g., Panchen and Smithson, 1988; Smithson, 1994; Lebedev and Coates, 1995; Paton et al., 1999; Clack and Carroll, 2000) and was explicitly defined as the name of the amniote total clade by Vallin and Laurin (2004).

The taxon *Anthracosauria* Säve-Söderbergh 1934 originally included only embolomeres, and it retained this delimitation for a few decades (Panchen, 1970: 2). Its composition subsequently expanded to include proterogyrinids, and gephyrostegids (Panchen, 1975: 609), and later, seymouriamorphs (Panchen, 1980: 343). This expansion culminated in Carroll's (1988: 613) inclusion in *Anthracosauria* of all other taxa that were thought to be stem-amniotes, namely *Solenodonsaurus janenschi* and diadectomorphs (in addition to other taxa that are now considered to be amniotes), and the name was explicitly defined as designating the amniote total clade by Gauthier et al. (1988) and Laurin (1998a).

Following the proposal of Gauthier and de Queiroz (2001; see also de Queiroz, 2007), Rowe (2004) used the name *Pan-Amniota* for the amniote total clade. Because the *PhyloCode* (Art. 10.3; Cantino and de Queiroz, 2020) promotes the use of panclade names for total clades, we are converting the name *Pan-Amniota.* Although the names *Reptiliomorpha* and *Anthracosauria* have been used more frequently than *Pan-Amniota*, neither name is very widely used (88 and 321 respective results returned by a Google Scholar search on May 12, 2014). Moreover, many uses of the name *Anthracosauria* are not for the amniote total clade (or even for a paraphyletic group originating in the same ancestor), and the name *Reptiliomorpha* could easily be misinterpreted by non-specialists as applying to the total clade of *Reptilia* (see entry in this volume) rather than that of *Amniota.* Thus, we decided to convert the name *Pan-Amniota*, even though the names *Reptiliomorpha* and *Anthracosauria* appear more often in the literature.

Literature Cited

Broili, F. 1924. Ein Cotylosaurier aus der oberkarbonischen Gaskohle von Nürschan in Böhmen. *Sber. Bayer. Akad. Wiss.* 1924:3–11.

Boy, J. A., and K. Bandel. 1973. *Bruktererpeton fiebigi* n. gen. n. sp. (*Amphibia: Gephyrostegida*) der erste Tetrapode aus dem Rheinisch-Westfälischen Karbon (Namur B; W.-Deutschland). *Palaeontogr. Abt. A* 145:39–77.

Cantino, P. D., and K. de Queiroz. 2020. *International Code of Phylogenetic Nomenclature (PhyloCode).* Version 6. CRC Press, Boca Raton, FL.

Carroll, R. L. 1988. *Vertebrate Paleontology and Evolution.* W. H. Freeman, New York.

Clack, J. A., and R. L. Carroll. 2000. Early Carboniferous tetrapods. Pp. 1030–1043 in *Amphibian Biology* (H. Heatwole and R. L. Carroll, eds.). Surrey Beatty & Sons, Chipping Norton, Australia.

Cope, E. D. 1884. The *Batrachia* of the Permian period of North America. *Am. Nat.* 18:26–39.

Gauthier, J. A., and K. de Queiroz. 2001. Feathered dinosaurs, flying dinosaurs, crown dinosaurs, and the name "*Aves.*" Pp. 7–41 in *New*

Perspectives on the Origin and Early Evolution of Birds: Proceedings of the International Symposium in Honor of John H. Ostrom (J. A. Gauthier and L. F. Gall, eds.). Peabody Museum of Natural History, Yale University, New Haven, CT.

Gauthier, J. A., A. G. Kluge, and T. Rowe. 1988. The early evolution of the *Amniota*. Pp. 103–155 in *The Phylogeny and Classification of the Tetrapods. Vol. 1: Amphibians, Reptiles, Birds* (M. J. Benton, ed.). Clarendon Press, Oxford.

Jaekel, O. 1902. Ueber *Gephyrostegus bohemicus* n.g. n.sp. *Z. Dtsch. Geol. Ges. 54*:127–132.

Laurin, M. 1998a. The importance of global parsimony and historical bias in understanding tetrapod evolution. Part I. Systematics, middle ear evolution, and jaw suspension. *Ann. Sci. Nat., Zool., 13 Ser.* 19:1–42.

Laurin, M. 1998b. The importance of global parsimony and historical bias in understanding tetrapod evolution. Part II. Vertebral centrum, costal ventilation, and paedomorphosis. *Ann. Sci. Nat., Zool., 13 Ser.* 19:99–114.

Laurin, M., O. Lapauze, and D. Marjanović 2019. What do ossification sequences tell us about the origin of extant amphibians? *PCI Paleontol.* doi:10.1101/352609.

Laurin, M., and R. R. Reisz. 1999. A new study of *Solenodonsaurus janenschi*, and a reconsideration of amniote origins and stegocephalian evolution. *Can. J. Earth Sci.* 36:1239–1255.

Lebedev, O. A., and M. I. Coates. 1995. The postcranial skeleton of the Devonian tetrapod *Tulerpeton curtum* Lebedev. *Zool. J. Linn. Soc.* 114:307–348.

Linnaeus, C. 1758. *Systema Naturae.* 10th edition. Laurentii Salvii, Holmiae (Stockholm).

Lombard, R. E., and S. S. Sumida. 1992. Recent progress in understanding early tetrapods. *Amer. Zool.* 32:609–622.

Marjanović, D., and M. Laurin. 2019. Phylogeny of Paleozoic limbed vertebrates reassessed through revision and expansion of the largest published relevant data matrix. *PeerJ* 6:e5565.

Österdam, A. 1766. *Siren lacertina.* PhD thesis, Uppsala University, Uppsala.

Panchen, A. L. 1970. *Anthracosauria.* Pp. 1–83 in *Encyclopedia of Paleoherpetology* (O. Kuhn, ed.). Gustav Fischer, Stuttgart.

Panchen, A. L. 1975. A new genus and species of anthracosaur amphibian from the Lower Carboniferous of Scotland and the status of *Pholidogaster pisciformis* Huxley. *Philos. Trans. R. Soc. Lond. B Biol. Sci.* 269:581–637.

Panchen, A. L. 1980. The origin and relationships of the anthracosaur *Amphibia* from the late Paleozoic. Pp. 319–350 in *The Terrestrial Environment and the Origin of Land Vertebrates* (A. L. Panchen, ed.). Academic Press, London.

Panchen, A. L., and T. R. Smithson. 1988. The relationships of the earliest tetrapods. Pp. 1–32 in *The Phylogeny and Classification of the Tetrapods. Vol. 1: Amphibians, Reptiles, Birds* (M. J. Benton, ed.). Clarendon Press, Oxford.

Paton, R. L., T. R. Smithson, and J. A. Clack. 1999. An amniote-like skeleton from the Early Carboniferous of Scotland. *Nature* 398:508–513.

Pyron, R. A., and J. J. Wiens 2011. A large-scale phylogeny of *Amphibia* including over 2800 species, and a revised classification of extant frogs, salamanders, and caecilians. *Mol. Phyl. Evol.* 61:543–583.

de Queiroz, K. 2007. Toward an integrated system of clade names. *Syst. Biol.* 56:956–974.

Rowe, T. 2004. Chordate phylogeny and development. Pp. 384–409 in *Assembling the Tree of Life* (J. Cracraft and M. J. Donoghue, eds.). Oxford University Press, Oxford.

Ruta, M., M. I. Coates, and D. D. L. Quicke. 2003. Early tetrapod relationships revisited. *Biol. Rev.* 78:251–345.

Säve-Söderbergh, G. 1934. Some points of view concerning the evolution of the vertebrates and the classification of this group. *Ark. Zool.* 26A:1–20.

Smithson, T. R. 1994. *Eldeceeon rolfei*, a new reptiliomorph from the Viséan of East Kirkton, West Lothian, Scotland. *Trans. R. Soc. Edinb.* 84:377–382.

Vallin, G., and M. Laurin. 2004. Cranial morphology and affinities of *Microbrachis*, and a reappraisal of the phylogeny and lifestyle of the first amphibians. *J. Vertebr. Paleontol.* 24:56–72.

Watson, D. M. S. 1917. A sketch classification of the pre-Jurassic tetrapod vertebrates. *Proc. Zool. Soc. Lond.* 1917:167–186.

Authors

Michel Laurin; CR2P, CNRS/MNHN/SU; Muséum National d'Histoire Naturelle; CP 48; 43 rue Buffon, 75005 Paris, France. Email: michel.laurin@mnhn.fr.

Tim Smithson; University Museum of Zoology; Downing Street; Cambridge; CB2 3EJ, UK. Email: ts556@cam.ac.uk.

Date Accepted: 3 June 2014

Editors: Philip Cantino, Jacques A. Gauthier, Kevin de Queiroz

Amniota E. Haeckel 1866 [M. Laurin and R. R. Reisz], converted clade name

Registration Number: 7

Definition: The smallest crown clade containing *Homo sapiens* Linnaeus 1758 (*Synapsida*), *Testudo graeca* Linnaeus 1758 (*Testudines*), and *Crocodylus* (originally *Lacerta*) *niloticus* Laurenti 1768 (*Diapsida*; see Reference Phylogeny). This is a minimum-crown-clade definition. Abbreviated definition: min crown ∇ (*Homo sapiens* Linnaeus 1758 & *Testudo graeca* Linnaeus 1758 & *Crocodylus niloticus* Laurenti 1768).

Etymology: Derived from "amnion," the name of a structure within which amniote embryos develop (Haeckel, 1866: 132), itself a Greek word that means "membrane around a fetus."

Reference Phylogeny: The primary reference phylogeny is Modesto (1999: Fig. 4A), which expands upon that of Laurin and Reisz (1995: Fig. 2). Both place turtles outside diapsids, as sister group of procolophonoids among parareptiles (which would then be stem-turtles; see Joyce et al. this volume). However, the above definition applies to a taxon that is stable in composition, despite debate over the position of turtles outside or within diapsids (see Comments). The specifier *Crocodylus niloticus*, which is not included in the reference phylogeny, is most closely related to *Younginiformes* in that phylogeny (Rieppel and Reisz, 1999: Fig. 1, in which *Crocodylus* is part of *Archosauriformes*).

Composition: *Amniota* encompasses mammals, turtles, lepidosaurs, archosaurs (including birds), and all extinct forms that derive from their most recent common ancestor. This includes, among other groups: stem-diapsids such as captorhinids and "protorothyridids," the latter of which appear to be paraphyletic (Müller and Reisz, 2006); parareptiles (also called anapsids by some authors, and might also be a paraphyletic assemblage) such as procolophonoids, pareiasaurs, lanthanosuchids, bolosaurs, millerettids, and mesosaurs; and the many extinct clades of synapsids, such as caseasaurs, varanopids, ophiacodontids, edaphosaurids, "haptodontines" (paraphyletic, as shown by Laurin, 1993), sphenacodontines, dinocephalians, dicynodonts, gorgonopsians, therocephalians, and several extinct clades of cynodonts.

Diagnostic Apomorphies: Skeletal apomorphies that distinguish amniotes from their closest known extinct relatives include (Gauthier et al., 1988a,b; Laurin and Reisz, 1995; deBraga and Rieppel, 1997; Laurin, 1998a,b; Rieppel and Reisz, 1999; Vallin and Laurin, 2004): frontal bone enters orbital margin; occipital flange of squamosal gently convex; squamosal forms part of post-temporal fenestra; large tooth row on posterior edge of transverse flange of pterygoid (lost in extant and several extinct amniotes); coronoid denticles present (lost in extant and several extinct amniotes); axial neural spine extends anterodorsally; cleithrum restricted to anterior edge of scapulocoracoid (the cleithrum is lost in most extant and several extinct amniotes; Lyson et al. 2013b); presence of an astragalus formed by fusion of three elements (possibly diagnostic of amniotes and diadectomorphs). Soft tissue apomorphies that distinguish amniotes from lissamphibians include: internal fertilization; membranes of the amniotic egg, including the shell, amnion, chorion and allantois; a fairly waterproof skin comprising lipid layers held in place by keratin; horny claws; the absence of aquatic larvae;

and the absence of lateral line neuromasts at all ontogenetic stages; however, the distribution of these features is difficult or impossible to assess in most extinct vertebrates.

Synonyms: *Reptilia* Laurenti 1768 (approximate, partial); *Anapsida* Williston 1917 and *Cotylosauria* sensu Cope (1896) (both approximate, partial).

Comments: Historically, the taxon *Amniota* has been conceptualized in two main ways. Many authors emphasized reproductive mode; for example, Watson (1917: 171) stated that "The essential feature of a reptile [amniote] is that it can carry out the whole of its life-history on dry land, not producing a gill-breathing larva"; similarly, Romer (1966: 102) argued that "The major definitive character of the class Reptilia [*Amniota*] is the fact that, in contrast to amphibians, they have developed a type of egg which can be laid on land–the amniote egg." More recently, a similar point of view has been discussed by Lee and Spencer (1997). This conceptualization is more similar to an apomorphy-based definition based on a reproductive or life cycle character than to a crown-clade definition. However, *Amniota* has also been conceptualized as the crown clade originating in the most recent common ancestor of the extant amniotes (e.g., Gauthier et al. 1988b; Berman et al., 1992; Laurin and Reisz, 1995). Given that *Amniota* is the name that has most commonly been applied to that crown clade, it should be applied to the amniote crown in accordance with *PhyloCode* Recommendation 10.1B (Cantino and de Queiroz, 2020). Furthermore, defining *Amniota* based on reproductive apomorphies would generate uncertainty about the delimitation of *Amniota* in the fossil record, because life cycle and reproductive data are lacking for most extinct taxa. For example, the presence of the relevant characters could not be determined for the two clades present on the amniote stem under the phylogeny of Vallin and Laurin (2004: Fig. 6), namely *Diadectomorpha* and *Solenodonsaurus* (see below). Gauthier et al. (1988b) proposed a crown-clade definition of *Amniota*, which has the advantage that its application in a palaeontological context is fairly straightforward in that osteological characters are sufficient to apply it, provided that they resolve the taxonomic affinities of the relevant fossils. Applying the name to the crown also maximizes the veracity of statements about *Amniota* in the literature (i.e., statements about characters, including an amniotic egg, that can be inferred to have been present in the most recent common ancestor of extant amniotes, but not necessarily in stem amniotes). This is the solution adopted here.

Ever since *Amniota* was named by Haeckel (1866), it has included *Aves, Mammalia,* and (paraphyletic) *Reptilia,* among extant taxa; however, the extinct composition of *Amniota* has changed in two ways over time. Discoveries of additional extant members of these taxa, as well as members of their stem groups, resulted in the compositional expansion of this taxon (e.g., Cope, 1878). Some of the taxa that were at some point added to *Amniota,* such as seymouriamorphs and diadectomorphs (Watson, 1917), are now generally excluded. For *Seymouriamorpha,* this has been uncontroversial since larvae with external gills were discovered (Romer, 1947: 281). For *Diadectomorpha,* there has been less consensus; some studies recently suggested that diadectomorphs should be considered amniotes, either for phylogenetic reasons (Berman et al., 1992: Fig. 13), or because of inferred reproductive mode (Lee and Spencer, 1997). However, arguments based on the reproductive mode have not generally been accepted because the critical data for diadectomorphs are missing (e.g., Laurin and Reisz, 1999). The possible phylogenetic justification consisted of inferring diadectomorphs to be the sister-group of all other known synapsids (stem-mammals) (Berman et al., 1992); however, this result has not been upheld by more recent investigations (Laurin, 1998a,b; Laurin and Reisz, 1999; Vallin and Laurin, 2004).

Evidence for the monophyly of *Amniota* is strong both from gene sequences (e.g., Hugall et al., 2007; Fong et al., 2012; Lu et al., 2013) and from morphology (Gauthier et al., 1988a,b; Vallin and Laurin, 2004; Hill, 2005). By contrast, phylogenetic relationships within *Amniota* are controversial, and several alternative phylogenies are plausible. Among these, many morphological studies place turtles outside diapsids (archosaurs plus lepidosaurs), often among the "parareptiles" as sister-group of pareiasaurs (Lee, 1995: Fig. 22) or among pareiasaurs (Lee, 1996: Fig. 1), as sister-group of procolophonoids (Laurin and Reisz, 1995: Fig. 1), or of *Eunotosaurus africanus* (Lyson et al., 2010: Fig. 2; Lyson et al. 2013a,b: Fig 4). Turtles have also been grouped with captorhinids (e.g., Gauthier et al., 1988a,b), an early clade off the diapsid stem, but that hypothesis has not been supported by more recent studies. Certain morphological studies place turtles with lepidosaurs, and thus in *Diapsida* (Rieppel and Reisz, 1999: Fig 1; Hill, 2005: Fig. 5; Lyson et al., 2012: Fig. 2). Some recent studies even suggest that turtles fit within parareptiles, and that the latter fit within diapsids (Laurin and Piñeiro, 2017; Ford and Benson, 2020). Most phylogenies based on gene sequences place turtles within diapsids, but related to archosaurs (Zardoya and Meyer, 1998; Hedges and Poling, 1999; Janke et al., 2001: Fig. 1; Iwabe et al., 2005; Hugall et al., 2007: Fig. 1). Lee (2001: Fig. 2) and Frost et al. (2006: Fig. 50) are among the few recent phylogenies that consider sequence data and place turtles outside *Diapsida*. Lu et al. (2013) have recently attributed this conflict to systematic error from gene heterogeneity masked by concatenation,

and inferred gene trees in which turtles are either sister to diapsids, sister to archosaurs, or sister to lepidosaurs, but all with poor bootstrap support. The inclusion of a turtle as a specifier contributes to the compositional stability of *Amniota* because in some older phylogenies (e.g., Gaffney, 1980: Fig. 1A), turtles were inferred to be the sister-group of all other extant amniotes. This idea has not been upheld recently, but in view of the continuing controversy about turtle affinities, we have included a turtle among the internal specifiers.

The list of apomorphies of *Amniota* was obtained by cross-checking several recent papers because the recent suggestion (Marjanović and Laurin, 2019) that lissamphibians are more closely related to amniotes than are embolomeres, seymouriamorphs and gephyrostegids, which were previously thought to be stem-amniotes, changes inferences concerning the origin of characters just outside *Amniota* (Laurin, 1998a: Fig. 3; Laurin, 1998b: Fig. 1; Laurin and Reisz, 1999: Fig. 5; Vallin and Laurin, 2004: Fig. 6).

Literature Cited

Berman, D. S., S. S. Sumida, and R. E. Lombard. 1992. Reinterpretation of the temporal and occipital regions in *Diadectes* and the relationships of diadectomorphs. *J. Paleont.* 66:481–499.

deBraga, M., and O. Rieppel. 1997. Reptile phylogeny and the interrelationships of turtles. *Zool. J. Linn. Soc.* 120:281–354.

Cantino, P. D., and K. de Queiroz. 2020. *International Code of Phylogenetic Nomenclature (PhyloCode)*, Version 6. CRC Press, Boca Raton, FL.

Cope, E. D. 1878. Descriptions of extinct *Batrachia* and *Reptilia* from the Permian formation of Texas. *Proc. Am. Philos. Soc.* 17:505–530.

Cope, E. D. 1896. The reptilian order *Cotylosauria*. *Proc. Am. Philos. Soc.* 34:436–457.

Fong, J. J., J. M. Brown, M. K. Fujita, and B. Boussau. 2012. A phylogenomic approach to vertebrate phylogeny supports a turtle-archosaur affinity and a possible paraphyletic *Lissamphibia*. *PLOS ONE* 7:e48990, 1–14.

Ford, D. P., and R. B. Benson. 2020. The phylogeny of early amniotes and the affinities of *Parareptilia* and *Varanopidae*. *Nat. Ecol. Evol* 4:57–65.

Frost, D. R., T. Grant, J. Faivovich, R. H. Bain, A. Haas, C. F. B. Haddad, R. O. de Sá, A. Channing, M. Wilkinson, S. C. Donnellan, C. J. Raxworthy, J. A. Campbell, B. Blotto, P. Moler, R. C. Drewes, R. A. Nussbaum, J. D. Lynch, D. M. Green, and W. C. Wheeler. 2006. The amphibian tree of life. *Bull. Am. Mus. Nat. Hist.* 297:1–370.

Gaffney, E. S. 1980. Phylogenetic relationships of the major groups of amniotes. Pp. 593–610 in *The Terrestrial Environment and the Origin of Land Vertebrates* (A. L. Panchen, ed.). Academic Press, London.

Gauthier, J. A., A. G. Kluge, and T. Rowe. 1988a. Amniote phylogeny and the importance of fossils. *Cladistics* 4:105–209.

Gauthier, J. A., A. G. Kluge, and T. Rowe. 1988b. The early evolution of the *Amniota*. Pp. 103–155 in *The Phylogeny and Classification of the Tetrapods. Vol. 1: Amphibians, Reptiles, Birds* (M. J. Benton, ed.). Clarendon Press, Oxford.

Haeckel, E. 1866. *Generelle Morphologie der Organismen*. Reimer, Berlin.

Hedges, S. B., and L. L. Poling 1999. A molecular phylogeny of reptiles. *Science* 283:998–1001.

Hill, R. V. 2005. Integration of morphological data sets for phylogenetic analysis of *Amniota*: the importance of integumentary characters and increased taxonomic sampling. *Syst. Biol.* 54:530–547.

Hugall, A. F., R. Foster, and M. S. Y. Lee. 2007. Calibration choice, rate smoothing, and the pattern of tetrapod diversification according to the long nuclear gene RAG-1. *Syst. Biol.* 56:543–563.

Iwabe, N., Y. Hara, Y. Kumazawa, K. Shibamoto, Y. Saito, T. Miyata, and K. Katoh. 2005. Sister group relationship of turtles to the bird-crocodilian clade revealed by nuclear DNA-coded proteins. *Mol. Biol. Evol.* 22:810–813.

Janke, A., D. Erpenbeck, M. Nilsson, and U. Arnason. 2001. The mitochondrial genomes of the Iguana (*Iguana iguana*) and the caiman (*Caiman crocodylus*): implications for amniote phylogeny. *Proc. R. Soc. Lond. B Biol. Sci.* 268:623–631.

Laurenti, J. N. 1768. *Specimen Medicum, Exhibens Synopsin Reptilium Emendatam cum Experimentis Circa Venena et Antidota Reptilium Austriacorum*. PhD thesis, University, Vienna.

Laurin, M. 1993. Anatomy and relationships of *Haptodus garnettensis*, a Pennsylvanian synapsid from Kansas. *J. Vertebr. Paleontol.* 13:200–229.

Laurin, M. 1998a. The importance of global parsimony and historical bias in understanding tetrapod evolution. Part I. Systematics, middle ear evolution, and jaw suspension. *Ann. Sci. Nat., Zool., 13 Ser.* 19:1–42.

Laurin, M. 1998b. The importance of global parsimony and historical bias in understanding tetrapod evolution. Part II. Vertebral centrum, costal ventilation, and paedomorphosis. *Ann. Sci. Nat., Zool., 13 Ser.* 19:99–114.

Laurin, M., and G. Piñeiro. 2017. A Reassessment of the taxonomic position of mesosaurs, and a surprising phylogeny of early amniotes. *Front. Earth Sci.* 5:1–13.

Laurin, M., and R. R. Reisz. 1995. A reevaluation of early amniote phylogeny. *Zool. J. Linn. Soc.* 113:165–223.

Laurin, M., and R. R. Reisz. 1999. A new study of *Solenodonsaurus janenschi*, and a reconsideration of amniote origins and stegocephalian evolution. *Can. J. Earth Sci.* 36:1239–1255.

Lee, M. S. Y. 1995. Historical burden in systematics and the interrelationships of 'parareptiles'. *Biol. Rev.* 70:459–547.

Lee, M. S. Y. 1996. Correlated progression and the origin of turtles. *Nature* 379:812–815.

Lee, M. S. Y. 2001. Molecules, morphology, and the monophyly of diapsid reptiles. *Contr. Zool.* 70:1–18.

Lee, M. S. Y., and P. S. Spencer 1997. Crown-clades, key characters and taxonomic stability: When is an amniote not an amniote? Pp. 61–84 in *Amniote Origins—Completing the Transition to Land* (S. Sumida and K. Martin, eds.). Academic Press, London.

Linnaeus, C. 1758. *Systema Naturae*. 10th edition. Laurentii Salvii, Holmiae (Stockholm).

Lu, B., W. Yang, Q. Dai, and J. Fu. 2013. Using genes as characters and a parsimony analysis to explore the phylogenetic position of turtles. *PLOS ONE* 8:1–14.

Lyson, T. R., G. S. Bever, B.-A. S. Bhullar, W. G. Joyce, and J. A. Gauthier. 2010. Transitional fossils and the origin of turtles. *Biol. Lett.* 6:830–833.

Lyson, T. R., G. S. Bever, T. M. Scheyer, A. Y. Hsiang, and J. A. Gauthier. 2013a. Evolutionary origin of the turtle shell. *Curr. Biol.* 23:1113–1119.

Lyson, T. R., B. A. S. Bhullar, G. S. Bever, W. G. Joyce, K. de Queiroz, A. Abzhanov, and J. A. Gauthier. 2013b. Homology of the enigmatic nuchal bone reveals novel reorganization of the shoulder girdle in the evolution of the turtle shell. *Evol. Dev.* 15:317–325.

Lyson, T. R., E. A. Sperling, A. M. Heimberg, J. A. Gauthier, B. L. King, and K. J. Peterson. 2012. MicroRNAs support a turtle + lizard clade. *Biol. Lett.* 8:104–107.

Marjanović, D., and M. Laurin. 2019. Phylogeny of Paleozoic limbed vertebrates reassessed through revision and expansion of the largest published relevant data matrix. *PeerJ* 6:e5565.

Modesto, S. P. 1999. Observations on the structure of the Early Permian reptile *Stereosternum temidum* Cope. *Palaeont. Afr.* 35:7–19.

Müller, J., and R. R. Reisz. 2006. The phylogeny of early eureptiles: comparing parsimony and Bayesian approaches in the investigation of a basal fossil clade. *Syst. Biol.* 55:503–511.

Rieppel, O., and R. R. Reisz. 1999. The origin and early evolution of turtles. *Annu. Rev. Ecol. Syst.* 30:1–22.

Romer, A. S. 1947. Review of the *Labyrinthodontia*. *Bull. Mus. Comp. Zool.* Harv. 99:1–368.

Romer, A. S. 1966. *Vertebrate Paleontology*. 3rd edition. University of Chicago Press, Chicago, IL.

Vallin, G., and M. Laurin. 2004. Cranial morphology and affinities of *Microbrachis*, and a reappraisal of the phylogeny and lifestyle of the first amphibians. *J. Vertebr. Paleontol.* 24:56–72.

Watson, D. M. S. 1917. A sketch classification of the pre-Jurassic tetrapod vertebrates. *Proc. Zool. Soc. Lond.* 1917:167–186.

Williston, S. W. 1917. The phylogeny and classification of reptiles. *Contrib. Walker Mus.* 2:61–71.

Zardoya, R., and A. Meyer. 1998. Complete mitochondrial genome suggests diapsid affinities of turtles. *Proc. Natl. Acad. Sci. USA* 95:14226–14231.

Authors

Michel Laurin; CR2P, CNRS/MNHN/SU; Muséum National d'Histoire Naturelle; CP 48; 43 rue Buffon, 75005 Paris, France. Email: michel.laurin@mnhn.fr.

Robert R. Reisz; Department of Biology; University of Toronto Mississauga; 3359 Mississauga Rd. N. Mississauga, ON, L5L 1C6, Canada. Email: rreisz@utm.utoronto.ca.

Date Accepted: 30 March 2014

Editors: Philip Cantino, Jacques A. Gauthier, Kevin de Queiroz

SECTION 7

Pan-Mammalia T. B. Rowe 2004 [T. B. Rowe], converted clade name

Registration Number: 224

Definition: The total clade of the crown clade *Mammalia* Linnaeus 1758. This is a crown-based total-clade definition. Abbreviated definition: total ∇ of *Mammalia*.

Etymology: Derived from the Greek *pan-* (all, in reference to a total clade) and *Mammalia* (this volume).

Reference Phylogeny: The primary reference phylogeny is Figure 3 in Gauthier et al. (1988a), where *Pan-Mammalia* corresponds approximately to the clade labeled *Synapsida*. See also Gauthier et al. (1988b), where *Pan-Mammalia* corresponds approximately to the clade labeled *Mammalia*.

Composition: *Pan-Mammalia* includes all extant and extinct *Mammalia* (this volume), plus all other extinct species and clades that are closer to *Mammalia* than to *Reptilia*, according to current hypotheses (Romer and Price, 1940; Romer, 1956, 1966; Carroll, 1988, Gauthier et al., 1988a; Reisz, 1980, 1986, 2014; Laurin, 1993; Benton, 2015). These include *Varanopidae* Romer and Price 1940 (= *Varanopsidae*, *Varanopseidae*; possibly paraphyletic), *Casesauria* Williston 1912; *Ophiacodontidae* Nopsca 1923; *Edaphosauridae* Cope 1882, *Haptodontidae* Romer and Price 1940, *Sphenacodontidae* Marsh 1878, plus the extant and extinct species and clades of *Therapsida* (this volume).

A number of incomplete pan-mammalian fossils from the Early Permian of North America and China have been referred to *Therapsida*; however, they probably do not lie within *Therapsida* and warrant brief mention here. Olson and Beerbower (1953) described several of these, including *Dimacrodon, Steppesaurus, Tappenosaurus,* and *Knoxosaurus*. Olson (1962) later named *Gorgodon, Eosyodon, Driveria,* and *Mastersonia*, followed by *Watongia* (Olson, 1974). *Tetraceratops insignis* Matthew 1908 and *Raranimus dashankouensis* Liu et al. 2009 have also been considered as the "oldest therapsids", but recent analysis (Amson and Laurin, 2011) indicates that they also lie outside that clade (see *Therapsida*, this volume).

Diagnostic Apomorphies: *Pan-Mammalia*, like all total clades, need not have any apomorphies *per se* (Gauthier and de Queiroz, 2001), but as is generally true, fossils can be referred to *Pan-Mammalia* based on the possession of one or more characters hypothesized as derived within the clade (Rowe, 1988). That said, the diagnostic feature most often associated with the mammalian total clade is a lower temporal fenestra that faces laterally and is surrounded by the jugal, squamosal, and postorbital (Williston, 1925*); these bones form what is known as the 'synapsid' arch. The quadratojugal may have originally contributed to the arch, but it is excluded from the arch in most descendant taxa and is entirely absent in *Mammalia*.

At some point, we may expect to find an early pan-mammal without the lower temporal fenestra. In light of what is currently known, such a fossil might be recognizable as a pan-mammal if it preserved at least one of the following osteological

---

* Samuel W. Williston died in 1918. His classic work, *The Osteology of the Reptiles*, was posthumously arranged and edited by William King Gregory (Williston, 1925). In Williston's penultimate classification of 'reptiles' *Sauropterygia* and *Placodontia* are included under *Synapsida* (Williston, 1917: 420), and in his phylogenetic diagram, *Procolophonia, Archosauria, Parasuchia, Rhynchocephalia,* and *Sauropterygia* all descend from *Synapsida*. Hence, the credit sometimes given Williston for restricting the name *Synapsida* exclusively to taxa now thought to belong to the mammalian stem probably belongs instead to W. K. Gregory.

features: the septomaxilla has a broad base that contacts the premaxilla and maxilla and is perforated by the septomaxillary foramen. The nasal bone is longer than the frontal, and there is a single postparietal bone (the interparietal) in mature individuals. The craniomandibular joint lies posterior to a forward-sloping occiput. The supraoccipital is expanded laterally over the top of the post-temporal fenestra. The angular has a ventrally keeled edge (later modified to support the tympanum in an impedance-matching ear). There are at least three sacral vertebrae in mature individuals. Discussion of these and other potentially diagnostic characters can be found in Romer (1956, 1966), Romer and Price (1940), Reisz (1980, 1986), Kemp (1982, 1988, 2005), Brinkman and Eberth (1983), Rowe (1986), Gauthier et al. (1988a, appendix B; 1988b, appendix C), Berman et al. (1992), Laurin (1993), Laurin and Reisz (1995, 1999), and Kammerer et al. (2014).

Synonyms: *Synapsida* (sensu Rowe, 1986; de Queiroz and Gauthier, 1992; Rubidge and Sidor, 2001; Kemp, 2006) is an unambiguous synonym. Approximate synonyms include *Mammalia* Linnaeus 1758 (sensu Hennig, 1983; Ax, 1987; Loconte, 1990), *Synapsida* (sensu Gauthier et al., 1988a; Rowe, 1988; and Sidor, 2001), and *Theropsida* Goodrich 1916. *Synapsida* sensu Laurin and Reisz (1995) refers to a slightly less inclusive clade (see Comments).

Partial synonyms include *Promammalia* Haeckel 1877, *Pelycosauria* Cope 1878a, *Theromorpha* Cope 1878b, *Hypotheria* Huxley 1880, *Theromora* Cope 1888, *Sauromammalia* Baur 1887, *Anomodontia* Owen 1860a (sensu Seeley, 1889; Lydekker, 1889; Watson, 1917), *Theriodontia* Owen 1876 (sensu Haeckel, 1895), and *Synapsida* Osborn 1903a (sensu Williston, 1925; Romer, 1956, 1966; Carroll, 1988; Sidor and Hopson, 1998).

Comments: In the wake of publication of *On the Origin of Species* (Darwin, 1859), Ernest Haeckel, Thomas Huxley, and Edward Cope were among the first to understand that palaeontology had taken on an entirely new meaning as transitional fossils were discovered that connected disparate living taxa in deep time. How to classify fossils in the context of preexisting classifications of living taxa soon became problematic. As Huxley observed, "The root of the matter appears to me to be that the palaeontological facts which have come to light in the course of the last ten or fifteen years have completely broken down existing taxonomical conceptions, and that attempts to construct fresh classifications upon the old model are necessarily futile" (Huxley, 1880: 652).

Some of the first attempts are embodied in the names *Promammalia* Haeckel 1877 and *Hypotheria* Huxley 1880. Huxley arranged living mammals into a "scala mammalium", as he called it, for which he created several names that reflected stages in evolution. He divided living mammals into *Eutheria* (= *Placentalia*), *Metatheria* (= *Marsupialia*), and *Prototheria* (= *Monotremata*). He observed that "Our existing classifications have no place for the 'submammalian' stage of evolution (already indicated by Haeckel under the name *Promammale* [sic]) ... I propose to term the representatives of this stage *Hypotheria*; and I do not doubt that when we have a fuller knowledge of the terrestrial Vertebrata of the later Palaeozoic epochs, forms belonging to this stage will be found among them" (Huxley, 1880: 660). Huxley's prediction proved true, although *Promammalia* and *Hypotheria* were effectively treated as theoretical constructs, and no fossils were ever assigned to them. None was conceived as including *Mammalia*, and both names have lapsed.

By this time, however, the first fossils that would eventually be recognized as members of the mammalian stem group had already been collected from Permian sediments in the western Ural Mountains in Russia, followed quickly by discoveries from Permo-Triassic terrestrial sediments of South Africa, India, Europe, and America. Richard Owen and Huxley were among the first to study them. They both recognized mammal-like features, but in classifying them, they were more impressed with the many 'reptilian' (plesiomorphic) features that they retained, leading these authors to place them in *Reptilia*. Owen confronted the unprecedented combination of mammalian and 'reptilian' features that these fossils presented, but he would never reconcile them as indicating a relationship with *Mammalia*, rejecting as he did the Lamarckian and Darwinian hypotheses (Owen, 1876: 76).

Cope (1878a) was the first to postulate a phylogenetic connection between *Mammalia* and fossils representing the oldest known and most plesiomorphic members of what we now call the mammalian stem group. Cope's insight was all the more remarkable in that his fossils were even older and more plesiomorphic than those mentioned above, and he forged a link between specimens from North America, Europe, Russia, India, and Africa. Cope's material was collected from Carboniferous and Early Permian terrestrial sediments, and he established the name *Pelycosauria* Cope 1878a, based primarily on *Clepsydrops* and *Dimetrodon*, for nearly a dozen genera from North America and Europe. He recognized a view, still held, that *Dimetrodon* is allied to *Deuterosaurus* and *Eurosaurus* (now seen as dinocephalian therapsids) from Russia, and an even closer resemblance exists between *Deuterosaurus* and *Lycosaurus*, which is now viewed as a gorgonopsid therapsid, a member of Owen's *Theriodontia* from South Africa. Cope went on to state that "Prof Owen has named a group of Triassic and Permian reptiles the *Theriodontia*, characterized by the mammal-like differentiation of the incisor and canine teeth. The animals thus referred by Prof. Owen probably enter my suborder *Pelycosauria*" (Cope, 1878a: 529).

Cope considered *Pelycosauria* to have close affinities to *Rhynchocephalia* (*Reptilia*) in accord with the widespread view of his time that rhynchocephalians represented the least modified living reptiles. However, he quickly recognized his *Pelycosauria* and Owen's *Theriodontia* as closely related but distinct taxa, believing that they diverged from a common ancestor. He united them in *Theromorpha* Cope 1878b (later renamed *Theromora* Cope 1888), explaining that: "the order *Theromorpha* [is] regarded … as approximating the *Mammalia* more closely than any other division of *Reptilia*, and as probably the ancestral group from which the latter were derived" (Cope, 1878b: 829–830).

Cope initially used *Pelycosauria* (Cope, 1878a) and then *Theromorpha* (Cope, 1878b, 1880a) in reference solely to taxa that we now recognize as belonging to the mammalian stem group. However, in subsequent publications, Cope (1880b, 1881, 1885) and his contemporaries included within both taxa various forms that are now considered members of *Reptilia* or its stem group (e.g., *Bolosauridae*, *Pareiasauridae*, *Procolophonidae*), or the stem group of *Amniota* (e.g., *Cotylosauria*, *Diadectidae*), or that of *Lissamphibia* (*Diplocaulidae*) (see reviews by E. C. Case, 1907; Romer and Price, 1940; and Gauthier et al., 1988b, 1989). Cope maintained that *Theromorpha* was "structurally nearer to both the Batrachia and the Mammalia than any other" but added that it "present[s] characters which render it probable that all other reptiles, with possibly the exception of the Ichthyopterygia, derived their being from them" (Cope, 1885: 246–247). *Theromorpha* had become, in effect, the paraphyletic stem group from which all amniotes, not just mammals, evolved. Neither *Pelycosauria* nor *Theromorpha* (= *Theromora*) ever included *Mammalia*.

In the early decades of the twentieth century, it became evident that the various taxa included in Cope's expanded *Theromorpha* were not each other's closest relatives. Some authors (e.g., Broom, 1910, 1915, 1932; Williston, 1925) continued to use the name in a restricted sense, as an approximate synonym for Cope's (1878a) original *Pelycosauria*. Moreover, Broom had at last removed the various extraneous taxa that today are no longer considered to be members of the mammalian stem group (see *Therapsida*, this volume). But others held that its varied use rendered *Theromorpha* "invalid" (Osborn, 1903b: 453) or "ambiguous" (Romer and Price, 1940: 14), and it was discarded in favor of *Pelycosauria* at a time when naming paraphyletic ancestral groups was a widely-accepted practice (e.g., Romer, 1956, 1966, 1968; Romer and Price, 1940; Carroll, 1988).

Both *Pelycosauria* and *Theromorpha* are available for conversion as names for the total clade of *Mammalia*, but there is little merit in either case. *Theromorpha* has all but disappeared from the literature. *Pelycosauria* has generally been used for the paraphyletic early part of the mammal stem group, one long interpreted as a grade of evolution rather than as a clade (e.g., Olson, 1959, 1962, 1974; Boonstra, 1971, 1972; Chudinov, 1983, Ivakhnenko, 2002, 2003). *Pelycosauria* survives today in the vernacular, where a number of authors find utility in referring to 'pelycosaur-grade' members of the mammalian stem group (e.g., Rubidge and Sidor, 2001; Reisz, 2014; Sumida et al., 2014).

Two of Richard Owen's names, *Anomodontia* Owen 1860a and *Theriodontia* Owen 1876, are also available for conversion for the total clade of *Mammalia*, but their histories are perhaps the most complicated of all. *Anomodontia* originally included two families of dicynodonts (*Dicynodontia*, *Cryptodontia*) and the archosauromorph *Rhynchosaurus* (Owen, 1860a, b). The following year, Owen (1861) named *Cynodontia*, which he also included within *Anomodontia*. Owen later restricted *Anomodontia* to the herbivorous dicynodonts and substituted *Theriodontia* for all of the carnivorous therapsids, while abandoning *Cynodontia* (Owen, 1876).

Subsequent authors grouped heterogeneous assemblages of taxa under both names, but none included *Mammalia*. For example, Seeley (1889, 1894a, 1895), Lydekker (1889, 1890), and Watson (1917) effectively substituted *Anomodontia* for *Theromorpha* (or even *Amniota*) by including within it *Pareiasauria*, *Procolophonia*, *Dicynodontia*, *Gennetotheria*, *Pelycosauria*, *Theriodontia*, *Cotylosauria*, *Placodontia*, *Mesosauria*, and *Nothosauria*. Later, Osborn (1903a, b) included the known therapsids as well as placodonts within *Anomodontia*. Boonstra (1963; see also Watson and Romer, 1956) used *Anomodontia* in reference to the dicynodonts and herbivorous dinocephalians (a polyphyletic assemblage according to current hypotheses), and later recommended abandonment of the name altogether owing to its varied and contradictory use (Boonstra, 1972). Broom (1905) and Williston (1925) restricted its use exclusively to dicynodonts; while not unanimous, this has been the prevailing view in recent decades.

Seeley (1892) and Haeckel (1895) used (order) *Theriodontia* for a heterogeneous assemblage that included *Pelycosauria* and several taxa now considered as *Reptilia*. With these complex histories and a tradition of excluding *Mammalia*, no one has proposed either *Anomodontia* or *Theriodontia* as a name for the total clade of *Mammalia*.

Baur (1886) proposed *Sauromammalia* first as an approximate synonym of *Amniota*, and the following year as approximating the mammalian total clade (Baur, 1887: 104). In the second paper he presented a diagram that, if read as a cladogram, linked *Pelycosauria* and *Anomodontia* as sister taxa, with *Mammalia* as their sister taxon branching from a node named *Sauromammalia*. This name can be read as designating (approximately) the total clade of *Mammalia*, although it is not certain that *Mammalia* was included and it never gained popularity.

By the start of the twentieth century, a considerable diversity of fossils had been identified as belonging to the mammalian stem group.

Henry Fairfield Osborn established the "primary division of the *Reptilia* into two subclasses, *Synapsida* and *Diapsida*" (Osborn, 1903a: 275), based on the structure of the temporal arches of the skull. *Diapsida* has two temporal arches (and fenestrae), and Osborn included within it both paraphyletic stem taxa and some (but not all) crown clades of *Reptilia*, as well as *Pelycosauria*, which he believed to have the diapsid condition. Like Cope, Osborn subscribed to the notion that "rhynchocephalians" (composed at the time of what are now regarded as crown lepidosaurs, living *Sphenodon punctatus*, and stem group archosaurs called rhynchosaurs) represented the most generalized 'reptiles' and were ancestral to all other amniotes, hence he considered the presence of two arches the ancestral state.

In deriving the name *Synapsida*, Osborn translated the Greek prefix 'syn' as meaning "together" (1903b: 455). He evidently believed the synapsid arch corresponded to both of the arches present in *Diapsida*, but on a single page he referred to the synapsid arch variously as "single", "compound", and "undivided" (Osborn, 1903b: 455). Within *Synapsida*, Osborn included several paraphyletic stem groups plus the extant *Testudines* (his *Testidudinata*). Osborn also included *Anomodontia* (as a superorder), and within it he placed *Theriodontia* (comprising *Cynodontia* and *Gomphodontia*), *Dicynodontia*, and *Placodontia*. Osborn considered *Synapsida* as "giving rise to the Mammalia" (Osborn, 1903a: 276; Osborn, 1903b). He also considered Cope's *Theromorpha* as invalid, and his concepts of *Anomodontia* and *Theriodontia* were idiosyncratic. From today's perspective, this first application of the name *Synapsida*, a name that later became popular first in reference to the stem group and then to the total clade of *Mammalia*, was in considerable disarray.

Harry G. Seeley (1894a,b) and Robert Broom (1901) had by this time already articulated a modern view of the structure of the temporal arches. As Broom explained, "In the Dicynodonts, as in the Theriodonts and Mammals, there is but a single arch formed by the jugal and squamosal" and "as it is moderately certain that the reptiles with two arches have been derived from ancestral forms which had the temporal region completely roofed, by quite a different line from that by which the Dicynodonts have arisen, the single arch in the latter cannot be homologous with either of the arches in the more typical reptile" (Broom, 1901: 181–182).

Broom (1905) soon named *Therapsida* for certain Late Permian and Triassic fossils from South Africa. His purpose was to redistribute heterogeneous taxa that various authors had lumped together in *Theriodontia* and in *Anomodontia*. In doing so, he contributed significantly to current hypotheses of amniote phylogeny by excluding extraneous taxa that now are considered members of *Reptilia* or its stem group, and by restricting *Therapsida* exclusively to taxa now viewed as relatively crownward members of the mammalian stem group. For most of the twentieth century *Therapsida* designated an extinct, paraphyletic stem group, as Broom intended. Broom (1907, 1910, 1932) also reaffirmed the resemblances of *Therapsida* to Cope's original (paraphyletic) *Pelycosauria* Cope 1878a, but he maintained that they diverged from a common ancestor.

In the next decades, with the discovery of more intact fossils, Watson (1921), Williston (1925), and Gregory (e.g., 1929, 1953) emphasized the many morphological transitions that could be traced from *Mammalia* back through a sequence of fossils that dated to the Carboniferous, more than 300 million years ago. The mammalian stem was conveniently divided into three paraphyletic groups. The oldest was *Pelycosauria*, which included the most primitive Carboniferous and Early Permian taxa from North America and Europe. Before becoming extinct, it 'gave rise' to *Therapsida*, which included the Middle to Late Permian and Triassic taxa best known from South Africa and Russia. *Cynodontia* was viewed as the extinct

Triassic group within *Therapsida* from which mammals ultimately evolved (Broom, 1932; Romer and Price, 1940; Gregory, 1953; Romer, 1956, 1966, 1968; Carroll, 1988).

Edwin S. Goodrich was perhaps the first to name the total clade of *Mammalia* when he coined the name *Theropsida* Goodrich 1916. He explained that "These synapsidan reptiles and the Mammalia make up a monophyletic offshoot to which the name Theropsidan branch may be applied" (Goodrich, 1916: 264). In emphasizing an extant taxon (*Mammalia*) and its extinct relatives, *Theropsida* is virtually synonymous with *Pan-Mammalia*. However, the total-clade concept had yet to be clearly articulated, not to mention widely appreciated. *Theropsida* never gained wide acceptance, and in its place most twentieth-century authors preferred the formal taxonomic separation of crown *Mammalia* from its paraphyletic stem group(s).

Despite early confusion over the structure and evolution of the amniote temporal region and early amniote relationships, Osborn's name *Synapsida* gained wide acceptance in the following decades for a taxon that joined *Pelycosauria* with *Therapsida* (including *Cynodontia*) as the so-called 'mammal-like reptiles,' a popular term approximating the entire stem group of *Mammalia*. *Synapsida* had the advantage that it had never been used for a less-inclusive group (as had *Pelycosauria*, *Theromorpha*, and *Anomodontia*), and from the start it was used for taxa believed to be involved in the ancestry of mammals. Once *Testudines* and *Sauropterygia* were removed, *Synapsida* referred exclusively to taxa viewed (under current hypotheses) as comprising the mammalian stem group, although not without controversy regarding its 'upper' limit and the corresponding 'lower' limit of *Mammalia* (see *Mammaliamorpha*, *Mammaliaformes*, and *Mammalia*, this volume).

As the concept of monophyly became clearly understood in reference to an ancestor and all its descendants, *Synapsida* was extended to include *Mammalia*, thus approximating the mammalian total clade. However, even this held a measure of ambiguity in early monophyletic taxonomies. For example, some authors used a maximum-clade definition in which *Synapsida* comprises *Mammalia* and all extinct taxa closer to *Mammalia* than to *Reptilia*. With this definition, *Synapsida* was applied unambiguously to the total clade (Rowe, 1986, 1988; de Queiroz and Gauthier, 1992; Rubidge and Sidor, 2001). However, others used minimum-clade definitions with fossils as internal specifiers, potentially restricting *Synapsida* to a subset of the total clade of *Mammalia*. For example, Laurin and Reisz (1995: 179) defined *Synapsida* as "The last common ancestor of *Eothyris*, *Varanops* and mammals, and all its descendants." By adopting such a definition, they preclude membership of more stemward pan-mammals (should such be recognized) within *Synapsida*, and consequently that name does not designate the total clade.

Other publications using the name *Synapsida* offered no explicit definition, and its meaning can only be inferred from published cladograms, with differing results. For example, *Synapsida* was labeled at the node representing the most recent common ancestor that *Mammalia* shares with *Casea* (*Casesauria*) by Gauthier et al. (1988a: Fig. 3) and Kemp (2005: Fig. 4.1, where *Eothyris* represents *Casesauria*). In contrast, Kemp (1988: Fig.1.1) and Sidor and Hopson (1998: Fig. 2) placed *Synapsida* at the node representing the most recent common ancestor shared by *Mammalia* and *Ophiacodontia*. A third definition can be inferred in Laurin (1993: Fig. 22), in which *Synapsida* is labeled at the node representing the most recent common ancestor of therapsids (including *Mammalia*) and *Varanopidae* (= *Varanopsidae*). These differences may be subtle, but nevertheless they underscore the desirability of an unambiguous definition when, even among post-Hennigian authors who largely agree about the phylogeny, the name *Synapsida* has been associated with different ancestors.

As an alternative to *Synapsida*, Hennig (1983), Ax (1987), and Loconte (1990) proposed extending the composition of *Mammalia* to include its stem group, approximating the total clade. Whereas this suggestion has the merit of naming the total clade, those authors proposed no alternative name for the crown, a clade about which most biologists will want to communicate (Rowe and Gauthier, 1992).

Of the available names, the strongest case in terms of popularity can be made for converting *Synapsida* to denominate the total clade of *Mammalia*. However, the proposal set forth here to adopt the name *Pan-Mammalia* rather than *Synapsida* for the total clade of *Mammalia* differs in that *Pan-Mammalia* is explicitly formulated as the name of the mammalian total clade. This is accomplished using an explicit total-clade definition (*ICPN* Art. 9.10) that (implicitly) uses the internal specifiers of the name of the corresponding crown clade (*Mammalia*) as its internal specifiers (Note 9.10.2)—in this case, the extant species *Homo sapiens* (*Placentalia*), *Didelphis marsupialis* (*Marsupialia*), and *Tachyglossus aculeatus* (*Monotremata*). *Pan-Mammalia* refers to a clade that explicitly and necessarily includes *Mammalia*. Both *Pan-Mammalia* and *Synapsida* could be utilized by converting *Synapsida* either to an apomorphy-based name, or as a node-based name for a clade within *Pan-Mammalia*, consistent with its use by Gauthier et al. (1988a), Laurin (1993), Laurin and Reisz (1995), Kemp (1988, 2005), and Sidor and Hopson (1998).

Pan-Mammalia also has the pragmatic virtue of employing the panclade convention (Gauthier and de Queiroz, 2001; de Queiroz, 2007), in which the names of total clades are formed by adding the prefix *Pan-* to the names of their corresponding crowns, instead of inventing altogether new names or converting specialized names used principally by palaeontologists,

such as *Pelycosauria* or *Synapsida*. General adoption of the panclade convention promises to facilitate communication among biologists who study distantly related clades, to enhance digital indexing and retrieval of phylogenetic knowledge for the total clade, and to promote a clear distinction between crown and total clades. Additionally, the name *Mammalia* Linnaeus 1758 predates both *Pelycosauria* and *Synapsida* by more than a century. By restricting the name *Mammalia* to the crown (see *Mammalia* in this volume) and using the panclade convention to designate the total clade of *Mammalia* as *Pan-Mammalia*, this nomenclature communicates most efficiently and unambiguously the evolutionary meanings of these names to non-specialists by integrating living organisms and their extinct relatives under a unified system of names (de Queiroz, 2007).

Literature Cited

Amson, E., and M. Laurin. 2011. On the affinities of *Tetraceratops insignis*, an Early Permian synapsid. *Acta Palaeontol. Pol.* 56:301–312.

Ax, P. 1987. *The Phylogenetic System: The Systematization of Organisms on the Basis of their Phylogenesis.* John Wiley & Sons, Toronto.

Baur, G. 1886. Ueber die Kanale im Humerus der Amnioten. *Morphol. Jahrb.* 12:299–305.

Baur, G. 1887. On the phylogenetic arrangement of the *Sauropsida*. *J. Morphol.* 1:93–104.

Benton, M. 2015. *Vertebrate Palaeontology.* 4th edition. Wiley Blackwell, West Sussex.

Berman, D. S., S. S. Sumida, and R. E. Lombard. 1992. Reinterpretation of the temporal and occipital regions in *Diadectes* and the relationships of diadectomorphs. *J. Paleontol.* 66:481–499.

Boonstra, L. D. 1963. Early dichotomies in the therapsids. *S. Afr. J. Sci.* 59:176–195.

Boonstra, L. D. 1971. The early therapsids. *Ann. S. Afr. Mus.* 59:17–46.

Boonstra, L. D. 1972. Discard the names *Theriodontia* and *Anomodontia*: a new classification of the *Therapsida*. *Ann. S. Afr. Mus.* 59:315–338.

Brinkman, D., and D. A. Eberth. 1983. The interrelationships of pelycosaurs. *Mus. Comp. Zool. Breviora* 473:1–35.

Broom, R. 1901. On the structure and affinities of *Udenodon*. *Proc. Zool. Soc. Lond.* 1901:162–190.

Broom, R. 1905. On the use of the term *Anomodontia*. *Rec. Albany Mus. Grahamstown S. Afr.* 4:266–269.

Broom, R. 1907. The origins of the mammal-like reptiles. *Proc. Zool. Soc. Lond.* 1907:1047–1061.

Broom, R. 1910. A comparison of the Permian reptiles of North America with those of South Africa. *Bull. Am. Mus. Nat. Hist.* 28:197–234.

Broom, R. 1915. On the origin of mammals. *Philos. Trans. R. Soc. Lond. B Biol. Sci.* 206:1–48.

Broom, R. 1932. *The Mammal-Like Reptiles of South Africa and the Origin of Mammals*. H. F. & G. Witherby, London.

Carroll, R. L. 1988. *Vertebrate Paleontology and Evolution*. W. H. Freeman & Co., New York.

Case, E. C. 1907. *Revision of the* Pelycosauria *of North America*. Publ. Carnegie Instit., Washington, DC 55:1–176.

Chudinov, P. K. 1983. Early therapsids. Academy of Sciences of the USSR. *Trans. Palaeontol. Inst. USSR* 202:3–229 (in Russian).

Cope, E. D. 1878a. Descriptions of extinct *Batrachia* and *Reptilia* from the Permian Formation of Texas. *Proc. Am. Philos. Soc.* 17:505–530.

Cope, E. D. 1878b. The theromorphous *Reptilia*. *Am. Nat.* 12:829–830.

Cope, E. D. 1880a. Contributions to the history of *Vertebrata* of the Permian formation of Texas. *Proc. Am. Philos. Soc.* 19:38–58.

Cope, E. D. 1880b. The skull of *Empidocles*. *Am. Nat.* 14:304.

Cope, E. D. 1881. Catalogue of *Vertebrata* of the Permian Formation of the United States. *Am. Nat.* 15:162–164.

Cope, E. D. 1882. Third contribution to the history of *Vertebrata* of the Permian formation of Texas. *Proc. Am. Philos. Soc.* 20:447–461.

Cope, E. D. 1885. On the evolution of the *Vertebrata*, progressive and retrogressive. *Am. Nat.* 19:234–247.

Cope, E. D. 1888. On the shoulder-girdle and extremities of *Eryops*. *Trans. Am. Philos. Soc.* 16:362–367.

Darwin, C. 1859. *On the Origin of Species by Means of Natural Selection*. Murray, London.

Gauthier, J. A., A. G. Kluge, and T. B. Rowe. 1988a. Amniote phylogeny and the importance of fossils. *Cladistics* 4:105–209.

Gauthier, J. A., A. G. Kluge, and T. B. Rowe. 1988b. The early evolution of the *Amniota*. Pp. 103–155 in *The Phylogeny and Classification of the Tetrapods. Vol. 1: Amphibians, Reptiles and Birds* (M. Benton, ed.). Systematics Association Special Volume No. 35a. Clarendon Press, Oxford.

Gauthier, J. A., D. Cannatella, K. de Queiroz, A. G. Kluge, and T. Rowe. 1989. Tetrapod phylogeny. Pp. 337–353 in *The Hierarchy of Life [Nobel Symposium]* (B. Fernholm, H. Bremer, and H. Jornvall, eds.). Nobel Symposium 70. Excerpta Medica, Amsterdam.

Gauthier, J., and K. de Queiroz. 2001. Feathered dinosaurs, flying dinosaurs, crown dinosaurs, and the name "*Aves*". Pp. 7–41 in *New Perspectives on the Origin and Evolution of Birds* (J. Gauthier and L. F. Gail, eds.). Yale University Press, New Haven, CT.

Goodrich, E. S. 1916. On the classification of the *Reptilia*. *Proc. R. Soc. Lond. B Biol. Sci.* 89:261–276.

Gregory, W. K. 1929. *Our Face From Fish to Man*. G. P. Putnam's Sons, New York.

Gregory, W. K. 1953. *Evolution Emerging*. Macmillan, New York.

Haeckel, E. 1877. *Anthropogenie oder Entwickelungsgeschichte des Menschen*. Wilhelm Engelmann, Leipzig.

Haeckel, E. 1895. *Systematische Phylogenie der Wirbeltiere*. Verlag von Georg Reimer, Berlin.

Hennig, W. 1983. *Stammesgeschichte der Chordaten. Fortschritte in der zoologischen Systematik und Evolutionsforschung*. Verlag Paul Parey, Hamburg and Berlin.

Huxley, T. H. 1880. On the application of the laws of evolution to the arrangement of the *Vertebrata*, and more particularly of the *Mammalia*. *Proc. Zool. Soc. Lond.* 1880:649–662.

Ivakhnenko, M. F. 2002. The origin and early divergence of therapsids. *Paleontol. J.* 36:168–175.

Ivakhnenko, M. F. 2003. Eotherapsids from the east European placket (Late Permian). *Paleontol. J.* 37:S339–S465.

Kammerer, C. F., K. D. Angielczyk, and J. Fröbisch, eds. 2014. *Early Evolutionary History of the Synapsida*. Springer Dordrecht, Heidelberg.

Kemp, T. S. 1982. *Mammal-Like Reptiles and the Origin of Mammals*. Academic Press, London.

Kemp, T. S. 1988. Interrelationships of the *Synapsida*. Pp. 1–22 in *The Phylogeny and Classification of the Tetrapods. Vol. 2: Mammals* (M. J. Benton, ed.). Systematics Association Special Volume No. 35B. Clarendon Press, Oxford.

Kemp, T. S. 2005. *The Origin and Evolution of Mammals*. Oxford University Press, Oxford.

Kemp, T. S. 2006. The origin and early radiation of the therapsid mammal-like reptiles: a palaeobiological hypothesis. *J. Evol. Biol.* 19:1231–1247.

Laurin, M. 1993. Anatomy and relationships of *Haptodus garnettensis*, a Pennsylvanian synapsid from Kansas. *J. Vertebr. Paleontol.* 13:200–229.

Laurin, M., and R. R. Reisz. 1995. A reevaluation of early amniote phylogeny. *Zool. J. Linnean Soc.* 113:165–223.

Laurin, M., and R. R. Reisz. 1999. A new study of *Solenodonsaurus janenschi*, and a reconsideration of amniote origins and stegocephalian evolution. *Can. J. Earth Sci.* 36:1239–1255.

Linnaeus, C. 1758. *Systema Naturae Per Regna Tria Naturae, Secundum Classes, Ordines, Genera, Species, Cum Characteribus, Differentiis, Synonymis, Locis*. Tomus I, Editio decima, reformata. Laurentius Salvius, Holmiae (Stockholm).

Liu, J., B. Rubidge, and J. Li. 2009. New basal synapsid supports Laurasian origin for therapsids. *Acta Palaeontol. Pol.* 54:393–400.

Loconte, H. 1990. Cladistic classification of *Amniota*: a response to Gauthier et al. *Cladistics* 6:187–190.

Lydekker, R. 1889. *Palaeozoology: Vertebrata. A Manual of Palaeontology for the Use of Students with General Introduction on the Principles of Palaeontology*, Vol. 2. W. Blackwood and Sons, London.

Lydekker, R. 1890. *Catalogue of the Fossil Reptilia and Amphibia in the British Museum (Natural History)*. Part IV. Trustees of the British Museum (Natural History), London.

Marsh, O. C. 1878. Notice of new fossil reptiles. *Am. J. Sci.* 89:409–411.

Matthew, W. D. 1908. A four-horned pelycosaurian from the Permian of Texas. *Bull. Am. Mus. Nat. Hist.* 24:183–185.

Nopsca, F. 1923. Die Familien der Reptilien. *Fortschritte Geol. Paläontol.* 2:1–210.

Olson, E. C. 1959. The evolution of mammalian characters. *Evolution* 13:344–353.

Olson, E. C. 1962. Late Permian terrestrial vertebrates, USA and USSR. *Trans. Am. Philos. Soc.* 52:1–224.

Olson, E. C. 1974. On the source of therapsids. *Ann. S. Afr. Mus.* 64:27–46.

Olson, E. C., and J. R. Beerbower. 1953. The San Angelo Formation, Permian of Texas, and its vertebrates. *J. Geol.* 61:389–423.

Osborn, H. F. 1903a. On the primary divisions of the *Reptilia* into two sub-classes, *Synapsida* and *Diapsida*. *Science* 17:275–276.

Osborn, H. F. 1903b. The reptilian subclasses *Diapsida* and *Synapsida* and the early history of the *Diaptosauria*. *Mem. Am. Mus. Nat. Hist.* 1:451–507.

Owen, R. 1860a. On the orders of fossil and recent *Reptilia* and their distribution in time. *Rep. Br. Assoc. Adv. Sci.* 29(for 1859):153–166.

Owen, R. 1860b. *Paleontology or a Systematic Summary of Extinct Animals and their Geological Relations*. 1st edition. Adams and Charles Black, Edinburgh.

Owen, R. 1861. *Paleontology or a Systematic Summary of Extinct Animals and their Geological Relations*. 2nd edition. Adams and Charles Black, Edinburgh.

Owen, R. 1876. *Descriptive and Illustrated Catalogue of the Fossil Reptilia of South Africa in the Collection of the British Museum*. Trustees of the British Museum, London.

de Queiroz, K. 2007. Toward an integrated system of clade names. *Syst. Biol.* 56:956–974.

de Queiroz, K., and J. A. Gauthier. 1992. Phylogenetic taxonomy. *Annu. Rev. Ecol. Syst.* 1992:449–480.

Reisz, R. R. 1980. The *Pelycosauria*: a review of phylogenetic relationships. Pp. 553–591 in *The Terrestrial Environment and the Origin of Land Vertebrates* (A. L. Panchen and M. Benton, eds.). Systematics Association Special Volume No. 15. Academic Press, London and New York.

Reisz, R. R. 1986. *Pelycosauria. Encyclopedia of Paleoherpetology, 17A.* Gustav Fischer Verlag, Stuttgart.

Reisz, R. R. 2014. 'Pelycosaur'-grade synapsids: introduction. Pp. 3–5 in *Early Evolutionary History of the Synapsida* (C. F. Kammerer, K. D. Angielczyk, and J. Fröbisch, eds.). Springer Dordrecht, Heidelberg.

Romer, A. S. 1956. *Osteology of the Reptilia.* University of Chicago Press, Chicago, IL.

Romer, A. S. 1966. *Vertebrate Paleontology.* University of Chicago Press, Chicago, IL.

Romer, A. S. 1968. *Notes and Comments on Vertebrate Paleontology.* University of Chicago Press, Chicago, IL.

Romer, A. S., and L. W. Price. 1940. Review of the *Pelycosauria. Geol. Soc. Am. Spec. Pap.* 28:1–534.

Rowe, T. B. 1986. Osteological diagnosis of *Mammalia*, L. 1758, and its relationships to extinct *Synapsida*. PhD dissertation, University of California, Berkeley, CA.

Rowe, T. B. 1988. Definition, diagnosis and origin of *Mammalia. J. Vertebr. Paleontol.* 8:241–264.

Rowe, T. B. 2004. Chordate phylogeny and development. Pp. 384–409 in *Assembling the Tree of Life* (J. Cracraft and M. J. Donoghue, eds.). Oxford University Press, Oxford and New York.

Rowe, T. B., and J. A. Gauthier. 1992. Ancestry, paleontology, and definition of the name *Mammalia. Syst. Biol.* 41:372–378.

Rubidge, B. S., and C. A. Sidor. 2001. Evolutionary patterns among Permo-Triassic therapsids. *Annu. Rev. Ecol. Syst.* 32:449–480.

Seeley, H. G. 1889. Researches on the structure, organization, and classification of the fossil *Reptilia*. Part IV. On the anomodont *Reptilia* and their allies. *Philos. Trans. R. Soc. Lond. B Biol. Sci.* 180:215–296.

Seeley, H. G. 1892. Researches on the structure, organization, and classification of the fossil *Reptilia*. Part VI. Further observations of *Pareiasaurus. Philos. Trans. R. Soc. Lond. B Biol. Sci.* 183:311–370.

Seeley, H. G. 1894a. Researches on the structure, organization, and classification of the fossil *Reptilia*. Part IX, Section 1. On the *Therosuchia. Proc. R. Soc. Lond. B Biol. Sci.* 55:224–226.

Seeley, H. G. 1894b. Researches on the structure, organization, and classification of the fossil *Reptilia*. Part IX, Section 5. On the skeleton in new *Cynodontia. Proc. R. Soc. Lond. B Biol. Sci.* 56:291–294.

Seeley, H. G. 1895. Researches on the structure, organization and classification of the fossil Reptilia. IX, Section 1. On the *Therosuchia. Philos. Trans. R. Soc. Lond. B Biol. Sci.* 185:987–1018.

Sidor, C. A. 2001. Simplification as a trend in synapsid cranial evolution. *Evolution* 55:1419–1442.

Sidor, C. A., and J. A. Hopson. 1998. Ghost lineages and 'mammalness': assessing the temporal pattern of character acquisition in the *Synapsida. Paleobiology* 24:254–273.

Sumida, S. S., V. Pelletier, and D. S. Berman. 2014. New information on the basal pelycosaurian-grade synapsid *Oedaleops*. Pp. 6–23 in *Early Evolutionary History of the Synapsida* (C. F. Kammerer, K. D. Angielczyk, and J. Fröbisch, eds.). Springer Dordrecht, Heidelberg.

Watson, D. M. S. 1917. A sketch classification of the pre-Jurassic tetrapod vertebrates. *Proc. Zool. Soc. Lond.* 1917:167–186.

Watson, D. M. S. 1921. The bases of classification of the *Theriodontia. Proc. Zool. Soc. Lond.* 1921:35–98.

Watson, D. M. S., and A. S. Romer. 1956. A classification of therapsid reptiles. *Bull. Mus. Comp. Zool.* 114:37–89.

Williston, S. W. 1912. Primitive reptiles—a review. *J. Morphol.* 23:637–666.

Williston, S. W. 1917. The phylogeny and classification of reptiles. *J. Geol.* 25:411–421.

Williston, S. W. 1925. *The Osteology of the Reptiles* (Arranged and edited by William King Gregory). Harvard University Press, Cambridge, MA.

Author

Timothy B. Rowe; Jackson School of Geosciences; The University of Texas at Austin; C1100, Austin, TX 78712. Email rowe@utexas.edu.

Date Accepted: 12 June 2017

Primary Editors: Jacques A. Gauthier, Kevin de Queiroz

Synapsida H. F. Osborn 1903 [M. Laurin and R. R. Reisz], converted clade name

Registration Number: 290

Definition: The largest clade that includes *Cynognathus crateronotus* Seeley 1895 but not *Testudo graeca* Linnaeus 1758, *Iguana iguana* (Linnaeus 1758), and *Crocodylus niloticus* Laurenti 1768. This is a maximum-clade definition. Abbreviated definition: max ∇ (*Cynognathus crateronotus* Seeley 1895 ~ *Testudo graeca* Linnaeus 1758 & *Iguana iguana* (Linnaeus 1758) & *Crocodylus niloticus* Laurenti 1768).

Etymology: Derived from the Greek *syn-* (together) and *apsis* (arch).

Reference Phylogeny: The main reference phylogeny, at least for the Palaeozoic and early Mesozoic members of *Synapsida*, is Sidor (2001: Fig. 3), which is a supertree built from many sources in the primary literature (Sigogneau, 1970; Clark and Hopson, 1985; Sues, 1985; Hopson and Barghusen, 1986; King, 1988; Rubidge, 1991, 1994; Reisz et al., 1992; Laurin, 1993; van den Heever, 1994; Hopson, 1994; Luo, 1994, Berman et al., 1995; Modesto, 1995; Laurin and Reisz, 1996; Rubidge and van den Heever, 1997; Reisz et al., 1998; Modesto et al., 1999, 2001; Rubidge and Sidor, 2001). None of the external specifiers are included in the reference phylogeny, but they are clearly outside *Synapsida* as defined above, as shown by amniote phylogenies (Gauthier et al., 1988; Laurin and Reisz, 1995).

Composition: *Synapsida* includes many extinct clades, including caseasaurs, possibly varanopids, ophiacodontids, edaphosaurids, "haptodontines" (paraphyletic, as shown by Laurin, 1993), sphenacodontines, dinocephalians, dicynodonts, gorgonopsians, therocephalians, and several extinct clades of cynodonts in addition to extinct and extant mammals.

Diagnostic Apomorphies: Skeletal synapomorphies of the most inclusive node for currently known taxa within *Synapsida* include (Reisz, 1986): a lower (inferior) temporal fenestra bordered by the jugal, squamosal and postorbital; a broad, posterodorsally facing occipital plate formed by the supraoccipital and the paroccipital process of the opisthotic; posttemporal fenestra bounded above and below by the lateral flange of the supraoccipital and the paroccipital process of the opisthotic; and a septomaxilla composed of a large dorsal process and a broad base that straddles on the premaxilla and the maxilla.

Synonyms: *Pan-Mammalia* Gauthier et al. 2004 (unambiguous); *Theropsida* Goodrich 1916 (unambiguous); *Mammalia* Linnaeus 1758 (approximate); *Pelycosauria* Cope 1878 (approximate, partial).

Comments: Earlier in its history, the composition of *Synapsida* was variable. Osborn (1903: 456) included in *Synapsida* several clades that are now considered reptiles, such as turtles, sauropterygians and cotylosaurs (a group that then included various anapsid amniotes such as pareiasaurs and procolophonids, as well as diadectomorphs), and excluded all the then known Permo-Carboniferous taxa that are now considered synapsids. Indeed, *Pelycosauria* was then part of *Diapsida* (Osborn, 1903: 456). The composition of *Synapsida* changed slowly. Williston (1917) included *Pelycosauria* and excluded *Cotylosauria* and turtles from *Synapsida*. Sauropterygians were removed from *Synapsida* slightly later (Broom, 1924; Goodrich, 1930), and from then on, *Synapsida* included only extinct relatives of mammals (e.g., Romer, 1966). In the last few decades, the composition of *Synapsida* has been nearly invariant. A recent challenge to established views is the suggestion that diadectomorphs are synapsids (Berman et al., 1992). That suggestion, however, has not been upheld by more recent investigations (Laurin, 1998a,b; Laurin and Reisz, 1999; Vallin and Laurin, 2004). A more recent proposed change in composition is the suggestion that varanopids may be early reptiles (Laurin and Piñeiro, 2018; MacDougall et al., 2018; Ford and Benson, 2020).

In most cases, definitions that explicitly designate total clades use extant specifiers. Thus, the first proposed phylogenetic definition of *Synapsida* was "mammals and their extinct allies" (Gauthier et al., 1989). A similar definition was represented by the taxonomy given by de Queiroz and Gauthier (1992). Because *Synapsida* traditionally included only extinct taxa (the paraphyletic *Pelycosauria* and *Therapsida*), it is preferable to use extinct taxa as internal specifiers. Laurin and Reisz (1995) proposed a node-based (minimum clade) definition that incorporated extinct and extant specifiers ("The last common ancestor of *Eothyris*, *Varanops* and mammals, and all its descendants"). The internal specifier proposed here is a species that was included in *Synapsida* when the taxon was erected (Osborn, 1903: 461). For the external specifiers, we have used extant reptiles because for a long time, *Synapsida* has included extinct forms thought to be more closely related to mammals than to extant reptiles. This also insures stability in content. The proposed definition designates the total clade that includes mammals and all extinct taxa that are more closely related to mammals than to extant reptiles under all phylogenies that have been published in the last 30 years.

We preferred the name *Synapsida* instead of *Pan-Mammalia* because *Synapsida* has been extensively used in the literature; 33 papers in the library of the first author (M. L.) include "*Synapsida*" in their title, and a Google Scholar search reveals about 857 web pages (search conducted on May 1, 2008); these numbers are 57 papers and 888 web pages, respectively, for the vernacular equivalent "synapsid". Our search suggests that this name has been most frequently used after 1980, and defined explicitly with respect to a clade since 1989 (Gauthier et al., 1989), so it has not fallen into disuse. The alternative *Theropsida* has been available for more than 90 years, but it is rarely used, and in this respect is not a viable alternative. No papers in the library of M. L. had any variant of this word in its title, and a GoogleScholar search yielded only 52 pages for "*Theropsida*" and only 47 for "theropsid". Not surprisingly, because the name "*Pan-Mammalia*" is only a few years old, it occurs relatively rarely in the literature (no title in M. L.'s library; 7 hits in Google Scholar; or no hits for the vernacular "panmammal").

Mammalia has occasionally been applied to the clade here called *Synapsida*. However, this usage has been limited to a few studies (e.g., Ax, 1987; Loconte, 1990), and *Mammalia* is usually applied to a much less inclusive clade, which coincides with the crown of synapsids, or a slightly more inclusive clade (Rowe and Gauthier, 1992).

Similarly, *Pelycosauria* has been applied to a paraphyletic group which has included all Permo-Carboniferous synapsids for most of its history. However, because *Pelycosauria* has always excluded therapsids and mammals, and because it has always included taxa derived from the same ancestor as *Synapsida*, it seems better not to use *Pelycosauria* for any clade.

Literature Cited

Ax, P. 1987. *The Phylogenetic System. The Systematization of Organisms on the Basis of their Phylogenesis.* 1st edition. John Wiley & Sons, Toronto.

Berman, D. S., R. R. Reisz, J. R. Bolt, and D. Scott. 1995. The cranial anatomy and relationships of the synapsid *Varanosaurus* (*Eupelycosauria*: *Ophiacodontidae*) from the Early Permian of Texas and Oklahoma. *Ann. Carnegie Mus. Nat. Hist.* 64:99–133.

Berman, D. S., S. S. Sumida, and R. E. Lombard. 1992. Reinterpretation of the temporal and occipital regions in *Diadectes* and the relationships of diadectomorphs. *J. Paleontol.* 66:481–499.

Broom, R. 1924. On the classification of the reptiles. *Bull. Am. Mus. Nat. Hist.* 51:39–65.

Clark, J. M., and J. A. Hopson. 1985. Distinctive mammal-like reptile from Mexico and its bearing on the phylogeny of the *Tritylodontidae*. *Nature* 315:398–400.

Cope, E. D. 1878. A new fauna. *Am. Nat.* 12:327–328.

Ford, D. P., and R. B. Benson. 2020. The phylogeny of early amniotes and the affinities of *Parareptilia* and *Varanopidae*. *Nat. Ecol. Evol.* 4:57–65.

Gauthier, J., D. C. Cannatella, K. de Queiroz, A. G. Kluge, and T. Rowe. 1989. Tetrapod phylogeny. Pp 337–353 in *The Hierarchy of Life* (B. Fernholm, K. Bremer, and H. Jornvall, eds.). Elsevier Science Publishers B. V. (Biomedical Division), New York.

Gauthier, J., A. G. Kluge, and T. Rowe. 1988. Amniote phylogeny and the importance of fossils. *Cladistics* 4:105–209.

Gauthier, J. A., K. de Queiroz, W. G. Joyce, J. F. Parham, T. Rowe, and J. Clarke. July 2004. Phylogenetic nomenclature for the major clades of *Amniota* Haeckel 1866, with emphasis on non-avian *Reptilia* Laurentus 1768. Abstract in *First International Phylogenetic Nomenclature Meeting* (M. Laurin, ed.). Paris, France.

Goodrich, E. S. 1916. On the classification of the *Reptilia*. *Proc. R. Soc. Lond. B Biol. Sci.* 89:261–276.

Goodrich, E. S. 1930. *Studies on the Structure and Development of Vertebrates*. Macmillan, London.

Hopson, J. A. 1994. Synapsid evolution and the radiation of non-eutherian mammals. Pp. 190–219 in *Major Features of Vertebrate Evolution* (D. B. Prothero and R. M. Schoch, eds.). Paleontological Society, Knoxville, TN.

Hopson, J. A., and H. R. Barghusen. 1986. An analysis of therapsid relationships. Pp. 83–106 in *The Ecology and Biology of Mammal-Like Reptiles* (N. Hotton III, P. D. Maclean, J. J. Roth, and E. C. Roth, eds.). Smithsonian Institution Press, Washington, DC.

King, G. M. 1988. Anomodontia. Encyclopedia of Paleoherpetology, Vol. 17C. Gustav Fischer, Stuttgart and New York.

Laurenti, J. N. 1768. *Specimen Medicum, Exhibens Synopsin Reptilium Emendatum cum Experimentis Circa Venena et Antidota Reptilium Austriacorum*. Wien.

Laurin, M. 1993. Anatomy and relationships of *Haptodus garnettensis*, a Pennsylvanian synapsid from Kansas. *J. Vertebr. Palaeontol.* 13:200–229.

Laurin, M. 1998a. The importance of global parsimony and historical bias in understanding tetrapod evolution. Part I. Systematics, middle ear evolution, and jaw suspension. *Ann. Sci. Nat. Zool. 13 Ser.* 19:1–42.

Laurin, M. 1998b. The importance of global parsimony and historical bias in understanding tetrapod evolution. Part II. Vertebral centrum, costal ventilation, and paedomorphosis. *Ann. Sci. Nat. Zool. 13 Ser.* 19:99–114.

Laurin, M., and G. Piñeiro. 2018. Response: Commentary: a reassessment of the taxonomic position of mesosaurs, and a surprising phylogeny of early amniotes. *Front. Earth Sci.* 6:220.

Laurin, M., and R. R. Reisz. 1995. A reevaluation of early amniote phylogeny. *Zool. J. Linn. Soc.* 113:165–223.

Laurin, M., and R. R. Reisz. 1996. The osteology and relationships of *Tetraceratops insignis*, the oldest known therapsid. *J. Vertebr. Paleontol.* 16:95–102.

Laurin, M., and R. R. Reisz. 1999. A new study of *Solenodonsaurus janenschi*, and a reconsideration of amniote origins and stegocephalian evolution. *Can. J. Earth Sci.* 36:1239–1255.

Linnaeus, C. 1758. *Systema Naturae*. 10th edition. Laurentius Salvius, Holmiae (Stockholm).

Loconte, H. 1990. Cladistic classification of *Amniota*: a response to Gauthier et al. *Cladistics* 6:187–190.

Luo, Z. 1994. Sister-group relationships of mammals and transformations of diagnostic mammalian characters. Pp. 98–128 in *In the Shadow of the Dinosaurs* (N. C. Fraser and H.-D. Sues, eds.). Cambridge University Press, Cambridge, UK.

MacDougall, M. J., S.P. Modesto, N. Brocklehurst, A. Verrière, R. R. Reisz, and J. Fröbisch. 2018. Response: a reassessment of the taxonomic position of mesosaurs, and a surprising phylogeny of early amniotes. *Front. Earth Sci.* 6:99.

Modesto, S. P. 1995. The skull of the herbivorous synapsid *Edaphosaurus boanerges* from the Lower Permian of Texas. *Palaeontology* 38:213–239.

Modesto, S., B. Rubidge, and J. Welman. 1999. The most basal anomodont therapsid and the primacy of Gondwana in the evolution of the anomodonts. *Proc. R. Soc. Lond. B Biol. Sci.* 266:331–337.

Modesto, S., C. A. Sidor, B. S. Rubidge, and J. Welman. 2001. A second varanopseid skull from the Upper Permain of South Africa: implications for Late Permian 'pelycosaur' evolution. *Lethaia* 34:249–259.

de Queiroz, K., and J. Gauthier. 1992. Phylogenetic taxonomy. *Annu. Rev. Ecol. Syst.* 23:449–480.

Osborn, H. F. 1903. The reptilian subclasses *Diapsida* and *Synapsida* and the early history of the *Diaptosauria. Mem. Am. Mus. Nat. Hist.* 1:265–270.

Reisz, R. R. 1986. Pelycosauria. Encyclopedia of paleoherpetology, Vol. 17A. (P. Wellnhofer, ed.). Gustav Fischer Verlag, Stuttgart.

Reisz, R. R., D. S. Berman, and D. Scott. 1992. The cranial anatomy and relationships of *Secodontosaurus*, an unusual mammal-like reptile (*Synapsida*: Sphenacodontidae) from the early Permian of Texas. *Zool. J. Linn. Soc.* 104:127–184.

Reisz, R. R., D. W. Dilkes, and D. S. Berman. 1998. Anatomy and relationships of *Elliotsmithia longiceps* Broom, a small synapsid (*Eupelycosauria*: *Varanopseidae*) from the Late Permian of South Africa. *J. Vertebr. Paleontol.* 18:602–611.

Romer, A. S. 1966. *Vertebrate Paleontology*. 3rd edition. University of Chicago Press, Chicago, IL.

Rowe, T., and J. Gauthier. 1992. Ancestry, paleontology, and definition of the name *Mammalia. Syst. Biol.* 41:372–378.

Rubidge, B. S. 1991. A new primitive dinocephalian mammal-like reptile from the Permian of Southern Africa. *Palaeontology* 34:547–559.

Rubidge, B. S. 1994. *Australosyodon*, the first primitive anteosaurid dinocephalian from the Upper Permian of Gondwana. *Palaeontology* 37:579–594.

Rubidge, B. S., and C. A. Sidor. 2001. Evolutionary patterns among Permo-Triassic therapsids. *Annu. Rev. Ecol. Syst.* 32:449–480.

Rubidge, B. S., and J. A. Van den Heever. 1997. Morphology and systematic position of the dinocephalian *Styracocephalus platyrhynchus. Lethaia* 30:157–168.

Seeley, H. G. 1895. Researches on the structure, organization, and classification of the fossil *Reptilia*. Part IX, Section 5. On the skeleton in new *Cynodontia* from the Karroo rocks. *Philos. Trans. R. Soc. Lond. B Biol. Sci.* 186:59–148.

Sidor, C. A. 2001. Simplification as a trend in synapsid cranial evolution. *Evolution* 55:1419–1442.

Sigogneau, D. 1970. Révision systématique des gorgonopsiens Sud-Africains. *Cah. Paléontol.*: 1–416.

Sues, H.-D. 1985. The relationships of the *Tritylodontidae* (Synapsida). *Zool. J. Linn. Soc.* 5:205–217.

Vallin, G., and M. Laurin. 2004. Cranial morphology and affinities of *Microbrachis*, and a reappraisal of the phylogeny and lifestyle of the first amphibians. *J. Vertebr. Paleontol.* 24:56–72.

Van den Heever, J. A. 1994. The cranial anatomy of the early *Therocephalia* (Amniota: Therapsida). *Ann. Univ. Stellenbosch* 1991:1–59.

Williston, S. W. 1917. The phylogeny and classification of reptiles. *J. Geol.* 2:61–71.

Authors

Michel Laurin; CR2P, CNRS/MNHN/SU; Muséum National d'Histoire Naturelle; CP 48; 43 rue Buffon, 75005 Paris, France. Email: michel.laurin@ccr.jussieu.fr.

Robert R. Reisz; Department of Biology; University of Toronto at Mississauga; 3359 Mississauga Rd. N. Mississauga, ON, L5L 1C6, Canada. Email: rreisz@utm.utoronto.ca.

Date Accepted: 5 October 2020

Primary editor: Jacques A. Gauthier

Therapsida R. Broom 1905 [T. B. Rowe], converted clade name

Registration Number: 226

Definition: The smallest clade containing *Thrinaxodon liorhinus* Seeley 1894a (*Cynodontia*, this volume) and *Biarmosuchus tener* Chudinov 1960 (*Biarmosuchia*). This is a minimum-clade definition. Abbreviated definition: min ∇ (*Thrinaxodon liorhinus* Seeley 1894a & *Biarmosuchus tener* Chudinov 1960).

Etymology: Broom never explained the etymology of *Therapsida*. Apparently, it is derived from the Greek *therios* (wild beast) and the Latin *apsis* (arch), in reference to the zygomatic arch composed of jugal and squamosal bones that encloses the single temporal fenestra, as in *Mammalia*.

Reference Phylogeny: Figure 3 in Gauthier et al. (1988) is the primary reference phylogeny.

Composition: *Therapsida* includes all extant and extinct *Mammalia* (this volume) plus a subset of extinct species and clades that are viewed under current hypotheses as being closer to mammals than are 'pelycosaur-grade' pan-mammals (Kuhn, 1965; Romer, 1966; Gauthier et al., 1988; Kemp, 2005, 2006). These include: *Biarmosuchidae* Olson 1962 (= *Biarmosuchia*, includes *Phthinosuchia*, *Phthinosauridae*, *Hipposauridae*); *Dinocephalia* Seeley 1895a (includes *Brithopodidae*, *Deuterosauria*, *Eotitanosuchia*, *Titanosuchia*); *Gorgonopsia* Seeley 1894a,b; *Dicynodontia* Owen 1860a (= *Betatherapsida* Boonstra 1972, = *Anomodontia* [sensu Broom 1905, 1932]; = *Therochelonia* Seeley 1894a,b; includes *Dromasauria* Broom 1907a); *Therocephalia* Broom 1903a (sensu Huttenlocker, 2009); plus the extant and extinct species of *Cynodontia* Owen 1861 (this volume).

Therapsida unquestionably includes *Bauridae* Broom 1911 (= *Bauriidae*, = *Bauriamorpha*, = *Bauriasauria*; see Huttenlocker, 2009: Fig. 3), but whether this taxon lies within *Cynodontia* (this volume) is unresolved.

Diagnostic Apomorphies: Phylogenetic analyses differ somewhat in details, but they agree broadly that characters of the skull related to enhanced macro-predation and characters in the postcranial skeleton related to more agile locomotion separate the earliest members of *Therapsida* from 'pelycosaur-grade' pan-mammals (Hopson and Barghusen, 1986; Rowe 1986, 1988; Rowe and van den Heever, 1986; Gauthier et al., 1988; Laurin, 1993; Sidor and Hopson, 1998; Rubidge and Sidor, 2001; Sidor, 2003; Sidor and Welman, 2003; Kemp, 2005, 2006).

In the face, the internasal process of the premaxilla is elongated and extends between the nasals to a level behind the nares (loss of this process is apomorphic of *Mammalia*). From the septomaxilla, a projecting facial process forms the rear margin of the naris as it separates the maxilla and nasal (septomaxilla loss occurs within *Mammalia*). The choanae are elongated posteriorly, possibly indicating expansion of the nasal capsule and enhanced olfaction.

The upper canine is exceedingly long and robust (in multiple reversals, the canine, and in some cases the entire dentition, fails to develop in various therapsid subclades). To accommodate the canine root, the maxilla increases in height to meet the prefrontal, thus eliminating contact between the nasal and lacrimal. The long, trenchant canine also separates specialized anterior incisors from posterior sharp, unicuspid, compressed, recurved postcanine teeth.

The palate is remodeled by losing its mobile ball-in-socket joint with the braincase (the basipterygoid articulation), as the pterygoids become closely joined against the sphenoid in a non-kinetic articulation. The transverse process of the pterygoid shifts forward to a level beneath the orbit and is oriented vertically; it resists rotation of the mandible when the jaws are open (this process is reduced or lost in *Mammalia* in association with complex new chewing and swallowing behaviors). The palatal processes of pterygoids meet on the midline in front of the transverse processes, and the interpterygoid vacuity is shifted posteriorly and diminishes in size or closes altogether during ontogeny. The craniomandibular joint is placed forward, lying approximately level with the fenestra vestibuli and occipital condyle. A reduced quadrate and quadratojugal are supported by a broad descending flange of the squamosal. Just in front of the craniomandibular joint, the temporal arch is emarginated from below, possibly signaling differentiation and migration of a small part of the *M. massetericus* from within the temporal fossa onto the ventral edge of the zygomatic arch.

Two additional transformations are related to the evolution of an impedance-matching middle ear: the reflected lamina of the angular (which separates from the mandible to become the ectotympanic bone in *Mammalia*) is deeply incised along its dorsal margin; and the stapes no longer contacts the paroccipital process of the opisthotic.

In the vertebral column, pleurocentra and intercentra of the trunk coosify, the notochordal canal through the vertebral centra ossifies and closes in mature individuals, and the dorsal ribs articulate entirely with the pleurocentrum. There is a single (fused), ossified sternum (which later becomes segmented at inter-costal junctions into a series of sternebrae; see *Mammaliamorpha*, this volume).

The glenoid (shoulder socket) is formed predominantly by the scapula and posterior coracoid, while the contribution by the anterior coracoid is diminished or lost altogether. The glenoid faces posteroventrally (instead of laterally); only the head of the humerus, which is now dorsally inflected and smoothly convex (instead of strap-like), articulates with the glenoid. At the elbow, the radius and ulna have separate, distinct, articular surfaces with the humerus. The manual intermedium is much smaller than manual centrale I, and the manual phalangeal count is reduced to 2-3-4-4-3, or further.

In the hindlimb, the iliac blade is expanded upwards and forwards; the acetabulum is circular with an expanded supra-acetabular buttress. The femur has a smoothly convex, rounded head that is inflected medially, and the fourth trochanter is lost as a distinct crest in mature individuals. The tibia articulates against the proximal (posterior) face of the astragalus, while the fibula articulates with its dorsal surface. In the ankle, there is a small tuberosity ('heel') on the calcaneum; the astragalus broadly overlaps the calcaneum; a single navicular bone ossifies in the place of separate medial and lateral centralia, and the pedal phalangeal formula is reduced to 2-3-4-4-3, or further. Collectively, these changes mark an early, incremental shift in the direction of parasagittal gait, as the hands turn forward and the elbows turns backward while the knees and feet turn forward.

Synonyms: Partial synonyms include *Anomodontia* Owen 1860a (sensu Seeley, 1889; Osborn, 1903a,b) and *Therosuchia* Seeley 1894a.

Mammalia Linnaeus 1758 (sensu Van Valen [1960]) is an approximate synonym.

Comments: A brief review of the history of palaeontological discoveries and their classification may help to underscore subtle changes in the evolution of taxonomic concepts that have led to the current recommendation to convert *Therapsida* as a node-based name that includes crown *Mammalia* plus a subset of its stem group specified by the most recent common ancestor shared with *Biarmosuchus tener* Chudinov 1960 from the early Late Permian.

The first fossils that would eventually be recognized as early therapsids were all based on fragmentary specimens collected during mining operations in the Permian of the western Ural Mountains in Russia. These include *Brithopus priscus* Kutorga 1838, *Syodon biamricum* Kutorga 1838, and *Rhopalodon wangenheimi* Fischer 1841. Based on only a broken humerus, Kutorga regarded *Brithopus* to be a mammal, but it was so fragmentary that opinion varied widely as to its identity (Seeley, 1889: 279). At the time, these fossils were all classified as "saurian" (~scaly reptiles with four sprawling legs and long tails). Casts of some of these specimens quickly found their way into the hands of Richard Owen, who tentatively identified them as "belonging to the Crocodilian division of Sauria" based on their 'thecodont' dentition, in which the roots are implanted into sockets (quoted in Murchison et al., 1845: 637; see also Eichwald, 1860).

An ensuing flood of discoveries from Permo-Triassic rocks in South Africa and India fell into the hands of Richard Owen, Thomas Huxley, and Edward Cope. The first of these that Owen described was a weathered, isolated skull of a highly derived herbivore, one that had lost all teeth save a pair of tusk-like canines and that had a beak like a turtle. Its name, *Dicynodon* Owen 1844, acknowledged a unique resemblance to mammals, which Owen later noted when explaining its etymology as combining the Greek prefix *dyo* for two, with *kuvikos*, "a term applied by Hippocrates to the canine teeth, and expressing the same idea as their common English denomination. The two teeth, which are so largely and exclusively developed in the present most extraordinary Reptilia of ancient Africa, answer to the canines of Mammalia" (Owen, 1845: 61). Elaborating on this he explained, "The most remarkable character in these fossils is the presence of two long curved tusks, which, like those of the Walrus, descend one from each superior maxillary bone, and pass on the outside of the fore part of the lower jaw, a character rare even in Mammals, and hitherto only met in that class; but in these specimens combined with a structure of the cranium, proving that the animals belonged to the class Reptilia, but were members neither of the Crocodilian nor Chelonian orders" (Owen, 1844: 500). "The principal difference in the microscopic texture of the tusks of the Dicynodon, as compared with the teeth of the crocodile, consists in the closer and more compact arrangement of the calcigerous tubes of the dentine; by which character it makes a closer approach to the intimate texture of that tissue in the canine teeth of the carnivorous Mammalia ... The Dicynodons not only manifest the higher type of free implantation of the base of the tooth in a deep and complete socket, common to Crocodilians, Megalosaurs, and Thecodonts, but make an additional and much more important step towards the Mammalian type of dentition by maintaining the serviceable state of the tusk by virtue of constant renovation of the substance of one and the same matrix, according to the principle manifested in the long-lived and ever-growing tusks and scalpriform incisors of the Mammalia" (Owen, 1844: 502–503).

Owen soon classified the mounting numbers of dicynodont species between two "families" (*Dicynodontia* and *Cryptodontia*). To contain these families plus *Rhynchosaurus* (now considered a stem archosaur), he created the name *Anomodontia* Owen 1860a for an order within *Reptilia* (Owen 1860a, b). The following year, he created the name *Cynodontia* Owen 1861 for two carnivorous taxa. Owen later restricted (order) *Anomodontia* to the herbivorous dicynodonts, he abandoned *Cynodontia* without explanation, and he created the new name (order) *Theriodontia* Owen 1876 for the carnivorous taxa, including his former cynodonts and the extraneous *Procolophon trigoniceps* (currently believed to be a member of *Reptilia* or its stem group). Owen classified *Anomodontia* and *Theriodontia* as separate orders under *Reptilia*. He also classified *Tapinocephalus atherstonei* Owen 1876 within *Dinosauria*; it would soon be recognized as a dinocephalian therapsid. Whereas Owen acknowledged the mammal-like features of the dentition in descriptive anatomy and even in the names he created for some of these fossils, his anti-Darwinian world-view (Rupke, 1994) never permitted him to reconcile this seeming conflict by viewing those resemblances as evidence of common ancestry.

Unlike Owen, Thomas Huxley was looking for phylogenetic patterns. Huxley's contribution to understanding the evolutionary history of *Mammalia* was more theoretical in nature, in helping to establish that a 'proto-mammalian' stem group made up of fossils should be expected (Huxley, 1880), and in redirecting the efforts of palaeontologists toward the new enterprise of reconstructing evolutionary history (see *Pan-Mammalia*, this volume). Huxley had opportunities only to study specimens of the 'anomalous' *Dicynodontia* from India and South Africa. He conceded that the detailed organization of their skulls could not be explained

by analogy with any extant or extinct *Reptilia* (Huxley, 1859) and noted in passing that the pelvis in *Dicynodon* resembled that of mammals in certain points (Huxley, 1879). But Huxley failed to recognize their transitional nature in the evolution of *Mammalia*, and he classified them as *Reptilia*. Cope (1878a,b) was the first to see their significance (see *Pan-Mammalia*, this volume) but years would pass before his insight was widely appreciated.

Harry G. Seeley (1894a–f, 1895a,b) was the next to make progress in mapping fossil therapsids to their proper places (as we see it today) within the mammalian stem group, although not without introducing new confusion. Seeley's major impact was to correctly recognize that some of Owen's theriodonts approached *Mammalia* more closely than others in having a secondary palate. For those with a secondary palate, Seeley restored Owen's name *Cynodontia* (this volume), and he later detailed similarities in all parts of the skeleton that cynodonts share with mammals (Seeley, 1895a,b). Based on Owen's theriodont *Gorgonops torvus*, Seeley named *Gorgonopsia* for those theriodonts that lacked a secondary palate, and Seeley removed *Gorgonopsia* from *Theriodontia*, treating it as a taxon of equal rank (Seeley, 1894a). Being characterized only by plesiomorphic characters, *Gorgonopsia* quickly grew to contain a large number of taxa that would later prove to be divisible into several distinct clades, including *Biarmosuchia*.

In Seeley's classifications (Seeley, 1894a, b), *Theriodontia* contained *Cynodontia*, plus *Gomphodontia* and *Lycosauria*. However, *Theriodontia* was ranked equally with *Gorgonopsia* under his new (order) *Therosuchia* Seeley 1894a. Seeley also included *Pareiasauria* and *Procolophonia* among the therosuchians (taxa currently believed to be members of *Reptilia* or its stem group). He excluded dicynodonts, which were placed in the separate (order) *Therochelonia*. *Therosuchia* was quickly seen as an artificial (paraphyletic) assemblage and lapsed, while *Therochelonia* was abandoned in favor of *Dicynodontia*. Thus, although it could be argued that *Therosuchia* (ignoring the procolophonians and pariesaurs) is associated with approximately the same ancestor as is *Therapsida*, there is no reason to prefer the former name.

By Seeley's time the name *Anomodontia* had become used for rather different assemblages, and at one time or another it included all of the therapsids then known plus other taxa now viewed as members of the reptilian stem group (Seeley, 1894a; Osborn, 1903a; see *Pan-Mammalia*, this volume). It was also used as a redundant name at a higher rank that contained only *Dicynodontia* and was imposed to convey the wide anatomical differences separating dicynodonts from other taxa (e.g., Cope, 1889, 1898). Moreover, *Anomodontia* is in current use for a therapsid subclade that includes dicynodonts plus newly discovered fossils believed to be most closely related to dicynodonts (e.g., Liu et al., 2009a). While technically available, there is no merit in substituting *Anomodontia* for *Therapsida*.

Robert Broom was the next to make great strides in understanding the relationships among the Permo-Triassic members of the mammalian stem group. As did Seeley, Broom (1900) observed that some of Owen's theriodonts had a secondary palate, like mammals, while others did not. He initially left the former group in *Theriodontia*. For the latter group, he ignored Seeley's *Gorgonopsia*, and created the name *Therocephalia* Broom 1903a (based on *Scylacosaurus sclateri*, *Ictidosaurus angusticeps*, and *Scymnosaurus ferox*). From the published descriptions, Broom initially believed that *Gorgonops* had temporal fenestrae that were

both closed and roofed by the parietals. In this respect it differed from his therocephalians, "but in most points the resemblance to the typical Therocephalians are so marked as to render it not improbable that it should be placed in the same group" (Broom, 1903b: 155). By this time, the name *Theriodontia* had been used for different assemblages of carnivorous therapsids, and both *Gorgonopsia* and *Therocephalia* were defined (diagnosed) in terms of plesiomorphic characters.

In 1905, Broom confronted the confusion that had by this time enveloped both *Theriodontia* and *Anomodontia*. He advocated restricting the membership of *Anomodontia* to the families of dicynodonts, a practice that many, but by no means all, palaeontologists subsequently followed (see *Pan-Mammalia*). He then argued that the carnivorous *Theriodontia* contained two very different groups and was therefore unnatural. He followed Seeley in reinstating *Cynodontia* for those taxa with a well-developed secondary palate, a double-occipital condyle, a rudimentary quadrate, and a jaw formed almost entirely by the dentary. The other group lacks a secondary palate, has a single occipital condyle, and has "a large number of other primitive characters" (Broom, 1905: 268). This group he had already named *Therocephalia* (Broom, 1903a, b). Broom proposed the name *Therapsida* to contain only the four mammal-like groups that he now recognized, viz., *Therocephalia*, *Dinocephalia*, *Anomodontia*, and *Cynodontia*. "These four groups, of which the most primitive is *Therocephalia*, are all closely related" (Broom, 1905: 269).

Broom never explained the etymology of *Therapsida*, nor did he offer a character-based diagnosis. He conveyed the meaning of *Therapsida* (ranked as a superorder) by enumeration of its constituent taxa, and he did not list excluded taxa. In this approach, Broom's concept of *Therapsida* resembles the minimum-clade (node-based) definition proposed herein more closely than it does a maximum clade (branch-based) or apomorphy-based definition. In naming *Therapsida*, Broom contributed much to the modern concept of that taxon by restricting its membership to fossils that are part of the mammalian stem group (according to current hypotheses), and he at last removed a number of extraneous taxa now considered members of *Reptilia* or its stem group (turtles, plesiosaurs, placodonts), or the stem group of *Amniota* ("cotylosaurs"). Broom treated (superorder) *Therapsida* as a paraphyletic group that was ancestral to (class) *Mammalia* and following the practice of the day *Mammalia* was formally excluded. Therapsids became the so-called 'mammal-like reptiles' and Broom left little room to doubt that mammals had evolved from them (e.g., Broom, 1907b, 1910, 1915, 1932).

In 1910, Broom had the chance to personally inspect *Gorgonops torvus* at the British Museum (Natural History). He recognized that in fact it has a large temporal opening, but one that faces laterally and is roofed by a broad parietal region. He conceded, "Still, *Gorgonops* one has always felt differed from the typical Therocephalians" (Broom and Haughton, 1913: 26). From then on, the name *Gorgonopsia* was widely applied to carnivorous therapsids that lacked a secondary palate, possessed a broad temporal region, and were otherwise characterized by retaining many plesiomorphic features.

During the first half of the twentieth century, Broom and his colleagues described several new gorgonopsian taxa that would eventually be recognized as biarmosuchians. At that time, *Biarmosuchus tener* had yet to be discovered, but these new taxa were recognized as highly plesiomorphic *Gorgonopsia*. They include *Ictidorhinus martinsi* Broom 1913, *Burnetia mirabilis* Broom 1923, *Hipposaurus boonstri* Haughton 1929,

Rubidgina angusticeps Broom 1942, see also Broom, 1943, and *Lemurosaurus pricei* Broom 1949. They were all based on rather poor material, but as better specimens were discovered, other authors recognized their plesiomorphic nature within *Gorgonopsia*, and some or all were sequestered under *Ictidorhinidae* Broom 1932, which was considered to be the most primitive family of *Gorgonopsia* (Haughton and Brink, 1954; Sigogneau, 1968, 1970; Mendrez-Carroll, 1975).

During the Soviet era, palaeontological expeditions into the Russian Permian recovered some of the most complete and best-preserved Permian therapsids in the world, including a complete skull and skeleton of *Biarmosuchus tener* Chudinov 1960. Thanks in part to its completeness, Olson used it as the basis for erecting *Biarmosuchidae* Olson 1962. Eventually, the most plesiomorphic South African gorgonopsians and the biarmosuchids were assigned to *Biarmosuchia* (Hopson and Barghusen, 1986; Rowe, 1986; Gauthier et al., 1988; Sigogneau-Russell, 1989; Rubidge and Sidor, 2001; Sidor, 2003; Sidor and Welman, 2003). The fact that Broom named several biarmosuchians (as they are now recognized), and unhesitatingly classified them as therapsids, is an important rationale in choosing *Biarmosuchus tener* as an internal specifier in the phylogenetic definition of *Therapsida* proposed here.

Kemp summarized the current situation: "It has always been recognized that therapsids are in a general way more 'advanced', or 'progressive' in their biology than their pelycosaurian forebears" (Kemp, 2006: 1237). Whether viewed as a grade or a clade, therapsids "... had evolved a higher rate of food assimilation and of ventilatory capacity, a more agile, faster, more energetic mode of locomotion, more elaborate and therefore more sensitive olfaction and hearing, and an increased growth rate" (Kemp,

2006: 1237). Osteological characters upon which most of Kemp's physiological interpretations are based are listed in the diagnosis above.

By the first decades of the twentieth century, Permo-Triassic therapsids had been discovered in Europe, Asia, India, Africa, South America, and North America. In the years preceding the emergence of plate tectonic theory, this geographical distribution was difficult to explain, and the idea emerged that *Therapsida* represented a polyphyletic grade of adaptation achieved independently several times on different continents (Olson, 1944, 1959, 1962, 1974, 1986; Boonstra, 1963, 1972; Chudinov, 1983). As geophysical evidence of plate tectonics began to grow, the discovery of the dicynodont *Lystrosaurus* in Antarctica (Kitching et al., 1972; Colbert, 1974) offered corroborating palaeobiogeographic evidence that helped to establish the theory of Plate Tectonics in its modern form (Colbert, 1973). In turn, plate tectonics helped to dispel objections to therapsid monophyly (but see Ivakhnenko, 2002, 2003).

One might see a challenge to a minimum-clade (node-based) definition of *Therapsida* in Olson's career-long work in the North American Permian. Based on exceedingly fragmentary fossils, some of which are now lost, Olson described several "carnivorous synapsids intermediate between pelycosaurian and therapsid levels of organization, as these terms are commonly used, and carnivorous synapsids at an unprogressive level of therapsid organization" (Olson, 1962: 48). Olson and Beerbower (1953) described *Dimacrodon hottoni*, *Steppesaurus gurleyi*, *Tappenosaurus magnus*, and *Knoxosaurus niteckii*, and Olson (1962) later named *Gorgodon minutus*, *Eosyodon hudsoni*, *Driveria ponderosa*, and *Mastersonia driverensis*. At first, Olson classified all of these as members of *Therapsida*, and although he viewed therapsids as comprising a grade, his proposal

might otherwise be interpreted as reflecting a maximum-clade concept of *Therapsida*. Romer (1966) and Chudinov (1983) accepted these as primitive therapsids, but Romer (1968) and most later researchers disputed the idea of a polyphyletic origin of therapsids. Olson himself later removed all of these fossils from *Therapsida* and informally designated them as 'eotherapsids' (Olson, 1986). Subsequent workers have either ignored them (e.g., Hopson and Barghusen, 1986; Tatarinov, 1974; Rowe, 1986; Gauthier et al., 1988; Kemp, 2005; Benton, 2015), or rejected their inclusion within *Therapsida* on the grounds that severe incompleteness renders them indeterminate (Kemp, 1982, 2006; Parrish et al., 1986; King, 1988), or argued that they were not of "therapsid-grade" (Sidor and Hopson, 1995: 53A).

Olson later named *Watongia meieri* (Olson, 1974) based on a fragmentary skeleton from the Middle Permian of Oklahoma, arguing that "*Watongia* is a such an excellent intermediate between the sphenacodontines and gorgonopsids that it could in fact represent the actual transitional stage" (Olson, 1974:41). Chudinov (1983) disregarded *Watongia* as an early therapsid, while Reisz and Laurin (2004) argued that it is instead a varanopsid "pelycosaur", and "Further evidence that *Watongia* is not a therapsid is provided by its primitive bone histology ... which is entirely "pelycosaurian" in nature and unlike that of any known therapsid" (Reisz and Laurin, 2004: 384). So far as can be ascertained, Olson's taxa are members of *PanMammalia*, but not *Therapsida*.

Van Valen (1960) proposed that tetrapod classes should be based on their major adaptive differences and that the mammalian grade was largely reached by the early therapsids. He concluded that therapsids should therefore be included in *Mammalia*. While this suggestion has the merit of acknowledging the relationship between crown mammals and part of their stem group, it was not adopted by subsequent authors, in part because no alternative name was proposed for the mammalian crown (Rowe and Gauthier, 1992).

A major shift in the meaning of the name *Therapsida* occurred as Hennig's (1966) concept of monophyly was applied, and it came to be recognized as applying to a clade that includes *Mammalia* as its crown (Rowe 1986, 1988; Gauthier et al., 1988; Hopson, 1994; Laurin and Reisz, 1996; Kemp, 2005, 2006). During the decades leading up to this shift and afterwards, the composition of *Therapsida* was remarkably stable in terms of its Permo-Triassic members. As Kemp explained: "Most contemporary commentators accept the monophyly of the taxon *Therapsida*, defined as the last common ancestor of biarmosuchians and mammals, plus all its descendants" (Kemp, 2006: 1232).

The only explicit alternative to a minimum-clade (node-based) definition of a monophyletic *Therapsida* was proposed by Laurin and Reisz (1990, 1996), who advocated a maximum-clade definition: "... we define Therapsida as mammals and all other synapsids that share a more recent common ancestor with them than with sphenacodontids" (Laurin and Reisz, 1996: 100; see also Amson and Laurin, 2011). Under this definition, they recognized the Early Permian fossil *Tetraceratops insignis* Matthew 1908 as the "oldest therapsid" based on a single, crushed, partial skull and their interpretation of the construction of the medial wall of the temporal fenestra and its possible relationship to jaw adductor musculature. Their proposal was met with criticism that the incompleteness and poor preservation of the only known specimen of *Tetraceratops* left the interpretation of its anatomy and its phylogenetic position equivocal (Sidor and Hopson, 1998). Conrad and Sidor (2001) and Liu et al.

(2009b) concluded that *Tetraceratops* is a member of *Sphenacodontidae*. Rubidge and Sidor (2001) and Kemp (2005, 2006) accepted the hypothesized transitional phylogenetic position of *Tetraceratops insignis* but they rejected Laurin and Reisz's definition and argued instead that *Tetraceratops* was the sister to, rather than a member of, *Therapsida*.

More recently, Liu et al. (2009b) described a Middle Permian fossil from China, *Raranimus dashankouensis*, calling it "the oldest therapsid" with no further discussion of how the name *Therapsida* was defined. *Raranimus* is known from a single specimen consisting of an isolated, broken snout. It preserves the diagnostic therapsid features of the premaxilla and septomaxilla mentioned above but preserves no characters hypothesized as derived within *Therapsida*. According to their analysis, *Raranimus* would be considered the sister taxon to *Therapsida,* or as sister taxon to *Therapsida* + *Tetraceratops insignis* (Amson and Laurin, 2011), under the minimum-clade definition proposed here.

The definition of *Therapsida* proposed by Laurin and Reisz (1996) used mammals as an internal specifier. A subsequent recommendation of the *ICPN* (Rec. 11A) advises against using specifiers that traditionally were excluded from the taxon designated by the defined name. In accordance with that recommendation, the Triassic fossil *Thrinaxodon liorhinus* Seeley 1894a is used here as one of the internal specifiers and, as explained above, the Permian *Biarmosuchus tener* Chudinov 1960 as the other. Of many potential specifiers, *Thrinaxodon liorhinus* was chosen because it has always been regarded as a cynodont (which under monophyletic hypotheses now includes *Mammalia*), it is well-represented in museum collections, and its skull (Fourie, 1974; Rowe et al., 1995) and postcranial skeleton (Jenkins,

1971) are among the best known of the non-mammalian cynodonts traditionally included within *Therapsida*.

Literature Cited

Amson, E., and M. Laurin. 2011. On the affinities of *Tetraceratops insignis*, an Early Permian synapsid. *Acta Palaeontol. Polonica* 56:301–312.

Benton, M. 2015. *Vertebrate Palaeontology*. 4th edition. Wiley Blackwell, West Sussex.

Boonstra, L. D. 1963. Early dichotomies in the therapsids. *S. Afr. J. Sci.* 59:176–195.

Boonstra, L. D. 1972. Discard the names *Theriodontia* and *Anomodontia*: a new classification of the *Therapsida*. *Ann. S. Afr. Mus.* 59:315–38.

Broom, R. 1900. On the structure of the palate in *Dicynodon* and its allies. *Trans. S. Afr. Philos. Soc.* 11:177–184.

Broom, R. 1903a. On the classification of the theriodonts and their allies. *Rept. S. Afr. Assoc. Adv. Sci.* 1:286–294.

Broom, R. 1903b. On some new primitive theriodonts in the South African Museum. *Ann. S. Afr. Mus.* 4:147–158.

Broom, R. 1905. On the use of the term *Anomodontia. Rec. Albany Mus. Grahamstown S. Afr.* 4:266–269.

Broom, R. 1907a. On some new fossil reptiles from the Karroo beds of Victoria West, South Africa. *Trans. S. Afr. Philos. Soc.* 18:31–42.

Broom, R. 1907b. The origins of the mammal-like reptiles. *Proc. Zool. Soc. Lond.* 1907:1047–1061.

Broom R. 1910. A comparison of the Permian reptiles of North America with those of South Africa. *Bull. Am. Mus. Nat. Hist.* 28:197–234.

Broom, R. 1911. On the structure of the skull in cynodont reptiles. *Proc. Zool. Soc. Lond. B* 81:893–925.

Broom, R. 1913. On some new carnivorous therapsids. *Bull. Am. Mus. Nat. Hist.* 32:557–561.

Broom, R. 1915. On the origin of mammals. *Philos. Trans. R. Soc. Lond. B Biol. Sci.* 206:1–48.

Broom, R. 1923. On the structure of the skull in the carnivorous dinocephalian reptiles. *Proc. Zool. Soc. Lond. B* 93:661–684.

Broom, R. 1932. *The Mammal-Like Reptiles of South Africa and the Origin of Mammals.* H. F. & G. Witherby, London.

Broom, R. 1942. Evidence of a new suborder of mammal-like reptiles. *Bull. S. Afr. Mus. Assoc.* 2:386.

Broom, R. 1943. Some new types of mammal-like reptiles. *Proc. Zool. Soc. Lond. B* 113:17–24.

Broom, R. 1949. New fossil reptile genera from the Bernard Price collection. *Ann. Transvaal Mus.* 21:187–194.

Broom, R., and S. H. Haughton. 1913. On a new species of *Scymnognathus* (*S. tigriceps*). *Ann. S. Afr. Mus.* 12:26–35.

Chudinov, P. K. 1960. Upper Permian therapsids from the Ezhovo locality. *Paleontol. J.* 4:81–94, (in Russian).

Chudinov, P. K. 1983. Early therapsids. Academy of Sciences of the USSR. *Trans. Palaeontol. Inst. USSR* 202:3–229, (in Russian).

Colbert, E. H. 1973. *Wandering Lands and Animals.* E. P. Dutton, New York.

Colbert, E. H. 1974. *Lystrosaurus* from Antarctica. *Am. Mus. Nat. Hist. Novitates* 2535:1–44.

Conrad, J., and C. A. Sidor. 2001. Re-evaluation of *Tetraceratops insignis* (Synapsida, Sphenacodontia). *J. Vertebr. Paleontol.* 21:42A.

Cope, E. D. 1878a. Descriptions of extinct *Batrachia* and *Reptilia* from the Permian formation of Texas. *Proc. Am. Philos. Soc.* 17:505–530.

Cope, E. D. 1878b. The theromorphous *Reptilia*. *Am. Nat.* 12:829–830.

Cope, E. D. 1889. Synopsis of the families of *Vertebrata. Am. Nat.* 23:849–877.

Cope, E. D. 1898. *Syllabus of lectures on the Vertebrata, with an introduction by Henry Fairfield Osborn.* Philadelphia, PA. Published for the University of Pennsylvania.

Eichwald, E. 1860. *Lethaea Rossica ou Paléontologie de la Russie. Vol. 1: Seconde Section de l'ancienne Périod.* E. Schweizerbart, Stuttgart.

Fischer, W. G. 1841. Notice sur le *Rhopalodon*, nouveau genre de sauriens fossils du versant occidental de l'Oural. *Bull. Moscow Soc. Naturalists* 14:460–464.

Fourie, S. 1974. The cranial morphology of *Thrinaxodon liorhinus* Seeley. *Ann. S. Afr. Mus.* 65:337–400.

Gauthier, J. A., A. G. Kluge, and T. B. Rowe. 1988. Amniote phylogeny and the importance of fossils. *Cladistics* 4:105–209.

Haughton, S. H. 1929. On some new therapsid genera. *Ann. S. Afr. Mus.* 28:55–78.

Haughton, S. H., and A. S. Brink. 1954. A bibliographic list of *Reptilia* from the Karroo beds of Africa. *Palaeontol. Afr.* 2:1–187.

Hennig, W. 1966. *Phylogenetic Systematics*, translated by D. Davis and R. Zangerl. University of Illinois Press, Urbana, IL.

Hopson, J. A. 1994. Synapsid evolution and the radiation of non-eutherian mammals. Pp.190–219 in *Major Features of Vertebrate Evolution* (D. R. Prothero and R. M. Schoch, eds.). Paleontological Society Short Course No. 7.

Hopson, J. A., and H. R. Barghusen. 1986. An analysis of therapsid relationships. Pp. 83–106 in *Ecology and Biology of Mammal-Like Reptiles* (N. Hotton, III, P. D. McLean, J. J. Roth, and E. C. Roth, eds.). Smithsonian Institution and National Institute of Mental Health, Washington, DC.

Huttenlocker, A. 2009. An investigation into the cladistic relationships and monophyly of therocephalian therapsids (*Amniota: Synapsida*). *Zool. J. Linn. Soc.* 157:865–891.

Huxley, T. H. 1859. On a new species of *Dicynodon* (*D. murrayi*) from near Colesberg, South Africa; and the structure of the skull in the dicynodonts. *Q. J. Geol. Soc. Lond.* 15:649–658.

Huxley, T. H. 1879. On the characteristics of the pelvis in the *Mammalia*, and the conclusions respecting the origin of mammals which may be based on them. *Proc. R. Soc. Lond. B Biol. Sci.* 28:395–405.

Huxley, T. H. 1880. On the application of the laws of evolution to the arrangement of the *Vertebrata*, and more particularly of the *Mammalia*. *Proc. Zool. Soc. Lond.* 1880:649–662.

Ivakhnenko, M. F. 2002. The origin and early divergence of therapsids. *Paleontol. J.* 36:168–175.

Ivakhnenko, M. F. 2003. Eotherapsids from the east European placket (Late Permian). *Paleontol. J.* 37:S339–S465.

Jenkins, Jr., F. A. 1971. The postcranial skeleton of African cynodonts. *Bull. Peabody Mus. Nat. Hist.* 36:1–216.

Kemp, T. S. 1982. *Mammal-Like Reptiles and the Origin of Mammals*. Academic Press, London.

Kemp, T. S. 2005. *The Origin and Evolution of Mammals*. Oxford University Press, Oxford.

Kemp, T. S. 2006. The origin and early radiation of the therapsid mammal-like reptiles: a palaeobiological hypothesis. *J. Evol. Biol.* 19:1231–1247.

King, G.M. 1988. *Anomodontia*. Gustav Fischer Verlag, Stuttgart.

Kitching, J. W., J. W. Collinson, D. H. Elliot, and E. H. Colbert. 1972. *Lystrosaurus* zone (Triassic) fauna from Antarctica. *Science* 175:524–527.

Kuhn, O. 1965. *Fossilium Catalogus. 1: Animalia, part 110. Therapsida*. Uitgeverij dr W. Junk, Gravenhage.

Kutorga, S. S. 1838. Beitrag zur Kenntniss der organisched Überreste des Kupfersandstein am westlichen Abhange des Urals. Verhandlungen dk mineralogische Gesellschaft, St. Petersburg.

Laurin, M. 1993. Anatomy and relationships of *Haptodus garnettensis*, a Pennsylvanian synapsid from Kansas. *J. Vertebr. Paleontol.* 13:200–229.

Laurin, M., and R. R. Reisz. 1990. *Tetraceratops* is the oldest known therapsid. *Nature* 345:249–250.

Laurin, M., and R. R. Reisz. 1996. The osteology and relationships of *Tetraceratops insignis*, the oldest known therapsid. *J. Vertebr. Paleontol.* 16:95–102.

Linnaeus, C. 1758. *Systema Naturae Per Regna Tria Naturae, Secundum Classes, Ordines, Genera, Species, cum Characteribus, Differentiis, Synonymis, Locis*. Tomus I, Editio decima, reformata. Laurentius Salvius, Holmiae (Stockholm).

Liu, J., R. Riubidge, and J. Li. 2009a. A new specimen of *Biseridens quilianicus* indicates its position as the most basal anomodont. *Proc. R. Soc. Lond. B Biol. Sci.* 227:285–292.

Liu, J., B. Rubidge, and J. Li. 2009b. New basal synapsid supports Laurasian origin for therapsids. *Acta Palaeontol. Polonica* 54:393–400.

Matthew, W. D. 1908. A four-horned pelycosaurian from the Permian of Texas. *Bull. Am. Mus. Nat. Hist.* 24:183–185.

Mendrez-Carroll, C. H. 1975. Comparaison du palais chez les Thérocéphales primitifs, les Gorgonopsiens et les Ictidorhinidae. *C. R. Acad. Sci.* 280:17–20.

Murchison, R. I., E. de Verneuil, and A. von Keyserling. 1845. *The Geology of Russia in Europe and the Ural Mountains.* John Murray, London.

Olson, E. C. 1944. Origin of mammals based upon cranial morphology of the therapsid suborders. *Geol. Soc. Am. Spec. Pap.* 55:1–130.

Olson, E. C. 1959. The evolution of mammalian characters. *Evolution* 13:344–353.

Olson, E. C. 1962. Late Permian terrestrial vertebrates, USA and USSR. *Trans. Am. Philos. Soc.* 52:1–224.

Olson, E. C. 1974. On the source of therapsids. *Ann. S. Afr. Mus.* 64:27–46.

Olson, E. C. 1986. Relationships and ecology of the early therapsids and their predecessors. Pp. 47–60 in *The Ecology and Biology of the Mammal-Like Reptiles* (N. Hotton, III, P. D. MacLean, J. J. Roth and E. C. Roth, eds.). Smithsonian Institution Press, Washington, DC.

Olson, E. C., and J. R. Beerbower. 1953. The San Angelo formation, Permian of Texas, and its vertebrates. *J. Geol.* 61:389–423.

Osborn, H. F. 1903a. On the primary divisions of the *Reptilia* into two sub-classes, *Synapsida* and *Diapsida*. *Science* 17:275–276.

Osborn, H. F. 1903b. The reptilian subclasses *Diapsida* and *Synapsida* and the early history of the *Diaptosauria*. *Mem. Am. Mus. Nat. Hist.* 1:451–507.

Owen, R. 1844. Description of certain fossil crania, discovered by A. G. Bain, Esq., in sandstone rocks at the south-eastern extremity of Africa, referable to different species of an extinct Genus of *Reptilia* (*Dicynodon*), and indicative of a new tribe of suborder *Sauria*. *Proc. Geol. Soc. Lond.* 4:500–504.

Owen, R. 1845. Description of certain fossil crania, discovered by A. G. Bain, Esq., in sandstone rocks at the south-eastern extremity of Africa, referable to different species of an extinct Genus of *Reptilia* (*Dicynodon*), and indicative of a new tribe or suborder of *Sauria*. *Trans. Geol. Soc. Lond.* 7:59–84.

Owen, R. 1860a. On the orders of fossil and recent *Reptilia* and their distribution in time. *Rept. Br. Assoc. Adv. Sci.* 29 1859:153–166.

Owen, R. 1860b. *Paleontology or a Systematic Summary of Extinct Animals and Their Geological Relations.* 1st edition. Adams and Charles Black, Edinburgh.

Owen, R. 1861. *Paleontology or a Systematic Summary of Extinct Animals and Their Geological Relations.* 2nd edition. Adams and Charles Black, Edinburgh.

Owen, R. 1876. *Descriptive and illustrated catalogue of the fossil* Reptilia *of South Africa in the collection of the British Museum.* Trustees of the British Museum (Natural History), London.

Parrish, J. M., J. T. Parrish, and A. M. Zeigler. 1986. Permian-Triassic paleogeography and implications for therapsid distribution. Pp. 109–131 in *The Ecology and Biology of Mammal-Like Reptiles* (N. Hotton, III, P. D. MacLean, J. J. Roth, and E. C. Roth, eds.). Smithsonian Institution Press, Washington, DC.

Reisz, R. R., and M. Laurin. 2004. A reevaluation of the enigmatic Permian synapsid *Watongia* and of its stratigraphic significance. *Can. J. Earth Sci.* 41:377–386.

Romer, A. S. 1966. *Vertebrate Paleontology.* University of Chicago Press, Chicago, IL.

Romer, A. S. 1968. *Notes and Comments on Vertebrate Paleontology.* University of Chicago Press, Chicago, IL.

Rowe, T. B. 1986. *Osteological Diagnosis of* Mammalia, *L. 1758, and Its Relationship to Extinct* Synapsida. PhD dissertation, University of California, Berkeley, CA.

Rowe, T. B. 1988. Definition, diagnosis and origin of *Mammalia*. *J. Vertebr. Palentol.* 8:241–264.

Rowe, T. B., and J. A. Gauthier. 1992. Ancestry, paleontology, and definition of the name *Mammalia*. *Syst. Biol.* 41:372–378.

Rowe, T. B., and J. van den Heever. 1986. The hand of *Anteosaurus magnificus* (*Therapsida, Dinocephalia*) and its bearing on the origin of the mammalian manual phalangeal formula. *S. Afr. J. Sci.* 82:641–645.

Rowe, T., W. Carlson, and W. Bottorff. 1995. *Thrinaxodon: Digital Atlas of the Skull.* CD-ROM. 2nd edition, for Windows and Macintosh platforms. University of Texas Press.

Rubidge, B. S., and C. A. Sidor. 2001. Evolutionary patterns among Permo-Triassic therapsids. *Annu. Rev. Ecol. Syst.* 32:449–480.

Rupke, N. A. 1994. *Richard Owen—Victorian Naturalist.* Yale University Press, New Haven and London.

Seeley, H. G. 1889. Researches on the structure, organization, and classification of the fossil *Reptilia.* Pt. IV. On the anomodont *Reptilia* and their allies. *Philos. Trans. R. Soc. Lond. B Biol. Sci.* 180:215–296.

Seeley, H. G. 1894a. Researches on the structure, organization, and classification of the fossil *Reptilia.* Part IX, Section 1. On the *Therosuchia. Proc. R. Soc. Lond. B Biol. Sci.* 55:224–226.

Seeley, H. G. 1894b. Researches on the structure, organization and classification of the fossil *Reptilia.* IX, Section 1. On the *Therosuchia. Philos. Trans. R. Soc. Lond. B Biol. Sci.* 185:987–1018.

Seeley, H. G. 1894c. Researches on the structure, organization, and classification of the fossil *Reptilia.* Part VIII. Further evidences of the skeleton in *Deuterosaurus* and *Rhopalodon,* from the Permian rocks of Russia. *Philos. Trans. R. Soc. Lond. B Biol. Sci.* 185:663–718.

Seeley, H. G. 1894d. Researches on the structure, organization, and classification of the fossil *Reptilia.* Part IX. Section 4. On the *Gomphodontia. Proc. R. Soc. Lond. B Biol. Sci.* 56:288–291.

Seeley, H. G. 1894e. Researches on the structure, organization, and classification of the fossil *Reptilia.—* Part IX, Section 3. On *Diademodon. Philos. Trans. R. Soc. Lond. B Biol. Sci.* 185:1029–1041.

Seeley, H. G. 1894f. Researches on the structure, organization, and classification of the fossil *Reptilia.—*Part IX, Section 2. The reputed mammals from the Karroo formation of Cape Colony. *Philos. Trans. R. Soc. Lond. B Biol. Sci.* 185:1019–1028.

Seeley, H. G. 1895a. Researches on the structure, organization, and classification of the fossil *Reptilia.* – Part IX, Section 5. On the skeleton of new *Cynodontia* from the Karoo rocks. *Philos. Trans. R. Soc. Lond. B Biol. Sci.* 186:59–148.

Seeley, H. G. 1895b. Researches on the structure, organization, and classification of the fossil *Reptilia.* Part IX, Section 6. Associated remains of two small skeletons from Klipfontein, Fraserburg. *Philos. Trans. R. Soc. Lond. B Biol. Sci.* 185:149–162.

Sidor, C. A. 2003. The naris and palate of *Lycaenodon longiceps* (*Therapsida: Biarmosuchia*), with comments on their early evolution in the *Therapsida. J. Paleontol.* 77:977–984.

Sidor, C. A., and J. A. Hopson. 1995. The taxonomic status of the Upper Permian eotheriodont therapsids of the San Angelo Formation (Guadalupian), Texas. *J. Vertebr. Paleontol.* 15:53A.

Sidor, C. A., and J. A. Hopson. 1998. Ghost lineages and 'mammalness': assessing the temporal pattern of character acquisition in the *Synapsida. Paleobiology* 24:254–273.

Sidor, C. A., and J. Welman. 2003. A second specimen of *Lemurosaurus pricei* (*Therapsida: Burnetiamorpha*). *J. Vertebr. Paleontol.* 23:631–642.

Sigogneau, D. 1968. On the classification of the *Gorgonopsia. Palaeontol. Afr.* 11:33–46.

Sigogneau, D. 1970. Contribution a la Connaissance des Ictidorhinidés (*Gorgonopsia*). *Palaeontol. Afr.* 13:25–38.

Sigogneau-Russell, D. 1989. *Theriodontia* I. Pp. 1–127 in *Encyclopedia of Paleoherpetology, Part 17B* (P. Wellnhofer, ed.). Gustav Fischer, Stuttgart.

Tatarinov, L. P. 1974. Theriodonts of the USSR. Academy of Sciences of the USSR. *Trans. Palaeontol. Inst. USSR.* 143:5–250, (in Russian).

Van Valen, L. 1960. Therapsids as mammals. *Evolution* 14:304–313.

Author

Timothy B. Rowe; Jackson School of Geosciences; The University of Texas at Austin; C1100, Austin, TX 78712, USA. E-mail: rowe@utexas. edu.

Date Accepted: 12 June 2017

Primary editors: Jacques A. Gauthier, Kevin de Queiroz

Cynodontia R. Owen 1861 [T. B. Rowe], converted clade name

Registration Number: 219

Definition: The clade possessing the apomorphy occlusal triconodont molariform postcanine teeth—i.e., the crowns of the molariform postcanine teeth have three principal cusps aligned longitudinally (or further modifications of this condition) and the upper and lower molariform teeth occlude, at least irregularly—as inherited by *Thrinaxodon liorhinus* Seeley 1894a. This is an apomorphy-based definition. The apomorphy is complex, and it is intended that all components must be present to satisfy the definition. Abbreviated definition: ∇ apo occlusal triconodont molariform postcanine teeth [*Thrinaxodon liorhinus* Seeley 1894a].

Etymology: As explained by Owen (1861), *Cynodontia* is derived from the Greek *kyon* (dog) and *odontos* (tooth), although it referred to the general form of the entire dentition rather than to the specific characters used to define the name here (see Comments).

Reference Phylogeny: Figure 3 in Gauthier et al. (1988a) is the primary reference phylogeny.

Composition: *Cynodontia* includes *Mammalia* (this volume) plus a subset of extinct species and clades viewed since the time of Seeley (Seeley, 1894a,b, 1895), and under current hypotheses, as the closest extinct relatives of *Mammalia* within *Therapsida* (Gauthier et al., 1988a; Rubidge and Sidor, 2001; Kemp, 2005). These include *Procynosuchidae* Broom 1948 (includes *Silphilestidae*); *Dvinia prima* Amalitzky 1922; *Galesauridae* Lydekker 1890 (includes *Thrinaxodontidae, Nythosauridae, Cynosuchidae,* and *Cynosauridae*); *Ecteninion*

lunensis Martinez et al. 1996; *Parathrinaxodon proops* (Parrington, 1936); *Cynognathidae* (Watson, 1917); *Platycraniellus elegans* (van Hoepen, 1917; Abdala, 2007); *Diademodontidae* (Haughton, 1924); *Chiniquodontidae* (von Huene, 1936) (sensu Abdala and Giannini, 2002); *Trirachodontidae* Crompton 1955; *Traversodontidae* (Lehman, 1961); *Nanocynodon seductus* (Tatarinov, 1968); *Probainognathus jenseni* (Romer, 1970); *Lumkuia fuzzi* (Hopson and Kitching, 2001); *Prozostrodon parvus* Bonaparte and Barberena 2001; plus the extant and extinct species of *Mammaliamorpha* (this volume).

Tritheledontidae (= *Ictidosauria*; see Sidor and Hancox, 2006; Martinelli and Rougier, 2007) and *Brasilodontidae* (Bonaparte et al., 2003, 2005) are unquestionably cynodonts, but whether they lie inside or just outside *Mammaliamorpha* (this volume) is ambiguous.

Under the definition presented above, the Early-Middle Triassic *Bauridae* Broom 1911 (= *Bauriidae,* = *Bauriamorpha,* = *Bauriasauria*) may belong to *Cynodontia*, based on possession of a secondary palate, an occlusal dentition, and possibly a divided occipital condyle. This taxon traditionally included *Microgomphodon oligocynus* Seeley 1895, *Melinodon simus* Broom 1905a, *Sesamodon browni* Broom 1905a, *Aelurosuchus browni* Broom 1906, *Bauria cynops* Broom 1909, *Sesamodontoides pauli* Broom 1950, and *Herpetogale marsupialis* Keyser and Brink 1979; only *Microgomphodon* and *Bauria* were recognized as valid by Abdala et al. (2014). Seeley (1895) and Broom (1911) initially assigned these taxa to *Cynodontia*, but others have considered them part of its sister group, *Therocephalia*. This hypothesis requires that baurids convergently evolved a cynodont-like secondary palate and an occlusal dentition (Hopson and Barghusen,

1986). Huttenlocker's (2009) analysis yielded a monophyletic *Therocephalia*, with *Bauria cynops* as its most deeply nested and youngest member. However, that analysis sampled only craniodental characters and only three undisputed early cynodonts, a strategy that may be inadequate to resolve the position of *Bauridae*.

Diagnostic Apomorphies: In early cynodonts, the rear postcanine teeth are 'molariform' as they develop complex crowns in which there are generally three principal cups aligned longitudinally, with the middle cusp the tallest, and with a row of smaller cuspules on a narrow shelf at the base of the inner surface (Crompton, 1963; Rowe et. al., 1995; Botha et al., 2007). In this arrangement, an occlusal relationship developed between upper and lower molariform teeth in which outer (buccal) surfaces of lower molariform teeth occlude against the inner (lingual) surfaces of the upper molariforms, producing irregular wear facets. The evolution of an occlusal dentition suggests that cynodonts had begun to masticate food items, a new behavior that produces a faster and enriched caloric return (Crompton, 1963, 1972, 1989; Rowe and Shepherd, 2016). The occlusal 'triconodont' pattern set the stage for an unprecedented diversification of molariform teeth into the mammalian 'dental array' (Rowe, 2017; see Comments).

Associated changes in the mandible indicate a shift in the force vectors of the adductor musculature. In early therapsids, the greatest bite forces were exerted at the front of the mouth for grasping prey. In cynodonts, the mandible has a masseteric fossa on the lateral surface of the coronoid process and dentary ramus, and the zygomatic arch has a ventral facet; these indicate that the temporalis and masseter muscles were differentiated, and that the largest mandibular forces had shifted toward the back of the tooth row for mastication (Gregory, 1953; Crompton, 1989; Kemp, 2005).

Numerous additional characters of the skull and postcranial skeleton diagnose *Cynodontia*. In the skull, a secondary palate arose to separate the oral cavity from the nasal cavity as shelves of the maxillae and palatines grew toward the midline. In the Late Permian cynodont *Procynosuchus delaharpeae*, the maxillae and palatines form a pair of shelves that extend two-thirds the length of the snout, but they remain separated on the midline, and a narrow channel separates them from the vomer (Kemp, 1979). In all other cynodonts, the maxillae and palatines meet on the midline beneath the vomer to form a complete secondary palate (reversed in many *Cetacea*). A double occipital condyle is formed by the right and left exoccipitals, which are positioned at the ventrolateral quadrants of the foramen magnum, increasing mobility of the head on the neck (Jenkins, 1969; Gauthier et al., 1988a).

Ossification of the lateral walls of the braincase is expanded by ventral sheets from the frontal and parietal bones, and forward expansion of the neurocranial prootic. In addition, the newly formed alisphenoid bone arises as a compound element joining the endochondral ala temporalis of the epipterygoid with a new, membranous ossification in the spheno-obturator membrane (Gauthier et al., 1988a; Rowe, 1996a,b). Collectively, these new ossifications enclose a larger endocranial volume, and indicate an increase in the relative size of the brain, as measured using computed tomography (Rowe et al., 1995, 1997). This was the first of several measurable pulses in encephalization among stem-mammals that would eventually culminate in the characteristically huge brain of living mammals (Northcutt, 2011; Rowe et al., 2011).

In the ontogeny of living mammals, ossification of the alisphenoid occurs in response to growth and inflation of the olfactory (piriform) cortex, and it may be an epigenetic response of connective tissues surrounding the brain to increased tension produced by rapid growth of a larger brain (Rowe, 1996b). The onset of odorant receptor gene expression in the olfactory epithelium of the nose induces the development and controls the size of the olfactory bulb through the number of odorant receptor axons that reach the olfactory bulb. Projections from the olfactory bulb in turn determine the size and number of layers of the olfactory cortex, a division of the telencephalon that is invariably enclosed by the alisphenoid bone in living mammals (Rowe and Shepherd, 2016). The correlated appearance of an enlarged endocranial cavity and the ossified alisphenoid suggests that the origin of *Cynodontia* was tied to a pulse in encephalization that was driven specifically by enhanced olfaction, a sensory modality developed to a higher level in living mammals than in any other vertebrates (Rowe et al., 2011; Rowe and Shepherd, 2016).

In the postcranium, the thoracic and lumbar regions become differentiated such that long ribs persist on the anterior thoracic vertebrae, while the posterior three to five ribs form attenuated processes that fuse to their respective neural arches (i.e., lumbar ribs). Differentiation of separate thoracic and lumbar regions marked the development of a muscular diaphragm, which separated the thoracic and abdominal cavities and initiated onset of the vacuum-chamber or bellows-like tidal diaphragmatic ventilation of extant *Mammalia* (Jenkins, 1971; Gauthier et al., 1988a; Rowe and Shepherd, 2016; Rowe, 2017).

Synonyms: Partial synonyms include *Theriodontia* Owen 1876 sensu Seeley (1889: 292); and *Gomphodontia* Seeley 1894b. *Mammalia* Linnaeus 1758 sensu Lucas and Hunt (1990) is an approximate synonym.

Comments: Richard Owen was impressed by the mammal-like dentitions he observed in the weathered skull of *Galesaurus planiceps* Owen

1859 and the broken snout of *Cynochampsa laniaria* Owen 1859. In the second edition of his textbook on Palaeontology, Owen grouped them under the name *Cynodontia*, which he characterized as "having a pair of teeth in each jaw resembling in shape, position, and relative size to the other teeth, the canines of carnivorous mammals, and dividing the incisors from the molars" (Owen, 1861: 267). Our modern phylogenetic framework identifies most of these modifications of the dentition as apomorphic of the more inclusive clade *Therapsida* (this volume), and an enlarged caniniform region of the dentition can be traced to early amniotes (Gauthier et al., 1988b).

Owen grouped (family) *Cynodontia* under (order) *Anomodontia* (which contained two dicynodont families plus the stem archosaur *Rhynchosaurus articeps*), and all were placed within (class) *Reptilia*. In this way, he chose to emphasize the many plesiomorphic characters distancing cynodonts and other "anomodonts" from mammals. Later, in his influential *Catalogue of the Fossil Reptilia of South Africa* (Owen, 1876), he abandoned the name *Cynodontia* without explanation, and substituted the name *Theriodontia* for his (former) cynodonts plus a diversity of recently discovered carnivorous forms. The taxa that Owen lumped under *Theriodontia* would later be distributed among several different clades of early *Therapsida* (this volume) now known as *Gorgonopsia*, *Therocephalia*, and the resurrected name *Cynodontia*. Owen was keenly aware that his theriodonts had lost a number of plesiomorphic characters as they gained unique resemblances to *Mammalia*. However, his anti-Darwinian worldview (Rupke, 1994) prevented him from accepting this unprecedented combination of characters as transitional in an evolutionary sense (Owen, 1876: 76).

Harry G. Seeley initiated a reconceptualization of *Cynodontia* by recognizing that some of Owen's theriodonts approached *Mammalia* more closely than others in having a secondary palate. For those with a secondary palate, Seeley restored Owen's term *Cynodontia*, and he later detailed similarities in all parts of the skeleton that cynodonts share with mammals (Seeley, 1894a,b, 1895). Robert Broom agreed with Seeley that Owen's *Theriodontia* contained two very different groups and was therefore unnatural. One group is characterized by "a very mammal-like form with a well-developed secondary palate ... two occipital condyles, a rudimentary quadrate, and a jaw formed almost entirely by the dentary" (Broom, 1905b: 268) and for this group he too chose to apply the name *Cynodontia*. Both Seeley (1895) and Broom (1905b, 1907, 1910, 1932) included only taxa that are viewed as cynodonts under current hypotheses. They recognized that the immediate ancestry of mammals was to be found among extinct taxa they referred to *Cynodontia*. Following the common practice of that era, however, they preferred to regard *Cynodontia* as an extinct, paraphyletic subgroup within an extinct, paraphyletic *Therapsida*. *Cynodontia* and other extinct therapsids were classified as *Reptilia*, while *Mammalia* remained in its own class.

Nevertheless, the idea of an evolutionary linkage between extinct therapsids, particularly the cynodonts, and mammals was solidly forged. Wide recognition of this relationship grew through the work of Osborn (1903a,b), and more emphatically by the prodigious efforts of Robert Broom, who nicknamed therapsids the "mammal-like-reptiles" (Broom, 1907, 1932). This term solidified their evolutionary link, but it obscured what would eventually be seen as a fallacy in most twentieth century classifications, the practice of categorizing the extinct, paraphyletic stem group of mammals in class *Reptilia* while retaining *Mammalia* in a class of its own (Broom, 1932; Romer, 1956, 1966;

Hopson and Kitching, 1972; Kemp, 1982; Carroll, 1988).

As acceptance of Hennig's (1966) concept of monophyly grew, *Cynodontia* was again reconceptualized. Today *Cynodontia* is generally regarded as a clade that includes *Mammalia* (this volume) plus its closest extinct relatives (Hopson and Barghusen, 1986; Rowe, 1986, 1993, 2004; Gauthier et al., 1988a; Kemp, 2005, Benton, 2015) and as being nested within the more inclusive clade *Therapsida* (this volume) and the mammalian total clade (see *Pan-Mammalia,* this volume).

Other names are potentially available for the clade here named *Cynodontia*. These include Owen's *Theriodontia*, which Seeley (1889) used as a synonym of *Cynodontia*. However, the name *Theriodontia* is now commonly used for a subset of *Therapsida* (this volume) that includes gorgonopsids and therocephalians as well as cynodonts (Hopson and Barghusen, 1986; Sidor and Hopson, 1998). *Gomphodontia* (Seeley, 1894b) has always excluded *Mammalia*, and the name is commonly used to refer to an extinct subset of cynodonts of doubtful monophyly (e.g., Romer, 1967; Liu and Olsen, 2010; Benton, 2015). *Mammalia* Linnaeus 1758 sensu Lucas and Hunt (1990) is an approximate synonym, but those authors proposed no alternative name for the mammalian crown (Rowe and Gauthier, 1992).

The implications of a monophyletic *Cynodontia* are still being explored, but already they can be felt in sweeping new generalizations about trends and hypothesized mechanisms that shaped the history of the clade, including those involved in the origin of its crown clade, the subsequent diversification of mammalian lineages, and even in the origin of our own species (Rowe and Shepherd, 2016).

Comparative and developmental anatomy of living mammals offer much insight into what the ancestral mammal must have been like and what shaped its history. Such studies have postulated numerous evolutionary driving factors, including increased encephalization and origin of the neocortex and innovations involving hearing, feeding, taste, olfaction, miniaturization, parental care, endothermy, elevated metabolism, nocturnality, and others. Although deeply informative, this approach reveals little about timing or sequences of historical events that preceded the origin of mammals, nor does it speak to the relative importance of the various proposed driving mechanisms at particular points in pan-mammalian history.

Clarification of many of these matters is emerging from a succession of phylogenetic analyses conducted over the last three decades that explored the place of *Mammalia* within more inclusive clades containing its closest extinct relatives among *Cynodontia, Therapsida* (this volume), and *Pan-Mammalia* (this volume) (Rowe, 1986, 1988, 1993; Gauthier et al., 1988a; Sidor and Hopson, 1998; Rubidge and Sidor, 2001; Kielan-Jaworowska et al., 2004; Meng et al., 2006; Ji et al., 2006; Liu and Olson, 2010). The more powerful analyses combed the entire skeleton of fossil and Recent taxa for phylogenetically informative characters without assuming *a priori* that any subsystem, for example the dentition or postcranium, is more or less susceptible to homoplasy. This approach has illuminated unrecognized and unexpected correlations between morphological characters, as well as nuanced patterns in the historical sequence of morphological transitions in the skeleton, based on the most strongly supported trees.

The occlusal 'triconodont' molariform tooth found in the earliest cynodonts set the stage for a remarkable diversification of tooth structure. In more crownward cynodonts (e.g., *Kuehneotherium praecursoris*), the two minor cups become 'circumducted' or rotated relative to the taller principal cusp, by moving to the outer side in the upper teeth, and to the inner side in the lowers, thus establishing cheek teeth that not only occluded but also interlocked to produce an elongated shearing edge for mastication. This condition is termed the 'tritubercular' molar. A further step was taken later, as the tritubercular molariform tooth became elaborated into the dual-action 'tribosphenic' molar along the stem of therian mammals (Osborn, 1907; Gregory, 1916, 1922, 1953). Opposing upper and lower crowns have taken on the form of reversed-triangular interlocking molars, where the principal cusps support long crests that shear past one another, and where the principal upper molar cusp (protocone) bites into the lower talonid basin with a crushing action (e.g., *Didelphis*).

Edward Drinker Cope (1883, 1888) was the first to recognize the significance of the tritubercular pattern, as he called it, hypothesizing that this was the basic form from which most or all mammalian teeth were derived. This came to be known as Cope's Tritubercular Theory (Wortman, 1886, 1902; Gidley, 1906). Henry Fairfield Osborn (1888, 1907[*]) traced the evolution of mammalian tooth crowns in considerable detail, and the theory came to be known as the Cope-Osborn theory (e.g., Gregory, 1910, 1916, 1922). Although the detailed homologies of various cusps and cuspules in different mammalian sub-clades have been debated for decades, it is generally true that the tritubercular pattern represents the morphological primordium from which an immense diversity of crown morphologies ultimately evolved (Cope, 1883; Osborn, 1907; Gregory 1922; Romer, 1966; Carroll,

---

[*] Osborn (1907) was edited by William King Gregory. Whereas Henry Fairfield Osborn is credited with refining the 'Cope-Osborn' theory of tritubercular molar evolution, much of the leg-work was performed by W. K. Gregory, and many of the successful insights in Osborn (1907) are probably Gregory's as well.

1988; Benton, 2015). The tritubercular theory is focused on evolution of the crowns of mammalian (mostly therian) molars, but its antecedent states and the trend toward elaboration of molariform tooth crowns can now be traced back to the earliest cynodonts (Crompton, 1971, 1972, 1974; Crompton and Jenkins, 1968; Kielan-Jaworowska et al., 2004; Davis, 2011). Additional insights can be gained from considering the evolutionary history of the larger integrated system of which the tooth crowns are a part, and their correlated and historical sequences of transformation.

In early cynodonts (e.g., *Thrinaxodon liorhinus*), each molariform tooth was implanted into a shallow socket via a single, short root that was not much longer than the height of the crown. It was held in place by a thin ring of bone, and the teeth were regularly replaced throughout life. Soon, however, the root became elongated, it was anchored into a deep socket by a periodontal ligament, the so-called thecodont gomphosis (Rowe, 1993), and this innovation in implantation was correlated with reduction in the rate and mode of tooth replacement (Cifelli et al., 1996; Luo et al., 2004). This innovation applies to *Eucynodontia*, the clade originating in the last common ancestor that mammals share with the Early Triassic *Cynognathus platyceps* (Rowe, 1993). At the same time, the molariform crowns increased in occlusal surface area and complexity, and soon thereafter the roots became 'incipiently divided', that is, they now have two root canals, each with its own dental nerve, but the mineralized roots were still connected by a web of dentine (e.g., *Diademodon*; Rowe, 2017). This innovation is correlated with expansion of the temporal fossa and strengthening of the masticatory musculature and the skeleton that supports it (viz., heightened coronoid process of dentary, more robust zygomatic arch, tall sagittal crest). One mechanical effect of deep tooth implantation was a stronger bite force applied to a broader occlusal surface (Kemp, 1982, 2005; Rowe, 1986, 1993).

Generally overlooked is the doubled (or more) innervation of the molariform teeth indicated by their multiple root canals, as well as the probable innervation of the periodontal ligament in these early cynodonts. Whereas the teeth in vertebrates are generally innervated, the degree of innervation and the size of their projection to somatosensory areas of the neocortex are unique to mammals (Rowe and Shepherd, 2016). This trend traces back to the earliest cynodonts, whose occluding molariform teeth were retrieving new levels of textural information about what they were masticating from differential loading of various regions of each molariform crown. With occlusion and mastication, the cynodont dentition was becoming integrated into a new mechanosensory system whose emergence correlates with the first measurable pulse in encephalization to take place among pan-mammals (Rowe et al., 2011; Rowe, 2017).

The degree of morphological disparity found in the dentition is perhaps best expressed by noting that nearly all species of *Cynodontia* (living and extinct) can be identified by their molariform teeth alone (for reviews of *Mammalia*, see Gregory, 1910, 1916, 1922, 1953; Hillson, 2009; Ungar, 2010). This is not true for other clades of extinct pan-mammals. Beginning with the first cynodonts, a factor now seen as a driving force across the entire history of the clade involves a new variational modality (character identity network; Wagner, 2014) in dental evolution that involved cusp numbers, shapes, crown sizes and geometry, occlusal relationships, replacement rates, numbers and shapes of roots, and a corresponding increase in evolutionary rate of the dentition as a system or module.

To frame this in a simple quantitative perspective, consider the basis of a series of phylogenetic analyses that scored cynodont dental characters. Gauthier et al. (1988a) scored 207 characters across 29 taxa that characterized skeletal variation across the major amniote clades, including the entire stem groups of *Mammalia* and *Reptilia*. Only 8% (17) of these characters reflect dental variation. At this scale of analysis, most of the observed dental variation among pan-mammals presented itself as autapomorphy and was phylogenetically uninformative. But for a data matrix designed to capture variation specifically among extinct cynodonts and early mammals, Meng et al. (2006) scored 435 total osteological characters for 58 taxa, and 25% (108 characters) reflect dental variation. Building on that matrix, Ji et al. (2006) scored 445 characters for 103 taxa, of which 39% (173) describe dental variation. In a matrix of 3,660 osteological characters for 86 fossil and living *Mammaliaformes*, O'Leary et al. (2013) found 40% of total skeletal variation (1,450 characters) to reside in the dentition. Additionally, in a study of cynodont (including early mammalian) relationships, Rowe (1993) found in a matrix of 151 characters for 24 taxa that there was 30% more homoplasy in cynodont dental characters than in either the skull, the postcranium, or the combined skeleton. Whereas these numbers all beg systematic questions about character independence and weighting (Harjunmaa et al., 2014), they illustrate the general trend of accelerated rate in dental evolution and increased homoplasy that began with the first cynodonts and can be traced into many clades within *Mammalia* (Rowe and Shepherd, 2016).

A number of the correlated diagnostic features of cynodonts contributed to a larger system of integrated senses, powered by an enlarged and elaborated brain. In the earliest pan-mammals, the choanae were located at the front of the mouth. These macropredators had simple, conical teeth and used their mouths to secure and dismember prey, employing a mode of inertial swallowing that was aided by teeth on the palate (Kemp, 2005). The appearance of the secondary palate marked a profound reorganization in feeding behavior and olfaction. Palatal teeth and the inertial mode of swallowing were lost in cynodonts, and the tongue became the major guide of food around the mouth during mastication and toward the esophagus in swallowing (Barghusen, 1986; Crompton, 1989). The secondary palate also created a new passageway through the nose, the nasopharyngeal passage, which lies above the secondary palate (its floor) and beneath the nasal capsule (its roof, whether cartilaginous or ossified), further isolating the olfactory epithelium from the mouth, and displacing the choanae to the back of the oral cavity.

With an occlusal dentition and secondary palate, the new feeding behaviors of early cynodonts became the basis of a new form of olfaction, known as ortho-retronasal olfaction (Rowe and Shepherd, 2016). Mammals retain the ancient tetrapod mode of sniffing known as 'othonasal' olfaction, in which airborne environmental odorant molecules are drawn through the nares to activate olfactory epithelium during inhalation. Mammals are unique in also having a counterpart to orthonasal olfaction known as 'retronasal' olfaction (Rozin, 1982), in which air exhaled from the lungs carries with it an entirely new information domain of volatile odor molecules liberated in the mouth through the breakdown of food by chewing, saliva, and actions of the tongue. These molecules pass forward from the oropharynx across the olfactory epithelium before being expelled through the nares. In retronasal olfaction, *smell* combines with *taste* and other senses (somatosensation, vision, and hearing) to generate our sense of *flavor* (Shepherd, 2004, 2006, 2012). Orthonasal smell, retronasal smell, taste, and somatosensory signals from the lips, tongue, and teeth all converge on single neurons in the neocortical area known as the orbitofrontal cortex (de Araujo et al., 2003; Small et al., 2007; Rolls and Grabenhorst, 2008). That flavor is a multisensory map in which distinct classes of information are integrated is evident in clinical data from patients who lost olfactory sensation following nasal infection or cranial trauma (Cullen and Leopold, 1999; Franselli et al., 2004; Bonfils et al., 2005) and from laboratory experiments (Heilmann and Humel, 2004; Sun and Halpern, 2005; Gautam and Verhagen, 2012). The discovery of cortical integration of multiple sensory modalities, which in other tetrapods remain independent, suggested the emergence of a new sensory system of "ortho-retronasal olfaction" in *Mammalia*. Osteological evidence traces the origin of this new system (but not the full development seen in the mammalian crown) to the earliest cynodonts (Rowe and Shepherd, 2016).

Also apomorphic of *Cynodontia* is a modified craniovertebral joint, in which the basioccipital recedes from the joint, and a 'double occipital condyle' is formed by the right and left exoccipitals that are positioned at the ventrolateral edges of the foramen magnum. The new articulation expanded the degree of stable dorsoventral and lateral excursion of the head on the neck without impairing passage of the spinal cord through the foramen magnum and along the cervical neural canal (Jenkins, 1969, 1971). The ventrolateral position of the condyles also suggests that the head was habitually held at a tilt with the nose toward the ground. Many mammals target their noses towards the ground and move their heads rapidly from side to side in scent-tracking and scent-guided navigation. More agile head movement potentially enabled cynodont olfaction to assert its importance in tracking and navigation (Rowe and Shepherd, 2016).

Augmenting this function is the evidence of a shift from a plesiomorphic buccal pump ventilation system to diaphragmatic ventilation in the earliest cynodonts. Differentiation of separate thoracic and lumbar regions marks the development of a muscular diaphragm, which separated the thoracic and abdominal cavities and initiated the onset of the stereotyped vacuum-chamber or bellows-like tidal diaphragmatic ventilation of mammals. While stationary or at rest, ventilation in living mammals is driven by the diaphragm (Bramble, 1989; Alexander, 2003). Chewing food is mostly a stationary action, thus ortho-retronasal olfaction is driven by diaphragmatic ventilation. The rapid sniffing so characteristic of many mammals in exploring their environments (Stoddart, 1980; Shepherd, 2012) is also driven by the diaphragm. In early cynodonts, the proximal ends of the thoracic ribs are flattened and imbricate in a condition unknown in other extinct therapsids or in living mammals. Hence their modes of breathing and locomotion were not entirely modern, but the important new capacity of diaphragmatic ventilation was introduced with the origin of *Cynodontia*. All of this is consistent with the view that olfactory ecology and dietary diversification had become primary influences on cynodont diversification (Hayden et al., 2010). This trend can be traced across approximately ~250 million years of evolution, from the first cynodonts to their living descendants.

In summary, conceptualizing *Cynodontia* as a clade including *Mammalia*, and understanding the correlation of characters traced to its origin, gives added insight into the significance of the dental apomorphy proposed here as the defining characteristic of its name.

Literature Cited

Abdala, F. 2007. Redescription of *Platycraniellus elegans* (*Therapsida, Cynodontia*) from the Lower Triassic of South Africa, and the cladistic relationships of eutheriodonts. *Palaeontology* 50:591–618.

Abdala, F., and N. P. Giannini. 2002. Chiniquodontid cynodonts: systematic and morphometric considerations. *Palaeontology* 4:1151–1170.

Abdala, F. T., B. S. Jashashvili, B. Rubidge, and J. van den Heever. 2014. New material of *Microgomphodon oligocynus* (*Eutherapsida, Therocephalia*) and the taxonomy of southern African *Bauriidae*. Pp. 209–231 in *Early Evolutionary History of the Synapsida* (C. F. Kammerer, K. D. Angielczyk, and Jörg Fröbisch, eds.). Springer, Netherlands.

Alexander, R. M. 2003. *Principles of Animal Locomotion*. Princeton University Press, Princeton, NJ.

Amalitzky, V. 1922. Diagnoses of the new forms of vertebrates and plants from the Upper Permian of North Dvina. *Bull. Acad. Imp. Sci. St. Petersburg* 16:329–340.

Barghusen, H. R. 1986. On the evolutionary origin of the therian tensor veli palatini and tensor tympani muscles. Pp. 253–262 in *The Ecology and Biology of Mammal-Like Reptiles* (N. H. Hotton, III, P. D. MacLean, J. J. Roth, and E. C. Roth, eds.). Smithsonian Institution Press, Washington, DC.

Benton, M. J. 2015. *Vertebrate paleontology*. 4th edition. Wiley Blackwell, West Sussex.

Bonaparte, J. F., A. G. Martinelli, and C. L. Schultz. 2005. New information on *Brasilodon* and *Brasilitherium* (*Cynodontia, Probainognathia*) from the late Triassic of southern Brazil. *Rev. Bras. Paleontol.* 8:25–46.

Bonaparte, J. F., A. G. Martinelli, C. L. Schultz, and R. Rubert. 2003. The sister group of mammals: small cynodonts from the Late Triassic of southern Brazil. *Rev. Bras. Paleontol.* 5:5–27.

Bonaparte, J. F., and M. C. Barberena. 2001. On two advanced carnivorous cynodonts from the Late Triassic of southern Brazil. *Bull. Mus. Comp. Zool.* 156(1):59–80.

Bonfils, P., P. Avan, P. Faulcon, and D. Malinvaud. 2005. Distorted odorant perception—analysis of a series of 56 patients with parosmia. *Arch. Otolaryngol. Head Neck Surg.* 131:107–112.

Botha, J., F. Abdala, and R. Smith. 2007. The oldest cynodont: new clues on the origin and early diversification of the *Cynodontia. Zool. J. Linn. Soc.* 149:477–492.

Bramble, D. M. 1989. Axial-appendicular dynamics and the integration of breathing and gait in mammals. *Am. Zool.* 29:171–186.

Broom, R. 1905a. Preliminary notice of some new fossil reptiles collected by Mr. Alfred Brown at Aliwal North, South Africa. *Rec. Albany Mus. Grahamstown S. Afr.* 4:269–275.

Broom, R. 1905b. On the use of the term *Anomodontia. Rec. Albany Mus. Grahamstown S. Afr.* 4:266–269.

Broom, R. 1906. On a new cynodont reptile (*Aelurosuchus browni*). *Trans. S. Afr. Philos. Soc.* 16:376–378.

Broom, R. 1907. The origins of the mammal-like reptiles. *Proc. Zool. Soc. Lond.* 1907:1047–1061.

Broom, R. 1909. Notice of some new South African fossil amphibians and reptiles. *Ann. S. Afr. Mus.* 7:270–278.

Broom, R. 1910. A comparison of the Permian reptiles of North America with those of South Africa. *Bull. Am. Mus. Nat. Hist.* 28:197–234.

Broom, R. 1911. On the structure of the skull in cynodont reptiles. *Proc. Zool. Soc. Lond.* 81:893–925.

Broom, R. 1932. *The Mammal-Like Reptiles of South Africa and the Origin of Mammals*. H. F. & G. Witherby, London.

Broom, R. 1948. A contribution to our knowledge of the vertebrates of the Karroo beds of South Africa. *Trans. R. Soc. Edinb.* 61:577–629.

Broom, R. 1950. Some fossil reptiles form the Karroo beds of Lady Frere. *S. Afr. J. Sci.* 47:86–88.

Carroll, R. L. 1988. *Vertebrate Paleontology and Evolution*. W. H. Freeman & Co., New York.

Cifelli, R. L., T. B. Rowe, W. P. Luckett, J. Banta, R. Reyes, and R. I. Howes. 1996. Fossil evidence for the origin of the marsupial pattern of tooth replacement. *Nature* 379:715–718.

Cope, E. D. 1883. On the trituberculate type of molar tooth in the *Mammalia. Proc. Am. Philos. Soc.* 1883:324–326.

Cope, E. D. 1888. On the tritubercular molar in human dentition. *J. Morphol.* 2:7–23.

Crompton, A. W. 1955. On some Triassic cynodonts from Tanganyika. *Proc. Zool. Soc. Lond.* 125:617–669.

Crompton, A. W. 1963. Tooth replacement in the cynodont *Thrinaxodon liorhinus. Ann. S. Afr. Mus.* 46:479–521.

Crompton, A. W. 1971. The origin of the tribosphenic molar. Pp. 65–87 in *Early Mammals* (D. M. Kermack and K. A. Kermack, eds.). *Zool. J. Linn. Soc.* 50(suppl. 1).

Crompton, A. W. 1972. Postcanine occlusion in cynodonts and tritylodontids. *Bull. Br. Mus. Nat. Hist. Geol.* 21:27–71.

Crompton, A. W. 1974. The dentitions and relationships of the Southern African Triassic mammals *Erythrotherium parringtoni*, and *Megazostrodon rudneri. Bull. Br. Mus. Nat. Hist. Geol.* 24:399–443.

Crompton, A. W. 1989. The evolution of mammalian mastication. Pp. 23–40 in *Complex Organismal Functions: Integration and Evolution in Vertebrates* (D. B. Wake and G. Roth, eds.). John Wiley, New York.

Crompton, A. W., and F. A. Jenkins, Jr. 1968. Molar occlusion in Late Triassic mammals. *Biol. Rev.* 43:427–458.

Cullen, M. M., and D. A. Leopold. 1999. Disorders of smell and taste. *Med. Clin. N. Am.* 83:57–74.

Davis, B. M. 2011. Evolution of the tribosphenic molar pattern in early mammals, with comments on the "dual-origin" hypothesis. *J. Mamm. Evol.* 18:227–244.

de Araujo, I. E., E. T. Rolls, M. L. Kringelbach, F. McGlone, and N. Phillips. 2003. Taste-olfactory convergence, and the representation of the pleasantness of flavor, in the human brain. *Eur. J. Neurosci.* 18:2059–2068.

Franselli, J., B. N. Landis, S. Heilmann, B. Hauswald, K. B. Huttenbrink, J. S. Lacroix, D. A. Leopold, and T. Hummel. 2004. Clinical presentation of qualitative olfactory dysfunction. *Eur. Arch. Otorhinolaryngol.* 261:411–415.

Gautam, S. H., and J. V. Verhagen. 2012. Retronasal odor representations in the dorsal olfactory bulb of rats. *J. Neurosci.* 32:7949–7959.

Gauthier, J. A., A. G. Kluge, and T. B. Rowe. 1988a. Amniote phylogeny and the importance of fossils. *Cladistics* 4:105–209.

Gauthier, J. A., A. G. Kluge, and T. B. Rowe. 1988b. The early evolution of the *Amniota*. Pp. 103–155 in *The Phylogeny and Classification of the Tetrapods. Vol. 1: Amphibians, Reptiles and Birds* (M. Benton, ed.). Systematics Association Special Volume No. 35a. Clarendon Press, Oxford.

Gidley, J. W. 1906. Evidence bearing on tooth-cusp development. *Proc. Wash. Acad. Sci.* 8:91–110.

Gregory, W. K. 1910. The orders of mammals. *Bull. Am. Mus. Nat. Hist.* 27:1–524.

Gregory, W. K. 1916. Studies on the evolution of the primates. Part I. The Cope-Osborn "theory of trituberculy" and the ancestral molar patterns of the primates. Part II. Phylogeny of recent and extinct anthropoids, with special reference to the origin of man. *Bull. Am. Mus. Nat. Hist.* 35:239–355.

Gregory, W. K. 1922. *The Origin and Evolution of the Human Dentition.* Williams and Wilkins, Baltimore, MD.

Gregory, W. K. 1953. *Evolution Emerging.* Macmillan, New York.

Harjunmaa, E., K. Seidel, T. Häkkinen, E. Renvoisé, I. J. Corfe, A. Kallonen, Z. Q. Zhang, A. R. Evans, M. L. Mikkola, I. Salazar-Ciudad, and O. D. Klein. 2014. Replaying evolutionary transitions from the dental fossil record. *Nature* 512:44–48.

Haughton, S. H. 1924. A bibliographic list of Pre-Stormberg Karroo *Reptilia*, with a table of horizons. *Trans. R. Soc. S. Afr.* 12:51–104.

Hayden, S., M. Bekaert, T. A. Crider, S. Mariani, W. J. Murphy, and E. C. Teeling. 2010. Ecological adaptation determines functional mammalian olfactory subgenomes. *Genome Res.* 20:1–9.

Heilmann, S., and T. Hummel. 2004. A new method for comparing orthonasal and retronasal olfaction. *Behav. Neurosci.* 118:412–419.

Hennig, W. 1966. *Phylogenetic Systematics.* Translated by D. D. Davis and R. Zangerl. University of Illinois Press, Urbana, IL.

Hillson, S. 2009. *Teeth.* 2nd edition. Cambridge University Press, Cambridge, UK.

van Hoepen, E. C. N. 1917. Note on *Myriodon* and *Platycranium. Ann. Transv. Mus.* 5:217.

Hopson, J. A., and H. R. Barghusen. 1986. An analysis of therapsid relationships. Pp. 83–106 in *Ecology and Biology of Mammal-Like Reptiles* (N. Hotton, III, P. D. McLean, J. J. Roth, and E. C. Roth, eds.). Smithsonian Institution and National Institute of Mental Health, Washington, DC.

Hopson, J. A., and J. W. Kitching. 1972. A revised classification of cynodonts (*Reptilia*; *Therapsida*). *Palaeontol. Afr.* 14:71–85.

Hopson, J. A., and J. W. Kitching. 2001. A probainognathian cynodont from South Africa and the phylogeny of non-mammalian cynodonts. *Bull. Mus. Comp. Zool.* 156:5–35.

von Huene, F. 1936. Die fossilen Reptilien des Südamerikanischen Gondwanalandes. Pp. 93–159 in *Lieferung 2* (Franz R. Heine, ed.). Tübingen. Reprinted in Huene, 1944, C. H. Beck, Munich, 1935–1942.

Huttenlocker, A. 2009. An investigation into the cladistic relationships and monophyly of therocephalian therapsids (*Amniota: Synapsida*). *Zool. J. Linn. Soc.* 157:865–891.

Jenkins, Jr., F. A. 1969. The evolution and development of the dens of the mammalian axis. *Anat. Rec.* 164:173–184.

Jenkins, Jr., F. A. 1971. The postcranial skeleton of African cynodonts. *Bull. Peabody Mus. Nat. Hist.* 36:1–216.

Ji, Q., Z.-X. Luo, C. Yuan, and A. R. Tabrum. 2006. A swimming mammaliaform from the Middle Jurassic and ecomorphological diversification of early mammals. *Science* 311:1123–1127.

Kemp, T. S. 1979. The primitive cynodont *Procynosuchus*: functional anatomy of the skull and relationships. *Philos. Trans. R. Soc. Lond. B Biol. Sci.* 285:73–122.

Kemp, T. S. 1982. *Mammal-Like Reptiles and the Origin of Mammals*. Academic Press, London.

Kemp, T. S. 2005. *The Origin and Evolution of Mammals*. Oxford University Press, Oxford.

Keyser, A. W., and A. S. Brink. 1979. A new bauriamorph (*Herpetogale marsupialis*) from the Omigonde Formation (Middle Triassic) of South West Africa. *Ann. Geol. Surv. S. Afr.* 12:91–105.

Kielan-Jaworowska, Z., R. L. Cifelli, and Z.-X. Luo. 2004. *Mammals from the Age of Dinosaurs*. Columbia University Press, New York.

Lehman, J.-P. 1961. Cynodontia. Pp. 140–191 in *Traité de Paleontologie*, Tome 6 (J. Piveteau, ed.). Masson, Paris.

Linnaeus, C. 1758. *Systema Naturae Per Regna Tria Naturae, Secundum Classes, Ordines, Genera, Species, cum Characteribus, Differentiis, Synonymis, Locis*. Editio decima. Laurentii Salvii, Holmiae (Stockholm).

Liu, J., and P. Olsen. 2010. The phylogenetic relationships of *Eucynodontia* (*Amniota: Synapsida*). *J. Mamm. Evol.* 17:151–176.

Lucas, S. G., and A. P. Hunt. 1990. The oldest mammal. *New Mexico J. Sci.* 30:41–49.

Luo, Z.-X., Z. Kielan-Jaworowska, and R. L. Cifelli. 2004. Evolution of dental replacement in mammals. *Bull. Carnegie Mus. Nat. Hist.* 36:159–175.

Lydekker, R. 1890. *Catalogue of the Fossil* Reptilia *and* Amphibia *in the British Museum (Natural History)*. Part IV. Trustees of the British Museum (Natural History), London.

Martinelli, A. G., and G. W. Rougier. 2007. On *Chaliminia musteloides* (*Eucynodontia: Tritheledontidae*) from the Late Triassic of Argentina, and a phylogeny of *Ictidosauria*. *J. Vertebr. Paleontol.* 27:442–460.

Martinez, R. N., C. L. May, and C. A. Forster. 1996. A new carnivorous cynodont from the Ischigualasto Formation (Late Triassic, Argentina), with comments on eucynodont phylogeny. *J. Vertebr. Paleontol.* 16:271–284.

Meng, J., Y. Hu, Y. Wang, X. Wang, and C. Li. 2006. A Mesozoic gliding mammal from northeastern China. *Nature* 444:889–893.

Northcutt, G. L. 2011. Evolving large and complex brains. *Science* 332:926.

O'Leary, M. A., Bloch, J. I., J. J. Flynn, T. J. Gaudin, A. Giallombardo, N. P. Giannini, S. L. Goldberg, B. P. Kraatz, Z.-X. Luo, J. Meng, X. Ni, M. J. Novacek, F. A. Perini, Z. Randall, G. W. Rougier, E. J. Sargis, M. T. Silcox, N. B. Simmons, M. Spaulding, P. M. Velazco, M. Weksler, J. R. Wible, and A. L. Cirranello. 2013. The placental mammal ancestor and the post-K-Pg radiation of placentals. *Science* 339:662–667.

Osborn, H. F. 1888. The evolution of mammalian molars to and from the tritubercular type. *Am. Nat.* 22:1067–1079.

Osborn, H. F. 1903a. On the primary divisions of the *Reptilia* into two sub-classes, *Synapsida* and *Diapsida*. *Science* 17:275–276.

Osborn, H. F. 1903b. The reptilian subclasses *Diapsida* and *Synapsida* and the early history of the *Diaptosauria. Mem. Am. Mus. Nat. Hist.* 1:451–507.

Osborn, H. F. 1907. *The Evolution of Mammalian Molar Teeth to and from the Triangular Type* (William K. Gregory, ed.). Macmillan, New York.

Owen, R. 1859. On some reptilian fossils from South Africa. *Q. J. Geol. Soc. Lond.* 16:49–63.

Owen, R. 1861. *Paleontology or a Systematic Summary of Extinct Animals and Their Geological Relations.* 2nd edition. Adams and Charles Black, Edinburgh.

Owen, R. 1876. *Descriptive and Illustrated Catalogue of the Fossil Reptilia of South Africa in the Collection of the British Museum.* Trustees of the British Museum (Natural History), London.

Parrington, F. R. 1936. On the tooth replacement in theriodont reptiles. *Philos. Trans. R. Soc. Lond. B Biol. Sci.* 226:121–142.

Rolls, E. T., and F. Grabenhorst. 2008. The orbitofrontal cortex and beyond: from affect to decision-making. *Prog. Neurobiol.* 86:216–244.

Romer, A. S. 1956. *Osteology of the Reptilia.* University of Chicago Press, Chicago, IL.

Romer, A. S. 1966. *Vertebrate Paleontology.* University of Chicago Press, Chicago, IL.

Romer, A. S. 1967. The Chãnares (Argentina) Triassic reptile fauna: III. Two new gomphodonts, *Massetognathus pascuali* and *M. teruggii. Mus. Comp. Zool. Breviora* 264:1–25.

Romer, A. S. 1970. The Chãnares (Argentina) Triassic Reptile Fauna: VI. A chiniquodontid cynodont with an incipient squamosal-dentary jaw articulation. *Mus. Comp. Zool. Breviora* 344:1–18.

Rowe, T. B. 1986. Osteological diagnosis of *Mammalia,* L. 1758, and its relationships to extinct *Synapsida.* PhD dissertation, University of California, Berkeley, CA.

Rowe, T. B. 1988. Definition, diagnosis and origin of *Mammalia. J. Vertebr. Paleontol.* 8:241–264.

Rowe, T. B. 1993. Phylogenetic systematics and the early history of mammals. Pp. 129–145 in *Mammalian Phylogeny* (F. S. Szalay, M. J. Novacek, and M. C. McKenna, eds.). Springer-Verlag, New York.

Rowe, T. B. 1996a. Coevolution of the mammalian middle ear and neocortex. *Science* 273:651–654.

Rowe, T. B. 1996b. Brain heterochrony and evolution of the mammalian middle ear. Pp. 71–96 in *New Perspectives on the History of Life* (M. Ghiselin and G. Pinna, eds.). California Academy of Sciences, Memoir 20.

Rowe, T. B. 2004. Chordate phylogeny and development. Pp. 384–409 in *Assembling the Tree of Life* (M. J. Donoghue and J. Cracraft, eds.). Oxford University Press, Oxford and New York.

Rowe, T. B. 2017. The emergence of mammals. Pp. 1–52 in *Evolution of Nervous Systems.* 2nd edition, Vol. 2 (Jon Kaas, ed.). Oxford: Elsevier.

Rowe, T. B., W. Carlson, and W. Bottorff. 1995. Thrinaxodon: *Digital Atlas of the Skull.* CD-ROM. 2nd edition, for Windows and Macintosh platforms. University of Texas Press.

Rowe, T. B., and J. A. Gauthier. 1992. Ancestry, paleontology, and definition of the name *Mammalia. Syst. Biol.* 41:372–378.

Rowe, T. B., J. Kappelman, W. D. Carlson, R. A. Ketcham, and C. Denison. 1997. High-Resolution Computed Tomography: a breakthrough technology for Earth scientists. *Geotimes* 42:23–27.

Rowe, T. B., T. E. Macrini, and Z.-X. Luo 2011. Fossil evidence on origin of the mammalian brain. *Science* 332:955–957.

Rowe, T. B., and G. M. Shepherd. 2016. The role of ortho-retronasal olfaction in mammalian cortical evolution. *J. Comp. Neurol.* 524:471–495.

Rozin, P. 1982. Taste-smell confusions and the duality of the olfactory sense. *Atten. Percept. Psycho.* 31:397–401.

Rubidge, B. S., and C. A. Sidor. 2001. Evolutionary patterns among Permo-Triassic therapsids. *Annu. Rev. Ecol. Syst.* 32:449–480.

Rupke, N. 1994. *Richard Owen: Victorian Naturalist.* Yale University Press, New Haven, CT.

Seeley, H. G. 1889. Researches on the structure, organization, and classification of the fossil

Reptilia. Pt. IV. On the anomodont *Reptilia* and their allies. *Philos. Trans. R. Soc. Lond. B Biol. Sci.* 180:215–296.

Seeley, H. G. 1894a. Researches on the structure, organization, and classification of the fossil *Reptilia*. Part IX. Section 5. On the skeleton in new *Cynodontia* from the Karroo rocks. *Proc. R. Soc. Lond. B Biol. Sci.* 56:291–294.

Seeley, H. G. 1894b. Researches on the structure, organization, and classification of the fossil *Reptilia*. Part IX, Section 4. On the *Gomphodontia*. *Proc. R. Soc. Lond. B Biol. Sci.* 56: 288–291.

Seeley, H. G. 1895. Researches on the structure, organization, and classification of the fossil *Reptilia*. Part IX, Section 5. On the skeleton of new *Cynodontia* from the Karoo rocks. *Philos. Trans. R. Soc. Lond. B Biol. Sci.* 186:59–148.

Shepherd, G. M. 2004. The human sense of smell: are we better than we think? *PLOS Biol.* 2:e146.

Shepherd, G. M. 2006. Smell images and the flavour system in the human brain. *Nature* 444:316–321.

Shepherd, G. M. 2012. *Neurogastronomy: How the Brain Creates Flavor and Why It Matters.* Columbia University Press, New York.

Sidor, C. A., and J. A. Hopson. 1998. Ghost lineages and 'mammalness': assessing the temporal pattern of character acquisition in the *Synapsida. Paleobiology* 24:254–273.

Sidor, C. A., and P. J. Hancox. 2006. *Elliotherium kersteni*, a new tritheledontid from the Lower Elliot Formation (Upper Triassic) of South Africa. *J. Paleontol.* 80:333–342.

Small, D. M., G. Bender, M. G. Veldhuizen, K. Rudenga, D. Nachtigal, and J. Felsted. 2007. The role of the human orbitofrontal cortex in taste and flavor processing. *Ann. N. Y. Acad. Sci.* 1121:136–151.

Stoddart, D. M. 1980. *Olfaction in Mammals.* Academic Press, London.

Sun, B. C., and B. P. Halpern. 2005. Identification of air phase retronasal and orthonasal odorant pairs. *Chem. Senses* 30:693–706.

Tatarinov, L. P. 1968. New theriodonts from the Upper Permian of USSR. Pp. 32–46 in *Upper Paleozoic and Mesozoic Amphibia and Reptilia from the U.S.S.R.* Academy of Science of the USSR, Moscow (in Russian).

Ungar, P. S. 2010. *Mammal Teeth: Origin, Evolution, and Diversity.* Johns Hopkins University Press, Baltimore, MD.

Wagner, G. 2014. *Homology, Genes, and Evolutionary Innovation.* Princeton University Press, Princeton, NJ.

Watson, D. M. S. 1917. A sketch classification of the pre-Jurassic tetrapod vertebrates. *Proc. Zool. Soc. Lond.* 1917:167–186.

Wortman, J. L. 1886. Comparative anatomy of the teeth in *Vertebrata. Am. Syst. Dent.* 1886:351–503.

Wortman, J. L. 1902. Origin of the tritubercular molar. *Am. J. Sci.* 8:93–98.

Author

Timothy B. Rowe; Jackson School of Geosciences; The University of Texas at Austin; C1100, Austin, TX 78712. Email rowe@utexas.edu.

Date Accepted: 12 June 2017

Primary Editors: Jacques A. Gauthier, Kevin de Queiroz

Mammaliamorpha T. B. Rowe 1988 [T. B. Rowe], converted clade name

Registration Number: 222

Definition: The smallest clade containing *Homo sapiens* Linnaeus 1758 (*Mammalia*) and *Tritylodon longaevus* Owen 1884 (*Tritylodontidae*). This is a minimum-clade definition. Abbreviated definition: min ∇ (*Homo sapiens* Linnaeus 1758 & *Tritylodon longaevus* Owen 1884).

Etymology: Combines *Mammalia* (this volume) with the Greek *morphe* (shape, form).

Reference Phylogeny: Figure 3 in Rowe (1988) is the primary reference phylogeny. See also Figure 1 in Luo (2007).

Composition: *Mammaliamorpha* includes *Mammalia* (this volume) plus a subset of the extinct species and clades that, under current hypotheses (Gauthier et al., 1988; Rowe, 1988, 1993; Kielan-Jaworowska et al., 2004; Luo, 2007; Ruta et al., 2013; Luo et al., 2015), are viewed as the closest extinct relatives of *Mammalia* within *Cynodontia* (this volume). At one time or another, most of the extinct taxa mentioned below were considered to be members of the mammalian crown, or were at least referred to as 'mammals'. Included in *Mammaliamorpha* are *Adelobasileus cromptoni* Lucas and Hunt 1990 (see also Lucas and Luo, 1993), *Sinoconodon rigneyi* Patterson and Olson 1961 (see also Crompton and Sun, 1985), *Tritylodontidae*, and the extant and extinct members of *Mammaliaformes* (this volume).

Possibly included within *Mammaliamorpha* are some or all of the taxa assigned to *Brasilodontidae*. This is a problematic taxon of doubtful monophyly, based on a few, small, incomplete specimens, including *Brasilodon tetragonus*, *Brasilitherium riograndensis*, *Minicynodon maieri*, and *Protheriodon estudianti*, from the Middle and Late Triassic of Brazil and Argentina; and *Panchetocynodon damodarensi* from the Early Triassic of India (Bonaparte, 2003, 2013; Bonaparte et al., 2013). Bonaparte (2013) considered *Brasilodontidae* to be monophyletic and to be the sister taxon to *Mammaliaformes* (his *Mammalia*), but his taxon sampling omitted *Tritylodontidae*, leaving his results equivocal with respect to membership in *Mammaliamorpha*. Analyses by Luo (2007, scoring only *Brasilitherium*), and Liu and Olson (2010, scoring *Brasilodon*, and considering *Brasilitherium* as its junior synonym), found them to lie within *Mammaliamorpha*, but outside of *Mammaliaformes*. Abdala (2007) found the group to be paraphyletic, with *Brasilitherium* as the sister taxon to *Mammaliaformes*, and *Brasilodon* the sister taxon to *Mammaliamorpha* + *Tritheledontidae* (represented by *Pachygenelus* Watson 1913).

Additional potential members of *Mammaliamorpha* are some or all of the Late Triassic to Early Jurassic species referred to *Tritheledontidae* (= *Ictidosauria* Broom 1929; = *Diarthrognathidae*, = *Dromatheriidae* + *Tritheledon riconoi*, sensu Haughton and Brink, 1954). Recent reviews of *Tritheledontidae* have all found weak support for its monophyly (Martinelli et al., 2005; Martinelli and Rougier, 2007; Sidor and Hancox, 2006). All of the known specimens are very small and incomplete, and the few known postcranial elements are largely undescribed. The name has been used for different sets of taxa by different authors, and the phylogenetic results of different studies are difficult to compare owing to different taxon sampling. With that caveat, some authors infer *Tritheledontidae* to lie within *Mammaliamorpha* (e.g., Bonaparte et al., 2005; Hopson and Barghusen, 1986; Hopson and Kitching, 2001; Rubidge and Sidor, 2001; Luo, 2007; Luo et al., 2015), while others have placed it just outside (e.g., Rowe, 1993; Martinez et al., 1996; Abdala, 2007). Lucas and Luo (1993) obtained both results, with *Tritheledontidae* just inside or just outside of *Mammaliamorpha*. Sidor and Hancox (2006) and Martinelli and Rougier (2007) addressed relationships among tritheledontids but neither study included *Tritylodontidae*, hence their results are uninformative with respect to inclusion in *Mammaliamorpha*. Liu and Olsen (2010) inferred tritheledontids to be paraphyletic, with *Pachygenelus monus* and *Riograndia guaibensis* as successive outgroups to *Mammaliamorpha*.

Diagnostic Apomorphies: In early mammaliamorphs, the entire skeleton takes on many detailed, if subtle, resemblances to the skeleton in *Mammalia*, and the node to which the name *Mammaliamorpha* is attached represents the approximate point in history in which members of the mammalian stem lineage became miniaturized (Rowe, 1993). Especially noteworthy are apomorphies that involve the nose, orbital wall, ear, and palate. An ascending lateral process of the premaxilla forms the rear margin of the naris. There is also the first hint of ossification of the rear parts of the nasal capsule (Kielan-Jaworowska et al., 2004). Around the orbit, the prefrontal and postorbital bones are absent (still present in brasilodontids), and the postorbital arch no longer separates the orbit from the temporal fenestra. Sheets of bone from the orbitosphenoid, frontal, and palatine join to form an extensively ossified medial orbital wall and a rigid anchor for the extrinsic ocular musculature (a complex feature absent in tritheledontids and brasilodontids).

The otic capsule and adjacent regions including the craniomandibular joint underwent multiple transformations that collectively indicate an ear that was increasingly sensitive to a broader range of higher frequencies. The internal auditory meatus is walled medially with separate foramina for the vestibular and cochlear nerves, and the cochlea is now elongated (also present in some brasilodonts). The middle ear is still bound to the mandible, however, and the articular has *de novo* dorsal and ventral processes, the ventral being the homolog of the *manubrium mallei* in *Mammalia*. The postdentary bones are reduced to a collective narrow rod that lies deeply set into a trough beneath an apomorphically elongated, posteriorly-directed dentary condylar process. The surangular is no longer involved in the craniomandibular joint, which now lies primarily between the quadrate and articular.

Adjacent to the otic capsule, the prootic, alisphenoid, and quadrate ramus of the pterygoid join to form a laterally directed flange near the rear edge of the trigeminal foramen. The paroccipital process is directed laterally (instead of ventrolaterally) and is bifurcated distally, with one distal process forming a separate condyle for a kinetic articulation with the quadrate, and the other apparently articulating with the hyoid.

The palate and basicranium present several apomorphic transformations that may have affected the suspension of the musculature for swallowing. The parabasisphenoid and pterygoid no longer form a single continuous ventral parasagittal ridge, and instead form parallel parasagittal ridges separated by a shallow trough. The basicranium is broadly expanded to widely separate the pterygoid transverse processes. The parasphenoid wings (alae) are believed to form broad ventrolateral flanges (Sues, 1985, 1986); if so, they fuse indistinguishably to the basicranium at an early stage in ontogeny (the parasphenoid disappears as a separate ossification in

Mammalia). In addition, a partial floor forms beneath the cavum epiptercycum and indicates expansion of the trigeminal ganglion (Rowe, 1988, 1993).

In the dentition, the postcanine teeth have two or more fully divided roots, each with its own dental nerve canal, and tooth crowns have complex occlusal patterns (roots are undivided in brasilodontids and tritheledontids).

In the postcranium, numerous transformations suggest improved agility. The atlantal post-zygapophyses are absent, the atlas centrum is flattened, and the dens is present on the anterior face of the axis centrum (collectively affording greater stability and flexibility of the skull at the cranioversal joint). The caudal vertebrae are graded in length with elongated centra in the distal-most vertebrae. The sternum is segmented to form sternebrae (facilitating parasagittal flexion-extension of the trunk). The anterior coracoid (procoracoid) is widely excluded from participation in the glenoid. A prominent olecranon process is present on the ulna, clasping the distal humerus in a hemicylindrical articulation. The iliac blade is low with a flat dorsal margin, and divided into dorsal and ventral components by a longitudinal ridge, giving this bone a triangular shape in coronal cross section. The ischium, pubis and acetabulum are rotated posteriorly to lie entirely behind the sacrum where they enclose an obturator foramen that exceeds the acetabulum in diameter. The epipubic bones are present (lost in *Placentalia*). The femoral head is nearly spherical with a distinct fovea for attachment of the *ligamentum capitis femoris*. The calcaneum articulates with the navicular (see Kemp, 1983, 2005; Gauthier et al., 1988; Rowe, 1986, 1988, 1993).

Synonyms: *Mammalia* Linnaeus 1758 (sensu Gow, 1986; Clemens, 1986; Crompton and Sun, 1985; Hopson and Barghusen, 1986; Kemp, 1988; Kielan-Jaworowska et al., 2004;

Lucas, 1992; Lucas and Hunt, 1990; Miao, 1991; Owen, 1871, 1884) is an approximate synonym. *Ictidosauria* Broom 1929 is a partial synonym.

Comments: When Richard Owen (1884) described *Tritylodon longaevus* from the Late Triassic Stormberg beds of Lesotho (Basutoland), he considered it to be the oldest known species of *Mammalia*. Owen had already encountered one other tritylodont fossil, *Stereognathus ooliticus,* based on isolated teeth from the Middle Jurassic Stonesfield Slate of England, which he unhesitatingly identified as a mammal. In their expansive occlusal surfaces, Owen found these teeth most comparable to artiodactyls (Owen, 1857) and Tertiary perissodactyls (Owen, 1871). The occlusal nature of its molariform teeth and their implantation into the jaw by fully divided roots were sufficient justification in Owen's opinion for assigning tritylodonts to *Mammalia*.

In *Tritylodontidae*, the triconodontid tooth crown pattern is elaborated into parallel ridges of three or more tubercles; the two ridges of the lower molariforms occlude into slots between the three ridges of the upper molariforms, and occlusion occurs in propalinal movement of the mandible (Luo et al., 2015). Nowhere among *Therapsida* or *Reptilia*, as then conceived by Owen, could he find a comparable occlusal dentition, one effective in both cutting soft and breaking hard food items, and with fully divided roots. Furthermore, he stated that this "preliminary act of digestion should be a necessary correlation, or be in harmony, with a more complete conversion of the food into chyle and blood—and that such more efficient type of the whole digestive machinery should be correlated and necessarily so, with the hot blood, quick-beating heart and quick-breathing lungs, with the higher instincts, and more vigorous and varied acts, of a Mammal" (Owen, 1857:5). Seeley argued that skull structure indicated *Tritylodon longaevus* to be a theriodont ('reptile'), but one that approached *Mammalia* more closely than any of the others in having double-rooted teeth, and that "the character may be important enough to place *Tritylodon* in a group of animals intermediate between Mammals and Theriodonts" (Seeley, 1895: 1028), but he never named such a group.

When Robert Broom described the first tritheledont, *Karoomys browni* (Broom, 1903), from Late Triassic beds in the Karoo basin, he thought at first that it was a mammal. He later discovered that it retained the articular-quadrate craniomandibular joint and reclassified it as a non-mammalian cynodont ('reptile'). As other specimens came to light, he named *Ictidosauria* Broom 1929 for a new sub-order intermediate between *Cynodontia* and *Mammalia*. He considered *Ictidosauria* as the group that included the immediate ancestors of mammals. Watson and Romer (1956: 67) included within *Ictidosauria* tritylodontids plus some "poorly known forms [that] may be either advanced therapsids or archaic mammals... merely for the sake of completeness."

Ictidosauria is therefore potentially available as a name for the clade designated by *Mammaliamorpha*. However, *Mammaliamorpha* contrasts with *Ictidosauria* in explicitly designating a clade and using an extant internal specifier (*Homo sapiens*) representing the crown. *Mammaliamorpha* refers to a clade that explicitly and necessarily includes crown *Mammalia*. It also bears mention that the name *Mammaliamorpha* was originally defined as comprising the last common ancestor of *Tritylodontidae* and *Mammalia*, and all its descendants (Rowe, 1988). The definition proposed here differs in using species as specifiers, following the *ICPN* (Art. 11.1).

One might have choosen *Tritheledontidae* rather than *Tritylodontidae* as the second internal specifier for *Mammaliamorpha*. Because both were initially identified as mammals, and both lie very close to the crown clade, the choice between them may seem arbitrary, but *Tritylodontidae* has the pragmatic advantage of being much better known. Based on multiple specimens from different parts of the world, the entire skull, dentition, and postcranial skeleton of tritylodontids are known, facilitating more informative phylogenetic analyses of taxa that are proximate extinct relatives of crown *Mammalia*.

Historically, opinions on the phylogenetic position of tritylodonts have varied widely. Much of the debate centered on the *a priori* selection of phylogenetically informative anatomical systems. Because most Mesozoic mammals are known only from teeth, the dentition became central to nearly all classifications (e.g., Simpson, 1928, 1945). Under this system, tritylodonts were grouped with other extinct taxa having molariform teeth with multiple cusps aligned in rows (*Haramiyidae*, *Multituberculata*). More recently, they were grouped in a paraphyletic assemblage of herbivorous non-mammalian "gomphodont" cynodonts that have tooth crowns with broad crushing surfaces (e.g., Hopson, 1969; Crompton and Jenkins, 1973; Sues, 1986; Carroll, 1988; Hopson and Kitching, 2001). This latter position was defended in an analysis of cynodont relationships by Hopson and Barghusen (1986) that represents an early attempt to identify apomorphic characters. However, they examined only cranial characters, and support for an herbivorous clade was very weak. Later, using a similar argument, Sues and Jenkins (2006) defended tritylodonts as gomphodonts, asserting that most of the postcranial similarities uniquely shared by *Tritylodontidae* and *Mammalia* are only superficial in nature. These arguments are rooted in older views of the polyphyletic origins for *Therapsida* (this volume) and *Mammalia* (this volume), and cynodont classifications that were overly focused on teeth.

As systematists moved away from a dentition-centered system of classification and began to comb the entire skeleton for phylogenetically informative variation, a large body of data rapidly accumulated in support of a proposal first made by Kemp—that *Tritylodontidae* is a proximal relative of *Mammalia,* rather than a gomphodont (Kemp, 1983; Rowe, 1986, 1988, 1993; Gauthier et al., 1988; Luo, 1994, 2007, 2015). It is also clear that *Tritylodontidae* retains numerous plesiomorphic states that exclude it from the crown (e.g., middle ear attached to the lower jaw; quadrate-articular jaw articulation; skeleton lacking epiphyseal ossifications). It was this combination of apomorphic resemblances to *Mammalia* and retained plesiomorphies that led to the recognition and naming of the more inclusive clade *Mammaliamorpha* (Rowe, 1988, 1993; Gauthier et al., 1988).

In a recent detailed analysis of the dentition of *Haramiyidae* and other mammaliamorphs possessing 'multituberculate' teeth (*Tritylodontidae, Multituberculata*), Luo et al. (2015) provided compelling evidence that the multi-cusped molariform teeth in each of these clades is highly autapomorphic, that each occludes and functions in its own distinctive way in each clade, and that other character discordance separates them widely across the mammaliamorph tree. When all available osteological and dental data were considered, they concluded that 'multituberculate' teeth must have evolved independently several times.

The improved phylogenetic precision that places *Tritylodontidae* only a short distance outside of *Mammalia,* and that breaks apart the different 'multituberculate' clades, reflects a shift from a system of taxonomy centered on dental morphology to one more directly concerned with common ancestry and the analysis of all available data. Viewing *Mammaliamorpha* as nested within *Cynodontia,* and as a clade whose crown is *Mammalia,* adds new insights into the assembly and correlation of its many diagnostic apomorphies. The origin of *Cynodontia* (this volume) corresponded to the emergence of an occlusal dentition and a radically new feeding behavior. The remarkable diversity of cynodont tooth crown patterns was carried to special degree in tritylodontids and leaves little doubt that dietary diversification continued to drive cynodont evolution as *Mammaliamorpha* originated and diversified. The ancestry of *Mammaliamorpha* is marked by the appearance of the epipubis (possibly supporting an external pouch), and remodeled palate (possibly for a neonatal liquid diet). Ossification of the orbital wall may have been a response to increased encephalization, although the endocast of tritylodonts is not yet known. It may also reflect a more solid attachment site for the extrinsic eye musculature, and enhanced mobility of the eyeballs. More straightforward in their interpretations are modifications of the middle and inner ear for wider spectrum and higher-frequency audition. Partial ossification of the nasal capsule extends a trend in the evolution of ortho-retronasal olfaction, and possibly scent-tracking, and suggests that olfactory ecology was becoming paramount as the miniaturized early mammaliamorphs were entering a world governed to an increasing degree by olfactory information (Rowe and Shepherd, 2016).

Literature Cited

Abdala, F. 2007. Redescription of *Platycraniellus elegans* (*Therapsida, Cynodontia*) from the Lower Triassic of South Africa, and the cladistics relationships of eutheriodonts. *Palaeontology* 50:591–618.

Bonaparte, J. F. 2013. Evolution of the *Brasilodontidae* (*Cynodontia-Eucynodontia*). *Hist. Biol.* 25:643–653.

Bonaparte, J. F., A. G. Martinelli, and C. L. Schultz. 2005. New information on *Brasilodon* and *Brasilitherium* (*Cynodontia, Probainognathia*) from the late Triassic of southern Brazil. *Rev. Bras. Paleontol.* 8:25–46.

Bonaparte, J. F., A. G. Martinelli, C. L. Schultz, and R. Rubert. 2003. The sister group of mammals: small cynodonts from the Late Triassic of southern Brazil. *Rev. Bras. Paleontol.* 5:5–27.

Bonaparte, J. F., M. B. Soares, and A. G. Martinelli. 2013. Discoveries in the Late Triassic of Brazil improves knowledge on the origin of mammals. *Hist. Nat., 3rd Ser.* 2:5–30.

Broom, R. 1903. On the lower jaw of a small mammal From the Karoo Beds of Ariwal North, South Africa. *Geol. Mag.* 10:345.

Broom, R. 1929. On some recent new light on the origin of mammals. *Proc. Linn. Soc. N.S.W.* 54:688–694.

Carroll, R. L. 1988. *Vertebrate Paleontology and Evolution.* W. H. Freeman, New York.

Clemens, W. A. 1986. On Triassic and Jurassic mammals. Pp. 237–246 in *The Beginning of the Age of Dinosaurs* (K. Padian, ed.). Cambridge University Press, Cambridge, UK.

Crompton, A. W., and F. A. Jenkins, Jr. 1973. Mammals from reptiles: a review of mammalian origins. *Annu. Rev. Earth Planet. Sci.* 1:131–155.

Crompton, A. W., and A. L. Sun. 1985. Cranial structure and relationships of the Liassic mammal *Sinoconodon. Zool. J. Linn. Soc.* 85(2):99–119.

Gauthier, J. A., A. G. Kluge, and T. B. Rowe. 1988. Amniote phylogeny and the importance of fossils. *Cladistics* 4:105–209.

Gow, C. E. 1986. A new skull of *Megazostrodon* (*Mammalia: Triconodonta*) from the Elliot Formation (Lower Jurassic) of southern Africa. *Palaeontol. Afr.* 26:13–23.

Haughton, S. H., and A. S. Brink. 1954. A bibliographic list of *Reptilia* from the Karoo beds of Africa. *Palaeontol. Afr.* 2:1–187.

Hopson, J. A. 1969. The origin and adaptive radiation of mammal-like reptiles and nontherian mammals. *Ann. N. Y. Acad. Sci.* 167(1):199–216.

Hopson, J. A., and H. R. Barghusen. 1986. An analysis of therapsid relationships. Pp. 83–106 in *Ecology and Biology of Mammal-Like Reptiles* (N. Hotton, III, P. D. McLean, J. J. Roth, and E. C. Roth, eds.). Smithsonian Institution and National Institute of Mental Health, Washington, DC.

Hopson, J. A., and J. W. Kitching. 2001. A probainognathian cynodont from South Africa and the phylogeny of nonmammalian cynodonts. *Bull. Mus. Comp. Zool.* 156:5–35.

Kemp, T. S. 1983. The relationships of mammals. *Zool. J. Linn. Soc.* 77:353–384.

Kemp, T. S. 1988. A note on the Mesozoic mammals, and the origin of therians. Pp. 23–29 in *The Phylogeny and Classification of the Tetrapods. Vol. 2: Mammals* (M. J. Benton, ed.). Systematics Association Special Volume No. 35B. Clarendon Press, Oxford.

Kemp, T. S. 2005. *The Origin and Evolution of Mammals.* Oxford University Press, Oxford.

Kielan-Jaworowska, Z., R. L. Cifelli, and Z.-X. Luo. 2004. *Mammals from the Age of Dinosaurs.* Columbia University Press, New York.

Linnaeus, C. 1758. *Systema Naturae Per Regna Tria Naturae, Secundum Classes, Ordines, Genera, Species, cum Characteribus, Differentiis, Synonymis, Locis.* Tomus I, Editio decima, reformata. Laurentius Salvius, Holmiae (Stockholm).

Liu, J., and P. Olsen. 2010. The phylogenetic relationships of *Eucynodontia* (*Amniota: Synapsida*). *J. Mamm. Evol.* 17:151–176.

Lucas, S. G. 1992. Extinction and the definition of the Class *Mammalia. Syst. Zool.* 41:370–371.

Lucas, S. G., and A. P. Hunt. 1990. The oldest mammal. *New Mexico J. Sci.* 30:41–49.

Lucas, S. G., and Z.-X. Luo. 1993. *Adelobasileus* from the Upper Triassic of West Texas: the oldest mammal. *J. Vertebr. Paleontol.* 13:309–334.

Luo, Z.-X. 1994. Sister-group relationships of mammals and transformations of diagnostic mammalian characters. Pp. 98–128 in *In the Shadow of the Dinosaurs: Early Mesozoic Tetrapods* (N. C. Fraser and H.-D. Sues, eds.). Cambridge University Press, Cambridge and New York.

Luo, Z.-X. 2007. Transformation and diversification in early mammal evolution. *Nature* 450:1011–1019.

Luo Z.-X., S. M. Gatesy, F. A. Jenkins, Jr., W. W. Amaral, and N. H. Shubin. 2015. Mandibular and dental characteristics of Late Triassic mammaliaform *Haramiyavia* and their ramifications for basal mammal evolution. *Proc. Natl. Acad. Sci. USA* 112:E7101–E7109.

Martinelli, A. G., and G. W. Rougier. 2007. On *Chaliminia musteloides* (*Eucynodontia*: *Tritheledontidae*) from the Late Triassic of Argentina, and a phylogeny of *Ictidosauria*. *J. Vertebr. Paleontol.* 27:442–460.

Martinelli, A. G., J. F. Bonaparte, C. L. Schultz and R. Rubert. 2005. A new tritheledontid (*Therapsida, Eucynodontia*) from the Late Triassic of Rio Grande do Sul (Brazil) and its phylogenetic relationships among carnivorous non-mammalian eucynodonts. *Amegheniana* 42:191–208.

Martinez, R. N., C. L. May, and C. A. Forster. 1996. A new carnivorous cynodont from the Ischigualasto Formation (Late Triassic, Argentina), with comments on eucynodont phylogeny. *J. Vertebr. Paleontol.* 16:271–284.

Miao, D. 1991. On the origin of mammals. Pp. 579–597 in *Origins of the Higher Groups of Tetrapods* (H.-P. Schultze and L. Treub, eds.). Cornell University Press, Ithaca, NY and London.

Owen, R. 1857. On the affinities of *Stereognathus ooliticus* (Charlesworth), a mammal from the oolitic slate of Stonesfield. *Q. J. Geol. Soc. Lond.* 13:1–11.

Owen, R. 1871. *Monograph of the Fossil Mammalia of the Mesozoic Formations*. The Palæontographical Society, London.

Owen, R. 1884. On the skull and dentition of a Triassic mammal (*Tritylodon longaevus* Owen) from South Africa. *Q. J. Geol. Soc. Lond.* 40:146–152.

Patterson, B., and E. C. Olson. 1961. A triconodontid mammal from the Triassic of Yunnan. Pp. 129–191 in *International Colloquium on the Evolution of Lower and Non-Specialized Mammals* (G. Vandebroek, ed.). Koninklijke Vlaamse Academie voor Wetenschappen, Lettern en Schoene Kunsten van Belgie, Brussels.

Rowe, T. B. 1986. *Osteological Diagnosis of Mammalia, L. 1758, and Its Relationship to Extinct* Synapsida. PhD dissertation, University of California, Berkeley, CA.

Rowe, T. B. 1988. Definition, diagnosis and origin of *Mammalia*. *J. Vertebr. Paleontol.* 8:241–264.

Rowe, T. B. 1993. Phylogenetic systematics and the early history of mammals. Pp. 129–145 in *Mammalian Phylogeny* (F. S. Szalay, M. J. Novacek, and M. C. McKenna, eds.). Springer-Verlag, New York.

Rowe, T. B., and G. M. Shepherd. 2016. The role of ortho-retronasal olfaction in mammalian cortical evolution. *J. Comp. Neurol.* 524(3):471–495.

Rubidge, B. S., and C. A. Sidor. 2001. Evolutionary patterns among Permo-Triassic therapsids. *Annu. Rev. Ecol. Syst.* 32:449–480.

Ruta, M., J. Botha-Brink, S. A. Mitchell, and M. J. Benton. 2013. The radiation of cynodonts and the ground plan of mammalian morphological diversity. *Proc. R. Soc. Lond. B Biol. Sci.* 280:20131865.

Seeley, H. G. 1895. Researches on the structure, organization, and classification of the fossil *Reptilia*. Part IX, Section 2. The reputed mammals from the Karroo Formation of Cape Colony. *Philos. Trans. R. Soc. Lond. B Biol. Sci.* 185:1019–1028.

Sidor, C. A., and P. J. Hancox. 2006. *Elliotherium kersteni*, a new tritheledontid from the Lower Elliot Formation (Upper Triassic) of South Africa. *J. Vertebr. Paleontol.* 80:333–342.

Simpson, G. G. 1928. *A Catalogue of the Mesozoic Mammalia in the Geological Department of the British Museum*. Trustees of the British Museum, London.

Simpson, G. G. 1945. The principles of classification and a classification of mammals. *Bull. Am. Mus. Nat. Hist.* 85:1–350.

Sues, H.-D. 1985. The relationships of the *Tritylodontidae* (*Synapsida*). *Zool. J. Linnean Soc.* 85:205–217.

Sues, H.-D. 1986. The skull and dentition of two tritylodont synapsids from the Lower Jurassic of western North America. *Bull. Mus. Comp. Zool.* 151:217–268.

Sues, H.-D., and F. A. Jenkins, Jr. 2006. The postcranial skeleton of *Kayentatherium wellesi* from the Lower Jurassic Kayenta Formation of Arizona and the phylogenetic significance of postcranial features. Pp. 114–152 in *Amniote Paleobiology* (M. T. Carrano, T. J. Gaudin, R. W. Blob, and J. R. Wible, eds.). University of Chicago Press, Chicago, IL.

Watson, D. M. S. 1913. On a new cynodont from the Stormberg. *Geol. Mag.* 10:145–148.

Watson, D. M. S., and A. S. Romer. 1956. A classification of therapsid reptiles. *Bull. Mus. Comp. Zool.* 114:37–89.

Author

Timothy B. Rowe; Jackson School of Geosciences; University of Texas at Austin; C1100, Austin, TX 78712, USA. E-mail: rowe@utexas.edu.

Date Accepted: 12 June 2017

Primary Editors: Jacques A. Gauthier, Kevin de Queiroz

Mammaliaformes T. B. Rowe 1988 [T. B. Rowe], converted clade name

Registration Number: 221

Definition: The smallest clade containing *Homo sapiens* Linnaeus 1758 (*Mammalia*) and *Morganucodon oehleri* Rigney 1963 (*Morganucodontidae*). This is minimum-clade definition. Abbreviated definition: min ∇ (*Homo sapiens* Linnaeus 1758 & *Morganucodon oehleri* Rigney 1963).

Etymology: Combines *Mammalia* (this volume) with the Latin *forma* (form).

Reference Phylogeny: Figure 3 in Rowe (1988) is the primary reference phylogeny. See also Figure 1 in Luo (2007).

Composition: *Mammaliaformes* includes *Mammalia* (this volume) plus a subset of extinct species and clades that are viewed under current hypotheses as the closest extinct relatives of *Mammalia* within *Mammaliamorpha* (Gauthier et al., 1988; Rowe, 1988, 1993; Luo, 2007; Luo et al., 2015). At one time or another, the extinct taxa mentioned below were considered to be members of the mammalian crown, or at least referred to as 'mammals'. These include *Morganucodonta* Kermack et al. 1973 (sensu Kielan-Jaworowska et al., 2004); *Haramiyidae* Simpson 1947 (sensu Luo et al., 2015), *Megazostrodontidae* Gow 1986 (sensu Kielan-Jaworowska et al., 2004) and *Docodonta* Kretzoi 1946 (sensu Kielan-Jaworowska et al., 2004).

Owing to incompleteness, the positions of some additional fossils are equivocal, and some listed below may prove to be members of *Mammalia* (this volume). They include *Eutriconodontidae* Kermack et al., 1973 (sensu Kielan-Jaworowska et al., 2004); *Hadrocodium wui* Luo et al. 2001; *Fruitafossor windschiffeli* Luo and Wible 2005; and members of *Kuehneotheria* McKenna 1975 (possibly paraphyletic).

In addition, a number of fossils have been hypothesized as members of *Pan-Monotremata* (this volume), but character conflicts indicate that they may lie outside of the mammalian crown clade as members of *Mammaliaformes*. These include: *Ambondro mahabo* Flynn et al. 1999; *Asfaltomylos patagonicus* Rauhut et al. 2002; *Ausktribosphenos nyktos* Rich et al. 2001; *Bishops whitmorei* Rich et al. 2001; *Henosferus molus* Rougier et al. 2007; *Kollikodon ritchiei* Flannery et al. 1995; *Shuotheriidae* Chow and Rich 1982; and *Pseudotribos robustus* Luo et al. 2007.

Diagnostic Apomorphies: The most striking feature of *Mammaliaformes* is that the parietals, basioccipital, and basisphenoid are expanded dorsally, posteriorly and laterally. The endocranial cavity thus was enlarged and, except for the lack of a cribriform plate (and for neural and vascular penetrations), it is entirely enclosed by bone. The inner surfaces of these bones are embossed with the external surface of the brain, and cranial endocasts made digitally using computed tomography offer details of how the brain was transformed in the ancestral mammaliaform (Rowe et al., 1995, 2011; Carlson et al., 2003).

Brain volume measured relative to estimated body mass produces an encephalization quotient (EQ), based on an empirically derived equation that affords a quantitative comparison of relative brain size between different taxa (Jerison, 1973; Eisenberg, 1981). The mammaliaform brain is nearly twice as large as the brain in early cynodonts such as *Thrinaxodon liorhinus*. The

EQ of early cynodonts ranges from ~0.16 to 0.23, whereas the EQ of *Morganucodon oehleri* is ~0.32, reflecting an increase of 30 to 50%. The olfactory bulb and olfactory (piriform) cortex are the regions of by far the greatest expansion. A deep annular fissure encircles the olfactory tract, marking a distinctive external division of the brain between the inflated olfactory bulbs and the cortex. The cortex is inflated and wider than the cerebellum, covering the midbrain and the pineal stalk. The cerebellum is also enlarged, implying expansion of the basal nuclei, thalamus, and medulla, and the spinal cord is thicker. In shape and proportions, the brain resembles the brain in a living mammal more than it resembles the brain in early cynodonts (Rowe et al., 2011).

Differentiation of the neocortex can be inferred from integumentary evidence preserved in the Middle Jurassic (~165 MA) fossil docodont *Castorocauda lutrasimilis* (Ji et al., 2006). This remarkable fossil preserves evidence of a pelt of modern aspect, with guard hairs and velus underfur. Because docodonts are either members of *Morganucodonta* (Rowe, 1988), or just slightly closer to crown *Mammalia* (Luo, 2007), the presence of a pelt is diagnostic of *Mammaliaformes*, or an unnamed node just within. In extant mammals, guard hairs are equipped with at least three different kinds of mechanoreceptors that induce somatosensory maps on the neocortex. In the relatively smallest and least modified mammalian brains (e.g., *Didelphis*), the neocortex is dominated by a single primary somatosensory field whose primary function is to map sensations from mechanoreceptors in the skin, hair follicles, muscle spindles, and joint receptors. Together with endocast size and shape, the pelt in *Castorocauda lutrasimilis* allows the inference that a small neocortex had differentiated from the telencephalic pallium (Rowe et al., 2011).

Additional diagnostic apomorphies include expansion of the petrosal to provide a bony floor beneath the cavum epiptericum, and presence of the petrosal promontorium (Gauthier et al., 1988; Rowe, 1988, 1993). The secondary palate extends to the caudal margin of the tooth row. Tooth replacement has a diphyodont pattern (Luo et al., 2004). The dentary has an expanded condyle that articulates with a well-developed glenoid fossa on the squamosal; these are now the primary load-bearing bones in the craniomandibular joint. The dentition is differentiated into anterior bicuspid premolariform and posterior molariform teeth with three to five main cusps. Novel dental wear facets on the molariform teeth indicate that they occluded in a complex unilateral pattern in which rotational movement of the mandibles occurred during mastication (Kielan-Jaworowska et al., 2004; McKenna and Bell, 1997).

In the postcranial skeleton, the fibular fabellum is present as a small bone that fuses with the proximal posterior end of the fibula. Often said to be a 'sesamoid', this endochondral element does not form within a tendon, nor does it share any of the temporal ontogenetic characteristics of sesamoids, which tend to ossify in complete synchrony in the forelimbs and then in the hindlimbs. Sesamoid bones arose later, as an apomorphy of *Mammalia* (this volume).

The lumbar vertebrae are markedly differentiated from the thoracic vertebrae, with vertical and anticlinal lumbar neural spines, and lumbar centra with articular faces that are inclined instead of vertical, such that the lumbar region is arched dorsally (Jenkins and Parrington, 1976).

Synonyms: Approximate synonyms include *Mammalia* Linnaeus 1758 sensu Benton (1990, 2015), Carroll (1988), Clemens (1986), Crompton and Jenkins (1968, 1973, 1979), Gow (1986), Hopson and Barghusen (1986), Kemp (1982, 1983, 1988, 2005), Kermack (1963); Kermack et al. (1973, 1981); Kermack and Kermack (1984), Kielan-Jaworowska et al. (2004), Lucas (1992), Lucas and Hunt (1990), Miao (1991), and Parrington (1941, 1947, 1967, 1971, 1973, 1978). Partial synonyms include *Eotheria* Kermack and Mussett 1958, *Prototheria* Gill 1872 (sensu Hopson, 1970, but not Huxley, 1880), and *Atheria* Kermack et al. 1973.

Comments: Since *Mammalia* was named (Linnaeus, 1758), neontologists have faced only a few uncertainties in determining whether a particular species or group is, or is not, a mammal (see *Monotremata*, this volume). But when mammalian stem group fossils were discovered, the circumscription of *Mammalia* became a persistent problem. The first specimens of *Tritylodontidae* (Owen, 1857, 1871, 1884) and then *Tritheledontidae* (Broom, 1903) were unhesitatingly classified within *Mammalia* (Osborn, 1907; Simpson, 1931, 1945). Later, with the recognition of certain plesiomorphic features (Broom, 1932; Kühne, 1943, 1958), both taxa were reassigned to a paraphyletic *Cynodontia* (see *Mammaliamorpha*, this volume). Many subsequent fossil discoveries would be the subject of similar equivocation.

For much of the twentieth century, controversy over what was or was not considered a mammal was fueled by diagreement over its defining characters. By the 1980s, numerous opinions had been published, but not a single character was common to them all (see Rowe, 1988: Table 1). In practice, the name *Mammalia* had taken on multiple meanings in terms of its composition and consequently its distribution in time and space. As a conceptually clouded taxonomic construct, the name *Mammalia* presented a failure of nomenclature, the goal of which should be precise and efficient communication (Rowe, 1988; de Queiroz and Gauthier, 1990).

A popular solution to the problem has been to restrict the name *Mammalia* (this volume) to

the crown clade (Jefferies, 1979) comprising the most recent common ancestor of living mammals and all of its descendants (Rowe, 1988; de Queiroz and Gauthier, 1992; de Queiroz, 1994; McKenna and Bell, 1997). To the extent that this proposal is adopted, it has the advantage of stabilizing the meaning of the name. Moreover, the question with regard to the classification of fossils can be answered quite simply by determining whether they are descended from the most recent common ancestor of the Recent species. From this perspective, a number of fossils that were once considered the oldest (Triassic) and least modified mammals are now generally recognized to lie just outside of the crown (Kemp, 1983), that is, of *Mammalia*, and the name *Mammaliaformes* was formulated to designate the more inclusive clade containing those fossils (Rowe, 1988).

Mammaliaformes was originally defined as the name of the clade originating in the last common ancestor of *Morganucodontidae* (= *Morganucodonta*) and *Mammalia* (Rowe, 1988). The definition adopted herein differs slightly from the original in using the species *Homo sapiens*, rather than the clade *Mammalia*, as one of the internal specifiers (following *ICPN*, Art. 11.1). In either case, *Mammaliaformes* refers to a clade that explicitly and necessarily includes living *Mammalia*. The current definition also differs from the original in using the species *Morganucodon oehleri* Rigney 1963, rather than the clade *Morganucodontidae* as the other internal specifier. The taxonomy of *Morganucodon* and the various species assigned to *Morganucodontidae* and *Morganucodonta* is complex. The incompleteness of most species assigned to these taxa, some consisting only of isolated teeth, suggests that a measure of uncertainty will continue to surround its precise composition (for details, see Mills, 1971; Kermack et al., 1973, 1981; Clemens, 1979, 1986). The advantage of using *Morganucodon oehleri* (e.g., rather than the partial jaw of *M. watsoni*, which is the type of *Morganucodon*) is that the type specimen of *M. oehleri* is a fairly complete skull and dentition, and other referable specimens subsequently recovered from the type locality include postcranial elements (Kielan-Jaworowska et al., 2004).

Conceptualizing *Mammalia* as nested within the more inclusive clades *Mammaliaformes, Mammaliamorpha, Cynodontia, Therapsida,* and *Pan-Mammalia* brings a sharper focus to the phylogenetic context, sequence, and timing of the acquisition of mammalian characters. Traditional "defining" mammalian characters such as hair and a large brain with a neocortex (Northcutt, 2011), which distinguish mammals from other living organisms, can now be seen to have deeper histories that can be traced in the fossil record to the ancestral mammaliaform, and beyond. Apomorphies of *Cynodontia* (this volume) include a first measurable pulse in encephalization that was tied to an occlusal dentition, the new feeding behavior of mastication, and probably to the origin of the novel system of ortho-retronasal olfaction (Rowe and Shepherd, 2016). Some 50 million years later, following miniaturization (see *Mammaliamorpha*, this volume), the skull in early mammaliaforms underwent a second pulse in encephalization, in which the mammaliaform brain nearly doubled in relative size. The olfactory bulb and olfactory (pyriform) cortex are the regions of by far the greatest expansion and indicate that an integrated system of ortho-retronasal olfaction continued to grow in importance.

The discovery of a pelt of modern aspect in an early mammaliaform (see Diagnostic Apomorphies) suggests that the novel six-layered neocortex had differentiated, given that the mechanoreceptors that innervate hair follicles induce somatosensory maps onto the outer layer of the neocortex in all living mammals (Rowe et al., 2011). In living mammals with small brains (e.g., *Didelphis*), the small neocortex is dominated by a single primary somatosensory field whose conscious component involves tactile exploration of the environment and body surface monitoring. Peripheral somatosensory input is mapped to the neocortex as an 'animunculus'. A parallel, underlying neocortical motor map contains pyramidal neurons whose axons form the pyramidal tract that projects via the brainstem directly to the spinal column to program and execute skilled movements requiring precise control of distal musculature. Increased agility and coordination are correlative behaviors. During its ontogeny in living mammals, hair performs first as a tactile organ and only later does it insulate as underfur thickens and matures. The body temperature of newborns is initially regulated by their mothers. This sequence in the ontogeny of the function of hair in *Mammalia* in conjunction with the pelage of modern aspect in early mammaliaforms implies that parental care and endothermy were present in *Mammaliaformes* ancestrally (if not before: see *Mammaliamorpha*, this volume). Endothermy may have been a consequence of mammaliaform encephalization because a large brain operates properly only within narrow thermal tolerances, and it is metabolically the most expensive organ to maintain. Metabolism is under hormonal control that does not command large cerebral regions and thus did not itself drive encephalization.

The emergence of these features, so distinctive of mammals among living species, now appears to predate the origin of (crown) *Mammalia* by ~40 million years and can be traced to the last common ancestor of *Mammaliaformes*. They reflect the continuation of behaviors begun nearly 250 million years ago as cynodont dentitions and diets diversified. The brain continued its more ancient trend toward increase in relative size as mammaliaforms further exploited the information of odors and scents in their microhabitats, along with a new measure of gnostic somatosensory input provided by hair (Rowe et al., 2011; Rowe and Shepherd, 2016).

Literature Cited

Benton, M. J. 1990. Phylogeny of the major tetrapod groups: morphological data and divergence dates. *J. Mol. Evol.* 30:409–424.

Benton, M. J. 2015. *Vertebrate Palaeontology.* 4th edition. Wiley Blackwell, West Sussex.

Broom, R. 1903. On the lower jaw of a small mammal from the Karroo beds of Aliwal North, South Africa. *Geol. Mag.* 10:345.

Broom, R. 1932. *The Mammal-Like Reptiles of South Africa and the Origin of Mammals.* H. F. & G. Witherby, London.

Carlson, W. D., T. Rowe, R. A. Ketcham, and M. W. Colbert. 2003. Geological applications of high-resolution X-ray computed tomography in petrology, meteoritics and palaeontology. Pp. 7–22 in *Applications of X-ray Computed Tomography in the Geosciences* (F. Mees, R. Swennen, M. Van Geet, and P. Jacobs, eds.). Geological Society, London. Special Publication 215.

Carroll, R. L. 1988. *Vertebrate Paleontology and Evolution.* W. H. Freeman, New York.

Chow, M., and T. H. Rich. 1982. *Shuotherium dongi,* n. gen. and sp., a therian with pseudotribosphenic molars from the Jurassic of Sichuan, China. *Aust. Mammal.* 5:127–142.

Clemens, W. A. 1979. A problem in morganucodontid taxonomy (*Mammalia*). *Zool. J. Linn. Soc.* 66:1–14.

Clemens, W. A. 1986. On Triassic and Jurassic mammals. Pp. 237–246 in *The Beginning of the Age of Dinosaurs* (K. Padian, ed.). Cambridge University Press, Cambridge, UK.

Crompton, A. W., and F. A. Jenkins, Jr. 1968. Molar occlusion in Late Triassic mammals. *Biol. Rev.* 43:427–458.

Crompton, A. W., and F. A. Jenkins, Jr. 1973. Mammals from reptiles: a review of mammalian origins. *Ann. Rev. Earth Planet. Sci.* 1:131–155.

Crompton, A. W., and F. A. Jenkins, Jr. 1979. Origin of mammals. Pp. 59–73 in *Mesozoic Mammals: The First Two-Thirds of Mammalian History* (J. A. Lillegraven, Z. Kielan-Jaworowska, and W. A. Clemens, eds.). University of California Press, Berkeley, CA and Los Angeles, CA.

Eisenberg, J. F. 1981. *The Mammalian Radiations: An Analysis of Trends in Evolution, Adaptation, and Behavior.* University of Chicago Press, Chicago, IL.

Flannery, T. F., M. Archer, T. H. Rich, and R. Jones. 1995. A new family of monotremes from the Cretaceous of Australia. *Nature* 377:418–420.

Flynn, J. J., J. M. Parrish, B. Rakotosamimanana, W. F. Simpson, and A. R. Wyss. 1999. A middle Jurassic mammal from Madagascar. *Nature* 401:57–60.

Gauthier, J. A., A. G. Kluge, and T. B. Rowe. 1988. Amniote phylogeny and the importance of fossils. *Cladistics* 4:105–209.

Gill, T. 1872. Arrangement of the families of mammals. *Smithson. Misc. Collect.* 11:1–98.

Gow, C. E. 1986. A new skull of *Megazostrodon* (*Mammalia: Triconodonta*) from the Elliot Formation (Lower Jurassic) of southern Africa. *Palaeontol. Afr.* 26:13–23.

Hopson, J. A. 1970. The classification of nontherian mammals. *J. Mammal.* 51:1–9.

Hopson, J. A., and H. R. Barghusen. 1986. An analysis of therapsid relationships. Pp. 83–106 in *Ecology and Biology of Mammal-Like Reptiles* (N. Hotton III, P. D. McLean, J. J. Roth, and E. C. Roth, eds.). Smithsonian Institution and National Institute of Mental Health, Washington, DC.

Huxley, T. H. 1880. On the application of the laws of evolution to the arrangement of the *Vertebrata* and more particularly of the *Mammalia*. *Proc. Zool. Soc. Lond.* 1880:649–662.

Jefferies, R. P. S. 1979. The origin of chordates—a methodological essay. Pp. 443–477 in *The Origin of Major Invertebrate Groups* (M. R. House, ed.). Academic Press, London.

Jenkins, Jr., F. A., and F. R. Parrington. 1976. The postcranial skeleton of the Triassic mammals *Eozostrodon, Megazostrodon,* and *Erythrotherium. Philos. Trans. R. Soc. Lond. B Biol. Sci.* B 273:387–431.

Jerison, H. J. 1973. *Evolution of the Brain and Intelligence.* Academic Press, New York.

Ji, Q., Z.-X. Luo, C. Yuan, and A. R. Tabrum. 2006. A swimming mammaliaform from the Middle Jurassic and ecomorphological diversification of early mammals. *Science* 311:1123–1127.

Kemp, T. S. 1982. *Mammal-Like Reptiles and the Origin of Mammals.* Academic Press, London.

Kemp, T. S. 1983. The relationships of mammals. *Zool. J. Linn. Soc.* 77:353–384.

Kemp, T. S. 1988. A note on the Mesozoic mammals, and the origin of therians. Pp. 23–29 in *The Phylogeny and Classification of the Tetrapods. Vol. 2: Mammals* (M. J. Benton, ed.). Systematics Association Special Volume No. 35B. Clarendon Press, Oxford.

Kemp, T. S. 2005. *The Origin and Evolution of Mammals.* Oxford University Press, Oxford.

Kermack, D. M., and K. A. Kermack. 1984. *The Evolution of Mammalian Characters.* Croom Helm, London, Sydney, and Kapitan Szabo Publishers, Washington, DC.

Kermack, K. A. 1963. The cranial structure of the triconodonts. *Philos. Trans. R. Soc. Lond. B Biol. Sci.* 240:95–133.

Kermack, K. A., and F. Mussett. 1958. The jaw articulation of the *Docodonta* and the classification of Mesozoic mammals. *Proc. R. Soc. Lond. B Biol. Sci.* 149:204–215.

Kermack, K. A., F. Mussett, and H. W. Rigney. 1973. The lower jaw of *Morganucodon*. *Zool. J. Linn. Soc.* 53:87–175.

Kermack, K. A., F. Mussett, and H. W. Rigney. 1981. The skull of *Morganucodon*. *Zool. J. Linn. Soc.* 71:1–158.

Kielan-Jaworowska, Z., R. L. Cifelli, and Z.-X. Luo. 2004. *Mammals from the Age of Dinosaurs.* Columbia University Press, New York.

Kretzoi, M. 1946. On *Docodonta*, a new order of Jurassic mammals. *Ann. Hist.-Nat. Mus. Natn. Hung.* 39:108–111.

Kühne, W. G. 1943. The dentary of *Tritylodon* and the systematic position of the *Tritylodontidae*. *J. Nat. Hist.* 10:589–601.

Kühne, W. G. 1958. Rhaetische Triconodonten aus Glamorgan, ihre Stellung zwischen den Klassen *Reptilia* und *Mammalia* und ihre Bedeutung für die Reichart'sche Theorie. *Paläontolog. Zeitschr.* 32:197–235.

Linnaeus, C. 1758. *Systema Naturae Per Regna Tria Naturae, Secundum Classes, Ordines, Genera, Species, cum Characteribus, Differentiis, Synonymis, Locis.* Tomus I, Editio decima. Laurentius Salvius, Holmiae (Stockholm).

Lucas, S. G. 1992. Extinction and the definition of the class *Mammalia*. *Syst. Zool.* 41:370–371.

Lucas, S. G., and A. P. Hunt. 1990. The oldest mammal. *New Mexico J. Sci.* 30:41–49.

Luo, Z.-X. 2007. Transformation and diversification in early mammal evolution. *Nature* 450:1011–1019.

Luo, Z.-X., A. W. Crompton, and A.-L. Sun. 2001. A new mammal from the Early Jurassic and evolution of mammalian characteristics. *Science* 292:1535–1540.

Luo Z.-X., S. M. Gatesy, F. A. Jenkins, Jr., W. W. Amaral, and N. H. Shubin. 2015. Mandibular and dental characteristics of Late Triassic mammaliaform *Haramiyavia* and their ramifications for basal mammal evolution. *Proc. Natl. Acad. Sci. USA* 112:E7101–E7109.

Luo, Z.-X., Z. Kielan-Jaworowska, and R. L. Cifelli. 2004. Evolution of dental replacement in mammals. *Bull. Carnegie Mus. Nat. Hist.* 36:159–175.

Luo, Z.-X., J. Quang, and C.-X. Yuan. 2007. Convergent dental adaptations in pseudotribosphenic and tribosphenic mammals. *Nature* 450:93–97.

Luo, Z.-X., and J. R. Wible. 2005. A Late Jurassic digging mammal and early mammalian diversification. *Science* 308:103–107.

McKenna, M. C. 1975. Toward a phylogenetic classification of the *Mammalia*. Pp. 21–46 in *Phylogeny of the Primates* (W. P. Luckett and F. S. Szalay, eds.). Plenum Publishing Company, New York.

McKenna, M. C., and S. K. Bell. 1997. *Classification of Mammals above the Species Level.* Columbia University Press, New York.

Miao, D. 1991. On the origin of mammals. Pp. 579–597 in *Origins of the Higher Groups of Tetrapods* (H.-P. Schultze and L. Treub, eds.). Cornell University Press, Ithaca, NY and London.

Mills, J. R. E. 1971. The dentition of *Morganucodon*. Pp. 29–63 in *Early Mammals* (D. M. Kermack and K. A. Kermack, eds.). *Zool. J. Linn. Soc.* 50(suppl. 1).

Northcutt, G. L. 2011. Evolving large and complex brains. *Science* 332:926.

Osborn, H. F. 1907. *The Evolution of Mammalian Molar Teeth to and from the Triangular Type.* (William K. Gregory, ed.) Macmillan, New York.

Owen, R. 1857. On the affinities of *Stereognathus ooliticus* (Charlesworth), a mammal from the oolitic slate of Stonesfield. *Q. J. Geol. Soc. Lond.* 13:1–11.

Owen, R. 1871. *Monograph of the Fossil Mammalia of the Mesozoic Formations.* The Palæontographical Society, London.

Owen, R. 1884. On the skull and dentition of a Triassic mammal (*Tritylodon longaevus* Owen) from South Africa. *Q. J. Geol. Soc. Lond.* 40:146–152.

Parrington, F. R. 1941. On two mammalian teeth from the Lower Rhaetic of Somerset. *Ann. Mag. Nat. Hist.* 8:140–144.

Parrington, F. R. 1947. On the collection of Rhaetic mammalian teeth. *Proc. Zool. Soc. Lond.* (for 1946) 116:707–728.

Parrington, F. R. 1967. The origins of mammals. *Adv. Sci. Lond.* 24:1–9.

Parrington, F. R. 1971. On the Upper Triassic mammals. *Philos. Trans. R. Soc. Lond. B Biol. Sci.* 261:231–272.

Parrington, F. R. 1973. The dentitions of the earliest mammals. *Zool. J. Linn. Soc.* 52:85–95.

Parrington, F. R. 1978. A further account of the Triassic mammals. *Philos. Trans. R. Soc. B Biol. Sci.* 282:177–204.

de Queiroz, K. 1994. Replacement of an essentialistic perspective on taxonomic definitions as exemplified by the definition of "*Mammalia*". *Syst. Biol.* 43:497–510.

de Queiroz, K., and J. A. Gauthier. 1990. Phylogeny as a central principle in taxonomy: phylogenetic definitions of taxon names. *Syst. Biol.* 39:307–322.

de Queiroz, K., and J. A. Gauthier. 1992. Phylogenetic taxonomy. *Ann. Rev. Ecol. System.* 1992:449–480.

Rauhut, O. W. M., T. Martin, E. Oritz-Jaureguizar, and P. Puerta. 2002. A Jurassic mammal from South America. *Nature* 416:165–168.

Rich, T. H., T. F. Flannery, P. Trusler, A. Constantine, L. Kool, P. Vickers-Rich, A. Kool, and N. van Klaveren. 2001. An advanced ausktribosphenid from the Early Creteaceous of Australia. *Rec. Queen Victoria Mus.* 110:1–9.

Rigney, H. W. 1963. A specimen of *Morganucodon* from Yunnan. *Nature* 197:1122–1123.

Rougier, G. W., A. G. Martinelli, A. M. Forasiepi, and M. J. Novacek. 2007. New Jurassic mammals from Patagonia, Argentina: a reappraisal of australosphenidan morphology and interrelationships. *Am. Mus. Nat Hist. Novitates* 3566:1–54.

Rowe, T. B. 1988. Definition, diagnosis and origin of *Mammalia. J. Vertebr. Paleontol.* 8:241–264.

Rowe, T. B. 1993. Phylogenetic systematics and the early history of mammals. Pp. 129–145 in *Mammalian Phylogeny* (F. S. Szalay, M. J. Novacek, and M. C. McKenna, eds.). Springer-Verlag, New York.

Rowe, T. B., W. Carlson, and W. Bottorff. 1995. Thrinaxodon: *Digital Atlas of the Skull.* CD-ROM. 2nd edition, for Windows and Macintosh platforms. University of Texas Press.

Rowe, T. B., T. E. Macrini, and Z.-X. Luo. 2011. Fossil evidence on origin of the mammalian brain. *Science* 332:955–957.

Rowe, T. B., and G. M. Shepherd. 2016. The role of ortho-retronasal olfaction in mammalian cortical evolution. *J. Comp. Neurol.* 524:471–495.

Simpson, G. G. 1931. A new classification of mammals. *Bull. Am. Mus. Nat. Hist.* 59:259–293.

Simpson, G. G. 1945. The principles of classification and a classification of mammals. *Bull. Amer. Mus. Nat. Hist.* 85:1–350.

Simpson, G. G. 1947. *Haramiya*, new name, replacing *Microcleptes* Simpson, 1928. *J. Paleontol* 23:497–497.

Author

Timothy B. Rowe; Jackson School of Geosciences; University of Texas at Austin; C1100, Austin, TX 78712, USA. E-mail rowe@utexas.edu.

Date Accepted: 12 June 2017

Primary Editors: Jacques A. Gauthier, Kevin de Queiroz

Mammalia C. Linnaeus 1758 [T. B. Rowe], converted clade name

Registration Number: 220

Definition: The smallest crown clade containing *Homo sapiens* Linnaeus 1758 (*Placentalia*), *Didelphis marsupialis* Linnaeus 1758 (*Marsupialia*), and *Tachyglossus aculeatus* (Shaw 1792) (*Monotremata*). This is a minimum-crown-clade definition. Abbreviated definition: min crown ∇ (*Homo sapiens* Linnaeus 1758 & *Didelphis marsupialis* Linnaeus 1758 & *Tachyglossus aculeatus* Shaw 1792).

Etymology: A name made in analogy with the familiar Latin word "animal" to designate those animals that possess mammae (Gill, 1902).

Reference Phylogeny: Figure 3 in Rowe (1988).

Composition: *Mammalia* includes the total clades of *Monotremata* (this volume) and *Theria*. Knowledge of living mammalian diversity has grown by nearly 25% in just the last 30 years, implying that much remains to be learned about its composition. A recent tabulation lists 5,416 living mammal species (Wilson and Reeder, 2005), whereas only 4,154 species were listed in the most authoritative summary of the preceding generation of taxonomists (Nowak and Paradiso, 1983). Many thousands of extinct mammal species have been named from fossils; the most exhaustive tabulation above the species level is by McKenna and Bell (1997).

There remains some uncertainty regarding which Triassic and Jurassic mammaliamorph fossils belong to *Mammalia*. One of the most important of these equivocal mammals is the Chinese Jurassic fossil *Hadrocodium wui*, which is known from a single isolated skull (Luo et al., 2001). Computed tomography indicated that its encephalization quotient is ~0.5 (compared to ~0.32 in *Morganucodon*), which places the brain of *Hadrocodium* within the size range of crown *Mammalia*. On the other hand, *Hadrocodium* lacks the mammalian apomorphy of ossified ethmoid and maxillary turbinals (below), but these often fail to be preserved in unquestionable fossil (crown) mammals. Different analyses place *Hadrocodium* within (Rowe et al., 2008) or just outside the mammalian crown (Luo et al., 2015).

Eutriconodontia Kermack et al. 1973 (sensu Kielan-Jaworowska et al., 2004) is a clade known from numerous species, a few of which are represented by rather complete skulls and skeletons from the Cretaceous of North American and Asia. None of these fossils is known to preserve the mammalian apomorphy of epiphyseal ossifications in the limb bones (see Diagnostic Apomorphies). Most analyses place *Eutriconodontia* within the crown (Rowe, 1993; Ji et al., 2006; Meng et al., 2006; Luo, 2007; Luo et al., 2015); however, Rougier et al. (2007) inferred this group to be paraphyletic with its members placed as successive sister taxa positioned outside of the crown.

Diagnostic Apomorphies: Compared to other living animals, *Mammalia* is distinguishable by features of virtually all anatomical systems, from genes and cells to organ systems, and from molecular sequences to behavior. Among its most distinctive features is a relatively large brain in which the central region of the telencephalic pallium is the six-layered neocortex (isocortex). It receives afferents from the spinal cord, thalamus, and other parts of the brain that project to restricted neocortical regions dominated by two or more layers of pyramidal cells whose association projections remain almost exclusively within the cortex and basal ganglia (i.e., intratelencephalic). Specialized corticothalamic pyramidal cells, and pyramidal tract motor output cells project from neocortex throughout the brainstem and directly to the spinal cord (i.e., the corticospinal tract). Well-developed specific motor nuclei are situated rostrally in the ventral half of the thalamus. Mammalian ependymal cells fail to reach the periphery of the brain in adults, and three meningeal layers enclose the brain (Rowe and Shepherd, 2016).

The increase in brain size diagnostic of *Mammalia* represents a continuation of the trend first established as diagnostic of *Cynodontia* (this volume), and further elaborated in *Mammaliaformes* (this volume), in which successive increases in encephalization were driven primarily (but not exclusively) by enhanced olfaction. The olfactory system in *Mammalia* is unique in its degree of elaboration. The olfactory sub-genome comprises ~1,000 or more different odorant receptor genes, the largest single gene subfamily in the mammalian genome (Niimura, 2009, 2012). Although some secondarily aquatic mammals have reduced olfactory capacities (cetaceans, platypus), their large olfactory genomes have converted to pseudogenes to varying degrees, reflecting their descent from hyperosmic ancestors (McGowen et al., 2008; Rowe et al., 2005).

The mammalian heart is completely divided into four chambers by a compact myocardium, it has a single left aortic arch, and a combination of three tricuspid valves and a single semilunar valve. The mammalian heart is activated via the auriculo-ventricular node and Purkinje fibers. Mammalian erythrocytes lack nuclei, and thrombocytes take the form of blood platelets.

The basic skeletal armature of a muscular diaphragm is diagnostic of *Cynodontia* (this volume), and in *Mammalia* its complete expression can be seen as a fundamental component of the mammalian system of ventilation. The diaphragm encloses the pleural cavities, marking the transition from thoracic to lumbar vertebrae, and it drives diaphragmatic breathing. The lungs are complex and divided into lobes, bronchioles and alveoli; and they expand ventrally, surrounding the heart and almost meeting on the ventral midline, leaving only a ventral mediastinum connecting the pericardial sac with the ventral body wall.

The *musculus panniculus carnosus* forms a continuous sheath of muscle wrapping the trunk and neck, and mobile facial musculature is differentiated into groups associated with the eye, ear, and snout (Huber, 1930a,b; Lightoller, 1942).

The pancreas has vascularized islets of Langerhans. The kidney includes the renal macula densa and loops of Henle, and the adrenal gland has a unique histology.

The mammalian integument differentiates into sebaceous glands, sweat glands, and mammary glands, and supports hair that grows from dermal papillae equipped with erector muscles. Thanks to the remarkably preserved Chinese Jurassic fossil *Castorocauda lutrasimilis* (Ji et al., 2006), a pelt of modern aspect with guard hairs and velus underfur is now known to diagnose a more inclusive clade than *Mammalia* (see *Mammaliaformes*, this volume).

Even when fossils are considered, *Mammalia* can be distinguished from its closest extinct relatives by many features of the skeleton. Experts differ on diagnostic details, but there is broad agreement that *Mammalia* can be distinguished from even its closest known extinct relatives among Mesozoic fossils by a suite of attributes that involve 'repackaging' of the skull around an enlarged brain and elaborated sensory systems (see Kielan-Jaworowska et al., 2004; Luo et al., 2007, 2015; Rowe, 1988, 1993, 1996a,b; Rowe et al., 2011; Rowe and Shepherd, 2016). The internasal process of the premaxilla is absent, leaving the external nares confluent.

An intricate internal skeleton develops as the cartilaginous nasal capsule ossifies to form the ethmoid complex. It forms an elaborate, delicate scaffolding of thin, interfolded bony sheets known as ethmoid turbinals (or turbinates). In *Mammalia* ancestrally, their appearance afforded a 10-fold increase in the surface area of olfactory epithelium that could be deployed inside the nasal cavity (Rowe et al., 2005). The ethmoid turbinals coalesce around the olfactory nerve fascicles to form the bony cribriform plate, a compound structure that separates the olfactory recess from the cavum cranii. The turbinals grow rostrally from the cribriform plate as the olfactory epithelium matures, and their mature geometry is highly variable among mammals (secondarily lost in some cetaceans). Also ossifying in the nose is the maxillary turbinal, which increases the epithelial surface area by nearly an order of magnitude and is involved in regulating respiratory moisture and heat exchange (Rowe et al., 2005).

The mammalian middle ear is a highly distinctive chain of three tiny ossicles that transmit vibrations received by the tympanic membrane to the fenestra vestibuli of the inner ear. These bones become detached from their embryonic and ancestral connection to the mandible as the brain grows, becoming suspended from a new position beneath the cranium during maturation. As a result, the dentary is the only bone in the adult mandible and it articulates only with the squamosal, leaving the dentary and squamosal as the sole elements of the mature craniomandibular joint. Whether this is an autapomorphy of *Mammalia* or represents convergent evolution within mammals is controversial (Rowe, 1988, 1996a,b; Bever et al., 2005; Meng et al., 2006, 2011; Luo et al., 2007; Ji et al., 2009).

A double occipital condyle arose in *Cynodontia* (this volume), and in *Mammalia* the condyles expand to surround the entire ventral half of the foramen magnum. Correspondingly, the mammalian atlas, or first vertebra, is highly distinctive in that it forms a bony ring via the ontogenetic fusion of three separate ossification centers that remain separate throughout life in all other pan-mammals. The cervical ribs are also apomorphic in fusing to their respective vertebrae in early ontogeny. The limbs and girdles develop secondary ossification centers, the most obvious of which are in the cartilaginous epiphyses of the long bones. Sesamoid bones form in tendons of the flexor muscles of the hands and feet, and in the hindlimb a single large sesamoid forms the patella (Rowe, 1988).

Synonyms: Approximate synonyms include *Mastodia* Rafinesque 1814, *Opthalmozoa* Oken 1847, *Thricozoa* Oken 1847, and *Aistheseozoa* Oken 1847. *Prototheria* Gill 1872 sensu Huxley (1880) and Hopson (1970) is a partial synonym.

Comments: With the exception of *Prototheria*, the synonyms of *Mammalia* (*Mastoidea*, *Opthalmozoa*, *Thricozoa*, and *Aistheseozoa*) have almost never been used, and no compelling reason has been advanced to consider converting any of them to designate the mammalian crown clade. Huxley (1880) used *Prototheria* to designate a primitive grade of mammalian evolution that was represented by *Monotremata* among extant organisms, and it might therefore be considered a partial synonym of *Mammalia*. However, nearly all subsequent authors used *Prototheria* to designate a taxon containing *Monotremata* plus a rather variable assemblage of extinct taxa once thought to be part of the monotreme total clade (see *Pan-Monotremata*, this volume).

Among living organisms, there is little ambiguity in determining what is, or is not, a member of *Mammalia*. However, in the nineteenth and twentieth centuries that distinction became blurred by the discovery of fossils, some of which were close relatives of extant mammals within the mammalian stem group while others represented the oldest extinct members of the crown (see *Mammaliaformes*, *Mammaliamorpha*, this volume). During the twentieth century, the composition and basic concept of *Mammalia* as a taxon became clouded, as preeminent palaeontologists (e.g., Olson, 1959; Simpson, 1959) argued that the name designated a grade of evolution, rather than a clade united by shared ancestry (see *Therapsida*, this volume). Even when *Mammalia* was viewed as a clade, arguments were advanced that assigned the name *Mammalia* to virtually every clade more inclusive than the crown ever named by palaeontologists along the mammalian stem (Rowe and Gauthier, 1992). Most of those proposals recommended applying the name *Mammalia* to a clade originating in the same ancestor as one of its named paraphyletic stem groups, for example, *Therapsida* or *Synapsida*, because one or more 'putative defining' characters could be traced to the base of those clades. The intention seems to have been to conceptualize *Mammalia* as part (*Therapsida*) or nearly all (*Synapsida*) of its total clade. But this was not explicit, and no substitute name for the crown was ever proposed. In any event, the focus eventually shifted from establishing arbitrary 'defining' characters of *Mammalia* to resolving the phylogenetic relationships among the living and extinct species of the mammalian total clade (Gauthier et al., 1988; Rowe, 1987, 1988, 1993; de Queiroz, 1994; de Queiroz and Gauthier, 1992, 1994).

Restricting *Mammalia* to the crown clade as proposed by Rowe (1988) has been criticized as breaking from a palaeontological tradition, and some palaeontologists still prefer to apply *Mammalia* to the node designated in this volume as *Mammaliaformes* or to a slightly more inclusive clade (e.g., Lee, 1996; Benton, 2015). They contend that their proposal is consistent with widespread and traditional usage, that the taxon

is relatively stable in terms of membership, and that it is diagnosed by biologically significant characters (Kielan-Jaworowska et al., 2004).

In fact, different palaeontologists include different fossils as the internal specifiers in their definitions of the name *Mammalia*, which highlights that the palaeontological tradition is one of instability. The most popular palaeontological definition expands *Mammalia* beyond the crown to include the Late Triassic *Morganucodonta* (e.g., Crompton and Jenkins, 1973; Kermak et al., 1973, 1981; Kemp, 1983, 1988a,b). Under this definition, the temporal span of *Mammalia* extends into the Triassic, minimally to the Rhaetian (~ 201–208 million years ago). Others expand *Mammalia* to include the more distant *Sinoconodon rigneyi*, which expands composition if not temporal duration (Kielan-Jaworowska et al., 2004). Still others expand *Mammalia* to include *Adelobasileus cromptoni* (Lucas and Luo, 1993; Benton, 2015), which expands composition and extends the minimum age of *Mammalia* further back in time, into the Carnian (~ 228–235 million years ago). None of these propositions is defensible in the absence of a substitute name for the crown clade.

It may be true that within the larger hierarchy of *Pan-Mammalia*, where one chooses to assign the name *Mammalia* is arbitrary. But that is not to say that a system that accepts or even encourages multiple meanings for a single name is meritorious. Instability in the meaning of *Mammalia* is naturally accompanied by instability in the properties associated with that name, including its composition, its diagnostic characters, and its distribution in time and space (Rowe, 1988). This was the situation in the latter part of the twentieth century, when the meaning of the name *Mammalia* came to be specified in terms of the names of individual researchers, for example "*Mammalia* sensu Kermack" or "*Mammalia* sensu Simpson" or

"*Mammalia* sensu Benton" (Rowe, 1986, 1988). Granted, this practice has not entirely disappeared, but one hopes that the practice of articulating phylogenetic definitions such as the one proposed here will put an end to it.

Employing *Mammalia* as the name of a crown clade stabilizes its meaning in terms of the most recent common ancestor of living species. Conceptualizing *Mammalia* as a crown clade nested within a series of more inclusive clades (see *Mammaliaformes*, *Mammaliamorpha*, *Cynodontia*, *Therapsida*, and *Pan-Mammalia* in this volume) highlights the entire mammalian stem lineage and thus helps to identify the processes of evolution of which *Mammalia* is a product. By restricting the name *Mammalia* to the crown clade, by conceptualizing that clade as the least inclusive in a series of nested clades from total to crown, and by using the pan-clade convention to designate the total clade of *Mammalia* as *Pan-Mammalia* (this volume), this nomenclature communicates most efficiently and unambiguously the evolutionary meanings of these names to non-specialists, using a single nomenclature in which living and extinct members of a total clade are integrated into a unified system of taxon names (de Queiroz, 2007).

Finally, in considering the long-standing question of the origin of mammals, new benefits can be seen from conceptualizing *Mammalia* as crown clade nested within the more inclusive clades *Mammaliaformes* (this volume), *Mammaliamorpha* (this volume), *Cynodontia* (this volume), *Therapsida* (this volume), and *Pan-Mammalia* (this volume). Numerous mechanisms have been postulated as driving the origin of mammals, including nocturnality, an enlarged brain with a neocortex, fur, endothermy, parental care of young, miniaturization, enhanced olfaction and hearing, and others. While these features distinguish mammals from other living organisms, we have now seen that at least some of them can be traced into the mammalian stem. For example, successive olfactory enhancements occurred first in *Cynodontia*, followed by additional enhancements that were apomorphic of *Mammaliaformes* and *Mammalia*. Miniaturization and the first potential indications of parental care are diagnostic of *Mammaliamorpha*. The presence of fur and probably also endothermy and a neocortex are diagnostic of *Mammaliaformes*. The discovery of three opsin genes in platypus (Davies et al., 2007) suggests that the ancestral crown mammal was diurnal with trichromatic vision, and that dichromatic nocturnality evolved later, and only within therian mammals. Considering *Mammalia* as the last in a series of nested clades, from total to crown, has already begun to provide a nuanced sequential framework in which to answer many enduring questions surrounding the emergence of mammalian characters and the origin of mammals.

Literature Cited

Bever, G. S., T. B. Rowe, E. G. Ekdale, T. E. Macrini, M. W. Colbert, and A. M. Balanoff. 2005. Comment on: Independent origins of middle ear bones in monotremes and therians. *Science* 309:1492a.

Benton, M. 2015. *Vertebrate Palaeontology*. 4th edition. Wiley Blackwell, West Sussex.

Crompton, A. W., and F. A. Jenkins, Jr. 1973. Mammals from reptiles: a review of mammalian origins. *Annu. Rev. Earth Planet. Sci.* 1:131–155.

Davies, W. L., L. S. Carvalho, J. A. Cowing, L. D. Beazley, D. M. Hunt, and C. A. Arrese. 2007. Visual pigments of the platypus: a novel route to mammalian colour vision. *Curr. Biol.* 17:R161–R163.

Gauthier, J. A., A. G. Kluge, and T. B. Rowe. 1988. Amniote phylogeny and the importance of fossils. *Cladistics* 4:105–209.

Gill, T. 1872. Arrangement of the families of mammals. *Smithson. Misc. Collect.* 11:1–98.

Gill, T. 1902. The story of a word—mammal. *Pop. Sci. Monthly* 61:434–438.

Hopson, J. A. 1970. The classification of nontherian mammals. *J. Mammal.* 51:1–9.

Huber, E. 1930a. Evolution of facial musculature and cutaneous field of Trigeminus. Part I. *Q. Rev. Biol.* 5:133–188.

Huber, E. 1930b. Evolution of facial musculature and cutaneous field of Trigeminus. Part II. *Q. Rev. Biol.* 5:389–437.

Huxley, T. H. 1880. On the application of the laws of evolution to the arrangement of the *Vertebrata*, and more particularly of the *Mammalia. Proc. Zool. Soc. Lond.* 1880:649–662.

Ji, Q., Z.-X. Luo, C. Yuan, and A. R. Tabrum. 2006. A swimming mammaliaform from the Middle Jurassic and ecomorphological diversification of early mammals. *Science* 311:1123–1127.

Ji, Q., Z.-X. Luo, X. Zhang, C.-X. Yuan, and L. Xu. 2009. Evolutionary development of the middle ear in Mesozoic therian mammals. *Science* 326:278–281.

Kemp, T. S. 1983. The relationships of mammals. *Zool. J. Linn. Soc.* 77:353–384.

Kemp, T. S. 1988a. Interrelationships of the *Synapsida*. Pp. 1–22 in *The Phylogeny and Classification of the Tetrapods. Vol. 2: Mammals* (M. J. Benton, ed.). Systematics Association Special Volume No. 35B. Clarendon Press, Oxford.

Kemp, T. S. 1988b. A note on the Mesozoic mammals, and the origin of therians. Pp. 23–29 in *The Phylogeny and Classification of the Tetrapods. Vol. 2: Mammals* (M. J. Benton, ed.). Systematics Association Special Volume No. 35B. Clarendon Press, Oxford.

Kermack, K. A., F. Mussett, and H. W. Rigney. 1973. The lower jaw of *Morganucodon. Zool. J. Linn. Soc.* 53:87–175.

Kermack, K. A., F. Mussett, and H. W. Rigney. 1981. The skull of *Morganucodon. Zool. J. Linn. Soc.* 71:1–158.

Kielan-Jaworowska, Z., R. L. Cifelli, and Z.-X. Luo. 2004. *Mammals from the Age of Dinosaurs.* Columbia University Press, New York.

Lee, M. S. 1996. Stability in meaning and content of taxon names: an evaluation of crown-clade definitions. *Proc. R. Soc. Lond. B Biol. Sci.* 263: 1103–1109.

Lightoller, G. S. 1942. Matrices of the facialis musculature: homologization of the musculature in monotremes with that of marsupials and placentals. *J. Anat.* 76:258–269.

Linnaeus, C. 1758. *Systema Naturae Per Regna Tria Naturae, Secundum Classes, Ordines, Genera, Species, cum Characteribus, Differentiis, Synonymis, Locis.* Tomus I, Editio decima. Laurentius Salvius, Holmiae (Stockholm).

Lucas, S. G., and Z.-X. Luo. 1993. *Adelobasileus* from the Upper Triassic of west Texas: the oldest mammal. *J. Vertebr. Paleontol.* 13:309–334.

Luo, Z.-X. 2007. Transformation and diversification in early mammal evolution. *Nature* 450:1011–1019.

Luo, Z.-X., A. W. Crompton, and A.-L. Sun. 2001. A new mammal from the Early Jurassic and evolution of mammalian characteristics. *Science* 292:1535–1540.

Luo, Z.-X., P. Chen, G. Li, and M. Chen. 2007. A new eutriconodont mammal and evolutionary development in early mammals. *Nature* 446:288–293.

Luo Z.-X., S. M. Gatesy, F. A. Jenkins, Jr., W. W. Amaral, and N. H. Shubin. 2015. Mandibular and dental characteristics of Late Triassic mammaliaform *Haramiyavia* and their ramifications for basal mammal evolution. *Proc. Natl. Acad. Sci. USA* 112:E7101–E7109.

McGowen, M. R., C. Clark, and J. Gatesy. 2008. The vestigial olfactory receptor subgenome of odontocete whales: phylogenetic congruence between gene-tree reconciliation and supermatrix methods. *Syst. Biol.* 57:574–590.

McKenna, M. C., and S. K. Bell. 1997. *Classification of Mammals above the Species Level.* Columbia University Press, New York.

Meng, J., Y. Hu, Y. Wang, X. Wang, and C. Li. 2006. A Mesozoic gliding mammal from northeastern China. *Nature* 444:889–893.

Meng, J., Y. Q. Wang, and C. K. Li. 2011. Transitional mammalian middle ear from a new Cretaceous Jehol eutriconodont. *Nature* 472:181–185.

Niimura, Y. 2009. On the origin and evolution of vertebrate olfactory receptor genes: comparative genome analysis among 23 chordate species. *Genome Biol. Evol.* 1:34–44.

Niimura, Y. 2012. Olfactory receptor multigene family in vertebrates: from the viewpoint of evolutionary genomics. *Curr. Genom.* 13:103–114.

Nowak, R. M., and J. L. Paradiso. 1983. *Walker's Mammals of the World.* Johns Hopkins University Press, Baltimore, MD.

Oken, L. 1847. *Elements of Physiophilosophy* (English Translation by A. Tulk), Vol. 10. John Ray Society, London.

Olson, E. C. 1959. The evolution of mammalian characters. *Evolution* 13:344–353.

de Queiroz, K. 1994. Replacement of an essentialistic perspective on taxonomic definitions as exemplified by the definition of "*Mammalia*". *Syst. Biol.* 43:497–510.

de Queiroz, K. 2007. Toward an integrated system of clade names. *Syst. Biol.* 56:956–974.

de Queiroz, K., and J. A. Gauthier. 1992. Phylogenetic taxonomy. *Annu. Rev. Ecol. Syst.* 23:449–480.

de Queiroz, K., and J. A. Gauthier. 1994. Toward a phylogenetic system of biological nomenclature. *Trends Ecol. Evol.* 9:27–31.

Rafinesque, C. S. 1814. Précis des Découvertes et Travaux Somiologiques de mr CS Rafinesque-Schmaltz Entre 1800 et 1814 ou Choix Raisonné de ses Principales Découvertes en Zoologie et en Botanique. Royale Typographie Militaire, aux Dépens de L'auteur.

Rougier, G. W., A. G. Martinelli, A. M. Forasiepi, and M. J. Novacek. 2007. New Jurassic mammals from Patagonia, Argentina: a reappraisal of australosphenidan morphology and interrelationships. *Am. Mus. Nat. Hist. Novitates* 3566:1–54.

Rowe, T. B. 1986. *Osteological diagnosis of* Mammalia, *L. 1758, and its relationship to extinct* Synapsida. PhD dissertation, University of California, Berkeley, CA.

Rowe, T. B. 1987. Definition and diagnosis in the phylogenetic system. *Syst. Zool.* 36:208–211.

Rowe, T. B. 1988. Definition, diagnosis and origin of *Mammalia. J. Vertebr. Paleontol.* 8:241–264.

Rowe, T. B. 1993. Phylogenetic systematics and the early history of mammals. Pp. 129–145 in *Mammalian Phylogeny* (F. S. Szalay, M. J. Novacek, and M. C. McKenna, eds.). Springer-Verlag, New York.

Rowe, T. B. 1996a. Coevolution of the mammalian middle ear and neocortex. *Science* 273:651–654.

Rowe, T. B. 1996b. Brain heterochrony and evolution of the mammalian middle ear. Pp. 71–96 in *New Perspectives on the History of Life* (M. Ghiselin and G. Pinna, eds.). California Academy of Sciences, Memoir 20.

Rowe, T B., T. P. Eiting, T. E. Macrini, and R. A. Ketcham. 2005. Organization of the olfactory and respiratory skeleton in the nose of the gray short-tailed opossum *Monodelphis domestica. J. Mamm. Evol.* 12:303–336.

Rowe, T. B., and J. A. Gauthier. 1992. Ancestry, paleontology, and definition of the name *Mammalia. Syst. Biol.* 41:372–378.

Rowe, T. B., T. E. Macrini, and Z.-X. Luo. 2011. Fossil evidence on origin of the mammalian brain. *Science* 332:955–957.

Rowe, T. B., T. H. Rich, P. Vickers-Rich, M. Springer, and M. O. Woodburne. 2008. The oldest platypus, and its bearing on divergence timing of the platypus and echidna clades. *Proc. Natl. Acad. Sci. USA* 105:1238–1242.

Rowe, T. B., and G. M. Shepherd. 2016. The role of ortho-retronasal olfaction in mammalian cortical evolution. *J. Comp. Neurol.* 524:471–495.

Shaw, G. 1792. *The Porcupine Ant-Eater. The Naturalist's Miscellany, or Coloured Figures of Natural Objects Drawn and Described Immediately from Nature.* Plate 109. London. 7 Vols. (1790–1813).

Simpson, G. G. 1959. Mesozoic mammals and the polyphyletic origin of mammals. *Evolution* 13:405–414.

Wilson, D. E., and D. M. Reeder, eds. 2005. *Mammal Species of the World.* Johns Hopkins University Press, Baltimore, MD.

Author

Timothy B. Rowe; Jackson School of Geosciences; C1100, The University of Texas at Austin; Austin, TX 78712, USA. Email rowe@utexas.edu.

Date Accepted: 12 June 2017

Primary Editors: Jacques A. Gauthier, Kevin de Queiroz

Pan-Monotremata T. B. Rowe, B.-A. Bhullar, and R. V. S. Wallace, new clade name

Registration Number: 225

Definition: The total clade of the crown clade *Monotremata*. This is a crown-based total-clade definition. Abbreviated definition: total ∇ of *Monotremata*.

Etymology: Derived from the Greek *pan-* (all, in reference to a total clade) and *Monotremata* (this volume).

Reference Phylogeny: Figure 9 in Rougier et al. (2007), where *Pan-Monotremata* ≈ *Australosphenida*, is the primary reference phylogeny. See also Figure 4 in Luo et al. (2015).

Composition: *Pan-Monotremata* includes *Monotremata* (this volume), plus all extinct species more closely related to monotremes than to therian mammals. Whether any known fossils are unequivocally members of the stem group of *Monotremata* has been a matter of debate for more than a century. There is broad agreement that extinct *Obdurodon tharalkooschild*, *Obdurodon dicksoni*, *Steropodon galmani*, *Monotrematum sudamericanum*, and *Teinolophos trusleri* are part of *Pan-Monotremata*, but controversy surrounds whether they belong within crown *Monotremata* as members of the total clade of *Ornithorhynchus*, or lie outside of the crown as members of the monotreme stem group (see *Monotremata*, this volume).

Several fragmentary Jurassic and Cretaceous fossils from the Southern Hemisphere may be stem monotremes. They include *Ausktribosphenos nyktos* Rich et al. 1997, *Ambondro mahabo* Flynn et al. 1999, *Bishops whitmorei* Rich et al. 2001a, *Asfaltomylos patigonicus* Rauhut et al. 2002, *Henosferus molus*

Rougier et al. 2007, *Shuotheriidae* Chow and Rich 1982 (sensu Luo et al., 2007), *Kryoryctes cadburyi* Pridmore et al. 2005, and *Kollikodon ritchiei* Flannery et al. 1995 (see also Pian et al., 2016). See Comments for details.

Diagnostic Apomorphies: *Pan-Monotremata*, like all total clades, need not have any apomorphies *per se* (Gauthier and de Queiroz, 2001); however, as is generally true, fossils can be referred to *Pan-Monotremata* based on the possession of one or more characters derived within the clade (Rowe, 1988). In light of the fossils mentioned above, several apomorphies have been proposed for the total clade of *Monotremata*: the presence of a sharp morphological break between penultimate and ultimate premolars, the last lower premolar (where known) has a strongly molarized trigonid, with a fully formed, obtusely triangulated protoconid, paraconid, and a metaconid, little or no talonid, and a strongly developed mesial cingulid. The lower molars are apomorphic in having a strongly developed mesial cingulid that tends to wrap around to the lingual side of the tooth (Kielan-Jaworowska et al., 2004).

Synonyms: *Prototheria* sensu Sereno (2006) is an unambiguous synonym. Approximate synonyms include *Prototheria* Gill 1872 sensu Cope (1889, 1898, but not Huxley, 1880), *Atheria* Kermack et al. 1973, *Yinotheria* Chow and Rich 1982 sensu Kielan-Jaworowska et al. (2004), and *Australosphenida* Luo et al. 2001 sensu Kielan-Jaworowska et al. (2004: 17) and Rougier et al. (2007).

Comments: By the end of the nineteenth century, *Monotremata* and *Theria* were recognized as comprising the two fundamental divisions of extant *Mammalia* (see *Monotremata*, this volume). Fossils and molecular divergence-time analyses broadly agree that these two clades had diverged by the Middle Jurassic (Rowe, 1988, 1993; Rowe et al., 2008; Woodburne et al., 2003; Benton and Donoghue, 2007; Bininda-Emonds et al., 2007). Whereas the stem group of *Theria* is represented by a very dense fossil record (albeit comprising highly incomplete specimens that consist mostly of broken jaws and teeth), the fossil record of the stem group of *Monotremata* is mostly, if not entirely, unknown. The few uncontested pan-monotreme fossils may all belong to crown *Monotremata* (this volume). All other fossils proposed as members of the stem group of *Monotremata* are controversial (see Composition).

E. D. Cope (1889) may have been the first to include fossils within a classification of living mammals. He recognized Gill's (1872) basic dichotomy between *Prototheria* and *Eutheria* (= *Theria* of contemporary authors), and under *Prototheria* Cope included living monotremes (his *Ornithostomi*) along with the fossil groups *Protodonta* and *Multituberculata* (which included *Tritylodontidae*). To these, Osborn (1907) and Gregory (1910) later added *Allotheria* (= *Multituberculata* + *Tritylodontidae* + *Haramiyidae*). *Prototheria* eventually came to contain the extinct taxa *Morganucodonta*, *Triconodontidae*, *Amphilestidae*, *Docodonta*, *Multituberculata*, and *Haramiyidae*, in addition to living monotremes. None of the fossils just mentioned is considered a pan-monotreme today.

During the twentieth century, palaeontologists led by G. G. Simpson (1928, 1931, 1945) questioned whether classifications of living taxa were adequate to accommodate fossils, and in many subsequent conceptualizations *Prototheria* was treated as a grade of organization that acquired certain "essential" mammalian features (e.g., endothermy, hair, mammary glands), but not the "defining" features of therians (e.g., vibrissae, nipples, vivipary). Unlike the early conceptualizations of *Therapsida*, *Pelycosauria*, and *Synapsida* (this volume), the protetherians were not considered to be an extinct, ancestral stem group because living monotremes were always included among them. Rather, Huxley's (1880) concept of *Prototheria* as the most primitive grade of mammalian evolution was augmented by including fossils along with monotremes (see *Mammalia*, this volume). *Prototheria* was conceived as a grade of organization that may have evolved to a "mammalian level of organization" independently of therians, and they became known in the vernacular as the "non-therian mammals" (e.g., Kermack and Kielan-Jaworowska, 1971). It was doubted that the two grades shared a common ancestor that would be considered a mammal, *Mammalia* was viewed as a polyphyletic grade, and *Prototheria* became a holding tank for problematic fossils that retained plesiomorphic features that were lost in therians. The ancestry of *Monotremata* and its relationship to *Theria* became the most problematic quandary of mammalian evolution (Simpson, 1959, 1971; Broom, 1932; Olson, 1944, 1959; Romer, 1966, 1968; Clemens et al., 1979; Kermack and Kermack, 1984; Carroll, 1988; Jenkins, 1989).

Hopson (1970) attempted to diagnose *Prototheria*: "an anterior extension of the prootic, rather than the alisphenoid, forms the greater part of the orbitotemporal region of the braincase. Principle cusps in molar teeth aligned in an antero-posterior row. This group is also characterized by the retention of many primitive features that are lost early in the subclass Theria" (Hopson, 1970: 6). But all of the characters proposed as diagnosing ("defining") *Prototheria* eventually proved to be plesiomorphic resemblances. Kermack et al. (1973) argued that the name *Prototheria* should be abandoned owing to the uncertainty surrounding the taxon's composition and diagnosis, and in its place they proposed *Atheria* to contain *Monotremata* plus the extinct *Multituberculata, Docodonta, Triconodontida,* and *Morganucodonta*; they excluded tritylodontids and ignored the highly incomplete *Protodonta*. But the name *Atheria* quickly lapsed.

Ever since it was articulated that *Monotremata* and *Theria* were the primary sister clades comprising *Mammalia* (Rowe, 1988), a steady debate has surrounded which fossils, if any, belong to the monotreme stem group. Recent phylogenetic analyses are in general agreement that *Morganucodonta, Docodonta, Tritylodontidae,* and *Haramiyidae* lie outside of crown *Mammalia* (see *Mammaliamorpha, Mammaliaformes,* this volume). The members of *Protodonta* are considered to be non-mammalian cynodonts of uncertain affinities (Sues, 2001). *Multituberculata,* which once included tritylodontids and haramiyids, continues to be whittled down in its composition, and it various components (*Paulchoffatia, Taeniolabioidea, Ptilodontoidea*) may form separate clades within the stem group of *Theria* (Rowe, 1988, 1993; but see Kielan-Jaworowska et al., 2004, for an alternative view). By the end of the twentieth century, it seemed that there were no unequivocal members of the stem group of *Monotremata,* and the name *Prototheria* came to be restricted to the monotremes.

Soon, however, new discoveries in the southern hemisphere yielded fossils that have been proposed as early members of the monotreme stem group. They include *Ambondro mahabo* (Flynn et al., 1999), known only from a jaw from Middle Jurassic rocks of Madagascar (Bathonian, ~165 million years old). *Ambondro* was originally assigned to *Tribosphenida* (i.e., the stem group of *Theria*), based on dental wear facets in a well-developed talonid that suggest occlusion by the protocone and a functionally

tribosphenic occlusal condition. It also possesses a strong distal metacristid that tends to place it within *Tribosphenida* rather than on the monotreme stem (Davis, 2011: 237). In contrast, an analysis by Luo et al. (2015) concluded that it is a stem-monotreme, based on characters listed in the diagnosis of pan-monotremes (see Diagnostic Apomorphies).

The next potential stem monotreme discoveries included an assemblage of jaws representing three taxa from the Early Cretaceous Flat Rocks locality of Victoria, Australia (~ Albian-Aptian, or 121–112.5 million years old; Rich et al. 1997, 1999, 2001a,b). These include *Teinolophos trusleri, Ausktribosphenos nyktos,* and *Bishops whitmorei*. Each of these taxa was initially regarded as a tribosphenic (i.e., stem therian) mammal (Rich et al., 1999, 2001a,b). Rich et al. (2001c) subsequently argued that *Teinolophos* is a monotreme, and the only point of contention with this newer conclusion is whether *Teinolophos* is a member of crown *Monotremata* or of its stem group (see *Monotremata,* this volume).

More recently, *Asfaltomylos patagonicus* Rauhut et al. 2002 (see also Martin and Rauhut, 2005) and *Henosferus molus* Rougier et al. 2007 were described from the same Middle to Late Jurassic locality in Argentina (~158 million years old). Both were considered to belong to the monotreme stem group, and phylogenetic analyses of these fossils added strength to the idea that *Ambondro, Ausktribosphenos,* and *Bishops* are also members of the monotreme total clade. *Shuotheriidae* Chow and Rich 1982 (sensu Luo et al., 2007) from the Jurassic of China has also been proposed as a member of the total clade of *Monotremata*.

With the new discoveries of possible stem-monotremes, two new names became associated with taxa approximating the monotreme total clade. Chow and Rich (1982) proposed (legion) *Yinotheria* containing (order) *Shuotheridia* and (family) *Shuotheriidae*, which at the time contained only *Shoutherium dongi*, based on an isolated jaw and later on a partial skeleton (Luo et al., 2007). Chow and Rich (1982) and McKenna and Bell (1997) initially considered *Shoutherium* to be a stem therian. Arguing instead that *Shuotherium* lies on the monotreme stem, Kielan-Jaworowska et al. (2004) and Luo et al. (2007) used *Yinotheria* for a clade approximating the total clade of *Monotremata*. Luo et al. (2001) also named *Australosphenida* for a subset of the total clade that excluded *Shuotherium* (see also Martin and Rauhut, 2005). In their encyclopedic book on *Mammals from the Age of Dinosaurs*, Kielan-Jaworowska et al. (2004) equivocated on the relationships of *Shuotherium*. Their Table 1.2 treated the relationships of *Shuotheridia* as uncertain, did not recognize the taxon *Yinotheria*, and used *Australosphenida* for a taxon approximating the monotreme total clade. However, elsewhere in the book (e.g., p. 204), *Shuotheridia* and *Australosphenida* were grouped together under *Yinotheria* (see also Luo et al., 2007). Rougier et al. (2007) found *Shuotherium* to lie outside the mammalian crown, and they used the name *Australosphenida* for a clade approximating the total clade of *Monotremata*.

A good deal of criticism has been directed at the inclusion of these fossils within the total clade of *Monotremata*. It implies either a dual origin for the tribosphenic molar or its secondary loss along the monotreme stem, and dual origins of the mammalian middle ear. These characters were interpreted as unique transformations in phylogenetic analyses based on taxa that are relatively complete (Rowe, 1988, 1993), whereas the evidence that places the relevant fossils in the monotreme stem group is largely limited to dental and mandibular characters. Other objections have been advanced that include contested homologies between designated premolars and molars, and between individual cusps, contested decisions on character weighting and character ordering, and disputes over rooting and transformation polarity (Rich et al., 2001c; Woodburne, 2003; Woodburne et al., 2003; Rougier et al., 2007; Rowe et al., 2008; Davis, 2011). More complete fossils are needed to resolve the phylogenetic uncertainty and to enable the unequivocal assignment of any taxon to the stem group of *Monotremata*.

The names *Prototheria*, *Atheria*, *Yinotheria*, and *Australosphenida* are all potentially available for the total clade of *Monotremata*, although each is problematic. *Prototheria* and *Atheria* were always used for groups that included *Monotremata*, but none of the extinct taxa traditionally assigned to them are considered part of the total clade of *Monotremata* under current hypotheses. *Atheria* quickly lapsed. *Yinotheria* has been used for groups that excluded monotremes, while *Australosphenida* was originally applied to a subset of the total clade of monotremes. *Prototheria* is deeply entrenched in the literature, and in recent years it has generally been employed as a redundant "supernomen" (subclass) for *Monotremata*. Of all the historical names available for the monotreme total clade, the strongest case can be made for *Prototheria* sensu Sereno (2006).

The proposal set forth here, to adopt the name *Pan-Monotremata* for the total clade of *Monotremata*, reflects a subtle distinction in being based on extant internal specifiers, principally *Tachyglossus aculeatus* and *Ornithorhynchus anatinus*. *Pan-Monotremata* refers to a clade that explicitly and necessarily applies to the total clade of *Monotremata* (this volume), whether or not any of the fossils mentioned above prove to lie within it. *Pan-Monotremata* also has the pragmatic virtue of employing the panclade convention (Gauthier and de Queiroz, 2001; de Queiroz, 2007; *ICPN*, Art. 10.3), in which the names of total clades are formed by adding the prefix *Pan-* to the names of their corresponding crowns, instead of inventing altogether new names or converting specialized names used principally by palaeontologists, such as *Australosphenida* or *Yinotheria*. General adoption of the panclade convention promises to facilitate communication among biologists who study distantly related clades, to enhance digital indexing and retrieval of phylogenetic knowledge for the total clade, and to promote a clear distinction between crown and total clades. Additionally, the name *Monotremata* Fischer 1823 (cf. Geoffroy, 1803) predates all others by decades and is the most widely used of the names. By restricting the name *Monotremata* to the crown (see *Monotremata*, this volume) and using the panclade convention to designate the total clade of *Monotremata* as *Pan-Monotremata*, this nomenclature communicates most efficiently and unambiguously the evolutionary meanings of these names to non-specialists by integrating living organisms and their extinct relatives under a unified system of clade names (de Queiroz, 2007).

Literature Cited

Benton, M. J., and P. C. J. Donoghue. 2007. Paleontological evidence to date the tree of life. *Mol. Biol. Evol.* 24:26–53.

Bininda-Emonds, O. R. P., M. Cardillo, K. E. Jones, R. D. MacPhee, R. M. Beck, R. Grenyer, S. A. Price, R. A. Vos, J. L. Gittleman, and A. Purvis. 2007. The delayed rise of present-day mammals. *Nature* 446:507–512.

Broom, R. 1932. *The Mammal-Like Reptiles of South Africa and the Origin of Mammals*. H. F. & G. Witherby, London.

Carroll, R. L. 1988. *Vertebrate Paleontology and Evolution*. W. H. Freeman & Co., New York.

Chow, M., and T. H. Rich. 1982. *Shuotherium dongi*, n. gen. and sp., a therian with pseudotribosphenic molars from the Jurassic of Sichuan, China. *Aust. Mammal.* 5:127–142.

Clemens, W. A., J. A. Lillegraven, and Z. Kielan-Jaworowska. 1979. *Mesozoic Mammals—The First Two-Thirds of Mammalian Evolution*. University of California Press, Berkeley, CA.

Cope, E. D. 1889. Synopsis of the families of Vertebrata. *Am. Natural.* 23:849–877.

Cope, E. D. 1898. *Syllabus of Lectures on the Vertebrata, with an Introduction by Henry Fairfield Osborn*. University of Pennsylvania, Philadelphia, PA.

Davis, B. M. 2011. Evolution of the tribosphenic molar pattern in early mammals, with comments on the "dual-origin" hypothesis. *J. Mammal. Evol.* 18:227–244.

Fischer, G. 1823. *Enchiridion Generum Animalium*. Moscow.

Flannery, T. F., M. Archer, T. H. Rich, and R. Jones. 1995. A new family of monotremes from the Cretaceous of Australia. *Nature* 377:418–420.

Flynn, J. J., J. M. Parrish, B. Rakotosamimanana, W. F. Simpson, and A. R. Wyss. 1999. A middle Jurassic mammal from Madagascar. *Nature* 401:57–60.

Gauthier, J. A., and K. de Queiroz. 2001. Feathered dinosaurs, flying dinosaurs, crown dinosaurs, and the name "*Aves*". Pp. 7–41 in *New Perspectives on the Origin and Evolution of Birds* (J. Gauthier and L. F. Gail, eds.). Yale University Press, New Haven, CT.

Geoffroy St. Hillaire, E. 1803. Extrait des observations anatomiques de M. Home, sur l'echidné. *Bull. Sci. Soc. Philom. Tome* 3:225–227.

Gill, T. 1872. Arrangement of the families of mammals. *Smithson. Misc. Collect.* 11:1–98.

Gregory, W. K. 1910. The orders of mammals. *Bull. Am. Mus. Nat. Hist.* 27:1–524.

Hopson, J. A. 1970. The classification of nontherian mammals. *J. Mammal.* 51:1–9.

Huxley, T. H. 1880. On the application of the laws of evolution to the arrangement of the *Vertebrata*, and more particularly of the *Mammalia*. *Proc. Zool. Soc. Lond.* 1880:649–662.

Jenkins, Jr., F. A. 1989. Monotremes and the biology of Mesozoic mammals. *Neth. J. Zool.* 40:5–31.

Kermack, D. M., and K. A. Kermack. 1984. *The Evolution of Mammalian Characters*. Croom Helm, London, Sydney, and Kapitan Szabo Publishers, Washington, DC.

Kermack, K. A., and Z. Kielan-Jaworowska. 1971. Therian and non-therian mammals.

Pp. 103–115 in *Early Mammals* (D. M. Kermack and K. A. Kermack, eds.). *Zool. J. Linn. Soc.* 50(suppl. 1).

Kermack, K. A., F. Mussett, and H. W. Rigney. 1973. The lower jaw of *Morganucodon. Zool. J. Linn. Soc.* 53:87–175.

Kielan-Jaworowska, Z., R. L. Cifelli, and Z.-X. Luo. 2004. *Mammals from the Age of Dinosaurs.* Columbia University Press, New York.

Luo, Z.-X., R. L. Cifelli, and Z. Kielan-Jaworowska. 2001. Dual origin of tribosphenic mammals. *Nature* 409:53–57.

Luo, Z.-X., J. Quang, and C.-X. Yuan. 2007. Convergent dental adaptations in pseudotribosphenic and tribosphenic mammals. *Nature* 450:93–97.

Luo Z.-X., S. M. Gatesy, F. A. Jenkins, Jr., W. W. Amaral, and N. H. Shubin. 2015. Mandibular and dental characteristics of Late Triassic mammaliaform *Haramiyavia* and their ramifications for basal mammal evolution. *Proc. Natl. Acad. Sci. USA* 112:E7101–E7109.

Martin, T., and O. W. Rauhut. 2005. Mandible and dentition of *Asfaltomylos patagonicus* (*Australosphenida, Mammalia*) and the evolution of tribosphenic teeth. *J. Vertebr. Paleontol.* 25:414–425.

McKenna, M. C., and S. K. Bell. 1997. *Classification of Mammals above the Species Level.* Columbia University Press, New York.

Olson, E. C. 1944. Origin of mammals based upon cranial morphology of the therapsid suborders. *Geol. Soc. Am. Spec. Pap.* 55:1–130.

Olson, E. C. 1959. The evolution of mammalian characters. *Evolution* 13:344–353.

Osborn, H. F. 1907. *The Evolution of Mammalian Molar Teeth to and from the Triangular Type.* (William K. Gregory, ed.). Macmillan, New York.

Pian, R., M. Archer, S. J. Hand, R. M Beck, and A. Cody. 2016. The upper dentition and relationships of the enigmatic Australian Cretaceous mammal *Kollikodon ritchiei. Mem. Mus. Vict.* 74:97–105.

Pridmore, P. A., T. H. Rich, P. Vickers-Rich, and P. P. Gambaryan. 2005. A tachyglossid-like humerus from the Early Cretaceous of southeastern Australia. *J. Mamm. Evol.* 12:359–378.

de Queiroz, K. 2007. Toward an integrated system of clade names. *Syst. Biol.* 56:956–974.

Rauhut, O. W., T. Martin, E. Ortiz-Jaureguizar, and P. Puerta. 2002. A Jurassic mammal from South America. *Nature* 416:165–168.

Rich, T. H., P. Vickers-Rich, A. Constantine, T. A. Flannery, L. Kool, and N. von Klaveren. 1997. A tribosphenic mammal from the Mesozoic of Australia. *Science* 278:1438–1442.

Rich, T. H., P. Vickers-Rich, A. Constantine, T. F. Flannery, L. Kool, and N. van Klaveren. 1999. Early Cretaceous mammals from Flat Rocks, Victoria, Australia. *Rec. Queen Victoria Mus.* 106:1–34.

Rich, T. H., T. F. Flannery, P. Trusler, A. Constantine, L. Kool, P. Vickers-Rich, and N. van Klaveren. 2001a. An advanced ausktribosphenid from the Early Cretaceous of Australia. *Rec. Queen Victoria Mus.* 110:1–9.

Rich, T. H., T. F. Flannery, P. Trusler, L. Kool, N. A. van Klaveren, and P. Vickers-Rich. 2001b. A second tribosphenic mammal from the Mesozoic of Australia. *Rec. Queen Victoria Mus.* 110:1–9.

Rich, T. H., P. Vickers-Rich, P. Trusler, T. F. Flannery, R. L. Cifelli, A. Constantine, L. Kool, and N. Van Klaveren. 2001c. Monotreme nature of the Australian Early Cretaceous mammal *Teinolophos. Acta Palaeontol. Pol.* 46:113–118.

Romer, A. S. 1966. *Vertebrate Paleontology.* University of Chicago Press, Chicago, IL.

Romer, A. S. 1968. *Notes and Comments on Vertebrate Paleontology.* University of Chicago Press, Chicago, IL.

Rougier, G. W., A. G. Martinelli, A. M. Forasiepi, and M. J. Novacek. 2007. New Jurassic mammals from Patagonia, Argentina: a reappraisal of australosphenidan morphology and interrelationships. *Am. Mus. Nat. Hist. Novit.* 3566:1–54.

Rowe, T. B. 1988. Definition, diagnosis and origin of *Mammalia. J. Vertebr. Paleontol.* 8:241–264.

Rowe, T. B. 1993. Phylogenetic systematics and the early history of mammals. Pp. 129–145 in *Mammalian Phylogeny* (F. S. Szalay, M. J. Novacek, and M. C. McKenna, eds.) Springer-Verlag, New York.

Rowe, T. B., T. H. Rich, P. Vickers-Rich, M. Springer, and M. O. Woodburne. 2008. The oldest platypus, and its bearing on divergence timing of the platypus and echidna clades. *Proc. Natl. Acad. Sci. USA* 105:1238–1242.

Sereno, P. 2006. Shoulder girdle and forelimb in multituberculates: evolution of parasagittal forelimb posture in mammals. Pp. 315–366 in *Amniote Paleobiology* (M. T. Carrano, T. J. Gaudin, R. W. Blob, and J. R. Wible, eds.). University of Chicago Press, Chicago, IL.

Simpson, G. G. 1928. *A Catalogue of the Mesozoic Mammalia in the Geological Department of the British Museum*. Trustees of the British Museum (Natural History), London.

Simpson, G. G. 1931. A new classification of mammals. *Bull. Am. Mus. Nat. Hist.* 59:259–293.

Simpson, G. G. 1945. The principles of classification and a classification of mammals. *Bull. Am. Mus. Nat. Hist.* 85:1–350.

Simpson, G. G. 1959. Mesozoic mammals and the polyphyletic origin of mammals. *Evolution* 13:405–414.

Simpson, G. G. 1971. Concluding remarks: Mesozoic mammals revisited. Pp. 181–198 in *Early Mammals* (D. M. Kermack and K. A. Kermack, eds.). *Zool. J. Linn. Soc.* 50(suppl. 1).

Sues, H. D. 2001. On *Microconodon*, a Late Triassic cynodont from the Newark Supergroup of eastern North America. *Bull. Mus. Comp. Zool.* 156:37–48.

Woodburne, M. O. 2003. Monotremes as pre-tribosphenic mammals. *J. Mamm. Evol.* 10:195–248.

Woodburne, M. O., T. H. Rich, and M. S. Springer. 2003. The evolution of tribospheny and the antiquity of mammalian clades. *Mol. Phylogenet. Evol.* 28:360–385.

Authors

Timothy B. Rowe; Jackson School of Geosciences; C1100, University of Texas at Austin; Austin, TX 78712, USA. E-mail: rowe@utexas.edu.

Rachel V. S. Wallace; Jackson School of Geosciences; C1100, University of Texas at Austin; Austin, TX 78712, USA. E-mail: rvsimon@utexas.edu.

Bhart-Anjan Bhullar; Department of Geology and Geophysics; Yale University; 210 Whitney Ave., New Haven, CT 06520-8109; USA. E-mail: bhart-anjan.bhullar@yale.edu.

Date Accepted: 12 June 2017

Primary Editors: Jacques A. Gauthier, Kevin de Queiroz

Monotremata G. Fischer 1823 [T. B. Rowe, B.-A. S. Bhullar, and R. V. S. Wallace], converted clade name

Registration Number: 223

Definition: The crown clade originating in the most recent common ancestor of *Tachyglossus* (originally *Myrmecophaga*) *aculeatus* (Shaw 1792) and *Ornithorhynchus* (originally *Platypus*) *anatinus* (Shaw 1799). This is a minimum-crown-clade definition. Abbreviated definition: min crown ∇ (*Ornithorhynchus anatinus* Shaw 1799 & *Tachyglossus aculeatus* Shaw 1792).

Etymology: From the Greek *monos*, single, plus *trema*, orifice, in reference to the (plesiomorphic) organization of the excretory and reproductive systems, which collectively open into a common chamber known as the cloaca (Gill, 1903).

Reference Phylogeny: Figure 5 in Rowe et al. (2008).

Composition: *Monotremata* comprises two extant sister clades: the spiny anteaters or echidnas, *Tachyglossidae* Gill 1872, and the duck-billed platypus, *Ornithorhynchus anatinus* (Shaw 1799). Extant echidnas include the short-beaked echidna, *Tachyglossus* (*Echidna* = *Echinopus* = *Syphomia*) *aculeatus* Shaw 1792, and the long-beaked echidna, *Zaglossus* Gill 1877 (= *Pröechidna* = *Acanthoglossus* = *Bruynia* = *Bruijnia* = *Prozaglossus*). *Zaglossus* includes three extant species: *Zaglossus bartoni*, *Zaglossus bruijni*, and *Zaglossus attenboroughi* (the latter possibly extinct; see Flannery and Groves, 1998; Helgen et al., 2012).

Also included in *Monotremata* are a number of named fossils from Australia and Tasmania that belong to the tachyglossid total clade (Murray, 1978; Griffiths et al., 1991; Musser, 2003). They include *Zaglossus owenii* (Krefft 1868), *Zaglossus ramsayi* (Owen 1884), *Zaglossus robustus* (Dun 1895), and *Zaglossus hacketti* Glauert 1914. Various synonymies among these names have been proposed and at present most or all of these fossils are considered conspecific with *Megalibgwilia ramsayi* (see Murray, 1978; Woodburne et al., 1985; Griffiths et al., 1991; Musser, 2003; Helgen et al., 2012; Simon, 2013).

Extinct members of the *Ornithorhynchus* total clade (see Musser, 2003; Simon, 2013) include *Ornithorhynchus agilis* de Vis 1885 (Pliocene), based on a tibia and edentulous dentary, and *Ornithorhynchus maximus* Dun 1895 (Pliocene), based on a large humerus from New South Wales. Also considered part of the platypus total clade is *Obdurodon insignis* Woodburne and Tedford 1975, based on isolated teeth (Oligocene or Miocene), *Obdurodon dicksoni* Archer et al. 1992, based on a complete skull, partial mandible, and partial dentition (Miocene), and *Obdurodon tharalkooschild* Pian et al. 2013 (Late Miocene or Pliocene), based on isolated teeth. While it is generally assumed that *Obdurodon* is monophyletic, it is possible that its named species are consecutive outgroups to *Ornithorhynchus anatinus*, nor can the possibility be ruled out that they are conspecific with one or more of the extinct species currently referred to *Ornithorhynchus*.

Three fragmentary taxa from Cretaceous and Palaeocene sediments of the southern hemisphere are widely acknowledged as members of the monotreme total clade, but whether they belong within the monotreme crown is disputed (Rowe et al., 2008; Phillips et al., 2009). These are *Steropodon galmani* Archer et al. 1985 and *Teinolophos trusleri* Rich et al. 1999, both from the Early Cretaceous of Australia, and *Monotrematum sudamericanum* Pascual et al. 1992a, from the Palaeocene of Argentina. A fourth taxon, *Kollikodon ritchiei* Flannery et al. 1995, was initially thought to be a crown monotreme but was more recently inferred to be a member of the monotreme stem group (Pian et al., 2016).

Diagnostic Apomorphies: In pre-cladistic literature, the grouping of echidnas and platypuses was based on dual criteria. First was their retention of plesiomorphic features that were lost in therian mammals, including ovipary, an interclavicle, procoracoid, and epipubic bones, and the egg tooth. The second criterion was the absence of derived features that diagnose therian mammals, such as whiskers (vibrissae), a rhinarium, and viviparity (Huxley, 1880; Flower and Lydekker, 1891; Gregory, 1910). In the last three decades, a large assemblage of apomorphic features has accumulated that strongly corroborates the monophyly of *Monotremata*. They include approximately 180 miRNAs now known to be shared uniquely by platypuses and echidnas (Murchison et al., 2008), multiple sex chromosomes, with the pattern X1Y1X2Y2X3Y3X4Y4X5Y5 in males (Rens et al., 2004), and unique duplications of the beta-casein genes associated with lactation (Lefèvre et al., 2009).

The sequence and timing of events in monotreme brain and cranial nerve development is unique and produces a relatively large brain compared to other mammals with unique forebrain architecture (Rowe et al., 2011; Ashwell, 2012; Ashwell et al., 2012). Cytoarchitecture of the olfactory bulb is derived in projections of cell somata, which spread throughout its external plexiform layer (Ashwell, 2006), and in the absence of mitral cells in monolayers in the olfactory bulb (Switzer and Johnson, 1977). Cyto- and chemoarchitecture in the sensory trigeminal nucleus is also unique (Ashwell et al.,

2006), as is the developmental path of the trigeminal peripheral nerves (Macrini et al., 2006; Ashwell et al., 2011).

Monotremes are virtually unique among amniotes in possessing an electroreception system mediated by hundreds (*Tachyglossus, Zaglossus*) to tens of thousands (*Ornithorhynchus*) of electroreceptor cells located in the skin of the snout (Manger et al., 1997; Proske et al., 1998; Pettigrew, 1999). This system evolved *de novo* along the monotreme stem. It is innervated by cranial nerve V, in contrast to the diverse electroreceptive systems in aquatic anamniotic vertebrates, which are generally innervated by cranial nerve VIII. Electroreception was recently discovered in paedomorphic primordia of whiskers in the Guiana dolphin (*Sotalia guianensis*), but the wide phylogenetic separation and the differing embryological derivation argue that this is a case of homoplasy (Czech-Damal et al., 2012). Monotremes also share derived developmental pathways for sensory neurons associated with pressure reception ("push rods") in and around the oral cavity (Ashwell et al., 2011).

The mature pattern of monotreme facial musculature and its pathway of embryological differentiation are unique in many ways (Huber, 1930a,b; Lightoller, 1942). Among these features is the remarkable *m. detrahens mandibulae*, a mandibular depressor that opens the jaws. It evolved from the *m. adductor mandibulae*, which in other tetrapods elevates and closes the jaws (Edgeworth, 1935; Rowe, 1986). Throughout the monotreme body, unique skeletomuscular traits underlie unique behaviors involving feeding and locomotion (Winge, 1941).

The osteological apomorphies of *Monotremata* described below were compiled from personal observation and literature (Gregory, 1910, 1947; Gauthier et al., 1988; Rowe, 1986, 1988, 1993; Rowe et al., 2008; Zeller, 1989 a,b, 1993) and from an unpublished thesis aimed specifically at testing monotreme monophyly, providing a revised osteological diagnosis, and testing whether the controversial fossils mentioned above lie within crown *Monotremata* or along its stem (Simon, 2013).

The monotreme snout is apomorphic in being greater than half the length of the skull. Other mammals have long snouts, such as myrmecophagid anteaters, pangolins (*Manidae*), numerous cetaceans, and certain nectar-feeding bats (*Platalina*), but these are universally recognized as homoplastic resemblances. We note that the stem tachyglossid *Megalibgwilia ramsayi* and the extinct platypus *Obdurodon dicksoni*, both known from complete skulls, have exceedingly long faces, and their phylogenetic positions suggest that a secondary shortening of the face has independently affected *Tachyglossus* and *Ornithorhynchus* (Simon, 2013).

Although platypuses have broad bills and echidnas have narrow beaks, they share unique aspects of facial architecture. The narial aperture is circular and is directed dorsally, rather than anteriorly or laterally. Ventrally, a pointed process of the premaxilla extends far posteriorly along the lateral margin of the rostrum, but the premaxilla does not contribute to the secondary palate. The secondary palate comprises only palatal processes of the maxilla, palatine, and pterygoid, and it extends far posteriorly, to beneath the basisphenoid-basioccipital suture where the choana opens in a unique position at the rear of the skull. The long secondary palate contributes to an apomorphic pattern of mastication shared by echidnas and platypuses, in which the rear part of the tongue bears keratinous "tongue teeth" (Owen, 1868) that masticate prey items by crushing them against the back of the secondary palate, where the tongue can exert its greatest mechanical force (Grant, 2007).

Monotremes lack a lacrimal bone, and the foramen for the canal of the lacrimal gland is bordered by the maxilla and frontal. Medial processes of each frontal are wedged between the two nasals on the skull roof. The orbital process of the frontal contacts the maxilla within the orbit, and there is no contact between the frontal and alisphenoid. A facial process of the large parietal extends anteriorly and closely approaches or actually contacts the nasal.

The monotreme cochlea is elongate and partly coiled to about 270° in contrast to the 70° coil inferred as present ancestrally in crown *Mammalia* (Kielan-Jaworowska et al., 2004). A full coil of 360° (or more) occurred independently in *Theria*. The fenestra ovalis faces ventrally from the ventral surface of the skull, and the ectotympanic and tympanum lie horizontally beneath the skull, in a condition that was convergently acquired in some therians with large brains. The hypoglossal nerve exits the endocranial cavity via the jugular foramen (the embryonic metotic fissure), instead of through its own distinct foramen. Both the enlarged brain and reduced masticatory musculature affect the surface topography of the skull such that the lambdoidal and sagittal crests never develop. Instead, the masticatory muscles attaching to the surface of the sub-spherical cranium leave distinctive muscle scarring only on the parietal and dorsal extremity of the supraoccipital. The cranial sutures of monotremes fuse very early in postnatal development.

The glenoid fossa on the squamosal is apomorphic in its geometry and the range of movement that it affords the mandible. It is anteroposteriorly elongate, mediolaterally narrowed and shallow, imposing relatively little constraint on either propalinal or rotational movements of the lower jaw. The dentary peduncle is gracile, elevated above the dentary ramus, and vertically directed where it supports a condyle that is rounded about an antero-posterior axis of rotation. The coronoid process is reduced and its dorsal tip is directed laterally. The mandibular foramen is located in the anterior apex of the pterygoideus fossa. The mandibular angle is reduced. The rear edge of the mandible bears a spiral scar or ridge for the insertion of the *m. detrahens mandibulae*.

Monotreme spinal nerves pass through foramina perforating the lamina of the neural arches along the entire pre-sacral column, instead of issuing from between the vertebrae. Monotreme ribs have only a single head (capitulum) that articulates with its respective centrum. Calcified ventral ribs are broad and imbricating, in a unique pattern associated with the ability of monotremes to flatten their bodies while squeezing through confined spaces (Winge, 1941).

In mature individuals, the cranial margin of the interclavicle fuses to the medial half of each clavicle and to the manubrium of the sternum. A large distinctive fossa for the *m. teres major* is present on the lateral aspect of the scapular blade. The articulation of the humerus with the radius and ulna is unique in that the elbow joint is not aligned with the long axis of the humerus. The capitulum and ulnar trochlea form a continuous synovial surface that is anteroposteriorly cylindrical. The ulnar olecranon process is divided into dual processes, which diverge posteromedially and anterolaterally. The radius and ulna are straight and appressed along their entire lengths, virtually eliminating pronation-supination of the forearm and hand. A substantial portion of the wrist articulates to the ulna where a deep trochlea is formed on the distal surface of the ulna; this contrasts to most other mammals, in which the radius has the broadest contact with the wrist, and the ulna contributes only marginally to the wrist with a pointed styloid process. Dual concave facets are present on the distal end of the radius for the proximal carpals.

The dorsal margin of the ischium is deeply concave, and its ventral edge bears a hypertrophied tuberosity. The dorsal margin of the acetabulum is closed with a complete rim that lacks the cotyloid notch. A preacetabular tubercle is present on the ilium for the *m. rectus femoris*. The lesser trochanter of the femur is present as a continuous ridge connected to the greater trochanter. The fibular flabellum (see *Mammaliaformes*, this volume) is an elaborated protuberance extending far caudally from the proximal end of the fibula in monotremes. The calcaneum has a distinct, long, and laterally projecting peroneal process and a distinct, deep peroneal groove. The calcaneal tubercle is gracile, conical, and laterally divergent, rather than being robust, square, and ventrally inflected. Metatarsal V and the peroneal process of the calcaneum contact side-by-side. The cuboid is skewed to the medial side of the long axis of the calcaneum. The sesamoid bones in the digital flexor tendons are unpaired; whether this last feature is the plesiomorphic condition for *Mammalia* or apomorphic of *Monotremata* is equivocal, as they are generally paired in therians.

Other unique cranial developmental patterns (Zeller, 1989a,b, 1993) and unique skeletal ossification sequences have been described that unite *Ornithorhynchus* with *Tachyglossidae* in a monophyletic *Monotremata* (Weisbecker, 2011; Werneburg and Sánchez-Villagra, 2011).

Synonyms: Approximate synonyms include *Reptantia* Illiger 1811, *Monotremia* Rafinesque 1815, *Didelphes Anomaux* de Blainville 1816, *Metataxymeria* Fischer 1823, *Ornithodelphes* de Blainville 1834 (published in Gervais, 1836), *Amasta* Haeckel 1866, *Ornithodelphia* Huxley 1864, *Prototheria* Gill 1872 (sensu Flower, 1883; Gregory, 1910), *Sauropsidelphia* Roger 1887, and *Ornithostomi* Cope 1889.

Comments: Monotremes are distributed on the mainland and surrounding islands of Australia, an area otherwise known as the "Greater Australian continent," or as "Meganesia" or the "Sahul" region. *Ornithorhynchus anatinus*, the semiaquatic duck-billed platypus, is distributed along the sub-tropical eastern edge of Australia, as well as on Tasmania, King Island, and Kangaroo Island (Griffiths, 1978; Grant, 1992). *Tachyglossus aculeatus*, the short-beaked echidna, is distributed over a wide range of habitats across much of Australia, Tasmania, and the larger neighboring islands in the Bass Strait, including King Island, Kangaroo Island, and Flinders Island, and the island of New Guinea (Griffiths, 1968). All species of *Zaglossus* are currently restricted to New Guinea, although *Zaglossus bruijni* may have been present during historic times in the Northern Territory of Australia (Helgen et al., 2012)

Following the discovery of *Tachyglossus* (Shaw, 1792) and *Ornithorhynchus* (Shaw, 1799), a relationship between the two was instantly proposed (e.g., Blumenbach, 1800; Home, 1802a, b; Geoffroy, 1803; de Blainville, 1816; Cuvier, 1817; Burrell, 1927). Everard Home even referred to them both under the name *Ornithorhynchus*, as *O. paradoxus* (Home, 1802a) and *O. hystrix* (Home, 1802b). However, the relationship of *Monotremata* to other mammals and even to other vertebrates quickly became controversial. Home had raised anatomical suspicions that both monotremes either laid eggs or were "oviviviparous", and he considered them as "an intermediate link between the classes Mammalia, Aves, and Amphibia" (Home, 1802b: 360). Geoffroy vacillated. He recognized their affinity, based chiefly on their cloaca, and proposed the name Monotremes along with a brief diagnosis for what he first considered to be a new mammalian order (Geoffroy, 1803; see Gill, 1903: 434). But later he and others sided with Lamarck (1809) in considering monotremes so far anatomically removed from mammals as to belong in a class of their own (Geoffroy, 1822). Illiger (1811) grouped the platypus and echidna together along with "*Pamphractus*" (possibly a

pangolin) under the name *Reptantia*, which he believed represented an intermediate division between reptiles and mammals. Fischer (1823) also placed *Monotremata* in a class by itself, under the name *Metataxymeria*. Both *Reptantia* and *Metataxymeria* quickly lapsed.

Meckel's (1826) discovery of mammary glands in *Ornithorhynchus*, "the outstanding characteristic of the mammalian class" (Burrell, 1927: 29), in an animal suspected of ovipary, provoked further controversy. Since monotremes lack nipples, the question of homology of the mammary glands in monotremes and other mammals arose. The question of whether monotremes should be classified as *Mammalia* eventually embroiled most of the zoological luminaries of the century. The details are recounted by Lydekker (1894), Burrell (1927), and Gould (1985). The controversy was fueled partly by anatomy, partly by field biology, and it typifies the essentialist debates over the relative importance of certain characters in defining the meanings of taxonomic names, a style of debate that would carry well into the twentieth century.

Over a long career, de Blainville classified monotremes as mammals, but he vacillated on their relationships to other mammals. His 1816 classification of *Mammalia* consisted of two primary divisions, *Monodelphes* and *Didelphes*, and in the latter he placed *Ornithorhynchus* and *Tachyglossus*, to the exclusion of other taxa, as *Didelphes Anomaux*. Many of de Blainville's taxa were subdivided into *Normaux* ('normal') and *Anomaux* ('anomalous'). Based on brain characters, Bonaparte (1837) also produced a two-fold classification in which *Placentalia* comprised the first series, and the second series was named *Ovovivipera* for *Marsupialia* + *Monotremata*.

In a footnote, de Blainville (1816) added the opinion that monotremes constituted a distinct subclass. Gregory (1910: 76) called this a "great step, perhaps the most important one in the history of the classification of mammals" in that it identified the three major clades of *Mammalia*, viz., *Placentalia*, *Marsupialia*, and *Monotremata* to use their modern names. The classification used by de Blainville in his lecture course of 1834 (published by Gervais in 1836; see Gregory, 1910: 82) later formalized his three-fold ("ternary") classification by giving monotremes the name (subclass) *Ornithodelphes* and separating it from (subclass) *Didelphes* and (subclass) *Monodelphes*. Huxley (1864) converted de Blainville's names to *Ornithodelphia*, *Didelphia*, and *Monodelphia*, in a ternary system that was widely used (e.g., Owen, 1868; Huxley, 1872; Gill, 1870). Even though Huxley (1880) and others (e.g., Flower and Lydekker, 1891) soon substituted the names *Prototheria*, *Metatheria*, and *Eutheria*, the ternary system was popular because it was unclear that placentals and marsupials are each other's closest relatives. Marsupials had a combination of primitive and derived characters, and the ternary system forestalled a decision on how to balance conflicting criteria. Gill (1872) grouped marsupials and placentals under *Eutheria*, but with little explanation. Parker and Haswell (1897) finally settled the issue by joining marsupials and placentals under the name *Theria* with an extensive list of diagnostic features.

The name (order) *Monotremata* Fischer 1823 gained almost universal recognition, albeit along with a variety of redundant "supernomina" associated with its ranking as an order contained within its own subclass. Gill was the next to introduce a new supernomen for *Monotremata* when he listed "Subclass (Prototheria) Ornithodelphia" (Gill, 1872: vi), which contained only *Monotremata*. This is the first time that *Prototheria* appeared in print, and by setting it in parentheses next to *Ornithodelphia*, Gill evidently acknowledged their synonymy. Flower (1883) also equivocated over the two redundant names by placing *Monotremata* alone in "Subclass Prototheria or *Ornithodelphia*".

Fortunately, this practice of double-redundancy quickly faded away.

In the meantime, Haeckel (1877) and Huxley (1880) began to confront the problem of classifying fossils. Haeckel (1877) proposed the name *Promammalia* for the extinct stem group of *Mammalia*, but at this stage it was only a theoretical construct and no fossils that might qualify as members had been identified (Rieppel, 2011). Huxley re-conceptualized *Prototheria* (containing only *Monotremata*) to represent the most primitive stage of evolution within *Mammalia* (see *Pan-Mammalia*, this volume, for more on Huxley's system of 1880). Roger (1887) proposed *Sauropsidelphia* and Cope (1889) proposed *Ornithostomi* as additional redundant (subclass) names to contain (order) *Monotremata*.

The names *Monotremia*, *Metataxymeria*, *Ornithodelphes*, *Amasta*, and *Ornithodelphia* quickly lapsed. But toward the end of the nineteenth century *Prototheria* surged in popularity as the name for a group that included monotremes plus varying assemblages of fossils (e.g., Flower and Lydekker, 1891; Cope, 1898). In all of its early implementations, *Prototheria* proved to be merely a holding tank for problematic fossils that preserved recognizably plesiomorphic features (see *Pan-Monotremata*, this volume). McKenna and Bell (1997) are alone in promoting the name *Prototheria* over *Monotremata*, the latter being adopted here owing to its priority (Fischer, 1823), and to its traditional and nearly universal use over the last century.

Two systematic issues confronted *Monotremata* in recent decades, with implications for the meaning of its name in terms of its composition and spatio-temporal distribution. One involved the question of monotreme monophyly. For most of the nineteenth and twentieth centuries, monophyly of *Monotremata* (*Ornithorhynchus* + *Tachyglossidae*) was widely accepted. However, in the late twentieth century the early application of molecular systematics arrived at the surprising conclusion that either *Tachyglossus* or *Ornithorhynchus* was phylogenetically nested within, or sister taxon to, *Marsupialia*. Evidence came from sequence analyses of both mitochondrial (Janke et al., 1996, 1997, 2002; Penny et al., 1999; Nilsson et al., 2004) and nuclear (Janke et al., 2002; Nowak et al., 2004) genes.

These findings led to resurrection of Gregory's (1947) '*Marsupionta*' hypothesis. Gregory arrived at the novel conclusion that monotremes were secondarily primitive in many of the features that united them, including such seemingly profound characters as ovipary. While Gregory argued that *Ornithorhynchus* and tachyglossids grouped together, he posited that monotremes were the closest relatives of *Marsupialia*, and that monotremes and marsupials constituted the taxon *Marsupionta*, which was sister to *Placentalia*. A few morphologists endorsed the idea (e.g., Kühne, 1973, 1974), but most others rejected it on a variety of grounds (e.g., Hopson, 1970; Parrington, 1974; Rowe, 1988; McKenna and Bell, 1997).

More recent work indicates that support for *Marsupionta* is a sampling artifact (aka homoplasy). When both monotreme clades are sampled simultaneously, all recent molecular and morphological analyses support the conventional view that *Tachyglossidae* and *Ornithorhynchus* are sister clades in a monophyletic *Monotremata*, and that *Monotremata* is the sister clade of *Theria* (Toyosawa et al., 1998; Reyes et al., 2004; van Rheede et al., 2006; Bininda-Emonds et al., 2007; Kullberg et al., 2008).

The second issue concerns the systematic position of four fragmentary Cretaceous and Palaeogene fossils from Gondwanan localities that have been proposed as either members of crown *Monotremata* or members of the monotreme stem group. The first to be discovered was *Steropodon galmani* Archer et al. 1985, based on an opalized partial dentary

with three teeth from the Early Cretaceous of Australia (middle Albian, or ~112.99 million years old; Flannery et al., 1995). The second fossil in this sequence was *Kollikodon ritchiei* Flannery et al. 1995, also proposed as a Cretaceous monotreme from the same locality as *Steropodon*, based on an opalized dentary fragment with three molariform teeth. It reportedly has an enlarged dentary canal that might place it within the stem group of *Ornithorhynchus*. The discovery of additional material, including a maxilla with highly apomorphic 'multituberculate' teeth cast doubt on whether *Kollikodon* is a member of the monotreme total clade (Musser, 2005); but the new material was later used as a rationale to place *Kollikodon* in the monotreme stem group (Pian et al., 2016). McKenna and Bell (1997) considered it to be a non-mammalian mammaliaform, while Kielan-Jaworowska et al. (2004) considered it a monotreme "*sedis mutabilis*".

Next to be discovered was *Monotrematum sudamericanum* Pascual et al. 1992a, from the Early Palaeocene (Danian, or ~63 million years old) of Patagonia based at first on a single tooth, and later on other teeth and isolated postcranial material (Pascual et al., 1992b, 2002). The fourth was *Teinolophos trusleri* Rich et al. 1999, from the Early Cretaceous of Australia, (Aptian, or between 112.5–121 million years old; Rich et al., 2001a; Rich et al., 2001b), based on numerous jaws and teeth.

With the exception of *Kollikodon*, each of these fossils has been widely recognized as 'a platypus' based on unique dental similarities shared with the Oligo-Miocene *Obdurodon dicksoni*, which is known from a complete skull that preserved an unmistakable platypus rostrum (which supports the bill), along with a partial dentary, and most of the dentition, which is fully mineralized (Archer et al., 1992; Musser and Archer, 1998; Musser, 2005). Whether they are members of crown *Monotremata* or its stem group has substantial consequences for how the history of *Monotremata* is interpreted.

Before these fossils were known, molecular clock models were used to estimate the timing of divergence between the echidna and platypus clades. The estimates broadly agreed on a split in the mid- to late-Cenozoic, although there was little overlap between the early estimates, and none of the early estimates allowed for molecular evolutionary rate heterogeneities (reviewed in Rowe et al., 2008; Phillips et. al., 2009). Given this molecular context and the antiquity of the fossils, most authors placed these fossils on the monotreme stem rather than as members of the crown, in spite of their dental resemblances to the unequivocal stem platypus *Obdurodon dicksoni*. Pascual et al., (1992a,b, 2002) were the first to articulate that if these fossils are indeed stem monotremes, then the aquatic electropredatory ecomorph of *Ornithorhynchus* is ancestral for crown *Monotremata* and that the terrestrial echidnas evolved from an aquatic ancestor.

McKenna and Bell (1997) recognized the unique resemblances of *Ornithorhynchus*, *Obdurodon*, *Steropodon*, and *Monotrematum* and included them together in *Ornithorhynchidae*, and thus implicitly as members of crown *Monotremata*. One implication of this view is that the total clades of *Ornithorhynchus* and *Tachyglossidae* must have diverged from their last common ancestor by the Early Cretaceous. Another implication of McKenna and Bell's larger classification of *Mammalia* and *Mammaliaformes* (McKenna and Bell, 1997: Appendix B) is that monotremes most likely had a terrestrial ancestor, that the terrestrial ecomorph of echidnas is primary, and that the aquatic ecomorph of platypuses is derived within *Monotremata*.

After examining several of specimens of *Teinolophos* using high-resolution computed tomographic (CT) scanning (Rowe et al., 1997), and also scanning the putative stem monotremes *Ausktribosphenos* and *Bishops* (see *Pan-Monotremata*, this volume), Rowe et al. (2008) found data that favored the McKenna-Bell hypothesis that *Teinolophos*, *Obdurodon*, *Steropodon*, and *Monotrematum* were crown monotremes and members of the total clade of *Ornithorhynchus*.

This was challenged in favor of the alternative hypothesis, which placed *Teinolophos*, *Steropodon* and *Monotrematum* on the monotreme stem (Phillips et al., 2009, 2010; Phillips, 2014). The main impetus for this argument was the multiple molecular estimates placing the platypus versus echidna divergence in the Cenozoic. Phillips et al. (2009) also modified the dataset used by Rowe et al. (2008) to arrive at a phylogeny that placed the fossils in question on the monotreme stem. Disagreement regarding character conceptualization and scoring of a few characters is at the heart of this phylogenetic controversy. We also note that four of seven 'constraint tests' conducted by Phillips et al. (2009) yielded results consistent with the McKenna-Bell hypothesis; the others excluded this hypothesis by design.

Subsequently, two unpublished MS theses have discovered additional data favoring the McKenna-Bell hypothesis (Simon, 2013; Latimer, 2014). Simon (2013) found nineteen dental characters and seven mandibular characters uniting *Teinolophos*, *Steropodon*, and *Obdurodon*, and *Ornithorhynchus*. These dental resemblances have never been disputed, but the fact that all currently known members of the total clade of *Tachyglossidae* are edentulous (except for retaining the egg tooth) renders their level of generality equivocal. In other words, the dental apomorphies might be diagnostic of either *Monotremata* or the total clade of *Ornithorhynchus*.

Less equivocation surrounds features of the dentary, which can be compared among all relevant taxa. The expanded mandibular canal is a feature unique to the platypus among living mammals, and indeed among all known members of *Pan-Mammalia* (this volume) including other putative stem monotremes. In *Ornithorhynchus* the enlarged canal conveys a thick mandibular nerve (cranial nerve V₃) that carries large axon bundles from the lower bill (electroreceptors and pressure receptors) as well as from non-mineralized keratinized cheek teeth. Many authors have mistakenly claimed that teeth are absent in the platypus (e.g., Geoffroy, 1803; Gregory, 1910). However, *Ornithorhynchus* has a fully mineralized milk dentition consisting of three or four molariform cheek teeth that are covered by crystalline enamel (Poulton, 1888a,b, 1894; Rowe, 1993; Latimer, 2014). At weaning, these teeth are shed and replaced by molariform teeth whose keratin framework develops, but they subsequently fail to mineralize, and the roots are severely stunted (paedomorphic). CT scanning revealed that the mature molariform teeth are *each* perforated by ~2,000–4,000 canals (Latimer, 2014) and that each canal held a pressure receptor known as a push-rod (Poulton, 1888a). Each mineralized molariform tooth of *Obdurodon* has four to six roots, indicating an extraordinary degree of innervation compared to other mammals, and its mandibular canal is correspondingly of large diameter, as in *Ornithorhynchus*. A large mandibular canal is present in the dentaries of *Teinolophos* and *Steropodon* (unknown in *Monotrematum*), and the scans of *Teinolophos* revealed molariform roots that are reduced in length, as in *Obdurodon*, and mediolaterally widened and ovoid (Rowe et al., 2008). Thus, *Teinolophos* can be inferred to reflect an early stage in hyper-innervation of the mandible, teeth, and cutaneous covering known only in *Ornithorhynchus* among extant species (Archer et al., 1985; Rowe et al., 2008).

CT scans and enlarged 3D printouts of the tiny original specimens also showed in *Teinolophos* modifications of the coronoid, condylar, angular processes, as well as modifications of the mandible involving the pterygoideus fossa and masseteric fossa that uniquely resemble those of living platypuses. These are tied to transformations of the masticatory musculature and a unique feeding pattern observed in *Ornithorhynchus* that involves rapid fluttering (opening and closure) of the mouth to flush prey from the substrate while hunting in the water, followed by oral food processing in which the cheek teeth remove the chitin shells of consumed arthropods and the bones of vertebrates, which they spit out before swallowing the residue or feeding it to young offspring (Grant, 2007). All of these distinctive features of the ornithorhynchid mandible differ from the pattern of mandibular reduction observed in tachyglossids (Simon, 2013).

It is also difficult to reconcile the extreme hypertrophy of the tachyglossid main olfactory system, which may exceed that of all other mammals, as an adaptation to a secondarily terrestrial existence (Camens, 2010; Ashwell, 2013; Ashwell et al., 2014). The main olfactory system receives airborne odorant molecules, and tachyglossid olfactory turbinals have far greater surface area than any other mammal measured to date and relatively larger olfactory bulbs (Macrini et al., 2006; Ashwell et al., 2014). In contrast, there is evidence suggesting that the main olfactory system (olfactory bulb, piriform cortex, and gross encephalization quotient) in *Ornithorhynchus* is reduced compared to *Obdurodon dicksoni* and in contrast to what is inferred as the ancestral state for *Mammalia* and for *Theria* (Macrini et al., 2006; Rowe et al., 2005, 2011; Rowe and Sheppard, 2016). This suggests a trend towards increased aquatic adaptation that is confined to the total clade of *Ornithorhynchus*. Additionally, all available fossil evidence, including the many recent discoveries of non-mammalian mammaliamorphs and early stem-therians, suggests that the ancestral mammal was a terrestrial scratch-digger (Kielan-Jaworowska et al., 2004; Luo, 2007). There are no documented instances where any member of a committed, aquatic mammalian clade includes a member that has re-emerged onto land, as Phillips et al. (2009, 2010) hypothesize for *Tachyglossidae*.

In summary, the controversy over whether *Steropodon*, *Monotrematum*, and *Teinolophos* are members of the monotreme crown or its stem group is driven by a lack of unambiguous evidence. Its resolution is important because it fundamentally impacts how the age of *Monotremata* and divergence of its two living sister clades is calibrated. It also affects how one reconstructs the ancestral ecology of *Monotremata* and interprets the subsequent histories of the platypus and echidna total clades. Overall, molecular estimates tend to favor a Cenozoic divergence, but they differ in methodological detail, and several analyses are now compatible with Early Cretaceous divergence (Woodburne et al., 2003; van Rheede et al., 2006; Rowe et al., 2008; Meredith et al., 2011; Phillips, 2014). It is difficult to see a decisive answer to this controversy using molecular divergence time estimates, given their inevitably wide confidence intervals. Additionally, because known fossils are so incomplete, the parsimony analyses yielding the two alternative hypotheses are only a few steps apart, and they depend on different views of how a very small number of characters are conceptualized and scored. The debate may not be decisively settled until more complete fossils are discovered (Gauthier et al., 1988; Donoghue et al., 1989).

Literature Cited

Archer, M., T. F. Flannery, A. Ritchie, and R. E. Molnar. 1985. First Mesozoic mammal from Australia—an Early Cretaceous monotreme. *Nature* 318:363–366.

Archer, M., F. A. Jenkins, Jr., S. J. Hand, P. Murray, and H. Godthelp. 1992. Description of the skull and non-vestigial dentition of a Miocene platypus (*Obdurodon dicksoni* n. sp.) from Riversleigh, Australia, and the problem of monotreme origins. Pp. 15–27 in *Platypus and Echidnas* (M. Augee, ed.). Royal Society of New South Wales, Sydney.

Ashwell, K. W. S. 2006. Chemoarchitecture of the monotreme olfactory bulb. *Brain Behav. Evol.* 67:69–84.

Ashwell, K. W. S. 2012. Development of the cerebellum in the platypus (*Ornithorhynchus anatinus*) and short-beaked echidna (*Tachyglossus aculeatus*). *Brain Behav. Evol.* 79:237–251.

Ashwell, K. W. S., ed. 2013. *Neurobiology of Monotremes.* CSIRO Publishing, Collingwood, Victoria.

Ashwell, K. W. S., C. D. Hardman, and G. Paxinos. 2006. Cyto-and chemoarchitecture of the sensory trigeminal nuclei of the echidna, platypus, and rat. *J. Chem. Neuroanat.* 31:81–107.

Ashwell, K. W. S., C. D. Hardman, and P. Giere. 2011. Distinct development of peripheral trigeminal pathways in the platypus (*Ornithorhynchus anatinus*) and short-beaked echidna (*Tachyglossus aculeatus*). *Brain Behav. Evol.* 79:113–127.

Ashwell, K. W. S., C. D. Hardman, and P. Giere. 2012. Distinct development of peripheral trigeminal pathways in the platypus (*Ornithorhynchus anatinus*) and short-beaked echidna (*Tachyglossus aculeatus*). *Brain Behav. Evol.* 79:113–127.

Ashwell, K. W. S., C. D. Hardman, and A. M. Musser. 2014. Brain and behaviour of living and extinct echidnas. *Zoology* 117:349–361.

Blumenbach, J. F. 1800. Über das Schnabelthier (*Ornithorhynchus paradoxus*) ein neu entdecktes Geschlecht von Säugethieren des fünften Welttheils. *Magazine für die neueste Zuständig Naturkunde* 2:205–214.

Bininda-Emonds, O. R. P., M. Cardillo, K. E. Jones, R. D. MacPhee, R. M. Beck, R. Grenyer, S. A. Price, R. A. Vos, J. L. Gittleman, and A. Purvis. 2007. The delayed rise of present-day mammals. *Nature* 446:507–512.

de Blainville, H. M. D. 1816. Prodrome d'une nouvelle distribution sytématique du régne animal. *Bull. Sci. Soc. Philomath. Paris* 8:113–124.

Bonaparte, C. L. 1837. A new systematic arrangement of vertebrated animals. *Trans. Linn. Soc. Lond.* 18:247–304.

Burrell, H. 1927. *The Platypus.* Angus & Robertson Limited, Sydney.

Camens, A. B. 2010. Were early Tertiary monotremes really all aquatic? Inferring paleobiology and phylogeny from a depauperate fossil record. *Proc. Natl. Acad. Sci. USA.* 107:E12.

Cope, E. D. 1889. Synopsis of the families of *Vertebrata. Am. Nat.* 23:849–877.

Cope, E. D. 1898. *Syllabus of Lectures on the Vertebrata*, with an introduction by Henry Fairfield Osborn. Published for the University of Pennsylvania, Philadelphia, PA.

Cuvier, G. 1817. *Le Règne Animal.* A. Belin, Paris.

Czech-Damal, N. U., A. Liebschner, L. Miersch, G. Klauer, F. D. Hanke, C. Marshall, G. Dehnhardt, and W. Hanke. 2012. Electroreception in the Guiana dolphin (*Sotalia guianensis*). *Proc. R. Soc. Lond. B Biol. Sci.* 279:663–668.

Donoghue, M. J., J. Doyle, J. A. Gauthier, A. G. Kluge, and T. Rowe. 1989. Importance of fossils in phylogeny reconstruction. *Annu. Rev. Ecol. Syst.* 20:431–460.

Dun, W. S. 1895. Notes on the occurrence of monotreme remains in the Pliocene of New South Wales. *Rec. Geol. Surv. N.S.W.* 4:118–126.

Edgeworth, F. H. 1935. *The Cranial Muscles of Vertebrates.* Cambridge University Press, Cambridge, UK.

Fischer, G. 1823. *Enchiridion Generum Animalium.* Moscow.

Flannery, T. F., M. Archer, T. H. Rich, and R. Jones. 1995. A new family of monotremes from the Cretaceous of Australia. *Nature* 377:418–420.

Flannery, T. F., and C. P. Groves. 1998. A revision of the genus *Zaglossus* (*Monotremata, Tachyglossidae*), with description of new species and subspecies. *Mammalia* 62:367–396.

Flower, W. H. 1883. On the arrangement of the orders and families of existing *Mammalia. Proc. Zool. Soc. Lond.* 1883:178–186.

Flower, W. H., and R. Lydekker. 1891. *An Introduction to the Study of Mammals: Living and Extinct.* Adam and Charles Black, London.

Gauthier, J. A., A. G. Kluge, and T. B. Rowe. 1988. Amniote phylogeny and the importance of fossils. *Cladistics* 4:105–209.

Geoffroy St. Hillaire, E. 1803. Extrait des observations anatomiques de M. Home, sur l'echidné. *Bull. Sci. Soc. Philomath.* 3:125–127. [Gill (1903) correctly observed these printed page numbers to be incorrect and that in actual sequence the pages number 225–227].

Geoffroy St. Hillaire, E. 1822. Monotremes ovipares. *Bull. Sci. Soc. Philomath. Paris, pour l'année* 1822:95–96.

Gervais, P. 1836. Mammalogie ou mastologie. *Dictionnaire Pittoresque d'Histoire Naturelle et des Phenomènes de la Nature* 4:614–640.

Gill, T. 1870. On the relations of the orders of mammals. *Proc. Am. Assoc. Adv. Sci.* 1870:267–270.

Gill, T. 1872. Arrangement of the families of mammals with analytical tables. *Smithson. Misc. Collect.* 11:i–iv, 1–98.

Gill, T. 1877. Vertebrate zoology. Pp. 171–172 in *Annual Record of Science and Industry for 1876* (S. F. Baird, ed.). Harper and Brothers, New York.

Gill, T. 1903. Origin of the name monotremes. *Science* 17:433–434.

Glauert, L. 1914. The mammoth cave. *Rec. W. Austral. Mus.* 6:11–38.

Gould, S. J. 1985. To be a platypus. *Nat. Hist. Mag.* 94:10–15.

Grant, T. R. 1992. Historical and current distribution of the platypus, *Ornithorhynchus anatinus*, in Australia. Pp. 232–254 in *Platypus and Echidna* (M. L. Augee, ed.). Royal Zoological Society of New South Wales, Mosman.

Grant, T. R. 2007. *Platypus.* 4th edition. CSIRO Publishing, Collingwood, Victoria.

Gregory, W. K. 1910. The orders of mammals. *Bull. Am. Mus. Nat. Hist.* 27:1–524

Gregory, W. K. 1947. Monotremes and the palimpsest theory. *Bull. Am. Mus. Nat. Hist.* 88:1–52.

Griffiths, M. 1968. *Echidnas. International Series of Monographs in Pure and Applied Biology, Zoology Division*, Vol. 38. Pergamon Press Ltd., London.

Griffiths, M. 1978. *The Biology of the Monotremes.* Academic Press, New York.

Griffiths, M., R. T. Wells, and D. J. Barrie. 1991. Observations on the skulls of fossil and extant echidnas (*Monotremata: Tachyglossidae*). *Austral. Mamm.* 14:97–101.

Haeckel, E. 1866. *Generelle Morphologie der Organismen. Allgemeine Grundzüge der organischen Formen-Wissenschaft, Mechanisch Begründet Durch die von Charles Darwin Reformirte Descendenz-Theorie.* Georg Reimer, Berlin.

Haeckel, E. 1877. *Anthropogenie oder Entwickelungsgeschichte des Menschen.* Wilhelm Engelmann, Leipzig.

Helgen, K. M., R. P. Miguez, J. L. Kohen, and L. E. Helgen. 2012. Twentieth century occurrence of the Long-Beaked Echidna *Zaglossus bruijni* in the Kimberley region of Australia. *ZooKeys* 255:103–132.

Home, E. 1802a. A description of the anatomy of the *Ornithorhynchus paradoxus. Philos. Trans. R. Soc. Lond. B Biol. Sci.* 92:67–84.

Home, E. 1802b. Description of the anatomy of *Ornithorhynchus hystrix. Philos. Trans. R. Soc. Lond. B Biol. Sci.* 92:348–364.

Hopson, J. A. 1970. The classification of nontherian mammals. *J. Mamm.* 51:1–9.

Huber, E. 1930a. Evolution of facial musculature and cutaneous field of Trigeminus. Part I. *Q. Rev. Biol.* 5:133–188.

Huber, E. 1930b. Evolution of facial musculature and cutaneous field of Trigeminus. Part II. *Q. Rev. Biol.* 5:389–437.

Huxley, T. H. 1864. *Lectures on the Elements of Comparative Anatomy—on the Classification of Animals and on the Vertebrate Skull.* Churchill and Sons, London.

Huxley, T. H. 1872. *A Manual of the Anatomy of Vertebrate Animals.* Appleton and Company, New York.

Huxley, T. H. 1880. On the application of the laws of evolution to the arrangement of the *Vertebrata*, and more particularly of the *Mammalia. Proc. Zool. Soc. Lond.* 1880:649–662.

Illiger, C. 1811. *Prodromus Systematis Mammalium et Avium Additis Terminis Zoographicis Utriusque Classis.* C. Salfeld, Berlin.

Janke, A., N. J. Gemmell, G. Feldmaier-Fuchs, A. von Haeseler, and S. Pääbo. 1996. The mitochondrial genome of a monotreme—the platypus (*Ornithorhynchus anatinus*). *J. Mol. Evol.* 42:153–159.

Janke, A., X. Xu, and U. Arnason. 1997. The complete mitochondrial genome of the wallaroo (*Macropus robustus*) and the phylogenetic relationship among *Monotremata, Marsupialia,* and *Eutheria. Proc. Natl. Acad. Sci. USA.* 94:1276–1281.

Janke, A., O. Magnell, G. Wieczorek, M. Westerman, and U. Arnason. 2002. Phylogenetic analysis of 18S rRNA and the mitochondrial genomes of the wombat, *Vombatus ursinus,* and the spiny anteater, *Tachyglossus aculeatus:* increased support for the *Marsupionta* hypothesis. *J. Mol. Evol.* 54:71–80.

Kielan-Jaworowska, Z., R. L. Cifelli, and Z.-X. Luo. 2004. *Mammals from the Age of Dinosaurs.* Columbia University Press, New York.

Krefft, G. 1868. On the discovery of a new and gigantic fossil species of *Echidna* in Australia. *Ann. Mag. Nat. Hist.* 1:113–114.

Kühne, W. G. 1973. The systematic position of monotremes reconsidered (*Mammalia*). *Z. Morphol. Tiere* 75:59–64.

Kühne, W. G. 1974. On the *Marsupionta,* a reply to Dr. Parrington. *J. Nat. Hist.* 11:225–228.

Kullberg M., B. M. Hallström, U. Arnason, and A. Janke. 2008. Phylogenetic analysis of 1.5 Mbp and platypus EST data refute the *Marsupionta* hypothesis and unequivocally support *Monotremata* as sister group to *Marsupialia/Placentalia. Zool. Scr.* 37:115–127.

Lamarck, J. B. P. A. 1809. *Philosophie Zoologique,* Vol. 1. J. B. Baillière, Paris.

Latimer, A. 2014. *Redescription of the Teeth and Epithelial Plates from the Platypus (Ornithorhynchus anatinus): Morphological and Evolutionary Implications.* MS thesis, University of Texas at Austin.

Lefèvre, C. M., J. A. Sharp, and K. R. Nicholas. 2009. Characterization of monotreme caseins reveals lineage-specific expansion of an ancestral casein locus in mammals. *Reprod. Fertil. Dev.* 21:1015–1027.

Lightoller, G. S. 1942. Matrices of the facialis musculature: homologization of the musculature in monotremes with that of marsupials and placentals. *J. Anat.* 76:258–269.

Luo, Z.-X. 2007. Transformation and diversification in early mammal evolution. *Nature* 450:1011–1019.

Lydekker, R. 1894. *A Hand-Book to the* Marsupialia *and* Montremata. W. H. Hale & Co., London.

Macrini, T. E., T. B. Rowe, and M. Archer. 2006. Description of a cranial endocast from a fossil platypus, *Obdurodon dicksoni* (*Monotremata, Ornithorhynchidae*), and the relevance of endocranial characters to monotreme monophyly. *J. Morph.* 267:1000–1015.

Manger, P. R., R. Collins, and J. D. Pettigrew. 1997. Histological observations on presumed electroreceptors and mechanoreceptors in the beak skin of the long-beaked echidna, *Zaglossus bruijni. Proc. R. Soc. Lond. B Biol. Sci.* 264:165–172.

McKenna, M. C., and S. K. Bell. 1997. *Classification of Mammals above the Species Level.* Columbia University Press, New York.

Meckel, I. F. 1826. Ornithorhynchi Paradoxi *Descriptuo Anatomica.* Gerhardum Fleischerum, Leipzig.

Meredith, R. W., J. E. Janecka, J. Gatesy, O. A. Ryder, C. A. Fisher, E. C. Teeling, A. Goodbla, E. Eizirik, T. L. L. Simão, T. Stadler, D. L. Rabosky, R. L. Honeycutt, J. J. Flynn, C. M. Ingram, C. Steiner, T. L. Williams, T. J. Robinson, A. Burk-Herrick, M. Westerman, N. A. Ayoub, M. S. Springer, and W. J. Murphy. 2011. Impacts of the Cretaceous Terrestrial Revolution and KPg extinction on mammal diversification. *Science* 334:521–524.

Murchison, E. P., P. Kheradpour, R. Sachidanandam, C. Smith, E. Hodges, Z. Xuan, M. Kellis, F. Grützner, A. Stark, and G. J. Hannon. 2008. Conservation of small RNA pathways in platypus. *Genome Res.* 18:995–1004.

Murray, P. F. 1978. Late Cenozoic monotreme anteaters. *Austral. Zool.* 20:29–55.

Musser, A. M. 2003. Review of the monotreme fossil record and comparison of palaeontological and molecular data. *Comp. Biochem. Physiol. Part A Mol. Integr. Physiol.* 136:927–942.

Musser, A. M. 2005. *Investigations into the Evolution of Australian Mammals with a Focus on* Monotremata. PhD dissertation, University of New South Wales, Sydney, Australia.

Musser, A. M., and M. Archer. 1998. New information about the skull and dentary of the Miocene platypus *Obdurodon dicksoni*, and a discussion of ornithorhynchid relationships. *Philos. Trans. R. Soc. Lond. B Biol. Sci.* 353:1063–1079.

Nilsson, M. A., U. Arnason, P. B. Spencer, and A. Janke. 2004. Marsupial relationships and a timeline for marsupial radiation in South Gondwana. *Gene* 340:189–196.

Nowak, M. A., Z. E. Parra, L. Hellman, and R. D. Miller. 2004. The complexity of expressed kappa light chains in egg-laying mammals. *Immunogenetics* 56:555–563.

Owen, R. 1868. *On the Anatomy of Vertebrates. Vol. III: Mammals.* Longmans, Green, and Co., London.

Owen, R. 1884. Evidence of a large extinct monotreme (*Echidna Ramsayi*, Ow.) from the Wellington Breccia Cave, New South Wales. *Philos. Trans. R. Soc. Lond. B Biol. Sci.* 185:273–274.

Parker, T. J., and W. A. Haswell. 1897. *A Text-Book of Zoology.* 1st edition, 2 Vols. Macmillan and Co., Ltd., London.

Parrington, F. R. 1974. The problem of the origin of the monotremes. *J. Nat. Hist.* 8:421–426.

Pascual, R., M. Archer, E. Ortiz-Jaureguizar, J. L. Prado, H. Godthelp, and S. J. Hand. 1992a. First discovery of monotremes in South America. *Nature* 356:704–706.

Pascual, R., M. Archer, E. Oritz-Jaureguizar, J. L. Prado, H. Godthelp, and S. J. Hand. 1992b. The first non-Australian monotreme: an early Paleocene South American platypus (*Monotremata, Ornithorhynchidae*). Pp. 22–15 in *Platypus and Echidnas* (M. L. Augee, ed.). Royal Society of New South Wales, Sydney.

Pascual, R., F. J. Goin, L. Balarino, and D. U. Sauthier. 2002. New data on the Paleocene monotreme *Monotrematum sudamericanum*, and the convergent evolution of triangulate molars. *Acta Palaeontol. Polon.* 47:487–492.

Penny, D., M. Hasegawa, P. J. Wadell, and M. D. Hendy. 1999. Mammalian evolution: timing and implications from using log-determinant transformation for proteins of differing amino acid composition. *Syst. Biol.* 48:76–93.

Pettigrew, J. D. 1999. Electroreception in monotremes. *J. Exp. Biol.* 202:1447–1454.

Phillips, M. J. 2014. Four mammal fossil calibrations: balancing competing palaeontological and molecular considerations. *Palaeontol. Electron.* 18:1–16.

Phillips, M. J., T. H. Bennett, and M. S. Lee. 2009. Molecules, morphology, and ecology indicate a recent, amphibious ancestry for echidnas. *Proc. Natl. Acad. Sci. USA* 106:17089–17094.

Phillips, M. J., T. H. Bennett, and M. S. Lee. 2010. Reply to Camens: how recently did modern monotremes diversify? *Proc. Natl. Acad. Sci. USA* 107:E13.

Pian, R., M. Archer, and S. J. Hand. 2013. A new, giant platypus, *Obdurodon tharalkooschild*, sp. nov. (*Monotremata, Ornithorhynchidae*), from the Riversleigh World Heritage Area, Australia. *J. Vertebr. Paleontol.* 33:1255–1259.

Pian, R., M. Archer, S. J. Hand, R. M Beck, and A. Cody. 2016. The upper dentition and relationships of the enigmatic Australian Cretaceous mammal *Kollikodon ritchiei. Mem. Mus. Vic.* 74:97–105.

Poulton, E. B. 1888a. True teeth and horny plates of *Ornithorhynchus. Q. J. Microsc. Sci.* 29:9–48.

Poulton, E. B. 1888b. True teeth in the young *Ornithorhynchus paradoxus. Proc. R. Soc. Lond. B Biol. Sci.* 43:353–356.

Poulton, E. B. 1894. The structure of the bill and hairs of *Ornithorhynchus paradoxus*; with a discussion of the homologies and origin of mammalian hair (with plates 14, 80 15, 15a). *Q. J. Microsc. Sci.* 36:143–200.

Proske, U., J. E. Gregory, and A. Iggo. 1998. Sensory receptors in monotremes. *Philos. Trans. R. Soc. Lond. B Biol. Sci.* 353:1187–1198.

Rafinesque, C. S. 1815. *Analyse de la Nature.* Palerme.

Rens, W., F. Grützner, P. Ferguson-Smith, and M. Ferguson-Smith. 2004. Resolution and evolution of the duck-billed platypus karyotype with an X1Y1X2Y2X3Y3X4Y4X5Y5 male sex chromosome constitution. *Proc. Natl. Acad. Sci. USA* 101:16257–16261.

van Rheede, T., T. Bastiaans, D. N. Boone, S. Blair Hedges, W. W. de Jong, and O. Madsen. 2006. The platypus is in its place: nuclear genes and indels confirm the sister group relation of monotremes and therians. *Mol. Biol. Evol.* 23:587–597.

Reyes, A., C. Gissi, F. Catzeflis, E. Nevo, G. Pesole, and C. Sacconell. 2004. Congruent mammalian trees from mitochondrial and nuclear genes using Bayesian methods. *Mol. Biol. Evol.* 21:397–403.

Rich, T. H., P. Vickers-Rich, A. Constantine, T. F. Flannery, L. Kool, and N. van Klaveren. 1999. Early Cretaceous mammals from Flat Rocks, Victoria, Australia. *Rec. Queen Vic. Mus.* 106:1–35.

Rich, T. H., T. F. Flannery, P. Trusler, L. Kool, N. van Klaveren, and P. Vickers-Rich. 2001a. A second tribosphenic mammal from the Mesozoic of Australia. *Rec. Queen Vic. Mus.* 110:1–9.

Rich, T. H., P. Vickers-Rich, P. Trusler, T. F. Flannery, R. L. Cifelli, A. Constantine, L. Kool, and N. Van Klaveren. 2001b. Monotreme nature of the Australian Early Cretaceous mammal *Teinolophos. Acta Palaeontol. Polon.* 46:113–118.

Rieppel, O. 2011. Ernst Haeckel (1834–1919) and the monophyly of life. *J. Zool. Syst. Evol. Res.* 49:1–5.

Roger, O. 1887. Verzeichniss der bisher bekannten fossilen Säugethiere. *Bericht des naturwissenschaftliches Vereines für Schwaben und Neuburg* 29:1–162.

Rowe, T. B. 1986. *Osteological Diagnosis of Mammalia, L. 1758, and Its Relationship to Extinct* Synapsida. PhD dissertation, University of California, Berkeley, CA.

Rowe, T. B. 1988. Definition, diagnosis and origin of *Mammalia. J. Vertebr. Paleontol.* 8:241–264.

Rowe, T. B. 1993. Phylogenetic systematics and the early history of mammals. Pp. 129–145 in *Mammalian Phylogeny* (F. S. Szalay, M. J. Novacek, and M. C. McKenna, eds.). Springer-Verlag, New York.

Rowe, T B., T. P. Eiting, T. E. Macrini, and R. A. Ketcham. 2005. Organization of the olfactory and respiratory skeleton in the nose of the gray short-tailed opossum *Monodelphis domestica. J. Mamm. Evol.* 12:303–336.

Rowe, T. B., J. Kappelman, W. D. Carlson, R. A. Ketcham, and C. Denison. 1997. High-resolution computed tomography: a breakthrough technology for Earth scientists. *Geotimes* 42:23–27.

Rowe, T. B., T. E. Macrini, and Z.-X. Luo. 2011. Fossil evidence on origin of the mammalian brain. *Science* 332:955–957.

Rowe, T. B., T. H. Rich, P. Vickers-Rich, M. Springer, and M. O. Woodburne. 2008. The oldest platypus, and its bearing on divergence timing of the platypus and echidna clades. *Proc. Natl. Acad. Sci. USA* 105:1238–1242.

Rowe, T. B., and G. M. Shepherd. 2016. The role of ortho-retronasal olfaction in mammalian cortical evolution. *J. Comp. Neurol.* 524:471–495.

Shaw, G. 1792. *The Porcupine Ant-Eater. The Naturalist's Miscellany, or Colored Figures of Natural Objects Drawn and Described Immediately from Nature.* Plate 109. London. 7 Vols. (1790–1813).

Shaw, G. 1799. *The Duck-Billed Platypus. The Naturalist's Miscellany, or Colored Figures of Natural Objects Drawn and Described Immediately from Nature.* Plate 385. London. 7 Vols. (1790–1813).

Simon, R. V. 2013. *Cranial Osteology of the Long-beaked Echidna, and the Definition, Diagnosis, and Origin of* Monotremata *and Its Major Subclades.* MS thesis, University of Texas at Austin.

Switzer, III, R. C., and J. I. Johnson, Jr. 1977. Absence of mitral cells in monolayer in monotremes. *Cells Tissues Organs* 99:36–42.

Toyosawa, S., C. O'Huigin, F. Figueroa, H. Tichy, and J. Klein. 1998. Identification and characterization of amelogenin genes in monotremes, reptiles, and amphibians. *Proc. Natl. Acad. Sci. USA* 95:13056–13061.

de Vis, C. W. 1885. On an extinct monotreme *Ornithorhynchus agilis. Proc. R. Soc. Qld.* 2:35–38.

Weisbecker, V. 2011. Monotreme ossification sequences and the riddle of mammalian skeletal development. *Evolution* 65:1323–1335.

Werneburg, I., and M. R. Sánchez-Villagra. 2011. The early development of the echidna, *Tachyglossus aculeatus* (*Mammalia: Monotremata*), and patterns of mammalian development. *Acta Zool.* 92:75–88.

Winge, H. 1941. *The Interrelationships of the Mammalian Genera,* Vol. 1. Reitzel, Copenhagen.

Woodburne, M. O., and R. H. Tedford. 1975. The first Tertiary monotreme from Australia. *Am. Mus. Nat. Hist. Novitates* 2588:1–11.

Woodburne, M. O., R. H. Tedford, M. Archer, W. D. Turnbull, M. D. Plane, and E. L. Lundelius. 1985. Biochronology of the continental mammal record of Australia and New Guinea. *Spec. Publ. (South Australia Dept. Mines Energy)* 5:347–363.

Woodburne, M. O., T. H. Rich, and M. S. Springer. 2003. The evolution of tribospheny and the antiquity of mammalian clades. *Mol. Phylogenet. Evol.* 28:360–385.

Zeller, U. 1989a. Die Entwicklung und Morphologie des Schädels von *Ornithorhynchus anatinus* (*Mammalia: Prototheria: Monotremata*). *Abh. Senckenb. Naturforsch. Ges.* 545:1–188.

Zeller, U. 1989b. The braincase of *Ornithorhynchus. Fortschr. Zool.* 35:386–391.

Zeller, U. 1993. Ontogenetic evidence for cranial homologies in monotremes and therians, with special reference to *Ornithorhynchus.* Pp. 95–107 in *Mammalian Phylogeny* (F. S. Szalay, M. J. Novacek, and M. C. McKenna, eds.). Springer-Verlag, New York.

Authors

Timothy B. Rowe; Jackson School of Geosciences; C1100, University of Texas at Austin; Austin, TX 78712, USA. Email rowe@utexas.edu.

Rachel V. S. Wallace; Jackson School of Geosciences; C1100, University of Texas at Austin; Austin, TX 78712, USA. Email rvsimon@utexas.edu.

Bhart-Anjan Bhullar; Department of Geology and Geophysics; Yale University; 210 Whitney Ave., New Haven, CT 06520-8109, USA. Email bhart-anjan.bhullar@yale.edu.

Date Accepted: 12 June 2017

Primary Editors: Jacques A. Gauthier, Kevin de Queiroz

Pan-Xenarthra B. J. Shockey, new clade name

Registration Number: 256

Definition: The total clade of the crown clade *Xenarthra*. This is a crown-based total-clade definition. Abbreviated definition: total ∇ of *Xenarthra*.

Etymology: Derived from the Greek *pan-* (all, the whole), indicating that the name designates a total clade (or pan-monophyllum), and *Xenarthra*, the name of the corresponding crown clade. See *Xenarthra* (this volume) for the etymology of that name.

Reference Phylogeny: Gibb et al. (2016: Figs. 1, 2).

Composition: As the total clade of *Xenarthra* (this volume), *Pan-Xenarthra* is composed of *Dasypus novemcinctus* Linnaeus 1758 (*Dasypodoidea* armadillos) and all mammals that are more closely related to it than to *Afrotheria*, *Laurasiatheria*, and *Euarchontaglires* under current hypotheses of mammalian phylogeny (Madsen et al., 2001; Murphy et al., 2001; Delsuc et al., 2004). No unambiguous stem xenarthrans are currently recognized. Extant xenarthrans include the *Dasypodoidea* Gray 1821 (armadillos), two distinct clades of *Folivora* Delsuc et al. 2001 (sloths, *Bradypus* and *Choloepus*) and two clades of *Vermilingua* Illiger 1811 (anteaters, *Myrmecophagidae* and *Cyclopedidae*) (Vizcaíno and Loughry, 2008). Although many extinct taxa of *Pan-Xenarthra* are known (e.g., rigidly armored glyptodonts, the armored and horned peltephilids, as well as a huge diversity of extinct sloths [Gaudin and McDonald, 2008]), they are probable members of crown *Xenarthra* (Gaudin and Wible, 2006; Shockey, 2017).

Diagnostic Apomorphies: See *Xenarthra* (this volume) and Gaudin and McDonald (2008) for a summary of the numerous apomorphies of the crown. However, because stem xenarthrans are unknown, the sequence in which those apomorphies arose is likewise unknown.

Synonyms: The first two names (below) could be approximate synonyms of *Pan-Xenarthra* as defined here. The third, *Americatheria*, is synonymous.

1. *Xenarthra* Cope 1889. For the want of a name for the total-clade of *Xenarthra*, authorities have generally applied the name of the crown-clade (*Xenarthra*) to the total-clade (e.g., Madsen et al., 2001; Murphy et al., 2001; Kriegs et al., 2006).
2. *Edentata* Cuvier 1798. Simpson (1945) used *Edentata* as a potential equivalent to total-clade *Xenarthra*, expressing his belief that the extinct palaeanodonts were "a primitive offshoot from the ancestry of the *Xenarthra*" (Simpson, 1945: p. 191).
3. *Americatheria* Shockey 2017. This pseudo-Linnaean name was explicitly given for the total clade *Xenarthra*.

Comments: Highly modified crown xenarthrans are remnants of one of the earliest divergences among crown placental mammals (Madsen et al., 2001; Murphy et al., 2001). This major clade is usually referred to as "*Xenarthra*" without any explicit distinction between the total-clade and the crown-clade (e.g., Madsen et al., 2001; Murphy et al., 2001; Kriegs et al., 2006). However, the major temporal disparity between the divergence of the total-clade (here named *Pan-Xenarthra*) from other placentals (≈100–108 Ma; Springer et al., 2003; Delsuc et al., 2004) and the basal divergence within crown *Xenarthra* (~65 Ma; Springer et al., 2003; Delsuc et al., 2004), combined with the numerous and distinctive modifications that occurred in the stem lineage leading up to crown *Xenarthra* (Gaudin and McDonald, 2008), illustrates the problem with using the name "*Xenarthra*" for both crown and total clades (Shockey, 2017). This practice is undesirable, not just because it is imprecise, but because it misleads; i.e., there is no reason to expect that early members of *Pan-Xenarthra* (the so-called "ghosts" sensu Delsuc et al., 2004) would have shared many apomorphic traits with crown xenarthrans. For example, from the first conception of the group, Cope (1889) recognized that the ancestors of xenarthrans would have had enamel-covered dentitions. It also is likely that early diverging members of *Pan-Xenarthra*, unlike crown-clade members, had tribosphenic molars, as well as the general eutherian dental formula, and that they would have lacked most of the apomorphies that distinguish crown xenarthrans from the members of other crown clades of placental mammals, such as the extra articulations of posterior thoracic and lumbar vertebrae, and ischial contact with the axial skeleton.

One could fairly regard the number of apomorphies for a clade to be inversely proportional to our knowledge of that clade's origin. Given the large number of synapomorphies for crown *Xenarthra* (Gaudin and McDonald, 2008), we can conclude that we know very little regarding the history of its stem lineage. No taxa are currently recognized as stem xenarthrans, though it is possible that early diverging fossils of *Pan-Xenarthra* are already in collections but have yet to be recognized as such (Shockey, 2017).

Literature Cited

Cope, E. D. 1889. The *Edentata* of North America. *Am. Nat.* 23:657–664.

Cuvier, G. L. C. F. D. 1798. *Tableau Élémentaire de L'Histoire Naturelle Des Animaux.* J. B. Bailière, Paris.

Delsuc, F., F. M. Catzeflis, M. J. Stanhope, and E. J. P. Douzery. 2001. The evolution of armadillos, anteaters and sloths depicted by nuclear and mitochondrial phylogenies: implications for the status of the enigmatic fossil *Eurotamandua*. *Proc. R. Soc. Lond. B Biol. Sci.* 268:1605–1615.

Delsuc F., S. Vizcaíno, and E. J. P. Douzery. 2004. Influence of Tertiary paleoenvironmental changes on the diversification of South American mammals: a relaxed molecular study within xenarthrans. *BMC Evol. Biol.* 4:11

Gaudin, T., and H. G. McDonald. 2008. Morphology-based investigations of the phylogenetic relationships among extant and fossil xenarthrans. Pp. 24–36 in *The Biology of the Xenarthra* (S. Vizcaíno and W. J. Loughry, eds.). University Press of Florida, Gainesville, FL.

Gaudin, T. J., and J. R. Wible. 2006. The phylogeny of the living and extinct armadillos (*Mammalia, Xenarthra, Cingulata*): a craniodental analysis. Pp. 153–198 in *Amniote Paleobiology: Perspectives on the Evolution of Mammals, Birds, and Reptiles* (M. T. Carrano, T. J. Gaudin, R. W. Blob, and J. R. Wible, eds.). University of Chicago Press, Chicago, IL.

Gibb, G. C., F. L. Condamine, M. Kuch, J. Enk, N. Moraesbarros, M. Superina, H. N. Poinar, and F. Delsuc. 2016. Shotgun mitogenomics provides a reference phylogenetic framework and timescale for living xenarthrans. *Mol. Biol. Evol.* 33:621–642.

Gray, J. E. 1821. On the natural arrangement of vertebrose animals. *Lond. Med. Repos.* 15:296–310.

Illiger, C. D. 1811. *Prodromus Systematis Mammalium et Avium Additis Terminis Zoographicis Uttriusque Classis.* Salfeld, Berlin.

Kriegs, J. O., G. Churakov, M. Kiefman, U. Jordan, J. Brosius, and J. Schmitz. 2006. Retroposed elements as archives for the evolutionary history of placental mammals. *PLOS Biol.* 4:e91.

Linnaeus, C. 1758. *Systema Naturæ Per Regna Tria Naturæ, Secundum Classes, Ordines, Genera, Species, cum Characteribus, Differentiis, Synonymis, Locis.* Tomus I, Editio decima, reformata. Laurentii Salvii, Holmiae (Stockholm).

Madsen, O., M. Scally, C. J. Douady, D. J. Kao, R. W. DeBry, R. Adkins, H. M. Amrine, M. J. Stanhope, W. W. de Jong, and M. S. Springer. 2001. Parallel adaptive radiations in two major clades of placental mammals. *Nature* 409:610–614.

Murphy, W. J., E. Eizirik, W. E. Johnson, Y. P. Zhang, O. A. Ryder, S. J. O'Brien. 2001. Molecular phylogenies and the origins of placental mammals. *Nature* 409:614–618.

Shockey, B. J. 2017. New early diverging cingulate (*Xenarthra*: *Peltephilidae*) from the late Oligocene of Bolivia and considerations regarding the origin of crown Xenarthra. *Bull. Peabody Mus. Nat. Hist.* 58:371–396.

Simpson, G. G. 1945. The principles of classification and a classification of mammals. *Bull. Am. Mus. Nat. Hist.* 85:1–350.

Springer, M. S., W. J. Murphy, E. Eizirik, and S. J. O'Brien. 2003. Placental mammal diversification and the Cretaceous-Tertiary boundary. *Proc. Natl. Acad. Sci. USA* 100:1056–1061.

Vizcaíno, S., and W. J. Loughry. 2008. *The Biology of the* Xenarthra. University Press of Florida, Gainesville, FL.

Author

Bruce J. Shockey; Biology Department; Manhattan College; Manhattan College Parkway; Riverdale, NY; Department of Vertebrate Paleontology; Division of Vertebrate Paleontology, Yale Peabody Museum of Natural History, 170 Whitney Ave. Yale University, New Haven, CT 06511. American Museum of Natural History; Central Park West at 79th St.; New York, NY 10024. Email: bshockey@amnh.org.

Date Accepted: 5 October 2017

Primary Editor: Jacques A. Gauthier

Xenarthra E. D. Cope 1889 [B. J. Shockey], converted clade name

Registration Number: 257

Definition: The crown clade originating in the most recent common ancestor of *Dasypus novemcinctus* Linnaeus 1758 (*Dasypodoidea*; armadillos), *Choloepus didactylus* Linnaeus 1758 (*Folivora*; sloths), and *Myrmecophaga tridactyla* Linnaeus 1758 (*Vermilingua*; anteaters). This is a minimum-crown-clade definition. Abbreviated definition: min crown ∇ (*Dasypus novemcinctus* Linnaeus 1758 & *Choloepus didactylus* Linnaeus 1758 & *Myrmecophaga tridactyla* Linnaeus 1758).

Etymology: From the Greek, ξένος (*xenos*; foreign or strange) combined with ἀρθρωση (*arthros*; joint), in reference to extra zygapophyses on vertebrae of the lumbar region (Cope, 1889; Engelmann, 1985).

Reference Phylogeny: Gibb et al. (2017: Figs 1, 2).

Composition: The extant xenarthrans include 21 species of *Dasypodoidea* Gray 1821 (armadillos = Dasypodidae Gray 1821 + Chlamyphoridae Gray 1869), six species of *Folivora* Delsuc et al. 2001 (sloths) in two distinct clades (*Bradypus* and *Choloepus*), and four species of *Vermilingua* Illiger 1811 (anteaters) in two clades (*Myrmecophagidae* and *Cyclopedidae*) (Vizcaíno and Loughry, 2008). In terms of species richness, range in body size, morphological variation, and habitats occupied, the greatest diversity of crown *Xenarthra* occurred earlier in the Cenozoic, when numerous extinct clades arose that are thought to be nested within the crown. Such clades include the large-bodied, solid-shelled glyptodonts nested within the Chlamyphoridae (Mitchell et al. 2016) as well as numerous mylodontoid and megatherioid *Folivora* (Gaudin and McDonald, 2008; Shockey and Anaya, 2011).

Diagnostic Apomorphies: Compared to other crown clades within placental mammals, crown *Xenarthra* is diagnosed by more than 50 putative synapomorphies (Gaudin and McDonald, 2008). The most conspicuous of these include the complete or nearly complete loss of dental enamel associated with an ever-growing "root analog" (sensu Tummers and Thesleff, 2008), reduction in the number of teeth (McDonald, 2003), non-replacement of dentition (McDonald, 2003), enlarged teres fossa of the scapula resulting in a second scapular spine (Engelmann, 1985), extra ("xenarthrus") joints in the lumbar vertebrae (Cope, 1889), and a secondary pelvic attachment to the axial skeleton via the ischium (Engelmann, 1985).

Synonyms: *Edentata* Cuvier 1798 (as "*Des Edentés*") is an approximate synonym. With the removal of most of taxa originally included within *Edentata* (below), the name has been occasionally used as a synonym for *Xenarthra* as defined here (e.g., McKenna, 1975; Novacek, 1992). However, *Edentata* Cuvier 1798 has been more generally used as an inclusive group to mean "*Xenartha*-plus-" (Simpson, 1945: 191), with the non-xenarthran components varying greatly (see Comments).

Comments: Cope's concept of *Xenarthra* is equivalent to that discussed here and included the same general taxonomic groups, i.e., folivoran sloths (with "*Bradypodidae*" and "*Megatheriidae*" specified by Cope, 1889: 658), vermilinguan anteaters ("*Myrmecophagidae*") and armored cingulates ("*Dasypodidae*" and "*Glyptodontidae*")

(Cope, 1889: 658). As in Cope's classification, *Xenarthra* has long been grouped among the "edentates", a more inclusive, but unstable and artificial assemblage that variously contained the Old World *Orycteropodidae* Gray 1821 (aardvarks), *Manidae* Gray 1821 (pangolins), and a variety of extinct taxa (e.g., palaeanodonts, taeniodonts, e.g., Simpson, 1945). The evidence for this inclusive group, has however, long been doubted (e.g., Huxley, 1872; Scott, 1903–4), and recent molecular studies (e.g., Madsen et al., 2001; Murphy et al., 2001) have confirmed the polyphyletic nature of extant taxa formerly assigned to *Edentata*: *Orycteropodidae* is now considered part of *Afrotheria*, *Manidae* part of *Laurasiatheria* (most closely related to *Carnivora*), and *Xenarthra* comprises a distinct placental clade that is part of neither *Afrotheria* nor *Laurasiatheria*. The use of *Edentata* as a synonym for *Xenarthra* has been infrequent (e.g., McKenna, 1975; Novacek, 1992) and has been generally abandoned (e.g., McKenna and Bell, 1997; Vizcaíno and Loughry, 2008). Rose et al. (2005) recommended that the term be discarded, and since 2000, *Xenarthra* has been almost exclusively used, excepting for some publications in the newsletter of the *Anteater, Sloth and Armadillo Specialist Group*, which is titled "*Edentata*".

The monophyly of crown *Xenarthra* is not seriously doubted (e.g., Simpson, 1945; Delsuc et al., 2004; Gaudin and Wible, 2006; Gibb et al., 2016; Mitchell et al., 2016). They are the only extant placental mammals that are not contained in either the major clades *Afrotheria* or *Boreoeutheria* (Madsen et al., 2001; Murphy et al., 2001). It consists of two primary crown clades: *Dasypodoidea* Gray 1821 and *Pilosa* Flower 1883. Morphological and recent molecular phylogenies suggest that the extinct glyptodonts are nested within crown *Dasypodoidea* (Engelmann, 1978, 1985; Gaudin and Wible, 2006; Mitchell et al., 2016) and that the *Folivora* and *Vermilingua* are more closely related to one another than either is to dasypodoidians, forming the clade *Pilosa* Flower 1883 (Gaudin and Wible, 2006). *Xenarthra* is primarily South American, with various late Cenozoic migrations into North America and the West Indies (Vizcaíno and Loughry, 2008).

Literature Cited

Cope, E. D. 1889. The *Edentata* of North America. *Am. Nat.* 23:657–664.

Cuvier, G. L. C. F. D. 1798. *Tableau Élémentaire de L'histoire Naturelle des Animaux*. J. B. Bailière, Paris.

Delsuc, F., F. M. Catzeflis, M. J. Stanhope, and E. J. P. Douzery. 2001. The evolution of armadillos, anteaters and sloths depicted by nuclear and mitochondrial phylogenies: implications for the status of the enigmatic fossil *Eurotamandua*. *Proc. R. Soc. Lond. B Biol. Sci.* 268:1605–1615.

Delsuc F., S. Vizcaíno, and E. J. P. Douzery. 2004. Influence of Tertiary paleoenvironmental changes on the diversification of South American mammals: a relaxed molecular study within xenarthrans. *BMC Evol. Biol.* 4:11.

Engelmann, G. 1978. *The Logic of Phylogenetic Analysis and the Phylogeny of the* Xenarthra. PhD dissertation, Columbia University, New York.

Engelmann, G. 1985. The phylogeny of the Xenarthra. Pp. 51–64 in *The Evolution and Ecology of Armadillos, Sloths, and Vermilinguas* (G. G. Montgomery, ed.). Smithsonian Institution Press, Washington, DC.

Flower, W. H. 1883. On the mutual affinities of the animals composing the order *Edentata*. *Proc. Zool. Soc. Lond.* 25:358–367.

Gaudin, T., and H. G. McDonald. 2008. Morphology-based investigations of the phylogenetic relationships among extant and fossil xenarthrans. Pp. 24–36 in *The Biology of the Xenarthra* (S. Vizcaíno and W. J. Loughry, eds.). University Press of Florida, Gainesville, FL.

Gaudin, T. J., and J. R. Wible. 2006. The phylogeny of the living and extinct armadillos (*Mammalia, Xenarthra, Cingulata*): a craniodental analysis. Pp. 153–198 in *Amniote Paleobiology: Perspectives on the Evolution of Mammals, Birds, and Reptiles* (M. T. Carrano,

T. J. Gaudin, R. W. Blob and J. R. Wible, eds.). University of Chicago Press, Chicago, IL.

Gibb, G. C., F. L. Condamine, M. Kuch, J. Enk, N. Moraesbarros, M. Superina, H. N. Poinar, and F. Delsuc. 2016. Shotgun mitogenomics provides a reference phylogenetic framework and timescale for living xenarthrans. *Mol. Biol. Evol.* 33:621–642.

Gray, J. E. 1821. On the natural arrangement of vertebrose animals. *Lond. Med. Repos.* 15:296–310.

Huxley, T. H. 1872. *A Manual of the Anatomy of Vertebrated Animals.* D. Appleton and Co., New York.

Illiger, C. D. 1811. *Prodromus Systematis Mammalium et Avium Additis Terminis Zoographicis Uttriusque Classis.* Salfeld, Berlin.

Linnaeus, C. 1758. *Systema Naturæ Per Regna Tria Naturæ, Secundum Classes, Ordines, Genera, Species, cum Characteribus, Differentiis, Synonymis, Locis.* Tomus I, Editio decima, reformata. Laurentii Salvii, Holmiae (Stockholm).

Madsen, O., M. Scally, C. J. Douady, D. J. Kao, R. W. DeBry, R. Adkins, H. M. Amrine, M. J. Stanhope, W. W. de Jong, and M. S. Springer. 2001. Parallel adaptive radiations in two major clades of placental mammals. *Nature* 409:610–614.

McDonald, H. G. 2003. Xenarthran skeletal anatomy: primitive or derived? (*Mammalia, Xenarthra*). *Senckenb. Biol.* 83:5–17.

McKenna, M. C. 1975. Toward a phylogenetic classification of the *Mammalia*. Pp. 21–46 in *Phylogeny of the Primates: A Multidisciplinary Approach* (W. P. Luckett and F. Szalay, eds.). Plenum Press, New York.

McKenna, M. C., and S. K. Bell. 1997. *Classification of Mammals above the Species Level.* Columbia University Press, New York.

Mitchell, K. J., A. Scanferla, E. Soibelzon, R. Bonni, J. Ochoa, and A. Cooper. 2016. Ancient DNA from the extinct South American giant glyptodont Doedicurus sp. (Xenarthra: Glyptodontidae) reveals that glyptodonts evolved from Eocene armadillos. *Molecular Ecology* 14:3499–3508.

Murphy, W. J., E. Eizirik, W. E. Johnson, Y. P. Zhang, O. A. Ryder, S. J. O'Brien. 2001. Molecular phylogenies and the origins of placental mammals. *Nature* 409:614–618.

Novacek, M. J. 1992. Mammalian phylogeny: shaking the tree. *Nature* 356:121–125.

Rose, K. D., R. J. Emry, T. J. Gaudin, and G. Storch. 2005. *Xenarthra* and *Pholidota*. Pp. 106–126 in *The Rise of Placental Mammals* (K. D. Rose and J. D. Archibald, eds.). Johns Hopkins University Press, Baltimore , MD and London.

Scott, W. B. 1903–4. *Mammalia of the Santa Cruz Beds.* Part 1: *Edentata.* Reports of the Princeton University Expeditions to Patagonia, 1896–1899, Vol. 5. Princeton University, Princeton, NJ.

Shockey, B. J., and F. Anaya. 2011. Grazing in a new late Oligocene mylodontid sloth and a mylodontid radiation as a component of the Eocene—Oligocene faunal turnover and the early spread of grasslands/savannas in South America. *J. Mamm. Evol.* 18:101–118.

Simpson, G. G. 1945. The principles of classification and a classification of mammals. *Bull. Am. Mus. Nat. Hist.* 85:1–350.

Tummers, M., and I. Thesleff. 2008. Observations on continuously growing roots of the sloth and the K14-Eda transgenic mice indicate that epithelial stem cells can give rise to both the ameloblast and root epithelium cell lineages creating distinct tooth patterns. *Evol. Dev.* 10:187–195.

Vizcaíno, S., and W. J. Loughry. 2008. *The Biology of the Xenarthra.* University Press of Florida, Gainesville, FL.

Author

Bruce J. Shockey; Biology Department; Manhattan College; Manhattan College Parkway; Riverdale, New York; Division of Vertebrate Paleontology, Yale Peabody Museum of Natural History, 170 Whitney Ave., Yale University, New Haven, CT 06511; Department of Vertebrate Paleontology; American Museum of Natural History; Central Park West at 79th St., New York, NY 10024. Email: bshockey@amnh.org.

Date Accepted: 7 October 2017

Primary editor: Jacques A. Gauthier

Scandentia J. A. Wagner 1855 [E. J. Sargis], converted clade name

Registration Number: 291

Definition: The crown clade originating in the most recent common ancestor of *Ptilocercus lowii* Gray 1848 and *Tupaia* (originally *Sorex*) *glis* (Diard 1820). This is a minimum-crown-clade definition. Abbreviated definition: min crown ∇ *Ptilocercus lowii* Gray 1848 & *Tupaia glis* (Diard 1820).

Etymology: Derived from the Latin *scandens* (climbing).

Reference Phylogeny: Roberts et al. (2011: Fig. 2).

Composition: *Scandentia* contains 23 currently recognized extant species (Helgen, 2005; Sargis et al., 2013a,b; 2014a,b; 2017) and seven described extinct species of treeshrews (Sargis, 2004; Ni and Qiu, 2012; Li and Ni, 2016). All species are divided into two sub-clades (traditionally ranked as families), *Ptilocercidae* and *Tupaiidae* (Helgen, 2005). Six fossil species are included in *Tupaiidae* and one is included in *Ptilocercidae*.

Diagnostic Apomorphies: *Scandentia* is diagnosed among extant euarchontans by the shared presence of the following apomorphies (Silcox et al., 2005: 132) (authorship credited below to first explicit identification as an apomorphy): (1) enclosure of the intratympanic portion of the internal carotid artery in a bony canal floored proximally and distally by the entotympanic, and by the petrosal in between (Wible and Zeller, 1994); (2) enclosure of the intratympanic portion of the stapedial artery by the petrosal in a canal on the promontorium, and within the epitympanic crest beneath the tympanic roof (Wible and Zeller, 1994); (3) absence of an exit for the arteria diploëtica magna (Wible and Zeller, 1994); (4) alisphenoid canal present (Wible and Zeller, 1994); (5) maxillary artery passes medial to the mandibular nerve, beneath the foramen ovale (Wible and Zeller, 1994); (6) laryngeopharyngeal artery present (Wible and Zeller, 1994); (7) scaphoid (radiale) and lunate (intermedium) fused into the scapholunate (Sargis, 2002a); (8) articular facet for the medial malleolus on the posterior side of the sustentaculum (Szalay and Lucas, 1996; Sargis, 2002b).

Synonyms: The following are approximate synonyms: *Scandentiformes* Kinman 1994; *Tupaii* Broers 1963; *Tupaioidea* Straus 1949; *Tupayae* Peters 1864.

Comments: The taxon name *Scandentia* was first used by Wagner (1855) at the traditional family level. The first use of *Scandentia* at the traditional ordinal level was by Butler (1972). Considerations of rank aside, this name is more widely used both in general and (implicitly) for the crown clade than any of its synonyms (e.g., Olson et al., 2004, 2005; Helgen, 2005; Roberts et al., 2011) and was selected for that reason. All currently known members of *Scandentia* are included in one of two sub-clades (viz., *Pan-Ptilocercus* and *Pan-Tupaiidae*) often referred to as *Ptilocercidae* and *Tupaiidae* (e.g., Helgen, 2005; Olson et al., 2004, 2005).

Some phylogenetic analyses have inferred a close relationship between the Oligocene fossil *Anagale gobiensis* and *Tupaia*, but *Ptilocercus lowii* was not included in these analyses (Asher et al., 2003). Although Simpson (1945) considered extinct anagalids as closely related to scandentians, McKenna (1963) provided evidence against such a relationship. Other than *Anagale gobiensis*, no fossils have been proposed to be stem scandentians. Miocene fossils such as *Palaeotupaia sivalicus*, *Prodendrogale yunnanica*, *Prodendrogale engesseri*, *Tupaia miocenica*, and *Tupaia storchi* all appear to be crown *Tupaiidae*, and thus crown *Scandentia* (Sargis, 2004; Ni and Qiu, 2012; Li and Ni, 2016). This is also implied for the earliest treeshrew fossil from the middle Eocene of China, *Eodendrogale parvum* (Sargis, 2004), though its scandentian relationships have been questioned (Ni and Qiu, 2012; Li and Ni, 2016). *Prodendrogale* has been included in a phylogenetic analysis and was inferred to be the closest relative of the extant *Dendrogale* (Li and Ni, 2016). *Tupaia miocenica*, *Tupaia storchi*, and *Palaeotupaia sivalicus* were originally described as sharing apomorphies with extant *Tupaia* (Mein and Ginsburg, 1997; Ni and Qiu, 2012; Chopra and Vasishat, 1979), though this has not been tested in a phylogenetic analysis; Luckett and Jacobs (1980) proposed that fossils attributed to *Palaeotupaia* are indistinguishable from *Tupaia* and should be included in this clade, but again, this proposal has not yet been assessed in a phylogenetic study. There may be Miocene fossils related to *Ptilocercus* at Yuanmou in China (Ni and Qiu, 2002, 2012), and *Ptilocercus kylin*, which has been inferred to be the closest relative of the extant *Ptilocercus lowii*, was recently described from the earliest Oligocene (~34 Ma) at Lijiawa in Yunnan, China (Li and Ni, 2016).

The origin of *Scandentia* (i.e., the divergence between *Ptilocercus lowii* and *Tupaiidae*) is estimated at 63.4 (67.5–62.2) Ma (Janečka et al., 2007), 55.9 (63.9–45.0) Ma (Meredith et al., 2011), or 60.2 Ma (Roberts et al., 2011). The earliest divergence in *Scandentia* thus joins those of many other crown clades of *Placentalia* radiating in the wake of the K-Pg mass extinction (O'Leary et al., 2013).

Literature Cited

Asher, R. J., M. J. Novacek, and J. H. Geisler. 2003. Relationships of endemic African mammals and their fossil relatives based on morphological and molecular evidence. *J. Mamm. Evol.* 10:131–194.

Broers, C. J. 1963. La position taxonomique de *Tupaia* parmi les primates, basée entre autres sur la structure de sa caisse du tympan. *C. R. Assoc. Anat.* 116:361–375.

Butler, P. M. 1972. The problem of insectivore classification. Pp. 253–265 in *Studies in Vertebrate Evolution* (K. A. Joysey and T. S. Kemp, eds.). Oliver and Boyd, Edinburgh.

Chopra, S. R. K., and R. N. Vasishat. 1979. Sivalik fossil tree shrew from Haritalyangar, India. *Nature* 281:214–215.

Diard, P. M. 1820. Report of a meeting of the Asiatic Society for March 10. *Asiatic J. Monthly Regis.* 10:477–478.

Gray, J. E. 1848. Description of a new genus of insectivorous *Mammalia*, or *Talpidae*, from Borneo. *Proc. Zool. Soc. Lond.* 1848:23–24.

Helgen, K. M. 2005. Order *Scandentia*. Pp. 104–109 in *Mammal Species of the World: A Taxonomic and Geographic Reference.* 3rd edition (D. E. Wilson and D. M. Reeder, eds.). The Johns Hopkins University Press, Baltimore, MD.

Janečka, J., W. Miller, T. H. Pringle, F. Wiens, A. Zitzmann, K. M. Helgen, M. S. Springer, and W. J. Murphy. 2007. Molecular and genomic data identify the closest living relative of primates. *Science* 318:792–794.

Kinman, K. E. 1994. *The Kinman System: Toward a Stable Cladisto-Eclectic Classification of Organisms* (living and extinct; 48 phyla, 269 classes, 1,719 orders). Hays, Kansas.

Li, Q., and X. Ni. 2016. An early Oligocene fossil demonstrates treeshrews are slowly evolving "living fossils." *Sci. Rep.* 6:18627.

Luckett, W. P., and L. L. Jacobs. 1980. Proposed fossil tree shrew genus *Palaeotupaia*. *Nature* 288:104.

McKenna, M. C. 1963. New evidence against tupaioid affinities of the mammalian family *Anagalidae*. *Am. Mus. Novit.* 2158:1–16.

Mein, P., and L. Ginsburg. 1997. Les mammifères du gisement miocène inférieur de Li Mae Long, Thaïlande: systématique, biostratigraphie et paléoenvironnement. *Geodiversitas* 19:783–844.

Meredith, R. W., J. E. Janečka, J. Gatesy, O. A. Ryder, C. A. Fisher, E. C. Teeling, A. Goodbla, E. Eizirik, T. L. Simao, T. Stadler, D. L. Rabosky, R. L. Honeycutt, J. J. Flynn, C. M. Ingram, C. Steiner, T. L. Williams, T. J. Robinson, A. Burk-Herrick, M. Westerman, N. A. Ayoub, M. S. Springer, and W. J. Murphy. 2011. Impacts of the Cretaceous Terrestrial Revolution and KPg extinction on mammal diversification. *Science* 334:521–524.

Ni, X., and Z. Qiu. 2002. The micromammalian fauna from the Leilao, Yuanmou hominoid locality: implications for biochronology and paleoecology. *J. Hum. Evol.* 42:535–546.

Ni, X., and Z. Qiu. 2012 Tupaiine tree shrews (*Scandentia, Mammalia*) from the Yuanmou *Lufengpithecus* locality of Yunnan, China. *Swiss J. Palaeontol.* 131:51–60.

O'Leary, M. A., J. I. Bloch, J. J. Flynn, T. J. Gaudin, A. Giallombardo, N. P. Giannini, S. L. Goldberg, B. P. Kraatz, Z.-X. Luo, J. Meng, X. Ni, M. J. Novacek, F. A. Perini, Z. Randall, G. W. Rougier, E. J. Sargis, M. T. Silcox, N. B. Simmons, M. Spaulding, P. M. Velazco, M. Weksler, J. R. Wible, and A. L. Cirranello. 2013. The placental mammal ancestor and the post-K-Pg radiation of placentals. *Science* 339:662–667.

Olson, L. E., E. J. Sargis, and R. D. Martin. 2004. Phylogenetic relationships among treeshrews (*Scandentia*): a review and critique of the morphological evidence. *J. Mamm. Evol.* 11:49–71.

Olson, L. E., E. J. Sargis, and R. D. Martin. 2005. Intraordinal phylogenetics of treeshrews (*Mammalia: Scandentia*) based on evidence from the mitochondrial 12S rRNA gene. *Mol. Phylogenet. Evol.* 35:656–673.

Peters, W. 1864. Uber die Säugethiergattung *Solenodon. Abhandl. Konig. Akad. Wissensch. Berlin* 1863/1864:1–22.

Roberts, T. E., H. C. Lanier, E. J. Sargis, and L. E. Olson. 2011. Molecular phylogeny of treeshrews (*Mammalia: Scandentia*) and the time-scale of diversification in Southeast Asia. *Mol. Phylogenet. Evol.* 60:358–372.

Sargis, E. J. 2002a. Functional morphology of the forelimb of tupaiids (*Mammalia, Scandentia*) and its phylogenetic implications. *J. Morphol.* 253:10–42.

Sargis, E. J. 2002b. The postcranial morphology of *Ptilocercus lowii* (*Scandentia, Tupaiidae*): an analysis of primatomorphan and volitantian characters. *J. Mamm. Evol.* 9:137–160.

Sargis, E. J. 2004. New views on tree shrews: the role of tupaiids in primate supraordinal relationships. *Evol. Anthropol.* 13:56–66.

Sargis, E. J., K. K. Campbell, and L. E. Olson. 2014a. Taxonomic boundaries and craniometric variation in the treeshrews (*Scandentia, Tupaiidae*) from the Palawan faunal region. *J. Mamm. Evol.* 21:111–123.

Sargis, E. J., N. Woodman, A. T. Reese, L. E. Olson. 2013a. Using hand proportions to test taxonomic boundaries within the *Tupaia glis* species complex (*Scandentia, Tupaiidae*). *J. Mamm.* 94:183–201.

Sargis, E. J., N. Woodman, N. C. Morningstar, A. T. Reese, L. E. Olson. 2013b. Morphological distinctiveness of Javan *Tupaia hypochrysa* (*Scandentia, Tupaiidae*). *J. Mamm.* 94:183–201.

Sargis, E. J., N. Woodman, N. C. Morningstar, A. T. Reese, L. E. Olson. 2014b. Island history affects faunal composition: the treeshrews (*Mammalia: Scandentia: Tupaiidae*) from the Mentawai and Batu Islands, Indonesia. *Biol. J. Linn. Soc.* 111:290–304.

Sargis, E. J., N. Woodman, N. C. Morningstar, T. N. Bell, and L. E. Olson. 2017. Skeletal variation and taxonomic boundaries among mainland and island populations of the common treeshrew (*Mammalia: Scandentia: Tupaiidae*). *Biol. J. Linn. Soc.* 120:286–312.

Silcox, M. T., J. I. Bloch, E. J. Sargis, and D. M. Boyer. 2005. *Euarchonta* (*Dermoptera, Scandentia, Primates*). Pp. 127–144 in *The Rise of Placental Mammals: Origins and Relationships of the Major Extant Clades* (K. D. Rose and J. D. Archibald, eds.). The Johns Hopkins University Press, Baltimore, MD.

Simpson, G. G. 1945. The principles of classification and a classification of mammals. *Bull.
Am. Mus. Nat. Hist.* 85:1–350.

Straus, W. L. 1949. The riddle of man's ancestry. *Q.
Rev. Biol.* 24:200–223.

Szalay, F. S., and S. G. Lucas. 1996. The postcranial morphology of Paleocene *Chriacus* and
Mixodectes and the phylogenetic relationships
of archontan mammals. *Bull. New Mexico
Mus. Nat. Hist. Sci.* 7:1–47.

Wagner, J. A. 1855. *Die Säugethiere in Abbildungen
nach der Natur.* Weiger, Leipzig.

Wible, J. R., and U. A. Zeller. 1994. Cranial circu-
lation of the pen-tailed tree shrew *Ptilocercus
lowii* and relationships of *Scandentia. J.
Mamm. Evol.* 2:209–230.

Author

Eric J. Sargis; Department of Anthropology; Yale
University; P.O. Box 208277; New Haven, CT
06520; Divisions of Vertebrate Zoology and
Vertebrate Paleontology; Peabody Museum of
Natural History. Email: eric.sargis@yale.edu.

Date Accepted: 08 August 2017; updated 13
November 2017

Primary Editor: Jacques A. Gauthier

Pan-Primates S. G. B. Chester and E. J. Sargis, new clade name

Registration Number: 258

Definition: The total clade of the crown clade *Primates*. This is a crown-based total-clade definition. Abbreviated definition: total ∇ of *Primates*.

Etymology: *Pan-*, from the Greek *pan-* or *pantos* (all, the whole), indicating that the name refers to a total clade, and *Primates*, from the Latin *primas* (of the highest rank, principal).

Reference Phylogeny: Figure 4 in Bloch et al. (2007).

Composition: *Pan-Primates* includes *Primates* (this volume) and all extinct mammals that are more closely related to that crown clade than to any other extant mammals. Currently, the closest extant relatives of *Primates* are thought to be either *Dermoptera* (e.g., Janečka et al., 2007; Ni et al., 2013; Meredith et al., 2011; Mason et al., 2016) or *Sundatheria* (*Scandentia* + *Dermoptera*) (e.g., Murphy et al., 2001; Bloch et al., 2007; Sargis, 2007; O'Leary et al., 2013). "Plesiadapiforms" are a diverse, likely paraphyletic or possibly polyphyletic group of mammals known from the Palaeocene and Eocene of North America, Europe, and Asia, which have long been considered close fossil relatives of *Primates* (Simpson, 1935; Szalay and Delson, 1979). The most recent comprehensive analyses either support "plesiadapiforms" as members of *Pan-Primates* (e.g., Bloch et al., 2007; Silcox et al., 2010; Chester et al., 2015; Chester et al., 2017) or as members of the stem group of *Primates* + *Dermoptera* (*Primatomorpha*) (Ni et al., 2013; Ni et al., 2016). "Plesiadapiforms" are here considered members of *Pan-Primates* following the traditional view that they are "archaic primates" (e.g., Gidley, 1923; Simpson, 1935; Szalay, 1975; Gingerich, 1975; Szalay and Delson, 1979; Silcox, 2007) and recent phylogenetic results that place them closer to *Primates* than to other euarchontan mammals (Bloch et al., 2007; Silcox et al., 2010; Chester et al., 2015; Chester et al., 2017).

Eleven taxa traditionally ranked as families are currently considered "plesiadapiforms": *Purgatoriidae, Micromomyidae, Palaechthonidae, Paromomyidae, Picromomyidae, Picrodontidae, Saxonellidae* (= *Saxonella*), *Carpolestidae, Plesiadapidae, Toliapinidae,* and *Microsyopidae* (Silcox et al., 2017). Exclusion of *Microsyopidae* from *Pan-Primates* has been proposed mainly based on their retention of a relatively plesiomorphic basicranium that lacks an osseous auditory bulla (Szalay, 1969; Szalay and Delson, 1979).

Diagnostic Apomorphies: The clade *Primates* is generally diagnosed based on traits associated with grasping (longer phalanges, divergent pollex and hallux, nails rather than claws on digits), leaping (relatively long hind limbs, elongated tarsus), an improved visual system with reduced reliance on olfaction (convergent orbits with postorbital bar or septum, less prognathic snout), dental traits for herbivory (low crowned teeth with bunodont cusps, broad talonid basins), and a petrosal bulla (Silcox et al., 2015). "Plesiadapiforms" have some, but not all, of these traits. Certain "plesiadapiforms" such as plesiadapids and carpolestids have been considered to share a petrosal bulla with *Primates* (e.g., Russell, 1964; Szalay et al., 1987; Bloch and Silcox, 2006), and the carpolestid *Carpolestes simpsoni* has an opposable hallux with a nail considered to be homologous with that of *Primates* (Bloch and Boyer, 2002; Bloch et al., 2007; Sargis et al., 2007).

Some of the features that apparently arose early in the history of the primate stem are particularly useful for referring fossils to *Pan-Primates*, including upper molars with a post-protocingulum, a lower third molar that is longer and has a relatively larger hypoconulid than the other lower molars, and elongate manual phalanges (Bloch et al., 2007).

Synonyms: Palaeontologists have often used the name *Primates* to refer to the total clade, with the crown clade—or at least taxa with 'classic' primate apomorphies—referred to as *Euprimates* Hoffstetter 1977, and the stem group as paraphyletic "*Plesiadapiformes*" (Simons and Tattersall, in Simons, 1972) or "*Paromomyiformes*" (Szalay, 1973; Szalay et al., 1987) (see *Primates*, this volume). *Primates* of some authors is therefore an approximate synonym of *Pan-Primates*, and "*Plesiadapiformes*" and "*Paromomyiformes*" are approximate and partial synonyms.

Comments: As discussed in *Primates* (this vol.), "plesiadapiforms" have long been considered closely related to *Primates* based on traits such as teeth with broad talonid basins and bunodont cusps (e.g., Gidley, 1923; Simpson, 1935; Szalay, 1975; Gingerich, 1975; Szalay et al., 1987). These Palaeogene mammals have also been considered most closely related to extant colugos (*Dermoptera*) (e.g., Beard, 1993; Kay et al., 1992) or as members of the stem group of *Primates* + *Dermoptera* (*Primatomorpha*) (Ni et al., 2013; Ni et al., 2016). Gingerich (1975, 1976) considered "plesiadapiforms" to represent the stem group of *Tarsiiformes* ("*Plesitarsiiformes*"), but that hypothesis has not been corroborated by subsequent analyses.

The origin of *Pan-Primates* is estimated at ~65 Ma (Bloch et al., 2007; O'Leary et al., 2013), 79.6 (84.7–77.8) Ma (Janečka et al., 2007), or 71.5 (78.4–64.3) Ma (Meredith et al., 2011).

Literature Cited

Beard, K. C. 1993. Phylogenetic systematics of the *Primatomorpha*, with special reference to *Dermoptera*. Pp. 129–150 in *Mammal Phylogeny: Placentals* (F. S. Szalay, M. C. McKenna, and M. J. Novacek, eds.). Springer-Verlag, New York.

Bloch, J. I., and D. M. Boyer. 2002. Grasping primate origins. *Science* 298:1606–1610.

Bloch, J. I., and M. T. Silcox. 2006. Cranial anatomy of the Paleocene plesiadapiform *Carpolestes simpsoni* (*Mammalia, Primates*) using ultra high-resolution X-ray computed tomography, and the relationships of plesiadapiforms to *Euprimates*. *J. Hum. Evol.* 50:1–35.

Bloch, J. I., M. T. Silcox, D. M. Boyer, and E. J. Sargis. 2007. New Paleocene skeletons and the relationship of plesiadapiforms to crown-clade primates. *Proc. Natl. Acad. Sci. USA* 104:1159–1164.

Chester, S. G. B., J. I. Bloch, D. M. Boyer, and W. A. Clemens. 2015. Oldest known euarchontan postcrania and affinities of Paleocene *Purgatorius* to *Primates*. *Proc. Natl. Acad. Sci. USA* 112:1487–1492.

Chester, S. G. B., T. E. Williamson, J. I. Bloch, M. T. Silcox, and E. J. Sargis. 2017. Oldest skeleton of a plesiadapiform provides additional evidence for an exclusively arboreal radiation of stem primates in the Palaeocene. *R. Soc. Open Sci.* 4:170329.

Gidley, J. W. 1923. Paleocene primates of the Fort Union, with discussion of relationships of Eocene primates. *Proc. US Natl. Mus.* 63:1–38.

Gingerich, P. D. 1975. Systematic position of *Plesiadapis*. *Nature* 253:111–113.

Gingerich, P. D. 1976. Cranial anatomy and evolution of early tertiary *Plesiadapidae* (*Mammalia, Primates*). *Univ. Mich. Pap. Paleontol.* 15:1–141.

Hoffstetter, R. 1977. Phylogenie des primates. Confrontation des resultats obtenus par les diverses voies d'approche du problem. *Bull. Mem. Soc. Anthropol. Paris* t.4, serie XIII:327–346.

Janečka, J., W. Miller, T. H. Pringle, F. Wiens, A. Zitzmann, K. M. Helgen, M. S. Springer, and

W. J. Murphy. 2007. Molecular and genomic data identify the closest living relative of primates. *Science* 318:792–794.

Kay, R. F., J. G. M. Thewissen, and A. D. Yoder. 1992. Cranial anatomy of *Ignacius graybullianus* and the affinities of the *Plesiadapiformes*. *Am. J. Phys. Anthropol.* 89:477–498.

Mason, V. C., G. Li, P. Minx, J. Schmitz, G. Churakov, L. Doronina, A. D. Melin, N. J. Dominy, N. T-L. Lim, M. S. Springer, R. K. Wilson, W. C. Warren, K. M. Helgen, and W. J. Murphy. 2016. Genomic analysis reveals hidden biodiversity within colugos, the sister group to primates. *Sci. Adv.* 2:e1600633.

Meredith, R. W., J. E. Janečka, J. Gatesy, O. A. Ryder, C. A. Fisher, E. C. Teeling, A. Goodbla, E. Eizirik, T. L. L. Simão, T. Stadler, D. L. Rabosky, R. L. Honeycutt, J. J. Flynn, C. M. Ingram, C. Steiner, T. L. Williams, T. J. Robinson, A. Burk-Herrick, M. Westerman, N. A. Ayoub, M. S. Springer, W. J. Murphy. 2011. Impacts of the Cretaceous Terrestrial Revolution and KPg extinction on mammal diversification. *Science* 334:521–524.

Murphy, W. J., E. Eizirik, S. J. O'Brien, O. Madsen, M. Scally, C. J. Douady, E. Teeling, O. A. Ryder, M. J. Stanhope, W. W. de Jong, M. S. Springer. 2001. Resolution of the early placental mammal radiation using Bayesian phylogenetics. *Science* 294:2348–2351.

Ni, X., D. L. Gebo, M. Dagosto, J. Meng, P. Tafforeau, J. J. Flynn, and K. C. Beard. 2013. The oldest known primate skeleton and early haplorhine evolution. *Nature* 498:60–64.

Ni, X., Q. Li, K. Li, and K. C. Beard. 2016. Oligocene primates from China reveal divergence between African and Asian primate evolution. *Science* 352:673–677.

O'Leary, M. A., J. I. Bloch, J. J. Flynn, T. J. Gaudin, A. Giallombardo, N. P. Giannini, S. L. Goldberg, B. P. Kraatz, Z-X Luo, J. Meng, X. Ni, M. J. Novacek, F. A. Perini, Z. Randall, G. W. Rougier, E. J. Sargis, M. T. Silcox, N. B. Simmons, M. Spaulding, P. M. Velazco, M. Weksler, J. R. Wible, and A. L. Cirranello. 2013. The placental mammal ancestor and the post-K-Pg radiation of placentals. *Science* 339:662–667.

Russell, D. E. 1964. Les mammifères paléocènes d'Europe. *Mém. Mus. Natl. Hist. Nat., Série C* 13:1–324.

Sargis, E. J. 2007. The postcranial morphology of *Ptilocercus lowii* (*Scandentia, Tupaiidae*) and its implications for primate supraordinal relationships. Pp. 51–82 in *Primate Origins: Adaptations and Evolution* (M. J. Ravosa and M. Dagosto, eds.). Springer, New York.

Sargis, E. J., D. M. Boyer, J. I. Bloch, and M. T. Silcox. 2007. Evolution of pedal grasping in primates. *J. Hum. Evol.* 53:103–107.

Silcox, M. T. 2007. Primate taxonomy, plesiadapiforms, and approaches to primate origins. Pp. 143–178 in *Primate Origins: Adaptations and Evolution* (M. J. Ravosa and M. Dagosto, eds.). Springer, New York.

Silcox, M. T., J. I. Bloch, D. M. Boyer, S. G. B. Chester, and S. López-Torres. 2017. The evolutionary radiation of plesiadapiforms. *Evol. Anthropol.* 26:74–94.

Silcox, M. T., J. I. Bloch, D. M. Boyer, and P. Houde. 2010. Cranial anatomy of Paleocene and Eocene *Labidolemur kayi* (*Mammalia: Apatotheria*), and the relationships of the *Apatemyidae* to other mammals. *Zool. J. Linn. Soc.* 160:773–825.

Silcox, M. T., E. J. Sargis, J. I. Bloch, and D. M. Boyer. 2015. Primate origins and supraordinal relationships: morphological evidence. Pp. 1053–1081 in *Handbook of Paleoanthropology. Vol. 2: Primate Evolution and Human Origins.* 2nd edition (W. Henke and I. Tattersall, eds.). Springer, Heidelberg.

Simons, E. L. 1972. *Primate Evolution: An Introduction to Man's Place in Nature.* MacMillan, New York.

Simpson, G. G. 1935. The Tiffany fauna, upper Paleocene. II. Structure and relationships of *Plesiadapis*. *Am. Mus. Novit.* 816:1–30.

Szalay, F. S. 1969. *Mixodectidae, Microsyopidae*, and the insectivore-primate transition. *Bull. Am. Mus. Nat. Hist.* 140:193–330.

Szalay, F. S. 1973. New Paleocene primates and a diagnosis of the new suborder *Paromomyiformes*. *Folia Primatol.* 19:73–87.

Szalay, F. S. 1975. Where to draw the non-primate-primate taxonomic boundary. *Folia Primatol.* 23:158–163.

Szalay, F. S., and E. Delson. 1979. *Evolutionary History of the* Primates. Academic Press, New York.

Szalay, F. S., A. L. Rosenberger, and M. Dagosto. 1987. Diagnosis and differentiation of the order *Primates. Yearb. Phys. Anthropol.* 30:75–105.

Authors

Stephen G. B. Chester; Department of Anthropology and Archaeology; Brooklyn College, City University of New York; 2900 Bedford Avenue; Brooklyn, NY 11210; Department of Anthropology; The Graduate Center; City University of New York; 365 Fifth Avenue; New York, NY 10016; New York Consortium in Evolutionary Primatology; New York, NY 10024. Email: stephenchester@brooklyn.cuny.edu

Eric J. Sargis; Department of Anthropology; Yale University; P.O. Box 208277; New Haven, CT 06520; Divisions of Vertebrate Paleontology and Vertebrate Zoology; Peabody Museum of Natural History. Email: eric.sargis@yale.edu

Date Accepted: 10 July 2017

Primary Editor: Jacques A. Gauthier

Primates C. Linnaeus 1758 [G. F. Gunnell and A. D. Yoder], converted clade name

Registration Number: 259

Definition: The smallest crown clade containing *Lemur catta* Linnaeus 1758 (*Lemuridae*), *Loris* (originally *Lemur*) *tardigradus* (Linnaeus 1758) (*Lorisidae*), *Tarsius* (originally *Lemur*) *tarsier* (Erxleben 1777) (*Tarsiidae*), and *Homo sapiens* Linnaeus 1758 (*Hominidae*). This is a minimum-crown-clade definition. Abbreviated definition: min crown ∇ (*Lemur catta* Linnaeus 1758 & *Loris tardigradus* Linnaeus 1758 & *Tarsius tarsier* (Erxleben 1777) & *Homo sapiens* Linnaeus 1758.)

Etymology: Derived from the Latin *primus* (of the highest rank, principal).

Reference Phylogeny: Perelman et al. (2011: Fig. 1); also see Bloch et al. (2007: Fig. 4) and Springer et al. (2012: Fig. 1).

Composition: *Primates* encompasses at least 380 extant species (Groves, 2005; Fleagle, 2013), plus at least 376 extinct species (Fleagle, 2013) included in crown *Primates* as defined above. Palaeontologists often include at least 135 additional extinct species of "plesiadapiforms" in *Primates*. But they are not part of the crown in any current phylogenies, although they may be stem *Primates* (see *Pan-Primates*, this volume and Comments below). We note here that the number of extant primate species recognized today exceeds that of Groves (2005) almost entirely due to the addition of cryptic species since 2005. At least 216 of the extinct species belong to Palaeogene taxa known as *Adapiformes* (= *Adapoidea*, *Adapidae*) and *Omomyiformes* (= *Omomyoidea*, *Omomyidae*). These potentially paraphyletic taxa, plus those necessarily

included in crown *Primates* (= *Strepsirrhini* + *Haplorhini*; note that the correct original spelling of the latter name Pocock, 1918 is with one "r"), have traditionally been referred to collectively as *Euprimates* (Hofstetter, 1977; see Comments).

Diagnostic Apomorphies: *Primates* is readily diagnosed among extant mammals by several apomorphies, including the presence of nails instead of claws on all digits, with the exception that the second (*Strepsirrhini*), or the second and third (*Tarsius*), pedal digits bear specialized grooming claws (that could also be an apomorphy of Primates (see Maiolino et al., 2012). Given relationships among South American platyrrhine anthropoids (Perez and Rosenberger, 2014; Schneider and Sampaio, 2015), *Callitrichinae* appears to have secondarily developed claw-like nails and *Aotus* a grooming claw (Maiolino et al., 2011). Additional apomorphies include a hallux that is opposable and has an enlarged peroneal process; relatively elongate tarsal bones; an enhanced visual system including convergent orbits; a complete postorbital bar or septum; low-crowned, bunodont teeth; an enlarged and complex brain including an expanded frontal cortex; and an ossified auditory bulla formed by the petrosal bone (Cartmill, 1974, 1992; Wible and Covert, 1987; Silcox, 2007). Of these characteristics, the fossil record currently indicates that most of them are plausible apomorphies of crown *Primates* (or at least *Euprimates*; see below), though a few may have originated earlier in potential stem forms (i.e., some "plesiadapiforms"), thus being plesiomorphic for crown *Primates*. Among these potentially earlier-arising apomorphies are: a divergent and opposable hallux with a nail (known for certain to occur among plesiadapiforms only in a single species of *Carpolestes* [Bloch and Boyer, 2002]); low-crowned, bunodont teeth (common to many "plesiadapiforms" [Bloch et al., 2007]); and possibly an ossified auditory bulla formed by the petrosal bone (potentially present in some *Plesiadapoidea* [Bloch and Silcox, 2006; Bloch et al., 2007], but clearly absent in other known "plesiadapiforms" [Bloch and Silcox, 2006]). The composition of the bulla can, however, be difficult to assess in fossils because the only reliable method for identifying the bony elements that comprise the bulla is via longitudinal ontogenetic analysis (MacPhee et al., 1983; MacPhee and Cartmill, 1986; Kay et al., 1992; Silcox, 2007). Moreover, possession of low-crowned, bunodont teeth offers only weak evidence for primate relationships, as that tooth form is not exclusive to primates, being generally typical of frugivorous mammals.

Synonyms: *Euprimates* Hofstetter 1977 (approximate).

Comments: *Primates* has been employed as a taxon name in a number of ways. It has most typically been associated with the crown, viz., the clade composed of taxa bracketed by extant strepsirrhine and haplorhine species (e.g., Springer et al., 2007, 2012; Meredith et al., 2011). Alternatively, it has been applied to a more inclusive clade composed of the crown plus all fossils thought to be more closely related to extant *Primates* than to *Dermoptera* and *Scandentia* (e.g., Hofstetter, 1977; Bloch et al., 2007; Silcox, 2007; Chester et al., 2015; also see *Pan-Primates,* this volume). Thus, if *Primates* is used for the total clade, a different name would be needed for the crown. No one doubts the monophyly of the clade composed of extant strepsirrhines and haplorhines (crown *Primates*). Likewise, palaeontologists have long agreed that

potential primate relatives among extinct "plesiadapiforms" are not part of that clade, even if they are related to it (which may account for why Hofstetter [1977] proposed a new name - *Euprimates* – for the crown clade). Because in the literature *Primates*, rather than *Euprimates*, has been by far the most commonly used name associated with the crown (confirmed by an extensive internet search), we elect to apply the name *Primates* to the crown clade.

The situation is complicated because *Euprimates* has also been associated with a well-supported taxon bearing all of the "classic" primate apomorphies (see Diagnostic Apomorphies) that are absent in (most) extinct "plesiadapiforms" but present in crown *Primates* and——so far as they are preserved—in taxa included in extinct "adapiforms" and "omomyiforms" (Gunnell et al., 2008). If the name *Euprimates* were to be defined with respect to these apomorphies, then any taxon possessing them would be safely referred to that clade regardless of whether they are crown primates. However, those "classic" apomorphies likely antedate the crown. Instead, they probably diagnose somewhat more inclusive clade(s) containing not only the crown, but also some taxa that may be close to, but still stemward of, the last ancestor shared by strepsirrhines and haplorhines (= crown *Primates*). (Apomorphy- and node-based definitions may be alternative ways of defining the same names, but they are unlikely to delimit the same clades, even if one is nested within the other; Gauthier and de Queiroz, 2001). But defining the name *Euprimates* with respect to that set of apomorphies could have a potentially undesirable consequence. If the "classic" primate apomorphies originated before the crown did, then that would reverse a commonly understood hierarchical relationship—instead of *Euprimates* being nested within *Primates* (sensu lato), *Euprimates* would contain *Primates* as that name is defined here.

Defining the name *Euprimates* by composition, without regard to particular defining apomorphies, would not necessarily avoid this outcome. For example, *Euprimates* could be defined to include not only extant specifiers, but also every extinct species currently referred to "adapiforms" and "omomyiforms". Or *Euprimates* could be defined by extant specifiers plus extinct *Adapis parisiensis* Cuvier 1812 and *Omomys carteri* Leidy 1869. Should any of these taxa prove to lie outside the crown, however, then *Primates* would still be a subclade of *Euprimates*, just the opposite of traditional understanding. Alternatively, if (under the first approach) all "adapiforms" and "omomyiforms" are nested within the crown, or if (under the second) just *Adapis* and *Omomys* are crown primates, then *Euprimates* would be a junior synonym of *Primates* as defined here.

In order to be referred to crown-clade *Primates*, an extinct taxon must not only possess the "classic" primate apomorphies traditionally associated with *Euprimates*, but also have at least one apomorphy that arose within the crown (Gauthier and de Queiroz, 2001). For example, whether the Eocene "omomyiform" *Teilhardina belgica* is a (crown) primate rests on whether one accepts the apomorphies it shares with extant tarsiers as homologous (i.e., anatomical details in the femoral head, position of the 3rd trochanter, distal elongation of the tarsals, and elongation of the fingers and toes; Gebo et al., 2015). If one accepts these apomorphies as homologous, then *T. belgica* would be a stem tarsier and thus a (crown) primate; otherwise, even with all the "classic" primate apomorphies indicating that *T. belgica* is a euprimate (and a pan-primate), that would be inadequate justification for including it in (crown) *Primates*.

A lingering issue of contention relates to the relationship of crown *Primates* to the large group of fossil taxa traditionally known as *Plesiadapiformes* (Szalay, 1969; Gunnell, 1989;

Bloch et al., 2007; Silcox, 2007; Ni et al., 2013). Those fossils have been placed in a broadly defined (~ total-clade) *Primates* (e.g., Hofstetter, 1977; Bloch et al., 2007; Chester et al., 2015), or as a group called *Proprimates* (ranked as a traditional order separate from *Primates*; Gingerich, 1989, 1990; see also Beard, 1990), or as ecological vicars of, but perhaps not closely related to, *Primates* (Cartmill, 1972; Gunnell, 1989; Kay et al., 1992; Clemens, 2004). Indeed, Ni et al. (2013) argued that plesiadapiforms have no exclusive connection to primates, and instead represent an extinct clade that is sister to *Dermoptera* + *Primates* (= *Primatomorpha* Beard 1991) within *Euarchonta*. In contrast, Chester and Sargis (*Pan-Primates*, this volume) consider "*Plesiadapiformes*" a paraphyletic assemblage of stem primates. Despite these disagreements, plesiadapiforms are universally regarded to be outside of the crown, and thus are not *Primates* as that name is defined here.

Another point of contention is the phylogenetic relationship between crown *Primates* and other mammalian crown clades. Some weak morphological evidence favors a *Primates* + *Scandentia* relationship (Godinot, 2007), but the weight of evidence supports *Primates* as sister to a clade that includes *Dermoptera* and *Scandentia* (= *Sundatheria* Olson et al. 2005) (e.g., Murphy et al., 2001; Silcox et al., 2005; Bloch et al., 2007; O'Leary et al., 2013). In contrast, most molecular evidence tends to favor a *Primates* + *Dermoptera* clade (*Primatomorpha* Beard 1991) (e.g., Janecka et al., 2007; Perelman et al., 2011; Meredith et al., 2011; Pozzi et al., 2014). Some molecular studies suggest that treeshrews (*Scandentia*) may be more closely related to *Glires* than to *Primates* or *Dermoptera* (e.g., Meredith et al., 2011; Xu et al., 2012). Rapid speciation at the base of *Euarchontoglires* suggests that this problem might be especially difficult to resolve (Zhou et al., 2015). Nevertheless, the majority of analyses support a *Scandentia-Dermoptera-Primates* clade (= *Euarchonta* Waddell et al. 1999) (see also Lin et al., 2014; Chester et al., 2015) which is most closely related to a clade including *Rodentia* and *Lagomorpha* (= *Glires* Linnaeus 1758) and together constitute the more inclusive clade *Euarchontoglires* Murphy et al. 2001, which is sister to *Laurasiatheria* Waddell et al. 1999 within *Boreoeutheria* Murphy et al. 2001.

Relationships within crown *Primates* are generally well resolved, at least at the more inclusive levels (Perelman et al., 2011; Springer et al., 2012). Extant *Primates* is composed of two primary crown clades: *Strepsirrhini* and *Haplorhini*. Strepsirrhines include all Malagasy lemuriforms including *Daubentonia* plus Afro-Asian lorisiforms, with haplorhines encompassing all other living primates. Among haplorhines, *Tarsiidae* is sister to a large clade containing all other living haplorhines, i.e., crown *Anthropoidea* (= *Simiiformes* sensu Perelman et al., 2011). *Anthropoidea* encompasses two crown clades, *Platyrrhini*, which includes all extant New World monkeys, and *Catarrhini*, which includes all extant Old World monkeys and apes (including humans). The preferred strepsirrhine-haplorhine taxonomy is generally considered to be phylogenetically accurate, as opposed to the "grade-based" prosimian-anthropoid taxonomy.

Fossils indicate both crown-primates and crown-haplorhines had already diverged by the Early Eocene. At least some well-preserved "omomyiforms" from that time (e.g., *Archicebus achilles*, Asia) are inferred to be stem tarsiers, and thus crown haplorhines (Ni et al., 2013; Gebo et al., 2015). And at least some well-preserved "adapiforms" (e.g., *Darwinius masillae* Franzen et al. 2009, Middle Eocene, Europe) are inferred to be stem strepsirrhines (Rose et al., 2011; López-Torres et al., 2015). Reasonably well-known haplorhine stem anthropoids, such as *Catophithecus browni* (Simons and Rasmussen, 1996), are known from the Late

Eocene of North Africa. If these relationships are accepted, then these species would rank among the earliest well-founded examples of (crown) *Primates* (and crown *Haplorhini*). The fossil record indicates a remarkably diverse and widespread early pan-primate fauna across Laurasia in Hot House conditions during the Palaeogene (Rose, 1995). Although the teeth and jaws by which most of these species are known have proven a rich character system, much remains to be discovered about other details of their anatomy and phylogenetic relationships.

Literature Cited

Beard, K. C. 1990. Do we need the newly proposed order *Proprimates*? *J. Hum. Evol.* 19:817–820.

Beard, K. C. 1991. Vertical postures and climbing in the morphotype of *Primatomorpha*: implications for locomotor evolution in primate history. Pp. 79–87 in *Origine(s) de la Bipedie chez les Hominides* (Y. Coppens and B. Senut, eds.). Cahiers de Paleoanthropologie, Editions du CNRS, Paris.

Bloch, J. I., D. M. Boyer. 2002. Grasping primate origins. *Science* 298:1606–1610.

Bloch, J. I., and M. T. Silcox. 2006. Cranial anatomy of the Paleocene plesiadapiform *Carpolestes simpsoni* (*Mammalia, Primates*) using ultra high-resolution X-ray computed tomography, and the relationships of plesiadapiforms to *Euprimates*. *J. Hum. Evol.* 50:1–35.

Bloch, J. I., M. T. Silcox, D. M. Boyer, and E. J. Sargis. 2007. New Paleocene skeletons and the relationship of plesiadapiforms to crown-clade primates. *Proc. Natl. Acad. Sci. USA* 104:1159–1164.

Cartmill, M. 1972. Arboreal adaptations and the origin of the order *Primates*. Pp. 97–122 in *The Functional and Evolutionary Biology of Primates* (R. Tuttle, ed.). Aldine-Atherton, Chicago, IL.

Cartmill, M. 1974. Rethinking primate origins. *Science* 184:436–443.

Cartmill, M. 1992. New views on primate origins. *Evol. Anthropol.* 1:105–111.

Chester, S. G. B., J. I. Bloch, D. M. Boyer, and W. A. Clemens. 2015. Oldest known euarchontan tarsals and affinities of Paleocene *Purgatorius* to *Primates*. *Proc. Natl. Acad. Sci. USA* 112(5):1487–1492.

Clemens, W. A. 2004. *Purgatorius* (*Plesiadapiformes, Primates?, Mammalia*), a Paleocene immigrant into northeastern Montana: stratigraphic occurrences and incisor proportions. *Bull. Carnegie Mus. Nat. Hist.* 36:3–13.

Cuvier, G. 1812. *Recherches sur les Ossemens Fossiles des Quadrupeds*. 1st edition. Chez Deterville, Paris.

Erxleben, J. C. P. 1777. *Systema Regni Animalis per Classes, Ordines, Genera, Species, Varietates, cum Synonymia et Historia Animalium. Classis I.* Mammalia. Weygand, Leipzig.

Fleagle, J. G. 2013. *Primate Adaptation and Evolution*. 3rd edition. Academic Press, New York.

Franzen, J. L., P. D. Gingerich, J. Habersetzer, J. H. Hurum, W. Von Koenigswald, and B. H. Smith. 2009. Complete primate skeleton from the Middle Eocene of Messel in Germany: morphology and paleobiology. *PLOS ONE* 4(5):e5723.

Gauthier, J., and K. de Queiroz. 2001. Feathered dinosaurs, flying dinosaurs, crown dinosaurs and the name *"Aves"*. Pp. 7–41 in *New Perspectives on the Origin and Early Evolution of Birds: Proceedings of the International Symposium in Honor of John H.Ostrom* (J. Gauthier and L. F. Gall, eds.). Peabody Museum of Natural History, Yale University, New Haven, CT.

Gebo, D. L., R. Smith, M. Dagosto, and T. Smith. 2015. Additional postcranial elements of *Teilhardina belgica*: the oldest European primate. *Am. J. Phys. Anthropol.* 156:388–406.

Gingerich, P. D. 1989. New earliest Wasatchian mammalian fauna from the Eocene of northwestern Wyoming: composition and diversity in a rarely sampled high-floodplain assemblage. *Papers Paleontol. Univ. Michigan* 28:1–97.

Gingerich, P. D. 1990. Mammalian order *Proprimates*—response to Beard. *J. Hum. Evol.* 19:821–822.

Godinot, M. 2007. Primate origins: a reappraisal of historical data favoring tupaiid affinities.

Pp. 83–142 in *Primate Origins—Adaptations and Evolution* (M. J. Ravosa and M. Dagosto, eds.). Springer, New York.

Groves, C. P. 2005. Order *Primates*. Pp. 111–184 in *Mammal Species of the World: A Taxonomic and Geographic Reference*. 3rd edition (D. E. Wilson and D. M. Reeder, eds.). Johns Hopkins University Press, Baltimore, MD.

Gunnell, G. F. 1989. Evolutionary history of *Microsyopoidea* (*Mammalia, ?Primates*) and the relationship between *Plesiadapiformes* and *Primates*. *Papers Paleontol. Univ. Michigan* 27:1–157.

Gunnell, G. F., K. D. Rose, and D. T. Rasmusssen. 2008. *Euprimates*. Pp. 239–261 in *Evolution of Tertiary Mammals of North America*, Volume 2 (C. M. Janis, G. F. Gunnell, and M. D. Uhen, eds.). Cambridge University Press, Cambridge, UK.

Hofstetter, R. 1977. Phylogénie des primates. *Bull. Mém. Soc. Anthrop. Paris* 4:327–346.

Janecka, J. E., W. Miller, T. H. Pringle, F. Wiens, A. Zitzmann, K. M. Helgen, M. S. Springer, and W. J. Murphy. 2007. Molecular and genomic data identify the closest living relative of primates. *Science* 318(5851):792–794.

Kay, R. F., J. G. M. Thewissen, and A. D. Yoder. 1992. Cranial anatomy of *Ignacius graybullianus* and the affinities of *Plesiadapiformes*. *Am. J. Phys. Anthropol.* 89:477–498.

Leidy, J. 1869. Notice of some extinct vertebrates from Wyoming and Dakota. *Proc. Acad. Nat. Sci. Phila.* 21:63–67.

Lin, J., G. Chen, L. Gu, Y. Shen, M. Zheng, W. Zheng, X. Hu, X. Zhang, Y. Qiu, X. Liu, and C. Jiang. 2014. Phylogenetic affinity of tree shrews to *Glires* is attributed to fast evolution rate. *Mol. Phylogenet. Evol.* 71:193–200.

Linnaeus, C. 1758. *Systema Naturae Per Regna Tria Naturae, Secundum Classes, Ordines, Genera, Species, cum Characteribus, Differentiis, Synonymis, Locis*. Tomus I, Editio decima, reformata. Laurentii Salvii, Holmiae (Stockholm).

López-Torres, S., M. A. Schillaci, and M. T. Silcox. 2015. Life history of the most complete fossil primate skeleton: exploring growth models for *Darwinius*. *R. Soc. Open Sci.* 2(9):e150340.

MacPhee, R. D. E., and M. Cartmill. 1986. Basicranial structures and primate systematics.

Pp. 219–275 in *Comparative Primate Biology, Vol. 1: Systematics, Evolution, and Anatomy* (D. R. Swisher and J. Erwin, eds.). Alan R. Liss, New York.

MacPhee, R. D. E, M. Cartmill, and P. D. Gingerich. 1983. New Paleogene primate basicrania and the definition of the order *Primates*. *Nature* 301:509–511.

Maiolino, S., D. M. Boyer, A. R. Rosenberger. 2011. Morphological correlates of the grooming claw in distal phalanges of platyrrhines and other primates: a preliminary study. *Anat Record* 294(12):1975–1990.

Maiolino, S., D. M. Boyer, J. I. Bloch, C. C. Gilbert, and J. Groenke. 2012. Evidence for a grooming claw in a North American adapiform primate: implications for anthropoid origins. *PLOS ONE* 7(1):e29135.

Meredith, R. W., J. E. Janecka, J. Gatesy, O. A. Ryder, C. A. Fisher, E. C. Teeling, A. Goodbla, E. Eizirik, T. L. L. Simão, T. Stadler, D. L. Rabosky, R. L. Honeycutt, J. J. Flynn, C. M. Ingram, C. Steiner, T. L. Williams, T. J. Robinson, A. Burk-Herrick, M. Westerman, N. A. Ayoub, M. S. Springer, and W. J. Murphy. 2011. Impacts of the Cretaceous Terrestrial Revolution and KPg extinction on mammal diversification. *Science* 334:521–524.

Murphy, W. J., E. Eizerik, S. J. O'Brien, O. Madsen, M. Scally, C. J. Douady, E. Teeling, O. A. Ryder, M. J. Stanhope, W. W. de Jong, M. S. Springer. 2001. Resolution of the early placental mammal radiation using Bayesian phylogenetics. *Science* 294:2348–2351.

Ni, X., D. L. Gebo, M. Dagosto, J. Meng, P. Tafforeau, J. J. Flynn, and K. C. Beard. 2013. The oldest known primate skeleton and early haplorhine evolution. *Nature* 498:60–64.

O'Leary, M. A., J. I. Bloch, J. J. Flynn, T. J. Gaudin, A. Giallombardo, N. P. Giannini, S. L. Goldberg, B. P. Kraatz, Z.-X. Luo, J. Meng, X. Ni, M. J. Novacek, F. A. Perini, Z. Randall, G. W. Rougier, E. J. Sargis, M. T. Silcox, N. B. Simmons, M. Spaulding, P. M. Velazco, M. Weksler, J. R. Wible, A. L. Cirranello. 2013. The placental mammal ancestor and the post-K radiation of placentals. *Science* 339:662–667.

Olson, L. E., E. J. Sargis, and R. D. Martin. 2005. Intraordinal phylogenetics of treeshrews (*Mammalia: Scandentia*) based on evidence from the mitochondrial 12s rRNA gene. *Mol. Phylogenet. Evol.* 35:656–673.

Perelman, P., W. E. Johnson, C. Roos, H. N. Seuánez, J. E. Horvath, M. A. M. Moreira, B. Kessing, J. Pontius, M. Roelke, Y. Rumpler, M. P. C. Schneider, A. Silva, S. J. O'Brien, J. Pecon-Slattery. 2011. A molecular phylogeny of living primates. *PLOS Genet.* 7(3):e1001342.

Perez, S. I., and A. L. Rosenberger. 2014. The status of platyrrhine phylogeny: a meta-analysis and quantitative appraisal of topological hypotheses. *J. Human Evol.* 76:177–187.

Pocock, R. I. 1918. On the external characters of the lemurs and of *Tarsius*. *Proc. Zool. Soc. Lond.* 88:19–53.

Pozzi L., J. A. Hodgson, A. S. Burrell, K. N. Sterner, R. L. Raaum, and T. R. Disotell. 2014. Primate phylogenetic relationships and divergence dates inferred from complete mitochondrial genomes. *Mol. Phylogenet. Evol.* 75:165–183.

Rose, K. D. 1995. The earliest *Primates*. *Evol. Anthro.* 3:159–173.

Rose, K. D, S. G. B. Chester, R. H. Dunn, D. M. Boyer, and J. I. Bloch. 2011. New fossils of the oldest North American euprimate *Teilhardina brandti* (*Omomyidae*) from the Paleocene–Eocene thermal maximum. *Am. J. Phys. Anthropol.* 146:281–305.

Schneider, H., and I. Sampaio. 2015. The systematics and evolution of New World primates—a review. *Mol. Phylogenet. Evol.* 82:348–357.

Silcox, M. T. 2007. Primate taxonomy, plesiadapiforms, and approaches to primate origins. Pp. 143–178 in *Primate Origins—Adaptations and Evolution* (M. J. Ravosa and M. Dagosto, eds.). Springer, New York.

Silcox, M. T., J. I. Bloch, E. J. Sargis, and D. M. Boyer. 2005. *Euarchonta* (*Dermoptera, Scandentia, Primates*). Pp. 127–144 in *The Rise of Placental Mammals: Origins and Relationships of the Major Extant Clades* (K. D. Rose and J. D. Archibald, eds.). The Johns Hopkins University Press, Baltimore, MD.

Simons, E. L., and D. T. Rasmussen. 1996. Skull of *Catopithecus browni*, an early Tertiary catarrhine. *Am. J. Phys. Anthropol.* 100(2):261–292.

Springer, M. S., R. W. Meredith, J. Gatesy, C. A. Emerling, J. Park, D. L. Rabosky, T. Stadler, C. Steiner, O. A. Ryder, J. E. Janečka, C. A. Fisher, and W. J. Murphy. 2012. Macroevolutionary dynamics and historical biogeography of primate diversification inferred from a species supermatrix. *PLOS ONE* 7(11):e49521.

Springer, M. S., W. J. Murphy, E. Eizirik, O. Madsen, M. Scally., C. J. Douady, E. C. Teeling, M. J. Stanhope, W. de Jong, and S. J. O'Brien. 2007. A molecular classification for the living orders of placental mammals and the phylogenetic placement of *Primates*. Pp. 1–28 in *Primate Origins—Adaptations and Evolution* (M. J. Ravosa and M. Dagosto, eds.). Springer, New York.

Szalay, F. S. 1969. *Mixodectidae, Microsyopidae,* and the insectivore-primate transition. *Bull. Am. Mus. Nat. Hist.* 140:195–330.

Wadell, P. J., N. Okada, and M. Hasegawa. 1999. Towards resolving the interordinal relationships of placental mammals. *Syst. Biol.* 48:1–5.

Wible, J. R., and H. H. Covert. 1987. *Primates*: cladistic diagnosis and relationships. *J. Hum. Evol.* 16:1–22.

Xu, L., S.-Y.Chen, W.-H. Nie, X.-L. Jiang, Y.-G. Yao. 2012. Evaluating the phylogenetic position of the Chinese Tree Shrew (*Tupaia belangeri chinensis*) based on complete mitochondrial genome: implications for using tree shrew as an alternative experimental animal to primates in biomedical research. *J. Genet. Genom.* 39:131–137.

Zhou, X., F. Sun, S. Xu, G. Yang, and L. Ming. 2015. The position of tree shrews in the mammalian tree: comparing multi-gene analyses with phylogenomic results leaves monophyly of *Euarchonta* doubtful. *Integr. Zool.* 10:186–198.

Authors

Gregg F. Gunnell[†]; Division of Fossil Primates; Duke University Lemur Center; 1013 Broad Street; Durham, NC 27705, USA. Email: gregg.gunnell@duke.edu.

Yoder, A. D., Department of Biology; 315 Biological Sciences Building; Duke University; Durham, NC 27708, USA. Email: anne.yoder@duke.edu.

Date Accepted: 11 February 2016

Primary editor: Jacques A. Gauthier

[†] Deceased.

Apo-Chiroptera A. L. Cirranello, N. B. Simmons, and G. F. Gunnell, new clade name

Registration Number: 237

Definition: The clade for which the unique modifications of the hand, forearm, humerus, scapula, hip, and ankle (see Diagnostic Apomorphies) associated with flapping flight, as inherited by *Vespertilio murinus* Linnaeus 1758, are apomorphies. This is an apomorphy-based definition. Abbreviated definition: ∇ apo hand-wing [*Vespertilio murinus* Linnaeus 1758].

Etymology: Derived from the Greek *apo* (from, away), in reference to the apomorphy-based definition of the name, and *Chiroptera* (see entry in this volume for etymology).

Reference Phylogeny: Figure 4 of Simmons et al. (2008), in which flight and associated apomorphies are inferred to have arisen on the branch subtending the clade (labeled "*Chiroptera*") that includes all of the ingroup taxa, should be treated as the primary reference phylogeny.

Composition: *Apo-Chiroptera* comprises *Chiroptera* (this volume) as well as the Early to Middle Eocene fossil taxa *Onychonycteridae* (*Onychonycteris finneyi, Aegina tobieni, Honrovits tsuwape, ?Honrovits joeli, Eppsinycteris anglica, Marnenycteris michauxi*), *Icaronycteridae* (*Icaronycteris index, I. menui, I. sigei*), *Archaeonycteridae* (*Archaeonycteris trigonodon, A. pollex, A. brailloni, A. relicta; ?Archaeonycteris praecursor, ?Archaeonycteris storchi, Protonycteris gunnelli*), *Hassianycteridae* (*Hassianycteris messelensis, H. magna, H. revilliodi, H. kumari; Cambaya complexus*), and *Palaeochiropterygidae* (*Palaeochiropteryx tupaiodon, P. spiegeli; Stehlinia*

gracilis, S. pusilla, S. rutimeyeri S. quercyi, S. minor, S. bonisi, S. revilliodi, S. alia; Cecilionycteris prisca; Matthesia germanica, ?M. insolita; Lapichiropteryx xiei; Microchiropteryx folieae). Three other extinct taxa that originated in the Eocene, *Philisidae* (*Dizzya exsultans, Philisis sphingis, P. sevketi; Witwatia eremicus, W. schlosseri, W. sigei*), *Mixopterygidae* (*Carcinipteryx trassounius, C. maximinensis, C. liaudae; Mixopteryx dubia* [known only from Oligocene deposits], *M. perrierensis, M. weithoferi*), and *Tanzanycteridae* (*Tanzanycteris mannardi*), also belong to *Apo-Chiroptera* although they probably nest within (crown) *Chiroptera* (Smith et al., 2012). In the reference phylogeny, *Palaeochiropteryx* (*Palaeochiropterygidae*), *Hassianycteris* (*Hassianycteridae*), *Archaeonycteris* (*Archaeonycteridae*), *Icaronycteris* (*Icaronycteridae*), and *Onychonycteris* (*Onychonycteridae*), appear as stem taxa that are successively more distantly related to the crown clade.

Diagnostic Apomorphies: There is one craniodental character that is unambiguously diagnostic of *Apo-Chiroptera*; all other known or putative apomorphies are postcranial features that are part of the flight apparatus (Vaughan, 1959; Hill and Smith, 1984; Simmons, 1995). Although some of these traits may have evolved in a stepwise fashion, the order of acquisition remains unknown. Should these features eventually be shown to have evolved at different levels in the tree, modification of the manus (apomorphy 2 below) is the definitive apomorphy necessary for membership in *Apo-Chiroptera*. This list is based on the phylogeny of Simmons et al. (2008; matrix download available from MorphoBank [O'Leary and Kaufman, 2007]) and draws on the work of Simmons (1994,

1995) and Simmons and Geisler (1998; see also *Chiroptera*). The list of apomorphies follows that of Simmons and Geisler (1998).

1. Fenestra rotunda faces directly posteriorly: The fenestra rotunda (= fenestra cochlea) faces directly posteriorly in some lagomorphs, bats, sirenians, and dermopterans (Novacek, 1986). The primitive condition appears to be a fenestra rotunda that faces posterolaterally (Novacek, 1986). Although this feature was considered an apomorphy of *Volitantia* (*Chiroptera* + *Dermoptera*; Wible and Novacek, 1988; Simmons 1994, 1995), recent systematic work indicates that *Dermoptera* is part of *Euarchontaglires*, while *Apo-Chiroptera* is part of *Laurasiatheria* (Meredith et al., 2011; O'Leary et al., 2013). Given these relationships, the presence of a fenestra rotunda that faces directly posteriorly appears to be derived within *Laurasiatheria* and represents an apomorphy of *Apo-Chiroptera*.

2. Modification of the manus (hand): Digits II-V of forelimb elongate (Simmons et al., 2008: character 117), with complex carpometacarpal and intermetacarpal joints and interdigital patagia (wing membranes). In contrast to all other mammals, the metacarpals and proximal two phalanges on digits II-V of the forelimb of all species of *Apo-Chiroptera* are remarkably elongate, combining to make these digits longer than the forearm. In extant taxa, the elongated fingers support interdigital patagia, which are continuously attached between the digits of the manus. Among mammals, the only other group with patagia continuously attached between the digits is *Dermoptera* (sensu Simmons,

1994), which has much shorter digits. Given the distant relationships between *Chiroptera* and *Dermoptera* (Meredith et al., 2011; O'Leary et al., 2013), this feature appears to have arisen convergently in these groups. Patagia are inferred for most fossil members of this clade based on elongation of digits, and are confirmed by visible soft tissue impressions of patagia in many Eocene fossils from Messel, Germany (e.g., *Archaeonycteris*, *Palaeochiropteryx*, *Hassianycteris*). Although *Onychonycteris*, the most stemward member of this clade, has overall forelimb proportions that fall between those of gliding mammals and all other known *Apo-Chiroptera* species (see Simmons et al., 2008), the digits are longer than the forelimb in this taxon. Simmons and Geisler (1998) included the absence of claws on wing digits III-V with this suite of modifications; however, the presence of claws on all the wing digits of *Onychonycteris* suggests that claws were present primitively in *Apo-Chiroptera*.

3. Manus locked into fixed position, with axis of flexion of the proximal wrist joint at an angle 90 degrees to that of the elbow (Wible and Novacek, 1988). Among quadrupedal mammals, the axis of flexion of the wrist joint lies largely parallel to the axis of flexion of the elbow. In most mammals, the manus is locked into this position, and rotation is prevented by modifications of the wrist and elbow. This condition appears to be primitive for mammals (Wible and Novacek, 1988). In contrast, in *Apo-Chiroptera*, the distal end of the radius is "twisted" from its primitive condition so that the axis of flexion of the radiocarpal facets is perpendicular to

(instead of parallel to) the axis of flexion of the elbow. This rotation of the manus 90 degrees from the primitive condition is also found in *Dermoptera* and was considered an apomorphy of *Volitantia* by Simmons (1994, 1995). However, this feature now appears to have arisen convergently in these groups. Rotation of the manus 90 degrees from the primitive condition should therefore be considered an apomorphy of *Apo-Chiroptera* as it appears to be derived within *Laurasiatheria*.

4. Modification of the elbow: Olecranon process of ulna small and with a greatly reduced humeral articular surface (Simmons et al., 2008: character 110). Humerus lacks an olecranon fossa (Simmons et al., 2008: character 110). These modifications are necessary for forelimb-powered flight, as they allow the forearm to extend in a broad arc, while controlling the rotation at the distal end of the limb (see Simmons, 1995). In extant cetaceans the olecranon process is also reduced and the olecranon fossa is lost; however, this feature seems to have evolved independently in bats and cetaceans (Simmons, 1994). Although Simmons and Geisler (1998) originally included the presence of the ulnar patella as an apomorphy of this clade, this feature appears to be an apomorphy of (crown) *Chiroptera*, as an ulnar patella apparently does not occur in any Eocene stem chiropterans.

5. Absence of supinator ridge on humerus: The supinator ridge provides a wide surface of origin for the supinator muscles of the forearm (Flower and Gadow, 1885). Absence of this ridge seems to have evolved independently in bats, cetaceans, and glires (Simmons, 1994).

6. Modification of the scapula: Scapular spine originates at the posterior edge of the glenoid fossa. Long axis of scapular spine offset 20–30 degrees from axis of rotation of the humeral head. Scapular spine reduced in height—acromion process appears more strongly arched and less well supported than in other mammals. Presence of at least two facets in infraspinous fossa (Simmons et al., 2008: character 94). These features are unique among mammals. Modification of the scapular spine likely reflects the different stresses applied to the scapulohumeral and acromioclavicular joints during flight (Simmons, 1995), while changes in the size and shape of the infraspinous fossa may reflect the increased importance of the subscapularis muscle in producing the downstroke of the wing and the infraspinatus and teres major, which function to elevate the wing and stabilize the humerus during the upstroke (Vaughan, 1959).

7. Modification of hip joint: 90-degree rotation of hindlimbs effected by reorientation of acetabulum and shaft of femur (Simmons et al., 2008: character 168). Neck of femur reduced. Ischium tilted dorsolaterally. Anterior pubes widely flared and pubic spine present (Simmons et al., 2008: character 165; the pubic spine may potentially be absent in *Onychonycteris*). Absence of obturator internus muscle. Modification of the hip joint in *Apo-Chiroptera* is also related to flight, allowing the femur to project laterally rather than ventrally, which results in a streamlined profile for both the posterior plagiopatagium and uropatagium, a broad upstroke, and more efficient flight (Altenbach, 1979; Simmons, 1995).

8. Modification of ankle joint: Reorientation of upper ankle joint facets on calcaneum and astragalus. Trochlea of astragalus convex (Wible and Novacek, 1988), lacks medial and lateral guiding ridges (Wible and Novacek, 1988), tuber of calcaneum projects in plantolateral direction away from ankle and foot. Peroneal process of calcaneum absent (Wible and Novacek, 1988). Sustentacular process of calcaneum reduced, calcaneoastragalar and sustentacular facets on calcaneum and astragalus coalesced. Absence of groove on astragalus for tendon of the flexor digitorum fibularis. Most of the features in this suite of characters are unique to bats among mammals. The reorganization of the ankle permits almost full extension of the foot and may contribute to the ability of these species to roost hanging by their feet for long periods (see Simmons, 1994, 1995).

9. Presence of a calcar and the depressor ossis styliformis (Simmons et al., 2008: character 117): The calcar, a spur of cartilage and/or bone that supports the trailing edge of the uropatagium (tail membrane), is unique among mammals. With the exception of *Craseonycteridae* and *Rhinopomatidae*, a calcar is found in all extant bats (in *Pteropodidae* a cartilaginous spur is present; Schutt and Simmons, 1998). The depressor ossis styliformis, which runs from the calcaneus and fifth metatarsal to insert on the calcar, controls the spread and tautness of the uropatagium by pulling the calcar laterad. This muscle is present in extant species with a calcar, and it is inferred to be present in the stem Eocene fossils for which we have postcranial remains that indicate the presence of a calcar. The calcar

of the type specimen of *Onychonycteris* exhibits an irregular and porous surface texture, which is unlike the smooth surfaces of the bones of the rest of the skeleton (Simmons et al., 2008). From this, as well as the apparent absence of a calcar in the paratype, Simmons et al. (2008) concluded that the calcar was cartilaginous in *Onychonycteris*. The calcar is cartilaginous, rather than bony, in many extant bats (Schutt and Simmons, 1998). The calcar also appears in fossils of *Palaeochiropteryx* and *Hassianycteris*—its absence in *Icaronycteris* and *Archaeonycteris* may be related to preservation. Simmons (1994, 1995) considered the calcar an apomorphy of (crown) *Chiroptera*, but Simmons and Geisler (1998), prior to the discovery of *Onychonycteris*, and given the relationships of the Eocene taxa that appeared to lack a calcar, did not.

Simmons and Geisler (1998: Table 7) listed 33 apomorphies of a clade equivalent to *Apo-Chiroptera*. Many of these are apomorphies of *Chiroptera* (see entry in this volume). However, the presence of posterior laminae on ribs (Simmons et al., 2008: character 81) is an unambiguous synapomorphy of *Icaronycteris* + *Archaeonycteris* [inferred] + *Hassianycteris* + *Palaeochiropteryx* + *Chiroptera*. The presence of a deciduous dentition with hook-like anterior teeth (Simmons et al., 2008: character 1) currently appears to be a synapomorphy of *Palaeochiropteryx* + *Chiroptera*, but these deciduous teeth are unknown due to preservation in the other Eocene fossil taxa. Hook-like deciduous teeth are unique among mammals (Simmons, 1994), and this may represent another apomorphy of *Apo-Chiroptera*. Although the termination of the sacrum posterior to midpoint of acetabulum (Simmons et al., 2008: character 161) appears to be a synapomorphy for the clade *Palaeochiropteryx* + *Chiroptera,* this feature has an equivocal reconstruction for other fossil taxa, two of which lack data. Outgroup relationships also impact the interpretation of this character on the tree.

Other cranial and postcranial characters that Simmons and Geisler considered apomorphies have equivocal reconstructions on the Simmons et al. (2008) tree or are so widespread among mammals that their interpretation changes depending on the phylogeny used for reconstruction: palatal process of premaxilla reduced (Simmons et al., 2008: character 14; see Giannini and Simmons, 2007); postpalatine torus absent; jugal reduced, and jugolacrimal contact lost (Simmons et al., 2008: character 19), and presence of a baculum (Simmons et al., 2008: 175). Finally, two of these characters appear to be primitive retentions for *Apo-Chiroptera* based on reconstructions on the Simmons et al. (2008) tree (and see discussion in Simmons, 1994): enlarged fenestra rotunda (Simmons et al., 2008: character 26) and absence of the entepicondylar foramen (Simmons et al., 2008: character 109).

Given the developing consensus that dermopterans *(in Euarchontaglires)* and bats *(in Laurasiatheria)* are only distantly related (see Meredith et al. 2011; O'Leary et al., 2013), we wished to evaluate features that had been previously suggested as apomorphies of *Volitantia* (see Simmons 1994, 1995) in light of these new relationships. Of the 17 characters discussed by Simmons (1994, 1995), we found that many can be interpreted as apomorphies of either *Chiroptera* (see entry in this volume) or *Apo-Chiroptera* (see above). However, several features – low to absent spines on the cervical vertebrae, the flattened condition of the ribs near their vertebral ends, displacement of the insertions of the pectoral and deltoid muscles, the fusion of the distal radius and ulna, the transverse widening

of the distal radius, the presence of deep grooves for carpal extensors on the distal radius, presence of a scaphocentrolunate, the presence of ungual phalanges that are proximally and distally deep and mediolaterally compressed, and elongation of the fourth and fifth pedal rays (Wible and Novacek, 1988; Szalay and Lucas, 1993; see Simmons 1994)—are difficult to assess as putative apomorphies. These features appear to have been principally compared across archontan taxa when they were designated as apomorphies of *Volitantia.* Given our current understanding of mammalian relationships, these features need to be more broadly examined, especially across *Laurasiatheria,* before we can determine if they represent apomorphies of *Apo-Chiroptera* or *Chiroptera.* Presence of the humeropatagialis muscle does not appear to be apomorphic for *Chiroptera* as most yinpterochiropterans, and many yangochiropterans, lack this muscle.

Synonyms: Nearly all authors who have discussed a clade comprising extant and extinct bats have applied the name *Chiroptera* to that group. *Ptética* Ameghino 1889 is the only additional synonym for this clade that we are aware of. Major references that use *Chiroptera* to refer to the clade here named *Apo-Chiroptera* include Gregory (1910), Simpson (1945), McKenna and Bell (1997), Simmons and Geisler (1998), Gunnell and Simmons (2005), and Simmons et al. (2008). These uses are approximate. However, note that the internal structure of the clade often differs dramatically in some of these works. For instance, although McKenna and Bell (1997) included all fossil bats in *Chiroptera,* the Eocene fossil taxa *Archaeonycteridae, Palaeochiropterygidae,* and *Hassianycteridae* were placed within *Microchiroptera,* reflecting then-current phylogenetic hypotheses.

Comments: No definitive pre-Eocene stem or crown chiropterans are yet known, although some potential Palaeocene candidates have been mentioned (Gingerich, 1987; Hooker, 1996). Among the Eocene fossil taxa, only *Onychonycteris, Icaronycteris, Archaeonycteris, Palaeochiropteryx,* and *Hassianycteris* are known from relatively complete skeletons, and consequently the relationships of these taxa are better understood. With the exception of one fossil that may fall within the crown clade (*Tachypteron,* a putative emballonurid; Storch et al., 2002), all other Eocene taxa are principally known only by teeth and dentitions, so their definitive taxonomic placement remains to be determined (see *Chiroptera* in this volume for fossils referred to the crown clade).

The relationships of *Archaeonycteris, Hassianycteris, Icaronycteris,* and *Palaeochiropteryx* have been problematic (see Simmons and Geisler, 1998). Morphological studies that have included these Eocene taxa have shown them as either successive outgroups to *Microchiroptera* (Simmons and Geisler, 1998), or with *Icaronycteris* and *Onychonycteris* (= "New Green River Bat") as successive outgroups to the crown clade, which includes the other Eocene fossil bats (Gunnell and Simmons, 2005; see also analysis 3b of Hermsen and Hendricks, 2008). The most recent results using a molecular scaffold and the full morphological data set from Simmons and Geisler (1998; e.g., Teeling et al., 2005), or an updated version of this data set (Simmons et al., 2008), indicate that the Eocene fossils *Onychonycteris, Icaronycteris, Archaeoncyteris, Hassianycteris,* and *Palaeochiropteryx* all lie outside the bat crown clade (but see Springer et al., 2001; Hermsen and Hendricks, 2008). We prefer to use the Simmons et al. (2008) topology as it has associated bootstrap and decay values for the morphological data set.

For a discussion concerning the restriction of *Chiroptera* to the crown clade, see the *Chiroptera* entry in this volume. We have chosen to use the name *Apo-Chiroptera* for the clade including *Chiroptera* and all stem fossils possessing the unique apomorphies related to flight; at present, we refrain from naming the total clade because all currently known stem fossils possess these apomorphies. However, we reserve *Pan-Chiroptera* for the clade that includes *Apo-Chiroptera* and all stem chiropterans that lack these unique limb apomorphies (i.e., the total clade); we anticipate being able to apply this name in the future, once earlier-diverging non-flying stem members of this clade are discovered.

Literature Cited

Ameghino, F. 1889. *Contribucion al Conocimiento de los Mamiferos Fosiles de la República Argentina.* P. E. Coni é Hijos, Buenos Aires.

Altenbach, J. S. 1979. Locomotor morphology of the vampire bat, *Desmodus rotundus. Spec. Publs. Am. Soc. Mammal.* 6:1–137.

Flower, W. H, and H. Gadow. 1885. *An Introduction to the Osteology* of the *Mammalia.* MacMillan, London.

Giannini, N., and N. B. Simmons. 2007. The chiropteran premaxilla: a reanalysis of morphological variation and its phylogenetic interpretation. *Am. Mus. Novit.* 3585:1–44.

Gingerich, P. D. 1987. Early Eocene bats (*Mammalia, Chiroptera*) and other vertebrates in freshwater limestones of the Willwood Formation, Clarks Fork Basin, Wyoming. *Contrib. Mus. Paleontol. Univ. Mich.* 27:275–320.

Gregory, W. K. 1910. The orders of mammals. *Bull. Am. Mus. Nat. Hist.* 27:1–524.

Gunnell, G. F., and N. B. Simmons. 2005. Fossil evidence and the origin of bats. *J. Mammal. Evol.* 12:209–246.

Hermsen, E. J., and J. R. Hendricks. 2008. W(h)ither fossils? Studying morphological character evolution in the age of molecular sequences. *Ann. Mo. Bot. Gard.* 95:72–100.

Hill, J. E., and J. D. Smith. 1984. *Bats: A Natural History.* University Texas Press, Austin.

Hooker, J. J. 1996. A primitive emballonurid bat (*Chiroptera, Mammalia*) from the earliest Eocene of England. *Palaeovertebrata* 25:287–300.

Linnaeus, C. 1758. *Systema Naturae Per Regna Tria Naturae, Secundum Classes, Ordines, Genera, Species, cum Characteribus, Differentiis, Synonymis, Locis.* Editio decima, reformata. Laurentii Salvii, Holmiae (Stockholm).

McKenna, M. C., and S. K. Bell. 1997. *Classification of Mammals above the Species Level.* Columbia University Press, New York.

Meredith, R. W., J. E. Janečka, J. Gatesy, O. A. Ryder, C. A. Fisher, E. C. Teeling, A. Goodbla, E. Eizirik, T. L. L. Simão, T. Stadler, D. L. Rabosky, R. L. Honeycutt, J. J. Flynn, C. M. Ingram, C. Steiner, T. L. Williams, T. J. Robinson, A. Burk-Herrick, M. Westerman, N. A. Ayoub, M. S. Springer, and W. J. Murphy. 2011. Impact of the Cretaceous Tertiary Revolution and KPg Extinction on mammal diversification. *Science* 344(6055):521–524.

Novacek, M. J. 1986. The skull of leptictid insectivorans and the higher-level classification of eutherian mammals. *Bull. Am. Mus. Nat. Hist.* 183:1–112.

O'Leary, M. A., J. I. Bloch, J. J. Flynn, T. J. Gaudin, A. Giallombardo, N. P. Giannini, S. L. Goldberg, B. P. Kraatz, Z.-X. Luo, J. Meng, X. Ni, M. J. Novacek, F. A. Perini, Z. S. Randall, G. W. Rougier, E. J. Sargis, M. T. Silcox, N. B. Simmons, M. Spaulding, P. M. Velazco, M. Weksler, J. R. Wible, and A. L. Cirranello. 2013. The placental mammal ancestor and the post K-Pg radiation of placentals. *Science* 339:662–667.

O'Leary, M. A., and S. G. Kaufman. 2007. MorphoBank 2.5: Web application for morphological phylogenetics and taxonomy. Available at http://www.morphobank.org.

Schutt, W. A., and N. B. Simmons. 1998. Morphology and homology of the chiropteran calcar, with comments on the phylogenetic relationships of *Archaeopteropus. J. Mamm. Evol.* 5:1–32.

Simmons, N. B. 1994. The case for chiropteran monophyly. *Am. Mus. Novit.* 3103:1–54.

Simmons, N. B. 1995. Bat relationships and the origin of flight. Pp. 27–43 in *Ecology, Evolution and Behavior of Bats* (P. A. Racey and S. M. Swift, eds.). Symposia of the Zoological Society of London 67. Oxford University Press, London.

Simmons, N. B., and J. H. Geisler. 1998. Phylogenetic relationships of *Icaronycteris, Archaeonycteris, Hassianycteris,* and *Palaeochiropteryx* to extant bat lineages, with comments on the evolution of echolocation and foraging strategies in *Microchiroptera. Bull. Am. Mus. Nat. Hist.* 235:1–182.

Simmons, N. B, K. L. Seymour, J. Habersetzer, and G. F. Gunnell. 2008. Primitive Early Eocene bat from Wyoming and the evolution of flight and echolocation. *Nature* 451:818–821.

Simpson, G. G. 1945. The principles of classification and a classification of mammals. *Bull. Am. Mus. Nat. Hist.* 85:i–xvi, 1–350.

Smith, T., J. Habersetzer, N. B. Simmons, and G. F. Gunnell. 2012. Systematics and paleobiogeography of early bats. Pp. 23–66 in *Evolutionary History of Bats: Fossils, Molecules, and Morphology* (G. F. Gunnell and N. B. Simmons, eds.). Cambridge University Press, New York.

Springer, M. S., E. C. Teeling, O. Madsen, M. J. Stanhope, and W. W. de Jong. 2001. Integrated fossil and molecular data reconstruct bat echolocation. *Proc. Natl. Acad. Sci. USA* 98:6241–6246.

Storch, G., B. Sigé, and J. Habersetzer. 2002. *Tachypteron franzeni* n. gen., n. sp., earliest emballonurid bat from the Middle Eocene of Messel (*Mammalia, Chiroptera*). *Palaontol. Z.* 76(2):189–199.

Szalay, F. S., and S. G. Lucas. 1993. Cranioskeletal morphology of archontans, and diagnoses of *Chiroptera, Volitantia,* and *Archonta.* Pp. 187–226 in *Primates and Their Relatives in Phylogenetic Perspective* (R. D. E. MacPhee, ed.). Plenum, New York.

Teeling, E. C., M. S. Springer, O. Madsen, P. Bates, S. J. O'Brien, and W. J. Murphy. 2005. A molecular phylogeny for bats illuminates biogeography and the fossil record. *Science* 307:580–584.

Vaughan, T. A. 1959. Functional morphology of three bats: *Eumops, Myotis, Macrotus. Univ. Kans. Publ. Mus. Nat. Hist.* 12:1–153.

Wible, J., and M. J. Novacek. 1988. Cranial evidence for the monophyletic origin of bats. *Am. Mus. Novit.* 2911:1–19.

Authors

Andrea L. Cirranello; Research Associate; Department of Mammalogy; Division of Vertebrate Zoology, American Museum of Natural History; 72nd Street at Central Park West; New York, NY 10024; USA. Email: andreacirranello@gmail.com.

Nancy B. Simmons; Curator-in-Charge; Department of Mammalogy; Division of Vertebrate Zoology; American Museum of Natural History; 72nd Street at Central Park West; New York, NY 10024; USA. E-mail: simmons@amnh.org.

Gregg F. Gunnell[†]; Division of Fossil Primates; Duke University Lemur Center; 1013 Broad Street; Durham, NC 27705, USA. Email: gregg.gunnell@duke.edu.

Date Accepted: 17 February 2016

Primary Editors: Jacques A. Gauthier, Kevin de Queiroz

[†] Deceased

Chiroptera J. F. Blumenbach 1779 [A. L. Cirranello, N. B. Simmons, and G. F. Gunnell], converted clade name

Registration Number: 238

Definition: The crown clade originating in the most recent common ancestor of *Pteropus vampyrus* (Linnaeus 1758) (*Pteropodidae*), *Rhinolophus ferrumequinum* (Schreber 1774) (*Rhinolophoidea*), and *Vespertilio murinus* Linnaeus 1758 (*Vespertilionoidea*). This is a minimum-crown-clade definition. Abbreviated definition: min crown ∇ (*Pteropus vampyrus* (Linnaeus 1758) & *Rhinolophus ferrumequinum* (Schreber 1774) & *Vespertilio murinus* Linnaeus 1758).

Etymology: Derived from the Greek *kheir* (hand) and *pterón* (wing).

Reference Phylogeny: Figure 4 of Simmons et al. (2008) is selected as the primary reference phylogeny, because it includes *Onychonycteris*, an important stem bat, incorporates both morphological and molecular data (as a molecular scaffold), and illustrates relationships among all major clades of extant bats and several Eocene fossil stem taxa.

Composition: *Chiroptera* is composed of two primary subclades: *Yinpterochiroptera* and *Yangochiroptera* (see entries in this volume). *Chiroptera* currently comprises more than 1,300 extant species (Tsang et al., 2016). More than 65 extant taxa traditionally ranked as genera are represented by fossils, and more than 70 extinct taxa traditionally ranked as genera are also known (see Gunnell and Simmons, 2005; Smith et al., 2012 for reviews). Three extinct taxa that originated in the Eocene, *Philisidae* (*Dizzya exsultans*, *Philisis sphingis*, *P. sevketi*; *Witwatia*

eremicus, *W. schlosseri*, *W. sigei*), *Mixopterygidae* (*Carcinipteryx trassounius*, *C. maximinensis*, *C. liaudae*; *Mixopteryx dubia M. perrierensis*, *M. weithoferi*), and *Tanzanycteridae* (*Tanzanycteris mannardi*), may nest within *Chiroptera* (Smith et al., 2012). Early and Middle Eocene taxa that appear to belong to extant clades within *Chiroptera* include the potential emballonurids *Vespertiliavus* (*V. lapradensis*, *V. gracilis*, *V. schlosseri*, *V. disjunctus*, *V. lizierensis*) and *Tachypteron franzeni*, the hipposiderid *Palaeophyllophora* (*P. quercyi*) and *Hipposideros* (*Pseudorhinolophus*) (*H. p. morloti*, *H. p. salemensis*, *H. p. tenuis*), and the possible molossid *Wallia scalopidens* (Smith et al., 2012). Three additional Eocene taxa of uncertain affiliation may also belong to *Chiroptera*: *Necromantis* (*N. adichaster*, *N. marandati*, *N. gezei*), *Jaegeria cambayensis*, and *Australonycteris clarkae* (Smith et al., 2012). With the exception of *Tachypteron*, these Eocene taxa are known only from dentitions and skull fragments, so their definitive taxonomic placement remains to be determined.

Diagnostic Apomorphies: Simmons (1994, 1995) presented a list of chiropteran apomorphies. Simmons and Geisler (1998: Table 7) revised the list based on their findings that four Eocene fossil taxa (*Palaeochiropteryx*, *Hassianycteris*, *Archaeonycteris*, and *Icaronycteris*) nested within the crown clade as successive sister taxa to *Microchiroptera*. However, a more recent hypothesis presented by Simmons et al. (2008), which constrained the morphological data with the molecular tree, indicated that these Eocene fossil taxa and *Onychonycteris* may actually be successive outgroups to the bat crown clade (*Chiroptera*). We therefore present this revised apomorphy list for *Chiroptera* based on the phylogeny of Simmons et al. (2008) and drawing on the work of Simmons (1994, 1995) and Simmons and Geisler (1998; see also *Apo-Chiroptera* in this volume).

The list of apomorphies follows that of Simmons and Geisler (1998). The matrix from Simmons et al. (2008) is available for download from MorphoBank (O'Leary and Kaufman, 2007). An asterisk indicates that the reconstruction is equivocal only for fossil taxa. In most cases, features marked with an asterisk cannot be observed in fossils (e.g., soft tissue morphology). We list these features below because they diagnose *Chiroptera* relative to other crown clades, even if they were present in stem fossils.

1. Posterior deciduous premolars styliform or hooklike, not molariform (Simmons et al., 2008: character 2). Unique among mammals (Simmons, 1994).

2. Small rostral entotympanic element associated with internal carotid artery (Simmons et al., 2008: character 31—part). The original description reads "Two entotympanic elements in floor of middle-ear cavity: a large caudal element and a small rostral element associated with the internal carotid artery" (Simmons, 1994, 1995; Simmons and Geisler, 1998). Two entotympanics (rostral and caudal) occur in all extant bats studied; the distribution of this feature is unknown in Eocene stem bats. Presence of a large caudal entotympanic and a small rostral entotympanic in association with the internal carotid artery is found only in bats and carnivorans, and appears to be absent in the sister taxon of *Carnivora*. This distribution of character states suggests that this feature may have evolved independently in bats and carnivorans (see Simmons, 1994).

3. Tegmen tympani tapers to an elongate process that projects into the middle ear cavity medial to the epitympanic recess and does not form roof over the mallear-incudal articulation or the entire ossicle chain; the anteroventral part of the tegmen tympani remains cartilaginous forming the posterior part of the cartilage of the auditory tube (Wible and Novacek, 1988). This feature cannot be observed in fossil taxa due to preservation; accordingly, we list it here but recognize that it may characterize a more inclusive clade. Unique among mammals (Wible and Novacek, 1988; Simmons, 1994).

4. Proximal stapedial artery enters cranial cavity medial to the tegmen tympani: ramus inferior passes anteriorly dorsal to the tegmen tympani. Unique among mammals (Simmons, 1994).

5. Ramus infraorbitalis of the stapedial artery passes through the cranial cavity dorsal to the alisphenoid. This intracranial passage is found in many bats, all dermopterans, and some eulipotyphlans (Wible and Novacek, 1988). In all other placentals, the ramus infraorbitalis takes an extracranial course, with passage ventral to the alisphenoid apparently being primitive for *Placentalia* (Wible and Novacek, 1988). Although this feature was considered an apomorphy of *Volitantia* (= *Chiroptera* + *Dermoptera*; Wible and Novacek, 1988; Simmons, 1994, 1995), recent systematic work indicates that *Dermoptera* is part of *Euarchontoglires*, while *Chiroptera* is part of *Laurasiatheria*. Given these relationships, an intracranial course for the ramus infrorbitalis appears to be an apomorphy of *Chiroptera* that is convergent in *Dermoptera*, with secondary reversal in some chiropterans.

6. Accessory olfactory bulb absent (Simmons et al., 2008: character 193). The accessory olfactory bulb (AOB) is part of the chemosensory vomeronasal system, which is served by the vomeronasal organ (Meisami and Bhatnagar, 1998). Among mammals, the AOB is absent in many marine mammals (cetaceans, sirenians, and some pinnipeds), catarrhine primates, and most bats. Given the relationships of the clades in question, the absence of the AOB in *Chiroptera* may be an apomorphy of this clade, with some secondary reversals.

7. Ulnar patella present and ossified (Simmons et al., 2008: character 111). An ulnar patella is a large sesamoid distal to the olecranon process and associated with the insertional tendons of the forearm extensor muscles. The ulnar patella appears to be absent in *Onychonycteris*, *Icaronycteris*, *Archaeonycteris*, and *Palaeochiropteryx*. An ulnar patella, which is unique among mammals, is present and ossified only in *Chiroptera*.

8. Occipitopollicalis muscle present in leading edge of propatagium; individual muscle bellies of the occipitopollicalis receive dual innervation of cranial nerve VII and cervical spinal nerves (Simmons et al., 2008: character 128)*. While other gliding mammals (e.g., dermopterans, gliding rodents and marsupials) have a similar arrangement of muscles in the leading edge of the propatagium, the occipital origin for the muscle is restricted to bats, as is the dual innervation of the individual muscle bellies. Originally the description of this apomorphy (see Simmons, 1994) included the presence of the cephalic vein in the leading edge of the propatagium; however, this may be an apomorphy of a larger clade as it is present in *Palaeochiropteryx*, an Eocene stem species (see *Apo-Chiroptera*, this volume; Thewissen and Babcock, 1993). Presence of a cephalic vein in the leading edge is also found in *Petaurus*, a gliding marsupial.

9. Caput breve of the biceps brachii muscle present and one-third the size of the caput longum (Simmons et al., 2008: character 148)*. The biceps brachii is a single muscle (the caput breve is absent) in most members of *Laurasiatheria*: most carnivorans, pinnipeds, perrisodactyls, and artiodactyls (Howell, 1937). Other mammalian taxa that have a single biceps brachii muscle include paenungulates and some primates (Howell, 1937). Among the former members of *Insectivora* and among rodents, some species have two muscle bellies and others do not (Howell, 1937). Distribution of these character states and relationships among these taxa suggests that the presence of a caput breve of the biceps brachii is an apomorphy for *Chiroptera*.

10. Ascending process of ilium does not extend dorsally beyond the iliosacral articulation; iliac fossa relatively small and poorly defined (Simmons et al., 2008: character 163). The ascending process of the ilium extends dorsally beyond the iliosacral articulation in *Icaronycteris*, *Hassianycteris*, *Paleochiropteryx*, and *Hipposideridae + Rhinolophidae* (Simmons and Geisler, 1998). In these taxa the iliac fossa is large and well defined. However, in all other members of *Chiroptera* the ascending process does not extend beyond the iliosacral articulation, and the iliac fossa is small and poorly defined (Simmons and Geisler, 1998). This condition is an

apomorphy of *Chiroptera* with a subsequent reversal in *Hipposideridae + Rhinolophidae*.

11. Sartorius muscle absent (Simmons et al., 2008: character 156)*. Absence of this muscle appears to have evolved independently in bats, macroscelidans, and eulipotyphlans (Simmons, 1994).

12. Vastus muscle complex undifferentiated and present as a single muscle rather than multiple muscles (Simmons et al., 2008: character 155)*. Unique among mammals (Simmons, 1994).

13. Entocuneiform tarsal bone proximodistally shortened, with flat triangular distal facet. Unique among mammals (Simmons, 1994). Although this feature may be present in fossil taxa that lie outside the crown clade, it is impossible to see in a fully articulated skeleton.

14. Elongation of proximal phalanx of digit I of foot (Simmons et al., 2008: character 172)*. The digits of the foot are subequal in length in *Chiroptera* due to the elongation of the proximal phalanx of pedal digit I. However, the proximal phalanx of digit I is not elongate in *Onychonycteris* and *Icaronycteris*. The condition of the proximal phalanx of pedal digit I is unknown in the other three Eocene stem taxa, but this may be an apomorphy of *Chiroptera*.

15. Presence of a tendon locking mechanism (TLM) on the proximal phalanges of the digits of the pes (Simmons et al., 2008: character 174; see Simmons and Quinn, 1994). Among mammals, a TLM has been described in *Dermoptera* and most *Chiroptera* but appears to be absent in other mammals (Simmons and Quinn, 1994). Presence of this feature in bats and dermopterans was previously interpreted as an apomorphy

of *Volitantia* by Simmons (1994, 1995). The appearance of this feature in *Chiroptera* (*Laurasiatheria*) and *Dermoptera* (*Euarchontoglires*) suggests that the TLM evolved independently in their ancestral lineages; accordingly, this feature should be interpreted as an apomorphy of *Chiroptera*, with secondary loss in many noctilionoids and some vespertilionoids.

16. Transverse vulval opening (Simmons et al., 2008: character 178). A transverse vulval opening is derived among placental mammals (Carter and Mess, 2008). Although noctilionoid bats have a vulval opening that is anteroposteriorly elongate, this appears to represent a single reversal (Carter and Mess, 2008), and a transverse vulval opening appears to be a diagnostic apomorphy of *Chiroptera*.

17. Definitive yolk sac gland-like with hypertrophied endodermal cells (Simmons et al., 2008: character 186)*. Although there is considerable variation in yolk sac development among bats, and some optimizations are equivocal (Simmons, 1994; Carter and Mess, 2008), presence of a hypertrophied gland-like definitive yolk sac is unique among mammals and may represent an apomorphy of *Chiroptera*.

18. Cortical somatosensory representation of forelimb reverse of that in other mammals. The map of the somatosensory cortex of the brain, which processes the sense of touch, shows the areas of the brain that receive input from specific areas of the body. In bats, the forelimb in the mapped area of the brain points caudally relative to the trunk. In all other mammals, the forelimb points rostrally. Unique among mammals (Simmons, 1994).

Synonyms: Based on stated compositions (i.e., they included all extant bats then described), the following named groups are approximate synonyms of (crown clade) *Chiroptera: Cheiroptera* Gray 1821; *Vespertilionidae* sensu Gray (1825); *Chiropteriformes* Kinman 1994.

Comments: Chiropteran monophyly is strongly supported by both morphological and molecular data (summarized by Simmons, 1994, 1995; see also Kirsch, 1996; Miyamoto, 1996; Murphy et al., 2001; Arnason et al., 2002; Teeling et al., 2002, 2005; Van Den Bussche et al., 2002; Miller-Butterworth et al., 2007; Meredith et al., 2011; O'Leary et al., 2013). By contrast, relationships of bats to other mammalian clades remain poorly understood. Morphological data support the placement of bats within *Archonta* (with *Dermoptera*, *Primates*, and *Scandentia*; e.g., Wible and Novacek, 1988; Simmons, 1995; Miyamoto, 1996), and some support exists for *Volitantia*, a hypothesized clade composed of *Dermoptera* and *Chiroptera* (e.g., Simmons, 1993; Simmons and Geisler, 1998; Szalay and Lucas, 1993, 1996). In numerous molecular studies based on both nuclear and mitochondrial genes, however, *Chiroptera* never groups with *Primates* and their relatives. Instead *Chiroptera* uniformly falls within *Laurasiatheria*, usually grouping with either the ferungulates (pholidotans, carnivorans, perissodactyls, and artiodactyls—including cetaceans, see *Artiodactyla* this volume), or eulipotyphlans (minimally including soricids and talpids; see e.g., Miyamoto et al., 2000; Murphy et al., 2001; Arnason et al., 2002; Douady et al., 2002; Van Den Bussche et al., 2002; Van Den Bussche and Hoofer, 2004; Meredith et al., 2011; O'Leary et al., 2013). Because a laurasiatherian origin for bats is now widely accepted, we have used this relationship in evaluating potential apomorphies of *Chiroptera* and *Apo-Chiroptera* (this volume).

Within *Chiroptera*, there are substantial disagreements concerning relationships of several higher-level clades. The traditional arrangement of bats into two groups traditionally ranked as suborders (i.e., *Megachiroptera* and *Microchiroptera*) has prevailed in classification for more than 150 years (e.g., Gray, 1821; Gill, 1872; Dobson, 1875; Miller, 1907; Simpson, 1945; McKenna and Bell, 1997) and precedes the discovery of echolocation. Phylogenetic studies based on morphological data support this view (e.g., Simmons, 1998; Simmons and Geisler, 1998; Gunnell and Simmons, 2005). Molecular data, however, strongly support a conflicting hypothesis, identifying two groups of bats: *Yinpterochiroptera* (this volume) comprises the non-echolocating *Pteropodidae* (formerly *Megachiroptera*) plus the echolocating *Rhinolophoidea* (sensu Teeling et al., 2003; formerly in *Microchiroptera*), while *Yangochiroptera* (this volume) includes the remaining echolocating taxa (formerly in *Microchiroptera*) (Hutcheon et al., 1998; Teeling et al., 2000, 2002, 2005; Hulva and Horacek, 2002; Van Den Bussche and Hoofer, 2004; Van Den Bussche et al., 2002, 2003; Miller-Butterworth et al., 2007; Meredith et al., 2011). However, using a supermatrix approach to analyze both the morphological and molecular data sets, Hermsen and Hendricks (2008) found "secondary support" in the molecular data set for a reciprocally monophyletic *Megachiroptera* and *Microchiroptera*. O'Leary et al. (2013), in a study that used both morphological and molecular data, also found support for *Megachiroptera* and *Microchiroptera*, suggesting that some work remains before we arrive at a consensus view of relationships within *Chiroptera*.

Although the name *Chiroptera* has been widely applied to the group including all (i.e., stem and crown) bats, especially in mammalian classifications or literature pertaining to fossil bats (e.g., McKenna and Bell, 1997; Simmons and Geisler, 1998), the vast majority of the biological literature on bats focuses on extant species within the crown clade. Selecting an alternative name for the crown clade would have caused substantial confusion among most biologists. Other names applied to the crown clade have not been widely used (*Cheiroptera* Gray 1821, *Chiropteriformes* Kinman 1994) or have been restricted to clades within *Chiroptera* (*Vespertilionidae* sensu Gray, 1825). We have therefore chosen to restrict the reference of *Chiroptera* to the crown clade and define the name in such a way that, even if support is found for *Megachiroptera* and *Microchiroptera* rather than *Yinpterochiroptera* and *Yangochiroptera*, this name will still be applied to a clade of identical composition. Given uncertainty regarding placement of several Eocene fossils (see *Apo-Chiroptera* in this volume), some additional fossil taxa may be placed within *Chiroptera* in the future.

Literature Cited

Arnason, U., J. A. Adegoke, K. Bodin, E. W. Born, Y. B. Esa, A. Gullberg, M. Nilsson, R. V. Short, X. Xu, and A. Janke. 2002. Mammalian mitogenomic relationships and the root of the eutherian tree. *Proc. Natl. Acad. Sci. USA* 99:8151–8156.

Blumenbach, J. F. 1779. *Handbuch der Naturgeschichte. Mit Kupfern.* [Erster Theil], pp. 1–13, 1–448, Tabs. I–II. J. C. Dieterich, Göttingen.

Carter, A. M., and A. Mess. 2008. Evolution of the placenta and associated reproductive characters in bats. *J. Exp. Zool. (Mol. Dev. Evol.)* 310B:428–449.

Dobson G. E. 1875. Conspectus of the suborder, families, and genera of *Chiroptera* arranged according to their natural affinities. *Ann. Mag. Nat. Hist., Ser. 4* 16:345–357.

Douady, C. J., P. I. Chatelier, O. Madsen, W. W. de Jong, F. Catzeflis, M. S. Springer, and M. J. Stanhope 2002. Molecular phylogenetic

evidence confirming the *Eulipotyphla* concept and in support of hedgehogs as the sister group to shrews. *Mol. Phylogenet. Evol.* 25:200–209.

Gill, T. 1872. *Arrangement of the Families of Mammals.* Smithsonian Institution, Washington, DC.

Gray, J. E. 1821. On the natural arrangement of the vertebrose animals. *London Med. Repos.* 15:296–310.

Gray, J. E. 1825. An outline of an attempt at the disposition of *Mammalia* into tribes and families, with a list of the genera apparently appertaining to each tribe. *Ann. Philos.* 10:337–344.

Gunnell, G. F., and N. B. Simmons. 2005. Fossil evidence and the origin of bats. *J. Mamm. Evol.* 12:209–246.

Hermsen, E. J., and J. R. W. Hendricks. 2008. W(h)ither fossils? Studying morphological character evolution in the age of molecular sequences. *Ann. Mo. Bot. Gard.* 95:72–100.

Howell, A. B. 1937. Morphogenesis of the shoulder architecture. Part VI. Therian *Mammalia.* *Q. Rev. Biol.* 12:440–463.

Hulva, P., and I. Horacek. 2002. *Craseonycteris thonglongyai* (*Chiroptera: Craseonycteridae*) is a rhinolophoid: molecular evidence from cytochrome *b*. *Acta Chiropterol.* 4:107–120.

Hutcheon, J. M., J. A. W. Kirsch, and J. D. Pettigrew. 1998. Base compositional biases and the bat problem. III. The question of microchiropteran monophyly. *Philos. Trans. R. Soc. Lond. B Biol. Sci.* 353:607–617.

Kinman, K. E. 1994. *The Kinman System: Towards a Stable Cladisto-Eclectic Classification of Organisms (Living and Extinct; 48 phyla, 269 classes, 1,719 orders)*, pp. ii + 88pp. Kinman, Hays, KS.

Kirsch, J. A. W. 1996. Bats are monophyletic; megabats are monophyletic; but are microbats also? *Bat Res. News* 36:78.

Linnaeus, C. 1758. *Systema Naturae Per Regna Tria Naturae, Secundum Classes, Ordines, Genera, Species, cum Characteribus, Differentiis, Synonymis, Locis.* Editio decima, reformata. Laurentii Salvii, Holmiae (Stockholm).

McKenna, M. C., and S. K. Bell. 1997. *Classification of Mammals above the Species Level.* Columbia University Press, New York.

Meisami, E., and K. P. Bhatnagar. 1998. Structure and diversity in mammalian accessory olfactory bulb. *Micros. Res. Tech.* 43:476–499.

Meredith, R. W., J. E. Janečka, J. Gatesy, O. A. Ryder, C. A. Fisher, E. C. Teeling, A. Goodbla, E. Eizirik, T. L. L. Simão, T. Stadler, D. L. Rabosky, R. L. Honeycutt, J. J. Flynn, C. M. Ingram, C. Steiner, T. L. Williams, T. J. Robinson, A. Burk-Herrick, M. Westerman, N. A. Ayoub, M. S. Springer, and W. J. Murphy. 2011. Impact of the Cretaceous Terrestrial Revolution and KPg extinction on mammal diversification. *Science* 344(6055): 521–524.

Miller, G. S. 1907. The families and genera of bats. *Bull. U.S. Natl. Mus.* 57:1–282.

Miller-Butterworth, C. M., W. J. Murphy, S. J. O'Brien, D. S. Jacobs, M. S. Springer, and E. C. Teeling. 2007. A family matter: conclusive resolution of the taxonomic position of the long-fingered bats, *Miniopterus. Mol. Biol. Evol.* 24:1553–1561.

Miyamoto, M. M. 1996. A congruence study of molecular and morphological data for eutherian mammals. *Mol. Phylogenet. Evol.* 6:373–390.

Miyamoto, M. M., C. Porter, and M. Goodman. 2000. cMyc gene sequences and the phylogeny of bats and other eutherian mammals. *Syst. Biol.* 49:501–514.

Murphy, W. J., E. Eizirik, W. E. Johnson, Y. P. Zhang, O. A. Ryder, and S. J. O'Brien. 2001. Molecular phylogenetics and the origin of placental mammals. *Nature* 409:614–618.

O'Leary, M. A., J. I. Bloch, J. J. Flynn, T. J. Gaudin, A. Giallombardo, N. P. Giannini, S. L. Goldberg, B. P. Kraatz, Z.-X. Luo, J. Meng, X. Ni, M. J. Novacek, F. A. Perini, Z. S. Randall, G. W. Rougier, E. J. Sargis, M. T. Silcox, N. B. Simmons, M. Spaulding, P. M. Velazco, M. Weksler, J. R. Wible, and A. L. Cirranello. 2013. The placental mammal ancestor and the post K-Pg radiation of placentals. *Science* 339:662–667.

O'Leary, M. A., and S. G. Kaufman. 2007. MorphoBank 2.5: Web application for morphological phylogenetics and taxonomy. Available at http://www.morphobank.org.

Schreber, J. C. D. 1774. *Die Säugthiere in Abbildungen nach der Natur mit Beschreibungen 1776–1778*. Wolfgang Walther, Erlangen.

Simmons, N. B. 1993. The importance of methods: archontan phylogeny and cladistic analysis of morphological data. Pp. 1–61 in *Primates and Their Relatives in Phylogenetic Perspective* (R. D. E. MacPhee, ed.). Advances in Primatology Series, Plenum Publ. Co., New York.

Simmons, N. B. 1994. The case for chiropteran monophyly. *Am. Mus. Novit.* 3103:1–54.

Simmons, N. B. 1995. Bat relationships and the origin of flight. Pp. 27–43 in *Ecology, Evolution and Behavior of Bats* (P. A. Racey and S. M. Swift, eds.). Symposia of the Zoological Society of London 67. Oxford University Press, London.

Simmons, N. B. 1998. A reappraisal of interfamilial relationships of bats. Pp. 3–26 in *Bat Biology and Conservation* (T. H. Kunz and P. A. Racey, eds.). Smithsonian Institution Press, Washington, DC.

Simmons, N. B., and J. H. Geisler. 1998. Phylogenetic relationships of *Icaronycteris, Archaeonycteris, Hassianycteris*, and *Palaeochiropteryx* to extant bat lineages, with comments on the evolution of echolocation and foraging strategies in *Microchiroptera. Bull. Am. Mus. Nat. Hist.* 235:1–182.

Simmons, N. B., and T. H. Quinn. 1994. Evolution of the digital tendon locking mechanism in bats and dermopterans: a phylogenetic perspective. *J. Mamm. Evol.* 2:231–254.

Simmons, N. B., K. L. Seymour, J. Habersetzer, and G. F. Gunnell. 2008. Primitive Early Eocene bat from Wyoming and the evolution of flight and echolocation. *Nature* 451:818–821.

Simpson, G. G. 1945. The principles of classification and a classification of mammals. *Bull. Am. Mus. Nat. Hist.* 85:i–xvi, 1–350.

Smith, T., J. Habersetzer, N. B. Simmons, and G. F. Gunnell. 2012. Systematics and paleobiogeography of early bats. Pp. 23–66 in *Evolutionary History of Bats: Fossils, Molecules, and Morphology* (G. F. Gunnell and N. B. Simmons, eds.). Cambridge University Press, New York.

Szalay, F. S., and S. G. Lucas. 1993. Cranioskeletal morphology of archontans, and diagnoses of *Chiroptera, Volitantia, and Archonta.* Pp. 187–226 in *Primates and their Relatives in Phylogenetic Perspective* (R. D. E. MacPhee, ed.). Plenum, New York.

Szalay, F. S., and S. G. Lucas. 1996. The postcranial morphology of Paleocene *Chriacus* and *Mixodectes* and the phylogenetic relationships of archontan mammals. *Bull. New Mexico Mus. Nat. Hist. Sci.* 7:1–47.

Teeling, E. C., M. Scully, D. J. Kao, M. L. Romagnoli, M. S. Springer, and M. J. Stanhope. 2000. Molecular evidence regarding the origin of echolocation and flight in bats. *Nature* 403:188–192.

Teeling, E. C., O. Madsen, R. A. Van Den Bussche, W. W. de Jong, M. J. Stanhope, and M. S. Springer. 2002. Microbat paraphyly and the convergent evolution of a key innovation in Old World rhinolophid microbats. *Proc. Natl. Acad. Sci. USA* 99:1431–1436.

Teeling, E. C., M. S. Springer, O. Madsen, P. Bates, S. J. O'Brien, and W. J. Murphy. 2005. A molecular phylogeny for bats illuminates biogeography and the fossil record. *Science* 307:580–584.

Teeling, E. C., O. Madsen, W. J. Murphy, M. S. Springer, and S. J. O'Brien. 2003. Nuclear gene sequences confirm an ancient link between New Zealand's short-tailed bat and South American noctilionoid bats. *Mol. Phylogenet. Evol.* 28:308–319.

Thewissen, J. G. M., and S. K. Babcock. 1993. Propatagial muscles in archontan systematics. Pp. 91–109 in *Primates and Their Relatives in Phylogenetic Perspective* (R. D. E. MacPhee, ed.). Plenum, New York.

Tsang, S. M., A. L. Cirranello, P. J. Bates, and N. B. Simmons. 2016. The roles of taxonomy and systematics in bat conservation. Pp. 503–538 in *Bats in the Anthropocene: Conservation of Bats in a Changing World* (C. C. Voight and T. Kingston, eds.). Springer International Publishing.

Van Den Bussche, R. A., and S. R. Hoofer. 2004. Phylogenetic relationships among recent chiropteran families and the importance of choosing appropriate out-group taxa. *J. Mammal.* 85:321–330.

Van Den Bussche, R. A., S. R. Hoofer, and E. W. Hansen. 2002. Characterization and phylogenetic utility of the mammalian protamine P1 gene. *Mol. Phylogenet. Evol.* 22:333–341.

Van Den Bussche, R. A., S. A. Reeder, E. W. Hansen, and S. R. Hoofer. 2003. Utility of the dentin matrix protein 1 (DMP1) gene for resolving mammalian intraordinal relationships. *Mol. Phylogenet. Evol.* 26:89–101.

Wible, J. R., and M. J. Novacek. 1988. Cranial evidence for the monophyletic origin of bats. *Am. Mus. Novit.* 2911:1–19.

Authors

Andrea L. Cirranello; Department of Mammalogy; Division of Vertebrate Zoology; American Museum of Natural History; 72nd Street at Central Park West; New York, NY 10024, USA. Email: acirranello@amnh.org.

Nancy B. Simmons; Division of Vertebrate Zoology; American Museum of Natural History; 72nd Street at Central Park West; New York, NY 10024, USA. E-mail: simmons@amnh.org.

Gregg F. Gunnell[†]; Division of Fossil Primates; Duke University Lemur Center; 1013 Broad Street; Durham, NC 27705, USA. Email: gregg.gunnell@duke.edu.

Date Accepted: 17 February 2016

Primary Editors: Jacques A. Gauthier, Kevin de Queiroz

[†] Deceased

Yinpterochiroptera M. S. Springer et al. 2001 [A. L. Cirranello, N. B. Simmons, and G. F. Gunnell], converted clade name

Registration Number: 240

Definition: The crown clade originating in the most recent common ancestor of *Pteropus vampyrus* (Linnaeus 1758) (*Pteropodidae*) and all extant species that share a more recent common ancestor with *Pteropus vampyrus* Linnaeus 1758 (*Pteropodidae*) and *Rhinolophus ferrumequinum* (Schreber 1774) (*Rhinolophoidea*) than with *Vespertilio murinus* Linnaeus 1758 (*Vespertilionoidea*). This is a maximum-crown-clade definition with two internal specifiers. Abbreviated definition: max crown ∇ (*Pteropus vampyrus* (Linnaeus 1758) & *Rhinolophus ferrumequinum* (Schreber 1774) ~ *Vespertilio murinus* Linnaeus 1758).

Etymology: The name *Yinpterochiroptera* is based in part on the Chinese concept of yin-yang: universal complementary opposites, with yin being an active element in any pair of concepts. This name is a combination of *Yinochiroptera* (coined by Koopman in 1985 for a higher-level clade including several bat families) and *Pteropodidae*, thus, *Yinpterochiroptera*.

Reference Phylogeny: Simmons et al. (2008: Fig. 4) should be treated as the primary reference phylogeny.

Composition: *Yinpterochiroptera* includes *Pteropodidae* and *Rhinolophoidea* (sensu Teeling et al., 2003: *Rhinolophidae*, *Hipposideridae*, *Megadermatidae*, *Rhinopomatidae*, and *Craseonycteridae*). As such, *Yinpterochiroptera* comprises at least 350 extant species (Simmons, 2005) and 8 extinct taxa traditionally ranked as genera (McKenna and Bell, 1997).

Diagnostic Apomorphies: Support for *Yinpterochiroptera* comes almost entirely from molecular data (see Comments); there are few compelling synapomorphies from other systems. Springer et al. (2001) discussed a unique amino acid apomorphy for *Yinpterochiroptera*. Specifically, in the RAG2 gene they found a replacement of a serine residue (found in yangochiropterans and all outgroups) by alanine in the eight yinpterochiropterans they examined. Using G-banding and chromosome painting, Ao et al. (2007) discovered a karyotypic inversion that appears to be an apomorphy for *Yinpterochiroptera*. In *Yangochiroptera* (*Vespertilionidae* and *Molossidae*), human-homologous chromosome segments 4/8 and 13 lie adjacent to each other. By contrast, in *Pteropodidae*, *Rhinolophidae*, and *Hipposideridae*, an inversion has taken place and segment 13 is split into flanking regions (i.e., 13/8/4/13). This appears to be a unique condition among all bats studied, although the majority of bat families have yet to be sampled.

Although *Yinochiroptera* (Koopman, 1985) was originally proposed based on the morphology of the premaxilla, yinpterochiropteran bats do not unequivocally share any derived conditions of the premaxilla. Absence of both flanges of the palatine process of the premaxilla may be an apomorphy of *Yinpterochiroptera* (Simmons et al., 2008: character 14; see Giannini and Simmons, 2007), but this condition might also be primitive for *Chiroptera*—the reconstruction is equivocal. While there are other features that may be apomorphies of *Yinpterochiroptera* (e.g., character 46: insertion of the mylohyoideus muscle onto the basihyal and thyrohyal; character 48: presence of the mandibulohyoideus muscle; character 143: reduction in the origin of serratus anterior muscle to 1 or 2 ribs; and character 162: sacral lamellae absent or small), all these have equivocal reconstructions on the tree of Simmons et al. (2008).

The presence of two incisors in each dentary (Simmons et al., 2008: character 4) is the single unequivocal morphological apomorphy of *Yinpterochiroptera* (three lower incisors on each side is the primitive condition in *Chiroptera*). However, this character is homoplastic; two lower incisors also occur within *Yangochiroptera* in some *Emballonuridae* and in the clades *Mormoopidae* + *Phyllostomidae* and *Molossidae*, but this condition appears to have evolved convergently in these taxa. Still further reduction to only a single lower incisor on each side is seen in all *Noctilionidae* and *Mystacinidae*, and in some *Phyllostomidae* and *Molossidae*. Given this distribution, it is difficult to interpret this character as providing strong support for any particular grouping of bat taxa.

Synonyms: *Pteropodiformes* Hutcheon and Kirsch 2006 (approximate).

Comments: Traditionally, non-echolocating *Pteropodidae* has been placed in the higher taxon *Megachiroptera*. All echolocating bats have traditionally been placed in *Microchiroptera*, including the members of *Rhinolophoidea* (sensu Teeling et al., 2002: *Rhinopomatidae*, *Rhinolophidae*, *Hipposideridae*, *Megadermatidae*, and *Craseonycteridae*; *Nycteridae*, previously included in *Rhinolophoidea* is now placed in *Yangochiroptera*; see Teeling et al., 2002 and *Yangochiroptera* entry in this volume). The traditional arrangement, with one group of echolocating bats and one group of non-echolocators, has prevailed in classification for more than 150 years (e.g., Gray, 1821; Gill, 1872; Dobson, 1875; Miller, 1907; Simpson, 1945; McKenna and Bell, 1997), although the division predates the discovery of echolocation. Phylogenetic studies based

on morphological data support this view (e.g., Simmons, 1998; Simmons and Geisler, 1998; Gunnell and Simmons, 2005). *Yinpterochiroptera* has never been inferred in analyses of morphological data, which instead support *Megachiroptera* and *Microchiroptera* as reciprocally monophyletic groups, with monophyly of *Microchiroptera* moderately well supported as indicated by bootstrap or Bremer support values.

Support for *Yinpterochiroptera* comes from molecular data. Porter et al. (1996) first inferred a clade that included two pteropodids and a rhinolophoid in a sequence analysis of exon 28 of the von Willebrand factor gene, but this clade did not have strong support. Kirsch and Pettigrew (1998) presented data that indicated an association between *Pteropodidae* and *Rhinolophidae* based on DNA hybridization data. However, sampling issues and a problem with AT bias in the DNA of rhinolophoid bats made this novel result somewhat suspect (Hutcheon et al., 1998). Hutcheon et al. (1998) further analyzed DNA hybridization data based on comparisons of representatives of 10 chiropteran taxa traditionally ranked as families, and again found a strongly supported *Pteropodidae* + *Rhinolophoidea* clade. Subsequently, many studies of sequence data have collectively analyzed more than 13 kilobases of DNA sequence from 20 nuclear and 4 mitochondrial genes (Teeling et al., 2000, 2002, 2003, 2005; Springer et al., 2001; Hulva and Horacek, 2002; Van Den Bussche and Hoofer, 2004; Van Den Bussche et al., 2002; Miller-Butterworth et al., 2007; Meredith et al., 2011), and all have supported *Yinpterochiroptera*, often with very high support values (e.g., bootstrap, Bremer values, posterior probabilities). The sole exception, a recent phylogeny using only the cytochrome *b* gene (Agnarsson et al., 2011), weakly rejects *Yinpterochiroptera* when all 648 species are used but supports *Yinpterochiroptera* monophyly under a pruned data set used for a divergence time analysis.

Although this large body of molecular evidence is increasingly compelling, the conflicts between the morphological and molecular data have not been fully explored. Some investigators (e.g., Springer et al., 2001; Simmons et al., 2008) have used a constraint tree consistent with the molecular results (i.e., a molecular scaffold) in order to fit fossil taxa, which lack molecular data, into the tree. Obviously, this type of analysis does not allow the characters within the morphological and molecular data sets to interact. Intriguingly, Hermsen and Hendricks (2008) used a supermatrix approach to analyze both the morphological and molecular data sets simultaneously, and they found "secondary support" in the molecular data set for a reciprocally monophyletic *Megachiroptera* and *Microchiroptera*. O'Leary et al. (2013) also found support for *Megachiroptera* and *Microchiroptera* in an analysis that combined morphological and molecular data, although with a very small sample of bats (8 taxa). Future work should focus on exploring conflicts between and within these data sets to address this issue.

Should future analyses lead to the conclusion that *Microchiroptera* is monophyletic, then the name *Yinpterochiroptera* as defined here would not apply to any clade, as it designates a clade that includes both *Pteropus vampyrus* and *Rhinolophus ferrumequinum* but not *Vespertilio murinus*, and there is no such clade on phylogenies in which *Microchiroptera* is monophyletic. We have chosen to convert the name *Yinpterochiroptera* (Springer et al., 2001) rather than *Pteropodiformes* (Hutcheon and Kirsch, 2006) because the former has been in use for a longer period of time and is more prevalent in the literature.

Literature Cited

Agnarsson, I., C. M. Zambrana-Torrelio, N. P. Flores-Saldana, and L. J. May-Collado. 2011. A time-calibrated species-level phylogeny of bats (*Chiroptera, Mammalia*). *PLOS Curr. Tree Life.* 4 February 2011. 1st edition. doi:10.1371/currents.RRN1212.

Ao, L., X. Mao, W. Nie, X. Gu, Q. Feng, J. Wang, W. Su, Y. Wang, M. Volleth, and F. Yang. 2007. Karyotypic evolution and phylogenetic relationships in the order *Chiroptera* as revealed by G-banding comparison and chromosome painting. *Chromosome Res.* 15:257–267.

Dobson, G. E. 1875. Conspectus of the suborder, families, and genera of *Chiroptera* arranged according to their natural affinities. *Ann. Mag. Nat. Hist., Ser. 4* 16:345–357.

Giannini, N. P., and N. B. Simmons. 2007. The chiropteran premaxilla: a reanalysis of morphological variation and its phylogenetic interpretation. *Am. Mus. Novit.* 3585:1–44.

Gill, T. 1872. *Arrangement of the Families of Mammals.* Smithsonian Institution, Washington, DC.

Gray, J. E. 1821. On the natural arrangement of the vertebrose animals. *London Med. Repos.* 15:296–310.

Gunnell, G. F., and N. B. Simmons. 2005. Fossil evidence and the origin of bats. *J. Mamm. Evol.* 12:209–246.

Hermsen, E. J., J. R. W. Hendricks. 2008. W(h)ither fossils? Studying morphological character evolution in the age of molecular sequences. *Ann. Mo. Bot. Gard.* 95:72–100.

Hulva, P., and I. Horacek. 2002. *Craseonycteris thonglongyai* (*Chiroptera: Craseonycteridae*) is a rhinolophoid: molecular evidence from cytochrome *b. Acta Chiropterol.* 4:107–120.

Hutcheon, J. M., and J. A. W. Kirsch. 2006. A moveable face: deconstructing the *Microchiroptera* and a new classification of extant bats. *Acta Chiropterol.* 8:1–10.

Hutcheon, J. M., J. A. W. Kirsch, and J. D. Pettigrew. 1998. Base compositional biases and the bat problem. III. The question of microchiropteran monophyly. *Philos. Trans. R. Soc. Lond. B Biol. Sci.* 353:607–617.

Kirsch, J. A. W., and J. D. Pettigrew. 1998. Base-compositional biases and the bat problem. II. DNA hybridization trees based on AT- and GC-enriched tracers. *Philos. Trans. R. Soc. Lond. B Biol. Sci.* 353:381–388.

Koopman, K. F. 1985. A synopsis of the families of bats, Part VII. *Bat Res. News* 25:25–27.

Linnaeus, C. 1758. *Systema Naturae Per Regna Tria Naturae, Secundum Classes, Ordines, Genera, Species, cum Characteribus, Differentiis, Synonymis, Locis.* Editio decima, reformata. Laurentii Salvii, Holmiae (Stockholm).

McKenna, M. C., and S. K. Bell. 1997. *Classification of Mammals above the Species Level.* Columbia University Press, New York.

Meredith, R. W., J. E. Janečka, J. Gatesy, O. A. Ryder, C. A. Fisher, E. C. Teeling, A. Goodbla, E. Eizirik, T. L. L. Simão, T. Stadler, D. L. Rabosky, R. L. Honeycutt, J. J. Flynn, C. M. Ingram, C. Steiner, T. L. Williams, T. J. Robinson, A. Burk-Herrick, M. Westerman, N. A Ayoub, M. S. Springer, W. J. Murphy. 2011. Impact of the Cretaceous Terrestrial Revolution and KPg extinction on mammal diversification. *Science* 334(6055):521–524.

Miller, G. S. 1907. The families and genera of bats. *Bull. U.S. Natl. Mus.* 57:1–282.

Miller-Butterworth, C. M., W. J. Murphy, S. J. O'Brien, D. S. Jacobs, M. S. Springer, and E. C. Teeling. 2007. A family matter: conclusive resolution of the taxonomic position of the long-fingered bats, *Miniopterus. Mol. Biol. Evol.* 24:1553–1561.

O'Leary, M. A., J. I. Bloch, J. J. Flynn, T. J. Gaudin, A. Giallombardo, N. P. Giannini, S. L. Goldberg, B. P. Kraatz, Z.-X. Luo, J. Meng, X. Ni, M. J. Novacek, F. A. Perini, Z. S. Randall, G. W. Rougier, E. J. Sargis, M. T. Silcox, N. B. Simmons, M. Spaulding, P. M. Velazco, M. Weksler, J. R. Wible, and A. L. Cirranello. 2013. The placental mammal ancestor and the post K-Pg radiation of placentals. *Science* 339:662–667.

Porter, C. A., M. Goodman, and M. J. Stanhope. 1996. Evidence on mammalian phylogeny from sequences of exon 28 of the von Willebrand factor gene. *Mol. Phylogenet. Evol.* 5:89–101.

Schreber, J. C. D. 1774. *Die Säugthiere in Abbildungen nach der Natur mit Beschreibungen 1776–1778.* Wolfgang Walther, Erlangen.

Simpson, G. G. 1945. The principles of classification and a classification of mammals. *Bull. Am. Mus. Nat. Hist.* 85:i–xvi, 1–350.

Simmons, N. B. 1998. A reappraisal of interfamilial relationships of bats. Pp. 3–26 in *Bat Biology and Conservation* (T. H. Kunz and P. A. Racey, eds.). Smithsonian Institution Press, Washington, DC.

Simmons, N. B. 2005. Order *Chiroptera.* Pp. 312–529 in *Mammal Species of the World: A Taxonomic and Geographic Reference.* 3rd edition, Vol. 1 (D. E. Wilson and D. M. Reeder, eds.). Smithsonian Institution Press, Washington, DC.

Simmons, N. B., and J. H. Geisler. 1998. Phylogenetic relationships of *Icaronycteris, Archaeonycteris, Hassianycteris,* and *Palaeochiropteryx* to extant bat lineages, with comments on the evolution of echolocation and foraging strategies in *Microchiroptera. Bull. Am. Mus. Nat. Hist.* 235:1–182.

Simmons, N. B., K. L. Seymour, J. Habersetzer, and G. F. Gunnell. 2008. Primitive Early Eocene bat from Wyoming and the evolution of flight and echolocation. *Nature* 451:818–821.

Springer, M. S., E. C. Teeling, O. Madsen, M. J. Stanhope and W. W. de Jong. 2001. Integrated fossil and molecular data reconstruct bat echolocation. *Proc. Natl. Acad. Sci. USA* 98:6241–6246.

Teeling, E. C., M. Scully, D. J. Kao, M. L. Romagnoli, M. S. Springer, and M. J. Stanhope. 2000. Molecular evidence regarding the origin of echolocation and flight in bats. *Nature* 403:188–192.

Teeling, E. C., O. Madsen, R. A. Van Den Bussche, W. W. de Jong, M. J. Stanhope, and M. S. Springer. 2002. Microbat paraphyly and the convergent evolution of a key innovation in Old World rhinolophid microbats. *Proc. Natl. Acad. Sci. USA* 99:1431–1436.

Teeling, E. C., O. Madsen, W. J. Murphy, M. S. Springer, and S. J. O'Brien. 2003. Nuclear gene sequences confirm an ancient link between New Zealand's short-tailed bat and South American noctilionoid bats. *Mol. Phylogenet. Evol.* 28:308–319.

Teeling, E. C., M. S. Springer, O. Madsen, P. Bates, S. J. O'Brien, and W. J. Murphy. 2005. A molecular phylogeny for bats illuminates biogeography and the fossil record. *Science* 307:580–584.

Van Den Bussche, R. A., and S. R. Hoofer. 2004. Phylogenetic relationships among recent chiropteran families and the importance of choosing appropriate out-group taxa. *J. Mammal.* 85:321–330.

Van Den Bussche, R. A., Hoofer, S. R., and Hansen, E. W. 2002. Characterization and phylogenetic utility of the mammalian protamine P1 gene. *Mol. Phylogenet. Evol.* 22:333–341.

Authors

Andrea L. Cirranello; Department of Mammalogy; Division of Vertebrate Zoology; American Museum of Natural History; 72nd Street at Central Park West; New York, NY 10024, USA. E-mail: andreacirranello@gmail.com.

Nancy B. Simmons; Division of Vertebrate Zoology; American Museum of Natural History; 72nd Street at Central Park West; New York, NY 10024, USA. Email: simmons@amnh.org.

Gregg F. Gunnell[†]; Division of Fossil Primates; Duke University Lemur Center; 1013 Broad Street; Durham, NC 27705, USA. Email: gregg.gunnell@duke.edu.

Date Accepted: 17 February 2016

Primary Editors: Jacques A. Gauthier, Kevin de Queiroz

[†] Deceased

Yangochiroptera K. F. Koopman 1985 [A. L. Cirranello, N. B. Simmons, and G. F. Gunnell], converted clade name

Registration Number: 239

Definition: The crown clade originating in the most recent common ancestor of *Vespertilio murinus* Linnaeus 1758 (*Vespertilionoidea*) and all extant species that share a more recent common ancestor with *Vespertilio murinus* Linnaeus 1758 than with *Pteropus vampyrus* (Linnaeus 1758) (*Pteropodidae*) and *Rhinolophus ferrumequinum* (Schreber 1774) (*Rhinolophoidea*). This is a maximum-crown-clade definition. Abbreviated definition: max crown ∇ (*Vespertilio murinus* Linnaeus 1758 ~ *Pteropus vampyrus* (Linnaeus 1758) & *Rhinolophus ferrumequinum* (Schreber 1774)).

Etymology: The name *Yangochiroptera* is based on the Chinese concept of yin-yang—universal complementary opposites. This clade originally was paired with *Yinochiroptera* by Koopman (1985). *Yinochiroptera* was named for a clade with a "moveable" premaxilla (a ligamentous attachment to the maxilla), while *Yangochiroptera* was originally named for a clade (*Noctilionoidea* + *Vespertilionoidea*) with a premaxilla fused to the maxilla. Koopman (1985: 27) noted that "The names for the two bat infraorders are particularly apt since in Confucian philosophy, Yin is an active, Yang is a passive element in any pair of concepts."

Reference Phylogeny: Simmons et al. (2008: Fig. 4) should be treated as the primary reference phylogeny.

Composition: *Yangochiroptera* includes *Emballonuroidea* (*Emballonuridae* and *Nycteridae*), *Noctilionoidea* (*Noctilionidae*, *Mormoopidae*, *Phyllostomidae*, *Furipteridae*, *Thyropteridae*, *Mystacinidae*, and *Myzopodidae*), and *Vespertilionoidea* (*Vespertilionidae*, *Miniopteridae*, *Molossidae*, and *Natalidae*). *Yangochiroptera* comprises more than 700 extant species (Simmons, 2005) and 22 extinct taxa traditionally ranked as genera (McKenna and Bell, 1997).

Diagnostic Apomorphies: Support for *Yangochiroptera* (as described above) comes almost entirely from molecular data (see below); there are few compelling apomorphies from other systems. As originally proposed by Koopman (1985; comprising just *Noctilionoidea* + *Vespertilionoidea*), *Yangochiroptera* included only taxa in which the premaxilla is solidly fused to the maxilla. In a review of the morphology of the chiropteran premaxilla, Giannini and Simmons (2007) found that in all *Noctilionoidea* and *Vespertilionoidea* (*Yangochiroptera* sensu Koopman, 1985) the premaxilla is complete—the body, palatine and nasal processes are all present—and the palatine and nasal processes are fused to the maxilla (see Giannini and Simmons, 2007). This contrasts with conditions seen in other bats in which one or more parts of the premaxilla are absent and there is a ligamentous attachment of the processes to the maxilla. However, with the inclusion of *Emballonuroidea* in *Yangochiroptera* (see below), the name *Yangochiroptera* is now applied to a more inclusive clade for which this condition of the premaxilla is no longer diagnostic. Instead, it remains an apomorphy of the less inclusive clade to which the name was originally applied (i.e., *Noctilionoidea* + *Vespertilionoidea*).

There are some unequivocal apomorphies of *Yangochiroptera* (Simmons, 2008); however, none of these provide strong support for this clade. A small to moderately sized fenestra rotunda (Simmons et al., 2008: character 26) appears to have evolved convergently in *Yangochiroptera*, *Hipposideridae* + *Rhinolophidae*, and *Megadermatidae* (the primitive state for *Chiroptera* is an enlarged fenestra rotunda). Within *Yangochiroptera*, there may be as many as three reversals to the primitive state within *Noctilionoidea* (the reconstruction is equivocal). Similarly, the presence of a xiphisternal keel (Simmons et al., 2008: character 87) appears as an unequivocal apomorphy of *Yangochiroptera*. However, this character evolved three times within *Chiroptera* (it also appears in *Rhinolophidae* and *Craseonycteridae* + *Megadermatidae*), and within *Yangochiroptera* there is subsequent loss of the keel in four families (*Myzopodidae*, *Noctilionidae*, *Phyllostomidae*, and *Molossidae*). Finally, presence of a tragus (Simmons et al., 2008; character 205) appears as another unequivocal apomorphy of *Yangochiroptera* on the tree of Simmons et al. (2008); however, this structure occurs in nearly all echolocating bats (i.e., *Microchiroptera*) except *Rhinolophidae* + *Hipposideridae*, which makes this a questionable apomorphy of *Yangochiroptera*. Given their widespread distribution among echolocating bat taxa, it is not surprising that Simmons and Geisler (1998) listed these three features as apomorphies of *Microchiroptera*.

A clavicular origin of the pectoralis profundus muscle (Simmons et al., 2008: character 125) also appears as an apomorphy of *Yangochiroptera*; however, within *Yangochiroptera* there is variation within *Vespertilionidae* (costal-only origin, both costal and clavicular origins, and clavicular-only origins occurring in different taxa), *Phyllostomidae* (costal and clavicular, and clavicular-only origins), and *Thyropteridae*

(a costal and clavicular origin). While the primitive condition for *Chiroptera* is a costal and clavicular origin for the pectoralis profundus, the clavicular origin that characterizes much of *Yangochiroptera* also appears in *Yinpterochiroptera*, in *Megadermatidae* and within *Pteropodidae*.

Synonyms: *Vespertilioniformes* Hutcheon and Kirsch 2006 (approximate).

Comments: In a key to the microchiropteran families, Miller (1907: 79) first used the condition of the premaxillaries to recognize a division within *Microchiroptera*. Koopman (1985) formalized this division taxonomically by naming *Yinochiroptera* for those bats with a premaxilla that is "moveable" (i.e., not fused or sutured) in relation to the maxillaries (*Emballonuridae*, *Rhinopomatidae*, and *Craseonycteridae*, *Rhinolophidae* [including *Hipposiderinae*], *Megadermatidae*, and *Nycteridae*), and *Yangochiroptera* for those bats with a premaxilla fused to the maxilla (*Noctilionidae*, *Mormoopidae*, *Phyllostomidae*, *Vespertilionidae*, *Molossidae*, *Furipteridae*, *Natalidae*, *Thyropteridae*, *Mystacinidae* and *Myzopodidae*). Subsequent morphological work has strongly supported monophyly of *Noctilionoidea* + *Vespertilionoidea* (*Yangochiroptera* sensu Koopman, 1985) (Simmons, 1998; Simmons and Geisler, 1998; Gunnell and Simmons, 2005).

The name *Yangochiroptera* has subsequently been applied to a larger clade (including *Emballonuroidea*: *Emballonuridae* + *Nycteridae*) by molecular researchers following Teeling et al. (2002). Support for this more inclusive *Yangochiroptera*, as here conceptualized and now widely recognized, comes primarily from molecular data. DNA hybridization data based on comparisons of representatives of 10 chiropteran families support *Yangochiroptera* (Hutcheon et al., 1998). Many studies using molecular data have collectively analyzed more than 13 kilobases of DNA sequence from 20 nuclear and 4 mitochondrial genes (Teeling et al., 2000, 2002, 2003, 2005; Springer et al., 2001; Hulva and Horacek, 2002; Van Den Bussche and Hoofer, 2004; Van Den Bussche et al., 2002; Miller-Butterworth et al., 2007; Agnarsson et al., 2011, Meredith et al., 2011), and all have inferred *Yangochiroptera*, often with very high support values (e.g., bootstrap, Bremer values, posterior probabilities). *Yangochiroptera* is also supported by a 15-base-pair (bp) deletion in the BRCA1 gene and a 7-bp deletion in the PLCB4 gene (Teeling et al., 2005). These two deletions are found in all members of *Yangochiroptera*, and are absent in all yinpterochiropteran and outgroup taxa (Teeling et al., 2005). In a simultaneous analysis of morphological and molecular data, Hermsen and Hendricks (2008) also inferred *Yangochiroptera* including *Emballonuroidea*. While these authors did not present support values, the inference of this clade from a combined data analysis is noteworthy.

Given the maximum-crown-clade definition with the internal and external specifiers that we have used here, we expect the application of the name *Yangochiroptera* to remain stable. *Emballonuroidea* (*Emballonuridae* + *Nycteridae*), previously thought to be closely related to *Rhinolophoidea*, is strongly supported as closer to *Vespertilionoidea* and *Noctilionoidea* than to *Rhinolophoidea* (e.g., O'Leary et al., 2013: jacknife value of 100; Meredith et al., 2011: posterior probability of 1.00). We have also constructed the definition so that even if *Microchiroptera* is monophyletic, the name *Yangochiroptera* will still apply to the crown clade of species that share a more recent common ancestor with *Vespertilio murinus* than with *Rhinolophus ferrumequinum* (i.e., the definition will recognize the split between *Yinochiroptera* and *Yangochiroptera* within *Microchiroptera*,

although the compositions of those taxa might be different than originally proposed). We use *Yangochiroptera* (Koopman, 1985) instead of *Vespertilioniformes* (Hutcheon and Kirsch, 2006), because the latter has been less widely used to refer to this clade.

Literature Cited

Agnarsson, I., C. M. Zambrana-Torrelio, N. P. Flores-Saldana, and L. J. May-Collado. 2011. A time-calibrated species-level phylogeny of bats (*Chiroptera, Mammalia*). *PLOS Curr. Tree Life*. 4 February 2011. 1st edition. doi:10.1371/currents.RRN1212.

Giannini, N. P., and N. B. Simmons. 2007. The chiropteran premaxilla: a reanalysis of morphological variation and its phylogenetic interpretation. *Am. Mus. Novit.* 3585:1–44.

Gunnell, G. F., and N. B. Simmons. 2005. Fossil evidence and the origin of bats. *J. Mammal. Evol.* 12:209–246.

Hermsen, E. J., and J. R. W. Hendricks. 2008. W(h)ither fossils? Studying morphological character evolution in the age of molecular sequences. *Ann. Mo. Bot. Gard.* 95:72–100.

Hoofer, S. R., S. Solari, P. A. Larsen, R. D. Bradley, and R. J. Baker 2008. Phylogenetics of the fruit-eating bats (*Phyllostomidae: Artibeina*) inferred from mitochondrial DNA sequences. *Occas. Pap. Mus. Tex. Tech Univ.* 227:1–15.

Hulva, P., and I. Horacek. 2002. *Craseonycteris thonglongyai* (*Chiroptera: Craseonycteridae*) is a rhinolophoid: molecular evidence from cytochrome *b*. *Acta Chiropterol.* 4:107–120.

Hutcheon, J. M., and J. A. W. Kirsch. 2006. A moveable face: deconstructing the *Microchiroptera* and a new classification of extant bats. *Acta Chiropterol.* 8:1–10.

Hutcheon, J. M., J. A. W. Kirsch, and J. D. Pettigrew. 1998. Base compositional biases and the bat problem. III. The question of microchiropteran monophyly. *Philos. Trans. R. Soc. Lond. B Biol. Sci.* 353:607–617.

Koopman, K. F. 1985. A synopsis of the families of bats, Part VII. *Bat Res. News* 25:25–27. [Dated

1984 but published in 1985 (see e.g., Hoofer et al. 2008).]

Linnaeus, C. 1758. *Systema Naturae Per Regna Tria Naturae, Secundum Classes, Ordines, Genera, Species, cum Characteribus, Differentiis, Synonymis, Locis.* Editio decima, reformata. Laurentii Salvii, Holmiae (Stockholm).

McKenna, M. C., and S. K. Bell. 1997. *Classification of Mammals Above the Species* Level. Columbia University Press, New York.

Meredith, R. W., J. E. Janečka, J. Gatesy, O. A. Ryder, C. A. Fisher, E. C. Teeling, A. Goodbla, E. Eizirik, T. L. L. Simão, T. Stadler, D. L. Rabosky, R. L. Honeycutt, J. J. Flynn, C. M. Ingram, C. Steiner, T. L. Williams, T. J. Robinson, A. Burk-Herrick, M. Westerman, N. A. Ayoub, M. S. Springer, W. J. Murphy. 2011. Impact of the Cretaceous Terrestrial Revoluition and KPg extinction on mammal diversification. *Science* 334:521–524.

Miller, G. S. 1907. The families and genera of bats. *Bull. U.S. Natl. Mus.* 57:1–282.

Miller-Butterworth, C. M., W. J. Murphy, S. J. O'Brien, D. S. Jacobs, M. S. Springer, and E. C. Teeling. 2007. A family matter: conclusive resolution of the taxonomic position of the long-fingered bats, *Miniopterus. Mol. Biol. Evol.* 24:1553–1561.

O'Leary, M. A., J. I. Bloch, J. J. Flynn, T. J. Gaudin, A. Giallombardo, N. P. Giannini, S. L. Goldberg, B. P. Kraatz, Z.-X. Luo, J. Meng, X. Ni, M. J. Novacek, F. A. Perini, Z. S. Randall, G. W. Rougier, E. J. Sargis, M. T. Silcox, N. B. Simmons, M. Spaulding, P. M. Velazco, M. Weksler, J. R. Wible, and A. L. Cirranello. 2013. The placental mammal ancestor and the post K-Pg radiation of placentals. *Science* 339:662–667.

Schreber, J. C. D. 1774. *Die Säugthiere in Abbildungen nach der Natur mit Beschreibungen 1776–1778.* Wolfgang Walther, Erlangen.

Simmons, N. B. 1998. A reappraisal of interfamilial relationships of bats. Pp. 3–26 in *Bat Biology and Conservation* (T. H. Kunz and P. A. Racey, eds.). Smithsonian Institution Press, Washington, DC.

Simmons, N. B. 2005. Order *Chiroptera.* Pp. 312–529 in *Mammal Species of the World: A Taxonomic and Geographic Reference.* 3rd edition, Vol. 1 (D. E. Wilson and D. M. Reeder, eds.). Smithsonian Institution Press, Washington, DC.

Simmons, N. B., and J. H. Geisler. 1998. Phylogenetic relationships of *Icaronycteris, Archaeonycteris, Hassianycteris,* and *Palaeochiropteryx* to extant bat lineages, with comments on the evolution of echolocation and foraging strategies in *Microchiroptera. Bull. Am. Mus. Nat. Hist.* 235:1–182.

Simmons, N. B, K. L. Seymour, J. Habersetzer, and G. F. Gunnell. 2008. Primitive Early Eocene bat from Wyoming and the evolution of flight and echolocation. *Nature* 451:818–821.

Springer, M. S., E. C. Teeling, O. Madsen, M. J. Stanhope, and W. W. de Jong. 2001. Integrated fossil and molecular data reconstruct bat echolocation. *Proc. Natl. Acad. Sci. USA* 98:6241–6246.

Teeling, E. C., O. Madsen, W. J. Murphy, M. S. Springer, and S. J. O'Brien. 2003. Nuclear gene sequences confirm an ancient link between New Zealand's short-tailed bat and South American noctilionoid bats. *Mol. Phylogenet. Evol.* 28:308–319.

Teeling, E. C., O. Madsen, R. A. Van Den Bussche, W. W. de Jong, M. J. Stanhope, and M. S. Springer. 2002. Microbat paraphyly and the convergent evolution of a key innovation in Old World rhinolophid microbats. *Proc. Natl. Acad. Sci. USA* 99:1431–1436.

Teeling, E. C., M. Scully, D. J. Kao, M. L. Romagnoli, M. S. Springer, and M. J. Stanhope. 2000. Molecular evidence regarding the origin of echolocation and flight in bats. *Nature* 403:188–192.

Teeling, E. C., M. S. Springer, O. Madsen, P. Bates, S. J. O'Brien, and W. J. Murphy. 2005. A molecular phylogeny for bats illuminates biogeography and the fossil record. *Science* 307:580–584.

Van Den Bussche, R. A., and S. R. Hoofer. 2004. Phylogenetic relationships among recent chiropteran families and the importance of choosing appropriate out-group taxa. *J. Mammal.* 85:321–330.

Van Den Bussche, R. A., S. R. Hoofer, and E. W. Hansen. 2002. Characterization and phylogenetic utility of the mammalian protamine P1 gene. *Mol. Phylogenet. Evol.* 22:333–341.

Authors

Andrea L. Cirranello; Department of Mammalogy; Division of Vertebrate Zoology; American Museum of Natural History; 72nd Street at Central Park West; New York, NY 10024, USA. Email: andreacirranello@gmail.com.

Nancy B. Simmons; Division of Vertebrate Zoology; American Museum of Natural History; 72nd Street at Central Park West; New York, NY 10024, USA. Email: simmons@amnh.org.

Gregg F. Gunnell[†]; Division of Fossil Primates; Duke University Lemur Center; 1013 Broad Street; Durham, NC 27705, USA. Email: gregg.gunnell@duke.edu.

Date Accepted: 17 February 2016

Primary Editors: Jacques A. Gauthier, Kevin de Queiroz

[†] Deceased

Ungulata C. Linnaeus 1766 [J. D. Archibald], converted clade name

Registration Number: 292

Definition: The least inclusive crown clade containing *Bos primigenius* Bojanus 1827 (= *Bos taurus* Linnaeus 1758) (*Artiodactyla*) and *Equus ferus* Boddaert 1785 (= *Equus caballus* Linnaeus 1758) (*Perissodactyla*), provided that this clade does not include *Felis silvestris* Schreber 1777 (= *Felis catus* Linnaeus 1758) (*Carnivora*), *Manis pentadactyla* Linnaeus 1758 (*Pholidota*), *Vespertilio murinus* Linnaeus 1758 (*Chiroptera*), or *Erinaceus europaeus* Linnaeus 1758 (*Lipotyphla*). This is a minimum-crown-clade definition with a qualifying clause. Abbreviated definition: min crown ∇ (*Bos primigenius* Bojanus 1827 & *Equus ferus* Boddaert 1785 ~ (*Felis silvestris* Schreber 1778 ∨ *Manis pentadactyla* Linnaeus 1758 ∨ *Vespertilio murinus* Linnaeus 1758 ∨ *Erinaceus europaeus* Linnaeus 1758).

Etymology: "*Ungulata*" derives from the Latin *ungula* (hoof), alluding to the presence of hooves in many included taxa, including all extant species except *Cetacea*, possibly even in their most recent common ancestor.

Reference Phylogeny: For the purposes of applying this definition, Figure 1 in O'Leary et al. (2013) serves as the primary reference phylogeny. It depicts relationships among internal and external specifiers, with the exception of *Vespertilio murinus*, which is part of *Chiroptera* on that tree. *Ungulata* as defined here corresponds to *Euungulata* in O'Leary et al. (2013). See also Figure 5 (A and B) in Asher et al. (2003), Figures 1 and 2 in Beck et al. (2006), Figure 3 in Hou et al. (2009), Figure 2 in Spaulding et al. (2009), and Figure 2 in Zhou et al. (2012).

Composition: The taxa here considered to belong to *Ungulata* are currently thought to be arrayed in 341 living species grouped in 135 more inclusive taxa traditionally ranked as genera (Wilson and Reeder, 2005). McKenna and Bell (1997) list some 1,045 extinct ungulate genera. Of the two primary included crown clades, *Perissodactyla* has shown the greatest decline since its appearance in the Early Eocene, with only about 7% of the "genera" now extant. In the sister crown clade, *Artiodactyla* (this volume, including *Cetacea*), 16% of the "genera" are extant. It is also first known from the Early Eocene.

Diagnostic Apomorphies: There are no unambiguous non-molecular apomorphies for *Ungulata*. "Hooves"—thick, short, blunt and heavily keratinized structures homologous to claws and nails—have figured prominently in the traditional diagnosis of this taxon. They are present in all extant terrestrial members of both primary subclades, albeit borne on mesaxonic (*Perissodactyla*) or paraxonic (*Artiodactyla*) digits. That difference need not contradict homology, which only requires that "hooves" arose prior to these alternative modifications to the digital arch (see, for example, mesaxonic hands and paraxonic feet in the stem artiodactyl *Diacodexis pakistanensis*; Thewissen and Hussain, 1990). Hoof-like structures in afrotherian *Paenungulata*, which led some to associate them with *Ungulata* historically, are apparently non-homologous with those in ungulates.

Synonyms: Ongulogrades Normaux de Blainville 1816 (partial and approximate); *Diplarthra* Cope 1883 (partial and approximate); *Ungulata Vera* Flower and Lydekker 1891 (partial and approximate); *Euungulata* Waddell et al. 2001 (unambiguous).

Comments: Of the various names that have been used for a taxon approximating the named clade, Ongulogrades Normaux, *Diplarthra*, and *Ungulata Vera* are no longer used. A Google Scholar search on August 18, 2014, found 14,100 hits for "*Ungulata*" vs. 44 for "*Euungulata*". Waddell et al. (2001) proposed the latter as a new name for this clade because whales were not traditionally included in *Ungulata*. In the phylogenetic era, whales were initially thought to be sister to artiodactyls based on fossils and morphology, but are now inferred to nest inside *Artiodactyla* based on molecular and combined datasets. *Ungulata* is clearly the name most often associated with the clade originating in the last common ancestor of crown perissodactyls and artiodactyls, and is therefore the name to be preferred. For the names of specifiers, I follow *ICZN* Opinion 2027 (2003) in using specific names first applied to wild forms despite earlier publication of specific names for domesticated forms, viz., for cows, horses, and cats.

Gregory (1910) provided the most comprehensive history of the use of the name *Ungulata*, at least until the early twentieth century. He stated, "the so called '*Ungulata Vera*,' embracing the *Artiodactyla* and *Perissodactyla*, is almost as unnatural an assemblage as 'the Pachydermes' of Cuvier" (pp. 342–343). According to Gregory, Ray (1693) was the first to recognize *Ungulata* much as used here (with obvious exclusion of *Cetacea* from *Artiodactyla*). Starting with the 1735 (first) edition of his *Systema Naturae*, Linnaeus began and continued altering the names and composition of Ray's *Ungulata*.

Many of the classifications of the nineteenth century included taxa other than artiodactyls and perissodactyls within *Ungulata*. Considering only extant taxa, notable exceptions

were de Blainville (1816; except that he referred to them as "Ongulogrades Normaux"), Owen (1868; except for the inclusion of an unspecified hyrax), Gill (1872), Flower and Lydekker (1891; except that they called it *Ungulata Vera* within an *Ungulata* that also included *Subungulata*), Cope (1883; except that he called it *Diplarthra* within an *Ungulata* that also included a variety of other extant and extinct taxa).

Simpson coined *Ferungulata* in 1945 to include the following extant clades traditionally ranked as orders: *Carnivora*, *Proboscidea*, *Hyracoidea*, *Sirenia*, *Perissodactyla*, and *Artiodactyla*. In 1997, McKenna and Bell did not recognize *Ferungulata*, but grouped *Proboscidea*, *Hyracoidea*, *Sirenia*, *Perissodactyla*, and *Artiodactyla* under *Ungulata* with the addition of *Tubulidentata* and *Cetacea*. *Cetacea* was linked to *Artiodactyla* by these authors within *Eparctocyona* to the exclusion of other extant placental clades. Molecular studies (see *Artiodactyla* in this volume for references) have strongly supported that *Cetacea* is not simply sister to *Artiodactyla* but is closer to *Hippopotamidae* within *Artiodactyla* (e.g., Gatesy, 1997). *Proboscidea*, *Hyracoidea*, and *Sirenia*, which have long been recognized as a clade (e.g., Simpson, 1945), usually known as *Paenungulata*, have been linked with strong support to other African clades including *Tubulidentata* (viz., Afrotheria).

The earliest known crown perissodactyls and crown artiodactyls are confidently identified from the Early Eocene onwards. The age of origin of their respective total clades, as well as the total clade of *Ungulata*, is unclear. The stems of all three clades have traditionally been sought among a diverse assemblage of Early Cenozoic placentals collectively known as "condylarths" or "archaic ungulates" (Archibald, 1998). While no clear candidates for stem *Ungulata* have emerged, possible ancestry for *Perissodactyla* may be among phenacodontid "condylarths" (see Hooker, 2005, for a review) and for *Artiodactyla* among arctocyonid, hyopsodontid, or mioclaenid "condylarths" (see Theodor et al., 2005, for a review). A possible stem relationship to *Ungulata* was proposed for the Late Cretaceous zhelestids by Archibald (1996), but more recent phylogenetic analyses (Wible et al., 2007 and Wible et al., 2009) indicate that zhelestids are more likely stem rather than crown placentals.

Whatever else it may have contained over the years, *Ungulata* has always contained crown clades traditionally included in *Perissodactyla* and *Artiodactyla*. The definition of "*Ungulata*" proposed here is thus rooted explicitly in the *Perissodactyla* + *Artiodactyla* hypothesis, which has been supported in a recent study considering both fossil and Recent species and data from disparate morphological systems and diverse gene sequences (O'Leary et al., 2013). However, this hypothesis has been disputed by most studies based on gene sequences alone (e.g., Murphy et al., 2001; Nishihara et al., 2006; Springer et al., 2007; Matthee et al., 2007; Prasad et al., 2008). These hypotheses nest one or more major clades (traditional mammalian orders) inside the smallest clade containing all artiodactyls and perissodactyls. For this reason, the proposed definition includes external specifiers representing clades—crown chriopterans, lipotyphlans, pholidotans, and carnivorans—that, should they prove to be descended from the last ancestor of crown perissodactyls and artiodactyls, will render the name *Ungulata* inapplicable to any clade.

Literature Cited

Archibald, J. D. 1996. Fossil evidence for a Late Cretaceous origin of "hoofed" mammals. *Science* 272:1150–1153.

Archibald, J. D. 1998. Archaic ungulates ("*Condylartha*"). Pp. 292–331 in *Evolution of Tertiary Mammals of North America*, Vol. 1 (C. Janis, K. Scott, and L. Jacobs, eds.). *Terrestrial Carnivores, Ungulates, and Ungulate like Mammals*. Cambridge University Press, Cambridge, UK.

Asher, R. J., M. J. Novacek, and J. H. Geisler. 2003. Relationships of endemic African mammals and their fossil relatives based on morphological and molecular evidence. *J. Mamm. Evol.* 10:131–194.

Beck, R. M. D., O. R. P. Bininda-Emonds, M. Cardillo, F.-G. R. Liu, and A. Purvis. 2006. A higher-level MRP supertree of placental mammals. *BMC Evol. Biol.* 6:93.

de Blainville, H. M. D. 1816. Prodome d'une nouvelle distribution systématique du règne animal. *Bull. Soc. Philomath., pour l'année* 1816:105.

Boddaert, P. 1785. *Elenchus Animalium, Volumen 1: Sistens Quadrupedia huc usque nota, Eorumque Varietates*. C. R. Hake, Rotterdam.

Bojanus, L. H. 1827. De uro nostrate eiusque sceleto commentatio. Scripsit et Bovis primigenii scelto auxit. *Nova Acta Leopoldina* 13: 413–467.

Cope, E. D. 1883. The classification of the *Ungulata*. *Proc. Am. Assoc. Adv. Sci., 31st Meeting, Montreal*, 1882:477–479.

Flower, W. H., and R. Lydekker. 1891. *An Introduction to the Study of Mammals Living and Extinct*. Adam and Charles Black, London.

Gatesy, J. 1997. More support for a *Cetacea/Hippopotamidae* clade: the blood clotting protein gene g-fibrinogen. *Mol. Biol. Evol.* 14:537–543.

Gill, T. N. 1872. Arrangement of the families of mammals with analytical tables. *Smithson. Misc. Collect.* 230:1–98.

Gregory W. K. 1910. The orders of mammals. *Bull. Am. Mus. Nat. Hist.* 27:1–524.

Hooker, J. J. 2005. *Perissodactyla*. Pp. 199–214 in *The Rise of Placental Mammals: Origin and Relationships of the Major Extant Clades* (K. D. Rose and J. D. Archibald, eds.). Johns Hopkins University Press, Baltimore, MD.

Hou, Z., R. Romero, and D. E. Wildman. 2009. Phylogeny of the *Ferungulata* (*Mammalia: Laurasiatheria*) as determined from phylogenomic data. *Mol. Phylogenet. Evol.* 52:660–664.

International Commission on Zoological Nomenclature. 2003. Opinion 2027 (Case 3010) Usage of 17 specific names based on wild species which are pre-dated by or contemporary with those based on domestic animals (Lepidoptera, Osteichthyes, Mammalia): conserved. *Bull. Zool. Nomen.* 60:81–84.

Linnaeus, C. 1735. *Systema Naturae, sive Regna Tria, Secundum Classes, Ordines Naturae, Secundum Classes, Ordines, Genera, & Species.* Fol. Lugduni Batavorum.

Linnaeus, C. 1758. *Systema Naturae Per Regna Tria Naturae, Secundum Classes, Ordines, Genera, Species, cum Charcateribus, Differentiis, Synonymis, Locis.* Editio X. Laurentii Salvii, Holmiae (Stockholm).

Linnaeus, C. 1766. *Systema Naturae Per Regna Tria Naturae, Secundum Classes, Ordines, Genera, Species, cum Charcateribus, Differentiis, Synonymis, Locis.* Editio XII. Laurentii Salvii, Holmiae (Stockholm).

Matthee, C. A., G. Eick, S. Willows-Munro, C. Montgelard, A. T. Pardini, and T. J. Robinson. 2007. Indel evolution of mammalian introns and the utility of non-coding nuclear markers in eutherian phylogenetics. *Mol. Phylogenet. Evol.* 42:827–837.

McKenna, M. C., and S. K. Bell. 1997. *Classification of Mammals above the Species Level.* Columbia University Press, New York.

Murphy, W. J., E. Eizirik, S. J. O'Brien, O. Madsen, M. Scally, C. J. Douady, E. Teeling, O. A. Ryder, M. J. Stanhope, W. W. de Jong, and M. S. Springer. 2001. Resolution of the early placental mammal radiation using Bayesian phylogenetics. *Science* 294: 2348–2351.

Nishihara, H., M. Hasegawa, and N. Okada. 2006. *Pegasoferae*, an unexpected mammalian clade revealed by tracking ancient retroposon insertions. *Proc. Natl. Acad. Sci. USA* 103:9929–9934.

O'Leary, M. A., J. I. Bloch, J. J. Flynn, T. J. Gaudin, A. Giallombardo, N. P. Giannini, S. L. Goldberg, B. P. Kraatz, Z.-X. Luo, J. Jin Meng, X. Ni, M. J. Novacek, F. A. Perini, Z. Randall, G. W. Rougier, E. J. Sargis, M. T. Silcox, N. B. Simmons, M. Spaulding, P. M. Velazco, M. Weksler, J. R. Wible, and A. L. Cirranello. 2013. The placental mammal ancestor and the post-K radiation of placentals. *Science* 339:662–667.

Owen, R. 1868. *The Anatomy of Vertebrates, Vol. 3. Mammals.* Longmans, Green, and Co., London.

Prasad, A. B., M. W. Allard, and E. D. Green. 2008. Confirming the phylogeny of mammals by use of large comparative sequence data sets. *Mol. Biol. Evol.* 25:1795–1808.

Ray, J. 1693. *Synopsis Methodica Animalium Quadrupedum et Serpentini Generis.* Smith and Walford, London.

Schreber, J. C. D. 1778. *Die Säugethiere in Abbildungen nach der Natur mit Beschreibungen, Dritter Theil.* Wolfgang Walther, Erlangen. [According to C. D. Sherborn (*Proc. Zool. Soc. London* 1891:587–592), the part containing the description of *Felis silvestris* was published in 1777.]

Simpson, G. G. 1945. The principles of classification and a classification of mammals. *Bull. Am. Mus. Nat. Hist.* 85:1–350.

Spaulding, M., M. A. O'Leary, and J. Gatesy. 2009. Relationships of *Cetacea* (*Artiodactyla*) among mammals: increased taxon sampling alters interpretations of key fossils and character evolution. *PLOS ONE* 4:1–14.

Springer, M. S., A. Burk-Herrick, R. Meredith, E. Eizirik, E. Teeling, S. J. O'Brien, and W. J. Murphy. 2007. The adequacy of morphology for reconstructing the early history of placental mammals. *Syst. Biol.* 56: 673–684.

Theodor, J. M., K. D. Rose, and J. Erfurt. 2005. *Artiodactyla*. Pp. 215–233 in *The Rise of Placental Mammals: Origin and Relationships of the Major Extant Clades* (K. D. Rose and J. D. Archibald, eds.). Johns Hopkins University Press, Baltimore, MD.

Waddell, P. J., H. Kishino, and R. Ota. 2001. A phylogenetic foundation for comparative mammalian genomics. *Genome Inf. Ser.* 12:141–154.

Wible, J. R., G. W. Rougier, M. J. Novacek, and R. J. Asher. 2007. Cretaceous eutherians and Laurasian origin for placental mammals near the K/T boundary. *Nature* 447:1003–1006.

Wible, J. R., M. J. Novacek, G. W. Rougier, and R. J. Asher. 2009. The eutherian mammal *Maelestes gobiensis* from the Late Cretaceous of Mongolia and the phylogeny of Cretaceous Eutheria. *Bull. Am. Mus. Nat. Hist.* 327:1–123.

Wilson, D. E., and D. M. Reeder, eds. 2005. *Mammal Species of the World.* Johns Hopkins University Press, Baltimore, MD.

Zhou, X., S. Xu, J. Xu, B. Chen, K. Zhou, and G. Yang. 2012. Phylogenomic analysis resolves the interordinal relationships and rapid diversification of the laurasiatherian mammals. *Syst. Biol.* 61:150–164.

Author

J. David Archibald; Department of Biology; San Diego State University; San Diego, CA 92182-4614, USA. Email: darchibald@sdsu.edu

Date Accepted: 12 September 2014

Primary Editor: Jacques A. Gauthier

Artiodactyla R. Owen 1848 [M. A. O'Leary, M. J. Orliac, M. Spaulding, and J. Gatesy], converted clade name

Registration Number: 293

Definition: The least inclusive crown clade containing *Camelus dromedarius* Linnaeus 1758 (*Camelidae*), *Hippopotamus amphibius* Linnaeus 1758 (*Hippopotamidae*), *Sus scrofa* Linnaeus 1758 (*Suidae*), and *Bos primigenius* Bojanus 1827 (*Bovidae*). This is a minimum-crown-clade definition. Abbreviated definition: min crown ∇ (*Camelus dromedarius* Linnaeus 1758 & *Hippopotamus amphibius* Linnaeus 1758 & *Sus scrofa* Linnaeus 1758 & *Bos primigenius* Bojanus 1827).

Etymology: Derived from the Greek *artio* (even-numbered) and *daktylos* (finger).

Reference Phylogeny: We use Spaulding et al. (2009: Fig. 2) as the primary reference phylogeny upon which this definition is based because it is the most comprehensive analysis of *Artiodactyla* to date in terms of characters, extinct taxa, and extant taxa. *Camelus dromedarius* is represented by *Camelus* in that tree. Geisler et al. (2007: Fig. 3.2), Marcot (2007: Fig. 2.1), O'Leary and Gatesy (2008: Figs. 4 and 5), Gatesy et al. (2013: Figs. 7, 8, and 9), and Hassanin et al. (2012: Fig. 1) are additional large-scale phylogenetic studies of *Artiodactyla* and infer a clade of similar composition.

Composition: *Artiodactyla* includes approximately 314 extant species (Oliver, 1993; Reeves et al., 2003; Wilson and Reeder, 2005). *Ruminantia* constitutes most of this diversity with more than 210 extant species (Wilson and Reeder, 2005), and *Cetacea* is composed of approximately 80 extant species (Reeves et al., 2003); *Camelidae* (six species), *Hippopotamidae* (two species) and *Suina* (16 species) combined represent less than 8% of the extant diversity of *Artiodactyla*. There is no explicit count of the number of extinct species of *Artiodactyla*; however, there are 40 extinct families and more than 900 extinct genera (McKenna and Bell, 1997; Prothero and Foss, 2007), indicating that most species in the clade are extinct. The inclusion of a few fossil taxa (e.g., *Gobiohyus*) within *Artiodactyla* is uncertain (Spaulding et al., 2009).

Diagnostic Apomorphies: Spaulding et al. (2009) found 20 most parsimonious trees; in 16 of these, *Gobiohyus* falls within *Artiodactyla*, whereas in four of the trees, *Gobiohyus* falls outside *Artiodactyla*. Six morphological synapomorphies are present for *Artiodactyla* in 16 shortest trees, however, these become synapomorphies of *Artiodactyla* + *Gobiohyus* in the remaining four trees. These synapomorphies are: (1) an astragalus sustentacular facet equal to 70% of the width of the astragalus; (2) contact of the distal astragalus with the cuboid; (3) lateral orientation of the astragalus ectal facet; (4) lateral edges of proximal and distal astragalar trochleae aligned; (5) dorsoplantar arc of the astragalar head with wide arc (200 degrees); and (6) astragalar neck as wide as tibial trochlea. See Comments for discussion of additional potential synapomorphies.

Synonyms: *Cetartiodactyla* Montgelard et al. 1997 (approximate).

Comments: *Artiodactyla* has a long history of study from both phylogenetic and evolutionary perspectives (e.g., Owen, 1848; Kowalevsky, 1873; Cope, 1887; Matthew, 1929), which is intertwined with attempts to position *Cetacea* (this volume) in the overall mammalian tree. Historically, placing *Cetacea* among mammals has been a challenging problem (e.g., Simpson, 1945; Novacek, 1992) because cetaceans are highly transformed anatomically relative to other mammals. The nesting of *Cetacea* within *Artiodactyla* was first supported by molecular data (e.g., Miyamoto and Goodman, 1986; Irwin et al., 1991; Graur and Higgins, 1994; Irwin and Arnason, 1994; Gatesy et al., 1996; Gatesy, 1997, 1998). This result prompted Montgelard et al. (1997) to propose a new name for the combined group: *Cetartiodactyla*. Despite the subsequent widespread use of this name (e.g., Geisler and Uhen, 2005; Theodor and Foss, 2005; O'Leary and Gatesy, 2008), placing *Cetacea* within *Artiodactyla* was never grounds to retire the name *Artiodactyla*, rather than simply revising its hypothesized composition to include *Cetacea*, according to rules of phylogenetic nomenclature (see discussion in Spaulding et al., 2009).

Subsequent documentation of a paraxonic tarsus and a typical double pulley ankle joint in early Eocene whales (Gingerich et al., 2001; Thewissen et al., 2001) provided even more support for a grouping of *Cetacea* with *Artiodactyla*, a result that had already been demonstrated by Geisler and Luo (1998), and again by O'Leary and Geisler (1999), in cladistic analyses of morphological characters. In taxonomically and anatomically comprehensive studies that include such ankle data, morphology alone generally does not place *Cetacea* within *Artiodactyla* (Theodor and Foss, 2005; Thewissen et al., 2007; O'Leary and Gatesy, 2008; Spaulding et al., 2009), but oftentimes as the sister taxon of *Artiodactyla*. This result suggests that many anatomically intermediate fossils remain to be discovered to explain the transition to aquatic life within *Artiodactyla*.

Multiple nuclear gene sequences (Gatesy, 1998; Matthee et al., 2001; Murphy et al., 2001a,b), mitochondrial genomes (Arnason et al., 2004) and transposons (Nikaido et al., 1999) support a close relationship between *Cetacea* and *Hippopotamidae*. These data reject the traditional grouping of *Suiformes* (*Suina* + *Hippopotamidae*) that had been supported by morphological data (O'Leary and Geisler, 1999; Thewissen et al., 2001; Theodor and Foss, 2005; O'Leary and Gatesy, 2008; Spaulding et al., 2009: Fig. 5). Analyses that combine the molecular data with morphological characters (Geisler and Uhen, 2005; Geisler et al., 2007; O'Leary and Gatesy, 2008, Spaulding et al., 2009; Gatesy et al., 2013) support relationships that are congruent with molecule-based topologies and nest *Cetacea* within *Artiodactyla*.

The relationship of *Artiodactyla* to other mammalian clades traditionally ranked as orders remains unresolved (e.g., Novacek, 1992; Gatesy, 1998; Murphy et al., 2001b; Nishihara et al., 2006), as is the placement of some Eocene groups whose position within or outside of the crown clade is still not well established (e.g., "*Dichobunoidea*" and *Choeropotamidae* in Geisler et al., 2007: Fig. 3.2). *Mesonychia* was nested within *Artiodactyla* in the combined morphological and molecular analysis of O'Leary and Gatesy (2008: Fig. 4), but it fell outside *Artiodactyla* with the inclusion of additional extant and extinct species (Spaulding et al., 2009: Fig. 2).

Identification of morphological synapomorphies supporting *Artiodactyla* is complicated by the unstable placement of extinct taxa relative to the crown clade. Synapomorphies identified are primarily osteological and may, therefore, potentially be identified in all members of the clade, both extinct and extant. Unambiguous synapomorphies noted above describe features of the ankle, and the evolution of ankle features has been widely discussed for *Artiodactyla* (e.g., Schaeffer, 1948; Prothero et al., 1988; Thewissen, 1994; Rose, 1996). Spaulding et al. (2009; see also O'Leary and Gatesy, 2008) described this bone by its constituent parts so that the separate evolutionary transformations of these parts could be traced.

Many other features have been described historically as synapomorphies of *Artiodactyla*, and ongoing research continues to test exactly where on the tree these transform. For example, *Artiodactyla* has been described as including ungulates with a paraxonic pes in which digits three and four are of equal length (Prothero et al., 1988; Thewissen, 1994; Geisler and Luo, 1998; Spaulding et al., 2009). This condition implies that the weight-bearing axis of symmetry of the foot passes between digits three and four. *Artiodactyla* has also previously been supported by the presence of a relatively small third trochanter of the femur (e.g., Luckett and Hong, 1998; O'Leary and Geisler, 1999). Some artiodactyls also exhibit a peculiar milk dentition pattern with a trilobate deciduous lower fourth premolar that has six cusps, including an additional neomorphic cusp on the paracristid (e.g., Gentry and Hooker, 1988; Luckett and Hong, 1998; Spaulding et al., 2009). Soft tissue characters such as sparse cavernous tissue in the penis, the presence of three primary bronchi in the lungs and a plurilocular stomach are likely to be upheld as synapomorphies of *Artiodactyla* (Thewissen, 1994; Gatesy et al., 1996; Langer, 2001) in future studies. Many of these features require further investigation in living species as they are ambiguous synapomorphies of *Artiodactyla* in Spaulding et al. (2009).

Varied types of molecular evidence favor the monophyly of *Artiodactyla* (including *Cetacea*). Individual nuclear loci (e.g., Gatesy, 1998), mitochondrial genomes (e.g., Kjer and Honeycutt, 2007; Hassanin et al., 2012), and long concatenations of nuclear and mitochondrial genes (e.g., Murphy et al., 2001a), robustly support the group. As an example, the supermatrix of 22 genes from Murphy et al. (2001b) that sampled all major clades of *Artiodactyla* and representatives of all extant placental clades traditionally ranked as orders showed 100% bootstrap support for *Artiodactyla*. Parsimony analysis of this matrix reveals 207 unequivocally optimized molecular synapomorphies for this clade, with the supporting nucleotide substitutions distributed among 17 nuclear loci and mitochondrial DNA.

The definition of the taxon name *Artiodactyla* is based on four specifiers: *Camelus dromedarius, Hippopotamus amphibius, Sus scrofa* and *Bos primigenius*. These species represent the four major subgroups of traditional *Artiodactyla* (*Camelidae, Hippopotamidae, Suina*, and *Ruminantia*, respectively). In combination with *Cetacea*, the above four clades encompass the extant diversity of "*Artiodactyla*" as defined here. Molecular data strongly support the monophyly of each of these subclades (Marcot, 2007; Hassanin et al., 2012).

Literature Cited

Arnason, U., A. Gullberg, and A. Janke. 2004. Mitogenomic analyses provide new insights into cetacean origin and evolution. *J. Mol. Evol.* 333:27–34.

Cope, E. D. 1887. The classification and phylogeny of the *Artiodactyla. Proc. Am. Philos. Soc.* 24:377–400.

Gatesy, J. 1997. More DNA support for a *Cetacea-Hippopotamidae* clade: the blood-clotting protein gene gamma-fibrinogen. *Mol. Biol. Evol.* 14:537–543.

Gatesy, J. 1998. Molecular evidence for the phylogenetic affinities of *Cetacea*. Pp. 63–111 in *The Emergence of Whales* (J. G. M. Thewissen, ed.). Plenum Press, New York.

Gatesy, J., J. H. Geisler, J. Chang, C. Buell, A. Berta, R. W. Meredith, M. S. Springer, and M. R. McGowen. 2013. A phylogenetic blueprint for a modern whale. *Mol. Phylogenet. Evol.* 66(2):479–506.

Gatesy, J., C. Hayashi, M. A. Cronin, and P. Arctander. 1996. Evidence from milk casein genes that cetaceans are close relatives of hippopotamid artiodactyls. *Mol. Biol. Evol.* 13:954–963.

Geisler, J. H., and Z. Luo. 1998. Relationships of *Cetacea* to terrestrial ungulates and the evolution of cranial vasculature in *Cete*. Pp. 163–212 in *The Emergence of Whales: Evolutionary Patterns in the Origin of Cetacea* (J. G. M. Thewissen, ed.). Plenum Press, New York.

Geisler, J. H., and M. D. Uhen. 2005. Phylogenetic relationships of extinct cetartiodactyls: results of simultaneous analyses of molecular, morphological, and stratigraphic data. *J. Mamm. Evol.* 12:145–160.

Geisler, J.H., J. M. Theodor, M. D. Uhen, and S. E. Foss. 2007. Phylogenetic relationships of cetaceans to terrestrial artiodactyls. Pp. 19–31 in *The Evolution of Artiodactyla* (D. R. Prothero and S. E. Foss, eds.). Johns Hopkins University Press, Baltimore, MD.

Gentry, A. W., and J. J. Hooker. 1988. The phylogeny of the *Artiodactyla*. Pp. 235–272 in *The Phylogeny and Classification of the Tetrapods* (M. J. Benton, ed.). Systematics Association Special Volume No. 35B. Clarendon Press, Oxford.

Gingerich, P. D., M. U. Haq, I. S. Zalmout, I. H. Khan, M. S. Malkani. 2001. Origin of whales from early artiodactyls: hands and feet of Eocene *Protocetidae* from Pakistan. *Science* 293:2239–2242.

Graur, D., and D. G. Higgins. 1994. Molecular evidence for the inclusion of cetaceans within the order *Artiodactyla*. *Mol. Biol. Evol.* 11:357–364.

Hassanin, A., F. Delsuc, A. Ropiquet, C. Hammer, B van Vuuren, C. Matthee, M. Ruiz-Garcia, F. Catzeflis, V. Areskoug, T. Nguyen, and A. Couloux. 2012. Pattern and timing of diversification of *Cetartiodactyla* (*Mammalia*, *Laurasiatheria*), as revealed by a comprehensive analysis of mitochondrial genomes. *C. R. Biologies* 335:32–50.

Irwin, D. M., and I. Arnason. 1994. Cytochrome b gene of marine mammals: phylogeny and evolution. *J. Mamm. Evol.* 2:37–55.

Irwin, D. M., T. D. Kocher, and A. C. Wilson. 1991. Evolution of the cytochrome b gene of mammals. *J. Mamm. Evol.* 32:128–144.

Kjer, K. M., and R. L. Honeycutt. 2007. Site specific rates of mitochondrial genomes and the phylogeny of *Eutheria*. *BMC Evol. Biol.* 7:8. doi:10.1186/1471-2148-7-8.

Kowalevsky, W. 1873. Monographie der Gattung *Anthracotherium* Cuv. und Versuch einer natürlichen Classification der fossilen Huftiere. *Palaeontographica* 22:131–347.

Langer, P. 2001. Evidence from the digestive tract on phylogenetic relationships in ungulates and whales. *J. Zool. Syst. Evol. Res.* 39:77–90.

Luckett, W. P., and N. Hong. 1998. Phylogenetic relationships between the orders *Artiodactyla* and *Cetacea*: a combined assessment of morphological and molecular evidence. *J. Mamm. Evol.* 5(2):127–182.

Marcot, J. D. 2007. Molecular phylogeny of terrestrial artiodactyls: conflicts and resolution. Pp. 4–18 in *The Evolution of Artiodactyla* (D. R. Prothero and S. E. Foss, eds.). Johns Hopkins University Press, Baltimore, MD.

Matthee, C. A., J. D. Burzlaff, J. F. Taylor, and S. K. Davis. 2001. Mining the mammalian genome for artiodactyl systematics. *Syst. Biol.* 50:367–390.

Matthew, W. D. 1929. Reclassification of the artiodactyl families. *Bull. Geol. Soc. Am.* 40:403–408.

McKenna, M. C., and S. K. Bell. 1997. *Classification of Mammals above the Species Level.* Columbia University Press, New York.

Miyamoto, M. M., and M. Goodman. 1986. Biomolecular systematics of eutherian mammals: phylogenetic patterns and classification. *Syst. Zool.* 35:230–240.

Montgelard, C., F. Catzeflis, and E. Douzery. 1997. Phylogenetic relationships of artiodactyls and cetaceans as deduced from the comparison of cytochrome b and 12S rRNA mitochondrial sequences. *Mol. Phylogenet. Evol.* 14(5):550–559.

Murphy, W. J., E. Eizirik, W. E. Johnson, Y. P. Zhang, O. A. Ryder, and S. J. O'Brien. 2001a. Molecular phylogenetics and the origins of placental mammals. *Nature* 409:614–618.

Murphy W. J., E. Eizirik, S. J. O'Brien, O. Madsen, M. Scally, C. J. Douady, E. Teeling, O. A. Ryder, M. J. Stanhope, W. W. DeJong, and M. S. Springer. 2001b. Resolution of the early placental mammal radiation using Bayesian phylogenetics. *Science* 294:2348–2351.

Nikaido, M., A. P. Rooney, and N. Okada. 1999. Phylogenetic relationships among cetartiodactyls based on insertions of short and long interpersed elements: hippopotamuses are the closest extant relatives of whales. *Proc. Natl. Acad. Sci. USA* 96:10261–10266.

Nishihara, H., M. Hasegawa, and N. Okada. 2006. *Pegasoferae*, an unexpected mammalian clade revealed by tracking ancient retroposon insertions. *Proc. Natl. Acad. Sci. USA* 103:9929–9934.

Novacek, M. J. 1992. Mammalian phylogeny: shaking the tree. *Nature* 356:121–125.

O'Leary, M. A., and J. Gatesy. 2008. Impact of increased character sampling on the phylogeny of *Cetartiodactyla* (*Mammalia*): combined analysis including fossils. *Cladistics* 23:1–46.

O'Leary, M. A., and J. H. Geisler. 1999. The position of *Cetacea* within *Mammalia*: phylogenetic analysis of extant and extinct taxa. *Syst. Biol.* 48(3):455–490.

Oliver, W. L. R., ed. 1993. *Pigs, Peccaries, and Hippos: Status Survey and Conservation Action Plan*. IUCN, Gland, Switzerland.

Owen, R. 1848. Description of teeth and portions of jaws of two extinct anthracotherioid quadrupeds (*Hyopotamus vectianus* and *Hyop. bovinus*) discovered by the Marchioness of Hastings in the Eocene deposits on the NW coast of the Isle of Wight: with an attempt to develope Cuvier's idea of the classification of pachyderms by the number of their toes. *Q. J. Geol. Soc. Lond.* 4:103–141.

Prothero, D. R., E. M. Manning, and M. Fischer. 1988. The phylogeny of the ungulates. Pp. 201–234 in *The Phylogeny and Classification of the Tetrapods*, Vol. 2 (M. J. Benton, ed.). Clarendon Press, London.

Prothero, D., and S. E. Foss. 2007. *The Evolution of Artiodactyls*. Johns Hopkins University Press, Baltimore, MD.

Reeves, R. R., B. D. Smith, E. A. Crespo, and G. Notarbartolo di Sciara (compilers). 2003. *Dolphins, Whales and Porpoises: 2002–2010 Conservation Action Plan for the World's Cetaceans*. IUCN/SSC Cetacean Specialist Group. IUCN, Gland, Switzerland and Cambridge, UK.

Rose, K. 1996. On the origin of the order *Artiodactyla*. *Proc. Natl. Acad. Sci. USA* 93:1705–1709.

Schaeffer, B. 1948. The origin of a mammalian ordinal character. *Evolution* 2:164–175.

Simpson, G. G. 1945. The principles of classification and a classification of mammals. *Bull. Am. Mus. Nat. Hist.* 85:1–350.

Spaulding, M., M. A. O'Leary, and J. Gatesy. 2009. Relationships of *Cetacea* (*Artiodactyla*) among mammals: increased taxon sampling alters interpretations of key fossils and character evolution. *PLOS ONE* 4(9):1–14.

Theodor, J. M., and S. Foss. 2005. Deciduous dentition of Eocene cebochoerid artiodactyls and cetartiodactylan relationships. *J. Mamm. Evol.* 12:161–181.

Thewissen, J. G. M. 1994. Phylogenetic aspects of cetacean origins: a morphological perspective. *J. Mamm. Evol.* 2:157–184.

Thewissen, J. G. M., L. N. Cooper, M. T. Clementz, S. Bajpai, and B. N. Tiwari. 2007. Whales originated from aquatic artiodactyls in the Eocene epoch of India. *Nature* 450:1190–1195.

Thewissen, J. G. M., E. M. Williams, L. J. Roe, and S. T. Hussain. 2001. Skeletons of terrestrial cetaceans and the relationships of whales to artiodactyls. *Nature* 413:277–281.

Wilson, D. E., and D. M. Reeder, eds. 2005. *Mammal Species of the World: A Taxonomic and Geographic Reference*. 3rd edition. Johns Hopkins University Press, Baltimore, MD.

Authors

Maureen A. O'Leary; Department of Anatomical Sciences; HSC T-8 (040); Stony Brook University; Stony Brook, NY 11794-8081, USA. Email: maureen.oleary@stonybrook.edu.

Maeva J. Orliac; Institut des Sciences de l'Evolution; Université Montpellier; Place Eugène Bataillon, 34095 Montpellier cedex 05, France. Email: maeva.orliac@umontpellier.fr.

Michelle Spaulding; Department of Biological Sciences; Purdue University Northwest, Westville Campus, 1401 S. U.S. 421; Westville, IN 46391, USA. Email: mspauldi@pnw.edu.

John Gatesy; Sackler Institute for Comparative Genomics and Division of Vertebrate Zoology; American Museum of Natural History; New York, NY 10024, USA. Email: jgatesy@amnh.org.

Date Accepted: 07 March 2013

Primary Editor: Jacques A. Gauthier

Pan-Cetacea T. A. Demére, new clade name

Registration number: 260

Definition: The total clade of the crown-clade *Cetacea*. This is a crown-based total-clade definition. Abbreviated definition: total ∇ of *Cetacea*.

Etymology: *Pan-*, from the Greek *pan-* or *pantos* (all, the whole), indicates that the name refers to a total clade, and *Cetacea*, from the Latin *cetus* (m. nominative) meaning "any large sea creature, such as whale, shark, or dolphin." It is a Latinized form of the Ancient Greek κῆτος (*kêtos*, "any sea-monster or huge fish").

Reference Phylogeny: The primary reference phylogeny is Figure 9 of Gatesy et al. (2013). *Pan-Cetacea* corresponds to the clade originating in the branch (Branch D) separating *Pakicetidae*, the *Indohyus-Khirtharia* clade, and the rest of the cetacean total clade from the total clade of *Hippopotamidae*. See also Figure 2 of Spaulding et al. (2009).

Composition: See *Cetacea* (this volume) for extinct and extant crown cetaceans. Early Eocene through Late Eocene taxa currently regarded as members of *Pan-Cetacea* that are outside of crown *Cetacea* include terrestrial *Helohyus* and *Diacodexis* (Spaulding et al., 2009), the semi-aquatic *Raoellidae* (Thewissen et al., 2007) (or the individual taxa *Indohyus* and *Khirtharia*, as *Raoellidae* may be non-monophyletic [see Geisler and Theodor, 2009]), the freshwater aquatic *Pakicetidae* (or the individual taxa *Himalyacetus*, *Ichthyolestes*, *Nalacetus*, and *Pakicetus*, as *Pakicetidae* may be non-monophyletic [see Thewissen and Hussain, 1998; Gingerich, 2005]), the

amphibious marine *Ambulocetidae* (Thewissen and Williams, 2002; Gingerich, 2005), *Remingtonocetidae* (Thewissen and Bajpai, 2009), and *Protocetidae* (or the individual taxa *Aegyptocetus*, *Artiocetus*, *Babiacetus*, *Carolinacetus*, *Crenatocetus*, *Dhedacetus*, *Eocetus*, *Gaviacetus*, *Georgiacetus*, *Indocetus*, *Kharodacetus*, *Maiacetus*, *Makaracetus*, *Natchitochia*, *Pappocetus*, *Pontobasileus*, *Protocetus*, *Qaisracetus*, *Rodhocetus*, *Takracetus*, and *Togocetus*, as *Protocetidae* may be non-monophyletic [see Fordyce and Muizon, 2001; Uhen, 2008; Martínez-Cáceres et al., 2018]), and fully marine taxa traditionally assigned to "*Basilosauridae*" (e.g., *Basilosaurus*, *Chrysocetus*, *Dorudon*, *Saghacetus*, and *Zygorhiza*) are generally considered paraphyletic with respect to crown *Cetacea* [see Uhen, 2008, and discussions in Fordyce and Muizon, 2001 and Uhen, 2010; but see Martínez-Cáceres et al., 2018 for a different topology indicating basilosaurid monophyly]). The majority of stem cetaceans have traditionally been referred to as "archaeocetes" (Kellogg, 1936), a taxonomic grouping that today is widely regarded as paraphyletic (Luo and Gingerich, 1999; Fordyce and Muizon, 2001; Uhen, 2004, 2010). See McKenna and Bell (1997) and the Palaeobiology Database (paleobiod.org) for a more inclusive list of pan-cetaceans.

Diagnostic Apomorphies: Apomorphies inferred to have arisen near the base of *Pan-Cetacea* were listed by Luo and Gingerich, 1999: 60), Thewissen et al. (2007: 1190), and Gatesy et al. (2013: 484): thickened medial lip of ectotympanic bulla (involucrum); presence of thin splinter-like plate of tympanic ring (sigmoid process); anterior dentition (incisors and canines) in line with cheek teeth; and high crowns on posterior premolars. Luo and Gingerich (1999: 74), Thewissen et al. (2007: 1190), and Uhen (2010: 199) listed additional apomorphies for a smaller aquatic clade within *Pan-Cetacea* (Branch E in reference phylogeny): postorbital and temporal region of skull long and narrow; presence of distally-enlarged posterior process of ectotympanic bulla, direct articulation of bulla with exoccipital, presence of platform for bullar process of squamosal; long external auditory meatus; double-rooted P3; lack of P4 protocone; lower molars lacking trigonid and talonid basins; upper molars with very small trigon basin; M1-2 metacones present but small; and lack of M1-2 hypocone. The taxonomic distribution of some of these character states is still subject to debate (e.g., Thewissen at al., 2007). See also Uhen (2010) and Martínez-Cáceres et al. (2018) for various clades within *Pan-Cetacea* that are more inclusive than the crown clade *Cetacea*, but more closely related to *Cetacea* than to *Raoellidae* (*Indohyus* and *Khirtharia*). Among these is *Pelagiceti* (Uhen, 2008, 2010), a clade of pan-cetaceans (Branch M in reference phylogeny) that includes basilosaurids and more crownward stem cetaceans (e.g., *Kekenodon*) sharing the following apomorphies: multiple accessory denticles on cheek teeth; greatly reduced size of pelvis and pes; rotation of pelvis; high number of lumbar vertebrae (>10); and rectangular, short, and dorsoventrally compressed posterior caudal vertebrae (caudal ball vertebrae). The latter feature is considered an osteological proxy for the presence of soft-tissue tail flukes, a critical adaptation for an obligate, fully marine life style (Uhen, 2004). Although based on an incomplete sampling of stem cetaceans, Martínez-Cáceres et al. (2018, Fig. 96 and Appendix 1) identified an additional 19 potentially unambiguous cranial, dental, and postcranial synapomorphies for *Pelagiceti* (branch uniting *Basilosauridae* [Clade B] and *Cetacea* [Clade N] of their Fig. 96), including:

external bony nares posterior edge lies between level of P1 and P2; frontal supraorbital process posterior edge posterolaterally oriented and forming acute angle with midline in temporal region; pterygoid sinus fossa extends well anteriorly with deep trough; basioccipital with well-developed and transversely thin falcate process; supraoccipital nearly vertical and posteriorly facing; vertex almost dorsal to foramen magnum; exoccipital lateral margin less than or equal to width of mastoid process of periotic; periotic anterior process length between 59% and 94% of promontorial length; deep groove for tensor tympani muscle clearly visible in ventral view; Eustachian tube external opening at anterior end of tympanic bulla; tympanic bulla median furrow forming broad embayment at posterior edge of bulla; tympanic bulla posterior edge not contacting paroccipital process of exoccipital; seven teeth in maxilla; well-developed accessory denticles in mesial and/or distal edge of crown, in upper and lower premolars and molars; loss of protocone in upper dentition; two anteroposteriorly aligned roots in upper molars; pelvis greatly reduced and shorter than first sacral vertebra; obturator foramen smaller in diameter than acetabulum; and pubis with well-developed ventromedial expansion extending ventromedial to obturator foramen.

Synonyms: *Mutica* Linnaeus 1758 sensu Simpson (1945) (approximate); *Zeugloceta* Haekel 1866 (partial and approximate); *Zeuglodontia* Gill 1871 (partial and approximate); *Archaeoceti* Flower 1883 (partial and approximate); *Cetacea* sensu Simpson (1945) (approximate); *Cete* sensu McKenna and Bell (1997) (approximate); *Cetaceamorpha* Spaulding et al. 2009 (unambiguous).

Comments: Simpson (1945) used Brisson's (1762) circumscription of *Cetacea* by including crown-clade whales but deviated from it by adding stem taxa that he termed *"Archaeoceti"* that were unknown in Brisson's time. McKenna and Bell (1997) used the name *Cete* for the total clade, which they considered to contain mesonychians in addition to aquatic stem and crown cetaceans. The latter two groups represented their concept of *Cetacea*. *Cetaceamorpha* Spaulding et al. 2009 (see also Luo and Gingerich, 1999) was explicitly tied to the total clade including the crown plus all extinct taxa more closely related to extant members of *Cetacea* than to any other living species. As thus defined, the little-used *Cetaceamorpha* is equivalent to *Pan-Cetacea*, as both include crown *Cetacea* plus all fully marine, amphibious, semi-aquatic, and terrestrial stem taxa that are successive outgroups to crown *Cetacea*. In spite of this synonymy, however, and in the interests of establishing an integrated system of clade names, the *Pan-* convention is adopted here to form the name of the cetacean total clade, viz., *Pan-Cetacea*.

Current evidence indicates that *Pan-Cetacea* is nested within (crown) *Artiodactyla* (Spaulding et al., 2009; Gatesy et al., 2013). The extant sister to *Cetacea* among crown placentals is not clearly established based on morphological evidence, though cetaceans have generally been grouped with crown 'ungulates' (artiodactyls and perissodactyls) (e.g., Novacek, 1986). However, most molecular evidence (e.g., McGowen et al., 2007) and several combined molecular and morphological analyses including fossils (Spaulding et al., 2009; Gatesy et al., 2013) indicate that *Hippopotamidae* is the extant sister to *Cetacea*, forming a clade named *Whippomorpha* (e.g., Waddell et al., 1999; Asher and Helgen, 2010) or *Cetancodonta* (e.g., Arnason et al., 2000; Spaulding et al., 2009).

Uhen (2010) applied methods described by Strauss and Sadler (1989) to stratigraphic data on stem cetaceans to estimate an Early Eocene (Ypresian; ~54 Ma) origin for the cetacean total clade.

Literature Cited

Arnason, U., A. Gullberg, S. Gretarsdottir, B. Ursing, and A. Janke. 2000. The mitochondrial genome of the sperm whale and a new molecular reference for estimating eutherian divergence. *J. Mol. Evol.* 50(6): 569–578.

Asher, R. J., and K. M. Helgen. 2010. Nomenclature and placental mammal phylogeny. *BMC Evol. Biol.* 10:102. doi:10.1186/1471-2148-10-102.

Brisson, M. J. 1762. *Regnum Animale in Classes IX Distributum Sive Synopsis Methodica.* Editio altera auctior. Theodorum Haak, Leiden.

Fordyce, R. E., and C. de Muizon. 2001. Evolutionary history of the cetaceans: a review. Pp. 169–233 in *Secondary Adaptations of Tetrapods to Life in the Water* (J. M. Mazin and V. de Buffrénil, eds.). Verlag Dr. Friedrich Pfeil, Munich.

Gatesy, J., J. H. Geisler, J. Chang, C. Buell, A. Berta, R. W. Meredith, M. S. Springer, and M. R. McGowen. 2013. A phylogenetic blueprint for a modern whale. *Mol. Phylogenet. Evol.* 66:479–506.

Geisler, J. H., and J. M. Theodor. 2009. *Hippopotamus* and whale phylogeny. *Nature* 458:E1–4.

Gingerich, P. D. 2005. *Cetacea.* Pp. 234–252 in *The Rise of Placental Mammals: Origins and Relationships of the Major Extant Clades* (K. D. Rose and J. D. Archibald, eds.). Johns Hopkins University Press, Baltimore, MD.

Kellogg, R. 1936. A review of the *Archaeoceti. Carnegie Instit. Wash. Publ.* 482:1–366.

Linnaeus, C. 1758. *Systema Naturæ Per Regna Tria Naturæ, Secundum Classes, Ordines, Genera, Species, cum Characteribus, Differentiis, Synonymis, Locis.* Tomus I, Editio decima, reformata. Laurentii Salvii, Holmiae (Stockholm).

Luo, Z., and P. D. Gingerich. 1999. Terrestrial *Mesonychia* to aquatic *Cetacea*: transformation of the basicranium and evolution of hearing in whales. *Univ. Mich. Pap. Paleontol.* 31:1–98.

Martínez-Cáceres, M., O. Lambert, and C. de Muizon. 2018. The anatomy and phylogenetic affinities of *Cynthiacetus peruvianus*, a large *Dorudon*-like basilosaurid (*Cetacea, Mammalia*) from the late Eocene of Peru. *Geodiversitas* 39:7–163.

McGowen, M. R., M. Spaulding, and J. Gatesy. 2007. Divergence date estimation and a comprehensive molecular tree of extant cetaceans. *Mol. Phylogenet. Evol.* 53:891–906.

McKenna, M. C., and S. K. Bell. 1997. *Classification of Mammals above the Species Level*. Columbia University Press, New York.

Novacek, M. J. 1986. The skull of leptictid insectivorans and the higher-level classification of eutherian mammals. *Bull. Am. Mus. Nat. Hist.* 183(1):1–112.

Simpson, G. G. 1945. The principles of classification and a classification of mammals. *Bull. Am. Mus. Nat. Hist.* 85:1–350.

Spaulding, M., M. A. O'Leary, and J. Gatesy. 2009. Relationships of *Cetacea* (*Artiodactyla*) among mammals: increased taxon sampling alters interpretations of key fossils and character evolution. *PLOS ONE* 4(9):e7062.

Strauss, D., and P. M. Sadler. 1989. Classical confidence intervals and Bayesian probability estimates for ends of local taxon ranges. *Math. Geol.* 21(4):411–427.

Thewissen, J. G. M., and S. Bajpai. 2009. New skeletal material for *Andrewsiphius* and *Kutchicetus*, two Eocene cetaceans from India. *J. Paleontol.* 83:635–663.

Thewissen, J. G. M., L. N. Cooper, M. T. Clementz, S. Bajpai, and B. N. Tiwari. 2007. Whales originated from aquatic artiodactyls in the Eocene epoch of India. *Nature* 450:1190–1194.

Thewissen, J. G. M., and S. T. Hussain. 1998. Systematic review of the *Pakicetidae*, early and middle Eocene *Cetacea* (*Mammalia*) from Pakistan and India. Pp. 220–238 in *Dawn of the Age of Mammals in Asia* (K. C. Beard and M. R. Dawson, eds.). Bulletin Carnegie Museum Natural History, 34.

Thewissen, J. G. M., and E. M. Williams. 2002. The early radiations of *Cetacea* (*Mammalia*): evolutionary pattern and developmental correlations. *Annu. Rev. Ecol. Syst.* 33:73–90.

Waddell, P. J., N. Okada, and M. Hasegawa. 1999. Towards resolving interordinal relationships of placental mammals. *Syst. Biol.* 48:1–5.

Uhen, M. D. 2004. Form, function, and anatomy of *Dorudon atrox* (*Mammalia, Cetacea*): an archaeocete from the middle to late Eocene of Egypt. *Univ. Mich. Pap. Paleontol.* 34:1–222.

Uhen, M. D. 2008. New protocetid whale from Alabama and Mississippi, and a new cetacean clade, *Pelagiceti*. *J. Vertebr. Paleontol.* 28(3):589–593.

Uhen, M. D. 2010. The origin(s) of whales. *Annu. Rev. Earth Planet. Sci.* 38:189–219.

Author

Thomas A. Deméré; Department of Paleontology; San Diego Natural History Museum; San Diego, CA 92112, USA. Email: tdemere@ sdnhm.org.

Date Accepted: 19 July 2018

Primary Editor: Jacques A. Gauthier

Cetacea M. J. Brisson 1762 [T. A. Demèré], converted clade name

Registration Number: 261

Definition: The smallest crown clade containing *Tursiops* (originally *Delphinus*) *truncatus* (Montagu 1821) (*Odontoceti/Delphinidae*) and *Balaena mysticetus* Linnaeus 1758 (*Mysticeti/ Balaenidae*). This is a minimum-crown-clade definition. Abbreviated definition: min crown ∇ (*Tursiops truncatus* (Montagu 1821) & *Balaena mysticetus* Linnaeus 1758).

Etymology: "*Cetacea*" derives from the Latin *cetus* (m. nominative) meaning "any large sea creature, such as whale, shark, or dolphin" and is a Latinized form of Ancient Greek κῆτος (*kêtos*, "any sea-monster or huge fish").

Reference Phylogeny: Figure 5 in Gatesy et al. (2013) is the primary reference phylogeny. *Cetacea* corresponds to the clade originating at the node separating *Odontoceti* (*Monodontidae* + *Phocoenidae* + *Delphinidae* + *Iniidae* + *Pontoporiidae* + *Ziphiidae* + *Platanistidae* + *Kogiidae* + *Physteridae*) from *Mysticeti* (*Balaenopteridae* + *Eschrichtiidae* + *Neobalaenidae* + *Balaenidae*).

Composition: *Cetacea* includes more than 89 extant species, listings of which are given by Rice (1998) and the Committee on Taxonomy (2017). An account of 193 extinct crown cetaceans is provided by McKenna and Bell (1997), while 219 extinct crown cetaceans are listed by Berta (2017).

Diagnostic Apomorphies: Apomorphies relative to other crown clades include (references do not necessarily indicate where the features in question were originally described but rather are chosen to provide clear descriptions in a phylogenetic context): multiple dorsal maxillary foramina on posterior portion of maxilla (Fordyce and Muizon, 2001; Geisler and Sanders, 2003); antorbital notch for facial nerve (Fordyce and Muizon, 2001); partly open mesorostral groove (Lambert et al., 2017); posterior position of ascending process of premaxilla (Geisler and Sanders, 2003); posterior (mastoid) process of petrosal not exposed laterally on skull wall (Geisler and Sanders 2003); absence of contact between anterior part of tympanic and squamosal (Luo and Gingerich, 1999); epitympanic recess entirely on petrosal (Luo and Gingerich, 1999); large pterygoid sinus (Luo and Gingerich, 1999); supraoccipital shield anterodorsally inclined (Lambert et al., 2017); apex of zygomatic process of squamosal nearly contacting post-orbital process of frontal (Lambert et al., 2017); monophyodonty (Thewissen and Hussain, 1998); joints between humerus and radius and between humerus and ulna are both flat and meet at an obtuse angle (non-rotating elbow joint; Fordyce and Muizon, 2001; Gatesy et al., 2013); and pes lost and hindlimbs completely internal (Gatesy et al., 2013).

While all of these apomorphies distinguish *Cetacea* from other crown clades, some of them may diagnose more inclusive clades within *Pan-Cetacea* (this volume). According to Gatesy et al. (2013), apomorphies that diagnose *Cetacea* within *Pan-Cetacea* include monophyodonty, pes lost and hindlimbs completely internal, and non-rotating elbow joint. This low number of apomorphies uniquely diagnosing crown *Cetacea* partly reflects uncertainty surrounding the taxonomic distribution of several of the other candidate apomorphies listed above. This uncertainty stems largely from incomplete preservation of many fossil cetaceans.

Synonyms: *Cete* Linnaeus 1758 (approximate), *Autoceta* Haeckel 1866 (approximate); *Neoceti* Fordyce and Muizon 2001 (approximate). *Cete* as used by some recent authors is not a synonym (see Comments).

Comments: The name *Cetacea* is here selected from among its synonyms based on frequency of use. The clade to which it refers has long been studied, beginning with the early works of Brisson (1762), Lacépède (1804), and Cuvier (1817), and expanded on by Flower (1864, 1867), Gill (1871), Winge (1921), Miller (1923), Kellogg (1928), and Slijper (1936). Although not based on explicit phylogenies, Simpson (1945) and McKenna and Bell (1997) provided comprehensive taxonomies for extant and extinct cetaceans, while Hershkovitz (1966) and Rice (1998) provided taxonomies for extant cetaceans only. Linnaeus' (1758) *Cete* included taxa of both primary crown subclades—the mysticete *Balaena* and the odontocetes *Monodon*, *Physeter*, and *Delphinus*—bracketing *Cetacea* as that name is defined here. More recently, *Cete* has been used to refer to a more inclusive taxon composed of crown and undisputed stem cetaceans as well as extinct mesonychians (McKenna and Bell, 1997; Luo and Gingerich, 1999, O'Leary and Gatesy, 2008). However, more recent work strongly questions the monophyly of *Cete* in that sense, and instead suggests that likely stem ungulate mesonychians are more distantly related to cetaceans (Spaulding et al., 2009). Flower (1864, 1867) probably did more than anyone else to establish the longstanding primary division of extant *Cetacea* into *Odontoceti* and *Mysticeti*, and his recognition of "archaeocetes" as ancient whales (Flower and Lydekker, 1891) approximated the concept of stem cetaceans. For a time, there were serious debates about the monophyly of *Cetacea*, with several workers suggesting a diphyletic origin—that is, separate origins of cetacean characters by *Odontoceti* and *Mysticeti* (Slijper, 1936, 1979; Yablokov, 1964). The discovery and description of numerous fossil cetaceans has, however, filled many of the major morphological gaps between crown members of *Odontoceti* and *Mysticeti* as well as the even more pronounced gap between crown and total-clades of *Cetacea* (e.g., Fordyce, 2008; see *Pan-Cetacea* in this volume). Today, there is a surprising degree of agreement concerning the monophyly and composition of crown clade *Cetacea*, even as many newly recognized fossil taxa, and even a few new extant taxa, have been added to the clade (e.g., Fitzgerald, 2006; Racicot et al., 2014; Boessenecker and Fordyce, 2015; Sanders and Geisler, 2015; Wada et al., 2003). One notable exception to this general agreement concerns *Kekenodontidae*, a near-crown stem taxon that has occasionally been proposed as an early diverging stem mysticete clade within crown *Cetacea* (see McKenna and Bell, 1997; Fordyce and Muizon, 2001). Fossil evidence indicates that the divergence between *Odontoceti* and *Mysticeti* (= origin of crown *Cetacea*) likely occurred by the Late Eocene (Priabonian; ~36.5 Ma) based on the discovery of the stem mysticete *Mystacodon selenensis* from the Yumaque Formation of Peru (Martínez-Cáceres et al., 2018).

Spaulding et al. (2009) proposed initial phylogenetic definitions for major artiodactyl clades including *Cetacea*, and linked this name to the crown employing the same specifiers used here. Recent phylogenies for or including *Cetacea* have been published by Slater et al. (2010), Geisler et al. (2011), and Gatesy et al. (2013). The clade here named *Cetacea* is robustly supported in these and other studies.

Literature Cited

Berta, A. 2017. *The Rise of Marine Mammals-50 Million Years of Evolution.* John Hopkins University Press, Baltimore, MD.

Boessenecker, R. W., and R. E. Fordyce. 2015. Anatomy, feeding ecology, and ontogeny of a transitional baleen whale: a new genus and species of *Eomysticetidae* (*Mammalia: Cetacea*) from the Oligocene of New Zealand. *PeerJ,* 3:e1129.

Brisson, M. J. 1762. *Regnum Animale in Classes IX Distributum sive Synopsis Methodica. Editio Altera Auctior.* Theodorum Haak, Leiden.

Committee on Taxonomy. 2017. List of marine mammal species and subspecies. *Society for Marine Mammalogy.* Available at www.marinemammalscience.org, consulted on 17 August 2017.

Cuvier, G. 1817. *Le Règne Animal, Distribué d'Aprés son Organisation, pour Servir de Base à l'Histoire Naturelle des Animaux et d'Introduction à l'Anatomie Comparée,* Tome I. Deterville, Paris.

Fitzgerald, E. M. G. 2006. A bizarre new toothed mysticete (*Cetacea*) from Australia and the early evolution of baleen whales. *Proc. Biol. Sci.* 273:2955–2963.

Flower, W. H. 1864. Notes on the skeletons of whales in the principal museums of Holland and Belgium, with descriptions of two species apparently new to science. *Proc. Zool. Soc. London* 1864:384–420.

Flower, W. H. 1867. Description of the skeleton of *Inia geoffrensis* and of the skull of *Pontoporia blainvillii* with remarks on the systematic position of these animals in the Order *Cetacea. Trans. Zool. Soc. Lond.* 6:87–116.

Flower, W. H., and R. Lydekker. 1891. *An Introduction to the Study of Mammals Living and Extinct.* Adam and Charles Black, London.

Fordyce, R. E. 2008. *Neoceti.* Pp. 758–763 in *Encyclopedia of Marine Mammals.* 2nd edition (W. F. Perrin, B. Würsig, and J. G. M. Thewissen, eds.). Elsevier, Amsterdam.

Fordyce, R. E., and C. de Muizon. 2001. Evolutionary history of the cetaceans: a review. Pp. 169–233 in *Secondary Adaptations of Tetrapods to Life in the Water* (J. M. Mazin and V. de Buffrénil, eds.). Verlag Dr. Friedrich Pfeil, Munich.

Gatesy, J., J. H. Geisler, J. Chang, C. Buell, A. Berta, R. W. Meredith, M. S. Springer, and M. R. McGowen. 2013. A phylogenetic blueprint for a modern whale. *Mol. Phylogenet. Evol.* 66:479–506.

Geisler, J. H., and A. E. Sanders. 2003. Morphological evidence for the phylogeny of *Cetacea. J. Mamm. Evol.* 10:23–129.

Geisler, J. H., M. R. McGowen, G. Yang, and J. Gatesy. 2011. A supermatrix analysis of genomic, morphological, and paleontological data from crown *Cetacea. BMC Evol. Biol.* 11, Contrib. 112. doi:10.1186/1471-2148-11-112.

Gill, T. 1871. Synopsis of the primary subdivisions of the cetaceans. *Commun. Essex Inst.* 6:121–126.

Haeckel, E. 1866. *Generelle Morphologie der Organismen,* Vol. 2. Georg Reimer, Berlin.

Hershkovitz, P. 1966. Catalog of living whales. *Bull. U. S. Nat. Mus., Smithson. Inst.* 246:1–259.

Kellogg, R. 1928. The history of whales—their adaptation to life in the water. *Q. Rev. Biol.* 3:29–76, 174–208.

de Lacépède, B. G. 1804. *Histoire naturelle des cétacées.* Chez Plassan, Imprimeur Libraire, L'an XII de la Republique, Paris.

Lambert, O., M. Martínez-Cáceres, G. Bianucci, C. Di Celma, R. Salas-Gismondi, E. Steurbaut, M. Urbina, and C. de Muizon. 2017. Earliest mysticete from the Late Eocene of Peru sheds new light on the origin of baleen whales. *Curr. Biol.* 27:1535–1541.

Linnaeus, C. 1758. *Systema Naturae Per Regna Tria Naturae, Secundum Classes, Ordines, Genera, Species, cum Characteribus, Differentiis, Synonymis, Locis.* Tomus I, Editio decima, reformata. Laurentius Salvius, Holmiae (Stockholm).

Luo, Z., and P. D. Gingerich. 1999. Terrestrial *Mesonychia* to aquatic *Cetacea*: transformation of the basicranium and evolution of hearing in whales. *Univ. Michigan Papers Paleontol.* 31:1–98.

Martínez-Cáceres, M., O. Lambert, and C. de Muizon. 2018. The anatomy and phylogenetic affinities of *Cynthiacetus peruvianus,* a large *Dorudon*-like basilosaurid (*Cetacea, Mammalia*) from the late Eocene of Peru. *Geodiversitas* 39:7–163.

McKenna, M. C., and S. K. Bell. 1997. *Classification of Mammals above the Species Level*. Columbia University Press, New York.

Miller, G. S. 1923. The telescoping of the cetacean skull. *Smithson. Misc. Coll.* 76:1–70.

Montagu, G. 1821. Description of a species of *Delphinus*, which appears to be new. *Mem. Wern. Nat. Hist. Soc.* 3:75–82

O'Leary, M. A., and J. Gatesy. 2008. Impact of increased character sampling on the phylogeny of *Cetartiodactyla* (*Mammalia*): combined analysis including fossils. *Cladistics* 24:397–442.

Racicot, R. A., T. A. Deméré, B. L. Beatty, and R. W. Boessenecker. 2014. Unique feeding morphology in a new prognathous extinct porpoise from the Pliocene of California. *Curr. Biol.* 24:774–779.

Rice, D. W. 1998. Marine mammals of the world. Systematics and distribution. *Soc. Mar. Mamm. Spec. Publ.* 4.

Sanders, A. E., and J. H. Geisler. 2015. A new basal odontocete from the upper Rupelian of South Carolina, USA, with contributions to the systematics of *Xenorophus* and *Mirocetus* (*Mammalia, Cetacea*). *J. Vertebr. Paleontol.* 35:e890107.

Simpson, G. G. 1945. The principles of classification and a classification of mammals. *Bull. Am. Mus. Nat. Hist.* 85:1–350.

Slater, G. J., S. A. Price, F. Santini, and M. E. Alfaro. 2010. Diversity versus disparity and the radiation of modern cetaceans. *Proc. R. Soc. Lond. B Biol. Sci.* 277:3097–3104.

Slijper, E. J. 1936. Die Cetaceen. Vergleichenanatomisch und systematisch. *Capita Zool.* 7:1–590.

Slijper, E. J. 1979. *Whales*. Hutchison, London.

Spaulding, M., M. A. O'Leary, and J. Gatesy. 2009. Relationships of *Cetacea* (*Artiodactyla*) among mammals: increased taxon sampling alters interpretations of key fossils and character evolution. *PLOS ONE* 4(9):e7062.

Thewissen, J. G. M., and S. T. Hussain. 1998. Systematic review of the *Pakicetidae*, early and middle Eocene *Cetacea* (*Mammalia*) from Pakistan and India. Pp. 22–238 in *Dawn of the Age of Mammals in Asia* (K. C. Beard and M. R. Dawson, eds.). *Bull. Carnegie Mus. Nat. Hist.* 34.

Wada, S., M. Oishi, and T. K. Yamada. 2003. A newly discovered species of living baleen whale. *Nature* 426:278–281.

Winge, H. 1921. A review of the interrelationships of the *Cetacea*. *Smithson. Misc. Coll.* 72 (8):1–97.

Yablokov, A. V. 1964. Convergence or parallelism in the evolution of cetaceans. *Int. Geol. Rev.* 7:1461–1468.

Author

Thomas A. Deméré; Department of Paleontology; San Diego Natural History Museum; San Diego, CA 92112, USA. Email: tdemere@sdnhm.org.

Date Accepted: 19 July 2018

Primary editor: Jacques A. Gauthier

Pan-Bovidae F. Bibi and E. S. Vrba, new clade name

Registration Number: 262

Definition: The total clade of the crown clade *Bovidae*. This is a crown-based total-clade definition. Abbreviated definition: total ∇ of *Bovidae*.

Etymology: Derived from the Greek *pan-* (all, in reference to a total clade) and *Bovidae*, the name of a crown clade (see entry in this volume for etymology).

Reference Phylogeny: Figure 1 in Hassanin et al. (2012).

Composition: The crown clade of *Bovidae* and all extinct organisms that are more closely related to *Bovidae* than to any other extant pecorans. For taxa contained in the crown, and for potential stem species, see *Bovidae* and *Cavicornia* in this volume.

Diagnostic Apomorphies: Although extant species can easily be distinguished from other crown pecorans, no unambiguous stem bovids are known. The problem is that the same apomorphy—un-branched, non-deciduous cranial appendages covered with a permanent keratin sheath (= bovid horns sensu Janis and Scott, 1987)—that diagnoses *Cavicornia* relative to all other fossil and Recent pecorans also diagnoses *Bovidae* relative to all other crown pecorans. Furthermore, "bovid horns" are unlikely to diagnose *Pan-Bovidae* even though all currently known extinct taxa that might qualify as stem bovids possess such horns (e.g., *Eotragus* spp.). These three nested taxa have different theoretical compositions, and their diagnoses are expected to eventually differ accordingly. As a practical matter,

however, it is currently impossible to differentiate *Pan-Bovidae* from either *Cavicornia* or *Bovidae*.

Synonyms: *Cavicornia* Illiger 1811 (approximate); *Bovidae* Gray 1821 (approximate); *Bovina* Gray 1825 (approximate—included *Antilocapra*); *Tubicornes* Lesson 1827 (approximate—included *Antilocapra* and *Pudu*); *Cavicornidae* Reichenow 1886 (approximate).

Comments: Although *Bovidae* is arguably the name most commonly associated with this clade, that is largely a consequence of having failed to distinguish clearly between the crown clade (*Bovidae*; this volume), its total clade (*Pan-Bovidae*), and the origin of its distinctive horns (*Cavicornia*; this volume). *Pan-Bovidae* is proposed to make a distinction between names applied to crown vs. total clades in keeping with the broader goals of the *ICPN* to develop a general nomenclatural system for all biologists. Divergence-time estimates place the split between the stem lineages of *Bovidae* and *Moschidae*—i.e., the origin of *Pan-Bovidae*—at around 25–20 Ma (Hassanin et al., 2012) or 19–16 Ma (Bibi, 2013). Currently, the earliest and least modified pan-bovids and cavicornians are about 18–17 Ma in age. These include *Eotragus* spp. from Europe and Asia, an unnamed species from the Vihowa Formation in Pakistan, and *Namacerus gariepensis* from Namibia (Ginsburg and Heintz, 1968; Solounias et al., 1995; Morales et al., 2003).

Literature Cited

Bibi, F. 2013. A multi-calibrated mitochondrial phylogeny of extant *Bovidae* (*Artiodactyla, Ruminantia*) and the importance of the fossil record to systematics. *BMC Evol. Biol.* 13:166. Available at http://www.biomedcentral.com/1471-2148/13/166.

Ginsburg, L., and E. Heintz. 1968. La plus ancienne antilope d'Europe, *Eotragus artenensis* du Burdigalien d'Artenay. *Bull. Mus. Hist. Nat., 2e Sér.* 40:837–842.

Gray, J. E. 1821. On the natural arrangement of vertebrose animals. *Lond. Med. Reposit.* 15:296–310.

Gray, J. E. 1825. An outline of an attempt at the disposition of *Mammalia* into tribes and families, with a list of the genera apparently appertaining to each tribe. *Ann. Philos., N.S.* 10:337–344.

Hassanin, A., F. Delsuc, A. Ropiquet, C. Hammer, B. Jansen van Vuuren, C. Matthee, M. Ruiz-Garcia, F. Catzeflis, V. Areskoug, T. T. Nguyen, and A. Couloux. 2012. Pattern and timing of diversification of *Cetartiodactyla* (*Mammalia, Laurasiatheria*), as revealed by a comprehensive analysis of mitochondrial genomes. *C. R. Biol.* 335:32–50.

Illiger, C. 1811. *Prodromus Systematis Mammalium et Avium: Additis Terminis Zoographicis Utriusque Classis.* Sumptibus C. Salfield, Berolini (Berlin).

Janis, C. M., and K. M. Scott. 1987. The interrelationships of higher ruminant families with special emphasis on the members of the *Cervoidea. Am. Mus. Novit.* 2893:1–85.

Lesson, R. P. 1827. *Manuel de Mammalogie.* Roret, Paris.

Morales, J., D. Soria, M. Pickford, and M. Nieto. 2003. A new genus and species of *Bovidae* (*Artiodactyla, Mammalia*) from the early Middle Miocene of Arrisdrift, Namibia, and the origins of the family *Bovidae.* Pp. 371–384 in *Geology and Palaeontology of the Central Southern Namib*, Vol. 2 (M. Pickford and B. Senut, eds.). *Palaeontology of the Orange River Valley, Namibia.* Geological Survey of Namibia, Ministry of Mines and Energy, Windhoek, Namibia.

Reichenow, A. 1886. Bericht über die Leistungen in der Naturgeschichte der Säugethiere während des Jahres 1885. *Arch. Naturgesch.* 52:97–144.

Solounias, N., J. C. Barry, R. L. Bernor, E. H. Lindsay, and S. M. Raza. 1995. The oldest bovid from the Siwaliks, Pakistan. *J. Vertebr. Paleontol.* 15:806–814.

Authors

Faysal Bibi; Museum für Naturkunde; Leibniz Institute for Evolution and Biodiversity Science; Invalidenstrasse 43; 10115 Berlin, Germany. Email: faysal.bibi@mfn-berlin.de.

Elisabeth S. Vrba; Department of Geology & Geophysics; P.O. Box 208109; Yale University; New Haven, CT 06520-8109, USA. Email: elisabeth.vrba@yale.edu.

Date Accepted: 07 July 2016

Primary editor: Jacques A. Gauthier

Cavicornia C. Illiger 1811 [F. Bibi and E. S. Vrba], converted clade name

Registration Number: 263

Definition: The artiodactyl clade for which frontal bones with permanent, unbranched, and paired horn cores, as inherited by *Bos taurus* Linnaeus 1758, is an apomorphy. This is an apomorphy-based definition. Abbreviated definition: ∇ apo frontal bones with permanent, unbranched, and paired horn cores [*Bos taurus* Linnaeus 1758].

Etymology: Derived from Latin *cavus* (hollow) plus *cornu* (horn).

Reference Phylogeny: Figure 9 in Sánchez et al. (2015), where the name *Cavicornia* applies to a clade originating somewhere along the branch subtending the node uniting *Namacerus* (a possible stem bovid) and *Eudorcas* (a crown bovid).

Composition: All (crown) bovids and members of the bovid stem that bear horns as specified in the definition, including Miocene fossils not certainly in the crown, such as *Eotragus* spp. (Azanza and Morales, 1994; Solounias et al., 1995), *Namacerus* (Morales et al., 2003), and taxa referred to *Hypsodontinae* (Köhler, 1987).

Diagnostic Apomorphies: *Cavicornia* could be diagnosed by a type of horn traditionally associated with living *Bovidae*, viz., "cranial appendages that are unbranched, non-deciduous, and covered with a permanent keratin sheath" (Janis and Scott, 1988: 279). However, because *Cavicornia* is defined here to include extinct stem members of the bovid total clade in which the sheath might not be preserved, we have omitted apomorphies related to the keratin sheath (see Comments). We focus instead on the morphology of the readily preserved bony horn core, which palaeontologists have generally considered sufficient for assignment to *Bovidae* (e.g., Solounias et al., 1995). In many species of *Cavicornia*, horns are found only in males, and females are hornless. It is not known whether the earliest members of *Cavicornia* had horned or hornless females; we intend for the name to be applied to the clade characterized by presence of bovid-type horn cores, whether in one or both sexes.

Synonyms: *Bovidae* Gray 1821 (approximate); *Bovina* Gray 1825 (approximate—included *Antilocapra*); *Tubicornes* Lesson 1827 (approximate—included *Antilocapra* and *Pudu*); *Cavicornidae* Reichenow 1886 (approximate).

Comments: The presence of paired, unbranched, permanent horn cores on the frontal bones encased in permanent horn sheaths was noted as a taxonomically significant feature associated with the crown as early as Illiger (1811: 106): "*Cornua persistentia frontalia*". Taxa conceptualized in terms of apomorphies vs. composition, even if inter-nested, are unlikely to specify exactly the same clades (Gauthier and de Queiroz, 2001). As such, *Cavicornia* refers to a more inclusive clade than *Bovidae*, as bovid horns almost certainly appeared before the crown first diversified. The palaeontological literature has generally not distinguished between the crown clade (*Bovidae* as conceptualized in this volume) and the potentially more inclusive clade composed of those artiodactyls with bovid-type horns (*Cavicornia*). Therefore, numerous fossils referred to *Bovidae* in the literature cannot be assumed to lie within the crown; such a referral requires at least one apomorphy placing a fossil within some bovid subclade in addition to the presence of bovid-type horn cores in that fossil.

We apply the far more commonly used name *Bovidae* to the crown clade (see entry in this volume). We select the name *Cavicornia* from among the names previously applied (see Synonyms) to a group approximating the crown because it has priority and is suitably descriptive of the likely more inclusive clade of pan-bovids containing the crown (see *Pan-Bovidae* in this volume).

Although arguably integral to traditional conceptualizations of the 'bovid' horn, we have omitted apomorphies associated with the keratinized horn sheath, such as its permanent and unbranched nature, and indeed the sheath itself, from our definition. Keratin is unlikely to be preserved in fossils, although there can be morphological correlates preserved on the bony horn cores by which its presence can be inferred (e.g., Webb, 1973). Exceptional depositional environments could conceivably preserve horn sheaths, or their impressions, or perhaps even geochemical signatures of their presence. So, it might be possible in exceptional circumstances to determine whether the keratin sheath is branched or not. Nevertheless, it would be highly unlikely to establish if that sheath was permanent or shed annually. The definition could include just the presence of a keratinized sheath enclosing the horn core, as that seems the safest inference based on horn-core surface texture. But the bony horn core itself is by far the easiest feature to observe in the broadest range of fossils. And it is possible that the cavicornian horn first arose as a bony protuberance without an enclosing keratinous sheath (e.g., as evolved independently in *Giraffa camelopardalis*), or that it had a keratinous sheath that was shed annually (e.g., *Antilocapra americana*).

Antilocapra americana (the pronghorn) presents a special challenge in this regard as it also has permanent and unbranched horn cores, although the horn sheaths are branched, at least in males, and they are not permanent, but shed annually. Molecular and morphological datasets disagree on

its position among extant ruminant artiodactyls (e.g., Davis et al., 2011). Molecular analyses previously placed *Antilocapra* at the base of the pecoran radiation (Hassanin et al., 2012), while morphological analyses (e.g., Sánchez et al., 2015) often placed the total clade of *Antilocapra* close to or sister to the total clade of *Bovidae*. A recent whole-genome analysis placed *Antilocapra* as the sister taxon to *Giraffidae* (Chen et al., 2019). The fossil record also suggests that the horns of antilocaprids and bovids are unlikely to be homologous. The (branched) horn cores in deep-stem *Antilocapridae* (e.g., *Ramoceros osborni*) were apparently covered by skin (Heffelfinger et al., 2004), with a keratinized sheath evolving only later within the clade (Webb, 1973; Janis and Manning, 1998).

The earliest known examples of *Cavicornia* are from about 18-17 Ma (Early Miocene) and include *Eotragus* spp. from Europe and Asia, an unnamed species from the Vihowa Formation in Pakistan, and *Namacerus* from Namibia (Ginsburg and Heintz, 1968; Solounias et al., 1995; Morales et al., 2003). Although some relationships to bovid subclades have been proposed, it is not obvious which of these fossil species, if any, belong within the crown clade (*Bovidae*). Apomorphies used previously to assign some *Eotragus* species to the crown-clade, such as horn keels (e.g., Azanza and Morales, 1994: Fig. 7), are considerably homoplastic and therefore unreliable on their own.

Literature Cited

Azanza, B., and J. Morales. 1994. *Tethytragus* nov. gen et *Gentrytragus* nov. gen. deux nouveaux bovidés (*Artiodactyla, Mammalia*) du Miocène moyen. *Proc. K. Ned. Akad. Wet., Ser. B* 97:249–282.

Chen, L., Q. Qiu, Y. Jiang, K. Wang, Z. Lin, Z. Li, F. Bibi, Y. Yang, J. Wang, W. Nie, W. Su, G. Liu, Q. Li, W. Fu, X. Pan, C. Liu, J. Yang, C. Zhang, Y. Yin, Y. Wang, Y. Zhao, C. Zhang, Z. Wang, Y. Qin, W. Liu, B. Wang, Y. Ren,

R. Zhang, Y. Zeng, R. R. da Fonseca, B. Wei, R. Li, W. Wan, R. Zhao, W. Zhu, Y. Wang, S. Duan, Y. Gao, Y. E. Zhang, C. Chen, C. Hvilsom, C. W. Epps, L. G. Chemnick, Y. Dong, S. Mirarab, H. R. Siegismund, O. A. Ryder, M. T. P. Gilbert, H. A. Lewin, G. Zhang, R. Heller, and W. Wang. 2019. Large-scale ruminant genome sequencing provides insights into their evolution and distinct traits. Science 364(6446):eaav6202.

Davis, E. B., K. A. Brakora, and A. H Lee. 2011. Evolution of ruminant headgear: a review. *Proc. R. Soc. Lond. B Biol. Sci.* 278:2857–2865.

Gauthier, J. A. , and K. de Queiroz. 2001. Feathered dinosaurs, flying dinosaurs, crown dinosaurs and the name "*Aves*". Pp. 7–41 in *New Perspectives on the Origin and Early Evolution of Birds: Proceedings of the International Symposium in Honor of John H. Ostrom* (J. Gauthier and L. F. Gall, eds.). Peabody Museum of Natural History, Yale University, New Haven, CT.

Ginsburg, L., and E. Heintz. 1968. La plus ancienne antilope d'Europe, *Eotragus artenensis* du Burdigalien d'Artenay. *Bull. Mus. Natl. Hist. Nat.* 40:837–842.

Gray, J. E. 1821. On the natural arrangement of vertebrose animals. *Lond. Med. Reposit.* 15:296–310.

Gray, J. E. 1825. An outline of an attempt at the disposition of *Mammalia* into tribes and families, with a list of the genera apparently appertaining to each tribe. *Ann. Philos., N.S.* 10:337–344.

Hassanin, A., F. Delsuc, A. Ropiquet, C. Hammer, B. Jansen van Vuuren, C. Matthee, M. Ruiz-Garcia, F. Catzeflis, V. Areskoug, T. T. Nguyen, and A. Couloux. 2012. Pattern and timing of diversification of *Cetartiodactyla* (*Mammalia, Laurasiatheria*), as revealed by a comprehensive analysis of mitochondrial genomes. *C. R. Biol.* 335:32–50.

Heffelfinger, J. R., B. W. O'Gara, C. M. Janis, and R. Babb. 2004. A bestiary of ancestral antilocaprids. *Proceedings of the 20th Biennial Pronghorn Workshop* 20:87–111.

Illiger, C. 1811. *Prodromus Systematis Mammalium et Avium: Additis Terminis Zoographicis Utriusque Classis.* Sumptibus C. Salfield, Berolini (Berlin).

Janis, C. M., and E. Manning. 1998. *Antilocapridae.* Pp. 491–507 in *Evolution of Tertiary Mammals of North America. Vol. 1: Terrestrial Carnivores, Ungulates, and Ungulatelike Mammals.* (C. M. Janis, K. M. Scott, and L. L. Jacobs, eds.). Cambridge University Press, Cambridge, UK.

Janis, C. M., and K. M. Scott. 1988. The phylogeny of the *Ruminantia* (*Artiodactyla, Mammalia*). Pp. 273–282 in *The Phylogeny and Classification of Tetrapods. Vol. 2: Mammals.* (M. Benton, ed.). Academic Press [for the] Systematics Association, London–New York, International.

Köhler, M. 1987. Boviden des turkischen Miozäns (Känozoikum und Braunkohlen der Türkei). *Paleontol. Evol.* 21:133–246.

Lesson, R. P. 1827. *Manuel de Mammalogie.* Roret, Paris.

Linnaeus, C. 1758. *Systema Naturae Per Regna Tria Naturae, Secundum Classes, Ordines, Genera, Species, Cum Characteribus, Differentiis, Synonymis, Locis.* 10th edition, Tomes 1–2. Laurentii Salvii, Holmiae (Stockholm).

Morales, J., D. Soria, M. Pickford, and M. Nieto. 2003. A new genus and species of *Bovidae* (*Artiodactyla, Mammalia*) from the early middle Miocene of Arrisdrift, Namibia, and the origins of the family *Bovidae.* Pp. 371–384 in *Geology and Palaeobiology of the Central Southern Namib. Vol. 2: Palaeontology of the Orange River Valley, Namibia.* (M. Pickford and B. Senut, eds.). Geological Survey of Namibia, Ministry of Mines and Energy. Windhoek, Namibia.

Reichenow, A. 1886. Bericht über die Leistungen in der Naturgeschichte der Säugethiere während des Jahres 1885. *Arch. Naturgesch.* 52:97–144.

Sánchez, I. M., J. L. Cantalapiedra, M. Ríos, V. Quiralte, and J. Morales. 2015. Systematics and evolution of the Miocene three-horned palaeomerycid ruminants (*Mammalia, Cetartiodactyla*). *PLOS ONE* 10:e0143034.

Solounias, N., J. C. Barry, R. L. Bernor, E. H. Lindsay, and S. M. Raza. 1995. The oldest bovid from the Siwaliks, Pakistan. *J. Vertebr. Paleontol.* 15:806–814.

Webb, S. D. 1973. Pliocene pronghorns of Florida. *J. Mammal.* 54(1):203–221.

Authors

Faysal Bibi; Museum für Naturkunde; Leibniz Institute for Evolution and Biodiversity Science; Invalidenstrasse 43; 10115 Berlin, Germany. Email: faysal.bibi@mfn-berlin.de.

Elisabeth S. Vrba; Department of Geology & Geophysics; P.O. Box 208109; Yale University; New Haven, CT 06520-8109, USA. Email: elisabeth.vrba@yale.edu.

Date Accepted: 27 July 2016

Primary Editor: Jacques A. Gauthier

Bovidae J. E. Gray 1821 [F. Bibi and E. S. Vrba], converted clade name

Registration Number: 264

Definition: The smallest crown clade containing *Bos taurus* Linnaeus 1758 (*Bovini*) and *Antilope* (originally *Capra*) *cervicapra* (Linnaeus 1758) (*Antilopini*). This is a minimum-crown-clade definition. Abbreviated definition: min crown ∇ (*Bos taurus* Linnaeus 1758 & *Antilope cervicapra* (Linnaeus 1758)).

Etymology: Derived from the Latin *bovis* (cow).

Reference Phylogeny: Figure 1 in Hassanin et al. (2012).

Composition: *Bovidae* contains around 139 living species (Vrba and Schaller, 2000; IUCN, 2016) and hundreds of described fossil species. Extant species are divided between two subclades (subfamilies), *Bovinae* and *Antilopinae* (see entries in this volume), and further among ~10 lower-level clades (tribes) (Hassanin et al., 2012; Bibi, 2013).

Diagnostic Apomorphies: *Bovidae* is differentiated from all other living pecorans by the presence of permanent "cranial appendages that are unbranched, non-deciduous, and covered with a permanent keratin sheath" (Janis and Scott, 1988: 279). This morphological apomorphy— or at least those aspects determinable in the fossil record—is used to define the name of the more inclusive clade *Cavicornia* (this volume). It is also the only character that allows for the placement of early fossils within *Pan-Bovidae* (this volume). The monophyly of *Bovidae* among crown *Artiodactyla* is otherwise strongly supported by gene sequence data (e.g., Hassanin et al., 2012).

Synonyms: *Cavicornia* Illiger 1811 (approximate); *Bovina* Gray 1825 (approximate—included *Antilocapra*); *Tubicornes* Lesson 1827 (approximate—included *Antilocapra* and *Pudu*); *Cavicornidae* Reichenow 1886 (approximate).

Comments: We prefer the widely used name *Bovidae* to its rarely used approximate synonyms. The name was first used by Gray (1821) for a group that most closely approximates the one now called *Bovini* (he included only *Bos taurus* and excluded antelopes, goats, sheep, nilgais, and elands). Gervais (1855) used the French name "Bovidés" for a group of similar composition to the clade here named *Bovidae*. Wallace, adopting a classification he credited to Victor Brooke, described *Bovidae* as "all the animals commonly known as oxen, buffaloes, antelopes, sheep, and goats" (1876: 221). Brooke's classification of *Bovidae* included all then-known species currently referred to *Bovidae* with the addition of *Antilocapra*. The two species we use as specifiers, *Bos taurus* and *Antilope cervicapra*, were explicitly and implicitly included by Gray (1821) under his "*Bovidae*" and "*Antilopidae*", respectively; they are also the type species of *Bos* and *Antilope* (Wilson and Reeder, 2005). The presence of paired, unbranched, permanent horn cores on the frontal bones enclosed in permanent horn sheaths was noted as a taxonomically significant feature as early as Illiger (1811: 106): "*Cornua persistentia frontalia*".

Divergence-time estimates put the origin of *Bovidae* at about 25.4Ma (Hernández Fernández and Vrba, 2005), 22–18 Ma (Hassanin et al., 2012), or as young as 17-15 Ma (Bibi, 2013). It has been suggested that the basal split in *Bovidae* might reflect an early biogeographic separation, with *Bovinae* evolving in Eurasia and *Antilopinae* in Africa (Kingdon, 1982: 12). This hypothesis finds some support in phylogenetic work (Hassanin and Douzery, 1999; Hassanin and Ropiquet, 2004).

Literature Cited

Bibi, F. 2013. A multi-calibrated mitochondrial phylogeny of extant *Bovidae* (*Artiodactyla, Ruminantia*) and the importance of the fossil record to systematics. *BMC Evol. Biol.* 13:166.

Gervais, P. 1855. *Mammifères*. Curmer, Paris.

Gray, J. E. 1821. On the natural arrangement of vertebrose animals. *Lond. Med. Reposit.* 15:296–310.

Gray, J. E. 1825. An outline of an attempt at the disposition of *Mammalia* into tribes and families, with a list of the genera apparently appertaining to each tribe. *Ann. Philos., N.S.* 10:337–344.

Hassanin, A., F. Delsuc, A. Ropiquet, C. Hammer, B. Jansen van Vuuren, C. Matthee, M. Ruiz-Garcia, F. Catzeflis, V. Areskoug, T. T. Nguyen, and A. Couloux. 2012. Pattern and timing of diversification of *Cetartiodactyla* (*Mammalia, Laurasiatheria*), as revealed by a comprehensive analysis of mitochondrial genomes. *C. R. Biol.* 335:32–50.

Hassanin, A., and E. J. P. Douzery. 1999. The tribal radiation of the family *Bovidae* (*Artiodactyla*) and the evolution of the mitochondrial cytochrome b gene. *Mol. Phylogenet. Evol.* 13:227–243.

Hassanin, A., and A. Ropiquet. 2004. Molecular phylogeny of the tribe *Bovini* (*Bovidae, Bovinae*) and the taxonomic status of the Kouprey, *Bos sauveli* Urbain 1937. *Mol. Phylogenet. Evol.* 33:896–907.

Hernández Fernández, M., and E. S. Vrba. 2005. A complete estimate of the phylogenetic relationships in *Ruminantia*: a dated species-level supertree of the extant ruminants. *Biol. Rev. Camb. Philos. Soc.* 80:269–302.

Illiger, C. 1811. *Prodromus Systematis Mammalium et Avium: Additis Terminis Zoographicis Utriusque Classis*. Sumpibus C. Salfild, Berolini (Berlin).

IUCN. 2016. IUCN *Red List of Threatened Species*, Version 2015-3. Available at www.iucnredlist.org, accessed on 1 April 2016.

Janis, C. M., and K. M. Scott. 1988. The phylogeny of the *Ruminantia* (*Artiodactyla, Mammalia*). Pp. 273–282 in *The Phylogeny and Classification of Tetrapods. Vol. 2: Mammals*. (M. J. Benton, eds.). Academic Press [for the] Systematics Association, London–New York, International.

Kingdon, J. 1982. *East African Mammals: An Atlas of Evolution in Africa*, Vol. IIIC. Academic Press, London.

Lesson, R. P. 1827. *Manuel de Mammalogie*. Roret, Paris.

Linnaeus, C. 1758. *Systema Naturae Per Regna Tria Naturae, Secundum Classes, Ordines, Genera, Species, cum Characteribus, Differentiis, Synonymis, Locis*. 10th edition, Tomes 1–2. Laurentii Salvii, Holmiae (Stockholm).

Reichenow, A. 1886. Bericht über die Leistungen in der Naturgeschichte der Säugethiere während des Jahres 1885. *Arch. Naturgesch.* 52:97–144.

Vrba, E. S., and G. Schaller, eds. 2000. *Antelopes, Deer, and Relatives*. Yale University Press, New Haven, CT.

Wallace, A. R. 1876. *The Geographical Distribution of Animals*, Vol. 2. Harper & Brothers, New York.

Wilson, D. E., and D. M. Reeder, eds. 2005. *Mammal Species of the World*. 3rd edition. Johns Hopkins University Press. Baltimore, MD.

Authors

Faysal Bibi; Museum für Naturkunde; Leibniz Institute for Evolution and Biodiversity Science; Invalidenstrasse 43; 10115 Berlin, Germany. Email: faysal.bibi@mfn-berlin.de.

Elisabeth S. Vrba; Department of Geology & Geophysics; P.O. Box 208109; Yale University; New Haven, CT 06520-8109, USA. Email: elisabeth.vrba@yale.edu.

Date Accepted: 07 July 2016

Primary editor: Jacques A. Gauthier

Bovinae T. Gill 1872 [F. Bibi and E. S. Vrba], converted clade name

Registration Number: 265

Definition: The crown clade originating in the most recent common ancestor of *Bos taurus* Linnaeus 1758 (*Bovini*) and all extant taxa that share a more recent common ancestor with *Bos taurus* than with *Antilope* (originally *Capra*) *cervicapra* (Linnaeus 1758) (*Antilopini*). This is a maximum-crown-clade definition. Abbreviated definition: max crown ∇ (*Bos taurus* Linnaeus 1758 ~ *Antilope cervicapra* (Linnaeus 1758)).

Etymology: Derived from the Latin *bovis* (cow).

Reference Phylogeny: Figure 1 in Hassanin et al. (2012).

Composition: *Bovinae* contains around 23 living species (Vrba and Schaller, 2000; IUCN, 2016), including buffaloes and oxen (*Bovini*), spiral horned antelopes (*Tragelaphini*), the nilgais and chousinghas (*Boselaphini*), and the saola (*Pseudoryx nghetinhensis*).

Diagnostic Apomorphies: *Bovinae* may be distinguished from *Antilopinae* (this volume) by many characters. However, with the current lack of resolution of phylogenetic relationships of early *Pan-Bovidae* (this volume), it is not clear which (if any) of these are truly apomorphic, rather than retained primitive features of all *Bovidae*: braincase box-like with flattened dorsal, lateral, and posterior faces (probably primitive for all *Bovidae*); face only weakly flexed on braincase; cranial interfrontal and frontoparietal sutures simple, never extremely convoluted, and often obliterated by fusion; horns keeled, primitively 2 or 3 keels (1 anterior and 1 or 2 posterior keels; entirely lost in some *Bovini*; lost,

or perhaps never present, in *Pseudoryx nghetinhensis*); horn cores never annulated, horn sheath typically without annulation (transverse ridging) but weak if present (e.g., some *Bubalus*); horns with torsion that is anticlockwise on the right side (lost in some deeply nested taxa; lost, or perhaps never present, in *Pseudoryx nghetinhensis*; Gatesy and Arctander, 2001a); postcornual fossa usually absent, or very shallow and weakly developed if present; thermoregulation by sweating and not nasal panting (Kingdon, 1982:12); interdigital pedal glands absent (Kingdon, 1982; Pocock, 1910; Pocock, 1918); territorial behavior absent (except *Boselaphus tragocamelus*).

Synonyms: *Boodontia* Schlosser in Zittel 1911 sensu Vrba and Schaller (2000) (approximate); *Tragelaphoidea* Pilgrim (1939:23) (approximate); "*Boselaphini*" (partial, as used in most palaeontological literature see Bibi et al., 2009).

Comments: Monophyly of *Bovinae* relative to other crown *Artiodactyla* (this volume) has been supported by molecular and behavioral data (e.g., Buckland and Evans, 1978; Kingdon, 1982; Allard et al., 1992; Gatesy et al., 1997; Hassanin and Douzery, 1999a,b, 2003; Gatesy and Arctander, 2000b; Vrba and Schaller, 2000; Matthee and Davis, 2001; Hernández Fernández and Vrba, 2005; Bibi, 2013). We prefer using '*Bovinae*' for this clade to preserve its long association with *Bos taurus* and because potential alternatives are much less commonly used. Older uses of *Bovinae* referred to the taxon now named *Bovini* (e.g., Pilgrim, 1939; Wallace, 1876:222). While traditionally attributed to Gray (1821) because of the principle of coordination in rank-based nomenclature (e.g., Grubb, 2001), the first author to use the term

Bovinae in its current sense was Simpson (1945). The name itself was coined by Gill (1872; in Simpson, 1945).

Eotragus sansaniensis (14.5 Ma, France) and an unnamed taxon from the Vihowa Formation (ca. 17 Ma, Pakistan) have been proposed as early *Bovinae* (Azanza and Morales, 1994; Solounias et al., 1995). Such assignments are, however, based on characters that are highly homoplastic, or the polarity of which is not known with certainty.

Many fossils have traditionally been assigned to "*Boselaphini*" (e.g., Pilgrim, 1939; Thomas, 1984), a paraphyletic assemblage of fossil taxa from which crown *Bovini*, crown *Boselaphini*, and probably crown *Tragelaphini*, all descended (Bibi, 2007; Bibi et al., 2009).

Divergence estimates put the origin of crown *Bovinae* at 20.5 Ma (Hernández Fernández and Vrba, 2005), 18-14 Ma (Hassanin et al., 2012), or 15-13 Ma (Bibi, 2013).

Literature Cited

Allard, M. W., M. M. Miyamoto, L. Jarecki, F. Kraus, and M. R. Tennant. 1992. DNA systematics and evolution of the artiodactyl family *Bovidae*. *Proc. Natl. Acad. Sci. USA* 89:3972–3976.

Azanza, B., and J. Morales. 1994. *Tethytragus* nov. gen et *Gentrytragus* nov. gen. Deux nouveaux bovidés (*Artiodactyla, Mammalia*) du Miocène moyen. *Proc. K. Ned. Akad. Van Wet. B* 97:249–282.

Bibi, F. 2007. Origin, paleoecology, and paleobiogeography of early *Bovini*. *Palaeogeogr. Palaeoclimatol. Palaeoecol.* 248:60–72.

Bibi, F. 2013. A multi-calibrated mitochondrial phylogeny of extant *Bovidae* (*Artiodactyla, Ruminantia*) and the importance of the fossil record to systematics. *BMC Evol. Biol.* 13:166.

Bibi, F., M. Bukhsianidze, A. W. Gentry, D. Geraads, D. S. Kostopoulos, and E. S. Vrba. 2009. The fossil record and evolution of *Bovidae*: state of the field. *Palaeontol. Electron.* 12,10A:11p.

Buckland, R. A., and H. J. Evans. 1978. Cytogenetic aspects of phylogeny in the *Bovidae* I. G-banding. *Cytogenet. Cell Genet.* 21:42–63.

Gatesy, J., G. Amato, E. Vrba, G. Schaller, and R. DeSalle. 1997. A cladistic analysis of mitochondrial ribosomal DNA from the *Bovidae*. *Mol .Phylogenet. Evol.* 7:303–319.

Gatesy, J., and P. Arctander. 2000a. Hidden morphological support for the phylogenetic placement of *Pseudoryx nghetinhensis* with bovine bovids: a combined analysis of gross anatomical evidence and DNA sequences from five genes. *Syst. Biol.* 49:515–538.

Gatesy, J., and P. Arctander. 2000b. Molecular evidence for the phylogenetic affinities of *Ruminantia*. Pp. 143–155 in *Antelopes, Deer, and Relatives* (E. S. Vrba and G. B. Schaller, eds.). Yale University Press, New Haven, CT.

Gill, T. 1872. Arrangement of the families of mammals. *Smithson. Misc. Collect.* 11:1–98.

Gray, J. E. 1821. On the natural arrangement of vertebrose animals. *Lond. Med. Reposit.* 15:296–310.

Grubb, P. 2005. Family *Bovidae*. Pp. 673–722 in *Mammals Species of the World: A Taxonomic and Geographic Reference* (D. E. Wilson and D. A. M. Reeder, eds.). Johns Hopkins University Press, Baltimore, MD.

Hassanin, A., F. Delsuc, A. Ropiquet, C. Hammer, B. Jansen van Vuuren, C. Matthee, M. Ruiz-Garcia, F. Catzeflis, V. Areskoug, T. T. Nguyen, and A. Couloux. 2012. Pattern and timing of diversification of *Cetartiodactyla* (*Mammalia, Laurasiatheria*), as revealed by a comprehensive analysis of mitochondrial genomes. *C. R. Biol.* 335:32–50.

Hassanin, A., and E. J. P. Douzery. 1999a. Evolutionary affinities of the enigmatic saola (*Pseudoryx nghetinhensis*) in the context of the molecular phylogeny of *Bovidae*. *Proc. R. Soc. Lond. B Biol. Sci.* 266:893–900.

Hassanin, A., and E. J. P. Douzery. 1999b. The tribal radiation of the family *Bovidae* (*Artiodactyla*) and the evolution of the mitochondrial cytochrome b gene. *Mol. Phylogenet. Evol.* 13:227–243.

Hassanin, A., and E. J. P. Douzery. 2003. Molecular and morphological phylogenies of *Ruminantia* and the alternative position of the *Moschidae*. *Syst. Biol.* 52:206–228.

Hernández Fernández, M., and E. S. Vrba. 2005. A complete estimate of the phylogenetic relationships in *Ruminantia*: a dated species-level supertree of the extant ruminants. *Biol. Rev. Camb. Philos. Soc.* 80:269–302.

IUCN. 2016. *IUCN Red List of Threatened Species*, Version 2015-3. Available at www.iucnredlist.org, accessed on 1 April 2016.

Kingdon, J. 1982. *East African Mammals: An Atlas of Evolution in Africa*. Academic Press, London.

Linnaeus, C. 1758. *Systema Naturae Per Regna Tria Naturae, Secundum Classes, Ordines, Genera, Species, cum Characteribus, Differentiis, Synonymis, Locis.* 10th edition, Tomes 1–2. Laurentii Salvii, Holmiae (Stockholm).

Matthee, C. A., and S. K. Davis. 2001. Molecular insights into the evolution of the family *Bovidae*: a nuclear DNA perspective. *Mol. Biol. Evol.* 18:1220–1230.

Pilgrim, G. E. 1939. The fossil *Bovidae* of India. *Palaeontol. Indica* NS 26:1–356.

Pocock, R. I. 1910. On the specialized cutaneous glands of ruminants. *Proc. Zool. Soc. Lond.* (1910):840–986.

Pocock, R. I. 1918. On some external characters of ruminant *Artiodactyla*—Part V. The *Tragelaphinae*, Part VI. The *Bovinae*. *Ann. Mag. Nat. Hist.* s9, 2:440–459.

Simpson, G. G. 1945. The principles of classification and a classification of mammals. *Bull. Am. Mus. Nat. Hist.* 85:1–350.

Solounias, N., J. C. Barry, R. L. Bernor, E. H. Lindsay, and S. M. Raza. 1995. The oldest bovid from the Siwaliks, Pakistan. *J. Vertebr. Paleontol.* 15:806–814.

Thomas, H. 1984. Les *Bovidae* (*Artiodactyla*; *Mammalia*) du Miocène du sous-continent indien, de la péninsule arabique et de l'Afrique: biostratigraphie, biogéographie et écologie. *Palaeogeogr. Palaeoclimatol. Palaeoecol.* 45:251–299.

Vrba, E. S., and G. Schaller. 2000. Phylogeny of *Bovidae* based on behavior, glands, skulls, and postcrania. Pp. 203–222 in *Antelopes, Deer, and Relatives* (E. S. Vrba and G. Schaller, eds.). Yale University Press, New Haven, CT.

Wallace, A. R. 1876. *The Geographical Distribution of Animals*. Harper & Brothers, New York.

von Zittel, K. A. 1911. *Grunzüge der Paläontologie (Paläozoologie), II. Abteilung* Vertebrata. Oldenbourg, Munich.

Authors

Faysal Bibi; Museum für Naturkunde; Leibniz Institute for Evolution and Biodiversity Science; Invalidenstrasse 43; 10115 Berlin, Germany. Email: faysal.bibi@mfn-berlin.de.

Elisabeth S. Vrba; Department of Geology & Geophysics; P.O. Box 208109; Yale University; New Haven, CT 06520-8109, USA. Email: elisabeth.vrba@yale.edu.

Date Accepted: 07 July 2016

Primary Editor: Jacques A. Gauthier

Antilopinae S. F. Baird 1857 [F. Bibi and E. S. Vrba], converted clade name

Registration Number: 266

Definition: The crown clade originating in the most recent common ancestor of *Antilope* (originally *Capra*) *cervicapra* (Linnaeus 1758) (*Antilopini*) and all extant bovids that share a more recent common ancestor with *Antilope cervicapra* than with *Bos taurus* Linnaeus 1758 (*Bovini*). This is a maximum-crown-clade definition. Abbreviated definition: max crown ∇ (*Antilope cervicapra* (Linnaeus 1758) ~ *Bos taurus* Linnaeus 1758).

Etymology: Derived from the Greek *antholops* (horned animal).

Reference Phylogeny: Figure 1 in Hassanin et al. (2012).

Composition: *Antilopinae* includes the majority of living bovids, with around 115 extant species (Vrba and Schaller, 2000; IUCN, 2016) divided among the following clades (tribes): *Aepycerotini*, *Antilopini*, *Cephalophini*, *Reduncini*, *Hippotragini*, *Alcelaphini*, and *Caprini*.

Diagnostic Apomorphies: With the current lack of resolution of phylogenetic relationships of early *Pan-Bovidae* (this volume) it is not clear which of the following characters are truly apomorphic, rather than retained primitive features. Nonetheless, the following character states distinguish—if not diagnose—members of *Antilopinae* from those of its sister clade *Bovinae* (this volume). Braincase rounded, especially dorsally such that the transition between dorsal, occipital, and lateral faces is often smooth; cranial interfrontal and frontoparietal sutures typically very convoluted, not obliterated by fusion; horn core sheaths often annulated with transverse ridges (Kingdon, 1982; Vrba and Schaller, 2000); postcornual fossa present; thermoregulation by nasal panting rather than sweating (Kingdon, 1982); interdigital pedal glands present (lost in *Redunca*, *Kobus*, and *Aepyceros*) (Pocock, 1910, 1918; Kingdon, 1982); territorial behavior common.

Synonyms: *Aegodontia* Schlosser in Zittel 1911 sensu Vrba and Schaller (2000) (approximate).

Comments: *Antilopinae* is well supported (e.g., Hassanin et al., 2012) even if missing data and some character discordance in fossils introduce uncertainty in morphological analyses (e.g., Vrba and Schaller, 2000). The clade contains most of the species diversity and exhibits much of the morphological disparity among extant bovids. The name *Antilopinae* is far more commonly used in association with this clade than its approximate synonym, *Aegodontia*. Attributed to Gray (1821) under the principle of coordination under rank-based nomenclature, the actual name *Antilopinae* was coined by Baird (1857; Simpson, 1945). Baird included the American pronghorn (*Antilocapra americana*, not a bovid) and the American mountain goat (*Aplocerus montanus*, junior synonym of *Oreamnos americanus*) within *Antilopinae*, meaning his name can now be regarded as a partial synonym of *Pecora*. Since Simpson (1945), *Antilopinae* has been used by many authors to designate a smaller clade comprising only "*Antilopini*" (gazelles and close relatives) and "*Neotragini*" (dwarf antelopes). "*Neotragini*" has since been shown to be a polyphyletic assemblage of small antelopes, many of which belong in a monophyletic *Antilopini* (e.g., Hassanin and Douzery, 1999; Matthee and Robinson, 1999; Rebholz and Harley, 1999; Hassanin et al., 2012). *Antilopinae* was first used in its current sense by Kingdon (1982) then by Hassanin and Douzery (1999, 2003).

Extinct *Pseudoeotragus seegrabensis* from Austria might be the earliest known member of the total clade of *Antilopinae* (Azanza and Morales, 1994: Figs. 7–8) (viz., *Pan-Antilopinae*, a name not defined here), with a maximum age of 16.6 or 17.0 Ma (Mammal Neogene Unit (MN); Made, 1989; Agustí et al., 2001). By 14 Ma, at sites from the Mediterranean, Africa, Arabia, and southern Asia, numerous taxa are present that may be referred to the total clade of *Antilopinae* or even to crown *Antilopinae* (Azanza and Morales, 1994; Gentry, 2000, and references therein). See Bibi et al. (2009) and Bibi (2013) for a review.

Divergence-time estimates variously place the origin of *Antilopinae* (as defined here) at around 23 Ma (Hernández Fernández and Vrba, 2005), 17 Ma (Hassanin et al., 2012), or 16–13 Ma (Hassanin and Douzery, 1999, Hassanin and Ropiquet, 2004; Bibi, 2013). Based on the distribution of its extant species, the origin of *Antilopinae* is thought to lie in Africa (Kingdon, 1982; Gentry, 1994; Hassanin and Douzery, 1999).

Literature Cited

Agustí, J., L. Cabrera, M. Garces, W. Krijgsman, O. Oms, and J. M. Pares. 2001. A calibrated mammal scale for the Neogene of Western Europe. *State Art Earth-Sci. Rev.* 52:247–260.

Azanza, B., and J. Morales. 1994. *Tethytragus* nov. gen et *Gentrytragus* nov. gen. Deux nouveaux bovidés (*Artiodactyla, Mammalia*) du Miocène moyen. *Proc. K. Ned. Akad. Van Wet. B* 97:249–282.

Baird, S. F. 1857. Mammals. *Explorations and Surveys for a Railroad Route from the Mississippi River to the Pacific Ocean* 8:1–757.

Bibi, F. 2013. A multi-calibrated mitochondrial phylogeny of extant *Bovidae* (*Artiodactyla, Ruminantia*) and the importance of the fossil record to systematics. *BMC Evol. Biol.* 13:166.

Bibi, F., M. Bukhsianidze, A. W. Gentry, D. Geraads, D. S. Kostopoulos, and E.S. Vrba. 2009. The fossil record and evolution of *Bovidae*: state of the field. *Palaeontol. Electron.* 12(3):10A,11pp.

Gentry, A. W. 1994. The Miocene differentiation of Old World *Pecora* (*Mammalia*). *Hist. Biol.* 7:115–158.

Gentry, A. W. 2000. *Caprinae* and *Hippotragini* (*Bovidae, Mammalia*) in the Upper Miocene. Pp. 65–83 in *Antelopes, Deer, and Relatives* (E. S. Vrba and G. Schaller, eds.). Yale University Press, New Haven, CT.

Gray, J. E. 1821. On the natural arrangement of vertebrose animals. *Lond. Med. Reposit.* 15:296–310.

Hassanin, A., F. Delsuc, A. Ropiquet, C. Hammer, B. Jansen van Vuuren, C. Matthee, M. Ruiz-Garcia, F. Catzeflis, V. Areskoug, T. T. Nguyen, and A. Couloux. 2012. Pattern and timing of diversification of *Cetartiodactyla* (*Mammalia, Laurasiatheria*), as revealed by a comprehensive analysis of mitochondrial genomes. *C. R. Biol.* 335:32–50.

Hassanin, A., and E. J. P. Douzery. 1999. The tribal radiation of the family *Bovidae* (*Artiodactyla*) and the evolution of the mitochondrial cytochrome b gene. *Mol. Phylogenet. Evol.* 13:227–243.

Hassanin, A., and E. J. P. Douzery. 2003. Molecular and morphological phylogenies of *Ruminantia* and the alternative position of the *Moschidae*. *Syst. Biol.* 52:206–228.

Hassanin, A., and A. Ropiquet. 2004. Molecular phylogeny of the tribe *Bovini* (*Bovidae, Bovinae*) and the taxonomic status of the kouprey, *Bos sauveli* Urbain 1937. *Mol. Phylogenet. Evol.* 33:896–907.

Hernández Fernández, M., and E. S. Vrba. 2005. A complete estimate of the phylogenetic relationships in *Ruminantia*: a dated species-level supertree of the extant ruminants. *Biol. Rev. Camb. Philos. Soc.* 80:269–302.

IUCN. 2016. *IUCN Red List of Threatened Species,* Version 2015–3. Available at www.iucnredlist. org, accessed on 1 April 2016.

Kingdon, J. 1982. *East African Mammals: An Atlas of Evolution in Africa,* Volume IIIC. Academic Press, London.

Linnaeus, C. 1758. *Systema Naturae Per Regna Tria Naturae, Secundum Classes, Ordines, Genera, Species, cum Characteribus, Differentiis, Synonymis, Locis.* 10th edition, Tomes 1–2. Laurentii Salvii, Holmiae (Stockholm).

Made, J. v. d. 1989. The bovid *Pseudoeotragus seegrabensis* nov. gen., nov. sp. from the Aragonia (Miocene) of Seegraben near Leoben (Austria). *Proc. K. Ned. Akad. Van Wet. Ser. B* 92:215–240.

Matthee, C. A., and T. J. Robinson. 1999. Cytochrome b phylogeny of the family *Bovidae*: resolution within the *Alcelaphini, Antilopini, Neotragini,* and *Tragelaphini. Mol. Phylogenet. Evol.* 12:31–46.

Pocock, R. I. 1910. On the specialized cutaneous glands of ruminants. *Proc. Zool. Soc. Lond.* 1910:840–986.

Pocock, R. I. 1918. On some external characters of ruminant *Artiodactyla*—Part IV. The *Reduncinae (Cervicaprinae)* and *Aepycerinae. Ann. Mag. Nat. Hist.* s9, 2:367–374.

Rebholz, W., and E. Harley. 1999. Phylogenetic relationships in the bovid subfamily *Antilopinae* based on mitochondrial DNA sequences. *Mol. Phylogenet. Evol.* 12:87–94.

Simpson, G. G. 1945. The principles of classification and a classification of mammals. *Bull. Am. Mus. Nat. Hist.* 85:1–350.

Vrba, E. S., and G. Schaller. 2000. Phylogeny of *Bovidae* based on behavior, glands, skulls, and postcrania. Pp. 203–222 in *Antelopes, Deer, and Relatives* (E. S. Vrba and G. Schaller, eds.). Yale University Press, New Haven, CT.

von Zittel, K. A. 1911. *Grunzüge der Paläontologie (Paläozoologie), II. Abteilung* Vertebrata. Oldenbourg, Munich.

Authors

Faysal Bibi; Museum für Naturkunde; Leibniz Institute for Evolution and Biodiversity Science; Invalidenstrasse 43; 10115 Berlin, Germany. Email: faysal.bibi@mfn-berlin.de.

Elisabeth S. Vrba; Department of Geology & Geophysics; P.O. Box 208109; Yale University; New Haven, CT 06520-8109, USA. Email: elisabeth.vrba@yale.edu.

Date Accepted: 07 July 2016

Primary Editor: Jacques A. Gauthier

Pan-Carnivora J. J. Flynn, A. R. Wyss, and M. Wolsan, new clade name

Registration Number: 267

Definition: The total clade of the crown clade *Carnivora*. This is a crown-based total-clade definition. Abbreviated definition: total ∇ of *Carnivora*.

Etymology: *Pan-*, from the Greek *pan-* or *pantos* (all, the whole), indicating that the name refers to a total clade, and *Carnivora*, from the Latin *carnis* (genitive) meaning "flesh" and *vorare* meaning "to devour," i.e., "flesh eating."

Reference Phylogeny: The primary reference phylogeny is Figure 3 of Wesley-Hunt and Flynn (2005). *Pan-Carnivora* corresponds to the clade originating in the branch separating *Thinocyon* and all taxa to the left from *Leptictis*, *Erinaceus*, and *Echinosorex* in that figure. See also Flynn and Wesley-Hunt (2005: Fig. 12.2).

Composition: *Carnivora* (this volume; reference phylogeny in Flynn et al., 2005, Fig. 5) and all extinct mammals more closely related to *Carnivora* than to any other mutually exclusive crown clade. Late Palaeocene through middle Miocene taxa currently regarded as members of *Pan-Carnivora* that are outside of crown-clade *Carnivora* include *Creodonta* (or the individual taxa *Hyaenodontidae* and *Oxyaenidae*, as *Creodonta* may be non-monophyletic; see Polly, 1996, and Flynn and Wesley-Hunt, 2005), *Viverravidae,* and species traditionally assigned to "*Miacidae*" (now considered paraphyletic with respect to crown-clade *Carnivora*; see Spaulding and Flynn, 2009, 2012, and discussion in Flynn and Wesley-Hunt, 2005). It is conceivable that certain Late Cretaceous taxa are also members of *Pan-Carnivora* (Hunt and Tedford, 1993), but the phylogenetic placement of these taxa (e.g., *Cimolestes*) remains controversial (Flynn and Wesley-Hunt, 2005). See McKenna and Bell (1997) for a more detailed list of referred stem carnivorans.

Diagnostic Apomorphies: Apomorphies were listed by Wesley-Hunt and Flynn (2005: 11, Table 1, and accompanying appendix; see also Flynn and Wesley-Hunt, 2005, for apomorphies of various clades within *Pan-Carnivora* that are more inclusive than the crown, but more closely related to crown-clade *Carnivora* than to *Creodonta* [*Hyaenodontidae*, *Oxyaenidae*], including *Carnivoraformes* and *Carnivoramorpha*). The study of Flynn et al. (1988), although not employing a matrix-based analysis, proposed seven potential synapomorphies for a clade approximating *Pan-Carnivora* (*Carnivora* plus *Creodonta*; called *Ferae* by most workers [e.g., Simpson, 1945; Flynn and Wesley-Hunt, 2005; O'Leary et al., 2013], although *Ferae* is used by some authors for a more inclusive clade that also contains *Pholidota* [e.g., Asher et al., 2009]): restriction of carnassial shear to a small part of the postcanine dentition; well-developed osseous tentorium, with a strong medial projection, in the cranial cavity; strong processus hyoideus ("mastoid tubercle" of Novacek, 1986) of the petrosal, arching toward and closely approaching the promontorium; entotympanic contribution to the bulla; grooves on the petrosal indicating a transpromontorial path of the internal carotid artery; medial border of the astragalar tibial trochlea much longer than the lateral border; calcaneal anterior plantar tubercle relatively far posterior, and peroneal process reduced. Although sampling only two crown and three stem carnivorans, O'Leary et al. (2013: Fig. 1 and Appendix S2) identified

31 potential unambiguous cranial, dental and postcranial synapomorphies for these five taxa (*Creodonta* + *Carnivoramorpha*), a clade approximating *Pan-Carnivora* (Clade H of Fig. 1; Node 118 of Fig. S15 and Appendix S2), including presence of maxilla palatal process embrasure pits, presence of a squamosal post-tympanic crest, presence of a postorbital process, a moderately wide mandibular symphysis (≥ 15% but < 20% of maximum mandibular length), absence of a deltoid tubercle on the humerus, presence of an anterolateral ridge on the ulna olecranon, small ischial tuberosity in the pelvis, presence of a shelf connecting the third and greater trochanters of the femur, and astragalus trochlear articular surface extends onto astragalar neck. No unambiguous dental synapomorphies are known for this clade, as the nearest outgroup sampled (a pangolin) is edentulous, thus precluding unambiguous optimization of any dental characters at the base of *Pan-Carnivora*. Many additional, ambiguously optimized, potential morphological synapomorphies are provided by O'Leary et al. (2013, Appendices S3 [Deltran] and S4 [Acctran] parsimony optimizations) for *Ferae* (≈ *Pan-Carnivora*; "node 128 to 126"), including classic feraean dental features such as carnassial shear present at one or more cheek tooth loci and presence of a "V" (notch) or slit on the metastylar blade of the upper 5th premolar ["P4" of traditional studies] (Characters 2646 and 1975, respectively, under Deltran optimization).

Synonyms: *Ferae* and *Carnivora* sensu Simpson 1945 (approximate).

Comments: Simpson (1945) used Linnaeus's (1758) name *Ferae* to refer to *Creodonta*, *Fissipeda*, and *Pinnipedia* collectively, and *Carnivora* to include both crown-clade *Carnivora* plus fossil stem taxa he termed "*Miacoidea*". We have applied the name *Carnivora* to the crown clade (in this volume) and, in the interest of developing an integrated system of clade names, here apply the name *Pan-Carnivora*, rather than *Ferae*, to its total clade. *Carnivoramorpha* Wyss and Flynn 1993 (see also Bryant, 1996; and Flynn and Wesley-Hunt, 2005) was explicitly tied to a clade including all species more closely related to (crown) *Carnivora* than to *Creodonta* (*Hyaenodontidae* and *Oxyaenidae*), thus encompassing more crownward stem pan-carnivorans (including "miacoids" [*Viverravidae* and "*Miacidae*"]) (Flynn et al., 2010), and the name *Carnivoramorpha* will be formally linked to this less inclusive *Pan-Carnivora* clade at a later date. The extant sister to *Carnivora* among crown placentals is not clearly established based on morphological evidence; however, most molecular evidence (e.g., Murphy et al., 2001; Meredith et al., 2011) and a recent combined molecular and morphological analysis including fossils (O'Leary et al., 2013) indicate that *Pholidota* is the extant sister to *Carnivora*, forming a clade named *Ostentoria* (e.g., Meredith et al., 2011; O'Leary et al., 2013; called *Ferae* by some authors [e.g., McKenna and Bell, 1997; Asher et al., 2009]).

Literature Cited

Asher, R. J., N. Bennett, and T. Lehmann. 2009. The new framework for understanding placental mammal evolution. *BioEssays* 31(8):853–864.

Bryant, H. N. 1996. Explicitness, stability, and universality in the phylogenetic definition and usage of taxon names: a case study of the phylogenetic taxonomy of the *Carnivora* (*Mammalia*). *Syst. Biol.* 45:174–189.

Flynn, J. J., J. A. Finarelli, and M. Spaulding. 2010. Phylogeny of the *Carnivora* and *Carnivoramorpha*, and the use of the fossil record to enhance understanding of evolutionary transformations. Pp. 25–63 in Carnivora *Evolution* (A. Goswami and A. Friscia, eds.). Cambridge University Press, Cambridge, UK.

Flynn, J. J., J. A. Finarelli, S. Zehr, J. Hsu, and M. A. Nedbal. 2005. Molecular phylogeny of the *Carnivora* (*Mammalia*): assessing the impact of increased sampling on resolving enigmatic relationships. *Syst. Biol.* 54:317–337.

Flynn, J. J., N. A. Neff, and R. H. Tedford. 1988. Phylogeny of the *Carnivora*. Pp. 73–116 in *The Phylogeny and Classification of the Tetrapods. Vol. 2: Mammals* (M. J. Benton, ed.). The Systematics Association Special Volume No. 35B. Clarendon Press, Oxford.

Flynn, J. J., and G. D. Wesley-Hunt. 2005. *Carnivora*. Pp. 175–198 in *The Rise of Placental Mammals: Origins and Relationships of the Major Extant Clades* (K. D. Rose and J. D. Archibald, eds.). Johns Hopkins University Press, Baltimore, MD.

Hunt, Jr., R. M., and R. H. Tedford. 1993. Phylogenetic relationships within the aeluroid *Carnivora* and implications of their temporal and geographic distribution. Pp. 53–74 in *Mammal Phylogeny: Placentals*, Vol. 2. (F. S. Szalay, M. J. Novacek and M. C. McKenna, eds.). Springer-Verlag, New York.

Linnaeus, C. 1758. *Systema Naturae Per Regna Tria Naturae, Secundum Classes, Ordines, Genera, Species, cum Characteribus, Differentiis, Synonymis, Locis.* Tomus I, Editio decima, reformata. Laurentius Salvius, Holmiae (Stockholm).

McKenna, M. C., and S. K. Bell. 1997. *Classification of Mammals above the Species Level.* Columbia University Press, New York.

Meredith, R. W., J. E. Janecka, J. Gatesy, O. A. Ryder, C. A. Fisher, E. C. Teeling, A. Goodbla, E. Eizirik, T. L. L. Simão, T. Stadler, D. L. Rabosky, R. L. Honeycutt, J. J. Flynn, C. M. Ingram, C. Steiner, T. L. Williams, T. J. Robinson, A. Burk, M. Westerman, N. A. Ayoub, M. S. Springer, and W. J. Murphy. 2011. Impacts of the Cretaceous Terrestrial Revolution and KPg extinction on extant mammal diversification. *Science* 334: 521–524.

Murphy, W. J., E. Eizirik, S. J. O'Brien, O. Madsen, M. Scally, C. J. Douady, E. Teeling, O. A. Ryder, M. J. Stanhope, W. W. de Jong, and M. S. Springer. 2001. Resolution of the early placental mammal radiation using Bayesian phylogenetics. *Science* 294:2348–2351.

Novacek, M. J. 1986. The skull of leptictid insectivorans and the higher-level classification of eutherian mammals. *Bull. Am. Mus. Nat. Hist.* 183:1–112.

O'Leary, M. A., J. I. Bloch, J. J. Flynn, T. J. Gaudin, A. Giallombardo, N. P. Giannini, S. L. Goldberg, B. P. Kraatz, Z.-X. Luo, J. Meng, X. Ni, M. J. Novacek, F. A. Perini, Z. Randall, G. W. Rougier, E. J. Sargis, M. T. Silcox, N. B. Simmons, M. Spaulding, P. M. Velazco, M. Weksler, J. R. Wible, and A. L. Cirranello. 2013. The placental mammal ancestor and the post—K-Pg radiation of placentals. *Science* 339:662–667.

Polly, P. D. 1996. The skeleton of *Gazinocyon vulpeculus* gen. et comb. nov. and the cladistic relationships of *Hyaenodontidae* (*Eutheria, Mammalia*). *J. Vertebr. Paleontol.* 16:303–319.

Simpson, G. G. 1945. The principles of classification and a classification of mammals. *Bull. Am. Mus. Nat. Hist.* 85:1–350.

Spaulding, M., and J. J. Flynn. 2009. Anatomy of the postcranial skeleton of '*Miacis*' *uintensis* (*Mammalia*: *Carnivoramorpha*). *J. Vertebr. Paleontol.* 29 (4):1212–1223.

Wesley-Hunt, G. D., and J. J. Flynn. 2005. Phylogeny of the *Carnivora*: basal relationships among the carnivoramorphans, and assessment of the position of "*Miacoidea*" relative to crown-clade *Carnivora*. *J. Syst. Palaeontol.* 3(1):1–28.

Wyss, A. R., and J. J. Flynn. 1993. A phylogenetic analysis and definition of the *Carnivora*. Pp. 32–52 in *Mammal Phylogeny: Placentals* (F. Szalay, M. Novacek, and M. McKenna, eds.). Springer-Verlag, New York.

Authors

John J. Flynn; Division of Paleontology and Richard Gilder Graduate School; American Museum of Natural History; Central Park West at 79th Street; New York, NY 10024-5192, USA. Email: jflynn@amnh.org.

André R. Wyss; Department of Earth Science; University of California at Santa Barbara; Webb Hall; Santa Barbara, CA 93106-9630, USA. Email: wyss@geol.ucsb.edu.

Mieczysław Wolsan; Museum and Institute of Zoology; Polish Academy of Sciences; Wilcza 64; 00-679 Warszawa; Poland. Email: wolsan@miiz.waw.pl.

Date Accepted: 04 January 2016

Primary Editor: Jacques A. Gauthier

Carnivora T. Bowdich 1821 [J. J Flynn, A. R. Wyss, and M. Wolsan], converted clade name

Registration Number: 268

Definition: The least inclusive crown clade containing *Panthera leo* Linnaeus 1758 (*Felidae*), *Canis lupus* Linnaeus 1758 (*Canidae*), and *Phoca vitulina* Linnaeus 1758 (*Phocidae*). This is a minimum-crown-clade definition. Abbreviated definition: min crown ∇ (*Panthera leo* Linnaeus 1758 & *Canis lupus* Linnaeus 1758 & *Phoca vitulina* Linnaeus 1758).

Etymology: *Carnivora* derives from the Latin *carnis* (genitive) meaning "flesh" and *vorare* meaning "to devour," i.e., "flesh eating. "

Reference Phylogeny: Figure 5 in Flynn et al. (2005) is the primary reference phylogeny.

Composition: *Carnivora* includes some 260 extant species, listings of which are given in Wozencraft (2005). An account of extinct carnivorans is provided by McKenna and Bell (1997).

Diagnostic Apomorphies: Apomorphies relative to other crown clades include (references point to clear descriptions of the features in question in a phylogenetic framework, not necessarily the original sources of those features): first upper molar metastyle projects no further labially than parastyle (Wesley-Hunt and Flynn, 2005); broad first upper molar parastyle (Wesley-Hunt and Flynn, 2005); fourth upper premolar over first lower molar carnassial shear, lacking a migratory locus for the carnassial in the permanent dentition (Flynn and Galiano, 1982; Flynn et al., 1988; Wyss and Flynn, 1993); second lower molar talonid not elongate and without enlarged hypoconulid (Wesley-Hunt and Flynn, 2005); fourth upper premolar protocone positioned anterior of paracone (Flynn and Galiano, 1982; Flynn et al 1988; Wyss and Flynn, 1993); loss of third upper molar (Wesley-Hunt and Flynn, 2005); infraorbital foramen round (Wesley-Hunt and Flynn, 2005); mastoid process blunt, rounded, does not protrude significantly (Wesley-Hunt and Flynn, 2005); surface of anteromedial promontorium rugose or rostral entotympanic present (Wesley-Hunt and Flynn, 2005); fossa for tensor tympani muscle defined and deep (Wesley-Hunt and Flynn, 2005); flange on basioccipital lateral edge well-developed (entotympanic attachment) (Wesley-Hunt and Flynn, 2005); expanded braincase, fronto-parietal suture anteriorly located (Wesley-Hunt and Flynn, 2005); scaphoid and lunate fused (Wesley-Hunt and Flynn, 2005); depressions on squamosal and alisphenoid formed by the anterior expansion of the ectotympanic (Wesley-Hunt and Flynn, 2005).

While all of these apomorphies distinguish *Carnivora* from other crown clades, some of them diagnose more inclusive clades within *Pan-Carnivora* (this volume). According to Wesley-Hunt and Flynn (2005; see also Flynn and Wesley-Hunt, 2005), the most recent analysis that broadly sampled fossil taxa in addition to crown clade *Carnivora*, apomorphies of *Carnivora* within *Pan-Carnivora* include the flange on basioccipital lateral edge well-developed (entotympanic attachment); loss of third upper molar; expanded braincase, fronto-parietal suture anteriorly located; and scaphoid and lunate fused. The modest number of features uniquely diagnosing crown-clade *Carnivora* partly reflects the relatively recent recognition that many Palaeocene and Eocene taxa previously considered to be early diverging members of the two basal subclades of *Carnivora* (*Feliformia*, *Caniformia*) are instead excluded from the crown (e.g., Flynn and Galiano, 1982 and Flynn et al., 1988; versus Wyss and Flynn, 1993; Wesley-Hunt and Flynn, 2005; and Flynn and Wesley-Hunt, 2005). Although sampling only two crown and three stem carnivorans, O'Leary et al. (2013: Fig. 1 and Appendix S2) identified nearly 70 potentially unambiguous cranial, dental and postcranial synapomorphies of *Carnivora* (Node 59 of Fig. 1; Node 116 of Fig. S15 and Appendix S2) and another 35 for a clade of *Carnivora* plus two stem-carnivorans (Node 60 of Fig. 1; Node 117 of Fig. S15 and Appendix S2; the four sampled *Carnivoramorpha* [Wyss and Flynn, 1993] in that study, incorrectly equated to *Pan-Carnivora* in Appendix S2). Many additional, ambiguously optimized, potential molecular and morphological synapomorphies are provided by O'Leary et al. (2013, Appendices S3 [Deltran] and S4 [Acctran] parsimony optimizations) for *Carnivora* ("node 125 to 123") and the "carnivoramorphan" clade ("node 126 to 125").

Synonyms: *Ferae* Linnaeus 1758 (approximate); *Fissipeda* Blumenbach 1791 (partial and approximate), spelled "*Fissipedia*" by many later authors, see discussion in Simpson (1945); *Carnaria* Haeckel 1866 (approximate); *Carnivora Vera* Flower and Lydekker 1891 (approximate); *Carnassidentia* Wortman 1901 (approximate).

Comments: The name *Carnivora* is here selected from among its synonyms based on frequency of use. The clade to which it refers has long been studied, beginning with the works of Bowdich (1821) and Haeckel (1866), and expanded upon

by Flower (1869, 1883) and Matthew (1909). Simpson (1945) and McKenna and Bell (1987) provided comprehensive classifications of the clade, although not based on explicit phylogenies. Linnaeus' (1758) *Ferae* corresponded roughly to *Carnivora*, although in later editions of *Systema Naturae* it included some non-carnivoran taxa. *Ferae* has been used by most later workers (e.g., Simpson, 1945; Flynn and Wesley-Hunt, 2005; O'Leary et al., 2013) to refer to a more inclusive clade composed of *Carnivora* and closely related fossil taxa such as the "creodonts" *Oxyaenidae* and *Hyaenodontidae*, although some have used *Ferae* to refer to a more inclusive clade of *Carnivora* plus *Pholidota* (e.g., McKenna and Bell, 1997; Asher et al., 2009), which is named *Ostentoria* by others (e.g., Amrine-Madsen et al., 2003; Meredith et al., 2011; O'Leary et al., 2013). Haeckel (1866) used *Carnivora* for a group consisting of the terrestrial (i.e., non-flipper footed) extant species. Blumenbach (1791) used *Fissipeda* for essentially the same grouping, but this was amended by Simpson (1945) to include fossil taxa that he assigned to *Miacoidea* (now considered stem carnivorans). Haeckel (1866) united under *Pinnipedia* ("flipper-footed") the taxa *Otaria*, *Cystophora*, and *Phoca*, along with the manatee *Trichechus*. Haeckel united his *Carnivora* and *Pinnipedia* within *Carnaria*. This helped establish the longstanding primary division of *Carnivora* into *Fissipedia* (*Fissipeda*; or *Carnivora Vera* of Flower and Lydekker, 1891; or *Carnassidentia* of Wortman, 1901) and *Pinnipedia* (e.g., see Flower, 1869; Mivart, 1885; Simpson, 1945). There is now universal agreement from both morphological and molecular data that *Pinnipedia* is nested within the group once named *Fissipeda*, *Fissipedia*, *Carnivora Vera*, or *Carnassidentia*, which referred to a paraphyletic group originating in the same ancestor as *Carnivora*, thus rendering the latter names obsolete.

Wyss and Flynn (1993) and Wolsan (1993) provided phylogenetic definitions of *Carnivora*. Wyss and Flynn (1993) linked this name to the crown, an action emended by Bryant (1996) mainly to avoid use of the word "living" (see also Flynn and Wesley-Hunt, 2005). Recent phylogenies of or including *Carnivora* have been published by Tedford (1976), Flynn and Galiano (1982), Flynn et al. (1988), Wyss and Flynn (1993), Vrana et al. (1994; arctoids), Flynn and Nedbal (1998), Bininda-Emonds et al. (1999; supertree), Yoder et al. (2003; feliforms), Yu et al. (2004), Wesley-Hunt and Flynn (2005), Wesley-Hunt and Werdelin (2005), Flynn et al. (2005), Fulton and Strobeck (2006; arctoids), Yu and Zhang (2006; caniforms), Arnason et al. (2007; caniforms), Finarelli (2008; arctoids), Tomiya (2011), Meredith et al. (2011), Spaulding and Flynn (2012), and O'Leary et al. (2013; *Carnivora* within *Mammalia*). Flynn and Wesley-Hunt (2005) and Flynn et al. (2010) provided the most recent reviews of phylogenetic work for *Carnivora* and relationships within the crown clade. The monophyly of *Carnivora* is robustly supported in these and other studies.

Literature Cited

Amrine-Madsen, H., K. P. Koepfli, R. K. Wayne, and M. S. Springer. 2003. A new phylogenetic marker, apolipoprotein B, provides compelling evidence for eutherian relationships. *Mol. Phylogenet. Evol.* 28:225–240.

Arnason, U., A. Gullberg, A. Janke, and M. Kullberg. 2007. Mitogenomic analyses of caniform relationships. *Mol. Phylogenet. Evol.* 45:863–874.

Asher, R. J., N. Bennett, and T. Lehmann. 2009. The new framework for understanding placental mammal evolution. *Bioessays* 31(8):853–864.

Bininda-Emonds, O. R. P., J. L. Gittleman, and A. Purvis. 1999. Building large trees by combining phylogenetic information: a complete phylogeny of the *Carnivora* (*Mammalia*). *Biol. Rev. Camb. Philos. Soc.* 74:143–175.

Blumenbach, J. F. 1791. *Handbuch der Naturgeschichte. Vierte auflage.* Johann Christian Dieterich, Göttingen.

Bowdich, T. E. 1821. *An Analysis of the Natural Classifications of* Mammalia *for the Use of Students and Travelers.* J. Smith, *Paris.*

Bryant, H. N. 1996. Explicitness, stability, and universality in the phylogenetic definition and usage of taxon names: a case study of the phylogenetic taxonomy of the *Carnivora* (*Mammalia*). *Syst. Biol.* 45:174–189.

Finarelli, J. A. 2008. A total evidence phylogeny of the *Arctoidea* (*Carnivora*: *Mammalia*): relationships among basal taxa. *J. Mamm. Evol.* 15:231–259.

Flower, W. H. 1869. On the value of the characters of the base of the cranium in the classification of the Order *Carnivora*, and on the systematic position of *Bassaris* and other disputed forms. *Proc. Zool. Soc. Lond.* 1869:4–37.

Flower, W. H. 1883. On the arrangement of the orders and families of existing *Mammalia*. *Proc. Zool. Soc. Lond.* 1883:178–186.

Flower, W. H., and R. Lydekker. 1891. *An Introduction to the Study of Mammals Living and Extinct.* Adam and Charles Black, London.

Flynn, J. J., J. A. Finarelli, and M. Spaulding. 2010. Phylogeny of *Carnivora* and *Carnivoramorpha*, and the use of the fossil record to enhance understanding of evolutionary transformations. Pp. 25–63 in Carnivora *Evolution* (A. Goswami and A. Frisca, eds.). Cambridge University Press, Cambridge, UK.

Flynn, J. J., J. A. Finarelli, S. Zehr, J. Hsu, and M. A. Nedbal. 2005. Molecular phylogeny of the *Carnivora* (*Mammalia*): assessing the impact of increased sampling on resolving enigmatic relationships. *Syst. Biol.* 54:317–337.

Flynn, J. J., and H. Galiano. 1982. Phylogeny of Early Tertiary *Carnivora*, with a description of a new species of *Protictis* from the Middle Eocene of Northwestern Wyoming. *Am. Mus. Novit.* 2725:1–64.

Flynn, J. J., and M. A. Nedbal. 1998. Phylogeny of the *Carnivora* (*Mammalia*): congruence vs incompatibility among multiple data sets. *Mol. Phylogen. Evol.* 9(3):414–426.

Flynn, J. J., N. A. Neff, and R. H. Tedford. 1988. Phylogeny of the *Carnivora*. Pp. 73–116 in *The Phylogeny and Classification of the Tetrapods. Vol. 2: Mammals* (M. J. Benton, ed.). The Systematics Association Special Volume No. 35B. Clarendon Press, Oxford.

Flynn, J. J., and G. D. Wesley-Hunt. 2005. *Carnivora*. Pages 175–198 in *The Rise of Placental Mammals: Origins and Relationships of the Major Extant Clades* (K. D. Rose and J. D. Archibald, eds.). Johns Hopkins Univ. Press, Baltimore, MD.

Fulton, T. L., and C. Strobeck. 2006. Molecular phylogeny of the *Arctoidea* (*Carnivora*): effect of missing data on supertree and supermatrix analyses of multiple gene data sets. *Mol. Phylogen. Evol.* 41:165–181.

Haeckel, E. 1866. *Generelle Morphologie der Organismen*, Vol. 2. Georg Reimer, Berlin.

Linnaeus, C. 1758. *Systema Naturae Per Regna Tria Naturae, Secundum Classes, Ordines, Genera, Species, cum Characteribus, Differentiis, Synonymis, Locis*. Tomus I, Editio decima, reformata. Laurentius Salvius, Holmiae (Stockholm).

Matthew, W. D. 1909. The *Carnivora* and *Insectivora* of the Bridger Basin, middle Eocene. *Mem. Am. Mus. Nat. Hist.* 9:289–567.

McKenna, M. C., and S. K. Bell. 1997. *Classification of Mammals above the Species Level*. Columbia University Press, New York.

Meredith, R. W., J. E. Janecka, J. Gatesy, O. A. Ryder, C. A. Fisher, E. C. Teeling, A. Goodbla, E. Eizirik, T. L. L. Simão, T. Stadler, D. L. Rabosky, R. L. Honeycutt, J. J. Flynn, C. M. Ingram, C. Steiner, T. L. Williams, T. J. Robinson, A. Burk, M. Westerman, N. A. Ayoub, M. S. Springer, and W. J. Murphy. 2011. Impacts of the Cretaceous Terrestrial Revolution and KPg extinction on extant mammal diversification. *Science* 334:521–524.

Mivart, S. G. 1885. Notes on the *Pinnipedia*. *Proc. Zool. Soc. Lond.* 1885:484–501.

O'Leary, M. A., J. I. Bloch, J. J. Flynn, T. J. Gaudin, A. Giallombardo, N. P. Giannini, S. L. Goldberg, B. P. Kraatz, Z.-X. Luo, J. Meng, X. Ni, M. J. Novacek, F. A. Perini, Z. Randall, G. W. Rougier, E. J. Sargis, M. T. Silcox, N. B. Simmons, M. Spaulding, P. M. Velazco, M. Weksler, J. R. Wible, and A. L. Cirranello. 2013. The placental mammal ancestor and the post–K-Pg radiation of placentals. *Science* 339:662–667.

Simpson, G. G. 1945. The principles of classification and a classification of mammals. *Bull. Am. Mus. Nat. Hist.* 85:1–350.

Spaulding, M., and J. J. Flynn. 2012. Phylogeny of the *Carnivoramorpha*: the impact of postcranial characters. *J. Syst. Palaeont.* 10(4):653–677.

Tedford, R. H. 1976. Relationship of pinnipeds to other carnivores (*Mammalia*). *Syst. Zool.* 25(4):363–374.

Tomiya, S. 2011. A new basal caniform (*Mammalia: Carnivora*) from the middle Eocene of North America and remarks on the phylogeny of early carnivorans. *PLOS ONE* 6(9):e24146

Vrana, P. B., M. C. Milinkovitch, J. R. Powell, and W. C. Wheeler. 1994. Higher level relationships of the arctoid *Carnivora* based on sequence data and total evidence. *Mol. Phylogen. Evol.* 3:47–58.

Wesley-Hunt, G. D., and J. J. Flynn. 2005. Phylogeny of the *Carnivora*: basal relationships among the carnivoramorphans, and assessment of the position of "*Miacoidea*" relative to crown-clade *Carnivora. J. Syst. Palaeontol.* 3(1):1–28.

Wolsan, M. 1993. Phylogeny and classification of early European *Mustelida* (*Mammalia: Carnivora*). *Acta Theriol.* 38:345–384.

Wortman, J. L. 1901. Studies of Eocene *Mammalia* in the Marsh collection, Peabody Museum. *Am. J. Sci.* 4(11):333–348 (and figs. 1–6, pl. 5), 4(11):437–450 (and figs. 7–17, pl. 6); 4 (12):143–154 (and figs. 18–30), 4(12):193–206 (and figs. 31–43), 4(12):281–296 (and fig. 44, pls. 1–4), 4(12):377–382 (and figs. 45–48), 4(12):421–432 (and figs. 49–60, pls. 8–9).

Wozencraft, W. C. 2005. Order *Carnivora*. Pp. 532–628 in *Mammal Species of the World: A Taxonomic and Geographic Reference*

(D. E. Wilson and D. M. Reeder, eds.). 3rd edition, Vol. 1. Johns Hopkins University Press, Baltimore, MD.

Wyss, A. R., and J. J. Flynn. 1993. A phylogenetic analysis and definition of the *Carnivora*. Pp. 32–52 in *Mammal Phylogeny: Placentals* (F. Szalay, M. Novacek, and M. McKenna, eds.). Springer-Verlag, New York.

Yoder, A. D., M. M. Burns, S. Zehr, T. Delefosse, G. Veron, S. M. Goodman, and J. J. Flynn. 2003. Single origin of Malagasy *Carnivora* from an African ancestor. *Nature* 421:734–737.

Yu, L., Q. Li, O. A. Ryder, and Y. Zhang. 2004. Phylogenetic relationships within mammalian order *Carnivora* indicated by sequences of two nuclear DNA genes. *Mol. Phylogen. Evol.* 33:694–705.

Yu, L., and Y. P. Zhang. 2006. Phylogeny of the caniform *Carnivora*: evidence from multiple genes. *Genetica* 127:65–79.

Authors

John J. Flynn; Division of Paleontology and Richard Gilder Graduate School; American Museum of Natural History; Central Park West at 79th Street; New York, NY 10024-5192, USA. Email: jflynn@amnh.org.

André R. Wyss; Department of Earth Science; University of California at Santa Barbara; Webb Hall; Santa Barbara, CA 93106-9630, USA. Email: wyss@geol.ucsb.edu.

Mieczysław Wolsan; Museum and Institute of Zoology; Polish Academy of Sciences; Wilcza 64; 00-679 Warszawa, Poland. Email: wolsan@miiz.waw.pl.

Date Accepted: 15 December 2015

Primary Editor: Jacques A. Gauthier

Pan-Feliformia M. Wolsan, J. J. Flynn, and A. R. Wyss, new clade name

Registration Number: 79

Definition: The total clade of the crown clade *Feliformia*. This is a crown-based total-clade definition. Abbreviated definition: total ∇ of *Feliformia*.

Etymology: Derived from the Greek *pan-* (all, in reference to a total clade) and the Latin *felis* (cat) and *forma* (form).

Reference Phylogeny: The primary reference phylogeny is Figure 1 of Spaulding and Flynn (2012), where *Pan-Feliformia* corresponds to the clade originating at the base of the branch leading to the crown *Feliformia* (see also Eizirik et al., 2010: Fig. 2).

Composition: For the composition of the crown, see the chapter on *Feliformia* in this volume. The composition of the stem is not known with confidence. Although some phylogenetic analyses place *Nimravidae* on the stem of *Feliformia* (Hunt, 1987; Bryant, 1991; Wyss and Flynn, 1993; Wesley-Hunt and Flynn, 2005; Tomiya, 2011), support for placement of this taxon outside *Pan-Feliformia* is stronger (Spaulding and Flynn, 2012; see also Neff, 1983; Werdelin, 1996; Wesley-Hunt and Werdelin, 2005; Solé et al., 2014). Also, *Viverravidae*, which was considered within the feliform total clade in the past (e.g., Matthew, 1909; Flynn and Galiano, 1982), is currently placed outside of *Carnivora* as defined in this volume (Wesley-Hunt and Flynn, 2005; Wesley-Hunt and Werdelin, 2005; Polly et al., 2006; Spaulding and Flynn, 2009, 2012; Spaulding et al., 2010; Tomiya, 2011; Solé et al., 2014; see also Flynn and Wesley-Hunt, 2005).

Diagnostic Apomorphies: The diagnosis of *Pan-Feliformia* is currently the same as for *Feliformia* (this volume) because no clear stem feliforms are known.

Synonyms: *Feliformia* sensu Wyss and Flynn (1993: 38), Wolsan (1993: 351), Bryant (1996: 184), and Wolsan and Bryant (2004) is an unambiguous synonym. *Herpestoidei* Winge 1895: 46, *Epimycteri* Cope 1882: 473, *Aeluroidea* sensu Trouessart (1885: 6), *Feloidae* Hay 1930: 538, *Feloidea* sensu Simpson (1945: 115), and *Aeluroida* sensu Ginsburg (1999: 129) may be considered approximate synonyms.

Comments: Because the unambiguous synonym *Feliformia* is applied to the crown, we select the new name *Pan-Feliformia* over other synonyms to designate the total clade in agreement with the convention that the name of a total clade be formed by adding the prefix *Pan-* to the name of its corresponding crown (de Queiroz, 2007). The potential synonyms *Aeluroidea*, *Feloidea*, and *Aeluroida* have been applied to other clades (for the first two names, see Comments on *Feliformia* in this volume; for *Aeluroida*, see Flynn and Galiano, 1982), and the remaining potential synonyms (*Herpestoidei*, *Epimycteri*, and *Feloidae*) have not been used for more than 80 years.

Literature Cited

Bryant, H. N. 1991. Phylogenetic relationships and systematics of the *Nimravidae* (*Carnivora*). *J. Mamm.* 72:56–78.

Bryant, H. N. 1996. Explicitness, stability, and universality in the phylogenetic defnition and usage of taxon names: a case study of the phylogenetic taxonomy of the *Carnivora* (*Mammalia*). *Syst. Biol.* 45:174–189.

Cope, E. D. 1882. On the systematic relations of the *Carnivora Fissipedia. Palaeontol. Bull.* 35:471–475.

de Queiroz, K. 2007. Toward an integrated system of clade names. *Syst. Biol.* 56:956–974.

Eizirik, E., W. J. Murphy, K.-P. Koepfli, W. E. Johnson, J. W. Dragoo, R. K. Wayne, and S. J. O'Brien. 2010. Pattern and timing of diversification of the mammalian order *Carnivora* inferred from multiple nuclear gene sequences. *Mol. Phylogenet. Evol.* 56:49–63.

Flynn, J. J., and H. Galiano. 1982. Phylogeny of early Tertiary *Carnivora*, with a description of a new species of *Protictis* from the middle Eocene of northwestern Wyoming. *Am. Mus. Novit.* 2725:1–64.

Flynn, J. J., and G. D. Wesley-Hunt. 2005. *Carnivora.* Pp. 175–198 in *The Rise of Placental Mammals: Origins and Relationships of the Major Extant Clades* (K. D. Rose and J. D. Archibald, eds.). Johns Hopkins University Press, Baltimore, MD.

Ginsburg, L. 1999. Order *Carnivora.* Pp. 109–148 in *The Miocene Land Mammals of Europe* (G. E. Rössner and K. Heissig, eds.). Verlag Dr. Friedrich Pfeil, Munich.

Hay, O. P. 1930. *Second Bibliography and Catalogue of the Fossil Vertebrata of North America,* Vol. II. Carnegie Institution of Washington, Washington, DC.

Hunt, Jr., R. M. 1987. Evolution of the aeluroid *Carnivora*: significance of auditory structure in the nimravid cat *Dinictis. Am. Mus. Novit.* 2886:1–74.

Matthew, W. D. 1909. The *Carnivora* and *Insectivora* of the Bridger Basin, middle Eocene. *Mem. Am. Mus. Nat. Hist.* 9:289–567.

Neff, N. A. 1983. *The Basicranial Anatomy of the* Nimravidae (Mammalia: Carnivora): *Character Analyses and Phylogenetic Inferences.* PhD dissertation, City University of New York, New York.

Polly, P. D., G. D. Wesley-Hunt, R. E. Heinrich, G. Davis, and P. Houde. 2006. Earliest known carnivoran auditory bulla and support for a recent origin of crown-group *Carnivora* (*Eutheria, Mammalia*). *Palaeontology* 49:1019–1027.

Simpson, G. G. 1945. The principles of classification and a classification of mammals. *Bull. Am. Mus. Nat. Hist.* 85:I–XVI + 1–350.

Solé, F., R. Smith, T. Coillot, E. de Bast, and T. Smith. 2014. Dental and tarsal anatomy of "*Miacis*" *latouri* and a phylogenetic analysis of the earliest carnivoraforms (*Mammalia, Carnivoramorpha*). *J. Vertebr. Paleontol.* 34:1–21.

Spaulding, M., and J. J. Flynn. 2009. Anatomy of the postcranial skeleton of "*Miacis*" *uintensis* (*Mammalia*: *Carnivoramorpha*). *J. Vertebr. Paleontol.* 29:1212–1223.

Spaulding, M., and J. J. Flynn. 2012. Phylogeny of the *Carnivoramorpha*: the impact of postcranial characters. *J. Syst. Palaeontol.* 10:653–677.

Spaulding, M., J. J. Flynn, and R. K. Stucky. 2010. A new basal carnivoramorphan (*Mammalia*) from the "Bridger B" (Black's Fork Member, Bridger Formation, Bridgerian NALMA, middle Eocene) of Wyoming, USA. *Palaeontology* 53:815–832.

Tomiya, S. 2011. A new basal caniform (*Mammalia*: *Carnivora*) from the middle Eocene of North America and remarks on the phylogeny of early carnivorans. *PLOS ONE* 6(e24146): 1–24.

Trouessart, E.-L. 1885. Catalogue des Mammifères vivants et fossiles (Carnivores). *Bull. Soc. Études Sci. Angers* 1884(suppl.):1–108.

Werdelin, L. 1996. Carnivoran ecomorphology: a phylogenetic perspective. Pp. 582–624 in *Carnivore Behavior, Ecology, and Evolution,* Vol. 2 (J. L. Gittleman, ed.). Cornell University Press, Ithaca, NY.

Wesley-Hunt, G. D., and J. J. Flynn. 2005. Phylogeny of the *Carnivora*: basal relationships among the carnivoramorphans, and assessment of the position of "*Miacoidea*" relative to *Carnivora. J. Syst. Palaeontol.* 3:1–28.

Wesley-Hunt, G. D., and L. Werdelin. 2005. Basicranial morphology and phylogenetic position of the upper Eocene carnivoramorphan *Quercygale. Acta Palaeontol. Pol.* 50:837–846.

Winge, H. 1895. Jordfundne og nulevende Rovdyr (*Carnivora*) fra Lagoa Santa, Minas Geraes, Brasilien. Med Udsigt over Rovdyrenes indbyrdes Slaegtskab. Pp. 1–130 in *E Museo Lundii. En Samling af Afhandlinger Om de i Det Indre Brasiliens Kalkstenshuler af Professor Dr. Peter Vilhelm Lund Udgravede Og i Den Lundske palaeontologiske Afdeling af Kjøbenhavns Universitets Zoologiske Museum Opbevarede Dyre- Og Menneskeknogler.* Bind 2, Halvbind 2, Afhandling 4 (C. F. Lütken, ed.). H. Hagerups Boghandel, Copenhagen.

Wolsan, M. 1993. Phylogeny and classification of early European *Mustelida* (*Mammalia*: *Carnivora*). *Acta Theriol.* 38:345–384.

Wolsan, M., and H. N. Bryant. 2004. Phylogenetic nomenclature of carnivoran mammals. p. 32 in *First International Phylogenetic Nomenclature Meeting, Paris, Muséum National d'Histoire Naturelle, 6–9 July 2004: Abstracts* (M. Laurin, ed.). Muséum National d'Histoire Naturelle, Paris.

Wyss, A. R., and J. J. Flynn. 1993. A phylogenetic analysis and definition of the *Carnivora*. Pp. 32–52 in *Mammal Phylogeny: Placentals* (F. S. Szalay, M. J. Novacek, and M. C. McKenna, eds.). Springer-Verlag, New York.

Authors

Mieczysław Wolsan; Museum and Institute of Zoology; Polish Academy of Sciences; Wilcza 64; 00-679 Warszawa, Poland. Email: wolsan@miiz.waw.pl.

John J. Flynn; Division of Paleontology; American Museum of Natural History; Central Park West at 79th Street; New York, NY 10024-5192, USA. Email: jflynn@amnh.org.

André R. Wyss; Department of Earth Science; University of California at Santa Barbara; Webb Hall; Santa Barbara, CA 93106-9630, USA. Email: wyss@geol.ucsb.edu.

Date Accepted: 17 August 2015

Primary Editor: Philip Cantino

Feliformia M. Kretzoi 1945 [M. Wolsan, J. J. Flynn, and A. R. Wyss], converted clade name

Registration Number: 44

Definition: The largest crown clade containing *Felis silvestris* Schreber 1777 (*Felidae*) but not *Canis lupus* Linnaeus 1758 (*Canidae*). This is a maximum-crown-clade definition. Abbreviated definition: max crown ∇ (*Felis silvestris* Schreber 1777 ~ *Canis lupus* Linnaeus 1758).

Etymology: Derived from the Latin *felis* (cat) and *forma* (form).

Reference Phylogeny: Figure 1 (maximum parsimony) in Flynn et al. (2005).

Composition: Wozencraft (2005) lists 121 extant species, which represent *Felidae* (cats), *Prionodon* (linsangs), *Viverridae* (civets, genets, and oyans), *Hyaenidae* (hyenas), *Herpestidae* (mongooses), *Eupleridae* (Malagasy carnivorans), and *Nandinia binotata* (African palm civets). An extensive list of extinct supraspecific taxa is provided by McKenna and Bell (1997). Note that McKenna and Bell (1997) applied the name *Feliformia* to a clade larger than the crown, in which they included the extinct *Viverravidae* and *Nimravidae*, both of which are now thought to lie outside of *Carnivora* as defined in this volume (Spaulding and Flynn, 2012; Solé et al., 2014).

Diagnostic Apomorphies: Apomorphies of *Feliformia* relative to its sister crown clade *Caniformia*, as defined in this volume, include 13 cranial, dental, postcranial, and soft-anatomical features (Spaulding and Flynn, 2012: Table 1): (1) lacrimal facial process not present on the face (Wesley-Hunt and Flynn, 2005: character 1); (2) anterior loop of the internal carotid artery absent (Wesley-Hunt and Flynn, 2005: character 23); (3) carotid artery entering the auditory bulla anteriorly and without an enclosing tube (Wyss and Flynn, 1993: character 16); (4) promontorium with a ventral process positioned medially (Wesley-Hunt and Flynn, 2005: character 27); (5) paroccipital process flattened laterally (Wesley-Hunt and Flynn, 2005: character 10); (6) protocone and paracone of the first upper molar equal or subequal in size (Wesley-Hunt and Flynn, 2005: character 42); (7) transverse processes of the atlas extended posteriorly at a sharp angle (Spaulding and Flynn, 2012: character 208); (8) capitular eminence of the radial head very small (Spaulding and Flynn, 2012: character 142); (9) unciform as wide as the magnum (Spaulding and Flynn, 2012: character 146); (10) cuneiform articulating with the ulna on the medial (radial) margin (Spaulding and Flynn, 2012: character 150); (11) lesser trochanter of the femur reduced in size (Spaulding and Flynn, 2012: character 163); (12) Cowper's gland present (Wozencraft, 1989: character 91; Wyss and Flynn, 1993: character 58); and (13) prostate gland large, with ampulla, and bilobed (Wozencraft, 1989: character 92; Wyss and Flynn, 1993: character 59).

Synonyms: *Feloidea* sensu Wyss and Flynn (1993: 38) and Bryant (1996: 184) is an unambiguous synonym. *Aeluroidea* Flower 1869: 22, *Herpestoidea* Weber 1904: 528, *Mungotoidea* Pocock 1919: 515, and *Feloidea* Simpson 1931: 277 are approximate synonyms.

Comments: *Feliformia* is selected here from among its synonyms based on its frequency of use.

This name was erected by Kretzoi (1945: 62) to refer to a group consisting of *Felidae*, *Nimravidae*, and two other extinct taxa that are now included in *Felidae*. Tedford (1976) adopted *Feliformia* for the sister group of *Caniformia*. Most subsequent publications have either explicitly or implicitly applied *Feliformia* to the largest crown clade containing *Felidae* but not *Canidae* (e.g., Flynn and Nedbal, 1998; Veron and Heard, 2000; Zehr et al., 2001; Gaubert and Veron, 2003; Gaubert et al., 2004, 2005; Veron et al., 2004; Yu et al., 2004; Flynn et al., 2005; Wozencraft, 2005; Gaubert and Cordeiro-Estrela, 2006; Koepfli et al., 2006; Chaveerach et al., 2008; Eizirik and Murphy, 2009; Agnarsson et al., 2010; Eizirik et al., 2010; Veron, 2010; Meredith et al., 2011: Supporting Online Material; Stankowich et al., 2011; Abramov and Khlyap, 2012; Cornelis et al., 2012; Nyakatura and Bininda-Emonds, 2012; Spaulding and Flynn, 2012; Solé et al., 2014). *Feliformia* has sometimes been applied to the total clade of this crown or other clades larger than the crown (e.g., Flynn and Galiano, 1982; Wozencraft, 1989; Wolsan, 1993; Wyss and Flynn, 1993; Bryant, 1996; Flynn, 1996; Hunt, 1996; Werdelin, 1996; McKenna and Bell, 1997; Janis et al., 1998; Wolsan and Bryant, 2004; Flynn and Wesley-Hunt, 2005; Wesley-Hunt and Flynn, 2005; Tomiya, 2011), while *Feloidea* has sometimes been used for the crown (e.g., Flynn and Galiano, 1982; Wyss and Flynn, 1993; Bryant, 1996; Flynn, 1996; Werdelin, 1996; Janis et al., 1998; Schreiber et al., 1998; Ginsburg, 1999; Flynn and Wesley-Hunt, 2005; Wesley-Hunt and Flynn, 2005; Wesley-Hunt and Werdelin, 2005; Tomiya, 2011). Similarly *Aeluroidea* has been used for the crown (e.g., Flynn et al., 1988; Bryant, 1991; Hunt and Tedford, 1993; Veron and Catzeflis, 1993; Veron, 1995; Ivanoff, 2000, 2001; Hunt, 2001). *Feloidea* and *Aeluroidea* are used for this clade with less frequency than *Feliformia*, and both of the former names have recently been applied to other clades: *Feloidea* to a smaller clade excluding at least *Nandinia* (e.g., Wiig, 1985; Aristov and Baryshnikov, 2001; Flynn et al., 2010), and *Aeluroidea* to a larger clade containing the crown (e.g., Hunt, 1987, 1989, 1991, 1996). Although *Herpestoidea* was once applied to the crown or its near equivalent (e.g., Weber, 1904; Kükenthal, 1913; Matthey, 1946), this name has recently been applied to a clade containing only *Herpestidae*, *Eupleridae*, and *Hyaenidae* (Flynn et al., 2010; Meredith et al., 2011: Supporting Online Material). *Mungotoidea* (Pocock, 1919) has not been used in almost a century.

Molecular and morphological data strongly support the close relationship of *Felidae*, *Prionodon*, *Viverridae*, *Hyaenidae*, *Herpestidae*, *Eupleridae*, and *Nandinia binotata* to the exclusion of other extant carnivorans (e.g., Flynn and Nedbal, 1998; Yoder et al., 2003; Flynn et al., 2005; Eizirik et al., 2010; Meredith et al., 2011; Spaulding and Flynn, 2012).

Literature Cited

Abramov, A. V., and L. A. Khlyap. 2012. Order *Carnivora* Bowdich, 1821. *Sb. Tr. Zool. Muz. MGU* 52:313–382.

Agnarsson, I., M. Kuntner, and L. J. May-Collado. 2010. Dogs, cats, and kin: a molecular species-level phylogeny of *Carnivora*. *Mol. Phylogenet. Evol.* 54:726–745.

Aristov, A. A., and G. F. Baryshnikov. 2001. *The Mammals of Russia and Adjacent Territories. Carnivores and Pinnipeds*. Russian Academy of Sciences Zoological Institute, St. Petersburg.

Bryant, H. N. 1991. Phylogenetic relationships and systematics of the *Nimravidae* (Carnivora). *J. Mammal.* 72:56–78.

Bryant, H. N. 1996. Explicitness, stability, and universality in the phylogenetic defnition and usage of taxon names: a case study of the phylogenetic taxonomy of the *Carnivora* (*Mammalia*). *Syst. Biol.* 45:174–189.

Chaveerach, A., N. Srisamoot, S. Nuchadomrong, N. Sattayasai, P. Chaveerach, A. Tanomtong, and K. Pinthong. 2008. Phylogenetic relationships of wildlife order *Carnivora* in Thailand inferred from the internal transcribed spacer region. *J. Biol. Sci.* 8:278–287.

Cornelis, G., O. Heidmann, S. Bernard-Stoecklin, K. Reynaud, G. Veron, B. Mulot, A. Dupressoir, and T. Heidmann. 2012. Ancestral capture of *syncytin-Car1*, a fusogenic endogenous retroviral envelope gene involved in placentation and conserved in *Carnivora*. *Proc. Natl. Acad. Sci. USA* 109:E432–E441.

Eizirik, E., and W. J. Murphy. 2009. Carnivores (*Carnivora*). Pp. 504–507 in *The Timetree of Life* (S. B. Hedges and S. Kumar, eds.). Oxford University Press, New York.

Eizirik, E., W. J. Murphy, K.-P. Koepfli, W. E. Johnson, J. W. Dragoo, R. K. Wayne, and S. J. O'Brien. 2010. Pattern and timing of diversification of the mammalian order *Carnivora* inferred from multiple nuclear gene sequences. *Mol. Phylogenet. Evol.* 56:49–63.

Flower, W. H. 1869. On the value of the characters of the base of the cranium in the classification of the order *Carnivora*, and on the systematic position of *Bassaris* and other disputed forms. *Proc. Zool. Soc. Lond.* 1869:4–37.

Flynn, J. J. 1996. Carnivoran phylogeny and rates of evolution: morphological, taxic, and molecular. Pp. 542–581 in *Carnivore Behavior, Ecology, and Evolution*, Vol. 2 (J. L. Gittleman, ed.). Cornell University Press, Ithaca, NY.

Flynn, J. J., J. A. Finarelli, and M. Spaulding. 2010. Phylogeny of the *Carnivora* and *Carnivoramorpha*, and the use of the fossil record to enhance understanding of evolutionary transformations. Pp. 25–63 in *Carnivoran Evolution: New Views on Phylogeny, Form, and Function* (A. Goswami and A. Friscia, eds.). Cambridge University Press, Cambridge, UK.

Flynn, J. J., J. A. Finarelli, S. Zehr, J. Hsu, and M. A. Nedbal. 2005. Molecular phylogeny of the *Carnivora* (*Mammalia*): assessing the impact of increased sampling on resolving enigmatic relationships. *Syst. Biol.* 54:317–337.

Flynn, J. J., and H. Galiano. 1982. Phylogeny of early Tertiary *Carnivora*, with a description of a new species of *Protictis* from the middle Eocene of northwestern Wyoming. *Am. Mus. Novit.* 2725:1–64.

Flynn, J. J., and M. A. Nedbal. 1998. Phylogeny of the *Carnivora* (*Mammalia*): congruence vs incompatibility among multiple data sets. *Mol. Phylogenet. Evol.* 9:414–426.

Flynn, J. J., N. A. Neff, and R. H. Tedford. 1988. Phylogeny of the *Carnivora*. Pp. 73–115 in *The Phylogeny and Classification of the Tetrapods. Vol. 2: Mammals* (M. J. Benton, ed.). Clarendon Press, Oxford, UK.

Flynn, J. J., and G. D. Wesley-Hunt. 2005. *Carnivora*. Pp. 175–198 in *The Rise of Placental Mammals: Origins and Relationships of the Major Extant Clades* (K. D. Rose and J. D. Archibald, eds.). Johns Hopkins University Press, Baltimore, MD.

Gaubert, P., and P. Cordeiro-Estrela. 2006. Phylogenetic systematics and tempo of evolution of the *Viverrinae* (*Mammalia, Carnivora, Viverridae*) within feliformians: implications for faunal exchanges between Asia and Africa. *Mol. Phylogenet. Evol.* 41:266–278.

Gaubert, P., M. Tranier, A.-S. Delmas, M. Colyn, and G. Veron. 2004. First molecular evidence for reassessing phylogenetic affinities between genets (*Genetta*) and the enigmatic genet-like taxa *Osbornictis*, *Poiana* and *Prionodon* (*Carnivora, Viverridae*). *Zool. Scr.* 33:117–129.

Gaubert, P., and G. Veron. 2003. Exhaustive sample set among *Viverridae* reveals the sister-group of felids: the linsangs as a case of extreme morphological convergence within *Feliformia*. *Proc. R. Soc. Lond. B Biol. Sci.* 270:2523–2530.

Gaubert, P., W. C. Wozencraft, P. Cordeiro-Estrela, and G. Veron. 2005. Mosaics of convergences and noise in morphological phylogenies: what's in a viverrid-like carnivoran? *Syst. Biol.* 54:865–894.

Ginsburg, L. 1999. Order *Carnivora*. Pp. 109–148 in *The Miocene Land Mammals of Europe* (G. E. Rössner and K. Heissig, eds.). Verlag Dr. Friedrich Pfeil, Munich.

Hunt, Jr., R. M. 1987. Evolution of the aeluroid *Carnivora*: significance of auditory structure in the nimravid cat *Dinictis*. *Am. Mus. Novit.* 2886:1–74.

Hunt, Jr., R. M. 1989. Evolution of the aeluroid *Carnivora*: significance of the ventral promontorial process of the petrosal, and the origin of basicranial patterns in the living families. *Am. Mus. Novit.* 2930:1–32.

Hunt, Jr., R. M. 1991. Evolution of the aeluroid *Carnivora*: viverrid affinities of the Miocene carnivoran *Herpestides*. *Am. Mus. Novit.* 3023:1–34.

Hunt, Jr., R. M. 1996. Biogeography of the order *Carnivora*. Pp. 485–541 in *Carnivore Behavior, Ecology, and Evolution*, Vol. 2 (J. L. Gittleman, ed.). Cornell University Press, Ithaca, NY.

Hunt, Jr., R. M. 2001. Basicranial anatomy of the living linsangs *Prionodon* and *Poiana* (*Mammalia, Carnivora, Viverridae*), with comments on the early evolution of aeluroid carnivorans. *Am. Mus. Novit.* 3330:1–24.

Hunt, Jr., R. M., and R. H. Tedford. 1993. Phylogenetic relationships within the aeluroid *Carnivora* and implications of their temporal and geographic distribution. Pp. 53–73 in *Mammal Phylogeny: Placentals* (F. S. Szalay, M. J. Novacek, and M. C. McKenna, eds.). Springer-Verlag, New York.

Ivanoff, D. V. 2000. Origin of the septum in the canid auditory bulla: evidence from morphogenesis. *Acta Theriol.* 45:253–270.

Ivanoff, D. V. 2001. Partitions in the carnivoran auditory bulla: their formation and significance for systematics. *Mamm. Rev.* 31:1–16.

Janis, C. M., J. A. Baskin, A. Berta, J. J. Flynn, G. F. Gunnell, R. M. Hunt, Jr., L. D. Martin, and K. Munthe. 1998. Carnivorous mammals. Pp. 73–90 in *Evolution of Tertiary Mammals of North America. Vol. 1: Terrestrial Carnivores, Ungulates, and Ungulatelike Mammals* (C. M. Janis, K. M. Scott, and L. L. Jacobs, eds.). Cambridge University Press, Cambridge, UK.

Koepfli, K.-P., S. M. Jenks, E. Eizirik, T. Zahirpour, B. Van Valkenburgh, and R. K. Wayne. 2006. Molecular systematics of the *Hyaenidae*: relationships of a relictual lineage resolved by a molecular supermatrix. *Mol. Phylogenet. Evol.* 38:603–620.

Kretzoi, M. 1945. Bemerkungen über das Raubtiersystem. *Ann. Hist.-Nat. Mus. Natl. Hung.* 38:59–83.

Kükenthal, W. 1913. Säugetiere (*Mammalia*). Pp. 633–695 in *Handwörterbuch der Naturwissenschaften, Achter Band: Quartärformation–Sekretion* (E. Korschelt, G. Linck, F. Oltmanns, K. Schaum, H. T. Simon, M. Verworn, and E. Teichmann, eds.). Verlag von Gustav Fischer, Jena.

Linnaeus, C. 1758. *Systema Naturae Per Regna Tria Naturae, Secundum Classes, Ordines, Genera, Species, cum Characteribus, Differentiis, Synonymis, Locis.* Tomus I, Editio decima, reformata. Laurentius Salvius, Holmiae (Stockholm).

Matthey, R. 1946. Le Chien domestique et son origine. *Bull. Soc. Vaud. Sci. Nat.* 63:251–268.

McKenna, M. C., and S. K. Bell. 1997. *Classification of Mammals above the Species Level.* Columbia University Press, New York.

Meredith, R. W., J. E. Janečka, J. Gatesy, O. A. Ryder, C. A. Fisher, E. C. Teeling, A. Goodbla, E. Eizirik, T. L. L. Simão, T. Stadler, D. L. Rabosky, R. L. Honeycutt, J. J. Flynn, C. M. Ingram, C. Steiner, T. L. Williams, T. J. Robinson, A. Burk-Herrick, M. Westerman, N. A. Ayoub, M. S. Springer, and W. J. Murphy. 2011. Impacts of the Cretaceous Terrestrial Revolution and KPg extinction on mammal diversification. *Science* 334:521–524.

Nyakatura, K., and O. R. P. Bininda-Emonds. 2012. Updating the evolutionary history of *Carnivora* (*Mammalia*): a new species-level supertree complete with divergence time estimates. *BMC Biol.* 10(12):1–31.

Pocock, R. I. 1919. The classification of the mongooses (*Mungotidae*). *Ann. Mag. Nat. Hist.* (9) 3:515–524.

Schreber, J. C. D. 1777. Die wilde Kaze. P. 397 in *Die Säugthiere in Abbildungen nach der Natur mit Beschreibungen, Dritter Theil, Heft 23* (J. C. D. Schreber). Walthersche Kunst- und Buchhandlung, Erlangen.

Schreiber, A., K. Eulenberger, and K. Bauer. 1998. Immunogenetic evidence for the phylogenetic sister group relationship of dogs and bears (*Mammalia, Carnivora: Canidae* and *Ursidae*): a comparative determinant analysis of carnivoran albumin, C3 complement and immunoglobulin μ-chain. *Exp. Clin. Immunogenet.* 15:154–170.

Simpson, G. G. 1931. A new classification of mammals. *Bull. Am. Mus. Nat. Hist.* 59:259–293.

Solé, F., R. Smith, T. Coillot, E. de Bast, and T. Smith. 2014. Dental and tarsal anatomy of *"Miacis" latouri* and a phylogenetic analysis of the earliest carnivoraforms (*Mammalia, Carnivoramorpha*). *J. Vertebr. Paleontol.* 34:1–21.

Spaulding, M., and J. J. Flynn. 2012. Phylogeny of the *Carnivoramorpha*: the impact of postcranial characters. *J. Syst. Palaeontol.* 10:653–677.

Stankowich, T., T. Caro, and M. Cox. 2011. Bold coloration and the evolution of aposematism in terrestrial carnivores. *Evolution* 65:3090–3099.

Tedford, R. H. 1976. Relationship of pinnipeds to other carnivores (*Mammalia*). *Syst. Zool.* 25:363–374.

Tomiya, S. 2011. A new basal caniform (*Mammalia: Carnivora*) from the middle Eocene of North America and remarks on the phylogeny of early carnivorans. *PLOS ONE* 6(e24146):1–24.

Veron, G. 1995. La position systématique de *Cryptoprocta ferox* (*Carnivora*). Analyse cladistique des caractères morphologiques de carnivores *Aeluroidea* actuels et fossiles. *Mammalia* 59:551–582.

Veron, G. 2010. Phylogeny of the *Viverridae* and "viverrid-like" feliforms. Pp. 64–91 in *Carnivoran Evolution: New Views on Phylogeny, Form, and Function* (A. Goswami and A. Friscia, eds.). Cambridge University Press, Cambridge, UK.

Veron, G., and F. M. Catzeflis. 1993. Phylogenetic relationships of the endemic Malagasy carnivore *Cryptoprocta ferox* (*Aeluroidea*): DNA/DNA hybridization experiments. *J. Mamm. Evol.* 1:169–185.

Veron, G., M. Colyn, A. E. Dunham, P. Taylor, and P. Gaubert. 2004. Molecular systematics and origin of sociality in mongooses (*Herpestidae, Carnivora*). *Mol. Phylogenet. Evol.* 30:582–598.

Veron, G., and S. Heard. 2000. Molecular systematics of the Asiatic *Viverridae* (*Carnivora*) inferred from mitochondrial cytochrome *b* sequence analysis. *J. Zool. Syst. Evol. Res.* 38:209–217.

Weber, M. 1904. *Die Säugetiere. Einführung in Die Anatomie und Systematik der Recenten und Fossilen* Mammalia. Gustav Fischer, Jena.

Werdelin, L. 1996. Carnivoran ecomorphology: a phylogenetic perspective. Pp. 582–624 in *Carnivore Behavior, Ecology, and Evolution*, Vol. 2 (J. L. Gittleman, ed.). Cornell University Press, Ithaca, New York.

Wesley-Hunt, G. D., and J. J. Flynn. 2005. Phylogeny of the *Carnivora*: basal relationships among the carnivoramorphans, and assessment of the position of *"Miacoidea"* relative to *Carnivora*. *J. Syst. Palaeontol.* 3:1–28.

Wesley-Hunt, G. D., and L. Werdelin. 2005. Basicranial morphology and phylogenetic position of the upper Eocene carnivoramorphan *Quercygale. Acta Palaeontol. Pol.* 50:837–846.

Wiig, Ø. 1985. Relationship of *Nandinia binotata* (Gray) to the superfamily *Feloidea* (*Mammalia, Carnivora*). *Zool. Scr.* 14:155–159.

Wolsan, M. 1993. Phylogeny and classification of early European *Mustelida* (*Mammalia: Carnivora*). *Acta Theriol.* 38:345–384.

Wolsan, M., and H. N. Bryant. 2004. Phylogenetic nomenclature of carnivoran mammals. P. 32 in *First International Phylogenetic Nomenclature Meeting, Paris, Muséum National d'Histoire Naturelle, 6–9 July 2004: Abstracts* (M. Laurin, ed.). Muséum National d'Histoire Naturelle, Paris.

Wozencraft, W. C. 1989. The phylogeny of the Recent *Carnivora*. Pp. 495–535 in *Carnivore Behavior, Ecology, and Evolution* (J. L. Gittleman, ed.). Cornell University Press, Ithaca, NY.

Wozencraft, W. C. 2005. Order *Carnivora*. Pp. 532–628 in *Mammal Species of the World: A Taxonomic and Geographic Reference*. 3rd edition, Vol. 1 (D. E. Wilson and D. M. Reeder, eds.). Johns Hopkins University Press, Baltimore, MD.

Wyss, A. R., and J. J. Flynn. 1993. A phylogenetic analysis and definition of the *Carnivora*. Pp. 32–52 in *Mammal Phylogeny: Placentals* (F. S. Szalay, M. J. Novacek, and M. C. McKenna, eds.). Springer-Verlag, New York.

Yoder, A. D., M. M. Burns, S. Zehr, T. Delefosse, G. Veron, S. M. Goodman, and J. J. Flynn. 2003. Single origin of Malagasy *Carnivora* from an African ancestor. *Nature* 421:734–737.

Yu, L., Q. Li, O. A. Ryder, and Y. Zhang. 2004. Phylogenetic relationships within mammalian order *Carnivora* indicated by sequences of two nuclear DNA genes. *Mol. Phylogenet. Evol.* 33:694–705.

Zehr, S. M., M. A. Nedbal, and J. J. Flynn. 2001. Tempo and mode of evolution in an orthologous *Can* SINE. *Mamm. Genome* 12: 38–44.

Authors

Mieczysław Wolsan; Museum and Institute of Zoology; Polish Academy of Sciences; Wilcza 64; 00-679 Warszawa, Poland. Email: wolsan@miiz.waw.pl.

John J. Flynn; Division of Paleontology; American Museum of Natural History; Central Park West at 79th Street; New York, NY 10024-5192, USA. Email: jflynn@amnh.org.

André R. Wyss; Department of Earth Science; University of California at Santa Barbara; Webb Hall; Santa Barbara, CA 93106-9630, USA. Email: wyss@geol.ucsb.edu.

Date Accepted: 13 August 2015

Primary Editor: Philip Cantino

Pan-Caniformia M. Wolsan, J. J. Flynn, and A. R. Wyss, new clade name

Registration Number: 76

Definition: The total clade of the crown clade *Caniformia*. This is a crown-based total-clade definition. Abbreviated definition: total ∇ of *Caniformia*.

Etymology: Derived from the Greek *pan-* (all, in reference to a total clade) and the Latin *caniformis* (having the form of a dog, dog-shaped), a compound adjective derived from *canis* (dog) and *forma* (form).

Reference Phylogeny: The primary reference phylogeny is Figure 1 of Spaulding and Flynn (2012), where *Pan-Caniformia* corresponds to the clade originating at the base of the branch leading to the crown *Caniformia*. See also Wesley-Hunt and Flynn (2005: Fig. 3) and Eizirik et al. (2010: Fig. 2); in both cases, *Pan-Caniformia* is the clade originating at the base of the branch leading to what is labeled as *Caniformia*.

Composition: For taxa contained in the crown, see the chapter on *Caniformia* in this volume. The composition of the stem is not known with confidence. Some recent phylogenetic analyses inferred *Amphicyonidae* (beardogs) to be a member of the stem (Wesley-Hunt and Flynn, 2005; Tomiya, 2011), but a more recent analysis using a more comprehensive data set (Spaulding and Flynn, 2012) supported the traditional view that *Amphicyonidae* is part of the crown.

Diagnostic Apomorphies: Although the crown clade has many apomorphies relative to its sister crown clade (see *Caniformia*, this volume), no clear stem caniforms are known, so it is currently impossible to differentiate the diagnosis of *Pan-Caniformia* from that of *Caniformia*.

Synonyms: *Caniformia* sensu Wyss and Flynn (1993: 38), Wolsan (1993: 363), Bryant (1996: 184), and Wolsan and Bryant (2004) is an unambiguous synonym.

Comments: As the synonym *Caniformia* is applied to the crown, we introduce a new pan-clade name, which is in accord with the convention that the name of a total clade be formed by adding the prefix *Pan-* to the name of its corresponding crown (de Queiroz, 2007).

Literature Cited

Bryant, H. N. 1996. Explicitness, stability, and universality in the phylogenetic definition and usage of taxon names: a case study of the phylogenetic taxonomy of the *Carnivora* (*Mammalia*). Syst. Biol. 45:174–189.

Eizirik, E., W. J. Murphy, K.-P. Koepfli, W. E. Johnson, J. W. Dragoo, R. K. Wayne, and S. J. O'Brien. 2010. Pattern and timing of diversification of the mammalian order *Carnivora* inferred from multiple nuclear gene sequences. *Mol. Phylogenet. Evol.* 56:49–63.

de Queiroz, K. 2007. Toward an integrated system of clade names. *Syst. Biol.* 56:956–974.

Spaulding, M., and J. J. Flynn. 2012. Phylogeny of the *Carnivoramorpha*: the impact of postcranial characters. *J. Syst. Palaeontol.* 10:653–677.

Tomiya, S. 2011. A new basal caniform (*Mammalia*: *Carnivora*) from the middle Eocene of North America and remarks on the phylogeny of early carnivorans. *PLOS ONE* 6:e24146.

Wesley-Hunt, G. D., and J. J. Flynn. 2005. Phylogeny of the *Carnivora*: basal relationships among the carnivoramorphans, and assessment of the position of "*Miacoidea*" relative to *Carnivora*. *J. Syst. Palaeontol.* 3:1–28.

Wolsan, M. 1993. Phylogeny and classification of early European *Mustelida* (*Mammalia*: *Carnivora*). *Acta Theriol.* 38:345–384.

Wolsan, M., and H. N. Bryant. 2004. Phylogenetic nomenclature of carnivoran mammals. P. 32 in *First International Phylogenetic Nomenclature Meeting, Paris, Muséum National d'Histoire Naturelle, 6–9 July 2004: Abstracts* (M. Laurin, ed.). Muséum National d'Histoire Naturelle, Paris.

Wyss, A. R., and J. J. Flynn. 1993. A phylogenetic analysis and definition of the *Carnivora*. Pp. 32–52 in *Mammal Phylogeny: Placentals* (F. S. Szalay, M. J. Novacek, and M. C. McKenna, eds.). Springer-Verlag, New York.

Authors

Mieczysław Wolsan; Museum and Institute of Zoology; Polish Academy of Sciences; Wilcza 64; 00-679 Warszawa, Poland. Email: wolsan@miiz.waw.pl.

John J. Flynn; Division of Paleontology; American Museum of Natural History; Central Park West at 79th Street; New York, NY 10024-5192, USA. Email: jflynn@amnh.org.

André R. Wyss; Department of Earth Science; University of California at Santa Barbara; Webb Hall; Santa Barbara, CA 93106-9630, USA. Email: wyss@geol.ucsb.edu.

Date Accepted: 14 August 2015

Primary Editor: Philip Cantino

Caniformia M. Kretzoi 1943 [M. Wolsan, J. J. Flynn, and A. R. Wyss], converted clade name

Registration Number: 23

Definition: The largest crown clade containing *Canis lupus* Linnaeus 1758 (*Canidae*) but not *Felis silvestris* Schreber 1777 (*Felidae*). This is a maximum-crown-clade definition. Abbreviated definition: max crown ∇ (*Canis lupus* Linnaeus 1758 ~ *Felis silvestris* Schreber 1777).

Etymology: Derived from the Latin *caniformis* (having the form of a dog, dog-shaped), a compound adjective derived from *canis* (dog) and *forma* (form).

Reference Phylogeny: Figure 1 (maximum parsimony) in Flynn et al. (2005).

Composition: *Caniformia* contains the crown clade *Arctoidea* as defined in this volume and its stem (both compose *Pan-Arctoidea*, this volume) as well as the crown clade *Canidae* (dogs) and its stem. Wozencraft (2005) lists 127 living and three recently extinct species of *Arctoidea* (distributed among *Mustelidae*, *Procyonidae*, *Ailurus fulgens*, and *Mephitidae*, which are parts of *Musteloidea*; *Otariidae*, *Odobenus rosmarus*, and *Phocidae*, which are parts of *Pinnipedia*; and *Ursidae*) and 35 extant species of *Canidae*. For references regarding extinct arctoids, see the chapters on *Arctoidea*, *Pinnipedia*, and *Pan-Pinnipedia* in this volume. For recent phylogenies including extinct canids, see Tedford et al. (2009) and Prevosti (2010). The canid stem contains a paraphyletic array of the *Leptocyon* species (Tedford et al., 2009: Figs. 65 and 66), *Borophaginae* (Wang et al., 1999), a paraphyletic grouping of hesperocyonines (Wang, 1994), and may also include *Amphicyonidae*

(beardogs; Flynn et al., 1988; Wyss and Flynn, 1993; Spaulding and Flynn, 2012). Partial lists of amphicyonid species can be found in Viranta (1996) and Hunt (1998, 2011).

Here we tentatively follow the view that *Amphicyonidae* is closely related to *Canidae* (Spaulding and Flynn, 2012). This hypothesis revived a traditional view (de Blainville, 1837; Trouessart, 1885; Matthew, 1924; Simpson, 1945) that had prevailed until the hypotheses of ursid (Hough, 1948; Olsen, 1960; Ginsburg, 1966, 1977; Hunt, 1977; Flynn et al., 1988; Wyss and Flynn, 1993; Vrana et al., 1994) or more general arctoid (Flynn et al., 1988; Wolsan, 1993; Tedford et al., 1994; Baskin and Tedford, 1996) affinities of *Amphicyonidae* came into vogue. An alternative is that *Amphicyonidae* is a stem caniform (within *Pan-Caniformia* as defined in this volume; Wesley-Hunt and Flynn, 2005; Tomiya, 2011).

Diagnostic Apomorphies: Apomorphies of *Caniformia* with respect to its sister crown clade *Feliformia*, as defined in this volume, include 14 cranial, dental, and postcranial features: (1) maxilloturbinals large and branching, excluding nasoturbinals from the narial opening (Wyss and Flynn, 1993: character 4); (2) primary anterior opening of the palatine canal placed at the maxilla–palatine suture (Spaulding and Flynn, 2012: character 6); (3) posterior lacerate foramen a defined foramen, not a vacuity (Wesley-Hunt and Flynn, 2005: character 17); (4) internal carotid artery medial, extrabullar, inside a bony canal formed by the entotympanic (Wesley-Hunt and Flynn, 2005: character 25); (5) promontorium anteriorly elongate with a broad flat extension (Wesley-Hunt and

Flynn, 2005: character 28); (6) shelf between the mastoid and paroccipital processes laterally wide, with a rugose or bulbous surface (Wesley-Hunt and Flynn, 2005: character 33); (7) posterior lingual cingular shelf at the base of the protocone on the first upper molar larger than the anterior cingulum (Wesley-Hunt and Flynn, 2005: character 47); (8) ulnar collateral ligament insertion site on the humerus very large, forming a distinct circular pit (Spaulding and Flynn, 2012: character 120); (9) metacarpal IV not overlapped by metacarpal III proximally (Spaulding and Flynn, 2012: character 148); (10) medial phalanx with a symmetrical distal articular surface (Spaulding and Flynn, 2012: character 153); (11) medial phalanx without an excavation on the lateral margin (Spaulding and Flynn, 2012: character 155); (12) astragalus without an anterior ventral expansion on the lateral margin (Spaulding and Flynn, 2012: character 179); (13) cuboid articulation with metatarsal V occupies at least 40% of the distal surface (Spaulding and Flynn, 2012: character 195); and (14) baculum long and elaborated (Wozencraft, 1989: character 83; Wyss and Flynn, 1993: character 52).

Synonyms: *Canoidea* sensu Bryant (1996: 184), but not sensu Wolsan (1993: 351), is an unambiguous synonym. *Arctoidea* sensu Wolsan (1993: 363) is an unambiguous synonym under the hypothesis that *Amphicyonidae* is closely related to *Canidae*. *Hypomycteri* Cope 1882: 473, *Arctoidei* Winge 1895: 47, *Arctoidea* sensu Weber (1904: 528), *Ursoidae* Hay 1930: 488, *Canoidea* Simpson 1931: 276, and *Cynoidea* sensu Scott and Jepsen (1936: 54) are approximate synonyms.

Comments: The name *Caniformia* is here chosen from among its synonyms based on frequency of use. This name was coined by Kretzoi (1943: 194) for a group, now regarded as polyphyletic, composed of various taxa of *Pan-Carnivora* (as defined in this volume). Tedford (1976) refined the meaning of *Caniformia* by adopting this name for a clade containing only *Canidae* and *Arctoidea*. This more restricted meaning of *Caniformia* has since been universally followed in terms of taxon composition, although there have been differences in the concept of the clade. Most authors have either explicitly or implicitly applied *Caniformia* to the smallest crown clade containing *Canidae* and *Arctoidea* (e.g., Vrana et al., 1994; Flynn and Nedbal, 1998; Flynn et al., 2000, 2005; Zehr et al., 2001; Yu et al., 2004, 2011; Delisle and Strobeck, 2005; Finarelli and Flynn, 2006; Fulton and Strobeck, 2006; Yu and Zhang, 2006; Árnason et al., 2007; Chaveerach et al., 2008; Eizirik and Murphy, 2009; Sato et al., 2009; Schröder et al., 2009; Agnarsson et al., 2010; Eizirik et al., 2010; Meredith et al., 2011: Supporting Online Material; Nyakatura and Bininda-Emonds, 2012; Spaulding and Flynn, 2012; Luan et al., 2013), whereas others have applied *Caniformia* to the corresponding total clade (e.g., Flynn and Galiano, 1982; Wolsan, 1993; Wyss and Flynn, 1993; Wang and Tedford, 1994; Bryant, 1996; Wolsan and Bryant, 2004; Flynn and Wesley-Hunt, 2005; Wang et al., 2005; Wesley-Hunt and Flynn, 2005; Flynn et al., 2010; Tomiya, 2011), and some have instead used *Canoidea* for the crown (e.g., Coltman and Wright, 1994; Bryant, 1996; Schreiber et al., 1998; Flynn and Wesley-Hunt, 2005; Wesley-Hunt and Flynn, 2005; Flynn et al., 2010; Tomiya, 2011). However, others have applied *Canoidea* to a more restricted clade containing canids but not arctoids (e.g., Wozencraft, 1989; Wolsan, 1993; Flynn, 1996; Flynn and Nedbal, 1998; Ginsburg, 1999; Delisle and Strobeck, 2005; Fulton and Strobeck, 2006). Similarly, *Cynoidea* has in most cases been applied to a clade containing canids but not arctoids, in agreement with Flower's (1869) original concept (e.g., Tedford, 1976; Flynn et al., 1988; Wang and Tedford, 1994; McKenna and Bell, 1997; Janis et al., 1998; Delisle and Strobeck, 2005; Wang et al., 2005; Fulton and Strobeck, 2006; Árnason et al., 2007; Eizirik and Murphy, 2009; Eizirik et al., 2010). The synonym *Hypomycteri* has not been used for more than 100 years, and the synonyms *Arctoidei* and *Ursoidae* have seen little use. *Arctoidea* is applied to a subclade of *Caniformia*.

A close relationship between *Canidae*, *Ursidae*, *Pinnipedia*, and *Musteloidea* to the exclusion of other non-nested extant clades is well grounded on both morphological and genetic evidence (e.g., Flynn et al., 1988, 2000, 2005; Wyss and Flynn, 1993; Flynn and Nedbal, 1998; Wesley-Hunt and Flynn, 2005; Sato et al., 2009; Eizirik et al., 2010; Meredith et al., 2011; Spaulding and Flynn, 2012).

Literature Cited

Agnarsson, I., M. Kuntner, and L. J. May-Collado. 2010. Dogs, cats, and kin: a molecular species-level phylogeny of *Carnivora*. *Mol. Phylogenet. Evol.* 54:726–745.

Árnason, Ú., A. Gullberg, A. Janke, and M. Kullberg. 2007. Mitogenomic analyses of caniform relationships. *Mol. Phylogenet. Evol.* 45:863–874.

Baskin, J. A., and R. H. Tedford. 1996. Small arctoid and feliform carnivorans. Pp. 486–497 in *The Terrestrial Eocene–Oligocene Transition in North America* (D. R. Prothero and R. J. Emry, eds.). Cambridge University Press, Cambridge, UK.

de Blainville, H. M. D. 1837. Rapport sur un nouvel envoi de fossiles provenant du dépôt de Sansan. *C. R. Acad. Sci. Paris* 5:417–427.

Bryant, H. N. 1996. Explicitness, stability, and universality in the phylogenetic defnition and usage of taxon names: a case study of the phylogenetic taxonomy of the *Carnivora* (*Mammalia*). *Syst. Biol.* 45:174–189.

Chaveerach, A., N. Srisamoot, S. Nuchadomrong, N. Sattayasai, P. Chaveerach, A. Tanomtong, and K. Pinthong. 2008. Phylogenetic relationships of wildlife order *Carnivora* in Thailand inferred from the internal transcribed spacer region. *J. Biol. Sci.* 8:278–287.

Coltman, D. W., and J. M. Wright. 1994. *Can* SINEs: a family of tRNA-derived retroposons specific to the superfamily *Canoidea*. *Nucleic Acids Res.* 22:2726–2730.

Cope, E. D. 1882. On the systematic relations of the *Carnivora Fissipedia*. *Palaeontol. Bull.* 35:471–475.

Delisle, I., and C. Strobeck. 2005. A phylogeny of the *Caniformia* (order *Carnivora*) based on 12 complete protein-coding mitochondrial genes. *Mol. Phylogenet. Evol.* 37:192–201.

Eizirik, E., and W. J. Murphy. 2009. Carnivores (*Carnivora*). Pp. 504–507 in *The Timetree of Life* (S. B. Hedges and S. Kumar, eds.). Oxford University Press, New York.

Eizirik, E., W. J. Murphy, K.-P. Koepfli, W. E. Johnson, J. W. Dragoo, R. K. Wayne, and S. J. O'Brien. 2010. Pattern and timing of diversification of the mammalian order *Carnivora* inferred from multiple nuclear gene sequences. *Mol. Phylogenet. Evol.* 56:49–63.

Finarelli, J. A., and J. J. Flynn. 2006. Ancestral state reconstruction of body size in the *Caniformia* (*Carnivora, Mammalia*): the effects of incorporating data from the fossil record. *Syst. Biol.* 55:301–313.

Flower, W. H. 1869. On the value of the characters of the base of the cranium in the classification of the order *Carnivora*, and on the systematic position of *Bassaris* and other disputed forms. *Proc. Zool. Soc. Lond.* 1869:4–37.

Flynn, J. J. 1996. Carnivoran phylogeny and rates of evolution: morphological, taxic, and molecular. Pp. 542–581 in *Carnivore Behavior, Ecology, and Evolution*, Vol. 2 (J. L. Gittleman, ed.). Cornell University Press, Ithaca, NY.

Flynn, J. J., J. A. Finarelli, and M. Spaulding. 2010. Phylogeny of the *Carnivora* and *Carnivoramorpha*, and the use of the fossil record to enhance understanding of evolutionary transformations. Pp. 25–63 in *Carnivoran*

Evolution: New Views on Phylogeny, Form, and Function (A. Goswami and A. Friscia, eds.). Cambridge University Press, Cambridge, UK.

Flynn, J. J., J. A. Finarelli, S. Zehr, J. Hsu, and M. A. Nedbal. 2005. Molecular phylogeny of the *Carnivora* (*Mammalia*): assessing the impact of increased sampling on resolving enigmatic relationships. *Syst. Biol.* 54:317–337.

Flynn, J. J., and H. Galiano. 1982. Phylogeny of early Tertiary *Carnivora*, with a description of a new species of *Protictis* from the middle Eocene of northwestern Wyoming. *Am. Mus. Novit.* 2725:1–64.

Flynn, J. J., and M. A. Nedbal. 1998. Phylogeny of the *Carnivora* (*Mammalia*): congruence vs incompatibility among multiple data sets. *Mol. Phylogenet. Evol.* 9:414–426.

Flynn, J. J., M. A. Nedbal, J. W. Dragoo, and R. L. Honeycutt. 2000. Whence the red panda? *Mol. Phylogenet. Evol.* 17:190–199.

Flynn, J. J., N. A. Neff, and R. H. Tedford. 1988. Phylogeny of the *Carnivora*. Pp. 73–115 in *The Phylogeny and Classification of the Tetrapods. Vol. 2: Mammals* (M. J. Benton, ed.). Clarendon Press, Oxford.

Flynn, J. J., and G. D. Wesley-Hunt. 2005. *Carnivora*. Pp. 175–198 in *The Rise of Placental Mammals: Origins and Relationships of the Major Extant Clades* (K. D. Rose and J. D. Archibald, eds.). Johns Hopkins University Press, Baltimore, MD.

Fulton, T. L., and C. Strobeck. 2006. Molecular phylogeny of the *Arctoidea* (*Carnivora*): effect of missing data on supertree and supermatrix analyses of multiple gene data sets. *Mol. Phylogenet. Evol.* 41:165–181.

Ginsburg, L. 1966. Les Amphicyons des Phosphorites du Quercy. *Ann. Paléontol. (Vertébr.)* 52:23–64.

Ginsburg, L. 1977. *Cynelos lemanensis* (Pomel), carnivore ursidé de l'Aquitanien d'Europe. *Ann. Paléontol. (Vertébr.)* 63:57–104.

Ginsburg, L. 1999. Order *Carnivora*. Pp. 109–148 in *The Miocene Land Mammals of Europe* (G. E. Rössner and K. Heissig, eds.). Verlag Dr. Friedrich Pfeil, Munich.

Hay, O. P. 1930. *Second Bibliography and Catalogue of the Fossil Vertebrata of North America*, Vol. II. Carnegie Institution of Washington, Washington, DC.

Hough, J. R. 1948. The auditory region in some members of the *Procyonidae, Canidae*, and *Ursidae*. Its significance in the phylogeny of the *Carnivora. Bull. Am. Mus. Nat. Hist.* 92:67–118.

Hunt, Jr., R. M. 1977. Basicranial anatomy of *Cynelos* Jourdan (*Mammalia: Carnivora*), an Aquitanian amphicyonid from the Allier Basin, France. *J. Paleontol.* 51:826–843.

Hunt, Jr., R. M. 1998. *Amphicyonidae*. Pp. 196–227 in *Evolution of Tertiary Mammals of North America. Vol. 1: Terrestrial Carnivores, Ungulates, and Ungulatelike Mammals* (C. M. Janis, K. M. Scott, and L. L. Jacobs, eds.). Cambridge University Press, Cambridge, UK.

Hunt, Jr., R. M. 2011. Evolution of large carnivores during the mid-Cenozoic of North America: the temnocyonine radiation (*Mammalia, Amphicyonidae*). *Bull. Am. Mus. Nat. Hist.* 358:1–153.

Janis, C. M., J. A. Baskin, A. Berta, J. J. Flynn, G. F. Gunnell, R. M. Hunt, Jr., L. D. Martin, and K. Munthe. 1998. Carnivorous mammals. Pp. 73–90 in *Evolution of Tertiary Mammals of North America. Vol. 1: Terrestrial Carnivores, Ungulates, and Ungulatelike Mammals* (C. M. Janis, K. M. Scott, and L. L. Jacobs, eds.). Cambridge University Press, Cambridge, UK.

Kretzoi, M. 1943. *Kochictis centennii* n. g. n. sp., ein altertümlicher Creodonte aus dem Oberoligozän Siebenbürgens. *Földt. Közlöny* 73:180–195.

Linnaeus, C. 1758. *Systema Naturae Per Regna Tria Naturae, Secundum Classes, Ordines, Genera, Species, cum Characteribus, Differentiis, Synonymis, Locis*. Tomus I, Editio decima, reformata. Laurentius Salvius, Holmiae (Stockholm).

Luan, P., O. A. Ryder, H. Davis, Y. Zhang, and L. Yu. 2013. Incorporating indels as phylogenetic characters: impact for interfamilial relationships within *Arctoidea* (*Mammalia: Carnivora*). *Mol. Phylogenet. Evol.* 66:748–756.

Matthew, W. D. 1924. Third contribution to the Snake Creek fauna. *Bull. Am. Mus. Nat. Hist.* 50:59–210.

McKenna, M. C., and S. K. Bell. 1997. *Classification of Mammals above the Species Level.* Columbia University Press, New York.

Meredith, R. W., J. E. Janečka, J. Gatesy, O. A. Ryder, C. A. Fisher, E. C. Teeling, A. Goodbla, E. Eizirik, T. L. L. Simão, T. Stadler, D. L. Rabosky, R. L. Honeycutt, J. J. Flynn, C. M. Ingram, C. Steiner, T. L. Williams, T. J. Robinson, A. Burk-Herrick, M. Westerman, N. A. Ayoub, M. S. Springer, and W. J. Murphy. 2011. Impacts of the Cretaceous Terrestrial Revolution and KPg extinction on mammal diversification. *Science* 334:521–524.

Nyakatura, K., and O. R. P. Bininda-Emonds. 2012. Updating the evolutionary history of *Carnivora* (*Mammalia*): a new species-level supertree complete with divergence time estimates. *BMC Biol.* 10(12):1–31.

Olsen, S. J. 1960. The fossil carnivore *Amphicyon longiramus* from the Thomas Farm Miocene. Part II. Postcranial skeleton. *Bull. Mus. Comp. Zool.* 123:1–45.

Prevosti, F. J. 2010. Phylogeny of the large extinct South American canids (*Mammalia, Carnivora, Canidae*) using a "total evidence" approach. *Cladistics* 26:456–481.

Sato, J. J., M. Wolsan, S. Minami, T. Hosoda, M. H. Sinaga, K. Hiyama, Y. Yamaguchi, and H. Suzuki. 2009. Deciphering and dating the red panda's ancestry and early adaptive radiation of *Musteloidea. Mol. Phylogenet. Evol.* 53:907–922.

Schreber, J. C. D. 1777. Die wilde Kaze. P. 397 in *Die Säugthiere in Abbildungen nach der Natur mit Beschreibungen, Dritter Theil,* Heft 23 (J. C. D. Schreber). Walthersche Kunst- und Buchhandlung, Erlangen.

Schreiber, A., K. Eulenberger, and K. Bauer. 1998. Immunogenetic evidence for the phylogenetic sister group relationship of dogs and bears (*Mammalia, Carnivora: Canidae* and *Ursidae*): a comparative determinant analysis of carnivoran albumin, C3 complement and immunoglobulin μ-chain. *Exp. Clin. Immunogenet.* 15:154–170.

Schröder, C., C. Bleidorn, S. Hartmann, and R. Tiedemann. 2009. Occurrence of *Can*-SINEs and intron sequence evolution supports robust phylogeny of pinniped carnivores and their terrestrial relatives. *Gene* 448:221–226.

Scott, W. B., and G. L. Jepsen. 1936. The mammalian fauna of the White River Oligocene—Part I. *Insectivora* and *Carnivora. Trans. Am. Philos. Soc. New Ser.* 28:1–153.

Simpson, G. G. 1931. A new classification of mammals. *Bull. Am. Mus. Nat. Hist.* 59:259–293.

Simpson, G. G. 1945. The principles of classification and a classification of mammals. *Bull. Am. Mus. Nat. Hist.* 85:I–XVI + 1–350.

Spaulding, M., and J. J. Flynn. 2012. Phylogeny of the *Carnivoramorpha*: the impact of postcranial characters. *J. Syst. Palaeontol.* 10:653–677.

Tedford, R. H. 1976. Relationship of pinnipeds to other carnivores (*Mammalia*). *Syst. Zool.* 25:363–374.

Tedford, R. H., L. G. Barnes, and C. E. Ray. 1994. The early Miocene littoral ursoid carnivoran *Kolponomos*: systematics and mode of life. *Proc. San. Diego Soc. Nat. Hist.* 29:11–32.

Tedford, R. H., X. Wang, and B. E. Taylor. 2009. Phylogenetic systematics of the North American fossil *Caninae* (*Carnivora: Canidae*). *Bull. Am. Mus. Nat. Hist.* 325:1–218.

Tomiya, S. 2011. A new basal caniform (*Mammalia: Carnivora*) from the middle Eocene of North America and remarks on the phylogeny of early carnivorans. *PLOS ONE* 6(e24146):1–24.

Trouessart, E.-L. 1885. Catalogue des Mammifères vivants et fossiles (Carnivores). *Bull. Soc. Études Sci. Angers* 1884(suppl.):1–108.

Viranta, S. 1996. European Miocene *Amphicyonidae*—taxonomy, systematics and ecology. *Acta Zool. Fenn.* 204:1–61.

Vrana, P. B., M. C. Milinkovitch, J. R. Powell, and W. C. Wheeler. 1994. Higher level relationships of the arctoid *Carnivora* based on sequence data and "total evidence". *Mol. Phylogenet. Evol.* 3:47–58.

Wang, X. 1994. Phylogenetic systematics of the *Hesperocyoninae* (*Carnivora: Canidae*). *Bull. Am. Mus. Nat. Hist.* 221:1–207.

Wang, X., M. C. McKenna, and D. Dashzeveg. 2005. *Amphicticeps* and *Amphicynodon* (*Arctoidea, Carnivora*) from Hsanda Gol

Formation, central Mongolia and phylogeny of basal arctoids with comments on zoogeography. *Am. Mus. Novit.* 3483:1–57.

Wang, X., and R. H. Tedford. 1994. Basicranial anatomy and phylogeny of primitive canids and closely related miacids (*Carnivora: Mammalia*). *Am. Mus. Novit.* 3092:1–34.

Wang, X., R. H. Tedford, and B. E. Taylor. 1999. Phylogenetic systematics of the *Borophaginae* (*Carnivora: Canidae*). *Bull. Am. Mus. Nat. Hist.* 243:1–391.

Weber, M. 1904. *Die Säugetiere. Einführung in die Anatomie und Systematik der Recenten und Fossilen* Mammalia. Gustav Fischer, Jena.

Wesley-Hunt, G. D., and J. J. Flynn. 2005. Phylogeny of the *Carnivora*: basal relationships among the carnivoramorphans, and assessment of the position of "*Miacoidea*" relative to *Carnivora. J. Syst. Palaeontol.* 3:1–28.

Winge, H. 1895. Jordfundne og nulevende Rovdyr (*Carnivora*) fra Lagoa Santa, Minas Geraes, Brasilien. Med Udsigt over Rovdyrenes indbyrdes Slaegtskab. Pp. 1–130 in *E Museo Lundii. En Samling af Afhandlinger Om de i Det indre Brasiliens Kalkstenshuler af Professor Dr. Peter Vilhelm Lund Udgravede og i Den Lundske Palaeontologiske Afdeling af Kjøbenhavns Universitets Zoologiske Museum Opbevarede Dyre- Og Menneskeknogler.* Bind 2, Halvbind 2, Afhandling 4 (C. F. Lütken, ed.). H. Hagerups Boghandel, Copenhagen.

Wolsan, M. 1993. Phylogeny and classification of early European *Mustelida* (*Mammalia: Carnivora*). *Acta Theriol.* 38:345–384.

Wolsan, M., and H. N. Bryant. 2004. Phylogenetic nomenclature of carnivoran mammals. P. 32 in *First International Phylogenetic Nomenclature Meeting, Paris, Muséum National d'Histoire Naturelle, 6–9 July 2004: Abstracts* (M. Laurin, ed.). Muséum National d'Histoire Naturelle, Paris.

Wozencraft, W. C. 1989. The phylogeny of the Recent *Carnivora*. Pp. 495–535 in *Carnivore Behavior, Ecology, and Evolution* (J. L. Gittleman, ed.). Cornell University Press, Ithaca, NY.

Wozencraft, W. C. 2005. Order *Carnivora*. Pp. 532–628 in *Mammal Species of the World: A Taxonomic and Geographic Reference.* 3rd edition, Vol. 1 (D. E. Wilson and D. M. Reeder, eds.). Johns Hopkins University Press, Baltimore, MD.

Wyss, A. R., and J. J. Flynn. 1993. A phylogenetic analysis and definition of the *Carnivora*. Pp. 32–52 in *Mammal Phylogeny: Placentals* (F. S. Szalay, M. J. Novacek, and M. C. McKenna, eds.). Springer-Verlag, New York.

Yu, L., Q. Li, O. A. Ryder, and Y. Zhang. 2004. Phylogenetic relationships within mammalian order *Carnivora* indicated by sequences of two nuclear DNA genes. *Mol. Phylogenet. Evol.* 33:694–705.

Yu, L., P.-T. Luan, W. Jin, O. A. Ryder, L. G. Chemnick, H. A. Davis, and Y. Zhang. 2011. Phylogenetic utility of nuclear introns in interfamilial relationships of *Caniformia* (order *Carnivora*). *Syst. Biol.* 60:175–187.

Yu, L., and Y. Zhang. 2006. Phylogeny of the caniform *Carnivora*: evidence from multiple genes. *Genetica* 127:65–79.

Zehr, S. M., M. A. Nedbal, and J. J. Flynn. 2001. Tempo and mode of evolution in an orthologous *Can* SINE. *Mamm. Genome* 12:38–44.

Authors

Mieczysław Wolsan; Museum and Institute of Zoology; Polish Academy of Sciences; Wilcza 64; 00-679 Warszawa, Poland. Email: wolsan@miiz.waw.pl.

John J. Flynn; Division of Paleontology; American Museum of Natural History; Central Park West at 79th Street; New York, NY 10024-5192; USA. Email: jflynn@amnh.org.

André R. Wyss; Department of Earth Science; University of California at Santa Barbara; Webb Hall; Santa Barbara, CA 93106-9630, USA. Email: wyss@geol.ucsb.edu.

Date Accepted: 13 August 2015

Primary Editor: Philip Cantino

Pan-Arctoidea M. Wolsan, H. N. Bryant, J. J. Flynn, and A. R. Wyss, new clade name

Registration Number: 269

Definition: The total clade of the crown clade *Arctoidea*. This is a crown-based total-clade definition. Abbreviated definition: total ∇ of *Arctoidea*.

Etymology: Derived from the Greek *pan-* (all, in reference to a total clade), *arctos* (bear), and *oeidēs* (like, resembling).

Reference Phylogeny: Figure 1 of Sato et al. (2009), where *Pan-Arctoidea* corresponds to the clade originating at the base of the branch leading to *Arctoidea*. See also Flynn et al. (2005: Fig. 1), where *Pan-Arctoidea* is the clade originating at the base of the branch that separates *Ailuropoda melanoleuca* and all taxa below it in that figure from *Canis lupus* and all taxa above it.

Composition: For the composition of the crown, see *Arctoidea* (this volume). The composition of the stem is not known with confidence. *Amphicyonidae* (beardogs) has been considered part of the arctoid crown (Hough, 1948; Olsen, 1960; Ginsburg, 1966, 1977, 1982; Hunt, 1977; Flynn et al., 1988; Wyss and Flynn, 1993; Vrana et al., 1994) or its stem (Flynn et al., 1988; Wolsan, 1993; Tedford et al., 1994; Baskin and Tedford, 1996) in the past, but at present there is stronger support for the placement of this taxon outside *Pan-Arctoidea* (Wesley-Hunt and Flynn, 2005; Tomiya, 2011; Spaulding and Flynn, 2012).

Diagnostic Apomorphies: At their root, total clades are unlikely to have any diagnostic apomorphies. No clear stem arctoids are known, however, so the diagnosis of *Pan-Arctoidea* is currently the same as for *Arctoidea* (this volume).

Synonyms: *Arctomorpha* Wolsan 1993: 363 is an unambiguous synonym under the hypothesis of a close relationship between *Amphicyonidae* and *Canidae* (dogs; Spaulding and Flynn, 2012).

Comments: The preexisting name *Arctomorpha*, proposed by Wolsan (1993) for a clade more inclusive than crown *Arctoidea*, but less inclusive than its total clade, applies to the total clade under the hypothesis that *Amphicyonidae* is closer to canids than to arctoids (Spaulding and Flynn, 2012); however, that name has rarely been used, and its application to the total clade of *Arctoidea* under Wolsan's (1993) definition depends on the phylogenetic position of *Amphicyonidae*, which remains uncertain. In addition, the name *Arctoidea*, applied by Wolsan (1993) to the total clade of the crown clade here named *Arctoidea* under the hypothesis that *Amphicyonidae* is closer to arctoids than to canids (Flynn et al., 1988; Wolsan, 1993; Tedford et al., 1994; Baskin and Tedford, 1996), is a synonym of *Caniformia* (this volume) under the hypothesis that *Amphicyonidae* is closer to canids (Spaulding and Flynn, 2012). For these reasons, and given the advantages of an integrated system of clade names in which the names of total clades are formed by adding the prefix *Pan-* to the names of their corresponding crowns (de Queiroz, 2007), we apply the preexisting name *Arctoidea* to the crown clade (in this volume) and introduce a new panclade name to designate the total clade of *Arctoidea*.

Literature Cited

Baskin, J. A., and R. H. Tedford. 1996. Small arctoid and feliform carnivorans. Pp. 486–497 in *The Terrestrial Eocene–Oligocene Transition in North America* (D. R. Prothero and R. J. Emry, eds.). Cambridge University Press, Cambridge, UK.

Flynn, J. J., J. A. Finarelli, S. Zehr, J. Hsu, and M. A. Nedbal. 2005. Molecular phylogeny of the *Carnivora* (*Mammalia*): assessing the impact of increased sampling on resolving enigmatic relationships. *Syst. Biol.* 54:317–337.

Flynn, J. J., N. A. Neff, and R. H. Tedford. 1988. Phylogeny of the *Carnivora*. Pp. 73–115 in *The Phylogeny and Classification of the Tetrapods. Vol. 2: Mammals* (M. J. Benton, ed.). Clarendon Press, Oxford.

Ginsburg, L. 1966. Les amphicyons des phosphorites du quercy. *Ann. Paléontol. (Vertébr.)* 52:23–64.

Ginsburg, L. 1977. *Cynelos lemanensis* (Pomel), carnivore ursidé de l'Aquitanien d'Europe. *Ann. Paléontol. (Vertébr.)* 63:57–104.

Ginsburg, L. 1982. Sur la position systématique du petit panda, *Ailurus fulgens* (*Carnivora*, *Mammalia*). *Geobios Mém. Spéc.* 6:247–258.

Hough, J. R. 1948. The auditory region in some members of the *Procyonidae, Canidae*, and *Ursidae*. Its significance in the phylogeny of the *Carnivora. Bull. Am. Mus. Nat. Hist.* 92:67–118.

Hunt, Jr., R. M. 1977. Basicranial anatomy of *Cynelos* Jourdan (*Mammalia: Carnivora*), an Aquitanian amphicyonid from the Allier Basin, France. *J. Paleontol.* 51:826–843.

Olsen, S. J. 1960. The fossil carnivore *Amphicyon longiramus* from the Thomas Farm Miocene. Part II. Postcranial skeleton. *Bull. Mus. Comp. Zool.* 123:1–45.

de Queiroz, K. 2007. Toward an integrated system of clade names. *Syst. Biol.* 56:956–974.

Sato, J. J., M. Wolsan, S. Minami, T. Hosoda, M. H. Sinaga, K. Hiyama, Y. Yamaguchi, and H. Suzuki. 2009. Deciphering and dating the red panda's ancestry and early adaptive radiation of *Musteloidea. Mol. Phylogenet. Evol.* 53:907–922.

Spaulding, M., and J. J. Flynn. 2012. Phylogeny of the *Carnivoramorpha*: the impact of postcranial characters. *J. Syst. Palaeontol.* 10:653–677.

Tedford, R. H., L. G. Barnes, and C. E. Ray. 1994. The early Miocene littoral ursoid carnivoran *Kolponomos*: systematics and mode of life. *Proc. San Diego Soc. Nat. Hist.* 29:11–32.

Tomiya, S. 2011. A new basal caniform (*Mammalia: Carnivora*) from the middle Eocene of North America and remarks on the phylogeny of early carnivorans. *PLOS ONE* 6(e24146):1–24.

Vrana, P. B., M. C. Milinkovitch, J. R. Powell, and W. C. Wheeler. 1994. Higher level relationships of the arctoid *Carnivora* based on sequence data and "total evidence". *Mol. Phylogenet. Evol.* 3:47–58.

Wesley-Hunt, G. D., and J. J. Flynn. 2005. Phylogeny of the *Carnivora*: basal relationships among the carnivoramorphans, and assessment of the position of "*Miacoidea*" relative to *Carnivora. J. Syst. Palaeontol.* 3:1–28.

Wolsan, M. 1993. Phylogeny and classification of early European *Mustelida* (*Mammalia: Carnivora*). *Acta Theriol.* 38:345–384.

Wyss, A. R., and J. J. Flynn. 1993. A phylogenetic analysis and definition of the *Carnivora*. Pp. 32–52 in *Mammal Phylogeny: Placentals* (F. S. Szalay, M. J. Novacek, and M. C. McKenna, eds.). Springer-Verlag, New York.

Authors

Mieczysław Wolsan; Museum and Institute of Zoology; Polish Academy of Sciences; Wilcza 64; 00-679 Warszawa, Poland. Email: wolsan@miiz.waw.pl.

Harold N. Bryant; Royal Saskatchewan Museum; 2445 Albert Street; Regina, SK, S4P 4W7, Canada. Email: bryants@sasktel.net.

John J. Flynn; Division of Paleontology; American Museum of Natural History; Central Park West at 79th Street; New York, NY 10024-5192, USA. Email: jflynn@amnh.org.

André R. Wyss; Department of Earth Science; University of California at Santa Barbara; Webb Hall; Santa Barbara, CA 93106-9630, USA. Email: wyss@geol.ucsb.edu.

Date Accepted: 01 August 2015

Primary Editor: Jacques A. Gauthier

Arctoidea W. H. Flower 1869 [M. Wolsan, H. N. Bryant, J. J. Flynn, and A. R. Wyss], converted clade name

Registration Number: 294

Definition: The smallest crown clade containing *Ursus arctos* Linnaeus 1758 (*Ursidae*), *Mustela erminea* Linnaeus 1758 (*Mustelidae*), *Procyon lotor* (Linnaeus 1758) (*Procyonidae*), *Ailurus fulgens* F. Cuvier 1825, and *Mephitis mephitis* (Schreber 1776) (*Mephitidae*). This is a minimum-crown-clade definition. Abbreviated definition: min crown ∇ (*Ursus arctos* Linnaeus 1758 & *Mustela erminea* Linnaeus 1758 & *Procyon lotor* (Linnaeus 1758) & *Ailurus fulgens* F. Cuvier 1825 & *Mephitis mephitis* (Schreber 1776)).

Etymology: Derived from the Greek *arctos* (bear) and *oeidēs* (like, resembling).

Reference Phylogeny: Figure 1 in Sato et al. (2009) is the primary reference phylogeny. See also Flynn et al. (2005: Fig. 1).

Composition: Wozencraft (2005) listed 127 living and three recently extinct species distributed among *Ursidae* (bears), *Phocidae* (seals), *Odobenus rosmarus* (walruses), *Otariidae* (sea lions and fur seals), *Mustelidae* (weasels, otters, martens, badgers, and relatives), *Procyonidae* (raccoons), *Ailurus fulgens* (red pandas), and *Mephitidae* (skunks and stink badgers). *Mustelidae, Procyonidae, Ailurus fulgens*, and *Mephitidae* are parts of *Musteloidea* (e.g., Flynn et al., 2005; Fulton and Strobeck, 2006; Sato et al., 2006, 2009, 2012), whereas *Phocidae, Odobenus rosmarus*, and *Otariidae* are parts of *Pinnipedia* (e.g., Wyss and Flynn, 1993; Flynn and Nedbal, 1998; Flynn et al., 2005; Fulton and Strobeck, 2006). For references regarding extinct species of the pinniped crown and the composition of its stem, see the chapters on *Pinnipedia* and *Pan-Pinnipedia* in this volume. McKenna and Bell (1997) provided an extensive list of extinct supraspecific taxa thought to be arctoids. This list includes *Amphicyonidae* (beardogs). However, even though some features suggest a close relationship of this taxon to extant bears (*Ursidae*; Hough, 1948; Olsen, 1960; Ginsburg, 1966, 1977; Hunt, 1977; Flynn et al., 1988; Wyss and Flynn, 1993; Vrana et al., 1994), there is stronger contradictory evidence that indicates a position outside the arctoid total clade (*Pan-Arctoidea* as that name is defined in this volume; Wesley-Hunt and Flynn, 2005; Tomiya, 2011; Spaulding and Flynn, 2012).

Diagnostic Apomorphies: Apomorphies of *Arctoidea* relative to *Canidae* (dogs, the sister crown clade of *Arctoidea*; e.g., Wyss and Flynn, 1993; Flynn et al., 2005; Sato et al., 2009; Eizirik et al., 2010) include four features in the ear region of the skull (mastoid process of petrosal prominent; external auditory meatus tubular and prolonged laterally; hypotympanic space extended into base of auditory tube; suprameatal fossa present) and two soft-anatomical traits (caecum absent; prostate glands situated within wall of urethra; Flynn et al., 1988).

Synonyms: *Ursidae* sensu Turner (1848: 86) is an approximate synonym.

Comments: The name *Arctoidea* was introduced by Flower (1869: 15) to encompass a group of extant *Carnivora* taxa, namely, *Ursidae, Mustelidae, Procyonidae, Ailurus fulgens*, and what is currently referred to as *Mephitidae*.

Nineteenth-century and early twentieth-century users of the name *Arctoidea* either followed its original circumscription (e.g., Mivart, 1885a) or also included *Canidae* (e.g., Weber, 1904). Since the later decades of the twentieth century it has become widely recognized, based on both morphological and genetic evidence, that pinnipeds are arctoids and that arctoids and canids are sister crown clades (e.g., Tedford, 1976; Flynn et al., 1988, 2000, 2005; Wolsan, 1993; Wyss and Flynn, 1993; Flynn and Nedbal, 1998; Delisle and Strobeck, 2005; Wang et al., 2005; Flynn and Wesley-Hunt, 2005; Wesley-Hunt and Flynn, 2005; Fulton and Strobeck, 2006; Árnason et al., 2007; Sato et al., 2009; Schröder et al., 2009; Eizirik et al., 2010; Meredith et al., 2011). Although some palaeontological studies hypothesized that *Amphicyonidae* is a close relative of extant arctoids and used the name *Arctoidea* to encompass this extinct taxon (e.g., Wolsan, 1993; Wyss and Flynn, 1993; Tedford et al., 1994; note that *Arctoidea* sensu Wolsan [1993: 363] is a synonym of *Caniformia* under the hypothesis that amphicyonids are closely related to canids [Spaulding and Flynn, 2012]), the prevailing explicit or implicit use of the name *Arctoidea* in recent studies has been in reference to the smallest crown clade containing *Ursidae*, *Pinnipedia*, and *Musteloidea* (e.g., Bryant, 1996; Flynn and Nedbal, 1998; Flynn et al., 2000, 2005; Delisle and Strobeck, 2005; Fulton and Strobeck, 2006; Sato et al., 2006, 2009; Árnason et al., 2007; Schröder et al., 2009; Eizirik et al., 2010; Luan et al., 2013). *Arctoidea* is therefore the most suitable choice for designating this crown clade. The name *Ursidae*, applied to a group of similar composition by Turner (1848), has generally been applied to a more limited clade including only bears (e.g., Simpson, 1945; McKenna and Bell, 1997; Wozencraft, 2005).

Recent molecular investigations strongly support that *Mustelidae*, *Procyonidae*, *Ailurus fulgens*, and *Mephitidae* form a clade (*Musteloidea*);

this clade plus *Pinnipedia* forms a clade to the exclusion of *Ursidae* (Flynn et al., 2000, 2005; Fulton and Strobeck, 2006; Sato et al., 2006, 2009; Eizirik et al., 2010). Alternative hypotheses, although less well supported or not based on a phylogenetic analysis of a taxon/character dataset, postulate either a close relationship between *Pinnipedia* and *Ursidae* (Weber, 1904; Wyss and Flynn, 1993; Yu et al., 2011; Luan et al., 2013; for more references, see Sato et al., 2006) or pinniped diphyly, with *Otariidae* allied with *Ursidae* and *Phocidae* related to *Musteloidea* (Mivart, 1885b; McLaren, 1960; Tedford, 1976; Repenning et al., 1979). Five internal specifiers (an ursid, *Ursus arctos*; a mustelid, *Mustela erminea*; a procyonid, *Procyon lotor*; *Ailurus fulgens*; and a mephitid, *Mephitis mephitis*) reflect the spirit of Flower's (1869) original concept of *Arctoidea* and also unambiguously associate this taxon name with the intended clade under all of these conflicting hypotheses.

Literature Cited

Árnason, Ú., A. Gullberg, A. Janke, and M. Kullberg. 2007. Mitogenomic analyses of caniform relationships. *Mol. Phylogenet. Evol.* 45:863–874.

Bryant, H. N. 1996. Explicitness, stability, and universality in the phylogenetic defnition and usage of taxon names: a case study of the phylogenetic taxonomy of the *Carnivora* (*Mammalia*). *Syst. Biol.* 45:174–189.

Cuvier, F. 1825. Panda. Pp. 1–3 in *Histoire Naturelle des Mammifères, Avec des Figures Originales, Coloriées, Dessinées d'Après des Animaux Vivans, Tome Troisième, Livraison L* (É. Geoffroy Saint-Hilaire and F. Cuvier, eds.). A. Belin, Paris.

Delisle, I., and C. Strobeck. 2005. A phylogeny of the *Caniformia* (order *Carnivora*) based on 12 complete protein-coding mitochondrial genes. *Mol. Phylogenet. Evol.* 37:192–201.

Eizirik, E., W. J. Murphy, K.-P. Koepfli, W. E. Johnson, J. W. Dragoo, R. K. Wayne, and S. J. O'Brien. 2010. Pattern and timing of diversification of the mammalian order *Carnivora* inferred from multiple nuclear gene sequences. *Mol. Phylogenet. Evol.* 56:49–63.

Flower, W. H. 1869. On the value of the characters of the base of the cranium in the classification of the order *Carnivora*, and on the systematic position of *Bassaris* and other disputed forms. *Proc. Zool. Soc. Lond.* 1869:4–37.

Flynn, J. J., J. A. Finarelli, S. Zehr, J. Hsu, and M. A. Nedbal. 2005. Molecular phylogeny of the *Carnivora* (*Mammalia*): assessing the impact of increased sampling on resolving enigmatic relationships. *Syst. Biol.* 54:317–337.

Flynn, J. J., and M. A. Nedbal. 1998. Phylogeny of the *Carnivora* (*Mammalia*): congruence vs incompatibility among multiple data sets. *Mol. Phylogenet. Evol.* 9:414–426.

Flynn, J. J., M. A. Nedbal, J. W. Dragoo, and R. L. Honeycutt. 2000. Whence the red panda? *Mol. Phylogenet. Evol.* 17:190–199.

Flynn, J. J., N. A. Neff, and R. H. Tedford. 1988. Phylogeny of the *Carnivora*. Pp. 73–115 in *The Phylogeny and Classification of the Tetrapods. Vol. 2: Mammals* (M. J. Benton, ed.). Clarendon Press, Oxford, UK.

Flynn, J. J., and G. D. Wesley-Hunt. 2005. *Carnivora*. Pp. 175–198 in *The Rise of Placental Mammals: Origins and Relationships of the Major Extant Clades* (K. D. Rose and J. D. Archibald, eds.). Johns Hopkins University Press, Baltimore, MD.

Fulton, T. L., and C. Strobeck. 2006. Molecular phylogeny of the *Arctoidea* (*Carnivora*): effect of missing data on supertree and supermatrix analyses of multiple gene data sets. *Mol. Phylogenet. Evol.* 41:165–181.

Ginsburg, L. 1966. Les Amphicyons des Phosphorites du Quercy. *Ann. Paléontol. (Vertébr.)* 52:23–64.

Ginsburg, L. 1977. *Cynelos lemanensis* (Pomel), carnivore ursidé de l'Aquitanien d'Europe. *Ann. Paléontol. (Vertébr.)* 63:57–104.

Hough, J. R. 1948. The auditory region in some members of the *Procyonidae, Canidae,* and *Ursidae.* Its significance in the phylogeny of the *Carnivora. Bull. Am. Mus. Nat. Hist.* 92:67–118.

Hunt, Jr., R. M. 1977. Basicranial anatomy of *Cynelos* Jourdan (*Mammalia: Carnivora*), an Aquitanian amphicyonid from the Allier Basin, France. *J. Paleontol.* 51:826–843.

Linnaeus, C. 1758. *Systema Naturae Per Regna Tria Naturae, Secundum Classes, Ordines, Genera, Species, cum Characteribus, Differentiis, Synonymis, Locis.* Tomus I, Editio decima, reformata. Laurentius Salvius, Holmiae (Stockholm).

Luan, P., O. A. Ryder, H. Davis, Y. Zhang, and L. Yu. 2013. Incorporating indels as phylogenetic characters: impact for interfamilial relationships within *Arctoidea* (*Mammalia: Carnivora*). *Mol. Phylogenet. Evol.* 66:748–756.

McKenna, M. C., and S. K. Bell. 1997. *Classification of Mammals above the Species Level.* Columbia University Press, New York.

McLaren, I. A. 1960. Are the *Pinnipedia* biphyletic? *Syst. Zool.* 9:18–28.

Meredith, R. W., J. E. Janečka, J. Gatesy, O. A. Ryder, C. A. Fisher, E. C. Teeling, A. Goodbla, E. Eizirik, T. L. L. Simão, T. Stadler, D. L. Rabosky, R. L. Honeycutt, J. J. Flynn, C. M. Ingram, C. Steiner, T. L. Williams, T. J. Robinson, A. Burk-Herrick, M. Westerman, N. A. Ayoub, M. S. Springer, and W. J. Murphy. 2011. Impacts of the Cretaceous Terrestrial Revolution and KPg extinction on mammal diversification. *Science* 334:521–524.

Mivart, S. G. 1885a. On the anatomy, classification, and distribution of the *Arctoidea. Proc. Zool. Soc. Lond.* 1885:340–404.

Mivart, S. G. 1885b. Notes on the *Pinnipedia. Proc. Zool. Soc. Lond.* 1885:484–501.

Olsen, S. J. 1960. The fossil carnivore *Amphicyon longiramus* from the Thomas Farm Miocene. Part II. Postcranial skeleton. *Bull. Mus. Comp. Zool.* 123:1–45.

Repenning, C. A., C. E. Ray, and D. Grigorescu. 1979. Pinniped biogeography. Pp. 357–369 in *Historical Biogeography, Plate Tectonics, and the Changing Environment* (J. Gray and A. J. Boucot, eds.). Oregon State University Press, Corvallis, OR.

Sato, J. J., M. Wolsan, S. Minami, T. Hosoda, M. H. Sinaga, K. Hiyama, Y. Yamaguchi, and H. Suzuki. 2009. Deciphering and dating the red panda's ancestry and early adaptive radiation of *Musteloidea*. *Mol. Phylogenet. Evol.* 53:907–922.

Sato, J. J., M. Wolsan, F. J. Prevosti, G. D'Elía, C. Begg, K. Begg, T. Hosoda, K. L. Campbell, and H. Suzuki. 2012. Evolutionary and biogeographic history of weasel-like carnivorans (*Musteloidea*). *Mol. Phylogenet. Evol.* 63:745–757.

Sato, J. J., M. Wolsan, H. Suzuki, T. Hosoda, Y. Yamaguchi, K. Hiyama, M. Kobayashi, and S. Minami. 2006. Evidence from nuclear DNA sequences sheds light on the phylogenetic relationships of *Pinnipedia*: single origin with affinity to *Musteloidea*. *Zool. Sci.* 23:125–146.

Schreber, J. C. D. 1776. *Viverra Mephitis* Linn. Plate CXXI in *Die Säugthiere in Abbildungen nach der Natur mit Beschreibungen, Dritter Theil, Heft 17* (J. C. D. Schreber). Walthersche Kunst- und Buchhandlung, Erlangen.

Schröder, C., C. Bleidorn, S. Hartmann, and R. Tiedemann. 2009. Occurrence of *Can*-SINEs and intron sequence evolution supports robust phylogeny of pinniped carnivores and their terrestrial relatives. *Gene* 448:221–226.

Simpson, G. G. 1945. The principles of classification and a classification of mammals. *Bull. Am. Mus. Nat. Hist.* 85:I–XVI + 1–350.

Spaulding, M., and J. J. Flynn. 2012. Phylogeny of the *Carnivoramorpha*: the impact of postcranial characters. *J. Syst. Palaeontol.* 10:653–677.

Tedford, R. H. 1976. Relationship of pinnipeds to other carnivores (*Mammalia*). *Syst. Zool.* 25:363–374.

Tedford, R. H., L. G. Barnes, and C. E. Ray. 1994. The early Miocene littoral ursoid carnivoran *Kolponomos*: systematics and mode of life. *Proc. San Diego Soc. Nat. Hist.* 29:11–32.

Tomiya, S. 2011. A new basal caniform (*Mammalia*: *Carnivora*) from the middle Eocene of North America and remarks on the phylogeny of early carnivorans. *PLOS ONE* 6(e24146):1–24.

Turner, Jr., H. N. 1848. Observations relating to some of the foramina at the base of the skull in *Mammalia*, and on the classification of the order *Carnivora*. *Proc. Zool. Soc. Lond.* 16:63–88.

Vrana, P. B., M. C. Milinkovitch, J. R. Powell, and W. C. Wheeler. 1994. Higher level relationships of the arctoid *Carnivora* based on sequence data and "total evidence". *Mol. Phylogenet. Evol.* 3:47–58.

Wang, X., M. C. McKenna, and D. Dashzeveg. 2005. *Amphicticeps* and *Amphicynodon* (*Arctoidea*, *Carnivora*) from Hsanda Gol Formation, central Mongolia and phylogeny of basal arctoids with comments on zoogeography. *Am. Mus. Novit.* 3483:1–57.

Weber, M. 1904. *Die Säugetiere. Einführung in die Anatomie und Systematik der recenten und fossilen* Mammalia. Gustav Fischer, Jena.

Wesley-Hunt, G. D., and J. J. Flynn. 2005. Phylogeny of the *Carnivora*: basal relationships among the carnivoramorphans, and assessment of the position of "*Miacoidea*" relative to *Carnivora*. *J. Syst. Palaeontol.* 3:1–28.

Wolsan, M. 1993. Phylogeny and classification of early European *Mustelida* (*Mammalia*: *Carnivora*). *Acta Theriol.* 38:345–384.

Wozencraft, W. C. 2005. Order *Carnivora*. Pp. 532–628 in *Mammal Species of the World: A Taxonomic and Geographic Reference*. 3rd edition, Vol. 1 (D. E. Wilson and D. M. Reeder, eds.). Johns Hopkins University Press, Baltimore, MD.

Wyss, A. R., and J. J. Flynn. 1993. A phylogenetic analysis and definition of the *Carnivora*. Pp. 32–52 in *Mammal Phylogeny: Placentals* (F. S. Szalay, M. J. Novacek, and M. C. McKenna, eds.). Springer-Verlag, New York.

Yu, L., P.-T. Luan, W. Jin, O. A. Ryder, L. G. Chemnick, H. A. Davis, and Y. Zhang. 2011. Phylogenetic utility of nuclear introns in interfamilial relationships of *Caniformia* (order *Carnivora*). *Syst. Biol.* 60:175–187.

Authors

Mieczysław Wolsan; Museum and Institute of Zoology; Polish Academy of Sciences; Wilcza 64; 00-679 Warszawa, Poland. Email: wolsan@miiz.waw.pl.

Harold N. Bryant; Royal Saskatchewan Museum; 2445 Albert Street; Regina, SK, S4P 4W7, Canada. Email: bryants@sasktel.net.

John J. Flynn; Division of Paleontology; American Museum of Natural History; Central Park West at 79th Street; New York, NY 10024-5192, USA. Email: jflynn@amnh.org.

André R. Wyss; Department of Earth Science; University of California at Santa Barbara; Webb Hall; Santa Barbara, CA 93106-9630, USA. Email: wyss@geol.ucsb.edu.

Date Accepted: 01 August 2015

Primary editor: Jacques A. Gauthier

Pan-Pinnipedia M. Wolsan, A. R. Wyss, A. Berta, and J. J. Flynn, new clade name

Registration Number: 270

Definition: The total clade of the crown clade *Pinnipedia*. This is a crown-based total-clade definition. Abbreviated definition: total ∇ of *Pinnipedia*.

Etymology: Derived from the Greek *pan-* (all, in reference to a total clade), Latin *pinna* (fin or feather), and Latin *pedis* (the genitive of *pes*, foot).

Reference Phylogeny: Figure 2 in Fulton and Strobeck (2006), where *Pan-Pinnipedia* is the clade originating at the base of the branch leading to crown *Pinnipedia*. The scientific equivalents of common names used in this phylogeny are in Table 1 of Fulton and Strobeck (2006).

Composition: In addition to the crown (*Pinnipedia*; this volume), *Pan-Pinnipedia* contains all other *Pinnipedimorpha* (such as *Enaliarctos*, *Pinnarctidion*, *Pacificotaria*, and *Pteronarctos*; Deméré et al., 2003: Fig. 3.3, Table 3.2) and may also contain other extinct species such as *Puijila darwini* (Rybczynski et al., 2009), *Potamotherium valletoni* (Kohno, 1994; Rybczynski et al., 2009), *Amphicticeps shackelfordi* (Wolsan, 1993; Rybczynski et al., 2009), *Kolponomos clallamensis* (Wolsan, 1993; Tedford et al., 1994), *Allocyon loganensis*, *Pachycynodon crassirostris*, and *Amphicynodon velaunus* (Tedford et al., 1994), as well as their fossil relatives. In contrast to pinnipedimorphs, *Puijila* (Rybczynski et al., 2009), *Potamotherium* (Savage, 1957), and *Kolponomos* (Tedford et al., 1994) did not have flippers, and the status of this feature is unknown from other putative pan-pinnipeds. Some evidence, however, suggests that *Potamotherium* (Baskin, 1998; Wang et al., 2005; Finarelli, 2008; Sato et al., 2009), *Amphicticeps* (Wang et al., 2005; Finarelli, 2008), *Allocyon* and *Pachycynodon* (Rybczynski et al., 2009), as well as *Amphicynodon* (Wang et al., 2005; Finarelli, 2008; Rybczynski et al., 2009), are not pan-pinnipeds.

Diagnostic Apomorphies: An enlarged infraorbital foramen, an abbreviated infraorbital canal, and a shortened metastyle on the fourth upper premolar are thought to be among the earliest apomorphies to have evolved in the pinniped stem and are present in the earliest-branching taxa referred to *Pan-Pinnipedia* (Tedford et al., 1994: characters 12 to 14). Features thought to be synapomorphic of slightly smaller nested clades (but larger than *Pinnipedimorpha*) include a greatly reduced postglenoid foramen and the loss of the supraorbital process, supra-meatal fossa, and third lower molar (Tedford et al., 1994: characters 2, 21, 23, and 26), as well as a lingually displaced second upper molar (Wolsan, 1993: Fig. 5, character state 19b; Wolsan in Tedford et al., 1994: character 24). Some of these apomorphies have been modified or lost in later-diverging pan-pinnipeds, including some or all extant species of *Pinnipedia*. For more traits hypothesized to have evolved early in the pan-pinniped lineage leading to *Pinnipedimorpha*, see Tedford et al. (1994: Fig. 15). For synapomorphies of *Pinnipedimorpha* and the nested clade *Pinnipediformes* (which includes crown *Pinnipedia*), see Berta and Wyss (1994).

Synonyms: *Pinnipedimorpha* sensu Flynn and Wesley-Hunt (2005: 181), but not sensu Berta (1991: 4), is an unambiguous synonym.

Phocoidea sensu McKenna and Bell (1997: 252) may be an approximate synonym, and *Amphicynodontidae* Tedford et al. (1994: 12) may be a partial synonym. The uncertainty as to whether the last two names are synonyms of *Pan-Pinnipedia* arises from the fact that some phylogenetic analyses (Wang et al., 2005; Finarelli, 2008; Rybczynski et al., 2009) have inferred that *Allocyon*, *Pachycynodon*, or *Amphicynodon* (all of these extinct taxa are included in both *Phocoidea* sensu McKenna and Bell (1997) and *Amphicynodontidae*) are outside of *Pan-Pinnipedia*. Moreover, pinniped affinities have never been tested for several extinct taxa included in *Phocoidea* sensu McKenna and Bell (1997), namely *Plesiocyon*, *Parictis*, *Nothocyon*, *Adelpharctos*, and *Drassonax*.

Comments: We use a new panclade name to designate the total clade of *Pinnipedia* for three reasons. First, although Flynn and Wesley-Hunt (2005) defined *Pinnipedimorpha* to refer to this total clade, *Pinnipedimorpha* was originally defined as the name of a less inclusive clade (Berta, 1991) and has mostly been applied, either explicitly or implicitly, to such a clade (e.g., Berta et al., 1989; Berta and Ray, 1990; Berta, 1991; Berta and Wyss, 1994). Here, we follow the prevailing usage of *Pinnipedimorpha*, and accordingly apply this name to a clade less inclusive than *Pan-Pinnipedia*. Second, it is uncertain whether all taxa included in *Phocoidea* sensu McKenna and Bell (1997) and *Amphicynodontidae* are pan-pinnipeds. Third, the name *Phocoidea* has mostly been applied to refer to a clade less inclusive than *Pinnipedia* (e.g., Wyss and Flynn, 1993; Berta and Wyss, 1994; Deméré et al., 2003).

The phylogenetic relationships of pan-pinnipeds were long controversial but are currently largely resolved. Even though the composition of the most-basal part of *Pan-Pinnipedia* (prior to the origin of *Pinnipedimorpha*) remains equivocal, the status of this total clade as part of *Arctoidea* (this volume), as well as the monophyly of its crown (*Pinnipedia*, this volume), are strongly supported on both morphological and molecular grounds (Wyss, 1988; Wyss and Flynn, 1993; Berta and Wyss, 1994; Flynn and Nedbal, 1998; Flynn et al., 2000, 2005; Delisle and Strobeck, 2005; Fulton and Strobeck, 2006; Sato et al., 2006, 2009; Árnason et al., 2007; Schröder et al., 2009; Eizirik et al., 2010; Meredith et al., 2011; Yu et al., 2011; Luan et al., 2013). Furthermore, the relationship of *Pinnipedia* to other major crown clades of *Arctoidea*, namely, *Ursidae* (bears) and *Musteloidea* (weasels, raccoons, red pandas, skunks, and allies), though problematic until recently, has been clarified with multigene sequence data (Flynn and Nedbal, 1998; Flynn et al., 2000, 2005; Fulton and Strobeck, 2006; Sato et al., 2006, 2009; Schröder et al., 2009; Eizirik et al., 2010). These studies strongly argue for musteloid affinities of *Pinnipedia*.

Under the hypothesis of pinniped diphyly (Mivart, 1885; McLaren, 1960; Tedford, 1976; Repenning et al., 1979), *Pan-Pinnipedia* would become synonymous with *Pan-Arctoidea* as defined in the chapter on *Pan-Arctoidea* in this volume. Even though this diphyly hypothesis has no support in any recent computer-assisted phylogenetic analysis based on a taxon/character dataset, in the event that it should ever regain favor, we recommend that the name *Pan-Arctoidea* take precedence over *Pan-Pinnipedia*.

Literature Cited

Árnason, Ú., A. Gullberg, A. Janke, and M. Kullberg. 2007. Mitogenomic analyses of caniform relationships. *Mol. Phylogenet. Evol.* 45:863–874.

Baskin, J. A. 1998. *Mustelidae*. Pp. 152–173 in *Evolution of Tertiary Mammals of North America. Vol. 1: Terrestrial Carnivores,*

Ungulates, and Ungulatelike Mammals (C. M. Janis, K. M. Scott, and L. L. Jacobs, eds.). Cambridge University Press, Cambridge, UK.

Berta, A. 1991. New *Enaliarctos** (*Pinnipedimorpha*) from the Oligocene and Miocene of Oregon and the role of "enaliarctids" in pinniped phylogeny. *Smithson. Contrib. Paleobiol.* 69:i–iv + 1–33.

Berta, A., and C. E. Ray. 1990. Skeletal morphology and locomotor capabilites of the archaic pinniped *Enaliarctos mealsi*. *J. Vertebr. Paleontol.* 10:141–157.

Berta, A., C. E. Ray, and A. R. Wyss. 1989. Skeleton of the oldest known pinniped, *Enaliarctos mealsi*. *Science* 244:60–62.

Berta, A., and A. R. Wyss. 1994. Pinniped phylogeny. *Proc. San Diego Soc. Nat. Hist.* 29:33–56.

Delisle, I., and C. Strobeck. 2005. A phylogeny of the *Caniformia* (order *Carnivora*) based on 12 complete protein-coding mitochondrial genes. *Mol. Phylogenet. Evol.* 37:192–201.

Deméré, T. A., A. Berta, and P. J. Adam. 2003. Pinnipedimorph evolutionary biogeography. *Bull. Am. Mus. Nat. Hist.* 279:32–76.

Eizirik, E., W. J. Murphy, K.-P. Koepfli, W. E. Johnson, J. W. Dragoo, R. K. Wayne, and S. J. O'Brien. 2010. Pattern and timing of diversification of the mammalian order *Carnivora* inferred from multiple nuclear gene sequences. *Mol. Phylogenet. Evol.* 56:49–63.

Finarelli, J. A. 2008. A total evidence phylogeny of the *Arctoidea* (*Carnivora: Mammalia*): relationships among basal taxa. *J. Mamm. Evol.* 15:231–259.

Flynn, J. J., J. A. Finarelli, S. Zehr, J. Hsu, and M. A. Nedbal. 2005. Molecular phylogeny of the *Carnivora* (*Mammalia*): assessing the impact of increased sampling on resolving enigmatic relationships. *Syst. Biol.* 54:317–337.

Flynn, J. J., and M. A. Nedbal. 1998. Phylogeny of the *Carnivora* (*Mammalia*): congruence vs incompatibility among multiple data sets. *Mol. Phylogenet. Evol.* 9:414–426.

Flynn, J. J., M. A. Nedbal, J. W. Dragoo, and R. L. Honeycutt. 2000. Whence the red panda? *Mol. Phylogenet. Evol.* 17:190–199.

Flynn, J. J., and G. D. Wesley-Hunt. 2005. *Carnivora*. Pp. 175–198 in *The Rise of Placental Mammals: Origins and Relationships of the Major Extant Clades* (K. D. Rose and J. D. Archibald, eds.). Johns Hopkins University Press, Baltimore, MD.

Fulton, T. L., and C. Strobeck. 2006. Molecular phylogeny of the *Arctoidea* (*Carnivora*): effect of missing data on supertree and supermatrix analyses of multiple gene data sets. *Mol. Phylogenet. Evol.* 41:165–181.

Kohno, N. 1994. Cranial evidence for affinity of pinnipeds to the musteloids. *J. Vertebr. Paleontol.* 14(3, suppl.):32A.

Luan, P., O. A. Ryder, H. Davis, Y. Zhang, and L. Yu. 2013. Incorporating indels as phylogenetic characters: impact for interfamilial relationships within *Arctoidea* (*Mammalia: Carnivora*). *Mol. Phylogenet. Evol.* 66:748–756.

McKenna, M. C., and S. K. Bell. 1997. *Classification of Mammals above the Species Level*. Columbia University Press, New York.

McLaren, I. A. 1960. Are the *Pinnipedia* biphyletic? *Syst. Zool.* 9:18–28.

Meredith, R. W., J. E. Janečka, J. Gatesy, O. A. Ryder, C. A. Fisher, E. C. Teeling, A. Goodbla, E. Eizirik, T. L. L. Simão, T. Stadler, D. L. Rabosky, R. L. Honeycutt, J. J. Flynn, C. M. Ingram, C. Steiner, T. L. Williams, T. J. Robinson, A. Burk-Herrick, M. Westerman, N. A. Ayoub, M. S. Springer, and W. J. Murphy. 2011. Impacts of the Cretaceous Terrestrial Revolution and KPg extinction on mammal diversification. *Science* 334:521–524.

Mivart, S. G. 1885. Notes on the *Pinnipedia*. *Proc. Zool. Soc. Lond.* 1885:484–501.

Repenning, C. A., C. E. Ray, and D. Grigorescu. 1979. Pinniped biogeography. Pp. 357–369 in *Historical Biogeography, Plate Tectonics, and the Changing Environment* (J. Gray and A. J. Boucot, eds.). Oregon State University Press, Corvallis, OR.

Rybczynski, N., M. R. Dawson, and R. H. Tedford. 2009. A semi-aquatic Arctic mammalian carnivore from the Miocene epoch and origin of *Pinnipedia*. *Nature* 458:1021–1024.

Sato, J. J., M. Wolsan, S. Minami, T. Hosoda, M. H. Sinaga, K. Hiyama, Y. Yamaguchi, and H. Suzuki. 2009. Deciphering and dating the red panda's ancestry and early adaptive radiation of *Musteloidea*. *Mol. Phylogenet. Evol.* 53:907–922.

Sato, J. J., M. Wolsan, H. Suzuki, T. Hosoda, Y. Yamaguchi, K. Hiyama, M. Kobayashi, and S. Minami. 2006. Evidence from nuclear DNA sequences sheds light on the phylogenetic relationships of *Pinnipedia*: single origin with affinity to *Musteloidea*. *Zool. Sci.* 23: 125–146.

Savage, R. J. G. 1957. The anatomy of *Potamotherium* an Oligocene lutrine. *Proc. Zool. Soc. Lond.* 129:151–244.

Schröder, C., C. Bleidorn, S. Hartmann, and R. Tiedemann. 2009. Occurrence of *Can*-SINEs and intron sequence evolution supports robust phylogeny of pinniped carnivores and their terrestrial relatives. *Gene* 448:221–226.

Tedford, R. H. 1976. Relationship of pinnipeds to other carnivores (*Mammalia*). *Syst. Zool.* 25:363–374.

Tedford, R. H., L. G. Barnes, and C. E. Ray. 1994. The early Miocene littoral ursoid carnivoran *Kolponomos*: systematics and mode of life. *Proc. San Diego Soc. Nat. Hist.* 29:11–32.

Wang, X., M. C. McKenna, and D. Dashzeveg. 2005. *Amphicticeps* and *Amphicynodon* (*Arctoidea, Carnivora*) from Hsanda Gol Formation, central Mongolia and phylogeny of basal arctoids with comments on zoogeography. *Am. Mus. Novit.* 3483:1–57.

Wolsan, M. 1993. Phylogeny and classification of early European *Mustelida* (*Mammalia*: *Carnivora*). *Acta Theriol.* 38:345–384.

Wyss, A. R. 1988. Evidence from flipper structure for a single origin of pinnipeds. *Nature* 334:427–428.

Wyss, A. R., and J. J. Flynn. 1993. A phylogenetic analysis and definition of the *Carnivora*. Pp. 32–52 in *Mammal Phylogeny: Placentals* (F. S. Szalay, M. J. Novacek, and M. C. McKenna, eds.). Springer-Verlag, New York.

Yu, L., P.-T. Luan, W. Jin, O. A. Ryder, L. G. Chemnick, H. A. Davis, and Y. Zhang. 2011. Phylogenetic utility of nuclear introns in interfamilial relationships of *Caniformia* (order *Carnivora*). *Syst. Biol.* 60:175–187.

Authors

Mieczysław Wolsan; Museum and Institute of Zoology; Polish Academy of Sciences; Wilcza 64; 00-679 Warszawa, Poland. Email: wolsan@miiz.waw.pl.

André R. Wyss; Department of Earth Science; University of California at Santa Barbara; Webb Hall; Santa Barbara, CA 93106-9630, USA. Email: wyss@geol.ucsb.edu.

Annalisa Berta; Department of Biology; San Diego State University; 5500 Campanile Drive; San Diego, CA 92182-0063, USA. Email: aberta@sunstroke.sdsu.edu.

John J. Flynn; Division of Paleontology; American Museum of Natural History; Central Park West at 79th Street; New York, NY 10024-5192, USA. Email: jflynn@amnh.org.

Date Accepted: 28 July 2015

Primary Editor: Jacques A. Gauthier

Pinnipedia C. Illiger 1811 [M. Wolsan, A. R. Wyss, A. Berta, and J. J. Flynn], converted clade name

Registration Number: 271

Definition: The smallest crown clade containing *Phoca vitulina* Linnaeus 1758 (*Phocidae*), *Odobenus* (originally *Phoca*) *rosmarus* (Linnaeus 1758), and *Otaria* (originally *Phoca*) *byronia* (de Blainville 1820) (*Otariidae*). This is a minimum-crown-clade definition. Abbreviated definition: min crown ∇ (*Phoca vitulina* Linnaeus 1758 & *Odobenus rosmarus* (Linnaeus 1758) & *Otaria byronia* (de Blainville 1820)).

Etymology: Derived from the Latin *pinna* (fin or feather) and *pedis* (the genitive of *pes*, foot).

Reference Phylogeny: Figure 2 in Fulton and Strobeck (2006), where *Phoca vitulina* = Harbor seal, *Odobenus rosmarus* = Walrus, and *Otaria byronia* = South American sea lion.

Composition: The known species of *Pinnipedia* are contained in four mutually exclusive subordinate clades, three extant (*Phocidae*, *Otariidae*, and *Odobenus rosmarus*) and one extinct (*Desmatophocidae*) (Deméré et al., 2003: Fig. 3.3). Berta and Churchill (2012) listed 32 extant and one recently extinct species. Deméré (1994: 113–121), Barnes and Hirota (1995: 357), Kohno et al. (1995: 305), Koretsky (2001: 85–87), and Deméré et al. (2003: Fig. 3.3, Table 3.2) offered partial lists of most other extinct species known to date.

Diagnostic Apomorphies: Apomorphies diagnostic of *Pinnipedia* with respect to other non-nested crown clades of *Arctoidea* (i.e., *Musteloidea* and *Ursidae*; see the chapter on *Arctoidea* in this volume) include at least 19 skeletal features: (1) maxilla contributing significantly to the orbital wall to form part of the anterior and lateral walls of the orbit (Wyss, 1987; Berta and Wyss, 1994: character 9); (2) lacrimal with no contact to the jugal, fused early in ontogeny to the maxilla and frontal (Wyss, 1987; Berta and Wyss, 1994: character 10); (3) pit for the insertion of the tensor tympani muscle (in the petrosal anterior to the oval window) lost (Wyss, 1987; Berta and Wyss, 1994: character 30); (4) round window of petrosal enlarged with the round window fossula developed (Wyss, 1987; Berta and Wyss, 1994: character 25); (5) basal cochlear whorl of scala tympani enlarged (Wyss, 1987; Berta and Wyss, 1994: character 27); (6) cochlear aqueduct enlarged (Wyss, 1987; Berta and Wyss, 1994: character 31); (7) processus gracilis and anterior lamina of malleus reduced (Wyss, 1987; Berta and Wyss, 1994: character 48); (8) humerus shortened and robust (Wyss, 1988; Berta and Wyss, 1994: character 90); (9) greater and lesser tubercles of humerus enlarged (Wyss, 1988; Berta and Wyss, 1994: character 87); (10) deltopectoral crest of humerus strongly developed (Wyss, 1988; Berta and Wyss, 1994: character 88); (11) olecranon process of ulna flattened laterally and expanded posteriorly (Wyss, 1988; Berta and Wyss, 1994: character 94); (12) radius flattened anteroposteriorly with the distal half expanded (Wyss, 1988; Berta and Wyss, 1994: character 95); (13) digit I of manus enlarged and elongated, digits II to V progressively smaller and shorter (Wyss, 1987; Berta and Wyss, 1994: character 98); (14) ilium shortened relative to ischium and pubis (Berta and Ray, 1990; Berta and Wyss, 1994: character 110); (15) pubic symphysis remaining unfused in adults (Berta and Ray, 1990; Berta and Wyss,

1994: character 109); (16) greater trochanter of femur enlarged and flattened (Wyss, 1989; Berta and Wyss, 1994: character 117); (17) fovea for teres femoris ligament greatly reduced or absent (Wyss, 1988; Berta and Wyss, 1994: character 115); (18) medial inclination of femoral condyles increased (Berta and Ray, 1990; Berta and Wyss, 1994: character 118); and (19) digits I and V of pes enlarged and elongated (Wyss, 1987; Berta and Wyss, 1994: character 105). Apomorphies distinguishing *Pinnipedia* from its stem (i.e., other *Pan-Pinnipedia*; see *Pan-Pinnipedia* in this volume) include at least the loss of the pit for tensor tympani and a suppressed trigonid on the lower carnassial (Berta and Wyss, 1994: characters 30 and 71, respectively). Berta and Wyss (1994) offered a complete list of skeletal and soft-anatomical features considered pinniped apomorphies relative to other non-nested arctoid crown clades.

Synonyms: *Amphibiae* Gray 1821: 302, *Phocidae* (sensu Turner, 1848: 88), *Pinnigrades* Owen 1857: 31, *Pinnigrada* Owen 1857: 37, *Amphibia* (sensu Trouessart, 1878: 109), and *Phocae* Trouessart 1878: 109 are approximate synonyms.

Comments: The name *Pinnipedia* seemingly derives from Storr's (1780) *Pinnipedum* (applied to a group composed of what are now considered pinnipeds and sirenians) and was first used by Illiger (1811: 138) to refer to a group of living "paddle-footed" mammals, namely harbor seals (*Phoca vitulina*), walruses (*Odobenus rosmarus*), Steller sea lions (*Eumetopias jubatus*), northern fur seals (*Callorhinus ursinus*), and Cape fur seals (*Arctocephalus pusillus*). *Phoca vitulina* has been recognized (e.g., Higdon et al., 2007; Agnarsson et al., 2010) as part of *Phocidae* (seals), while *Eumetopias jubatus*, *Callorhinus ursinus*, and *Arctocephalus pusillus* are parts of *Otariidae* (sea lions and fur seals). The overwhelming majority of the literature citations of *Pinnipedia*, both in the past two centuries (e.g., Flower, 1869; Mivart, 1885; Gregory, 1910; Thenius, 1969; Berta and Wyss, 1994) and more recently (e.g., Deméré et al., 2003; Flynn et al., 2005; Berta et al., 2006; Fulton and Strobeck, 2006; Eizirik et al., 2010), are consistent with this use. Even though some palaeontologists have employed *Pinnipedia* to also encompass taxa outside the crown clade (e.g., McKenna and Bell, 1997; Wang et al., 2005; Rybczynski et al., 2009), other palaeontologists have explicitly applied this name to the crown (e.g., Berta et al., 1989; Berta, 1991; Wolsan, 1993a; Wyss and Flynn, 1993; Berta and Wyss, 1994; Bryant, 1996; Deméré et al., 2003; Flynn and Wesley-Hunt, 2005). All previous phylogenetic definitions of *Pinnipedia* have applied this name to the crown clade (Wolsan, 1993a,b; Wyss and Flynn, 1993; Berta, 1994; Bryant, 1996; Berta and Sumich, 1999; Cantino and de Queiroz, 2000, 2007; Wolsan and Bryant, 2004; Sereno, 2005). In contrast to the name *Pinnipedia*, which has been widely and consistently applied to encompass *Phocidae*, *Odobenus rosmarus*, and *Otariidae* for most of the past 200 years, none of its synonyms (except for *Phocidae*, which is now applied to a less inclusive clade) has gained widespread acceptance or use, making *Pinnipedia* the obvious choice for designating the crown clade.

It was long debated whether pinnipeds are monophyletic or diphyletic. Considerable morphological and genetic evidence favoring the monophyletic origin hypothesis has accumulated in recent decades. For a recent discussion and extensive bibliography surrounding the issue of pinniped monophyly versus diphyly, see Sato et al. (2006). For phylogenetic studies testing monophyly of taxa here referred to *Pinnipedia*, see Wyss and Flynn (1993), Berta and Wyss (1994), Flynn and Nedbal (1998), Flynn et al. (2005), Árnason et al. (2006, 2007), Fulton and Strobeck (2006),

Sato et al. (2006, 2009), Yu and Zhang (2006), Peng et al. (2007), Yonezawa et al. (2007), Yu et al. (2008, 2011), Schröder et al. (2009), Agnarsson et al. (2010), Eizirik et al. (2010), Meredith et al. (2011), Luan et al. (2013), and Scheel et al. (2014).

Some prior phylogenetic definitions of *Pinnipedia*, whether intended as nomenclatural acts (Wolsan and Bryant, 2004) or only as hypothetical examples (Cantino and de Queiroz, 2000, 2007; Sereno, 2005), have been worded such that this name would not be applicable to any clade in the context of the hypothesis of pinniped diphyly. Given the robustness of current support for the monophyly of the pinnipeds, such a complicated form of the definition seems no longer needed or justified. Under the hypothesis of pinniped diphyly, *Pinnipedia* would become synonymous with *Arctoidea* as defined in the chapter on *Arctoidea* in this volume. Should this hypothesis ever find favor, we recommend that the name *Arctoidea* take precedence over *Pinnipedia*.

Literature Cited

Agnarsson, I., M. Kuntner, and L. J. May-Collado. 2010. Dogs, cats, and kin: a molecular species-level phylogeny of *Carnivora*. *Mol. Phylogenet. Evol.* 54:726–745.

Árnason, Ú., A. Gullberg, A. Janke, and M. Kullberg. 2007. Mitogenomic analyses of caniform relationships. *Mol. Phylogenet. Evol.* 45:863–874.

Árnason, Ú., A. Gullberg, A. Janke, M. Kullberg, N. Lehman, E. A. Petrov, and R. Väinölä. 2006. Pinniped phylogeny and a new hypothesis for their origin and dispersal. *Mol. Phylogenet. Evol.* 41:345–354.

Barnes, L. G., and K. Hirota. 1995. Miocene pinnipeds of the otariid subfamily *Allodesminae* in the North Pacific Ocean: systematics and relationships. *Island Arc* 3(1994):329–360.

Berta, A. 1991. New *Enaliarctos** (*Pinnipedimorpha*) from the Oligocene and Miocene of Oregon and the role of "enaliarctids" in pinniped phylogeny. *Smithsonian Contrib. Paleobiol.* 69:i–iv + 1–33.

Berta, A. 1994. A new species of phocoid pinniped *Pinnarctidion* from the early Miocene of Oregon. *J. Vertebr. Paleontol.* 14:405–413.

Berta, A., and M. Churchill. 2012. Pinniped taxonomy: review of currently recognized species and subspecies, and evidence used for their description. *Mamm. Rev.* 42:207–234.

Berta, A., and C. E. Ray. 1990. Skeletal morphology and locomotor capabilites of the archaic pinniped *Enaliarctos mealsi*. *J. Vertebr. Paleontol.* 10:141–157.

Berta, A., C. E. Ray, and A. R. Wyss. 1989. Skeleton of the oldest known pinniped, *Enaliarctos mealsi*. *Science* 244:60–62.

Berta, A., and J. L. Sumich. 1999. *Marine Mammals: Evolutionary Biology*. Academic Press, San Diego, CA.

Berta, A., J. L. Sumich, and K. M. Kovacs. 2006. *Marine Mammals: Evolutionary Biology*. 2nd edition. Elsevier, Amsterdam.

Berta, A., and A. R. Wyss. 1994. Pinniped phylogeny. *Proc. San Diego Soc. Nat. Hist.* 29:33–56.

de Blainville, H. D. 1820. Sur quelques crânes de phoques. *J. Phys. Chim. Hist. Nat. Arts* 91:286–300.

Bryant, H. N. 1996. Explicitness, stability, and universality in the phylogenetic defnition and usage of taxon names: a case study of the phylogenetic taxonomy of the *Carnivora* (*Mammalia*). *Syst. Biol.* 45:174–189.

Cantino, P. D., and K. de Queiroz. 2000. *PhyloCode: A Phylogenetic Code of Biological Nomenclature*. Available at http://www.ohio.edu/phylocode/PhyloCode.pdf.

Cantino, P. D., and K. de Queiroz. 2007. *International Code of Phylogenetic Nomenclature*, Version 4a. Available at http://www.ohio.edu/phylocode/PhyloCode4a.pdf.

Deméré, T. A. 1994. The family *Odobenidae*: a phylogenetic analysis of fossil and living taxa. *Proc. San Diego Soc. Nat. Hist.* 29:99–123.

Deméré, T. A., A. Berta, and P. J. Adam. 2003. Pinnipedimorph evolutionary biogeography. *Bull. Am. Mus. Nat. Hist.* 279:32–76.

Eizirik, E., W. J. Murphy, K.-P. Koepfli, W. E. Johnson, J. W. Dragoo, R. K. Wayne, and S. J. O'Brien. 2010. Pattern and timing of diversification of the mammalian order *Carnivora* inferred from multiple nuclear gene sequences. *Mol. Phylogenet. Evol.* 56:49–63.

Flower, W. H. 1869. On the value of the characters of the base of the cranium in the classification of the order *Carnivora*, and on the systematic position of *Bassaris* and other disputed forms. *Proc. Zool. Soc. Lond.* 1869:4–37.

Flynn, J. J., J. A. Finarelli, S. Zehr, J. Hsu, and M. A. Nedbal. 2005. Molecular phylogeny of the *Carnivora* (*Mammalia*): assessing the impact of increased sampling on resolving enigmatic relationships. *Syst. Biol.* 54:317–337.

Flynn, J. J., and M. A. Nedbal. 1998. Phylogeny of the *Carnivora* (*Mammalia*): congruence vs incompatibility among multiple data sets. *Mol. Phylogenet. Evol.* 9:414–426.

Flynn, J. J., and G. D. Wesley-Hunt. 2005. *Carnivora*. Pp. 175–198 in *The Rise of Placental Mammals: Origins and Relationships of the Major Extant Clades* (K. D. Rose and J. D. Archibald, eds.). Johns Hopkins University Press, Baltimore, MD.

Fulton, T. L., and C. Strobeck. 2006. Molecular phylogeny of the *Arctoidea* (*Carnivora*): effect of missing data on supertree and supermatrix analyses of multiple gene data sets. *Mol. Phylogenet. Evol.* 41:165–181.

Gray, J. E. 1821. On the natural arrangement of vertebrose animals. *London Med. Repository* 15:296–310.

Gregory, W. K. 1910. The orders of mammals. *Bull. Am. Mus. Nat. Hist.* 27:i–vi + 1–524.

Higdon, J. W., O. R. P. Bininda-Emonds, R. M. D. Beck, and S. H. Ferguson. 2007. Phylogeny and divergence of the pinnipeds (*Carnivora*: *Mammalia*) assessed using a multigene dataset. *BMC Evol. Biol.* 7(216):1–19.

Illiger, C. 1811. *Prodromus Systematis Mammalium et Avium Additis Terminis Zoographicis Utriusque Classis, Eorumque Versione Germanica.* C. Salfeld, Berlin.

Kohno, N., L. G. Barnes, and K. Hirota. 1995. Miocene fossil pinnipeds of the genera *Prototaria* and *Neotherium* (*Carnivora*; *Otariidae*; *Imagotariinae*) in the North Pacific Ocean: evolution, relationships and distribution. *Island Arc* 3(1994):285–308.

Koretsky, I. A. 2001. Morphology and systematics of Miocene *Phocinae* (*Mammalia*: *Carnivora*) from Paratethys and the North Atlantic region. *Geol. Hung. Palaeontol.* 54:1–109.

Linnaeus, C. 1758. *Systema Naturae Per Regna Tria Naturae, Secundum Classes, Ordines, Genera, Species, cum Characteribus, Differentiis, Synonymis, Locis.* Tomus I, Editio decima, reformata. Laurentius Salvius, Holmiae (Stockholm).

Luan, P., O. A. Ryder, H. Davis, Y. Zhang, and L. Yu. 2013. Incorporating indels as phylogenetic characters: impact for interfamilial relationships within *Arctoidea* (*Mammalia*: *Carnivora*). *Mol. Phylogenet. Evol.* 66:748–756.

McKenna, M. C., and S. K. Bell. 1997. *Classification of Mammals above the Species Level.* Columbia University Press, New York.

Meredith, R. W., J. E. Janečka, J. Gatesy, O. A. Ryder, C. A. Fisher, E. C. Teeling, A. Goodbla, E. Eizirik, T. L. L. Simão, T. Stadler, D. L. Rabosky, R. L. Honeycutt, J. J. Flynn, C. M. Ingram, C. Steiner, T. L. Williams, T. J. Robinson, A. Burk-Herrick, M. Westerman, N. A. Ayoub, M. S. Springer, and W. J. Murphy. 2011. Impacts of the Cretaceous Terrestrial Revolution and KPg extinction on mammal diversification. *Science* 334:521–524.

Mivart, S. G. 1885. Notes on the *Pinnipedia*. *Proc. Zool. Soc. Lond.* 1885:484–501.

Owen, R. 1857. On the characters, principles of division, and primary groups of the class *Mammalia*. *J. Proc. Linn. Soc. Lond. Zool.* 2:1–37.

Peng, R., B. Zeng, X. Meng, B. Yue, Z. Zhang, and F. Zou. 2007. The complete mitochondrial genome and phylogenetic analysis of the giant panda (*Ailuropoda melanoleuca*). *Gene* 397:76–83.

Rybczynski, N., M. R. Dawson, and R. H. Tedford. 2009. A semi-aquatic Arctic mammalian carnivore from the Miocene epoch and origin of *Pinnipedia*. *Nature* 458:1021–1024.

Sato, J. J., M. Wolsan, S. Minami, T. Hosoda, M. H. Sinaga, K. Hiyama, Y. Yamaguchi, and H. Suzuki. 2009. Deciphering and dating the red panda's ancestry and early adaptive radiation of *Musteloidea. Mol. Phylogenet. Evol.* 53:907–922.

Sato, J. J., M. Wolsan, H. Suzuki, T. Hosoda, Y. Yamaguchi, K. Hiyama, M. Kobayashi, and S. Minami. 2006. Evidence from nuclear DNA sequences sheds light on the phylogenetic relationships of *Pinnipedia*: single origin with affinity to *Musteloidea. Zool. Sci.* 23:125–146.

Scheel, D.-M., G. J. Slater, S.-O. Kolokotronis, C. W. Potter, D. S. Rotstein, K. Tsangaras, A. D. Greenwood, and K. M. Helgen. 2014. Biogeography and taxonomy of extinct and endangered monk seals illuminated by ancient DNA and skull morphology. *Zookeys* 409:1–33.

Schröder, C., C. Bleidorn, S. Hartmann, and R. Tiedemann. 2009. Occurrence of Can-SINEs and intron sequence evolution supports robust phylogeny of pinniped carnivores and their terrestrial relatives. *Gene* 448:221–226.

Sereno, P. C. 2005. The logical basis of phylogenetic taxonomy. *Syst. Biol.* 54:595–619.

Storr, G. C. C. 1780. *Prodromus Methodi Mammalium.* Reiss, Tübingen.

Thenius, E. 1969. Stammesgeschichte der Säugetiere (einschließlich der Hominiden). Pp. I–VIII + 1–722 in *Handbuch der Zoologie*, Band 8, Teil 2, Beitrag 1 (J.-G. Helmcke, D. Starck, and H. Wermuth, eds.). Walter de Gruyter, Berlin.

Trouessart, E. L. 1878. Catalogue des mammifères vivants et fossiles. *Rev. Mag. Zool.* (3) 6:108–140.

Turner, Jr., H. N. 1848. Observations relating to some of the foramina at the base of the skull in *Mammalia*, and on the classification of the order *Carnivora. Proc. Zool. Soc. London* 16:63–88.

Wang, X., M. C. McKenna, and D. Dashzeveg. 2005. *Amphicticeps* and *Amphicynodon* (*Arctoidea, Carnivora*) from Hsanda Gol Formation, central Mongolia and phylogeny of basal arctoids with comments on zoogeography. *Am. Mus. Novit.* 3483:1–57.

Wolsan, M. 1993a. Phylogeny and classification of early European *Mustelida* (*Mammalia*: *Carnivora*). *Acta Theriol.* 38:345–384.

Wolsan, M. 1993b. Definitions, diagnoses, and classification of higher-level taxa of the *Mustelida* (*Carnivora*: *Arctoidea*). *Z. Säugetierkd.* 58(suppl.):79–80.

Wolsan, M., and H. N. Bryant. 2004. Phylogenetic nomenclature of carnivoran mammals. P. 32 in *First International Phylogenetic Nomenclature Meeting, Paris, Muséum National d'Histoire Naturelle, 6–9 July 2004: Abstracts* (M. Laurin, ed.). Muséum National d'Histoire Naturelle, Paris.

Wyss, A. R. 1987. The walrus auditory region and the monophyly of pinnipeds. *Am. Mus. Novit.* 2871:1–31.

Wyss, A. R. 1988. Evidence from flipper structure for a single origin of pinnipeds. *Nature* 334:427–428.

Wyss, A. R. 1989. Flippers and pinniped phylogeny: has the problem of convergence been overrated? *Mar. Mamm. Sci.* 5:343–360.

Wyss, A. R., and J. J. Flynn. 1993. A phylogenetic analysis and definition of the *Carnivora.* Pp. 32–52 in *Mammal Phylogeny: Placentals* (F. S. Szalay, M. J. Novacek, and M. C. McKenna, eds.). Springer-Verlag, New York.

Yonezawa, T., M. Nikaido, N. Kohno, Y. Fukumoto, N. Okada, and M. Hasegawa. 2007. Molecular phylogenetic study on the origin and evolution of *Mustelidae. Gene* 396:1–12.

Yu, L., J. Liu, P. Luan, H. Lee, M. Lee, M. Min, O. A. Ryder, L. Chemnick, H. Davis, and Y. Zhang. 2008. New insights into the evolution of intronic sequences of the β-fibrinogen gene and their application in reconstructing mustelid phylogeny. *Zool. Sci.* 25:662–672.

Yu, L., P.-T. Luan, W. Jin, O. A. Ryder, L. G. Chemnick, H. A. Davis, and Y. Zhang. 2011. Phylogenetic utility of nuclear introns in interfamilial relationships of *Caniformia* (order *Carnivora*). *Syst. Biol.* 60:175–187.

Yu, L., and Y. Zhang. 2006. Phylogeny of the caniform *Carnivora*: evidence from multiple genes. *Genetica* 127:65–79.

Authors

Mieczysław Wolsan; Museum and Institute of Zoology; Polish Academy of Sciences; Wilcza 64; 00-679 Warszawa, Poland. Email: wolsan@miiz.waw.pl.

André R. Wyss; Department of Earth Science; University of California at Santa Barbara; Webb Hall; Santa Barbara, CA 93106-9630, USA. Email: wyss@geol.ucsb.edu.

Annalisa Berta; Department of Biology; San Diego State University; 5500 Campanile Drive; San Diego, CA 92182-0063, USA. Email: aberta@sunstroke.sdsu.edu.

John J. Flynn; Division of Paleontology; American Museum of Natural History; Central Park West at 79th Street; New York, NY 10024-5192, USA. Email: jflynn@amnh.org.

Date Accepted: 20 July 2015

Primary Editor: Jacques A. Gauthier

SECTION 8

Reptilia C. Linnaeus 1758 [M. Laurin and R. R. Reisz], converted clade name

Registration Number: 88

Definition: The smallest crown clade containing *Testudo graeca* Linnaeus 1758 (*Testudines*), *Iguana iguana* Linnaeus 1758 (*Lepidosauria*), and *Crocodylus* (originally *Lacerta*) *niloticus* Laurenti 1768 (*Archosauria*). This is a minimum-crown-clade definition. Abbreviated definition: min crown ∇ (*Testudo graeca* Linnaeus 1758 & *Iguana iguana* Linnaeus 1758 & *Crocodylus* (originally *Lacerta*) *niloticus* Laurenti 1768).

Etymology: From Latin *reptilis* (creeping, crawling).

Reference Phylogeny: The primary reference phylogeny is Modesto (1999: Fig. 4A), which expands upon that of Laurin and Reisz (1995: Fig. 2). On that tree, *Iguana iguana* Linnaeus 1758 and *Crocodylus niloticus* Linnaeus 1758 are closest to *Younginiformes* (Gauthier, 1994: Figs. 6–8). Both place turtles outside diapsids, as sister-group of procolophonoids among "parareptiles". However, higher-level relationships within *Reptilia* are controversial (see Comments).

Composition: *Reptilia* is composed of turtles, lepidosaurs, archosaurs, and all extinct forms that derive from their most recent common ancestor. Under several recently published phylogenies, this includes, among other groups, relatives of diapsids, such as captorhinids and "protorothyridids", which appear to be paraphyletic (Müller and Reisz, 2006), and "parareptiles" such as procolophonoids, pareiasaurs, lanthanosuchids, bolosaurs, millerettids, and mesosaurs. The position of mesosaurs is weakly supported (Modesto, 1999); in the context of an alternative hypothesis of early amniote relationships (Laurin and Reisz, 1995: Fig. 2; Laurin and Piñeiro, 2017: Fig. 5), mesosaurs are not reptiles, but stem reptiles.

Diagnostic Apomorphies: Skeletal apomorphies include (Gauthier et al., 1988a,b; Laurin and Reisz, 1995; deBraga and Rieppel, 1997; Laurin, 1998a,b; Modesto, 1999; Rieppel and Reisz, 1999; Vallin and Laurin, 2004): posttemporal fenestrae large; only one coronoid present in each mandible; supinator process parallel to humeral shaft and separated from the distal head by a groove; medial pedal centrale absent (only one pedal centrale is present). This list rests on the assumption that the primary reference phylogeny is correct. If turtles are considered to be diapsids, several of these apomorphies have to be deleted and new ones have to be added. Other (i.e., non-skeletal) apomorphies whose distribution cannot be established in extinct taxa include, among others (Gauthier et al., 1988b): phi keratins; uric acid enzyme cycle in the liver; cricoid cartilage consisting of two or more tracheal rings; ornithuric acid produced in benzoic acid metabolism; external nasal grand lies outside nasal capsule; iris and ciliary muscles composed of striated muscle fibers; fully developed and highly mobile third eyelid; dorsoventricular ridge of telencephalon; fewer than three postmandibular branchial arches; masticatory muscle plate divides into m. constrictor dorsalis and m. adductor mandibulae.

Synonyms: *Sauroids* Huxley 1864a (approximate); *Sauropsida* Huxley 1864b (approximate) (see also Goodrich, 1916).

Comments: Linnaeus (1758) and Laurenti (1768) are both often credited for having erected the taxon *Reptilia*. This may be because Linnaeus (1758) ranked *Reptilia* as an order rather than a class, also used the spelling "*Reptiles*", and did not include snakes. However, Laurenti (1768) and other early authors (e.g., Brongniart, 1800; Sonnini and Latreille, 1802) used the name *Reptilia* or *Reptiles* for a group that included both amphibians and reptiles in the paraphyletic sense that excluded birds and mammals. Amphibians were first taken out of *Reptilia* by de Blainville (1816), so that *Reptilia* became the name for the paraphyletic group of non-avian, non-mammalian amniotes. Much later, birds were added to *Reptilia* and non-mammalian synapsids were removed from *Reptilia* to make it monophyletic (Gauthier, 1984).

Our phylogenetic definition of *Reptilia* results in the taxon having the same composition as under most previous phylogenetic definitions (but not under earlier gradistic circumscriptions), which explicitly designated a crown-clade. The first was by Gauthier (1984): "*Reptilia* is here defined as the taxon encompassing all the descendants of the most recent common ancestor of *Testudines* and *Diapsida*", and it was formulated in the context of a phylogeny where turtles were not nested within diapsids. Gauthier et al. (1988a,b) and Laurin and Reisz (1995) used similar definitions. The taxonomy of de Queiroz and Gauthier (1992) also implied a crown-group definition of *Reptilia* and placed turtles outside diapsids.

Two previous phylogenetic definitions did not apply the name *Reptilia* to a crown clade. DeBraga and Rieppel (1997) published the following definition: "The most recent common ancestor of diapsids and all its descendants", which makes *Diapsida* and *Reptilia* synonyms. The clade to which this definition refers, under all recent phylogenies (e.g., Reisz et al., 2011), is not a crown-clade because it includes stem-diapsids such as araeoscelidians. This definition was proposed in the context of a phylogeny in which turtles are nested within crown diapsids; thus, their definition excludes parareptiles from *Reptilia* under their inferred phylogeny. However, this definition is inconsistent with their taxonomy (p. 286 and Fig. 1), in which *Reptilia* includes both *Parareptilia* and *Eureptilia*, so it is unclear how they really meant to define this name. Modesto and Anderson (2004) proposed a maximum-clade definition of *Reptilia* that makes it the name of a total clade: "the most inclusive clade containing *Lacerta agilis* Linnaeus 1758 and *Crocodylus niloticus* Laurenti 1768, but not *Homo sapiens* Linnaeus 1758."

In all these previous works, *Reptilia* included all extant amniotes except mammals. However, the extinct taxa included in *Reptilia* are slightly less stable; e.g., Gauthier et al. (1988b) excluded *Parareptilia* from *Reptilia*, while Laurin and Reisz (1995) excluded mesosaurs. Given the instability of the position of turtles, a crown-group definition of *Reptilia* means that the diverse "parareptiles" will remain in *Reptilia* if they are stem turtles (Laurin and Reisz, 1995; Lee, 2001; Lyson et al., 2010, 2013a,b) and thus part of *Pan-Testudines* (Joyce et al., 2004; Joyce et al., this volume) but not if they are stem reptiles. Similarly, captorhinids, "protorothyridids", and the earliest diapsids (e.g., araeoscelidians and possibly younginiforms) which are all basal eureptiles and have nearly always been considered reptiles (e.g., Romer, 1966; Carroll, 1988; Benton, 2005), will be excluded from *Reptilia* if turtles turn out to be diapsids, as suggested by most gene-sequence phylogenies (e.g., Janke et al., 2001; Iwabe et al., 2005; Hugall et al., 2007; Field et al., 2014; but see Lu et al., 2013) and some palaeontological trees (Laurin and Piñeiro, 2017: Fig. 5; Ford and Benson, 2020). This crown clade may have the same composition as the diapsid crown group (*Sauria* of Gauthier, 1984) if turtles are crown diapsids (deBraga and Rieppel, 1997; Rieppel and Reisz, 1999; Janke et al., 2001; Hill, 2005), and in that case, we suggest that *Reptilia* should be given precedence, because *Sauria*, when phylogenetically defined (e.g., Gauthier, 1984, 1994), was conceptualized as excluding turtles, while in contrast *Reptilia* was conceptualized as including them.

The composition of *Reptilia* is uncertain mostly because of the controversial position of turtles. Most morphological studies agree on the "parareptilian" relationships of turtles, but disagree on where exactly they fit within that group: either as sister-group of pareiasaurs (Lee, 1995: Fig. 22), among pareiasaurs (Lee, 1996: Fig. 1), or as sister-group of procolophonids (Laurin and Reisz, 1995) or of *Eunotosaurus africanus* (Lyson et al., 2013b: Fig. 4). The "diapsid affinities" alternative represents the other major class of hypotheses regarding the position of turtles: morphologists have grouped turtles either with stem diapsids as sister to captorhinids (e.g., Gauthier et al., 1988a,b) or, more often, within crown diapsids as the extant sister-group of lepidosaurs (e.g., Rieppel and Reisz, 1999: Fig 1; Hill, 2005: Fig. 5). In contrast, most phylogenies based on gene sequences place turtles within crown diapsids, but as the extant sister-group of archosaurs (Zardoya and Meyer 1998; Hedges and Poling, 1999; Janke et al., 2001: Fig. 1; Iwabe et al., 2005; Hugall et al., 2007; Field et al., 2014). Lee (2001: Fig. 2) and Frost et al. (2006) are among the few recent phylogenies that consider molecular sequence data and that found turtles to lie outside *Diapsida* Osborn 1903. To add to the dilemma, Lu et al. (2013) used a genes-as-characters approach to indicate that among the several thousands of genes they studied, there are data to ally turtles with archosaurs, with lepidosaurs, or as sister-group to both; however, all three alternatives were only weakly supported. The position of turtles has been, and continues to be, the single most intractable problem in early amniote phylogeny.

Literature Cited

Benton, M. J. 2005. *Vertebrate Palaeontology.* 3rd edition. Blackwell, Oxford.

Brongniart, A. 1800. Essai d'une classification naturelle des reptiles. 1ère partie. Établissement des ordres. *Bull. Sci. Soc. Philom.* 2:81–82.

Carroll, R. L. 1988. *Vertebrate Paleontology and Evolution.* W. H. Freeman, New York.

de Blainville, M. H. 1816. Prodrome d'une nouvelle distribution systematique du règne animal. *Bull. Soc. Philomath. Paris, Ser. 3,* 8:105–124.

deBraga, M., and O. Rieppel. 1997. Reptile phylogeny and the interrelationships of turtles. *Zool. J. Linn. Soc.* 120:281–354.

Field, D. J., J. A. Gauthier, B. L. King, D. Pisani, T. R. Lyson, and K. J. Peterson. 2014. Toward consilience in reptile phylogeny: microRNAs support an archosaur, not lepidosaur, affinity for turtles. *Evol. Dev.* 16(4):189–196.

Ford, D. P., and R. B. Benson. 2020. The phylogeny of early amniotes and the affinities of *Parareptilia and Varanopidae. Nat. Ecol. Evol.* 4:57–65.

Frost, D. R., T. Grant, J. Faivovich, R. H. Bain, A. Haas, C. F. B. Haddad, R. O. de Sá, A. Channing, M. Wilkinson, S. C. Donnellan, C. J. Raxworthy, J. A. Campbell, B. Blotto, P. Moler, R. C. Drewes, R. A. Nussbaum, J. D. Lynch, D. M. Green, and W. C. Wheeler. 2006. The amphibian tree of life. *Bull. Am. Mus. Nat. Hist.* 297:1–370.

Gauthier, J. 1984. *A Cladistic Analysis of the Higher Systematic Categories of the Diapsida.* PhD thesis, University of California, Berkeley, CA.

Gauthier, J. 1994. The diversification of the amniotes. Pp. 129–159 in *Major Features of Vertebrate Evolution* (D. R. Prothero and R. M. Schoch, eds.). The Paleontological Society, Knoxville, TN.

Gauthier, J., A. G. Kluge, and T. Rowe. 1988a. The early evolution of the *Amniota.* Pp. 103–155 in *The Phylogeny and Classification of the Tetrapods. Vol. 1: Amphibians, Reptiles, Birds* (M. J. Benton, ed.). Clarendon Press, Oxford.

Gauthier, J., A. G. Kluge, and T. Rowe. 1988b. Amniote phylogeny and the importance of fossils. *Cladistics* 4:105–209.

Goodrich, E. S. 1916. On the classification of the *Reptilia*. *Proc. R. Soc. Lond. B Biol. Sci.* 89:261–276.

Hedges, S. B., and L. L. Poling 1999. A molecular phylogeny of reptiles. *Science* 283:998–1001.

Hill, R. V. 2005. Integration of morphological data sets for phylogenetic analysis of *Amniota*: the importance of integumentary characters and increased taxonomic sampling. *Syst. Biol.* 54:530–547.

Hugall, A. F., R. Foster, and M. S. Y. Lee. 2007. Calibration choice, rate smoothing, and the pattern of tetrapod diversification according to the long nuclear gene RAG-1. *Syst. Biol.* 56:543–563.

Huxley, T. H. 1864a. *Lectures on the Elements of Comparative Anatomy*. J. Churchill, London.

Huxley, T. H. 1864b. The structure and classification of the *Mammalia* in *Med. Times Gazette*. Available at http://aleph0.clarku.edu/huxley/UnColl/Gazettes/Mamma.html.

Iwabe, N., Y. Hara, Y. Kumazawa, K. Shibamoto, Y. Saito, T. Miyata, and K. Katoh. 2005. Sister group relationship of turtles to the bird-crocodilian clade revealed by nuclear DNA-coded proteins. *Mol. Biol. Evol.* 22:810–813.

Janke, A., D. Erpenbeck, M. Nilsson, and U. Arnason. 2001. The mitochondrial genomes of the iguana (*Iguana iguana*) and the caiman (*Caiman crocodylus*): implications for amniote phylogeny. *Proc. R. Soc. Lond. B Biol. Sci.* 268:623–631.

Joyce, W. G., J. F. Parham, and J. A. Gauthier. 2004. Developing a protocol for the conversion of rank-based taxon names to phylogenetically defined clade names, as exemplified by turtles. *J. Paleontol.* 78(5):989–1013.

Laurenti, J. N. 1768. *Specimen Medicum, Exhibens Synopsin Reptilium Emendatam cum Experimentis circa venena et Antidota Reptilium Austriacorum*. PhD thesis, University of Vienna.

Laurin, M. 1998a. The importance of global parsimony and historical bias in understanding tetrapod evolution. Part I. Systematics, middle ear evolution, and jaw suspension. *Ann. Sci. Nat., Zool., 13 Ser.* 19:1–42.

Laurin, M. 1998b. The importance of global parsimony and historical bias in understanding tetrapod evolution. Part II. Vertebral centrum, costal ventilation, and paedomorphosis. *Ann. Sci. Nat., Zool., 13 Ser.* 19:99–114.

Laurin M., and G. Piñeiro. 2017. A reassessment of the taxonomic position of mesosaurs, and a surprising phylogeny of early amniotes. *Front. Earth Sci.* 5:88.

Laurin, M., and R. R. Reisz. 1995. A reevaluation of early amniote phylogeny. *Zool. J. Linn. Soc.* 113:165–223.

Lee, M. S. Y. 1995. Historical burden in systematics and the interrelationships of "Parareptiles". *Biol. Rev.* 70:459–547.

Lee, M. S. Y. 1996. Correlated progression and the origin of turtles. *Nature* 379:812–815.

Lee, M. S. Y. 2001. Molecules, morphology, and the monophyly of diapsid reptiles. *Contrib. Zool.* 70:1–18.

Linnaeus, C. 1758. *Systema Naturae*. 10th edition. Laurentii Salvii, Holmiae (Stockholm).

Lu, B., W. Yang, Q. Dai, and J. Fu. 2013. Using genes as characters and a parsimony analysis to explore the phylogenetic position of turtles. *PLOS ONE* 8:e79348.

Lyson, T. R., G. S. Bever, B.-A. S. Bhullar, W. G. Joyce, and J. A. Gauthier. 2010. Transitional fossils and the origin of turtles. *Biol. Lett.* 6:830–833.

Lyson, T. R., G. S. Bever, T. M. Scheyer, A. Y. Hsiang, and J. A. Gauthier. 2013a. Evolutionary origin of the turtle shell. *Curr. Biol.* 23:1–7.

Lyson, T. R., B.-A. S. Bhullar, G. S. Bever, W. G. Joyce, K. de Queiroz, A. Abzhanov, and J. A.Gauthier. 2013b. Homology of the enigmatic nuchal bone reveals a novel reorganization of the shoulder girdle in the evolution of the turtle shell. *Evol. Dev.* 15(5):317–325.

Modesto, S. P. 1999. Observations on the structure of the early Permian reptile *Stereosternum temidum* Cope. *Palaeont. Afr.* 35:7–19.

Modesto, S. P., and J. S. Anderson. 2004. The phylogenetic definition of *Reptilia*. *Syst. Biol.* 53:815–821.

Müller, J., and R. R. Reisz. 2006. The phylogeny of early eureptiles: comparing parsimony and Bayesian approaches in the investigation of a basal fossil clade. *Syst. Biol.* 55:503–511.

Osborn, H. F. 1903. The reptilian subclasses *Diapsida* and *Synapsida* and the early history of the *Diaptosauria*. *Mem. Am. Mus. Nat. Hist.* 1:265–270.

de Queiroz, K., and J. Gauthier. 1992. Phylogenetic taxonomy. *Annu. Rev. Ecol. Syst.* 23:449–480.

Reisz, R. R., S. P. Modesto, and D. M. Scott. 2011. A new early Permian reptile and its significance in early diapsid evolution. *Proc. R. Soc. Lond. B Biol. Sci.* 278:3731–3737.

Rieppel, O., and R. R. Reisz. 1999. The origin and early evolution of turtles. *Annu. Rev. Ecol. Syst.* 30:1–22.

Romer, A. S. 1966. *Vertebrate Paleontology.* 3rd edition. University of Chicago Press, Chicago, IL.

Sonnini, C. S., and P. A. Latreille. 1802. *Histoire Naturelle des Reptiles: Avec Figures Dessinées d'Apres Nature.* Crapelet, Paris.

Vallin, G., and M. Laurin. 2004. Cranial morphology and affinities of *Microbrachis*, and a reappraisal of the phylogeny and lifestyle of the first amphibians. *J. Vertebr. Paleontol.* 24:56–72.

Zardoya, R., and A. Meyer. 1998. Complete mitochondrial genome suggests diapsid affinities of turtles. *Proc. Natl. Acad. Sci. USA* 95:14226–14231.

Authors

Michel Laurin; CR2P, CNRS/MNHN/SU; Muséum National d'Histoire Naturelle; CP 48; 43 rue Buffon, 75005 Paris, France. Email: michel.laurin@mnhn.fr.

Robert R. Reisz; Department of Biology, University of Toronto at Mississauga; 3359 Mississauga Rd. N. Mississauga, ON, L5L 1C6, Canada. Email: rreisz@utm.utoronto.ca.

Date Accepted: 27 May 2014

Editors: Philip Cantino, Jacques A. Gauthier, Kevin de Queiroz

Diapsida H. F. Osborn 1903 [J. A. Gauthier and K. de Queiroz], converted clade name

Registration Number: 120

Definition: The clade characterized by the apomorphy 'upper and lower temporal fenestrae' (see Diagnostic Apomorphies and Comments below), as inherited by *Sphenodon* (*Hatteria*) *punctatus* (Gray 1842) (*Rhynchocephalia*). This is an apomorphy-based definition. Abbreviated definition: ∇ apo upper and lower temporal fenestrae [*Sphenodon punctatus* (Gray 1842)].

Etymology: Derived from the Greek *di-*, two, and the Latin *apsis*, arch.

Reference Phylogeny: The primary reference phylogeny is deBraga and Rieppel (1997: Fig. 1), in which the defining apomorphy of *Diapsida* originates along branch F. See also Reisz (1981: Fig. 26), Gauthier (1984: Figs. 23–24), Gauthier et al. (1988b: Fig. 3), Müller and Reisz (2006: Fig. 2), Bever et al. (2015: Fig. 4), and Simões et al. (2018: Fig. 2).

Composition: See *Sauria* in this volume for composition of the crown if it does not include turtles. Diapsids outside of crown *Sauria* include the following taxa: *Araeoscelidia* Williston 1913, *Weigeltisauridae* Kuhn 1939, *Younginiformes* Romer 1947, *Claudiosaurus germaini* Carroll 1981, *Lanthanolania ivakhnenkoi* Modesto and Reisz 2003, and *Orovenator mayorum* Reisz et al. 2011 (see Comments). See Comments in *Sauria* (this volume) for discussion of the relationships of drepanosaurs, turtles, ichthyosaurs, sauropterygians, thalattosaurs, and choristoderans.

Diagnostic Apomorphies: Upper and lower fenestrae in the temporal region of the skull bordered by a relatively slender scaffolding of dermocranial bones, the upper and lower temporal arches (see Comments for details concerning the ancestral morphology and further modifications).

Synonyms: *Diaptosauria* Osborn 1903 (partial and approximate).

Comments: Heaton's (1979) landmark study of captorhinid cranial anatomy illustrates the ancestral condition in stem reptiles. The dermal bones forming the skull behind the orbit—the parietal, postorbital, squamosal, jugal, and quadratojugal—completely cover the skull roof and cheek regions. In the ancestral diapsid, however, this continuous bony surface was replaced by a temporal region with two conspicuous fenestrae, the margins of which were smooth and continuous and formed distinctive emarginations in adjacent dermal bones (e.g., Reisz, 1981; Heaton and Reisz, 1986). These fenestrae are not merely the incompletely ossified, irregular-margined gaps between bones growing together during ontogeny, as seen, for example, in the cheek region of some *Milleretta* (Gow, 1972) and *Mesosaurus* (Laurin and Piñeiro, 2017), although such gaps may represent developmental (and evolutionary) precursors of temporal fenestrae (e.g., Haridy et al., 2016). On the contrary, diapsid fenestration appears to reflect reorganization of the jaw-closing muscles, both in terms of their fibrous internal frameworks as well as by concentrating their areas of origin, to yield an open, frame-like skull optimized to ensure adequate bite forces while minimizing musculoskeletal volume, weight, and maintenance costs (Curtis et al., 2011).

The upper temporal fenestra largely separates skull-roofing bones that were originally in broad contact, leaving distinct emarginations along adjacent parietal, postorbital, and squamosal bones. On the posterior skull roof once covered by the squamosal bone, for example, only a narrow ascending process of the squamosal remains; this process, along with the supratemporal bone, forms what is here termed the post-temporal arch, which bounds the upper temporal fenestra posteriorly. The distinctive diapsid post-temporal arch was originally formed mainly by the supratemporal bone, which connected the squamosal laterally to the parietal medially (i.e., the parietal supratemporal process did not participate in the upper temporal fenestra as it does in *Sauria*; see entry in this volume and Gauthier et al., 2012). In the upper temporal arch unique to diapsids, the articulation between the postorbital and squamosal is equally diagnostic. Rather than being received in a posterodorsally sloping recess along the anterodorsal margin of the squamosal as in, for example, *Milleretta* (Gow, 1972), or in a broader and irregular-margined recess in the squamosal that is variably exposed above, rather than being entirely below, the postorbital overlap as in *Captorhinus* (Heaton, 1979; Kissel et al., 2002), the slender diapsid postorbital tapers into a narrow and smooth-margined, triangular recess that is enclosed entirely in the lateral face of the squamosal (Gauthier et al., 2012; see, e.g., *Araeoscelis* in Fig. 2 of Reisz et al., 1984, in which the postorbital is displaced to reveal this diapsid recess on the squamosal).

In amniotes ancestrally, the cheek (lower temporal) region was also completely covered by dermal bones, mainly by the squamosal, and to a lesser extent by the postorbital and jugal, with the posteroventral corner covered by the quadratojugal bone (Heaton, 1979). In contrast, diapsids possess a large lower temporal fenestra that deeply emarginates the squamosal in particular, but also the postorbital and jugal, and to a lesser extent the quadratojugal (e.g., Reisz, 1981). The squamosal, for example, still extends the full length of the suspensorial arch (which borders the lower temporal fenestra posteriorly) to attach to the quadratojugal. But instead of contacting broadly in the posterior cheek region as in amniotes ancestrally, the quadratojugal-squamosal contact is confined to the posteroventral corner of the lower temporal fenestra in diapsids (see, e.g., *Spinoaequalis schultzei* in deBraga and Reisz, 1995: Fig. 5).

The presence of upper and lower temporal fenestrae can often be inferred from the shape of their surrounding bones, even in disarticulated remains. A triradiate postorbital bone is perhaps the single most characteristic element in this character complex, as it acquires that shape via conspicuous posterodorsal and posteroventral emarginations reflecting, respectively, formation of the upper and lower temporal fenestrae. The same holds for the triradiate jugal, with its slender maxillary, postorbital and quadratojugal processes, the latter two reflecting deep emargination of the jugal posterodorsally by the lower temporal fenestra. Finally, although the supratemporal process of the parietal is barely developed in early amniotes and diapsids (Heaton, 1979; Reisz, 1981), it is prominently developed in later diapsids (Gauthier et al., 2012), largely supplanting the supratemporal bone in the post-temporal arch (e.g., *Proterosuchus fergusi*, Ezcurra and Butler, 2015: Fig. 3; *Gambelia wislizenii*, Gauthier et al., 2012: Fig. 175).

These two fenestrae were originally bordered by slender upper and lower temporal arches, the former composed of the postorbital and squamosal and the latter by the jugal and quadratojugal (e.g., Reisz, 1981). The lower temporal arch has had a complex history. Indeed, several Permo-Triassic diapsids crownward of *Araeoscelidia* appear to lack a complete bar, at least in bone (e.g., *Claudiosaurus germaini*; Carroll, 1981), while it is clearly intact in other early-diverging diapsids (e.g., *Champsosaurus* spp.; Gao and Fox, 1998). It may be noteworthy that, even in extinct taxa in which it is known to be present, such as *Youngina capensis* (Gow, 1975), a complete bony arch cannot be observed in most specimens owing to imperfect preservation (Gauthier et al., 1988a). The lower temporal arch displays a remarkable degree of homoplasy regardless; it is, for example, currently thought to have been lost early in diapsid evolution (Müller, 2003), and then to have re-evolved in *Youngina capensis*, and several more times within *Sauria* (e.g., Dilkes, 1998; Ezcurra and Butler, 2015; Jones et al., 2013). Among extant diapsids, the upper and lower temporal arches can vary significantly in terms of which bones predominate in them (e.g., Gauthier et al., 2012). Sometimes the bony arches can be replaced by ligaments (e.g., the lower temporal arch in *Squamata*; Broom, 1925), and can even be lost entirely (e.g., the upper temporal ligament within *Serpentes*; e.g., Rieppel, 1980). Even if only a ligament remains, we follow Broom (1925) in considering that structure to represent the arch, albeit in a transformed state. A ligamentous lower temporal arch can be present even in the absence of a quadratojugal process on the jugal in squamates (Oelrich, 1956); if there is any indication of that process on the jugal, however, a robust lower temporal ligament is invariably present in squamates, enabling the jugal to re-ossify down that ligament, nearly to the quadrate in at least one instance (e.g., Mo et al., 2009).

In addition to modifications in the continuity and composition of the bounding arches, the temporal fenestrae have been modified further in other ways. Thus, the upper temporal fenestra has in some cases been covered over secondarily (e.g., Gauthier et al., 2012; Bever et al., 2015). The lower temporal fenestra has also been closed secondarily several times, as seen in some araeoscelidians (e.g., Reisz et al., 1984), choristoderans (e.g., Gao and Fox, 2005), early archosauromorphs (e.g., Gregory, 1945), and

squamates (e.g., Savage, 1963). Early in the history of amniote phylogenetics, the presence of this fenestra in synapsids and diapsids, and its absence in fully shelled turtles from the Late Triassic to the Recent, was part of the argument for turtles being sister to all other amniotes (e.g., Gaffney, 1980). However, a lower temporal fenestra (presumably bordered ventrally by a ligamentous arch) is now known to have been present in Permian stem turtles (e.g., Lyson et al., 2010; Bever et al., 2015). Thus, *Testudines* appears to present yet another example of secondary closure among diapsids (see *Sauria*, this volume).

To complicate matters, a lower temporal fenestra—or at least an incompletely ossified cheek region—is also common among parareptiles, albeit in varying degrees of differentiation and situated between somewhat different bones (e.g., Haridy et al., 2016). Moreover, *Synapsida* also has a lower temporal fenestra, and it resembles that of *Diapsida* in being bordered by postorbital, squamosal, and jugal bones. But the lower temporal fenestra of *Diapsida* differs from that ancestral for synapsids in that it was also bordered by the quadratojugal. (There are only four bones in the cheek region, and their borders near the center of the cheek only ossify fully during post-hatching ontogeny; see, e.g., *Araeoscelis* in Reisz et al., 1984, and *Delorhynchus cifellii* in Haridy et al., 2016.) The matter is further complicated by mesosaurs, highly modified aquatic amniotes from the Early Permian of Gondwana. They are widely thought to have diverged near the base of the reptilian tree (e.g., Gauthier et al., 1988c), although their precise relationships to other reptiles and, indeed, whether they have a lower temporal fenestra, have long been debated (e.g., MacDougall et al., 2018). So it is possible that a lower temporal fenestra was ancestral for *Amniota* (e.g., Piñeiro et al., 2012). Nevertheless, the inferred nearest extinct relatives of *Diapsida* lack both lower and upper temporal fenestrae (e.g., Gauthier, 1994; Müller and Reisz, 2006; Bickelmann et al., 2009; Reisz et al., 2011; MacDougall et al., 2018).

The Late Pennsylvanian araeoscelidians *Petrolacosaurus kansensis* (Reisz, 1977) and *Spinoaequalis schultzei* (deBraga and Reisz, 1995), both of which are known from fairly complete skeletons, are universally considered to be the earliest-diverging diapsids. There has also been general agreement that the Early Permian *Orovenator mayorum* (Reisz et al., 2011) from Oklahoma represents the next earliest divergence from the saurian stem (e.g., Pritchard and Nesbitt, 2017). As for Late Permian diapsids, *Lanthanolania ivakhnenkoi* (Modesto and Reisz, 2003) from the early Late Permian of Russia is both the earliest and least complete, although it has some of the same saurian apomorphies seen in *Orovenator mayorum*; because of poor preservation, neither species otherwise adds much to our knowledge of Permian *Diapsida*. Apart from some of the end-Permian diapsids from South Africa, the late-surviving araeoscelidian *Araeoscelis casei* from the Early Permian of Texas is by far the best-known Permian diapsid (Reisz et al., 1984).

The globally distributed, gliding *Weigeltisauridae* poses another challenge, less from non-preservation than from a high degree of morphological modification, which renders some anatomical details difficult to interpret. Nevertheless, they have been inferred to be outside of crown *Sauria* since Evans and Haubold (1987; see also Evans, 1988; Laurin, 1991). The same is true of the long-necked aquatic species *Claudiosaurus germaini* of Madagascar (e.g., Carroll, 1981; Evans, 1988; Gauthier, 1994). There seems little agreement, however, regarding the relative positions of weigeltisaurs and *Claudiosaurus* on the saurian stem (e.g., Pritchard and Nesbitt [2017] vs. Simões et al. [2018]).

All of these Permian taxa are like *Younginiformes* (whether as a clade or a paraphylum) in having at least some saurian apomorphies, including upper and lower temporal fenestrae (whether the lower arch is entirely ossified or the lower fenestra nearly closed). Nevertheless, most lack the osteological correlates of an impedance-matching auditory system characteristic of crown *Sauria* (e.g., Laurin, 1991; Gauthier, 1994; deBraga and Rieppel, 1997; Dilkes, 1998; Müller, 2004; Senter, 2004; Ezcurra et al., 2014; Nesbitt et al., 2015; Pritchard and Nesbitt, 2017; Li et al., 2018).

As noted by Reisz et al. (2010), diapsids are relatively common in the fossil record from the early Mesozoic onward. Diapsid stem saurians are, however, relatively rare during the Palaeozoic, when other amniotes, particularly captorhinids but also parareptiles and diverse clades among the earliest stem mammals (e.g., *Varanops brevirostris*), predominated in terrestrial ecosystems. The reptile-mammal split marking the origin of *Amniota* is not particularly well constrained temporally (Müller and Reisz, 2005). It is estimated to have taken place roughly 312 million years ago (Benton and Donoghue, 2007), and the earliest known diapsids, such as *Petrolacosaurus kansensis* (Reisz, 1981) and *Spinoaequalis schultzei* (deBraga and Reisz, 1995), are estimated to have lived around 302 million years ago (Falcon-Lang et al., 2007). Taken at face value, these data suggest that the initial diversification of *Amniota* and the origin of *Diapsida* took place in the latter part of the Carboniferous. Simões et al. (2018) used a relaxed-clock Bayesian analysis to infer that *Diapsida* originated much earlier, in the Devonian, approximately 70 million years before any amniotes, let alone diapsids, are known in the fossil record.

Since Osborn (1903) coined the name *Diapsida*, palaeontologists have always associated that name with the clade bearing two temporal fenestrae/arches. Indeed, as soon as this apomorphy was confirmed in a fossil, they promptly shifted it from a less inclusive clade (e.g., the *Youngina* node) to a more inclusive clade (e.g., the *Petrolacosaurus* node) (e.g., Reisz, 1977, 1981; Gaffney 1979, 1980; Thulborn, 1980; Evans, 1980, 1982, 1984, 1988; Benton, 1982, 1983, 1984, 1985; Gauthier, 1984; Heaton and Reisz, 1986; Gauthier et al., 1988a,b,c, 1989). We have accordingly chosen to continue a tradition that has persisted from Osborn (1903) to the present day (e.g., Pritchard et al., 2018), by proposing an apomorphy-based definition for this taxon name. Consequently, our definition ties the name *Diapsida* to a potentially more inclusive clade than do previous explicitly stated (minimum-clade) phylogenetic definitions that tied the name to the clade originating in the most recent common ancestor of araeoscelidians, lepidosaurs and archosaurs (e.g., Laurin, 1991). Selection of the name *Diapsida* over its approximate synonym *Diaptosauria* is straightforward, as the former has been used much more frequently (particularly in the recent literature) and the match in composition to that of the named clade has always been much closer (*Diaptosauria* was originally conceptualized as the ancestral group from which other diapsids were derived).

Literature Cited

Benton, M. J. 1982. The *Diapsida*: a revolution in reptile relationships. *Nature* 296:306–307.

Benton, M. J. 1983. The Triassic reptile *Hyperodapedon* from Elgin: functional morphology and relationships. *Philos. Trans. R. Soc. Lond. B Biol. Sci.* 302(B):605–717.

Benton, M. J. 1984. The relationships and early evolution of the *Diapsida. Symp. Zool. Soc. Lond.* 575–596.

Benton, M. J. 1985. Classification and phylogeny of the diapsid reptiles. *Zool. J. Linn. Soc.* 84:97–164.

Benton, M. J., and P. C. J. Donoghue. 2007. Paleontological evidence to date the tree of life. *Mol. Biol. Evol.* 24(1):26–53.

Bever, G. S., T. R. Lyson, D. J. Field, and B.-A. S. Bhullar. 2015. Evolutionary origin of the turtle skull. *Nature* 525:239–242.

Bickelmann, C., J. Müller, and R. R. Reisz. 2009. The enigmatic diapsid *Acerodontosaurus piveteaui* (*Reptilia: Neodiapsida*) from the Upper Permian of Madagascar and the paraphyly of "younginiform" reptiles. *Can. J. Earth Sci.* 46(9):651–661.

deBraga, M., and R. R. Reisz. 1995. A new diapsid reptile from the uppermost Carboniferous (Stephanian) of Kansas. *Palaeontology* 38(1): 199–212.

deBraga, M., and O. Rieppel. 1997. Reptile phylogeny and the interrelationships of turtles. *Zool. J. Linn. Soc.* 120(3):281–354.

Broom, R. 1925. On the origin of lizards. *Proc. Zool. Soc. Lond.* 1926:487–492.

Carroll, R. L. 1981. Plesiosaur ancestors from the Upper Permian of Madagascar. *Philos. Trans. R. Soc. Lond. B Biol. Sci.* 293:315–383.

Curtis, N., M. E. H. Jones, J. Shi, P. O'Higgins, S. E. Evans, and M. J. Fagan. 2011. Functional relationship between skull form and feeding mechanics in *Sphenodon*, and implications for diapsid skull development. *PLOS ONE* 6(12): e29804.

Dilkes, D. W. 1998. The Early Triassic rhynchosaur *Mesosuchus browni* and the interrelationships of basal archosauromorph reptiles. *Philos. Trans. R. Soc. Lond. B Biol. Sci.* 353:501–541.

Evans, S. E. 1980. The skull of a new eosuchian reptile from the Lower Jurassic of South Wales. *Zool. J. Linn. Soc.* 70:203–264.

Evans, S. E. 1982. The gliding reptiles of the Upper Permian. *Zool. J. Linn. Soc.* 76(2):97–123.

Evans, S. E. 1984. The classification of the *Lepidosauria. Zool. J. Linn. Soc.* 82:87–100.

Evans, S. E. 1988. The early history and relationships of the *Diapsida*. Pp. 221–260 in *The Phylogeny and Classification of the Tetrapods. Vol. 1: Amphibians, Reptiles, Birds* (M. J. Benton, ed.). Systematics Association Special Volume No. 35A. Clarendon Press, Oxford.

Evans, S. E., and H. Haubold. 1987. A review of the Upper Permian genera *Coelurosauravus, Weigeltisaurus* and *Gracilisaurus* (*Reptilia: Diapsida*). *Zool. J. Linn. Soc.* 90(3):275–303.

Ezcurra, M. D., and R. J. Butler. 2015. Taxonomy of the proterosuchid archosauriforms (*Diapsida: Archosauromorpha*) from the earliest Triassic of South Africa, and implications for the early archosauriform radiation. *Palaeontology* 58(1):141–170.

Ezcurra, M. D., T. M. Scheyer, and R. J. Butler. 2014. The origin and early evolution of *Sauria*: reassessing the Permian saurian fossil record and the timing of the crocodile-lizard divergence. *PLOS ONE* 9(2):e89165.

Falcon-Lang, H.J., M. J. Benton, and M. Stimson. 2007. Ecology of early reptiles inferred from Lower Pennsylvanian trackways. *J. Geol. Soc. Lond.* 164(6):1113–1118.

Gaffney, E. S. 1979. Tetrapod monophyly: a phylogenetic analysis. *Bull. Carnegie Mus. Nat. Hist.* 13:92–105.

Gaffney, E. S. 1980. Phylogenetic relationships of the major groups of amniotes. Pp. 593–610 in *The Terrestrial Environment and the Origin of Land Vertebrates* (A. L. Panchen, ed.). Academic Press, London.

Gao, K.-Q., and R. C. Fox. 1998. New choristoderes (*Reptilia: Diapsida*) from the Upper Cretaceous and Palaeocene, Alberta and Saskatchewan, Canada, and phylogenetic relationships of *Choristodera*. *Zool. J. Linn. Soc.* 124:303–353.

Gao, K.-Q., and R. C. Fox. 2005. A new choristodere (*Reptilia: Diapsida*) from the Lower Cretaceous of western Liaoning Province, China, and phylogenetic relationships of *Monjurosuchidae*. *Zool. J. Linn. Soc.* 145(3):427–444.

Gauthier, J. A. 1984. *A Cladistic Analysis of the Higher Systematic Categories of the Diapsida*. University Microfilms International, Ann Arbor, #85-12825, vii + 564 pp.

Gauthier, J. A. 1994. The diversification of the amniotes. Pp. 129–159 in *Major Features of Vertebrate Evolution: Short Courses in Paleontology* (D. Prothero, ed.). The Paleontological Society, Knoxville, TN.

Gauthier, J. A., D. Cannatella, K. de Queiroz, A. Kluge, and T. Rowe. 1989. Tetrapod phylogeny. Pp. 337–353 in *The Hierarchy of Life: Proceedings of the 70th Nobel Symposium* (B. Fernholm, K. Bremer, and H. Jornvall, eds.). Elsevier Science Publishers B.V., Amsterdam.

Gauthier, J., R. Estes, and K. de Queiroz, 1988a. A phylogenetic analysis of *Lepidosauromorpha*. Pp. 15–98 in *Phylogenetic Relationships of the Lizard Families* (R. Estes and G. Pregill, eds.). Stanford University Press, Palo Alto, CA.

Gauthier, J. A., M. Kearney, J. A. Maisano, O. Rieppel, and A. Behlke. 2012. Assembling the squamate tree of life: perspectives from the phenotype and the fossil record. *Bull. Peabody Mus. Nat. Hist.* 53(1):3–308.

Gauthier, J., A. Kluge, and T. Rowe. 1988b. Amniote phylogeny and the importance of fossils. *Cladistics* 4(2):105–209.

Gauthier, J., A. Kluge, and T. Rowe. 1988c. The early evolution of the *Amniota*. Pp. 103–155 in *The Phylogeny and Classification of the Tetrapods. Vol. I: Amphibians, Reptiles and Birds* (M. J. Benton, ed.). Systematics Association Special Volume No. 35A. Clarendon Press, Oxford.

Gow, C. E. 1972. The osteology and relationships of the *Millerettidae* (*Reptilia: Cotylosauria*). *J. Zool.* 167:219–264.

Gow, C. E. 1975. The morphology and relationships of *Youngina capensis* Broom and *Prolacetla broomi* Parrington. *Palaeontol. Afr.* 18: 89–131.

Gray, J. E. 1842. Descriptions of two hitherto unrecorded species of reptiles from New Zealand; presented to the British Museum by Dr. Dieffenbach. *The Zoological Miscellany*. Privately published, London.

Gregory, J. T. 1945. Osteology and relationships of *Trilophosaurus*. *Univ. Texas Publ.* 4401:273–359.

Haridy, Y., M. J. Macdougall, D. Scott, and R. R. Reisz. 2016. Ontogenetic change in the temporal region of the Early Permian parareptile *Delorhynchus cifellii* and the implications for closure of the temporal fenestra in amniotes. *PLOS ONE* 11(12):e0166819.

Heaton, M. J. 1979. Cranial anatomy of primitive captorhinid reptiles from the Late Pennsylvanian and Early Permian, Oklahoma and Texas. *Bull. Okla. Geol. Surv.* 127:1–84.

Heaton, M. J., and R. R. Reisz. 1986. Phylogenetic relationships of captorhinomorph reptiles. *Can. J. Earth Sci.* 23(3):402–418.

Jones, M. E. H., C. L. Anderson, C. A. Hipsley, J. Müller, S. E. Evans, and R. R. Schoch. 2013. Integration of molecules and new fossils supports a Triassic origin for *Lepidosauria* (lizards, snakes, and tuatara). *BMC Evol. Biol.* 13:208.

Kissel, R. A., D. W. Dilkes, and R. R. Reisz. 2002. *Captorhinus magnus*, a new captorhinid (*Amniota*: *Eureptilia*) from the Lower Permian of Oklahoma, with new evidence on the homology of the astragalus. *Can. J. Earth Sci.* 39:1363–1372.

Kuhn, O. 1939. Schädelbau und systematische Stellung von *Weigeltisaurus*. *Paläont. Z.* 21(3):163–167.

Laurin, M. 1991. The osteology of a Lower Permian eosuchian from Texas and a review of diapsid phylogeny. *Zool. J. Linn. Soc.* 101(1):59–95.

Laurin, M., and G. H. Piñeiro. 2017. A reassessment of the taxonomic position of mesosaurs, and a surprising phylogeny of early amniotes. *Front. Earth Sci.* 5:88.

Li, C., N. E. Fraser, O. Rieppel, and X.-C. Wu. 2018. A Triassic stem turtle with an edentulous beak. *Nature* 560:476–479.

Lyson, T. R., G. S. Bever, B.-A. S. Bhullar, W. G. Joyce, and J. A. Gauthier. 2010. Transitional fossils and the origin of turtles. *Biol. Lett.* 6:830–833.

MacDougall, M. J., S. P. Modesto, N. Brocklehurst, A. Verrière, R. R. Reisz, and J. Fröbisch. 2018. Response: a reassessment of the taxonomic position of mesosaurs, and a surprising phylogeny of early amniotes. *Front. Earth Sci.* 6:99.

Mo, J.-Y., X. Xu, and S. E. Evans. 2009. The evolution of the lepidosaurian lower temporal bar: new perspectives from the Late Cretaceous of South China. *Proc. R. Soc. Lond. B Biol. Sci.* 277:331–36.

Modesto, S. P., and R. R. Reisz. 2003. An enigmatic new diapsid reptile from the Upper Permian of Eastern Europe. *J. Vertebr. Paleontol.* 22:851–855.

Müller, J. 2003. Early loss and multiple return of the lower temporal arcade in diapsid reptiles. *Naturwissenschaften* 90:473–476.

Müller, J. 2004. The relationships among diapsid reptiles and the influence of taxon selection. Pp. 379–408 in *Recent Advances in the Origin and Early Radiation of Vertebrates* (G. Arratia and R. Cloutier, eds.). Verlag Dr. Friedrich Pfeil, München, Germany.

Müller, J., and R. R. Reisz. 2005. Four well-constrained calibration points from the vertebrate fossil record for molecular clock estimates. *BioEssays* 27:1069–1075.

Müller, J., and R. R. Reisz. 2006. The phylogeny of early eureptiles: comparing parsimony and Bayesian approaches in the investigation of a basal fossil clade. *Syst. Biol.* 55:503–511.

Nesbitt, S. J., J. J. Flynn, A. C. Pritchard, J. M. Parrish, L. Ranivoharimanana, and A. R. Wyss. 2015. Postcranial osteology of *Azendohsaurus madagaskarensis* (Middle to Upper Triassic, Isalo Group of Madagascar) and its systematic position among stem archosaurs. *Bull. Am. Mus. Nat. Hist.* 398:1–126.

Oelrich, T. M. 1956. *The Anatomy of the Head of* Ctenosaura pectinata (Iguanidae). Miscellaneous Publications, Museum of Zoology, University of Michigan, Number 94.

Osborn, H. F. 1903. On the primary division of the *Reptilia* into two sub-classes, *Synapsida* and *Diapsida*. *Science* 17(424):275–276.

Piñeiro, G., J. Ferigolo, A. Ramos, and M. Laurin. 2012. Cranial morphology of the Early Permian mesosaurid *Mesosaurus tenuidens* and the evolution of the lower temporal fenestration reassessed. *C. R. Palevol* 11(5):379–391.

Pritchard, A. C., J. A. Gauthier, M. Hanson, G. S. Bever, and B.-A. S. Bhullar. 2018. A tiny Triassic saurian from Connecticut and the early evolution of the diapsid feeding apparatus. *Nat. Commun.* 9:1213.

Pritchard, A. C., and S. J. Nesbitt. 2017. A bird-like skull in a Triassic diapsid reptile increases heterogeneity of the morphological and phylogenetic radiation of *Diapsida*. *R. Soc. Open Sci.* 4:170499.

Reisz, R. R. 1977. *Petrolacosaurus*, the oldest known diapsid reptile. *Science* 196(4294):1091–1093.

Reisz, R. R. 1981. A diapsid reptile form the Pennsylvanian of Kansas. *Univ. Kans. Publ., Mus. Nat. Hist.* 7:1–74.

Reisz, R. R., D. S. Berman, and D. Scott. 1984. The anatomy and relationships of the Lower Permian reptile *Araeoscelis. J. Vertebr. Paleontol.* 4(1):57–67.

Reisz, R. R., M. Laurin, and D. Marjanovic. 2010. *Apsisaurus witteri* from the Lower Permian of Texas: yet another small varanopid synapsid, not a diapsid. *J. Vertebr. Paleontol.* 30:1628–1631.

Reisz, R. R., S. P. Modesto, and D. M. Scott. 2011. A new Early Permian reptile and its significance in early diapsid evolution. *Proc. R. Soc. Lond. B Biol. Sci.* 278(1725):3731–3737.

Rieppel, O. 1980. The trigeminal jaw adductors of primitive snakes and their homologies with the lacertilian jaw adductors. *J. Zool.* 190:447–471.

Romer, A. S. 1947. *Vertebrate Paleontology.* 2nd edition. University of Chicago Press, Chicago, IL.

Savage, J. M. 1963. Studies on the lizard family Xantusiidae. IV. The genera. *Contrib. Sci. (Los Angeles Co. Mus. Nat. Hist.)* 71:1–38.

Senter, P. 2004. Phylogeny of *Drepanosauridae* (*Reptilia: Diapsida*). *J. Syst. Palaeontol.* 2:257–268.

Simões, T. R., M. W. Caldwell, M. Tałanda, M. Bernardi, A. Palci, O. Vernygora, F. Bernardini, L. Mancini, and R. L. Nydam. 2018. The origin of squamates revealed by a Middle Triassic lizard from the Italian Alps. *Nature* 557:706–709.

Thulborn, R. A. 1980. The ankle joints of archosaurs. *Alcheringa* 4:241–261.

Williston, S. J. 1913. An ancestral lizard from the Permian of Texas. *Science* 38:825–826.

Authors

Jacques A. Gauthier; Department of Geology and Geophysics; Yale University; 210 Whitney Ave.; New Haven, CT 06520-8109, USA. Email: jacques.gauthier@yale.edu.

Kevin de Queiroz; Department of Vertebrate Zoology; National Museum of Natural History; Smithsonian Institution; Washington, DC 20560-0162, USA. Email: dequeirozk@si.edu.

Date Accepted: 22 October 2018

Primary Editor: Philip Cantino

Pan-Testudines W. G. Joyce, J. F. Parham, and J. A. Gauthier 2004 [W. G. Joyce, J. F. Parham, J. Anquetin, J. Claude, I. G. Danilov, J. B. Iverson, B. Kear, T. R. Lyson, M. Rabi, and J. Sterli], converted clade name

Registration Number: 272

Definition: The total clade of the crown clade *Testudines*. This is a crown-based total-clade definition. Abbreviated definition: total ∇ of *Testudines*.

Etymology: *Pan*, referring to the total clade or pan-monophylum, plus *Testudines*, the Latin vernacular for 'turtles.'

Reference Phylogeny: For the purposes of applying our definition, the tree presented in Figure 1 of Lyson et al. (2010) should be treated as the primary reference phylogeny. We arbitrarily chose this tree composed of Palaeozoic species because a single example must be selected for the reference phylogeny; nevertheless, we explicitly refrain from endorsing it as our favored topology for the relationship of turtles among amniotes (see Comments).

Composition: All *Testudinata* (this volume) plus *Odontochelys semitestacea,* and very likely *Pappochelys rosinae* and *Eunotosaurus africanus.* Otherwise, the hypothesized composition of *Pan-Testudines* varies widely because the placement of turtles within *Amniota* remains uncertain (see Comments).

Diagnostic Apomorphies: The characters that are considered diagnostic for *Pan-Testudines* vary widely depending on the favored topology (see Comments). We do not list any apomorphies here so as to avoid unduly favoring a particular topology.

Synonyms: *Anapsida* (sensu Gauthier 1994) (unambiguous); *Parareptilia* (sensu Laurin and Reisz 1995) (unambiguous); *Pantestudines* Joyce et al. 2004 (unambiguous).

Comments: The origin of turtles remains a highly contentious issue and the hypothesized composition of *Pan-Testudines* varies accordingly. Hypotheses based on morphological data originally positioned turtles as sister to other crown amniotes (e.g., Gaffney, 1980), but later hypotheses considering wider samples of morphology and fossils have placed turtles either outside of crown *Diapsida* (e.g., Gauthier, 1994; Laurin and Reisz, 1995; Lee, 1995; Lyson et al., 2010) or within crown *Diapsida* as sister to sauropterygians (e.g., DeBraga and Rieppel, 1997; Li et al., 2008; Schoch and Sues, 2018), or in an unresolved polytomy with sauropterygians and crown diapsids (= *Sauria*, this volume) (e.g., Bever et al., 2015; Schoch and Sues, 2018). Molecular data have consistently placed turtles inside crown *Diapsida* as sister to *Archosauria* (e.g., Chiari et al. 2012; Crawford et al., 2012; Fong et al., 2012; Field et al., 2014; Lu et al., 2013; Shaffer et al., 2013; Wang et al., 2013; but see Frost et al., 2006), although Brown and Thomson (2017) concluded that the phylogenetic signal encoded in the available data is too weak to sustain meaningful results.

Whereas hypotheses placing turtles outside of crown *Diapsida* typically consider various Palaeozoic amniotes as part of *Pan-Testudines*, most notably parareptiles such as milleretids, procolophonids, and pareiasaurs, those favoring a placement within crown *Diapsida* typically consider *Sauropterygia* as part of *Pan-Testudines*. The Permian *Eunotosaurus africanus*

(Lyson et al., 2010, 2013, 2014; Bever et al., 2015) and the Middle Triassic *Pappochelys rosinae* (Schoch and Sues, 2015) have been proposed to be close relatives of *Testudinata* (this volume), and thus parts of *Pan-Testudines*. At the very least, *Pan-Testudines* contains all representatives of *Testudinata* plus the toothed species with what everyone agrees is a partial shell, *Odontochelys semitestacea* (Li et al., 2008).

Prior to Joyce et al. (2004), the names *Anapsida* Williston 1917 and *Parareptilia* Olson 1947 were adapted to serve as names for the total clade of *Testudines* by Gauthier (1994) and Laurin and Reisz (1995), respectively. They are, therefore, unambiguous synonyms of *Pan-Testudines* as defined here. Nevertheless, to retain consistency with the nomenclature of Joyce et al. (2004), and in order to promote an integrated taxonomic system, we chose to use the name *Pan-Testudines* for the total clade of turtles.

Literature Cited

Bever, G. S., T. R. Lyson, D. J. Field, and B.-A. S. Bhullar. 2015. Evolutionary origin of the turtle skull. *Nature* 525:239–242.

de Braga, M., and O. Rieppel. 1997. Reptile phylogeny and the interrelationships of turtles. *Zool. J. Linn. Soc.* 120:281–354.

Brown, J. M., and R. C. Thomson. 2017. Bayes factors unmask highly variable information content, bias, and extreme influence in phylogenomic analyses. *Syst. Biol.* 66:517–530.

Chiari, Y., V. Cahais, N. Galter, and F. Delsuc. 2012. Phylogenomic analyses support the position of turtles as the sister group of birds and crocodiles (*Archosauria*). *BMC Biol.* 10:65.

Crawford, N. G., B. C. Faircloth, J. E. McCormack, R. T. Brumfield, K. Winker, and T. C. Glenn 2012. More than 1000 ultraconserved elements provide evidence that turtles are the sister group of archosaurs. *Biol. Lett.* 8:783–786.

Field, D. J., J. A. Gauthier, B. L. King, D. Pisani, T. R. Lyson, and K. J. Peterson. 2014. Toward consilience in reptile phylogeny: miRNAs

support an archosaur, not lepidosaur, affinity for turtles. *Evol. Dev.* 16:189–196.

Fong, J. J., J. M. Brown, B. Boussau, and M. K. Fujita. 2012. A phylogenomic approach to vertebrate phylogeny supports a turtle-crocodilian affinity and a paraphyletic *Lissamphibia*. *PLOS ONE* 7:e48990.

Frost, D. R., T. Grant, J. N. Faivovich, R. H. Bain, A. Haas, C. F. B. Haddad, R. O. de Sá, A. Channing, M. Wilkinson, S. C. Donnellan, C. J. Raxworthy, J. A. Campbell, B. L. Blotto, P. Moler, R. C. Drewes, R. A. Nussbaum, J. D. Lynch, D. M. Green, and W. C. Wheeler. 2006. The amphibian tree of life. *Bull. Am. Mus. Nat. Hist.* 297:1–370.

Gaffney, E. S. 1980. Phylogenetic relationships of the major groups of amniotes. Pp. 593–610 in *The Terrestrial Environment and the Origin of Land Vertebrates* (A. J. Panchen, ed.). Academic Press, London.

Gauthier, J. 1994. The diversification of the amniotes. Pp. 129–159 in *Major Features of Vertebrate Evolution: Short Courses in Paleontology* (D. Prothero and R. M. Schoch, eds.). The Paleontological Society, Knoxville, TN.

Joyce, W. G., J. F. Parham, and J. A. Gauthier. 2004. Developing a protocol for the conversion of rank-based taxon names to phylogenetically defined clade names, as exemplified by turtles. *J. Paleontol.* 78:989–1013.

Laurin, M., and R. R. Reisz. 1995. A reevaluation of early amniote phylogeny. *Zool. J. Linn. Soc.* 113:165–223.

Lee, M. S. Y. 1995. Historical burden in systematics and the interrelationships of "parareptiles." *Biol. Rev.* 70:459–547.

Li, C., X.-C. Wu, O. Rieppel, L.-T. Wang, and L.-J. Zhao. 2008. An ancestral turtle from the Late Triassic of southwestern China. *Nature* 456:497–501.

Lu, B., W. Yang, Q. Dai, J. Fu. 2013. Using genes as characters and a parsimony analysis to explore the phylogenetic position of turtles. *PLOS ONE* 8:e79348.

Lyson, T. R., G. S. Bever, B.-A. S. Bhullar, W. G. Joyce, and J. A. Gauthier. 2010. Transitional fossils and the origin of turtles. *Biol. Lett.* 6:830–833.

Lyson, T. R., G. S. Bever, T. M. Scheyer, A. Y. Hsiang, and J. A. Gauthier. 2013. Evolutionary origin of the turtle shell. *Curr. Biol.* 23:1–7.

Lyson, T. R., E. R. Schachner, J. Botha-Brink, T. M. Scheyer, M. Lambertz, G. S. Bever, B. S. Rubidge, and K. de Queiroz. 2014. Origin of the unique ventilatory apparatus of turtles. *Nat. Commun.* 5:5211.

Olson, E. C. 1947. The family *Diadectidae* and its bearing on the classification of reptiles. *Fieldiana: Geol.* 11:1–53.

Schoch, R. R., and H.-D. Sues. 2015. A Middle Triassic stem-turtle and the evolution of the turtle body plan. *Nature* 523:584–587.

Schoch, R. R., and H.-D. Sues. 2018. Osteology of the Middle Triassic stem-turtle *Pappochelys rosinae* and the early evolution of the turtle skeleton. *J. Syst. Palaeontol.* 16:927–965.

Shaffer, H., P. Minx, D. Warren, A. M. Shedlock, R. C. Thomson, N. Valenzuela, J. Abramyan, D. Badenhorst, K. K. Biggar, G. M. Borchert, C. W. Botka, R. M. Bowden, E. L. Braun, A. M. Bronikowski, B. G. Bruneau, L. T. Buck, B. Capel, T. A. Castoe, M. Czerwinski, K. D. Delehaunty, S. V. Edwards, C. C. Fronick, M. K. Fujita, L. Fulton, T. A. Graves, R. E. Green, W. Haerty, R. Hariharan, L. H. Hillier, A. K. Holloway, D. Janes, F. J. Janzen, C. Kandoth, L. Kong, J. de Koning, Y. Li, R. Literman, E. R. Mardis, S. E. McGaugh, P. Minx, L. Mork, M. O'Laughlin, R. T. Paitz, D. D. Pollock, C. P. Ponting, S. Radhakrishnan, B. J. Raney, J. M. Richman, J. St John, T. Schwartz, A. Sethuraman, B. Shaffer, P. Q. Spinks, K. B. Storey, N. Thane, T. Vinar, D. E. Warren, W. C. Warren, R. K. Wilson, L. M. Zimmerman, O. Hernandez, and C. T. Amemiya. 2013. The western painted turtle genome, a model for the evolution of extreme physiological adaptations in a slowly evolving lineage. *Genome Biol.* 14:R28.

Wang, Z., J. Pascual-Anaya, A. Zadissa, W. Li, Y. Niimura, Z. Huang, C. Li, S. White, Z. Xiong, D. Fang, B. Wang, Y. Ming, Y. Chen, Y. Zheng, S. Kuraku, M. Pignatelli, J. Herrero, K. Beal, M. Nozawa, Q. Li, J. Wang, H. Zhang, L. Yu, S. Shigenobu, J. Wang, J. Liu, P. Flicek, S. Searle, J. Wang, S. Kuratani, Y. Yin, B. Aken, G. Zhang, and N. Irie. 2013. The draft genomes of soft-shell turtle and green sea turtle yield insights into the development and evolution of the turtle-specific body plan. *Nat. Genet.* 45:701–706.

Williston, S. W. 1917. The phylogeny and classification of reptiles. *J. Geol.* 25:411–421.

Authors

International Turtle Nomenclature Committee followed by list of authors

Walter G. Joyce; Departement für Geowissenschaften; Universität Freiburg; 1700 Fribourg, Switzerland. Email: walter.joyce@unifr.ch.

James F. Parham; Department of Geological Sciences; California State University; Fullerton, CA 92834, USA. Email: jparham@fullerton.edu.

Jérémy Anquetin; Jurassica Museum; 2900 Porrentruy, Switzerland. Email: jeremy.anquetin@jurassica.ch.

Juliana Sterli; CONICET – Museo Egidio Feruglio; 9100 Trelew, Chubut, Argentina. Email: jsterli@mef.org.ar.

Julien Claude; Institut des Sciences de l'Evolution de Montpellier; Université de Montpellier, CNRS, IRD, EPHE; 34095 Montpellier, France. Email: julien.claude@umontpellier.fr.

Igor G. Danilov; Department of Herpetology; Zoological Institute of Russian Academy of Sciences; 199034 St. Petersburg, Russia. Email: turtle@zin.ru.

John B. Iverson; Department of Biology; Earlham College; Richmond, IN 47374, USA. Email: johni@earlham.edu.

Benjamin P. Kear; Museum of Evolution; Uppsala University; 752 36 Uppsala, Sweden. Email: benjamin.kear@em.uu.se.

Tyler R. Lyson; Department of Earth Sciences; Denver Museum of Nature and Science; Denver, CO 80205, USA. Email: tyler.lyson@dmns.org.

Date Accepted: 16 October 2017

Primary Editor: Jacques A. Gauthier

Testudinata I. T. Klein 1760 [W. G. Joyce, J. F. Parham, J. Anquetin, J. Claude, I. G. Danilov, J. B. Iverson, B. Kear, T. R. Lyson, M. Rabi, and J. Sterli], converted clade name

Registration Number: 273

Definition: The clade for which a complete turtle shell, as inherited by *Testudo graeca* Linnaeus 1758, is an apomorphy. A 'complete turtle shell' is herein defined as a composite structure consisting of a carapace with interlocking costals, neurals, peripherals, and a nuchal, together with the plastron comprising interlocking epi-, hyo-, meso- (lost in *Testudo graeca*), hypo-, xiphiplastra and an entoplastron that are articulated with one another along a bridge. This is an apomorphy-based definition. Abbreviated definition: ∇ apo complete turtle shell [*Testudo graeca* Linnaeus 1758].

Etymology: Derived from *Testudo*, the Latin vernacular for 'turtle' + -*ata* (L), having the nature of, pertaining to.

Reference Phylogeny: For the purpose of applying our definition, the phylogeny provided in Joyce et al. (2016: Fig. 8) should be treated as the primary reference.

Composition: In addition to crown *Testudines* (this volume), current phylogenetic hypotheses refer a broad set of fossil turtles with fully formed shells to *Testudinata*. These include, among others, *Proganochelys* spp., *Proterochersis* spp., and *Palaeochersis talampayensis* from the Late Triassic, *Australochelys africanus*, *Condorchelys antiqua*, *Eileanchelys waldmani*, *Heckerochelys romani*, *Indochelys spatula*, and *Kayentachelys aprix* from the Early to Middle Jurassic, and representatives of the clades *Helochelydridae*, *Meiolaniformes*, *Sichuanchelyidae*, and *Kallokibotion bajazidi* and *Spoochelys ormondea* from the Late Jurassic to Pleistocene (Rougier et al., 1995; Sukhanov, 2006; Gaffney et al., 2007; Joyce, 2007; Sterli, 2008; Joyce et al., 2009, 2014, 2016; Anquetin et al., 2009; Tong et al., 2012; Smith and Kear, 2013; Szczygielski and Sulej, 2016).

Diagnostic Apomorphies: As an apomorphy-based clade, *Testudinata* has a single diagnostic character, the turtle shell (see Definition above). The turtle shell is, however, a composite structure composed of both endochondral and dermal bones, including vertebrae (neurals), ribs (costals), fused cleithra (nuchal), clavicles (epiplastra), an interclavicle (entoplastron), and gastralia (hyo-, meso-, hypo-, and xiphiplastra), all of which may be extensively augmented by metaplastic bone. Carapace and plastron are articulated with one another along a bridge. Additional elements may be present as well (but are not part of the definition). The entire structure is encased in heavily keratinized scales. Some of these features are present in proto-shells known to have arisen prior to *Testudinata* as defined here (see Comments below). We do not know which of these apomorphies was the last one added to the turtle shell. Until all of them are present, however, a fossil should not be referred to *Testudinata*.

Synonyms: *Testudo* Linnaeus 1758 (approximate); *Testudines* Batsch 1788 (approximate); *Chelonii* Latreille 1800 (approximate); *Chelonia* Ross and Macartney 1802 (approximate). Many additional synonyms exist (see Joyce et al., 2004, for an exhaustive list), but these have not been extensively used beyond their initial publication.

Comments: Gaffney and Meylan (1988) provided the first comprehensive and explicit phylogeny of *Testudinata*, but much of that topology was derived by grafting parsimony-based subtrees. Many later morphology-based phylogenetic analyses have since assessed testudinate ingroup relationships with increasing confidence (e.g., Gaffney et al., 2007; Joyce, 2007; Anquetin, 2012; Sterli et al., 2013; Joyce et al., 2014, 2016; Zhou and Rabi, 2015). Molecular phylogenies are not considered here, because they do not include members of the turtle stem. We chose the morphology-based phylogeny of Joyce et al. (2016: Fig. 8) as our reference, as this is the most recently published global turtle phylogeny that incorporates a broad sample of extinct taxa.

Given that a number of disparate vertebrate clades developed a bony dermal shell during their evolution (e.g., armadillos, placodonts, anguid lizards, derived ankylosaurs), Linnaeus's (1758) original diagnosis (*corpus tetrapodum, caudatum test obtectum*—body four-legged, with a tail, covered by a shell) is somewhat insufficient for turtles. However, the detailed composition and structure of the turtle shell is unique among vertebrates and thus serves to diagnose *Testudinata* unambiguously. In particular, the turtle shell (carapace and plastron) is a composite structure that consists of an internal layer of metaplastic bone (usually) reinforcing a heavily keratinized external epidermis. It incorporates the ribs and the dermal components of the pectoral girdle, and fully surrounds the remaining elements of the pectoral and the pelvic girdles (see Lyson et al., 2013, for a recent summary). Historically, the application of this definition was straightforward because no amniote was known to manifest a "partial" turtle shell. However, this has changed with the recent findings of mid-Triassic *Odontochelys semitestacea* Li et al. 2008 that possess a well-developed plastron resembling those of basally branching testudinates. Nevertheless, the carapace lacks peripherals and interlocking neurals and costals. *Odontochelys semitestacea*, therefore, only incompletely fulfills the definition criteria for *Testudinata*, and is alternatively considered a representative of *Pan-Testudines*, just outside of *Testudinata*. A number of fossil and extant turtles, in particular leatherback turtles (*Dermochelyidae*) and soft-shelled turtles (*Pan-Trionychidae*), lack a complete shell as well, but as these turtles derive from an ancestor with a complete shell, they are included in *Testudinata*. See *Testudines* for an explanation of the name choice for this clade.

Literature Cited

Anquetin, J. 2012. Reassessment of the phylogenetic interrelationships of basal turtles (*Testudinata*). *J. Syst. Palaeontol.* 10:3–45.

Anquetin, J., P. M. Barrett, M. E. H. Jones, S. Moore-Fay, and S. E. Evans. 2009. A new stem turtle from the Middle Jurassic of Scotland: new insights into the evolution and palaeoecology of basal turtles. *Proc. R. Soc. Lond. B Biol. Sci.* 276:879–886.

Batsch, A. J. G. C. 1788. *Versuch einer Anleitung, zur Kenntniß und Geschichte der Thiere und Mineralien.* Akademische Buchhandlung, Jena.

Gaffney, E. S., and P. A. Meylan. 1988. A phylogeny of turtles. Pp. 157–219 in *The Phylogeny and Classification of the Tetrapods, Volume 1: Amphibians, Reptiles, Birds* (M. J. Benton, ed.). Clarendon Press, Oxford.

Gaffney, E. S., T. H. Rich, P. Vickers-Rich, A. Constantine, R. Vacca, and L. Kool. 2007. *Chubutemys*, a new eucryptodiran turtle from the Early Cretaceous of Argentina, and the relationships of the *Meiolaniidae*. *Am. Mus. Novit.* 3599:1–35.

Joyce, W. G. 2007. Phylogenetic relationships of Mesozoic turtles. *Bull. Peabody Mus. Nat. Hist.* 48:3–102.

Joyce, W. G., S. G. Lucas, T. Scheyer, A. Heckert, and A. P. Hunt. 2009. A thin-shelled reptile from the Late Triassic of North America and the origin of the turtle shell. *Proc. R. Soc. Lond. B Biol. Sci.* 276:507–513.

Joyce, W. G., J. F. Parham, and J. A. Gauthier. 2004. Developing a protocol for the conversion of rank-based taxon names to phylogenetically defined clade names, as exemplified by turtles. *J. Paleontol.* 78:989–1013.

Joyce, W. G., M. Rabi, J. M. Clark, and X. Xu. 2016. A toothed turtle from the Late Jurassic of China and the global biogeographic history of turtles. *BMC Evol. Biol.* 16:236.

Joyce, W. G., J. Sterli, and S. D. Chapman. 2014. The skeletal morphology of the solemydid turtle *Naomichelys speciosa* from the Early Cretaceous of Texas. *J. Paleontol.* 88:1257–1287.

Klein, I. T. 1760. *Klassification und kurze Geschichte der Vierfüßigen Thiere* (translation by F. D. Behn). Jonas Schmidt, Lübeck.

Latreille, P. A. 1800. *Histoire naturelle des salamandres de France.* Villier, Paris.

Li, C., X.-C. Wu, O. Rieppel, L.-T. Wang, and L.-J. Zhao. 2008. An ancestral turtle from the Late Triassic of southwestern China. *Nature* 456:497–501.

Linnaeus, C. 1758. *Systema Naturae Per Regna Tria Naturae, Secundum Classes, Ordines, Genera, Species, cum Characteribus, Differentiis, Synonymis, Locis.* 10th edition, Vol. 1. Laurentii Salvii, Holmiae (Stockholm).

Lyson, T. R., G. S. Bever, T. M. Scheyer, A. Y. Hsiang, and J. A. Gauthier. 2013. Evolutionary origin of the turtle shell. *Curr. Biol.* 23:1113–1119.

Ross, W., and J. Macartney. 1802. *Lectures on Comparative Anatomy, Translated from the French of G. Cuvier*, Vol. 1. Oriental Press, London.

Rougier, G. W., M. S. Fuente, and A. B. Arcucci. 1995. Late Triassic turtles from South America. *Science* 268:855–858.

Smith, E. T., and B. P. Kear. 2013. *Spoochelys ormondea* gen. et sp. nov., an archaic meiolaniid-like turtle from the Early Cretaceous of Lightning Ridge, Australia. Pp. 121–146 in *Morphology and Evolution of Turtles* (D. Brinkman, P. Holroyd, and J. Gardner, eds.). Springer, Dordrecht.

Sterli, J. 2008. A new, nearly complete stem turtle from the Jurassic of South America with implications for turtle evolution. *Biol. Lett.* 4:286–289.

Sterli, J., M. S. de la Fuente, and I. A. Cerda. 2013. A new species of meiolaniform turtle and a revision of the Late Cretaceous *Meiolaniformes* of South America. *Ameghiniana* 50: 240–256.

Sukhanov, V. B. 2006. An archaic turtle, *Heckerochelys romani* gen. et sp. nov., from the Middle Jurassic of Moscow region, Russia. *Russ. J. Herpetol. (Fossil Turtle Research 1)* 13:104–111.

Szczygielski, T., and T. Sulej. 2016. Revision of the Triassic European turtles *Proterochersis* and *Murrhardtia (Reptilia, Testudinata, Proterochersidae)*, with the description of new taxa from Poland and Germany. *Zool. J. Linn. Soc.* 177:395–427.

Tong, H., I. Danilov, Y. Ye, H. Ouyang, and G. Peng. 2012. Middle Jurassic turtles from the Sichuan Basin, China: a review. *Geol. Mag.* 149:675–695.

Zhou, C.-F., and M. Rabi. 2015. A sinemydid turtle from the Jehol Biota provides insights into the basal divergence of crown turtles. *Sci. Rep.* 5:16299.

Authors

International Turtle nomenclature Committee followed by the author sequence for this chapter

Walter G. Joyce; Departement für Geowissenschaften; Universität Freiburg; 1700 Fribourg, Switzerland. Email: walter.joyce@unifr.ch.

James F. Parham; Department of Geological Sciences; California State University; Fullerton, CA 92834, USA. Email: jparham@fullerton.edu.

Jérémy Anquetin; Jurassica Museum; 2900 Porrentruy, Switzerland. Email: jeremy.anquetin@jurassica.ch.

Julien Claude; Institut des Sciences de l'Evolution de Montpellier; Université de Montpellier;

CNRS, IRD, EPHE; 34095 Montpellier; France. E-mail: julien.claude@umontpellier.fr.

Igor G. Danilov; Department of Herpetology; Zoological Institute of Russian Academy of Sciences; 199034 St. Petersburg; Russia. E-mail: turtle@zin.ru.

John B. Iverson; Department of Biology; Earlham College; Richmond, IN 47374, USA. Email: johni@earlham.edu.

Benjamin P. Kear; Museum of Evolution; Uppsala University; 752 36 Uppsala, Sweden. Email: benjamin.kear@em.uu.se.

Tyler R. Lyson; Department of Earth Sciences; Denver Museum of Nature and Science; Denver, CO 80205, USA. Email: tyler.lyson@dmns.org.

Márton Rabi; Zentralmagazin Naturwissenschaftlicher Sammlungen; Martin-Luther-Universität Halle-Wittenberg; 06108 Halle (Saale), Germany. Email: iszkenderun@gmail.com.

Juliana Sterli; CONICET – Museo Egidio Feruglio; 9100 Trelew, Chubut, Argentina. Email: jsterli@mef.org.ar.

Date Accepted: 16 October 2017

Primary Editor: Jacques A. Gauthier

Testudines A. J. G. C. Batsch 1788 [W. G. Joyce, J. F. Parham, J. Anquetin, J. Claude, I. G. Danilov, J. B. Iverson, B. Kear, T. R. Lyson, M. Rabi, and J. Sterli], converted clade name

Registration Number: 274

Definition: The smallest crown clade containing the pleurodire *Chelus* (originally *Testudo*) *fimbriatus* (Schneider 1783), the trionychian *Trionyx* (originally *Testudo*) *triunguis* (Forskål 1775), the americhelydian *Chelonia* (originally *Testudo*) *mydas* (Linnaeus 1758), and the testudinoid *Testudo graeca* Linnaeus 1758. This is a minimum-crown-clade definition. Abbreviated definition: min crown ∇ (*Chelus fimbriatus* (Schneider 1783) & *Trionyx triunguis* (Forskål 1775) & *Chelonia mydas* (Linnaeus 1758) & *Testudo graeca* Linnaeus 1758).

Etymology: The Latin vernacular for 'turtles.'

Reference Phylogeny: For the purposes of applying our definition, the phylogeny provided in Figure 1 of Pereira et al. (2017) should be treated as the primary reference.

Composition: *Testudines* is currently believed to contain 356 extant species (Turtle Taxonomy Working Group, 2017) in two primary subclades, *Pan-Pleurodira* and *Pan-Cryptodira* (see entries in this volume). A synoptic overview of fossil *Testudines* is provided in Młynarski (1976). The oldest fossil that can be referred unambiguously to *Testudines* is a pan-pleurodire from the Late Jurassic (Oxfordian) of Cuba (de la Fuente and Iturralde-Vinent, 2001; Joyce et al., 2013). However, if future analyses confirm the pancryptodiran relationship of xinjiangchelyid turtles, the age of crown *Testudines* can be backdated to the Middle Jurassic (Zhou and Rabi, 2015).

Diagnostic Apomorphies: *Testudines* can be distinguished from all other crown clades by the presence of a turtle shell (see *Testudinata*), lack of teeth, the presence of a horny beak, the lack of lacrimal, tabular, postparietal, postfrontal, and supratemporal skull bones, a completely fused basicranial joint, fused vomers, a unique impedance-matching ear in which the quadrate guides the stapes and subdivides the middle ear cavity into outer and inner chambers, and the presence of a trochlear system in which the jaw musculature curves around the otic cavity (Gaffney and Meylan, 1988; Joyce, 2007).

Synonyms: *Testudo* Linnaeus 1758 (approximate); *Testudinata* Klein 1760 (approximate); *Chelonii* Latreille 1800 (approximate); *Chelonia* Ross and Macartney 1802 (approximate); *Casichelydia* Gaffney 1975 (approximate); *Chelonia* Gauthier et al. 1988 (unambiguous); see Joyce et al. (2004) for a review of additional, much less commonly used, synonyms.

Comments: Gaffney and Meylan (1988) presented the most comprehensive morphology-based tree of *Testudinata* (this volume) and *Testudines*. Numerous studies have otherwise analyzed molecular data (e.g., Shaffer et al., 1997; Fujita et al., 2004; Krenz et al., 2005; Crawford et al., 2015), and collectively sampled nearly every currently recognized living turtle species (Thomson and Shaffer, 2010; Pereira et al., 2017). We chose the phylogenetic hypotheses of Pereira et al. (2017: Fig 1) as our reference phylogeny, because it contains the most densely sampled set of extant turtles and is based on the largest DNA sequence-based dataset assembled to date.

Turtles have never been, and still are not, referred to by a single name. Of the more than 17 names that have been used, the most common names include *Testudines, Chelonia, Testudinata, Chelonii* and *Casichelydia*, ordered herein in descending popularity as determined by simple publication searches. We follow the recommendation of Joyce et al. (2004) by attaching the most commonly used name, *Testudines*, to the crown clade and the third most commonly used name, *Testudinata*, to the apomorphy-based, shell-bearing clade (see *Testudinata*). The second most commonly used name, *Chelonia,* is omitted from consideration because it is a homonym of a name of a clade of marine turtles that is traditionally ranked as a genus. For an extensive discussion of historical name usage, including prior application of the names *Testudines* and *Testudinata* to taxa approximating the clades to which they are here being applied, see Joyce et al. (2004).

Literature Cited

Batsch, A. J. G. C. 1788. *Versuch Einer Anleitung, zur Kenntniß und Geschichte der Thiere und Mineralien.* Akademische Buchhandlung, Jena.

Crawford, N. G., J. F. Parham, A. B. Sellas, B. C. Faircloth, T. C. Glenn, T. J. Papenfuss, J. B. Henderson, M. H. Hansen, and W. B. Simison. 2015. A phylogenomic analysis of turtles. *Mol. Phylogenet. Evol.* 83:250–257.

Forskål, P. 1775. *Descriptiones Animalium Avium, Amphibiorum, Piscium, Insectorum, Vermium; Quae in Itinere Orientali Observavit.* Möllerus, Haunia.

de la Fuente, M. S., and M. Iturralde-Vinent. 2001. A new pleurodiran turtle from the Jagua Formation (Oxfordian) of Western Cuba. *J. Paleontol.* 75:860–869.

Fujita, M. K., T. N. Engstrom, D. E. Starkey, and H. B. Shaffer. 2004. Turtle phylogeny: insights from a novel nuclear intron. *Mol. Phylogenet. Evol.* 31:1031–1040.

Gaffney, E. S. 1975. A phylogeny and classification of the higher categories of turtles. *Bull. Am. Mus. Nat. Hist.* 155:389–436.

Gaffney, E. S., and P. A. Meylan. 1988. A phylogeny of turtles. Pp. 157–219 in *The Phylogeny and Classification of the Tetrapods. Vol. 1: Amphibians, Reptiles, Birds* (M. J. Benton, ed.). Clarendon Press, Oxford.

Gauthier, J., R. Estes, and K. de Queiroz. 1988. A phylogenetic analysis of *Lepidosauromorpha.* Pp. 15–98 in *Phylogenetic Relationships of the Lizard Families* (R. Estes and G. Pregill, eds.). Stanford University Press, Stanford, CA.

Joyce, W. G. 2007. Phylogenetic relationships of Mesozoic turtles. *Bull. Peabody Mus. Nat. Hist.* 48:3–102.

Joyce, W. G., J. F. Parham, and J. A. Gauthier. 2004. Developing a protocol for the conversion of rank-based taxon names to phylogenetically defined clade names, as exemplified by turtles. *J. Paleontol.* 78:989–1013.

Joyce, W. G., J. F. Parham, T. R. Lyson, R. C. M. Warnock, and P. C. J. Donoghue. 2013. A divergence dating analysis of turtles using fossil calibrations: an example of best practices. *J. Paleontol.* 87:612–634.

Klein, I. T. 1760. *Klassification und kurze Geschichte der Vierfüßigen Thiere* (translation by F. D. Behn). Jonas Schmidt, Lübeck.

Krenz, J. G., G. J. P. Naylor, H. B. Shaffer, and F. J. Janzen. 2005. Molecular phylogenetics and evolution of turtles. *Mol. Phylogenet. Evol.* 37:178–191.

Latreille, P. A. 1800. *Histoire Naturelle des Salamandres de France.* Villier, Paris.

Linnaeus, C. 1758. *Systema Naturae Per Regna Tria Naturae, Secundum Classes, Ordines, Genera, Species, cum Characteribus, Differentiis, Synonymis, Locis.* 10th edition, Vol. 1. Laurentii Salvii, Holmiae (Stockholm).

Młynarski, M. 1976. *Testudines. Handbuch der Paläoherpetologie.* Part 7. Fischer Verlag, Stuttgart.

Pereira, A. G., J. Sterli, F. R. R. Moreira, and C. G. Schrago. 2017. Multilocus phylogeny and statistical biogeography clarify the evolutionary history of major lineages of turtles. *Mol. Phylogenet. Evol.* 113:59–66.

Ross, W., and J. Macartney. 1802. *Lectures on Comparative Anatomy, Translated from the French of G. Cuvier*, Vol. 1. Oriental Press, London.

Schneider, J. G. 1783. *Allgemeine Naturgeschichte der Schildkröten, Nebst Einem Systematischen Verzeichnisse der Einzelnen Arten und Zwey Kupfern.* Johann Gotfried Müllersche Buchhandlung, Leipzig.

Shaffer, H. B., P. Meylan, and M. L. McKnight. 1997. Tests of turtle phylogeny: molecular, morphological, and paleontological approaches. *Syst. Biol.* 46:235–268.

Thomson, R. C., and H. B. Shaffer. 2010. Sparse supermatrices for phylogenetic inference: taxonomy, alignment, rogue taxa, and the phylogeny of living turtles. *Syst. Biol.* 59:42–58.

Turtle Taxonomy Working Group [Rhodin, A. G. J., J. B. Iverson, U. Fritz, A. Georges, H. B. Shaffer, and P. P. van Dijk]. 2017. Turtles of the world: annotated checklist and atlas of taxonomy, synonymy, distribution, and conservation status (8th edition). *Chelonian Res. Monogr.* 7:1–292.

Zhou, C.-F., and M. Rabi. 2015. A sinemydid turtle from the Jehol Biota provides insights into the basal divergence of crown turtles. *Sci. Rep.* 5:16299.

Authors

International Turtle Nomenclature Committee then list the authors for this chapter.

Walter G. Joyce; Departement für Geowissenschaften; Universität Freiburg; 1700 Fribourg, Switzerland. Email: walter.joyce@unifr.ch.

James F. Parham; Department of Geological Sciences; California State University; Fullerton, CA 92834, USA. Email: jparham@fullerton.edu.

Jérémy Anquetin; Jurassica Museum; 2900 Porrentruy, Switzerland. Email: jeremy.anquetin@jurassica.ch.

Julien Claude; Institut des Sciences de l'Evolution de Montpellier; Université de Montpellier, CNRS, IRD, EPHE; 34095 Montpellier, France. Email: julien.claude@umontpellier.fr.

Igor G. Danilov; Department of Herpetology; Zoological Institute of Russian Academy of Sciences; 199034 St. Petersburg, Russia. Email: turtle@zin.ru.

John B. Iverson; Department of Biology; Earlham College; Richmond, IN 47374, USA. Email: johni@earlham.edu.

Benjamin P. Kear; Museum of Evolution; Uppsala University; 752 36 Uppsala, Sweden. Email: benjamin.kear@em.uu.se.

Tyler R. Lyson; Department of Earth Sciences; Denver Museum of Nature and Science; Denver, CO 80205, USA. Email: tyler.lyson@dmns.org.

Márton Rabi; Zentralmagazin Naturwissenschaftlicher Sammlungen; Martin-Luther-Universität Halle-Wittenberg; 06108 Halle (Saale), Germany. Email: iszkenderun@gmail.com.

Juliana Sterli; CONICET – Museo Egidio Feruglio; 9100 Trelew, Chubut, Argentina. Email: jsterli@mef.org.ar.

Date Accepted: 16 October 2017

Primary Editor: Jacques A. Gauthier

Pan-Pleurodira W. G. Joyce, J. F. Parham, and J. A. Gauthier 2004 [W. G. Joyce, J. F. Parham, J. Anquetin, J. Claude, I. G. Danilov, J. B. Iverson, B. Kear, T. R. Lyson, M. Rabi, and J. Sterli], converted clade name

Registration Number: 275

Definition: The total clade of the crown clade *Pleurodira*. This is a crown-based total-clade definition. Abbreviated definition: total ∇ of *Pleurodira*.

Etymology: *Pan-*, with reference to the total clade or pan-monophylum, plus *Pleurodira* (see *Pleurodira* in this volume for etymology).

Reference Phylogeny: For the purposes of applying our definition, we use the topology provided in Figure 8 of Joyce et al. (2016) because it is the most recent global turtle tree incorporating stem pleurodires.

Composition: *Pleurodira* and all extinct turtles that share more recent common ancestry with *Pleurodira* rather than *Cryptodira* (this volume). This includes turtles of the extinct clade *Platychelyidae* (Cadena and Joyce, 2016; López-Conde et al., 2017) and possibly also *Dortokidae* (Lapparent de Broin, 2001; Gaffney et al., 2006; Cadena and Joyce, 2016).

Diagnostic Apomorphies: see *Pleurodira* (this volume).

Synonyms: *Pleurodira* Cope 1865 (approximate); *Pleurodiromorpha* Lee 1995 (unambiguous).

Comments: The name *Pleurodira* is an approximate synonym of *Pan-Pleurodira* and is still sometimes used that way to denote the total clade of pleurodires (e.g., Gaffney et al., 2006), but this is rejected here to retain the most commonly used names for crown clades. *Pleurodiromorpha* is an unambiguous synonym of *Pan-Pleurodira* but we follow Joyce et al. (2004) in using the prefix "*Pan-*" when referring to total clades.

Literature Cited

Cadena E., and W. G. Joyce. 2015. A review of the fossil record of turtles of the clades *Platychelyidae* and *Dortokidae*. *Bull. Peabody Mus. Nat. Hist.* 56:3–20.

Cope, E. D. 1865. Third contribution to the herpetology of tropical America. *Proc. Acad. Nat. Sci. Phila.* 1865:185–198.

Gaffney, E. S., H. Tong, and P. A. Meylan. 2006. Evolution of the side-necked turtles: the families *Bothremydidae, Euraxemydidae*, and *Araripemydidae. Bull. Am. Mus. Nat. Hist.* 300:1–698.

Joyce, W. G., J. F. Parham, and J. A. Gauthier. 2004. Developing a protocol for the conversion of rank-based taxon names to phylogenetically defined clade names, as exemplified by turtles. *J. Paleontol.* 78:989–1013.

Joyce, W. G., M. Rabi, J. M. Clark, and X. Xu. 2016. A toothed turtle from the Late Jurassic of China and the global biogeographic history of turtles. *BMC Evol. Biol.* 16:236.

de Lapparent de Broin, F. 2001. The European turtle fauna from the Triassic to the Present. *Dumerilia* 4:155–217.

Lee, M. S. Y. 1995. Historical burden in systematics and the interrelationships of "parareptiles." *Biol. Rev.* 70:459–547.

López-Conde, O. A., J. Sterli, J. Alvarado-Ortega, and M. L. Chavarría-Arellano. 2017. A new platychelyid turtle (*Pan-Pleurodira*) from the Late Jurassic (Kimmeridgian) of Oaxaca, Mexico. *Pap. Palaeontol.* 3:161–174.

Authors

International Turtle Nomenclature Committee followed by the list of authors for this chapter.

Walter G. Joyce; Departement für Geowissenschaften; Universität Freiburg; 1700 Fribourg, Switzerland. Email: walter.joyce@unifr.ch.

James F. Parham; Department of Geological Sciences; California State University; Fullerton, CA 92834, USA. Email: jparham@fullerton.edu.

Jérémy Anquetin; Jurassica Museum; 2900 Porrentruy, Switzerland. Email: jeremy.anquetin@jurassica.ch.

Julien Claude; Institut des Sciences de l'Evolution de Montpellier; Université de Montpellier, CNRS, IRD, EPHE; 34095 Montpellier, France. Email: julien.claude@umontpellier.fr.

Igor G. Danilov; Department of Herpetology; Zoological Institute of Russian Academy of Sciences; 199034 St. Petersburg, Russia. Email: turtle@zin.ru.

John B. Iverson; Department of Biology; Earlham College; Richmond, IN 47374, USA. Email: johni@earlham.edu.

Benjamin P. Kear; Museum of Evolution; Uppsala University; 752 36 Uppsala, Sweden. Email: benjamin.kear@em.uu.se.

Tyler R. Lyson; Department of Earth Sciences; Denver Museum of Nature and Science; Denver, CO 80205, USA. Email: tyler.lyson@dmns.org.

Márton Rabi; Zentralmagazin Naturwissenschaftlicher Sammlungen; Martin-Luther-Universität Halle-Wittenberg; 06108 Halle (Saale), Germany. Email: iszkenderun@gmail.com.

Juliana Sterli; CONICET – Museo Egidio Feruglio; 9100 Trelew, Chubut, Argentina. Email: jsterli@mef.org.ar.

Date Accepted: 16 October 2017

Primary Editor: Jacques A. Gauthier

Pleurodira E. D. Cope 1865 [W. G. Joyce, J. F. Parham, J. Anquetin, J. Claude, I. G. Danilov, J. B. Iverson, B. Kear, T. R. Lyson, M. Rabi, and J. Sterli], converted clade name

Registration Number: 276

Definition: The smallest crown clade containing the chelid *Chelus* (originally *Testudo*) *fimbriatus* (Schneider 1783) the pelomedusid *Pelomedusa* (originally *Testudo*) *subrufa* (Bonnaterre 1789) and the podocnemid *Podocnemis* (originally *Emys*) *expansa* (Schweigger 1812). This is a minimum-crown-clade definition. Abbreviated definition: min crown ∇ (*Chelus fimbriatus* (Schneider 1783) & *Pelomedusa subrufa* (Bonnaterre 1789) & *Podocnemis expansa* (Schweigger 1812)).

Etymology: Derived from the Greek words πλευρο (*pleura*) for 'side' and δειρη (*deire*) for 'neck' or 'throat' in reference to sideways retraction of the neck into the shell.

Reference Phylogeny: For the purposes of applying our definition, we use Figure 1 of Pereira et al. (2017) because it was based on the most extensive molecular dataset compiled to date.

Composition: *Pleurodira* is currently believed to contain 93 extant species (Turtle Taxonomy Working Group, 2017). Numerous extinct taxa are known from the fossil record (e.g., Gaffney et al., 2006, 2011; Maniel and de la Fuente, 2016). While the oldest remains referable to *Pan-Pleurodira* (this volume) are dated Late Jurassic (Oxfordian), the oldest fossils referable to crown *Pleurodira* are Early Cretaceous (Barremian) (de la Fuente and Iturralde-Vinent, 2001; Joyce et al., 2013; Romano et al., 2014).

Diagnostic Apomorphies: *Pleurodira* can be diagnosed relative to other testudine crown clades by the absence of a contact of the prefrontal with the vomer and palatine, presence of a solid posterior orbit wall due to the contact of the postorbital with the palatine, wide closure of the incisura stapes, loss of the epipterygoids, presence of a processus trochlearis pterygoidei, presence of formed central articulations between the cervical vertebrae combined with closely placed zygapophyses that allow for extreme lateral movements of the neck, presence of a well-developed medial anal notch between the xiphiplastra, medial fusion of the gulars, loss of the inframarginals, and sutural articulation of the pelvis with the shell (Gaffney and Meylan, 1988; Gaffney et al., 2007; Joyce, 2007).

Synonyms: *Eupleurodira* Gaffney and Meylan 1988 (approximate). Joyce et al. (2004) provided an exhaustive list of other infrequently used historical synonyms.

Comments: Gaffney (1988) and Gaffney and Meylan (1988) provided the first comprehensive morphology-based phylogenetic trees for *Pleurodira*, but these were compiled from sub-trees rather than analysis of comprehensive taxon/character datasets. The phylogeny of pleurodires as a whole was only recently investigated in global turtle phylogenies that sampled nearly every recognized living turtle taxon (Thomson and Shaffer, 2010; Pereira et al., 2017). For the purposes of applying our definition, we use the topology presented in figure 1 of Pereira et al. (2017) because it is based on the most extensive molecular dataset compiled to date.

The name *Pleurodira* has been applied variously to groups ranging from the total clade to the crown clade of side-necked turtles (Joyce et al., 2004). For consistency, we follow the recommendations of Gauthier and de Queiroz (2001) by associating the most commonly used name with the crown clade and by forming the name of the total clade through the addition of the prefix "*Pan-*".

Literature Cited

Bonnaterre, P. J. 1789. *Tableau Encyclopédique et Méthodique des Trois Règnes de la Nature. Erpétologie.* Panckoucke, Paris.

Cope, E. D. 1865. Third contribution to the herpetology of tropical America. *Proc. Acad. Nat. Sci. Phila.* 1865:185–198.

de la Fuente, M. S., and M. Iturralde-Vinent. 2001. A new pleurodiran turtle from the Jagua Formation (Oxfordian) of Western Cuba. *J. Paleontol.* 75:860–869.

Gaffney, E. S. 1988. A cladogram of the pleurodiran turtles. *Acta Zool. Crac.* 31:487–492.

Gaffney, E. S., and P. A. Meylan. 1988. A phylogeny of turtles. Pp. 157–219 in *The Phylogeny and Classification of the Tetrapods. Vol. 1: Amphibians, Reptiles, Birds* (M. J. Benton, ed.). Clarendon Press, Oxford.

Gaffney, E. S., P. A. Meylan, R. C. Wood, E. Simons, and D. de Almeida Campos. 2011. Evolution of the side-necked turtles: the family *Podocnemididae. Bull. Am. Mus. Nat. Hist.* 350:1–237.

Gaffney, E. S., T. H. Rich, P. Vickers-Rich, A. Constantine, R. Vacca, and L. Kool. 2007. *Chubutemys*, a new eucryptodiran turtle from the Early Cretaceous of Argentina, and the relationships of the *Meiolaniidae. Am. Mus. Novit.* 3599:1–35.

Gaffney, E. S., H. Tong, and P. A. Meylan. 2006. Evolution of the side-necked turtles: the families *Bothremydidae, Euraxemydidae,* and *Araripemydidae. Bull. Am. Mus. Nat. Hist.* 300:1–698.

Gauthier, J., and K. de Queiroz. 2001. Feathered dinosaurs, flying dinosaurs, crown dinosaurs, and the name "*Aves*". Pp. 7–41 in *New Perspectives on the Origin and Early Evolution of Birds: Proceedings of the International Symposium in Honor of John H. Ostrom* (J. Gauthier and L. F. Gall, eds.). Peabody Museum of Natural History, New Haven, CT.

Joyce, W. G. 2007. Phylogenetic relationships of Mesozoic turtles. *Bull. Peabody Mus. Nat. Hist.* 48:3–102.

Joyce, W. G., J. F. Parham, and J. A. Gauthier. 2004. Developing a protocol for the conversion of rank-based taxon names to phylogenetically defined clade names, as exemplified by turtles. *J. Paleontol.* 78:989–1013.

Joyce, W. G., J. F. Parham, T. R. Lyson, R. C. M. Warnock, and P. C. J. Donoghue. 2013. A divergence dating analysis of turtles using fossil calibrations: an example of best practices. *J. Paleontol.* 87:612–634.

Maniel, I. J., and M. S. de la Fuente. 2016. A review of the fossil record of turtles of the clade *Pan-Chelidae. Bull. Peabody Mus. Nat. Hist.* 57:191–227.

Pereira, A. G., J. Sterli, F. R. R. Moreira, and C. G. Schrago. 2017. Multilocus phylogeny and statistical biogeography clarify the evolutionary history of major lineages of turtles. *Mol. Phylogenet. Evol.* 113:59–66.

Romano, P. S. R., V. Gallo, R. R. C. Ramos, and L. Antonioli. 2014. *Atolchelys lepida*, a new side-necked turtle from the Early Cretaceous of Brazil and the age of crown *Pleurodira. Biol. Lett.* 10:20140290.

Schneider, J. G. 1783. *Allgemeine Naturgeschichte der Schildkröten, Nebst Einem Systematischen Verzeichnisse der Einzelnen Arten und Zwey Kupfern.* Johann Gotfried Müllersche Buchhandlung, Leipzig.

Schweigger, A. F. 1812. Prodromus monographiae Cheloniorum, Part 1. *Königsberger Arch. Naturwiss. Math.* 1812:271–458.

Thomson, R. C., and H. B. Shaffer. 2010. Sparse supermatrices for phylogenetic inference: taxonomy, alignment, rogue taxa, and the phylogeny of living turtles. *Syst. Biol.* 59:42–58.

Turtle Taxonomy Working Group (Rhodin, A. G. J., J. B. Iverson, U. Fritz, A. Georges, H. B. Shaffer, and P. P. van Dijk). 2017. Turtles of the world: annotated checklist and atlas of taxonomy, synonymy, distribution, and conservation status (8th edition). *Chelonian Res. Monogr.* 7:1–292.

Authors

International Turtle Nomenclature Committee followed by the author sequence for this chapter.

Walter G. Joyce; Departement für Geowissenschaften; Universität Freiburg; 1700 Fribourg, Switzerland. Email: walter.joyce@unifr.ch.

James F. Parham; Department of Geological Sciences; California State University; Fullerton, CA 92834, USA. Email: jparham@fullerton.edu.

Jérémy Anquetin; Jurassica Museum; 2900 Porrentruy, Switzerland. Email: jeremy.anquetin@jurassica.ch.

Julien Claude; Institut des Sciences de l'Evolution de Montpellier; Université de Montpellier, CNRS, IRD, EPHE; 34095 Montpellier, France. Email: julien.claude@umontpellier.fr.

Igor G. Danilov; Department of Herpetology; Zoological Institute of Russian Academy of Sciences; 199034 St. Petersburg, Russia. Email: turtle@zin.ru.

John B. Iverson; Department of Biology; Earlham College; Richmond, IN 47374, USA. Email: johni@earlham.edu.

Benjamin P. Kear; Museum of Evolution; Uppsala University; 752 36 Uppsala, Sweden. Email: benjamin.kear@em.uu.se.

Tyler R. Lyson; Department of Earth Sciences; Denver Museum of Nature and Science; Denver, CO 80205, USA. Email: tyler.lyson@dmns.org.

Márton Rabi; Zentralmagazin Naturwissenschaftlicher Sammlungen; Martin-Luther-Universität Halle-Wittenberg; 06108 Halle (Saale), Germany. Email: iszkenderun@gmail.com.

Juliana Sterli; CONICET – Museo Egidio Feruglio; 9100 Trelew, Chubut, Argentina. Email: jsterli@mef.org.ar.

Date Accepted: 16 October 2017

Primary Editor: Jacques A. Gauthier

Pan-Cryptodira W. G. Joyce, J. F. Parham, and J. A. Gauthier 2004
[W. G. Joyce, J. F. Parham, J. Anquetin, J. Claude, I. G. Danilov, J. B. Iverson, B. Kear, T. R. Lyson, M. Rabi, and J. Sterli], converted clade name

Registration Number: 277

Definition: The total clade of crown clade *Cryptodira*. This is a crown-based total-clade definition. Abbreviated definition: total ∇ of *Cryptodira*.

Etymology: *Pan-*, referring to the total clade or pan-monophylum, plus *Cryptodira* (see *Cryptodira* in this volume for etymology).

Reference Phylogeny: For the purposes of applying our definition, we use the topology presented in Figure 8 of Joyce et al. (2016), because this is the most recent global phylogenetic analysis that highlights numerous fossil taxa as being stem cryptodires.

Composition: *Cryptodira* (this volume) and all extinct turtles that share more recent common ancestry with *Cryptodira* rather than with *Pleurodira* (this volume). Extinct clades likely placed within *Pan-Cryptodira* include stem taxa such as *Thalassochelydia* and *Xinjiangchelyidae* (e.g., Gaffney, 1996; Hirayama et al., 2000; Gaffney et al., 2007; Joyce, 2007; Rabi et al., 2014).

Diagnostic Apomorphies: See *Cryptodira* (this volume).

Synonyms: *Cryptodira* Cope 1868 (approximate); *Cryptodiromorpha* Lee 1995 (unambiguous).

Comments: The name *Cryptodira* is an approximate synonym of *Pan-Cryptodira* and has been used to denote this total clade (e.g., Gaffney et al., 2006). However, this is rejected here to retain the most commonly used names for crown clades. *Cryptodiromorpha* is an unambiguous synonym of *Pan-Cryptodira*, but we follow Joyce et al. (2004) in using the prefix "*Pan-*" for total clades to promote an integrated system of nomenclature (Gauthier and de Queiroz, 2001; de Queiroz, 2007).

Literature Cited

Cope, E. D. 1868. On the origin of genera. *Proc. Acad. Nat. Sci. Phila.* 1868:242–300.

Gaffney, E. S. 1996. The postcranial morphology of *Meiolania platyceps* and a review of the Meiolaniidae. *Bull. Am. Mus. Nat. Hist.* 229:1–166.

Gaffney, E. S., T. H. Rich, P. Vickers-Rich, A. Constantine, R. Vacca, and L. Kool. 2007. *Chubutemys*, a new eucryptodiran turtle from the Early Cretaceous of Argentina, and the relationships of the Meiolaniidae. *Am. Mus. Novit.* 3599:1–35.

Gaffney, E. S., H. Tong, and P. A. Meylan. 2006. Evolution of the side-necked turtles: the families Bothremydidae, Euraxemydidae, and Araripemydidae. *Bull. Am. Mus. Nat. Hist.* 300:1–698.

Gauthier, J., and K. de Queiroz. 2001. Feathered dinosaurs, flying dinosaurs, crown dinosaurs, and the name "*Aves*". Pp. 7–41 in *New Perspectives on the Origin and Early Evolution of Birds: Proceedings of the International Symposium in Honor of John H. Ostrom* (J. Gauthier and L. F. Gall, eds.). Peabody Museum of Natural History, Yale University, New Haven, CT.

Hirayama, R., D. B. Brinkman, and I. G. Danilov. 2000. Distribution and biogeography of nonmarine Cretaceous turtles. *Russ. J. Herpetol.* 7:181–198.

Joyce, W. G. 2007. Phylogenetic relationships of Mesozoic turtles. *Bull. Peabody Mus. Nat. Hist.* 48:3–102.

Joyce, W. G., J. F. Parham, and J. A. Gauthier. 2004. Developing a protocol for the conversion of rank-based taxon names to phylogenetically defined clade names, as exemplified by turtles. *J. Paleontol.* 78:989–1013.

Joyce, W. G., M. Rabi, J. M. Clark, and X. Xu. 2016. A toothed turtle from the Late Jurassic of China and the global biogeographic history of turtles. *BMC Evol. Biol.* 16:236.

Lee, M. S. Y. 1995. Historical burden in systematics and the interrelationships of "parareptiles." *Biol. Rev.* 70:459–547.

de Queiroz, K. 2007. Toward an integrated system of clade names. *Syst. Biol.* 56:956–974.

Rabi, M., V. Sukhanov, V. Egerova, I. G. Danilov, and W. G. Joyce. 2014. Osteology, relationships, and ecology of *Annemys* (*Testudines, Eucryptodira*) from the Late Jurassic of Shar Teg, Mongolia, and phylogenetic definitions for *Xinjiangchelyidae*, *Sinemydidae*, and *Macrobaenidae*. *J. Vertebr. Paleontol.* 34:327–352.

Authors

International Turtle Nomenclature Committee.

Walter G. Joyce; Departement für Geowissenschaften; Universität Freiburg; 1700 Fribourg, Switzerland. Email: walter.joyce@unifr.ch.

James F. Parham; Department of Geological Sciences; California State University; Fullerton, CA 92834, USA. Email: jparham@fullerton.edu.

Jérémy Anquetin; Jurassica Museum; 2900 Porrentruy, Switzerland. Email: jeremy.anquetin@jurassica.ch.

Julien Claude; Institut des Sciences de l'Evolution de Montpellier; Université de Montpellier, CNRS, IRD, EPHE; 34095 Montpellier, France. Email: julien.claude@umontpellier.fr.

Igor G. Danilov; Department of Herpetology; Zoological Institute of Russian Academy of Sciences; 199034 St. Petersburg, Russia. Email: turtle@zin.ru.

John B. Iverson; Department of Biology; Earlham College; Richmond, IN 47374, USA. Email: johni@earlham.edu.

Benjamin P. Kear; Museum of Evolution; Uppsala University; 752 36 Uppsala, Sweden. Email: benjamin.kear@em.uu.se.

Tyler R. Lyson; Department of Earth Sciences; Denver Museum of Nature and Science; Denver, CO 80205, USA. Email: tyler.lyson@dmns.org.

Márton Rabi; Zentralmagazin Naturwissenschaftlicher Sammlungen; Martin-Luther-Universität Halle-Wittenberg; 06108 Halle (Saale), Germany. Email: iszkenderun@gmail.com.

Juliana Sterli; CONICET – Museo Egidio Feruglio; 9100 Trelew, Chubut, Argentina. Email: jsterli@mef.org.ar.

Date Accepted: 16 October 2017

Primary editor: Jacques A. Gauthier

Cryptodira E. D. Cope 1868 [W. G. Joyce, J. F. Parham, J. Anquetin, J. Claude, I. G. Danilov, J. B. Iverson, B. Kear, T. R. Lyson, M. Rabi, and J. Sterli], converted clade name

Registration Number: 278

Definition: The smallest crown clade containing the testudinoid *Testudo graeca* Linnaeus 1758, the chelonioid *Chelonia* (originally *Testudo*) *mydas* (Linnaeus 1758), the trionychian *Trionyx* (originally *Testudo*) *triunguis* (Forskål 1775), the kinosternoid *Kinosternon* (originally *Testudo*) *scorpioides* (Linnaeus 1766), and the chelydrid *Chelydra* (originally *Testudo*) *serpentina* (Linnaeus 1758). This is a minimum-crown-clade definition. Abbreviated definition: min crown ∇ (*Testudo graeca* Linnaeus 1758 & *Chelonia mydas* (Linnaeus 1758) & *Trionyx triunguis* (Forskål 1775) & *Kinosternon scorpioides* (Linnaeus 1766) & *Chelydra serpentina* (Linnaeus 1758).

Etymology: Derived from the Greek κρυπτω (*kryptos*) for 'hidden' and δειρη (*deire*) for 'neck' or 'throat' in reference to the neck being hidden from view when the head is retracted into the shell.

Reference Phylogeny: We designate the topology presented in Figure 1 of Pereira et al. (2017) as our reference phylogeny.

Composition: *Cryptodira* is currently believed to contain approximately 263 extant species (Turtle Taxonomy Working Group, 2017). Many fossil taxa traditionally considered 'cryptodires' (e.g., *Xinjiangchelyidae, Thalassochelydia*) are now placed outside of the crown clade and are thus members of *Pan-Cryptodira* (this volume). A number of even more basally branching taxa once thought to be cryptodires (e.g., *Kayentachelys aprix, Helochelydridae, Meiolaniformes, Sichuanchelyidae*) are now recognized as stem testudines (e.g., Rougier et al., 1995; Joyce, 2007; Sterli, 2008; Anquetin, 2012; Joyce et al., 2014, 2016; Zhou and Rabi, 2015). Current phylogenies suggest that Middle Jurassic xinjiangchelyids are the oldest known pan-cryptodires and that Late Jurassic adocusians are the oldest known crown cryptodires (Zhou and Rabi, 2015; Joyce et al., 2016).

Diagnostic Apomorphies: A number of characters that were formerly thought to be distinct apomorphies of *Cryptodira*, such as the development of a trochlear process on the otic capsule or a contact of the pterygoid with the basisphenoid (Gaffney and Meylan, 1988), may actually diagnose a more inclusive clade (Joyce, 2007). Nevertheless, cryptodires are unique among vertebrates in exhibiting a series of modifications enabling extreme retraction of the neck within the shell. These include the formation of formed cervical central articulations, which are often biconvex, the presence of strongly developed ventral processes on the posterior cervicals, and the shortening of the eighth cervical centrum relative to the seventh cervical centrum (Gaffney and Meylan, 1988; Gaffney et al., 2007; Joyce, 2007).

Synonyms: *Polycryptodira* Gaffney and Meylan 1988 (approximate); see Joyce et al. (2004) for other synonyms.

Comments: Gaffney and Meylan (1988) produced the most comprehensive morphological phylogeny of *Cryptodira*, but their topology was based on grafted sub-trees. Numerous other densely sampled morphology-based phylogenies have since been produced (e.g., Gaffney et al., 2007; Joyce, 2007; Sterli et al., 2013; Joyce et al., 2016; Anquetin, 2012; Zhou and Rabi, 2015), but these conflict with the emerging molecular consensus (Shaffer et al., 1997; Fujita et al., 2004; Krenz et al., 2005; Thomson and Shaffer, 2010; Crawford et al., 2015; Pereira et al., 2017) in regards to the basal arrangement of the primary subclades. We here use the molecular phylogeny provided in Figure 1 of Pereira et al. (2017) as our reference phylogeny as it was based on the most extensive molecular dataset compiled to date.

The name *Cryptodira* has been consistently applied to groups that range from the total clade to the crown clade of hidden-necked turtles (Joyce et al., 2004). For consistency, we follow the recommendations of Gauthier and de Queiroz (2001; see also de Queiroz, 2007) by associating the most commonly used name with the crown clade and by forming the name of the total clade through the addition of the prefix "*Pan-*".

Literature Cited

Anquetin, J. 2012. Reassessment of the phylogenetic interrelationships of basal turtles (*Testudinata*). *J. Syst. Palaeontol.* 10:3–45.

Cope, E. D. 1868. On the origin of genera. *Proc. Acad. Nat. Sci. Phila.* 1868:242–300.

Crawford, N. G., J. F. Parham, A. B. Sellas, B. C. Faircloth, T. C. Glenn, T. J. Papenfuss, J. B. Henderson, M. H. Hansen, and W. B. Simison. 2015. A phylogenomic analysis of turtles. *Mol. Phylogenet. Evol.* 83:250–257.

Forskål, P. 1775. *Descriptiones Animalium Avium, Amphiborum, Piscium, Insectorum, Vermium; Quae in Itinere Orientali Observavit.* Möllerus, Haunia.

Fujita, M. K., T. N. Engstrom, D. E. Starkey, and H. B. Shaffer. 2004. Turtle phylogeny: insights from a novel nuclear intron. *Mol. Phylogenet. Evol.* 31:1031–1040.

Gaffney, E. S., and P. A. Meylan. 1988. A phylogeny of turtles. Pp. 157–219 in *The Phylogeny and Classification of the Tetrapods. Vol. 1: Amphibians, Reptiles, Birds* (M. J. Benton, ed.). Clarendon Press, Oxford.

Gaffney, E. S., T. H. Rich, P. Vickers-Rich, A. Constantine, R. Vacca, and L. Kool. 2007. *Chubutemys*, a new eucryptodiran turtle from the Early Cretaceous of Argentina, and the relationships of the *Meiolaniidae*. *Am. Mus. Novit.* 3599:1–35.

Gauthier, J., and K. de Queiroz. 2001. Feathered dinosaurs, flying dinosaurs, crown dinosaurs, and the name "*Aves*". Pp. 7–41 in *New Perspectives on the Origin and Early Evolution of Birds: Proceedings of the International Symposium in Honor of John H. Ostrom* (J. Gauthier and L. F. Gall, eds.). Peabody Museum of Natural History, Yale University, New Haven, CT.

Joyce, W. G. 2007. Phylogenetic relationships of Mesozoic turtles. *Bull. Peabody Mus. Nat. Hist.* 48:3–102.

Joyce, W. G., J. F. Parham, and J. A. Gauthier. 2004. Developing a protocol for the conversion of rank-based taxon names to phylogenetically defined clade names, as exemplified by turtles. *J. Paleontol.* 78:989–1013.

Joyce, W. G., M. Rabi, J. M. Clark, and X. Xu. 2016. A toothed turtle from the Late Jurassic of China and the global biogeographic history of turtles. *BMC Evol. Biol.* 16:236.

Joyce, W. G., J. Sterli, and S. D. Chapman. 2014. The skeletal morphology of the solemydid turtle *Naomichelys speciosa* from the Early Cretaceous of Texas. *J. Paleontol.* 88:1257–1287.

Krenz, J. G., G. J. P. Naylor, H. B. Shaffer, and F. J. Janzen. 2005. Molecular phylogenetics and evolution of turtles. *Mol. Phylogenet. Evol.* 37:178–191.

Linnaeus, C. 1758. *Systema Naturae Per Regna Tria Naturae, Secundum Classes, Ordines, Genera, Species, cum Characteribus, Differentiis, Synonymis, Locis.* 10th edition, Vol. 1. Laurentii Salvii, Holmiae (Stockholm).

Linnaeus, C. 1758. *Systema Naturae Per Regna Tria Naturae, Secundum Classes, Ordines, Genera, Species, cum Characteribus, Differentiis,*

Synonymis, Locis. 12th edition, Vol. 1. Laurentii Salvii, Holmiae (Stockholm).

Pereira, A. G., J. Sterli, F. R. R. Moreira, and C. G. Schrago. 2017. Multilocus phylogeny and statistical biogeography clarify the evolutionary history of major lineages of turtles. *Mol. Phylogenet. Evol.* 113:59–66.

de Queiroz, K. 2007. Toward an integrated system of clade names. *Syst. Biol.* 56:956–974.

Rougier, G. W., M. S. Fuente, and A. B. Arcucci. 1995. Late Triassic turtles from South America. *Science* 268:855–858.

Shaffer, H. B., P. Meylan, and M. L. McKnight. 1997. Tests of turtle phylogeny: molecular, morphological, and paleontological approaches. *Syst. Biol.* 46:235–268.

Sterli, J. 2008. A new, nearly complete stem turtle from the Jurassic of South America with implications for turtle evolution. *Biol. Lett.* 4:286–289.

Sterli, J., M. S. de la Fuente, and I. A. Cerda. 2013. A new species of meiolaniform turtle and a revision of the Late Cretaceous *Meiolaniformes* of South America. *Ameghiniana* 50:240–256.

Thomson, R. C., and H. B. Shaffer. 2010. Sparse supermatrices for phylogenetic inference: taxonomy, alignment, rogue taxa, and the phylogeny of living turtles. *Syst. Biol.* 59:42–58.

Turtle Taxonomy Working Group (Rhodin, A. G. J., J. B. Iverson, U. Fritz, A. Georges, H. B. Shaffer, and P. P. van Dijk). 2017. Turtles of the world: annotated checklist and atlas of taxonomy, synonymy, distribution, and conservation status (8th edition). *Chelonian Res. Monogr.* 7:1–292.

Zhou, C.-F., and M. Rabi. 2015. A sinemydid turtle from the Jehol Biota provides insights into the basal divergence of crown turtles. *Sci. Rep.* 5:16299.

Authors

International Turtle Nomenclature Committee.

Walter G. Joyce; Departement für Geowissenschaften; Universität Freiburg; 1700 Fribourg, Switzerland. Email: walter.joyce@unifr.ch.

James F. Parham; Department of Geological Sciences; California State University; Fullerton, CA 92834, USA. Email: jparham@fullerton.edu.

Jérémy Anquetin; Jurassica Museum; 2900 Porrentruy, Switzerland. Email: jeremy.anquetin@jurassica.ch.

Julien Claude; Institut des Sciences de l'Evolution de Montpellier; Université de Montpellier, CNRS, IRD, EPHE; 34095 Montpellier, France. Email: julien.claude@umontpellier.fr.

Igor G. Danilov; Department of Herpetology; Zoological Institute of Russian Academy of Sciences; 199034 St. Petersburg, Russia. Email: turtle@zin.ru.

John B. Iverson; Department of Biology; Earlham College; Richmond, IN 47374, USA. Email: johni@earlham.edu.

Benjamin P. Kear; Museum of Evolution; Uppsala University; 752 36 Uppsala, Sweden. Email: benjamin.kear@em.uu.se.

Tyler R. Lyson; Department of Earth Sciences; Denver Museum of Nature and Science; Denver, CO 80205, USA. Email: tyler.lyson@dmns.org.

Márton Rabi; Zentralmagazin Naturwissenschaftlicher Sammlungen; Martin-Luther-Universität Halle-Wittenberg; 06108 Halle (Saale), Germany. Email: iszkenderun@gmail.com.

Juliana Sterli; CONICET – Museo Egidio Feruglio; 9100 Trelew, Chubut, Argentina. Email: jsterli@mef.org.ar.

Date Accepted: 16 October 2017

Primary editor: Jacques A. Gauthier

Sauria J. Macartney 1802 [J. A. Gauthier and K. de Queiroz], converted clade name

Registration Number: 114

Definition: The smallest crown clade containing *Alligator* (originally *Crocdilus*) *mississippiensis* (Daudin 1802) (*Crocodylia*), *Sphenodon* (originally *Hatteria*) *punctatus* (Gray 1842), and *Lacerta agilis* Linnaeus 1758 (*Squamata*), but neither *Testudo graeca* Linnaeus 1758 (*Testudines*) nor *Homo sapiens* Linnaeus 1758 (*Mammalia*). This is a minimum-crown-clade definition with external specifiers. Abbreviated definition: min crown ∇ (*Alligator mississippiensis* (Daudin 1802) & *Sphenodon punctatus* (Gray 1842) & *Lacerta agilis* Linnaeus 1758 ~ *Testudo graeca* Linnaeus 1758 ∨ *Homo sapiens* Linnaeus 1758).

Etymology: Derived from the Greek *sauros* ("lizard").

Reference Phylogeny: The primary reference phylogeny is Figure 3 in Gauthier et al. (1988b), in which *Alligator mississippiensis* is part of *Pseudosuchia* (and *Archosauria*), *Lacerta agilis* and *Sphenodon punctatus* are in *Lepidosauromorpha* (= *Pan-Lepidosauria*), *Testudo graeca* is part of *Testudines*, and *Homo sapiens* is in *Mammalia*. See also Ezcurra et al. (2014: Fig. 1), Pritchard and Nesbitt (2017: Fig. 9), and Simões et al. (2018: Fig. 2).

Composition: *Pan-Archosauria* and *Pan-Lepidosauria*, which include the extant crocodilians, birds, tuatara, and lizards (including snakes) (see entries for *Pan-Archosauria*, *Archosauria*, *Pan-Lepidosauria*, and *Lepidosauria* in this volume for details concerning included taxa). *Sauria* is represented in the extant biota by more than 20,000 currently recognized species.

Diagnostic Apomorphies: Gauthier et al. (1988b) listed more than 35 apomorphies for *Sauria* relative to other crown amniotes. Comparison to stem taxa shortens that list considerably (e.g., Gauthier, 1994; Dilkes, 1998; Gottmann-Quesada and Sander, 2009; Ezcurra et al., 2014; Pritchard and Nesbitt, 2017; Simões et al., 2018; Li et al., 2018), partly from scant knowledge of soft anatomy in fossils, but also because of incomplete knowledge of hard anatomy in stem saurians that are closer to the crown than is *Petrolacosaurus kansensis* (*Araeoscelidia*) (e.g., Reisz, 1981). However, the primary challenge to diagnosing *Sauria* is that it remains unclear exactly which Permo-Triassic diapsids, most especially the highly modified aquatic forms (e.g., *Ichthyopterygia*), lie inside or outside of the crown (see Comments).

Relative to *Petrolacosaurus kansensis*, (crown) saurians can be diagnosed by possessing the following "hard" apomorphies that can be ascertained in fossils (see references above): (1) reduced anterior extent of lacrimal bone, accompanied by an enlarged maxilla facial process contacting the nasal bone, excluding lacrimal from naris; (2) loss of caniniform maxillary teeth; (3) origin of temporal adductor muscles spreads onto dorsal surface of lateral edge of parietal table; (4) reduction of supratemporal bone in post-temporal arch and concomitant increase in size of parietal supratemporal process to closely approach the squamosal; (5) loss of tabular bone in skull; (6) quadrate exposed behind squamosal in lateral view; (7) quadrate head hemispherical and received in fossa beneath squamosal; (8) an impedance-matching ear, including a tympanum, a quadrate shaft bowed anteriorly and jaw-opening muscles displaced posteriorly onto a prominent mandibular retroarticular process, to facilitate passage of middle ear cavity traversed by a slender stapes that enables transmission of high frequency, low-intensity airborne sounds from tympanum to inner ear; (9) loss of posterior coracoid bone in endochondral shoulder girdle (= scapulocoracoid cartilage); (10) loss of cleithrum in dermal shoulder girdle; (11) larger pelvic muscles indicated by an enlarged, subtriangular iliac blade with a horizontally oriented dorsal margin; (12) hindlimbs much longer than forelimbs (femur > 40% longer than humerus); (13) distal femoral condyles in approximately the same plane; (14) short and broad-based 5th metatarsal.

Exactly which, if any, of these apomorphies is diagnostic of the crown relative to all known (or suspected) diapsids that are closer to the crown than is *Petrolacosaurus kansensis* remains unclear.

Synonyms: *Diapsida* Osborn 1903 (approximate, partial); *Diaptosauria* Osborn 1903 (approximate, partial), *Eosuchia* Broom 1914 (approximate, partial); *Neodiapsida* Benton 1985 (approximate).

Comments: *Sauria* has been a conspicuous component of terrestrial ecosystems since the dawn of the Mesozoic. It has long been a globally distributed clade, the members of which crawl, run, climb, and slither through disparate habitats on land (and burrow in it), swim in freshwaters and on the high seas, and twice took to the skies in powered flight. It has suffered mass extinctions but re-radiated in their wake, and is still represented by at least 20,000 living species (nearly all of which are either birds or lizards, including snakes). Its members span several orders of magnitude in size—from tiny bee hummingbirds (i.e., *Mellisuga hellenae*) to enormous sauropods (e.g., *Argentinosaurus huinculensis*)—and are just as spectacularly disparate ecologically and morphologically— from head-first burrowing threadsnakes (e.g.,

Rena humilis), to amphibious crocodilians (e.g., *Gavialis gangeticus*), to arboreal chameleons (e.g., *Trioceros jacksonii*), to birds that fly in the air (e.g., *Upupa epops*) or underwater (e.g., *Aptenodytes patagonicus*), or are bipedal cursors unable to fly at all (e.g., *Struthio camelus*), as well as more "lizard-like" forms, including the unusually cold-adapted, nocturnal tuatara (*Sphenodon punctatus*).

Several early naturalists recognized a group composed of four-legged, long-tailed, non-shelled, and (mostly) oviparous vertebrates, including "lizards" (non-serpentiform squamates) and crocodilians (e.g., Linnaeus, 1758; Laurenti, 1768; Blumenbach, 1779), although the scale-less salamanders were soon removed from the group (e.g., Brongniart, 1800; Cuvier, 1800). It was for such a group that Cuvier's (1800) "Les Sauriens" was Latinized to *Sauria* by Macartney (1802). *Crocodylia* was subsequently removed (e.g., Blainville, 1816, 1822; Merrem, 1820), and "*Sauria*" became associated with "lizards" alone (as indicated by standard English translations of the Ancient Greek term "saurian"). A close relationship between *Serpentes* and "lizards" was proposed by some of the same authors (e.g., Oppel, 1811; Blainville, 1816, 1822; Merrem, 1820), although the nesting of snakes deep within "lizards" was not recognized until relatively recently (e.g., Estes et al., 1988; see *Squamata*, this volume).

Osborn's (1903) *Diapsida*, which refers to the presence of two temporal arches (and fenestrae; see entry in this volume), largely supplanted *Sauria* as the favored name for a taxon that included only *Sphenodon*, *Squamata* and *Crocodylia* among extant "reptiles". Although Osborn (1903) acknowledged an evolutionary relationship between birds and his "diapsid reptiles", *Aves* was not explicitly included in *Diapsida* until much later (e.g., Gauthier, 1986), in the context of a revised concept of monophyly (Hennig, 1966). *Youngina capensis* from the Late Permian of South Africa lent key early support to Osborn's concept of "*Diapsida*". Broom (1914) relied on this very early and relatively unmodified two-arched species as the basis of his "*Eosuchia*", a taxon that he conceptualized as the ancestral group from which all other "diapsid reptiles" would later emerge (see also Romer, 1966).

Growing knowledge of the Late Pennsylvanian *Petrolacosaurus kansensis* (Lane, 1945, 1946; Peabody, 1952; Reisz, 1977, 1981), most particularly that it possessed a two-arched skull, led palaeontologists to apply the name *Diapsida* to a more inclusive clade originating in the Carboniferous (see *Diapsida*, this volume). That left the crown (which emerged much later in the Permian) without a name, so Gauthier et al. (1988a) proposed using *Sauria* for that clade. *Neodiapsida* (Benton, 1985) is sometimes considered a synonym of *Sauria* (e.g., Evans, 1988). However, Benton did not define the name explicitly in terms of ancestry. Thus, when *Younginiformes*, one of the extinct groups he included in *Neodiapsida*, was later inferred to be outside of the crown (e.g., Laurin, 1991), *Neodiapsida* became associated with clade(s) more inclusive than the crown (e.g., Senter, 2004; Reisz et al., 2011). Like Benton (1985), Gauthier et al. (1988a) also regarded younginiforms to be part of *Sauria*, as they were thought to represent the deepest divergence on the lepidosaur stem at the time (see also Benton, 1982, 1983; Gauthier, 1984; Evans, 1988). But because Gauthier et al. (1988a) explicitly defined *Sauria* as applying to the crown, when younginiforms were inferred to be outside of that clade, they were accordingly excluded from it by subsequent workers (e.g., Laurin, 1991; Clark et al., 1993; Gauthier, 1994; deBraga and Rieppel, 1997; Müller, 2004; Senter, 2004; Ezcurra et al., 2014; Nesbitt et al., 2015; Pritchard and Nesbitt, 2017; Pritchard et al., 2018). Although younginiforms are now widely thought to be near-crown stem saurians, it is not entirely clear that they are all related to one another (see, e.g., Müller [2004] and Bickelmann et al. [2009] vs. Gauthier et al. [1988a] and Pritchard and Nesbitt [2017]).

The highly modified drepanosaurs, a radiation of chameleon-like arboreal diapsids with sharp-snouted skulls, have had a more checkered taxonomic history (summarized in Pritchard and Nesbitt, 2017). They were often thought to be early diverging stem archosaurs (e.g., Evans, 1988; Gauthier, 1994; Renesto, 1994; Merck, 1997; Dilkes, 1998; Renesto et al., 2010; and see *Pan-Archosauria*, this volume). But sometimes they were inferred to be outside of crown *Sauria* (e.g., Senter, 2004; Müller, 2004), albeit in variable positions on its stem. Pritchard and Nesbitt's (2017) comprehensive study firmly placed Late Triassic drepanosaurs deep on the saurian stem, with only the Early Permian *Orovenator mayorum* and the Late Pennsylvanian *Araeoscelidia* being more basally branching among diapsids. This requires a long ghost lineage in keeping with their many morphological modifications. According to Pritchard and Nesbitt (2017), drepanosaurs show no trace of the apomorphies associated with the saurian ear, nor do they possess the peg-in-socket quadrate-squamosal articulation of the crown (Gauthier et al., 2012).

The situation is murkier for diapsids that are highly modified for life underwater: the marine *Sauropterygia*, *Ichthyopterygia*, and *Thalattosauria*, and the freshwater *Choristodera*. They must have diverged in the Palaeozoic based on their positions on most trees of early saurians (e.g., Gauthier, 1994; Simões et al., 2018). But even the earliest examples of the marine clades, which first appear in the Early Triassic, already display significant adaptations to marine environments (e.g., Nakajima et al., 2014). Choristoderans are not known with certainty before the Middle Jurassic (Gao et al., 2013). They have variously been allied to either lepidosaurs or archosaurs, and thus inside the crown (e.g., Gauthier et al., 1988b; Rieppel, 1993; Gauthier, 1994; Merck, 1997). The most recent analysis to consider their relationships among diapsids is that of Simões et al. (2018). Their preferred relaxed-clock Bayesian tree based on combined morphological and molecular data inferred strong support for Merck's (1997) marine clade *Euryapsida* Colbert 1945 (= thalattosaurs, sauropterygians, and ichthyosaurs). Euryapsids and choristoderans formed a polytomy just below a weakly supported crown node, so it is not yet clear if they are crown saurians.

Turtles (*Pan-Testudines*, see entry this volume) present yet another challenge. Crown turtles (*Testudines*, see entry this volume) possess an impedance-matching auditory system reminiscent of that in saurians (and some parareptiles; Lyson et al., 2010). But they appear to have acquired many of the associated apomorphies convergently, as most of them are absent in stem turtles (Bever et al., 2015). Nevertheless, even stem turtles have a saurian-like ear in some respects, such as, for example, in having a slenderer stapes that is more horizontally disposed, rather than being thicker, shorter, and more ventrolaterally oriented as in reptiles ancestrally (Gaffney, 1990). Absence of the full suite of saurian ear apomorphies is consistent with other evidence, such as retention of the cleithrum in the shoulder girdle (Lyson et al., 2013b), indicating that turtles lie outside of *Sauria*, all of whose members lack this bone (e.g., Lyson et al., 2010; Lyson et al., 2013a). Likewise, saurians have conspicuously long hindlimbs compared to those of even the earliest stem turtles, which retain the more even limb proportions of the earliest diapsid, *Petrolacosaurus kansensis* (Gauthier et al., 2011).

Nevertheless, there is growing evidence that turtles are diapsids (*contra* e.g., Gaffney, 1980; Gauthier, 1984; Gauthier et al., 1988a,b,c; Gauthier et al., 1989; Reisz and Laurin, 1991; Gauthier, 1994; Lee, 1996), at least in the sense of having descended from an ancestor with upper and lower temporal fenestrae (see *Diapsida*, this volume). For example, in addition to a lower temporal fenestra, there appears to be an upper temporal fenestra, albeit covered over secondarily by an enlarged supratemporal, in the earliest and least-modified stem turtle currently known, *Eunotosaurus africanus* (Bever et al., 2015). This fenestra is unique to diapsids among amniotes. That, coupled with weak morphological support linking them with lepidosaurs (e.g., deBraga and Rieppel, 1997; Bever et al., 2015), and strong molecular support allying them to archosaurs (e.g., Field et al., 2014; Crawford et al., 2015), suggests that turtles might well be crown diapsids (see also Rest et al. [2003], Hugall et al. [2007], and Crawford et al. [2012]; but see Frost et al. [2006], and especially Lu et al. [2013], for an alternative view). However, at least one of these inferences must be mistaken, as turtles cannot simultaneously be sister to archosaurs and sister to lepidosaurs. To further complicate matters, Simões et al. (2018) inferred moderate support from a combined morphological and molecular dataset for turtles being outside of, if near to, *Sauria* (see also Li et al., 2018).

Provided that turtles are not included, the name *Sauria* is the most appropriate name for the clade in question, which is convenient given that it is the root word from which the names of its primary subclades, *Archosauria* and *Lepidosauria,* are formed. It has been defined explicitly as applying to this clade (Gauthier et al., 1988a), and that practice has generally been followed in subsequent phylogenetic studies (e.g., Laurin, 1991; Clark et al., 1993; Gauthier, 1994; deBraga and Rieppel, 1997; Müller, 2004; Senter, 2004; Ezcurra et al., 2014; Nesbitt et al., 2015; Pritchard and Nesbitt, 2017; Pritchard et al., 2018). By contrast, the approximate synonyms *Diapsida* and *Neodiapsida* are commonly applied to more inclusive clades (e.g., Reisz et al., 2011). The name *Eosuchia* has also been defined as applying to a more inclusive clade (Laurin, 1991), and it has traditionally been applied to a group long understood to be a wastebasket of miscellaneous early diapsids without clear connections to either archosaurs or lepidosaurs (e.g., Broom, 1926; Kuhn, 1952; Romer, 1956, 1966; Tatarinov, 1964; Cruickshank, 1972; Gow, 1972, 1975; Carroll, 1976, 1978, 1981; Sigogneau-Russell and Russell, 1978; Olsen, 1978; Evans, 1980).

We here update the definition of *Sauria* proposed by Gauthier et al. (1988a) by using species as specifiers that are consistent with the composition of the taxon as originally circumscribed by Macartney (1802). Although not originally included in *Reptilia* (and thus not listed among our internal specifiers), birds (and snakes) are unambiguously part of this clade (see e.g., *Archosauria*, *Dinosauria* and *Aves* in this volume). We have included a turtle species as an external specifier in our definition so that the name *Sauria* will not apply to any clade in the context of phylogenies in which turtles are descended from the most recent common ancestor of lepidosaurs and archosaurs. In that case, *Reptilia* would be the name of the crown composed of all three clades. These clades appear to have diversified rapidly around 260 million years ago in the early Late Permian, based on the stem turtle *Eunotosaurus africanus* (Capitanian; Lyson et al., 2010) and the stem archosaur *Aenigmastropheus parringtoni* (Guadalupian; Ezcurra et al., 2014). However, Simões et al. (2018) estimated that the primary saurian divergence occurred much earlier, in the Carboniferous.

Literature Cited

Benton, M. J. 1982. The *Diapsida*: a revolution in reptile relationships. *Nature* 296:306–307.

Benton, M. J. 1983. The Triassic reptile *Hyperodapedon* from Elgin: functional morphology and relationships. *Philos. Trans. R. Soc. Lond. B Biol. Sci.* 302(B):605–717.

Benton, M. J. 1985. Classification and phylogeny of the diapsid reptiles. *Zool. J. Linn. Soc.* 84:97–164.

Bever, G. S., T. R. Lyson, D. J. Field, and B.-A. S. Bhullar. 2015. Evolutionary origin of the turtle skull. *Nature* 525:239–242.

Bickelmann, C., J. Müller, and R. R. Reisz. 2009. The enigmatic diapsid *Acerodontosaurus piveteaui* (*Reptilia*: *Neodiapsida*) from the Upper Permian of Madagascar and the paraphyly of "younginiform" reptiles. *Can. J. Earth Sci.* 46(9):651–661.

de Blainville, H.-M. D. 1816. Prodrome d'une nouvelle distribution systématique du règne animal. *Bull. Soc. Philom. Paris* 8:113–124.

de Blainville, H.-M. D. 1822. *De L'organisation des Animaux ou Principes D'anatomie Comparée*. F. G. Levrault, Paris.

Blumenbach, J. F. 1779. *Handbuch der Naturgeschichte. Mit Kupfern*. Johann Christian Dieterich, Göttingen.

Brongniart, A. 1800. Essai d'une classification naturelle des reptiles. 1ère Partie. Etablissement des ordres. *Bull. Soc. Philom. Paris* 2(35):81–82.

Broom, R. 1914. A new thecodont reptile. *Proc. Zool. Soc. Lond. B* 84(4):1072–1077.

Broom, R. 1926. On a nearly complete skeleton of a new eosuchian reptile (*Palaeagama vielhaueri*. gen. et sp. nov.). *Proc. Zool. Soc. Lond.* 1926: 487–492.

Carroll, R. L. 1976. Eosuchians and the origin of archosaurs. Pp. 58–79 in *Athlon Essays on Paleontology in Honour of Loris Shano Russell* (C. Churcher, ed.). Miscellaneous Publications of the Life Sciences Division of the Royal Ontario Museum, Toronto.

Carroll, R. L. 1978. Permo-Triassic "lizards" from the Karroo, Part 2. A gliding reptile from the Upper Permian of Madagascar. *Palaeontol. Afr.* 21:143–159.

Carroll, R. L. 1981. Plesiosaur ancestors from the Upper Permian of Madagascar. *Philos. Trans. R. Soc. Lond. B Biol. Sci.* 293:315–383.

Clark, J. M., J. Welman, J. A. Gauthier, and J. M. Parrish. 1993. The laterosphenoid bone of early archosauriforms. *J. Vertebr. Paleontol.* 13(1):48–57.

Colbert, E. H. 1945. *The Dinosaur Book.* American Museum of Natural History, New York.

Crawford, N. G., B. C. Faircloth. J. E. McCormack, R. T. Brumfield, K. Winker, and T. C. Glenn. 2012. More than 1000 ultraconserved elements provide evidence that turtles are the sister group of archosaurs. *Biol. Lett.* 8(5):783–786.

Crawford, N. G., J. F. Parham, A. B. Sellas, B. C. Faircloth, T. C. Glenn, T. J. Papenfuss, J. B. Henderson, M. H. Hansen, and W. B. Simison. 2015. A phylogenomic analysis of turtles. *Mol. Phylogenet. Evol.* 83:250–257.

Cruickshank, A. R. 1972. The proterosuchian thecodonts. Pp. 89–119 in *Studies in Vertebrate Evolution* (K. A. Joysey and T. S. Kemp, eds.). Oliver and Boyd, Edinburgh.

Cuvier, G. 1800. *Leçons D'anatomie Comparée,* Tome I. Baudouin, Paris.

Daudin, F. M. 1802. *Histoire Naturelle, Générale et Particulière, des Reptiles: Ouvrage Faisant Suite à l'istoire Naturelle Générale et Particulière, Composée par Leclerc de Buffon, et Rédigée par C.S. Sonnini,* Vol. 2. F. Dufart, Paris.

deBraga, M., and O. Rieppel. 1997. Reptile phylogeny and the interrelationships of turtles. *Zool. J. Linn. Soc.* 120(3):281–354.

Dilkes, D. W. 1998. The Early Triassic rhynchosaur *Mesosuchus browni* and the interrelationships of basal archosauromorph reptiles. *Philos. Trans. R. Soc. Lond. B Biol. Sci.* 353:501–541.

Estes, R., K. de Queiroz, and J. Gauthier. 1988. Phylogenetic relationships within *Squamata.* Pp. 119–281 in *Phylogenetic Relationships of the Lizard Families* (R. Estes and G. Pregill, eds.). Stanford University Press, Stanford, CA.

Evans, S. E. 1980. The skull of a new eosuchian reptiles from the Lower Jurassic of South Wales. *Zool. J. Linn. Soc.* 70:203–264.

Evans, S. E. 1988. The early history and relationships of the *Diapsida.* Pp. 221–260 in *The Phylogeny and Classification of the Tetrapods. Vol. 1: Amphibians, Reptiles, Birds* (M. J. Benton, ed.). Systematics Association Special Volume No. 35A. Clarendon Press, Oxford.

Ezcurra, M. D., T. M. Scheyer, and R. J. Butler. 2014. The origin and early evolution of *Sauria*: reassessing the Permian saurian fossil record and the timing of the crocodile-lizard divergence. *PLOS ONE* 9(2):e89165.

Field, D. J., J. A. Gauthier, B. L. King, D. Pisani, T. R. Lyson, and K. J. Peterson. 2014. Toward consilience in reptile phylogeny: miRNAs support an archosaur, not lepidosaur, affinity for turtles. *Evol. Dev.* 16(4):189–196.

Frost, D. R., T. Grant, J. N. Faivovich, R. H. Bain, A. Haas, C. F. B. Haddad, R. O. de Sá, A. Channing, M. Wilkinson, S. C. Donnellan, C. J. Raxworthy, J. A. Campbell, B. L. Blotto, P. Moler, R. C. Drewes, R. A. Nussbaum, J. D. Lynch, D. M. Green, and W. C. Wheeler. 2006. The amphibian tree of life. *Bull. Am. Mus. Nat. Hist.* 297:1–370.

Gaffney, E. S. 1980. Phylogenetic relationships of the major groups of amniotes. Pp. 593–610 in *The Terrestrial Environment and the Origin of Land Vertebrates* (A. L. Panchen, ed.). Academic Press, London.

Gaffney, E. S. 1990. The comparative osteology of the Triassic turtle *Proganochelys. Bull. Am. Mus. Nat. Hist.* 194:1–263.

Gao, K.-Q., C.-F. Zhou, L. Hou, and R. C. Fox. 2013. Osteology and ontogeny of Early Cretaceous *Philydrosaurus* (*Diapsida*: *Choristodera*) based on new specimens from Liaoning, China. *Cretac. Res.* 45:91–102.

Gauthier, J. A. 1984. *A Cladistic Analysis of the Higher Systematic Categories of the* Diapsida. University Microfilms International, Ann Arbor, MI, #85-12825.

Gauthier, J. A. 1986. Saurischian monophyly and the origin of birds. Pp. 1–55 in *The Origin of Birds and the Evolution of Flight* (K. Padian, ed.). *Mem. Cali. Acad. Sci.* 8. California Academy of Sciences, San Francico, CA.

Gauthier, J. A. 1994. The diversification of the amniotes. Pp. 129–159 in *Major Features of Vertebrate Evolution: Short Courses in Paleontology* (D. Prothero, ed.). The Paleontological Society.

Gauthier, J. A., D. Cannatella, K. de Queiroz, A. G. Kluge, and T. Rowe. 1989. Tetrapod phylogeny. Pp. 337–353 in *The Hierarchy of Life* (B. Fernholm, K. Bremer, and H. Jörnvall, eds.).

Elsevier Science Publishers B. V. (Biomedical Division).

Gauthier, J., R. Estes, and K. de Queiroz. 1988a. A phylogenetic analysis of *Lepidosauromorpha*. Pp. 15–98 in *Phylogenetic Relationships of the Lizard Families* (R. Estes and G. Pregill, eds.). Stanford University Press, Stanford, CA.

Gauthier, J. A., M. Kearney, J. A. Maisano, O. Rieppel, and A. Behlke. 2012. Assembling the squamate tree of life: perspectives from the phenotype and the fossil record. *Bull. Peabody Mus. Nat. Hist.* 53(1):3–308.

Gauthier, J., A. Kluge and T. Rowe. 1988b. Amniote phylogeny and the importance of fossils. *Cladistics* 4(2):105–209.

Gauthier, J., A. Kluge, and T. Rowe. 1988c. The early evolution of the *Amniota*. Pp. 103–155 in *The Phylogeny and Classification of the Tetrapods. Vol. I: Amphibians, Reptiles and Birds* (M. J. Benton, ed.). Systematics Association Special Volume No. 35A. Clarendon Press, Oxford.

Gauthier, J. A., S. J. Nesbitt, E. R. Schachner, G. S. Bever, and W. G. Joyce. 2011. The bipedal stem crocodilian *Poposaurus gracilis*: inferring function in fossils and innovation in archosaur locomotion. *Bull. Peabody Mus. Nat. Hist.* 52(1):107–126.

Gottmann-Quesada, A., and P. M. Sander. 2009. A redescription of the early archosauromorph *Protorosaurus speneri* Meyer, 1832 and its phylogenetic relationships. *Palaeontogr. Abt. A* 287:123–220.

Gow, C. E. 1972. The osteology and relationships of the *Millerettidae* (*Reptilia: Cotylosauria*). *J. Zool.* 167:219–264.

Gow, C. E. 1975. The morphology and relationships of *Youngina capensis* Broom and *Prolacetla broomi* Parrington. *Palaeontol. Afr.* 18:89–131.

Gray, J. E. 1842. Descriptions of two hitherto unrecorded species of reptiles from new zealand; presented to the british museum by dr. Dieffenbach. *The Zoological Miscellany.* Privately published, London.

Hennig, W. 1966. *Phylogenetic Systematics.* Translated by D. Davis and R. Zangerl. University of Illinois Press, Urbana, IL.

Hugall, A. F, R. Foster, and M. S. Y. Lee. 2007. Calibration choice, rate smoothing, and the pattern of tetrapod diversification according to the long nuclear gene RAG-1. *Syst. Biol.* 56:543–563.

Kuhn, E. 1952. *Askeptosaurus italicus* Nopcsa. *Schweiz. Paläontol. Abh.* 69:6–73.

Lane, H. H. 1945. New Mid-Pennsylvanian reptiles from Kansas. *Trans. Kansas Acad. Sci.* 47:381–390.

Lane, H. H. 1946. A survey of fossil vertebrates of Kansas: part III: the reptiles. *Trans. Kansas Acad. Sci.* 49:289–332.

Laurenti, J. N. 1768. *Specimen Medicum, Exhibens Synopsin Reptilium Emendatam cum Experimentis Circa Venena et Antidota Reptilium Austriacorum.* Joan. Thom. Nob. de Trattnern, Viennae.

Laurin, M. 1991. The osteology of a Lower Permian eosuchian from Texas and a review of diapsid phylogeny. *Zool. J. Linn. Soc.* 101(1): 59–95.

Lee, M. S. Y. 1996. Pareiasaur phylogeny and the origin of turtles. *Zool. J. Linn. Soc.* 3(1):197–280.

Li, C., N. E. Fraser, O. Rieppel, and X.-C. Wu. 2018. A Triassic stem turtle with an edentulous beak. *Nature* 560:476–479.

Linnaeus, C. 1758. *Systema Naturae Per Regna Tria Naturae, Secundum Classes, Ordines, Genera, Species, cum Characteribus, Differentiis, Synonymis, Locis.* 10th edition, Tome 1. Laurentius Salvius, Holmiae (Stockholm).

Lu, B., W. Yang, Q. Dai, and J. Fu. 2013. Using genes as characters and a parsimony analysis to explore the phylogenetic position of turtles. *PLOS ONE* 8(11):e79348.

Lyson T. R., G. S. Bever, B.-A. S. Bhullar, W. G. Joyce, and J. A. Gauthier. 2010. Transitional fossils and the origin of turtles. *Biol. Lett.* 6:830–833.

Lyson T. R., G. S. Bever, T. M. Scheyer, A. Y. Hsiang, and J. A. Gauthier. 2013a. Evolutionary origin of the turtle shell. *Curr. Biol.* 23:1113–1119.

Lyson T. R., B.-A. S. Bhullar, G. S. Bever, W. G. Joyce, K. de Queiroz, A. Abzhanov, and J. A. Gauthier. 2013b. Homology of the enigmatic nuchal bone reveals novel reorganization of the shoulder girdle in the evolution of the turtle shell. *Evol. Dev.* 15(5):317–325.

Macartney, J. 1802. Tables of classification. In *Lectures on Comparative Anatomy* (G. Cuvier ed.). Translated by William Ross under the inspection of James Macartney. T. N. Longman and O. Rees, London.

Merck, J. W. 1997. *A Phylogenetic Analysis of the Euryapsid Reptiles*. PhD thesis, University of Texas, Austin, TX.

Merrem, B. 1820. *Versuch eines Systems der Amphibien*. Johann Christian Krieger, Marburg.

Müller, J. 2004. The relationships among diapsid reptiles and the influence of taxon selection. Pp. 379–408 in *Recent Advances in the Origin and Early Radiation of Vertebrates* (G. Arratia and R. Cloutier, eds.). Verlag Dr. Friedrich Pfeil, Munich.

Nakajima, Y., A. Houssaye, and H. Endo. 2014. Osteohistology of the Early Triassic ichthyopterygian reptile *Utatsusaurus hataii*: implications for early ichthyosaur biology. *Acta Palaeontol. Pol.* 59(2):343–352.

Nesbitt, S. J., J. J. Flynn, A. C. Pritchard, J. M. Parrish, L. Ranivoharimanana, and A. R. Wyss. 2015. Postcranial osteology of *Azendohsaurus madagaskarensis* (Middle to Upper Triassic, Isalo Group of Madagascar) and its systematic position among stem archosaurs. *Bull. Am. Mus. Nat. Hist.* 398:1–126.

Olsen, P. E. 1978. A new aquatic eosuchian from the Newark Supergroup (Late Triassic-Early Jurassic) of North Carolina and Virginia. *Postilla* 176:1–30.

Oppel, M. 1811. *Die Ordnungen, Familien, und Gattungen der Reptilien, als Prodrom Einer Naturgeschichte Derselben*. Joseph Lindauer, Munich.

Osborn, H. F. 1903. On the primary division of the *Reptilia* into two sub-classes, *Synapsida* and *Diapsida*. *Science* 17(424):275–276.

Peabody, F. E. 1952. *Petrolacosaurus kansensis* Lane, a Pennsylvanian reptile from Kansas. *Univ. Kansas Paleontol. Contr., Vertebr.* 1:1–41.

Pritchard, A. C., J. A. Gauthier, M. Hanson, G. S. Bever, and B.-A. S. Bhullar. 2018. A tiny Triassic saurian from Connecticut and the early evolution of the diapsid feeding apparatus. *Nat. Commun.* 9:1213.

Pritchard, A. C., and S. J. Nesbitt. 2017. A bird-like skull in a Triassic diapsid reptile increases heterogeneity of the morphological and phylogenetic radiation of *Diapsida*. *R. Soc. Open Sci.* 4:170499.

Reisz, R. R. 1977. *Petrolacosaurus*, the oldest known diapsid reptile. *Science* 196(4294):1091–1093.

Reisz, R. R. 1981. A diapsid reptile form the Pennsylvanian of Kansas. *Univ. Kansas Publ., Mus. Nat. Hist.* 7:1–74.

Reisz, R. R., and M. Laurin. 1991. *Owenetta* and the origin of turtles. *Nature* 349:324–326.

Reisz, R. R., S. P. Modesto, and D. M. Scott. 2011. A new Early Permian reptile and its significance in early diapsid evolution. *Proc. R. Soc. Lond. B Biol. Sci.* 278:3731–3737.

Renesto, S. 1994. *Megalancosaurus*, a possibly arboreal archosauromorph (*Reptilia*) from the Upper Triassic of northern Italy. *J. Vertebr. Paleontol.* 14:38–52.

Renesto, S., J. A. Spielmann, S. G. Lucas, and G. T. Spagnoli. 2010. The taxonomy and paleobiology of the Late Triassic (Carnian-Norian: Adamanian-Apachean) drepanosaurs (*Diapsida: Archosauromorpha: Drepanosauromorpha*). *Bull. New Mexico Mus. Nat. Hist.* 46:1–81.

Rest, J. S., J. C. Ast, C. C. Austin, P. J. Waddell, E. A. Tibbetts, J. M. Hay, and D. P. Mindell. 2003. Molecular systematics of primary reptilian lineages and the tuatara mitochondrial genome. *Mol. Phylogenet. Evol.* 29:289–297.

Rieppel, O. 1993. Euryapsid relationships: a preliminary analysis. *Neues Jahrb. Geol. Paläontol.* 188(2):241–264.

Romer, A. S. 1956. *The Osteology of the Reptiles*. University of Chicago Press, Chicago, IL.

Romer, A. S. 1966. *Vertebrate Paleontology*. 3rd edition. University of Chicago Press, Chicago, IL.

Senter, P. 2004. Phylogeny of *Drepanosauridae* (*Reptilia*: *Diapsida*). *J. Syst. Palaeontol.* 2(3):257–268.

Sigogneau-Russell, D., and D. E. Russell. 1978. Etude osteologique du reptile *Simoedosaurus* (*Choristodera*). *Ann. Paléontol.* (*Vertébrés*) 64:1–84.

Simões, T. R., M. W. Caldwell, M. Tałanda, M. Bernardi, A. Palci, O. Vernygora, F. Bernardini, L. Mancini, and R. L. Nydam.

2018. The origin of squamates revealed by a Middle Triassic lizard from the Italian Alps. *Nature* 557:706–709.

Tatarinov, L. P. 1964. *Lepidosauria.* Pp. 439–492 in *Osnovy Paleontologii* (5) (J. A. Orlov, ed.). Nauka, Moscow.

Authors

Jacques A. Gauthier; Department of Geology and Geophysics; Yale University; 210 Whitney Ave.; New Haven, CT 06520-8109, USA. Email: jacques.gauthier@yale.edu.

Kevin de Queiroz; Department of Vertebrate Zoology; National Museum of Natural History; Smithsonian Institution; Washington, DC 20560-0162, USA. Email: dequeirozk@si.edu.

Date Accepted: 22 October 2018

Primary Editor: Philip Cantino

Pan-Lepidosauria J. A. Gauthier and K. de Queiroz, new clade name

Registration Number: 118

Definition: The total clade of the crown clade *Lepidosauria*. This is a crown-based total-clade definition. Abbreviated definition: total ∇ of *Lepidosauria*.

Etymology: Derived from *pan* (Greek), here referring to "pan-monophylum," another term for "total clade," and *Lepidosauria*, the name of the corresponding crown clade; hence, "the total clade of *Lepidosauria*."

Reference Phylogeny: Gauthier et al. (1988) Figure 13, where the clade in question is named *Lepidosauromorpha* and is hypothesized to include *Younginiformes*, which is no longer considered part of the clade (see Composition).

Composition: *Pan-Lepidosauria* is composed of *Lepidosauria* (*Pan-Sphenodon* plus *Pan-Squamata*, see entry in this volume) and all extinct species (stem lepidosaurs) that share a more recent common ancestor with the species of *Lepidosauria* than they do with any other extant amniotes (*Aves*, *Testudines*, *Crocodylia*, and *Mammalia*). The Late Permian fossils *Saurosternon bainii* and *Lanthanolania ivakhnenkoi*, as well as the Permo-Triassic fossils *Paliguana whitei* and *Palaeagama vielhaueri*, have sometimes been inferred to be stem lepidosaurs (e.g., Gauthier et al., 1988; Modesto and Reisz, 2002; Evans and Jones, 2010) but other times to be stem saurians (e.g., Modesto and Reisz, 2002; Müller, 2004). The Upper Triassic *Kuehneosauridae* and its inferred Early Triassic relative *Pamelina polonica* (but see Simões et al., 2018) have often been inferred to be stem lepidosaurs (Gauthier et al., 1988; Evans, 1988,

2009; Evans and Borsuk-Bialylicka, 2009; Evans and Jones, 2010; but see Müller, 2004; Jones et al., 2013). More recently, kuehneosaurs have been inferred to be stem archosaurs close to *Trilophosaurus buettneri* by Pritchard and Nesbitt (2017), although Simões et al. (2018) placed them as either stem saurians or stem archosaurs depending on the analysis, but in either case only distantly related to *T. buettneri*. The Early Triassic *Sophineta cracoviensis* and the Middle Jurassic *Marmoretta oxoniensis* have been more consistently regarded as stem lepidosaurs, although that inference depends upon correct association among disarticulated remains (Evans, 1991; Waldman and Evans, 1994; Evans, 2009; Evans and Borsuk-Bialylicka, 2009; Evans and Jones, 2010; Jones et al., 2013; Renesto and Bernardi, 2014). Depending on the analysis, Simões et al. (2018) inferred *S. cracoviensis* to be either a stem lepidosaurian or a stem squamatan, but they consistently inferred *M. oxoniensis* as sister to *Megachirella wachtleri* on the squamate stem. Renesto and Bernardi (2014) previously placed the Middle Triassic *Megachirella wachtleri* either inside or outside of the lepidosaurian crown. The Early or Middle Jurassic *Tamaulipasaurus morenoi* may be a stem or a crown lepidosaur, but its relationships are difficult to assess because of its high degree of modification for head-first burrowing (Clark and Hernandez, 1994). The highly modified Late Triassic drepanosaurs have sometimes been inferred to be stem archosaurs (Laurin, 1991; Simões et al., 2018) or stem lepidosaurs (Evans, 2009), but most authors have inferred drepanosaurs to be stem saurians that must accordingly have diverged much earlier than the fossil record currently indicates (i.e., by the Mid-Permian; Guadalupian) (e.g.,

Müller, 2004; Senter, 2004; Renesto et al., 2010; Pritchard and Nesbitt, 2017; Simões et al., 2018). The authors of several early phylogenetic analyses inferred that *Younginiformes* are stem lepidosaurs (e.g., Benton, 1985; Evans, 1988; Gauthier et al., 1988); however, more recent analyses have placed *Younginiformes*, either as a clade or a paraphylum, on the stem of a more inclusive crown clade composed of (in addition to lepidosaurs) archosaurs and sometimes also turtles (e.g., Laurin, 1991; Gauthier, 1994; Caldwell, 1996; deBraga and Rieppel, 1997; Müller, 2004; Evans and Borsuk-Bialylicka, 2009; Renesto and Bernardi, 2014; Bever et al., 2015; Pritchard and Nesbitt, 2017; Simões et al., 2018). See Evans and Jones (2010) for other taxa that are no longer considered stem lepidosaurs.

Diagnostic Apomorphies: As a total clade (and therefore also a maximum clade), *Pan-Lepidosauria* is not necessarily expected to have diagnostic apomorphies (characters that arose simultaneously with the split of lepidosaur stem lineage from that of archosaurs; see de Queiroz, 2007); however, any of the apomorphies that originated along the lepidosaur stem lineage (see Diagnostic Apomorphies for *Lepidosauria*, this volume), as well as those diagnosing various side branches, would allow referral of a specimen or species to *Pan-Lepidosauria*. Uncertainty in placement of several incomplete, and often highly modified, fossils further complicates the question. If, as we suspect, the Early Triassic *Sophineta cracoviensis* is a stem lepidosaur, it shares the following apomorphies with the crown: (1) maxilla participates broadly in ventral orbital margin; (2) postfrontal wraps around fronto-parietal suture; (3) lacrimal reduced, confined largely to orbital rim; (4) marginal teeth attached superficially to lingual margin of jaw (pleurodont dentition); and (5), a weak zygosphene-zygantrum accessory intervertebral articulation.

Synonyms: *Lepidosauria* Haeckel 1866 of various authors (e.g., Romer, 1933, 1945, 1956, 1966; Gardiner, 1982) and *Lepidosauromorpha* Gauthier in Benton 1983 of Benton (1983, 1985) and Evans (1984) are approximate synonyms. *Eosuchia* (e.g., of Romer, 1933, 1945, 1956, 1966), characterized as being composed of "primitive lepidosaurians" and "direct ancestors of the lizards and snakes", and *Holapsida* Underwood 1957, composed of "*Eosuchia* and *Rhynchocephalia*", are partial (and approximate) synonyms. *Eolacertilia* Romer 1966 (often incorrectly attributed to Robinson, 1967) sensu Carroll (1975, 1977) and Estes (1983) is a partial (and approximate) synonym in the context of phylogenies that place all of the included taxa in the lepidosaur stem group (see Composition). *Lepidosauromorpha* Gauthier in Benton 1983 of Gauthier et al. (1988) is an unambiguous synonym.

Comments: Although the name *Lepidosauromorpha* has previously been defined phylogenetically as applying to the total clade of *Lepidosauria* (Gauthier et al., 1988), we have chosen instead to apply the name *Pan-Lepidosauria* to that clade in the interest of promoting a standardized form for the names of total clades (e.g., de Queiroz and Gauthier, 1992; Gauthier and de Queiroz, 2001; de Queiroz, 2007). If the name *Lepidosauromorpha* is to be retained, it could be applied to a deep node within *Pan-Lepidosauria*, although that would require careful formulation of the (non-standard) definition, given the uncertain status of many of its potential early members (see Composition). Alternatively, the name could be applied to a clade coinciding with the origin of one of the early-evolving lepidosaur apomorphies.

In the context of phylogenies in which *Sphenodon punctatus* is more closely related to turtles, crocodilians, and birds than to *Squamata* (e.g., Hedges and Poling, 1999; Zardoya and Meyer, 2000), the names *Lepidosauria* (this volume) and *Reptilia* (this volume) are synonyms, and *Reptilia* is to be granted precedence (see *Lepidosauria* in this volume). In the context of such phylogenies, the name *Pan-Lepidosauria* should not be applied to any clade (*ICPN* Art. 14.5; Cantino and de Queiroz, 2020).

Sophineta cracoviensis and possibly *Pamelina polonica* (see Composition) are the earliest known pan-lepidosaurs (~245Ma, late Olenekian, Early Triassic; Evans, 2009; Evans and Borsuk-Białynicka, 2009). But the initial divergence of *Pan-Lepidosauria* from *Pan-Archosauria* (this volume) is likely older (~265 Ma), based on the age of the earliest-known putative pan-archosaur, *Aenigmastropheus parringtoni* (Capitanian, mid-Late Permian; Ezcurra et al., 2014).

Literature Cited

Benton, M. 1983. The Triassic reptile *Hyperodapedon* from Elgin: functional morphology and relationships. *Philos. Trans. R. Soc. Lond. B Biol. Sci.* 302:605–717.

Benton, M. 1985. Classification and phylogeny of the diapsid reptiles. *Zool. J. Linn. Soc.* 84:97–164.

Bever, G. S., T. R. Lyson, D. J. Field, and B.-A. S. Bhullar. 2015. Evolutionary origin of the turtle skull. *Nature* 525:239–242.

deBraga, M., and O. Rieppel. 1997. Reptile phylogeny and the interrelationships of turtles. *Zool. J. Linn. Soc.* 120:281–354.

Caldwell, M. 1996. *Ichthyosauria*: a preliminary phylogenetic analysis of diapsid affinities. *Neues Jahrb. Geol. Paläontol. Abh.* 200:361–386.

Cantino, P. D., and K. de Queiroz. 2020. *International Code of Phylogenetic Nomenclature (PhyloCode)*, Version 6. CRC Press, Boca Raton, FL.

Carroll, R. L. 1975. Permo-Triassic "lizards" from the Karroo. *Palaeontol. Afr.* 18:17–87.

Carroll, R. L. 1977. The origin of lizards. Pp. 359–396 in *Problems in Vertebrate Evolution* (S. Andrews, R. Miles and A. Walker, eds.). Linn. Soc. London, Symp. Ser. 4. Academic Press, London.

Clark, J. M., and R. Hernandez. 1994. A new burrowing diapsid from the Jurassic La Boca Formation of Tamaulipas, Mexico. *J. Vertebr. Paleontol.* 14:180–195.

Estes, R. 1983. Sauria *terrestria*, Amphisbaenia. *Handbuch der Paläoherpetologie. Teil 10A.* Gustav Fischer, Stuttgart.

Evans, S. E. 1984. The classification of the *Lepidosauria. Zool. J. Linn. Soc.* 82:87–100.

Evans, S. E. 1988. The early history and relationships of the *Diapsida*. Pp. 221–260 in *The Phylogeny and Classification of the Tetrapods. Vol. 1: Amphibians, Reptiles, Birds* (M. J. Benton, ed.). Systematics Association Special Volume No. 35A. Clarendon Press, Oxford.

Evans, S. E. 1991. A new lizard-like reptile (*Diapsida: Lepidosauromorpha*) from the Middle Jurassic of England. *Zool. J. Linn. Soc.* 103:391–412.

Evans, S. E. 2009. An early kuehneosaurid reptile (*Reptilia: Diapsida*) from the Early Triassic of Poland. *Palaeontol. Pol.* 65:145–178.

Evans, S. E., and M. Borsuk-Białynicka. 2009. A small lepidosauromorph reptile from the Early Triassic of Poland. *Palaeontol. Pol.* 65:179–202.

Evans, S. E., and M. E. H. Jones. 2010. The origin, early history and diversification of lepidosauromorph reptiles. Pp. 27–44 in *New Aspects of Mesozoic Biodiversity* (S. Bandyopadhyay, ed.). Lecture Notes in Earth Sciences 132.

Ezcurra, M. D., T. M. Scheyer, and R. J. Butler. 2014. The origin and early evolution of *Sauria*: reassessing the Permian saurian fossil record and the timing of the crocodile-lizard divergence. *PLOS ONE* 9(2):e89165.

Gardiner, B. G. 1982. Tetrapod classification. *Zool. J. Linn. Soc.* 74:207–232.

Gauthier, J. A. 1994. The diversification of the amniotes. Pp. 129–159 in *Major Features of Vertebrate Evolution: Short Courses in Paleontology* (D. Prothero, ed.). Paleontological Society.

Gauthier, J. A., and K. de Queiroz. 2001. Feathered dinosaurs, flying dinosaurs, crown dinosaurs, and the name "*Aves*". Pp. 7–41 in *New Perspectives on the Origin and Early Evolution of Birds: Proceedings of the International Symposium in Honor of John H. Ostrom* (J. Gauthier and L. F. Gall, eds.). Peabody Mus. Nat. Hist., Yale Univ., New Haven, CT.

Gauthier, J. A., R. Estes, and K. de Queiroz. 1988. A phylogenetic analysis of *Lepidosauromorpha*. Pp. 15–98 in *Phylogenetic Relationships of the Lizard Families* (R. Estes and G. Pregill, eds.). Stanford University Press, Stanford, CA.

Haeckel, E. 1866. *Generelle Morphologie der Organismen. Band 2: Allgemeine Entwicklungsgeschichte der Organismen.* George Reimer, Berlin.

Hedges, S. B., and L. L. Poling. 1999. A molecular phylogeny of reptiles. *Science* 283:998–1001.

Jones, M. E. H., C. L. Anderson, C. A. Hipsley, J. Müller, S. E. Evans, and R. R. Schoch. 2013. Integration of molecules and new fossils supports a Triassic origin for *Lepidosauria* (lizards, snakes, and tuatara). *BMC Evol. Biol.* 13:208.

Laurin, M. 1991. The osteology of a Lower Permian eosuchian from Texas and a review of diapsid phylogeny. *Zool. J. Linn. Soc.* 101:59–95.

Modesto, S., and R. R. Reisz. 2002. An enigmatic new diapsid reptile from the Upper Permian of Eastern Europe. *J. Vertebr. Palontol.* 22:851–855.

Müller, J. 2004. The relationships among diapsid reptiles and the influence of taxon selection. Pp. 379–408 in *Recent Advances in the Origin and Early Radiation of Vertebrates* (G. Arratia, M. V. H. Wilson, and R. Cloutier, eds.). Friedrich Pfeil, Munich.

Pritchard, A. C., and S. J. Nesbitt. 2017. A bird-like skull in a Triassic diapsid reptile increases heterogeneity of the morphological and phylogenetic radiation of *Diapsida*. *R. Soc. Open Sci.* 4:170499.

de Queiroz, K. 2007. Toward an integrated system of clade names. *Syst. Biol.* 56:956–974.

de Queiroz, K., and J. Gauthier. 1992. Phylogenetic taxonomy. *Annu. Rev. Ecol. Syst.* 23:449–480.

Renesto, S., and M. Bernardi. 2014. Redescription and phylogenetic relationships of *Megachirella wachtleri* Renesto et Posenato, 2003 (*Reptilia, Diapsida*). *Paläontol. Z.* 88:197.

Renesto, S., J. A. Spielmann, S. G. Lucas, and G. T. Spagnoli. 2010. The taxonomy and paleobiology of the Late Triassic (Carnian-Norian: Adamanian-Apachean) drepanosaurs (*Diapsida: Archosauromorpha: Drepanosauromorpha*). *Bull. New Mexico Mus. Nat. Hist. Sci.* 46.

Robinson, P. 1967. Triassic vertebrates from lowland and upland. *Sci. Cult.* 33:169–173.

Romer, A. S. 1933. *Vertebrate Paleontology.* University of Chicago Press, Chicago, IL.

Romer, A. S. 1945. *Vertebrate Paleontology.* 2nd edition. University of Chicago Press, Chicago, IL.

Romer, A. S. 1956. *Osteology of the Reptiles.* University of Chicago Press, Chicago, IL.

Romer, A. S. 1966. *Vertebrate Paleontology.* 3rd edition. University of Chicago Press, Chicago, IL.

Senter, P. 2004. Phylogeny of *Drepanosauridae* (*Reptilia: Diapsida*). *J. Syst. Palaeontol.* 2:257–268.

Simões, T. R., M. W. Caldwell, M. Tałanda, M. Bernardi, A. Palci, O. Vernygora, F. Bernardini, L. Mancini, and R. L. Nydam. 2018. The origin of squamates revealed by a Middle Triassic lizard from the Italian Alps. *Nature* 557:706–709.

Underwood, G. 1957. On lizards of the family *Pygopodidae*: a contribution to the morphology and phylogeny of the *Squamata*. *J. Morphol.* 100:207–268.

Waldman, M., and S. E. Evans. 1994. Lepidosauromorph reptiles from the Middle Jurassic of Skye. *Zool. J. Linn. Soc.* 112:135–150.

Zardoya, R., and A. Meyer. 2000. Mitochondrial evidence on the phylogenetic position of caecilians (*Amphibia: Gymnophiona*). *Genetics* 155:765–775.

Authors

Jacques A. Gauthier; Department of Geology and Geophysics; Yale University; 210 Whitney Avenue; New Haven, CT 06511, USA. Email: jacques.gauthier@yale.edu.

Kevin de Queiroz; Department of Vertebrate Zoology; National Museum of Natural History; Smithsonian Institution; Washington, DC 20560-0162, USA. Email: dequeirozk@si.edu.

Date Accepted: 11 August 2018

Primary Editor: Philip Cantino

Lepidosauria E. Haeckel 1866 [K. de Queiroz and J. A. Gauthier], converted clade name

Registration Number: 61

Definition: The smallest crown clade containing *Lacerta agilis* Linnaeus 1758 (*Squamata*) and *Sphenodon* (originally *Hatteria*) *punctatus* (Gray 1842) (*Rhynchocephalia*). This is a minimum-crown-clade definition. Abbreviated definition: min crown ∇ (*Lacerta agilis* Linnaeus 1758 & *Sphenodon punctatus* (Gray 1842)).

Etymology: Derived from the Greek *lepidos*, scale, plus *sauros*, lizard, reptile.

Reference Phylogeny: Gauthier et al. (1988a: Fig. 13), where *Lacerta agilis* is part of *Squamata* and *Sphenodon punctatus* is part of *Rhynchocephalia*. See also Gauthier (1984: Figs. 32–33), Evans (1984: Fig. 2, 1988: Figs. 6.1–6.2), Rest et al. (2003: Fig. 3), Hill (2005: Fig. 3), Evans and Jones (2010: Fig. 2.1), Crawford et al. (2012: Fig. 2), Jones et al. (2013: Figs. 3–4), and Simões et al. (2018: Fig. 2).

Composition: *Lepidosauria* is composed of two primary crown clades: the New Zealand endemic *Sphenodon* with one currently recognized extant species (Hay et al., 2010) and the globally distributed *Squamata* with approximately 10,078 currently recognized extant species (Uetz, 2016). See *Squamata* and *Pan-Squamata* (this volume) for references regarding extinct species in those clades. Reviews of disparate and diverse fossil rhynchocephalians can be found in Jones et al. (2013), Apesteguía et al. (2014), and Bever and Norell (2017).

Diagnostic Apomorphies: According to Gauthier et al. (1988a), *Lepidosauria* has at least 35 apomorphies relative to other extant amniotes (only some of which are diagnostic relative to different stem lepidosaurs). These apomorphies derive from disparate anatomical systems, are apparently unrelated functionally and developmentally, and have persisted for hundreds of millions of years among lepidosaurs with remarkably divergent ecologies. They include the following apomorphies (those lacking citations are from Gauthier et al., 1988a): (1) transversely oriented external opening of cloaca and loss of amniote penis (hemipenes of squamatans are neomorphic); (2) kidney in tail base and adrenal gland suspended in gonadal mesentery (Gabe, 1970); (3) lingual prey prehension; (4) sagittal crest of scales projecting from neck, body and tail; (5) scales composed of superimposed, rather than juxtaposed, amniote alpha keratin and reptilian phi keratin layers (Maderson, 1985); (6) skin shed regularly in its entirety; (7) prefrontal braces skull roof on palate (Gauthier, 1994); (8) lacrimal bone largely confined to orbital rim; (9) maxilla broadly contributes to ventral orbital margin (Gauthier, 1994); (10) marginal teeth attached superficially to lingual surface of jaw (rather than in shallow sockets; further modified to more apical attachment in some taxa); (11) teeth lost from transverse process of pterygoid and from sphenoid bones (Gauthier et al., 1988b); (12) zygosphene-zygantrum accessory intervertebral joints formed from dorsal extensions of zygapophysial surfaces (see Petermann and Gauthier, 2018); (13) autotomic and regenerable tail (autotomy, but not regeneration, has also been reported in some captorhinids; LeBlanc et al., 2018); (14) all non-ossifying cartilaginous parts of skeleton calcify during postnatal ontogeny; (15) neomorphic ossification centers in limb bone epiphyses that fuse to diaphyses near maximum adult size; (16) medial centrale larger than lateral central in wrist (Gauthier et al., 1988b); (17) radiale contacts 1st distal carpal or 1st metacarpal in hand (Gauthier et al., 2012); (18) 4th metacarpal shorter than 3rd (symmetrical metacarpals); (19) fenestrate pelvic girdle; (20) posterodorsally sloping ilium; (21) embryonic fusion between anlage of lateral centrale and astragalus in tarsus; (22) astragalus and calcaneum fused in adult (and lack a perforating foramen between them as neurovascular system passes between tibia and fibula proximal to tarsus; Rieppel and Reisz, 1999); (23) absence of separate 1st distal tarsal in foot; (24) hooked 5th metatarsal—and absence of discrete 5th distal tarsal presumably incorporated into it—modified to act as both a 'heel' and a grasping 'thumb' enabling the 5th toe to rotate 90° with respect to rest of the foot according to Robinson (1975).

Synonyms: All synonyms are approximate (not phylogenetically defined). The names that follow were used after *Sphenodon punctatus* was first recognized (Gray, 1831, 1842) for taxa that explicitly included *Sphenodon* (sometimes as *Hatteria* or *Rhynchocephalus*) and squamatans: *Squamata* of Gray (1845), partial (amphisbaenians excluded); *Saura* of Gray (1845), partial (amphisbaenians and snakes excluded); *Lacertia* of Owen (1845), partial (snakes excluded); *Squamata* of Cope (1864) and Günther (1867); *Lacertilia* of Cope (1864) and Huxley (1886), partial (snakes excluded); *Saurii* of Gegenbaur (1874, 1878), partial (snakes excluded). In most of these cases (except for Günther, 1867), *Sphenodon punctatus* was considered nested within the taxon corresponding to *Squamata*.

Comments: Changing ideas about the relationship between *Sphenodon punctatus* and *Squamata* have come nearly full circle. Originally

described as an agamid "lizard" (Gray, 1831, 1842), *S. punctatus* was later separated from agamids as *Hatteriidae* (Cope, 1864), and later still from "lizards" (and snakes) as *Rhynchocephalia* within *Squamata* (Günther, 1867). Cope (1875) furthered the separation, first, by not recognizing *Squamata*, second, by including the extinct protorosaurs (i.e., *Protorosaurus speneri*) and rhynchosaurs (i.e., *Rhynchosaurus articeps*) along with extant *Sphenodon punctatus* in his *Rhynchocephalia*, and third, by interposing turtles (*Testudines*) between *Rhynchocephalia* and "lizards" (and snakes) in his taxonomy. Cope (1889) later continued this trend by separating *Rhynchocephalia* and *Squamata* as taxa of equal rank (although the taxa were adjacent in his list). Distancing them further still, Cope (1900) placed *Rhynchocephalia* and *Squamata* in separate higher taxa, *Archosauria* and *Streptostylica*, respectively, although his phylogenetic diagram (p. 160) had them closely related. In general, many late nineteenth and early to mid twentieth century authors treated *Sphenodon* and *Squamata* (not always using that name) as relatively distantly related among reptiles (e.g., Cope, 1875; Haeckel, 1895; Williston, 1917, 1925), or at least considered *Sphenodon* closer to rhynchosaurs (and sometimes also to choristoderans) than to *Squamata* (e.g., Gadow, 1898, 1901; Jaekel, 1911; Nopcsa, 1923; Romer, 1933, 1945, 1956, 1966; Underwood, 1957; Kuhn, 1966). Dissolution of Cope's idea that rhynchosaurs and "protorosaurs" are closely related to *Sphenodon* began with Hughes' (1968) study of the rhynchocephalian tarsus and culminated in Carroll's (1975) argument that characters traditionally used to unite *Sphenodon punctatus* (and its legitimate fossil relatives) with rhynchosaurs and "protorosaurs" are erroneous. Subsequent authors working in an explicitly phylogenetic framework (Benton, 1982; Evans, 1984; Gauthier, 1984; Carroll, 1985) presented evidence that rhynchosaurs and "protorosaurs" (including *Protorosaurus speneri* and *Prolacerta broomi*) are related to archosaurs while *Sphenodon punctatus* is closer to squamatans. Gauthier et al. (1988a) provided extensive morphological evidence for a close relationship between *Sphenodon punctatus* and *Squamata* based on a computer-assisted analysis of 171 characters in 13 taxa, and this result has been corroborated by subsequent studies based on morphology (e.g., Evans, 1988; Hill, 2005; Evans and Jones, 2010; Gauthier et al., 2012), molecules (e.g., Rest et al., 2003; Crawford et al., 2012; Pyron et al., 2013), and combined morphological and molecular data (e.g., Jones et al., 2013; Reeder et al., 2015; Simões et al., 2018).

The name *Lepidosauria* was proposed by Haeckel (1866) for what is now known as *Squamata*—that is, "lizards" (a paraphyletic group) and snakes. Although *Sphenodon punctatus* was not explicitly included, members of this taxon were originally thought to be "lizards" (e.g., Gray, 1831, 1842). After Günther's (1867) study demonstrating major anatomical differences between *Sphenodon* (as *Hatteria*) and "lizards", and the subsequent increasing taxonomic separation of these groups (see previous paragraph), the name *Lepidosauria* was seldom used, or it was used for the taxon now called *Squamata* (e.g., Zittel, 1887–1890; Williston, 1904; Jaekel, 1911). Romer (1933) resurrected the name *Lepidosauria* for a taxon designated "validity doubtful" composed of *Eosuchia* (also considered of doubtful validity), *Rhynchocephalia* (choristoderans, rhynchosaurs, and sphenodontians), and *Squamata*. This arrangement became more solidly established in Romer's subsequent publications (e.g., 1945, 1956, 1966), although choristoderans (= champsosaurids) were transferred from *Rhynchocephalia* to *Eosuchia*. Despite *Sphenodon* having been considered more closely related to (extinct) rhynchosaurs than to squamatans, it should be noted that *Sphenodon* and *Squamata*

were considered most closely related among extant taxa. Therefore, when rhynchosaurs were allied to *Archosauria* and evidence was presented for an exclusive relationship between *Rhynchocephalia* and *Squamata*, the name *Lepidosauria* was applied to the group including the latter two taxa (e.g., Benton, 1982, 1983, 1985; Gardiner, 1982; Evans, 1984; Gauthier, 1984; Gauthier et al., 1988a; Pritchard and Nesbitt, 2017).

Selection of the name *Lepidosauria* for the *Sphenodon* + *Squamata* clade is relatively straightforward. Other names that have been applied to a taxon composed of *Sphenodon* and squamatans (see Synonyms) either have been little used (*Saura, Lacertia, Saurii*) or have been more commonly associated with a less inclusive clade (*Squamata*) or a paraphyletic group originating in the same ancestor (*Lacertilia* as well as the seldom-used names *Saura, Lacertia,* and *Saurii*).

The name *Lepidosauria* was first defined phylogenetically by Gauthier et al. (1988a: 34) as "the most recent common ancestor of *Sphenodon* and squamates and all of its descendants." We have updated that definition by using species as specifiers. In the context of phylogenies in which *Sphenodon punctatus* is more closely related to turtles, crocodilians, and birds than to *Squamata* (e.g., Hedges and Polling, 1999; Zardoya and Meyer, 2000), the names *Lepidosauria* and *Reptilia* (this volume) are synonyms, and *Reptilia* is to be granted precedence.

Until very recently, the earliest known unambiguous crown lepidosaurs were extinct relatives of *Sphenodon punctatus* extending back at least ~220 Ma (late Carnian; early Late Triassic; e.g., Pritchard and Nesbitt, 2017, and references therein). Simões et al. (2018) have, however, recently concluded that *Megachirella wachtleri* represents a stem squamatan (~240 Ma; Anisian; Middle Triassic), making it the only pan-squamatan currently known from the Triassic or Early Jurassic (Evans, 2003; see

Pan-Squamata, this volume). Moreover, some of their analyses placed *Sophineta cracoviensis* on the squamatan stem as well, which would extend the lepidosaurian crown into the Late Permian (see *Pan-Lepidosauria* this volume). Early and Middle Triassic terrestrial sediments are rare, and small fossil vertebrates are difficult to find; these factors doubtless contribute to a poor early lepidosaurian record (Jones et al., 2013). Nevertheless, rhynchocephalian fossils from the Late Triassic are reasonably common and already well-diversified phylogenetically (e.g., Whiteside, 1986; Bever and Norell, 2017). Rhynchocephalians had spread across Pangaea by the early Mesozoic, and have been in Gondwana since the Triassic, persisting in South America until at least the Palaeocene (Apesteguía et al., 2014). They remain common well into the Jurassic of Laurasia (e.g., Cocude-Michel, 1963), but are rare thereafter, with the last known occurrence a single species from the Early Cretaceous of southern Mexico (Reynoso, 2000). Extinct taxa closer to *Sphenodon punctatus* have been on New Zealand since at least the Early Miocene (Jones et al., 2009). Subfossil Quaternary remains of an extinct relative closely resembling *Sphenodon* were named *S. diversum* by Colenso (1886).

Clevosaur rhynchocephalians are distributed worldwide by the Late Triassic and are relatively abundant in early Mesozoic localities that have yet to produce any representatives of the squamatan total clade (e.g., Hsiou et al., 2015). Even granting the inadequacies of the early lepidosaur record, stem squamatans must have been present, which suggests a disparity in relative abundance between the two primary lepidosaurian subclades during the Triassic and Jurassic that is exactly the opposite of today. Squamatans do not appear to surpass rhynchocephalians in relative abundance until the Early Cretaceous (e.g., Evans and Matsumoto, 2015; and references therein).

Literature Cited

Apesteguía, S., R. O. Gómez, and G. W. Rougier. 2014. The youngest South American rhynchocephalian, a survivor of the K/Pg extinction. *Proc. R. Soc. Lond. B Biol. Sci.* 281:20140811.

Benton, M. J. 1982. The *Diapsida*: a revolution in reptile relationships. *Nature* 296:306–307.

Benton, M. J. 1983. The Triassic reptile *Hyperodapedon* from Elgin: functional morphology and relationships. *Philos. Trans. R. Soc. Lond. B Biol. Sci.* 302:605–717.

Benton, M. J. 1985. Classification and phylogeny of the diapsid reptiles. *Zool. J. Linn. Soc.* 84:97–164.

Bever, G. S., and M. A. Norell. 2017. A new rhynchocephalian (*Reptilia*: *Lepidosauria*) from the Late Jurassic Solnhofen (Germany) and the origin of the marine *Pleurosauridae*. *R. Soc. Open Sci.* 4(11):170570.

Carroll, R. L. 1975. The early differentiation of diapsid reptiles. *Colloq. Internat. CNRS* 218:433–449.

Carroll, R. L. 1985. A pleurosaur from the Lower Jurassic and the taxonomic position of the *Sphenodontida*. *Palaeontographica* 189:1–28.

Cocude-Michel, M. 1963. Les rhynchocephales et les sauriens des calcaires lithographiques (Jurassique superieur) d'Europe occidentale. *Nouv. Arch. Mus. Hist. Nat. Lyon* 7:1–187.

Colenso, W. 1886. Notes on the bones of a species of *Sphenodon*, (*S. diversum*, Col.,) apparently distinct from the species already known. *Trans. Proc. R. Soc. NZ* 18:118–128.

Cope, E. D. 1864. On the characters of the higher groups of *Reptilia Squamata*—and especially of the *Diploglossa*. *Proc. Acad. Nat. Sci. Phila.* 16:224–231.

Cope, E. D. 1875. Check-list of North American *Batrachia* and *Reptilia* with a systematic list of the higher groups and an essay on geographical distribution. *Bull. U.S. Natl. Mus.* 1:1–104.

Cope, E. D. 1889. Synopsis of the families of *Vertebrata*. *Am. Nat.* 23:849–877.

Cope, E. D. 1900. The crocodilians, lizards, and snakes of North America. *Ann. Rep. U. S. Natl. Mus.* 1898:155–1270 + 36 plates.

Crawford, N. G., B. C. Faircloth, J. E. McCormack, R. T. Brumfield, K. Winker, and T. C. Glenn. 2012. More than 1000 ultraconserved elements provide evidence that turtles are the sister group of archosaurs. *Biol. Lett.* 8:783–786.

Evans, S. E. 1984. The classification of the *Lepidosauria. Zool. J. Linn. Soc.* 82:87–100.

Evans, S. E. 1988. The early history and relationships of the *Diapsida.* Pp. 221–260 in *The Phylogeny and Classification of the Tetrapods. Vol. 1: Amphibians, Reptiles, Birds* (M. J. Benton, ed.). Systematics Association Special Volume No. 35A. Clarendon Press, Oxford.

Evans, S. E. 2003. At the feet of dinosaurs: the early history and radiation of lizards. *Biol. Rev.* 78:513–551.

Evans, S. E., and M. E. H. Jones. 2010. The origin, early history and diversification of lepidosauromorph reptiles. Pp. 27–44 in *New Aspects of Mesozoic Biodiversity* (S. Bandyopadhyay, ed.). Lecture Notes in Earth Sciences 132.

Evans, S. E., and R. Matsumoto. 2015. An assemblage of lizards from the Early Cretaceous of Japan. *Palaeontol. Electron.* 18.2.36A:1–36.

Gabe, M. 1970. The adrenal. Pp. 263–318 in *Biology of the Reptilia* (Morphology C). Vol. 3 (C. Gans and T. S. Parsons, eds.). Academic Press, London.

Gadow, H. 1898. *A Classification of* Vertebrata *Recent and Extinct.* Adam and Charles Black, London.

Gadow, H. 1901. *Amphibia and Reptiles. The Cambridge Natural History*, Vol. 8. Macmillan, London.

Gardiner, B. G. 1982. Tetrapod classification. *Zool. J. Linn. Soc.* 74:207–232.

Gauthier, J. A. 1984. *A Cladistic Analysis of the Higher Systematic Categories of the* Diapsida. PhD dissertation, University of California, Berkeley, CA.

Gauthier, J. A. 1994. The diversification of the amniotes. Pp. 129–159 in *Major Features of Vertebrate Evolution: Short Courses in Paleontology* (D. Prothero, ed.). Paleontological Society.

Gauthier, J. A., R. Estes, and K. de Queiroz. 1988a. A phylogenetic analysis of *Lepidosauromorpha.* Pp. 15–98 in *Phylogenetic Relationships of the Lizard Families* (R. Estes and G. Pregill, eds.). Stanford University Press, Stanford, CA.

Gauthier, J. A., M. Kearney, J. A. Maisano, O. Rieppel, and A. Behlke. 2012. Assembling the squamate tree of life: perspectives from the phenotype and the fossil record. *Peabody Mus. Nat. Hist. Bull.* 53(1):3–308.

Gauthier, J., A. Kluge, and T. Rowe. 1988b. Amniote phylogeny and the importance of fossils. *Cladistics* 4(2):105–209.

Gegenbaur, C. 1874. *Grundriss der Vergleichenden Anatomie.* Wilhelm Engelmann, Leipzig.

Gegenbaur, C. 1878. *Elements of Comparative Anatomy.* 2nd edition, English translation. MacMillan, London.

Gray, J. E. 1831. Note on a peculiar structure in the head of an *Agama. Zool. Misc.* 1831:13–14.

Gray, J. E. 1842. Descriptions of two hitherto unrecorded species of reptiles fro New Zealand; presented to the British Museum by Dr. Dieffenbach. *Zool. Misc.* 1842:72.

Gray, J. E. 1845. *Catalogue of the Specimens of Lizards in the Collection of the British Museum.* Trustees [of the British Museum], London.

Günther, A. 1867. Contribution to the anatomy of *Hatteria (Rhynchocephalus*, Owen). *Philos. Trans. R. Soc. Lond. B Biol. Sci.* 157:595–629.

Haeckel, E. 1866. *Generelle Morphologie der Organismen. Band 2: Allgemeine Entwicklungsgeschichte der Organismen.* George Reimer, Berlin.

Haeckel, E. 1895. *Systematicshe Phylogenie. Dritter Theil: Systematicshe Phylogenie der Wirbelthiere* (Vertebrata). Georg Reimer, Berlin.

Hay, J. M., S. D. Sarre, D. M. Lambert, F. W. Allendorf, and C. H. Daugherty. 2010. Genetic diversity and taxonomy: a reassessment of species designation in tuatara (*Sphenodon: Reptilia*). *Conserv. Genet.* 11:1063–1081.

Hedges, S. B., and L. L. Polling. 1999. A molecular phylogeny of reptiles. *Science* 283:998–1001.

Hill, R. V. 2005. Integration of morphological data sets for phylogenetic analysis of *Amniota*: the importance of integumentary characters and increased taxonomic sampling. *Syst. Biol.* 54:530–547.

Hsiou, A. S., M. A. G. De França, and J. Ferigolo. 2015. New data on the *Clevosaurus* (*Sphenodontia: Clevosauridae*) from the Upper Triassic of Southern Brazil. *PLOS ONE* 10(9):e0137523.

Hughes, B. 1968. The tarsus of rhynchocephalian reptiles. *J. Zool.* 156:457–481.

Huxley, T. H. 1886. *A Manual of the Anatomy of Vertebrated Animals.* D. Appleton, New York.

Jaekel, O. 1911. *Die Wirbeltiere; Eine Übersicht Über Die Fossilen und Lebenden Formen.* Gebrüder Borntraeger, Berlin.

Jones, M. E. H., C. L. Anderson, C. A. Hipsley, J. Müller, S. E. Evans, and R. R. Schoch. 2013. Integration of molecules and new fossils supports a Triassic origin for *Lepidosauria* (lizards, snakes, and tuatara). *BMC Evol. Biol.* 13:208.

Jones, M. E. H., A. J. D. Tennyson, J. P. Worthy, S. E. Evans, and T. H. Worthy. 2009. A sphenodontine (*Rhynchocephalia*) from the Miocene of New Zealand and palaeobiogeography of the tuatara (*Sphenodon*). *Proc. R. Soc. Lond. B Biol. Sci.* 276:1385–1390.

Kuhn, O. 1966. *Die Reptilien. System und Stammesgeschichte.* Oeben, Krailling.

LeBlanc, A. R. H., M. J. MacDougall, Y. Haridy, D. Scott, and R. R. Reisz. 2018. Caudal autotomy as anti-predatory behaviour in Palaeozoic reptiles. *Sci. Rep.* 8:3328. doi:10.1038/s41598-018-21526-3.

Linnaeus, C. 1758. *Systema Naturae Per Regna Tria Naturae, Secundum Classes, Ordines, Genera, Species, cum Characteribus, Differentiis, Synonymis, Locis.* 10th edition. Laurentii Salvii, Holmiae (Stockholm).

Maderson, P. F. A. 1985. Some developmental problems of the reptilian integument. Pp. 525–598 in *Biology of the Reptilia*, Vol. 14. (P. F. A. Maderson, C. Gans, and F. Billett, eds.). John Wiley & Sons, New York.

Nopcsa, F. 1923. *Die Familien der Reptilien.* Gebrüder Borntraeger, Berlin.

Owen, R. 1845. Description of certain fossil crania, discovered by A. G. Bain, Esq., in sandstone rocks at the south-eastern extremity of Africa, referable to a different species of an extinct genus of *Reptilia* (*Dicynodon*), and indicative of a new tribe or sub-order of *Sauria. Trans. Geol. Soc. Lond.* 7:59–84 + 4 plates.

Petermann, H., and J. A. Gauthier. 2018. Fingerprinting snakes: paleontological and paleoecological implications of zygantral growth rings in *Serpentes. Peer J.* 6:e4819. doi:10.7717/peerj.481.

Pritchard, A. C., and S. J. Nesbitt. 2017. A bird-like skull in a Triassic diapsid reptile increases heterogeneity of the morphological and phylogenetic radiation of *Diapsida. R. Soc. Open Sci.* 4:170499.

Pyron, R. A., F. T. Burbrink, and J. J. Wiens. 2013. A phylogeny and revised classification of *Squamata*, including 4161 species of lizards and snakes. *BMC Evol. Biol.* 13:93.

Reeder, T. W., T. M. Townsend, D. G. Mulcahy, B. P. Noonan, P. L. Wood, J. W. Sites, and J. J. Wiens. 2015. Integrated analyses resolve conflicts over squamate reptile phylogeny and reveal unexpected placments for fossil taxa. *PLOS ONE* 10(3):e0118199.

Rest, J. S., J. C. Ast, C. C. Austin, P. J. Waddell, E. A. Tibbetts, J. M. Hay, and D. P. Mindell. 2003. Molecular systematics of primary reptilian lineages and the tuatara mitochondrial genome. *Mol. Phylogenet. Evol.* 29:289–297.

Reynoso, V.-H. 2000. An unusual aquatic sphenodontian (*Reptilia: Diapsida*) from the Tlayua Formation (Albian), Central Mexico. *J. Paleontol.* 74:133–148.

Rieppel, O., and R. Reisz. 1999. The origin and early evolution of turtles. *Annu. Rev. Ecol. Syst.* 30:1–22.

Robinson, P. L. 1975. The functions of the hooked fifth metatarsal in lepidosaurian reptiles. *Coll. Internat. CNRS* 218:461–83.

Romer, A. S. 1933. *Vertebrate Paleontology.* University of Chicago Press, Chicago, IL.

Romer, A. S. 1945. *Vertebrate Paleontology.* 2nd edition. University of Chicago Press, Chicago, IL.

Romer, A. S. 1956. *Osteology of the Reptiles.* University of Chicago Press, Chicago, IL.

Romer, A. S. 1966. *Vertebrate Paleontology.* 3rd edition. University of Chicago Press, Chicago, IL.

Simões, T. R., M. W. Caldwell, M. Tałanda, M. Bernardi, A. Palci, O. Vernygora, F.

Bernardini, L. Mancini, and R. L. Nydam. 2018. The origin of squamates revealed by a Middle Triassic lizard from the Italian Alps. *Nature* 557:706–709.

Uetz, P. 2016. Species numbers (as of August 2016). The reptile database. Available at http://www.reptile-database.org/.

Underwood, G. 1957. On lizards of the family *Pygopodidae*: a contribution to the morphology and phylogeny of the *Squamata*. *J. Morphol.* 100:207–268.

Whiteside, D. 1986. The head skeleton of the Rhaetian sphenodontid *Diphydontosaurus avonis* gen. et sp. nov. and the modernizing of a living fossil. *Philos. Trans. R. Soc. Lond. B Biol. Sci.* 312:379–430.

Williston, S. W. 1904. The relationships and habits of the mosasaurs. *J. Geol.* 12:43–51.

Williston, S. W. 1917. The phylogeny and classification of reptiles. *J. Geol.* 25:411–421.

Williston, S. W. 1925. *The Osteology of the Reptiles*. Harvard University Press, Cambridge, MA.

Zardoya, R., and A. Meyer. 2000. Mitochondrial evidence on the phylogenetic position of caecilians (*Amphibia*: *Gymnophiona*). *Genetics* 155:765–775.

Zittel, K. A. 1887–1890. *Handbuch der Palæontologie. I Abtheilung. Palæozoologie. III Band*. Vertebrata (Pices, Amphibia, Reptilia, Aves). R. Oldenbourg, Munich.

Authors

Kevin de Queiroz; Department of Vertebrate Zoology; National Museum of Natural History; Smithsonian Institution; Washington, DC 20560-0162, USA. Email: dequeirozk@si.edu.

Jacques A. Gauthier; Department of Geology and Geophysics; Yale University; 210 Whitney Avenue, New Haven, CT 06511, USA. Email: jacques.gauthier@yale.edu.

Date Accepted: 11 August 2018

Primary Editor: Philip Cantino

Pan-Squamata J. A. Gauthier and K. de Queiroz, new clade name

Registration Number: 122

Definition: The total clade of the crown clade *Squamata*. This is a crown-based total-clade definition. Abbreviated definition: total ∇ of *Squamata*.

Etymology: Derived from the Greek *pan* (all, every), here referring to "pan-monophylum," another term for "total clade," and *Squamata*, the name of the corresponding crown (for etymology, see *Squamata* in this volume); hence, "the total clade of *Squamata*."

Reference Phylogeny: Figure 1 of Gauthier et al. (2012) is the primary reference phylogeny (see also Evans, 1984: Fig. 3; Gauthier, 1984: Fig. 32; Benton, 1985: Fig. 10; Evans, 1988: Fig. 6.2; Gauthier et al., 1988: Fig. 13; Reynoso, 1998: Fig. 10; Evans and Barbadillo, 1998: Fig. 10; Lee, 1998: Fig. 1; Evans and Barbadillo, 1999: Fig. 6; Evans et al., 2005: Fig. 18B; Conrad, 2008: Fig. 56; Evans and Wang, 2010: Fig. 11; Bolet and Evans, 2010: Fig. 6; Simões et al., 2018: Fig. 2). On the primary reference phylogeny, *Squamata* includes *Anolis carolinensis* and all taxa below it in the figure, while *Pan-Squamata* includes those taxa plus *Huehuecuetzpalli mixtecus*.

Composition: *Squamata* and its stem group— that is, *Squamata* and all extinct species that are more closely related to that crown clade than they are to *Sphenodon punctatus*. Although several extinct taxa have, at one time or another, been considered stem squamatans, the best candidate for a stem squamatan is the Early Cretaceous *Huehuecuetzpalli mixtecus* (Reynoso, 1998; Gauthier et al., 2012). See Comments for further discussion of *H. mixtecus* and other potential stem squamatans.

Diagnostic Apomorphies: Possession of any of the putative synapomorphies of *Squamata* (this volume), or those diagnosing its subclades, permit referral of fossils to *Pan-Squamata*. Some of the most obvious characters that are likely to be preserved in fossils are included in the Diagnostic Apomorphies for *Squamata* (this volume). According to Gauthier et al. (2012), these include the following apomorphies that the stem squamatan *Huehuecuetzpalli mixtecus* shares with *Squamata*: (1) frontoparietal suture roughly transverse; (2) jugal closely approaches prefrontal below orbit; (3) jugal entirely exposed above orbital margin of maxilla; (4) jugal quadratojugal process absent; (5) quadrate head pivots on slender tapering tip of squamosal; (6) quadratojugal absent; (7) pterygoid only narrowly overlaps quadrate; (8) epipterygoid columelliform; (9) processus ascendens of synotic tectum present; (10) angular does not reach level of mandibular condyle; (11) coronoid eminence formed entirely by coronoid; (12) coronoid arches over dorsal margin of mandible to reach lateral face of surangular; (13) coronoid posteromedial process present; (14) scapulocoracoid emargination present; (15) anterior coracoid emargination present; (16) pubis symphysial process tapered distally; (17) penultimate phalanges in hand longer than antepenultimate phalanges; (18) fibula-astragalar joint involves most of distal end of fibula; (19) tibia and fibula only narrowly separated on ankle.

Synonyms: All of the names listed as approximate synonyms of *Squamata* (this volume) can also be interpreted as approximate synonyms of *Pan-Squamata* because the authors of those names did not explicitly distinguish between crown and total clades. In addition, *Lacertilia* of de Queiroz and Gauthier (1992) is an unambiguous synonym.

Comments: See *Squamata* (this volume) for historical information concerning the recognition of a group corresponding to one or more of the clades in the squamatan total-crown series. The authors of early phylogenetic analyses corroborating the existence of this clade either did not distinguish nomenclaturally between crown and total clades (e.g., Rage, 1982; Evans, 1984, 1988) or applied the name *Squamata* to the crown (Gauthier, 1984; Gauthier et al., 1988; Estes et al., 1988). Because no known taxa were inferred to be stem squamatans, there was not then a pressing need to name the total clade and, in any case, it was not named. De Queiroz and Gauthier (1992) proposed using the name *Lacertilia* for the squamatan total clade; however, that proposal was ignored when both previously known and newly discovered taxa were referred to the squamatan stem (e.g., Reynoso, 1998; Evans and Barbadillo, 1998, 1999; Simões et al., 2018). The name *Lacertilia* is a less appropriate choice for a clade containing *Squamata* because it has commonly been applied to a (paraphyletic) subgroup of *Squamata* (e.g., Williston, 1925; Romer, 1956, 1966; Carroll, 1988). Among the names that have been applied ambiguously to this clade (see Synonyms for *Squamata*, this volume), the best-known names (e.g., *Sauria*) have disadvantages similar to those of *Lacertilia* or are more appropriately applied to different clades (e.g., *Reptilia, Lepidosauria, Pholidota*), while the remaining names (e.g., *Saurophidia, Streptostylica, Lyognathi*) have been used so infrequently that there would be little advantage to selecting one of them as the name of the total clade. Consequently, use of a pan-clade name in this case seems uncontroversial, and because of the advantages of basing the name of a total clade on that of its corresponding crown (de Queiroz, 2007), we have chosen to name the total clade *Pan-Squamata*.

The assignment of fossils to the squamatan stem is disputed. *Pan-Squamata*, which encompasses that stem, must extend deep into the Triassic (if not the Permian; Simões et al., 2018), as its sister clade (represented by *Sphenodon punctatus* in the extant biota) was already diverse, disparate and widespread by the Late Triassic (e.g., Fraser and Benton, 1989). There is also a substantial set of apomorphies diagnosing the squamatan crown relative to rhynchocephalians, suggesting a stem lineage of substantial duration. Thus, it is surprising that so few potential stem squamatans have been identified, although this situation results in part from disagreements concerning relationships within *Squamata* inferred from morphological versus molecular data (see Comments for *Squamata*, this volume).

Caldwell et al. (2015) referred the Upper Jurassic/Lower Cretaceous *Parviraptor estesi*, originally described by Evans (1994b) as related to varanoid anguimorphs, to *Serpentes*. However, while tooth form and implantation are indeed snake-like, the rest of the associated elements—including notochordal vertebrae, an anteriorly bowed fronto-parietal suture, and paired parietals—suggest that this species could instead represent a stem squamatan with a snake-like dentition.

Reynoso (1998) inferred that the Early Cretaceous Mexican fossil *Huehuecuetzpalli mixtecus* was a stem squamatan, although he noted that it possessed a few iguanian apomorphies. In Conrad's (2008) analysis, *H. mixtecus* was part of a trichotomy at the squamatan crown node in a strict consensus tree (Fig. 54), but part of the iguanian stem group in an Adams consensus (Fig. 60; see also Evans and Wang, 2010). Gauthier et al. (2012) inferred *H. mixtecus* to be a stem squamatan in both parsimony and Bayesian analyses. In an analysis combining morphological and molecular characters (Wiens et al., 2010), *H. mixtecus* again appears on the iguanian stem, although that tree has *Iguania* deeply nested within *Squamata*, far from its position inferred from morphological characters as one of the two primary squamatan subclades (e.g., Estes et al., 1988; Conrad, 2008; Gauthier et al., 2012). In a more recent combined analysis in which *Iguania* was similarly nested within *Squamata*, however, *H. mixtecus* was consistently inferred to be a stem squamatan (Simões et al., 2018).

Evans and Barbadillo (1998, 1999) presented results suggesting that the extinct taxa *Ardeosaurus brevipes, Eichstaettisaurus schroederi, Scandensia ciervensis, Hoyalacerta sanzi,* and *Bavarisaurus macrodactylus* are representatives of the squamatan stem group—that is, members of *Pan-Squamata* but not *Squamata*. *A. brevipes, E. schroederi,* and *B. macrodactylus* had previously been interpreted as related to gekkotans and thus nested within *Squamata* (e.g., Hoffstetter, 1962, 1964, 1967; Estes, 1983; Evans, 1994a). By contrast, when added to Lee's (1998) dataset, these species were placed inside the squamatan crown (Evans et al., 2005). Similarly, Conrad's (2008) analysis placed all of these fossils, the putative stem gekkotans as well as *H. sanzi* and *S. ciervensis,* within the crown, a result confirmed by others adding new Mesozoic fossil taxa, as well as new material of *S. ciervensis,* to Conrad's dataset (e.g., Evans and Wang, 2010; Bolet and Evans, 2010, 2011). Gauthier et al. (2012) found strong support for placement of *E. schroederi* within the crown as part of the gekkotan stem, and Simões et al. (2017) placed both *E. schroederi* and *Ardeosaurus digitalellus* as stem gekkotans.

The highly-modified burrowing form *Tamaulipasaurus morenoi* has been interpreted as a possible stem squamatan, although the data are ambiguous (Clark and Hernandez, 1994): shortest trees placed it on either the lepidosaurian or the squamatan stem. In either case, its paired premaxillae, primitive quadrate suspension, complete lower temporal bar (including a quadratojugal), and the large size of its jugular foramen—indicating passage of the jugular vein and an undivided metotic fissure—suggest that it is outside the squamatan crown. Derived states that *T. morenoi* shares with crown squamatans, such as procoelous vertebrae lacking intercentra, are absent in other potential stem squamatans (e.g., *Huehuecuetzpalli mixtecus*) that share other derived states with crown squamatans (most notably a mobile peg-and-socket squamosal-quadrate articulation, for which *T. morenoi* retains the plesiomorphic condition (for *Diapsida*) in which the quadrate head sits in a fossa below the squamosal), suggesting that the resemblances of *T. morenoi* to squamatans may be homoplastic.

Megachirella wachtleri, from the Middle Triassic of Italy, was originally inferred to be a stem squamatan in some analyses but a stem lepidosaur in others (Renesto and Bernardi, 2014). Renesto and Bernardi (2014) considered its placement within the total clade of *Lepidosauria* well supported but noted that more data were needed to assess its relationships to *Squamata*. Simões et al. (2018) have recently inferred this species to be a stem squamatan closely related to the Middle Jurassic *Marmoretta oxoniensis,* a taxon previously regarded as a stem lepidosaur (e.g., Evans, 1991), in several analyses using either morphology only or morphology in combination with DNA-sequence data. Relationships of *M. oxoniensis* appear somewhat unstable, however, as the relaxed-clock Bayesian analysis of Simões et al. (2018) placed it with *Huehuecuetzpalli mixtecus;* all three species were nevertheless inferred to be stem squamatans in all their analyses.

Gephyrosaurus bridensis, from the Early Jurassic of Wales, was placed by Evans (1984: Fig. 3) as a stem squamatan. However, she noted that other characters suggested a closer relationship to *Sphenodon,* and most subsequent analyses have placed this taxon as the earliest-diverging member of *Rhynchocephalia*, and thus part of the stem group of *Sphenodon* (e.g., Evans, 1988, 2003; Gauthier et al., 1988, 2012; Fraser, 1988; Fraser and Benton, 1989; Bever and Norell, 2017).

Kuroyuriella mikikoi, from the Early Cretaceous of Japan, was placed on the squamatan stem in some analyses but on the stem of *Scincidae* (i.e., within crown *Squamata*) in others (Evans and Matsumoto, 2015). Evans and Matsumoto (2015) considered placement on the squamatan stem "problematic and probably artifactual" and treated the relationships of *K. mikkoi* as *incertae sedis*, presumably within the total clade of *Squamata*.

Thus, there are currently no undisputed representatives of the squamatan stem group. In our view, *Huehuecuetzpalli mixtecus* is the best candidate for a stem squamatan. It retains a number of ancestral features that are inferred to have been modified prior to the crown node, including paired premaxillae, parietals that fuse late in post-hatching ontogeny, a long supratemporal and short parietal supratemporal process, a plesiomorphic postorbital-squamosal relationship, as well as unicuspid teeth, amphicoelous vertebrae, persistent trunk intercentra, and a second distal tarsal in the foot (Reynoso, 1998; Gauthier et al., 2012). However, the deep nesting of this fossil within crown *Squamata* (as a stem iguanian) as inferred from some analyses of combined morphological and molecular data (Wiens et al., 2010) suggests that its status as a stem versus crown squamatan must await resolution of the current incongruence between trees inferred from morphological versus molecular data (see Losos et al., 2012; McMahan et al., 2015; Reeder et al., 2015). On the other hand, recent combined analyses (Simões et al., 2018) inferred a similarly nested position

for *Iguania*, but still had *H. mixtecus* as a stem squamatan.

Based on their relaxed-clock Bayesian estimate, Simões et al. (2018) inferred that the squamatan stem originated in the latest Permian (~255 Ma). But with the possible exception of *Megachirella wachtleri*, no other pan-squamatans are known from anywhere in the world from throughout the Permian, Triassic or Early Jurassic (Evans, 2003); they must have been present in the early Mesozoic, however, as rhynchocephalians are diverse, disparate and distributed world-wide by at least ~220 Ma (late Carnian; early Late Triassic; e.g., Hsiou et al., 2015), and are known from the Middle Triassic of Germany (~240 Ma; Anisian; Jones et al., 2013). Microvertebrate-producing localities usually yield pan-squamatan remains from the latter part of the Jurassic to the present, and the disparity in relative abundance between the two primary subclades of *Lepidosauria* is currently attributed to the early ecological dominance of rhynchocephalians (see *Pan-Lepidosauria*, this volume).

Literature Cited

Benton, M. J. 1985. Classification and phylogeny of the diapsid reptiles. *Zool. J. Linn. Soc.* 84:97–164.

Bever, G. S., and M. A. Norell. 2017. A new rhynchocephalian (*Reptilia: Lepidosauria*) from the Late Jurassic of Solnhofen (Germany) and the origin of the marine *Pleurosauridae. R. Soc. Open Sci.* 4:170570..

Bolet, A., and S. E. Evans. 2010. A new lizard from the Early Cretaceous of Catalonia (Spain), and the Mesozoic lizards of the Iberian Peninsula. *Cretac. Res.* 31:447–457.

Bolet, A., and S. E. Evans. 2011. New material of the enigmatic *Scandensia*, an Early Cretaceous lizard from the Iberian Peninsula. *Spec. Pap. Palaeontol.* 86:99–108.

Caldwell, M. W., R. L. Nydam, A. Palci, and S. Apesteguía. 2015. The oldest known snakes from the Middle Jurassic-Lower Cretaceous provide insights on snake evolution. *Nat. Commun.* 6:5996.

Carroll, R. L. 1988. *Vertebrate Paleontology and Evolution*. W. H. Freeman, New York.

Clark, J. M., and R. Hernandez. 1994. A new burrowing diapsid from the Jurassic La Boca Formation of Tamaulipas, Mexico. *J. Vertebr. Paleontol.* 14:180–195.

Conrad, J. L. 2008. Phylogeny and systematics of *Squamata* (*Reptilia*) based on morphology. *Bull. Am. Mus. Nat. Hist.* 310:1–182.

Estes, R. 1983. Sauria *Terrestria*, Amphisbaenia. *Handbuch der Paläoherpetologie*. Teil 10A. Gustav Fischer, Stuttgart.

Estes, R., K. de Quieroz, and J. Gauthier. 1988. Phylogenetic relationships within *Squamata*. Pp. 119–281 in *Phylogenetic Relationships of the Lizard Families* (R. Estes and G. Pregill, eds.). Stanford University Press, Stanford, CA.

Evans, S. E. 1984. The classification of the *Lepidosauria*. *Zool. J. Linn. Soc.* 82:87–100.

Evans, S. E. 1988. The early history and relationships of the *Diapsida*. Pp. 221–260 in *The Phylogeny and Classification of the Tetrapods. Vol. 1: Amphibians, Reptiles, Birds* (M. J. Benton, ed.). Systematics Association Special Volume No. 35A. Clarendon Press, Oxford.

Evans, S. E. 1991. A new lizard-like reptile (*Diapsida: Lepidosauromorpha*) from the Middle Jurassic of England. *Zool. J. Linn. Soc.* 103:391–412.

Evans, S. E. 1994a. Jurassic lizard assemblages. *Rev. Paléobiol. Vol. Spec.* 7:55–65.

Evans, S. E. 1994b. A new anguimorph lizard from the Jurassic and Lower Cretaceous of England. *Palaeontology* 37:33–49.

Evans, S. E. 2003. At the feet of dinosaurs: the early history and radiation of lizards. *Biol. Rev.* 78:513–551.

Evans, S. E., and L. J. Barbadillo. 1998. An unusual lizard (*Reptilia: Squamata*) from the Early Cretaceous of Las Hoyas, Spain. *Zool. J. Linn. Soc.* 124:235–265.

Evans, S. E., and L. J. Barbadillo. 1999. A short-limbed lizard from the Lower Cretaceous of Spain. *Spec. Pap. Palaeontol.* 60:73–85.

Evans, S. E., and R. Matsumoto. 2015. An assemblage of lizards from the Early Cretaceous of Japan. *Palaeontol. Elect.* 18.2.36A:1–36..

Evans, S. E., and Y. Wang. 2010. A new lizard (*Reptilia: Squamata*) with exquisite preservation of soft tissue from the Lower Cretaceous of Inner Mongolia, China. *J. Syst. Paleontol.* 8:81–95.

Evans, S. E., Y. Wang, and C. Li. 2005. The Early Cretaceous lizard genus *Yabeinosaurus* from China: resolving an enigma. *J. Syst. Paleontol.* 3:319–335.

Fraser, N. C. 1988. The osteology and relationships of *Clevosaurus* (*Reptilia: Sphenodontida*). *Philos. Trans. R. Soc. Lond. B Biol. Sci.* 321:125–178.

Fraser, N. C., and M. J. Benton. 1989. The Triassic reptiles *Brachyrhinodon* and *Polysphenodon* and the relationships of the sphenodontids. *Zool. J. Linn. Soc.* 96:413–445.

Gauthier, J. A. 1984. *A Cladistic Analysis of the Higher Systematic Categories of the* Diapsida. PhD dissertation, University of California, Berkeley, CA.

Gauthier, J. A., R. Estes, and K. de Queiroz. 1988. A phylogenetic analysis of *Lepidosauromorpha*. Pp. 15–98 in *Phylogenetic Relationships of the Lizard Families* (R. Estes and G. Pregill, eds.). Stanford University Press, Stanford, CA.

Gauthier, J. A., M. Kearney, J. A. Maisano, O. Rieppel, and A. D. B. Behlke. 2012. Assembling the squamate tree of life: perspectives from the phenotype and the fossil record. *Bull. Peabody Mus. Nat. Hist.* 53:3–308.

Hoffstetter, R. 1962. Révue des récentes acquisitions concernant l'histoire et la systématique des squamates. Pp. 243–279 in *Problèmes Actuels de Paléontologie* (évolution des vertébrés) (J. P. Lehman, ed.). Coll. Int. CNRS (104), Paris.

Hoffstetter, R. 1964. Les *Sauria* du Jurassique Supérieur et specialement les *Gekkota* de Bavière et de Mandchourie. *Senckenb. Biol.* 45:281–324.

Hoffstetter, R. 1967. A propos des genres *Ardeosaurus* et *Eichstaettisaurus* (*Reptilia, Sauria, Gekkonoidea*) du Jurassique Supérieur de Franconie. *Bull. Soc. Géol. France* 7:592–595.

Hsiou, A. S., M. A. G. De França, and J. Ferigolo. 2015. New data on the *Clevosaurus* (*Sphenodontia: Clevosauridae*) from the Upper Triassic of Southern Brazil. *PLOS ONE* 10(9):e0137523.

Jones, M. E. H., C. L. Anderson, C. A. Hipsley, J. Müller, S. E. Evans, and R. R. Schoch. 2013. Integration of molecules and new fossils supports a Triassic origin for *Lepidosauria* (lizards, snakes, and tuatara). *BMC Evol. Biol.* 13: 208.

Lee, M. S. Y. 1998. Convergent evolution and character correlation in burrowing reptiles: towards a resolution of squamate relationships. *Biol. J. Linn. Soc.* 65:369–453.

Losos, J. B., D. M. Hillis, and H. W. Greene. 2012. Who speaks with a forked tongue? *Science* 338:1428–1429.

McMahan, C. D., L. R. Freeborn, W. C. Wheeler, and B. I. Crother. 2015. Forked tongues revisited: molecular apomorphies support morphological hypotheses of squamate evolution. *Copeia* 103:525–529.

de Queiroz, K. 2007. Toward an integrated system of clade names. *Syst. Biol.* 56:956–974.

de Queiroz, K., and J. A. Gauthier. 1992. Phylogenetic taxonomy. *Annu. Rev. Ecol. Syst.* 23:449–480.

Rage, J.-C. 1982. La phylogenie des lepidosauriens (*Reptilia*): une approche cladistique. *C. R. Acad. Sci. Paris* 284:1765–1768.

Reeder, T. W., T. M. Townsend, D. G. Mulcahy, B. P. Noonan, P. L. Wood, Jr., J. W. Sites, Jr., and J. J. Wiens. 2015. Integrated analyses resolve conflicts over squamate reptile phylogeny and reveal unexpected placements for fossil taxa. *PLOS ONE* 10(3):e0118199.

Renesto, S., and M. Bernardi. 2014. Redescription and phylogenetic relationships of *Megachirella wachtleri* Renesto et Posenato, 2003 (*Reptilia, Diapsida*). *Paläontol. Z.* 88:197.

Reynoso, V.-H. 1998. *Huehuecuetzpalli mixtecus* gen. et sp. nov: a basal squamate (*Reptilia*) from the Early Cretaceous of Tepexi de Rodríguez, central México. *Philos. Trans. R. Soc. Lond. B Biol. Sci.* 353:477–500.

Romer, A. S. 1956. *Osteology of the Reptiles.* University of Chicago Press, Chicago, IL.

Romer, A. S. 1966. *Vertebrate Paleontology.* 3rd edition. University of Chicago Press, Chicago, IL.

Simões, T. R., M. W. Caldwell, R. L. Nydam, and P. Jiménez-Huidobro. 2017. Osteology, phylogeny, and functional morphology of two Jurassic lizard species and the early evolution of scansoriality in geckoes. *Zool. J. Linn. Soc.* 180:216–241.

Simões, T. R., M. W. Caldwell, M. Tałanda, M. Bernardi, A. Palci, O. Vernygora, F. Bernardini, L. Mancini, and R. L. Nydam. 2018. The origin of squamates revealed by a Middle Triassic lizard from the Italian Alps. *Nature* 557:706–709.

Wiens, J. J., C. A. Kuczynski, T. Townsend, T. W. Reeder, D. G. Mulcahy, and J. W. Sites, Jr. 2010. Combining phylogenomics and fossils in higher-level squamate reptile phylogeny: molecular data change the placement of fossil taxa. *Syst. Biol.* 59:674–688.

Williston, S. W. 1925. *The Osteology of the Reptiles.* Harvard University Press, Cambridge, MA.

Authors

Jacques A. Gauthier; Department of Geology and Geophysics; Yale University; 210 Whitney Avenue; New Haven, CT 06511, USA. Email: jacques.gauthier@yale.edu.

Kevin de Queiroz; Department of Vertebrate Zoology; National Museum of Natural History; Smithsonian Institution; Washington, DC 20560-0162, USA. Email: dequeirozk@si.edu.

Date Accepted: 18 July 2018

Primary Editor: Philip Cantino

Squamata M. Oppel 1811 [K. de Queiroz and J. A. Gauthier], converted clade name

Registration Number: 101

Definition: The largest crown clade containing *Lacerta agilis* Linnaeus 1758 but not *Sphenodon* (originally *Hatteria*) *punctatus* (Gray 1842). This is a maximum-crown-clade definition. Abbreviated definition: max crown ∇ (*Lacerta agilis* Linnaeus 1758 ~ *Sphenodon punctatus* (Gray 1842)).

Etymology: Derived from the Latin *squama* (scale) + *-ata* (provided with), thus, "scaled."

Reference Phylogeny: Figure 1 of Gauthier et al. (2012) is the primary reference phylogeny (see also Estes et al. 1988: Fig. 6; Lee, 1998: Figs. 1–2; Conrad, 2008: Fig. 56); however, our definition has been formulated so that it will apply to a clade of identical composition in the context of phylogenies based on molecular (e.g., Townsend et al., 2004: Fig. 8; Vidal and Hedges, 2005: Fig. 1; Wiens et al., 2010: Fig. 4; Pyron et al., 2013: Fig. 1) and combined morphological and molecular data (Reeder et al., 2015: Fig. 1; Simões et al., 2018: Fig. 2) that differ regarding relationships within *Squamata* (see Composition and Comments). Although *Lacerta agilis* is not included in the primary reference phylogeny, it is most closely related to *Lacerta viridis* of the taxa included in that tree (see Baeckens et al., 2015).

Composition: *Squamata* was hypothesized by Estes et al. (1988; see also Gauthier, 1982) to be composed of two primary extant subclades, *Iguania* and *Scleroglossa*, and those subclades have been corroborated by subsequent analyses based on morphological data (e.g., Lee, 1998; Conrad, 2008; Gauthier et al., 2012). By contrast, analyses based on DNA-sequence data place the root of the squamatan tree within "*Scleroglossa*," usually with either *Dibamidae* or *Gekkota*, or a clade composed of both taxa, as the sister to all other squamatans (e.g., Townsend et al., 2004; Vidal and Hedges, 2005; Wiens et al., 2010; Pyron et al., 2013; Reeder et al., 2015). In the context of these alternative hypotheses, the two primary extant subclades are different and in one case have been given the names *Dibamia* and *Bifurcata* (Vidal and Hedges, 2005). *Squamata* is composed of 10,078 currently recognized extant species (Uetz, 2016), thus rivaling *Aves* and surpassing its sister clade, *Sphenodon* (with only one currently recognized extant species), by four orders of magnitude. Somewhat dated lists of extinct species are available for *Serpentes* (Rage, 1984) and for non-*Serpentes*, non-mosasaurian *Squamata* (Estes, 1983).

Diagnostic Apomorphies: Gauthier et al. (1988) listed 69 + 5 (addendum) putative synapomorphies of *Squamata* relative to other extant diapsids, which are also diagnostic relative to most other extinct diapsids given that so few potential stem squamatans have been identified (see *Pan-Squamata* in this volume). Estes et al. (1988) added 18 potential synapomorphies that have reversed in some squamatans. Gauthier et al. (2012) listed only 59, including 20 that are shared with the putative stem squamatan *Huehuecuetzpalli mixtecus* (see *Pan-Squamata*, this volume), but this lower number reflects the near-absence of soft anatomical characters in their study. Some of the most obvious diagnostic apomorphies are paired male intromittent organs (hemipenes), loss of caruncle (false egg tooth), mobile fronto-parietal joint (modified in fossorial forms and mosasauroids), embryonic fusion of parietals (reversed in some gekkotans, some xantusiids, and *Sineoamphisbaena hexatabularis*), mobile quadrate with peg-in-socket squamosal-quadrate articulation (lost in chameleons and modified in snakes), pterygoids and vomers separated by palatines (reversed in some amphisbaenians and some polyglyphanodontians), very slender stapes (reversed in some fossorial forms), slender epipterygoid, subdivision of embryonic metotic fissure into vagus (jugular) foramen posteriorly and recessus scalae tympani anteriorly, embryonic or early post-embryonic (dibamids and some xantusiids) fusion of exoccipitals and opisthotics, coronoid eminence formed by coronoid bone only (reversed in some snakes and some amphisbaenians), keeled cervical intercentra, all ribs (including cervicals) single-headed, procoelous vertebrae (reversed in some gekkotans), absence of trunk intercentra (reversed in some gekkotans and full-grown xantusiids), emarginated scapulocoracoid (reversed in *Heloderma*, amphisbaenians, chameleons, some dolichosaurs, some dibamids, some pygopodids), and absence of gastralia.

Synonyms: Numerous approximate synonyms of *Squamata* match the modern concept to various degrees in terms of composition. *Squamosa* of Latreille (1825, except for the inclusion of caecilians), *Saurophidia* of Blainville (1835), *Reptilia* of Bonaparte (1840, 1841), *Streptostylica* of Stannius (1856) and Cope (1900), *Lepidosauria* and *Diplophalli* of Haeckel (1866, inclusion of *Sphenodon* unclear), *Pholidota* of Haeckel (1895), *Sauria* of Gadow (1898, 1901), and *Lyognathi* and *Lyognatha* of Jaekel (1911) refer to a taxon of more or less identical composition.

Sauria of MaCartney (1802) and others (some of whom included crocodilians), *Saurii* of Kuhl (1820) and others, *Gradientia* of Merrem

(1820) and Haworth (1825), *Sauri* of Gray (1825), *Pedata*, *Tetrapoda*, and *Communipedis* of Haworth (1825), *Saurae* of Gray (1827) and Wagler (1828), *Lacertae* of Wagler (1830) and others, *Lacertiformes* of Bonaparte (1831), *Lacertina* of Müller (1831), *Squamati* of Wiegmann (1834), *Saures* of Swainson (1839), *Lacertilia* of Owen (1842 [including *Rhynchosaurus articeps*], 1866) and others, *Saura* of Gray (1845), *Lacertia* of Owen (1845), *Kionocrania* of Stannius (1856) and Huxley (1886), *Lepidota* of Jan (1857), *Autosaurii* of Gegenbaur (1859), *Autosauria* of Haeckel (1866), *Lacertilia vera* of Boulenger (1884, 1885–7), *Eusauri* of Gadow (1898), *Autosauri* of Gadow (1898, 1901), *Lacerti* of Jaekel (1911), *Lacertosauria* of Tornier (1913), and *Lacertidae* of Nopcsa (1923) are partial as well as approximate synonyms that refer to taxa that exclude various highly modified squamatans, such as snakes, amphisbaenians, other serpentiform species, and chameleons (in various combinations).

Other partial (and approximate) synonyms were used for a paraphyletic group made up of *Iguania* and *Gekkota*, which is associated with the same ancestor as *Squamata* on the reference phylogeny (but not on some of those based on molecular data—see Reference Phylogeny). These include *Pachyglossi* of Bonaparte (1840), *Amblyglossae* of Fitzinger (1843), *Pachyglossae* of Gray (1845), *Ascalabota* of Camp (1923), and *Iguaniformes* of Hay (1930), although Gray's (1845) *Pachyglossae* might alternatively be considered a partial synonym of *Lepidosauria* given that it included *Sphenodon punctatus*. Still other partial synonyms were used for a doubly paraphyletic group that excluded both *Autarchoglossa* and *Chamaeleonidae*; these include *Ascalabotae* of Merrem (1820), *Ascolabata* of Gray (1825), *Inextensilinguis* of Haworth (1825), *Pachyglossae* of Weigmann (1834), and *Pachyglossa* of Strauch (1887; who also included *Xenosaurus* and *Heloderma*).

Comments: Many eighteenth and early nineteenth century naturalists based their primary groupings within ectothermic tetrapods (*Amphibia* of some authors, *Reptilia* of others) on mode of locomotion, separating the slitherers with small or no limbs (*Serpentes*, *Ophidia*, and variants of those names) from the limbed-propelled walkers and crawlers (*Gradientia*, *Reptilia*, and variants of those names). Thus, squamatans with well-developed limbs ("typical lizards") were commonly grouped with crocodilians and sometimes with turtles and even salamanders and frogs, rather than with snakes, while those with long bodies and reduced limbs (e.g., *Anguis*, *Amphisbaena*, and sometimes even the amphibian *Caecilia*) were considered "snakes" (e.g., Linnaeus, 1758; Laurenti, 1768; Scopoli, 1777). The "lizards" (under various names) were first separated from turtles (e.g., Brongniart, 1800; Cuvier, 1800; Daudin, 1802–1803) and later from crocodilians (e.g., Blainville, 1816, 1822; Merrem, 1820; Gray, 1825). Derivation of snakes from within "lizards" was inferred by several authors (e.g., Haeckel, 1866; Camp, 1923; McDowell and Bogert, 1954), and the formal abandonment of the paraphyletic group "lizards", by that time usually referred to as either *Sauria* or *Lacertilia*, was proposed by Estes et al. (1988). Although the paraphyly of "lizards" (relative to both snakes and amphisbaenians) is now highly corroborated and widely acknowledged, the group continues to be treated as a taxonomic unit, although no longer given a formal name (e.g., Vitt and Caldwell, 2014; Pough et al., 2016; Crother, 2017).

Despite the long history of treating "lizards" and snakes as separate taxa, a close relationship between them has also been recognized for a long time. The first author to recognize and name a taxon composed exclusively of "lizards" (including amphisbaenians) and snakes appears to have been Blainville (1816), who used the French vernacular name *Bispeniens* in reference to the hemipenes, which are still considered diagnostic (see Diagnostic Apomorphies). The taxon was recognized inconsistently under a variety of names by nineteenth century authors (e.g., Blainville, 1822, 1835; Latreille, 1825; Haworth, 1825; Fitzinger, 1826; Gray, 1831a; Bonaparte, 1840, 1841, 1850; Stannius, 1856; Zittel, 1887–1890; Cope, 1889; Haeckel, 1895), but during the twentieth century it was recognized more consistently under the name *Squamata* (e.g., Cope, 1900; Hay, 1902; Williston, 1917, 1925; Camp, 1923; Nopcsa, 1923; Romer, 1933, 1945, 1956, 1966; Gans, 1978; Estes, 1983; Estes et al., 1988; Lee, 1998), a use that continues in the present century (e.g., Conrad, 2008; Wiens et al., 2010; Gauthier et al., 2012). Several early explicit phylogenetic analyses based on morphology supported squamatan monophyly (e.g., Rage, 1982; Benton, 1985; Gauthier et al., 1988; Evans, 1988), a result that has been corroborated by subsequent molecular and combined morphological and molecular studies (e.g., Townsend et al., 2004; Hugall et al., 2007; Albert el al., 2009; Reeder et al., 2015; Simões et al., 2018).

The name *Squamata* was first used by Oppel (1811), who applied it to a group composed of "lizards" (including crocodilians) and snakes. Crocodilians (*Loricata*), were removed from *Squamata* by Merrem (1820; see also Blainville, 1816), thus more closely approximating the modern composition of the taxon, and this proposal was followed by most subsequent authors who recognized a taxon designated by that name (e.g., Haworth, 1825; Fitzinger, 1826; Gray, 1831a; Bonaparte, 1841 [as a synonym]; Gravenhorst, 1843 [though he excluded snakes]; Gray, 1844, 1845 [though he excluded amphisbaenians]; Bonaparte, 1850). When *Sphenodon punctatus* (tuatara) was first described (Gray, 1831b), it was considered an agamid "lizard" and thus implicitly part of *Squamata*.

Cope (1864) separated *Sphenodon* (= *Hatteria*) from *Agamidae* as *Hatteriidae*, and Günther (1867) further separated it from all "lizards" ("*Lacertilia*") as *Rhynchocephalia*, though both authors retained *Sphenodon* within *Squamata*. Cope (1875) increased the separation further still by including extinct taxa (*Protorosauridae* and *Rhynchosauridae*) within *Rhynchocephalia*, not recognizing *Squamata*, and placing turtles between *Rhynchocephalia* and "lizards" (and snakes) in his taxonomy. However, he later (Cope, 1889) recognized *Squamata* as a group including "lizards" and snakes but excluding *Sphenodon*, and this arrangement has been adopted by most authors since the beginning of the twentieth century (e.g., Cope, 1900; Williston, 1917, 1925; Nopcsa, 1923; Romer, 1933, 1945, 1956, 1966; McDowell and Bogert, 1954; Gans, 1978; Estes, 1983; Estes et al., 1988; Lee, 1998, 2005; Conrad, 2008).

Squamata is clearly the most appropriate name for the clade in question. Other candidate names (see Synonyms) either have seldom been used since the eighteenth century (e.g., *Saurophidia*, *Streptostylica*, *Lyognathi*), are more appropriately applied to different clades (e.g., *Pholidota*, *Reptilia*, *Lepidosauria*), or were used (most commonly) for a paraphyletic group associated with the same ancestor (e.g., *Sauria*, *Lacertilia*) and thus do not correspond as closely in terms of composition (*Squamata* itself was used in this way by Gravenhorst, 1843, and Haeckel, 1866). Moreover, the name *Squamata* has previously been defined phylogenetically as designating the same (crown) clade to which it is applied here (Estes et al., 1988; Lee, 1998), although it has sometimes been applied implicitly to a more inclusive clade (see Conrad, 2008; Simões et al., 2018). We have defined the name using a maximum crown clade definition in the interest of presenting a simple definition (one with few specifiers) that will nevertheless apply to the same clade in the context of major differences between the internal relationships in phylogenies based on morphological (e.g., Estes et al., 1988; Lee, 1998; Conrad, 2008; Gauthier et al., 2012; but see Simões et al., 2018) versus molecular (e.g., Townsend et al., 2004; Vidal and Hedges, 2005; Wiens et al., 2010; Pyron et al., 2013; see also Reeder et al., 2015) data (see Composition).

Despite these striking disagreements, data from both DNA sequences and phenotypic sources strongly support many of the same crown squamatan subclades: traditional *Anguimorpha* plus *Serpentes*, *Dibamidae*, *Gekkota*, *Iguania*, *Lacertoidea* (including *Amphisbaenia*), and *Scincoidea*. Estimating the timing of their divergence is complicated because their interrelationships are in dispute and their early fossil records are sparse. A single partial, but articulated, specimen from the mid-Triassic (~240 Ma), *Megachirella wachtleri* (Renesto and Posenato, 2003), has recently been inferred to be a stem squamatan (Simões et al., 2018). It is the only potential pan-squamatan known from the Triassic or Early Jurassic. Otherwise, the oldest pan-squamatan fossils that might be referrable to the total clades of any of the crown squamatan subclades are mainly partial jawbones known from ~167 Ma (Bathonian, Middle Jurassic; e.g., Caldwell et al., 2015). Relatively intact, or at least associated, remains of stem-members of three of them (*Gekkota*, *Scincoidea*, *Anguimorpha*) are known from ~150 Ma (Kimmeridgian, Late Jurassic; e.g., Evans, 2003). Apart from *Dibamidae*, which has no fossil record, most of the rest are diverse and disparate by ~75 Ma (Campanian, Late Cretaceous; e.g., Gao and Norell, 2000). Crown squamatans appear to have been the dominant lepidosaurs in terrestrial ecosystems since the Early Cretaceous (e.g., Evans and Matsumoto, 2015; and references therein), while rhynchocephalians predominated during the early Mesozoic (e.g., Cocude-Michel, 1963; Hsiou et al., 2015; and references therein; see also *Lepidosauria*, this volume).

Literature Cited

Albert, E. M., D. San Mauro, M. Garcia-Paris, L. Ruber, and R. Zardoya. 2009. Effect of taxon sampling on recovering the phylogeny of squamate reptiles based on complete mitochondrial genome and nuclear gene sequence data. *Gene* 441:12–21.

Baeckens, S., S. Edwards, K. Huyghe, and R. Van Damme. 2015. Chemical signaling in lizards: an interspecific comparison of femoral pore numbers in *Lacertidae*. *Biol. J. Linn. Soc.* 114:44–57.

Benton, M. J. 1985. Classification and phylogeny of the diapsid reptiles. *Zool. J. Linn. Soc.* 84:97–164.

de Blainville, H. M. D.. 1816. Prodrome d'une nouvelle distribution systématique du règne animal. *Bull. Sci. Soc. Philom. Paris* 8:113–124.

de Blainville, H. M. D.. 1822. *De L'organisation des Animaux ou Principes d'anatomie Comparée.* F. G. Levrault, Paris.

de Blainville, H. M. D.. 1835. Description de quelques espèces de reptiles de la Californie, précédée de l'analyse d'un système général d'erpétologie et d'amphibiologie. *Nouv. Ann. Mus. Hist. Nat. Paris* 4:233–296 + 4 plates.

Bonaparte, C. L. 1831. *Saggio di Una Distribuzione Metodica Degli Animali Vertebrati.* Antonio Boulzaler, Roma.

Bonaparte, C. L. 1840. Prodromus systematis herpetologiae. *Nuovi Ann. Sci. Nat. Bologna* 4:90–101.

Bonaparte, C. L. 1841. A new systematic arrangement of vertebrated animals. *Trans. Linn. Soc. Lond.* 18:247–304.

Bonaparte, C. L. 1850. *Conspectus Systematum Herpetologiae et Amphibiologiae.* E. J. Brill, Lugduni Batavorum.

Boulenger, G. A. 1884. Synopsis of the families of existing *Lacertilia. Ann. Mag. Nat. Hist., Ser.* 5 14:117–122.

Boulenger, G. A. 1885–1887. *Catalogue of the Lizards in the British Museum (Natural History)*, Vols. 1–3. Trustees of the British Museum, London.

Brongniart, A. 1800. Essai d'une classification naturelle des reptiles. Iere Partie. Etablissement des ordres. *Bull. Sci. Soc. Philom. Paris* 2(35):81–82.

Caldwell, M. W., R. L. Nydam, A. Palci, and S. Apesteguía. 2015. The oldest known snakes from the Middle Jurassic—Lower Cretaceous provide insights on snake evolution. *Nat. Commun.* 6:5996.

Camp, C. L. 1923. Classification of the lizards. *Bull. Amer. Mus. Nat. Hist.* 48:289–481.

Cocude-Michel, M. 1963. Les rhynchocephales et les sauriens des calcaires lithographiques (Jurassique superieur) d'Europe occidentale. *Nouv. Arch. Mus. Hist. Nat. Lyon* 7:1–187.

Conrad, J. L. 2008. Phylogeny and systematics of *Squamata* (*Reptilia*) based on morphology. *Bull. Am. Mus. Nat. Hist.* 310:1–182.

Cope, E. D. 1864. On the characters of the higher groups of *Reptilia Squamata*—and especially of the *Diploglossa. Proc. Acad. Nat. Sci. Phila.* 16:224–231.

Cope, E. D. 1875. Check-list of North American *Batrachia* and *Reptilia*; with a systematic list of the higher groups, and an essay on geographical distribution. *Bull. U.S. Natl. Mus.* 1:1–104.

Cope, E. D. 1889. Synopsis of the families of *Vertebrata. Am. Nat.* 23:849–877.

Cope, E. D. 1900. The crocodilians, lizards, and snakes of North America. *Annu. Rep. Board Regents Smithson. Inst.* 1898:155–1270 + 36 plates.

Crother, B., ed. 2017. *Scientific and Standard English Names of Amphibians and Reptiles of North America North of Mexico, Which Comment Regarding Confidence in Our Understanding.* Society for the Study of Amphibians and Reptiles, Herpetological Circular No. 43.

Cuvier, G. 1800. *Leçons D'anatomie Comparée.* Tome I. Baudouin, Paris.

Daudin, F. M. 1802–1803. *Histoire Naturelle, Générale et Particulière des Reptiles.* F. Dufart, Paris.

Estes, R. 1983. Sauria *terrestria*, Amphisbaenia. *Handbuch der Paläoherpetologie.* Teil 10A. Gustav Fischer, Stuttgart.

Estes, R., K. de Queiroz, and J. Gauthier. 1988. Phylogenetic relationships within *Squamata*. Pp. 119–281 in *Phylogenetic Relationships of the Lizard Families* (R. Estes and G. Pregill, eds.). Stanford University Press, Stanford, CA.

Evans, S. E. 1988. The early history and relationships of the *Diapsida*. Pp. 221–260 in *The Phylogeny and Classification of the Tetrapods. Vol. 1: Amphibians, Reptiles, Birds* (M. J. Benton, ed.). Systematics Association Special Volume No. 35A. Clarendon Press, Oxford.

Evans, S. E. 2003. At the feet of the dinosaurs: the early history and radiation of lizards. *Biol. Rev.* 78(4):513–551.

Evans, S. E., and R. Matsumoto. 2015. An assemblage of lizards from the early cretaceous of Japan. *Palaeontol. Electron.* 18.2.36A:1–36.

Fitzinger, L. I. 1826. *Neue Classification der Reptilien Nach Ihren Natürlichen Verwandtschaften. Nebst Einer Verwandtschafts-Tafel Und Einem Verzeichnisse der Reptilien-Sammlung des K. K. Zoologischen Museum's zu Wien*. J. G. Heubner, Wien.

Fitzinger, L. I. 1843. *Systema Reptilium. Fasiculus Primus*, Amblyglossae. Braumüller et Seidel, Vindobonae.

Gadow, H. 1898. *A Classification of* Vertebrata *Recent and Extinct*. Adam and Charles Black, London.

Gadow, H. 1901. *Amphibia and Reptiles. The Cambridge Natural History*, Vol. 8. Macmillan, London.

Gans, C. 1978. The characteristics and affinities of the *Amphisbaenia*. *Trans. Zool. Soc. London* 34:347–416.

Gao, K., and M. A. Norell. 2000. Taxonomic composition and systematics of Late Cretaceous lizard assemblages from Ukhaa Tolgod and adjacent localities, Mongolian Gobi Desert. *Bull. Am. Mus. Nat. Hist.* 249:1–118.

Gauthier, J. A. 1982. Fossil *Xenosauridae* and *Anguidae* from the Lower Eocene Wasatch Formation, southcentral Wyoming, and a revision of the *Anguioidea*. *Contr. Geol. Univ. Wyoming* 21:7–54.

Gauthier, J., R. Estes, and K. de Queiroz. 1988. A phylogenetic analysis of *Lepidosauromorpha*. Pp. 15–98 in *Phylogenetic Relationships of the Lizard Families* (R. Estes and G. Pregill, eds.). Stanford University Press, Stanford, CA.

Gauthier, J. A., M. Kearney, J. A. Maisano, O. Rieppel, and A. D. B. Behlke. 2012. Assembling the squamate tree of life: perspectives from the phenotype and the fossil record. *Bull. Peabody Mus. Nat. Hist.* 53:3–308.

Gegenbaur, C. 1859. *Grundzüge der Vergleichenden Anatomie*. Wilhelm Englemann, Leipzig.

Gravenhorst, J. L. 1843. *Vergleichende Zoologie*. Breslau, Graf, Barth und Comp.

Gray, J. E. 1825. A synopsis of the genera of *Reptiles* and *Amphibia*, with a description of some new species. *Ann. Philos., Ser. 2* 10:193–217.

Gray, J. E. 1827. A synopsis of the genera of saurian reptiles in which some new genera are indicated, and the others reviewed by actual examination. *Philos. Mag., Ser. 2* 2:54–58.

Gray, J. E. 1831a. A synopsis of the species of the class *Reptilia*. Pp. 1–110 in *The Animal Kingdom Arranged in Conformity with Its Organization, by the Baron Cuvier, with Additional Descriptions of all the Species Hitherto Named, and of Many Not Before Noticed, by Edward Griffith and Others. Vol. 9: The Class Reptilia*. Appendix (E. Griffith, ed.). Whittaker, Treacher, and Co., London.

Gray, J. E. 1831b. Note on a peculiar structure in the head of an *Agama. Zool. Misc.* 1831:13–14.

Gray, J. E. 1842. Descriptions of two hitherto unrecorded species of reptiles fro New Zealand; presented to the British Museum by Dr. Dieffenbach. *Zool. Misc.* 1842:72.

Gray, J. E. 1844. *Catalogue of the Tortoises, Crocodiles, and Amphisbaenians in the Collection of the British Museum*. Trustees of the British Museum, London.

Gray, J. E. 1845. *Catalogue of the Specimens of Lizards in the Collection of the British Museum*. Trustees [of the British Museum], London.

Günther, A. 1867. Contribution to the anatomy of *Hatteria* (*Rhynchocephalus*, Owen). *Philos. Trans. R. Soc. Lond. B Biol. Sci.* 157:595–629.

Haeckel, E. 1866. *Generelle Morphologie der Organismen: Allgemeine Grundzüge der Organischen Formen-Wissenschaft, Mechanisch*

Begründet Durch die von Charles Darwin Reformirte Descendenz-Theorie. Band 2: Allgemeine Entwickelungsgeschichte der Organismen. G. Reimer, Berlin.

Haeckel, E. 1895. *Systematicshe Phylogenie. Dritter Theil: Systematicshe Phylogenie der Wirbelthiere* (Vertebrata). Georg Reimer, Berlin.

Haworth, A. H. 1825. A binary arrangement of the class *Amphibia. Philos. Mag. J.* 65:372–373.

Hay, O. P. 1902. *Bibliography and Catalogue of the Fossil* Vertebrata *of North America.* United States Geological Survey, Washington, DC.

Hay, O. P. 1930. *Second Bibliography and Catalogue of the Fossil* Vertebrata *of North America.* Carnegie Institution, Washington, DC.

Hsiou, A. S., M. A. G. De França, and J. Ferigolo. 2015. New data on the *Clevosaurus* (*Sphenodontia: Clevosauridae*) from the Upper Triassic of southern Brazil. *PLOS ONE* 10(9):e0137523.

Hugall, A. F., R. Foster, and M. S. Y. Lee. 2007. Calibration choice, rate smoothing, and the pattern of tetrapod diversification according to the long nuclear gene RAG-1. *Syst. Biol.* 56:543–563.

Huxley, T. H. 1886. *A Manual of the Anatomy of Vertebrated Animals.* D. Appleton, New York.

Jaekel, O. 1911. *Die Wirbeltiere: Eine Übersicht Über Die Fossilen und Lebenden Formen.* Gebrüder Borntraeger, Berlin.

Jan, G. 1857. *Cenni sul Museo Civico di Milano ed Indice Sistematico dei Rettili ed Anfibi Esposti nel Medesimo.* Luigi di Giacomo Priola, Milano.

Kuhl, H. 1820. Beiträge zur Kenntnifs der Amphibien. Pp. 75–131 in *Beiträge zur Zoologie Vergleichenden Anatomie.* Hermannschen Buchhandlung, Frankfurt am Main.

Latreille, P. A. 1825. *Familles Naturelles du Règne Animal, Exposées Succinctement et dans un Ordre Analytique, avec L'indication de Leurs Genres.* J. B. Baillière, Paris.

Laurenti, J. N. 1768. *Specimen Medicum, Exhibens Synopsin Reptilium Emendatam cum Experimentis circa Venena et Antidota Reptilium Austriacorum.* Joan. Thom. Nob. de Trattnern, Viennae.

Lee, M. S. Y. 1998. Convergent evolution and character correlation in burrowing reptiles: towards a resolution of squamate relationships. *Biol. J. Linn. Soc.* 65:369–453.

Lee, M. S. Y. 2005. Molecular evidence and marine snake origins. *Biol. Lett* 1(2):227–230.

Linnaeus, C. 1758. *Systema Naturae Per Regna Tria Naturae, Secundum Classes, Ordines, Genera, Species, cum Characteribus, Differentiis, Synonymis, Locis.* 10th edition. Laurentii Salvii, Holmiae (Stockholm).

MaCartney, J. 1802. Scientific names [see p. vi] in Cuvier, G. 1802. *Lectures on Comparative Anatomy.* Translated by William Ross under the inspection of James MaCartney. T. N. Longman and O. Rees, London.

McDowell, Jr., S. B., and C. M. Bogert. 1954. The systematic position of *Lanthanotus borneensis,* and the affinities of the anguinomorphan lizards. *Bull. Am. Mus. Nat. Hist.* 105:1–142.

Merrem, B. 1820. *Versuch Eines Systems der Amphibien.* Johann Christian Krieger, Marburg.

Müller, J. 1831. Beiträge zur Anatomie und Naturgeschichte der Amphibien. *Z. Physiol.* 4:190–275.

Nopcsa, F. 1923. *Die Familien der Reptilien. Fortschritte der Geologie und Palaeontologie, Heft 2.* Gebrüder Borntraeger, Berlin.

Oppel, M. 1811. *Die Ordnungen, Familien, und Gattungen der Reptilien, als Prodrom Einer Naturgeschichte Derselben.* Joseph Lindauer, München.

Owen, R. 1842. Report on British fossil reptiles, part II. *Rep. Br. Assoc. Adv. Sci.* 11:60–204.

Owen, R. 1845. Description of certain fossil crania, discovered by A. G. Bain, Esq., in sandstone rocks at the south-eastern extremity of Africa, referable to a different species of an extinct genus of *Reptilia* (*Dicynodon*), and indicative of a new tribe or sub-order of *Sauria. Trans. Geol. Soc. Lond.* 7:59–84 + 4 plates.

Owen, R. 1866. *On the Anatomy of Vertebrates. Vol. 1: Fishes and Reptiles.* Longmans, Green, and Co., London.

Pough, F. H., R. M. Andrews, M. L. Crump, A. D. Savitsky, K. D. Wells, and M. C. Brandley. 2016. *Herpetology.* 4th edition. Sinauer, Sunderland, MA.

Pyron, R. A., F. T. Burbrink, and J. J. Wiens. 2013. A phylogeny and revised classification of *Squamata*, including 4161 species of lizards and snakes. *BMC Evol. Biol.* 13:93.

Rage, J.-C. 1982. La phylogenie des lepidosauriens (*Reptilia*): une approche cladistique. *C. R. Acad. Sci. Paris* 284:1765–1768.

Rage, J.-C. 1984. Serpentes. *Handbuch der Paläoherpetologie*, Teil 11. Gustav Fischer, Stuttgart.

Reeder, T. W., T. M. Townsend, D. G. Mulcahy, B. P. Noonan, P. L. Wood, Jr., J. W. Sites, Jr., and J. J. Wiens. 2015. Integrated analyses resolve conflicts over squamate reptile phylogeny and reveal unexpected placements for fossil taxa. *PLOS ONE* 10: e0118199.

Renesto, S., and R. Posenato. 2003. A new lepidosauromorph reptile from the Middle Triassic of the Dolomites. *Riv. Ital. Paleo. Strat.* 109:463–474.

Romer, A. S. 1933. *Vertebrate Paleontology*. University of Chicago Press, Chicago, IL.

Romer, A. S. 1945. *Vertebrate Paleontology*. 2nd edition. University of Chicago Press, Chicago, IL.

Romer, A. S. 1956. *Osteology of the Reptiles*. University of Chicago Press, Chicago, IL.

Romer, A. S. 1966. *Vertebrate Paleontology*. 3rd edition. University of Chicago Press, Chicago, IL.

Scopoli, I. A. 1777. *Introductio ad Historiam Naturalem Sistens Genera Lapidum, Plantarum, et Animalium Hactenus Detecta, Caracteribus Essentialibus Donata, in Tribus Divisa, Subinde ad Leges Naturae*. Wolfgangum Gerle, Pragae.

Simões, T. R., M. W. Caldwell, M. Tałanda, M. Bernardi, A. Palci, O. Vernygora, F. Bernardini, L. Mancini, and R. L. Nydam. 2018. The origin of squamates revealed by a Middle Triassic lizard from the Italian Alps. *Nature* 557:706–709.

Stannius, H. 1856. *Handbuch der Zootomie. Zweiter Theil. Die Wirbelthiere. Zweites Buch. Zootomie der Amphibien*. Veit and Comp., Berlin.

Strauch, A. 1887. Bemerkungen über die Geckoniden-Sammlung im zoologischen Museum der Kaiserlichen Akademie der Wissenshaften zu St. Petersburg. *Mem. Acad. Imp. Sci. St.-Petersb. Ser. 7* 35:1–72.

Swainson, W. W. 1839. *The Natural History of Fishes, Amphibians, & Reptiles, or Monocardian Animals. Vol. II: The Cabinet Cyclopædia; Natural History; Conducted by Dionysius Lardner*. Longman, Orme, Brown, Green, and Longmans, and John Taylor, London.

Tornier, G. 1913. *Reptilia*. Pp. 315–376 in *Handwörterbuch der Naturwissenshaften, Achter Band* (E. Korschelt, ed.). Gustav Fischer, Jena.

Townsend, T. M., A. Larson, E. Louis, and J. R. Macey. 2004. Molecular phylogenetics of *Squamata*: the position of snakes, amphisbaenians and dibamids, and the root of the squamate tree. *Syst. Biol.* 53:735–757.

Uetz, P., ed. 2016. Species numbers (as of August 2016). The Reptile Database. Available at http://www.reptile-database.org/.

Vidal, N., and S. B. Hedges. 2005. The phylogeny of squamate reptiles (lizards, snakes, and amphisbaenians) inferred from nine nuclear protein-coding genes. *C. R. Biol.* 328:1000–1008.

Vitt, L. J., and J. P. Caldwell. 2014. *Herpetology, an Introductory Biology of Amphibians and Reptiles*. 4th edition. Academic Press, London.

Wagler, J. 1828. Vorläufige Übersicht des Gerüstes, so wie Ankündigung seines Systema Amphibiorum. *Isis von Oken* 21:859–861.

Wagler, J. 1830. *Natürliches System der Amphibien, mit Vorangehender Classification der Säugethiere und Vögel. Ein Beitrag zur Vergleichenden Zoologie*. J. G. Cotta'schen, München, Stuttgart und Tübingen.

Wiegmann, A. F. A. 1834. *Herpetologia Mexicana, seu Descriptio Amphibiorum Novae Hispaniae, quae Itineribus Comitis de Sack, Ferdinandi Deppe et Chr. Guil. Schiede in Museum Zoologicum Berolinense Pervenerunt. Pars Prima, Saurorum Species Amplectens, Adiecto Systematis Saurorum Prodromo, Additisque Multis in Hunc Amphibiorum Ordinem Observationibus*. C. G. Lüderitz, Berolini.

Wiens, J. J., C. A. Kuczynski, T. Townsend, T. W. Reeder, D. G. Mulcahy, and J. W. Sites, Jr. 2010. Combining phylogenomics and fossils in higher-level squamate reptile phylogeny: molecular data change the placement of fossil taxa. *Syst. Biol.* 59:674–688.

Williston, S. W. 1917. The phylogeny and classification of reptiles. *J. Geol.* 25:411–421.

Williston, S. W. 1925. *The Osteology of the Reptiles.* Harvard University Press, Cambridge, MA.

Zittel, K. A. 1887–1890. *Handbuch der Palaeontologie. I. Abtheilung. Palaeozoologie. III. Band.* Vertebrata (Pices, Amphibia, Reptilia, Aves). R. Oldenbourg, München und Leipzig.

Authors

Kevin de Queiroz; Department of Vertebrate Zoology; National Museum of Natural History; Smithsonian Institution; Washington, DC 20560-0162, USA. Email: dequeirozk@si.edu.

Jacques A. Gauthier; Department of Geology and Geophysics; Yale University; 210 Whitney Avenue, New Haven, CT 06520-8109, USA. Email: jacques.gauthier@yale.edu.

Date Accepted: 18 July 2018

Primary Editor: Philip Cantino

Mosasauridae P. Gervais 1853 [D. Madzia and J. L. Conrad], converted clade name

Registration Number: 200

Definition: The least inclusive clade containing *Mosasaurus hoffmanni* Mantell 1829, *Tylosaurus* (originally *Macrosaurus*) *proriger* (Cope 1869a), and *Halisaurus platyspondylus* Marsh 1869. This is a minimum-clade definition. Abbreviated definition: min ∇ (*Mosasaurus hoffmanni* Mantell 1829 & *Tylosaurus proriger* (Cope 1869a) & *Halisaurus platyspondylus* Marsh 1869).

Etymology: Derived from the stem of the name *Mosasaurus* Conybeare 1822, the name of an included taxon, which combines the words *Mosa* (the Latin name of the river Meuse) and *sauros* (the Greek equivalent of "lizard" or "reptile").

Reference Phylogeny: As the data of Bell and Polcyn (2005), modified from the set published by Bell (1997), represent the primary source for most phylogenetic analyses of *Mosasauridae*, Figure 7 of Bell and Polcyn (2005) is treated here as the primary reference phylogeny. For updated versions see Dutchak and Caldwell (2006: Fig. 6), Caldwell and Palci (2007: Figs. 7, 8), Cuthbertson et al. (2007: Fig. 10), Dutchak and Caldwell (2009: Figs. 4, 5), and Leblanc et al. (2012: Figs. 12, 13, 14). Although *Mosasaurus hoffmanni* is not included in the reference phylogeny, it is inferred to be closely related to *Plotosaurus* (*P. bennisoni* [misspelled in the reference phylogeny], *P. tuckeri*) and to other species of *Mosasaurus* (e.g., *M. conodon*, *M. missouriensis*) based on the results of Schulp (2006) and Leblanc et al. (2012).

Composition: Three major groups within *Mosasauridae* are usually distinguished:

Halisaurinae, *Mosasaurinae*, and a clade comprising *Plioplatecarpinae* and *Tylosaurinae* (none of the subgroup names is defined phylogenetically in this volume). For taxa included within these groups as well as those usually placed within the sister taxon to *Mosasaurinae*, but outside the *Plioplatecarpinae* plus *Tylosaurinae* clade, see for example Bell (1997), Dortangs et al. (2002), Bardet et al. (2005), Bell and Polcyn (2005), Konishi and Caldwell (2011), and Leblanc et al. (2012). See Comments for the possible inclusion of additional taxa within *Mosasauridae*. There is no recently published list of mosasaurid species that are considered to be taxonomically valid; however, the total number is likely more than 40 but less than 100.

Diagnostic Apomorphies: Historically, mosasaurids were characterized as large to very large carnivorous squamates that were partially to fully adapted to marine environments, with numerous derived features thought to have been associated with an aquatic lifestyle, including elongate skulls, short paddle-shaped limbs, and compressed tails (Bell, 1993). They were also characterized as having clawless digits with hyperphalangy, mandibles with a ligamentous (as opposed to sutural) mandibular symphysis, no sacrum, seven cervical vertebrae, cylindrical (as opposed to depressed) centra, reduced or absent clavicles and interclavicle, teeth set in sockets, etc. (see e.g., Gadow, 1898; Camp, 1923). However, the discovery of additional fossils has resulted in alternative phylogenetic hypotheses of mosasaurid interrelationships and the unstable position of that taxon within *Squamata* (this volume), so that no unambiguous diagnostic apomorphies are currently accepted universally. Nevertheless, several character states have commonly been considered synapomorphic for *Mosasauridae*. For instance, as unequivocal character transformations Bell (1997) listed a relatively narrowly arcuate or anteriorly acute shape of the premaxilla and no scapula-coracoid fusion at any life stage. However, the premaxilla condition is unknown in the early mosasaurine *Dallasaurus turneri* (Bell and Polcyn, 2005), and the absence of scapula-coracoid fusion cannot be confirmed in many early mosasaurids or in the important taxon *Aigialosaurus dalmaticus* (Bell and Polcyn, 2005; Dutchak and Caldwell, 2006). More recently, Conrad (2008) diagnosed *Mosasauridae* as lacking a functional sacrum (for definition see Conrad, 2008: 62) and having a flattened "hourglass-shaped" humerus. However, the evolution of pelvic anatomy is still a matter of debate, and these states may have evolved several times independently (see e.g., Bell and Polcyn, 2005; Caldwell and Palci, 2007).

Synonyms: The following taxon names are approximate synonyms: *Natantia* Owen 1851; *Pythonomorpha* Cope 1869b; *Mosasauria* Huxley 1871; *Mosasauri* Gadow 1898; *Mosasauromorpha* Fürbringer 1900; *Mosasauroidea* Camp 1923; *Mosasauriformes* Hay 1930 (see Comments for discussion of their historical usage).

Comments: *Mosasauridae* is a relatively species-rich clade of semi- to fully aquatic squamates of the Late Cretaceous. Mosasaurids have traditionally been understood as a group that includes *Mosasaurinae*, *Tylosaurinae*, *Plioplatecarpinae*, and *Halisaurinae*, and in most cases excludes *Aigialosaurus dalmaticus*, *Carsosaurus marchesettii*, and *Opetiosaurus bucchichi* (e.g., Russell, 1967; Bell, 1993; Bell and Polcyn, 2005). We used a phylogenetic definition based on traditionally included taxa that is equivalent to those of Bell (1993) and Conrad (2008), the only phylogenetic definitions proposed for the name *Mosasauridae* in the literature. Although an unambiguous diagnosis of this clade is currently

problematical (see Diagnostic Apomorphies), the monophyly of *Mosasauridae*, as traditionally circumscribed, has been supported in numerous studies (e.g., Bell, 1997; Bell and Polcyn, 2005; Cuthbertson et al., 2007); however, some studies (e.g., Dutchak and Caldwell, 2006; Caldwell and Palci, 2007; Leblanc et al., 2012) have placed certain taxa traditionally considered to be "aigialosaurids" within the mosasaurid clade.

By contrast, the interrelationships of the above-mentioned taxa and their close relatives are a matter of debate (compare Bell and Polcyn, 2005: Fig. 7; Dutchak and Caldwell, 2006: Fig. 6; Caldwell and Palci, 2007: Figs. 7, 8; Cuthbertson et al., 2007: Fig. 10; Dutchak and Caldwell, 2009: Figs. 4, 5; Leblanc et al., 2012: Figs. 12, 13, 14). Two major clades have consistently been inferred within the *Mosasauridae* (one named *Mosasaurinae*, and another consisting of the subclades *Plioplatecarpinae*, *Tylosaurinae*, and their close relatives); however, a more precise resolution of mosasaurid phylogeny including forms outside of those two clades is problematic. For instance, the species of *Halisaurus*, together with their closest relatives, *Phosphorosaurus ortliebi* and *Eonatator sternbergii*, form a clade named *Halisaurinae*. Although the composition of this clade is rather stable, its position within *Mosasauridae* varies considerably. Halisaurines have been inferred to be (1) sister taxon to a clade consisting of all other mosasaurids (e.g., Bell, 1993; Caldwell, 1999; Christiansen and Bonde, 2002; Bardet et al., 2005; Conrad, 2008; Dutchak and Caldwell, 2009), (2) sister taxon to *Plioplatecarpinae*, *Tylosaurinae* and their close relatives (Dutchak and Caldwell, 2006; Cuthbertson et al., 2007), (3) sister taxon to the "russellosaur group" (= *Russellosaurus coheni*, *Yaguarasaurus columbianus*, and *Tethysaurus nopcsai*), the putative close relatives of *Plioplatecarpinae* and *Tylosaurinae* (Caldwell and Palci, 2007; Leblanc et al., 2012), which may not be monophyletic

(Polcyn and Bell, 2005), and (4) sister taxon to *Komensaurus carrolli* (= "Trieste aigialosaur") (Bell and Polcyn, 2005).

Several taxon names have been used more or less synonymously with *Mosasauridae*. For example, *Mosasauri*, *Mosasauromorpha*, or *Mosasauroidea* all originally contained a single subtaxon, the *Mosasauridae*, and thus referred to the same group in terms of known composition. The situation is similar with *Mosasauria*. Although Huxley (1871) did not specify the content of *Mosasauria* (except that he mentioned *Mosasaurus* as a member), he excluded *Dolichosaurus*, a taxon generally considered related to, but not a member of, *Mosasauridae*. *Natantia* was proposed to include *Mosasaurus* and *Leiodon* (= *Liodon*), but even though this name was introduced in 1851, two years before the name *Mosasauridae*, it was the latter that gained widespread use. *Pythonomorpha* and *Mosasauriformes* were initially divided into *Mosasauridae* and *Clidastidae* (Cope, 1869b), and *Mosasauridae* and *Globidentidae* (Hay, 1930), respectively. Because *Clidastes propython*, *Globidens alabamaensis* and their relatives are now considered to be a part of *Mosasauridae* (e.g., Bell, 1993; Bell and Polcyn, 2005), *Pythonomorpha* and *Mosasauriformes* can be viewed as approximate synonyms of *Mosasauridae* as well. It is worth noting, however, that most taxon names listed here as approximate synonyms of *Mosasauridae*, except *Mosasauri* and *Mosasauromorpha*, have recently been applied to more or less inclusive clades (e.g., *Natantia* of Bell [1993, 1997], *Pythonomorpha* of Lee [1997] and Caldwell [1999], *Mosasauria* and *Mosasauriformes* of Conrad [2008], and *Mosasauroidea* of Bell [1993, 1997], Bell and Polcyn [2005] and many others). *Mosasauri* and *Mosasauromorpha* were not selected because *Mosasauridae* is much more commonly used for this clade.

Historically, the name *Mosasauridae* was used for a group thought to be related to but excluding *Aigialosauridae*, a group of semi-aquatic squamatans from the early Late Cretaceous that can roughly be characterized by the presence of plesiomorphic (presumably not obligatorily aquatic) limb and pelvic girdle morphology. *Mosasauridae* and *Aigialosauridae* were then united as *Mosasauroidea* (e.g., Bell, 1993). However, *Aigialosauridae* may be paraphyletic relative to *Mosasauridae* (for review see Dutchak, 2005); moreover, some of the "aigialosaurids" might be nested within the mosasaurid clade (e.g., *Aigialosaurus dalmaticus* as found by Dutchak and Caldwell [2006], *Carsosaurus marchesettii* as found by Caldwell and Palci [2007] and Leblanc et al. [2012]).

Given that "aigialosaurids" likely represent a non-monophyletic assemblage close to mosasaurids (and some possibly belong to *Mosasauridae*), and that the name *Mosasauroidea* traditionally covers all the taxa of the "aigialosaur morphotype" (or, "plesiopedal mosasaurs/mosasauroids" sensu Bell and Polcyn, 2005), it would be appropriate to apply the name *Mosasauroidea* to a larger taxon including both "aigialosaurids" and *Mosasauridae* with a phylogenetic definition. Although it is beyond the scope of this contribution, one possibility would be to implement the definition of *Mosasauroidea* offered by Conrad (2008), wherein the name *Mosasauroidea* was defined as a name for the clade of *Mosasaurus hoffmanni* and all taxa more closely related to it than to *Dolichosaurus longicollis*. This maximum-clade definition would maintain the inclusion of taxa traditionally referred to as "aigialosaurids" and taxa to be discovered that possess a similar morphotype (though it might also end up including some taxa not possessing that morphotype), as well as the traditional exclusion of "dolichosaurs" (and potentially, "adriosaurs").

Literature Cited

Bardet, N., X. P. Suberbiola, M. Iarochene, F. Bouyahyaoui, B. Bouya, and M. Amaghzaz. 2005. A new species of *Halisaurus* from the Late Cretaceous phosphates of Morocco, and the phylogenetical relationships of the Halisaurinae (*Squamata: Mosasauridae*). *Zool. J. Linn. Soc.* 143:447–472.

Bell, G. L. 1993. *A Phylogenetic Revision of* Mosasauroidea (Squamata). Unpublished doctoral dissertation, University of Texas, Austin, TX.

Bell, G. L. 1997. A phylogenetic revision of North American and Adriatic *Mosasauroidea*. Pp. 293–332 in *Ancient Marine Reptiles* (J. M. Callaway, and E. L. Nicholls, eds.). Academic Press, San Diego, CA.

Bell, G. L., and M. J. Polcyn. 2005. *Dallasaurus turneri*, a new primitive mosasauroid from the Middle Turonian of Texas and comments on the phylogeny of *Mosasauridae* (*Squamata*). *Neth. J. Geosci.* 84:177–194.

Caldwell, M. W. 1999. Squamate phylogeny and the relationships of snakes and mosasauroids. *Zool. J. Linn. Soc.* 125:115–147.

Caldwell, M. W., and A. Palci. 2007. A new basal mosasauroid from the Cenomanian (U. Cretaceous) of Slovenia with a review of mosasauroid phylogeny and evolution. *J. Vertebr. Paleontol.* 27:863–880.

Camp, C. L. 1923. Classification of the lizards. *Bull. Am. Mus. Nat. Hist.* 48:289–480.

Christiansen, P., and N. Bonde. 2002. A new species of gigantic mosasaur from the Late Cretaceous of Israel. *J. Vertebr. Paleontol.* 22:629–644.

Conrad, J. L. 2008. Phylogeny and systematics of *Squamata* (*Reptilia*) based on morphology. *Bull. Am. Mus. Nat. Hist.* 310:1–182.

Conybeare, W. D. 1822. Fossil crocodiles and other saurian animals. Pp. 284–304 in *Outlines of Oryctology: An Introduction to the Study of Fossil Organic Remains; Especially of Those Found in the British Strata* (J. Parkinson, ed.). Printed for the author, London.

Cope, E. D. 1869a. Remarks on *Holops brevispinus*, *Ornithotarsus immanis* and *Macrosaurus proriger*. *Proc. Acad. Nat. Sci. Phila.* 11:1–123.

Cope, E. D. 1869b. On the reptilian orders *Pythonomorpha* and *Streptosauria*. *Proc. Boston Soc. Nat. Hist.* 12:250–266.

Cuthbertson, R. S., J. C. Mallon, N. E. Campione, and R. B. Holmes. 2007. A new species of mosasaur (*Squamata: Mosasauridae*) from the Pierre Shale (lower Campanian) of Manitoba. *Can. J. Earth Sci.* 44:593–606.

Dortangs, R. W., A. S. Schulp, E. W. A. Mulder, J. W. M. Jagt, H. H. G. Peeters, and D. Th. de Graaf. 2002. A large new mosasaur from the Upper Cretaceous of the Netherlands. *Neth. J. Geosci.* 81:1–8.

Dutchak, A. R. 2005. A review of the taxonomy and systematics of aigialosaurs. *Neth. J. Geosci.* 84:221–229.

Dutchak, A. R., and M. W. Caldwell. 2006. Redescription of *Aigialosaurus dalmaticus* Kramberger, 1892, a Cenomanian mosasauroid lizard from Hvar Island, Croatia. *Can. J. Earth Sci.* 43:1821–1834.

Dutchak, A. R., and M. W. Caldwell. 2009. A redescription of *Aigialosaurus* (= *Opetiosaurus*) *bucchichi* (Kornhuber, 1901) (*Squamata: Aigialosauridae*) with comments on mosasauroid systematics. *J. Vertebr. Paleontol.* 29:437–452.

Fürbringer, M. 1900. Zur vergleichenden Anatomie des Brustschulterapparates und der Schultermuskeln. *Jenaische Zeitschr. Naturw.* 34:215–718.

Gadow, H. 1898. *A Classification of* Vertebrata *Recent and Extinct*. Adam and Charles Black, London.

Gervais, P. 1853. Observations relatives aux Reptiles fossiles de France (deuxième partie). *C. R. Hebd. Seanc. Acad. Sci. Paris* 36:470–474.

Hay, O. P. 1930. *Second Bibliography and Catalogue of the Fossil* Vertebrata *of North America*. Carnegie Institution of Washington, Washington, DC.

Huxley, T. H. 1871. *A Manual of the Anatomy of Vertebrated Animals*. J. & A. Churchill, London.

Konishi, T., and M. W. Caldwell. 2011. Two new plioplatecarpine (*Squamata, Mosasauridae*) genera from the Upper Cretaceous of North

America, and a global phylogenetic analysis of plioplatecarpines. *J. Vertebr. Paleontol.* 31:754–783.

Leblanc, A. R. H., M. W. Caldwell, and N. Bardet. 2012. A new mosasaurine from the Maastrichtian (Upper Cretaceous) phosphates of Morocco and its implications for mosasaurine systematics. *J. Vertebr. Paleontol.* 32:82–104.

Lee, M. S. Y. 1997. The phylogeny of varanoid lizards and the affinities of snakes. *Philos. Trans. R. Soc. Lond. B Biol. Sci.* 352:53–91.

Mantell, G. A. 1829. A tabular arrangement of the organic remains of the county of Sussex. *Trans. Geol. Soc.* 2:201–216.

Marsh, O. C. 1869. Notice of some new mosasauroid reptiles from the Greensand of New Jersey. *Am. J. Sci.* 48:392–397.

Owen, R. 1851. *A History of British Fossil Reptiles. Vol. I, Section II: The Fossil* Reptilia *of the Cretaceous Formations.* Cassell & Company Limited, London: 155–274.

Polcyn, M. J., and G. L. Bell. 2005. *Russellosaurus coheni* n. gen., n. sp., a 92 million-year-old mosasaur from Texas (USA), and the definition of the parafamily *Russellosaurina. Neth. J. Geosci.* 84:321–333.

Russell, D. A. 1967. Systematics and morphology of American mosasaurs. *Bull. Peabody Mus. Nat. Hist.* 23:1–241.

Schulp, A. S. 2006. A comparative description of *Prognathodon saturator* (*Mosasauridae, Squamata*), with notes on its phylogeny. Pp. 19–56 in *On Maastricht Mosasaurs* (A. S. Schulp, ed.). Publicaties van het Natuurhistorisch Genootschap in Limburg 45.

Authors

Daniel Madzia; Institute of Paleobiology; Polish Academy of Sciences; Twarda 51/55; PL-00-818 Warsaw, Poland. Email: dmadzia@twarda.pan.pl.

Jack L. Conrad[†]; Department of Anatomy; New York College of Osteopathic Medicine; Old Westbury, NY 11568-8000, USA. Email: jack.conrad@gmail.com.

Date Accepted: 10 December 2012

Primary Editor: Kevin de Queiroz

[†] Deceased.

Pan-Gekkota A. M. Bauer, new clade name

Registration Number: 236

Definition: The total clade of the crown clade *Gekkota*. This is a crown-based total-clade definition. Abbreviated definition: total ∇ of *Gekkota*.

Etymology: Derived from the Greek *pantos* (all), indicating that the name refers to a total clade or pan-monophyllum, plus *Gekkota*, the name of the corresponding crown clade (see the entry for *Gekkota* in this volume for the etymology of that name).

Reference Phylogeny: For the purposes of applying the definition above, Figure 54B of Conrad (2008) should be treated as the primary reference phylogeny.

Composition: As currently recognized, *Gekkota* (this volume) and the Jurrasic/Lower Cretaceous *Parviraptor* spp. Evans 1994, the Upper Cretaceous *Gobekko cretacicus* Borsuk-Białynicka 1990, and AMNH FR 21444 (unnamed taxon) (Conrad, 2008). However, the referral of *Parviraptor* to *Pan-Gekkota* has not been hypothesized by any previous authors and is relatively weakly supported, as fossils of this taxon can be scored for few of the diagnostic characters of the gekkotan total clade. See Comments for additional discussion of taxa that have been previously or tentatively assigned to *Pan-Gekkota*.

Diagnostic Apomorphies: Conrad (2008) identified six skeletal synapomorphies uniting *Gekkota* and the fossil taxa he allocated to its total clade (his *Gekkonomorpha* = *Pan-Gekkota*): (1) posteromedial parietal flange, (2) occipital condyle bipartite and constructed primarily of the exoccipital portions of the otooccipitals, (3) presence of a distinct supratrigeminal process that anteriorly closes the trigeminal foramen, (4) sphenoid enclosing the lateral head vein, (5) anterior location of the sphenooccipital tubercle (this character may be size related and its diagnostic value is, therefore, questionable), and (6) retroarticular process posteriorly expanded.

Synonyms: *Gekkonomorpha* sensu Conrad and Norell (2006) and Conrad (2008) is an unambiguous synonym (Fürbringer's (1900) original use of *Geckonomorpha* excluded pygopodids). *Gekkonomorpha* sensu Kluge (1987) and Rösler (2000), *Gekkonoidea* sensu Kluge (1967, 1976) and Underwood (1971) and *Gekkota* sensu Kluge (1967), Estes (1983), Evans (1993, 1994, 1995, 2003), Gao and Norell (2000), and Bauer et al. (2005) all include extinct taxa no longer considered part of the gekkotan total clade, but are consistent with use of the name for that clade in light of phylogenetic knowledge of those times. *Gekkonoidea* and *Gekkota* sensu Hoffstetter (1962, 1964, 1967) are partial synonyms of *Pan-Gekkota* as they exclude *Pygopodidae*. *Nyctisauria* sensu Romer (1956) and *Gekkonidae* sensu Vidal and Hedges (2004) and Townsend et al. (2004) are approximate synonyms in that the authors do not clearly associate their respective names with the crown rather than the total clade.

Comments: Hoffstetter (1962, 1964, 1967) and Estes (1983) included the Jurassic and Early Cretaceous taxa *Bavarisaurus*, *Ardeosaurus*, *Eichstaettisaurus*, *Paleolacerta* and *Yabeinosaurus* in their respective implied total gekkotan clades, but more recent work has suggested that these taxa are outside of *Pan-Gekkota* (Evans, 1993, 1994, 1995, 2003; Evans et al., 2004, 2005; Conrad and Norell 2006). The unresolved positions of *Bavarisaurus*, *Ardeosaurus*, *Eichstaettisaurus* and *Scandensia ciervensis* in Conrad's (2008) phylogeny leaves open the possible inclusion of these taxa in the total clade of *Gekkota*, although he did not include any of them in his *Gekkonomorpha* (= *Pan-Gekkota*). *Myrmecodaptria micropha-gosa* from the Upper Cretaceous Djadokhta Formation of the Mongolian Gobi Desert was considered by Gao and Norell (2000) to be *incertae sedis* within the *Gekkota*, but Conrad and Norell (2006) considered this fossil, as well as all the Jurassic taxa above, to be outside of the gekkotan total clade. The Cretaceous *Hoburogekko suchanovi* Alifanov 1989, considered by Conrad and Norell (2006) as of uncertain affinities within the gekkotan total clade, has been subsequently regarded as a crown gekkotan by Conrad (2008). *Cretaceogekko burmae*, a 100-million-year-old lizard in amber, has been allocated to the *Pan-Gekkota* (≈ their *Gekkota*) by Arnold and Poinar (2008), but they were equivocal with respect to their interpretation of it as a member of the crown versus total clade.

Although the name *Gekkonomorpha* has been defined as designating the gekkotan total clade (Conrad and Norell, 2006; Conrad, 2008), the name *Pan-Gekkota* was selected for this clade in the interest of developing an integrated system of crown and total clade names and to avoid confusion with earlier uses of *Gekkonomorpha*. In contrast to the total clade definition adopted here for *Pan-Gekkota*, Conrad and Norell (2006) and Conrad (2008) provided a maximum clade definition for *Gekkonomorpha* using explicit external specifiers. In the context of the reference phylogeny, those two names are heterodefinitional synonyms (*PhyloCode* Art. 14.1).

Literature Cited

Arnold, E. N., and G. Poinar. 2008. A 100-million-year-old gecko with sophisticated adhesive toe pads, preserved in amber from Myanmar. *Zootaxa* 1847:62–68.

Bauer, A. M., W. Böhme, and W. Weitschat. 2005. A Lower Eocene gecko from Baltic amber and the evolution of the gekkonid adhesive system. *J. Zool. Lond.* 265:327–332.

Conrad, J. L. 2008. Phylogeny and systematics of *Squamata* (*Reptilia*) based on morphology. *Bull. Am. Mus. Nat. Hist.* 310:1–182.

Conrad, J. L., and M. A. Norell. 2006. High resolution X-ray computed tomography of an Early Cretaceous gekkonomorph (*Squamata*) from Öösh (Övörkhangai; Mongolia). *Hist. Biol.* 18:405–431.

Estes, R. 1983. *Handbuch der Paläoherpetologie. Teil 10A*. Sauria *terrestria*, Amphisbaenia. Gustav Fischer Verlag, Stuttgart.

Evans, S. E. 1993. Jurassic lizard assemblages. *Rev. Paléobiol., Vol. Spec.* 7:55–65.

Evans, S. E. 1994. The Solnhofen (Jurassic: Tithonian) lizard genus *Bavarisaurus*: new skull material and a reinterpretation. *N. Jb. Geol. Paläont. Abh.* 192:37–52.

Evans, S. E. 1995. Lizards: evolution, early radiation and biogeography. Pp. 51–55 in *Sixth Symposium on Mesozoic Terrestrial Ecosystems and Biota, Short Papers* (A. Sun, and Y. Wang, eds.). China Ocean Press, Beijing.

Evans, S. E. 2003. At the feet of the dinosaurs: the early history and radiation of lizards. *Biol. Rev. Camb. Philos. Soc.* 78:513–551.

Evans, S. E., P. Raia, and C. Barbera. 2004. New lizards and rhynchocephalians from the Lower Cretaceous of southern Italy. *Acta Palaeontol. Pol.* 49:393–408.

Evans, S. E., Y. Wang, and C. Li. 2005. The early Cretaceous lizard genus *Yabeinosaurus* from China: resolving an enigma. *J. Syst. Palaeontol.* 4:319–335.

Fürbringer, M. 1900. Zur vergleichenden Anatomie des Brustschulterapparates und der Schultermuskeln. *Jena. Z. F. Naturwiss.* 34:215–718, pls. 13–17.

Gao, K., and M. A. Norell. 2000. Taxonomic composition and systematics of Late Cretaceous lizard assemblages from Ukhaa Tolgod and adjacent localities, Mongolian Gobi Desert. *Bull. Am. Mus. Nat. Hist.* 249:1–118.

Hoffstetter, R. 1962. Révue des récentes acquisitions concernant l'histoire et la systématique des squamates. Pp. 243–279 in *Problèmes actuels de paléontologie (évolution des vertébrés)* (J.-P. Lehman, ed.). Colloque International du Centre National de la Recherche Scientifique (104), Paris.

Hoffstetter, R. 1964. Les *Sauria* du Jurassique supérieur et spécialement les *Gekkota* de Bavière et de Mandchourie. *Senckenberg. Biol.* 45:281–324.

Hoffstetter, R. 1967. A propos des genres *Ardeosaurus* et *Eichstaettisaurus* (*Reptilia*: *Sauria*: *Gekkonoidea*) du Jurassique supérieur de Franconie. *Bull. Soc. Géol. Fr.* 8 (1966):592–595.

Kluge, A. G. 1967. Higher taxonomic categories of gekkonid lizards and their evolution. *Bull. Am. Mus. Nat. Hist.* 135:1–60, pls. 1–5.

Kluge, A. G. 1976. Phylogenetic relationships in the lizard family *Pygopodidae*: an evaluation of theory, methods and data. *Misc. Publ. Mus. Zool. Univ. Mich.* 152:i–iv + 1–72.

Kluge, A. G. 1987. Cladistic relationships in the *Gekkonoidea* (Squamata, Sauria). *Misc. Publ. Mus. Zool. Univ. Mich.* 173:i–iv, 1–54.

Romer, A. S. 1956. *Osteology of the Reptiles.* University of Chicago Press, Chicago, IL.

Rösler, H. 2000. Kommentierte Liste der rezent, subrezent und fossil bekannten Geckotaxa (*Reptilia*: *Gekkonomorpha*). *Gekkota* 2:28–153.

Townsend, T., A. Larson, E. Louis, and J. R. Macey. 2004. Molecular phylogenetics of *Squamata*: the position of snakes, amphisbaenians, and dibamids, and the root of the squamate tree. *Syst. Biol.* 53:735–757.

Underwood, G. L. 1971. A modern appreciation of Camp's "Classification of the lizards." Pp. vii–xvii in *Camp's Classification of the Lizards* (K. Adler, ed.). Society for the Study of Amphibians and Reptiles, Athens, OH.

Vidal, N., and S. B. Hedges. 2004. Molecular evidence for a terrestrial origin of snakes. *Proc. R. Soc. Lond. B Biol. Sci.* 271(suppl.):226–229.

Author

Aaron M. Bauer; Department of Biology; Villanova University; Villanova, PA 19085, USA. Email: aaron.bauer@villanova.edu.

Date accepted: 8 February 2012

Primary Editor: Kevin de Queiroz

Gekkota C. L. Camp 1923 [A. M. Bauer], converted clade name

Registration Number: 148

Definition: The crown clade originating in the most recent common ancestor of *Gekko* (originally *Lacerta*) *gecko* Linnaeus 1758, *Eublepharis* (originally *Cyrtodactylus*) *macularius* Blyth 1854 [1855], and *Pygopus* (originally *Cryptodelma*) *nigriceps* Fischer 1882. This is a minimum-crown-clade definition. Abbreviated definition: min crown ∇ (*Gekko gecko* Linnaeus 1758 & *Eublepharis macularius* Blyth 1854 [1855] & *Pygopus nigriceps* Fischer 1882).

Etymology: Derived from the vernacular name "gecko" which is itself most likely a variant of either the Malay (behasa Melayu) "gēkoq" – of onomatopoeic origin, or the Elu (the precursor of modern Sinhala) "gēgo," meaning "house lizard."

Reference Phylogeny: For the purposes of applying the definition above, Figure 2 of Gamble et al. (2008a) should be treated as the primary reference phylogeny.

Composition: *Gekkota* is currently believed to contain approximately 1,470 described extant species based on recent updates and corrections to species lists provided by Rösler (2000), Kluge (2001), and Uetz (2012). The major extant subclades are *Eublepharidae, Diplodactylidae, Carphodactylidae, Pygopodidae, Gekkonidae, Phyllodactylidae,* and *Sphaerodactylidae* as recognized by Han et al. (2004: Table 3) and subsequently modified by Gamble et al. (2008a: Table 8, 2008b: 363). Hoffstetter (1946), Estes (1983), Rösler (2000), and Bauer et al. (2005) provided reviews of extinct gekkotans. The Cretaceous *Hoburogekko suchanovi* Alifanov 1989 and *Gobekko cretacicus* Borsuk-Białynicka 1990 have been considered to be gekkotans by Evans (1993, 1995, 2003), whereas Conrad and Norell (2006) considered the relationships of the former uncertain within the *Gekkonomorpha* (= *Pan-Gekkota*) but included the latter species within the *Gekkota*, and Conrad (2008) accepted Alifanov's (1989, 2000) interpretation of *Hoburogekko* as a gekkotan but considered *Gobekko* as the sister taxon to *Gekkota*. *Cretaceogekko burmae*, a 100-million-year-old lizard in amber, has been referred to *Gekkota* by Arnold and Poinar (2008). Although the precise conceptualization of *Gekkota* by those authors is not entirely clear, their questionable assignment of *Cretaceogekko* to the *Gekkonidae* suggests that they considered its inclusion in the gekkotan crown clade possible.

Diagnostic Apomorphies: Underwood (1957), Kluge (1967, 1987), and Rieppel (1984a), amongst others, reviewed apomorphic features of *Gekkota*. Estes et al. (1988) provided a list of 46 apomorphies diagnosing *Gekkota*, Lee (2000) provided 24 soft anatomical characters, and Conrad and Norell (2006) identified 12 unambiguous skeletal synapomorphies of the *Gekkota* (although not all are unique to gekkotans among squamates): (1) jugal reduced to a small splint or absent, (2) postorbital/postfrontal ramus of supratemporal arch absent, (3) squamosal not in contact with postorbital/postfrontal (questionably independent of character 2), (4) ectopterygoid contacts palatine along anterior border of suborbital fenestra, (5) processus ascendens tecti synotici absent, (6) occipital recess not hidden by spheno-occipital tubercle in ventral view, (7) posterior mylohyoid foramen located posterior to the level of the coronoid apex, (8) Meckel's canal closed and fused, (9) splenial extends anteriorly for less than one half length of dentary tooth row, (10) dentary extends posteriorly beyond the level of coronoid apex, (11) retroarticular process medially offset with a lateral notch, (12) aperture of the recessus scalae tympani subdivided. In a more recent analysis of all major squamate clades, Conrad (2008) considered fused frontal bones and short parietal supratemporal processes as unambiguous synapomorphies of *Gekkota*.

Synonyms: *Geckotii* Latrielle 1825, *Geckonomorpha* Fürbringer 1900, *Nyctisaura* Cope 1900 (and implicitly Cope 1864), *Geckones* Gadow 1901, *Gekkonoideae* Deraniyagala 1932, and *Gekkonoidea* Underwood 1954 are partial synonyms of *Gekkota* in that *Pygopodidae* was excluded. *Gekkonoidea* sensu Kluge 1987 is a partial synonym in that *Eublepharidae* was excluded. *Nyctisauria* sensu Romer (1956), *Gekkonoidea* sensu Kluge (1967, 1976a,b, 1982, 1983, 1991, 1993), and *Gekkonidae* sensu Vidal and Hedges (2004) and Townsend et al. (2004) are approximate synonyms in that the authors do not clearly associate their respective names with the crown rather than the total clade (extinct taxa are not explicitly referred to, or their inferred relationships are not specified).

Comments: The taxon name *Gekkota* is frequently attributed to Cuvier (1817) (e.g., Kluge, 1987; Estes et al., 1988; Alifanov, 2000; Gao and Norell, 2000; Augé, 2003; Conrad, 2008) or Cuvier (1807) [sic] (Borsuk-Białynicka, 1990). Cuvier (1817), however, used the French vernacular name "Geckotiens." Other authors (e.g., Hoffstetter, 1946) attribute the name to Latreille (1825), who employed the term *Geckotii*, which was subsequently modified by Camp (1923) to *Gekkota*.

Recognition of the *Gekkota* as a clade including both the limbed geckos and the "limbless"

pygopodids originated with McDowell and Bogert (1954); thus the earlier names *Nyctisaura* Cope 1900, *Geckonomorpha* Fürbringer 1900, and *Geckones* Gadow 1901, as well as pre-1954 usages of *Gekkota* are not fully inclusive of the living diversity of *Gekkota* as here defined in that they excluded pygopodids. McDowell and Bogert (1954) also included *Xantusiidae* in *Gekkota* as have several more recent authors (e.g., Hoffstetter, 1962; Underwood, 1971; Kluge, 1976b; Macey et al., 1997), and Underwood (1957) questionably included *Dibamidae* and *Anelytropsidae* (his terminology) but not *Xantusiidae*. Romer (1956) was the first author to use the name *Gekkota* (treated by him as a synonym of *Nyctisauria*) as conceptualized here. Kluge (1967) used a similar concept for *Gekkota*, as have most subsequent authors (e.g., Moffat, 1973; Hecht, 1976; Rieppel, 1984a,b; Kluge, 1987; Estes et al,. 1988; Grismer 1988; Donnellan et al. 1999; Lee 1998, 2000; Russell and Bauer 2002). However, in those cases in which extinct taxa are not discussed explicitly, or their inferred relationships are not specified, it is not possible to determine if these authors intended a more inclusive use of the name (i.e., including some or all members of the stem group). *Gekkota* of Alifanov (2000) included his *Nyctisauria* (incorporating all extant crown gekkotans and the late Cretaceous and Tertiary fossil geckos then known as well as *Yabeinosauridae*) and *Paraamphisbaenia* (*Sineoamphisbaenidae*, *Rhineuridae, Hyporhinidae*).

Several previous authors have provided or adopted explicit phylogenetic definitions of the name *Gekkota* (Estes et al., 1988; Lee, 1998; Gao and Norell, 2000; Conrad, 2008). Although those definitions differ in details of specifiers and wording, all of those authors applied the name to the same crown clade to which it is applied here.

Hypotheses similar to the reference phylogeny (Gamble et al., 2008a) have been published by Kluge (1987: Fig. 11), Grismer (1988: Fig. 4), Donnellan et al. (1999: Fig.1A–B), Harris et al. (1999: Figs. 1–2, 2001: Fig. 1), Han et al. (2004: Fig. 2); Townsend et al. (2004: Fig. 3); Vidal and Hedges (2005: Fig. 1, 2009: Fig. 2), Lee (2005: Fig. 1), Gamble et al. (2008b: Fig. 2), and Gamble et al. (2011: Fig. 1). Although differing in suggested patterns of relationship among constituent taxa, all are fully consistent with the definition of *Gekkota* provided here with regard to composition.

Literature Cited

Alifanov, V. R. 1989. The oldest gecko (*Lacertilia, Gekkonidae*) from the lower Cretaceous of Mongolia. *Paleontol. Zhur.* 1:124–126 (in Russian).

Alifanov, V. R. 2000. Macrocephalosaurs and the early evolution of lizards of Central Asia. *Trans. Paleontol. Inst. Russ. Acad. Sci.* 272:1–126 (in Russian).

Arnold, E. N., and G. Poinar. 2008. A 100 million year old gecko with sophisticated adhesive toe pads, preserved in amber from Myanmar. *Zootaxa* 1847:62–68.

Augé, M. 2003. La faune de *Lacertilia* (*Reptilia, Squamata*) de l'Éocène inférieur de Prémontré (Bassin de Paris, France). *Geodiversitas* 25:539–574.

Bauer, A. M., W. Böhme, and W. Weitschat. 2005. A Lower Eocene gecko from Baltic amber and the evolution of the gekkonid adhesive system. *J. Zool. Lond.* 265:327–332.

Blyth, E. 1854 [1855]. Report of the Curator, Zoological Department, for September, 1854. *Proc. Asiatic Soc. Bengal* 23:729–740.

Borsuk-Białynicka, M. 1990. *Gobekko cretacicus* gen. et sp. n., a new gekkonid lizard from the Cretaceous of the Gobi Desert. *Acta Palaeontol. Pol.* 35:67–76, pls. 17–18.

Camp, C. L. 1923. Classification of the lizards. *Bull. Amer. Mus. Nat. Hist.* 48:289–481.

Conrad, J. L. 2008. Phylogeny and systematics of *Squamata* (*Reptilia*) based on morphology. *Bull. Amer. Mus. Nat. Hist.* 310:1–182.

Conrad, J. L., and M. A. Norell. 2006. High resolution X-ray computed tomography of an Early Cretaceous gekkonomorph (*Squamata*) from Öösh (Övörkhangai; Mongolia). *Hist. Biol.* 18:405–431.

Cope, E. D. 1864. On the characters of the higher groups of *Reptilia Squamata*—and especially of the *Diploglossa*. *Proc. Acad. Nat. Sci. Philad.* 16:224–231.

Cope, E. D. 1900. The crocodilians, lizards, and snakes of North America. *Ann. Rept. U. S. Natl. Mus.* 1898:153–1270, pls. 1–36.

Cuvier, G. 1817. *Le Règne Animal Distribué d'après son Organisation, pour Servir de Base a l'Histoire Naturelle des Animaux et d'Introduction a l'Anatomie Comparée. Tome II, Contenant les Reptiles, les Poissons, les Mollusques et les Annélides*. Chez Deterville, Paris. xviii + 532 pp.

Deraniyagala, P. E. P. 1932. The *Gekkonoideae* of Ceylon. *Spolia Zeylanica* 16:291–310, pls. LVIII–LXIV.

Donnellan, S. C., M. N. Hutchinson, and K. M. Saint. 1999. Molecular evidence for the phylogeny of Australian gekkonoid lizards. *Biol. J. Linn. Soc.* 67:97–118.

Estes, R. 1983. *Handbuch der Paläoherpetologie. Teil 10A. Sauria terrestria*, Amphisbaenia. Gustav Fischer Verlag, Stuttgart.

Estes, R., K. de Queiroz, and J. Gauthier. 1988. Phylogenetic relationships within *Squamata*. Pp. 119–281 in *Phylogenetic Relationships of the Lizard Families* (R. Estes and G. Pregill, eds.). Stanford University Press, Stanford, CA.

Evans, S. E. 1993. Jurassic lizard assemblages. *Rev. Paléobiol., Vol. Spec.* 7:55–65.

Evans, S. E. 1995. Lizards: evolution, early radiation and biogeography. Pp. 51–55 in *Sixth Symposium on Mesozoic Terrestrial Ecosystems and Biota, Short Papers* (A. Sun, and Y. Wang, eds.). China Ocean Press, Beijing.

Evans, S. E. 2003. At the feet of the dinosaurs: the early history and radiation of lizards. *Biol. Rev. Camb. Philos. Soc.* 78:513–551.

Fischer, J. G. 1882. Herpetologische Bemerkungen. *Arch. f. Naturgesch.* 48:281–302, pls. XVI–XVII.

Fürbringer, M. 1900. Zur vergleichenden Anatomie des Brustschulterapparates und der Schultermuskeln. *Jena. Z. F. Naturwiss.* 34:215–718, pls. 13–17.

Gadow, H. 1901. *Amphibia and Reptiles.* Macmillan and Co., London. xiii + 668 pp., map.

Gamble, T., A. M. Bauer, E. Greenbaum, and T. R. Jackman. 2008a. Evidence for Gondwanan vicariance in an ancient clade of gecko lizards. *J. Biogeogr.* 35:88–104.

Gamble, T., A. M. Bauer, E. Greenbaum, and T. R. Jackman. 2008b. Out of the blue: cryptic higher level taxa and a novel, trans-Atlantic clade of gecko lizards (*Gekkota, Squamata*). *Zool. Scr.* 37:355–366.

Gamble, T., A. M. Bauer, G. R. Colli, E. Greenbaum, T. R. Jackman, L. J. Vitt, and A. M. Simons. 2011. Coming to America: multiple origins of New World geckos. *J. Evol. Biol.* 24:231–244.

Gao, K., and M. A. Norell. 2000. Taxonomic composition and systematics of Late Cretaceous lizard assemblages from Ukhaa Tolgod and adjacent localities, Mongolian Gobi Desert. *Bull. Am. Mus. Nat. Hist.* 249:1–118.

Grismer, L. 1988. Phylogeny, taxonomy, classification and biogeography of eublepharid geckos. Pp. 369–469 in *Phylogenetic Relationships of the Lizard Families* (R. Estes, and G. Pregill, eds.). Stanford University Press, Stanford, CA.

Han, D., K. Zhou, and A. M. Bauer. 2004. Phylogenetic relationships among the higher taxonomic categories of gekkotan lizards inferred from C-*mos* nuclear DNA sequences. *Biol. J. Linn. Soc.* 83:353–368.

Harris, D. J., J. C. Marshall, and K. A. Crandall. 2001. Squamate relationships based on C-*mos* nuclear DNA sequences: increased taxon sampling improves bootstrap support. *Amphib.-Reptil.* 22:235–242.

Harris, D. J., E. A. Sinclair, N. L. Mercader, J. C. Marshall, and K. A. Crandall. 1999. Squamate relationships based on C-*mos* nuclear DNA sequences. *Herpetol. J.* 9:147–151.

Hecht, M. K. 1976. Phylogenetic inference and methodology as applied to the vertebrate record. *Evol. Biol.* 9:335–363.

Hoffstetter, R. 1946. Sur les *Gekkonidae* fossiles. *Bull. Mus. Natn. Hist. Nat., 2ᵉ Sér.* 18:195–203.

Hoffstetter, R. 1962. Révue des récentes acquisitions concernant l'histoire et la systématique des squamates. Pp. 243–279 in *Problèmes Actuels de Paléontologie (Evolution des Vertébrés)* (J.-P. Lehman, ed.). Colloque International du Centre National de la Recherche Scientifique (104), Paris.

Kluge, A. G. 1967. Higher taxonomic categories of gekkonid lizards and their evolution. *Bull. Am. Mus. Nat. Hist.* 135:1–60, pls. 1–5.

Kluge, A. G. 1976a. A reinvestigation of the abdominal musculature of gekkonoid lizards and its bearing on their phylogenetic relationships. *Herpetologica* 32:295–298.

Kluge, A. G. 1976b. Phylogenetic relationships in the lizard family *Pygopodidae*: an evaluation of theory, methods and data. *Misc. Publ. Mus. Zool. Univ. Mich.* 152:i–iv + 1–72.

Kluge, A. G. 1982. Cloacal bones and sacs as evidence of gekkonoid lizard relationships. *Herpetologica* 38:348–355.

Kluge, A. G. 1983. Epidermal gland evolution in gekkonoid lizards. *J. Herpetol.* 17:89–90.

Kluge, A. G. 1987. Cladistic relationships in the *Gekkonoidea (Squamata, Sauria)*. *Misc. Publ. Mus. Zool. Univ. Mich.* 173:i–iv + 1–54.

Kluge, A. G. 1991. Checklist of gekkonoid lizards. *Smithson. Herpetol. Inform. Serv.* 85:1–35.

Kluge, A. G. 1993. *Gekkonoid Lizard Taxonomy.* International Gecko Society, San Diego, CA.

Kluge, A. G. 2001. Gekkotan lizard taxonomy. *Hamadryad* 26:1–209.

Latreille, P. A. 1825. *Familles Naturelles du Règne Animal, Exposées Succinctement et dans un Ordre Analytique, avec l'Indication de Leurs Genres.* J.-B. Baillière, Paris. [6] + 570 pp.

Lee, M. S. Y. 1998. Convergent evolution and character correlation in burrowing reptiles: towards a resolution of squamate relationships. *Biol. J. Linn. Soc.* 65:369–453.

Lee, M. S. Y. 2000. Soft anatomy, diffuse homoplasy, and the relationships of lizards and snakes. *Zool. Scr.* 29:101–130.

Lee, M. S. Y. 2005. Squamate phylogeny, taxon sampling, and data congruence. *Org. Div. Evol.* 5:25–45.

Linnaeus, C. 1758. *Systema Naturæ Per Regna Tria Naturæ, Secundum Classes, Ordines, Genera,*

Species, cum Characteribus, Differentiis, Synonymis, Locis. Tomus I, Editio decima. Laurenti Salvi, Holmiæ (Stockholm). [4], 823, [1] pp.

Macey, J. R., A. Larson, N. B. Ananjeva, and T. J. Papenfuss. 1997. Replication slippage may cause parallel evolution in the secondary structures of mitochondrial transfer RNAs. *Mol. Biol. Evol.* 14:30–39.

McDowell, S. B., and C. M. Bogert. 1954. The systematic position of *Lanthanotus* and the affinities of the anguinomorphan lizards. *Bull. Amer. Mus. Nat. Hist.* 105:1–142, pls. 1–16.

Moffat, L. A. 1973. The concept of primitiveness and its bearing on the phylogenetic classification of the *Gekkota. Proc. Linn. Soc. New South Wales* 97:275–301.

Rieppel, O. 1984a. The structure of the skull and jaw adductor musculature in the *Gekkota*, with comments on the phylogenetic relationships of the *Xantusiidae* (*Reptilia*: *Lacertilia*). *Zool. J. Linn. Soc.* 82:291–318.

Rieppel, O. 1984b. The cranial morphology of the fossorial lizard genus *Dibamus* with a consideration of its phylogenetic relationships. *J. Zool. Lond.* 204:289–327.

Romer, A. S. 1956. *Osteology of the Reptiles.* University of Chicago Press, Chicago, IL. xxi + 772 pp.

Rösler, H. 2000. Kommentierte Liste der rezent, subrezent und fossil bekannten Geckotaxa (*Reptilia*: *Gekkonomorpha*). *Gekkota* 2:28–153.

Russell, A. P., and A. M. Bauer. 2002. Underwood's classification of the geckos: a 21st century appreciation. *Bull. Nat. Hist. Mus. Lond.* (*Zool.*) 68:113–121.

Townsend, T., A. Larson, E. Louis, and J. R. Macey. 2004. Molecular phylogenetics of *Squamata*: the position of snakes, amphisbaenians, and dibamids, and the root of the squamate tree. *Syst. Biol.* 53:735–757.

Uetz, P. 2012. The Reptile Database. Available at http://www.reptile-database.org, accessed on 1 February 2012.

Underwood, G. 1954. On the classification and evolution of geckos. *Proc. Zool. Soc. Lond.* 124:469–492.

Underwood, G. 1957. On lizards of the family *Pygopodidae*: a contribution to the morphology and phylogeny of the *Squamata. J. Morphol.* 100:207–268.

Underwood, G. L. 1971. A modern appreciation of Camp's "Classification of the lizards." Pp. vii–xvii in *Camp's Classification of the Lizards.* Society for the Study of Amphibians and Reptiles, Athens, OH.

Vidal, N., and S. B. Hedges. 2004. Molecular evidence for a terrestrial origin of snakes. *Proc. R. Soc. Lond. B Biol. Sci.* 271(suppl.):226–229.

Vidal, N., and S. B. Hedges. 2005. The phylogeny of squamate reptiles (lizards, snakes, and amphisbaenians) inferred from nine nuclear protein-coding genes. *C. R. Biol.* 328:1000–1008.

Vidal, N., and S. B. Hedges. 2009. The molecular evolutionary tree of lizards, snakes, and amphisbaenians. *C. R. Biol.* 332:129–139.

Author

Aaron M. Bauer; Department of Biology; Villanova University; Villanova, PA 19085, USA. Email: aaron.bauer@villanova.edu

Date Accepted: 8 February 2012

Primary Editor: Kevin de Queiroz

Pan-Amphisbaenia M. Kearney and K. de Queiroz, new clade name

Registration Number: 115

Definition: The total clade of the crown clade *Amphisbaenia*. This is a crown-based total-clade definition. Abbreviated definition: total ∇ of *Amphisbaenia*.

Etymology: Derived by adding the prefix *Pan-* (for Pan-Monophylum = total clade) to *Amphisbaenia* (see entry for *Amphisbaenia*, this volume, for the etymology of that name).

Reference Phylogeny: Figure 2b of Longrich et al. (2015) is the primary reference phylogeny. On that tree, *Pan-Amphisbaenia* applies to the clade that includes *Rhineuridae*, *Chthonophidae†*, *Oligodontosauridae†*, *Bipedidae*, *Blanidae*, *Cadeidae*, *Trogonophidae*, and *Amphisbaenidae*, as well as all extinct taxa (of which there are none on that tree) that are more closely related to them than to *Teius*, *Tachydromus*, and *Lacerta*. However, the definition stipulates that the name is to be applied to the total clade of *Amphisbaenia* regardless of which extant taxa are inferred to be its closest relatives.

Composition: *Amphisbaenia* (applied to a crown clade in this volume) and the members of its stem group. See entry for *Amphisbaenia* (in this volume) for the composition of the crown clade. Wu et al. (1993, 1996) considered the fossil taxon *Sineoamphisbaena hexatabularis* from the Upper Cretaceous of China to belong to the amphisbaenian stem group and thus part of the clade that is here named *Pan-Amphisbaenia*; however, subsequent analyses by Kearney (2003a), Lee (2005), and Gauthier et al. (2012) indicated that *Sineoamphisbaena* is not part of that clade. Müller et al. (2011) inferred the fossil

taxon *Cryptolacerta hassiaca* from the Eocene of Germany to be a stem amphisbaenian; however, a more recent analysis places *Cryptolacerta* closer to *Lacertidae* than to *Amphisbaenia* (Longrich et al., 2015). As noted in the Composition section for *Amphisbaenia* (this volume), some North American fossil amphisbaenians possess seemingly ancestral characters (e.g., a complete postorbital bar) not seen in any living amphisbaenians (see e.g., Berman, 1972, 1973, 1976); however, current phylogenetic inferences (e.g., Kearney, 2003a; Hembree, 2007; Gauthier et al., 2012) place those fossils as crown rather than stem amphisbaenians.

Diagnostic Apomorphies: As a maximum (and total) clade, *Pan-Amphisbaenia* may not have any apomorphies (de Queiroz, 2007); however, possession of any of the apomorphies of *Amphisbaenia* (see Kearney, 2003b) constitutes evidence for inclusion of a species or specimen within *Pan-Amphisbaenia*. Those apomorphies of *Amphisbaenia* that are most likely to be preserved in fossils include: enlarged median premaxillary tooth, elongated postorbital region of skull, braincase enclosed anteriorly by enlarged orbitosphenoid(s), absence of suborbital fenestra, absence of epipterygoid, reduction of hind limbs (and possibly fore limbs), short tail with caudal autotomy septa confined to a single vertebra. In the absence of well-supported assignments of fossils to the amphisbaenian stem group, the order of evolution of these apomorphies is currently unknown. Müller et al. (2011) inferred 19 characters to be synapomorphies of *Cryptolacerta* and *Amphisbaenia*, including a tongue-and-groove articulation of the frontals, transversely widened frontal downgrowths, thickened frontals and maxillae, the absence of a tympanic crest on the quadrate, very low vertebral neural spines, and a sutural prefrontal-postorbitofrontal contact. If *Cryptolacerta* is a stem amphisbaenian, then this list likely includes some of the earliest characters to evolve along the amphisbaenian stem lineage.

Synonyms: Most, if not all, of the names listed as approximate (and partial) synonyms of *Amphisbaenia* (see entry in this volume) are also approximate (and partial) synonyms of *Pan-Amphisbaenia*, given that authors who used those names rarely, if ever, clearly distinguished between the total clade and the crown. Wu et al. (1993) clearly applied both the names *Amphisbaenia* (see also Wu et al., 1996) and *Annulata* (treating them as synonyms) to a clade more inclusive than the crown, but it is not clear if they applied those names to the total clade.

Comments: See the Comments in the entry for *Amphisbaenia* (this volume) for historical information concerning recognition of the named group and support for its status as a clade (given that earlier authors rarely distinguished between the crown and total clades). The names *Amphisbaenia* and *Pan-Amphisbaenia* were selected for the crown and total clades, respectively, in the interest of developing an integrated system of names for those categories of clades (e.g., Meier and Richter, 1992; Gauthier and de Queiroz, 2001; de Queiroz, 2007). In this context, *Amphisbaenia* was selected for the crown clade because it appears to be the most widely used name (see Comments in the entry for *Amphisbaenia*, this volume), and the name of the total clade was then formed by adding the prefix *Pan-* to the name of the crown clade (*Amphisbaenia*). Moreover, none of the synonyms appear to have been used unambiguously for the total clade, and they remain available for application to nested clades between the total

clade and the crown. For example, the name *Annulata* could be applied to the clade originating in the first ancestor of *Amphisbaena fuliginosa* Linnaeus 1758 that possessed the apomorphy of having the body scales arranged in rings.

Literature Cited

Berman, D. S. 1972. *Hyporhina tertia*, new species (*Reptilia*: *Amphisbaenia*) from the early Oligocene (Chadronian) White River Formation of Wyoming. *Ann. Carnegie Mus.* 44:1–10.

Berman, D. S. 1973. *Spathorhynchus fossorium*, a middle Eocene amphisbaenian (*Reptilia*) from Wyoming. *Copeia* 1973(4):704–721.

Berman, D. S. 1976. A new amphisbaenian (*Reptilia*: *Amphisbaenia*) from the Oligocene-Miocene John Day Formation, Oregon. *J. Vertebr. Paleontol.* 50:165–174.

Gauthier, J. A., and K. de Queiroz. 2001. Feathered dinosaurs, flying dinosaurs, crown dinosaurs, and the name *"Aves."* Pp. 7–41 in *New Perspectives on the Origin and Early Evolution of Birds: Proceedings of the International Symposium in Honor of John H. Ostrom* (J. Gauthier and L. F. Gall, eds.). Peabody Museum of Natural History, Yale University, New Haven, CT.

Gauthier, J. A., M. Kearney, J. A. Maisano, O. Rieppel, and A. D. B. Behlke. 2012. Assembling the squamate tree of life: perspectives from the phenotype and the fossil record. *Bull. Peabody Mus. Nat. Hist.* 53:3–308.

Hembree, D. I. 2007. Phylogenetic revision of *Rhineuridae* (*Reptilia*: *Squamata*: *Amphisbaenia*) from the Eocene to Miocene of North America. *Univ. Kansas Paleontol. Contrib.* 15:1–20.

Kearney, M. 2003a. The phylogenetic position of *Sineoamphisbaena hexatabularis* reexamined. *J. Vertebr. Paleontol.* 23:394–403.

Kearney, M. 2003b. Systematics and evolution of the *Amphisbaenia*: a phylogenetic hypothesis based on morphological evidence from fossil and recent forms. *Herpetol. Monogr.* 17:1–75.

Lee, M. S. Y. 2005. Squamate phylogeny, taxon sampling, and data congruence. *Org. Divers. Evol.* 5:25–45.

Linnaeus, C. 1758. *Systema Naturae, Per Regna Tria Naturae, Secundum Classes, Ordines, Genera, Species cum Characteribus, Differentiis, Synonymis, Locis.* 10th edition. Laurentii Salvii, Holmiae (Stockholm).

Longrich, N. R., J. Vinther, R. A. Pyron, D. Pisani, and J. A. Gauthier. 2015. Biogeography of worm lizards (*Amphisbaenia*) driven by end-Cretaceous mass extinction. *Proc. R. Soc. Lond. B Biol. Sci.* 282: 20143034.

Meier, R., and S. Richter. 1992. Suggestions for a more precise usage of proper names of taxa: ambiguities related to the stem lineage concept. *Z. Zool. Syst. Evolutionsforsch.* 30:81–88.

de Queiroz, K. 2007. Toward an integrated system of clade names. *Syst. Biol.* 56:956–974.

Müller, J., C. A. Hipsley, J. J. Head, N. Kardjilov, A. Hilger, M. Wuttke, and R. R. Reisz. 2011. Eocene lizard from Germany reveals amphisbaenian origins. *Nature* 473:364–367.

Wu, X.-C., D. B. Brinkman, and A. P. Russell. 1996. *Sineoamphisbaena hexatabularis*, an amphisbaenian (*Diapsida: Squamata*) from the Upper Cretaceous redbeds at Bayan Mandahu (Inner Mongolia, People's Republic of China), and comments on the phylogenetic relationships of the *Amphisbaenia. Can. J. Earth Sci.* 33:541–577.

Wu, X.-C., D. B. Brinkman, A. P. Russell, Z.-M. Dong, P. J. Currie, L.-H. Hou, and G.-H. Cui. 1993. Oldest known amphisbaenian from the Upper Cretaceous of Chinese Inner Mongolia. *Nature* 366:57–59.

Authors

Maureen Kearney; Office of Science, Policy, and Society Programs, American Association for the Advancement of Science, 1200 New York Avenue NW, Washington, DC 20005, USA. Email: mkearney@aaas.org.

Kevin de Queiroz; Department of Vertebrate Zoology; National Museum of Natural History; Smithsonian Institution; Washington, DC 20560-0162, USA. Email: dequeirozk@si.edu.

Date Accepted: 1 November 2013; updated 3 November 2017

Primary Editor: Philip Cantino

Amphisbaenia J. E. Gray 1844 [M. Kearney and K. de Queiroz], converted clade name

Registration Number: 10

Definition: The largest crown clade containing *Amphisbaena fuliginosa* Linnaeus 1758 (*Amphisbaenidae*) but not *Lacerta agilis* Linnaeus 1758 (*Lacertidae*) and *Teius* (*Lacerta*) *teyou* (Daudin 1802) (*Teioidea*) and *Dibamus novaeguineae* Duméril and Bibron 1839 (*Dibamidae*) and *Coluber constrictor* Linnaeus 1758 (*Serpentes*). This is a maximum-crown-clade definition. Abbreviated definition: max crown ∇ (*Amphisbaena fuliginosa* Linnaeus 1758 ~ *Lacerta agilis* Linnaeus 1758 & *Teius teyou* (Daudin 1802) & *Dibamus novaeguineae* Duméril and Bibron 1839 & *Coluber constrictor* Linnaeus 1758).

Etymology: Derived from the Greek *amphi-* (on both sides, double) + *baeno* (walk, go, pass), thus, "moving both backwards and forwards."

Reference Phylogeny: Figure 1 of Gauthier et al. (2012) should be treated as the primary reference phylogeny, although similar relationships have been supported by molecular phylogenetic analyses of both amphisbaenians (Macey et al., 2004; Vidal et al., 2008; Longrich et al., 2015) and squamatans (Townsend et al., 2004; Vidal and Hedges, 2005; Pyron et al., 2013). Although not included in the reference phylogeny, the specifier *Lacerta agilis* is most closely related to *Lacerta viridis* among the included taxa (e.g., Godinho et al., 2005).

Composition: *Amphisbaenia* contains 193 currently recognized extant species (Uetz, 2017), which have been referred to the following mutually exclusive clades: *Rhineuridae, Bipedidae, Blanidae, Cadeidae, Trogonophidae,* and *Amphisbaenidae* (Kearney, 2003a; Vidal et al., 2008). A list of extinct species was provided by Estes (1983), though several taxa have been described subsequently (e.g., Charig and Gans, 1990; Gans and Montero, 1998; Smith, 2009; Augé, 2012; Longrich et al., 2015). Apart from *Sineoamphisbaena hexatabularis* and *Crythiosaurus mongoliensis*, all fossils previously referred to *Amphisbaenia* that were included in the analysis of Kearney (2003a; *Listromycter leakeyi, Lophocranion rusingense,* and various rhineurids) were inferred to be part of the crown clade. Only a few of the currently known fossils (e.g., *Chthonophis subterraneus, Oligodontosaurus*) are outside of the total clades of the previously mentioned amphisbaenian subgroups (Longrich et al., 2015). Although some North American fossil amphisbaenians possess seemingly ancestral characters (e.g., a complete postorbital bar) not seen in any living amphisbaenians (see e.g., Berman, 1972, 1973, 1976), current phylogenetic inferences (e.g., Kearney, 2003a; Hembree, 2007; Gauthier et al., 2012; Longrich et al., 2015) place those fossils in the stem group of *Rhineura* and thus within the crown clade *Amphisbaenia*. In the context of some molecular phylogenies (e.g., Vidal and Hedges, 2004; Wiens et al., 2012), in which *Lacerta* is as or more closely related to *Amphisbaena* than is *Rhineura*, *Rhineura* and its fossil relatives are not part of *Amphisbaenia* according to our definition.

Diagnostic Apomorphies: Kearney (2003a) listed 17 putative synapomorphies of *Amphisbaenia* relative to other extant and extinct squamatans (see also Gans, 1978; Estes et al., 1988; Gauthier et al., 2012). Some of the most obvious diagnostic apomorphies are: eyes reduced or absent, body scales arranged in rings (annuli), enlarged median premaxillary tooth, elongated postorbital region of skull, braincase enclosed anteriorly by enlarged orbitosphenoid(s), absence of suborbital fenestra, absence of epipterygoid, anteriorly elongated extracolumella with distal connections to the dermis, loss of external hind limbs (vestigial femora present internally in *Bipes* and *Blanus*), reduction of right lung (the left lung is reduced in most other elongate squamatans), short tail with caudal autotomy septum confined to a single vertebra and lacking regeneration.

Synonyms: *Amphisbaenoidea* (e.g., of Fitzinger, 1826; Müller, 1831; Stannius, 1856; Günther, 1867; Camp, 1923), approximate.

Amphisboena of Gray (1831), approximate.

Annulati (e.g., of Wiegmann, 1834; Fitzinger, 1843; Cope, 1900), approximate.

Amphisbaenae (e.g., of Wiegmann, 1834; Parker, 1868; Huxley, 1886), approximate.

Saurophidii of Bonaparte (1840) but not of Bonaparte (1841), which included *Chalcides*, approximate.

Cancellata of Gravenhorst (1843), partial (*Bipes* excluded).

Annulata of Gravenhorst (1843), partial (*Amphisbaena* and *Blanus* excluded).

Ophiosauri of Cope (1864), approximate.

Glyptodermata of Haeckel (1866), approximate.

Annulata (e.g., of Haeckel, 1866, as synonym; Gegenbaur, 1878, as synonym; Romer, 1956, 1966), approximate.

Opheosauri of Cope (1875, 1889), approximate.

Ophisauri of Cope (1875; attributed to Merrem), approximate.

Amphisbaenida (e.g., of Gegenbaur, 1878; Strauch, 1887), approximate.

Amphisbaenidae (e.g., of Boulenger, 1884, 1885–1887; Fürbringer, 1900; Gadow, 1901; Hay, 1902, 1930; Camp, 1923; Williston, 1925; Romer, 1933, 1945, 1956, 1966; Vanzolini, 1951; McDowell and Bogert, 1954), approximate.

Glyptoderma of Haeckel (1895), approximate.

Amphisbaenoida of Huxley (1886), approximate.

Amphisbaenomorpha of Fürbringer (1900), approximate.

Amphisbenidae of Nopcsa (1908), approximate.

Comments: Many early naturalists used the presence versus absence of limbs as the basis of major groups within *Reptilia*, and consequently, they placed the limbless amphisbaenians (e.g., *Amphisbaena*) within the serpents ≈ ophidians (e.g., Linnaeus, 1758; Shaw, 1802; Merrem, 1820). *Bipes* (*Chirotes*), which possesses forelimbs but not hindlimbs, was placed either within the saurians ≈ lacertilians (e.g., Daudin, 1802–1803; Oppel, 1811; Cuvier, 1817), within the serpents ≈ ophidians (e.g., Blainville, 1816, 1822), or in a separate taxon (e.g., *Saurophidii*, *Ophiosauri*) that was apparently considered intermediate between the two (e.g., La Cépède, 1788–1789; Wagler, 1828, 1830; Gray, 1831). However, even those early authors (e.g., Blainville, 1816, 1822) who considered both *Bipes* and *Amphisbaena* to be snakes did not group them together within the snake taxon. The first author to recognize a taxon composed only of species that are now considered amphisbaenians appears to have been Fitzinger (1826), who placed *Amphisbaena* and *Bipes*, along with the recently described *Leposternon*, in his (Familia) *Amphisbaenoidea*. This grouping was soon followed by other authors (e.g., Gray, 1831, 1844; Müller, 1831; Wiegmann, 1834), though often under different names (see Synonyms), and by the end of the nineteenth century, most authors recognized a group similar to the modern concept of *Amphisbaenia* (e.g., Haeckel, 1895; Gadow, 1898, 1901; Cope, 1900; Fürbringer, 1900).

Amphisbaenians are highly modified for burrowing and share several associated derived features supporting monophyly of the group (see Diagnostic Apomorphies), including several that are not shared by other burrowing squamatans (e.g., the arrangement of scales in annuli, a solid skull with bony enclosure of the braincase anteriorly, unique modifications of the extracolumella). Moreover, monophyly of *Amphisbaenia* has been supported by the results of explicit phylogenetic analyses based on both morphological (Kearney, 2003a; Gauthier et al., 2012) and molecular data (Kearney and Stuart, 2004; Townsend et al., 2004; Pyron et al., 2013). Nevertheless, some molecular studies have indicated that *Amphisbaenia* may be paraphyletic or polyphyletic, with *Rhineura* more closely related to *Lacerta* than to *Amphisbaena* (Vidal and Hedges, 2004) or outside of a clade containing both *Amphisbaena* and *Lacerta* (Wiens et al., 2012).

Although the amphisbaenian taxon has, over the years, been treated by various authors as separate from "lizards" (e.g., Wiegmann, 1834; Bonaparte, 1840; Gray, 1844; Hoffstetter, 1955; Kuhn, 1961; Halstead Tarlo, 1968; Gans, 1978), explicit phylogenetic analyses based on both morphological (e.g., Estes et al., 1988; Lee, 1998; Kearney, 2003a; Gauthier et al., 2012) and molecular (e.g., Kearney and Stuart, 2004; Townsend et al., 2004; Vidal and Hedges, 2005; Pyron et al., 2013) data strongly support the nesting of amphisbaenians within "lizards". The relationships of amphisbaenians to other "lizards" inferred from morphological data have been considered suspect because of the repeated evolution of body elongation and limb loss within *Squamata* (e.g., Estes et al., 1988; Lee, 1998; Kearney, 2003a; Gauthier et al., 2012). Those

inferred from molecular data indicate deep nesting of amphisbaenians within *Squamata* close to lacertids (e.g., Townsend et al., 2004; Vidal and Hedges, 2005; Pyron et al., 2013).

Selection of the name *Amphisbaenia* for the clade in question is relatively straightforward, as this name has been applied most frequently to that clade during roughly the last 60 years (e.g., McDowell and Bogert, 1954; Gans, 1967, 1978, 2005; Underwood, 1971; Estes, 1983; Estes et al., 1988; Kearney, 2003a). Other names that have been applied to this group (see Synonyms) have been used rarely after 1900, with the exceptions of *Amphisbaenidae* (e.g., Vanzolini, 1951; McDowell and Bogert, 1954; Romer, 1956) and *Annulata* (e.g., Romer, 1956; Wu et al., 1993). The name *Amphisbaenidae*, however, is now most commonly applied to a subgroup of amphisbaenians (e.g., Gans, 1967, 1978, 2005; Underwood, 1971; Kearney, 2003a). The name *Annulata* is almost certainly less commonly used than is *Amphisbaenia* (though frequency of use is difficult to assess using automated searches because *annulata* is also commonly used as a species name); moreover, it has been applied by Lee (1998) to a more inclusive putative clade composed of amphisbaenians, dibamids, and the extinct *Sinoamphisbaena hexatabularis*.

The name *Amphisbaenia* was defined phylogenetically by Estes et al. (1988) as the name of a crown clade, and our proposed definition is intended to associate the name with the same clade. Our definition has been formulated so that it will refer to a taxon of identical composition under an alternative hypothesis of relationships (Kearney, 2003a: Fig. 31), which differs in placing *Bipes* rather than *Rhineura* as the sister group to other extant amphisbaenians. It has also been formulated as a maximum-crown-clade definition so that it will restrict the name to the larger of the two extant clades (and the one that includes *Amphisbaena*) in the context of phylogenetic hypotheses in which the species traditionally included in *Amphisbaenia* form a paraphyletic or polyphyletic group (e.g., Vidal and Hedges, 2004; Wiens et al., 2012). Our definition differs in the second respect from the minimum-clade (node-based) definition of Estes et al. (1988), which would result in the fully limbed *Lacerta* (and presumably other lacertids) being included in *Amphisbaenia* in the context of the aforementioned phylogenetic hypotheses. We consider that outcome less consistent with the historical concept of the taxon than the exclusion of *Rhineura*. Because the inferred closest extant relatives of amphisbaenians differ between analyses based on morphological characters (e.g., Kearney, 2003a,b; Gauthier et al., 2012), which infer dibamids and snakes, and those based on molecular sequences (e.g., Townsend et al., 2004; Vidal and Hedges, 2004; Wiens et al., 2012), which infer lacertids and teioids, we have included external specifiers representing each of those taxa in our definition.

Literature Cited

Augé, M. L. 2012. Amphisbaenians from the European Eocene: a biogeographical review. *Palaeobio. Palaeoenv.* 92:425–443.

Berman, D. S. 1972. *Hyporhina tertia*, new species (*Reptilia: Amphisbaenia*) from the early Oligocene (Chadronian) White River Formation of Wyoming. *Ann. Carnegie Mus.* 44:1–10.

Berman, D. S. 1973. *Spathorhynchus fossorium*, a middle Eocene amphisbaenian (*Reptilia*) from Wyoming. *Copeia* 1973(4):704–721.

Berman, D. S. 1976. A new amphisbaenian (*Reptilia: Amphisbaenia*) from the Oligocene-Miocene John Day Formation, Oregon. *J. Vertebr. Paleontol.* 50:165–174.

de Blainville, H. D. 1816. Prodrome d'une nouvelle distribution systématique du règne animal. *Bull. Sci. Soc. Philom. Paris* 1816:113–132.

de Blainville, H. D. 1822. *De L'organisation des Animaux ou Principes D'anatomie Comparée.* F. G. Levrault, Paris.

Bonaparte, C. L. 1840. Prodromus systematis herpetologiae. *Nuov. Ann. Sci. Nat. Bologna* 4:90–101.

Bonaparte, C. L. 1841. *A New Systematic Arrangement of Vertebrated Animals.* Taylor, London.

Boulenger, G. A. 1884. Synopsis of the families of existing *Lacertilia. Ann. Mag. Nat. Hist., Ser.* 5 14:117–122.

Boulenger, G. A. 1885–1887. *Catalogue of the Lizards in the British Museum (Natural History).* 2nd edition. Trustees of the British Museum, London.

Camp, C. L. 1923. Classification of the lizards. *Bull. Am. Mus. Nat. Hist.* 48:289–481.

Charig, A. J., and C. Gans. 1990. Two new amphisbaenians from the Lower Miocene of Kenya. *Bull. Br. Mus. Nat. Hist. Geol.* 46:19–36.

Cope, E. D. 1864. On the Characters of the higher Groups of *Reptilia Squamata*—and especially of the *Diploglossa. Proc. Acad. Nat. Sci. Phila.* 16:224–231.

Cope, E. D. 1875. Checklist of North American *Batrachia* and *Reptilia. Bull. U.S. Natl. Mus.* 1:1–104.

Cope, E. D. 1889. Synopsis of the families of *Vertebrata. Am. Nat.* 23:849–877.

Cope, E. D. 1900. The crocodilians, lizards and snakes of North America. *Annu. Rep. U.S. Natl. Mus.* 1898:153–1270 + 36 plates.

Cuvier, G. 1817. *La Règne Animal, Distribué D'après son Organisation.* 1st edition. Deterville, Paris.

Daudin, F. M. 1802–1803. *Histoire Générale et Particulière des Reptiles,* 8 Vols. F. Dufart, Paris.

Duméril, A. M., and G. Bibron. 1839. *Erpétologie Générale ou Histoire Naturelle Complète des Reptiles,* Tome 5. Roret, Paris.

Estes, R. 1983. Sauria *Terrestria,* Amphisbaenia. Handbuch der Paläoherpetologie, Teil 10A. Gustav Fischer Verlag, Stuttgart.

Estes, R., K. de Queiroz, and J. A. Gauthier. 1988. Phylogenetic relationships within *Squamata.* Pp. 119–282 in *Phylogenetic Relationships of the Lizard Families. Essays Commemorating Charles L. Camp* (R. Estes and G. K. Pregill, eds.). Stanford University Press, Stanford, CA.

Fitzinger, L. J. 1826. *Neue Classification der Reptilien Nach Ihren Natürlichen Verwandtschaften.* J. G. Heubner, Vienna.

Fitzinger, L. J. 1843. *Systema Reptilium. Fasciculus Primus.* Amblyglossae. Baumüller and Seidel, Vienna.

Fürbringer, M. 1900. Zur vergleichenden Anatomie des Brustschulterapparates und der Schultermuskeln. *Jen. Zeit. Naturwissen.* 34:215–718.

Gadow, H. 1898. *A Classification of* Vertebrata *Recent and Extinct.* Adam and Charles Black, London.

Gadow, H. 1901. *Amphibia and Reptiles. Vol. 8: The Cambridge Natural History* (S. F. Harmer and A. E. Shipley, eds.). Macmillan, London.

Gans, C. 1967. A check list of Recent amphisbaenians (*Amphisbaenia, Reptilia*). *Bull. Am. Mus. Nat. Hist.* 119:129–204.

Gans, C. 1978. The characteristics and affinities of the *Amphisbaenia. Trans. Zool. Soc. Lond.* 34:347–416.

Gans, C. 2005. Checklist and bibliography of the *Amphisbaenia* of the world. *Bull. Am. Mus. Nat. Hist.* 289:1–130.

Gans, C., and R. Montero. 1998. Two new fossil amphisbaenids (*Reptilia: Squamata*) from the Pleistocene of Lagoa Santa (Minas Gerais, Brasil). *Steenstrupia* 24(1):9–22.

Gauthier, J. A., M. Kearney, J. A. Maisano, O. Rieppel, and A. D. B. Behlke. 2012. Assembling the squamate tree of life: perspectives from the phenotype and the fossil record. *Bull. Peabody Mus. Nat. Hist.* 53:3–308.

Gegenbaur, C. 1878. *Elements of Comparative Anatomy.* (Translated by F. J. Bell.) Macmillan, London.

Godinho, R., E. G. Crespo, N. Ferrand, and D. J. Harris. 2005. Phylogeny and evolution of the green lizards, *Lacerta* spp. (*Squamata: Lacertidae*) based on mitochondrial and nuclear DNA sequences. *Amphibia-Reptilia* 26:271–285.

Gravenhorst, J. L. C. 1843. *Vergleichende Zoologie.* Graß, Barth & Comp., Breslau.

Gray, J. E. 1831. A synopsis of the species of the class *Reptilia*. Appendix. Pp. 1–110 in *The Animal Kingdom Arranged in Conformity with*

Its Organization, by the Baron Cuvier, Vol. 9: The class Reptilia (E. Griffith and E. Pidgeon, eds.). Whittaker, Treacher and Co., London.

Gray, J. E. 1844. *Catalogue of the Tortoises, Crocodiles, and Amphisbaenians in the Collection of the British Museum.* Trustees of the British Museum, London.

Günther, A. 1867. Contribution to the anatomy of *Hatteria* (*Rhynchocephalus*, Owen). *Philos. Trans. R. Soc. Lond. B Biol. Sci.* 157:595–629.

Haeckel, E. 1866. *Generelle Morphologie der Organismen.* Georg Reimer, Berlin.

Haeckel, E. 1895. *Systematische Phylogenie. Vol. 3: Systematische Phylogenie der Wirbelthiere* (Vertebrata). Georg Reimer, Berlin.

Halstead Tarlo, L. B. 1968. An outline classification of the squamates. *Br. J. Herpetol.* 4:32–35.

Hay, O. P. 1902. Bibliography and catalogue of the fossil *Vertebrata* of North America. *Bull. U.S. Geol. Surv.* 179:1–868.

Hay, O. P. 1930. *Second Bibliography and Catalogue of the Fossil* Vertebrata *of North America.* Carnegie Institution of Washington, Washington, D.C.

Hembree, D. I. 2007. Phylogenetic revision of *Rhineuridae* (*Reptilia: Squamata: Amphisbaenia*) from the Eocene to Miocene of North America. *Univ. Kansas Paleontol. Contrib.* 15:1–20.

Hoffstetter, R. 1955. Squamates de type moderne. Pp. 606–662 in *Traité de Paléontology*, Vol. 5. (J. Piveteau, ed.). Masson et Cie, Paris.

Huxley, T. H. 1886. *A Manual of the Anatomy of Vertebrated Animals.* D. Appleton and Co., New York.

Kearney, M. 2003a. Systematics of the *Amphisbaenia* (*Lepidosauria: Squamata*) based on morphological evidence from recent and fossil forms. *Herpetol. Monogr.* 17:1–74.

Kearney, M. 2003b. The phylogenetic position of *Sineoamphisbaena hexatabularis* reexamined. *J. Vertebr. Paleontol.* 23:394–403.

Kearney, M., and B. Stuart. 2004. Repeated evolution of limblessness and digging heads in worm lizards revealed by DNA from old bones. *Proc. R. Soc. Lond. B Biol. Sci.* 271(1549):1677–1683.

Kuhn, O. 1961. *Die Reptilien. System und Stammesgeschichte.* Verlag Oeben, Munich.

La Cépède, B. G. É. 1788–1789. *Histoire Naturelle des Quadrupèdes Ovipares et des Serpens*, 2 Vols. Hôtel de Thou, Paris.

Lee, M. S. Y. 1998. Convergent evolution and character correlation in burrowing reptiles: towards a resolution of squamate relationships. *Biol. J. Linn. Soc.* 65:369–453.

Linnaeus, C. 1758. *Systema Naturae, Per Regna Tria Naturae, Secundum Classes, Ordines, Genera, Species cum Characteribus, Differentiis, Synonymis, Locis.* 10th edition. Laurentii Salvii, Holmiae (Stockholm).

Longrich, N. R., J. Vinther, R. A. Pyron, D. Pisani, and J. A. Gauthier. 2015. Biogeography of worm lizards (*Amphisbaenia*) driven by end-Cretaceous mass extinction. *Proc. R. Soc. Lond. B Biol. Sci.* 282:20143034.

McDowell, Jr., S. B., and C. M. Bogert. 1954. The systematic position of *Lanthanotus* and the affinities of the anguinomorphan lizards. *Bull. Am. Mus. Nat. Hist.* 105(1):1–142.

Macey, J. R., T. J. Papenfuss, J. V. Kuehl, H. M. Fourcade, and J. L. Boore. 2004. Phylogenetic relationships among amphisbaenian reptiles based on complete mitochondrial genomic sequences. *Mol. Phylogenet. Evol.* 33:22–31.

Merrem, B. 1820. *Versuch eines Systems der Amphibien.* Johann Christian Krieger, Marburg.

Müller, J. 1831. Beiträge zur Anatomie und Naturgeschichte der Amphibien. *Z. Physiol.* (F. Tiedemann, G. R. Treviranus, and L. C. Treviranus, eds.) 4:190–275.

Nopcsa, F. 1908. Zur Kenntnis der fossilen Eidechsen. *Beitr. Paläontol. Geol. Österr.-Ung. Orients* 21:33–62.

Oppel, M. 1811. *Die Ordnungen, Familien und Gattungen der Reptilien als Prodrom einer Naturgeschichte Derselben.* J. Lindauer, Munich.

Parker, W. K. 1868. *A Monograph on the Structure and Development of the Shoulder-Girdle in the Vertebrata.* Ray Society, London.

Pyron, R. A., F. T. Burbrink, and J. J. Wiens. 2013. A phylogeny and revised classification of *Squamata*, including 4161 species of lizards and snakes. *BMC Evol. Biol.* 13:93.

Romer, A. S. 1933. *Vertebrate Paleontology.* University of Chicago Press, Chicago, IL.

Romer, A. S. 1945. *Vertebrate Paleontology.* 2nd edition. University of Chicago Press, Chicago, IL.

Romer, A. S. 1956. *Osteology of the Reptiles.* University of Chicago Press, Chicago, IL.

Romer, A. S. 1966. *Vertebrate Paleontology.* 3rd edition. University of Chicago Press, Chicago, IL.

Shaw, G. 1802. *General Zoology or Systematic Natural History. Vol. III: Amphibia.* G. Kearsley, London.

Smith K. T. 2009. A new lizard assemblage from the earliest Eocene (zone WAO) of the Bighorn Basin, Wyoming, USA. Biogeography during the warmest interval of the Cenozoic. *J. Syst. Palaeont.* 7:299–358.

Stannius, H. 1856. *Handbuch der Anatomie der Wirbelthiere. Zweites Buch: Zootomie der Amphibien. Handbuch der Zootomie.* 2nd edition (C. T. E. von Siebold and H. Stannius, eds.). Veit & Comp., Berlin.

Strauch, A. 1887. Bemerkungen über die Geckoniden-Sammlung im zoologischen Museum der Kaiserlichen Akademie der Wissenschaften zu St. Petersburg. *Mem. Acad. Sci. St. Petersb., VIIE Ser.* 35(2):1–72.

Townsend, T. M., A. Larson, E. Louis, and J. R. Macey. 2004. Molecular phylogenetics of *Squamata*: the position of snakes, amphisbaenians, and dibamids, and the root of the squamate tree. *Syst. Biol.* 53:735–757.

Uetz, P., ed. 2017. The Reptile Database, Version 15 October 2017. Available at http://www.reptile-database.org.

Underwood, G. L. 1971. A modern appreciation of Camp's "Classification of the lizards". Pp. vii–xvii in *Camp's Classification of the Lizards.* Facsimile reprint by the Society for the Study of Amphibians and Reptiles.

Vanzolini, P. E. 1951. A systematic arrangement of the family *Amphisbaenidae (Sauria)*. *Herpetologica* 7(3):113–123.

Vidal, N., A. Azvolinsky, C. Cruaud, and S. B. Hedges. 2008. Origin of tropical American burrowing reptiles by transatlantic rafting. *Biol. Lett.* 4:115–118.

Vidal, N., and S. B. Hedges. 2004. Molecular evidence for a terrestrial origin of snakes. *Proc. R. Soc. Lond. B Biol. Sci.* 271(suppl.):S226–S229.

Vidal, N., and S. B. Hedges. 2005. The phylogeny of squamate reptiles (lizards, snakes, and amphisbaenians) inferred from nine nuclear protein-coding genes. *C. R. Biol.* 328:1000–1008.

Wagler, J. G. 1828. Vorläufige Übersicht des Gerüstes, so wie Ankündigung seines Systema Amphibiorum. *Isis von Oken* 21:859–861.

Wagler, J. G. 1830. *Natürliches System der Amphibien, Mit Vorangehender Classification der Säugthiere und Vögel.* J. G. Cotta, Munich.

Wiegmann, A. F. A. 1834. *Herpetologia Mexicana, seu Descriptio Amphibiorum Novae Hispaniae. Pars Prima, Saurorum Species.* C. G. Lüderitz, Berlin.

Wiens, J. J., C. R. Hutter, D. G. Mulcahy, B. P. Noonan, T. M. Townsend, J. W. Sites, Jr., and T. W. Reeder. 2012. Resolving the phylogeny of lizards and snakes (*Squamata*) with extensive sampling of genes and species. *Biol. Lett.* 8:1043–1046.

Williston, S. W. 1925. *Osteology of the Reptiles.* Harvard University Press, Cambridge, UK.

Wu, X.-C., D. B. Brinkman, A. P. Russell, Z.-M. Dong, P. J. Currie, L.-H. Hou, and G.-H. Cui. 1993. Oldest known amphisbaenian from the Upper Cretaceous of Chinese Inner Mongolia. *Nature* 366:57–59.

Authors

Maureen Kearney; Office of Science, Policy, and Society Programs, American Association for the Advancement of Science, 1200 New York Avenue NW, Washington, DC 20005, USA. Email: mkearney@aaas.org.

Kevin de Queiroz; Department of Vertebrate Zoology; National Museum of Natural History; Smithsonian Institution; Washington, DC 20560-0162, USA. Email: dequeirozk@si.edu.

Date Accepted: 1 November 2013; updated 3 November 2017

Primary Editor: Philip Cantino

Pan-Serpentes J. Head, K. de Queiroz and H. Greene, new clade name

Registration Number: 121

Definition: The total clade of the crown clade *Serpentes*. This is a crown-based total-clade definition. Abbreviated definition: total ∇ of *Serpentes*.

Etymology: *Pan* (prefix indicating that the name refers to a total clade; derived from the Greek *Pantos*, all, the whole) + *Serpentes* (see etymology for *Serpentes* in this volume).

Reference Phylogeny: The reference phylogeny is Figure 1 of Gauthier et al. (2012), where *Pan-Serpentes* includes *Najash rionegrina*† and all taxa below it (p. 12). This is the result of the first morphological phylogenetic analysis that: (1) tests monophyly of *Serpentes* and constituent subclades via extensive taxon sampling; (2) infers at least some fossil taxa as stem snakes; and (3) places the snake total clade within a comprehensive phylogenetic hypothesis of *Squamata*. Previous analyses have satisfied some, but not all, of these criteria (e.g., Lee, 1997, 1998; Tchernov et al., 2000; Rieppel and Zaher, 2000; Scanlon and Lee, 2000; Lee and Scanlon, 2002; Apesteguía and Zaher, 2006; Scanlon, 2006; Conrad, 2008; Wilson et al., 2010; Müller et al., 2011; Zaher and Scanferla, 2012).

Composition: *Serpentes* and all taxa sharing a more recent ancestry with that crown clade than with the most closely related crown group. See Comments for a discussion of taxa that have been hypothesized to be members of the snake stem group.

Diagnostic Apomorphies: As a total clade, *Pan-Serpentes* may not have any apomorphies (de Queiroz, 2007); however, possession of any of the apomorphies of *Serpentes* (this volume) constitutes evidence for inclusion of a species or specimen within *Pan-Serpentes*. In addition, Apesteguía and Zaher (2006) and Longrich et al. (2012) inferred characters to be synapomorphies of *Serpentes* and the putative stem snakes *Najash rionegrina* and *Coniophis precedens*. Well-delineated characters shared by these fossils and crown snakes include interdental ridges forming partial alveoli in marginal tooth-bearing elements (present in some other squamatans); parietal descending process contacts dorsal and anterior margins of prootic and dorsal margin of parasphenoid rostrum, laterally enclosing braincase; loss of upper temporal arch and squamosal (also in some other squamatans); subcentral paralymphatic fossae present on posterior precloacal vertebrae; lymphapophyses present on cloacal vertebrae (also in some other long-bodied limb-reduced squamatans); haemapophyses and pleuropophyses present on caudal vertebrae; and pleurocentral hypapophyses on (at least) anterior precloacal vertebrae (also in some other squamatans).

Synonyms: Approximate (some of which are also partial) synonyms are the same as those listed for *Serpentes* (this volume). *Serpentes* of Estes et al. (1988) is an unambiguous synonym.

Comments: Although the concept of the snake total clade is straightforward, there is considerable disagreement concerning both the closest extant relatives of snakes and the members of the snake stem group. Various fossil taxa have been proposed to be stem snakes, including *Dinilysia patagonica* (Scanlon and Lee, 2000; Lee and Scanlon, 2002, Zaher and Scanferla,

2012), *Najash rionegrina* (Apesteguía and Zaher, 2006), *Coniophis precedens* (Longrich et al., 2012) the pachyophiids *Pachyrhachis problematicus* (Lee, 1998; Lee and Caldwell, 2000; Lee and Scanlon, 2002), *Pachyophis woodwardi* (Lee et al., 1999; Lee and Caldwell, 2000), *Haasiophis terrasanctus* (Lee and Scanlon, 2002; Scanlon, 2006; Lee, 2009), and *Eupodophis descouensi* (Rage and Escuille, 2000; Palci et al., 2013), and madtsoiids such as *Wonambi naracoortensis* and *Yurlunggur spp.* (Scanlon and Lee, 2000; Lee and Scanlon, 2002; Scanlon, 2006). All of the above taxa have highly elongated bodies and highly reduced limbs, based on specimens preserving articulated or directly associated cranial and postcranial elements, except for *Coniophis*, whose postcranial body form is inferred from the hypothesized association of isolated, disarticulated elements, and all have been alternatively considered to be crown snakes (Hecht, 1959; Rieppel and Zaher, 2000; Tchernov et al., 2000; Rieppel et al., 2002; Apesteguía and Zaher, 2006; Conrad, 2008; Wilson et al., 2010; Gauthier et al., 2012; Longrich et al., 2012; Vasile et al., 2013). Other taxa exhibiting minimal to moderate body elongation and fully formed (if sometimes modified) limbs have also sometimes been inferred to be members of the snake stem group, specifically "aigialosaurs", mosasaurids (highly aquatic forms), dolichosaurids, *Aphanizocnemus*, and *Adriosaurus* (Lee, 1997, 1998, 2009; Lee and Caldwell, 2000). However, more recent and more comprehensive morphological phylogenetic analyses of *Squamata* (Conrad, 2008; Gauthier et al., 2012) and analyses using combined morphological and molecular data (Wiens et al., 2010; Müller et al., 2011) do not support dolichosaurids, *Adriosaurus,* "aigialosaurs", or mosasaurids as stem snakes.

Estes et al. (1988) applied the name *Serpentes* to the total clade of snakes using an explicit phylogenetic definition. However, that proposal contradicted their own stated adoption of the conventions proposed by Gauthier et al. (1988), which include using the best-known names for crown clades. We have adopted that convention and thus apply the name *Serpentes* (in this volume) to the snake crown clade. Conrad (2008) mentioned the possibility of using the name *Ophidia* for the snake total clade but did not formally adopt that idea. Because *Ophidia* has been defined explicitly as applying to a less inclusive clade (see below), and because we support the convention of forming the name of the total clade by adding a standard prefix to the name of the crown (see de Queiroz, 2007, and references therein), we use the name *Pan-Serpentes* for the snake total clade. De Queiroz (2007) used the name *Pan-Serpentes* to illustrate some of the advantages of an integrated system of crown and total clade names; however, we are not treating *Pan-Serpentes* as a preexisting name because it was used as a hypothetical example rather than as a nomenclatural proposal (see *ICPN*, Art. 7.2b; Cantino and de Queiroz, 2020).

Lee (1997) defined the name *Pythonomorpha* Cope 1869 as applying to the clade originating in the most recent common ancestor of snakes and mosasauroids (see also Lee, 1998; Lee and Caldwell, 1998, 2000). However, given that snakes were explicitly excluded from *Pythonomorpha* by Cope (1869), it would be more appropriate (*ICPN*, Rec. 11A; Cantino and de Queiroz, 2020) to define that name so that the inclusion of snakes is permitted but not required (e.g., as the smallest clade containing both *Mosasaurus hoffmanni* and *Dolichosaurus longicollis*, which would include snakes in the context of the phylogeny of Lee and Caldwell [2000]). Lee and Caldwell (1998) defined the name *Ophidia* Macartney 1802 as applying to the clade originating in the most recent common ancestor of *Serpentes* and *Pachyrachis problematicus*, to which Lee (1998) added a qualifying clause that would prevent use of the name *Ophidia* in the context of phylogenies in which any of 25 taxa not normally considered snakes are part of that clade. Those definitions are inappropriate given that the authors equated the name *Ophidia* with the English vernacular name "snakes", which they intended to refer to a clade composed of organisms that possess most of the diagnostic characters of crown-group snakes (Lee, 2001). According to the stated definition, an extinct species very similar in appearance to *Pachyrachis* but slightly more distantly related to extant snakes would not be part of *Ophidia*, seemingly contrary to the authors' intent. A more appropriate definition of *Ophidia* would specify the reference of that name using an apomorphy or set of apomorphies.

Literature Cited

Apesteguía, S., and H. Zaher. 2006. A Cretaceous terrestrial snake with robust hindlimbs and a sacrum. *Nature* 440:1037–1040.

Cantino, P. D., and K. de Queiroz. 2020. *International Code of Phylogenetic Nomenclature (PhyloCode)*, Version 6. CRC Press, Boca Raton, FL.

Conrad, J. L. 2008. Phylogeny and systematics of *Squamata (Reptilia)* based on morphology. *Bull. Am. Mus. Nat. Hist.* 310:1–182.

Cope, E. D. 1869. On the reptilian orders, *Pythonomorpha* and *Streptosauria*. *Proc. Boston Soc. Nat. Hist.* 12:250–266.

Estes, R., K. de Queiroz, and J. A. Gauthier. 1988. Phylogenetic relationships within *Squamata*. Pp. 119–281 in *Phylogenetic Relationships of the Lizard Families* (R. Estes and G. K. Pregill, eds.). Stanford University Press, Stanford, CA.

Gauthier, J. A., R. Estes, and K. de Queiroz. 1988. A phylogenetic analysis of *Lepidosauromorpha*. Pp. 15–98 in *Phylogenetic Relationships of the Lizard Families* (R. Estes and G. K. Pregill, eds.). Stanford University Press, Stanford, CA.

Gauthier, J. A., M. Kearney, J. A. Maisano, O. Rieppel, and A. D. B. Behlke. 2012. Assembling the squamate tree of life: perspectives from the phenotype and the fossil record. *Bull. Peabody Mus. Nat. Hist.* 53:3–308.

Hecht, M. K. 1959. Amphibians and reptiles. Pp. 130–146 in *The Geology and Paleontology of the Elk Mountain and Tabernacle Butte Area, Wyoming* (P. O. McGrew, ed.). *Bull. Am. Mus. Nat. Hist.* 117:1–176.

Lee, M. S. Y. 1997. The phylogeny of varanoid lizards and the affinities of snakes. *Philos. Trans. R. Soc. Lond. B Biol. Sci.* 352:53–91.

Lee, M. S. Y. 1998. Convergent evolution and character correlation in burrowing reptiles: towards a resolution of squamate phylogeny. *Biol. J. Linn. Soc.* 65:369–453.

Lee, M. S. Y. 2001. Snake origins and the need for scientific agreement on vernacular names. *Paleobiology* 27:1–6.

Lee, M. S. Y. 2009. Hidden support from unpromising data sets strong unites snakes with anguimorph "lizards". *J. Evol. Biol.* 22:1308–1316.

Lee, M. S. Y., and M. W. Caldwell. 1998. Anatomy and relationships of *Pachyrhachis problematicus*, a primitive snake with legs. *Philos. Trans. R. Soc. Lond. B Biol. Sci.* 353:1521–1552.

Lee, M. S. Y., and M. W. Caldwell. 2000. *Adriosaurus* and the affinities of mosasaurs, dolichosaurs, and snakes. *J. Paleontol.* 74:915–937.

Lee, M. S. Y., M. W. Caldwell, and J. D. Scanlon. 1999. A second primitive marine snake: *Pachyophis woodwardi* from the Cretaceous of Bosnia-Herzegovina. *J. Zool. Lond.* 248:509–520.

Lee, M. S. Y., and J. D. Scanlon. 2002. Snake phylogeny based on osteology, soft anatomy and ecology. *Biol. Rev.* 77:333–401.

Longrich, N. R., B.-A. Bhullar, and J. A. Gauthier. 2012. A transitional snake from the Late Cretaceous period of North America. *Nature* 488:205–208.

Macartney, J. 1802. *Lectures on Comparative Anatomy. Vol. 1: Translated from the French of G. Cuvier by William Ross Under the Inspection of James Macartney*. Wilson and Co., London.

Müller, J., C. Hipsley, J. J. Head, N. Kardjilov, A. Hilger, M. Wuttke, and R. R. Reisz. 2011. Limbed lizard from the Eocene of Germany reveals amphisbaenian origins. *Nature* 473:364–367.

Palci, A., M. W. Caldwell, and R. L. Nydam. 2013. Reevaluation of the anatomy of the Cenomanian (Upper Cretaceous) hind-limbed marine fossil snakes *Pachyrhachis*, *Haasiophis*, and *Eupodophis*. *J. Vertebr. Paleontol.* 33:1328–1342.

de Queiroz, K. 2007. Toward an integrated system of clade names. *Syst. Biol.* 56:956–974.

Rage, J.-C., and F. Escuille. 2000. Un nouveau serpent bipède du Cénomanien (Crétacé). Implications phylétiques. *C. R. Acad. Sci. Ser. IIA Sci. Ter.* 330:513–520.

Rieppel, O., A. G. Kluge, and H. Zaher. 2002. Testing the phylogenetic relationships of the Pleistocene snake *Wonambi naracoortensis* Smith. *J. Vertebr. Paleontol.* 23:812–829.

Rieppel, O., and H. Zaher. 2000. The intramandibular joint in squamates, and the phylogenetic relationships of the fossil snake *Pachyrhachis problematicus* Haas. *Fieldiana Geol.* 43:1–69.

Scanlon, J. D. 2006. Skull of the large non-macrostomatan snake *Yurlunggur* from the Australian Oligo-Miocene. *Nature* 439:839–842.

Scanlon, J. D., and M. S. Y. Lee. 2000. The Pleistocene serpent *Wonambi* and the early evolution of snakes. *Nature* 403:416–420.

Tchernov, E., O. Rieppel, H. Zaher, M. J. Polcyn, and L. L. Jacobs. 2000. A fossil snake with limbs. *Science* 287:2010–2012.

Vasile, S., Csiki-Sava, Z., and Venczel, M. 2013. A new madtsoiid snake from the upper Cretaceous of the Hateg Basin, western Romania. *J. Vertebr. Paleontol.* 33:1100–1119.

Wiens, J. J., C. A. Kuczynski, T. Townsend, T. W. Reeder, D. G. Mulcahy, and J. W. Sites, Jr. 2010. Combining phylogenomics and fossils in higher-level squamate reptile phylogeny: molecular data change the placement of fossils. *Syst. Biol.* 59:674–688.

Wilson, J. A., D. Mohabey, S. Peters, and J. J. Head. 2010. Predation upon hatchling sauropod dinosaurs by a new basal snake from the Late Cretaceous of India. *PLOS Biol.* 8:e1000322.

Zaher, H., and C. A. Scanferla. 2012. The skull of the upper Cretaceous snake *Dinilysia patagonica* Smith-Woodward, 1901, and its phylogenetic position revisited. *Zool. J. Linn. Soc.* 164:194–238.

Authors

Jason J. Head; Department of Zoology and University Museum of Zoology; University of Cambridge; Cambridge UK, CB2 3EJ; Email: jjh71@cam.ac.uk.

Kevin de Queiroz; Department of Vertebrate Zoology; National Museum of Natural History; Smithsonian Institution; Washington, DC 20560-0162, USA. Email: dequeirozk@si.edu.

Harry W. Greene; Department of Ecology and Evolutionary Biology; Cornell University; E251 Corson Hall, Ithaca, NY 14853, USA. Email: hwg5@cornell.edu.

Date Accepted: 18 December 2014

Primary Editor: Philip Cantino

Registration Number: 96

Definition: The crown clade originating in the most recent common ancestor of *Anomalepis mexicanus* Jan 1860 (*Scolecophidia/Anomalepididae*), *Rena dulcis* Baird and Girard 1853 (*Scolecophidia /Leptotyphlopidae*), *Typhlops linneolatus* Jan 1863 (*Scolecophidia/Typhlopidae*), *Anilius* (originally *Anguis*) *scytale* (Linnaeus 1758) (*Alethinophidia/ Anilioidea*), and *Coluber constrictor* Linnaeus 1758 (*Alethinophidia/Caenophidia*). This is a minimum-crown-clade definition. Abbreviated definition: min crown ∇ (*Anomalepis mexicanus* Jan 1860 & *Rena dulcis* Baird and Girard 1853 & *Typhlops linneolatus* Jan 1863 & *Anilius scytale* (Linnaeus 1758) & *Coluber constrictor* Linnaeus 1758).

Etymology: Derived from the Latin vernacular for "snake" (*serpens*, singular; *serpentis*, plural).

Reference Phylogeny: The reference phylogeny is Figure 4 of Kluge (1991), where *Coluber constrictor* is a representative of "higher snakes." This is the first phylogeny of snakes based on an explicitly phylogenetic analysis of a taxon × character matrix incorporating all major extant snake clades, including those containing the specifiers used here. Phylogenies based on subsequent analyses of morphological (e.g., Cundall et al., 1993; Lee and Scanlon, 2002; Gauthier et al., 2012), molecular (e.g., Slowinski and Lawson, 2002; Wilcox et al., 2002; Wiens et al., 2012), and combined data (e.g., Lee et al., 2007; Wiens et al., 2010) result in application of the name *Serpentes* to a clade of identical composition.

Composition: *Serpentes* consists of more than 3,400 currently recognized extant (Uetz, 2012) and more than 100 currently recognized fossil species (e.g., Rage, 1984; Holman, 2000). It consists of two primary subclades: the total clades of *Scolecophidia* and *Alethinophidia* (although the former is inferred to be paraphyletic in many molecular analyses; see Comments).

Diagnostic Apomorphies: Diagnosing *Serpentes* relative to purported sister taxa is difficult because there is no consensus on the position of *Serpentes* relative to other extant *Squamata* or to putative stem-group fossils (see Comments). Despite these problems, extant snakes possess numerous morphological characters that optimize unambiguously as apomorphies relative to other extant taxa. The following list (which is not comprehensive) is derived from anatomical reviews and recent morphological phylogenetic analyses (Bellairs and Underwood, 1951; Underwood, 1967; Tchernov et al., 2000; Lee and Scanlon, 2002; Conard, 2008; Cundall and Irish, 2008; Gauthier et al., 2012).

1) Subolfactory process of frontal contacts parasphenoid (also in some other fossorial squamatans).
2) Anterolateral margin of frontal subolfactory process includes prefrontal process that extends into socket of prefrontal.
3) Reduction in number of posterior orbital bones (also in some other squamatans).
4) Crista circumfenestralis formed from crista prootica, crista interfenestralis, and crista tuberalis partially encloses stapedial footplate.
5) Platytrabic braincase.
6) Palatoquadrate (pterygoquadrate) cartilage reduced, forms only the quadrate.

This character additionally describes the absence of an epipterygoid in snakes (Bellairs and Kamal, 1981).

7) Processus ascendens tectum synoticum absent (also in some other squamatans).

8) Scleral ossicles absent.

9) Mandibles connected only by soft tissues; bony symphysis absent.

10) Splenial reduced so that Meckel's canal is broadly open anteriorly for ~ 50% the length of the dentary tooth row (also in some other squamatans).

11) Posterior terminus of splenial does not extend posteriorly beyond coronoid apex (also in some other squamatans).

12) Intramandibular joint in which the angular receives the splenial (absent in typhlopids, anomalepidids, and many viperids).

13) Vertebrae with well-developed zygosphenes, including straight to convex anterior margins and dorsolaterally angled articular facets, and zygantra with corresponding ventromedially angled articular facets (although these accessory articulations are present in some other squamatans, the strait to convex anterior margin of the zygosphenes is unique to snakes).

14) Loss of pectoral girdle and forelimb (also in some other squamatans).

15) Loss of sternal skeleton (also in some other squamatans)

16) Increase in precloacal vertebral numbers beyond 100 (also in some other squamatans).

Soft-tissue apomorphies of extant snakes (but unknown in fossils) relative to other extant squamatans include:

17) Gall bladder separated from liver (posteriorly) by a distinct gap.

18) Left lung reduced or absent (also in some other elongate squamatans).

19) Kidneys positioned well anterior of cloaca.

20) Soft secondary palate formed by the closure of each choanal groove to form a ductus nasopharyngeus (Groombridge, 1979). Absent in typhlopids and leptotyphlopids, but not anomalepidids (McDowell, 1972). A secondary palate is also present in some other squamatans, although often with contributions from the maxillary and palatal bones (hard secondary palate).

21) Eye covered by brille (also in some other squamatans).

22) Tympanum lost (also in some other squamatans).

The osteological apomorphies described above (characters 1–16) are diagnostic for *Serpentes* if *Dinilysia patagonica*, *Pachyophiidae* and *Madtsoiidae* are nested within the snake crown (Tchernov et al., 2000; Rieppel et al., 2002; Apesteguía and Zaher, 2006; Conrad, 2008; Wilson et al., 2010). If some or all of these taxa are outside of the crown (Lee and Caldwell, 1998; Lee, 1998; Scanlon and Lee, 2000; Scanlon, 2006; Longrich et al., 2012; Zaher and Scanferla, 2012; Palci et al., 2013), then some of the listed characters may diagnose slightly larger clades.

Synonyms: Approximate (and sometimes partial—see Comments) synonyms include *Serpentia* (e.g., Linnaeus, 1735, 1740, 1748, 1759; Laurenti, 1768; Scopoli, 1777), *Serpens* (e.g., Aldrovandi, 1640; Klein, 1755; Lacépède, 1788, 1790; Sonnini and Latreille, 1801—see Harper, 1940 for publication date information); *Ophidia* (Macartney, 1802—see Gill, 1900); *Ophidii* (e.g., Latreille, 1804, 1825; Oppel, 1811; Gray, 1825); *Epalpebrata* (Haworth, 1825); *Ophes* (Wagler, 1828); *Idiophides* (Latreille, 1825; Bonaparte, 1831); and *Ophides* (Swainson, 1838–1839).

Comments: Recognition of a group called *Serpentes*, *Ophidia*, or a variant of one of those names can be found in some of the earliest writings on natural history (e.g., Aristotle's *Historia Animalium*; Aldrovandi, 1640; Ray, 1693); however, early concepts seem to have been based primarily on the elongate, limb-reduced body form and therefore included several taxa now thought to have independently evolved this body form that are no longer considered snakes, such as *Anguis*, *Amphisbaena*, and *Caecilia* (e.g., Linnaeus, 1758; Laurenti, 1768; Scopoli, 1777; Gmelin, 1788; Daudin, 1801–3; Rafinesque, 1815). During the early nineteenth century, *Caecilia* was removed and placed among the amphibians (e.g., Oppel, 1811; Blainville, 1816; Merrem, 1820); *Anguis* (and *Ophisaurus*) was removed and placed among the "lizards" or in a separate group for limbless squamatans other than snakes (e.g., Oppel, 1811; Gray, 1825; Bonaparte, 1831); and *Amphisbaena* was removed and placed among the "lizards", in a separate group for limbless squamatans other than snakes, or in a separate amphisbaenian group (e.g., Gray, 1825; Bonaparte, 1831; Wiegmann, 1834). By the middle of the nineteenth century, the composition of the snake taxon had become congruent with contemporary concepts (e.g., Wagler, 1830; Duméril and Bibron, 1834–1854; Fitzinger, 1843; Bonaparte, 1850; Duméril and Duméril, 1851; Stannius, 1856; Jan, 1857; Cope, 1864). Monophyly of the taxon is supported by numerous morphological characters (see Diagnostic Apomorphies) and has been inferred by most authors in the evolutionary era, often with strong support (e.g., Cope, 1900; Estes et al., 1988; Lee, 1998; Vidal and Hedges, 2009; Gauthier et al., 2012; Wiens et al., 2012; but see McDowell and Bogert, 1954).

Cope (1900) presented one of the first phylogenetic hypotheses regarding relationships within snakes. He recognized an ancestral taxon *Peropoda* (boas, pythons, dwarf boas) from which were derived two groups, one composed of *Scolecophidia* (typhlopids and leptotyphlopids) and *Tortricina* (*Anilius*, *Cylindrophis*, uropeltids) and the other composed of *Aglyphodonta* (*Xenopeltis*, caenophidians without grooved teeth), *Opisthoglypha* (caenophidians with posterior grooved teeth on the maxilla), *Proteroglypha* (caenophidians with an anterior tubular tooth on the maxilla), and *Solenoglypha* (caenophidians with only an anterior tubular tooth on the maxilla*). Nopcsa (1923) presented a phylogeny of snakes in which an ancestral snake taxon *Cholophidia* (fossil forms) gave rise to two groups: *Angiostomata* (typhlopids and leptotyphlopids) and *Alethinophidia* ("true snakes"). Hoffstetter (1939) proposed a taxonomy that recognized four major mutually exclusive taxa of snakes: *Cholophidia*, *Scolecophidia*, *Henophidia* ("relatively ancient" snakes), and *Caenophidia* ("new or modern" snakes), the last two corresponding to Nopcsa's *Alethinophidia*. Subsequent phylogenetic analyses have demonstrated the paraphyly of *Cholophidia* and *Henophidia* but supported the monophyly of both *Scolecophidia* and *Alethinophidia*, including those of Rieppel (1988), Kluge (1991), Tchernov et al. (2000), Scanlon and Lee (2000), Lee and Scanlon (2002), Apesteguía and Zaher (2006), Scanlon (2006), Gauthier et al. (2012), and Longrich et al. (2012) based on morphology, and Slowinski and Lawson (2002), Wilcox et al. (2002), Vidal and Hedges (2002, 2004), and Vidal and David (2004) based on molecular sequence data. Heise et al. (1995), Conrad (2008), Wiens et al. (2012), and Pyron et al. (2013) inferred a monophyletic *Serpentes*, but not a monophyletic *Scolecophidia*.

---

* Duméril and Bibron (1834–1854), who proposed the taxon *Solenoglyphes*, characterized its members as having a single fang in each maxilla. In reality, there are two tooth positions in each maxilla, although for most of the tooth replacement cycle only one position or the other bears a tooth that is firmly attached to the jaw (Klauber, 1972; Gauthier et al., 2012; Nagy et al., 2013).

The first explicit phylogenetic definition of the name *Serpentes* is that of Estes et al. (1988), who, contrary to their convention of using widely known names for crown clades (Gauthier et al., 1988: 62), defined the name *Serpentes* as referring to a total clade. However, most subsequent authors have used *Serpentes* for the crown, either implicitly (e.g., Rieppel, 1988; Lee, 1998; Palci and Caldwell, 2007) or explicitly (Lee, 1997; Greene and Cundall, 2000; Conrad, 2008; Gauthier et al., 2012). We have used the name *Serpentes* for the snake crown clade because (1) when defined explicitly, that is the clade to which the name is most commonly applied, (2) it is by far the most widely used of existing names that have been at least loosely associated with that clade (roughly five times more commonly used than *Ophidia*, the most widely used of the approximate synonyms), (3) there are advantages to using the most widely known name for the crown clade (de Queiroz and Gauthier, 1992), and (4) the name refers etymologically only to a vague character (< Latin *serpere*, to creep, crawl) rather than a specific apomorphy (see de Queiroz, 2007). Based on the commonly inferred basal split between the stems of *Scolecophidia* and *Alethinophidia*, as well as the fact that *Scolecophidia* is not always inferred to be monophyletic, we selected distantly related species within each of those clades as internal specifiers.

Despite possessing a highly modified body form and unique feeding morphologies relative to other squamatans, determination of unambiguous synapomorphies for the snake crown clade is complicated by the disparate phylogenetic hypotheses concerning the relationships of *Serpentes* both to other extant *Squamata* and to fossil taxa. Phylogenies derived from molecular sequence data place *Serpentes* closest to *Anguimorpha* and/or *Iguania* (Townsend et al., 2004; Fry et al., 2006; Wiens et al., 2006; Vidal and Hedges, 2009; Lee, 2009; Wiens et al.,

2012), those derived from morphology place the clade closest to *Amphisbaenia* or *Amphisbaenia* + *Dibamidae* (Evans and Barbadillo, 1998; Kearney, 2003a, b; Evans and Wang, 2005; Evans et al., 2005; Conrad, 2008; Gauthier et al., 2012), and combined analyses place *Serpentes* closest to either *Anguimorpha* (Wiens et al., 2010; Müller et al., 2011) or *Anguimorpha* + *Iguania* (Wiens et al., 2010; Wiens et al., 2012). The fossil clades *Pachyophiidae* and *Madtsoiidae* have been inferred to be either stem snakes (Lee and Caldwell, 1998; Lee, 1998; Scanlon and Lee, 2000; Lee and Scanlon, 2002; Scanlon, 2006; Palci et al., 2013) or crown snakes nested within or on the stem of *Alethinophidia* (Tchernov et al., 2000; Rieppel et al., 2002; Rieppel and Head, 2004; Apesteguía and Zaher, 2006; Wilson et al., 2010; Zaher and Scanferla, 2012; Gauthier et al., 2012; Longrich et al., 2012, for *Pachyophiidae*) or on the alethinophidian stem (Longrich et al., 2012; for *Madtsoiidae*). Similarly, the fossil taxon *Dinilysia patagonica* has been considered either a stem snake (Zaher and Scanferla, 2012; Gauthier et al., 2012) or a stem alethinophidian (Wilson et al., 2010; Longrich et al., 2012). As a result, the diagnosis of both the snake crown and its total clade are considerably more difficult than expected given the distinctiveness of snakes.

Literature Cited

Aldrovandi, U. 1640. *Historia Serpentum et Draconum Libri Duo*. Bononiæ, C. Ferronium, Bologna.

Apesteguía, S., and H. Zaher. 2006. A Cretaceous terrestrial snake with robust hindlimbs and a sacrum. *Nature* 440:1037–1040.

Baird, S. F., and C. Girard. 1853. *Catalogue of North American Reptiles in the Museum of the Smithsonian Institution. Part I, Serpents*. Smithsonian Institution, Washington, DC.

Bellairs, A. d'A., and A. M. Kamal. 1981. The chondrocranium and the development of the skull in recent reptiles. Pp. 1–263 in *Biology of the* Reptilia. *Vol. 11: Morphology F* (C. Gans, and T. S. Parsons, eds.). Academic Press, New York.

Bellairs, A. d'A., and G. Underwood. 1951. The origin of snakes. *Biol. Rev. Camb. Philos. Soc.* 26:193–237.

de Blainville, H. 1816. Prodrome d'une nouvelle distribution systématique du règne animal. *Bull. Sci. Soc. Philom. Paris* 8:105–112.

Bonaparte, C. L. 1831. *Saggio di una Distribuzione Metodica Degli Animali Vertebrati.* Presso Antonio Bouzaler, Roma.

Bonaparte, C. L. 1850. *Conspectus Systematicum Herpetologicae et Amphiologicae. Editio Altera Reformata.* E. J. Brill, Lugduni Batavorum (Leiden).

Conrad, J. L. 2008. Phylogeny and systematics of *Squamata* (*Reptilia*) based on morphology. *Bull. Am. Mus. Nat. Hist.* 310:1–182.

Cope, E. D. 1864. On the characters of the higher groups of *Reptilia—Squamata*—and especially of the *Diploglossa. Proc. Acad. Nat. Sci. Phila.* 16:224–231.

Cope, E. D. 1900. The crocodilians, lizards, and snakes of North America. *Annu. Rep. U.S. Natl. Mus.* 1898:153–1270 + 36 plates.

Cundall, D., and F. Irish. 2008. The snake skull. Pp. 349–692 in *Biology of the* Reptilia, *vol. 20, Morphology H* (C. Gans, A. S. Gaunt, and K. Adler, eds.). Society for the Study of Amphibians and Reptiles, Ithaca, NY.

Cundall, D., V. Wallach, and D. A. Rossman. 1993. The systematic relationships of the snake genus *Anomochilus. Zool. J. Linn. Soc.* 109:275–299.

Daudin, F. M. 1801–1803. *Histoire Naturelle, Generale et Particuliere, Des Reptiles. Tome Premier-Tome Huitieme.* F. Dufart, Paris.

Duméril, A. M. C., and G. Bibron. 1834–1854. *Erpétologie Générale ou Histoire Naturelle Complète des Reptiles.* 1–10. Librairie Encyclopédique de Roret, Paris.

Duméril, A. M. C., and A. H. A. Duméril. 1851. *Catalogue Méthodique de la Collection des Reptiles du Muséum d'Histoire Naturelle de Paris.* Gide et Baudry/Roret, Paris.

Estes, R., K. de Queiroz, and J. A. Gauthier. 1988. Phylogenetic relationships within *Squamata.*

Pp. 119–281 in *Phylogenetic Relationships of the Lizard Families* (R. Estes and G. K. Pregill, eds.). Stanford University Press, Stanford, CA.

Evans, S. E., and L. J. Barbadillo. 1998. An unusual lizard from the Early Cretaceous of Las Hoyas, Spain. *Zool. J. Linn. Soc.* 124:235–265.

Evans, S. E., and Y. Wang. 2005. The Early Cretaceous lizard *Dalinghosaurus* from China. *Acta Palaeontol. Pol.* 50:725–742.

Evans, S. E., Y. Wang, and C. Li. 2005. The early Cretaceous lizard genus *Yabeinosaurus* from China: resolving an enigma. *J. Syst. Palaeontol.* 4:319–335.

Fitzinger, L. J. 1843. *Systerma Reptilium. Fasiculus Primus.* Amblyglossae. Braumüller et Seidel, Vindobonae (Vienna).

Fry, B. G., N. Vidal, J. A. Norman, F. J. Vonk, H. Scheib, S. F. R. Ramjan, S. Kuruppu, K. Fung, S. B. Hedges, M. K. Richardson, W. C. Hodgson, V. Ignjatovic, R. Summerhayes, and E. Kochva. 2006. Early evolution of the venom system in lizards and snakes. *Nature* 439:584–588.

Gauthier, J. A., R. Estes, and K. de Queiroz. 1988. A phylogenetic analysis of *Lepidosauromorpha.* Pp. 15–98 in *Phylogenetic Relationships of the Lizard Families* (R. Estes and G. K. Pregill, eds.). Stanford University Press, Stanford, CA.

Gauthier, J. A., M. Kearney, J. A. Maisano, O. Rieppel, and A. D. B. Behlke. 2012. Assembling the squamate tree of life: perspectives from the phenotype and the fossil record. *Bull. Peabody Mus. Nat. Hist.* 53:3–308.

Gill, T. 1900. The earliest use of the names *Sauria* and *Batrachia. Science* 12:730.

Gmelin, J. F. 1788. *Systema Naturae Per Regna Tria Naturae, Secundum Classes, Ordines, Genera, Species; cum Characteribus, Differentiis Synonymis, Locis,* Vol. 1. Georg. Emanuel Beer, Lipsiae. i–xii + 1–500.

Gray, J. E. 1825. A synopsis of the genera of reptiles and amphibia, with a description of some new species. *Ann. Philo., Ser. 2,* 10:193–217.

Greene, H. W., and D. Cundall. 2000. Limbless tetrapods and snakes with legs. *Science* 287:1939–1941.

Groombridge, B. C. 1979. Variations in morphology of the superficial palate of henophidian snakes and some systematic implications. *J. Nat. Hist.* 13:447–475.

Harper, F. 1940. Some works of Bartram, Daudin, Latreille, and Sonnini, and their bearing upon North American herpetological nomenclature. *Am. Midl. Nat.* 23:692–723.

Haworth, A. H. 1825. A binary arrangement of the Class *Amphibia. Philos. Mag. J.* 65:372–373.

Hoffstetter, R. 1939. Contribution a l'étude des *Elapidae* actuels et fossiles et de l'ostéologie des ophidians. *Arch. Mus. Hist. Nat. Lyon* 15:1–78.

Holman, J. A. 2000. *The Fossil Snakes of North America.* Indiana University Press, IBloomington, IN.

Heise, P. J., L. R. Maxson, H. G. Dowling, and S. B. Hedges. 1996. Higher-level snake phylogeny inferred from mitochondrial DNA sequences of 12S rRNA and 16S rRNA genes. *Mol. Phylogenet. Evol.* 12:259–265.

Jan, G. 1857. *Cenni sul Museo Civico di Milano ed Indice Sistematico dei Rettili ed Anfibi Esposti nel Medesimo.* Milan, Pirola.

Jan, G. 1860. In Jan, G., and F. Sordelli. 1860 ("1864"). *Iconographie Générale des Ophidiens,* Tome I. Chez l'auteur, Milan.

Jan, G. 1863. *Elenco Sistematico Degli Ofidi Descritti e Disegnati per L'iconografia Generale Edita dal Prof. G. Jan.* Tipografia di. A. Lombardi, Milano.

Kearney, M. 2003a. Systematics of the *Amphisbaenia* (*Lepidosauria*: *Squamata*) based on morphological evidence from Recent and fossil forms. *Herpetol. Monogr.* 17:1–74.

Kearney, M. 2003b. The phylogenetic position of *Sineoamphisbaena hexatabularis* reexamined. *J. Vertebr. Paleontol.* 23:394–403.

Klauber, L. M. 1972. *Rattlesnakes. Their Habits, Life Histories, and Influence on Mankind.* 2nd edition. University of California Press, Berkeley, CA and Los Angeles, CA.

Klein, J. T. 1755. *Tentamen Herpetologiae.* Eliam Luzac, Leidae and Gottingae.

Kluge, A. G. 1991. Boine snake phylogeny and research cycles. *Misc. Publ. Mus. Zool. Univ. Mich.* 178:1–58.

Lacépède, B. G. E. 1788. *Histoire Naturelle des Quadrupèdes Ovipares et des Serpens. Tome Premier.* Hôtel de Thou, Paris.

Lacépède, B. G. E. 1790. *Histoire Naturelle des Serpens.* Tome Troisieme. Hotel de Thou, Paris.

Latreille, P. A. 1804. *Tableau Méthodique des Reptiles. Nouveau Dictionnaire D'histoire Naturelle, Appliquée aux Arts, à L'agriculture, à L'économie Rurale et Domestique, à la Médecine, etc. xxiv.* Deterville, Paris.

Latreille, P. A. 1825. *Familles Naturelles du Règne Animal, Exposées Succinctement et Dans un Ordre Analitique, Avec L'indication de Leurs Genres.* J.B. Baillière, Paris.

Laurenti, J. N. 1768. *Specimen Medicum, Exhibens Synopsin Reptilium Emendatus cum Experimentis Circa Venea et Antidota Reptilium Austriacorum.* J. T. de Trattern, Vienna.

Lee, M. S. Y. 1997. The phylogeny of varanoid lizards and the affinities of snakes. *Philos. Trans. R. Soc. Lond. B Biol. Sci.* 352:53–91.

Lee, M. S. Y. 1998. Convergent evolution and character correlation in burrowing reptiles: towards a resolution of squamate phylogeny. *Biol. J. Linn. Soc.* 65:369–453.

Lee, M. S. Y. 2009. Hidden support from unpromising data sets strong unites snakes with anguimorph "lizards". *J. Evol. Biol.* 22:1308–1316.

Lee, M. S. Y., and M. W. Caldwell. 1998. Anatomy and relationships of *Pachyrhachis problematicus,* a primitive snake with legs. *Philos. Trans. R. Soc. Lond. B Biol. Sci.* 353:1521–1552.

Lee, M. S. Y., A. F. Hugall, R. Lawson, and J. D. Scanlon. 2007. Phylogeny of snakes (*Serpentes*): combining morphological and molecular data in likelihood, Bayesian and parsimony analyses. *Syst. Biodivers.* 5:371–389.

Lee, M. S. Y., and J. D. Scanlon. 2002. Snake phylogeny based on osteology, soft anatomy and ecology. *Biol. Rev.* 77:333–401.

von Linnaeus, C. 1735. *Systema Naturae Sive Regna Tria Naturae Systematice Proposita per Classes, Ordines, Genera, & Species.* Haak, Lugduni Batavorum (Leiden).

von Linnaeus, C. 1740. *Systema Naturae in Quo Naturae Regna Tria, Secundum. Classes, Ordines, Genera, Species, Systematice Proponuntur.* Kiesewetter, Stockholm.

von Linnaeus, C. 1748. *Systema Naturae Sistens Regna Tria Naturae, in Classes et Ordines*

Genera et Species Redacta Tabulisque Aeneis Illustrata. Kiesewetter, Stockholm.

von Linnaeus, C. 1758. *Systema Naturae Per Regna Tria Naturae, Secundum Classes, Ordines, Genera, Species, cum Characteribus, Differentiis, Synonymis, Locis.* 10th edition, Tomes 1–2. Laurentii Salvii, Holmiae (Stockholm).

von Linnaeus, C. 1759. *Systema Naturae Per Regna Tria Naturae, Secundum Classes, Ordines, Genera, Species, cum Characteribus, Differentiis, Synonymis, Locis.* Tomus II, Editio decima, reformata. Laurentii Salvii, Holmiae (Stockholm).

Longrich, N. R., A.-B. S. Bhullar, and J. A. Gauthier. 2012. A transitional snake from the Late Cretaceous period of North America. *Nature* 488:205–208.

Macartney, J. 1802. *Lectures on Comparative Anatomy,* Vol. 1. Translated from the French of G. Cuvier by William Ross under the inspection of James Macartney. Wilson and Co., London.

McDowell, Jr., S. B. 1972. The evolution of the tongue in snakes and its bearing on snake origins. Pp. 191–273 in *Evolutionary Biology,* Vol. 6 (T. Dobzhansky, M. K. Hecht, and W. C. Steere, eds.). Appleton-Century-Crofts, New York.

McDowell, Jr., S. B., and C. M. Bogert. 1954. The systematic position of *Lanthanotus borneensis,* and the affinities of the anguinomorphan lizards. *Bull. Am. Mus. Nat. Hist.* 105(1):1–142.

Merrem, B. 1820. *Versuch eines Systems der Amphibien I (Tentamen Systematis Amphibiorum).* J. C. Kriegeri, Marburg.

Müller, J., C. Hipsley, J. J. Head, N. Kardjilov, A. Hilger, M. Wuttke, and R. R. Reisz. 2011. Limbed lizard from the Eocene of Germany reveals amphisbaenian origins. *Nature* 473:364–367.

Nagy, Z. T., D. Adriaens, E. Pauwels, L. Van Hoorebeke, J. Kielgast, C. Kusamba, and K. Jackson. 2013. 3D reconstruction of fang replacement in the venomous snakes *Dendroaspis jamesoni* (*Elapidae*) and *Bitis arietans* (*Viperidae*). *Salamandra* 49:109–113.

Nopcsa, F. 1923. *Eidolosaurus* und *Pachyophis.* Zwei neue Neocom-Reptilien. *Palaeontographica* 65:99–154.

Oppel, M. 1811. *Die Ordnungen, Familien und Gattungen der Reptilien als Prodom Einer Naturgeschichte Derselben.* Joseph Lindauer Verlag, München.

Palci, A., and M. W. Caldwell. 2007. Vestigial forelimbs and axial elongation in a 95-million-year-old non-snake squamate. *J. Vertebr. Paleontol.* 27:1–7.

Palci, A., M. W. Caldwell, and R. L. Nydam. 2013. Reevaluation of the anatomy of the Cenomanian (Upper Cretaceous) hind-limbed marine fossil snakes *Pachyrhachis, Haasiophis,* and *Eupodophis. J. Vertebr. Paleontol.* 33: 1328–1342.

Pyron, R. A., F. T. Burbrink, and J. J. Wiens. 2013. A phylogeny and revised classification of *Squamata,* including 4161 species of lizards and snakes. *BMC Evol. Biol.* 13:93.

de Queiroz, K. 2007. Toward an integrated system of clade names. *Syst. Biol.* 56:956–974.

de Queiroz, K., and J. Gauthier. 1992. Phylogenetic taxonomy. *Ann. Rev. Ecol. Syst.* 23:449–480.

Rafinesque, C. S. 1815. *Analyse de la Nature ou Tableau de L'univers et des Corps Organisés.* Palerme.

Rage, J.-C. 1984. *Serpentes. Handbuch der Paläoherpetologie,* Teil 11 (P. Wellnhofer, ed.). G. Fischer, Stuttgart.

Ray, J. 1693. *Synopsis Methodica Animalium Quadrupedum et Serpentini Generis.* Smith and Walford, London.

Rieppel, O. 1988. A review of the origin of snakes. *Evol. Biol.* 22:37–130.

Rieppel, O., and J. J. Head. 2004. New specimens of the fossil snake genus *Eupodophis* Rage and Escuillié, from the mid-Cretaceous of Lebanon. *Mem. Soc. Ital. Sci. Nat. Mus. Civ. Stor. Nat. Milano* 23(2):1–26.

Rieppel, O., A. G. Kluge, and H. Zaher. 2002. Testing the phylogenetic relationships of the Pleistocene snake *Wonambi naracoortensis* Smith. *J. Vertebr. Paleontol.* 23:812–829.

Scanlon, J. D. 2006. Skull of the large non-macrostomatan snake *Yurlunggur* from the Australian Oligo-Miocene. *Nature* 439:839–842.

Scanlon, J. D., and M. S. Y. Lee. 2000. The Pleistocene serpent *Wonambi* and the early evolution of snakes. *Nature* 403:416–420.

Scopoli, G. A. 1777. *Introductio ad Historiam Naturalem, Sistens Genera Lapidum, Plantarum et Animalium Hactenus Detecta, Caracteribus Essentialibus Donata, In Tribus Divisa, Subinde ad Leges Naturae.* Wolfgangum Gerle, Prague.

Slowinski, J. B., and R. Lawson. 2002. Snake phylogeny: evidence from nuclear and mitochondrial genes. *Mol. Phylogenet. Evol.* 24:194–202.

Sonnini, C. S., and P. A. Latreille. 1801. *Histoire Naturelle des Reptiles, avec Figures Dissinees Dápres Nature*, 4 Vols. Chez Deterville, Paris.

Stannius, 1856. *Handbuch der Zootomie, Zweiter Theil.* Wirbelthiere, Berlin.

Swainson, W. 1838–1839. *On the Natural History and Classification of Fishes, Amphibians and Reptiles or Monocardian Animals*, 2 Vols.: 1, (1838); 2 (1839). Longman, Orme, Brown, Green and Longman, and J. Taylor, London.

Tchernov, E., O. Rieppel, H. Zaher, M. J. Polcyn, and L. L. Jacobs. 2000. A fossil snake with limbs. *Science* 287:2010–2012.

Townsend, T. M., A. Larson, E. Louis, and J. R. Macey. 2004. Molecular phylogenetics of *Squamata*: the position of snakes, amphisbaenians, and dibamids, and the root of the squamate tree. *Syst. Biol.* 53:735–757.

Uetz, P. 2012. The Reptile Database, Version 24 December 2012. Available at www.reptile-database.org.

Underwood, G. 1967. *A Contribution to the Classification of Snakes.* British Museum (Natural History), London.

Vidal, N., and P. David. 2004. New insights into the early history of snakes inferred from two nuclear genes. *Mol. Phylogenet. Evol.* 31:783–787.

Vidal, N., and S. B. Hedges. 2002. Higher-level relationships of snakes inferred from four nuclear and mitochondrial genes. *C. R. Biol.* 325:977–985.

Vidal, N., and S. B. Hedges. 2004. Molecular evidence for a terrestrial origin of snakes. *Proc. R. Soc. Lond. B Biol. Sci.* 271(suppl.):S226–S229.

Vidal, N., and S. B. Hedges. 2009. The molecular evolutionary tree of lizards, snakes, and amphisbaenians. *C. R. Biol.* 332:129–139.

Wagler, J. G. 1828. Auszüge aus einem systema amphibiorem. *Isis von Oken* 21:740–744.

Wagler, J. G. 1830. *Natürliches System der Amphibien, mit Vorangehender Classification der Säugthiere und Vögel.* J. G. Cotta'schen, München.

Wiegmann, A. F. A. 1834. *Herpetologia Mexicana, seu Descriptio Amphibiorum Novae Hispaniae, Quae Itineribus Comitis de Sack, Ferdinandi Deppe et Chr. Guil. Schiede in Museum Zoologicum Berolinense Pervenerunt.* C. G. Lüderitz, Berolini (Berlin).

Wiens, J. J., M. C. Brandley, and T. W. Reeder. 2006. Why does a trait evolve multiple times within a clade? Repeated evolution of snake-like body form in squamate reptiles. *Evolution* 60:123–141.

Wiens, J. J., C. A. Kuczynski, T. Townsend, T. W. Reeder, D. G. Mulcahy, and J. W. Sites, Jr. 2010. Combining phylogenomics and fossils in higher-level squamate reptile phylogeny: molecular data change the placement of fossils. *Syst. Biol.* 59:674–688.

Wiens, J. J., C. R. Hutter, D. G. Mulcahy, B. P. Noonan, T. M. Townsend, J. W. Sites, Jr., and T. W. Reeder. 2012. Resolving the phylogeny of lizards and snakes (*Squamata*) with extensive sampling of genes and species. *Biol. Lett.* 8:1043–1046.

Wilson, J. A., D. Mohabey, S. Peters, and J. J. Head. 2010. Predation upon hatchling sauropod dinosaurs by a new basal snake from the Late Cretaceous of India. *PLOS Biol.* 8:e1000322.

Wilcox, T. P., D. J. Zwickl, T. A. Heath, and D. M. Hillis. 2002. Phylogenetic relationships of the dwarf boas and a comparison of Bayesian and bootstrap measures of phylogenetic support. *Mol. Phylogenet. Evol.* 25:361–371.

Zaher, H., and C. A. Scanferla. 2012. The skull of the upper Cretaceous snake *Dinilysia patagonica* Smith-Woodward, 1901, and its phylogenetic position revisited. *Zool. J. Linn. Soc.* 164:194–238.

Authors

Jason J. Head; Department of Zoology and University Museum of Zoology; University of Cambridge; Cambridge, UK. CB2 3EJ. Email: jjh71@cam.ac.uk.

Kevin de Queiroz; Department of Vertebrate Zoology; National Museum of Natural History; Smithsonian Institution; Washington, DC 20560-0162, USA. Email: dequeirozk@si.edu.

Harry W. Greene; Department of Ecology and Evolutionary Biology, Cornell University; E251 Corson Hall, Ithaca, NY 14853, USA. Email: hwg5@cornell.edu.

Date Accepted: 18 December 2014

Primary Editor: Philip Cantino

Pan-Iguania Krister T. Smith, nomen cladi novum

Registration Number: 180

Definition: The total clade of the crown clade *Iguania*. This is a crown-based total-clade definition. Abbreviated definition: total ∇ of *Iguania*.

Etymology: *Pan-*, Gr. "all, total" indicating reference to a total clade + *Iguania* (see *Iguania*, this volume).

Reference Phylogeny: Estes et al. (1988: Fig. 6).

Composition: *Iguania* (this volume) and all extinct lizards that share more recent ancestry with it than with its extant sister-group, which according to the reference phylogeny is *Scleroglossa* (see Comments for alternative hypotheses). A number of very well-preserved taxa from the Campanian (Late Cretaceous; Dashzeveg et al., 2005) of Mongolia and China—namely *Anchaurosaurus gilmorei*, *Xihaina aquilonia*, *Ctenomastax parva*, *Temujinia ellisoni*, *Zapsosaurus sceliphros*, *Polrussia mongoliensis*, *Igua minuta*, and *Saichangurvel davidsoni* (Borsuk-Białynicka and Alifanov, 1991; Gao and Hou, 1995; Gao and Norell, 2000; Conrad and Norell, 2007)—are clearly pan-iguanians. Conrad (2008) found these taxa to be within *Iguania* (the crown) scattered among the members of a paraphyletic *Iguanidae*; on the other hand, Conrad and Norell (2007) found many of these taxa to form a clade, *Gobiguania*, within a monophyletic *Iguanidae*. Of the above taxa that they included, Gauthier et al. (2012) also found them to be in crown *Iguania*, but either on the stem of *Iguanidae* or of *Acrodonta*. Some of these taxa, particularly *T. ellisoni* and *S. davidsoni*, have alternatively been suggested to lie on the stem of *Iguania* (Smith, 2009). Unnamed taxa

represented by fragmentary remains from the Cenomanian (early Late Cretaceous; Archibald and Averianov, 2005) of Uzbekistan (Gao and Nessov, 1998), the Campanian of Spain (Rage, 1999; see also Narváez and Ortega, 2010), and the Maastrichtian (Late Cretaceous) of France (Sigé et al., 1997) are pan-iguanians of uncertain affinity. Conrad (2008) interpreted *Hoyalacerta sanzi* Evans and Barbadillo 1999 from the Early Cretaceous of Spain as a pan-iguanian.

Diagnostic Apomorphies: Interpretation of diagnostic apomorphies of parts of *Pan-Iguania* depends on the phylogenetic position of taxa potentially on the stem of *Iguania*. If Smith (2009) is correct that *Temujinia ellisoni* and *Saichangurvel davidsoni* (Gao and Norell, 2000; Conrad and Norell, 2007) are stem iguanians, then the following features that diagnose *Iguania* among living taxa (Estes et al., 1988) arose prior to the divergence of those fossil taxa: frontals fused, frontals constricted between orbits, parietal foramen (i.e., parietal organ) located at frontoparietal suture, angular process of prearticular present, and teeth distinctively tricuspid (Smith, 2006). If Conrad and Norell (2007), Conrad (2008) and Gauthier et al. (2012) are correct that these species belong (somewhere) in the iguanian crown, then there is no direct evidence about when the features mentioned above arose along the iguanian stem.

Synonyms: *Iguania* (approximate): several authors (e.g., Reynoso, 1998; Gao and Norell, 2000) have used *Iguania* to refer to species not necessarily in the crown.

Iguanidae (partial): many authors (e.g., Augé, 2007; Borsuk-Białynicka and Alifanov, 1991; Gao and Hou, 1995; Rage, 1999) have used *Iguanidae* in approximately the following sense.

Traditionally (e.g., Boulenger, 1885) *Iguanidae* was applied to a group of living squamates based on plesiomorphy (non-acrodont iguanians). When extended to include fossil taxa, that concept would include parts of the iguanian stem (as well as the stems of *Acrodonta* and *Iguanidae* as defined in this volume).

Iguanomorpha Fürbringer 1900 (partial): Fürbringer (1900) included *Iguanidae* and *Agamidae* but not *Chamaeleonidae*.

Iguanomorpha sensu Sukhanov (1961) (approximate or partial): Sukhanov appears to have been inconsistent, applying this name both to *Iguania* (p. 8, translation) and to *Iguania* exclusive of *Chamaeleonidae* (p. 1, translation). Based on his discussion Sukhanov probably would have included at least some members of the iguanian stem in this taxon.

Iguanomorpha sensu Conrad (2008) (unambiguous).

Comments: The choice to coin the new name *Pan-Iguania* for the total clade of *Iguania*, rather than using the existing name *Iguanomorpha*, contributes to a more uniform or integrated nomenclatural system. The name *Iguanomorpha* has only rarely been used since it was first proposed by Fürbringer (1900) and then seemingly independently by Sukhanov (1961). As far as I am aware, there is only one appropriate use of *Iguanomorpha* sensu Conrad (2008) outside of the context in which it was proposed. Thus, stability of nomenclature is not greatly infringed.

Oddly, unambiguous representatives of *Pan-Iguania* are unknown in the fossil record prior to the Late Cretaceous (but see Comments under *Pan-Acrodonta*, this volume, on early Mesozoic taxa referred to *Acrodonta*), despite the fact that its supposed sister-taxon, *Pan-Scleroglossa*, was already diversifying in the Late Jurassic (Rieppel, 1994; Evans, 2003). The earliest known remains clearly referable to *Pan-Iguania* are from the Cenomanian (earliest Late Cretaceous) of Uzbekistan (Nessov, 1988: pl. 17, Fig. 1a,b; Gao and Nessov, 1998). These facts could be interpreted as at least weak support for recent phylogenies based on genetic data, according to which *Iguania* is deeply nested within *Squamata* (e.g., Vidal and Hedges, 2004; Townsend et al., 2004; Fry et al., 2006). The name *Pan-Iguania* (defined as referring to the total clade of *Iguania*) should apply to a clade of similar composition regardless of whether that clade is the sister of all other *Squamata* (as it is in the reference phylogeny) or is deeply nested within the larger clade.

Literature Cited

Archibald, J. D., and A. O. Averianov. 2005. Mammalian faunal succession in the Cretaceous of the Kyzylkum Desert. *J. Mammal. Evol.* 12:9–22.

Augé, M. 2007. Past and present distribution of iguanid lizards. *Arq. Mus. Nac. Rio de Janeiro* 65:403–416.

Borsuk-Białynicka, M., and V. R. Alifanov. 1991. First Asiatic "iguanid" lizards in the Late Cretaceous of Mongolia. *Acta Palaeontol. Pol.* 36:325–342.

Boulenger, G. A. 1885. *Catalogue of the Lizards in the British Museum (Natural History).* 2nd edition, Vol. II. Taylor & Francis, London.

Conrad, J. L. 2008. Phylogeny and systematics of *Squamata* (*Reptilia*) based on morphology. *Bull. Am. Mus. Nat. Hist.* 310:1–182.

Conrad, J. L., and M. A. Norell. 2007. A complete Late Cretaceous iguanian (*Squamata, Reptilia*) from the Gobi and identification of a new iguanian clade. *Am. Mus. Novit.* 3584:1–47.

Dashzeveg, D., L. Dingus, D. B. Loope, C. C. Swisher, III, T. Dulam, and M. R. Sweeney. 2005. New stratigraphic subdivision, depositional environment, and age estimate for the Upper Cretaceous Djadokhta Formation, southern Ulan Nur Basin, Mongolia. *Am. Mus. Novit.* 3498:1–31.

Estes, R., K. de Queiroz, and J. A. Gauthier. 1988. Phylogenetic relationships within *Squamata*.

Pp. 119–281 in *Phylogenetic Relationships of the Lizard Families* (R. Estes and G. K. Pregill, eds.). Stanford University Press, Stanford, CA.

Evans, S. E. 2003. At the feet of the dinosaurs: the early history and radiation of lizards. *Biol. Rev.* 78:513–551.

Evans, S. E., and J. Barbadillo. 1999. A short-limbed lizard from the Early Cretaceous of Spain. *Spec. Pap. Palaeontol.* 60:73–85.

Fry, B. G., N. Vidal, J. A. Norman, F. J. Vonk, H. Scheib, S. F. Ryan Ramjan, S. Kuruppu, K. Fung, S. B. Hedges, M. K. Richardson, W. C. Hodgson, V. Ignjatovic, R. Summerhayes, and E. Kochva. 2006. Early evolution of the venom system in lizards and snakes. *Nature* 439:584–588.

Fürbringer, M. 1900. Zur vergleichenden Anatomie des Brustschulterapparates und der Schultermuskeln. *Jena. Zs. f. Naturwiss.* 34:215–718.

Gao, K., and L. H. Hou. 1995. Iguanians from the Upper Cretaceous Djadochta Formation, Gobi Desert, China. *J. Vertebr. Paleontol.* 15:57–78.

Gao, K., and L. A. Nessov. 1998. Early Cretaceous squamates from the Kyzylkum Desert, Uzbekistan. *Neues Jahrb. Geol. Paläontol. Abh.* 207:289–309.

Gao, K., and M. A. Norell. 2000. Taxonomic composition and systematics of Late Cretaceous lizard assemblages from Ukhaa Tolgod and adjacent localities, Mongolian Gobi Desert. *Bull. Am. Mus. Nat. Hist.* 249:1–118.

Gauthier, J. A., M. Kearney, J. A. Maisano, O. Rieppel, and A. Behlke. 2012. Assembling the squamate tree of life: perspectives from the phenotype and the fossil record. *Bull. Peabody Mus. Nat. Hist.* 53:3–308.

Narváez, I., and F. Ortega. 2010. Análisis preliminar de los restos de *Iguanidae* indet. del Cretácico Superior de Lo Hueco (Fuentes, Cuenca). *Cidaris* 30:205–209.

Nessov, L. A. 1988. Late Mesozoic amphibians and lizards of Soviet Middle Asia. *Acta Zool. Cracoviensia* 31:475–486.

Rage, J.-C. 1999. Squamates (*Reptilia*) from the Upper Cretaceous of Laño (Basque Country, Spain). *Est. Mus. Cienc. Nat. Alava* 14(número especial 1):121–133.

Reynoso, V. H. 1998. *Huehuecuetzpalli mixtecus* gen. et sp. nov: a basal squamate (*Reptilia*) from the Early Cretaceous of Tepexi de Rodríguez, central México. *Philol. Trans. R. Soc. Lond. B Biol. Sci.* 353:477–500.

Rieppel, O. 1994. The *Lepidosauromorpha*: an overview with special emphasis on the *Squamata*, Pp. 23–37 in *In the Shadow of the Dinosaurs: Early Mesozoic Tetrapods* (N. C. Fraser and H. D. Sues, eds.). Cambridge University Press, Cambridge, UK.

Sigé, B., A. D. Buscalioni, S. Duffaud, M. Gayet, B. Orth, J.-C. Rage, and J. L. Sanz. 1997. Etat des données sur le gisement Crétacé supérieur continental de Champ-Garimond (Gard, Sud de la France). *Münchner Geowiss. Abh., A* 34:111–130.

Smith, K. T. 2006. A diverse new assemblage of late Eocene squamates (*Reptilia*) from the Chadron Formation of North Dakota, U.S.A. *Palaeontol. Electron.* 9(2): Art. 5A, 44 pp.

Smith, K. T. 2009. Eocene lizards of the clade *Geiseltaliellus* (*Squamata*: *Iguania*) from Messel and Geiseltal, Germany, and the early radiation of *Iguanidae*. *Bull. Peabody Mus. Nat. Hist.* 50:219–306.

Sukhanov, V. B. 1961. Some problems of the phylogeny and systematics of *Lacertilia* (seu *Sauria*) *Zool. Zh.* 40:73–83. (in Russian; trans. G. Jacobs).

Townsend, T. M., A. Larson, E. Louis, and J. R. Macey. 2004. Molecular phylogenetics of *Squamata*: the position of snakes, amphisbaenians, and dibamids, and the root of the squamate tree. *Syst. Biol.* 53:735–757.

Vidal, N., and S. B. Hedges. 2004. Molecular evidence for a terrestrial origin of snakes. *Proc. R. Soc. Lond. B Biol. Sci.* 271:S226–S229.

Author

Krister T. Smith; Faculty of Biological Sciences; Goethe University; and Senckenberg Museum; Senckenberganlage 25, 60325 Frankfurt am Main, Germany. Frankfurt am Main, Germany. Email: krister.smith@senckenberg.de.

Date Accepted: 17 January 2012; revised 19 May 2014

Primary Editor: Kevin de Queiroz

Iguania E. D. Cope 1864 [K. de Queiroz, O. Torres-Carvajal and J. A. Schulte, II], converted clade name

Registration Number: 51

Definition: The largest crown clade containing *Iguana* (*Lacerta*) *iguana* (Linnaeus 1758) (*Iguanidae*), *Agama* (*Lacerta*) *agama* (Linnaeus 1758) (*Agamidae*), and *Chamaeleo* (*Lacerta*) *chamaeleon* (Linnaeus 1758) (*Chamaeleonidae*), but not *Lacerta agilis* Linnaeus 1758 (*Lacertidae*) and *Scincus* (*Lacerta*) *scincus* (Linnaeus 1758) (*Scincidae*) and *Anguis fragilis* Linnaeus 1758 (*Anguidae*) and *Coluber constrictor* Linnaeus 1758 (*Serpentes*). This is a maximum-crown-clade definition. Abbreviated definition: max crown ∇ (*Iguana iguana* (Linnaeus 1758) & *Agama agama* (Linnaeus 1758) & *Chamaeleo chamaeleon* (Linnaeus 1758) ~ *Lacerta agilis* Linnaeus 1758 & *Scincus scincus* (Linnaeus 1758) & *Anguis fragilis* Linnaeus 1758 & *Coluber constrictor* Linnaeus 1758).

Etymology: Derived from *Iguana*, the name of one of its subclades, which is based on the Spanish "Iguana", which is in turn derived from the Carib "iwana" (Burghardt and Rand, 1982).

Reference Phylogeny: The primary reference phylogeny is Figure 6 of Estes et al. (1988).

Composition: *Iguania* is composed of two primary crown clades, *Iguanidae* (*Pleurodonta* of some authors) and *Acrodonta* (Macey et al., 1997; Schulte et al., 2003; Gauthier et al., 2012), the latter composed of *Agamidae* and *Chamaeleonidae*, and approximately 1,876 currently recognized extant species (Uetz, 2017). Estes (1983) presented a compilation of the lizard fossil record including *Iguania*, and subsequent additions can

be found in Evans (2003), Conrad and Norell (2007), Conrad (2008, 2015), Longrich et al. (2012), Smith and Gauthier (2013), Simões et al. (2015), Apesteguía et al. (2016), and possibly DeMar et al. (2017).

Diagnostic Apomorphies: Estes et al. (1988) listed 14 apomorphies of *Iguania*, including the following:

1. Frontals fuse embryonically (Estes et al., 1988).
2. Frontals strongly constricted between orbits (Estes et al., 1988).
3. Broad frontal shelf underlying nasals (Estes et al., 1988).
4. Postfrontal reduced (or absent), not forked, and confined to orbital rim (Estes et al., 1988; Presch, 1988).
5. Parietal foramen on frontoparietal suture or within frontal (Estes et al., 1988).
6. *M. intercostalis ventralis* absent (Camp, 1923).
7. Tongue mucocytes mostly serous and sero-mucous (Gabe and Saint Girons, 1969; Schwenk, 1988).

Some of these characters exhibit reversals within *Iguania* (3,5) or independent (convergent or parallel) origin in other *Squamata* (1,2,5) and therefore are diagnostic only when used in combination with other characters. Gauthier et al. (2012) identified additional iguanian apomorphies (40 total, 21 unambiguous) including a novel kinetic joint between the postorbital and squamosal bones (reversed in some *Acrodonta*).

Synonyms: *Iguanomorpha* sensu Sukhanov (1961) corresponds in terms of composition but was not defined phylogenetically and is therefore an approximate synonym. The following names are partial (and approximate) synonyms of *Iguania* in the paraphyletic sense that excludes chamaeleonids (see Comments), though some included taxa no longer considered iguanians, such as *Cordylus* and even *Pterodactylus*: *Iguaniens* of Cuvier (1817, 1829), Duméril and Bibron (1834–1854), and Duméril and Duméril (1851); *Crassilingues* of Wiegmann (1834); *Eunotes* of Duméril and Bibron (1834–1854) and Duméril and Duméril (1851); *Iguanii* of Latreille (1825); *Pachyglossae* of Wagler (1828, 1830); *Stellionidae* of Bonaparte (1831); *Crassilinguia* of Gravenhorst (1843); *Strobilosaura* of Gray (1845); *Iguanidae* of Jones (1847–1849); *Pachyglossa* of Stannius (1856), Cope (1900), and Fürbringer (1900); *Eunota* of Huxley (1886) and Furbringer (1900); *Iguanomorpha* of Fürbringer (1900); and *Iguanoidae* of Hay (1930).

Comments: Nineteenth and early twentieth century authors commonly recognized a group composed primarily or exclusively of species here referred to *Iguania* based on their short, thick tongues and thus excluding the highly modified *Chamaeleonidae* (e.g., Cuvier, 1817; Latreille, 1825; Wagler, 1830; Gray, 1845; Cope, 1900; Camp, 1923). (The tendency to separate chamaeleonids reached an extreme in the systems of Boulenger [1884, 1885–1887] and Hay [1902], who placed them in a taxon, *Rhiptoglossa* or *Rhiptoglossi*, separate from and of equal rank to a taxon containing all other lizards.) Some authors from this time period grouped chameleons with other iguanians in a larger taxon that also included the similarly broad-tongued gekkotans (e.g., Gray, 1831; Fitzinger, 1843; Camp, 1923; Hay, 1930). Others implied a close relationship by listing the taxa sequentially (e.g., Laurenti, 1768; Blainville, 1822, 1835; Bonaparte, 1841; Cope, 1889, 1900; Romer, 1945), though it is unclear whether they considered all taxa now considered iguanians to be more closely related to one another than to any other lizards (a paraphyletic group). Fitzinger (1826) recognized a group, based on a dilatable throat, composed almost exclusively of iguanians (his *Chamaeleonoidea*, *Pneustoidea*, *Draconoidea*, and *Agamoidea*), though he also included distantly related pterosaurs with gliding iguanian lizards (in his *Draconoidea*). Fitzinger did not name this group, nor did he recognize it in his later major work on the classification of reptiles (Fitzinger, 1843). Camp (1923) proposed that iguanids, agamids, and chamaeleonids shared an exclusive common ancestry, though he did not place all three together in an exclusive taxon. McDowell and Bogert (1954) placed chamaeleonids with agamids and iguanids in *Iguania*, and the exclusive common ancestry of these three taxa has been corroborated by subsequent explicit phylogenetic analyses based on both morphological (e.g., Estes et al., 1988; Lee, 1998; Reynoso, 1998; Gauthier et al., 2012) and molecular (e.g., Townsend et al., 2004; Vidal and Hedges, 2004; Pyron et al., 2013) data. Phylogenetic analyses based on morphological data place *Iguania* as one of the two primary crown clades within *Squamata* (e.g., Estes et al., 1988; Schwenk, 1988; Lee, 1998; Gauthier et al., 2012); by contrast, those based on molecular sequence data have *Iguania* more deeply nested within *Squamata* (e.g., Townsend et al., 2004; Vidal and Hedges, 2004; Pyron et al., 2013).

The name *Iguania* has been attributed to Cuvier 1817 (e.g., Estes et al., 1988) and Latreille 1825 (e.g., Camp, 1923); however, Cuvier used the name *Iguaniens* and Latreille used the name *Iguanii*. According to Kuhn (1967), Cope (1864) was the first author to use the name *Iguania*, though he used it for his *Anolidae* plus *Iguanidae* (see also Cope, 1875, 1889) so that his *Iguania* corresponds to *Iguanidae* of other authors (e.g., Gray 1827, 1845; Boulenger, 1884, 1885–1887) and his own later work (Cope, 1900). Camp (1923; see also Romer, 1945) used the name *Iguania* for a taxon composed of *Iguanidae* and *Agamidae* but not *Chamaeleonidae* (*Rhiptoglossa*); nevertheless, he considered chamaeleonids to be derived from within *Iguania*, sharing a more recent common ancestor with *Agamidae* than with *Iguanidae* (see his Figure p. 333). McDowell and Bogert (1954:136) added *Chamaeleonidae* to *Iguania*, an arrangement that has been followed by most subsequent authors (e.g., Romer, 1956, 1966; Underwood, 1971; Estes et al., 1988; Frost and Etheridge, 1989; Lee, 1998; Zug et al., 2001; Pyron et al., 2013). Selection of the name *Iguania* for the crown clade in question is highly appropriate given that it is by far the most commonly used name for that clade and has previously been tied to it using explicit phylogenetic definitions (Estes et al., 1988; Lee, 1998). The main alternative name, *Iguanomorpha* Furbinger 1900, has been used at roughly 1/100 the frequency of *Iguania* and has been explicitly defined as the name of a more inclusive clade (Conrad, 2008).

Although molecular data support a basal dichotomy within *Iguania* leading to the crown clades *Iguanidae* and *Acrodonta* (Macey et al., 1997; Schulte et al., 2003; Pyron et al., 2013), the lack of morphological corroboration for the monophyly of *Iguanidae* (Etheridge and de Queiroz, 1988; Frost and Etheridge, 1989; Gauthier et al., 2012), and thus potential uncertainty concerning the basal relationships within *Iguania*, has led us to use a maximum-crown-clade definition rather than a simpler minimum-clade definition. Internal specifiers are the type species of the type genera of the three taxa within *Iguania* traditionally ranked as families (*Iguanidae*, *Agamidae*, and *Chamaeleonidae*); all three are included to

ensure that all three of those taxa are included in *Iguania*. Our definition designates a clade of identical composition in the context of alternative hypotheses (i.e., to the reference phylogeny) concerning the relationships of *Iguania* to other *Squamata* (e.g., Townsend et al., 2004; Vidal and Hedges, 2004; Pyron et al., 2013).

Literature Cited

Apesteguía, S., J. D. Daza, T. R. Simões, and J. C. Rage. 2016. The first iguanian lizard from the Mesozoic of Africa. *R. Soc. Open Sci.* 3:160462.

de Blainville, H. D. 1822. *De l'organisation des Animaux ou Principes D'anatomie Comparée*. F. G. Levrault, Paris.

de Blainville, H. D. 1835. Description de quelques espèces de reptiles de la Californie, précédée de l'analyse d'un système général d'erpétologie et d'amphibiologie. *Nouv. Ann. Mus. Hist. Nat. Paris* 4:1–64.

Bonaparte, C. L. 1831. *Saggio di una Distribuzione Metodica Degli Animali Vertebrati*. Antionio Boulzaler, Roma.

Bonaparte, C. L. 1841. *A New Systematic Arrangement of Vertebrated Animals*. Taylor, London.

Boulenger, G. A. 1884. Synopsis of the families of existing *Lacertilia*. *Ann. Mag. Nat. Hist.* 14:117–122.

Boulenger, G. A. 1885–1887. *Catalogue of the Lizards in the British Museum (Natural History)*. Taylor & Francis, London.

Burghardt, G. M., and A. S. Rand. 1982. *Iguanas of the World: Their Behavior, Ecology, and Conservation*. Noyes Publications, Park Ridge, NJ.

Camp, C. L. 1923. Classification of the lizards. *Bull. Am. Mus. Nat. Hist.* 48:289–481.

Conrad, J. L. 2008. Phylogeny and systematics of *Squamata* (*Reptilia*) based on morphology. *Bull. Am. Mus. Nat. Hist.* 310:1–182.

Conrad, J. L. 2015. A new Eocene casquehead lizard (*Reptilia*, *Corytophanidae*) from North America. *PLOS ONE* 10(7):e0127900.

Conrad, J. L., and M. A. Norell. 2007. A complete Late Cretaceous iguanian (*Squamata, Reptilia*) from the Gobi and identification of a new iguanian clade. *Am. Mus. Novit.* 3584:1–47.

Cope, E. D. 1864. On the characters of the higher groups of *Reptilia Squamata* and especially of the *Diploglossa*. *Proc. Acad. Nat. Sci. Phila.* 16:224–231.

Cope, E. D. 1875. Check-list of North American *Batrachia* and *Reptilia*; with a systematic list of the higher groups, and an essay on geographical distribution. Based on the specimens contained in the U.S. National Museum. *Bull. U.S. Natl. Mus.* 1:1–104.

Cope, E. D. 1889. Synopsis of the families of *Vertebrata. Am. Nat.* 23:849–877.

Cope, E. D. 1900. The crocodilians, lizards, and snakes of North America. *Annu. Rep. U.S. Natl. Mus.* 1898:153–270.

Cuvier, G. 1817. *Le Regne Animal Distribué D'après Son Organization, Pour Servir de Base a L'histoire Naturèlle Des Animaux et D'introduction a L'anatomie Comparée. Tome 2, les Reptiles, les Poissons, les Mollusques, et Les Annélides.* Déterville, Paris.

Cuvier, G. 1829. *Le Règne Animal Distribué D'après Son Organization, Pour Servir de Base a L'histoire Naturelle des Animaux et D'introduction a L'anatomie Comparée*, Tome 2. Déterville, Paris.

DeMar, D. G., J. L. Conrad, J. J. Head, D. J. Varricchio, and G. P. Wilson. 2017. A new Late Cretaceous iguanomorph from North America and the origin of New World *Pleurodonta* (*Squamata, Iguania*). *Proc. R. Soc. Lond. B Biol. Sci.* 284:20161902.

Duméril, A. M., and G. Bibron. 1834–1854. *Erpétologie Générale ou Histoire Naturelle Complète des Reptiles.* Roret, Paris.

Duméril, M. C., and A. M. Duméril. 1851. *Catalogue Méthodique de la Collection des Reptiles.* Gide et Baudry, Paris.

Estes, R. 1983. Sauria *terrestria*, Amphisbaenia. Handbuch der Paläoherpetologie, Vol. 10A. Gustav Fischer, Stuttgart.

Estes, R., K. de Queiroz, and J. Gauthier. 1988. Phylogenetic relationships within *Squamata*. Pp. 119–281 in *Phylogenetic Relationships of the Lizard Families* (R. Estes and G. Pregill, eds.). Stanford University Press, Stanford, CA.

Etheridge, R., and K. de Queiroz. 1988. A phylogeny of *Iguanidae*. Pp. 283–367 in *Phylogenetic Relationships of the Lizard Families* (R. Estes and G. Pregill, eds.). Stanford University Press, Stanford, CA.

Evans, S. 2003. At the feet of the dinosaurs: the early history and radiation of lizards. *Biol. Rev.* 78:513–551.

Fitzinger, L. I. 1826. *Neue Classification der Reptilien nach ihren natürlichen Verwandtschaften. Nebst einer Verwandtschafts-Tafel und einem Verzeichnisse der Reptilien-Sammlung des K. K. Zoologischen Museum's zu Wien.* J. G. Heubner, Vienna.

Fitzinger, L. I. 1843. *Systema Reptilium. Fasciculus Primus.* Amblyglossae. Baumüller and Seidel, Vienna.

Frost, D. R., and R. Etheridge. 1989. A phylogenetic analysis and taxonomy of iguanian lizards (*Reptilia: Squamata*). *Misc. Publ. Nat. Hist. Mus. Univ. Kans.* 81:1–65.

Fürbringer, M. 1900. Zur vergleichenden Anatomie des Brustschulterapparates und der Schultermuskeln. *Jenaischen Z. Natwiss.* 34:215–718. [Section D (pp. 597–682), titled Systematische und genealogische Schlüsse, was published as a separate reprint under the title Beitrag zur Systematik und Genealogie der Reptilien.]

Gabe, M., and H. Saint Girons. 1969. Données histologiques sur les glandes salivaires des lépidosauriens. *Mem. Mus. Natl. Hist. Nat.* 58:1–112.

Gauthier, J. A., M. Kearney, J. A. Maisano, O. Rieppel, and A. D. B. Behlke. 2012. Assembling the squamate tree of life: perspectives from the phenotype and the fossil record. *Bull. Peabody Mus. Nat. Hist.* 53:3–308.

Gravenhorst, J. L. C. 1843. *Vergleichende Zoologie.* Graß, Barth and Comp., Breslau.

Gray, J. E. 1827. A synopsis of the genera of saurian reptiles, in which some new genera are indicated, and the others reviewed by actual examination. *Philos. Mag.* 2:54–58.

Gray, J. E. 1831. A synopsis of the species of the class *Reptilia*. Pp. 1–110 in *The Animal Kingdom Arranged in Conformity with Its Organization, by the Baron Cuvier, with Additional Descriptions of all the Species Hitherto Named, and of Many Not Before Noticed, by Edward Griffith and Others. Vol. 9: The Class* Reptilia. Whittaker, Treacher, and Co., London.

Gray, J. E. 1845. *Catalogue of the Specimens of Lizards in the Collection of the British Museum.* Edward Newman, London.

Hay, O. P. 1902. Bibliography and catalogue of the fossil *Vertebrata* of North America. *Bull. U.S. Geol. Surv.* 179:1–868.

Hay, O. P. 1930. *Second Bibliography and Catalogue of the Fossil* Vertebrata *of North America,* Vol. 2. Carnegie Institution of Washington, Washington, DC.

Huxley, T. H. 1886. *A Manual of the Anatomy of Vertebrated Animals.* D. Appleton and Co., New York.

Jones, T. R. 1847–1849. *Reptilia.* Pp. 264–325 in *The Cyclopaedia of Anatomy and Physiology* (R. B. Todd, ed.). Longman, Brown, Green, Longmans, and Roberts, London.

Kuhn, O. 1967. *Amphibien und Reptilien. Katalog der Subfamilien und höheren Taxa mit Nachweis der ersten Auftretens.* Gustav Fischer, Stuttgart.

Latreille, M. 1825. *Familles Naturelles du Règne Animal, Exposées Succinctement et Dans un Ordre Analytique, Avec l'Indication de Leurs Genres.* J.-B. Baillière, Paris.

Laurenti, J. N. 1768. *Specimen Medicum, Exhibens Synopsin Reptilium Emendatam cum Experimentis Circa Venena et Antidota Reptilium Austriacorum.* Joan. Thomae, Vienna.

Lee, M. S. Y. 1998. Convergent evolution and character correlation in burrowing reptiles: towards a resolution of squamate relationships. *Biol. J. Linn. Soc.* 65:369–453.

Linnaeus, C. 1758. *Systema Naturae Per Regna Tria Naturae, Secundum Classes, Ordines, Genera, Species, cum Characteribus, Differentiis, Synonymis, Locis.* 10th edition. Laurentii Salvii, Holmiae (Stockholm).

Longrich, N. R., B.-A. S. Bhullar, and J. A. Gauthier. 2012. Mass extinction of lizards and snakes at the Cretaceous–Paleogene boundary. *Proc. Natl. Acad. Sci. USA* 109:21396–21401.

McDowell, S. B., and C. M. Bogert. 1954. The systematic position of *Lanthanotus* and the affinities of the anguinomorphan lizards. *Bull. Am. Mus. Nat. Hist.* 105:1–142.

Macey, J. R., A. Larson, N. B. Ananjeva, and T. J. Papenfuss. 1997. Evolutionary shifts in three major structural features of the mitochondrial genome among iguanian lizards. *J. Mol. Evol.* 44:660–674.

Presch, W. 1988. Phylogenetic relationships of the *Scincomorpha.* Pp. 471–492 in *Phylogenetic Relationships of the Lizard Families* (R. Estes and G. Pregill, eds.). Stanford University Press, Stanford, CA.

Pyron, R. A., F. T. Burbrink, and J. J. Wiens. 2013. A phylogeny and revised classification of *Squamata,* including 4161 species of lizards and snakes. *BMC Evol. Biol.* 13:93.

Reynoso, V.-H. 1998. *Huehuecuetzpalli mixtecus* gen. et sp. nov: a basal squamate (*Reptilia*) from the Early Cretaceous of Tepexi de Rodríguez, central México. *Philos. Trans. R. Soc. Lond. B Biol. Sci.* 353:477–500.

Romer, A. S. 1945. *Vertebrate Paleontology.* 2nd edition. University of Chicago Press, Chicago, IL.

Romer, A. S. 1956. *Osteology of the Reptiles.* University of Chicago Press, Chicago, IL.

Romer, A. S. 1966. *Vertebrate Paleontology.* 3rd edition. University of Chicago Press, Chicago, IL.

Schulte, II, J. A., J. P. Valladares, and A. Larson. 2003. Phylogenetic relationships within *Iguanidae* inferred using molecular and morphological data and a phylogenetic taxonomy of iguanian lizards. *Herpetologica* 59: 399–419.

Schwenk, K. 1988. Comparative morphology of the lepidosaur tongue and its relevance to squamate phylogeny. Pp. 569–598 in *Phylogenetic Relationships of the Lizard Families* (R. Estes and G. Pregill, eds.). Stanford University Press, Stanford, CA.

Simões, T. R., E. Wilner, M. W. Caldwell, L. C. Weinschütz, and A. W. A. Kellner. 2015. A stem acrodontan lizard in the Cretaceous of Brazil revises early lizard evolution in Gondwana. *Nat. Commun.* 6:8149.

Smith, K., and J. A. Gauthier. 2013. Early Eocene lizards of the Wasatch Formation near bitter creek, Wyoming: diversity and paleoenvironment during an interval of global warming. *Bull. Peabody Mus. Nat. Hist.* 54:135–230.

Stannius, H. 1856. *Handbuch der Zootomie (zweiter Theil). Die Wirbelthiere.* Veit and Comp., Berlin.

Sukhanov, V. B. 1961. Some problems of the phylogeny and systematics of *Lacertilia* (seu *Sauria*). *Zool. Zh.* 40:73–83. (English translation in *Smithsonian Herpetological Information Service* No. 38, 1976.)

Townsend. T., A. Larson, E. Louis, and J. R. Macey. 2004. Molecular phylogenetics of *Squamata*: the position of snakes, amphisbaenians, and dibamids, and the root of the squamate tree. *Syst. Biol.* 53:735–757.

Uetz, P., ed. 2017. The Reptile Database, Version 15 October 2017. Available at http://www.reptile-database.org.

Underwood, G. L. 1971. A modern appreciation of Camp's "Classification of the lizards". Pp. vii–xvii in *Camp's Classification of the Lizards.* Facsimile reprint by the Society for the Study of Amphibians and Reptiles.

Vidal, N., and S. B. Hedges. 2004. Molecular evidence for a terrestrial origin of snakes. *Proc. R. Soc. Lond. B Biol. Sci.* 271(suppl.):S226–S229.

Wagler, J. 1828. Vorläufige Übersicht des Gerüstes, so wie Ankündigung seines Systema Amphibiorum. *Isis von Oken* 21:859–861.

Wagler, J. 1830. *Naturliches System der Amphibien, Mit Vorangehender Klassification der Saügethiere und Vögel. Ein Beiträg zur vergleichenden Zoologie.* J. G. Cotta'schen Munich, Stuttgart, and Tübingen.

Wiegmann, A. F. A. 1834. *Herpetologia Mexicana, seu Descriptio Amphibiorum Novae Hispanae quae Itineribus Comitis de Sack, Ferdinandi Deppe et Chr. Guil. Schiede. Pars Prima, Saurorum Species Amplectens.* C. G. Lüderitz, Berlin.

Zug, G. R., L. J. Vitt, and J. P. Caldwell. 2001. *Herpetology: An Introductory Biology of Amphibians and Reptiles.* 2nd edition. Academic Press, San Diego, CA.

Authors

Kevin de Queiroz; Department of Vertebrate Zoology; National Museum of Natural History; Smithsonian Institution; Washington, DC 20560-0162, USA. Email: dequeirozk@si.edu.

Omar Torres-Carvajal; Museo de Zoología; Escuela de Ciencias Biológicas; Pontificia Universidad Católica del Ecuador; Av. 12 de Octubre 1076 y Roca, Quito, Ecuador. Email: omartorcar@gmail.com.

James A. Schulte, II; Division of Amphibians and Reptiles; National Museum of Natural History; Smithsonian Institution; Washington, DC 20560-0162, USA. Email: schulte.jim@gmail.com.

Date Accepted: 8 April 2014; updated 3 November 2017

Primary Editor: Philip Cantino

Pan-Iguanidae Krister T. Smith, nomen cladi novum

Registration Number: 181

Definition: The total clade of the crown clade *Iguanidae*. This is a crown-based total-clade definition. Abbreviated definition: total ∇ of *Iguanidae*.

Etymology: *Pan-*, Gr. "all, total" indicating reference to a total clade + *Iguanidae* (see *Iguanidae*, this volume).

Reference Phylogeny: Estes et al. (1988: Fig. 6).

Composition: *Iguanidae* (this volume) and all extinct lizards that share more recent ancestry with it than with its extant sister group, *Acrodonta* (this volume). There is no consensus on taxa on the stem of *Iguanidae* (see Comments). Gauthier et al. (2012) interpreted a clade comprising the Campanian (Late Cretaceous; Dashzeveg et al., 2005) taxa *Temujinia ellisoni* Gao and Norell (2000) and *Saichangurvel davidsoni* Conrad and Norell (2007) as pan-iguanids (see entry on *Pan-Iguania* in this volume for alternative interpretations). They called that clade *Temujiniidae*. Conrad and Norell (2007) called a clade of similar composition on their phylogeny *Gobiguania*. See Comments for a discussion of other potential pan-iguanids.

Diagnostic Apomorphies: Morphological evidence for iguanid monophyly is presently weak (Conrad and Norell, 2007; Smith, 2009; cf. Conrad, 2008; Gauthier et al., 2012), making it difficult to discuss character evolution on the iguanid stem. Gauthier et al. (2012) suggested that the following derived features could unite the extinct *Temujiniidae* with *Iguanidae*: autotomy plane located posterior to transverse process in caudal vertebrae, and Meckelian groove restricted. A character of the dentary-coronoid articulation, namely insertion of the anteromedial process of the coronoid into a sulcus beneath the posterior tooth-bearing border of the dentary, which was interpreted by Smith (2009) as an apomorphy of the clade here called *Pan-Iguania*, might instead be an apomorphy of *Pan-Iguanidae* if *Temujiniidae* is part of that clade (Gauthier et al. 2012, character 388). If *Temujiniidae* is not on the stem of *Iguanidae*, the evolution of characters on that stem cannot be discussed at present.

Synonyms: Palaeontological uses of *Iguanidae* and its synonyms (see Synonyms in the entry for *Iguanidae* in this volume) can be interpreted as approximate synonyms of *Pan-Iguanidae*. However, those names were traditionally applied to groups of extant species (e.g., Boulenger, 1885), and insofar as they were defined at all, the designated taxon concepts were clearly based on plesiomorphy ("non-acrodont iguanians"). Therefore, when extended to include fossils, those taxa would incorporate large portions of *Pan-Iguania* exclusive of *Acrodonta*. Thus, the names in question (particularly *Iguanidae*) can also (and arguably more appropriately) be interpreted as partial synonyms of *Pan-Iguania* (see Synonyms in the entry for *Pan-Iguania* in this volume).

Comments: Cretaceous and Palaeogene taxa have been referred to *Iguanidae* since the late nineteenth century. Most early reports of fossil iguanids from the Old World proved not to be, or are at any rate dubious (e.g., Filhol, 1876, 1877; Kuhn, 1940; reviewed in Estes, 1983a; Zerova and Chkhikvadze, 1984 [trans. A. Averianov, 2008]); this is also true of many New World fossil taxa (Marsh, 1872; Cope, 1873; Gilmore, 1928).

Because living members of *Iguanidae* are almost exclusively confined to the New World, the clade probably originated there (e.g., Estes and Price, 1973), and proximate stem-members of *Pan-Iguanidae* are also likely to be found there. Thus, although all proposed pan-iguanians from the Cretaceous and Palaeocene of the New World are represented by fragmentary material (mostly jaws), some of them potentially belong to *Pan-Iguanidae*, and it may be useful to list them here. Estes and Price (1973) described *Pristiguana brasiliensis* from the Late Cretaceous (probably Campanian–Maastrichtian: Bertini et al., 1993) of Brazil on the basis of several skeletal elements, which Daza et al. (2012) interpreted as a stem-iguanian. Nava and Martinelli (2011) described another possible pan-iguanian, *Brasiliguana prudentis*, from the Turonian–Santonian (Late Cretaceous) of the same country. Apesteguía et al. (2005) described a partial pan-iguanian frontal bone from the Cenomanian–Turonian (Late Cretaceous) of Argentina. Albino (2007) mentioned pan-iguanian jaws from the Campanian–Maastrichtian of Argentina. Other remains are known from the Palaeocene of Bolivia (de Muizon et al., 1983; Rage, 1991) and Brazil (Estes, 1970), some of which might belong in *Iguanidae* (Estes, 1983b). Gao and Fox (1996) attributed several species from the Late Cretaceous of Canada to *Iguanidae*, including two unnamed taxa; Longrich et al. (2012) considered instead that all of these are polyglyphanodontines. Longrich et al. (2012) described *Pariguana lancensis* from the Late Cretaceous (Maastrichtian) as either on the stem of crown *Iguanidae* or a member of the crown, and *Lamiasaura ferox* from the same horizon as an indeterminate pan-iguanian. Most recently DeMar et al. (2017) described a pan-iguanian, *Magnuviator ovimonsensis*, from the Campanian of Montana. Early pan-iguanians from the Old World that have been referred to *Iguanidae* were discussed under *Pan-Iguania* (this volume).

Literature Cited

Albino, A. 2007. *Lepidosauromorpha*. Pp. 87–115 in *Patagonian Mesozoic Reptiles* (Z. Gasparini, L. Salgado, and R. A. Coria, eds.). Indiana University Press, Bloomington, IN.

Apesteguía, S., F. L. Agnolin and G. L. Lio. 2005. An early Late Cretaceous lizard from Patagonia, Argentina. *C. R. Palevol* 4: 311–315.

Bertini, R. J., L. G. Marshall, M. Gayet and P. Brito. 1993. Vertebrate faunas from the Adamantina and Marília formations (Upper Baurú Group, late Cretaceous, Brazil) in their stratigraphic and paleobiogeographic context. *Neues Jahrb. Geol. Paläontol. Abh.* 188:71–101.

Boulenger, G. A. 1885. *Catalogue of the Lizards in the British Museum (Natural History)*. 2nd edition, Vol. II. Taylor & Francis, London.

Conrad, J. L. 2008. Phylogeny and systematics of *Squamata* (*Reptilia*) based on morphology. *Bull. Am. Mus. Nat. Hist.* 310:1–182.

Conrad, J. L., and M. A. Norell. 2007. A complete Late Cretaceous iguanian (*Squamata, Reptilia*) from the Gobi and identification of a new iguanian clade. *Am. Mus. Novit.* 3584:1–47.

Cope, E. D. 1873. Synopsis of new *Vertebrata* from the Tertiary of Colorado, obtained during the summer of 1873. *Annu. Rep. U.S. Geol. Surv. Territories* 7:3–19.

Dashzeveg, D., L. Dingus, D. B. Loope, C. C. Swisher, III, T. Dulam, and M. R. Sweeney. 2005. New stratigraphic subdivision, depositional environment, and age estimate for the Upper Cretaceous Djadokhta Formation, southern Ulan Nur Basin, Mongolia. *Am. Mus. Novit.* 3498:1–31.

Daza, J. D., V. Abdala, J. S. Arias, D. García-López, and P. Ortiz. 2012. Cladistic analysis of *Iguania* and a fossil lizard from the late Pliocene of northwestern Argentina. *J. Herpetol.* 46:104–119.

DeMar, Jr., D. G., J. L. Conrad, J. J. Head, D. J. Varricchio, and G. P. Wilson. 2017. A new Late Cretaceous iguanomorph from North America and the origin of New World Pleurodonta (*Squamata, Iguania*). *Proc. R. Soc. Lond. B Biol. Sci.* 284:20161902.

Estes, R. 1970. Origin of the Recent North American lower vertebrate fauna: an inquiry into the fossil record. *Forma et Functio* 3:139–163.

Estes, R. 1983a. Sauria *Terrestria*, Amphisbaenia (*Handbuch der Paläoherpetologie, v. 10A*). Gustav Fischer, Stuttgart.

Estes, R. 1983b. The fossil record and early distribution of lizards. Pp. 365–398 in *Advances in Herpetology and Evolutionary Biology* (A. G. J. Rhodin, and K. Miyata, eds.). Harvard University Press, Cambridge, MA.

Estes, R., K. de Queiroz, and J. A. Gauthier. 1988. Phylogenetic relationships within *Squamata*. Pp. 119–281 in *Phylogenetic Relationships of the Lizard Families* (R. Estes and G. K. Pregill, eds.). Stanford University Press, Stanford, CA.

Estes, R., and L. I. Price. 1973. Iguanid lizard from the Upper Cretaceous of Brazil. *Science* 180:748–751.

Filhol, H. 1876 (1877). Sur les Reptiles fossiles des phosphorites du Quercy. *Bull. Soc. Philom. Paris (Sér. 6)* 11:27–28.

Filhol, H. 1877. *Recherches sur les Phosphorites du Quercy; Étude des Fossiles qu'on y Rencontre et Spécialement des Mammifères*. G. Masson, Paris.

Gao, K., and R. C. Fox. 1996. Taxonomy and evolution of Late Cretaceous lizards (*Reptilia*: *Squamata*) from western Canada. *Bull. Carnegie Mus. Nat. Hist.* 33:1–107.

Gao, K., and M. A. Norell. 2000. Taxonomic composition and systematics of Late Cretaceous lizard assemblages from Ukhaa Tolgod and adjacent localities, Mongolian Gobi Desert. *Bull. Am. Mus. Nat. Hist.* 249:1–118.

Gauthier, J. A., M. Kearney, J. A. Maisano, O. Rieppel, and A. Behlke. 2012. Assembling the squamate tree of life: perspectives from the phenotype and the fossil record. *Bull. Peabody Mus. Nat. Hist.* 53:3–308.

Gilmore, C. W. 1928. Fossil lizards of North America. *Mem. Natl. Acad. Sci.* 22:1–201.

Kuhn, O. 1940. Crocodilier- und Squamatenreste aus dem oberen Paleocän von Walbeck. *Zentralb. Mineral., Geol. Paläontol. Abt. B* 1940:21–25.

Longrich, N. R., B.-A. S. Bhullar, and J. A. Gauthier. 2012. Mass extinction of lizards and snakes at the Cretaceous–Paleogene boundary. *Proc. Natl. Acad. Sci. USA* 109:21396–21401.

Marsh, O. C. 1872. Preliminary description of new Tertiary reptiles. *Am. J. Sci. (Ser. 3)* 4:298–309.

de Muizon, C., M. Gayet, A. Lavenu L. G. Marshall, B. Sigé and C. Villaroel. 1983. Late Cretaceous vertebrates, including mammals, from Tiupampa, southcentral Bolivia. *Geobios* 16:747–753.

Nava, W. R., and A. G. Martinelli. 2011. A new squamate lizard from the Upper Cretaceous Adamantina Formation (Bauru Group), São Paulo State, Brazil. *An. Acad. Bras. Ciênc.* 83:291–299.

Rage, J.-C. 1991. Squamate reptiles from the early Paleocene of the Tiupampa area (Santa Lucía Formation), Bolivia. Pp. 503–508 in *Fosiles y Facies de Bolivia. Vol. I: Vertebrados* (R. Suarez-Soruco, ed.) (Revista Técnica de YPFB 12/3–4). Yacimientos Petrolíferos Fiscales Bolivianos, Santa Cruz, Bolivia.

Smith, K. T. 2009. Eocene lizards of the clade *Geiseltaliellus* (*Squamata*: *Iguania*) from Messel and Geiseltal, Germany, and the early radiation of *Iguanidae*. *Bull. Peabody Mus. Nat. Hist.* 50:219–306.

Zerova, G. A., and V. M. Chkhikvadze. 1984. Review of Cenozoic lizards and snakes of the USSR *Izv. Akad. Nauk Gruz. SSR Ser. Biol.* 10:319–326. (in Russian, with brief English summary).

Author

Krister T. Smith; Faculty of Biological Sciences; Goethe University; and Senckenberg Museum; Senckenberganlage 25, 60325 Frankfurt am Main, Germany. Email: krister.smith@ senckenberg.de.

Date Accepted: 26 January 2012; revised 15 August 2018

Primary Editor: Kevin de Queiroz

Iguanidae T. Bell 1825 [O. Torres-Carvajal, K. de Queiroz and J. A. Schulte II], converted clade name

Registration Number: 52

Definition: The largest crown clade containing *Iguana* (*Lacerta*) *iguana* (Linnaeus 1758) (*Iguanidae*), but not *Agama* (*Lacerta*) *agama* (Linnaeus 1758) (*Agamidae*) and *Chamaeleo* (*Lacerta*) *chamaeleon* (Linnaeus 1758) (*Chamaeleonidae*). This is a maximum-crown-clade definition. Abbreviated definition: max crown ∇ (*Iguana iguana* (Linnaeus 1758) ~ *Agama agama* (Linnaeus 1758) & *Chamaeleo chamaeleon* (Linnaeus 1758)).

Etymology: Derived from *Iguana*, the name of one of its subclades, which is based on the Spanish "Iguana", which is in turn derived from the Carib "iwana" (Burghardt and Rand, 1982).

Reference Phylogeny: Figure 6 of Estes et al. (1988). For details concerning the composition and internal relationships see Etheridge and de Queiroz (1988), Frost and Etheridge (1989), Schulte et al. (2003), and Townsend et al. (2011), although some of those authors use the name *Iguanidae* for a smaller clade and call the clade in question *Pleurodonta* (see Comments).

Composition: *Iguanidae* contains approximately 1200 currently recognized extant species (Uetz, 2020) within 12 mutually exclusive clades (Schulte et al., 2003): *Anolis, Corytophaninae, Crotaphytinae, Hoplocercinae, Iguaninae, Leiocephalus, Leiosaurini, Liolaemini, Oplurinae, Phrynosomatinae, Polychrus, Tropidurini* (some authors [e.g., Townsend et al., 2011] use names that all end in *-idae* for the same clades). A compilation of fossil *Iguanidae* can be found in Estes (1982), with subsequent additions in

Alifanov (2000), Evans (2003), Conrad and Norell (2007), Conrad (2008, 2015), Longrich et al. (2012), and Smith and Gauthier (2013).

Diagnostic Apomorphies: Unambiguous morphological synapomorphies of *Iguanidae* have not been reported (Estes et al., 1988; Etheridge and de Queiroz, 1988; Frost and Etheridge, 1989; Gauthier et al., 2012); lists of characters that change along the relevant branches are provided by DeMar et al. (2017).

Synonyms: *Iguanoïdes* Blainville (1816, 1822), *Iguanae* (Spix, 1825), *Pachyglossae coelodontae* (Wagler, 1828), *Pachyglossae platycormae pleurodontes + Pachyglossae stenocormae pleurodontes* (Wagler, 1830), *Iguanina* Bonaparte (1831, 1840, 1841), *Dendrobatae Prosphyodontes + Humivagae Prosphyodontes* (Wiegmann, 1834), *Iguaniens Pleurodontes + Eunotes Pleurodontes* (Duméril and Bibron, 1834–1854; Duméril and Duméril, 1851), *Pleurodontes + Prosphyodontes* (Fitzinger, 1843), *Iguanoidea + Agamida Pleurodonta* (Stannius, 1856), *Iguania* (Cope, 1864, 1875, 1889), *Iguanida* Strauch (1887), and *Pleurodonta* (Frost et al., 2001) are all approximate synonyms.

Comments: Early authors used diverse characters to divide iguanian lizards (often excluding the highly modified chamaeleons) into two primary subgroups. Those characters included compressed versus depressed body form (e.g., Wagler, 1830; Bonaparte, 1841), toothed versus toothless palate (e.g., Gray, 1825; Cuvier, 1829; Bonaparte, 1831), tree climbing versus ground walking habits (e.g., Wiegmann, 1834; Fitzinger, 1843; Gravenhorst, 1843), and pleurodont versus "acrodont" tooth implantation (e.g., Wagler, 1828; Duméril and Bibron, 1834–1854; Bonaparte, 1831; Gray, 1845; see Estes et al. 1988 concerning tooth implantation in supposedly acrodont iguanians). The division based on tooth implantation ultimately prevailed, with the names *Iguanidae* (for the pleurodont forms) and *Agamidae* (for the "acrodont" forms) being widely adopted from the late 1800s to the late 1900s (e.g., Boulenger, 1885; Cope, 1900; Fürbringer, 1900; Gadow, 1901; Camp, 1923; Williston, 1925; Romer, 1933, 1945, 1956, 1966; McDowell and Bogert, 1954; Underwood, 1971; Estes, 1982; Estes et al., 1988). By the late 1900s, authors realized that pleurodonty is an ancestral character state and that morphological evidence for the monophyly of *Iguanidae* was lacking (e.g., Etheridge and de Queiroz, 1988; Estes et al., 1988), which led to a proposal to abandon the taxon as previously circumscribed (Frost and Etheridge, 1989). Nonetheless, subsequent phylogenetic analyses of DNA sequences have found strong support for the monophyly of the *Iguanidae* as traditionally circumscribed (e.g., Macey et al., 1997; Schulte et al., 1998, 2003; Harris et al., 2001; Townsend et al., 2011; Pyron et al., 2013).

Of the names previously applied to the clade in question (see Synonyms), most have been rarely used after the nineteenth century, and some of them are compound names and therefore unsuitable for conversion (*ICPN*, Art. 17.1; Cantino and de Queiroz, 2020). Only two names have been applied to the clade during the last 100 years, *Iguanidae* and *Pleurodonta*. Of these two names, the former has been used widely since the early 1900s, while the latter has been used for the clade in question only since Frost et al. (2001). Concomitant with this historical difference, *Iguanidae* has been used far more commonly for a taxon approximating the clade for which we are establishing it, including

effective application to the crown. Moreover, the name *Pleurodonta* has at least three undesirable properties. First, it describes an ancestral character state that is widely distributed outside of the clade in question. Second, when used in the nineteenth century, the name was applied either to a much less inclusive group (e.g., Stannius, 1856) or, in keeping with the ancestral status and widespread distribution of pleurodonty, to a much more inclusive one (e.g., Cope, 1864, 1875). Third, it is a homonym of *Pleurodonta* Beck 1837, a name, applied to a taxon of mollusks, that is not currently in use but is nevertheless available under the *ICZN* (*International Commission on Zoological Nomenclature*, 1999) and therefore could be converted under the *ICPN*. For all of these reasons, the name *Iguanidae* is the more appropriate name for the clade under consideration.

Related to the alternative names for the clade in question, as well as to disagreements about rank assignment under the *ICZN*, the name *Iguanidae* has been applied to two different clades in the recent literature. Thus, when Frost and Etheridge (1989; see also Frost et al., 2001) rejected the traditional *Iguanidae* because of its then-questionable monophyly, they restricted that name, and its associated rank of family, to one of eight subgroups recognized by themselves and previous authors (e.g., Etheridge and de Queiroz, 1988). By contrast, other authors have preferred to apply the name *Iguanidae* to the group with which it was traditionally associated, at first with an explicit acknowledgment of its uncertain status (e.g., Estes et al., 1988; see also Zug, 1993; Schwenk, 1994), and later under subsequently discovered support for its monophyly (e.g., Macey et al., 1997; Harris et al., 2001; Schulte et al., 2003; Gauthier et al., 2012). This traditional use became established in the nineteenth century (e.g., Gray, 1827, 1845; Boulenger, 1884, 1885), was almost universal for most of the twentieth century (e.g., Cope, 1900;

Gadow, 1901; Camp, 1923; Williston, 1925; Romer, 1933, 1945, 1956, 1966; McDowell and Bogert, 1954; Underwood, 1971; Estes et al., 1988; Etheridge and de Queiroz, 1988), and continues to be accepted by many authors (e.g., Macey et al., 1997; Harris et al., 2001; Schulte et al., 2003; Gauthier et al., 2012; Smith and Gauthier, 2013), including those of influential general works (e.g., Zug et al., 2001; Pianka and Vitt, 2003; Pough et al., 2004). This long-established use justifies applying the name to the more inclusive clade. Moreover, the smaller clade has been given a different name, *Iguaninae* Cope 1886, which is one of the earliest names to have been given an explicit phylogenetic definition (de Queiroz, 1987).

Under the *ICPN*, Thomas Bell is considered the author of *Iguanidae* because he was the first author to use that name (Bell, 1825). Under the *ICZN*, Oppel (1811) is considered the author of *Iguanidae* because his *Iguanoides* seems to have been the first name based on the name *Iguana* proposed at the rank of family.

Literature Cited

Alifanov, V. R. 2000. The fossil record of Cretaceous lizards from Mongolia. Pp. 368–389 in *The Age of Dinosaurs in Russia and Mongolia* (M. J. Benton, M. A. Shishkin, D. M. Unwin, and E. N. Kurochkin, eds.). Cambridge University Press, Cambridge, UK.

Beck, H. 1837. *Index Molluscorum Præsentis ævi Musei Principis Augustissimi Christiani Frederici*. Hafniæ.

Bell, T. 1825. On a new genus of *Iguanidae*. *Zool. J.* 2:204–207.

de Blainville, H. D. 1816. Prodrome d'une nouvelle distribution systématique du règne animal. *Bull. Sci. Soc. Philom. Paris* 1816:113–132.

de Blainville, H. D. 1822. *De L'organisation des Animaux ou Principes D'anatomie Comparée*. F. G. Levrault, Paris.

Bonaparte, C. L. 1831. *Saggio di Una Distribuzione Metodica Degli Animali Vertebrati*. Antonio Boulzaler, Roma.

Bonaparte, C. L. 1840. Prodromus systematis herpetologiae. *Nuovi Ann. Sci. Nat. Bologna* 4:90–101.

Bonaparte, C. L. 1841. *A New Systematic Arrangement of Vertebrated Animals*. Taylor, London.

Boulenger, G. A. 1884. Synopsis of the families of existing *Lacertilia*. *Ann. Mag. Nat. Hist.* 14:117–122.

Boulenger, G. A. 1885. *Catalogue of the Lizards in the British Museum* (*Natural History*), Vol. 2. Taylor & Francis, London.

Burghardt, G. M., and A. S. Rand. 1982. *Iguanas of the World: Their Behavior, Ecology, and Conservation*. Noyes Publications, Park Ridge, NJ.

Camp, C. L. 1923. Classification of the lizards. *Bull. Am. Mus. Nat. Hist.* 48:289–481.

Cantino, P. D., and K. de Queiroz. 2020. *International Code of Phylogenetic Nomenclature* (*PhyloCode*), Version 6. CRC Press, Boca Raton, FL.

Conrad, J. L. 2008. Phylogeny and systematics of *Squamata* (*Reptilia*) based on morphology. *Bull. Am. Mus. Nat. Hist.* 310:1–182.

Conrad, J. L. 2015. A new Eocene casquehead lizard (*Reptilia*, *Corytophanidae*) from North America. *PLOS ONE* 10(7):e0127900.

Conrad, J. L., and M. A. Norell. 2007. A complete Late Cretaceous iguanian (*Squamata*, *Reptilia*) from the Gobi and identification of a new iguanian clade. *Am. Mus. Novit.* 3584:1–47.

Cope, E. D. 1864. On the characters of the higher groups of *Reptilia Squamata* and especially of the *Diploglossa*. *Proc. Acad. Nat. Sci. Phila.* 16:224–231.

Cope, E. D. 1875. Check-list of North American *Batrachia* and *Reptilia*; with a systematic list of the higher groups, and an essay on geographical distribution. Based on the specimens contained in the U.S. National Museum. *Bull. U.S. Natl. Mus.* 1:1–104.

Cope, E. D. 1886. On the species of *Iguaninae*. *Proc. Am. Philos. Soc.* 23:261–271.

Cope, E. D. 1889. Synopsis of the families of *Vertebrata*. *Am. Nat.* 23:849–877.

Cope, E. D. 1900. The crocodilians, lizards, and snakes of North America. *Annu. Rep. U.S. Natl. Mus.* 1898:153–1270.

Cuvier, G. 1829. *Le Règne Animal Distribué d'après son Organisation, Pour Servir de Base a l'histoire Naturelle des Animaux et D'introduction a L'anatomie Comparée*, Tome 2. Chez Déterville Libraire, Paris.

DeMar, D. G., J. L. Conrad, J. J. Head, D. J. Varricchio, and G. P. Wilson. 2017. A new Late Cretaceous iguanomorph from North America and the origin of New World *Pleurodonta* (*Squamata, Iguania*). *Proc. R. Soc. Lond. B Biol. Sci.* 284:20161902.

de Queiroz, K. 1987. Phylogenetic systematics of iguanine lizards. *Univ. Calif. Publ. Zool.* 118:1–203.

Duméril, A. M., and G. Bibron. 1834–1854. *Erpétologie Générale ou Histoire Naturelle Complète des Reptiles*. Librairie Encyclopedique de Roret, Paris.

Duméril, M. C., and A. M. Duméril. 1851. *Catalogue Méthodique de la Collection des Reptiles*. Gide et Baudry, Libraires-Éditeurs, Paris.

Estes, R. 1982. Handbuch der Paläoherpetologie, Vol. 10, Sauria *Terrestria* Amphisbaenia. Gustav Fischer Verlag, Stuttgart.

Estes, R., K. de Queiroz, and J. Gauthier. 1988. Phylogenetic relationships within *Squamata*. Pp. 119–281 in *Phylogenetic Relationships of the Lizard Families* (R. Estes and G. Pregill, eds.). Stanford University Press, Stanford, CA.

Etheridge, R., and K. de Queiroz. 1988. A phylogeny of *Iguanidae*. Pp. 283–367 in *Phylogenetic Relationships of the Lizard Families* (R. Estes and G. Pregill, eds.). Stanford University Press, Stanford, CA.

Evans, S. 2003. At the feet of the dinosaurs: the early history and radiation of lizards. *Biol. Rev.* 78:513–551.

Fitzinger, L. I. 1843. *Systema Reptilium. Fasciculus primus*. Amblyglossae. Baumüller and Seidel, Vienna.

Frost, D. R., and R. Etheridge. 1989. A phylogenetic analysis and taxonomy of iguanian lizards (*Reptilia: Squamata*). *Misc. Publ. Nat. Hist. Mus. Univ. Kans.* 81:1–65.

Frost, D. R., R. Etheridge, D. Janies, and T. Titus. 2001. Total evidence, sequence alignment, evolution of polychrotid lizards, and a reclassification of the *Iguania* (*Squamata: Iguania*). *Am. Mus. Novit.* 3343:1–38.

Fürbringer, M. 1900. Zur vergleichenden Anatomie des Brustschulterapparates und der Schultermuskeln. *Jenaischen Z. Natwiss.* 34:215–718. [Section D (pp. 597–682), titled Systematische und genealogische Schlüsse, was published as a separate reprint under the title *Beitrag zur Systematik und Genealogie der Reptilien.*]

Gadow, H. 1901. *Amphibia and Reptiles. The Cambridge Natural History*, Vol. 8. MacMillan and Co., Limited, London.

Gauthier, J. A., M. Kearney, J. A. Maisano, O. Rieppel, and A. D. B. Behlke. 2012. Assembling the squamate tree of life: perspectives from the phenotype and the fossil record. *Bull. Peabody Mus. Nat. Hist.* 53:3–308.

Gravenhorst, J. L. C. 1843. *Vergleichende Zoologie*. Graß, Barth and Comp., Breslau.

Gray, J. E. 1825. A synopsis of the genera of reptiles and amphibia, with a description of some new species. *Ann. Philos.* 10:193–217.

Gray, J. E. 1827. A synopsis of the genera of saurian reptiles, in which some new genera are indicated, and the others reviewed by actual examination. *Philos. Mag.* 2:54–58.

Gray, J. E. 1845. *Catalogue of the Specimens of Lizards in the Collection of the British Museum*. Edward Newman, London.

Harris, D. J., J. C. Marshall, and K. A. Crandall. 2001. Squamate relationships based on C-mos nuclear DNA sequences: increased taxon sampling improves bootstrap support. *Amphibia-Reptilia* 22:235–242.

International Commission on Zoological Nomenclature. 1999. *International Code of Zoological Nomenclature*. 4th edition. International Trust for Zoological Nomenclature, London.

Linnaeus, C. 1758. *Systema Naturae Per Regna Tria Naturae, Secundum Classes, Ordines, Genera, Species, cum Characteribus, Differentiis, Synonymis, Locis*. 10th edition. Laurentii Salvii, Holmiae (Stockholm).

Longrich, N. R., B.-A. S. Bhullar, and J. A. Gauthier. 2012. Mass extinction of lizards and snakes at the Cretaceous–Paleogene boundary. *Proc. Natl. Acad. Sci. USA* 109: 21396–21401.

Macey, J. R., A. Larson, N. B. Ananjeva, and T. J. Papenfuss. 1997. Evolutionary shifts in three major structural features of the mitochondrial genome among iguanian lizards. *J. Mol. Evol.* 44:660–674.

McDowell, S. B., and C. M. Bogert. 1954. The systematic position of *Lanthanotus* and the affinities of the anguinomorphan lizards. *Bull. Am. Mus. Nat. Hist.* 105:1–142.

Oppel, M. 1811. *Die Ordnungen, Familien und Gattungen der Reptilien als Prodrom einer Naturgeschichte Derselben.* J. Lindauer, Munich.

Pianka, E., and L. J. Vitt. 2003. *Lizards: Windows to the Evolution of Diversity.* University of California Press, Berkeley, CA.

Pough, F. H., R. M. Andrews, J. E. Cadle, M. L. Crump, A. H. Savitzky, and K. D. Wells. 2004. *Herpetology.* 3rd edition. Pearson Prentice Hall, Upper Saddle River, NJ.

Pyron, R. A., F. T. Burbrink, and J. J. Wiens. 2013. A phylogeny and revised classification of *Squamata*, including 4161 species of lizards and snakes. *BMC Evol. Biol.* 13:93.

Romer, A. S. 1933. *Vertebrate Paleontology.* 1st edition. University of Chicago Press, Chicago, IL.

Romer, A. S. 1945. *Vertebrate Paleontology.* 2nd edition. University of Chicago Press, Chicago, IL.

Romer, A. S. 1956. *Osteology of the Reptiles.* University of Chicago Press, Chicago, IL.

Romer, A. S. 1966. *Vertebrate Paleontology.* 3rd edition. University of Chicago Press, Chicago, IL.

Schulte, II, J. A., J. R. Macey, A. Larson, and T. J. Papenfuss. 1998. Testing the monophyly of four iguanid subfamilies: a comparison of molecular and morphological data. *Mol. Phylogenet. Evol.* 10:367–376.

Schulte, II, J. A., J. P. Valladares, and A. Larson. 2003. Phylogenetic relationships within *Iguanidae* inferred using molecular and morphological data and a phylogenetic taxonomy of iguanian lizards. *Herpetologica* 59:399–419.

Schwenk, K. 1994. Systematics and subjectivity: the phylogeny and classification of iguanian lizards revisited. *Herpetol. Rev.* 25:53–57.

Smith, K., and J. A. Gauthier. 2013. Early Eocene lizards of the Wasatch Formation near Bitter Creek, Wyoming: diversity and paleoenvironment during an interval of global warming. *Bull. Peabody Mus. Nat. Hist.* 54:135–230.

von Spix, J. B. 1825. *Animalia Nova Sive Species Novae Lacertarum Quas in Itinere per Brasiliam Annis 1817–1820 Jusu et Auspicius Maximilian; Josephi I Bavariae Regis Suscepto Collegit Et Descripsit Dr. J. B. de Spix.* T. Weigel, Munich.

Stannius, H. 1856. *Handbuch der Zootomie (zweiter Theil).* Die Wirbelthiere. Verlag von Veit and Comp., Berlin.

Strauch, A. 1887. Bemerkungen über die Geckoniden-Sammlung im zoologischen Museum der Kaiserlichen Akademie der Wissenschaften zu St. Petersburg. *Mem. Acad. Impériale Sci. St. Petersbg. 7ème Ser.* 35(2):1–72.

Townsend, T. M., D. G. Mulcahy, B. P. Noonan, J. W. Sites, Jr., C. A. Kuczynski, J. J. Wiens, and T. W. Reeder. 2011. Phylogeny of iguanian lizards inferred from 29 nuclear loci, and a comparison of concatenated and species-tree approaches for an ancient, rapid radiation. *Mol. Phylogenet. Evol.* 61:363–380.

Uetz, P., ed. 2020. The Reptile Database, Version 14 January 2020. Available at http://www.reptile-database.org.

Underwood, G. L. 1971. A modern appreciation of Camp's "Classification of the lizards". Pp. vii–xvii in *Camp's Classification of the Lizards.* Facsimile reprint by the Society for the Study of Amphibians and Reptiles.

Wagler, J. 1828. Vorläufige Übersicht des Gerüstes, so wie Ankündigung seines Systema Amphibiorum. *Isis von Oken* 21:859–861.

Wagler, J. 1830. *Naturliches System der Amphibien, mit vorangehender Klassification der Saügethiere und Vögel. Ein Beiträg zur vergleichenden Zoologie.* München, Stuttgart, and Tübingen.

Wiegmann, A. F. A. 1834. *Herpetologia Mexicana, seu Descriptio Amphibiorum Novae Hispanae quae Itineribus Comitis de Sack, Ferdinandi Deppe et Chr. Guil. Schiede. Pars prima, Saurorum Species Amplectens.* C. G. Lüderitz, Berlin.

Williston, S. W. 1925. *The Osteology of the Reptiles.* Harvard University Press, Cambridge, MA.

Zug, G. R. 1993. *Herpetology: An Introductory Biology of Amphibians and Reptiles*. Academic Press, San Diego, CA.

Zug, G. R., L. J. Vitt, and J. P. Caldwell. 2001. *Herpetology: An Introductory Biology of Amphibians and Reptiles*. 2nd edition. Academic Press, San Diego, CA.

Authors

Omar Torres-Carvajal; Museo de Zoología; Escuela de Ciencias Biológicas; Pontificia Universidad Católica del Ecuador; Av. 12 de Octubre 1076 y Roca, Quito, Ecuador. Email: omartorcar@gmail.com.

Kevin de Queiroz; Department of Vertebrate Zoology; National Museum of Natural History; Smithsonian Institution; Washington, DC 20560-0162, USA. Email: dequeirozk@si.edu.

James A. Schulte, II; Department of Vertebrate Zoology; National Museum of Natural History; Smithsonian Institution; Washington, DC 20560-0162, USA. Email: schulte.jim@gmail.com.

Date Accepted: 10 April 2014; updated 6 November 2017

Primary Editor: Philip Cantino

Pan-Acrodonta Krister T. Smith, nomen cladi novum

Registration Number: 179

Definition: The total clade of the crown clade *Acrodonta*. This is a crown-based total-clade definition. Abbreviated definition: total ∇ of *Acrodonta*.

Etymology: *Pan-*, Gr. "all, total" indicating reference to a total clade + *Acrodonta* (see *Acrodonta*, this volume).

Reference Phylogeny: Conrad (2008: Fig. 55A), where the clade is named *Chamaeleontiformes*.

Composition: *Acrodonta* (this volume) and all extinct lizards that share more recent ancestry with it than with any other extant iguanians (see *Iguania* in this volume). *Tinosaurus* from the Palaeogene of Laurasia is a pan-acrodontan form-taxon (see Smith et al., 2011). Gilmore (1943) and Estes (1983) noted similarities between "agamid" acrodontans and *Arretosaurus ornatus* from the late Eocene (Meng and McKenna, 1998) of China, and Conrad (2008) suggested that it could be related to priscagamines; this problematic taxon, to which other Palaeogene species might be related (Alifanov, 2012), is in need of restudy. The earliest clear representatives of *Pan-Acrodonta* are well-preserved fossil taxa from the Campanian (Late Cretaceous; Dashzeveg et al., 2005) of Mongolia and China, including the priscagamines *Priscagama gobiensis*, *Phrynosomimus asper*, *Mimeosaurus crassus*, *M. tugrikinensis*, and *Flaviagama dzerzhinskii* (Gilmore, 1943; Alifanov, 1989, 1996; Borsuk-Białynicka and Moody, 1984; Gao and Hou, 1996; Gao and Norell, 2000). Gauthier et al. (2012) interpreted *Ctenomastax parva*, from these horizons, as a pan-acrodontan basal to *Priscagaminae*. Alifanov (1989) suggested that

Isodontosaurus gracilis Gilmore (1943), also from the Late Cretaceous of Mongolia, is a priscagamine, whereas Conrad (2008) found it to represent an earlier branch from the stem of *Acrodonta*, and Gauthier et al. (2012) placed it on the stem of *Iguanidae*. Nessov (1988: pl. XVII) noted the presence of a "new form of *Priscagaminae*" (sensu Borsuk-Białynicka and Moody, 1984) from the Bissekty Formation (Turonian, Late Cretaceous; Archibald and Averianov, 2005) of Uzbekistan. Nessov later illustrated this (and related) material again, applying to it the nomen nudum *Sheikhia priscagama* (Nessov, 1995: 146, pl. I, Fig. 25; Nessov, 1997: 205, pl. 22, Fig. 8) and explicitly designating a holotype (Nessov, 1995: 146) but apparently not satisfying other requirements for the name to be considered available (established) under the Zoological Code. Li et al. (2007) described a gliding diapsid from the Lower Cretaceous of China, *Xianglong zhaoi*, as a member of the total clade of *Acrodonta*; although based on a complete skeleton with soft-tissue remnants, it is difficult to be certain of the reported apomorphies because the bone is poorly preserved (pers. obs.). Two fossil species with acrodont dentition from the early Mesozoic of India, *Tikiguania estesi* and *Bharatagama rebbanensis*, were also referred to that clade (Datta and Ray, 2006; Evans et al., 2002). However, the latter authors noted that the taxon in question also shows some rhynchocephalian features, and Hutchinson et al. (2012) concluded that *T. estesi* is a late Cenozoic contaminant in the Triassic Tiki Formation. Two subsequent Gondwanan species have been inferred to pertain to *Pan-Acrodonta*: *Gueragama sulamericana* Simões et al. 2015 and *Jeddaherdan aleadonta* Apesteguía et al. 2016. The former was inferred to lie on the stem of *Acrodonta* between *Priscagamidae* and the crown (Simões et al., 2015). A later analysis using a different data-set concluded that both species are closely related to extant *Uromastyx* within crown *Acrodonta* (Apesteguía et al., 2016). Daza et al. (2016) reported crown representatives of *Acrodonta* in mid-Cretaceous amber from Myanmar, which would corroborate a Mesozoic divergence for the crown.

Diagnostic Apomorphies: Gauthier et al. (2012) gave the following as unambiguous synapomorphies of *Ctenomastax parva*, *Priscagaminae*, and *Acrodonta*: postorbital and postfrontal fused, postorbital firmly sutured to dermal roofing bones, quadrate vertical, palatine foramen absent, dentary does not restrict Meckelian groove, anterior inferior alveolar foramen entirely confined to splenial, median premaxillary tooth absent, enlarged anterior maxillary teeth ("caniniforms"), dentary tooth count between 10 and 20, and wear facets formed between dentary teeth by occlusion of maxillary teeth. They additionally gave the following as unambiguous synapomorphies of *Priscagaminae* and *Acrodonta*: reduced lateral process of premaxilla, nasals in dorsal contact over nasal process of premaxilla, maxilla posterior process (and tooth row) extend to posterior one-third of orbit, and tall dorsal process of ectopterygoid. Previously noted synapomorphies of *Priscagamidae* and *Acrodonta* are: "acrodont" dentition (see below), a meeting of the anteromedial processes of the maxillae behind the premaxilla (but see Gao and Norell, 2000; Conrad and Norell, 2007), and a reduced postfrontal (Gao and Norell, 2000: 33; but see the reinterpretation of Gauthier et al., 2012, above). Acrodonty was traditionally defined as the attachment of teeth to the apex of the jaw ramus (e.g., Romer, 1956). Fürbringer (1900), Estes et al. (1988), and Alifanov (1996) discussed variation in the degree of acrodonty in reptiles and especially squamates. In *Acrodonta* the more posterior teeth attach not only at the apex of the jaw but also partly on its medial surface. Evans (2003; Evans et al., 2002) called this intermediate condition "pleuroacrodonty". Alifanov (1996) developed a nomenclatural system according to which this might be called "hemipleurodonty". That which more specifically characterizes *Acrodonta* and part of its stem (as discussed above) is the absence of tooth replacement in the cheek-tooth series (Cooper et al., 1970) and the reduction of tooth bases (and hence acrodonty or at least reduced pleurodonty; Estes et al., 1988).

Synonyms: *Acrodonta* (approximate): some authors (e.g., Simões et al., 2015) have used the name *Acrodonta* for iguanians in which at least some teeth show more apical implantation, which probably evolved before the crown. *Chamaeleonidae* sensu Frost and Etheridge (1989) (approximate): these authors made no explicit distinction between the crown and total clades but included *Priscagama* in their *Chamaeleonidae*, indicating that their concept of that taxon was more inclusive than the crown. *Agamidae* sensu Borsuk-Białynicka and Moody (1984) (partial): these authors recognized that their osteological characters of *Agamidae* might be diagnostic of a larger group, in which case their *Agamidae* corresponds to *Chamaeleonidae* sensu Frost and Etheridge (1989) minus chameleons. *Chamaeleontiformes* sensu Conrad (2008) (unambiguous).

Comments: The choice to coin the new name *Pan-Acrodonta* for the total clade of *Acrodonta*, rather than using the existing name *Chamaeleontiformes*, was made based on two considerations. First, the new name contributes to a more uniform or integrated nomenclatural system. Second, if *Chamaeleontiformes* suggests a total clade at all, it is that of *Chamaeleonidae* and not *Acrodonta*, given that the proposed change to the content of *Chamaeleonidae* by Frost and Etheridge (1989) did not become established and moreover was later reversed by the same authors (Frost et al., 2001). To my knowledge, as of 15 August 2018 the name *Chamaeleontiformes* has been used three times for the total clade of *Acrodonta* outside of the context in which it was first proposed; thus, stability of nomenclature is not greatly infringed.

There is a history of mistaking acrodont rhynchocephalians for acrodont iguanians, as with species of the Jurassic taxon *Euposaurus* Jourdan (1862) (see Evans, 1994). Evans et al. (2002) discussed possible ways to distinguish isolated jaws of these groups. Although *Acrodonta* is today entirely confined to the Old World, fossil pan-acrodontans were first described, as *Chamaeleo pristinus* (Leidy, 1872, 1873), from the Eocene of North America. This species and (many) others are now placed in *Tinosaurus* Marsh (1872); their relations within *Pan-Acrodonta* and to each other are uncertain (Smith et al., 2011).

Literature Cited

Alifanov, V. R. 1989. New *Priscagamida* [sic] (*Lacertilia*) from the Upper Cretaceous of Mongolia and their systematic position among *Iguania*. *Paleontol. J.* 1989:68–80.

Alifanov, V. R. 1996. Lizards of the families *Priscagamidae* and *Hoplocercidae* (*Sauria, Iguania*): Phylogenetic position and new representatives from the Late Cretaceous of Mongolia. *Paleontol. J.* 30:466–483.

Alifanov, V. R. 2012. Lizards of the family *Arretosauridae* Gilmore, 1943 (*Iguanomorpha, Iguania*) from the Paleogene of Mongolia. *Paleontol. J.* 46:412–420.

Apesteguía, S., J. D. Daza, T. R. Simões, and J. C. Rage. 2016. The first iguanian lizard from the Mesozoic of Africa. *R. Soc. Open Sci.* 3:160462.

Archibald, J. D., and A. O. Averianov. 2005. Mammalian faunal succession in the Cretaceous of the Kyzylkum Desert. *J. Mammal. Evol.* 12:9–22.

Borsuk-Białynicka, M., and S. M. Moody. 1984. *Priscagaminae*, a new subfamily of the *Agamidae*

(*Sauria*) from the Late Cretaceous of the Gobi Desert. *Acta Palaeontol. Pol.* 29:51–81.

Conrad, J. L. 2008. Phylogeny and systematics of *Squamata* (*Reptilia*) based on morphology. *Bull. Am. Mus. Nat. Hist.* 310:1–182.

Conrad, J. L., and M. A. Norell. 2007. A complete Late Cretaceous iguanian (*Squamata, Reptilia*) from the Gobi and identification of a new iguanian clade. *Am. Mus. Novit.* 3584:1–47.

Cooper, J. S., D. F. G. Poole, and R. Lawson. 1970. The dentition of agamid lizards with special reference to tooth replacement. *J. Zool. Lond.* 162:85–98.

Dashzeveg, D., L. Dingus, D. B. Loope, C. C. Swisher, III, T. Dulam, and M. R. Sweeney. 2005. New stratigraphic subdivision, depositional environment, and age estimate for the Upper Cretaceous Djadokhta Formation, southern Ulan Nur Basin, Mongolia. *Am. Mus. Novit.* 3498:1–31.

Datta, P. M., and S. Ray. 2006. Earliest lizard from the Late Triassic (Carnian) of India. *J. Vertebr. Paleontol.* 26:795–800.

Daza, J. D., E. L. Stanley, P. Wagner, A. M. Bauer, and D. A. Grimaldi. 2016. Mid-Cretaceous amber fossils illuminate the past diversity of tropical lizards. *Sci. Adv.* 2:31501080.

Estes, R. 1983. Sauria *Terrestria*, Amphisbaenia (*Handbuch der Paläoherpetologie, v. 10A*). Gustav Fischer, Stuttgart.

Estes, R., K. de Queiroz, and J. A. Gauthier. 1988. Phylogenetic relationships within *Squamata*. Pp. 119–281 in *Phylogenetic Relationships of the Lizard Families* (R. Estes and G. K. Pregill, eds.). Stanford University Press, Stanford, CA.

Evans, S. E. 1994. A re-evaluation of the Late Jurassic (Kimmeridgian) reptile *Euposaurus* (*Reptilia: Lepidosauria*) from Cerin, France. *Geobios* 27:621–631.

Evans, S. E. 2003. At the feet of the dinosaurs: the early history and radiation of lizards. *Biol. Rev.* 78:513–551.

Evans, S. E., G. V. R. Prasad, and B. K. Manhas. 2002. Fossil lizards from the Jurassic Kota Formation of India. *J. Vertebr. Paleontol.* 22:299–312.

Frost, D. R., and R. Etheridge. 1989. A phylogenetic analysis and taxonomy of iguanian lizards (*Reptilia: Squamata*). *Misc. Publ., Univ. Kansas Mus. Nat. Hist.* 81:1–65.

Frost, D. R., R. Etheridge, D. Janies, and T. A. Titus. 2001. Total evidence, sequence alignment, evolution of polychrotid lizards, and a reclassification of the *Iguania* (*Squamata: Iguania*). *Am. Mus. Novit.* 3343:1–38.

Fürbringer, M. 1900. Zur vergleichenden Anatomie des Brustschulterapparates und der Schultermuskeln. *Jena. Z. f. Naturwiss.* 34:215–718.

Gao, K. Q., and L. H. Hou. 1996. Systematics and taxonomic diversity of squamates from the Upper Cretaceous Djadochta Formation, Bayan Mandahu, Gobi Desert, People's Republic of China. *Can. J. Earth Sci.* 33:578–598.

Gao, K. Q., and M. A. Norell. 2000. Taxonomic composition and systematics of Late Cretaceous lizard assemblages from Ukhaa Tolgod and adjacent localities, Mongolian Gobi Desert. *Bull. Am. Mus. Nat. Hist.* 249:1–118.

Gauthier, J. A., M. Kearney, J. A. Maisano, O. Rieppel, and A. Behlke. 2012. Assembling the squamate tree of life: perspectives from the phenotype and the fossil record. *Bull. Peabody Mus. Nat. Hist.* 53:3–308.

Gilmore, C. W. 1943. Fossil lizards of Mongolia. *Bull. Am. Mus. Nat. Hist.* 81:361–384.

Hutchinson, M. N., A. Skinner, and M. S. Y. Lee. 2012. *Tikiguania* and the antiquity of squamate reptiles (lizards and snakes). *Biol. Lett.* 8:665–669.

Jourdan, C. 1862. Géologie et Paléontologie du bassin du Rhône. *Rev. Soc. Savantes* 2:260–261, pls. I–X.

Leidy, J. 1872. Remarks on fossils from Wyoming. *Proc. Acad. Nat. Sci. Phila.* 1872:277.

Leidy, J. 1873. Contributions to the extinct vertebrate fauna of the Western Territories. *Ann. Rep. U.S. Geol. Surv. Terr.* 6:14–358.

Li, P.-P., K.-Q. Gao, L.-H. Hou, and X. Xu. 2007. A gliding lizard from the Early Cretaceous of China. *Proc. Natl. Acad. Sci. USA* 104:5507–5509.

Marsh, O. C. 1872. Preliminary description of new Tertiary reptiles. *Am. J. Sci. (Ser. 3)* 4:298–309.

Meng, J., and M. C. McKenna. 1998. Faunal turnovers of Palaeogene mammals from the Mongolian Plateau. *Nature* 394:364–367.

Nessov, L. A. 1988. Late Mesozoic amphibians and lizards of Soviet Middle Asia. *Acta Zool. Cracov.* 31:475–486.

Nessov, L. A. 1995. *Dinosaurs of Northern Eurasia: New Data about Assemblages, Ecology and Palaeobiogeography* University of St. Petersburg Institute of Earth Crust, St. Petersburg. 156 pp., 14 pls. (in Russian).

Nessov, L. A. 1997. *Cretaceous Nonmarine Vertebrates of Northern Eurasia* University of St. Petersburg Institute of Earth Crust, St. Petersburg. 218 pp., 60 pls. (in Russian).

Romer, A. S. 1956. *Osteology of the Reptiles.* University of Chicago Press, Chicago, IL.

Simões, T. R., E. Wilner, M. W. Caldwell, L. C. Weinschütz, and A. W. A. Kellner. 2015. A stem acrodontan lizard in the Cretaceous of Brazil revises early lizard evolution in Gondwana. *Nat. Commun.* 6:8149.

Smith, K. T., S. F. K. Schaal, W. Sun, and C.-T. Li. 2011. Acrodont iguanians (*Squamata*) from the middle Eocene of the Huadian Basin of Jilin Province, China, with a critique of the taxon "*Tinosaurus*". *Verteb. PalAsiat.* 49:69–84.

Authors

Krister T. Smith; Faculty of Biological Sciences; Goethe University; and Senckenberg Museum; Senckenberganlage 25, 60325 Frankfurt am Main, Germany.

Date Accepted: 25 January 2012; revised 15 August 2018

Primary Editor: Kevin de Queiroz

Acrodonta H. Stannius 1856 [J. A. Schulte, II, K. de Queiroz and O. Torres-Carvajal], converted clade name

Registration Number: 4

Definition: The crown clade originating in the most recent common ancestor of *Agama* (originally *Lacerta*) *agama* (Linnaeus 1758), *Chamaeleo* (originally *Lacerta*) *chamaeleon* (Linnaeus 1758), *Uromastyx* (originally *Lacerta*) *aegyptia* (Forskål 1775) and *Leiolepis guttata* Cuvier 1829. This is a minimum-crown-clade definition. Abbreviated definition: min crown ∇ (*Agama agama* (Linnaeus 1758) & *Chamaeleo chamaeleon* (Linnaeus 1758) & *Uromastyx aegyptia* (Forskål 1775) & *Leiolepis guttata* Cuvier 1829).

Etymology: Derived from the Greek *akros* (at the top) + *odontos* (tooth), referring to the fact that the marginal teeth of these lizards are attached to the jaws more apically than was the case ancestrally in squamatans.

Reference Phylogeny: For the purposes of applying our definition, Supplement Figure 1 of Schulte and Moreno-Roark (2010) should be treated as the primary reference phylogeny. *Leiolepis guttata* is not included in that phylogeny but is considered most closely related to *L. guentherpetersi* of the taxa that are included (Grismer and Grismer, 2010; Grismer et al., 2014). Application of our definition in the context of phylogenies inferred by Macey et al. (2000a), Conrad (2008), Gauthier et al. (2012), Pyron et al. (2013), and Reeder et al. (2015) result in the name being applied to a clade of identical composition.

Composition: *Acrodonta* is believed to contain at least 692 extant species (Uetz, 2017) distributed among the following 7 mutually exclusive clades: *Agaminae*, *Amphibolurinae*, *Chamaeleonidae*, *Draconinae*, *Hydrosaurus*, *Leiolepis*, and *Uromastycinae* (Macey et al., 2000b; Schulte et al., 2003). Estes (1982) presented a compilation of the lizard fossil record including *Acrodonta*, and Evans (2003) provided a more recent review; however, those authors did not distinguish clearly between crown and stem, and several of the taxa referred by them to *Acrodonta* appear to be outside of the crown (Conrad, 2008).

Diagnostic Apomorphies: The following is a list of some of the more obvious derived characters shared by the members of *Acrodonta* relative to other extant iguanians (though several exhibit homoplasy):

1. Dorsolateral portion of dentary extends well posterior to coronoid apex (Etheridge and de Queiroz, 1988).
2. Splenial reduced or absent (Estes et al., 1988).
3. Maxillae meet anteromedially below palatal portion of premaxilla (Cope, 1864), and are expanded dorsally at the contact (Gauthier et al., 2012).
4. Lacrimal foramen enlarged (Etheridge and de Queiroz, 1988).
5. Pterygoid teeth absent (Estes et al., 1988).
6. All but the most anterior maxillary and dentary teeth relatively apically attached (acrodont), with the spaces between them filled by bone of attachment; those teeth are not replaced (Estes et al., 1988; Gauthier et al., 2012).
7. Number of ossicles in scleral ring reduced to 11 or 12 (Estes et al., 1988).

8. Fracture planes in caudal vertebrae absent (Estes et al., 1988).

9. Dorsal muscles of lower leg innervated by interosseous nerve (Estes et al., 1988).

10. *M. mylohyoideus anterior* in two layers, superficial layer transverse or anteriorly oblique, profound layer directed transversely and obliquely backward (Camp, 1923).

11. Reticular papillae present on fore and hind tongue (Schwenk, 1988).

12. Absence of a recognizable origin for mitochondrial light-strand replication (O_L) between the tRNA^{Asn} and tRNA^{Cys} genes (Macey et al., 1997a, 2000a).

13. A rearrangement of the mitochondrial genome in which the positions of the tRNA^{Ile} and tRNA^{Gln} genes are switched in order (Macey et al., 1997a,b, 2000a).

14. D-stem replaced by a D-arm replacement loop in the mitochondrial tRNA^{Cys} gene (Macey et al., 1997a,c, 2000a).

Additional apomorphies of *Acrodonta* have been summarized by Estes et al. (1988), Conrad (2008), and Gauthier et al. (2012).

Synonyms:
The following are approximate synonyms:

Agamoïdes of Blainville (1822), partial (see Comments);

Acrodontes of Fitzinger (1843), partial (see Comments);

Chamaeleonidae of Frost and Etheridge (1989);

Chamaeleonoidea of Vidal and Hedges (2009).

If *Agamidae*, as traditionally circumscribed, were to be found to be paraphyletic relative to *Chamaeleonidae* (see Comments), then *Agamidae* and its synonyms would also be partial synonyms of *Acrodonta*.

Comments: From the early nineteenth to the mid twentieth centuries, many systematists separated the highly modified chamaeleonids from the agamids, often grouping the latter with iguanids and various other "lizards" (e.g., Cuvier, 1817, 1829; Merrem, 1820; Gray, 1825, 1845; Latreille, 1825; Wagler, 1828, 1830; Bonaparte, 1831, 1850; Wiegmann, 1834; Duméril and Bibron, 1834–1854; Gravenhorst, 1843; Jones, 1847–1849; Duméril and Duméril, 1851; Stannius, 1856; Strauch, 1887; Haeckel, 1895; Hay, 1902; Williston, 1904, 1925; Camp, 1923; Nopcsa, 1923; Romer, 1933, 1945). This tendency reached its extreme in the taxonomies of Haworth (1825) and Boulenger (1884; 1885–1887), who placed chamaeleons in a taxon separate from one including all other limbed squamatans (both taxa under various names). Nonetheless, at least some authors during this period grouped chamaeleonids with agamids (or at least some of them) to the exclusion (for the most part) of iguanids (e.g., Blainville, 1822, see also 1835; Fitzinger, 1826, 1843). Perhaps the closest early (i.e., nineteenth century) approximations (in terms of composition) to the clade here named *Acrodonta* were Blainville's (1822) taxon Agamoïdes and Gray's (1827) unnamed group for those lizards with slightly notched tongues and "Teeth simple, marginal, entire" (p. 57). Blainville's Agamoïdes contained *Agama*, *Chamaeleo*, and *Draco* but also the iguanid *Basiliscus*; Gray's unnamed group was made up of agamids and chamaeleonids, although the former included *Zonurus* (= *Cordylus*), which is now considered distantly related.

Camp (1923) presented an early phylogeny of lizards with a clade corresponding to *Acrodonta* as recognized here, although he did not recognize it taxonomically. Instead, he grouped *Agamidae* with *Iguanidae* in a paraphyletic *Iguania*, assigning *Chamaeleonidae* to the mutually exclusive taxon *Rhiptoglossa*. McDowell and Bogert (1954) placed chamaeleonids within *Iguania* but did not group them specifically with agamids. Based on an explicit phylogenetic analysis, Estes et al. (1988) inferred a clade composed of agamids and chamaeleonids, which has been corroborated by subsequent analyses based on morphological (e.g., Frost and Etheridge, 1989; Lee, 1998; Conrad, 2008; Gauthier et al., 2012), molecular (e.g., Macey et al., 2000b; Townsend et al., 2004, 2011; Hugall et al., 2007; Schulte and Moreno-Roark, 2010; Pyron et al., 2013), and combined morphological and molecular data (e.g., Macey et al., 1997a; Reeder et al., 2015).

The name *Acrodonta* was coined by Stannius (1856; see also Jan, 1857) for a group composed of what are now considered agamids but not chamaeleonids, although earlier authors had used a similar name (Acrodontes) for a group of similar composition (Duméril and Bibron, 1834–1854; Duméril and Duméril, 1851) or one composed of chameleons and arboreal (but not terrestrial) agamids (Fitzinger, 1843). Cope (1864) applied the name *Acrodonta* to a group composed of *Chamaeleonidae*, *Agamidae*, and *Hatteriidae* (*Sphenodon*, which is now considered distantly related), although he did not recognize that taxon in some subsequent works (e.g., Cope, 1875, 1900) and removed chamaeleonids from it in others (e.g., Cope, 1889). In the twentieth century, the name *Acrodonta* was seldom used until it was applied by Estes et al. (1988), using an explicit phylogenetic definition, to the crown clade composed of agamids and chamaeleonids, and it has been widely used for that clade in subsequent works (e.g., Macey et al., 1997a, 2000a,b; Lee, 1998; Frost et al., 2001; Schulte et al., 2003; Townsend et al., 2004, 2011; Conrad, 2008; Gauthier et al., 2012; Reeder et al., 2015). Because alternative names for this clade (see synonyms) have been used rarely, selection of the name *Acrodonta* for the clade in question would seem uncontroversial (except for the possibility of using that name for a more inclusive apomorphy-based or total clade). Use of *Chamaeleonidae* for the clade here named *Acrodonta* (Frost and Etheridge, 1989) is at odds with the most common uses of both names (*Acrodonta* for the clade in question and *Chamaeleonidae* for a less inclusive clade). Moreover, the authors of that proposal subsequently used *Acrodonta* for the clade in question (Frost et al., 2001). The name *Chamaeleonoidea* Fitzinger 1826 sensu Vidal and Hedges (2009), but not sensu Gauthier et al. (2012), seems to be a redundant and thus unnecessary name that refers to a clade of identical composition to *Acrodonta*. Given the possibility of agamid paraphyly relative to chamaeleonids (e.g., Estes et al., 1988; Frost and Etheridge, 1989; Conrad, 2008; Schulte and Moreno-Roark, 2010; Gauthier et al., 2012), if *Agamidae* were to be considered a synonym of *Acrodonta*, the latter name should have precedence based on a closer match in terms of hypothesized composition. However, that situation could be avoided by defining the name *Agamidae* using a definition analogous to the one used for the name *Iguanidae* (this volume)—that is, the most inclusive crown clade containing *Agama agama* (Linnaeus 1758) but not *Iguana iguana* (Linnaeus 1758) (*Iguanidae*) and *Chamaeleo chamaeleon* (Linnaeus 1758) (*Chamaeleonidae*). Under that definition, the names *Acrodonta* and *Agamidae* could never be considered synonyms, because *Acrodonta* refers to a clade that necessarily includes *Chamaeleo chamaeleon*, whereas *Agamidae* would refer to a clade that necessarily excludes that taxon.

Literature Cited

de Blainville, H. D.. 1822. *De L'organisation des Animaux ou Principes D'anatomie Comparée*. F. G. Levrault, Paris.

de Blainville, H. D.. 1835. Description de quelques espèces de reptiles de la Californie, précédée de l'analyse d'un système général d'erpétologie et d'amphibiologie. *Nouv. Ann. Mus. Hist. Nat. Paris* 4:1–64.

Bonaparte, C. L. 1831. *Saggio di una Distribuzione Metodica Degli Animali Vertebrati*. Antionio Boulzaler, Roma.

Bonaparte, C. L. 1850. *Conspectus Systematum Herpetologiae et Amphibiologiae*. E. J. Brill, Lugduni Batavorum (Leiden).

Boulenger, G. A. 1884. Synopsis of the families of existing *Lacertilia*. *Ann. Mag. Nat. Hist.* 14:117–122.

Boulenger, G. A. 1885–1887. *Catalogue of the Lizards in the British Museum* (*Natural History*). Taylor & Francis, London.

Camp, C. L. 1923. Classification of the lizards. *Bull. Am. Mus. Nat. Hist.* 48:289–481.

Conrad, J. L. 2008. Phylogeny and systematics of *Squamata* (*Reptilia*) based on morphology. *Bull. Am. Mus. Nat. Hist.* 310:1–182.

Cope, E. D. 1864. On the characters of the higher groups of *Reptilia Squamata* and especially of the *Diploglossa*. *Proc. Acad. Nat. Sci. Phila.* 16:224–231.

Cope, E. D. 1875. Check-list of North American *Batrachia* and *Reptilia*; with a systematic list of the higher groups, and an essay on geographical distribution. Based on the specimens contained in the U.S. National Museum. *Bull. U.S. Natl. Mus.* 1:1–104.

Cope, E. D. 1889. Synopsis of the families of *Vertebrata*. *Am. Nat.* 23:849–877.

Cope, E. D. 1900. The crocodilians, lizards, and snakes of North America. *Annu. Rep. U.S. Natl. Mus.* 1898:153–1270.

Cuvier, G. 1817. *Le Regne Animal Distribué D'après Son Organization, Pour Servir de Base a L'histoire Naturèlle des Animaux et D'introduction a L'anatomie Comparée, Tome 2: Les Reptiles, Les Poissons, Les Mollusques, et Les Annélides*. Déterville, Paris.

Cuvier, G. 1829. *Le Règne Animal Distribué D'après Son Organisation, Pour Servir De Base a L'histoire Naturelle des Animaux et D'introduction a L'anatomie Comparée, Tome 2: Chez Déterville, Paris*.

Duméril, A. M., and G. Bibron. 1834–1854. *Erpétologie Générale ou Histoire Naturelle Complète des Reptiles*. Roret, Paris.

Duméril, M. C., and A. M. Duméril. 1851. *Catalogue Méthodique de la Collection des Reptiles*. Gide et Baudry, Paris.

Estes, R. 1982. *Sauria* Terrestria, *Amphisbaenia*. *Handbuch der Paläoherpetologie, Vol. 10* Gustav Fischer, Stuttgart.

Estes, R., K. de Queiroz, and J. Gauthier. 1988. Phylogenetic relationships within *Squamata*. Pp. 119–281 in *Phylogenetic Relationships of the Lizard Families* (R. Estes and G. Pregill, eds.). Stanford University Press, Stanford, CA.

Etheridge, R., and K. de Queiroz. 1988. A phylogeny of *Iguanidae*. Pp. 283–367 in *Phylogenetic Relationships of the Lizard Families* (R. Estes and G. Pregill, eds.). Stanford University Press, Stanford, CA.

Evans, S. 2003. At the feet of the dinosaurs: the early history and radiation of lizards. *Biol. Rev.* 78:513–551.

Fitzinger, L. I. 1826. *Neue Classification der Reptilien Nach Ihren natürlichen Verwandtschaften. Nebst Einer Verwandtschafts-Tafel und Einem Verzeichnisse der Reptilien-Sammlung des K. K. Zoologischen Muesums zu Wien*. J. G. Heubner, Vienna.

Fitzinger, L. I. 1843. *Systema Reptilium. Fasciculus Primus*. Amblyglossae. Baumüller and Seidel, Vienna.

Forskål, P. 1775. *Descriptiones Animalium, Avium, Amphibiorum, Piscium, Insectorum, Vermium, Quae in Itinere Orientale Observavit Petrus Forsskål. Post Mortem Auctoris Editit C. Niebuhr, Adjuncta est Materia Medica Kahirina Atque Tabula Maris Rubri Geographica*. Heineck and Faber, Copenhagen.

Frost, D. R., and R. Etheridge. 1989. A phylogenetic analysis and taxonomy of iguanian lizards (*Reptilia*: *Squamata*). *Misc. Publ. Nat. Hist. Mus. Univ. Kans.* 81:1–65.

Frost, D. R., R. Etheridge, D. Janies, and T. Titus. 2001. Total evidence, sequence alignment, evolution of polychrotid lizards, and a reclassification of the *Iguania* (*Squamata*: *Iguania*). *Am. Mus. Novit.* 3343:1–38.

Gauthier, J. A., M. Kearney, J. A. Maisano, O. Rieppel, and A. D. B. Behlke. 2012. Assembling the squamate tree of life: perspectives from the phenotype and the fossil record. *Bull. Peabody Mus. Nat. Hist.* 53:3–308.

Gravenhorst, J. L. C. 1843. *Vergleichende Zoologie*. Graß, Barth & Comp., Breslau.

Gray, J. E. 1825. A synopsis of the genera of reptiles and amphibia, with a description of some new species. *Ann. Philos.* 10:193–217.

Gray, J. E. 1827. A synopsis of the genera of saurian reptiles, in which some new genera are indicated, and the others reviewed by actual examination. *Philos. Mag.* 2:54–58.

Gray, J. E. 1845. *Catalogue of the Specimens of Lizards in the Collection of the British Museum.* Edward Newman, London.

Grismer, J. L., A. M. Bauer, L. L. Grismer, K. Thirakhupt, A. Aowphol, J. R. Oaks, P. L. Wood, C. K. Onn, N. Thy, M. Cota, and T. Jackman. 2014. Multiple origins of parthenogenesis, and a revised species phylogeny for the Southeast Asian butterfly lizards, *Leiolepis. Biol. J. Linn. Soc.* 113:1080–1093.

Grismer, J. L., and L. L. Grismer. 2010. Who's your mommy? Identifying maternal ancestors of asexual species of *Leiolepis* Cuvier, 1829 and the description of a new endemic species of asexual *Leiolepis* Cuvier, 1829 from Southern Vietnam. *Zootaxa* 2433:47–61.

Haeckel, E. 1895. *Systematicshe Phylogenie der Wirbelthiere* (Vertebrata). Georg Reimer, Berlin.

Haworth, A. H. 1825. A binary arrangement of the Class *Amphibia. Philos. Mag.* 65:372–373.

Hay, O. P. 1902. Bibliography and catalogue of the fossil *Vertebrata* of North America. *Bull. U.S. Geol. Surv.* 179:1–868.

Hugall, A. F., R. Foster, and M. S. Y. Lee. 2007. Calibration choice, rate smoothing, and the pattern of tetrapod diversification according to the long nuclear gene *RAG-1. Syst. Biol.* 56:543–563.

Jan, G. 1857. *Indice Sistematico dei Rettili ed Anfibi Esposti nel Museo Civico di Milano.* Luigi di Giacomo Priola, Milano.

Jones, T. R. 1847–1849. *Reptilia.* Pp. 264–325 in *The Cyclopaedia of Anatomy and Physiology* (R. B. Todd, ed.). Longman, Brown, Green, Longmans, & Roberts, London.

Latreille, M. 1825. *Familles Naturelles du Règne Animal, Exposées Succinctement et dans un Ordre Analytique, avec l'indication de leurs Genres.* J.-B. Baillière, Paris.

Lee, M. S. Y. 1998. Convergent evolution and character correlation in burrowing reptiles: towards a resolution of squamate relationships. *Biol. J. Linn. Soc.* 65:369–453.

Linnaeus, C. 1758. *Systema Naturae Per Regna Tria Naturae, Secundum Classes, Ordines, Genera, Species, cum Characteribus, Differentiis, Synonymis, Locis.* 10th edition. Laurentii Salvii, Holmiae (Stockholm).

Macey, J. R., A. Larson, N. B. Ananjeva, and T. J. Papenfuss. 1997a. Evolutionary shifts in three major structural features of the mitochondrial genome among iguanian lizards. *J. Mol. Evol.* 44:660–674.

Macey, J. R., A. Larson, N. B. Ananjeva, Z. Fang, and T. J. Papenfuss. 1997b. Two novel gene orders and the role of light-strand replication in rearrangement of the vertebrate mitochondrial genome. *Mol. Biol. Evol.* 14:91–104.

Macey, J. R., A. Larson, N. B. Ananjeva, and T. J. Papenfuss. 1997c. Replication slippage may cause parallel evolution in the secondary structures of mitochondrial transfer RNAs. *Mol. Biol. Evol.* 14:30–39.

Macey, J. R., J. A. Schulte, II, and A. Larson. 2000a. Evolution and phylogenetic information content of mitochondrial genomic structural features illustrated with acrodont lizards. *Syst. Biol.* 49:257–277.

Macey, J. R., J. A. Schulte, II, A. Larson, N. B. Ananjeva, Y. Wang, R. Pethiyagoda, N. Rastegar-Pouyani, and T. J. Papenfuss. 2000b. Evaluating trans-Tethys migration: an example using acrodont lizard phylogenetics. *Syst. Biol.* 49:233–256.

McDowell, S. B., and C. M. Bogert. 1954. The systematic position of *Lanthanotus* and the affinities of the anguinomorphan lizards. *Bull. Am. Mus. Nat. Hist.* 105:1–142.

Merrem, B. 1820. *Versuch eines Systems der Amphibien. Tentamen Systematis Amphibiorum.* Johann Christian Krieger, Marburg.

Nopcsa, F. 1923. *Die Familien der Reptilien.* Gebrüder Borntraeger, Berlin.

Pyron, R. A., F. T. Burbrink, and J. J. Wiens. 2013. A phylogeny and revised classification of *Squamata*, including 4161 species of lizards and snakes. *BMC Evol. Biol.* 13:93.

Reeder, T. W., T. M. Townsend, D. G. Mulcahy, B. P. Noonan, P. L. Wood, J. W. Sites, and J. J. Wiens. 2015. Integrated analyses resolve conflicts over squamate reptile phylogeny and reveal unexpected placements for fossil taxa. *PLOS ONE* 10(3): e0118199.

Romer, A. S. 1933. *Vertebrate Paleontology*. 1st edition. University of Chicago Press, Chicago, IL.

Romer, A. S. 1945. *Vertebrate Paleontology*. 2nd edition. University of Chicago Press, Chicago, IL.

Schulte, II, J. A., and F. Moreno-Roark. 2010. Live birth among iguanian lizards predates Pliocene–Pleistocene glaciations. *Biol. Lett.* 6:216–218.

Schulte, II, J. A., J. P. Valladares, and A. Larson. 2003. Phylogenetic relationships within *Iguanidae* inferred using molecular and morphological data and a phylogenetic taxonomy of iguanian lizards. *Herpetologica* 59:399–419.

Schwenk, K. 1988. Comparative morphology of the lepidosaur tongue and its relevance to squamate phylogeny. Pp. 569–598 in *Phylogenetic Relationships of the Lizard Families* (R. Estes and G. Pregill, eds.). Stanford University Press, Stanford, CA.

Stannius, H. 1856. *Handbuch der Zootomie (zweiter Theil). Die Wirbelthiere. Zweites Buch, Zootomie der Amphibien*. Verlag von Veit & Comp., Berlin.

Strauch, A. 1887. Bemerkungen über die Geckoniden-Sammlung im zoologischen Museum der Kaiserlichen Akademie der Wissenschaften zu St. Petersburg. *Mem. Acad. Imp. Sci. St. Petersbg.* 7(35)2:1–72.

Townsend. T., A. Larson, E. Louis, and J. R. Macey. 2004. Molecular phylogenetics of *Squamata*: the position of snakes, amphisbaenians, and dibamids, and the root of the squamate tree. *Syst. Biol.* 53:735–757.

Townsend, T. M., D. G. Mulcahy, B. P. Noonan, J. W. Sites, Jr., C. A. Kuczynski, J. J. Wiens, and T. W. Reeder. 2011. Phylogeny of iguanian lizards inferred from 29 nuclear loci, and a comparison of concatenated and species-tree approaches for an ancient, rapid radiation. *Mol. Phylogenet. Evol.* 61:363–380.

Uetz, P., ed. 2017. The Reptile Database, Version October 2017. Available at http://www.reptile-database.org.

Vidal, N., and S. B. Hedges. 2009. The molecular evolutionary tree of lizards, snakes, and amphisbaenians. *C. R. Biol.* 332:129–139.

Wagler, J. 1828. Vorläufige Übersicht des Gerüstes, so wie Ankündigung seines Systema Amphibiorum. *Isis von Oken* 21:859–861.

Wagler, J. 1830. *Naturliches System der Amphibien, Mit Vorangehender Klassification der Saügethiere und Vögel. Ein Beiträg zur Vergleichenden Zoologie*. J. G. Cotta'schen, Munich, Stuttgart, and Tübingen.

Wiegmann, A. F. A. 1834. *Herpetologia Mexicana, seu Descriptio Amphibiorum Novae Hispanae quae Itineribus Comitis de Sack, Ferdinandi Deppe et Chr. Guil. Schiede. Pars Prima, Saurorum Species Amplectens.* C. G. Lüderitz, Berlin.

Williston, S. W. 1904. The relationships and habits of the mosasaurs. *J. Geol.* 12:43–51.

Williston, S. W. 1925. *The Osteology of the Reptiles*. Harvard University Press, Cambridge, MA.

Authors

James A. Schulte, II; Department of Vertebrate Zoology; National Museum of Natural History; Smithsonian Institution; Washington, DC 20560-0162, USA. Email: schulte.jim@gmail.com.

Kevin de Queiroz; Department of Vertebrate Zoology; National Museum of Natural History; Smithsonian Institution; Washington, DC 20560-0162, USA. Email: dequeirozk@si.edu.

Omar Torres-Carvajal; Museo de Zoología; Escuela de Ciencias Biológicas; Pontificia Universidad Católica del Ecuador; Av. 12 de Octubre 1076 y Roca, Quito, Ecuador. Email: omartorcar@gmail.com.

Date Accepted: 10 August 2014; updated 6 November 2017

Primary Editor: Philip Cantino

Pan-Archosauria J. A. Gauthier, new clade name

Registration Number: 176

Definition: The total clade of the crown clade *Archosauria*. This is a crown-based total-clade definition. Abbreviated definition: total ∇ of *Archosauria*.

Etymology: *Pan-*, from the Greek *pan-* or *pantos* (all, the whole), indicating that the name refers to a total clade, and *Archosauria* (see etymology for that name in this volume), the name of its crown clade.

Reference Phylogeny: Gauthier et al. (1988) Figure 3, where the total clade of *Archosauria* is called *Archosauromorpha* and is the sister group of *Lepidosauromorpha* (*Pan-Lepidosauria* in this volume).

Composition: See entries for *Archosauria*, *Archosauriformes* and *Archosauromorpha* (in this volume) for references regarding extant and extinct species referred to those clades, which are nested within *Pan-Archosauria*. Merck (1997) and Gauthier (1994) provided partial lists of even more remotely related extinct diapsids that may be parts of this clade, including the marine reptiles of the Mesozoic (e.g., *Euryapsida*; but see Comments).

Diagnostic Apomorphies: Although total clades need have no apomorphies, taxa inferred to be members of *Pan-Archosauria* share the apomorphy of distally tapering, rather than broad and flat, cervical ribs. Broad cervical ribs are retained in *Lepidosauria* (rhynchocephalians, most scleroglossans, but not iguanians; Gauthier et al., 2012). The derived condition is present in *Choristodera, Thalattosauria,* *Megalancosaurus, Longisquama, Sauropterygia, Ichthyosauria,* and *Archosauromorpha* (Gauthier, 1994). Enlarged premaxillae forming most of snout tip, though difficult to quantify owing to marked modifications to the rostrum across this wildly disparate clade, may also be diagnostic (Gauthier, 1994).

Synonyms: *Archosauromorpha* of Gauthier et al. (1988) (unambiguous); traditional *Archosauria* of Cope (1869) and *Archosauromorpha* of von Huene (1946) are partial and approximate synonyms.

Comments: The name *Pan-Archosauria* is proposed for the total clade of *Archosauria*, a clade that formerly bore the name *Archosauromorpha* (Gauthier et al., 1988, after Gauthier, 1984; see also Benton 1983, 1984, 1985), which is applied to a less inclusive clade in this volume (see *Archosauromorpha* entry for justification). This clade may include additional taxa, such as *Choristodera, Thalattosauria, Megalancosaurus, Sauropterygia* and *Ichthyosauria,* that might have diverged from the archosaur stem prior to the origin of *Archosauromorpha* (Gauthier, 1994; Merck, 1997). There are, however, several reasons why the relationships of these taxa to one another and to the archosaur stem are uncertain. Support is weak, as might be expected for the earliest members of any clade. But the main problem is that we have yet to find relatively little-modified examples of any of these clades. If correctly placed on the tree, all these clades must have diverged from the archosaur stem by the late Permian (e.g., given the Permian archosauriform *Archosaurus rossicus* and archosauromorph *Protorosaurus speneri*), and none of them is currently known from that time.

Even the earliest Mesozoic fossils attributable to these clades are already highly modified (most are aquatic specialists, though *Megalancosaurus* is a chameleon-like arboreal specialist). The potential for convergence is great owing to the demands of an aquatic lifestyle, a situation further exacerbated by significant missing data both from non-preservation and from profound morphological transformation, even in the earliest known examples of these clades. As a consequence, there is considerable variation in hypotheses about their phylogenetic relationships to other fossil and Recent amniotes (e.g., Merck, 1997; Gauthier, 1994; Rieppel and Reisz, 1999; Gottmann-Quesada and Sander, 2009; Ezcurra et al., 2014).

If turtles are crown diapsids closer to archosaurs than to lepidosaurs, as indicated by virtually all gene sequences (e.g., Field et al., 2014), if not when the genes themselves are treated as characters and their sequences as states (Lu et al., 2013), there is little indication in the phenotype that turtles nest within *Archosauria*, *Archosauriformes* or *Archosauromorpha*. Bhullar and Bever (2009) tentatively identified an archosaurian laterosphenoid ossification in the near-crown stem turtle *Proganochelys quenstedti* (but not in crown turtles). However, when that well-known stem turtle was added to a dataset designed specifically to address early archosauromorph phylogeny (Dilkes, 1998), the ossification enclosing the trigeminal region in *P. quenstedti* was inferred to be non-homologous with the laterosphenoid present universally among *Archosauriformes* (Clark et al., 1993). In Bhullar and Bever's (2009) analysis, turtles fell outside of *Archosauromorpha*, all the early-diverging members of which lack laterosphenoids but share many other apomorphies with archosauriforms that are absent in turtles. Nevertheless, turtles (*Pan-Testudines*) might still be sister to *Pan-Archosauria*. Some data, such as

retention of a cleithrum in the shoulder girdle, suggest that turtles lie outside of crown diapsids, all of which lack this ossification (e.g., Lyson et al., 2010; Lyson et al., 2013a,b). Nevertheless, there is growing evidence that turtles might still be diapsids, at least in the sense of having descended from an ancestor with upper and lower temporal fenestrae, even if they are not crown diapsids. For example, in addition to a lower temporal fenestra, there appears to be an upper temporal fenestra, albeit enclosed secondarily by an enlarged supratemporal, in the earliest and least-modified stem turtle currently known, *Eunotosaurus africanus* (Bever et al., 2014). This indicates that turtles share an affinity with total-clade diapsids that, coupled with weak support in morphological data linking them with lepidosaurs (Bever et al., 2014), and strong support in molecular data allying them to archosaurs (Field et al., 2014), suggest that turtles might be crown diapsids (though at least one of these inferences must be incorrect).

Literature Cited

Benton, M. J. 1983. The Triassic reptile *Hyperodapedon* from Elgin: functional morphology and relationships. *Philos. Trans. R. Soc. Lond. B Biol. Sci.* 302(B):605–717.

Benton, M. J. 1984. The relationships and early evolution of *Diapsida*. Pp. 575–596 in *The Structure, Development, and Evolution of Reptiles* (M. W. J. Ferguson, ed.). Academic Press, London.

Benton, M. J. 1985. Classification and phylogeny of the diapsid reptiles. *Zool. J. Linn. Soc.* 84:97–164.

Bever, G. S., T. Lyson, and B.-A. Bhullar. 2014. Fossil evidence for a diapsid origin of the anapsid turtle skull. *J. Vertebr. Paleontol., Program and Abstracts*, 2014:91.

Bhullar, B.-A. S., and G. S. Bever. 2009. An archosaur-like laterosphenoid in early turtles (*Reptilia: Pantestudines*). *Breviora* 518:1–11.

Clark, J. M., J. Welman, J. A. Gauthier, and J. M. Parrish. 1993. The laterosphenoid bone of early archosauriforms. *J. Vertebr. Paleontol.* 13:48–57.

Cope, E. D. 1869. Synopsis of the extinct *Batrachia* and *Reptilia* of North America, Part I. *Trans. Am. Philos. Soc., New Ser.* 14:1–235.

Dilkes, D. W. 1998. The Early Triassic rhynchosaur *Mesosuchus browni* and the interrelationships of basal archosauromorph reptiles. *Philos. Trans. R. Soc. Lond. B Biol. Sci.* 353:501–541.

Ezcurra, M. D., T. M. Scheyer, and R. J. Butler. 2014. The origin and early evolution of *Sauria*: reassessing the Permian saurian fossil record and the timing of the crocodile-lizard divergence. *PLOS ONE* 9(2):e89165.

Field, D. J., J. A. Gauthier, B. J. King, D. Pisani, T. R. Lyson, and K. J. Peterson. 2014. Toward consilience in reptile phylogeny: miRNAs support an archosaur, not lepidosaur, affinity for turtles. *Evol. Dev.* 16:189–196.

Gauthier, J. A., 1984. A cladistic analysis of the higher systematic categories of the *Diapsida*. University Microfilms International, Ann Arbor, MI. #85-12825, vii + 564 pp.

Gauthier, J. A., 1994. The diversification of the amniotes. Pp. 129–159 in *Major Features of Vertebrate Evolution: Short Courses in Paleontology*, Vol. 7 (D. Prothero, ed.). Paleontological Society.

Gauthier, J. A., M. Kearney, J. A. Maisano, O. Rieppel, and A. Behlke. 2012. Assembling the squamate tree of life: perspectives from the phenotype and the fossil record. *Bull. Yale Peabody Mus.* 53:3–308.

Gauthier, J., A. Kluge, and T. Rowe, 1988. Amniote phylogeny and the importance of fossils. *Cladistics* 4:105–209.

Gottmann-Quesada, A., and P. M. Sander. 2009. A redescription of the early archosauromorph *Protorosaurus speneri* Meyer, 1832 and its phylogenetic relationships. *Palaeontogr. Abt. A* 287:123–220.

Lu, B., W. Yang, Q. Dai, and J. Fu. 2013. Using genes as characters and a parsimony analysis to explore the phylogenetic position of turtles. *PLOS ONE* 8(11):e79348.

Lyson T. R., G. S. Bever, B.-A. S. Bhullar, W. G. Joyce, and J. A. Gauthier. 2010. Transitional fossils and the origin of turtles. *Biol. Lett.* 6:830–833.

Lyson T. R., G. S. Bever, T. M. Scheyer, A. Y. Hsiang, and J. A. Gauthier. 2013a. Evolutionary origin of the turtle shell. *Curr. Biol.* 23:1113–1119.

Lyson T. R., B.-A. S. Bhullar, G. S. Bever, W. G. Joyce, K. de Queiroz, A. Abzhanov, and J. A. Gauthier. 2013b. Homology of the enigmatic nuchal bone reveals novel reorganization of the shoulder girdle in the evolution of the turtle shell. *Evol. Dev.* 15:317–325.

Merck, J. W. 1997. *A Phylogenetic Analysis of the Euryapsid Reptiles*. PhD dissertation, Department of Geological Sciences, The University of Texas at Austin.

Rieppel, O., and R. R. Reisz. 1999. The origin and early evolution of turtles. *Annu. Rev. Ecol. Syst.* 30:1–22.

Author

Jacques A. Gauthier; Department of Geology and Geophysics; Yale University; 210 Whitney Ave; New Haven, CT 06520-8109, USA. Email: jacques.gauthier@yale.edu.

Date Accepted: 24 August 2015

Primary Editor: Kevin de Queiroz

Archosauromorpha F. von Huene 1946 [J. A. Gauthier], converted clade name

Registration Number: 175

Definition: The least inclusive clade containing *Gallus* (originally *Phasianus*) *gallus* (*Aves*) (Linnaeus 1758), *Alligator* (originally *Crocdilus*) *mississippiensis* (Daudin 1802) (*Crocodylia*), *Mesosuchus browni* Watson 1912 (*Rhynchosauria*), *Trilophosaurus buettneri* Case 1928 (*Trilophosauridae*), *Prolacerta broomi* Parrington 1935 (*Prolacertiformes*), and *Protorosaurus speneri* von Meyer 1830 (*Protorosauria*). This is a minimum-clade definition. Abbreviated definition: min ∇ (*Gallus gallus* (Linnaeus 1758) & *Alligator mississippiensis* (Daudin 1802) & *Mesosuchus browni* Watson 1912 & *Trilophosaurus buettneri* Case 1928 & *Prolacerta broomi* Parrington 1935 & *Protorosaurus speneri* von Meyer 1830).

Etymology: Derived from the Greek *archon* (ruler), *sauros* (lizard-like reptile), and *morpha*, shape.

Reference Phylogeny: Ezcurra et al. (2014) Figure 1. On the figured tree, *Euparkeria* represents the smallest clade containing the crown archosaurs (*Aves* and *Crocodylia*).

Composition: See the entries for *Archosauria* and *Archosauriformes* (this volume) for the composition of the included crown and the proximal part of its stem. See Rieppel et al. (2003), Gottmann-Quesada and Sander (2009), and Ezcurra et al. (2014) for more distal stem archosaurs, including such taxa as *Protorosaurus*, prolacertiforms, rhynchosaurs, and trilophosaurids, here included in *Archosauromorpha*.

Diagnostic Apomorphies: Gauthier et al. (1988) identified 17 putative synapomorphies for *Archosauromorpha*, including what Thulborn (1980) termed a "duplex" ankle joint (see Comments). If the definition adopted here is applied to the tree inferred by Ezcurra et al. (2014: Fig. 1) or Gottmann-Quesada and Sander (2009: Fig. 28A), that list shrinks to two unambiguous synapomorphies: a retroarticular process formed by co-ossified articular and prearticular bones, and three-headed cervical ribs with slender, tapering shafts paralleling the vertebral centra.

Synonyms: Traditional *Archosauria* Cope 1869 could be regarded as a partial and approximate synonym (see Comments on *Archosauria*, this volume).

Comments: The ankle provided palaeontologists with the first clues to the existence of this clade (Hughes, 1963; Cruickshank, 1972; Carroll, 1976; Thulborn, 1980; Brinkman, 1981; Gauthier, 1984, 1988; Benton, 1985). The "duplex" joint (sensu Thulborn, 1980) between the proximal tarsals is particularly characteristic, consisting of a pair of concavo-convex articular surfaces located proximal and distal to the perforating foramen passing between the astragalus and calcaneum. A gently concave surface lies on the proximal part of the astragalus and a gently convex surface sits distally, matching complementary convex and concave articular surfaces on the calcaneum (Brinkman, 1981). No extant archosaur conserves this morphology as such. The more distal of the two concavo-convex surfaces became the rotary ankle joint that arose in early *Archosauriformes* and was conserved, albeit in an elaborated form, in extant *Crocodylia* (Gauthier, 1994). In contrast, although early stem birds such as *Lagosuchus* *talampayensis* retain that joint, it was suppressed entirely in the simplified hinge-like ankle of dinosaurs (Novas, 1996). Extinct diapsids such as *Choristodera* (Gauthier et al., 1988) and *Euryapsida* (Merck, 1997) have sometimes been assigned to the archosaur stem. But *Prolacerta* and *Protorosaurus* (sometimes collectively known as *Protorosauria*) (Benton, 1983; Evans, 1984; Gauthier, 1984; Gauthier et al., 1988; Laurin, 1991; Senter, 2005; Modesto and Sues, 2004; Gottmann-Quesada and Sander, 2009; Ezcurra et al., 2014), rhynchosaurs (Dilkes, 1998) and *Trilophosaurus* (Mueller and Parker, 2006; Heckert et al., 2006), as well as early and relatively unmodified *Archosauriformes* such as *Proterosuchus*, have since Brinkman (1981) invariably been included in a clade here designated *Archosauromorpha*.

In his pioneering monograph on *Hyperodapedon gordoni*, Benton (1983) inferred that *Rhynchosauria* and *Prolacertiformes* represented very early (Late Permian) divergences from the archosaur stem. He resurrected von Huene's (1946) taxon name *Archosauromorpha* to include them in a clade along with *Archosauria*. The former name, as used by von Huene, was effectively a phylogenetic junior synonym of the latter—i.e., their proposed compositions and "defining" characters were identical—and "*Archosauromorpha*" was little used subsequently. Benton's resurrection of *Archosauromorpha* was suitable for a clade originating deep in the archosaur stem, as first Cope (1869) and later von Huene (1946, 1956) explicitly grouped rhynchosaurs with more traditional archosaurs in his "*Archosauria*" (albeit based on a symplesiomorphy).

As defined here, *Archosauromorpha* is affixed to a node (see also Laurin, 1991), although it was defined originally as a maximum-clade (branch-based) name for the archosaur total clade by Gauthier et al. (1988; see also Gauthier, 1984). Benton (1983, 1984, 1985) made no such

distinctions, but he often used *Archosauromorpha* as if it applied to the archosaur total clade. However, in order to achieve the larger goal of establishing an integrated system of clade names (de Queiroz, 2007), the archosaur total clade will here be referred to as *Pan-Archosauria* (see entry in this volume). *Archosauromorpha* is reserved for a well-founded pan-archosaur sub-clade, the one originating in the last ancestor that the archosaurs *Gallus gallus* and *Alligator mississippiensis* shared with *Protorosaurus speneri*, *Prolacerta broomi*, *Mesosuchus browni*, and *Trilophosaurus buettneri*.

For most of the time since it was first proposed, *Archosauria* Cope (1869) never included *Aves*. Ignoring that tradition by using an avian as a specifier in the definition of *Archosauria* would violate *ICPN* Recommendation 11A. However, this is not the case with *Archosauromorpha* von Huene (1946). Like *Archosauria*, it also excluded *Aves* when first proposed. But unlike *Archosauria*, *Archosauromorpha* was almost never used by palaeontologists (except von Huene) until it was resurrected by Benton (1983). Since then, however, *Archosauromorpha* has always included *Aves*, either implicitly (e.g., Benton, 1983) or explicitly (e.g., Gauthier et al., 1988). So it seems that in this case, unlike that of *Archosauria*, using both avian and crocodilian specifiers in the definition of *Archosauromorpha* is consistent with preserving traditional usage.

Literature Cited

Benton, M. J. 1983. The Triassic reptile *Hyperodapedon* from Elgin: functional morphology and relationships. *Philos. Trans. R. Soc. Lond. B Biol. Sci.* 302(B):605–717.

Benton, M. J. 1984. The relationships and early evolution of *Diapsida*. Pp. 575–596 in *The Structure, Development, and Evolution of Reptiles* (M. W. J. Ferguson, ed.). Academic Press, London.

Benton, M. J. 1985. Classification and phylogeny of the diapsid reptiles. *Zool. J. Linn. Soc.* 84:97–164.

Brinkman, D. 1981. The origin of the crocodiloid tarsi and the interrelationships of thecodontian reptiles. *Breviora* 464:1–23.

Carroll, R. L. 1976. Eosuchians and the origin of archosaurs. Pp. 58–79 in *Athlon: Essays on Paleontology in Honor of Loris Shano Russell* (C. S. Churcher, ed.). Miscellaneous Publications of the Royal Ontario Museum, Toronto.

Cope, E. D. 1869. Synopsis of the extinct *Batrachia* and *Reptilia* of North America, Part I. *Trans. Amer. Philos. Soc. New Ser.* 14:1–235.

Cruickshank, A. R. I., 1972. The proterosuchian thecodonts. Pp. 89–119 in *Studies in Vertebrate Evolution* (K. A. Joysey and T. S. Kemp, eds.). Winchester Press, New York.

Dilkes, D. W. 1998. The Early Triassic rhynchosaur *Mesosuchus browni* and the interrelationships of basal archosauromorph reptiles. *Philos. Trans. R. Soc. Lond. B Biol. Sci.* 353:501–541.

Evans, S. E. 1984. The classification of the *Lepidosauria. Zool. J. Linn. Soc.* 82:87–100.

Ezcurra, M. D., T. M. Scheyer, and R. J. Butler. 2014. The origin and early evolution of *Sauria*: reassessing the Permian saurian fossil record and the timing of the crocodile-lizard divergence. *PLOS ONE* 9(2):e89165.

Gauthier, J. A. 1984. A cladistic analysis of the higher systematic categories of the *Diapsida*. University Microfilms International, Ann Arbor, MI. #85-12825, vii + 564 pp.

Gauthier, J. A. 1988. Evolution of archosaur ankle joints: A reply to Cruickshank and Benton. *Nature* 331:218.

Gauthier, J. A. 1994. The diversification of the amniotes. Pp. 129–159 in *Major Features of Vertebrate Evolution. Short Courses in Paleontology*, Vol. 7 (D. Prothero, ed.). Paleontological Society.

Gauthier, J., A. Kluge, and T. Rowe. 1988. Amniote phylogeny and the importance of fossils. *Cladistics* 4:105–209.

Gottmann-Quesada, A., and P. M. Sander. 2009. A redescription of the early archosauromorph *Protorosaurus speneri* Meyer, 1832 and its

phylogenetic relationships. *Palaeontogr. Abt. A* 287:123–220.

Heckert, A. B., S. G. Lucas, L. F. Rinehart, J. A. Spielmann, A. P. Hunt, and R. Kahle. 2006. Revision of the archosauromorph reptile *Trilophosaurus*, with a description of the first skull of *Trilophosaurus jacobsi*, from the Upper Triassic Chinle Group, West Texas, USA. *Palaeontology* 49:621–640.

von Huene, F. 1946. Die Grossen Stämme der Tetrapoden in den geologischen. *Biol. Zent.* 65:268–275.

von Huene, F. 1956. *Paläontologie und Phylogenie der Niederen Tetrapoden.* G. Fischer Verlag, Jena, Berlin.

Hughes, B. 1963. The earliest archosaurian reptiles. *S. Afr. J. Sci.* 59:221–241.

Laurin, M. 1991. The osteology of a Lower Permian eosuchian from Texas and a review of diapsid phylogeny. *Zool. J. Linn. Soc.* 101:59–95.

Merck, J. W. 1997. *A Phylogenetic Analysis of the Euryapsid Reptiles.* PhD dissertation, Department of Geological Sciences, The University of Texas at Austin.

Mueller, B. D., and W. G. Parker. 2006. A new species of *Trilophosaurus* (*Diapsida: Archosauromorpha*) from the Sonsela Member (Chinle Formation) of Petrified Forest National Park, Arizona. Pp. 119–125 in *A Century of Research at Petrified Forest National Park 1906–2006: Geology and Paleontology* (W. G. Parker, S. R. Ash, and R. B. Irmis, eds.). Museum of Northern Arizona Bulletin 62.

Modesto, S. P., and H.-D. Sues. 2004. The skull of the Early Triassic archosauromorph reptile *Prolacerta broomi* and its phylogenetic significance. *Zool. J. Linn. Soc.* 140:335–351.

Novas, F. E. 1996. Phylogenetic relationships of the basal dinosaurs, the *Herrerasauridae. J. Vertebr. Paleontol.* 16:723–741.

de Queiroz, K. 2007. Toward an integrated system of clade names. *Syst. Biol.* 56:956–974.

Rieppel, O. C., N. C. Fraser, and S. Nosotti. 2003. The monophyly of *Protorosauria* (*Reptilia, Archosauromorpha*): a preliminary analysis. *Atti Soc. Ital. Sci. Nat.* 144:359–382.

Senter, P. 2005. Phylogenetic taxonomy and the names of the major archosaurian (*Reptilia*) clades. *PaleoBios* 25:1–7.

Thulborn, R. A. 1980. The ankle joints of archosaurs. *Alcheringa* 4:241–261.

Author

Jacques A. Gauthier; Department of Geology and Geophysics; Yale University; 210 Whitney Ave.; New Haven, CT 06520-8109, USA. Email: jacques.gauthier@yale.edu.

Date Accepted: 24 August 2015

Primary Editor: Kevin de Queiroz

Archosauriformes J. A. Gauthier 1994 [J. A. Gauthier], converted clade name

Registration Number: 174

Definition: The least inclusive clade containing *Gallus* (originally *Phasianus*) *gallus* (*Aves*) (Linnaeus 1758), *Alligator* (originally *Crocdilus*) *mississippiensis* (Daudin 1802) (*Crocodylia),* and *Proterosuchus fergusi* Broom 1903 (*Proterosuchidae*). This is a minimum-clade definition. Abbreviated definition: min ∇ (*Gallus gallus* (Linnaeus 1758) & *Alligator mississippiensis* (Daudin 1802) & *Proterosuchus fergusi* Broom 1903).

Etymology: Derived from the Greek *archon* (ruler), *sauros* (lizard-like reptile), and the Latin *forma* (shape).

Reference Phylogeny: Gauthier et al. (1988) Figure 3, where *Gallus gallus* is part of *Ornithosuchia*, *Alligator mississippiensis* is part of *Pseudosuchia*, and *Proterosuchus fergusi* is part of *Proterosuchidae*.

Composition: See *Archosauria* (this volume) for references regarding crown species. The more inclusive clade *Archosauriformes* also contains some extinct clades outside of the crown, including taxa (listed in order of decreasing relationship to *Archosauria*) such as *Phytosauria* (Gauthier et al., 2011; Nesbitt, 2011), *Euparkeria capensis* (Ewer, 1965; Gower and Weber, 1998; Senter, 2003), *Proterochampsidae* (including *Doswellia*; Dilkes and Sues, 2009*),* *Erythrosuchidae* (Parrish, 1992; Gower, 2003), and *Proterosuchidae* (Cruickshank, 1972; Ezcurra, 2014).

Diagnostic Apomorphies: This clade can currently be diagnosed by a number of "classic" archosaurian features (see Comments), including serrated teeth, a caniniform dentary tooth, an antorbital fenestra, a mandibular fenestra, and a neomorphic laterosphenoid ossification (Clark et al., 1993), which is probably why they have traditionally been considered to *be* archosaurs.

Synonyms: *Proterosuchia* Broom 1906 (partial and approximate); *Thecodontia* Owen 1859 (partial and approximate); *Archosauria* Cope 1869 (partial and approximate).

Comments: *Thecodontia* Owen 1859 was one of the original groups placed by Cope (1869) in his *Archosauria*. He appeared to equate it with what is now called *Phytosauria*, as he mentions only *Belodon* (a *nomen dubium* that is in any case a phytosaur) and lists "external nostrils posterior" (p. 32)—a phytosaur synapomorphy—among its distinguishing characteristics. Nevertheless, "*Thecodontia*" later came to be applied to a paraphyletic assemblage including part of the archosaur stem as well as some early crown archosaurs that were neither dinosaurs nor pterosaurs nor crocodilians (e.g., Williston, 1917). The earliest-diverging and in many ways most plesiomorphic "thecodonts" have traditionally been grouped as "*Proterosuchia*" (Broom, 1903, 1906; Charig and Reig, 1970; Cruickshank, 1972; Bonaparte, 1982), a paraphyletic subgroup within a more inclusive paraphyletic group ("*Thecodontia*"), within an even more inclusive paraphyletic group "*Archosauria*" (if *Aves* and *Pterosauria* are excluded). The concept of "*Thecodontia*" as an "ancestral group" has underlain virtually every major controversy in archosaur phylogeny for more than a century (Gauthier and Padian, 1985; Gauthier, 1986), especially those involving the origin of dinosaurs (e.g., Charig, 1976) and the origin of birds (e.g., Feduccia et al., 2007).

When Gauthier and Padian (1985) formally restricted *Archosauria* to a crown clade (see also Gauthier, 1984; 1986), the more inclusive clade containing the remaining "thecodonts" that were previously included in *Archosauria* needed a new name. Using "*Thecodontia*" for that clade should probably be avoided for reasons outlined above. I therefore formally redefine *Archosauriformes* of Gauthier (1994) to apply to this widely recognized clade (see also Gauthier, 1984; Gauthier et al, 1988; Clark et al., 1993; Senter, 2003, 2005; Dilkes and Sues, 2009; Gottmann-Quesada and Sander, 2009; Gauthier et al., 2011; Nesbitt, 2011; Ezcurra et al., 2014). This definition differs from the previous version only in using species, rather than more inclusive taxa (i.e., *Proterosuchidae* + *Archosauria*), as specifiers (*ICPN* Art. 11.1). *Alligator mississippiensis* and *Gallus gallus* were selected to represent *Archosauria* in this definition because both species are well known, widely available, and have had their genomes sequenced. Unlike the case with *Archosauria*, birds have explicitly been considered part of *Archosauriformes* from the outset, so using *Gallus gallus* as a specifier does not violate *ICPN* Recommendation 11A.

Literature Cited

Bonaparte, J. F. 1982. Classification of the *Thecodontia*. *Geobios Mém. Spéc.* 6:99–112.

Broom, R. 1903. On a new reptile (*Proterosuchus fergusi*) from the Karroo Beds of Tarkastad, South Africa. *Ann. S. Afr. Mus.* 6:161–163.

Broom, R. 1906. On the South African diaptosaurian reptile *Howesia*. *Proc. Zool. Soc. Lond.* 1906:591–600.

Clark, J. M., J. Welman, J. A. Gauthier and J. M. Parrish. 1993. The laterosphenoid bone of early archosauriforms. *J. Vertebr. Paleontol.* 13:48–57.

Charig, A. J. 1976. "Dinosaur monophyly and a new class of vertebrates": a critical review. Pp. 65–104 in *Morphology and Biology of Reptiles* (A. d'A. Bellairs and C. B. Cox, eds.). Academic Press, London.

Charig, A. J., and Reig, O. A. 1970. The classification of the *Proterosuchia. Biol. J. Linn. Soc.* 2:125–171.

Cope, E. D. 1869. Synopsis of the extinct *Batrachia* and *Reptilia* of North America, Part I. *Trans. Am. Philos. Soc. New Ser.* 14:1–235.

Cruickshank, A. R. I. 1972. The proterosuchian thecodonts. Pp. 89–119 in *Studies in Vertebrate Evolution* (K. A. Joysey and T. S. Kemp, eds.). Winchester Press, New York.

Dilkes, D., and H.-D. Sues. 2009. Redescription and phylogenetic relationships of *Doswellia kaltenbachi* (*Diapsida: Archosauriformes*) from the Upper Triassic of Virginia. *J. Vertebr. Paleontol.* 29:58–79.

Ewer, R. F. 1965. The anatomy of the thecodont reptile *Euparkeria capensis* Broom. *Philos. Trans. R. Soc. Lond. B Biol. Sci.* 248:379–435.

Ezcurra, M. D. 2014. The osteology of the basal archosauromorph *Tasmaniosaurus triassicus* from the Lower Triassic of Tasmania, Australia. *PLOS ONE* 9(1):e86864.

Ezcurra. M. D., T. M. Scheyer, and R. J. Butler. 2014. The origin and early evolution of *Sauria*: reassessing the Permian saurian fossil record and the timing of the crocodile-lizard divergence. *PLOS ONE* 9:e89165.

Feduccia, A., L. D. Martin, and S. Tarsitano. 2007. *Archaeopteryx* 2007: Quo Vadis? *Auk* 124:373–380.

Gauthier, J. A. 1984. *A Cladistic Analysis of the Higher Systematic Categories of the* Diapsida. University Microfilms International, Ann Arbor, MI. #85–12825, vii + 564 pp.

Gauthier, J. A. 1986. Saurischian monophyly and the origin of birds. Pp 1–55 in *The Origin of Birds and the Evolution of Flight* (K. Padian, ed.). Mem. Calif. Acad. Sci. 8. California Academy of Sciences, San Francisco, CA.

Gauthier, J. A. 1994. The diversification of the amniotes. Pp. 129–159 in *Major Features of Vertebrate Evolution: Short Courses in Paleontology* (D. Prothero, ed.). Paleontological Society.

Gauthier, J. A., A. Kluge, and T. Rowe. 1988. Amniote phylogeny and the importance of fossils. *Cladistics* 4:105–209.

Gauthier, J. A., S. J. Nesbitt, E. R. Schachner, G. S. Bever, and W. G. Joyce. 2011. The bipedal stem crocodilian *Poposaurus gracilis*: inferring function in fossils and innovation in archosaur locomotion. *Bull. Peabody Mus. Nat. Hist.* 52:107–126.

Gauthier, J., and K. Padian. 1985. Phylogenetic, functional, and aerodynamic analyses of the origin of birds. Pp. 185–197 in *The Beginnings of Birds; Proceedings of the International* Archaeopteryx *Conference, Eichstätt 1984* (M. K. Hecht, J. H. Ostrom, G. Viohl, and P. Wellenhoffer, eds.). Freunde des Jura-Museums, Eichstätt, West Germany.

Gottmann-Quesada, A., and P. M. Sander. 2009. A redescription of the early archosauromorph *Protorosaurus speneri* Meyer, 1832 and its phylogenetic relationships. *Palaeontogr. Abt. A* 287:123–220.

Gower, D. J. 2003. Osteology of the early archosaurian reptile *Erythrosuchus africanus* Broom. *Ann. S. Afr. Mus.* 110:1–84.

Gower, D. J., and E. Weber. 1998. The braincase of *Euparkeria*, and the evolutionary relationships of birds and crocodilians. *Biol. Rev.* 73:367–411.

Nesbitt, S. J. 2011. The early evolution of archosaurs: relationships and the origin of major clades. *Bull. Am. Mus. Nat. Hist.* 352:1–292.

Owen, R. 1859. Palaeontology. Pp. 91–176 in *Encyclopaedia Britannica*. 8th edition, Vol. 17. A. & C. Black, Edinburgh.

Parrish, J. M. 1992. Phylogeny of the *Erythrosuchidae* (*Reptilia*: *Archosauriformes*). *J. Vertebr. Paleontol.* 12:93–102.

Senter, P. 2003. New information on cranial and dental features of the Triassic archosauriform reptile *Euparkeria capensis*. *Palaeontology* 46:613–621.

Senter, P. 2005. Phylogenetic taxonomy and the names of the major archosaurian (*Reptilia*) clades. *PaleoBios* 25:1–7.

Williston, S. W. 1917. The phylogeny and classification of reptiles. *J. Geol.* 25:411–421.

Author

Jacques A. Gauthier; Department of Geology and Geophysics; Yale University; 210 Whitney Ave.: New Haven, CT 06520-8109, USA. Email: jacques.gauthier@yale.edu.

Date Accepted: 24 August 2015

Primary Editor: Kevin de Queiroz

Archosauria E. D. Cope 1869 [J. A. Gauthier and K. Padian], converted clade name

Registration Number: 173

Definition: The smallest crown clade containing *Alligator* (originally *Crocdilus*) *mississippiensis* (Daudin 1802) (*Crocodylia*) and *Compsognathus longipes* Wagner 1859 (*Dinosauria*). This is a minimum-crown-clade definition. Abbreviated definition: min crown ∇ (*Alligator mississippiensis* (Daudin 1802) & *Compsognathus longipes* Wagner 1859).

Etymology: Derived from the Greek *archon* (ruler) plus *sauros* (lizard-like reptile).

Reference Phylogeny: Gauthier et al. (1988b: Fig. 3, where *Alligator mississippiensis* is part of a clade labeled *Pseudosuchia* and *Compsognathus longipes* is part of a clade called *Ornithosuchia*. See also Gauthier (1984: Figs. 22–29), Rest et al. (2003: Fig. 3), Irmis et al. (2007: Fig. 3), Hugall et al. (2007: Fig. 1), Gauthier et al. (2011: Fig. 4), Nesbitt (2011: Figs. 51 and 52), and Field et al. (2014: Fig. 1).

Composition: *Archosauria* is composed of two primary crown clades: *Crocodylia*, with 23 extant species (Uetz, 2008) and *Aves*, with about 9,700 extant species (Clements, 2000). Partial lists of extinct members of *Crocodylia* and part of its stem can be found in Brochu (2003, 2004) and McAliley et al. (2006). See the entry for *Aves* (this volume) for some extinct species of that crown clade, and Chiappe (2007) for many species of extinct winged dinosaurs from outside that crown, as well as some extinct species inside crown *Aves*. More distantly related Mesozoic stem crocodilians and stem avians can be found in Gauthier (1986), Benton and Clark (1988), Sereno (1991), Parrish (1993), Juul (1994), Novas (1996), Benton (1999), Rauhut (2003), Weishampel et al. (2004), Irmis et al. (2007), Nesbitt et al. (2007), Brusatte et al. (2010), and Nesbitt (2011).

Diagnostic Apomorphies: Gauthier et al. (1988b) listed 82 unambiguous synapomorphies of crown archosaurs relative to other extant amniotes. Supporting apomorphies have been reported from diverse character systems including DNA sequences, histology, gross anatomy, and even behavior. When fossils are considered, however, that list shrinks dramatically owing to a diverse array of Permo-Triassic stem archosaurs illuminating the sequential assembly of the crown's distinctive morphology (e.g., Gottmann-Quesada and Sander, 2009; Nesbitt, 2011; Ezcurra et al., 2014). As a consequence, only a few skeletal apomorphies remain unambiguously diagnostic of *Archosauria* relative to its near-crown stem relatives, including: an antorbital fossa completely surrounding the antorbital fenestra; maxillary vomerine (= palatal) processes that meet on the palate; and an *m. coracobrachialis* eminence on the coracoid (mistakenly called the "biceps tubercle" in stem avians and homologous with the "acrocoracoid process"—to which the *m. biceps brachii* now originates—in crown avians). That there is relatively low support in the skeleton for *Archosauria* indicates that we have learned a great deal about the origin of the crown from the rich fossil record of its stem.

Synonyms: *Neoarchosauria* Benton 1985 (approximate); *Avesuchia* Benton 1999 (unambiguous). The old but rarely used names *Ornithomorpha* Seeley 1892 and *Ornithocrania* Haeckel 1895 are noteworthy for referring to a taxon that explicitly included crocodilians, pterosaurs, ornithischians, "saurischians", and avians, qualifying them as approximate junior synonyms of *Archosauria* Cope 1869 as defined here, albeit with decidedly more modern conceptions than embraced by Cope of the relationship between phylogeny and taxonomy. A few obscure names coined in the nineteenth century for patently paraphyletic taxa might be added to this list as partial synonyms.

Comments: *Aves* and *Crocodylia* were included explicitly within a monophyletic *Archosauria* by Wiley (1979; see also Gaffney, 1980), and the name *Archosauria* was defined explicitly with respect to the last common ancestor of crown birds and crown crocodilians by Gauthier and Padian (1985; see also Gauthier 1984, 1986). Moreover, an implicit genealogical connection between birds and crocodilians among crown amniotes runs deep into the history of palaeontology. Even Cope (1869: p. 2) observed that one of his archosaurian groups, the *Dinosauria*, "present[s] a graduated series of approximations to the birds…standing between [*Aves*] and the *Crocodilia*."

Nevertheless, following traditional practice, Cope did not include the Class *Aves* in the Class *Reptilia*, much less in his Order *Archosauria*. His original composition included *Crocodylia*, *Thecodontia*, *Dinosauria*, *Sauropterygia*, *Anomodontia*, and *Rhynchocephalia* (but not *Pterosauria*, which he regarded as a separate taxon of equal categorical rank). Thus, if one were to view Cope's *Archosauria* in terms of his proposed composition, *Archosauria* might qualify as a junior partial synonym of *Amniota* under current phylogenies (e.g., Gauthier et al., 1988a,b; Rest et al., 2003). That is to say, although he explicitly excluded turtles from *Archosauria*, Cope mistakenly included some then-poorly-known stem mammals (*Anomodontia*) in the group. To make

matters worse, Cope's "important feature" for *Archosauria* is a plesiomorphy likely ancestral for all gnathostomes (i.e., a fixed quadrate firmly sutured to the skull, in contrast to the derived mobile quadrate of squamates).

Cope's (1869) *Archosauria* included better-known crown species (i.e., crocodilians and dinosaurs) that would come to be thought of as "archosaurian" from Williston (1917) on. But it also included an assortment of stem archosaurs that may be more (e.g., *Belodon* [*Phytosauria*] Nesbitt, 2011) or less (e.g., *Nothosaurus* [*Sauropterygia*] Merck, 1997) closely related to the crown. By the early twentieth century (e.g., Broom, 1913; Williston, 1917), most of what are now considered non-archosaurian amniotes had been pruned away, and *Pterosauria* added, to yield what became the traditional composition of *Archosauria* throughout most of the twentieth century.

Traditional ideas about archosaur composition always encompassed a group called "*Thecodontia*" Owen 1859. The term "thecodont" was used originally by Owen (1845) to refer to a dentition in which the teeth were set in distinct sockets, in contrast to the superficially attached pleurodont implantation (sensu Estes et al., 1988) seen in some other *Sauria* (a name which Owen used for what we would regard as a paraphyletic group, crocodilians and "lizards", partially approximating crown *Diapsida*; see Gauthier, 1984). Even then, however, Owen appeared to have also had a taxon in mind, for he excluded several taxa, such as the dinosaur *Megalosaurus*, that he described as having thecodont teeth. Indeed, when he formally coined the taxon name "*Thecodontia*" in 1859, Owen specified the same composition, viz., *Thecodontosaurus*, *Palaeosaurus*, *Cladyodon*, *Belodon*, *Protosaurus*, and *Bathygnathus*. Currently, *Thecodontosaurus* is a well-known sauropodomorph saurischian dinosaur. *Palaeosaurus* (a preoccupied name) is a chimaera (e.g., Galton, 2007). *Cladyodon* (also known as *Cladeiodon*) appears to be a rauisuchian stem crocodilian. Both *Protorosaurus* and *Belodon* are, respectively, distal and proximal stem archosaurs. *Protorosaurus*, a Permian species, is now thought to rank among the stratigraphically lowest, and anatomically least modified, unambiguous stem-archosaurs currently known. In contrast, *Belodon*, one of the familiar *Phytosauria* of the Triassic, is so close to the crown that it was commonly included within *Archosauria* until very recently (Gauthier et al., 2011; Nesbitt et al., 2011). Indeed, phytosaurs embodied the concept of *Thecodontia* for a very long time (because they were aquatic, they are relatively well represented in the fossil record). Finally, although *Bathygnathus* does indeed have socketed teeth, it is not even remotely related to archosaurs, but is instead a very early and relatively unmodified stem mammal (also known as a synapsid "pelycosaur"). Part of the problem here relates to the meaning of the anatomical term "thecodont", as shallowly socketed (= subthecodont) teeth appear ancestral for *Amniota* (Gauthier et al., 1988a,b). Within the archosaur total clade, the teeth become more deeply implanted and secured via a periodontal ligament, so that the teeth typically fall out when the skull is macerated or desiccated (Gauthier, 1984). In contrast, all vestiges of these shallow sockets are lost in the lepidosaur line leading to squamates (Gauthier et al., 1988a; Estes et al., 1988).

As stem members of the crocodilian total clade (e.g., aetosaurs such as *Staganolepis*) became better known, they were also grouped with the phytosaurs by von Huene (1902), although under "*Parasuchia*", an old name for the phytosaurs implying that they were "near to, but not" crocodilians (whose Egyptian name was "*souchus*"). Boulenger (1903) resurrected Owen's *Thecodontia* when describing Triassic fossils from Scotland. For a while, there was

confusion about whether to use the word as a taxon name or a dental condition, but Watson (1917) clarified it to refer to "archosaurians with clavicles and an interclavicle. Pelvis platelike." Because all these states are also plesiomorphic for archosaurs, that led to further complications. For example, Watson also included some "eosuchians" (including near-crown stem diapsids) and rhynchosaurs (early diverging stem archosaurs) among the thecodonts. Von Huene's later work (e.g., 1956) helped to develop the concept of "*Thecodontia*" that endured until the 1980s. The taxon was obviously paraphyletic because it explicitly excluded dinosaurs, pterosaurs, and crocodilians, even though all of these clades were widely thought to have evolved from a "thecodontian stock" (e.g., Owen, 1860a,b; Broom, 1903; von Huene, 1946, 1956; Romer, 1956; Cruickshank, 1972; Bonaparte, 1975; Charig, 1976).

Following Gauthier (1984), Gauthier and Padian (1985) proposed a solution to this muddle by tying the name *Archosauria* to a crown clade, and Gauthier (1994) defined a new name—*Archosauriformes* (see entry in this volume)—for the more inclusive clade traditionally associated with *Archosauria*, at least among vertebrate palaeontologists. Benton, however, preferred to conserve the traditional palaeontological circumscription of *Archosauria*, and instead coined two new names—*Neoarchosauria* Benton 1985 and *Avesuchia* Benton 1999—for the crown, although only the latter was defined explicitly in phylogenetic terms. A Google Scholar search (January 9, 2015) revealed just 31 results for *Avesuchia*, whereas a search for *Archosauria* found about 4,150; the former much-less-commonly used name benefits from its explicit reference to the crown; the latter name is far more commonly used, but is more ambiguous in its reference as it could apply either to the crown or to the more inclusive clade that also contains part of the stem (= *Archosauriformes*; which returned about 600 results, implying at least implicit acceptance of using *Archosauria*, instead of *Avesuchia*, for the crown). Because they generally sample only Recent species, the larger community of systematists using gene-sequence data is implicitly referring to the crown when they use the name *Archosauria*. But even among palaeontologists, the name *Archosauria* is often used for the crown rather than a more inclusive clade. Another Google Scholar search (January 18, 2015) for the undisputed stem taxon *Proterosuchus*, for example, yielded 10 instances in which that species was explicitly referred to *Archosauria* as opposed to 28 papers in which it was referred to *Archosauriformes* (based on articles published since 2000 that were listed in the first ten pages returned in this search). Accordingly, we prefer *Archosauria* to *Avesuchia* as the name for the crocodilian + avian crown clade, a common practice among neontologists since Wiley (1979) and among palaeontologists since Gauthier and Padian (1985).

Prior to the acceptance of Hennig's (1966) definition of monophyly, *Aves* was never part of traditional reptilian, diapsid, or archosaurian taxonomy. In the post-Hennigian era, however, birds have often been used as specifiers in formal definitions of *Archosauria* and associated clade names (e.g., Senter, 2005; Sereno, 2005). However, in order to apply *Archosauria* to a crown clade while remaining faithful to traditional ideas about its composition (*ICPN* Rec. 11A), we avoid using any avians (or avialans), and instead select internal specifiers familiar to Cope from taxa that he explicitly included in *Archosauria*: *Crocodylia* (i.e., *Alligator mississippiensis*) and *Dinosauria* (i.e., *Compsognathus longipes*). The former species is a particularly suitable specifier because it is well known, including a sequenced genome, and widely available for study. The latter species is also well known, and is of special historical significance as one of the first articulated non-avian dinosaurs ever found

(Wagner, 1859), and for its role in inspiring the dinosaur/bird hypothesis (e.g., Gegenbaur, 1863). Even the few remaining researchers who reject the hypothesis that extinct dinosaurs are actually stem birds have never doubted that *Compsognathus longipes* is a dinosaur, and thus also an archosaur. Of course, in the context of the proposed definition, *Aves* would also be part of *Archosauria* according to all dataset-driven phylogenetic analyses published since Gauthier (1986).

The single possible exception to the idea that birds are archosaurs stems from the ambiguous results obtained by James and Pourtless (2009). Their "alternative analysis" of the origin of birds was burdened by a number of shortcomings, not the least of which was a narrowly focused set of characters, 86% of which derived from a dataset developed specifically to assess the position of birds among a subset of Mesozoic theropods (i.e., maniraptor coelurosaurs; Clark et al., 2002), rather than the phylogeny of *Archosauria* as a whole. As a consequence, they were only able to infer a large polytomy among a set of taxa normally considered to be crown archosaurs (James and Pourtless, 2009: Fig. 14; with the conspicuous exception of the inclusion of the distal stem-archosaur *Longisquama insignis*). The specifiers *Compsognathus* and *Alligator*—archosaurs by definition—are represented in their 50% majority-rule consensus tree, but limited character sampling otherwise renders the composition of *Archosauria* ambiguous on that tree (viz., based on their consensus tree, it is not clear if, for example, *Tyrannosaurus rex*, is an archosaur). Nevertheless, birds would still be inside *Archosauriformes* (see entry in this volume) on that tree, with 97% of the bootstrap replicates placing them in a clade including *Compsognathus* and *Alligator*, but excluding Early Triassic near-crown stem archosaurs ("thecodonts") such as *Proterosuchus*, *Erythrosuchus*, and *Euparkeria*.

Note that *Archosauria* was not based on the later-proposed name *Archosaurus,* so excluding *Archosaurus rossicus* (Tatarinov, 1960) from the list of internal specifiers—and indeed from the clade *Archosauria*—does not violate *ICPN* Article 11.10. Although our definition uses an extinct species as an internal specifier, it is intended to apply to a crown clade.

Literature Cited

Benton, M. J. 1985. Classification and phylogeny of the diapsid reptiles. *Zool. J. Linn. Soc.* 84:97–164.

Benton, M. J. 1999. *Scleromochlus taylori* and the origin of dinosaurs and pterosaurs. *Philos. Trans. R. Soc. B Biol. Sci.* 354:1423–1446.

Benton, M. J., and J. M. Clark. 1988 Archosaur phylogeny and the relationships of the *Crocodylia.* Pp. 289–332 in *The Phylogeny and Classification of Tetrapods,* Vol. 1 (M. J. Benton, ed.). Clarendon Press, Oxford.

Bonaparte. J. F. 1975. Nuevos materiales de *Lagosuchus talampayensis* Romer (*Thecodontia – Pseudosuchia*) y su significado en el origen de los *Saurischia,* Chañarense Inferior, Triasico Medio de Argentina *Acta Geol. Lillo.* 13:5–90.

Boulenger, G. A. 1903. On reptilian remains from the Trias of Elgin. *Geol. Mag.* 10:354–357.

Brochu, C. A. 2003. Phylogenetic approaches toward crocodylian history. *Annu. Rev. Earth Planet. Sci.* 31:357–397.

Brochu, C. A. 2004. Alligatorine phylogeny and the status of *Allognathosuchus* Mook, 1921. *J. Vertebr. Paleontol.* 24:856–872.

Broom, R. 1903. On a new reptile (*Proterosuchus fergusi*) from the Karroo Beds of Tarkastad, South Africa. *Ann. S. Afr. Mus.* 6: 161–163.

Broom, R. 1913. On the South African pseudosuchian *Euparkeria* and allied genera. *Proc. Zool. Soc. Lond.* 1913:619–633.

Brusatte, S. L., M. J. Benton, J. B. Desoja, and M. C. Langer. 2010. The higher-level phylogeny of *Archosauria* (*Tetrapoda: Diapsida*). *J. Syst. Palaeontol.* 8:3–47.

Charig, A. J. 1976. "Dinosaur monophyly and a new class of vertebrates": a critical review. Pp. 65–104 in *Morphology and Biology of Reptiles* (A. d'A. Bellairs and C. B. Cox, eds.). Academic Press, London.

Chiappe, L. M. 2007. *Glorified Dinosaurs: The Origin and Early Evolution of Birds.* John Wiley & Sons, Hoboken, NJ.

Clark, J. M., M. A. Norell, and P. J. Makovicky. 2002. Cladistic approaches to the relationships of birds to other theropod dinosaurs. Pp. 31–60 in *Mesozoic Birds: Above the Heads of Dinosaurs* (L. M. Chiappe and L. M. Witmer, eds.). University of California Press, Berkeley, CA.

Clements, J. F. 2000. *Birds of the World: A Checklist.* 5th edition. Ibis, Vista, CA.

Cope, E. D. 1869. Synopsis of the extinct *Batrachia* and *Reptilia* of North America, Part I. *Trans. Am. Philos. Soc. New Ser.* 14:1–235.

Cruickshank, A. R. I. 1972. The proterosuchian thecodonts. Pp. 89–119 in *Studies in Vertebrate Evolution* (K. A. Joysey and T. S. Kemp, eds.). Winchester Press, New York.

Estes, R., K. de Queiroz, and J. Gauthier. 1988. Phylogenetic relationships within *Squamata.* Pp. 119–281 in *Phylogenetic Relationships of the Lizard Families* (R. Estes and G. Pregill, eds.). Stanford University Press, Palo Alto, CA.

Ezcurra, M. D., T. M. Scheyer, and R. J. Butler. 2014. The origin and early evolution of *Sauria*: reassessing the Permian saurian fossil record and the timing of the crocodile-lizard divergence. *PLOS ONE* 9(2):e89165.

Field, D. J., J. A. Gauthier, B. J. King, D. Pisani, T. R. Lyson, and K. J. Peterson. 2014. Toward consilience in reptile phylogeny: miRNAs support an archosaur, not lepidosaur, affinity for turtles. *Evol. Dev.* 16:189–196.

Gaffney, E. 1980. Phylogenetic relationships of the major groups of amniotes. Pp. 593–610 in *The Terrestrial Environment and the Origin of Land Vertebrates* (A. Panchen, ed.). Academic Press, London/New York.

Galton, P. 2007. Notes on the remains of archosaurian reptiles, mostly basal sauropodomorph dinosaurs, from the 1834 fissure fill (Rhaetian, Upper Triassic) at Clifton in Bristol, southwest England. *Rev. Paléobiol.* 26:505–591.

Gauthier, J. A. 1984. *A Cladistic Analysis of the Higher Systematic Categories of the Diapsida.* University Microfilms International, Ann Arbor, MI. #85–12825, vii + 564 pp.

Gauthier, J. A. 1986. Saurischian monophyly and the origin of birds. Pp. 1–55 in *The Origin of Birds and the Evolution of Flight* (K. Padian, ed.). Mem. Calif. Acad. Sci. 8, California Academy of Sciences, San Francisco, CA.

Gauthier, J. A. 1994. The diversification of the amniotes. Pp. 129–159 in *Major Features of Vertebrate Evolution: Short Courses in Paleontology* (D. Prothero, ed.). Paleontological Society.

Gauthier, J. A., R. Estes, and K. de Queiroz. 1988a. A phylogenetic analysis of *Lepidosauromorpha.* Pp. 15–98 in *Phylogenetic Relationships of the Lizard Families* (R. Estes and G. Pregill, eds.). Stanford University Press, Palo Alto, CA.

Gauthier, J. A., A. Kluge, and T. Rowe, 1988b. Amniote phylogeny and the importance of fossils. *Cladistics*, 4:105–209.

Gauthier, J. A., S. J. Nesbitt, E. R. Schachner, G. S. Bever, and W. G. Joyce. 2011. The bipedal stem crocodilian *Poposaurus gracilis*: inferring function in fossils and innovation in archosaur locomotion. *Bull. Peabody Mus. Nat. Hist.* 52:107–126.

Gauthier, J. A., and K. Padian. 1985. Phylogenetic, functional, and aerodynamic analyses of the origin of birds. Pp. 185–197 in *The Beginnings of Birds*; *Proceedings of the International* Archaeopteryx *Conference, Eichstätt 1984* (M. K. Hecht, J. H. Ostrom, G. Viohl, and P. Wellenhoffer, eds.). Freunde des Jura-Museums, Eichstätt, West Germany.

Gegenbaur, C. 1863. Vergeichendend-anatomische Bermerkungen uber das Fusskelet dr Vogel. *Arch. Anat. Physiol. Wiss. Med.* 1863:450–472.

Gottmann-Quesada, A., and P. M. Sander. 2009. A redescription of the early archosauromorph *Protorosaurus speneri* Meyer, 1832 and its phylogenetic relationships. *Palaeontogr. Abt. A* 287:123–220.

Haeckel, E. 1895. *Systematische Phylogenie der Wirbelthiere* (Vertebrata). *Dritter Theil des Entwurfs Einer Systematischen Stammesgeschichte.* G. Reimer, Berlin.

Hennig, W. 1966. *Phylogenetic Systematics.* Translated by D. Davis and R. Zangerl. University of Illinois Press, Urbana, IL.

von Huene, F. 1902. Übersicht über die Reptilien der Trias. *Geol. Paläontol. Abh. Neue Folge* 6:1–84.

von Huene, F. 1946. Die Grossen Stämme der Tetrapoden in den geologischen. *Biol. Zent.* 65:268–275.

von Huene, F. 1956. *Paläontologie und Phylogenie der Niederen Tetrapoden.* G. Fischer Verlag, Jena, Berlin.

Hugall, A. F., R. Foster, and M. S. Y. Lee. 2007. Calibration choice, rate smoothing, and the pattern of tetrapod diversification according to the long nuclear gene RAG-1. *Syst. Biol.* 56:543–563.

Irmis, R. B., S. J. Nesbitt, K. Padian, N. D. Smith, A. H. Turner, D. Woody, and A. Downs. 2007. A late Triassic dinosauromorph assemblage from New Mexico and the rise of dinosaurs. *Science* 317:358–361.

James, F. C., and J. A. Pourtless. 2009. Cladistics and the origin of birds: a review and two new analyses. *Ornithol. Monogr.* 66:1–78.

Juul, L. 1994. The phylogeny of basal archosaurs. *Palaeontol. Afr.* 31:1–38.

McAliley, L. R., R. E. Willis, D. A. Ray, P. S. White, C. A. Brochu, and L. D. Densmore. 2006. Are crocodiles really monophyletic? – evidence for subdivisions from sequence and morphological data. *Mol. Phylogenet. Evol.* 39:16–32.

Merck, J. W. 1997. *A Phylogenetic Analysis of the Euryapsid Reptiles.* PhD dissertation, Department of Geological Sciences, The University of Texas at Austin.

Nesbitt, S. J., R. B., Irmis, and W. G., Parker. 2007. A critical re-evaluation of the Late Triassic dinosaur taxa of North America. *J. Syst. Palaeontol.* 5:209–243.

Nesbitt, S. J. 2011. The early evolution of archosaurs: relationships and the origin of major clades. *Bull. Am. Mus. Nat. Hist.* 352:1–292.

Novas, F. E. 1996. Phylogenetic relationships of the basal dinosaurs, the *Herrerasauridae. J. Vertebr. Paleontol.* 16:723–741.

Owen, R. 1845. *Odontography: Or, a Treatise on the Comparative Anatomy of the Teeth; Their*

Physiological Relations, Mode of Development, and Microscopic Structure, in the Vertebrate Animals. Bailliere, London.

Owen, R. 1859. Palaeontology. Pp. 91–176 in *Encyclopaedia Britannica.* 8th edition, Vol. 17. A. & C. Black, Edinburgh.

Owen, R. 1860a. On the orders of fossil and recent *Reptilia*, and their distribution in time. *Rep. Twenty-Ninth Meet. Brit. Assoc. Adv. Sci.* 1859:153–166.

Owen, R. 1860b. *Palaeontology, or a Systematic Summary of Extinct Animals and Their Geological Relations.* 1st edition. A. & C. Black, Edinburgh.

Parrish, J. M. 1993. Phylogeny of the *Crocodylotarsi*, with reference to archosaurian and crurotarsan monophyly. *J. Vertebr. Paleontol.* 13:287–308.

Rauhut, O. W. M. 2003. The interrelationships and evolution of basal theropod dinosaurs. *Spec. Pap. Palaeontol.* 69:1–214.

Rest, J. S., J. C. Ast, C. C. Austin, P. J. Waddell, E. A. Tibbetts, J. M. Hay, and D. P. Mindell. 2003. Molecular systematics of primary reptilian lineages and the tuatara mitochondrial genome. *Mol. Phylogenet. Evol.* 29:289–297.

Romer, A. S. 1956. *Osteology of the Reptiles.* University of Chicago Press, Chicago, IL.

Seeley, H. G. 1892. Researches on the structure, organization, and classification of the fossil *Reptilia.* VII. Further observations on *Pareiasaurus. Philos. Trans. R. Soc. Lond. B Biol. Sci.* 183:311–370

Senter, P. 2005. Phylogenetic taxonomy and the names of the major archosaurian (*Reptilia*) clades. *PaleoBios* 25:1–7.

Sereno, P. C. 1991. Basal archosaurs: phylogenetic relationships and functional implications. *Soc. Vertebr. Paleontol. Mem.* 2:1–53.

Sereno, P. C. 2005. The logical basis of phylogenetic taxonomy. *Syst. Biol.* 54:595–619.

Tatarinov, L. P. 1960. Discovery of pseudosuchians in the Upper Permian of the USSR. *Paleontol. Zh.* 1960:74–80.

Uetz, P. 2008. Species numbers (as of February 2008). The TIGR Reptile Database. Available at http://www.reptile-database.org/.

Wagner, J. A. 1859. Über einige im lithographischen Schiefer neu aufgefundene Schildkröten und Saurier. *Gel. Anz. Bayer. Akad. Wiss.* 49:553.

Watson, D. M. S. 1917. A sketch-classification of the pre-Jurassic tetrapod vertebrates. *Proc. Zool. Soc. Lond.* 1917:167–86.

Weishampel, D. B., P. Dodson, and H. Osmolska. 2004. The Dinosauria. 2nd edition. University of California Press, Berkeley, CA.

Wiley, E. O. 1979. Ventral gill arch muscles and the interrelationships of gnathostomes, with a new classification of the *Vertebrata. Zool. J. Linn. Soc.* 67:149–179.

Williston, S. W. 1917. The phylogeny and classification of reptiles. *J. Geol.* 25:411–421.

Authors

Jacques A. Gauthier; Department of Geology and Geophysics; Yale University; 210 Whitney Ave.; New Haven, CT 06520-8109, USA. Email: jacques.gauthier@yale.edu.

Kevin Padian; Department of Integrative Biology and Museum of Paleontology; University of California, Berkeley, CA 94720-3140, USA. Email kpadian@berkeley.edu.

Date Accepted: 24 August 2015

Primary Editor: Kevin de Queiroz

Pterosauromorpha E. Kuhn-Schnyder and H. Rieber 1986
[B. Andres and K. Padian], converted clade name

Registration Number: 149

Definition: The clade consisting of *Pterodactylus* (originally *Ornithocephalus*) *antiquus* (Sömmerring 1812) (*Pterosauria*) and all organisms or species that share a more recent common ancestor with it than with *Alligator* (originally *Crocdilus*) *mississippiensis* (Daudin 1802) (*Suchia*) and *Compsognathus longipes* Wagner 1859 (*Dinosauromorpha*). This is a maximum-clade definition. Abbreviated definition: max ∇ (*Pterodactylus antiquus* (Sömmerring 1812) ~ *Alligator mississippiensis* (Daudin 1802) & *Compsognathus longipes* Wagner 1859).

Etymology: Derived from the Greek *ptero-* (wing), *sauros* (reptile), and *morphe* (shape); see *Pterosauria* (Comments) in this volume for historical details.

Reference Phylogeny: Figure 26A of Sereno (1991), which delineates the probable relationships of *Pterosauria* (including *Pterodactylus antiquus*) and *Scleromochlus taylori* to *Dinosauromorpha* (including *Compsognathus longipes*) and *Suchia* (including *Alligator mississippiensis*).

Composition: Includes *Pterosauria* (see *Pterosauria*, this volume, for included taxa) and possibly *Scleromochlus taylori* Woodward 1907 (Huene, 1914; Padian, 1984, 1997; Novas, 1996; Sereno, 1991; Brusatte et al., 2010).

Diagnostic Apomorphies: As a maximum clade ("branch-based" taxon), *Pterosauromorpha* may not have any diagnostic apomorphies. Possible apomorphies for the deepest currently known node within *Pterosauromorpha* (uniting *Pterosauria* and *Scleromochlus taylori*) include: skull length more than half of presacral column length but lacking an elongate prenarial rostrum (after Sereno, 1991); four sacral and nine cervical vertebrae if the first vertebra bearing a rib that articulates with the sternum is interpreted as the first thoracic (after Benton, 1999, and Bennett, 2007, 2014); ulna and/or radius longer than the humerus, both longer than scapula (Padian, 1997) with deltopectoral crest extending less than one-quarter of the way down the humeral shaft (after Bennett, 1996); bowed femur (Huene, 1914); fourth trochanter of the femur extremely reduced or absent (after Sereno, 1991); tibia longer than the femur with splint-like fibula (Padian, 1997); four elongated, closely appressed metatarsals with the fifth angled posterolaterally (Padian, 1997), and metatarsal I at least 85 percent of metatarsal III in length (Sereno, 1991).

Synonyms: None.

Comments: Pterosaurs appear in the fossil record of the Late Triassic as fully formed flying organisms with no known unequivocal transitional forms to give insight into their origin. Although their monophyly is uncontroversial, questions about the relationships of pterosaurs to other animals began at their discovery (Wellnhofer, 1978, 1991; Taquet and Padian, 2004; Unwin, 2006). Identification of a pterosaur specimen as a "reptile" dates to the description of the first known specimen by Cuvier (1800, 1809), but the hypothesized relationships of pterosaurs within *Reptilia* have had a more varied history. Haeckel was the first to place *Pterosauria* in a tree diagram and, in doing so, allied it with a group composed of birds and turtles (Haeckel, 1866: Plate 7). He later depicted *Pterosauria* as sister to a group including "*Dinosauria*" and *Aves* in his 1876 *The History of Creation* (Haeckel, 1876: Plate 14), and both *Pterosauria* and *Aves* as stemming from *Dinosauria* in his 1895 *Systematische Phylogenie der Wirbelthiere* (Haeckel, 1895: Fig. 328). Seeley (1901) regarded pterosaurs (his "*Ornithosauria*") as showing "affinities" and "relations" to both "dinosaurs" and birds.

Huene (1914, 1956) first classified pterosaurs in both a treelike fashion and within *Archosauria*. Later authors have generally followed this classification. Wild (1983) suggested an alternative hypothesis of pterosaurs as close relatives of *Prolacertiformes*, which are currently considered stem archosaurs only distantly related to crown *Archosauria* (e.g., Gauthier, 1984; Nesbitt, 2011), but he provided no phylogenetic analysis; only Peters (2000) has supported this view, but his analysis of previous work is controversial and has not been supported by independent analyses of other workers. Most phylogenetic analyses have corroborated the archosaurian hypothesis, placing *Pterosauria* as sister to all other known pan-avians (Gauthier, 1984, 1986; Gauthier and Padian, 1985; Sereno and Arcucci, 1990; Sereno and Novas, 1990; Sereno, 1991, 1996; Juul, 1994; Padian et al., 1995; Bennett, 1996, in part; Novas, 1996; Benton, 1999, 2004; Irmis et al., 2007; Brusatte et al., 2008, 2010; Nesbitt et al., 2009; Nesbitt, 2011). Some analyses have placed *Pterosauria* among stem archosaurs, but not as a sister taxon of *Prolacertiformes*; Benton (1984, 1985) did, but revised these preliminary analyses in later works (i.e., 1999, 2004); Bennett (1996) obtained this phylogenetic position in one of his analyses by excluding characters of the hindlimb, and by reformulating them in Bennett (2013); Renesto and Binelli (2006) placed the pterosaur *Eudimorphodon ranzii*

as the sister to *Drepanosauridae* but did not place *Drepanosauridae* within *Prolacertiformes*. Reanalyses of previous work have found support for pterosaurs both within and outside of crown *Archosauria*, but have rejected the prolacertiform hypothesis (Hone and Benton, 2007, 2008).

The long list of the apomorphies diagnosing *Pterosauria* illustrates the high degree of modification of the pterosaur lineage before its first known appearance in the fossil record, which underlies the difficulty in attempting to link it to other clades. Unambiguous transitional fossils linking less modified archosaurs to pterosaurs have yet to be found (or recognized as such). The only non-pterosaur that has been repeatedly associated with *Pterosauria* is *Scleromochlus taylori* Woodward 1907 from the Late Triassic of Scotland. Huene (1914) first suggested that *Scleromochlus* may be a candidate for a pterosaur ancestor, and its close relationship to pterosaurs has been supported by subsequent studies (Padian, 1984; Gauthier, 1986; Padian et al., 1995; Sereno, 1991; Bennett, 1996, in part; Novas, 1996; Brusatte et al., 2010). However, Sereno (1991) suggested that *Scleromochlus* might be closer to his *Dinosauromorpha* (= *Lagerpeton chanarensis*, *Lagosuchus talampayensis*, *Pseudolagosuchus major*, *Dinosauria*, and all descendants of their common ancestor), and Benton (1999, 2004) inferred *Scleromochlus* as the sister taxon of the clade containing *Pterosauria* and *Dinosauromorpha*. The sister relationship of *Pterosauria* and *Scleromochlus* remains in question largely because some workers have put forward that the synapomorphies listed above are convergent or symplesiomorphic, or have conceptualized them differently than we do here (e.g., Sereno, 1991). Furthermore, this discussion has often been confused with the open questions of whether *Scleromochlus* and the origin of pterosaur flight were arboreal (Huene, 1914) or terrestrial (Padian, 1984).

The name *Pterosauromorpha* was coined by Kuhn-Schnyder and Rieber (1986) as that of a subclass in order to remove *Pterosauria* from the Subclass *Archosauria* because the authors thought it was impossible for pterosaurs to be derived from the same ancestral "thecodonts" that "gave rise" to the archosaurs. Padian (1997:617) defined the name phylogenetically as "*Pterosauria* and all ornithodiran archosaurs closer to them than to dinosaurs" in order to accommodate pterosaur relatives outside of *Pterosauria*. Although considerable support has been found for pterosaurs as pan-avian archosaurs and the sister taxon to *Scleromochlus*, alternative placements for these taxa have been put forward (even though they are not in general agreement), and uncertainty concerning the relationships of pterosaurs to other archosauriforms is often mentioned in the literature (e.g., Unwin, 2006). The name *Pterosauromorpha* is established here as a maximum clade name using *Pterodactylus antiquus* as an internal specifier and using the two internal specifiers of *Archosauria* (*Alligator mississippiensis* and *Compsognathus longipes*) as external specifiers. So defined, *Pterosauromorpha* will include *Pterosauria* and all taxa closer to it than to birds and crocodilians regardless of whether it is placed inside or outside of *Archosauria* and regardless of the relationships of *Scleromochlus taylori*.

Literature Cited

Bennett, S. C. 1996. The phylogenetic position of the pterosaurs within the *Archosauromorpha*. *Zool. J. Linn. Soc.* 118:261–308.

Bennett, S. C. 2007. A second specimen of the pterosaur *Anurognathus ammoni*. *Palaeontol. Z.* 81(4):376–398.

Bennett, S. C. 2013. The phylogenetic position of the *Pterosauria* within the *Archosauromorpha* re-examined. *Hist. Biol.* 25(5–6):545–563.

Bennett, S. C. 2014. A new specimen of the pterosaur *Scaphognathus crassirostris*, with comments on constraint of cervical vertebrae number in pterosaurs. *N. Jb. Geol. Paläont. Abh.* 271(3):327–348.

Benton, M. J. 1984. The relationships and early evolution of the *Diapsida*. *Symp. Zool. Soc. Lond.* 52:575–596.

Benton, M. J. 1985. Classification and phylogeny of the diapsid reptiles. *Zool. J. Linn. Soc.* 84:97–164.

Benton, M. J. 1999. *Scleromochlus taylori* and the origin of dinosaurs and pterosaurs. *Philos. Trans. R. Soc. Lond. B Biol. Sci.* 354:1423–1446.

Benton, M. J. 2004. Origin and relationships of *Dinosauria*. Pp. 7–19 in *The* Dinosauria (D. B. Weishampel, P. Dodson, and H. Osmólska, eds.). University of California Press, Berkeley, CA.

Brusatte, S. L., M. J. Benton, M. Ruta, and G. T. Lloyd. 2008. Evolutionary radiation of dinosaurs: superiority, competition, and opportunism in the evolutionary radiation of dinosaurs. *Science* 321:1485–1488.

Brusatte, S. L., M. J. Benton, J. B. Desojo, and M. C. Langer. 2010. The higher-level phylogeny of *Archosauria* (*Tetrapoda*: *Diapsida*). *J. Syst. Palaeontol.* 8(1):3–47.

Cuvier, G. 1800. *Extrait d'un ouvrage Sur les Espèces de Quadrupèdes Dont on a Trouvé les Ossemens Dans L'intérieur de la Terre*. Baudouin, Paris.

Cuvier, G. 1809. Mémoire sur le squelette fossile d'un reptile volant des environs d'Aichstedt, que quelques naturalistes ont prispour un oiseau, et dont nous formons un genre de sauriens, sous le nom de Petro-Dactyle [sic]. *Ann. Mus. Hist. Nat.* 13:424–437.

Daudin, F. M. 1802. *Histoire Naturelle, Générale et Particulière des Reptiles; Ouvrage Faisant Suit a L'histoire Naturelle Générale et Particulière, Composée par Leclerc de Buffon, et Rédigee par C. S. Sonnini, Membre de Plusieurs Sociétés Savantes*, Dufart, Paris. 2.

Gauthier, J. 1984. *A Cladistic Analysis of the Higher Systematic Categories of the* Diapsida. PhD dissertation, University of California, Berkeley, CA.

Gauthier, J. 1986. Saurischian monophyly and the origin of birds. Pp. 1–55 in *The Origin of Birds and the Evolution of Flight* (K. Padian, ed.). Mem. Calif. Acad. Sci. 8. California Academy of Sciences, San Francisco, CA.

Gauthier, J., and K. Padian. 1985. Phylogenetic, functional, and aerodynamic analyses of the origins of birds and their flight. Pp. 185–197 in *The Beginnings of Birds: Proceedings of the International Archaeopteryx Conference Eichstätt 1984*. Freunde des Jura-Museum Eichstätt, Eichstätt.

Haeckel, E. 1866. *Generelle Morphologie der Organismen. Allegemeine Grundzuge der Organischen Formen Wissenschaft, Mechanisch Begrundet Durch Die von Charles Darwin Reformierte Daszendenz-Theorie. II. Allgemeine Entwicklungsgeschicte der Organismen. Kritische Grundzuge der Mechanischen Wissenschaft von Dan Entstehenden Formen der Organismen, Begrundet Durch Die Deszendenz-Theorie.* Verlag von Georg Reimer, Berlin.

Haeckel, E. 1876. *The History of Creation: Or the Development of the Earth and Its Inhabitants by the Action of Natural Causes. II.* D. Appleton and Company, New York.

Haeckel, E. 1895. *Systematische Phylogenie der Wirbelthiere.* Verlag von Georg Reimer, Berlin.

Hone, D. W. E., and M. J. Benton. 2007. An evaluation of the phylogenetic relationships of the pterosaurs among archosauromorph reptiles. *J. Syst. Palaeontol.* 5:465–469.

Hone, D. W. E., and M. J. Benton. 2008. Contrasting supertree and total-evidence methods: the origin of pterosaurs. *Zitteliana, Ser. B* 28:35–60.

von Huene, F. R.. 1914. Beiträge zur Geschichte der Archosaurier. *Geol. Palaeontol. Abh., NF.* 13:1–53.

von Huene, F. R.. 1956. *Paläontologie und Phylogenie der Niederen Tetrapoden.* Gustav Fischer Verlag, Jena.

Irmis, R. B., S. J. Nesbitt, K. Padian, N. D. Smith, A. H. Turner, D. Woody, and A. Downs. 2007. A Late Triassic dinosauromorph assemblage from New Mexico and the rise of dinosaurs. *Science* 317:358–361.

Juul, L. 1994. The phylogeny of basal archosaurs. *Palaeontol. Afr.* 31:1–38.

Kuhn-Schnyder, E., and H. Rieber. 1986. *Handbook of Paleozoology.* The John Hopkins University Press, Baltimore, MD.

Nesbitt, S. J. 2011. The early evolution of archosaurs: relationships and the origin of major clades. *Bull. Am. Mus. Nat. Hist.* 352:1–292.

Nesbitt, S. J., R. B. Irmis, W. G. Parker, N. D. Smith, A. H. Turner, and T. Rowe. 2009. Hindlimb osteology and distribution of basal dinosauromorphs from the Late Triassic of North America. *J. Vertebr. Paleontol.* 29(2):498–516.

Novas, F. E. 1996. Dinosaur monophyly. *J. Vertebr. Paleontol.* 16:723–741.

Padian, K. 1984. The origins of pterosaurs. Pp. 163–168 in *Third Symposium on Mesozoic Terrestrial Ecosystems* (W.-E. Reif and F. Westphal, eds.). Attempto Verlag, Tübingen.

Padian, K. 1997. *Pterosauromorpha.* Pp. 617–618 in *Encyclopedia of Dinosaurs* (P. J. Currie and K. Padian ed.). Academic Press, San Diego, CA.

Padian, K., J. A. Gauthier, and N. C. Fraser. 1995. *Scleromochlus taylori* and the early evolution of pterosaurs. *J. Vertebr. Paleontol.* 15(suppl. 3):47A.

Peters, D. 2000. A reexamination of four prolacertiforms with implications for pterosaur phylogenesis. *Riv. Ital. Paleontol. S.* 105(3):293–335.

Renesto, S., and G. Binelli. 2006. *Vallesaurus cenensis* Wild, 1991, a drepanosaurid (*Reptilia, Diapsida*) from the Late Triassic of northern Italy. *Riv. Ital. Paleontol. S.* 112(1):77–94.

Seeley, H. G. 1901. *Dragons of the Air.* Methuen and Co., London.

Sereno, P. C. 1991. Basal archosaurs: phylogenetic relationships and functional implications. *Soc. Vertebr. Paleontol. Memoir 2, J. Vertebr. Paleontol.* 11(suppl. 4):1–53.

Sereno, P. C. 1996. The phylogenetic position of pterosaurs within archosaurs. *J. Vertebr. Paleontol.* 16(suppl. to 3):65A.

Sereno, P. C., and A. Arcucci. 1990. The monophyly of crurotarsal archosaurs and the origin of bird and crocodile ankle joints. *N. Jb. Geol. Paläontol. Abh.* 180(1):21–52.

Sereno, P. C., and F. E. Novas. 1990. Dinosaur origins and the phylogenetic position of pterosaurs. *J. Vertebr. Palaeontol.* 9(suppl. 3):42A.

von Sömmerring, S. T. 1812. Über einen *Ornithocephalus. Denkschr. Kgl. Bayer. Akad. Wiss., Math.-Phys. Cl.* 3:89–158.

Taquet, P., and K. Padian. 2004. The earliest known restoration of a pterosaur and the philosophical origins of Cuvier's *Ossemens Fossiles. C. R. Palevol.* 3:157–175.

Unwin, D. M. 2006. *Pterosaurs from Deep Time.* Pi Press, New York.

Wagner, J. A. 1859. Über einige im lithographischen Schiefer neu aufgefundene Schildkröten und Saurier. *Gelehrte Anz. Bayer. Akad. Wiss,.*49:553.

Wellnhofer, P. 1978. Pterosauria. *Handbuch der Paläontologie*, Teil 19. Gustav Fischer Verlag, Stuttgart.

Wellnhofer, P. 1991. *The Illustrated Encyclopedia of Pterosaurs.* Salamander Books, London.

Wild, R. 1983. Über den Ursprung der Flugsaurier. Pp. 231–238 in *Erwin Rutte-Festschrift*, Weltenberger Akademie, Kelheim/Weltenburg.

Woodward, A. S. 1907. On a new dinosaurian reptile (*Scleromochlus taylori* gen. et sp. nov.) from the Trias of Lossiemouth, Elgin. *Q. J. Geol. Soc. Lond.* 63:140–146.

Authors

Brian Andres; Department of Animal & Plant Sciences; University of Sheffield; Sheffield, S10 2TN, UK. Email: brian.andres@aya.yale.edu.

Kevin Padian; Department of Integrative Biology and Museum of Paleontology; University of California, Berkeley, CA 94720-4780, USA. Email: kpadian@berkeley.edu.

Date Accepted: January 27, 2016

Primary Editors: Jacques A. Gauthier, Kevin de Queiroz

Pterosauria R. Owen 1842 [B. Andres and K. Padian], converted clade name

Registration Number: 157

Definition: The clade characterized by the apomorphy fourth manual digit hypertrophied to support a wing membrane, as inherited by *Pterodactylus* (originally *Ornithocephalus*) *antiquus* (Sömmerring 1812). This is an apomorphy-based definition. Abbreviated definition: ∇ apo fourth manual digit hypertrophied to support wing membrane [*Pterodactylus antiquus* (Sömmerring 1812)].

Etymology: Derived from the Greek *ptero-* (wing) and *sauros* (lizard).

Reference Phylogeny: Figure 3 in Andres et al. (2014), where the hypertrophied fourth manual digit is inferred to have arisen along the branch subtending the clade labeled "*Pterosauria*".

Composition: *Pterosauria* currently contains about 205 extinct species of Mesozoic flying archosaurs (B. Andres, pers. obs.), including *Pterodactyloidea* (this volume).

Diagnostic Apomorphies: Romer's (1956:639–640) diagnosis still largely serves to differentiate pterosaurs from other reptiles, though it has been modified by later authors (Wellnhofer, 1978; Padian, 1984; Sereno, 1991; Bennett, 1996; Kellner, 1996; Dalla Vecchia, 2014; this entry): pneumatic and hollow skeleton; large skull 50–90 percent of torso length, with correspondingly large fenestrae; elongate median dorsal process of the premaxillae extends posterior to the external naris and at least partially separates the nasals; maxilla forms about one-third of the border of the external naris; quadratojugal located at anterior margin of the infratemporal fenestra and does not contact squamosal; otic notch absent; internal naris located posterior to palatal bones; basipterygoid processes elongate and rod-shaped; slender jaw lacking external mandibular fenestra; procoelous presacral vertebrae (also in *Squamata*); postaxial cervical vertebrae considerably longer than thoracic vertebrae; elongate middle and distal caudal centra more than five times longer than high (reversed in short-tailed anurognathids and pterodactyloids); expanded sternum with a prominent anterior cristospine; ossified coracosternal joint present dorsolateral to base of cristospine; scapula and coracoid both long and slender, meeting at an acute angle, and subequal in length, or with a longer scapula; coracoid slants ventromedially and slightly posteriorly to cristospine; scapula oriented anteroposteriorly and slightly medially on dorsum (connects to fused dorsal spines, or notarium, in some pterodactyloids); coracoid foramen absent; preacetabular process of ilium low and elongate; ischia plate-like, fused to short, cylindrical pubes, which are connected to long slender prepubes with spatulate or forked distal ends (the prepubes are considered the pubes by some authors, but no evidence of identity is definitive); both ischiadic plates and distal prepubes have medial symphyses in adults; humerus with a bowed shaft, subequal in length to the femur; enlarged deltopectoral crest and saddle-shaped proximal articulation on humerus; carpals fused into proximal and distal syncarpals, with a separate medial distal carpal and sesamoid that subtend a long, attenuated, medially directed pteroid bone; manual digit V absent; metacarpals I–III slender and elongated, appressed to the robust metacarpal IV; extensor tendon process on the proximal articulation of the first phalanx of the wing digit; manual digit IV with flattened, cuplike concave-convex interphalangeal joints; phalangeal formula 2–3–4–4–0; manual digit IV lacks an ungual; femoral head hemispherical and offset from the shaft by constriction; femoral shaft bowed dorsally; proximal tarsals reduced to astragalus and calcaneum, and fused to each other and to tibia in adults, calcaneum extremely reduced to less than one-fourth the size of the astragalus; pedal digit V with a short, twisted, medially hooked metatarsal and two elongate phalanges but no ungual (the entire digit and metatarsal V are absent or reduced in the pterodactyloids); pelage of epidermal filaments; wing membrane supported internally by a subcutaneous network of radiating stiffening-fibers. The foregoing list is secondary to the apomorphic condition listed in the definition; it is expected that as earlier or earlier-diverging pterosauromorphs (this volume) are discovered, the distribution of these synapomorphies will shift with respect to this apomorphy.

Synonyms: Ptero-dactyle (informal) Cuvier 1809, *Pterosaurii* Kaup 1834, *Podoptera* Fischer 1834, *Ornithosauri* Fitzinger 1836, *Ornithosaurii* Bonaparte 1838, *Saurornia* Seeley 1864, *Ornithosauria* Seeley 1869, *Patagiosauria* Fürbringer 1888, *Pterodracones* Haeckel 1895, and *Pterodactylia* Haeckel 1895, are all approximate synonyms.

Comments: Young (1964: Fig. 9), Kuhn (1967: Fig. 24d), and Wellnhofer (1975: Fig. 45; 1978: Fig. 32) provided early evolutionary trees of pterosaur relationships. Kellner (2003: Fig. 1) and Unwin (2003: Fig. 6) provided the first, albeit disparate, published phylogenetic analyses of the relationships among all major pterosaur groups. Bennett (2007) and Andres et al. (2010) revised basal pterosaur relationships proposed in these previous analyses. Andres et al. (2014) was the largest, most comprehensive, most detailed, and most recent phylogenetic analysis of pterosaur relationships and was therefore selected as the reference phylogeny. This analysis was updated by Longrich et al. (2018) but presented in less detail.

The first known scientific report of a pterosaur is in a description by Collini (1784) of a small specimen from the Solnhofen Limestone of Bavaria. He correctly described many aspects of its anatomy but was stymied by the possible function of its elongated finger and the classification of the creature, which he did not name. He concluded that its "originals" (a pre-evolutionary concept of antecedents) must be sought among unknown marine animals, because living terrestrial animals were well-enough known to exclude all of them from affinity to the new find (Taquet and Padian, 2004). Georges Cuvier did not learn of the specimen until 1800, but working only from drawings, he recognized that it was a "reptile" and that it flew, reporting it as such in Cuvier (1800) (Taquet and Padian, 2004). In his 1809 description, he gave it the name "ptéro-dactyle" or "wing-finger," although the name first appeared misprinted as "pétro-dactyle" or "rock finger." Since then, the term "pterodactyl" has been used as an informal name for a subset of the nested groups of Mesozoic flying archosaurs to which this specimen belongs (see *Pterodactyloidea*). Cuvier did not initially provide a formal scientific name for this specimen, but he later (Cuvier, 1819) assigned it to *Pterodactylus longirostris*. In the meantime, Sömmerring (1812) named *Ornithocephalus antiquus* for the same specimen. Historical usage, later validated by the *International Code of Zoological Nomenclature*, has awarded priority to Cuvier for the genus *Pterodactylus* and to Sömmerring for the species *antiquus*; therefore, the first named pterosaur taxon is *Pterodactylus antiquus* (Sömmerring 1812).

Authorship of the name *Pterosauria* is often attributed to Kaup (1834), who was the first to suggest a name at the Linnaean ordinal rank that the group has maintained since. He used the name "*Pterosaurii*" instead, writing only that it included the "*Pterodactyli*" and could certainly be subdivided into several genera. *Pterosauria* was coined by Sir Richard Owen (1842). Owen also gave this name an ordinal rank, and specified *Pterodactylus* as its type, and characterized these organisms as "reptiles" that achieved flight "by the modification of their pectoral extremity" (Owen, 1842:156), and listed a number of species in the group. The other approximate synonyms listed above have long been abandoned.

A phylogenetic analysis of relationships within *Pterosauria* was not published until Howse's (1986) analysis of *Pterodactyloidea*, and a phylogenetic analysis of both basal pterosaur and pterodactyloid relationships was not published until 2003. In the same year and edited volume, Kellner (2003) and Unwin (2003) put forward two disparate hypotheses of relationships, taxonomy, and definitions of *Pterosauria*. Because both authors used minimum-clade definitions and they differed in their schemes of basal pterosaur relationships, using one of these definitions with the other phylogeny could result in the exclusion from *Pterosauria* of some taxa traditionally considered pterosaurs. The primary reason for this conflict in circumscription is that alternative interrelationships were proposed for *Anurognathidae*, a clade of highly derived Upper Jurassic pterosaurs. Furthermore, new species of flying reptiles that diverged earlier than the last common ancestor of the taxa specified in those definitions would also be excluded from *Pterosauria*. This problem would also be encountered with Sereno's (1991) minimum-clade definition of *Pterosauria*, which used almost all of the early-diverging pterosaur genera known at the time as internal specifiers.

The name *Pterosauria* is established here with an apomorphy-based definition using the hypertrophied wing digit found in *Pterodactylus antiquus* (Sömmerring 1812) because this definition is consistent with historical and current usage. This definition encapsulates Owen's (1842) original concept of *Pterosauria*. He used our internal specifier as the type and denoted this as a group of flying "reptiles". Although he listed a number of species that he included in the group, history has placed emphasis on the diagnostic apomorphy of wings rather than on the composition specified by Owen. In that regard, the tradition has been to refer to

Pterosauria all reptiles inferred to have diverged earlier than previously known species as long as they possess a wing-supporting digit. Currently there is no clear consensus on basal pterosaur relationships. However, the proposed apomorphy-based definition would include all current pterosaur species regardless of tree topology (including both Kellner's and Unwin's), as well as any future discoveries of earlier-diverging wing-fingered pterosauromorphs. Furthermore, the name *Pterosauria* itself refers to the diagnostic apomorphy of "wings."

Literature Cited

Andres, B., J. M. Clark, and X. Xu. 2010. A new rhamphorhynchid pterosaur from the Upper Jurassic of Xinjiang, China, and the phylogenetic relationships of basal pterosaurs. *J. Vertebr. Paleontol.* 30:163–187.

Andres, B., J. M. Clark, and X. Xu. 2014. The earliest pterodactyloid and the origin of the group. *Curr. Biol.* 24:1011–1016.

Bennett, S. C. 1996. The phylogenetic position of the pterosaurs within the *Archosauromorpha*. *Zool. J. Linn. Soc.* 118:261–308.

Bennett, S. C. 2007. A second specimen of the pterosaur *Anurognathus ammoni*. *Palaeontol. Z.* 81(4):376–398.

Bonaparte, C. L. 1838. Synopsis vertebratorum systematis. Amphibiorum. *Nuovi Ann. Sci. Nat. Bologna* 1:391–397.

Collini, C. A. 1784. Sur quelques zoolithes du cabinet d'histoire naturelle de S.A.S.E. Palatine *& de Bavière, à Mannheim. Historia et Commentationes Academiae Electoralis Scientiarum et Elegantiorum Litterarum Theodoro-Palatinae, Manheim* 5:58–103.

Cuvier, G. 1800. *Extrait d'un Ouvrage sur les Espèces de Quadrupèdes Dont on a Trouvé Les Ossemens dans L'intérieur de la Terre.* Baudouin, Paris.

Cuvier, G. 1809. Mémoire sur squelette fossile d'un reptil volant des environs d'Aischstedt, que quelques naturalistes ont pris pour un oiseau, et dont nous formons un genre de Sauriens, sous le nom de Petro-Dactyle [sic]. *Ann. Mus. Hist. Nat.* 13:424–437.

Cuvier, G. 1819. *Pterodactylus longirostris. Isis von Oken* 2:1126,1788.

Dalla Vecchia, F. M. 2014. *Gli Pterosauri Triassici.* Comune di Udine, Udine.

Fitzinger, L. 1836. Entwurf einer systematischen Anordnung der Schildkroten. *Ann. Wien. Mus.* 1:105–128.

Fürbringer, M. 1888. Untersuchungen zur Morphologie und Systematik der Vögel, zugleich ein Beitrag zur Anatomie der Stützund Bewegungsorgane. *Bijd. Dierkunde.* 15:1–834.

Haeckel, E. 1895. *Systematische Phylogenie der Wirbelthiere.* Verlag von Georg Reimer, Berlin.

Howse, S. C. B. 1986. On the cervical vertebrae of the *Pterodactyloidea* (*Reptilia: Archosauria*). *Zool. J. Linn. Soc.* 88:307–328.

Kaup, J. J. 1834. Versuch einer Eintheilung der Saugethiere in 6 Stämme und der Amphibien in 6 Ordnung. *Isis von Oken* 3:311–316.

Kellner, A. W. A. 1996. *Description of new material of* Tapejaridae *and* Anhangueridae *(Pterosauria, Pterodactyloidea) and discussion of pterosaur phylogeny.* Thesis, Columbia University, New York.

Kellner, A. W. A. 2003. Pterosaur phylogeny and comments on the evolutionary history of the group. Pp. 105–137 in *Evolution and Palaeobiology of Pterosaurs* (E. Buffetaut and J.-M. Mazin, eds.). Geological Society, London, Special Publications 217.

Kuhn, O. 1967. *Die Fossile Wirbeltierklasse* Pterosauria. Verlag Oeben, Krailling bei München.

Longrich, N. R., D. M. Martill, and B. Andres. 2018. Late Maastrichtian pterosaurs from North Africa and mass extinction of *Pterosauria* at the Cretaceous-Paleogene boundary. *PLOS Biol.* 16:e2001663.

Owen, R. 1842. Report on British fossil reptiles, Part II. *Rep. Br. Ass. Advmt. Sci.* 60–204.

Padian, K. 1984. The origins of pterosaurs. Pp. 163–168 in *Third Symposium on Mesozoic Terrestrial Ecosystems* (W.-E. Reif and F. Westphal, eds.). Attempto Verlag, Tübingen.

Romer, A. S. 1956. *Osteology of the Reptiles.* University of Chicago Press, Chicago, IL.

Seeley, H. G. 1864. On the pterodactyle as evidence of a new subclass of *Vertebrata* (*Saurornia*). *Rep. Br. Ass. Advmt. Sci.* 34:69.

Seeley, H. G. 1869. *Index to the Fossil Remains of Aves,* Ornithosauria *and* Reptilia *in the Woodwardian Museum Cambridge.* Deighton, Bell, and Co., Cambridge, UK.

Sereno, P. C. 1991. Basal archosaurs: phylogenetic relationships and functional implications. *Soc. Vert. Paleontol. Memoir 2, J. Vertebr. Paleontol.* 11(suppl. #4):1–53.

von Sömmerring, S. T. 1812. Über einen *Ornithocephalus. Denkschr. Kgl. Bayer. Akad. Wiss., Math.-Phys. Cl.* 3:89–158.

Taquet, P., and K. Padian. 2004. The earliest known restoration of a pterosaur and the philosophical origins of Cuvier's *Ossemens Fossiles. C. R. Palevol.* 3:157–175.

Unwin, D. M. 2003. On the phylogeny and evolutionary history of pterosaurs. Pp. 138–190 in *Evolution and Palaeobiology of Pterosaurs* (E. Buffetaut and J-M. Mazin, eds.). Geological Society, London, Special Publications 217.

von Waldheim, G. F. 1834. *Bibliotheca Palaeontologica* Animalium Systematica. Typis Universitatis Caesareae, Moscow.

Wellnhofer, P. 1975. Die *Rhamphorhynchoidea* (*Pterosauria*) der Oberjura-Plattenkalke Süddeutschlands. III: Palökologie und Stammesgeschichte. *Palaeontogr. Abt. A Palaeozool.-Stratigr.* 149:1–30.

Wellnhofer, P. 1978. Pterosauria. *Handbuch der Palaeoherpetologie,* Teil 19. Gustav Fischer Verlag, Stuttgart.

Young, C. C. 1964. On a new pterosaurian from Sinkiang, China. *Vertebr. PalAs.* 8:221–256.

Authors

Brian Andres; Department of Animal & Plant Sciences; University of Sheffield; Sheffield, S10 2TN, UK. Email: brian.andres@aya.yale.edu

Kevin Padian; Department of Integrative Biology and Museum of Paleontology; University of California, Berkeley, CA 94720-4780, USA. Email: kpadian@berkeley.edu

Date Accepted: January 27, 2016

Primary Editors: Jacques A. Gauthier, Kevin de Queiroz

Pterodactyloidea F. Plieninger 1901 [B. Andres and K. Padian], converted clade name

Registration Number: 158

Definition: The clade characterized by the apomorphy elongate wing metacarpal that is at least 80 percent as long as the humerus (at adult stage), as inherited by *Pterodactylus* (originally *Ornithocephalus*) *antiquus* (Sömmerring 1812). This is an apomorphy-based definition. Abbreviated definition: ∇ apo fourth metacarpal at least 80% as long as humerus [*Pterodactylus antiquus* (Sömmerring 1812)].

Etymology: Derived from the Greek *ptero-* (wing), *daktylos* (finger), and *-eides* (like).

Reference Phylogeny: Figure 3 in Andres et al. (2014: Fig. 3), where the elongate wing metacarpal that is at least 80 percent as long as the humerus is inferred to have arisen along the branch subtending the clade labeled "*Pterodactyloidea*".

Composition: *Pterodactyloidea* contains about 150 extinct species of pterosaurs (B. Andres, pers. obs.), most of which are currently referred to the subclades *Archaeopterodactyloidea* and *Ornithocheiroidea* (Andres et al., 2014).

Diagnostic Apomorphies: In addition to the fourth metacarpal at least 80% as long as humerus, apomorphies relative to other pterosaurs include elongate skull more than four times longer than the height at the jaw articulation (after Romer, 1956); basipterygoids united to form an elongate median bar of bone (Unwin, 2003); occipital condyles oriented posteroventrally (instead of posteriorly); deltopectoral crest of humerus proximally placed, does not recurve proximally, and end of crest blunt and rounded rather than expanded (after Kellner, 2003); fifth pedal digit reduced to a small metatarsal with no more than a single phalanx (after Romer, 1956).

Synonyms: *Ornithocheirae* Seeley 1870, *Ornithocheiroidea* Seeley 1891, *Dracochira* Haeckel 1895, *Pterodactyloidei* Khozatskii et al. 1964, and *Ornithocheiria* Kuhn 1967 are all approximate synonyms.

Comments: Howse (1986: Figs. 11, 12), Bennett (1989: Fig. 3, 1994: Fig. 8), Kellner (2003: Fig. 1), and Unwin (2003: Fig. 6) provided early, albeit disparate, published phylogenetic analyses of *Pterodactyloidea*. Andres and Ji (2008) and Lü et al. (2008) revised and increased the scope of these previous works. Andres et al. (2014) was the largest, most comprehensive, most detailed, and most recent phylogenetic analysis and was therefore selected as the reference phylogeny. This analysis was updated by Longrich et al. (2018) but presented in less detail.

Early-discovered pterosaurs were generally placed in the genus *Pterodactylus* or one of its synonyms. It was not until about 60 years after the first description of a pterosaur that *Pterosauria* was given a second genus, *Rhamphorhynchus*, based on its long tail, short wing metacarpal, and long fifth toe. As new pterosaurs were described, they were aligned with either *Rhamphorhynchus* or *Pterodactylus* almost dichotomously, depending on alternative states of these features (viz., long vs. short tail, shorter vs. longer wing metacarpal, long vs. reduced fifth toe). Haeckel (1895) named these groupings (ranked as traditional orders) *Draconura* and *Dracochira*, respectively, based on a number of features (e.g., his Figure 342), but he used only the length of the metacarpus in his Figure 327. However, these names have not been used subsequently. Plieninger (1901) renamed those groups as the suborders "*Rhamphorhynchoidea*" and *Pterodactyloidea*. Other approximate synonyms have long been abandoned. Prior to the use of phylogenetic methods, there were few attempts to develop evolutionary trees of pterosaurs beyond this traditional "subordinal" division (see Young, 1964; Kuhn 1967; Wellnhofer, 1978). Traditional "family" and "sub-family" names were erected at one time or another for nearly every genus, more to emphasize their distinctiveness from each other than to unite them evolutionarily (Bennett, 1996). "*Rhamphorhynchoidea*" was traditionally considered more "primitive" than *Pterodactyloidea*, because its members were exclusively Jurassic and Late Triassic, whereas pterodactyloids encompassed Late Jurassic and Cretaceous forms and often showed derived features such as loss of teeth, notaria, large cranial crests, and great size. With the advent of the explicit methods of phylogenetic analysis, "*Rhamphorhynchoidea*" was discovered to be paraphyletic with respect to *Pterodactyloidea* (Padian, 1984). A better term for this paraphyletic grouping is "non-pterodactyloid pterosaurs".

Pterodactyloidea is the best-known and best-supported clade within *Pterosauria*. It contains most of the known pterosaur species, and its monophyly has never been seriously questioned. Although phylogenetic analyses agree on its composition, they disagree on the relationships within *Pterodactyloidea*, especially regarding the basal branches. Kellner (2003) and Unwin (2003) put forward such disparate topologies and node-based definitions for *Pterodactyloidea* that, were the definition of one to be applied to the other's topology, they would exclude taxa that both consider pterodactyloids. In the original circumscription of *Pterodactyloidea*,

Plieninger (1901) listed two diagnostic characters, a short tail and elongate wing metacarpal, and included four species in two families within the group. Subsequently, tradition has looked to the diagnostic characters to determine whether a newly discovered taxon was to be placed within or outside *Pterodactyloidea*. When a pterosaur (*Anurognathus ammoni* Döderlein 1923) was described that had both a short tail (derived) and short metacarpus (ancestral), it was excluded from *Pterodactyloidea* based on the short metacarpus, as well as a long fifth pedal digit with two phalanges (also ancestral). In this manner, historical usage has placed an emphasis on the length of the metacarpus for determining inclusion in *Pterodactyloidea*.

For these reasons, the name *Pterodactyloidea* is established here as an apomorphy-based name founded on the elongate wing metacarpal (homologous/synapomorphic with that in *Pterodactylus antiquus*) using the character description from Unwin (2003). This apomorphy was used in the original circumscription of *Pterodactyloidea* and by later authors to refer species to this group. So defined, this name will apply to all taxa currently considered pterodactyloids regardless conflicting ideas about relationships within the clade. Furthermore, the wing metacarpal is a robust bone that has a higher preservation potential than many other elements of the pterosaur skeleton and is likely to be preserved in fragmentary specimens.

Literature Cited

Andres, B., and Q. Ji. 2008. A new pterosaur from the Liaoning Province of China, the phylogeny of the *Pterodactyloidea*, and convergence in their cervical vertebrae. *Palaeontology* 51:453–470.

Andres, B., J. M. Clark, and X. Xu. 2014. The earliest pterodactyloid and the origin of the group. *Curr. Biol.* 24:1011–1016.

Bennett, S. C. 1989. A pteranodontid pterosaur from the early Cretaceous of Peru, with comments on the relationships of Cretaceous pterosaurs. *J. Paleontol.* 63(5):669–677.

Bennett, S. C. 1994. Taxonomy and systematics of the Late Cretaceous pterosaur *Pteranodon* (*Pterosauria, Pterodactyloidea*). *Occas. Pap. Mus. Nat. Hist.* (Lawrence, KS) 169:1–70.

Bennett, S. C. 1996. The phylogenetic position of the pterosaurs within the *Archosauromorpha*. *Zool. J. Linn. Soc.* 118:261–308.

Döderlein, L. 1923. *Anurognathus Ammoni*, ein neuer Flugsaurier. *Sber. Bayer. Akad. Wiss. Math.-Phys. Kl.* 117–164.

Haeckel, E. 1895. *Systematische Phylogenie der Wirbelthiere* (Vertebrata). Verlag von Georg Reimer, Berlin.

Howse, S. C. B. 1986. On the cervical vertebrae of the *Pterodactyloidea* (*Reptilia: Archosauria*). *Zool. J. Linn. Soc.* 88:307–328.

Kellner, A. W. A. 2003. Pterosaur phylogeny and comments on the evolutionary history of the group. Pp. 105–137 in *Evolution and Palaeobiology of Pterosaurs* (E. Buffetaut and J.-M. Mazin, eds.). Geological Society, London, Special Publications, 217.

Kuhn, O. 1967. *Die fossile Wirbeltierklasse* Pterosauria. Verlag Oeben, Denkschr. Kgl. Bayer. Akad. Wiss., Math.-Phys. Cl.

Longrich, N. R., D. M. Martill, and B. Andres. 2018. Late Maastrichtian pterosaurs from North Africa and mass extinction of Pterosauria at the Cretaceous-Paleogene boundary. *PLOS Biol.* 16:e2001663.

Lü, J. D., M. Unwin, L. Xu, and X. Zhang. 2008. A new azhdarchoid pterosaur from the Lower Cretaceous of China and its implications for pterosaur phylogeny and evolution. *Naturwissenschaften* 95(9):891–897.

Padian, K. 1984. The origins of pterosaurs. Pp. 163–168 in *Third Symposium on Mesozoic Terrestrial Ecosystems* (W.-E. Reif and F. Westphal, eds.). Attempto Verlag, Tübingen.

Plieninger, F. 1901. Beiträge zur Kenntnis der Flugsaurier. *Palaeontographica* 48:65–90.

Romer, A. S. 1956. *Osteology of the Reptiles*. University of Chicago Press, Chicago, IL.

Seeley, H. G. 1870. *The* Ornithosauria: *An Elementary Study of the Bones of Pterodactyles*. Cambridge, UK.

Seeley, H. G. 1891. On the shoulder girdle in Cretaceous *Ornithosauria*. *Ann. Mag. Nat. Hist.* 6(7):438–445.

von Sömmerring, S. T. 1812. Über einen *Ornithocephalus*. *Denkschr. Kgl. Bayer. Akad. Wiss., Math.-Phys. Cl.* 3:89–158.

Unwin, D. M. 2003. On the phylogeny and evolutionary history of pterosaurs. Pp. 138–190 in *Evolution and Palaeobiology of Pterosaurs* (E. Buffetaut and J-M. Mazin, eds.). Geological Society, London, Special Publications, 217.

Wellnhofer, P. 1978. Pterosauria. *Handbuch der Paläoherpetologie*, Teil 19. Gustav Fischer Verlag, Stuttgart.

Young, C. C. 1964. On a new pterosaurian from Sinkiang, China. *Vertebr. PalAs.* 8:221–256.

Authors

Brian Andres; Department of Animal & Plant Sciences; University of Sheffield; Sheffield, S10 2TN, UK. Email: brian.andres@aya.yale.edu.

Kevin Padian; Department of Integrative Biology and Museum of Paleontology; University of California, Berkeley, CA 94720-4780, USA. Email: kpadian@berkeley.edu.

Date Accepted: January 27, 2016

Primary Editors: Jacques A. Gauthier, Kevin de Queiroz

Dinosauria R. Owen 1842 [M. C. Langer, F. E. Novas, J. S. Bittencourt, M. D. Ezcurra, and J. A. Gauthier], converted clade name

Registration Number: 194

Definition: The smallest clade containing *Iguanodon bernissartensis* Boulenger in Beneden 1881 (*Ornithischia/Euornithopoda*) *Megalosaurus bucklandii* Mantell 1827 (*Theropoda/Megalosauroidea*) and *Cetiosaurus oxoniensis* Phillips 1871 (*Sauropodomorpha*). This is a minimum-clade definition. Abbreviated definition: min ∇ (*Iguanodon bernissartensis* Boulenger in Beneden 1881 & *Megalosaurus bucklandii* Mantell 1827 & *Cetiosaurus oxoniensis* Phillips 1871).

Etymology: Derived from the ancient Greek: δεινός = fearful, terrible; σαύρος = lizard-like (Liddell and Scott, 1882); seemingly in reference to the large size of the fossils originally assigned to this group. Indeed, Owen (1842) translated δεινός as "fearfully great," conceivably employing the term as "the quality of objects which, from their vastness, magnitude, etc., inspire fear, awe, reverence, power, etc." (Pickering, 1873; in Creisler, 1996).

Reference Phylogeny: There are a plethora of phylogenies depicting relationships within and among various parts of the host of extinct clades that diverged from along the avian stem during the Mesozoic. More comprehensive studies (e.g., Sereno, 1999) are rare. No one has yet attempted to simultaneously analyze a broad range of species from across the avian total clade, which includes more than 11,000 species ranging from mid-Triassic to Recent (see *Aves*, this volume). Because it contains the largest sample of non-avian dinosaurs, the phylogeny derived from the supertree analysis of Lloyd et al. (2008; supplementary Fig. 2) was selected as the primary reference phylogeny. Note that the definition uses clade addresses (e.g., *Theropoda: Megalosauroidea*) from Lloyd et al. (2008) to facilitate finding specifiers (e.g., *Megalosaurus bucklandii*) on this densely branched supertree. Only species, not more inclusive taxa, can be specifiers in this kind of definition. We accordingly do not regard these additional taxon names as parts of the formal definition, nor do we mean to endorse their use for these clades.

Composition: According to most phylogenetic hypotheses, *Dinosauria* is composed of two primary subclades—*Ornithischia* and *Saurischia* (this volume)—with the latter composed of *Sauropodomorpha* (this volume) and *Theropoda* (including *Aves*, this volume) (e.g., Gauthier et al., 1989; Sereno, 1999; Ezcurra, 2006; Langer and Benton, 2006; Irmis et al., 2007; Lloyd et al., 2008; Apaldetti et al., 2011; Martínez et al., 2011; Sereno et al., 2013; Niedźwiedzki et al., 2014; Pretto et al., 2015; Cabreira et al., 2016). There is some contention over relations of a few very early species—e.g., some have proposed that herrerasaurs (e.g., Gauthier et al., 1989; Novas, 1992; Fraser et al., 2002) and silesaurids (e.g., Langer and Ferigolo, 2013) are, respectively, outside and within *Dinosauria* as defined here. The composition and relations of those primary dinosaurian subclades have otherwise been remarkably stable, and alternative topologies proposed among basalmost branches (e.g., Cooper, 1985; Baron et al., 2017; but see Langer et al., 2017) do not alter the circumscription of *Dinosauria* as that taxon name is defined here. Starting with the compilation of Weishampel et al. (2004), Benton (2008) identified 726 valid dinosaur species of Mesozoic age. This count includes early winged dinosaurs such as *Archaeopteryx lithographica*, with early avialans comprising about 10% of Mesozoic dinosaur species diversity. The International Ornithological Union recognizes 10,672 extant species of *Aves* (Gill and Donsker, 2017); hence, there are accordingly at least 11,398 species currently assigned to *Dinosauria* as that taxon name is defined here.

Starrfelt and Liow (2016) used a Poisson sampling model to more accurately estimate species richness among Mesozoic dinosaurs, more than doubling the estimated diversity to around 2,000 species. Considering the episodic nature of deposition on land, the dearth of dinosaur-bearing beds of Mesozoic age around the world, and the uneven geographic distribution of palaeontologists prospecting for them, this is still likely to be an underestimate. Extant dinosaurs (*Aves*) may be far more diverse in terms of species, doubtless reflecting their relatively small body size, but large-bodied stem avians (traditional "dinosaurs") are far more disparate morphologically. Among the astonishing variety of dinosaurs alive at the end of the Cretaceous, only the very earliest branches of crown *Aves* survived the Chicxulub impact at 66 Ma (Longrich et al., 2011; see *Aves* this volume).

Diagnostic Apomorphies: With respect to other "reptiles", Owen (1842) characterized his three species of *Dinosauria*—now recognized as two ornithischians and a theropod saurischian—by their large size and unusual combination of osteological traits, including a distinctive sacral construction and an upright limb posture resembling those of "bulky terrestrial mammals." Owen's conception was developed in the pre-Darwinian era and was burdened further by there being so few extinct dinosaurs known in the mid-nineteenth century. Nevertheless,

pelvic and hind-limb modifications thought to reflect acquisition of an upright striding bipedal stance and gait (e.g., Gauthier et al., 2011) figured prominently in the diagnosis of *Dinosauria* from Owen's time to the dawn of the Hennigian (phylogenetic) era (e.g., Bakker and Galton, 1974; Gauthier, 1986). This collection of apomorphies remains diagnostic of *Dinosauria* relative to the last ancestor it shared with *Alligator mississippiensis* (*Archosauria,* this volume). Nevertheless, these distinctly bird-like "dinosaurian" synapomorphies likely originated before the divergence amongst *Ornithischia,* *Theropoda,* and *Sauropodomorpha* (= *Dinosauria* as defined here). A growing understanding of species diversity and anatomical disparity of Triassic dinosaurs and their close kin reveals that many of these traditional "dinosaurian" apomorphies did indeed evolve earlier on the avian stem (e.g., Novas, 1996; Sereno, 1999; Langer and Benton, 2006; Ezcurra, 2006; Irmis et al., 2007; Nesbitt, 2011; Cabreira et al., 2016; Nesbitt et al., 2017).

Several authors have reviewed the synapomorphies of *Dinosauria* (e.g., Brusatte et al., 2010; Langer et al., 2010; Nesbitt et al., 2010; Nesbitt, 2011; Nesbitt et al., 2012; Cabreira et al., 2016; Baron et al., 2017). They indicate that the precise diagnosis of the clade turns on whether it contains silesaurids (Dzik, 2003; Langer and Ferigolo, 2013; Niedźwiedzki et al., 2014; Cabreira et al., 2016) or not. In addition to some uncertainty from character conflict, incomplete preservation in Triassic taxa (e.g., *Nyasasaurus parringtoni*; Nesbitt et al., 2012) remains a serious problem in that important characters of the skull and manus are very poorly known among non-dinosaur dinosauromorphs (Langer et al., 2013).

A list of traditional dinosaur synapomorphies based on reviews noted above is presented below. It is mainly composed of uncontroversial dinosaur apomorphies, but it also includes some apomorphies with uncertain distributions among a few non-dinosaur dinosauromorphs from the Triassic (indicated by *). Others are not present in all early dinosaurs, but might be synapomorphic for the clade depending on assumptions about the frequency of independent origin vs. secondary loss in evolution (indicated by **). Finally, some are also known in *Nyasaurus parringtoni* (indicated by †) and in some silesaurids (indicated by ‡), taxa that are either sister to, or just inside of, *Dinosauria*. They nevertheless suffice to distinguish dinosaurs from more distant outgroups near the base of the avian stem such as *Lagosuchus talampayensis (= Marasuchus lilloensis)* and *Lagerpeton chanarensis* (authorship credited to first explicit identification as an apomorphy): (1) postfrontal absent* (Gauthier, 1986); (2) jugal with bifurcated quadratojugal process* (Sereno and Novas, 1992); (3) supratemporal fossa extends onto frontal rostral to supratemporal fenestra (Gauthier, 1986); (4) post-temporal opening much smaller than foramen magnum (Sereno and Novas, 1993); (5) exoccipitals fail to meet on the floor of endocranial cavity* (Nesbitt, 2011); (6) epipophyses present on cranial cervical vertebrae ‡ (Gauthier, 1986); (7) forelimb (humerus+radius) to hindlimb (femur+tibia) ratio less than 0.55** (Gauthier, 1986); (8) deltopectoral crest extends for more than 30% the humeral length† ‡ (Bakker and Galton, 1974); (9) humerus with proximal articulation separated by gap from deltopectoral crest‡ (Nesbitt, 2011); (10) radius shorter than 80% of humerus** (Irmis et al., 2007); (11) reduced postaxial digits of hand* (Gauthier and Padian, 1985); (12) more than two sacral vertebrae** † ‡ (Galton, 1976); (13) concave ventral margin of the acetabulum in the ilium** (Bakker and Galton, 1974); (14) ilium with well-developed brevis fossa** ‡ (Gauthier and Padian, 1985); (15) pubis longer than 70% of femur length‡ (Novas, 1996); (16) ischium articular surfaces with ilium and pubis separated by broad concave surface** (Irmis et al., 2007); (17) ischia with extensive contact between their shafts‡ (Nesbitt, 2011); (18) femoral head distinctly off-set at a sharp angle from the shaft‡ (Bakker and Galton, 1974); (19) femur with dorsolateral trochanter‡ (Gauthier, 1986); (20) fourth trochanter with distal margin forming steeper angle relative to femoral shaft** (Langer and Benton, 2006); (21) tibia cnemial crest arcs craniolaterally‡ (Benton and Clark, 1988); (22) caudal surface of tibia with proximodistally oriented ridge** (Nesbitt, 2011); (23) flange on distal portion of tibia overlapping caudally the ascending process of the astragalus‡ (Novas, 1992); (24) cranial edge of proximal portion of fibula tapers to point and arches medially‡ (Nesbitt, 2011); (25) broad astragalar ascending process‡ (Gauthier, 1986); (26) astragalus with fibular facet occupying less than 1/3 of transverse width‡ (Langer and Benton, 2006); (27) concave articular surface of the fibula in the calcaneum‡ (Novas, 1996); (28) calcaneal tuber absent** (Gauthier, 1986); (29) distal tarsal IV with flat proximal surface (Nesbitt, 2011); (30) metatarsal IV subequal in length to metatarsal II‡ (Gauthier, 1986).

Synonyms: *Pachypodes* Meyer 1845, *Harpagosauria* Haeckel 1866, and *Ornithoscelida* Huxley 1870 are approximate (not defined phylogenetically) and partial (did not include birds) synonyms; *Eudinosauria* Novas 1992 is an unambiguous synonym. In the context of the tree of Baron et al. (2017), *Saurischia* Seeley 1888 as traditionally understood is a partial and approximate synonym, and *Eusaurischia* Padian et al. 1999 is an unambiguous synonym.

Comments: After Owen (1842) coined the name *Dinosauria*, a few more taxon names entered the literature to designate a similar group of organisms (see Synonyms), but none gained much traction. Note that *Ornithoscelida* Huxley 1870 has recently been resurrected for a proposed clade composed only of *Ornithischia* + *Theropoda* (Baron et al., 2017), and excluding *Sauropodomorpha*, thus rendering *Eusaurischia* a junior synonym of *Dinosauria* and *Saurischia* a junior synonym of *Theropoda* sensu Gauthier (1986); (see *Saurischia* and *Theropoda*, this volume.)) *Dinosauria* was used in most early classification schemes (e.g., Cope, 1866; Huxley, 1870; Marsh, 1882), including that of Seeley (1888) who, although questioning dinosaur monophyly, divided "dinosaurs" into ornithischians and saurischians, thereby establishing a nomenclature used in all future works on dinosaur taxonomy. Indeed, most influential works of the twentieth century (Huene, 1956; Colbert, 1964; Romer, 1966) accepted *Dinosauria* as composed of *Saurischia* plus *Ornithischia* even though they were generally considered to have arisen independently from "thecodont" ancestors (e.g., Thulborn, 1975; see *Archosauria*, this volume). Dinosaur monophyly in the Hennigian sense was first proposed by Bakker and Galton (1974), was firmly established by Gauthier (1986; see also Gauthier, 1984), and that hypothesis now represents the consensus view emerging from extensive subsequent study (e.g., Novas, 1996; Sereno, 1999; Langer and Benton, 2006; Ezcurra, 2006; Irmis et al., 2007; Brusatte et al., 2010; Langer et al., 2010; Nesbitt, 2011; Nesbitt et al., 2012; Apaldetti et al., 2011; Martínez et al., 2011; Sereno et al., 2013; Niedźwiedzki et al., 2014; Bittencourt et al., 2014; Pretto et al., 2015; Cabreira et al., 2016; Baron et al., 2017).

Following the taxonomic orthodoxy of the time (Gauthier, 1984, 1986; Brinkman and Sues, 1987; Novas, 1989; Benton, 1990), Novas (1992) proposed the first phylogenetic definition of *Dinosauria* to include saurischians, ornithischians, and herrerasaurs (but see Gauthier et al., 1989), but not near-dinosaur "thecodonts"

such as *Lagosuchus talampayensis* (Romer, 1971; Bonaparte, 1975; Sereno and Arcucci, 1994; Agnolin & Ezcurra, 2019). Because Novas (1992) inferred herrerasaurs to be outside of the *Ornithischia-Saurischia* dichotomy, he also coined the name *Eudinosauria* for that less inclusive clade. Most subsequent analyses have, however, consistently placed herrerasaurs within the saurischian radiation (Novas, 1996; Sereno, 1999; Ezcurra, 2006, 2010; Langer and Benton, 2006; Irmis et al., 2007; Nesbitt, 2011; Nesbitt et al., 2012; Apaldetti et al., 2011; Martínez et al., 2011; Sereno et al., 2013; Niedźwiedzki et al., 2014; Pretto et al., 2015; Cabreira et al., 2016). In that context, *Eudinosauria* and *Dinosauria* (as defined by Novas, 1992) are currently synonymous phylogenetically. Herrerasaurs have recently been inferred to lie at the base of *Sauropodomorpha*, with *Ornithischia* and *Theropoda* as sisters (Baron et al., 2017) (see *Saurischia*, this volume), but that still places them within *Dinosauria* as that taxon name is defined here.

A surprising number of phylogenetic definitions for the name *Dinosauria* were proposed subsequent to Novas (1992). Although they agreed on the circumscription of the clade, tying it to the *Ornithischia-Saurischia* dichotomy proposed by Seeley (1888), they often differed in their choice of internal specifiers (e.g., Padian and May, 1993; Novas, 1996; Holtz in Padian, 1997; Sereno, 1998, 2005a,b; Olshevsky, 2000; Fraser et al., 2002; Kischlat, 2002; Clarke, 2004). Various workers selected clades rather than species as specifiers, often using avians to represent the saurischian branch of *Dinosauria*.

There are two problems with some of these early efforts in phylogenetic nomenclature: instead of more inclusive taxa, valid specifiers must be species, specimens or apomorphies (Cantino and de Queiroz, 2020; *ICPN* Art. 11.1); and those specifiers should preserve traditional ideas about the taxon's composition (Cantino and de Queiroz,

2020; *ICPN* Rec. 11A). Instead of using any avians as specifiers—an approach introduced by Gauthier (1986; see also Gauthier, 1984) but hardly consistent with the scope of Owen's *Dinosauria* (Cantino and de Queiroz, 2010)—we selected three traditional "dinosaurs" to bracket this clade, including the two best-known species explicitly included by Owen (1842), as suggested previously by Holtz (in Padian, 1997), Olshevsky (2000), and Clarke (2004).

Unfortunately, Owen (1842) used only generic names for the three taxa he included in his *Dinosauria* (*Megalosaurus, Iguanodon* and *Hylaeosaurus*). We accordingly follow Benson et al. (2008) in considering *M. bucklandii* the type of *Megalosaurus* (*contra* Lloyd et al., 2008) and adopting the subsequently designated *I. bernissartensis* as the type of *Iguanodon* (*ICZN*, 2000). Owen (1842) possessed scant knowledge of sauropodomorph dinosaurs, mistaking the remains of the sauropod dinosaur *Cetiosaurus* for a gigantic, marine, crocodilian-like "reptile". But with more specimens he later accepted that *Cetiosaurus* belonged with the dinosaurs (Owen, 1875). Because Owen coined both names, *Dinosauria* and *Cetiosaurus*, and because *C. oxoniensis* Phillips 1871 ranks among the earliest discovered, and is by far the best known, sauropod from England (Upchurch and Martin, 2003), we have added that species as a specifier to the definition to preserve the traditional inclusion of sauropodomorphs in *Dinosauria*. Although no bird species is used as an internal specifier, according to nearly every phylogeny published since Gauthier (1986; see also Gauthier, 1984), *Aves* is part of *Dinosauria* as defined here (but see James and Pourtless, 2009). Archetypical dinosaurs such as the horned ornithischian *Triceratops horridus,* no less than the winged theropod *Archaeopteryx lithographica,* can accordingly be regarded as stem birds, even though the former species is far more distantly related to *Aves* than is the latter.

Literature Cited

Agnolin, F., and M. D. Ezcurra. 2019. The validity of *Lagosuchus talampayensis* Romer, 1971 (*Archosauria, Dinosauriformes*), from the Late Triassic of Argentina. *Breviora* 565 (1):1–21.

Apaldetti, C., R. N. Martinez, O. A. Alcober, and D. Pol. 2011. A new basal sauropodomorph (*Dinosauria: Saurischia*) from Quebrada del Barro formation (Marayes-El Carrizal Basin), northwestern Argentina. *PLOS ONE* 6(11):1–19.

Bakker, R. T., and P. M. Galton. 1974. Dinosaur monophyly and a new class of vertebrates. *Nature* 248:168–172.

Baron, M. G., D. B. Norman, and P. M. Barrett. 2017. A new hypothesis of dinosaur relationships and early dinosaur evolution. *Nature* 543:501–506.

Beneden, P. 1881. Sur l'arc pelvien chez les dinosauriens de Bernissart. *Bull. Acad. Roy. Belg., 3 sér.* 1(5):600–608.

Benson, R. B. J., P. M. Barrett, P. Powell, and D. B. Norman. 2008. The taxonomic status of *Megalosaurus bucklandii* (*Dinosauria, Theropoda*) from the Middle Jurassic of Oxfordshire, UK. *Palaeontology* 51:419–424.

Benton, M. J. 1990. Origin and interrelationships of dinosaurs. Pp. 11–30 in *The* Dinosauria (D. B. Weishampel, P. Dodson and H. Osmólska, eds.). University of California Press, Berkeley, CA.

Benton, M. J. 2008. How to find a dinosaur, and the role of synonymy in biodiversity studies. *Paleobiology* 34:516–533.

Benton, M. J., and J. Clark. 1988. Archosaur phylogeny and the relationships of the *Crocodylia*. Pp. 289–332 in *The Phylogeny and Classification of Tetrapods*, Vol. 1, Amphibians, Reptiles, Birds (M. J. Benton, ed.). Clarendon Press, Oxford.

Bever, G. S., J. A. Gauthier, and G. P. Wagner. 2011. Finding the frame shift: digit loss, developmental variability, and the origin of the avian hand. *Evol. Dev.* 13:269–279.

Bittencourt, J. S., A. B. Arcucci, C. A. Marsicano, and M. C. Langer. 2014. Osteology of the Middle Triassic archosaur *Lewisuchus admixtus* Romer (Chañares Formation, Argentina), its inclusivity, and relationships amongst early dinosauromorphs. *J. Syst. Palaeontol.* 13:189–219.

Bonaparte, J. F. 1975. Nuevos materiales de *Lagosuchus talampayensis* Romer (*Thecodontia – Pseudosuchia*) y su significado en el origen de los *Saurischia*, Chañarense Inferior, Triásico Medio de Argentina. *Acta Geol. Lillo.* 13:5–90.

Boulenger, G. A. 1881. *Iguanodon bernissartensis.* P. 606 in P.-J. Beneden, Sur l'arc pelvien chex les dinosauiens de Bernissart. *Bull. Acad. Roy. Belg., 3 Sér.* 1(5):600–608.

Brinkman, D. B., and H.-D. Sues. 1987. A staurikosaurid dinosaur from the Upper Triassic Ischigualasto Formation of Argentina and the relationships of the *Staurikosauridae. Palaeontology* 30:493–503.

Brusatte, S., S. J. Nesbitt, R. B. Irmis, R. Butler, M. J. Benton, and M. A. Norell. 2010. The origin and early radiation of dinosaurs. *Earth Sci. Rev.* 101:68–100.

Cabreira, S. F., A. W. A. Kellner, S. Dias-da-Silva, L. R. da Silva, M. Bronzati, J. C. de A. Marsola, R. T. Müller, J. de S. Bittencourt, B. J. Batista, T. Raugust, R. Carrilho, A. Brodt, and M. C. Langer. 2016. A unique Late Triassic dinosauromorph assemblage reveals dinosaur ancestral anatomy and diet. *Curr. Biol.* 26:3090–3095.

Cantino, P. D., and K. de Queiroz. 2010. International code of phylogenetic nomenclature, Version 4c. Available at http://www.ohio.edu/phylocode/.

Cantino, P. D., and K. de Queiroz. 2020. *International Code of Phylogenetic Nomenclature (PhyloCode)*, Version 6. CRC Press, Boca Raton, FL.

Clarke, J. A. 2004. Morphology, phylogenetic taxonomy, and systematics of *Ichthyornis* and *Apatornis* (*Avialae: Ornithurae*). *Bull. Am. Mus. Nat. Hist.* 286:1–179.

Colbert, E. H. 1964. Relationships of the saurischian dinosaurs. *Am. Mus. Novit.* 2181:1–24.

Cooper, M. R. 1985. A revision of the ornithischian dinosaur *Kangnasaurus coetzeei* Haughton, with a classification of the *Ornithischia. Ann. S. Afr. Mus.* 95:281–317.

Cope, E. D. 1866. Remarks on dinosaur remains from New Jersey. *Proc. Acad. Nat. Sci. Phila.* 1866:275–279.

Creisler, B. 1996. *Dinosauria* translation and pronunciation guide. Available at http://www.dinosauria.com/dml/dmlf.htm, accessed on 16 January 2009.

Dzik, J. 2003. A beaked herbivorous archosaur with dinosaur affinities from the early Late Triassic of Poland. *J. Vertebr. Paleontol.* 23:556–574.

Ezcurra, M. D. 2006. A review of the systematic position of the dinosauriform archosaur *Eucoelophysis baldwini* from the Upper Triassic of New Mexico, USA. *Geodiversitas* 28:649–684.

Ezcurra, M. D. 2010. A new early dinosaur (*Saurischia: Sauropodomorpha*) from the Late Triassic of Argentina: a reassessment of dinosaur origin and phylogeny. *J. Syst. Palaeontol.* 8:371–425.

Fraser, N. C., K. Padian, G. M. Walkden, and A. L. M. Davis. 2002. Basal dinosauriform remains from Britain and the diagnosis of the *Dinosauria. Palaeontology* 45:78–95.

Galton, P. M. 1976. Prosauropod dinosaurs (*Reptilia: Saurischia*) of North America. *Postilla* 169:1–98.

Gauthier, J. A., 1984. *A Cladistic Analysis of the Higher Systematic Categories of the* Diapsida. University Microfilms International, Ann Arbor, MI, #85-12825, vii + 564 pp.

Gauthier, J. A. 1986. Saurischian monophyly and the origin of birds. Pp. 1–55 in *The Origin of Birds and the Evolution of Flight* (K. Padian, ed.). California Academy of Sciences, San Francisco, CA.

Gauthier, J. A., D. Cannatella, K. de Queiroz, A. G. Kluge, and T. Rowe. 1989. Tetrapod phylogeny. Pp. 337–353 in *The Hierarchy of Life: Molecules and Morphology in Phylogenetic Analysis* (B. Fernholm, K. Bremer, and H. Jornvall, eds.). Elsevier, Amsterdam.

Gauthier, J. A., and K. de Queiroz, 2001. Feathered dinosaurs, flying dinosaurs, crown dinosaurs, and the name "*Aves*." Pp. 7–41 in *New Perspectives on the Origin and Early Evolution of Birds: Proceeding of the International Symposium in Honor of John H. Ostrom* (J. A. Gauthier and L. F. Gall, eds.). Spec. Pub. Peabody Mus. Nat. Hist., New Haven, CT.

Gauthier, J. A., and K. Padian. 1985. Phylogenetic, functional, and aerodynamic analyses of the origin of birds and their flight. Pp. 185–197 in *The Beginnings of Birds: Proceedings of the International* Archaeopteryx *Conference,*

Eichstätt, 1984 (M. K. Hecht, J. H. Ostrom, G. Viohl and P. Wellnhofer, eds.). Freunde des Jura-Museum, Eichstätt.

Gauthier, J. A., S. J. Nesbitt, E. R. Schachner, G. S. Bever, and W. G. Joyce. 2011. The bipedal stem crocodilian *Poposaurus gracilis*: inferring function in fossils and innovation in archosaur locomotion. *Bull. Peabody Mus. Nat. Hist.* 52(1):107–126.

Gill, F., and D. Donsker, eds. 2017. *IOC World Bird List* (v 7.1). doi:10.14344/IOC.ML.7.1.

Haeckel, E. 1866. *Generelle Morphologie der Organismen, Allgemeine Grundzüge der Organischen Formen-Wissenschaft, Mechanisch Begründet Durch die von Charles Darwin Reformirte Descendez-Theorie.* Verlag von Georg Reimer, Berlin.

von Huene, F.. 1956. *Paläontologie und Phylogenie der Niederen Tetrapoden.* Gustav Fischer Verlag, Jena.

Huxley, T. H. 1870. On the classification of the *Dinosauria* with observations on the *Dinosauria* of the Trias. *Q. J. Geol. Soc. Lond.* 26:32–51.

ICZN. 2000. Opinion 1947. *Iguanodon* Mantel, 1825 (*Reptilia, Ornithischia*): *Iguanodon bernissartensis* Boulenger in Beneden, 1881 designated as the type species and a lectotype designated. *Bull. Zool. Nomencl.* 57:61–62.

Irmis, R. B., S. J., Nesbitt, K. Padian, N. D. Smith, A. H. Turner, D. Woody, and A. Downs. 2007. A Late Triassic dinosauromorph assemblage from New Mexico and the rise of dinosaurs. *Science* 317:358–361.

James, F. C., and J. A. Pourtless. 2009. Cladistics and the origin of birds: a review and two new analyses. *Ornithol. Monogr.* 66:1–78.

Kischlat, E.-E. 2002. Tecodôncios: a aurora dos arcossáurios no Triássico. Pp. 273–316 in *Paleontologia do Rio Grande do Sul* (M. Holz and L. F. de Ros, eds.). 2nd edition. CIGO/UFRGS, Porto Alegre.

Langer, M. C., and M. J. Benton. 2006. Early dinosaurs: a phylogenetic study. *J. Syst. Palaeontol.* 4:309–358.

Langer, M. C., M. D. Ezcurra, J. S. Bittencourt, and F. E. Novas. 2010 The origin and early evolution of dinosaurs. *Biol. Rev.* 85:55–110.

Langer, M. C., M. D. Ezcurra, O. W. M. Rauhut, M. J. Benton, F. Knoll, B. W. McPhee, F. E. Novas, D. Pol, and S. L. Brusatte. 2017. Untangling the dinosaur family tree. *Nature* 551:E1–E3.

Langer, M. C., and J. Ferigolo. 2013. The Late Triassic dinosauromorph *Sacisaurus agudoensis* (Caturrita Formation; Rio Grande do Sul, Brazil): anatomy and affinities. *Geol. Soc. Lond. Spec. Publ.* 379:353–392.

Langer, M. C., S. Nesbitt, J. S. Bittencourt, and R. Irmis. 2013. Non-dinosaurian Dinosauromorpha. *Geol. Soc. Lond. Spec. Publ.* 379:157–186.

Liddell, H. G., and R. Scott. 1882. *Greek-English Lexicon.* 7th edition. Harper & Brothers, New York.

Lloyd, G. T., K. E. Davis, D. Pisani, J. E. Tarver, M. Ruta, M. Sakamoto, D. W. E. Hone, R. Jennings, and M. J. Benton. 2008. Dinosaurs and the Cretaceous Terrestrial Revolution. *Proc. R. Soc. Lond. B Biol. Sci.* 275:2483–2490.

Longrich, N. R., T. Tokaryk, and D. J. Field. 2011. Mass extinction of birds at the Cretaceous-Paleogene (K-Pg) boundary. *Proc. Natl. Acad. Sci. USA* 108:15253–15257.

Mantell, G. A. 1827. *Illustrations of the Geology of Sussex: A General View of the Geological Relations of the South-Eastern Part of England, with Figures and Descriptions of the Fossils of Tilgate Forest.* Lupton Relfe, London.

Marsh, O. C. 1882. Classification of the *Dinosauria. Am. J. Sci.* 23:81–86.

Martinez, R. N., P. C. Sereno, O. A. Alcober, C. E. Colombi, P. R. Renne, I. P. Montañez, and B. S. Currie. 2011. A basal dinosaur from the dawn of the dinosaur era in southwestern Pangaea. *Science* 331:206–210.

Meyer, H. v. 1845. System der fossilen Saurier. *Neues Jahrb. Mineral. Geogn. Geol. Petrefakten-Kunde* 1845:278–285.

Mongiardino Koch, N., and J. A. Gauthier. 2018 Noise and biases in genomic data may underlie radically different hypotheses for the position of *Iguania* within *Squamata. PLOS ONE* 13(8):e0202729.

Nesbitt, S. J., C. A. Sidor, R. B. Irmis, K. D. Angielczyk, R. M. H. Smith, and L. A. Tsuji. 2010. Ecologically distinct dinosaurian sister group shows early diversification of Ornithodira. *Nature* 464(7285):95–98.

Nesbitt, S. J. 2011. The early evolution of *Archosauria*: relationships and the origin of major clades. *Bull. Am. Mus. Nat. Hist.* 352:1–292.

Nesbitt, S. J., P. M. Barrett, S. Werning, C. A. Sidor, and A. J. Charig. 2012. The oldest dinosaur? A Middle Triassic dinosauriform from Tanzania. *Biol. Lett.* 9:e20120949.

Nesbitt, S. J., R. J. Butler, M. D. Ezcurra, P. M. Barrett, M. R. Stocker, K. D. Angielczyk, R. M. H. Smith, C. A. Sidor, G. Niedźwiedzki, A. Sennikov, and A. J. Charig. 2017. The earliest bird-line archosaurs and the assembly of the dinosaur body plan. *Nature* 544:484–487.

Niedźwiedzki, G., S. L. Brusatte, T. Sulej, and R. J. Butler. 2014. Basal dinosauriform and theropod dinosaurs from the mid-late Norian (Late Triassic) of Poland: implications for Triassic dinosaur evolution and distribution. *Palaeontology* 57(6):1121–1142.

Novas, F. E. 1989. The tibia and tarsus in *Herrerasauridae* (*Dinosauria, incertae sedis*) and the origin and evolution of the dinosaurian tarsus. *J. Paleontol.* 63:677–690.

Novas, F. E. 1992. Phylogenetic relationships of the basal dinosaurs, the *Herrerasauridae*. *Palaeontology* 35:51–62.

Novas, F. E. 1996. Dinosaur monophyly. *J. Vertebr. Paleontol.* 16:723–741.

Olshevsky, G. 2000. An annotated checklist of dinosaur species by continent. *Mesozoic Meanderings* 3:1–157.

Owen, R. 1842. Report on British fossil reptiles. Part II. *Rep. Br. Assoc. Adv. Sci.* 11:60–204.

Owen, R. 1875. Monograph of the Mesozoic *Reptilia*, part 2: monograph on the genus *Cetiosaurus*. *Palaeontogr. Soc. Monogr.* 29:27–43.

Padian, K. 1997. *Dinosauria*: definition. Pp. 175–179 in *Encyclopedia of Dinosaurs* (P. J. Currie and K. Padian, eds.). Academic Press, San Diego, CA.

Padian, K., and C. L. May. 1993. The earliest dinosaurs. *New Mexico Mus. Nat. Hist. Sci.* Bull. 3:379–381.

Padian, K., J. R. Hutchinson, and T. R., Holtz, Jr. 1999. Phylogenetic definitions and nomenclature of the major taxonomic categories of the carnivorous *Dinosauria* (*Theropoda*). *J. Vertebr. Paleontol.* 19:69–80.

Phillips, J. 1871. *Geology of Oxford and the Valley of the Thames*. Clarendon Press, Oxford.

Pretto, F. A., C. L. Schultz, and M. C. Langer. 2015. New dinosaur remains from the Late Triassic of southern Brazil (Candelária Sequence, *Hyperodapedon* Assemblage Zone). *Alcheringa* 39:1–10.

Riedl, R. 1978. *Order in Living Organisms: A Systems Analysis of Evolution*. Wiley, New York.

Romer, A. S. 1966. *Vertebrate Paleontology*. University of Chicago Press, Chicago, IL.

Romer, A. S. 1971. The Chañares (Argentina) Triassic reptile fauna. X. Two new but incompletely known long-limbed pseudosuchians. *Breviora* 378:1–10.

Seeley, H. G. 1888. On the classification of the fossil animals commonly named *Dinosauria*. *Proc. R. Soc. Lond. B Biol. Sci.* 43:165–171.

Sereno, P. C. 1998. A rationale for phylogenetic definitions, with application to the higher-level taxonomy of *Dinosauria*. *Neues Jahrb. Geol. Paläontol. Abh.* 210(1):41–83.

Sereno, P. C. 1999. The evolution of dinosaurs. *Science* 284:2137–2147.

Sereno, P. C. 2005a. The logical basis of phylogenetic taxonomy. *Syst. Biol.* 54(4):595–619.

Sereno, P. C. 2005b. TaxonSearch: Database for Suprageneric Taxa and Phylogenetic Definitions. Available at http://www.taxonsearch.org/, accessed on 16 January 2009.

Sereno, P. C., and Arcucci, A. B. 1994. Dinosaurian precursors from the Middle Triassic of Argentina: *Marasuchus lilloensis* gen. nov. *J. Vertebr. Paleontol.* 14:53–73.

Sereno, P. C., Martinez, R. N., and Alcober, O. A. 2013. Osteology of *Eoraptor lunensis* (*Dinosauria: Sauropodomorpha*). Pp. 83–179 in *Basal Sauropodomorphs and the Vertebrate Fossil Record of the Ischigualasto Formation (Late Triassic: Carnian–Norian) of Argentina* (P. C. Sereno, ed.). Society of Vertebrate Paleontology Memoir 12.

Sereno, P. C., and F. E. Novas. 1992. The complete skull and skeleton of an early dinosaur. *Science* 258:1137–1140.

Sereno, P. C., and F. E. Novas. 1993. The skull and neck of the basal theropod *Herrerasaurus ischigualastensis. J. Vertebr. Paleontol.* 13:451–476.

Starrfelt, J., and L. H Liow. 2016. How many dinosaur species were there? Fossil bias and true richness estimated using a Poisson sampling model. *Philos. Trans. R. Soc. Lond. B Biol. Sci.* 371: 20150219

Stephanic, C. M., and S. J. Nesbitt. 2019. The evolution and role of the hyposphene-hypantrum articulation in Archosauria: phylogeny, size and/ or mechanics? *Roy. Soc. Open Sci.* 6:190258.

Thulborn, R. A. 1975. Dinosaur polyphyly and the classification of archosaurs and birds. *Aust. J. Zool.* 23:249–270.

Upchurch, P., and J. Martin. 2003. The anatomy and taxonomy of *Cetiosaurus* (*Saurischia, Sauropoda*) from the Middle Jurassic of England. *J. Vertebr. Palaeontol.* 23:208–231.

Wagner, G. P. 2014. *Homology, Genes, and Evolutionary Innovation*. Princeton University Press, Princeton, NJ.

Weishampel, D. B., P. Dodson, and H. Osmolska, eds. 2004. *The* Dinosauria. 2nd edition. University of California Press, Berkeley/Los Angeles/London.

Authors

Max Cardoso Langer; Departamento de Biologia, FFCLRP; Universidade de São Paulo; Ribeirão Preto-SP, 14040-901, Brazil. Email: mclanger@ffclrp.usp.br.

Fernando E. Novas; Laboratorio de Anatomía Comparada y Evolución de los Vertebrados; Museo Argentino de Ciencias Naturales "Bernardino Rivadavia"; Buenos Aires, C1405DJR, Argentina. Email: fernovas@yahoo.com.ar.

Jonathas S. Bittencourt; Departamento de Geologia, Instituto de Geociências; Universidade Federal de Minas Gerais; Belo Horizonte-MG, 31270-901, Brazil. Email: bittencourt.paleo@gmail.com.

Martín D. Ezcurra; Sección Paleontología de Vertebrados; Museo Argentino de Ciencias Naturales "Bernardino Rivadavia"; Buenos Aires, C1405DJR, Argentina. Email: martindezcurra@yahoo.com.ar.

Jacques A. Gauthier; Department of Geology and Geophysics; Yale University; New Haven, CT 06520-8109, USA. E-mail: jacques.gauthier@yale.edu.

Date Accepted: 27 July 2018

Primary Editor: Kevin de Queiroz

Saurischia H. G. Seeley 1888 [J. A. Gauthier, M. C. Langer, F. E. Novas, J. Bittencourt, and M. D. Ezcurra], converted clade name

Registration Number: 195

Definition: The largest clade containing *Allosaurus fragilis* Marsh 1877 (*Theropoda/Carnosauria*) and *Camarasaurus supremus* Cope 1877 (*Sauropodomorpha*), but not *Stegosaurus stenops* Marsh 1887 (*Ornithischia /Stegosauridae*). This is a maximum-clade definition. Abbreviated definition: max ∇ (*Allosaurus fragilis* Marsh 1877 & *Camarasaurus supremus* Cope 1877 ~ *Stegosaurus stenops* Marsh 1887).

Etymology: Derived from the Greek: σαύρος = saurian and ισχίον = hip joint (Liddell and Scott, 1882). Seeley (1888) translated ισχίον as "pelvis".

Reference Phylogeny: Because it contains the largest sample of non-avian saurischians, the tree derived from the supertree analysis of Lloyd et al. (2008: supplementary Fig. 2) is selected as the primary reference phylogeny. Note that the definition uses clade addresses (e.g., *Theropoda*: *Carnosauria*) from Lloyd et al. (2008) to facilitate finding specifiers (e.g., *Allosaurus fragilis*) on their densely branched supertree. Taxa more inclusive than species cannot be specifiers; we do not regard these additional taxon names as parts of the formal definition, nor do we mean to endorse their use for these clades.

Composition: According to most phylogenetic hypotheses, *Saurischia* is composed of two primary subclades: *Sauropodomorpha* (this volume) and *Theropoda* (including *Aves*, this volume) (e.g., Gauthier, 1984, 1986; Gauthier et al., 1989; Novas, 1989, 1992; Sereno, 1999; Rauhut, 2003; Ezcurra, 2006; Langer and Benton,

2006; Irmis et al., 2007; Lloyd et al., 2008; Nesbitt et al., 2010; Nesbitt, 2011; Apaldetti et al., 2011; Martínez et al., 2011; Novas et al., 2011; Sereno et al., 2013; Niedźwiedzki et al., 2014; Pretto et al., 2015; Cabreira et al., 2016; Langer et al., 2017). Relations of a few Triassic species, most notably *Eoraptor lunensis* and *Herrerasauridae*, remain unclear even though they are represented by reasonably complete specimens (e.g., contrast Fraser et al., 2002, vs. Martínez and Alcober, 2009, vs. Nesbitt et al., 2010). Somewhat dated lists of included taxa and relationships among them were reviewed in Weishampel et al. (2004). Based on their review of the Paleobiology Database, Starrfelt and Liow (2016) found 762 species of saurischians described from the Mesozoic; as they argued, that number likely underestimates their true diversity (e.g., their estimate was about 1,628 spp.; see *Dinosauria*, this volume). That is certainly the case for avian saurischians of the Cenozoic given their poor preservation potential (see *Aves*, this volume). Nevertheless, with the addition of the 10,672 surviving species of *Aves* (Gill and Donsker, 2017), there are at least 11,434 described species currently assigned to *Saurischia*. In the recently proposed rearrangement of relationships among major dinosaur clades (Baron et al., 2017), *Saurischia* (as redefined by those authors) would include only *Sauropodomorpha* and *Herrerasauridae*, which have no living representatives.

Diagnostic Apomorphies: The earliest species belonging to *Saurischia*, as a maximum clade, may not have apomorphies that arose just as it diverged from its sister clade (see de Queiroz, 2007). In practice, any one of the apomorphies arising along that deepest branch will suffice for taxon assignment (Gauthier and de Queiroz, 2001). In that circumstance, however, support for one relationship rather than another may not be strong. Nevertheless, a clade composed of sauropodomorphs and theropods (including *Aves*) has been a constant feature in dinosaur phylogenetics for the past few decades. Apomorphies shared by saurischians have been reviewed extensively (e.g., Gauthier, 1986; Gauthier et al., 1989; Novas, 1994; Sereno, 1999; Langer, 2004; Ezcurra, 2006; Langer and Benton, 2006; Yates, 2007; Irmis et al., 2007; Nesbitt et al., 2010; Apaldetti et al., 2011; Martínez et al., 2011; Sereno et al., 2013; Pretto et al., 2015; Cabreira et al., 2016).

According to Nesbitt (2011), for example, there may be as many as 25 synapomorphies for *Saurischia*, including the following (authorship credited to first explicit identification as an apomorphy, not the current optimization of that apomorphy): (1) subnarial foramen on border between premaxilla and maxilla (Benton and Clark, 1988); (2) foramen in ventral part of splenial (Rauhut, 2003); (3) epipophyses on posterior cervical vertebrae (Gauthier, 1986); (4) hyposphene-hypantrum accessory joint in trunk vertebrae (Gauthier, 1986); (5) manus comprises more than 30% of humerus + radius length (Gauthier, 1986); (6) distal carpal V absent (Sereno, 1999); (7) 1st phalanx in digit I longest non-ungual phalanx in manus (Gauthier, 1986); (8) pointed posterior prong on distal tarsal IV (Langer and Benton, 2006); (9) distinct medial process projecting from middle of distal tarsal IV (Nesbitt, 2011); (10) markedly rimmed, elliptical fossa posterior to ascending process on astragalus (Langer and Benton, 2006).

If theropods are sister to ornithischians, and not to sauropodomorphs (Baron et al., 2017), then these apomorphies either apply to a more inclusive clade (i.e., *Dinosauria*), or they arose convergently in sauropodomorphs and theropods.

Synonyms: In trees in which *Sauropodomorpha* and *Theropoda* are sisters, there are no synonyms for the taxon name *Saurischia*. In the context of the Baron et al. (2017) tree, however, *Theropoda* as defined by Gauthier (1986) is an unambiguous synonym of *Saurischia*. As the name *Saurischia* is defined here, however, that name would not apply to any clade on the Baron et al. (2017) tree (see Comments).

Comments: The name *Saurischia* was coined by Seeley (1888) to contain Marsh's (1882) *Theropoda* and *Sauropoda*. It was to be contrasted with Seeley's second cardinal contribution to dinosaur nomenclature published in that same article: *Ornithischia* Seeley 1888. *Saurischia* and *Ornithischia* collectively yield nearly 26,000 records in a Google Scholar search (June 29, 2018), and Seeley (1888) was also the nineteenth most-cited "Proceedings B" article during that same month according to the Royal Society of London. Although Seeley distinguished *Ornithischia* on the basis of some characters now regarded as apomorphies, he differentiated (and named) *Saurischia* exclusively on the basis of plesiomorphies in pelvic anatomy, most notably that the pubis was still directed anteroventrally as in other reptiles. Like other taxonomists of his era, Seeley did not appreciate this distinction between character states. Nor did he explicitly state whether his taxa were to be associated more with a set of taxa or a set of characters (although we suspect the latter given taxonomic practices of the time). As Joyce et al. (2004) observed in their review of the history of turtle nomenclature, in such cases we can only be sure that Seeley coined the taxon names *Ornithischia* and *Saurischia* and not how he conceptualized these taxa.

Despite continued use of *Saurischia* as a taxonomic unit in the dinosaur literature since the late nineteenth century (e.g., Matthew and Brown, 1922; Huene, 1932; Romer, 1956, 1966; Colbert, 1964; Steel, 1970), the implied close relationship between theropods and sauropodomorphs continued to be questioned. During the mid-late twentieth century, for example, saurischians were either thought to be polyphyletic (evolved from different groups of "thecodonts", e.g., Charig et al., 1965; Thulborn, 1975), or paraphyletic (instead of being related to meat-eating theropods, at least some herbivorous sauropodomorphs were thought related to herbivorous ornithischians, e.g., Bakker and Galton, 1974; Sereno, 1984; Cooper, 1985). Indeed, the monophyly of *Saurischia* was not proposed until nearly a century after Seeley coined that name (Gauthier, 1986; and see corroborating references in Composition and Diagnostic Apormorphies above). Bonaparte's (1975) landmark study of *Lagosuchus* presented a tree (Fig. 2) that appears to depict *Ornithischia* as sister to *Saurischia* (= herrerasaurs, "prosauropods", "coelurosaurs", and "carnosaurs" according to Bonaparte). Nevertheless, the structure of Bonaparte's (1975) argument in the text demonstrates that he thought of "saurischians" in their traditional sense, that is, as those dinosaurs that are not ornithischians, rather than as sister to *Ornithischia*. Despite being repeatedly corroborated, *Saurischia* could fall out of use, as theropods have recently been inferred to be related to ornithischians instead of sauropodomorphs (Baron et al., 2017). If that tree is correct, then the name *Saurischia* would apply to no clade at all by our definition, viz., there would be no clade including the theropod *Allosaurus fragilis* and the sauropodomorph *Camarasaurus supremus* that excludes the ornithischian *Stegosaurus stenops*.

Gauthier (1986) proposed the first phylogenetic definition of *Saurischia*: "birds and all dinosaurs that are closer to birds than they are to *Ornithischia*" (p.15). A variety of alternative definitions were proposed subsequently differing mainly in their choice of internal and external specifiers (e.g., Padian and May, 1993; Padian, 1997; Padian et al., 1999; Sereno, 1998, 1999; Holtz and Osmólska, 2004; Langer, 2004). Gauthier's (1986) definition of *Saurischia* violates the *PhyloCode* in two important respects: it uses clades rather than species as specifiers (*ICPN* Art. 11.1; Cantino and de Queiroz, 2020); and it uses an internal specifier (birds) that was not traditionally included in the taxon (*ICPN* Rec. 11A; Cantino and de Queiroz, 2020). Instead of using any avians as specifiers—an approach introduced by Gauthier (1986; see also Gauthier, 1984)—we selected three extinct dinosaurs: two species of saurischians—a theropod and a sauropod sensu Marsh (1882)—as internal specifiers, plus an ornithischian species as an external specifier (Sereno, 2005a, b). At least by reference to their genus names, all of these species were mentioned and figured as examples of their respective groups by Seeley (1888). If early Late Triassic (Carnian) taxa such as *Eodromaeus murphi* and *Buriolestes schultzi* represent early saurischians, then the clade is at least 228 million years old (Cabreira et al., 2016).

Although no avian species is used in our definition as an internal specifier, *Aves* is part of *Saurischia* (as defined here) according to nearly every phylogenetic analysis published since Gauthier (1986). Although the name *Saurischia* (as defined here) would not apply to any clade in Baron et al.'s tree, birds would still be part of *Dinosauria* and *Theropoda* (see entries in this volume). Instead of being saurischians, they would be ornithoscelidans sensu Baron et al. (2017).

Literature Cited

Apaldetti, C., R. N. Martinez, O. A. Alcober, and D. Pol. 2011. A new basal sauropodomorph (*Dinosauria: Saurischia*) from Quebrada del Barro formation (Marayes-El Carrizal Basin), northwestern Argentina. *PLOS ONE* 6(11):1–19.

Bakker, R. T., and P. M. Galton. 1974. Dinosaur monophyly and a new class of vertebrates. *Nature* 248:168–172.

Baron, M. G., D. B. Norman, and P. M. Barrett. 2017. A new hypothesis of dinosaur relationships and early dinosaur evolution. *Nature* 543:501–506.

Benton, M. J., and J. Clark. 1988. Archosaur phylogeny and the relationships of the *Crocodylia*. Pp. 289–332 in *The Phylogeny and Classification of Tetrapods. Vol. 1: Amphibians, Reptiles, Birds* (M. J. Benton, ed.). Clarendon Press, Oxford.

Bonaparte, J. F. 1975. Nuevos materiales de *Lagosuchus talampayensis* Romer (*Thecodontia-Pseudosuchia*) y su significado en el origen de los *Saurischia. Acta Geol. Lillo.* 13:5–90.

Cabreira, S. F., A. W. A. Kellner, S. Dias-da-Silva, L. R. da Silva, M. Bronzati, J. C. de Almeida Marsola, R. T. Müller, J. de Souza Bittencourt, B. J. Batista, T. Raugust, R. Carrilho, A. Brodt, M. C. Langer. 2016. A unique Late Triassic dinosauromorph assemblage reveals dinosaur ancestral anatomy and diet. *Curr. Biol.* 26:3090–3095.

Cantino, P. D., and K. de Queiroz. 2020. *International Code of Phylogenetic Nomenclature (PhyloCode)*, Version 6. CRC Press, Boca Raton, FL.

Charig, A. J., J. Attridge, and A. W. Crompton. 1965. On the origin of the sauropods and the classification of the *Saurischia. Proc. Linn. Soc. Lond.* 176:197–221.

Colbert, E. H. 1964. Relationships of the saurischian dinosaurs. *Am. Mus. Novit.* 2181:1–24.

Cooper, M. R. 1985. A revision of the ornithischian dinosaur *Kangnasaurus coetzeei* Haughton, with a classification of the *Ornithischia. Ann. S. Afr. Mus.* 95:281–317.

Cope, E. D. 1877. On a gigantic saurian from the Dakota eopoc (sic) of Colorado. *Palaeontol. Bull.* 25:5–10.

Ezcurra, M. D. 2006. A review of the systematic position of the dinosauriform archosaur *Eucoelophysis baldwini* from the Upper Triassic of New Mexico, USA. *Geodiversitas* 28:649–684.

Fraser, N. C., K. Padian, G. M. Walkden, and A. L. M. Davis. 2002. Basal dinosauriform remains from Britain and the diagnosis of the *Dinosauria. Palaeontology* 45:78–95.

Gauthier, J. A., 1984. *A Cladistic Analysis of the Higher Systematic Categories of the Diapsida.* University Microfilms International, Ann Arbor, MI, #85-12825, vii + 564 pp.

Gauthier, J. A. 1986. Saurischian monophyly and the origin of birds. Pp. 1–55 in *The Origin of Birds and the Evolution of Flight* (K. Padian, ed.). Mem. Calif. Acad. Sci. 8. California Academy of Sciences, San Francisco, CA.

Gauthier, J. A., D. Cannatella, K. de Queiroz, A. G. Kluge, and T. Rowe. 1989. Tetrapod phylogeny. Pp. 337–353 in *The Hierarchy of Life: Molecules and Morphology in Phylogenetic Analysis* (B. Fernholm, K. Bremer, and H. Jornvall, eds.). Elsevier, Amsterdam.

Gauthier, J. A., and K. de Queiroz. 2001. Feathered dinosaurs, flying dinosaurs, crown dinosaurs, and the name *"Aves."* Pp. 7–41 in *New Perspectives on the Origin and Early Evolution of Birds: Proceeding of the International Symposium in Honor of John H. Ostrom* (J. A. Gauthier and L. F. Gall, eds., Spec. Publ. Peabody Mus. Nat. Hist., New Haven, CT.

Gill, F., and D. Donsker, eds. 2017. *IOC World Bird List* (v 7.1). doi:10.14344/IOC.ML.7.1.

Holtz, T. R., and H. Osmólska. 2004. *Saurischia.* Pp. 21–24 in *The* Dinosauria (D. B. Weishampel, P. Dodson and H. Osmólska, eds.). University of California Press, Berkley, CA.

von Huene, F.. 1932. Die fossile Reptil-Ordnung *Saurischia*, ihre Entwicklung und Geschichte. *Monogr. Geol. Palaeontol. Ser. 1.* 4:1–361.

Irmis, R. B., S. J. Nesbitt, K. Padian, N. D. Smith, A. H. Turner, D. Woody, and A. Downs. 2007. A Late Triassic dinosauromorph assemblage from New Mexico and the rise of dinosaurs. *Science* 317:358–361.

Joyce, W. G., J. F. Parham, and J. A. Gauthier. 2004. Developing a protocol for the conversion of rank-based taxon names to phylogenetically defined clade names, as exemplified by turtles. *J. Paleontol.* 78:989–1013.

Langer, M. C. 2004. Basal *Saurischia*. Pp. 25–46 in *The* Dinosauria (D. B. Weishampel, P. Dodson and H. Osmólska, eds.). University of California Press, Berkeley, CA.

Langer, M. C., and M. J. Benton. 2006. Early dinosaurs: a phylogenetic study. *J. Syst. Palaeontol.* 4:309–358.

Langer, M. C., M. D. Ezcurra, O. W. M. Rauhut, M. J. Benton, F. Knoll, B. W. McPhee, F. E. Novas, D. Pol, and S. L. Brusatte. 2017. Untangling the dinosaur family tree. *Nature* 551:E1–E3.

Liddell, H. G., and R. Scott. 1882. *Greek-English Lexicon.* 7th edition. Harper & Brothers, New York, CA.

Lloyd, G. T., K. E. Davis, D. Pisani, J. E. Tarver, M. Ruta, M. Sakamoto, D. W. E. Hone, R. Jennings, and M. J. Benton. 2008. Dinosaurs and the Cretaceous Terrestrial Revolution. *Proc. R. Soc. Lond. B Biol. Sci.* 275:2483–2490.

Marsh, O. C. 1877. Notice of new dinosaurian reptiles from the Jurassic formations. *Am. J. Sci.* 3:541–516.

Marsh, O. C. 1882. Classification of the *Dinosauria*. *Am. J. Sci.* 23:81–86.

Marsh, O. C. 1887. Principal characters of American Jurassic dinosaurs. Part IX: The skull and dermal armor of *Stegosaurus*. *Am. J. Sci.* 3:413–417.

Martínez, R. N., and O. A. Alcober. 2009. A basal sauropodomorph (*Dinosauria: Saurischia*) from the Ischigualasto Formation (Triassic, Carnian) and the early evolution of *Sauropodomorpha*. *PLOS ONE* 4(2):1–12.

Martínez, R. N., P. C. Sereno, O. A. Alcober, C. E. Colombi, P. R. Renne, I. P. Montañez, and B. S. Currie. 2011. A basal dinosaur from the dawn of the dinosaur era in southwestern Pangaea. *Science* 331:206–210.

Matthew, W. D., and B. Brown. 1922. The family *Deinodontidae*, with notice of a new genus from the Cretaceous of Alberta. *Bull. Am. Mus. Nat. Hist.* 46:367–385.

Niedźwiedzki, G., S. L. Brusatte, T. Sulej, and R. J. Butler. 2014. Basal dinosauriform and theropod dinosaurs from the mid-late Norian (Late Triassic) of Poland: implications for Triassic dinosaur evolution and distribution. *Palaeontology* 57(6):1121–1142.

Nesbitt, S. J. 2011. The early evolution of *Archosauria*: relationships and the origin of major clades. *Bull. Am. Mus. Nat. Hist.* 352:1–292.

Nesbitt, S. J., C. A. Sidor, R. B. Irmis, K. D. Angielczyk, R. M. Smith, and L. A Tsuji. 2010. Ecologically distinct dinosaurian sister group shows early diversification of *Ornithodira*. *Nature* 464:95–98.

Novas, F. E. 1989. The tibia and tarsus in *Herrerasauridae* (*Dinosauria, incertae sedis*) and the origin and evolution of the dinosaurian tarsus. *J. Paleontol.* 63:677–690.

Novas, F. E. 1992. Phylogenetic relationships of the basal dinosaurs, the *Herrerasauridae*. *Palaeontology* 35:51–62.

Novas, F. E. 1994. New information on the systematics and postcranial skeleton of *Herrerasaurus ischigualastensis* (*Theropoda: Herrerasauridae*) from the Ischigualasto Formation (Upper Triassic) of Argentina. *J. Vertebr. Paleontol.* 13:400–423.

Novas, F. E., M. D. Ezcurra, S. Chatterjee, and T. S. Kutty. 2011. New dinosaur species from the Upper Triassic Upper Maleri and Lower Dharmaram formations of Central India. *Earth Environ. Sci. Trans. R. Soc. Edinb.* 101:333–349.

Padian, K. 1997. *Saurischia*. Pp. 647–653 in *Encyclopedia of Dinosaurs* (P. J. Currie and K. Padian, eds.). Academic Press, San Diego, CA.

Padian, K., J. R. Hutchinson, and T. R., Holtz, Jr. 1999. Phylogenetic definitions and nomenclature of the major taxonomic categories of the carnivorous *Dinosauria* (*Theropoda*). *J. Vertebr. Paleontol.* 19:69–80.

Padian, K., and C. L. May. 1993. The earliest dinosaurs. *Bull. New Mexico Mus. Nat. Hist. Sci.* 3:379–381.

Pretto, F. A., C. L. Schultz, and M. C. Langer. 2015. New dinosaur remains from the Late Triassic of southern Brazil (Candelária Sequence, *Hyperodapedon* Assemblage Zone). *Alcheringa* 39:1–10.

de Queiroz, K. 2007. Toward an integrated system of clade names. *Syst. Biol.* 56:956–974.

Rauhut, O. W. M. 2003. The interrelationships and evolution of basal theropod dinosaurs. *Spec. Pap. Palaeontol.* 69:1–213.

Romer, A. S. 1956. *Osteology of the Reptiles.* University of Chicago Press, Chicago, IL.

Romer, A. S. 1966. *Vertebrate Paleontology.* 3rd edition. University of Chicago Press, Chicago, IL and London.

Seeley, H. G. 1888. On the classification of the fossil animals commonly named *Dinosauria. Proc. R. Soc. Lond. B Biol. Sci.* 43:165–171.

Sereno, P. C. 1984. The phylogeny of *Ornithischia*: a reappraisal. Pp. 219–227 in *Third Symposium on Mesozoic Terrestrial Ecosystems, Short Papers* (W.-E. Reif and F. Westphal, eds.). Attempto Verlag, Tübingen.

Sereno, P. C. 1998. A rationale for phylogenetic definitions, with application to the higher-level taxonomy of *Dinosauria. Neues Jahrb. Geol. Paläontol. Abh.* 210:41–83.

Sereno, P. C. 1999. The evolution of dinosaurs. *Science* 284:2137–2147.

Sereno, P. C. 2005a. The logical basis of phylogenetic taxonomy. *Syst. Biol.* 54:595–619.

Sereno, P. C. 2005b. TaxonSearch: Database for Suprageneric Taxa and Phylogenetic Definitions. Available at http://www.taxonsearch.org/, accessed on 16 January 2009.

Sereno, P. C., R. N., Martinez, and O. A., Alcober. 2013. Osteology of *Eoraptor lunensis* (*Dinosauria: Sauropodomorpha*). Pp. 83–179 in *Basal Sauropodomorphs and the Vertebrate Fossil Record of the Ischigualasto Formation (Late Triassic: Carnian–Norian) of Argentina* (P. C. Sereno, ed.). Society of Vertebrate Paleontology Memoir 12.

Starrfelt, J., and L. H Liow. 2016. How many dinosaur species were there? Fossil bias and true richness estimated using a Poisson sampling model. *Philos. Trans. R. Soc. Lond. B Biol. Sci.* 371:20150219.

Steel, R. 1970. *Saurischia. Handbuch der Paläoherpetologie.* Teil 14. Gustav Fischer, Stuttgart.

Thulborn, R. A. 1975. Dinosaur polyphyly and the classification of archosaurs and birds. *Aust. J. Zool.* 23:249–270.

Weishampel, D. B., P. Dodson, and H. Osmólska. 2004. *The* Dinosauria. 2nd edition. University of California Press, Berkeley, CA.

Yates, A. M. 2007. Solving a dinosaurian puzzle: the identity of *Aliwalia rex* Galton. *Hist. Biol.* 19:93–123.

Authors

Jacques A. Gauthier; Department of Geology and Geophysics; Yale University; 210 Whitney Ave.; New Haven, CT 06520-8109, USA. Email: jacques.gauthier@yale.edu.

Max Cardoso Langer; Departamento de Biologia, FFCLRP; Universidade de São Paulo; Ribeirão Preto-SP, 14040-901, Brazil. Email: mclanger@ffclrp.usp.br.

Fernando E. Novas; Laboratorio de Anatomia Comparada y Evolución de los Vertebrados; Museo Argentino de Ciencias Naturales "Bernardino Rivadavia"; Buenos Aires, C1405DJR, Argentina. Email: fernovas@yahoo.com.ar.

Jonathas Bittencourt; Departamento de Biologia, FFCLRP; Universidade de São Paulo; Ribeirão Preto-SP, 14040-901, Brazil. Email: sigmaorionis@yahoo.com.br.

Martin D. Ezcurra; Sección Paleontología de Vertebrados; Museo Argentino de Ciencias Naturales "Bernardino Rivadavia"; Buenos Aires, C1405DJR, Argentina. Email: martindezcurra@yahoo.com.ar.

Date Accepted: 11 August 2018

Primary Editor: Kevin de Queiroz

Sauropodomorpha F. R. von Huene 1932 [M. Fabbri, E. Tschopp, B. McPhee, S. Nesbitt, D. Pol, and M. Langer], converted clade name

Registration Number: 295

Definition: The largest clade containing *Saltasaurus loricatus* Bonaparte and Powell 1980 (*Sauropodomorpha*) but not *Allosaurus fragilis* Marsh 1877 (*Theropoda*) and *Iguanodon bernissartensis* Boulenger in Beneden 1881 (*Ornithischia*). This is a maximum-clade definition. Abbreviated definition: max ∇ (*Saltasaurus loricatus* Bonaparte and Powell 1980 ~ *Allosaurus fragilis* Marsh 1877 & *Iguanodon bernissartensis* Boulenger in Beneden 1881).

Etymology: Derived from the Ancient Greek *sauros* (lizard), *podos* (foot), and *morphe* (form), presumably in reference to the general resemblance between Huene's Triassic "prosauropods" and Marsh's Jurassic *Sauropoda*, which are both included in *Sauropodomorpha*.

Reference Phylogeny: Figure 17 in Otero et al. (2015) is selected as the primary reference phylogeny because it considers a broad range of out- and in-group taxa (including a dense sample of *Sauropoda*). Updated versions of this phylogeny with additional taxa and characters are presented in Cerda et al. (2017: Fig. 1) and Apaldetti et al. (2018: suppl. Fig. 9b). The primary reference phylogeny is itself an update of previous datasets from Yates (2007: Fig. 19), Otero and Pol (2013: Fig. 19), and McPhee et al. (2015: Fig. 12), and represents a general consensus in the field regarding phylogenetic relationships within *Sauropodomorpha*; see also the supertree of Lloyd et al. (2008: suppl. Fig. 2), which depicts detailed relationships of the specifier species within *Dinosauria*. The specifiers in our maximum-clade definition of *Sauropodomorpha* are not shown in Otero et al. (2015: Fig. 17). Nevertheless, *Saltasaurus loricatus* Bonaparte and Powell 1980, *Allosaurus fragilis* Marsh 1877, and *Iguanodon bernissartensis* Boulenger in Beneden 1881 are represented by the supraspecific (terminal) taxa *Neosauropoda*, *Neotheropoda*, and *Ornithischia*, respectively, in that tree. The chosen specifiers ensure that *Sauropodomorpha* will apply to the intended maximum clade regardless of the ultimate resolution of current controversies about basal divergences within *Dinosauria* (see Comments, and *Dinosauria* and *Saurischia* contributions in this volume).

Composition: With the possible exception of some Triassic species, such as *Guaibasaurus candelariensis* and *Eoraptor lunesis*, and a few mistaken referrals of isolated fragmentary remains (see Comments), the composition of *Sauropodomorpha* has been agreed upon (e.g., Gauthier, 1986; Galton and Upchurch, 2004; Yates, 2007, 2010; Otero et al., 2015; Apaldetti et al., 2018). Since *Sauropodomorpha* was first proposed, it has always included *Sauropoda*, a group whose monophyly has never been questioned, and "*Prosauropoda*", whether as a clade or as a paraphylum (see Comments). Dated lists of included species can be found in Galton and Upchurch (2004) and Upchurch et al. (2004). An updated list from the Paleobiology Database (https://paleobiodb.org/#/) records 246 species within *Sauropodomorpha* (Starrfelt and Liow, 2016).

Diagnostic Apomorphies: Maximum clades are unlikely to have apomorphies that arose simultaneously as they split from their sister clades (Gauthier and de Queiroz, 2001; de

Queiroz, 2007). To further complicate matters, recently described Late Triassic taxa attributed to *Sauropodomorpha* with relatively few, and sometimes conflicting, apomorphies play key roles in current controversies regarding the earliest divergences within *Dinosauria* (see entry this volume). The early phylogeny of *Dinosauria* and *Sauropodomorpha* are accordingly in flux, complicating the diagnosis of maximum-clade *Sauropodomorpha*.

There is wide agreement that some of the earliest known Triassic dinosaurs, such as *Saturnalia tupiniquim*, *Panphagia protos*, *Pampadromaeus barberenai*, and *Buriolestes schultzi*, are sauropodomorphs (e.g., Langer et al., 1999; Yates, 2007; Upchurch et al., 2007; Ezcurra, 2010; Pol et al., 2011; Martinez and Alcober, 2009; Martinez et al., 2012; Otero et al., 2015; Cabreira et al., 2016; Baron et al., 2017; Langer et al., 2017; Pretto et al., 2018; Müller et al., 2018; and Apaldetti et al., 2018; but see Bonaparte et al., 2007) (see Comments below). That being said, the diagnosis of *Sauropodomorpha* appears to depend on the sample of taxa and characters included in each analysis. In the context of the analysis of Apaldetti et al. (2018), for example, *Sauropodomorpha* is diagnosed by the following apomorphies: (1) supraoccipital semilunate and wider than tall; (2) postzygodiapophyseal lamina in neural arches of cervical vertebrae 4–8 absent; (3) remaining laminae on neural arches of cervical vertebrae 4–8 weakly developed; (4) interpostzygapophyseal notch absent between postzygapophyses in proximal caudal vertebrae; (5) humerus length 55%–65% of femur length; (6) transverse width of distal humerus more than 33% of humerus length; (7) preacetabular process of ilium with pointed, cranioventrally projecting corner and rounded dorsum; (8) partially open medial wall of acetabulum with straight ventral margin between peduncles; (9) length of pubic peduncle of ilium more than twice craniocaudal width distally; (10) triangular outline of ischial shaft in transverse section.

With a different sample of characters and taxa (e.g., the inclusion of *Buriolestes schultzi*), apomorphies diagnosing *Sauropodomorpha* in Cabreira et al. (2016) are the following: (1) medial wall of maxillary antorbital fossa with one or more foramen- or fenestra-sized perforations close to base of facial process; (2) ventral margins of suborbital and quadratojugal rami of jugal straight or forming an angle of more than 180°; (3) suborbital process of jugal excluded from internal antorbital fenestra by lacrimal or maxilla; (4) length versus height of dentary more than 0.2; (5) lateral surface of surangular with prominent horizontal shelf; (6) caudal dorsal centra nearly as long or longer than tall; (7) cranial margin of femur in distal view straight or convex; (8) tibia cnemial crest absent or just a slight bump; (9) astragalus proximal articular facet for fibula less than 0.3 times transverse width.

Cabreira et al. (2016) also identified a minimum clade including *Saturnalia tupiniquim* Langer et al. 1999 and some other Late Triassic/Early Jurassic taxa universally recognized as sauropodomorphs—e.g., *Pantydraco caducus* Galton et al. (2007), *Efraasia minor* (Huene, 1908), *Thecodontosaurus antiquus* (Morris, 1843), *Plateosaurus engelhardti* (Meyer, 1837), and *Massospondylus carinatus* (Owen, 1854)—that share the following apomorphies: (1) skull length less than half of femur length; (2) ventrally curved dorsal margin of dentary tip (usually with an inset first tooth); (3) ventrally displaced mandibular condyle relative to tooth row; (4) prevalence of teeth with leaf-shaped crowns with enlarged denticles (four per millimeter); (5) more distal teeth in jaws with significantly lower crowns than mesial teeth; (6) pubic peduncle of ilium at least twice as long as craniocaudal width.

Synonyms: "*Prosauropoda*" sensu Huene 1932 (approximate, partial); *Saurischia* (unambiguous, but only as redefined by Baron et al. [2017] and in the context of their tree).

Comments: The origin and early divergences within *Sauropodomorpha* remain controversial owing to recently discovered early dinosaurs and new phylogenetic analyses focused on the major branches of *Dinosauria* (see below and entries for *Dinosauria* and *Saurischia* in this volume). Nevertheless, the composition of *Sauropodomorpha* has, with few exceptions, otherwise been remarkably stable over the years. The early sauropodomorph *Efraasia minor* was once referred to *Teratosaurus*, a carnivorous stem crocodilian, but Galton (1985; see also Benton, 1986) corrected that error. Additionally, *Segnosaurus galbinensis* from the Cretaceous was briefly thought to be a relatively early diverging sauropodomorph (Paul, 1984; Gauthier, 1986; Olshevsky, 1991). More material referable to that species and the discovery of closely related taxa later showed that *Segnosaurus galbinensis* is part of the *Therizinosauria*, a highly modified clade of non-avian coelurosaurs (Clark et al., 2004; Zanno, 2010). Thus, the apomorphies shared by Triassic sauropodomorphs and Cretaceous therizinosaurs are now inferred to result from convergence related to herbivorous diets.

Sauropodomorpha Huene 1932 was proposed to encompass "*Prosauropoda*" Huene 1920 and *Sauropoda* Marsh 1878. Huene (1932) conceptualized "*Prosauropoda*" as an ancestral supraspecific taxon (see also Charig et al., 1965). Regardless of how it was originally conceptualized, the first actual evidence proposed for "prosauropod" paraphyly did not emerge until much later (e.g., Gauthier, 1986), following Raath's (1972) description of *Vulcanodon karibaensis* as a "prosauropod" convergent on sauropods. After a period during which some researchers proposed that (most) of the remaining early sauropodomorphs comprised a clade (= *Prosauropoda* sensu Galton, 1990; Salgado et al., 1997; Wilson and Sereno, 1998; Sereno, 1999, 2007; Benton et al., 2000; and Upchurch et al., 2007), they are now widely seen as a paraphylum—albeit including some smaller clades such as *Massospondylidae* (e.g., Smith and Pol, 2007; Apaldetti et al., 2014; Otero et al., 2015), *Plateosauridae* (e.g., Otero et al., 2015; Apaldetti et al., 2018), and *Riojasauridae* (e.g., Otero et al., 2015; McPhee and Choiniere, 2017; Apaldetti et al., 2018)—composed of early sauropodomorphs more or less closely related to *Sauropoda* (e.g., Gauthier, 1986; Langer et al., 1999; Upchurch et al, 2007; Yates, 2007, 2010; Ezcurra, 2010; Yates et al., 2011; Pol et al., 2011; McPhee et al., 2014; Cabreira et al., 2016; Langer et al., 2017).

Early sauropodomorphs (traditional "prosauropods") of the Late Triassic and Early Jurassic include such familiar long-necked, herbivorous dinosaurs as *Pantydraco*, *Efraasia*, *Thecodontosaurus*, *Plateosaurus*, and *Massospondylus*. These taxa, along with the gigantic *Sauropoda*, are universally thought to constitute a clade regardless of whether herrerasaurids (e.g., Baron et al, 2017) or theropods (e.g., Langer et al., 2017) are their closest relatives. Several early and relatively unmodified dinosaurs described more recently may also belong to the herbivorous *Sauropodomorpha*, including *Eoraptor lunensis* (Sereno et al., 1993), *Panphagia protos* (Martinez and Alcobar, 2009), *Chromogisaurus novasi* (Ezcurra, 2010), and *Pampadromaeus barberenai* (Cabreira et al., 2011), as well as the still-carnivorous *Buriolestes schultzi* (Cabreira et al., 2016). According to a recent phylogeny (Baron et al., 2017), *Herrerasauridae* would also be considered *Sauropodomorpha* according to our definition; however, a revised and corrected version of this dataset placed *Herrerasauridae* in its more traditional position among early *Saurischia* (Langer et al., 2017).

We reject Baron et al.'s (2017) proposed redefinition of the name *Sauropodomorpha* to explicitly exclude *Herrerasauridae*. Although it would be consistent with tradition in one sense—conserving most of the traditional composition *Sauropodomorpha*—it would be inconsistent in another—*Sauropodomorpha* would no longer be associated with one of the fundamental branches of the dinosaurian tree (Baron et al. [2017] proposed renaming that major clade with a redefined version of *Saurischia*). Moreover, instead of being sister to a remarkably diverse, disparate, long-lived, and ecologically dominant clade—whether that is *Theropoda* or *Ornithoscelida*—the sister to *Sauropodomorpha* would only comprise a fairly homogenous and short-lived clade composed of less than a half-dozen species known only from a brief interval during the latter part of the Triassic (in Baron et al.'s [2017] tree).

The tenuousness of current ideas about the early evolution of *Dinosauria*, hence the precise composition and diagnosis of *Sauropodomorpha*, can be illustrated by the impact of the relatively well-preserved *Eodromaeus murphi* from the early Late Triassic of Argentina (Martinez et al., 2011) on phylogenetic analysis. Most of the phylogenetic studies including this species infer *Eoraptor lunensis* to be a sauropodomorph (e.g., Martinez et al., 2011; Cabreira et al., 2011; modified dataset from Smith et al., 2008, in Novas et al., 2015; Cabreira et al., 2016; Langer et al., 2017; Pretto et al., 2018; Müller et al., 2018), contrary to few recent analyses (Baron et al., 2017; Müller et al., 2018). Yet studies lacking *Eodromaeus murphi* more frequently infer *Eoraptor lunensis* to be either a theropod (Nesbitt et al., 2009; Ezcurra, 2010; Nesbitt et al., 2010; Nesbitt, 2011; Sues et al., 2011; Cabreira et al., 2011; Carrano et al., 2012; Novas et al., 2015; Pretto et al., 2015; Martill et al., 2016),

or a basal saurischian (e.g., Langer and Benton, 2006; Martinez and Alcober, 2009; Cabreira et al., 2011; Apaldetti et al., 2011; Otero and Pol, 2013; McPhee et al., 2014; Novas et al., 2015), or occasionally as a basal sauropodomorph (Martinez et al., 2012; Müller et al., 2018; Pretto et al., 2018). Moreover, in the tree of Baron et al. (2017), in which *Sauropodomorpha* is sister to *Herrerasauridae* rather than to *Theropoda*, both *Eoraptor lunensis* and *Eodromaeus murphi* are basal theropods rather than basal saurischians (e.g., Cabreira et al., 2016; Langer et al., 2017; Müller et al., 2018; Pretto et al., 2018).

In addition, the phylogenetic affinities of *Guiabasaurus candelariensis* and its inclusion within *Sauropodomorpha* also remain a topic of debate. This species has been regarded as a basal saurischian (e.g., Bonaparte et al., 1999; Langer et al., 2017; Pretto et al., 2018), an early theropod (Langer, 2004; Yates, 2007; Langer et al., 2007a, 2010; Müller et al., 2018; Pretto et al., 2018), or an early sauropodomorph (Bonaparte et al., 2007; Ezcurra and Novas, 2007; Ezcurra, 2008; McPhee et al., 2015; Baron et al., 2017; McPhee and Choiniere, 2017; Pretto et al., 2018).

Several phylogenetic definitions were previously proposed for the name *Sauropodomorpha*. These include minimum-clade and maximum-clade definitions and differ in the proposed internal and external specifiers. Salgado et al. (1997) defined the name *Sauropodomorpha* as "the clade including the most recent common ancestor of Prosauropoda and Sauropoda and all of its descendants". Sereno (1998) proposed using genera as specifiers instead of more inclusive clades: "*Plateosaurus, Saltasaurus*, their most recent common ancestor and all descendants". The first maximum clade definition for *Sauropodomorpha* was proposed by Upchurch (1997): "Sauropoda + Prosauropoda and all saurischians closer to them than to birds". The definition was changed by Galton and Upchurch (2004), who used a (monotypic) genus plus a more inclusive clade as specifiers: "all taxa more closely related to *Saltasaurus* than to Theropoda". Sereno (2007) improved the maximum-clade definition by using species as specifiers, including a third specifier explicitly excluding the ornithischian branch: "the most inclusive clade containing *Saltasaurus loricatus* Bonaparte and Powell 1980 but not *Passer domesticus* Linnaeus 1758 and *Triceratops horridus* Marsh 1889". Sereno's definition has the advantage of conserving the traditional composition of *Sauropodomorpha* against potential revisions to hypothesized basal relationships within *Dinosauria* (i.e., Baron et al., 2017). Nevertheless, the inclusion of *Passer domesticus* as a specifier in Sereno's (2007) definition remains problematic: even though *Aves* is unambiguously part of *Theropoda* (this volume), *Passer domesticus* was not generally considered part of the traditional composition of that group (e.g., Marsh, 1881), thus contravening Recommendation 11A of the *PhyloCode*.

Although both minimum- and maximum-clade definitions are acceptable, the latter kind of definition has been preferred in most studies (e.g., Yates 2007: 31; Langer et al., 2007b: 115). Because the name has long been associated with one of the three primary branches of the dinosaurian tree, the name *Sauropodomorpha* would most appropriately be defined using a maximum-clade (branch-based) definition (see de Queiroz, 2013). We build on these early efforts at phylogenetic nomenclature by modifying previously proposed maximum-clade definitions so that they comply with the current rules of the *PhyloCode* (Cantino and de Queiroz, 2020). Individual species, rather than more inclusive taxa, are used as specifiers. These specifiers are chosen from among the species traditionally included in their respective clades. We also buffer our definition against potential changes in ideas about phylogenetic relationships by using deeply nested species (following Sereno, 1998),

and by using an additional external specifier to ensure that *Sauropodomorpha* applies to the intended clade regardless of current uncertainty in relationships among the major clades of *Dinosauria*. We anchor our definition to the internal specifier *Saltasaurus loricatus* because it is a deeply nested titanosaur sauropod often used in phylogenetic definitions (e.g., Wilson and Sereno, 1998; Wilson, 2002; Upchurch et al., 2004, 2007). Fossil specimens referred to *Saltasaurus loricatus* are abundant, the species is well known, and its phylogenetic position is consistent among phylogenetic analyses. We selected our external specifiers, *Allosaurus fragilis* and *Iguanodon bernissartensis*, because they are deeply nested within their respective clades (*Theropoda* and *Ornithischia*, respectively) (see Sereno, 1998), and they are represented by multiple specimens that have been studied extensively. The oldest potential sauropodomorphs are from the early Late Triassic (Cabreira et al., 2016). *Sauropoda* completely replaced other sauropodomorphs in Mesozoic ecosystems during the Jurassic. No sauropodomorph is known to have survived the asteroid impact marking the end of the Cretaceous (e.g., Weishampel et al., 2004).

Literature Cited

Apaldetti, C., R. N. Martinez, D. Pol, and T. Souter. 2014. Redescription of the skull of *Coloradisaurus brevis* (*Dinosauria, Sauropodomorpha*) from the Late Triassic Los Colorados Formation of the Ischigualasto-Villa Union Basin, northwestern Argentina. *J. Vertebr. Paleontol.* 34:1113–1132.

Apaldetti, C., R. N. Martínez, I. A. Cerda, D. Pol, and O. Alcober. 2018. An early trend towards gigantism in Triassic sauropodomorph dinosaurs. *Nat. Ecol. Evol.* 2:1227.

Apaldetti, C., R. N. Martinez, O. A. Alcober, and D. Pol. 2011. A new basal sauropodomorph (*Dinosauria: Saurischia*) from Quebrada del Barro Formation (Marayes-El Carrizal Basin), northwestern Argentina. *PLOS ONE* 6(11):e26964.

Baron, M. G., D. B. Norman, and P. M. Barrett. 2017. A new hypothesis of dinosaur relationships and early dinosaur evolution. *Nature* 543:501–506.

Benton, M. J. 1986. The late Triassic reptile *Teratosaurus*—a rauisuchian, not a dinosaur. *Paleontology* 29:293–301.

Benton, M. J., L. Juul, G. W. Storrs, and P. M. Galton. 2000. Anatomy and systematics of the prosauropod dinosaur *Thecodontosaurus antiquus* from the Upper Triassic of southwest England. *J. Vertebr. Paleontol.* 20:77–108.

Bonaparte, J. F., G. Brea, C. L. Schultz and A. G. Martinelli. 2007. A new specimen of *Guiabasaurus candelariensis* (basal *Saurischia*) from the Late Triassic Caturrita Formation of southern Brazil. *Hist. Biol.* 19(1):73–82.

Bonaparte, J. F., J. Ferigolo, and A. M. Ribeiro. 1999. A new early Late Triassic saurischian dinosaur from Rio Grande do Sul state, Brazil. *Natl. Sci. Mus. Monogr.* 15:89–109.

Bonaparte, J. F., and J. E. Powell. 1980. A continental assemblage of tetrapods from the Upper Cretaceous beds of El Brete, northwestern Argentina (*Sauropoda-Coelurosauria-Carnosauria-Aves*). *Mem. Geol. Soc. Fr.* 139:19–28.

Boulenger, G. A. 1881. *Iguanodon bernissartensis*. P. 606 in P.-J. Beneden, Sur l'arc pelvien chex les dinosauriens de Bernissart. *Bull. Acad. R. Sci. Belg. 3 Sér.* 1(5):600–608.

Cabreira, S. F., A.W.A. Kellner, S. Dias-da-Silva, L. R. da Silva, M. Bronzati, J. C. de Almeida Marsola, R. T. Müller, J. de Souza Bittencourt, B. J. Batista, T. Raugust, R. Carrilho, A. Brodt, and M. C. Langer. 2016. A unique Late Triassic dinosauromorph assemblage reveals dinosaur ancestral anatomy and diet. *Curr. Biol.* 26:3090–3095.

Cabreira, S. F., C. L. Schultz, J. S. Bittencourt, M. B. Soares, D. C. Fortier, L. R. Silva, and M. C. Langer. 2011. New stem-sauropodomorph (*Dinosauria, Saurischia*) from the Triassic of Brazil. *Naturwissenschaften* 98:1035–1040.

Cantino, P. D., and K. de Queiroz. 2020. *International Code of Phylogenetic Nomenclature (PhyloCode)*, Version 6. CRC Press, Boca Raton, FL.

Carrano, M. T., R. B. Benson, and S. D. Sampson. 2012. The phylogeny of *Tetanurae* (*Dinosauria: Theropoda*). *J. Syst. Palaeontol.* 10:211–300.

Cerda, I. A., A. Chinsamy, D. Pol, C. Apaldetti, A. Otero, J. E. Powell, and R. N. Martínez. 2017. Novel insight into the origin of the growth dynamics of sauropod dinosaurs. *PLOS ONE* 12:e0179707.

Charig, A. J., J. Attridge, and A. W. Crompton. 1965. On the origin of the sauropods and the classification of the *Saurischia*. *Proc. Linn. Soc. Lond.* 176:197–221.

Clark, J. M., T. Maryanska, and R. Barsbold. 2004. *Therizinosauroidea*. Pp. 151–164 in *The Dinosauria*. 2nd edition (D. B. Weishampel, P. Dodson, and H. Osmólska, eds.). University of California Press, Berkeley, CA and Los Angeles, CA.

Ezcurra, M. D. 2008. A new early dinosaur from the Carnian Ischigualasto Formation (NW Argentina) and the origin of dinosaurs. In *Libro de Resúmenes, III Congreso Latinoamericano de Paleontologia de Vertebrados*, 87. (J. O. Calvo, R. J. Valieri, J. D. Porfiri, and D. dos Santos, eds.). Universidade Nacional del Comahue, Neuquen, AR.

Ezcurra, M. D. 2010. A new early dinosaur (*Saurischia: Sauropodomorpha*) from the Late Triassic of Argentina: a reassessment of dinosaur origin and phylogeny. *J. Syst. Palaeontol.* 8:371–425.

Ezcurra, D. M., and F. E. Novas. 2007. Phylogenetic relationships of the Triassic theropod *Zupaysaurus rougieri* from NW Argentina. *Hist. Biol.* 19:35–72.

Galton, P. M. 1985. The poposaurid thecodontian *Teratosaurus suevicus* Meyer, plus referred specimens mostly based on prosauropod dinosaurs. *Stuttg. Beitr. Naturkd. B* 116:1–29.

Galton, P. M. 1990. Basal *Sauropodomorpha*—*Prosauropoda*. Pp. 320–344 in *The* Dinosauria (D. B. Weishampel, P. Dodson, and H. Osmólska, eds.). University of California Press, Berkeley, CA.

Galton, P. M., A. M. Yates, and D. Kermack. 2007. *Pantydraco* n. gen. for *Thecodontosaurus caducus* Yates, 2003, a basal sauropodomorph dinosaur from the Upper Triassic or Lower Jurassic of South Wales, UK. *Neues Jahrb. Geol. Paläontol.* 243:119–125.

Galton, P. M., and P. Upchurch. 2004. *Prosauropoda.* Pp. 232–258 in *The* Dinosauria. 2nd edition (D. B. Weishampel, P. Dodson, and H. Osmólska, eds.). University of California Press, Berkeley, CA and Los Angeles, CA.

Gauthier, J. 1986. Saurischian monophyly and the origin of birds. Pp. 1–55 in The Origin of Birds and the Evolution of Flight (K. Padian, ed.). *Mem. Calif. Acad. Sci.* 8. California Academy of Sciences, San Francisco, CA.

Gauthier, J., and K. de Queiroz. 2001. Feathered dinosaurs, flying dinosaurs, crown dinosaurs, and the name "*Aves*". Pp. 7–41 in *New Perspectives on the Origin and Early Evolution of Birds: Proceedings of the International Symposium in Honor of John H. Ostrom* (J. Gauthier and L. F. Gall, eds.). Peabody Museum of Natural History, Yale University, New Haven, CT.

Huene, F. v. 1908. Die Dinosaurier der Europäischen Triasformation mit Berücksichtigung der Aussereuropäischen Vorkommnisse. *Geol. Palaeontol. Abhand. Suppl.* 1:1–419.

Huene, F. v. 1920. Bemerkungen zur Systematik und Stammesgeschichte einiger Reptilien. *Z. Indukt. Abstamm. Ver.* 24:162–166.

Huene, F. v. 1932. Die fossile Reptile-Ordnung *Saurischia*, ihre Entwicklung und Geschichte. *Monogr. Geol. Palaeontol (Ser. 1)* 4:1–361.

Langer, M. C. 2004. Basal *Saurischia*. Pp. 25–46 in *The* Dinosauria. 2nd edition (D. B. Weishampel, P. Dodson, and H. Osmólska, eds.). University of California Press, Berkeley, CA.

Langer, M. C., and M. J. Benton. 2006. Early dinosaurs: a phylogenetic study. *J. Syst. Palaeontol.* 4(4):309–358.

Langer, M. C., F. Abdala, M. Richter, and M. J. Benton. 1999. A sauropodomorph dinosaur from the Upper Triassic (Carnian) of southern Brazil. *C. R. Acad. Sci., Paris (Sciences de la Terre et des Planètes)* 329:511–517.

Langer, M. C., J. S. Bittencourt, and C. L. Schultz. 2007a. The inclusivity and phylogenetic position of *Guiabasaurus candelariensis*: a basal dinosaur from the Late Triassic of Brazil. *J. Vertebr. Paleontol.* 27(3, suppl.):103A–104A.

Langer, M. C., M. Ezcurra, J. S. Bittencourt, and F. Novas. 2010. The origin and early evolution of dinosaurs. *Biol. Rev.* 85:55–110.

Langer, M. C., M. A. G. França, and S. Gabriel. 2007b. The pectoral girdle and forelimb anatomy of the stem-sauropodomorph *Saturnalia tupiniquim* (Upper Triassic, Brazil). Pp. 113–137 in *Special Papers in Palaeontology 77: Evolution and Palaeobiology of Early Sauropodomorph Dinosaurs* (P. M. Barrett and D. J. Batten, eds.). The Palaeontological Association, UK.

Langer, M. C., M. D. Ezcurra, O. W. M. Rauhut, M. J. Benton, F. Knoll, B. W. McPhee, F. E. Novas, D. Pol, and S. L. Brusatte. 2017. Untangling the dinosaur family tree. *Nature* 551:E1–E3.

Linnaeus, C. 1758. *Systema Naturae Per Regnum Tria Naturae Secundum Classes, Ordines, Genera, Species, Cum Characteribus, Differentiis, Synonimis, Loci.* 10th edition, Vol. 1. Laurentius Salvius, Holmiae (Stockholm).

Lloyd, G. T., K. E. Davis, D. Pisani, J. E. Tarver, M. Ruta, M. Sakamoto, D. W. E. Hone, R. Jennings, and M. J. Benton. 2008. Dinosaurs and the Cretaceous terrestrial revolution. *Proc. R. Soc. Lond. B Biol. Sci.* 275:2483–2490.

Marsh, O. C. 1877. Notice of new dinosaurian reptiles from the Jurassic formation. *Am. J. Sci. Arts (Ser. 3)* 14:514–516.

Marsh, O. C. 1878. Principal characters of American Jurassic dinosaurs. Part I. *Am. J. Sci. (Ser. 3)* 16:411–416.

Marsh, O. C. 1881. Principal characters of American Jurassic dinosaurs. Part V. *Am. J. Sci. (Ser. 3)* 21:417–423.

Marsh, O. C. 1889. Notice of new American dinosaurs. *Am. J. Sci. (Ser. 3)* 37:331–336.

Martill, D. M., S. U. Vidovic, C. Howells, and J. R. Nudds. 2016. The oldest Jurassic dinosaur: a basal neotheropod from the Hettangian of Great Britain. *PLOS ONE* 11:e0145713.

Martínez, R. N., and O. A. Alcober. 2009. A basal sauropodomorph (*Dinosauria: Saurischia*) from the Ischigualasto Formation (Triassic, Carnian) and the early evolution of *Sauropodomorpha*. *PLOS ONE* 4:e4397.

Martínez, R. N., C. Apaldetti, and D. Abelin. 2012. Basal sauropodomorphs from the Ischigualasto Formation. *J. Vertebr. Paleontol.* 32(suppl. 3):51–69.

Martínez, R. N., P. C. Sereno, O. A. Alcober, C. E. Colombi, P. R. Renne, I. P. Montañez, and B. S. Currie. 2011. A basal dinosaur from the dawn of the dinosaur era in southwestern Pangaea. *Science* 331:206–210.

McPhee, B. W., A. M. Yates, J. N. Choiniere, and F. Abdala. 2014. The complete anatomy and phylogenetic relationships of *Antetonitrus ingenipes* (*Sauropodiformes, Dinosauria*): implications for the origins of *Sauropoda*. *Zool. J. Linn. Soc.* 171:151–205.

McPhee, B. W., and J. N. Choiniere. 2017. The osteology of *Pulanesaura eocollum*: implications for the inclusivity of *Sauropoda* (*Dinosauria*). *Zool. J. Linn. Soc.* 182:830–861.

McPhee, B. W., J. N. Choiniere, A. M. Yates, and P. A. Viglietti. 2015. A second species of *Eucnemesaurus* Van Hoepen, 1920 (*Dinosauria, Sauropodomorpha*): new information on the diversity and evolution of the sauropodomorph fauna of South Africa's lower Elliot Formation (latest Triassic). *J. Vertebr. Paleontol.* 35:e980504.

Meyer, H. v. 1837. Mitteilung an Prof. Bronn (*Plateosaurus engelhardti*). *Neues Jahrb. Geol. Paläontol.* 1837:316.

Morris, J. 1843. *A Catalogue of British Fossils*. British Museum, London.

Müller, R. T., M. C. Langer, M. Bronzati, C. P. Pacheco, S. F. Cabreira, and S. Dias-Da-Silva. 2018. Early evolution of sauropodomorphs: anatomy and phylogenetic relationships of a remarkably well-preserved dinosaur from the Upper Triassic of southern Brazil. *Zool. J. Linn. Soc.* 184(4):1187–1248.

Nesbitt, S. J. 2011. The early evolution of archosaurs: relationships and the origin of major clades. *Bull. Am. Mus. Nat. Hist.* 352:1–292.

Nesbitt, S. J., C. A. Sidor, R. B. Irmis, K. D. Angielczyk, R. M. Smith, and L. A. Tsuji. 2010. Ecologically distinct dinosaurian sister group shows early diversification of *Ornithodira*. *Nature* 464:95–98.

Nesbitt, S. J., N. D. Smith, R. B. Irmis, A. H. Turner, A. Downs, and M. A. Norell. 2009. A complete skeleton of a Late Triassic saurischian and the early evolution of dinosaurs. *Science* 326:1530–1533.

Novas, F. E., L. Salgado, M. Suárez, F. L Agnolín, M. D. Ezcurra, N. R. Chimento, R. De La Cruz, M. P. Isasi, A. O. Vargas, and D. Rubilar-Rogers. 2015. An enigmatic plant-eating theropod from the Late Jurassic period of Chile. *Nature* 522:331–334.

Olshevsky, G. 1991. A revision of the parainfraclass *Archosauria* Cope, 1869, excluding the advanced *Crocodylia*. *Mesozoic Meanderings* 2:1–196.

Otero, A., and D. Pol. 2013. Postcranial anatomy and phylogenetic relationships of *Mussaurus patagonicus* (*Dinosauria, Sauropodomorpha*). *J. Vertebr. Paleontol.* 33:1138–1168.

Otero, A., E. Krupandan, D. Pol, A. Chinsamy, and J. Choiniere. 2015. A new basal sauropodiform from South Africa and the phylogenetic relationships of basal sauropodomorphs. *Zool. J. Linn. Soc.* 174:589–634.

Owen, R. 1854. *Descriptive Catalogue of the Fossil Organic Remains of Reptilia and Pisces Contained in the Museum of the Royal College of Surgeons of England*. Taylor & Francis, London.

Paul, G. S. 1984. The segnosaurian dinosaurs: relics of the prosauropod-ornithischian transition? *J. Vertebr. Paleontol.* 4:507–515.

Pol, D., A. Garrido, and I. A. Cerda. 2011. A new sauropodomorph dinosaur from the Early Jurassic of Patagonia and the origin and evolution of the sauropod-type sacrum. *PLOS ONE* 6:e14572.

Pretto, F. A., C. L. Schultz, and M. C. Langer. 2015. New dinosaur remains from the Late Triassic of southern Brazil (Candelária Sequence, *Hyperodapedon* Assemblage Zone). *Alcheringa* 39:264–273.

Pretto, F. A., M. C. Langer and C. L. Schultz. 2018. A new dinosaur (*Saurischia*: *Sauropodomorpha*) from the Late Triassic of Brazil provides insights on the evolution of sauropodomorph body plan. *Zool. J. Linn. Soc.* 185(2):388–416.

de Queiroz, K. 2007. Toward an integrated system of clade names. *Syst. Biol.* 56:956–974.

de Queiroz, K. 2013. Nodes, branches, and phylogenetic definitions. *Syst. Biol.* 62:625–632.

Raath, M. A. 1972. Fossil vertebrate studies in Rhodesia: a new dinosaur (*Reptilia*: *Saurischia*) from near the Triassic–Jurassic boundary. *Arnoldia* 5:1–37.

Salgado, L., R. A. Coria, and J. O. Calvo. 1997. Evolution of titanosaurid sauropods. I: Phylogenetic analysis based on the postcranial evidence. *Ameghiniana* 34:3–32.

Sereno, P. C. 1998. A rationale for phylogenetic definitions, with application to the higher-level taxonomy of *Dinosauria*. *Neues Jahrb. Geol. Paläontol.* 210:41–83.

Sereno, P. C. 1999. The evolution of dinosaurs. *Science* 284:2137–2147.

Sereno, P. C. 2007. Basal *Sauropodomorpha*: historical and recent phylogenetic hypotheses, with comments on *Ammosaurus major* (Marsh, 1889). Pp. 261–289 in *Special Papers in Palaeontology 77: Evolution and Palaeobiology of Early Sauropodomorph Dinosaurs* (P. M. Barrett and D. J. Batten, eds.). The Palaeontological Association, UK.

Sereno, P. C., C. A. Forster, R. R. Rogers, and A. M. Monetta. 1993. Primitive dinosaur skeleton from Argentina and the early evolution of *Dinosauria*. *Nature* 361:64–66.

Smith, N. D., and D. Pol. 2007. Anatomy of a basal sauropodomorph dinosaur from the Early Jurassic Hanson Formation of Antarctica. *Acta Palaeontol. Pol.* 52:657–674.

Smith, N. D., P. J. Makovicky, F. L. Agnolin, M. D. Ezcurra, D. F. Pais, and S. W. Salisbury. 2008. A *Megaraptor*-like theropod (*Dinosauria*: *Tetanurae*) in Australia: support for faunal exchange across eastern and western Gondwana in the Mid-Cretaceous. *Proc. R. Soc. Lond. B Biol. Sci.* 275:2085–2093.

Starrfelt, J., and L. H. Liow. 2016. How many dinosaur species were there? Fossil bias and true richness estimated using a Poisson sampling model. *Philos. Trans. R. Soc. Lond. B Biol. Sci.* 371:20150219.

Sues, H. D., S. J. Nesbitt, D. S. Berman, and A. C. Henrici. 2011. A late-surviving basal theropod dinosaur from the latest Triassic of North America. *Proc. R. Soc. Lond. B Biol. Sci.* 278:3459–3464.

Upchurch, P. 1997. *Sauropodomorpha*. Pp. 658–660 in *The Encyclopedia of Dinosaurs* (P. J. Currie and K. Padian, eds.). Academic Press, San Diego, CA.

Upchurch, P., P. M. Barrett, and P. Dodson. 2004. *Sauropoda*. Pp. 259–322 in *The* Dinosauria. 2nd edition (D. B. Weishampel, P. Dodson, and H. Osmólska, eds.). University of California Press, Berkeley, CA and Los Angeles, CA.

Upchurch, P., P. M. Barrett, and P. M. Galton. 2007. A phylogenetic analysis of basal sauropodomorph relationships: implications for the origin of sauropod dinosaurs. Pp. 57–90 in *Special Papers in Palaeontology 77: Evolution and Palaeobiology of Early Sauropodomorph Dinosaurs* (P. M. Barrett and D. J. Batten, eds.). The Palaeontological Association, UK.

Weishampel, D. B., P. Dodson, and H. Osmólska, eds. 2004. *The* Dinosauria. 2nd edition. University of California Press, Berkeley, CA.

Wilson, J. A. 2002. Sauropod dinosaur phylogeny: critique and cladistic analysis. *Zool. J. Linn. Soc.* 136:217–276.

Wilson, J. A., and P. C. Sereno. 1998. Early evolution and higher-level phylogeny of sauropod dinosaurs. *Soc. Vertebr. Paleontol. Mem.* 5:1–68.

Yates, A. M. 2007. The first complete skull of the Triassic dinosaur *Melanorosaurus* Haughton (*Sauropodomorpha*: *Anchisauria*). Pp. 9–55 in *Special Papers in Palaeontology 77: Evolution and Palaeobiology of Early Sauropodomorph Dinosaurs* (P. M. Barrett and D. J. Batten, eds.). The Palaeontological Association, UK.

Yates, A. M. 2010. A revision of the problematic sauropodomorph dinosaurs from Manchester, Connecticut and the status of *Anchisaurus* Marsh. *Palaeontology* 53:739–752.

Yates, A. M., M. F. Bonnan, and J. Neveling. 2011. A new basal sauropodomorph dinosaur from the Early Jurassic of South Africa. *J. Vertebr. Paleontol.* 31(3):610–625.

Zanno, L. E. 2010. A taxonomic and phylogenetic re-evaluation of *Therizinosauria* (*Dinosauria*: *Maniraptora*). *J. Syst. Palaeontol.* 8:503–543.

Authors

Matteo Fabbri; Department of Geology and Geophysics; Yale University, New Haven, CT, 06520, USA. Email: matteo.fabbri@yale.edu.

Emanuel Tschopp; Division of Paleontology; American Museum of Natural History, New York, NY, 10024, USA. Email: etschopp@amnh.org.

Blair W. McPhee; Departamento de Biologia, FFCLRP; Universidade de São Paulo; Ribeirão Preto, São Paulo 14040-901, Brazil. Email: blair.mcphee@gmail.com.

Sterling J. Nesbitt; Department of Geosciences; Virginia Polytechnic Institute and State University, Blacksburg, VA, 24061, USA. Email: sjn2104@vt.edu.

Diego Pol; CONICET-Museo Paleontológico Egidio Feruglio; Trelew, Argentina. Email: cacopol@gmail.com.

Max C. Langer; Laboratório de Paleontologia de Ribeirão Preto, FFCLRP; Universidade de São Paulo; Av. Bandeirantes 3900, 14040-901 Ribeirão Preto, São Paulo, Brazil. Email: mclanger@ffclrp.usp.br.

Date Accepted: 15 October 2018

Primary Editor: Jacques A. Gauthier

Theropoda O. C. Marsh 1881 [D. Naish, A. Cau, T. R. Holtz, Jr., M. Fabbri, and J. A. Gauthier], converted clade name

Registration Number: 216

Definition: The largest clade containing *Allosaurus fragilis* Marsh 1877 (*Theropoda*) but neither *Plateosaurus engelhardti* Meyer 1837 (*Sauropodomorpha*) nor *Heterodontosaurus tucki* Crompton and Charig 1962 (*Ornithischia*). This is a maximum-clade definition. Abbreviated definition: max ∇ (*Allosaurus fragilis* Marsh 1877 ~ *Plateosaurus engelhardti* Meyer 1837 & *Heterodontosaurus tucki* Crompton and Charig 1962).

Etymology: "Beast-foot". From the Greek *ther* (beast) and *podos* (foot).

Reference Phylogeny: Figure 3 of Cau (2018) is the primary reference phylogeny. It contains an extensive sample of non-avian dinosaurs usually referred to *Theropoda*, including our internal specifier, *Allosaurus fragilis*, as well as a reasonable sample of potential outgroups to *Theropoda* (see Comments), including our external specifiers *Plateosaurus* and *Heterodontosaurus*. Phylogenies in Nesbitt (2011: Fig. 52) and Langer et al. (2017: Fig 1), or based on successive iterations of their datasets (e.g., Langer and Ferigolo, 2013: Fig. 22; Parry et al., 2017: Fig. 5), are based on a broad sample of Mesozoic pan-avians including both our internal and external specifiers. Phylogenies in Sereno (1999a: Fig. 2), Holtz and Osmólska (2004: Figs. S.1, S.2), Ezcurra (2006: Fig. 3), Nesbitt et al. (2009: Fig. S.7), and Hendrickx et al. (2015: Figs. 4 and 5) also considered an extensive range of Mesozoic stem avians, including our internal specifier, *Allosaurus fragilis*.

Composition: Weishampel et al. (2004) provided somewhat dated lists of extinct, non-avialan theropods, including reviews of many major clades diverging from the avian stem. Chiappe and Qingjin (2016) presented a lavishly illustrated overview of Mesozoic flying dinosaurs (*Avialae*), and Del Hoyo et al. (2013) supplied a comprehensive list of extant members of the crown clade.

From therizinosaurs to hummingbirds, theropods exhibit striking morphological disparity. They are also exceptionally diverse in terms of number of species, although this is likely influenced by severe constraints on body mass imposed by flight (Field et al., 2013), and the correlation between smaller body size and higher diversity (e.g., Martin, 2016), in a major subclade of theropods. Starrfelt and Liow (2016) reported 516 extinct species of Mesozoic theropods and estimated a total species richness of 1,115 species (confidence range: 780–1,653). These are likely to be underestimates, given the limited sample of geographic and temporal ranges represented by known "dinosaur"-bearing sediments. In contrast, there are 10,711 species of avian theropods alive today (Gill and Donsker, 2018; see *Aves*, this volume). A higher number (ca. 18,043) has been extrapolated via application of the phylogenetic species concept to a sample of 200 "conventionally" recognized species (Barrowclough et al., 2016). Some authors (e.g., Brodkorb, 1960) estimate that the avian crown alone might contain more than 1.5 million species if missing extinct species were to be added. Nevertheless, reliable estimates for the total number of extinct species within *Theropoda* (both in and outside of the avian crown) are unavailable (Nesbitt, 2011).

Diagnostic Apomorphies: More than 99% of species referred to *Theropoda* are uncontroversial (see Composition). As might be expected, however, that is not the case for a few relatively unmodified Triassic dinosaurs (and perhaps the Late Jurassic *Chilesaurus diegosuarezi* Novas et al. 2015; see Baron and Barrett, 2017, and Cau, 2018), and the list of apomorphies shared by *Theropoda* accordingly varies among studies (see Comments). For example, lists of shared apomorphies were proposed by Novas (1993), Sereno et al. (1993), and Sereno (1999a) to unite *Eoraptor lunensis* and *Herrerasauridae* with *Neotheropoda* Bakker 1986 (see below), including the following (author citations are for first explicit identification as apomorphies): lacrimal broadly exposed on skull roof (Gauthier, 1986); prong-shaped epipophyses on cervical vertebrae (Novas, 1993); strap-shaped scapular blade (Gauthier, 1986); boot-like expansion at the distal end of pubis (Gauthier, 1986); obturator process on the ischium (Gauthier, 1986); arched brevis fossa (Gauthier, 1986); distal femoral depression (Sereno, 1999a); and trochanteric shelf (Gauthier, 1986).

Langer (2004) argued, however, that all of these apomorphies are present in other saurischians and do not support a clade that includes only *Eoraptor lunensis*, *Herrerasauridae* and *Neotheropoda*. Furthermore, *Eoraptor lunensis* and *Herrerasauridae* were inferred by Langer (2004) to be early saurischians, diverging in some analyses before the clade composed of *Theropoda* + *Sauropodomorpha* (= *Eusaurischia* Padian et al. 1999; see below). If correct, this topology requires that the apomorphies proposed by Novas (1993), Sereno et al. (1993), and Sereno (1999a) to diagnose *Theropoda* are in fact saurischian apomorphies later lost in *Sauropodomorpha*. Alternatively, if either *Eoraptor lunensis* (Sereno et al., 2012) or *Herrerasauridae* (Baron et al., 2017a) are early diverging *Sauropodomorpha,* then at least some of these apomorphies could be ancestral for *Dinosauria* or might have evolved independently in *Eoraptor lunensis* and/or *Herrerasauridae*.

Because of this uncertainty, we propose a definition of *Theropoda* that ensures only that unambiguous neotheropods (e.g., *Allosaurus*) are part of this clade, and that ornithischians (e.g., *Heterdontosaurus*) and sauropodomorphs (e.g., *Plateosaurus*) are not. We are content to let the controversial taxa fall where they may, and their taxonomic designations change accordingly, in future studies.

Despite these questions, there are still several synapomorphies diagnosing uncontroversial theropods, now known as *Neotheropoda* Bakker 1986 (sensu Sereno, 2005a,b = min ∇ (*Coelophysis bauri* Cope 1887 & *Passer domesticus* Linnaeus 1758)), which approximates the traditional composition of *Theropoda* prior to the discovery of species (e.g., *Tawa hallae* Nesbitt et al. 2009) stemward of their last common ancestor (or outside of *Ornithoscelida* on Cau's 2018 tree). According to Cau (2018), the following combination of apomorphies is unique to *Neotheropoda* (sensu Sereno, 2005b): narrow snout with subparallel maxillae (Cau, 2018); elongate and subvertical orbital ramus of lacrimal (Cau, 2018); higher degree of antorbital and braincase pneumaticity (Sereno et al., 2012); furcula (fused clavicles) (Gauthier, 1986); loss of distal carpal 5 (Cau, 2018); metacarpal V vestigial/absent (Gauthier, 1986); enlarged preacetabular process of ilium (Gauthier, 1986); proximally elongate fibular crest on tibia (Cau, 2018); slender fibula adpressed along entire length to tibia (Gauthier, 1986); astragalus with anterior platform and horizontal sulcus (Cau, 2018); foot functionally tridactyl (Gauthier, 1986) with pedal digit I failing to contact both the substrate and the tarsus (Gauthier, 1986), and with metatarsal V half or less as long as metatarsal IV (Gauthier, 1986).

Synonyms: *Goniopoda* Cope 1866; *Harpagosauria* Haeckel 1866 (consistently misspelled *Harpagmosauria* since Cope, 1870);

Carnosauriformes Cooper 1985; all are partial, approximate synonyms. In the context of Baron et al.'s (2017a) tree, *Saurischia* as defined by Gauthier (1986:15)—"birds and all dinosaurs that are closer to birds than they are to Ornithischia"—is an unambiguous synonym of *Theropoda*.

Comments: The first-named Mesozoic theropod (and also the first-named non-avian dinosaur) was *Megalosaurus bucklandii* Buckland 1824 from the British Middle Jurassic. The type material of this species, the basis of Huxley's (1869) traditional taxon name *Megalosauridae*, is diagnosable, despite its fragmentary nature (Benson et al., 2008). Cope (1866) coined *Goniopoda* for *Megalosaurus* and its supposed North American relative *Laelaps aquilunguis* (= *Dryptosaurus*). *Goniopoda* was not strictly equivalent to *Theropoda* of modern usage as it explicitly excluded small-bodied forms such as *Compsognathus longipes*, which were included in a distinct taxon, *Symphypoda*, which was thought to be especially closely related to birds (Cope, 1867). *Goniopoda* was used only occasionally by later authors (e.g., Matthew and Brown, 1922) and has generally been ignored. *Harpagosauria* was coined for a group comprising *Megalosaurus*, the sauropodomorphs *Plateosaurus* and *Pelorosaurus*, and the ornithischian *Hylaeosaurus* (but not *Iguanodon*). Thus, Haeckel's (1866) taxon is in modern understanding a paraphyletic version of *Dinosauria* missing at least some ornithischians and birds. However, Haeckel described the harpagosaurs as "*Carnivore Lindwürmer*" (literally, "carnivorous dragons"), and thus Cope (1870) described this taxon as comparable to his *Goniopoda*. (Cope misspelled Haeckel's taxon as "*Harpagmosauria*", a misspelling repeated in the few subsequent references mentioning this name.) None of these taxon names have been used by working scientists in recent decades, in contrast to *Theropoda*.

Marsh's (1881) initial concept of *Theropoda* included only his family *Allosauridae*, then thought to include *Allosaurus, Creosaurus* and *Labrosaurus*. Like *Goniopoda*, *Theropoda* as originally conceived excluded such small-bodied taxa as *Coelurus fragilis*. Marsh (1884) later expanded *Theropoda* to include *Megalosauridae, Ceratosauridae, Labrosauridae, Zanclodontidae, Amphisauridae, Coeluria* and *Compsognatha*. This combination of taxa, with the exception of those now known to be sauropodomorphs (*Amphisauridae*) or stem crocodylians (*Zanclodontidae*), essentially set the stage for the traditional composition of *Theropoda* followed by most authors thereafter.

Theropoda as conceived by Marsh (1884) did not include *Aves* and would be regarded as paraphyletic in modern taxonomies. Von Huene (1923, 1926, 1932) thought that *Theropoda* was diphyletic and classified its members in two suborders: *Coelurosauria* and *Pachypodosauria*. *Coelurosauria* mostly contained small-bodied taxa such as *Coelophysis bauri* and *Ornitholestes hermanni* (but also the large bodied *Ceratosaurus nasicornis* and *Tyrannosaurus rex*, among others) while *Pachypodosauria* included large-bodied *Carnosauria* and *Sauropodomorpha*. Von Huene's main reason for regarding carnosaurs and sauropodomorphs as close relatives rested on the erroneous association of predatory stem archosaurs (e.g., *Phytosauria*) or stem crocodylians (e.g., *Rauisuchia*) with herbivorous sauropodomorph remains (Benton, 1986). Matthew and Brown (1922) rejected von Huene's classification and resurrected Cope's *Goniopoda*, including *Megalosauridae, Spinosauridae, Coeluridae, Deinodontidae* (= *Tyrannosauridae*) and *Ornithomimidae* within this group. Their use of the name *Goniopoda* has not been followed. Galton and Jensen (1979:10) suggested that *Megalosauridae* (and perhaps other large theropods included by them in *Carnosauria*) might be "descendants of the

Prosauropoda rather than of another theropod currently included in the Coelurosauria", and therefore endorsed diphyly of *Theropoda*. Numerous derived characters shared by megalosaurids and other theropods (see Diagnostic Apomorphies) demonstrate that any similarities shared by megalosaurids and sauropodomorphs are either convergent or plesiomorphic. It has also been suggested by some authors that *Therizinosauroidea* (= segnosaurs), *Avialae*, and even the whole of *Maniraptora* might not be part of *Theropoda* (Paul, 1984; Feduccia et al., 2007), but these suggestions are contradicted by nearly all available evidence.

Cooper (1985:283) proposed the name *Carnosauriformes* for the group of dinosaurs "retaining the primitive condition of recurved thecodontian dentition with finely serrated cutting edges". While no constituent taxa were listed, discussed in the text, or marked on Cooper's phylogenetic diagram, it appears that this name was intended as a replacement for *Theropoda* (at least in part). The name has not, to our knowledge, been used subsequently.

Finally, Olshevsky (1991:89) proposed the name *Theropodomorpha* for "the smallest clade including *Megalancosaurus*, *Longisquama*, and (possibly) *Cosesaurus*, together with all the well-known theropod dinosaurs, *minus* the clade Aves (or Carinatae)." The name *Theropodomorpha* does not correspond to any clade as it is paraphyletic by definition. Furthermore, the hypothesis that *Megalancosaurus*, *Longisquama*, and *Cosesaurus* might be close to *Theropoda* (or even *Archosauria*) is heterodox and lacks support in any trees including a range of stem archosaurs (e.g., Gottmann-Quesada and Sander, 2009).

The first phylogenetic definition proposed for the name *Theropoda* was Gauthier's (1986:18) maximum-clade definition: "birds and all other saurischians closer to birds than they are to sauropodomorphs." Subsequent authors have continued to use this style of definition but selected different specifiers. For example, in place of *Sauropodomorpha*, Currie (1997) used *Plateosaurus* and *Diplodocus* as external specifiers, Sereno (1998) used *Saltasaurus*, Padian et al. (1999) used *Cetiosaurus*, and Kischlat (2000) used *Morosaurus*. Note that all of these definitions used names of "higher" taxa rather than of species and are thus in violation of *ICPN* Article 11.1 (Cantino and de Queiroz, 2020). Only Holtz and Osmólska (2004) used a species, *Cetiosaurus oxoniensis*, to represent *Sauropodomorpha*. Following Nesbitt (2011), we use relatively well-known species, the sauropodomorph *Plateosaurus engelhardti* and the ornithischian *Heterodontosaurus tucki*, as our external specifiers. These two species partly buffer our definition of *Theropoda* from current debates regarding early dinosaur phylogeny (e.g., Baron et al., 2017a,b; Langer et al., 2017; Cau, 2018), ensuring preservation of (most) of its traditional composition should either *Ornithischia* or *Sauropodomorpha* turn out to be closer to *Theropoda* as defined here.

The use of the avian species *Passer domesticus* as the internal specifier for *Theropoda* (e.g., Sereno, 1998; Holtz and Osmólska, 2004) contradicts Recommendation 11A of the *ICPN*. Although *Theropoda* certainly includes *Passer domesticus* (*Aves*), the use of this species as a specifier does not conform to traditional ideas about its composition. Indeed, although they acknowledged an evolutionary connection, neither Marsh nor Ostrom included birds among theropod dinosaurs. Ostrom's ground-breaking studies of *Deinonychus* and *Archaeopteryx* inspired the Dinosaur Renaissance in which "dinosaurs" were reimagined as active, hot-blooded, stem birds rather than sluggish, cold-blooded, overgrown "lizards". Justly famous for inspiring this revolution in dinosaurology, Ostrom nevertheless firmly believed that "dinosaurs are dinosaurs, and birds are birds" (pers. comm. to J. A. Gauthier in 1996). We have accordingly opted to use *Allosaurus fragilis* as the internal specifier for *Theropoda* (as proposed previously by Kischlat, 2000). This species is well represented in collections with dozens of skeletons, has an excellent and substantially complete topotype specimen (USNM 4734), is comparatively well described (e.g., Madsen, 1976), is often included in phylogenies of *Theropoda* (e.g., Gauthier, 1986; Novas, 1996; Sereno, 1999a; Holtz and Osmólska, 2004; Brusatte and Sereno, 2008; Carrano and Sampson, 2008; Carrano et al., 2012; Smith et al., 2008; Benson et al., 2010; Lee et al., 2014; Cau, 2018), its ontogeny is well known both from morphological and osteohistological perspectives (e.g., Smith, 1998; Bybee et al., 2006; Lee and Werning, 2008), its pathologies are well documented (e.g., Hanna, 2002; Foth et al., 2015), and this species served as the basis of Marsh's (1881) taxon when he coined the name *Theropoda*.

The overall composition of *Theropoda* has been remarkably stable in the phylogenetic era despite fundamentally different topologies proposed for relationships at the very base of *Dinosauria, Saurischia* and *Theropoda* (compare, e.g., Sereno, 1999a, with Baron et al., 2017a). Nevertheless, character discordance and incomplete preservation complicate identification of the putatively earliest-diverging theropods, especially the Triassic taxa *Eoraptor lunensis* (Sereno et al., 1993), *Herrerasauridae* (Reig, 1963; Novas, 1993, 1996), and *Guiabasaurus candelariensis* (Bonaparte et al., 1999; Bonaparte et al., 2007). For example, while herrerasaurids have conventionally been regarded as basal dinosaurs (e.g., Novas, 1992), the discovery of *Eoraptor lunensis* and of new specimens of *Herrerasaurus ischigualastensis* led Novas (1993, 1996), Sereno et al. (1993), and Sereno (1999a,b) to regard these taxa as the earliest-diverging theropods. However, the relationships relative to *Theropoda* (see below), *Sauropodomorpha* and *Dinosauria* are uncertain

(e.g., Padian and May, 1993; Sereno, 1999a,b, 2007; Langer, 2004; Langer and Benton, 2006; Yates, 2007; Smith et al., 2008; Martínez and Alcober, 2009; Nesbitt et al., 2009; Ezcurra, 2010; Langer et al., 2010; Nesbitt et al., 2010; Apaldetti et al., 2011; Kammerer et al., 2011; Martínez et al., 2011; Nesbitt, 2011; Sues et al., 2011; Martínez et al. 2012; Sereno et al., 2012; Otero and Pol, 2013; Cabriera et al., 2016; Baron et al., 2017a,b; Müller et al., 2018). These early and relatively unmodified taxa will only be included in *Theropoda* if they are closer to *Allosaurus* than to both *Sauropodomorpha* (Gauthier, 1986) and *Ornithischia* (Baron et al., 2017a). Among Triassic dinosaurs, *Tawa hallae* (Nesbitt et al., 2009), *Daemonosaurus chauliodus* (Sues et al., 2011), and *Eodromaeus murphi* (Martínez et al., 2011) may potentially be regarded among the earliest and least-modified species that are outside of *Neotheropoda* and still inside *Theropoda* as that name is defined here. These taxa display the "blind-pouch" stage of pleurocoel evolution (Nesbitt, 2011), in which the pulmonary air sac system is beginning to invade the cervical vertebrae, recapitulating the earliest stages of the development of this system in avian ontogeny (Gauthier, 1986; Britt et al., 1998; Rauhut, 2003; Nesbitt, 2011). Under our preferred topology, *Theropoda* thus stems from at least the early-middle Norian (Late Triassic), more than 220 million years ago (Ezcurra and Brusatte, 2011).

Gauthier's (1986) *Theropoda* was hypothesized to contain two major subclades: *Ceratosauria* and *Tetanurae*. (Gauthier conceptualized *Theropoda* as a maximum clade; some later authors applied the name *Neotheropoda* Bakker 1986 to the minimum clade containing both *Coelophysis,* representing Gauthier's *Ceratosauria,* and *Allosaurus,* representing Gauthier's *Tetanurae*). Those two subclades were supported by some later authors (Holtz, 2000; Holtz and Osmólska, 2004; Tykoski and

Rowe, 2004; Allain et al., 2007). The oldest formal phylogenetic definition of *Ceratosauria* (Sereno, 1998: 64) was "all theropods closer to *Coelophysis* than to *Neornithes*", coined in light of this topology. However, others rejected that tree, arguing instead that some of Gauthier's ceratosaurs, most notably *Ceratosaurus nasicornis*, are more closely related to *Tetanurae* than are others, such as *Coelophysis bauri* and *Dilophosaurus wetherelli*. As the earlier definition lacked the eponymous taxon as a specifier, Padian et al. (1999) defined the taxon as *Ceratosaurus* and all taxa sharing a more recent common ancestor with it than with *Neornithes*. In that context, *Ceratosauria* encompasses only *Ceratosaurus nasicornis* and *Abelisauridae* (and their close relatives) (Carrano et al., 2002; Rauhut, 2003; Wilson et al., 2003; Ezcurra, 2006; Carrano and Sampson, 2008; Smith et al., 2008; Nesbitt et al., 2009; Carrano et al., 2012). The name *Tetanurae* Gauthier 1986 was defined as referring to a maximum clade incorporating birds and all theropods closer to birds than to *Ceratosauria* (see also Padian et al., 1999). It was subsequently redefined to apply to a less inclusive clade, viz., "all neotheropods closer to *Neornithes* than to *Torvosaurus*" by Sereno (1998:64) [according to Wilson et al. (2003), in which Sereno was a co-author, this "redefinition" was a typographical error]. These early efforts at phylogenetic definitions violate *ICPN* Article 11.1 (Cantino and de Queiroz, 2020), which requires that valid specifiers must either be species, specimens, or apomorphies.

For current perspectives on theropod ingroup relationships in addition to those cited above, see: Clarke et al. (2005); Livezey and Zusi (2007); Senter (2007); Clarke and Middleton (2008); Hackett et al. (2008); Balanoff and Norell (2012); Hu et al. (2012); Turner et al. (2012); Agnolín and Novas (2013); Brusatte et al. (2013); Godefroit et al. (2013); Brusatte et al. (2014); Lamanna et al. (2014); Lee et al. (2014); Nesbitt and Ezcurra (2015); Prum et al. (2015); Xu et al. (2015); Brusatte and Carr (2016); Huang et al. (2016); Nesbitt and Clarke (2016); Cau et al. (2017); and Wang et al. (2017).

Literature Cited

Agnolín, F. L., and F. E. Novas. 2013. *Avian Ancestors: a Review of the Phylogenetic Relationships of the Theropods* Unenlagiidae, Microraptoria, Anchiornis *and* Scansoriopterygidae. Springer, Dordrecht.

Allain, R., R. Tykoski, N. Aquesbi, N.-E. Jalil, M. Monbaron, D. Russell, and P. Taquet. 2007. An abelisauroid (*Dinosauria: Theropoda*) from the Early Jurassic of the High Atlas Mountains, Morocco, and the radiation of ceratosaurs. *J. Vertebr. Paleontol.* 27:610–624.

Apaldetti, C., R. N. Martínez, O. A. Alcober, and D. Pol. 2011. A new basal sauropodomorph (*Dinosauria: Saurischia*) from Quebrada del Barro Formation (Marayes-El Carrizal Basin), northwestern Argentina. *PLOS ONE* 6(11):e26964.

Bakker, R. T. 1986. *The Dinosaur Heresies*. William Morrow, New York.

Balanoff, A. M., and M. A. Norell. 2012. Osteology of *Khaan mckennai* (*Oviraptorosauria: Theropoda*). *Bull. Am. Mus. Nat. Hist.* 372:1–77.

Baron, M. G., and P. M. Barrett. 2017. A dinosaur missing-link? *Chilesaurus* and the early evolution of ornithischian dinosaurs. *Biol. Lett.* 13:1–5.

Baron, M. G., D. B. Norman, and P. M. Barrett. 2017a. A new hypothesis of dinosaur relationships and early dinosaur evolution. *Nature* 543:501–506.

Baron, M. G., D. B. Norman, and P. M. Barrett. 2017b. Baron et al. reply. *Nature* 551:E4–E5.

Barrowclough, G. F., J. Cracraft, J. Klicka, and R. M. Zink. 2016. How many kinds of birds are there and why does it matter? *PLOS ONE* 11(11):e0166307.

Benson, R. B. J., P. M. Barrett, H. P. Powell, and D. B. Norman. 2008. The taxonomic status of *Megalosaurus bucklandii* (*Dinosauria, Theropoda*) from the Middle Jurassic of Oxfordshire, UK. *Palaeontology* 51:419–424.

Benson, R. B. J., M. T. Carrano, and S. L. Brusatte. 2010. A new clade of archaic large-bodied predatory dinosaurs (*Theropoda: Allosauroidea*) that survived to the latest Mesozoic. *Naturwissenschaften* 97:71–78.

Benton, M. J. 1986. The Late Triassic reptile *Teratosaurus*—a rauisuchian, not a dinosaur. *Palaeontology* 29:293–301.

Bonaparte, J. F., G. Brea, C. L. Schultz, and A. G. Martinelli. 2007. A new specimen of *Guiabasaurus candelariensis* (basal *Saurischia*) from the Late Triassic Caturrita Formation of southern Brazil. *Hist. Biol.* 19:73–82.

Bonaparte, J. F., J. Ferigolo, and A. M. Ribeiro. 1999. A new early Late Triassic saurischian dinosaur from Rio Grande do Sol state, Brazil. *Proceedings of the Second Gondwanan Dinosaur Symposium, National Science Museum Monographs* 15:89–109.

Britt, B. B., P. J. Makovicky, J. Gauthier, and N. Bonde. 1998. Postcranial pneumatization in *Archaeopteryx*. *Nature* 6700:374–376.

Brodkorb, P. 1960. How many bird species have existed? *Bull. Florida State Mus. Biol. Sci.* 5:41–56.

Brusatte, S. L., and T. D. Carr. 2016. The phylogeny and evolutionary history of tyrannosauroid dinosaurs. *Sci. Rep.* 6:20252.

Brusatte S. L., G. Lloyd, S. Wang, and M. A. Norell. 2014. Gradual assembly of avian body plan culminated in rapid rates of evolution across the dinosaur–bird transition. *Curr. Biol.* 24:2386–2392.

Brusatte, S. L., and P. C. Sereno. 2008. Phylogeny of *Allosauroidea* (*Dinosauria*: *Theropoda*): comparative analysis and resolution. *J. Syst. Palaeontol.* 6:155–182.

Brusatte, S. L., M. Vremir, Z. Csiki-Sava, A. H. Turner, A. Watanabe, G. M. Erickson, and M. A. Norell. 2013. The osteology of *Balaur bondoc*, an island-dwelling dromaeosaurid (*Dinosauria: Theropoda*) from the Late Cretaceous of Romania. *Bull. Am. Mus. Nat. Hist.* 374:1–100.

Buckland, W. 1824. Notice on the *Megalosaurus* or great fossil lizard of Stonesfield. *Trans. Geol. Soc. Lond. (Ser. 2)* 1:390–396.

Bybee, P. J., A. H. Lee, and E. T. Lamm. 2006. Sizing the Jurassic theropod dinosaur *Allosaurus*: assessing growth strategy and evolution of ontogenetic scaling of limbs. *J. Morphol.* 267:347–359.

Cabreira, S. F., A. W. A. Kellner, S. Dias-da-Silva, L. R. da Silva, M. Bronzati, J. C. de Almeida Marsola, R. T. Müller, J. de Souza Bittencourt, B. J. Batista, T. Raugust, R. Carrilho, A. Brodt, and M. C. Langer. 2016. A unique Late Triassic dinosauromorph assemblage reveals dinosaur ancestral anatomy and diet. *Curr. Biol.* 26:3090–3095.

Cantino, P. D., and K. de Queiroz. 2020. *International Code of Phylogenetic Nomenclature (PhyloCode)*, Version 6. CRC Press, Boca Raton, FL.

Carrano, M. T., R. B. J. Benson, and S. D. Sampson. 2012. The phylogeny of *Tetanurae* (*Dinosauria*: *Theropoda*). *J. Syst. Paleontol.* 10:211–300.

Carrano, M. T., and S. D. Sampson. 2008. The phylogeny of *Ceratosauria* (*Dinosauria*: *Theropoda*). *J. Syst. Palaeontol.* 6:183–236.

Carrano, M. T., S. D. Sampson, and C. A. Forster. 2002. The osteology of *Masiakasaurus knopfleri*, a small abelisauroid (*Dinosauria: Theropoda*) from the Late Cretaceous of Madagascar. *J. Vertebr. Paleontol.* 22:510–534.

Cau, A. 2018. The assembly of the avian body plan: a 160-million-year-long process. *Boll. Soc. Paleontol. Ital.* 57:1–25.

Cau, A., V. Beyrand, D. F. A. E. Voeten, V. Fernandez, P. Tafforeau, K. Stein, R. Barsbold, K. Tsogtbaatar, P. J. Currie, and P. Godefroit. 2017. Synchrotron scanning reveals amphibious ecomorphology in a new clade of bird-like dinosaurs. *Nature* 552:395–399.

Chiappe, L. M., and M. Qingjin. 2016. *Birds of Stone*. Johns Hopkins University Press, Baltimore, MD.

Clarke, J. A., and K. M. Middleton. 2008. Mosaicism, modules, and the evolution of birds: results from a Bayesian approach to the study of morphological evolution using discrete character data. *Syst. Biol.* 57:185–201.

Clarke, J. A., C. P. Tambussi, J. I. Noriega, G. M. Erickson, and R. A. Ketcham. 2005. Definitive fossil evidence for the extant avian radiation in the Cretaceous. *Nature* 433:305–308.

Cooper, M. R. 1985. A revision of the ornithischian dinosaur *Kangnasaurus coetzeei* Haughton, with a classification of the *Ornithischia. Ann. S. Afr. Mus.* 95:281–317.

Cope, E. D. 1866. On the anomalous relations existing between the tibia and fibula in certain of the *Dinosauria. Proc. Acad. Nat. Sci. Phila.* 18:316–317.

Cope, E. D. 1867. An account of the extinct reptiles which approached the birds. *Proc. Acad. Nat. Sci. Phila.* 19:234–235.

Cope, E. D. 1870. Synopsis of the extinct *Batrachia* and *Reptilia* of North America. *Trans. Am. Philos. Soc.* 14:1–252.

Cope, E. D. 1887. The dinosaurian genus *Coelurus. Am. Nat.* 21:367–369.

Crompton, A. W., and A. J. Charig. 1962. A new ornithischian from the Upper Triassic of South Africa. *Nature* 196:1074–1077.

Currie, P. J. 1997. *Theropoda.* Pp. 731–737 in *Encyclopedia of Dinosaurs* (P. J. Currie and K. Padian, eds.). University of Californian Press, San Diego, CA.

Del Hoyo, J., A. Elliott, J. Sargatal, and D. A. Christie, eds. 2013. *Handbook of the Birds of the World.* Special volume. New species and global index. Lynx Edicions, Barcelona.

Ezcurra, M. D. 2006. A review of the systematic position of the dinosauriform archosaur *Eucoelophysis baldwini* Sullivan & Lucas, 1999 from the Upper Triassic of New Mexico, USA. *Geodiversitas* 28:649–684.

Ezcurra, M. D. 2010. A new early dinosaur (*Saurischia: Sauropodomorpha*) from the Late Triassic of Argentina: a reassessment of dinosaur origin and phylogeny. *J. Syst. Palaeontol.* 8:371–425.

Ezcurra, M. D., and S. L. Brusatte. 2011. Taxonomic and phylogenetic reassessment of the early neotheropod dinosaur *Camposaurus arizonensis* from the Late Triassic of North America. *Palaeontology* 54:763–772.

Feduccia, A., L. D. Martin, and S. Tarsitano. 2007. *Archaeopteryx* 2007: quo vadis? *Auk* 124:373–380.

Field, D. J., C. Lynner, C. Brown, and S. A. F. Darroch. 2013. Skeletal correlates of body mass in modern and fossil birds. *PLOS ONE* 8(11):e82000.

Foth, C., S. W. Evers, B. Pabst, O. Mateus, A. Flisch, M. Patthey, and O. W. Rauhut. 2015. New insights into the lifestyle of *Allosaurus* (*Dinosauria: Theropoda*) based on another specimen with multiple pathologies. *PeerJ* 3:e940.

Galton, G. M., and J. A. Jensen. 1979. A new large theropod dinosaur from the Upper Jurassic of Colorado. *Brigham Young Univ. Geol. Stud.* 26:1–12.

Gauthier, J. A. 1986. Saurischian monophyly and the origin of birds. Pp. 1–55 in *The Origin of Birds and the Evolution of Flight* (K. Padian, ed.). Memoirs of the California Academy of Sciences 8. California Academy of Sciences, San Francisco, CA.

Gill, F., and D. Donsker, eds. 2018. *IOC World Bird List (v8.2).* doi:10.14344/IOC.ML.8.2.

Godefroit, P., A. Cau, H. Dong-Yu, F. Escuillié, W. Wenhao, and G. Dyke. 2013. A Jurassic avialan dinosaur from China resolves the early phylogenetic history of birds. *Nature* 498:359–362.

Gottmann-Quesada, A., and P. M. Sander. 2009. A redescription of the early archosauromorph *Protorosaurus speneri* Meyer, 1832 and its phylogenetic relationships. *Palaeontogr. Abt. A* 287:123–220.

Hackett, S. J., R. T. Kimball, S. Reddy, R. C. K. Bowie, E. L. Braun, M. J. Braun, J. L. Chojnowski, W. A. Cox, K.-L. Han, J. Harshman, C. J. Huddleston, B. D. Marks, K. J. Miglia, W. S. Moore, F. H. Sheldon, D. W. Steadman, C. C. Witt, and T. Yuri. 2008. A phylogenomic study of birds reveals their evolutionary history. *Science* 320:1763–1768

Haeckel, E. 1866. *Generelle Morphologie der Organismen. Band 2: Allgemeine Entwicklungsgeschichte der Organismen.* George Reimer, Berlin.

Hanna, R. R. 2002. Multiple injury and infection in a sub-adult theropod dinosaur *Allosaurus fragilis* with comparisons to allosaur pathology in the Cleveland-Lloyd dinosaur quarry collection. *J. Vertebr. Paleontol.* 22:76–90.

Hendrickx, C., S. A. Hartman, and O. Mateus. 2015. An overview of non-avian theropod discoveries and classification. *PalArch's J. Vertebr. Palaeontol.* 12:1–73.

Holtz, Jr., T. R. 2000. A new phylogeny of the carnivorous dinosaurs. *Gaia* 15:5–61.

Holtz, Jr., T. R., and H. Osmólska. 2004. *Saurischia.* Pp. 21–24 in *The* Dinosauria. 2nd edition (D. B. Weishampel, P. Dodson, and H. Osmólska, eds.). University of California Press, Berkeley, CA.

Hu, D., X. Xu, L. Hou, and C. Sullivan. 2012. A new enantiornithine bird from the Lower Cretaceous of Western Liaoning, China, and its implications for early avian evolution. *J. Vertebr. Paleontol.* 32:639–645.

Huang, J., X. Wang, Y. Hu, J. Liu, J. A. Peteya, and J. A. Clarke. 2016. A new ornithurine from the Early Cretaceous of China sheds light on the evolution of early ecological and cranial diversity in birds. *PeerJ* 4:e1765.

Huene, F. v. 1923. Carnivorous *Saurischia* in Europe since the Triassic. *Bull. Geol. Soc. Am.* 34:449–458.

Huene, F. v. 1926. The carnivorous *Saurischia* in the Jura and Cretaceous formations, principally in Europe. *Rev. Mus. La Plata* 29:35–127.

Huene, F. v. 1932. Die fossile Reptile Ordnung *Saurischia*, ihre Entwicklung und Geschichte. *Monogr. Geol. Palaeontol. (Ser. 1)* 4:1–361.

Huxley, T. H. 1869. On the upper jaw of *Megalosaurus. Q. J. Geol. Soc. Lond.* 25:311–314.

Kammerer, C. F., S. J. Nesbitt, and N. H. Shubin. 2011. The first silesaurid dinosauriform from the Late Triassic of Morocco. *Acta Palaeontol. Pol.* 57:277–284.

Kischlat, E.-E. 2000. Tecodôncios: a aurora dos arcossáurios no Triássico. Pp. 273–316 in *Paleontologia do Rio Grande do Sul* (M. Holz and L. F. De Ros, eds.). Centro de Investigação do Gonduana/Universidade Federal do Rio Grande do Sul.

Lamanna, M. C., H. D. Sues, E. R. Schachner, and T. R. Lyson. 2014. A new large-bodied oviraptorosaurian theropod dinosaur from the latest Cretaceous of western North America. *PLOS ONE* 9(3):e92022.

Langer, M. C. 2004. Basal *Saurischia.* Pp. 25–46 in *The* Dinosauria. 2nd edition (D. B. Weishampel, P. Dodson, and H. Osmólska, eds.). University of California Press, Berkeley, CA.

Langer, M. C., and M. J. Benton. 2006. Early dinosaurs: a phylogenetic study. *J. Syst. Palaeontol.* 4:309–358.

Langer, M. C., M. D. Ezcurra, J. S. Bittencourt, and F. E. Novas. 2010. The origin and early evolution of dinosaurs. *Biol. Rev.* 85:55–110.

Langer, M. C., M. D. Ezcurra, O. W. M. Rauhut, M. J. Benton, F. Knoll, B. W. McPhee, F. E. Novas, D. Pol, and S. L. Brusatte. 2017. Untangling the dinosaur family tree. *Nature* 551:E1–E3.

Langer, M. C., and J. Ferigolo. 2013. The Late Triassic dinosauromorph *Sacisaurus agudoensis* (Caturrita Formation; Rio Grande do Sul, Brazil): anatomy and affinities. *Geol. Soc. Lond. Spec. Publ.* 379:353–392.

Lee, A. H., and S. Werning 2008. Sexual maturity in growing dinosaurs does not fit reptilian growth models. *Proc. Natl. Acad. Sci. USA* 105:582–587.

Lee, M. S. Y., A. Cau, D. Naish, and G. J. Dyke. 2014. Sustained miniaturization and anatomical innovation in the dinosaurian ancestors of birds. *Science* 345:562–566.

Livezey, B. C., and R. L. Zusi. 2007. Higher-order phylogeny of modern birds (*Theropoda, Aves: Neornithes*) based on comparative anatomy. II. Analysis and discussion. *Zool. J. Linn. Soc.* 149:1–95.

Madsen, J. H. 1976. Allosaurus fragilis*: A Revised Osteology.* Bulletin 109, Utah Geological Survey.

Marsh, O. C. 1877. Notice of new dinosaurian reptiles from the Jurassic formation. *Am. J. Sci. Arts (Ser. 3)* 14:514–516.

Marsh, O. C. 1881. Principal characters of American Jurassic dinosaurs. Part V. *Am. J. Sci. (Ser. 3)* 21:417–423.

Marsh, O. C. 1884. Principal characters of American Jurassic dinosaurs. Part VIII. The order *Theropoda. Am. J. Sci. (Ser. 3)* 27:329–340.

Martin, R. A. 2016. Body size in (mostly) mammals: mass, speciation rates and the translation from gamma to alpha diversity on evolutionary timescales. *Hist. Biol.* 29:576–593.

Martínez, R. N., and O. A. Alcober. 2009. A basal sauropodomorph (*Dinosauria*: *Saurischia*) from the Ischigualasto Formation (Triassic, Carnian) and the early evolution of *Sauropodomorpha*. *PLOS ONE* 4(2):e4397.

Martínez, R. N., C. Apaldetti and D. Abelin. 2012. Basal sauropodomorphs from the Ischigualasto Formation. *J. Vertebr. Paleontol.* 32(s1):51–69.

Martínez, R. N., P. C. Sereno, O. A. Alcober, C. E. Colombi, P. R. Renne, I. P. Montañez, and B. S. Currie. 2011. A basal dinosaur from the dawn of the dinosaur era in southwestern Pangaea. *Science* 331:206–210.

Matthew, W. D., and B. Brown. 1922. The family *Deinodontidae*, with notice of a new genus from the Cretaceous of Alberta. *Bull. Am. Mus. Nat. Hist.* 46:367–385.

Meyer, H. v. 1837. Mitteilung an Prof. Bronn (*Plateosaurus engelhardti*). *Neues Jahrb. Geol. Paläontol.* 1837:316.

Müller, R. T., M. C. Langer, M. Bronzati, C. P. Pacheco, S. F. Cabreira, and S. Dias-da-Silva. 2018. Early evolution of sauropodomorphs: anatomy and phylogenetic relationships of a remarkably well-preserved dinosaur from the Upper Triassic of southern Brazil. *Zool. J. Linn. Soc.* 184:1187–1248.

Nesbitt, S. J. 2011. The early evolution of archosaurs: relationships and the origin of major clades. *Bull. Am. Mus. Nat. Hist.* 352:1–292.

Nesbitt, S. J., and J. A. Clarke. 2016. The anatomy and taxonomy of the exquisitely preserved Green River Formation (Early Eocene) lithornithids (*Aves*) and the relationships of *Lithornithidae*. *Bull. Am. Mus. Nat. Hist.* 406:1–91.

Nesbitt, S. J., and M. D. Ezcurra. 2015. The early fossil record of dinosaurs in North America: a new neotheropod from the base of the Upper Triassic Dockum Group of Texas. *Acta Palaeontol. Pol.* 60:513–526.

Nesbitt, S. J., C. A. Sidor, R. B. Irmis, K. D. Angielczyk, R. M. Smith, and L. A. Tsuji. 2010. Ecologically distinct dinosaurian sister group shows early diversification of *Ornithodira*. *Nature* 464:95–98.

Nesbitt, S. J., N. D. Smith, R. B. Irmis, A. H. Turner, A. Downs, and M. A. Norell. 2009. A complete skeleton of a Late Triassic saurischian and the early evolution of dinosaurs. *Science* 326:1530–1533.

Novas, F. E. 1992. Phylogenetic relationships of the basal dinosaurs, the *Herrerasauridae*. *Palaeontology* 35:51–62.

Novas, F. E. 1993. New information on the systematics and postcranial skeleton of *Herrerasaurus ischigualastensis* (*Theropoda*: *Herrerasauridae*) from the Ischigualasto Formation (Upper Triassic) of Argentina. *J. Vertebr. Paleontol.* 13:400–423.

Novas, F. E. 1996. Dinosaur monophyly. *J. Vertebr. Paleontol.* 16:723–741.

Novas, F. E., L. Salgado, M. Suárez, F. L. Agnolín, M. N. D. Ezcurra, N. S. R. Chimento, R. de la Cruz, M. P. Isasi, A. O. Vargas, and D. Rubilar-Rogers. 2015. An enigmatic plant-eating theropod from the Late Jurassic period of Chile. *Nature* 522:331–334.

Olshevsky, G. 1991. A revision of the parainfraclass *Archosauria* Cope, 1869, excluding the advanced *Crocodylia*. *Mesozoic Meanderings* 2:1–196.

Otero, A., and Pol, D. (2013). Postcranial anatomy and phylogenetic relationships of *Mussaurus patagonicus* (*Dinosauria*, *Sauropodomorpha*). *J. Vertebr. Paleontol.* 33:1138–1168.

Padian, K., J. R. Hutchinson, and T. R. Holtz, Jr. 1999. Phylogenetic definitions and nomenclature of the major taxonomic categories of the carnivorous *Dinosauria* (*Theropoda*). *J. Vertebr. Paleontol.* 19:69–80.

Padian, K., and C. L. May. 1993. The earliest dinosaurs. Pp. 379—381 in *The Nonmarine Triassic* (S. G. Lucas and M. Morales, eds.). Bulletin of the New Mexico Museum of Natural History and Science 3. New Mexico Museum of Natural History and Science, Albuquerque, NM.

Parry, L. A., M. G. Barron, and J. Vinther. 2017. Multiple optimality criteria support *Ornithoscelida*. *R. Soc. Open Sci.* 4(10):170833.

Paul, G. S. 1984. The segnosaurian dinosaurs: relics of the prosauropod-ornithischian transition? *J. Vertebr. Paleontol.* 4:507–515.

Prum, R. O., J. S. Berv, A. Dornburg, D. J. Field, J. P. Townsend, E. M. Lemmon, and A. R. Lemmon. 2015. A comprehensive phylogeny of birds (*Aves*) using targeted next-generation DNA sequencing. *Nature* 526:569–573.

Rauhut, O. W. M. 2003. The interrelationships and evolution of basal theropod dinosaurs. *Spec. Pap. Palaeontol.* 69:1–213.

Reig, O. A. 1963. La presencia de dinosaurios saurisquios en los "Estratos de Ischigualasto" (Mesotriasico Superior) de las provincias de San Juan y La Rioja (República Argentina). *Ameghiniana* 3:3–20.

Senter, P. 2007. A new look at the phylogeny of *Coelurosauria* (*Dinosauria: Theropoda*). *J. Syst. Palaeontol.* 5:429–463.

Sereno, P. C. 1998. A rationale for phylogenetic definitions, with application to the higher-level taxonomy of *Dinosauria. Neues Jahrb. Geol. Paläontol. Abh.* 20:41–83.

Sereno, P. C. 1999a. The evolution of dinosaurs. *Science* 284:2137–2147.

Sereno, P. C. 1999b. A rationale for dinosaurian taxonomy. *J. Vertebr. Paleontol.* 19:788–790.

Sereno, P. C. 2005a. The logical basis of phylogenetic taxonomy. *Syst. Biol.* 54:595–619.

Sereno, P. C. 2005b. TaxonSearch: Database for Suprageneric Taxa and Phylogenetic Definitions. Available at http://www.taxonsearch.org/, accessed on 16 January 2009.

Sereno, P. C. 2007. The phylogenetic relationships of early dinosaurs: a comparative report. *Hist. Biol.* 19:145–155.

Sereno, P. C., C. A. Forster, R. R. Rogers, and A. M. Monetta. 1993. Primitive dinosaur skeleton from Argentina and the early evolution of *Dinosauria. Nature* 361:64–66.

Sereno, P. C., R. N. Martínez, and O. A. Alcober. 2012. Osteology of *Eoraptor lunensis* (*Dinosauria, Sauropodomorpha*). *J. Vertebr. Paleontol.* 32:83–179.

Smith, D. K. 1998. A morphometric analysis of *Allosaurus. J. Vertebr. Paleontol.* 18:126–142.

Smith, N. D., P. J. Makovicky, F. L. Agnolín, M. D. Ezcurra, D. F. Pais, and S. W. Salisbury. 2008. A *Megaraptor*-like theropod (*Dinosauria: Tetanurae*) in Australia: support for faunal exchange across eastern and western Gondwana in the Mid-Cretaceous. *Proc. R. Soc. B Biol. Sci.* 275:2085–2093.

Starrfelt, J., and L. H. Liow. 2016. How many dinosaur species were there? Fossil bias and true richness estimated using a Poisson sampling model. *Philos. Trans. R. Soc. Lond. B Biol. Sci.* 371:20150219 (1–10).

Sues, H. D., S. J. Nesbitt, D. S. Berman, and A. C. Henrici. 2011. A late-surviving basal theropod dinosaur from the latest Triassic of North America. *Proc. R. Soc. Lond. B Biol. Sci.* 278:3459–3464.

Turner, A. H., P. J. Makovicky, and M. A. Norell. 2012. A review of dromaeosaurid systematics and paravian phylogeny. *Bull. Am. Mus. Nat. Hist.* 371:1–206.

Tykoski, R. S., and T. Rowe. 2004. *Ceratosauria.* Pp. 47–70 in *The* Dinosauria. 2nd edition (D. B. Weishampel, P. Dodson, and H. Osmólska, eds.). University of California Press, Berkeley, CA.

Xu, X., X. Zheng, C. Sullivan, X. Wang, L. Xing, Y. Wang, X. Zhang, J. K. O'Connor, F. Zhang, and Y. Pan. 2015. A bizarre Jurassic maniraptoran theropod with preserved evidence of membranous wings. *Nature* 521:70–73.

Wang, S., J. Stiegler, R. Amiot, X. Wang, G.-H. Du, J. M. Clark, and X. Xu. 2017. Extreme ontogenetic changes in a ceratosaurian theropod. *Curr. Biol.* 27:144–148.

Weishampel, D. B., P. Dodson, and H. Osmólska. 2004. *The* Dinosauria. 2nd edition. University of California Press, Berkeley, CA.

Wilson, J. A., P. C. Sereno, S. Srivastava, D. K. Bhatt, A. Khosla, and A. Sahni. 2003. A new abelisaurid (*Dinosauria, Theropoda*) from the Lameta Formation (Cretaceous, Maastrichtian) of India. *Contr. Mus. Paleontol. Univ. Michigan* 31:1–42.

Yates, A. M. 2007. Solving a dinosaurian puzzle: the identity of *Aliwalia rex* Galton. *Hist. Biol.* 19:93–123.

Authors

Darren Naish; Ocean and Earth Science; National Oceanography Centre, Southampton; University of Southampton; Southampton SO14 3ZH, UK. Email: eotyrannus@gmail.com.

Andrea Cau; Museo Geologico Giovanni Capellini; Alma Mater Università di Bologna; Via Zamboni 63; 40126 Bologna, Italy. Email: cauand@gmail.com.

Thomas R. Holtz, Jr.; Department of Geology; University of Maryland; 8000 Regents Drive; College Park, MD 20742-4211; USA, and Department of Paleobiology; National Museum of Natural History; Smithsonian Institution; Washington, DC 20013-7012. Email: tholtz@umd.edu.

Matteo Fabbri; Department of Geology and Geophysics; Yale University; New Haven, CT 06520-8109, USA. Email: matteo.fabbri@yale.edu.

Jacques A. Gauthier; Department of Geology and Geophysics; Yale University; New Haven, CT 06520-8109, USA. Email: jacques.gauthier@yale.edu.

Date Accepted: 15 August 2018

Primary Editor: Kevin de Queiroz

Aves C. Linnaeus 1758 [J. A. Clarke, D. P. Mindell, K. de Queiroz, M. Hanson, M. A. Norell, and J. A. Gauthier], converted clade name

Registration Number: 113

Definition: The smallest crown clade containing *Struthio camelus* Linnaeus 1758 (*Palaeognathae*), *Tinamus* (originally *Tetrao*) *major* (Gmelin 1789) (*Palaeognathae/Tinamidae*), *Phasianus colchicus* Linnaeus 1758 (*Neognathae/Galloanserae/Galliformes*), and *Vultur gryphus* Linnaeus 1758 (*Neognathae/Neoaves/Accipitriformes*). This is a minimum-crown-clade definition. Abbreviated definition: min crown ∇ (*Struthio camelus* Linnaeus 1758 & *Tinamus major* (Gmelin 1789) & *Phasianus colchicus* Linnaeus 1758 & *Vultur gryphus* Linnaeus 1758).

Etymology: Derived from the Latin vernacular for "birds."

Reference Phylogeny: Figure 1 in the comprehensive molecular analysis of Prum et al. (2015). We selected specifier species from among those originally used by Linnaeus to represent uncontroversial avian subclades in order to bracket the crown clade. Because some of these specifier species are not included in the reference phylogeny, more inclusive taxa containing them are listed parenthetically in the definition to facilitate its application.

Composition: *Aves* currently contains more than 10,000 described species, but could include as many as 5,000–10,000 more depending upon reassessment of currently recognized subspecies (Barrowclough et al., 2016). These include all those listed in Brodkorb (1963, 1964, 1967, 1971, 1978), Unwin (1993), Mlíkovsky (2002), Clements (2007), Mayr (2009), del Hoyo et al. (2013), and Gill and Donsker (2018), provided lists of many extinct members of the crown, as well as a few extinct species that are here regarded to have diverged from the avian stem.

Diagnostic Apomorphies: With respect to other extant amniotes, Huxley's (1867: 416–417) diagnosis of *Aves* still serves (see Comments below):

1. "[E]pidermal appendages developed in sacs of the dermis, and having the structure of feathers.
2. [A] remarkably large sacrum, the vertebrae, through the intervertebral foramina of which the roots of the sacral plexus (and, consequently, of the great sciatic nerve) pass, are not provided with expanded ribs abutting against the ilium externally, and against the bodies of these vertebrae by their inner ends. [Instead, these vertebrae are connected to the ilia via] slender transverse processes, which seem to answer to those which unite with the tubercles of the ribs in the dorsal region [in other reptiles].
3. The broad and expanded part of the sternum, which immediately follows the coracoidal articular surfaces, receives all the sternal ribs.
4. The ischia never unite in a median ventral symphysis; and both pubes and ischia are directed backwards, approximately parallel with one another and with the spinal column.
5. The proximal constituent of the tarsus is anchylosed [sic] with the tibia into

one tibio-tarsal bone; the distal element of the tarsus similarly unites with the second, third, and fourth metatarsal bones, and gives rise to the tarso-meta-tarsal bone. The metatarsal of the hallux is shorter than the others, and does not reach the tarsus.

6. [H]ot blood ... a single aortic arch, and remarkably modified respiratory organs."

Additional apomorphies have been listed by Cracraft (1988) and Kurochkin (1995). Currently, there are no unambiguous synapomorphies of *Aves* that will distinguish members of the crown from all known members of its stem (see Comments).

Synonyms: *Ornithurae* Haeckel 1866 (approximate); *Neornithes* Gadow 1892 (approximate); see review in Gauthier and de Queiroz (2001) and Comments below.

Comments: The ease with which *Aves* (the crown) could be diagnosed in the mid-nineteenth century depended entirely upon missing data, as does its ease of diagnosis relative to other extant taxa today. However, due both to the discovery of an array of intermediate forms (e.g., Cau, 2018), and to the fact that extinct species closest to the crown are so incompletely preserved, it is now much more difficult to distinguish members of the crown (*Aves*) from the nearest members of its stem. The poorly known extinct species most closely related to the crown were placed outside the crown when considered individually (e.g., Clarke, 2004). Nevertheless, when considered together, there was so little overlap among their preserved remains that there were no longer any unambiguously optimized morphological apomorphies for *Aves*. This remains true today.

Linnaeus proposed the name *Aves* for a group composed entirely of extant species representing

both branches of the basal split within the crown clade (Gauthier, 1986). A century later, discovery of the stem bird *Archaeopteryx lithographica* (von Meyer, 1861; Owen, 1863) with its mosaic of ancestral (e.g., teeth, long bony tail) and derived (e.g., feathers, wings) characters, engendered controversy regarding the circumscription of *Aves* (reviewed in de Beer, 1954). A consensus eventually emerged that *Archaeopteryx* should be considered part of *Aves,* as first proposed explicitly by Haeckel (1866), who also proposed *Ornithurae* to distinguish the "living ... true birds" (p. 140) from *Archaeopteryx*. A loose association between *Aves* and at least part of its stem lineage remained in steady use for more than a century following Haeckel (1866), although the exact clade to which the name applied became more varied with the discovery of additional fossil intermediates, increasing knowledge of phylogenetic relationships, and changing taxonomic philosophies. More specifically, the name *Aves* became associated with at least five different nested clades (see Gauthier and de Queiroz, 2001, for a review): (1) the total clade of birds (e.g., Patterson, 1993); (2) the clade characterized by pinnate feathers (e.g., James and Pourtless, 2009; Feduccia, 2013); (3) the clade characterized by flight (e.g., Ji and Ji, 2001; Xu et al., 2009); (4) the clade stemming from the *Archaeopteryx* node (e.g., Padian and Chiappe, 1998); and (5), the crown clade (e.g., Gauthier, 1986).

Some contemporary systematists (those adopting one of the first four uses of *Aves* listed above) prefer the name *Neornithes* Gadow 1892 for the crown. This name was proposed by Gadow (see also Gadow, 1893) as an explicit replacement name for *Ornithurae* to distinguish extant birds as well as the Cretaceous toothed birds (Marsh, 1880), which were unknown to Haeckel (1866), from *Archaeopteryx*. Subsequently, however, the associations of these names diverged. *Ornithurae* came to be applied to a more inclusive clade than the crown, i.e., to those stem avians sharing apomorphies of the tail to which the name implicitly refers (e.g., Gauthier, 1986; Elzanowski, 1995; Gauthier and de Queiroz, 2001; Clarke, 2004; Zhou and Zhang, 2006). *Neornithes* was first applied explicitly to the crown by Walker (1981), and that practice has been followed frequently in the palaeontological literature (e.g., Thulborn, 1984; Chiappe, 1995; Dyke and van Tuinen, 2004). But palaeontologists have also used *Neornithes* for more inclusive clades, such as for the crown plus *Ichthyornithes* (e.g., Martin, 1983), or the crown and all stem birds closer to the crown than to *Enantiornithes* (e.g., Elzanowski, 1995).

Given that most published papers on birds deal only with extant species, however, the preference for *Neornithes* for the crown does not accord well with the most widespread use of the name *Aves* (Gauthier and de Queiroz, 2001). Google Scholar (Aug. 2018) yields more than one million records for searches on *"Aves"* or "avian"; by contrast, searches on *"Neornithes"* or "neornithine" (sic) yield approximately 2,500 and 800 records, respectively. Contrasting usage seems to follow disciplinary boundaries, with palaeontological authors being more likely to use *Neornithes* than ornithologists focused on extant birds. Unfortunately, some authors who have explicitly chosen to use the name *Neornithes* for the crown have not been entirely successful at avoiding inconsistent use of the name *Aves*. For example, consider the title of the paper "Phylogenetic Relationships among Modern Birds (Neornithes)" with the subtitle "Toward an Avian Tree of Life," and one section of that paper entitled "The Challenge of Resolving Avian Relationships" (Cracraft et al., 2004). Similarly, authors who analyzed the "Higher-order phylogeny of modern birds (Theropoda, Aves: Neornithes)" refer to "Broad affinities of long standing among the avian orders ... that were not supported by the present analysis,"

even though they were addressing relationships within the crown (Livezey and Zusi, 2007: 42).

Despite efforts to promote the use of *Neornithes*, a preference for using *Aves* when referring to the crown remains strong. Recent landmark molecular phylogenetic studies, for example, which were necessarily restricted to the crown clade, have not used *Neornithes* or "neornithine" but referred to their topic as "avian" diversification and their target clade as *Aves* (Hackett et al., 2008; Jetz et al., 2012; Burleigh et al., 2015; Prum et al., 2015). Jarvis et al. (2014) used both *Neornithes* and "avian" when referring to the crown clade. Other studies have avoided using either *Aves* or *Neornithes* but continued to use "avian" (e.g., McCormack et al., 2013; Reddy et al., 2017).

None of the uses of the name *Aves* is optimal in all respects. Exclusion of *Archaeopteryx lithographica* from *Aves* is disruptive given that this species was included in that taxon by many authors for more than 150 years. Associating *Aves* with pinnate feathers would be even more disruptive, as such feathers are now known to diagnose a much larger clade than the *Archaeopteryx* node that includes more traditional "non-avian" theropods such as *Velociraptor* (e.g., Clarke, 2013). Applying the name to the total clade requires including even more distantly related extinct forms not traditionally included in *Aves* such as sauropodomorphs (e.g., *Brontosaurus*), ornithischians (e.g., *Triceratops*), and pterosaurs (e.g., *Rhamphorhynchus*). Linking the name *Aves* to the origin of "flight" is hard to maintain in the face of abundant discoveries that show this complex character was not a simple apomorphy diagnosing a single node (e.g., Feo et al., 2015). Additionally, the fraction of the scientific community that needs to distinguish between the origin of volant dinosaurs and the crown clade is exceeding small.

In sum, most authors continue to use *"Aves"* or "avian" to discuss aspects of bird biology—such as their genomes—that have been documented only in extant species. Authors even use *"Aves"* or "avian" when discussing features that are known to be absent in early members of the clade to which they apply those terms (see examples in de Queiroz, 2007: 968). *Neornithes* is rarely used when discussing extant birds alone, even though that name was proposed more than a century ago. Reluctance to do so is understandable, given that precision regarding the name of the crown-clade would require that the familiar name *Aves* be supplanted by the obscure name *Neornithes*. For reasons discussed at length in Gauthier and de Queiroz (2001) and de Queiroz (2007), we have applied the best known and most frequently used name (*Aves*) to the best known and most frequently discussed clade (the crown). For those who are concerned about the status of *Archaeopteryx*, it is still a (stem) bird even if it is not part of *Aves*.

Literature Cited

Barrowclough, G. F., J. Cracraft, J. Klicka, and R. M. Zink. 2016. How many kinds of birds are there and why does it matter? *PLOS ONE* 11(11):e0166307.

de Beer, G. 1954. Archaeopteryx lithographica: *A Study Based on the British Museum Specimen*. Trustees of the British Museum (Natural History), London. No. 224, 68pp. [Reprinted 1967 Waterford, UK: Taylor Garnett Evans.].

Brodkorb, P. 1963. Catalogue of fossil birds. Part 1 (*Archaeopterygiformes* through *Ardeiformes*). *Bull. Fl. State Mus. Biol. Sci.* 7(4):179–293.

Brodkorb, P. 1964. Catalogue of fossil birds. Part 2 (*Anseriformes* through *Galliformes*). *Bull. Fl. State Mus. Biol. Sci.* 8(3):195–335.

Brodkorb, P. 1967. Catalogue of fossil birds. Part 3 (*Ralliformes, Ichthyornithiformes, Charadriiformes*). *Bull. Fl. State Mus. Biol. Sci.* 11(3):99–220.

Brodkorb, P. 1971. Catalogue of fossil birds. Part 4 (*Columbiformes* through *Piciformes*). *Bull. Fl. State Mus. Biol. Sci.* 15(4):163–266.

Brodkorb, P. 1978. Catalogue of fossil birds. Part 5 (*Passeriformes*). *Bull. Fl. State Mus. Biol. Sci.* 23(2):139–228.

Burleigh, J. G., R. T. Kimball, and E. L. Braun. 2015. Building the avian tree of life using a large-scale, sparse supermatrix. *Mol. Phylogenet. Evol.* 84:53–63.

Cau, A. 2018. The assembly of the avian body plan: a 160-million-year-long process. *Boll. Soc. Paleontol. Ital.* 57(1):1–25.

Chiappe, L. M. 1995. The first 85 million years of avian evolution. *Nature* 378:349–355.

Clarke, J. A. 2004. The morphology, phylogenetic taxonomy and systematics of *Ichthyornis* and *Apatornis* (*Avialae: Ornithurae*). *Bull. Am. Mus. Nat. Hist.* 286:1–179.

Clarke, J. A. 2013. Feathers before flight. *Science* 340(6133):690–692.

Clements, J. F. 2007. *Clements Checklist of Birds of the World*. Comstock Pub. Associates, Cornell University Press.

Cracraft, J. 1988. The major clades of birds. Pp. 339–361 in *The Phylogeny and Classification of the Tetrapods. Vol. 1: Amphibians, Reptiles, Birds* (M. J. Benton, ed.). Systematics Association Special Volume No. 35A. Clarendon Press, Oxford.

Cracraft, J, F. K. Barker, M. Braun, J. Harshman, G. J. Dyke, J. Feinstein, S. Stanley, A. Cibois, P. Schikler, P. Beresford, J. Garcia-Moreno, M. D. Sorenson, T. Yuri, and D. P. Mindell. 2004. Phylogenetic relationships among modern birds (*Neornithes*): toward an avian tree of life. Pp. 468–489 in *Assembling the Tree of Life* (J. Cracraft and M. J. Donoghue, eds.). Oxford University Press, Oxford, UK.

Dyke, G. J., and M. van Tuinen. 2004. The evolutionary radiation of modern birds (*Neornithes*): reconciling molecules, morphology, and the fossil record. *Zool. J. Linn. Soc.* 141(2):153–177.

Elzanowski, A. 1995. Cretaceous birds and avian phylogeny. *Cour. Forsch. Inst. Senckenberg* 181:37–53.

Feduccia, A. 2013. Bird origins anew. *Auk* 130:1–12.

Feo, T. J., D. J. Field, and R. O. Prum. 2015. Comparison of barb geometry in modern and Mesozoic asymmetrical flight feathers reveals a transitional morphology during the evolution of avian flight. *Proc. R. Soc. Lond. B Biol. Sci.* 282 (1803):20142864.

Gadow, H. 1892. On the classification of birds. *Proc. Zool. Soc. Lond.* 17:229–256.

Gadow, H. 1893. Vogel. II. Systematischer Theil. Pp. 1–304 in *Dr. H. G. Bronn's Klassen und Ordnungen des Thier-Reichs, Wissenschaftlich Dargestellt in Wort und Bild*, Band 6, Abt. 4. C. F. Winter, Leipzig.

Gauthier, J. A. 1986. Saurischian monophyly and the origin of birds. Pp. 1–55 in *The Origin of Birds and the Evolution of Flight* (K. Padian, ed.). Mem. Calif. Acad. Sci. 8. California Academy of Sciences, San Francisco, CA.

Gauthier, J. A., and K. de Queiroz. 2001. Feathered dinosaurs, flying dinosaurs, crown dinosaurs, and the name *"Aves."* Pp. 7–41 in *New Perspectives on the Origin and Early Evolution of Birds: Proceedings of the International Symposium in Honor of John H. Ostrom* (J. Gauthier and L. F. Gall, eds.). Peabody Museum of Natural History, Yale University, New Haven, CT.

Gill, F., and D. Donsker, eds. 2018. *IOC World Bird List (v8.2).* doi:10.14344/IOC.ML.8.2.

Gmelin, J. E. 1788–1789. *Caroli a Linné, Systema Naturae Per Regna Tria Naturae, Secundum Classes, Ordines, Genera, Species, cum Characteribus, Differentiis, Synonymis, Locis.* 13th edition, Tome 1, Pars 1–2. Impensis Georg Emanuel Beer, Lipsiae.

Hackett, S. J., R. T. Kimball, S. Reddy, R. C. K. Bowie, E. L. Braun, M. J. Braun, J. L. Chojnowski, W. A. Cox, K.-L. Han, J. Harshman, C. J. Huddleston, B. D. Marks, K. J. Miglia, W. S. Moore, F. H. Sheldon, D. W. Steadman, C. C. Witt, and T. Yuri. 2008. A phylogenomic study of birds reveals their evolutionary history. *Science* 320:1763–1768.

Haeckel, E. 1866. *Generelle Morphologie der Organismen. Allegemeine Grundzuge der Organischen Formen Wissenschaft, Mechanisch Begrundet Durch die von Charles Darwin reformierte Daszendenz-Theorie. II. Allgemeine Entwicklungsgeschicte der Organismen. Kritische Grundzuge der Mechanischen Wissenschaft von dan Entstehenden Formen der Organismen, Begrundet Durch die Deszendenz-Theorie.* Georg Reimer, Berlin.

del Hoyo, J., A. Elliott, J. Sargatal, and D. A. Christie, eds. 2013. *Handbook of the Birds of the World. Special Volume. New Species and Global Index.* Lynx Edicions, Barcelona.

Huxley, T. H. 1867. On the classification of birds; and on the taxonomic value of the modifications of certain of the cranial bones observable in that class. *Proc. Zool. Soc. Lond.* 1867:415–472.

James, F. C., and J. A. Pourtless. 2009. Cladistics and the origin of birds: a review and two new analyses. *Ornithol. Monogr.* 66.

Jarvis E. D., S. Mirarab, A. J. Aberer, B. Li, P. Houde, C. Li, S. Y. W. Ho, B. C. Faircloth, B. Nabholz, J. T. Howard, A. Suh, C. C. Weber, R. R. da Fonseca, J. Li, F. Zhang, H. Li, L. Zhou, N. Narula, L. Liu, G. Ganapathy, B. Boussau, M. S. Bayzid, V. Zavidovych, S. Subramanian, T. Gabaldon, S. Capella-Gutierrez, J. Huerta-Cepas, B. Rekepalli, K. Munch, M. Schierup, B. Lindow, W. C. Warren, D. Ray, R. E. Green, M. W. Bruford, X. Zhan, A. Dixon, S. Li, N. Li, Y. Huang, E. P. Derryberry, M. F. Bertelsen, F. H. Sheldon, R. T. Brumfield, C. V. Mello, P. V. Lovell, M. Wirthlin, F. Cruz Schneider, J. A. Prosdocimi, A. M. Samaniego, A. M. Vargas, A. M. Velazquez, A. Alfaro-Nunez, P. F. Campos, B. Petersen, T. Sicheritz-Ponten, A. Pas, T. Bailey, P. Scofield, M. Bunce, D. M. Lambert, Q. Zhou, P. Perelman, A. C. Driskell, B. Shapiro, Z. Xiong, Y. Zeng, S. Liu, Z. Li, B. Liu, K. Wu, J. Xiao, X. Yinqi, Q. Zheng, Y. Zhang, H. Yang, J. Wang, L. Smeds, F. E. Rheindt, M. Braun, J. Fjeldså, L. Orlando, F. K. Barker, K. A. Jonsson, W. Johnson, K.-P. Koepfli, S. O'Brien, D. Haussler, O. A. Ryder, C. Rahbek, E. Willerslev, G. R. Graves, T. C. Glenn, J. McCormack, D. Burt, H. Ellegren, P. Alström, S. V. Edwards, A. Stamatakis, D. P. Mindell, J. Cracraft, E. L. Braun, T. Warnow, W. Jun, M. T. P. Gilbert, and G. Zhang. 2014.

Whole-genome analyses resolve early branches in the tree of life of modern birds. *Science* 346:1320–1331.

Jetz, W. W., G. H. G. Thomas, J. B. J. Joy, K. K. Hartmann, and A. O. A. Mooers. 2012. The global diversity of birds in space and time. *Nature* 491(7424):444–448.

Ji, Q., and S.-A. Ji. 2001. How can we define a feathered dinosaur as a bird? Pp. 43–46 in *New Perspectives on the Origin and Early Evolution of Birds: Proceedings of the International Symposium in Honor of John H. Ostrom* (J. Gauthier and L. F. Gall, eds.). Peabody Museum of Natural History, Yale University, New Haven, CT.

Kurochkin, E. N. 1995. Synopsis of Mesozoic birds and early evolution of class *Aves. Archaeopteryx* 13:47–66.

Linnaeus, C. 1758. *Systema Naturae Per Regna Tria Naturae, Secundum Classes, Ordines, Genera, Species, cum Characteribus, Differentiis, Synonymis, Locis.* 10th edition, Tome 1. Laurentii Salvii, Holmiae (Stockholm).

Livezey, B. C., and R. L. Zusi. 2007. Higher-order phylogeny of modern birds (*Theropoda, Aves: Neornithes*) based on comparative anatomy. II. Analysis and discussion. *Zool. J. Linn. Soc.* 149:1–95.

Marsh, O. C. 1880. Odontornithes: *A Monograph on the Extinct Toothed Birds of North America.* U. S. Government Printing Office, Washington, DC.

Martin, L. D. 1983. The origin and early radiation of birds. Pp. 291–337 in *Perspectives in Ornithology: Essays Presented for the Centennial of the American Ornithologists' Union* (A. H. Brush and G. A. Clark, Jr., eds.). Cambridge University Press, Cambridge, UK.

Mayr, G. 2009. *Paleogene Fossil Birds.* Springer, Heidelberg.

McCormack, J. E., M. G., Harvey, B. C. Faircloth, N. G. Crawford, T. C. Glenn, and R. T. Brumfield. 2013. A phylogeny of birds based on over 1,500 loci collected by target enrichment and high-throughput sequencing. *PLOS ONE* 8(1):e54848.

Mlíkovsky, J. 2002. *Cenozoic Birds of the World. Part 1: Europe.* Ninox Press, Praha.

Owen, R. 1863. On the *Archaeopteryx* of von Meyer, with a description of the fossil remains of a long-tailed species from the lithographic slate of Solnhofen. *Philos. Trans. R. Soc. Lond. B Biol. Sci.* 153:33–47.

Padian, K., and L. M. Chiappe. 1998. The origin and early evolution of birds. *Biol. Rev.* 73:1–42.

Patterson, C. 1993. Naming names. *Nature* 366:518.

Prum, R. O., J. S. Berv, A. Dornburg, D. J. Field, J. P. Townsend, E. M. Lemmon, and A. R. Lemmon. 2015. A comprehensive phylogeny of birds (*Aves*) using targeted next-generation DNA sequencing. *Nature* 526:569–573.

de Queiroz, K. 2007. Toward an integrated system of clade names. *Syst. Biol.* 56:956–974.

Reddy, S., R. T. Kimball, A. Pandey, P. A. Hosner, M. J. Braun, S. J. Hackett, K.-L. Han, J. Harshman, C. J. Huddleston, S. Kingston, B. D. Marks, K. J. Miglia, W. S. Moore, F. H. Sheldon, C. C. Witt, T. Yuri, and E. L. Braun. 2017. Why do phylogenomic datasets yield conflicting trees? Data type influences the avian tree of life more than taxon sampling. *Syst. Biol.* 66:857–879.

Thulborn, R. A. 1984. The avian relationships of *Archaeopteryx*, and the origin of birds. *Zool. J. Linn. Soc.* 82:119–158.

Unwin, D. M. 1993. *Aves.* Pp. 717–737 in *The Fossil Record*, Vol. 2 (M. J. Benton, ed.). Chapman and Hall, London.

von Meyer, H. 1861. *Archaeopteryx lithographica* (Vogel-Feder) und *Pterodactylus* von Solnhofen. *Neues Jahrb. Geol. Paläontol.* 1861:679–679.

Walker, C. A. 1981. New subclass of birds from the Cretaceous of South America. *Nature* 292:51–53.

Xu, X., O. Zhao, M. A. Norell, C. Sullivan, D. Hone, C. Erickson, X. Wang, and F. Han. 2009. A new feathered maniraptoran dinosaur fossil that fills a morphological gap in avian origin. *Chin. Sci. Bull.* 54:430–435.

Zhou, Z., and F. Zhang. 2006. A beaked basal ornithurine bird (*Aves, Ornithurae*) from the Lower Cretaceous of China. *Zool. Scr.* 35:363–373.

Authors

Julia Clarke; Department of Geological Sciences; Jackson School of Geosciences, The University of Texas, Austin, TX 78712-1722. Email: Julia_Clarke@jsg.utexas.edu.

David P. Mindell; Museum of Vertebrate Zoology; University of California; Berkeley, CA 94720-3160, USA. Email: dpmindell@gmail.com.

Kevin de Queiroz; Department of Vertebrate Zoology; National Museum of Natural History; Smithsonian Institution; Washington, DC 20560-0162, USA. Email: dequeirozk@si.edu.

Michael Hanson; Department of Geology and Geophysics; Yale University; New Haven, CT 06520-8109, USA. Email: michael.hanson@yale.edu.

Mark A. Norell; Division of Paleontology; American Museum of Natural History; New York, NY 10024, USA. Email: norell@anmh.org.

Jacques A. Gauthier; Department of Geology and Geophysics; Yale University; 210 Whitney Ave, New Haven, CT 06520-8109, USA. Email: jacques.gauthier@yale.edu.

Date Accepted: 14 October 2018

Primary Editor: Philip Cantino

Galloanserae C. Sibley, J. E. Ahlquist, and B. L. Monroe 1988 [D. P. Mindell], converted clade name

Registration Number: 279

Definition: The smallest crown clade containing *Gallus* (originally *Phasianus*) *gallus* (Linnaeus 1758) (*Galliformes*) and *Anser* (originally *Anas*) *anser* (Linnaeus 1758) (*Anseriformes*). This is a minimum-crown-clade definition. Abbreviated definition: min crown ∇ (*Gallus gallus* (Linnaeus 1758) & *Anser anser* (Linnaeus 1758)).

Etymology: "*Galloanserae*" derives from a combination of the names of its two primary subclades, *Galliformes* and *Anseriformes*, which are derived in turn from the Latin vernacular names for "chicken" (*Gallus*) and "goose" (*Anser*), respectively.

Reference Phylogeny: Figure 1 in Prum et al. (2015) should be treated as the primary reference phylogeny. See also Sorenson et al. (2003: Figs. 1, 2) and Jarvis et al. (2014: Fig. 1).

Composition: *Galliformes* (i.e., all pheasants, grouse, partridges, quail, chachalacas, guans, curassows, megapodes, and guineafowl) and *Anseriformes* (i.e., Magpie Goose, whistling ducks, all other ducks, swans, geese, and screamers) comprise *Galloanserae*. See Dickinson (2003) for constituent species of *Galliformes* and *Anseriformes*.

Diagnostic Apomorphies: Morphological synapomorphies uniting *Galliformes* and *Anseriformes* include from Cracraft (1988): ectethmoid weakly developed or absent; rostropterygoid articulation present; inflated, rounded and broad sphenoid plate; well-marked depression on posterolateral side of sphenoid plate for nerves and arteries; palatines long, thin, and widely separated anteriorly; well-developed process on quadrate anterior to quadrate-prootic articulation; external mandibular condyle of quadrate large and oriented lateromedially; mandibular articular surface with single anteroposterior ridge; articular surface of mandible lacking posteromedial and lateral walls; long, dorsally oriented, internal articular process of mandible; retroarticular process long, curving, strongly compressed lateromedially; and from Livezey (1997): strongly bilobate occipital condyle; rostral basipterygoid process; insertion of m. adductor mandibulae elongated, marked by substantial separation of rostral and caudal termini by (typically) crest-like prominence; lateral mandibular process laterally prominent, squared, perpendicular to retroarticular process; maxillary process of premaxilla dorsoventrally deep and lateromedially compressed; pneumatic pores present on cranial margin of ribs; pneumatic foramen present on ventral face of sternum; hypotarsal ridges of tarsometatarsus reduced in number and displaced laterally; metatarsal I (hallux) fossa distinct; and acrotarsium of podotheca reticulate.

Synonyms: There are no synonyms.

Comments: Early evidence for a close phylogenetic relationship between *Galliformes* and *Anseriformes* was presented by Garrod (1874) and Seebohm (1889). More recent evidence for the sister relationship between these two taxa is based on immunological distances (Ho et al., 1976), DNA hybridization analyses (Sibley and Ahlquist, 1990), mitochondrial and nuclear gene sequences (e.g., Caspers et al.,

1997; Mindell et al., 1997; 1999; Groth and Barrowclough, 1999; van Tuinen et al., 2000; Paton et al., 2002; Garcia-Moreno et al., 2003; Sorenson et al., 2003; Ericson et al., 2006; Slack et al., 2006; Hackett et al., 2008; Jarvis et al., 2014; Prum et al., 2015), morphological characters (Cracraft, 1988; Dzerhinsky, 1995; Livezey, 1997; Cracraft and Clarke, 2001; Mayr and Clarke, 2003; Livezey and Zusi, 2007), and combined molecular and morphological datasets (Cracraft and Mindell, 1989; Cracraft et al., 2004).

Literature Cited

Caspers, G. J., D. Uit de Weerd, J. Wattel, and W. W. de Jong. 1997. Alpha-crystallin sequences support a galliform/anseriform clade. *Mol. Phylogenet. Evol.* 7:185–188.

Cracraft, J. 1988. The major clades of birds. Pp. 333–355 in *The Phylogeny and Classification of the Tetrapods* (M. J. Benton, ed.). Systematics Assoc. Special Vol. No. 35A, Clarendon Press, Oxford.

Cracraft, J., F. K. Barker, M. J. Braun, J. Harshman, G. Dyke, J. Feinstein, S. Stanley, A. Cibois, P. Schikler, P. Beresford, J. García-Moreno, M. D. Sorenson, T. Yuri, and D. P. Mindell. 2004. Phylogenetic relationships among modern birds (*Neornithes*): toward an avian tree of life. Pp. 468–489 in *Assembling the Tree of Life* (J. Cracraft and M. J. Donoghue, eds.). Oxford University Press, New York.

Cracraft, J., and J. Clarke. 2001. The basal clades of modern birds. Pp. 143–156 in *New Perspectives on the Origin and Early Evolution of Birds* (J. Gauthier and L. F. Gall, eds.). Peabody Museum of Natural History, Yale University, New Haven, CT.

Cracraft, J., and D. P. Mindell. 1989. The early history of modern birds: a comparison of molecular and morphological evidence. Pp. 389–403 in *The Hierarchy of Life*. (B. Fernholm, K. Bremer, and H. Jörnvall, eds.). *Proc. Nobel Symposia*. Elsevier Science Publishers, Amsterdam.

Dickinson, E. C. 2003. *The Howard and Moore Complete Checklist of the Birds of the World.* Christopher Helm, London

Dzerhinsky, R. Y. 1995. Evidence for common ancestry of *Galliformes* and *Anseriformes*. *Courier Forsch. Senckenberg* 181:325–336.

Ericson, P. G. P., C. L. Anderson, T. Britton, A. Elzanowski, U. S. Johansson, M. Källersjö, J. I. Ohlson, T. J. Parsons, D. Zuccon, and G. Mayr. 2006. Diversification of *Neoaves*: integration of molecular sequence data and fossils. *Biol. Lett.* 2(4):543–547.

García-Moreno, J., M. D. Sorenson, and D. P. Mindell. 2003. Congruent avian phylogenies inferred from mitochondrial and nuclear DNA sequences. *J. Mol. Evol.* 57:27–37.

Garrod, A. H. 1874. On certain muscles of birds and their value in classification. *Proc. Zool. Soc. Lond.* 1874:339–348.

Groth, J. G., and G. F. Barrowclough. 1999. Basal divergences in birds and the phylogenetic utility of the nuclear RAG-1 gene. *Mol. Phylog. Evol.* 12:115–123.

Hackett, S. J., R. T. Kimball, S. Reddy, R. C. K. Bowie, E. L. Braun, M. J. Braun, J. L. Chojnowski, W. A. Cox, K.-L. Han, J. Harshman, C. J. Huddleston, B. D. Marks, K. J. Miglia, W. S. Moore, F. H. Sheldon, D. W. Steadman, C. C. Witt, and T. Yuri. 2008. A phylogenomic study of birds reveals their evolutionary history. *Science* 320:1763–1768.

Ho, C. Y.-K., E. M. Prager, A. C. Wilson, D. T. Osuga, and R. E. Feeney. 1976. Penguin evolution: protein comparisons demonstrate phylogenetic relationship to flying aquatic birds. *J. Mol. Evol.* 8:271–282.

Jarvis, E. D., S. Mirarav, A. J. Aberer, B. Li, P. Houde, C. Li, S. Y. W. Ho, B. C. Faircloth, B. Nabholz, J. T. Howard, A. Suh, C. C. Weber, R. R. da Fonseca, J. Li, F. Zhang, H. Li, L. Zhou, N. Narula, L. Liu, G. Ganaopathy, B. Boussau, Md. S. Bayzid, V. Zavidovych, S. Subramanian, T. Gabaldón, S. Capella-Gutierrez, J. Huerta-Cepas, B. Rekepalli, K. Munch, M. Schierup, B. Lindow, W. C. Warren, D. Ray, R. E. Green, M. W. Bruford, X. Zhan, A. Dixon, S. Li, N. Li, Y. Huang, E. P. Derryberry, M. F. Bertelsen, F. H. Sheldon,

R. T. Brumfield, C. V. Mello, P. V. Lovell, M. Wirthlin, M. P. C. Schneider, F. Prosdocimi, J. A. Samaniego, A. M. V. Velazquez, A. Alfaro-Núñez, P. F. Campos, B. Petersen, T. Sicheritz-Ponten, A. Pas, T. Bailey, P. Scofield, M. Bunce, D. M. Lambert, Q. Zhou, P. Perelman, A. C. Driskell, B. Shapiro, Z. Xiong, L. Zeng, S. Liu, Z. Li, B. Liu, K. Wu, J. Xiao, X. Yinqi, Q. Zheng, Y. Zhang, H. Yang, J. Wang, L. Smeds, F. E. Rheindt, M. Braun, J. Fjeldsa, L. Orlando, F. K. Barker, K. A. Jønsson, W. Johnson, K.-P. Koepfli, S. O'Brien, D. Haussler, O. A. Ryder, C. Rahbek, E. Wilerslev, G. R. Graves, T. C. Glenn, J. McCormack, D. Burt, H. Ellegren, P. Alström, S. V. Edwards, A. Stamatakis, D. P. Mindell, J. Cracraft, E. L. Braun, T. Warnow, W. Jun, M. T. P. Gilbert, and G. Zhang. 2014. Whole-genome analyses resolve early branches in the tree of life of modern birds. *Science* 346(6215):1320–1331.

Linnaeus, C. 1758. *Systema Naturae Per Regna Tria Naturae, Secundum Classes, Ordines, Genera, Species, cum Characteribus, Differentiis, Synonymis, Locis.* Editio decima, reformata. Laurentii Salvii, Holmiae (Stockholm).

Livezey, B. C. 1997. A phylogenetic analysis of basal *Anseriformes*, the fossil *Presbyornis*, and the interordinal relationships of waterfowl. *Zool. J. Linn. Soc.* 121:361–428.

Livezey, B. C., and R. L. Zusi. 2007. Higher-order phylogeny of modern birds (*Theropoda, Aves: Neornithes*) based on comparative anatomy. II. Analysis and discussion. *Zool. J. Linn. Soc.* 149:1–95.

Mayr, G., and J. Clarke. 2003. The deep divergences of neornithine birds: a phylogenetic analysis of morphological characters. *Cladistics* 19(6):527–553.

Mindell, D. P., M. D. Sorenson, D. E. Dimcheff, M. Hasegawa, J. C. Ast, and T. Yuri. 1999. Interordinal relationships of birds and other reptiles based on whole mitochondrial genomes. *Syst. Biol.* 48:138–152.

Mindell, D. P., M. D. Sorenson, C. J. Huddleston, H. C. Miranda, Jr., A. Knight, S. J. Sawchuk, and T. Yuri. 1997. Phylogenetic relationships among and within select avian orders based on mitochondrial DNA. Pp. 213–247 in *Avian Molecular Evolution and Systematics* (D. P. Mindell, ed.). Academic Press, San Diego, CA.

Paton, T., O. Haddrath, and A. J. Baker. 2002. Complete mitochondrial DNA genome sequences show that modern birds are not descended from transitional shorebirds. *Proc. R. Soc. Lond. B Biol. Sci.* 269B:839–846.

Prum, R. O., J. S. Berv, A. Dornburg, D. J. Field, J. P. Townsend, E. M. Lemmon, and A. R. Lemmon. 2015. A comprehensive phylogeny of birds (*Aves*) using targeted next-generation DNA sequencing. *Nature* 526(7574):569–573.

Seebohm, H. 1889. An attempt to diagnose the suborders of the ancient Ardeino-Anserine assemblage of birds by the aid of osteological characters alone. *Ibis* 31:92–104.

Sibley, C. G., and J. E. Ahlquist. 1990. *Phylogeny and Classification of Birds: A Study in Molecular Evolution.* Yale University Press, New Haven, CT.

Sibley, C. G., J. E. Ahlquist, and B. L. Monroe. 1988. A classification of the living birds of the world based on DNA-DNA hybridization studies. *Auk* 105:409–423.

Slack, K. E., F. Delsuc, P. A. Mclenachan, U. Arnason, and D. Penny. 2006. Resolving the root of the avian mitogenomic tree by breaking up long branches. *Mol. Phylogenet. Evol.* 42(1):1–13.

Sorenson, M. D., E. Oneal, J. García-Moreno, and D. P. Mindell. 2003. More taxa, more characters: the hoatzin problem is still unresolved. *Mol. Biol. Evol.* 20:1484–1499.

van Tuinen, M., C. G. Sibley, and S. B Hedges. 2000. The early history of modern birds inferred from DNA sequences of nuclear and mitochondrial ribosomal genes. *Mol. Biol. Evol.* 17:451–457.

Author

David P. Mindell; Museum of Vertebrate Zoology; University of California; Berkeley, CA 94720-3160, USA. Email: dpmindell@gmail.com.

Date Accepted: 15 February 2016

Primary Editor: Jacques A. Gauthier

Cuculidae W. E. Leach 1820 [G. Sangster], converted clade name

Registration Number: 280

Definition: The smallest crown clade containing *Carpococcyx* (originally *Calobates*) *radiceus* (Temminck 1832) and *Centropus* (originally *Cuculus*) *senegalensis* (Linnaeus 1766) and *Crotophaga ani* Linnaeus 1758 and *Cuculus canorus* Linnaeus 1758 and *Neomorphus* (originally *Coccyzus*) *geoffroyi* (Temminck 1820) and *Piaya cayana* (originally *Cuculus cayanus*) (Linnaeus 1758). This is a minimum-crown-clade definition. Abbreviated definition: min crown ∇ (*Carpococcyx radiceus* (Temminck 1832) & *Centropus senegalensis* (Linnaeus 1766) & *Crotophaga ani* Linnaeus 1758 & *Cuculus canorus* Linnaeus 1758 & *Neomorphus geoffroyi* (Temminck 1820) & *Piaya cayana* (Linnaeus 1758)).

Etymology: Derived from the Latin *cuculus*, meaning a cuckoo.

Reference Phylogeny: For the purpose of applying the definition of the taxon name *Cuculidae*, Figure 5.1 in Payne (2005) should be regarded as the primary reference phylogeny. The reference phylogeny does not indicate the species studied; the specifiers used here are type species of genera included in that phylogeny.

Composition: *Cuculidae* is currently believed to contain about 140 Recent species (Payne, 2005; Dickinson and Remsen, 2013) in the following groups (sensu Payne, 2005): *Crotophaginae* (Guira Cuckoo and anis), *Neomorphinae* (New World ground-cuckoos, roadrunners and allies), *Centropus* (coucals), *Couinae* (Old World ground-cuckoos and couas), and *Cuculinae* ("true" cuckoos). The phylogenetic relationships of *Opisthocomus hoazin* (Hoatzin) remain controversial; a proposed position among cuckoos (Sibley and Ahlquist, 1990), or a close relationship to cuckoos (Avise et al., 1994; Hedges et al., 1995; Livezey and Zusi, 2007), has not been corroborated by recent molecular phylogenetic studies (Hughes and Baker, 1999; Sorenson et al., 2003; Ericson et al., 2006; Brown et al., 2008; Hackett et al., 2008).

Diagnostic Apomorphies: Hughes (2000) identified 14 potential apomorphies of *Cuculidae* but Payne (2005) rejected most of these because he noted some homoplasy in apomorphies identified by Hughes. Livezey and Zusi (2006, 2007) identified eight apomorphies of *Cuculidae*, of which six are unique and unreversed (consistency index = 1). Because no unambiguous members of the stem-group of *Cuculidae* are known (Mayr, 2006; Mayr, 2009), these apomorphies are relative to other crown clades. Diagnostic apomorphies are (characters and states are indicated by their number-letter combination in Livezey and Zusi, 2006): (1) extremitas caudalis synsacri, vertebra synsacri ultima, facies ventralis, processes transverses distinctly paddle-shaped (937b); (2) extremitas distalis femoris, condylae lateralis et medialis, pronounced subangular orientation of extremitas distalis caudal to axis majoris present (2034b); (3) extremitas distalis femoris, epicondylus medialis, forma as distinctly raised, medially smooth, irregularly subovate eminentia present (2046b); (4) extemitas proximalis tibiotarsi, caput tibiotarsi, facies articularis medialis narrowly crescentiform, medially constricted (2061b); (5) corpus fibulae, tuberculum insertii musculus iliofibularis, situs proximodistalis relative to crista fibularis tibiotarsi opposite (lateral) to terminus distalis cristae (2200c); (6) extremitas distalis tarsometatarsi, trochleae accessoriae group III, trochleae accessoriae metatarsalia II (secundus) et IV (quartus), i.e., primarily or solely involving differential modifications of trochleae metatarsalia II et IV and infrequently also trochlea metatarsale I, present and comprising a deep fossa supratrochlearis plantaris (perforated by comparatively large, distal foramen vasculare distale), a jugum separating fossa metatarsale I from basis trochlearis metarsale II, extreme medial displacement and hamulate shape of trochlea metatarsi II, a variably plantomediad-oriented spinus trochlearis metatarsi IV, and associated with a form of zygodactyly (2334e).

Synonyms: The following are approximate synonyms (see Comments): *Cuculiformes* Coues 1884 (sensu Wolters, 1975–1982; Sibley and Ahlquist, 1990; and Cracraft, 2013); *Cuculi* Wagler 1830 (sensu Wetmore, 1930; Storer, 1960; Brodkorb, 1971; Sibley, 1996; del Hoyo et al., 1997; and Livezey and Zusi, 2007); *Cuculimorphae* Sibley, Ahlquist, and Monroe 1988 (sensu Cracraft, 2013).

Comments: Monophyly of the cuckoos is supported by morphological (Hughes, 2000; Posso and Donatelli, 2006) and molecular (Payne, 2005; Hackett et al., 2008) studies. The clade was also inferred in a supertree analysis (Davis and Page, 2014).

Use of the name *Cuculidae* preserves current usage (e.g., Payne, 2005; Cracraft, 2013), and has priority over other possible names for this clade. Some authors have used the name *Cuculiformes* for the cuckoo clade and have restricted *Cuculidae* to a less inclusive clade (Wolters, 1975–1982; Sibley and Ahlquist, 1990). However, this use is not followed here in view of the long application of the name *Cuculiformes* to a more inclusive group

formed by cuckoos (*Cuculidae*) and turacos (*Musophagidae*) (e.g., Wetmore, 1930; Storer, 1960; Brodkorb, 1971; Cracraft, 1981; Bock, 1982; Livezey and Zusi, 2007). The names *Cuculimorphae*, *Cuculiformes* and *Cuculi* are included in some classifications as redundant names for the more widely used name *Cuculidae* (Wetmore, 1930; Storer, 1960; Livezey and Zusi, 2007; Cracraft, 2013).

A minimum-crown-clade definition was selected because there is no strongly supported, congruent evidence for the sister-group relationship of the cuckoos (van Tuinen et al., 2000; Mayr and Clarke, 2003; Sorenson et al., 2003; Chubb, 2004; Cracraft et al., 2004; Fain and Houde, 2004; Mayr and Ericson, 2004; Ericson et al., 2006; Hackett et al., 2008; Pacheco et al., 2011; Yuri et al., 2013). The basal subdivision of *Cuculidae* differs among various morphological, behavioral and molecular phylogenetic studies (Hughes, 1996, 2000; Aragón et al., 1999; Johnson et al., 2000; Payne, 2005; Posso and Donatelli, 2006; Livezey and Zusi, 2007). Therefore, the internal specifiers include representatives of the two basal subclades inferred in each of these studies.

The name *Cuculidae* is often attributed to Vigors (1825) (e.g., Brodkorb, 1971; Livezey and Zusi, 2007) but was first used by Leach (1820) (see Bock, 1994).

Literature Cited

Aragón, S., A. P. Møller, J. J. Soler, and M. Soler. 1999. Molecular phylogeny of cuckoos supports a polyphyletic origin of brood parasitism. *J. Evol. Biol.* 12:495–506.

Avise, J. C., W. S. Nelson, and C. G. Sibley. 1994. Why onekilobase sequences from mitochondrial DNA fail to solve the Hoatzin phylogenetic enigma. *Mol. Phylogen. Evol.* 3:175–184.

Bock, W. J. 1982. *Aves*. Pp. 967–1015 *in Synopsis and Classification of Living Organisms*, Vol. 2 (S. P. Parker, ed.). McGraw-Hill, New York.

Bock, W. J. 1994. History and nomenclature of avian family-group names. *Bull. Am. Mus. Nat. Hist.* 222:1–281.

Brodkorb, P. 1971. Catalogue of fossil birds: Part 4 (*Columbiformes* through *Piciformes*). *Bull. Fl. State Mus., Biol. Sci.* 15(4):163–266.

Brown, J. W., J. S. Rest, J. Garcia-Moreno, M. D. Sorenson, and D. P. Mindell. 2008. Strong mitochondrial DNA support for a Cretaceous origin of modern avian lineages. *BMC Biol.* 6:6.

Chubb, A. L. 2004. New nuclear evidence for the oldest divergence among neognath birds: the phylogenetic utility of ZENK. *Mol. Phylogen. Evol.* 30:140–151.

Coues, E. 1884. *Key to North American Birds*, Vol. 2. Estes and Lauriat, Boston, MA.

Cracraft, J. 1981. Toward a phylogenetic classification of the recent birds of the world (Class *Aves*). *Auk* 98:681–714.

Cracraft, J. 2013. Avian higher-level relationships and classification: nonpasseriforms. Pp. xxi–xliii in *The Howard and Moore Complete Checklist of the Birds of the World*. 4th edition. Vol 1: *Non-passerines* (E. C. Dickinson and J. V. Remsen, Jr., eds.). Aves Press, London.

Cracraft, J., F. K. Barker, M. Braun, J. Harshman, G. J. Dyke, J. Feinstein, S. Stanley, A. Cibois, P. Schikler, P. Beresford, J. García-Moreno, M. D. Sorenson, T. Yuri, and D. P. Mindell. 2004. Phylogenetic relationships among modern birds (*Neornithes*): towards an avian tree of life. Pp. 468–489 in *Reconstructing the Tree of Life* (J. Cracraft and M. J. Donoghue, eds.). Oxford University Press, New York.

Davis, K. E., and R. D. M. Page. 2014. Reweaving the tapestry: a supertree of birds. *PLOS Curr. Tree Life.* 9 June 2014. 1st edition. doi: 10.1371/currents.tol.c1af68dda7c999ed9f1e4b2d2df7a08e.

Dickinson, E. C., and J. V. Remsen, Jr., eds. 2013. *The Howard and Moore Complete Checklist of the Birds of the World*. 4th edition. Vol. 1: *Non-Passerines*. Aves Press, London.

Ericson, P. G. P., C. L. Anderson, T. Britton, A. Elzanowski, U. S. Johansson, M. Källersjö, J. I. Ohlson, T. J. Parsons, D. Zuccon, and G. Mayr. 2006. Diversification of *Neoaves*: integration of molecular sequence data and fossils. *Biol. Lett.* 2:543–547.

Fain, M. G., and P. Houde. 2004. Parallel radiations in the primary clades of birds. *Evolution* 58:2558–2573.

Hackett, S. J., R. T. Kimball, S. Reddy, R. C. K. Bowie, E. L. Braun, M. J. Braun, J. L. Chojnowski, W. A. Cox, K.-L. Han, J. Harshman, C. J. Huddleston, B. D. Marks, K. J. Miglia, W. S. Moore, F. H. Sheldon, D. W. Steadman, C. C. Witt, and T. Yuri. 2008. A phylogenomic study of birds reveals their evolutionary history. *Science* 320:1763–1768.

Hedges, S. B., M. D. Simmons, M. A. M. van Dijk, G. J. Caspers, W. W. de Jong, and C. G. Sibley. 1995. Phylogenetic relationships of the Hoatzin, an enigmatic South American bird. *Proc. Natl. Acad. Sci. USA* 92:11662–11665.

del Hoyo, J., A. Elliot, and J. Sargatal. 1997. *Handbook of the Birds of the World, Vol. 4. Sandgrouse to Cuckoos*. Lynx Edicions, Barcelona.

Hughes, J. M. 1996. Phylogenetic analysis of the *Cuculidae* (*Aves, Cuculiformes*) using behavioral and ecological characters. *Auk* 113:10–22.

Hughes, J. M. 2000. Monophyly and phylogeny of cuckoos (*Aves, Cuculidae*) inferred from osteological characters. *Zool. J. Linn. Soc.* 130:263–307.

Hughes, J. M., and A. J. Baker. 1999. Phylogenetic relationships of the enigmatic Hoatzin (*Opisthocomus hoazin*) resolved using mitochondrial and nuclear gene sequences. *Mol. Biol. Evol.* 16:1300–1307.

Johnson, K. P., S. M. Goodman, and S. M. Lanyon. 2000. A phylogenetic study of the Malagasy couas with insights into cuckoo relationships. *Mol. Phylogen. Evol.* 14:436–444.

Leach, W. E. 1820. Eleventh room. In *Synopsis of the Contents of the British Museum*. 17th edition. British Museum (Natural History), London.

Linnaeus, C. 1758. *Systema Naturae Per Regna Tria Naturae, Secundum Classes, Ordines, Genera, Species, cum Characteribus, Differentiis, Synonymis, Locis*. Editio decima, reformata. Laurentii Salvii, Holmiae (Stockholm).

Linnaeus, C. 1766. *Systema Naturae Per Regna Tria Naturae, Secundum Classes, Ordines, Genera, Species, cum Characteribus, Differentiis, Synonymis, Locis*. Edito duodecima, reformata. Laurentii Salvii, Holmiae (Stockholm).

Livezey, B. C., and R. L. Zusi. 2006. Higher-order phylogeny of modern birds (*Theropoda, Aves: Neornithes*) based on comparative anatomy. I. Methods and characters. *Bull. Carnegie Mus. Nat. Hist.* 37:1–556.

Livezey, B. C., and R. L. Zusi. 2007. Higher-order phylogeny of modern birds (*Theropoda, Aves: Neornithes*) based on comparative anatomy. II. Analysis and discussion. *Zool. J. Linn. Soc.* 149:1–95.

Mayr, G. 2006. A specimen of *Eocuculus* Chandler, 1999 (*Aves, ? Cuculidae*) from the early Oligocene of France. *Geobios* 39:865–872.

Mayr, G. 2009. *Paleogene Fossil Birds*. Springer-Verlag, Berlin and Heidelberg.

Mayr, G., and J. Clarke. 2003. The deep divergences of neornithine birds: a phylogenetic analysis of morphological characters. *Cladistics* 19:527–553.

Mayr, G., and P. G. P. Ericson. 2004. Evidence for a sister group relationship between the Madagascan mesites (*Mesitornithidae*) and cuckoos (*Cuculidae*). *Senckenb. Biol.* 84:119–135.

Pacheco, M. A., F. U. Battistuzzi, M. Lentino, R. Aguilar, S. Kumar, and A. A. Escalante. 2011. Evolution of modern birds revealed by mitogenomics: timing the radiation and origin of major orders. *Mol. Biol. Evol.* 28:1927–1942.

Payne, R. B. 2005. *Bird Families of the World: Cuckoos*. Oxford University Press, Oxford.

Posso, S. R., and R. J. Donatelli. 2006. Análise filogenética e implicações sistemáticas e evolutivas nos *Cuculiformes* (*Aves*) com base na osteologia, comportamento e ecologia. *Rev. Bras. Zool.* 23:608–629.

Sibley, C. G. 1996. *Birds of the World*. [CD-ROM] Version 2.0. Thayer Birding Software, Cincinnati, OH.

Sibley, C. G., and J. E. Ahlquist. 1990. *Phylogeny and Classification of Birds*. Yale University Press, New Haven, CT.

Sibley, C. G., J. E. Ahlquist, and B. L. Monroe, Jr. 1988. A classification of the living birds of the world based on DNA-DNA hybridization studies. *Auk* 105:409–423.

Sorenson, M. D., E. Oneal, J. García-Moreno, and D. P. Mindell. 2003. More taxa, more characters: the Hoatzin problem is still unresolved. *Mol. Biol. Evol.* 20:1484–1498.

Storer, R. W. 1960. The classification of birds. Pp. 57–93 in *Biology and Comparative Physiology of Birds*, Vol. 1 (A. J. Marshall, ed.). Academic Press, New York.

Temminck, C. J. 1820. *Nouveau Recueil de Planches Coloriées d'Oiseaux, Pour Servir de Suite et de Complément aux Planches Enluminées de Buffon. Imprimerie Royale, 1778*. II. G. Dufour & E. d'Ocagne, Paris.

Temminck, C. J. 1832. *Nouveau Recueil de Planches Coloriées d'Oiseaux*, III. G. F. Levrault, Paris.

van Tuinen, M., C. G. Sibley, and S. B. Hedges. 2000. The early history of modern birds inferred from DNA sequences of nuclear and mitochondrial ribosomal genes. *Mol. Biol. Evol.* 17:451–457.

Wagler, J. 1830. *Natürliches System der Amphibien, mit Vorangehender Classification der Säugthiere und Vogel. Ein Beitrag zur vergleichenden Zoologie*. J. G. Cotta, München, Stuttgart and Tübingen.

Wetmore A. 1930. A systematic classification for the birds of the world. *Proc. US Natl. Mus.* 76(24):1–8.

Wolters, H. E. 1975–1982. *Die Vogelarten der Erde*. Paul Parey, Hamburg.

Yuri, T., R. T. Kimball, J. Harshman, R. C. K. Bowie, M. J. Braun, J. L. Chojnowski, K.-L. Han, S. J. Hackett, C. J. Huddleston, W. S. Moore, S. Reddy, F. H. Sheldon, D. W. Steadman, C. C. Witt, and E. L. Braun. 2013. Parsimony and model-based analyses of indels in avian nuclear genes reveal congruent and incongruent phylogenetic signals. *Biology* 2:419–444.

Author

George Sangster; Department of Bioinformatics and Genetics; Swedish Museum of Natural History; P. O. Box 50007, SE–104 05 Stockholm; Sweden; Naturalis Biodiversity Center; Darwinweg 2, P. O. Box 9517, 2333 CC Leiden; Netherlands. Email: g.sangster@planet.nl.

Date Accepted: 11 December 2014

Primary Editor: Jacques A. Gauthier

Mirandornithes G. Sangster 2005 [G. Sangster], converted clade name

Registration Number: 281

Definition: The smallest crown clade containing *Phoenicopterus chilensis* Molina 1782 and *Podiceps* (originally *Colymbus*) *auritus* (Linnaeus 1758). This is a minimum-crown-clade definition. Abbreviated definition: min crown ∇ (*Phoenicopterus chilensis* Molina 1782 & *Podiceps auritus* (Linnaeus 1758)).

Etymology: Derived from the Latin adjective *mirandus*, meaning wonderful, and the Greek noun ορνίς (*ornis*), meaning bird.

Reference Phylogeny: For the purpose of applying the definition of *Mirandornithes*, Figure 2 in Hackett et al. (2008) should be regarded as the primary reference phylogeny. The left tree in Figure 2 of van Tuinen et al. (2001) and Figure 1 of Mayr (2004) may be regarded as secondary reference phylogenies.

Composition: Based on current evidence, *Mirandornithes* includes, among extant taxa, only *Phoenicopteriformes* (flamingos; 6 Recent species) and *Podicipediformes* (grebes; 22 Recent species).

Diagnostic Apomorphies: Mayr (2004) and Manegold (2006) identified 14 apomorphies of *Mirandornithes*, but their studies did not include members of its stem-group. Some of these characters were subsequently found to be apomorphies of a larger clade that also includes the mid-Eocene stem species *Juncitarsus merkeli* (Mayr, 2014). Derived morphological characters unique to *Mirandornithes* (relative to *Juncitarsus*) include (1) at least four thoracic vertebrae fused and forming a notarium; (2) ulna, tuberculum ligamenti collateralis ventralis greatly elongated; (3) phalanx proximalis digiti majoris (manual digit II) craniocaudally narrow (ratio length to craniocaudal width more than 4.5); (4) patella enlarged and cristae cnemiales of tibiotarsus projecting markedly beyond proximal articular surfaces; (5) tarsometatarsus, proximal portion of plantar surface of trochlea metatarsi III elongate and of subtriangular shape; (6) absence of fossa metatarsi I and reduction of hallux; (7) ungual phalanges dorsoventrally flattened and widened (Mayr, 2014). Suh et al. (2012) identified two retroposon insertions that are apomorphic relative to other crown clades.

Synonyms: *Phoenicopterimorphae* Cracraft 2013 and *Phoenicopteriformes* C. L. Bonaparte 1831 (sensu Dickinson and Remsen, 2013) are approximate synonyms; see Comments.

Comments: This clade was first discovered by van Tuinen et al. (2001) and first named by Sangster (2005). The clade represents one of best-supported higher-level relationships among birds and is now corroborated by studies based on mitochondrial and nuclear DNA sequences (Chubb, 2004; Cracraft et al., 2004; Ericson et al., 2006; Brown et al., 2008; Hackett et al., 2008; Morgan-Richards et al., 2008; Pratt et al., 2009; Pacheco et al., 2011; McCormack et al., 2013; Mahmood et al. 2014), nuclear DNA insertions and deletions (Suh et al., 2012; Yuri et al., 2013), and morphology (Mayr and Clarke, 2003; Mayr, 2004; but see Livezey and Zusi, 2007). The clade was also recovered in a recent supertree analysis (Davis and Page, 2014).

The name *Mirandornithes* has been used in several recent publications (e.g., Morgan-Richards et al., 2008; Suh et al., 2012; Ksepka et al.,

2013; Yuri et al., 2013; Davis and Page, 2014). Two alternative names (*Phoenicopterimorphae, Phoenicopteriformes*) have been used for this clade by different authors within a single publication (Cracraft, 2013; Dickinson and Remsen, 2013). These result from different opinions about the appropriate rank of this clade and the desire to apply standardized endings to clade names not governed by the *ICZN* (International Commission on Zoological Nomenclature, 1999). The name *Phoenicopteriformes* has long been associated with flamingos (reviewed by Sangster, 2005), so its application to a more inclusive group that also includes grebes goes against a fundamental principle of phylogenetic nomenclature—that is, maintaining the association between a name and a clade. The association of the name *Phoenicopteriformes* with flamingos, and the priority of the name *Mirandornithes* for the flamingo + grebe clade, are the reasons for selecting the name *Mirandornithes* for the latter clade.

A minimum-crown-clade definition was selected because the basal subdivision of *Mirandornithes* is well supported whereas its extant sister clade is unresolved or poorly supported (Ericson et al., 2006; Brown et al., 2008; Hackett et al., 2008; McCormack et al., 2013; Yuri et al., 2013). The extinct *Juncitarsus merkeli* was identified by Mayr (2014) as a member of the stem-group of *Mirandornithes*.

Literature Cited

Brown, J. W., J. S. Rest, J. Garcia-Moreno, M. D. Sorenson, and D. P. Mindell. 2008. Strong mitochondrial DNA support for a Cretaceous origin of modern avian lineages. *BMC Biol.* 6:6.

Chubb, A. L. 2004. New nuclear evidence for the oldest divergence among neognath birds: the phylogenetic utility of ZENK. *Mol. Phylogen. Evol.* 30:140–151.

Cracraft, J. 2013. Avian higher-level relationships and classification: nonpasseriforms. Pp. xxi–xliii in *The Howard and Moore Complete Checklist of the Birds of the World*. 4th edition. Vol. 1: *Non-passerines* (E. C. Dickinson and J. V. Remsen, Jr., eds.). Aves Press, London.

Cracraft, J., F. K. Barker, M. Braun, J. Harshman, G. J. Dyke, J. Feinstein, S. Stanley, A. Cibois, P. Schikler, P. Beresford, J. García-Moreno, M. D. Sorenson, T. Yuri, and D. P. Mindell. 2004. Phylogenetic relationships among modern birds (*Neornithes*): toward an avian tree of life. Pp. 468–489 in *Reconstructing the Tree of Life* (J. Cracraft and M. J. Donoghue, eds.). Oxford University Press, New York.

Davis, K. E., and R. D. M. Page. 2014. Reweaving the tapestry: a supertree of birds. *PLOS Curr. Tree Life.* 9 June 2014. 1st edition. doi:10.13 71/currents.tol.c1af68dda7c999ed9f1e4b2d 2df7a08e.

Dickinson, E. C., and J. V. Remsen, Jr., eds. 2013. *The Howard and Moore Complete Checklist of the Birds of the World*. 4th edition. Vol. 1: *Non-passerines*. Aves Press, London.

Ericson, P. G. P., C. L. Anderson, T. Britton, A. Elzanowski, U. S. Johansson, M. Källersjö, J. I. Ohlson, T. J. Parsons, D. Zuccon, and G. Mayr. 2006. Diversification of *Neoaves*: integration of molecular sequence data and fossils. *Biol. Lett.* 2:543–547.

Hackett, S. J., R. T. Kimball, S. Reddy, R. C. K. Bowie, E. L. Braun, M. J. Braun, J. L. Chojnowski, W. A. Cox, K.-L. Han, J. Harshman, C. J. Huddleston, B. D. Marks, K. J. Miglia, W. S. Moore, F. H. Sheldon, D. W. Steadman, C. C. Witt, and T. Yuri. 2008. A phylogenomic study of birds reveals their evolutionary history. *Science* 320:1763–1768.

International Commission on Zoological Nomenclature. 1999. *International Code of Zoological Nomenclature*. 4th edition. International Trust for Zoological Nomenclature, London.

Ksepka, D. T., A. M. Balanoff, M. A. Bell, and M. D. Houseman. 2013. Fossil grebes from the Truckee Formation (Miocene) of Nevada and a new phylogenetic analysis of *Podicipediformes* (Aves). *Palaeontology* 56:1149–1169.

Linnaeus, C. 1758. *Systema Naturae Per Regna Tria Naturae, Secundum Classes, Ordines, Genera, Species, cum Characteribus, Differentiis, Synonymis, Locis.* Editio decima, reformata. Laurentii Salvii, Holmiae (Stockholm).

Livezey, B. C., and R. L. Zusi. 2007. Higher-order phylogeny of modern birds (*Theropoda, Aves: Neornithes*) based on comparative anatomy. II. Analysis and discussion. *Zool. J. Linn. Soc.* 149:1–95.

Mahmood, M. T., P. A. McLenachan, G. C. Gibb, and D. Penny. 2014. Phylogenetic position of avian nocturnal and diurnal raptors. *Gen. Biol. Evol.* 6:326–332.

Manegold, A. 2006. Two additional synapomorphies of grebes *Podicipedidae* and flamingos *Phoenicopteridae. Acta Ornithol.* 41:79–82.

Mayr, G. 2004. Morphological evidence for sister group relationship between flamingos (*Aves: Phoenicopteridae*) and grebes (*Podicipedidae*). *Zool. J. Linn. Soc.* 140:157–169.

Mayr, G. 2014. The Eocene *Juncitarsus*—its phylogenetic position and significance for the evolution and higher-level affinities of flamingos and grebes. *C. R. Palevol.* 13: 9–18.

Mayr, G., and J. Clarke. 2003. The deep divergences of neornithine birds: a phylogenetic analysis of morphological characters. *Cladistics* 19:527–553.

McCormack, J. E., M. G. Harvey, B. C. Faircloth, N. G. Crawford, T. C. Glenn, and R. T. Brumfield. 2013. A phylogeny of birds based on over 1,500 loci collected by target enrichment and high-throughput sequencing. *PLOS ONE* 8(1):e54848.

Morgan-Richards, M., S. A. Trewick, A. Bartosch-Härlid, O. Kardialsky, M. J. Phillips, P. A. McLenachan, and D. Penny. 2008. Bird evolution: testing the *Metaves* clade with six new mitochondrial genomes. *BMC Evol. Biol.* 8:20.

Pacheco, M. A., F. U. Battistuzzi, M. Lentino, R. Aguilar, S. Kumar, and A. A. Escalante. 2011. Evolution of modern birds revealed by mitogenomics: timing the radiation and origin of major orders. *Mol. Biol. Evol.* 28:1927–1942.

Pratt, R. C., G. C. Gibb, M. Morgan-Richards, M. J. Phillips, M. D. Hendy, and D. Penny. 2009. Towards resolving deep *Neoaves* phylogeny: data, signal enhancement and priors. *Mol. Biol. Evol.* 26:313–326.

Sangster, G. 2005. A name for the flamingo-grebe clade. *Ibis* 147:612–615.

Suh, A., J. O. Kriegs, S. Donnellan, J. Brosius, and J. Schmitz. 2012. A universal method for the study of CR1 retroposons in nonmodel bird genomes. *Mol. Biol. Evol.* 29:2899–2903.

van Tuinen, M., D. B. Butvill, J. A. W. Kirsch, and S. B. Hedges. 2001. Convergence and divergence in the evolution of aquatic birds. *Proc. R. Soc. Lond. B Biol. Sci.* 268:1345–1350.

Yuri, T., R. T. Kimball, J. Harshman, R. C. K. Bowie, M. J. Braun, J. L. Chojnowski, K.-L. Han, S. J. Hackett, C. J. Huddleston, W. S. Moore, S. Reddy, F. H. Sheldon, D. W. Steadman, C. C. Witt, and E. L. Braun. 2013. Parsimony and model-based analyses of indels in avian nuclear genes reveal congruent and incongruent phylogenetic signals. *Biology* 2:419–444.

Author

George Sangster; Department of Bioinformatics and Genetics; Swedish Museum of Natural History; P. O. Box 50007, SE–104 05 Stockholm; Sweden; and Naturalis Biodiversity Center; Darwinweg 2, P. O. Box 9517, 2333 CC Leiden; the Netherlands. Email: g.sangster@planet.nl.

Date Accepted: 11 December 2014

Primary Editor: Jacques A. Gauthier

Charadriiformes A. H. Garrod 1874 [G. Sangster], converted clade name

Registration Number: 282

Definition: The smallest crown clade containing *Charadrius hiaticula* Linnaeus 1758 (*Charadrii*), *Scolopax minor* Gmelin 1789 (*Scolopaci*) and *Larus marinus* Linnaeus 1758 (*Lari*). This is a minimum-crown-clade definition. Abbreviated definition: min crown ∇ (*Charadrius hiaticula* Linnaeus 1758 & *Scolopax minor* Gmelin 1789 & *Larus marinus* Linnaeus 1758).

Etymology: The Greek χαραδριός (*charadrios*) referred to a "bird dwelling in clefts or gullies". The name is found in the writings of Aristotle, Aristophanes, and Suidas without further identification (Marchant and Higgins, 1993). The suffix *-formes* is derived from the Latin *forma* (having the form of).

Reference Phylogeny: For the purpose of applying the definition of *Charadriiformes*, Figure 1 of Baker et al. (2007) should be regarded as the primary reference phylogeny. Baker et al. (2007) did not include *Chariadrius hiaticula*, which is the type species of *Charadrius* and is thus required as a specifier for *Charadriiformes*. Instead, they sampled *Charadrius vociferus*. These two species are closely related (Barth et al., 2013).

Composition: The crown clade here named *Charadriiformes* contains about 375 Recent species (Dickinson and Remsen, 2013). Colloquially known as 'shorebirds', *Charadriiformes* includes *Charadrii* (Magellanic Plover, sheathbills, Egyptian Plover, thick-knees, avocets, stilts, Ibisbill, oystercatchers, plovers and lapwings), *Scolopaci* (Plains-wanderer, seedsnipes, painted-snipes, jacanas, curlews, godwits, snipes, sandpipers and phalaropes) and *Lari* (Crab Plover, pratincoles, coursers, skuas, auks, terns, skimmers and gulls). Recent studies indicate that the Plains-wanderer (*Pedionomus torquatus*) (Olson and Steadman, 1981; Sibley and Ahlquist, 1990; Paton et al., 2003; Paton and Baker, 2006; Baker et al., 2007; Hackett et al., 2008; Mayr, 2011), and the buttonquails (*Turnicidae*) (Paton et al., 2003; Fain and Houde, 2004; Paton and Baker, 2006; Baker et al., 2007; Fain and Houde, 2007; Hackett et al., 2008), which were previously included in *Gruiformes*, are instead parts of *Charadriiformes*.

Diagnostic Apomorphies: Cracraft (1988) listed one apomorphy of *Charadriiformes* relative to other crown-clades: a derived mobility pattern of malate dehydrogenase (Kitto and Wilson, 1966). Subsequent morphological studies of charadriiform relationships did not identify any further apomorphies (Strauch, 1978; Mickevich and Parenti, 1980; Björklund, 1994; Chu, 1995; Dove, 2000). Livezey and Zusi (2006, 2007), in the largest phylogenetic study based on morphological characters to date, however, identified a single 'supportive' apomorphy (0.5 < consistency index < 1) relative to other major crown-clades: Musculus adductor mandibulae externus, caput zygomaticus, origo, inclusion of arcus suborbitalis et arcus jugalis present (Livezey and Zusi, 2006: character state 2462b). Two recent morphological studies of *Charadriiformes* did not list apomorphies for this clade (Livezey, 2010; Mayr, 2011). No unambiguous members of the stem-group of *Charadriiformes* are known (Mayr, 2009); therefore, all known apomorphies are relative to other crown clades.

Synonyms: *Charadriomorphae* Huxley 1867 (partial); *Limicolae* (*Laro-Limicolae*) Mayr and Amadon 1951 (approximate); *Charadriides* Sibley, Ahlquist and Monroe 1988 (approximate); *Charadriimorphae* auct. sensu Cracraft (2013) (approximate); *Charadriia* Cracraft 2013 (approximate).

Comments: Monophyly of the shorebirds is supported by morphological (Livezey, 2010; Mayr, 2011) and molecular studies (Sibley and Ahlquist, 1990; Ericson et al., 2003; Baker et al., 2007; Fain and Houde, 2007; Hackett et al., 2008; Yuri et al., 2013), and by a recent supertree analysis (Davis and Page, 2014).

The name *Charadriiformes* is widely accepted for this clade. The name *Charadriides* has only been used by C. G. Sibley and co-workers (Sibley et al., 1988; Sibley and Ahlquist, 1990), and the names *Charadriimorphae* and *Charadriia* appear to have been used for this clade only as redundant names. The name *Charadriiformes* has also been applied to a smaller taxon that does not include the sheathbills, gulls, terns, skimmers, skuas, and auks (Wolters, 1975–1982); this taxon is now known to be paraphyletic (e.g., Baker et al., 2007).

A minimum-crown-clade definition was selected because the sister-taxon of *Charadriiformes* remains unclear (Cracraft et al., 2004; Brown et al., 2008; Hackett et al., 2008; Pacheco et al., 2011; McCormack et al., 2013; Yuri et al., 2013), whereas the subdivision of *Charadriiformes* into *Charadrii*, *Scolopaci* and *Lari* is well-supported (Sibley and Ahlquist, 1990; Christian et al., 1992; Ericson et al., 2003; Paton et al., 2003; Thomas et al., 2004; Paton and Baker, 2006; Baker et al., 2007; Fain and Houde, 2007; Hackett et al., 2008). The sandgrouse (*Pteroclidae*) are sometimes considered closely related to, or part of, *Charadriiformes* (Fjeldså, 1976; Sibley and Ahlquist, 1990) but this is not supported by recent morphological

or nucleotide sequence studies (Ericson et al., 2003; Mayr and Clarke, 2003; Paton et al., 2003; Cracraft et al., 2004; Ericson et al., 2006; Livezey and Zusi, 2007; Hackett et al., 2008).

Under rank-based nomenclature, Huxley (1867) is considered the author of the name *Charadriiformes*, even though he spelled the name *Charadriomorphae*, because he was the first person to publish a name based on *Charadrius* at the rank of order. The first author to use the name *Charadriiformes* appears to have been Garrod (1874) (see Brodkorb, 1967), and he is accordingly regarded as the nominal author of the name selected for this clade.

Literature Cited

Baker, A. J., S. L. Pereira, and T. A. Paton. 2007. Phylogenetic relationships and divergence times of *Charadriiformes* genera: multigene evidence for the Cretaceous origin of at least 14 clades of shorebirds. *Biol. Lett.* 3: 205–209.

Barth, J. M. I., M. Matschiner, and B. C. Robertson. 2013. Phylogenetic position and subspecies divergence of the endangered New Zealand Dotterel (*Charadrius obscurus*). *PLOS ONE* 8(10):e78068.

Björklund, M. 1994. Phylogenetic relationships among *Charadriiformes*: reanalysis of previous data. *Auk* 111:825–832.

Brodkorb, P. 1967. Catalogue of fossil birds: part 3 (*Ralliformes, Ichthyornithiformes, Charadriiformes*). *Bull. Fl. State Mus., Biol. Sci.* 11(3):99–220.

Brown, J. W., J. S. Rest, J. Garcia-Moreno, M. D. Sorenson, and D. P. Mindell. 2008. Strong mitochondrial DNA support for a Cretaceous origin of modern avian lineages. *BMC Biol.* 6:6.

Christian, P. D., L. Christidis, and R. Schodde. 1992. Biochemical systematics of the *Charadriiformes* (shorebirds): relationships between the *Charadrii, Scolopaci* and *Lari*. *Aust. J. Zool.* 40:291–302.

Chu, P. C. 1995. Phylogenetic reanalysis of Strauch's osteological data set for the *Charadriiformes*. *Condor* 97:174–196.

Cracraft, J. 1988. The major clades of birds. Pp. 339–361 in *The Phylogeny and Classification of the Tetrapods. Vol. 1: Amphibians, Reptiles, Birds* (M. J. Benton, ed.). Systematics Association Special Volume No. 35A. Clarendon Press, Oxford.

Cracraft, J. 2013. Avian higher-level relationships and classification: nonpasseriforms. Pp. xxi–xliii in *The Howard and Moore Complete Checklist of the Birds of the World*. 4th edition. Vol. 1: *Non-passerines* (E. C. Dickinson and J. V. Remsen, Jr., eds.). Aves Press, London.

Cracraft, J., F. K. Barker, M. Braun, J. Harshman, G. J. Dyke, J. Feinstein, S. Stanley, A. Cibois, P. Schikler, P. Beresford, J. García-Moreno, M. D. Sorenson, T. Yuri, and D. P. Mindell. 2004. Phylogenetic relationships among modern birds (*Neornithes*): toward an avian tree of life. Pp. 468–489 in *Reconstructing the Tree of Life* (J. Cracraft and M. J. Donoghue, eds.). Oxford University Press, New York.

Davis, K. E., and R. D. M. Page. 2014. Reweaving the tapestry: a supertree of birds. *PLOS Curr. Tree Life*. 9 June 2014. 1st edition. doi:10.13 71/currents.tol.c1af68dda7c999ed9f1e4b2d 2df7a08e.

Dickinson, E. C., and J. V. Remsen, Jr., eds. 2013. *The Howard and Moore Complete Checklist of the Birds of the World*. 4th edition. Vol. 1: *Non-passerines*. Aves Press, London.

Dove, C. J. 2000. A descriptive and phylogenetic analysis of plumulaceous feather characters in *Charadriiformes*. *Ornithol. Monogr.* 51:1–163.

Ericson, P. G. P., C. L. Anderson, T. Britton, A. Elzanowski, U. S. Johansson, M. Källersjö, J. I. Ohlson, T. J. Parsons, D. Zuccon, and G. Mayr. 2006. Diversification of *Neoaves*: integration of molecular sequence data and fossils. *Biol. Lett.* 2:543–547.

Ericson, P. G. P., I. Envall, M. Irestedt, and J. A. Norman. 2003. Inter-familial relationships of the shorebirds (*Aves: Charadriiformes*) based on nuclear DNA sequence data. *BMC Evol. Biol.* 3 (16):1–14.

Fain, M. G., and P. Houde. 2004. Parallel radiations in the primary clades of birds. *Evolution* 58:2558–2573.

Fain, M. G., and P. Houde. 2007. Multilocus perspectives on the monophyly and phylogeny of the order *Charadriiformes* (*Aves*). *BMC Evol. Biol.* 7 (1):35.

Fjeldså, J. 1976. The systematic affinities of the sandgrouse, *Pteroclididae*. *Vidensk Medd. Dansk Naturh. Foren.* 139:179–243.

Garrod, A. H. 1874. On certain muscles of birds and their value in the classification. Part II. *Proc. Zool. Soc. Lond.* 1874:111–123.

Gmelin, J. F. 1789. *Caroli a Linné. Systema Naturae Per Regna Tria Naturae: Secundum Classes, Ordines, Genera, Species, cum Characteribus, Differentiis, Synonymis, Locis*. Tomus I: Georg Emanuel Beer, 1788–1793, Leipzig.

Hackett, S. J., R. T. Kimball, S. Reddy, R. C. K. Bowie, E. L. Braun, M. J. Braun, J. L. Chojnowski, W. A. Cox, K.-L. Han, J. Harshman, C. J. Huddleston, B. D. Marks, K. J. Miglia, W. S. Moore, F. H. Sheldon, D. W. Steadman, C. C. Witt, and T. Yuri. 2008. A phylogenomic study of birds reveals their evolutionary history. *Science* 320:1763–1768.

Huxley, T. H. 1867. On the classification of birds; and on the taxonomic value of the modifications of certain of the cranial bones observable in that class. *Proc. Zool. Soc. Lond.* 1867:415–472.

Kitto, G. B., and A. C. Wilson. 1966. Evolution of malate dehydrogenase in birds. *Science* 153:1408–1410.

Linnaeus, C. 1758. *Systema Naturae Per Regna Tria Naturae, Secundum Classes, Ordines, Genera, Species, cum Characteribus, Differentiis, Synonymis, Locis*. Editio decima, reformata. Laurentii Salvii, Holmiae (Stockholm).

Livezey, B. C. 2010. Phylogenetics of modern shorebirds (*Charadriiformes*) based on phenotypic evidence: analysis and discussion. *Zool. J. Linn. Soc.* 160:567–618.

Livezey, B. C., and R. L. Zusi. 2006. Higher-order phylogeny of modern birds (*Theropoda, Aves: Neornithes*) based on comparative anatomy. I. Methods and characters. *Bull. Carnegie Mus. Nat. Hist.* 37:1–556.

Livezey, B. C., and R. L. Zusi. 2007. Higher-order phylogeny of modern birds (*Theropoda, Aves: Neornithes*) based on comparative anatomy. II. Analysis and discussion. *Zool. J. Linn. Soc.* 149:1–95.

Marchant, S., and P. Higgins. 1993. *Handbook of Australian, New Zealand & Antarctic Birds,* Vol. 2. Oxford University Press, Melbourne.

Mayr, E., and D. Amadon. 1951. A classification of recent birds. *Am. Mus. Novit.* 1496:1–42.

Mayr, G. 2009. *Paleogene Fossil Birds.* Springer-Verlag, Berlin and Heidelberg.

Mayr, G. 2011. The phylogeny of charadriiform birds (shorebirds and allies) – reassessing the conflict between morphology and molecules. *Zool. J. Linn. Soc.* 161:916–934.

Mayr, G., and J. Clarke. 2003. The deep divergences of neornithine birds: a phylogenetic analysis of morphological characters. *Cladistics* 19:527–553.

McCormack, J. E., M. G. Harvey, B. C. Faircloth, N. G. Crawford, T. C. Glenn, and R. T. Brumfield. 2013. A phylogeny of birds based on over 1,500 loci collected by target enrichment and high-throughput sequencing. *PLOS ONE* 8(1):e54848.

Mickevich, M. F., and L. R. Parenti. 1980. Review of Strauch (1978): The phylogeny of the *Charadriiformes* (*Aves*): a new estimate using the method of character compatibility analysis. *Syst. Zool.* 29:108–113.

Olson, S. L., and D. W. Steadman. 1981. The relationships of the *Pedionomidae* (*Aves: Charadriiformes*). *Smithson. Contr. Zool.* 337.

Pacheco, M. A., F. U. Battistuzzi, M. Lentino, R. Aguilar, S. Kumar, and A. A. Escalante. 2011. Evolution of modern birds revealed by mitogenomics: timing the radiation and origin of major orders. *Mol. Biol. Evol.* 28:1927–1942.

Paton, T. A., and A. J. Baker. 2006. Sequences from 14 mitochondrial genes provide a well-supported phylogeny of the charadriiform birds congruent with the nuclear RAG-1 tree. *Mol. Phylogen. Evol.* 39:657–667.

Paton, T. A., A. J. Baker, J. G. Groth, and G. F. Barrowclough. 2003. RAG-1 sequences resolve phylogenetic relationships within charadriiform birds. *Mol. Phylogen. Evol.* 29:268–278.

Sibley, C. G., and J. E. Ahlquist. 1990. *Phylogeny and Classification of Birds.* Yale University Press, New Haven, CT.

Sibley, C. G., J. E. Ahlquist, and B. L. Monroe. 1988. A classification of the living birds of the world based on DNA-DNA hybridization studies. *Auk* 105:409–423.

Strauch, J. G. 1978. The phylogeny of the *Charadriiformes* (*Aves*): a new estimate using the method of character compatibility analysis. *Trans. Zool. Soc. Lond.* 34:263–345.

Thomas, G. H., M. A. Wills, and T. Székely. 2004. Phylogeny of shorebirds, gulls, and alcids (*Aves: Charadrii*) from the cytochrome-b gene: parsimony, Bayesian inference, minimum evolution, and quartet puzzling. *Mol. Phylogen. Evol.* 30:516–526.

Wolters, H. E. 1975–82. *Die Vogelarten der Erde.* Paul Parey, Hamburg.

Yuri, T., R. T. Kimball, J. Harshman, R. C. K. Bowie, M. J. Braun, J. L. Chojnowski, K.-L. Han, S. J. Hackett, C. J. Huddleston, W. S. Moore, S. Reddy, F. H. Sheldon, D. W. Steadman, C. C. Witt, and E. L. Braun. 2013. Parsimony and model-based analyses of indels in avian nuclear genes reveal congruent and incongruent phylogenetic signals. *Biology* 2:419–444.

Author

George Sangster; Department of Bioinformatics and Genetics; Swedish Museum of Natural History; P. O. Box 50007, SE–104 05 Stockholm; Sweden; and Naturalis Biodiversity Center; Darwinweg 2, P. O. Box 9517, 2333 CC Leiden; Netherlands.

Date Accepted: 11 December 2014

Primary Editor: Jacques A. Gauthier

Procellariiformes M. Fürbringer 1888 [G. Sangster], converted clade name

Registration Number: 283

Definition: The smallest crown clade containing *Procellaria aequinoctialis* Linnaeus 1758, *Diomedea exulans* Linnaeus 1758, *Pelecanoides* (originally *Procellaria*) *urinatrix* (Gmelin 1789), *Hydrobates* (originally *Procellaria*) *pelagicus* (Linnaeus 1758) and *Oceanites* (originally *Procellaria*) *oceanicus* (Kuhl 1820). This is a minimum-crown-clade definition. Abbreviated definition: min crown ∇ (*Procellaria aequinoctialis* Linnaeus 1758 & *Diomedea exulans* Linnaeus 1758 & *Pelecanoides urinatrix* (Gmelin 1789) & *Hydrobates pelagicus* (Linnaeus 1758) & *Oceanites oceanicus* (Kuhl 1820)).

Etymology: Derived from the taxon name *Procellaria* which is "a modern adjectival form of the Latin word for a storm (*procella*) and gives the meaning of 'creatures of the storm' as is appropriate" (Marchant and Higgins, 1990: 557). The suffix *-formes* is derived from the Latin *forma* (having the form of).

Reference Phylogeny: For the purpose of applying the definition of *Procellariiformes*, Figure 2A of Kennedy and Page (2002) should be regarded as the primary reference phylogeny. Figure 2 in Nunn and Stanley (1998) and Figure 2 in Penhallurick and Wink (2004) are regarded as secondary reference phylogenies.

Composition: *Procellariiformes* is currently believed to contain 120–125 Recent species (Brooke, 2004; Dickinson and Remsen, 2013). These are often (e.g., Bock, 1982; Dickinson and Remsen, 2013) grouped into four main, mutually exclusive subtaxa: *Diomedeidae* (albatrosses), *Procellariidae* (petrels and shearwaters), *Pelecanoididae* (diving-petrels) and *Hydrobatidae* (storm-petrels). The last of these may not be monophyletic and is sometimes split, under rank-based nomenclature, into *Oceanitidae* and *Hydrobatidae* (e.g., Dickinson and Remsen, 2013).

Diagnostic Apomorphies: Mayr (2003) identified six morphological apomorphies of *Procellariiformes*: (1) external nostrils tubular; (2) fossae glandulae nasales very marked and situated on dorsal surface of supraorbital margin of orbitae; (3) coracoid, extremas sternalis, processus lateralis greatly elongated; (4) humerus, large and strongly protruding processus supracondylaris dorsalis present; (5) tibiotarsus, proximal end, cristae cnemialis strongly proximally protruding; (6) hallux greatly reduced and consisting of a single phalanx only. In a much larger study, Livezey and Zusi (2006, 2007) identified five apomorphies of *Procellariiformes*, of which four are unique (consistency index = 1) (characters and states are indicated by their number-letter combination in Livezey and Zusi, 2006): (1) extremitas sternalis coracoidei, processus lateralis, minimal craniocaudal width subequal to lateromedial width of facies articularis sternalis (1305b); (2) extremitas sternalis coracoidei, processus lateralis, terminus processi, facies ancorae ligamentosa present (1306b); (3) proventriculus gastris, regio glandularis, plicae proventriculi, glandulae proventriculares productive of odoriferous excretory products subject to voluntary (defensive) emesis (2847b); (4) remiges, ecdysis remigum primariae synchronous (2819b).

Mayr and Smith (2012) identified 4 or 5 apomorphies (depending on which outgroups were selected), of which most were not previously identified by Mayr (2003) or Livezey and Zusi (2006, 2007).

The studies by Mayr (2003) and Livezey and Zusi (2006, 2007) did not include any members of the stem-group of *Procellariiformes*. Therefore, apomorphies listed in these studies are relative to other crown clades. Mayr and Smith (2012) included the extinct *Rupelornis definitas* but they could not resolve whether this species is a member of the *Procellariiformes* crown.

Synonyms: The following are approximate synonyms (see Comments): *Tubinares* Illiger 1811 (sensu Seebohm, 1890, and Mayr and Amadon, 1951); *Procellariidae* Boie 1826 (sensu Sibley and Ahlquist, 1990, and Brooke, 2004).

Comments: Monophyly of *Procellariiformes* is supported by morphological (Livezey and Zusi, 2007; Mayr and Smith, 2012) and molecular (Ericson et al., 2006; Hackett et al., 2008; Yuri et al., 2013) studies. The clade was also inferred in a recent supertree analysis (Davis and Page, 2014). A few anomalous trees have been published in which albatrosses, petrels and shearwaters, diving-petrels and storm-petrels did not form a monophyletic group (Siegel-Causey, 1997: Fig. 6.2; Johnson, 2001; Kennedy and Page, 2002: Fig. 3; Chubb, 2004: Fig. 5), but none of these trees can be regarded as a rigorous test of monophyly, and node support was either low or not specified.

Two names are currently in use for this clade. *Procellariiformes* is the most commonly used name (e.g., Wetmore, 1960; Storer, 1971; Jouanin and Mougin, 1979; del Hoyo et al., 1992; Livezey and Zusi, 2007; Cracraft, 2013) and is adopted here. The name *Procellariidae* is used by some authors, but is more commonly applied to a subclade formed by petrels, fulmars, and shearwaters. The name *Tubinares* is no longer in use.

A minimum-crown-clade definition was selected because it preserves current use of the name (i.e., as a crown clade). Multiple internal specifiers are used to define the name *Procellariiformes* because internal relationships of the clade are not well supported, although most studies have identified the storm-petrels (if monophyletic) or a subclade (if storm-petrels are not monophyletic) as the sister of all remaining procellariiforms (Kuroda et al., 1990, Sibley and Ahlquist, 1990, Nunn and Stanley, 1998; Kennedy and Page, 2002, Cracraft et al., 2004, Penhallurick and Wink, 2004, Mayr, 2005, Hackett et al., 2008; but see Livezey and Zusi, 2007). Because storm-petrels formed a non-monophyletic group in some of these studies, representatives of both groups of storm-petrels are included as internal specifiers in the definition, plus representatives of *Diomedeidae*, *Procellariidae* and *Pelecanoididae*.

The closest relatives of *Procellariiformes* are sometimes considered to be the loons (*Gaviidae*; e.g., Sibley and Ahlquist, 1990; Mayr, 2005) or the penguins (*Sphenisciformes*; e.g., Livezey and Zusi, 2007; Hackett et al., 2008), but the evidence does not clearly favor one of these hypotheses, and several results are inconsistent with both (e.g., Mayr and Clarke, 2003; Fain and Houde, 2004; Slack et al., 2006; Watanabe et al., 2006; Brown et al., 2008; Pratt et al., 2009; McCormack et al., 2013).

Literature Cited

Bock, W. J. 1982. *Aves*. Pp. 967–1015 in *Synopsis and Classification of Living Organisms*, Vol. 2 (S. P. Parker, ed.). McGraw-Hill, New York.

Brooke, M. de L. 2004. *Albatrosses and Petrels Across the World*. Oxford University Press, Oxford.

Brown, J. W., J. S. Rest, J. Garcia-Moreno, M. D. Sorenson, and D. P. Mindell. 2008. Strong mitochondrial DNA support for a Cretaceous origin of modern avian lineages. *BMC Biol.* 6:6.

Chubb, A. L. 2004. New nuclear evidence for the oldest divergence among neognath birds: the phylogenetic utility of ZENK. *Mol. Phylogen. Evol.* 30:140–151.

Cracraft, J. 2013. Avian higher-level relationships and classification: nonpasseriforms. Pp. xxi–xliii in *The Howard and Moore Complete Checklist of the Birds of the World*. 4th edition. Vol. 1: *Non-passerines* (E. C. Dickinson and J. V. Remsen, Jr., eds.). Aves Press, London.

Cracraft, J., F. K. Barker, M. Braun, J. Harshman, G. J. Dyke, J. Feinstein, S. Stanley, A. Cibois, P. Schikler, P. Beresford, J. García-Moreno, M. D. Sorenson, T. Yuri, and D. P. Mindell. 2004. Phylogenetic relationships among modern birds (*Neornithes*): toward an avian tree of life. Pp. 468–489 in *Reconstructing the Tree of Life* (J. Cracraft and M. J. Donoghue, eds.). Oxford University Press, New York.

Davis, K. E., and R. D. M. Page. 2014. Reweaving the tapestry: a supertree of birds. *PLOS Curr. Tree Life*. 9 June 2014. 1st edition. doi:10.1371/currents.tol.c1af68dda7c999ed9f1e4b2d2df7a08e.

Dickinson, E. C., and J. V. Remsen, Jr., eds. 2013. *The Howard and Moore Complete Checklist of the Birds of the World*. 4th edition. Vol. 1: *Non-passerines*. Aves Press, London.

Ericson, P. G. P., C. L. Anderson, T. Britton, A. Elzanowski, U. S. Johansson, M. Källersjö, J. I. Ohlson, T. J. Parsons, D. Zuccon, and G. Mayr. 2006. Diversification of *Neoaves*: integration of molecular sequence data and fossils. *Biol. Lett.* 2:543–547.

Fain, M. G., and P. Houde. 2004. Parallel radiations in the primary clades of birds. *Evolution* 58:2558–2573.

Fürbringer, M. 1888. *Untersuchungen zur Morphologie und Systematik der Vögel zugleich ein Beitrag zur Anatomie der Stütz- und Bewegungsorgane* (Vol. 2). T. J. Van Holkema, Amsterdam.

Gmelin, J. F. 1789. *Caroli a Linné. Systema Naturae Per Regna Tria Naturae: Secundum Classes, Ordines, Genera, Species, cum Characteribus, Differentiis, Synonymis, Locis*. Tomus I: Georg Emanuel Beer, Leipzig.

Hackett, S. J., R. T. Kimball, S. Reddy, R. C. K. Bowie, E. L. Braun, M. J. Braun, J. L. Chojnowski, W. A. Cox, K.-L. Han, J. Harshman, C. J. Huddleston, B. D. Marks, K. J. Miglia, W. S. Moore, F. H. Sheldon, D. W. Steadman, C. C. Witt, and T. Yuri. 2008. A phylogenomic study of birds reveals their evolutionary history. *Science* 320:1763–1768.

del Hoyo, J., A. Elliot, and J. Sargatal. 1992. *Handbook of the Birds of the World. Vol. 1: Ostrich to Ducks*. Lynx Edicions, Barcelona.

Johnson, K. P. 2001. Taxon sampling and the phylogenetic position of *Passeriformes*: evidence from 916 avian cytochrome b sequences. *Syst. Biol.* 50:128–136.

Jouanin, C., and J.-L. Mougin. 1979. Order Procellariiformes. Pp. 48–121 in *Check-list of Birds of the World*. 2nd edition, Vol. 1 (E. Mayr and G. W. Cottrell, eds.). Mus. Comp. Zool., Cambridge, MA.

Kennedy, M., and R. D. M. Page. 2002. Seabird supertrees: combining partial estimates of procellariiform phylogeny. *Auk* 119:88–108.

Kuhl, H. 1820. *Beiträge zur Zoologie und Vergleichenden Anatomie. 1. Abtheilung Beiträge zur Zoologie*. Verlag der Hermannschen Buchhandlung.

Kuroda, N., R. Kakizawa, and M. Watada. 1990. Genetic divergence and relationships in fifteen species of *Procellariiformes*. *J. Yamashina Inst. Ornithol.* 22:114–123.

Linnaeus, C. 1758. *Systema Naturae Per Regna Tria Naturae, Secundum Classes, Ordines, Genera, Species, cum Characteribus, Differentiis, Synonymis, Locis*. Editio decima, reformata. Laurentii Salvii, Holmiae (Stockholm).

Livezey, B. C., and R. L. Zusi. 2006. Higher-order phylogeny of modern birds (*Theropoda, Aves: Neornithes*) based on comparative anatomy. I. Methods and characters. *Bull. Carnegie Mus. Nat. Hist.* 37:1–556.

Livezey, B. C., and R. L. Zusi. 2007. Higher-order phylogeny of modern birds (*Theropoda, Aves: Neornithes*) based on comparative anatomy. II. Analysis and discussion. *Zool. J. Linn. Soc.* 149:1–95.

Marchant, S., and P. Higgins. 1990. *Handbook of Australian, New Zealand & Antarctic Birds*, Vol. 1. Oxford University Press, Melbourne.

Mayr, E., and D. Amadon. 1951. A classification of recent birds. *Am. Mus. Novit.* 1496:1–42.

Mayr, G. 2003. The phylogenetic affinities of the Shoebill (*Balaeniceps rex*). *J. Ornithol.* 144:157–175.

Mayr, G. 2005. Tertiary plotopterids (*Aves, Plotopteridae*) and a novel hypothesis on the phylogenetic relationships of penguins (*Spheniscidae*). *J. Zool. Syst. Evol. Res.* 43:61–71.

Mayr, G., and J. Clarke. 2003. The deep divergences of neornithine birds: a phylogenetic analysis of morphological characters. *Cladistics* 19:527–553.

Mayr, G., and T. Smith. 2012. Phylogenetic affinities and taxonomy of the Oligocene *Diomedeoididae*, and the basal divergences amongst extant procellariiform birds. *Zool. J. Linn. Soc.* 166:854–875.

McCormack, J. E., M. G. Harvey, B. C. Faircloth, N. G. Crawford, T. C. Glenn, and R. T. Brumfield. 2013. A phylogeny of birds based on over 1,500 loci collected by target enrichment and high-throughput sequencing. *PLOS ONE* 8(1):e54848.

Nunn, G. B., and S. E. Stanley. 1998. Body size and rates of cytochrome b evolution in tube-nosed seabirds. *Mol. Biol. Evol.* 15:1360–1371.

Penhallurick, J., and M. Wink. 2004. Analysis of the taxonomy and nomenclature of the *Procellariiformes* based on complete nucleotide sequences of the mitochondrial cytochrome b gene. *Emu* 104:125–147.

Pratt, R. C., G. C. Gibb, M. Morgan-Richards, M. J. Phillips, M. D. Hendy, and D. Penny. 2009. Toward resolving deep *Neoaves* phylogeny: data, signal enhancement and priors. *Mol. Biol. Evol.* 26:313–326.

Seebohm, H. 1890. *Classification of Birds: An Attempt to Diagnose the Subclasses, Orders, Suborders and Some of the Families of Existing Birds*. R. H. Porter, London.

Sibley, C. G., and J. E. Ahlquist. 1990. *Phylogeny and Classification of Birds*. Yale University Press, New Haven, CT.

Siegel-Causey, D. 1997. Phylogeny of the *Pelecaniformes*: molecular systematics of a privative group. Pp. 159–171 in *Avian Molecular Evolution and Systematics* (D. P. Mindell, ed.). Academic Press, San Diego, CA.

Slack, K. E., C. M. Jones, T. Ando, G. L. Harrison, E. Fordyce, U. Arnason, and D. Penny. 2006. Early penguin fossils, plus mitochondrial genomes, give a firm calibration point for avian evolution. *Mol. Biol. Evol.* 23:1144–1155.

Storer, R. W. 1971. Classification of birds. *Avian Biol.* 1:1–18.

Watanabe, M., M. Nikaido, T. T. Tsuda, T. Kobayashi, D. P. Mindell, Y. Cao, N. Okada, and M. Hasegawa. 2006. New candidate species most closely related to penguins. *Gene* 378:65–73.

Wetmore, A. 1960. A classification for the birds of the world. *Smithson. Misc. Coll.* 139(11):1–37.

Yuri, T., R. T. Kimball, J. Harshman, R. C. K. Bowie, M. J. Braun, J. L. Chojnowski, K.-L. Han, S. J. Hackett, C. J. Huddleston, W. S. Moore, S. Reddy, F. H. Sheldon, D. W. Steadman, C. C. Witt, and E. L. Braun. 2013. Parsimony and model-based analyses of indels in avian nuclear genes reveal congruent and incongruent phylogenetic signals. *Biology* 2:419–444.

Author

George Sangster; Department of Bioinformatics and Genetics; Swedish Museum of Natural History; P. O. Box 50007, SE–104 05 Stockholm; Sweden; and Naturalis Biodiversity Center; Darwinweg 2, P. O. Box 9517, 2333 CC Leiden; Netherlands.

Date Accepted: 11 December 2014

Primary Editor: Jacques A. Gauthier

Strigiformes R. B. Sharpe 1899 [G. Sangster], converted clade name

Registration Number: 284

Definition: The smallest crown clade containing *Tyto* (originally *Strix*) *alba* (Scopoli 1769) and *Strix aluco* Linnaeus 1758. This is a minimum-crown-clade definition. Abbreviated definition: min crown ∇ (*Tyto alba* (Scopoli 1769) & *Strix aluco* Linnaeus 1758).

Etymology: Derived from the Latin *strix*, a screaming nightbird (Brodkorb, 1971), plus the Latin *forma* (having the form of).

Reference Phylogeny: For the purpose of applying the definition of *Strigiformes*, Figure 1 in Wink et al. (2008) should be regarded as the primary reference phylogeny.

Composition: *Strigiformes*, as defined here, includes all 213–250 species of Recent owls (König and Weick, 2008; Dickinson and Remsen, 2013). These are often grouped into two main subtaxa: *Tytonidae* and *Strigidae* (König and Weick, 2008; Livezey and Zusi, 2007; Dickinson and Remsen, 2013).

Diagnostic Apomorphies: Livezey and Zusi (2006, 2007) identified 24 apomorphies of *Strigiformes* relative to other crown clades. Because putative members of the stem-group of *Strigiformes* (Mayr, 2009) have not been included in any phylogenetic analysis, some of these characters are likely to be apomorphies of more inclusive clades containing stem owls. Eleven of the apomorphies identified by Livezey and Zusi (2006, 2007) are unique and unreversed apomorphies (consistency index = 1) (characters and states are indicated by their number-letter combination in Livezey and

Zusi, 2006): (1) regio ossium frontales et parietales (frons), pronounced tickening produced by multi-layered cellulae pneumaticae (os spongiosum) present (13b); (2) cotyla quadratica otici, foramen pneumaticum magnum present (154b); (3) os frontale, facies orbitalis, margo supraorbitalis, variably prominent, rounded, highly pneumatic, hornlike processus supraocularis present (174b); (4) os lacrymale, tumulus pneumaticus present (193b); (5) os quadratum, processus oticus quadrati, capitulum (condylus) oticum, marked elongation and uniquely caudimedial orientation, directed approximately perpendicularly to processus oticus present (548b); (6) corpus radii, margo interosseus (caudalis), arcus origii musculus extensor longus digiti majoris, pars proximalis present (1549b); (7) phalanx proximalis digiti II (majus, secundus) manus, (especially phalanx proximalis), margo cranialis phalangis, pila cranialis phalangis distinctly concave, resulting in sulcus on facies cranialis pilae (1714b); (8) extremitas proximalis tibiotarsi, caput tibiotarsi, crista cnemialis cranialis, facies lateralis with distinct and approximately uniform concavity of surface (2072b); (9) corpus fibulae, tuberculum insertii musculus iliofibularis, situs proximodistalis relative to crista fibularis tibiotarsi opposite (lateral) to terminus proximalis cristae (2200a); (10) hypotarsus (plantar perspective), sulci et/aut canales magnus hypotarsis present, bounded asymmetrically by lamina medialis (2286b); (11) corpus tarsometatarsi, facies plantaris, sulcus flexorius, crista medioplantaris, situs lateromedialis shifted distinctly laterad throughout, underlying (in part) the virtually complete concavity of facies plantaris tarsometatarsi (2300b).

An apomorphic indel of 1 base pair was inferred for this clade in a phylogenetic study based on sequences of beta-fibrinogen intron 7 (Prychitko and Moore, 2003).

Synonyms: The following are approximate synonyms (see Comments): *Strigidae* Leach 1820 (sensu Mayr and Amadon, 1951, and Cracraft, 1981, 1988); *Striges* Wagler 1830; *Strigoidea* Cracraft 1981; *Strigi* Sibley and Ahlquist 1990; *Strigimorphae* Cracraft 2013.

Comments: Monophyly of the owls is supported by analyses based on morphology (Livezey and Zusi, 2007), DNA-DNA hybridization (Sibley and Ahlquist, 1990) and DNA sequences (Prychitko and Moore, 2003; Hackett et al., 2008; Wink et al., 2008; Mahmood et al., 2014; Yuri et al., 2013). The clade was also inferred in a supertree analysis (Davis and Page, 2014).

In the current literature, several names are used for the owl clade due to differences in opinion among authors over its proper taxonomic rank. *Strigiformes* is the most widely used name associated with this clade (e.g., del Hoyo et al., 1999; Livezey and Zusi, 2007; König and Weick, 2008; Cracraft, 2013) and is adopted here. Some authors have used *Strigidae* but this name is more commonly applied to a sub-clade (i.e., the owls that are more closely related to *Strix* than to *Tyto*). The names *Strigoidea* and *Strigimorphae* have each been used in at least one classification but as redundant, rank-related names. Sibley and Ahlquist (1990) introduced the name *Strigi* for the owl clade and used the name *Strigiformes* for a more inclusive putative clade that also includes the nightjars (*Caprimulgiformes*). Wagler's (1830) taxon *Striges* appears to be no longer in use.

The name *Strigiformes* is sometimes applied to a larger clade that also includes extinct taxa outside the crown clade (e.g., Mourer-Chauviré, 1987; Mayr, 2005). This use of the name *Strigiformes* could be formalized with a maximum-clade definition. In the interest of establishing a general system of taxonomy (de Queiroz, 2007), *Pan-Strigiformes* would be the preferable name for the total clade. Recognizing fossils as belonging to the strigiform total clade could be complicated because of contradictory evidence for the extant sister to *Strigiformes* (Brown et al., 2008; Hackett et al., 2008; Pratt et al., 2009; Pacheco et al., 2011; McCormack et al., 2013; Yuri et al., 2013). A minimum-crown-clade definition was selected because (i) the long-recognized basal subdivision of *Strigiformes* into *Tytonidae* and *Strigidae* is well-supported (e.g., Sibley and Ahlquist, 1990; Randi et al., 1991; Wink et al., 2008) and (ii) this definition preserves current use of the name (i.e., for the crown clade).

Under rank-based nomenclature, Wagler (1830) is considered the author of the name *Strigiformes*, even though he spelled the name *Striges*, because he was the first person to publish a name based on *Strix* at the rank of order. The first author to use *Strigiformes* appears to have been Sharpe (1899) (see Brodkorb, 1971), and he is accordingly regarded as the nominal author for this taxon name.

Literature Cited

Brodkorb, P. 1971. Catalogue of fossil birds: part 4 (*Columbiformes through Piciformes*). *Bull. Fl. State Mus., Biol. Sci.* 15(4):163–266.

Brown, J. W., J. S. Rest, J. Garcia-Moreno, M. D. Sorenson, and D. P. Mindell. 2008. Strong mitochondrial DNA support for a Cretaceous origin of modern avian lineages. *BMC Biology* 6:6.

Cracraft, J. 1981. Toward a phylogenetic classification of the recent birds of the world (Class *Aves*). *Auk* 98:681–714.

Cracraft, J. 1988. The major clades of birds. Pp. 339–361 in *The Phylogeny and Classification of the Tetrapods. Vol. 1: Amphibians, Reptiles, Birds* (M. J. Benton, ed.). Systematics Association Special Volume No. 35A. Clarendon Press, Oxford.

Cracraft, J. 2013. Avian higher-level relationships and classification: nonpasseriforms. Pp. xxi–xliii in *The Howard and Moore Complete Checklist of the Birds of the World*. 4th edition. Vol. 1: *Non-passerines* (E. C. Dickinson and J. V. Remsen, Jr., eds.). Aves Press, London.

Davis, K. E., and R. D. M. Page. 2014. Reweaving the tapestry: a supertree of birds. *PLOS Curr. Tree Life*. 9 June 2014. 1st edition. doi:10.13 71/currents.tol.c1af68dda7c999ed9f1e4b2d 2df7a08e.

Dickinson, E. C., and J. V. Remsen, Jr. 2013. *The Howard and Moore Complete Checklist of the Birds of the World*. 4th edition. Vol. 1: *Non-passerines*. Aves Press, London.

Hackett, S. J., R. T. Kimball, S. Reddy, R. C. K. Bowie, E. L. Braun, M. J. Braun, J. L. Chojnowski, W. A. Cox, K.-L. Han, J. Harshman, C. J. Huddleston, B. D. Marks, K. J. Miglia, W. S. Moore, F. H. Sheldon, D. W. Steadman, C. C. Witt, and T. Yuri. 2008. A phylogenomic study of birds reveals their evolutionary history. *Science* 320:1763–1768.

del Hoyo, J., A. Elliot, and J. Sargatal. 1999. *Handbook of the Birds of the World. Vol. 5: Barn-Owls to Hummingbirds*. Lynx Edicions, Barcelona.

König, C., and F. Weick. 2008. *Owls of the World*. 2nd edition. Pica Press, Robertsbridge.

Leach, W. E. 1820. "Eleventh Room". Pp. 65–70, *Synopsis of the Contents of the British Museum*. 17th edition. British Museum, London.

Linnaeus, C. 1758. *Systema Naturae Per Regna Tria Naturae, Secundum Classes, Ordines, Genera, Species, cum Characteribus, Differentiis, Synonymis, Locis*. Editio decima, reformata. Laurentii Salvii, Holmiae (Stockholm).

Livezey, B. C., and R. L. Zusi. 2006. Higher-order phylogeny of modern birds (*Theropoda, Aves: Neornithes*) based on comparative anatomy. I. Methods and characters. *Bull. Carnegie Mus. Nat. Hist.* 37:1–556.

Livezey, B. C., and R. L. Zusi. 2007. Higher-order phylogeny of modern birds (*Theropoda, Aves: Neornithes*) based on comparative anatomy. II. Analysis and discussion. *Zool. J. Linn. Soc.* 149:1–95.

Mahmood, M. T., P. A. McLenachan, G. C. Gibb, and D. Penny. 2014. Phylogenetic position of avian nocturnal and diurnal raptors. *Gen. Biol. Evol.* 6:326–332.

Mayr, E., and D. Amadon. 1951. A classification of Recent birds. *Am. Mus. Novit.* 1496:1–42.

Mayr, G. 2005. The Paleogene fossil record of birds in Europe. *Biol. Rev.* 80:1–28.

Mayr, G. 2009. *Paleogene Fossil Birds*. Springer-Verlag, Berlin and Heidelberg.

McCormack, J. E., M. G. Harvey, B. C. Faircloth, N. G. Crawford, T. C. Glenn, and R. T. Brumfield. 2013. A phylogeny of birds based on over 1,500 loci collected by target enrichment and high-throughput sequencing. *PLOS ONE* 8(1):e54848.

Mourer-Chauviré, C. 1987. Les *Strigiformes* (*Aves*) des Phosphorites du Quercy (France): Systématique, biostratigraphie et paléobiogéographie. *Doc. Lab. Géol. Lyon* 99:89–135.

Pacheco, M. A., F. U. Battistuzzi, M. Lentino, R. Aguilar, S. Kumar, and A. A. Escalante. 2011. Evolution of modern birds revealed by mitogenomics: timing the radiation and origin of major orders. *Mol. Biol. Evol.* 28:1927–1942.

Pratt, R. C., G. C. Gibb, M. Morgan-Richards, M. J. Phillips, M. D. Hendy, and D. Penny. 2009. Toward resolving deep *Neoaves* phylogeny: data, signal enhancement and priors. *Mol. Biol. Evol.* 26:313–326.

Prychitko, T. M., and W. S. Moore. 2003. Alignment and phylogenetic analysis of ß-fibrinogen intron 7 sequences among avian orders reveal conserved regions within the intron. *Mol. Biol. Evol.* 20:762–771.

de Queiroz, K. 2007. Toward an integrated system of clade names. *Syst. Biol.* 56:956–974.

Randi, E., G. Fusco, R. Lorenzini, and F. Spina. 1991. Allozyme divergence and phylogenetic relationships within the *Strigiformes*. *Condor* 93:295–301.

Scopoli, G. A. 1769. *Annus I. Historico-Naturalis*. Hilscher, Leipzig.

Sharpe, R. B. 1899. *A Hand-list of the Genera and Species of Birds*, Vol. 1. British Museum, London.

Sibley, C. G., and J. E. Ahlquist. 1990. *Phylogeny and Classification of Birds.* Yale University Press, New Haven, CT.

Wagler, J. G. 1830. *Natürliches System der Amphibien mit vorangehender Classification der Säugethiere und Vögel.* J. G. Cotta Buchhandlung, München, Stuttgart und Tübingen.

Wink, M., P. Heidrich, H. Sauer-Gürth, A.-A. Elsayed, and J. Gonzalez. 2008. Molecular phylogeny and systematics of owls (*Strigiformes*). Pp. 42–63 in *Owls: A Guide to the Owls of the World.* 2nd edition (C. König and F. Weick, eds.). Pica Press, Robertsbridge.

Yuri, T., R. T. Kimball, J. Harshman, R. C. K. Bowie, M. J. Braun, J. L. Chojnowski, K.-L. Han, S. J. Hackett, C. J. Huddleston, W. S. Moore, S. Reddy, F. H. Sheldon, D. W. Steadman, C. C. Witt, and E. L. Braun. 2013. Parsimony and model-based analyses of indels in avian nuclear genes reveal congruent and incongruent phylogenetic signals. *Biology* 2:419–444.

Author

George Sangster; Department of Bioinformatics and Genetics; Swedish Museum of Natural History; P. O. Box 50007, SE–104 05 Stockholm; Sweden; and Naturalis Biodiversity Center; Darwinweg 2, P. O. Box 9517, 2333 CC Leiden; Netherlands.

Date Accepted: 11 December 2014

Primary Editor: Jacques A. Gauthier

Picidae W. E. Leach 1820 [G. Sangster], converted clade name

Registration Number: 285

Definition: The smallest crown clade containing *Jynx torquilla* Linnaeus 1758 and *Picus viridis* Linnaeus 1758. This is a minimum-crown-clade definition. Abbreviated definition: min crown ∇ (*Jynx torquilla* Linnaeus 1758 & *Picus viridis* Linnaeus 1758).

Etymology: Derived from the Latin *picus*, meaning a woodpecker.

Reference Phylogeny: For the purpose of applying the definition of *Picidae*, Figure 1 in Benz et al. (2006) should be regarded as the primary reference phylogeny. Figure 4 in Webb and Moore (2005) may be regarded as a secondary reference phylogeny.

Composition: Based on current knowledge (Swierczewski and Raikow, 1981; Webb and Moore, 2005; Benz et al., 2006), *Picidae* comprises approximately 217 Recent species of woodpeckers, piculets and wrynecks (del Hoyo et al., 2002; Dickinson and Remsen, 2013).

Diagnostic Apomorphies: Two apomorphies were listed by Swierczewski and Raikow (1981): Musculus femorotibialis internus with one belly, and M. flexor perforatus digiti III heads separate. Manegold and Töpfer (2013) identified five apomorphies: (1) apparatus hyobranchialis, paraglossum short with either greatly or even completely reduced cornua; (2) apparatus hyobranchialis, basihyale elongated; (3) apparatus hyobranchialis, urohyale completely reduced; (4) apparatus hyobranchialis, epibranchials strongly elongated, at least twice as long as ceratobranchials; (5) mandibula, pars caudalis, sulcus intercotylaris without caudal margin. Two synapomorphic indels of 14-bp each were inferred for this clade in a phylogenetic study of the beta-fibrinogen intron 7 (Prychitko and Moore, 2003). These studies did not include any members of the stem-group of *Picidae*. In fact, no unambiguous members of the stem-group are known (Mayr, 2009). Therefore, the apomorphies listed here are relative to other crown clades.

Synonyms: *Pici* Meyer and Wolf 1810 (sensu Wetmore, 1930 and Storer, 1960) (approximate).

Comments: Monophyly of the woodpeckers, piculets and wrynecks is supported by analyses based on morphological data (Swierczewski and Raikow, 1981; Manegold and Töpfer, 2013) and DNA sequence data (Webb and Moore, 2005; Benz et al., 2006; Winkler et al., 2013). This clade was also inferred in a supertree analysis (Davis and Page, 2014).

The name *Picidae* has long been in use for the woodpeckers, piculets and wrynecks (Seebohm, 1890; Wetmore, 1930; Mayr and Amadon, 1951; Storer, 1971; Cracraft, 1981; Bock, 1982; Short, 1982; Sibley and Ahlquist, 1990; Livezey and Zusi, 2007; Dickinson and Remsen, 2013) and is therefore selected for this clade. *Picidae* also has been used for a less inclusive clade comprising the piculets and woodpeckers but not the wrynecks (Wolters, 1975–1982; Fain and Houde, 2004). This minority view stems from a difference in opinion over which traditional taxonomic rank should be accorded to wrynecks.

Older classifications (e.g., Wetmore, 1930; Storer, 1960) used *Pici* as a redundant name for *Picidae*. The name *Pici* is currently used for a more inclusive clade formed by the woodpeckers

(*Picidae*), honeyguides (*Indicator*) and toucans and barbets (*Ramphastidae*) (e.g., Simpson and Cracraft, 1981; Swierczewski and Raikow, 1981; Mayr, 1998; Johansson and Ericson, 2003; Livezey and Zusi, 2007).

A minimum-crown-clade definition was selected because (i) the basal subdivision of this clade into *Jynx* vs. all other *Picidae* is strongly supported by morphological data (Swierczewski and Raikow, 1981; Manegold and Töpfer, 2013) and mitochondrial and nuclear DNA sequence data (Webb and Moore, 2005; Benz et al., 2006; Fuchs et al., 2007; Winkler et al., 2013) and (ii) this preserves current usage of the name (i.e., for the crown clade).

The name *Picidae* is often attributed to Vigors 1825 (e.g., Brodkorb, 1971; Livezey and Zusi, 2007), but it was first used by Leach (1820) (see Bock, 1994), who is regarded here as the nominal author.

Literature Cited

Benz, B. W., M. B. Robbins, and A. T. Peterson. 2006. Evolutionary history of woodpeckers and allies (*Aves: Picidae*): placing key taxa on the phylogenetic tree. *Mol. Phylogen. Evol.* 40:389–399.

Bock, W. J. 1982. *Aves*. Pp. 967–1015 in *Synopsis and Classification of Living Organisms*, Vol. 2 (S. P. Parker, ed.). McGraw-Hill, New York.

Bock, W. J. 1994. History and nomenclature of avian family-group names. *Bull. Am. Mus. Nat. Hist.* 222:1–281.

Brodkorb, P. 1971. Catalogue of fossil birds: part 4 (*Columbiformes through Piciformes*). *Bull. Fl. State Mus., Biol. Sci.* 15(4):163–266.

Cracraft, J. 1981. Toward a phylogenetic classification of the Recent birds of the world (Class *Aves*). *Auk* 98:681–714.

Davis, K. E., and R. D. M. Page. 2014. Reweaving the tapestry: a supertree of birds. *PLOS Curr. Tree Life*. 9 June 2014. 1st edition. doi:10.1371/currents.tol.c1af68dda7c999ed9f1e4b2d2df7a08e.

Dickinson, E. C., and J. V. Remsen, Jr. 2013. *The Howard and Moore Complete Checklist of the Birds of the World*. 4th edition. Vol. 1: *Non-passerines*. Aves Press, London.

Fain, M. G., and P. Houde. 2004. Parallel radiations in the primary clades of birds. *Evolution* 58:2558–2573.

Fuchs, J., J. I. Ohlson, P. G. P. Ericson, and E. Pasquet. 2007. Synchronous intercontinental splits between assemblages of woodpeckers suggested by molecular data. *Zool. Scr.* 36:11–25.

del Hoyo, J., A. Elliot, and J. Sargatal, eds. 2002. *Handbook of the Birds of the World. Vol. 7: Jacamars to Woodpeckers*. Lynx Edicions, Barcelona.

Johansson, U. S., and P. G. P. Ericson. 2003. Molecular support for a sister group relationship between *Pici* and *Galbulae* (*Piciformes sensu* Wetmore 1960). *J. Avian Biol.* 34:185–197.

Leach, W. E. 1820. Eleventh room. In *Synopsis of the Contents of the British Museum*. 17th edition. British Museum (Natural History), London.

Livezey, B. C., and R. L. Zusi. 2007. Higher-order phylogeny of modern birds (*Theropoda, Aves: Neornithes*) based on comparative anatomy. II. Analysis and discussion. *Zool. J. Linn. Soc.* 149:1–95.

Manegold, A., and T. Töpfer. 2013. The systematic position of *Hemicircus* and the stepwise evolution of adaptations for drilling, tapping and climbing up in true woodpeckers (*Picinae, Picidae*). *J. Zool. Syst. Evol. Res.* 51:72–82.

Mayr, G. 1998. "Coraciiforme" und "piciforme" Kleinvögel aus dem Mittel-Eozän der Grube Messel (Hessen, Deutschland). *Courier Forschungsinst. Senckenb.* 205:1–101.

Mayr, G. 2009. *Paleogene Fossil Birds*. Springer-Verlag, Berlin and Heidelberg.

Mayr, E., and D. Amadon. 1951. A classification of Recent birds. *Am. Mus. Novit.* 1496:1–42.

Prychitko, T. M., and W. S. Moore. 2003. Alignment and phylogenetic analysis of ß-fibrinogen intron 7 sequences among avian orders reveal conserved regions within the intron. *Mol. Biol. Evol.* 20:762–771.

Seebohm, H. 1890. *Classification of Birds. An Attempt to Diagnose the Subclasses, Orders, Suborders and Some of the Families of Existing Birds.* R. H. Porter, London.

Short, L. L. 1982. *Woodpeckers of the World.* Delaware Museum of Natural History, Greenville, Del.

Sibley, C. G., and J. E. Ahlquist. 1990. *Phylogeny and Classification of Birds.* Yale University Press, New Haven, CT.

Simpson, S. F., and J. Cracraft. 1981. The phylogenetic relationships of the *Piciformes* (Class *Aves*). *Auk* 98:481–494.

Storer, R. W. 1960. The classification of birds. Pp. 57–93 *in Biology and Comparative Physiology of Birds*, Vol. 1 (A. J. Marshall, ed.). Academic Press, New York.

Storer, R. W. 1971. Classification of birds. *Avian Biol.* 1:1–18.

Swierczewski, E. V., and R. J. Raikow. 1981. Hind limb morphology, phylogeny, and classification of the *Piciformes*. *Auk* 98:466–480.

Webb, D. M., and W. S. Moore. 2005. A phylogenetic analysis of woodpeckers and their allies using 12S, Cyt b, and COI nucleotide sequences (class *Aves*; order *Piciformes*). *Mol. Phylogen. Evol.* 36:233–248.

Wetmore, A. 1930. A systematic classification for the birds of the world. *Proc. US Natl. Mus.* 76(24):1–8.

Winkler, H., A. Gamauf, F. Nittinger, and E. Haring. 2013. Relationships of Old World woodpeckers (*Aves*: *Picidae*)—new insights and taxonomic implications. *Ann. Naturhist. Mus. Wien B* 116:69–86.

Wolters, H. E. 1975–82. *Die Vogelarten der Erde.* Paul Parey, Hamburg.

Author

George Sangster; Department of Bioinformatics and Genetics; Swedish Museum of Natural History; P. O. Box 50007, SE–104 05 Stockholm; Sweden; and Naturalis Biodiversity Center; Darwinweg 2, P. O. Box 9517, 2333 CC Leiden; Netherlands.

Date Accepted: 11 December 2014

Primary Editor: Jacques A. Gauthier

Psittaciformes M. Fürbringer 1888 [G. Sangster], converted clade name

Registration Number: 286

Definition: The smallest crown clade containing *Strigops habroptilus* Gray 1845 and *Psittacus erithacus* Linnaeus 1758. This is a minimum-crown-clade definition. Abbreviated definition: min crown ∇ (*Strigops habroptilus* Gray 1845 & *Psittacus erithacus* Linnaeus 1758).

Etymology: Derived from the Latin *psittacus*, meaning a parrot, plus the Latin *forma* (having the form of).

Reference Phylogeny: For the purpose of applying the definition of *Psittaciformes*, Figure 3 in Schirtzinger et al. (2012) should be regarded as the primary reference phylogeny. Figure 1 in Wright et al. (2008) may be regarded as a secondary reference phylogeny.

Composition: *Psittaciformes* is estimated to contain about 376 Recent species (Dickinson and Remsen, 2013). This crown clade includes all extant parrots, parakeets, parrotlets, lories, lovebirds, cockatoos, and macaws.

Diagnostic Apomorphies: Livezey and Zusi (2006, 2007) identified 35 apomorphies of *Psittaciformes* of which 22 are unique (consistency index = 1) but their study did not include any members of the stem. Ksepka et al. (2011) noted that when extinct stem taxa are considered, many putative synapomorphies of *Psittaciformes* are optimized as synapomorphies of more inclusive groups including both stem and crown taxa.

Apomorphies identified by Livezey and Zusi (2006, 2007) that are diagnostic for *Psittaciformes* relative to other extant birds are as follows (characters and states are indicated by their number-letter combination in Livezey and Zusi, 2006): (1) os parasphenoidale, lamina paraspenoidalis, ala paraspenoidalis, continuity as broad, slightly concave, sloping planum with os exoccipitale, processus paroccipitalis, and demarcated medially by conspicuous carina from lamina basa paraspenoidalis present (118b); (2) paries medialis orbitae, large cavum ethmomandibularis lateral to septum interorbitale and caudodorsal in orientation present (246b); (3) conchae nasalis oseae, caudiomedial terminus, conformation as laminar surface perpendicular to zona flexoria craniofacialis and in which nares internae are separated by thick osseus septum present (354b); (4) os maxillare, facies ventralis, margo caudalis medial to lateral limit of sutura maxillaropalatinus, conformation as distinct plica, continuous across the midline, and distinctly ventral to os palatinum, processus rostralis present (410b); (5) os palatinum (especially pars lateralis), marked ventral angling relative to rostrum parasphenoidale present (429b); (6) syndesmosis jugo-maxillaris present (593b); (7) zona flexoria palatina present, zona simplex, restricted by ligamentum palatomaxillare (605b); (8) rostrum (symphysis) mandibulae, "scoop-like" conformation involving truncate apical margin, flattened and angular facies ventralis present (650b); (9) ramus mandibulare, angulus dorsalis mandibulae, distinctly deeper dorsoventrally than remainder of ramus mandibulae, emphasizing sharp angulus ventralis of pars rostralis cristae tomialis (margo caudalis of pars symphysialis) present (679b); (10) ramus mandibulae, pars caudalis, fossa caudalis aut recessus insertii m. depressor mandibulae present (703b); (11) os urohyale (basibranchiale caudale), facies ventralis, nodulus (sesamoideum ligamenti noduloceratobranchiale) present (761b); (12) costae vertebralis (especially longest, most robust elements), corpus costae, processus uncinatus, augmentation by diminutive, acuminate spinus uncinatus dorsally on each costae present (1060b); (13) carina sterni, margo cranialis carinae, pila carinae, orientation relative to margo cranialis carinae distinctly caudoventral (1210b); (14) extremitas distalis tarsometatarsi, trochleae accessoriae group III , trochleae accessoriae metatarsalia II (secundus) et IV (quartus), i.e., primarily or solely involving differential modifications of trochleae metatarsalia II et IV and infrequently also trochlea metatarsale I, present and comprising a medial tuberculum trochlearis metatarsi I in articulation with eminentia medialis trochlea metatarsi II, and expanded lobus lateralis and approximately 90° plantar rotation of trochlea metatarsi IV (2334d); (15) digiti III-IV pedis, phalanges exclusive of phalanges unguales, hypertruncate, discoid phalanges—characterized by maximal width being at least twice the length of the phalanx and foveae ligamentorum collaterium obsolete—present, in moderate form in digitus IV (phalanges proximalis et one-two intermediate) (2405d); (16) m. ethmomandibularis present (2490b); (17) m. extensor hallucis longus vestigial or absent (2703b); (18) sulcus infraorbitalis (sinus antorbitalis) present (2712b); (19) torus palatinus present, thick and highly vascularized (2832b); (20) fenestra membrana tracheosyringealis present, dorsoventral (2876c); (21) membrana tympaniformis lateralis present (2878b); (22) diverticula cervicocephala, recessus pneumatici cervicocephalica, extension as paired recessi around columna vertebralis, enclosing ingluvies, and resting sellariform on regio omalis present (2884b).

Synonyms: The following are approximate synonyms: *Psittacidae* Vigors 1825 (sensu Wetmore, 1940; Storer, 1960; Sibley and Ahlquist, 1990; Mayr and Daniels, 1998; and Dickinson, 2003); Psittaci Wagler 1830 (sensu Seebohm, 1890; Mayr and Amadon, 1951; and Stresemann, 1959); Psittacimorphae Huxley 1867 (sensu Sibley and Ahlquist, 1990).

Comments: Monophyly of the parrots (including parakeets, parrotlets, lories, lovebirds, cockatoos and macaws) is supported by studies based on morphology (Ksepka et al., 2011), DNA sequences (Schweizer et al., 2011; Schirtzinger et al., 2012) and supertree analysis (Davis and Page, 2014).

The name *Psittaciformes* is widely used for the parrots, including parakeets, parrotlets, lories, lovebirds, cockatoos and macaws (e.g., Forshaw, 1973; Christidis et al., 1991; del Hoyo et al., 1997; de Kloet and de Kloet, 2005; Livezey and Zusi, 2007; Wright et al., 2008; Ksepka et al., 2011; Cracraft, 2013) and either implicitly or explicitly refers to the crown. The names *Psittacidae* and *Psittaci* are less often used for this clade, and the latter seems no longer in use. In view of the large number of species in this clade, the name *Psittacidae* is best used for a less-inclusive clade (e.g., Joseph et al., 2012; Dickinson and Remsen, 2013).

The name *Psittaciformes* is sometimes applied to a larger clade that also includes extinct taxa outside the crown-clade (e.g., Mayr and Daniels, 1998; Dyke and Cooper, 2000; Mayr, 2005). However, recent authors have found it useful to distinguish between the parrot crown and total clades by using *Pan-Psittaciformes* for the total clade and *Psittaciformes* for the crown (Ksepka et al., 2011; Mayr, 2011).

Phylogenetic studies of mitochondrial DNA, nuclear DNA and morphology provide overwhelming evidence that a clade comprising the large New Zealand parrots *Strigops* and *Nestor* is the sister of all other extant *Psittaciformes* (Miyaki et al., 1998; Nothwang, 2000; de Kloet and de Kloet, 2005; Tavares et al., 2006; Tokita et al., 2007; Wright et al., 2008; Schweizer et al.,

2011; Schirtzinger et al., 2012). Thus, members of these two clades have been selected as internal specifiers in the definition. A minimum-crown-clade definition was selected because this preserves the most common current use of the name (i.e., for the crown clade).

According to rank-based nomenclature, Wagler (1830) is considered the author of the name *Psittaciformes*, even though he spelled the name *Psittaci*, because he was the first person to publish a name based on *Psittacus* at the rank of order. The first author who used the name *Psittaciformes* appears to have been Fürbringer (1888; see Brodkorb, 1971).

Literature Cited

Brodkorb, P. 1971. Catalogue of fossil birds: part 4 (*Columbiformes* through *Piciformes*). *Bull. Fl. State Mus., Biol. Sci.* 15(4):163–266.

Christidis, L., R. Schodde, D. D. Shaw, and S. F. Maynes. 1991. Relationships among the Australo-Papuan parrots, lorikeets, and cockatoos (*Aves: Psittaciformes*): protein evidence. *Condor* 93:302–317.

Cracraft, J. 2013. Avian higher-level relationships and classification: nonpasseriforms. Pp. xxi–xliii in *The Howard and Moore Complete Checklist of the Birds of the World*. 4th edition. Vol. 1: *Non-passerines* (E. C. Dickinson and J. V. Remsen, Jr., eds.). Aves Press, London.

Davis, K. E., and R. D. M. Page. 2014. Reweaving the tapestry: a supertree of birds. *PLOS Curr. Tree Life*. 9 June 2014. 1st edition. doi:10.13 71/currents.tol.c1af68dda7c999ed9f1e4b2d 2df7a08e.

Dickinson, E. C. 2003. *The Howard and Moore Complete Checklist of the Birds of the World*. 3rd edition. Christopher Helm, London.

Dickinson, E. C., and J. V. Remsen, Jr. 2013. *The Howard and Moore Complete Checklist of the Birds of the World*. 4th edition. Vol. 1: *Non-passerines*. Aves Press, London.

Dyke, G. J., and J. Cooper. 2000. A new psittaciform bird from the London Clay (Lower Eocene) of England. *Palaeontology* 43:271–285.

Forshaw, J. M. 1973. *Parrots of the World*. Lansdowne Press, Sydney.

Fürbringer, M. 1888. *Untersuchungen zur Morphologie und Systematik der Vögel zugleich ein Beitrag zur Anatomie der Stütz- und Bewegungsorgane*, Vol. 2. T. J. Van Holkema, Amsterdam.

Gray, G.R. 1845. *The Genera of Birds Comprising Their Generic Characters, a Notice of the Habits of Each Genus, and an Extensive List of Species Referred to Their Several Genera*, Vol. 2. Longman, Brown, Green and Longmans, London.

del Hoyo, J., A. Elliot, and J. Sargatal. 1997. *Handbook of the Birds of the World*, Vol. 4. *Sandgrouse to cuckoos*. Lynx Edicions, Barcelona.

Joseph, L., A. Toon, E. E. Schirtzinger, T. F. Wright, and R. Schodde. 2012. A revised nomenclature and classification for family-group taxa of parrots (*Psittaciformes*). *Zootaxa* 3205:26–40.

de Kloet, R. S., and S. R. de Kloet. 2005. The evolution of the spindlin gene in birds: sequence analysis of an intron of the spindlin W and Z gene reveals four major divisions of the *Psittaciformes*. *Mol. Pylogen. Evol.* 36:706–721.

Ksepka, D. T., J. A. Clarke, and L. Grande. 2011. Stem parrots (*Aves, Halcyornithidae*) from the Green River Formation and a combined phylogeny of *Pan-Psittaciformes*. *J. Paleontol.* 85:835–852.

Linnaeus, C. 1758. *Systema Naturae Per Regna Tria Naturae, Secundum Classes, Ordines, Genera, Species, cum Characteribus, Differentiis, Synonymis, Locis*. Editio decima, reformata. Laurentii Salvii, Holmiae (Stockholm).

Livezey, B. C., and R. L. Zusi. 2006. Higher-order phylogeny of modern birds (*Theropoda, Aves: Neornithes*) based on comparative anatomy. I. Methods and characters. *Bull. Carnegie Mus. Nat. Hist.* 37:1–556.

Livezey, B. C., and R. L. Zusi. 2007. Higher-order phylogeny of modern birds (*Theropoda, Aves: Neornithes*) based on comparative anatomy. II. Analysis and discussion. *Zool. J. Linn. Soc.* 149:1–95.

Mayr, E., and D. Amadon. 1951. A classification of Recent birds. *Am. Mus. Novit.* 1496:1–42.

Mayr, G. 2005. The postcranial osteology and phylogenetic position of the Middle Eocene *Messelastur gratulator* Peters, 1994 – a morphological link between owls (*Strigiformes*) and falconiform birds? *J. Vert. Paleontol.* 25:635–645.

Mayr, G. 2011. Well-preserved new skeleton of the middle Eocene *Messelastur* substantiates sister group relationship between *Messelasturidae* and *Halcyornithidae* (*Aves*, ?*Pan- Psittaciformes*). *J. Syst. Palaeontol.* 9:159–171.

Mayr, G., and M. Daniels. 1998. Eocene parrots from Messel (Hessen, Germany) and the London Clay of Walton-on-the-Naze (Essex, England). *Senckenberg. Lethaea* 78:157–177.

Miyaki, C. Y., S. R. Matioli, T. Burke, and A. Wajntal. 1998. Parrot evolution and paleogeographical events: mitochondrial DNA evidence. *Mol. Biol. Evol.* 15:544–551.

Nothwang, U. 2000. *Rekonstruktion der Phylogenie der Papageien* (Aves: Psittaciformes) *anhand osteologischer Merkmale.* Dipl.-Arb. J. W. Goethe-University, Frankfurt am. Meine.

Schirtzinger, E. E., E. S. Tavares, L. A. Gonzales, J. R. Eberhard, C. Y. Miyaki, J. J. Sanchez, A. Hernandez, H. Mueller, G. R. Graves, R. C. Fleischer, and T. F. Wright. 2012. Multiple independent origins of mitochondrial control region duplications in the order *Psittaciformes*. *Mol. Phylogen. Evol.* 64:342–356.

Schweizer, M., O. Seehausen, and S. T. Hertwig. 2011. Macroevolutionary patterns in the diversification of parrots: effects of climate change, geological events and key innovations. *J. Biogeogr.* 38:2176–2194.

Seebohm, H. 1890. *Classification of Birds. An Attempt to Diagnose the Subclasses, Orders, Suborders and Some of the Families of Existing Birds.* R. H. Porter, London.

Sibley, C. G., and J. E. Ahlquist. 1990. *Phylogeny and Classification of Birds.* Yale University Press, New Haven, CT.

Storer, R. W. 1960. The classification of birds. Pp. 57–93 in *Biology and Comparative Physiology of Birds*, Vol. 1 (A. J. Marshall, ed.). Academic Press, New York.

Stresemann, E. 1959. The status of avian systematics and its unresolved problems. *Auk* 76:269–280.

Tavares, E. S., A. J. Baker, S. L. Pereira, and C. Y. Miyaki. 2006. Phylogenetic relationships and historical biogeography of Neotropical parrots (*Psittaciformes*: *Psittacidae*: *Arini*) inferred from mitochondrial and nuclear DNA Sequences. *Syst. Biol.* 55:454–470.

Tokita, M., T. Kiyoshi, and K. N. Armstrong. 2007. Evolution of craniofacial novelty in parrots through developmental modularity and heterochrony. *Evol. Dev.* 9:590–601.

Wagler, J. G. 1830. *Natürliches System der Amphibien mit Vorangehender Classification der Säugethiere und Vögel.* J. G. Cotta Buchhandlung, München, Stuttgart und Tübingen.

Wetmore, A. 1940. A systematic classification for the birds of the world. *Smithson. Misc. Coll.* 99(7):1–11.

Wright, T. F., E. E. Schirtzinger, T. Matsumoto, J. R. Eberhard, G. R. Graves, J. J. Sanchez, S. Capelli, H. Muller, J. Scharpegge, G. K. Chambers, and R. C. Fleischer. 2008. A multilocus molecular phylogeny of the parrots (*Psittaciformes*): support for a Gondwanan origin during the Cretaceous. *Mol. Biol. Evol.* 25:2141–2156.

Author

George Sangster; Department of Bioinformatics and Genetics; Swedish Museum of Natural History; P. O. Box 50007, SE–104 05 Stockholm; Sweden; and Naturalis Biodiversity Center; Darwinweg 2, P. O. Box 9517, 2333 CC Leiden; Netherlands.

Date Accepted: 11 December 2014

Primary Editor: Jacques A. Gauthier

Daedalornithes G. Sangster 2005 [G. Sangster], converted clade name

Registration Number: 287

Definition: The smallest crown clade containing *Aegotheles* (originally *Caprimulgus*) *cristatus* (Shaw 1790) (*Aegothelidae*) and *Apus* (originally *Hirundo*) *apus* (Linnaeus 1758) (*Apodidae*). This is a minimum-crown-clade definition. Abbreviated definition: min crown ∇ (*Aegotheles cristatus* (Shaw 1790) & *Apus apus* (Linnaeus 1758)).

Etymology: The clade was named after Daedalus, the Greek mythological figure who fabricated wings that allowed him and his son Icarus to soar upwards into the air. The suffix -*ornithes* is derived from the Greek noun ορνις (*ornis*), meaning bird. The name refers to the agile flight capabilities that evolved within the clade (Sangster, 2005).

Reference Phylogeny: For the purpose of applying the definition of *Daedalornithes*, Figure 2 in Mayr (2002) should be regarded as the primary reference phylogeny.

Composition: Based on current knowledge (Mayr, 2002; Mayr et al., 2003; Ericson et al., 2006; Hackett et al., 2008; Nesbitt et al., 2011; Ksepka et al., 2013), *Daedalornithes* includes, among extant taxa, only *Aegothelidae* (owlet-nightjars, 11 species; Dickinson and Remsen, 2013), *Hemiprocnidae* (tree swifts, 4 species), *Apodidae* (swifts, 95 species), and *Trochilidae* (hummingbirds, 338 species).

Diagnostic Apomorphies: Mayr (2002) identified six morphological apomorphies of *Daedalornithes*: (1) os palatinum with greatly protruding angulus caudolateralis; (2) processus basipterygoidei reduced; (3) pneumatic foramina on the caudal surface of the processus oticus; (4) extremitas omalis of coracoid hooked and processus lateralis greatly reduced; (5) musculus splenius capitis with cruciform origin; (6) caeca absent. Mayr et al. (2003) identified two additional morphological synapomorphies: (7) processus terminalis ischii of pelvis very narrow and slender, touching pubis at an angle of 45–90°, fenestra ischiopubica very wide; (8) musculus fibularis longus absent. They identified character (5) as an unambiguous synapomorphy and characters (3), (4), and (6) to (8) as synapomorphies that, although not unique to *Daedalornithes*, are optimized in their phylogenetic analysis as independently derived in the common ancestor of this clade. Because no stem *Daedalornithes* are known, some of the characters listed above may actually be apomorphies of more inclusive clades.

Synonyms: The following are approximate synonyms (see Comments): *Apodiformes* Peters 1940 (sensu Mayr, 2005); Apodimorphae Sibley, Ahlquist and Monroe 1988 (sensu Mayr, 2010); and Trochiloidea Cracraft 2013.

Comments: This clade was discovered by Mayr (2002) and first named by Sangster (2005). It is supported by studies based on morphology (Mayr et al. 2003; Nesbitt et al., 2011; Ksepka et al., 2013) and molecular data (Barrowclough et al., 2006; Ericson et al., 2006; Hackett et al., 2008; Pacheco et al., 2011; Yuri et al., 2013). The clade was also inferred in a recent supertree analysis (Davis and Page, 2014). Unfortunately, despite complete agreement among these authors about the monophyly, composition, and basal relationships of this group, three additional names have been proposed by various authors, based on differences in opinion about the categorical rank of this clade and personal preferences for particular names, enabled by the lack of regulation of names above the "family-group level" by the *ICZN* (International Commission on Zoological Nomenclature, 1999). Thus, considerable nomenclatural instability has arisen within a little over a decade since the clade was discovered.

Mayr (2005) and Barrowclough et al. (2006) used the name *Apodiformes* for this clade. However, this use of the name *Apodiformes* is inadvisable for several reasons: (1) it is inconsistent with the traditional usage of the name *Apodiformes* and ignores the long association of that name with the swift-hummingbird clade (see *Apodiformes* in this volume); (2) it is unneccessary given the availability of the name *Daedalornithes*; and (3), it leaves the swift-hummingbird clade without a name.

Mayr (2010) did not accept priority of the name *Daedalornithes* because taxon names above the "family-group level" are not regulated by the *ICZN*. He preferred the name *Apodimorphae* because he believed that this more "appropriately reflects the actual content of this clade". Use of the name *Apodimorphae* for this clade is inadvisable because: (1) the name was originally proposed by Sibley et al. (1988) for the clade formed by swifts and hummingbirds, and thus should be treated as a junior synonym of *Apodiformes*; and (2) it is unneccessary given the availability of the name *Daedalornithes*.

Cracraft (2013) placed the owlet-nightjars, swifts, and hummingbirds in a 'superfamily' named *Trochiloidea*. However, this nomenclatural change was based on an opinion about the appropriate rank of this clade under rank-based nomenclature, and the name is thus inherently unstable; it is also unnecessary given the priority of the name *Daedalornithes*.

A minimum-crown-clade definition was selected because the basal divergence within *Daedalornithes* is well supported, whereas evidence for its sister-group is contradictory (Hackett et al., 2008; Mayr, 2010; Nesbitt et al., 2011; Pacheco et al., 2011; Ksepka et al., 2013; Yuri et al., 2013).

Literature Cited

Barrowclough, G. F., J. G. Groth, and L. A. Mertz. 2006. The RAG-1 exon in the avian order *Caprimulgiformes*: phylogeny, heterozygosity, and base composition. *Mol. Phylogen. Evol.* 41:238–248.

Cracraft, J. 2013. Avian higher-level relationships and classification: nonpasseriforms. Pp xxi–xliii in *The Howard and Moore Complete Checklist of the Birds of the World*. 4th edition. Vol. 1: *Non-passerines* (E. C. Dickinson and J. V. Remsen, Jr., eds.). Aves Press, London.

Davis, K. E., and R. D. M. Page. 2014. Reweaving the tapestry: a supertree of birds. *PLOS Curr. Tree Life.* 9 June 2014. 1st edition. doi:10.1371/currents.tol.c1af68dda7c999ed9f1e4b2d2df7a08e.

Dickinson, E. C., and J. V. Remsen, Jr. 2013. *The Howard and Moore Complete Checklist of the Birds of the World*. 4th edition. Vol. 1: *Non-passerines*. Aves Press, London.

Ericson, P. G. P., C. L. Anderson, T. Britton, A. Elzanowski, U. S. Johansson, M. Källersjö, J. I. Ohlson, T. J. Parsons, D. Zuccon, and G. Mayr. 2006. Diversification of *Neoaves*: integration of molecular sequence data and fossils. *Biol. Lett.* 2:543–547.

Hackett, S. J., R. T. Kimball, S. Reddy, R. C. K. Bowie, E. L. Braun, M. J. Braun, J. L. Chojnowski, W. A. Cox, K.-L. Han, J. Harshman, C. J. Huddleston, B. D. Marks, K. J. Miglia, W. S. Moore, F. H. Sheldon, D. W. Steadman, C. C. Witt, and T. Yuri. 2008. A phylogenomic study of birds reveals their evolutionary history. *Science* 320: 1763–1768.

International Commission on Zoological Nomenclature. 1999. *International Code of Zoological*

Nomenclature. 4th edition. International Trust for Zoological Nomenclature, London.

Ksepka, D. T., J. A. Clarke, S. J. Nesbitt, F. B. Kulp, and L. Grande. 2013. Fossil evidence of wing shape in a stem relative of swifts and hummingbirds (*Aves, Pan-Apodiformes*). *Proc. R. Soc. Biol. Sci.* 280:20130580.

Linnaeus, C. 1758. *Systema Naturae Per Regna Tria Naturae, Secundum Classes, Ordines, Genera, Species, cum Characteribus, Differentiis, Synonymis, Locis.* Editio decima, reformata. Laurentii Salvii, Holmiae (Stockholm).

Mayr, G. 2002. Osteological evidence for paraphyly of the avian order *Caprimulgiformes* (nightjars and allies). *J. Ornithol.* 143:82–97.

Mayr, G. 2005. The Paleogene fossil record of birds in Europe. *Biol. Rev.* 80:1–28.

Mayr, G. 2010. Phylogenetic relationships of the paraphyletic 'caprimulgiform' birds (nightjars and allies). *J. Zool. Syst. Evol. Res.* 48:126–137.

Mayr, G., A. Manegold, and U. S. Johansson. 2003. Monophyletic groups within 'higher land birds'-comparison of morphological and molecular data. *J. Zool. Syst. Evol. Res.* 41: 233–248.

Nesbitt, S. J., D. T. Ksepka, and J. A. Clarke. 2011. Podargiform affinities of the enigmatic *Fluvioviridavis platyrhamphus* and the early diversification of *Strisores* ("*Caprimulgiformes*" + *Apodiformes*). *PLOS ONE* 6(11):e26350.

Pacheco, M. A., F. U. Battistuzzi, M. Lentino, R. Aguilar, S. Kumar, and A. A. Escalante. 2011. Evolution of modern birds revealed by mitogenomics: timing the radiation and origin of major orders. *Mol. Biol. Evol.* 28:1927–1942.

Sangster, G. 2005. A name for the clade formed by owlet-nightjars, swifts and hummingbirds (*Aves*). *Zootaxa* 799:1–6.

Shaw, G. In White, J. 1790. *Journal of a Voyage to New South Wales.* J. Debrett, London.

Sibley, C. G., J. E. Ahlquist, and B. L. Monroe. 1988. A classification of the living birds of the world based on DNA-DNA hybridization studies. *Auk* 105:409–423.

Yuri, T., R. T. Kimball, J. Harshman, R. C. K. Bowie, M. J. Braun, J. L. Chojnowski, K.-L. Han, S. J. Hackett, C. J. Huddleston, W. S. Moore, S. Reddy, F. H. Sheldon, D. W. Steadman, C. C. Witt, and E. L. Braun. 2013. Parsimony and model-based analyses of indels in avian nuclear genes reveal congruent and incongruent phylogenetic signals. *Biology* 2:419–444.

Author

George Sangster; Department of Bioinformatics and Genetics; Swedish Museum of Natural History; P. O. Box 50007, SE–104 05 Stockholm; Sweden; and Naturalis Biodiversity Center; Darwinweg 2, P. O. Box 9517, 2333 CC Leiden; Netherlands. Email: g.sangster@planet.nl.

Date Accepted: 11 December 2014

Primary Editor: Jacques A. Gauthier

Apodiformes J. L. Peters 1940 [G. Sangster], converted clade name

Registration Number: 288

Definition: The smallest crown clade containing *Phaethornis griseogularis* Gould 1851 and *Apus* (originally *Hirundo*) *apus* (Linnaeus 1758). This is a minimum-crown-clade definition. Abbreviated definition: min crown ∇ (*Phaethornis griseogularis* Gould 1851 & *Apus apus* (Linnaeus 1758)).

Etymology: Derived from the taxonomic name *Apus* Scopoli 1777, which is based on the Greek *a-* (without) and *pous* (foot), referring to the small feet of the members of this taxon, plus the Latin *forma* (having the form of).

Reference Phylogeny: For the purpose of applying the definition of *Apodiformes*, Figure 2 in Hackett et al. (2008) should be regarded as the primary reference phylogeny. Hackett et al. did not sample *Apus apus*, which is the type species of *Apus* and thus required as a specifier for the name *Apodiformes*. In the primary reference phylogeny, *Apus apus* is most closely related to *Aerodramus vanikorensis* and *Streptoprocne zonaris* (Price et al., 2005; Davis and Page, 2014).

Composition: Based on current knowledge (Hackett et al., 2008; Mayr, 2010), *Apodiformes* includes, among extant taxa, only *Hemiprocnidae* (tree swifts), *Apodidae* (swifts), and *Trochilidae* (hummingbirds). This clade includes approximately 437 Recent species (Dickinson and Remsen, 2013).

Diagnostic Apomorphies: Livezey and Zusi (2006, 2007) identified 13 apomorphies of *Apodiformes* relative to other crown clades. Eight of these are unambiguously diagnostic

(consistency index = 1; characters and states are indicated by their number-letter combination in Livezey and Zusi, 2006): (1) corpus sterni, margo cranialis sterni, depressio (sulcus) articularis coracoideus, facies articularis coracoideus weakly sellariform or convex with rectangular facies (1125b); (2) extremitas proximalis humeri, crista deltopectoralis (oblique caudal perspective), distal displacement and variably conspicuous dorsocranial elongation and hamulation present, crista extremely elongated, deeply hamulate processus (1375b); (3) extremitas proximalis humeri, crista bicipitalis, terminus on corpus humeri, margo ventralis abuptly discontinued proximally on corpus humeri, margo ventralis (1413b); (4) extremitas distalis humeri, facies cranialis, incisura intercondylaris, lamina mediana sulci serving as boundary for sulcus ligamenti musculus scapulotriceps present and small (1455b); (5) extremitas distalis humeri et corpus humeri, epicondylus dorsalis (ectepicondylus), ancora origii m. deltoideus pars propatagialis, pars cranialis subdistal, on cristula on margo dorsalis corporis shared with ancorae origii m. extensor digitorum communis (1465b); (6) extremitas distalis humeri et corpus humeri, epicondylus dorsalis (ectepicondylus), ancora origii m. extensor digitorum communis subdistal, a tuberculum proximal to epicondylus dorsalis (1466b); (7) juncturae interphalangeales longitudinis digiti II-III pedis present, prosynostosis phalangiorum proximales et intermedia digitorum III-IV apparent by extreme truncation and reduced mobility of articulationes (2449c); (8) musculus caudofemoralis, pars caudalis present but poorly developed (2661b).

Livezey and Zusi (2006, 2007) did not include any members of the stem (*Eocypselus rowei*, *E. vincenti*; Ksepka et al., 2013), and therefore it is likely that some of the apomorphies of *Apodiformes* listed above are actually present in more inclusive clades that also include stem-group species. Ksepka et al. (2013) identified six apomorphies of *Apodiformes* relative to *Eocypselus*: (1) a sternum with weakly saddle-shaped or convex articular surfaces for the coracoids; (2) a dorsoventrally thick and finger-like crista deltopectoralis; (3) a well-projected tubercle or process placed well proximal to the distal end of the humerus for attachment of m. tensor propatagialis pars brevis (modified in some *Apodidae*); (4) a well-developed processus internus indicis (convergently present in *Fluvioviridavis*, *Steatornis*, *Nyctibius*, and *Caprimulgus*); (5) distal expansion of manual phalanx II-2; (6) prominent projection of the tuberositas musculi tibialis cranialis of the tarsometatarsus.

Synonyms: The following are approximate synonyms: *Macrochires* Nitzsch 1840 (sensu Mayr and Amadon, 1951); *Cypseliformes* Garrod 1874; *Micropodii* Knowlton 1909; *Micropodiiformes* Wetmore and Miller 1926; *Micropodiformes* Wetmore 1930; *Apodimorphae* Sibley and Ahlquist 1990.

Comments: There is strong support for a clade formed by swifts, tree swifts, and hummingbirds, including studies based on morphological data (Cracraft, 1988; Mayr 2002, Mayr et al., 2003, Nesbitt et al., 2011; Ksepka et al., 2013), molecular data (Sibley and Ahlquist, 1990; Bleiweiss et al., 1994; van Tuinen et al., 2000; Ericson et al., 2006; Hackett et al., 2008; Pacheco et al., 2011), and supertree analysis (Davis and Page, 2014).

A few authors have restricted the name *Apodiformes* to the swifts and tree swifts (Brodkorb, 1971; Sibley and Ahlquist, 1990; Dyke and van Tuinen, 2004), whereas others (Mayr, 2005; Barrowclough et al., 2006) have expanded it to the clade of owlet-nightjars, swifts, tree swifts, and hummingbirds (see comments under *Daedalornithes*, this volume). However, the name *Apodiformes* has been most commonly applied to the swifts, tree swifts, and hummingbirds (e.g., Peters, 1940, 1945; Wetmore, 1951, 1960; Storer, 1960, 1971; Cracraft, 1981; Bock, 1982; del Hoyo et al., 1999; Cracraft et al., 2003; Livezey and Zusi, 2007; Nesbitt et al., 2011; Ksepka et al., 2013; Yuri et al., 2013) and is therefore retained for this clade. Sibley and Ahlquist (1990) used the name *Apodimorphae* for this clade. This minority view stems from a difference in opinion over the taxonomic rank of the clade formed by swifts, tree swifts, and hummingbirds. The name *Macrochires* is no longer in use. Other names that have been used in the past are based on generic names (*Micropus*, *Cypselus*) that are synonyms of *Apus* (see Peters, 1940; Wetmore, 1947) and were replaced prior to 1961 (the date adopted by the *ICZN* (International Commission on Zoological Nomenclature, 1999) to allow maintenance of substitute family-group names in prevailing use that were proposed because of synonymy of the type genus).

A minimum-crown-clade definition was selected because this maintains traditional use of the name (i.e., for the crown clade; Nesbitt et al., 2011; Ksepka et al., 2013).

Literature Cited

Barrowclough, G. F., J. G. Groth, and L. A. Mertz. 2006. The RAG-1 exon in the avian order *Caprimulgiformes*: phylogeny, heterozygosity, and base composition. *Mol. Phylogen. Evol.* 41:238–248.

Bleiweiss, R., J. A. Kirsch, and F. J. Lapointe. 1994. DNA-DNA hybridization-based phylogeny for "higher" nonpasserines: reevaluating a key portion of the avian family tree. *Mol. Phylogen. Evol.* 3:248–255.

Bock, W. J. 1982. Aves. Pp. 967–1015 in *Synopsis and Classification of Living Organisms*, Vol. 2 (S. P. Parker, ed.). McGraw-Hill, New York.

Brodkorb, P. 1971. Catalogue of fossil birds: Part 4 (*Columbiformes* through *Piciformes*). *Bull. Fl. State Mus., Biol. Sci.* 15(4):163–266.

Cracraft, J. 1981. Toward a phylogenetic classification of the Recent birds of the world (Class *Aves*). *Auk* 98:681–714.

Cracraft, J. 1988. The major clades of birds. Pp. 339–361 in *The Phylogeny and Classification of the Tetrapods. Vol. 1: Amphibians, Reptiles, Birds* (M. J. Benton, ed.). Systematics Association Special Volume No. 35A. Clarendon Press, Oxford.

Cracraft, J., F. K. Barker, and A. Cibois. 2003. Avian higher-level phylogenetics and the Howard and Moore checklist of birds. Pp. 16–21 in *The Howard and Moore Complete Checklist of the Birds of the World*. 3rd edition (E. C. Dickinson, ed.). Christopher Helm, London.

Davis, K. E., and R. D. M. Page. 2014. Reweaving the tapestry: a supertree of birds. *PLOS Curr. Tree Life*. 9 June 2014. 1st edition. doi:10.13 71/currents.tol.c1af68dda7c999ed9f1e4b2d 2df7a08e.

del Hoyo, J., A. Elliot, and J. Sargatal. 1999. *Handbook of the Birds of the World. Vol. 5: Barn-Owls to Hummingbirds*. Lynx Edicions, Barcelona.

Dickinson, E. C., and J. V. Remsen, Jr. 2013. *The Howard and Moore Complete Checklist of the Birds of the World*. 4th edition. Vol. 1: *Non-passerines*. Aves Press, London.

Dyke, G. J., and M. van Tuinen. 2004. The evolutionary radiation of modern birds (*Neornithes*): reconciling molecules, morphology and the fossil record. *Zool. J. Linn. Soc.* 141:153–177.

Ericson, P. G. P., C. L. Anderson, T. Britton, A. Elzanowski, U. S. Johansson, M. Källersjö, J. I. Ohlson, T. J. Parsons, D. Zuccon, and G. Mayr. 2006. Diversification of *Neoaves*: integration of molecular sequence data and fossils. *Biol. Lett.* 2:543–547.

Garrod, A. H. 1874. On certain muscles of birds and their value in the classification. Part II. *Proc. Zool. Soc. Lond.* 1874:111–123.

Gould, J. 1851. *A Monograph of the* Trochilidae, *or Family of Humming-Birds*. Published by the author, London.

Hackett, S. J., R. T. Kimball, S. Reddy, R. C. K. Bowie, E. L. Braun, M. J. Braun, J. L. Chojnowski, W. A. Cox, K.-L. Han, J. Harshman, C. J. Huddleston, B. D. Marks, K. J. Miglia, W. S. Moore, F. H. Sheldon, D. W. Steadman, C. C. Witt, and T. Yuri. 2008. A phylogenomic study of birds reveals their evolutionary history. *Science* 320:1763–1768.

International Commission on Zoological Nomenclature. 1999. *International Code of Zoological Nomenclature*. 4th edition. International Trust for Zoological Nomenclature, London.

Knowlton, F. H. 1909. *Birds of the World*. Henry Holt and Co., New York.

Ksepka, D. T., J. A. Clarke, S. J. Nesbitt, F. B. Kulp, and L. Grande. 2013. Fossil evidence of wing shape in a stem relative of swifts and hummingbirds (*Aves, Pan-Apodiformes*). *Proc. R. Soc. Lond. B Biol. Sci.* 280: 20130580.

Linnaeus, C. 1758. *Systema Naturae Per Regna Tria Naturae, Secundum Classes, Ordines, Genera, Species, cum Characteribus, Differentiis, Synonymis, Locis*. Editio decima, reformata. Laurentii Salvii, Holmiae (Stockholm).

Livezey, B. C., and R. L. Zusi. 2006. Higher-order phylogeny of modern birds (*Theropoda, Aves: Neornithes*) based on comparative anatomy. I. Methods and characters. *Bull. Carnegie Mus. Nat. Hist.* 37:1–556.

Livezey, B. C., and R. L. Zusi 2007. Higher-order phylogeny of modern birds (*Theropoda, Aves: Neornithes*) based on comparative anatomy. II. Analysis and discussion. *Zool. J. Linn. Soc.* 149:1–95.

Mayr, E., and D. Amadon. 1951. A classification of recent birds. *Am. Mus. Novit.* 1496:1–42.

Mayr, G. 2002. Osteological evidence for paraphyly of the avian order *Caprimulgiformes* (nightjars and allies). *J. Ornithol.* 143:82–97.

Mayr, G. 2005. The Paleogene fossil record of birds in Europe. *Biol. Rev.* 80:1–28.

Mayr, G. 2010. Phylogenetic relationships of the paraphyletic 'caprimulgiform' birds (nightjars and allies). *J. Zool. Syst. Evol. Res.* 48:126–137.

Mayr, G., A. Manegold, and U. S. Johansson. 2003. Monophyletic groups within 'higher land birds'-comparison of morphological and molecular data. *J. Zool. Syst. Evol. Res.* 41:33–248.

Nesbitt, S. J., D. T. Ksepka, and J. A. Clarke. 2011. Podargiform affinities of the enigmatic *Fluvioviridavis platyrhamphus* and the early diversification of *Strisores* ("*Caprimulgiformes*" + *Apodiformes*). *PLOS ONE* 6(11):e26350.

Pacheco, M. A., F. U. Battistuzzi, M. Lentino, R. Aguilar, S. Kumar, and A. A. Escalante. 2011. Evolution of modern birds revealed by mitogenomics: timing the radiation and origin of major orders. *Mol. Biol. Evol.* 28:1927–1942.

Peters, J. L. 1940. *Check-list of Birds of the World*, Vol. 4. Mus. Comp. Zool., Cambridge, MA.

Peters, J. L. 1945. *Check-list of Birds of the World*, Vol. 5. Mus. Comp. Zool., Cambridge, MA.

Price, J. J., K. P. Johnson, S. E. Bush, and D. H. Clayton. 2005. Phylogenetic relationships of the Papuan Swiftlet *Aerodramus papuensis* and implications for the evolution of avian echolocation.

Scopoli, G. A. 1777. *Introductio ad historiam naturalem*. Apud Wolfgangum Gerle, Prague.

Sibley, C. G., and J. E. Ahlquist. 1990. *Phylogeny and Classification of Birds*. Yale University Press, New Haven, CT.

Storer, R. W. 1960. The classification of birds. Pp. 57–93 in *Biology and Comparative Physiology of Birds*, Vol. 1 (A. J. Marshall, ed.). Academic Press, New York.

Storer, R. W. 1971. Classification of birds. *Avian Biol.* 1:1–18.

van Tuinen, M., C. G. Sibley, and S. B. Hedges. 2000. The early history of modern birds inferred from DNA sequences of nuclear and mitochondrial ribosomal genes. *Mol. Biol. Evol.* 17:451–457.

Wetmore, A. 1930. A systematic classification for the birds of the world. *Proc. US Natl. Mus.* 76(24):1–8.

Wetmore, A. 1947. Nomenclature of the higher groups in swifts. *Wilson Bull.* 59:211–212.

Wetmore, A. 1951. A revised classification for the birds of the world. *Smithson. Misc. Coll.* 117(4):1–22.

Wetmore, A. 1960. A classification for the birds of the world. *Smithson. Misc. Coll.* 139(11):1–37.

de Wetmore, A., and W. Miller 1926. The revised classification for the fourth edition of the A.O.U. check-list. *Auk* 43:337–346.

Yuri, T., R. T. Kimball, J. Harshman, R. C. K. Bowie, M. J. Braun, J. L. Chojnowski, K.-L. Han, S. J. Hackett, C. J. Huddleston, W. S. Moore, S. Reddy, F. H. Sheldon, D. W. Steadman, C. C. Witt, and E. L. Braun. 2013. Parsimony and model-based analyses of indels in avian nuclear genes reveal congruent and incongruent phylogenetic signals. *Biology* 2:419–444.

Author

George Sangster; Department of Bioinformatics and Genetics; Swedish Museum of Natural History; P. O. Box 50007, SE–104 05 Stockholm; Sweden; and Naturalis Biodiversity Center; Darwinweg 2, P. O. Box 9517, 2333 CC Leiden; Netherlands.

Date Accepted: 11 December 2014

Primary Editor: Jacques A. Gauthier

Index

Note: Page numbers in **bold** denotes the starting page of chapter.